Introduction to Electronic Circuit Design

Richard R. Spencer
University of California, Davis

Mohammed S. Ghausi
University of California, Davis

Pearson Education, Inc.,
Upper Saddle River, New Jersey 07458

Library of Congress Cataloging-in-Production Data

Spencer, Richard
Introduction to electronic circuit design / Richard Spencer, Mohammed Ghausi–1st ed.
p. cm.
ISBN 0-201-36183-3
1. Electronic circuit design. I. Ghausi, Mohammed Shuaib. II. Title.
TK7867.S817 2001 2001021251
621.3815—dc21 CIP

Vice President and Editorial Director, ECS: *Marcia J. Horton*
Vice President and Director of Production and Manufacturing, ESM: *David W. Riccardi*
Publisher: *Tom Robbins*
Executive Managing Editor: *Vince O'Brien*
Production Editor: *Scott Disanno*
Director of Creative Services: *Paul Belfanti*
Creative Director: *Carole Anson*
Art Editor: *Xiaohong Zhu*
Manager of Electronic Composition & Digital Content: *Jim Sullivan*
Electronic Production Specialist: *Allyson Graesser*
Electronic Composition: *Karen Stephens, Jacqueline Ambrosius, Terrance Cummings, Teresa Dillon, DonnaMarie Paukovits*
Manufacturing Manager: *Trudy Pisciotti*
Manufacturing Buyer: *Lisa McDowell*
Senior Marketing Manager: *Holly Stark*
Editorial Assistant: *Jody McDonnell*

Pearson Education, Inc.
Upper Saddle River, NJ 07458

Printed in the United States of America

10 9 8 7 6 5 4 3 2 1

ISBN 0-201-36183-3

Pearson Education Ltd., *London*
Pearson Education Australia Pty. Ltd., *Sydney*
Pearson Education Singapore, Pte. Ltd.
Pearson Education North Asia Ltd., *Hong Kong*
Pearson Education Canada, Inc., *Toronto*
Pearson Educacíon de Mexico, S.A. de C.V.
Pearson Education—Japan, *Tokyo*
Pearson Education Malaysia, Pte. Ltd.
Pearson Education, Inc., *Upper Saddle River, New Jersey*

LDD	Lightly-doped drain
LED	Light-emitting diode
LO	Local oscillator
LOS	Line of symmetry
LPCVD	Low-pressure CVD
LPF	Low-pass filter
LSB	Least significant bit
LSI	Large-scale integrated circuit
LTI	Linear time invariant
MESFET	Metal-semiconductor field-effect transistor
MODFET	Modulation-doped field-effect transistor
MOS	Metal-oxide semiconductor
MOSFET	Metal-oxide semiconductor field-effect transistor
MSB	Most significant bit
MSI	Medium-scale integrated circuit
MTBF	Mean time between failures
NAND	Not AND (logic gate)
NMOS	N-type MOS
NOR	Not OR (logic gate)
npn	Negative-positive-negative (a type of BJT)
PCB	Printed circuit board
PDP	Power-delay product
PIV	Peak inverse voltage
PLL	Phase-locked loop
PMOS	P-type MOS
pnp	Positive-negative-positive (a type of BJT)
pot	Potentiometer
ppm	Parts per million
PROM	Programmable read-only memory
PSRR	Power-supply rejection ratio
RAM	Random-access memory
RC	Resistor and capacitor
RF	Radio frequency
RLC	Resistor, inductor, and capacitor
ROM	Read-only memory
RTL	Resistor-transistor logic or register transfer level
SAR	Successive approximation register
SCR	Silicon controlled rectifier
SNR	Signal-to-noise ratio
SOA	Safe operating area
SPICE	Simulation Program with Integrated Circuit Emphasis
SPDT	Single pole double throw (a kind of switch)
SPST	Single pole single throw (a kind of switch)
SR	Slew rate or set-reset (a type of flip flop)
SRAM	Static RAM
STTL	Schottky TTL
TC	Temperature coefficient
TTL	Transistor-transistor logic
VCCS	Voltage-controlled current source
VCO	Voltage-controlled oscillator
VCR	Voltage-controlled resistance
VCVS	Voltage-controlled voltage source
VLSI	Very-large-scale integrated circuit
VTC	Voltage transfer characteristic
WL	Word line
$R_1 \parallel R_2$	the notation means that R_1 is in parallel with R_2

The standard engineering prefixes used in this book are shown below.

Prefix	Symbol	Value
femto	f	10^{-15}
pico	p	10^{-12}
nano	n	10^{-9}
micro	μ	10^{-6}
milli	m	10^{-3}
kilo	k	10^{3}
mega	M	10^{6}
giga	G	10^{9}

I am eternally grateful to my Lord Jesus Christ
and pray that this text will in some small way
bring honor to his name.—R.R.S.

*"Unless the Lord builds the house,
its builders labor in vain."*—from Psalm 127

Table of Contents

Preface

In the forward to the first issue of the *IEEE Journal of Solid-State Circuits* in September 1966, Dr. James D. Meindl wrote

> *"Within the past two decades, perhaps no sector of electronics has developed more rapidly than solid-state circuits. The nature of this development has imposed an expanding set of requirements on the breadth of knowledge one must possess in order to design a circuit well. – Most recently, the uniquely interdependent material, device, circuit, and system design considerations of large scale integration have again extended the scope of the problem of circuit design."*

It is remarkable that such broad statements, written over 30 years ago, are still accurate today. Given the complexity of present-day integrated circuits, circuit designers need to have a greater breadth of knowledge than ever before. This text is intended to provide an introduction to this important and rapidly changing discipline.

Many engineers who will never design an electronic circuit need to have a basic understanding of the characteristics of electronic circuits because they fabricate, test, or use these circuits, or they design systems that eventually have to be implemented using these circuits. In addition, there are many techniques and principles used in the design of electronic circuits that find widespread use outside of this discipline (e.g., small-signal linearity and feedback). Therefore, for those of you who will not become circuit designers this text still has much to offer that will be important in your careers.

The field of electrical engineering changes very rapidly. What can you expect to learn from this book that will still be useful in ten or twenty years? A great deal I hope. Although the devices, the economics of which components you favor, (e.g., resistors used to be cheaper than transistors, but the opposite is true in integrated circuits), and the computer-aided design tools will definitely change, there is still much that will stay the same. I can't predict exactly what will remain useful, but it seems unlikely that the *concept* of how to analyze a circuit so that you can see how to improve it will change, or that the *concept* of small-signal linearity will become unimportant, and certainly the ability you develop to solve problems will always be useful; after all, that is what engineers do.

I have tried to concentrate on helping you learn the *concepts* in this book. To be good at circuit design and many other engineering disciplines requires a healthy dose of intuition. While you certainly need to know how to write nodal equations and solve them, no one is going to pay you to do that because you can't possibly compete with a computer program. As soon as any area of engineering is well enough understood that we can write rigid procedures guaranteed to produce a correct answer, a computer program will take over. Therefore, it is true that you will always be working with systems that you don't completely understand and systems that cannot be analyzed exactly. Design will always require an ability to model the real system with a model that is simultaneously simple enough to allow you to "see" what is going on and think of ways to improve the performance, while being complete enough to adequately model the salient characteristics of the system. In addition, design requires that we do our analyses in a different way; we aren't just seeking the *answer*, but an understanding of how we can *modify* the system and/or *choose* the component values to achieve a desired result. That is, at least in part, what it takes to do design.

Therefore, I have focused on providing explanations and examples of how different circuits work and have focused on the underlying principles more than "rules of thumb" or design procedures, although some of these are certainly given. If you are searching for a cookbook approach, you won't find it here. However, remember that no one will pay you to follow set procedures. A designer is only valuable if he or she understands the problem well enough to come up with a good solution, even when no "procedure" exists.

This textbook came together when the second author, who was looking for someone to revise an existing text, met with the first author, who was contemplating writing a new book. After discussion, it was agreed that a new textbook was needed, but that material from the older text could be used. The result is the book you have in your hands; most of the material is completely new and written by the first author, but some of it is adapted from *Electronic Devices and Circuits: Discrete and Integrated*, by Mohammed S. Ghausi.

Organization and Features of the Text

The material in this text is organized logically by topic, rather than sequentially in the order I would present it. Therefore, I *do not expect* that you will read this book linearly. Rather, I expect that you will at times, jump around a bit. Organizing the book in this way has two advantages; first, I do not dictate the order of presentation and second, it emphasizes the different types of analyses that must be used in the design process.

Not specifying the order of presentation is important, because it allows each instructor more flexibility in choosing the topics to be covered and the depth of coverage of each topic. For example, you may cover field-effect transistor (FET) circuits first, or bipolar junction transistor (BJT) circuits first. This flexibility is partly brought about by placing the material on small-signal linearity in a separate chapter, and partly through the use of a generic transistor to present certain information that is common to both FETs and BJTs.

Emphasizing the different types of analyses used in design is important, because students are frequently confused about when and why a particular analysis or model should be used. For example, why is a capacitor modeled as an open or short circuit for some analyses while it is retained for others? And how do I know which model I should use for the transistors in my circuit? By covering DC bias point analysis, small-signal midband AC analysis, frequency response, large-signal AC performance, and digital circuits in different chapters, I emphasize the models, methods, and motivation for each type of analysis. Where possible, one example will be used throughout several chapters so, for example, you can learn about the DC biasing, midband gain, frequency response and large-signal swing of a common-emitter amplifier using a single example circuit. But the distinctions between the different types of analyses are emphasized by having each of them in a different chapter.

One feature of this text is the use of a generic transistor to present many of the basic principles that are common to FET and BJT circuits. While this transistor is fictitious, the terminal names used focus attention on the *functionality* of the device and the models used are the same as real transistors. The advantages of using this slight fiction are:

1. It helps to develop intuition about the operation of a given type of circuit without reference to the active device used (for example, a common-emitter BJT amplifier, common-source FET amplifier and even a common-cathode vacuum tube amplifier have much in common).
2. It allows common information to be presented once without dictating which type of transistor is covered first.
3. It helps to foster a modern device-independent way of thinking about circuits.

With this mode of thinking, the designer first considers the functionality of the circuit and then considers which type of device is best suited to a given application.

Another feature of this text is that it often breaks complex topics up into different levels of coverage to enable an instructor to decide how much detail to cover on a given topic at a given time. For example, most sections in Chapter 2 describing the operation of solid-state devices have an intuitive description followed by a more detailed derivation of the equations. The intuitive description can be used by students who have already had—or will have—a more detailed course in device physics. The detailed derivations can be used by students who have only one course covering electronic circuits and devices. As another example, the frequency response chapter presents both first-order methods for estimating the bandwidth of simple circuits (e.g., the Miller effect) and the more general zero-value time constant method. The more advanced zero-value time constant method may be left out of an introductory course taken by general Electrical and Computer Engineering students and can then be added in during a second-term course for students specializing in the circuits area.

One other feature of this text is that full solutions are provided for the exercises, rather than just numerical answers. It is my belief that if an exercise is involved enough to be of any real use to the student, simply providing the numerical answer is insufficient.

The final major feature of this text is the use of Asides. Asides are used for two different purposes. Sometimes, they are used to present material that is essential, but the student may already know from a prerequisite course, or may want to refer to later. Having the material in a separate Aside allows students to skip it or easily refer to it later. Asides are also sometimes used to present optional material that may expand on or further explain the material covered in the section. A separate index is provided to the Asides.

The CD that accompanies this text is also important. In addition to containing an evaluation version of the MicroSim DesignLab 8 software, which includes PSPICE, a postprocessor (PROBE), and a schematic capture program, the CD also includes all of the simulation files for over 100 exercises, examples, figures, and comments in the text. There are indexes on the CD for these files so that you can find, for example, a simulation file that shows how to use PSPICE to find the input or output resistance of an amplifier. In addition, the CD contains companion sections for a few places in the text where material was not printed for the sake of brevity (including two appendixes). I felt that some students would want this material.

There is more than enough material in this text for a two semester or three quarter sequence in electronic circuit design. Several different instructors have used drafts of this text for a two-quarter sequence at the University of California at Davis for several years. The first quarter of that sequence is required of all Electrical and Computer Engineering majors and covers Chapters 1 and 6, parts of Chapters 2, 7, and 8, the introductory material in Chapter 9, Chapter 14 and part of Chapter 15. The second quarter then adds the zero-value time constant method in Chapter 9, covers all of Chapter 10 and some of Chapters 4, 12, and 13. The material in Chapters 5 and 11 is used as reference material in other courses and Chapter 3 is left for the students to read, if they are interested.

Acknowledgments

Writing this book has been a *very* long process, and there are so many people to thank that I am sure I will forget someone. So, whoever you are, I ask your forgiveness in advance. To begin with, I thank my wife, Patti, and my children, David, Andrew, and Elizabeth, for putting up with hearing "when the book gets done" far too many times. I would next like to thank the many students who almost universally exhibited great patience while using early versions of the manuscript. I owe a debt of gratitude to the excellent teachers I have had the pleasure of learning from. I particularly want to thank Professor Artice M. Davis of San Jose State University and James D. Plummer of Stanford University, they are the finest two teachers I had when I was a student, and I am privileged to call them both friends. I would also like to thank Dr. Bruce M. Fleischer of IBM and Professor Thomas W. Matthews of Sacramento State University, who as students and friends have contributed to my own understanding of electronic circuits. It is a joy to work with Professors Stephen H. Lewis and Paul J. Hurst of UC Davis, they have both helped to make the Solid-State Circuits Research Laboratory an enjoyable place to work and learn. I also thank Dave Crook, Graham Baskerville, and John Steininger of National Semiconductor, and Ron Guly of HP, who were all helpful in providing data, photographs and other information that helped with this project. Dwight Morejohn helped tremendously with the cover art and Zoe Marlowe took the photograph for the Chapter Four opening figure. Professor James D. Plummer, Dr. Michael D. Deal and Dr. Peter B. Griffin were kind enough to allow me to extract material from their book, *Silicon VLSI Technology; Fundamentals, Practice, and Modeling*, to write Chapter 3—thanks! Professor Venkatesh Akella provided some of the information on digital CAD tools in Chapter Four. Charles Blas wrote the installer for the CD. Many students helped with problem solutions and reviewing the text; Salma Begum, Stephen Bruss, Efram Burlingame, Nick Chang, Chieu Yin Chia, Michael Collins, Ozan Erdogan, Alex Gros-Balthazard, Royce Higashi, Tunde Gyurics, Chris Holm, Yardley Ip, Jessi Johnson, Frank Lau, Tom McDonald, and Sophia Tang. In addition, Professor Thomas Matthews provided some of the problems and solutions. I am also grateful to the many reviewers who took the time to constructively comment on the text; Alok K. Berry from George Mason Universiy, Amir Farhat from Northeastern University, Samuel J. Garret from the University of South Florida, Rhett T. George from Duke University, Can E. Korman from the George Washington University, Sam Kozaitis from Florida Institute of Technology, Thomas Matthews from California State University Sacramento, Venkata Rao Mulpuri from George Mason University, Dennis Polla from the University of Minnesota, B. Song from San Jose State University, Karl A. Spuhl from Washington University, and John Uyemura from the Georgia Institute of Tehnology. I want to thank Tom Robbins, Scott Disanno, and the fine staff at Prentice Hall for putting up with me, keeping a sense of humor, and working hard to make the book as good as possible.

RICHARD SPENCER,
Davis, CA

About the Authors

Richard R. Spencer received the B.S.E.E. degree from San Jose State University in 1978 and the M.S. and Ph.D. degrees in Electrical Engineering from Stanford University in 1982 and 1987, respectively. He is a senior member of the IEEE.

He has been with the Department of Electrical and Computer Engineering at the University of California, Davis, since 1986, where is he is currently the Vice Chair for Undergraduate Studies and the Child Family Professor of Engineering. His research focuses on analog and mixed-signal circuits for signal processing and digital communication. He is an active consultant to the IC design industry.

Professor Spencer has won the UCD-IEEE Outstanding Undergraduate Teaching Award three times. He served on the IEEE International Solid-State Circuits Conference program committee for nine years, has been a guest editor of the IEEE Journal of Solid-State Circuits and has been an organizer and session chair for various IEEE conferences and workshops.

Mohammed S. Ghausi received the B.S.E.E., M.S. and Ph.D. degrees in Electrical Engineering from the University of California at Berkeley. He is a Professor Emeritus of Electrical and Computer Engineering as well as Dean Emeritus of the College of Engineering at UC Davis. His research interests are in electronics circuits and systems, and network theory, and he is the author or co-author of six textbooks. He was formerly a Professor of Electrical Engineering at New York University and later John F. Dodge Professor and Dean of Engineering and Computer Science at Oakland University. He is a recipient of the Alexander von Humboldt Prize, the IEEE Centennial Medal, the Circuits and Systems Society's 1991 Education Award, and the 1988 Outstanding Alumnus award of the Department of Electrical Engineering and Computer Science at the University of California at Berkeley.

List of Asides

CHAPTER 1

Electronic Circuit Design

The image illustrates different levels of abstraction used for analysis and design of electronic circuits. Starting with the actual circuit, we move up to device equations, then transistor-level schematics and then block-diagram level schematics.

INTRODUCTION

This chapter is written with the assumption that you have not previously been exposed to design in any systematic way, but have had one or more courses on circuit analysis. Therefore, one of the goals of the chapter is to explain the differences between analysis and design. Although the process of design is different for everyone who does it, there are certain practices and techniques that are almost universally accepted. In addition, although the process of design involves a great deal of analysis, the philosophy and methodology used for analysis is different when the goal is to enable the process of design, rather than to arrive at the answer.

Although we will certainly seek systematic approaches in this book, analysis for design does not generally follow any set procedure that can be used universally. When a set procedure can be described, computers will be programmed to do the work, and no one will pay you to do it. Therefore, although it may be frustrating at times to have each problem be a little different, that is the nature of real engineering.

While describing the process of design, we illustrate the distinction between analog and digital systems, discuss block diagrams and different levels of abstraction, and present two examples of sophisticated electronic systems. In addition, several useful results are derived in self-contained asides so that you can refer to them later. The last section of the chapter presents the notation that will be used throughout the text.

Hopefully, this introduction will provide motivation for further study and will help in making the mental shift necessary to go from analyzing the performance of electronic circuits and systems to designing them. Even if you do not become a circuit designer, you will develop useful problem-solving skills while studying this material. Many of the issues within the area of circuit design are common to almost every area of engineering—for example, the need to choose the appropriate level of modeling of a physical system for the given task.

1.1 THE PROCESS OF DESIGN

The process of design usually starts with a problem statement. For example, suppose you want to design a new electric car that can go farther on one charge and be less expensive to produce than those currently available. Developing a clear statement of the design requirements is not always a simple task. Sometimes, a lot of effort must be put into this first step in the design process. It is never a good idea to proceed with a design when the objectives are not clearly understood.

Once a clear problem statement is completed, a team of designers must develop a plan of action. They usually start by enumerating the goals of the project in more detail, perhaps including priorities so that any trade-offs that must be made later can be properly evaluated (e.g., how important is cost relative to range on one charge?). The next step is to identify possible approaches to achieving the stated goals. Often, this step is facilitated by a brainstorming[1] session.

After the possibilities have been listed, each is critically reviewed to see if there are any obvious reasons for dismissing it from further consideration. The resulting—usually shorter—list is then subjected to even closer scrutiny. At this point, the design will usually need to be broken up into different subsystems (e.g., motors, body), and subgroups of the designers will repeat the process for each particular subsystem. The subsystems may then be broken up further.

At each step there is a four-step process to be followed:

- First, the goals of the design are enumerated and, perhaps, priorities are established.
- Second, a list is generated containing as many ways of achieving the stated goals as possible.
- Third, this initial list of alternatives is subjected to critical review to select choices that appear to be feasible.
- Fourth, the remaining options are subjected to longer and more thoughtful analysis; note that the design may be partitioned into smaller subsystems at this point.

The process may be repeated many times, with the subgroups coming back together and modifying the partitioning or dealing with trade-offs in the different subassemblies (e.g., if the body is made more aerodynamically efficient, the motor doesn't need to be as powerful). Eventually, one or two approaches are settled on, and work begins in earnest. However, by this time, each of the groups has done a substantial amount of design, since ***design*** *to a large extent is an exploration of alternatives, followed by analysis, followed by choosing the alternative that seems best*

[1] "Brainstorming" is the process of listing *all* of the different ways you can conceive of solving a given problem *without* consideration of whether or not the given solutions are practical. The critical review of the proposed solutions should be postponed to a later step [1.1].

suited to the task at hand. Sometimes the process of design includes what most people think of when you mention design, namely, synthesis.

Synthesis *is the process of creating something new,* at least to the person who creates it. Although one can certainly get better at synthesis through practice and by critically examining the process used, there isn't much that can be done to directly teach you how to synthesize entirely new electronic circuits. In general, the process is similar to design in that alternatives are listed and then subjected to different levels of analysis. However, the difference is in the listing of alternatives. Rather than just enumerating all the known ways of performing a function, you are interested in figuring out how to perform a function in an entirely new way, either because no way currently exists or because all existing techniques are for some reason inadequate for the application.

Although synthesis is less systematic than design[2] and arguably not something that can be taught, much of what you learn about the process of design will help you with synthesis as well. To be good at synthesis, you must first be familiar with as many different circuits and functions as possible. You must also learn to think about circuits at several different levels of abstraction so that you can see how to modify circuits to improve them or alter their performance. The study of circuit design, as presented in this text, should help you gain knowledge and skill that will help you with synthesis as well as design.

As a final note, synthesis is more likely to consume a large amount of time to complete and is also riskier than using known methods in that your great new idea may never work properly! Therefore, a good designer will carefully research previous efforts at solving the same and similar problems to discover as many approaches as possible. From a business perspective, it usually isn't a good idea to repeat the work of others.

1.2 ANALYSIS FOR DESIGN

How does the method of analysis you are familiar with differ from the type of analysis that is used in the design process? A couple of examples will help to explain the difference. We begin with circuits containing only resistors and sources and then generalize to circuits containing complex impedances.

1.2.1 Frequency-Independent Analysis for Design

First, consider the circuit shown in Figure 1-1; it is typical of the kinds of circuits that will be analyzed many times later in this text. The circuit uses a ***Thévenin*** model of the input source and a ***controlled***, or ***dependent***, source. If you are not familiar with Thévenin equivalents or controlled sources, you should review Asides A1.1 and A1.2.

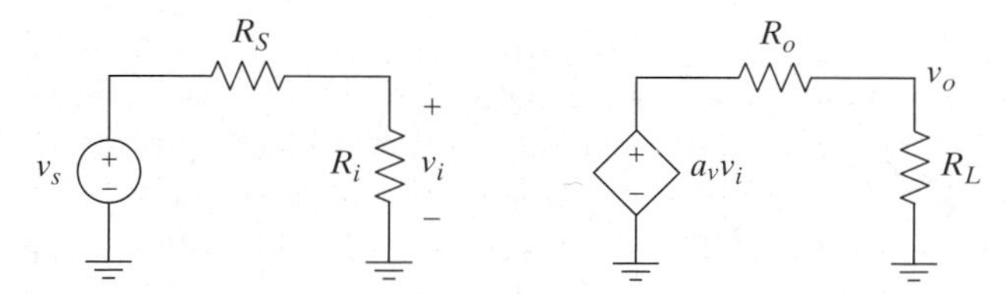

Figure 1–1 An amplifier circuit.

[2] There are well established very systematic approaches to synthesizing networks that have desired transfer functions. But we are referring to synthesizing a broader range of functions where even the basic topology may be entirely new.

The circuit in Figure 1-1 can be analyzed using either loop or nodal analysis. We choose nodal analysis and note that there are only two independent nodes in the circuit: the input, v_i, and the output, v_o. Our choice of using nodal analysis is not as arbitrary as it might seem, although for this circuit it turns out to not make a very big difference. Aside A1.3 compares loop and nodal analysis, and it shows why we usually prefer nodal analysis when we have a choice.

Using Kirchhoff's current law (KCL) at v_i in Figure 1-1 leads to

$$v_i\left(\frac{1}{R_S}+\frac{1}{R_i}\right)-\frac{v_s}{R_S}=0. \tag{1.1}$$

Writing KCL for v_o results in

$$v_o\left(\frac{1}{R_L}+\frac{1}{R_o}\right)-a_v v_i\frac{1}{R_o}=0. \tag{1.2}$$

ASIDE A1.1 SOURCE MODELS: THÉVENIN AND NORTON

Electronic circuits may be driven by many different signal sources. For example, the source may be a phonograph cartridge (not too common anymore!), a microphone, a guitar pickup, a pressure sensor, or any one of countless other examples. How are we going to model these sources? In some applications it is necessary to use a very complex model of the source, but we can often adequately model the source using one of two standard models for a linear time-invariant (LTI) ***one-port network***. Therefore, we use either a Thévenin or a Norton equivalent as shown in Figures A1-1(a) and (b), respectively. The Thévenin resistance shown in these figures may need to be a complex impedance but often is not. We will use these models for the sources available to us in this text unless specifically stated otherwise. These two models are equivalent so long as we use the same Thévenin impedance in each case, ignore the extreme cases where R_{Th} is zero or infinite, and set

$$i_N = v_{Th}/R_{Th}. \tag{A1.1}$$

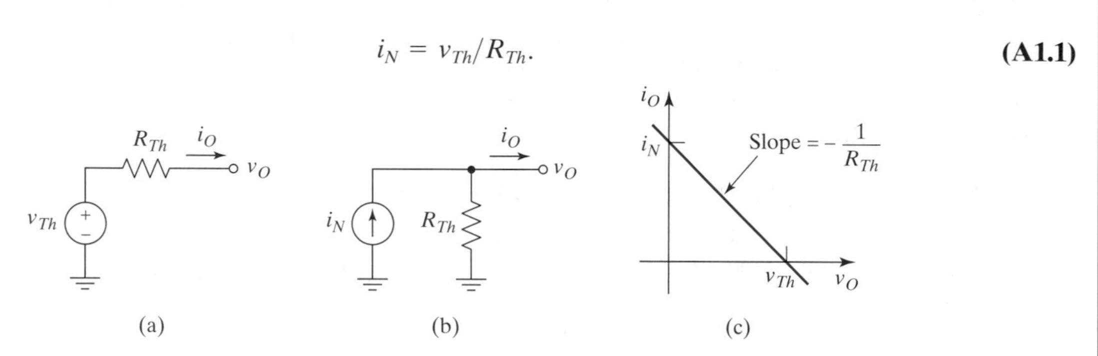

Figure A1-1 (a) A Thévenin equivalent source model, (b) a Norton equivalent source model, and (c) the *I-V* characteristic for these models.

Since these two models are equivalent, the question arises as to why we would choose one over the other in a given application. There are three possible answers to this question: (1) Either we don't really care, and we choose one arbitrarily, or (2) we choose the one that matches the way we think about the source (i.e., is the physical mechanism generating a voltage or a current?), or (3) we choose the one that yields the simplest analysis.

As a general guideline, if the source impedance is low relative to the impedances in our circuit, we probably want to use the Thévenin equivalent. If the source impedance is large relative to the impedances in our circuit, we probably want to use the Norton equivalent.

ASIDE A1.2 CONTROLLED, OR DEPENDENT, SOURCES

In modeling electronic systems, we often need to use ***controlled***, or ***dependent***, sources. These sources provide either a voltage or a current as their output. As the names imply, the output is controlled by some other quantity (i.e., is dependent on that quantity). The controlling quantity can also be a voltage or a current, so there are four types of controlled sources; namely, the voltage-controlled voltage source (VCVS), the voltage-controlled current source (VCCS), the current-controlled current source (CCCS), and the current-controlled voltage source (CCVS), as shown in Figure A1-2. Although it is possible to have nonlinear controlled sources (i.e., where the output is a nonlinear function of the input), we restrict ourselves at this point to linear controlled sources. Note that these sources are two-port networks.

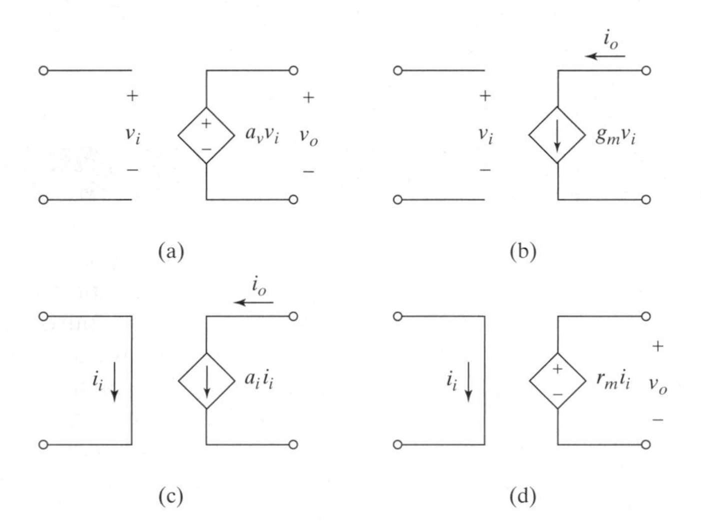

Figure A1-2 The four different types of controlled sources: (a) voltage-controlled voltage source, (b) voltage-controlled current source, (c) current-controlled current source, (d) current-controlled voltage source.

In the case of the voltage-controlled voltage source, the gain relating v_o to v_i is dimensionless (i.e., $a_v = v_o/v_i$). For the voltage-controlled current source, however, the gain has the units of a conductance (i.e., amps/volt, or siemens).[1] Since this conductance relates, or transfers, the input voltage to the output current, we call it a ***transconductance*** (a combination of the words *transfer* and *conductance*) and denote it by g_m. The g is used since the quantity has units of conductance, and the m subscript refers to it being a *mutual* conductance, an old terminology that has been replaced by the term transconductance, but which, unfortunately, still sets the accepted standard subscript. Similarly, the current-controlled current source has a dimensionless current gain a_i, whereas the current-controlled voltage source has a ***transresistance*** with units of ohms.

So long as a controlled source is linear, we can analyze a circuit containing it in the same way we would a circuit containing any other source. The only difference is that the output is a function of some other voltage or current, so the resulting equations will be more complicated than they would be for an independent source. If the controlled source is not linear, we need to deal with it in a completely different way.

[1] You may still occasionally come across the obsolete unit *mho* as well, but we will not use it.

ASIDE A1.3 NODAL VERSUS LOOP ANALYSIS

Although there are notable exceptions, nodal analysis is usually a better option than loop analysis when either one will work. There are at least two reasons that this is true: First, the node voltages are real measurable quantities, whereas the loop currents often are not; second, it is usually straightforward to see the minimum number of nodal equations required (i.e., the number of independent nodes), whereas it is often difficult to see how many independent loops there are. To make these points clear, examine Figure A1-3.

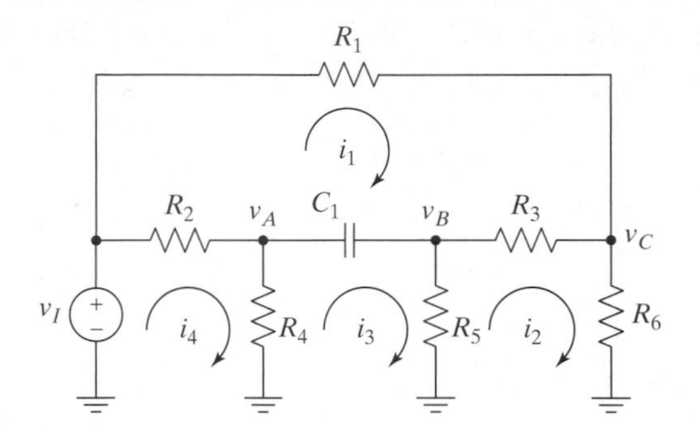

Figure A1-3 A circuit for demonstrating the differences between loop and nodal analysis.

If loop analysis is used to analyze this circuit, four equations in four unknowns (i.e., i_1, i_2, i_3, and i_4) must be solved. If nodal analysis is used, however, only three equations in three unknowns (i.e., v_A, v_B, and v_C) need to be solved. Also, the node voltages are measurable quantities, whereas the current i_3 by itself does not flow through any element in the circuit.

Solving (1.1) for v_i yields

$$v_i = \frac{1}{R_S\left(\frac{1}{R_S} + \frac{1}{R_i}\right)} v_s = \left(\frac{R_i}{R_i + R_S}\right) v_s, \tag{1.3}$$

and plugging this result into (1.2) produces

$$v_o\left(\frac{1}{R_L} + \frac{1}{R_o}\right) = \frac{a_v}{R_o}\left(\frac{R_i}{R_i + R_S}\right) v_s, \tag{1.4}$$

from which the final transfer function is obtained;

$$\begin{aligned}\frac{v_o}{v_s} &= \frac{a_v}{R_o}\left(\frac{R_i}{R_i + R_S}\right)\left(\frac{1}{\frac{1}{R_L} + \frac{1}{R_o}}\right) \\ &= \frac{a_v}{R_o}\left(\frac{R_i}{R_i + R_S}\right)\left(\frac{R_o R_L}{R_L + R_o}\right) \\ &= a_v\left(\frac{R_i}{R_i + R_S}\right)\left(\frac{R_L}{R_L + R_o}\right).\end{aligned} \tag{1.5}$$

We have deliberately included most of the steps in the derivation of (1.5) to illustrate that this brute-force approach to analyzing the circuit, while correct, tends to focus concentration on the algebra, rather than the circuit. An alternate analysis technique—and the one that all experienced designers would use—is to break the problem into a sequence of smaller problems and use a few well-known results to write the answer by inspection. The method starts by noticing that the overall transfer function can be written as the product of two simpler transfer functions:

$$\frac{v_o}{v_s} = \frac{v_i}{v_s}\frac{v_o}{v_i}. \tag{1.6}$$

The overall transfer function of a complex network can always be broken down as shown in (1.6), but doing so is useful only if the individual transfer functions can be written by inspection, or at least with significantly less work. For example, if the individual networks are all ***unilateral***, their transfer functions can usually be derived with minimal effort. A unilateral network is a network in which signals flow in only one direction. In other words, for this example, v_i does not depend on v_o. Each of the individual transfer functions on the right-hand side of (1.6) can be written by inspection, since they are both resistive voltage dividers. Examining the output of the circuit, we get

$$v_o = \left(\frac{R_L}{R_L + R_o}\right)a_v v_i, \tag{1.7}$$

or

$$\frac{v_o}{v_i} = a_v\left(\frac{R_L}{R_L + R_o}\right). \tag{1.8}$$

Examining the input side, we get

$$v_i = \left(\frac{R_i}{R_i + R_S}\right)v_s, \tag{1.9}$$

or

$$\frac{v_i}{v_s} = \frac{R_i}{R_i + R_S}. \tag{1.10}$$

Substituting these two transfer functions back into (1.6), or substituting (1.9) into (1.7) directly, leads to (1.5) again, but with less effort and more importantly, a clear realization on our part of where each term in the equation comes from.

We have chosen a very simple example to begin with, so the benefits may not be obvious at this stage. However, in more complicated circuits, the difference in the difficulty of solving simultaneous equations versus writing a sequence of simpler transfer functions can be quite large. We will do more complicated examples later, but we first want to introduce a notation that will be used throughout this text and that helps to cultivate the proper style of analysis.

In analyzing a circuit like this, it is often convenient to define an ***input attenuation factor***

$$\alpha_i \equiv \frac{v_i}{v_s} = \left(\frac{R_i}{R_i + R_S}\right) \tag{1.11}$$

and an ***output attenuation factor***

$$\alpha_o \equiv \frac{v_o(\text{with } R_L)}{v_o(\text{without } R_L)} = \left(\frac{R_L}{R_L + R_o}\right). \tag{1.12}$$

These attenuation factors will be different for different circuits, but the concept is always the same and leads to simplified analyses. The general definition of an attenuation factor is given in Aside A1.4.

ASIDE A1.4 ATTENUATION FACTORS

In analyzing unilateral circuits, we will often encounter situations like those illustrated in Figure A1-4, wherein a Thévenin equivalent is seen driving a load in part (a) and a Norton equivalent is seen driving a load in part (b). The sources may be independent as shown here, or they may be dependent. The output voltage or current is always given by a voltage or current divider. For the Thévenin case, we get

$$v_L = \left(\frac{R_L}{R_L + R_{Th}}\right) v_{Th}, \tag{A1.2}$$

and for the Norton equivalent, we have

$$i_L = \left(\frac{G_L}{G_L + G_{Th}}\right) i_N = \left(\frac{R_{Th}}{R_{Th} + R_L}\right) i_N. \tag{A1.3}$$

Because the signal being considered in either case is attenuated (i.e., reduced in amplitude) by the divider, we call the ratio of the actual output to the no-load output the ***attenuation factor*** and denote it by α. For the voltage divider, we have the attenuation factor

$$\alpha \equiv \frac{v_L}{v_{Th}} = \left(\frac{R_L}{R_L + R_{Th}}\right), \tag{A1.4}$$

and for the current divider, we have

$$\alpha \equiv \frac{i_L}{i_N} = \left(\frac{R_{Th}}{R_{Th} + R_L}\right). \tag{A1.5}$$

Since we often find this kind of divider network at the input and output of an amplifier, we will call the associated attenuation factors the input attenuation factor (α_i) and output attenuation factor (α_o), respectively. Attenuation factors are always strictly less than one.

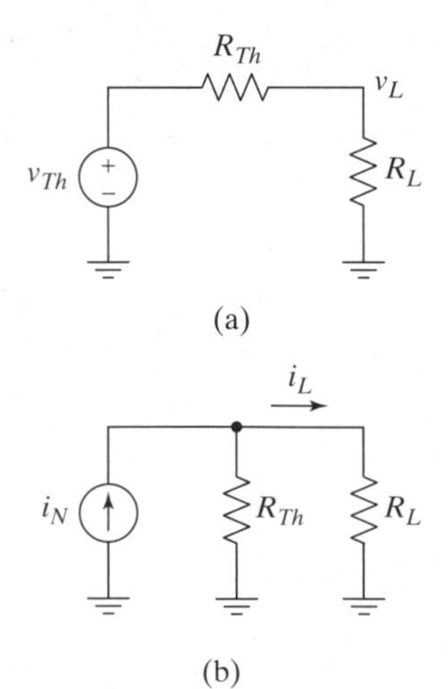

Figure A1-4 (a) A voltage divider and (b) a current divider.

In designing amplifiers and many other circuits, we often desire to make the attenuation factors as close to one as possible. If the input attenuation factor is too small, the signal is reduced prior to any amplification or processing. Additive noise, which is always present, can then be a significant problem. On the other hand, if the output attenuation is too small, it means we are wasting some of our signal in driving the load, and we have to increase the gain in the system to make up for the loss. Note that it is usually the attenuation factor, not the power transfer, that should be maximized. You may have previously worked out how to maximize the *power* transferred from a source to a load, but Aside A1.5 explains why this is not usually what we want to do.

Exercise 1.1[3]

Consider the circuit shown in Figure 1-2. Find the overall transfer function from v_S to v_O (i.e., v_O/v_S) using (a) nodal analysis and (b) repeated application of a voltage divider. Which is easier? Which lends the most insight?

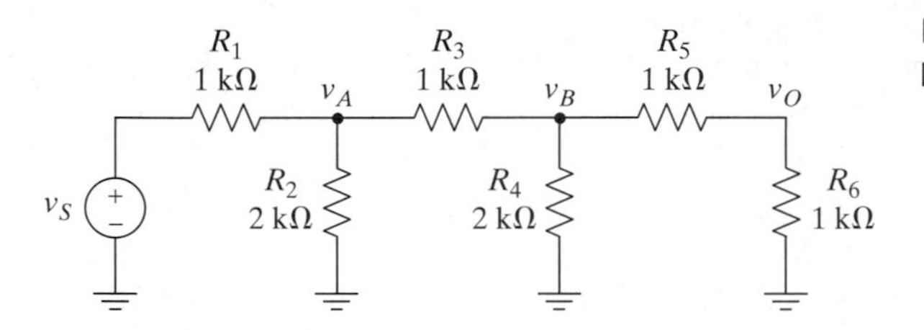

Figure 1–2 The circuit for Exercise 1.1.

ASIDE A1.5 MAXIMUM POWER AND SIGNAL TRANSFER

Consider the circuit shown in Figure A1-5, which consists of a Thévenin equivalent source driving a load resistor. Assuming that the source resistance is fixed, we wish to find the value of R_L that will maximize the power delivered to the load.

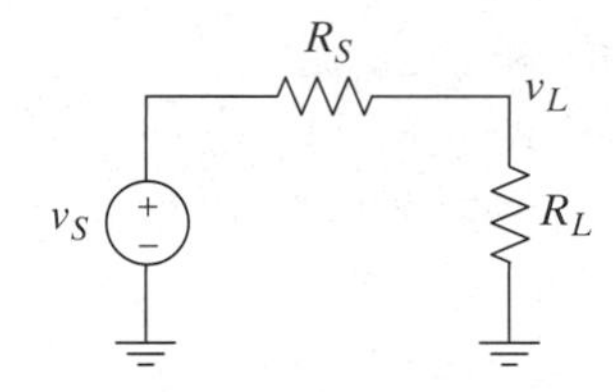

Figure A1-5 A Thévenin equivalent source driving a load resistor.

The load voltage in Figure A1-5 is given by

$$v_L = \frac{R_L}{R_L + R_S} v_S, \tag{A1.6}$$

and the corresponding power in the load is

$$P_L = \frac{v_L^2}{R_L} = \frac{R_L v_S^2}{(R_L + R_S)^2}. \tag{A1.7}$$

[3] The solutions to exercises are found at the end of the chapter in which they appear.

By differentiating (A1.7) with respect to R_L and setting the result equal to zero, we find that the power is maximized when $R_L = R_S$. If the source has a complex impedance Z_S, the load impedance that maximizes the power delivered to the load is the complex conjugate, Z_S^*.

This result is classical and is repeated in many introductory circuit texts. However, it is often not an important result from a practical point of view, since we typically do not want to adjust the load impedance to maximize the power transferred to it from a fixed source.

In electronic circuit design, it is far more common to deal with one of two situations: either we are designing the input stage of some system and are driven by a fixed source, or we are designing the output stage to drive a fixed load.

In the first case, we are usually not concerned with optimizing the power delivered to our circuits; rather, we are concerned about getting the largest amplitude signal possible (i.e., voltage or current) into the circuit. Usually the signal amplitude at the input of an electronic system is small (e.g., the output of an antenna), and if the signal amplitude drops too much, the additive noise that is always present will reduce the signal-to-noise ratio and degrade the overall performance.

In the second case, we are designing an output stage for a fixed load (e.g., the output of an audio amplifier to drive an 8 Ω speaker), and we are usually concerned about the efficiency of the amplifier. In other words, we don't want to waste power in the amplifier; we want to deliver it all to the load. In this case, R_S in Figure A1-5 would model the output resistance of our amplifier and R_L would model the speaker. We now want to adjust R_S to maximize the power delivered to a fixed R_L. The solution is to set $R_S = 0$, as you can verify by examining (A1.7). Therefore, we design our circuits to achieve the lowest output resistance possible.

In both of these cases, it is the attenuation factor that has been maximized rather than the power transfer. The only situation commonly encountered where it is desirable to match impedances occurs when transmission lines are used. But even then, maximum power transfer is not the goal. For a transmission line, it is desirable to have both the impedance of the source driving the line and the load impedance terminating the line matched to the line's characteristic impedance (*see* Section 14.4 for a discussion of transmission lines). If the impedances are not matched, there will be reflections on the line that can cause problems. For example, the ghost images that are sometimes seen on a TV connected to a cable are usually caused by reflections on the cable, which are caused by improper termination.

Example 1.1

Problem:
For the circuit shown in Figure 1-3(a), determine the Thévenin and the Norton equivalent circuits looking into the open terminals as indicated.

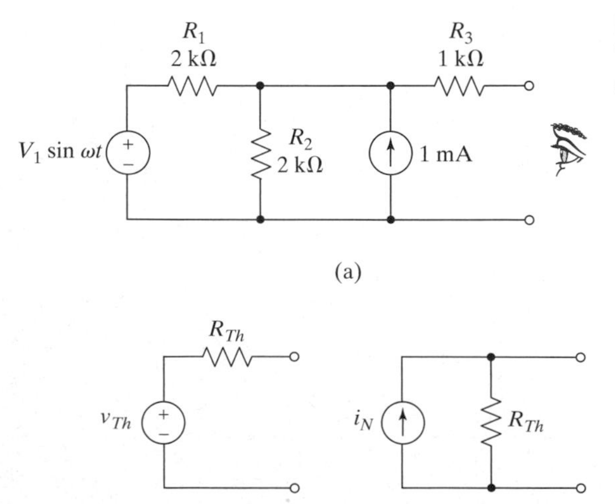

Figure 1–3 (a) Example circuit, (b) Thévenin equivalent, and (c) Norton equivalent.

Solution:
Note that we have only independent sources in the circuit and R_{Th} is determined by open-circuiting the current source and short-circuiting the voltage source. By inspection, $R_{Th} = (R_1 \| R_2) + R_3 = 2\text{ k}\Omega$ (the notation $R_1 \| R_2$ means that R_1 is in parallel with R_2). To find v_{Th} we can use superposition, since the network is linear. Recall that the superposition principle allows us to determine the response due to each source separately and then find the total response as the sum of the individual responses. Hence, v_{Th} due to the current source is obtained by setting the voltage source equal to zero and determining the open-circuit voltage at the port.

This process yields $v_{Th1} = (R_1 \| R_2)(1\text{ mA}) = 1\text{V}$. Similarly, v_{Th} due to the voltage source is obtained by open-circuiting the current source and determining the open-circuit voltage. This process yields

$$v_{Th2} = \left(\frac{R_2}{R_1 + R_2}\right)V_1 \sin\omega t = \frac{V_1}{2}\sin\,\omega t\text{ V}.$$

By superposition, we have

$$v_{Th} = v_{Th1} + v_{Th2} = 1 + \frac{V_1}{2}\sin\,\omega t\text{ V}.$$

For the Norton current, we obtain

$$i_N = \frac{v_{Th}}{R_{Th}} = 0.5 + \frac{V_1}{4 \times 10^3}\sin\,\omega t\text{ mA}.$$

The resulting Thévenin and Norton equivalent circuits are shown in Figures 1-3(b) and (c), respectively.

■

Exercise 1.2

Suppose a temperature sensor has an open-circuit output voltage of 1 mV/°C and the output is 0.2 mV/°C when a 1-kΩ load is driven. Find the Thévenin and Norton models for the source.

□

Exercise 1.3

Find the Thévenin and Norton equivalents for the circuit shown in Figure 1-4.

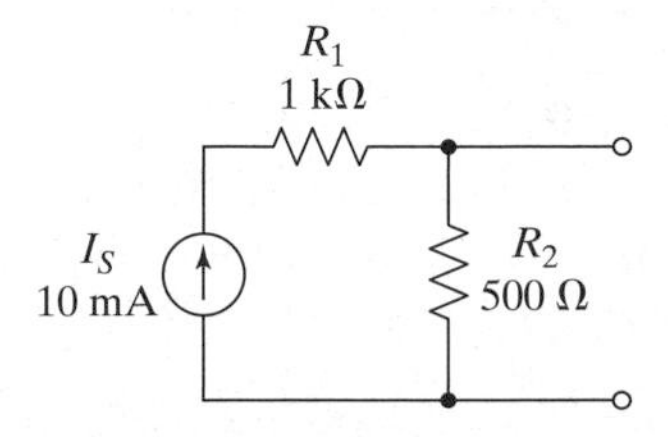

Figure 1–4 The circuit for Exercise 1.3.

□

Example 1.2

Problem:
Find the Thévenin equivalent for the circuit shown in Figure 1-5(a). Note that in this case we have a dependent source.

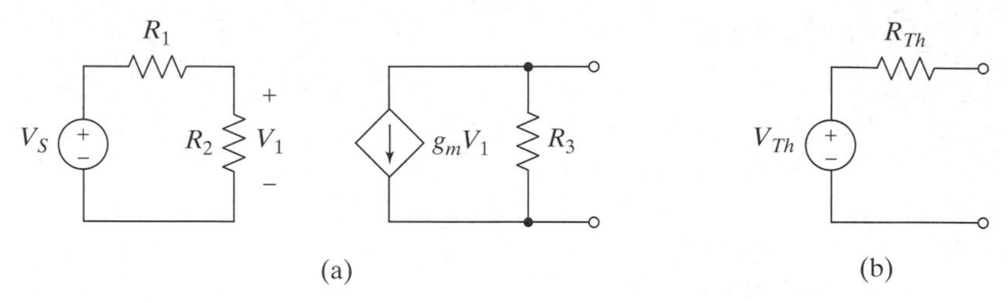

Figure 1–5 (a) Example circuit and (b) Thévenin equivalent.

Solution:
Important Note: When finding a Thévenin or Norton equivalent for a circuit containing controlled sources, the controlled sources are not set equal to zero, although they may certainly turn out to have zero value.

The open-circuit output voltage is given by

$$V_{Th} = -g_m V_1 R_3 = -g_m R_3 \frac{R_2}{R_1 + R_2} V_S,$$

the short-circuit output current is given by

$$I_N = -g_m V_1 = -g_m \frac{R_2}{R_1 + R_2} V_S,$$

and the Thévenin resistance is

$$R_{Th} = \frac{V_{Th}}{I_N} = R_3.$$

The resulting Thévenin equivalent is shown in part (b) of the figure.

■

Exercise 1.4

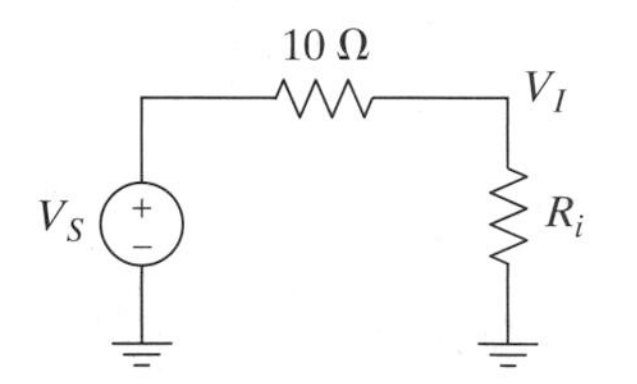

Figure 1–6 The circuit for Exercise 1.4.

Consider the circuit in Figure 1-6, which models a 10 Ω source driving the input to an amplifier. (a) Find the value of input resistance, R_i, necessary to achieve maximum power transfer into the amplifier. What is the power? (b) Find the power transferred into the amplifier, assuming that $V_S = 3$ V and $R_i = 100$ Ω, and express it as a percentage of the maximum power that could be transferred. (c) Calculate the input attenuation factors for both cases. (d) Which of these values for R_i is a better choice? Why?

□

1.2.2 Frequency-Dependent Analysis for Design

Now consider a somewhat more complicated example to illustrate the difference between analysis for its own sake and analysis for design.[4] If we want to analyze the circuit shown in Figure 1-7 using nodal analysis techniques, we first use Kirchhoff's current law (KCL) at v_1 to get

[4] This section assumes some knowledge of frequency-domain analysis techniques and may be omitted without loss of continuity.

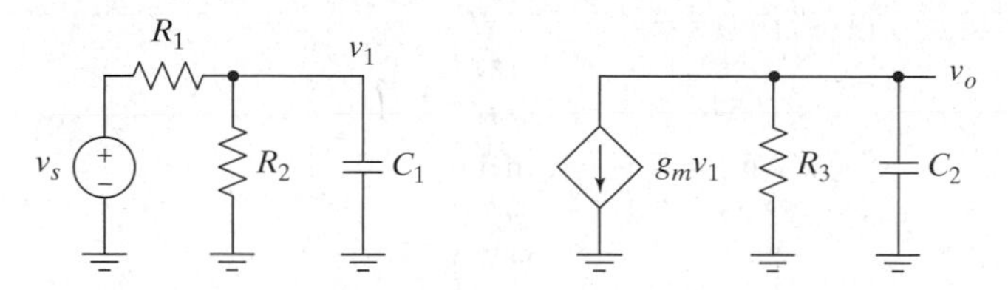

Figure 1–7 A circuit for demonstrating analysis for design.

$$V_1(j\omega)\left(\frac{1}{R_1} + \frac{1}{R_2} + j\omega C_1\right) - V_s(j\omega)\frac{1}{R_1} = 0, \tag{1.13}$$

where we have shifted to uppercase variables because we are now dealing with functions of the complex frequency $j\omega$.[5] The notation used in this text is completely described at the end of this chapter, but is not critical at this stage. Writing KCL at v_o leads to

$$V_o(j\omega)\left(\frac{1}{R_3} + j\omega C_2\right) = -g_m V_1(j\omega). \tag{1.14}$$

Solving (1.14) for $V_1(j\omega)$ and substituting into (1.13) leads to

$$\frac{V_o(j\omega)\left(\frac{1}{R_3} + j\omega C_2\right)}{g_m}\left(\frac{1}{R_1} + \frac{1}{R_2} + j\omega C_1\right) = \frac{-V_s(j\omega)}{R_1}, \tag{1.15}$$

or, after a little algebra,

$$\frac{V_o(j\omega)}{V_s(j\omega)} = \frac{-g_m}{R_1\left(\frac{1}{R_3} + j\omega C_2\right)\left(\frac{1}{R_1} + \frac{1}{R_2} + j\omega C_1\right)}. \tag{1.16}$$

This result is certainly correct and could be simplified with a little work, but neither the form of the answer nor the process of deriving it lends any insight into the operation of the circuit.

Now let's see how to analyze this circuit in a way that is more useful for design. We first recognize that the overall transfer function can be written as the product of two simpler transfer functions:

$$\frac{V_o(j\omega)}{V_s(j\omega)} = \frac{V_1(j\omega)}{V_s(j\omega)}\frac{V_o(j\omega)}{V_1(j\omega)}. \tag{1.17}$$

Each of the individual transfer functions in (1.17) can be written by inspection, provided that we know the standard form for a single-pole low-pass transfer function. Aside A1.6 reviews the standard forms for single-pole transfer functions. The first term in the product on the right-hand side of (1.17) is

$$\frac{V_1(j\omega)}{V_s(j\omega)} = \frac{\frac{R_2}{R_1 + R_2}}{1 + j\omega\frac{R_1 R_2 C_1}{R_1 + R_2}} = \frac{H_1(j0)}{1 + j\frac{\omega}{\omega_{p1}}}. \tag{1.18}$$

---

[5] Readers whose memory of linear network analysis is shaky might profit from reviewing Appendix D on the CD.

Similarly, the second transfer function is

$$H_2(j\omega) \equiv \frac{V_o(j\omega)}{V_1(j\omega)} = \frac{H_2(j0)}{1 + j\dfrac{\omega}{\omega_{p2}}}. \tag{1.19}$$

$H_2(j0)$ is found by open-circuiting C_2, which leads to

$$V_o(j0) = -g_m V_1(j0) R_3, \tag{1.20}$$

and therefore,

$$H_2(j0) = \frac{V_o(j0)}{V_1(j0)} = -g_m R_3. \tag{1.21}$$

To find the pole frequency, we need to set v_s to zero. Doing so also guarantees that v_1 is zero. Therefore, the controlled current source has zero value and can be represented by an open circuit. The equivalent resistance seen by C_2 is then R_3, and the pole frequency is given by $\omega_{p2} = 1/R_3C_2$. Combining all of these intermediate results leads to

$$H(j\omega) = \frac{V_o(j\omega)}{V_s(j\omega)} = H_1(j\omega)H_2(j\omega) = \frac{\dfrac{R_2}{R_1 + R_2}}{1 + \dfrac{j\omega R_1 R_2 C_1}{R_1 + R_2}}\left(\frac{-g_m R_3}{1 + j\omega R_3 C_2}\right). \tag{1.22}$$

This transfer function may look imposing, but it has the distinct advantage that it is the product of two standard forms, and it can be readily seen what the effect of changing any of the circuit elements will be on the overall performance of the circuit. Of course, the equation looks much simpler if it is written in terms of $H_1(j0)$, $H_2(j0)$, and the two pole frequencies. With this form of result, we can, for example, see just what to change if we must change the bandwidth or the gain of the circuit, as demonstrated in the following example.

Example 1.3

Problem:

Consider the amplifier shown in Figure 1-8. This is the same circuit we had in Figure 1-7, but we have now included numerical values for all the components. Find the gain and bandwidth of this amplifier. Which components dominate the bandwidth?

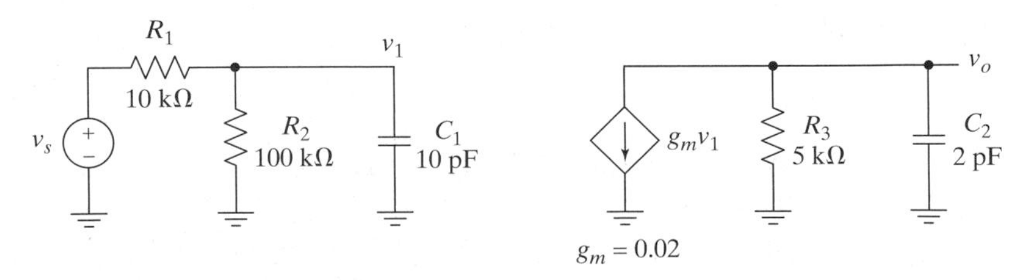

Figure 1–8 An amplifier.

Solution:
Using (1.22), we find the voltage gain:

$$A_v(j\omega) = \frac{V_o(j\omega)}{V_s(j\omega)} = \frac{0.91}{1 + j\omega/1.1 \times 10^7}\left(\frac{-100}{1 + j\omega/1 \times 10^8}\right).$$

The low-frequency gain of the amplifier is, therefore, −91. Since the second pole frequency is much higher than the first, the −3 dB frequency of the amplifier is approximately equal to the first pole:

$$\omega_{p1} = \frac{1}{(R_1 \| R_2)C_1} = 1.1 \times 10^7.$$

But, because $R_2 \gg R_2$, this pole is approximately equal to $1/R_1C_1$. Therefore, we see that the bandwidth of this amplifier is dominated by R_1 and C_1. If we want to change the bandwidth, changing the values of other components will have little or no effect.

■

ASIDE A1.6 SINGLE-POLE TRANSFER FUNCTIONS

Since we often deal with single-pole networks, it is helpful to develop standard forms for the transfer functions so that we can write them with less effort. Any circuit with a single reactive element, either a capacitor or an inductor, is a single-pole network. Of course, there can be multiple capacitors or inductors, but as long as they can be reduced to a single equivalent capacitor or inductor (i.e., they are in series and/or parallel), the network will have just one pole. There are two types of single-pole transfer functions: low pass and high pass. We will deal with low-pass networks first.

Single-Pole Low-Pass Transfer Functions

A low-pass transfer function is one that passes low-frequency signals with less attenuation or more gain than it does signals with frequencies above the cutoff, or pole, frequency. The magnitude of a low-pass transfer function is plotted versus frequency in Figure A1-6.

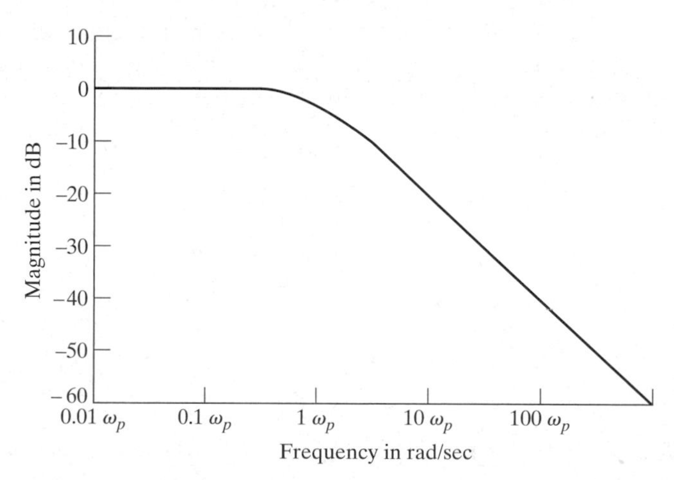

Figure A1-6 Magnitude plot of a low-pass transfer function with DC gain of one and pole at ω_p.

The transfer function shown has one pole, no finite zeros, and a DC gain of one (i.e., the gain at $\omega = 0$ is one). The transfer function can be written as

$$H(j\omega) = \frac{1}{1 + j\omega/\omega_p}. \tag{A1.8}$$

More generally, a single-pole low-pass transfer function can always be written in the form

$$H(j\omega) = \frac{H(j0)\left(1 + j\dfrac{\omega}{\omega_z}\right)}{1 + j\dfrac{\omega}{\omega_p}}, \tag{A1.9}$$

or, in the Laplace domain,[1]

$$H(s) = \frac{H(0)(1 + s\tau_z)}{1 + s\tau_p}, \tag{A1.10}$$

where $\tau_p = 1/\omega_p$, $\tau_z = 1/\omega_z$, and it is entirely possible for the zero to occur at infinity (i.e., $\omega_z = \infty$). The form used for the pole and zero in (A1.9)—that is, the form $(1 + j\omega/\omega_{pz})$, where ω_{pz} is a pole or zero—is convenient for a low-pass function even when there are more than one pole and zero, because the term is approximately one for $\omega << \omega_{pz}$.

The DC transfer function of a single-pole capacitive network, $H(j0)$, is found by open-circuiting the capacitor, since $Z_C(j0)$ is infinite. For example, for the circuit in Figure A1-7, the DC transfer function from v_s to v_1 is given by the resistive voltage divider

$$H_1(j0) \equiv \frac{V_1(j0)}{V_s(j0)} = \frac{R_2}{R_1 + R_2}. \tag{A1.11}$$

For an inductive single-pole low-pass circuit, the inductor is short-circuited to find $H(j0)$, since $Z_L(j0)$ is zero.

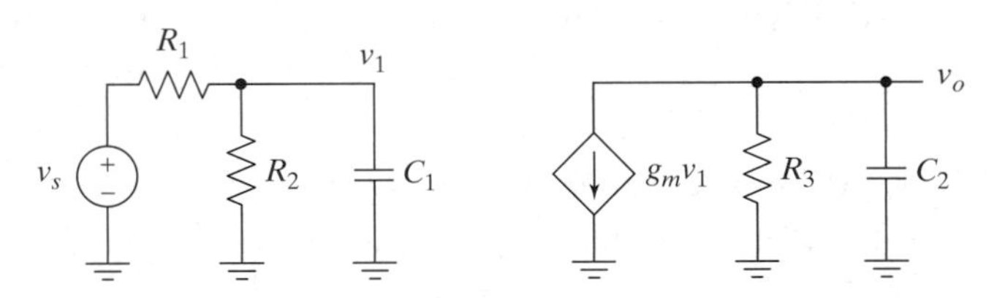

Figure A1-7 A circuit for demonstrating analysis for design.

For a single-pole network, the pole is the reciprocal of the time constant, and the time constant is given by $R_{eq}C$ (or L/R_{eq} if the reactive element is an inductor), where R_{eq} is the resistance "seen" by the capacitor (or inductor) when all independent sources have been set equal to zero. The resistance R_{eq} is called the ***driving-point resistance***. For the circuit in Figure A1-7, if v_s is set to zero and C_1 is removed, the equivalent resistance seen looking back into the terminals where C_1 was connected is calculated as shown in Figure A1-8.

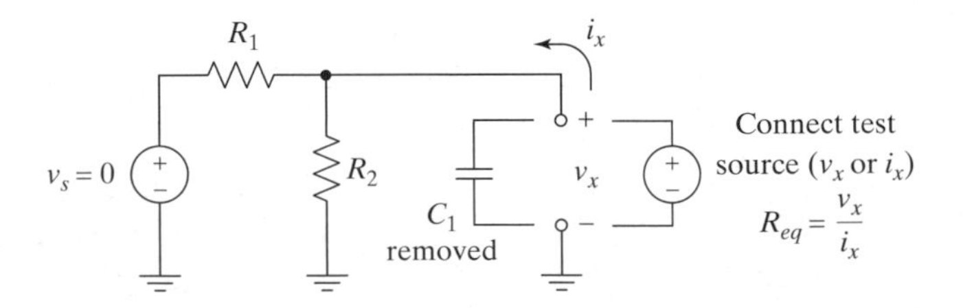

Figure A1-8 Finding the resistance seen by C_1 in the circuit of Figure 1-7.

The equivalent resistance is $R_{eq} = R_1 \| R_2$. Therefore, the time constant for the transfer function from v_s to v_1 is

$$\tau_p = R_{eq}C = (R_1 \| R_2)C_1 = \frac{R_1 R_2 C_1}{R_1 + R_2}. \tag{A1.12}$$

[1] Readers not familiar with Laplace transforms may ignore (A1.10) and (A1.15) without loss of understanding.

The frequency of the zero is infinity for this example. We see this by considering what happens in the circuit as the frequency of the source is increased beyond ω_p. As ω increases, the impedance of the capacitor steadily decreases and causes the voltage across it, and therefore the magnitude of the transfer function, V_1/V_s, to decrease without limit, only going to zero as ω goes to infinity. Unfortunately, there isn't a simple procedure for finding a zero that is finite and not at DC. Nevertheless, the frequency of the zero is often infinity, or at least high enough in frequency to be ignored, and this can often be determined by inspection, as was just demonstrated. Therefore, the complete transfer function from v_s to v_1 is

$$H_1(j\omega) = \frac{V_1(j\omega)}{V_s(j\omega)} = \frac{H_1(j0)\left(1 + j\dfrac{\omega}{\omega_z}\right)}{1 + j\dfrac{\omega}{\omega_p}} = \frac{\dfrac{R_2}{R_1 + R_2}}{1 + \dfrac{j\omega R_1 R_2 C_1}{R_1 + R_2}}. \tag{A1.13}$$

Single-Pole High-Pass Transfer Functions

A high-pass transfer function is a transfer function that passes high-frequency signals with significantly less attenuation (or more gain) than it does frequencies below the cutoff, or pole, frequency. The magnitude of a high-pass transfer function is plotted versus frequency in Figure A1-9.

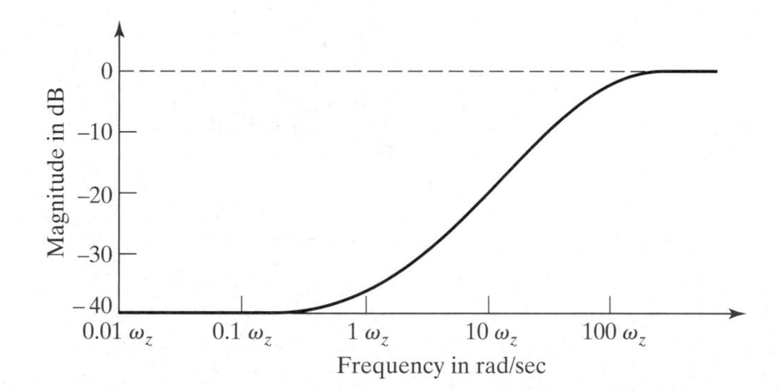

Figure A1-9 Magnitude plot for a single-pole high-pass transfer function with a high-frequency gain of one, a zero at ω_z, and a pole at $\omega_p = 100\,\omega_z$.

A single-pole high-pass transfer function can always be written in the form

$$H(j\omega) = \frac{H(j\infty)(j\omega + \omega_z)}{j\omega + \omega_p}, \tag{A1.14}$$

or, in the Laplace domain,

$$H(s) = \frac{H(j\infty)(s + \omega_z)}{s + \omega_p} = \frac{H(j\infty)(s + 1/\tau_z)}{s + 1/\tau_p}. \tag{A1.15}$$

The form used for the pole and zero in (A1.14)—that is, the form $(j\omega + \omega_{pz})$, where ω_{pz}, is a pole or zero—is a convenient form for a high-pass function even when there are more than one pole and zero, because the magnitude is approximately equal to ω for $\omega \gg \omega_{pz}$ and the terms in the numerator and denominator cancel.

The value of a high-pass transfer function at $\omega = \infty$ is found by replacing an inductor by an open circuit or a capacitor by a short circuit. The pole frequency is found in exactly the same manner as was described for the single-pole low-pass transfer function.

The frequency of the zero is again more troublesome, since no simple procedure exists for finding it. Nevertheless, it is not unusual for the zero to be at DC, or at a frequency that is low enough that it can be approximated as DC. Whether or not the zero is at DC can be determined by inspecting the circuit while letting a capacitor approach an open circuit or an inductor approach a short circuit. If the resulting transfer function approaches zero, then the zero is at DC.

Exercise 1.5

Using the analysis procedure just presented, derive a transfer function for the circuit in Figure 1-9.

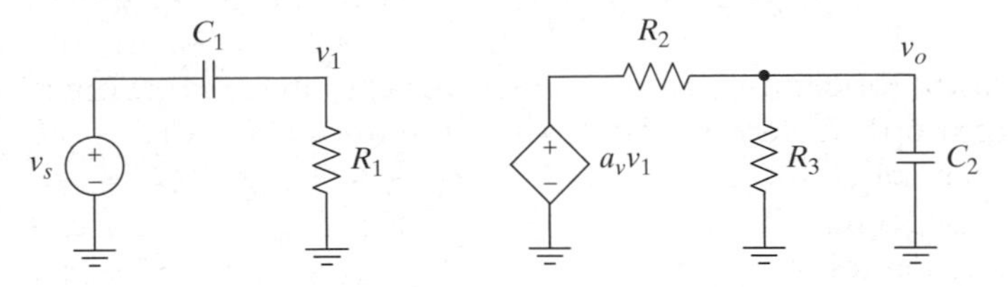

Figure 1–9 Circuit for Exercise 1.5.

The previous example dealt entirely with single-pole networks, but even when the circuits are more complicated, the process of design will be simplified if the analysis is done in a way that leads to a clear correspondence between the different terms in the equation and the elements in the circuit. For example, consider the circuit of Figure 1-10(a). This circuit can again be analyzed directly by writing the nodal equation at v_o:

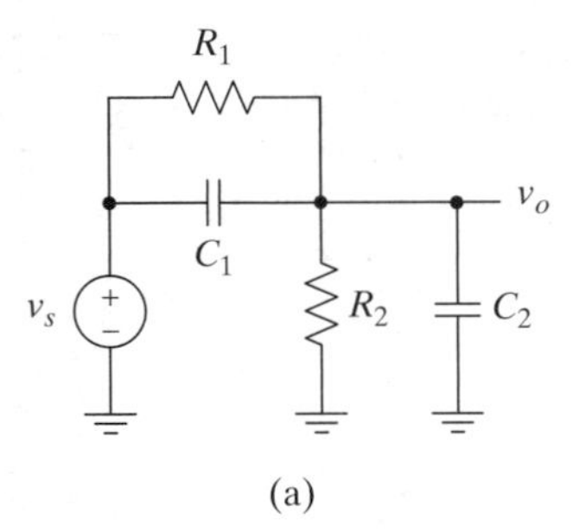

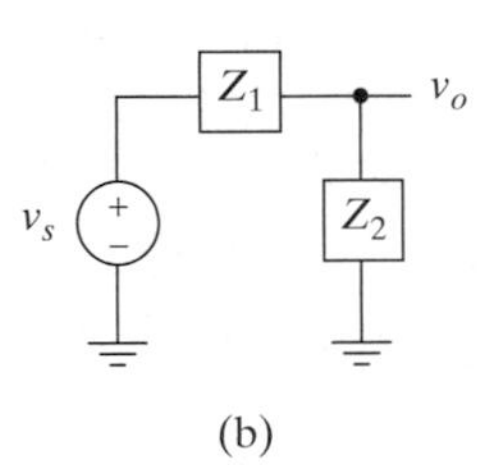

Figure 1–10 A more complicated circuit for demonstrating analysis for design.

$$V_o(j\omega)\left(\frac{1}{R_1} + \frac{1}{R_2} + j\omega(C_1 + C_2)\right) - V_s(j\omega)\left(\frac{1}{R_1} + j\omega C_1\right) = 0. \quad \textbf{(1.23)}$$

This leads to

$$\frac{V_o(j\omega)}{V_s(j\omega)} = \frac{R_2 + j\omega R_1 R_2 C_1}{R_2 + R_1 + j\omega R_1 R_2 (C_1 + C_2)}. \quad \textbf{(1.24)}$$

However, a much simpler result is obtained if the circuit is thought of as an impedance divider, as shown in Figure 1-10(b). In this case we can write by inspection that

$$\frac{V_o(j\omega)}{V_s(j\omega)} = \frac{Z_2(j\omega)}{Z_1(j\omega) + Z_2(j\omega)}. \quad \textbf{(1.25)}$$

A circuit designer would often stop at this point unless there was a particular reason to expand (1.25) further. Two advantages to this analysis technique are apparent at this point: First, it was much simpler to derive (1.25) than (1.24); second, the form of (1.25) is much simpler than (1.24), and it is much easier to establish a clear one-to-one correspondence between the terms in the equation and the components (or groups of components) in the original circuit.

Take this particular example one step further. The impedance of a resistor in parallel with a capacitor shows up frequently in circuit analysis and is given by

$$Z_{R\|C}(j\omega) = \frac{R}{1 + j\omega RC}. \quad \textbf{(1.26)}$$

Using this result in (1.25) leads to

$$\frac{V_o(j\omega)}{V_s(j\omega)} = \frac{\dfrac{R_2}{1 + j\omega R_2 C_2}}{\dfrac{R_1}{1 + j\omega R_1 C_1} + \dfrac{R_2}{1 + j\omega R_2 C_2}}. \quad \textbf{(1.27)}$$

Examination of (1.27) reveals that if we set $R_1C_1 = R_2C_2$, the transfer function reduces to a frequency-independent voltage divider:

$$\frac{V_o(j\omega)}{V_s(j\omega)} = \frac{R_2}{R_1 + R_2}. \tag{1.28}$$

This circuit has many uses. For example, if the impedance Z_2 represents the input impedance of an oscilloscope, Z_1 can represent a scope probe, and we have the basic structure for a compensated scope probe. By adjusting the value of C_1 so that $R_1C_1 = R_2C_2$ (the other component values are fixed), the probe provides a frequency-independent attenuation. The same idea can be used to extend the usable bandwidth of other circuits as well. It is very difficult to see from (1.24) that the transfer function can be made constant by setting $R_1C_1 = R_2C_2$; again, we see the power of doing analysis in a more intuitive way.

In this section, we have seen that analyzing circuits by writing nodal equations (or loop equations) is often not the best approach for design, even though it is sometimes necessary. The nodal equation approach is certainly systematic and will always yield a correct result, but it is often difficult to establish the one-to-one correspondence between terms in the equation and components in the circuit that is essential to the design process. In addition, the nodal equation approach is often more work than breaking the circuit up into a series of transfer functions and viewing these transfer functions as either single-pole circuits or impedance dividers. Of course, with more complicated circuits, we may need to use a combination of nodal (or loop) analysis and the methods illustrated here. The penalty we pay for using the techniques presented here in lieu of nodal or loop analysis is that the techniques presented here are less systematic; we cannot define a step-by-step procedure that will always produce the desired result. That is why computer programs still use nodal analysis! However, for the purpose of doing design, the advantages of our approach overwhelm this disadvantage.

1.3 ELECTRONIC SYSTEMS

Having presented a basic idea of what constitutes design and of how analysis for the sake of design differs from analysis for the sake of determining performance, we next discuss electronic systems. There are three reasons for this discussion: First, it will provide motivation for the material that follows; second, it will help to present an overall framework for the upcoming material so that it is easier to see how it fits into a broader picture of the field of electrical engineering; and finally, it will provide an opportunity to indicate clearly what it is about this material that makes it different from the material presented in the linear circuits and signals courses you have had before.

1.3.1 Electronic versus Electric Circuits

Strictly speaking, an electric circuit is a complete loop in which electrons can flow and that includes circuits containing transistors. Nevertheless, most people (but not all) would distinguish between an electric circuit and an electronic circuit—hence the title of this text. An electric circuit is a circuit that contains resistors, capacitors, inductors, transformers, switches and the like, but no transistors, integrated circuits, or tubes (yes, they *do* still get used once in a while). Ignoring tubes, and recognizing that integrated circuits contain transistors, resistors, capacitors, and, occasionally inductors, the real distinction between an electric circuit and an electronic circuit boils down to whether or not it contains transistors. So what is it about transistors that make them special? The short answer is that they are ***active*** devices, while resistors, capacitors, and similar components are ***passive*** devices. A passive device is a device that can neither provide energy nor direct the flow of energy. An active device can control the flow of energy or can supply energy.[6]

A mechanical analogy may be useful. If you want to lift a heavy box, you can use a lever, as shown in Figure 1-11(a), or you can use a forklift, as shown in part (b) of the figure. The lever is a passive device: You are directly supplying the required energy. The lever simply lets you exert a small force over a large distance to produce a larger force over a shorter distance. On the other hand, the forklift is an active device: You direct its motion, but it uses another source of power to accomplish the task.

An electric analog to the lever is a transformer, as shown in Figure 1-12(a): The transformer can boost the voltage supplied to a given load (analogous to force in Figure 1-11), but the current supplied to the load is proportionately lower than the current drawn from the source (analogous to distance in Figure 1-11). The electric analog of the forklift is the transistor circuit shown in Figure 1-12(b): It uses an input voltage to control an output voltage, but the input source does not need to supply the power to the output. The circuit includes the input source, the transistor, the load, and a separate supply of energy—a battery in this case. The battery is analogous to the motor in the forklift. The transistor is a device that has a current flowing from the output to ground that is controlled by the voltage appearing from the input to ground. The voltage across the load is equal to the current times the load resistance, and this voltage can be much larger than the source voltage.

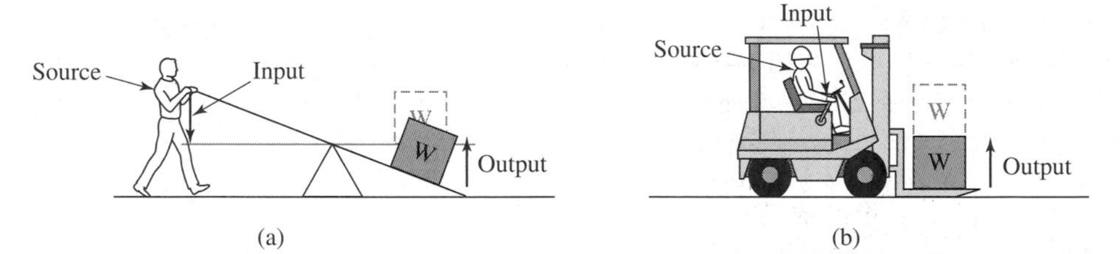

Figure 1–11 Mechanical analogy of (a) passive and (b) active devices.

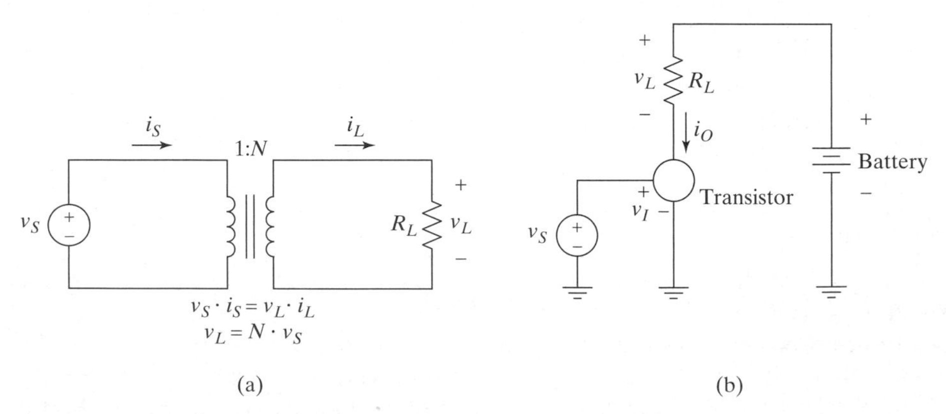

Figure 1–12 Electrical examples of (a) passive and (b) active voltage gain.

---

[6] Strictly speaking, active devices supply energy and passive devices absorb energy. Batteries and generators are examples of active devices. In the case of electronic circuits, however, we think of circuits that can increase the power of a signal as active circuits. The key function required is the ability to control the flow of energy.

Unlike the transformer, however, the output current from the transistor can also be much larger than the input current, which is ideally zero.

At this point an important difference between active and passive circuits can be pointed out. Since mechanical power is force times distance per unit time (integrated if the force is not constant), the power delivered to the load in Figure 1-11(a) is no more than the power obtained from the source (i.e., the person pushing on the lever). In reality, the output power is lower because of friction losses and other effects. But in Figure 1-11(b), the power delivered to the load can be significantly larger than that provided by the source (i.e., the driver), which is negligible in this case, since all the "source" has to do is push the lever. The power in part (b) of the figure is supplied by the motor in the forklift.

In the electrical case, power is equal to voltage times current. For the ideal transformer in Figure 1-12(a), the power out is exactly equal to the power in. However, the ideal transformer can only be approximated in practice. There will always be some loss so that the power out will actually be lower than the power in and the power gain of the system is strictly less than one. ALL **passive circuits have power gains less than one**.

For the transistor circuit shown in Figure 1-12(b), however, the signal power out can be much larger than the power delivered by the signal source, so the power gain for the signal can be greater than one. Energy is not produced in the circuit, of course; it simply uses the energy provided by the power supply (the battery in the figure) to drive the load. In other words, the transistor uses the input signal as a template to produce an output that looks like the input, but may have greater power. Because they control or direct the flow of energy from some power supply (e.g., a battery), **active circuits can have power gains greater than one**.

1.3.2 Analog and Digital Electronic Circuits

To gain a better understanding of electronic systems, let's briefly examine two common electronic systems: an FM receiver and a computer. Before we go into any detail, however, we need to understand a fundamental difference between these two systems: The FM receiver is an *analog* system (although modern receivers certainly contain digital components as well, the transmitted signal is analog), whereas the computer is a *digital* system (computers do contain analog circuits as well, most notably in the display, power supply, sound reproduction, and mass storage devices). The different kinds of signals we will encounter are further explained in Aside A1.7.

The input to an FM receiver is a *frequency-modulated* (FM) signal. That is, the frequency of a high-frequency *carrier* is changed in response to the information to be transmitted (the music in this case). There are two main reasons for not transmitting the audio signal itself: First, there wouldn't be any way to separate one station from another, and, second, the antennas would need to be huge. Antennas need to be a reasonable fraction of a wavelength to work well and, as an example, the wavelength of a 1 kHz wave propagating in air is about 186 miles.[7]

To have practical systems, therefore, we use high-frequency carriers, commonly called *radio-frequency* (RF) waves. To transmit the signal, we *modulate* (i.e., change) some parameter of the carrier wave (e.g., amplitude or frequency) with the signal. The receiver must then *demodulate* the carrier to recover the original signal.

In the case of a stereo receiver, the situation is complicated further by the desire to send both left- and right-channel signals and to have old-fashioned monaural receivers still work (not important now, but it was when FM was first introduced).

---

[7] Very low frequency electromagnetic waves are used to communicate with submarines because the attenuation increases with increasing frequency in salt water. Nevertheless, even though the wavelengths are shorter in water, the antennas are extremely long.

ASIDE A1.7 SIGNAL CLASSIFICATION

Consider the signals shown in Figure A1-10. The signal shown in part (a) of the figure is a ***continuous-time analog signal***; that is, it is a continuous function of time and can have any amplitude. If we sample this signal at certain times as shown in part (b) of the figure, the resulting signal is a ***discrete-time analog signal***. If we then go one step further and ***quantize*** (i.e., round off to one of a discrete set of levels) the amplitude as shown in part (c) of the figure, the resulting signal is a ***digital signal***. The discrete sample values shown in part (c) of the figure are each held until the new sample arrives. If the different levels are represented by binary numbers, another way of representing the digital signal would be to use a serial stream of binary digits with one three-bit binary word per sample time, as shown in part (d) of the figure; this is a ***binary*** digital signal, commonly called just a binary signal. It may be necessary to use a different number of bits in each binary word, of course; three bits is just an example.

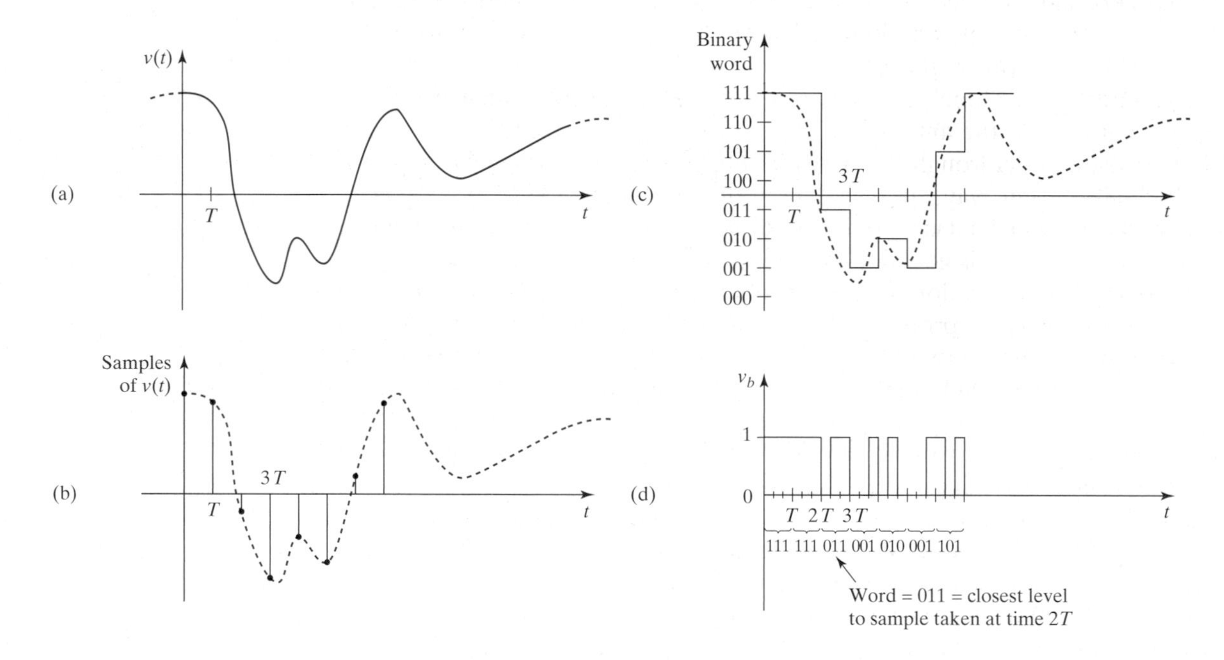

Figure A1–10 Classification of signals: (a) continuous-time analog, (b) discrete-time analog, (c) digital, and (d) binary digital.

All four of the signal types shown in Figure A1-10 get used in modern electronic systems. For example, an FM receiver uses continuous-time analog signals, modems often use discrete-time analog signal processing, and compact disc (CD) players store the music as binary words, convert those words to digital signals as in part (c) of the figure, and then smooth the resulting output to get a continuous-time signal prior to sending it to an amplifier. Often, continuous-time analog signals are just called analog signals if there is no possibility of confusion. Similarly, binary digital signals are often just referred to as digital signals, especially if digital logic is being discussed. In addition to binary digital systems, multivalued logic does exist, but it is uncommon enough that people usually explicitly say if something is multivalued.

Therefore, if you simply demodulate an FM signal, you get the left- and the right-channel signals combined, denoted $(L + R)$; this is what a monaural receiver uses. In addition, there is a separate subcarrier modulating the RF wave that contains the difference between the left- and right-channel signals $(L - R)$. This subcarrier is ignored by monaural receivers but is demodulated by stereo receivers. By adding and subtracting the two demodulated signals, the original stereo signal can be recovered; that is, $(L + R) + (L - R) = 2L$ and $(L + R) - (L - R) = 2R$. We will ignore this complication and show only a monaural receiver in the discussion that follows.

A block diagram of a monaural FM receiver is shown in Figure 1-13. This block diagram shows a ***superheterodyne*** receiver, which simply means that the RF wave is first brought down to an ***intermediate frequency*** (IF), goes through some signal processing there, and is then demodulated or converted to the audio signal. Heterodyning is the process of combining two signals to produce a beat tone and is accomplished by multiplying the signals.

Returning now to the block diagram, we note that the receiver first filters the RF signal coming from the antenna to choose the one station you want to listen to. The diagonal arrow through the RF filter block is a common way of showing that the block is in some way variable; in this case, you can tune the filter to select different stations. This signal is amplified and sent to a ***mixer***, which multiplies the signal by a sine wave that is generated locally by the ***local oscillator*** (LO).

To understand the operation of the mixer, consider a single sinusoidal input. Multiplying two sine waves produces sine waves at the sum of the two frequencies and at the difference of the two frequencies.[8] If the two frequencies are close, the difference will be at a much lower frequency where it is less expensive to perform further signal processing. It is also possible to use an LO frequency equal to the nominal RF frequency, in which case the resulting receiver is called a ***direct-conversion***, or ***homodyne***, receiver.

In the case of broadcast-band FM, the IF is at 10.7 MHz, so to receive a station at 100.1 MHz, the LO is set to 110.8 MHz. The output of the mixer, which is the intermediate frequency, is then sent to the IF filter and amplifier. Most of the filtering is done in the IF section, since it is less expensive to implement good filters at that frequency. These filters often use crystals, as will be discussed in Chapter 11. The RF filter is only a preselect filter. The LO frequency is changed to mix the desired signal frequency down to the IF frequency. The IF amplifier is often an ***automatic gain control*** (AGC) amplifier that changes its gain so that the output amplitude is held constant. The output of the IF section then goes into a demodulator that converts the frequency changes into voltage changes to produce the audio output.

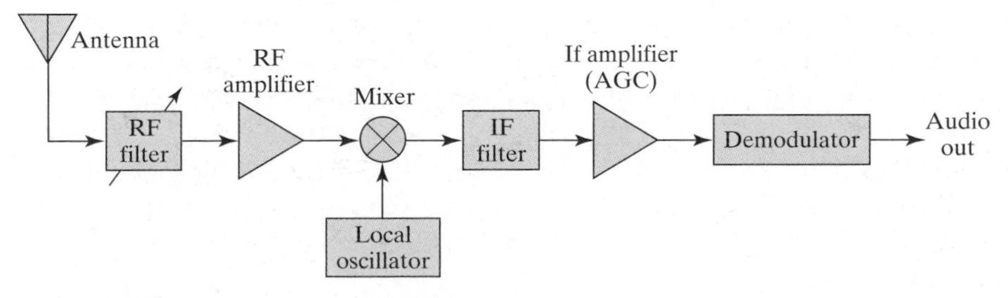

Figure 1–13 Block diagram of a monaural receiver.

[8] Remember that $\cos A \cos B = (\cos(A + B) + \cos(A - B))/2$.

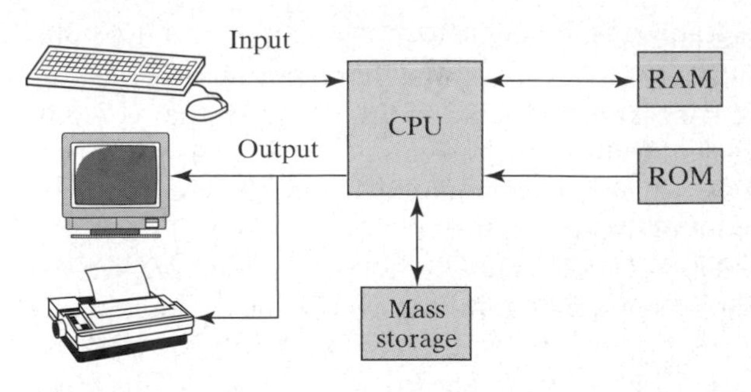

Figure 1–14 Block diagram of a basic computer.

A block diagram for a basic computer is shown in Figure 1-14. The main element in the computer is the ***central processing unit*** (CPU). When the computer is turned on, the CPU initially reads some basic instructions stored in the nonvolatile ***read-only memory*** (ROM). Using these basic instructions, and perhaps automatically loading additional information from nonvolatile ***mass storage*** (e.g., disk drive), the computer is ready to run. Information that must be accessed quickly is stored in ***random-access memory*** (RAM), and information that is not needed immediately is stored on some slower mass storage device, most commonly a disk drive. Input from the user is accepted from a keyboard, a mouse, and perhaps other devices; output can be directed to the screen, a printer, or other peripheral devices.

It is beyond the scope of the present discussion to go into more detail. For example, there are many ways of organizing the CPU, and the computer can include many more devices than have been shown here and can be constructed using many different circuit technologies. For our purposes, it will suffice at this point to say that the circuits in the computer are mostly digital, and the information is manipulated in binary words composed of sequences of binary digits, as shown in Figure A1-10(d).

The Process Of Design Revisited We are now ready to revisit the topic of design with reference to the two specific example systems described above. The process of design proceeds at many different levels of abstraction as noted in Section 1.1.

In the case of a communication system (i.e., transmitter and receiver), one must first decide what form of modulation to use—for example, ***amplitude modulation*** (AM) or FM. The characteristics of the modulation must also be determined. For example, how far will we allow the frequency of our carrier to be changed in the FM case? These decisions were made long ago for standard broadcast systems, but even if they are already determined, many choices remain for the designer of the receiver itself. To start with, we have shown only one type of receiver; it is possible to use different approaches (e.g., the direct conversion receiver mentioned briefly before). Assuming our block diagram, however, there are still plenty of choices to make. How much filtering should be done in the RF section? How large must the AGC control range be? Do we even want to use an AGC amplifier? (The performance will be degraded with simpler schemes like amplitude limiting, but they are cheaper.) Do we use a local oscillator that is analog and tuned by varying some component value, or do we use a single-frequency oscillator and a programmable digital divider circuit to make a "digital" FM receiver? How many stages of IF filtering do we use and what type? What kind of demodulator do we use? We see there are many options at the block diagram level alone.

Similar choices exist for the computer. How large a digital word should be used (e.g., 16-bit, 32-bit) for data? How large a digital word should be used for memory addressing? What kind of CPU architecture should be used? Should we use parallel processors? Should there be a local cache memory? There are again many more choices that could be listed, even at the block diagram level.

All of these choices involve trade-offs between cost and performance, or one performance specification over another. Some of these trade-offs are quite complex, and many depend on the details of how each block is implemented. There are similar lists of choices and trade-offs to be evaluated in the design of each individual block as well; there are, for example, several good ways to make an AGC amplifier. That is why the design process bounces back and forth between different levels of abstraction, from system-level block diagrams to the implementation of the individual blocks and back again. These individual blocks are themselves often made of many identifiable blocks, and the same kind of discussion applies to them individually.

It is important to realize that *optimum* choices are very rarely identifiable. Designers should do the best they can to consider the alternatives and make an informed choice, but sometimes the choice boils down to an arbitrary one. The important point is to make sure that where clear-cut differences exist, we understand them and make informed choices. Any arbitrary choices that are made should not have an impact on the overall system performance or cost. (If they do, they really can't be considered arbitrary.)

Note that neither of the systems discussed is entirely analog or entirely digital. Therefore, it is important for a circuit designer to know something about both kinds of design, even though most designers will specialize in one or the other. It is also important for the circuit designer to understand some of the system issues involved in the design. Both of these systems are too complicated for any one person to know everything from the system level down through the transistor-level circuit design and on to the solid-state device design and fabrication. Nevertheless, for the overall design team to be able to go through the kind of process described, each member must know something about the disciplines that are adjacent to their own. For the circuit designer, this means understanding the system design to some degree, as well as understanding solid-state device physics and fabrication to some degree. For the communication system engineer, it means understanding both the mathematics underlying the system analysis techniques and the transistor-level circuits to some degree. For the engineer involved with solid-state devices, it means understanding circuits as well as the physics and fabrication.

Exercise 1.6

In the late 1800s, there was a debate about whether it is better to use AC or DC power for homes and businesses. Consider this problem and list some arguments either in favor of or against each method.[9]

□

1.3.3 Modeling Electronic Systems

In the preceding sections, the idea of modeling a system at many different levels of abstraction was discussed a great deal. As an example, a communication theorist might draw a single block labeled "receiver" while considering different models for the communication channel and how to choose the best possible modulation scheme for a given application. The receiver block could then be further broken down as shown in Figure 1-13. The AGC amplifier could then be shown as a block diagram, using a variable-gain amplifier and some other blocks to indicate how

[9] Thomas Edison and the Edison Electric Light Company were in favor of DC distribution and had the largest systems in place at the time, while George Westinghouse and his Westinghouse Electric and Manufacturing Company pushed for the AC system. Obviously, Westinghouse eventually won.

the gain is varied. The variable-gain amplifier can then be shown as a block diagram, involving some variable element and some amplifying stages. The transistor-level schematics can then be drawn for these stages. Of course, we can then draw more detailed circuit models for the transistors, or we could represent them by drawings of the physical structures, and so on. This process can be compared to peeling an onion: We take off one layer at a time. These different levels of drawing a system are also present in the mathematics used to represent a system; we can represent the receiver with a single equation, or we can go all the way down to the equations describing the motions of the electrons in the transistors.

One of the most difficult aspects of design is choosing the appropriate level of abstraction with which to examine a given problem. If the communication theorist thinks about the motions of the electrons in one transistor in the middle of the AGC amplifier, he or she won't get very far in figuring out what kind of modulation scheme to use! Unfortunately, all designers fall into less obvious versions of this same trap far too often. We must be able to bounce back and forth between the different levels of abstraction as we work on the design of any system, and we must be aware of the trap of staying stuck at the wrong level.

A similar problem exists in deciding which representation to use at any given level of abstraction. For example, signals can be represented in the time domain or the frequency domain, and circuits can be represented by schematics or by equations. No one representation is always the best.

A common problem encountered in circuit design is what level of model to use for a given device. To analyze circuits containing transistors, we must have a model for how the transistor works. The problem is that there are many models for the transistor. Using the most complete model available at all times is not appropriate, because it makes the analysis so difficult that it hinders the process of design. Therefore, **we choose the simplest model that will retain the necessary accuracy for the task at hand**. We will return to this subject many times throughout this text, but you should be aware at the outset that there isn't just one model for any component. You must always ask yourself, When is it appropriate to use this model?

As an example of models for electronic components, consider the amplifier circuit shown in Figure 1-12(b), which is repeated in Figure 1-15(a). In the discussion regarding that circuit, we simply said that the transistor "has a current flowing from output to ground that is controlled by the voltage appearing from the input to ground." We can come up with a simple model for the transistor based on this statement; it is a voltage-controlled current source, as discussed in Aside A1.2. If we assume that the output is a linear function of the input and call the constant of proportionality G, we arrive at the model shown in the schematic in Figure 1-15(b).

Using this model, we can analyze the circuit quantitatively, whereas before we examined it only qualitatively. From Figure 1-15(b), we see that

$$v_L = Gv_I R_L = Gv_S R_L, \tag{1.29}$$

so that the gain from v_S to v_L is

$$\frac{v_L}{v_S} = GR_L. \tag{1.30}$$

We have used the simplest possible model for the transistor in this analysis, and it shows the basic operation of the device as a voltage-controlled current source. Of course, the model is overly simplistic. For example, the output current of a real transistor is not a linear function of the input voltage, and the input impedance is not infinite. Nevertheless, this simple model allows us to see that the circuit is an amplifier, and it gives us some indication of the gain. We will use more sophisticated models of the device to gain a more complete understanding later.

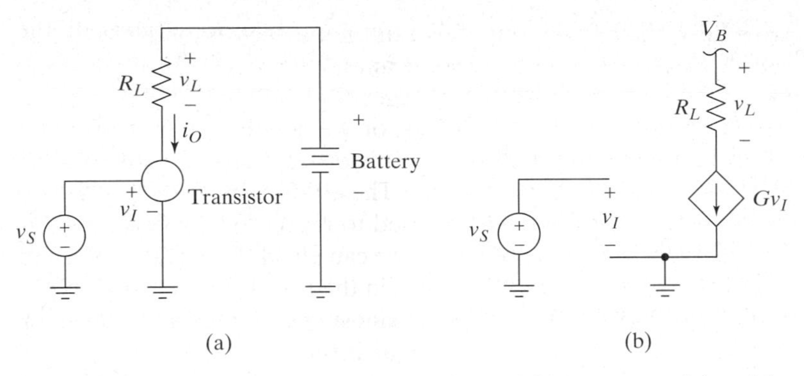

Figure 1–15 (a) A simple transistor amplifier and (b) a circuit model used for the transistor.

Virtually everything we do in circuit design is based on models. There are no pure resistors, capacitors, inductors, batteries, or other components available to us. Every real resistor has some inductance and capacitance. Every real capacitor has some inductance and resistance, and so on for every other component we use. In many cases, we can satisfactorily model elements as being just a resistor or capacitor, but we must always remember that there will be times when those models fail to accurately predict the system's performance (e.g., at very high frequencies). Sometimes more than one equivalent model is available. In such cases, we choose either the model that makes the most sense in terms of how we think about the operation of the element or the one that yields the simplest analysis.

Consider a complicated amplifier like the LM1558 shown in Figure 1-16. We certainly don't know how to analyze a circuit like this yet, but we can discuss the different levels at which it might be useful to model the circuit.

The schematic shown in Figure 1-16, although certainly complex, is itself only a model for the real device and does not tell us everything we need to know to build

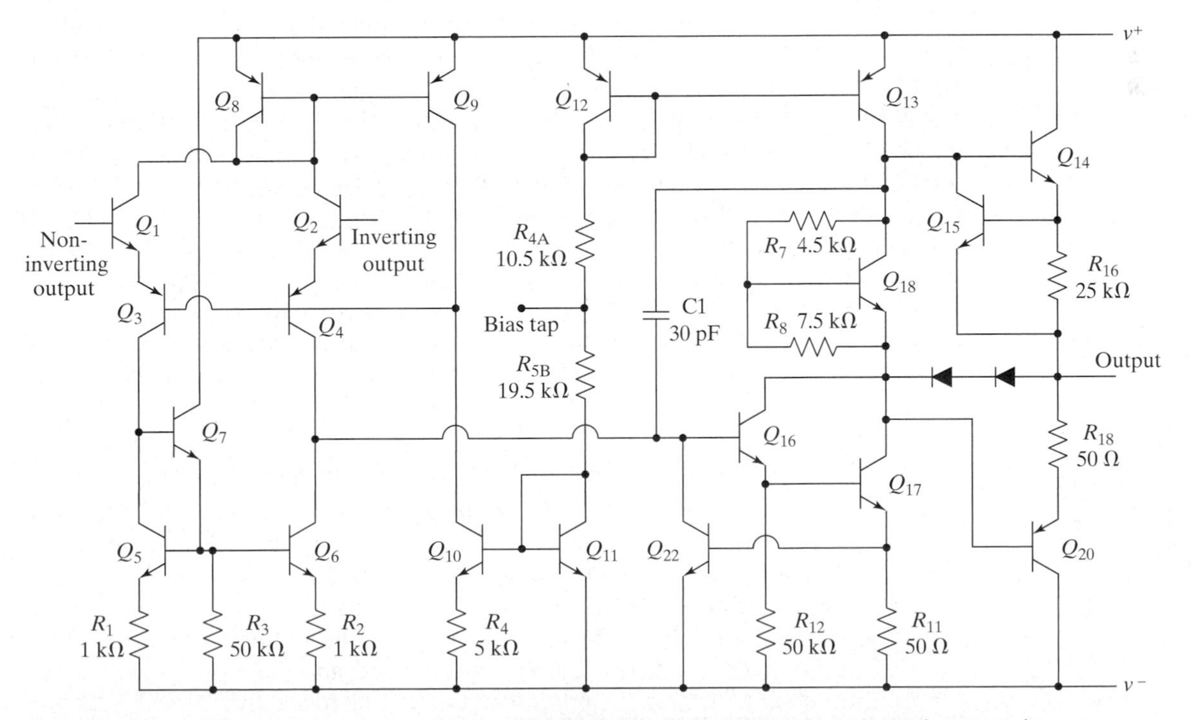

Figure 1–16 The schematic of one op amp in an LM1558. (The part contains four identical op amps.) (This schematic was adapted from an image from National Semiconductor.)

the circuit. For example, we need to know the exact process used to fabricate the circuit and the exact layout of all of the devices. Nevertheless, the figure does show the lowest level of detail that we will typically deal with in this text.

When an LM1558 is used in the design of some larger circuit, the designer does not usually look at the detailed schematic. The highest level of abstraction for dealing with the overall circuit is to draw an operational amplifier symbol as shown in Figure 1-17(a). This symbol allows us to get away from all of the details embodied in the schematic of Figure 1-16 and focus on the global performance of the op amp. Of course, there are equations that go along with the symbol shown. For example, we must know that the output voltage is equal to the gain times the difference in the input voltages:

$$v_O = a(v_P - v_N) = av_{ID}. \quad \textbf{(1.31)}$$

There are many equations we could show that describe other aspects of the amplifier's performance, but for some applications this equation is sufficient. The symbol shown in Figure 1-17(a) is not a complete representation of the op amp; in fact, it doesn't even show the power supply connections. However, it is a very useful model.

A lower level model for the circuit is shown in part (b) of the figure and includes the differential input resistance R_{id}, a controlled source to model the gain, and the output resistance R_o. This model is still incomplete, but it provides more detail than the schematic shown in part (a) and can be used to perform more detailed analyses.

There are many more models that could be shown for the op amp. Each one is useful for some analyses and not for others. The job of the circuit designer is to choose the appropriate model for the job at hand. An important question to ask, whether you are a student or an experienced designer, is whether the model you are using is appropriate for the analysis you are doing. If the model is too complicated, your analysis may take a long time, yield results that are difficult or impossible to interpret, or be impossible. If the model is too simple, your results are very likely to be wrong.

Figure 1-18 demonstrates three different levels of abstraction for an amplifier design. In part (a) of the figure, we show a block diagram of some source driving an amplifier, which then drives a load. In part (b) of the figure, we use a Thévenin model for the source, a high-level circuit model for the amplifier, and a resistive model for the load. In part (c) of the figure, we expand the amplifier one step further to show the transistors and individual resistors used to construct the device. Much of our effort in this text will focus on taking a schematic like the one in part (c) of the figure and finding the model elements for the model shown in part (b) of the figure. The advantage of modeling the circuit at a higher level of abstraction

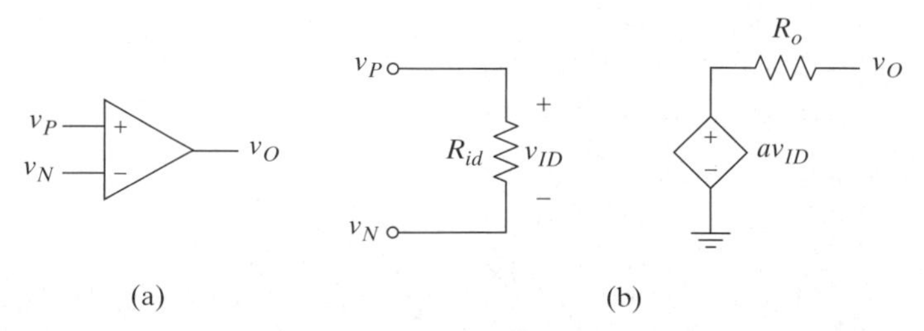

Figure 1–17 (a) The highest level of schematic abstraction for the op amp in Figure 1-16. (b) A lower level of abstraction for the op amp.

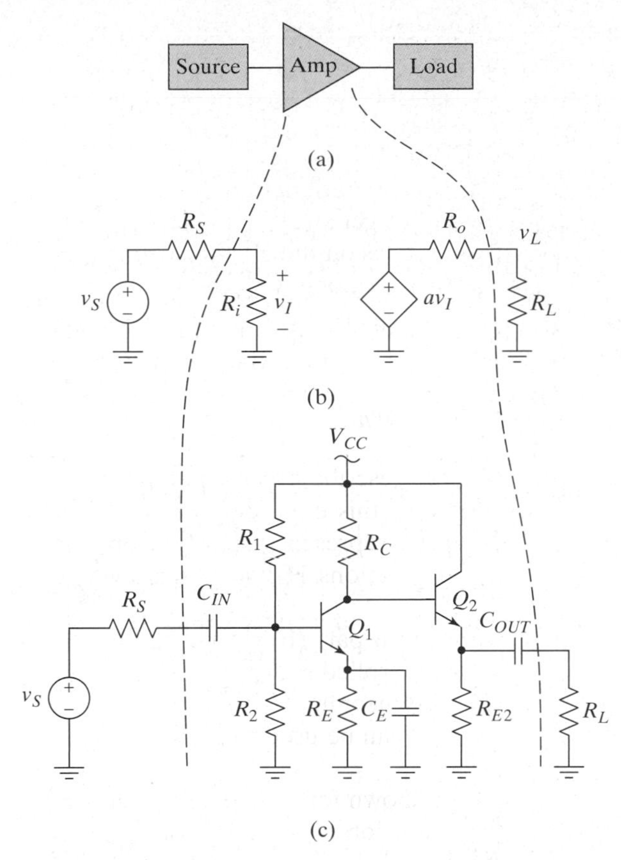

Figure 1–18 (a) Block diagram view of an amplifier, (b) a high-level circuit model for the amplifier, and (c) a transistor-level schematic for the amplifier.

is that the analyses are simpler and our understanding of the system is greater. Even when we don't draw the circuit shown in part (b) of the figure, we have it in mind. For example, we know what the output resistance of our amplifier needs to be if we want to be able to drive a given load without having the output attenuation factor be too small. We can also use more complex models for the source and load if that is necessary.

Exercise 1.7

Suppose the amplifier shown in Figure 1-19 is driven by the temperature sensor of Exercise 1.2. Would a Thévenin or a Norton source model be most convenient? Why?

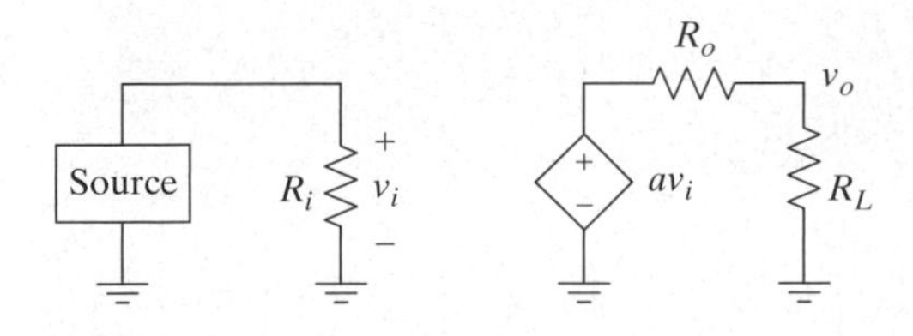

Figure 1–19 A temperature sensor driving an amplifier.

1.3.4 Discrete, Integrated, and Hybrid Circuits

A *discrete* circuit is constructed using separate components, usually on a *printed circuit board* (PCB). For example, a circuit constructed using individual resistors, capacitors, transistors, integrated circuits, and so on, is considered a discrete circuit. Figure 1-20 shows a picture of one such circuit.

An *integrated circuit* (IC) is fabricated on a single substrate. The substrate is most commonly silicon and the fabrication methods are discussed in Chapter 3. Figure 1-21 shows a microphotograph of one such circuit. In this case, all of the components, which are usually restricted to being transistors, resistors, and capacitors, are fabricated at the same time and on the same silicon wafer. As a result, the cost of the circuit is determined by the cost of processing the wafer and the number of circuits on each wafer. In other words, the cost is a function of the area consumed. Consequently, the relative cost is determined by the area used for individual components. This fact is in stark contrast to the discrete circuit, where the cost of a component is a function of the manufacturing process used, and costs vary widely from one component to the next.

There are many differences in how discrete and integrated circuits are designed. For example, in discrete circuits, resistors tend to be inexpensive and are used extensively, whereas in ICs, resistors are expensive because they consume a lot of area. Therefore, circuit techniques have been developed that use fewer resistors. Also, components that are close to each other on an IC have almost identical characteristics and temperatures. Integrated circuits are designed to take advantage of this good matching and thermal tracking, which are not present in discrete components. On the other hand, it is easy to get discrete resistors that are accurate to within ±1%, but such an absolute tolerance is unattainable on an IC without expensive trimming steps being added to the process. These differences imply that the design of ICs is quite different from the design of discrete circuits. Nevertheless, there is a great deal of common material as well. Some of the differences will be pointed out in more detail later in this text.

There are also *hybrid* circuits—that is, circuits that are not truly discrete and not fully integrated. Usually these circuits comprise various ICs and discrete components mounted on a ceramic substrate, as shown in Figure 1-22.

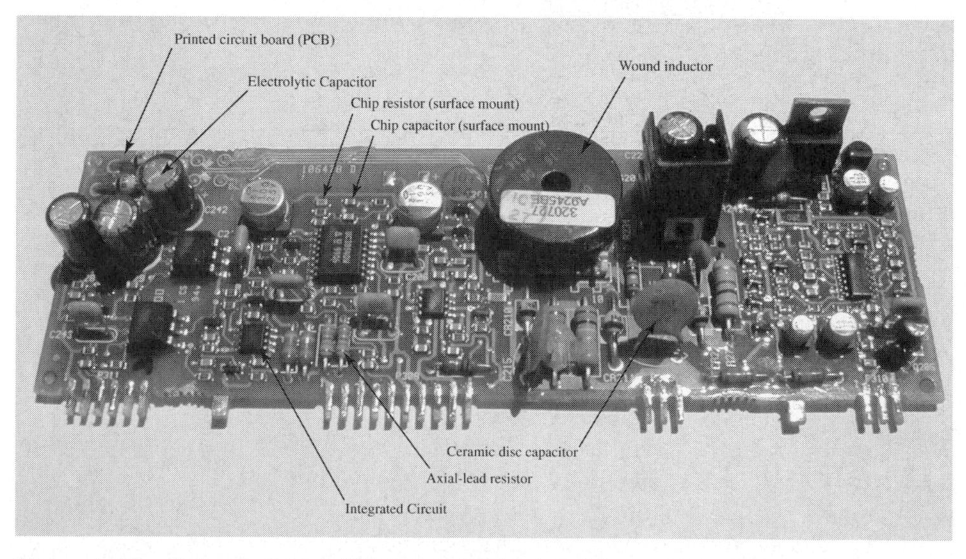

Figure 1–20 A discrete circuit. This is a controller board from a Hewlett-Packard power supply. (Photo by Ron Guly, courtesy of Agilent.)

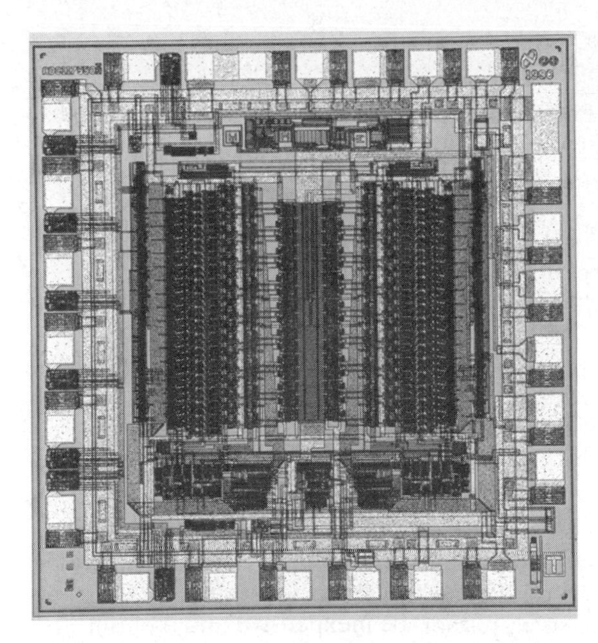

Figure 1–21 An integrated circuit. This particular circuit is an 8-bit 50-MSample/second analog-to-digital converter (ADC1175). The circuit has a minimum feature size of 0.65 μm. (Photo courtesy of National Semiconductor.)

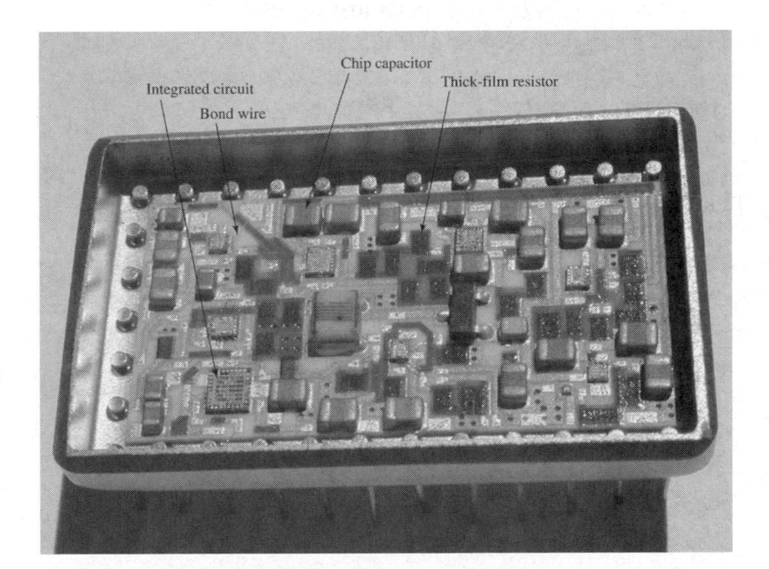

Figure 1–22 A hybrid circuit. (Photo by Ron Guly.)

1.4 NOTATION

The notation we will use throughout this text is a *de facto* standard used in most journals and books on the subject in the United States. Suppose, for example, we want to describe the input voltage of some circuit. The total instantaneous value of the voltage at time t is denoted by $v_I(t)$; the variable name is lowercase and the subscript is uppercase. We will often drop the explicit reference to time when no confusion will result; therefore, we will often write just v_I. This voltage can be described as the sum of its average and time-varying components; that is, $v_I(t) = V_I + v_i(t)$, where V_I is the average, or DC, value and $v_i(t)$ is the time-varying, or AC, value. If v_I

Table 1.1 Summary of Notation.

Notation	Explanation
$v_I(t)$	Total instantaneous quantities are represented by lowercase variables with uppercase subscripts. (The explicit function of time is often dropped.)
V_I	Average, or DC, quantities are represented by uppercase variables with uppercase subscripts.
$v_i(t)$	Time varying, or AC, quantities are represented by lowercase variables with lowercase subscripts. (The explicit function of time is often dropped.)
V_i, $v_I(t) = V_i \sin\omega t$	Sine wave peak values, phasors, and Fourier and Laplace transforms are represented by uppercase variables with lowercase subscripts.

is a sinusoidal function of time, we can represent it by a phasor (i.e., the magnitude and phase angle of the sine wave). We denote the phasor quantity and the peak value of the sine wave by V_i. The same phasor notation is used for Fourier and Laplace transforms. The notation is summarized in Table 1.1.

In addition to the standards just summarized, we adopt a number of other conventions: When a voltage is given with a single subscript (e.g., V_A), it is assumed that the voltage is specified with respect to ground (i.e., $V_A = V_A - 0$). If two subscripts are used, the voltage is the difference; for example, $V_{AB} = V_A - V_B$. When a single subscript is repeated twice (e.g., V_{CC}), it indicates a DC power supply. Other notational conventions will be introduced as needed in the book.

SOLUTIONS TO EXERCISES

1.1 Either way the answer is $v_O/v_S = 1/8$. The repeated application of the voltage divider, starting from the right and moving back to the left, is easier and lends more insight. To apply the voltage divider, first find

$$\frac{v_O}{v_B} = \frac{R_6}{R_5 + R_6} = \frac{1}{2},$$

then find

$$\frac{v_B}{v_A} = \frac{(R_5 + R_6)\|R_4}{(R_5 + R_6)\|R_4 + R_3} = \frac{1}{2},$$

and similarly get $v_A/v_S = 1/2$. Finally,

$$\frac{v_O}{v_S} = \frac{v_O}{v_B}\frac{v_B}{v_A}\frac{v_A}{v_S} = \left(\frac{1}{2}\right)^3 = \frac{1}{8}.$$

1.2 The Thévenin equivalent circuit is shown driving the 1-kΩ load in Figure 1–23. The output voltage is given by

$$V_O = \frac{R_L}{R_{Th} + R_L} V_{Th},$$

where the Thévenin voltage is equal to the open-circuit output voltage given in the exercise (1 mV/°C). Since the output voltage with the load present is 0.2 mV/°C, we know that

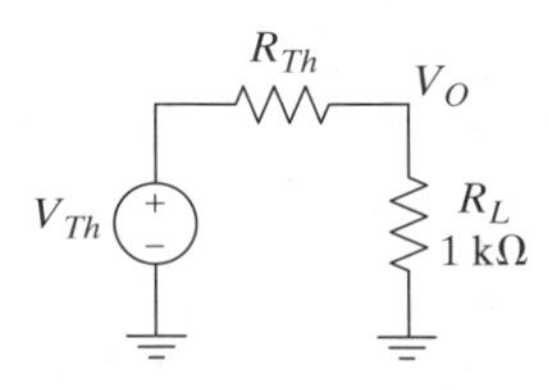

Figure 1–23 Thévenin equivalent driving the load.

$$0.2 = \frac{R_L}{R_{Th} + R_L},$$

which leads to $R_{Th} = 4\ \text{k}\Omega$. The corresponding Norton equivalent will have a short-circuit current given by

$$I_N = \frac{V_{Th}}{R_{Th}} = 250\ \text{nA/°C}.$$

1.3 We first find the open-circuit output voltage:

$$V_{Th} = I_s R_2 = 5\ \text{V}.$$

The Thévenin resistance is found by setting the independent current source to zero. The result is $R_{Th} = R_2 = 500\ \Omega$. The short-circuit current is then given by

$$I_N = \frac{V_{Th}}{R_{Th}} = 10\ \text{mA},$$

which could also be found by direct analysis. Note that R_1 does not affect the operation of this circuit in *any way*. Resistance in series with a perfect current source has no effect.

1.4 **(a)** 10 Ω and 0.225 W; **(b)** 0.0744 W and 33%; **(c)** 0.5 for case (a) and 0.91 for case (b); and **(d)** $R_i = 100\ \Omega$ would typically be a better choice because of the larger input attenuation factor (even though P_i is about three times smaller than when $R_i = 10\ \Omega$!).

1.5 We get

$$\frac{V_o(j\omega)}{V_s(j\omega)} = \frac{V_o(j\omega)}{V_1(j\omega)}\frac{V_1(j\omega)}{V_s(j\omega)},$$

where

$$\frac{V_o(j\omega)}{V_1(j\omega)} = \frac{a_v H_o(j0)}{1 + j\omega/\omega_{po}}$$

with

$$H_o(j0) = \frac{R_3}{R_2 + R_3} \text{ and } \omega_{po} = \frac{1}{(R_2 \| R_3)C_2},$$

and

$$\frac{V_1(j\omega)}{V_s(j\omega)} = \frac{H_1(j\infty)(j\omega + \omega_{z1})}{j\omega + \omega_{p1}}$$

with

$$H_1(j\infty) = 1, \omega_{z1} = 0, \text{ and } \omega_{p1} = \frac{1}{R_1 C_1}.$$

1.6 First we recognize that it is desirable to keep the voltage supplied to the end user relatively low and nearly constant. If it is too high or varies too much, it will be dangerous, and electric appliances will be difficult to design and expensive to build. Then we note that the power delivered to the end user is proportional to the current if the voltage is constant (since $P = IV$). Therefore, the power wasted in delivering the current to the end user over a transmission line of resistance R is $P_{wasted} = I^2R$. This loss can be reduced either by reducing the resistance or by distributing the power using higher voltages so that a smaller current is used. Reducing the resistance means using bigger transmission lines or lower resistivity material for the lines. Either of these solutions is expensive to implement. That leaves the option of distributing the power at higher voltages and then dropping the voltage down before providing it to the end user.

AC systems can use transformers to change the voltage with very low power loss. Therefore, AC systems can use very high voltages to transmit the power over long distances and then use transformers to bring the voltage down to a more reasonable level for the end users. Only recently have practical techniques been developed to convert a large DC voltage into a lower DC voltage without wasting power, and even now, these systems are more complicated and expensive than the transformers used in AC distribution systems. Therefore, DC systems still cannot distribute the power at a voltage much higher than that provided to the end user without adding extra cost and, at the time the original discussion took place on this topic, high-voltage DC distribution was simply not possible in any practical way. There are other issues as well, but this is the most important one. Interested readers should consult [1.2].

1.7 Since the input voltage is the variable of interest to the amplifier, it is most natural to use a Thévenin model for the source.

CHAPTER SUMMARY

- The process of design begins with a clear problem statement, followed by an iterative process. At each iteration, four steps are usually followed: (1) Enumerate goals, (2) list possible solutions, (3) examine possible solutions, and (4) work on the most promising approaches.
- As a result of the process listed above, most of the design effort involves analysis.
- Synthesis is the process of creating something new. Your ability to synthesize new circuits is developed by experience in *properly* analyzing many circuits.
- Analysis for design is very different from analysis performed for the sake of the final answer. Analysis for design focuses on obtaining results that can be readily understood, have a clear connection to the original circuit, and lend insight into the circuit's operation. It is acceptable if these results are only approximately correct.
- We are often interested in knowing how much a signal is attenuated as it enters or exits an amplifier or other circuit. We define the input and output attenuation factors to help with intuitive analysis.
- Single-pole transfer functions show up often, and we have standard forms for their transfer functions. The pole can always be found by evaluating the driving point resistance seen by the reactive element.

- Electronic circuits contain *active* elements. An active element can control the flow of energy or provide energy.
- All passive circuits have power gains less than one. Active elements are necessary to achieve power gain.
- Signals are classified as either continuous- or discrete-time analog, or digital.
- We must choose the level of modeling appropriate for each situation. In general, we choose the simplest model that will retain the necessary accuracy for the task at hand.
- Circuits are classified as either discrete, integrated, or hybrid.

REFERENCES

[1.1] J.L. Adams, *Conceptual Blockbusting: A Guide to Better Ideas*, 3rd ed. Reading, MA. Addison-Wesley, 1986.

[1.2] J.D. Ryder and D.G. Fink, *Engineers & Electrons*, New York: IEEE Press, 1984.

PROBLEMS

1.1 The Process of Design

D P1.1 Describe two different ways to make a dimmer switch (i.e., a control that allows you to continuously vary the intensity of a light from off to full brightness) for a 110-VAC light operating from a US standard 60-Hz sinusoidal source. You may use blocks that describe functions you don't know how to implement; just describe the required function of each block. You may use a combination of verbal descriptions, block diagrams, and schematics. What are the advantages and disadvantages of each method you propose?

D P1.2 Suppose you are to design a new porch light that has three modes of operation. In the first mode the light is off. In the second mode, it is on continuously, and in the third mode, the light turns on and off, based on its own enclosed motion detector (such lights are readily available). You desire to have this light work with existing house wiring. Describe two different ways that you could make the light work with an existing single-pole single-throw switch as shown in Figure 1-24. Is one of your methods better than the other? Why?

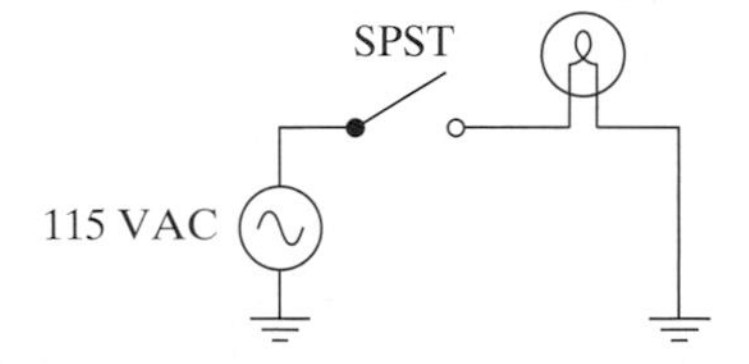

Figure 1–24 Standard wiring for a porch light showing the existing SPST switch.

D P1.3 Show how to make a single 2-wire connection work for a telephone. You must come up with the block diagrams for the phone and the switching office as well as the protocol used (i.e., what is done to the line to signal an incoming call?). The system you devise must support the following functions: (1) it must cause the phone to ring when there is an incoming call; (2) it must provide a dial tone when the phone is picked up; (3) it must allow the phone to send a sequence of dial tones to the switching office; (4) the wires must provide DC power to the phone at all times.

D P1.4 Suppose you want to connect to a remote temperature sensor using only two wires. The sensor produces a pulse output with a period of 1 ms and a pulse width that varies with the temperature. The wires need to supply the sensor with DC power and also need to carry the sensor output signal. How can you accomplish this?

D P1.5 We wish to design an integrated circuit that is capable of performing DC-to-DC conversion without using any inductors or transformers. That is, we want to be able to use a single DC voltage input and produce a DC voltage out that is either bigger than the input or of the opposite polarity. Draw a block diagram showing at least one way to accomplish this.

1.2 Analysis for Design

1.2.1 Frequency Independent Analysis for Design

P1.6 A potentiometer (called a "pot" for short) has three terminals: one terminal at each end of a fixed resistor and one terminal for a wiper that slides along the length of the resistor. In a rotary pot, the wiper is moved by rotating a shaft. Usually there is a control knob on the shaft for easy manual operation (e.g., the volume control on a radio).

The circuit of Figure 1-25 shows a pot used as a variable voltage divider. The total resistance of the pot is denoted R_P. The voltage V_O can be adjusted from 0 V to 10 V, depending on the position of the wiper. Assume that the control knob of the pot points to a linear scale running from 0 V to 10 V. It is desired that the user be able to set the value of V_O by setting the pot according to the scale; V_O need not be measured. The accuracy of V_O will, of course, depend on the linearity of the pot. However, even if a perfectly linear pot is assumed, an additional error is introduced when the output of the circuit is loaded. This homework problem will investigate the loading error.

The pot can be modeled as two separate resistors, as shown in Figure 1-26. The parameter x represents the position of the wiper as a fraction of its total travel. When the wiper is contacting the grounded end of the pot, $x = 0$, and when the wiper is at the top of the pot, $x = 1$.

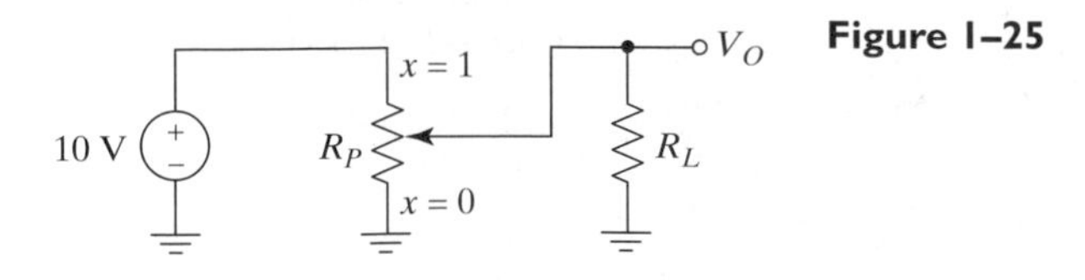

Figure 1–25

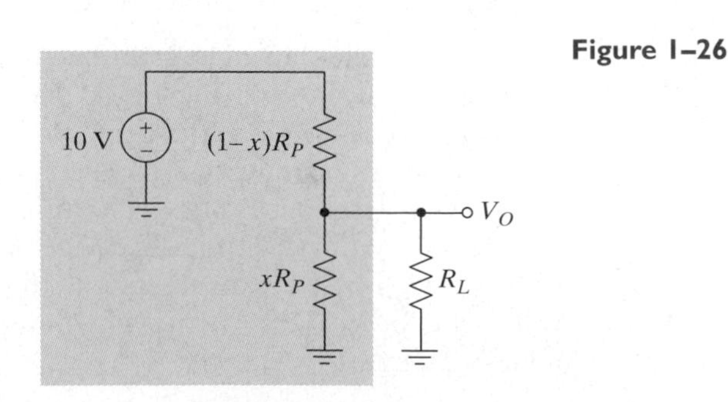

Figure 1–26

(a) Find V_{Th} and R_{Th} so that the circuit of Figure 1-27 is equivalent to that of Figure 1-26.

(b) Show that V_O is a linear function of x if R_L is infinite.

(c) If R_L is not infinite, V_O will not be equal to V_{Th}. Let the relative error be defined by

$$\text{Relative error} = \frac{V_O - V_{Th}}{V_{Th}} \times 100\%.$$

The maximum relative error in V_O will occur when the voltage divider formed by R_{Th} and R_L has its maximum attenuation. To what value of x does this correspond? What is R_{Th} in terms of R_P at this value of x?

(d) What is the relative error in terms of R_L and R_P when its magnitude is greatest?

(e) What R_L/R_P ratio is required for a 5% (magnitude) maximum relative error?

(f) What value of R_P would you use to keep the magnitude of the relative error below 10% while minimizing power dissipation in the pot, given that $R_L = 1000\ \Omega$? Assume that the value of R_P must be a multiple of 1.0, 2.5, or 5.0 (e.g., 250 Ω, 500 Ω, 1 kΩ, and 2.5 kΩ are all acceptable values).

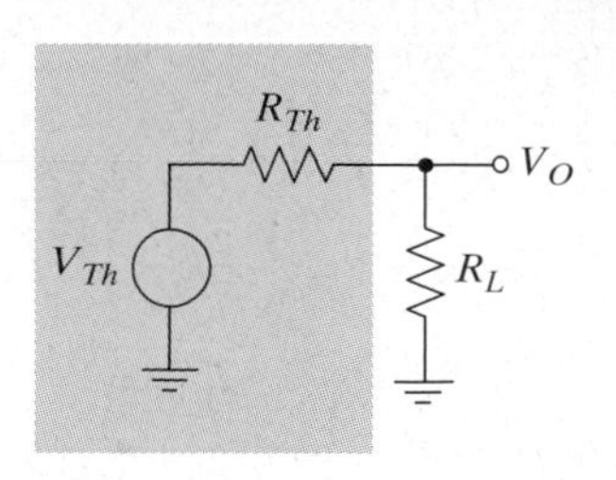

Figure 1–27

P1.7 Find the overall transfer function, v_O/v_S, of the circuit in Figure 1-28.

P1.8 Find the overall transfer function, i_O/v_S, of the circuit in Figure 1-28.

P1.9 Find the overall transfer function, v_O/i_S, of the circuit in Figure 1-29.

P1.10 Find the overall transfer function, i_O/i_S, of the circuit in Figure 1-29.

P1.11 Find the overall transfer function, v_O/i_S, of the circuit in Figure 1-30.

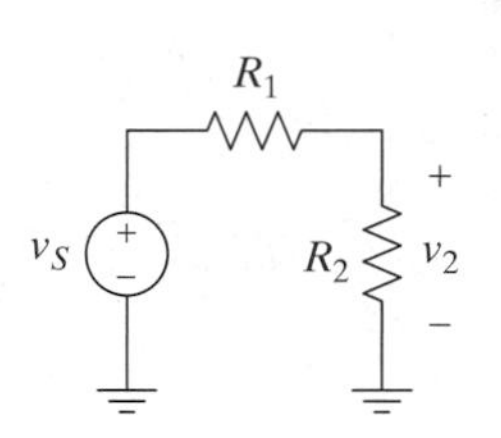

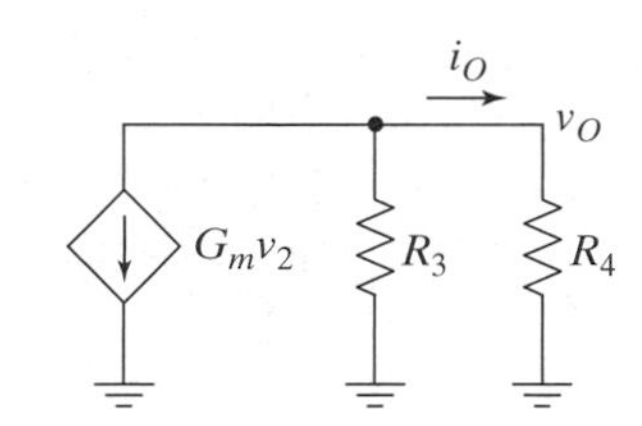

Figure 1–28

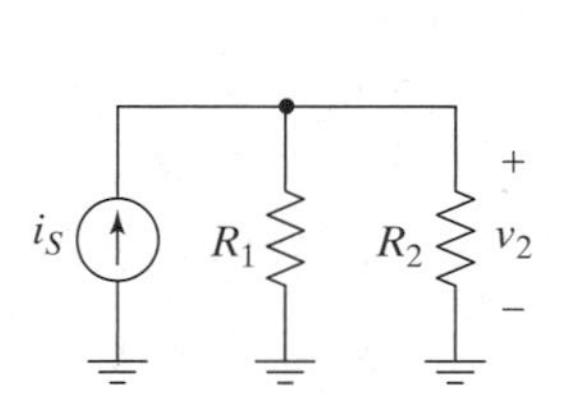

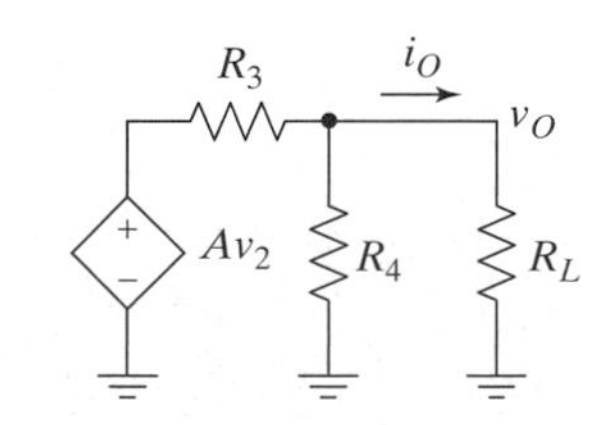

Figure 1–29

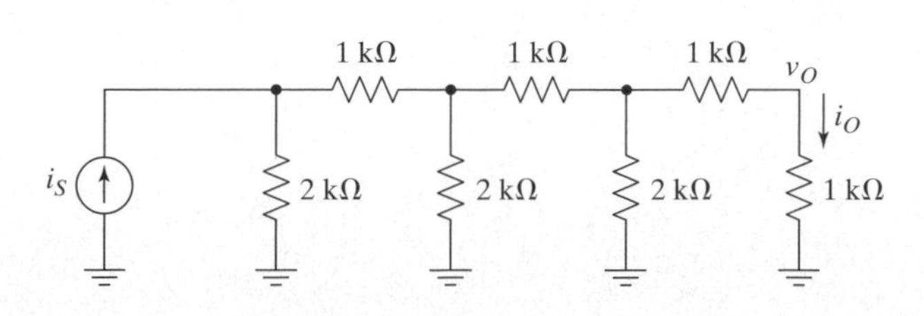

Figure 1–30

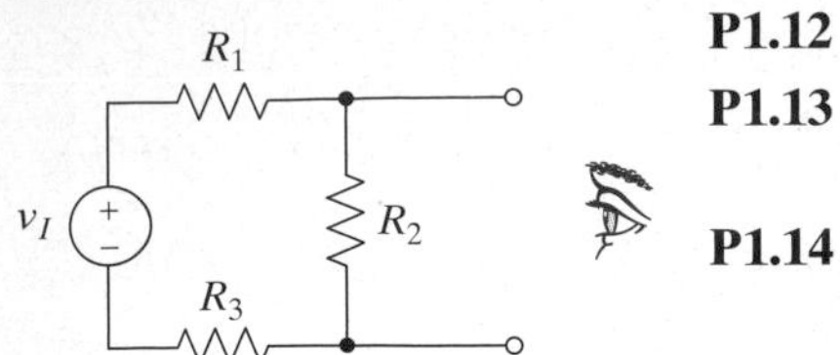
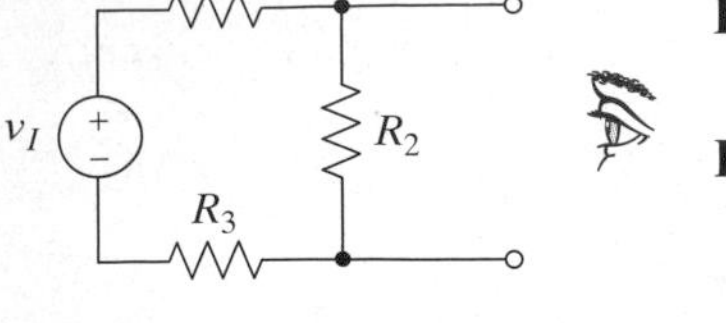

Figure 1–31

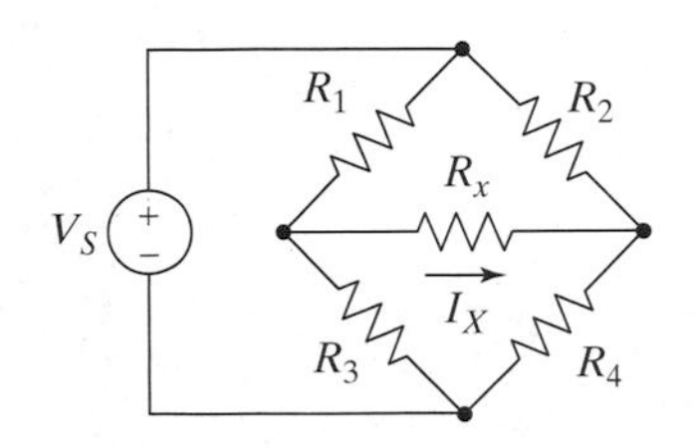

Figure 1–32

P1.12 Find the overall transfer function, i_O/i_S, of the circuit in Figure 1-30.

P1.13 Find the Thévenin and Norton equivalents looking into the terminals shown for the circuit in Figure 1-31.

P1.14 **(a)** Derive an equation for the current I_X in the circuit shown in Figure 1-32. **(b)** Assume that V_S, R_1, R_2, R_3, and R_4 are known, and write an equation for R_x as a function of I_X. (*Hint*: Use a Thévenin equivalent.)

P1.15 Find the Thévenin and Norton equivalents looking into the terminals shown for the circuit in Figure 1-33.

P1.16 Find the Thévenin and Norton equivalents looking into the terminals shown for the circuit in Figure 1-34.

P1.17 Find the Thévenin and Norton equivalents looking into the terminals shown for the circuit in Figure 1-35.

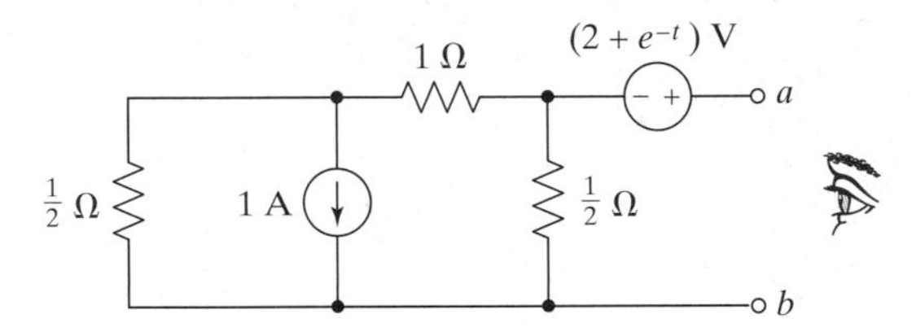

Figure 1–33

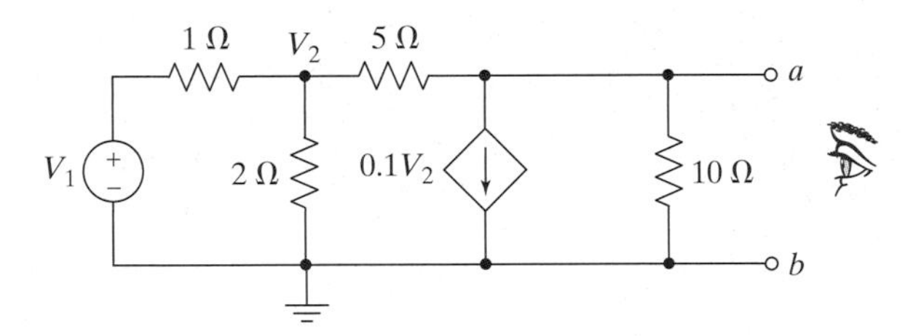

Figure 1–34

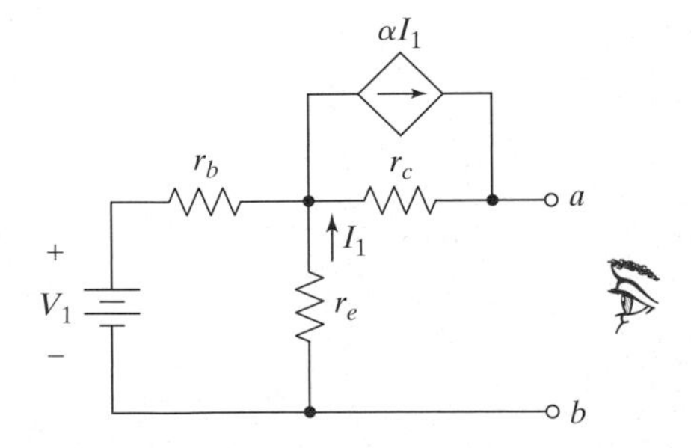

Figure 1–35

1.2.2 Frequency-Dependent Analysis for Design

P1.18 Find the overall transfer function, $V_o(j\omega)/V_s(j\omega)$, of the circuit in Figure 1-36.

D P1.19 **(a)** Design a compensated 10X probe for an oscilloscope that has an input impedance of 1 MΩ in parallel with 10 pF. A 10X probe is a probe that attenuates the signal by a factor of 10 so, for example, a 1-V signal at the probe tip is only 100 mV at the scope input. A compensated probe is one that has a flat frequency response, *see* (1.28). **(b)** If the capacitor in the scope probe is variable, what must its range of adjustment be to allow the user to always be able to adjust the transfer function of the probe to be independent of frequency if the scope's input resistance and capacitance can both vary by ±10%? Note that the attenuation of the probe will vary if the input resistance of the scope varies; only the frequency response can be adjusted using the variable capacitor.

P1.20 Prove that for a voltage source with a complex source impedance Z_S, maximum power will be delivered to the load when it has an impedance equal to Z_S^*.

P1.21 Find the overall transfer function, $V_o(j\omega)/V_s(j\omega)$, of the circuit in Figure 1-37.

P1.22 Find the overall transfer function, $V_o(j\omega)/V_s(j\omega)$, of the circuit in Figure 1-38.

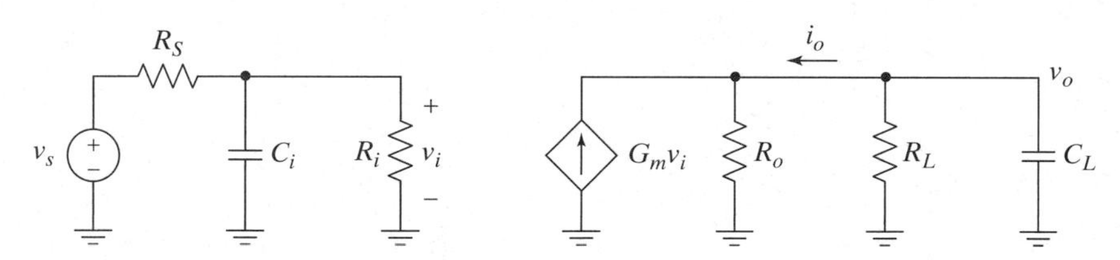

Figure 1–36

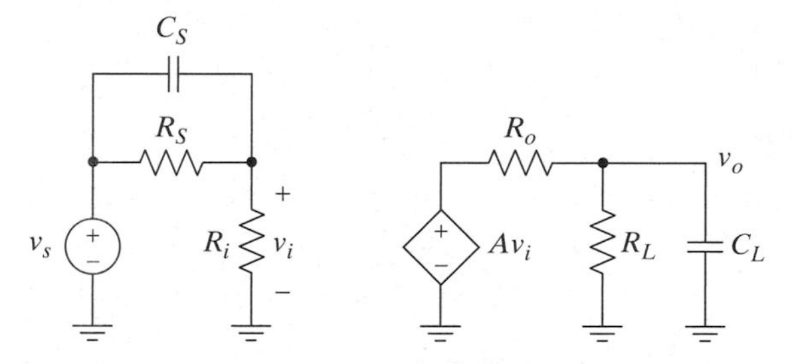

Figure 1–37

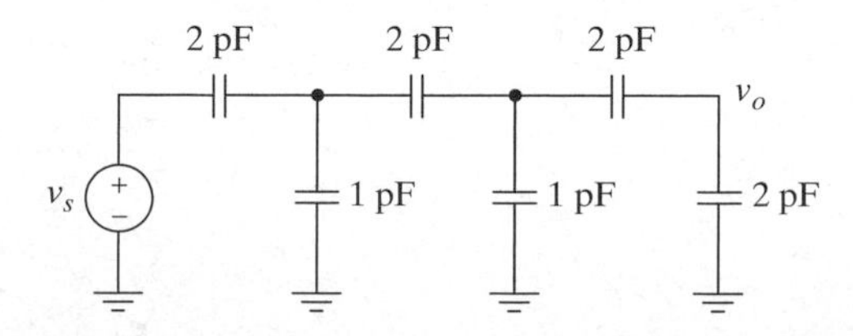

Figure 1–38

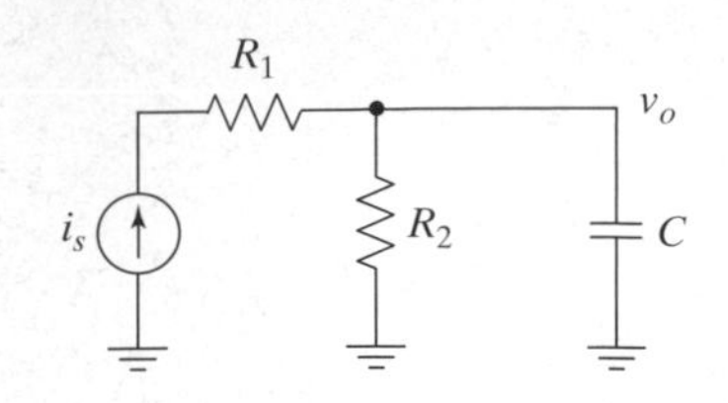

Figure 1–39

P1.23 Find the overall transfer function, $V_o(j\omega)/I_s(j\omega)$, of the circuit in Figure 1-39.

P1.24 Find the overall transfer function, $V_o(j\omega)/V_s(j\omega)$, of the circuit in Figure 1-40.

P1.25 Consider the circuits shown in Figure 1-41. For each circuit, **(a)** identify whether it is a low-pass filter or a high-pass filter, **(b)** find the pole frequency, and **(c)** state whether the zero is at $\omega = 0$, $\omega = \infty$, or some finite nonzero frequency.

1.3 Electronic Systems

1.3.1 Electronic versus Electric Circuits

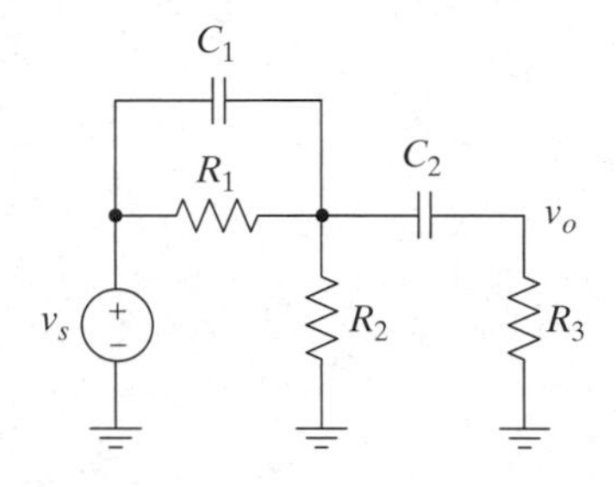

Figure 1–40

P1.26 Consider a series RLC circuit driven by a sinusoidal AC voltage source as shown in Figure 1-42(a). Assume the voltage across the inductor is taken to be the output. **(a)** At what frequency does this voltage have its maximum value if the input amplitude is held constant? **(b)** What is the voltage gain of the circuit (v_o/v_i) at the frequency found in part (a)? **(c)** Now consider connecting a load impedance to the circuit (in parallel with L). Explain why it is not possible to get a power gain larger than one from the circuit. (Power gain is defined to be the power in the load divided by the power supplied by v_i). *Hint*: Think of the circuit at a higher level of abstraction as shown in part (b) of the figure, and consider the second law of thermodynamics (which implies that perpetual motion is impossible).

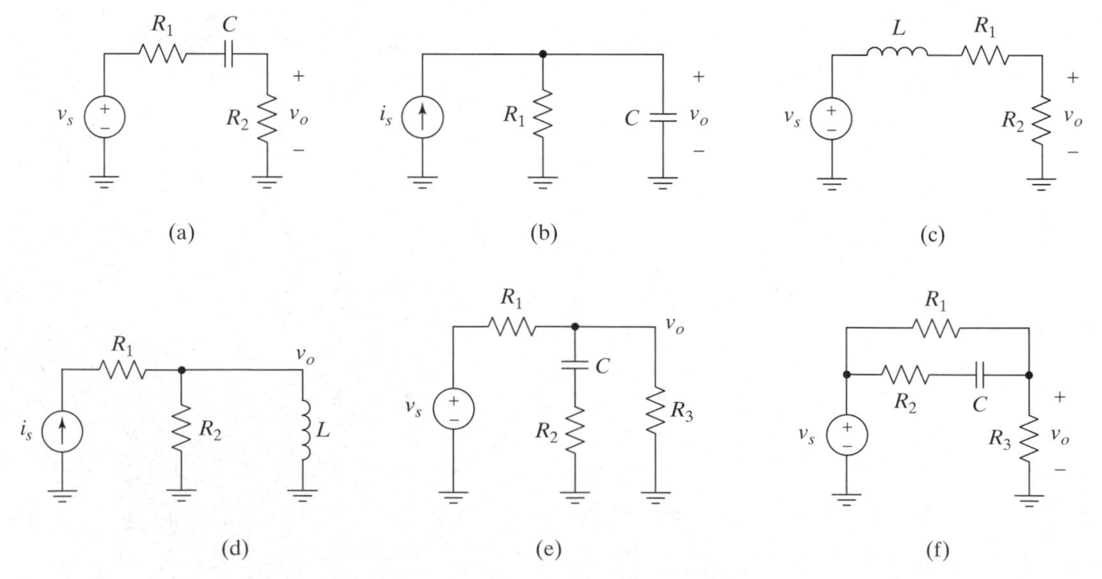

Figure 1–41

1.3.2 Analog and Digital Electronic Circuits

P1.27 Suppose an analog signal has been sampled and quantized and that each sample has been converted to a four-bit binary word. Assume that 0000 corresponds to zero volts, 1111 corresponds to one volt, and the sequence of binary words obtained is the following five-word pattern repeated: [0111, 1111, 1100, 0011, 0000]. **(a)** What is the corresponding sequence of sample values? **(b)** Draw the resulting digital waveform (voltage versus time) if each sample value is held and the sample time is 200 μs.

S P1.28 Use the digital waveform found in Problem P1.27 to define a piecewise linear voltage source in SPICE. (Use 1 μs transition times between values.) Use this source to drive a single-pole RC lowpass filter with a cutoff frequency of 2 kHz. Perform a transient simulation for 2 ms. Plot both the input and output waveforms, and explain what you see. Also, perform a fast Fourier transform (FFT) and plot the result.

P1.29 An analog pressure sensor has an output voltage of 2 V when the pressure is 500 torr and the output increases by 1 mV/torr as the pressure increases. If the sensor is used in an application where the pressure varies between 500 and 3000 torr, and we wish to digitize it's output, how many bits are required if we want a minimum resolution of 0.5 torr?

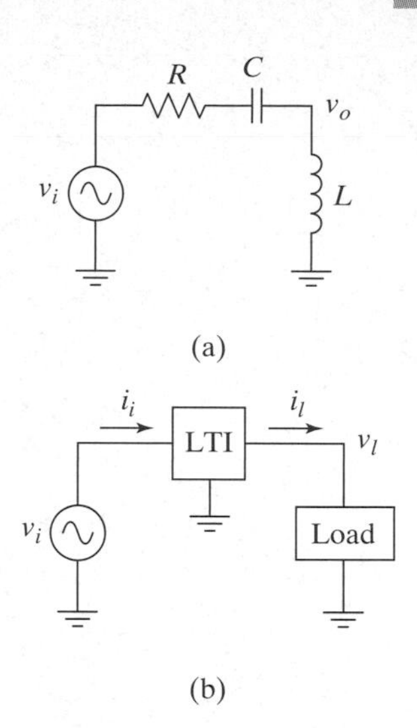

Figure 1–42 (a) A series RLC circuit. (b) The circuit viewed as an arbitrary LTI network with a load attached.

1.3.3 Modeling Electronic Systems

P1.30 Consider the circuit shown in Figure 1-43. Assume that the switch has been open for a long time and that $V_1 = 5$ V and $V_2 = 0$ V. If the switch is closed and we wait a long time, **(a)** find the final voltage across the two capacitors, assuming that energy is conserved, and **(b)** find the final voltage across the two capacitors, assuming that charge is conserved. **(c)** Compare the two answers from parts (a) and (b). Can you explain why they are different? *Hint*: Think of the models used. Is there anything you could do to more accurately model the operation of a real circuit?

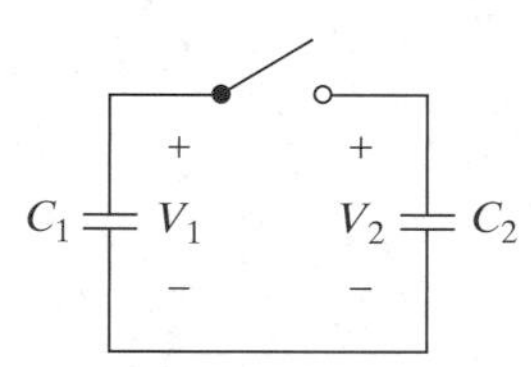

Figure 1–43

P1.31 Consider the circuits shown in Figure 1-44. Can you find the value of V_1 in either circuit? If not, why not? What is wrong with the models used?

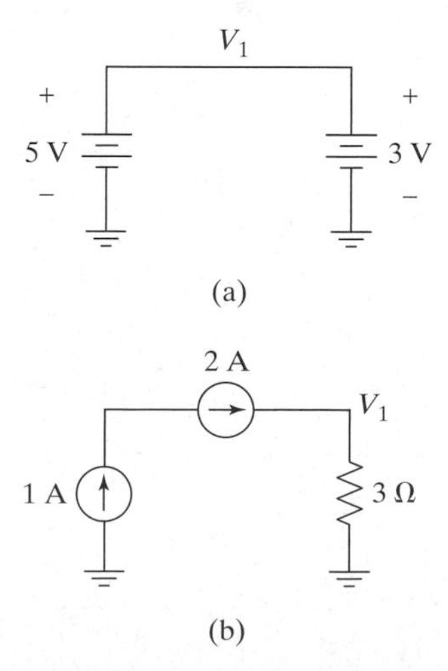

Figure 1–44

CHAPTER 2

Semiconductor Physics and Electronic Devices

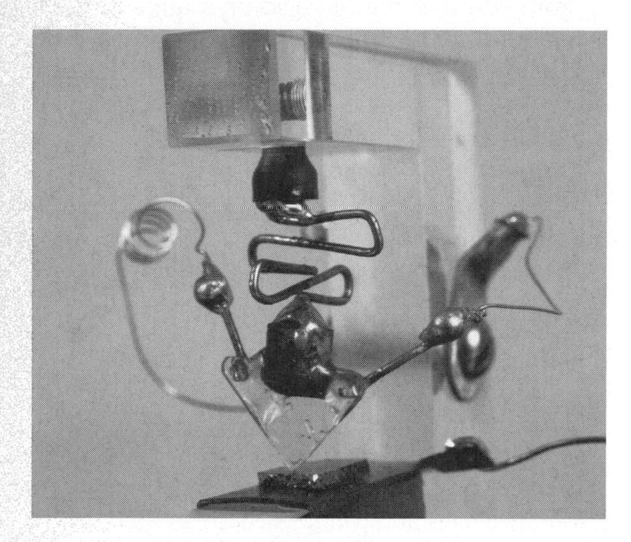

The first transistor. Invented at Bell Telephone Laboratories in 1947 by Bardeen, Brattain and Schockley.

INTRODUCTION

For the electronic-circuit designer, device physics forms one of the boundary disciplines that must be understood to do their job well. While circuit designers do not need to be experts in device physics, they do need to thoroughly understand the operation of solid-state devices. In this chapter, we focus on developing an intuitive understanding of device operation, backed up by an understanding of the underlying physical processes.

Most solid-state devices are fabricated in silicon, although germanium and various compound materials are also used. Common compound semiconductors include combinations of group-III and group-V elements[1] (e.g., gallium arsenide).

The main reason underlying the dominance of silicon is the ease with which we can fabricate silicon devices—in particular, the ease with which a high-quality insulating oxide can be grown. The importance of this oxide layer to the fabrication of silicon devices is covered in Chapter 3, but it also has properties that have a significant impact on device performance. In particular, it is the quality of the interface between silicon and silicon dioxide that has made ***insulated-gate field-effect transistors*** (IGFETs) possible.

[1] The *group* an element belongs to corresponds to the column in which it appears in the periodic table of the elements and is equal to the number of valence electrons.

IGFETs are commonly called ***metal-oxide semiconductor FETs***, or MOSFETs, even though the top conducting layer is heavily doped polysilicon rather than metal in modern processes. MOSFETs are currently the dominant transistor used in integrated circuits, although the ***bipolar junction transistor*** remains important, especially in analog design, and circuits manufactured with both types of transistors (called BiCMOS circuits, for **Bi**polar **C**omplementary **MOS**) are becoming quite common.

2.1 MATERIAL PROPERTIES

Before studying the physical operation of solid-state devices, you must understand some of the properties of the materials used to construct them. All materials are made up of atoms, and atoms in turn consist of electrons orbiting nuclei. The nucleus contains protons and neutrons (except for the hydrogen atom, which doesn't have any neutrons). Neutrons account for most of the mass of the atom. Protons are positively charged particles, electrons are negatively charged particles, and neutrons are electrically neutral, as the name suggests. The charge on an electron is equal in magnitude and opposite in polarity to the charge on a proton, and atoms usually have the same number of electrons as protons, so that they are charge neutral.

The electrons are arranged in shells, and each shell is broken into orbits. The distinction between shells and orbits is unimportant for our purposes. The electrons in the orbits closest to the nucleus are tightly bound to the nucleus by the Coulomb force, whereas those in the outer orbits are less tightly bound. The outermost shell is called the ***valence shell*** and is largely responsible for the way in which atoms chemically react with each other. If the outermost shell is full, the atoms are inert and will not readily interact with other elements.

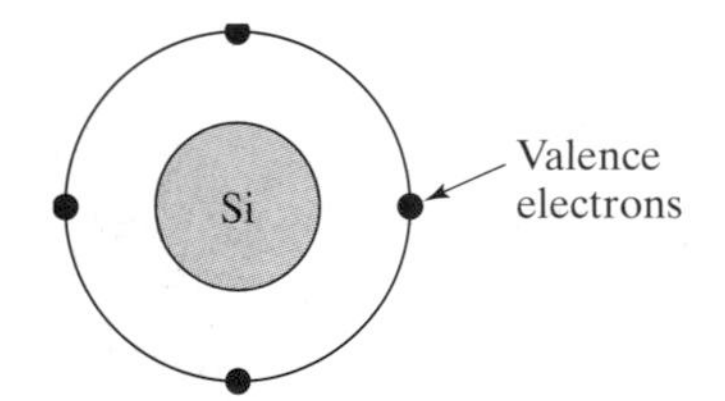

Figure 2–1 A diagram of an atom of silicon.

Since the chemical properties depend mostly on the number of electrons in the valence shell, the atom is frequently drawn as demonstrated for silicon in Figure 2-1. The central shaded region represents the nucleus *and* all the inner shells of electrons. The valence shell of electrons is shown as a circle with the electrons indicated as small black circles. Since these outer electrons are in the valence shell, they are called valence electrons. In the case of silicon, there are 14 protons and 14 neutrons in the nucleus and 14 electrons overall, with 4 in the valence shell.

The atoms in semiconductors form ***covalent bonds*** that hold the material together. Covalent bonding occurs when atoms share their valence electrons in such a way that all of the atoms appear to have a stable outer shell, which requires eight electrons. This bonding arrangement is illustrated for silicon in Figure 2-2. The center atom in the figure can be seen to share one electron with each of its four nearest neighbors. In this way, each atom appears to have a full outer shell and is stable.

Semiconductors can be either group-IV elements, or compounds of elements in groups III and V, or II and VI. For our purposes, it is sufficient to know that group-IV elements have four electrons in their valence shell and group-III and group-V elements have three and five electrons in their valence shells, respectively. In either silicon or groups III–V or II–VI compounds, the average number of electrons per atom is four, so that a stable bond results when valence electrons are shared.

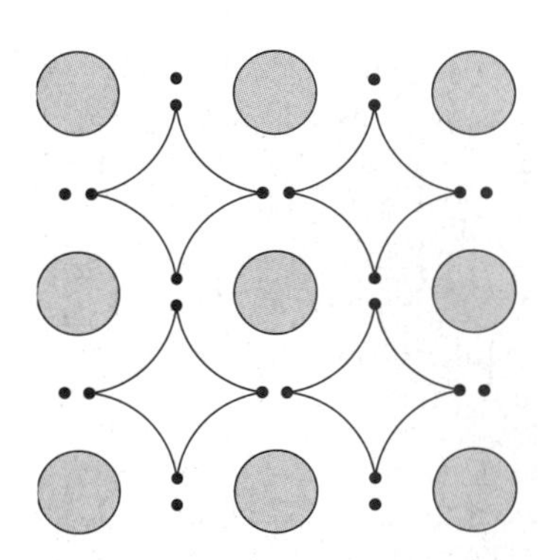

Figure 2–2 Covalent bonding of silicon.

2.1.1 Crystal Structure

Solid materials can be ***crystalline***, wherein the atoms are arranged in a set pattern; ***polycrystalline***, in which there are contiguous regions of crystalline material with irregular boundaries and no relationship between the orientation of the crystals in each region; or ***amorphous***, wherein there isn't any global order to the arrangement of the atoms. Most semiconductor devices manufactured today use crystalline materials for the basic substrate. It is quite common to use polycrystalline materials for some added layers, but polycrystalline devices are rare and will not be discussed here. There are also some applications, such as large-area flat-panel displays, in which amorphous materials are used; these applications will not be covered here either. Silicon forms a diamond lattice structure, but for our purposes, it is only important to recognize that a regular structure exists, so that the physical properties of the material are consistent.

2.1.2 Conductors, Insulators, and Semiconductors

The atomic structure discussed at the beginning of this section is the result of certain laws governing the behavior of subatomic particles. For example, no two electrons may occupy exactly the same ***state*** in an atom. (This is called the Pauli exclusion principle.) The state of an electron depends on its energy and spin, but we will ignore the spin, since it won't affect our understanding of device operation. There are many different possible states for electrons in a crystal. Each state represents a particular set of possible properties of an electron (e.g., position and momentum).

An analogy may be helpful in understanding what a state is and how it can have properties (e.g., energy) even though it isn't a physical object. If we think of cars traveling on a crowded freeway, we can say that there are different possible states for the cars. Only one car can occupy each state, and the car must match the speed and direction of a state to be able to move into it. In this crude example, the properties of the state (i.e., its direction and speed) are determined by the vehicles surrounding it.

When a number of silicon atoms come together and form a crystal, the electrons all interact, and no two may occupy exactly the same state. The result is that the energy levels that electrons are allowed to reside in are different for the crystal than they are for the isolated atoms. To accommodate all the neighboring electrons, each allowed energy level for an isolated atom must split into multiple levels when the atoms come together in a crystal. The result is that we end up with a large number of allowed energy levels that form what are called ***energy bands***.

The effects of neighboring atoms on electrons close to the nucleus of an atom are small. On the other hand, the electrons in the valence shells interact strongly. Since each isolated atom can support eight electrons in a stable outer shell, but only has four when it is part of the crystal, only half of the states corresponding to the valence shells are used by electrons participating in the covalent bonding of the crystal. The band of energies that corresponds to all of the possible states for electrons participating in the covalent bonding process is called the ***valence band***. The other half of the states from the valence shells split to represent states available to electrons that are *not* participating in the bonding process. Electrons that are *not* participating in the bonding process and are not in an inner shell of an atom can be made to move about in the material by adding a small amount of energy to them and are in the ***conduction band***.

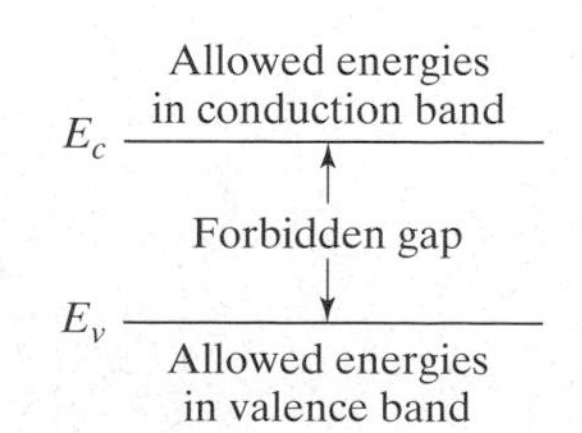

Figure 2–3 Energy band diagram. The vertical scale is energy, and the horizontal scale is distance in one dimension. These scales are understood and are *not* usually shown in band diagrams.

A common way of showing these energy bands as a function of position is the ***band diagram*** illustrated in Figure 2-3. In this figure, only the lower edge of the conduction band, E_C, and the upper edge of the valence band, E_V, are shown. The energies are usually given in electron volts, $E = qV$. (Remember that voltage is defined as electrical potential energy per unit charge.)

Materials can be characterized on the basis of the structure of their energy bands. If the valence and conduction bands overlap or are only partially full, so that electrons can move about freely in the material, the material is called a ***conductor***, because it can conduct electricity relatively easily. On the other hand, if there is a gap between the valence and conduction bands, that means there is a range of energies that electrons are not allowed to occupy. Therefore, unless we somehow excite electrons from the valence band into the conduction band, the electrons will remain a part of the bonding process and will not be free to participate in the conduction of electric current. Such materials are either ***insulators*** or ***semiconductors***, depending on the size of the energy gap.

The gap between the valence and conduction bands is called the forbidden energy gap—or simply, ***energy gap***. If this energy gap is small enough, at room temperature the normal thermal agitation of the atoms in the crystal will supply sufficient energy to move a significant number of electrons from the valence to the conduction band. If this number is large enough to carry an appreciable electric current, the material is called a semiconductor. If the gap is too large to allow a significant number of electrons to move into the conduction band, the material will not conduct electricity well, and it is called an insulator.

Since the electrons that reside in the conduction band, are able to move, they may carry an electric current and are called ***conduction***, ***free***, or ***mobile*** electrons. Each time an electron is taken from the valence band, it leaves behind a ***hole*** in the valence band. Electrons in the valence band that are adjacent to a hole can move into it relatively easily, since the move does not involve gaining the energy necessary to jump out of the valence band. In this way, the hole itself is mobile and able to contribute to an electric current. Electric currents can be carried by mobile electrons, holes,[2] or both.

We can alter the number of electrons or holes available for conduction in silicon by ***doping*** it with impurities. For example, if we add some atoms of boron to the silicon lattice, we will create one hole in the valence band for each atom of boron. The hole results because boron has only three electrons in its valence band and cannot fully satisfy the covalent bonding of the silicon crystal. As soon as an adjacent electron moves into this available slot, the net charge of the boron atom becomes negative, and it is ***ionized*** (i.e., it becomes a charged atom, or ***ion***). Owing to the random thermal motion of the valence band electrons, virtually all of the boron atoms will be ionized at room temperature in silicon. Since this atom is not free to move about in the lattice, it represents ***fixed***, or ***bound***, charge, in contrast to the mobile charge that the hole represents.

Similarly, if we add phosphorus to silicon, we have one extra electron for each phosphorus atom since phosphorus has five electrons in its valence band. Although this electron is bound to the phosphorus atom, it is not held very strongly, since it is not needed to satisfy the bonding arrangement. Therefore, it does not take much energy for that electron to move away from the phosphorus atom and ionize it. The ionized phosphorus atom then constitutes fixed positive charge in the silicon lattice, and the electron that moved away is mobile charge.

Pure silicon is called ***intrinsic*** silicon, and silicon with impurities added is called ***extrinsic***. The impurities that are added are called ***dopants***. Group-III elements, like boron, are called ***acceptors***, because they can accept an extra electron from the crystal, whereas group-V elements, like phosphorus, are called ***donors***, because they donate an extra electron to the lattice.

---

[2] Unlike electrons, which may be tightly bound to a particular atom, *all* holes are mobile, so there isn't any need to add the adjective mobile in referring to holes.

Silicon that has been doped with acceptor impurities will have an excess of positive mobile charge (holes) and is therefore called *p-type* silicon. Similarly, silicon that has been doped with donor impurities will have an excess of negative mobile charge (electrons) and is therefore called *n-type* silicon. As a result, acceptor impurities are also called p-type dopants, and donor impurities are also called n-type dopants. The concentration of donor impurities is denoted N_D, and the concentration of acceptor impurities is denoted N_A. (Concentrations are typically measured in atoms per cubic centimeter.) As stated before, these impurities are virtually all ionized at room temperature in silicon. Therefore, each donor impurity produces one mobile electron and one bound positive ion, and each acceptor impurity produces one hole and one bound negative ion. The *overall* material is charge neutral;[3] that is, the number of negative charges present equals the number of positive charges present.

2.1.3 Generation and Recombination

Two processes that affect the number of electrons and holes available for conduction are generation and recombination. As mentioned in the previous section, the normal thermal agitation of the atoms in the crystal will supply sufficient energy to move some of the electrons from the valence band to the conduction band, even at room temperature. Whenever an electron moves from the valence band to the conduction band, it leaves behind a hole in the valence band, and an *electron–hole pair* (EHP) is created. In other words, we now have a mobile electron and a hole available to carry electric current. The rate at which EHPs are generated is called the ***generation rate***, G, and is, to first order, only a function of temperature.

The complementary process of EHP ***recombination*** also occurs. Recombination can occur in different ways. However, for our purposes, the details are unimportant, and we can simply envision an electron in the conduction band losing energy and falling into a hole in the valence band. This recombination process can occur only if both mobile electrons and holes are available. The rate at which recombination occurs is, in fact, proportional to the product of the density of mobile electrons, n, and the density of holes, p. (These densities are usually measured in number of electrons or holes per cubic centimeter.) In other words, the ***recombination rate***, R, is proportional to np. All electrons and holes in intrinsic silicon are created as part of an EHP, and therefore, $n = p$. The intrinsic concentrations of carriers are denoted by adding a subscript i to produce n_i and p_i, so we have $n_i = p_i$.

The generation and recombination rates must be equal at thermodynamic equilibrium. If they were not, then the net number of EHPs would be increasing or decreasing, and we would not yet be at equilibrium. Using the facts that $R \propto np$, G is a function of temperature alone, $n_i = p_i$, and $R = G$ leads to the *mass–action law* for semiconductors; namely,

$$np = n_i^2, \qquad \textbf{(2.1)}$$

which is true at equilibrium whether or not the material is intrinsic. The intrinsic carrier concentration in silicon is a strong function of temperature; near room temperature, it approximately doubles for every 11°C rise in temperature. At room temperature (300 K), the intrinsic carrier concentration in silicon is about 1.45×10^{10} carries/cm^3.

---

[3] There can be regions within the material where the charges do not balance (e.g., a pn junction as discussed in Section 2.4.1).

Exercise 2.1

Derive (2.1). Assume that $R = npf_R(T)$ *and* $G = f_G(T)$.

If we have an n-type piece of silicon at room temperature in which virtually all of the donor impurities are ionized, the total number of mobile electrons will be equal to N_D plus the number of EHPs that have been generated. If N_D is much larger than the number of EHPs, as is usually the case, then

$$n \approx N_D, \tag{2.2}$$

and using (2.1), we have

$$p_n \approx \frac{n_i^2}{N_D}, \tag{2.3}$$

where the subscript n indicates that the hole concentration is in n-type material. Similarly, for reasonably doped p-type material, we have

$$p \approx N_A \tag{2.4}$$

and

$$n_p \approx \frac{n_i^2}{N_A}. \tag{2.5}$$

Since the number of electrons in n-type silicon is much larger than the number of holes, the electrons are called the ***majority carriers*** and the holes are called the ***minority carriers***. Similarly, in p-type silicon, the holes are the majority carriers and the electrons are the minority carriers.

Exercise 2.2

If a silicon crystal is doped with phosphorus with $N_D = 3 \times 10^{16}$ *atoms/cm*^{3}*, what is* p_n *at 300 K?*

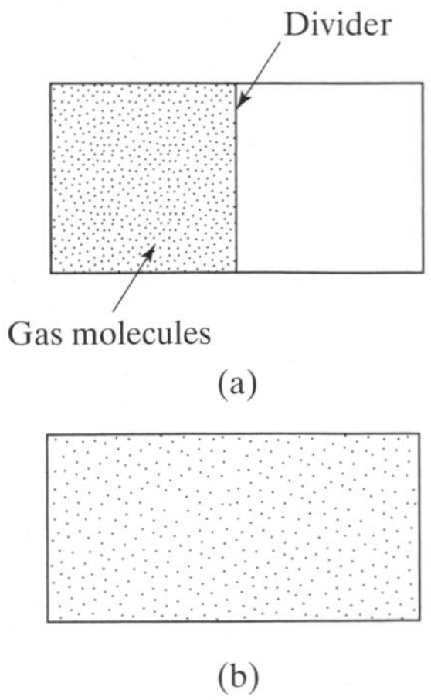

2.2 CONDUCTION MECHANISMS

Before we discuss the operation of electronic devices, we need to understand two mechanisms that cause electric currents to flow: drift and diffusion. Both of these mechanisms are involved in semiconductor device operation. We will not discuss the two phenomena in detail, but rather will be satisfied with a heuristic explanation. The interested reader is referred to [2.1], which has a very good discussion of both of these conduction mechanisms.

2.2.1 Diffusion

Picture a box with a divider in the middle, and imagine that half of the box is filled with molecules of some gas and the other half of the box is empty, as shown in Figure 2-4(a). Due to thermal energy, the gas molecules will be randomly moving around in their half of the box. If the divider is removed, we know intuitively that

Divider

Gas molecules

(a)

(b)

Figure 2–4 A thought experiment to demonstrate diffusion. In (a), half the box is filled with some gas. In (b), the barrier is removed and the gas diffuses to fill the entire box.

the gas molecules will spread evenly through the box as shown in part (b) of the figure. This phenomenon is called ***diffusion*** and is simply the net motion of particles from a place of high concentration to a place of lower concentration. Whenever there is a concentration gradient of some type of particle, *and* the particles have some random motion, *and* they are free to move to a region of lower concentration, diffusion will result. In particular, in a solid containing electrons that are free to move about, diffusion tends to distribute the electrons evenly throughout the material, unless there is some force to counteract the diffusion (e.g., drift, which is discussed in the next section).

Diffusion results in a net flux of particles from the region of higher concentration to the region of lower concentration. If the particles are electrically charged, like electrons, this flux constitutes an electric current called the ***diffusion current***. We intuitively expect that the magnitude of the current is proportional to the difference in the number of particles in the two regions; in other words, it should be proportional to the gradient, or derivative, of the particle distribution. In addition, the electric current must be proportional to the charge on an individual particle.

It is often convenient to find the current per unit area instead of the absolute magnitude of the current in a given device. The current per unit area is called the ***current density*** and is denoted by J, with typical units of amperes/cm^2. A detailed analysis of the diffusion process reveals that the diffusion current density in the x direction for electrons in a solid is given by

$$J_{nxdiff}(x) = qD_n\frac{dn(x)}{dx}, \quad \textbf{(2.6)}$$

where $n(x)$ is the density of *mobile* electrons per unit volume, D_n is the ***diffusion coefficient*** for electrons, and q is the charge on an electron. If the derivative is positive, $J_{nx\text{diff}}$ is positive, indicating current flow from a place of lower concentration to a place of higher concentration. This direction may seem incorrect at first, but it is correct, since $J_{nx\text{diff}}$ represents conventional current flow, which is opposite to electron current flow.

Exercise 2.3

Suppose an n-type sample of silicon has a doping density that changes linearly from 10^{15} atoms/cm^3 to 10^{12} atoms/cm^3 over a 1-µm distance in the x direction. The sample measures 10 µm by 10 µm in the dimensions perpendicular to x. The diffusion coefficient is 35 cm^2/s at these densities and at 300 K. What is the diffusion current in this sample?

□

Exercise 2.4

Excluding a concentration gradient, what else can cause a diffusion current to flow?

□

2.2.2 Drift

Let's consider again the box filled with molecules of some gas from Figure 2-4(b). We ignored the effect of gravity on the gas molecules for our picture in Figure 2-4, but let's now assume the particles to be massive enough that gravity will have a noticeable effect. With gravity included, the molecules have a constant force pulling downward on them. We know from Newton's second law ($F = ma$) that a mass m with a constant force F applied will undergo a constant acceleration equal to F/m. In the case of the gas, there will also be collisions between individual molecules and between molecules and the sides of the box. These collisions are what keep the molecules scattered about, but gravity does introduce a preferred direction of motion, and there will, on average, be more molecules near the bottom of the box than near the top of the box.

Now let one end of the box be elevated slightly, as shown in Figure 2-5; what will happen to the distribution of the gas molecules in the box? Lifting one end of the box causes one component of the gravitational acceleration to line up with the length of the box, so that the particles have a constant acceleration toward one end of the box. We can describe the associated force by saying that the molecules ***drift*** toward the lower end of the box. The final distribution of particles seen is the result of a balance between the two phenomena: drift and diffusion. As more molecules move down in response to gravity, they increase the magnitude of the concentration gradient and, therefore, increase the magnitude of the diffusion current opposing the drift. At equilibrium, the drift and diffusion currents must balance each other.

In the case of electrons in a solid, the force that causes them to drift can be due to an electric or a magnetic field. The force on an electron is

$$\boldsymbol{F} = -q(\boldsymbol{\mathcal{E}} + \boldsymbol{v} \times \boldsymbol{B}), \tag{2.7}$$

where q is the charge on the electron, $\mathcal{E}$ is the electric field strength,[4] $\boldsymbol{v}$ is the velocity, $\boldsymbol{B}$ is the magnetic field strength, and '×' denotes the cross product. We will ignore the magnetic field and concentrate on the electric field. Newton's second law says that an electron in a constant electric field will have a constant acceleration equal to $-q\mathcal{E}/m$, where m is the mass of the electron. However, mobile electrons in solids do not accelerate indefinitely when an electric field is applied. Owing to collisions with the atoms in the material, there is an average ***drift velocity*** given by

$$\boldsymbol{v}_d = -\mu_n \boldsymbol{\mathcal{E}}, \tag{2.8}$$

where μ_n is the electron ***mobility*** and is a function of the solid (the subscript n is used because electrons have negative charge). These collisions are the electrical equivalent of mechanical friction and are the cause of electrical resistance. Each individual electron is accelerated by the field in between collisions, so the electrons are not moving at a constant velocity, nor are they all moving at the same velocity. The drift velocity is the average velocity of all of the electrons contributing to the current flow. The drift velocity is typically much smaller than the velocity due to thermal motion, which is randomly directed, so each electron continues to move about randomly even in the presence of a drift current. One result of this fact is that electrons will still diffuse if there is a concentration gradient, even if there is an average drift velocity present.

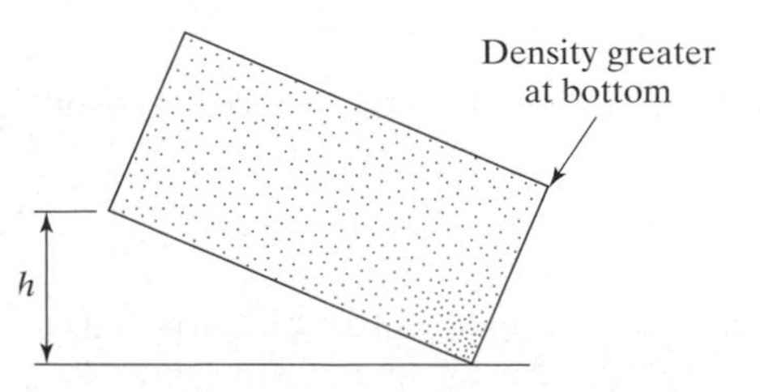

Figure 2–5 A box filled with an imaginary gas made up of molecules with significant mass and having one end elevated.

[4] Vector quantities are denoted by **boldface**.

When electrons in a material move under the influence of an electric field, they constitute a ***drift current***. We expect the magnitude of this drift current to be proportional to the drift velocity, the charge per electron, and the number of free electrons present. A detailed analysis shows that the drift current density in the x direction is given by

$$J_{nx\text{drift}}(x) = q\mu_n n(x)\mathcal{E}(x), \tag{2.9}$$

where all of the terms have already been defined. If the electric field is positive (i.e., net positive charge on the left and net negative charge on the right), then the electrons move to the left, resulting in a conventional current flowing to the right, which is defined to be positive.

The total electron current in one dimension in a solid is the sum of the drift and diffusion currents found in (2.9) and (2.6):

$$J_{nx}(x) = q\mu_n n(x)\mathcal{E}(x) + qD_n\frac{dn(x)}{dx}. \tag{2.10}$$

Example 2.1

Problem:
What is the conductivity of doped silicon?

Solution:
The electrical resistance of a piece of material is given by

$$R = \frac{\rho L}{A}, \tag{2.11}$$

where ρ is the resistivity in ohm-cm, L is the length in cm, and A is the cross-sectional area in cm^2. The conductivity, σ, is the reciprocal of the resistivity. Using Ohm's law and (2.11), we can write

$$\sigma = \frac{1}{\rho} = \frac{IL}{VA} = \frac{J}{\mathcal{E}},$$

where we have assumed that the electric field is constant throughout the length of the sample. If we have a piece of n-type silicon that is uniformly doped, then the diffusion current is zero, and (2.10) gives us the relationship between J and $\mathcal{E}$. Using this, we find that

$$\sigma = q\mu_n n,$$

but this accounts only for mobile electrons. Since there will always be holes present as well, the conductivity is given by

$$\sigma = q\mu_n n + q\mu_p p. \tag{2.12}$$

For extrinsic silicon, the majority carriers will typically dominate the conductivity, and the minority carriers can be ignored.

■

Exercise 2.5

Consider again the n-type sample from Exercise 2.3. What voltage would need to be applied to the sample to cause a drift current equal in amplitude to the diffusion current?

Table 2.1 presents some useful material properties for silicon and silicon dioxide, and Table 2.2 presents some useful physical constants.

2.3 CONDUCTOR-TO-SEMICONDUCTOR CONTACTS

We must have some way of making electrical connections to semiconductors in order to use them. In this section, we will examine the physics of conductor-to-semiconductor contacts by examining what happens when metal contacts silicon. If a metal—say, aluminum—is deposited on a clean silicon surface, the resulting contact may be ***ohmic***, or it may be ***rectifying***. An ohmic contact creates a linear resistance to the flow of electric current that obeys Ohm's law, whereas a rectifying contact allows substantial current flow only in one direction. To determine which kind of contact is produced when metal is brought into intimate contact with silicon (e.g., by evaporation onto a clean silicon surface), we need to understand the concept of the ***Fermi level*** and how to use it in drawing band diagrams and determining carrier concentrations. Aside A2.1 presents an introductory view of Fermi diagrams.

Table 2.1 Material Properties of Silicon and Silicon Dioxide. Sources: [2.1], [2.2], [2.11]

Property	Silicon	SiO_2
Atoms (or molecules)/cm^3	5.0×10^{22}	2.2×10^{22}
Electric field at breakdown	3×10^5 V/cm	6 to 9 MV/cm
Electron mobility,* μ_n	1350 cm^2/(V s)	20 cm^2/(V s)
Energy gap at 300 K, E_g	1.12 eV	N/A
Hole mobility,* μ_p	480 cm^2/(V s)	$\approx 10^{-8}$ cm^2/(V s)
Intrinsic carrier concentration, n_i	1.45×10^{10} cm^{-3}	N/A
Relative permittivity, ϵ_r	11.7	3.9

*Value quoted for low doping levels.

Table 2.2 Physical Constants. Source: [2.11]

Physical Constant	Symbol	Value
Boltzmann constant	k	1.38066×10^{-23} J/K
Electron charge	q	1.60218×10^{-19} C
Electron volt	eV	1 eV $= 1.60218 \times 10^{-19}$ J
Permittivity in vacuum	ϵ_o	8.85418×10^{-14} F/cm

ASIDE A2.1 BAND DIAGRAMS, THE FERMI LEVEL, AND CARRIER CONCENTRATIONS

A band diagram like the one in Figure A2-1 shows the ranges of energies of states that an electron may occupy, but it does not provide information about either the density of states available or the probability of occupancy at a given energy. Therefore, we cannot use that figure to arrive at any quantitative statements about carrier concentrations. This aside examines the electron and hole concentrations in more detail.

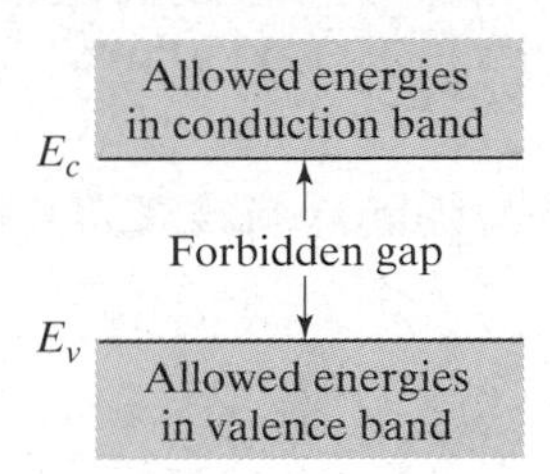

Figure A2-1 Energy band diagram. The vertical scale is energy, and the horizontal scale is distance in one dimension. These scales are understood and are *not* usually shown in band diagrams.

The probability of occupancy of a state with energy E is given by the ***Fermi-Dirac*** distribution[1], sometimes called the ***Fermi function***,

$$f(E) = \frac{1}{1 + e^{(E-E_f)/kT}}, \quad \textbf{(A2.1)}$$

where E_f is the Fermi level, k is Boltzmann's constant (1.38×10^{-23} J/K), and T is absolute temperature (i.e., measured in Kelvin). At a temperature of absolute zero, we see that $f(E) = 1$ for $E < E_f$ and that $f(E) = 0$ for $E > E_f$; in other words, all of the available states below E_f are filled, and all of those above E_f are empty. At any temperature, we notice that when $E = E_f$, the probability of occupancy is one-half. We show how to find the Fermi level later in this aside.

The density of states available at a given energy, $N(E)$, can be determined from quantum mechanical considerations, but it is beyond the scope of the present treatment (one good elementary presentation is given in [2.1]). If $N(E)$ is known, the number of mobile electrons in equilibrium can be determined by integrating the product of $N(E)$ and the Fermi function over the entire conduction band:

$$n = \int_{E_c}^{E_{top}} N(E)f(E)dE. \quad \textbf{(A2.2)}$$

Here E_{top} is the energy at the top of the conduction band. So long as E_f is several kT below E_c, the Fermi function can be approximated by

$$f(E) \approx \exp\left(-\frac{E - E_f}{kT}\right), \quad \textbf{(A2.3)}$$

and the integral in (A2.2) evaluates to

$$n = n_c \exp\left(-\frac{E_c - E_f}{kT}\right), \quad \textbf{(A2.4)}$$

where n_c is the ***effective density of states*** at the conduction band edge. Similarly, using $1 - f(E)$ and integrating over the valence band yields

$$p = n_v \exp\left(\frac{E_v - E_f}{kT}\right), \quad \textbf{(A2.5)}$$

where n_v is the effective density of states at the valence band edge. Figure A2-2 illustrates the density of states, the Fermi function, and the carrier concentrations for intrinsic silicon at equilibrium.

---

[1] This function applies to states that are subject to the Pauli exclusion principle.

Figure A2-2 The density of states, the Fermi function, and the carrier concentrations for intrinsic silicon at equilibrium.

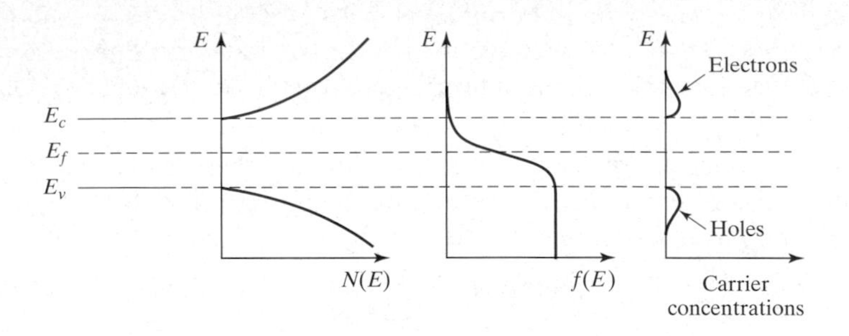

For intrinsic silicon, we know that $n = p = n_i$, so we can set (A2.4) equal to (A2.5) and solve for the ***intrinsic Fermi level***, E_i:

$$E_i = \frac{E_c + E_v}{2} + \frac{kT}{2}\ln\left(\frac{n_v}{n_c}\right). \qquad \textbf{(A2.6)}$$

E_i is very near, but slightly below, the center of the bandgap, due to the fact that n_c is slightly greater than n_v. We can use E_i to obtain useful equations for the carrier concentrations. Using (A2.4) for intrinsic silicon yields

$$n = n_c \exp\left(-\frac{E_c - E_i}{kT}\right). \qquad \textbf{(A2.7)}$$

If we solve (A2.7) for n_c and plug the result back into (A2.4), we obtain

$$n = n_i \exp\left(\frac{E_f - E_i}{kT}\right). \qquad \textbf{(A2.8)}$$

Similarly, for holes, we find that

$$p = n_i \exp\left(\frac{E_i - E_f}{kT}\right). \qquad \textbf{(A2.9)}$$

These equations show that the difference between E_f and E_i determines the carrier concentrations. We will make use of these relations, which are sometimes called the ***Boltzmann relations***, in later sections.

Let's now consider extrinsic silicon. For n-type silicon, there are more mobile electrons than are present in intrinsic material. The probability of occupancy of a given state in the conduction band must therefore be higher. Consequently, the Fermi level is closer to E_c. For p-type silicon, on the other hand, there are more holes than are present in intrinsic material, and therefore, the probability of occupancy for a state in the valence band is lower, and the Fermi level is closer to the valence band. These three situations are shown in Figure A2-3.

Figure A2-3 Band diagrams for (a) intrinsic, (b) n-type, and (c) p-type silicon.

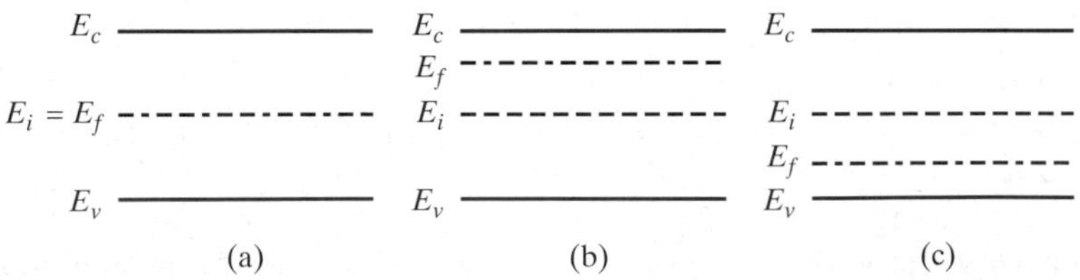

(a): E_c, $E_i = E_f$, E_v (b): E_c, E_f, E_i, E_v (c): E_c, E_i, E_f, E_v

Including the Fermi level in our band diagrams is essential because it will help us see what happens when the doping is varied or different materials are in contact. The Fermi level is helpful because *it is constant in a material at equilibrium*. To see why the Fermi level must be constant, consider states with identical energies that are physically adjacent to each other in the material. If these states did *not* have the same probability of occupancy, there would be a net diffusion of electrons from the region of higher concentration (i.e., higher probability of occupancy) to the region of lower concentration (i.e., lower probability of occupancy), and the material would not be in equilibrium. Therefore, the Fermi level must be constant at equilibrium.

We found the intrinsic Fermi level in (A2.6). Finding the Fermi level for extrinsic silicon is also straightforward. For example, if we have n-type silicon, the concentration of mobile electrons is about equal to N_D at equilibrium (assuming room temperature and $N_A = 0$). Setting (A2.4) equal to N_D and solving yields

$$E_f = E_c - kT\ln\left(\frac{n_c}{N_D}\right). \quad \textbf{(A2.10)}$$

Similarly, for p-type material, we obtain (with similar assumptions)

$$E_f = E_v + kT\ln\left(\frac{n_v}{N_A}\right). \quad \textbf{(A2.11)}$$

In essence, the Fermi level adjusts itself so as to maintain charge neutrality in the material (i.e., when a material is doped n-type and the donors are ionized, the Fermi level must be higher so that there are enough mobile electrons to compensate for the ionized donors).

2.3.1 Rectifying Contacts

When a metal is brought into intimate contact with silicon, as shown in Figure 2-6(a), and some path exists for electrons to flow from one material to the other (there will always be a path in the devices we consider), the Fermi level in the metal must be the same as the Fermi level in the silicon when the system is in equilibrium. The Fermi levels may not initially be the same, however. For them to equalize, there must be a momentary flux of electrons from the silicon to the metal or vice versa, depending on the kind of metal and the doping of the silicon. The band diagram shown in part (b) of the figure assumes that the silicon is n-type and that electrons had to flow from the silicon into the metal to achieve equilibrium. The net result is that the probability of occupancy of states in the conduction band of the silicon decreases as we approach the surface of the metal; this shows up as the increasing difference between E_f and E_c in the diagram.

Because there are fewer mobile electrons in the silicon near the interface, there are ionized donor impurities that do not have mobile electrons nearby to compensate for their positive charge. In other words, the material is no longer locally charge neutral. We say that the ionized donor impurities in this region are ***uncompensated***. The result is that we have a region of net positive charge. A region of net charge (positive or negative) is often called a ***space-charge region***. At the same time, because the electrons that diffused into the metal are attracted by this positive charge in the silicon, there exists a sheet of negative charge at the surface of the metal. The total net charge on each side of the interface must be equal in magnitude and opposite in sign.

These charge profiles are shown in part (c) of Figure 2-6, where, for simplicity, we have assumed that the doping in the silicon is uniform and that the material is completely depleted of mobile carriers up to a distance x_d from the interface and is completely charge neutral beyond that distance (i.e., has no field). The second assumption is called the ***depletion approximation***. We call the region in the silicon from the interface to x_d the ***depletion region***.

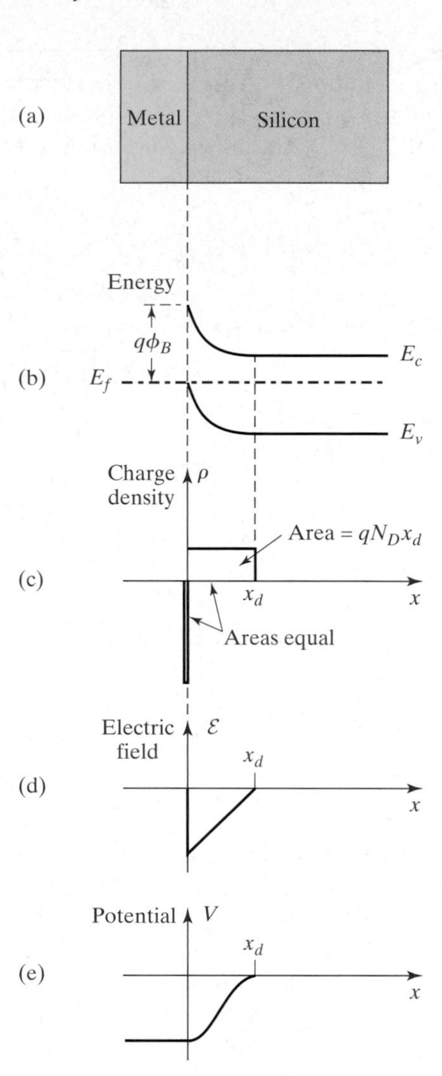

Figure 2–6 (a) A metal-and-silicon system, (b) the corresponding band diagram, (c) the charge density, (d) the electric field, and (e) the electric potential.

Because the metal is a good conductor, the carriers it acquired from the silicon are mobile, and the net charge in the metal exists as a sheet of charge at the surface. Because the charge profile resulted from diffusion and the device was initially charge neutral, the total charge on each side of the junction must be equal. In other words, the total sheet charge at the surface of the metal, Q_s, is (in coulombs/cm^2)

$$Q_s = qN_Dx_d. \tag{2.13}$$

Remember from electrostatics that the electric field caused by a spatial distribution of charge is given by the integral of the charge distribution. This fact is also expressed by Gauss' law, which, in point form, says that the divergence of the electric flux density, $\boldsymbol{D}$, at a point is equal to the charge density at that point,

$$\nabla \cdot \boldsymbol{D} = \rho. \tag{2.14}$$

Substituting the definition of electric flux, $\boldsymbol{D} = \epsilon\mathcal{E}$, into (2.14) and also recalling that the electric field is the negative of the gradient of the electric potential leads to Poisson's equation,

$$\nabla \cdot \nabla V = -\frac{\rho}{\epsilon}, \tag{2.15}$$

which, in one dimension, reduces to

$$\frac{d^2V}{dx^2} = -\frac{\rho}{\epsilon}. \tag{2.16}$$

Integrating both sides of (2.16) for the charge distribution in the silicon in part (c) of Figure 2-6, we get

$$\mathcal{E} = -\frac{dV}{dx} = \frac{qN_D}{\epsilon}x + C_1; 0 < x < x_d \tag{2.17}$$

where C_1 is a constant that we find by applying the boundary condition that $\mathcal{E} = 0$ outside the depletion region (i.e., at $x = x_d$). The resulting field is

$$\mathcal{E} = \frac{qN_D}{\epsilon}(x - x_d) \; ; \quad 0 < x < x_d \tag{2.18}$$

and is plotted in part (d) of the figure (the field in the metal is assumed to be zero, because the charge exists in an infinitely thin sheet at the surface). Integrating both sides of (2.18) and applying the boundary condition that $V = 0$ at $x = x_d$ leads to

$$V(x) = \frac{qN_D}{\epsilon}\left(x_d x - \frac{x^2 + x_d^2}{2}\right); \quad 0 < x < x_d \; , \tag{2.19}$$

which is plotted in part (e) of the figure.

Notice in Figure 2-6(b) that there is a potential-energy barrier of height $q\phi_B$ that a typical electron in the metal must surmount to move from the metal into the silicon.[5] If this barrier is large, very few electrons will possess enough energy to get over it, and therefore, the electron current from the metal to the silicon will be extremely small. Similarly, there is a barrier that electrons trying to move from the silicon into the metal must overcome. This barrier height varies with the applied bias and leads to an asymmetric *i–v* characteristic for the junction, as we shall now see.

Applying an external potential disturbs the equilibrium situation, and we can no longer expect the Fermi levels in the metal and silicon to be the same. In fact, the Fermi level in the metal will go down relative to that in the silicon when we apply a positive voltage to the metal and will go up when we apply a negative potential (the energy moves in the opposite direction from what you might expect because electrons have negative charge; remember, $E = qV$). In either case, the barrier for electron flow from the metal to the silicon ($q\phi_B$) stays the same; what changes is the barrier for electrons moving from the silicon to the metal, as seen in Figure 2-7.

For $V_A < 0$, as in part (b) of the figure, the barrier for electron flow from silicon to metal increases, and the only current that flows is a very small ***reverse saturation current***, due to electrons surmounting the barrier from the metal into the silicon.[6] This current is denoted I_S, and a typical value would be around 10^{-9} amperes. If we are not dealing with the external current, it is sometimes conve-

[5] We cannot, in general, say that E_f is the average energy of an electron. But the probability of occupancy of a state with $E > E_f$ decreases exponentially with $(E - E_f)$. Therefore, it is useful to think of an electron with energy E_f as "typical" in some sense.

[6] This statement is meant to impart intuition, but should not be read too literally. The "barrier" to electron movement is not like a physical barrier in classical physics; for example, *see* Aside A2.2.

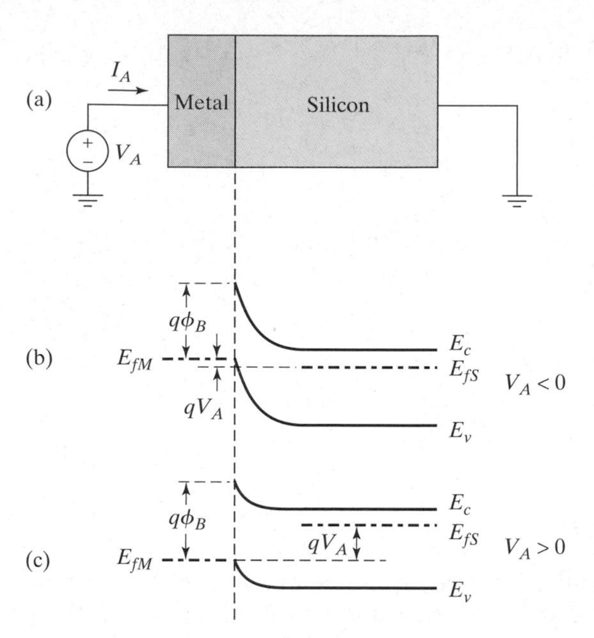

Figure 2–7 The band diagrams for the system of Figure 2-6 with applied bias.

nient to deal with the current density instead; we denote the saturation current density by J_S, which is equal to I_S divided by the area of the junction. Having $V_A < 0$ is called the ***reverse bias*** condition, and the current that flows is, to first order, independent of the magnitude of the reverse bias voltage. Since the current initially increases from zero to I_S, and after that it is relatively constant as the reverse bias is increased, the current is said to have saturated—hence the name.

When $V_A > 0$, as in part (c) of the figure, the device is said to be ***forward biased***, the barrier for electron flow from silicon to metal is reduced, and it becomes possible to have significant current flow. The net result is that significant terminal current can flow only when the device is forward biased. When significant current does flow, it is made up of majority carriers from the silicon—electrons in this case—diffusing into the metal. The electrons diffuse because electrons in the conduction band in the silicon possess more energy than the electrons in the metal. In other words, if you consider the probability of occupancy of a state at the conduction band edge in the silicon right at the metal–silicon interface, you will find that it is higher than the probability of occupancy of a state at the same energy in the metal, and we would therefore expect a net diffusion of carriers. This electron flow results in a positive current directed as shown in the figure (I_A). A detailed analysis reveals that the terminal current is given by

$$I_A = I_S(e^{qV_A/kT} - 1), \tag{2.20}$$

where I_S is a constant that depends on the individual device and is equal to the junction area times J_S, k is Boltzmann's constant, T is temperature in Kelvin, and q is the charge on an electron. Equation (2.20) is plotted in normalized form in Figure 2-8. I_S is called the ***reverse saturation current***, because when the device is reverse biased (i.e., $V_A < 0$), the current increases rapidly until it nearly reaches $-I_S$, at which point it saturates and cannot increase further.

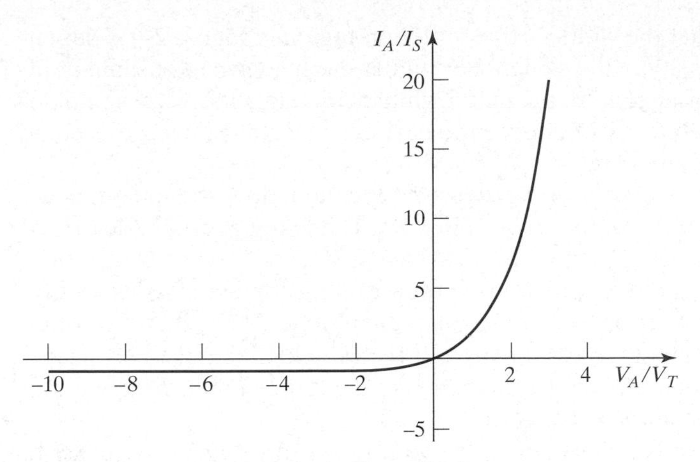

Figure 2–8 The diode equation from (2.20).

Equation (2.20) is the classic diode equation.[7] The rectifying metal-semiconductor junction forms what is called a ***Schottky diode***. We will come back to the Schottky diode after we have discussed *pn*-junction diodes.

Although we will not pursue a detailed derivation of (2.20) here, we can justify the form of the equation. When the diode is strongly reverse biased, essentially no electrons can surmount the barrier from the silicon to the metal, so the net current is just the reverse saturation current made up of the electrons that can make it over the barrier from the metal to the silicon. This barrier is fixed at $q\phi_B$, as noted earlier. The number of electrons with sufficient energy to surmount the barrier is proportional to the probability of occupancy of a state at that energy, as given by the Fermi function, (A2.1). Therefore, we expect the saturation current to be proportional to the Fermi function:

$$I_S \propto \frac{1}{1 + e^{q\phi_B/kT}} \approx e^{-q\phi_B/kT}. \quad \textbf{(2.21)}$$

When the device is forward biased, a significant number of electrons can surmount the reduced barrier from the silicon to the metal, and they make up the net current flow. The number of electrons able to surmount this barrier is again proportional to the probability of occupancy of a state at that energy, as given by the Fermi function. Therefore, the current in forward bias should be

$$I_A \propto \frac{1}{1 + e^{q(\phi_B - V_A)/kT}} \approx e^{-q\phi_B/kT} e^{qV_A/kT} \propto I_S e^{qV_A/kT}, \quad \textbf{(2.22)}$$

which agrees with (2.20) in forward bias.

2.3.2 Ohmic Contacts

The Schottky diode is a useful device, but it does not provide us with an ohmic connection to silicon. In this section, we examine two ways in which ohmic contacts can be produced.

---

[7] In device physics texts, this equation is sometimes called the *ideal* diode equation. We will not use that terminology, so as to avoid confusion with what circuit designers commonly call the ideal diode (which is presented elsewhere). It is also sometimes called the Shockley equation in honor of William Shockley, who developed the theory of junction devices.

We start by noting that the width of the depletion region in Figure 2-6 is dependent on the doping density in the silicon. For higher doping levels, the same number of electrons can be supplied to the metal while depleting a narrower region of the silicon. For very high levels of doping, the barrier to electron flow can become very narrow, as illustrated in Figure 2-9.

This barrier can become so narrow that a new conduction phenomenon becomes important: ***quantum mechanical tunneling***. Tunneling is briefly described in Aside A2.2.

Tunneling can occur in either direction, so the resulting contact loses its rectifying property and becomes an ohmic contact. In practice, the silicon is often ***degenerately*** doped; in other words, the doping level is so high that the Fermi level is no longer in the bandgap, but rather is in the conduction band for n-type material or the valence band for p-type material.

A second kind of ohmic contact occurs if the transfer of electrons required to bring the Fermi levels of the metal and silicon together is in the opposite direction to what was assumed when drawing Figure 2-6. The result for n-type silicon is shown in Figure 2-10, where we see that electrons accumulate at the surface of the silicon and there is a corresponding sheet of net positive charge at the surface of the metal. The result is a reduction in the barrier height for electron flow into the silicon and no barrier for electron flow from the silicon into the metal. If the barrier, $q\phi_B$, is small enough, current can flow in either direction with little difficulty.

Similar band diagrams can be drawn for ohmic Schottky contacts to p-type silicon. Most of the ohmic contacts made to n-type silicon devices are of the tunneling type.

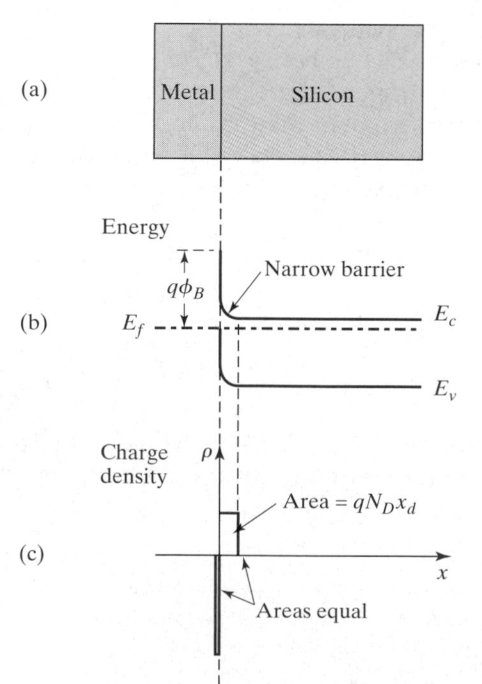

Figure 2–9 Band diagram for a metal–silicon system with very high doping in the silicon (n-type shown).

ASIDE A2.2 QUANTUM MECHANICAL TUNNELING

To understand quantum mechanical tunneling, we must be careful with our interpretation of band diagrams. The conduction band edge in these plots represents the energy associated with the lowest energy state in the conduction band. If there is a bump or step in this energy, we say that there is a "barrier" to electron flow. This terminology is not entirely appropriate, however, since this change in E_C is not completely analogous to a physical barrier to particle flow in classical physics. There is some finite probability of an electron penetrating a barrier (so long as the barrier doesn't have infinite height) without acquiring all of the energy necessary to surmount the barrier. Remember that the energy of an individual electron may be quite different from the statistical average. This phenomenon is beyond the scope of the text to discuss in detail, but the probability of finding an electron past the barrier drops off exponentially with distance. If the barrier becomes very narrow, however, then there is some reasonable probability of an electron appearing on the other side. When an electron successfully traverses the barrier without having sufficient energy to surmount it, we say it has ***tunneled*** through the barrier.

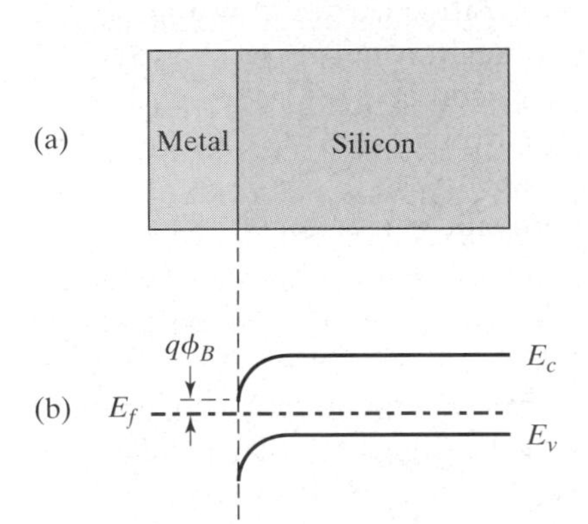

Figure 2–10 Band diagram for a metal–silicon system in which electrons must be transferred to the metal to achieve equilibrium.

2.4 *PN*-JUNCTION DIODES

The most common solid-state devices are *pn*-junction diodes. This section covers *pn*-junction diode operation and compares it with the operation of the Schottky diode already covered as a part of the section on metal-to-semiconductor contacts.[8] We then move on to cover the bipolar transistor, which is constructed using *pn* junctions, in Section 2.5.

A *pn* junction is formed by doping one part of a semiconductor *p* type and an adjacent part *n* type, as shown in Figure 2-11. The figure also shows contacts to

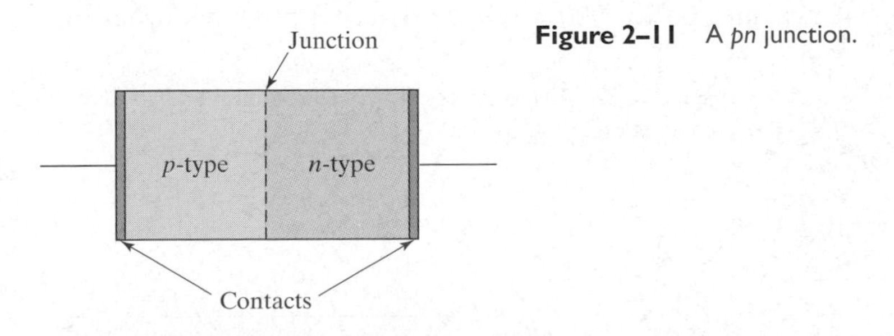

Figure 2–11 A *pn* junction.

[8] Readers who have skipped Section 2.3 will not be at a disadvantage; they can safely skip the references to that section.

each of the regions, so that we can connect the device in a circuit. To understand how this device works, we will start by considering the equilibrium situation with no external connections. We will also restrict ourselves to the so-called ***step junction***, where the doping changes abruptly from N_A acceptors/cm^3 on the p side to N_D donors/cm^3 on the n side. The doping is also assumed to be uniform within each region.

2.4.1 Intuitive Treatment

Suppose that we could start out with the n-type and p-type materials separated and then bring them into contact[9] at time $t = 0$. Although both the n- and p-type sides are globally charge neutral, they do have different types of mobile carriers. The n-type material has electrons as majority carriers with a concentration approximately given by $n = N_D$ and has holes as minority carriers with a concentration approximately given by $p_{no} = n_i^2/N_D$, where the subscript n reminds us that we are talking about holes in the n-type material (i.e., minority carriers) and the subscript o tells us it is the equilibrium value in the material. The situation is reversed in the p-type material and it has $p = N_A$ and $n_{po} = n_i^2/N_A$. It is important to remember that charge neutrality in these regions is maintained because every mobile carrier is either part of an electron–hole pair (EHP) or the result of ionizing a dopant atom, in which case there is the ion itself left in a fixed position in the lattice.

When the n and p regions are brought together, diffusion currents will be created because of the concentration gradients of the different types of mobile carriers. The ionized dopant atoms, however, are fixed in the crystal lattice and will not diffuse. Therefore, there will be a diffusion of mobile electrons from the n-type material into the p-type material and a diffusion of holes from the p-type material into the n-type material, as shown in Figure 2-12.

As majority carriers diffuse out of each side of the junction, they leave behind dopant ions that no longer have mobile charge of the opposite polarity nearby to keep the region charge neutral. We say that these dopant ions are ***uncompensated***, and the resulting region with net charge is called a ***space-charge region***. The uncompensated dopant ions are positively charged donor ions (N_D^+) on the n-type side and negatively charged acceptor ions (N_A^-) on the p-type side. The separation of charge creates an electric field in the space-charge region, which attracts any negative charge in the region toward the n-type side and any positive charge in the region toward the p-type side. In other words, the electric field causes drift currents to flow that oppose the original diffusion currents. Thermodynamic equilibrium is achieved when the drift and diffusion currents balance so that there is no net movement of charge. The resulting charge density profile, electric field, and current components are indicated in Figure 2-13.

A couple of additional comments need to be made about Figure 2-13 before we proceed. First, a balance must be reached for *each* carrier type at equilibrium, not

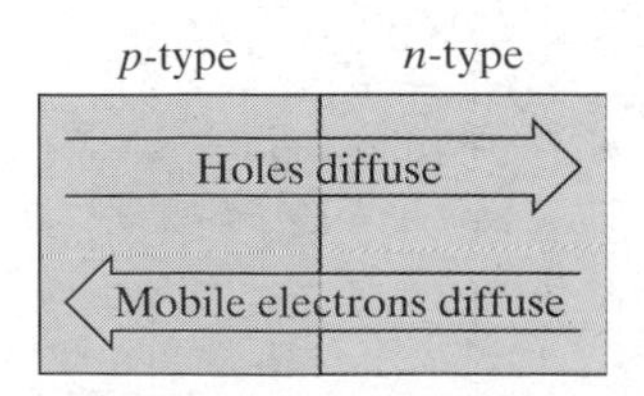

Figure 2–12 Diffusion of carriers to establish equilibrium in a *pn* junction.

---

[9] We cannot really do this, because the crystals would not match up and surface effects at the interface would dominate the behavior. In reality, the n and p regions must be diffused into the same piece of semiconductor, but this is a useful thought experiment nonetheless.

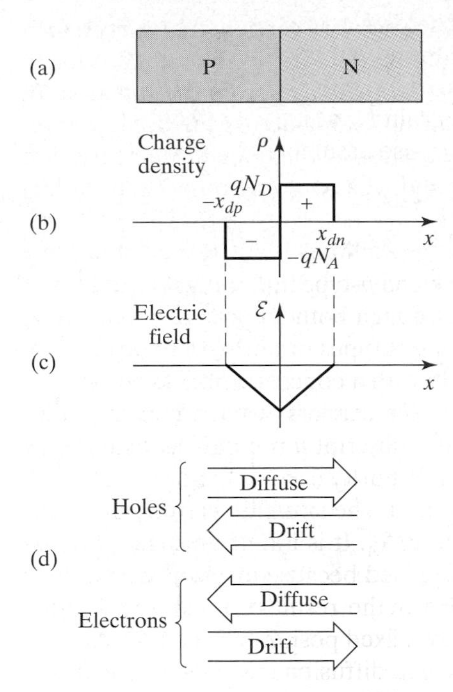

Figure 2–13 A *pn* junction at equilibrium: (a) the junction, (b) charge density profile, (c) electric field, and (d) current components.

just the total. In other words, electrons and holes must achieve a balance of drift and diffusion individually, not just collectively. If this were not so, electrons or holes would be accumulating somewhere, and the junction would not be at equilibrium. Second, in drawing part (b) of the figure, we made two assumptions that together make up what is typically called the ***depletion approximation***. First, we assumed that the region near the junction is completely depleted of mobile carriers. This region is called either the ***depletion region*** or the ***space-charge region***. Similarly, we assumed that the regions outside the depletion region are entirely charge neutral and therefore do not have any electric fields. We will call these the ***neutral regions***. In Figure 2-13 the depletion region extends from x_{dp} to x_{dn}, and we have used the assumption that it is depleted of mobile carriers to draw the charge density profile.

If we integrate the electric field across the *pn* junction shown in Figure 2-13(c), we find that there is a potential difference, just as was found in Figure 2-6(e) for the Schottky contact. The potential difference is called the ***potential barrier***, or ***built-in potential***. This potential difference cannot be measured directly, however, because the contact potentials created by contacting the *n*- and *p*-type materials will exactly cancel the junction potential. This fact may seem amazing at first, but notice that if we could measure the potential barrier, then we could connect a resistor externally from the *n* side to the *p* side, and we would see a current flow at equilibrium. This current would dissipate energy in the resistor in the form of heat, and we would be extracting energy from a closed system at equilibrium; in other words, we could make a perpetual-motion machine. However, perpetual motion is a violation of the second law of thermodynamics (not to mention common sense), so we conclude that it must not be possible to measure the built-in potential directly.

Before we proceed, it will help if we examine the balance between drift and diffusion a bit more. In the discussion that follows, we ignore generation and recombination in the depletion region for simplicity. We also define the phrase "***net***

diffusion current" to refer to the current due to majority carriers that *successfully* diffuse across the depletion region and the phrase "***net drift current***" to refer to the current due to minority carriers that drift *all* the way across the junction. In other words, we do not count carriers that start diffusing across the depletion region, but never make it across due to the acceleration of the field. Using these simplified definitions makes it easier to understand what is going on without hiding any essential characteristics of device operation.

The electric field in the depletion region does two things. First, it accelerates any minority carriers that enter this region (e.g., electrons from the p side) and tends to sweep them across the junction. Second, it opposes majority carriers diffusing across the junction. As a result, the magnitudes of the net diffusion currents are functions of the electric field. On the other hand, the magnitudes of the two net drift currents that flow, due to holes and electrons, turn out to be independent of the magnitude of the built-in potential because they are limited by the availability of minority carriers, which are supplied by EHP generation.

Let's now examine the electron current in more detail. All electrons that drift across the junction from the p side came from EHPs generated in or near the depletion region.[10] Since the EHP generation rate is small, the current is determined by the available supply of carriers, rather than the magnitude of the accelerating potential. This situation is analogous to the flow of cars on a freeway. If the cars are traveling bumper to bumper, then the average number of cars passing a given point in a given time (i.e., the "car current") is proportional to the speed they are all traveling. If the cars are all being released from a stoplight one at a time, however, and are able to go as fast as necessary on the freeway once they are past the light, the average current depends on how many cars are released in a given time rather than how fast they are traveling. In the case of the minority carriers, it is the rate at which they are supplied to the depletion region that determines the current.

The net diffusion currents crossing the pn junction, however, are supplied by the majority carriers on either side of the junction and are not, therefore, limited by the available supply of carriers. These currents do depend on the magnitude of the built-in potential. The larger the potential, the fewer the number of carriers with sufficient energy to completely diffuse across the junction. In other words, some of the carriers that start to diffuse across the junction never make it; the acceleration caused by the field stops them and turns them around.

We are now in a position to examine what happens when an external bias is applied to the diode, as shown in Figure 2-14(a), where we have defined the applied bias to be positive on the p side relative to the n side and have defined the current as being positive into the p side.

When the external voltage is set to zero, it does not affect the pn junction in any way; remember that this voltage *is* zero in equilibrium anyway. When the external voltage is positive, however, it is in opposition to the built-in potential (which is positive on the n side and negative on the p side) and thereby reduces the magnitude of the potential across the depletion region. The reduced potential allows a larger net diffusion current to flow, but does not, to first order, affect the net drift current. The result is that a large terminal current can flow, as indicated for $V_A > 0$ in part (b) of the figure. This situation is called ***forward bias***.

When the external voltage is negative, the voltage across the depletion region is increased, the net diffusion currents are reduced, and the terminal current that can flow is no larger than the sum of the electron and hole net drift currents. Since

---

[10] Only the EHPs generated outside the depletion region contribute to the *net* drift current as we have defined it.

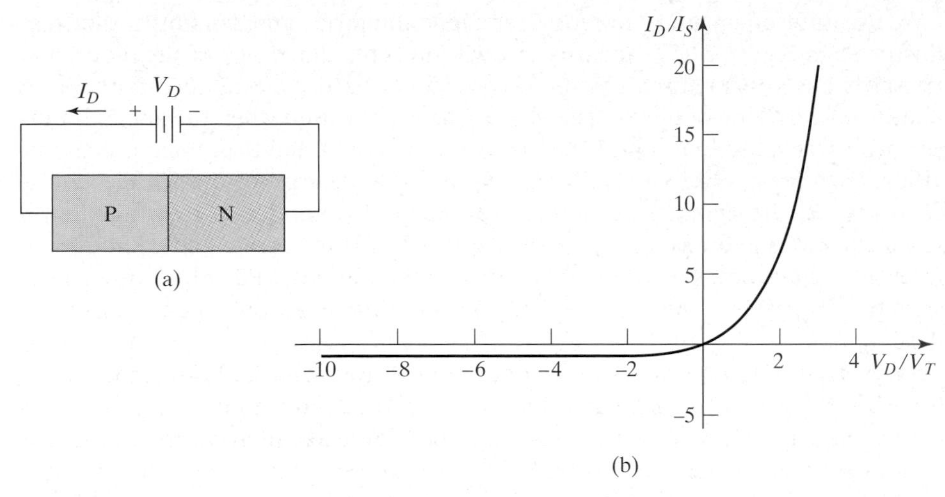

Figure 2–14 (a) A *pn*-junction diode with external bias applied and (b) the resulting *i–v* characteristic.

these currents are supplied by the thermal generation of EHPs near the depletion region, they are quite small. The result is that the external current that flows is very small (a typical value at room temperature is 10^{-14} amps or less), as shown in the figure. This situation is called ***reverse bias***, and the small external current that flows is called the ***reverse saturation current***[11] or just the ***saturation current*** and is denoted I_S. The resulting terminal current of the diode is given by

$$I_D = I_S(e^{qV_D/kT} - 1). \tag{2.23}$$

The term kT/q occurs so frequently that we give it a special name, the ***thermal voltage***, and denote it by

$$V_T = \frac{kT}{q}, \tag{2.24}$$

where k is again Boltzmann's constant, T is temperature in Kelvin, q is the charge on an electron, and the resulting value for V_T is about 26 mV at room temperature. Using (2.24) for the thermal voltage, (2.23) can be rewritten as

$$I_D = I_S(e^{V_D/V_T} - 1). \tag{2.25}$$

It is sometimes important to know that the saturation current increases rapidly with increasing temperature, typically doubling for every 6°C rise in temperature.

We are now in a position to contrast the Schottky diode with the *pn*-junction diode. (If you skipped the material on Schottky diode, you don't need to feel left out: The details of its operation are not necessary for the present discussion.) The Schottky diode's *i–v* relationship is given by (2.20), which is identical to the equation we have developed for the *pn*-junction, (2.25). The saturation current of the Schottky diode is, however, typically about five orders of magnitude larger than the saturation current of a *pn*-junction. Because of this difference in saturation currents, the forward voltage across a Schottky diode is smaller than that across a *pn* junction diode with the same forward current. The voltage across a forward-biased *pn* junction diode is typically about 0.7 V, and that across a Schottky diode is typically about 0.4 V.

---

[11] The name comes from the fact that the current quickly reaches a maximum value of I_S as V_D becomes more negative and then saturates (i.e., it won't increase with further increases in reverse bias).

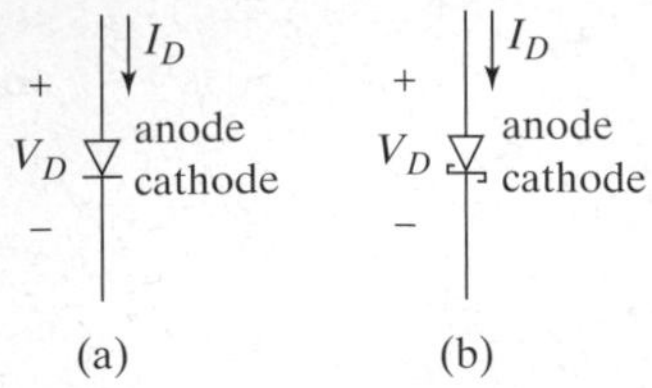

Figure 2–15 Schematic symbols for (a) a junction diode and (b) a Schottky diode.

We use different symbols to represent a junction diode and a Schottky diode, as illustrated in Figure 2-15. The large arrow represents the ***anode*** of the diode, and the short line represents the ***cathode***. For the Schottky diode, the short line is made to look like the letter 'S.' The diode is forward biased when the anode is positive with respect to the cathode and is reverse biased when the anode is negative with respect to the cathode. The direction of positive current flow is indicated by the arrow on the anode, so this serves as a visual reminder of the direction in which current is allowed to flow. A simple mnemonic for anode and cathode is to note that when they are listed alphabetically (i.e., anode first), their order indicates the direction of current flow through the device (i.e., from anode to cathode).

It is often useful to think about the operation of the diode as approximating that of an ***ideal diode***. Aside A2.3 describes the ideal diode in detail.

Both the *pn*-junction diode and the Schottky diode are reasonable approximations to the ideal diode. The most notable difference is that the voltage across either of these diodes is not zero in forward bias.

2.4.2 Detailed Analysis of Current Flow

We begin the more detailed treatment of current flow by considering again the depletion approximation made earlier. This approximation seems more plausible if we draw a band diagram for the *pn* junction, as shown in Figure 2-16. Part (a) of the figure shows the isolated *p*- and *n*-type semiconductors, and part (b) of the figure shows them together at equilibrium, where the Fermi levels must be equal (*see* Aside A2.1). Notice that the Fermi level must cross the intrinsic Fermi level. Also, remember that the Fermi level determines the probability of occupancy of the states that electrons might occupy. When $E_f = E_i$, we know that the material appears intrinsic and the mobile carrier concentrations are very low, equal to n_i.

ASIDE A2.3 IDEAL DIODE

The ***ideal diode*** can be defined as a device that has the *i–v* characteristic shown in Figure A2-4; the schematic symbol is also shown in the figure. An ideal diode does not allow current to flow at all in the negative direction (as defined in the figure); in other words, $i_D = 0$ for all $v_D < 0$. On the other hand, the voltage drop across the device is zero for *any* magnitude of current in the positive direction. There is a minor problem with using the ideal diode model to analyze the operation of some circuits: The model is not a true function, since there isn't a one-to-one mapping of voltage to current. One can always correct this problem by thinking of the characteristic for $v_D \geq 0$ as a straight line with an arbitrarily large, but finite, slope.

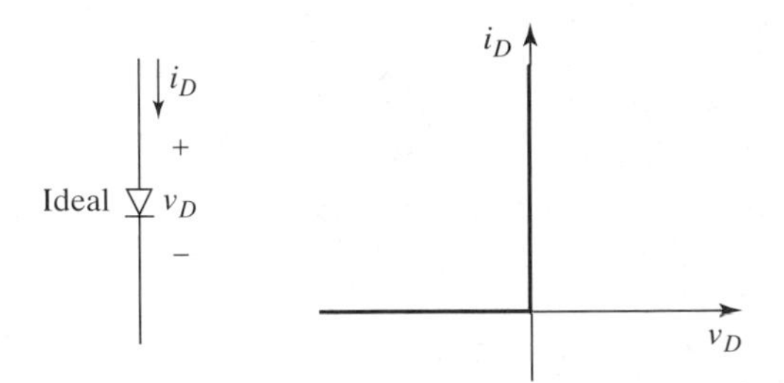

Figure A2-4 The ideal diode schematic symbol and *i–v* characteristic.

The ideal diode has two distinct regions of operation: When $v_D < 0$, the diode is said to be ***reverse biased*** (or off), and it looks like an open circuit, since the current through it is zero; when $v_D = 0$ and $i_D > 0$, the diode is said to be ***forward biased*** (or on), and it looks like a short circuit, since the voltage across it is zero for any current.

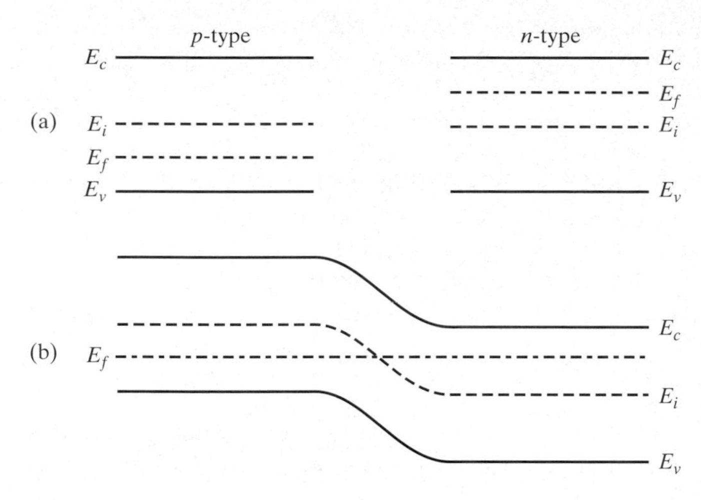

Figure 2–16 Energy band diagrams for (a) the separate *n* and *p* regions and (b) the *pn* junction at equilibrium.

Since the probability of occupancy is an exponential function of energy, if we plotted the density of mobile carriers versus position, we would find that it is much smaller near the junction, where $|E_f - E_i|$ is small, than it is in either the *n*- or *p*-type material away from the junction, where $|E_f - E_i|$ is larger. In other words, the region near the junction is depleted of mobile carriers. Of course, the variation in mobile carrier density is smooth rather than abrupt, as we assume, but the approximation significantly simplifies both the analysis and our thinking and is surprisingly accurate in practice.

Before examining what happens with an applied bias, it will be useful to take a moment and consider the built-in potential of the *pn* junction. Figure 2-13 showed the charge profile and field of a *pn* junction; in Figure 2-17, we repeat this figure and add to it a plot of the potential difference across the junction. Because the charge profile resulted from diffusion and the device was initially charge neutral, the total charge on each side of the junction must be equal. In other words,

$$N_A x_{dp} = N_D x_{dn}. \tag{2.26}$$

We will now calculate the magnitude of the built-in potential that results from this charge distribution.

Remember from electrostatics that the electric field caused by a spatial distribution of charge is given by the integral of the charge distribution. This fact is also expressed by Gauss' law, which, in point form, says that the divergence of the electric flux density, $\boldsymbol{D}$, at a point is equal to the charge density at that point:

$$\nabla \cdot \boldsymbol{D} = \rho. \tag{2.27}$$

Substituting the definition of electric flux, $\boldsymbol{D} = \epsilon \mathcal{E}$, into (2.27) and also recalling that the electric field is the negative of the gradient of the electric potential leads to Poisson's equation,

$$\nabla \cdot \nabla V = -\frac{\rho}{\epsilon}, \tag{2.28}$$

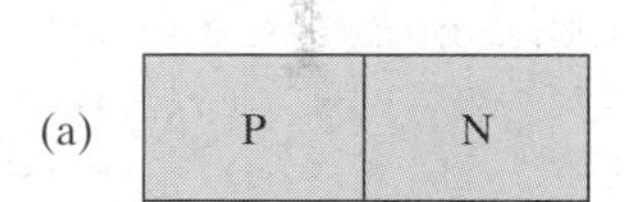

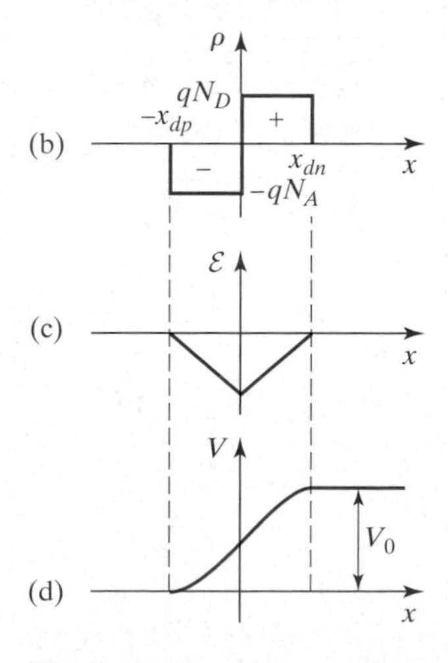

Figure 2–17 (a) A *pn* junction at equilibrium, (b) the charge profile, (c) the electric field, and (d) the potential.

which, in one dimension, reduces to

$$\frac{d^2V}{dx^2} = -\frac{\rho}{\epsilon}. \tag{2.29}$$

Integrating both sides of (2.29) for the charge distribution in part (b) of Figure 2-17, we get

$$\mathcal{E} = -\frac{dV}{dx} = \begin{cases} -\left(\frac{qN_A}{\epsilon}x + C_1\right); & -x_{dp} < x < 0 \\ \left(\frac{qN_D}{\epsilon}x + C_2\right); & 0 < x < x_{dn} \end{cases} \tag{2.30}$$

where C_1 and C_2 are constants that we find by applying the boundary conditions $\mathcal{E} = 0$ outside the depletion region (i.e., at $x = x_{dn}$ and $x = x_{dp}$). The resulting field is

$$\mathcal{E} = \begin{cases} -\frac{qN_A}{\epsilon}(x_{dp} + x); & -x_{dp} < x < 0 \\ \frac{qN_D}{\epsilon}(x - x_{dn}); & 0 < x < x_{dn} \end{cases} \tag{2.31}$$

and is plotted in part (c) of the figure. Integrating both sides of (2.31) and applying the arbitrary boundary condition that $V = 0$ at $x = -x_{dp}$ leads to (Problem P2.21)

$$V(x) = \begin{cases} \frac{qN_A}{\epsilon}\left(\frac{x_{dp}^2}{2} + x_{dp}x + \frac{x^2}{2}\right); & -x_{dp} < x < 0 \\ \frac{qN_D}{\epsilon}\left(x_{dn}x - \frac{x^2}{2}\right) + \frac{qN_A}{\epsilon}\frac{x_{dp}^2}{2}; & 0 < x < x_{dn} \end{cases}, \tag{2.32}$$

which is plotted in part (d) of the figure. The resulting built-in potential is

$$V_0 = V(x_{dn}) = \frac{q}{2\epsilon}(N_A x_{dp}^2 + N_D x_{dn}^2). \tag{2.33}$$

The built-in potential is rarely expressed as in (2.33), since a more convenient form can be derived by setting the drift and diffusion current densities equal to each other (*see* Problem P2.15). The result of this alternative calculation is

$$V_0 = \frac{kT}{q}\ln\frac{N_A N_D}{n_i^2} = V_T \ln\frac{N_A N_D}{n_i^2}. \tag{2.34}$$

We will find it useful later to have equations for the depths of penetration of the depletion region into each side of the junction, x_{dp} and x_{dn}. Combining (2.33) with (2.26) leads to

$$x_{dn} = \sqrt{\frac{2\epsilon V_0}{qN_D(1 + N_D/N_A)}} \tag{2.35}$$

and

$$x_{dp} = \sqrt{\frac{2\epsilon V_0}{qN_A(1 + N_A/N_D)}}. \tag{2.36}$$

Exercise 2.6

Consider an abrupt, uniformly doped pn junction at room temperature with $N_D = 10^{18}$ atoms/cm^3 and $N_A = 10^{15}$ atoms/cm^3. (a) Calculate V_0. (b) Find x_{dn} and x_{dp}.

We now turn our attention to calculating the terminal current that results from an applied bias. To do this, we first need to develop equations for the excess minority-carrier concentrations at the edges of the depletion region. We restrict ourselves to considering the minority-carrier electrons, but analogous equations result if the minority-carrier holes are considered. We will find it convenient to use the Fermi level, which is constant at equilibrium, as our reference potential. We then define the potentials on each side of the junction as the difference $E_i - E_f$, as shown in Figure 2-18(b). Using this figure and (A2.8), we can write

$$n(-x_{dp}) = n_i \exp\left(\frac{E_f - E_i(-x_{dp})}{kT}\right) = n_i e^{-q\phi_p/kT} \tag{2.37}$$

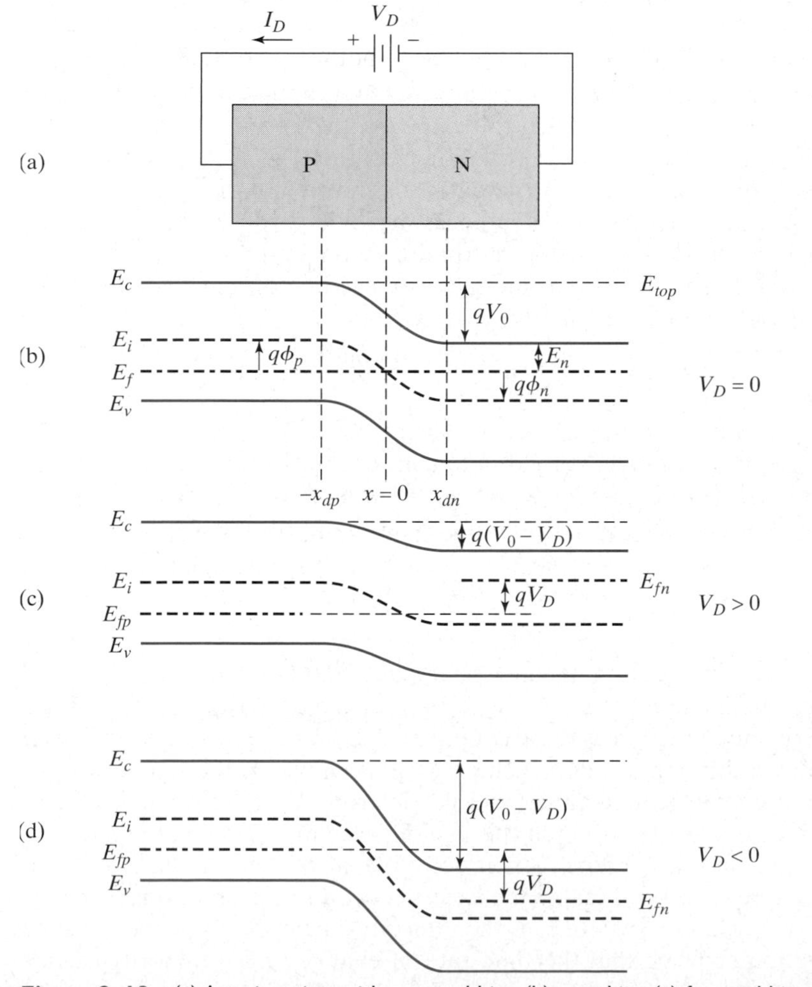

Figure 2–18 (a) A *pn* junction with external bias, (b) zero bias, (c) forward bias, and (d) reverse bias.

and

$$n(x_{dn}) = n_i \exp\left(\frac{E_f - E_i(x_{dn})}{kT}\right) = n_i e^{-q\phi_n/kT}. \quad \textbf{(2.38)}$$

Taking the ratio of these two concentrations yields

$$\frac{n(-x_{dp})}{n(x_{dn})} = e^{-q(\phi_p - \phi_n)/kT} = e^{-qV_0/kT}. \quad \textbf{(2.39)}$$

Assuming equilibrium and room temperature, we can say that $n(x_{dn}) = N_D$. We then find the equilibrium concentration of minority carriers on the p side to be

$$n_{p0} = n(-x_{dp}) = N_D e^{-qV_0/kT}. \quad \textbf{(2.40)}$$

Now, consider applying a bias V_D as shown in the figure. Remember that the contact potentials exactly cancel the built-in potential, so that $V_D = 0$ corresponds to the equilibrium condition. If V_D is not zero, the thermodynamic equilibrium of the junction will be disturbed. Remember also that we have assumed that both the n- and p-type materials *outside the depletion region* are neutral. In other words, we are assuming that these regions can conduct currents reasonably well without any significant voltage drops across them. Therefore, *all* of the applied voltage will show up across the depletion region. When this happens, the Fermi levels on the two sides will no longer be the same: They will differ by exactly qV_D. If V_D is positive, it *reduces* the magnitude of the internal potential, as shown in part (c) of the figure, and if it is negative, it *increases* the potential, as shown in part (d) of the figure.

If we restrict our analysis to small perturbations from equilibrium (called ***low-level injection***), we can assume that the majority-carrier concentrations will not change appreciably, and we can modify (2.40) to obtain

$$n_p(-x_{dp}) = N_D e^{-q(V_0 - V_D)/kT} = n_{p0} e^{V_D/V_T}. \quad \textbf{(2.41)}$$

We will find it convenient in what follows to consider the excess minority-carrier concentration—that is, the amount by which the minority-carrier concentration exceeds its equilibrium value (we denote the excess concentration with a Δ):

$$\Delta n_p(-x_{dp}) = n_p(-x_{dp}) - n_{p0} = n_{p0}(e^{V_D/V_T} - 1). \quad \textbf{(2.42)}$$

Now that we have this equation in hand, we need to find how this excess concentration varies with position before we can find the current. To do this, we change the x–axis origin and direction as shown in Figure 2-19 to make the equations simpler. Examining a differential slice of the region as shown in the figure, we can write an equation that accounts for all mobile electrons. Because we have an elevated number of minority carriers in this region, recombination will greatly exceed generation, and we ignore generation. We also use the fact that the recombination rate is proportional to the excess concentration and express the recombination rate as $\Delta n/\tau_n$, where τ_n is the minority-carrier lifetime. We can now write an equation showing that the time rate of change of the minority-carrier concentration in the differential slice is equal to the difference in the number of carriers entering and leaving, minus the recombination rate:

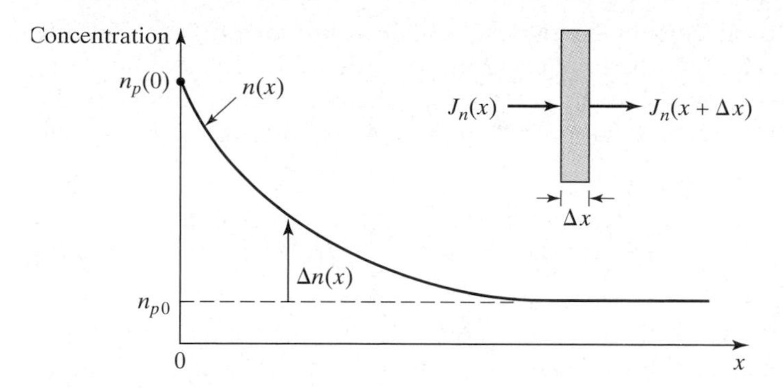

Figure 2–19 Minority-carrier concentration versus position. The total concentration, $n(x)$, excess concentration, $\Delta n(x)$, and equilibrium concentration, n_{p0}, are all shown. Also shown is a small slice of the material, indicating the currents entering and leaving the slice.

$$\frac{\partial n(x)}{\partial t} = \frac{1}{q}\frac{J_n(x) - J_n(x + \Delta x)}{\Delta x} - \frac{\Delta n(x)}{\tau_n}. \tag{2.43}$$

In the limit as Δx goes to zero, we obtain the ***continuity equation*** for electrons:

$$\frac{\partial n(x)}{\partial t} = \frac{1}{q}\frac{\partial J_n(x)}{\partial x} - \frac{\Delta n(x)}{\tau_n}. \tag{2.44}$$

We can now solve this equation for the steady-state condition. In the steady state, the time derivative is, by definition, zero. Also, recall that the depletion approximation includes assuming that the neutral regions cannot support an electric field (i.e., they are highly conductive due to the number of majority carriers present). Therefore, any minority-carrier current that flows will be due exclusively to diffusion, and we can use (2.6) for the current. Using these facts, we find that (2.44) becomes

$$\Delta n(x) = D_n \tau_n \frac{d^2 n(x)}{dx^2}. \tag{2.45}$$

Remembering that $n(x) = n_{p0} + \Delta n(x)$ and assuming that the region is long enough to allow the concentration to decay away naturally, we see that the solution to this equation is

$$\Delta n(x) = \Delta n(0) e^{-x/\sqrt{D_n \tau_n}}, \tag{2.46}$$

which is plotted in Figure 2-19. We can now again make use of (2.6) and determine the magnitude of the diffusion current. We obtain

$$J_n(x) = qD_n \frac{dn(x)}{dx} = qD_n \frac{d\Delta n(x)}{dx} = -q\Delta n(x)\sqrt{\frac{D_n}{\tau_n}}. \tag{2.47}$$

The current is negative because the electrons are diffusing to the right in this figure and that corresponds to conventional current flow to the left. The magnitude of the current decreases with increasing x because the electrons are recombining with holes. The hole current necessary to supply this recombination will increase as the electron diffusion current decreases, but the total current will stay the same.[12] Therefore, we can find the total current due to the electrons that have diffused across the forward-biased junction by finding the current at the edge of the depletion region before recombination has occurred. Making use of (2.47) and (2.42), we obtain

$$J_{n\text{total}} = J_n(0) = -q\sqrt{\frac{D_n}{\tau_n}}\Delta n(0) = -q\sqrt{\frac{D_n}{\tau_n}}n_{p0}(e^{V_D/V_T} - 1). \quad \textbf{(2.48)}$$

A similar equation results for the holes on the n side, and the total current is the sum of the two diffusion currents, (The electrons and holes diffuse in opposite directions, but have opposite signs of charge.) Combining the results and denoting the area of the junction in the plane perpendicular to the current flow A results in

$$\begin{aligned} I_D &= A(J_p - J_n) = qA\left(\sqrt{\frac{D_n}{\tau_n}}n_{p0} + \sqrt{\frac{D_p}{\tau_p}}p_{n0}\right)(e^{V_D/V_T} - 1) \\ &= I_S(e^{V_D/V_T} - 1) \end{aligned} \quad \textbf{(2.49)}$$

where the reverse saturation current, I_S, is implicitly given by this equation and our result is the same as was given in (2.25). The fact that I_S and, therefore, I_D are proportional to the junction area is significant and should be noted. The diode characteristic is plotted in Figure 2-20.

Since the current is an exponential function of the voltage, the voltage across a forward-biased diode does not change much with changing current. For example, at room temperature, the voltage across the diode will change by about 18 mV for every doubling (or halving) of the current and by about 60 mV for every factor-of-ten change in the current (*see* Problem P2.20). In other words, there is an 18-mV-per-octave change, or a 60-mV-per-decade change, in voltage. Since the logarithm of a product is the sum of the logarithms, if the current changes by a factor of 20, we see a change in voltage of about $60 + 18 = 78$ mV. Similarly, for a four-order-of-magnitude change in current, the voltage varies only by about $4 \times 60 = 240$ mV.

The numbers just given illustrate why it is often reasonable to assume that the drop across a forward-biased *pn*-junction diode is between about 0.6 and 0.8V. If the other voltages in a circuit are much larger than this, we can reasonably approximate the diode as having a constant 0.7 V drop across it. Making this approx-

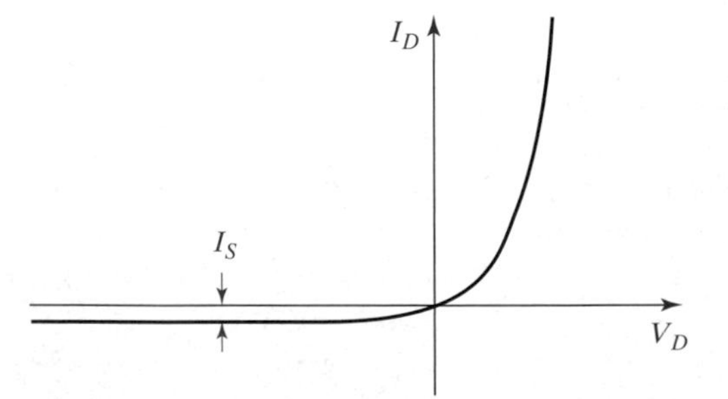

Figure 2–20 A typical diode *i–v* characteristic.

[12] This hole current obviously requires a small electric field, but the depletion approximation assumes the majority-carrier density to be so high that the field is small enough not to affect the minority-carrier diffusion current.

imation and neglecting the small reverse saturation current leads to the common piecewise linear diode model shown in Figure 2-21. When this model is used, engineers frequently talk about the *cut-in*, or *turn-on*, voltage of the diode; in this case, it is 0.7 V.

It is occasionally *very important* to remember that a real diode does *not* exhibit the discontinuous behavior shown in Figure 2-21. A real diode follows an exponential characteristic and does not have any turn-on voltage. The diode is forward biased for *any* $V_D > 0$, and the current changes by about a factor of 10 every time the voltage increases by 60 mV. To illustrate the vague nature of talking about a cut-in, or turn-on, voltage, consider Figure 2-22, where we show the *same* diode characteristic plotted with three different scales for the vertical axis. Note that you would say that this diode has three distinct turn-on voltages, depending on which scale you used!

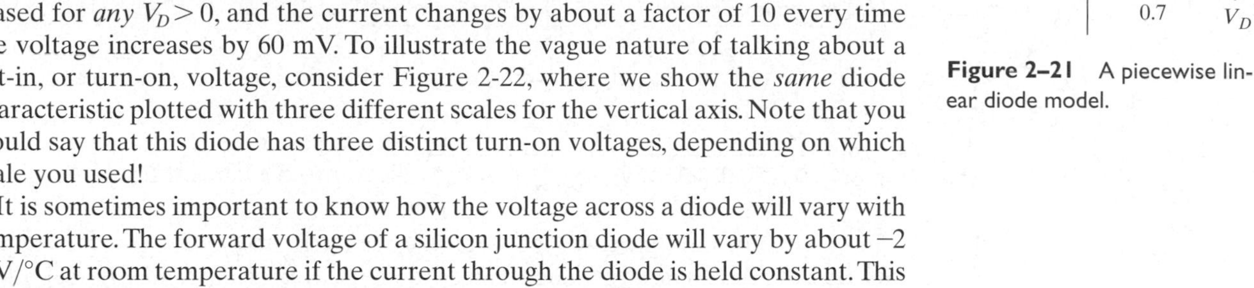

Figure 2–21 A piecewise linear diode model.

It is sometimes important to know how the voltage across a diode will vary with temperature. The forward voltage of a silicon junction diode will vary by about −2 mV/°C at room temperature if the current through the diode is held constant. This rate of change is an example of a *temperature coefficient* (TC), and we would say that the diode forward voltage has a TC = −2 mV/°C. We also sometimes need to know how I_S varies with temperature. Roughly, I_S doubles for every 6°C increase in temperature, as stated earlier.

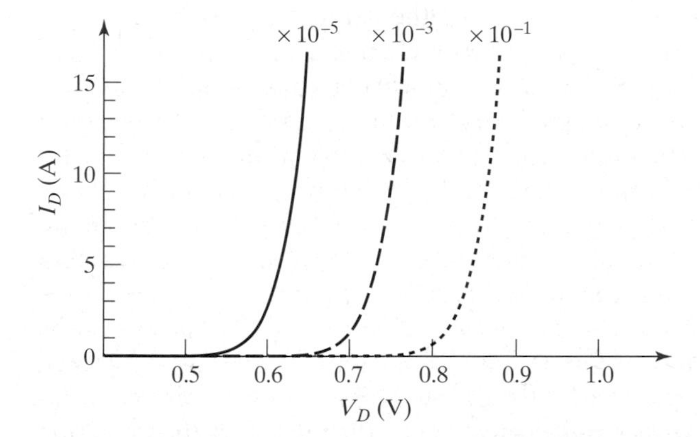

Figure 2–22 Diode characteristic plotted with different scales for the current axis.

2.4.3 Minority-Carrier Profiles

In preparation for describing the operation of the bipolar transistor in the next section, we need to examine minority carriers in a *pn* junction in more detail. The minority-carrier densities for a *pn* junction are shown as a function of position in Figure 2-23. In part (a) of the figure, we show the densities with no applied bias. This is the equilibrium case, and except within the depletion region, the densities have the values they would have in equilibrium in isolated *n*- or *p*-type material. The drift currents are being supplied by EHP generation both in and near the depletion region, and the minority carriers that drift away are exactly compensated for by the diffusion of carriers from the other side. For example, electrons generated as part of electron–hole pairs near the depletion region edge in the *p*-side may move to the edge of the depletion region and be swept across by the field. But this loss of electrons is compensated for by electrons diffusing from the *n* side to the *p* side, so the density remains at the equilibrium value.

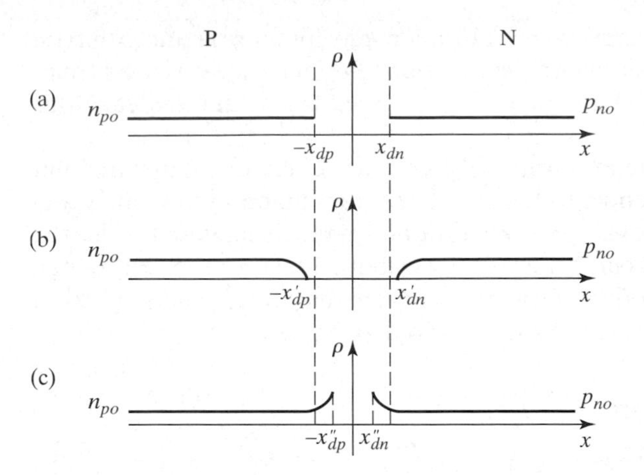

Figure 2–23 Minority-carrier charge density profiles: (a) no applied bias, (b) reverse bias, and (c) forward bias.

In part (b) of the figure, we see what happens with an applied reverse bias. In this case, the field gets larger and reduces the net diffusion currents. The net effect on the minority-carrier densities is that the equilibrium values can no longer be maintained, and the densities are reduced at the edges of the depletion region, as predicted by (2.41) for electrons when $V_D < 0$. For any substantial applied reverse bias the densities essentially go to zero as indicated in the figure. The external terminal current in this case will be due to the drift current.

This reverse drift current is supplied by minority carriers diffusing toward the junction and then being swept across due to the field (these minority carriers are supplied by EHP generation, as mentioned before) and, to a lesser extent, by EHP generation in the depletion region itself.[13] There is a net generation of EHPs in these regions, because wherever the minority-carrier densities are below their equilibrium values, the generation rate exceeds the recombination rate (since there are fewer carriers available to recombine). In addition, because the applied reverse bias increases the potential drop across the junction, there must be more uncompensated dopant ions on each side of the junction to support this change in potential, and the result is that the depletion region gets wider, as indicated in the figure.

In part (c) of the figure, we see what happens when a forward bias is applied. In this case, the net diffusion currents increase in proportion to $\exp(V_D/V_T)$, as noted earlier. Since the majority carriers diffusing across the junction end up as minority carriers on the other side (for example, electrons diffusing across the depletion region from the n side become minority carriers on the p side), the minority-carrier densities are elevated at the edges of the depletion region, as shown by (2.41) for electrons. These elevated minority-carrier concentrations then lead to increased recombination, so they decline as you move further away from the depletion region. As seen in (2.41), the minority-carrier densities right at the depletion region boundaries can be expressed as

$$n_p(x''_{dp}) = n_{po}e^{V_D/V_T} \tag{2.50}$$

and

$$p_n(x''_{dn}) = p_{no}e^{V_D/V_T}, \tag{2.51}$$

where x''_{dp} and x''_{dn} are as defined in the figure. We also note that the edges of the depletion region have again moved relative to the case with no applied bias. This

[13] EHPs generated within the depletion region do contribute to the drift current, but not to the *net* drift current as we have defined it.

time, the depletion region gets narrower, since the applied voltage reduces the potential, and we must have fewer uncompensated dopant ions on each side to support the field. We will use (2.50) and (2.51) in our derivation of bipolar transistor operation in Section 2.5.

Finally, we point out that the minority-carrier profiles shown here assume that the n and p regions are long enough that the concentrations can decay naturally back to their equilibrium bulk values (i.e., n_{po} and p_{no}). This is typically called the ***long-base diode***. If the regions are shorter, then the border (either another junction or a metal contact) will define a new boundary condition. This other possibility, called the ***short-base diode***, is implicitly covered in Section 2.5.

2.4.4 Summary of Current Flow

At this point, it is worthwhile to complete the picture we have of steady-state current flow in a pn junction. By "steady-state," we simply mean that this is not a transient situation, but that some constant external current is flowing and has been for some time. We know that majority carriers diffuse across the depletion region and minority carriers drift across it in the opposite direction. To complete the picture, we need to understand what happens in the neutral regions on either side of the depletion region.

Consider the case of a forward-biased junction diode, and ignore the small drift currents across the depletion region for simplicity. We saw in Figure 2-23 that the minority-carrier densities are elevated at the edges of the depletion region. Let's consider one carrier type to simplify the discussion. The net diffusion current of electrons from the n side to the p side, is supplied by electrons drifting toward the depletion region in the neutral n side, as shown by the first current component J_{n1}. in Figure 2-24. Because the electron density in the n-type material is very high, the field necessary to support this drift current is quite small and does not invalidate our assumption of neutrality.

Notice that the magnitude of J_{nl} is reduced as you move toward the junction, due to electrons recombining with excess holes diffusing away from the junction, which constitute the current J_{p2}. The total current at any point in the device is the sum of the electron and hole currents, since they are going in opposite directions, and must always be constant; otherwise, charge would accumulate somewhere, and this would not be a steady-state situation. The electrons that reach the depletion region edge then diffuse across it, where they become excess minority carriers on the p side.

Because of the excess of minority electrons near the depletion region in the p-type material, there is a concentration gradient, and these electrons diffuse away from the edge of the depletion region. This diffusion current component is shown

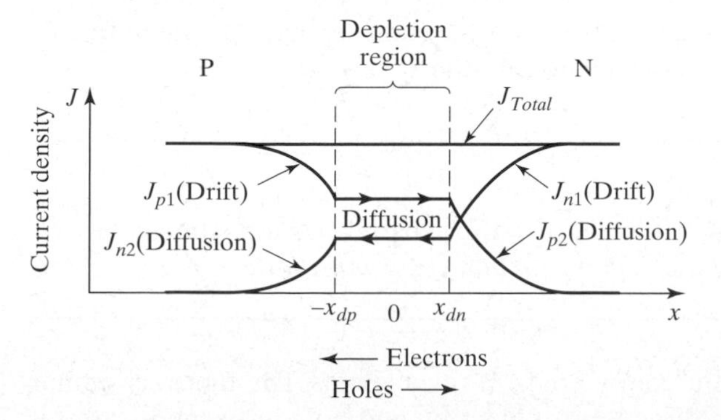

Figure 2–24 Current components in a forward-biased pn junction. Note that the net diffusion currents are only constant in the depletion region because of our definition; see the text for details.

as J_{n2} in the figure. Also, because of the elevated concentration, the rate of EHP recombination exceeds the rate of EHP generation in this region, and the concentration decreases as you move away from the junction. Therefore, the magnitude of J_{n2} decreases to zero as you move away from the junction, as shown in Figure 2-24. These same arguments apply to holes drifting toward the junction on the p side (J_{p1}), diffusing across, and then diffusing away from the junction as minority carriers on the n side (J_{p2}).

It is extremely important to realize that the net diffusion currents in the depletion region are constant only *because we have defined them that way*! We defined the "net diffusion current" to be made up of those carriers which successfully diffuse *all the way* across the depletion region. If we instead define the diffusion current as the number of carriers per unit time diffusing across an imaginary plane, we see that it is a strong function of position in the depletion region, due to carriers that start diffusing across, but are turned around by the field and drift back to the side they started from.

If second-order effects are included, the situation is not as clear cut as we have implied up to this point. For example, the components J_{n2} and J_{p2} are not entirely due to diffusion. The small fields in the neutral regions are in the correct direction to aid these diffusion currents. Nevertheless, since drift currents are proportional to the carrier density (*see* [2.9]), the drift components in these currents are quite small, and the currents are mostly due to diffusion. Similarly, charge neutrality requires that the majority-carrier densities in the neutral regions have the same shape as the minority-carrier densities. Therefore, J_{n1} and J_{p1} will contain small diffusion components as well.

The example shown in Figure 2-24 assumes that $N_A > N_D$, so that $x_{dn} > x_{dp}$ and the hole current injected into the n side is larger than the electron current injected into the p side. Real junction diodes are typically asymmetrical, as illustrated here.

Finally, it should be noted that Figure 2-24 ignores generation and recombination within the depletion region. Generation is ignored because it contributes to the drift currents, which play an inconsequential role in forward bias. If the recombination component is included, it leads to more majority carriers on each side of the depletion region drifting toward this region and then recombining. In practical diodes at very low bias, recombination in the depletion region is often the dominant term.

Recombination in the depletion region varies with bias as $\exp(V_D/2V_T)$, rather than $\exp(V_D/V_T)$. Therefore, to the extent that it affects the diode current, the exponent in our equation for the diode current will change. This effect is usually modeled by adding the ***ideality factor***, n, to the equation; that is,

$$I_D = I_S(e^{V_D/nV_T} - 1), \tag{2.52}$$

where $n = 1$ for an ideal diode and n approaches 2 as recombination in the depletion region becomes more important. Because recombination in the depletion region is more important at low currents, n is a function of bias. Throughout this text, we will use $n = 1$ unless specifically noted otherwise.

Exercise 2.7

Measurements on a given diode show that $I_D = 731\ \mu A$ when $V_D = 0.65$ V and $I_D = 2.26$ mA when $V_D = 0.7$ V. Find the values of n and I_S for this diode.

Figure 2-25 shows the current components in reverse bias. The figure is similar to Figure 2-24, except that the directions of electron and hole movements are re-

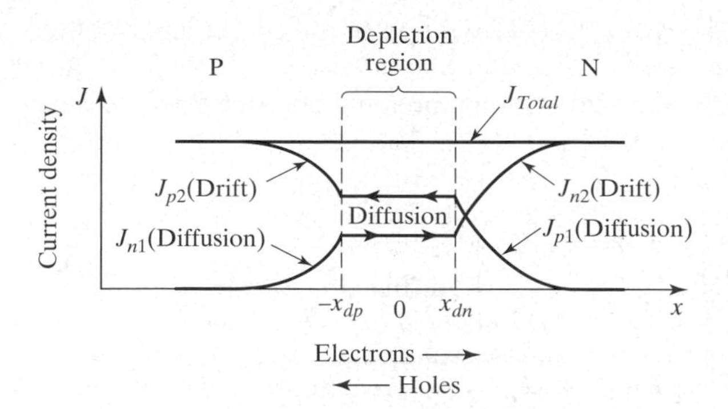

Figure 2–25 Current components in a reverse-biased *pn* junction. Note that the net drift currents are only constant in the depletion region because of our definition; see the text for details.

versed, the magnitude of the resulting current is much smaller, and the carriers drift across the depletion region. In the reverse bias case, the minority-carrier densities go to zero at the edges of the depletion region, so the minority carriers diffuse *toward* the depletion region (components J_{n1} and J_{p1} in the figure), with EHP generation supplying the needed carriers. The generation rate exceeds the recombination rate in the neutral regions near the depletion region because the number of minority carriers is below the equilibrium concentration. The carriers then drift across the depletion region and on through the neutral regions (components J_{n2} and J_{p2}). The magnitudes of the drift currents increase as they move away from the depletion region because of the EHP generation that is supplying the minority carriers diffusing the other way.

This picture ignores EHP generation and recombination in the depletion region. Recombination is ignored because there are very few carriers to recombine, and it does not, therefore, have a significant effect. If generation is included, it simply increases the magnitudes of the drift currents as they traverse the depletion region. EHP generation in the depletion region can be the dominant term in I_S for real diodes. Also, the total number of EHPs contributing to I_S will increase if the width of the depletion region increases. Since the depletion-region width increases with increasing reverse bias, the saturation current will also increase. This increase in I_S is a second-order effect that we will ignore.

We should again note, as we did for Figure 2-24, that the current components in the depletion region in Figure 2-25 are constant only because we have defined the net drift current to be made up of those carriers which drift *all* the way across the region. We could also define drift current as the number of carriers per unit time drifting across an imaginary plane. We would then see that it is a strong function of position in the depletion region, due to carriers that start diffusing across, but are turned around by the field and drift back to the side they started from.

2.4.5 Charge Storage and Varactor Diodes

In discussing the effect of applied bias on the minority-carrier charge densities, we noted that the amount of charge stored in the depletion region changes with applied bias. In other words, the junction exhibits capacitive behavior. Remember that for a linear capacitor, the charge stored is given by $Q = CV$. In the case of a junction diode, however, the capacitance is not linear; in other words, the charge is not simply proportional to the applied voltage. As a result, we must consider the *change* in the stored charge for a given *change* in the applied voltage.

Therefore, we must use a derivative and find the slope of the Q_{DR}-versus-V_D characteristic as a function of V_D, where Q_{DR} is the charge stored in the depletion region. Since this slope is a function of V_D, the capacitance is not constant. By evaluating the capacitance at a particular applied voltage, however, we can find a capacitance that will be approximately constant for small changes about the nominal applied voltage. A model that is valid only for small changes about the bias point is called a ***small-signal*** or ***incremental*** model; it is useful in small-signal analysis, the major topic of Chapter 6.

When the required analysis is carried out for a junction diode, one finds that the ***junction capacitance*** can be expressed by

$$C_j \equiv \frac{dQ_{DR}}{dV_D} = \begin{cases} \dfrac{C_{j0}}{(1 - V_D/V_0)^{m_j}} & ; \quad \textit{in reverse bias} \\ 2C_{j0} & ; \quad \textit{in forward bias} \end{cases}, \tag{2.53}$$

where C_{j0} is the zero bias value of the junction capacitance, V_0 is the built-in potential of the junction, and m_j is an exponent that is a function of the doping profile in the device and is typically between 0.33 and 0.5. The value given for C_j in forward bias is only an approximation. The equation given for reverse bias is actually accurate in forward bias as well, so long as the forward current is small. As V_D approaches V_0, however, the equation predicts that the capacitance goes to infinity. At that point, the assumptions used in deriving the equation are no longer valid. Equation (2.53) is useful from a circuit designer's viewpoint, since it shows how C_j varies with applied bias. Example 2.2 illustrates the calculation of a small-signal capacitance.

Example 2.2

Problem:
Consider equations (2.35) and (2.36) giving the depths of penetration of the depletion region in an abrupt *pn* junction with uniform doping. These equations were derived with no applied bias. Modify the equations to account for an applied bias by replacing the built-in potential with $(V_0 - V_D)$ and derive the junction capacitance per unit area.

Solution:
Making use of the definition of small-signal junction capacitance given by (2.53), along with the charge stored on either side of the junction as given by (2.26), we have

$$C_j' \equiv \frac{dQ_{DR}'}{dV_D} = qN_D\frac{dx_{dn}}{dV_D} = qN_A\frac{dx_{dp}}{dV_D} \tag{2.54}$$

where the primes indicate that we are dealing with charge and capacitance per unit area. Using $(V_0 - V_D)$ in (2.35) and then using (2.54) allows us to calculate the small-signal junction capacitance of an abrupt, uniformly doped *pn*-junction diode. We have

$$C_j' \equiv qN_D\frac{dx_{dn}}{dV_D} = qN_D\frac{d}{dV_D}\sqrt{\frac{2\epsilon(V_0 - V_D)}{qN_D(1 + N_D/N_A)}}, \tag{2.55}$$

which leads to

$$C_j' = \sqrt{\frac{\epsilon q}{2(V_0 - V_D)(1/N_D + 1/N_A)}}. \tag{2.56}$$

We can equate (2.56) with the formula given in (2.53) and discover that $m_j = 0.5$ and

$$C'_{j0} = \sqrt{\frac{\epsilon q}{2V_0(1/N_D + 1/N_A)}}. \tag{2.57}$$

We now note that the junction capacitance is similar to a parallel-plate capacitor where the depletion region acts as the dielectric and the two neutral regions act as the conducting plates. This view is practically useful, since it correctly predicts the way the capacitance changes with applied bias. Increasing the magnitude of the reverse bias increases the depletion region width, which means that the parallel-plate capacitor has a larger plate spacing, which reduces the capacitance.

There are applications wherein the voltage dependence of the junction capacitance can be quite useful. For example, by varying the DC bias on a diode, the capacitance seen by a small AC signal can be varied and the diode can be used to electronically tune some filter or oscillator. Special diodes optimized for use as voltage-dependent capacitors (operating in reverse bias) are called ***varactor diodes***. The schematic symbol for a varactor diode is shown in Figure 2-26. Varactor diodes typically have capacitances in the range of a few picofarads to a few hundred picofarads, and the ratio of maximum to minimum capacitance is typically about three, although there are some varactors with much larger ranges. The reverse voltages are usually in the range of a few tenths of a volt to a few tens of volts.

Figure 2–26 A varactor diode symbol.

The junction capacitance is the only significant capacitance present in reverse bias, but in forward bias there is another charge storage mechanism to consider. We have already seen that a forward-biased junction diode has minority-carrier densities that are elevated above the equilibrium values, *see* Figure 2-23(c). The number of excess minority carriers (i.e., $n_p - n_{po}$ or $p_n - p_{no}$) changes with applied bias and constitutes a voltage-dependent charge storage, or capacitance. Because this extra charge is caused by the diffusion of majority carriers across the depletion region, the capacitance is called the ***diffusion capacitance***. Ignoring the small drift currents across the depletion region, we see that the external current that flows in a forward-biased diode is caused by the net diffusion currents alone.

Now, define the average ***minority-carrier lifetime*** to be the ratio of the total excess minority-carrier charge present on one side of the junction to the current caused by that charge. For example, if we consider just the electrons on the *p* side of the junction and denote the current due to these electrons by I_n, we have

$$\tau_n = Q_n/I_n, \tag{2.58}$$

where Q_n is the total excess charge stored (i.e., the integral of $n_p - n_{po}$ throughout the *p* region) and τ_n is the average minority-carrier lifetime for electrons on the *p* side. Similarly,

$$\tau_p = Q_p/I_p, \tag{2.59}$$

where the notation is analogous. These minority-carrier lifetimes are functions of the material, the doping density, and the temperature. For a given device at a fixed temperature, they can therefore be taken to be constants, and we can express the currents in terms of the charge profiles and the lifetimes. Remembering that the total current in the device is the sum of the electron and hole net diffusion currents, we can write

$$I_D = I_n + I_p = Q_n/\tau_n + Q_p/\tau_p. \tag{2.60}$$

For simplicity, we now restrict our attention to what is often called a ***one-sided*** diode—that is, a diode with the doping on one side much heavier than on the other side, so that one of the two terms on the right side of (2.60) can be safely ignored. Suppose that we have a diode made with a heavily doped n-type region and a relatively lightly doped p-type region. Such a diode is often referred to as an n^+–p diode. In this case, the hole current in (2.60) will be negligible. Therefore, we can write

$$Q_n \approx I_D \tau_n \approx \tau_n I_S e^{V_D/V_T}. \tag{2.61}$$

The diffusion capacitance can now be found by differentiating (2.61):

$$C_d \equiv \frac{dQ_n}{dV_D} = \tau_n \frac{dI_D}{dV_D} = \tau_n \frac{I_S e^{V_D/V_T}}{V_T} = \tau_n \frac{I_D}{V_T}. \tag{2.62}$$

This capacitance is again only approximately constant for small deviations from the nominal applied voltage.

Schottky diodes do have junction capacitance, and the value can be predicted by treating the diode as an abrupt, one-sided pn-junction diode with the highly doped side corresponding to the metal in the Schottky diode. Schottky diodes do not have diffusion capacitance, however, since the current is made up of majority carriers, as pointed out in Section 2.3.1; therefore, there isn't any significant minority-carrier charge storage.

Exercise 2.8

Consider an abrupt, uniformly doped pn junction with $N_A = 10^{14}$ atoms/cm^3, $N_D = 10^{17}$ atoms/cm^3, $\tau_n = 0.1$ ns, and an area of 10,000 μm^2. (a) Find the built-in potential for this diode. (b) Find the zero-bias junction capacitance, C_{j0}. (c) Find the total capacitance of this diode when it is reverse biased with $V_D = -5V$ and (d) when it is forward biased with $I_D = 2$ mA.

□

2.4.6 Breakdown and Zener Diodes

When the reverse bias on a junction diode is increased, the current will not stay the same forever. Eventually, the junction will ***break down***, and the current will increase rapidly. Contrary to what one might expect from the name, reverse breakdown is not a destructive mechanism by itself. A junction diode can be operated safely in this region so long as the power dissipation and terminal current are kept within acceptable limits. In fact, there is a wide range of applications for what are called ***zener diodes***. A zener diode is a diode with a specified reverse breakdown voltage. Zener diodes are often used as voltage references. As we will soon see, calling these diodes zeners is frequently a misnomer, but it is the common terminology. There are really two different breakdown mechanisms that must be understood: ***avalanche breakdown*** and ***zener breakdown***.

We first discuss avalanche breakdown. In a reverse-biased junction, the minority carriers drift across the depletion region. On their way across this region, they occasionally have collisions with atoms in the lattice. With a large enough field, a carrier drifting across the depletion region is accelerated to the point where it has enough energy to knock a valence electron free from its host atom during a collision. The field then separates the electron and hole of this newly created EHP, and we now have three mobile carriers instead of one. This process is called ***avalanche multiplication***. The multiplication can become quite large if the carriers generated by this collision also acquire enough energy to create more carriers, thereby initiating a chain reaction. Once the process starts, the number of multiplications that can occur from a single collision increases rapidly with further increases in the reverse bias, so the terminal current grows rapidly, and we say that the junction "breaks down." This is the process of avalanche breakdown.

As the doping levels in a *pn* junction increase, the depletion region gets narrower. Therefore, at a fixed reverse bias, the electric field will be larger for more highly doped junctions, as is evident from evaluating (2.31) at $x = 0$, which is where the field is largest. The result is that the critical field necessary to induce avalanche breakdown occurs at a lower terminal voltage. In other words, the breakdown voltage of a *pn*-junction diode decreases as the doping levels increase. The geometry of the device can also have a large influence on the voltage at which avalanche breakdown occurs, since sharp corners produce significantly larger electric fields than straight lines do (because more charge per unit distance is located on the inside of a curve than is located on a straight line).

Example 2.3

Problem:

Assuming that a *pn*-junction diode will begin to have avalanche breakdown when the electric field reaches a critical value of $\mathcal{E}_{crit} = 3 \times 10^5$ V/cm, what is the breakdown voltage of a diode that has $N_A = 10^{16}$ atoms/cm^3 and $N_D = 10^{18}$ atoms/cm^3?

Solution:

Using (2.31), we find that the maximum field in the depletion region occurs at $x = 0$ and is given by

$$\mathcal{E}(0) = \frac{-qN_A x_{dp}}{\epsilon}. \tag{2.63}$$

Setting (2.63) equal to the critical field (ignoring the minus sign in the equation since it is the magnitude of the field that matters) and using the given values along with $q = 1.6 \times 10^{-19}$ coulombs, $\epsilon_0 = 8.854 \times 10^{-14}$ F/cm, and $\epsilon_r = 11.7$ for silicon, we find that breakdown will occur when the depletion region edge in the p region is

$$x_{dp} = \frac{\mathcal{E}_{crit}\epsilon}{qN_A} = 1.9\ \mu\text{m}. \tag{2.64}$$

We can now modify (2.36) and use it to find the breakdown voltage. Equation (2.36) was derived for the equilibrium condition, but if we replace V_0 by $(V_0 - V_D)$, we can also apply it when a bias is present. Making the substitution and setting (2.36) equal to (2.64) yields

$$V_0 - V_D = \frac{qN_A}{2\epsilon}\left(1 + \frac{N_A}{N_D}\right)x_{dp}^2 = 28.2\ \text{V}. \tag{2.65}$$

Finally, we use (2.34) and the fact that $n_i = 1.45 \times 10^{10}$ cm^{-3} to find that

$$V_0 = V_T \ln\frac{N_A N_D}{n_i^2} = 0.82\ \text{V} \tag{2.66}$$

and combine this result with (2.65) to find that the junction breaks down when $V_D = -27.4$ V. The negative sign is usually left off when specifying a breakdown voltage, since reverse bias is assumed. Therefore, we conclude that the breakdown voltage is 27.4 V.

■

We now turn our attention to zener breakdown. If the doping levels of a junction get high enough, then when a reverse voltage is applied, the energy bands quickly reach the point shown in Figure 2-27(b) where the conduction band edge in the n-type material is actually at a lower energy than the valence band edge in the p-type material. Under these conditions, a significant current can flow by quantum mechanical tunneling, as was described in Aside A2.2. This tunneling is the zener effect. An alternative explanation of this phenomenon is to say that in very highly doped devices the depletion region is so narrow that the electric field can get large enough to simply rip valence band electrons free from their parent atoms. This process is called ***field ionization***.

In silicon junction diodes, zener breakdown dominates for breakdown voltages less than about 5 V, while avalanche breakdown is the dominant mechanism for voltages above about 7 V. For breakdown voltages around 5–7 V, both mechanisms occur simultaneously.

Exercise 2.9

Consider an abrupt pn junction diode with uniform doping, $N_A = 10^{17}$ atoms/cm^3 and $N_D = 10^{18}$ atoms/cm^3. What would the reverse breakdown voltage of this junction be if avalanche breakdown could occur? Does the junction exhibit avalanche breakdown or zener breakdown?

□

As mentioned earlier, diodes intended for use in the reverse breakdown region are commonly called zener diodes, independent of which mechanism is dominant. The symbol for a zener diode is shown in Figure 2-28(a) and the i–v characteristic of a typical zener (a 1N3999) is shown in part (b) of the figure. Since the device is

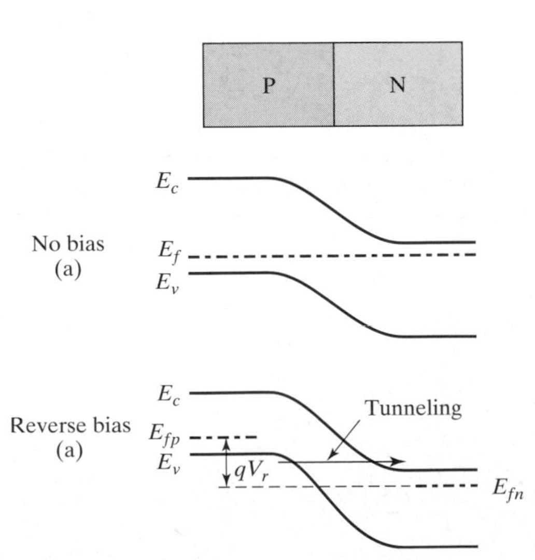

Figure 2–27 Band diagram of a heavily doped pn junction (a) with no applied bias and (b) with an applied reverse bias of magnitude V_r.

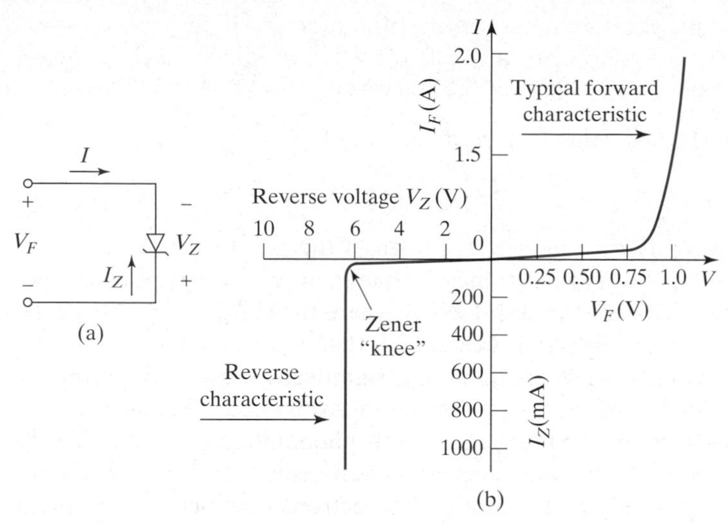

Figure 2–28 (a) Zener diode schematic symbol and (b) the *i*–*v* characteristic of a 1N3999. (Note the different scales used in each quadrant.)

intended to be operated in reverse breakdown, the directions of the voltage and current are reversed with respect to a conventional diode. To make the *i*–*v* characteristic more understandable, however, we have plotted I and V using the more familiar polarities for a regular diode as shown on the figure. A wide range of zener diodes are commercially available; breakdown voltages from 2 to 200 V are typical, with power ratings from a fraction of a watt to 100 watts.

Since the reverse characteristic of a zener diode is the region of interest, let us examine this region more closely. A typical zener diode *i*–*v* characteristic, which exhibits the key parameters of the device, is shown in Figure 2-29. I_{ZK} and V_{ZK} denote the zener "knee" current and voltage, respectively, while I_{ZT} and V_{ZT} denote the test current and voltage, respectively. The resistance of the diode at the knee is R_{ZK} and is given by

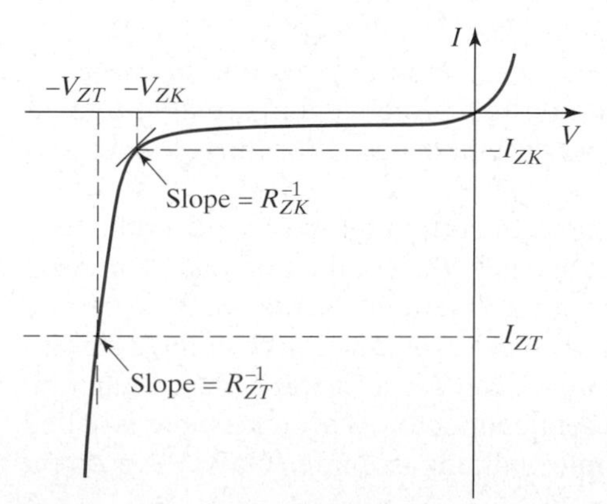

Figure 2–29 A zener diode characteristic in the region of interest with key parameters identified.

$$R_{ZK} = \left.\frac{dV_Z}{dI_Z}\right|_{\text{at knee}}, \tag{2.67}$$

and the resistance at the test value is R_{ZT} and is given by

$$R_{ZT} = \left.\frac{dV_Z}{dI_Z}\right|_{\text{at test point.}} \tag{2.68}$$

R_{ZK} is much larger than R_{ZT}. Typical values of R_{ZK} for 1-W zener diodes are between a few hundred and a few thousand ohms, whereas for R_{ZT}, these are in the range of a few ohms to a few hundred ohms.

For zener breakdown, the breakdown voltage decreases as the temperature increases because the energy of the valence band electrons goes up, and they are more likely to tunnel through the barrier. For avalanche breakdown, the breakdown voltage increases as the temperature increases because the lattice is vibrating and the electrons have more frequent collisions with it. With more frequent collisions, the electrons are not able to pick up as much energy, due to the acceleration of the field, so it takes a higher field to induce breakdown. One important result of these mechanisms having opposite temperature coefficients is that a zener diode with a breakdown voltage near 6 V, where both mechanisms are about equally active, can have a temperature coefficient near zero.

2.4.7 Other Types of Diodes

So far in this section, we have discussed junction diodes, Schottky diodes, zener diodes, and varactor diodes. There are many other kinds of diodes in use; we will mention a few of the more important and common ones here before going on to the bipolar junction transistor.

Two very common diodes make use of the fact that photons of light can be absorbed by semiconductor diodes and, if the light has sufficient energy, can produce extra EHPs. The ***optical generation rate*** is a function of the optical power, the wavelength of the light, and the bandgap of the semiconductor. Only photons with energies greater than or equal to the bandgap are absorbed; lower energy light is transmitted through the material or reflected.

If extra EHPs are generated within the depletion region of a reverse-biased diode, the electrons and holes will be separated by the electric field, and the reverse saturation current of the junction will increase. Therefore, the reverse current will be proportional to the optical power incident on the diode. Diodes that are designed to use this effect are called ***photodiodes***.

The second kind of diode that makes use of the optical generation of EHPs is the ***solar cell***. If a junction diode is open circuited and illuminated by light of an appropriate wavelength, a voltage is produced across the junction; this is called the ***photovoltaic effect***. Solar cells operate on this principle and are able to convert light into electrical power. The schematic symbols used for photodiodes and solar cells are shown in Figure 2-30.

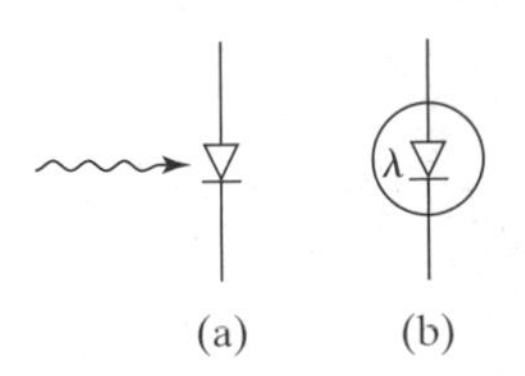

Figure 2–30 Photodiode and solar cell symbols (a) used for photodiodes and (b) used for photodiodes and solar cells. (λ is the symbol used for the wavelength of light.)

Diodes are also capable of generating light. In certain types of semiconductors, called ***direct bandgap*** semiconductors, electrons in the conduction band can drop directly into the valence band, and the energy liberated in the transfer can be given off as a photon of light. Since forward-biased *pn* junctions can have significant recombination in the depletion region, we can use a forward-biased junction diode constructed from a direct bandgap semiconductor to produce what is called a ***light-emitting diode*** (LED). For example, gallium arsenide (GaAs) is a direct bandgap semiconductor and can be used to produce LEDs. Silicon and germanium are examples of ***indirect bandgap*** semiconductors. For indirect bandgap semiconductors, electrons cannot typically combine directly with holes in the valence band, but must go through a recombination site (e.g., certain impurities provide a two-step recombination process) and cannot emit light. Aside A2.4 gives a very casual description of the difference between direct bandgap and indirect bandgap semiconductors. The schematic symbol used for an LED is shown in Figure 2-31.

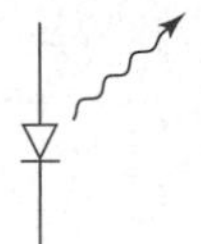

Figure 2–31 The schematic symbol for an LED.

2.5 BIPOLAR JUNCTION TRANSISTORS (BJTs)

There are two general types of transistors in common use: bipolar and field-effect transistors. Field-effect transistors are described in Sections 2.6, 2.7, and 2.8. The operation of the ***bipolar junction transistor***, or BJT, is described in this section. The BJT was discovered[14] at Bell Laboratories in the late 1940s, and the 1956 Nobel prize in physics went to John Bardeen, Walter Brattain, and William Shockley for the accomplishment. Although there are certainly more field-effect transistors fabricated than BJTs nowadays, the BJT remains an important device in both discrete and integrated circuit design.

In Section 1.3.1, we showed that active devices control the flow of energy and are necessary to achieve power gain. BJTs are active devices and do provide a reasonable approximation to a controlled source. What characterizes an ideal controlled source? Two things: first, the output current (or voltage) is constant, independent of the load connected to it. Second, the output current (or voltage) is controlled by some input voltage (or current) that is not required to provide any power.

ASIDE A2.4 DIRECT BANDGAP AND INDIRECT BANDGAP SEMICONDUCTORS

In our previous view of energy bands, we talked only about the total energies of the allowed states for an electron. A more detailed description would require treating the electrons as waves propagating in the crystal. The crystal then establishes a periodic potential in space in which these waves must exist. The energy can then be plotted as a function of the ***wave vector*** for the electron. (The wave vector is proportional to the momentum of the electron.) If it is possible for an electron to make a minimum energy jump from the conduction band to the valence band without changing its wave vector (i.e., while conserving its momentum), the material is a direct bandgap semiconductor. If an electron must change its wave vector to make a minimum energy jump, then the material is an indirect bandgap semiconductor. In direct bandgap semiconductors, an electron can drop from the bottom of the conduction band to the top of the valence band and give off a photon of light in the process. In the case of indirect bandgap semiconductors, however, the electron must also change its momentum in order to drop from the conduction band to the valence band, and this usually entails giving up energy to the lattice in the form of heat instead of giving off a photon of light.

A very crude analogy for the indirect bandgap semiconductor would be to say that if you want to enter the freeway, it isn't sufficient for there to be a space available going the same speed you are going, the space must also be in the lane closest to the on ramp. The only way for you to get to an available "state" in the middle lane is to make an indirect, and perhaps painful, transition from the on ramp to the lane!

[14] We say discovered rather than invented, because Bardeen and Brattain were actually trying to make a field-effect transistor when they produced the first BJT. After its operation was demonstrated, Shockley's contribution was to understand and describe the BJT's theory of operation.

2.5.1 Intuitive Treatment

If we examine the *i–v* characteristic of a junction diode and ask how we might produce a controlled source using it, two possibilities come to mind: The voltage is nearly constant in forward bias, and the current is nearly constant in reverse bias. Therefore, if we can electronically control the forward voltage drop, we can make a controlled voltage source, and if we can electronically control the reverse saturation current, we can make a controlled current source. Further thought reveals that the forward voltage drop is a function of the doping profiles and material properties of the device and is not electronically variable (at least not in any practical way we can conceive—maybe you can come up with a new device!). However, the reverse saturation current is a different matter.

Remember that the reverse saturation current is limited by the rate at which minority carriers can be supplied to the depletion region edge. All minority carriers that arrive there are swept across the junction by the applied bias (ignoring recombination). So, if we can somehow control the availability of minority carriers to a reverse-biased *pn* junction, we will be able to change the reverse saturation current.

In considering the possibilities for controlling minority-carrier densities, we remember (hopefully) that the minority-carrier densities near a forward-biased *pn* junction are elevated above the equilibrium values and are controlled by the bias voltage across the junction, *see* (2.50) and (2.51). Therefore, if we place a forward-biased junction close enough to a reverse-biased junction, the reverse saturation current of the reverse-biased junction will be controlled by the voltage applied across the forward-biased junction. Since the excess minority-carrier densities of a forward-biased junction drop off rapidly as you move away from the depletion region edge, we must place the reverse-biased junction very near to the forward-biased junction in order to see the desired effect. We have just described the operation of a BJT. The device comprises two *pn* junctions very close to each other and works like a voltage-controlled current source if one junction is forward biased while the other is reverse biased.

The region in the middle of the device is common to both junctions and is called the ***base*** of the transistor. The other side of the forward-biased junction is called the ***emitter***, since it is the source of the minority carriers injected (or emitted) into the base. The carriers then diffuse across the narrow base until they reach the reverse-biased junction, where they are swept across the depletion region by the electric field. The other side of the reverse-biased junction is called the ***collector***, since it finally collects the carriers that have been swept across the reverse-biased depletion region.

BJTs can be constructed with the *p*-type material common to both junctions, in which case an ***npn*** structure results, or they can be made with the *n*-type material common to both junctions, in which case a ***pnp*** structure results. A prototypical *npn* BJT is shown in Figure 2-32, along with the appropriate biasing, the names of

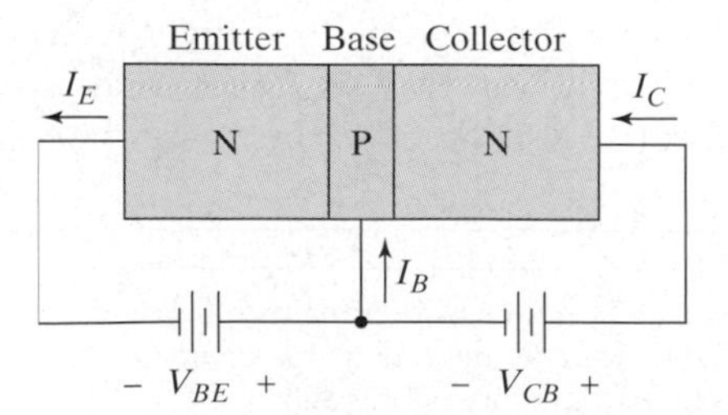

Figure 2–32 A prototypical *npn* BJT.

the regions, and the defined directions for positive terminal currents. Remember from Section 1.4 that V_{BE} represents the voltage from the base to the emitter and that V_{CB} similarly represents the voltage from the collector to the base. We have chosen to define the emitter current, I_E, as being positive coming out of the device, since this is the direction it will actually flow in an *npn* transistor. In network theory (e.g., in two-port networks), currents are usually defined as being positive into a device, but that convention is unnecessarily awkward here.

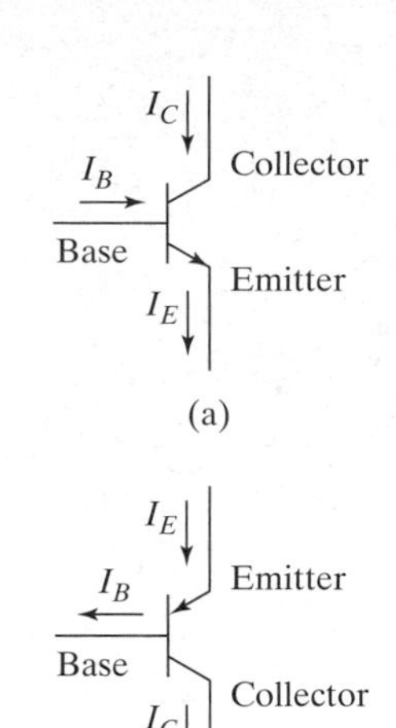

Figure 2–33 Schematic symbols and current directions for (a) an *npn* transistor and (b) a *pnp* transistor.

Figure 2-33 shows the schematic symbols and current directions for both *npn* and *pnp* transistors. In the next section, we develop a more detailed picture of the current flow in a BJT. As a result, we will derive the basic equation describing the operation of the device and discuss some of the important nonideal characteristics. For now, we simply state that the collector current of an *npn* transistor biased as shown in Figure 2-32 is given by

$$I_C = I_S e^{V_{BE}/V_T}, \qquad \textbf{(2.69)}$$

where I_S is a constant and V_T is the thermal voltage introduced in Section 2.4.4.

2.5.2 Detailed Analysis of Current Flow

In this section, we use the preceding qualitative description of transistor operation, along with the equations we have for the operation of *pn* junctions, to derive the basic *i–v* characteristic of the BJT. Since there are two back-to-back *pn* junctions in a BJT, there are four regions of operation to consider, corresponding to each of the junctions being either forward or reverse biased. The region of operation we have described qualitatively, where the base-emitter junction is forward biased and the base-collector junction is reverse biased, is called the ***forward-active*** region of operation. We will consider this region first and then return to the other possibilities.

Figure 2-34 shows an *npn* transistor structure and the corresponding minority-carrier densities when the BJT is operating in the forward-active region. Because the base-emitter junction is forward biased, we can apply (2.50) and (2.51) to write (note that $x = 0$ has been set to the base edge of the base-emitter depletion region)

$$n_p(0) = n_{po} e^{V_{BE}/V_T} \qquad \textbf{(2.70)}$$

and

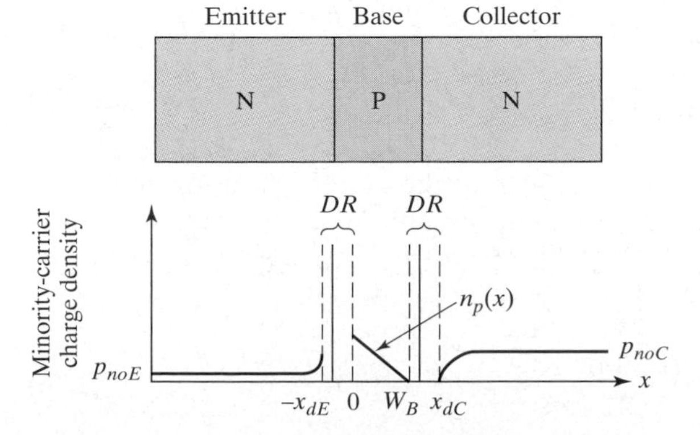

Figure 2–34 Minority carrier densities for a BJT biased in the forward-active region of operation.

$$p_n(-x_{dE}) = p_{noE}{}^{V_{BE}/V_T}, \tag{2.71}$$

where $-x_{dE}$ is the emitter edge of the base-emitter depletion region, as shown in the figure. In addition, we know that the minority-carrier densities at the edges of the base-collector depletion region will be zero due to the reverse bias on the junction; that is,

$$n_p(W_B) = 0 \tag{2.72}$$

and

$$p_n(x_{dC}) = 0, \tag{2.73}$$

where W_B and x_{dC} are the edges of the base-collector depletion region, as shown in the figure.

If we assume that the base region (i.e., from $x = 0$ to $x = W_B$) is narrow enough that carriers can diffuse across it without a significant chance of recombining, the diffusion current through the base will be constant, and we can draw a straight line between the two boundary values for the minority carrier density in the base, as shown in the figure. (Remember that the diffusion current is proportional to the slope of the carrier density.)

The transistor operates in the following manner (an *npn* transistor is used for illustration): Electrons are injected into the base from the emitter, they diffuse across the base because of the density gradient established by the boundary conditions imposed by the junctions, and they are swept up by the field at the collector-base depletion region. The magnitude of the current that flows in the collector lead is determined by the diffusion current traversing the base, since it provides the minority carriers to the reverse-biased base-collector junction. The diffusion current in the base can be found by using (2.6) and the geometry of the minority carrier density in the base (shown in Figure 2-34):

$$J_n(\text{base}) = qD_n \frac{dn(x)}{dx}\bigg|_{\text{in base}} = qD_n \frac{n_p(0) - n_p(W_B)}{0 - W_B} = -qD_n \frac{n_p(0)}{W_B}. \tag{2.74}$$

The minus sign in (2.74) results from the fact that conventional current goes to the left in Figure 2-34, but positive current is defined as going to the right. When we switch to dealing with the collector current, the minus sign will disappear, since we have defined I_C as being positive flowing into the collector. Substituting (2.70) into (2.74) yields

$$J_n(\text{base}) = -\frac{qD_n n_{po}}{W_B} e^{V_{BE}/V_T}. \tag{2.75}$$

The collector current is equal to the negative of the current diffusing across the base, which is equal to the emitter-base junction area times the current density given by (2.75). Using this fact and substituting $n_{po} = n_i^2/N_A$ from (2.5) yields the equation for the collector current:

$$I_C = \frac{qD_n A n_i^2}{N_A W_B} e^{V_{BE}/V_T} = I_S e^{V_{BE}/V_T}. \tag{2.76}$$

Here, I_S is the ***linking saturation current***, so called because it appears in the same location in the equation as the saturation current of a *pn* junction, but it represents the linking between the emitter and the collector. It is important to note that I_S, which is commonly just called the saturation current, is proportional to the area of the emitter-base junction. Also, except at very low currents, recombination in the depletion region plays a much smaller role here than it does for the diode (*see* Section 2.4.4), so the ideality factor is usually quite close to unity and is often ignored, as we have done here. It is also important to notice that I_S is highly temperature dependent; typically, I_S approximately doubles for every 5°C rise in temperature. For circuit design, it is usually more useful to express this temperature dependence in terms of the base-emitter voltage. From (2.76), we see that the base-emitter voltage is given by

$$V_{BE} = V_T \ln (I_C/I_S). \quad \textbf{(2.77)}$$

The base-emitter voltage decreases by about 2 mV for every 1°C rise in temperature if the collector current is held constant. In other words, V_{BE} has a TC of −2 mV/°C, the same as that of a *pn*-junction diode (assuming that $n = 1$ for the diode).

Since our original intent was to produce a voltage-controlled current source, it is worthwhile to examine whether or not we have succeeded. From (2.76), we can see that the collector current is controlled by the base-emitter voltage, but it is not a linear function as we would like. Nevertheless, in Chapter 6 we will see that it is possible to limit ourselves to signals that are small compared with the average values present and thereby achieve approximately linear operation. In the case of small perturbations about the average, the bipolar transistor can be described as a controlled source with a transconductance found by differentiating (2.76):

$$g_m \equiv \frac{dI_C}{dV_{BE}} = \frac{I_S e^{V_{BE}/V_T}}{V_T} = \frac{I_C}{V_T}. \quad \textbf{(2.78)}$$

We can now summarize what we know about bipolar transistor operation in the forward-active region and then move on to consider the other regions. When the collector-base junction is reverse biased, the current flowing across it depends on the available supply of minority carriers. By forward biasing the nearby emitter-base junction, we supply a controlled number of minority carriers and bring the current across the collector-base junction under the control of the emitter-base voltage.

We now consider the other possible regions of operation. If both junctions are reverse biased, then neither junction will have enough minority carriers available to support significant external current flow, and we say the device is ***cutoff***. If the collector-base junction is forward biased and the emitter-base junction is reverse biased, minority carriers will be supplied by the collector and swept up by the emitter; in other words the device is active, but in the reverse direction. We call this region of operation the ***reverse active***, or ***inverse active***, region. The final possibility to consider is what happens if both junctions are forward biased, which is called the ***saturation*** region.

In saturation, the minority-carrier profile in the base is still a straight line (if we continue to neglect recombination in the base) as shown in Figure 2-34, but it doesn't go to zero at either junction. Instead, the boundary values, and hence the slope of the minority-carrier density and the diffusion current, depend on the magnitudes of the forward biases applied to both junctions. This situation can be viewed as a superposition of the forward- and reverse-active regions of operation, as shown in Figure 2-35. In other words, the transistor works like two devices connected in parallel, as shown in Figure 2-36. Viewing the operation as a superposition of forward- and reverse-active devices is the basis for the ***Ebers–Moll*** model of the transistor. We will return to the topic of modeling the transistor in Chapter 4.

Because both junctions are forward biased in saturation, the collector and emitter are at nearly the same voltage and V_{CE} is very small. The collector current in this situation is determined mostly by the external circuit (i.e., the transistor looks like a small voltage source from collector to emitter).

A symmetric BJT would work equally well in either the forward- or reverse-active mode. In reality, however, the doping profile is usually set to optimize the forward operation of the device. This is done by setting the doping level in the emitter to be much larger than that in the base and the doping level in the base to be much larger than that in the collector. The net result is that the number of carriers injected into the base from the emitter is maximized, and both the numbers of carriers injected from the base into the emitter and the small drift current from collector to base are minimized. Therefore, most BJTs do not operate nearly as well in the reverse direction.

Table 2.3 shows the four regions of operation, and Figure 2-37 shows a typical family of $I_C - V_{CE}$ curves for different values of V_{BE}. (V_{CE} denotes the voltage from collector to emitter.) These curves are commonly called the ***collector characteristics***, and we will consider several features of them in more detail later. In analog circuits, the bipolar transistor is most often used in the forward-active region, where it approximates a voltage-controlled current source and can be used to make amplifiers. In digital circuits, the transistor is most commonly used either in the forward-active region or as a switch going back and forth between the cutoff and saturation regions of operation. The reverse-active region of operation is seldom used.

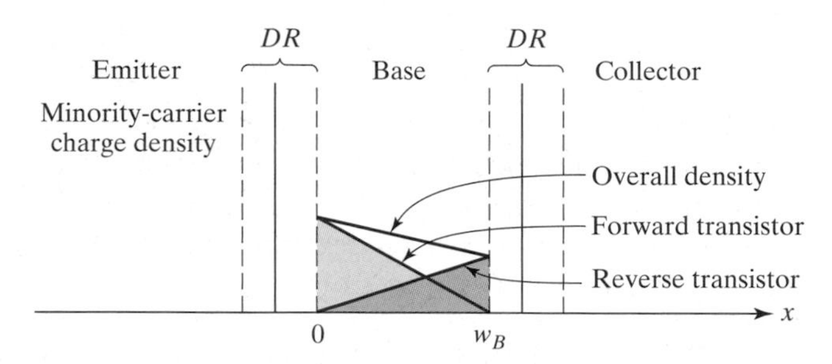

Figure 2–35 The minority-carrier profile in a saturated bipolar transistor. The profile is the superposition of the forward- and reverse-active profiles as shown.

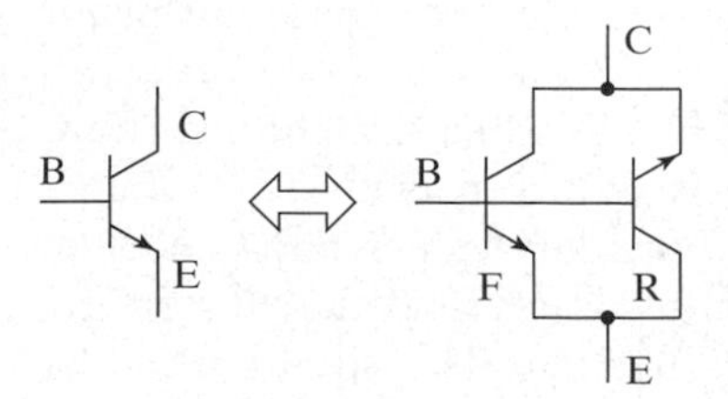

Figure 2–36 A single bipolar transistor represented as forward (*F*) and reverse (*R*) transistors connected in parallel.

Table 2.3 The Four Regions of Operation of a Bipolar Junction Transistor

Region of Operation	Emitter-Base Junction	Collector-Base Junction	Description
Forward active	forward biased	reverse biased	collector current given by (2.76)
Reverse active	reverse biased	forward biased	same as forward with appropriate substitutions; - usually does not work as well.
Cutoff	reverse biased	reverse biased	no significant current flow
Saturated	forward biased	forward biased	external currents given by superposition of forward- and reverse-active regions

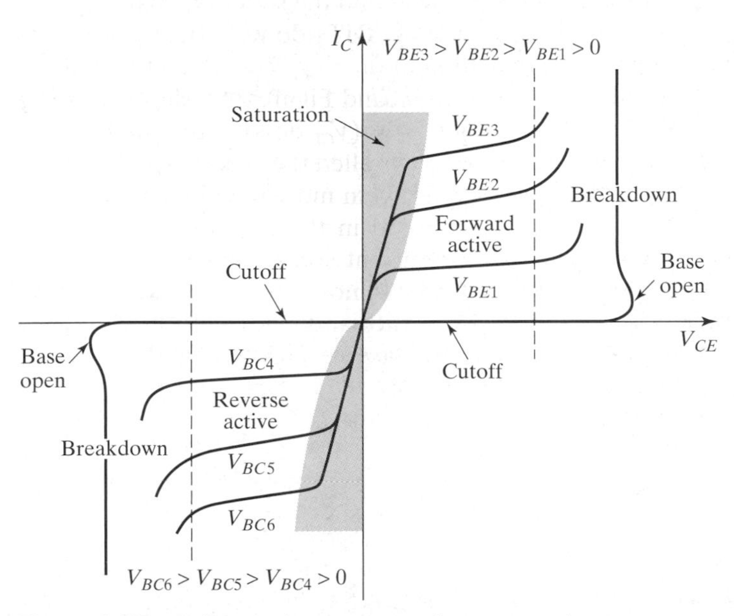

Figure 2–37 Collector characteristics of an *npn* transistor.

We have been implicitly assuming that the base-emitter voltage is the input and that the collector current is the output. In this case, the emitter is connected to both the input and output circuits, so this is called the ***common-emitter*** connection. When no configuration is specified, the common-emitter connection should be assumed. The curves in Figure 2-37 are also referred to as the ***common-emitter characteristics***.

Now that we have described the basic operation of the bipolar transistor, the rest of this section is devoted to the most important second-order effects. Up to this point in our treatment, the only nonideal characteristic observed has been the fact that the collector current is not a linear function of the base-emitter voltage. However, in addition, the base current of the transistor is not zero as we would like it to be, the collector current does depend on the collector voltage, there is capacitance present, the device will break down if the terminal voltages are too large, and the operation is different at very low or very high collector currents. We discuss each of these nonideal effects separately.

Exercise 2.10

Consider an npn transistor with emitter doping $N_D = 10^{18}$ atoms/cm^3, base doping $N_A = 10^{16}$ atoms/cm^3, and collector doping $N_D = 10^{14}$ atoms/cm^3. The metallurgical width of the base (i.e., ignoring the depletion regions) is 0.6 μm, and the junction area is 100 μm^2. Assume that $D_n = 35$ cm^2/s in the base. (a) Find V_0 for both junctions. (b) Find the depletion region widths and the width of the base region. (c) Find the saturation current.

□

2.5.3 Base Current

An ideal voltage-controlled current source would not require any current from the controlling voltage source. However, a bipolar transistor does require some base current, which is caused by two effects: recombination in the base and the base-emitter depletion region, and carriers injected into the emitter from the base (called ***reverse injection***). In modern transistors, the second term dominates at normal operating currents, but recombination in the depletion region dominates at very low currents. Notice from (2.70) and (2.71) that the density of minority carriers injected into one side of a *pn* junction is directly proportional to the nominal minority-carrier density on that side (e.g., p_n is proportional to p_{no}). Therefore, the reverse injection current can be reduced by increasing the doping in the emitter relative to that in the base.

It is also important to note from (2.70) and (2.71) that both minority-carrier densities depend on the applied base-emitter voltage in precisely the same way the collector current does. Therefore, we can express the base current that flows into the transistor as a fraction of the collector current. The ratio of the collector current to the base current is a parameter of fundamental importance and is called the ***beta*** of the transistor:

$$\beta \equiv \frac{I_C}{I_B}. \quad \textbf{(2.79)}$$

In an ideal transistor, β is infinite. β is also referred to as the common-emitter current gain, or just the current gain. Because the collector current is proportional to the base current, many books describe the bipolar transistor as a current-controlled current source. Although this is clearly an acceptable view in light of (2.79), it does not agree with the fundamental physics of operation of the device and does not allow us to make use of the logarithmic relationship between the base-emitter voltage and collector current. The relationship is sometimes very useful (e.g., in making log amps). The relationship expressed by (2.79) is extremely useful in bias analysis, since it is a linear equation (*see* Chapter 7). Although it is often reasonable to approximate β as a constant, it is dependent on collector current and temperature.

As the current in the device is reduced, recombination in the base-emitter depletion region becomes more significant, and there is a significant drop in β. At high currents, a situation called ***high-level injection*** exists such that the change in the majority-carrier concentration becomes important. The majority-carrier profile has the same basic shape as the minority-carrier profile since charge neutrality is maintained. The resulting increase in majority-carrier concentration on the base side of the emitter-base depletion region effectively makes the base appear to be doped higher and increases the magnitude of the reverse injection current, thereby reducing the magnitude of β. In addition, the increased number of majority carriers further increases recombination, also causing an increase in base current and a concomitant decrease in beta.

The changes that take place with increasing temperature are complex, but the net result is that β typically increases with increasing temperature.[15] This increase can lead to a problem called ***thermal runaway***, where the temperature is increased due to power dissipation in the device. This increase in temperature causes an increase in β, which in turn increases the collector current and power dissipation, thereby causing a further increase in temperature, and so on. Thermal runaway will destroy a transistor unless the current is limited.

Looking at the terminal currents as defined in Figure 2-33, we see that $I_E = I_B + I_C$. It is useful to define a new parameter, ***alpha***, that is the ratio of I_C to I_E:

$$\alpha \equiv \frac{I_C}{I_E} = \frac{I_C}{I_C + I_B} = \frac{\beta I_B}{(\beta+1)I_B} = \frac{\beta}{\beta+1}. \tag{2.80}$$

α is also called the common-base current gain, a name that will make more sense after the common-base connection has been introduced (in Chapter 8). In an ideal transistor, α is unity. We will see later that subscripts may be needed for alpha and beta, since they are functions of frequency and collector current, and we can define them for both the forward- and reverse-active regions of operation. Nevertheless, these distinctions are often not important, and we will usually dispense with the subscripts. When subscripts are not used, it is understood that we are talking about forward-active operation, and the values given are the low-frequency values.

Exercise 2.11

A typical BJT has a β of 100. What is the corresponding value of α? If the collector current is 1 mA, what are the base and emitter currents?

2.5.4 Base-width Modulation (the Early Effect)

The output current of an ideal voltage-controlled current source does not depend on the voltage across the source. Unfortunately, the collector current of a bipolar transistor operating in the forward-active region does depend on the collector voltage. The effect is known as the ***Early effect*** (after James Early, who first analyzed it) or, more descriptively, as ***base-width modulation***. When the collector voltage of a bipolar transistor is varied while keeping the base voltage constant, the reverse bias across the collector-base depletion region changes, and therefore, the width of the depletion region changes. As a result, the width of the active base changes with V_{CB}, as shown in Figure 2-38.

As the reverse bias on the collector-base depletion region is increased, the depletion region gets wider and the base width is reduced. The collector current is proportional to the slope of the minority-carrier profile in the base, as shown in (2.74), so as W_B decreases, the slope gets larger and the current increases. This effect is visible on the collector characteristics, as shown in Figure 2-37 and, in an

[15] *See* [2.1] for a short treatment.

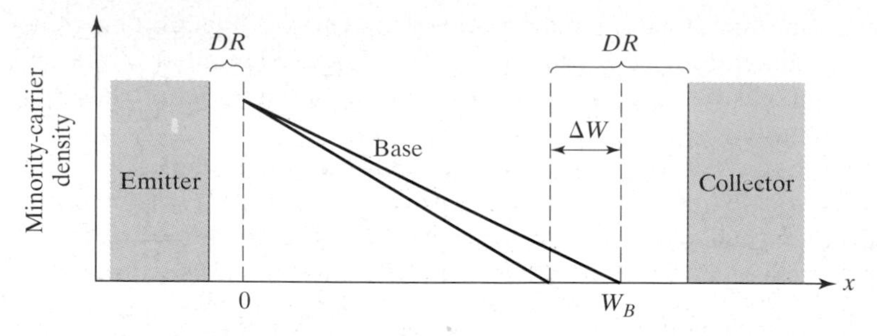

Figure 2–38 The width of the collector-base depletion region varies with the collector voltage, resulting in a variation in the active base width W_B. The slope of the minority-carrier density then changes, as shown by the two densities drawn.

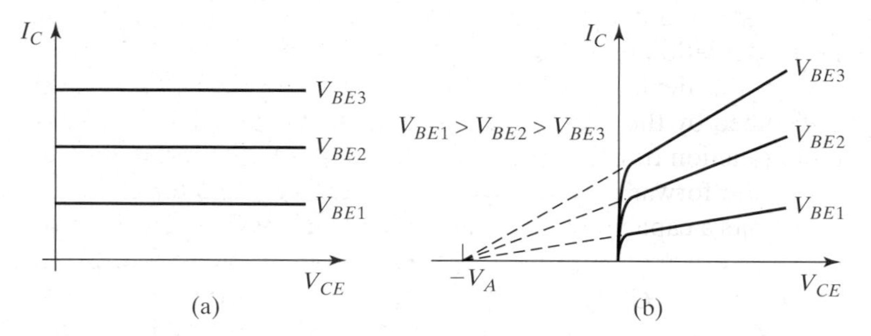

Figure 2–39 Collector characteristics of an *npn* transistor in forward-active operation (a) without base-width modulation and (b) with base-width modulation.

idealized form, in Figure 2-39(b). In part (a) of Figure 2-39, the characteristics are shown for an ideal transistor as described by (2.76), whereas part (b) of the figure shows what the characteristics look like if the Early effect is included. Notice that if the characteristics in the forward-active region of operation are extrapolated back to the V_{CE} axis, they all intersect it at the same point. The magnitude of this intercept is typically called the ***Early voltage*** and is denoted by V_A.[16] The equation for the collector current can also be modified to account for this effect:

$$I_C = I_S e^{V_{BE}/V_T}\left(1 + \frac{V_{CE}}{V_A}\right). \tag{2.81}$$

This equation is typically the one used, but it is occasionally important to recognize that it isn't correct, since it should yield (2.76) when the base-collector voltage is zero. Instead, (2.81) simplifies to (2.76) when V_{CE} is zero, at which time the transistor is not forward active. Since V_{CE} is usually much larger than V_{BE}, this correction is often unimportant (and, in fact, is often not recognized or mentioned). Nevertheless, it is possible to modify the equation to be more accurate by using V_{CB} instead of V_{CE} and an appropriately modified value for the Early voltage.

[16] Some people use the intercept itself, which is negative, as V_A. In that case, the magnitude of V_A needs to be used in (2.81). We will always assume that the Early voltage is positive and will simply adjust the signs as necessary to yield an increase in the collector current for an increase in the reverse bias on the collector-base junction. That way, we don't need to memorize signs for different types of transistors.

As the width of the base is decreased, the total minority charge stored in the base also decreases, and the base current decreases as a result. This is a minor effect in most transistors, since the base current is dominated by reverse injection into the emitter, but it does cause the base current to decrease, and therefore β to increase, as the reverse bias on the collector-base junction is increased.

Exercise 2.12

A typical BJT has an Early voltage of 50 V. How much is the collector current increased by base-width modulation when $V_{CE} = 20$ V?

2.5.5 Charge Storage

In Section 2.4.5, we discussed two charge storage terms present in *pn* junctions: charge storage in the depletion region and charge storage in the excess minority-carrier profiles when the device is forward biased. These two charge storage mechanisms are modeled by the junction capacitance and diffusion capacitance, respectively. Bipolar junction transistors also have these capacitors present.

When operating in the forward-active region, the collector-base junction is reverse biased and exhibits a capacitance of the form given by (2.53), namely,

$$C_{jc} = \frac{C_{jc0}}{\left(1 - \frac{V_{BC}}{V_0}\right)^{m_{jC}}} \tag{2.82}$$

where C_{jc0} is the zero-bias value of the base-collector junction capacitance, V_0 is the built-in potential of the junction, and m_{jC} is the ***base-collector junction exponent*** (sometimes called the ***grading factor***, since it depends on the doping profile). Equation (2.82) assumes an *npn* device. If a *pnp* device is used, V_{BC} needs to be replaced by V_{CB} or have the sign changed.[17]

In addition to C_{jc}, there is a junction capacitance associated with the base-emitter junction, C_{je}. Since the base-emitter junction is forward biased when the transistor is in the forward-active region of operation, its junction capacitance is approximately given by

$$C_{je} \approx 2C_{je0}, \tag{2.83}$$

as shown in (2.53).

Finally, there is also a diffusion capacitance caused by the excess minority-carrier charge stored in the base region and, to a lesser extent, in the emitter. We can again define an average minority-carrier lifetime, as was done for the diode. If we define the total excess minority charge stored in the forward-active region of operation by Q_F (this includes charge stored in the base and the excess charge injected into the emitter) and define a lifetime associated with this charge storage, τ_F, the collector current can be written as

$$I_C = \frac{Q_F}{\tau_F}, \tag{2.84}$$

[17] There isn't any need to memorize the signs. If you take the view that the capacitance across the depletion region is similar to that across a parallel-plate capacitor, then you know that the capacitance will decrease when the plate spacing increases. Therefore, whatever signs you use must be such that an increase in the reverse bias on the junction, which increases the depletion region width, will decrease the capacitance.

just as was done in (2.58) and (2.59). Finally, we can find the capacitance at a given operating point by differentiation:

$$C_d = \frac{dQ_F}{dV_{BE}} = \tau_F \frac{dI_C}{dV_{BE}} = \tau_F \frac{d(I_S e^{V_{BE}/V_T})}{dV_{BE}} = \frac{I_S \tau_F}{V_T} e^{V_{BE}/V_T} = \tau_F \frac{I_C}{V_T} = g_m \tau_F. \quad \textbf{(2.85)}$$

The diffusion capacitance given by (2.85) shows up in parallel with C_{je}, since both capacitors represent charge storage that depends on the base-emitter voltage. In (2.85), we implicitly assumed that τ_F was independent of I_C. This assumption is not correct at high currents. Since Q_F is approximately equal to the minority charge in the base, τ_F can loosely be thought of as the ***average base transit time***. At high currents (i.e., high-level injection), minority carriers flood the collector-base depletion region, and the effective base width increases, causing an increase in τ_F. This phenomenon is called either the ***Kirk effect*** or ***base pushout***.

All three capacitances—C_{je}, C_{jc}, and C_d—are only approximately constant for small deviations from the operating point. Finally, we note that if the device is operated in the reverse-active region, similar equations apply, but the junctions will reverse their roles. Also, if the device is operated in saturation, both junctions will be forward biased, so both will have diffusion capacitance and both junction capacitances will have the form of (2.83). If the device is cut off, both junctions are reverse biased; therefore, they do not have diffusion capacitance, but both have junction capacitance of the form given by (2.82).

Exercise 2.13

A particular BJT has a zero-bias base-emitter junction capacitance of 0.7 pF and a forward transit time of 500 ps. What is the total capacitance seen from base to emitter when the collector current is 500 μA? Assume room temperature.

□

2.5.6 Breakdown Voltages

The terminal voltages of a bipolar transistor may not be arbitrarily large. As with any *pn* junction, the junctions in a transistor will break down if the reverse voltage across them gets too large. The breakdown mechanisms are the same as those already discussed for *pn*-junction diodes, but the transistor action complicates the discussion. Nevertheless, if we consider either of the junctions in isolation, they will break down in precisely the way explained for diodes.

The doping of the base-emitter junction is usually large enough that tunneling is important; the breakdown voltage from base to emitter is typically on the order of 5 or 6 V and is denoted by BV_{EBO}, where the subscripts indicate the breakdown is from emitter to base with the collector open circuited. Another common notation is $V_{(BR)EBO}$. This parameter is not usually of great interest, since the base-emitter junction is rarely reverse biased; in fact, BV_{EBO}, is often not even specified.

Of greater importance in circuit design is the breakdown voltage of the collector-base junction, since this junction is reverse biased when the transistor either is operating in the forward-active region or is cutoff. The largest possible breakdown voltage for the collector-base junction is observed if the emitter is open circuited, in which case the breakdown is just the breakdown of the *pn* junction by itself. This breakdown voltage is denoted BV_{CBO} and varies from a few volts for high-speed integrated circuit transistors to hundreds of volts for common discrete power transistors.

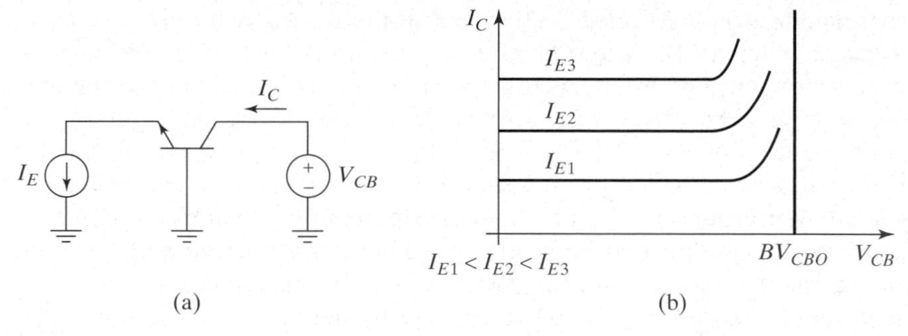

Figure 2–40 (a) The circuit for generating the common-base collector characteristics shown in (b).

The situation is more complicated if the emitter is not open circuited. To understand what happens, consider the circuit shown in Figure 2-40(a) and the corresponding plot of I_C versus V_{CB} in part (b). If we consider the current source to be the input and the collector voltage to be the output of this circuit, then the base of the transistor is common to both the input and output. This configuration is therefore called the ***common-base*** connection, and the curves in part (b) of the figure are the ***common-base characteristics***. As the value of the emitter current is increased from zero, which is the open-circuit condition that yields BV_{CBO}, the extra current traversing the collector-base depletion region causes avalanche multiplication to show up at lower values of V_{CB}. Therefore, the characteristics curve upward at values of V_{CB} below BV_{CBO}.

A more common situation is the common-emitter connection, where we are interested in the breakdown voltage from the collector to the emitter. This situation is depicted in Figure 2-41(a) for an *npn* transistor, and the resulting plot of I_C versus V_{CE} is shown in part (b) of the figure.

Let's first consider what happens when I_B is zero or, in other words, an open circuit. As V_{CE} is increased from zero, the base-emitter junction is forward biased, the base-collector junction is reverse biased, and there will be some small reverse leakage current flowing. This current comprises holes drifting from the *n*-type collector into the *p*-type base and electrons drifting from the base to the collector. The magnitude of this leakage current is limited by the available supplies of minority carriers, which in the base are supplied by injection from the forward-biased base-emitter junction.

The transistor does require base current, so where does this base current come from, given that the external base current is set to zero? The "base current" is supplied by the holes drifting from the collector to the base. This drift current acts as

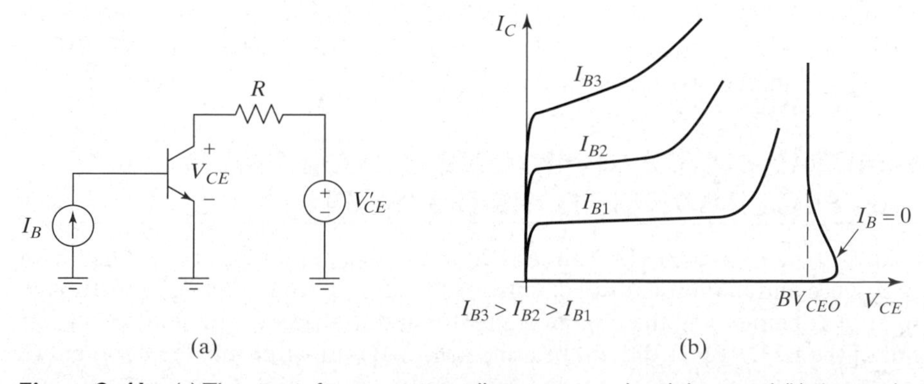

Figure 2–41 (a) The circuit for examining collector-emitter breakdown and (b) the resulting characteristics.

a base current and gets multiplied by the beta of the transistor to produce even more "collector" current. The external collector current consists of two terms: the internal "collector" current and holes that are supplying the needed base current. Since beta is small when the overall current is small, this effect is not initially all that significant. As soon as the voltage increases to the point where avalanche multiplication first starts to occur, however, the EHPs created in the collector-base depletion region get separated by the field. The resulting electrons get swept into the collector to contribute to the collector current directly, whereas the holes get swept into the base, where they increase the effective base current.

As the current increases, the beta will increase, and the device can actually supply the same external collector current with a *smaller* reverse voltage. This explains why the characteristic for $I_B = 0$ in Figure 2-41(b) curves back to the collector-to-emitter breakdown voltage with the base open, BV_{CEO}. Because the higher "breakdown" voltage that is reached momentarily cannot be sustained, BV_{CEO} is sometimes called the ***sustaining voltage*** and is then denoted $V_{(SUS)CEO}$. (This voltage is also sometimes denoted as $V_{(BR)CEO}$ or LV_{CEO}.) For reasonable values of base current, the value of beta does not change as much, and the characteristics do not fold back on themselves. In addition, due to the extra collector current flowing when I_B is not zero, the effects of avalanche multiplication are again observed at lower values of reverse bias, so the characteristics start to curve upward prior to reaching BV_{CEO}, as shown in the figure.

As was the case with the diode, breakdown is not inherently a destructive phenomenon. The device will be destroyed only if it is allowed to dissipate too much power or if the current exceeds the maximum level that can be supported.

2.5.7 Other Types of Junction Transistors

A common variation on the standard BJT is the ***phototransistor***. A phototransistor is simply a transistor that uses the optical generation of EHPs as the means for providing its base current. In other words, the optically generated EHPs provide the necessary carriers for recombination and reverse injection into the emitter. The phototransistor is better than a photodiode in some applications, because it yields a larger output current, due to the optically generated current being multiplied by the current gain of the transistor. The trade-off is that phototransistors are slower than photodiodes, since the base is electrically open circuited and recombination is the only mechanism available to remove the excess minority carriers from the base if the device is switched from forward active to cutoff (i.e., the light source is suddenly turned off).

Input Output

Figure 2–42 An optoisolator using a phototransistor as the detector. The base lead may or may not be provided as an external connection.

A number of applications exist for phototransistors, but perhaps the most common is in ***optoisolators***. An optoisolator is a device that contains both an LED and a photodetector of some sort (usually a phototransistor), as shown in Figure 2-42. These devices are very useful for isolating circuits electrically. Several applications with circuit examples are discussed in [2.4].

2.6 METAL-OXIDE SEMICONDUCTOR FIELD-EFFECT TRANSISTORS (MOSFETS)

The ***field-effect transistor*** (FET) is one of the two general types of transistors in common use and undoubtedly accounts for the largest volume of sales. Although simpler in principle than the bipolar transistor and first conceived almost 20 years ahead of the BJT, [18] FETs did not become practical until after the development of

[18] FET devices were described by patents in the United States and Germany in the 1930s [2.6].

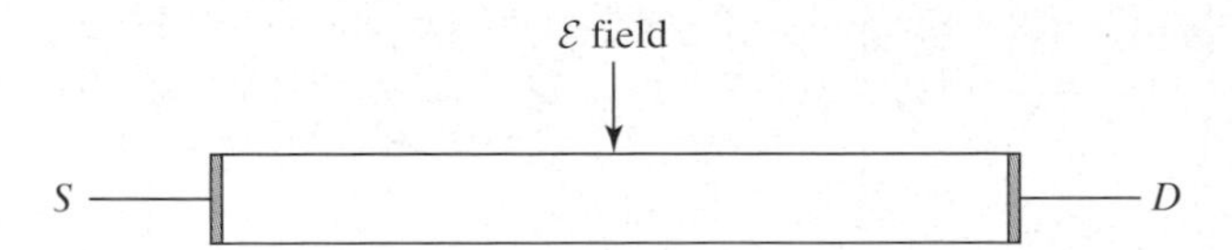

Figure 2–43 A conceptual generic field-effect transistor.

the BJT. In fact, the most popular FET, the *metal-oxide semiconductor* FET (MOSFET), did not become practical until the early 1960s and did not become commonplace until the 1970s. The operation of a MOSFET is adversely affected by poor control of the silicon–silicon-dioxide interface, so practical devices could not be reliably manufactured until silicon technology had matured sufficiently to produce a good-quality interface.

Although FETs are simpler in principle than BJTs, the second-order effects are far more significant in FETs than in BJTs. As a result, first-order hand calculations involving FETs are frequently less accurate than they are for BJTs, and the models used for FETs in circuit simulation programs tend to be more complicated and include more empirical elements.

The basic concept underlying the operation of any FET is illustrated in Figure 2-43. Starting with a semiconducting material with one connection at each end, an electric field is applied perpendicularly to the length of the sample as shown. The electric field modulates the conductivity of the material, so the device functions as a variable resistor. For a given electric field, the current through the device cannot increase beyond an upper bound, called the saturation current. When the current is saturated, the device functions like a controlled current source. The contacts at the ends of the semiconductor are called the *source* and the *drain*, since the carriers (electrons for an n-type channel and holes for an p-type channel) are supplied by the source and leave the material at the drain. The differences between FETs are mostly in how the $\mathcal{E}$ fields are generated, which is why Figure 2-43 leaves out that part of the device. Nevertheless, there is always another terminal, called the *gate*, that controls the field.

The MOSFET is discussed in this section and in the most detail, since it is by far the most common and commercially important. The junction FET (JFET) is covered in Section 2.7, but the discussion is much shorter, because the device is rapidly becoming obsolete. The metal–semiconductor FET is introduced in Section 2.8.

Although "MOSFET" is an acronym for "metal-oxide–semiconductor FET," almost all MOSFETs are made with heavily doped polysilicon gates, rather than metal gates. The more general term "*insulated-gate* FET" ("IGFET") is sometimes used, but most often "MOSFET" is used as the generic term. In Section 2.6.1, we provide an intuitive description of the operation of a MOSFET. Section 2.6.2 then provides a derivation of the first-order equations governing the operation of the device. In the remaining sections, we discuss some of the more important second-order effects.

2.6.1 Intuitive Treatment

Figure 2-44 shows a simplified MOSFET structure. The device shown is fabricated on a p-type substrate and has heavily-doped n-type regions for the source and drain. In addition, there is a contact to the *body*, or *bulk*, of the device, and a *gate* which is separated from the silicon by the oxide layer. The gate conductor is heavily-doped polysilicon in modern devices. As shown in the figure, we will assume that the source and bulk terminals are grounded and will examine the behavior as the gate and drain potentials are varied.

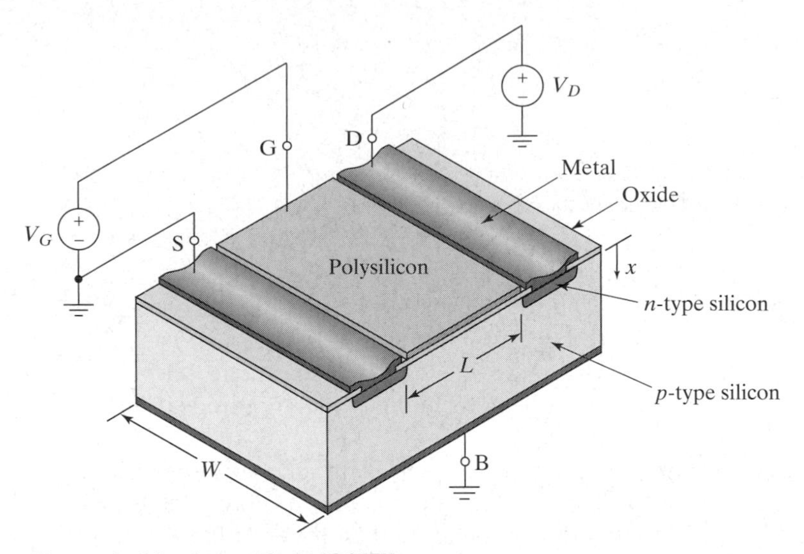

Figure 2–44 A simplified MOSFET structure.

When $V_G \leq 0$, the field perpendicular to the surface of the silicon (i.e., in the x direction in Figure 2-44) attracts holes to the surface of the silicon. Under this condition, no current can flow from source to drain, because either the source-to-bulk *pn* junction or the drain-to-bulk *pn* junction is guaranteed to be reverse biased for any value of V_D. In this case, we say the device is ***cutoff***.

On the other hand, as V_G is increased above zero, holes are repelled from the surface of the silicon under the gate, and mobile electrons are attracted. For a sufficiently large positive value of V_G, the surface under the gate will contain more mobile electrons than holes. In other words, the surface will appear to have been converted into *n*-type material, and a continuous *n*-type path will exist from source to drain. In this condition, a significant current can flow. Because the material starts out as *p*-type and gets converted to *n*-type by the vertical field, we say the surface has been ***inverted***, and the layer of mobile electrons is frequently referred to as the ***inversion layer***. Because the conducting channel in this case is *n*-type, the transistor shown here is an ***n-channel*** MOSFET.[19]

The gate voltage for which the density of mobile electrons in the channel equals the density of holes in the bulk material is defined to be the ***threshold voltage*** and is denoted V_{th}. (V_T is sometimes used for the threshold voltage, but we will reserve that notation for the thermal voltage, kT/q, to avoid confusion.) Once $V_G \geq V_{th}$, we say the device is operating in ***strong inversion***. When the surface has more mobile electrons than holes, but $V_G < V_{th}$, we say the device is operating in ***weak inversion***, or ***subthreshold***. If the device is simply said to be inverted, strong inversion is usually what is meant.

The simplest possible view (although a poor approximation) is to say that no inversion layer exists for $V_G < V_{th}$ and that one does exist for $V_G \geq V_{th}$. This approximation is frequently used, but can lead to trouble if you don't remember that substantial drain current can flow with $V_G < V_{th}$. Although it is occasionally convenient to talk and think in such terms, real devices do not have such sharp distinctions between the different regions of operation.

[19] "Channel" simply refers to the region under the gate between the source and drain, whether or not it is currently inverted.

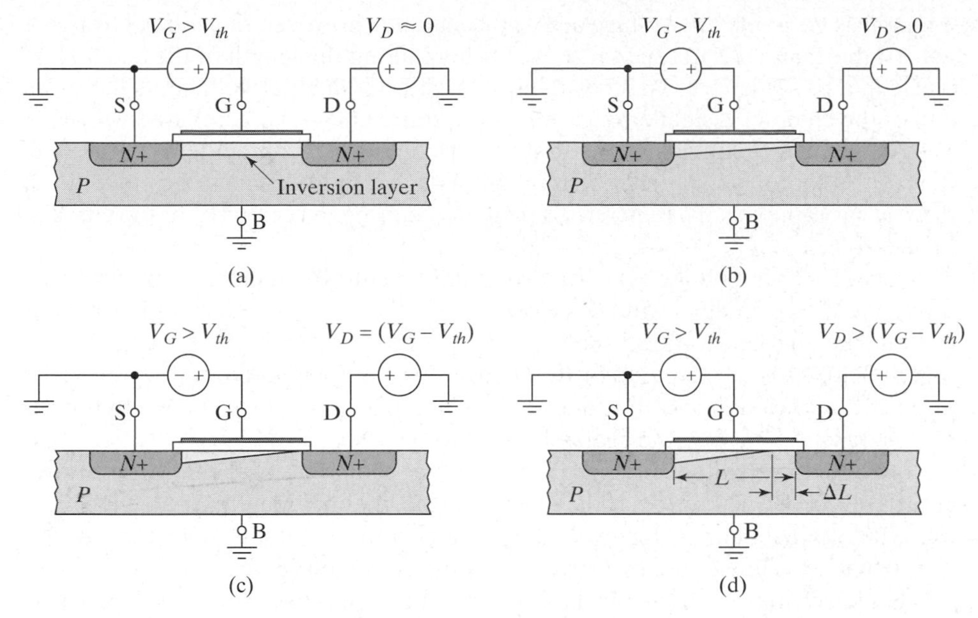

Figure 2–45 The charge in the inversion layer as a function of V_D: (a) small values of V_D, (b) larger values of V_D, (c) pinch-off, and (d) V_D greater than pinch-off.

Once an inversion layer has been formed, current can flow freely from the drain to the source if V_D is small, but not zero. When V_D is near zero, the device can be represented as shown in Figure 2-45(a). The corresponding region of the I_D-versus-V_D curve is also labeled (a) in Figure 2-46 and is called the ***linear***, ***triode***,[20] or ***voltage-controlled resistance*** (VCR), region. The drain current in this region is approximately given by

$$I_D \approx K'\frac{W}{L}(V_{GS} - V_{th})V_{DS} \tag{2.86}$$

where K' is a device-dependent constant, W and L are the width and length of the channel, respectively (*see* Figure 2-44), and we have allowed for the source not being grounded by using V_{GS} and V_{DS} in place of V_G and V_D respectively. Another commonly used notation is to define a new constant, $K = (K'/2)(W/L)$, so that

$$I_D \approx 2K(V_{GS} - V_{th})V_{DS}. \tag{2.87}$$

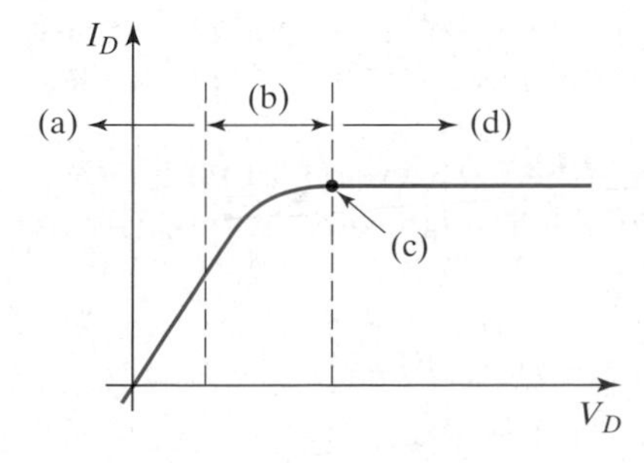

Figure 2–46 The drain current versus drain voltage for an *n*-channel MOSFET when $V_G \gtrsim V_{th}$. The regions shown correspond to the regions in Figure 2-45.

[20] This region of operation is referred to as the triode region because the characteristics look like those of a three-electrode vacuum tube which is called a triode.

As soon as we apply any voltage across the channel, however, the voltage from the gate to the channel is not the same everywhere along the length of the channel. In particular, as V_D is increased positively, the gate-to-channel voltage at the drain end of the channel is smaller than that at the source ($V_G - V_D$, compared with $V_G - V_S$). As a result, the inversion layer at the drain end will contain less charge than that at the source end, as shown in Figure 2-45(b), and the drain current will not increase as rapidly as V_D is increased further, resulting in region (b) of the curve in Figure 2-46.

When the drain voltage gets large enough, the gate-to-channel potential at the drain end of the channel will fall below V_{th}, and the inversion layer will disappear (taking the simple view), as shown in part (c) of Figure 2-45 and point (c) in Figure 2-46. As V_D is increased further above this value, the depletion region at the drain end of the channel will widen, as shown in part (d) of Figure 2-45. In this region (i.e., $V_D \geq V_G - V_{th}$), the channel is said to be ***pinched off***.

Notice that the voltage in the channel right at the point it pinches off is always equal to $V_G - V_{th}$. Therefore, if the change in the channel length, ΔL, is much less than the original channel length L, the drain current will not change appreciably. This situation is illustrated by region (d) of the curve in Figure 2-46.

When electrons reach the pinched-off end of the channel, they are swept across the depletion region by the lateral field. Since the drain current stops increasing with increasing drain voltage in this region, we say that the current has ***saturated***. This region of operation is, therefore, called the ***saturation region***. The region is also called the ***forward-active*** region of operation, because the device is operating in the forward direction (i.e., we have not interchanged the roles of the source and drain) and it is active (i.e., it can be used to provide amplification). The drain current in this region of operation is approximately given by

$$I_D \approx \frac{K'}{2}\frac{W}{L}(V_{GS} - V_{th})^2 = K(V_{GS} - V_{th})^2, \quad \textbf{(2.88)}$$

where all of the terms are the same as those used in (2.86) and (2.87). The major approximation made in deriving (2.88) is that we have ignored the change in channel length that occurs with increasing drain voltage. Therefore, that equation is accurate as long as $\Delta L \ll L$.

This discussion of device operation has distinguished between the source and drain of the device, but MOSFETs are usually made symmetrically, so that the source and drain are interchangeable. The discussion has also assumed that no inversion layer existed until a gate bias was applied; however, it is possible to fabricate devices that nominally have a conducting channel. In these devices, the channel must be depleted of mobile carriers to turn the device off. Such devices are called ***depletion-mode*** transistors. Devices in which the inversion layer must be created to turn them on, like the one we have described, are called ***enhancement-mode*** transistors.

The most common symbols used for n-type enhancement-mode MOSFETs are shown in Figure 2-47. The arrow on the bulk connection indicates the polarity of

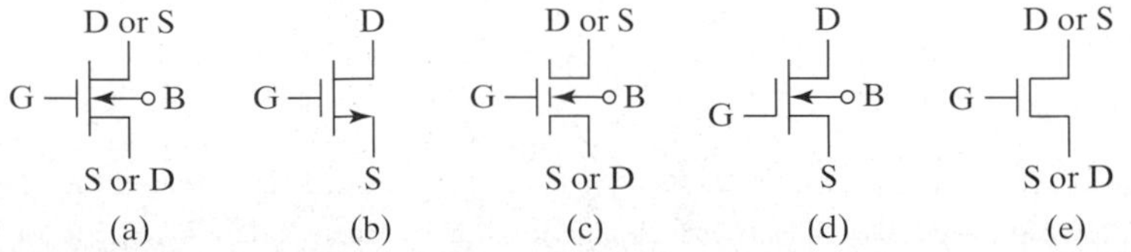

Figure 2–47 The symbols most commonly used for n-type enhancement-mode MOSFETs.

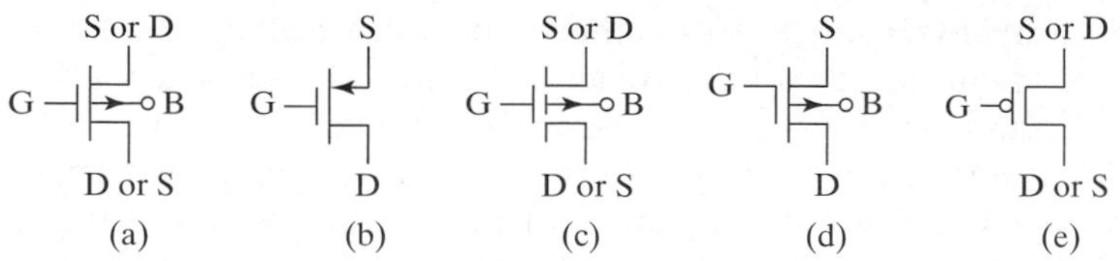

Figure 2–48 The symbols most commonly used for *p*-type enhancement-mode MOSFETs.

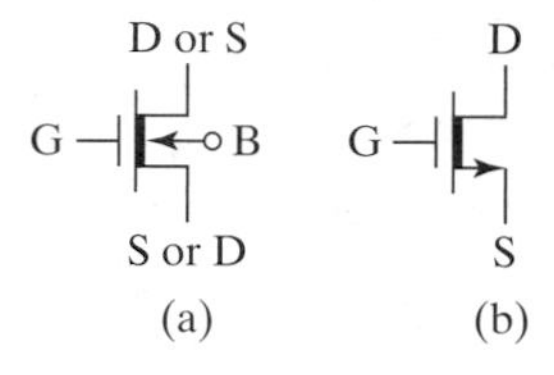

Figure 2–49 The symbols most commonly used for *n*-type depletion-mode MOSFETs.

the pseudo *pn* junction that is formed between the bulk and the inversion layer, with the arrow pointing from *p*-type to *n*-type, as it does for diodes and BJTs. The symbol in part (b) of the figure is probably the most popular, although it does not follow the established convention for having the arrow indicate a *pn* junction. In this case, the symbol is intended to look like a BJT, and the arrow indicates the direction of conventional current flow, as well as which lead is the source. Even with symmetric devices, it is nice to know which lead the designer was thinking of as the source, because that information makes analyzing the operation of a circuit easier. This symbol also assumes that the bulk connection is understood and does not explicitly show it.

The symbol in part (c) uses a broken line from source to drain to show that a channel does not exist unless a gate bias is applied (i.e., this MOSFET is an enhancement-mode device). The symbol in (d) is sometimes used, the gate is offset to show which of the other two terminals is the source. The symbol in (e) is sometimes used for simplicity when no confusion can result.

The most common symbols used for *p*-type enhancement-mode MOSFETs are shown in Figure 2-48 and are similar to their *n*-type counterparts. The symbol in part (e) uses a small circle on the gate to indicate that this MOSFET is a *p*-channel device. The circle is like the inversion dot used in logic circuits. In this text, we will use the symbols in parts (a) and (b) of Figures 2-47 and 2-48.

The most common symbols for *n*- and *p*-type depletion-mode MOSFETs are shown in Figures 2-49 and 2-50, respectively. The broad line from source to drain indicates that a channel exists if no bias is applied to the gate.

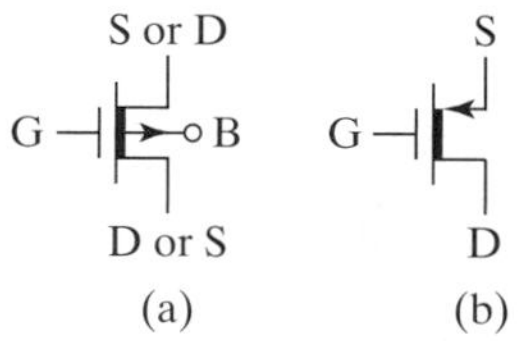

Figure 2–50 The symbols most commonly used for *p*-type depletion-mode MOSFETs.

Exercise 2.14

Assume that an n-channel enhancement-mode MOSFET has $V_{th} = 0.5$ *V and* $K = 100\ \mu A/V^2$. *(a) If the transistor is operated with* $V_{GS} = 1$ *V and* $V_{DS} = 0.2$ *V, what region of operation is the device in? What is the drain current? (b) What is* I_D *if* V_{GS} *is kept the same and* V_{DS} *is dropped to 0.1 V? (c) If the transistor is operated with* $V_{GS} = 0.7$ *V and* $V_{DS} = 1$ *V, what region of operation is the device in? What is the drain current? (d) What is* I_D *if* V_{DS} *is now increased to 2 V?*

□

2.6.2 Detailed Analysis of Current Flow

We now proceed with a detailed derivation of the first-order equations describing the operation of the MOSFET. We will proceed in the following way: First, we examine the energy band diagrams and approximate charge distributions for the MOS structure with varying gate bias. This analysis will allow us to derive an equation for the threshold voltage and improve our understanding of the MOS structure. Second, we examine current flow from drain to source through the inversion layer as a function of V_D.

Consider the metal-oxide–silicon structure shown in Figure 2-51(a). This structure is a vertical slice under the gate of the device in Figure 2-44 (or Figure 2-45). There will always be an electrical connection from the gate to the silicon for the structures we are interested in (e.g., through the circuit providing the gate bias), so the Fermi levels in the silicon and the metal must be the same.[21] The resulting band diagram is shown in part (b) of the figure, where we have assumed that electrons had to be transferred from the metal to the silicon in order to achieve equilibrium. Because of this charge transfer, the metal and silicon are no longer charge neutral, and an electric field will exist. The net positive charge in the metal is attracted by the net negative charge in the silicon, so the charges will be concentrated near the metal-oxide and silicon-oxide interfaces. The oxide is assumed to be free of charge, and the charge in the metal will exist as a very thin sheet right at the interface, since the metal is a good conductor.

In the example shown, the electrons that were transferred to the silicon have recombined with holes near the silicon-oxide interface, so that the region is depleted of holes. In other words, the charge in the silicon is caused by uncompensated acceptor ions. We invoke the depletion approximation to draw the charge profile shown in part (c) of the figure.

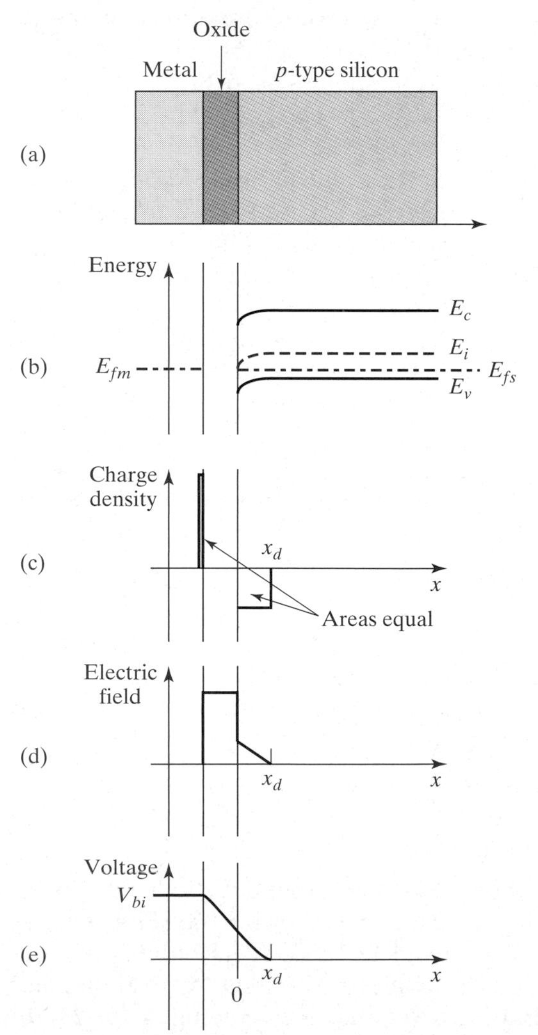

Figure 2–51 (a) A metal-oxide–silicon system with no applied bias, (b) the equilibrium band diagram, (c) the charge density profile, (d) the electric field, and (e) the electric potential, showing the resulting built-in potential V_{bi}.

[21] If you don't recall what a Fermi level is or why it must be constant, consult Aside A2.1.

The charge distribution can be integrated to find the electric field, as shown in part (d) of the figure (*see* Problem P2.35), and the field can then be integrated to find the potential, as shown in part (e) (the bulk of the silicon is arbitrarily taken to be at zero volts; *see* Problem P2.36). Notice that there is a built-in potential for the MOS system at equilibrium, just as was found for the metal–semiconductor contact and the *pn* junction. The field in the oxide is larger than the field at the surface of the silicon because of the difference in the permittivities of silicon ($\epsilon_r = 11.7$) and silicon dioxide ($\epsilon_r = 3.9$).[22]

The MOS system is essentially a parallel-plate capacitor with the oxide layer as the dielectric. The gate forms one plate and the silicon forms the other plate. For any capacitor (whether or not it is linear), the charges on the two plates must be equal and opposite; in other words, the charges must balance in any vertical slice through the MOS system, which is why the areas must be equal in Figure 2-51(c).

In the intuitive treatment of MOSFET operation, we compared the density of mobile carriers at the surface with the density in the bulk of the semiconductor. This comparison is a convenient reference point for quantitative analysis as well, and we therefore ask what gate potential must be applied to the device to force the band structure of Figure 2-51(b) to be flat, as shown in Figure 2-52. When the bands are flat, the carrier concentration is the same everywhere in the silicon. This condition is called the ***flat-band*** condition, and the gate voltage required to produce it is exactly the voltage required to *remove* the negative charge that was transferred to the silicon to cause the Fermi level to be constant as in Figure 2-51. In other words, if we apply a voltage equal in magnitude and opposite in sign to the built-in potential ($V_{FB} = -V_{bi}$), we will remove the charge and end up with the bands flat (but the Fermi levels different in the silicon and the metal). For the example at hand, we apply a negative voltage to the metal (relative to the silicon), which raises the Fermi level in the metal relative to the silicon.[23] The flat-band voltage for an *n*-channel MOSFET will typically be negative, as in this example.

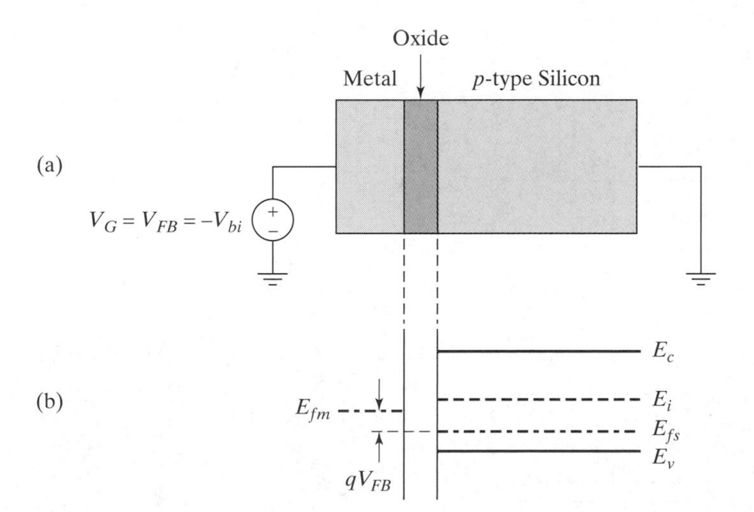

Figure 2–52 (a) A MOS system with the flat-band voltage applied and (b) the resulting band diagram.

[22] The electric flux (D) is given by $D = \epsilon\mathcal{E}$ and must be continuous at the interface.

[23] Since the charge on an electron is negative and the band diagram shows electron energy, a negative voltage will raise the Fermi level.

We can now describe the operation of the MOS system more accurately with reference to the flat-band voltage. We will describe the operation of an *n*-channel MOSFET, but the descriptions are all valid for *p*-channel devices if appropriate changes are made in the signs and carrier types.

Whenever $V_G < V_{FB}$ (i.e., more negative in our example), holes will be attracted to the surface of the silicon and will accumulate. This condition is known as ***accumulation***, and the resulting band diagram and charge distribution are shown in Figure 2-53. When the device is biased in the accumulation region, there are back-to-back *np*- and *pn*-junction diodes from source to drain. Therefore, no current can flow from the drain to the source, unless one of the *pn* junctions breaks down. Therefore, the MOSFET is cutoff in this region.

Now consider the case where $V_G > V_{FB}$, but the difference is small. In this case, holes are repelled from the surface of the silicon and the region is depleted of mobile charge. Therefore, a depletion region exists just like that at a *pn* junction. This region of operation is called the ***depletion region*** of operation. The band diagram and charge distribution for this condition are shown in Figure 2-54. We again invoke the depletion approximation for simplicity, so the charge distribution is drawn as a rectangle. The depth of the depletion region will increase with increasing V_G, since the charge distribution can be integrated twice to obtain the voltage, and this must increase as V_G increases. Since there isn't any mobile charge available to conduct current from the source to the drain, the MOSFET is cutoff when it is biased into depletion.

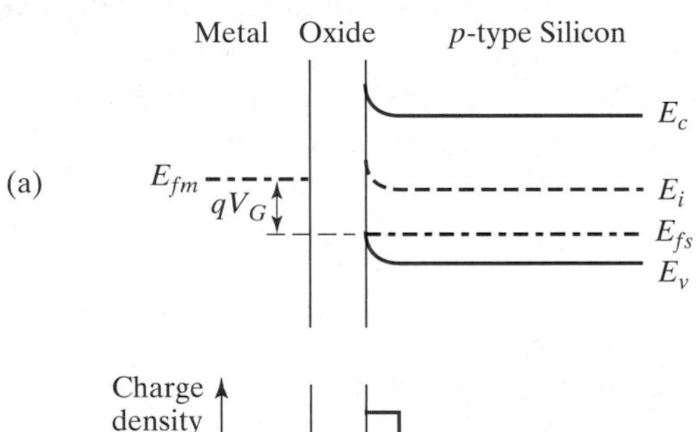

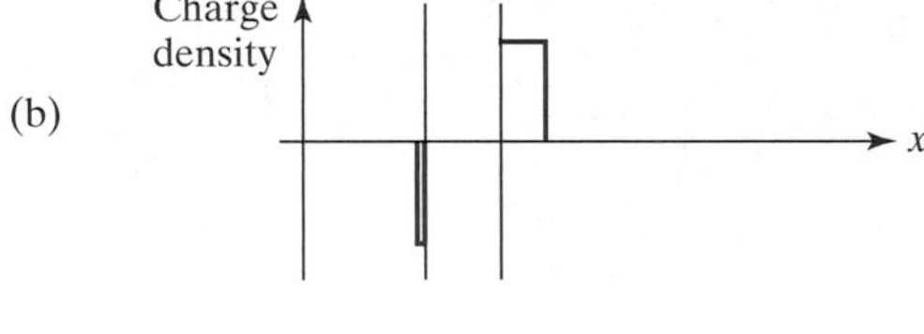

Figure 2–53 A MOS system biased in accumulation. (a) Band diagram and (b) charge distribution for an *n*-channel MOS device in accumulation.

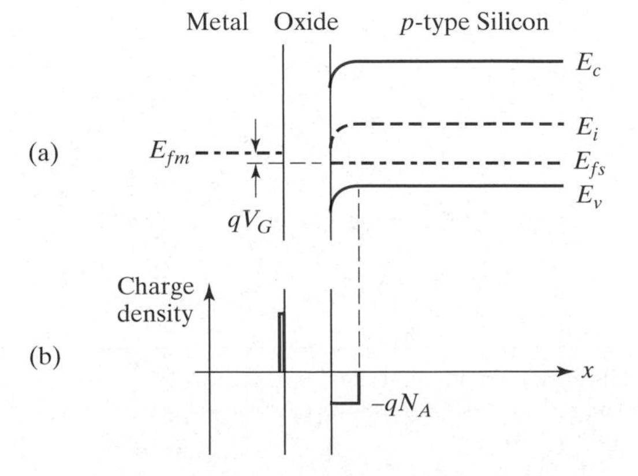

Figure 2–54 An *n*-channel MOS device biased in depletion. (a) Band diagram and (b) charge distribution.

For some value of V_G such that $V_{th} > V_G > V_{FB}$, the bands will bend just far enough that E_i will cross E_{fs}, and the surface will then be inverted; that is, the population of mobile electrons will exceed the population of holes. (Remember that in intrinsic material, where $E_i = E_f$, we have $n = p$.) At this point, it is possible to have current flow from the drain to the source, but the number of mobile carriers available is quite small and the resulting current will be small. This is the subthreshold, or weak-inversion, region of operation.

When V_G is increased so that $V_G = V_{th}$, the density of mobile electrons at the surface equals the density of holes in the bulk. In other words, the bands are bent so that E_{fs} is above E_i at the surface by the same amount that it is below E_i in the bulk. The band diagram and charge distribution for this situation are shown in Figure 2-55, where it is now necessary to distinguish between charge in the surface inversion layer and charge in the depletion region. This is the strong-inversion region of operation.

While the MOSFET is in weak inversion, the total-inversion layer charge is negligible in comparison with the charge in the depletion region. Therefore, the charge in the depletion region is approximately equal to the charge on the metal. As we increase V_G, the depletion region gets deeper, and the voltage at the surface of the silicon, $x = 0$ in Figure 2-51(e), increases. We know from (A2.1) that the probability of occupancy of a state is an exponential function of the surface potential, $E_{sur} = qV_{sur}$, so the charge in the inversion layer changes far more rapidly than the charge in the depletion region. Therefore, when the device is in strong inversion and the charge in the inversion layer dominates, we can say that the depletion region has approximately reached its maximum depth, denoted by $x_{d\max}$ in Figure 2-55, and further increases in V_G will be absorbed entirely by changing the inversion layer charge.

The maximum depth of the depletion region in modern devices is on the order of 0.1 to 0.3 μm. Figure 2-55 also introduces a new parameter, ϕ_f, which is implicitly defined in the figure; $q\phi_f = E_i - E_f$ in the bulk (i.e., away from the interface). This parameter is a function of the doping density in the silicon and can be shown to be given by (*see* Problem P2.38)

$$\phi_f = \begin{cases} V_T \ln \dfrac{N_A}{n_i}; & \text{p-type} \\ -V_T \ln \dfrac{N_D}{n_i}; & \text{n-type.} \end{cases} \tag{2.89}$$

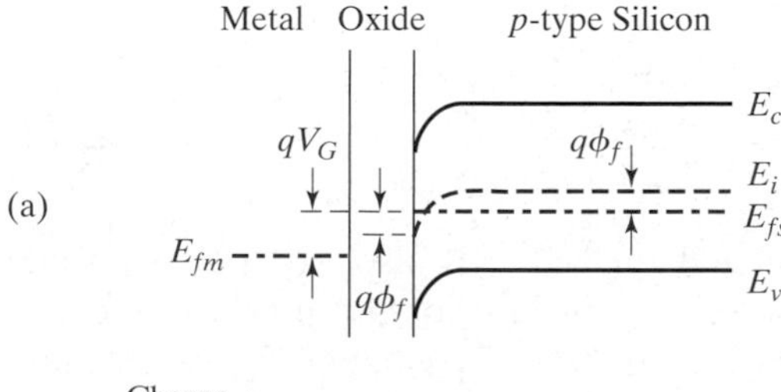

Figure 2–55 (a) Band diagram and (b) charge distribution for an *n*-channel MOS device biased at the onset of strong inversion.

It is worthwhile to note the similarity between (2.89) and (2.34) for the *pn* junction. We can rewrite (2.34) as

$$V_0 = V_T \ln \frac{N_A N_D}{n_i^2} = V_T\left(\ln\frac{N_A}{n_i} + \ln\frac{N_D}{n_i}\right), \tag{2.90}$$

which shows that the built-in potential of a *pn* junction can be expressed as the sum of the potentials developed across each side of the junction. We now recognize ϕ_f as being the potential developed across the single-sided depletion region present in the MOS system.

Using Figure 2-55, we can derive an equation for the threshold voltage. Consider starting with $V_G = V_{FB}$, so that we have the flat band condition. Then increase the gate voltage until the depletion region has attained its maximum depth. Since $Q = CV$ for any capacitor, the change in gate voltage necessary to compensate for the charge in the depletion region is $\Delta V_G = Q'_{dr}/C'_{ox}$, where C'_{ox} and Q'_{dr} represent the oxide capacitance per unit area and the depletion region charge per unit area, respectively. From the figure, we see that $Q'_{dr} = qN_A x_{d\max}$. Finally, increase V_G enough to bend the bands by a total of $2\phi_f$ to reach the onset of strong inversion. The resulting threshold voltage is the sum of all these terms:

$$V_{th} = V_{FB} + 2\phi_f + \frac{qN_A x_{d\max}}{C'_{ox}}. \tag{2.91}$$

A similar equation results for *p*-channel devices. It is useful to express (2.91) in a different way by plugging in an equation for $x_{d\max}$. $x_{d\max}$ can be found using Poisson's equation, (2.29), the charge distribution, and the fact that the potential at the surface is $2\phi_f$ at the onset of strong inversion if the bulk of the silicon is grounded[24] (*see* Problem P2.37). The result is

$$x_{d\max} = \sqrt{\frac{4\epsilon_{si}\phi_f}{qN_A}}, \tag{2.92}$$

where ϵ_{si} is the permittivity of silicon. Interestingly, (2.92) is exactly what you get for a one-sided ($N_A \gg N_D$) abrupt *pn* junction if the built-in potential is $2\phi_f$. Plugging (2.92) into (2.91) yields

$$\begin{aligned} V_{th} &= V_{FB} + 2\phi_f + \frac{\sqrt{2qN_A\epsilon_{si}}}{C'_{ox}}\sqrt{2\phi_f} \\ &= V_{FB} + 2\phi_f + \gamma\sqrt{2\phi_f} \end{aligned} \tag{2.93}$$

where we have implicitly defined γ.

As a final comment on V_{th}, we note that the value is also dependent on any charge trapped in the oxide or at the silicon/silicon-dioxide interface and any charge introduced by a threshold adjusting implant (*see* Chapter 3). All of these extra charges can be included by adding terms to (2.93) of the form Q/C'_{ox}, where Q is the total equivalent charge at the interface.

---

[24] At the onset of strong inversion, the total change in the energy bands from the bulk to the surface is $2q\phi_f$, as shown in Figure 2-55. Since $E = qV$ and the bands bend down, the surface potential is $2\phi_f$.

Exercise 2.15

Consider a MOS system made using p-type silicon with uniform doping and $N_A = 10^{15}$ atoms/cm^3. (a) What is the value of ϕ_f? (b) What is x_{dmax}? (c) If $V_{FB} = 1$ V and the oxide is 500 Å thick (Å stands for angstrom; 1 Å $= 10^{-8}$ cm), what is V_{th}?

We are now in a position to consider again the qualitative discussion of the previous section and derive equations for the drain current of the MOSFET in both weak and strong inversion. The drain current, I_D, flows from drain to source and is caused by both drift and diffusion—drift because of the lateral field set up when $V_D \neq 0$ and diffusion because the density of carriers in the inversion layer will vary along the length of the channel when $V_D \neq 0$. The variation in the density of carriers in the inversion layer is also caused by the lateral field, since this changes the voltage as a function of position along the length of the channel. The change in channel voltage then produces a change in the vertical field and, consequently, in the charge in the inversion layer. Unlike the *pn* junction, however, the drift and diffusion components of the drain current are in the same direction: Electrons move from the source to the drain for positive V_D.

Before we can find equations for the current, we need to know how the density of carriers in the inversion layer depends on the applied gate bias. That is a difficult problem to solve, but we will be able to find useful approximations without too much work. Consider Figure 2-56, which shows the band diagram, charge density, electric field, and electric potential plots for an MOS system biased in weak inversion.

We have already noted that for every increase in charge on the gate, there must be an equal increase in the total charge in the inversion layer and depletion region combined. Also, when the density of charge in the inversion layer is low compared with the density of charge in the depletion region, the charge in the inversion layer can be ignored, and changes in the applied gate voltage are almost entirely absorbed by increases in the depletion region depth. Since the charge in the inversion layer is an exponential function of the surface potential, V_{sur}, and V_{sur} increases as x_d increases, the charge in the inversion layer increases rapidly with increasing V_G. In fact, Q_{inv} grows exponentially with increasing V_G.

Once strong inversion is reached, the density of carriers in the inversion layer exceeds the density of charge in the depletion region, and further increases in gate charge are compensated for almost entirely by increases in the inversion layer charge density, resulting in the depletion region stopping at x_{dmax}, as mentioned earlier. At this point, the surface potential becomes almost independent of the applied gate bias, and any further increase in inversion layer charge can be found by considering the oxide to be a parallel-plate capacitor, so that[25] $\Delta Q_{inv} = C_{ox}\Delta V_G$. In sum, in weak inversion the inversion-layer charge density varies exponentially with gate bias, whereas in strong inversion it varies linearly with gate bias.

The preceding discussion and the plots shown in Figure 2-56 assume that the source and bulk are grounded and that $V_{DS} = 0$. When $V_{DS} > 0$, the discussion and

[25] This statement cannot be strictly true, since the surface potential *must* increase to cause the increase in charge density. Nevertheless, the point is that the change in potential is not absorbed entirely by the depletion region anymore, so the increase in surface potential is not as great. The net result is a linear dependence of Q_{inv} on V_G when the device is biased heavily into strong inversion. A good discussion and detailed treatment are provided in [2.5].

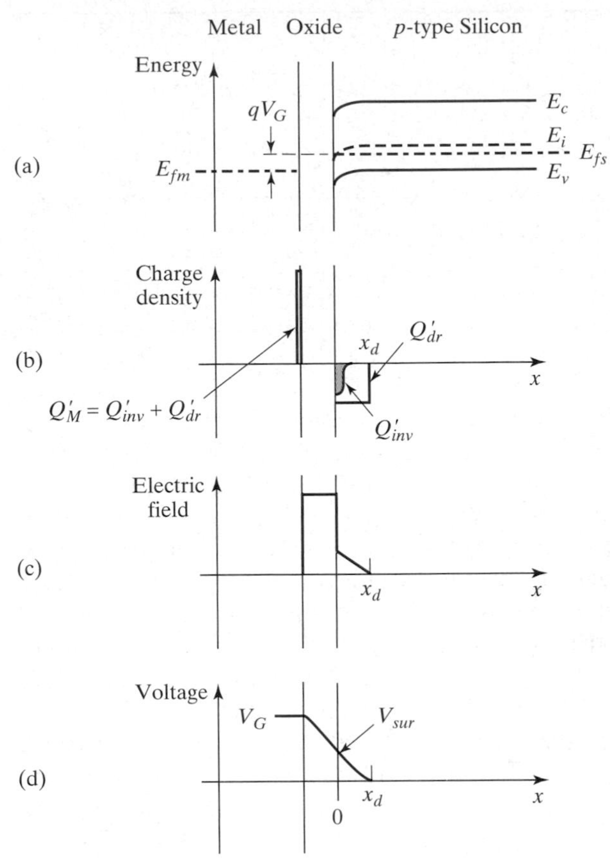

Figure 2–56 (a) Band diagram, (b) charge density, (c) electric field, and (d) electric potential plots for an MOS system biased in weak inversion.

figures are valid only at the source end of the channel. At the drain end, the surface potential is increased ($V_{sur} = V_D$), which implies that the depletion region is deeper[26] and, because the total charge density on the gate is constant, the inversion layer charge density must be correspondingly lower, so that Q'_m is still equal to $Q'_{inv} + Q'_{dr}$. An alternative view of this phenomenon is to note that as V_{sur} is raised, the Fermi level at the drain end of the channel is lowered in the silicon, which reduces Q_{inv}. In addition to causing a gradient in inversion-layer charge density along the length of the device, the variation in surface potential sets up a lateral field in the inversion layer that results in a drift current. In weak inversion the diffusion current will dominate, while in strong inversion the drift current will dominate. There is a transition region in which both components are important; in [2.5], this is called ***moderate inversion***, but we will use the more common simplification and divide the operation into weak and strong inversion only.

To understand which component of the current will dominate in each region of operation, remember that the diffusion current is proportional to the rate of change of carrier density, while drift current is proportional to the carrier density, *see* (2.6) and (2.9). In weak inversion, the number of carriers present is small, and the variation in inversion-layer carrier density will be large due to the exponential dependence on local potential, so the lateral diffusion current will dominate. In

[26] It will even be deeper than $x_{d\max}$ under some bias levels. Integrating the total charge density from x_d to $x = 0$ twice must yield V_{sur}, which at the drain end of the channel is equal to V_D.

strong inversion, on the other hand, the inversion-layer charge density is approximately a linear function of the surface potential, so the lateral gradient will be smaller. In addition, the number of carriers present is much larger. Therefore, the drift current dominates in strong inversion. In the following discussions, we will assume that the drain current is due entirely to diffusion in weak inversion and drift in strong inversion.

We know that the carrier density is an exponential function of the gate bias in weak inversion. We also know that the diffusion current is proportional to the derivative of the carrier density and that the derivative of an exponential is an exponential. Therefore, we expect the drain current to be an exponential function of the gate bias. This is in fact the case, and a more detailed approximate analysis reveals that the drain current in weak inversion is given by [2.5]

$$I_D \approx \frac{W}{L} I_X e^{(V_G - V_X)/nV_T}(1 - e^{-V_{DS}/V_T}), \tag{2.94}$$

where I_X and n are constants for a given device, V_T is the thermal voltage as used in the equation for the junction diode, W and L are the width and length of the device as defined in Figure 2-44, and V_X is slightly less than, but closely related to, V_{th}. Note that for values of V_{DS} greater than about $3V_T$, the equation can be simplified to one that closely resembles the equation for the collector current of a bipolar transistor:

$$I_D \approx \frac{W}{L} I_X e^{(V_G - V_X)/nV_T}. \tag{2.95}$$

Now consider the drain current of the MOSFET when $V_G > V_{th}$. Figure 2-57 is a repeat of part of Figure 2-45. Part (a) of Figure 2-57 depicts the situation where V_D is greater than zero, but less than $V_G - V_{th}$. To find an equation for the drain current in this case, we first remember that the drift component of the drain current will dominate. Starting from (2.9), which describes the one-dimensional drift-current density, and using l in place of x because we recognize that the drain current is due to drift along the length of the channel, we write

$$J_D(l) = -q\mu_n n(l)\mathcal{E}(l), \tag{2.96}$$

where we have used the fact that the mobility we want is the mobility of an electron, denoted μ_n, and where we have defined positive drain current as being *into* the drain, which is in the negative l direction. Note that J_D must be constant along the length of the channel and we want the current, not the current density. We can safely assume that $n(l)$ is constant across the width of the device, but it does vary

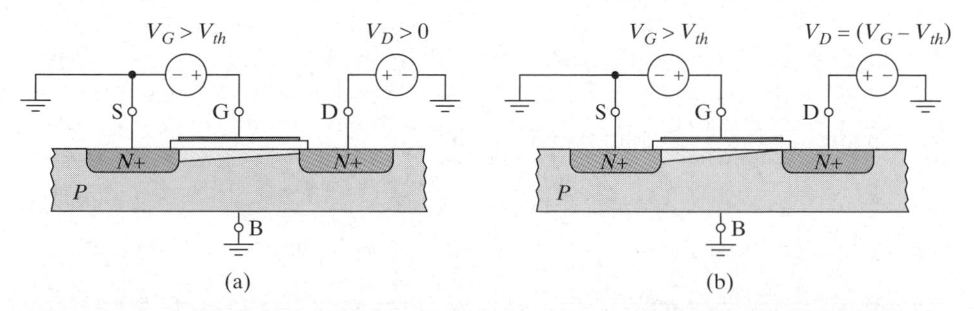

Figure 2–57 The charge in the inversion layer as a function of V_D: (a) values of V_D less than required for pinch-off and (b) pinch-off.

with depth x. Therefore, we would need to integrate the density in the x direction and multiply by W to get density per unit length. If we were to perform the integration, multiply by W, and further multiply by the charge per electron, q, the result would be the charge per unit length in the inversion layer;[27] call it $Q_{inv}(l)$. The drain current is then equal to

$$I_D = -\mu_n Q_{inv}(l). \qquad \textbf{(2.97)}$$

The charge per unit length in (2.97) can be found for the MOS system in strong inversion by using the capacitance of the oxide and the fact that $Q = CV$ for a capacitor. The capacitance of the oxide is given by the equation for a parallel-plate capacitor, namely,

$$C_{ox} = \frac{\epsilon_{ox}A}{t_{ox}} = \frac{\epsilon_r\epsilon_o A}{t_{ox}} = \frac{\epsilon_r\epsilon_o WL}{t_{ox}} = WLC'_{ox}, \qquad \textbf{(2.98)}$$

where t_{ox} is the oxide thickness, W and L are the width and length of the channel, respectively, and we have previously defined C'_{ox} to be the capacitance per unit area.

Performing a first-order analysis, we say that the total charge in the inversion layer is approximately zero when $V_G = V_{th}$, and all further increases in V_G result in increased channel charge (i.e., the depletion region does not change width). Therefore, when $V_D = 0$, the total charge in the inversion layer in strong inversion will be $Q_{inv} = C_{ox}(V_G - V_{th})$, and the charge per unit length will be

$$Q_{inv}(l) = \frac{Q_{inv}}{L} = WC'_{ox}(V_G - V_{th}). \qquad \textbf{(2.99)}$$

When the drain voltage is not zero, the voltage in the channel will be a function of position, and so will the charge density. We can express the charge density as

$$Q_{inv}(l) = WC'_{ox}(V_G - V_{ch}(l) - V_{th}). \qquad \textbf{(2.100)}$$

Substituting the charge per unit length from (2.100) into (2.97) and expressing the field as the negative of the derivative of the channel voltage, we get

$$I_D = \mu_n WC'_{ox}(V_G - V_{ch}(l) - V_{th})\frac{dV_{ch}(l)}{dl}. \qquad \textbf{(2.101)}$$

We now integrate both sides of (2.101) to get

$$\int_0^L I_D dl = \mu_n C'_{ox} W \int_0^L (V_G - V_{ch}(l) - V_{th})\frac{dV_{ch}}{dl}dl. \qquad \textbf{(2.102)}$$

As l varies from 0 to L, V_{ch} varies from 0 to V_D, so we can rewrite (2.102) as

$$I_D = \mu_n C'_{ox}\frac{W}{L}\int_0^{V_D}(V_G - V_{ch}(l) - V_{th})dV_{ch}, \qquad \textbf{(2.103)}$$

and performing the integration, we finally obtain

$$I_D = \mu_n C'_{ox}\frac{W}{L}\left[(V_G - V_{th})V_D - \frac{V_D^2}{2}\right]. \qquad \textbf{(2.104)}$$

---

[27] The charge in the depletion region is excluded because it is *bound* charge in the form of uncompensated dopant ions and is not free to move. Therefore, it does not contribute to the drift current.

Equation (2.104) provides a first-order equation for the drain current for a large[28] MOSFET with $V_G > V_{th}$ and V_D small enough that the inversion layer extends all the way from the source to the drain as shown in Figure 2-57(a). As noted earlier, this region of operation is known as the linear, triode, or voltage-controlled resistance (VCR) region. For small V_D, the squared term in (2.104) can be ignored, and I_D is approximately a linear function of V_D, with the resistance being a function of $V_G - V_{th}$. More accurate expressions for I_D can be derived (e.g., by including the variation in x_d with l, rather than assuming that $x_d = x_{dmax}$; *see* [2.3], Vol. IV, and [2.5]), but their complexity usually limits their usefulness for hand analysis and design, and they will not be pursued here.

As the drain voltage is increased, the gate-to-channel voltage at the drain end of the channel continues to decrease. When $V_G - V_D - V_{th} = 0$, the inversion layer at the drain end of the channel disappears, as shown in Figure 2-57(b). Electrons that drift from the source toward the drain still make it to the drain, however, since no barrier is established to their flow. Further increases in V_D are mostly absorbed across the small depletion region formed at the end of the channel and do not affect the voltage in the inversion layer. Since the drain current is assumed to be entirely due to the drift of carriers in the inversion layer, once the drain end of the channel pinches off, the drain current saturates and does not increase for further increases in V_D. The MOSFET is now said to be in saturation, and we can find the drain current by substituting $V_D = V_G - V_{th}$ into (2.104) to obtain

$$I_{D\text{sat}} = \frac{\mu_n C'_{ox}}{2}\frac{W}{L}(V_G - V_{th})^2. \quad \textbf{(2.105)}$$

Our analysis so far has concentrated on positive values of V_D, but the same arguments apply if V_D is negative. In that case, the inversion layer is pinched off at the source end of the channel, and the drain current is negative. Because MOSFETs are usually symmetric, it is common practice to redefine the source and drain terminals as necessary to keep the drain current positive with the device in saturation.

It will not always be true that the source and bulk of a MOSFET are grounded, so we should modify our equations to account for this fact. We will always assume that the source-to-bulk and drain-to-bulk *pn* junctions are reverse biased and the source-to-bulk voltage is constant, unless explicitly stated otherwise. Therefore, (2.104) and (2.105) can be modified by recognizing that V_D and V_G are referred to ground only because we have the source grounded. If the source is not grounded, the voltages we are interested in are V_{DS} and V_{GS}. Making this change yields the equations we will use for the MOSFET: In the linear region, the drain current is given by (we have again used K as in (2.87), but we now have a complete definition of the constant in terms of fundamental physical constants and device dimensions)

$$I_D = \mu_n C'_{ox}\frac{W}{L}\left[(V_{GS} - V_{th})V_{DS} - \frac{V_{DS}^2}{2}\right] = 2K\left[(V_{GS} - V_{th})V_{DS} - \frac{V_{DS}^2}{2}\right], \quad \textbf{(2.106)}$$

which is a modified version of (2.104), and in saturation the drain current is

$$I_{D\text{sat}} = \frac{\mu_n C'_{ox}}{2}\frac{W}{L}(V_{GS} - V_{th})^2 = K(V_{GS} - V_{th})^2, \quad \textbf{(2.107)}$$

---

[28] Second-order effects become very important in modern submicron devices, and this equation is not very accurate in those circumstances, although it still provides a serviceable approximation for hand analysis.

which is a modified version of (2.105). The electron mobility used in these equations is the *surface* mobility, not the bulk mobility. The surface mobility is typically about one-half of the bulk mobility, due to increased scattering of carriers at the interface between the silicon and the oxide. The transistor is saturated when $V_{DS} \geq V_{GS} - V_{th}$, which is a modified version of our earlier condition. Because the saturation region of a MOSFET is the region of operation used for amplification, it is referred to as the forward-active region of operation.

We are interested in seeing whether or not the MOSFET does a reasonable job of approximating a voltage-controlled current source in the forward-active region of operation. In Chapter 6, we will see that it is possible to limit ourselves to signals that are small compared with the average values present and still produce useful circuits. For small perturbations about the average, the MOSFET can be described as a controlled source with a transconductance given by differentiating (2.107):

$$\begin{aligned} g_m \equiv \frac{dI_D}{dV_{GS}} &= \mu_n C'_{ox}\frac{W}{L}(V_{GS} - V_{th}) = 2\sqrt{\frac{\mu_n C'_{ox}}{2}\frac{W}{L}I_D} \\ &= 2K(V_{GS} - V_{th}) = 2\sqrt{KI_D} \end{aligned} \quad \textbf{(2.108)}$$

The drain characteristics for an n-channel MOSFET are shown in Figure 2-58 and illustrate the different regions of operation. A key characteristic of a FET is that the drain characteristics all have different slopes for small V_{DS}. The different slopes are indicative of the fact that the MOSFET looks like a voltage-controlled resistor in this region. The effective resistance of a MOSFET operating in the linear region can be calculated from (2.106). Assuming that V_{DS} is small enough that the squared term can be ignored, you get

$$R_{DS} = \frac{V_{DS}}{I_D} = \frac{1}{\mu_n C'_{ox}\frac{W}{L}(V_{GS} - V_{th})}. \quad \textbf{(2.109)}$$

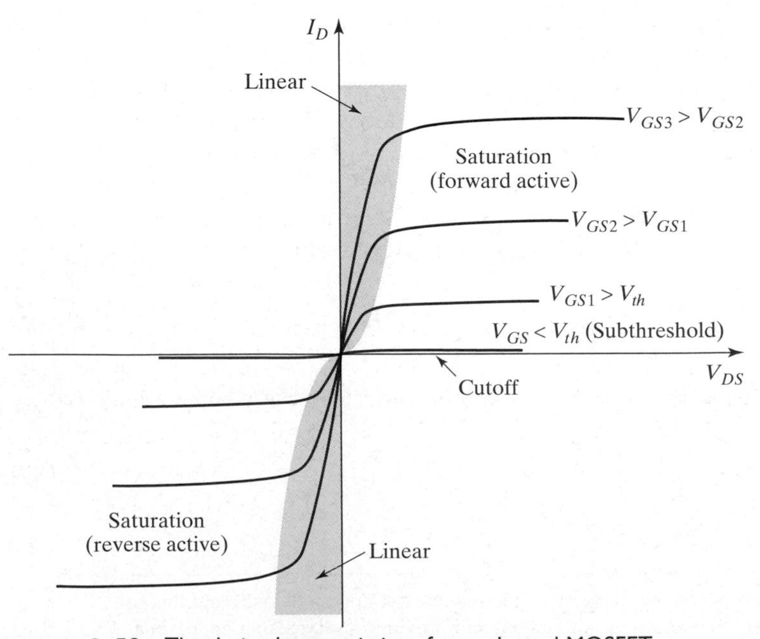

Figure 2–58 The drain characteristics of an n-channel MOSFET.

Exercise 2.16

Consider an n-channel MOSFET with $W = 1\ \mu m$, $L = 0.5\ \mu m$, an oxide thickness of 150 Å, and $V_{th} = 0.5$ V. (a) Assuming that the surface mobility is one-half of the bulk mobility, what is the value of K? (b) If $V_{GS} = 1$ V and $V_{DS} = 3$ V, what region of operation is the device in? (c) What is the corresponding drain current?

Now that we have a basic quantitative understanding of the operation of a MOSFET, we can investigate the more important second-order effects. One important second-order effect has already been covered: subthreshold conduction. In the next several sections, we discuss channel-length modulation and the effect of changing the source-to-bulk voltage.

2.6.3 Channel-Length Modulation

Figure 2-59 shows a picture of a MOSFET channel when $V_{DS} > V_{GS} - V_{th}$. As V_{DS} increases above the value required to saturate the device, there is a slight widening of the depletion region at the drain end, so that the effective channel length is $L_{eff} = L - \Delta L$. By definition, the channel voltage at the point where the channel just pinches off is

$$V_{ch}(L_{eff}) = V_G - V_{th}. \qquad \textbf{(2.110)}$$

Electrons in the channel that drift from the source to the point L_{eff} will be swept across the depletion region by the lateral field and will contribute to the drain current. Therefore, the drain current depends only on what happens from $l = 0$ to $l = L_{eff}$. We can find the new drain current by changing the limits of integration in (2.102) to get (*see* Problem P2.51)

$$I_{Dsat} = \frac{\mu_n C'_{ox}}{2} \frac{W}{L_{eff}} (V_G - V_{th})^2. \qquad \textbf{(2.111)}$$

This change in the drain current with changing L_{eff} is called ***channel-length modulation***. Comparing (2.111) and (2.107), we see that one reasonable way to modify the drain current in saturation to account for channel-length modulation is to write

$$I_{Dsat} = \frac{\mu_n C'_{ox}}{2} \frac{W}{L} (V_G - V_{th})^2 \left(\frac{L}{L_{eff}}\right). \qquad \textbf{(2.112)}$$

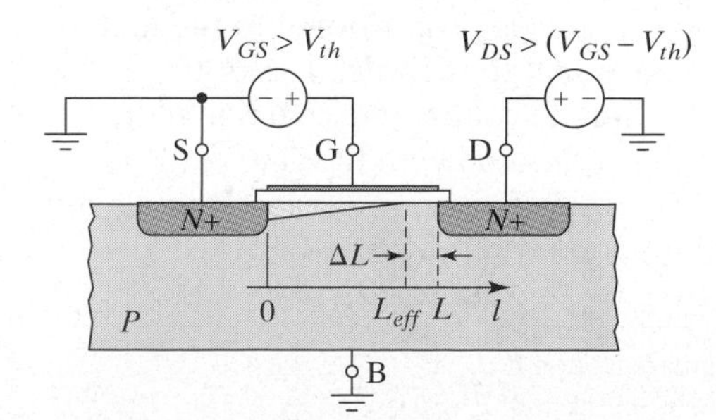

Figure 2–59 MOSFET channel when $V_{DS} > V_{GS} - V_{th}$.

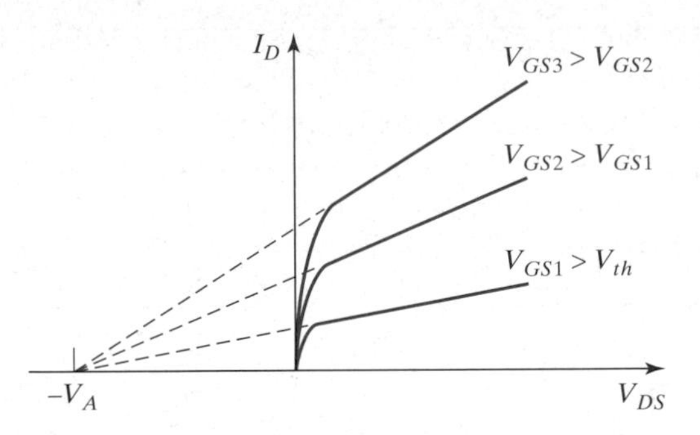

Figure 2–60 Drain characteristics of an *n*-channel MOSFET including channel-length modulation.

We now need to know how L_{eff} varies with V_{DS} to model the effect. A detailed analysis of this effect is quite complicated and must take into account the two-dimensional field distribution, especially for devices with short channel lengths and thin oxides. We will not pursue the analysis here, but will only present a simple model that is often used. Figure 2-60 shows the effect of channel-length modulation on the idealized drain characteristics of a MOSFET. The idealized drain characteristics in the forward-active region of operation have a finite slope, and they all extrapolate to a common intercept on the V_{DS} axis.

As a first-order model, it is often acceptable to write

$$I_{D\text{sat}} = \frac{\mu_n C'_{ox}}{2}\frac{W}{L}(V_G - V_{th})^2(1 + \lambda V_{DS}), \tag{2.113}$$

where λ is a device parameter. The same relationship is often expressed in terms of the Early voltage,[29] so that the equation can be written

$$I_{D\text{sat}} = \frac{\mu_n C'_{ox}}{2}\frac{W}{L}(V_G - V_{th})^2\left(1 + \frac{V_{DS}}{V_A}\right). \tag{2.114}$$

Although it is possible to find an equation for λ (or V_A), we will treat it as a constant that will be given or determined by measurement. It is important to note, however, that longer devices have smaller values of λ (or larger V_A), as one would expect (since the longer the device is, the less difference a small change makes).

Exercise 2.17

Consider an n-channel MOSFET that is measured to have $I_D = 100\ \mu A$ when $V_{GS} = 2V$ and $V_{DS} = 3\ V$ and has $I_D = 104\ \mu A$ when $V_{GS} = 2V$ and $V_{DS} = 4\ V$. (a) What is the value of λ? (b) If $L = 1\ \mu m$, what is L_{eff} when $V_{DS} = 4\ V$?

□

2.6.4 Charge Storage

The most important charge storage mechanism in a MOSFET is the gate oxide capacitance. From (2.98), we see that this capacitance is proportional to the area of the gate oxide (i.e., WL) and inversely proportional to the thickness of the oxide. The capacitance shows up between the gate and the conducting channel in the device. In addition, there are small capacitances associated with the gate oxide overlapping the source and drain regions. These are called ***overlap*** capacitances and can be highly significant, especially in small-geometry devices.

[29] In honor of James Early, who first described a similar effect in BJTs.

In modeling a MOSFET, we need to split the gate-to-channel capacitance into two terms: C_{gs} and C_{gd}. Each of these capacitors will contain an overlap capacitance, C_{ol}, but how we split the gate oxide capacitor depends on the region of operation the device is in. In the linear region of operation, a conducting channel (the inversion layer) exists all the way from the source to the drain, and each of these terminals affects the charge stored in the channel. Therefore, for the linear region of operation, we can split the gate oxide capacitance equally and obtain

$$C_{gs} = C_{gd} = \frac{WLC'_{ox}}{2} + C_{ol} \,. \tag{2.115}$$

When the device is saturated, however, V_{DS} has no effect on the charge stored in the channel, so the gate oxide capacitance does not contribute to C_{gd} at all. Therefore, in saturation, we have

$$C_{gd} = C_{ol}. \tag{2.116}$$

To find the value of C_{gs} in saturation, we can show that the total charge in the inversion layer is (*see* Problem P2.49)

$$Q_{inv} = \frac{2}{3}WLC'_{ox}(V_{GS} - V_{th}) \,. \tag{2.117}$$

Since this equation is not of the form $Q = CV$, it does not represent a linear capacitor. Nevertheless, as noted earlier, we can restrict ourselves to small deviations from a nominal operating point and use derivatives to find linear approximations (*see* Chapter 6 for complete details). The gate oxide capacitance in saturation is then

$$C_{ox} = \frac{dQ_{inv}}{dV_{GS}} = \frac{2}{3}WLC'_{ox} \,, \tag{2.118}$$

and the total gate-to-source capacitance in saturation is

$$C_{gs} = \frac{2}{3}WLC'_{ox} + C_{ol} \,. \tag{2.119}$$

In addition to the gate oxide capacitance, MOSFETs do have other ***parasitic*** capacitors. These are capacitors that are not theoretically necessary to the operation of the device, but nonetheless appear due to the method of fabrication. The major parasitic capacitors in a MOSFET are the source-to-bulk and drain-to-bulk junction capacitors. These are identical to the capacitance seen across a reverse-biased *pn*-junction diode, as discussed in Section 2.4.5, and will not be treated again here.

Exercise 2.18

Consider an n-channel MOSFET with a 200-Å-thick oxide, $W = 3$ μm, and $L = 1$ μm. Ignore overlap capacitance. (a) What are C_{gs} and C_{gd} when the device is operated in the linear region? (b) What are C_{gs} and C_{gd} when the device is saturated?

2.6.5 The Effect of Bulk Bias

Thus far in our examination of MOSFET operation, we have kept the source-to-bulk potential at zero volts. Keeping $V_{SB} = 0$ is not always possible in practice, so it is important to know what happens when this voltage varies. A first-order analysis of the effect of changing V_{SB} can be carried out by considering charge balance in the MOS structure once again.

Figure 2-61 shows the different charge terms present in the MOSFET structure when $V_{DS} = 0$. To maintain balance, it must be true that $Q_M = Q_{inv} + Q_{dr}$. However, notice that the inversion layer acts somewhat like the *n*-type side of a *pn* junction with respect to the *p*-type bulk material. In other words, if the voltage in the inversion layer changes relative to the bulk (i.e., V_{SB} changes), while keeping the gate-to-channel voltage constant (i.e., V_{GS} constant), the depth of the depletion region must change to compensate for the potential difference, just as it would if this were a *pn* junction. If V_{SB} increases, the depletion region will widen and Q_{dr} will increase. To maintain charge balance, there must be a concomitant decrease in the charge in the inversion layer, Q_{inv}. Therefore, Q_{inv} is under the control of both the gate voltage and the bulk voltage (with both taken relative to the source).

Since the drain current is a function of Q_{inv}, the bulk voltage can affect the drain current. Since the bulk is, in effect, acting like another gate, it is sometimes called the ***back gate***. The modulation of Q_{inv} by the bulk potential is usually called the ***body effect***, since the bulk is also referred to as the body. A common way to model the body effect for constant V_{SB} is to modify the value of V_{th}. Since increasing V_{SB} decreases Q_{inv}, a larger gate voltage will be required to reach strong inversion; in other words, V_{th} will increase. The threshold voltage with body effect is given by [2.5]

$$V_{th} = V_{th0} + \gamma\left[\sqrt{2\phi_f + V_{SB}} - \sqrt{2\phi_f}\right], \tag{2.120}$$

where ϕ_f was introduced in Figure 2-55, γ was introduced in (2.93), and V_{th0} is V_{th} when $V_{SB} = 0$ as given by (2.93). Equation (2.120) is derived from (2.93) by replacing the surface potential, $2\phi_f$, with $2\phi_f + V_{SB}$ to account for applied bias.

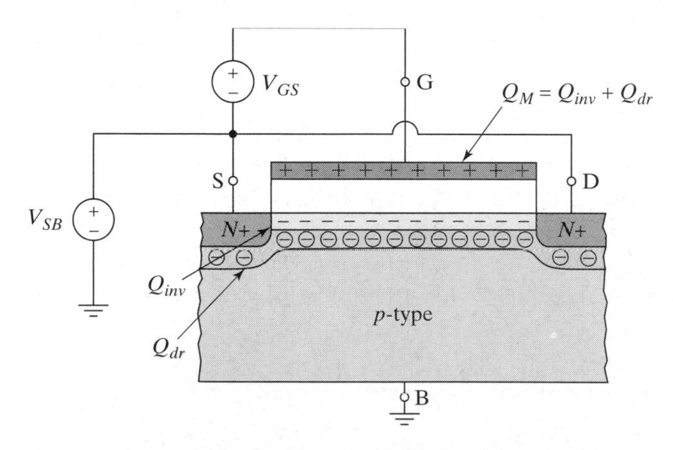

Figure 2–61 Charge present in a MOSFET biased with $V_{GS} > V_{th}$. The charge in the metal must balance the sum of the inversion layer charge and the charges in the depletion region under the gate. The charge in the inversion layer is shown as mobile charge, whereas the charges in the depletion region have circles around them to indicate that they are bound charges (ionized acceptors).

Since changing V_{SB} results in a change in the charge stored in the depletion region, there is a capacitance associated with the depletion region that exists from the channel to the bulk. We denote this capacitance by C_{DR} and note that it is a function of the doping profile in the bulk and the nominal value of V_{SB}. We can use the capacitance to arrive at a simple model for the effect of the back-gate bias on the drain current in saturation. In strong inversion, we approximated the change in inversion layer charge with gate voltage by $\Delta Q_{inv} = C_{ox}\Delta V_G$ (where V_S was assumed to be ground). With the source not grounded, we can revise this equation and say that $\Delta Q_{inv} = C_{ox}\Delta V_{GS}$. Now, if V_{SB} changes, we also see a change in Q_{inv} as described, but the charge decreases for increasing positive V_{SB}. Therefore, we can write $\Delta Q_{inv} = -C_{DR}\Delta V_{SB}$.

For small changes from the bias point, we can again define a transconductance that relates changes in V_{BS} to the resulting changes in I_D. This transconductance is called the ***bulk transconductance*** and is

$$g_{mb} \equiv \frac{dI_D}{dV_{BS}}, \tag{2.121}$$

where the result will be positive, since we are using V_{BS} in the definition instead of V_{SB}. Because I_D is proportional to Q_{inv} in strong inversion, *see* (2.97), and because we have $\Delta Q_{inv} = C_{ox}\Delta V_{GS}$ and $\Delta Q_{inv} = -C_{DR}\Delta V_{SB} = C_{DR}\Delta V_{BS}$, we can express g_{mb} as a function of g_m (given by (2.108));

$$g_{mb} = g_m \frac{C_{DR}}{C_{ox}} = g_m \chi, \tag{2.122}$$

where we have implicitly defined the parameter χ.

Exercise 2.19

Reconsider the MOSFET from Exercise 2.15. If $V_{SB} = 2$ V, what is the new value of V_{th}?

□

2.6.6 Breakdown

MOSFETs exhibit two different kinds of breakdown. First, there are several nondestructive breakdown mechanisms that are similar to those present in *pn* junctions and BJTs. Second, a destructive breakdown of the gate oxide can occur. If the gate-to-channel voltage exceeds the maximum field that the insulator can withstand, permanent damage will result. For silicon dioxide, this field is about 600 V/μm, so for typical modern devices, the breakdown voltages are on the order of a few volts or tens of volts. Protection circuits are almost always included at the inputs to MOS circuits to prevent static charge from destroying the devices. This is also why MOS integrated circuits are shipped in static-free containers and why special handling precautions are used.

The nondestructive breakdown mechanisms that are present are similar to those present in junction diodes. First, the source-to-bulk and drain-to-bulk *pn* junctions can be broken down in exactly the same way a *pn*-junction diode breaks down. Second, if the MOSFET is saturated, the carriers at the drain end are traveling the fastest (the lateral field is highest there) and may have enough energy to create new EHPs when they collide with atoms in the crystal; this is the same impact ionization phenomenon described for avalanche breakdown of *pn* junctions. In this case, the drain current can increase rapidly with increasing V_{DS}, as shown in Figure 2-62.

2.6.7 Short- and Narrow-Channel Effects

The analyses presented so far in this section have assumed that the MOSFET can be treated by analyzing the vertical fields and horizontal fields separately. For example, in deriving (2.91) for the threshold voltage, we ignored the fact that some charge in the depletion region under the gate is really part of the drain-to-bulk depletion region and is not under the control of the gate voltage alone. For large devices (say about $L > 3$ μm), this assumption is reasonable, but for modern devices with channel lengths on the order of 0.1 μm, it is not reasonable. In fact, as shown in Figure 2-63, the source-to-bulk and drain-to-bulk depletion regions can account for a significant fraction of the charge under the gate of a short-channel MOSFET. Several ***short-channel effects*** result from the fact that the drain, source, bulk, and gate terminals all have a significant influence on the charge in the inversion layer. We will briefly mention some short-channel effects here; for a more detailed treatment, you should consult [2.5] or the *Advanced MOS Devices* volume of [2.3].

The most important short-channel effect is a decrease in the threshold voltage of the device as the channel length is reduced. Because a portion of the charge in the depletion region under the gate is controlled by the drain-to-bulk and source-

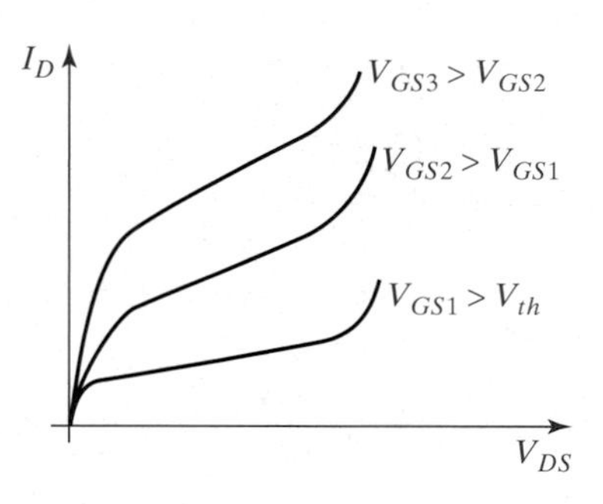

Figure 2–62 The drain characteristics of an *n*-channel MOSFET, showing breakdown.

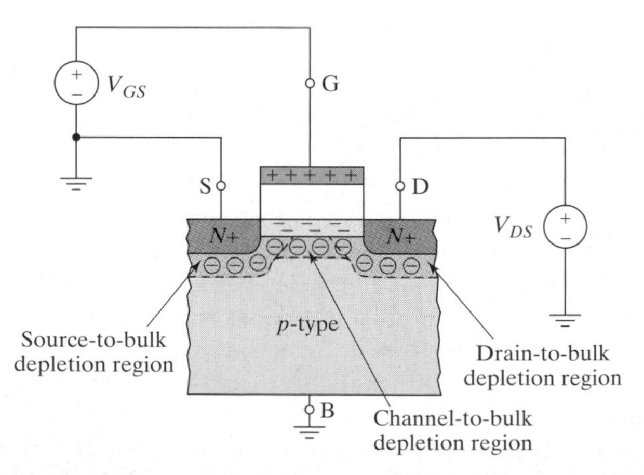

Figure 2–63 A short-channel MOSFET, showing the depletion region charge associated with the drain-to-bulk, source-to-bulk, and channel-to-bulk depletion regions.

to-bulk junctions, it does not need to be compensated for by charge on the gate. Therefore, when all of the charges are balanced, there is more charge in the inversion layer than our previous simple picture predicted, and to reach the point where the density of carriers in the inversion layer equals the density of mobile carriers in the bulk, the gate voltage will not have to be as high in a short-channel device. In other words, the threshold voltage will be reduced. V_{th} can be reduced by several tenths of a volt (compared with the value for a large device) for devices with channel lengths on the order of 1 μm [2.5].

Another important short-channel effect is that the output resistance of the device will be decreased. In other words, the drain characteristics (as in Figure 2-60) will have a steeper slope in the saturation region. This effect is due to the fact that an increasing drain voltage not only reduces the channel length, but also increases the percentage of the depletion region charge that is due to the drain-to-bulk junction, thereby reducing V_{th} and causing a further increase in drain current. In addition, the change in effective channel length is a larger fraction of the total channel length for short devices.

One bit of good news is that the body effect is less severe in short-channel devices. Since the channel-to-bulk depletion region charge is a smaller fraction of the total charge in the depletion region under the gate, changes in this charge due to changing bulk bias have a correspondingly smaller effect on device performance for short-channel devices.

The final short-channel effect we will mention is that a new breakdown mechanism is introduced. If the source-to-bulk and drain-to-bulk depletion regions meet, then significant current can flow from the drain to the source. This current flows below the surface and is not under the control of the gate [2.5]. The condition is known as ***punch-through*** and can lead to breakdown at voltages lower than those predicted by the mechanisms described in Section 2.6.6.

Devices made with very narrow channels also exhibit significant deviations from the theory we have presented. Figure 2-64 shows a cross-sectional view of the width of a MOSFET, indicating that there is extra charge in the sides of the depletion region that are not directly under the gate of the MOSFET. This charge is due to the fringing fields and is not accounted for in our simple vertical picture of the channel. For wide channel devices, the charge is not significant, but for narrow devices, it changes the characteristics. For narrow devices, there is *more* charge in the depletion region under the control of the gate than our simple analysis accounted for, so the inversion-layer charge density is *lower* than we expect. Therefore, the major ***narrow-width effects*** are in opposition to the short-channel effects. Narrow devices have larger threshold voltages and greater body effect than wider devices. Threshold voltage shifts of tenths of a volt are again observed for devices with widths on the order of 1 μm [2.5].

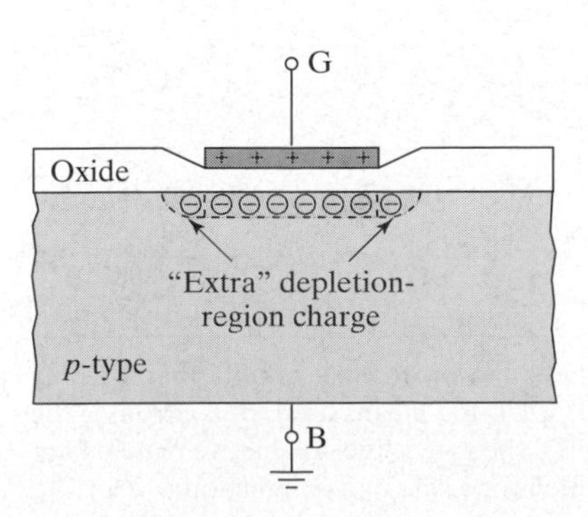

Figure 2–64 Cross section of the width of a MOSFET in the channel, showing the extra depletion-region charge not directly under the gate.

2.7 JUNCTION FIELD-EFFECT TRANSISTORS (JFETS)

The ***junction FET***, or JFET, has a normally conducting region connecting the ***source*** and ***drain***; this region is called the ***channel***. It also has a ***gate***, which is the contact used to control current flow through the channel. The gate and channel form a *pn* junction as in Figure 2-65, which shows a simplified *n*-channel JFET structure. As a bias is applied to the gate, the depletion region gets wider, and the extent to which this region depletes the channel of mobile carriers determines the overall conductance seen between the source and the drain.[30] For the JFET shown in Figure 2-65, the gate is the *p*-type region on top. Real JFETs often have another gate on the other side of the channel, with the top and bottom gates connected together so that there are two depletion regions impinging on the channel.

2.7.1 Intuitive Treatment

When the drain-to-source voltage is near zero, the depletion region formed between the *p*-type and *n*-type regions in the JFET of Figure 2-65 would have a nearly uniform width along the length of the channel, as shown in the figure. The width of this depletion region is varied by changing the voltage across it, which is equal to V_{GS} when $V_{DS} = 0$. JFETs are usually fabricated with the doping in the gate region much higher than the doping in the channel region so that the depletion region will extend further into the channel than it does into the gate. Since the JFET in Figure 2-65 is an *n*-channel device, V_{GS} must be negative to reverse bias the junction.

When V_{GS} is small (and negative) as shown in the figure (i.e., the depletion region does not consume the entire width of the conducting channel between source and drain), the JFET is in what is called the ***linear***, ***triode***, or ***voltage-controlled resistance*** (VCR) region of operation. The device functions as a nonlinear voltage-controlled resistance in this region of operation. For small values of V_{DS}, however, the drain current is approximately a linear function of V_{DS}. The resistance seen between the drain and the source is controlled by the voltage from the gate to the source, V_{GS}. When V_{DS} is small, the drain current in the linear region of operation is given by

$$I_D \approx G_0\left(1 - \frac{1}{X_{ch}}\sqrt{\frac{2\epsilon(V_0 - V_{GS})}{qN_D(1 + N_D/N_A)}}\right)V_{DS}, \tag{2.123}$$

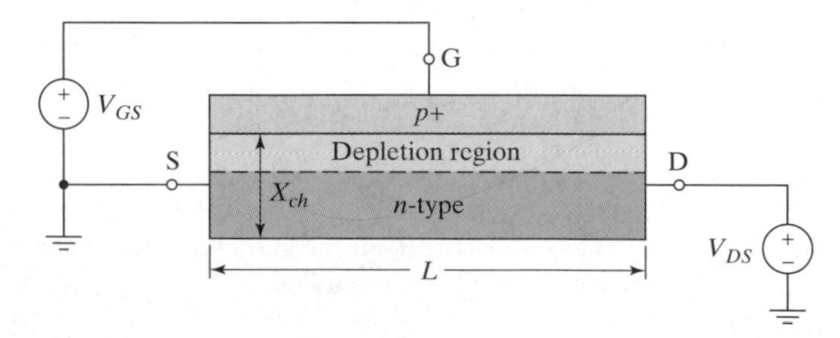

Figure 2–65 A simplified *n*-channel JFET structure.

[30] For those readers already familiar with MOSFET operation, it is worthwhile to note that the only difference between the depletion-mode MOSFET and the JFET is the means used for controlling the width of the depletion region. In the depletion-mode MOSFET, the gate is the conductive plate on top of the silicon-oxide sandwich. In the case of the JFET, the gate is one side of a *pn* junction.

with

$$G_0 = q\mu_n N_D X_{ch} \frac{W}{L}, \tag{2.124}$$

where X_{ch} and L are as defined in Figure 2-65, W is the width of the device (i.e., in the direction perpendicular to the cross section shown in the figure), N_D and N_A are the doping densities in the channel and gate, respectively, V_0 is the built-in potential of the pn junction, and the other terms have their normal meanings.

Examination of (2.123) reveals that the device does indeed look like a voltage-controlled resistor. As the magnitude of V_{GS} increases, while it remains negative, the conductance seen from drain to source decreases, or, in other words, the resistance increases. This behavior is evident on the plot of I_D versus V_{DS} shown in Figure 2-66 if you examine the region where V_{DS} is small.

When V_{DS} is positive and is increased, the width of the depletion region is no longer uniform along the length of the channel but, instead, it becomes wider at the drain end because the reverse bias on the gate-channel junction is increased to $V_{DS} - V_{GS}$. (Remember, V_{GS} is negative.) This situation is depicted in Figure 2-67. When the depletion region extends all the way through the height of the channel, we say that the channel is pinched off. The voltage present in the channel at the point it first pinches off is a function of the device and does not change. At that point, the drain current becomes approximately constant, and any further increases in V_{DS} result in a slight widening of the depletion region at the drain end (in the l direction), but otherwise do not disturb the operation of the device. We say that

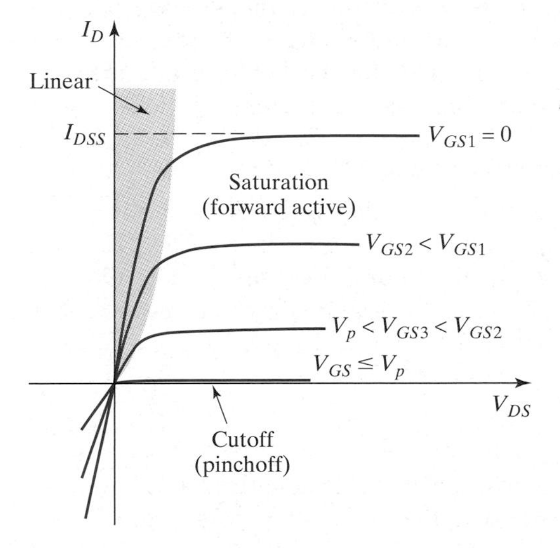

Figure 2–66 The drain characteristics of an n-channel JFET.

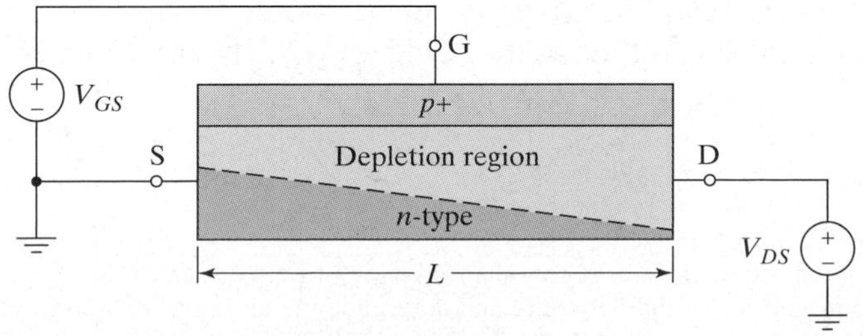

Figure 2–67 The JFET of Figure 2-65 when V_{DS} is larger.

the drain current in this region has ***saturated***; therefore, we call this region the ***saturation region*** of operation. The region is also shown in Figure 2-66. Since this region is also the region of operation used for amplification, it is also referred to as the ***forward-active region*** of operation.

If the effective length of the channel does not change significantly with V_{DS}, then the drain current in saturation is approximately given by

$$I_D = I_{DSS}\left(1 - \frac{V_{GS}}{V_p}\right)^2, \tag{2.125}$$

where V_p is called the pinch-off voltage (it is negative for an n-channel JFET) and I_{DSS} is a device-dependent constant that represents the maximum drain current available in saturation, which occurs for $V_{GS} = 0$.

If V_{GS} is made sufficiently negative, it is possible to pinch off the entire length of the channel. If this occurs, then no drain current can flow and the device is ***pinched off***, or ***cutoff***. The gate-to-source voltage required to pinch off the FET when $V_{DS} = 0$ is denoted V_p, is negative for an n-channel JFET (since V_{GS} is negative), and is the same V_p that is used in (2.125). V_p is sometimes called $V_{GS(off)}$, especially on data sheets for discrete JFETs. In saturation, it is only the drain end of the channel that is pinched off, because the reverse bias at that end is greater than the reverse bias at the source end of the channel.

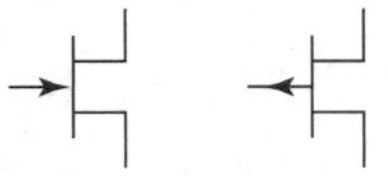

(a) (b)

Figure 2–68 The symbols most commonly used for (a) n-channel and (b) p-channel JFETs.

Figure 2-68 shows the symbols most commonly used for n- and p-channel JFETs. Both V_{GS} and V_p are negative for n-channel JFETs, while for p-channel JFETs they are both positive. I_{DSS} is positive for n-channel JFETs and is defined to be negative for p-channel JFETs *if the drain current is considered to be positive flowing into both devices*. If, on the other hand, you define I_D to be positive when flowing into an n-channel device and positive when flowing *out of* a p-channel device, which is the direction it really flows in saturation, I_{DSS} is positive for both types of device. We will always define I_{DSS} to be positive, but no confusion should result, since one can always determine from first principles which way the current will flow.

2.7.2 Detailed Analysis of Current Flow

Consider the JFET of Figure 2-65 again. The drain current that flows in the device is almost entirely due to drift. The drift current density is given by (2.9). The drain current is the current density in the channel, multiplied by the area of the conducting region. The conducting part of the channel is the part that has not been depleted of mobile electrons, and its area is $(X_{ch} - x_d)W$. The resulting drain current is[31]

$$I_D = [X_{ch} - x_d(l)]Wq\mu_n n(l)|\mathcal{E}(l)|, \tag{2.126}$$

where some quantities are functions of position along the length of the channel as indicated. If the channel is uniformly doped, then $n(l)$ is a constant and is very nearly equal to N_D.

Let's first consider the linear region, where V_{DS} is small. For very small V_{DS}, x_d is approximately constant, and the magnitude of the field is V_{DS}/L. The drain current then becomes

[31] To avoid confusion regarding the conventions used for what direction of field or current is considered to be positive, we use the magnitude of the field here and determine the current direction from first principles.

$$
\begin{aligned}
I_D &\approx q\mu_n N_D \frac{W}{L}(X_{ch} - x_d)V_{DS} \\
&= q\mu_n N_D X_{ch} \frac{W}{L}(1 - x_d/X_{ch})V_{DS} \\
&= G_0(1 - x_d/X_{ch})V_{DS}
\end{aligned}
\tag{2.127}
$$

where we have defined

$$
G_0 = q\mu_n N_D X_{ch} \frac{W}{L}. \tag{2.128}
$$

In the section on *pn*-junction diodes, we derived equations for how far the depletion region extends into each region. On the *n*-side of the junction, the depletion region width was found to be

$$
x_{dn} = \sqrt{\frac{2\epsilon V_0}{qN_D(1 + N_D/N_A)}}, \tag{2.129}
$$

which is just (2.35) repeated for convenience. This equation was derived with no externally applied bias on the junction, but still applies if we replace the built-in potential V_0 with the total junction potential. In the case of our *n*-channel JFET, the total junction potential is $V_0 - V_{GS}$, and the resulting depletion region width in the channel is (the subscript *n* has been dropped for simplicity)

$$
x_d = \sqrt{\frac{2\epsilon(V_0 - V_{GS})}{qN_D(1 + N_D/N_A)}}. \tag{2.130}
$$

Substituting (2.130) into (2.127) leads to

$$
I_D \approx G_0\left(1 - \frac{1}{X_{ch}}\sqrt{\frac{2\epsilon(V_0 - V_{GS})}{qN_D(1 + N_D/N_A)}}\right)V_{DS}, \tag{2.131}
$$

where the approximation is valid only for small V_{DS}.

Next, let's consider what happens as the value of V_{DS} is increased. We must now take into account the fact that x_d is a function of position along the channel. Noting that the gate-to-channel voltage is $V_{GS} - V_{ch}$, where V_{ch} is the voltage in the channel, and using (2.130), we can write

$$
x_d(l) = \sqrt{\frac{2\epsilon(V_0 - V_{GS} + V_{ch}(l))}{qN_D(1 + N_D/N_A)}}. \tag{2.132}
$$

Substituting this equation back into (2.126) while still recalling that $n(l) = N_D$ yields

$$
I_D = q\mu_n N_D X_{ch} W\left(1 - \frac{1}{X_{ch}}\sqrt{\frac{2\epsilon(V_0 - V_{GS} + V_{ch}(l))}{qN_D(1 + N_D/N_A)}}\right)|\mathcal{E}(l)|. \tag{2.133}
$$

We now integrate both sides of (2.133) along the length of the channel,

$$
\int_0^L I_D dl = q\mu_n N_D X_{ch} W \int_0^L \left(1 - \frac{1}{X_{ch}}\sqrt{\frac{2\epsilon(V_0 - V_{GS} + V_{ch}(l))}{qN_D(1 + N_D/N_A)}}\right)|\mathcal{E}(l)|dl, \tag{2.134}
$$

and, because I_D is constant, we get

$$I_D = G_0 \int_0^L \left(1 - \frac{1}{X_{ch}}\sqrt{\frac{2\epsilon(V_0 - V_{GS} + V_{ch}(l))}{qN_D(1 + N_D/N_A)}}\right)|\mathcal{E}(l)|\,dl, \tag{2.135}$$

where G_0 is given by (2.128). We can simplify the integration by using

$$|\mathcal{E}(l)| = \frac{dV_{ch}(l)}{dl} \tag{2.136}$$

and the fact that $V_{ch}(0) = 0$ and $V_{ch}(l) = V_{DS}$. The result is

$$I_D = G_0 \int_0^{V_{DS}} \left(1 - \frac{1}{X_{ch}}\sqrt{\frac{2\epsilon(V_0 - V_{GS} + V_{ch}(l))}{qN_D(1 + N_D/N_A)}}\right)dV_{ch}, \tag{2.137}$$

which can be integrated to obtain

$$I_D = G_0\left[V_{DS} - K(V_0 - V_{GS} + V_{DS})^{3/2} + K(V_0 - V_{GS})^{3/2}\right], \tag{2.138}$$

where

$$K = \frac{2}{3X_{ch}}\sqrt{\frac{2\epsilon}{qN_D(1 + N_D/N_A)}}. \tag{2.139}$$

Equation (2.138) provides a more accurate description of the drain current of the JFET in the linear region of operation and gives results equal to (2.131) when V_{DS} is small.

Now let us consider the saturation region of operation. The drain end of the channel will pinch off when the gate-to-channel voltage is equal to V_p, as shown in Figure 2-69. In this region of operation, there is a depletion region that extends from the drain to the point at which the channel pinches off (in the l dimension). Further increases in V_{DS} will widen this depletion region in the l direction and will cause the effective channel length, L_{eff}, to change. In the discussion that follows, we will ignore this effect and assume that $L_{eff} = L$. We will come back and address this issue in Section 2.7.3.

For our n-channel JFET, V_p is negative, and when the source is grounded as shown in Figure 2-65, the drain end will pinch off when

$$V_{Gch}(l) = V_{GS} - V_{DS} = V_p, \tag{2.140}$$

or, in other words, when

$$V_{DS} = V_{GS} - V_p. \tag{2.141}$$

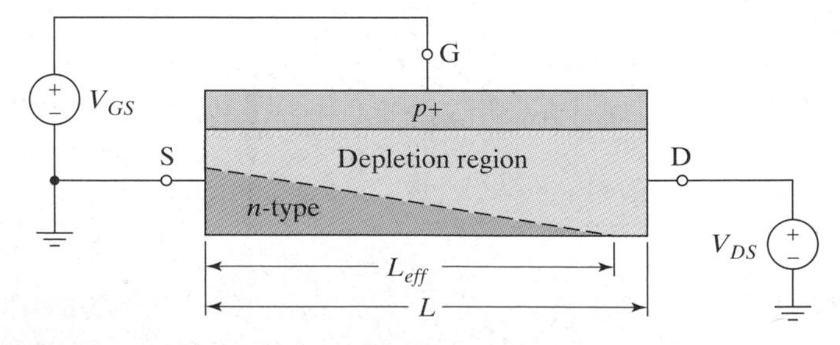

Figure 2–69 An n-channel JFET in saturation.

When the drain end of the channel pinches off (i.e., when the device is saturated), the current will not increase for further increases in V_{DS}, assuming that $L_{eff} = L$. Therefore, the drain current in saturation can be found by substituting (2.141) into (2.138). The result is

$$I_{Dsat} = G_0\left[V_{GS} - V_p - K(V_0 - V_p)^{3/2} + K(V_0 - V_{GS})^{3/2}\right]. \quad \textbf{(2.142)}$$

While this equation is reasonably accurate, it is also very cumbersome to use. In addition, we would need to rederive it, using more realistic doping profiles instead of the abrupt junction assumed in deriving (2.130). In practice, it is found that a much simpler square-law approximation to (2.142) is quite accurate, and designers typically use the equation

$$I_{Dsat} = I_{DSS}\left(1 - \frac{V_{GS}}{V_p}\right)^2, \quad \textbf{(2.143)}$$

where I_{DSS} is the maximum value of I_{Dsat} and occurs when $V_{GS} = 0$. I_{DSS} is temperature dependent and decreases with increasing temperature. We can find an equation for I_{DSS} by plugging $V_{GS} = 0$ into (2.142); the result is

$$I_{DSS} = G_0\left[-V_p - K(V_0 - V_p)^{3/2} + KV_0^{3/2}\right]. \quad \textbf{(2.144)}$$

Or, if we use (2.128), we find that

$$I_{DSS} = q\mu_n N_D X_{ch}\frac{W}{L}\left[-V_p - K(V_0 - V_p)^{3/2} + KV_0^{3/2}\right], \quad \textbf{(2.145)}$$

which shows explicitly how I_{DSS} is a function of the dimensions of the device.

We can also find an equation for the pinch-off voltage, V_p. The pinch-off voltage is, by definition, the value of V_{GS} required to make x_d, as given by (2.130), equal to X_{ch} when V_{DS} is small. Solving this equation, we find that

$$V_p = V_0 - \frac{qN_D(1 + N_D/N_A)X_{ch}^2}{2\epsilon}. \quad \textbf{(2.146)}$$

Exercise 2.20

What are the values of V_0, V_p, G_0, I_{DSS}, and K for an n-channel JFET made using an abrupt-junction, uniformly doped structure with $N_A = 10^{17}$ atoms/cm^3, $N_D = 10^{15}$ atoms/cm^3, $L = 10$ μm, $W = 2$ μm, and $X_{ch} = 2$ μm? Assume room temperature.

Utilizing the results of Exercise 2.20, we can plot I_{Dsat} versus V_{GS} using (2.142) and (2.143). The results are shown in Figure 2-70 and demonstrate that the square-law equation is a reasonable approximation.

2.7.3 Second-Order Effects

There are several second-order effects that are important in analyzing circuits containing JFETs. We will briefly address three of them in this section. The first of these is ***channel-length modulation***. As V_{DS} is increased above $V_{GS} - V_p$ (the value required to pinch off the channel), the depletion region at the drain end must get

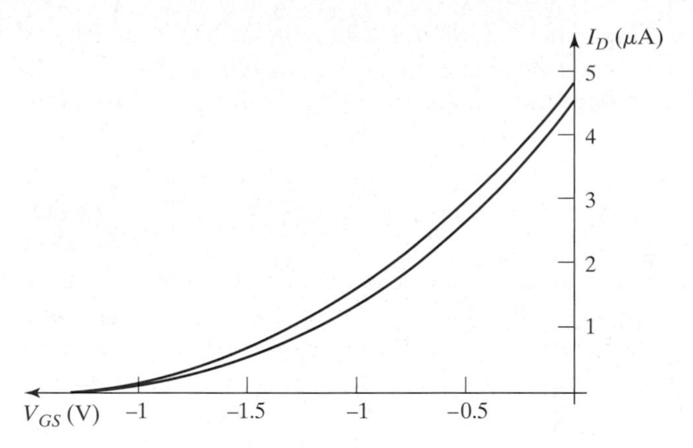

Figure 2–70 I_{Dsat} versus V_{GS}, using (2.142) and (2.143).

wider in order to support the larger field caused by the difference between V_{DS} and $V_{GS} - V_p$. Therefore, the point at which the channel pinches off moves toward the source end of the channel. We can account for this effect by using the effective channel length in place of L in (2.128) and then deriving an equation for how the effective channel length varies with V_{DS}. It is clear from the fact that L shows up in the denominator for G_0, *see* (2.128), that as V_{DS} is increased (and, therefore, the effective L decreases), I_D will increase as shown in the drain characteristics plotted in Figure 2-71.

Examination of the drain characteristics reveals that all of the curves in the saturation region can, at least approximately, be extrapolated back to a common point on the V_{DS} axis, as shown in the figure. This point is labeled $-V_A$ in the plot, and V_A is often called the Early voltage in honor of James Early, who first described a similar effect in bipolar transistors. By recognizing that the curves in the saturation region are nearly straight lines with a common x-axis intercept, we can modify (2.143) to account for channel-length modulation, obtaining

$$I_{Dsat} = I_{DSS}\left(1 - \frac{V_{GS}}{V_p}\right)^2\left(1 + \frac{V_{DS}}{V_A}\right), \tag{2.147}$$

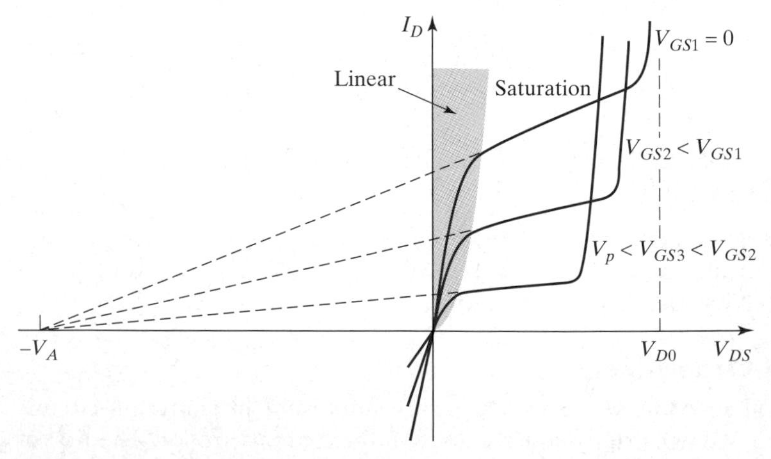

Figure 2–71 The drain characteristics of an *n*-channel JFET with channel-length modulation and breakdown included.

or

$$I_{Dsat} = I_{DSS}\left(1 - \frac{V_{GS}}{V_p}\right)^2(1 + \lambda V_{DS}), \quad \textbf{(2.148)}$$

where $\lambda = 1/V_A$.

Another second-order effect is the junction capacitance that appears between the gate and source and between the gate and drain. The capacitance of a reverse-biased *pn* junction was treated in Section 2.4.5, and we will use the formulas derived there in the discussion that follows.

The capacitance from the gate to the channel of a JFET is modeled by two capacitors: one from the gate to the source and one from the gate to the drain. The total capacitance from gate to channel results from a reverse-biased *pn* junction and so has the form given in (2.53), namely,

$$C_{gch} = \frac{C_{gch0}}{\left(1 + \frac{V_{Gch}}{V_0}\right)^{m_j}}, \quad \textbf{(2.149)}$$

where m_j is typically 1/3 in integrated JFETs. When a JFET is operated in the linear region, C_{gs} and C_{gd} are usually taken to be equal and are each given by one-half of the total capacitance of the junction. When the device is saturated, however, most of the channel is isolated from the drain by the pinched-off region, and so C_{gs} is larger than C_{gd}.

The final second-order effect we will discuss for JFETs is the breakdown voltage. Since the device is operated with a reverse-biased *pn* junction, this junction was subject to breakdown just like any other. The breakdown of a *pn* junction was described in detail in Section 2.4.6. In the case of a JFET, breakdown does not usually occur if the device is being operated in the linear region, since there is no advantage to increasing the reverse bias once the device is cut off. When the device is in saturation, breakdown may occur between the drain and the gate, since the reverse bias on the gate-to-channel junction is largest there. The breakdown voltage of a JFET is highest when $V_{GS} = 0$, and that maximum value is denoted V_{D0}. When V_{GS} is not zero, the breakdown voltage is reduced and is given by

$$BV_{DG} = V_{D0} + V_{GS}. \quad \textbf{(2.150)}$$

The maximum breakdown voltage, V_{D0}, is shown on the drain characteristics of Figure 2-71, and the fact that the breakdown voltage decreases as V_{GS} is made more negative is evident.

2.8 METAL-SEMICONDUCTOR FIELD-EFFECT TRANSISTORS (MESFETS)

Another type of FET that deserves brief mention is the ***metal-semiconductor FET***, or MESFET. In a MESFET, a rectifying Schottky contact is used to control the conductance of the channel. The first-order analysis is identical to that of the JFET, except that the depletion region that controls the width of the channel is formed by the Schottky diode rather than a *pn*-junction diode. MESFETs can be either enhancement- or depletion-mode devices, but depletion-mode devices are by far the most common, since the gate voltage of an enhancement-mode device cannot change very much without forward biasing the Schottky diode and causing a large gate current to flow. The drain current of a MESFET in saturation is given by

$$I_D = \gamma \frac{W}{L}(V_{GS} - V_{th})^2 \tag{2.151}$$

where W, L, and V_{th} have their usual meanings and γ is called the ***current-driving factor***.

When MESFETs are constructed using GaAs, the device operation can be improved in ways that are not possible with silicon devices. Specifically, techniques collectively referred to as ***bandgap engineering*** can be used.

In any transistor, it is desirable to have as large a transconductance as possible. In a MESFET, the transconductance will increase if the conductance of the channel is increased. One way to increase the conductance of the channel is to increase the doping in the channel. However, this method has the drawback that the increased number of impurities leads to a reduction in mobility. (The mobile charges that make up the current interact with the ionized impurities in the lattice in a mechanism called ***impurity scattering***, only one of a number of scattering mechanisms. *See* [2.1] for a brief treatment of this topic.)

Therefore, it is desirable to find a way to increase the channel conductance without decreasing the mobility. With compound semiconductors like GaAs, it is possible to make a device that has a sandwich structure, with, for example, AlGaAs layers on either side of a thin, undoped GaAs layer. What happens in this case is that mobile electrons from the AlGaAs layers diffuse into the undoped GaAs layer and get trapped there by the energy barrier that forms between the layers. (Due to the different bandgaps, AlGaAs has a wider bandgap than GaAs does.[32]) These trapped electrons, collectively called a ***two-dimensional electron gas*** (2-DEG), are free to move in the undoped layer if a field is present. Since the layer is undoped, it does not have impurity scattering and the mobility is very high. Transistors that make use of this 2-DEG for their conducting channel are called either ***modulation-doped FETs*** (MODFETs) or ***high electron mobility transistors*** (HEMTs).

Because of the high mobility attainable with MODFETs, the fact that the saturation velocities for carriers are higher in GaAs, and because the interfaces can be smoother than the Si–SiO_2 interface, MODFETs exhibit high-frequency performance superior to silicon MOSFETs. In addition, because impurity scattering is negligible in the channel of a MODFET, the dominant scattering mechanism is due to collisions with atoms in the lattice itself. Since the lattice vibrates less at low temperature, the lattice scattering decreases at low temperatures, and MODFETs work even better at such temperatures. If impurity scattering is present, as it is in MOSFETs, then at low temperature the impurity scattering dominates because the carriers move more slowly and are more significantly affected by ionized impurities. The fabrication processes for MODFETs are more complicated and less well developed, however, so their use at this point in time is limited.

2.9 SILICON-CONTROLLED RECTIFIER AND POWER-HANDLING DEVICES

A ***silicon-controlled rectifier***, or SCR, is a three-terminal device that has two stable modes of operation, or states, called the ***on state*** and the ***off state***. The three terminals are called the ***gate***, the ***anode***, and the ***cathode***. The device is normally in the off state and, when in that state, is able to block the flow of current from the

[32] This structure is an example of a heterojunction—that is, a junction formed between two materials with different bandgaps.

anode to the cathode, independent of the polarity of the voltage across the device. When triggered into the on state, the device allows current to flow freely from anode to cathode until the current either is momentarily interrupted or begins to change direction (in an AC circuit). When in the on state, the device does not need a gate input to sustain the state. Once the current stops or begins to change direction, the device again enters the off state. The device is triggered into the on state by the application of a control current on the gate electrode.

SCRs are commonly used in power control circuits—for example, in lighting control. After covering the basic operation of an SCR, we will briefly mention some other semiconductor devices used in power distribution and control. The use of integrated circuits in applications involving very large voltages and currents is expanding rapidly; [2.7] and [2.8] are good references.

SCRs are also sometime called ***thyristors***. This name is a carryover from the old gas thyratron tubes that performed a similar function.

An SCR is a *pnpn* structure, as shown in Figure 2-72(a). The device is best understood by considering the circuit model in part (b) of the figure. The model shows that the device can be considered to be two bipolar transistors connected together so that the *pnp* transistor can provide the base current to the *npn* transistor and vice versa.

Consider the circuit in part (b) of the figure, and assume for the moment that the gate is an open circuit, so that $I_G = 0$. As the anode-to-cathode voltage, V_{AC}, is changed, no current can flow, because the transistors need base current supplied to them *before* they can turn on. Examination of the structure in part (a) of the figure shows that in this state there is always at least one reverse-biased *pn* junction in series between the anode and the cathode, and therefore, the device can prevent current flow up to some breakdown voltage, as shown in part (c) of the figure. In the forward direction the breakdown voltage is denoted by BV_F in the figure, while in the reverse direction it is denoted by BV_R. If an external gate current pulse is supplied in the direction shown *when* $V_{AC} > 0$, the device will turn on (so long as the external gate current pulse is large enough and lasts long enough). When the external gate current is applied, it supplies base current to the *npn* transistor, which then starts to conduct. The collector current of the *npn* transistor supplies base current to the *pnp* transistor, which also starts to conduct. Once the *pnp* transistor is conducting, its collector current can supply the base current to the *npn* device, and the external gate current is no longer necessary. We will now derive the conditions necessary for the device to remain in the on state.

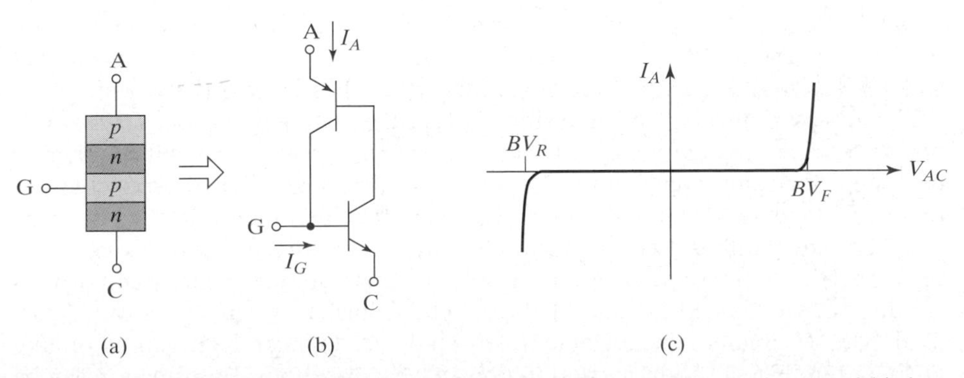

Figure 2–72 (a) The basic structure of an SCR, (b) a circuit model for the device, and (c) the resulting *i–v* characteristic.

In Section 2.5.3, we discovered that the current gain, β, of a BJT drops as the collector current drops. In fact, for very small values of collector current, the value of β can be substantially less than unity. Therefore, we will examine what the values of the current gains for the *pnp* and *npn* transistors must be for the circuit in Figure 2-72(b) to be able to sustain current flow from the anode to the cathode when no gate current is present.

When the gate current is zero, we can write by inspection that

$$I_{Cp} = \beta_p I_{Bp} = \beta_p I_{Cn} = \beta_p \beta_n I_{Cp}. \tag{2.152}$$

This equation is true only if I_{Cp} is zero or $\beta_p\beta_n = 1$. When the currents in the circuit are very small, the product of the current gains will be less than unity, and the circuit cannot maintain any external anode current without a steady gate current. If there is a pulse of gate current that is sufficiently large and long, however, the collector currents will become large enough that (2.152) is satisfied. In fact, the product of the current gains will quickly exceed unity. When this happens, the external gate current can be removed and the circuit will ***regenerate***.

Regeneration refers to a positive feedback process that can sustain itself with no outside help. In the case at hand, once we have $\beta_p\beta_n > 1$, the collector currents will continue to increase, until both transistors saturate so heavily that the gains are again reduced and (2.152) is satisfied. At this point, the SCR is in the on state, and the voltage from the anode to the cathode will be quite low. In terms of the circuit model, the voltage in the on state will be

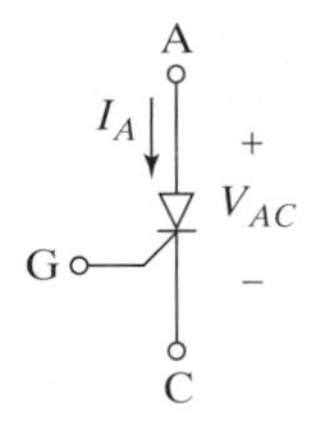

Figure 2–73 The circuit symbol for an SCR.

$$V_{AC} = V_{ECsat}(pnp) + V_{CEsat}(npn), \tag{2.153}$$

which is typically a few tenths of a volt. If the external current to the SCR falls low enough, the product of the current gains will fall below unity, and the device will turn off, as stated before. The minimum current necessary for the device to remain in the on state is called the ***holding current***. The circuit symbol for an SCR is shown in Figure 2-73.

The most common occurrence of an SCR in integrated circuits is not as a desired element, but as a parasitic element in CMOS circuits. Parasitic SCRs in CMOS circuits lead to a destructive mechanism known as ***latchup***. If no gate connection is present, a *pnpn* structure can still be turned on, due to a large rate of change of the voltage across it. Since each of the *pn* junctions has capacitance, if V_{AC} is changed rapidly, there will be a current induced equal to CdV/dt. If this current is sufficiently large, it can cause the product of the current gains to exceed unity, and the device will turn on. Once this state is initiated, the device will remain on until the current is interrupted or begins to change direction. In a CMOS integrated circuit, the parasitic SCRs may be turned on when the device is first powered up. The result is a very large current drawn from the supply. This current will not stop until either the supply is turned off or the device is destroyed.

The process of turning on a parasitic SCR is called latchup. It was a very serious problem in the early days of CMOS circuit design and is still a difficult problem to overcome for some applications. Nevertheless, for most CMOS circuit designs there are standard layout and design rules that effectively deal with the problem.

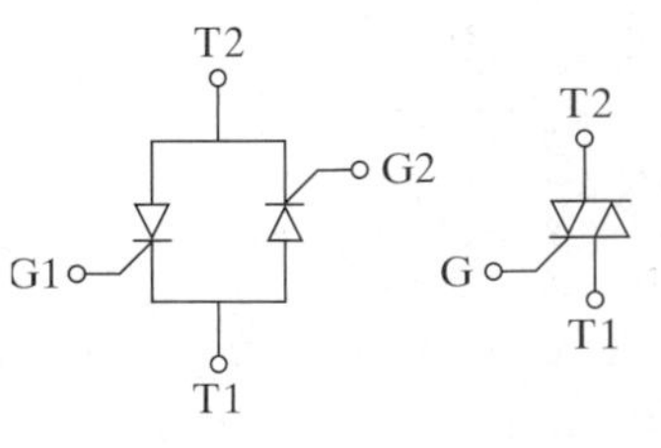

Figure 2–74 (a) Two SCRs in parallel and (b) a triac.

SCRs intended for power applications have doping profiles and device structures specifically tailored to allow large breakdown voltages and large current-handling capability and to inhibit triggering by voltage transients (due to Cdv/dt) [2.6]. For AC circuits, it is desirable to have a device that can be triggered on like an SCR, but that will allow current to flow in both directions. This function can be achieved by connecting two SCRs back-to-back as shown in Figure 2-74; the resulting device is called a ***triac***. The single gate of the triac may trigger the device with an input of either polarity, since a somewhat more complicated structure is used (e.g., *see* [2.9]).

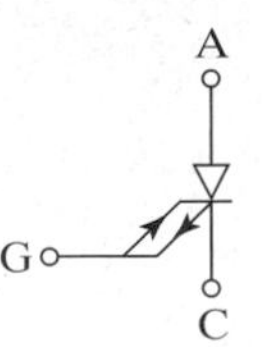

Figure 2–75 A gate-turnoff thyristor.

Another useful modification of an SCR is the ***gate-turnoff thyristor***, or GTO thyristor (sometimes just called a GTO). Since normal SCRs require the anode current to approach zero before they will turn off, they are not suitable for controlling power in DC systems. The GTO is a thyristor, or SCR, that can be turned off by the application of a large negative current pulse at the gate, and its schematic symbol is shown in Figure 2-75. This current pulse may need to be one-third the magnitude of the anode current. The negative gate current is used to remove charge from the base of the *npn* transistor in the SCR (much like taking a standard BJT out of saturation); as a result, the device is relatively slow to turn off. Since the device can be turned off without requiring the anode current to decrease, the GTO thyristor is useful for DC control applications.

Another relatively new device is the ***insulated gate bipolar transistor***, or IGBT. This device was developed because power BJTs have poor current gains and require large base currents, which complicates the drive circuitry, while power MOSFETs have high series resistance if they are made to have large breakdown voltages. In the IGBT, a MOSFET device is merged with a BJT so that the lower series resistance of the BJT is combined with the simpler drive requirements of the FET. There are also MOS-gated thyristor structures [2.7].

2.10 COMPARISON OF DEVICES

There are many differences in the operation of bipolar and field-effect transistors. In addition, there are significant variations in the specifications of individual transistors of each type. In this brief section, we intend only to point out some of the major differences. These differences, and others, will be noted again at other appropriate places in the text. It should be stated at the outset that neither bipolar nor field-effect transistors are inherently superior for all applications. A good circuit designer knows the characteristics of each and how to take advantage of their good properties while avoiding some of the problems. The main differences are summarized as follows:

- FETs are ***unipolar***; that is, the current flow is made up of majority carriers, whereas bipolar devices depend, as the name implies, on both majority and minority carriers. The lack of minority-carrier charge storage in the basic FET (it is still present in the parasitic junctions associated with the device) leads to higher switching speeds for FETs.
- FETs are usually smaller than BJTs if both devices are laid out at minimum size, so FET circuits tend to be more dense than bipolar ones.
- A key difference is that the input impedance of a FET is much higher than that of a bipolar transistor, especially at low frequencies.
- FETs have a negative temperature coefficient for the drain current; therefore, they do not exhibit thermal runaway, whereas BJTs do.
- FETs make better switches than BJTs do because in the linear region they can have essentially no voltage drop across them, whereas BJTs have a small drop across them even when they are saturated. In fact, the main difference between the drain characteristics of a MOSFET and the collector characteristics of a BJT is that the BJT characteristics all come together for small V_{CE} as the device saturates (*see* Figure 2-37) while the MOSFET drain characteristics all have different slopes for small V_{DS} (*see* Figure 2-58).

- BJTs achieve a much larger transconductance at a given bias current than FETs do. Therefore, bipolar amplifiers have more gain at a given current.
- The difference in gains is large enough that the gain-bandwidth product of BJTs is typically higher than that of FETs, even though FETs are inherently faster. This difference in gain-bandwidth product effectively negates the higher speed of FETs in many applications. In fact, for many analog applications, the higher gain-bandwidth product of bipolar transistors makes them the device of choice.
- The matching of bipolar devices is usually better than that of comparably sized FETs, which is again particularly useful in analog design.
- The logarithmic nature of the current-voltage relationship in a BJT is sometimes useful in nonlinear signal processing (e.g., log amps), but it also leads to BJTs having more distortion than FETs, which approximately obey a square law (i.e., $I_D = K(V_{GS} - V_{th})^2$).
- The fact that the base-emitter voltage of a BJT is nearly 0.7 volt and does not change much with current is useful in some circuit functions (e.g., as a simple voltage reference).

Finally, it is interesting to note the difference between how the term "saturation" is used in reference to BJTs and MOSFETs. In both cases, it means that a current stops depending on some parameter. The current in question is the collector current for the BJT and the drain current for the MOSFET. In the case of the BJT, "saturation" refers to the operating region where the collector current is no longer dependent on the base-emitter voltage (to first order), while for the MOSFET, "saturation" refers to the operating region where the drain current is no longer dependent on the drain-to-source voltage.

SOLUTIONS TO EXERCISES

2.1 For intrinsic silicon, we have $R = n_i\, p_i\, f_R(T) = G = f_G(T)$, from which we obtain $f_G(T) = n_i^2 f_R(T)$. For extrinsic silicon, we then have $npf_R(T) = f_G(T)$, and substituting for $f_G(T)$, we obtain $np = n_i^2$.

2.2 Using (2.3), we find that $p_n \approx (1.45 \times 10^{10})^2/3 \times 10^{16} = 7 \times 10^3$ holes/cm^3.

2.3 The concentration gradient is

$$\frac{dn}{dx} = \frac{10^{15} - 10^{12}}{1\ \mu\text{m}} \approx \frac{10^{15}}{10^{-4}\ \text{cm}} = 10^{19}\ \text{atoms/cm}^4 .$$

Using this in (2.6), along with $q = 1.6 \times 10^{-19}$ and the given diffusion coefficient, we find a current density of 56 A/cm^2. The area of the sample is 10 μm × 10 μm, which is 10^{-6} cm^2, so the diffusion current is 56 μA.

2.4 It is possible to have a diffusion current caused by a difference in average energy. If the particles in one region are more energetic than in another region (e.g., if they are being heated), they will be moving faster on average. This greater average velocity will cause a diffusion current to flow, even if there isn't a concentration gradient.

2.5 Since the doping is a linear function of length along the sample, the average doping is 5×10^{14} atoms/cm^3. Using this figure and ignoring the minority carriers, we find that $\sigma_{avg} = q\mu_n(5 \times 10^{14}) = 0.108$ (ohm-cm)$^{-1}$. The overall resistance of the sample is then $R = \rho L/A = L/(\sigma A) = 10^{-4}/(0.108 \times 10^{-6}) =$ 926 Ω. To have a 56 μA drift current, we need to apply a voltage $V =$ (926)56 μ = 52 mV.

2.6 Using the given data and (2.34), we find that $V_0 = 759$ mV. Plugging this into (2.35) with the given data yields $x_{dn} = 9.9 \times 10^{-8}$ cm. Similarly, (2.36) yields $x_{dp} = 9.9 \times 10^{-5}$ cm ≈ 1 μm. Since the n side is much more heavily doped than the p side, the depletion region extends almost entirely into the p side.

2.7 Using (2.52) and ignoring the −1, since $V_D >> V_T$, we take the ratio of the two currents to find $I_{D1}/I_{D2} = e^{(V_{D1}-V_{D2})/nV_T}$, which can be solved for n. We then plug in the given data and find that $n = 1.7$. We then use this value of n along with either set of data and (2.52) to find that $I_S = 3 \times 10^{-10}$.

2.8 **(a)** Using the given data and (2.34), we find that $V_0 = 639$ mV. **(b)** Using V_0 and the given data in (2.57) yields $C'_{j0} = 3.6$ nF/cm^2. Using the given area we then find that $C_{j0} = 0.36$ pF. **(c)** When this diode is reverse biased with $V_D = -5$ V, we can use (2.53) with $m_j = 0.5$, as noted in Example 2.2. The result is $C_j = 0.12$ pF. Since there isn't any diffusion capacitance in reverse bias, this is the total capacitance. **(d)** When the diode is forward biased, we get, from (2.53), $C_j \approx 2C_{j0} = 0.72$ pF. In addition, we find from (2.62) that the diffusion capacitance is $C_d = 7.7$ pF (note that (2.53) does apply, since this is a one-sided junction). Therefore, the total capacitance in forward bias is $C_D = 8.4$ pF.

2.9 Using (2.64), we find that breakdown will occur when $x_{dp} = 0.19$ μm. Plugging this value into (2.65) yields $V_0 - V_D = 1.6$ V. Finally, using (2.66), we obtain $V_0 =$ 0.879 V, and therefore, we predict that the diode will break down when $V_D =$ 0.72 V. Now that we see how low this number is, however, we know that the diode will actually exhibit zener breakdown, not avalanche breakdown, and our result for the breakdown voltage is, therefore, not correct.

2.10 **(a)** Using (2.34), we find that $V_0 = 819$ mV for the base-emitter junction and $V_0 = 579$ mV for the base-collector junction. **(b)** The electrical base width, W_B, is equal to the metallurgical base width minus the widths of the two depletion regions extending into the base. Using (2.36), we can find how far the base-emitter depletion region extends into the base: $x_{dpE} = 0.33$ μm. Similarly, we find how far the base-collector depletion region extends into the base: $x_{dpC} = 0.03$ μm. The electrical base width is then $W_B = 0.24$ μm. **(c)** We can now use this and the given data in (2.76) to find that $I_S = 4.9 \times 10^{-15}$ A. (Remember to use consistent units, since D_n is in cm^2/s, you should use W_B in cm.)

2.11 Using (2.80), we find that $\alpha = 0.99$, $I_B = I_C/\beta = 10$ μA, and $I_E = I_C + I_B =$ 1.01 mA.

2.12 Using (2.81), we find that I_C is increased by 40%.

2.13 Using (2.85), we find that the diffusion capacitance is 9.6 pF. Using (2.83), we find that the junction capacitance is about 1.4 pF, so the total is 11 pF.

2.14 **(a)** Since $V_{GS} > V_{th}$ and $V_{DS} < V_{GS} - V_{th}$, the device is in the linear region of operation. Using (2.87) with the given values yields $I_D = 20$ μA. **(b)** With V_{DS} dropped to half its previous value, the current drops to half: $I_D = 10$ μA. **(c)** Since $V_{GS} > V_{th}$ and $V_{DS} > V_{GS} - V_{th}$, the device is in saturation. Using (2.88) with the given values yields $I_D = 4$ μA. **(d)** Increasing V_{DS} does not have any effect when the device is saturated, so I_D is still 4 μA.

2.15 **(a)** Using (2.89), we obtain $\phi_f = 290$ mV. **(b)** Using (2.92), we find that $x_{d\max} =$ 0.9 μm. **(c)** We first use data from Tables 2.1 and 2.2 to find $C'_{ox} = \epsilon_{ox}/t_{ox} = (3.9)(8.854 \times 10^{-14})/5 \times 10^{-6} = 69$ nF/cm^2. Then we use (2.91) with the given data to obtain $V_{th} = 1.78$ V.

2.16 **(a)** Using (2.107), we find that $K = (\mu_n C'_{ox}/2)(W/L)$. We first calculate $C'_{ox} =$ 230 nF/cm^2. Then, using half of the mobility found in Table 2.1 along with the given data, we find that $K = 155\ \mu\text{A/V}^2$. **(b)** Since $V_{GS} > V_{th}$ and $V_{DS} > V_{GS} - V_{th}$, the device is in saturation. **(c)** Using (2.107), we find that $I_D = 39\ \mu\text{A}$.

2.17 **(a)** Using (2.113), we can write two equations in two unknowns: $100\ \mu\text{A} = C(1 + 3\lambda)$ and $104\ \mu\text{A} = C(1 + 4\lambda)$ where we have defined $C = K(V_{GS} - V_{th})^2$. Solving the equations yields $\lambda = 0.05$. **(b)** Equating (2.113) and (2.112), we find that $L_{eff} = L/(1 + \lambda V_{DS})$. Plugging in the given data, we obtain $L_{eff} = 0.83\ \mu\text{m}$ when $V_{DS} = 4$ V.

2.18 **(a)** We first find $C'_{ox} = 173$ nF/cm^2. Then we use (2.115) to find $C_{gs} = C_{gd} = 2.6$ fF. **(b)** Next, we use (2.119) to obtain $C_{gs} = 3.5$ fF and (2.116) to obtain $C_{gd} = 0$.

2.19 In Exercise 2.15 we found that $V_{th} = 1.78$ V, which is really the value of V_{th0}. We also found $\phi_f = 290$ mV and $C'_{ox} = 69$ nF/cm^2. Using (2.93) and the data supplied, we calculate $\gamma = 0.26$, and then, using (2.120), we obtain $V_{th} = 2$ V.

2.20 Using (2.34) and the data supplied, we find $V_0 = 699$ mV. Then, we use (2.146) to obtain $V_p = -2.39$ V, (2.128) to get $G_0 = 8.64\ \mu\text{A/V}$, (2.139) to find $K = 0.379\ \text{V}^{-1/2}$, and (2.144) to calculate $I_{DSS} = 4.79\ \mu\text{A}$.

CHAPTER SUMMARY

- Solid-state devices are made using crystalline materials that can be doped to control their conductivity and the polarity of the majority charge carriers.
- Doped semiconductors have both mobile charge in the form of holes and free electrons, and fixed charge in the form of uncompensated donor and acceptor impurities.
- Mobile charge carriers will diffuse from a region of high concentration (or energy) to a region with a lower concentration (or energy) if they are free to do so.
- The diffusion of carriers establishes a charge imbalance, which produces an electric field. Mobile charge carriers will drift in response to the field, and the drift and diffusion currents will balance each other at equilibrium.
- Metal–semiconductor contacts can be ohmic or rectifying. Rectifying contacts result from the fact that the barrier to diffusion in one direction is controlled by the applied bias, whereas the barrier to diffusion in the other direction is approximately constant.
- The asymmetrical *i–v* characteristics of *pn* junctions result because the diffusion currents are supplied by majority carriers and are controlled by the potential barrier, whereas the drift currents are supplied by minority carriers and are therefore limited by the number of carriers available and are not, to first order, a function of the applied bias.
- Bipolar junction transistors (BJTs) use a forward-biased *pn* junction to control the supply of minority carriers available to a reverse-biased *pn* junction. In this way, they form a voltage-controlled current source.
- Field-effect transistors (FETs) use an electric field to modulate the number of mobile charge carriers present in the channel. They can function as voltage-controlled resistors, or if the voltage at one end of the channel is changed so that the conducting layer is pinched off, they can function as voltage-controlled current sources.
- Metal-oxide semiconductor FETs (MOSFETs) use a MOS capacitor to modulate the number of mobile carriers in the channel.

- Junction FETs (JFETs) use a reverse-biased *pn* junction to modulate the number of mobile carriers in the channel.
- Metal–semiconductor FETs (MESFETs) use a metal–semiconductor junction to modulate the number of mobile carriers in the channel.
- Silicon-controlled rectifiers (SCRs) are four-layer *pn*-junction devices that have two states: a nonconducting off state and a highly conducting on state. They are used for power control and also appear as parasitic elements in CMOS circuits, where the highly conductive on state leads to a destructive mechanism called latchup.

REFERENCES

[2.1] B.G. Streetman, *Solid State Electronic Devices*, 3/E. Englewood Cliffs, NJ: Prentice Hall, 1990.

[2.2] R.S. Muller and T.I. Kamins, *Device Electronics for Integrated Circuits*, 2/E. New York: John Wiley & Sons, Inc., 1986.

[2.3] R.F. Pierret and G.W. Neudeck, eds. *The Modular Series on Solid-State Devices*, Ed., Reading, MA. Addison-Wesley Publishing Company, 1990. Volume I: R.F. Pierret, *Semiconductor Fundamentals*, 2nd Edition, 1989. Volume III: G.W. Neudeck, *The Bipolar Junction Transistor*, 2nd Edition, 1989. Volume IV: R.F. Pierret, *Field Effect Devices*, 2nd Edition, 1990. No volume given: Dieter K. Schroder, *Advanced MOS Devices*.

[2.4] P. Horowitz and W. Hill, *The Art of Electronics*, 2/E. Cambridge UK: Cambridge University Press, 1989.

[2.5] Y.P. Tsividis, *Operation and Modeling of the MOS Transistor*. New York: McGraw-Hill, 1987.

[2.6] C.T. Sah, "Evolution of the MOS transistor—from conception to VLSI," *Proc. of the IEEE*, Vol. 76, No. 10, Oct. 1988, pp 1280–1326.

[2.7] B. Jayant Baliga, *Power Semiconductor Devices*. Boston: PWS Publishing Company, 1996.

[2.8] N. Mohan, T.M. Undeland, and W.P. Robbins, *Power Electronics*, 2/E. New York: John Wiley & Sons, Inc., 1995.

[2.9] Teccor Application Note AN1002, *Gating, Latching and Holding of SCRs and TRIACs*. Teccor Electronics Inc., Irving Texas.

[2.10] R. Soares Ed., *GaAs MESFET Circuit Design*. Boston: Artech House, 1988.

[2.11] S.M. Sze, *Physics of Semiconductor Devices*, 2/E. New York: John Wiley & Sons, Inc., 1981.

PROBLEMS

2.1 Material Properties

P2.1 If a *p*-type silicon crystal has a doping density of $N_A = 2 \times 10^{18}$ atoms/cm^3 and we consider the sample at 300 K, **(a)** what is the density of holes, p_p? **(b)** What is the density of mobile electrons, n_p? **(c)** What would these densities be if the temperature increased to 344 K?

P2.2 Silicon has 5×10^{22} atoms/cm^3. If an *n*-type silicon crystal has a doping density of $N_D = 2 \times 10^{16}$ atoms/cm^3 and we consider the sample at 300 K, **(a)** what is the minority carrier density? **(b)** What is the majority carrier density? **(c)** How many silicon atoms are there in the lattice for every dopant atom?

P2.3 A sample of *p*-type silicon crystal has a doping density of $N_A = 5 \times 10^{14}$ atoms/cm^3. At what temperature would the intrinsic carrier concentration due to EHP generation equal the concentration due to ionized acceptor impurities?

2.2 Conduction Mechanisms

P2.4 Suppose you take a bar of doped silicon and heat the left end of the bar. (Perhaps you put a soldering iron it.) Answer each of the following questions and explain your answer: **(a)** Will any current flow in the material the first instant you apply the heat? **(b)** Will a steady-state current flow? **(c)** If a current does flow, whether or not it does in the steady state, what direction does the current flow for *n*-type silicon? for *p*-type?

P2.5 Consider a sample of silicon that is 10 μm long (in the x direction), 1 μm wide (in the y direction) and 4 μm deep (in the z direction). The sample starts out as intrinsic silicon but is doped in the z direction so that $n(z) = 10^{18}e^{-z/z_0}$ electrons/cm^3, where z is taken to be zero at the surface of the sample and $Z_0 = 0.6$ μm. Find the resistance of the sample when measured from end to end in the x direction. Assume that you can use an average mobility of 750 cm^2/Vs for the electrons. (The mobility varies with doping level, so this is a crude approximation.)

P2.6 Consider an intrinsic sample of silicon that has been doped in the x direction so that $N_D(x) = N_o e^{-x/x_0}$ atoms/cm^3. If the sample is at equilibrium, the total current flow must be zero (i.e., drift must balance diffusion). Using this fact and (2.10), derive an equation for the electric field as a function of x. Assume that we restrict our attention to a region where $n(x) \gg n_i$.

2.3 Conductor-to-Semiconductor Contacts

P2.7 An *n*-type Schottky diode has $\phi_B = 0.62$ V. **(a)** Find I_D if $T = 300$ K and $V_A = 0.3$ V. **(b)** if $T = 350$ K and $V_A = 0.3$ V. **(c)** if $T = 300$ K and $V_A = 0.4$ V.

P2.8 Assuming that $I_S = 4 \times 10^{-11}$ A, find the change in V_A necessary to double the current in a Schottky diode at room temperature. What ΔV_A is needed to increase the current by a factor of 10?

P2.9 Assuming that we measure a particular Schottky diode and find that $I_D =$ 1.5 mA and $V_A = 0.43$ V at room temperature, what is the value of I_S?

2.4 pn-Junction Diodes

2.4.1 Intuitive Treatment

P2.10 On linear axes, plot the *i–v* characteristic of a *pn*-junction diode that has $I_S = 1 \times 10^{-14}$ and $n = 1$ at room temperature . Plot the characteristic with three different vertical scales: **(a)** 0 to 200 μA, **(b)** 0 to 20 mA, and **(c)** 0 to 2 A. For each plot, extrapolate the approximately straight high-current part of the plot down to the voltage axis, and determine the turn-on voltage of the diode. Why are the three values different?

P2.11 Suppose you want to use two diodes in parallel to pass a large current. You want to be sure that the current is shared approximately equally by the two diodes. If they are identical in every way except for the areas of their junctions (measured perpendicular to the flow of current), how much different can the areas be if the currents must be the same within ±10 %? Now suppose the currents are identical, but one of the diodes is 10 degrees hotter than the other. If the voltage across them is 0.6 V, how different are the currents? Based on these results, do you think it would be practical to put discrete diodes in parallel to boost the power-handling capability? Why or why not?

P2.12 If the saturation current of a *pn*-junction diode increases 16 times due to an increase in the temperature, approximately how much has the temperature increased? Assume that the initial temperature was near room temperature.

P2.13 If a Schottky diode has a reverse saturation current that is five orders of magnitude larger than that of a *pn* junction diode, what is the difference in their forward voltage drops at room temperature when they both have the same forward current?

P2.14 Consider the circuit shown in Figure 2-76, where the Schottky diode has $I_S = 4 \times 10^{-10}$ and the *pn*-junction diode has $I_S = 1 \times 10^{-15}$. What is the voltage across the two diodes?

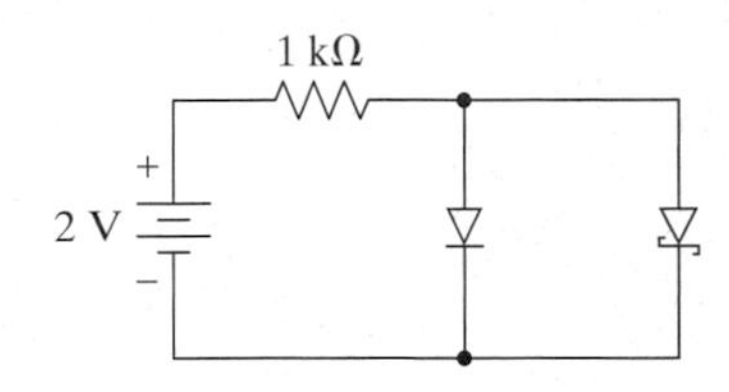

Figure 2–76

2.4.2 Detailed Analysis of Current Flow

P2.15 Consider an abrupt *pn* junction with uniform doping in each region. Consider just the electron current, and set the drift and diffusion components equal to each other in order to obtain an equation for the electric field across the junction. Integrate this equation and show that (2.34) is correct. You need to know that $D_n/\mu_n = KT/q$ (Einstein's Relation).

P2.16 How much do the excess minority-carrier densities of a *pn*-junction diode at the edge of the depletion region increase when the applied forward bias increases from 0.5 V to 0.7 V at room temperature?

P2.17 Consider a *pn* junction with $N_D = 10^{19}$ atoms/cm^3 and $N_A = 10^{16}$ atoms/cm^3. Find the built-in potential and the total width of the depletion region that is formed at equilibrium at room temperature.

P2.18 Consider two *pn*-junction diodes in parallel being driven by a 1 mA current source. If the diodes are identical in every way, except that one of them has four times the area of the other (measured in the plane perpendicular to the current flow), what are the two diode currents? If the smaller diode has $I_S = 10^{-15}$A, what is the voltage across the two diodes?

P2.19 Consider a *pn* junction with $N_D = 10^{17}$ atoms/cm^3, $N_A = 10^{15}$ atoms/cm^3, and an area of 5μm × 5μm. Assume that the density of mobile electrons varies linearly from its value in the *n*-type material to its value in the *p*-type material and that the transition takes place over a distance of 100 Å (1 Å = 10^{-10} m). If the diffusion constant for electrons is 35 cm^2/sec, what is the magnitude of the electron diffusion current?

P2.20 Show that at room temperature the voltage across a forward-biased diode changes **(a)** by 18 mV for every factor-of-2 change in the diode current and **(b)** by about 60 mV for every factor-of-10 change in the diode current. Assume that $V_T = 0.026$ V. **(c)** Show that if the current changes by a factor of 400, the voltage should change by about 60 + 60 + 18 + 18 = 156 mV.

P2.21 Derive (2.32) starting from (2.31).

2.4.5 Charge Storage and Varactor Diodes

P2.22 What is the zero-bias junction capacitance of the diode in Problem P2.17 at room temperature?

2.4.6 Breakdown and Zener Diodes

P2.23 For what value of applied reverse bias would the junction in P2.17 break down at room temperature? (Use the data in Table 2.1.)

P2.24 Consider an abrupt, uniformly doped *pn* junction diode with $N_D = 10^{19}$ atoms/cm^3. We wish to find out how high the doping on the *p*-side of the junction must be for this diode to exhibit both avalanche and zener breakdown and, therefore, have a breakdown voltage with a low temperature coefficient. Considering avalanche breakdown alone, what must the doping on the *p*-side be to produce a breakdown voltage of 5 V at room temperature?

2.5 The Bipolar Junction Transistor

2.5.1 Intuitive Treatment

P2.25 Given each of the following sets of terminal voltages, state which region of operation an *npn* transistor would be in: **(a)** $V_E = 0$ V, $V_B = 0.4$ V, and $V_C = 1$ V. **(b)** $V_E = 0$ V, $V_B = 0.8$ V, and $V_C = 3$ V. **(c)** $V_E = 0$ V, $V_B = -1$ V, and $V_C = 1$ V. **(d)** $V_E = -0.7$ V, $V_B = 0$ V, and $V_C = -0.5$ V. **(e)** $V_E = 2$ V, $V_B = 0$ V, and $V_C = -0.6$ V. **(f)** $V_E = 3$ V, $V_B = 2.6$ V, and $V_C = 3$ V. **(g)** $V_E = 3$ V, $V_B = 3.7$ V, and $V_C = 3.3$ V.

P2.26 Given each of the following sets of terminal voltages, state which region of operation a *pnp* transistor would be in: **(a)** $V_E = 6$ V, $V_B = 6.3$ V, and $V_C = 5$ V. **(b)** $V_E = 6$ V, $V_B = 5.3$ V, and $V_C = 3$ V. **(c)** $V_E = 0$ V, $V_B = -0.8$ V, and $V_C = -0.2$ V. **(d)** $V_E = 0$ V, $V_B = 1$ V, and $V_C = 1.6$ V. **(e)** $V_E = 0$ V, $V_B = 0$ V, and $V_C = 0$ V. **(f)** $V_E = 10$ V, $V_B = 9.2$ V, and $V_C = 4$ V. **(g)** $V_E = 10$ V, $V_B = 9.2$ V, and $V_C = 9.6$ V.

P2.27 How much must we increase the value of V_{BE} for a forward active *npn* transistor at room temperature if we desire to increase I_C by a factor of three?

P2.28 How much must we increase the value of V_{EB} for a forward active *pnp* transistor at room temperature if we desire to increase I_C by a factor of five?

P2.29 If we measure $I_C = 1.5$ mA when $V_{BE} = 0.65$ V, what is the corresponding value of I_S for an *npn* transistor? Assume that the transistor is in the forward-active region of operation and at room temperature.

P2.30 If we measure $I_C = 8$ mA when $V_{EB} = 0.6$ V, what is the corresponding value of I_S for a *pnp* transistor? Assume that the transistor is in the forward-active region of operation and at room temperature.

P2.31 Plot I_C versus V_{BE} on linear axes for an *npn* transistor with $I_S = 10^{-14}$ A. Use three different scales for I_C: **(a)** 0 to 200 μA, **(b)** 0 to 20 mA, and **(c)** 0 to 2 A. For each plot, extrapolate the approximately straight high-current part of the plot down to the voltage axis, and determine what is sometimes called the turn-on voltage of the transistor. Why are the three values different?

2.5.2 Detailed Analysis of Current Flow

P2.32 Consider an *npn* transistor with emitter doping $N_D = 4 \times 10^{18}$ atoms/cm^3, base doping $N_A = 3 \times 10^{15}$ atoms/cm^3, and collector doping $N_D = 10^{14}$ atoms/cm^3. The metallurgical width of the base (i.e., ignoring the depletion regions) is 1 μm, and the junction area is 100 μm^2. Assuming that $D_n = 35$ cm^2/s in the base, find the saturation current for this transistor at room temperature.

P2.33 Consider a *pnp* transistor with emitter doping $N_A = 2 \times 10^{18}$ atoms/cm^3, base doping $N_D = 10^{15}$ atoms/cm^3, and collector doping $N_A = 2 \times 10^{14}$ atoms/cm^3. The metallurgical width of the base (i.e., ignoring the depletion regions) is 1 μm, and the junction area is 100 μm^2. Assuming that $D_p = 17$ cm^2/s in the base, find the saturation current for this transistor at room temperature.

2.5.4 Base-Width Modulation (the Early Effect)

P2.34 Assuming the simplest possible model for channel-length modulation and a nominal base width of 0.4 μm, what is the active base width of an *npn* BJT that has $V_A = 30$ V when it is operated in the forward-active region with $V_{CE} = 15$ V?

2.6 Metal-Oxide Semiconductor Field-Effect Transistors

P2.35 Starting from Poisson's equation in one dimension as given by (2.29) and the charge distribution shown in Figure 2-51, derive equations for the electric field in each region. Assume that the charge in the metal exists in an infinitesimally thin sheet. [*Hint:* Look at how we derived (2.70).]

P2.36 Using the results from Problem P2.35, derive equations for the voltage in each region in Figure 2-51. Assume the voltage to be equal to zero in the bulk of the semiconductor. Write an equation for the built-in voltage.

P2.37 Starting from Poisson's equation in one dimension as given by (2.29) and the charge distribution shown in Figure 2-55, derive an equation for the electric field in the semiconductor if the charge in the inversion layer can be neglected. Using the field just derived, find an equation for the voltage as a function of position, assuming that the voltage in the bulk is zero. Then solve for the value of $x_{d\max}$ by setting the potential at the semiconductor surface equal to $2\phi_f$.

P2.38 Derive (2.89) for *p*-type silicon. [*Hint:* Look at Aside A2.1.]

P2.39 Consider a *p*-channel MOSFET with $V_{th} = -0.8$ V and $K = 50$ μA/V^2. **(a)** If $V_{GS} = -1.8$ V and $V_{DS} = -4$ V, what region of operation is the device in? What is the corresponding drain current? **(b)** If $V_{GS} = -1$V and $V_{DS} = -$V, what region of operation is the device in? What is the corresponding drain current?

P2.40 Consider a *p*-channel MOSFET with $V_{th} = -1$ V and $K = 250$ μA/V^2. **(a)** If $V_{GS} = -1.5$ V and $V_{DS} = -3$ V, what region of operation is the device in? What is the corresponding drain current? **(b)** If the width of the device is doubled while keeping everything else constant, what does the drain current become?

P2.41 Consider an n-channel MOSFET with $V_{th} = 0.8$ V and $K = 50\ \mu A/V^2$. **(a)** If $V_{GS} = 1.8$ V and $V_{DS} = 4$ V, what region of operation is the device in? What is the corresponding drain current? **(b)** If $V_{GS} = 1$ V and $V_{DS} = 1$ V, what region of operation is the device in? What is the corresponding drain current?

P2.42 Consider an n-channel MOSFET with $V_{th} = 1$ V and $K = 250\ \mu A/V^2$. **(a)** If $V_{GS} = 1.5$ V and $V_{DS} = 3$ V, what region of operation is the device in? What is the corresponding drain current? **(b)** If the width of the device is doubled while keeping everything else constant, what does the drain current become?

P2.43 Consider an n-channel MOSFET with $W = 2\ \mu m$, $L = 0.5\ \mu m$, an oxide thickness of 100 Å, and $V_{th} = 0.7$ V. **(a)** Assuming that the surface mobility is one-half of the bulk mobility, what is the value of K? **(b)** If $V_{GS} = 2$ V and $V_{DS} = 3$ V, what region of operation is the device in? **(c)** What is the corresponding drain current?

P2.44 **(a)** Repeat part (a) of Problem P2.43. **(b)** If $V_{GS} = 2$ V and $V_{DS} = 1$ V, what region of operation is the device in? **(c)** What is the corresponding drain current?

P2.45 Consider an n-channel MOSFET made using p-type silicon with uniform doping and $N_A = 10^{17}$ atoms/cm^3. The transistor has $W = 3\ \mu m$, $L = 1\ \mu m$, an oxide thickness of 100 Å, and $V_{FB} = -1$ V. **(a)** What is V_{th} at room temperature? **(b)** Assuming that the surface mobility is one-half of the bulk mobility, what is the value of K? **(c)** If the device is operated in saturation, what V_{GS} is required to achieve $I_D = 500\ \mu A$? **(d)** For the value of V_{GS} found in (c), what is the minimum V_{DS} for which the device will remain in saturation?

P2.46 Consider an n-channel MOSFET with $V_{th} = 0.7$ V, $K = 50\ \mu A/V^2$, and $\lambda = 0.02$. **(a)** What is I_D when $V_{GS} = 2$ V and $V_{DS} = 2$ V? **(b)** What is I_D when $V_{GS} = 2$ V and $V_{DS} = 4$ V? **(c)** If we model the drain-to-source port of the transistor by a Norton model for the conditions $V_{GS} = 2$ V and $2 \leq V_{DS} \leq 4$, what is the equivalent resistance in parallel with the current source (i.e., the Thévenin resistance)?

P2.47 Consider a p-channel MOSFET that is measured to have $I_D = 100\ \mu A$ when $V_{GS} = -2$ V and $V_{DS} = -3$ V, and $I_D = 103\ \mu A$ when $V_{GS} = -2$ V and $V_{DS} = -4$ V. **(a)** What is the value of λ? **(b)** If $L = 0.5\ \mu m$, what is L_{eff} when $V_{DS} = -4$ V?

P2.48 Consider an n-channel MOSFET with $V_{th} = 1$ V, $K = 150\ \mu A/V^2$, and $V_A = 50$ V. **(a)** If we model the drain-to-source port of the transistor by a Norton model when the device is saturated, what is the equivalent resistance in parallel with the current source (i.e., the Thévenin resistance)?

P2.49 Starting from (2.100) and replacing V_G by V_{GS}, integrate along the length of the channel to derive (2.117). Since the device is saturated, you want to integrate only from $l = 0$ to $l = L_{eff}$ (*see* Figure 2-59). [*Hint:* You want to change variables so that you integrate with respect to channel voltage, not l. Use (2.111) to help in the change of variables, and use (2.107) to replace I_D.]

P2.50 Consider an n-channel MOSFET made using p-type silicon with uniform doping, $N_A = 10^{17}$ atoms/cm^3 and $V_{FB} = -1.1$ V. The transistor has an oxide thickness of 200 Å. **(a)** What is V_{th0}? **(b)** If $V_{SB} = 3$ V, what is V_{th}?

P2.51 Derive (2.111).

2.7 Junction Field-Effect Transistors (JFETs)

P2.52 Consider an n-channel JFET made using a uniformly doped abrupt junction with $N_D = 10^{15}$ atoms/cm^3 in the channel and $N_A = 10^{17}$ atoms/cm^3 in the gate. The channel width is 3 μm, the length is 6 μm, and the depth (i.e., X_{ch}) is 2 μm. **(a)** What is the pinch-off voltage? **(b)** What is I_{DSS}?

P2.53 Consider a p-channel JFET made using a uniformly doped abrupt junction with $N_A = 10^{15}$ atoms/cm^3 in the channel and $N_D = 10^{17}$ atoms/cm^3 in the gate. The channel width is 4 μm, the length is 10 μm, and the depth (i.e., X_{ch}) is 1.5 μm. **(a)** What is the pinch-off voltage? **(b)** What is I_{DSS}?

P2.54 Consider the n-channel JFET from Exercise 2.20. Using whatever computer program is most readily available, plot a family of curves for I_D versus V_{DS} while varying V_{GS} in 0.5 V steps from zero to V_p.

P2.55 For a p-channel JFET operating in pinchoff with $V_p = 1$ V and $I_{DSS} =$ 250 μA, by how much do you have to change V_{GS} to double the drain current if it starts at **(a)** 25 μA? **(b)** 100 μA?

P2.56 Show that (2.143) for a JFET and (2.107) for a depletion-mode MOSFET are identical. Derive the relationships between V_{th} and V_p and between I_{DSS} and K.

CHAPTER 3

Solid-State Device Fabrication

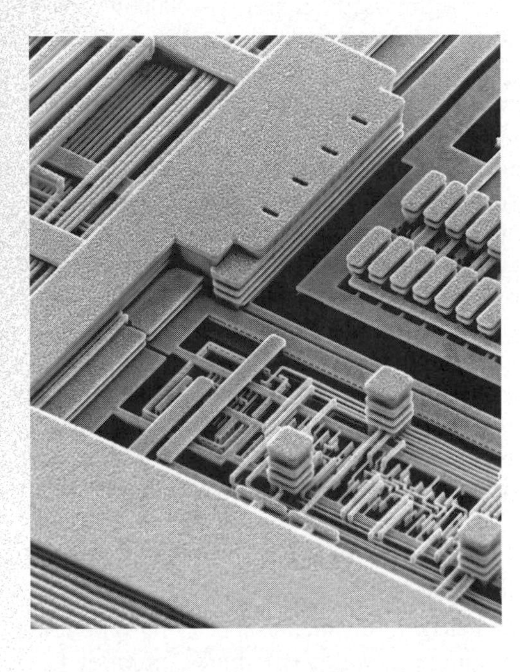

Cross section of six levels of copper metal in IBM's CMOS process with copper interconnect. (Photo courtesy of International Business Machines Corporation.)

INTRODUCTION

The fabrication of solid-state electronic devices is an amazing and rapidly changing field that has fundamentally altered the world we live in during the past 40 years. In this chapter, we will take a brief look at this technology to establish some terminology and develop a better understanding of the devices that are used throughout this text. It is important for a circuit designer to know the basic structures of the devices used, since these structures determine the intrinsic device performance and give rise to parasitic components that affect the operation of the devices and circuits. Since the processes change from year to year, so do the devices. A good designer must have at least a rudimentary understanding of how these changes affect circuit performance, rather than depending solely on detailed simulation models (which are usually not available right away).

The presentation in this chapter is necessarily brief, and we encourage you to examine the references for more detailed information on the processes introduced here. Almost all of the treatment presented here is extracted directly from the excellent text by Plummer, Deal, and Griffin [3.1], and we thank them for granting permission to use the material here.

We first give a brief description of the process flow required to fabricate a modern CMOS circuit. CMOS is chosen because it is the dominant technology in use today. We then discuss how the circuit models used in this text relate to elements in the physical structure of the device. After the basic process flow has been described, we show a typical cross section of a bipolar transistor fabricated using the same techniques and again relate the circuit models used to the physical structure.

3.1 CMOS TECHNOLOGY

Modern integrated circuits may contain tens of millions of transistors and be nearly 1 inch on a side. The minimum lateral feature size in the circuit, which is determined by a photolithographic process, is around a tenth of a micron (1 micron = 1 μm = 10^{-6} meter = 0.000039 inch). It is projected that by the year 2009, a microprocessor will contain 520 million transistors with minimum feature sizes of 70 nm (i.e., 0.07 micron) [3.2].

Silicon integrated circuits are fabricated using a photolithographic process. A modern CMOS process, may use two dozen or more repetitions, called mask layers, of this process interspersed with oxidation, film depositions, ion implantation, and other processes. The main reason for the dominance of silicon in IC technology is that the oxide that can be easily grown on silicon is stable, is reproducible, yields a high-quality surface interface, is an effective shield for most dopants (intended or otherwise), and can be selectively etched (i.e., with etchants that do not effectively etch silicon).

The description that follows of a modern CMOS process flow is a significantly shortened version of Chapter 2 in [3.1]. The end result of the process will produce a CMOS inverter like that shown in Figure 3-1 and described in detail in Chapter 15 of this text. This circuit is chosen for its simplicity. The physical structure of the transistors for the inverter is shown in Figure 3-2.

3.1.1 The Beginning: Choosing a Substrate

Before we begin actual wafer fabrication, we must of course choose the starting wafers. In general, this means specifying type (*N* or *P*), resistivity (doping level), crystal orientation, wafer size, and a number of other parameters having to do with wafer flatness, trace impurity levels, etc. The major choices are the type, resistivity, and orientation.

Figure 3-2 indicates that the final structure has a *P*-type substrate. In most CMOS integrated circuits, the substrate has a moderately high resistivity (25–50 Ω·cm), which corresponds to a doping level[1] on the order of 10^{15} cm^{-3}. As is apparent from Figure 3-2, the active devices are actually built in wells diffused into the surface of the wafer. The doping levels in these wells are chosen to optimize the electrical properties of the active devices, as we will see later in the chapter. Typically, the well doping levels are on the order of 10^{16}–10^{17} cm^{-3} near the wafer surface. In order to manufacture such wells reproducibly, the background doping (the substrate doping in this case) needs to be significantly less than the well doping.

The observant reader might notice that the NMOS device could actually be built directly in the *P* substrate without adding the *P* well near the surface. While some CMOS circuits actually are built this way today, the twin well process illustrated in Figure 3-2 is much more common because the doping process to produce

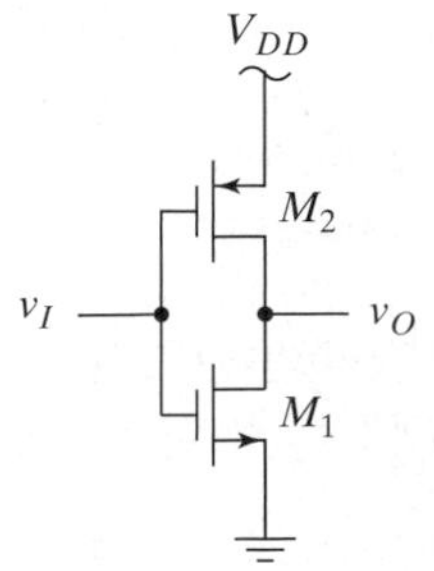

V_{DD} M_2 v_I v_O M_1

Figure 3–1 A CMOS inverter. This circuit is also shown in Figure 15-23, and its operation is described in Section 15.3.2. The bulk, or substrate, connections are not shown here because the bulk terminal of each device is connected to its source.

[1] Doping levels are specified in terms of the number of dopant atoms per cubic cm, see Chapter 2 for details.

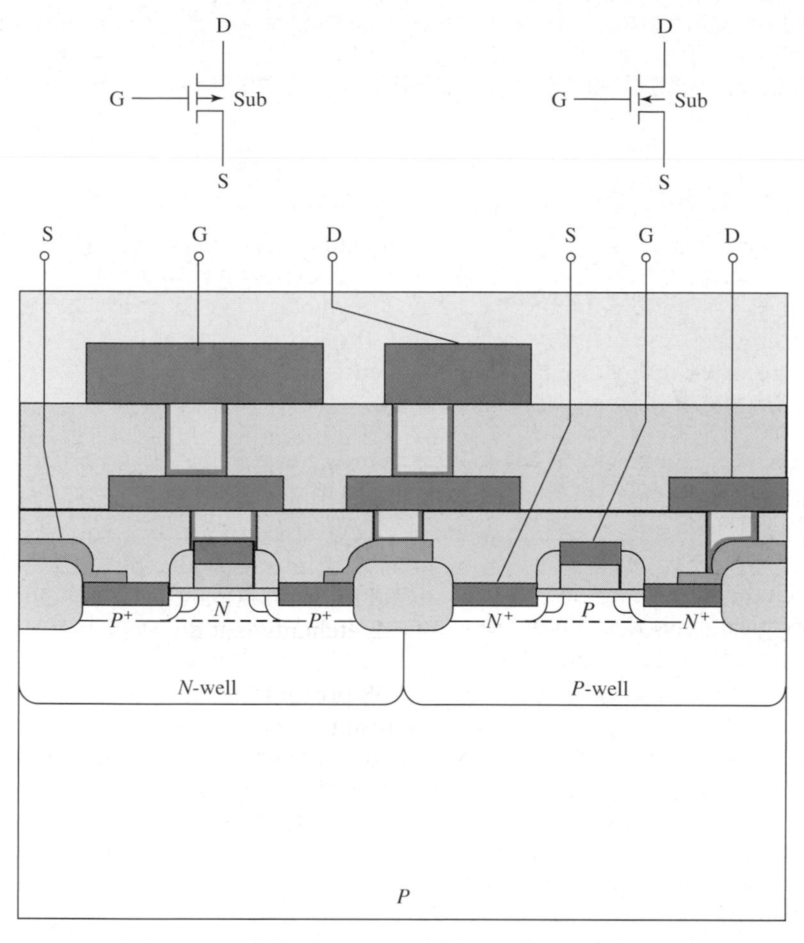

Figure 3–2 Cross section of the final CMOS integrated circuit. The PMOS transistor is shown on the left and the NMOS device is on the right. Remember that the bulk, or substrate, of each device is tied to its source (not shown in this cross section).

the *P* well (ion implantation) is much better controlled in manufacturing than is the substrate doping. Also, since the *P* well and *N* well doping concentrations are on the same order, it is easier to start with a much more lightly doped substrate and tailor the wells for the NMOS and PMOS devices individually.

3.1.2 Active Region Formation

Modern CMOS chips integrate millions of active devices (NMOS and PMOS) side by side in a common silicon substrate. Circuits are designed with these devices to implement complex logic or analog functions. In designing such circuits, it is usually assumed that the individual devices do not interact with each other except through their circuit interconnections. In other words, we need to make certain that the individual devices on the chip are electrically isolated from each other. This is accomplished most often by growing a fairly thick layer of silicon dioxide, SiO_2, in between each of the active devices. SiO_2 is essentially a perfect insulator and provides the needed isolation. This process of locally oxidizing the silicon substrate is known as the LOCOS process (*local oxidation* of *silicon*). The regions between these thick SiO_2 layers, where transistors will be built, are called the "active" regions of the substrate.

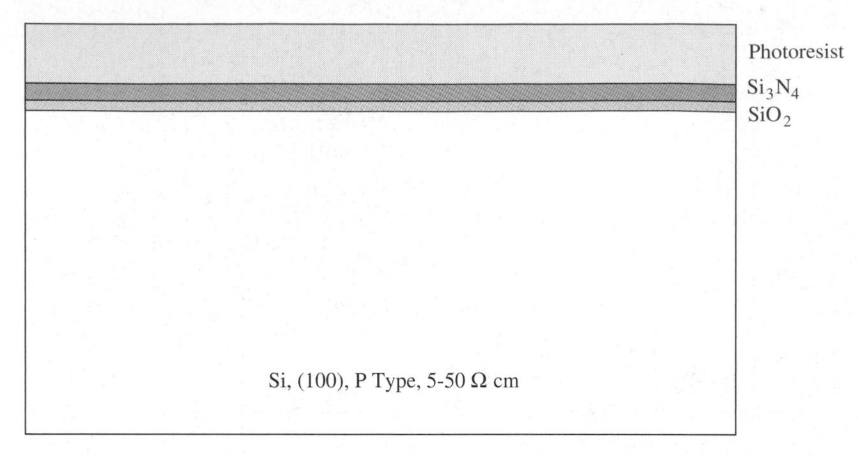

Figure 3–3 Following initial cleaning, an SiO_2 layer is thermally grown on the silicon substrate. An Si_3N_4 layer is then deposited by LPCVD. Photoresist is spun on the wafer to prepare for the first masking operation.

We begin with the steps shown in Figure 3-3. The wafers are first cleaned in a combination of chemical baths that remove any impurities from the surface. A thermal SiO_2 layer is then grown on the Si surface by placing the wafers in a high-temperature furnace with either an O_2 or H_2O ambient. A typical thickness would be about 40 nm (400 Å).

The wafers are then transferred to a second furnace, which is used to deposit a thin layer (typically 80 nm) of silicon nitride, Si_3N_4, usually just called nitride. Generally, this deposition is done below atmospheric pressure because this produces better uniformity over larger wafer lots in the deposited films. Systems in which such depositions are done are usually called low-pressure chemical vapor deposition (LPCVD) systems.

The nitride layers deposited by such machines are normally highly stressed, with the Si_3N_4 under tensile stress. This produces a large compressive stress in the underlying Si, substrate, which can lead to defects if it is not carefully controlled. In fact, the major purpose of the SiO_2 layer under the Si_3N_4 is to help relieve this stress. SiO_2 layers are under compressive stress when they are thermally grown on Si and if the thicknesses of the SiO_2 and Si_3N layers are properly chosen, the stresses in the two layers can partially compensate for each other, reducing the stresses in the Si substrate. The thicknesses chosen above do this.

The final step in Figure 3-3 is the deposition of a photoresist layer in preparation for masking. Since photoresists are liquids at room temperature, they are normally simply spun onto the wafers. The resist viscosity and the spin speed determine the final resist thickness, which is typically about 1 μm. (Note that the dimensions in all the drawings in this chapter are not exactly to scale, since the photoresist layer in Figure 3-3 is really more than 10 times the thickness of the oxide or nitride layers and the substrate is typically 500 times as thick as the photoresist layer. The liberties we take with scale in these drawings are intended to improve clarity.)

After the photoresist is spun onto the wafer, it is usually baked at about 100°C in order to drive off solvents from the layer. The resist is then exposed using a mask, which defines the pattern for the LOCOS regions. The machines that accomplish the exposure are often called "steppers," because they usually expose only a small area of the wafer during each exposure and then "step" to the next adjacent field to expose it. Such machines must be capable today of printing lines on the order of 250 nm (0.25 μm) and placing these patterns on the wafer with an accuracy that is less than 100 nm. They typically cost several million dollars.

The photoresists themselves are complex hydrocarbon mixtures. The actual UV-light-sensitive part of the resist is only a portion of the total mixture. In the case of a positive resist, which is the most common type today, the molecule in the resist which is sensitive to light, absorbs UV photons and changes its chemical structure in response to the light. The result is that the molecule and the resist itself then dissolve in the developing solution. Negative resists also respond to UV light, but become insoluble in the regions in which they are exposed. Figure 3-4 shows our CMOS wafer after the resist has been exposed and developed.

An additional step is also illustrated in Figure 3-4. After the pattern is defined in the resist, the Si_3N_4 is etched using dry etching, with the resist as a mask. This is usually accomplished in a fluorine plasma.

Once the Si_3N_4 etching is completed, we are through with the resist, and it can be chemically removed in sulfuric acid or stripped in an O_2 plasma, neither of which significantly attacks the underlying Si_3N_4 and SiO_2 layers. Following cleaning, the wafers are placed into a furnace in an oxidizing ambient. This grows a thick SiO_2 layer locally on the wafer surface. The Si_3N_4 layer on the surface prevents oxidation where it is present, because Si_3N_4 is a very dense material and prevents the H_2O or O_2 from diffusing to the Si surface, where oxidation takes place. This local oxide would typically be about 500 nm (0.5 μm) thick. The structure at this point is illustrated in Figure 3-5.

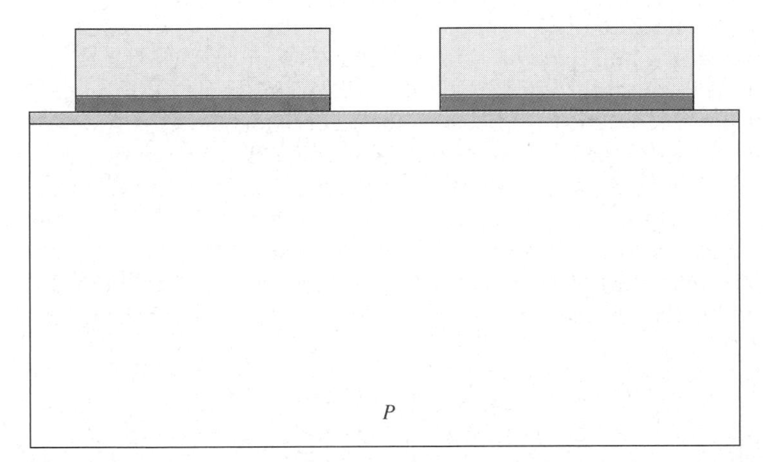

Figure 3–4 Mask #1 patterns the photoresist. The Si_3N_4 layer is removed by dry etching where it is not protected by the photoresist.

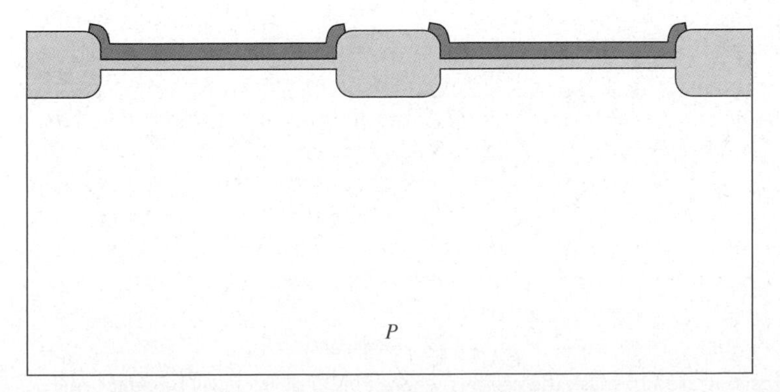

Figure 3–5 After photoresist stripping, the field oxide is grown in an oxidizing ambient.

After the furnace operation, the Si_3N_4 layer can be stripped. This is conveniently done in hot phosphoric acid, which is highly selective between Si_3N_4 and SiO_2.

3.1.3 *N* and *P* Well Formation

In the final device cross section in Figure 3-2, the active devices are shown in *p*- and *n*-type wells. These wells tailor the substrate doping locally to provide optimum device characteristics. The well doping affects device characteristics such as the MOS transistor threshold voltage, *iv* characteristics, and *pn*-junction capacitances.

In Figure 3-6, photoresist is spun onto the wafer and mask #2 is used to expose the resist and to define the regions where *P* wells are to be formed. The *P* regions are created by a process known as ion implantation. The machines that perform this step are really small linear accelerators. A source of the ion to be implanted (boron in this case) is provided, usually from a gas. Positively charged ions (B^+) are formed by exposing the source gas to an arc discharge. The ions are then accelerated in an electric field to some final energy, usually expressed in KeV. Since many types of ions may be formed in the source region, all ion implanters select the particular ion to be implanted by bending the ion beam through a magnetic field. Ions of different masses will bend at different rates in the magnetic field, allowing one type of ion to be selected at the output by adjustment of the field strength. Once the selection process is complete, final acceleration of the B^+ takes place, along with either electrostatic scanning of the beam or mechanical scanning of the wafer, to provide a uniform implant dose across the wafer.

In our case, we would need to pick an implant energy sufficiently large that the B^+ ions penetrated the thin and thick SiO_2 layers on the wafer surface, but not so large that the beam penetrated through the photoresist, which must mask against the implant. This is possible in this case because the field oxide is on the order of 0.5 μm and the photoresist is at least twice that thick. The B^+ implant needs to penetrate through the thin SiO_2 layer in order to form the *P* well. It also needs to penetrate through the field oxide although the reason is not so obvious in this case. As was pointed out earlier, the purpose of the field oxide is to provide lateral isolation between adjacent MOS transistors. If the doping is too light under the field oxide, it is possible that surface inversion can occur in these regions, provid-

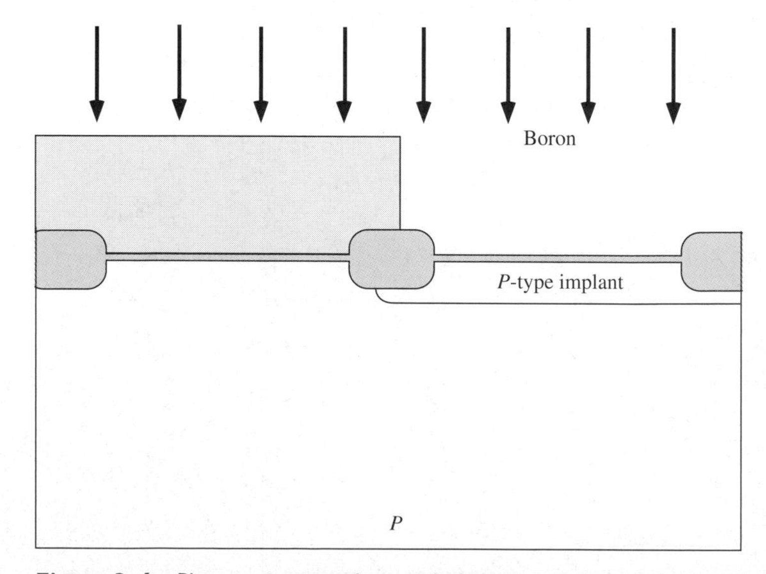

Figure 3–6 Photoresist is used to mask the regions where PMOS devices will be built using Mask #2. A boron implant provides the doping for the *P* wells for the NMOS devices.

ing electrical connections between adjacent devices through parasitic MOS devices (field oxide transistors). By ensuring that the well implants penetrate through the field oxide, the doping is increased under the field oxide, preventing this parasitic inversion problem. For the situation described here, an energy of 150–200 KeV would be typical.

The amount of B^+ we implant (or the dose) is determined by the device requirements. Here we are forming a P well whose concentration is required to be on the order of 5×10^{16} to 10^{17} cm^{-3} to provide correct device electrical characteristics. A dose on the order of 10^{13} cm^{-2} would be typical for this case. (Note that implant doses are expressed as an areal dose per cm^2, while doping concentrations are volume concentrations per cm^3.)

An important point about ion implantation has not been made to this point. Implantation of ions into a crystalline substrate causes damage. This is easy to visualize, since an incoming ion with an energy of perhaps 100 KeV can clearly collide with and dislodge silicon atoms in the substrate, which have a binding energy of only 4 Si–Si bonds (about 12 eV). Visualize a billiard-ball-like collision between the incoming 100 KeV B^+ ion and a stationary Si atom, and you can easily imagine that the silicon atom will likely be recoiled a significant distance from its original lattice site. In fact, many such recoils are produced as the B^+ ion gradually comes to rest. This damage must somehow be repaired, since the devices we want to end up with require virtually perfect crystalline substrates. Fortunately, the repair process is not as difficult as it might initially seem. A simple furnace step usually suffices. Heating the wafers allows the dislodged silicon atoms to diffuse and find a vacant lattice site, thus repairing the damage. Such a high-temperature step will soon occur in our process description and will accomplish the crystal repair function. This can be done in short times at high temperatures (e.g., 10 sec at 1000°C) or in longer times at lower temperatures (e.g., 30 min at 800°C).

Once the boron implant is complete, we are finished with the photoresist, and it is then stripped either chemically or in an O_2 plasma. Photoresist and mask #3 are now used as shown in Figure 3-7 to define the regions where *N* wells will be placed in the silicon. The process is identical to that just described for the *P* wells, except that in this case an *N*-type dopant, phosphorus, is implanted. The energy of the

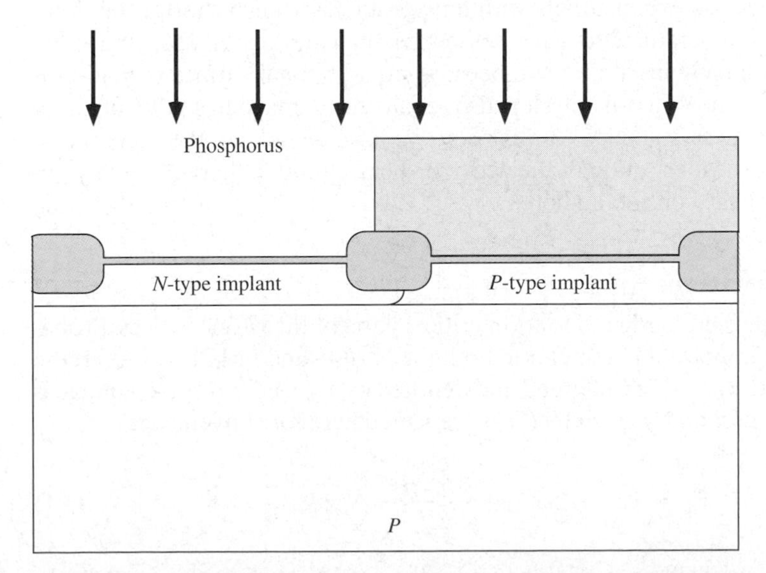

Figure 3–7 Photoresist is used to mask the regions where NMOS devices will be built using Mask #3. A phosphorus implant provides the doping for the *N* wells for the PMOS devices.

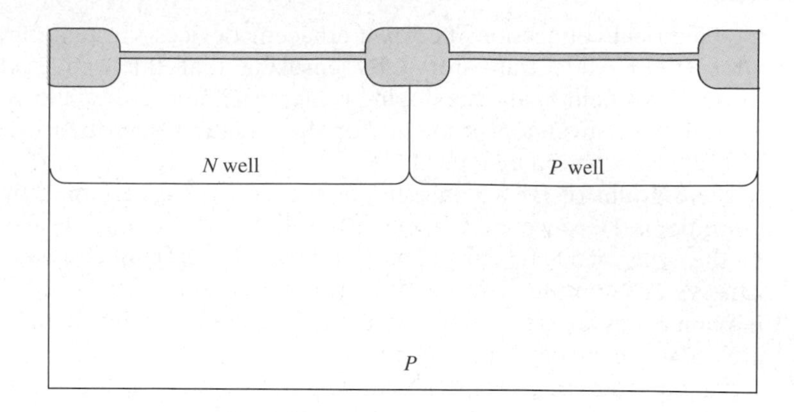

Figure 3–8 A high temperature drive-in completes the formation of the N and P wells.

phosphorus implant is again chosen to penetrate the oxide layers but not the photoresist. Phosphorus is a heavier atom than boron (atomic mass = 31 versus 11), so a higher energy is required to obtain an implant to the same depth into the silicon. In this situation, an energy of 300/400 KeV would be chosen. The dose of the phosphorus implant would typically be on the same order as the boron P well implant, since the purpose is similar in both cases.

There are several common N-type dopants available for use in silicon (phosphorus, arsenic, and antimony), and yet we specifically chose phosphorus in this case. The next step in the process is to diffuse the P and N wells into a junction depth of typically several microns, as illustrated in Figure 3-8. Boron and phosphorus have essentially matched diffusion coefficients, so they will produce wells with about the same junction depth when they are simultaneously diffused.

After the phosphorus implant, the photoresist is removed and the wafers are cleaned. They are next placed in a drive-in furnace, which diffuses the wells to a junction depth of 2–3 microns. (Actually, the depths they reach in this step will not be their "final" depths, because all subsequent high-temperature steps will continue to diffuse the dopants. However, later high temperature steps will generally be either at lower temperatures or for shorter times, so that most of the well diffusion occurs during this drive-in process.) A typical thermal cycle might be 4–6 hours at 1000–1100°C. Diffusion coefficients increase exponentially with temperature, so much shorter times are required at higher temperatures to achieve the same junction depth. This step could be performed in a largely inert ambient, because no additional surface oxidation is needed at this point. The well drive-in step also repairs the damage from the implants, restoring the substrate crystallinity. At this point, we have completed the preparation of the substrate for the fabrication of the active devices, although there are many options that have not been considered here.

3.1.4 Gate Formation

The next several steps are designed to form critical parts of the MOS devices. Probably the single most important parameter in both the NMOS and PMOS devices is the threshold voltage, discussed in Chapter 2 and denoted by V_{th}. The threshold voltage is derived in Section 2.6.2 and is given by (2.93), repeated here for convenience

$$V_{th} = V_{FB} + 2\phi_f + \frac{\sqrt{2qN_A\epsilon_{si}}}{C'_{ox}}\sqrt{2\phi_f}. \tag{3.1}$$

Notice that V_{th} is dependent on the doping under the gate, N_A, and on the oxide thickness, which sets C'_{ox}.

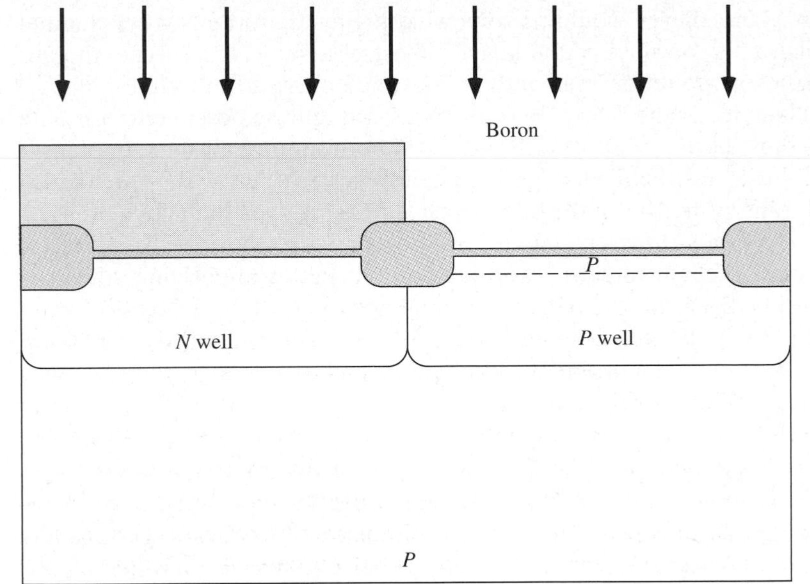

Figure 3–9 After photoresist is spun onto the wafer, Mask #4 is used to define the NMOS transistors. A boron implant adjusts the *N*-channel V_{th}.

In modern CMOS devices, the threshold voltages are adjusted by a separate ion implantation step. The target threshold voltages are generally around 0.5 to 0.8 V in magnitude.

Figure 3-9 illustrates the masking to adjust the NMOS V_{th}. Photoresist is applied, and Mask #4 is used to open the areas where NMOS devices are located. After development, a boron implant is used to adjust V_{th}. A dose of $1–5 \times 10^{12}$ cm^{-2} at an energy of 50–75 KeV might be used. The energy is chosen to be high enough to get the implant dose through the thin oxide, but low enough to keep the boron near the silicon surface.

Figure 3-10 illustrates the same process sequence, now applied to the PMOS device. Mask #5 is used. The required implant could be either *n*- or *p*-type, depending on the doping level in the *N* well and the required PMOS V_{th}. An *n*-type implant is illustrated in the figure. This would typically be arsenic with a dose of

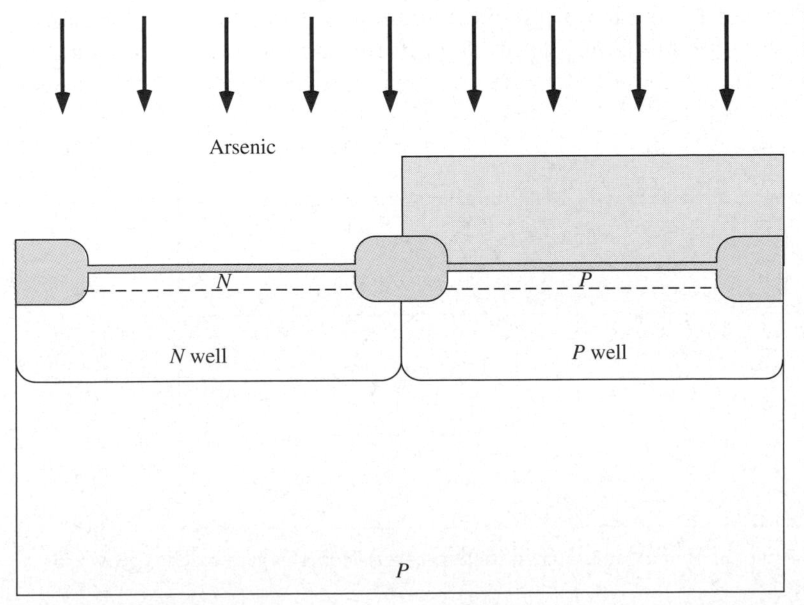

Figure 3–10 After photoresist is spun onto the wafer, Mask #5 is used to define the PMOS transistors. An arsenic implant adjusts the *P*-channel V_{th}.

$1–5 \times 10^{12}$ cm^{-2}. The energy would be somewhat higher than the NMOS channel implant in Figure 3-9, because of the heavier mass of arsenic. If a *p*-type implant were needed, boron would be used with a dose and energy in the same range as for the NMOS device in the figure. In some cases, it might be possible to use only one mask to adjust both NMOS and PMOS V_{th} if both require a *p*-type implant.

Figure 3-11 illustrates the next steps. We are now ready to grow the gate oxides for the MOS transistors. The thin oxide, which is present over the active areas of each transistor, is first stripped in a dilute solution of hydrofluoric acid, HF. HF is a highly selective etchant and will stop etching when the underlying silicon is reached. Note, however, that we will etch a small portion of the field oxide during this step, because the HF etch is unmasked. Since we are etching only 10 or 20 nm of oxide, this is usually not a problem, although the etch needs to be timed so that it does not etch too much of the field oxide.

The reason that the thin oxide is stripped and then regrown to form the MOS gate oxide is that the oxide on the silicon surface prior to stripping is too thick to serve as the device gate oxide. Stripping and regrowing this oxide results in a well-controlled final oxide thickness. The original thin oxide on the wafer surface has also been exposed to several implants at this stage in the process, which create damage in the SiO_2, so stripping and regrowing a new oxide produces a higher quality gate oxide. In state-of-the-art MOS devices today, the gate oxide is typically thinner than 10 nm. This oxide could be formed by a variety of processes (times and temperatures). For example, oxidation in O_2 at 800°C for 2 hours would produce about 10 nm of oxide. Similarly, oxidation in H_2O at 800°C for 25 min would produce a similar thickness of oxide. Figure 3-11 illustrates the devices after the gate oxide has been grown.

The next steps deposit and define the polysilicon gate electrodes for the MOS devices. By using LPCVD, a layer of polysilicon is deposited over the entire wafer surface, as illustrated in Figure 3-12. The deposited layer may be either amorphous or polycrystalline, depending on the deposition temperature.

The polysilicon is then doped *N* type by an unmasked ion implant. Either phosphorus or arsenic could be used here, since both are highly soluble in silicon (and polysilicon) and thus can produce low-sheet-resistance poly layers. The implant energy is not very critical here, provided that the phosphorus or arsenic does not penetrate through the poly and into the underlying gate oxide and substrate. Both dopants rapidly redistribute in poly at elevated temperatures, because diffusion is rapid along the grain boundaries in poly, so uniform doping of the poly will occur later in the process, when the wafers are next heated in a furnace. The N^+ dose is

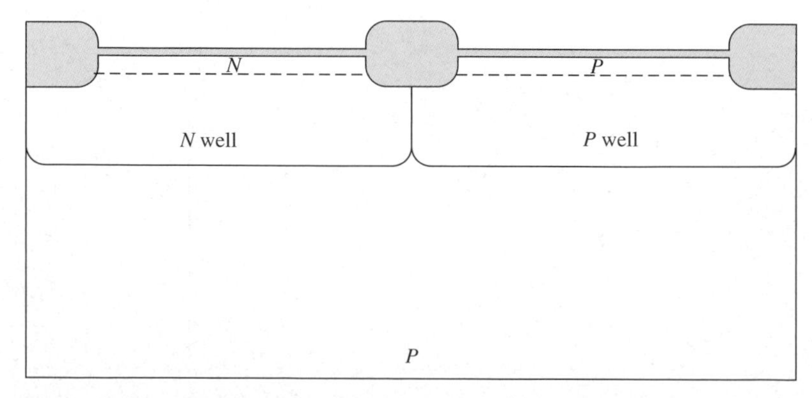

Figure 3–11 After the thin oxide is etched back to bare silicon, the gate oxide is grown for the MOS transistors.

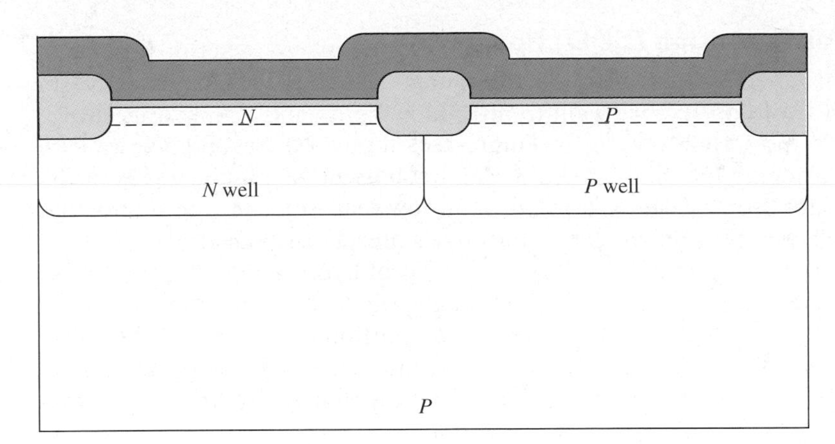

Figure 3–12 A layer of polysilicon is deposited. Ion implantation of phosphorus follows the deposition to heavily dope the poly.

not critical for the MOS gates other than the fact that we would like it to be as high as possible in order to obtain low poly sheet resistivity and hence low gate resistance. A dose of about 5×10^{15} cm^{-2} would be typical. In some polysilicon deposition systems, the poly can be doped while it is being deposited. This is referred to as "in-situ" doped poly. In this case, the ion implantation doping step would not be necessary.

The final step, illustrated in Figure 3-13, uses resist and Mask #6 to etch the poly away in regions where it is not needed. Photoresist is spun onto the wafer, baked, and then exposed and developed. The poly etching would again be done in a plasma etcher. Although it is not shown in the figure, the polysilicon layer can also be used to provide wiring between active devices on the chip (for example, to connect the NMOS and PMOS poly gates together). In this sense, it can serve as the first level of interconnect. Since the poly sheet resistance is relatively high compared with that of metal layers that will be deposited later ($\approx 10\ \Omega$/sq versus $< 0.1\ \Omega$/sq),[2] long interconnects are not made with the polysilicon. The RC delays associated with long poly lines would have a significant effect on circuit performance—hence the term "local interconnects" for these relatively short polysilicon wires.

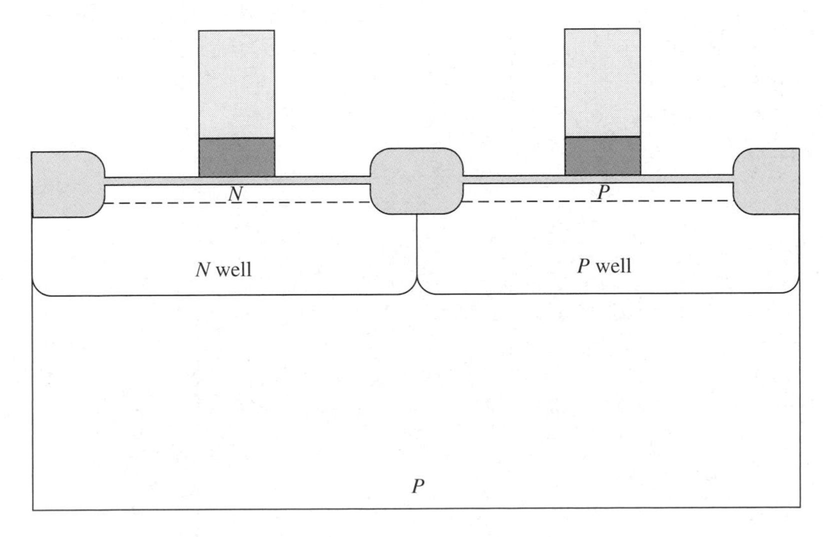

Figure 3–13 Photoresist is applied, and Mask #6 is used to define the regions where MOS gates are located. The polysilicon layer is then etched by means of plasma etching.

[2] Ω/sq means ohms per square. A rectangular pattern on the wafer with length L and width W has (L/W) squares. If it defines a region with resistivity ρ and thickness t, the resistance is: $R = \frac{\rho L}{tW} = \frac{\rho}{t}$(# of squares). Therefore, expressing ρ/t in Ω/sq is convenient.

3.1.5 Tip or Extension (LDD) Formation

The next several steps are illustrated in Figures 3-14 to 3-17. Our objective in these steps is twofold. First, we want to introduce the N^- and P^- implants shown in the NMOS and PMOS devices in Figures 3-14 and 3-15. Second, we want to place a thin oxide layer usually called a "sidewall spacer" along the edges of the polysilicon gates. Both of these steps are required because of scaling trends that have taken place in the semiconductor industry over the past decade.

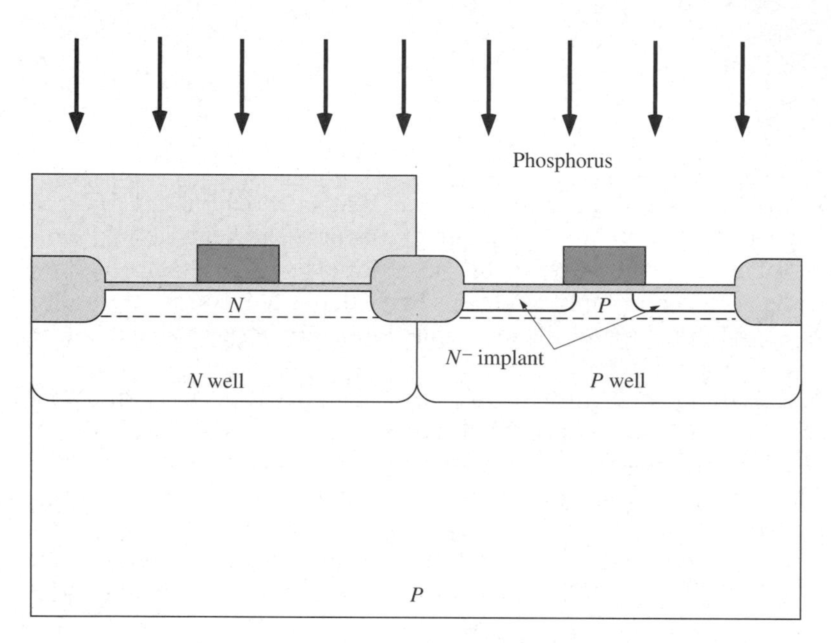

Figure 3–14 Mask #7 is used to cover the PMOS devices. A phosphorus implant is used to form the tip or extension (LDD) regions in the NMOS devices.

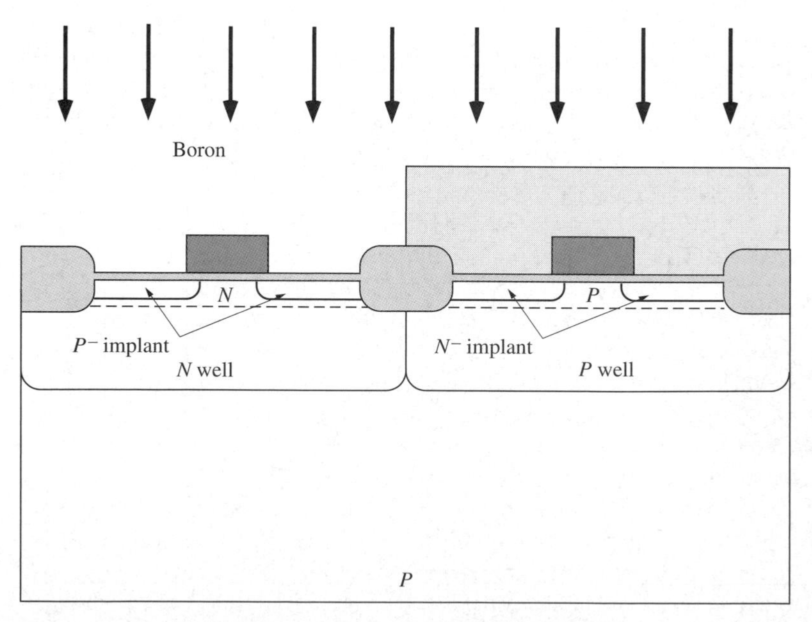

Figure 3–15 Mask #8 is used to cover the NMOS devices. A boron implant is used to form the tip or extension (LDD) regions in the PMOS devices.

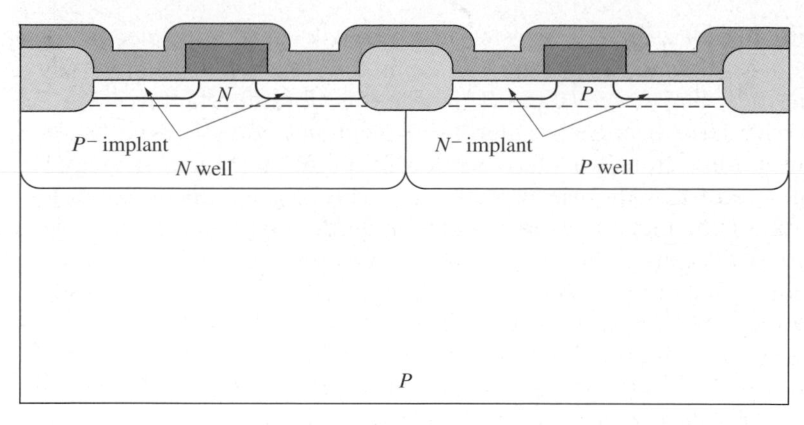

Figure 3–16 A conformal layer of SiO_2 is deposited on the wafer in preparation for sidewall spacer formation.

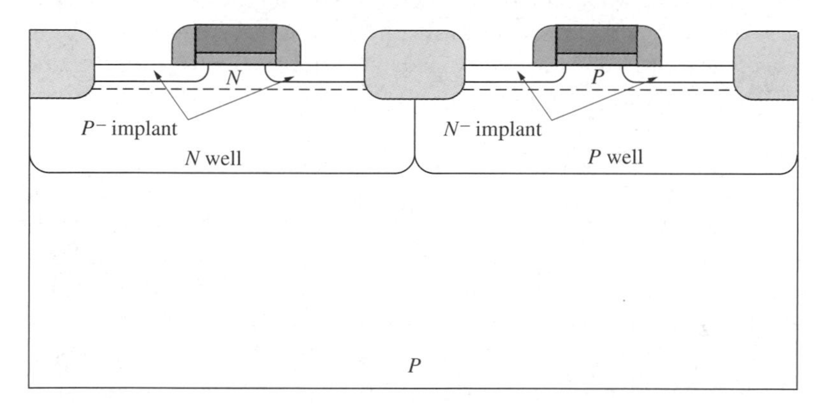

Figure 3–17 The deposited SiO_2 layer is etched back anisotropically, leaving sidewall spacers along the edges of the polysilicon.

Ten years ago, MOS devices used in ICs were built with minimum dimensions well above 1 micron and were operated in circuits with supply voltages of 5 V. Today, device dimensions have been reduced to 0.18 μm to improve performance. However, supply voltages in circuits have not been proportionally reduced. Some ICs still use 5-V power supplies, although many chips are now being designed with 3.3-, 2.5-, or 1.8-V supplies. There is great benefit at the system level to maintaining a standard power supply level, because then new ICs are compatible with older parts, system power supplies do not have to be redesigned, and circuit noise margins remain adequate. However, if device dimensions are reduced and voltage levels are not correspondingly scaled, electric fields inside the devices necessarily rise. The fields can easily become large enough to produce carriers, called "hot" carriers (electrons or holes), with sufficient energy to tunnel into the gate oxide and change the electrical characteristics of the device.

Because of these larger fields in scaled devices, considerable effort has gone into designing MOS device structures that can withstand high electric fields. One of the innovations that is almost universally used is the lightly doped drain, or LDD, device. The idea behind this structure is to grade the doping in the drain region to produce an N^+N^-P profile between the drain and channel in the NMOS devices and a corresponding P^+P^-N profile in the PMOS devices. This allows the drain voltage to be dropped over a larger distance than would be the case if an abrupt N^+P junction were formed. This reduces the peak value of the electric field in the near drain region and can make a significant difference in device reliability.

A final point regarding these N^- and P^- implants is also important to make. As device geometries have become smaller, "short-channel effects" have become very important in MOS transistors (*see* Section 2.6.7). An important strategy for minimizing these effects is the use of shallow junctions. Such junctions are less susceptible to short-channel effects essentially because their geometry minimizes the junction areas adjacent to the channel. The LDD structure also provides these shallow junctions, which in this context are often called the "tip" or "extension" regions, since they must be combined with deeper source and drain junctions away from the channel in order to make reliable contacts to the device. The N^- and P^- implants in Figures 3-14 and 3-15 and the sidewall spacers in Figure 3-17 are used to construct these tip or extension or LDD regions.

In Figure 3-14, photoresist is spun on the wafer, and Mask #7 is then used to protect all the devices except the NMOS transistors. A phosphorus implant is done to form the N^- region. The dose and the energy are carefully controlled in this implant to ultimately produce the desired graded drain junction. Typically, a dose of about 5×10^{13} to 5×10^{14} cm^{-2} at a low energy might be used. A similar sequence of steps is used for the PMOS devices in Figure 3-15 to produce the LDD regions in these devices. A similar implant would be performed, although boron would be used in this case.

The next step is the LPCVD deposition of a conformal spacer dielectric layer (SiO_2 or Si_3N_4) on the wafer surface, shown in Figure 3-16. The thickness of this layer will determine the width of the sidewall spacer region and would be chosen to optimize device characteristics. Typically, this might be a few hundred nanometers.

Notice in Figure 3-16 that the deposited SiO_2 layer is much thicker along the edges of the polysilicon than it is above flat regions of the wafer surface. This is because of the vertical edges of the polysilicon regions and the conformal deposition of the oxide. If we now etch back the deposited SiO_2 layer, using an etching technique that is highly anisotropic (one that etches vertically, but not horizontally), we will be left with the structure shown in Figure 3-17. Typically, this would be done in a fluorine-based plasma.

The deposited oxide is removed everywhere except along the edges of vertical steps in the underlying structures. Simply by a deposition and then an etchback, we have formed the sidewall spacers. We have also created lateral features on the chip surface that are smaller than the minimum feature size of the lithographic process. In fact, the width of the spacers is determined largely by the thickness of the deposited oxide. It should also be noted that there is nothing magic about the use of SiO_2 in this application. We could form such spacers by using almost any deposited thin film. In this particular case we need an insulating material, for reasons that will become apparent shortly, so SiO_2 is a convenient choice.

3.1.6 Source–Drain Formation

At this point, most of the doped regions in the structure have been formed, except for the MOS transistor source and drain regions. However, note that in Figure 3-17, the oxide was etched off the source and drain regions as part of the sidewall spacer formation process. Generally, implants are done through a thin "screen" oxide, so prior to doing the source–drain implants, a thin screen oxide of perhaps 10 nm is grown.

The first source–drain implant is illustrated in Figure 3-18. Photoresist and Mask #9 are used to define the regions where NMOS source–drain implants will be done. Arsenic would be the dopant of choice in modern processes, because of the need to keep junctions shallow in small-geometry devices. An implant of $2–4 \times 10^{15}$ cm^{-2} at an en-

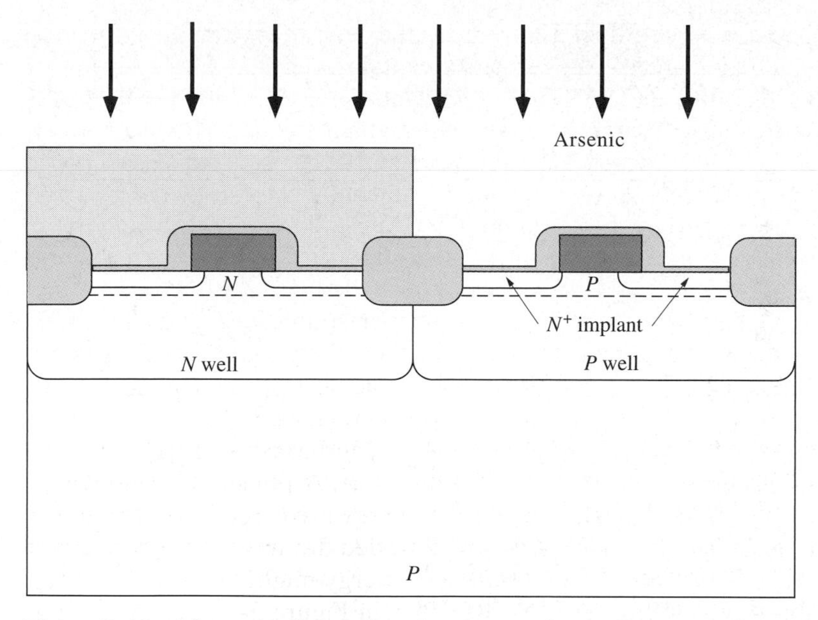

Figure 3–18 After a thin "screen" oxide is grown, photoresist is applied and Mask #9 is used to protect the PMOS transistors. An arsenic implant then forms the NMOS source and drain regions.

ergy of 75 KeV might be typical. This would allow the arsenic to penetrate the screen oxide in the implanted areas, but still be easily masked by the photoresist.

The final mask used for doping is illustrated in Figure 3-19. Photoresist and Mask #10 allow a boron implant to form the PMOS source–drain regions. This implant would also be a high-dose implant, on the order of $1–3 \times 10^{15}$ cm^{-2}, but at a lower energy of about 50–75 KeV because boron is much lighter than arsenic and therefore requires less energy to reach the same range. High-dose implants minimize the parasitic resistances associated with the source and drain regions in the MOS transistors. It is

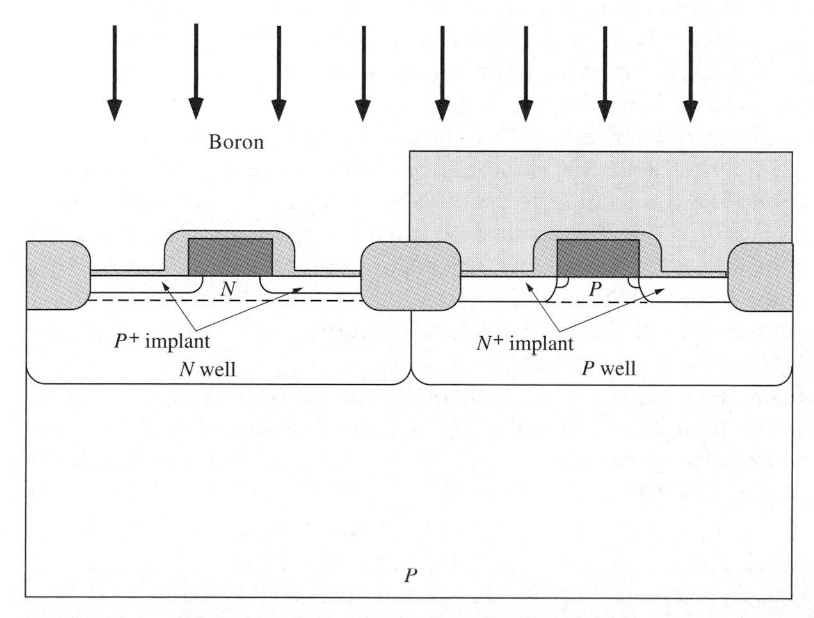

Figure 3–19 After photoresist is applied, Mask #10 is used to protect the NMOS transistors. A boron implant then forms the PMOS source and drain regions.

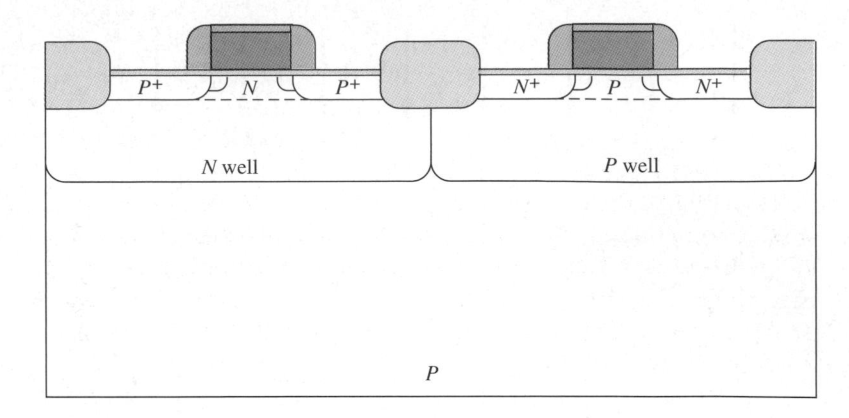

Figure 3–20 A final high temperature drive-in activates all the implanted dopants and diffuses junctions to their final depth.

also interesting to notice that the polysilicon gate regions receive at least two high dose implants in the process flow we have described. The first, which was N^+, occurred in connection with Figure 3-14 and initially heavily doped the poly N type. In the NMOS devices, a second N^+ implant goes into the poly in Figure 3-18. In the PMOS device, a P^+ implant dopes the poly in Figure 3-19. In most processes today, both the PMOS and NMOS gates are N type. If so, then the P^+ dose in Figure 3-19 needs to be smaller than the N^+ dose implanted in Figure 3-14. This is the case for the numbers we have used in this process flow.

The final step in active device formation is illustrated in Figure 3-20. A furnace anneal, typically at ≈ 900°C for 30 min, or perhaps a rapid thermal anneal for ≈ 1 min at 1000–1050°C, activates all the implants,[3] anneals implant damage, and drives the junctions to their final depths.

3.1.7 Contact and Local Interconnect Formation

All of the steps needed to form the active devices have now been completed. However, we obviously need to provide a means of interconnecting devices on the wafer to form circuits and a means to bring the input and output connections off the chip for packaging. The CMOS process we are describing will actually provide three levels of wiring to accomplish these objectives. The first or lowest level is often called the "local interconnect," and the steps needed to form it are shown in Figures 3-21 to 3-24. (Actually, as we pointed out earlier, the polysilicon gate level itself can also be used as a local interconnect.)

The first step, illustrated in Figure 3-21, removes the oxide from the areas the interconnect is to contact. Since this is the bottom level of the interconnect structure, it will provide the connections to essentially all doped regions in the silicon and to all polysilicon regions. This oxide etch can actually be unmasked, because the oxide is quite thin over the regions we wish to contact (the ≈10-nm screen oxide we grew just prior to source–drain formation). A short dip in a buffered HF etching solution will remove these oxide layers, without significantly reducing the thickness of the oxide layers elsewhere.

The next step, shown in Figure 3-22, involves the deposition of a thin layer (50–100 nm) of titanium, Ti, on the wafer surface. This is usually done by sputtering from a Ti target. In a sputtering system, atoms of the desired material (Ti in this case) are physically knocked off a solid target by bombarding the Ti target

[3] To "activate" an implant means to repair damage so that the dopant atoms substitute for silicon atoms in the lattice and can properly participate in the covalent bonding process discussed in Section 2.1. If the dopants are between normal lattice sites, rather than substituting for silicon atoms, they do not participate in the bonding in the normal way.

with Ar^+ ions. The Ti atoms then deposit on any substrates that are located nearby. This produces a continuous coating of Ti on the wafer, as shown in the figure.

The next step, shown in Figure 3-23, makes use of two chemical reactions. The wafers are heated in an N_2 ambient at about 600°C for a short time (about 1 minute). At this temperature, the Ti reacts with Si where they are in contact, to form titanium silicide, $TiSi_2$, consuming some silicon in the process. This is why deeper source and drain junctions are required outside the tip or extension regions. Silicide is an excellent conductor and forms low-resistance contacts to both N^+ and P^+ silicon or polysilicon. This material is shown in black in the figure. The Ti also reacts with N_2 to form TiN (the dark top layer in Figure 3-23). This material is also a conductor, although its conductivity is not as high as that of most metals. For this reason, it is used only for "local" or short-distance interconnects. The resistance of long lines made from TiN would cause unacceptable RC delays in most circuits.

Figure 3-24 illustrates the patterning of the TiN layer. Photoresist is applied, and Mask #11 protects the TiN where we want it to remain on the wafer. The remaining TiN is removed by etching. It is interesting to note at this point that the sidewall spacers we used earlier to provide graded N^+N^- or P^+P^- drain junctions also serve the function here of separating the $TiSi_2$ on the poly gates from contacting the silicon-doped regions. After photoresist removal, the wafer would typically be heated in a furnace in an Ar ambient at about 800°C for about 1 minute, to reduce the resistivity of the TiN and $TiSi_2$ layers to their final values (about 10 Ω/sq and 1 Ω/sq respectively). The photoresist would be removed prior to this last high-temperature anneal in an O_2 plasma or through chemical stripping, since photoresist cannot tolerate temperatures much above 100°C.

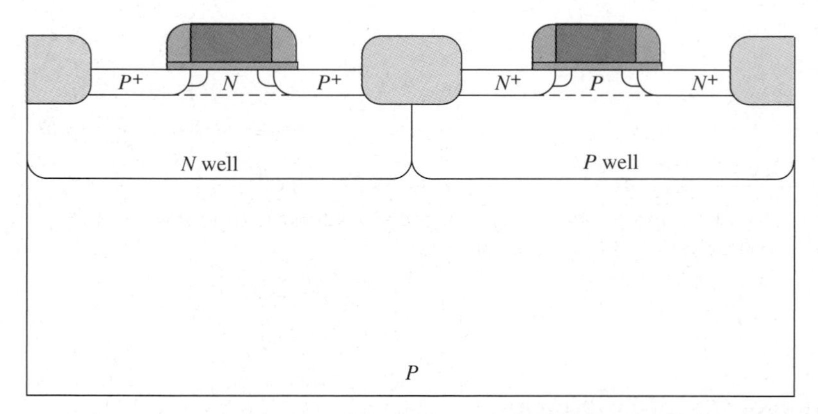

Figure 3–21 An unmasked oxide etch removes the SiO_2 from the device source and drain regions and from the top surface of the polysilicon.

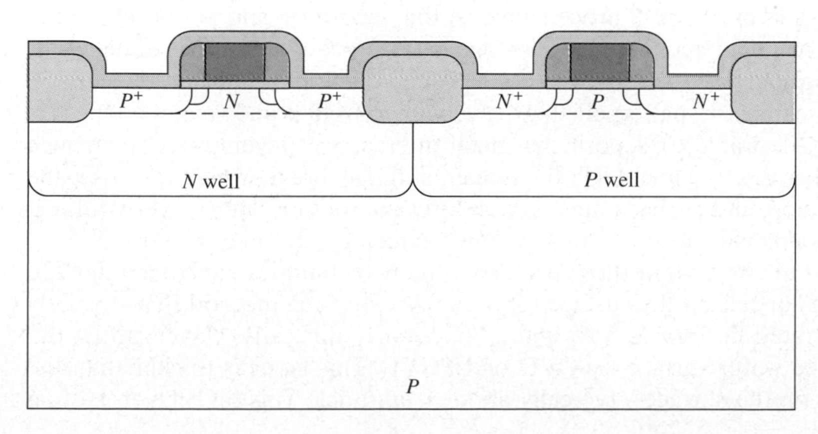

Figure 3–22 Titanium is deposited on the wafer surface by sputtering.

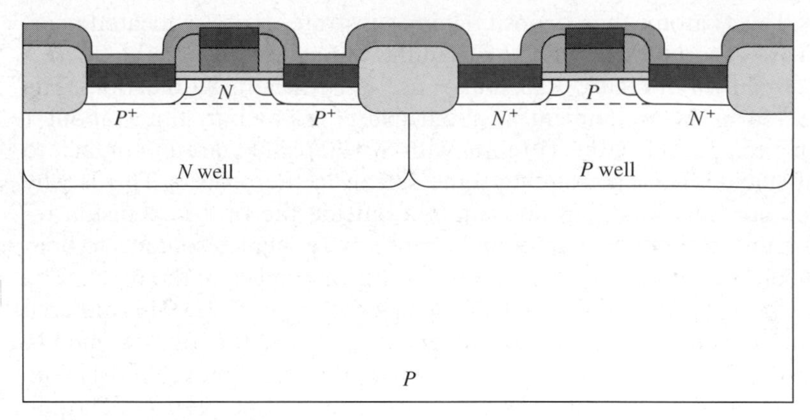

Figure 3–23 The titanium is reacted in an N_2 ambient, forming $TiSi_2$ where it contacts silicon or polysilicon (black regions in the figure) and TiN elsewhere.

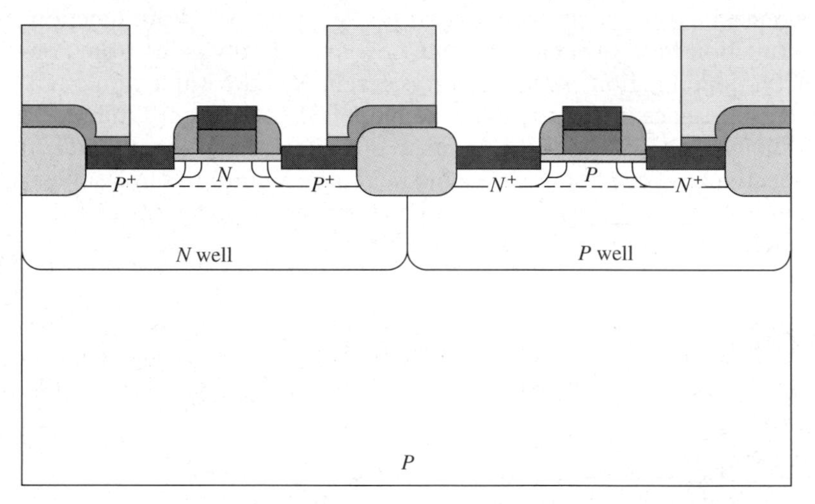

Figure 3–24 Photoresist is applied and Mask #11 is used to define the regions where TiN local interconnects will be used. The TiN is then etched.

3.1.8 Multilevel Metal Formation

The final steps in our CMOS process involve the deposition and patterning of the two layers of metal interconnect. At this stage in the process, the surface of the wafer is highly nonplanar. We have grown and deposited many thin films on the surface, and after these films are patterned, they leave numerous hills and valleys on the surface. It is not desirable to deposit the metal interconnect layers directly on such topography, because the metal will be thinner, and may break, where it crosses the steps. In addition, and perhaps most importantly, photolithography is very difficult with highly nonplanar substrates, especially when metal patterning is involved.

In an effort to circumvent these problems, many techniques have been devised to "planarize" or flatten the surface topography. One such method that is widely used is illustrated in Figures 3-25 and 3-26. A fairly thick SiO_2 layer is first deposited on the wafer surface by CVD or LPCVD. This layer is thicker than the largest steps on the surface—typically about 1 μm thick. This SiO_2 layer is often

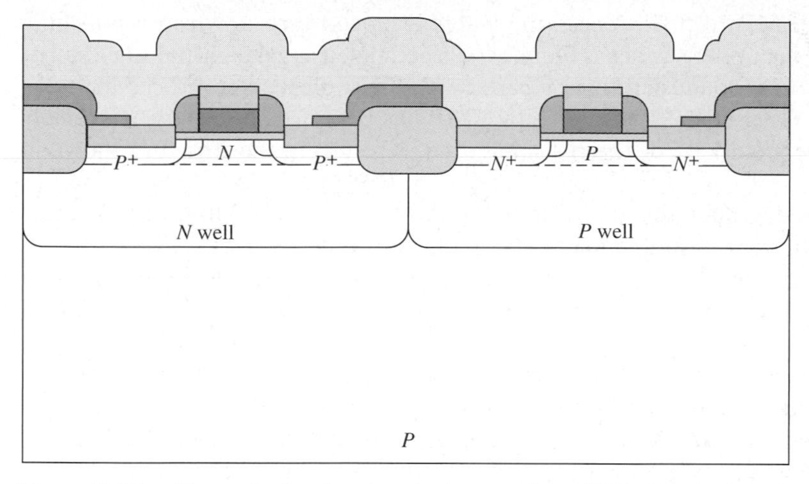

Figure 3–25 After stripping the photoresist, a conformal SiO_2 layer is deposited by LPCVD.

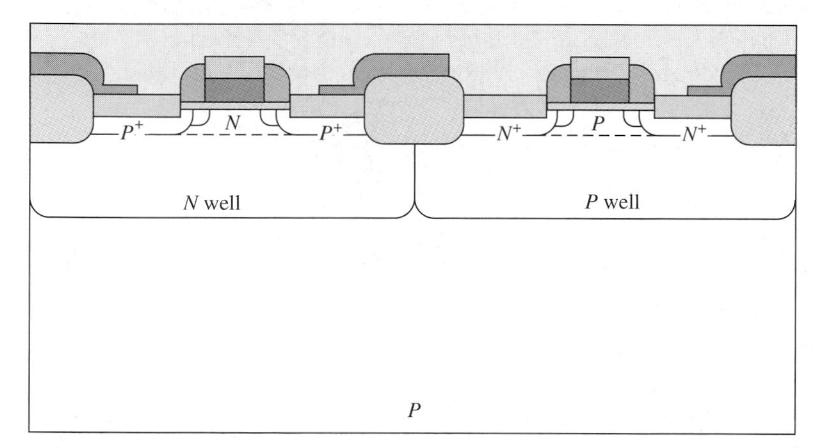

Figure 3–26 Chemical–mechanical polishing (CMP) is used to polish the deposited SiO_2 layer. This planarizes the wafer surface.

doped with phosphorus and sometimes with boron as well, in which cases the deposited oxide is known as PSG (phosphosilcate glass), or BPSG (borophosphosilicate glass), respectively. In some cases an undoped SiO_2 layer is added on top of the PSG or BPSG layer. The phosphorus provides some protection against mobile ions like Na^+ which can cause instabilities in MOS devices.[4] The addition of the boron reduces the temperature at which the deposited glass layer "flows." This is important because following the deposition, the wafer is often heated to a temperature of 800–900°C which allows the glass to flow and to smooth the surface topography. Adding the boron minimizes the heat treatment required to accomplish this reflow, which is an issue because of the limited temperature tolerance of some of the underlying films at this point in the process.

Reflowing the deposited PSG or BPSG layer is not sufficient to completely "planarize" the surface topography, so, generally, additional steps are required. In most modern processes, the final "planarization" is accomplished by chemical–mechanical

[4] Because these ions are mobile in SiO_2 and possess electric charge, as they move under the influence of the electric field that is present they shift the value of the threshold voltage.

polishing, or CMP. In this process, the wafer is placed face down in a polishing machine, and the upper surface is literally polished flat, using a high-pH silica slurry. While this process sounds crude compared with the sophisticated processing techniques generally used to fabricate chips, CMP has been found to work extremely well and is frequently used. The polishing process results in the structure shown in Figure 3-26.

The next step is again the application of photoresist. We use Mask #12 to define the regions where we want contact to be made between metal level #1 and underlying structures,[5] as shown in Figure 3-27. The SiO_2 layer would be etched in a plasma. After the contact holes are etched, the photoresist would be stripped off the wafer.

We wish to maintain the planar surface as we add metal layers to the structure. While there are a number of process flows that can achieve this, the particular process we will describe here, Figure 3-28, is one of the more common. The first step is a blanket deposition of a thin TiN layer or Ti–TiN bilayer by sputtering or CVD. This layer is typically only a few tens of nanometers thick. It provides good adhesion to the SiO_2 and other underlying materials present in the structure at this point. The TiN also acts as an effective barrier layer between the upper metal layers and the lower local interconnect layers that connect to the active devices. The next step is the deposition of a blanket W layer by CVD, as illustrated in Figure 3-28.

The next step again involves CMP to "planarize" the wafer. The polishing in this case removes the W and the TiN everywhere except in the contact holes and provides a planar surface on which the first level of metal can be deposited. This process flow we have described, in which contact holes are etched, filled, and planarized, is known as the *damascene process.*

Metal 1 is then deposited, usually by sputtering, and defined using resist and Mask #13, as shown in Figure 3-29. The metal is commonly Al with a small per-

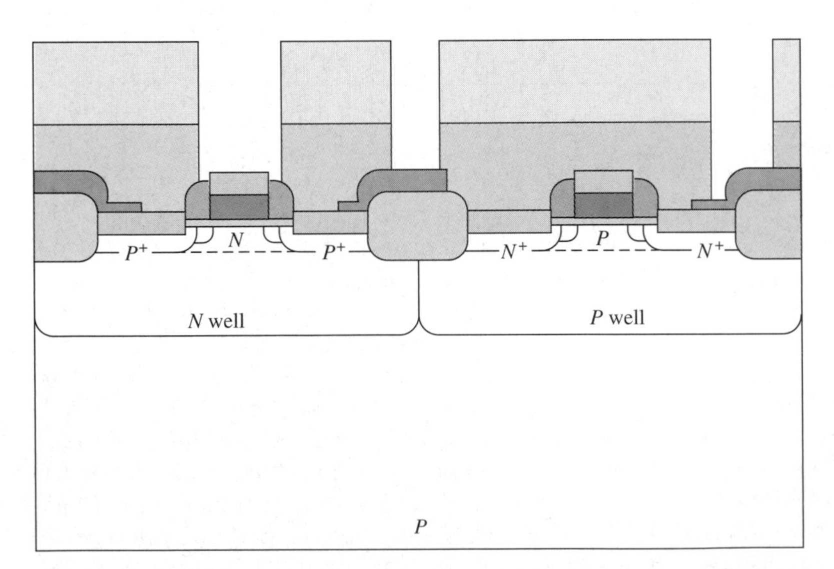

Figure 3–27 Photoresist is spun onto the wafer. Mask #12 is used to define the contact holes. The deposited SiO_2 layer is then etched to allow connections to the silicon, polysilicon, and local interconnect regions.

[5] Note that the contacts are not usually made over the active channel as shown in this figure. Since the poly layer defining the channel is usually the minimum feature size available in the process, the contacts are typically made off to the side of the transistor.

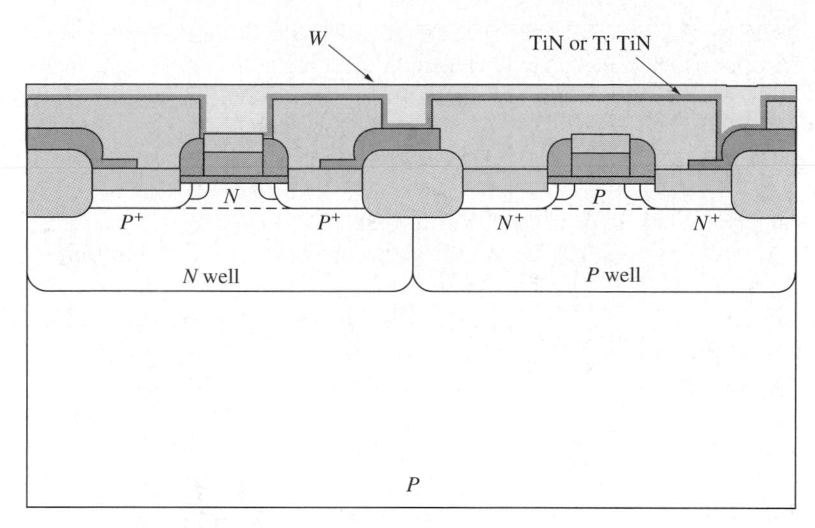

Figure 3–28 A thin TiN barrier–adhesion layer is deposited on the wafer by sputtering, followed by deposition of a *W* layer by CVD.

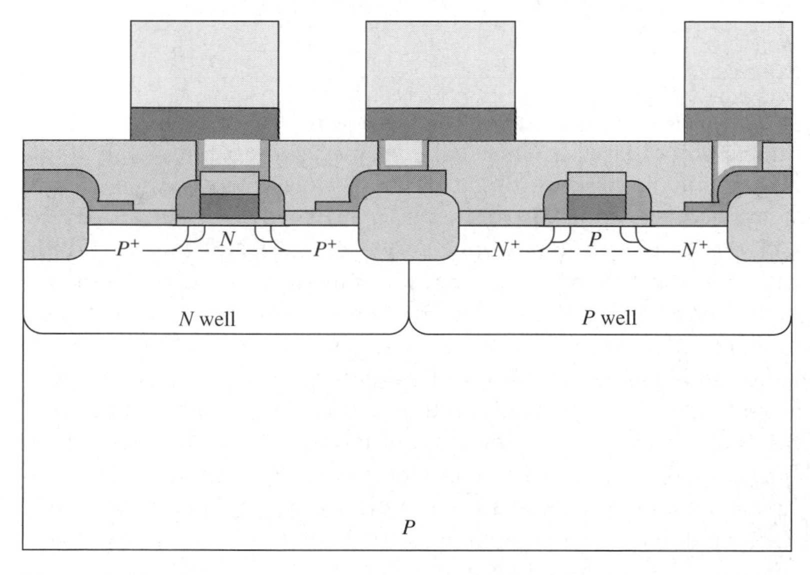

Figure 3–29 Aluminum is deposited on the wafer by sputtering. Photoresist is spun onto the wafer, and Mask #13 is used to define the first level of metal. The Al is then plasma etched.

centage of Si and Cu in it. The Si is used because Si is soluble in Al up to a few percent, and if the silicon is not already present in the Al, it may by absorbed by the Al from underlying silicon-rich layers. This can cause problems with contact resistance and contact reliability. The Cu is added because it helps to prevent a reliability problem known as electromigration in Al thin films. This phenomenon causes open circuits in Al interconnects after many hours of circuit operation, especially at elevated temperatures and high current densities, and is due to the formation of voids in the Al lines caused by the diffusion of Al atoms. The Cu helps to prevent this from occurring.

Because of its better electrical conductivity, Cu is now beginning to replace Al as the interconnect metal. Cu is deposited by using electroplating. Because Cu is quite difficult to etch by means of plasma etching, a somewhat different process flow is required than that described here for Al-based interconnects.

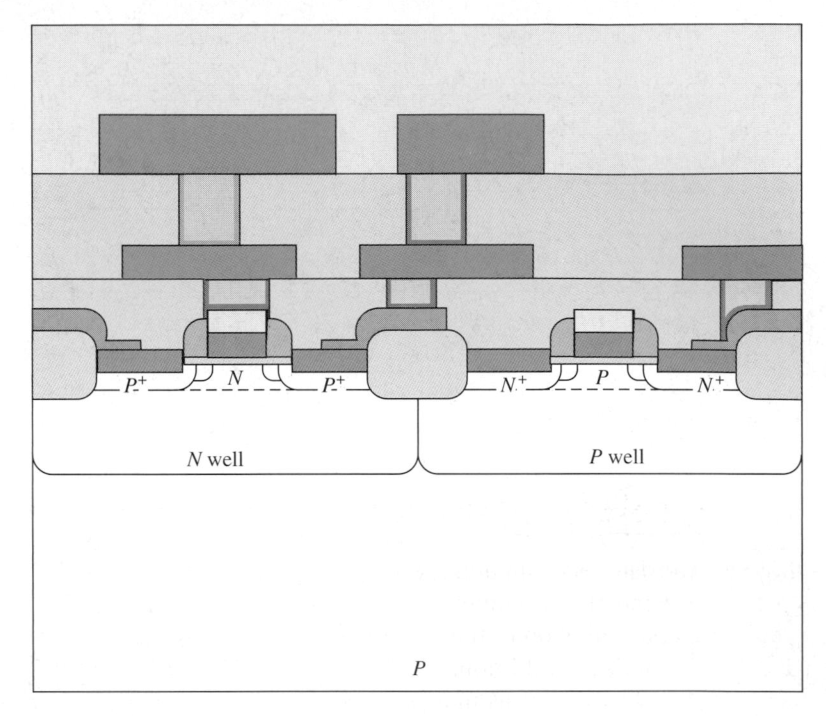

Figure 3–30 The steps to form the second level of Al interconnect follow those in Figs. 3-25 through 3-29. Mask #14 is used to define via holes between metal 2 and metal 1. Mask #15 is used to define metal 2. The last step in the process is the deposition of a final passivation layer, usually Si_3N_4 deposited by PECVD. The last mask (#16) is used to open holes in this mask over the bonding pads.

Most modern VLSI processes use more than one level of wiring on the wafer surface, because in complex circuits it is usually very difficult to completely interconnect all the devices in the circuit without multiple levels. The processes that are used to deposit and define each level are similar to those we described for level 1 and usually involve a "planarization" step. Figure 3-30 illustrates this for the dielectric between metal 1 and metal 2. The process would again involve depositing an oxide layer and using CMP to "planarize" the deposited oxide. the figure also illustrates the filling of the via holes between metal 1 and metal 2 with TiN and W, the deposition and etching of metal 2 and the final deposition of a top dielectric to protect the finished chip. This top layer could be either SiO_2 or Si_3N_4 and is designed to provide some protection for the chip during the mechanical handling it will receive during packaging, as well as to provide a final passivation layer to protect the chip against ambient contamination (by Na^+ or K^+). After the final processing steps are completed, an anneal and alloy step at a relatively low temperature (400–450°C) for about 30 minutes in forming gas (10% H_2 in N_2) is used to alloy the metal contacts in the structure and to reduce some of the electrical charges associated with the Si–SiO_2 interface. This finally brings us back to Figure 3-2, which is the completed CMOS chip that we started out to build.

3.1.9 Electrical Model Related to Physical Structure

The major elements of our electrical models can be related to the physical structure of a transistor as presented in Figure 3-30. The most important element in the circuit model, outside of the transconductance describing the fundamental operation, is the capacitance of the gate. The gate capacitance has three terms in general: the gate-to-source capacitance, C_{gs}; the gate-to-drain capacitance, C_{gd}; and the gate-to-substrate capacitance, C_{gsub}.

Both C_{gs} and C_{gd} have two components: some fraction of the capacitance from the gate to the channel and an overlap capacitance. The overlap capacitors result from the fact that the gate electrode overlaps the source and drain regions, as can be seen in the figure. The gate-to-channel capacitance splits equally between the source and drain when the device is in the linear region and appears mostly from the gate to the source in saturation, as explained in Chapter 2. The gate also overlaps the substrate on the sides of the transistor, which leads to a parasitic capacitance there as well.

There are also parasitic resistances in series with each terminal, due to the resistance of the conductor (metal or polysilicon), the contact to the silicon, and the bulk region leading to the active device.

Another result of the fabrication sequence is that the actual channel length of the device is shorter than the ***drawn channel length***. The drawn channel length is what the designer specifies when the circuit is laid out, but because the source and drain diffusions extend underneath the gate in the finished device, the active channel is shorter. The difference becomes a larger fraction of the drawn channel length as the drawn channel length becomes shorter. The difference must be taken into account by the designer.

3.2 BIPOLAR TECHNOLOGY

Bipolar transistors dominated electronic circuits until the 1970s. In the late 1970s, NMOS became the dominant process for large digital circuits. Starting in the 1980s and continuing on to the present, CMOS has been the dominant technology for digital circuits. In addition, there has been a tremendous incentive to use digital CMOS processes for as many analog functions as possible. The incentive derives mostly from two factors: First, the volume of digital CMOS circuits has brought down the cost and increased the availability of this process. Second, if the analog circuits can be fabricated in a digital CMOS process, then large mixed-signal chips can be built that implement the analog and digital functions together on one chip, thereby saving space on the printed circuit board, reducing manufacturing costs, and, in some cases, improving performance as well.

In spite of the dominance of CMOS technology, bipolar transistors remain an important element in electronic circuits, especially for analog functions. A quick perusal of the papers presented at the 2001 International Solid-state Circuits Conference, for example, shows that if microprocessors and memories are ignored, roughly 20% of the papers contain bipolar transistors [3.3].

3.2.1 Device Fabrication

Bipolar transistors can be fabricated using the same process technology described in Section 3.1. The individual transistors may be isolated from one another using reverse-biased *pn* junctions or with trenches filled with dielectric. They may be made in a dedicated process, or they may be part of a CMOS process. If the CMOS process has been modified to produce good-quality bipolar transistors, it is called a BiCMOS process.

Figure 3-31 shows the cross sections of typical junction-isolated bipolar transistors made using a traditional process (i.e., the one that has dominated bipolar technology since the 1960s). There are many modern variations on this basic process that will not be covered here. The *npn* transistor shown in part (a) of the figure is a ***vertical transistor***, so called because the carriers injected from the emitter flow vertically down to the collector region. The *pnp* transistor shown in part (b) of the figure is a ***lateral transistor***, so called because the carriers injected from the emitter flow laterally to the collector region. The main advantages of a vertical structure are that the active base region can be narrower than it is possible to define using lateral structures, which have dimensions limited by the resolution of the photolithographic process, and the active base region is in the bulk, rather than at the surface, which improves carrier lifetimes and mobilities.

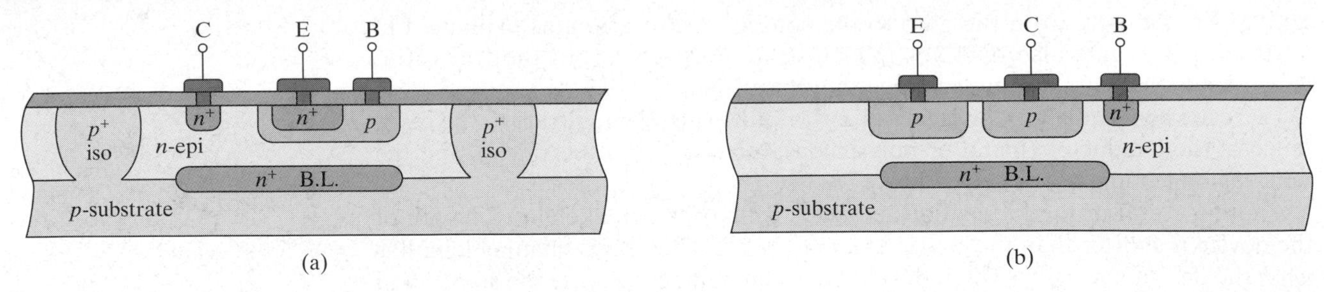

Figure 3–31 Junction-isolated bipolar transistors. (a) A vertical *npn* transistor. (b) A lateral *pnp* transistor.

The buried layer shown in this figure serves to lower the series collector resistance of the *npn* device and the series base resistance of the *pnp* device. Buried layers are almost always used in bipolar transistors, unless the devices are made in a well in a CMOS process, where a buried layer is not possible. (This is how most bipolar transistors in BiCMOS processes are made.) Many modern processes use dielectric isolation and allow the *pnp* transistors to be vertical devices as well.

3.2.2 Electrical Model Related to Physical Structure

Examining the device structures in Figure 3-31 reveals a number of important electrical features of the resulting devices. First, there is resistance in series with each terminal. The resistance is caused by the metal lines, the contact to the silicon, and the fact that the carriers must traverse a region of the silicon to reach the active area of the transistor. In addition, there are capacitances associated with each *pn* junction, including the substrate.

Many modern processes modify these structures to improve the devices by minimizing or eliminating some of the parasitic capacitances, making both *npn* and *pnp* devices vertical, reducing series resistance, and making the overall device dimensions smaller.

CHAPTER SUMMARY

- The main reason for the dominance of silicon technology is that the oxide which is easily grown is stable, is reproducible, yields a high-quality interface, is an effective dopant shield, and can be selectively etched.
- The patterns on an integrated circuit are formed by using a photolithographic process: A photoresist is deposited on the wafer, the resist is baked, the resist is exposed through a mask, the resist is developed, the desired process step is performed, and the resist is stripped.
- Dopants are introduced to the silicon by ion implantation.
- Threshold voltages are adjusted by separate implantation steps.
- Modern MOSFETs use lightly doped drain structures to minimize short-channel effects.
- The surface of the wafer needs to be "planarized" prior to depositing metal interconnects.
- Multiple levels of interconnect are used in modern processes.
- Bipolar transistors can be fabricated using a different sequence of the same kind of processing steps that are used to form CMOS circuits.

REFERENCES

[3.1] J.D. Plummer, M.D. Deal, and P.B. Griffin, *Silicon VLSI Technology: Fundamentals, Practice and Modeling*. Upper Saddle River, NJ, Prentice Hall, 2000.

[3.2] National Technology Roadmap for Semiconductors, SIA, 1997.

[3.3] *2001 Digest of Technical Papers, IEEE International Solid-State Circuits Conference*, IEEE, 2001.

PROBLEMS

Section 3.1.1

P3.1 Sketch a process flow that would result in the structure shown in Figure 3-32 by drawing a series of drawings similar to those in this chapter. You only need to describe the flow up through the stage at which active device formation starts since from that point on, the process is similar to that described in this chapter.

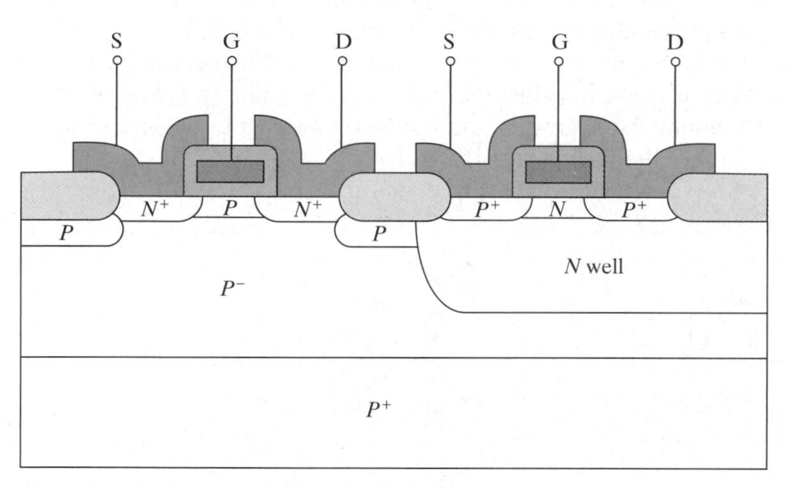

Figure 3–32 Technology typical of the 1980s.

P3.2 During the 1970s, the dominant logic technology was NMOS, as described briefly in Chapter 15. A cross-sectional view of this technology is shown in Figure 3-33. The depletion-mode device is identical to the enhancement-mode device except that a separate channel implant is done to create a negative threshold voltage. Design a plausible process flow to fabricate such a structure, following the ideas of the CMOS process flow in this chapter. You do not have to include any quantitative process parameters (times, temperatures, doses, etc.) Your answer should be given in terms of a series of sketches of the structure after each major process step, like the figures in the chapter. Briefly explain your reasoning for each step and the order in which you choose to do things.

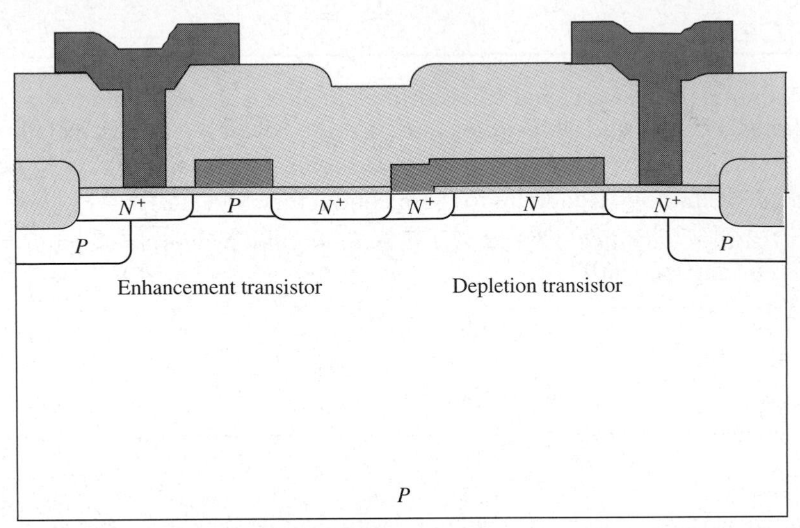

Figure 3–33 NMOS technology cross section.

P3.3 The cross section in Figure 3-31(a) shows a simple bipolar transistor fabricated as part of a silicon IC. Design a plausible process flow to fabricate such a structure, following the ideas of the CMOS process flow in this chapter. You do not have to include any quantitative process parameters (times, temperatures, doses, etc.) Your answer should be given in terms of a series of sketches of the structure after each major process step, like the figures in the chapter. Briefly explain your reasoning for each step and the order in which you choose to do things. Ignore the buried layer (n^+ B.L.).

CHAPTER 4

Computer-Aided Design: Tools and Techniques

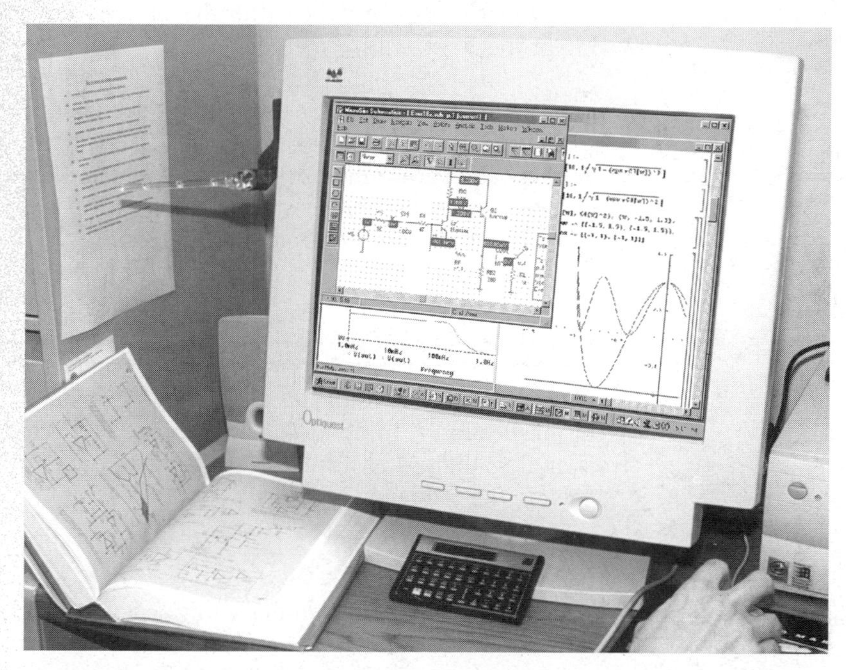

Modern circuit design relies heavily on computer-aided design tools. (Photo by Zoë Marlowe.)

INTRODUCTION

At this point in time (and for the foreseeable future!), computers do not design circuits or systems, people do. The fact that computers do not design circuits or systems explains why the tools designers use are generally referred to using the compound adjective computer-*aided*. Therefore, we refer to **computer-aided design** (CAD) tools or **computer-aided engineering** (CAE). Computers merely aid the process by performing tedious and complex calculations rapidly and providing useful graphical output. It is not necessary for the designer to know exactly *how* the tools work, although such knowledge is sometimes helpful. Nevertheless, the designer must completely understand *what* the tools are doing and must know in general what results are expected, otherwise, the simulation can produce incorrect results and the designer won't know it.

Before the advent of computer simulation, circuits could only be **breadboarded** in order to verify the design. A breadboard is a physical circuit constructed using one of a number of different techniques (e.g., soldering parts together with wires or prototype printed-circuit boards). The circuit can then be tested and modified as necessary until the designer is satisfied. At that point the design is released to production and many copies of the circuit are made. The first integrated circuits were designed this way, but there are significant problems in breadboarding integrated circuits. Unlike discrete circuits, ICs often rely on the very close matching of device properties and temperatures to work properly. It is not usually possible to build a breadboard that can mimic this matching to a sufficient degree. This problem, along with the increasing complexity of the circuits and the increasing capabilities of computers, drove the development of circuit simulation programs in the 1960s. Although breadboards are sometimes still used for discrete circuits, they have been replaced in most instances for integrated circuit design. Even for discrete circuits, it is much faster to simulate the circuit first and only build a breadboard (or prototype) after the simulations indicate that the design is working properly.

In the first section of this chapter we introduce the basic CAD tools most frequently used by transistor-level circuit designers and briefly mention some of the tools used for higher-level design as well. We discuss when it is appropriate to use each tool and how to incorporate the tool into the design process. In the rest of the chapter we focus on the simulation tool most often used by circuit designers; ***SPICE*** (Simulation Program with Integrated Circuit Emphasis).

4.1 OVERVIEW OF SIMULATION TECHNIQUES

CAD tools can be used at different levels in the design process. There is a distinct difference in the methodologies and tools used depending on whether the design is analog, digital, or a mixture of the two. We will discuss the methods used for each in the following sections.

4.1.1 Analog Systems

As an example, consider designing a complex communication channel as described in Section 1.3.2. The first simulation s may be done using mathematical representations of the system blocks. This level of simulation is often called ***block-diagram level simulation*** or ***system-level simulation***. This type of simulation can be performed using either custom-written code (e.g., in C++) or using general-purpose simulators or math packages like MATLAB®,[1] or by using system-level simulators written specifically for communication systems. System-level simulators often allow the user to connect blocks together in a graphic display to produce block diagrams. The individual blocks can then be described by time-domain or frequency-domain equations, by lookup tables, or by custom subroutines (MATLAB, for example, has the SIMULINK® environment that allows the blocks to link to the user's custom-written subroutines).

After the system-level simulations are carried out with ideal blocks in the diagram, the next step is often to see how the system will perform with realistic models for electronic implementations of the required functions. The more realistic models needed for this stage in the design process can again be implemented using equations, but the equations now include quantization errors, offsets, gain errors, phase errors, and other effects that are likely to be present in physical implementations. This level of simulation is frequently called ***behavioral simulation***. Behavioral simulation is often carried out using the same tools as block-diagram level simulation.

---

[1] MATLAB is a registered trademark of The Math Works, Inc.

When the behavioral simulations for a system are complete, the designers have a reasonable understanding of how the errors in the circuit implementations of each block will affect the performance of the overall system. Using this knowledge, the different blocks can be specified and designed. It is frequently necessary to go back and forth between behavioral simulation and lower-level design, since we cannot always know in advance whether or not we can meet the required specifications for a given block. There are usually tradeoffs that can be made at the system level to accommodate changes in the specifications of individual blocks in the systems, as discussed in Chapter 1.

The next level below behavioral simulation is circuit simulation and the most commonly used CAD tool is SPICE or one of its derivatives. SPICE was originally developed at the University of California at Berkeley in the early 1970s. Since that time SPICE has been continually upgraded to include more accurate device models, more robust and faster numerical techniques and more capabilities. Many versions of SPICE are now available that include some form of behavioral simulation capability, i.e., the user can define an element using an equation, so SPICE is increasingly being used for that kind of work as well.

Once the circuit has been simulated, it is laid out. *Layout* refers to the process of determining the physical locations and interconnections of the individual components that comprise a circuit. If the circuit is going to be constructed using discrete components, the layout is usually in the form of a *printed circuit board*, or PCB. A PCB is a thin board of some insulating material, usually some kind of epoxy and fiberglass mixture, with copper traces on it. The layout of the PCB is simply the physical description of the locations and dimensions of all components and connections. In the case of an integrated circuit, the layout consists of a set of masks; one for each of the different layers in the IC process (*see* Chapter 3).

In either case, there will be parasitic capacitances introduced by the lines connecting different elements and, in the case of an IC, there may be parasitic junction diodes and other structures as well. These parasitics can strongly influence the operation of a circuit, and it is common to go back and redo the simulations with estimates of these parasitics included. In fact, most IC layout programs have the ability to extract a SPICE netlist that includes these parasitic elements. This netlist can then be simulated to see the performance of the circuit as laid out. The process of adding parasitics from the layout back into the simulation program is called *back annotation*.

4.1.2 Digital Systems

There are four phases to the design of a digital system; *specification*, *functional simulation* (sometimes called *verification*), *synthesis*, and *timing simulation*.

The specification process of a digital system is analogous to the block-diagram level simulation of an analog system and may in fact use the same CAD tools. However, special-purpose *hardware-description languages* like Verilog or VHDL[2] are more frequently used. The system can be described at a behavioral level or at a *register-transfer level* (RTL). The RTL description is somewhat closer to the hardware and is the more popular of the two approaches.

Once the overall system is specified, the next step for digital blocks is functional simulation. Functional simulation is usually performed using *event-driven* simulators, and, circuits are described in a behavioral code or language at this level. Event-driven simulators take advantage of the fact that the voltages in most digital circuits only change rapidly at periodic intervals and are nearly constant the

[2] VHDL stands for VHSIC hardware description language. VHSIC stands for very high speed integrated circuit.

rest of the time. At this level of simulation, the designer is only interested in the overall input/output relationships and is not concerned with the detailed timing of the system.

The next step in the design process is called ***synthesis***.[3] If the initial model of the system was behavioral, this step is called ***behavioral synthesis***. If the initial model was described at the register-transfer level, then this step is called ***RTL synthesis***. For some functions, ***silicon compilers*** are used to produce the final design based on the behavioral description. More often, some intervention is required in the process of moving from a behavioral or RTL description to an actual circuit. This kind of digital design presupposes that the basic building blocks of digital circuits, the logic gates, shift registers and so on, are already designed.

These pre-designed ***cells*** exist in ***design libraries*** that include the behavioral specifications (e.g., delay time, setup, and hold times) of the individual cells. These libraries may be discrete ***small-scale integration*** (SSI) and ***large-scale integration*** (LSI) ICs; for example, CD4000 series CMOS logic gates. In this case, the next step is to layout a PCB using the netlist of gates generated by the synthesis step. In modern systems, it is common for the libraries to be cells designed in a specific IC process. These cells may be part of a ***gate array***, or they may be cells used to produce a custom ***application-specific integrated circuit*** (ASIC).

The final step in the digital design process is timing verification. In this step the netlist of gates generated at the synthesis step is re-simulated using the known performance of the individual cells and the expected parasitics owing to the layout of either a PCB, gate array, or custom chip. This level of simulation allows the designer to verify that the circuit will work with realistic delays and parasitics accounted for.

In this text, we are interested in transistor-level circuit design. In the case of the design process just described, transistor-level design is used to generate the library of digital cells. SPICE is the program most frequently used for transistor-level simulation of digital circuits. Although SPICE is often not fast enough to allow the simulation of large digital systems at the transistor level using the full nonlinear device models, it is often not necessary to do so. Instead, SPICE is used to simulate smaller functional blocks (the cells), and the other methods just described are used for the larger system.

4.1.3 Mixed Analog and Digital Systems

One area of simulation that has received a tremendous amount of attention is ***mixed-mode simulation***. A mixed-mode simulation is one that includes analog and digital circuits together. If the digital circuits are very small, an analog simulator could be used, and the digital circuits could all be described at the transistor level. If a substantial amount of digital logic is included, however, using full nonlinear transistor-level models in a SPICE-type simulator may not be a reasonable option. The problem results from the fact that SPICE must use a small time step to simulate the analog circuits, but because of the large number of transistors contained in the digital circuits, the simulation becomes very slow. Since the voltages in the digital circuits typically change only once in a while, it is not a good use of computing resources to perform all of the nonlinear computations with a very small time step for the digital portion of the circuit. This problem has been ad-

[3] This use of the word synthesis is distinct from our use in Chapter 1. In the present context, the term does not imply that something entirely new is created without precedent. The circuit that is created in this step has already been described at the behavioral or RTL level.

dressed in different ways. For example, a transistor-level description of the analog circuits can be simulated using embedded behavioral models for the digital sections. This kind of capability is built into many modern versions of SPICE.

4.2 CIRCUIT SIMULATION USING SPICE

There are advantages and disadvantages to using circuit simulation instead of a breadboard. The major differences are:

Whereas a breadboard uses real devices operating in real time, a simulation necessarily uses models of these devices and operates in discrete time steps. If inadequate models are used or the discrete-time nature of the simulation is ignored, meaningless results may be obtained.

Computer simulations can help us pinpoint why a particular design does not work like we think it should. Since we always use simplified models during hand analysis, we sometimes miss important features of device operation. If the simulations don't agree with our hand analysis, we have the ability to simplify the models the simulator uses. When the simulator uses the same models we did in our hand analysis, we should agree (unless, of course, we made a mistake). We can then add different elements back into the model until we find what it is that is causing the disagreement. This method will sometimes help us to see how to fix the problem as well.

Computer simulation allows us to monitor some voltages and currents that would be either extremely difficult or impossible to monitor on a breadboard.

It is easier to control parasitic components in a simulation so that they reflect those that will be present in the final circuit. This point is especially important for ICs.

Computer simulation makes it much easier to vary temperature and component values. In fact, many versions of SPICE allow us to specify tolerances and do ***Monte Carlo*** simulations[4] to see what the resulting spread in performance will be.

The remaining parts of this section briefly discuss the SPICE input format and the types of analyses that can be performed. The required syntax for each statement is provided in Appendix A to make it easier to use as a quick reference. This chapter will focus on understanding what the commands do and how to properly use the program. Examples of the input syntax will be presented only as needed and will not be exhaustive.

4.2.1 SPICE Input and Output

For SPICE to simulate a circuit, we must first describe our circuit in a way that the program can understand. The input file used by SPICE contains three types of statements; a description of the circuit called the ***netlist***, ***simulation commands***, and ***program control statements***. Circuits are often entered using graphic ***schematic capture*** programs. These programs allow the designer to place schematic symbols, connect them together and define attributes for the different

[4] A Monte Carlo simulation varies the parameters in a random fashion and reruns the simulation many times. The results can then be viewed to see the average performance and the expected deviations in performance.

components (e.g., resistor value, transistor model name). The schematic capture program outputs a netlist and, perhaps, the simulation commands and program control statements as well.[5] The input file can also be entered manually using a text editor. Figure 4-1 shows a low-pass filter and its corresponding SPICE input file.[6] We will explain the different elements in this file as they come up in the following discussion.

The first line in the file must always be a title. In our example, the title is **`LOW-PASS FILTER`**.[7] This line is repeated at the top of every page of output and is not part of the circuit description, program control, or the simulation commands. Lines that begin with an asterisk (or a semicolon) are comments and are ignored by the program. Alternatively, a semicolon can be placed after other information on a line, and material to the right of the semicolon will be parsed as a comment. It is a good idea to organize your input files in some reasonable fashion as shown here. As with any kind of programming, many simple mistakes can be avoided by adopting a standard organized approach. From here on out we will ignore the first line of the file and all comment lines in our discussion. For every other line in the file the first character on the line indicates what kind of statement it is. If the first character is a period, the line is providing program control or a simulation command. If it isn't a period, the line is describing a circuit element.

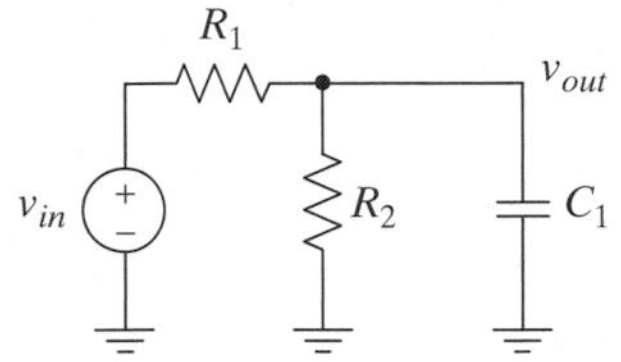

All circuit elements have their name in the first field, and the first character of the name indicates what kind of component is being described (e.g., **`v`** indicates a voltage source. The rest of the name is arbitrary, with certain restrictions depending on the version of SPICE. The following fields indicate what nodes the element is connected to, the element value or the name of the model to be used and other optional information as described in Appendix A. In Figure 4-1 for example, the line "**`r1 in out 1k`**" indicates that the resistor R_1 is connected between the nodes labeled "in" and "out," and it has a value of 1 kΩ.

Not all versions of SPICE allow nodes to be named (numbers must be used), and some versions do not allow ground to be named (it is always assigned the number 0). All circuits must have a ground node.

SPICE output can be obtained in several different ways. All versions of SPICE support the old **`.PRINT`** and **`.PLOT`** statements that produce line printer output. The **`.PRINT`** statement provides a tabular printout of the values requested (e.g., node voltages), whereas the **`.PLOT`** statement produces a line-printer plot of the requested data. In addition to these statements, most versions of SPICE have some kind of a *postprocessor* that provides a graphic interface for viewing the results of a simulation. These postprocessors are different for each version of SPICE, and the manual should be consulted for the details on how to use them. To use a postprocessor, the program needs to be told to save the data obtained during the simulation. There is a special command set aside to accomplish this task; for example, in PSPICE,[8] the **`.PROBE`** command is used as shown in Figure 4-1, whereas in HSPICE[9] the '**`.OPTIONS post`**' command is used. Usually, the pro-

R_1 v_{out} v_{in} R_2 C_1

```
LOW-PASS  FILTER
*
* Simulation  Control
*
.probe
.ac  dec  10  .1  1k
*
* Sources
*
vin  in  0  ac  1
*
* Passive  components
*
r1   in   out   1k
r2   out  0     1k
c1   out  0     1u
.end
```

Figure 4–1 (a) A low-pass filter and (b) its SPICE input file.

[5] The word netlist is sometimes used to refer to the entire input file, including simulation commands and program control statements.

[6] The input file is frequently referred to as an input *deck*. This terminology is a holdover from the days of punched cards, when the input file would truly have been a deck of cards.

[7] Bold Courier font will be used to represent all SPICE statements and model parameter names.

[8] PSPICE is a version of SPICE that runs on PCs and is sold by Cadence. The post processor is called PROBE.

[9] HSPICE is a version of SPICE that is sold by Avanti. The post processor most often used with HSPICE is called AWAVES.

gram will store all node voltages for every simulation point, but there are other variables that can be requested individually. If the program kept track of every terminal current, charge, and so on for every simulation step, the output would be too large to be practical for many circuits.

4.2.2 Simulation Modes and Types of Analysis

SPICE has three different ***simulation modes***: DC, AC, and transient. Within each of these modes of simulation there are different ***analysis types*** available.

In the DC simulation mode, SPICE first replaces all capacitors with open circuits and all inductors with short circuits and iteratively solves for the operating point of the circuit. The AC simulation mode uses the operating point found in the DC simulation and replaces all nonlinear elements by linearized models valid for small deviations from the operating point (the concept of small-signal analysis is covered in Chapter 6). If a DC simulation was not requested, the program will automatically run one to obtain the operating point. The program then solves the resulting linear circuit. The transient simulation mode computes the node voltages in the circuit as a function of time using the full nonlinear models.

These three simulation modes are described in more detail in the following sections. In Figure 4-1, the only simulation requested is an AC analysis. The input line "`.ac dec 10 .1 1k`" tells SPICE to perform an AC simulation while sweeping the frequency of the input. The frequency sweep is to be logarithmic, and the program is to store the results (for printing or plotting) ten times for each decade of frequency. The frequency is to change from 0.1 Hz to 1 kHz.

DC Simulation Mode To perform a DC simulation on a circuit, SPICE starts with a guess at an operating point, finds linear models for all nonlinear devices (valid at the assumed operating point), sets up modified nodal equations for the linear circuit and solves them.[10] The result is then compared with the original guess to see if it was correct. If it wasn't, a new guess is generated, and the process is repeated.

There are times when this iterative process will not converge to a solution; in these cases, SPICE issues an error statement. Options exist to allow the user to modify this process to help the program converge [4.1,4.2]. The simplest techniques are to increase the number of iterations allowed (using the `.OPTIONS` statement), increase the error tolerances (using the `.OPTIONS` statement), or supply SPICE with better initial guesses for the node voltages (using the `.NODESET` statement).

There are four different types of analysis that SPICE can perform within the DC simulation mode; an operating point analysis (`.OP`), a small-signal DC transfer function (`.TF`), a DC sweep (`.DC`), and a sensitivity analysis (`.SENS`). The DC operating-point analysis finds and prints the operating point. The transfer function analysis finds the DC transfer function from a specified source to a specified load. Be warned that this analysis is not often useful for discrete circuits, since bypass and coupling capacitors will be represented as open circuits. The DC sweep is used to find the operating point of the circuit, as the values of one or more DC sources are swept over a specified range. The sensitivity analysis finds the normalized sensitivity of the operating point to component values and model parameters.

---

[10] The nodal equations are modified to allow for voltage sources and inductors (which have been shorted for the DC analysis). *See* [4.1] for an introductory treatment of the procedure.

Example 4.1

Problem:
Create a SPICE input file to run a DC sweep on the bridge circuit shown in Figure 4-2. Sweep the source from 4.9 V to 5.1 V in 0.001 V steps. Examine the voltage across R_1 and the current through R_5.

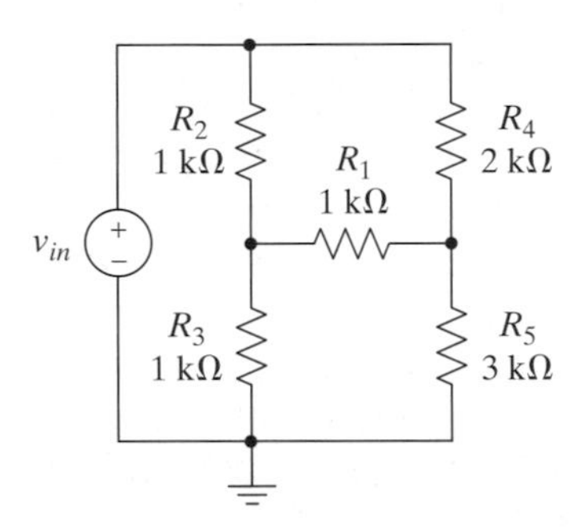

Figure 4–2 A bridge circuit.

Solution:
The SPICE input file necessary for running the requested simulation on PSPICE and viewing the results on the PROBE postprocessor is shown in Figure 4-3. Note that a zero-value DC voltage source has been inserted in series with R_5 to monitor the current through it. Although several versions of SPICE have ways of allowing you to access the current into a given terminal of a device, these methods vary from version to version and do not always work with the postprocessors; therefore, the technique shown here is often useful.

```
Bridge  Circuit
.probe  ;  create  a  probe  data  file
.dc  vin  4.9  5.1  .001
vin  1  0  dc  5
v_ir5  3  4  dc  0
r1  2  3  1k
r2  1  2  1k
r3  2  0  1k
r4  1  3  2k
r5  4  0  3k
.end
```

Figure 4–3 SPICE input listing.

A plot of the voltage across R_1 and the current through R_5 was obtained using PROBE; the results are shown in Figure 4-4. Note that two vertical axes have been used, since the values are so different.

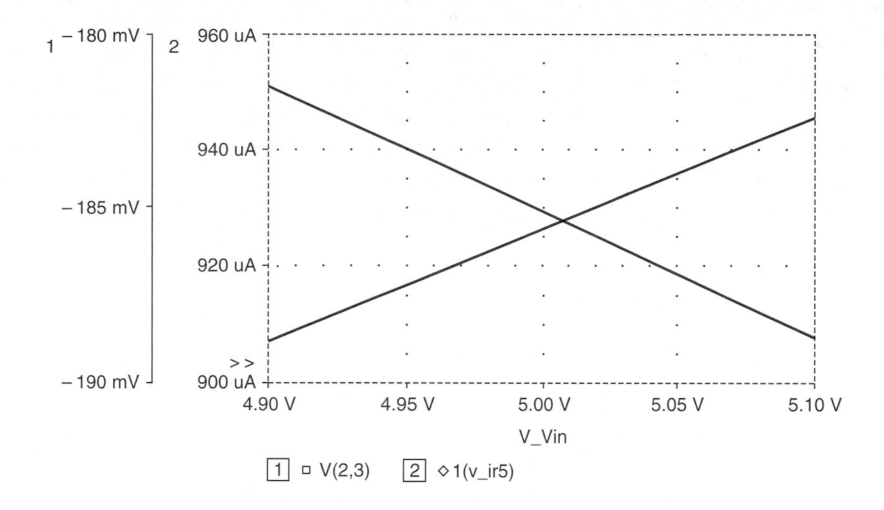

Figure 4–4 A plot of the voltage across R_1 and the current through R_5 versus V_I.

Exercise 4.1

Modify the SPICE file in Example 4.1 to find the sensitivity of the voltage across R_I to changes in the resistor values and input DC value when $V_I = 5$ V.

AC Simulation Mode In the AC simulation mode, SPICE represents the linear elements in the circuit as complex impedances and uses linear models evaluated at the DC operating point for all nonlinear devices. The DC values of all sources are set to zero during AC analysis. The program then uses the modified nodal equation method to calculate the frequency response of the resulting linear circuit. The input sources are all assumed to be sinusoidal for the AC analysis. Usually only one input source is included; otherwise, the results can be confusing. The only common exception to this rule is that two input sources are often used to simulate a differential input.

Depending of the version of SPICE used, there are several types of analysis available within the AC simulation mode. All versions of SPICE support the small-signal AC frequency response analysis (**`.AC`**) and the noise analysis (**`.NOISE`**). Some versions of SPICE also support the distortion analysis (**`.DISTO`**) and/or the pole–zero analysis (**`.PZ`**).

For example, the AC frequency response is used to generate plots of the magnitude and phase of the transfer function for a circuit. The magnitude of the input source is usually set to one, as shown in Figure 4-1(b). Since a linear analysis will be performed, it doesn't matter whether the input value is one volt, one million volts, or one millivolt. All that matters is the value of the other voltages in comparison to it, so one is a convenient value. The currents in the circuit depend on the magnitude of this voltage, but it is again only the relative magnitudes that are significant.

The noise analysis is used, as the name makes clear, to estimate the noise present in the circuit.[11] The distortion analysis is also self explanatory (the distortion analysis performed in the AC analysis mode is an approximation that is only valid for small distortion; therefore, many designers prefer to use the Fourier analysis included as a part of the transient simulation mode to examine distortion). The pole-zero analysis prints out the real and imaginary parts of all of the poles and zeros present for a specified input and output.

Example 4.2

Problem:
Create a SPICE input file to examine the magnitude of the transfer function of the low-pass filter shown in Figure 4-5. Sweep the frequency from 10 Hz to 1 MHz.

Figure 4–5 A low-pass filter.

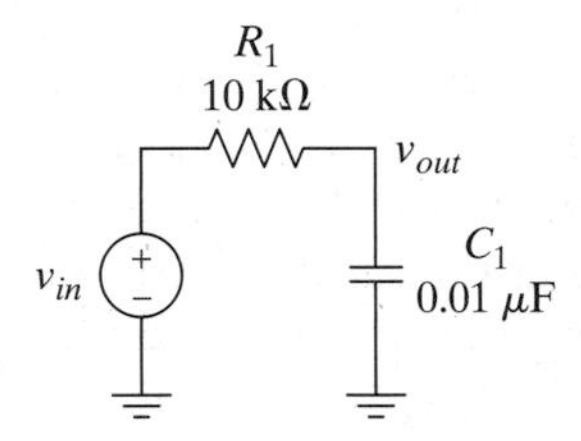

Solution:
The SPICE file for running the simulation requested is shown in Figure 4-6.

Figure 4–6 The SPICE input listing.

```
Low-pass  filter
.probe  ;  create  a  probe  data  file
.ac  dec  10  10  1meg
vin  in  0  ac  1
r1  in  out  10k
c1  out  0  .01u
.end
```

The PROBE output showing the magnitude of the transfer function is shown in Figure 4-7. Note that since v_i is equal to one, the magnitude of v_o is equal to the magnitude of the transfer function v_o/v_i.

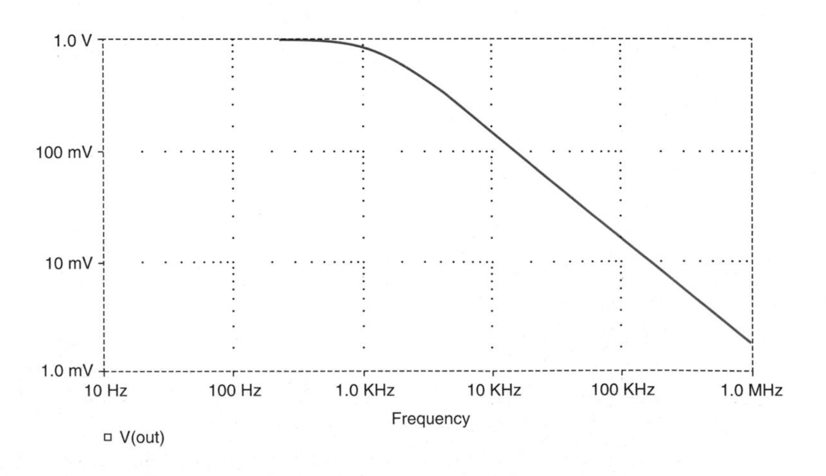

Figure 4–7 The magnitude of the transfer function for the low-pass filter.

[11] Only noise generated by devices in the circuit is included. SPICE cannot predict noise coupled into the circuit from outside sources, which is often very significant, unless you include models for the coupling mechanism and the source of the noise.

Exercise 4.2

Modify the SPICE file in Example 4.2 so that you plot the transfer function of the filter for three different values of C_1: 0.001 μF, 0.01 μF, and 0.1 μF.

Transient Simulation Mode In the transient simulation mode, SPICE solves the nonlinear equations describing the circuit using an iterative procedure similar to what is used in the DC simulation mode. The differences are that the inductors and capacitors are represented by their respective differential equations in the transient mode, and voltage and current sources can have a transient component as well (e.g., a sine wave, pulse, or other shape). The program then finds a new solution at each time step while keeping track of the charge accumulating on capacitors and the fields building up in coils. The time step will automatically be reduced if the voltages change too much from one time step to the next. However, there is a limit to how small the time step is allowed to be. There are again techniques that can be used if the program fails to converge during a transient simulation [4.1,4.2]. Transient simulations are the most time-consuming of all SPICE simulations and should be used with care, especially for large circuits.

There are two analysis types supported in the transient simulation mode: the transient analysis (`.TRAN`) and the Fourier analysis (`.FOUR`). The transient analysis calculates the voltages and currents in the circuit as a function of time. The Fourier analysis mode computes the magnitudes of the fundamental and harmonic components of a given frequency. (Warning: the results of the Fourier analysis are very sensitive to the fundamental frequency you specify—you must be accurate.)

Example 4.3

Problem:

Create a SPICE input file and run it to find the output voltage of the half-wave rectifier circuit in Figure 4-8 when the input is a 20-V peak 60-Hz sine wave.

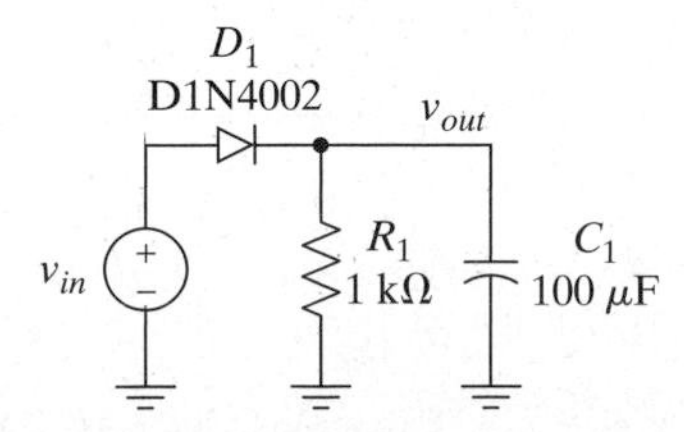

Figure 4–8 A half-wave rectifier circuit with filter capacitor.

Solution:
The SPICE file for running the simulation requested is shown in Figure 4-9 (the diode model is from the model library included with the student version of PSPICE).

```
Half-wave  rectifier  with  filter
.probe  ;  create  a  probe  data  file
.tran  .1m  30m
vin  in  0  sin(0  20  60)
d1  in  out  D1N4002
r1  out  0  1k
c1  out  0  100u
*  the  `+'  in  the  first  column,  as  in
*  some  of  the  lines  in  the following
*  model  indicates  to  SPICE  the  lines
*  are  a  continuation  of  the  previous
*  line.
.model  D1N4002  D(is=14.11e-9,  n=1.984,
+rs=33.89e-3,  ikf=94.81,  xti=3,
+eg=1.110,  cjo=51.17e-12,  m=.2762,
+vj=.3905,  fc=.5,  isr=100.0e-12,  nr=2,
+bv=100.1,  ibv=10,  tt=4.761e-6)
.end
```

Figure 4–9 The SPICE input listing.

The PROBE output showing both the input and output voltages is shown in Figure 4-10.

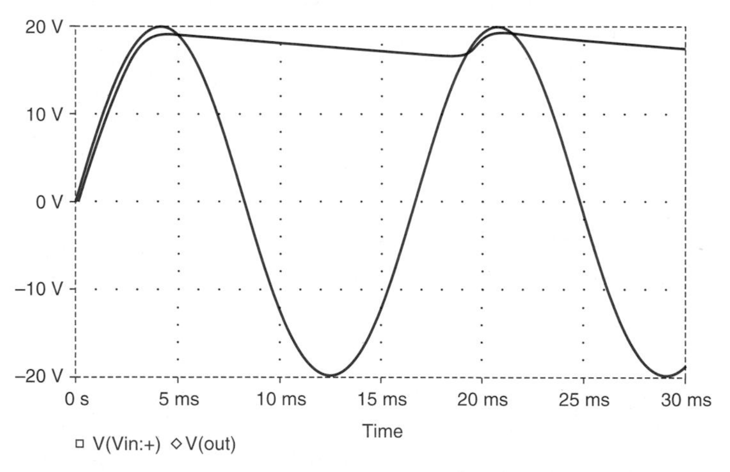

Figure 4–10 v_I and v_O for the half-wave rectifier circuit with filter.

■

Exercise 4.3

Re-simulate the half-wave rectifier from Example 4.3 while sweeping the filter capacitor value from 10 μF to 1 mF using two values per decade. Observe the change in the magnitude of the ripple on v_O.

□

4.3 CIRCUIT ELEMENTS AND MODELS FOR SPICE

Some elements, like resistors and capacitors, are typically specified entirely on one line in the SPICE input file, as is true for the example in Figure 4-1. Other elements, like transistors, usually require separate `.model` statements that supply SPICE with the different parameters needed to accurately model the component. Resistor and capacitor models can also be complicated enough to warrant separate model statements, and transistors may occasionally be simple enough to specify on one line, but these are unusual circumstances. When a separate model statement is used, the model is given a name and the input line for the element specifies the name of the model. For example, if a diode has its anode connected to node 1, its cathode connected to node two, and its model is called **`dsimple`**, the input line for the diode would be "**`d1 1 2 dsimple`**."

There are two fundamentally different ways of approaching the problem of modeling elements for simulation. ***Physically based*** models start from equations derived by considering the physics underlying the operation of a given device. In these models, the parameters that need to be specified correspond to physically meaningful quantities (e.g., doping density, oxide thickness) and can, at least in principle, be predicted for a given device based on knowledge of the fabrication details. ***Empirical*** models, on the other hand, use curve fitting to try and model the external *i–v* characteristics of a given device. In this case the model parameters are only curve-fitting parameters and do not typically have any underlying physical significance.

Physically based models are easier for a designer to correlate with the models used in hand analysis. They also make it easier to predict the effect a given change in the fabrication of a device will have on its performance. In addition, a designer can simplify the models by leaving out some second-order effects when needed to speed up simulation or to confirm a hand analysis. The main disadvantage of physically based models is that they can rapidly become very complex mathematically and therefore tend to make the simulations run slower. In addition, when the underlying physics of a device are not fully understood or are mathematically intractable, the model may not be as accurate as an empirical model. Conversely, the main advantage of empirical models is that using them leads to faster and sometimes more accurate simulations. SPICE models are typically combinations of physically based and empirical models.

4.3.1 Sources

SPICE supports independent voltage and current sources as well as all four types of controlled source. Independent voltage and current source names must begin with a **`V`** or an **`I`**, respectively. Voltage-controlled voltage source names begin with **`E`**, current-controlled voltage source names begin with **`H`**, voltage-controlled current source names begin with **`G`**, and current-controlled current source names begin with **`F`**. Each source can have a DC value, an AC value, a transient value, or any combination of the three. However, note that when a particular simulation mode is used (DC, AC, or transient), *all* of the sources in the netlist are included in the analysis. Those that have no value specified for that mode are set equal to zero, but all others are included. Therefore, except for representing differential signals or other special cases, only one source at a time should be given an AC or transient value. If two or more sources have an AC value, for example, it will not be possible to determine their separate contributions to some output.

The DC and AC values assigned to a source are numbers representing the magnitude of the voltage or current. For the transient mode, however, there are a number of different options and not all versions of SPICE offer the same functions. For example, the source can be a sinusoidal function of time, a pulse, or a frequency-modulated sine wave. *See* Appendix A or a SPICE manual for the details. The source listed in Figure 4-1 only has an AC value. SPICE will set it equal to zero for DC and transient simulations.

4.3.2 Passive Devices

All versions of SPICE include resistors, capacitors, inductors, transformers (i.e., coupled inductors), and ideal transmission lines. The names of these elements begin with **R**, **C**, **L**, **K**, and **T** respectively. Some versions also include lossy transmission lines, voltage- and current-controlled switches and uniformly distributed RC lines.

Resistors, capacitors, and inductors are most often only given a value and an optional temperature coefficient.[12] In some versions of SPICE, it is possible to include more detailed models (e.g., to model the parasitics associated with resistors on an integrated circuit). We will not discuss these models in this text and will assume that these elements are always described by a single value.

4.3.3 Diodes

Diodes, like all semiconductor devices, differ from resistors, capacitors, and inductors in that the model used is typically too complicated to specify on the same line as the element, although it can be, if desired. A separate **.model** statement is used instead and the element line specifies the name of the model to be used. All diode names in SPICE must begin with the letter **D**.

The diode model in SPICE is almost entirely physically based, although the parameters can be obtained by curve fitting instead. The model is based on the equation derived in Section 2.4.2,

$$I_D = I_S\left(e^{qV_D/nkT} - 1\right), \qquad (4.1)$$

where I_S is the saturation current, q is the charge on an electron, k is Boltzmann's constant, T is the temperature in Kelvin and n is the ideality factor. Only two model parameters are needed to specify the model given by (4.1): **IS** and **N**. In addition to these parameters, SPICE also models the reverse breakdown of the junction by specifying the voltage (**-BV**) and current (**-IBV**) at the knee of the characteristic as shown in Figure 4-11.

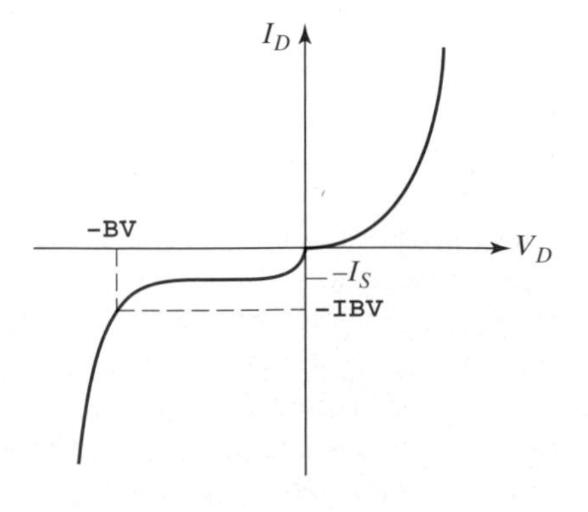

Figure 4–11 The diode characteristic showing the definition of the reverse breakdown voltage and current at the knee.

[12] The temperature variation of a resistor may be modeled by a second-order polynomial if desired. In this case, two coefficients are specified. When only one coefficient is given, a linear variation is assumed.

Two final model parameters complete the SPICE DC model for diodes: **RS** and **EG**. **RS** is the resistance in series with the diode and **EG** is the bandgap of the semiconductor used. The bandgap is used in the equation to allow for different semiconductors (e.g., silicon or germanium) and to model Schottky diodes (Schottky diodes can also be modeled by varying the value of I_S, as discussed in Chapter 2). The default value of **EG** is for silicon *pn*-junction diodes.

For AC analyses, the diode is modeled in forward bias by its small-signal conductance, found by differentiating (4.1) with respect to V_D:

$$g_d = \frac{dI_D}{dV_D} \approx \frac{qI_D}{nkT} = \frac{I_D}{nV_T}, \tag{4.2}$$

where V_T is the thermal voltage defined in Chapter 2. This conductance is calculated using the DC bias point and is then used as a linear element for AC analyses. Unless the default is overridden, SPICE performs all analyses assuming a temperature of 27°C (300 K). This temperature leads to a thermal voltage of $V_T = 25.86$ mV.

Charge storage effects in diodes are also modeled in SPICE. As discussed in Section 2.4.5, there are two charge storage mechanisms to consider: the depletion capacitance and the diffusion capacitance. Both of these capacitances are modeled in SPICE as capacitors in parallel with the diode and both are functions of the bias point. In the AC analysis mode, these capacitors are constant and are determined by the DC bias point, whereas in the transient analysis mode, the charge storage is handled using the nonlinear equations. Both capacitances are present in forward bias, but only the junction capacitance is present in reverse bias, as shown in Figure 4-12.

The depletion capacitance was given in Section 2.4.5 and is repeated here for convenience:

$$C_j = \begin{cases} \dfrac{C_{j0}}{(1 - V_D/V_0)^{m_j}}; & \text{in reverse bias} \\ 2C_{j0} \quad ; & \text{in forward bias} \end{cases} \tag{4.3}$$

where C_{j0} is the zero bias junction capacitance, V_0 is the built-in potential of the junction and m_j is the grading coefficient. The SPICE parameter names for these terms are **CJO**,[13] **VJ**, and **M** respectively. The value given for the capacitance in forward bias is an approximation used for hand analysis. SPICE handles the forward bias situation differently, but we will not concern ourselves with the details here, since the net difference is not very significant.

The second charge storage mechanism, the diffusion capacitance, is modeled by

$$C_d = \tau \frac{I_D}{nV_T} = g_d\tau, \tag{4.4}$$

where τ is the average minority-carrier lifetime (sometimes called the transit time); SPICE denotes τ by **TT**. The parameter **N** was introduced in (4.1).

Finally, we note that SPICE allows the relative junction area of a diode to be specified on the element definition line, and this area will scale some of the model parameters. The parameters **IS**, **CJO**, and **IBV** scale proportionally with **AREA**, and the parameter **RS** scales inversely with **AREA**. The diode model parameters are summarized in Table 4.1. An example of a simple diode circuit and the use of the model was given in Example 4.3.

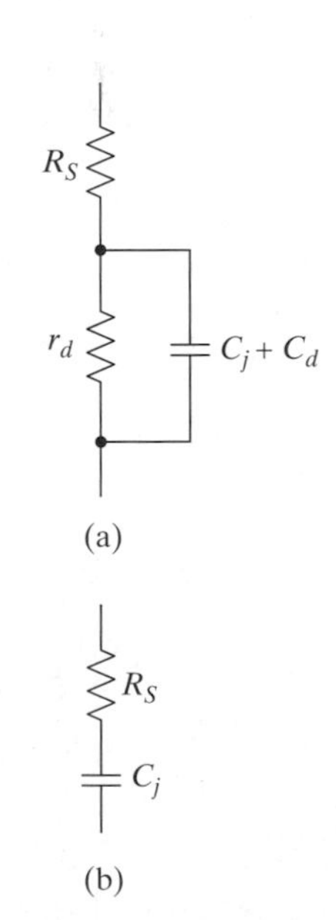

Figure 4–12 The SPICE small-signal AC diode models, (a) in forward bias and (b) in reverse bias.

[13] This is the letter 'O,' not the number 0.

Table 4.1 SPICE Diode Model Parameters[1]

Name	Parameter	Units	Default	Example
BV	Breakdown voltage	V	infinite	60
CJO[2]	Zero-bias junction capacitance	F	0	1E-12
EG	Bandgap	eV	1.11	1.11 Si 0.69 Schottky 0.67 Ge
IBV	Current at BV	A	10^{-10} (10^{-3})	1E-4
IS	Saturation current	A	10^{-14}	5E-15
M	Grading coefficient	none	0.5	0.33
N	Ideality factor	none	1	1.5
RS	Series resistance	Ω	0	5
TT	Transit time	s	0	1E-9
VJ	Built-in potential	V	1 (.8)	0.65

[1] The defaults are for PSPICE and HSPICE; when they are different, the HSPICE values are given in parentheses.

[2] The last character in the parameter name is the letter 'O', not the number zero.

4.3.4 Bipolar Junction Transistors

All bipolar junction transistor (BJT) names in SPICE begin with Q. There are two types of model available, **NPN** and **PNP**. The same parameters are used for each type of transistor. We will describe a discrete transistor model in this section. SPICE also allows the transistor to be entered as a four-terminal device, where the fourth terminal describes the substrate node that is present in integrated transistors; the model then includes the parameters necessary to describe the junction diode associated with the substrate.

The SPICE BJT model is complex, and we will only provide an introductory treatment here. It is predominantly a physically based model, although the way in which some characteristics are modeled is empirical. More detailed treatments can be found in SPICE manuals, the literature, and in [4.3]. An old but still very useful reference is [4.4]. The following discussion covers the model used in SPICE 2 and its derivatives.

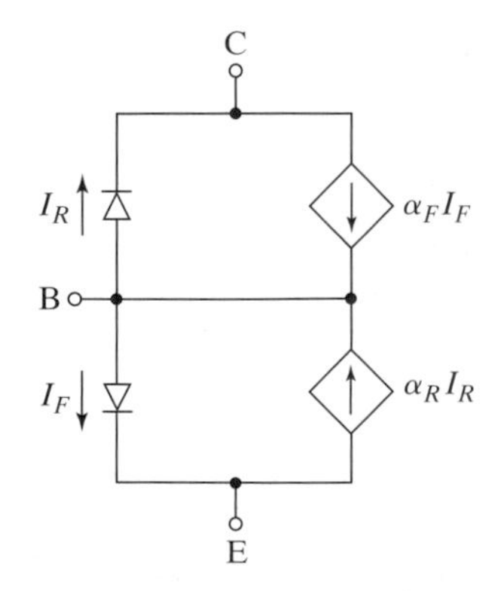

Figure 4–13 The Ebers–Moll model of a BJT.

DC Model for the BJT The model used by SPICE for a BJT depends on which model parameters are specified. The simplest model used for DC analysis is a modified form of the Ebers–Moll model briefly mentioned in Section 2.5.2. This model is based on representing the transistor as a superposition of a forward transistor and a reverse transistor. The model parameters describing the reverse transistor are only important when it is active, which usually only occurs when the transistor is saturated. Therefore, for much of analog design the reverse parameters described below can be allowed to take on their default values. As long as the transistors are not saturated, the results will be close to as accurate as the forward model will allow.

With the forward and reverse transistors each modeled by a *pn*-junction diode and a controlled source, the Ebers–Moll model takes the form shown in Figure 4-13, wherein the forward and reverse common-base current gains are denoted by α_F and α_R respectively. The diode currents are given by

$$I_F = I_{SF}\,(e^{V_{BE}/n_F V_T} - 1) \tag{4.5}$$

and

$$I_R = I_{SR}\,(e^{V_{BC}/n_R V_T} - 1), \tag{4.6}$$

where I_{SF} and I_{SR} are the forward and reverse saturation currents and n_F and n_R are the forward and reverse ***emission coefficients*** respectively (the emission coefficients are like the ideality factor for a diode; for most BJTs $n_F \approx 1$ and n_R is closer to two).

Using (4.5) and (4.6) and the figure, we can write the equations for the three terminal currents of the transistor (all defined as positive into the device);

$$I_C = \alpha_F I_{SF}\,(e^{V_{BE}/n_F V_T} - 1) - I_{SR}\,(e^{V_{BC}/n_R V_T} - 1), \tag{4.7}$$

$$I_E = \alpha_R I_{SR}\,(e^{V_{BC}/n_R V_T} - 1) - I_{SF}\,(e^{V_{BE}/n_F V_T} - 1), \tag{4.8}$$

and

$$I_B = (1 - \alpha_F) I_{SF}\,(e^{V_{BE}/n_F V_T} - 1) + (1 - \alpha_R) I_{SR}\,(e^{V_{BC}/n_R V_T} - 1). \tag{4.9}$$

Finally, we note that we will need to make use of the ***reciprocity*** condition [4.5]; namely, that

$$\alpha_F I_{SF} = \alpha_R I_{SR} = I_S \tag{4.10}$$

where I_S is the linking saturation current used in Chapter 2.

Now that we have shown the basic Ebers–Moll model, we are prepared to show the form used in SPICE. The motivation is to have a form for the model that is a nonlinear version of the standard hybrid-π small-signal model used for hand analysis (i.e., we want to have a single controlled source from collector to emitter). The resulting model, which is shown in Figure 4-14, is a modified version of the Ebers–Moll model that is sometimes referred to as a ***nonlinear hybrid-π model***.

The controlled source in this π model is given by (*see* Problem P4.12)

$$\begin{aligned} I_{C\pi} &= \alpha_F I_F - \alpha_R I_R \\ &= \alpha_F I_{SF}\,(e^{V_{BE}/n_F V_T} - 1) - \alpha_R I_{SR}\,(e^{V_{BC}/n_R V_T} - 1), \end{aligned} \tag{4.11}$$

and applying reciprocity leads to

$$\begin{aligned} I_{C\pi} &= I_S[(e^{V_{BE}/n_F V_T} - 1) - (e^{V_{BC}/n_R V_T} - 1)] \\ &= I_S\,(e^{V_{BE}/n_F V_T} - e^{V_{BC}/n_R V_T}), \end{aligned} \tag{4.12}$$

The diode currents must now be modified as well to keep the terminal currents the same as we had before. Since the diode currents now represent the base current of the transistor, they are given by

$$I_{F\pi} = \frac{\alpha_F I_F}{\beta_F} = \frac{I_S}{\beta_F}\,(e^{V_{BE}/n_F V_T} - 1) \tag{4.13}$$

and

$$I_{R\pi} = \frac{\alpha_R I_R}{\beta_R} = \frac{I_S}{\beta_R}\,(e^{V_{BC}/n_R V_T} - 1). \tag{4.14}$$

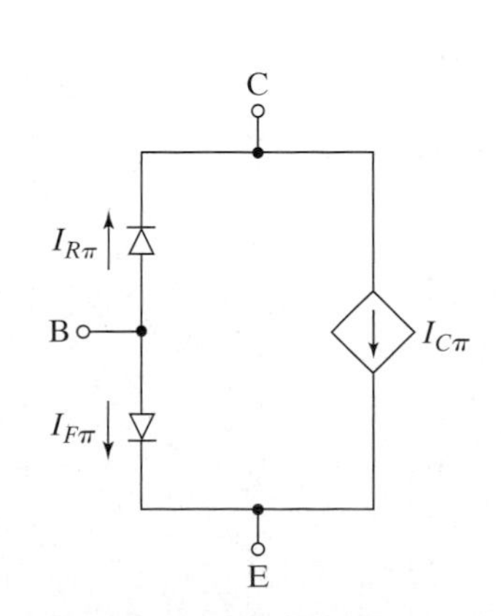

Figure 4-14 The modified Ebers–Moll model used by SPICE.

We can now express the terminal currents of the transistor in terms of the nonlinear hybrid-π model:

$$I_C = I_S\,(e^{V_{BE}/n_F V_T} - e^{V_{BC}/n_R V_T}) - \frac{I_S}{\beta_R}\,(e^{V_{BC}/n_R V_T} - 1), \tag{4.15}$$

$$I_E = I_S\,(e^{V_{BC}/n_R V_T} - e^{V_{BE}/n_F V_T}) - \frac{I_S}{\beta_F}\,(e^{V_{BE}/n_F V_T} - 1), \tag{4.16}$$

and

$$I_B = \frac{I_S}{\beta_F}\left(e^{V_{BE}/n_F V_T} - 1\right) + \frac{I_S}{\beta_R}\left(e^{V_{BC}/n_R V_T} - 1\right). \tag{4.17}$$

Equations (4.15), (4.16), and (4.17) present the DC SPICE model in a simplified form. Notice that these equations require only five model parameters: the saturation current, the forward and reverse betas, and the forward and reverse emission coefficients. In SPICE, these parameters are usually denoted by **IS**, **BF**, **BR**, **NF**, and **NR**, respectively. The utility of the Ebers–Moll model in modeling the saturation region of operation is illustrated by Example 4.4.

Example 4.4

Problem:
Find the value of $V_{CE\text{sat}}$ for a BJT as a function of I_C, I_B and the device model parameters.

Solution:
Since both the base-emitter and base-collector junctions are forward biased in saturation, we assume that (4.15) can be approximated by

$$\begin{aligned} I_C &= I_S\left(e^{V_{BE}/n_F V_T} - e^{V_{BC}/n_R V_T}\right) - \frac{I_S}{\beta_R} e^{V_{BC}/n_R V_T} \\ &= I_S e^{V_{BE}/n_F V_T} - I_S\left(1 + \frac{1}{\beta_R}\right) e^{V_{BC}/n_R V_T}. \end{aligned} \tag{4.18}$$

Similarly, we approximate (4.17) by

$$I_B = \frac{I_S}{\beta_F} e^{V_{BE}/n_F V_T} + \frac{I_S}{\beta_R} e^{V_{BC}/n_R V_T}. \tag{4.19}$$

We now define the ***forced beta***, β_{forced}, as the actual ratio of I_C to I_B when the device is saturated:

$$\beta_{\text{forced}} = \left.\frac{I_C}{I_B}\right|_{\text{in saturation}}. \tag{4.20}$$

Substituting (4.18) and (4.19) into (4.20) yields

$$\begin{aligned} &I_S e^{V_{BE}/n_F V_T} - I_S\left(1 + \frac{1}{\beta_R}\right) e^{V_{BC}/n_R V_T} \\ &= \frac{\beta_{\text{forced}} I_S}{\beta_F} e^{V_{BE}/n_F V_T} + \frac{\beta_{\text{forced}} I_S}{\beta_R} e^{V_{BC}/n_R V_T}. \end{aligned}$$

Gathering terms, we obtain

$$\left(1 - \frac{\beta_{\text{forced}}}{\beta_F}\right) e^{V_{BE}/n_F V_T} = \left(\frac{\beta_{\text{forced}} + \beta_R + 1}{\beta_R}\right) e^{V_{BC}/n_R V_T},$$

which we then solve to get the final result:

$$V_{CE\text{sat}} = V_T \ln\left[\frac{\beta_{\text{forced}} + \beta_R + 1}{\beta_R\left(1 - \beta_{\text{forced}}/\beta_F\right)}\right]. \tag{4.21}$$

Figure 4-15 is a plot of (4.21).

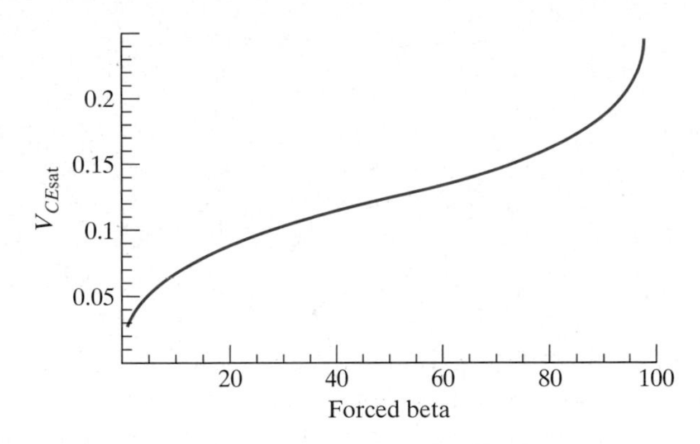

Figure 4–15 V_{CEsat} versus forced beta for a transistor with $\beta_F = 100$ and $\beta_R = 2$.

Exercise 4.4

*Consider the common-emitter transistor stage shown in Figure 4-16. Assuming that $I_B = 100\ \mu A$ and that the transistor is adequately described by just three parameters; **IS** = 1E-14, **BF** = 100, and **BR** = 5, find the value of R_C necessary to produce a forced beta of 20 (i.e., the actual ratio $I_C/I_B = 20$). What value of V_{CEsat} is predicted by (4.21) in this case? Plug the resulting circuit into SPICE and confirm your calculations.*

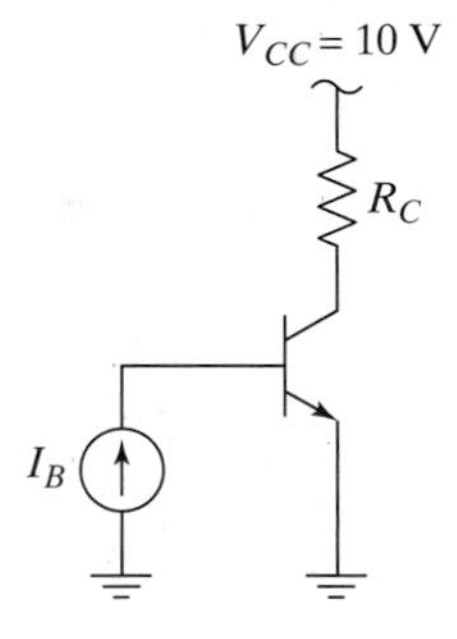

Figure 4–16 A common-emitter transistor stage.

AC Model for the BJT The small-signal AC model used by SPICE for forward-active operation is, in its simplest form, equivalent to the hybrid-π model used in hand analysis and is shown in Figure 4-17. The more complex form of the model is based on charge-control analysis and will not be presented here; we refer you to [4.5]. When SPICE linearizes the equations already presented for the DC model, it essentially finds values for the midband resistances r_π, r_μ, and r_o, as well as the transconductance g_m. For our purposes, it is sufficient to assume that SPICE will use the same values for these model parameters as we derive in Section 8.1.5:

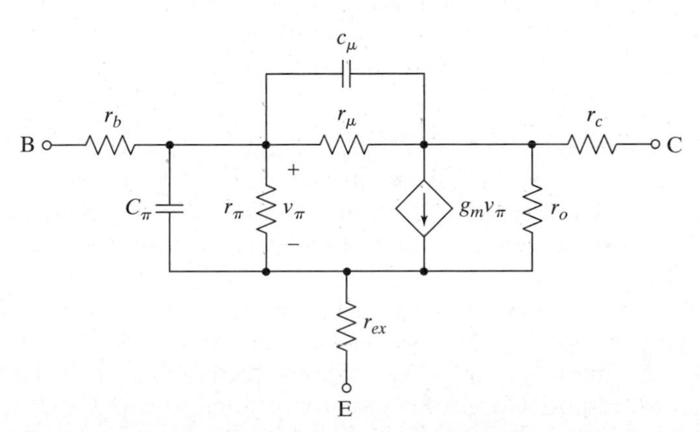

Figure 4–17 The small-signal AC hybrid-π model used by SPICE.

$$g_m = \frac{I_C}{V_T}, \tag{4.22}$$

$$r_\pi = \frac{\beta}{g_m}, \tag{4.23}$$

$$r_o = \frac{V_A}{I_C} \tag{4.24}$$

and

$$r_\mu \geq 10\beta r_o. \tag{4.25}$$

The only new SPICE parameter needed for these calculations is the Early voltage. SPICE again has both a forward and a reverse Early voltage, denoted **VAF** and **VAR** respectively. In addition to these model resistors, there are three real terminal resistors: r_b, r_c, and r_{ex} (discussed in Section 8.1.5) which are entered directly in the SPICE model as **RB**, **RC**, and **RE**, respectively. Finally, the charge storage mechanisms presented in Section 2.5.5 are modeled by the inclusion of two capacitors: C_π and C_μ.

The base-collector capacitor, C_μ, is the junction capacitance of the reverse-biased base-collector junction and is given by an equation identical to (4.3) for the diode. For C_μ, the model parameters are **CJC**, **VJC**, and **MJC** in place of **CJO**, **VJ**, and **M** for the diode. The base-emitter capacitor, C_π, models the base-emitter junction capacitance and the diffusion capacitance. The junction capacitance term is again modeled in the same way as the diode (and C_μ), but the corresponding model parameters are called **CJE**, **VJE**, and **MJE**. The diffusion capacitance was discussed in Section 2.5.5 and the equation given there was

$$C_d = g_m \tau_F, \tag{4.26}$$

where τ_F is the forward transit time. SPICE essentially uses this equation as well, although τ_F is modified at high currents as discussed later in this section. The nominal value of τ_F is a SPICE input parameter, **TF**. The equivalent parameter for the reverse transistor is **TR**.

Transient Model for the BJT In the transient simulation mode, SPICE uses the charge-control model of Gummel and Poon. We briefly presented the concept of charge-control analysis in (2.84) in Section 2.5.5, although we did not refer to it by that name. The charge control model that SPICE uses is based on the AC model parameters and essentially boils down to keeping track of all the charges stored in the different capacitances. Therefore, we will not discuss this model further and refer you to the introductory treatment in [4.5].

The first-order model parameters described in this section are summarized in Table 4.2.

Second-Order Effects in the SPICE BJT Model This section will present the SPICE models for a few of the most important second-order effects in BJTs. The most important second-order effects in the DC SPICE model deal with the change in beta with collector current and the effect of base-width modulation (the Early effect).

At very low currents, recombination in the base-emitter depletion region becomes a significant part of the base current, as explained in Section 2.5.3. This term is proportional to $\exp(V_{BE}/2V_T)$, whereas recombination in the base and reverse injection are both closer to being proportional to $\exp(V_{BE}/V_T)$, like the collector current. Therefore, as I_C decreases, the fraction of I_B caused by recombination in the depletion region increases faster than I_C and the effective

Table 4.2 SPICE BJT First-Order Model Parameters[1]

Name	Parameter	Units	Default	Example
`BF`	Forward beta	none	100	180
`BR`	Reverse beta	none	1	5
`CJC`	Zero-bias base-collector junction capacitance	F	0	1E-12
`CJE`	Zero-bias base-emitter junction capacitance	F	0	2E-12
`IS`	Saturation current	A	10^{-16}	3E-16
`MJC`	Base-collector grading coefficient	none	0.33	0.5
`MJE`	Base-emitter grading coefficient	none	0.33	0.33
`NF`	Forward emission coefficient	none	1	1.2
`NR`	Reverse emission coefficient	none	1	2
`RB`	Base resistance	Ω	0	50
`RC`	Collector resistance	Ω	0	100
`RE`	Emitter resistance	Ω	0	1
`TF`	Forward transit time	s	0	1E-9
`TR`	Reverse transit time	s	0	5E-8
`VAF`	Forward Early voltage	V	Infinite	60
`VAR`	Reverse Early voltage	V	Infinite	120
`VJC`	Base-collector built-in potential	V	.75	0.65
`VJE`	Base-emitter built-in potential	V	.75	0.65

[1] The defaults are for PSPICE and HSPICE.

current gain decreases. SPICE accounts for this increase in base current by adding two terms to (4.17), one term for forward operation and one for reverse:

$$I_B = \frac{I_S}{\beta_F}\left(e^{V_{BE}/n_F V_T} - 1\right) + \frac{I_S}{\beta_R}\left(e^{V_{BC}/n_R V_T} - 1\right)$$
$$+ I_{SE}\left(e^{V_{BE}/n_E V_T} - 1\right) + I_{SC}\left(e^{V_{BC}/n_C V_T} - 1\right), \qquad \textbf{(4.27)}$$

where I_{SE} and n_E model the recombination in the base-emitter depletion region, which is important in forward operation, and I_{SC} and n_C model the recombination in the base-collector depletion region, which is important in reverse operation. The corresponding SPICE parameters are **`ISE`**, **`NE`**, **`ISC`**, and **`NC`**, respectively.

The change in current gain at high collector currents (i.e., in high-level injection) and the Early effect will be discussed together. The Early effect was described in Section 2.5.4, and a model useful for hand analysis was presented, which is repeated here for convenience:

$$I_C = I_S e^{V_{BE}/V_T}\left(1 + \frac{V_{CE}}{V_A}\right). \qquad \textbf{(4.28)}$$

V_A is the Early voltage, which was already introduced in discussing the small-signal output resistance. Although this model works reasonably well, it is based on a straight-line approximation to the variation of I_C with V_{CE}. SPICE handles the Early effect in a more accurate (and more complex) way by modifying the value of the saturation current. Returning to (4.15), SPICE modifies the equation by dividing the first occurrence of I_S by a term q_b:

$$I_C = \frac{I_S}{q_b}\left(e^{V_{BE}/n_F V_T} - e^{V_{BC}/n_R V_T}\right) - \frac{I_S}{\beta_R}\left(e^{V_{BC}/n_R V_T} - 1\right). \tag{4.29}$$

The term q_b is a complicated function of the forward and reverse Early voltages, the terminal voltages of the transistor, and two new parameters: the ***forward knee current***, **IKF**, and its counterpart for the reverse transistor, **IKR**. Roughly, as I_C is increased from a midrange value, **IKF** is the current at which the value of **BF** drops to **BF/2**. **IKR** has a similar meaning for the transistor operating in the reverse active mode. The variation in q_b thereby handles both the Early effect and high-level injection. In high-level injection, the value of q_b is such that $I_C \propto I_C/2V_T$ instead of $I_C \propto I_C/V_T$ [*see* 4.3, pg. 87]. In addition, the value of g_m is reduced by about a factor of two in high-level injection.

Another important second-order effect in the AC SPICE model for a BJT is the variation of the forward transit time with collector current. As briefly noted in Section 2.5.5, at high collector currents the Kirk effect causes an effective widening of the base region, which causes the transit time to increase. This effect is the main cause of the degraded frequency response of BJTs at high currents. SPICE models this effect with three parameters, **XTF**, **VTF**, and **ITF**. The value of τ_F is modified by [*see* 4.2 and 4.3, pg. 100]

$$\tau_{Feffective} = \tau_F\left[1 + XTF\left(\frac{I_{CC}}{I_{CC} + ITF}\right)^2 e^{V_{BC}/1.44VTF}\right], \tag{4.30}$$

where I_{CC} is given by[14]

$$I_{CC} = I_S\left(e^{V_{BE}/n_F V_T} - 1\right). \tag{4.31}$$

The second-order SPICE model parameters presented in this section are summarized in Table 4.3.

Table 4.3 SPICE BJT Second-Order Model Parameters[1]

Name	Parameter	Units	Default	Example
IKF	Knee current for high-current forward beta rolloff	A	infinite	5E-2
IKR	Knee current for high-current reverse beta rolloff	A	infinite	5E-2
ISC	Base-collector junction leakage saturation current	A	0	1E-16
ISE	Base-emitter junction leakage saturation current	A	0	1E-16
ITF	TF high-current parameter	A	0	N/A
NC	Base-collector junction leakage emission coefficient	none	2	2
NE	Base-emitter junction leakage emission coefficient	none	1.5	1.5
VTF	TF base-collector voltage dependence coefficient	V	infinite	N/A
XTF	TF bias dependence coefficient.	none	0	N/A

[1] The defaults are the same for PSPICE and HSPICE.

[14] HSPICE uses a different equation, see the manuals.

4.3.5 MOS Field-Effect Transistors

All MOSFET names in SPICE begin with **M**. There are two types of model available, n-channel (**NMOS**) and p-channel (**PMOS**). For each of these types of model, the same parameters are used, although the signs of some of them change (e.g., V_{th}). MOSFETs are fundamentally four-terminal devices (gate, source, drain, bulk, or substrate) and even though we often treat them as three terminal devices in hand analysis by assuming $V_{SB} = 0$, SPICE includes all four terminals all the time.

The modeling of MOSFETs is more complicated than the modeling of BJTs and there are many models to choose from in SPICE. The most commonly used models are probably the level-1 and BSIM[15] models.The level-1 model is described here. It is physically based and is almost identical to our simplest equations for hand analysis. It is only used for large discrete transistors and quick checks of hand analyses, since it does not model small geometry effects or subthreshold conduction. Small geometry effects are important in almost all modern MOS integrated circuits.

Empirical models, like the BSIM model, will not be discussed here, although they are the most commonly used models in the IC industry. For a thorough treatment of MOS device modeling, you are referred to [4.3] and [4.8]. To simplify the models to obtain agreement with hand analysis as described in the introduction to Section 4.2, the level-1 model is a good choice, even though the final simulations to verify the performance of a circuit would almost always be done with a more complex model.

The Level-1 SPICE MOSFET Model SPICE models the source-to-bulk and drain-to-bulk diodes as well as the intrinsic FET. These diodes are modeled by the common parameters **IS**, **N**, **MJ**, and **TT**, which correspond to the parameters **IS**, **N**, **M**, and **TT** for the diode (*see* Table 4.1). In addition, the source-to-bulk diode capacitance is modeled by the parameter **CBS** and the drain-to-bulk diode capacitance by **CBD**. Both of these parameters correspond to **CJO** for the discrete diode model.[16]

The level-1 model is essentially based on the equations we derived in Chapter 2. We cover the DC, AC, and transient models separately.

The DC Level-1 MOSFET Model The DC level-1 MOSFET model is shown in Figure 4-18, wherein the function used for I_D depends on the region of operation.

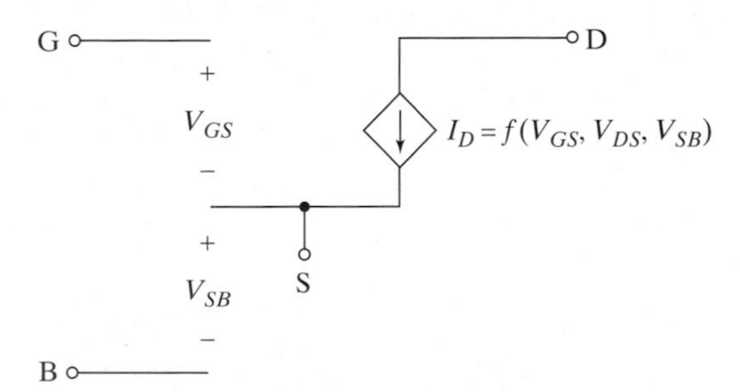

Figure 4–18 The DC level-1 MOSFET model.

---

[15] BSIM stands for Berkeley Short-channel IGFET model. IGFET stands for Insulated-Gate FET. The model is a semi-empirical model that uses physically meaningless parameters to do a curve fit to the device characteristics.

[16] The capacitances of the bulk diodes can also be specified by giving capacitance per unit area and sidewall capacitance per unit length. See the SPICE manual for details.

For operation in strong inversion, we found in Chapter 2 that the drain current in the linear region (i.e., for small values of V_{DS}) is:

$$I_D = \mu_n C'_{ox} \frac{W}{L}\left[(V_{GS} - V_{th})V_{DS} - \frac{V_{DS}^2}{2} \right], \qquad \textbf{(4.32)}$$

where μ_n is the electron mobility at the surface (an n-channel device is used for illustration), W and L are the channel width and length, respectively, C'_{ox} is the capacitance of the gate per unit area, and V_{th} is the threshold voltage. We also derived that when V_{DS} is large enough to saturate the device the drain current is given by

$$I_D = \frac{\mu_n C'_{ox}}{2} \frac{W}{L} (V_{GS} - V_{th})^2 \qquad \textbf{(4.33)}$$

when channel-length modulation is ignored. Finally, we concluded that in the cut-off region (i.e., when $V_{GS} < V_{th}$) the drain current is zero.

In the DC simulation mode, the level-1 SPICE model uses these same equations for I_D. In addition, the expression used for V_{th} includes the body effect and is

$$V_{th} = V_{th0} + \gamma\left[\sqrt{2\varphi_f + V_{SB}} - \sqrt{2\varphi_f} \right] \qquad \textbf{(4.34)}$$

where γ is a parameter modeling the body effect, ϕ_f is the Fermi potential in the bulk defined in Section 2.6.2, V_{SB} is the source-to-bulk voltage, and V_{th0} is given by

$$V_{th0} = V_{FB} + 2\varphi_f + \frac{qN_A x_{d\max}}{C'_{ox}} = V_{FB} + 2\varphi_f + \gamma\sqrt{2\varphi_f}, \qquad \textbf{(4.35)}$$

where V_{FB} is the flat-band voltage, $x_{d\max}$ is the maximum depletion depth, and N_A is the acceptor concentration in the bulk (an n-channel MOSFET is assumed). We can rewrite (4.34) as

$$V_{th} = V_{th0} - \gamma\sqrt{2\varphi_f} + \gamma\sqrt{2\varphi_f + V_{SB}} = V_{BI} + \gamma\sqrt{2\varphi_f + V_{SB}}, \qquad \textbf{(4.36)}$$

where we have implicitly defined what SPICE calls the built-in threshold V_{BI}. We can specify the parameters needed for V_{th} in different ways. If we supply SPICE with the substrate doping (N_A in (4.35) and **NSUB** in SPICE), the value of $2\phi_f$ will be calculated; alternatively, we can enter the SPICE parameter **PHI**, which is equal to $2\phi_f$. Secondly, the zero-bias threshold voltage can be entered directly as the SPICE parameter **VTO** (the letter O, not the number 0), or we can let SPICE calculate the value from the temperature, the value of $2\phi_f$ and the type of gate material (set by the parameter **TPG**). Finally, we can enter the parameter γ directly as **GAMMA**, or we can let SPICE calculate the value given **NSUB** and C'_{ox}. The capacitance C'_{ox} is calculated from the thickness of the oxide, **TOX**. The equations used by SPICE for these calculations are

$$\texttt{PHI} = 2V_T \ln\left(\frac{\texttt{NSUB}}{n_i}\right) \qquad \textbf{(4.37)}$$

and

$$\texttt{VTO} = \begin{cases} -\frac{E_g}{2} - type\frac{\texttt{PHI}}{2} - 105 + type\left(\texttt{GAMMA}\sqrt{\texttt{PHI}} + \texttt{PHI}\right); \texttt{TPG} = 0 \\ type\left(\texttt{TPG}\frac{E_g}{2} - \frac{\texttt{PHI}}{2}\right) + type\left(\texttt{GAMMA}\sqrt{\texttt{PHI}} + \texttt{PHI}\right); \texttt{TPG} = \pm 1 \end{cases} \qquad \textbf{(4.38)}$$

where *type* is +1 for n-channel devices and –1 for p-channel, and E_g is the energy gap (1.12 eV for silicon),

$$\mathbf{GAMMA} = \frac{\sqrt{2q\epsilon_s \mathbf{NSUB}}}{C'_{ox}}, \tag{4.39}$$

and

$$C'_{ox} = \frac{\epsilon_{ox}}{\mathbf{TOX}}. \tag{4.40}$$

We can simplify our equations a bit by defining the parameter K as[17]

$$K = \frac{\mu_n C'_{ox}}{2} \frac{W}{L}. \tag{4.41}$$

SPICE will either calculate K given the device dimensions, the mobility, μ_n, and the capacitance or we can override this calculation by entering the parameter **KP**. **KP** stands for K prime and is

$$\mathbf{KP} = \mu_n C'_{ox}. \tag{4.42}$$

The mobility can be set with the parameter **UO** (the letter O, not the number 0).

The level-1 model also accounts for channel-length modulation in exactly the way we did in Section 2.6.3, so that the overall drain current equations become

$$I_D = K\left[(V_{GS} - V_{th})V_{DS} - \frac{V_{DS}^2}{2}\right](1 + \lambda V_{DS}) \tag{4.43}$$

in the linear region and

$$I_D = K(V_{GS} - V_{th})^2 (1 + \lambda V_{DS}) \tag{4.44}$$

in saturation. The SPICE parameter **LAMBDA** must be specified to model channel-length modulation.

Since MOSFETs are usually used in integrated circuit design, SPICE allows us to specify the drawn channel length and width—that is, the dimensions from our mask design (*see* Chapter 3). SPICE then calculates the effective channel length allowing for lateral diffusion under the mask edges and channel-length modulation. The equation used is

$$L_{eff} = (L - 2\mathbf{LD}) \tag{4.45}$$

where **LD** is the SPICE parameter giving the amount of lateral diffusion present. HSPICE and PSPICE both include a similar calculation for the effective channel width, although it was not included in the level-1 model in the original SPICE program. For simplicity, we will continue to use the drawn width in our equations.

Example 4.5

Problem:
Derive a relationship between g_{mb} and g_m in terms of more fundamental device parameters.

Solution:
By definition, the bulk transconductance is given by

$$g_{mb} = \left.\frac{di_D}{dv_{BS}}\right|_{@\,\text{Q point}} = \frac{di_D}{dV_{th}}\left(\frac{-\,dV_{th}}{dV_{SB}}\right)$$

[17] Some texts use β instead of K.

which can be found using (4.33) and (4.34):

$$g_{mb} = \frac{2K(V_{GS} - V_{th})\gamma}{2\sqrt{2\phi_f + V_{SB}}} = \frac{g_m\gamma}{2\sqrt{2\phi_f + V_{SB}}}.$$

Therefore, we can write

$$\frac{g_{mb}}{g_m} \equiv \chi = \frac{\gamma}{2\sqrt{2\phi_f + V_{SB}}}.$$

The parameter χ (pronounced $k\bar{i}$) defined here is an important parameter for a MOSFET, since it indicates how seriously the body effect will affect the small-signal performance when v_{BS} is not zero.

The AC Level-1 MOSFET Model For the AC simulation mode, the SPICE level-1 MOSFET model is essentially identical to the model we use for hand analysis in Chapters 8 and 9 and is shown in Figure 4-19. For operation in saturation, the equation for the drain current is differentiated twice (at the bias point), once with respect to v_{GS} to find the transconductance and once with respect to v_{DS} to find the output resistance:

$$g_m = \left.\frac{di_D}{dv_{GS}}\right|_{@\text{ Q point}} = 2K(V_{GS} - V_{th}) = 2\sqrt{KI_D} \quad \textbf{(4.46)}$$

and

$$r_o = \left(\left.\frac{di_D}{dv_{DS}}\right|_{@\text{ Q point}}\right)^{-1} = \frac{1}{\lambda I_D}. \quad \textbf{(4.47)}$$

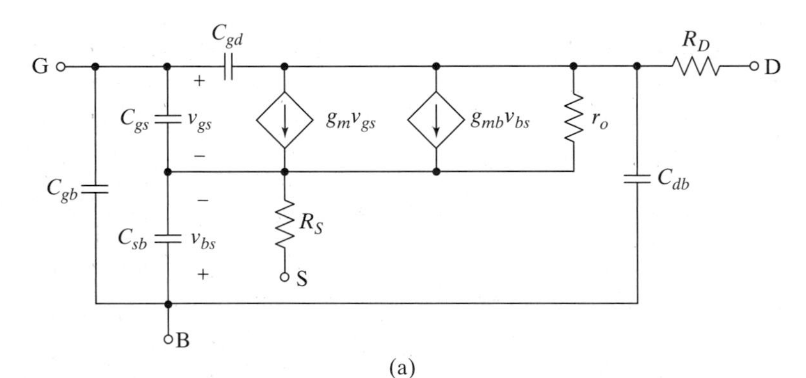

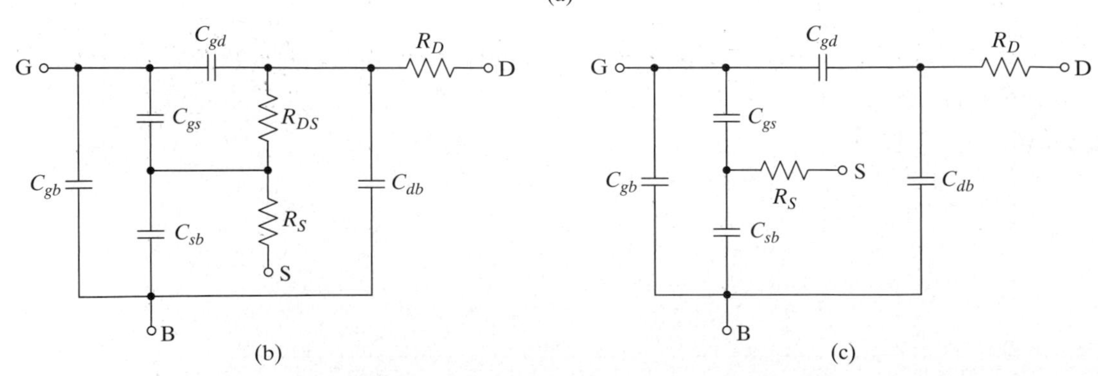

Figure 4–19 The AC level-1 MOSFET model: (a) in saturation, (b) in the linear region, and (c) in cutoff.

In addition, the bulk transconductance is found as illustrated in Example 4.5:

$$g_{mb} = \chi g_m. \tag{4.48}$$

For operation in the linear region, the drain-to-source resistance is found using (4.32) (i.e., $R_{DS} = V_{DS}/I_D$) and is assumed constant for the AC analysis. In addition, SPICE allows us to specify the resistances in series with the drain and source—`RD` and `RS`, respectively.

The small-signal capacitors in the MOSFET model are given by the capacitors of the drain-to-bulk and source-to-bulk diodes already mentioned, as well as the capacitors associated with the oxide. The oxide capacitance can be broken down into the capacitance associated with the intrinsic channel region and the ***overlap capacitance***. As shown in Chapter 3, the gate electrode overlaps the source, drain, and bulk regions slightly. As a result, there are small parallel-plate parasitic capacitances between the gate and each of these regions. These capacitors are called overlap capacitors and are given by

$$C_{GBO} = L_{eff} CGBO, \tag{4.49}$$

$$C_{GSO} = W \cdot CGSO, \tag{4.50}$$

and

$$C_{GDO} = W \cdot CGDO, \tag{4.51}$$

where `CGBO` is the gate-to-bulk overlap capacitance per unit length and `CGSO` and `CGDO` are the gate-to-source and gate-to-drain overlap capacitances per unit width, respectively.

How the intrinsic channel capacitance is modeled depends on the region of operation the device is in. The equations used by SPICE are cumbersome and not worth pursuing here. For our purposes it is sufficient to note that the capacitors can be reasonably approximated by the equations that we have for hand analysis. In other words, the total intrinsic gate capacitance when an inversion layer is present is given by

$$C_g = C'_{ox} W L_{eff}, \tag{4.52}$$

and we need to specify how it is divided between the source, drain, and bulk connections.

When the FET is cutoff, C_g is zero. When the FET is in the linear region of operation, C_g splits equally between the source and drain. When the FET is in the saturation region of operation, C_g is reduced to about 2/3 of its nominal value and is connected entirely to the source. No new model parameters are needed to specify these capacitors, since SPICE will determine C'_{ox} from (4.40), but `TOX` must be specified—SPICE will not use (4.42) to get C'_{ox} from `KP`. In summary, we have

$$C_{gb} = C_{GBO} \tag{4.53}$$

in all regions of operation,[18]

$$C_{gs} = \begin{cases} C_{GSO} + (2/3)C_g \text{ in saturation} \\ C_{GSO} + (1/2)C_g \text{ in the linear region,} \\ C_{GSO} \qquad\qquad\quad \text{in cutoff} \end{cases} \tag{4.54}$$

$$C_{gs} = \begin{cases} C_{GSO} + (2/3)C_g \text{ in saturation \& cutoff} \\ C_{GSO} + (1/2)C_g \text{ in the linear region} \end{cases} \tag{4.55}$$

---

[18] The SPICE output listing gives the overlap capacitors separately.

and both C_{sb} and C_{db} are given by the capacitance of the junction diodes formed by these regions as noted at the start of this section.

The Transient Level-1 MOSFET Model In the transient simulation mode, SPICE uses charge-control analysis and keeps track of all of the charge stored in the device using the nonlinear charge equations. The AC model parameters are used to specify the device.

The level-1 model parameters described in this section are summarized in Table 4.4.

4.3.6 Junction Field-Effect Transistors and MOSFETs

All JFET names in SPICE begin with `J`. There are both n-channel and p-channel models available, called **`NJF`** and **`PJF`**, respectively. The same parameters are used for both models. Junction FETs are inherently depletion-mode devices, so we expect the threshold voltage of an n-channel transistor to be negative, and it is. SPICE uses a strange convention for JFETs, however; *both* n- and p-channel depletion-mode devices have negative threshold voltages. If the threshold voltage is positive, it implies that the device is in enhancement mode.

The modeling of JFETs is identical to the level-1 MOSFET model for the DC drain current, but the equations are typically written in a somewhat different form. In the linear region we have

$$I_D = \beta_{DS}[2(V_{GS} - V_{th})V_{DS} - V_{DS}^2](1 + \lambda V_{DS}), \tag{4.56}$$

which is identical to (4.32) for the MOSFET except that we have included channel-length modulation and have used β in place of K. Similarly in saturation we have

$$I_D = \beta(V_{GS} - V_{th})^2 (1 + \lambda V_{DS}), \tag{4.57}$$

which is identical to (4.33) for the MOSFET. The drain current is zero when the device is cutoff, just like the level-1 MOSFET model.

The capacitances in the JFET model are different than in a MOSFET since a pn junction is used to modulate the channel conductance rather than a capacitor. We will not pursue the JFET SPICE model in more detail here, since JFETs are rapidly becoming obsolete.

4.4 MACRO MODELS IN SPICE

It is frequently advantageous to use ***macro models*** for some complex circuits, like operational amplifiers, when they are part of a larger simulation. A macro model is a simplified version of some circuit, and there are two main advantages to using a macro model instead of a detailed transistor-level schematic. First, the simulation may be much faster since the macro model only needs to produce the same external behavior and may be significantly less complex than the circuit it models. Second, we can generate a macro model for a circuit without having to know the details of its construction.

There are many possible levels of macro model that can be used, but the motivation is always to reproduce some aspects of the performance of a more complex circuit while simplifying the simulation. Many IC manufacturers now supply macro models for popular ICs, so that a board-level designer can use these models to simulate a circuit containing them.

We introduce the idea of macro modeling by presenting a few macro models for an operational amplifier. We could use the basic model shown in Figure 4-20. This

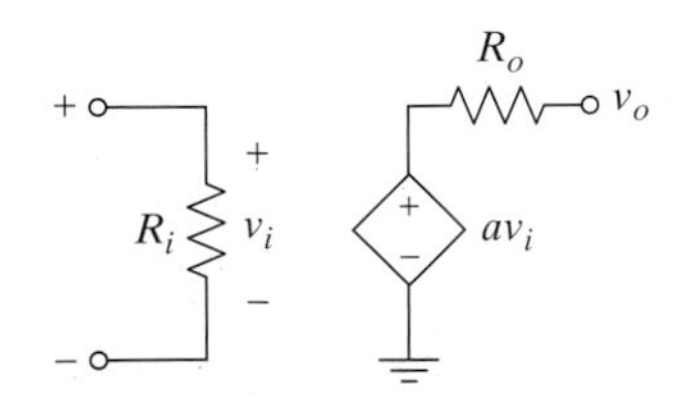

Figure 4–20 A simple macro model of an operational amplifier.

Table 4.4 SPICE Level-1 MOSFET Model Parameters[1]

Name	Parameter	Units	Default
`CBD`	Drain-to-bulk diode zero-bias capacitance	F	0
`CBS`	Source-to-bulk diode zero-bias capacitance	F	0
`CGBO`[2]	Gate-to-bulk overlap capacitance per unit length	F/m	0
`CGDO`[2]	Gate-to-drain overlap capacitance per unit width	F/m	0
`CGSO`[2]	Gate-to-source overlap capacitance per unit width	F/m	0
`GAMMA`	Bulk threshold parameter	$V^{0.5}$	calculated
`IS`	Source-to-bulk and drain-to-bulk diode saturation current	A	10^{-14}
`KP`	Transconductance parameter	A/V^2	2×10^{-5} (n-ch same, p-ch 8.6×10^{-6})
`L`	Drawn channel length (usually on element line)	m	**`DEFL`**(can be set by **`DEFL`** in **`.OPTIONS`**)
`LAMBDA`	Channel-length modulation parameter	V^{-1}	0
`LD`	Lateral diffusion length	m	0
`MJ`	Source-to-bulk and drain-to-bulk diode grading coefficient	none	0.5
`N`	Source-to-bulk and drain-to-bulk diode emission coefficient	none	1
`NSUB`	Substrate doping level	$1/cm^3$	none (1×10^{15})
`PHI`	Surface potential ($2\phi_f$)	V	0.6 (calculated)
`RD`	Series drain resistance	Ω	0
`RS`	Series source resistance	Ω	0
`TOX`	Oxide thickness	m	none (1×10^{-7})
`TPG`	Type of gate material ($0\Rightarrow$ Al, $+1\Rightarrow$ Poly with doping opposite polarity as substrate, -1 $\Rightarrow$ Poly with doping same polarity as substrate)	none	+1
`TT`	Source-to-bulk and drain-to-bulk diode transit time	s	0
`UO`[2]	Surface mobility	$cm^2/V{\cdot}s$	600
`VTO`	Zero-bias threshold voltage	V	0 (calculated)
`W`	Drain channel width (usually on element line)	m	**`DEFW`** (can be set by **`DEFW`** in **`.OPTIONS`**)

[1] The defaults are for PSPICE and HSPICE; the HSPICE values are shown in parentheses when different. Examples are not given because the values of some parameters vary tremendously from one device to another and for the other parameters the defaults are good examples.

[2] The last character in the parameter name is the letter 'O', not the number zero.

model only allows for the differential input resistance, the output resistance, and the differential-mode gain of the amplifier. The common-mode input resistance, input common-mode voltage range, output voltage limits, bandwidth, common-mode rejection ratio, and slew-rate are all implicitly assumed to be infinite, and many other second-order effects are ignored as well. Nevertheless, for some applications this model may be adequate.

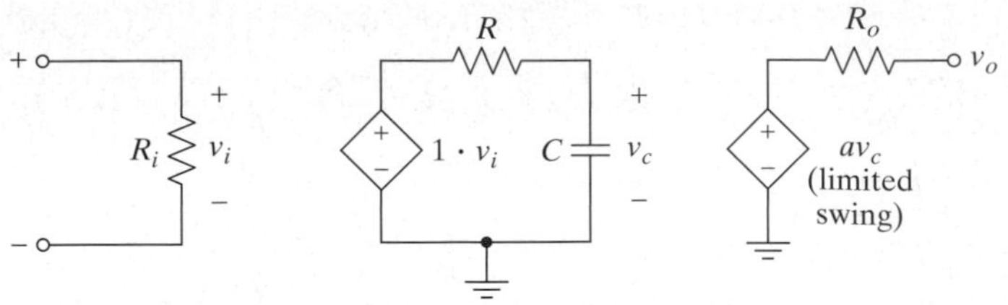

Figure 4–21 A more complex macro model of an operational amplifier.

A slightly more sophisticated model of the op amp is shown in Figure 4-21 and includes the small-signal bandwidth and limited output swing. This circuit provides an illustration of how complex systems can be modeled using controlled sources and relatively simple circuits.

In this model, the first controlled source drives an *RC* lowpass filter. The transfer function of this filter produces a single-pole approximation to the frequency response of the amplifier being modeled. The same effect could also be accomplished using an ***analog behavioral modeling*** block—for example, the **ELAPLACE** block in PSPICE.

The finite output swing of the amplifier is modeled by placing limits on the voltage-controlled voltage source output. How the limits are implemented depends on the version of SPICE being used; in the newer versions of PSPICE, you can do it using the **ETABLE** component and setting the limits in the lookup table, see the manual for details. One example use is shown in the solution to Exercise 4.5.

We can also extend this model to include the slew rate limit of the op amp. One way to do this is to use a current source in the single-pole filter section and use limits on the magnitude of this current. When the amplifier input changes too rapidly, the current will reach a limit and the voltage on the filter capacitor will slew.

Exercise 4.5

Show how you can use a voltage-controlled current source with limits on its swing to simultaneously model the small-signal bandwidth and the slew rate limit of an operational amplifier. Use a circuit similar to the one shown in Figure 4-21, but without output swing limits. Assume that the output voltage is limited by the external circuit to ± 10 *V. Produce a macro model that has an open-loop gain of 100,000, a small-signal bandwidth of 10 Hz,* $R_i = 100\ k\Omega$, $R_o = 10\ \Omega$, *and a slew rate of* $1\ V/\mu s$. *Confirm that your model works as expected.*

SOLUTIONS TO EXERCISES

4.1 The modified SPICE file is:

```
Bridge Circuit
.sens v(2,3)
vin 1 0 dc 5
r1 2 3 1k
r2 1 2 1k
r3 2 0 1k
r4 1 3 2k
r5 3 0 3k
.end
```

A summary of a portion of the output file is shown below.

DC Sensitivities Of Output V(2,3)			
Element Name	Element Value	Element Sensitivity (Volts/Unit)	Normalized Sensitivity (Volts/Percent)
r1	1.0000k	-116.5981u	-1.1660m
r2	1.0000k	-445.8162u	-4.4582m
r3	1.0000k	480.1097u	4.8011m
r4	2.0000k	246.9136u	4.9383m
r5	3.0000k	-137.1742u	-4.1152m
vin	5.0000	-37.0370m	-1.8519m

As an example of what these numbers mean, consider the sensitivity of $v(2,3)$ to changes in the value of R_4. $v(2,3)$ changes by 246.9 μV for every 1-Ω change in R_4. The normalized sensitivity is the change in $v(2,3)$ for a 1% change in R_4. A 1 Ω change in R_4 is 1/20%, so a 1% change in R_4 would lead to a change of 20(246.9 μV) = 4.94 mV, which agrees with the final number listed. Note that the sensitivity of $v(2,3)$ to changes in V_I (–37 mV/V) agrees with the slope of the line plotted in Example 4.1.

4.2 The modified SPICE file is:

```
Low-pass filter
.probe ; create a probe data file
.param cap .01u
.dc cap list .001u .01u .1u
.ac dec 10 10 1meg
vin in 0 ac 1
r1 in out 10k
c1 out 0 {cap} ;the curly braces are not
; required in all versions of SPICE
.end
```

The output plot is:

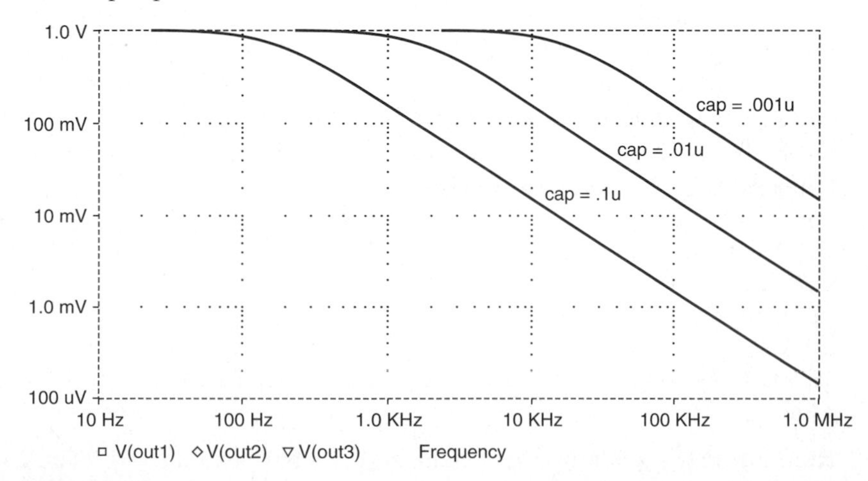

Figure 4–22

4.3 The modified SPICE file is:

```
Half-wave rectifier with filter
.probe ; create a probe data file
.param cap 100u
.dc dec cap 10u 1m 2
.tran .1m 30m
vin in 0 sin(0 20 60)
d1 in out D1N4002
r1 out 0 1k
c1 out 0 {cap} ;the curly braces are not
; required in all versions of SPICE
.model D1N4002 D(is=14.11e-9, n=1.984, rs=33.89e-
+3, ikf=94.81, xti=3, eg=1.110, cjo=51.17e-12,
+m=.2762, vj=.3905, fc=.5, isr=100.0e-12, nr=2,
+bv=100.1, ibv=10, tt=4.761e-6)
.end
```

The output plot is:

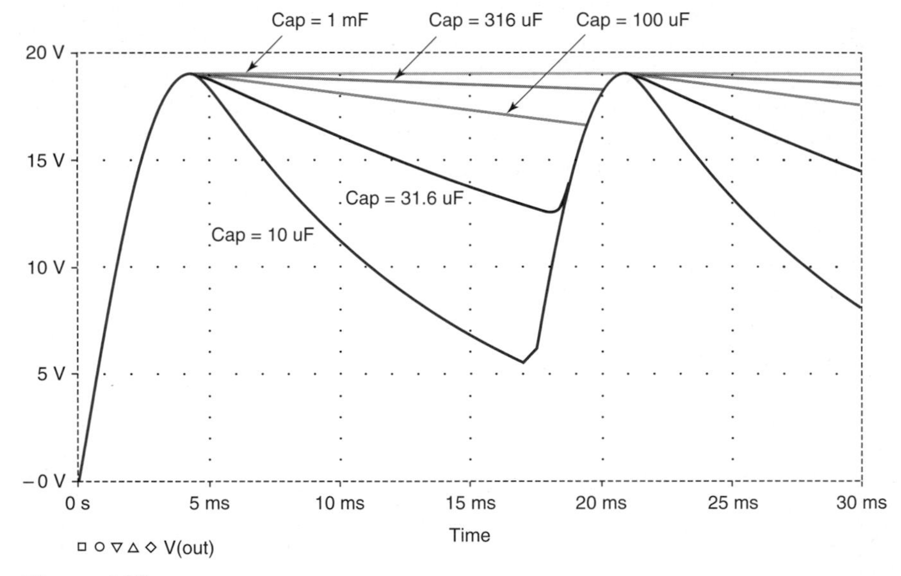

Figure 4-23

4.4 With $I_B = 100\ \mu\text{A}$, a forced beta of 20 requires $I_C = 2$ mA. Using (4.21) with the given parameters we find that $V_{CE\text{sat}} = 49$ mV at room temperature (i.e., $V_T = 26$ mV). With this value of V_{CE} and $I_C = 2$ mA, we find $R_C = (10\text{-}0.049)/0.002 = 4.98\ \text{k}\Omega$. Rounding off we use $R_C = 5\ \text{k}\Omega$. The resulting SPICE input file is:

```
Common-emitter stage saturation
.op
ib 0 b1 dc 100u
vcc vcc 0 dc 10
rc vcc c1 5k
q1 c1 b1 0 Nsimple
.model Nsimple npn(is=1e-14, bf=100, br=5)
.end
```

Running SPICE with this input yields $V_{CE\text{sat}} = 0.0483$ and $I_C = 1.99$ mA, very close to the predicted values.

4.5 The appropriate model is shown below, where the PSPICE schematics program has been used. The controlled-current source is produced using the **GTABLE** component. The output is equal to the negative of the input voltage difference as shown. (The negative difference is used because of the direction of the source.) The lookup table for the source is entered as: (−.159,−.159)(.159, .159). PSPICE uses these two points to define a transfer characteristic where $i_O = -.159$ A for $v_I \leq .159$ V, i_O (in A) = v_I (in V) for $-.159 < v_I < .159$ and $i_O = .159$ A for $v_I \geq .159$ V. Therefore, the current is limited to $\pm .159$ mA. Since the output voltage is limited by the external circuit to ± 10 V, the voltage on the capacitor is less than $\pm 10/100{,}000 = \pm 0.1$ mV. For a capacitor voltage this small, the current in R_A is insignificant, and virtually all of the current flows in the capacitor. Therefore, when the current limits, the voltage across the capacitor is approximately a linear ramp with a slope equal to $\pm I/C = 10$ V/s. When this slope is multiplied by the gain of 100,000, the output voltage has the desired slew rate of 1 V/μs. When the circuit is not slewing, the open–loop bandwidth is set by the RC time constant and is $BW = 1/(2\pi RC) = 10$ Hz.

To check the operation, we added an input resistor, R_1, a feedback resistor, R_2, and a load resistor, R_L, external to the op amp. An example simulation result is shown with the schematic in Figure 4-24.

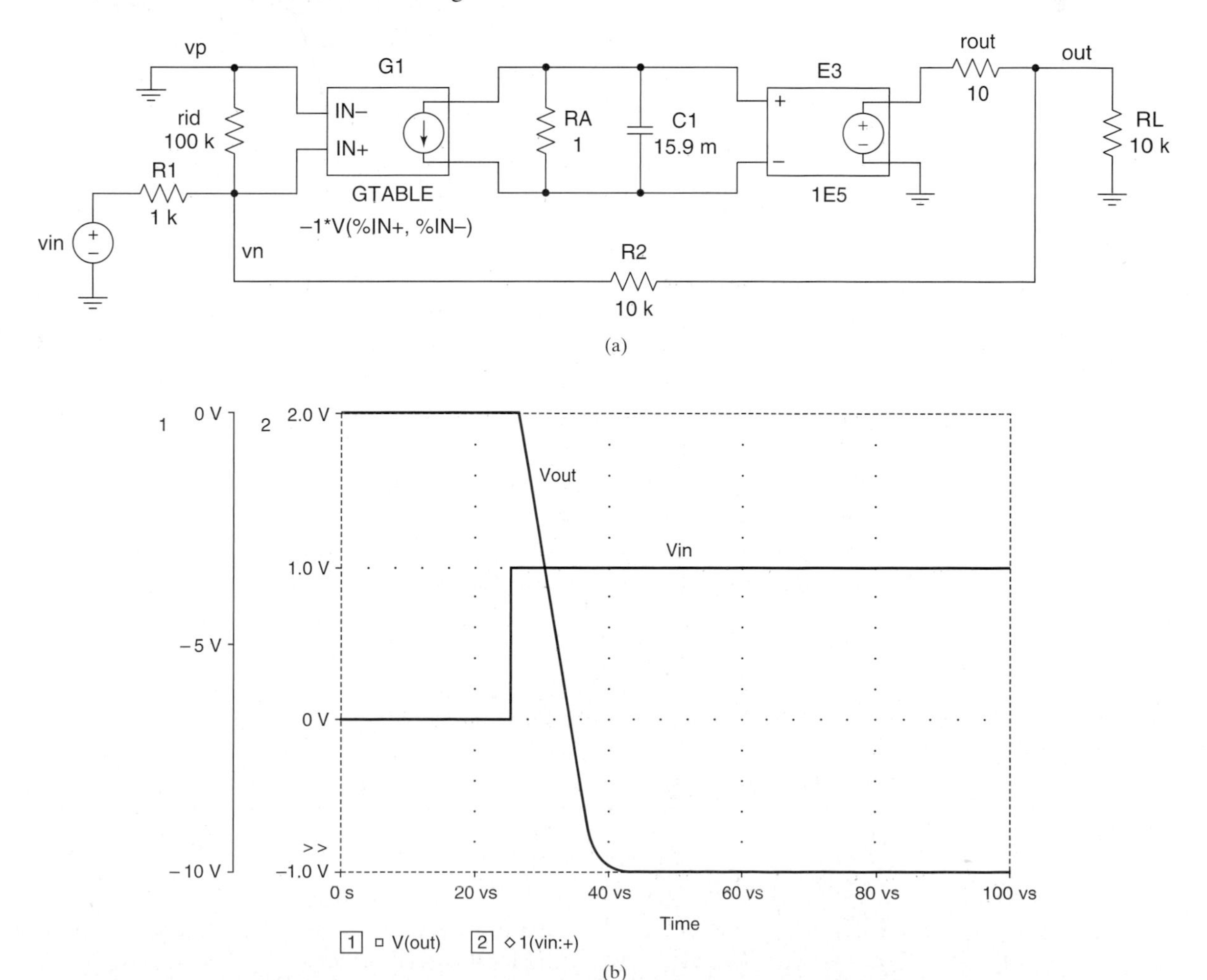

Figure 4–24

CHAPTER SUMMARY

- There are a number of different computer programs used to aid in the design of electronic systems. Many different tools are used for behavioral and other high-level simulations, especially for purely digital circuits. But, SPICE is the most commonly used transistor-level circuit simulation program.
- SPICE has many advantages over breadboards, but we must always remember the limitations of the program and realize that our results cannot be accurate if our models are not good enough, or we forget to allow for some parasitic effect.
- A SPICE input consists of a circuit description (the netlist), simulation commands, and program control statements.
- SPICE has three simulation modes: DC, AC, and transient.
- The DC simulation mode short-circuits inductors, open-circuits capacitors, and iteratively solves the nonlinear equations.
- The AC simulation mode uses complex impedances, but uses linear models evaluated at the DC bias point for all nonlinear elements.
- The transient simulation mode iteratively solves the full nonlinear equations, including reactive elements. Transient simulations take the longest time.
- There are different types of analyses available within each simulation mode.
- In the DC simulation mode, you can perform DC sweeps (`.DC`), examine a DC transfer function (`.TF`), look at the sensitivity of some voltage or current to parameter values (`.SENS`), or just ask for the operating point (`.OP`).
- In the AC simulation mode, you can perform AC sweeps (always versus frequency—`.AC`) and noise analyses (`.NOISE`). Some versions of SPICE also allow you to do a distortion analysis (`.DISTO`) and/or find the poles and zeros in a network (`.PZ`).
- In the transient simulation mode, you can perform a transient analysis (i.e., voltages and currents versus time—`.TRAN`) or perform a Fourier analysis to determine the spectral components for a given fundamental frequency (`.FOUR`).
- All circuit elements in SPICE specify the element by the first letter of the name; `Rxxx` is a resistor, `Cxxx` is a capacitor, `Lxxx` is an inductor, `Dxxx` is a diode, `Qxxx` is a BJT, `Mxxx` is a MOSFET, and `Jxxx` is a JFET.
- If the model to be used for a given element is too complex to fit on the element line, a separate `.model` statement is used.
- The models used by SPICE for semiconductor devices can be either physically based or empirical in nature. Physical models have certain advantages for the designer, but empirical models are usually more accurate and lead to faster simulations.
- Macro models are sometimes used for complex elements to speed up the simulation or to model a device for which the detailed circuit is not known.

REFERENCES

[4.1] A. Vladimirescu, *The SPICE Book*, New York: John Wiley & Sons, 1994.

[4.2] HSPICE manuals.

[4.3] G. Massobrio and P. Antognetti, *Semiconductor Device Modeling with SPICE*, 2E. New York: McGraw-Hill, 1993.

[4.4] I. Getreu, *Modeling the Bipolar Transistor*, Tektronix, Inc. part # 062-2841-00.

[4.5] R.S. Muller and T.I. Kamins, *Device Electronics for Integrated Circuits*, 2E. New York: John Wiley & Sons, 1986.

[4.6] A. Vladimirescu and S. Liu, *The Simulation of MOS Integrated Circuits Using SPICE2*, UCB/ERL Memorandum No. M80/7, Feb. 1980.

[4.7] P.R. Gray, P.J. Hurst, S.H. Lewis and R.G. Meyer, *Analysis and Design of Analog Integrated Circuits*, 4E, New York: John Wiley & Sons, 2001.

[4.8] D.P. Foty, *MOSFET Modeling with SPICE—Principles and Practice*, Englewood Cliffs, NY, Prentice Hall, 1997.

PROBLEMS

4.2 Circuit Simulation Using SPICE

4.2.2 Simulation Modes and Types of Analysis

S P4.1 Consider the bandpass filter shown in Figure 4-25. **(a)** Use SPICE to simulate and plot the magnitude and phase of the transfer function from 1 Hz to 100 kHz. Also, use SPICE to determine the magnitude of the input impedance of the filter from 100 Hz to 100 kHz. **(b)** Compare the low and high-frequency cutoff frequencies and the shape of the input impedance plot with what you would expect from rough hand approximations. (Hint: You can ignore each of the capacitors for some range of frequencies and use Aside A1.6.)

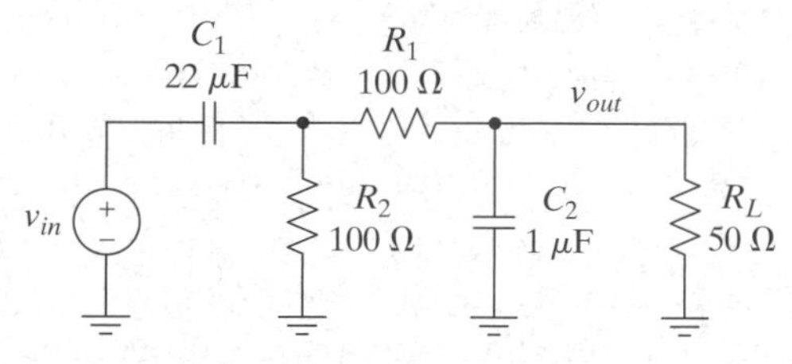

Figure 4–25 A bandpass filter.

S P4.2 Using the filter shown in Figure 4-25, run a SPICE simulation to examine what the output looks like when the input is a 4.0 kHz square wave from 0 to 1V. Hand in a plot of the output voltage and explain the results.

S P4.3 Consider the bipolar transistor amplifier shown in Figure 4-26. For this problem, model the transistor with only two parameters: `IS=1E-14` and `BF=100`. **(a)** Calculate the DC bias point by hand and then confirm your results using SPICE. **(b)** Use SPICE to find the sensitivity of the collector current to changes in all component values and model parameters. **(c)** Repeat part (b) with $R_1 = 6.2\text{ k}\Omega$, $R_2 = 3.9\text{ k}\Omega$, and $R_E = 3.3\text{ k}\Omega$ Explain the differences between the results in (b) and (c).

S P4.4 Create a SPICE model of an amplifier with an input resistance of 10 kΩ, an open-circuit voltage gain of 20 V/V, and an output resistance of 100 Ω. Connect a Thévenin model of a voltage source with an output resistance of 1 kΩ to the input of the amplifier, and also connect a load of 1 kΩ to the output. Use SPICE to find the overall voltage gain, the input resistance (seen by the Thévenin equivalent source), and the output resistance (seen by the load). Explain the methods used to find the impedances and show that the answers are what you expect.

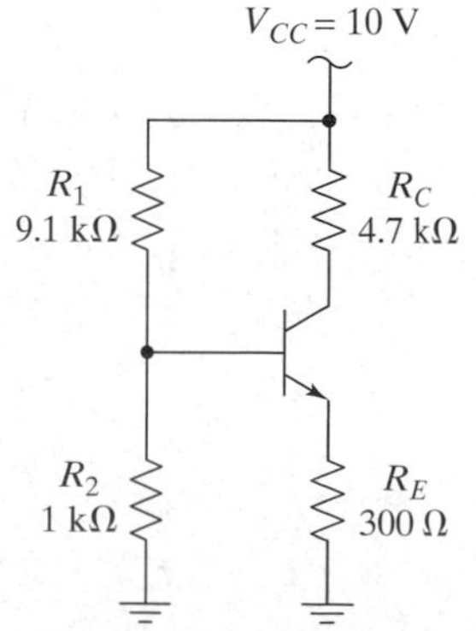

$V_{CC} = 10$ V; R_1 9.1 kΩ; R_C 4.7 kΩ; R_2 1 kΩ; R_E 300 Ω

Figure 4–26 A bipolar transistor amplifier stage.

S P4.5 Add an input capacitance of 10 pF (in parallel with R_i) to the amplifier model in Problem P4.4. Using the same source and load as in that problem, plot the magnitude of the overall voltage gain in dB versus frequency. What is the cutoff frequency? Does it agree with what you calculate by hand?

S P4.6 Simulate the operation of the full-wave rectifier circuit shown in Figure 4-27 and verify that it operates as you expect. Add a capacitor in parallel with the load resistor and sweep its value from 1 μF, to 100 μF, using a total of 3 different linearly-spaced values. Use the 1N4002 diode model and a 120 V peak 60 Hz sinusoidal input. Plot v_o versus time and explain the result.

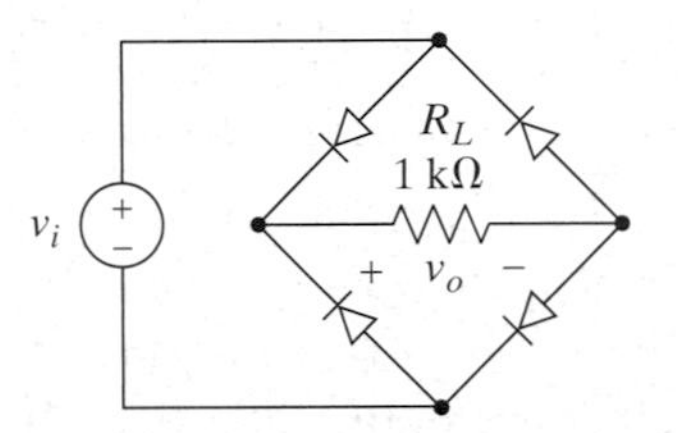

Figure 4–27 A full-wave rectifier.

4.3 Circuit Elements and Models for SPICE

4.3.3 Diodes

S P4.7 Given that $I_D = 2$ mA when $V_D = 630$ mV for a diode operating at room temperature, what is the value of **IS**?

S P4.8 Assume a diode has **IS** $= 3\times10^{-15}$ and **N** $= 1.8$. What is the current at room temperature if the voltage across the diode is 500 mV?

S P4.9 Given the room-temperature DC measurements presented below, derive a SPICE model for the diode using **IS** and **N**. Create a SPICE file that will allow you to confirm that your model produces the same results.

I_D	V_D
219 pA	0.3 V
37.0 nA	0.5 V
6.24 μA	0.7 V
1.05 mA	0.9

S P4.10 Assume that for a given diode, we already know that **N** $=1$, **RS** $=0$, **M** $=$ 0.33, and **VJ** $= 0.7$. We make the following measurements; the capacitance of the diode with $V_D = 4.9$ V is 1 pF, the capacitance in forward bias with $I_D = 2$ mA is 19.4 pF, and the corresponding forward voltage is $V_D = 705$ mV. All measurements are made at 300 Kelvin. Find values for **IS**, **CJO**, and **TT** for the SPICE model of this diode.

S P4.11 At 300 Kelvin, we measure $I_D = 32.9$ μA at $V_D = 600$ mV and $I_p = 641$ μA at $V_d = 700$ mV for a particular diode. **(a)** Find values for **IS** and **N** for this diode. **(b)** Using the values found in (a), set up a diode model in SPICE and create a test circuit in SPICE to confirm the currents and voltages given in the problem statement.

4.3.4 Bipolar Junction Transistors

P4.12 Start from Figure 4-13 and apply the split-source transformation from Aside A8.1 to both current sources to derive Figure 4-14 and Equations (4.11), (4.13), and (4.14). You will need to use:

$$\alpha_F = \frac{\beta_F}{\beta_F + 1}$$

and

$$\alpha_R = \frac{\beta_R}{\beta_R + 1}.$$

P4.13 Show that (4.21) can also be written as

$$V_{CEsat} = V_T \ln\left(\frac{1 + \beta_{\text{forced}}(1 - \alpha_R)}{\alpha_R\left[1 - \beta_{\text{forced}}\left(\frac{1 - \alpha_F}{\alpha_F}\right)\right]}\right),$$

which is the form found most often in the literature.

S P4.14 The plot of $\ln I_C$ versus V_{BE} shown in Figure 4-28 is called a *Gummel plot*. Using this plot and assuming $V_T = 28.6$ mV, derive a SPICE model for the transistor consisting of `IS`, `BF`, and `NF`. Use the straight portions of the plot and ignore the curvature at low and high currents. Create a SPICE file to reproduce the Gummel plot and confirm that your model works.

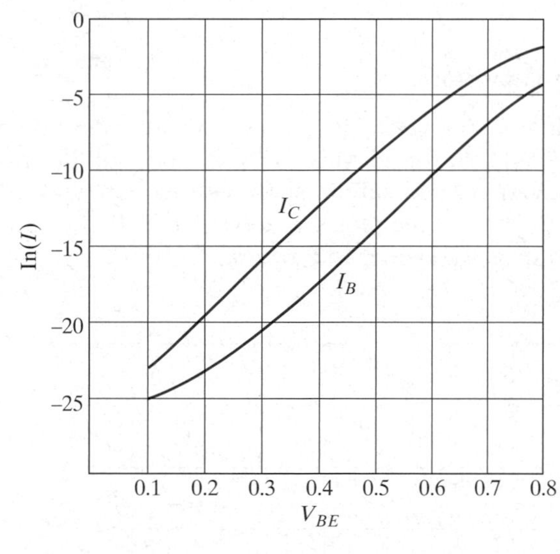

Figure 4-28 A Gummel plot for a BJT.

P4.15 If a BJT operating at room temperature has $I_C = 700$ μA when $V_{BE} =$ 650 mV, what is the value of `IS`?

DS P4.16 **(a)** Design the circuit shown in Figure 4-29 to obtain $I_C = 3$ mA, a voltage gain of about –10 and a stable operating point. Use standard ±5% resistor values from Aside A7.5. Assume that the coupling and bypass capacitors are large and use $R_L = 10$ kΩ. **(b)** Assuming that the transistor is modeled by the SPICE model given below, run a simulation to check your design. Set all three capacitor values equal to 100 μF for your simulation. First check the operating point and then the gain. Check the gain using the AC simulation mode and using the `.TF` command. Do the answers agree? If not, which one is correct and why? **(c)** Now calculate by hand the values you expect for C_π and C_μ at your operating point. Do they agree with the values given by SPICE? **(d)** Change the model by making I_S five times the default value and rerun your simulation. Did any of the results change significantly? Why or why not?

Use `.MODEL EXAMPLE NPN(CJC=1p, CJE=2p, TF=500p)`; all other parameters assume their default values.

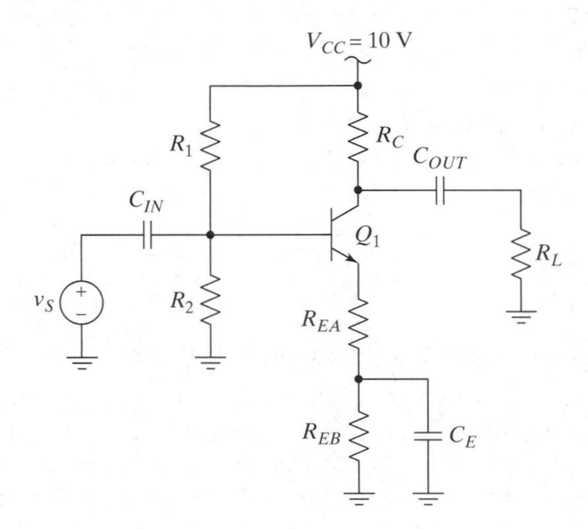

Figure 4–29 An AC-coupled common-emitter amplifier.

4.3.5 MOS Field-Effect Transistors

S P4.17 Given the plot of $\sqrt{I_D}$ versus V_{GS} shown in Figure 4-30, determine the values of **`KP`** and **`VTO`** for a level-1 SPICE model of the transistor. Ignore the body effect and subthreshold conduction, and only use the straight-line portion of the plot. Create a SPICE file to reproduce the figure and confirm that your model works.

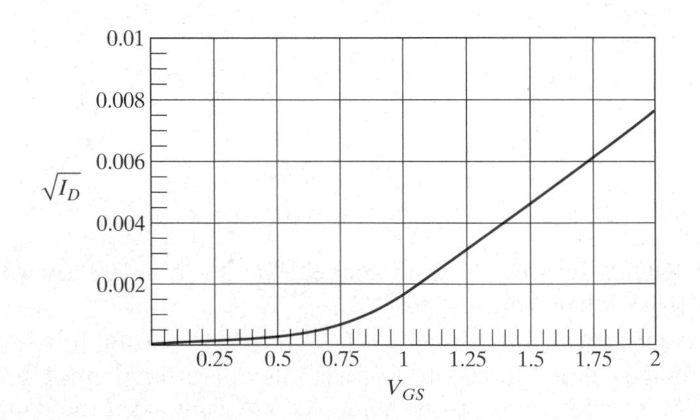

Figure 4–30 A plot of $\sqrt{I_D}$ versus V_{GS} for an n-channel MOSFET with $V_{GD} = 0$.

DS P4.18 **(a)** Design the circuit shown in Figure 4-31 to obtain $I_D = 0.5$ mA and a voltage gain of about –4. You are free to choose the values for W and L (assume $L_{min} = 1$ μm) and the DC component of the signal source. Assume that the transistor is modeled by the SPICE model given below and initially assume that $V_B = 0$ V. **(b)** Run a simulation to check your design. First, check the operating point and then the gain. **(c)** Now calculate by hand the values you expect for C_{gs} and C_{gd} at your operating point. Do they agree with the values given by SPICE? **(d)** Set the bulk bias at –0.5 V and rerun the simulation. Explain the results.

`.MODEL NMOS1 NMOS(level=1, vto=.5, tox=2.07E-8, kp=1E-4, lambda=0.04, gamma=0.5, LD=.1u, CGSO=1E-7, CGDO=1E-7, CGBO=2E-7)`; all other parameters assume their default values.

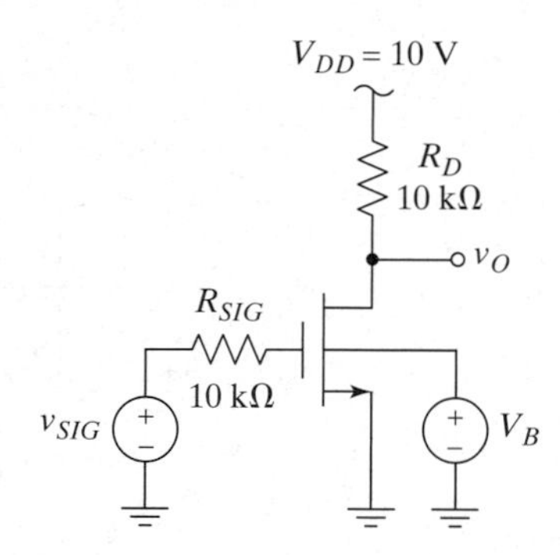

Figure 4–31 A common-source amplifier.

4.4 Macro Models in SPICE

S P4.19 We want to modify the op amp model shown in the solution to Exercise 4.5 to include positive and negative limits on the output voltage. **(a)** As a first try, use an ETABLE block for the final controlled source and assign it a gain of 100,000 and the lookup table (–5,–5)(5,5). Set the input voltage to be Vin = pulse (0 1 0 0) and run a 100 μs long transient analysis. Plot V(RA), V(vn) and V(out). Notice that V(RA) continues to ramp downward after the output limits. **(b)** Now set Vin = pulse(0 1 0 0 0 50u) and repeat the simulation from (a). Why doesn't V(out) start to change at 50 μs? **(c)** Now remove the limits on the output controlled source and come up with another way to limit the swing that will not have this problem. *Hint*: You can use voltage-controlled switches in SPICE, but the voltage difference between off and on can't be too small. Read the manual for details.

CHAPTER 5

Operational Amplifiers

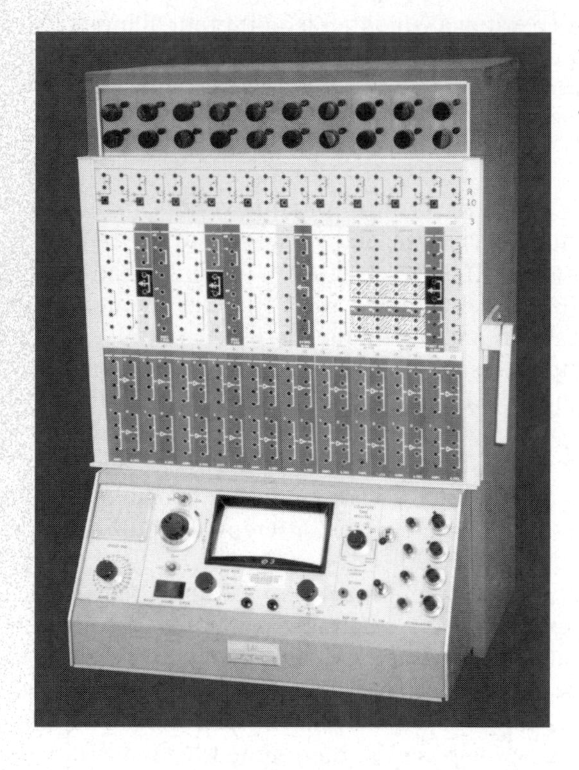

An analog computer from the 1960's. Operational amplifiers were used for gain, integration and summing.

INTRODUCTION

Operational amplifiers are one of the most important electronic building blocks of all time. Integrated op amps have been around since the mid-1960s, and one particular op amp, the LM741, has been in continuous production since it was first introduced in the late 1960s, a truly remarkable achievement in a field that changes as rapidly as the IC industry does.

Why are op amps so important? The answer lies in their versatility. Op amps can be used to perform mathematical operations like multiplication by a constant, addition, subtraction, integration, and differentiation, hence the name 'operational' amplifier. In addition, op amps can be used to produce comparators, waveform generators, current sources, voltage regulators, filters, and many other signal processing functions.

This chapter presents the analysis and design of circuits containing op amps. Although some of these circuits are undeniably digital in nature, the op amp is basically an analog building block. The chapter begins with the basic definition of an op amp, and the concept of an ideal op amp. Many circuits can be analyzed and designed using only the ideal op amp model. Later sections introduce many of the nonidealities present in real op amps and some of the ways in which they can be incorporated into analyses and compensated for by design. Along the way, many important and useful op amp circuits will be presented.

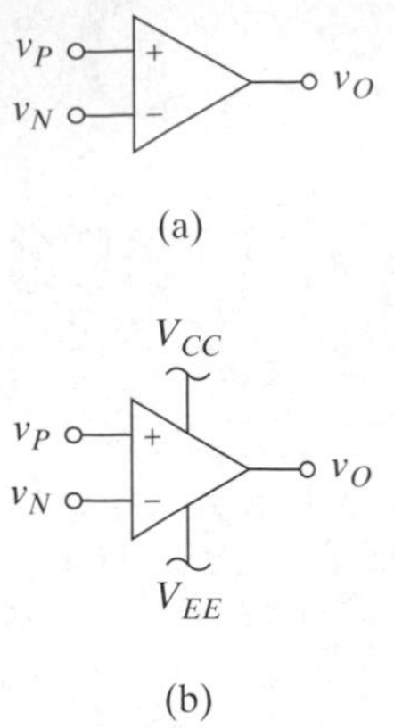

Figure 5–1 Op amp symbol (a) without power supply connections and (b) with power supply connections.

5.1 BASIC OP AMP CIRCUITS

An operational amplifier is a differential input, DC coupled[1] amplifier with a very large gain. The symbol is shown in Figure 5-1 and the input–output relationship is given by

$$v_O = a(v_P - v_N) \tag{5.1}$$

where v_P is the positive, or noninverting, input and v_N is the negative, or inverting, input, a is the gain, and v_O is the output voltage.[2] All terminal voltages are measured with respect to ground, but note that it is only the *difference* in the input voltages that matters in determining the output; hence, this is a ***differential input*** amplifier, and we frequently use $v_{ID} = v_P - v_N$ to denote the differential input voltage. Also notice that the output voltage is ***single-ended***; that is, it is measured with respect to ground. A single-ended output is not necessary, but it is by far the most common situation. Also note that if the supply voltages used with the op amp are not symmetric, there may be a significant DC component added to v_O. Nevertheless, the external circuit connection will almost always determine the DC output voltage when the differential input is zero, and we won't need to be bothered by this detail.

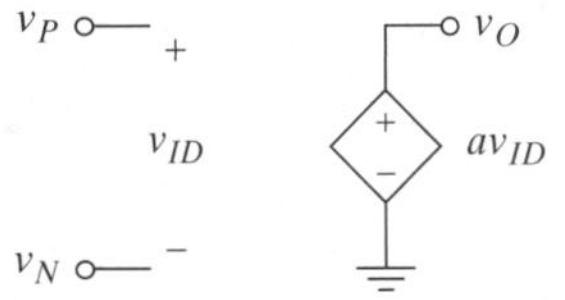

Figure 5–2 Circuit model of an op amp.

A real op amp has more connections than are shown in Figure 5-1(a), but the simplest possible configuration only adds power supply connections (usually a positive and a negative supply with no explicit ground connection to the op amp) as shown in part (b) of the figure, and for simplicity, these are frequently left off of a schematic during the analysis and design phase.

An op amp can be represented to first order by a voltage-controlled voltage source as shown in Figure 5-2. An ***ideal*** op amp is one that has zero input current, or equivalently, infinite input impedance (i.e., the open circuit between the inputs in the model), zero output impedance (i.e., no impedance in series with the controlled source in the model), infinite gain, and infinite bandwidth.

In the following sections we analyze a number of useful op amp circuits. Because op amps are invariably used with feedback, we must discuss feedback here as well. Feedback amplifiers, as the name suggests, take a sample of the output and feed it back to the input in some way. Feedback is discussed in detail in Chapter 10, but we will develop some intuition and basic concepts as we examine the noninverting amplifier in Section 5.1.1.

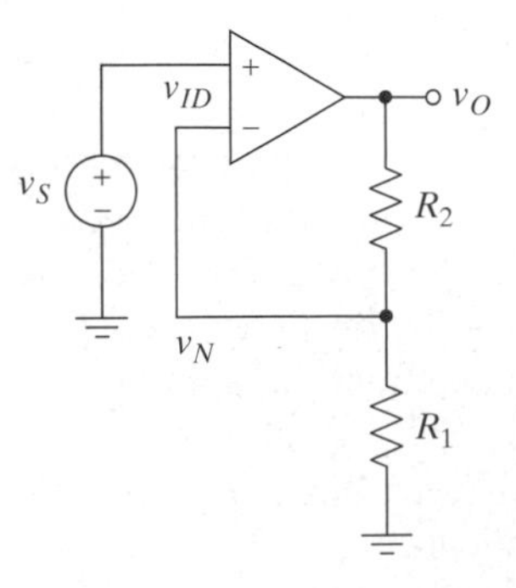

Figure 5–3 A noninverting op amp circuit.

5.1.1 The Noninverting Amplifier

A simple noninverting amplifier is shown in Figure 5-3. This connection provides ***negative feedback***, which is also sometimes called ***degenerative*** feedback. In other words, the feedback tends to reduce, or degenerate, the magnitude of any distur-

[1] Saying that an op amp is DC coupled simply means that (5.1) holds for DC voltages as well as AC voltages. AC and DC coupling of amplifiers is covered in depth in Chapter 6.

[2] We will use a lower case a for the gain of the basic op amp and an upper case A for the gain of amplifiers made using an op amp.

bance in the loop. To understand this statement, note that since the input current of an ideal op amp is zero, the voltage across R_1 is unaffected by the connection to the inverting input of the op amp, and we can write

$$v_{R_1} = v_N = \frac{R_1}{R_1 + R_2} v_O. \tag{5.2}$$

We also know that the output voltage of the op amp is given by (5.1), $v_P = v_S$, and v_N is given by (5.2). So, combining these equations we get

$$v_O = a(v_P - v_N) = a\left(v_S - \frac{R_1}{R_1 + R_2} v_O\right). \tag{5.3}$$

Equation (5.3) describes the operation of the feedback loop and can be solved for the transfer function from v_S to v_O. Before getting the transfer function, however, some intuition about the operation of the loop can be gleaned from this equation.

Suppose the loop is at a stable operating point, that v_O is finite, and that a is infinite. Now imagine that v_S increases; this causes the value of v_{ID}, which is nominally zero, to change to some small positive value, call it ε. As a result of the increase in v_{ID}, the value of v_O will increase positively in accordance with (5.1). The increase in v_O will, in turn, cause an increase in v_N as dictated by (5.2), and this change will reduce the magnitude of the original disturbance in v_{ID}. With infinite gain, the change will be compensated for entirely, and v_{ID} will again be zero. A similar line of reasoning works if the value of v_S is decreased.

Therefore, this particular connection will lead to a finite value for v_O and a zero value for v_{ID} so long as v_S is finite, a is infinite, and the ratio R_2/R_1 is finite. In other words, under these circumstances the op amp has a ***virtual short circuit***, or simply a virtual short between its inputs. It is 'virtual', because they aren't physically connected, and no current can flow. However, it acts like a short circuit in the sense that the two voltages must be the same.

In a sense, we just provided an operational definition of negative feedback. The phrase *negative feedback* means that the signal fed back from the output to the input of the ***forward*** amplifier (i.e., the op amp in Figure 5-3) opposes any change in v_{ID}. In the case of Figure 5-3, the op amp by itself is called the forward, or ***open-loop***, amplifier and R_1, R_2, and the op amp together comprise what is called the ***closed-loop*** amplifier. In this closed-loop amplifier, R_1 and R_2 comprise the ***feedback network***. The source, together with any load that is present (from v_O to ground), are outside the feedback loop.

We will consider in detail later what happens if the input terminals of the op amp are reversed, for now it is enough to state that it results in ***positive feedback***; that is, the signal fed back to the input ***increases*** the original change in v_{ID}. The result is that both v_{ID} and v_O increase without limit for the ideal op amp. For a realistic op amp there is some limit to how large v_O can be, but when that limit is reached, (5.1) no longer applies.

To summarize what we have learned; an ideal op amp will have a virtual short circuit between its input terminals *so long as it has a negative feedback connection*. This fact, combined with the assumption of no input current, makes the analysis of ideal op amp circuits particularly simple. In the case of Figure 5-3, these facts imply that we can write

$$v_S = v_P = v_N = \frac{R_1}{R_1 + R_2} v_O, \tag{5.4}$$

which leads directly to

$$v_O = \frac{R_1 + R_2}{R_1} v_S = \left(1 + \frac{R_2}{R_1}\right) v_S. \tag{5.5}$$

Notice that we get the same transfer function if we solve (5.3) for v_O and take the limit as a goes to infinity;

$$v_O = a v_S - \frac{aR_1}{R_1 + R_2} v_O, \tag{5.6}$$

or

$$v_O = \frac{a v_S}{1 + \dfrac{aR_1}{R_1 + R_2}}, \tag{5.7}$$

and, taking the limit yields

$$\lim_{a \to \infty} \frac{v_O}{v_S} = \lim_{a \to \infty} \frac{a}{1 + \dfrac{aR_1}{R_1 + R_2}} = \frac{R_1 + R_2}{R_1}, \tag{5.8}$$

which is the same as (5.5).

The noninverting amplifier is a useful building block in analog signal processing. It has a very high input impedance (ideally infinite), a low output impedance (ideally zero), and a well-controlled voltage gain set by a resistor ratio. The well-controlled gain is a demonstration of the main advantage of negative feedback; it allows us to trade gain for stability. The op amp by itself has a very large gain, but the magnitude of the gain is poorly controlled and can change with temperature and other factors. Using negative feedback, we are able to trade some of that large gain for stability in the sense that the closed-loop gain is lower than the open-loop gain, but is nearly independent of it. Note from (5.7) that so long as a is much larger than $(R_1 + R_2)/R_1$, the gain is nearly independent of a. Remember that a is a parameter of the forward amplifier and is *not affected by the feedback*; it is the closed-loop gain, not a, that is stabilized.

Exercise 5.1

Consider the circuit shown in Figure 5-4 and assume the op amp is ideal. (a) What is the gain v_L/v_S? (b) Does the gain depend on R_L at all? Why or why not?

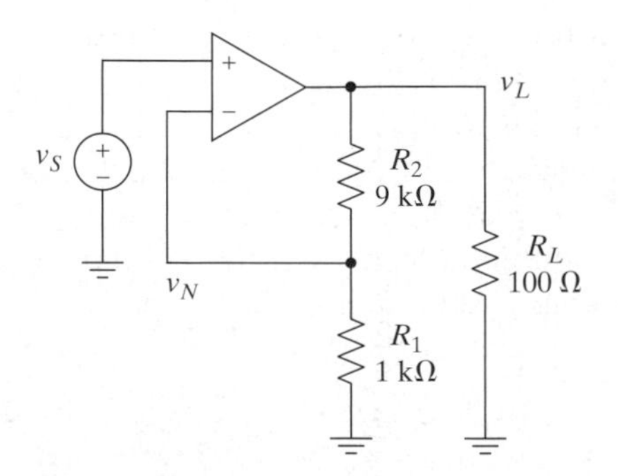

Figure 5–4 Circuit for Exercise 5.1.

□

Exercise 5.2

Consider the same circuit as in Exercise 5.1 only now model the nonzero output resistance of the op amp as shown in Figure 5-5. The op amp in the figure is still ideal. (a) What is the gain v_L/v_S now? (b) Explain why the gain still does not depend on R_L.

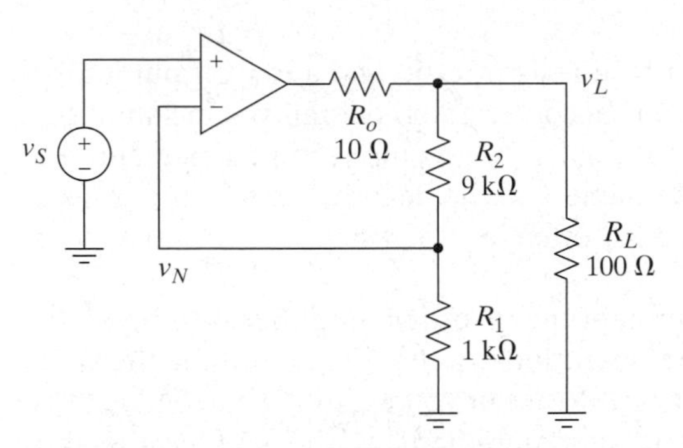

Figure 5–5 Circuit for Exercise 5.2.

5.1.2 The Inverting Amplifier

Figure 5-6 shows an inverting amplifier circuit. This configuration again uses negative feedback. In fact, by setting the source to zero (i.e., short it out) and redrawing the circuit, you can see that the feedback is the same as in the non-inverting amplifier of Figure 5-3 when its source is set to zero.

Since the circuit uses negative feedback, and we are assuming the op amp to be ideal, a virtual short will exist between the op amp inputs. With the noninverting input connected to ground, the virtual short implies that the inverting input will be a **virtual ground**. In other words, the voltage at the node is zero, but the current into the op amp is also zero, so we call it a *virtual* ground. Be careful to not misuse the concept of virtual ground! For example, if the noninverting input is connected to a 3 V supply, the inverting input is certainly *not* a virtual ground.

Having concluded that v_N is at ground in this case, the current through R_1 can be determined by Ohm's law;

$$i_{R_1} = \frac{v_S}{R_1}. \tag{5.9}$$

But where does this current go? We must remember that no current will flow into ground at the v_N node (it is only a *virtual* ground). In fact, since the op amp is assumed to be ideal, the current into the op amp is zero, and we must have

$$i_{R_2} = i_{R_1}. \tag{5.10}$$

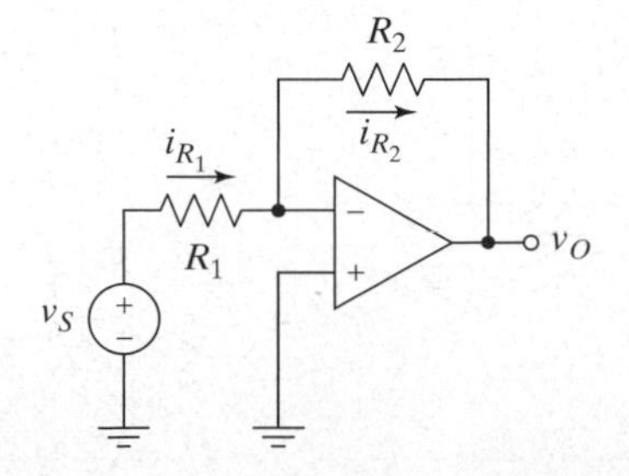

Figure 5–6 An inverting amplifier circuit.

The output voltage is then given by

$$v_O = v_N - i_{R_2}R_2 = -i_{R_2}R_2 = -i_{R_1}R_2, \tag{5.11}$$

or, substituting (5.9) into (5.11)

$$v_O = -\frac{R_2}{R_1}v_S. \tag{5.12}$$

The inverting amplifier has a gain set by a resistor ratio and a low output impedance just like the noninverting amplifier. In contrast to the noninverting amplifier, however, the inverting amplifier has a relatively low input impedance approximately equal to R_1. In addition, as the name and the negative sign in the transfer function imply, the output is inverted. In other words, when v_S is positive, v_O is negative and vice-versa.

A useful way to think of the operation of the inverting amplifier is to break the transfer function up into two separate operations. In the first operation, the virtual ground is used along with R_1 to convert the input voltage into a current, i_{R_1}. This current is then the input to a ***transresistance*** amplifier (i.e., the gain has the units of resistance, v_O/i_{R_2}) formed by the op amp and R_2 (*see* Problem P5.5).

This alternate view of the inverting amplifier operation leads directly to the idea of using the virtual ground as a summing node to form a summing amplifier as shown in Figure 5-7. There are three separate input currents shown in this figure, and they all sum together at the virtual ground to form the total input current for the transconductance stage;

$$i_{R_2} = i_{R_1} + i_{S2} + i_{R_3}. \tag{5.13}$$

Then we use (5.13) with (5.11), which still applies to this circuit, along with (5.9), which can be applied to both v_{S1} and v_{S2} with suitable changes in notation, to yield

$$v_O = -\frac{R_2}{R_1}v_{S1} - R_2 i_{S2} - \frac{R_2}{R_3}v_{S3}. \tag{5.14}$$

Figure 5–7 The inverting amplifier used as a summing amplifier.

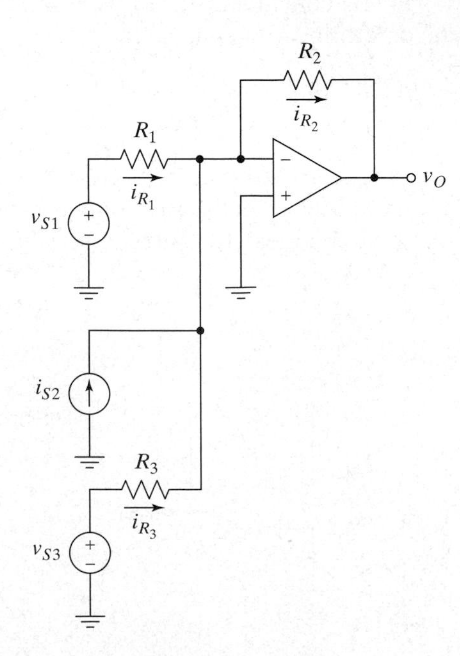

Example 5.1

Problem:
Find the input resistance of the inverting amplifier shown in Figure 5-6.

Solution:
The input resistance is the resistance seen by the source. Since R_1 is a part of the circuit, the source in this case has been modeled as being perfect; that is, it has zero output resistance. The resistance seen by the source is given by $R_i = v_S/i_S = v_S/i_{R_1}$. Substituting for i_{R_1} from (5.9) yields $R_i = R_1$.

■

Exercise 5.3

Consider the circuit in Figure 5-6. Determine values for R_1 and R_2 such that the closed-loop gain is −20, and the input resistance is 5 kΩ.

□

5.1.3 The Unity-Gain Amplifier, or Voltage Follower

The unity-gain amplifier, or voltage follower, is shown in Figure 5-8 and is a special case of the noninverting amplifier already covered (with $R_1 = 0$). Since the feedback is negative, and the op amp is ideal, there is a virtual short between the inputs. Therefore, we see by inspection that

$$v_O = v_I. \tag{5.15}$$

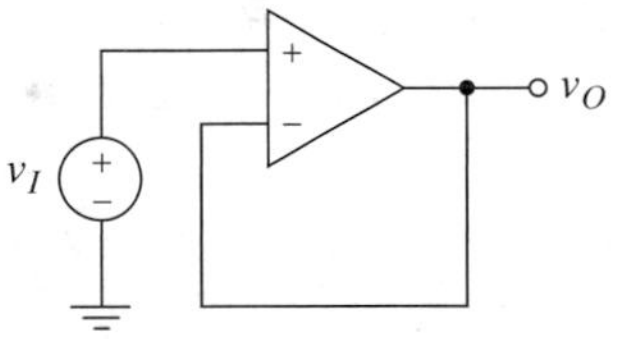

Figure 5–8 The unity-gain amplifier, or voltage follower, circuit.

The result is not as simple as (5.15) when a more practical model is used for the op amp (*see* Problem P5.8).

5.1.4 The Differential Amplifier

A differential amplifier circuit is shown in Figure 5-9. Since the ideal op amp is a linear circuit (consider the model in Figure 5-2, which has a single linear element), we can analyze this circuit using superposition. First, set v_{S2} equal to zero, and consider v_{S1} as the input. After doing this, we recognize that R_3 and R_4 are in parallel and the circuit can be redrawn as in Figure 5-10.

Since the input current to an ideal op amp is zero, both the current through and the voltage across the parallel combination of R_3 and R_4 in Figure 5-10 are zero. Therefore, the resistor $R_3 \| R_4$ does not affect the operation of the circuit and may be replaced by a short circuit; the result is the inverting amplifier of Figure 5-6. Therefore, the gain from v_{S1} to v_O is given by

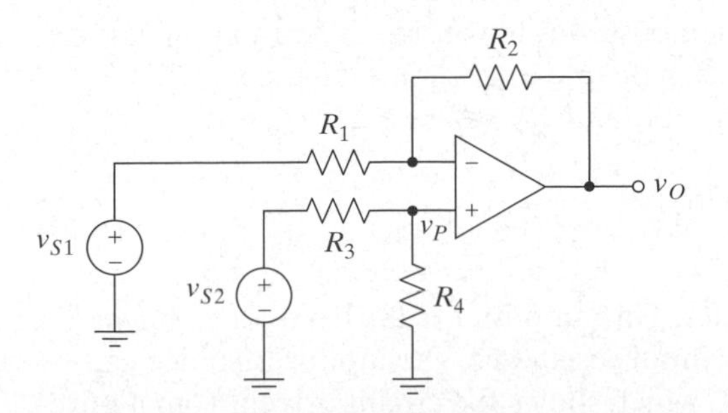

Figure 5–9 A differential amplifier circuit.

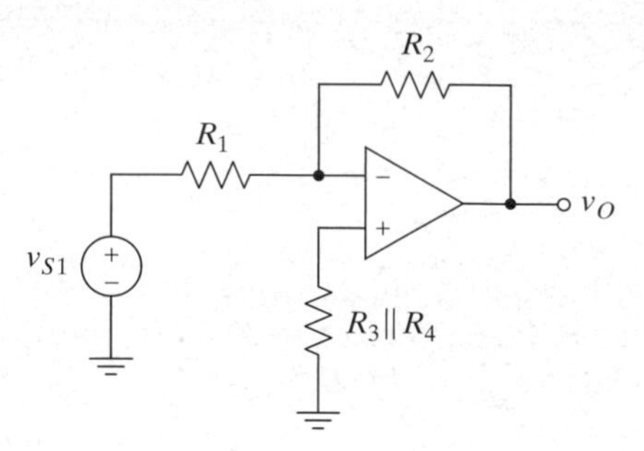

Figure 5–10 The circuit of Figure 5-9 with v_{S2} set to zero.

$$\frac{v_O}{v_{S1}} = -\frac{R_2}{R_1} \tag{5.16}$$

as in (5.12).

Similarly, when v_{S1} is shorted and we consider v_{S2} as the input, the circuit is identical to the noninverting amplifier of Figure 5-3, except for the presence of the voltage divider made up of R_3 and R_4. However, since the op amp input current is zero, the connection of the noninverting input of the op amp to this divider does not affect the operation of the divider in any way. Therefore, we can write

$$v_P = \frac{R_4}{R_3 + R_4} v_{S2}. \tag{5.17}$$

With v_{S1} shorted, the gain from v_P to v_O is given by the equation for the non-inverting amplifier, (5.5). Combining (5.5) and (5.17) while recognizing that v_S in (5.5) is equivalent to v_P here leads to

$$v_O = \left(\frac{R_4}{R_3 + R_4}\right)\left(1 + \frac{R_2}{R_1}\right) v_{S2}. \tag{5.18}$$

To finish up the analysis, we sum the results obtained for each input source individually:

$$v_O = -\frac{R_2}{R_1} v_{S1} + \left(\frac{R_4}{R_3 + R_4}\right)\left(1 + \frac{R_2}{R_1}\right) v_{S2}. \tag{5.19}$$

We are through with the analysis at this point, but it is worthwhile to consider how to make the output voltage a function of only the difference between v_{S1} and v_{S2}. In other words, we want to see how to make this circuit into a differential amplifier. Differential signals are extremely important and commonly used. They are dealt with in detail in Aside A5.1 and Aside A5.2 demonstrates how using differential signals can improve the performance of our circuits in the presence of interference.

Notice from (5.19) that if we set $R_3 = R_1$ and $R_4 = R_2$, we get

$$v_O = -\frac{R_2}{R_1} v_{S1} + \frac{R_2}{R_1} v_{S2} = \frac{R_2}{R_1}(v_{S2} - v_{S1}), \tag{5.20}$$

which describes a differential amplifier. This amplifier is far from ideal, however, mostly because of its low differential input resistance. The input resistance can be calculated with the aid of Figure 5-11, which shows the circuit driven from a purely differential source.

ASIDE A5.1 DIFFERENTIAL-MODE AND COMMON-MODE SIGNALS

A signal that is specified, generated, or measured as the difference in potential between some node and ground is called a ***single-ended*** or ***ground referenced*** signal.[1] It is sometimes convenient to express two single-ended signals as a differential signal. For example, consider the two single-ended voltage sources v_{S1} and v_{S2} shown in Figure A5-1(a). These two sources are connected to two nodes called v_A and v_B in the figure. We can also generate these two node voltages using the sources shown in part (b) of the figure.

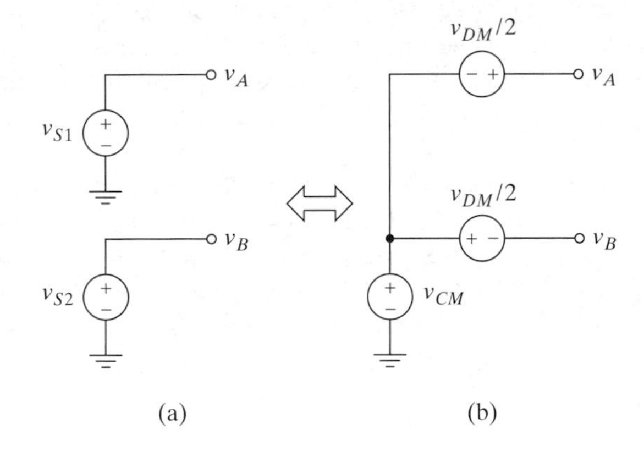

Figure A5-1 (a) Two single-ended voltage sources and (b) their representation as differential and common-mode sources.

Let's determine the values of the sources in part (b) of the figure necessary for the two circuits to generate the same values for v_A and v_B. From part (b) of the figure we have

$$v_A = v_{CM} + \frac{v_{DM}}{2} \tag{A5.1}$$

and

$$v_B = v_{CM} - \frac{v_{DM}}{2}. \tag{A5.2}$$

We can set $v_A = v_{S1}$ and $v_B = v_{S2}$ and solve to get

$$v_{DM} = v_{S1} - v_{S2}, \tag{A5.3}$$

which is the ***differential-mode component***, or the difference between the two single-ended signals, and

$$v_{CM} = \frac{v_{S1} + v_{S2}}{2}, \tag{A5.4}$$

which is the average or ***common-mode*** component of the two single-ended signals. This terminology makes good sense, since the common-mode source is common to both v_A and v_B as seen in both Figure A5-1(b) and Equations (A5.1) and (A5.2).

[1] Remember that a voltage is by definition the difference in potential between two points. Therefore, when we say a node is at some voltage (e.g., 2.4 V) what we mean is a voltage relative to some agreed upon reference, usually taken to be ground.

ASIDE A5.2 GROUNDING, DIFFERENTIAL-MODE SIGNALS AND INTERFERENCE REJECTION

In circuit analysis, the assumption is often made that all connections to ground represent a single node. However, in practice such a universal ground node does not exist, because connections are made with real conductors that have nonzero resistance and inductance. Therefore, the various "grounded" nodes in a practical circuit may not all be at the same potential.

In many cases, it is a reasonable and helpful approximation to ignore these differences in ground potential (e.g., for small blocks or circuitry where relatively low impedance connections to a *local ground* can be made). Nevertheless, it is sometimes necessary to recognize that the local grounds of two blocks of circuitry may not be the same. In this case, it is necessary to distinguish one "ground" from another. This aside illustrates a problem caused by a difference in ground potentials and introduces the differential amplifier as a way of solving it.

Suppose you are designing a control system for the heating, ventilating, and air conditioning (HVAC) system in some large building. Such systems use many sensors to measure airflow and temperature, and these sensors are often located a considerable distance from the control electronics. Consider just one of these sensors, and assume we want to connect it to a receiving amplifier in the control electronics.

Figure A5-2 shows the connection of the sensor to the receiving amplifier. The wires connecting the sensor to the amplifier are modeled by ideal conductors in series with resistances, R_{W1} and R_{W2}. The sensor is modeled as a voltage source v_s and its associated resistance R_S. Node v_{g1} is the local ground at the sensor and node v_{g2} is the local ground at the receiver. The amplifier is assumed to have an input resistance to local ground equal to R_i.

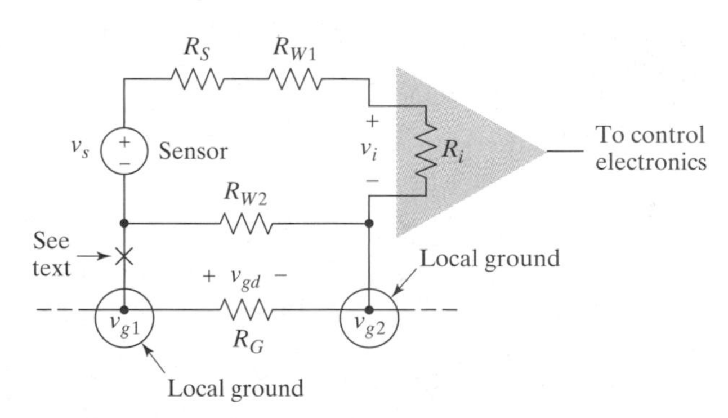

Figure A5-2 A remote sensor in an HVAC system connected to an amplifier in the control electronics.

The local grounds are connected together by the wires of the distributed ground system which have resistance R_G. Because there may be current flow in the ground connections, there is no reason to believe that the two local grounds are at the same potential. In fact, current from other electrical equipment flowing into the local ground nodes can create differences in the ground potentials that are quite large and also time-varying. For the purpose of analysis, the difference in potential between the local grounds will be denoted by v_{gd}. We do not know the precise value of v_{gd} and will not try to obtain an expression for it. This is acceptable, because our goal is to analyze the *effect* of the ground potential difference on the system performance, recognizing that the magnitude of that effect depends on the magnitude of v_{gd}.

Analysis of Figure A5-2 yields (since we assume we know v_{gd} we can treat it as a voltage source; therefore, R_{W2} does not enter into the equation)

$$v_i = (v_s + v_{gd})\frac{R_i}{R_i + R_S + R_{W1}}. \tag{A5.5}$$

From (A5.5) we see that v_i is not equal to v_s. For one thing, the signal is attenuated going from v_s to v_i, but if $R_i \gg (R_s + R_{W1})$, the loss can be neglected. A far more significant problem is that the unknown signal v_{gd} has been effectively added to the desired signal v_s, and there is no way to separate them. In other words, v_{gd} represents an error signal—or ***interference***—that has been added to the signal v_s.

One approach to solving the problem would be to use a single-point grounding scheme, which can be accomplished by disconnecting the sensor from its local ground at point "X" as shown in the figure. Analysis after this change yields:

$$v_i = v_s \frac{R_i}{R_i + R_S + R_{W1} + R_{W2}}, \tag{A5.6}$$

which does not show an interference component because of v_{gd}. Single-point grounding is often a good technique for avoiding problems caused by differing ground potentials, but it should be used with caution.[1]

Another solution to the problem is to use a differential amplifier for the receiver, as shown in Figure A5-3. To calculate the input voltage of the receiving amplifier, we will separately calculate the voltage at each of its input nodes, called v_P and v_N. To express these as single-ended voltages, they will be calculated with respect to the local ground at the receiving amplifier (i.e., assigning $v_{g2} = 0$). Using superposition, we get:

$$v_P = v_{gd} + v_s\left(\frac{R_{id} + R_{W2}}{R_{id} + R_{W2} + R_{W1} + R_S}\right) \tag{A5.7}$$

and

$$v_N = v_{gd} + v_s\left(\frac{R_{W2}}{R_{id} + R_{W2} + R_{W1} + R_S}\right). \tag{A5.8}$$

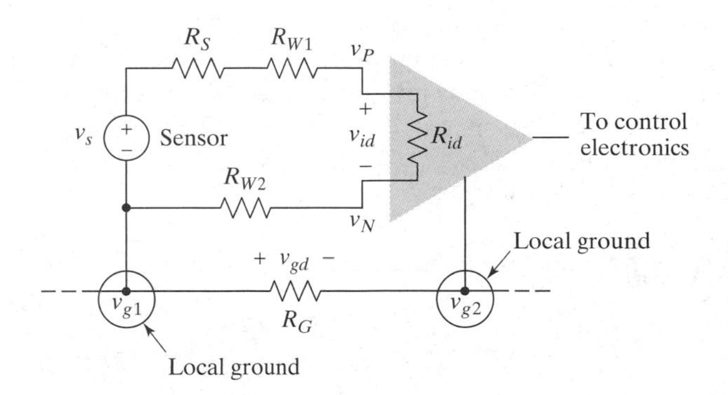

Figure A5-3 A remote sensor in an HVAC system connected to a differential amplifier in the control electronics.

Using (A5.7) and (A5.8) we find that the differential input voltage of the amplifier is:

$$v_{id} = v_P - v_N = v_s\left(\frac{R_{id}}{R_{id} + R_{W2} + R_{W1} + R_S}\right). \tag{A5.9}$$

This equation can also be derived by writing KVL around the loop containing the source, R_S, the wire resistances, and the input of the amplifier. From (A5.7) and (A5.8) we see that the same interference signal, v_{gd}, appears at both inputs of the differential amplifier; hence, it is a common-mode signal. The differential-mode input signal given by (A5.9) is only a function of v_s, so the common-mode interference will be rejected by the op amp and will not appear at the output. Practical op amps do respond to the common-mode input to a small degree, but the output is still predominately a function of v_s alone.

[1] Single-point grounding is fine at low frequencies for low-level signals, but it can cause problems at high frequencies. For example, the impedance of a wire increases rapidly with increasing frequency owing to the inductance and skin effect. Therefore, the chassis of a piece of electronic equipment may be a better high-frequency ground than a long wire is, even if it is a shared ground. Long wires may also radiate or pick up significant amounts of radio frequency interference.

A brief summary at this point will help to clarify the value of a differential amplifier. The input voltage of a single-ended amplifier is defined with respect to its own local ground. If the corresponding source voltage is generated with respect to a different ground, the difference in the ground potentials will be added to the signal. On the other hand, a differential amplifier, responds to the difference in voltage between its two inputs, so its own local ground does not have to be in the same loop as the desired signal. Therefore, a differential amplifier can be used to convert a signal referenced to one ground into the same signal referenced to another ground.

There are three other common ways in which interference can couple into this circuit

1 If there are other signal wires close enough so that the capacitance between those lines and the sensor leads is significant, the signals can couple capacitively (very small capacitances can be significant if the impedance of the node being coupled to is large),
2 If the sensor leads are part of a loop, changing magnetic fields like those around power lines can induce voltages in the loop, and
3 The sensor leads can act as antennas and pick up electromagnetic fields.

In case 2) the interference produces a current in the input loop and will generate a differential input. Therefore, differential input amplifiers will not help. In case 3), the situation can be quite complicated, but the interference can be a combination of differential and common-mode components. The common-mode components can be rejected by a differential amplifier. Differential amplifiers will also help with interference that capacitively couples into the circuit, as we now demonstrate.

We can model interference capacitively coupling into the circuit as shown in Figure A5-4. The figure does not show the resistance of the sensor leads R_{W1} and R_{W2}, since they are negligibly small compared to R_S and R_{id} and will not affect our conclusions in this case. The interfering signals are represented as current sources (i_{i1} and i_{i2}) connected to the signal wires. The local ground at the receiver is denoted by the symbol for ***chassis ground*** in this figure. A chassis ground symbol is used to represent the local ground in a system when it is important to distinguish it from earth ground. We are going to use the chassis ground as our reference in the following analysis (i.e., we define the voltage at that point to be zero volts).

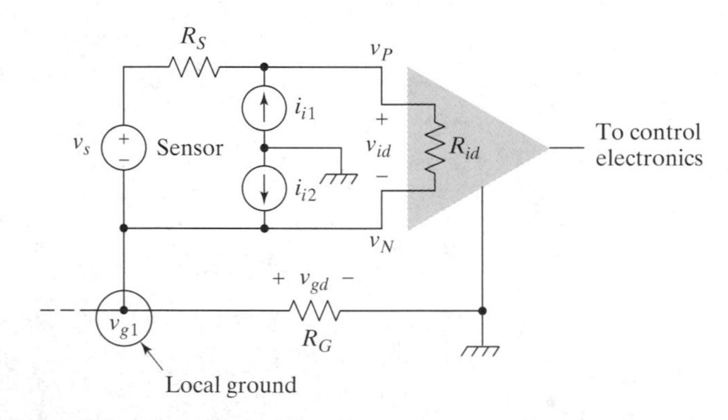

Figure A5-4 A model for analyzing the effect of interfering signals.

Once again, we will calculate the voltages at v_P and v_N with respect to the chassis ground. We treat the ground difference v_{gd} as if it were a voltage source in this analysis. Using superposition we find that the component at v_P due to i_{i1} alone is (treat v_{gd} as a voltage source, therefore set it to zero here)

$$v_P\Big|_{\text{due to } i_{i1}} = i_{i1}(R_S \| R_{id}), \quad \textbf{(A5.10)}$$

the component due to i_{i2} alone is

$$v_P\Big|_{\text{due to } i_{i2}} = 0, \quad \textbf{(A5.11)}$$

the component due to v_s is

$$v_P\big|_{\text{due to } v_s} = v_s\left(\frac{R_{id}}{R_{id} + R_S}\right) \qquad \textbf{(A5.12)}$$

and the component due to v_{gd} is

$$v_P\big|_{\text{due to } v_{gd}} = v_{gd}. \qquad \textbf{(A5.13)}$$

Combining these equations we find

$$v_P = i_{i1}(R_S \| R_{id}) + v_s\left(\frac{R_{id}}{R_{id} + R_S}\right) + v_{gd}. \qquad \textbf{(A5.14)}$$

Similarly, we find

$$v_N = v_{gd}, \qquad \textbf{(A5.15)}$$

so that

$$v_{id} = v_P - v_N = v_s\left(\frac{R_{id}}{R_{id} + R_S}\right) + i_{i1}(R_S \| R_{id}). \qquad \textbf{(A5.16)}$$

We see from (A5.16) that although the ground difference has been rejected, the interference can result in a differential-mode signal at the input of the receiving amplifier.

A solution to this problem is suggested by the fact that the interference to which each wire is exposed is very nearly the same (assuming that the two wires are physically close to each other, which is standard practice). Therefore, the two interference current sources produce nearly the same signal and we have $i_{i1} \approx i_{i2}$. If the circuit is altered so that the impedance to ground is the same for each sensor lead, the same interference signal will appear at each op amp input and the resulting common-mode interference will be rejected by the amplifier. Such a circuit is called **balanced** in reference to its symmetry. To balance the circuit of Figure A5-4, a resistor equal to R_S could be placed in series with the connection from the bottom of the signal source to v_N.[2]

The circuit can also be balanced by disconnecting the signal source from its local ground connection, as with a single-point grounding scheme. The simplified model of Figure A5-4 would then not show any path to ground for the interference current, and a more complicated model would be needed for analysis (one that includes the common-mode input resistance of the receiver and stray capacitance from the sensor leads to ground).

A few final comments are in order. The two leads used in a differential system like that in Figure A5-4 are often twisted together (commonly called a **twisted-pair**). Twisting the wires together reduces the area of the loop that is formed, decreasing the magnitude of the magnetically induced voltage. Each twist can be seen as a small-area loop, but the voltages induced in adjacent twists have opposite signs, so they tend to cancel. The twisting also ensures that the wires are physically adjacent, so that most capacitively coupled interference will be the same for each wire and consequently be a common-mode signal. In addition, the twisted pair may be put inside a shield to prevent the pickup of high-frequency common-mode interference. High-frequency interference can be a problem, since real amplifiers do respond to the common-mode component to some degree, and the response is larger at higher frequencies.

---

[2] We state this option to be complete, but we must remember that v_s and R_S comprise a Thévenin model of the source. It is usually not possible or necessary to add a resistor, as mentioned here. The common solution is to remove the local ground connection, as noted in the next paragraph.

Because of the purely differential source, the currents through R_1 and R_3 are equal[3]. Also, note that the ideal op amp and negative feedback connection force a virtual short between v_P and v_N. Therefore, we can write KVL around the loop formed by R_1, R_3, v_S, and the virtual short. Starting from the noninverting input and moving clockwise around the loop, we get

$$i_S R_3 - v_S + i_S R_1 = 0, \tag{5.21}$$

or,

$$v_S = i_S(R_3 + R_1). \tag{5.22}$$

We can solve (5.22) to get the differential input resistance;

$$R_{id} = \frac{v_S}{i_S} = R_3 + R_1 = 2R_1. \tag{5.23}$$

Since the gain is equal to R_2/R_1, and there are practical limits on how large R_2 can be, R_1 cannot be extremely large. Therefore, this circuit turns out to have lower R_{id} than we would like. In the next section, we discuss one possible solution to this problem.

Exercise 5.4

Consider the differential amplifier in Figure 5-11. Determine values for R_1, R_2, R_3, and R_4, so that the amplifier has a differential gain of 10 and a differential input resistance of 20 kΩ.

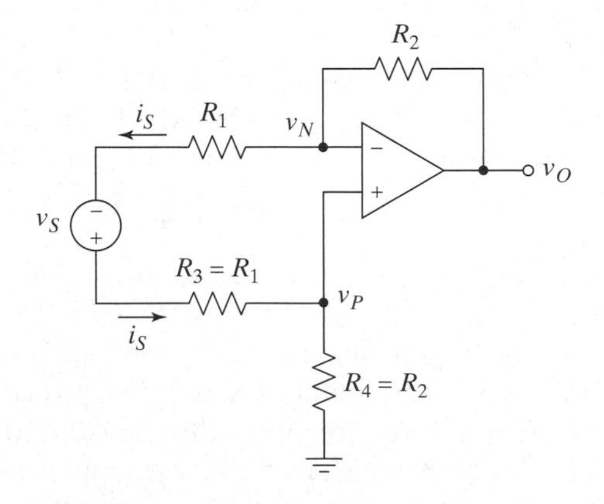

Figure 5–11 Finding the differential input resistance of the circuit in Figure 5-9.

□

[3] We cannot make the same statement for either source in the circuit in Figure 5-9, because the input is not *purely* differential.

5.1.5 The Instrumentation Amplifier

As the name implies, an ***instrumentation amplifier***, is one that is useful for different kinds of electronic measuring instruments. It is common in such situations to want to use differential inputs and to desire the results to be accurate and independent of the resistance of the leads connecting the source to the instrument. Therefore, we desire a differential amplifier with an input impedance much higher than the simple differential amplifier shown in Section 5.1.4.

Approaching this problem as a designer, we ask: what configuration do we know about that has a high input impedance? After some thought, and perhaps a look back at the previous sections, we remember that the noninverting amplifier stage has a high input impedance. So, one possible solution would be to use two noninverting amplifiers in front of the differential amplifier of Figure 5-9. If we do this, the noninverting amplifiers would be serving as ***buffers***. That is, they would isolate the source from the input of the differential amplifier. We don't need voltage gain in these buffers, so we could use the unity-gain configuration.

This approach would certainly meet our stated goal, but it has at least one drawback; the gain of the overall amplifier cannot be adjusted by varying a single resistive element. If we could adjust the gain with a single element, the input to our instrument could have a gain control that would be inexpensive to implement.

Let's begin our search for a solution by driving the standard differential amplifier by two noninverting amplifiers as shown in Figure 5-12. In this case, we have set it up so that both of the noninverting amplifiers have the same gain, $(1 + R_2/R_1)$, and so that the differential stage has the same magnitude of gain, R_4/R_3, from each input to the output. The overall gain of the amplifier is then given by

$$\frac{v_O}{v_S} = \left(1 + \frac{R_2}{R_1}\right)\frac{R_4}{R_3}. \tag{5.24}$$

The circuit as shown works, but it does not allow us to vary the gain with a single element, and it does not to reject common-mode signals as well as possible. We will analyze the common-mode rejection problem in Section 5.4.6 (*see* Example 5.7), but the circuit performs better if we remove the ground connection between the two R_1s in the figure and allow that node to float.

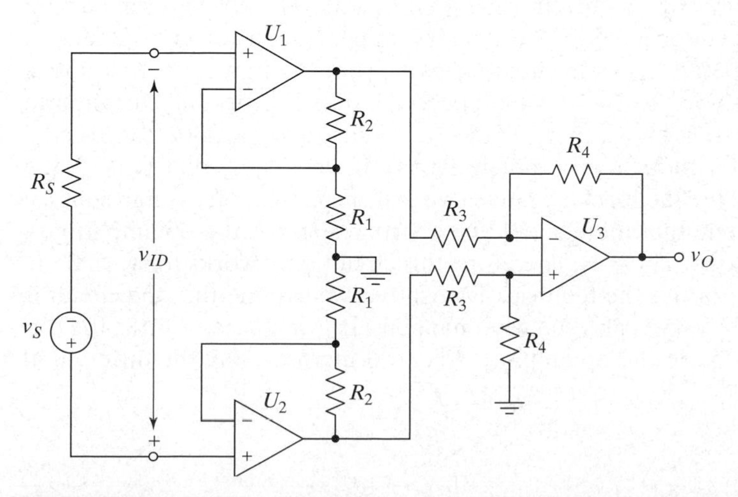

Figure 5–12 A differential amplifier with two noninverting amplifiers used to increase the input impedance.

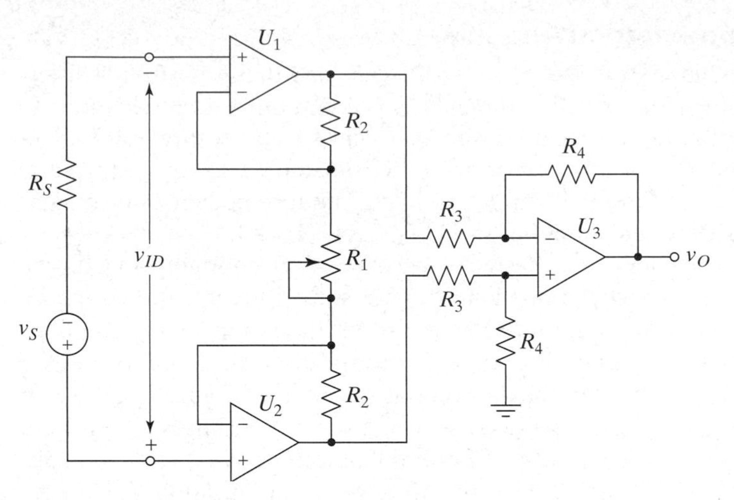

Figure 5–13 An instrumentation amplifier circuit.

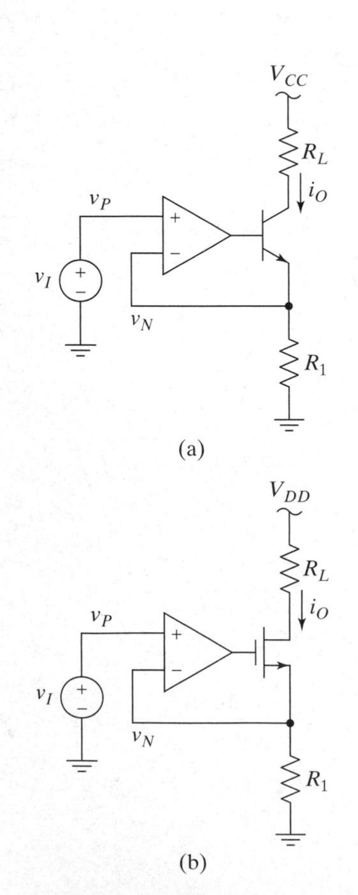

Figure 5–14 Op amp current sources; (a) using a bipolar transistor, and (b) using a MOSFET (other kinds of FETs can also be used).

Making this change also allows us to combine the two R_1s into a single variable resistor. This variable resistor will then set the gain of both sides at the same time, thereby allowing us to vary the overall differential gain with a single element. The resulting amplifier is a standard configuration known as an instrumentation amplifier and is shown in Figure 5-13. We encourage you to verify that the resulting gain is: $v_O/v_S = (1 + 2R_2/R_1)(R_4/R_3)$.

5.1.6 Current Sources[4]

An op amp can be used to construct a current source as shown in Figure 5-14. The transistor can either be a bipolar or field-effect transistor. The important points are that the voltage at the top of R_1 is controlled by the op amp output voltage, and the current into the transistor's base, or gate, is negligible in comparison to the output current. The circuits could more properly be called current *sinks,* since the output current is directed inward. Similar circuits can be constructed that have output currents that are directed out of the circuit. However, the distinction between a current source and a current sink is often ignored, and the term current source is used as a generic label for any circuit that provides a current output.

To analyze the operation of the circuit in part (a) of the figure, we first notice that v_N must be positive, or else we would need the transistor to pull current into its emitter. An *npn* transistor cannot do this (a similar source can be constructed using a *pnp* transistor though *see* Problem P5.14). If we assume that $v_N > 0$, the op amp is ideal, and the feedback negative (we will check this last assumption in a moment), we conclude that there should be a virtual short at the op amp inputs, and we can say that $v_N = v_P = v_I$. Therefore, this circuit only works for $v_I > 0$.

To check whether or not the feedback is negative, we assume that the circuit is functioning properly, vary v_I and see what happens. If v_I increases a little, the differential input voltage of the op amp, $v_P - v_N$, will increase. Say the differential

[4] This section requires some knowledge of transistor operation.

input increases by an amount ε. The op amp output voltage will then increase by $a\varepsilon$, where a is the gain of the op amp. The increase in op amp output voltage will in turn increase the v_{BE} of the transistor. Therefore, the current through the transistor will increase. Since the ideal op amp input current is zero, we note that $v_N = v_{R_1} = i_E R_1$; when the transistor current increases, so will v_N. Therefore, the feedback reduces the magnitude of ε and is negative. With an ideal op amp, a goes to infinity, and ε will be driven to zero (i.e., a virtual short).

Now that we know that $v_N = v_I$, we can immediately conclude from Ohm's law and the fact that the op amp input current is zero that

$$i_O = \alpha i_E = \alpha \frac{v_I}{R_1} \approx \frac{v_I}{R_1}, \tag{5.25}$$

where the approximation is valid, since the transistor's alpha is close to one, and we are ignoring any effect the collector voltage might have on i_O. It is interesting that the exact nature of the transistor's i_C versus v_{BE} characteristic does not matter. This independence occurs because the feedback forces the op amp output voltage; hence, the transistor's v_{BE} is whatever is required to set v_N equal to v_I. We must also remember that the circuit only works for $v_I > 0$.

The analysis of the FET circuit in part (b) of the figure is the same, except that the value of α *is* one for the MOSFET at low frequencies. There is a subtle difference in the two circuits though. In the case of the bipolar transistor, the voltage at the top of R_1 is controlled by the op amp output whether or not the collector of the transistor is connected to anything, since there is still a *pn* junction from base to emitter, and the current through R_1 can be supplied by the op amp directly. In the case of the FET, there must be a load connection, so that current can be supplied to R_1. Otherwise, the gate voltage does not control the voltage across R_1, and the negative feedback is lost.

5.1.7 Voltage Regulators[5]

A voltage regulator is a circuit that ideally provides a constant voltage independent of changes in the load, the temperature, the unregulated supply from which it is driven, or any other parameter. An op amp can be used to perform part of the function of the voltage regulator, as shown in Figure 5-15. The circuit shown is called a **series regulator**, since the regulator appears in series with the unregulated

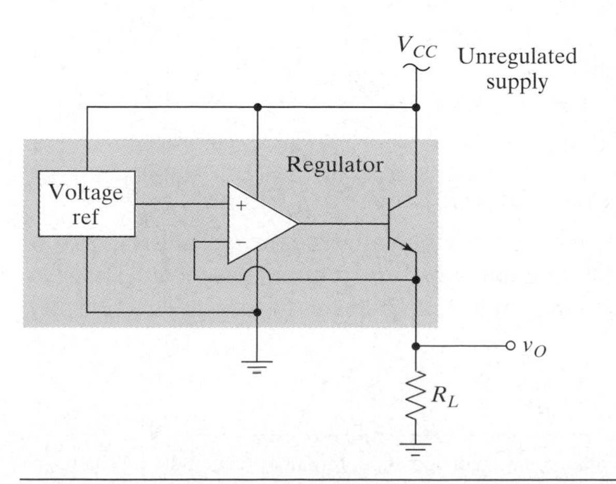

Figure 5–15 A series voltage regulator circuit.

[5] This section requires some knowledge of transistor operation.

supply, V_{CC}, and the load. The transistor present is called the *pass transistor*, since the load current passes through it. The figure also explicitly shows that the voltage reference and the op amp derive their power from the unregulated supply. The reference may derive its power from a separate supply to avoid variations in the supply caused by changes in the load current.

It is also possible to construct *shunt* regulators, where the regulator appears in parallel (i.e., in shunt) with the load, although we will not pursue that topic here (*see* Problem P5.16). For a series regulator, the unregulated supply voltage must be larger than the desired output. The minimum difference between the regulated and unregulated voltages for which the circuit will function is called the *dropout voltage* of the regulator.

This circuit is an example of a noninverting amplifier as covered in Section 5.1.1. The idea is that the voltage reference provides an output voltage that is stable with respect to changes in the unregulated supply, temperature, and other variables. The op amp and pass transistor together buffer the reference from the load.

It is much easier to design a stable reference, if it does not need to provide much output current and if its load is fairly constant. In other words, we have taken a divide-and-conquer approach to the design problem; we first generate a stable reference voltage and then use that reference along with negative feedback to provide a stable output voltage that can provide significant current to a potentially varying load. The pass transistor can be eliminated all together if the op amp can provide sufficient current on its own. However, the transistor is sometimes on a separate piece of silicon, even in packaged regulators, because it dissipates most of the power in the circuit and there can be significant problems with thermal feedback to the op amp's internal circuitry.

5.2 FREQUENCY-DEPENDENT OP AMP CIRCUITS

In this section we use op amps and reactive elements[6] to generate some useful and interesting circuits with frequency-dependent characteristics. Circuits that are deliberately designed to have frequency-dependent transfer functions are called *filters*.

Filters that include op amps are called *active filters* because they include active components, rather than being purely passive RLC filters. Unlike passive filters, active filters can have power gain and a low output impedance. In addition, active filters do not have to include inductors in order to generate transfer functions with complex poles (*see* Chapter 11), whereas passive RC circuits are incapable of producing complex poles. Complex poles are necessary to generate filters with transfer functions that roll off sharply (i.e., are very selective). We defer further discussion of this topic to Chapter 11.

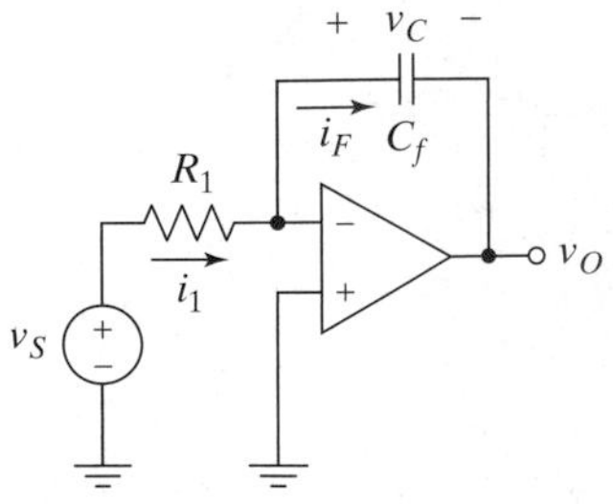

5.2.1 The Integrator and First-Order Low-Pass Filter

Consider the op amp circuit in Figure 5-16. We can analyze the circuit by first recognizing that it does have negative feedback for all frequencies except DC (i.e., zero frequency). At DC, the impedance of the feedback capacitor goes to infinity,

Figure 5–16 An integrator circuit.

[6] A reactive element is any element that has a nonzero imaginary component (called the reactance) to its impedance.

and the feedback is nonexistent. Ignoring DC operation for the moment, we can recognize that the negative feedback and ideal op amp will combine to produce a virtual ground at the inverting input. Therefore, we know that $i_1 = v_S/R_1$, and because the input current to the op amp is zero, we can also say that i_F is equal to i_1. Remembering that the voltage across a capacitor is proportional to the integral of the current into it, we can write

$$v_O(t) = -\frac{1}{C_f}\int_0^t \frac{v_S(\tau)}{R_1}d\tau - v_C(0). \quad \textbf{(5.26)}$$

If we assume that the capacitor is initially uncharged, the output voltage is proportional to the integral of the source voltage;

$$v_O(t) = -\frac{1}{R_1C_f}\int_0^t v_S(\tau)d\tau. \quad \textbf{(5.27)}$$

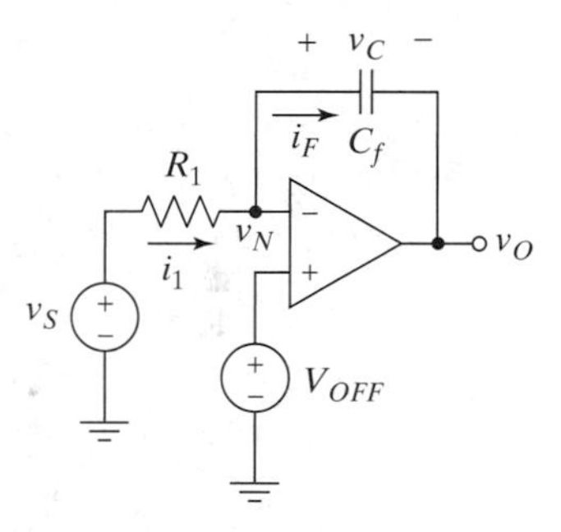

Figure 5–17 The integrator of Figure 5-16 with op amp offset included.

Now we need to examine what happens at DC. We will discover in Section 5.4.3 that all real op amps have some nonzero offset voltage that can be modeled as a voltage source in series with one of the inputs. In Figure 5-17 we have added this small source in series with the noninverting input of the op amp and have given it a magnitude V_{OFF}.

We only need to change our analysis slightly to handle this new situation. Instead of having the inverting input at ground, we now have $v_N = V_{OFF}$. This leads to

$$i_1(t) = \frac{v_S(t) - V_{OFF}}{R_1}, \quad \textbf{(5.28)}$$

and plugging this into the integral relationship while still assuming no initial charge storage leads to

$$v_O(t) = v_N - v_C = V_{OFF} - \frac{1}{C_f}\int_0^t \frac{v_S(\tau) - V_{OFF}}{R_1}d\tau \quad \textbf{(5.29)}$$

or,

$$v_O(t) = V_{OFF}\left(1 + \frac{t}{R_1C_f}\right) - \frac{1}{R_1C_f}\int_0^t v_S(\tau)d\tau. \quad \textbf{(5.30)}$$

Examination of (5.30) reveals that even if the signal source is set to zero, the output voltage will ramp towards infinity just because of a finite offset voltage. In other words, if we try to find the steady-state solution at DC, there isn't one; the output magnitude continues to increase without bound.

This action is a manifestation of the fact that the capacitor impedance goes to infinity at DC, so the DC gain of the inverting amplifier goes to infinity. With a real op amp, the gain of the circuit will not go to infinity, but the output will nevertheless ramp until it is limited by some other mechanism. Therefore, we can't really build an integrator as shown in Figure 5-16.

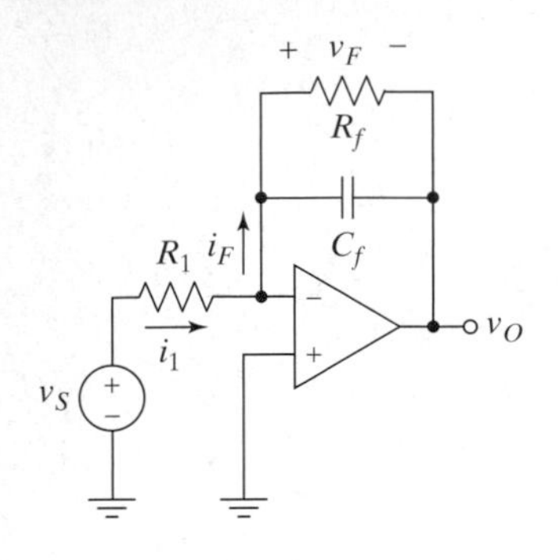

Figure 5–18 An integrator circuit with finite DC gain.

To correct this problem, we either need a switch in parallel with the feedback capacitor that will enable us to periodically reset the voltage across it to zero, or else we need some way of limiting the DC gain. In Figure 5-18 we show a modified integrator that has a finite DC gain.

For this circuit, we again have $i_1(t) = v_S(t)/R_1$ and $i_F = i_1$. If we define Z_f to be the feedback impedance (i.e., the parallel combination of R_f and C_f), then we have

$$Z_f(j\omega) = \frac{R_f}{1 + j\omega R_f C_f} \tag{5.31}$$

and

$$V_o(j\omega) = -I_f(j\omega)Z_f(j\omega) = \frac{-V_s(j\omega)}{R_1}Z_f(j\omega). \tag{5.32}$$

Plugging (5.31) into (5.32) and simplifying leads to

$$\frac{V_o(j\omega)}{V_s(j\omega)} = \frac{-R_f/R_1}{1 + j\omega R_f C_f}, \tag{5.33}$$

which is a single-pole, low-pass transfer function with a DC gain of $-R_f/R_1$ and a pole frequency of $\omega_p = 1/R_f C_f$. We can see that as R_f is increased, this circuit becomes a better approximation to the ideal integrator. The practical upper limit on R_f is set by the op amp offset and other factors we have not yet discussed.

Interestingly, this circuit is really a low-pass filter (LPF), as mentioned above. A true integrator has the transfer function $H(j\omega) = 1/j\omega$, which has a pole at DC (i.e., the magnitude of $H(j\omega)$ goes to infinity as ω approaches zero). Since we cannot build stable circuits that have infinite DC gain, we must approximate an integrator as a LPF with a very low pole frequency. The circuit only functions like an integrator for frequencies well above the pole frequency.

One final observation can be made about the integrator circuit. We could have written the transfer function (5.33) by using the formula we derived for the inverting amplifier circuit, (5.12), along with the complex impedances. In other words, (5.12) still holds even if the resistors are replaced by impedances. This statement is true of all the transfer functions we have derived for ideal op amp circuits so far. Nothing in the analyses performed would preclude their being used with complex impedances instead of resistances.

Example 5.2

Problem:
Suppose we have an op amp with a DC gain of 200,000 and we want to build a circuit that will be a reasonable approximation to an integrator for signals above 100 Hz. Further assume that we must limit the DC gain of the integrator to at most 2000 to avoid problems with DC offsets. If R_1 has already been chosen to be 1 kΩ for other reasons, find values for R_f and C_f.

Solution:
Using the maximum allowable DC gain we find that $R_f \leq 2000R_1 = 2\ \text{M}\Omega$. To do a reasonable job of approximating an integrator, assume that the pole frequency

must be at least two decades below the signal frequency.[7] Using this approximation implies that our pole must be no greater than 1 Hz. If we use the largest possible value for R_f, we must use $C_f \geq 80$ nF to achieve a pole less than 1 Hz. A reasonable standard value would be $C_f = 0.1$ μF. However, we might have trouble making this circuit work, since most capacitors as large as this have an equivalent parallel resistance that is not a lot larger than R_f, owing to the kind of dielectric materials that must be used to make a capacitor that large in a reasonable size package. Consequently, we might need to take this dielectric resistance into account.

Exercise 5.5

Design an integrator with a pole frequency of 100 Hz and a DC gain of −1000 using an ideal op amp and an inductor. Set the DC input impedance (i.e., the input resistance) equal to 10 Ω.

5.2.2 The Differentiator and First-Order High-Pass Filter

A differentiator circuit is shown in Figure 5-19. We have explicitly included a source resistance in this circuit, because it cannot be absorbed as in past sections (e.g., in the integrator circuit the source resistance could be considered to be a part of R_1). Because of the negative feedback and the ideal op amp we know that v_N is a virtual ground, and we can write

$$I_1(j\omega) = \frac{V_s(j\omega)}{R_S + 1/j\omega C_1}. \tag{5.34}$$

Using (5.34) along with the fact that $i_F = i_1$ leads to the transfer function of the differentiator;

$$\frac{V_o(j\omega)}{V_s(j\omega)} = \frac{-R_f}{R_S + \dfrac{1}{j\omega C_1}} = \frac{-(R_f/R_S)j\omega}{j\omega + 1/R_S C_1}. \tag{5.35}$$

This equation is that of a high-pass filter (HPF) with a DC zero, a high-frequency pole at $\omega_p = 1/R_S C_1$, and a high-frequency gain of $-R_f/R_S$. We could also have derived (5.35) by using the impedance of R_S in series with C_1 in place of R_1 in (5.12).

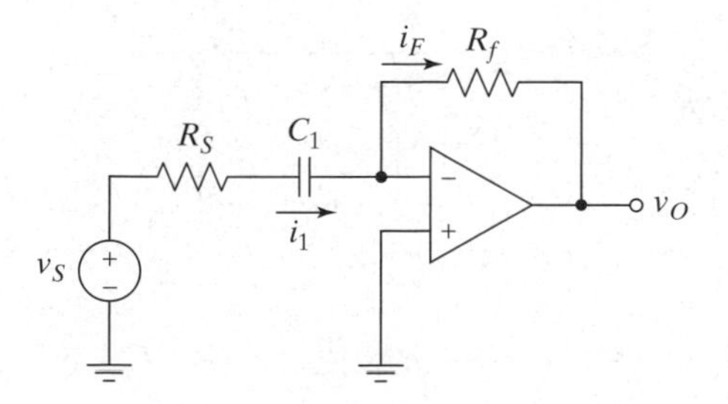

Figure 5–19 A differentiator circuit.

[7] Chosen so that the phase is nearly −90° at the signal frequency.

In contrast with the integrator circuit, we see that the differentiator circuit does not have a problem with DC stability, since even an ideal differentiator (i.e., $H(j\omega) = j\omega$) has a DC zero. Nevertheless, we also see that we cannot construct an ideal differentiator. It is unavoidable that the source will have some nonzero output resistance (R_S), and this causes the transfer function to have a high-frequency pole. Even if R_S were zero, any real op amp has a finite bandwidth. Therefore, the resulting circuit only functions like a differentiator for frequencies well below the pole frequency of the circuit.

We also notice that the only difference between a differentiator circuit and a single-pole HPF is our intent. In other words, a single-pole HPF *is* an approximation to a differentiator for frequencies well below the pole frequency. Therefore, we also see how to design a simple op amp HPF.

5.2.3 Second-Order Filters

We have seen examples of single-pole active filters in the preceding two sections. In this section, we want to see how to construct second-order filters using ideal op amps. A more detailed treatment of filters is deferred until Chapter 11, so it will suffice to say here that higher-order filters can be constructed by cascading first- and/or second-order filters together. We need to ensure that the filter characteristics of each individual stage are not changed when they are cascaded; however, in the case of ideal op amps, the output impedance of each stage is zero, so the output is unaffected by driving the input impedance of the next stage.

We can build many types of second-order filters, but all of their transfer functions can be written as a ratio of two quadratic functions of the complex-frequency[8] s. A transfer function of this type is called a ***biquadratic*** transfer function, or ***biquad***, and can be written as

$$H(s) = \frac{N(s)}{D(s)} = \frac{n_0 + n_1 s + n_2 s^2}{d_0 + d_1 s + s^2} \tag{5.36}$$

where the numerator and denominator polynomials are referred to as $N(s)$ and $D(s)$, respectively, and it is common to write the transfer function in such a way that the coefficient of s^2 in $D(s)$ is one as shown here.

One op amp circuit that is capable of producing transfer functions of the form given by (5.36) is the Sallen–Key[9] filter circuit shown in a generalized form in Figure 5-20. We first analyze this circuit in its general form and then will show how it can be made to perform different filtering operations by choosing the different impedances to be resistors or capacitors.

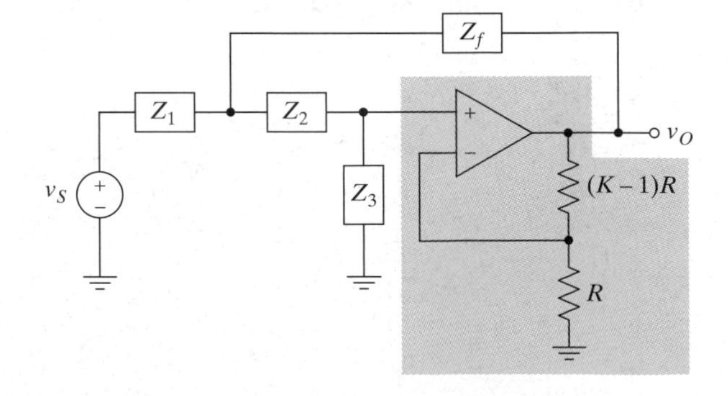

Figure 5–20 A generalized Sallen–Key filter circuit.

---

[8] For steady-state signals, $s = j\omega$.

[9] The circuit is named after the authors who originally published it. Filters of this form are also sometimes called controlled-source, or voltage-controlled voltage source (VCVS) filters [5.3].

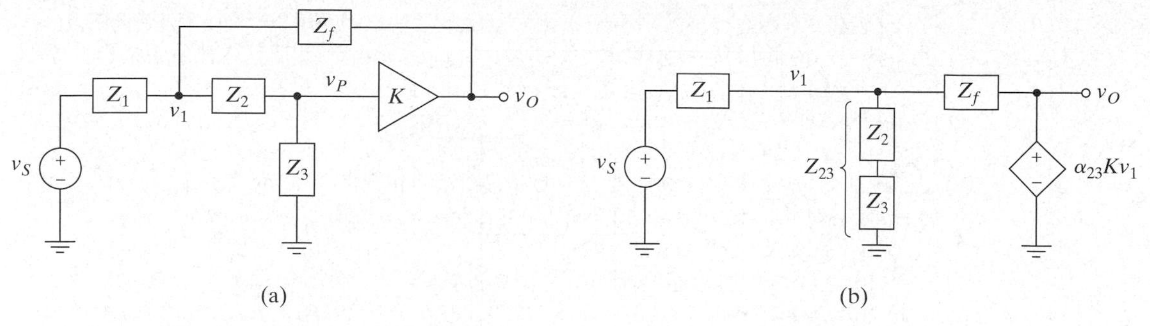

Figure 5–21 Simplifying the Sallen–Key filter circuit.

We could analyze this circuit by writing nodal or loop equations, but we can reduce the amount of algebra required (and get some useful practice with circuit manipulations) by simplifying the circuit first. We begin by recognizing that the op amp and resistors in the shaded box in Figure 5-20 comprise a noninverting amplifier with a gain of K. The circuit can then be drawn as in Figure 5-21(a). We next take advantage of the fact that the op amp input current is zero and write

$$v_P = \frac{Z_3}{Z_2 + Z_3} v_1 \equiv \alpha_{23} v_1. \tag{5.37}$$

Now we replace the op amp with a voltage-controlled voltage source (VCVS) as in Figure 5-2, but allow it to be controlled by v_1 instead of v_P through the use of (5.37). The resulting circuit is shown in Figure 5-21(b), where we have also defined $Z_{23} = Z_2 + Z_3$. This figure has only one independent node, v_1, and we can solve the circuit by writing the nodal equation at v_1 in standard form;

$$V_1(Y_1 + Y_{23} + Y_f) - V_o Y_f = V_s Y_1, \tag{5.38}$$

where the Ys are the admittances (i.e., the reciprocals of the impedances) of the elements and we use phasor notation for the voltages because we are using complex admittances. Or replacing V_o by the value of the controlled source, we get

$$V_1[Y_1 + Y_{23} + Y_f(1 - \alpha_{23}K)] = V_s Y_1, \tag{5.39}$$

from which we can derive the transfer function;

$$\frac{V_1}{V_s} = \frac{Y_1}{Y_1 + Y_{23} + Y_f(1 - \alpha_{23}K)}, \tag{5.40}$$

and finally,

$$\frac{V_o}{V_s} = \frac{V_o V_1}{V_1 V_s} = \frac{\alpha_{23} K Y_1}{Y_1 + Y_{23} + Y_f(1 - \alpha_{23}K)}. \tag{5.41}$$

We now show a few examples of filters made using the Sallen–Key circuit. In Figure 5-22 a LPF circuit is shown. By plugging in $R_1 = Z_1$, $R_2 = Z_2$, $C_1 = Z_f$, $C_2 = Z_3$, and manipulating the result a bit, you get

$$\frac{V_o(s)}{V_s(s)} = \frac{K/\tau_1\tau_2}{1 + s[1/\tau_1 + 1/R_2C_1 + (1 - K)/\tau_2] + s^2}, \tag{5.42}$$

where we have defined $\tau_1 = R_1 C_1$ and $\tau_2 = R_2 C_2$.

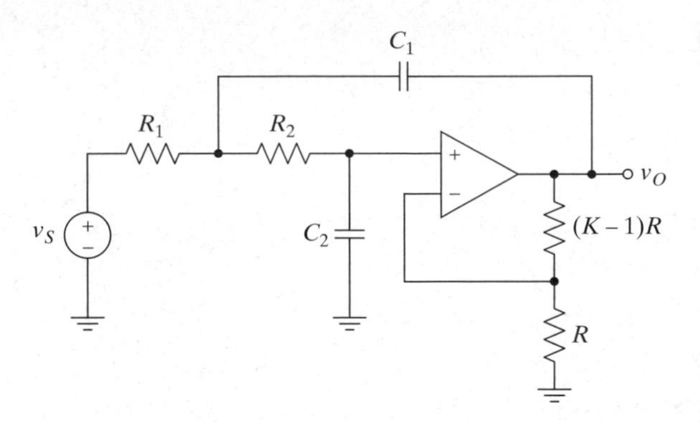

Figure 5–22 A low-pass Sallen–Key filter.

A HPF circuit is shown in Figure 5-23 and is derived by substituting $C_1 = Z_1$, $C_2 = Z_2$, $R_2 = Z_3$, and $R_1 = Z_f$ in the circuit in Figure 5-20. In this case, the transfer function is

$$\frac{V_o(s)}{V_s(s)} = \frac{Ks^2}{\dfrac{1}{\tau_1\tau_2} + s\left[\dfrac{(1-K)}{\tau_1} + \dfrac{1}{\tau_2} + \dfrac{1}{R_2C_1}\right] + s^2}, \tag{5.43}$$

where we have again used $\tau_1 = R_1C_1$ and $\tau_2 = R_2C_2$.

As a final example of the Sallen–Key filter structure, a bandpass filter is shown in Figure 5-24. As the name implies, a bandpass filter is one that readily passes signals within a given band of frequencies (called the pass band), while attenuating signals with frequencies outside the pass band. The bandpass filter is derived by substituting $R_1 = Z_1$, $C_1 = Z_2$, $R_3 \| C_2 = Z_3$, and $R_2 = Z_f$ in the circuit in Figure 5-20. The transfer function in this case is

$$\frac{V_o(s)}{V_s(s)} = \frac{K_SR_3C_1}{1 + \dfrac{R_1}{R_2} + s\left[R_3(C_1 + C_2) + \tau_1 + \dfrac{R_1R_3}{R_2}(C_2 + C_1(1-K))\right] + s^2\tau_1R_3C_2} \tag{5.44}$$

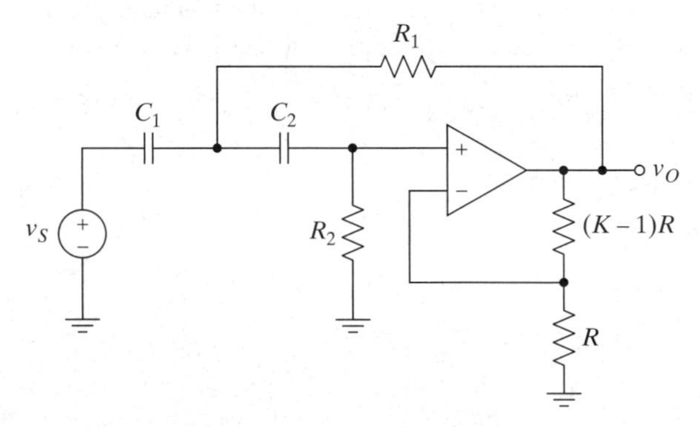

Figure 5–23 A high-pass Sallen–Key filter.

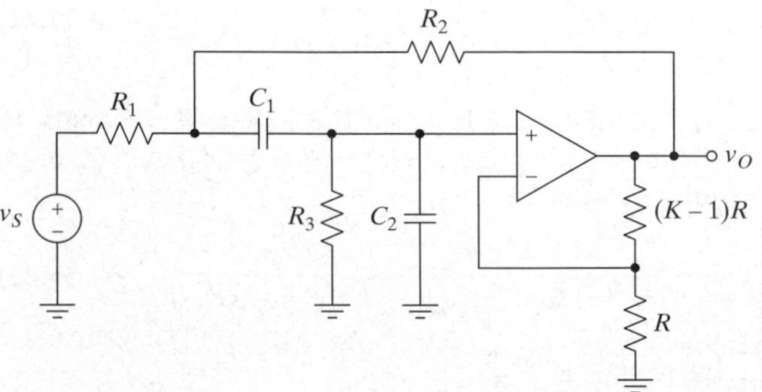

Figure 5–24 A bandpass Sallen–Key filter.

5.3 NONLINEAR OP AMP CIRCUITS

The previous sections have all concentrated on linear applications of op amps. In this section we present a few of the useful nonlinear circuits that can be constructed using op amps.

5.3.1 Comparators

A ***comparator*** is a device that compares the magnitude of two inputs and gives an output that indicates which of the two is larger. An ideal comparator is shown along with its transfer characteristic in Figure 5-25. The ideal comparator is a linear amplifier with infinite incremental gain *only* for $V_{O\max} \leq v_O \leq V_{O\min}$; once v_O reaches either of its limits, the gain drops to zero. The incremental gain is equal to the slope of the v_O versus v_I characteristic in Figure 5-25(c).

When the gain of an amplifier drops to zero because it has reached one of the limits, we say the amplifier output has ***saturated***. Notice that the ideal comparator is identical in operation to an ideal op amp, except that even an ideal comparator has swing limits; that is why the op amp symbol is usually used. It follows that the ideal comparator is also characterized by having zero output impedance and infinite input impedance. An ideal comparator also switches infinitely fast, which is equivalent to the assumption of an ideal op amp having infinite bandwidth.

In reality, an op amp and a comparator are usually designed differently, but an op amp can function as a comparator, and some comparators can function as op amps. The main difference conceptually is how the circuits are used; op amps are used with negative feedback, and comparators are usually used without negative feedback.

Comparators are sometimes called ***slicers***, although this term is really more general. A slicer is a circuit that compares a single input voltage (it may be differential) to one or more thresholds and provides an output that indicates the value of the input relative to the thresholds. Therefore, a comparator is a slicer with a single threshold.

Many comparators are ***strobed***, or ***clocked***; that is, they have a clock input that tells them when to make their comparison. In this case, the input voltage may be sampled, and the value present at that instant held. In addition, the output voltage may be prevented from changing until the next strobe occurs.

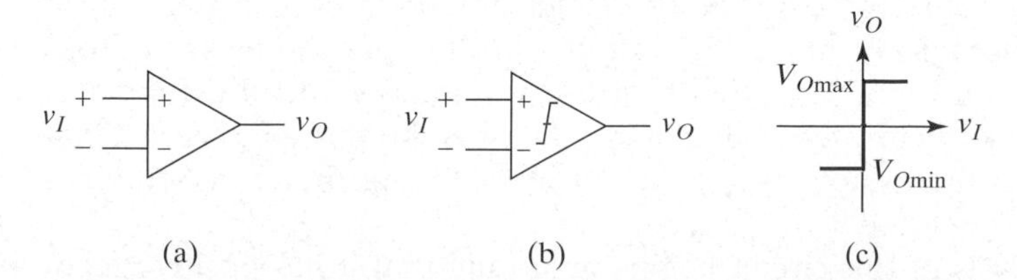

Figure 5–25 The ideal comparator; (a) the most common symbol used, (b) a less common symbol, and (c) the transfer characteristic.

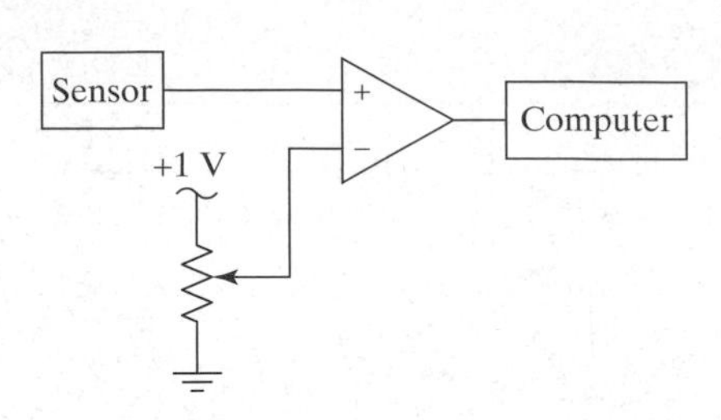

Figure 5–26 A comparator used to detect when a sensor output crosses a variable threshold.

An example comparator application is shown in Figure 5-26. The sensor could be a temperature or pressure sensor used in some manufacturing process. The comparator's job is to detect when the sensor output reaches a certain threshold and send a signal to the computer. In this case the threshold is adjustable by changing the setting of the potentiometer. The comparator may be needed for at least two reasons. First, the sensor output may not be large enough and/or powerful enough to drive the computer. Second, using the comparator allows us to have a continuously variable threshold yet provide the computer with a binary signal.

A common problem with circuits of the type shown in Figure 5-26 is that the output of the comparator may bounce rapidly back and forth between $V_{O\max}$ and $V_{O\min}$ if the input is very close to the threshold value. This action is caused by the noise that is invariably present in the system. If we model the noise as being added to the sensor output, we can see that if the *noiseless* sensor output is right at the threshold level, small amounts of noise can make the comparator input go back and forth across the threshold. In some applications, this behavior can cause serious problems and needs to be avoided.

We can prevent this problem by adding ***hysteresis*** to the comparator. Hysteresis means that an effect lags behind the cause of the effect and is associated with systems that have some memory of past events. In the case of our comparator, we would like the output to tend to stay at one level once it has changed. We can produce this behavior by applying some positive feedback as shown in Figure 5-27.

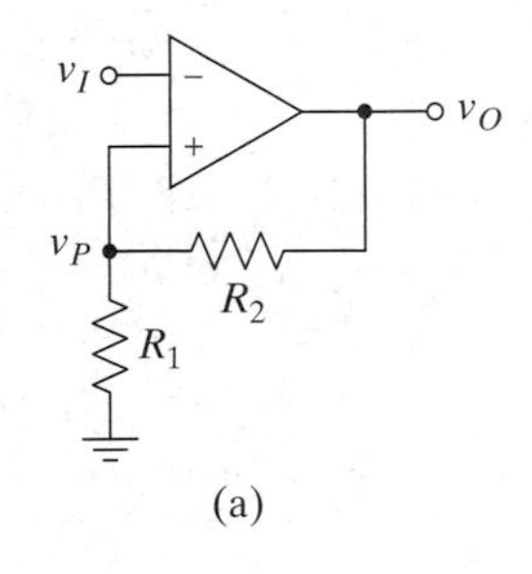

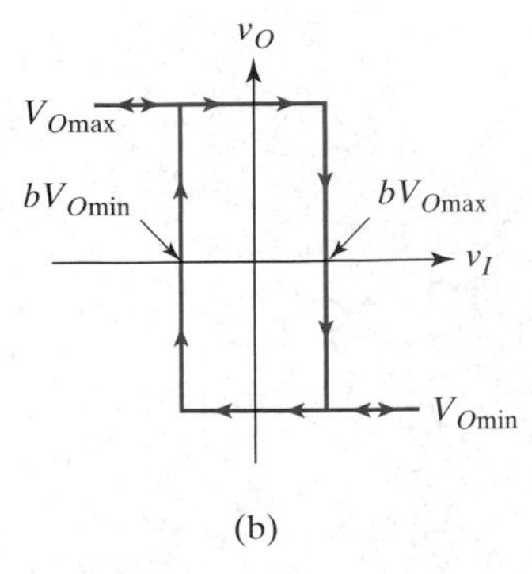

Figure 5–27 (a) A comparator with hysteresis and (b) the resulting transfer characteristic.

This particular circuit is an inverting comparator with hysteresis and as shown here the thresholds are on either side of ground. Problem P5.29 deals with a comparator with hysteresis and a variable threshold. Start the analysis by assuming that v_I is large and negative and that the output is at $V_{O\max}$. Notice that because the input current is zero,

$$v_P = \frac{R_1}{R_1 + R_2} v_O \equiv b v_O. \tag{5.45}$$

Therefore, with $v_O = V_{O\max}$, we know that $v_P > 0$ and, with v_I negative, our starting assumptions were consistent. Now let's see what happens as v_I is ramped positively (i.e., moved to the right in the transfer characteristic in Figure 5-27). So long as $v_I \le v_P = bV_{O\max}$ the output of the comparator will stay at $V_{O\max}$. As soon as v_I exceeds $bV_{O\max}$, the output will change to $V_{O\min}$. When the output changes, to $V_{O\min}$, the value of the threshold changes, too; $v_P = bV_{O\min}$. Therefore, if v_I decreases again because of noise, the output will not change unless v_I drops below the negative threshold. In other words, the circuit remembers the last change (as recorded in the present value of v_O) and sets the threshold accordingly. The magnitude of the hysteresis is adjusted by changing the resistor-divider feedback circuit.

An alternate analysis of this circuit begins by noting that it has positive feedback. We can determine that the feedback is positive by applying the same kind of test we did earlier when confirming that an amplifier had negative feedback. Assume that the loop is presently stable and consider what happens if the input voltage changes a little bit. If v_I is increased by ε, then v_O will decrease (i.e., become *more* negative) by $-a\varepsilon$ where a is the open-loop gain of the amplifier. As a result of the decrease in v_O, v_P will decrease as well. The decrease in v_P leads to a further reduction in v_O, so rather than opposing the original change, the feedback reinforces it. Therefore, the output of the comparator will continue to change until the output saturates. As a result of this positive feedback, the output has only two stable states; it is either saturated at $V_{O\max}$ or at $V_{O\min}$. We say that the circuit is ***bistable***.

It is worth pointing out that this circuit is a perfect example of where blindly applying the principle of a virtual short will fail. If you don't check to see that the feedback is positive, you would assume a virtual short and analyze the circuit in Figure 5-27 in exactly the same way as the noninverting amplifier circuit in Figure 5-3. You would conclude that the two circuits function exactly the same, which is certainly not correct.

5.3.2 Precision Rectification and Clipping[10]

Devices that can pass current in only one direction are very useful for signal processing. The diode is a two-terminal device that accomplishes this task. Chapter 12 shows a number of circuits that use the large-signal behavior of diodes. In this section we show how op amps can be used with diodes to improve the operation of some diode circuits.

One common application of diodes is demonstrated by the rectifier circuit shown in Figure 5-28(a). The resulting transfer characteristic if an ideal diode[11] is used is shown in part (b) of the figure, and the transfer characteristic attained with a real diode is shown in part (c) of the figure. The offset voltage of the real diode is unacceptable in many signal processing applications (e.g., if you need to rectify a very small signal).

An op amp can be used with a diode to approximate an ideal diode and do away with the offset apparent in Figure 5-28(c). Figure 5-29 shows a ***super diode*** used as a rectifier. The resulting rectifier characteristic looks essentially the same as that in Figure 5-28(b) for the ideal diode. We will now analyze this circuit to see how it functions. We will also see how the performance of the super diode differs from that of the ideal diode.

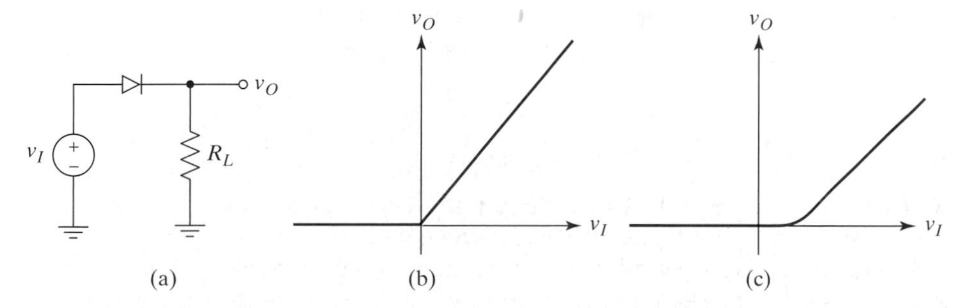

Figure 5–28 (a) A diode rectifier circuit. (b) The transfer characteristic if an ideal diode is used and (c) if a real diode is used.

[10] This section requires some knowledge of diodes.

[11] An ideal diode is one with no forward voltage and no reverse current. *See* Aside A2.3 if you are not familiar with the concept of an ideal diode.

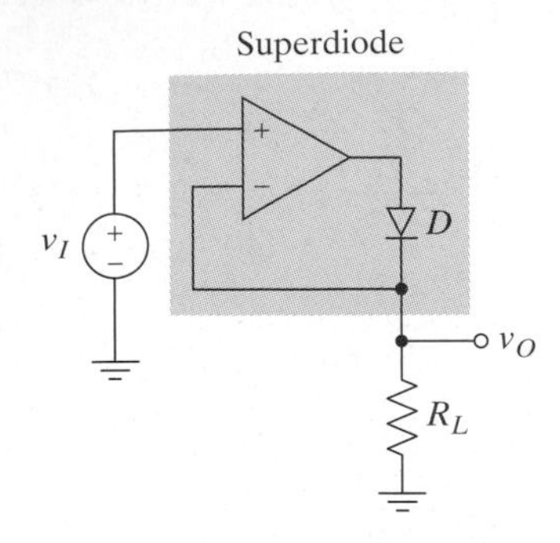

Figure 5–29 A super diode rectifier circuit. The superdiode is the circuit in the shaded box.

As with any feedback circuit, the first thing we want to do is determine whether or not the feedback is negative. In this case the diode makes that determination a bit more complicated. We have to check two different possibilities; what happens when the diode is off (i.e., reverse biased), and what happens when it is on (i.e., forward biased).

If the diode is off, the current through it is essentially zero. Since the input current of the op amp is also zero, the current through R_L is zero, and the output voltage must be zero. The diode will be off whenever the op amp output voltage is less than v_O. Since v_O is connected directly to the inverting input of the op amp, the diode will be off whenever $v_I < v_O = 0$. Because v_O is constant and independent of v_I while the diode is off, there isn't any feedback for $v_I < 0$.

However, for positive values of v_I the op amp differential input is positive, and its output will immediately go positive and turn the diode on. At this point the diode has a voltage of about 0.7 V across it and v_O is 0.7 V below the op amp output. As v_I increases, so will the op amp output, and so will v_O, thereby reducing the op amp differential input voltage. We now recognize this as negative feedback and know that there will be a virtual short between the inputs for an ideal op amp. Therefore, $v_O = v_I$ and the gain is one for any $v_I \geq 0$. This circuit in essence behaves like the ideal diode rectifier and has the transfer characteristic shown in Figure 5-28(b).

A common function that is similar to rectification is ***clipping***, or ***limiting***. It is sometimes useful to limit the amplitude of a signal as shown in Figure 5-30(b). We can accomplish this task using the circuit shown in part (a) of the same figure. The diodes shown in the figure are called ***zener diodes***. Zener diodes are normal diodes, except that they have a well defined reverse breakdown voltage denoted here by V_Z. Therefore, when they are reverse biased to a value V_Z they conduct significant current in the reverse direction. It is also possible to limit the signal for just positive or negative swings (*see* Exercise 5.6).

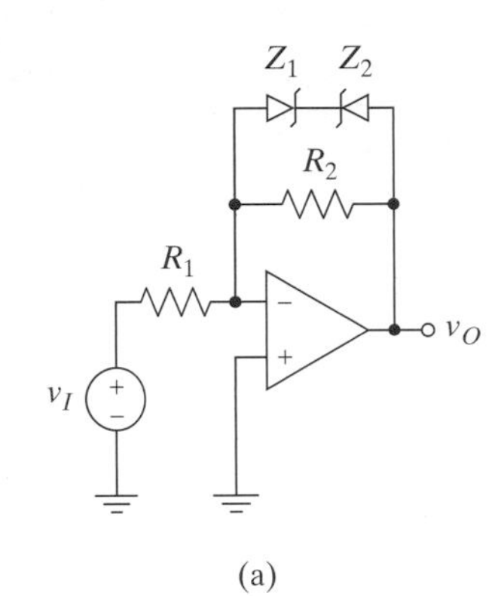

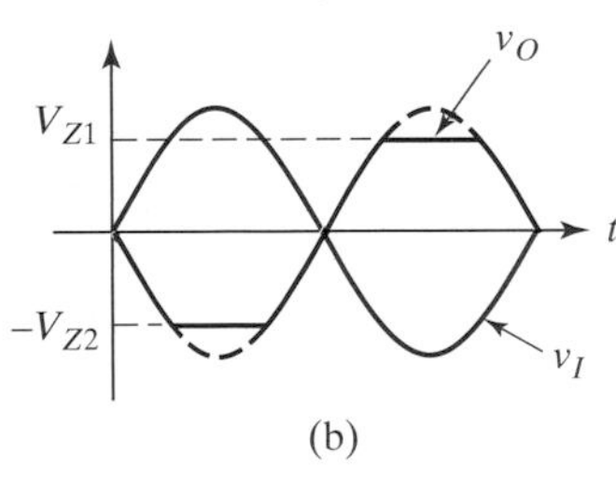

Figure 5–30 (a) A limiter circuit (b) the associated waveforms (we have assumed that the magnitudes of V_{Z1} and V_{Z2} are much larger than the forward drops, so we ignored the forward voltages in drawing the figure).

Exercise 5.6

Sketch the transfer characteristic for the circuit shown in Figure 5-31. Assume the zener diode has a reverse breakdown voltage of 3 V and is ideal in every way.

□

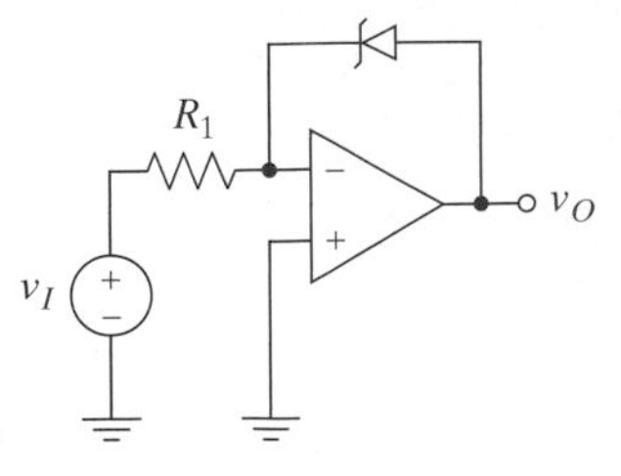

Figure 5–31 Circuit for Exercise 5.6.

5.3.3 Logarithmic Amplifiers

Amplifiers with logarithmic transfer functions are useful in a number of applications. For example, when an input covers a very large dynamic range, a log amp can compress the range to reduce the demands on any signal processing circuitry that follows. Log amps can also be used to produce instruments that respond to the logarithm of an input. Logarithmic responses are useful, since they are common in natural systems; for example, the human ear responds logarithmically. Log amps can also be used to produce multipliers, dividers, and other algebraic functions, since the log of a product is the sum of the logs (i.e., $\log AB = \log A + \log B$).

Figure 5-32 shows a log amp circuit. In this circuit we cannot consider the diode to be ideal or we will completely miss the main feature of the performance of the circuit. (We will return to this point after we finish the analysis.) Therefore, we use the diode equation in our analysis;

$$i_D = I_S(e^{v_D/nV_T} - 1). \tag{5.46}$$

If we restrict our attention to the case where $v_D \gg V_T$ and assume that $n = 1$, we have

$$i_D \approx I_S e^{v_D/V_T}. \tag{5.47}$$

and, therefore,

$$v_D = V_T \ln\left(\frac{i_D}{I_S}\right) = V_T \ln\left(\frac{v_I}{R_1 I_S}\right) \tag{5.48}$$

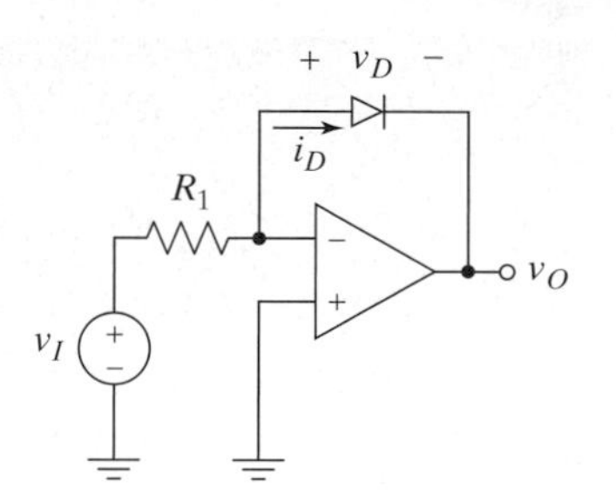

Figure 5–32 A logarithmic amplifier circuit.

In deriving this result, we have used two facts; 1) the input current of the op amp is zero, and 2) there is negative feedback so long as the diode is on. Therefore, when the diode is on, there will be a virtual ground at the inverting input of the op amp. Since current can only flow in one direction in the diode, this circuit only works for positive values of v_I. Because of the virtual ground, we can also write

$$v_O = -v_D = -V_T \ln\left(\frac{v_I}{R_1 I_S}\right) = -K \ln(v_I) \tag{5.49}$$

where we have implicitly defined the constant K.

A bipolar transistor can be used in place of the diode in Figure 5-32 and the circuit will perform better: this modification is considered in Problem P5.32.

The logarithmic amplifier circuit is a wonderful example of how important model selection is in circuit design. If you didn't know anything about the circuit, and you performed an analysis using the ideal diode model, you would conclude that the output was zero volts whenever the input was positive and that the output was saturated at the positive swing limit whenever the input was negative. If you then refined the model a bit and allowed for a constant 0.7 V drop across the diode when it is forward–biased, you would conclude that $v_O \approx -0.7$ V for positive inputs. This conclusion is essentially correct, but it still misses the essential feature of the operation of the circuit.

Therefore, we see that the model we use can have a dominant influence on the results of our analysis. We sometimes have to choose the appropriate model by repeating the analysis with progressively more complex models until the essential features of the circuit operation are evident. As we gain more experience, we can often choose the appropriate model based on our understanding of what the circuit is supposed to do. When uncertain as to what model you should use, it is always best to start with the simplest model possible.

5.4 NONIDEAL CHARACTERISTICS OF OP AMPS

In this section we examine some of the nonideal characteristics of op amps. Table 5.1 gives some of the specifications for some popular op amps. The different specifications listed in the table are all explained in this section. All of the op amps listed are internally compensated; that is, they are all designed to be stable (i.e., free of oscillations) for any reasonable resistive negative feedback connection. Uncompensated op amps are also available. We defer the topics of stability and compensation to Chapter 10.

From Table 5.1 and the definitions for the specifications as given in this section, it is clear that real op amps do a pretty good job of approximating an ideal op amp in some ways (e.g., gain), but not always in others (e.g., bandwidth). It is important in many applications to understand at least some of these limitations. The following sections discuss each of these second-order effects.

Table 5.1 Specifications of Some Common Op Amps

	Part #	Gain V/V	R_{in} Ω	C_{in} pF	Z_o[1] Ω	GBW MHz	I_B nA	V_{OS} mV	I_{OS} nA	CMRR dB	PSRR dB	SR V/μs	e_n nV/$\sqrt{\text{Hz}}$	i_n pA/$\sqrt{\text{Hz}}$
BIPOLAR	LM607	5M	2M	-	50	1.8	1	.015	.5	140	140	.7	6.5	.12
BIPOLAR	LF411[6]	.2M	10^{12}	-	10	4	.05	.8	.025	100	100	15	25	.01
BIPOLAR	LM741[2]	.2M	2M	-	-	1	80	1	20	90	96	.5	-	-
BIPOLAR	LM324[3]	.1M	-	-	-	1	45	≤9	5	85	100	-	-	-
CMOS	LMC6041	.5M[4]	$>10^{13}$	-	-	.075	.002pA	1	.001pA	75	75[5]	.02	83	.0002
CMOS	LMC6061	3M	$>10^{13}$	-	-	0.1	.01 pA	0.1	.005 pA	85	85[5]	.035	83	.0002
CMOS	LMC6081	.35M[4]	$>10^{13}$	-	-	1.3	.01 pA	.15	.005 pA	85	85[5]	1.5	22	.0002

All specifications are typical at room temperature, see the data sheets for specific notes. 1) Often given as a graph, in which case the value at 100 kHz and gain of 10 is shown. 2) The classic op amp but not recommended for new designs—substitute the LF411. 3) One of the cheapest op amps, but meant for single-supply operation and should not be used with dual supplies in general, it has a lot of "crossover" distortion. 4) Gain is quoted for sinking current, the gain is higher when sourcing. Also, the gain is a function of the load. 5) Different for positive and negative supplies, the lowest value is given here. 6) Has a JFET input stage.

5.4.1 Finite Gain

Although op amps have extremely large gains, they are not infinite. The true value of the gain can be included in the op amp model, as was shown in Figure 5-2. This finite gain does have an effect on the overall gain of an op amp circuit. As an example of how to analyze this effect, let's reconsider the noninverting amplifier configuration. Figure 5-33(a) shows a noninverting amplifier circuit, and we can analyze the effect of finite gain by using the op amp model shown in part (b) of the figure.

Using this model, we write

$$v_{ID} = v_S - \left(\frac{R_1}{R_1 + R_2}\right) a v_{ID} = v_S - ab v_{ID}, \tag{5.50}$$

where we have implicitly defined the ***feedback factor***, b. Solving (5.50) for v_{ID} leads to

$$v_{ID} = \frac{v_S}{1 + ab}, \tag{5.51}$$

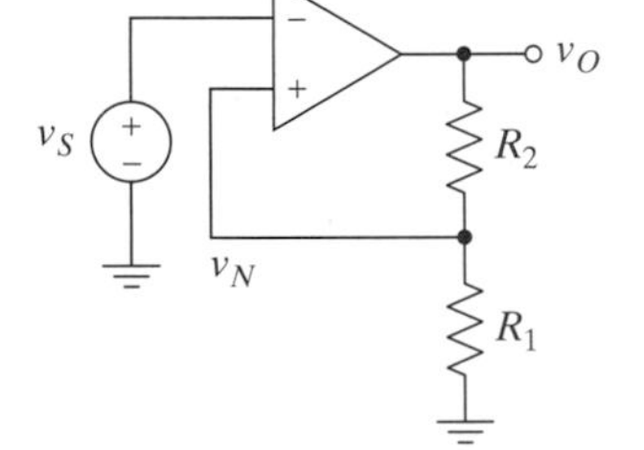

and finally

$$v_O = a v_{ID} = \frac{a}{1 + ab} v_S. \tag{5.52}$$

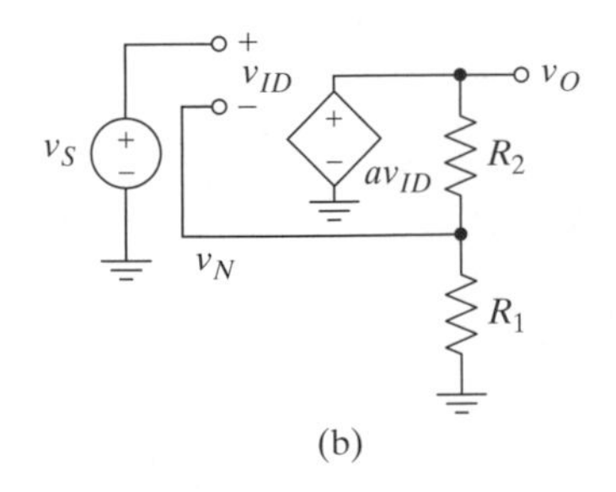

Figure 5–33 (a) The noninverting amplifier. (b) A model for the circuit.

Now we can see what effect the op amp gain, called the open-loop gain, has on the overall gain of the amplifier, which is called the closed-loop gain. First, notice that if a is infinite in (5.52), the closed-loop gain reduces to

$$\frac{v_O}{v_S} = \frac{1}{b} = \left(1 + \frac{R_2}{R_1}\right) \tag{5.53}$$

which is the same result we derived in Section 5.1.1. Therefore, if the open-loop gain is large enough, the closed-loop gain is roughly equal to $1/b$. Now say we are using an LF411 with a nominal gain of about 200,000, and we design our amplifier for a closed-loop gain of 10 by setting $b = 0.1$. Using (5.52) we get

$$\frac{v_O}{v_S} = \frac{200{,}000}{1 + 200{,}000 \times (0.1)} = 9.9995. \tag{5.54}$$

In other words, the closed-loop gain determined using the ideal op amp approximation was accurate to within 0.05%. We would be hard pressed to find resistors that accurate, so the ideal op amp approximation is certainly reasonable in this case. Note that if the closed-loop gain had been higher, the error would be greater. Also, remember that there are other nonideal characteristics that we have yet to take into account.

5.4.2 Input Bias and Offset Currents

The input currents of an op amp are not zero for two reasons: First, to function, the op amp's internal circuitry may demand that a small amount of DC bias current be supplied to, or drawn from, the input terminals. This current is called ***input bias current***. Second, the input impedance seen by the signal source is not infinite, so there will be some input current that is signal dependent. The input impedance will be treated later, we will concentrate on the bias current in this section.

We can model the input bias current by adding a current source from each input to ground as shown in Figure 5-34. Remember that these currents are directed *out* of the terminals for some op amps.

Nominally, the two bias currents are equal, but due to unavoidable variations in the processing of the integrated circuit, there is an unpredictable difference in the two currents. This difference is specified as the ***input offset current***. Therefore, we can say that the bias current specified in Table 5.1, I_B, is the average value of the two bias currents, and the offset current, I_{OS}, is the difference. That is,

$$I_B = \frac{I_{BP} + I_{BN}}{2}, \text{ and } I_{OS} = |I_{BP} - I_{BN}|. \tag{5.55}$$

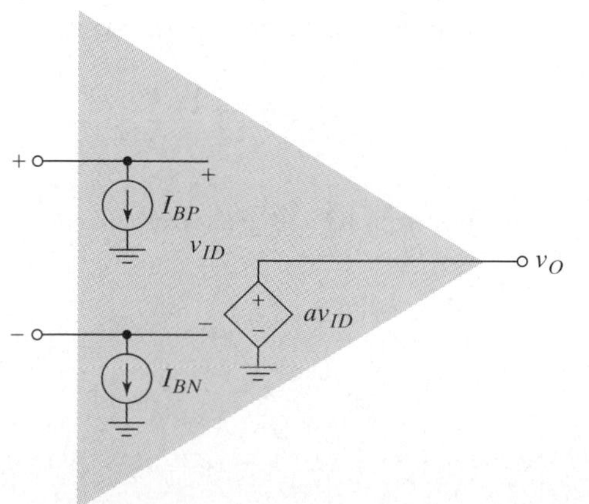

Figure 5–34 Op amp model including input bias currents.

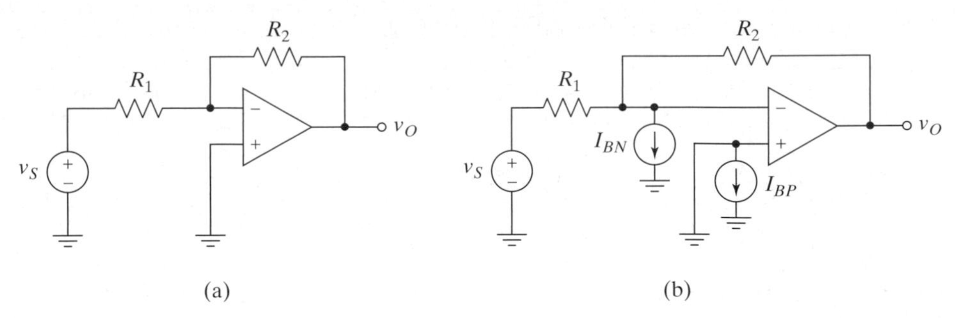

Figure 5–35 (a) Inverting amplifier circuit and (b) model for calculating the effect of input bias currents.

We sometimes need to allow for the input bias currents in our designs. For example, consider the inverting circuit shown in Figure 5-35(a), and the model for the circuit shown in part (b) of the figure, where we have assumed an ideal op amp, except for the input bias currents.

We first notice that I_{BP} will not affect the operation, since it drives a short circuit to ground and does not affect any of the voltages in the circuit. Next, we use superposition and set v_S to zero first to see the effect of I_{BN} alone. Because of the negative feedback and ideal op amp, the inverting input is a virtual ground, so the voltage across R_1 is zero, and it does not have any current through it. Therefore, all of I_{BN} flows through R_2, and we have

$$V_O\Big|_{\text{due to } I_{BN}} = I_{BN}R_2. \tag{5.56}$$

This is the output offset voltage that results from the input bias currents in this circuit.

We can cancel the output offset voltage caused by input bias currents by adding another resistor in series with the noninverting input as shown in Figure 5-36. This circuit is said to have input bias current compensation. The bias current I_{BP} will flow through R_3. If we chose the value of this resistor correctly, we can set $v_P = v_N$ when $v_S = 0$ and therefore make the output offset voltage go to zero.

To determine the correct value for R_3, we consider only the bias currents and set v_S to zero. We then set v_O to zero and notice that both R_1 and R_2 connect from v_N to a node at ground potential. Consequently, they appear in parallel with each other. Therefore, if we make $R_3 = R_1 \| R_2$, we will achieve our goal and have zero offset voltage at the output, assuming that $I_{BP} = I_{BN}$. In general, the two bias currents will not be equal (i.e., I_{OS} is not zero), but since the difference is random, the best we can do is cancel the average value.

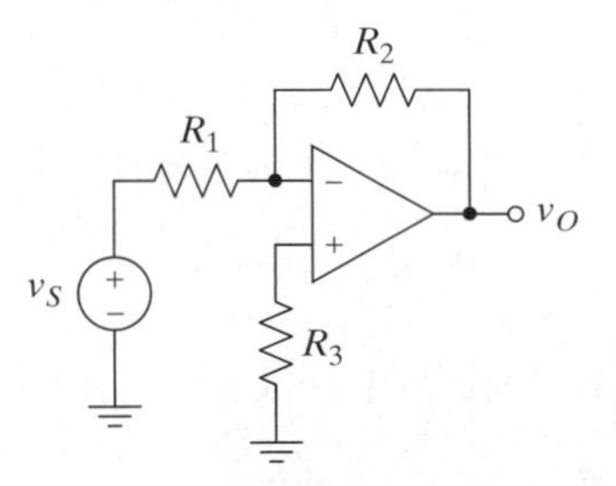

Figure 5–36 Inverting amplifier with input bias current compensation.

Example 5.3

Problem:
Using the model shown in this subsection, calculate the value of R_2 required to mitigate the effects of input bias current on the output offset voltage of the circuit in Figure 5-37.

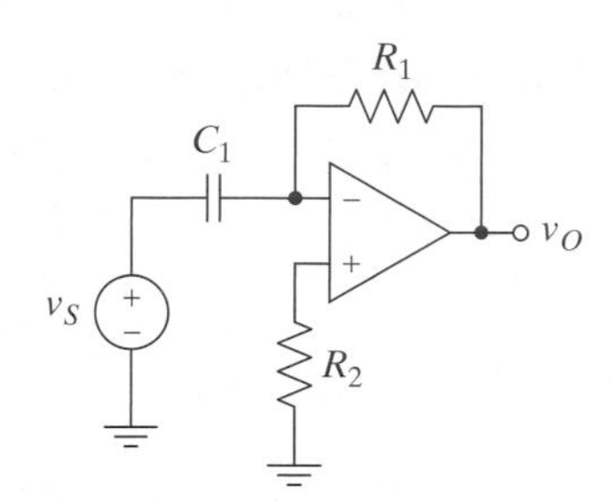

Figure 5–37 Figure for Example 5.3.

Solution:
Since the input bias current is a DC current, it cannot flow through the capacitor. Therefore, all of the current will flow through R_1, and we need to set $R_2 = R_1$ to guarantee that the input bias current does not contribute to an output offset voltage.

■

5.4.3 Input Offset Voltage

Due to unavoidable mismatches in the characteristics of the devices used to make op amps, the output voltage is not zero if the two inputs are physically tied together. Tying the inputs together sets the differential input voltage is zero, even if input bias and offset currents are accounted for. The parameter used to quantify and model this nonideality is the ***input offset voltage***, V_{OFF}.

The input offset voltage specifies the value of voltage that would need to be applied at the input to produce an output voltage of zero. The sign of this voltage has no significance, since the mismatches that cause it are random, and the sign may be different for different amplifiers of the same type. Figure 5-38 shows how V_{OFF} can be modeled; remember that the sign can be either way. In this figure, V_{OFF} is shown outside of the amplifier, so the amplifier shown in the figure does not have any offset voltage.

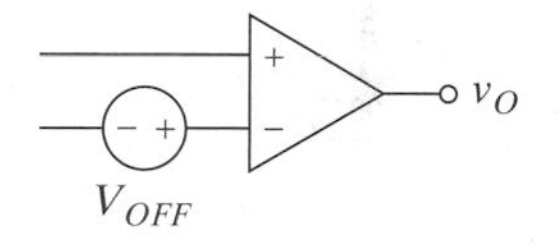

Figure 5–38 An op amp with an offset voltage source.

We already discussed one of the problems caused by offset voltage in the course of our investigation of the op amp integrator circuit (*see* Figure 5-17). But, there are many other places where the offset voltage presents a problem, as well. In general, it sets the minimum value of input voltage that can be reliably detected.

For example, consider the amplifier in Figure 5-39, and assume the op amp is ideal except for the offset shown. If we have an output voltage of zero, for an ideal op amp without offset we could also say that the input voltage was zero. But, with the offset present, we cannot draw the same conclusion. Note that with $v_O = 0$ the voltage at the junction of R_1 and R_2 is also zero. Since the op amp will force a virtual short between the inputs, the source voltage must be equal to V_{OFF}, which is not zero. Since V_{OFF} is unpredictable and changes with temperature, we don't know what v_S is. In general, we can't determine v_S any more accurately than $\pm V_{OFF}$ unless we have measured V_{OFF} for the particular device being used.

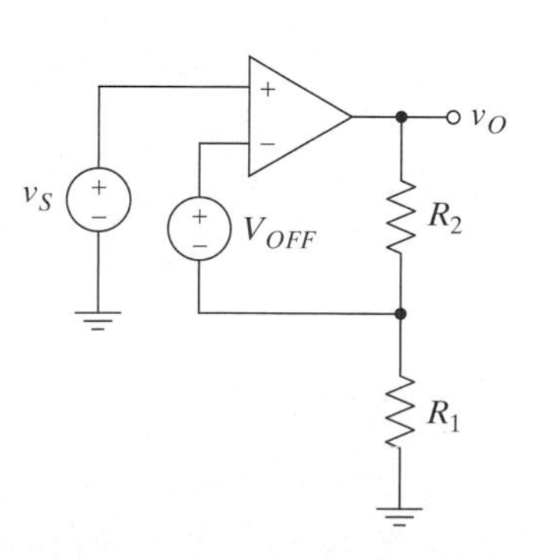

Figure 5–39 An amplifier with offset.

Because the offset voltage is random, there isn't any way of compensating for it with a fixed circuit. This situation is identical to that for the input offset current;

remember that we can only compensate for the bias current, not the offset. Nevertheless, many op amps include separate terminals to which we can connect a variable resistor (possibly also connected to one of the power supplies), which can be adjusted to cancel the offset voltage. There is an added side benefit to making this adjustment; the variation of offset with temperature is reduced when the nominal offset voltage has been canceled.

5.4.4 Finite Input and Output Impedances

In addition to the input bias currents discussed earlier, all op amps have finite input and output impedances. In the case of MOS-input op amps (i.e., op amps that have metal-oxide-semiconductor transistors in the input stage), the DC input impedance, or input resistance, is large enough to be essentially infinite for many applications. However, even then, there is input capacitance. The origins of amplifier input and output impedances will be discussed in Chapter 8. In this section, we simply present a model and illustrate how to use it. In the rest of this section, we refer to input and output *resistances*, but our statements apply to general impedances.

Since there are two inputs to an op amp, we need to consider both the **differential-mode input resistance** and the **common-mode input resistance**. The differential-mode input resistance, R_{id}, is the resistance seen between the two inputs, and the common-mode input resistance, R_{ic}, is that between the two inputs tied together and ground. Figure 5-40 shows a model of the input of an op amp, including both R_{ic} and R_{id}. The model assumes that $R_{ic} \gg R_{id}$, as is usually the case; otherwise the value of R_{id} would need to be modified in the model to account for the resistance $4R_{ic}$ in parallel with it. It is important to note that, since the common-mode resistance is usually much greater, it is the differential-mode input resistance that is specified unless noted otherwise. For example, the input resistances given in Table 5.1 are differential input resistances.

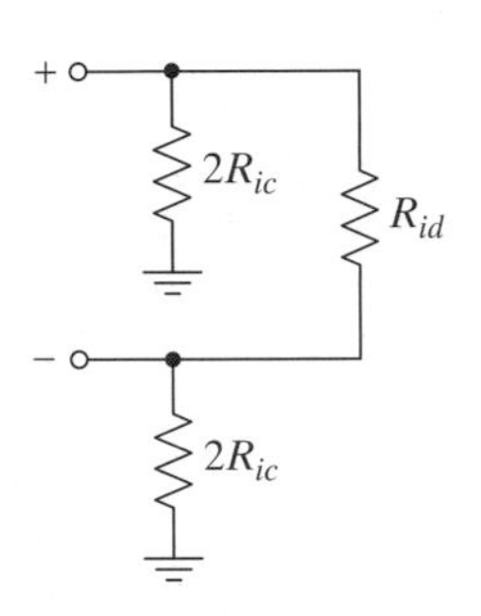

Figure 5–40 Model for the input resistances of an op amp.

Example 5.4

Problem:

Find the input impedance of the circuit shown in Figure 5-41. Assume that the op amp is ideal, except for having $R_{ic} = 1\ \text{M}\Omega$ and $R_{id} = 100\ \text{k}\Omega$.

Solution:

Start by replacing the op amp by its model, including the input resistances, as shown in Figure 5-42.

By definition, the input impedance is the impedance seen by the source driving the circuit; $R_i = v_S/i_S$. To simplify the algebra, notice that we can combine R_1 and the $2R_{ic}$ at the inverting input in parallel, resulting in the circuit shown in Figure 5-43. We then find the resistance R_x shown in the figure and combine it in parallel with the $2R_{ic}$ at the noninverting input yield

$$R_i = R_x \| 2R_{ic} = (v_S/i_x) \| 2R_{ic}. \quad \textbf{(5.57)}$$

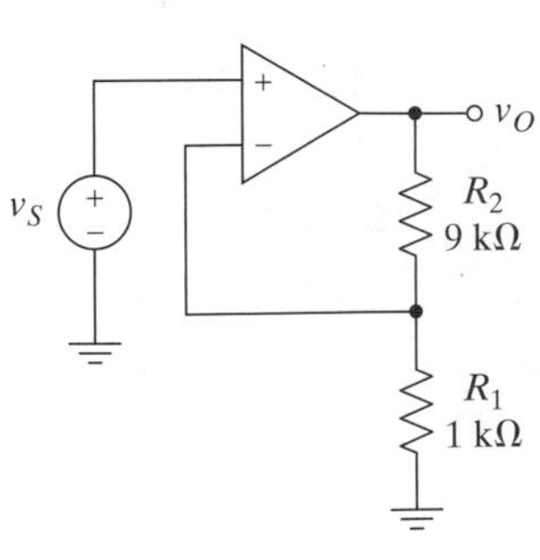

Figure 5–41 Circuit for Example 5.4.

An elegant way to solve this problem is presented in Chapter 10, but it can be solved by direct analysis. In fact, there is only one independent node in the circuit in Figure 5-43; v_N. The other two nodes are v_S, which is the driving source, and v_O, which is dependent on v_S and v_N. Therefore, writing the node equation at v_N we get

$$v_N(G_{id} + G_1' + G_2) - av_{ID}G_2 = v_SG_{id}. \quad \textbf{(5.58)}$$

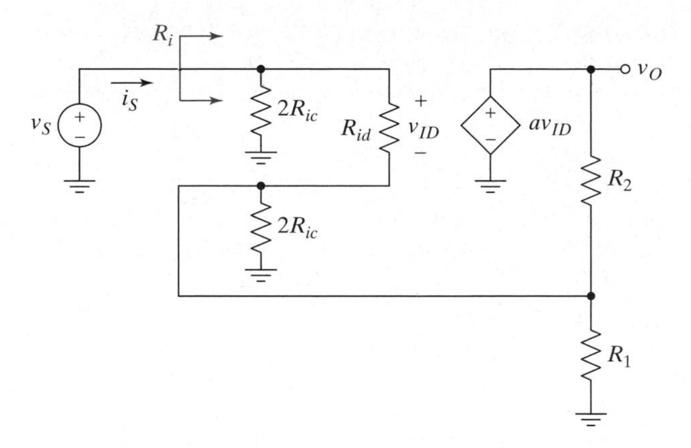

Figure 5–42 Circuit model to find input resistance of circuit in Figure 5-41.

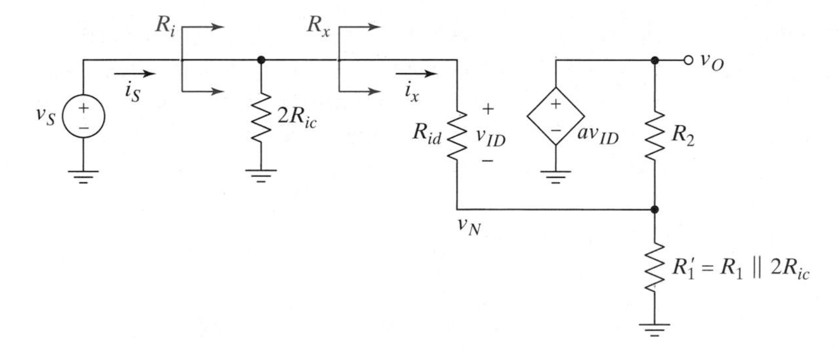

Figure 5–43 Simplified circuit for finding input impedance.

Substituting $v_{ID} = (v_S - v_N)$ and solving leads to

$$v_N = v_S \frac{G_{id} + aG_2}{G_{id} + G_1' + G_2(1 + a)}. \tag{5.59}$$

Now we can solve for i_x and, thereafter, R_x. First, i_x is given by

$$i_x = \frac{v_S - v_N}{R_{id}} = v_S G_{id}\left[1 - \frac{G_{id} + aG_2}{G_{id} + G_1' + G_2(1 + a)}\right], \tag{5.60}$$

and then R_x is (after several lines of algebra)

$$R_x = \frac{v_S}{i_x} = R_{id}\frac{G_{id} + G_1' + G_2(1 + a)}{G_1' + G_2}. \tag{5.61}$$

Simplifying a bit further, we get

$$R_x = R_{id}(1 + ab') + R_1' \| R_2, \tag{5.62}$$

where we have defined

$$b' \equiv \frac{R_1'}{R_1' + R_2} \tag{5.63}$$

in analogy to (5.50). Finally, the overall input resistance is

$$R_i = 2R_{ic} \| [R_{id}(1 + ab') + R_1' \| R_2]. \tag{5.64}$$

In other words, the input resistance of the basic op amp has been increased by the negative feedback. We will find in Chapter 10 that this is a general result for a connection of this type. Notice that if the op amp gain is large enough, the input resistance is entirely determined by the common-mode component; that is, $R_i \approx 2R_{ic}$. Therefore, the common-mode input resistance sets an upper limit on the input resistance we can achieve with this circuit.

■

The output impedance of a real op amp is nonzero and can be modeled as shown in Figure 5-44. This figure includes the input resistances as well.

5.4.5 Finite Bandwidth

We mentioned at the beginning of this section that many op amps are internally compensated to prevent oscillations. Compensation is covered in detail in Chapter 10. In this section we restrict our attention to internally compensated op amps. Internally compensated op amps can usually be modeled as having a single-pole low-pass transfer function (*see* Aside A1.6). Therefore, we can write

$$a(j\omega) = \frac{a_o}{1 + j\dfrac{\omega}{\omega_p}}, \tag{5.65}$$

where a_o is the DC gain and ω_p is the pole frequency. The magnitude plot for this response is shown in Figure 5-45 (note that we plot the gain in dB). Notice that for frequencies well above ω_p the magnitude can be approximated by

$$|a(j\omega)| \approx \frac{a_o\omega_p}{\omega}. \tag{5.66}$$

In other words, the gain is inversely proportional to ω for frequencies well above ω_p. An interesting observation can be made here; namely, the product of the frequency and the magnitude of the transfer function at that frequency is a constant called the ***gain-bandwidth product*** (GBW) on the high-frequency asymptote. In other words

$$\omega|a(j\omega)| = a_o\omega_p = \text{GBW} = \text{constant} \tag{5.67}$$

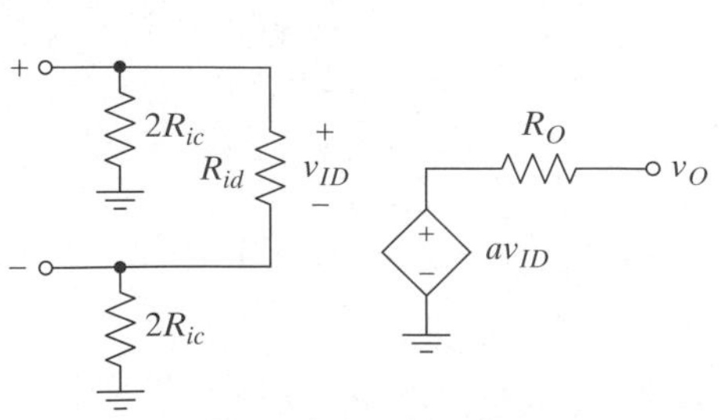

Figure 5–44 An op amp model incorporating the input and output resistances.

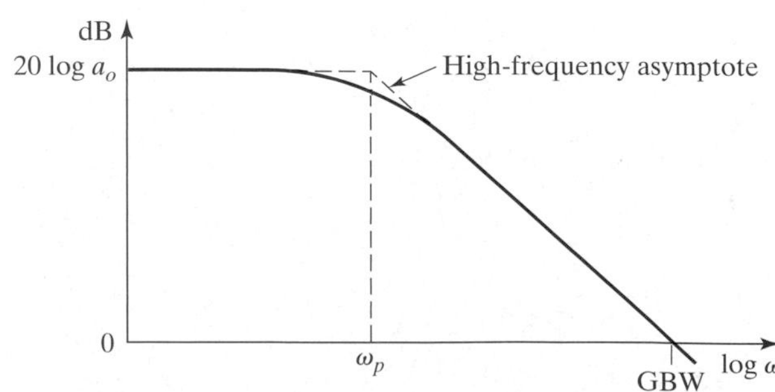

Figure 5–45 The magnitude response of (5.65).

on the high-frequency asymptote. Notice from (5.67) that the frequency at which the gain of the op amp falls to unity is equal to the GBW. This figure of merit only makes sense if the response of the amplifier can be modeled as a single-pole low-pass response. Notice that it is the GBW, not the pole frequency, that is specified in Table 5.1.

Example 5.5

Problem:

Find the transfer function of the noninverting amplifier circuit shown in Figure 5-46(a). Assume that the op amp is ideal, except that the gain is described by (5.65). Part (b) of the figure shows the equivalent circuit we will use for analysis.

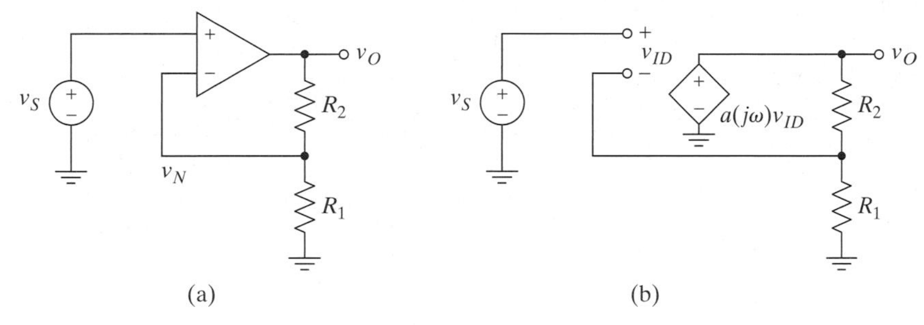

Figure 5–46 (a) A non-inverting amplifier circuit. (b) The model for determining the frequency response.

Solution:

From the figure we see that $V_o(j\omega) = a(j\omega)V_{id}(j\omega)$ and that

$$V_{id}(j\omega) = V_s(j\omega) - \frac{R_1}{R_1 + R_2}V_o(j\omega) = V_s(j\omega) - bV_o(j\omega), \tag{5.68}$$

where we have implicitly defined the feedback factor b. Therefore, we have

$$V_o(j\omega) = a(j\omega)(V_s(j\omega) - bV_o(j\omega)), \tag{5.69}$$

from which we obtain the transfer function

$$\frac{V_o(j\omega)}{V_s(j\omega)} = \frac{a(j\omega)}{1 + a(j\omega)b} \tag{5.70}$$

Plugging in $a(j\omega)$ from (5.65) we get

$$\frac{V_o(j\omega)}{V_s(j\omega)} = \frac{\dfrac{a_0}{1 + j\omega/\omega_p}}{1 + \dfrac{a_0 b}{1 + j\omega/\omega_p}}, \tag{5.71}$$

which simplifies to

$$\frac{V_o(j\omega)}{V_s(j\omega)} = \frac{\dfrac{a_0}{1 + a_0 b}}{1 + j\dfrac{\omega}{\omega_p(1 + a_0 b)}} \equiv \frac{A_{f0}}{1 + j\omega/\omega_{pf}}. \tag{5.72}$$

We call A_{f0} the DC gain with feedback and ω_{pf} the pole frequency with feedback. Notice that the transfer function is again a single-pole low-pass characteristic. The closed-loop gain is reduced by the factor $(1 + a_0 b)$ and the closed-loop pole frequency is increased by the same factor, so the GBW is the same for the circuit as it is for the op amp alone. We say that the GBW has been conserved in this case.

When the GBW is conserved, we can picture the tradeoff between gain and bandwidth as illustrated by Figure 5-47. Since the gain-bandwidth product is the same for any point lying on the high-frequency asymptote of the op amp transfer function, as was concluded in (5.67), the gain and bandwidth of the closed-loop amplifier also lie on this line. Therefore, we can see that the bandwidth of the closed-loop system can be readily determined from the closed-loop gain and the open-loop transfer function as demonstrated for the two possible closed-loop solutions shown in the figure. Similarly, the closed-loop gain attainable at a specified closed-loop bandwidth can be determined.

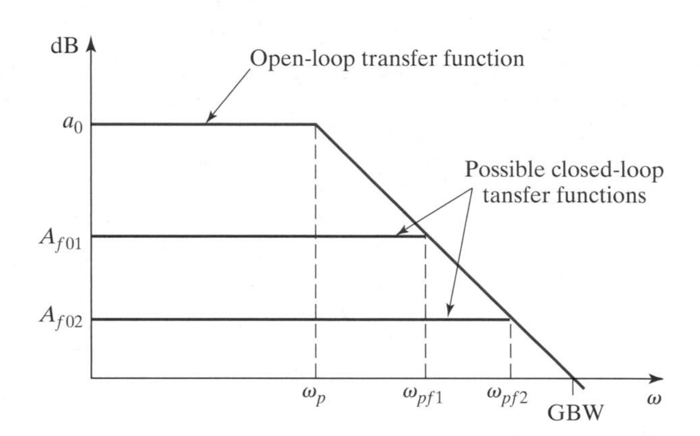

Figure 5–47 Plot showing the relationship between the open- and closed-loop transfer functions for a system that conserves GBW.

■

Example 5.6

Problem:
Find the transfer function of the inverting amplifier circuit shown in Figure 5-48(a). Assume the op amp is ideal, except that the gain is described by (5.65). Part (b) of the figure shows the equivalent circuit we will use for analysis.

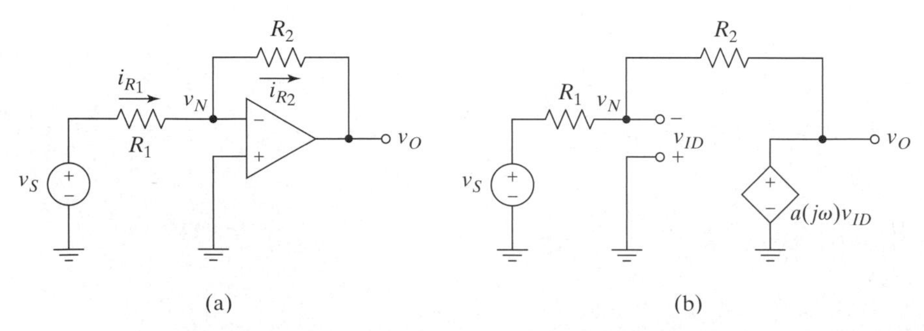

Figure 5–48 (a) An inverting amplifier circuit. (b) The model for determining the frequency response.

There is only one independent node in this circuit, v_N, so we can find the transfer function by writing a nodal equation at that node.

$$v_N(G_1 + G_2) - v_S G_1 - a v_{ID} G_2 = 0. \tag{5.73}$$

Substituting v_N for $-v_{ID}$ and simplifying leads to

$$v_N = v_s\frac{G_1}{G_1 + G_2(1 + a)} = v_s\frac{R_2}{R_2 + R_1(1 + a)}. \quad \textbf{(5.74)}$$

Using the fact that $v_O = av_{ID} = -av_N$ along with (5.74) and substituting (5.65) for $a(j\omega)$ leads to

$$\frac{V_o(j\omega)}{V_s(j\omega)} = \frac{\dfrac{-a_0R_2}{R_2 + R_1(1 + a_0)}}{1 + j\dfrac{\omega(R_1 + R_2)}{\omega_p(R_2 + R_1(1 + a_0))}} \equiv \frac{A_{f0}}{1 + j\dfrac{\omega}{\omega_{pf}}}. \quad \textbf{(5.75)}$$

where we have again defined the DC gain and pole frequency with feedback as A_{f0} and ω_{pf}, respectively. Since the transfer function is again a single-pole low-pass function, it makes sense to find the GBW;

$$\text{GBW} = a_0\omega_p\frac{R_2}{R_1 + R_2} = GBW_0\frac{R_2}{R_1 + R_2}. \quad \textbf{(5.76)}$$

Interestingly, the GBW of the inverting configuration is always lower than the GBW of the noninverting configuration. In fact, if the gain is set to one (so that $R_1 = R_2$), the GBW of the inverting stage is only half that of the noninverting stage. Because the GBW is not conserved in this case, we don't have a simple geometric way of determining the tradeoff between gain and bandwidth as was shown in Figure 5-47 for the noninverting case.

■

5.4.6 Common-Mode Rejection Ratio and Power-Supply Rejection Ratio

The output voltage of an ideal op amp is only a function of the differential input voltage. Unfortunately, real op amps also respond to the average, or common-mode, component of the input voltage. The ratio of the change in output voltage to the change in common-mode input voltage is called the ***common-mode gain***, a_{cm}, and should be as close to zero as possible. Real op amps have asymmetries caused by device mismatches that result in their common-mode gains having nonzero values, which may be positive or negative.

In most applications, the magnitude of the common-mode gain by itself is not that important. Rather, it is the ratio of the magnitude of the differential-mode gain, a_{dm}, to the magnitude of the common-mode gain, a_{cm} that is important.[12] This ratio is called the ***common-mode rejection ratio*** (CMRR) and is usually specified in decibels:

$$\text{CMRR}_{(\text{in dB})} = 20\log(a_{dm}/a_{cm}) = a_{dm\,(\text{in dB})} - a_{cm\,(\text{in dB})}. \quad \textbf{(5.77)}$$

[12] The differential-mode gain is often denoted by a (i.e., without the subscripts dm) when no confusion is likely. We have adopted that convention throughout this chapter, but in this section we use a_{dm} for clarity.

To realize the common-mode interference canceling capabilities of the op amp (as pointed out in Aside A5.2), the CMRR should be as large as possible (*see* Problem P5.47). Fortunately, CMRRs of over 100 dB are fairly common in commercial op amps.

Exercise 5.7

Using the data from Table 5.1, determine the common-mode gain of the LM607 op amp.

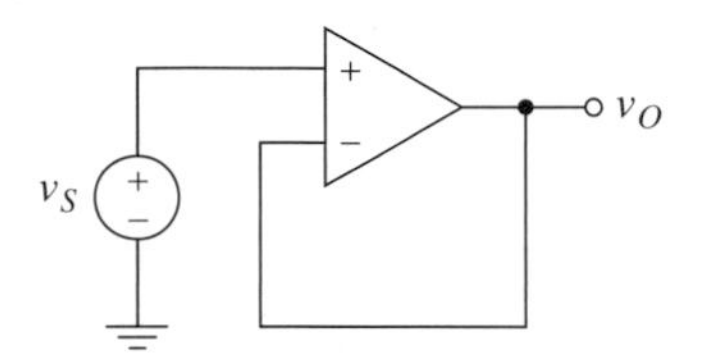

Figure 5–49 A unity-gain buffer amplifier.

The common-mode gain of an op amp can also affect its performance in applications where interference is not the main concern. For example, consider the effect of a finite CMRR on the performance of the buffer amplifier shown in Figure 5-49. If we include both the common-mode and differential-mode gains, the output voltage is given by

$$v_O = a_{dm}(v_S - v_O) + a_{cm}\left(\frac{v_S + v_O}{2}\right). \tag{5.78}$$

Solving this equation for the transfer function yields

$$\frac{v_O}{v_S} = \frac{a_{dm} + a_{cm}/2}{1 + a_{dm} - a_{cm}/2}. \tag{5.79}$$

Since we want a gain of one for the buffer, the nonzero common-mode gain will hurt our performance.

A common way of modeling the effect of a_{cm} on the operation of a circuit is to convert the change in output voltage caused by the common-mode input voltage into an equivalent differential input voltage. The change in output voltage caused by a_{cm} is given by

$$\Delta v_O = a_{cm} v_{Icm}. \tag{5.80}$$

The same change in v_O would be caused by a differential input voltage of

$$v_{Ideq} = \Delta v_O / a_{dm} = v_{Icm} a_{cm} / a_{dm} = v_{Icm}/\text{CMRR}. \tag{5.81}$$

If we place a voltage source in series with either input and give it the value specified by (5.81), it will produce the same change in v_O as a common-mode input equal to v_{Icm}. We can, therefore, model CMRR using exactly the same circuit we used to model V_{OFF}, but with the source equal to $V_{OFF} + v_{Ideq}$. Notice that this voltage is a function of the common-mode input voltage and is therefore *not* constant. Also notice that an inverting amplifier configuration with the noninverting input grounded has essentially no common-mode input. Therefore, the inverting configuration is, in general, not affected by finite CMRR.

Example 5.7

Problem:
Determine the CMRR of the first stages of the instrumentation amplifiers shown in Figure 5-50. For this analysis, assume that the op amps themselves are ideal. In Section 5.1.5 we stated that the circuit in Figure 5-50(b) has a higher CMRR than the circuit in part (a) of the figure. This example will prove that statement.

Solution:
Since the output of the first stage is a differential signal rather than a single-ended signal, we must first clarify our definition of common-mode gain for this case. To allow the common-mode gains of the first and second stages to multiply as we would expect them to, we must define the common-mode gain of the first stage to be v_{Ocm}/v_{Scm}. Similarly, the differential-mode gain is v_{Odm}/v_{Sdm}.

The circuit in part (a) of the figure is really two independent noninverting amplifiers. Amplifier A (on top) has an input of $v_{Scm} + v_{Sdm}/2$ and amplifier B has an input of $v_{Scm} - v_{Sdm}/2$. Both amplifiers have a gain of $(1 + R_2/R_1)$, where we assume that $R_{1A} = R_{1B} = R_1$ and $R_{2A} = R_{2B} = R_2$. Therefore, the output voltages are;

$$v_{OA} = (1 + R_2/R_1)(v_{Scm} + v_{Sdm}/2), \tag{5.82}$$

and

$$v_{OB} = (1 + R_2/R_1)(v_{Scm} - v_{Sdm}/2). \tag{5.83}$$

These equations lead to

$$v_{Ocm} = \frac{v_{OA} + v_{OB}}{2} = (1 + R_2/R_1)v_{Scm} \tag{5.84}$$

and

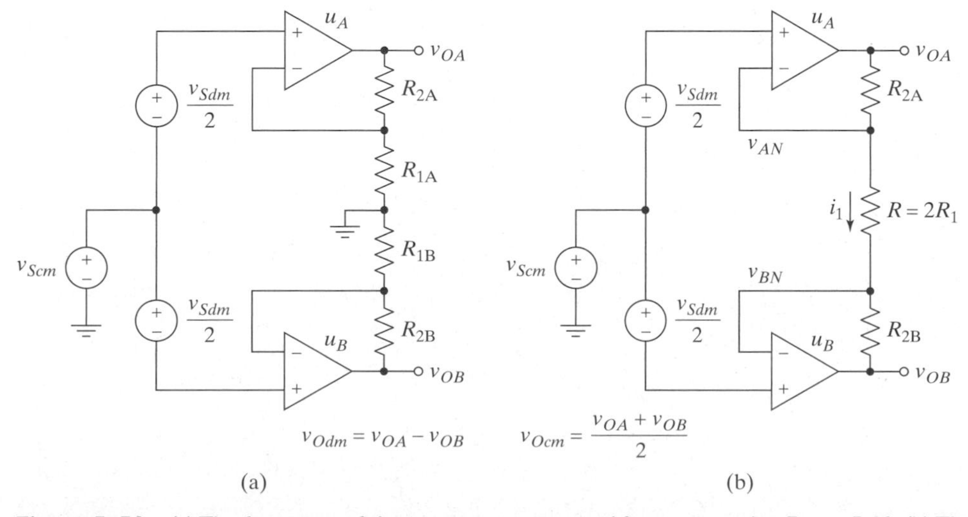

Figure 5–50 (a) The first stage of the instrumentation amplifier presented in Figure 5-12. (b) The first stage of the instrumentation amplifier in Figure 5-13.

$$v_{Odm} = v_{OA} - v_{OB} = (1 + R_2/R_1)v_{Sdm}, \tag{5.85}$$

so that we get

$$a_{cm} = \frac{v_{Ocm}}{v_{Scm}} = 1 + \frac{R_2}{R_1} \tag{5.86}$$

and

$$a_{dm} = \frac{v_{Odm}}{v_{Sdm}} = 1 + \frac{R_2}{R_1}. \tag{5.87}$$

Finally, we obtain the CMRR;

$$\text{CMRR} = \frac{a_{dm}}{a_{cm}} = 1 \tag{5.88}$$

or, in dB, CMRR = 20log(1) = 0 dB. This horrible CMRR would not be the disaster it might at first appear to be, since the second stage of the instrumentation amplifier still provides significant CMRR. Nevertheless, we should strive to do as well as possible in the first stage, so let's examine the circuit in part (b) of the figure.

In this circuit, the two op amps no longer operate independently. Because they each have negative feedback and are ideal however, we can assume virtual shorts at their inputs. Therefore, the voltages at the top and bottom of the resistor $2R_1$ are $v_{AN} = v_{Scm} + v_{Sdm}/2$ and $v_{BN} = v_{Scm} - v_{Sdm}/2$ respectively. The current through this resistor is

$$i_1 = \frac{v_{AN} - v_{BN}}{2R_1} = \frac{v_{Sdm}}{2R_1}. \tag{5.89}$$

Since the input currents of the op amps are zero, the output voltages are

$$v_{OA} = v_{AN} + i_1 R_{2A} = v_{Scm} + \frac{v_{Sdm}}{2} + \frac{R_2 v_{Sdm}}{2R_1}, \tag{5.90}$$

and

$$v_{OB} = v_{BN} - i_1 R_{2B} = v_{Scm} - \frac{v_{Sdm}}{2} - \frac{R_2 v_{Sdm}}{2R_1}. \tag{5.91}$$

Therefore, we get

$$v_{Ocm} = \frac{v_{OA} + v_{OB}}{2} = v_{Scm} \tag{5.92}$$

and

$$v_{Odm} = v_{OA} - v_{OB} = (1 + R_2/R_1)v_{Sdm}. \tag{5.93}$$

The common-mode gain is only one this time, and the differential-mode gain is the same as we had before. Therefore, the CMRR of the first stage is equal to the differential-mode voltage gain;

$$\text{CMRR}_{(\text{in dB})} = 20\log(1 + R_2/R_1). \quad \textbf{(5.94)}$$

We can now see that the assertion made in Section 5.1.5; namely, that removing the ground connection between R_{1A} and R_{1B} improves the CMRR, is correct. Also note that the differential gain can now be adjusted by varying a single resistor (R in Figure 5-50(b)).

■

In addition to responding to common-mode signals at the input, real op amps also respond to changes in the supply voltage(s). If we add a small AC signal to one of the supplies, we can measure the AC component in the output voltage and define a gain from the supply to the output; call it a_{sup}. This gain will not typically be the same from both supplies. However, we can average the results, or take the larger of the two and define a ***power-supply rejection ratio*** (PSRR) by

$$\text{PSRR}_{(\text{in dB})} \equiv 20\log(a_{dm}/a_{sup}). \quad \textbf{(5.95)}$$

The PSRR is important for noise rejection, since we can never eliminate all noise from the supply lines.

Both the CMRR and the PSRR are frequency dependent, decreasing with increasing frequency. We will not consider this effect here, but if you use an op amp at higher frequencies—say above about one twentieth of the GBW—you need to be aware that CMRR and PSRR are typically specified at DC.

5.4.7 Output Swing Revisited

The fact that real op amps have limited output swings was presented in Section 5.3.1, where the upper and lower swing limits were denoted by $V_{O\max}$ and $V_{O\min}$ respectively. However, these limits are not fixed for a given op amp. They depend on the power supplies used. Typically, the output voltage can only go to within a volt, or perhaps a few tenths of a volt, of either supply voltage. Therefore, if the supplies are set at ± 5 V, the limits are likely to be about ± 4 to ± 4.5 V. There are op amps with output swings that can go all the way to the supply voltages (called a *rail-to-rail* swing[13]), but that is not typical.

5.4.8 Slew Rate and Full-Power Bandwidth

In addition to the bandwidth limitation already discussed, op amps have another limitation on their speed of operation. The output voltage cannot change faster than an upper limit called the ***slew rate***. The slew rate limit occurs because there is only a finite current available to charge and discharge each node inside the circuit. In practice, one node in the circuit usually dominates the slew rate. If we model the dominant node as having only capacitance to ground, we can see that the voltage on the node will have a maximum rate of change given by $I_{\max}/C$, where $I_{\max}$ is the maximum current available to charge or discharge the node. This rate of change is the slew rate.

---

[13] This terminology results because the supply voltages are sometimes called the supply *rails*. This term probably refers to the fact that the positive and negative supply nodes are drawn on schematics, as horizontal lines at the top and bottom of the schematic, which looks a bit like the rails of a train track.

Note that slew-rate limiting is a nonlinear phenomenon, because once the limit is reached, increasing v_{ID} no longer increases v_O proportionally. Therefore, superposition does not apply. In contrast, the bandwidth limitation discussed previously is a linear characteristic of a circuit. Also, the slew rate is not generally the same in both directions (i.e., it is not symmetric). However, we will assume the slew rate is symmetric in the rest of this section for simplicity.

If the output voltage swing of the op amp is large, the slew rate can cause distortion to occur for frequencies well below the bandwidth. Therefore, it is of interest to determine the frequency at which we start to see slew-rate limiting. Suppose a sine wave of frequency ω is input to the op amp. The output voltage should be

$$v_O(t) = \hat{V}_O \sin \omega t, \tag{5.96}$$

from which the maximum rate of change of the output voltage can be determined to be

$$\max\left(\frac{dv_O(t)}{dt}\right) = \max(\hat{V}_O \omega \cos \omega t) = \hat{V}_O \omega. \tag{5.97}$$

If the output of the op amp is swinging completely from one limit to the other, we have $\hat{V}_O = (V_{O\max} - V_{O\min})/2$. Under this condition, the op amp is driving the load as hard as it possibly can. If we now set the maximum rate of change of the output equal to the slew rate, we find what is called the ***full-power bandwidth*** of the op amp:

$$\text{BW}_{\text{full power}} = \frac{\text{Slew Rate}}{\pi(V_{O\max} - V_{O\min})} \text{ Hz.} \tag{5.98}$$

To understand the full-power bandwidth better, consider Figure 5-51. The figure shows what an output sine wave should look like and what it really looks like if the maximum slope of the sine wave exceeds the slew rate. As mentioned earlier, slew-rate limiting is a nonlinear phenomenon. If we examine the frequency spectrum of the distorted sine wave, we find that components at harmonics of the fundamental frequency have been introduced. Because of the nonlinear nature of the slew-rate limitation, it cannot be analyzed using standard linear network techniques.

As an example, assume an LF411 op amp has $V_{O\max} = 5$ V and $V_{O\min} = -5$ V. Using the slew rate of 15 V/μs given in Table 5.1, we find a full-power bandwidth of 477 kHz—or only 12% of the unity-gain bandwidth. Therefore, if the LF411 is used as a unity-gain buffer with supply voltages leading to the limits assumed

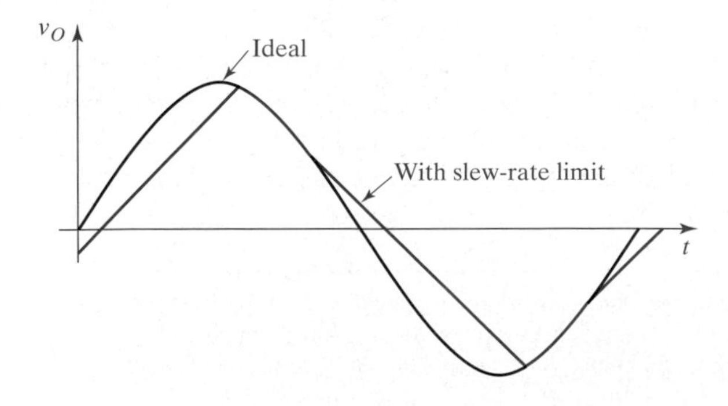

Figure 5–51 A sine wave and what happens if it is distorted by an op amp that cannot change its output voltage faster than its slew rate (see slew.sch on the CD).

above, the maximum frequency for which the output voltage will be undistorted is only 12% of the bandwidth. If we know that the output peak amplitude is always less than the saturation limits, the maximum frequency we can use without slew-rate limiting will be higher than the full-power bandwidth. The exact value for any peak output can be calculated by substituting the proper limits in the denominator of the right-hand side of (5.98).

Exercise 5.8

What slew rate is required of an op amp for it to produce a 20 kHz 10 V_{pp} sine wave at its output? Assume that the slew rate specification of the op amp is the same for positive-going and negative-going output signals.

5.4.9 Noise

In addition to the offsets discussed earlier, there is another problem that emerges when dealing with very small inputs: noise. In many situations (e.g., stereo preamplifiers) we need to maintain a large ***signal-to-noise ratio*** (SNR). Noise is always present in any system that dissipates energy, and the op amp is no exception. A good but brief treatment of noise in circuits is presented in [5.4].

The noise performance of op amps is characterized by specifying the equivalent input noise voltage and current. Usually, these are specified as spectral densities at several different frequencies. In Table 5.1 the values are given at a single frequency that is high enough that it indicates the white noise component. The flicker noise may be specified by providing values at other frequencies, or by giving the $1/f$ corner frequency. Sometimes, the value of the spectral density is integrated over a specified bandwidth, and the rms value is given.

Because the noise voltages present at both inputs of the op amp are independent, their mean-square values are summed prior to specifying the noise.[14] Summing the voltage noise terms allows a single noise voltage source to be used in noise analyses involving op amps; it is the magnitude of this *single* noise voltage generator that is specified on the data sheet. Unfortunately, the same technique cannot be applied to the current noise terms, since they may be working into different source impedances. The magnitude given for the noise current generator applies to each of the noise sources. The resulting noise model for an op amp is shown in Figure 5-52.

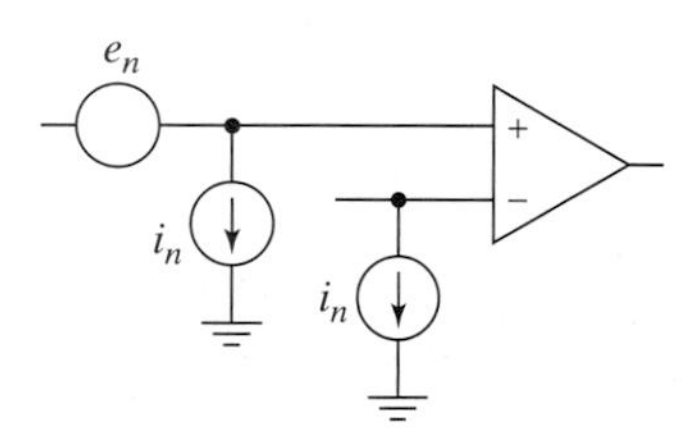

Figure 5–52 The standard noise model used for an op amp.

5.4.10 Op Amp Parameter Measurement

Although a number of op amp parameters can be measured without too much difficulty, some are extremely difficult to measure accurately. In some cases, what is measured is not the parameter as it is classically defined (e.g., CMRR). In this section, we examine how to measure two of the most important op amp parameters; gain and offset. More complete information can be found in industrial application notes and data books, and some basic information is contained in [5.1].

[14] This summation is only an approximation and depends on the CMRR being relatively high to be valid. However, the CMRR is virtually always high enough to make this approximation reasonable.

Open-Loop Gain Directly measuring the open-loop gain of an op amp is extremely difficult, because any small DC offset at the input will drive the output to one saturation limit or the other (see Problem P5.49). The offset could be the offset of the op amp itself, or it could be a slight imbalance in the external circuitry used to bias the op amp and inject the test signal.

Therefore, we are forced to consider methods of measuring the op amp gain using circuits with negative feedback. If we try the obvious thing and use a typical amplifier circuit like the noninverting amplifier of Section 5.1.1 to measure the open-loop gain, we find that the main advantage of negative feedback, which is to make the closed-loop gain insensitive to the open-loop gain, works against us. When we considered the gain of a noninverting amplifier with finite op amp gain in Section 5.4.1, we discovered that the output voltage was given by (5.52), from which we obtain

$$A \equiv \frac{v_O}{v_S} = \frac{a}{1 + ab} \quad \textbf{(5.99)}$$

where b is the feedback factor, as before. If we set $b = 1/1000$ so that the closed-loop gain will be close to 1000, we find that a 1% change in a, which is what we are trying to measure, only produces a 0.005% change in the closed-loop gain, A, for a nominal open-loop gain of $a = 200{,}000$. The sensitivity is even worse if we make the closed-loop gain smaller. Therefore, it would be extremely difficult to accurately determine the open-loop gain from measurements of the closed-loop gain.

We can get around this problem by using the op amp whose gain we are trying to measure as a part of the feedback loop of another op amp circuit. Then, since the closed-loop gain is almost entirely dependent on the feedback, we should be able to get a reasonable measure of the open-loop gain. The circuit shown in Figure 5-53 uses this principle. In this circuit, the op amp we are measuring the gain of is called the ***device under test*** (DUT). The other op amp forms an integrator from v_{OA} to v_O (i.e., consider v_{OA} to be the input and picture v_I as a ground). In this case, the integrator will be stable, since we are tying global DC negative feedback around it. There are two feedback paths in parallel; the path comprising R_f and C_f, and the path through R_2, R_1, and the DUT.

For ease of analysis, we assume the op amp in the integrator is ideal. Because the gain of the integrator is infinite at DC (negative infinity in this case), the only stable solution occurs when $v_{OA} = v_{BN} = v_I$, so that $i_1 = 0$ and v_O isn't changing. In other words, the feedback path containing the DUT must be driven to the point where the DC input to the integrator is zero. With this value of output voltage, the input to the DUT must be

$$v_{IDA} = \frac{v_{OA}}{a} = \frac{v_I}{a}. \quad \textbf{(5.100)}$$

We also know from the circuit that

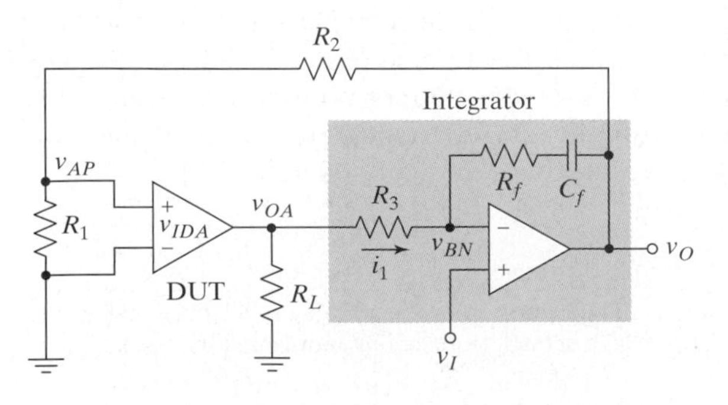

Figure 5–53 A circuit for measuring the open-loop gain of the DUT. Typical values for the elements would be; $R_1 = 51\ \Omega$, $R_2 = 51\ \text{k}\Omega$, $R_3 = 10\ \text{k}\Omega$, $R_L = 2\ \text{k}\Omega$, $R_f = 1\ \text{k}\Omega$, and $C_f = .01\ \mu\text{F}$ [5.1].

$$v_{IDA} = v_O \frac{R_1}{R_1 + R_2}. \tag{5.101}$$

Setting these two equations equal yields

$$a = \frac{v_I(R_1 + R_2)}{v_O R_1}. \tag{5.102}$$

To make v_O a reasonable value to measure, we typically set $R_2 \gg R_1$. Then, rather than making a single-point measurement, an average over a large range of inputs is made. To do this, first set v_I to some large negative value and measure the resulting value of v_O. Next, v_I is set to some large positive value, and the resulting value of v_O is again measured. The gain is determined from

$$a = \left(\frac{R_1 + R_2}{R_1}\right)\frac{\Delta v_I}{\Delta v_O} = \left(\frac{R_1 + R_2}{R_1}\right)\frac{v_{I\text{high}} - v_{I\text{low}}}{v_{O\text{high}} - v_{O\text{low}}}. \tag{5.103}$$

Offset Voltage The offset voltage of an op amp can also be measured using the circuit in Figure 5-53. If we set v_I to ground so that v_{OA} is also forced to ground, the voltage at the input of the DUT must then be V_{OFF}, and we have

$$V_{OFF} = v_O\left(\frac{R_1}{R_1 + R_2}\right). \tag{5.104}$$

SOLUTIONS TO EXERCISES

5.1 **(a)** $v_L/v_S = 10$, **(b)** The gain does not depend on R_L because the output resistance of an ideal op amp is zero.

5.2 We can analyze this circuit using the same procedure followed in (5.4) and (5.5), since there is still negative feedback, and the infinite gain still guarantees a virtual short. If you are unconvinced of this, repeat the analysis in (5.6) through (5.8) and allow for the voltage divider from v_O to v_L; the result is the same. Therefore, **(a)** $v_L/v_S = 10$. **(b)** The infinite gain causes a virtual short at the inputs. Therefore, since it is v_L that is fed back to the input, rather than v_O (which would be impossible, since it is modeling the *open-circuit* output voltage of the op amp), R_o has no effect on the gain. If the op amp gain is not infinite, the output resistance will affect the gain.

5.3 $R_1 = 5$ kΩ and $R_2 = 100$ kΩ.

5.4 $R_1 = R_3 = 10$ kΩ and $R_2 = R_4 = 100$ kΩ.

5.5 We know that the current through an inductor is the integral of the voltage across it. Therefore, if we use an inductor in place of R_1 in the standard inverting op amp gain stage, we will get an integrator. This circuit would have infinite DC gain (only if the inductor was perfect) and would only work if the source had no DC component. We can reduce the DC gain by including a resistor in series with the inductor, as shown in Figure 5-54.

We analyze this circuit by noting that the negative feedback will force a virtual short at the op amp inputs, therefore

$$I_1(j\omega) = \frac{V_s(j\omega)}{R_1 + j\omega L}. \tag{5.105}$$

We also know that i_2 must equal i_1, since the input current of the op amp is zero and that

$$V_o(j\omega) = -R_2 I_1(j\omega). \tag{5.106}$$

Figure 5–54 The solution to Exercise 5.5.

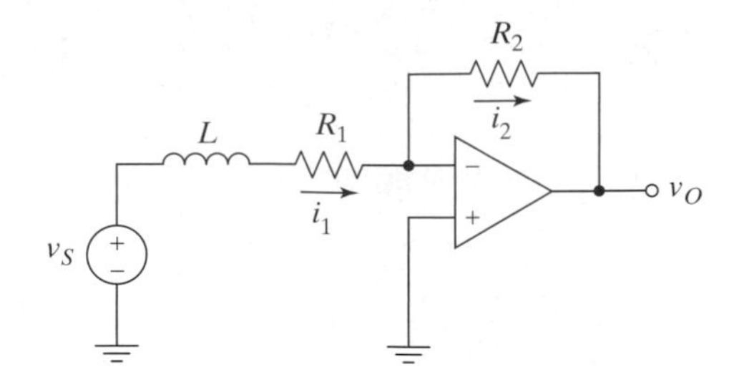

Combining these results we get the transfer function

$$\frac{V_o(j\omega)}{V_s(j\omega)} = \frac{-H(j0)}{1 + j\omega/\omega_p}, \tag{5.107}$$

where the DC gain is $H(j0) = -R_2/R_1$, and the pole is $\omega_p = R_1/L$. Now we can solve for the values requested in the exercise. The DC input resistance is just R_1, since the inductor has zero resistance. Therefore, we need $R_1 = 10\ \Omega$, $R_2 = 10\ \text{k}\Omega$, and $L = 16$ mH.

5.6 Whenever v_I is positive, the diode must be in reverse breakdown for any current to flow, and the resulting v_O is −3 V. Whenever v_I is negative, the diode is forward-biased, and v_O is 0 V (for an ideal diode).

5.7 We see from the table that the LM607 has a gain (i.e., differential-mode gain) of 5 million, or 134 dB. The CMRR is also given as 140 dB. Using the definition of the CMRR, we have;

$$a_{cm} = a \times 10^{-\text{CMRR}/20}, \tag{5.108}$$

where the CMRR is in dB, or

$$a_{cm\text{ in dB}} = a_{dm\text{ in dB}} - \text{CMRR}_{\text{in dB}}. \tag{5.109}$$

Plugging in the values for the LM607 leads to $a_{cm} = -6\text{ dB} = 0.5$.

5.8 The output waveform is given by $v_O = 5\sin(2\pi \times 20\text{ kHz} \times t)$ and

$$\frac{dv_O}{dt} = 6.283 \times 10^5 \cos(2\pi \times 20\text{ kHz} \times t)\text{ V/s}.$$

Therefore, we see that

$$\left|\frac{dv_O}{dt}\right|(\text{max}) = 6.283 \times 10^5\ \frac{\text{V}}{s} = 0.6283\ \frac{\text{V}}{\mu s}.$$

This is the minimum slew rate that the op-amp must have to reproduce the given output waveform.

CHAPTER SUMMARY

- Op amps are extremely versatile building blocks for board-level circuit design. The ideal op amp is a differential-input amplifier with infinite gain, infinite input impedance, infinite bandwidth, and zero output impedance.
- When negative feedback is used with an ideal op amp, we can analyze the circuit by using the concept of a virtual short between the op amp inputs.
- Op amps can be used to produce inverting amplifiers, noninverting amplifiers, buffer amplifiers, differential amplifiers, active filters, and a number of other linear and nonlinear functions. They can also be used to help make current sources, voltage regulators, and many other electronic subsystems.

- One of the main advantages of amplifiers with differential inputs and good common-mode rejection is that the system can reject many types of interfering signals and noise, because they show up as common-mode terms.
- Instrumentation amplifiers combine extremely high input impedance with good common-mode rejection and are very useful in many electronic systems.
- Active filters made with op amps have the advantages of providing power gain and a low output impedance. The low output impedance implies that these filters can be cascaded with minimal interaction.
- Although real op amps do an amazingly good job of approximating an ideal op amp for many applications, all real op amps do have nonideal characteristics that sometimes need to be taken into consideration. For example, finite gain, input bias and offset currents, offset voltage, finite input and output impedances, finite bandwidth, finite common-mode and power-supply rejection, finite output swing, finite slew rate, and noise.
- The slew rate limitation of op amps is a nonlinear effect that can limit the usable bandwidth significantly for large signal swings.
- Some op amp parameters are very difficult to measure directly and require special test circuits and techniques.

REFERENCES

[5.1] T.M. Frederiksen, *Intuitive Operational Amplifiers: From Basics to Useful Applications*, revised ed. New York: McGraw-Hill, 1988.

[5.2] J.K. Roberge, *Operational Amplifiers: Theory and Practice*, New York, NY: John Wiley & Sons, 1975.

[5.3] Horowitz & Hill, *The Art of Electronics*, 2/E, New York: Cambridge University Press, 1989.

[5.4] P.R. Gray, P.J. Hurst, S.H. Lewis and R.G. Meyer, *Analysis and Design of Analog Integrated Circuits*, 4E, New York: John Wiley & Sons, 2001.

PROBLEMS

5.1 Basic Op Amp Circuits

P5.1 Find the unknown voltages $V_1, V_2, \ldots$ for each circuit in Figure 5-55 (not every circuit has a V_1). Assume ideal opamps. In circuit (a), for example, it is possible to calculate V_3 by using superposition and (5.5) and (5.12). You should also recognize that the output of the opamp will go to whatever voltage is required to drive the inverting input equal to the noninverting input, and the circuit can be solved using that principle. Both methods will yield the same result if done correctly.

5.1.1 The Non-Inverting Amplifier

P5.2 The following questions pertain to a non-inverting amplifier built with an op amp. **(a)** Suppose that the closed-loop amplifier is to have a gain of 100 and a bandwidth of at least 20 kHz. What is the minimum gain-bandwidth product (GBW) required of the op amp? **(b)** The closed-loop gain of the amplifier is still 100, but it is required that the product $a(j\omega)b$, which is called the loop gain, be at least 20 dB at 20 kHz. What is the minimum GBW required of the op amp?

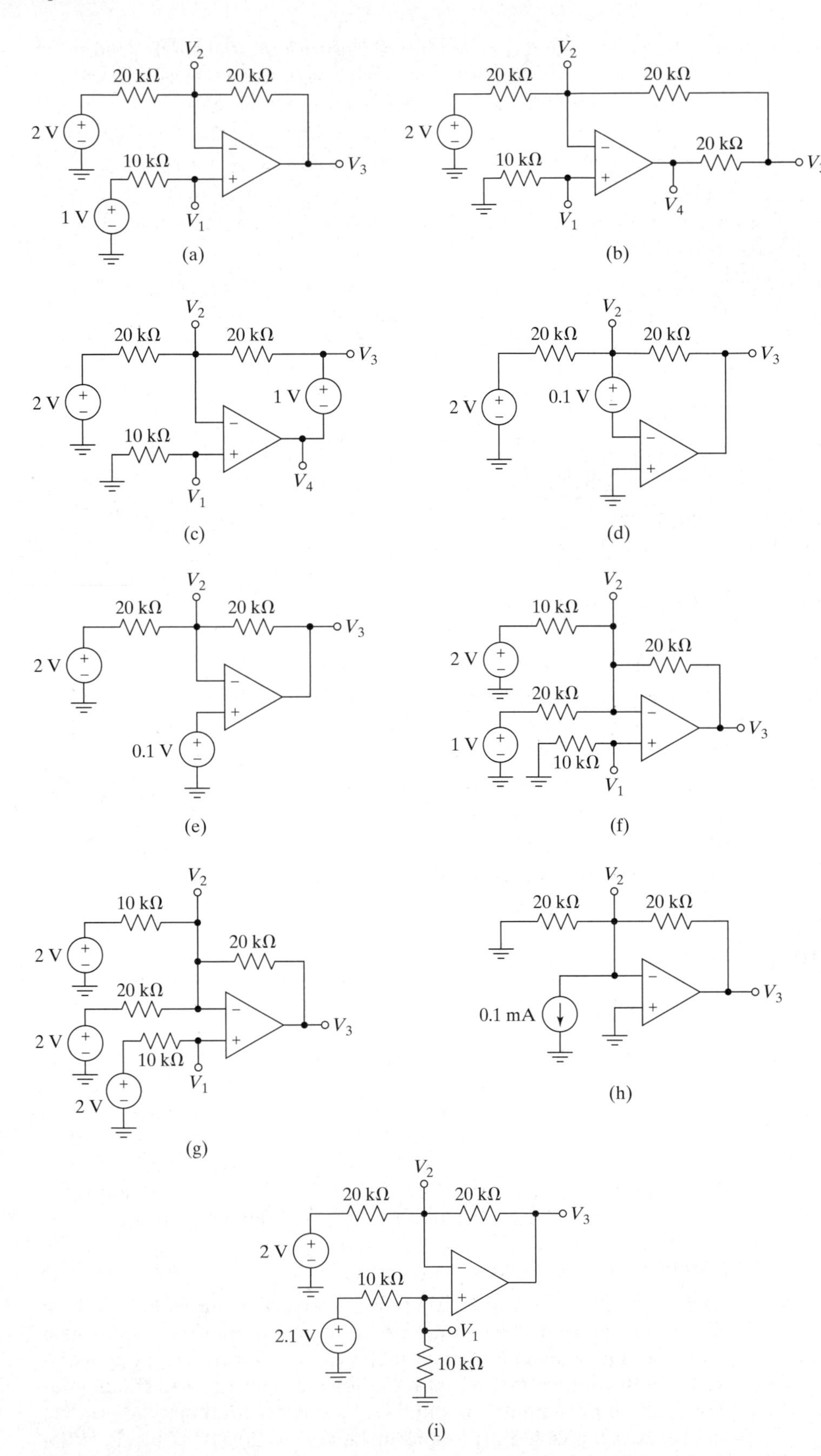

V2
20 kΩ
20 kΩ
2 V
10 kΩ
V3
1 V
V1
(a)
V2
20 kΩ
20 kΩ
2 V
20 kΩ
10 kΩ
V3
V1
V4
(b)
V2
20 kΩ
20 kΩ
V3
1 V
2 V
10 kΩ
V1
V4
(c)
V2
20 kΩ
20 kΩ
V3
2 V
0.1 V
(d)
V2
20 kΩ
20 kΩ
V3
2 V
0.1 V
(e)
V2
10 kΩ
2 V
20 kΩ
20 kΩ
V3
1 V
10 kΩ
V1
(f)
V2
10 kΩ
2 V
20 kΩ
20 kΩ
V3
2 V
10 kΩ
V1
2 V
(g)
V2
20 kΩ
20 kΩ
V3
0.1 mA
(h)
V2
20 kΩ
20 kΩ
V3
2 V
10 kΩ
V1
2.1 V
10 kΩ
(i)

Figure 5–55

P5.3 Consider a non-inverting amplifier as shown in Figure 5-3. If $(R_1 + R_2)/R_1 = 10$, then the closed-loop gain will be 10 if the op amp is ideal. **(a)** Assuming that the op amp is ideal except that the gain, a, is finite, determine the minimum value of a for which the closed-loop gain will be within 1% of the desired value of 10. **(b)** If $a = 200{,}000$, how accurate is the closed-loop gain? (Note: many op amps have low-frequency gains this high)

P5.4 Consider the non-inverting amplifier shown in Figure 5-5 and assume that $a = 100{,}000$. **(a)** Find the load voltage. **(b)** Find the load voltage with R_L removed. **(c)** From your results in (a) and (b), determine the output resistance of the closed-loop amplifier. (*Hint*: What is the output attenuation factor with R_L present?)

5.1.2 The Inverting Amplifier

P5.5 Consider the circuit in Figure 5-56. **(a)** Calculate v_O as a function of i_S and the two resistors in the circuit. **(b)** Suppose it is desired that $v_O = i_S R_1$. How should R_1 be chosen to make v_O approximately equal to its desired value with R_S present? What effect might this constraint on the value of R_1 have on the achievable transresistance gain of the circuit? **(c)** Draw a circuit using an ideal op amp as a transresistance amplifier for which the desired relationship in Part (b) is achieved (ignore the sign). **(d)** For the circuit which is the solution to Part (c), explain why R_S does not appear in the equation for the output voltage.

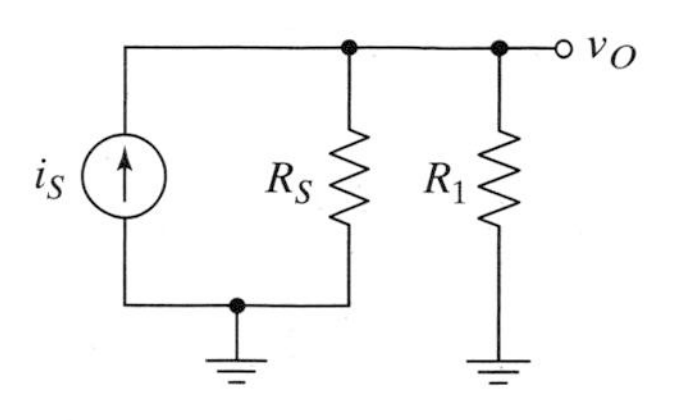

Figure 5–56

P5.6 Consider the inverting amplifier shown in Figure 5-6 and let $R_1 = 4.7\ \text{k}\Omega$, $R_2 = 100\ \text{k}\Omega$ and the op amp be ideal, except that its gain is $a = 50{,}000$. Find the closed-loop voltage gain of this circuit.

D P5.7 Using only inverting op amp stages, design a circuit that has three inputs and produces an output given by $v_O = 3v_{I1} - 2v_{I2} + 5v_{I3}$.

5.1.3 The Unity-Gain Amplifier, or Voltage Follower

P5.8 Derive an expression for the closed-loop gain of the voltage follower in Figure 5-8 if the op amp is ideal except for having a finite gain $a = 100{,}000$.

P5.9 Consider the voltage follower in Figure 5-8 and assume that the op amp is ideal, except for having a finite gain a. What is the minimum value of a required to produce a closed-loop gain greater than 0.99?

5.1.4 The Differential Amplifier

P5.10 Derive an expression for the closed-loop gain of the differential amplifier in Figure 5-9 if the op amp is ideal except for having a gain $a = 150{,}000$. Assume that $R_1 = R_3$ and $R_2 = R_4$.

D P5.11 **(a)** Design a differential amplifier with a gain of 30 and a differential input resistance of 50 kΩ. Assume the op amp is ideal. **(b)** If the amplifier is driven by two single-ended sources as shown in Figure 5-9, find the input resistance seen by each of these sources.

P5.12 Consider the circuit shown in Figure 5-57 where a sensor is connected to an op amp. The local grounds are different by an amount v_{gd}, $R_{ic} = 1\ \text{M}\Omega$, $R_{id} = 100\ \text{k}\Omega$, and $R_S = 100\ \Omega$. Assume that i_i may be as large as 50 μA but of unknown polarity. How large must v_S be for us to be sure we are seeing the signal and not interference?

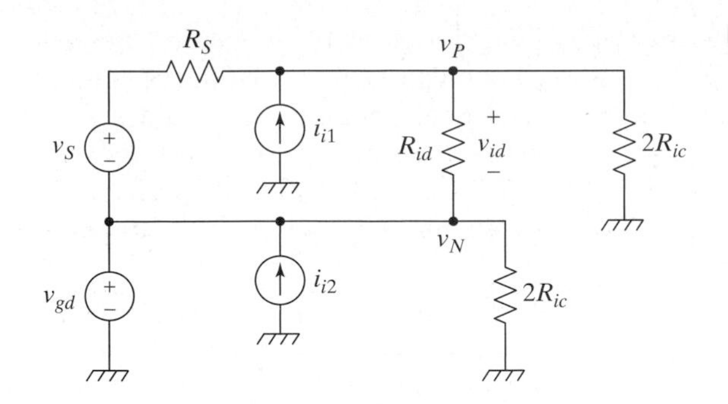

Figure 5–57

5.1.5 The Instrumentation Amplifier

D P5.13 Design an instrumentation amplifier with an overall voltage gain of 1000. Use the smallest spread in resistor values possible.

5.1.6 Current Sources

P5.14 The circuit of Figure 5-58 shows a current source using an op amp and a *pnp* transistor as active devices. **(a)** Express I_O in terms of circuit variables. These variables may include V_{CC}, V_{ref}, R_E and R_L. Assume that the transistor has infinite current gain (β) and infinite V_A. **(b)** Given that $V_{CC} = 10$ V, $V_{ref} = 8$ V, $R_E = 10$ kΩ, and $R_L = 20$ kΩ, what is I_O? What is the voltage at the collector of Q_1? **(c)** For the given conditions in part (b) but with R_L changed to 10 kΩ, what is I_O? What is the voltage at the collector of Q_1? **(d)** Because of changing current demands from other circuitry using the same power supply, the value of V_{CC} may change slightly from its nominal value of 10 V. From your answer to part (a), what is dI_O/dV_{CC} if V_{ref} remains constant? **(e)** Would I_O be better controlled if V_{ref} is regulated to be constant with respect to ground or with respect to V_{CC}? Justify your answer.

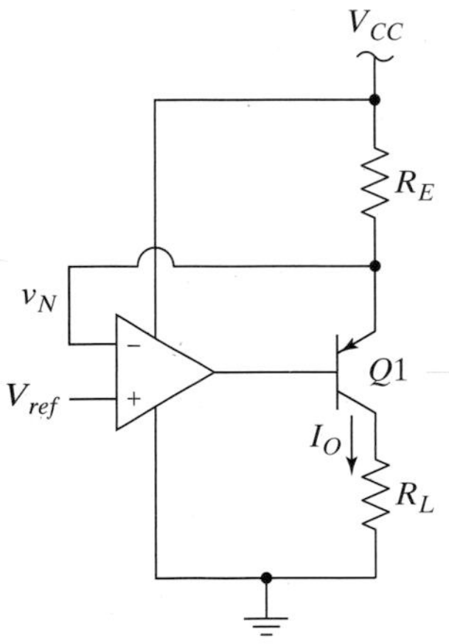

Figure 5–58

P5.15 Consider the current source of Figure 5-59. Assume that I_{ref} is adjusted so that $V_{D1} = 5$ V. **(a)** Assuming that M_1 is in saturation, what is the value of I_O? **(b)** What is the minimum voltage V_O for which M_1 is in saturation if M_1 has $I_D = K\,(V_{GS} - V_{th})^2$ with $K = 4$ mA/V^2?

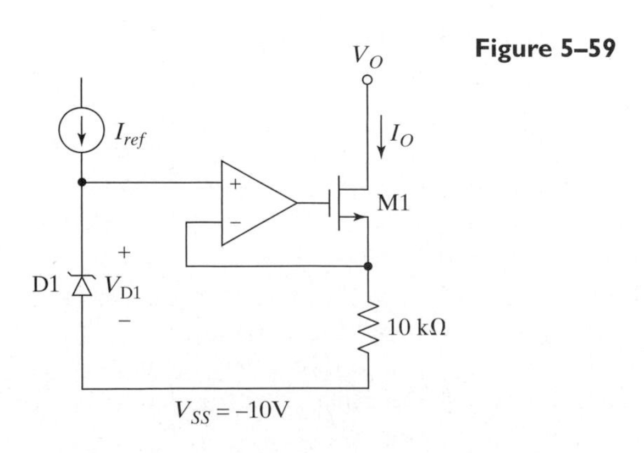

Figure 5–59

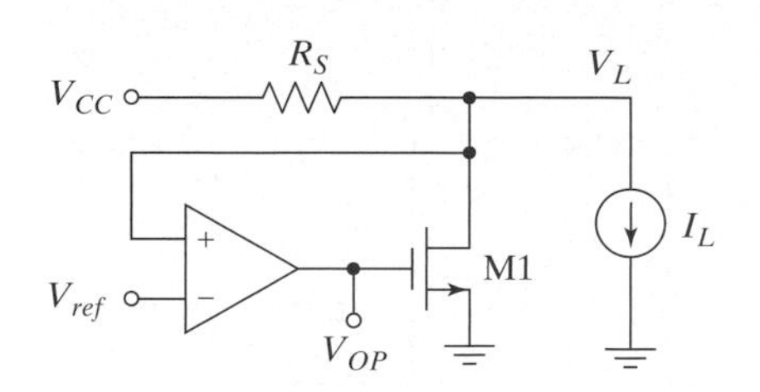

Figure 5–60

5.1.7 Voltage Regulators

P5.16 Consider the shunt voltage regulator of Figure 5-60. **(a)** Does this circuit have negative feedback? Explain. For the following parts, assume that $V_{CC} = 6$ V and it is desired that $V_L = 5$ V. It is also given that $0 \leq I_L \leq 50$ mA. **(b)** What value of V_{ref} should be used to obtain the desired output voltage? **(c)** What is the maximum allowable value of R_S? **(d)** Assume that $R_S = 18\ \Omega$. What is the maximum power dissipation in M_1 and at what value of I_L does it occur? **(e)** What is v_{OP} when $I_L = 30$ mA? Assume that M_1 has the following relationship: $I_{D1} = K_1(V_{GS1} - V_{th})^2$ where $K_1 = 50\ \text{mA/V}^2$ and $V_{th} = 1$V.

P5.17 Figure 5-61(a) represents a very simple form of voltage regulator that uses a single MOS transistor as a series-pass transistor. The voltage across the load is given by $V_L = V_{ref} - V_{GS}$, where V_{ref} is a constant reference voltage. A change in the load current I_L will cause a relatively small change in V_{GS} ; hence, a small change in V_L. This circuit is therefore useful as a voltage regulator. The small change in V_L as a function of I_L is characterized by the small-signal output resistance R_o, which is calculated from $R_o \equiv -dv_L/di_L = dv_{GS}/di_L$. Now consider the circuit in part (b) of the figure, in which the same MOSFET is placed in a feedback loop using an op amp. The op amp has open-loop gain a. **(a)** Application of (5.1) to this circuit yields $V_{OP} = a(V_{ref} - V_{GS})$. Eliminate V_{OP} from this equation by substituting $V_{OP} = V_L + V_{GS}$. Solve for V_L in terms of V_{ref} and V_{GS} . **(b)** What value does V_L approach as the open-loop gain a becomes larger? Contrast this with V_L in the circuit of Figure 5-61(a). **(c)** Find the small-signal output resistance with feedback, $R_{of} = -dv_L/di_L$, for the op amp circuit in terms of a and R_o. *Hint*: Find dv_L/dv_{GS} from your answer to part (a).

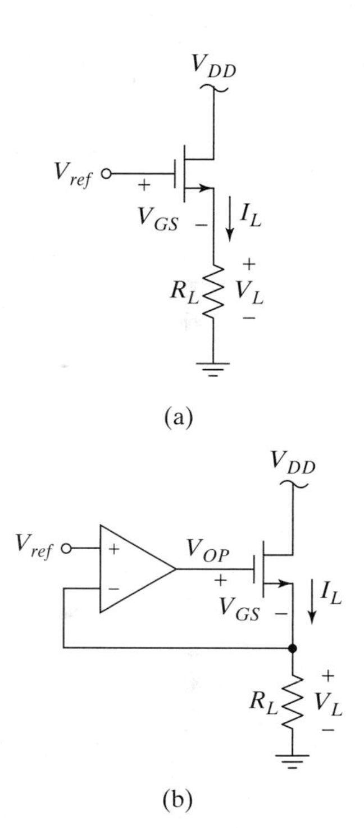

Figure 5–61

5.2 Frequency-Dependent Op Amp Circuits

5.2.1 The Integrator and First-Order Low-Pass Filter

P5.18 The circuit of Figure 5-62 shows an inverting integrator and its input voltage source. The voltage of v_S is graphed as a function of time in part (b) of the figure. The figure shows that v_S is equal to +2 V for some non-specified time T_H, and is equal to −1 V for time T_L. **(a)** If v_O is to have a constant long-term average value, what must be the ratio of T_H to $(T_H + T_L)$? This ratio is called the "duty cycle". **(b)** What is dv_O/dt when v_O is positive-going? When it is negative-going?

P5.19 Consider the integrator circuit in Figure 5-16 with $R_1 = 1\ \text{k}\Omega$ and $C_f = 0.1\ \mu\text{F}$. How long does it take for v_O to go from zero to −10 V if $v_I = 4$ mV DC? Assume $v_C(0) = 0$ V.

D P5.20 Design a LPF with an input resistance of 10 kΩ and a cutoff frequency of 25 kHz.

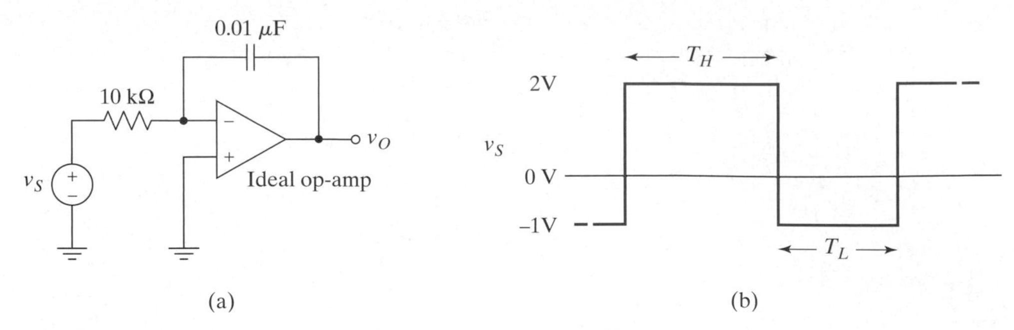

Figure 5–62

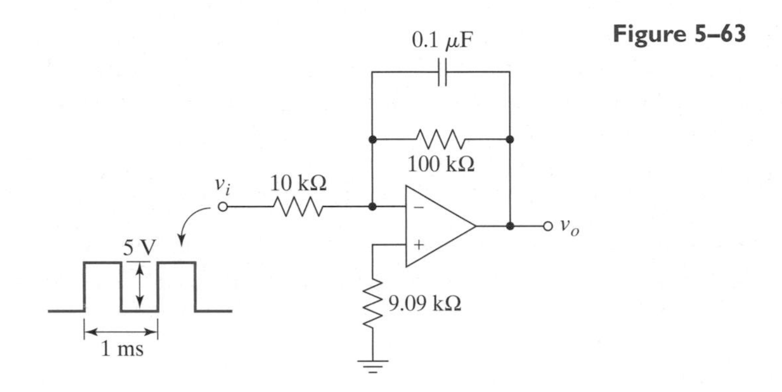

Figure 5–63

P5.21 For the integrator circuit shown in Figure 5-63, sketch the output signal for the square-wave input shown. Show the amplitude and time scale. The DC input voltage is zero.

5.2.2 The Differentiator and First-Order High-Pass Filter

D P5.22 **(a)** Design a HPF with a cutoff frequency of 100 kHz when driven by a source with $R_S = 1$ kΩ. **(b)** What is the cutoff frequency if R_S changes to 800 Ω? **(c)** Modify your design to provide a cutoff frequency that is independent of R_S.

P5.23 **(a)** Re-derive (5.35) if the op amp has a finite gain a. **(b)** If $R_f/R_S = 100$ and $a = 50{,}000$, how much does finite op amp gain change the pole frequency relative to an ideal op amp? **(c)** Repeat (b) with $R_f/R_S = 1000$.

P5.24 For the differentiator circuit shown in Figure 5-64, derive $V_{o(j\omega)}/V_{i(j\omega)}$ and sketch the output waveform for the given input waveform. The DC current input voltage is 1.5 V.

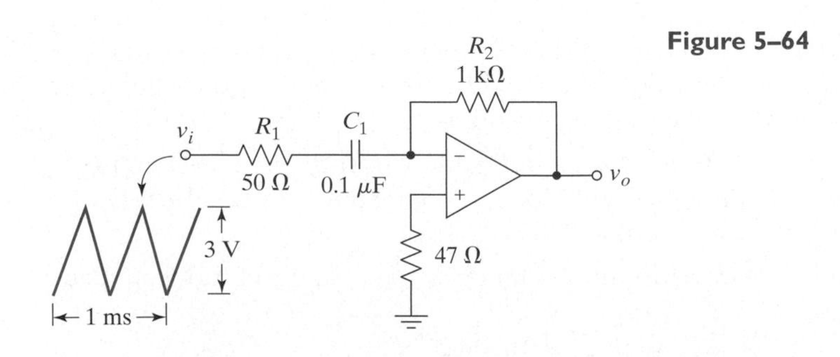

Figure 5–64

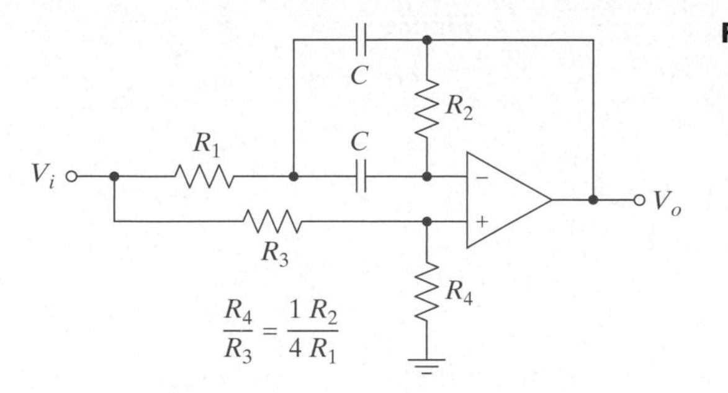

Figure 5–65

5.2.3 Second-Order Filters

P5.25 An all-pass filter is one that has a transfer function with a constant magnitude (i.e., only the phase changes with frequency). A second-order all-pass filter is shown in Figure 5-65. **(a)** Derive the transfer function and show that if $R_4/R_3 = R_2/(4R_1)$ it reduces to

$$\frac{V_o(s)}{V_i(s)} = K\frac{s^2 - (1/Q)s + 1}{s^2 + (1/Q)s + 1}$$

where $K = +R_4/(R_3 + R_4)$. What is Q? **(b)** Draw a plot of the poles and zeros in the complex plane.

5.3 Nonlinear Op Amp Circuits

P5.26 The op amp in the circuit of Figure 5-66 is ideal, except that its output is limited to $\pm$ 10 V. Find the values of v_O corresponding to the two stable states of the circuit. Assume that the diode has a voltage drop of 0.5 V when it is forward biased.

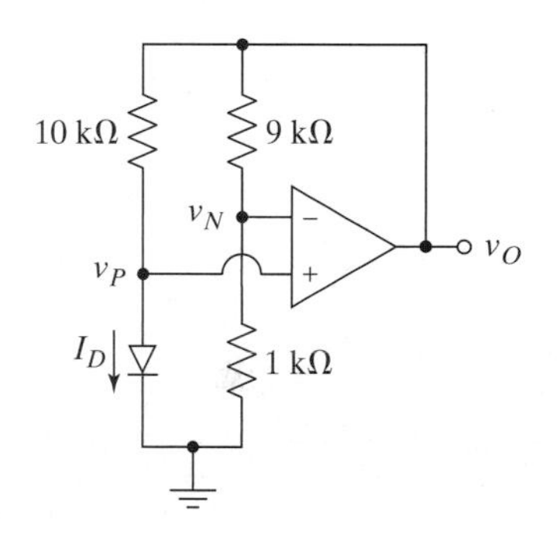

Figure 5–66

5.3.1 Comparators

P5.27 Assume the op amps in Figure 5-67 are ideal, except that their output voltage swings are limited to $\pm$10 V. **(a)** For the circuit of Figure 5-67(a), at what value of v_I does v_O change from -10 V to $+10$ V? At what value of v_I does v_O change from $+10$ V to -10 V? Draw the transfer characteristic from v_I to v_O. **(b)** For the circuit of Figure 5-67(b), assume that v_I is driven by an ideal voltage source. At what value of v_I does v_O change from -10 V to $+10$ V ? At what value of v_I does v_O change from $+10$ V to -10 V? Draw the transfer characteristic from v_I to v_O.

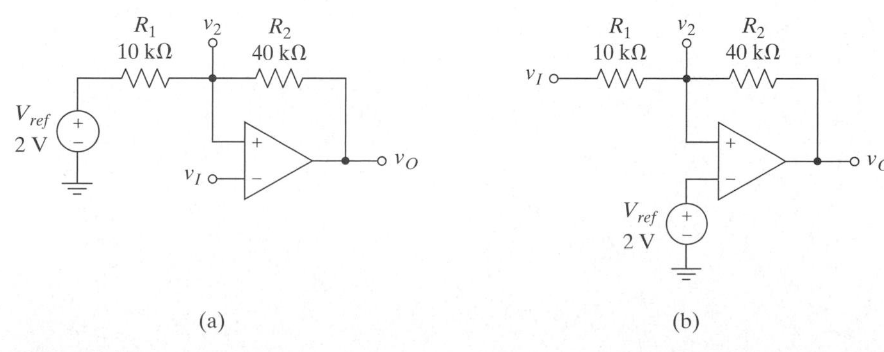

Figure 5–67

Figure 5–68

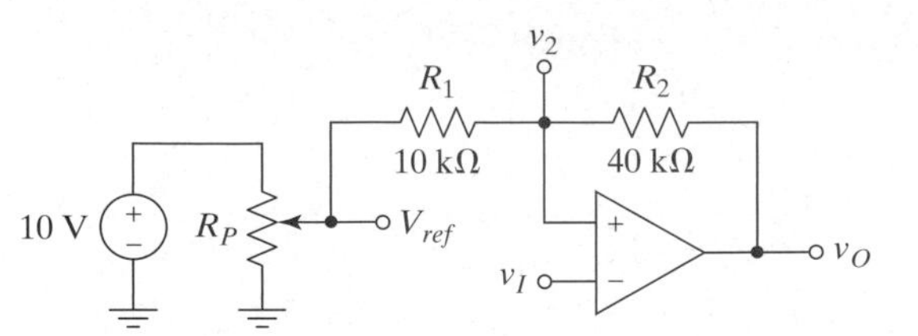

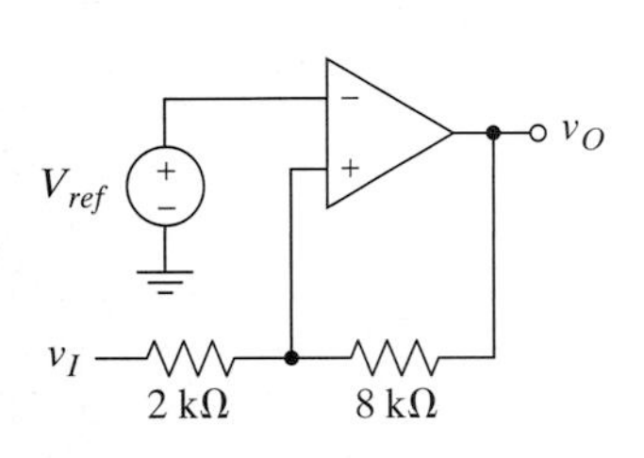

Figure 5–69

P5.28 It is desired that the V_{ref} source in Problem P5.27 be adjustable between 0 and +10 V by means of a pot as shown in Figure 5-68. How will this lead to difficulty in setting the amount of hysteresis in the comparator circuit? A brief qualitative answer will suffice. What circuit that uses a single op amp could be inserted between the pot and R_1 to make the amount of hysteresis independent of the pot position?

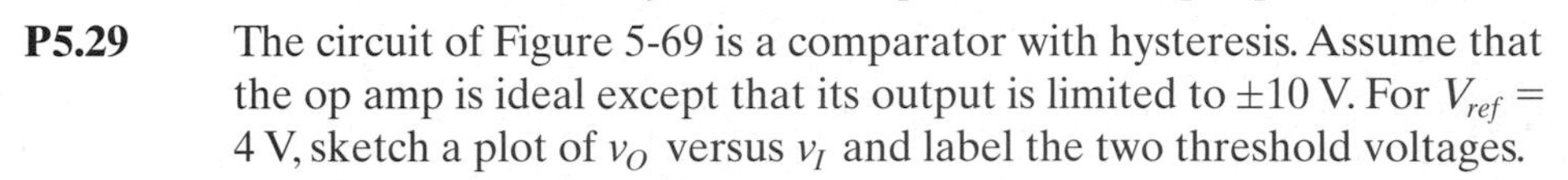

P5.29 The circuit of Figure 5-69 is a comparator with hysteresis. Assume that the op amp is ideal except that its output is limited to ±10 V. For $V_{ref} =$ 4 V, sketch a plot of v_O versus v_I and label the two threshold voltages.

5.3.2 Precision Rectification and Clipping

P5.30 Consider the circuit of Figure 5-70. Assume that the diodes have a constant forward voltage drop of 0.5 V when forward-biased and that they do not conduct any current when reverse biased. **(a)** For $|v_I| < 0.5$ V, what is the small-signal gain $|v_o/v_i|$? **(b)** For $v_I = 1$ V, what is v_O? **(c)** Sketch the DC transfer characteristic of the circuit. The horizontal axis should represent $-1\text{V} < v_I < 1\text{V}$, and the vertical axis should represent v_O. Label the coordinates of corners on the graph.

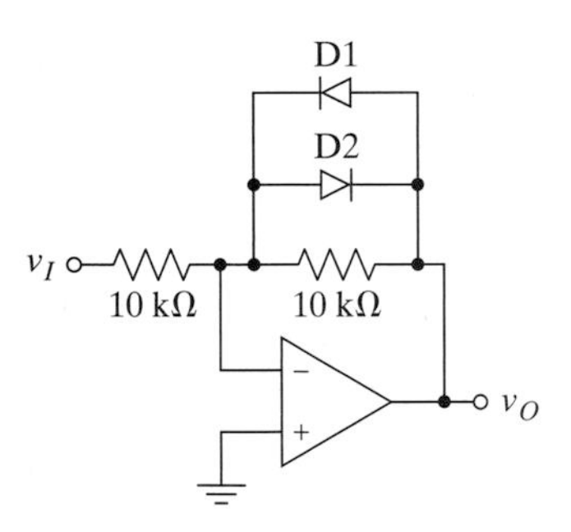

Figure 5–70

P5.31 For the circuit of Figure 5-71, assume that the diode has a constant forward voltage drop of 0.5 V when forward biased and that it does not conduct any current when reverse biased. The op amp is ideal except that it rails at ±5 V. **(a)** The bias condition of the diode (forward or reverse biased) depends on the relative values of v_I and v_O. What relationship between v_I and v_O will cause the diode to be reverse biased? **(b)** When the diode is forward biased, the circuit has negative feedback. When this is true, what is the relationship between v_I and v_O? **(c)** When the diode is reverse biased, what resistance is seen by C? What is the resulting time constant? **(d)** Using your answers to parts (b) and (c) above, find the values of v_I, v_{OP}, and v_O at $t = 0.5$ ms, 1 ms, and 2 ms. **(e)** Sketch the waveforms at v_I and v_O for $0 < t < 3$ ms. What would you call this circuit?

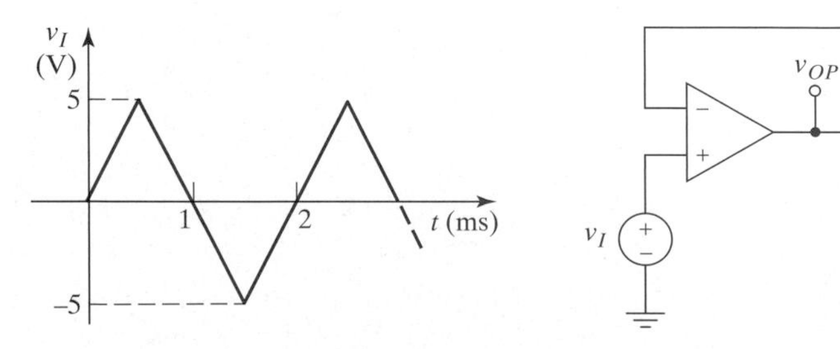

Figure 5–71

5.3.3 Logarithmic Amplifiers

P5.32 The circuit of Figure 5-72 is a logarithmic amp. Its operation is based on the relationship between the transistor's collector current I_C and its base-emitter voltage V_{BE}: $I_C = I_S \exp(V_{BE}/V_T)$, where I_S is a constant for the device, and V_T is the thermal voltage. For this problem, assume $I_S = 10^{-14}$ A and $V_T = 26$ mV. Also assume $R = 1$ kΩ. **(a)** For $v_I = 1$ V, find i_I. Assume a virtual short exists. **(b)** For the conditions of part (a), what must be the collector current of Q_1, and what is the corresponding V_{BE} at room temperature? What is the resulting v_O? **(c)** For $v_I = 10$ V what is the resulting v_O? **(d)** For this circuit, a ten-fold increase in v_I causes what incremental change in v_O? Does this relationship depend on the I_S of the transistor used?

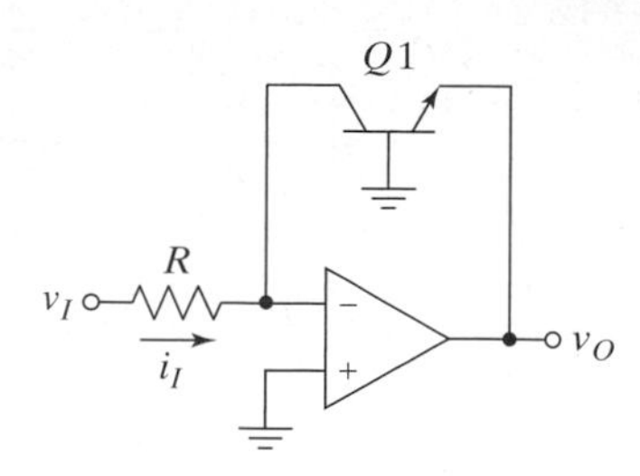

Figure 5–72

5.4 Nonideal Characteristics of Op Amps

5.4.1 Finite Gain

P5.33 Consider the circuit of Figure 5-33. Define

$$A_f \equiv \frac{a}{1 + ab}$$

where A_f represents the closed-loop gain of the amplifier, and define

$$A_f' \equiv \frac{1}{b},$$

where A_f' represents the closed-loop gain calculated under the assumption that $a = \infty$. Then the relative gain error can be defined:

$$\varepsilon \equiv \frac{A_f - A_f'}{A_f'} \times 100\%.$$

(a) Express the relative gain error ε as a function of ab. **(b)** Suppose a noninverting amplifier is to be operated at $A_f' = 100$. What open-loop gain a is required of the op amp to result in a relative gain error with magnitude less than 0.1%? If the amplifier is to be operated at $A_f' = 1000$, what open-loop gain a is required for the same gain error performance?

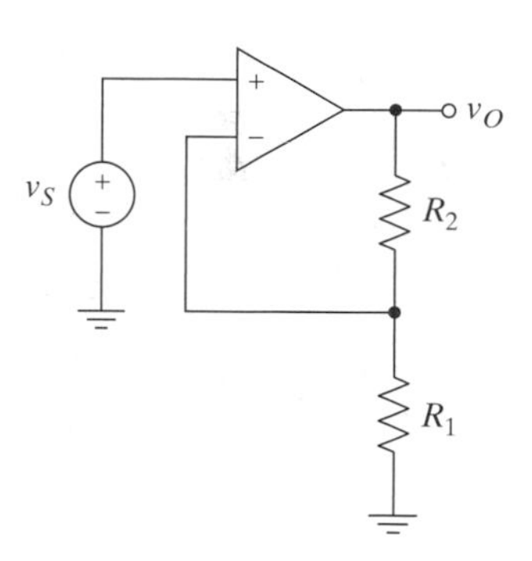

Figure 5–73

5.4.2 Input Bias and Offset Currents

P5.34 Calculate the output offset voltage caused by a bias current of I_B flowing into each input of the op amp for the circuit in Figure 5-35(a). Include the finite gain of the op amp in the analysis.

P5.35 Assume that the op amp shown in Figure 5-73 is ideal except for having nonzero input bias currents. Redraw the circuit using the appropriate model for the op amp and derive an equation for v_O resulting from the input bias currents alone. Assume the input offset current is zero.

DP5.36 Assume that the op amp shown in Figure 5-74 is ideal, except for having nonzero input bias currents. Redraw the circuit using the appropriate model for the op amp and then derive equations for R_1 and R_2 that would yield a gain of ten and no output offset due to the input bias currents. Assume the input offset current is zero, and the value of R_S is given.

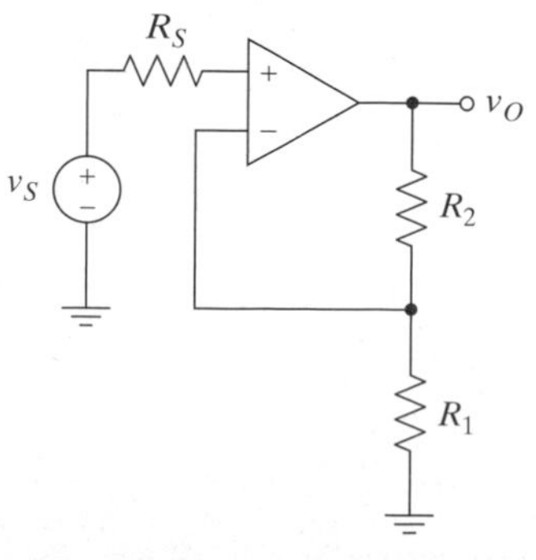

Figure 5–74

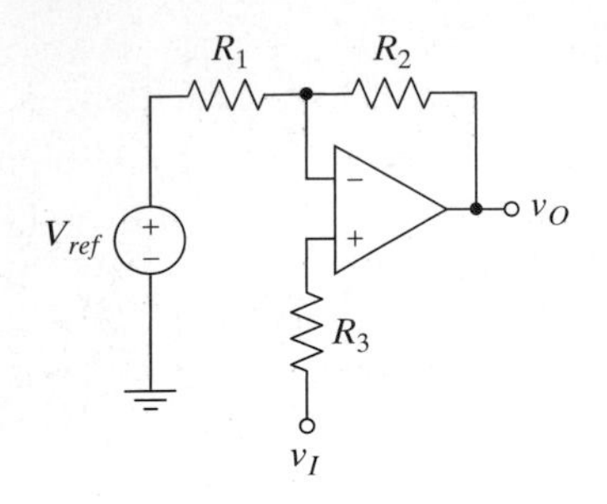

Figure 5–75

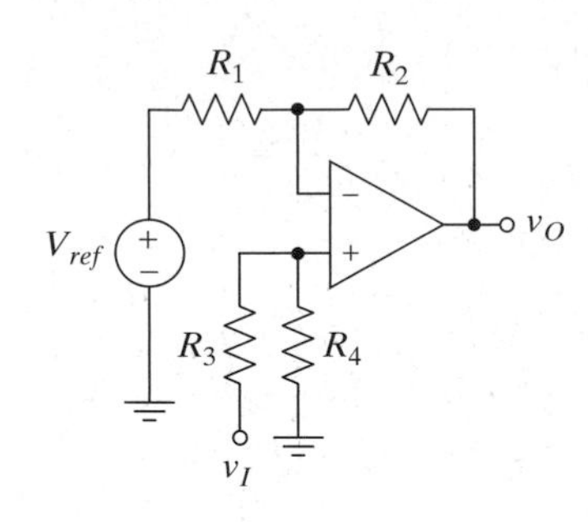

Figure 5–76

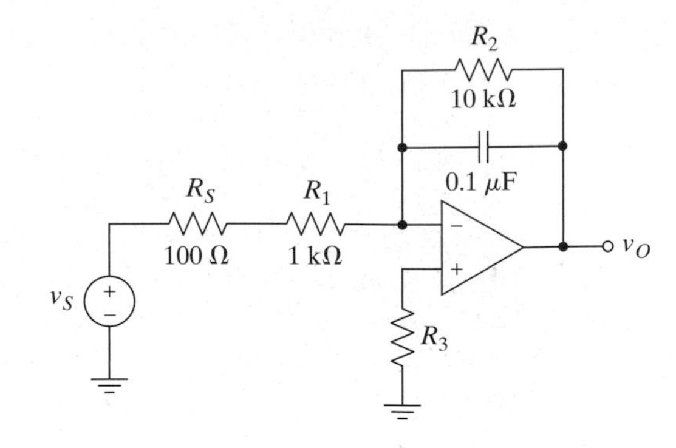

Figure 5–77

P5.37 A circuit is desired for which the following input-output relationships exist: $v_O = Av_I + V_X$, where A is a voltage gain, V_X is a constant, $v_O = 5$ V when $v_I = 0.5$ V and $v_O = 0$ V when $v_I = 0.3$ V. Find all of the resistor values and V_{ref} for the circuit in Figure 5-75 to meet the given specifications while limiting the current in the feedback network to no more than 4 mA and also minimizing the output offset voltage because of input bias currents.

P5.38 Repeat Problem P5.37 using the circuit in Figure 5-76.

P5.39 Determine the value of R_3 necessary to provide input bias current compensation for the LPF shown in Figure 5-77.

5.4.3 Input Offset Voltage

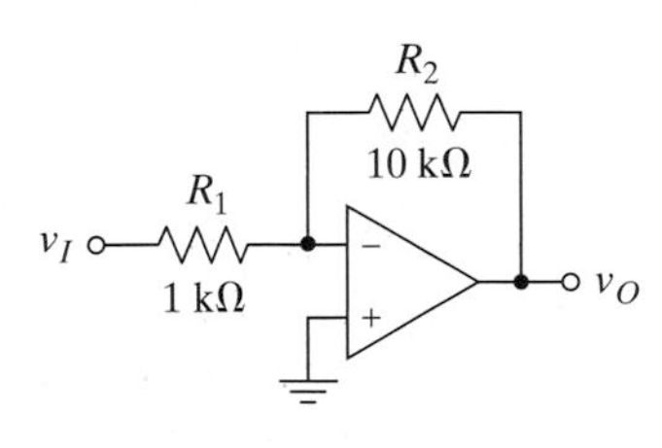

Figure 5–78

P5.40 Consider the circuit shown in Figure 5-78. **(a)** If the op amp is ideal except for having an offset voltage V_{OFF} where $|V_{OFF}| \leq 5$ mV, what is the smallest change in v_I that can be detected? **(b)** Now assume that the op amp has input bias currents equal to 5 μA (into the op amp inputs) in addition to V_{OFF}. What is the range of possible output voltages if $v_I = 0$?

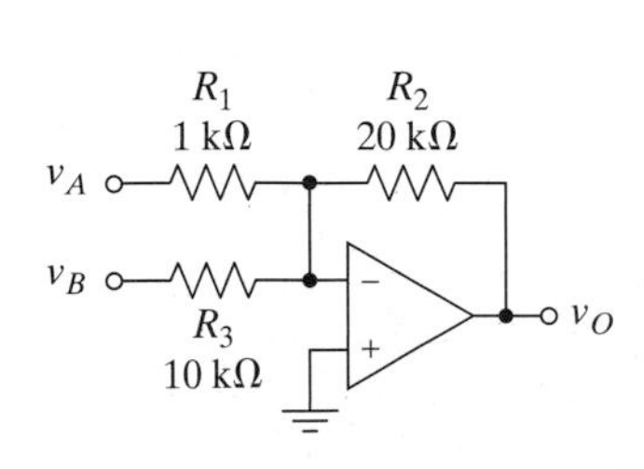

Figure 5–79

P5.41 Consider the summing circuit shown in Figure 5-79. Assume the op amp is ideal except that it has an offset voltage V_{OFF}. **(a)** Derive an equation for the output voltage in terms of the two inputs and the input offset voltage of the op amp, V_{OFF}. **(b)** If $|V_{OFF}| \leq 3$ mV, how small a value of v_A can we detect? How small a value of v_B can we detect?

5.4.4 Finite Input and Output Impedances

P5.42 Consider the transresistance amplifier circuit of Figure 5-80. **(a)** The op amp is ideal, except that its gain is finite and equal to a. What is the impedance seen looking into the input? **(b)** What is the input impedance if the op amp has finite R_{ic} and R_{id} ?

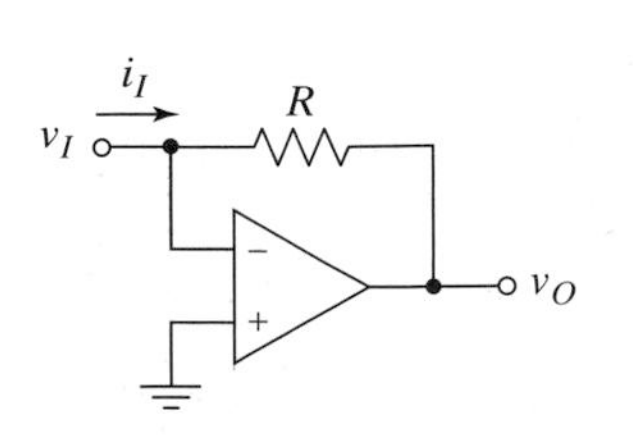

Figure 5–80

P5.43 **(a)** Derive an expression for the gain of the circuit in Figure 5-78 if the op amp is ideal except for having finite nonzero values of R_{id}, R_{ic}, R_o and a. **(b)** Repeat part (a) but allow the op amp gain to go to infinity.

P5.44 Derive expressions for the differential-mode and common-mode input resistances of the differential amplifier in Figure 5-11 if the op amp is ideal except for having finite values of R_{id} and R_{ic}.

5.4.5 Finite Bandwidth

P5.45 For an audio op amp circuit, it is desirable that the open-loop gain exceed the designed closed-loop gain by at least 20 dB at all frequencies of interest (there are other opinions on this!). **(a)** A non-inverting amplifier stage uses an op amp with a gain-bandwidth product of 2 MHz. What is the highest closed-loop gain for which the amplifier can be connected so that the open-loop gain exceeds the closed-loop gain by 20 dB at 20 kHz? **(b)** If the op amp is operated at the closed-loop gain calculated in part (a), what closed-loop bandwidth results?

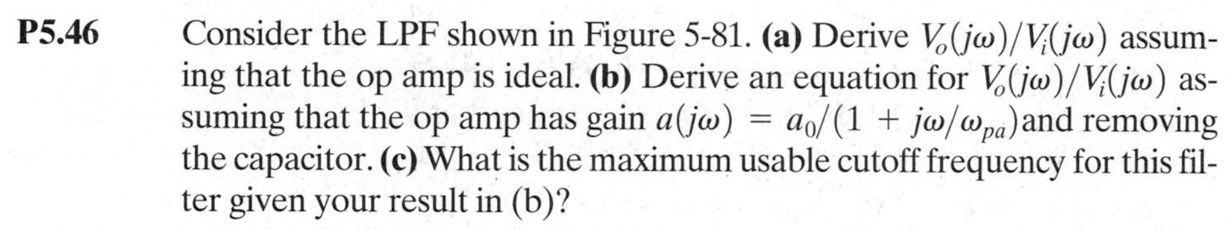

P5.46 Consider the LPF shown in Figure 5-81. **(a)** Derive $V_o(j\omega)/V_i(j\omega)$ assuming that the op amp is ideal. **(b)** Derive an equation for $V_o(j\omega)/V_i(j\omega)$ assuming that the op amp has gain $a(j\omega) = a_0/(1 + j\omega/\omega_{pa})$and removing the capacitor. **(c)** What is the maximum usable cutoff frequency for this filter given your result in (b)?

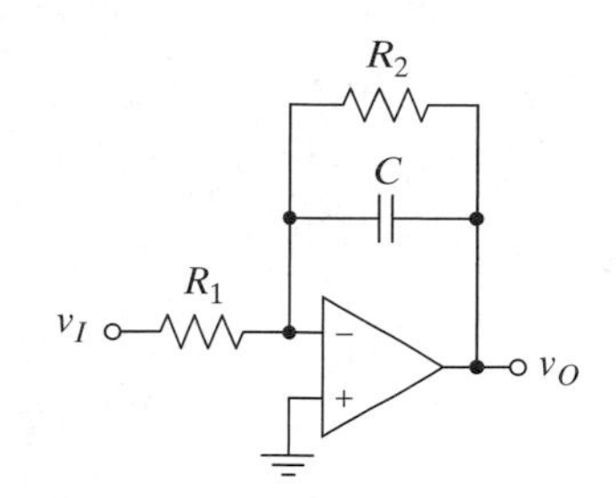

Figure 5–81

5.4.6 Common-Mode Rejection Ratio and Power-Supply Rejection Ratio

P5.47 The differential amplifier in Figure 5-82 uses an op amp with infinite CMRR. The CMRR of the circuit is not infinite, however, because of resistor mismatch. Solve for the CMRR of the circuit in terms of R and ΔR. **(a)**Use (5.19) to show that $a_{cm} = v_o/v_{cm} = \Delta R/(2R + \Delta R)$. **(b)** If $\Delta R \ll R$, $a_{dm} \approx v_O/(v_{S2} - v_{S1}) \approx 1$. Using this result and the answer to part (a), find the CMRR of the circuit for this condition. **(c)** If $R = 1\ \text{k}\Omega$ and $\Delta R = 10\ \Omega$, what is the CMRR of this circuit in dB?

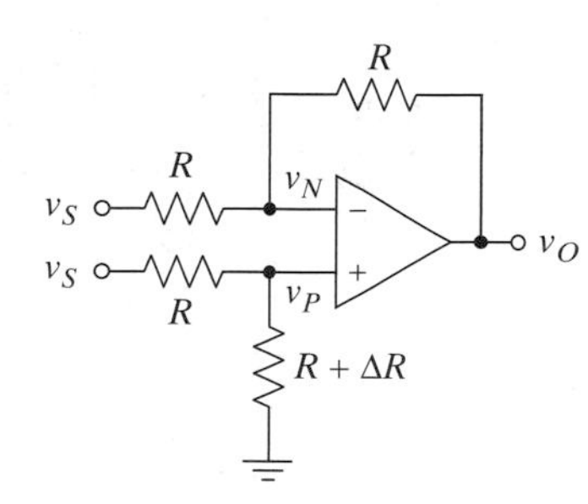

Figure 5–82

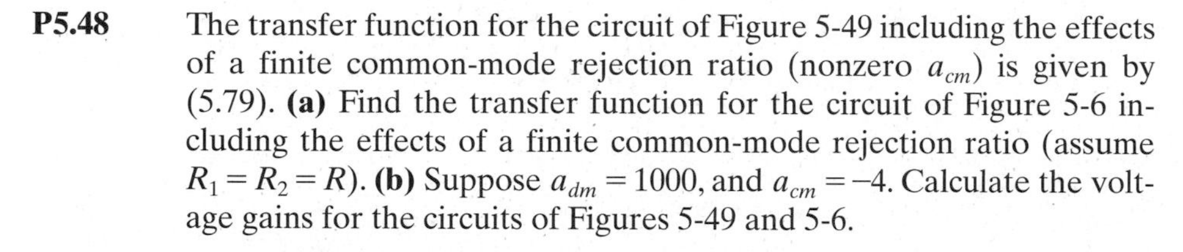

P5.48 The transfer function for the circuit of Figure 5-49 including the effects of a finite common-mode rejection ratio (nonzero a_{cm}) is given by (5.79). **(a)** Find the transfer function for the circuit of Figure 5-6 including the effects of a finite common-mode rejection ratio (assume $R_1 = R_2 = R$). **(b)** Suppose $a_{dm} = 1000$, and $a_{cm} = -4$. Calculate the voltage gains for the circuits of Figures 5-49 and 5-6.

5.4.8 Slew Rate and Full-Power Bandwidth

P5.49 The circuit of Figure 5-83 uses an op amp that is ideal, except that the output voltage cannot exceed the power supply voltages (supply rails) in either direction, and it has a symmetric slew rate of 2 V > µs. **(a)** Assume that v_S is a 2 V_{pp} ideal square wave at 20 kHz with zero DC offset. Sketch v_S and the resulting v_O on the same time axis for a period of at least one cycle. **(b)** What is the time delay between the zero crossing of v_S and the zero crossing of v_O?

P5.50 Consider the op amp circuit of Figure 5-3 and assume that an LM 607 is used. The specifications for the LM 607 are given in Table 5.1. Assume that when the amplifier is driven with a step input, the 10% to 90% rise time of the output is $0.35/BW_{CL}$ where BW_{CL} is the closed-loop bandwidth of the amplifier in Hertz. The rise time represents a limit on the amplifier's speed of response caused by to limited bandwidth and is a linear effect. The amplifier's response speed can also be limited because of slew rate, which is a nonlinear effect. **(a)** Let $R_2 = 9\ \text{k}\Omega$, $R_1 = 1\ \text{k}\Omega$, and v_s be a 1-V step (from 0 to 1 V). Calculate the amplifier's risetime and also the time it takes the amplifier's output

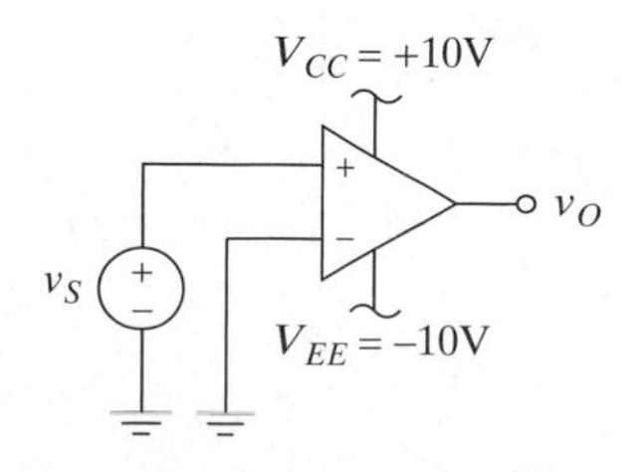

Figure 5–83

to slew from 0 to 10 V. Which effect dominates?[15] **(b)** Suppose that v_S only makes a 0.1 V step, and the amplifier's closed-loop gain is raised to 100 so that the output still goes from 0 to 10 V. Calculate the amplifier's rise time and also the time it takes the amplifier's output to slew from 0 to 10 V. Which effect dominates?

5.4.10 Op Amp Parameter Measurement

P5.51 Suppose the circuit shown in Figure 5-53 is used to measure the gain and offset of a particular op amp. The results of the measurements are given in Table 5.2. Find the values of the gain and offset for this op amp.

Table 5.2

V_I	V_O
−5 V	−0.0275 V
0 V	−3.2 V
5 V	0.0275 V

[15] Risetime is a measure of the time it takes the output to go from 10% to 90% of the way from its initial value to its final value. Why are we comparing it to the time it takes the output to slew *all the way* from its initial value to its final value? One certainly could multiply the slewing time by 0.8 to get a fairer comparison. However, if the times are comparable, the effects interact and the actual response is more difficult to calculate.

CHAPTER 6

Small-Signal Linearity and Amplification

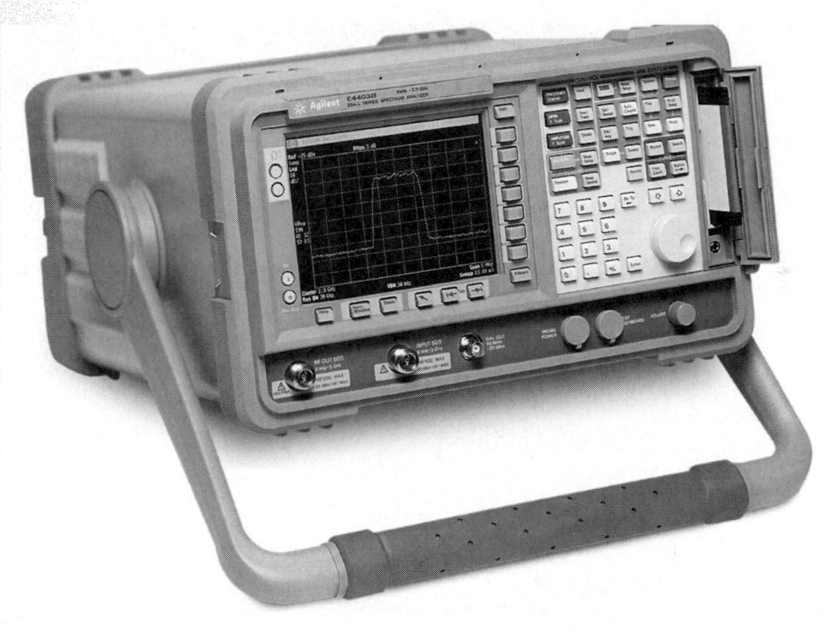

Nonlinearities in circuits result in harmonic distortion, which is measured using instruments like this Agilent distortion analyzer. (Photo courtesy of Agilent Technologies.)

INTRODUCTION

The vast majority of analog circuits perform functions that we desire to be linear; in other words, we want the output to be a linear function of the input (we will define a linear system precisely in Section 6.1). But, the solid-state electronic devices we have available are all nonlinear. Designing truly linear systems using devices with nonlinear characteristics is impossible. Nevertheless, we will be able to design circuits that behave approximately linearly by restricting the signals to be small enough that the nonlinear devices appear linear to a sufficient degree of accuracy. We will see in Chapter 10 that it is possible to extend the range of signal amplitudes over which a given device can be approximated as linear by using negative feedback, but no amount of feedback can overcome the fact that diodes and transistors are fundamentally nonlinear devices.

Nonlinear systems also pose a problem for the designer, because the mathematical tools available for dealing with them are not as well developed as those for linear systems and are frequently too complicated to effectively use in design. In fact, nonlinear analysis often requires that the problem be solved numerically rather than analytically. The way around these mathematical difficulties is again to approximate the devices as being linear for some set of circumstances.

The purpose of this chapter is to discuss the ***small-signal*** approximation in some detail without specifically tying the treatment to a particular device or circuit. Early sections use a diode in the examples for simplicity. Subsequent sections introduce a generic transistor model to discuss nonlinear active devices. The generic transistor is a generalized 3-terminal device that can represent the behavior of different types of transistors and even vacuum tubes. The application of the small-signal approximation to the design of small-signal amplifiers will be discussed later in this chapter. Finally, some different types of amplifiers will be presented and categorized.

6.1 LINEAR TIME-INVARIANT NETWORKS

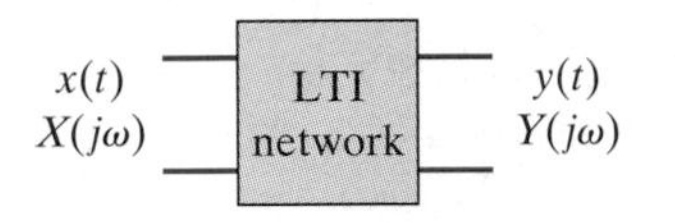

Figure 6–1 An electronic network.

Consider a network with input $x(t)$ and output $y(t)$ as shown in Figure 6-1. If an input $x_1(t)$, produces an output $y_1(t)$, and an input $x_2(t)$ produces an output $y_2(t)$, then the network is ***linear*** if and only if an input $x(t) = ax_1(t) + bx_2(t)$ produces an output $y(t) = ay_1(t) + by_2(t)$. If, in addition to being linear, the response of the network is independent of when an input is applied, the network is called a ***linear time-invariant*** (LTI) network. An LTI network is much easier to deal with mathematically than nonlinear or time-varying networks because it is completely described by its ***impulse response*** in the time domain. Given the impulse response, $h(t)$, of a network like the one in Figure 6-1, the output is given by the ***superposition integral*** (also known as the ***convolution integral***).[1]

$$y(t) = \int_{-\infty}^{\infty} x(\lambda)h(t - \lambda)d\lambda \tag{6.1}$$

Using (6.1), we are in essence breaking the input to the system up into an infinite summation of properly scaled and time-shifted impulses and then finding the output as an infinite summation of the responses to these individual impulses. In other words, we are using the definition of an LTI system in a direct manner. We do not need to recalculate the system response for each input. The impulse response is sufficient to allow us to calculate the output for any input.

If we can find a reasonable way to model nonlinear electronic components and circuits with linear elements, we will be able to use all of our LTI analysis tools to aid in the design process.

It is often more convenient to work in the frequency domain where the system is described by its transfer function. The transfer function of any real network can be expressed as either the Fourier transform of its impulse response, $H(j\omega)$, or the Laplace transform of its impulse response, $H(s)$. If the input to the network is described in the frequency domain, the output of the network is given by

$$Y(j\omega) = H(j\omega)X(j\omega) \tag{6.2}$$

where $X(j\omega)$ is the Fourier transform of $x(t)$ and $Y(j\omega)$ is the Fourier transform of $y(t)$.

The use of transforms combined with the complex exponential representation of signals and complex phasor analysis is a powerful tool for circuit analysis and design.

[1] If your memory of linear network analysis is shaky, you may want to read Appendix D on the CD.

These techniques depend on the circuit being linear and time invariant. Using these tools, we can frequently analyze a complex circuit with algebraic equations rather than having to resort to differential equations and convolution integrals.

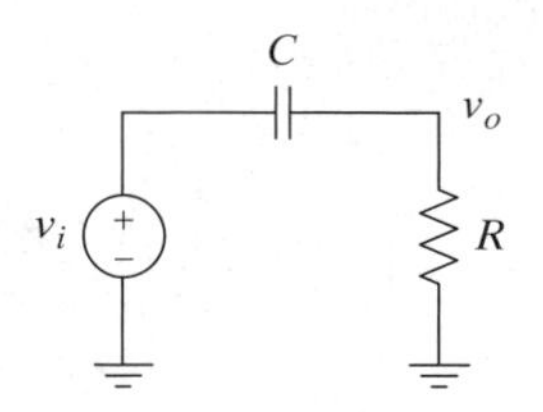

Figure 6–2 A single-pole high-pass filter.

Exercise 6.1

Prove that the circuit shown in Figure 6-2 has a linear transfer function for inputs applied after t = 0. Assume that the capacitor voltage is zero at t = 0.

6.2 NONLINEAR CIRCUIT ANALYSIS

Before considering linear models in detail, it is instructive to examine the methods available for analyzing a nonlinear circuit. Suppose we want to determine the output voltage for the circuit shown in Figure 6-3, where we assume that the frequency of the source is low enough that charge storage in the diode may be safely ignored. We will discuss three ways of solving this circuit; analytically using the equation for the diode, using graphical techniques, and using a suitable model for the diode.

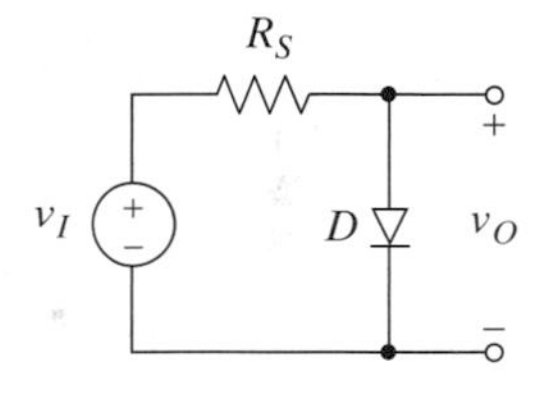

Figure 6–3 A nonlinear circuit.

6.2.1 Analytical Solution

Taking the first approach, we note that the output voltage is given by

$$v_O(t) = v_I(t) - i_D(t)R_S, \tag{6.3}$$

or, dropping the explicit reference to time as we will often do for notational simplicity,

$$v_O = v_I - i_D R_S. \tag{6.4}$$

Equation (6.4) represents the constraint placed on the solution by the circuit topology and Kirchoff's laws. The diode's characteristic equation for i_D versus v_D is another constraint that must be simultaneously satisfied;

$$i_D = I_S(e^{v_D/nV_T} - 1), \tag{6.5}$$

where I_S and n are constants associated with the diode, and V_T is a constant at a given temperature ($V_T \approx 26$ mV at room temperature). Substituting (6.5) into (6.4) leads to the final result (remember that $v_O = v_D$)

$$v_O = v_I - I_S R_S(e^{v_O/nV_T} - 1). \tag{6.6}$$

This equation cannot be algebraically solved for v_O in closed form; it is what is called a transcendental equation. We must solve it numerically. Although it is not difficult to solve the equation for a given value of v_I, the trouble is that a *general* analytical solution cannot be found. We must solve the equation numerically for *each* value of v_I. Another manifestation of this problem is that we cannot use superposition. In other words, we cannot represent a complex input by its Fourier series (or transform), find the output for each component of the input, and then sum (or integrate) the individual components of the output to find the total response.

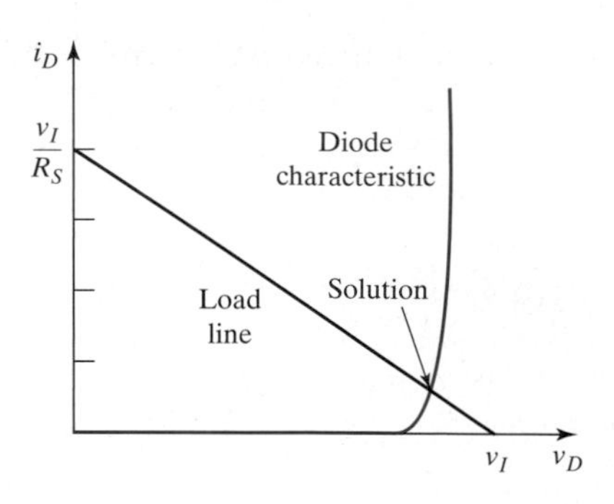

Figure 6–4 Graphical solution of (6.4) and (6.5) with $v_I > 0$ ($v_O = v_D$).

6.2.2 Graphical Solution

The second way to solve the problem is to simultaneously solve (6.4) and (6.5) graphically instead of algebraically. Figure 6-4 shows (6.4) and (6.5) plotted on the same axes for some $v_I > 0$. Since v_I is a function of time, we would need to redo the plot for each moment in time to get a complete solution. Therefore, the graphical method is only useful for DC and to help our intuition, although we can sometimes see how to superimpose small AC signals, as will be done in Section 6.3. The solution represented by (6.6) is the value of v_O where the two curves intersect (remember that $v_O = v_D$). The line representing (6.4) in the figure is labeled the load line. Aside A6.1 discusses the use of load lines in general.

6.2.3 Solving with Models

The third way to solve the circuit in Figure 6-3 is to use a model for the diode. In one sense, the first two solutions also used a model for the diode, since the equation is certainly not the diode itself and does not exactly represent the performance of the diode. Nevertheless, what we mean here is more accurately called a ***circuit model***. In other words, a collection of components that produce the same or similar characteristics. This third method would often be the method of choice, but *not always,* since the result is less accurate than solving (6.6) directly. The third method is introduced in Section 6.3 and is used extensively in the rest of this book.

Exercise 6.2

Consider the circuit shown in Figure 6-5(a). The glow tube has the v-i characteristic shown in part (b) of the figure.[2] *Plot the load line and the glow tube characteristic on the same set of axes and graphically determine the operating point.*

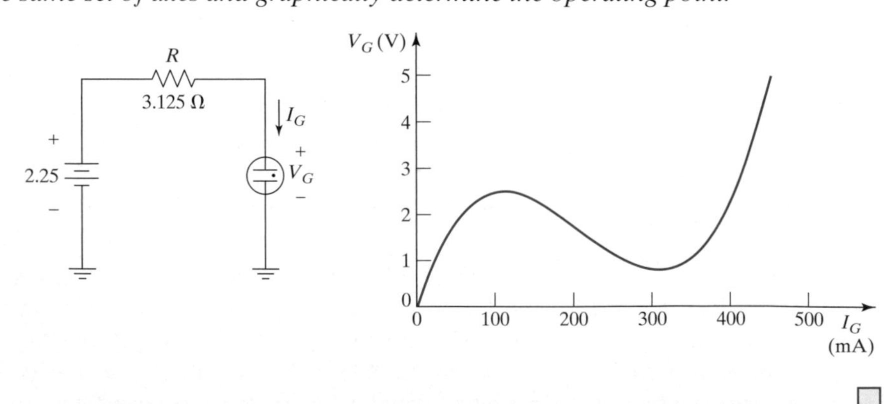

Figure 6–5 (a) A circuit with a glow tube and (b) the glow tube's characteristic.

□

[2] The characteristic *must* be plotted with voltage on the vertical axis, or it is not a true function (i.e., the current is not a single-valued function of the voltage). This same "trick" can be used with other problematic nonlinear functions, but they occur very rarely in electronic circuits.

ASIDE A6.1 LOAD LINES

In analyzing nonlinear circuits, we frequently encounter situations like the one illustrated in Figure A6-1, where the box represents a general nonlinear two-terminal device. To analyze this circuit, we desire to simultaneously solve two equations; the characteristic equation for the nonlinear device and the equation representing the rest of the circuit.

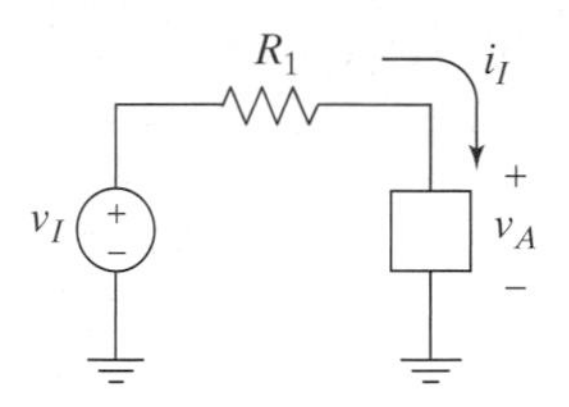

Figure A6-1 A circuit containing a two-terminal nonlinear device.

Writing KVL around the loop we get

$$v_I - i_I R_1 - v_A = 0, \tag{A6.1}$$

which can be rewritten as

$$i_I = \frac{v_I - v_A}{R_1}. \tag{A6.2}$$

We also have an equation describing the nonlinear device

$$i_I = f(v_A), \tag{A6.3}$$

and we must simultaneously solve (A6.2) and (A6.3).

We can solve these equations graphically, as illustrated in Figure A6-2. In this figure we have plotted the curve describing the nonlinear device and a straight line, called the ***load line***, that represents the constraint placed on the solution by the rest of the circuit as given by (A6.2). The terminology is imprecise, since the line does not always represent what would typically be called a load. Nevertheless, in simultaneously solving equations describing a nonlinear device and the circuit it is connected to, the line representing the circuit is usually called a load line, since it can be thought of as the load presented to the particular nonlinear element under consideration (a different name may occasionally be used when doing so makes sense). The two equations are simultaneously satisfied at any point where the load line and the device *i-v* characteristic intersect (there may be more than one solution).

The graphical method of solution is rarely used in practice, since we can solve the equations with a calculator or computer, if necessary. However, it does lend insight into the problem, and we will refer to similar pictures often.

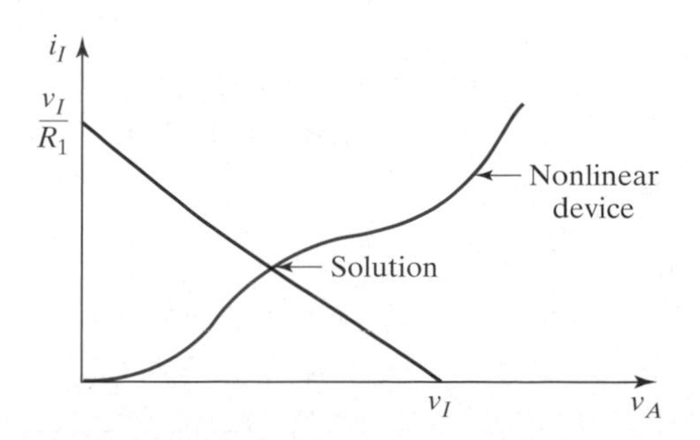

Figure A6-2 The graphical solution of the two equations for the circuit in Figure A6-1.

6.3 SMALL-SIGNAL ANALYSIS

It is now time to discuss how, when, and with what accuracy a nonlinear device can be modeled as linear. Consider again the diode circuit of Figure 6-3. If the input is just a DC voltage, the solution can be reasonably found using any of the methods of Section 6.2. If, on the other hand, the input has both a DC and an AC part, how can we model the diode with a linear circuit? The answer can be found by examining the graphical solution presented in Figure 6-4. Suppose the input voltage in Figure 6-3 is given by

$$v_I(t) = 3 + 0.005\cos(628t). \tag{6.7}$$

Consider just the DC part of the input, and solve the resulting equations for I_D and V_D.

The result can be presented graphically, as shown in Figure 6-6, which is the same as Figure 6-4, except that it is for the DC part of the specific example at hand where $V_I = 3.0$ V, and we have assumed $R_S = 9\ \text{k}\Omega$, $n = 1$, $I_S = 7.75 \times 10^{-16}$ and have used the room temperature value for V_T, $V_T \approx 26$ mV. The resulting solution is $I_D = 257\ \mu\text{A}$ and $V_D = 689.7$ mV; the voltage is given with this precision because it is necessary to get the correct current if you plug V_D into (6.5).

Notice in Figure 6-6 that adding the 5 mV peak AC input described by (6.7) to the 3 V DC input would correspond to moving the load line by an imperceptible amount. Figure 6-7 illustrates this point by enlarging a small part of the curve in Figure 6-6 along with three load lines; one for $v_I = V_I + 5$ mV, and one for $v_I = V_I - 5$ mV (the Q point shown in the figure will be defined shortly). The load lines appear to have a smaller slope in the enlarged portion of the figure because of the different scale used. At any point in time, the load line for the circuit with v_I given by (6.7) will lie somewhere between the outer two load lines in Figure 6-7. Figure 6-7 also shows a straight-line approximation to the diode characteristic that is tangent to it at the DC operating point. It can be seen from the figure that using the straight-line approximation in place of the actual curve would be quite accurate for load lines within the range seen in this example.

This example illustrates the idea behind **small-signal analysis**; that is, if the signal can be represented as a small perturbation about some operating point, the nonlinear characteristic can be modeled as a straight line tangent to the characteristic at the operating point. It is almost always true that the operating point is DC,[3] and it *is* always true for the circuits encountered in this book.

This DC operating point is usually chosen by the designer to help the circuit achieve a desired level of performance. In other words, the DC operating point is used to **bias** the nonlinear elements to operate in a certain way and is the point at

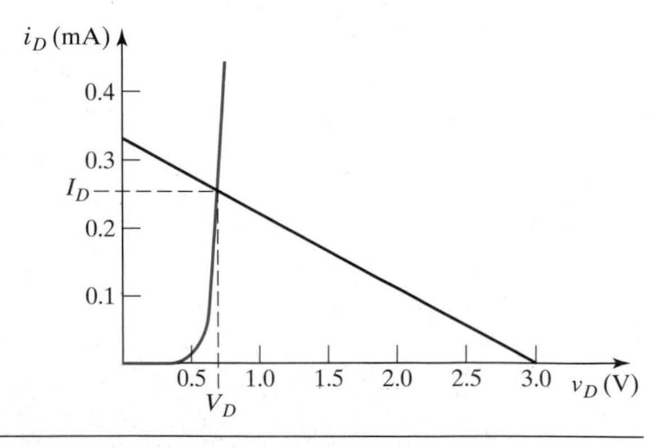

Figure 6–6 The DC load line and diode characteristic for the circuit of Figure 6-3 with $V_I = 3.0$ V, $R_S = 9\ \text{k}\Omega$, $n = 1$, and $I_S = 7.75 \times 10^{-16}$A ($V_T = .026$ V). The solution is $I_D = 257\ \mu\text{A}$ and $V_D = 689.7$ mV.

[3] An example of a circuit where it is *not* true is a microwave multiplier, where the circuit is deliberately "pumped" by a time-varying bias [6.1].

which the device will operate if no signal is present (i.e., it is "quiet"). Therefore, this DC operating point is also commonly referred to as the ***bias point***, or the ***quiescent point*** (Q point). The Q point is shown in Figure 6-7.

An alternate view of the small-signal approximation is presented in Figure 6-8 to further clarify the concept. This figure shows the *i*-*v* characteristic

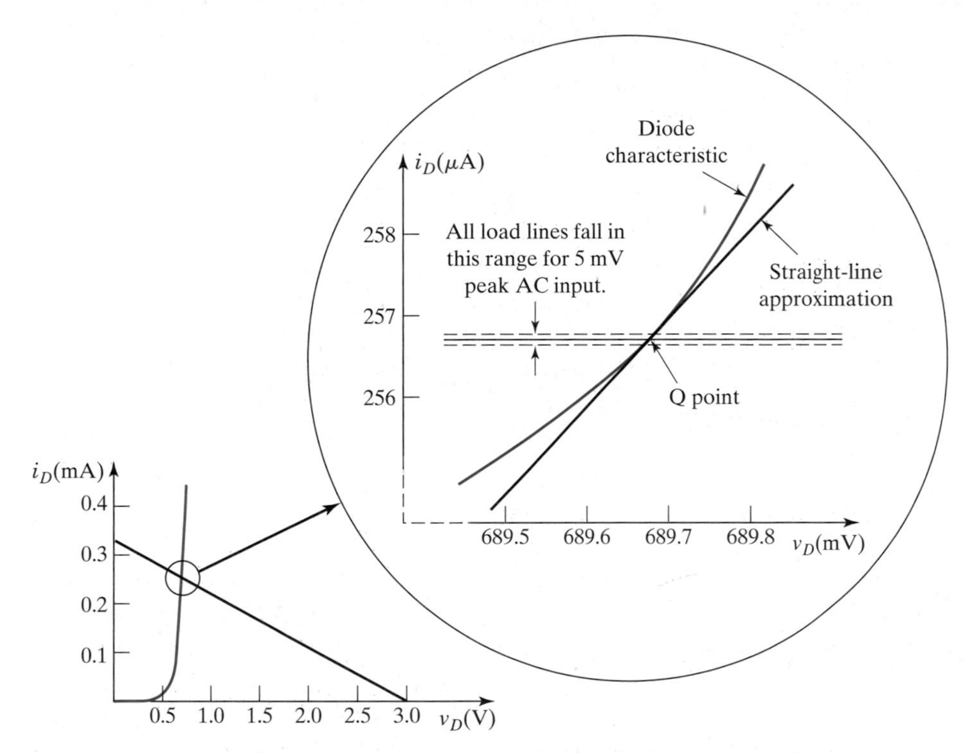

Figure 6–7 A close-up of the region around the operating point in Figure 6-6. The nominal and extreme load lines, the straight-line approximation, and the Q point (DC operating point) are also shown. The curve of the diode characteristic has been exaggerated for illustrative purposes. The exact value of V_D is 689.68 mV, and the resulting current is 256.7 μA.

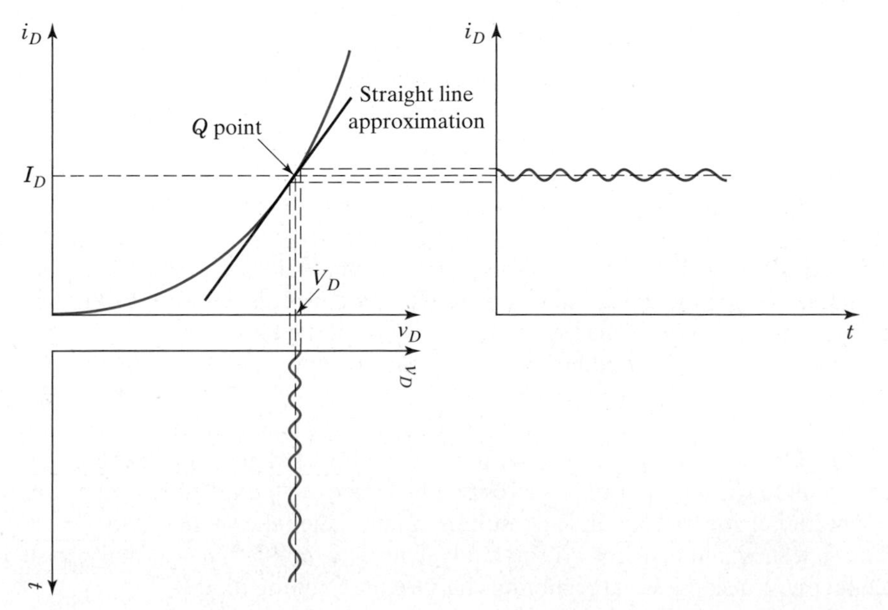

Figure 6–8 An alternate view of the small-signal approximation.

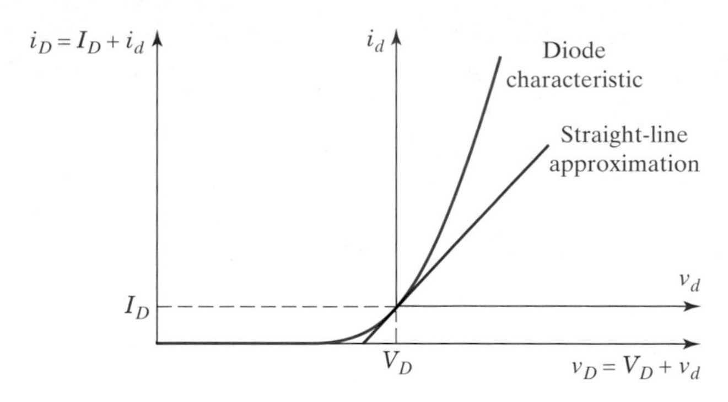

Figure 6–9 The axes for the time-varying components superimposed on the axes for the total instantaneous quantities.

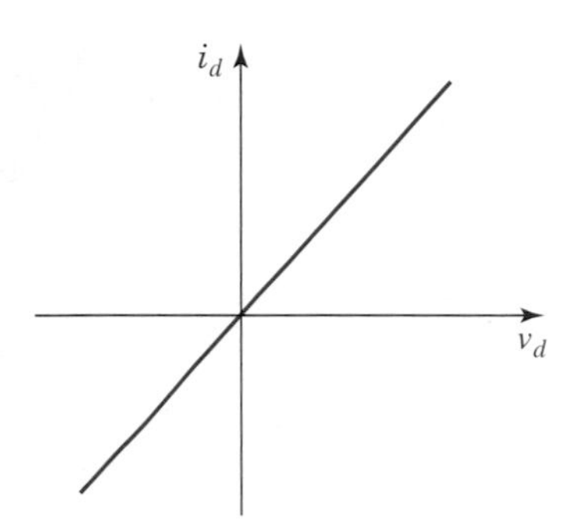

Figure 6–10 The axes for the time-varying components and the straight-line approximation shown alone.

of the diode, the straight-line approximation tangent to the characteristic at the Q point, and plots of v_D and i_D versus time. You can see from the figure that for small variations about the Q point, the straight-line approximation yields accurate results for the relationship between v_D and i_D.

Figure 6-9 is a repeat of the diode characteristic from Figure 6-6, but with a second set of axes shown with their origin at the Q point and the straight-line approximation to the diode characteristic also shown. Since the Q point is the solution when the time-varying component of the input is zero, the axes that have their origin at the Q point are the time-varying components of i_D and v_D (i.e., they are i_d and v_d). If we focus our attention on this set of axes, we see that the straight-line approximation to the diode characteristic goes through the origin as shown in Figure 6-10. Therefore, the straight-line approximation can be modeled as a resistor with a value equal to the inverse of the slope of the line. In other words, if we restrict our attention to small changes from the bias point, the diode can be replaced by a resistor r_d or a conductance g_d. We have

$$g_d = \left.\frac{di_D}{dv_D}\right|_{@\text{ Q point}} = \left.\frac{I_S e^{v_D/nV_T}}{nV_T}\right|_{@\ v_D = V_D} = \frac{I_D}{nV_T} \quad \textbf{(6.8)}$$

and the resistance is the reciprocal of g_d,

$$r_d = \frac{nV_T}{I_D}. \quad \textbf{(6.9)}$$

The elements r_d and g_d are called the ***small-signal*** (or ***incremental***) resistance and conductance respectively. For the example being considered, $r_d = 101\ \Omega$. This model for the diode is called either the small-signal or incremental model, since it is only valid for small or incremental variations from the Q point. It is important to notice that the model is a function of the bias point, since I_D sets the bias point.

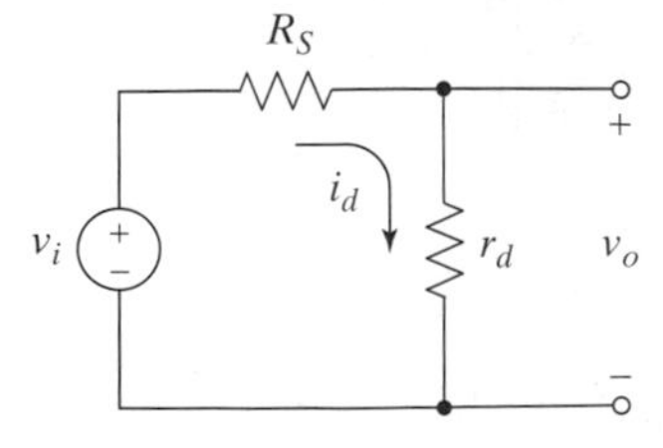

Figure 6–11 The low-frequency small-signal equivalent circuit for Figure 6-3.

Before we can draw a new equivalent circuit that is only valid for small changes about the bias point, we also need to determine appropriate models for the other elements in the original circuit (i.e., R_S and the input source). Since we are only considering changes from the bias point, we want to remove the average value of the input; that is, we want to use only v_i for the source, not v_I. However, the resistor R_S does not change, since it is a linear element. The resulting circuit, called the ***small-signal equivalent circuit***, is shown in Figure 6-11. This equivalent is the *low-frequency* small-signal equivalent circuit, because we are ignoring charge storage in the diode.

Using this equivalent circuit, we can solve for i_d and v_o directly. We have (remember that $v_i = 5$ mV peak, $R_S = 9$ kΩ and $r_d = 101$ Ω)

$$i_d = \frac{v_i}{R_S + r_d} = 0.55\cos(628t)\ \mu\text{A} \tag{6.10}$$

and

$$v_o = \frac{r_d}{r_d + R_S} v_i = 55\cos(628t)\ \mu\text{V} \tag{6.11}$$

by inspection.

The complete solution to the circuit is then found by adding these small-signal components to the DC bias solutions found in Figure 6-6;

$$i_D = [257 + 0.55\cos(628t)]\ \mu\text{A} \tag{6.12}$$

and

$$v_O = [689.7 + 0.055\cos(628t)]\ \text{mV}. \tag{6.13}$$

Summing the bias point and the small-signal solution as we have done in (6.12) and (6.13) is an application of the principle of superposition. Since we used a linear model, we can sum the DC solution, which represents just one point on the straight-line characteristic, with the small-signal AC solution.

To summarize this method of analyzing a nonlinear circuit;

1) Solve for the bias point using any convenient method.
2) Determine the small-signal model(s) of the nonlinear element(s) at the bias point.
3) Set the DC components of all independent sources in the circuit to zero, use small-signal model(s) for all nonlinear element(s), and draw the small-signal equivalent circuit.
4) Solve the resulting small-signal equivalent circuit using standard LTI analysis techniques.
5) Sum the small-signal and DC bias point solutions to get the complete solution. (This step is often omitted, since the small-signal quantities are usually the signals we are interested in. The DC bias is only of interest insofar as it affects the small-signal performance.)

It needs to be emphasized that the results we obtain with this procedure are only approximate, since the small-signal models used for the nonlinear elements are only approximate. We will consider the accuracy of this approximation in Section 6.3.2. In addition, note that even if we solve the equations to find the DC operating point, using the small-signal equivalent circuit has resulted in a significant reduction in the work required to obtain a general solution of the circuit. Finally, we note that we can—and do—allow the signal to have a DC component.

Exercise 6.3

Reconsider the glow tube circuit in Exercise 6.2. Assume that the source has a small AC component superimposed on the DC. Draw the small-signal equivalent circuit assuming the device is biased at point B (see the solution to Exercise 6.2) and find an approximate value for the component modeling the glow tube. Does the model make sense?

□

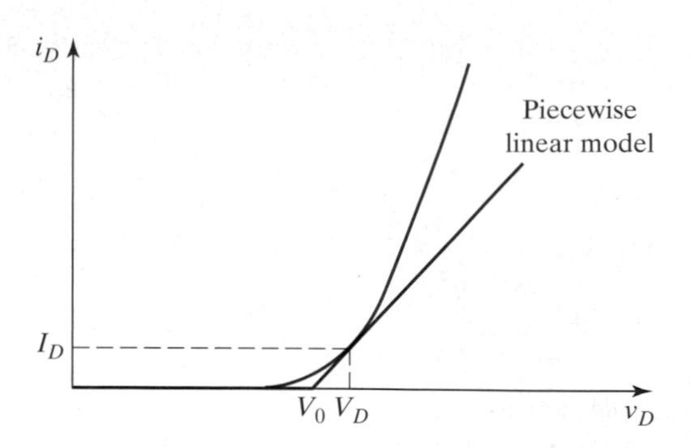

Figure 6–12 A piecewise-linear large-signal model for the diode.

It is possible to combine the small-signal and DC operating point analyses into one by using a linear large-signal model for the diode. Although the analyses would typically *not* be performed with them combined, going through an example this way will help to further illustrate the principle of small-signal analysis.

Looking back at Figure 6-9, we note that the diode could be represented by the piecewise-linear model shown in Figure 6-12, where one segment of the model is coincident with the v_D axis, and one segment is the same as the straight line from which the small-signal model was derived. The equivalent circuit for this piecewise-linear model depends on which segment is considered. When $v_D < V_0$, the model would simply be an open circuit. When $v_D > V_0$, the model is a battery of voltage $V_0 = 663.7$ mV in series with a resistor of value $r_d = 101\ \Omega$ as you can verify (*see* Problem P6.15). Using this model, the diode current is zero until the voltage across the device equals V_0. Therefore, this voltage is sometimes called the ***turn-on*** or ***cut-in*** voltage. However, note that the real device does *not* exhibit such discontinuous behavior!

Using the equivalent circuit for the piecewise-linear diode model when $v_D > V_0$ leads to the large-signal equivalent circuit shown in Figure 6-13. Solving the circuit of Figure 6-13 leads to

$$i_D = \frac{v_I - V_0}{R_S + r_d} = [257 + 0.55\cos(628t)]\ \mu\text{A} \tag{6.14}$$

and

$$v_O = V_0 + (v_I - V_0)\frac{r_d}{r_d + R_S} = [689.7 + 0.055\cos(628t)]\ \text{mV}. \tag{6.15}$$

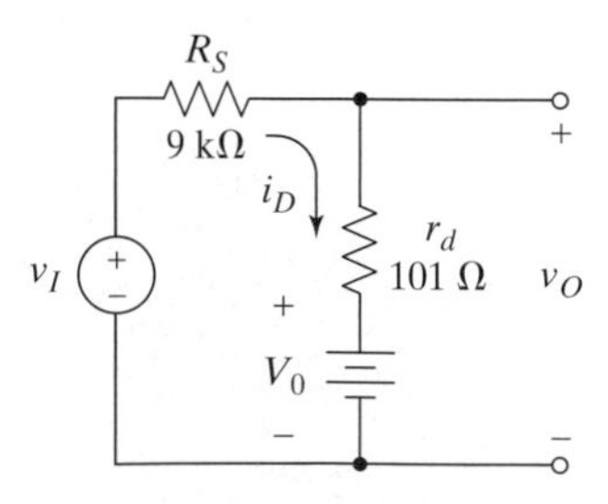

Figure 6–13 Large-signal equivalent circuit.

These final results are the same as were presented in (6.12) and (6.13).[4] Although this example helps us understand small-signal analysis, we can see why this method is not used in practice. Observe that the model used in this method depends on the bias point, since V_0 and r_d do. But, if the bias point has been determined, why bother to complicate the analysis by including it in the calculation again? In addition, although the model is now a "large-signal" model, it is *no more accurate* than the small-signal model used earlier! In fact, in the region of the bias point, which is the only region where this model is likely to be acceptably accurate, it is the *same* as the small-signal model.

---

[4] If you round off V_0 and r_d (as we have in presenting the numbers in the text) prior to performing all of the calculations, you will get slightly different values than presented here.

ASIDE A6.2 NOTATION FOR SMALL- AND LARGE-SIGNAL ANALYSIS

The notation for mixed small- and large-signal analysis is usually simplified by denoting the "signal" as AC only, even though it may contain a DC component. This point should not be overemphasized, but it is worth a few words of explanation to avoid confusion later on. It is entirely possible—in fact common—for the input signal to a linear system to include a DC component (i.e., have an average value other than zero). Nevertheless, we don't want to complicate our notation unnecessarily and will use the time-varying component to represent the signal and the DC component to represent the bias. Therefore, although writing $v_D = V_D + v_d$ means that V_D is the DC value and v_d is the AC value, we will use V_D to represent the bias value (which is always DC in this book) and v_d to represent the signal, which may contain DC. Allowing this minor ambiguity simplifies our notation significantly.

6.3.1 An Alternate View of the Small-Signal Approximation

In this section, we present an alternate view of the small-signal approximation. We show that the small-signal approximation is equivalent to using the first term of the Taylor-series expansion of the characteristic equation for a nonlinear element. This statement is true since finding the first term of a Taylor series expansion involves finding the first derivative of a function evaluated at a particular point, which is precisely what we do to find the small-signal model.

For the diode we had

$$i_D = I_S\,(e^{v_D/nV_T} - 1). \tag{6.16}$$

Assuming the diode is forward biased, the -1 can be neglected and we can write

$$i_D \approx I_S e^{v_D/nV_T} = I_S e^{(V_D+v_d)/nV_T} = I_S e^{V_D/nV_T} e^{v_d/nV_T} = I_D e^{v_d/nV_T}. \tag{6.17}$$

Now replace the exponential by its power series expansion

$$e^x = 1 + x + \frac{x^2}{2!} + \frac{x^3}{3!} + \cdots, \tag{6.18}$$

to get

$$i_D \approx I_D\left[1 + \frac{v_d}{nV_T} + \frac{1}{2}\left(\frac{v_d}{nV_T}\right)^2 + \frac{1}{6}\left(\frac{v_d}{nV_T}\right)^3 + \cdots\right]. \tag{6.19}$$

Finally, subtract the average value and examine just the AC component;

$$i_d = i_D - I_D \approx I_D\left[\left(\frac{v_d}{nV_T}\right) + \frac{1}{2}\left(\frac{v_d}{nV_T}\right)^2 + \frac{1}{6}\left(\frac{v_d}{nV_T}\right)^3 + \cdots\right]. \tag{6.20}$$

If we keep just the first term in (6.20) we have

$$i_d \approx \frac{I_D}{nV_T} v_d = g_d v_d = \frac{v_d}{r_d}, \tag{6.21}$$

which is the small-signal approximation used earlier in drawing Figure 6-11. Examining the other terms in the expansion allows us to compute the distortion produced by this nonlinear element (*see* Section 12.2.3).

6.3.2 Accuracy of the Small-Signal Approximation

To determine the accuracy of the small-signal approximation, we need to compare the values obtained for the AC component of i_D using the exact equation with those obtained using the small-signal approximation given by (6.21). Using the full equation for the diode current, and neglecting the −1, since we are interested in forward bias, we get

$$i_d = i_D - I_D \approx I_S e^{v_D/nV_T} - I_D = I_S e^{V_D/nV_T} e^{v_d/nV_T} - I_D = I_D\,(e^{v_d/nV_T} - 1). \quad \textbf{(6.22)}$$

Assume, as before, that the diode has $n = 1$ and $I_S = 7.75 \times 10^{-16}$, that $V_T = .026$ V, $r_d = 101\ \Omega$ and the bias point is $I_D = 257\ \mu$A. Table 6.1 gives values for i_d obtained from (6.22) and from (6.21), as well as the differences expressed as percentages of the exact values. From the table, it can be seen that the signal voltage across the diode must be about 5 mV or less for our analysis to be accurate within ±10%. Similar results apply to the other nonlinear devices discussed in this book, although MOS devices are more nearly linear than bipolar devices.

6.4 SMALL-SIGNAL AMPLIFIERS

The previous sections of this chapter have used a diode, which is a two-terminal nonlinear device, as the example. In this section we use the principle of small-signal linearity to show how a transistor, which is a 3-terminal nonlinear active device, can be used to construct an amplifier that has an approximately linear transfer characteristic. We first present some background material on small-signal modeling of transistors and proceed to an example application.

6.4.1 Small-Signal Models for Transistors

To avoid tying this presentation to any specific device and to encourage you to think about transistor circuits in a general way, we will use a device called a ***generic transistor*** in the following analysis. The generic transistor is introduced in Aside A6.3.

Now suppose the generic transistor shown in Figure A6-3 is described by the following equations (these equations apply *only* for this example; they do *not* always apply for the generic transistor);

$$i_O = \sum_{k=0}^{3} a_k \left(\frac{v_{CM}}{V_B}\right)^k = a_0 + a_1\left(\frac{v_{CM}}{V_B}\right) + a_2\left(\frac{v_{CM}}{V_B}\right)^2 + a_3\left(\frac{v_{CM}}{V_B}\right)^3 \quad \textbf{(6.23)}$$

Table 6.1 A Comparison of Exact AC Diode Currents with those Obtained from the Small-Signal Approximation

v_d in mV	Exact i_d from (6.22)	Approximate i_d from (6.21)	% difference
1	10.1 μA	9.9 μA	2
5	54.5 μA	49.5 μA	9.2
10	121 μA	99 μA	18.2
15	201 μA	149 μA	25.9

ASIDE A6.3 THE GENERIC TRANSISTOR — PART I

We will frequently find it useful to discuss transistor circuits in a general way without tying the discussion to any particular type of transistor. Since certain basic characteristics are common to bipolar and field-effect transistors (and many other transistors and vacuum tubes as well) we can use these characteristics to arrive at a **generic transistor** model that is useful. Figure A6-3(a) shows the schematic symbol we will use for an n-type generic transistor (we will introduce a p-type transistor when needed) and part (b) of the figure shows a model for the device. Notice on the schematic symbol that the triangle points in the direction of current flow.

The three terminals of the device are called **out**, **control**, and **merge** and will be denoted by O, C, and M, respectively. The merge terminal is the one at the end with the triangle. The names 'out' and 'control' describe the functions of the terminals and the name 'merge' implies that this is where the control and output currents merge. The output current of the transistor flows into the output terminal and is controlled by the voltage appearing between the control and merge terminals (positive on the control terminal). There is also a two-terminal nonlinear element between control and merge and there may in general be a nonzero input current i_C, but we will assume this current is small compared to i_O. The current out of the merge terminal is $i_M = i_C + i_O$.

If you are familiar with bipolar transistors, you will notice that the out terminal corresponds to the collector, control corresponds to the base, and merge to the emitter. For a field-effect transistor, out corresponds to the drain, control to the gate, and merge to the source. We will say more about the generic transistor as the need arises.

Figure A6–3 An n-type generic transistor. (a) The schematic symbol, and (b) a model.

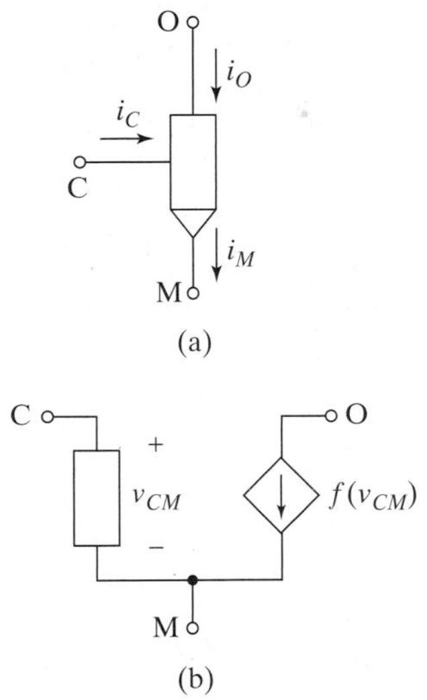

where V_B and the a_k are constants, and

$$v_{CM} = \sum_{k=0}^{1} b_k i_C^k = b_0 + b_1 i_C \quad \textbf{(6.24)}$$

where the b_k are constants. Note that the coefficients a_k in (6.23) all have units of amperes, but in (6.24) the coefficient b_0 has units of volts, and b_1 has units of ohms.

One important characteristic of the generic transistor model shown in Figure A6-3 is that it is unilateral; in other words, the relationship between v_{CM} and i_C does not depend on v_O or i_O, and we can use a unilateral two-port network for the small-signal model.[5] Therefore, we only need to find three component values for the small-signal model; the input and output resistances and the gain for the controlled source. This idea is explored further in Aside A6.4.

[5] A unilateral network is defined in Section 1.2.1.

ASIDE A6.4 UNILATERAL TWO-PORT MODELS

The transfer function of a network can be written as a product of the transfer functions of individual stages, as shown in Chapter 1. If the network is unilateral, these individual transfer functions can be found without solving simultaneous equations. This fact is frequently helpful in analysis and design, since it permits a straightforward method of analysis without writing and solving simultaneous equations, and because each term in the resulting equation can be uniquely associated with a particular portion of the circuit as illustrated in Chapter 1.

The purpose of this aside is to make the benefits of unilateral circuits clearer by example, and to introduce the ***unilateral two-port model*** of a two-port network. Two-port networks are covered in Appendix C on the CD. Many of the two-port networks we deal with can be modeled as unilateral, and the result will be a significant simplification in the resulting analyses.

Consider the amplifier shown in Figure A6-4, wherein a full two-port model has been used. The controlled source on the input side represents ***reverse transmission*** through the network and is typically undesirable in amplifiers. For almost every useful amplifier it will be true[1] that $a_v \gg a_R$, where a_v is the ***forward gain*** (i.e., the gain from input to output) of the amplifier and a_R is the reverse gain (i.e., from output to input) as shown.[2]

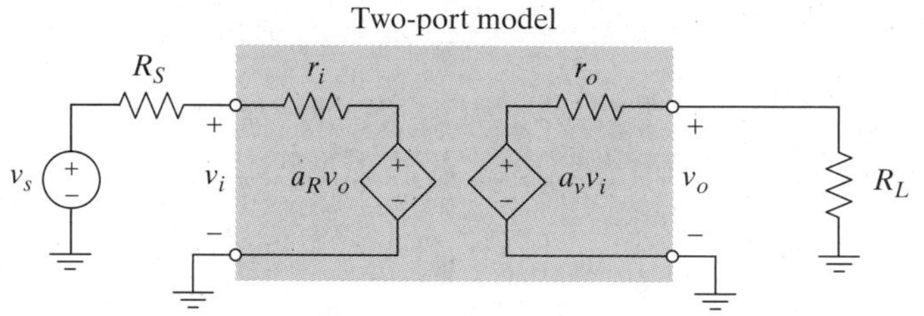

Figure A6-4 A voltage amplifier using a full two-port model.

We want to analyze this circuit to find the overall gain v_o/v_s. There are only two independent nodes in this circuit, v_i and v_o, so we can find our solution by writing nodal equations at these nodes;

$$v_i\left(\frac{1}{R_S} + \frac{1}{r_i}\right) - a_R v_o \frac{1}{r_i} - v_s \frac{1}{R_S} = 0, \tag{A6.4}$$

and

$$v_o\left(\frac{1}{R_L} + \frac{1}{r_o}\right) - a_v v_i \frac{1}{r_o} = 0. \tag{A6.5}$$

From (A6.5) we find that

$$v_i = \frac{v_o(R_L + r_o)}{R_L a_v} = \frac{v_o}{\alpha_o a_v}, \tag{A6.6}$$

where α_o is the output attenuation factor given by (attenuation factors are described in Aside A1.4)

$$\alpha_o = \frac{R_L}{R_L + r_o}. \tag{A6.7}$$

We could also have written (A6.6) by inspection, since $v_o = \alpha_o a_v v_i$. Plugging (A6.6) into (A6.4), solving for v_o and simplifying yields

$$\frac{v_o}{v_s} = \frac{\alpha_i a_v \alpha_o}{1 - \alpha_i a_v \alpha_o \dfrac{R_S a_R}{r_i}} \tag{A6.8}$$

[1] This statement is not always true at high frequencies, where the amplifier gain rolls off, and the signal may capacitively couple to the output, but then the gain is less than one.

[2] These gains are the *open-circuit* voltage gains; that is, they are the gain that is observed when no load is connected; therefore, no current flows, and there isn't a voltage drop across the resistors in series with them.

where the input attenuation factor is defined by

$$\alpha_i = \frac{r_i}{r_i + R_S}. \quad \textbf{(A6.9)}$$

Now consider what happens if we can reasonably ignore the reverse transmission given by a_R. In this case we arrive at the unilateral two-port model shown in Figure A6-5.

In this case we can write by inspection that

$$v_i = \frac{r_i}{r_i + R_S} v_s = \alpha_i v_s \quad \textbf{(A6.10)}$$

and

$$v_o = \frac{R_L}{R_L + r_o} a_v v_i = \alpha_o a_v v_i \quad \textbf{(A6.11)}$$

so that the transfer function becomes

$$\frac{v_o}{v_s} = \alpha_i a_v \alpha_o. \quad \textbf{(A6.12)}$$

This transfer function is much easier to derive and work with than (A6.8). The two transfer functions are equal if a_R is zero and (A6.12) is an accurate approximation so long as $a_R \ll \alpha_i a_v \alpha_o R_S / r_i$.

Any of the two-port networks we deal with can be made unilateral by setting the controlled source on the input side of the network equal to zero. We will frequently use these simplified two-port network models. We should point out that the approximation that a circuit is unilateral is often a very good one for a circuit without feedback, but for a circuit with feedback, the assumption is occasionally quite poor (*see* Chapter 10 for a detailed treatment of feedback).

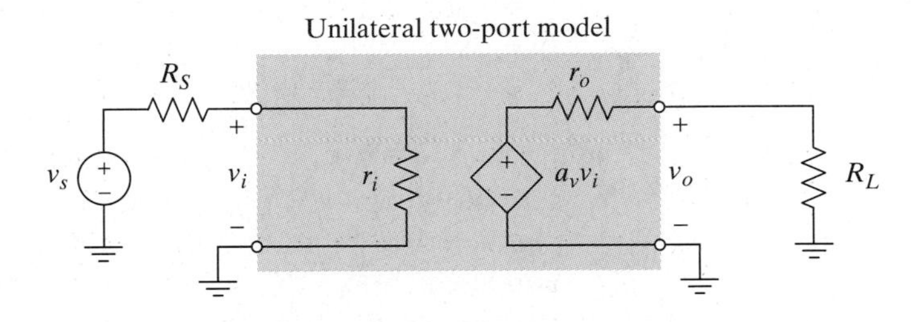

Figure A6-5 A voltage amplifier using a unilateral two-port model.

We now return to our previous analysis. Equation (6.23) indicates that the output *current* of the generic transistor is controlled by the input *voltage*. Consequently, it makes sense to use the Norton equivalent for the output port of our unilateral two-port model, as shown in Figure 6-14.

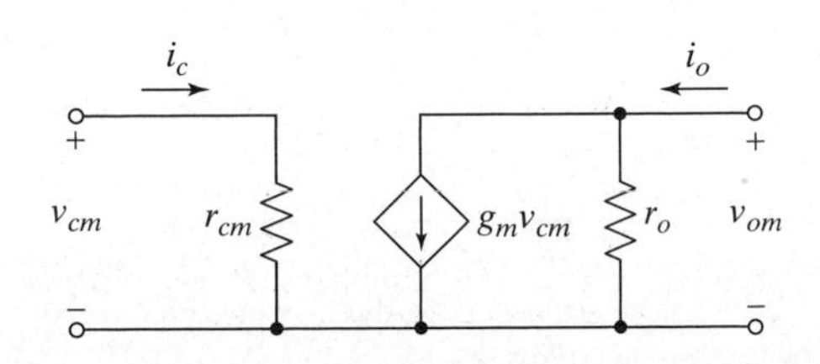

Figure 6–14 A small-signal model for the generic transistor.

Since the device is nonlinear, the small-signal input resistance is given by the slope of the v_{CM} versus i_C characteristic evaluated at the Q point (i.e., the ratio of v_{cm} to i_c). Assuming we know the Q point for the device, we can write (we use r_{cm} to denote the small-signal resistance from control to merge)

$$r_{cm} = \left.\frac{dv_{CM}}{di_C}\right|_{@\,Q} = b_1. \tag{6.25}$$

Although the Q point makes no difference in finding r_{cm} for this example, small-signal model parameters generally do depend on the Q point. The small-signal output resistance is given by the slope of the v_{OM} versus i_O characteristic at the Q point, but since i_O is independent of v_{OM}, we see that r_o is infinite in this example. Finally, the small-signal short-circuit transconductance is given by the slope of the i_O versus v_{CM} characteristic at the Q point (i.e., $v_{CM} = V_{CM}$) with the output shorted, so that $i_o = g_m v_{cm}$. Using (6.23) we find

$$g_m = \left.\frac{di_O}{dv_{CM}}\right|_{@\,Q \text{ with short}} = \frac{a_1}{V_B} + 2a_2\frac{V_{CM}}{V_B^2} + 3a_3\frac{V_{CM}^2}{V_B^3}. \tag{6.26}$$

We are now ready to see how to construct a linear amplifier from this nonlinear device.

6.4.2 Example Application

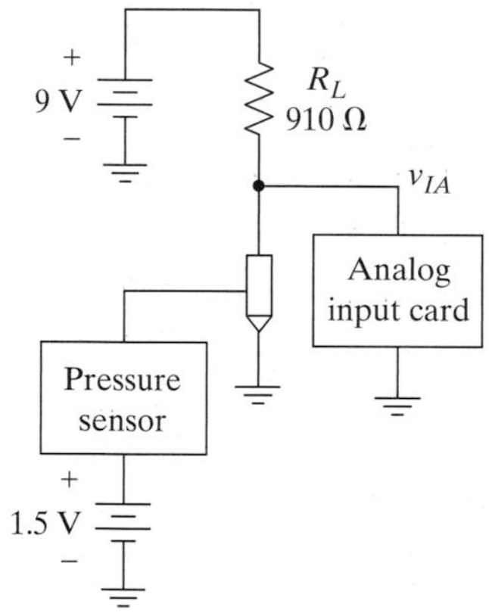

Let's consider a specific application as an example. Say we want to use the generic transistor to amplify the output of a pressure sensor so that it is large enough to drive an analog input card to a computer.[6] Suppose the sensor output is given by $v_s = .01(P - P_o)$ V, where P is the pressure in pounds per square inch (psi) and P_o is atmospheric pressure. Furthermore, suppose that we want to provide the analog input card with a signal that changes by 100 mV for every one psi change in pressure. Examining the equation for the sensor output reveals that the sensor output only changes by 10 mV/psi; therefore, we need a gain of ten. Figure 6-15 shows how we can achieve the desired gain by using a generic transistor. The 1.5V battery in the figure is used to bias the transistor at a particular point on its transfer characteristic. The 9V battery supplies the voltage necessary to achieve the gain and also supplies the power used by the transistor, the load resistor, and the input circuitry on the analog input card.

Figure 6–15 A pressure sensor and amplifier driving an analog input card.

Figure 6-16 shows the small-signal AC equivalent circuit with the transistor replaced by its small-signal model, the pressure sensor replaced by a Thévenin equivalent with an open-circuit voltage v_s and a resistance of 100 Ω, and the ana-

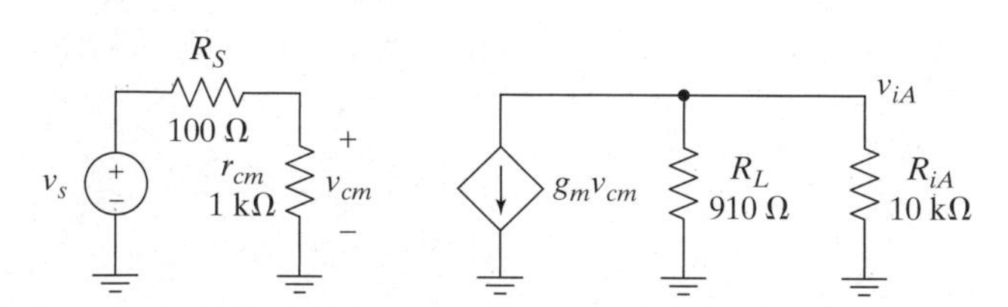

Figure 6–16 The small-signal AC equivalent circuit for Figure 6-15.

[6] The analog input card is a circuit board that would be plugged into an expansion slot on the computer and would have an analog-to-digital converter to convert the analog input voltage into a binary word for use by the computer.

log input card modeled by an input resistance of 10 kΩ. Because we are only interested in the *variations* from the bias point (i.e., the AC signal), the batteries are short-circuited (since the time-varying component of the battery voltage is zero, and zero volts implies a short circuit).

The small-signal model parameters for the transistor must be evaluated at the bias point. By definition, the bias point is the operating point that results when no signal is applied. In this case, that corresponds to atmospheric pressure, where the corresponding sensor output is zero volts. By convention, we use the average value to denote the bias, even though the signal may contain a DC component (*see* Aside A6.2). Therefore, we know that $V_{CM} = 1.5$ V. Assume that the device parameters are: $V_B = 3$ V, $b_0 = 1$ V, $b_1 = 1$ kV/A, $a_0 = 0$ A, $a_1 = 0.6$ mA, $a_2 = 2.4$ mA, and $a_3 = 48$ mA. Plugging into (6.23) yields $I_O = 6.9$ mA, (6.25) yields $r_{cm} = 1$ kΩ, and (6.26) yields $g_m = 13$ mA/V.

Because the circuit in Figure 6-16 is unilateral, we can perform the analysis one stage at a time. The output voltage is (the output of the amplifier is the input to the analog card)

$$v_{iA} = -g_m R_L' v_{cm}, \tag{6.27}$$

where we have defined $R'_L = R_L \| R_{iA}$. The amplifier's input voltage is

$$v_{cm} = \left(\frac{r_{cm}}{r_{cm} + R_S}\right) v_s = \alpha_i v_s, \tag{6.28}$$

where we have defined the input attenuation factor in the usual way. Substituting (6.28) into (6.27) yields the gain;

$$\frac{v_{iA}}{v_s} = -\alpha_i g_m R_L'. \tag{6.29}$$

Plugging in the given values leads to a gain of −9.86, very close to the desired gain of ten. The minus sign indicates that this amplifier inverts the input; that is, an increase in the input voltage produces a decrease in the output voltage and vice-versa. If the inversion is not acceptable for this application, a different circuit will be required (note that a positive gain would result from cascading two inverting stages). It can also be seen from (6.29) that different values for the gain can be achieved by changing R_L or the Q point of the transistor (changing the Q point will change the value of g_m).

The example presented in this section has shown how an active nonlinear device can be used to produce a circuit that operates in an approximately linear fashion for small signals. We have also shown how the nonlinear device can be modeled, resulting in a linear equivalent circuit. This equivalent circuit, called the small-signal equivalent circuit, can then be analyzed using standard LTI analysis techniques to yield an approximation to the circuit's performance. The resulting approximation is accurate as long as the operating point of the nonlinear device does not stray too far from the Q point.

Remember that the DC bias values need to be added to the small-signal solution to get the complete output. In this circuit, the small-signal v_o is added to a DC component of $V_O = 9 - I_O(910) = 2.7$ V (assuming no input current into the analog input card). In practice, a large capacitor would probably be used to couple the AC signal into the analog input card while isolating the DC voltages. The topic of coupling capacitors will be addressed later. Chapter 7 discusses how to find the operating point of a circuit using large-signal models for the nonlinear devices instead of solving the equations directly, as was done here. It also examines practical circuits for establishing the desired operating point. The small-signal analysis of nonlinear circuits is treated in more detail in Chapters 8 and 9.

Exercise 6.4

Find the unilateral two-port model for the amplifier in Figure 6-16 using a Thévenin output stage for the two port. What is the output attenuation factor?

6.5 TYPES OF AMPLIFIERS[7]

Before proceeding to the detailed treatment of amplifiers in Chapters 7 through 10, it is worthwhile to consider how amplifiers are modeled and classified. We first consider the different two-port representations used and move on to discuss how amplifiers are classified based on their frequency response.

6.5.1 Two-port Models for Amplifiers

There are four different two-port models that are commonly used for amplifiers. The four models result from the fact that the controlling variable may be a voltage or a current, and the most appropriate model for the output of the device may be a Thévenin or a Norton equivalent. The unilateral two-port models of the four different amplifier types are shown in Figure 6-17.

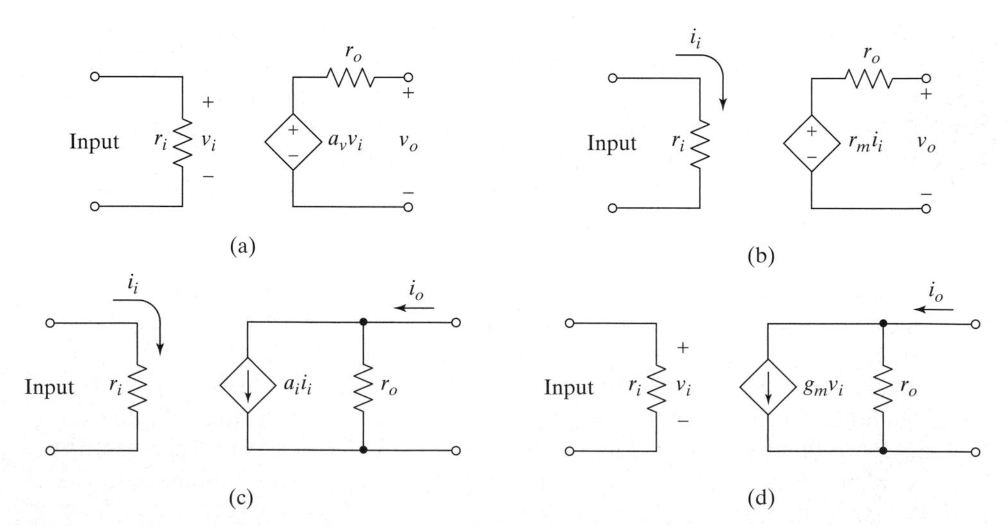

Figure 6–17 The four unilateral amplifier models. (a) Voltage, (b) transresistance, (c) current, and (d) transconductance.

[7] If you are not familiar with two-port networks, you should review Appendix C on the CD before proceeding.

Part (a) of the figure shows a ***voltage amplifier***, so called because the gain of the controlled source is a voltage gain. In this case, the input variable is usually taken to be the input voltage v_i, and the output is usually taken to be the output voltage v_o. Note that we can consider the currents instead, since we can relate them to the voltages. Therefore, our choice of the voltage amplifier model does not *require* us to consider only the voltages. Nevertheless, if we are interested in the currents, one of the other models is probably going to be more convenient. We will demonstrate this with an example later, but let's first briefly consider the other models shown in the figure.

Part (b) of the figure shows a ***transresistance amplifier***, so called because the gain of the controlled source is a transresistance. The term transresistance is a combination of *transfer* and *resistance*, as explained in Aside A1.2. In other words, the gain has the units of resistance, but it refers to the resistance that transfers the input to the output, rather than the resistance of a two-terminal device.

Part (c) of the figure shows a ***current amplifier***, so named because the gain of the controlled source is a current gain. Part (d) of the figure shows a ***transconductance amplifier***, so called because the gain of the controlled source is a transconductance, which means that the gain has units of conductance but relates the output current to the input voltage.

The notation used throughout this figure is lower-case variables with lower-case subscripts (e.g., r_i) and is intended only to remind us that these elements may be small-signal model parameters. In other words, they may represent the slope of some nonlinear function evaluated at a particular bias point rather than a real resistor or controlled source.

It is important to reiterate that the unilateral models shown in Figure 6-17 are, like all unilateral models, typically only approximately correct. The use of these models is often acceptable, however, as was pointed out in Aside A6.4. In the following examples we demonstrate the use of these models and also show that the models are usually interchangeable,[8] although the analysis is typically easier if we choose the proper model for a given situation.

Example 6.1

Problem:

Find the overall gain (i.e., v_o/v_s) of the voltage amplifier shown in Figure 6-18.

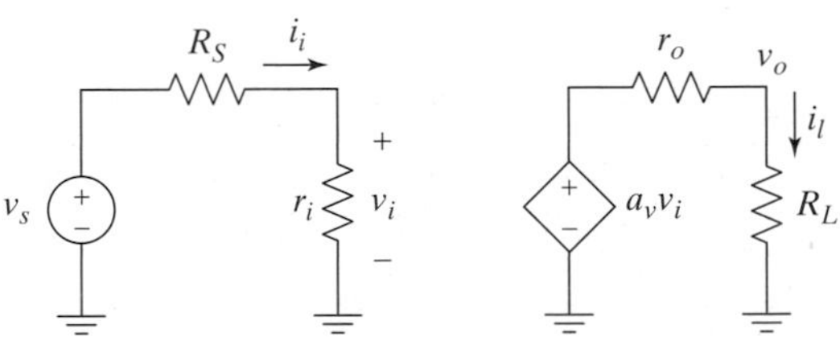

Figure 6–18 Voltage amplifier circuit.

Solution:

The input voltage is found using the voltage divider formed by R_S and r_i,

$$v_i = \left(\frac{r_i}{r_i + R_S}\right)v_S. \tag{6.30}$$

[8] The models are not always interchangeable. As a simple example, notice that if r_i is zero, we cannot use v_i as the controlling parameter for the controlled source.

Similarly, the output voltage is given by

$$v_o = \left(\frac{R_L}{R_L + r_o}\right) a_v v_i \tag{6.31}$$

and substituting for v_i leads to the desired transfer function;

$$\frac{v_o}{v_s} = \left(\frac{r_i}{r_i + R_S}\right) a_v \left(\frac{R_L}{R_L + r_o}\right). \tag{6.32}$$

Using the definitions of the input attenuation factor,

$$\alpha_i \equiv \left(\frac{r_i}{r_i + R_S}\right), \tag{6.33}$$

and the output attenuation factor,

$$\alpha_o \equiv \left(\frac{R_L}{R_L + r_o}\right), \tag{6.34}$$

allows the overall gain to be expressed as

$$\frac{v_o}{v_s} = \alpha_i a_v \alpha_o. \tag{6.35}$$

■

Example 6.2

Problem:
Find the overall gain (i.e., i_l/i_s) for the circuit in Figure 6-19.

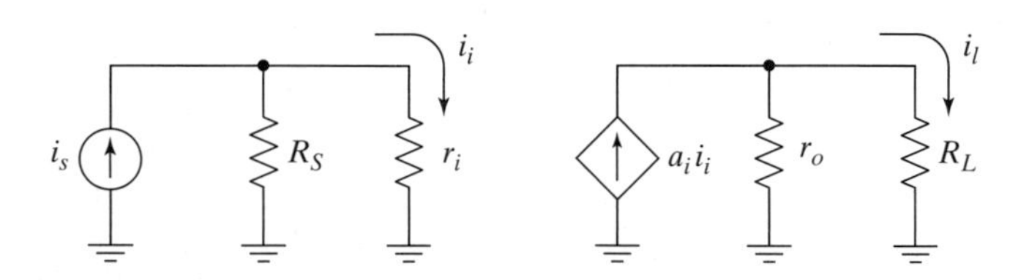

Figure 6–19 Current amplifier circuit.

Solution:
The output current is given by the current divider formed by r_o and R_L,

$$i_l = \left(\frac{r_o}{r_o + R_L}\right) a_i i_i. \tag{6.36}$$

Similarly, the input current is

$$i_i = \left(\frac{R_S}{R_S + r_i}\right) i_s. \tag{6.37}$$

Combining these two equations yields the desired transfer function;

$$\frac{i_l}{i_s} = \left(\frac{R_S}{R_S + r_i}\right) a_i \left(\frac{r_o}{r_o + R_L}\right). \tag{6.38}$$

We can again define input and output attenuation factors, but this time, they are both current dividers. The final result is

$$\frac{i_l}{i_s} = \alpha_i a_i \alpha_o, \tag{6.39}$$

where

$$\alpha_i \equiv \left(\frac{R_S}{R_S + r_i}\right), \tag{6.40}$$

and

$$\alpha_o \equiv \left(\frac{r_o}{r_o + R_L}\right). \tag{6.41}$$

Example 6.3

Problem:
Find the overall *current* gain (i.e., i_l/i_s where i_s is the short-circuit output current of the source) for the circuit of Figure 6-18 and find the equivalent current amplifier model. Compare the results with Example 6.2

Solution:
The load current in Figure 6-18 is given by

$$i_l = \frac{a_v v_i}{r_o + R_L} = \frac{a_v r_i i_i}{r_o + R_L}. \tag{6.42}$$

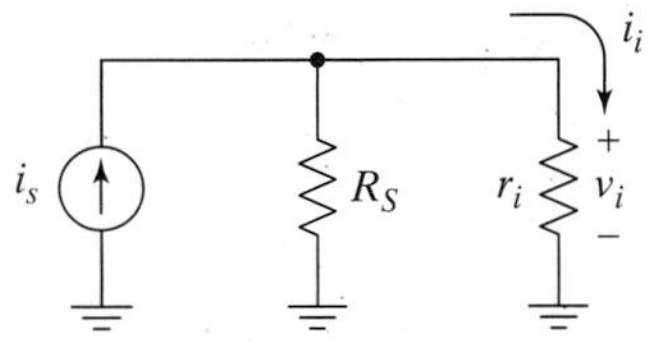

Figure 6–20 The input circuit of Figure 6-18 with the source replaced by its Norton equivalent.

Since we are examining current gain, let's replace the signal source by its Norton equivalent. After doing so, we can redraw the input of our circuit as shown in Figure 6-20. The short-circuit output current of the signal source is (i.e., the Norton equivalent current source)

$$i_s = \frac{v_s}{R_S} \tag{6.43}$$

and the resistance stays the same.

From this circuit we can find the relationship between i_i and i_s;

$$i_i = \left(\frac{R_S}{R_S + r_i}\right) i_s \tag{6.44}$$

Substituting this into our equation for i_l, (6.42), yields the desired transfer function;

$$\frac{i_l}{i_s} = \left(\frac{R_S}{R_S + r_i}\right)\frac{a_v r_i}{r_o + R_L}. \quad \textbf{(6.45)}$$

Although this equation is correct, it would sometimes be more useful to first convert the two-port model used here into a current amplifier model, as shown in Figure 6-19. If we do that, the short-circuit current gain is $a_i = a_v(r_i/r_o)$ and the output resistance is the same. You are encouraged to prove to yourself that this method yields the same results, but is somewhat more convenient when the current gain is what is desired.

Now relate this transfer function to the one obtained in Example 6.2 for a current amplifier. By examining (6.39), (6.40), and (6.41) we see that the gain in (6.45) is identical, except that we need to have $a_i = a_v(r_i/r_o)$. Similar relationships can be worked out for the other amplifier types (*see* Problems P6.22, P6.23, and P6.24).

■

To summarize the results of the preceding examples; if we choose the appropriate amplifier model, the overall gain will be given by the product of an input attenuation factor, the gain of the controlled source, and an output attenuation factor, wherein the attenuation factors were defined in Aside A1.4. If we choose a different two-port model, we can usually obtain the same results but with more work and a less pleasing form. The appropriate model to use for the amplifier is usually clear, but the basic guidelines are:

1) If the signal source model is a Norton equivalent, we should use a current-controlled amplifier model, whereas if the signal source is a Thévenin equivalent, we should use a voltage-controlled model for the amplifier.

2) If the output variable of interest is the voltage, we should model the output of the amplifier with a controlled voltage source (i.e., a Thévenin equivalent), whereas if the output current is desired, we should use a controlled current source (i.e., a Norton equivalent).

As noted previously, the different two-port models are usually interchangeable, so our choice is a matter of convenience most of the time. However, there will be cases in our idealized analyses where the models are not interchangeable; for example, if the input resistance is infinite, the input current will always be zero, and we *must* use a voltage controlled model.

6.5.2 AC- and DC-Coupled Transfer Functions[9]

In addition to the four amplifier types presented in the preceding section, amplifiers can be further classified as being ***DC-coupled*** or ***AC-coupled***. A DC-coupled amplifier is one that will allow a DC component from the source to pass through, whereas an AC-coupled amplifier will not pass DC signals. DC-coupled amplifiers are sometimes called ***direct-coupled*** amplifiers.

AC-coupled amplifiers have the advantage that the DC bias can be established independent of the signal being applied. In other words, we can establish the Q point for

[9] This section requires some knowledge of frequency-domain analysis.

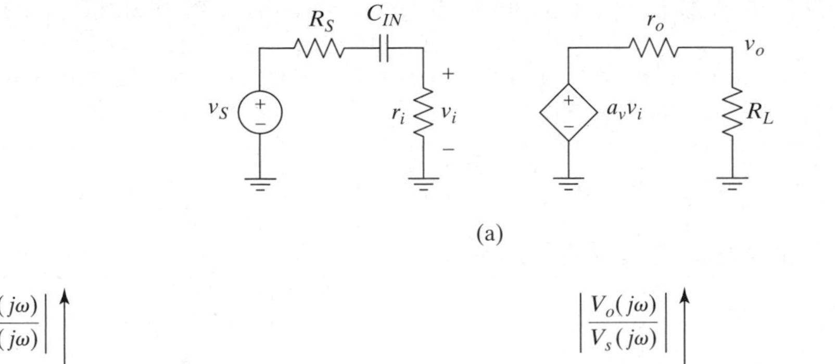

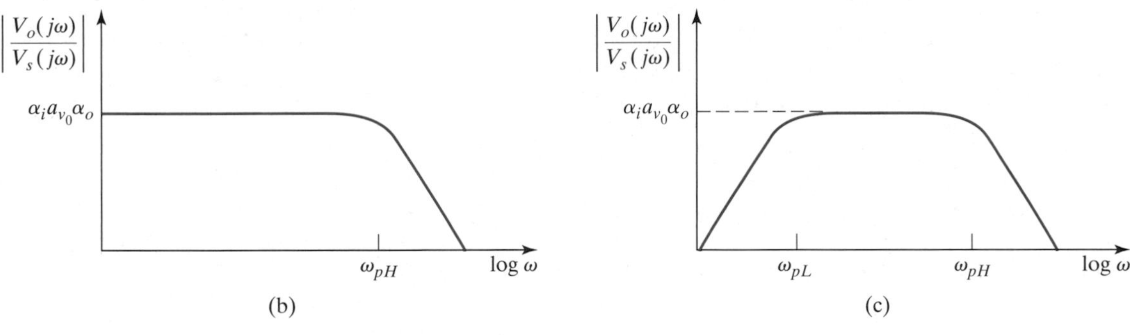

Figure 6–21 (a) An AC-coupled amplifier, (b) the log-magnitude response if the capacitor is shorted out, and (c) the log-magnitude response if the capacitor is not shorted out.

the nonlinear devices in the circuit using DC analysis alone without considering the signal, and we can then add the signal to this bias using capacitive coupling.

Consider the amplifier shown in Figure 6-21(a). If the capacitor were not present, this is precisely the circuit analyzed in Example 6.1, and the transfer function is given by (6.35). To make the analysis somewhat more realistic, however, let's assume that the amplifier's gain is a function of frequency;

$$a_v(j\omega) = \frac{a_{v0}}{1 + j\omega/\omega_{pH}}, \tag{6.46}$$

where ω_{pH} is used to denote the fact that this pole will turn out to be the high-frequency pole in the overall amplifier. Using (6.46) and (6.35), we can plot the magnitude response of the amplifier when it is DC coupled (i.e., the capacitor is shorted out as in Figure 6-18), this plot is shown in Figure 6-21(b), and we can see that the gain at DC will be $\alpha_i a_{v0} \alpha_o$.

Let's analyze what happens with the capacitor in place. The input voltage is now given by a single-pole high-pass transfer function (*see* Aside A1.6) with a high-frequency gain equal to α_i and a pole frequency given by

$$\omega_{pL} = \frac{1}{(R_S + r_i)C_{IN}} \tag{6.47}$$

where ω_{pL} is used to denote the fact that this will be the lower pole frequency in the overall amplifier. The resulting transfer function with the capacitor included is

$$\frac{V_o(j\omega)}{V_s(j\omega)} = \frac{\alpha_i a_{v0} \alpha_o j\omega}{(j\omega + \omega_{pL})(1 + j\omega/\omega_{pH})}, \tag{6.48}$$

wherein we note that we have used different forms for the high- and low-frequency poles for convenience as discussed in Aside A1.6. Figure 6-21(c) shows the magnitude versus frequency plot of (6.48), where we have assumed that C_{IN} has been chosen so that $\omega_{pH} \gg \omega_{pL}$.

Examining the DC-coupled amplifier characteristic in part (b) of the figure we see that it is a low-pass characteristic as defined in Aside 1.6. The characteristic for the AC-coupled amplifier shown in part (c) of the figure is frequently referred to as a ***bandpass*** characteristic since frequencies in the range $\omega_{pL} < \omega < \omega_{pH}$, referred to as the ***passband***, are passed with greater amplitude than those outside of this range.

SOLUTIONS TO EXERCISES

6.1 Writing KVL around the loop for $t > 0$ and remembering that the initial conditions are zero, we find

$$v_i(t) = v_o(t) + \frac{1}{C}\int_0^t i(\lambda)d\lambda.$$

But $i = v_o/R$, so we have

$$v_i(t) = v_o(t) + \frac{1}{RC}\int_0^t v_o(\lambda)d\lambda.$$

Now, if an input v_{i1} produces an output v_{o1} and an input v_{i2} produces an output v_{o2}, we have (the arguments have been dropped for simplicity)

$$v_{i1} = v_{o1} + \frac{1}{RC}\int v_{o1}d\lambda$$

and

$$v_{i2} = v_{o2} + \frac{1}{RC}\int v_{o2}d\lambda.$$

Multiplying the first equation by k_1, the second equation by k_2 and adding, we find

$$k_1v_{i1} + k_2v_{i2} = k_1v_{o1} + k_2v_{o2} + \frac{1}{RC}\int (k_1v_{o1} + k_2v_{o2})d\lambda.$$

This equation shows that an input

$$v_i = k_1v_{i1} + k_2v_{i2}$$

yields an output

$$v_o = k_1v_{o1} + k_2v_{o2},$$

so the transfer function is linear. We can also see that if an input is delayed,

$$v_{id}(t) = v_i(t - \tau),$$

the output will be delayed by the same amount, so the system is time invariant as well.

6.2 The equation for the load line is

$$V_G = 2.25 - 3.125I_G,$$

and the resulting plot is shown in Figure 6-22. There are three possible solutions for the operating point as indicated by points A, B, and C.

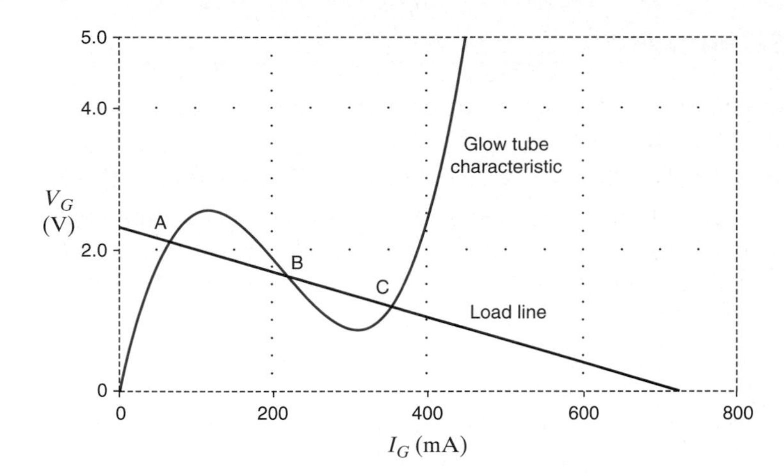

Figure 6–22 V_G versus I_G showing both the load line and the glow tube characteristic.

6.3 The small-signal equivalent circuit has the signal source (AC component only), linear resistor, and a resistor to model the glow tube as shown in Figure 6-23. The small-signal resistance of the glow tube is the slope of the line tangent to the *v-i* characteristic at point B in Figure 6-22 and is approximately −13 Ω. It makes sense that this value is negative, since at this bias point, a decrease in the voltage across the glow tube leads to an increase in the current through it. It should be noted that this operating point is not stable, since any small change (e.g., adding the AC component to the source) will tend to push the circuit in the direction of either operating point A or C.

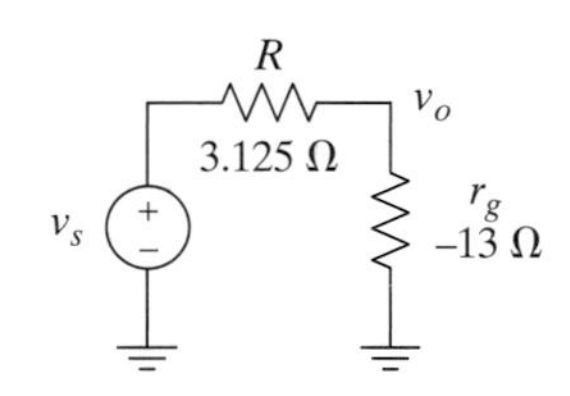

Figure 6–23 The small-signal equivalent for the glow tube circuit of Exercise 6.2 operating at point B in Figure 6-22.

6.4 The unilateral two-port model for the amplifier is shown in Figure 6-24. The input side is identical to Figure 6-16, so we see that $R_i = r_{cm}$. To make the outputs equivalent, we note that the open-circuit output voltage (i.e., with R_{iA} removed) from the circuit in Figure 6-16 is given by $v_{o.c.} = -g_m R_L v_{cm}$. The open-circuit output voltage in Figure 6-24 is $v_{o.c.} = a_v v_i$. So, recognizing that v_{cm} in Figure 6-16 is the same as v_i here, we find $a_v = -g_m R_L$. Also, the output resistance here is the same as the output resistance for the Norton circuit; $R_o = R_L$. The output attenuation factor is $\alpha_o = R_{iA}/(R_{iA} + R_o)$ so the overall gain is

$$\frac{v_o}{v_s} = \alpha_i a_v \alpha_o = \frac{R_i}{R_i + R_S}(-g_m R_L)\frac{R_{iA}}{R_{iA} + R_o}$$

$$= -\alpha_i g_m\left(\frac{R_L R_{iA}}{R_L + R_{iA}}\right) = -\alpha_i g_m R_L'$$

as in (6.29).

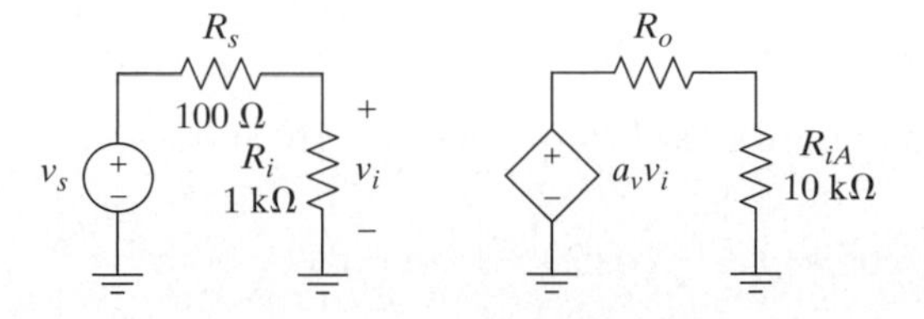

Figure 6–24 A unilateral two-port model for the amplifier in Figure 6-16.

CHAPTER SUMMARY

- Most analog circuits are designed to perform linear functions but must do so using nonlinear active components.
- Most of the analysis tools available to us only apply to LTI networks.
- Nonlinear devices can be approximated as linear for small deviations from some nominal operating point. If we consider only the deviations (i.e., we subtract out the bias values) and keep them small, we can use linearized small-signal models for the nonlinear devices.
- We can separately solve for the DC operating point, or Q point, and the small-signal performance of a circuit.
- The Q point can be found by direct solution of the nonlinear equations, by graphical techniques, or by using large-signal models.
- The small-signal performance is found using the small-signal equivalent circuit. All bias quantities are set to zero in this circuit, and all nonlinear devices are replaced by linear models evaluated at the bias point.
- Small-signal analysis is an approximation and the quality of the approximation rapidly degrades as the signal amplitudes increase.
- Unilateral two-port models are frequently used to simplify the analysis of amplifiers and other circuits. Real circuits are not usually unilateral, but the approximation is very good much of the time.
- There are four types of amplifiers: voltage, current, transresistance, and transconductance. They are usually, but not always, interchangeable, and the choice of which model to use is based on convenience.
- Amplifiers may be classified as either DC-coupled or as AC-coupled.

REFERENCES

[6.1] A.S. Sedra and K.C. Smith, *Microelectronic Circuits*, 4/E, New York: Oxford University Press, 1998.

[6.2] R.C. Jaeger, *Microelectronic Circuit Design*. New York: McGraw-Hill, 1997.

PROBLEMS

6.1 Linear Time-Invariant Networks

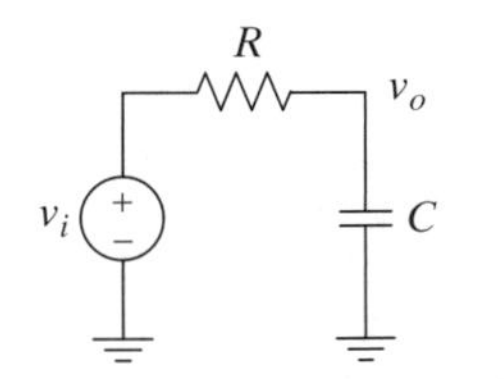

Figure 6–25 A low-pass filter.

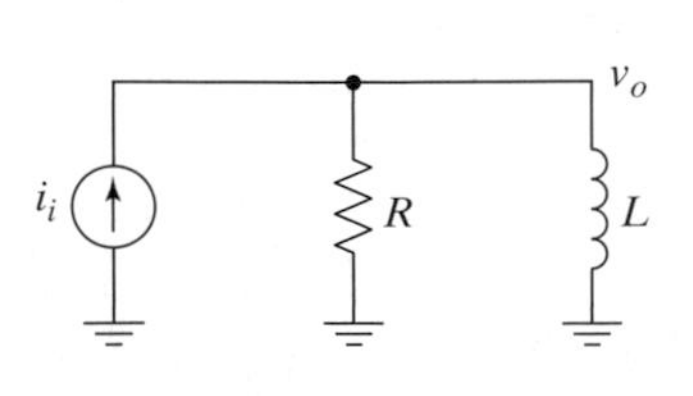

Figure 6–26

P6.1 Consider the LPF shown in Figure 6-25. Assuming that the capacitor voltage is zero at $t = 0$, **(a)** prove that the circuit is LTI for $t \geq 0$. **(b)** If we now apply an initial condition, $v_C(0) = 2$ V, is the system still LTI?

P6.2 Consider the circuit shown in Figure 6-26. Assuming that the inductor current is zero at $t = 0$, **(a)** prove that the circuit is LTI for $t \geq 0$. **(b)** If we now apply an initial condition, $i_L(0) = 0.5$ A, is the system still LTI?

P6.3 The Fourier series for a square wave of frequency f_0 is given by

$$v(t) = \frac{2}{\pi}\left(\cos\omega_o t + \frac{1}{3}\cos 3\omega_o t + \frac{1}{5}\cos 5\omega_o t + \cdots\right) \text{V},$$

where $\omega_o = 2\pi f_o$ rad/s. Suppose this square wave has $f_o = 1$ kHz and is passed through a circuit with the transfer function shown in Figure 6-27. What is the output of the circuit? (Ignore the phase shift produced by the circuit.)

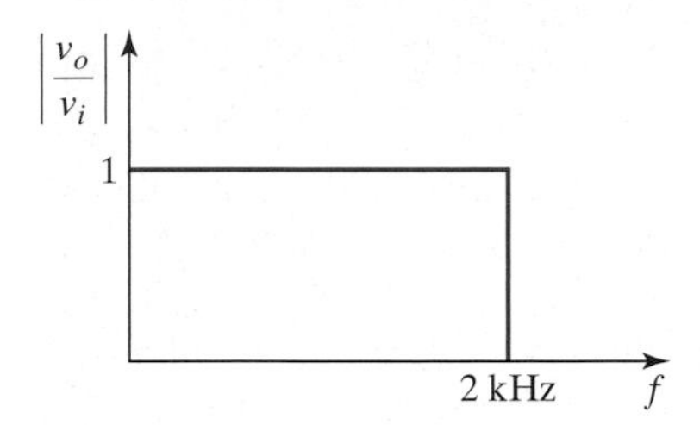

Figure 6–27

6.2 Nonlinear Circuit Analysis

P6.4 Consider the circuit in Figure 6-3. If $V_T = 26$ mV, $R_S = 1$ kΩ, $I_S = 10^{-14}$ A and $n = 1$, find v_O numerically when **(a)** $v_I = 0.3$ V, **(b)** $v_I = 0.6$ V, **(c)** $v_I = 0.9$ V, **(d)** $v_I = 2$ V.

P6.5 Consider the circuit in Figure 6-28 and assume that $V_T = 26$ mV, $I_S = 10^{-14}$ A and $n = 1$. **(a)** Find I_D graphically. **(b)** Find I_D numerically.

P6.6 Consider the circuit in Figure 6-29 and assume that $V_T = 26$ mV, $I_S = 10^{-15}$ A and $n = 1$. **(a)** Find I_D graphically. **(b)** Find I_D numerically.

P6.7 Consider the circuit in Figure 6-30 and assume that $V_T = 26$ mV, $I_S = 10^{-15}$ A and $n = 1$. **(a)** Find I_D graphically. **(b)** Find I_D numerically.

P6.8 Given a tunnel diode with the *i-v* characteristic shown in Figure 6-31(b), find the operating point of the circuit in part (a) of the figure. Ignore the point marked 'Q' for this problem.

DP6.9 Reconsider the circuit in Figure 6-31. Find a new value for R_1 so that the operating point of the circuit will be at the point marked 'Q' in the figure.

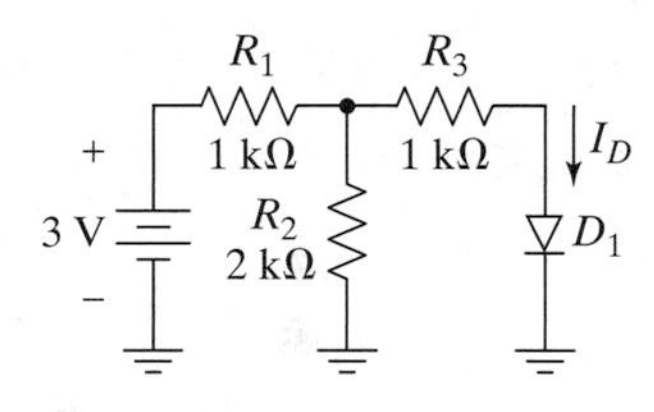

Figure 6–28

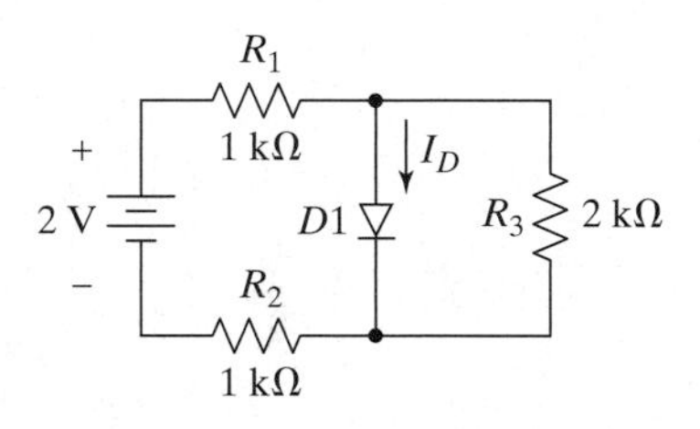

Figure 6–29

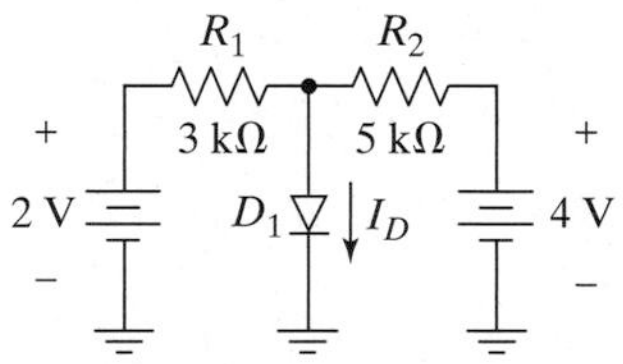

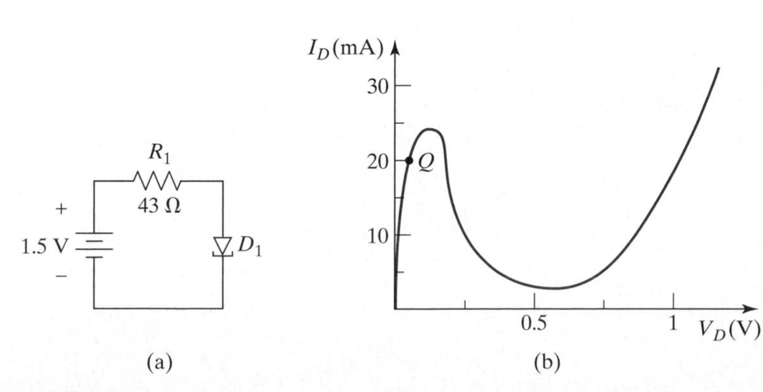

Figure 6–31

R_1 3 kΩ, R_2 5 kΩ, 2 V, 4 V, D_1, I_D

Figure 6–30

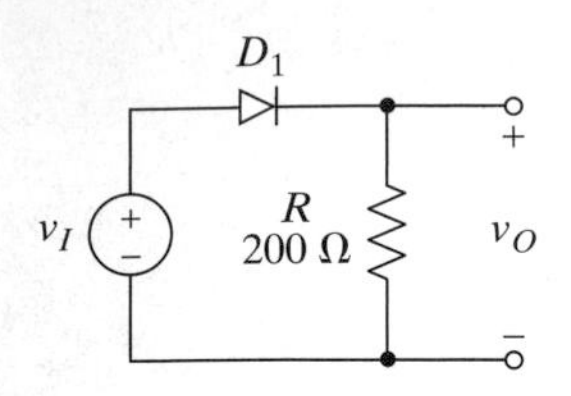

Figure 6–32

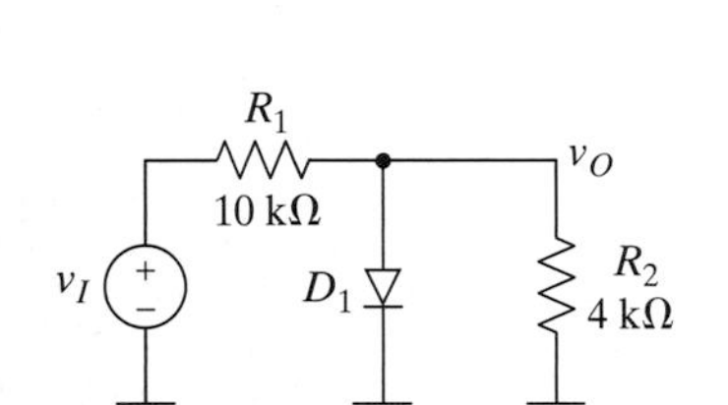

Figure 6–33

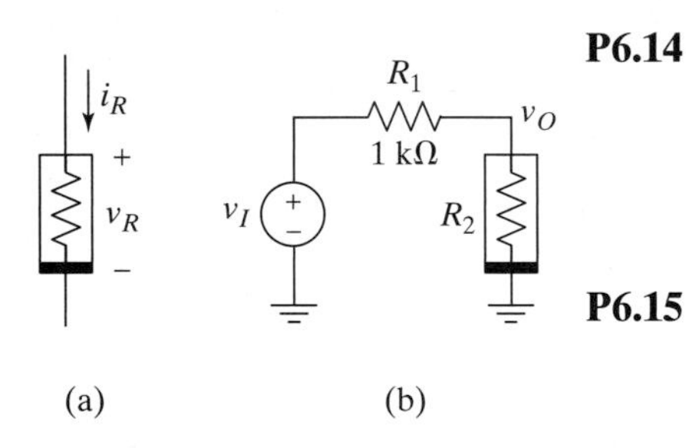

Figure 6–34

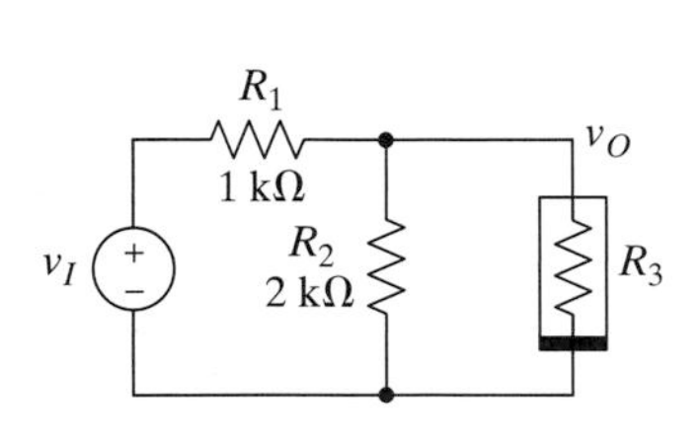

Figure 6–35

6.3 Small-Signal Analysis

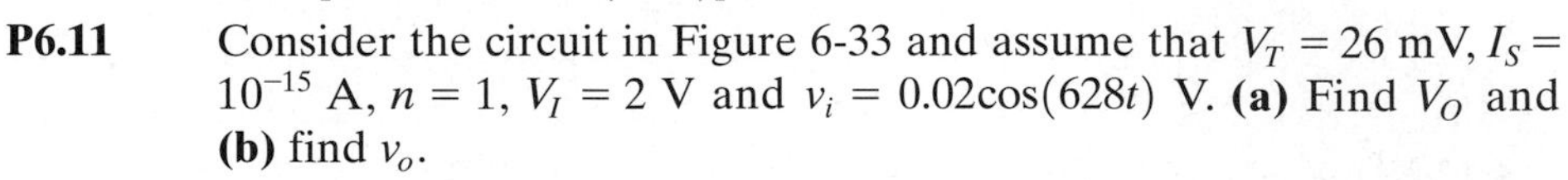

P6.10 Refer to the circuit of Figure 6-32. The diode has $I_S = 8 \times 10^{-14}$A and assume $V_T = 26$ mV. **(a)** Find the total solution for v_O if $v_I = [1 + .005\cos(628t)]$ V. **(b)** Find the total solution for v_O if $v_I = [-1 + .005\cos(628t)]$ V.

P6.11 Consider the circuit in Figure 6-33 and assume that $V_T = 26$ mV, $I_S = 10^{-15}$ A, $n = 1$, $V_I = 2$ V and $v_i = 0.02\cos(628t)$ V. **(a)** Find V_O and **(b)** find v_o.

P6.12 The schematic symbol for a nonlinear resistor is shown in Figure 6-34(a). Consider the circuit in part (b) of the figure and assume the resistor has $i_R = (v_R/R_0)(1 - 0.05\sqrt{v_R})$ and that $R_0 = 1$ kΩ, $V_I = 10$ V, and $v_i = 0.1\sin(628t)$ V. **(a)** Find I_R and V_O. **(b)** Find the small-signal model for R_2 evaluated at the operating point found in part (a). **(c)** Draw the resulting small-signal equivalent circuit and find v_o/v_i.

P6.13 Consider the circuit in Figure 6-35. The nonlinear resistor is described in Problem P6.12, except that in this case $R_0 = 2$ kΩ. **(a)** Find I_{R3} and V_O if $V_I = 3$ V. **(b)** Find the small-signal model for R_3 evaluated at the operating point found in part (a). **(c)** Draw the resulting small-signal equivalent circuit and find v_o/v_i.

P6.14 Find the piecewise linear model for a diode (as in Figure 6-12) for a diode with $I_S = 5 \times 10^{-15}$ A and $n = 1$; **(a)** when $V_D = 660$ mV and **(b)** when $V_D = 600$ mV. **(c)** Compare the two different cut-in voltages you obtained in parts (a) and (b). Explain how one diode can have two different cut-in voltages.

P6.15 Find the breakpoint, V_0, for the piecewise linear model illustrated in Figure 6-12 and discussed in the text and verify that the cut-in voltage given in the text is correct ($V_0 = 663.7$ mV). We used $n = 1$, $I_S = 7.75 \times 10^{-16}$ and had an operating point $I_D = 256.7$ μA and $V_D = 689.7$ mV.

P6.16 Find the breakpoint, V_0, for the piecewise linear model illustrated in Figure 6-12 for a diode operating at the Q point $V_D = 0.63$ V.

P6.17 Refer to the circuit of Figure 6-36(a) and assume that $I_S = 8 \times 10^{-14}$A and $V_T = 26$ mV. **(a)** Draw the DC equivalent circuit and solve for V_D and I_D. **(b)** Draw the small-signal equivalent circuit and solve for v_d and i_d. **(c)** Find r_d and V_0 to make a large-signal piecewise-linear model for the diode as shown in Figure 6-36(b). Select the model components so that the small-signal resistance of the model is equal to the small-signal resistance of the diode, and so that

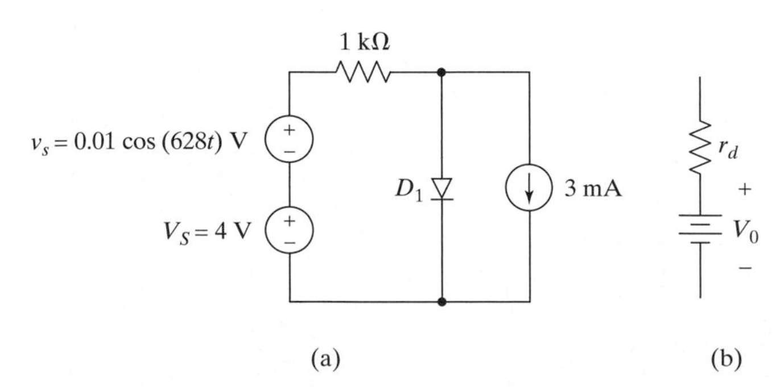

Figure 6–36

the DC voltage across the model is equal to V_D when I_D is flowing through the model. **(d)** Redraw the circuit of Figure 6-36(a) with the diode replaced by the model derived in part (c). Use the circuit with the model to solve for V_D, I_D, v_d, and i_d. Compare these answers with those of parts (a) and (b).

6.4 Small-Signal Amplifiers

P6.18 Consider the generic transistor amplifier in Figure 6-37. Assume that the DC component of the signal source establishes the DC bias so that $g_m = 30$ mA/V, $r_{cm} = 10$ kΩ and $r_o = 50$ kΩ. Given that $v_s = 2\cos(628t)$ mV, **(a)** draw the small-signal AC equivalent circuit, **(b)** find v_o, and **(c)** determine the signal power gain of the circuit (i.e., the signal power delivered to the load divided by the signal power provided by the source, remember that R_S is part of the source).

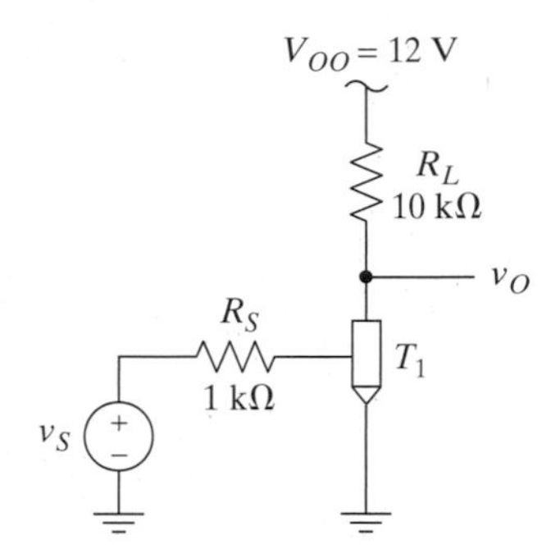

Figure 6–37

P6.19 Consider the generic transistor amplifier in Figure 6-38. **(a)** Draw the small-signal AC equivalent circuit for the amplifier, include r_o of the transistor. **(b)** Derive a unilateral two-port model for the basic amplifier (i.e., not including the source and load) using a Thévenin output stage for the model. **(c)** Write equations for the input and output attenuation factors and the open-circuit voltage gain of the two-port model, a_v.

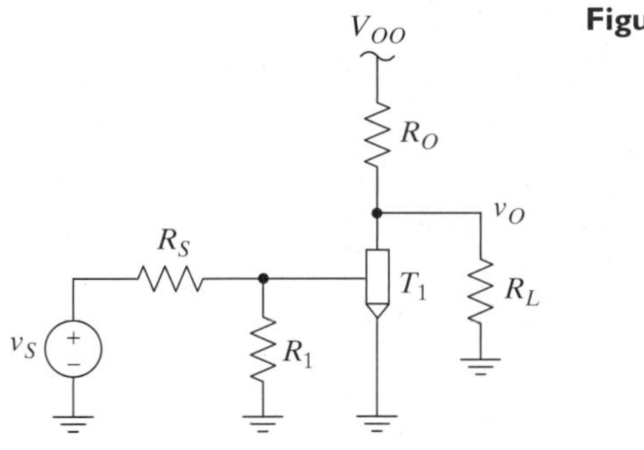

Figure 6–38

P6.20 Suppose a new transistor has been developed. It has three terminals; the output, the input, and a common. The current from the output to the common, i_O, is controlled by the voltage from the input to the common, v_I, and is given by

$$i_O = I_O + a_1 v_I + a_2 v_I^2.$$

(a) Derive an equation for the short-circuit transconductance of the device. **(b)** What must the units of the constants a_1 and a_2 be?

P6.21 Consider an amplifier with the small-signal AC equivalent circuit shown in Figure 6-39. Ignore the feedback resistor shown with dashed

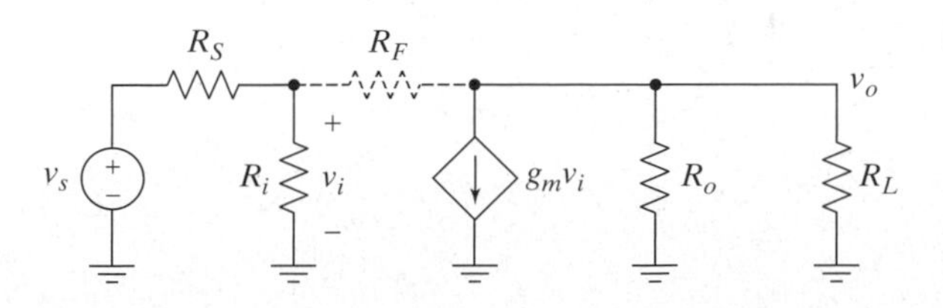

Figure 6–39

lines until part (c). **(a)** Draw the circuit using the unilateral two-port model with a Thévenin output stage for the basic amplifier (i.e., not including the source and load) and find equations for all of the components in the model in terms of the components in the figure. **(b)** Derive an equation for the overall voltage gain, v_o/v_s. **(c)** If R_f is included in the original circuit, would a unilateral two-port model for the basic amplifier still be a completely accurate representation of the original circuit? Why or why not?

6.5 Types of Amplifiers

P6.22 Derive relationships between the three parameters of a unilateral voltage amplifier and those of a unilateral transconductance amplifier.

P6.23 Derive relationships between the three parameters of a unilateral current amplifier and those of a unilateral transconductance amplifier.

P6.24 Derive relationships between the three parameters of a unilateral transresistance amplifier and those of a unilateral transconductance amplifier.

P6.25 Find the parameters r_i, r_o, and k for the two-port of Figure 6-40(a) so that it is electrically identical to the two-port of Figure 6-40(b).

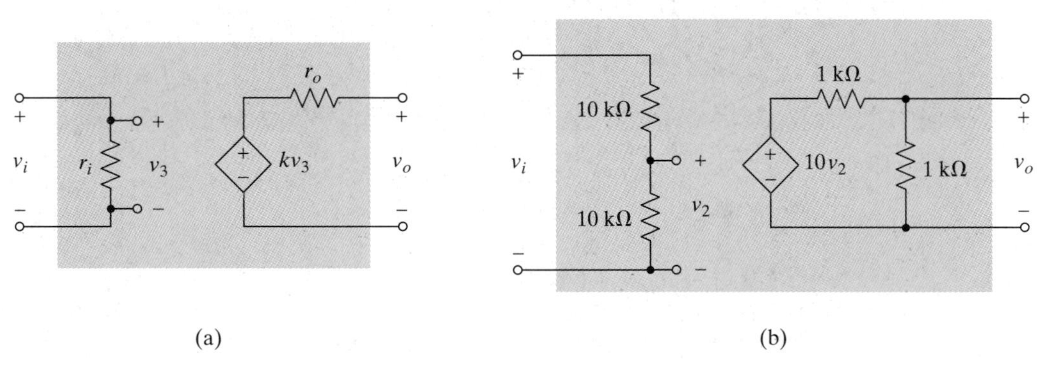

Figure 6–40

CHAPTER 7

DC Biasing

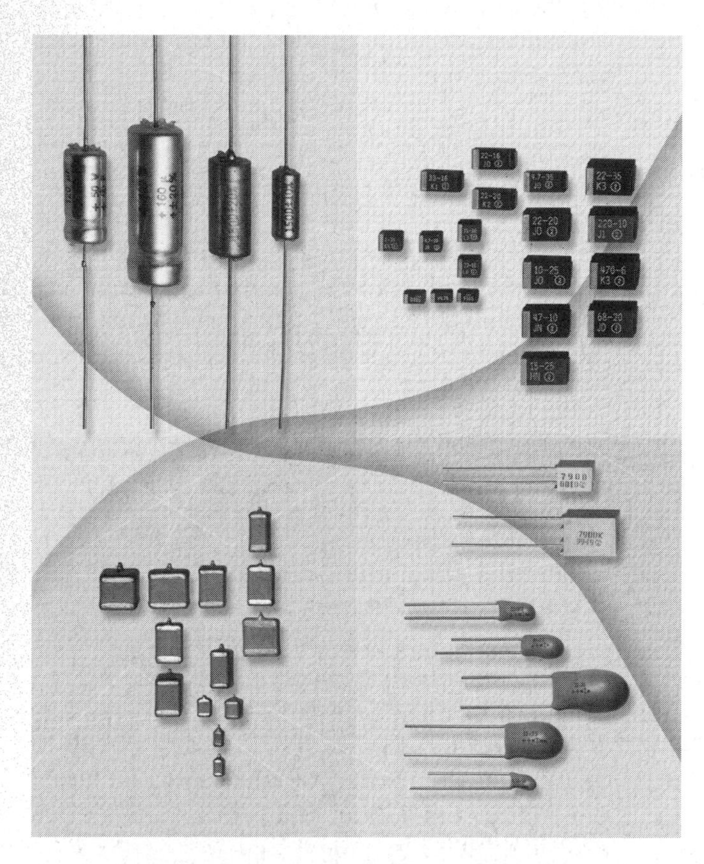

A collection of tantalum capacitors. Capacitors are often used to separate the DC biasing and AC signals in circuits. (Photo courtesy of Vishay Intertechnology.)

INTRODUCTION

The purpose of this chapter is to introduce the models, analysis techniques, and some of the circuits used to establish the desired operating points for transistor amplifiers. In Chapter 6, we learned how nonlinear devices can be used to produce amplifiers that are approximately linear for small signals. The basic idea is to pick an operating point (the Q point), design biasing circuitry to establish that operating point, and then superimpose the signals of interest. The operating point is chosen on the basis of the desired operating characteristics of the circuit; the gain, bandwidth, distortion, noise, and power dissipation may all enter into the decision. We initiate a discussion of how to select the appropriate bias point in this chapter, but we will return to the topic as we examine the small-signal performance of amplifiers at low and high frequencies in Chapters 8 and 9, and the midband large-signal performance in Chapter 12. In the current chapter, we will usually assume that the bias point has been chosen and will concentrate on how to establish it.

It is important to notice from the material in Chapter 6 that the small-signal characteristics of nonlinear devices in general (and transistors in particular) are functions of the bias point (the Q point). Therefore, we want to establish bias points that are not too sensitive to changes in temperature, poorly controlled device parameters, supply voltage, and other variables. In this chapter, we assume that the biasing is always accomplished with DC sources.

The majority of the chapter deals with resistive biasing and resistive loads—that is, circuits in which the bias network and the load are made up of resistors. However, integrated circuits (ICs) do not typically use resistive biasing and loads. IC biasing is covered in Sections 7.4 and 7.5.

7.1 DC AND LARGE-SIGNAL LOW-FREQUENCY MODELS FOR DESIGN

In this section, we are interested in deriving ***DC and large-signal low-frequency models*** suitable for hand analysis. Small-signal AC performance is discussed in Chapter 8 for low frequencies and in Chapter 9 for high frequencies, and large-signal low-frequency AC performance is treated in Chapter 12. Some of the models discussed here are also used in Chapters 12 and 15.

We begin by presenting DC models for independent sources; these will be the only models we develop that are used solely for DC analyses and that cannot also be applied to low-frequency AC analyses. We then present large-signal low-frequency models for linear passive elements (i.e., resistors, capacitors, and inductors) and nonlinear devices.

The models for the nonlinear devices are based directly on our understanding of the underlying device physics (*see* Chapter 2). All of these models will be derived assuming a particular operating condition for the device at hand (e.g., forward active), and it is up to the user to ensure that the choice of model is correct for a given situation. In other words, after performing an analysis, we always need to check that our solution leaves the devices in the operating regions we assumed.

If, for example, our solution indicates that a diode with a breakdown voltage of 10 V has a reverse bias of 20 V across it, then we know that our solution is incorrect, and we must modify our analysis by replacing the diode with a model that is appropriate in breakdown. It is far simpler to handle the different regions of operation in this way than it is to use models that are complicated enough to be appropriate in all regions of operation.

7.1.1 Independent Sources

DC analysis is only *part* of a complete analysis. We use the fact that superposition holds for LTI networks and break the complete analysis up into a DC part and an AC part. Although the DC analysis deals with large voltages and currents and the circuits often *cannot* be approximated as linear, we can find the DC operating point and then use linear small-signal models for small deviations from the bias point, as explained in Chapter 6. To use this analysis technique, however, we must separate all of our independent sources into DC and AC components. We set the AC components to zero for DC analysis, and we set the DC components to zero for the small-signal AC analysis. Therefore, for DC analyses, we deal with independent sources in the following way: First, break all independent sources into AC and DC parts. (The DC component is just the average value.) Second, set the AC components to zero (i.e., short circuit voltage sources and open circuit current sources). This procedure is illustrated in Figure 7-1 for both voltage and current sources.

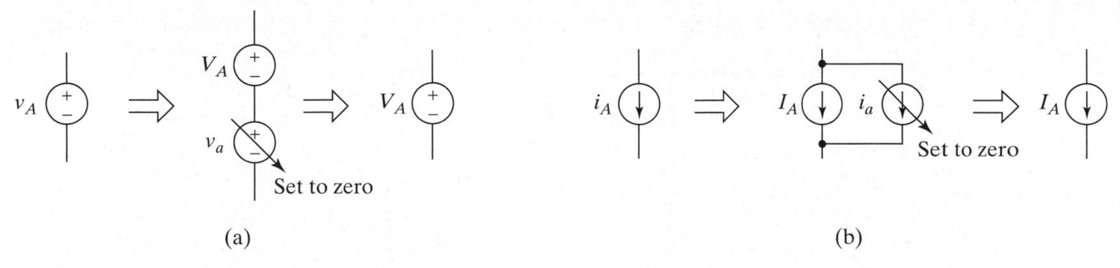

Figure 7–1 How to model independent sources for DC analysis. (a) A voltage source. (b) A current source.

7.1.2 Linear Passive Devices (*R*s, *L*s, and *C*s)

Although perfect resistors, capacitors, and inductors are not available, many real components can be modeled accurately by these ideal elements. Real resistors are never completely linear (i.e., they don't obey Ohm's law, $I = V/R$) for all signal levels.[1] Also, all real resistors have some inductance and capacitance. Similarly, all real capacitors have some resistance and inductance, and all real inductors have some resistance and capacitance.

These unwanted characteristics (e.g., the inductance of a capacitor) are called *parasitic elements*, because they are an unavoidable part of the host element, but are not a desired—or theoretically essential—part of the element. Frequently, we can safely ignore the parasitic elements in each of these models, but not always.

At DC, an ideal capacitor looks like an open circuit, since the current through it will be zero ($i_C = C \cdot dv_C/dt = 0$). Similarly, an ideal inductor looks like a short circuit (since $v_L = L \cdot di_L/dt = 0$), and an ideal resistor remains unchanged. The simplest possible DC models for R's, L's, and C's are therefore derived by ignoring all parasitic elements, open-circuiting all capacitors, and short-circuiting all inductors. The resulting models are shown in Figure 7-2 and are also approximately correct for low frequencies.

In this chapter, we will assume the models shown in Figure 7-2, unless specifically stated otherwise. At low power levels, where one can safely ignore the nonlinearity of resistors, the most important parasitics at DC are the series resistance of an inductor and the parallel resistance of a capacitor.

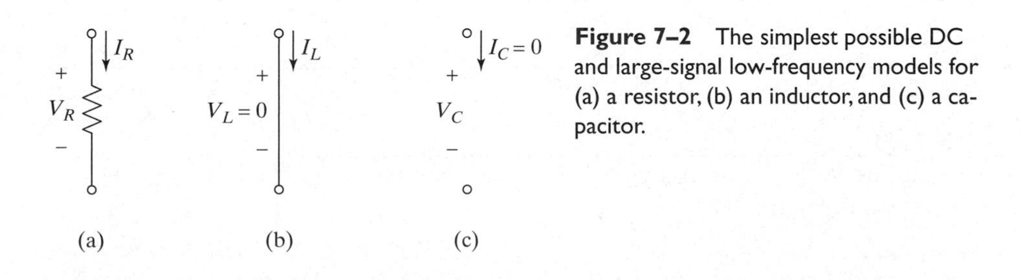

Figure 7–2 The simplest possible DC and large-signal low-frequency models for (a) a resistor, (b) an inductor, and (c) a capacitor.

[1] One example of nonlinear behavior is a result of the fact that resistance is usually temperature dependent. Since the temperature of a resistor increases as the current through it increases, its resistance will change. The result is that the resistance is not constant, and Ohm's law is not strictly valid.

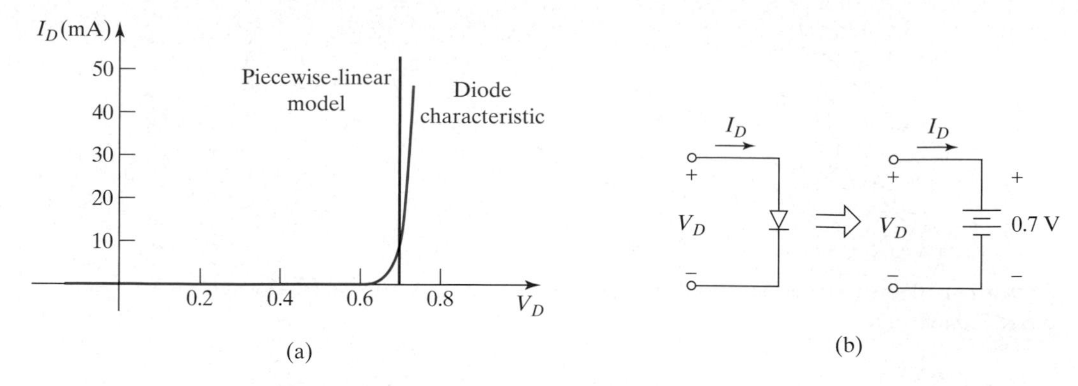

Figure 7–3 A plot of (7.1) for a diode with $I_S = 10^{-14}$ A and $n = 1$ and a piecewise-linear approximation to the characteristic. (b) The corresponding circuit model when the diode is forward biased. (A *pn*-junction diode symbol is used for illustration, but the same form of model applies for Schottky diodes.)

7.1.3 Diodes

At low frequencies, we can ignore all of the charge storage mechanisms in diodes. In this case, both the Schottky diode and the *pn*-junction diode can be modeled by an equation of the form

$$I_D = I_S(e^{V_D/nV_T} - 1) \tag{7.1}$$

if we ignore breakdown. A plot of (7.1) is shown in Figure 7-3(a) for a diode with $I_S = 10^{-14}$A and $n = 1$. The plot is made for room temperature, so $V_T \approx 26$ mV. When the diode is forward biased, the current is given approximately by

$$I_D = I_S e^{V_D/nV_T}, \tag{7.2}$$

and in reverse bias the current is approximately equal to $-I_S$. As was pointed out in Chapter 2, and as can be seen from Figure 7-3(a), the voltage drop across a forward-biased *pn*-junction diode is on the order of 0.6 V to 0.8 V for a wide range of currents. Similarly, the voltage drop across a forward-biased Schottky diode is on the order of 0.2 V to 0.4 V for a wide range of currents. Therefore, it is not unreasonable in many applications to use a simple piecewise-linear model for the diode, as shown in Figure 7-3(a), where V_D is about 0.3 V for a Schottky diode and about 0.7 V for a *pn*-junction diode. The corresponding circuit model in forward bias is shown in part (b) of the figure.

The model shown in Figure 7-3 will provide reasonably accurate results for the currents, and some voltages, in the circuit provided that the error in V_D is acceptably small.

Example 7.1

Problem:
Use the diode model from Figure 7-3 to find the voltages and currents in the circuit shown in Figure 7-4.

Solution:
Using the model, the circuit is redrawn in Figure 7-5. From this figure, we see immediately that the voltage across R_2 is 0.7 V; therefore, the current through it is 1 mA. We also see that the voltage across R_1 is 9.3 V, so the current through it is 10 mA. By KCL, we see that the diode current must be $I_D = I_1 - I_2 = 9$ mA.

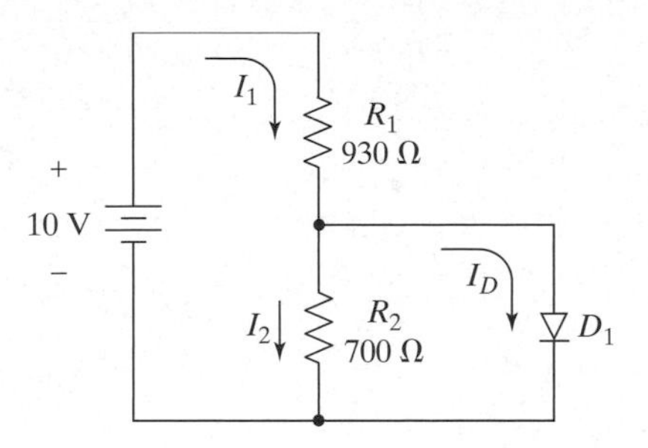

Figure 7–4 Circuit for Example 7.1.

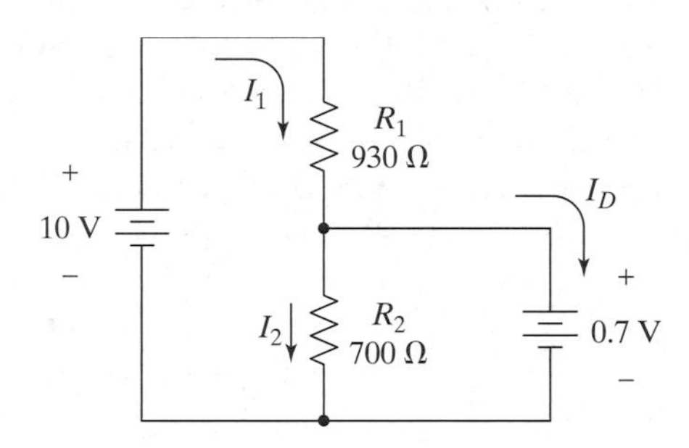

Figure 7–5 The circuit of Figure 7-4 redrawn using a model for the diode.

In Example 7.1, the voltage supplied to the circuit is much larger than the drop across the diode (i.e., $10 \gg 0.7$), so the values determined for V_{R1} and I_1 using the model are quite accurate, even if the diode voltage turns out to be 0.5 V (an extreme example). The diode voltage itself would be in error by 40% in this case, however, as would the current through R_2; therefore, we need to be careful in interpreting the results obtained using this simple model.

If the result we are interested in depends heavily on the accuracy of the model (e.g., I_2 or V_D in Example 7.1), it may be substantially in error. In the case of the diode model we are considering, as long as the other voltages in the circuit are much larger than the voltage drop across the diode (sometimes referred to as the *diode drop*), the currents determined using the model will be close to the correct values, except for elements in parallel with the diode (e.g., R_2). If the other voltages are not much larger than the diode drop, it is best to use the equation.

In performing quick mental calculations, and even for some real analyses, it is sometimes sufficient to assume that the diode is ideal,[2] so that $V_D = 0$. In Example 7.1, for instance, using $V_D = 0$ yields $I_1 = 10.8$ mA, which is only 8% higher than our previous estimate. In this example, the mental arithmetic is much easier with $V_D = 0.7$ V because the numbers were chosen to make that true. However, the idea is that, in general, one can estimate the values using whatever value of V_D (near 0.7 V) is most convenient numerically, and the answer will be close to right, *as long as the answer is not sensitive to the exact value chosen for V_D.*

Exercise 7.1

Find the correct values for I_1 and I_2 in the circuit of Figure 7-4 if the diode voltage is (a) 0.5 V and (b) 0.9 V.

[2] *See* Aside A2.3 on the ideal diode if this concept is unfamiliar.

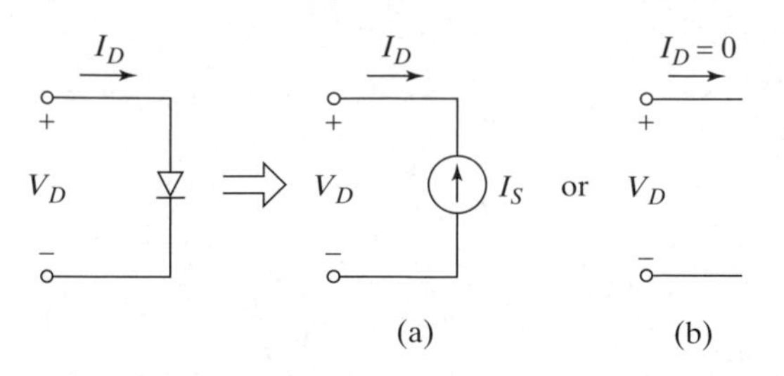

Figure 7–6 Two possible models for a reverse-biased diode: (a) constant current and (b) open circuit.

If the diode is reverse biased, there are two possible models, as shown in Figure 7-6. The first model, shown in part (a), models the current in the diode as having a constant value equal to $-I_S$. Since I_S is very small, however, the open-circuit model shown in part (b) is used most often. We will always assume that the open-circuit model is appropriate, unless specifically stated otherwise. We should point out that the value of the reverse current predicted by (7.1) is not accurate. Therefore, if the reverse current is significant, a better value needs to be obtained by measurement or more accurate modeling. The reverse current is usually much larger than I_S, due to surface leakage and other effects we have not allowed for. It is incumbent on the circuit designer to decide which model is appropriate in a given circumstance.

Example 7.2

Problem:

Find all the voltages and currents for the circuit in Figure 7-7.

Solution:

Suppose that we can't "see" which models to use immediately; how do we proceed? One way that will always work (although it may be very tedious in some cases) is to simply assume a particular region of operation for each nonlinear device, perform the analysis with the model for the assumed region of operation, and then check to see if the initial assumptions were correct. If we do this enough times, we will develop our intuition, and our initial guesses will be correct a larger percentage of the time. In this case, let's start by assuming that both diodes are forward biased. The resulting circuit, with the diodes replaced by the appropriate models, is shown in Figure 7-8.

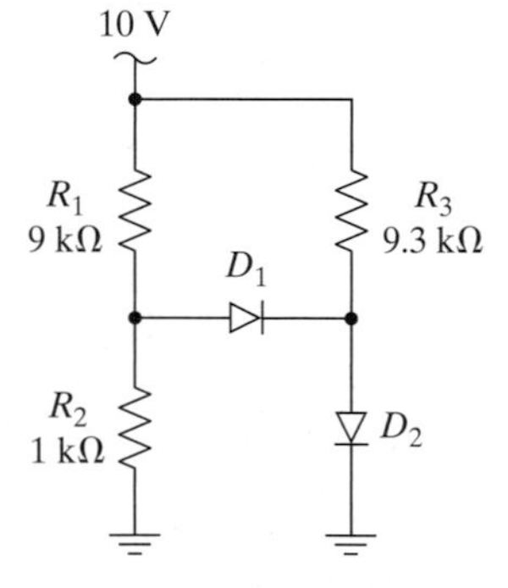

Figure 7–7 The circuit for Example 7.2.

Using this figure, we can see immediately that $V_B = 0.7$ V and $V_A = 1.4$ V. The resulting currents are found using Ohm's law and KCL and are $I_{R1} = 0.96$ mA, $I_{R2} = 1.4$ mA, $I_{R3} = 1$ mA, $I_{D1} = -0.44$ mA, and $I_{D2} = 0.56$ mA. In examining these currents, we see that we have a problem: The current through D_1 is going in the wrong direction (i.e., the value is negative). Therefore, our assumption that this diode is forward biased must be wrong. As a result, we redraw the circuit one more time, only this time assuming that D_1 is reverse biased and D_2 is forward biased. The resulting circuit is shown in Figure 7-9.

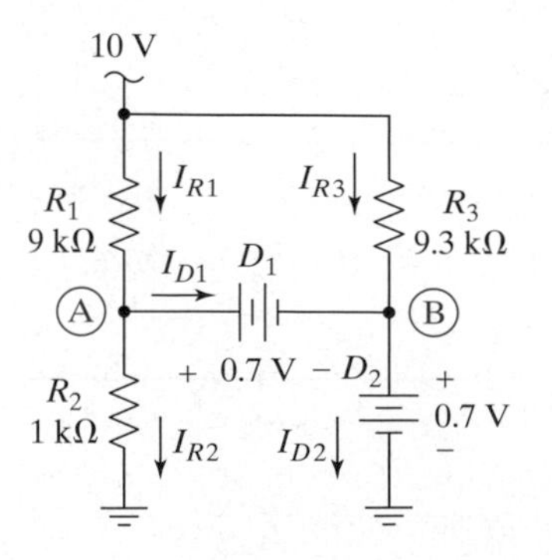

Figure 7–8 The circuit of Figure 7-7 with the diodes replaced by their models, assuming that they are both forward biased.

This circuit can be solved to obtain $V_A = 1$ V, $V_B = 0.7$ V, $I_{R1} = I_{R2} = 1$ mA, and $I_{R3} = I_{D2} = 1$ mA. The voltage across D_1 is $V_{D1} = V_A - V_B = 0.3$ V. Checking our models, we see that D_2 is forward biased as assumed (the current through it is positive) and D_1 is reverse-biased according to the piecewise-linear model we are using (i.e., $V_{D1} < 0.7$ V).

At this point, it is worth discussing the simplified model used here further. Since the voltage across D_1 is 0.3 V, this diode *is* forward biased, and we are *not* completely correct in modeling it as reverse biased. However, notice that the current through D_1 will be negligible in comparison with the other currents in this circuit if D_1 and D_2 have I_S's that are within a few orders of magnitude of each other. To see why this statement is true, remember from Chapter 2 that at room temperature the voltage across a *pn*-junction diode changes by about 60 mV for every decade change in the current through it. So, with $V_{D1} = 0.3$ V and $V_{D2} = 0.7$ V, the current through D_1 would be *more than six orders of magnitude smaller* than the current through D_2 if their I_S's were the same. Therefore, we are justified in using the reverse-biased model for D_1 in the analysis.

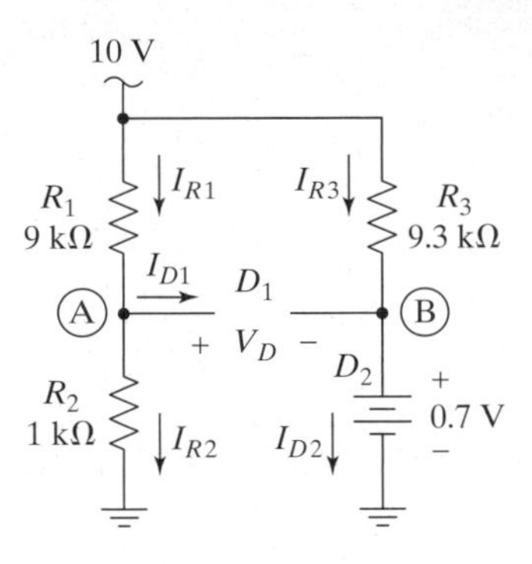

Figure 7–9 The circuit of Figure 7-7 with the diodes replaced by their models assuming that D_1 is reverse-biased and D_2 is forward-biased.

■

Let's reiterate the procedure we have been using to model nonlinear circuits:

1. Assume the region of operation for each nonlinear device and apply the appropriate models.
2. Perform the analysis.
3. Check the models.
4. Redo steps 2 and 3 with new models if necessary.

There will be times when the models we are using here are inadequate. It is again up to the designer to determine when that is true. One situation in which our model is inadequate is demonstrated in Example 7.3.

Example 7.3

Problem:

Determine both diode currents in the circuit shown in Figure 7-10.

Solution:

Start by assuming that both diodes are forward biased. The result is that the voltage across the diodes is 0.7 V and the current through the resistor is $I_{R1} = (5 - 0.7)/1\text{k} = 4.3$ mA. But how does this current divide between the two diodes? Our model is inadequate to answer this question, since the current through the diode can be anything from zero to infinity when the voltage is 0.7 V. If we resort to the equation for a forward-biased diode as given by (7.2) and assume that both diodes have $n = 1$, but allow for different saturation currents, we can show that (*see* Problem P7.6) $I_{D1} = (I_{S1}/I_{S2})I_{D2}$. Note that we cannot make any statement about the relative values of I_{D1} and I_{D2} based on the piecewise-linear model; to do so, we need to resort to the equations or derive a more accurate model.

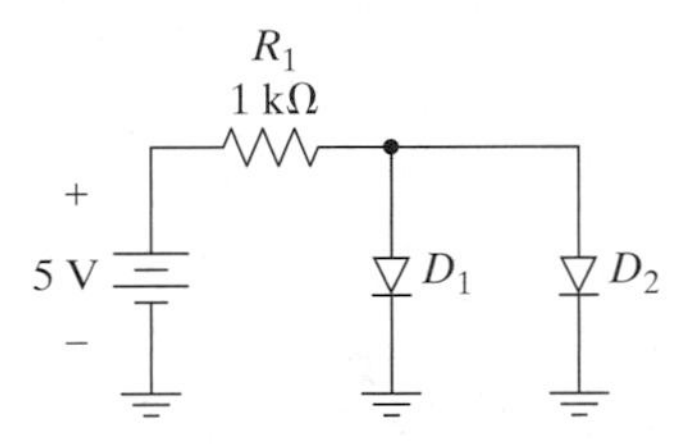

Figure 7–10 Circuit for Example 7.3.

■

The final region of operation for which we need a model is the breakdown region. If the reverse voltage across a diode exceeds the breakdown voltage, the voltage across the diode is, to first order, independent of the current through the device. The resulting model is shown in Figure 7-11, wherein V_{BR} is the breakdown

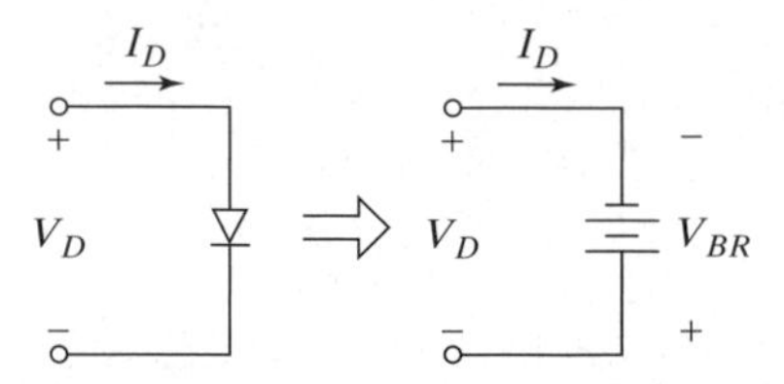

Figure 7–11 A model for a diode in the breakdown region of operation.

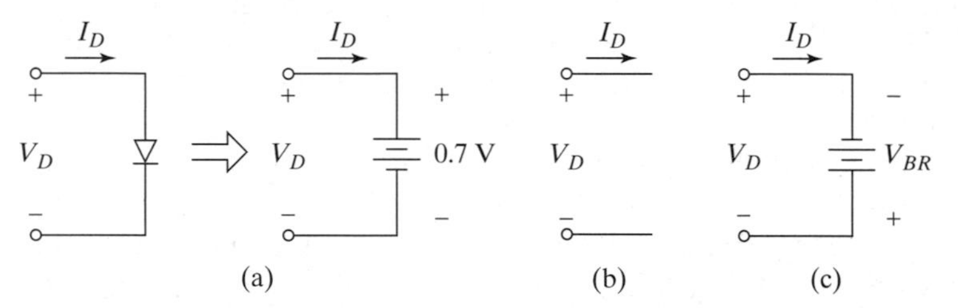

Figure 7–12 The large-signal low-frequency models used for a diode; (a) forward-biased, (b) reverse-biased, and (c) in breakdown.

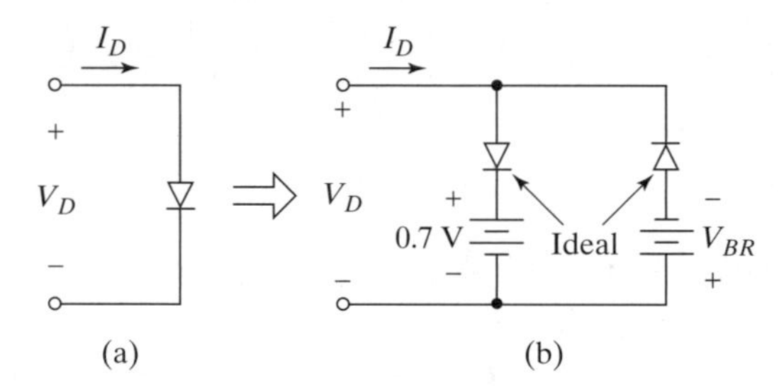

Figure 7–13 (a) A *pn*-junction diode and (b) a single model that works in all regions.

voltage. The diode current and voltage (I_D and V_D) are both negative when the diode is in breakdown.

The models that we will use for the diode are summarized in Figure 7-12. The models for a Schottky diode are the same, except that the forward drop is 0.3 V instead of 0.7 V. We could certainly improve on these models by adding series resistance, but that is unnecessary the vast majority of the time. Also, if we want the model to account for the slope of the *i–v* characteristic, calculating the required value for the series resistance requires that we already know the bias point, as pointed out in Section 6.3.

It also possible to construct a single model that will include all three of our models by using ideal diodes and sources. Consider the model in Figure 7-13, wherein the diodes shown in part (b) are all assumed to be ideal (i.e., with no forward drop, no breakdown, and infinite slope to the *i–v* characteristic at $V_D = 0$).

This model is identical in all three regions of operation to the models we have proposed in Figure 7-12. However, we would typically not want to use the model for hand analysis, because it is unnecessarily complicated.

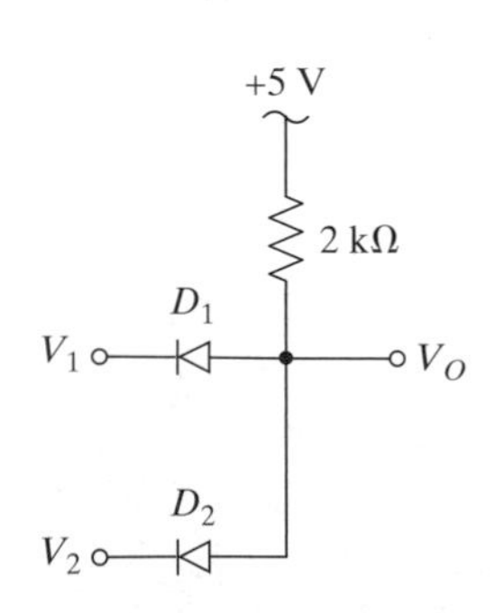

Figure 7–14

Exercise 7.2

For the circuit shown in Figure 7-14, determine V_O if (a) $V_1 = V_2 = 0$, (b) $V_1 = 0$ and $V_2 = 2$ V, and (c) $V_1 = V_2 = 5$ V.

7.1.4 Bipolar Junction Transistors

In this section, we first examine the models for an *npn* bipolar junction transistor (BJT) and then show the corresponding models for a *pnp* BJT. Since we are interested in low-frequency models, we will ignore all charge storage mechanisms. The collector characteristics of a BJT are shown in Figure 7-15. We only desire models for the device in the cutoff, saturation, and forward-active regions of operation.[3]

Since the collector current of a BJT is controlled by the voltage from the base to the emitter, it is useful to have two different, but equivalent, models for the device, one with the base as the input terminal and one with the emitter as the input terminal, as illustrated in Figure 7-16. The circuit in part (a) of the figure is called a ***common-emitter*** connection, and the circuit in part (b) of the figure is called a ***common-base*** connection.

In the common-emitter connection, the input is either V_B or I_B and the output is either I_C or V_C; note that the emitter is common to both the input and the output. For the common-base connection, the input is either V_E or I_E and the output is either I_C or V_C; note that the base is common to both the input and the output. We will not always ground the common terminal directly, as shown here. As a result, the input voltage will be V_{BE} instead of just V_B or V_E, but the models developed here will still apply.

The Forward-active Region of Operation Ignoring base-width modulation in the forward-active region of operation (i.e., assuming that the characteristics in the forward-active region in Figure 7-15 are flat), the collector current in this region is given by

$$I_C = I_S e^{V_{BE}/V_T} \tag{7.3}$$

or

$$I_C = \beta_F I_B, \tag{7.4}$$

where β is the ***common-emitter current gain*** (i.e., it is the ratio of the output current to the input current in the common-emitter connection). We have used the subscript to clearly identify β_F as the beta in forward-active operation. The subscript will be dropped if confusion is unlikely.

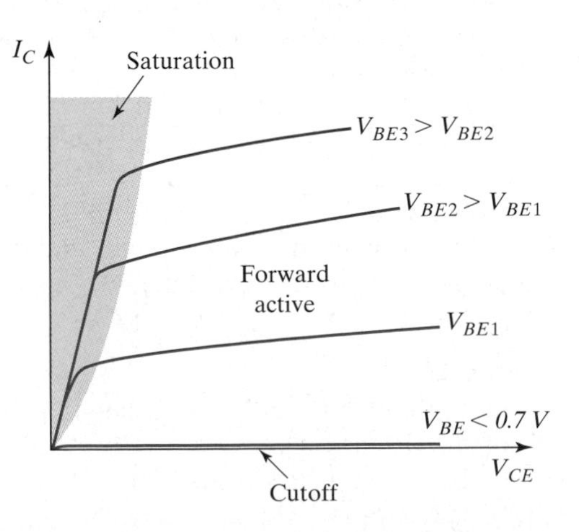

Figure 7–15 The collector characteristics of a BJT, showing the forward-active, cutoff, and saturation regions of operation.

---

[3] There are a few cases where the reverse-active region is used, but it can be modeled exactly like the forward-active region. The only differences are the specific values of the model parameters (e.g., the reverse β is typically much less than the forward β).

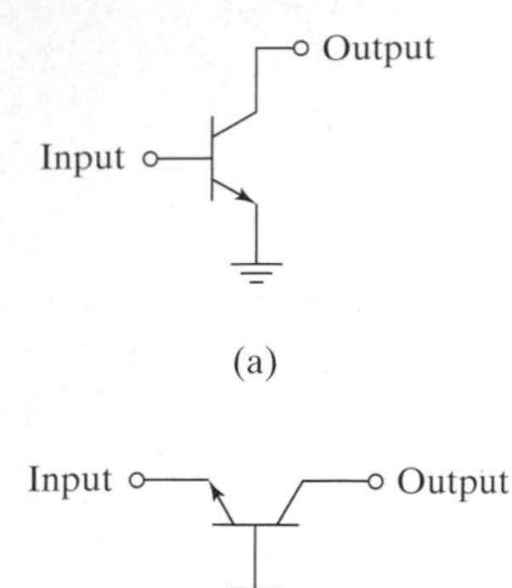

Figure 7–16 (a) The common-emitter connection and (b) the common-base connection for an *npn* BJT.

Turning our attention to the common-base configuration, we write,

$$I_E = I_B + I_C = (\beta_F + 1)I_B, \tag{7.5}$$

so that we can relate the emitter and collector currents by

$$I_C = \beta_F I_B = \frac{\beta_F}{\beta_F + 1} I_E = \alpha_F I_E, \tag{7.6}$$

where α_F is the ***common-base current gain***, since it is the ratio of the output current to the input current in the common-base connection. We have again used the subscript to clearly identify α_F as the α in forward-active operation.

Just as was the case for the *pn*-junction diode, we find that the base-emitter voltage of a bipolar transistor that is on is usually in the range of 0.6 V to 0.8 V for a wide range of reasonable collector currents; note that a plot of (7.3) would be just like a plot of (7.2). Therefore, if we use our diode model for the forward-biased base-emitter junction, the resulting base current will be accurate, as long as the other voltages in the loop are much larger than V_{BE}. Following this method, we can obtain a reasonable approximation to I_B or I_E, and using (7.4) or (7.6), we can then find a reasonable approximation to the collector current. The resulting models are shown in Figure 7-17 and are current-controlled models. The advantage of using these models is that the collector current is a *linear* function of I_B or I_E, whereas it is a *nonlinear* function of V_{BE}. Therefore, we use current-controlled large-signal low-frequency models, even though the underlying physics indicate that the BJT is a voltage-controlled device.

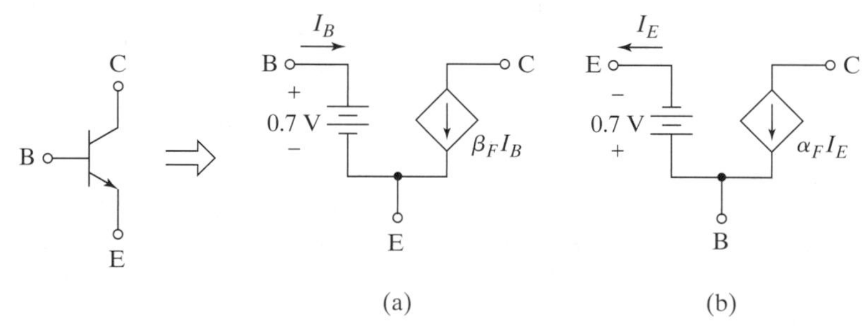

Figure 7–17 Large-signal low-frequency models for an *npn* BJT in forward-active operation: (a) the common-emitter model and (b) the common-base model.

Example 7.4

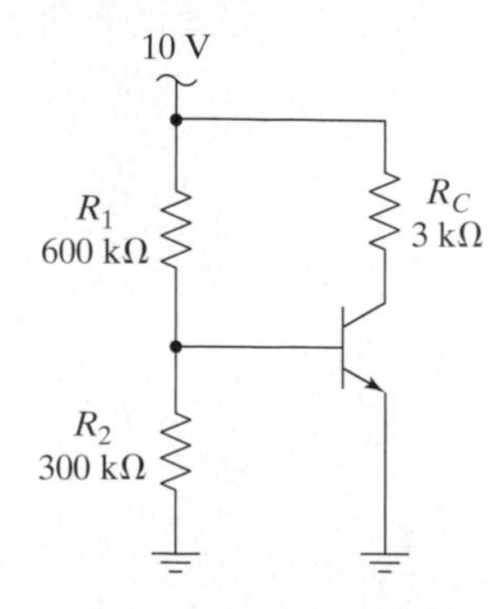

Figure 7–18 Circuit for Example 7.4.

Problem:
Find the currents and voltages in the circuit shown in Figure 7-18 by using the appropriate large-signal model for the BJT, assuming that it is forward active and has $\beta_F = 100$.

Solution:
First simplify the circuit by finding the Thévenin equivalent driving the base of the transistor. Referring to Figure 7-18, we see that the open-circuit base voltage is 3.33 V (i.e., disconnect the transistor from the middle of R_1 and R_2, and see what the voltage is). The Thévenin resistance is the parallel combination of R_1 and R_2. (The 10-V supply is an independent source and is set to zero to find the resistance—doing so shorts the top of R_1 to ground.) Because this Thévenin equivalent represents the DC biasing circuit driving the base, we will use the double-subscript notation that is used for power supplies; the voltage is V_{BB}, and the resistance is R_{BB}.

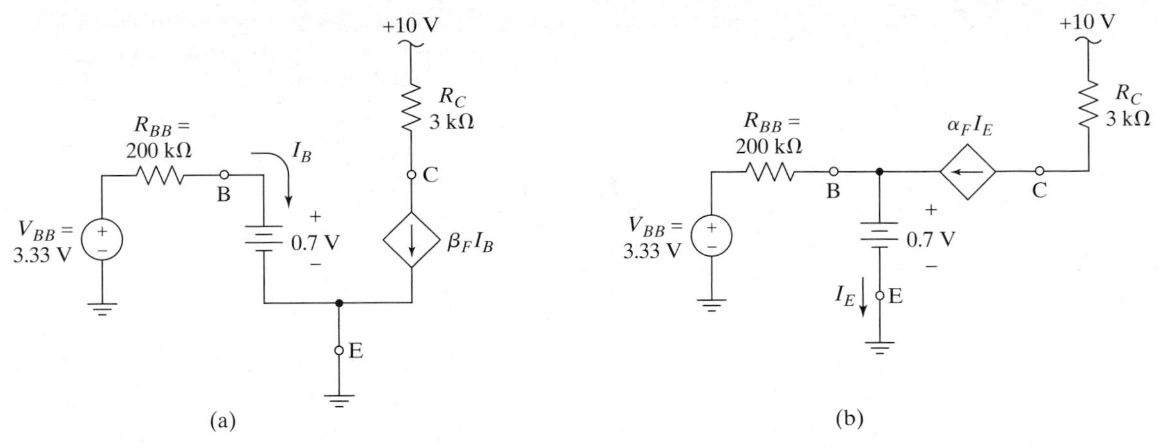

Figure 7–19 The circuit of Figure 7-18 with the transistor replaced by its model and the base circuit represented by its Thévenin equivalent. (a) Using the common-emitter model and (b) using the common-base model.

Now notice that the circuit is a common-emitter connection. Although our two models for the BJT are equivalent, one of them leads to a circuit that is easier to analyze. To ascertain that this statement is true, the circuits that result from replacing the transistor by both models are shown in Figure 7-19. (The Thévenin equivalent for the base circuit is also used.)

You are encouraged to solve the circuit shown in Figure 7-19(b), but will quickly discover that it is more difficult than the circuit in part (a) of the figure, due to that fact that you can't write a single loop equation to solve for the current. This is why it is convenient to have more than one model for the device.

Dealing with the circuit in Figure 7-19(a), we can now proceed with the solution in a straightforward manner. Examining the base-emitter loop, we see that

$$I_B = \frac{V_{BB} - 0.7}{R_{BB}} = 13.2 \quad \mu\text{A}.$$

Therefore, the collector current is $I_C = \beta_F I_B = 1.32$ mA. The collector voltage is then

$$V_C = V_{CC} - I_C R_C = 6 \text{ V}.$$

We can now check our initial assumption that the transistor is in forward-active operation. Since the base-emitter junction is forward biased and the base-collector junction is reverse biased, the device is in forward-active operation. ■

Example 7.5

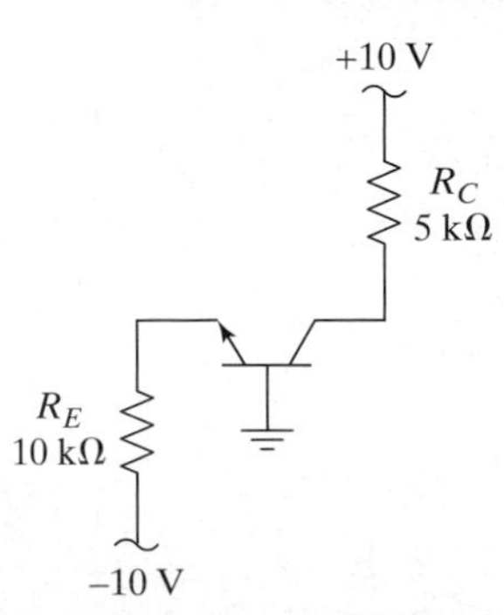

Figure 7–20 Circuit for Example 7.5.

Problem:
Find the currents and voltages in the circuit shown in Figure 7-20 by using the appropriate large-signal model for the BJT, assuming that it is forward active and has $\beta_F = 100$.

Solution:
This circuit is a common-base connection and will be easiest to analyze if we choose the common-base model. (You should draw out the circuit using the common-emitter model to convince yourself of this fact.) There isn't any need to simplify the circuitry external to the transistor, so we just plug in the model as shown in Figure 7-21 and solve.

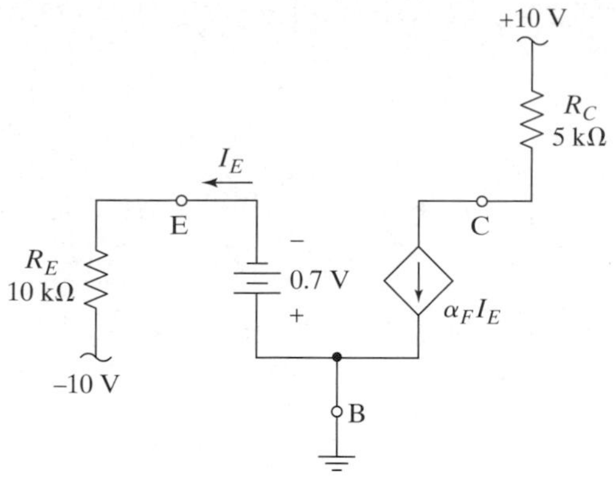

Figure 7–21 The circuit of Figure 7-20 with the transistor replaced by its common-base model.

Using (7.6) and the given value for β_F, the value of α_F is $\alpha_F = (100/101) = 0.99$. From the circuit we see that

$$I_E = \frac{10 - 0.7}{R_E} = 0.93 \text{ mA}.$$

Therefore, we have $I_C = 0.99(0.93 \text{ mA}) = 0.92$ mA, and the collector voltage turns out to be $V_C = 10 - I_C R_C = 5.4$ V. We should again check our initial assumption that the transistor is in forward-active operation. (It is.)

■

Exercise 7.3

If V_{BE} turned out to be 0.6 V in Example 7.4 instead of 0.7 V as assumed (a drop of 14%), how much lower than anticipated is I_B? Is the model a good one in this example?

□

The Cutoff Region of Operation As the base-emitter voltage of a BJT in forward-active operation is reduced, the collector current will decrease. The device will not technically be in cutoff until the base-emitter junction becomes reverse biased. However, because of the exponential dependence of the current on the voltage, the device will, for all intents and purposes, be in cutoff (sometimes just called *off*) whenever V_{BE} is substantially below 0.7 V. What constitutes "substantially below" depends on the currents in the circuit; but if the value of V_{BE} drops by 360 mV at room temperature, the collector current will decrease by about six orders of magnitude. For most circuits, this can certainly be considered to be off.

Therefore, assuming that the nominal V_{BE} is about 0.7 V, it is usually safe to say that a transistor with $V_{BE} \leq 0.3$ V is cut off. A similar argument applies to reverse operation, so we will adopt the convention (it is *only* a convention) that if *both* the base-emitter and base-collector junctions are either reverse biased, or forward biased by less than 0.3 V, the device is cut off.

When the device is cut off, the only currents that flow are the leakage currents associated with the junctions. The leakage currents are much larger in practice than is predicted from a first-order analysis, since they are dominated by surface leakage and other second-order effects. We could model these by current sources, as was done for the diode in Figure 7-6(a), but the resulting currents can almost always be ignored, and we will adopt the model shown in Figure 7-22 for a BJT operating in cutoff.

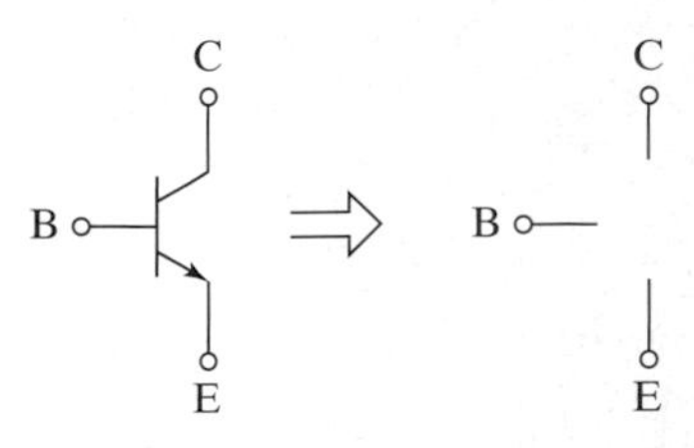

Figure 7–22 Model for a BJT in cutoff. This model is appropriate for either *npn* or *pnp* transistors.

The Saturation Region of Operation Saturation occurs when both *pn* junctions in a BJT become forward biased. To see how this can happen, consider the circuit shown in Figure 7-23.

Because of the direction of I_B, we know that the base-emitter junction is forward biased, and the transistor must be in either the forward-active or saturation region of operation. Let's start by supposing that R_C and I_B are chosen so that the transistor is forward active. What happens if we increase I_B? The collector current will initially increase in proportion, as dictated by (7.4). As I_C increases, the drop across R_C increases, and the voltage at the collector decreases. Eventually, the collector voltage will drop below V_B, the base-collector junction will become forward biased, and the transistor will be saturated.

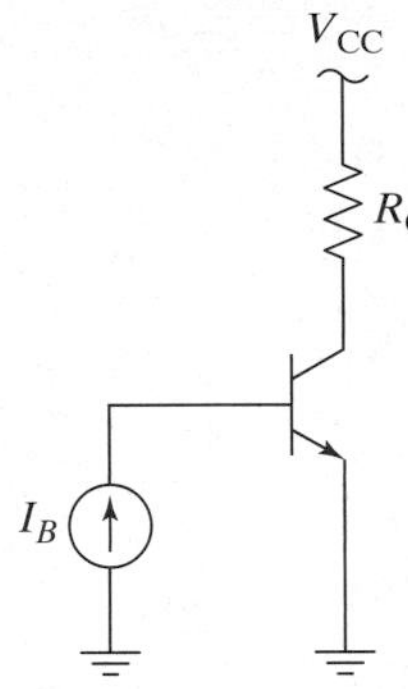

The collector voltage cannot continue to drop indefinitely, however. Certainly, it can't drop below ground. Therefore, the collector current can never be larger than V_{CC}/R_C, no matter how large I_B is. In fact, the collector current *saturates* before this value is reached and will not increase in proportion to I_B. In other words, the collector current is no longer controlled by I_B, but depends on the external circuit.

We can see this loss of control by noticing that all of the curves in Figure 7-15 collapse into a single line in the saturation region. The value of V_{CE} in saturation is denoted by V_{CEsat} and is a function of how hard we drive the transistor and the characteristics of the transistor. The value of V_{CEsat} is derived in Chapter 4, but for our present purposes, it is sufficient to note that it is typically about 0.2 V or 0.3 V.

The transistor is technically saturated as soon as the base-collector junction is forward biased (*see* Chapter 2). Nevertheless, even though this definition is precise, it is not very useful for design, since no real change in the performance of the device takes place when the base-collector junction goes from being slightly reverse biased to slightly forward biased. Therefore, we are forced to live with a more nebulous definition. We will assume that a BJT is saturated when V_{CE} drops to 0.2 V. The resulting model for a saturated *npn* BJT is shown in Figure 7-24.

V_{CC}
R_C
I_B

Figure 7–23 A BJT driven with a base current.

Exercise 7.4

Determine the maximum value that can be used for R_C in Figure 7-23 without saturating the transistor. Assume that $V_{CC} = 10$ V, $I_B = 15$ μA, and $\beta_F = 100$.

The Breakdown Region of Operation If a BJT enters breakdown, the appropriate model depends on which junction has broken down and in exactly what way. If the device is cut off and the reverse bias on either junction exceeds the breakdown voltage, either junction can be broken down just like a *pn*-junction diode, and the appropriate model is a battery of the proper polarity and value used in place of the junction that has broken down. The breakdown voltages in this case are the voltages with the third terminal open circuited (as denoted by the third subscript): BV_{EBO} and BV_{CBO} for the emitter-base and collector-base junctions, respectively. The appropriate models are shown in Figure 7-25.

Figure 7–24 The large-signal model for an *npn* transistor in saturation.

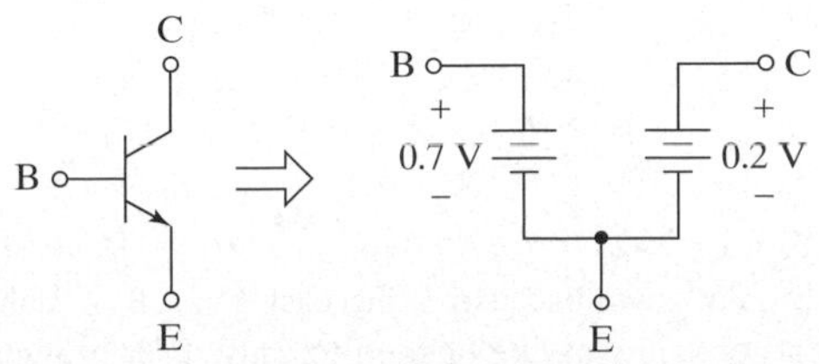

C
B
E
B
C
+
+
0.7 V
0.2 V
−
−
E

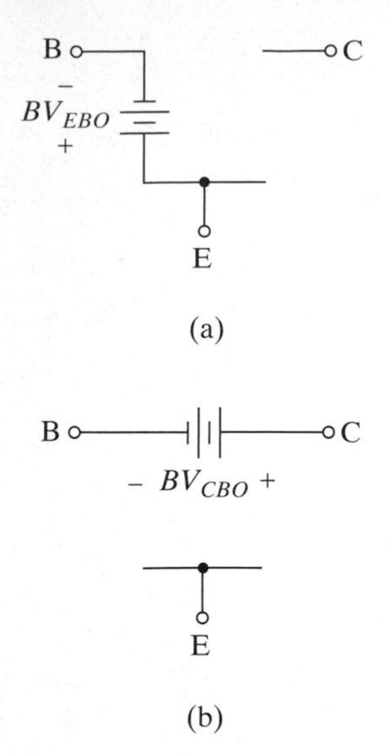

Figure 7–25 Models for an *npn* transistor with (a) the base-emitter junction broken down and (b) the base-collector junction broken down.

For the case where the transistor is forward active and the reverse bias on the base-collector junction exceeds the breakdown voltage, we need to consider the common-base and common-emitter circuits separately. Consider the common-base circuit first, as shown in Figure 7-26(a). In this case, the voltage at which the collector-base junction will break down depends on the value of I_E, as explained in Chapter 2. This voltage can never be greater than BV_{CBO}. The model that should be used in breakdown is shown in part (b) of the figure and the exact breakdown voltage would need to be calculated for the given I_E.

Now consider the common-emitter circuit of Figure 7-27(a). The voltage at which the collector-base junction will break down again depends on the current, this time set by I_B, as explained in Chapter 2. The largest breakdown voltage that can be sustained in this case is BV_{CEO}, but a voltage larger than that may be required to cause the breakdown if I_B is very small or zero. The model that should be used in the breakdown region is again shown in part (b) of the figure, and the exact breakdown voltage would again need to be calculated for each value of I_B.

Models for the* pnp *Transistor The large-signal low-frequency models we use for an *npn* transistor in forward-active operation, cutoff, and saturation are given in Figures 7-17, 7-22, and 7-24, respectively. The corresponding models for a *pnp* transistor are shown in Figure 7-28. The *pnp* models are obtained by reversing the polarities of all voltages and currents in the *npn* models.

The models used for *npn* transistors in breakdown are shown in Figures 7-25, 7-26(b), and 7-27(b) for the different types of breakdown possible. The corresponding models for the *pnp* transistor are shown in Figure 7-29. In this figure, BV_{BC} is a function of I_E, BV_{EC} is a function of I_B, and V_{EE} is a negative supply voltage. BV_{BC} is less than BV_{BCO} for any nonzero I_E. Similarly, BV_{EC} is less than BV_{ECO}; although a larger value may be needed to cause breakdown, it cannot be sustained.

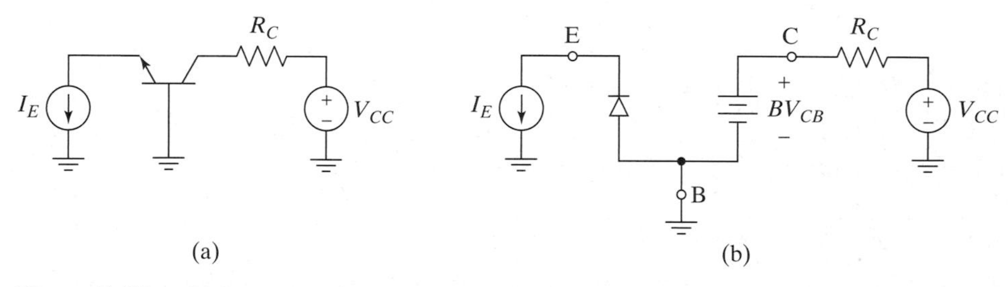

Figure 7–26 (a) A common-base circuit and (b) the circuit with the transistor modeled in the region where the collector-base junction breaks down. The breakdown voltage BV_{CB} is less than BV_{CBO} for any nonzero I_E.

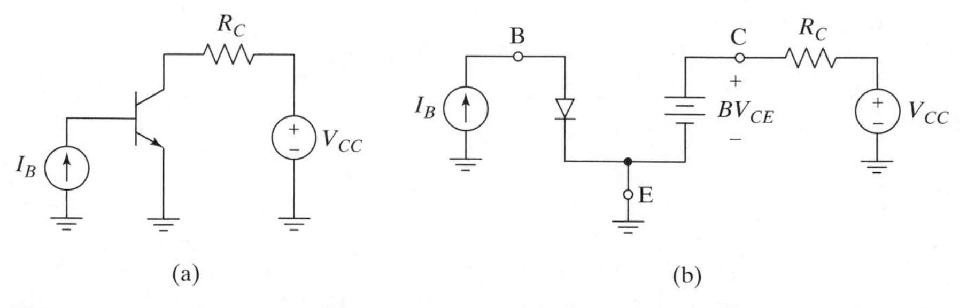

Figure 7–27 (a) A common-emitter circuit and (b) the model to be used when the transistor breaks down because of increasing V_{CC}. B_{VCE} is always less than or equal to BV_{CEO} (although a larger value may be needed to cause breakdown if $I_B \approx 0$, the larger value cannot be sustained).

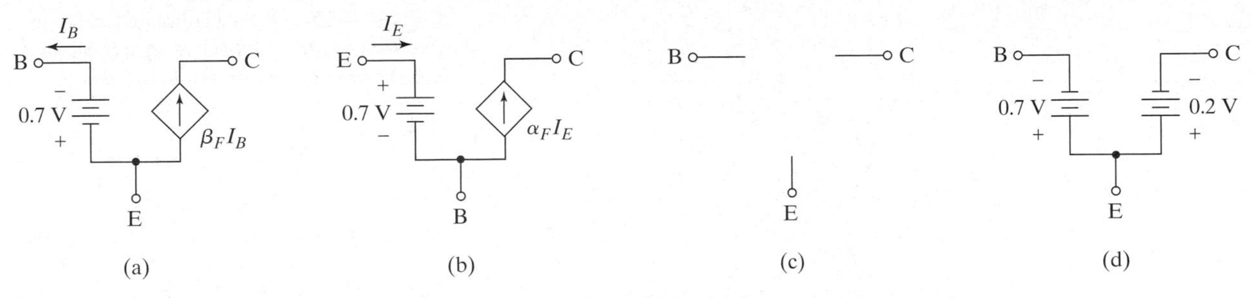

Figure 7–28 The large-signal low-frequency models for a *pnp* BJT in (a) forward-active operation in a common-emitter connection, (b) forward-active operation in a common-base connection, (c) cutoff, and (d) saturation.

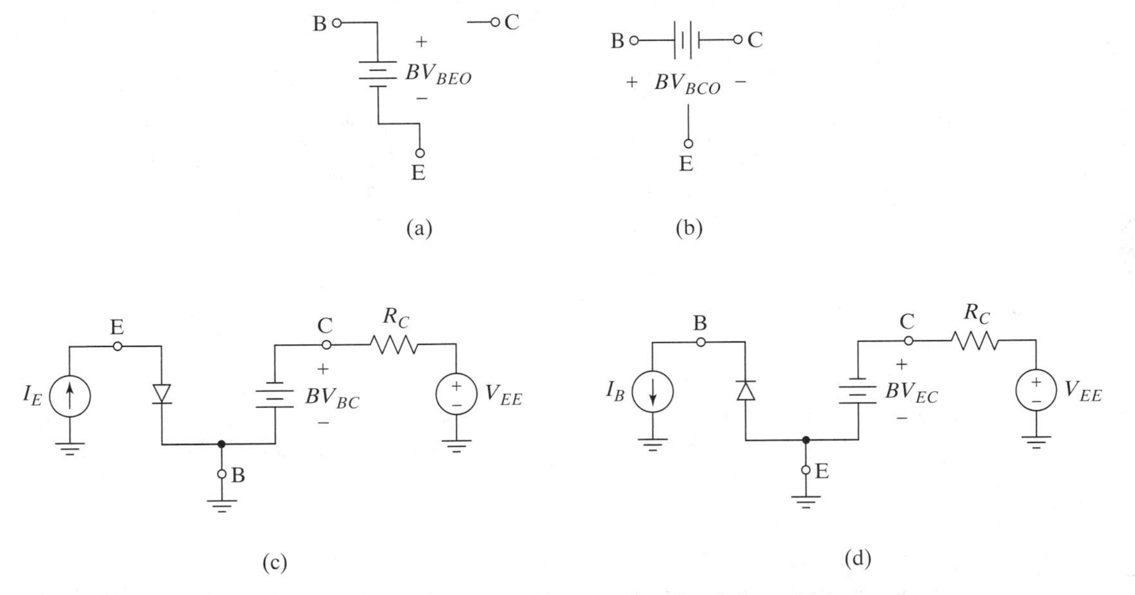

Figure 7–29 The breakdown models for a *pnp* BJT: (a) base-emitter breakdown, (b) base-collector breakdown when cutoff, (c) base-collector breakdown when operating in a common-base circuit, and (d) base-collector breakdown when operating in a common-emitter circuit. (V_{EE} is a negative power supply.)

7.1.5 MOS Field-Effect Transistors

In this section, we derive the large-signal low-frequency models for an n-channel enhancement-mode MOSFET. The models for a depletion-mode device will be the same, except for the value of the threshold voltage. The corresponding models for p-channel devices will be presented at the end of the section. Since we are interested only in low-frequency models here, we ignore all charge storage mechanisms. We also assume that the body effect has been accounted for by changing the value of the threshold voltage in accordance with (2.120).

There are three regions of operation we are concerned with: forward-active operation (i.e., saturation), the linear or voltage-controlled resistance (VCR) region,[4] and cutoff. The regions of operation are shown in Figure 7-30.

The Forward-Active Region of Operation Ignoring channel-length modulation and assuming that the device is in strong inversion, the drain current of an n-channel MOSFET in the forward active region is given by

[4] The linear regime is also sometimes referred to as the triode region, since the characteristics in this region are similar to those of a vacuum tube triode.

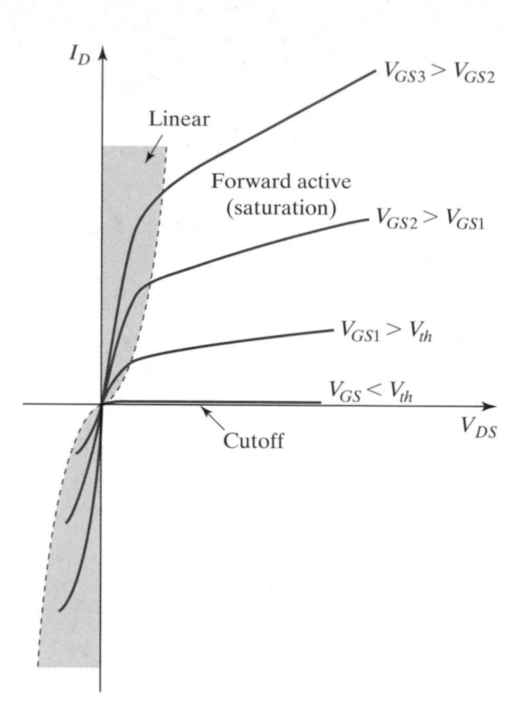

Figure 7–30 MOSFET drain characteristics, showing forward-active operation (i.e., saturation), the linear region, and cutoff.

$$I_D = \frac{\mu_n C'_{ox}}{2}\frac{W}{L}(V_{GS} - V_{th})^2, \tag{7.7}$$

where μ_n is the electron mobility at the surface, C'_{ox} is the oxide capacitance per unit area, V_{th} is the threshold voltage, and W and L are the width and length of the channel, respectively. If channel-length modulation is included, (7.7) is modified to

$$I_D = \frac{\mu_n C'_{ox}}{2}\frac{W}{L}(V_{GS} - V_{th})^2(1 + \lambda V_{DS}), \tag{7.8}$$

where λ is the channel-length modulation parameter presented in Chapter 2. We will ignore channel-length modulation for the rest of this discussion and also will ignore it in our analyses unless it is needed. Ignoring channel-length modulation implies that the curves in Figure 7-30 are approximately flat in the forward-active region of operation.

The value of V_{th} in these equations is a function of the source-to-bulk voltage, V_{SB}, as discussed in Chapter 2, and it is sometimes necessary to include this effect. When the bulk connection is included, the MOSFET is a four-terminal device. We will assume that the bulk connection can be safely ignored when it isn't shown.

The large-signal low-frequency model for a MOSFET is shown in Figure 7-31, where we have defined the constant K' in the usual way:

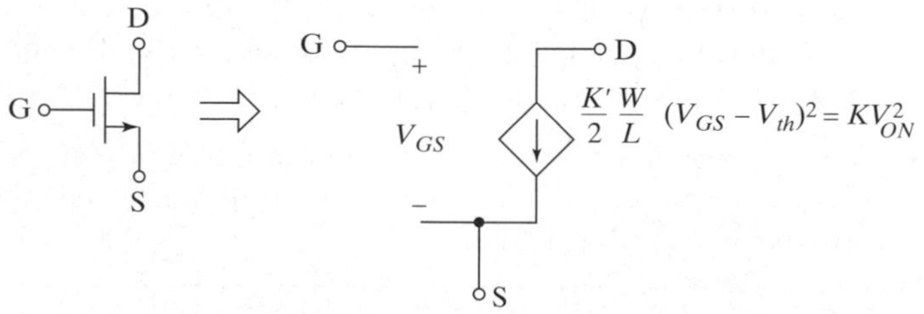

D G S ⇒ G + V_{GS} − S D $\frac{K'}{2}\frac{W}{L}(V_{GS} - V_{th})^2 = KV_{ON}^2$

Figure 7–31 A large-signal low-frequency model for an *n*-channel MOSFET in saturation.

$$K' \equiv \mu_n C'_{ox}. \tag{7.9}$$

Typical values for K' are in the range from 1.5×10^{-5} to 3.0×10^{-5} amperes per volt squared. Another common simplification is to define a new constant

$$K \equiv \frac{K'}{2}\frac{W}{L} = \frac{\mu_n C'_{ox}}{2}\frac{W}{L}, \tag{7.10}$$

which leads to

$$I_D = K(V_{GS} - V_{th})^2. \tag{7.11}$$

Because the term in parentheses in (7.11), $V_{GS} - V_{th}$, shows up frequently and is important in many equations involving MOSFET circuits and model parameters, it is given a special name, the **gate overdrive**. In other words, it is the amount by which the gate is driven beyond the threshold voltage required to turn the device on. Therefore, we will denote the gate overdrive by V_{ON}. It is also sometimes called the *effective* gate-to-source voltage [7.2] and is variously denoted by V_{eff} or ΔV.

For this model to be valid, the device must be in forward-active operation as stated prior to (7.7). In other words, we must have

$$V_{GS} > V_{th} \tag{7.12}$$

and

$$V_{DS} > (V_{GS} - V_{th}), \tag{7.13}$$

as was explained in Chapter 2. The minimum drain-to-source voltage required to keep a FET forward active (i.e., in saturation), as given by (7.13), is often called

$$V_{DS\text{sat}} = V_{GS} - V_{th} = V_{ON}. \tag{7.14}$$

It is also useful to rewrite (7.11) to obtain an equation for the gate-to-source voltage of a MOSFET. The result is[5]

$$V_{GS} = V_{th} + \sqrt{\frac{I_D}{K}} = V_{th} + V_{ON}. \tag{7.15}$$

Example 7.6

Problem:

Find the voltages and currents in the circuit shown in Figure 7-32. Assume that $K' = 2.0 \times 10^{-5}$ A/V², $W/L = 5$, and $V_{th} = 1$ V.

Solution:

We first simplify the circuit by finding the Thévenin equivalent for the circuitry driving the gate of the MOSFET. Because this Thévenin equivalent represents the DC biasing circuit driving the gate, we will use the double subscript notation that is used for power supplies; the voltage is V_{GG} and the resistance is R_{GG}. We have

$$V_{GG} = \frac{R_2}{R_2 + R_1} V_{DD} = 5\text{ V}.$$

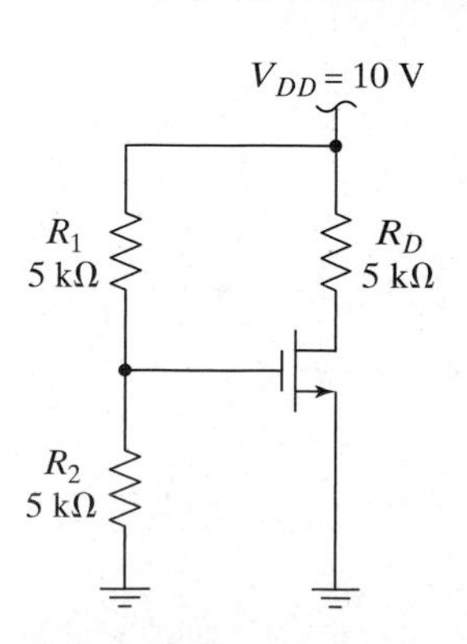

Figure 7–32 Circuit for Example 7.6.

---

[5] Note that we can choose the positive root because of (7.12).

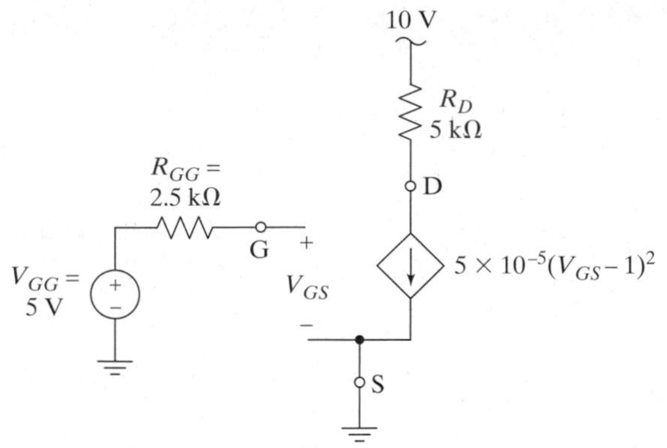

Figure 7–33 The circuit from Figure 7-32 simplified and with the MOSFET replaced by its large-signal model.

The Thévenin resistance is the parallel combination of R_1 and R_2 and is denoted by R_{GG}. (V_{DD} is an independent source and is set to zero to find the resistance. Doing so shorts the top of R_1 to ground.) Figure 7-33 shows the simplified circuit drawn using the model for the MOSFET. We are assuming that the MOSFET is forward active, but we'll come back and check this assumption at the end.

From this simplified circuit, we see that $V_{GS} = V_{GG} = 5$ V; therefore, the drain current is 0.8 mA. The resulting drain voltage is $V_{DD} - I_D R_D = 6$ V. We can now check whether or not the device is forward active as assumed. Since $V_{GS} > V_{th}$ and $V_{DS} > (V_{GS} - V_{th})$, the assumption was correct, and we are done with the analysis.

■

Example 7.7

Problem:

Find the voltages and currents in the circuit shown in Figure 7-34. Assume that $K' = 2.0 \times 10^{-5}$ A/V^2, $W/L = 5$, and $V_{th} = 1$ V.

Solution:

The Thévenin equivalent driving the gate of the MOSFET is the same as in Example 7.6, so the circuit can be redrawn as shown in Figure 7-35. We again assume the device is forward active.

Because of the feedback introduced by R_S (i.e., the input to the transistor, V_{GS}, now depends on its output, I_D), this circuit is more difficult to analyze than the one in Example 7.6. We note that

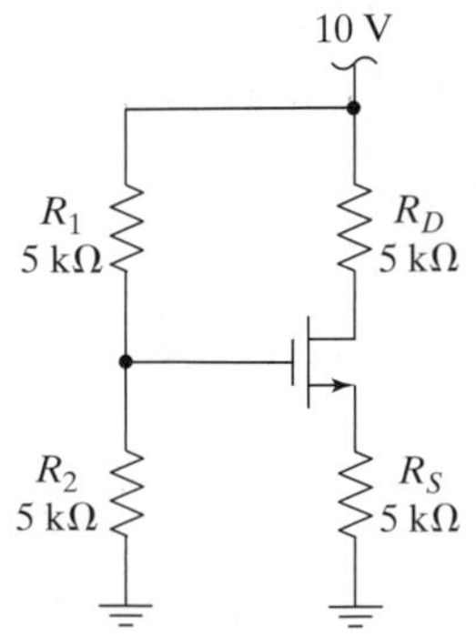

Figure 7–34 Circuit for Example 7.7.

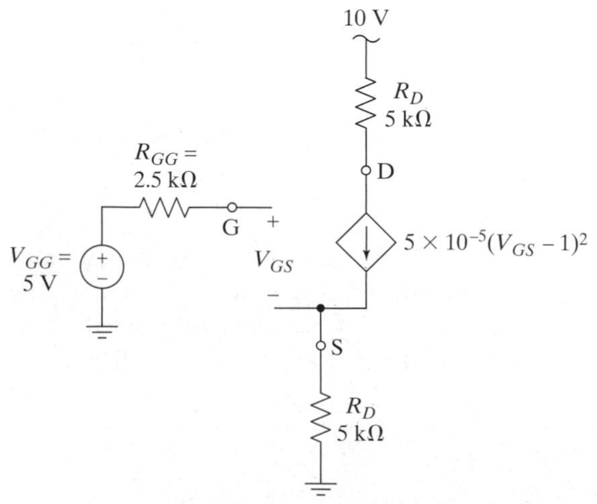

Figure 7–35 Large-signal DC equivalent of the circuit in Figure 7-34.

$$I_D = K(V_{GS} - V_{th})^2 = K(V_G - I_D R_S - V_{th})^2$$

and expand the square to get

$$I_D = K(V_G - V_{th})^2 - 2KI_D R_S(V_G - V_{th}) + KI_D^2 R_S^2.$$

We can rewrite this equation as a quadratic in standard form, namely,

$$KR_S^2 I_D^2 - [1 + 2KR_S(V_G - V_{th})]I_D + K(V_G - V_{th})^2 = 0$$

and solve using the quadratic formula to get

$$I_D = \frac{[1 + 2KR_S(V_G - V_{th})] \pm \sqrt{[1 + 2KR_S(V_G - V_{th})]^2 - 4K^2R_S^2(V_G - V_{th})^2}}{2KR_S^2}. \quad \textbf{(7.16)}$$

Plugging in the values for the present example leads to $I_D = 2.1$ mA or 0.3 mA. By examining the circuit, we recognize that the 2.1-mA solution cannot be correct, since it would lead to $V_S > 10$ V. Therefore, the solution must be 0.3 mA.

It is always possible to discard one of the two solutions arrived at using (7.16) by examining the original circuit and plugging in the values. Using $I_D = 0.3$ mA leads to $V_S = 1.5$ V, $V_{GS} = 3.5$ V, and $V_D = 8.5$ V. These values are all consistent with our assumption that the transistor is forward active. ■

Exercise 7.5

What is the maximum value of drain resistance that can be used in the circuit of Figure 7-32 while keeping the MOSFET in forward-active operation? □

The Linear Region of Operation If the drain resistor in Figure 7-32 is too large, the drain voltage of the MOSFET will drop far enough that the gate-to-channel voltage at the drain end of the channel is larger than V_{th} (i.e., $V_{GD} = V_G - V_D = V_{GS} - V_{DS} \geq V_{th}$). For drain voltages less than or equal to this value (i.e., $V_{DS} \leq (V_{GS} - V_{th})$), the inversion layer will extend all the way from the drain to the source and the device will be in the linear region. In this region of operation, the drain current is given by

$$I_D = \mu_n C'_{ox} \frac{W}{L}\left[(V_{GS} - V_{th})V_{DS} - \frac{V_{DS}^2}{2}\right], \quad \textbf{(7.17)}$$

as shown in Chapter 2. We can model the MOSFET in this region of operation by a nonlinear resistance. For small values of V_{DS}, the squared term in (7.17) can be ignored without significant error, and the MOSFET operates like a linear voltage-controlled resistor, which is why the region is called the *linear* region of operation. This approximation can also be seen in the curves in Figure 7-30; the curves are approximately straight lines going through the origin for small values of V_{DS}. The slope of each line is different and is a function of V_{GS}. We can derive the drain-to-source resistance of a MOSFET for small V_{DS} by ignoring the squared term in (7.17) and taking the ratio of V_{DS} to I_D:

$$R_{DS} \equiv \left.\frac{V_{DS}}{I_D}\right|_{\text{small } V_{DS}} = \frac{1}{\mu_n C'_{ox} \frac{W}{L}(V_{GS} - V_{th})}$$

$$= \frac{1}{2K(V_{GS} - V_{th})} = \frac{1}{2KV_{ON}}. \quad \textbf{(7.18)}$$

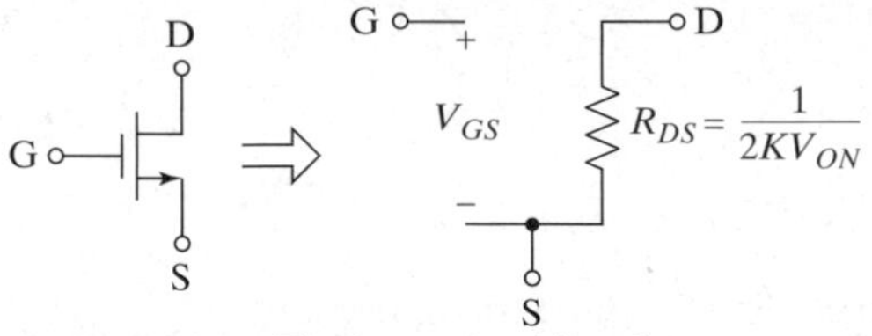

Figure 7–36 The large-signal low-frequency model for a MOSFET in the linear region of operation. The resistance is given by (7.18) for small V_{DS}.

The large-signal low-frequency model for the MOSFET in the linear region of operation is shown in Figure 7-36 where the resistance is given by (7.18). The same model will work in the nonlinear part of the region (i.e., when V_{DS} is not small) if the resistance is nonlinear and implicitly given by (7.17).

Since the MOSFET does a good job of approximating a linear voltage-controlled resistor for small values of drain-to-source voltage, it is sometimes used in this capacity in linear circuits.

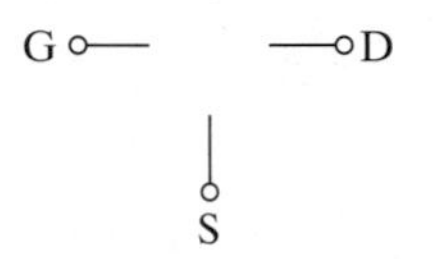

Figure 7–37 The low-frequency model for a MOSFET when cut off.

The Cutoff Region of Operation If the gate-to-source voltage of a MOSFET drops below V_{th}, the device is no longer in strong inversion. Although current can flow in this weak inversion region, we will use a simplistic model and assume that whenever $V_{GS} < V_{th}$ the device is in cutoff. (We frequently say the device is *off* rather than cutoff.) If the gate-to-channel voltage at either end of the channel exceeds the threshold, the device is in either the linear or forward-active region of operation. If the device is in the forward-active region of operation, the end of the channel that has the inversion layer present is acting as the source terminal.[6] When a MOSFET is cut off, the low-frequency model is simply an open circuit, as shown in Figure 7-37.

Example 7.8

Problem:
Solve for V_O in the circuit in Figure 7-38 when $V_{I_1} = 1$ V and $V_{I_2} = 1.5$ V. Assume that both FETs have $K = 10$ mA/V^2 and $V_{th} = 1.5$ V. Solve when (a) $V_{G1} = 5$ V and $V_{G2} = 0$ V and (b) $V_{G1} = 0$ V and $V_{G2} = 5$ V. The bulk connections have been shown to illustrate that the bulk must be tied to the most negative potential anywhere in the circuit (for an n-channel device) to prevent the source-to-bulk and drain-to-bulk junctions from becoming forward biased. In this example, we will assume that V_{SB} has already been accounted for in the value of V_{th} (although the magnitude of the term varies with the source voltage, so it can't be accounted for accurately until we know the source voltage). We will ignore that complication here. (*See* Problem P7.20.)

Solution:
To solve this problem, we must first determine what regions of operation the devices are in, and then replace them by the appropriate models, and perform the analysis. We must then check to see that we made the correct choices in determining the

[6] If we have defined the source and drain ends of the device externally, it is possible to operate a MOSFET backwards in what could be called the reverse-active region of operation. Nevertheless, the common usage is to simply redefine which terminal is the source and still refer to the device as being forward active or in saturation. Of course, this terminology will not work if the device is not symmetric.

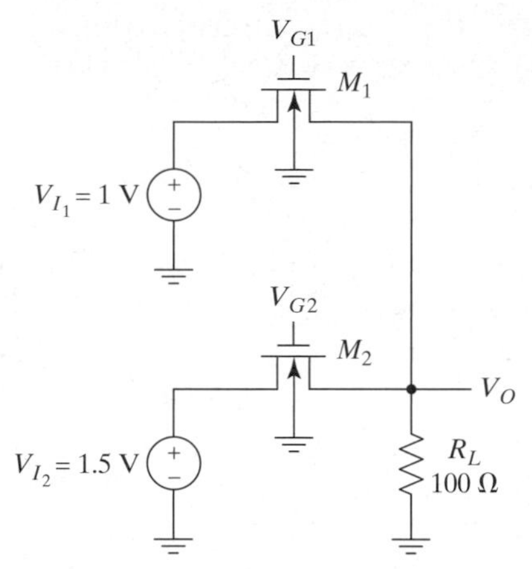

Figure 7–38 Circuit for Example 7.8.

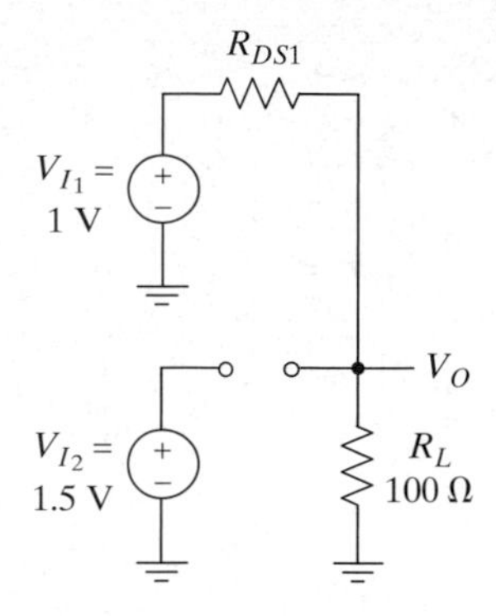

Figure 7–39 The circuit of Figure 7-38 with the transistors replaced by the models appropriate for part (a) of the problem—M_1 is in the linear region and M_2 is cut off.

regions of operation. Notice that both inputs are positive; therefore, whenever the gate voltage of one of the FETs is set to zero, there is no chance of having $V_{GS} > V_{th}$, and the device must be off. When the gate is set to 5 V, however, it is more than V_{th} above every other voltage in the circuit, so we can safely assume the device is in the linear region. Using these initial observations, we replace the transistors with the appropriate models to solve part (a) (M_1 in linear region and M_2 in cutoff) and show the resulting circuit in Figure 7-39.

The value of R_{DS} can be determined from (7.18) if we assume that V_{DS} is small. For small V_{DS}, it doesn't matter which terminal is the source, so we assume that the source is connected to V_{I_1} and find that $V_{GS_1} = 4$ V. The resulting value of resistance is $R_{DS} = 20\ \Omega$. Using this value we can solve for V_O to get

$$V_O = \frac{R_L}{R_L + R_{DS_1}} V_{I_1} = 0.83 \text{ V}.$$

Therefore, our assumption that V_{DS} is small is acceptable. We can also now check to see that M_1 is in the linear region and that M_2 is cut off. (They are.)

Figure 7-40 shows the equivalent circuit for part (b) of the problem. The solution is arrived at in exactly the same manner, and the result is that $R_{DS2} = 25\ \Omega$ and $V_O = 1.20$ V.

Notice that the transistors in this example are functioning as voltage-controlled switches. This is a common application for a MOSFET.

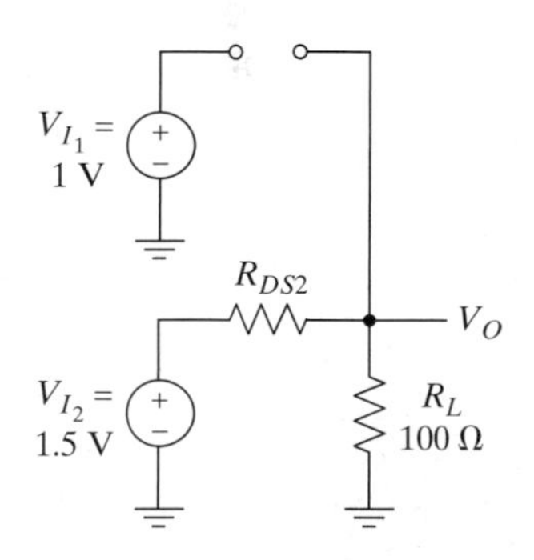

Figure 7–40 The circuit of Figure 7-38 with the transistors replaced by the models appropriate for part (b) of the problem—M_1 is cut off and M_2 is in the linear region.

■

The Breakdown Region of Operation The breakdown mechanisms of MOSFETs are discussed in Chapter 2. We will ignore the breakdown of the gate oxide in this section, since it is a destructive mechanism, and we do not, therefore, typically want to analyze circuits when the oxide is broken down. We also typically avoid breaking down the source-substrate and drain-substrate *pn* junctions, so we won't model that region of operation here either (although it can be modeled by adding two *pn*-junction diode models to the standard MOSFET models). The remaining breakdown mechanism is the drain-to-source breakdown. When drain-to-source breakdown occurs, the MOSFET can be modeled as shown in Figure 7-41. The breakdown voltage, BV_{DS}, depends on the details of the device fabrication.

P-Channel and Depletion-Mode MOSFETs The large-signal DC models for *p*-channel MOSFETs are the same as those of *n*-channel devices, except that the

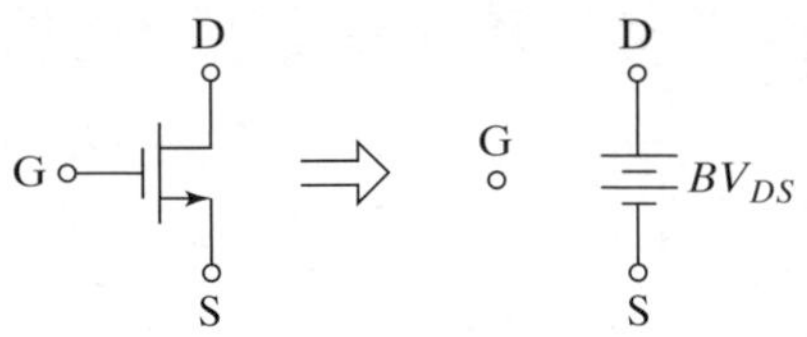

Figure 7–41 The model for a MOSFET when the drain-to-source breakdown voltage (BV_{DS}) has been reached.

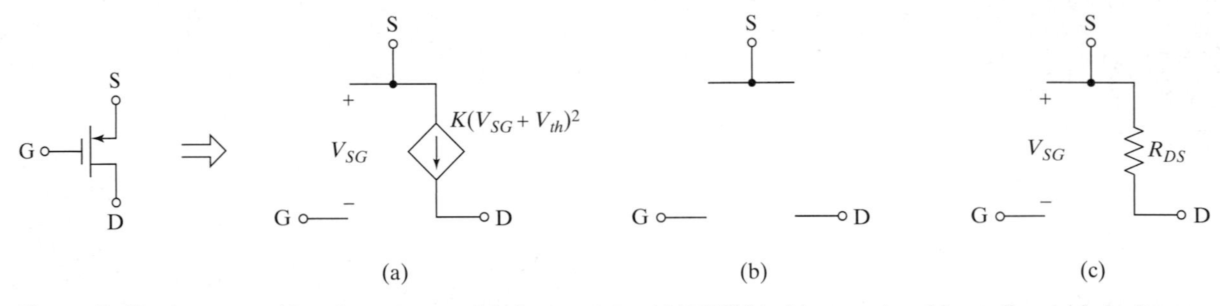

Figure 7–42 Large-signal low-frequency models for a *p*-channel MOSFET in (a) saturation, (b) cutoff, and (c) the linear region.

polarities of V_{GS}, V_{th}, and I_D are all opposite. Therefore, for *p*-channel devices, we define the drain current to be positive when flowing *out* of the device. The models are shown in Figure 7-42. The drain current in saturation is given by

$$I_D = \frac{\mu_p C'_{ox}}{2}\frac{W}{L}\left(-V_{GS} + V_{th}\right)^2$$

$$= \frac{K'}{2}\frac{W}{L}\left(V_{SG} + V_{th}\right)^2 = K(V_{SG} + V_{th})^2 = KV_{ON}^2, \tag{7.19}$$

where we use V_{SG} rather than $-V_{GS,}$ and the threshold voltage is negative. The device is in saturation so long as $V_{SD} > V_{SG} + V_{th}$ and $V_{SG} > |V_{th}|$.

Most books use V_{GS} instead of V_{SG} in this equation. The reason we have departed from the standard treatment is that most circuit designers *think* about the voltage from source to gate for a *p*-channel MOSFET.[7] The signs can get confusing, so rather than trying to memorize them, it is *strongly* recommended that you develop an understanding of how the device works, so that you can figure out the proper signs. For example, we know that the gate has to be more than $|V_{th}|$ *below* the source to turn on an enhancement-mode PMOS transistor. Although we do not recommend

Table 7.1 Signs for V_{GS} and V_{th} for *n*- and *p*-Channel Enhancement-and Depletion-Mode MOSFETs

	Enhancement mode		Depletion mode	
	n-channel	*p*-channel	*n*-channel	*p*-channel
V_{GS} (in sat.)	+	–	+ or –	+ or –
V_{th}	+	–	–	+

[7] We are tempted to change the sign of V_{th} for *p*-channel devices as well, but since threshold voltage is a frequently quoted parameter (and is used as a parameter in SPICE) and is *defined* to be negative for enhancement-mode PMOS devices, changing the sign on it would do you a disservice.

memorizing the signs, Table 7.1 presents them for V_{GS} and V_{th} for n- and p-channel enhancement- and depletion-mode MOSFETs.

Similarly, (7.17) is modified so that the drain current in the linear region is

$$I_D = \mu_p C'_{ox} \frac{W}{L}\left[(V_{SG} + V_{th})V_{SD} - \frac{V_{SD}^2}{2}\right], \tag{7.20}$$

where we note that V_{SD} has been used in place of V_{DS}, so that I_D is positive when flowing out of the drain. For small V_{SD}, we ignore the squared term in (7.20) as before and get

$$R_{DS} = \frac{1}{\mu_p C'_{ox} \frac{W}{L}(V_{SG} + V_{th})} = \frac{1}{2KV_{ON}}, \tag{7.21}$$

which is a modified version of (7.18).

The model for a PMOS transistor in cutoff is the same as for an NMOS transistor; all three terminals are open circuits. This model is accurate only as long as the terminal voltages on the device do not exceed the breakdown voltages.

The models for depletion-mode devices are the same as those for enhancement-mode devices, except for the sign of V_{th} (note, however, that many texts and data sheets supply parameters for depletion-mode MOSFETs using the notation we reserve for JFETs). The proper signs are given in Table 7.1.

7.1.6 Junction Field-Effect Transistors

JFETs are depletion-mode devices, and their large-signal low-frequency models are identical to the models for depletion-mode MOSFETs, except for the drain current equations. When operated in the linear region, JFETs are modeled as shown in Figure 7-36, but rather than using (7.18), the resistance is given by

$$R_{DS} \approx \frac{1}{G_0\left(1 - \frac{1}{X_{ch}}\sqrt{\frac{2\epsilon(V_0 - V_{GS})}{qN_D(1 + N_D/N_A)}}\right)}, \tag{7.22}$$

where

$$G_0 = q\mu_n N_D X_{ch} \frac{W}{L} \tag{7.23}$$

and all of the other terms are defined in Chapter 2. An n-channel JFET is in the linear region when $V_{GS} > V_p$ and $V_{DS} < V_{GS} - V_p$ (V_p is negative for n-channel JFETs).

When operated in saturation ($V_{GS} > V_p$ and $V_{DS} \geq V_{GS} - V_p$ for an n-channel JFET), JFETs are modeled as shown in Figure 7-43, which is the same model as Figure 7-31,

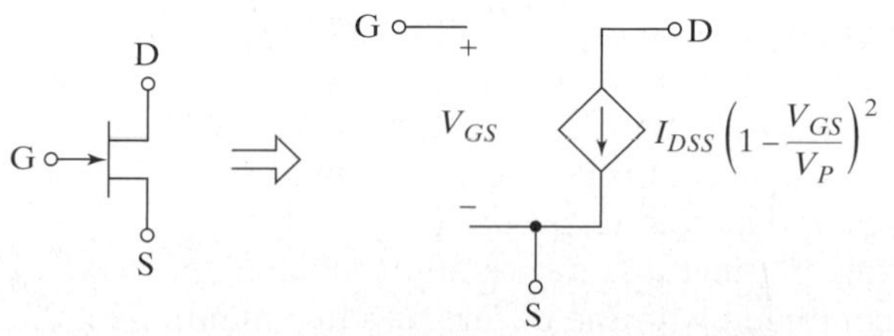

G D S ⇒ G D + V_{GS} − S $I_{DSS}\left(1 - \frac{V_{GS}}{V_P}\right)^2$

Figure 7–43 A large-signal low-frequency model for an n-channel JFET in saturation.

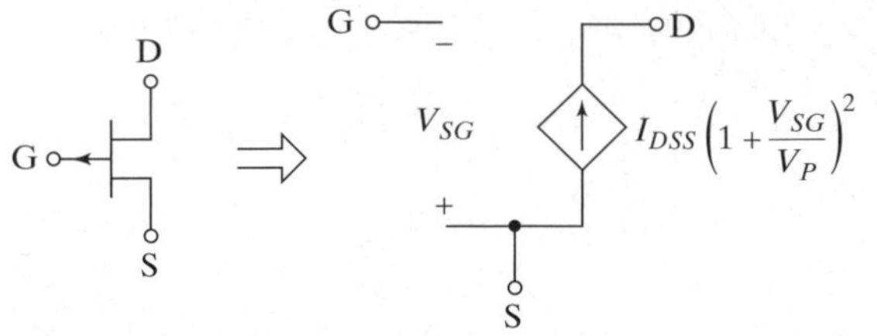

Figure 7–44 A large-signal low-frequency model for a *p*-channel JFET in saturation.

except for the form of the drain current equation. Although the form is different, the equation is the same as (7.11) for a MOSFET. Note that

$$I_D = I_{DSS}(1 - V_{GS}/V_p)^2 = \frac{I_{DSS}}{V_p^2}(V_{GS} - V_p)^2, \tag{7.24}$$

which is the same as (7.11) if we set $V_p = V_{th}$ and $K = I_{DSS}/V_p^2$. The pinch-off voltage is often denoted as $V_{GS(\text{off})}$ on data sheets for discrete JFETs.

The polarities of all terminal voltages and currents are reversed for *p*-channel JFETs. The resulting model is shown in Figure 7-44.

Example 7.9

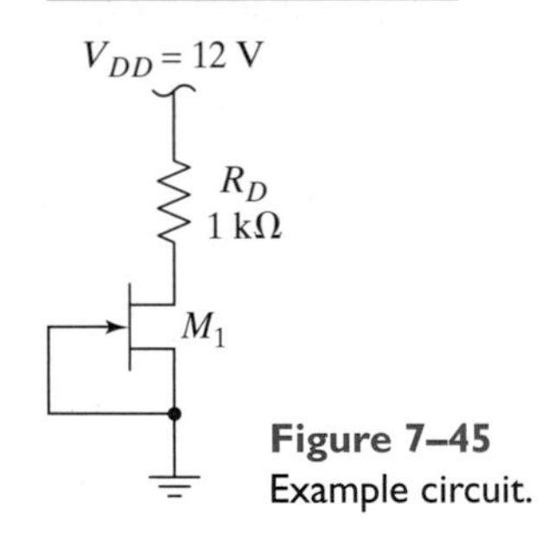

Figure 7–45 Example circuit.

Problem:

Determine the values of I_D and V_{DS} for the JFET circuit shown in Figure 7-45, given that $I_{DSS} = 5$ mA and $V_p = -3$ V.

Solution:

We first assume that the device is in the forward-active region of operation and then check our assumption. From the figure, we see that $V_{GS} = 0$, so using (7.24), we see that $I_D = I_{DSS} = 5$ mA. The drain voltage is then given by $V_D = V_{DD} - I_D R_D = 7$ V, and we also have $V_{DS} = V_D = 7$ V. We now check and see that $V_{GS} > V_p$ and $V_{DS} \geq V_{GS} - V_p$, so the transistor is forward active, as assumed.

7.1.7 Comparison of Bipolar and Field-Effect Transistor Biasing

The DC biasing of bipolar and field-effect transistors is quite different due to the very different large-signal characteristics of the devices. Because BJTs have an exponential relationship between the base-emitter voltage and the collector current, the value of V_{BE} does not vary much. This allows us to assume that it is about 0.7 V and proceed with our analysis. Combining the approximately constant V_{BE} with the fact that the BJT can be represented as a current-controlled device allows us to have a linear large-signal model for the device. The simplification of using $V_{BE} \approx 0.7$ V is usually justified and yields surprisingly accurate results.

In the case of the FET, however, the current is only proportional to the square of the gate overdrive, and, as a result, the value of V_{GS} can vary over a significant range. Therefore, we cannot approximate V_{GS} in advance by some "standard" value. Also, unlike the current-controlled model of the BJT, there is no transfer function for which the large-signal operation of FETs appears linear. Therefore, we end up solving quadratic equations to determine the bias points of FET circuits. However, the good news is that the gate of a FET appears as an open circuit at low frequencies, and this simplifies analysis. FETs also can be made as depletion-mode devices, and there isn't any bipolar counterpart to these.

The variations in bipolar device parameters are usually smaller than those of FET devices. When this fact is combined with the differences just mentioned, the result is that it is somewhat easier to establish accurate bias points in bipolar circuits than it is in FET circuits. Nevertheless, FETs have unique characteristics that make them superior to BJTs in some applications (e.g., high input impedance). We will compare these devices more in later chapters.

7.2 BIASING OF SINGLE-STAGE AMPLIFIERS

The small-signal performance of a circuit depends on the operating point; therefore, our biasing circuitry should provide an operating point that is predictable and stable with respect to variations in parameters that are either poorly controlled, or likely to change with temperature or time. In addition, we would like the biasing circuitry to be as simple as possible and not to consume any more power than necessary. Using these guidelines, we develop some rules in this section that allow us to design the biasing circuits for a given application. This section and the next one on multistage amplifiers deal with resistive biasing—that is, circuits in which the bias point is established using resistors. Resistive biasing is used mostly in discrete circuits. Integrated circuit biasing, which uses transistor current sources in place of most of the resistors, is covered in Section 7.4.

7.2.1 BJT Amplifiers

We have already noted that one major objective in designing a bias circuit is to obtain a bias point that is insensitive to poorly controlled or variable device parameters. In the case of BJTs, the parameters we are most concerned about are I_S and β, since they are both poorly controlled and temperature dependent. The base-emitter voltage of a BJT is given by

$$V_{BE} = V_T \ln I_C/I_S, \tag{7.25}$$

so we see that at a fixed collector current, a variation in I_S can be equated to a variation in V_{BE}. It is often more convenient to think in terms of V_{BE} varying; therefore, we think of V_{BE} and β as being the two parameters we are most concerned about.

Typical variations from one device to another for these parameters are presented in Table 7.2. Table 7.3 shows data for how these parameters vary with temperature. (I_{CO} is included here for reference, since its temperature dependence is sometimes important.)

This section presents the most popular resistive bias circuit for bipolar amplifiers. We show how to analyze the circuit and present some rules for designing the circuit to obtain a stable bias point. The companion section on the CD

Table 7.2 Matching Data for Discrete *npn* BJTs (2N2222 type) and Typical Medium-Size Integrated *npn* Transistors That Are Adjacent to Each Other on an IC

Parameter	Typical range for 2N2222		Typical matching for medium-size *npns* adjacent to each other on an IC
	Minimum	Maximum	
β	35	300	$\pm 0.001\%$
V_{BE} (@ a fixed I_C)	0.64 V	0.76 V	± 2 mV
I_S	2.0×10^{-15} A	2.0×10^{-13} A	$\pm 10\%$

Table 7.3 Variation of Large-Signal DC BJT Parameters with Temperature. All Temperatures Are in °C. For *pnp* Transistors, V_{BE} Should Be Replaced by V_{EB}

Parameter	Typical value at 25°C	Variation with Temperature	Typical Temperature Coefficient at 25°C
I_{CO}	1 nA	$I_{CO}(T) = I_{CO}(25) \cdot 2^{(T-25)/6}$ A	116 pA/°C
β	100	$\beta(T) = \beta(25)[1 + (T - 25)/80]$	12,500 ppm/°C
I_S	1×10^{-15} A	$I_S(T) = I_S(25) \times 10^{(T-25)/15}$	1.66×10^{-16} A/°C
V_{BE} (I_C constant)	0.7 V	$V_{BE}(T) = V_{BE}(25) - (T - 25)0.002$ V	−2 mV/°C

analyzes and compares three different biasing schemes, including the one presented here, in more detail. The more detailed section explains why the bias circuit shown here is the most popular, presents an analytical tool, the ***normalized small-signal sensitivity***, for quantitatively comparing the expected variation in a particular parameter, and uses that tool to derive the design rules presented here.

The circuit shown in Figure 7-46 is the most frequently used biasing arrangement for discrete single-transistor BJT amplifiers. We will show how we connect the signal input and where we take the output in Chapter 8. Since the small-signal parameters of the transistor are all dependent on the value of I_C, we are concerned mostly with it (i.e., the exact terminal voltages are not nearly as important, as long as the device is in the forward-active region of operation).

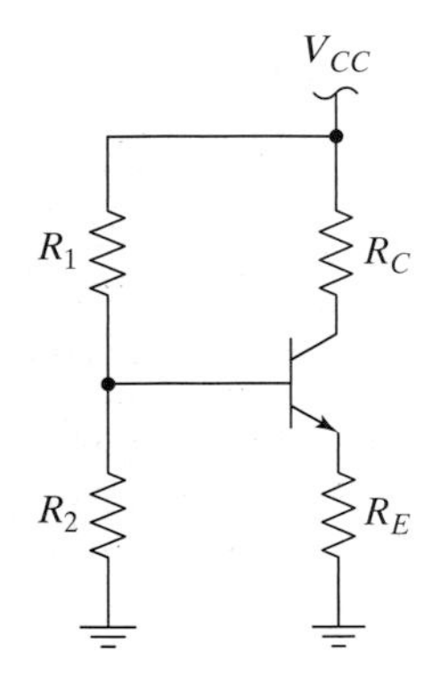

Figure 7–46 The most common single-transistor discrete BJT biasing circuit.

In this circuit, the presence of R_E stabilizes the operating point. If, for example, I_C increases due to a change in a component or the temperature, I_E will increase by nearly the same amount. The increase in I_E causes an increase in the voltage drop across R_E, which decreases V_{BE} and thereby tends to reduce I_C. Therefore, the circuit acts to oppose the initial change in I_C, and the initial disturbance is reduced. In other words, the circuit has ***negative feedback***, which opposes any change in the bias point. Negative feedback is discussed at length in Chapter 10. For now, it will suffice to understand the operation of the feedback qualitatively and to see that negative feedback tends to stabilize the bias point.

To analyze the circuit in Figure 7-46, we first draw the large-signal DC equivalent circuit shown in Figure 7-47. To draw this circuit, we replace every component by its large-signal DC model and then replace the resistor divider driving the base of the transistor with its Thévenin equivalent circuit. For the Thévenin equivalent, we get

$$V_{BB} = \frac{R_2}{R_1 + R_2} V_{CC} \tag{7.26}$$

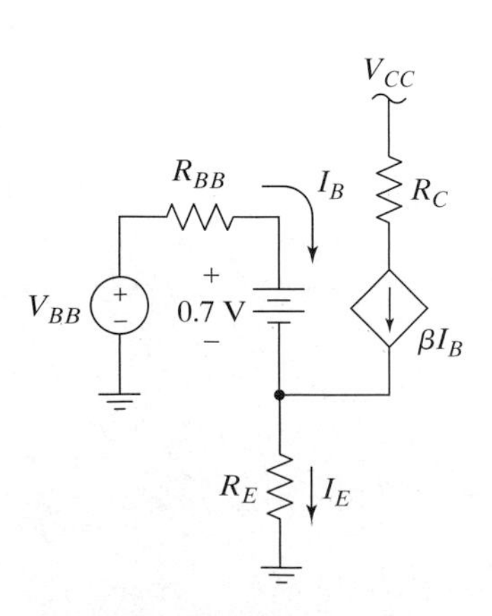

Figure 7–47 The large-signal DC equivalent circuit for Figure 7-46.

and

$$R_{BB} = R_1 \| R_2 = \frac{R_1 R_2}{R_1 + R_2}. \tag{7.27}$$

Writing KVL around the base-emitter loop yields

$$V_{BB} - I_B R_{BB} - 0.7 - I_E R_E = 0, \tag{7.28}$$

and replacing I_E with $(\beta + 1)I_B$ yields

$$V_{BB} - I_B R_{BB} - 0.7 - (\beta+1) I_B R_E = 0, \tag{7.29}$$

which can be solved to obtain

$$I_B = \frac{V_{BB} - 0.7}{R_{BB} + (\beta+1)R_E}. \quad \textbf{(7.30)}$$

If $V_{BB} \gg 0.7$, this equation will yield an accurate estimate for I_B, even though the value used for V_{BE} is only an approximation. If V_{BB} is not large enough, the value of V_{BE} can be expressed in terms of I_B and the resulting transcendental equation solved iteratively. Once I_B is known, the values of I_C and I_E can be readily determined. Notice in (7.30) that R_E shows up as if it is directly in series with R_{BB}, but multiplied by $(\beta+1)$. We say that R_E is *reflected into the base* by multiplying it by $(\beta+1)$. Impedance reflection is an important analysis tool and lends additional insight into the operation of bipolar transistor circuits; we discuss it further in Aside A7.1 at the end of this section.

Writing KVL in the loop containing the controlled source in Figure 7-47[8] (and recognizing that the voltage across the controlled source is V_{CE}) leads to

$$V_{CC} - I_C R_C - V_{CE} - I_E R_E = 0. \quad \textbf{(7.31)}$$

Solving for V_{CE} and assuming that $I_C \approx I_E$ (i.e., that β is large), we get

$$V_{CE} \approx V_{CC} - I_C(R_C + R_E). \quad \textbf{(7.32)}$$

We can now check that the transistor is in the forward-active region of operation as assumed. It is very important not to omit this final step, since the model we used for the BJT is wrong if the transistor is not in the forward-active region of operation.

Example 7.10

Problem:

Consider the circuit shown in Figure 7-46, and let $V_{CC} = 20$ V, $R_C = 5$ kΩ, $R_E = 1$ kΩ, $R_1 = 20$ kΩ, and $R_2 = 3$ kΩ. Determine the Q point if the transistor β is 100.

Solution:

From (7.26), we get

$$V_{BB} = \frac{3}{3 + 20}(20) = 2.6 \text{ V},$$

and using (7.27), we get $R_{BB} = (3\|20)\text{k} = 2.6$ kΩ. Using these values along with the other given values in (7.30) yields

$$I_B = \frac{2.6 - 0.7}{2.6\text{ k} + (101)1\text{ k}} = 18.3\ \mu\text{A}.$$

Therefore, $I_C = 1.83$ mA, and using (7.32), we find $V_{CE} = 9$ V. Checking these values, we see that the transistor is forward active, as assumed.

A SPICE simulation of this circuit using the 2N2222 model from Appendix B, but with β changed to 100, yields $I_C = 1.87$ mA, very close to our prediction.[9]

■

---

[8] To see the "loop" referred to, you must remember that V_{CC} is a fixed supply between ground and the connection labeled V_{CC}. If this supply is included in the drawing, the loop becomes evident.

[9] The SPICE files for all examples and exercises that use SPICE are included on the CD.

We now want to show that the bias point for this circuit is not too sensitive to changes in V_{BE} and β. In the next example, we let these values change and see what happens to the bias point calculated in Example 7.10.

Example 7.11

Problem:
Consider again the circuit shown in Figure 7-46, and let $V_{CC} = 20$ V, $R_C = 5$ kΩ, $R_E = 1$ kΩ, $R_1 = 20$ kΩ, and $R_2 = 3$ kΩ, as before. Determine the Q point if the transistor β is 50 and the V_{BE} is 0.6 V.

Solution:
Just as in Example 7.10, we get $V_{BB} = 2.6$ V and $R_{BB} = 2.6$ kΩ. Using these values along with the other given values in (7.30) now yields

$$I_B = \frac{2.6 - 0.6}{2.6\text{ k} + (51)1\text{ k}} = 37.3\ \mu\text{A} .$$

The resulting collector current is $I_C = 1.87$ mA, and again using (7.32), we find that $V_{CE} = 8.8$ V. The value of I_C is only 2% higher than before, and V_{CE} is only 4.4% below its previous value, so the Q point has not shifted very much, which demonstrates that the negative feedback has done a reasonable job of stabilizing the bias point.

Running the SPICE simulation from the previous example again, but sweeping β from 25 to 200, causes the collector current to vary from slightly over 1.7 mA to slightly less than 1.9 mA. Sweeping I_S from 1/100 to 100 times its nominal value, which corresponds to V_{BE} varying by about ±120 mV, we find that I_C varies from about 1.4 mA to 2 mA.

Problem P7.34 examines a more complete worst-case scenario for this topology.

■

We now shift our attention from analysis to design. The collector current is $I_C = \beta I_B$, so using (7.30), we can write (having used V_{BE} in place of its approximate value of 0.7 V)

$$I_C = \beta I_B = \frac{\beta(V_{BB} - V_{BE})}{R_{BB} + (\beta+1)R_E}. \tag{7.33}$$

By examining (7.33), we can see how to design the circuit so that I_C is not too sensitive to the values of β or V_{BE}. If we can make $(\beta+1)R_E \gg R_{BB}$, allowing for the fact that β is typically much larger than unity, we have

$$I_C \approx \frac{\beta(V_{BB} - V_{BE})}{(\beta+1)R_E} \approx \frac{V_{BB} - V_{BE}}{R_E}, \tag{7.34}$$

and I_C will be reasonably independent of β. Finally, if we can ensure that $(V_{BB} - V_{BE}) \gg \Delta V_{BE}$, where ΔV_{BE} is the anticipated possible variation in V_{BE}, then I_C will not be too sensitive to V_{BE}, either. These two conditions are met if we obey the following two ***rules of thumb for stable bias circuit design***:

1 Set the current in R_1 greater than or equal to 10 times the base current. Obeying this rule guarantees that V_B is approximately equal to V_{BB}, which is equivalent to saying that $(\beta+1)R_E \gg R_{BB}$. Consequently, the result is that we are not too sensitive to the value of β. It also guarantees that we can ignore I_B, so that the current in R_2 is about equal to the current in R_1. Since it is usually true that $\beta \geq 100$, we can restate this rule as follows: set $I_{R_1} \geq I_C/10$.

2 Set $V_{BB} \geq 2$ V. Given that $V_B \approx V_{BB}$ and that $V_{BE} \approx 0.7$ V, this rule is equivalent to saying we should set $V_E \geq 1.3$ V. If V_{BE} varies by no more than ± 0.13 V, the change in V_E will be less than 10%; and, therefore, so will the change in I_C. This rule guarantees that I_C is not too sensitive to changes in V_{BE} or, in other words, it is not too sensitive to changes in I_S.

It is very important to realize that these "rules" are derived on the basis of reasonable assumptions about how much the bias point should be allowed to shift and how much the transistor parameters are likely to vary. It is not uncommon to violate the rules, if necessary, nor is it uncommon to have to use stricter rules. What is most important is the logic that led to them.

We now have some guidelines to help us choose values for R_1, R_2, and R_E, given a desired value for the bias current I_C. Given a desired value for I_C, we say that $I_E \approx I_C$ and then choose R_E using $R_E = V_E/I_C$. We next use $I_{R2} \approx I_{R1} = 10I_B$ along with the value of V_{CC} to find the total resistance in the bias leg $(R_1 + R_2)$. Given the total resistance in the bias leg and the desired V_{BB}, we then determine the values for R_1 and R_2. The example that follows illustrates this procedure. We will discuss how to find R_C after this example.

Example 7.12

Problem:
Determine values for R_1, R_2, and R_E in Figure 7-46 to set $I_C = 2$ mA. Assume that $V_{CC} = 10$ V and $\beta = 50$.

Solution:
Following our rule of thumb for V_{BB}, we set it to 2 V and recognize that this is also going to be the base voltage. Therefore, V_E is about 1.3 V, and we find that

$$R_E = \frac{1.3}{I_E} \approx \frac{1.3}{I_C} = 650\ \Omega.$$

We next note that the base current will be 2 mA/50 = 40 μA. Using our rule of thumb for the current in R_1, and remembering that this is approximately the same as the current in R_2, we find the total resistance in the bias string:

$$R_1 + R_2 \approx \frac{V_{CC}}{I_{R1}} = \frac{V_{CC}}{10I_B} = 25\ \text{k}\Omega.$$

We next use the equation for V_{BB} along with its desired value of 2 V to get the individual values for R_1 and R_2; we obtain

$$V_{BB} = \frac{R_2}{R_1 + R_2}V_{CC},$$

$$R_2 = \frac{V_{BB}}{V_{CC}}(R_1 + R_2) = \frac{2}{10}25\ \text{k}\Omega = 5\ \text{k}\Omega,$$

and

$$R_1 = (25 - 5)\,\text{k}\Omega = 20\,\text{k}\Omega.$$

■

We now need to figure out how to determine the value of R_C. In the circuit we are examining, R_C does not affect the value of I_C; it only sets the collector voltage. The only requirement we currently have for the collector voltage is that the transistor must be kept in the forward-active region of operation. Therefore, V_C cannot be too low, or equivalently, R_C cannot be too large. The small-signal gain and bandwidth also turn out to be functions of R_C, so we have to defer a more complete discussion of what value to choose to Chapters 8 and 9. Nevertheless, one reasonable choice for V_C (as we will see in Chapter 8) is given in our ***third rule of thumb***.

3 Set V_C halfway between V_E and V_{CC}.

Once the desired value of V_C is known, R_C is determined by using Ohm's law.

Example 7.13

Problem:
Find the value of R_C required to set V_C halfway between V_E and V_{CC} in the circuit of Example 7.12.

Solution:
In Example 7.12, we had $I_C = 2$ mA, $V_{CC} = 10$ V, and $V_E = 1.3$ V. Therefore, V_C should be set to $(10 + 1.3)/2 = 5.65$ V, and we find that

$$R_C = \frac{V_{CC} - V_C}{I_C} = \frac{10 - 5.65}{2\,\text{mA}} = 2.2\,\text{k}\Omega\ .$$

■

Exercise 7.6

Determine the percentage change in the collector current for the circuit designed in Examples 7.12 and 7.13, given the following changes (take them one at a time): ***(a)*** $\beta = 100$, ***(b)*** $\beta = 25$, ***(c)*** $V_{BE} = 0.6$ V, ***(d)*** $V_{BE} = 0.8$ V, ***(e)*** $V_{CC} = 9.5$ V, *and* ***(f)*** $V_{CC} = 10.5$ V.

□

Exercise 7.7

Choose values for the resistors in the circuit in Figure 7-46 to establish $I_C = 500\ \mu$A, *using our rules of thumb and assuming that* $V_{CC} = 5$ V, $V_{BE} = 0.7$ V, *and* $\beta = 150$.

□

One final issue needs to be considered with regard to transistor biasing for amplification: We must not only guarantee that the transistor is in forward-active

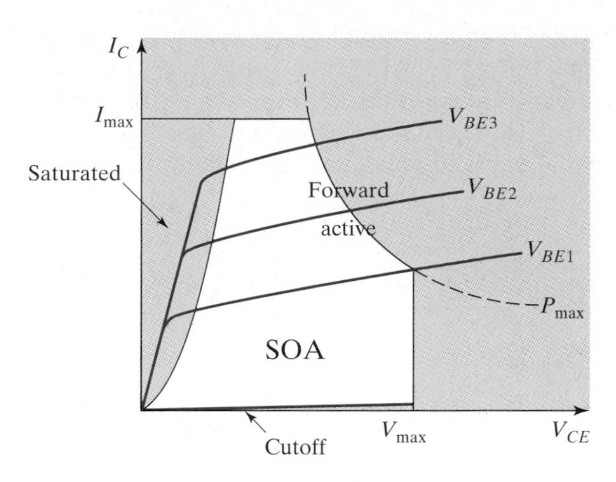

Figure 7–48 A family of collector characteristics and the safe operating area (SOA).

operation; we must also take care to not exceed the maximum limits for the terminal voltages, currents, or power. Figure 7-48 shows the collector characteristics (i.e., I_C versus V_{CE} for different values of I_B or V_{BE}) of a BJT. The safe portion of the forward-active region of operation is the unshaded region in the figure. The region shaded for very low V_{CE} is where the BJT is saturated, and the region shaded for very low I_C is where the device is cut off. Consequently, these regions can't be used for amplification. The figure also shows maximum limits for the current, voltage, and power, denoted by I_{max}, V_{max}, and P_{max}, respectively. The portion of the forward-active region below these limits is called the ***safe operating area*** (SOA). Since $P = VI$, the maximum power dissipation boundary is a hyperbola and is called the ***maximum-dissipation hyperbola***. The maximum allowable current, voltage, and power are listed on the data sheets for a given device. The maximum allowable power decreases as temperature increases.

ASIDE A7.1 DC IMPEDANCE REFLECTION

Consider the circuit shown in Figure A7-1(a), and write KVL around the loop containing R_{BB}. The result is

$$V_{BB} - I_B R_{BB} - V_{BE} - I_E R_E = 0. \qquad \textbf{(A7.1)}$$

Replacing I_E with $(\beta + 1)I_B$ in (A7.1) yields

$$V_{BB} - I_B R_{BB} - V_{BE} - (\beta+1) I_B R_E = 0, \qquad \textbf{(A7.2)}$$

and similarly, replacing I_B with $I_E/(\beta + 1)$ yields

$$V_{BB} - \frac{I_E}{\beta+1} R_{BB} - V_{BE} - I_E R_E = 0. \qquad \textbf{(A7.3)}$$

Notice that (A7.2) would also result from writing KVL around the loop shown in Figure A7-1(b), and (A7.3) would result from writing KVL around the loop shown in part (c) of the figure. In other words, these circuits are all equivalent in a limited way, namely, they all produce the same KVL equation. Consequently, the voltages around the loop are the same in all three loops.

In addition, the loop in part (b) of the figure has the same current as the *base* side of the transistor in part (a), so any resistances calculated on the base side will be correct. For example, the Thévenin resistance seen looking into the base of the transistor is $(\beta + 1)R_E$. Similarly, the loop in part (c) of the figure has the same current as the *emitter* side of the transistor in part (a), so any resistance calculated on the emitter side will be correct. For example, the Thévenin resistance seen looking into the emitter of the transistor is $R_{BB}/(\beta+1)$.

We say that the emitter resistance *reflected into the base* is equal to $R_E(\beta + 1)$. In other words, if we find the Thévenin equivalent seen looking in the base, we see that it consists of a constant V_{BE} in series with a resistance of value $R_E(\beta + 1)$ as shown in Figure A7-1(b). Similarly, the base resistance reflects into the emitter as $R_{BB}/(\beta + 1)$. Knowing this general result will allow us to simplify our equivalent circuits and makes the analyses easier.

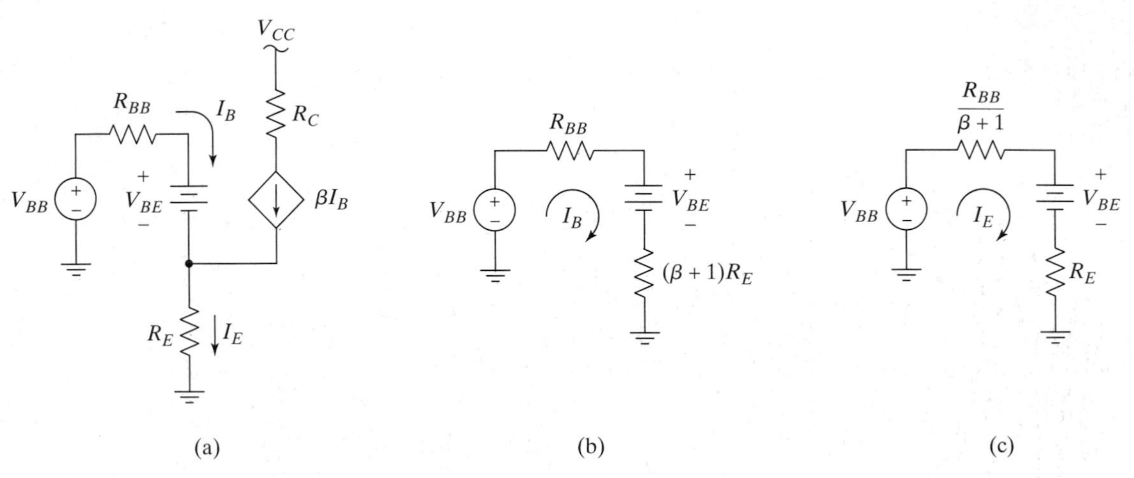

Figure A7-1 (a) A large-signal DC model of a BJT biasing circuit. (b) A circuit that yields the same loop equation as (a), but with the current equal to I_B everywhere in the loop. (c) A circuit that yields the same loop equation as (a), but with the current equal to I_E everywhere in the loop.

7.2.2 FET Amplifiers

Discrete FETs are seldom used for small-signal amplifiers because they don't offer any advantage over BJTs in most applications and because they have a lower gain at the same bias current and a wider spread in device characteristics. The major applications for discrete FETs are as switches, power devices, and high-frequency amplifiers (usually GaAs MESFETs, which are depletion-mode devices) and where high input impedance is needed. Nevertheless, this section presents some useful analysis examples and design guidelines and serves as a pedagogic stepping-stone on the way to integrated-circuit biasing techniques.

The examples in this section will use MOSFETs, but the circuits will work with JFETs and MESFETs as well. The only real differences for biasing are that the DC gate currents of JFETs and MESFETs are not zero. However, the gate currents are still quite small (so long as the gate-to-channel diodes are reverse biased), so they can usually be ignored.

A major objective in designing a bias circuit is to obtain a bias point that is insensitive to poorly controlled or variable device parameters. In the case of a FET, the values of K and V_{th} are poorly controlled and temperature dependent, so we will examine how to produce a bias point that is not too sensitive to these parameters.

Typical parameter variations from one device to another are presented in Table 7.4; the corresponding parameters for JFETs (V_p and I_{DSS}) are also listed. Table 7.5 shows data for how these parameters vary with temperature.

We start with a resistive bias circuit that is usable for both enhancement-mode and depletion-mode FET amplifiers. We show how to analyze the circuit and present some guidelines for designing the circuit to obtain a stable bias point. The companion section on the CD analyzes and compares different

Table 7.4 Matching Data for Discrete *n*-Channel FETs and Typical Medium-Size ($W = 10$ μm & $L = 2$ μm) Integrated *n*-Channel MOSFETs That Are Adjacent to Each Other on an IC

Parameter	Typical range for discrete transistors of the same type		Typical matching for medium-size FETs adjacent to each other on an IC
	Minimum	Maximum	
V_p (1)	−6	−1	±3 mV
V_{th}	0.5 V	2.5 V	±3 mV
K (2)	10 mA/V^2	150 mA/V^2	±1%
I_{DSS}	1 mA	5 mA	±4%

NOTES: (1) V_{th} applies to MOSFETs and V_p applies to JFETs. (2) K applies for MOSFETs and I_{DSS} applies for JFETs.

Table 7.5 Variation of Large-Signal DC MOSFET Parameters with Temperature. All Temperatures Are in Kelvin for the Equations. Data Taken from [7.3, pp. 121–122] and [7.4, pp. 357-358][1]

Parameter	Typical value at 25 °C	Variation with Temperature	Typical Temperature Coefficient at 25 °C
K	10^{-5} to 10^{-3} A/V^2	$K(T) = K(T_o)(T_o/T)^{1.5}$	−0.06 °C^{-1}
V_{th} (I_D constant) *n*-channel	0 to 5 V	$V_{th}(T) = V_{th}(T_o) - .002(T - T_o)$	−2 mV/°C
V_{th} (I_D constant) *p*-channel	0 to −5 V	$V_{th}(T) = V_{th}(T_o) + .002(T - T_o)$	2 mV/°C

[1] Note that (5.5-18) has a printing error in the copy we examined; it shows $(T/T_o)^{1.5}$, which is incorrect.

biasing schemes—including the one presented here—in more detail. This more detailed section presents an analytical tool, the ***normalized small-signal sensitivity***, for quantitatively comparing the expected variation in a particular parameter. The second part of this section presents a bias circuit that is useful only for depletion-mode devices, but that is the most popular circuit for those devices.

General Resistive Bias Circuit Consider the circuit shown in Figure 7-49. We will show how the signal is routed through this circuit in Chapter 8; for now, we just focus on establishing the bias point.

The presence of R_S in this circuit ensures that the bias current, I_D, is relatively insensitive to changes in the transistor parameters. For example, if the drain current increases, perhaps due to a change in temperature or component value, the voltage drop across R_S will also increase. The value of V_{GS} will then be reduced, since V_G is constant. This reduction in gate drive will tend to reduce I_D and partially compensate for the initial change. In other words, the change that would have been seen in I_D if V_{GS} were held constant has been decreased. This is an example of ***negative feedback***, which will be dealt with in detail in Chapter 10. For now, it will suffice to understand the operation of the feedback qualitatively and to see that negative feedback tends to stabilize the bias point.

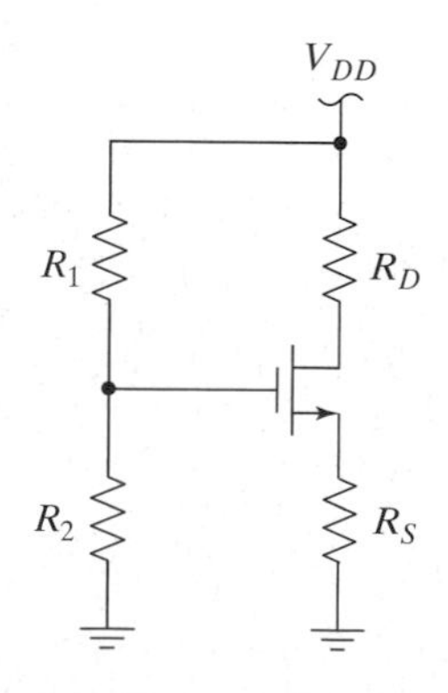

Figure 7–49 A discrete biasing circuit for FET amplifiers (shown with an *n*-channel enhancement-mode MOSFET).

To analyze the circuit of Figure 7-49, we first draw the large-signal DC equivalent circuit as shown in Figure 7-50. We have replaced every component by its large-signal DC model and have also replaced the resistor divider network driving the gate with its Thévenin equivalent. For the Thévenin equivalent, we get

$$V_{GG} = \frac{R_2}{R_1 + R_2} V_{DD} \tag{7.35}$$

and

$$R_{GG} = R_1 \| R_2 = \frac{R_1 R_2}{R_1 + R_2}. \tag{7.36}$$

Writing KVL around the gate-source loop yields

$$V_{GG} - V_{GS} - I_D R_S = 0, \tag{7.37}$$

where we have made use of the facts that $I_G = 0$ and $I_S = I_D$. We can now replace I_D in (7.37) by $K(V_{GS} - V_{th})^2$ and expand the square to obtain a quadratic equation that can be solved for V_{GS}:

$$KR_S V_{GS}^2 + (1 - 2KR_S V_{th})V_{GS} + KR_S V_{th}^2 - V_{GG} = 0. \tag{7.38}$$

Solving (7.38) yields

$$V_{GS} = \frac{2KR_S V_{th} - 1 \pm \sqrt{(1 - 2KR_S V_{th})^2 - 4KR_S(KR_S V_{th}^2 - V_{GG})}}{2KR_S}, \tag{7.39}$$

from which one solution can always be eliminated as being impossible. Once V_{GS} is known, I_D follows immediately. The final step is to find V_D and V_{DS} using the calculated I_D and the known R_D. From the drain loop[10] in Figure 7-50, we get

$$V_{DD} - I_D R_D - V_{DS} - I_D R_S = 0, \tag{7.40}$$

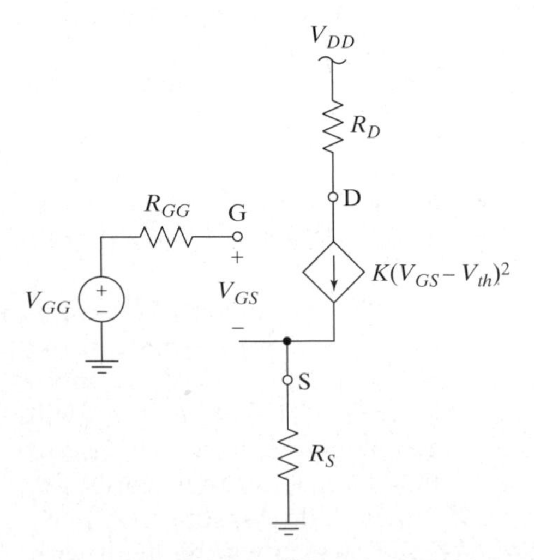

Figure 7–50 The large-signal DC equivalent for the circuit in Figure 7-49.

[10] To see the "loop" referred to, you must remember that V_{DD} is a fixed supply between ground and the connection labeled V_{DD}. If this supply is included in the drawing, the loop becomes evident.

which can be solved for V_{DS} to yield

$$V_{DS} = V_{DD} - I_D(R_D + R_S). \qquad \textbf{(7.41)}$$

After finding the bias point, we should always go back and check that the transistor is in the forward-active region of operation, as assumed. If it is not, our model is wrong, and we need to change the model and redo the analysis.

Exercise 7.8

Find the voltages and currents in the circuit of Figure 7-49 if $V_{DD} = 15$ V, $R_1 = 10$ MΩ, $R_2 = 5$ MΩ, $R_D = 4$ kΩ, $R_S = 800$ Ω, $V_{th} = 2$ V, *and* $K = 1$ mA/V^2.

We now turn our attention to designing the circuit of Figure 7-49 (i.e., how to calculate the component values necessary to achieve a particular bias point). As the following example shows, this task is somewhat less work than analysis, since we don't have to solve a quadratic equation.

Example 7.14

Problem:
Find values for the components in Figure 7-49 to establish a bias point of $I_D = 1$ mA and $V_{DS} = 3$ V. Assume that $V_{DD} = 15$ V and that the transistor has $K = 500$ μA/V^2 and $V_{th} = 0.5$ V.

Solution:
We must first find the value of V_{GS} necessary to provide the desired current. We have

$$I_D = K(V_{GS} - V_{th})^2,$$

from which we obtain

$$V_{GS} = V_{th} + \sqrt{\frac{I_D}{K}} = 0.5 + \sqrt{\frac{0.001}{0.0005}} = 1.91\text{ V}.$$

Since we want $V_{DS} = 3$ V, we have 12 V to drop across R_D and R_S. We have no guideline at this point for how to divide this voltage, let's choose 2 V across R_S and 10 V across R_D. This allows us to find the values of these resistors using Ohm's law; we obtain

$$R_D = \frac{10}{I_D} = \frac{10}{0.001} = 10\text{ k}\Omega,$$

and

$$R_S = \frac{2}{I_D} = \frac{2}{0.001} = 2\text{ k}\Omega.$$

All that remains is to find the values of R_1 and R_2. We start by finding the value of V_{GG}. From (7.37), we get

$$V_{GG} = V_{GS} + I_D R_S = 1.91 + 2 = 3.91\text{ V}.$$

The values of R_1 and R_2 can be determined once we decide how much current to put through them. Too much current will increase the power consumption, whereas too little current will lead to unrealistic values for the resistors and will make the circuit too sensitive to leakage currents and impossible to probe during testing. As a compromise, we choose a current of 1 μA through the two resistors. This is still a very small current; it would not be unreasonable to choose a current 10, or 100, times larger. Therefore,

$$R_1 + R_2 = \frac{15}{10^{-6}} = 15\,\text{M}\Omega.$$

Using this total resistance and (7.35), along with the desired V_{GG} of 3.91 V, yields

$$R_2 = (R_1 + R_2)\frac{V_{GG}}{V_{DD}} = (15M)\frac{3.9}{15} = 3.9\ \text{M}\Omega,$$

and therefore $R_1 = 11.1\ \text{M}\Omega$.

A SPICE simulation performed with the Nsimple_mos model described in Appendix B and with $W/L = 5$ confirms that the values calculated here yield $I_D = 1$ mA and $V_{DS} = 3$ V.

■

We now want to examine how the bias point established in Example 7.14 changes if the transistor parameters change. These changes could be the result of a change in temperature or our having designed the circuit using typical values for a given type of transistor, but then having built the circuit with a transistor whose parameters were not typical.[11]

Example 7.15

Problem:
Reconsider the design from Example 7.14, but let the transistor parameters change. Calculate the new values of I_D and V_{DS} if (a) V_{th} changes to 1 V and (b) K changes to 1 mA/V^2.

Solution:
(a) Using (7.39) with the given information leads to V_{GS} being 2.28 V, and the resulting drain current is $I_D = 0.81$ mA. Plugging the known values into (7.41) yields $V_{DS} = 4.9$ V. The transistor is still forward active in this case, so the answers are correct. A SPICE simulation of the circuit shows that as V_{th} is swept from 0.25 V to 1.5 V, the drain current decreases almost linearly from about 1.1 mA to about 640 μA.

(b) Repeating the same procedure as in part (a), but with $V_{th} = 0.5$ V and $K = 1$ mA/V^2 yields $V_{GS} = 1.58$ V, $I_D = 1.2$ mA, and $V_{DS} = 1.4$ V. The transistor is again forward active as assumed. A SPICE simulation shows that as K is varied from 250 μA/V^2 to 1.5 mA/V^2, the drain current varies from about 0.8 mA to about 1.2 mA.

■

[11] As you probably suspect, no device is ever "typical"; these values are averages.

Example 7.15 has demonstrated that our design from Example 7.14 is more sensitive to changes in device characteristics than we might desire. The sensitivity is not a fault of the topology chosen, but is a result of our specific component values. Looking back at Example 7.14, we see that the only free choices we made were how much voltage to drop across R_S and R_D and how much current we put through R_1 and R_2. The current through R_1 and R_2 has no effect on the stability of the bias point, since V_{GG} remains fixed for any set of transistor parameters. The voltage drops across R_S and R_D are a different story, however. Notice that we can solve (7.37) to obtain

$$I_D = \frac{V_{GG} - V_{GS}}{R_S}. \tag{7.42}$$

If the transistor is forward active, we know that

$$V_{GS} = V_{th} + \sqrt{I_D/K} = V_{th} + V_{ON}. \tag{7.43}$$

Notice from (7.43) that if the transistor is operated at a constant drain current, changes in K and V_{th} both lead to a change in V_{GS}. Now notice from (7.42) that if we make $V_{GG} - V_{GS}$ much larger than any expected variation in V_{GS}, the drain current will be approximately constant. Therefore, we can improve the stability of our design by increasing $V_{GG} - V_{GS}$, which (7.37) indicates is equal to the voltage drop across R_S. We are limited in how much we can increase this drop, however. From (7.40), we see that the drop across R_S is given by

$$V_{R_S} = I_D R_S = V_{DD} - V_{DS} - I_D R_D. \tag{7.44}$$

Therefore, with a given I_D, the only freedom we have is to decide how to partition the voltage drops across R_S, the transistor, and R_D. Consider what happens with a signal at the drain. The voltage cannot rise above V_{DD} nor can it go below $V_G - V_{th}$ (or the transistor leaves the forward-active region). Therefore, a reasonable choice is to center V_D between these extremes. Keeping this constraint in mind, we note that if the drop across R_S is at least 10 V, the design will usually be reasonably stable. But, we will not always be able to achieve such a large drop. Nevertheless, we can state a couple of ***rules of thumb for producing a stable bias design*** using the circuit of Figure 7-49:

1. We should make the DC voltage drop across $R_S \geq 10$ V or as large as possible for the given bias point and signal swing.
2. Set V_D halfway between V_{DD} and $V_G - V_{th}$.

Example 7.16

Problem:
Redo the design in Example 7.14 to make it as stable as possible, given the desired bias point and assuming that the minimum drop across R_D is 3 V.

Solution:
To produce a stable design, we want the drop across R_S to be as large as possible. With $V_{DS} = 3$ V and the drop across $R_D = 3$ V, the drop across R_S is 9 V. Therefore, $R_D = 3$ kΩ and $R_S = 9$ kΩ. We still have $V_{GS} = 1.9$ V, as calculated in Example 7.14, so we now want $V_{GG} = 10.9$ V. Therefore, we now get $R_1 = 4.1$ MΩ and $R_2 = 10.9$ MΩ.

■

Now let's examine how much better our new design works.

Example 7.17

Problem:
Consider again the design from Example 7.16, but let the transistor parameters change as in Example 7.15. Calculate the new values of I_D and V_{DS} if (a) V_{th} changes to 1 V and (b) K changes to 1 mA/V^2.

Solution:
(a) Using (7.39) with the given information leads to V_{GS} being 2.38 V, and the resulting drain current is $I_D = 0.95$ mA. Plugging the known values into (7.41) yields $V_{DS} = 3.2$ V, and the transistor is still forward active, as assumed.

(b) Repeating the same procedure as in part (a), but with $V_{th} = 0.5$ V and $K = 1$ mA/V^2, yields $V_{GS} = 1.52$ V, $I_D = 1.04$ mA, and $V_{DS} = 2.9$ V. The transistor is again forward active as assumed.

■

The results of this example show that the new design is significantly more robust than our first attempt. (Compare the results of Examples 7.17 and 7.15.)

One final issue needs to be considered with regard to FET biasing for amplification: We must not only guarantee that the transistor is in forward-active operation; we must also not exceed the maximum limits for the terminal voltages, currents, or power. Figure 7-51 shows the drain characteristics (i.e., I_D versus V_{DS} for different values of V_{GS}) of a FET. The safe portion of the forward-active region of operation is the unshaded region in the figure. The region shaded for very low V_{DS} is where the FET is in the linear region, and the region shaded for very low I_D is where the device is cut off, so these regions can't be used for amplification. The figure also shows maximum limits for the current, voltage, and power, denoted by I_{max}, V_{max}, and P_{max}, respectively. The portion of the forward-active region below these limits is called the ***safe operating area*** (SOA). Since $P = VI$, the maximum power dissipation boundary is a hyperbola and is called the ***maximum-dissipation hyperbola***. The maximum allowable current, voltage, and power are listed on the data sheets for a given device. The maximum allowable power decreases as temperature increases.

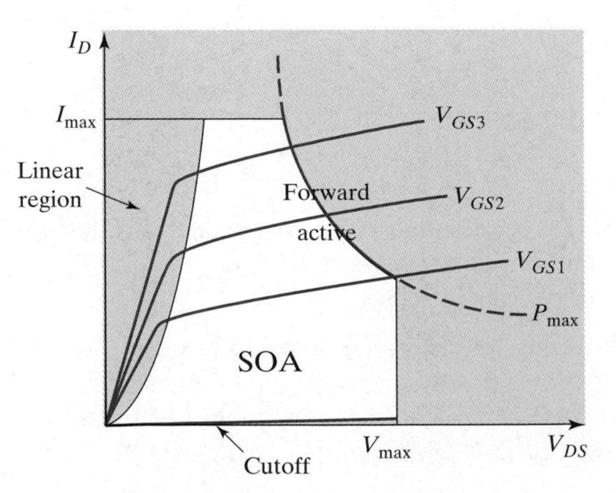

Figure 7–51 A family of drain characteristics and the SOA.

Exercise 7.9

Design the circuit of Figure 7-49 to achieve a stable bias point with $I_D = 100\ \mu A$, assuming that $K = 100\ \mu A/V^2$, $V_{th} = 0.5$ V, and $V_{DD} = 3.3$ V. Keep the drop across R_D and V_{DS} both greater than or equal to 1.2 V.

We end this section with an example using this bias circuit with a depletion-mode device.

Example 7.18

Problem:
Consider the circuit shown in Figure 7-52, where the transistor is a depletion-mode MOSFET. Find the operating point for the transistor, assuming that $V_p = -3$ V and $K = 167\ \mu A/V^2$. (These values are typical for a 2N3796 transistor.)

Solution:
We first note that $V_{GG} = 10$ V. We next assume that the transistor is forward active and use (7.39) to find that V_{GS} is either -0.5 V or -6 V. Since the transistor will be cut off if $V_{GS} \leq -3$ V, the correct answer must be $V_{GS} = -0.5$ V. Using this result, we find that

$$I_D = K(V_{GS} - V_{th})^2 = 167(-0.488 + 3)^2 \quad \mu A = 1.05 \text{ mA}.$$

Now we see that $V_S = I_D R_S = 10.5$ V, and the drain voltage is $V_D = V_{DD} - I_D R_D = 15$ V. Therefore, $V_{DS} = 4.5$ V. Finally, we check and see that the device is forward active as assumed, so our solution is correct.

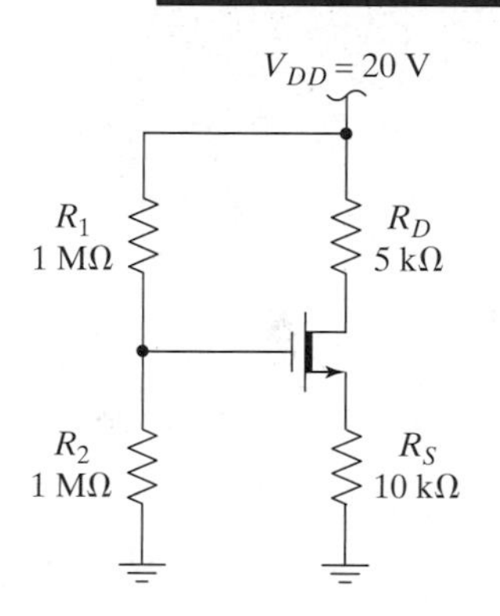

Figure 7–52 A depletion-mode MOSFET biasing circuit.

Depletion-mode Self-Biasing Because depletion-mode devices have a conducting channel when no bias is applied to the gate, a simpler biasing circuit can be used, as shown in Figure 7-53. An n-channel JFET is used in this example, but the circuit works equally well for any n-channel depletion-mode FET. (A similar circuit works for p-channel devices.)

When the circuit is first powered up, the FET has a conducting channel, so current flows in the drain circuit. The current flow then produces a voltage drop across R_S of a polarity that tends to turn the device off. The steady-state value of the drain current is determined by the simultaneous solution of two equations, one for the device and one for the circuit. Because the device is guaranteed to be on and has current flow without a positive bias voltage being applied to the gate, the circuit is called a ***self-biasing*** circuit. Assuming that the gate current is small enough that the drop across R_G can be neglected, we know that $V_G = 0$ V. Therefore, we can write

$$V_{GS} = V_G - V_S = -I_D R_S, \tag{7.45}$$

and this is the constraint placed on V_{GS} and I_D by the circuit. From (7.24), we know that

$$I_D = I_{DSS}\left(1 - \frac{V_{GS}}{V_p}\right)^2 \tag{7.46}$$

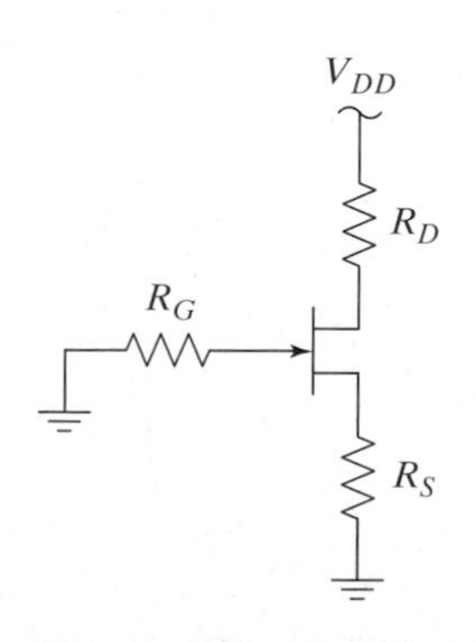

Figure 7–53 Self-biasing circuit for a depletion-mode n-channel FET.

so long as it is in saturation. We can solve (7.45) and (7.46) simultaneously to yield the bias point. Plugging (7.45) into (7.46) and expanding the square yields

$$I_D^2\left(\frac{I_{DSS}R_S^2}{V_p^2}\right) + I_D\left(\frac{2I_{DSS}R_S}{V_p} - 1\right) + I_{DSS} = 0, \tag{7.47}$$

which can be solved to get

$$I_D = \frac{1 - \dfrac{2I_{DSS}R_S}{V_p} \pm \sqrt{\left(\dfrac{2I_{DSS}R_S}{V_p} - 1\right)^2 - 4\dfrac{I_{DSS}^2R_S^2}{V_p^2}}}{\dfrac{2I_{DSS}R_S^2}{V_p^2}}. \tag{7.48}$$

We will always be able to reject one of the two solutions of (7.48). Once I_D is known, we get V_{GS} from (7.45) and then find V_{DS} by writing KVL in the drain loop,

$$V_{DD} = I_D(R_S + R_D) + V_{DS} \tag{7.49}$$

and solving:

$$V_{DS} = V_{DD} - I_D(R_S + R_D). \tag{7.50}$$

As a final step, we should always remember to check that the transistor is forward active, as we have assumed. If it is not, we need to use a different model and repeat the analysis.

Example 7.19

Problem:
Assume that a 2N5457 is used in the circuit in Figure 7-53, so that $I_{DSS} = 3$ mA and $V_p = -3$ V. Find the bias point if $R_G = 100$ kΩ, $R_S = 1.2$ kΩ, $R_D = 5$ kΩ, and $V_{DD} = 15$ V.

Solution:
Using (7.48), we find that I_D is either 6 mA or 1 mA. I_D cannot be 6 mA, because that would lead to $V_{GS} = -7.2$ V, which would cause the device to be pinched off. Therefore, $I_D = 1$ mA. Using this value of drain current along with (7.45) and (7.50), we find that $V_{GS} = -1.2$ V and $V_{DS} = 8.8$ V. With these values, the device is forward active, as assumed.

■

Designing the self-biasing circuit to achieve a particular operating point is straightforward. Given a desired I_D, we find the required V_{GS} by solving (7.46) to obtain

$$V_{GS} = V_p\left(1 - \sqrt{\frac{I_D}{I_{DSS}}}\right). \tag{7.51}$$

We then use (7.45) to find the required value of R_S (assuming that the drop across R_G can be safely ignored) as

$$R_S = \frac{V_{GS}}{I_D}, \tag{7.52}$$

and we use (7.49) along with the desired value of V_{DS} to find R_D:

$$R_D = \frac{V_{DD} - V_{DS} - I_D R_S}{I_D}. \tag{7.53}$$

Finally, we choose R_G so that the drop across it is negligible. Typical JFETs have gate currents in the range of 1 nA, so a 1-MΩ resistor produces only a 1-mV drop. Notice that this circuit does not provide the designer with any flexibility. For example, if the bias point turns out to be too sensitive to the values of I_{DSS} and V_p, there isn't anything we can do to reduce this sensitivity. If the stability of the bias point is unacceptable using this circuit, the designer would have to use a different topology. For example, the bias circuit shown in Figure 7-49 could be used.

7.3 BIASING OF MULTISTAGE AMPLIFIERS

In this section, we consider the analysis and design of biasing circuits for multistage amplifiers. By using coupling capacitors to isolate the DC biasing in each stage (i.e., AC coupling, as discussed in Chapter 6), it is possible to cascade individual amplifiers as shown in Figure 7-54, wherein the amplifier symbols represent complete amplifiers, including bias circuitry. The coupling capacitors, C_{IN}, C_{OUT}, C_{C1}, and C_{C2}, are all chosen to be large enough that their impedances are negligible at the frequencies of interest for the signal. They serve to couple the signal into and out of each amplifier without disturbing the DC bias. If a multistage amplifier is constructed in this way, the analysis and design of the biasing circuits for each individual stage are identical to what we have presented in previous sections.

Capacitive coupling is sometimes used in multistage amplifiers and it does provide the designer with the greatest degree of flexibility in choosing the bias point of each stage. Nevertheless, there are drawbacks to capacitive coupling. First of all, capacitively coupled amplifiers have zero gain at DC and therefore cannot be used with signals that contain a DC component. The second drawback to capacitive coupling is that the resulting amplifier contains more components than may be necessary and, as a result, will cost more, consume more power, and have a lower ***mean time between failures*** (MTBF).

Because of the problems associated with capacitive coupling, it is frequently desirable to directly couple amplifiers together (i.e., DC coupling as discussed in Chapter 6). The difficulty that arises with direct coupling is that the bias points of successive stages are not independent, and we have reduced freedom in setting the bias points of the individual transistors. Nevertheless, the analysis and design procedures are very much the same as for single-stage amplifiers if our rules of thumb are followed. In the sections that follow, we examine how our rules of thumb apply to multistage amplifiers.

7.3.1 Cascaded Bipolar Amplifiers

Consider the amplifier shown in Figure 7-55. This type of cascade is quite common. To analyze the circuit, we start with the first stage. If the base current of Q_2 is negligible in comparison with I_{C1}, the second stage does not affect the bias point of the first stage at all, and the analysis is exactly the same as for a single-stage amplifier. Once the value of V_{C1} is determined, we use that value as the DC base voltage of Q_2 and proceed with the analysis in the usual way.

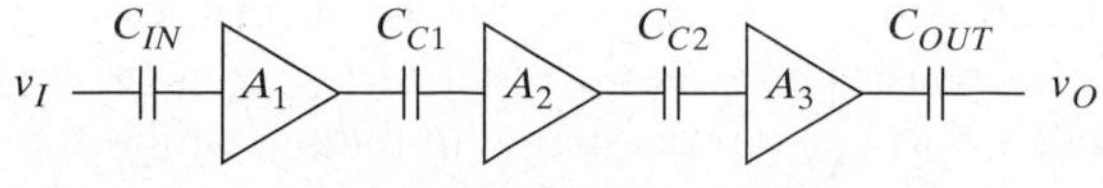

v_I — C_{IN} — A_1 — C_{C1} — A_2 — C_{C2} — A_3 — C_{OUT} — v_O

Figure 7–54 A capacitively coupled cascade of amplifiers.

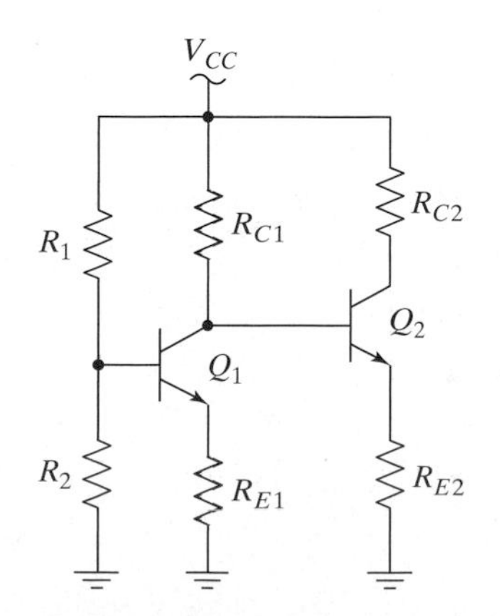

Figure 7–55 A common emitter-emitter follower cascade.

To design this circuit, we apply our previous rules of thumb directly to the first stage. We set the Thévenin voltage driving the base greater than 2 V and guarantee that the current in R_1 is at least 10 times the base current of Q_1. Then, in establishing the bias point of Q_2, we need to guarantee that $I_{C1} > 10I_{B2}$ so that if the base current of Q_2 changes (because of a change in β, component value, or temperature), the bias voltage at its base will not change too much. Since the emitter voltage of Q_2 is approximately 0.7 V below its base, a change in V_{B2} would show up as a change in V_{E2} and would cause a change in I_{C2} ($I_{C2} \approx V_{E2}/R_{E2}$).

If we set $I_{C1} = 10I_{B2}$ and if $\beta \approx 100$, then $I_{C2} \approx 10I_{C1}$. We continue to apply this rule of thumb as more stages are cascaded, so we have a general ***rule of thumb for cascaded bipolar amplifiers***:

> Set the bias currents in the transistors so that they increase by a factor of 10 as you move from left to right through the stages.

Finally, we note that if $V_{E1} > 1.3$ V and Q_1 is forward active, we are guaranteed that the base voltage of Q_2 will be high enough to make its bias current insensitive to V_{BE2}.

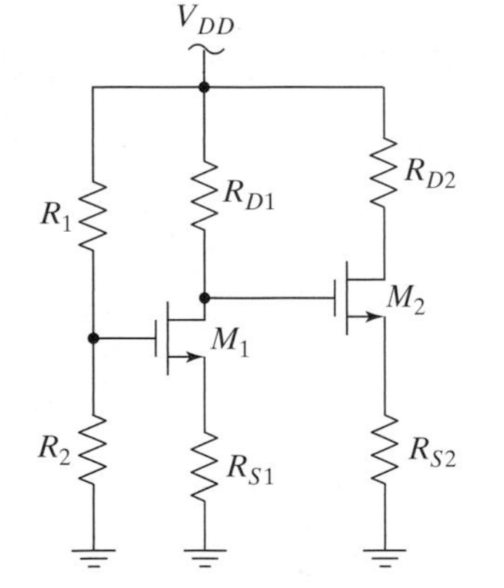

Figure 7–56 A common-source source follower cascade.

7.3.2 Cascaded FET Amplifiers

Consider the amplifier shown in Figure 7-56. Since the DC gate current of a MOSFET is zero, the second stage does not affect the DC bias point of the first stage at all. The only complication we have is that the DC drain voltage of the first stage is the DC gate voltage of the second stage, which significantly reduces our freedom in choosing the bias points of the two transistors.

The rule of thumb we had for producing a stable bias point in FET amplifiers said to keep the voltage across R_S greater than 10 V or as large as possible. If we follow that rule for M_1, and if M_1 is forward active, we are virtually guaranteed to follow the rule for later stages as well. Therefore, there aren't any new rules of thumb for biasing cascaded discrete MOS amplifiers.

7.4 BIASING FOR INTEGRATED CIRCUITS

The way we bias transistors on an integrated circuit (IC) is usually very different than in discrete circuits because the tolerances, matching, and relative costs of the components are significantly different. Since all of the components on an IC are fabricated at the same time on the same wafer, they tend to match each other very well. Also, all of the components on an IC are physically close to each other and mounted in the same package, so the differences in temperature from one device to another are much smaller than they are with discrete components. Finally, since the cost of fabricating a wafer with a given process is fixed, the cost of each individual element on the wafer is proportional to the area it consumes. Transistors and resistors vary quite a bit in size, but, in general, transistors are much smaller than resistors, which are much smaller than capacitors. Therefore, in designing an IC, we typically avoid using capacitors and resistors as much as possible.

Another factor to be considered is the quality of the components produced in a given process. Most IC resistors are not as good as discrete resistors. IC resistors typically vary more with temperature than discrete resistors do, and they are also voltage dependent (i.e., the value of resistance varies with the voltage applied), whereas discrete resistors are not, to first order, voltage dependent (although they do change value if the power dissipated in them changes their temperature). Resistors in an IC process also usually have a substantial amount of parasitic capacitance that is not present in discrete resistors. Similar statements apply to integrated capacitors, they are often voltage dependent and have more parasitics than discrete capacitors. Integrated inductors are rarely used, except at RF frequencies, especially in silicon. The reason is in part because the conductive substrate underneath the inductor allows large eddy currents to flow, which reduce the quality of the inductor. Another reason is that the lack of three-dimensional coils and high-permeability cores limits the amount of inductance attainable. All of these factors contribute to a preference for using transistors instead of resistors in IC design and for avoiding the use of capacitors and inductors whenever possible.

Capacitors are avoided in IC design by using DC-coupled amplifiers almost exclusively. To understand how we can use transistors instead of resistors for biasing, we first notice that the net result of the resistive bias circuits studied in the previous sections was to establish a particular operating point for each transistor in a circuit.

It is possible to bias a transistor using fewer resistors if we have current sources available in addition to the usual voltage sources. Figure 7-57 gives an example of a generic single-transistor amplifier biased using a current source and two voltage sources. This circuit establishes a bias current I_{BIAS} in the transistor by allowing the DC voltage at the merge node to be whatever is necessary to support the current. The bypass capacitor is an open circuit for DC-biasing purposes and is used to make sure that the merge node is an AC ground. In other words, the capacitor is chosen to be large enough that it will look essentially like a short circuit at the frequencies of interest for the signal. Without the capacitor present, the amplifier would not work. Since the current source has a constant value, the current out of the merge terminal on the transistor would be constant, and therefore, so would the other terminal currents. In other words, the merge node would be an open circuit for AC signals if the capacitor were not present. We will see how to eliminate this capacitor in Section 7.5, so we won't worry about it now, as it does not affect the biasing.

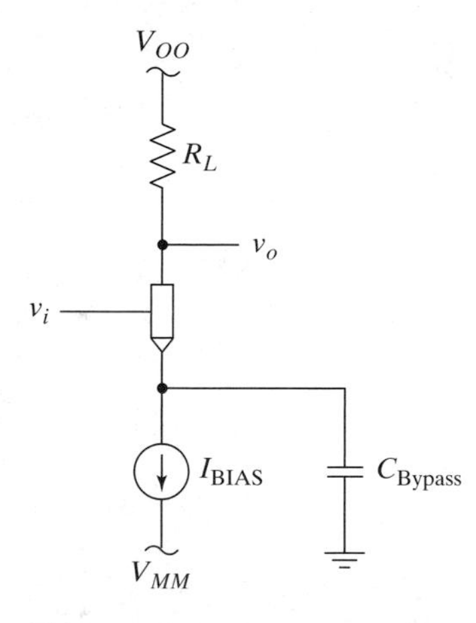

Figure 7–57 A generic single-transistor amplifier with current source biasing.

With the bias current established in the transistor, the only quantity remaining to determine is the output voltage. We have

$$V_O = V_{OO} - I_{BIAS}R_L, \tag{7.54}$$

which is identical to the equation for a resistively biased circuit. The examples that follow illustrate this technique with specific transistors. If you are unfamiliar with either bipolar or field-effect transistors at this point, you may safely ignore one of the examples.

Example 7.20

Problem:

Determine all of the DC voltages and currents for the circuit shown in Figure 7-58. The signal would be input at the base.

Solution:

By inspection, the emitter current of the transistor is 0.5 mA. The emitter voltage will be near −0.7 V, but the exact value has no effect on the operation of the circuit. Assuming that β is reasonable, we can conclude that the collector current is nearly 0.5 mA as well, so the collector voltage is 7.5 V, and the transistor is forward active, as implicitly assumed.

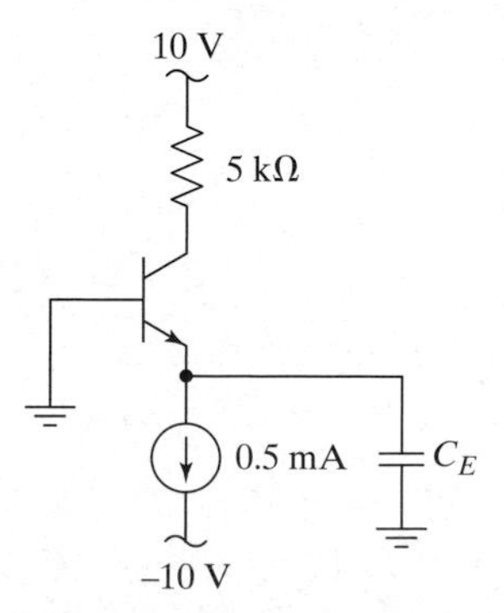

Figure 7–58 A bipolar gain stage using current source biasing.

■

Example 7.21

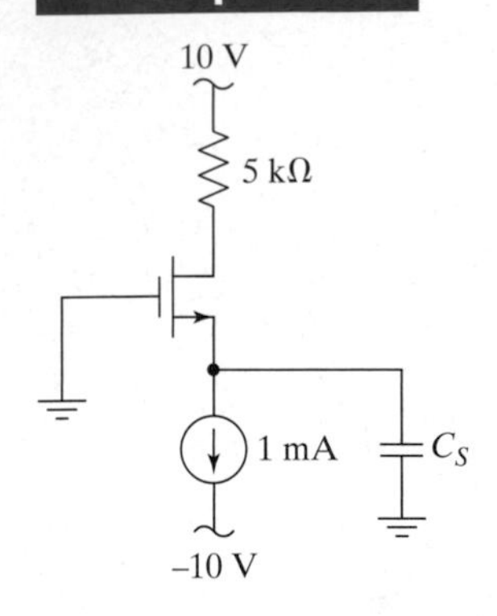

Figure 7–59 A FET gain stage using current source biasing.

Problem:
Determine all of the DC voltages and currents for the circuit shown in Figure 7-59. Assume that $V_{th} = 0.8$ V and that $K = 250\ \mu A/V^2$. The signal would be input at the gate.

Solution:
By inspection, the source current of the transistor is 1 mA. The source voltage will be

$$V_S = V_G - V_{GS} = -V_{GS} = -V_{th} - \sqrt{\frac{I_D}{K}},$$

which in this case yields –2.8 V, but the exact value has no effect on the operation of the circuit. Since the drain current is also 1 mA, the drain voltage is 5 V, and the transistor is forward active, as implicitly assumed.

■

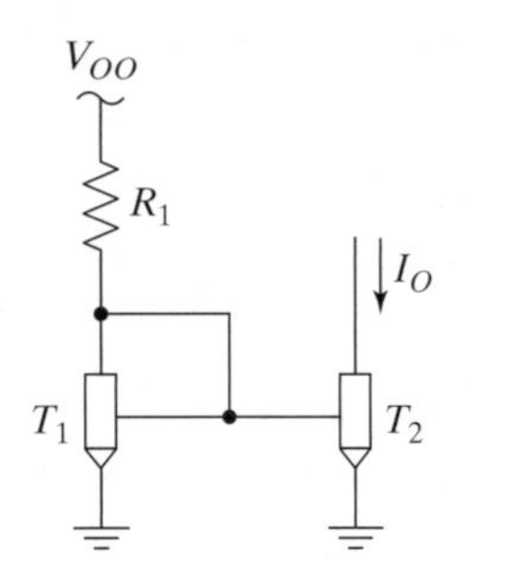

Figure 7–60 A transistor current source.

Now that we have seen how to bias a transistor by using a current source, the logical question is, how do we make the current source? One way of making a current source is shown in Figure 7-60. We have not indicated what the output is tied to, but the voltage at this node must be high enough to keep T_2 in the forward-active region. For this circuit to work, it is essential that the transistors be matched. *From here on out in this section, we will assume that all transistors are identical unless stated otherwise.*

The operation of the circuit can be understood by considering what happens when the power supply is first turned on (i.e., assume V_{OO} makes a step change from 0 to its normal value). Initially, T_1 and T_2 are off. But as V_{OO} increases, the control voltages of these transistors will increase and current will start to flow. The output current of T_1 will flow through R_1 and will reduce the magnitude of the control voltage. The circuit will find the operating point that simultaneously satisfies the equation for the transistor's operation and the equation for the rest of the circuit. For the sake of discussion, let's say that the final solution is known. Since the transistors are matched and have the same control voltages, they will also have the same output currents. If the currents into their control inputs are negligible compared with their output currents, the current through R_1 is the output current of T_1, which is approximately the output current of T_2 as well. In other words, the current through R_1 is *mirrored* to the output of T_2. As a result, this circuit is called a **current mirror**.

At this point, it may not be obvious that any real savings have been achieved relative to resistive biasing. Combining Figures 7-57 and 7-60, we see that we now need two resistors and three transistors to establish the bias current, whereas we previously needed four resistors. The real advantage of the current source biasing technique is not apparent if a single-transistor amplifier is considered, however. In ICs, we typically mirror the current to more than one location and, thereby, use a single resistor to establish many bias currents, as shown in Figure 7-61.

In the rest of this section, we treat field-effect and bipolar mirror circuits separately.

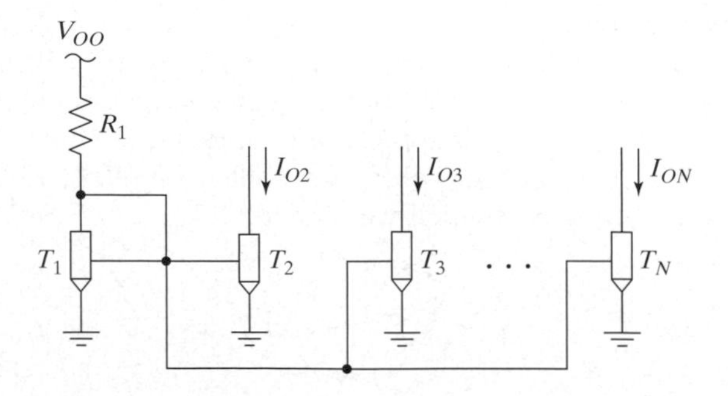

Figure 7–61 The repeated mirroring of the current through R_1.

7.4.1 Simple Bipolar Current Mirrors

A bipolar current mirror is shown in Figure 7-62 where the load has been represented by a voltage source. Q_1 in this circuit is often called a *diode-connected* transistor because it functions as a two-terminal diode.

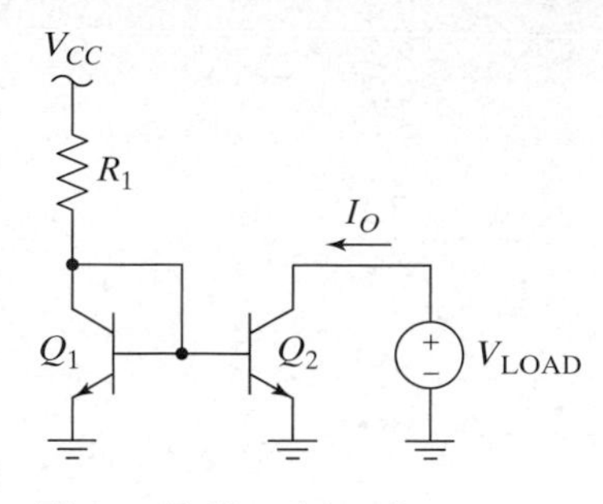

Figure 7–62 A bipolar current mirror.

Figure 7-63 shows the circuit with our standard models used for the transistors. Writing KCL at the collector of Q_1 leads to

$$I_{R_1} = \frac{V_{CC} - 0.7}{R_1} = (\beta_1 + 1)I_{B1} + I_{B2}, \tag{7.55}$$

which is one equation in two unknowns. If we ignore base-width modulation (i.e., assume that V_{CE} does not affect I_C), we can reasonably assume that $I_{B1} = I_{B2}$, since the devices are matched and have the same V_{BE}. Using this assumption yields (note: matched devices implies $\beta_1 = \beta_2 = \beta$)

$$I_{R1} = \frac{V_{CC} - 0.7}{R_1} = (\beta + 2)I_{B1}, \tag{7.56}$$

which implies that

$$I_{B2} = I_{B1} = \left(\frac{1}{\beta+2}\right)\frac{V_{CC} - 0.7}{R_1}. \tag{7.57}$$

Therefore,

$$I_O = I_{C2} = \beta I_{B2} = \left(\frac{\beta}{\beta+2}\right)\frac{V_{CC} - 0.7}{R_1}, \tag{7.58}$$

and if $\beta \gg 2$,

$$I_O \approx \frac{V_{CC} - 0.7}{R_1}. \tag{7.59}$$

We can get a more accurate answer by using the full equation for the collector current of a BJT, including base-width modulation. We first find the ratio of the collector currents. Using the full equations and setting $V_{BE2} = V_{BE1} = V_{BE}$, we find that

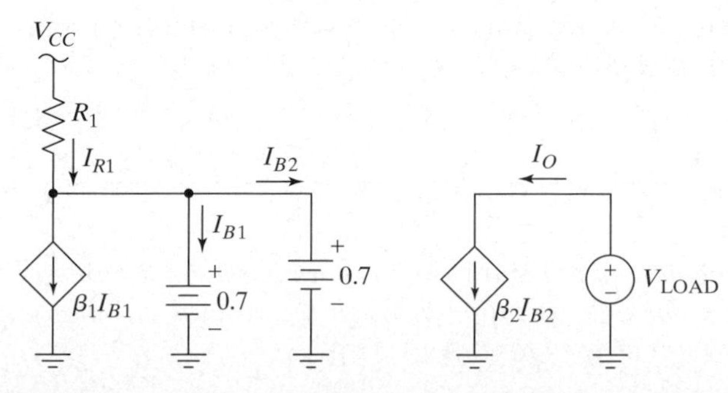

Figure 7–63 The mirror of Figure 7-62 with the transistors replaced by our standard large-signal DC model.

$$\frac{I_{C2}}{I_{C1}} = \frac{I_S e^{V_{BE}/V_T}(1 + V_{LOAD}/V_A)}{I_S e^{V_{BE}/V_T}(1 + V_{BE}/V_A)} = \frac{V_A + V_{LOAD}}{V_A + V_{BE}} \quad \textbf{(7.60)}$$

where we have assumed the transistors to be matched so that $I_{S1} = I_{S2} = I_S$, $V_{T1} = V_{T2} = V_T$, and $V_{A1} = V_{A2} = V_A$ and have also used the fact that $V_{CE1} = V_{BE}$ and $V_{CE2} = V_{LOAD}$. From (7.60), we see that so long as V_A is much larger than both V_{BE} and V_{LOAD}, we have $I_O = I_{C2} \approx I_{C1}$. So, if $\beta \gg 2$, we can ignore the base currents and write

$$I_O \approx I_{C1} \approx I_{R1} \approx \frac{V_{CC} - 0.7}{R_1}. \quad \textbf{(7.61)}$$

Exercise 7.10

Find the output current for the mirror in Figure 7-64. Ignore base-width modulation and assume that β is large.

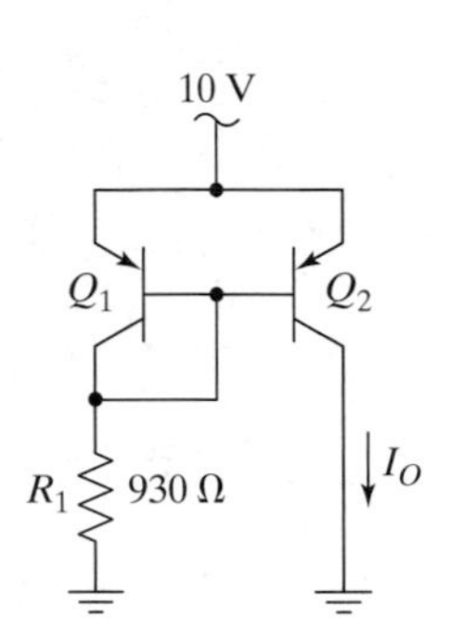

Figure 7–64 A *pnp* current mirror.

7.4.2 More Advanced Bipolar Current Mirrors

The current mirrors examined thus far have output currents approximately equal to the input current. However, it is very useful to be able to produce currents that are a fixed multiple or submultiple of the input current. Adjacent transistors on an IC are nearly identical in every way, but they do not have to be the same size. Since the linking saturation current of a BJT, I_S, is proportional to the emitter area of the device, adjacent transistors on an IC can have different values of I_S, whereas they remain matched in every other way. If we denote the emitter area of Q_N by A_N, we can write

$$\frac{I_{S1}}{I_{S2}} = \frac{A_1}{A_2}. \quad \textbf{(7.62)}$$

Using (7.62) in (7.60) and assuming that base-width modulation can be ignored and β is large, we find that

$$I_O \approx \frac{A_2}{A_1} I_{R_1}. \quad \textbf{(7.63)}$$

So, it is possible to scale a current at the same time as we mirror it. This fact is used quite frequently in IC design. If no special notation is made on the schematic, the transistors are all assumed to be the same size (usually the minimum size, denoted 1×). If a transistor is larger than minimum size, this is usually—but not always—indicated on the schematic by writing, for example, 4× next to the emitter for a transistor that has four times the minimum area.

Exercise 7.11

Find the output currents for the mirror in Figure 7-65. Ignore base-width modulation and assume that β is large. The line extending through the bases is a common way of showing that all of the bases are tied together.

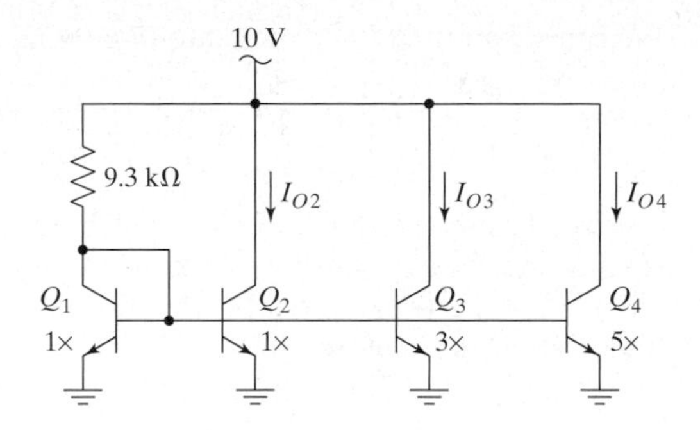

Figure 7–65 A multiple-output current mirror.

There are many variations and improvements possible to the simple current mirror. For example, if the base currents cause a significant error, a third transistor can be used, as shown in Figure 7-66. A helper transistor like this is usually used, especially if the diode-connected transistor is used to drive a number of other transistors as in Figure 7-65.

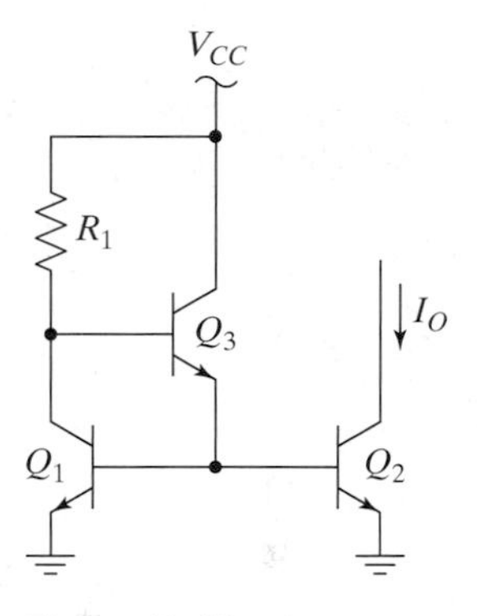

Figure 7–66 An improved current mirror.

Another common modification is to include one or more emitter resistors in the mirror, as demonstrated by R_E in Figure 7-67. The emitter resistor serves two purposes: It allows a much larger scaling than can be practically achieved by scaling the sizes of the transistors, and it increases the small-signal output resistance of the mirror (discussed in Chapter 8). This kind of mirror is frequently called a **Widlar current source** in honor of its inventor, Robert Widlar. This circuit allows us to produce very small currents without the use of large resistors, which require lots of area.

Notice that the voltage at the base of R_1 is still just the V_{BE} of Q_1, so we know that if we ignore base-width modulation and assume that β is large, we get

$$I_{C1} \approx \frac{V_{CC} - 0.7}{R_1}, \tag{7.64}$$

as before. To find I_O, we must write KVL around the base-emitter loop, namely,

$$V_{BE1} - V_{BE2} - I_O R_E = 0, \tag{7.65}$$

and solve to get

$$I_O = \frac{V_{BE1} - V_{BE2}}{R_E} = \frac{\Delta V_{BE}}{R_E}, \tag{7.66}$$

where we have implicitly defined a new quantity, ΔV_{BE}. If we assume that $V_{BE2} = 0.7$ V as usual, we arrive at the silly conclusion that I_O is zero. In other words, our simple approximation for the V_{BE} of a BJT cannot be used here. This circuit is a great illustration of a general principle: We *must* understand the limitations of our models, or our answers may be complete nonsense.

We can express ΔV_{BE} as a function of the two collector currents in the following way (we assume that the transistors are identical, so that $I_{S1} = I_{S2} = I_S$, although this assumption is not necessary):

$$\Delta V_{BE} = V_{BE1} - V_{BE2} = V_T \ln\frac{I_{C1}}{I_S} - V_T \ln\frac{I_{C2}}{I_S} = V_T \ln\frac{I_{C1}}{I_{C2}}. \tag{7.67}$$

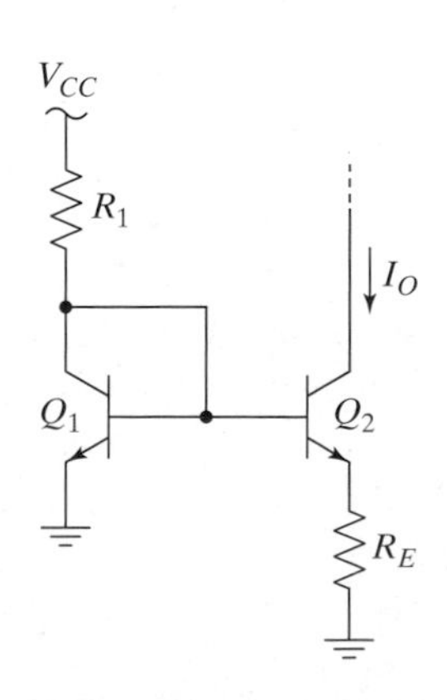

Figure 7–67 A Widlar current source.

We can substitute (7.67) back into (7.66) and note that $I_{C2} = I_O$ to get

$$I_O = \frac{V_T}{R_E}\ln\frac{I_{C1}}{I_O}. \quad \textbf{(7.68)}$$

This is a transcendental equation, which must be solved iteratively to find I_O. Designing the Widlar source is somewhat easier than analyzing it, since we don't need to solve a transcendental equation. If we want an output current that is one hundred times smaller than the input current, we use (7.67) to determine the required ΔV_{BE} and then use (7.66) to find the necessary value for R_E.

Example 7.22

Problem:
Determine the values required for R_1 and R_E for the circuit in Figure 7-67 to have $I_O = 10\ \mu A$ and $I_{R1} = 1$ mA. Neglect base-width modulation and assume that β is large. Use $V_{CC} = 10$ V.

Solution:
We first find R_1 by using (7.62) and get $R_1 = 9.3\ k\Omega$. We next use (7.67) to find $\Delta V_{BE} = 173$ mV. Finally, we use (7.66) to get

$$R_E = \frac{\Delta V_{BE}}{I_O} = \frac{173\ \text{mV}}{10\ \mu\text{A}} = 17.3\ \text{k}\Omega.$$

■

7.4.3 FET Current Mirrors

A MOSFET current mirror is shown in Figure 7-68, where the load has been represented by a voltage source. M_1 in this circuit is often called a ***diode-connected*** transistor because it functions as a two-terminal device with an *i–v* characteristic that looks like a diode.

We analyze the circuit in Figure 7-68 by writing KCL at the drain of M_1 to get

$$\frac{V_{DD} - V_{GS}}{R_1} = K_1(V_{GS} - V_{th})^2, \quad \textbf{(7.69)}$$

where we have assumed that $V_{th1} = V_{th2} = V_{th}$, but have allowed for different K's. This equation can be rewritten as a quadratic and solved for V_{GS} to obtain

$$V_{GS} = V_{th} - \frac{1}{2K_1R_1} \pm \frac{1}{2}\sqrt{\left(\frac{1}{K_1R_1} - 2V_{th}\right)^2 - 4\left(V_{th}^2 - \frac{V_{DD}}{K_1R_1}\right)}. \quad \textbf{(7.70)}$$

As always, we will be able to eliminate one of the two solutions given by (7.70) in a straightforward way. The output current can be found as soon as we know V_{GS}; we obtain

$$I_O = K_2(V_{GS} - V_{th})^2, \quad \textbf{(7.71)}$$

where we have ignored the correction for channel-length modulation.

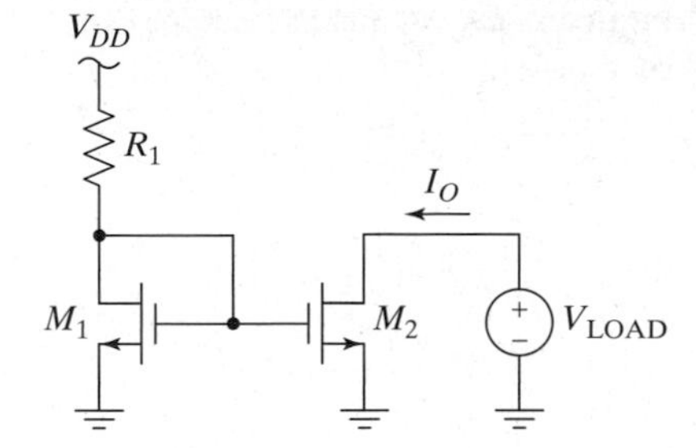

Figure 7–68 A MOSFET current mirror.

Example 7.23

Problem:
Find I_O for the circuit of Figure 7-68, assuming that $V_{DD} = 5$ V, $V_{th} = 0.7$ V, $R_1 = 10$ kΩ, and $K_1 = K_2 = 20$ μA/V².

Solution:
Using (7.70), we find that V_{GS} is either 3.5 V or −7 V. The answer must be 3.5 V, since nothing in the circuit can cause V_{GS} to be −7 V. Using 3.5 V in (7.71), we find that $I_O = 150$ μA.

Designing the current mirror of Figure 7-68 is easier than analyzing it, since we don't need to solve a quadratic. Notice that the ratio of the drain currents is (again ignoring channel-length modulation)

$$\frac{I_O}{I_{D1}} = \frac{I_{D2}}{I_{D1}} = \frac{K_2(V_{GS} - V_{th})^2}{K_1(V_{GS} - V_{th})^2} = \frac{K_2}{K_1} = \frac{(W/L)_2}{(W/L)_1}, \tag{7.72}$$

where we have assumed that the devices are perfectly matched except for the (W/L)'s. To design the mirror, we first use (7.72) to find the required ratio of (W/L)'s and then find the necessary V_{GS} from

$$V_{GS} = V_{th} + \sqrt{\frac{I_{D1}}{K_1}}. \tag{7.73}$$

Knowing V_{GS}, we find R_1 using Ohm's law:

$$R_1 = \frac{V_{DD} - V_{GS}}{I_{D1}}. \tag{7.74}$$

It is possible to use a single diode-connected transistor to drive a number of other transistors, as demonstrated in Exercise 7.12.

Exercise 7.12

Find the drain currents for the mirror in Figure 7-69. Ignore channel-length modulation, and assume that the devices are perfectly matched except for the (W/L)'s, which are given on the schematic. $V_{th} = 1$ V and $K' = 10$ μA/V². The line extending through the transistors is a common way of showing that all of the gates are tied together.

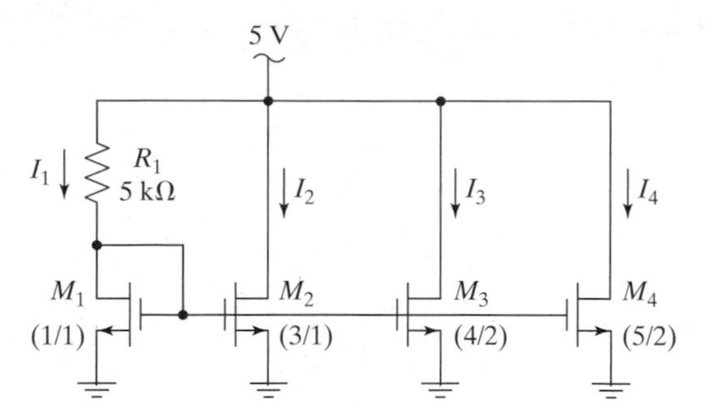

Figure 7–69 A multiple output current mirror.

7.5 BIASING OF DIFFERENTIAL AMPLIFIERS

All of the amplifiers discussed up to this point use ***single-ended*** signals; that is, the signal voltages appear between different nodes and ground. In this section, we want to consider how to construct amplifiers that have ***differential*** inputs; that is, the input voltage appears between two nodes other than ground. If you are not familiar with differential signals, you should review Aside A5.1 before proceeding. Aside A5.2 may also be useful to review, since it presents some justification for why differential signals are important. From this point on in this section, we will assume familiarity with the notation presented in Aside A5.1.

Differential amplifiers are ubiquitous in integrated circuit design, but rare in discrete circuits. The reason for this is that the transistors constituting the differential amplifier must be extremely well matched for the circuit to function. Therefore, differential amplifiers can be made with discrete components only if a matched pair of transistors (two transistors in one package, made on the same substrate) or a transistor array (more than two transistors in one package, made on the same substrate) is used, or if the differential pair is heavily ***degenerated***. We will define degeneration in Chapter 8. This section will be very brief, since it will discuss only the DC biasing of the differential amplifier. We defer a more complete discussion to Chapter 8.

A differential amplifier is shown in Figure 7-70, wherein the signal source has been represented by its differential and common-mode components and we have used generic transistors. The connection of two transistors in the manner shown is called a ***differential pair*** or, rarely, a ***long-tailed pair*** (an older terminology more common in Britain). The differential pair is the basic input stage used in virtually all integrated operational amplifiers, although resistive loads as shown in the figure are not usually used. The circuit is shown here as operating from a single supply voltage, but it is frequently used with both positive and negative supplies so that the DC common-mode input voltage can be zero. The DC current source labeled I_{MM} is often called the ***tail current*** of the differential pair and is usually implemented by using current mirrors.

To understand the biasing of a differential pair, let's work directly from the circuit in Figure 7-70. Assume that the differential signal is set to zero, but the DC part of the common-mode signal is large enough to turn on T_1 and T_2 and to keep the current source working. With v_{IDM} set to zero, we can see that the two transistors have the same control voltage. In addition, the merge terminals are tied together so that $v_{CM1} = v_{CM2}$, and if the transistors are perfectly matched, it follows that their output currents I_{O1} and I_{O2} are equal as well. We can also conclude that I_{M1} and I_{M2} are equal, so writing KCL at the merge node allows us to conclude that $I_{M1} = I_{M2} = I_{MM}/2$. Given the symmetry of the circuit, it is not surprising that the bias current splits equally between the two transistors when the differential input

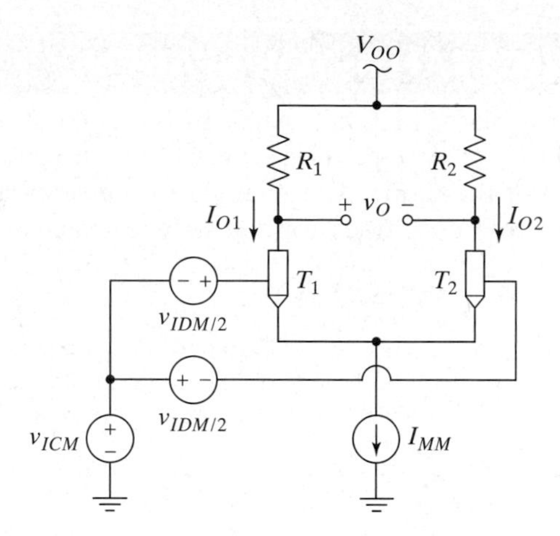

Figure 7–70 A differential amplifier.

voltage is zero. Given that I_{O1} and I_{O2} are equal, we can see that the differential output voltage will be zero if R_1 and R_2 are equal. The single-ended output voltage on either side (i.e., at the output of either T_1 or T_2) will be equal to $V_{OO} - I_{MM}R_1/2$.

Designing the bias circuit for a differential amplifier requires first setting the bias current to twice the desired operating currents of T_1 and T_2. We then choose the resistors R_1 and R_2 to provide the desired DC output voltages and check to see that the transistors are in the forward-active region of operation. We defer further discussion of the design to Chapter 8, since we need to understand the small-signal operation of the circuit.

As noted earlier in this section, most operational amplifiers do not use resistive loads as shown in Figure 7-70. Before we can show the loads actually used, however, we need to have a *p*-type generic transistor model, which is presented in Aside A7.2.

Figure 7-71 shows a differential pair with the load resistors replaced by a *p*-type current mirror. This circuit configuration is called an ***actively loaded*** differential pair, since the load is made up of active devices. (The load itself is referred to as an ***active load***.[12]) In this case, we have both positive and negative power supplies and have assumed that the common-mode input voltage is zero, so we have not bothered to use subscripts to indicate that the input shown is the differential-mode input. We have shown the amplifier driving a short-circuit for a load

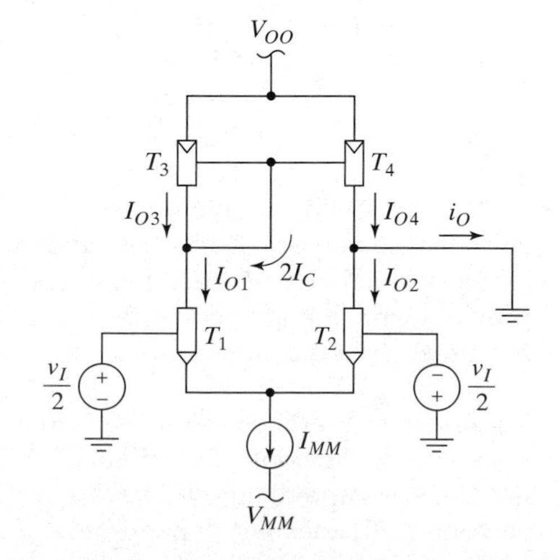

Figure 7–71 An actively loaded differential pair.

[12] It is also called a current-mirror load since T_3 and T_4 comprise a *p*-type current mirror.

ASIDE A7.2 THE GENERIC TRANSISTOR—PART II

We presented an n-type generic transistor in Aside A6.3. In the current aside, we add a p-type complement to it. The schematic symbol for a p-type generic transistor is shown in Figure A7-2(a), and a model for the device is shown in part (b) of the figure. The triangle in the schematic symbol again points in the direction of conventional current flow, as it does for the n-type transistor; the end with the triangle is again the merge terminal.

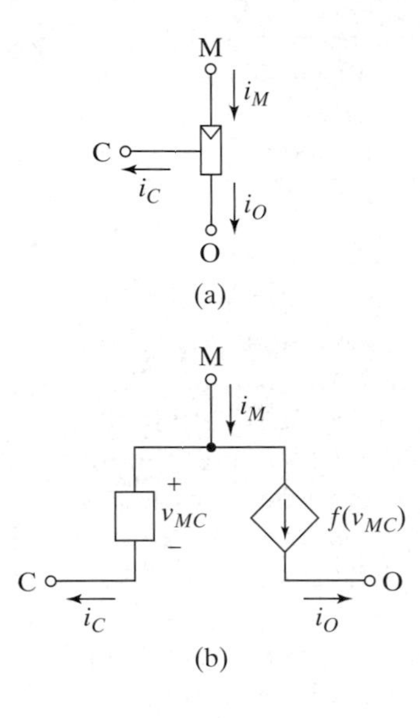

Figure A7-2 A p-type generic transistor: (a) the schematic symbol and (b) a model.

All of the currents and voltages in the p-type generic transistor have the opposite polarity of their counterparts in the n-type transistor. Therefore, in forward-active operation, v_{MC} and v_{MO} are both greater than zero, i_M flows into the device and i_O and i_C flow out of the device. We will assume that i_C is small compared with i_O, as before. The current into the merge terminal is $i_M = i_C + i_O$.

so that the output variable is i_O, but it is also possible to drive a resistive load and consider the output voltage. This configuration is representative of the first stage in the vast majority of operational amplifiers. The stage serves three purposes: It provides gain, it converts the differential input voltage into a single-ended output, and it rejects common-mode inputs.

The DC biasing of the differential pair in this amplifier is the same as was discussed in reference to Figure 7-70, so we know that $I_{O1} = I_{O2}$ if $V_I = 0$. If we assume that the control currents of T_3 and T_4 are negligible, then the current $2I_C$ can be ignored, and we have $I_{O3} \approx I_{O1}$. If we also assume that the output current of T_4 does not depend on V_{O4} at all, or that $V_{O1} \approx 0$ (so the voltages across T_3 and T_4 are the same), then we can further state that $I_{O4} \approx I_{O3}$. Therefore, $I_{O4} \approx I_{O1} = I_{O2}$ and $I_O \approx 0$.

7.6 WORST-CASE ANALYSIS AND PARAMETER VARIATION

Designing circuits that are to be mass produced requires that we consider a number of issues that have not yet been addressed. For example, only certain types and values of components are available, and we must allow for this in our design. Asides A7.4 and A7.5 provide some information on the types and values of discrete passive components that are available. In Chapter 9, we discuss some of the parasitic effects that are most important in passive components at high frequencies. A wonderful reference for general information on different types of capacitors, resistors, and so on is [7.3].

For the circuits we design to be useful in practice, we must be able to manufacture them with a good *yield*, which is the percentage of parts that meet the required specification. We must also know that they will continue to work with reasonable variations in temperature, supply voltage, and component values as the circuits age.

Worst-case analysis is the process of determining the worst possible variation expected in some variable (e.g., bias current) as the parameters in the circuit change. Aside A7.3 illustrates how to perform a worst-case analysis. One problem with doing worst-case analysis by hand is that it often grossly overestimates the variation that will occur.

A better way to approximate the variation is to use the probability distribution functions describing the individual parameters to randomly generate a large number of possible circuits and then look at the distribution of the output variable of interest and see whether or not it is acceptable. This kind of statistical analysis is performed using a computer and is called *Monte Carlo analysis* (a reference to the famous gambling capital).

Monte Carlo analyses can be performed with various different CAD tools, Aside A7.3 illustrates how to use a spreadsheet for this purpose. Many versions of the SPICE program will also allow you to perform Monte Carlo simulations. While these simulations can be useful in specific instances, they are not as widely used as you might suppose, because the simulations can take an extremely long time. In fact, for many complex circuits, a full simulation takes so long that running a Monte Carlo analysis is not practical. Therefore, the designer needs to use his or her judgment to decide which parts of a circuit are critical enough to warrant using this technique.

ASIDE A7.3 WORST-CASE AND MONTE CARLO ANALYSES

This aside illustrates how to perform worst-case analyses by hand and one way to perform Monte-Carlo analyses using a spreadsheet. Consider the two-resistor voltage divider shown in Figure A7-3.

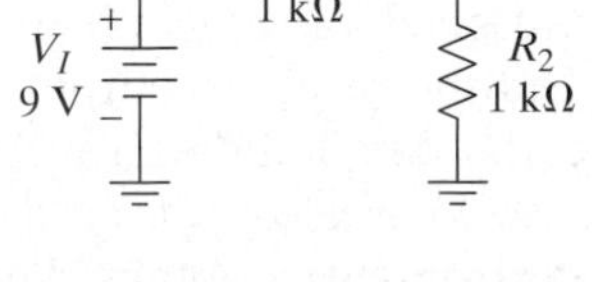

R_1 1 kΩ, V_O, V_I 9 V, R_2 1 kΩ

Figure A7-3 A two-resistor voltage divider.

We can determine by inspection that the nominal output voltage is 4.5 V; but if we know that V_I can vary by ±5% and the resistors both have a tolerance of ±10%, what is the maximum variation in V_O?

To answer this question, we first determine how changes in each of the parameters in the circuit will affect V_O. In this case, there are three parameters to consider: V_I, R_1, and R_2. Taking them one at a time, we see that; if V_I increases, V_O will increase; if R_1 increases, V_O will decrease; and if R_2 increases, V_O will increase. Therefore, the maximum possible value of V_O occurs when V_I and R_2 are maximum and R_1 is minimum; that is,

$$V_O(\max) = V_I(\max)\frac{R_2(\max)}{R_2(\max) + R_1(\min)}$$

$$= 9(1.05)\frac{1100}{1100 + 900} = 5.2 \text{ V}, \qquad \textbf{(A7.4)}$$

which is 15.6% above the nominal value. Similarly,

$$V_O(\min) = V_I(\min)\frac{R_2(\min)}{R_2(\min) + R_1(\max)}$$

$$= 9(0.95)\frac{900}{900 + 1100} = 3.85 \text{ V}, \qquad \textbf{(A7.5)}$$

which is 14.5% below the nominal value. These answers could be significantly worse if we included the variations in component values with temperature and age. Given that this example is a very simple circuit, it is easy to see that designing complex circuits to function well over time and temperature can be very difficult.

While the answers given in (A7.4) and (A7.5) are correct, these values of V_O are extremely unlikely to occur, since all three parameters have to be at or very near one extreme of their range. The more parameters there are involved, the less likely it will be that they will all simultaneously be near their extreme values. Therefore, this type of worst-case analysis may grossly overestimate the variation that is likely to occur in the parameter of interest.

If is often acceptable to have an occasional failure in manufacturing due to extreme component variations, as long as these failures are rare. Therefore, we examine the distribution to see what our yield will be. If you have never studied probability, you may not be able to follow the details in the following illustration of the Monte Carlo analysis technique, but the general idea can still be understood.

Consider again the divider of Figure A7-3, and assume that all parameters (V_I, R_1, and R_2) have uniform distributions[1] (i.e., all values within the given range are equally likely). The probability distribution for a resistor with a nominal value R_o and a uniform distribution is given by[2]

$$f_R(r) = \begin{cases} 1/(2\epsilon R_o) & ; R_o(1-\epsilon) \le r \le R_o(1+\epsilon) \\ 0 & ; \text{elsewhere} \end{cases}, \qquad \textbf{(A7.6)}$$

which is pictured in Figure A7-4(a). Part (b) of the figure shows the probability distribution for a standard random number, as generated by many computer programs (uniformly distributed over the interval from zero to one). We can use this random number, x, to produce R by scaling and shifting the distribution; in other words, given a value, x, from the random number generator, we create a random value for R using

$$R = R_o[1 + 2\epsilon(x - 0.5)]. \qquad \textbf{(A7.7)}$$

Using this form for the random parameter values, a spreadsheet was created. (It is on the CD.) Using the spreadsheet, 400 random parameter sets were generated and the resulting output voltages examined. A histogram of the results is shown in Figure A7-5. Examining the detailed data reveals that, for this particular group of random parameter sets, V_O was never less than 3.95 V or greater than 5.15 V and 97.5% of all the outputs were between 4 V and 5 V. These limits aren't too far from those obtained using our worst-case analysis (3.85 V and 5.2 V), but as the circuit being analyzed becomes more complex, the differences between the limits found using worst-case analysis and those found using Monte Carlo analysis get larger.

Figure A7-4 (a) Probability density function for a resistor and (b) the density function for a random number, as generated by many computer programs.

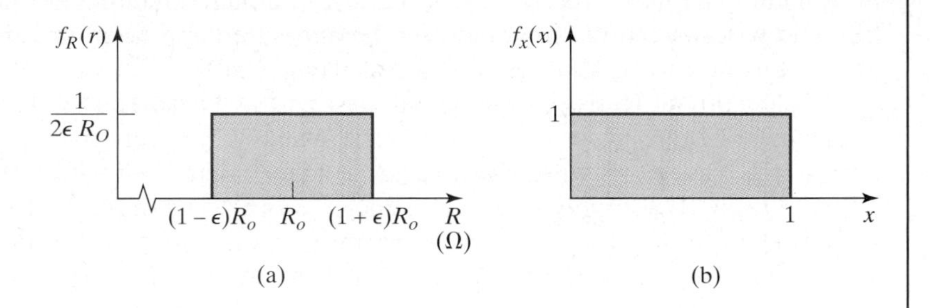

Figure A7-5 Histogram of output voltages for the circuit in Figure A7-3.

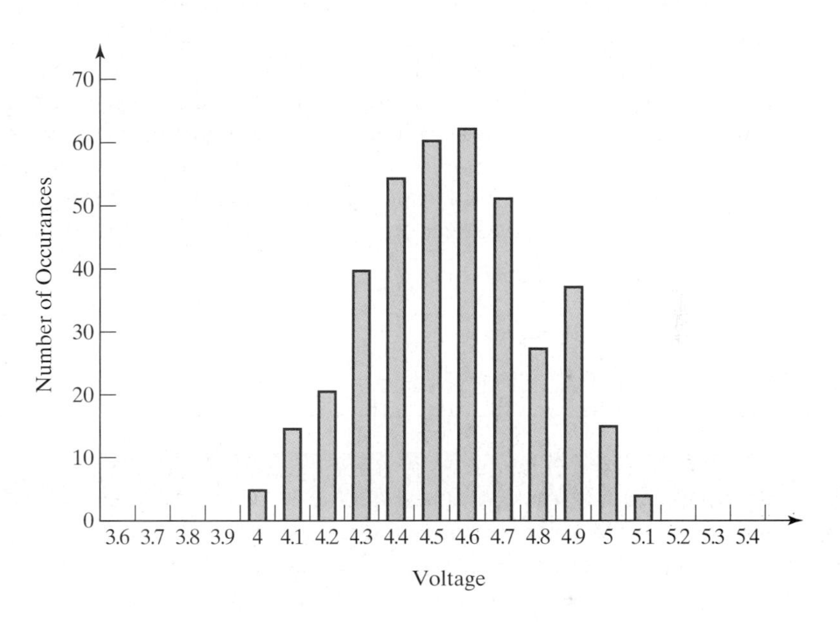

[1] It would be rare for these parameters to have uniform distributions, but it makes the example simpler to follow.

[2] The notation $f_R(r)$ denotes the probability density for the variable R having the value r. It is a probability *density*, rather than a probability, because the distribution is continuous. The probability of R taking on *any* value is zero, but the probability of it being in a given range is found by integrating $f_R(r)$ over that range. Since the variable must assume some value, the integral of $f_R(r)$ over all values must be one.

ASIDE A7.4 VARIATIONS IN DISCRETE PASSIVE COMPONENT VALUES

Resistors, capacitors, inductors, and other passive devices all have tolerances and temperature coefficients (TC's) associated with them. A ***temperature coefficient*** is the rate of change of some parameter with respect to temperature (e.g., a voltage may have a TC of 2 mV/°C). A TC may also be normalized with respect to the nominal value of the parameter, in which case the proper units are $°C^{-1}$. It is necessary to consult manufacturers' catalogs and data sheets for complete information, but some rough guidelines are presented here to give you a feeling for what is available. Only discrete passive components are considered in this aside.

Resistors: Resistors can be purchased in a wide variety of types, power ratings, tolerances, and temperature coefficients. The most common types of resistors are carbon composition or carbon film resistors. They are most commonly available in ±5% and ±10% tolerances (±2% and ±20% are also available, but less common), with no guaranteed specification on the temperature coefficient. Although it is not guaranteed, the temperature coefficient is typically on the order of 5000 ppm/°C (ppm = parts per million). The power ratings most commonly used are 1/8, 1/4, and 1/2 W, although much higher power ratings can be purchased. Metal film and wire-wound resistors can also be purchased and tolerances below ±1% are available. The TC's of metal film and wire-wound resistors are also much better than those of carbon resistors, typically on the order of 25–100 ppm/°C. Resistors can be purchased with guaranteed TC's lower than those quoted here.

Capacitors: Many types of capacitors are available. Crudely speaking, capacitors fall into one of two categories: large capacitors with poor tolerance, large leakage current, and low self-resonant frequencies[1] intended for filtering DC supply lines; and smaller, more accurate, capacitors with low leakage currents and high self-resonant frequencies. The large capacitors (about 1 μF to several farads) are most commonly electrolytic and have tolerances on the order of ±20% or worse. Smaller, higher quality, capacitors come in many types: ceramic, silver–mica, polystyrene, and others. Typical values fall in the range from tenths of a pF to several μF, and common tolerances range between ±2% and ±20%. The TC's of capacitors are often not specified, but are usually smaller than those of resistors. For capacitors that do specify the TC, the values can be positive, negative, or even zero. The most important considerations are usually leakage current and the self-resonant frequency.

Inductors: Inductors are available ranging from small molded inductors (similar in appearance to resistors) to large inductors wound on ferrite or iron cores. The range of inductance commonly available is from fractions of a μH to nearly 1 H, with tolerances on the order of ±10% (some ±5% and a few even better). The TC of inductors is almost never specified, but should be very low. The most important consideration is usually the series resistance.

[1] Capacitors always have some series inductance and therefore form a series resonant circuit. For frequencies above the self-resonant frequency, the impedance of a capacitor is inductive. For electrolytic capacitors, this frequency can be quite low (less than 1 MHz). *See* Section 9.1.2 for a discussion of capacitor parasitics and self resonance.

ASIDE A7.5 STANDARD COMPONENT VALUES

If resistors with a ±5% tolerance are used in a design, the values available are all multiples of the following set of numbers (the table is generated by starting at 1.0, increasing by 10% for each new number and then rounding), with standard ±10% resistor values given in boldface:

1.0	1.1	**1.2**	1.3	**1.5**	1.6	**1.8**	2.0	**2.2**	2.4	**2.7**	3.0
3.3	3.6	**3.9**	4.3	**4.7**	5.1	**5.6**	6.2	**6.8**	7.5	**8.2**	9.1

For example, you can buy the following values as ±5% resistors: 120 Ω, 1.2 kΩ, 12 kΩ, 120 kΩ, 51 Ω, 680 Ω, 8.2 kΩ, and so on. There is a different set of standard values for 1% resistors.

A similar situation exists for capacitors and inductors. In fact, many capacitors and inductors use the table of standard values presented here.

SOLUTIONS TO EXERCISES

7.1 **(a)** With the diode voltage equal to 0.5 V, we see by KVL that $V_{R1} = 9.5$ V, so $I_1 = 9.5/930 = 10.2$ mA. Also, $I_2 = 0.5/700 = 0.7$ mA. **(b)** Similarly, with $V_D = 0.9$ V, we find that $I_1 = 9.78$ mA and $I_2 = 1.3$ mA. The value of I_1 is not very sensitive to V_D and is, therefore, reasonably accurate. The value of I_2, on the other hand, is very sensitive to the value used for V_D and is not very accurate.

7.2 **(a)** With both V_1 and V_2 equal to zero, the diodes are forward biased and $V_O = 0.7$ V. This answer is equal to the voltage we *assume* as the forward bias on the diodes and therefore is not guaranteed to be accurate. **(b)** With $V_1 = 0$ and $V_2 = 2$ V, we find that D_1 is forward biased and D_2 is off. V_O is again equal to 0.7 V and not necessarily accurate. **(c)** With $V_1 = V_2 = 5$ V, both diodes are off, no current flows, and $V_O = 5$ V. This answer is accurate, since the resistor is so small that the reverse leakage currents of the diodes would not produce an appreciable voltage drop.

7.3 The new base current is $I_B = (V_{BB} - 0.6)/R_{BB} = 13.7\ \mu A$, which is about 4% higher than the predicted value. This error is not too large, so the model is reasonable in this case. Note that $V_{BB} - V_{BE}$ is nominally 2.6 V, so a ±0.1-V change in V_{BE} is only a 4% change as we saw; anytime $V_{BB} - V_{BE}$ is more than about 2 V, the error in finding I_B will not be too large.

7.4 Since the collector current is $I_C = \beta_F I_B = 1.5\,mA$ if the transistor is not saturated, and the transistor will saturate as soon as $V_{CE} = 0.2$ V, the maximum value for R_C is given by

$$R_{Cmax} = \frac{V_{CC} - V_{CEsat}}{I_C} = \frac{10 - 0.2}{1.5\ mA} = 6.5\,k\Omega.$$

7.5 Since the drain current is 0.8 mA when the transistor is forward active (*see* Example 7.6), and the transistor will enter the linear region as soon as $V_D = V_{DSsat} = V_{GS} - V_{th} = 4$ V, the maximum value for R_D is given by

$$R_{Dmax} = \frac{10 - V_{DSsat}}{I_D} = \frac{6}{0.0008} = 7.5\ k\Omega.$$

7.6 Start by noting that $V_{BB} = (R_2/(R_1 + R_2))V_{CC}$, which is nominally 2 V for the values found in the example. From (7.30), we obtain

$$I_C = \beta\frac{V_{BB} - V_{BE}}{R_{BB} + (\beta+1)R_E},$$

and we know that $R_{BB} = R_1 \| R_2 = 4\ k\Omega$. Plugging in the values for the nominal design, we find that $I_C = 1.75$ mA, which is 12.5% lower than the design goal. The error is caused by the low value of β. In the following answers, we compare the results with this true nominal I_C instead of the design goal: We find that: **(a)** with $\beta = 100$, $I_C = 1.87$ mA, which is 7% high; **(b)** with $\beta = 25$, $I_C = 1.56$ mA, which is 11% low; **(c)** with $V_{BE} = 0.6$ V, $I_C = 1.88$ mA, which is 7% high; **(d)** with $V_{BE} = 0.8$ V, $I_C = 1.62$ mA, which is 7% low; **(e)** with $V_{CC} = 9.5$ V, $V_{BB} = 1.9$ V and $I_C = 1.62$ mA, which is 7% low; and **(f)** with $V_{CC} = 10.5$ V, $V_{BB} = 2.1$ V, and $I_C = 1.88$ mA, which is 7% high. Except for parts (e) and (f), these deviations are all reasonably small, given the magnitudes of the parameter changes causing them. Changes in V_{CC} are not compensated for.

7.7 With $V_{BB} = 2$ V, we have $V_E = 1.3$ V and $R_E = 1.3/I_E \approx 1.3/I_C = 2.6\ k\Omega$. The base current will be $I_B = I_C/\beta = 3.33\ \mu A$, so, using our rule, we set the current in the base bias string to 33.3 μA, which leads to $(R_1 + R_2) = 5/33.3\ \mu = 150\ k\Omega$. To obtain $V_{BB} =$ 2 V as desired, we set $R_1 = 90\ k\Omega$ and $R_2 = 60\ k\Omega$. Finally, we set $V_C = 3.15$ V, which is halfway between V_E and V_{CC}. Therefore, $R_C = (V_{CC} - V_C)/I_C = 3.7\ k\Omega$.

7.8 From the given values, we find that $V_{GG} = 5$ V. Plugging the given values into (7.39) we find that $V_{GS} = 3.41$ or −0.66 V. The negative solution is impossible, so we have $V_{GS} = 3.41$ V. The corresponding I_D is 2 mA, which leads to $V_S = 1.6$ V and $V_D = 7$ V.

7.9 To achieve $I_D = 100\ \mu A$ requires that $V_{GS} = 1.5$ V. We can meet the required V_{DS} and V_{RD} by setting $V_S = \pm 0.5$ V and $V_D = 1.9$ V. Then $V_G = 2.0$. Setting the current in R_1 and R_2 to 1 μA implies that $R_1 = 1.3\ M\Omega$ and $R_2 = 2.0\ M\Omega$. Then $R_S = 5\ k\Omega$ and $R_D = 14\ k\Omega$. Simulating this circuit with the Nsimple_mos model in Appendix B verifies that the design works as expected.

7.10 The V_{BE} of Q_1 is approximately –0.7 V, so ignoring the base currents, we find that $I_{C1} \approx (10 - 0.7)/930 = 10$ mA. Since $V_{BE2} = V_{BE1}$ and the transistors are matched, we conclude that $I_{C2} \approx 10$ mA as well.

7.11 Assuming that the V_{BE} of Q_1 is 0.7 V and ignoring the base currents, since β is large, $I_{C1} = (10 - 0.7)/9.3\text{k} = 1$ mA. Because Q_2 is the same size as Q_1, we also have $I_{C2} = 1$ mA. Since Q_3 is three times as large as Q_1, its saturation current is three times as large, and, as shown in (7.62) and (7.63), we expect that $I_{C3} = 3$ mA. Similarly, $I_{C4} = 5$ mA.

7.12 Since the DC gate currents are all zero, the presence of M_3 and M_4 does not alter the equation for I_1, and (7.69) and (7.70) apply. Using the given values and (7.70), we find that $V_{GS} = 4.42$ V or –22.4 V. The answer must be 4.42 V, which corresponds to $I_1 = 117$ μA. Ignoring channel-length modulation and assuming that the devices are perfectly matched except for their (W/L)'s leads to $I_2 = 3I_1 = 351$ μA, $I_3 = 2I_1 = 234$ μA, and $I_4 = 2.5I_1 = 293$ μA.

CHAPTER SUMMARY

- When performing large-signal DC analyses, we set all AC sources to zero.
- We open-circuit all capacitors and short-circuit all inductors for DC bias analyses. Linear resistors are not changed for DC analysis.
- Diodes and other nonlinear devices must be replaced by suitable models for DC analysis. It is incumbent on the designer to choose the appropriate model for each circumstance. You should always check the final results to be sure that the model you used was the correct one. Some DC models are linear, and some are not. You must be aware of the approximations being made, or the analysis may be impossible or the results inaccurate.
- Bias circuits are used to establish a desired operating point for the devices in a circuit. The small-signal AC performance depends on the bias point, so we want it to be insensitive to changes that occur over time, or with changing temperature, or from one device to the next.
- We have rules of thumb for intelligent design that can help us design bias circuitry that is reasonably stable.
- Amplifiers are often cascaded. The stages may be DC coupled or AC coupled. AC coupling provides more degrees of freedom and somewhat easier design, but is more costly.
- Integrated circuit amplifiers are usually biased using current sources rather than resistive circuits because current sources typically use less area and are therefore less expensive.
- Differential amplifiers can be resistively loaded or have active loads. They can be easily DC coupled to a source and do not require bypass capacitors.
- Worst-case or Monte Carlo analyses are used to predict whether or not our designs will work when the expected component variations are taken into account. Worst-case analysis is safer and easier, but often overestimates the variation by a wide margin. Monte Carlo analyses are time consuming and are performed with computer simulations.

REFERENCES

[7.1] P.R. Gray, P.J. Hurst, S.H. Lewis and R.G. Meyer, *Analysis and Design of Analog Integrated Circuits*, 4E, New York: John Wiley & Sons, 2001.

[7.2] D.A. Johns and K. Martin, *Analog Integrated Circuit Design*. New York, NY: John Wiley & Sons, Inc., 1997.

[7.3] P. Horowitz and W. Hill, *The Art of Electronics*, 2/E. Cambridge, Mass: Cambridge University Press, 1989.

[7.4] R.L. Geiger, P.E. Allen and N.R. Strader, *VLSI Design Techniques for Analog and Digital Circuits*. New York, NY: McGraw-Hill, 1990.

PROBLEMS

7.1 Large-Signal DC and Low-Frequency Models for Design

7.1.1-3 Independent Sources, Rs, Ls, Cs, and Diodes

P7.1 Draw the large-signal DC equivalent for the circuit shown in Figure 7-72. Assuming that $v_I = 3 + 0.1\sin(628t)$ V, what is V_O?

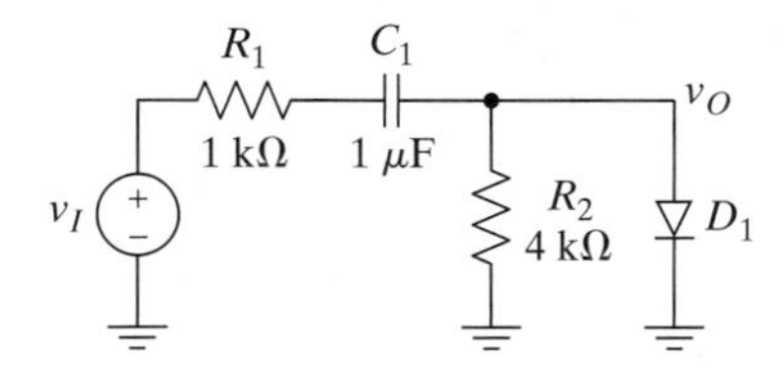

Figure 7–72

P7.2 Draw the large-signal DC equivalent for the circuit shown in Figure 7-73. Assuming that $v_I = 4 + 0.2\sin(628t)$ V, what is V_O?

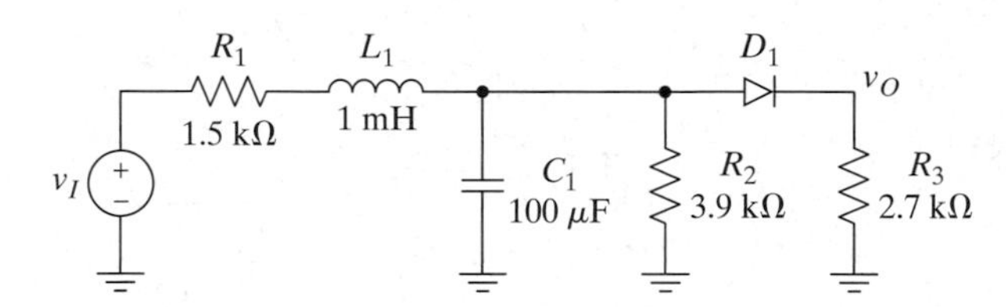

Figure 7–73

P7.3 Solve for all of the DC voltages and currents in the circuit shown in Figure 7-74 when **(a)** $V_A = 3$ V and **(b)** $V_A = 1$ V.

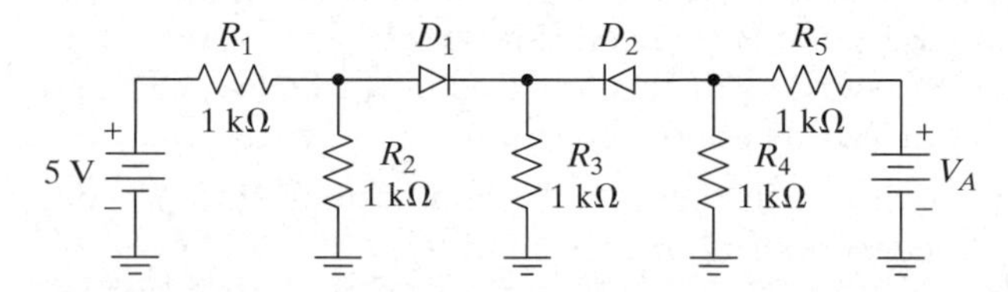

Figure 7–74

D P7.4 Consider the circuit shown in Figure 7-75. Diode D_2 is a light-emitting diode (LED) that has a forward-voltage drop of 2 V (assume that it has zero current for $V_D < 2$ V and can have any current for $V_D = 2$ V) and should nominally be biased at 20 mA of forward current. Find values for all three resistors so that D_2 turns on if V_2 exceeds 6 V and so that $I_{D2} = 20$ mA when $V_2 = 8$ V.

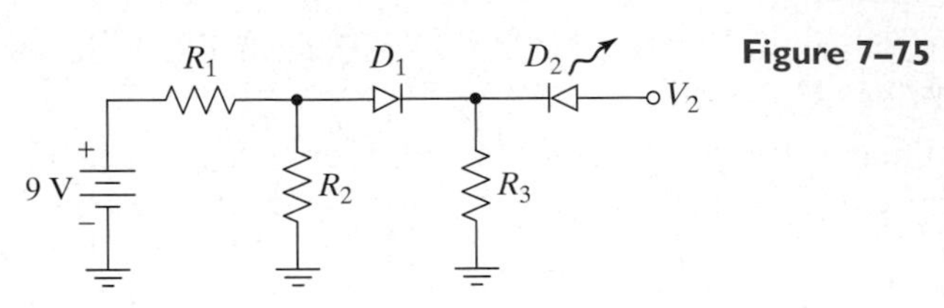

Figure 7–75

P7.5 For the circuit of Figure 7-76, make a sketch of V_O versus V_I for $-2V < V_I < 6V$. (Put V_I on the horizontal axis.) You may make the sketch using a simplified diode model. Check the accuracy of your sketch at $V_I = 0\text{V}$ and at $V_I = 2\text{V}$ by using the complete diode current characteristic equation with reasonable approximations. Use $I_S = 8 \times 10^{-14}$ and $V_T = 26$ mV.

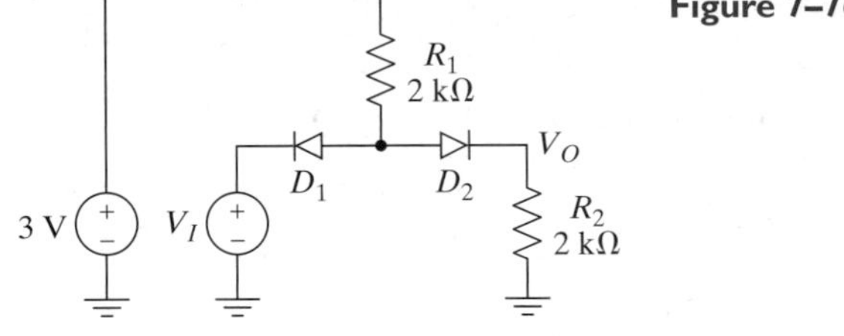
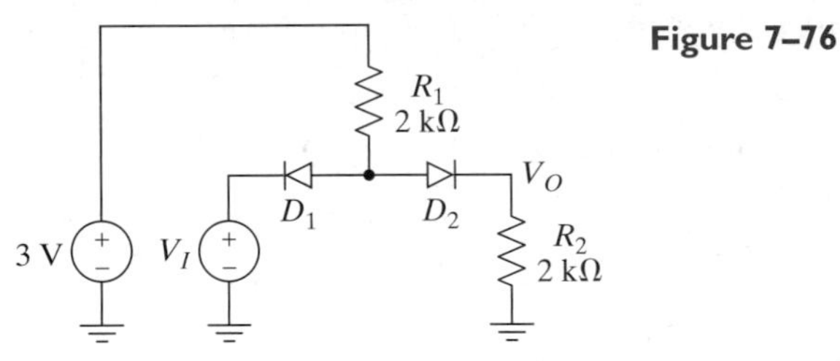

Figure 7–76

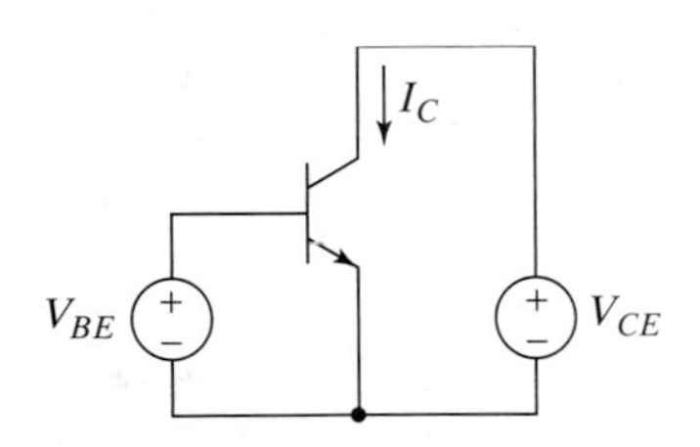

Figure 7–77

P7.6 Consider the circuit shown in Figure 7-77, and assume that the diodes are identical except that $I_{S1} \neq I_{S2}$. Using (7.2), show that $I_{D1} = (I_{S1}/I_{S2})I_{D2}$.

7.1.4 Bipolar Junction Transistors

Figure 7–78

P7.7 Refer to the circuit of Figure 7-78. The bipolar transistor has the following parameters: $I_S = 1.84 \times 10^{-14}$ A, $\beta = 203$, and $V_A = 65$ V. Remember from Chapter 2 that when base-width modulation is included, $I_C = I_S e^{V_{BE}/V_T}(1+V_{CE}/V_A)$. Assume that $V_T = 26$ mV. **(a)** If $V_{BE} = 0.6$ V and $V_{CE} = 0.6$ V, what is I_C? **(b)** What is the current supplied by the source V_{BE} under the conditions given in part (a)? **(c)** If $V_{BE} = 0.6$ V and $V_{CE} = 20$ V, what is I_C? **(d)** Suppose the transistor is replaced with one of identical design, except that its base-emitter junction has twice the area. If $V_{BE} = 0.6$ V and $V_{CE} = 0.6$ V, what is I_C?

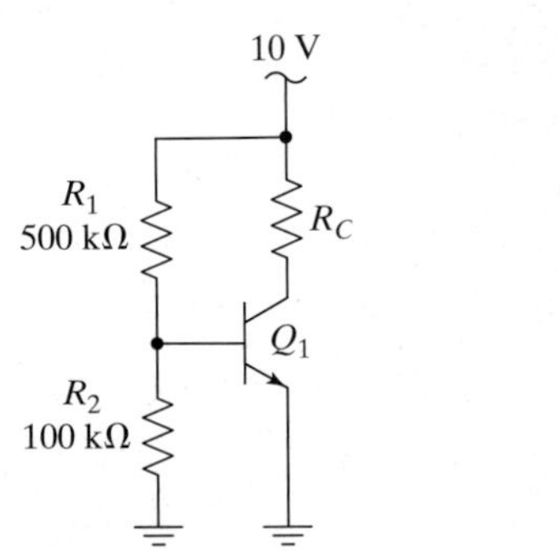

Figure 7–79

P7.8 Consider the circuit shown in Figure 7-79 and assume that $\beta = 100$. **(a)** Find I_C. **(b)** Find the value of R_C necessary to set $V_C = 5$ V. **(c)** Keeping all other values the same as in part (b), find new values for R_1 and R_2 that will cause Q_1 to barely enter saturation.

P7.9 Consider the circuit shown in Figure 7-80 and assume that $\beta = 50$. **(a)** What region of operation is Q_1 in? **(b)** Find V_B.

P7.10 Consider the circuit shown in Figure 7-81 and assume that $\beta = 100$. **(a)** Find I_C. **(b)** Find the value of R_C necessary to set $V_C = -3$ V. **(c)** Keeping all other values the same as in part (b), find new values for R_1 and R_2 that will just put Q_1 in saturation.

P7.11 Consider the circuit shown in Figure 7-82 and assume that $\beta = 50$. **(a)** What region of operation is Q_1 in? **(b)** Find V_B.

P7.12 Consider an *npn* BJT with $I_S = 3.5 \times 10^{-14}$ and assume that $V_T = 26$ mV. **(a)** If $V_{BE} = 0.65$ V and the transistor is forward active, what is I_C? **(b)** How large must V_{BE} be to make I_C 10 times larger? **(c)** 100 times larger? **(d)** 100 times smaller? **(e)** Redo (a) through (d) if the temperature is increased to 100 °C. (V_T is *not* the same in this case.)

P7.13 Consider the circuit shown in Figure 7-83, where the base biasing circuitry has been replaced by a Thévenin equivalent, $V_{BB} = 1$ V, and the transistor has $\beta = 100$. **(a)** If we model the transistor as usual by assuming that $V_{BE} = 0.7$ V, what is I_C? **(b)** If V_{BE} turns out to be 0.6 V instead, what does I_C become? Comment on the accuracy of the model in this case. Why is the error so large?

P7.14 Consider the circuit shown in Figure 7-84 and assume that $\beta = 50$ for both transistors. **(a)** Draw the large-signal DC equivalent circuit. **(b)** Find I_{C1} and I_{C2}. **(c)** Find V_C.

P7.15 Find the value of resistance needed to set the current in the circuit in Figure 7-85 equal to 1 mA.

P7.16 Assume that $\beta = 99$ for the transistor in Figure 7-86. **(a)** What region of operation is the device in? **(b)** What are I_B, I_C, and V_C? **(c)** How large can R_C become without saturating the transistor?

P7.17 Find the values of I_C and V_C for the circuit in Figure 7-87. Assume that $\beta = 100$.

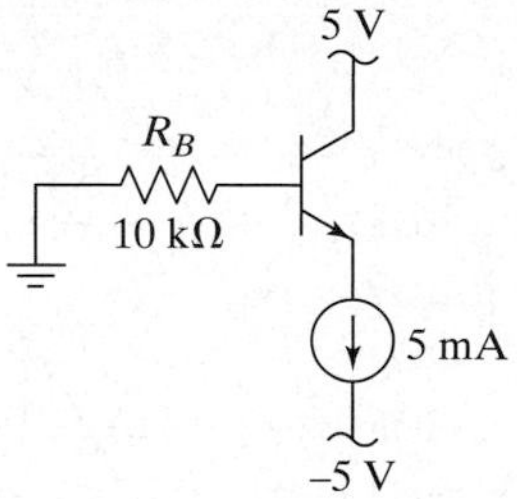

Figure 7–80

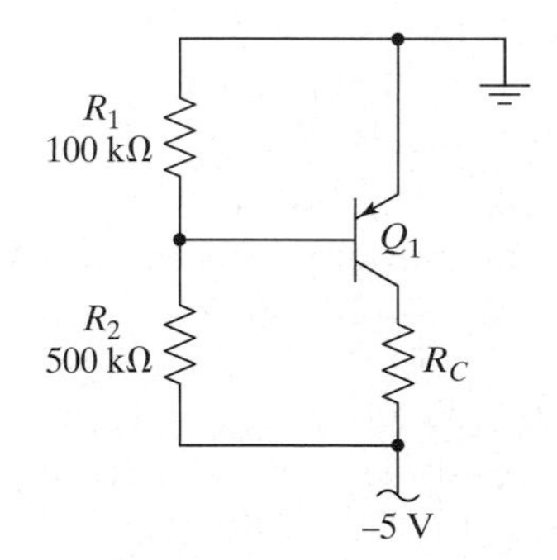

Figure 7–81

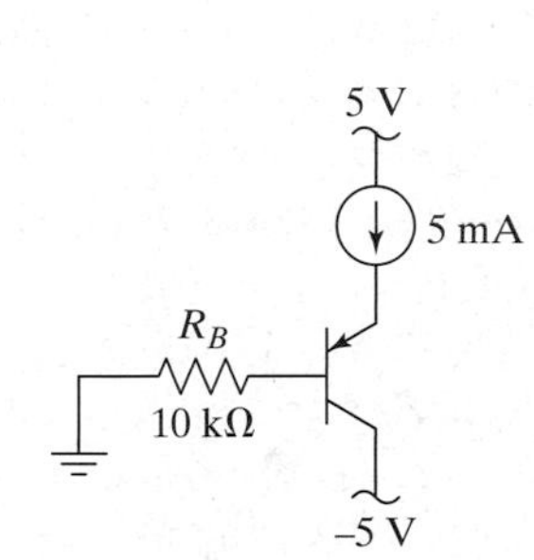

Figure 7–82

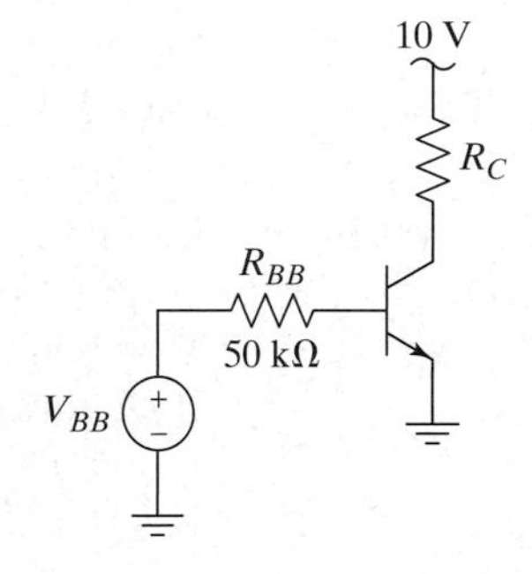

Figure 7–83

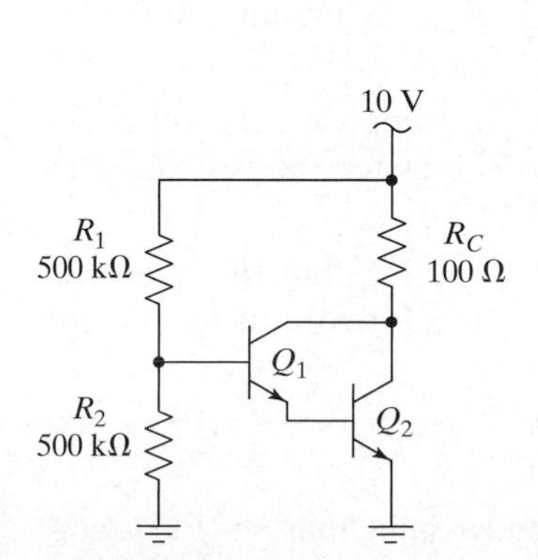

Figure 7–84

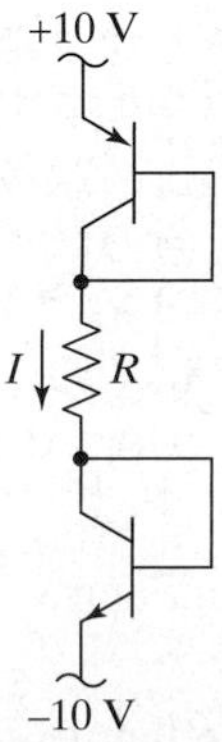

Figure 7–85

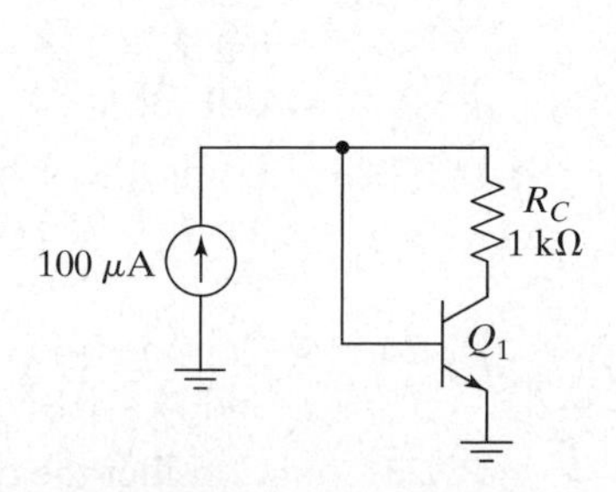

Figure 7–86

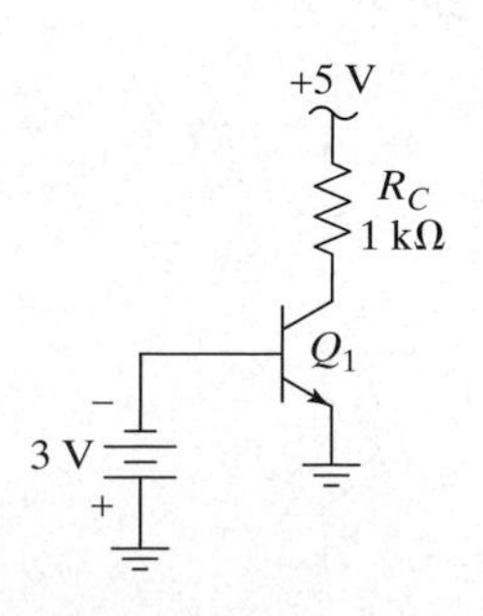

Figure 7–87

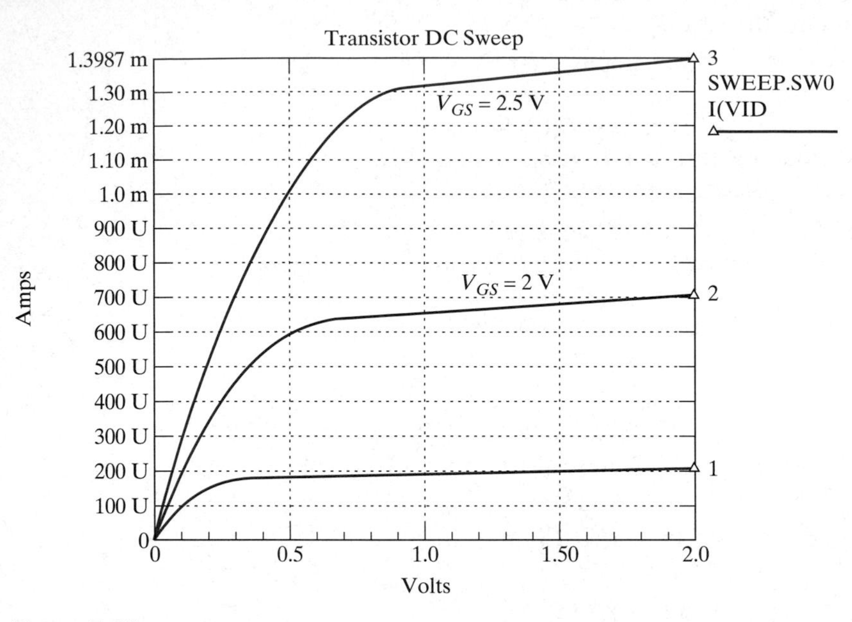

Figure 7–88

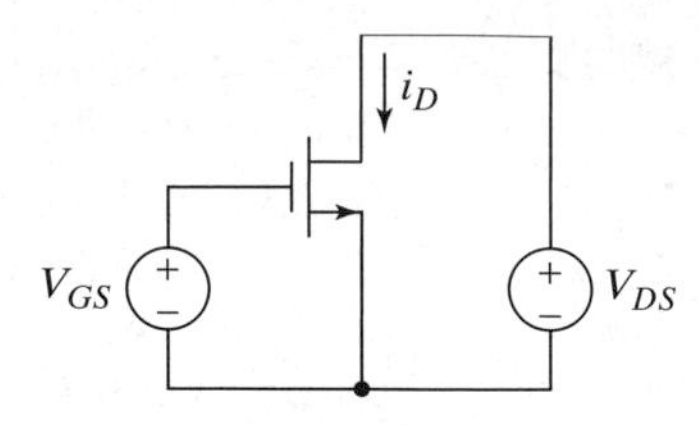

Figure 7–89

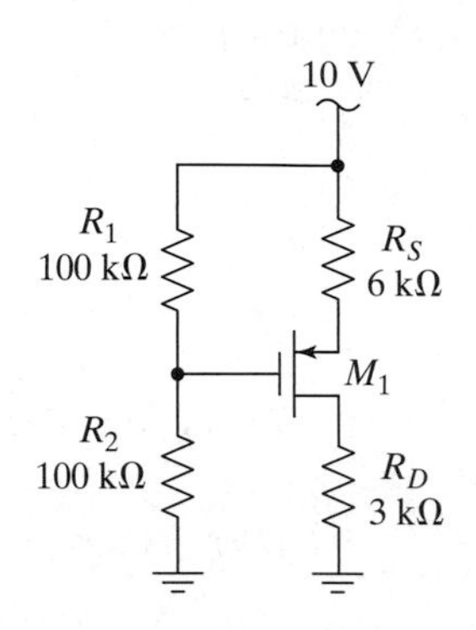

Figure 7–90

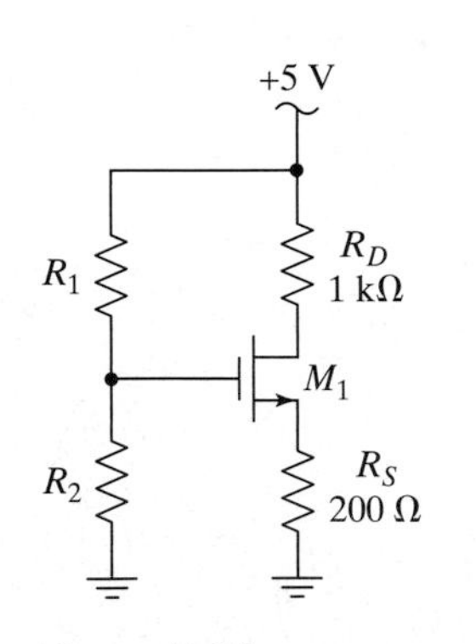

Figure 7–91

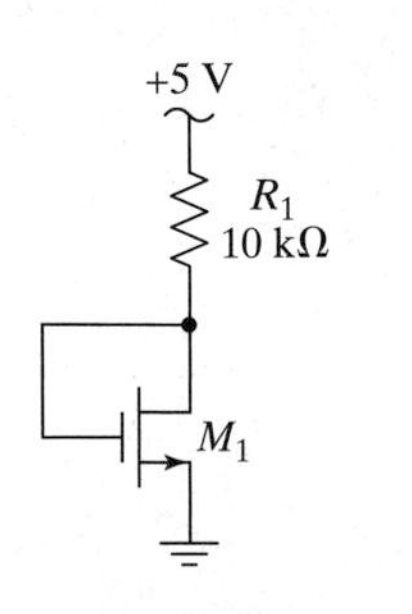

Figure 7–92

7.1.5 MOS Field-Effect Transistors

P7.18 Figure 7-88 shows measured drain characteristics for an *n*-channel MOSFET. "M" stands for 10^{-3} and "U" stands for 10^{-6}. **(a)** What is the on resistance of the transistor when it is in the linear region and $V_{GS} = 2.5$ V? **(b)** Derive a Norton model that has approximately the same *I–V* characteristics as the drain-source terminals of the transistor in the saturation region (forward active) when $V_{GS} = 2.0$ V.

P7.19 Refer to the circuit of Figure 7-89. The MOS transistor has the following parameters: $V_{th} = .978$ V, $K' = 5.31 \times 10^{-5}$ A/V^2, and $\lambda = 2.08 \times 10^{-2}$ V^{-1}. **(a)** Given $V_{GS} = 1.5$ V and $V_{DS} = 2$ V, what W/L ratio is required to make $I_D = 400\,\mu$A? **(b)** Given $V_{GS} = 1.5$ V, $V_{DS} = 5$ V, and $W/L = 60$, what is the resulting I_D? **(c)** Given $V_{GS} = 1.5$ V, $V_{DS} = 0.3$ V, and $W/L = 60$, what is the resulting I_D?

P7.20 Redo Example 7.8, including the body effect. Assume that V_{th} is given by $V_{th} = V_{th0} + \gamma[\sqrt{2\phi_f + V_{SB}} - \sqrt{2\phi_f}]$, which is (2.120), and that $V_{th0} = 1.5$, $\gamma = 0.53$, and $2\phi_f = 0.82$.

P7.21 Consider the circuit shown in Figure 7-90, and assume that $K = 75\ \mu$A/V^2 and $V_{th} = -0.8$ V. Find I_D, V_D, V_G, and V_S.

D P7.22 Using the schematic of Figure 7-90 and assuming that the transistor has $K = 125\ \mu$A/V^2 and $V_{th} = -0.6$ V, find values for the resistors that will set $I_D = 1$ mA and $V_{SD} = 3$ V.

D P7.23 Using the schematic of Figure 7-91 and assuming that the transistor has $K = 250\ \mu$A/V^2 and $V_{th} = 0.7$ V, find values for R_1 and R_2 that will set $I_D = 1$ mA.

P7.24 Consider the circuit shown in Figure 7-92, and assume that $K = 100\ \mu$A/V^2 and $V_{th} = 0.8$ V. Find I_D.

P7.25 Consider the circuit shown in Figure 7-93, and assume that $K = 150\ \mu$A/V^2 and $V_{th} = -0.8$ V. Find I_D.

P7.26 Consider the circuit shown in Figure 7-94, and assume that K = 250 μA/V² and V_{th} = 1 V. **(a)** If V_{DD} = 5 V, what region of operation is the transistor in? **(b)** With V_{DD} = 5 V, what is V_S? **(c)** If V_{DD} = −2 V, what region of operation is the transistor in? **(d)** With V_{DD} = −2 V, what is V_S?

P7.27 Consider the circuit shown in Figure 7-95, and assume that K = 75 μA/V² and V_{th} = −0.6 V. **(a)** If V_{SS} = −5 V, what region of operation is the transistor in? **(b)** With V_{SS} = −5 V, what is V_S? **(c)** If V_{SS} = 1 V, what region of operation is the transistor in? **(d)** With V_{SS} = 1 V, what is V_S?

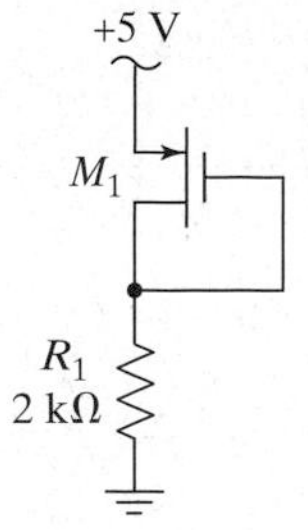

Figure 7–93

7.1.6 Junction Field-Effect Transistors

P7.28 How large can R_D be in the circuit in Example 7.9 before the transistor enters the linear region?

P7.29 The transistor in Figure 7-96 has I_{DSS} = 2 mA and V_p = 2 V. **(a)** Assuming that the transistor is operating in the forward-active region, find the value of R_D that will lead to V_{DS} = −4 V.

P7.30 For the circuit shown in Figure 7-97, the transistor has I_{DSS} = 5 mA and V_p = −2 V. Find the values of I_D and V_{DS}.

P7.31 The transistors in Figure 7-98 have I_{DSS} = 5 mA and V_p = −2 V. Find V_{DS1} and I_{D2}.

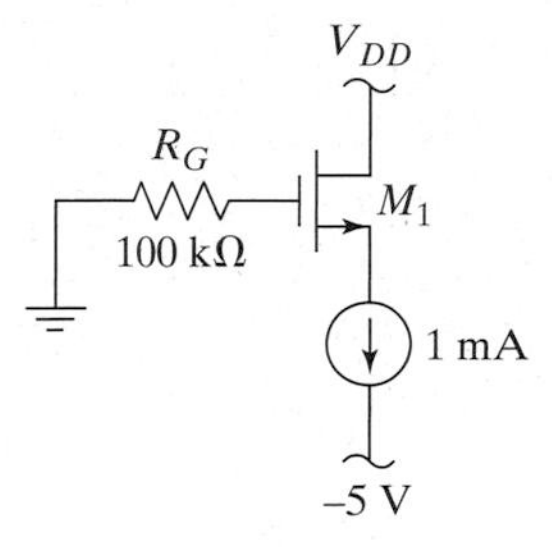

Figure 7–94

7.2 Biasing of Single-Stage Amplifiers

7.2.1 BJT Amplifiers

D P7.32 **(a)** Choose values for the resistors in the circuit in Figure 7-99 to establish I_C = 1.5 mA using our rules of thumb and assuming that V_{CC} = 9 V, V_{BE} = 0.7 V, and β = 75. **(b)** How much does I_C change if V_{BE} changes by 0.1 V? **(c)** If the change in I_C in part (b) is unacceptably large, what can we do to change the bias point so that the change in I_C is less?

D P7.33 Consider the circuit shown in Figure 7-99, and assume that β = 100 and V_{CC} = 10 V. Design the circuit to achieve a stable operating point with I_C = 500 μA and V_{CE} = 4 V.

P7.34 Consider the biasing arrangement shown in Figure 7-99. Assume that V_{CC} = 10 V, R_1 = 130 kΩ, R_2 = 33 kΩ, R_E = 3.3 kΩ, R_C = 10 kΩ, and β = 100. **(a)** Determine the value of I_C, including the base current in your calculation. **(b)** Perform a worst-case analysis, assuming that all of the resistors can vary by ±10%, V_{CC} can vary by ±5%, β can vary from 50 to 200, and V_{BE} can vary between 0.6 and 0.8 V.

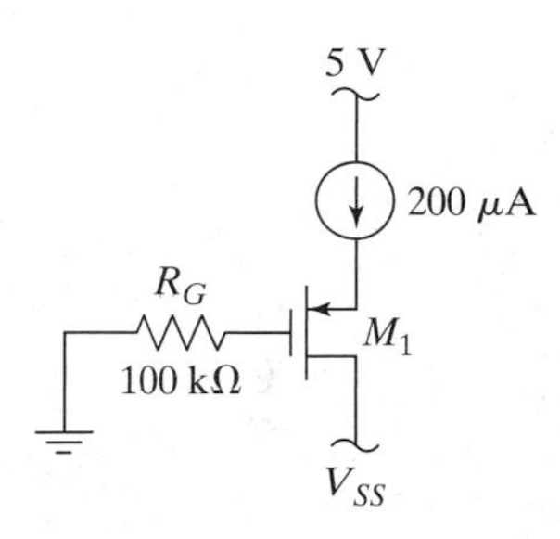

Figure 7–95

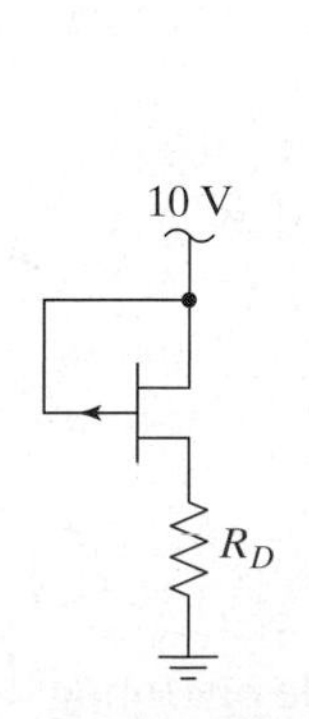

Figure 7–96

V_{DD} = 10 V
R_D
2 kΩ
M_1
R_S
1 kΩ

Figure 7–97

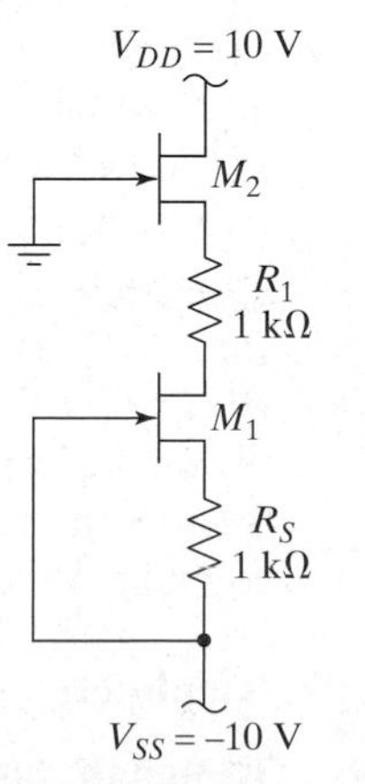

Figure 7–98

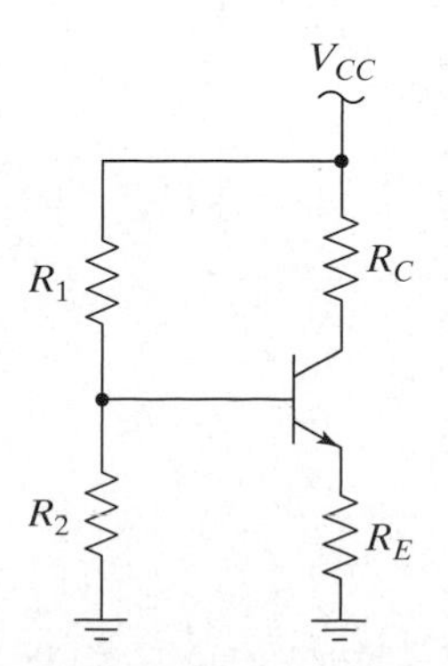

Figure 7–99

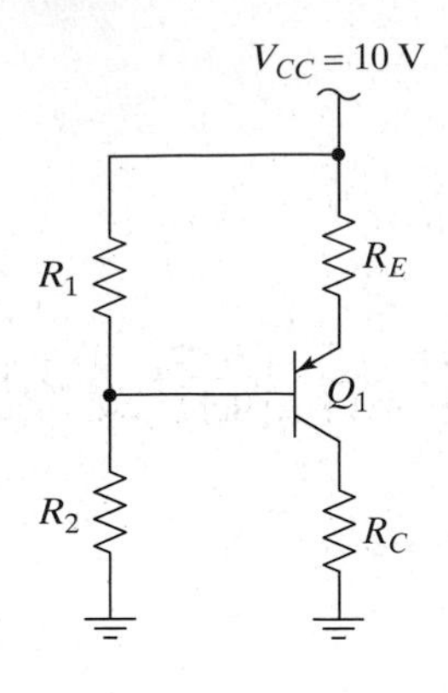

Figure 7–100

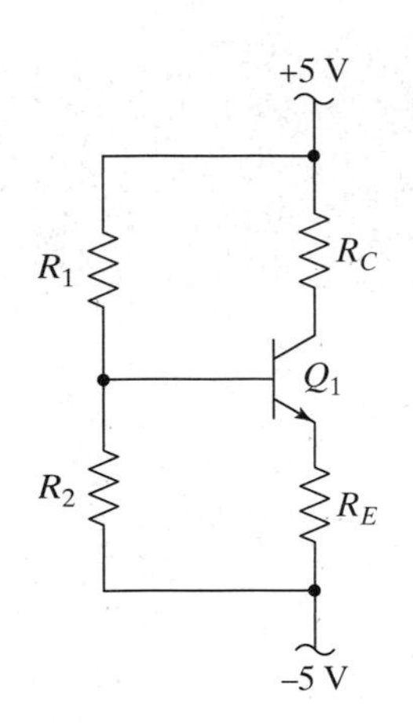

Figure 7–101

D P7.35 Consider the circuit shown in Figure 7-100, and assume that the transistor has $V_{BE} = -0.7$ V and $\beta = 50$. Find values for the resistors to establish a bias current of $I_C = 3$ mA. Follow the rules of thumb given in the text, but be careful to interpret them properly for a *pnp* transistor!

D P7.36 Consider the circuit shown in Figure 7-100, and assume that the transistor has $V_{BE} = -0.7$ V and $\beta = 75$. Find values for the resistors to establish a bias current of $I_C = 1$ mA. Follow the rules of thumb given in the text, but be careful to interpret them properly for a *pnp* transistor!

D P7.37 Consider the circuit shown in Figure 7-101 and assume that $\beta = 150$. Design the circuit to achieve $I_C = 500$ μA while maintaining our rules of thumb. You must modify the rules properly to account for the dual supplies.

D P7.38 Consider the circuit shown in Figure 7-101 and assume that $\beta = 100$. Design the circuit to achieve $I_C = 750$ μA, $V_{CE} = 2$ V, and $V_C = 1$ V. Follow our rules of thumb, but modify them properly for dual supplies.

D P7.39 Consider the circuit shown in Figure 7-99, and assume that $\beta = 100$ and $V_{CC} = 5$ V. Design the circuit to achieve a stable operating point with $I_C = 200$ μA and $V_{CE} = 2$ V.

S P7.40 Use the Nsimple model in Appendix B and SPICE to confirm that your design from Problem P7.39 achieves the desired bias point. Hand in a printout of your bias point.

S P7.41 Use the Nsimple model in Appendix B and SPICE to confirm that your design from Problem P7.39 has a stable operating point. Set up the file so that β and I_S in the transistor model are both parameters. Then print out the bias point when β is 50 and 200 and when I_S is 10 times smaller and 10 times larger than normal. Comment on your results.

D P7.42 Consider the circuit shown in Figure 7-99, and assume that $\beta = 100$ and $V_{CC} = 15$ V. Design the circuit to achieve a stable operating point with $I_C = 2$ mA and $V_{CE} = 6$ V.

P7.43 Assume that you are using the topology of Figure 7-99 to design an amplifier and that the transistor you are using has $V_{BE} = 0.7$ V at $T = 25$°C and $I_C = 1$ mA. If the V_{BE} changes by –2 mV for every 1°C rise in temperature, and if you desire to keep the change in I_C to less than 5% over the range 0°C to 70°C, what is the minimum V_{BB} you must use?

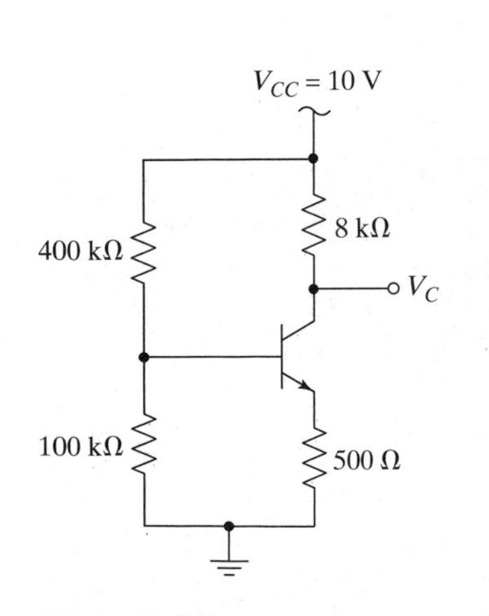

Figure 7–102

P7.44 Consider the circuit of Figure 7-102. The circuit has been designed assuming that $\beta = 50$ and $V_{BE} = 0.7$ V. **(a)** What is the designed value of V_C resulting from these assumptions? **(b)** The circuit is constructed as shown, but using a transistor that has $\beta = 100$. What is the resulting V_C? **(c)** What changes can be made in the circuit to make V_C less sensitive to changes in β?

P7.45 Consider the circuit of Figure 7-103. It has been designed assuming that $\beta = 50$ and $V_{BE} = 0.7$ V. **(a)** What is the designed value of V_C? **(b)** If the circuit is constructed as shown, but using a transistor that has $\beta = 100$, what is the resulting V_C? **(c)** If $\beta = 50$, but $V_{BE} = 0.6$ V, what is the resulting V_C? **(d)** What changes can be made in the circuit to make V_C less sensitive to changes in V_{BE}?

7.2.2 FET Amplifiers

D P7.46 Design the amplifier in Figure 7-104 to achieve the most stable operating point possible, given the following design goals: $I_D = 200$ μA, $V_{DS} = 3$ V, and drop across $R_D \geq 3$ V. Assume that $V_{DD} = 10$ V, $K = 100\ \mu A/V^2$, and $V_{th} = 0.5$ V.

S P7.47 Use the Nsimple_mos model in Appendix B and SPICE to confirm that your design from Problem P7.46 achieves the desired bias point. Hand in your output listing.

D P7.48 Draw the PMOS equivalent of the biasing circuit shown in Figure 7-104, and design the circuit to achieve the most stable bias point possible with $I_D = 500\ \mu A$, $V_{DS} \leq -3$ V, and a minimum drop of 3 V across R_D. Assume that you have both a positive and a negative supply available, that $V_{DD} = 5$ V and $V_{SS} = -5$ V, and that $K = 100\ \mu A/V^2$ and $V_{th} = -0.5$ V.

D P7.49 Consider the circuit shown in Figure 7-105, and assume that $K = 100\ \mu A/V^2$ and $V_{th} = 0.5$ V. Find the values of R_D and R_S necessary to achieve $I_D = 50\ \mu A$ and $V_{DS} = 3$ V.

S P7.50 Use the Nsimple_mos model from Appendix B and SPICE to confirm that your design from Problem P7.49 achieves the desired bias point. Hand in your output listing.

P7.51 Find the bias voltages and currents for the circuit in Example 7.18, assuming that the device has $K = 100\ \mu A/V^2$ and $V_{th} = -0.5$ V.

S P7.52 Use the Nsimpled_mos model from Appendix B and SPICE to confirm that your results from Problem P7.51 are correct. Hand in your output listing.

D P7.53 Consider using a depletion-mode device in the circuit in Figure 7-104. Assume that $V_{DD} = 10$ V, $K = 100\ \mu A/V^2$, and $V_{th} = -0.5$ V. Design the circuit to achieve a stable operating point of $I_D = 1$ mA and $V_{DS} = 3$ V with a minimum drop of 3 V across R_D.

S P7.54 Use the Nsimpled_mos model from Appendix B and SPICE to confirm that your design from Problem P7.53 achieves the desired bias point. Hand in your output listing.

D P7.55 Draw the PMOS equivalent of the depletion-mode biasing circuit shown in Example 7.18, and design the circuit to achieve $I_D = 500\ \mu A$, $V_{R_D} = 3$ V, and $V_{DS} = -3$ V. Assume that $V_{DD} = 10$ V, $K = 100\ \mu A/V^2$, and $V_{th} = 0.5$ V.

S P7.56 Use the Psimpled_mos model from Appendix B and SPICE to confirm that your design from Problem P7.55 achieves the desired bias point. Hand in your output listing.

D P7.57 Design the circuit shown in Figure 7-106 to achieve $I_D = 100\ \mu A$ and $V_D = 5$ V. Assume that $V_{DD} = 10$ V, $K = 500\ \mu A/V^2$, and $V_{th} = -0.5$ V.

S P7.58 Use the Nsimpled_mos model from Appendix B and SPICE to confirm that your design from Problem P7.57 achieves the desired bias point. Hand in your output listing.

D P7.59 Draw the PMOS equivalent of the circuit shown in Figure 7-106, using a single −10 V supply (called V_{SS}). Design the circuit to achieve $I_D = 100\ \mu A$ and $V_D = -5$ V, assuming that $K = 500\ \mu A/V^2$ and $V_{th} = 0.5$ V.

S P7.60 Use the Psimpled_mos model from Appendix B and SPICE to confirm that your design from Problem P7.59 achieves the desired bias point. Hand in your output listing.

D P7.61 Consider the circuit in Figure 7-104, and assume that $V_{DD} = 10$ V, $K = 100\ \mu A/V^2$ and $V_{th} = 0.5$ V. Design the circuit to have a stable operating point, $I_D = 250\ \mu A$, $V_{DS} = 3$ V, and $V_D \leq 7$ V.

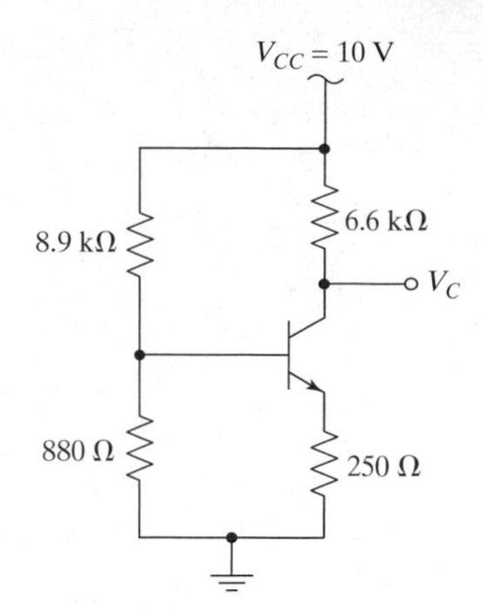

Figure 7–103

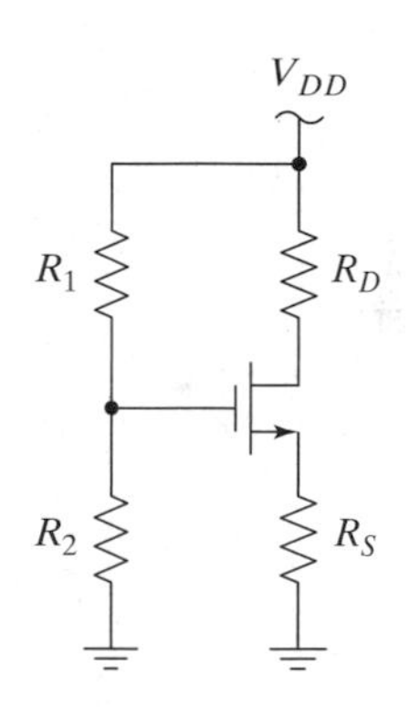

Figure 7–104

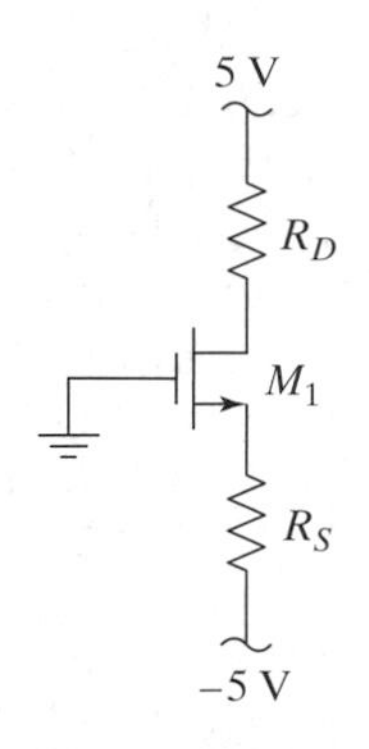

Figure 7–105

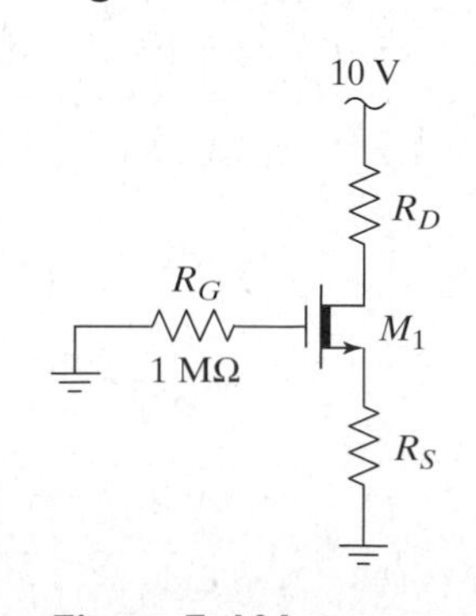

Figure 7–106

S P7.62 Use the Nsimple_mos model in Appendix B and SPICE to verify that your design in Problem P7.61 achieves the desired operating point. Hand in your output listing.

S P7.63 Use the Nsimple_mos model in Appendix B and SPICE to verify that your design in Problem P7.61 produces a stable bias point. **(a)** Change the value of K by ±25% and print out the resulting values of I_D (use the model parameter **KP**). **(b)** Now vary V_{th} by ±0.2 V and print out the resulting values of I_D. **(c)** Comment on the results.

D P7.64 Consider the circuit in Figure 7-104, and assume that $V_{DD} = 15$ V, $K = 100\ \mu A/V^2$, and $V_{th} = 0.5$ V. Design the circuit to have a stable operating point, $I_D = 500\ \mu A$, $V_{DS} = 5$ V, and $V_D \leq 10$ V.

S P7.65 Use the Nsimple_mos model in Appendix B and SPICE to verify that your design in Problem P7.64 achieves the desired operating point. Hand in your output listing.

S P7.66 Use the Nsimple_mos model in Appendix B and SPICE to verify that your design in Problem P7.64 produces a stable bias point. **(a)** Change the value of K by ±25% and print out the resulting values of I_D (use the model parameter **KP**). **(b)** Now vary V_{th} by ±0.2 V and print out the resulting values of I_D. **(c)** Comment on the results.

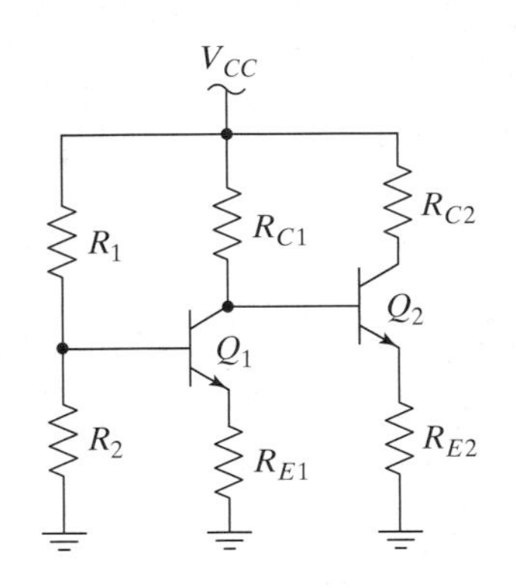

Figure 7–107 A cascade of bipolar amplifiers.

7.3 Biasing of Multistage Amplifiers

7.3.1 Cascaded Bipolar Amplifiers

P7.67 Consider the circuit in Figure 7-107. Assume that $V_{CC} = 10$ V, $R_1 = 80$ kΩ, $R_2 = 20$ kΩ, $R_{E1} = 2.6$ kΩ, $R_{C1} = 12$ kΩ, $R_{E2} = 3.3$ kΩ, $R_{C2} = 3.3$ kΩ, and $\beta = 100$. Find the DC bias points of both transistors.

D P7.68 Design the circuit of Figure 7-107 to achieve $I_{C1} = 1$ mA, $I_{C2} = 5$ mA, and $V_{CE1} = V_{CE2} = 3$ V and to have a stable bias point with both transistors forward active. Assume that $\beta = 100$ and $V_{CC} = 10$.

S P7.69 Using the *npn* Nsimple model presented in Appendix B, simulate the circuit from Problem P7.68 and verify your design.

D P7.70 Draw the *pnp* equivalent to the circuit of Figure 7-107 using a single −10 V supply. Design the circuit to achieve $I_{C1} = 2$ mA, $I_{C2} = 10$ mA, and $V_{CE1} = V_{CE2} = -3$ V and to have a stable bias point with both transistors forward active. Assume that $\beta = 100$.

S P7.71 Using the *pnp* Psimple model presented in Appendix B, simulate the circuit from Problem P7.70 and verify your design.

P7.72 Consider the circuit in Figure 7-107. Assume that $V_{CC} = 10$ V, $R_1 = 120$ kΩ, $R_2 = 30$ kΩ, $R_{E1} = 1.3$ kΩ, $R_{C1} = 5$ kΩ, $R_{E2} = 430$ Ω, $R_{C2} = 0$ Ω, and $\beta = 100$. Find the DC bias points of both transistors.

D P7.73 Design the circuit of Figure 7-107 to achieve $I_{C2} = 20$ mA, $V_{E2} = 5$ V, and a stable operating point with both transistors forward active. Use the minimum acceptable currents in Q_1 and R_1 and the minimum acceptable voltage at the base of Q_1. Assume that $\beta_1 = 100$ and $\beta_2 = 50$. Set $R_{C2} = 0$ Ω.

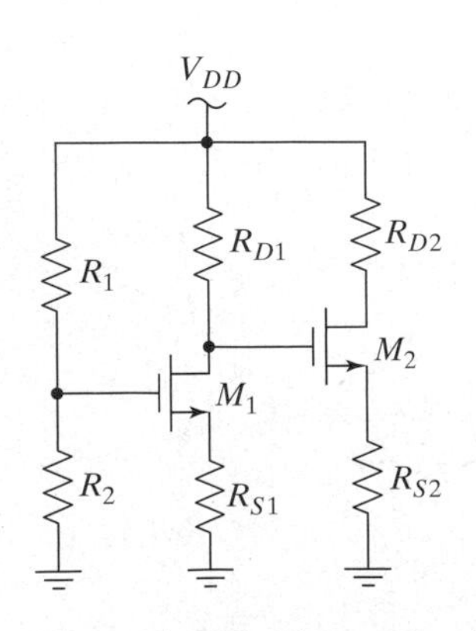

Figure 7–108 A cascade of common-emitter amplifiers.

7.3.2 Cascaded FET Amplifiers

P7.74 Consider the circuit in Figure 7-108. Assume that $V_{DD} = 10$ V, $R_1 = 560$ kΩ, $R_2 = 420$ kΩ, $R_{S1} = 15$ kΩ, $R_{D1} = 15$ kΩ, $R_{S2} = 3.3$ kΩ, $R_{D2} = 3$ kΩ, $K = 100\ \mu A/V^2$, and $V_{th} = 0.5$ V. Find the DC bias points of both transistors.

D P7.75 Design the circuit of Figure 7-108 to achieve $I_{D1} = 100\ \mu\text{A}$, $I_{D2} = 1$ mA, $V_{DS1} = 3$ V, and $V_{DS2} = 4$ V and to have a stable bias point with both transistors forward active. Assume that $V_{DD} = 10$ V, $K = 100\ \mu\text{A/V}^2$, and $V_{th} = 0.5$ V.

S P7.76 Using the Nsimple_mos model presented in Appendix B and SPICE, simulate the circuit from Problem P7.75 and verify your design. Hand in your output listing.

D P7.77 Draw the PMOS equivalent to the circuit of Figure 7-108, using a single −10 V supply. Design the circuit to achieve $I_{D1} = 200\ \mu\text{A}$, $I_{D2} = 1$ mA, $V_{DS1} = -3$ V, and $V_{DS2} = -4$ V and to have a stable bias point with both transistors forward active. Assume that $K = 100\ \mu\text{A/V}^2$ and $V_{th} = -0.5$ V.

S P7.78 Using the Psimple_mos model presented in Appendix B and SPICE, simulate the circuit from Problem P7.77 and verify your design. Hand in your output listing.

P7.79 Consider the circuit in Figure 7-108, but set $R_{D2} = 0\ \Omega$. Assume that $V_{DD} = 10$ V, $R_1 = 560$ kΩ, $R_2 = 420$ kΩ, $R_{S1} = 20$ kΩ, $R_{D1} = 7$ kΩ, $R_{S2} = 2$ kΩ, $K = 100\ \mu\text{A/V}^2$, and $V_{th} = 0.5$ V. Find the DC bias points of both transistors.

D P7.80 Design the circuit of Figure 7-108 to achieve $I_{D1} = 200\ \mu\text{A}$, $I_{D2} = 2$ mA, $V_{DS1} = 3$ V, and $V_{DS2} = 4$ V and to have a stable bias point with $R_{D2} = 0\ \Omega$ and both transistors forward active. Assume that $V_{DD} = 10$ V, $K = 100\ \mu\text{A/V}^2$, and $V_{th} = 0.5$ V.

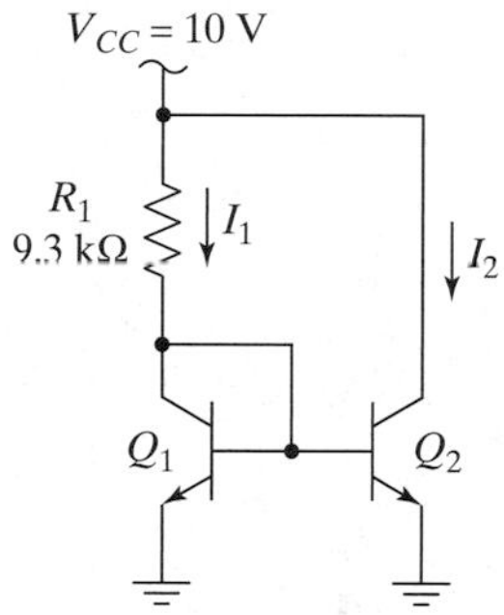

Figure 7–109

7.4 Biasing for Integrated Circuits

7.4.1 Simple Bipolar Current Mirrors

P7.81 Consider the mirror shown in Figure 7-109. The output current, which is the collector current of Q_2, is returned to V_{CC} for illustrative purposes; it would not be used this way in practice. **(a)** Find I_1 and I_2, assuming that the transistors are identical, have infinite β, and have no base-width modulation. **(b)** Find the currents, assuming that $V_A = 50$ V for Q_2. **(c)** Find the currents, assuming that $V_A = 50$ V for Q_2 and $\beta = 100$ for both transistors.

P7.82 Consider the mirror shown in Figure 7-109. The output current, which is the collector current of Q_2, is returned to V_{CC} for illustrative purposes; it would not be used this way in practice. **(a)** Find the currents, assuming that V_A and β are both infinite, but that the saturation current of Q_2 is 10% larger than the saturation current of Q_1. **(b)** Repeat part (a) with the saturation currents set equal, but with the two transistors at slightly different temperatures. Assume that Q_1 is at 300 K and Q_2 is at 305 K.

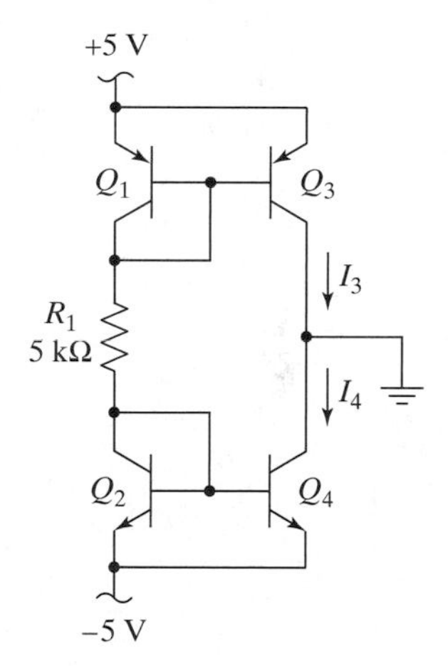

Figure 7–110

P7.83 Consider the circuit in Figure 7-109, but connect the collector of Q_2 to a separate voltage source called V_O. Sketch the output current-versus V_O characteristic if V_O is swept from zero to 20 V. Assume that $V_A = 50$ V for Q_2. Ignore the base currents.

P7.84 **(a)** Find I_3 and I_4 for the circuit in Figure 7-110, assuming that the *pnp* transistors are matched, the *npn* transistors are matched, and all the transistors have infinite β and V_A. **(b)** Repeat part (a), except set the β of the *pnp* transistors to 20 and the β of the *npn* transistors to 100.

7.4.2 More Advanced Bipolar Current Mirrors

P7.85 Find the value of I_O in the circuit of Figure 7-111, assuming that the

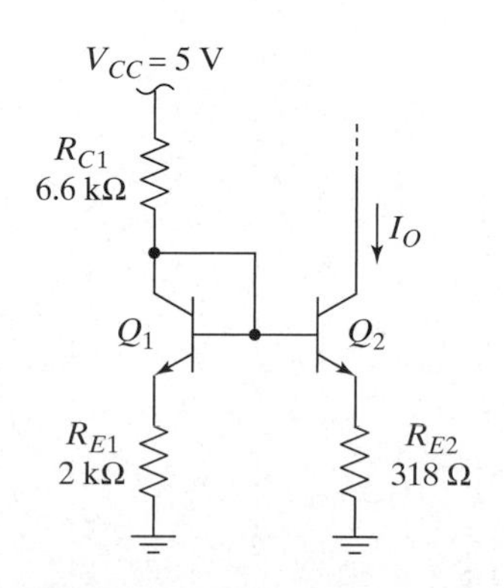

Figure 7–111

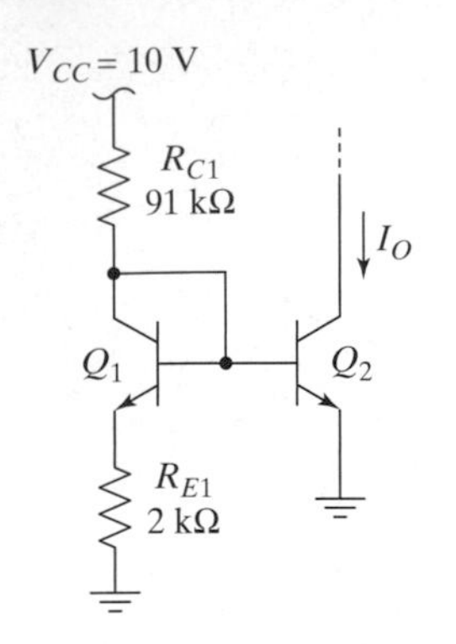

Figure 7–112

transistors are identical and that β is very large. Assume that the collector of Q_2 is connected to some other circuit capable of supplying the current and that Q_2 is forward active. Ignore base-width modulation.

P7.86 Find the value of I_O in the circuit of Figure 7-112, assuming that the transistors are identical and that β is very large. Assume that the collector of Q_2 is connected to some other circuit capable of supplying the current and that Q_2 is forward active. Ignore base-width modulation.

S P7.87 Use the Nsimple model presented in Appendix B and SPICE to simulate the mirror in Problem P7.86. Connect the collector of Q_2 to V_{CC}. How close are the currents to your prediction? Why is there a difference?

P7.88 Find the values of I_2 and I_4 in the circuit of Figure 7-113. Ignore base-width modulation, and assume that the *pnp* and *npn* transistor pairs are identical and β is very large.

S P7.89 Use the Nsimple model presented in Appendix B and SPICE to simulate the mirror in Problem P7.88. How close are the currents to your prediction? Why is there a difference?

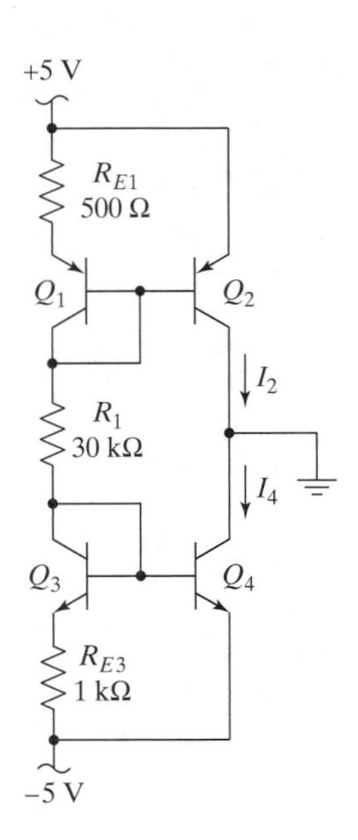

Figure 7–113

P7.90 Consider the mirror shown in Figure 7-114. This is called a ***peaking current source*** because the output current has a peak value as the input current is changed. **(a)** Derive an equation for I_O as a function of I_I. Assume that the transistors are perfectly matched, β and V_A are very large and Q_2 remains forward active. **(b)** Find the derivative of I_O with respect to I_I, and show that it is zero when $I_I R = V_T$.

S P7.91 Use the Nsimple model presented in Appendix B and SPICE to simulate the peaking current source from Figure 7-114. Drive the input with a DC current source, and connect the collector of Q_2 to a 1 V DC voltage source. Set $R = 26\ \Omega$, and then sweep I_I and observe I_O. Show that I_O does, in fact, peak at $I_I \approx 1$ mA (where $I_I R = V_T$).

P7.92 Consider the mirror shown in Figure 7-115, and assume that all of the transistors are matched except for having their emitter areas ratioed as shown in the figure. Assume that the collectors of Q_3 and Q_4 are tied to circuits that keep them in forward active operation while supplying their collector currents. **(a)** Assuming that β and V_A are very large, find I_3 and I_4. **(b)** Find the output currents if $\beta = 50$. (Still assume that V_A is very large.) **(c)** Remove Q_1, tie the collector of Q_2 to its base, and repeat part (b). **(d)** Comment on the purpose of Q_1.

7.4.3 FET Current Mirrors

P7.93 Consider the mirror shown in Figure 7-116. Assume that the transistors are perfectly matched and that $V_{DD} = 5$ V, $R_1 = 5$ kΩ, $K = 100\ \mu A/V^2$, $\lambda = 0$,

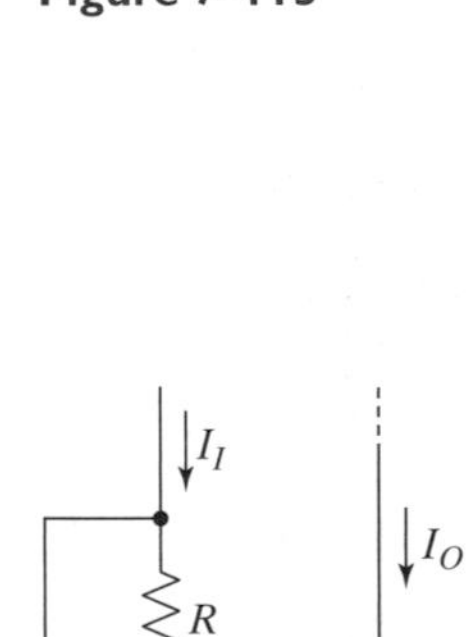

Figure 7–114

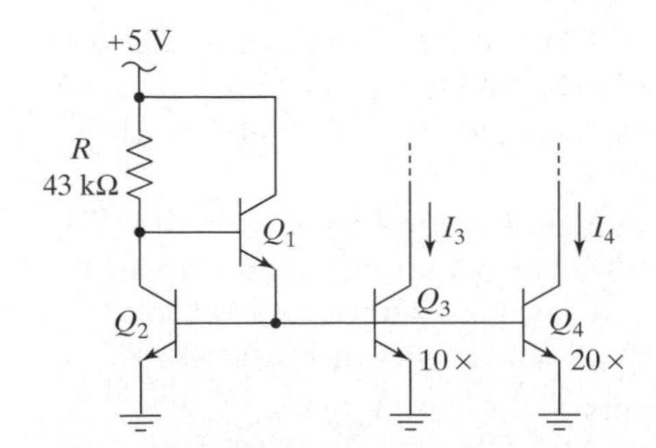

Figure 7–115

and $V_{th} = 0.5$ V. Furthermore, assume that the drain of M_2 is connected to a circuit that can supply the output current while keeping M_2 in saturation. What is the value of I_O?

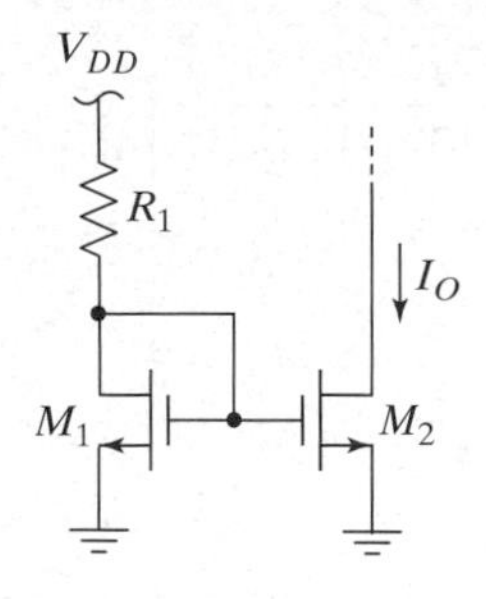

Figure 7–116

D P7.94 Consider the mirror shown in Figure 7-116. Assume that the transistors are perfectly matched and that $V_{DD} = 5$ V, $K = 100\ \mu\text{A/V}^2$, $\lambda = 0$, and $V_{th} = 0.5$ V. Assume further that the drain of M_2 is connected to a circuit that can supply the output current while keeping M_2 in saturation. Determine the value of R_1 necessary to set $I_O = 100\ \mu$A.

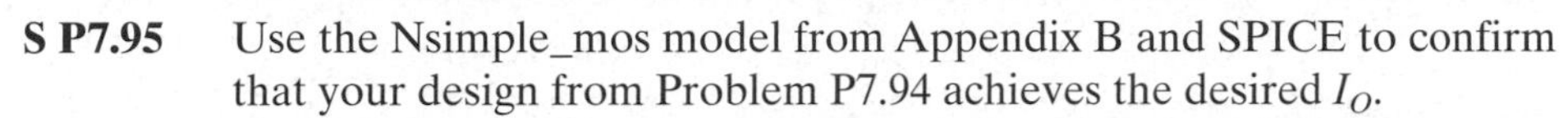

S P7.95 Use the Nsimple_mos model from Appendix B and SPICE to confirm that your design from Problem P7.94 achieves the desired I_O.

P7.96 Consider the mirror shown in Figure 7-117. Assume that the transistors are perfectly matched and that $V_{DD} = 5$ V, $R_1 = 5$ kΩ, $K = 100\ \mu\text{A/V}^2$, $\lambda = 0$, and $V_{th} = -0.5$ V. Assume further that the drain of M_2 is connected to a circuit that can supply the output current while keeping M_2 in saturation. What is the value of I_O?

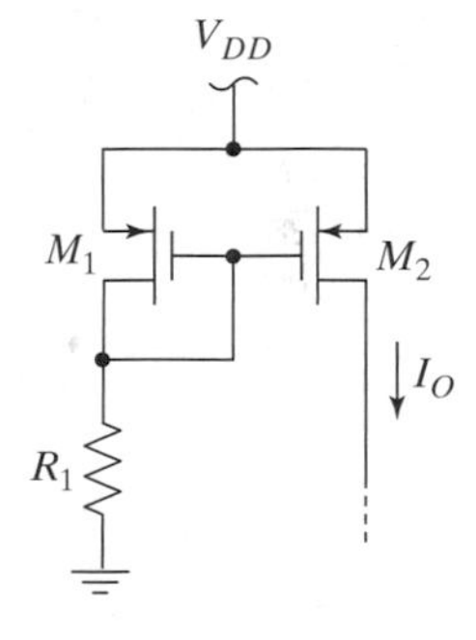

Figure 7–117

D P7.97 Consider the mirror shown in Figure 7-117. Assume that the transistors are perfectly matched and that $V_{DD} = 5$ V, $K = 100\ \mu\text{A/V}^2$, $\lambda = 0$, and $V_{th} = -0.5$ V. Assume further that the drain of M_2 is connected to a circuit that can supply the output current while keeping M_2 in saturation. Determine the value of R_1 necessary to set $I_O = 100\ \mu$A.

S P7.98 Use the Psimple_mos model from Appendix B and SPICE to confirm that your design from Problem P7.97 achieves the desired I_O.

P7.99 Consider the mirror shown in Figure 7-118. The PMOS transistors are perfectly matched and have $K = 100\ \mu\text{A/V}^2$, $\lambda = 0$, and $V_{th} = -0.5$ V. The NMOS transistors are also perfectly matched and have $K = 100\ \mu\text{A/V}^2$, $\lambda = 0$, and $V_{th} = 0.5$ V. Given $V_{DD} = 5$ V, $V_{SS} = -5$ V, and $R_1 = 10$ kΩ, what are the values of I_2, I_3, and I_5?

7.5 Biasing of Differential Amplifiers

P7.100 Consider the generic differential pair in Figure 7-119. If $V_{OO} = 10$ V, $V_{MM} = -10$ V, $I_{MM} = 1$ mA, $R_{O1} = R_{O2} = 5$ kΩ, and the transistors are perfectly matched, **(a)** what is the DC voltage at the output node of T_1 if $v_I = 0$? **(b)** What is the DC differential output voltage if $v_I = 0$? **(c)** If R_{O2} is 5% larger than R_{O1}, but everything else remains the same, what is the DC differential output voltage?

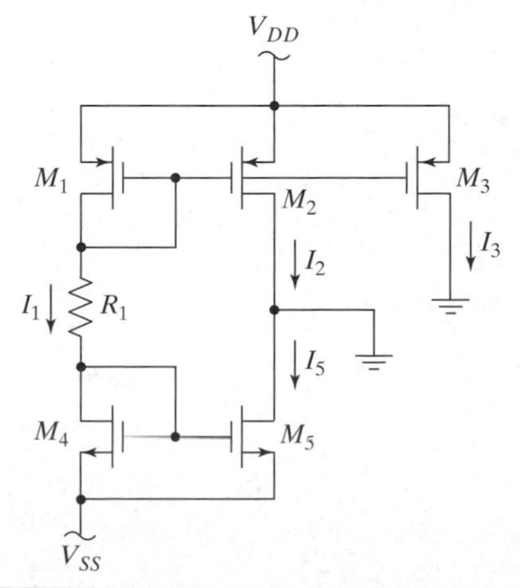

Figure 7–118

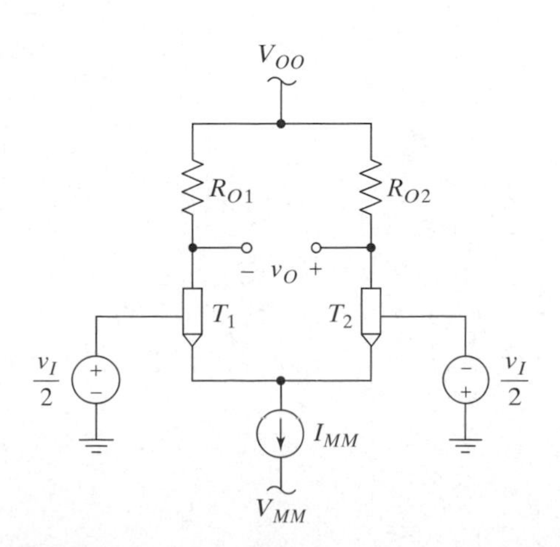

Figure 7–119

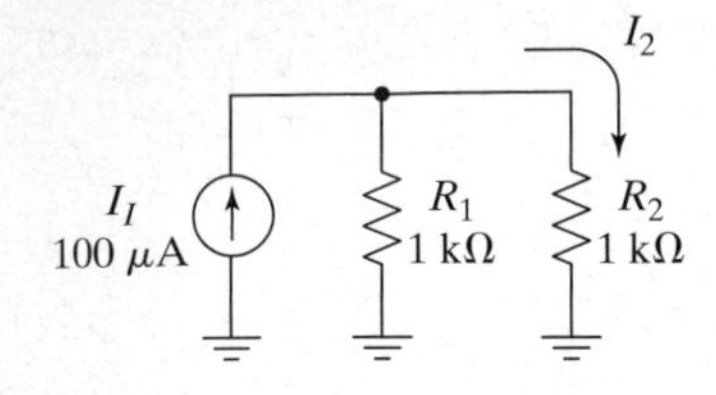

Figure 7–120

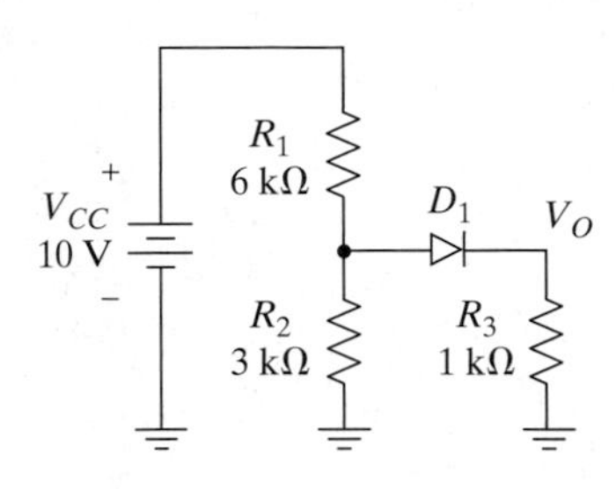

Figure 7–121

7.6 Worst-Case Analysis and Parameter Variation

P7.101 Consider the circuit shown in Figure 7-120. Assuming that both resistors and V_{CC} can vary by $\pm 10\%$, what are the worst-case limits for I_2?

P7.102 Consider the circuit shown in Figure 7-121. Assuming that all three resistors and V_{CC} can vary by $\pm 10\%$ and that the diode's forward voltage drop can vary from 0.6 to 0.8 V, what are the worst-case limits for V_O?

CHAPTER 8

Low-Frequency Small-Signal AC Analysis and Amplifiers

Musical instrument amplifiers are familiar to almost everyone, but amplifiers are used in almost every electronic system. (Photo courtesy of Marshall Amplification plc.)

INTRODUCTION

The concept of using nonlinear devices to make amplifiers that are approximately linear for small signals was presented in Chapter 6. In that chapter, we discovered that the nonlinear devices are biased at some nominal operating point, and then the signals are superimposed on the bias. In Chapter 7, we examined how to analyze and design the DC bias circuits necessary to establish the operating points (or Q points) for the nonlinear devices. In this chapter, we examine how to analyze the small-signal performance of amplifiers and how to design them to achieve a desired set of specifications.

In examining the small-signal AC performance of amplifiers, we must deal with the biasing circuitry as well, since these elements will always be present. In this chapter we concentrate on frequencies that are low enough that parasitic charge storage (and inductive effects) can be ignored. High-frequency models and analysis and design techniques are covered in Chapter 9.

We begin this chapter with an investigation of low-frequency small-signal AC models. We then go on to discuss the three types of single-transistor amplifiers that are possible. In each case the basic circuit is first presented using the generic transistor. After the basic circuit is understood, the following subsections give specific examples using bipolar and field-effect transistors.

8.1 LOW-FREQUENCY SMALL-SIGNAL MODELS FOR DESIGN

In this section, we derive circuit models suitable for hand analysis and design with small AC voltages and currents at low frequency. The resulting models are called ***low-frequency small-signal AC models***. In Chapter 9 we consider high-frequency AC performance, and in Chapter 12 we consider large-signal performance.

The first part of this section deals with models for independent sources. We treat linear passive elements (i.e., resistors, capacitors, and inductors) and then nonlinear devices. Except for the section on the generic transistor, the models for the nonlinear devices are directly based on our understanding of the underlying device physics (*see* Chapter 2).

8.1.1 Independent Sources

Small-signal AC analysis is only *part* of a complete analysis. We assume that the DC bias point has already been calculated and that during the small-signal analysis the circuit will be modeled as linear for incremental changes from the bias point. Therefore, superposition will apply for the small-signal AC analysis, and we can separate all of our independent sources into DC and AC components. When we perform the DC analysis, we set the AC components to zero, and when we perform the small-signal AC analysis, we set the DC components to zero. Therefore, we can state how we deal with independent sources for the small-signal AC analysis in the following way: First, break all independent sources into AC and DC parts. (The DC component is just the average value.) Second, set the DC components to zero (i.e., short out voltage sources and open-circuit current sources). This procedure is illustrated in Figure 8-1 for both voltage and current sources.

Notice that the models in Figure 8-1 are *not* appropriate for AC analysis with large signals, unless the circuit being analyzed is linear. Although you might think that it would still be reasonable to separate the AC and DC parts for non linear circuits, the separate solutions could not be added together to yield the total result, since superposition does not work for nonlinear circuits. When linear small-signal models are used for the nonlinear devices, however, the DC operating point corresponds to the point where the *signal* is zero. Therefore, the DC bias point *can* be added to the small-signal solution, *even though* the DC analysis may be performed using nonlinear models for the devices.

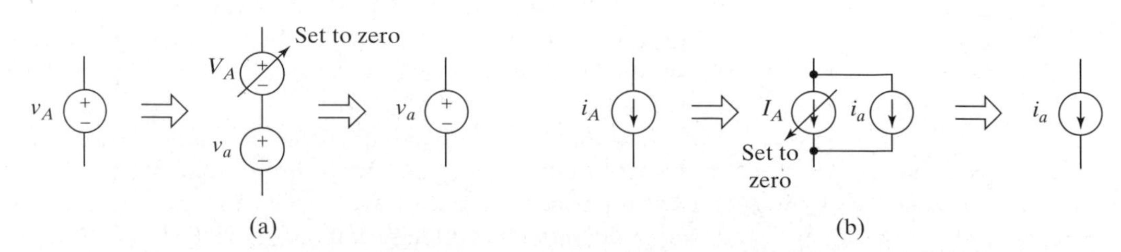

Figure 8–1 How to model independent sources for small-signal AC analysis. (a) A voltage source. (b) A current source.

8.1.2 Linear Passive Devices (Rs, Ls, & Cs)

No perfect resistors, capacitors, or inductors exist. All real elements must be modeled to the desired degree of accuracy. For low-frequency analysis and design, however, the parasitic elements can usually be ignored.

Since we are discussing amplifiers in this chapter, we will ignore signal-shaping filters for the moment.[1] (The topic of filters will be treated in Chapter 11.) In addition to making signal-shaping filters, inductors, and capacitors can be used to separate the DC biasing and AC operation of a circuit.

Inductors are most commonly used to filter noise from DC supply lines. Consider Figure 8-2, for example. In this case, an inductor and a capacitor are used to make sure that the supply voltage reaching our amplifier is really DC and does not contain noise. We have modeled the power supply as a battery in series with a noise source, but the noise may be a result of other circuits connected to the same supply (*see* Problem P8.1), or of extraneous signals getting coupled onto the supply line (*see* Problem P8.3). Independent of the cause of the added noise, it is frequently true that we need to filter the supply line locally to ensure that we have a clean DC supply to work with. The inductor is chosen to be large enough so that any alternating current will "see" a large impedance. In other words, we want the inductor to look like an open circuit to anything other than DC. The capacitor, on the other hand, is chosen to be large enough that any alternating current will see a small impedance and will be shunted to ground. In other words, we want the capacitor to look like a short circuit to anything other than DC.

At this point, we want to examine the function of coupling capacitors more closely. Consider the circuit shown in Figure 8-3, and examine the operation of the input capacitor first.

Since the DC current through a capacitor is zero, there can't be any DC current through the resistor R_S. Therefore, the DC voltage at the node labeled v_I' will be

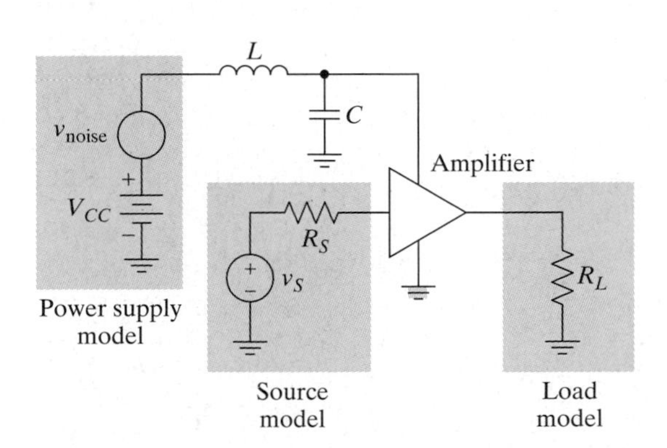

Figure 8–2 An LC power line filter used with an amplifier.

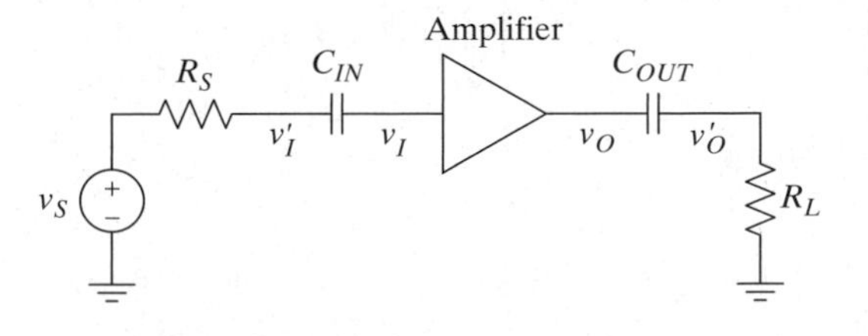

Figure 8–3 Input and output coupling capacitors.

[1] The qualifying phrase "signal-shaping" is used so that we can still discuss filters for our DC supply lines.

equal to V_S. At the same time, the DC voltage at the other side of the input capacitor will be V_I, which is not in general the same as V_S. In other words, the DC bias voltage at v_I is the same as it was without the signal source connected. Similarly, the DC voltage at v'_O is zero, but the voltage at v_O can be anything. Now consider the AC equivalent circuit shown in Figure 8-4, where we have replaced the amplifier by a unilateral two-port model and have allowed the amplifier's gain to be frequency dependent.

The input attenuation factor for this circuit at a given frequency is given by the impedance divider at the input:

$$\begin{aligned} \mathrm{A}_i(j\omega) &\equiv \frac{V_i(j\omega)}{V_s(j\omega)} = \frac{R_i}{R_i + R_S + Z_{C_{IN}}(j\omega)} \\ &= \frac{j\omega}{j\omega + \dfrac{1}{(R_i + R_S)C_{IN}}}\left(\frac{R_i}{R_i + R_S}\right) = \frac{\alpha_{i0} j\omega}{j\omega + \omega_{pi}}. \end{aligned} \tag{8.1}$$

Here, we have defined the DC input attenuation factor as

$$\alpha_{i0} = \frac{R_i}{R_i + R_S} \tag{8.2}$$

and the input pole frequency as

$$\omega_{pi} = \frac{1}{(R_i + R_S)C_{IN}}. \tag{8.3}$$

A similar analysis at the output reveals that

$$\mathrm{A}_o(j\omega) = \frac{\alpha_{o0} j\omega}{j\omega + \omega_{po}} \tag{8.4}$$

where

$$\alpha_{o0} = \frac{R_L}{R_L + R_o}, \tag{8.5}$$

and

$$\omega_{po} = \frac{1}{(R_L + R_o)C_{OUT}}. \tag{8.6}$$

The overall gain of the amplifier is then given by

$$A(j\omega) = \mathrm{A}_i(j\omega)a_v(j\omega)\mathrm{A}_o(j\omega). \tag{8.7}$$

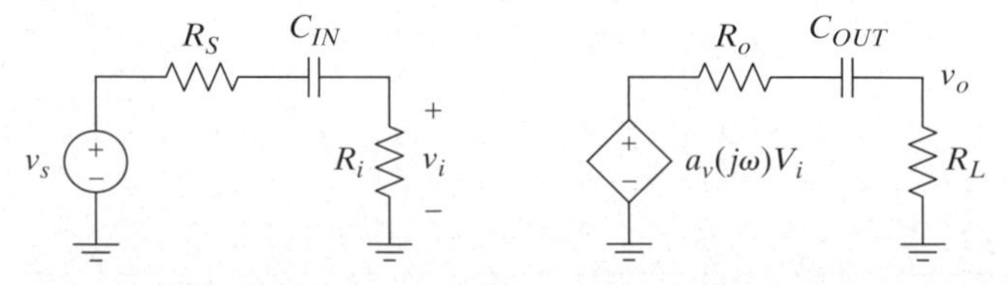

Figure 8–4 The AC equivalent circuit for Figure 8-3 with the amplifier replaced by a unilateral two-port model.

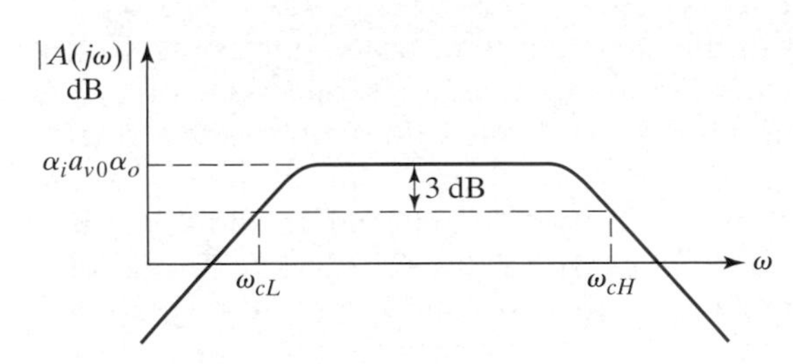

Figure 8–5 The magnitude of the voltage gain of the circuit in Figure 8-4.

If the amplifier's voltage gain has a low-pass characteristic, which is the usual case, the overall response will have a bandpass characteristic as shown in Figure 8-5. The response of this amplifier is fairly constant for a middle range of frequencies called the *midband* or the *passband*. At either end of this range, the magnitude of the gain falls off from its midband value. The frequencies where the gain has dropped by 3 dB from the midband gain are called the *high cutoff frequency*, ω_{cH}, and the *low cutoff frequency*, ω_{cL}. We use the term *cutoff frequency* instead of *pole frequency* to be general; the cutoff frequency is equal to a pole frequency when there is a single pole.

In this chapter we will be interested in the performance of an amplifier at frequencies above the low cutoff frequency, but below the high cutoff frequency. Analysis in this middle range of frequencies is called *midband analysis*.

For signals in this midband range, the coupling capacitors are so large that they look like short circuits. In other words, the magnitude of the capacitive reactance, $|1/j\omega C|$, is very small compared to the resistance in series with the coupling capacitor. We will see in Chapter 9 that the high cutoff frequency is caused by parasitic capacitances. These parasitic capacitances are usually quite small and effectively shunt the signal to ground. Consequently, for midband analysis, we can safely assume that they appear as open circuits. In other words, the magnitude of the capacitive reactance, $|1/j\omega C|$, is very large compared to the resistance of any path in parallel with the parasitic capacitor.

We can sum up what has been said so far with regard to inductors and capacitors in the following way. For small-signal midband analysis, we model coupling capacitors and shunt capacitors on the power line as short circuits. Similarly, we model inductors in series with the supply line as open circuits. We will soon see that we also use capacitors to bypass certain resistors in a circuit for midband frequencies. These capacitors are again chosen to be large enough that they are effective short circuits in midband.

Passive Elements in Integrated Circuits Resistors, capacitors, and inductors fabricated on integrated circuits have different parasitics than their discrete counterparts. We will briefly point out the differences in this section for those readers familiar with IC technology.

Resistors can be made in different ways on integrated circuits, but the most common and least expensive are diffused resistors. These devices frequently have relatively large parasitic capacitances associated with them. In addition, the resistance is a nonlinear function of the voltage applied, although the change is usually small enough that it can be modeled as a linear function (usually described by the *voltage coefficient of resistance*).

Capacitors are avoided when possible in IC design, because even small value capacitors consume large areas and are, therefore, very expensive. Unless very specialized technologies are used (e.g., so-called trench capacitors for memories), capacitors are usually limited to less than 100 pF. In addition, IC capacitors frequently have large parasitic capacitance from one of the plates to the substrate of the circuit and also have voltage-dependent capacitance.

Inductors are rarely used in ICs at this time, but their usage is growing. Inductors may be integrated as spiral shapes on the surface of a semiconductor, or the bond wires can be used as inductors. In either case, the values attainable are quite small (nH to tens of nH), because of the inability to form a large number of concentric loops and the lack of a high-permeability core material. In addition, when they are implemented on the semiconductor surface, there are large eddy currents induced in the semiconductor that lead to large losses and low *Q*s.

8.1.3 Diodes

The small-signal midband AC model for a diode can be derived by going back to the basic diode equation, as was done in Chapter 6. The diode equation is

$$i_D = I_S(e^{v_D/nV_T} - 1). \tag{8.8}$$

In reverse bias, the current is equal to I_S and is small enough to be ignored in almost all applications. Therefore, we will model the reverse-biased diode as an open circuit. When the diode is forward biased, the 1 in (8.8) can be safely ignored, and we find the small-signal conductance by differentiation:

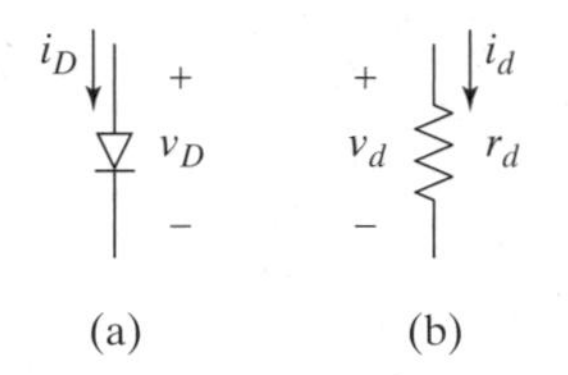

Figure 8–6 (a) A diode and (b) its small-signal midband AC model when forward biased.

$$g_d \equiv \left.\frac{di_D}{dv_D}\right|_{@\text{ Q point}} = \left.\frac{I_S e^{v_D/nV_T}}{nV_T}\right|_{@\text{ Q point}} = \frac{I_D}{nV_T}. \tag{8.9}$$

We will use $n = 1$ unless explicitly stated otherwise. For midband AC analysis, we ignore the junction capacitance and model the forward biased diode by its small-signal resistance as shown in Figure 8-6(b). Note that the capacitance would be in parallel with r_d, so most of the current flows through r_d in midband, and open-circuiting the capacitor is a reasonable approximation. The resistance is given by

$$r_d = \frac{1}{g_d} = \frac{nV_T}{I_D}. \tag{8.10}$$

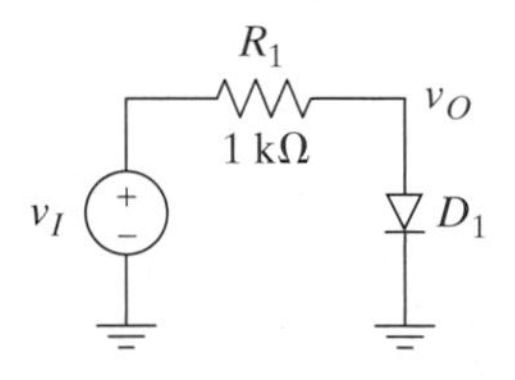

Figure 8–7

Since we are dealing strictly with amplifiers in this chapter, we won't need to deal with varactor or zener diodes. Nevertheless, it is worthwhile to note that both have the model shown in Figure 8-6(b) when they are forward biased. When reverse biased, a varactor is modeled as a voltage-controlled capacitor. The reverse-biased zener diode is modeled as an open circuit until the reverse voltage reaches the zener voltage, at which time it is modeled by a battery. These are only first-order models and can be improved upon if necessary.

Exercise 8.1

Consider the circuit shown in Figure 8-7. Find the small-signal midband model for the diode, draw the equivalent circuit, and find v_o. Assume that n = 1 and $v_I(t) = 3 + 0.02\cos(628t)$ V.

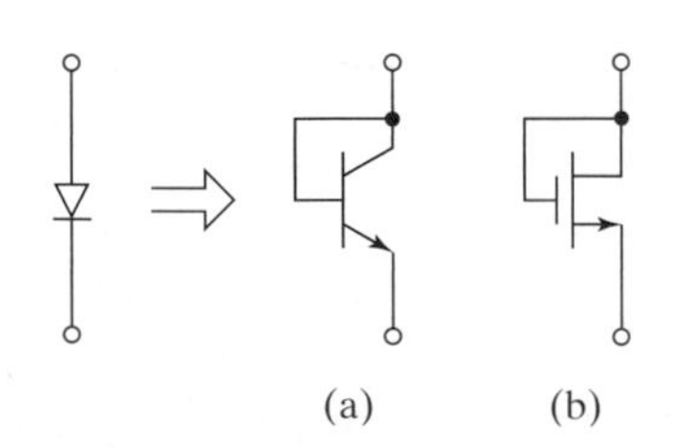

Figure 8–8 Diode-connected transistors; (a) a BJT and (b) a MOSFET.

Diodes in Integrated Circuits Diodes are not typically fabricated on an integrated circuit. A *diode-connected transistor*, as shown in Figure 8-8, is used instead. We show a diode-connected MOSFET for completeness, but even in a MOS process a parasitic bipolar transistor or *pn* junction will usually be used. Therefore, the transistor models presented in later sections are needed to accurately model an integrated diode. Transistors fabricated on integrated circuits have parasitic junction diodes associated with them, and these parasitic diodes are modeled as shown for discrete diodes in the preceding section.

8.1.4 The Generic Transistor

In Asides A6.3 and A7.2, we presented large-signal models for *n*- and *p*-type generic transistors and implicitly assumed they were operating in the forward active-region. We are interested in small-signal amplifiers in this chapter. Consequently, we are mostly concerned with forward-active operation. Nevertheless, we do need to briefly examine other regions of operation to determine the conditions under which our models apply.

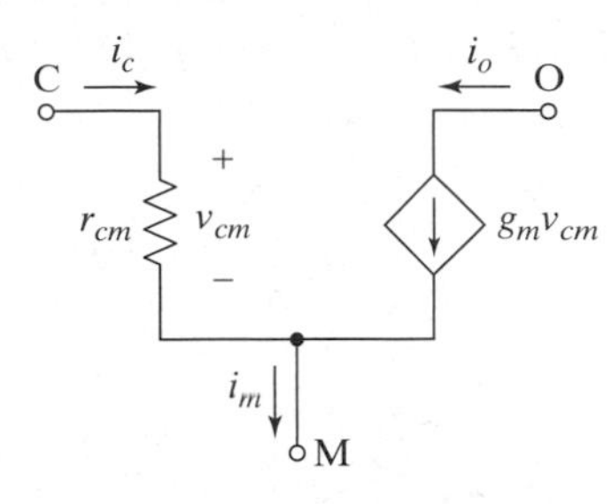

Figure 8–9 A generic transistor hybrid-π model for forward-active operation.

The low-frequency small-signal models of real transistors operating in the forward-active region all have a voltage-controlled current source and some input resistance (possibly infinite), as shown in Figure 8-9. Therefore, we will use this model for our generic transistor as well. For reasons that will be clear later, the model is called the ***hybrid-π model***. This model is most useful when the input is applied between the control and merge terminals and the output is taken between the output and merge terminals. Since the merge terminal is common to both the input and the output, such a connection is referred to as a ***common-merge*** connection.

The current i_o is much larger than the current i_c, and the current i_m must be the sum of the other two. Therefore, the generic transistor does have current gain, and we can see from the model that it is equal to

$$a_{im} = \frac{i_o}{i_c} = \frac{g_m v_{cm}}{i_c} = \frac{g_m i_c r_{cm}}{i_c} = g_m r_{cm}, \qquad \textbf{(8.11)}$$

where the subscript m indicates that this is the current gain for a common merge connection. Using this current gain, we could change the controlled source in Figure 8-9 into a current-controlled current source. This alternate representation is sometimes more convenient to use

We will have occasion to consider circuits in which the input is connected between the merge and control terminals, and the output is taken between the output and control terminals as shown in Figure 8-10. This connection is called a ***common-control*** circuit, since the control terminal is common to both the input and output. The hybrid-π model is not the most convenient to use in this situation; we will derive an alternate model after Aside A8.1, which presents a couple of network theorems we will use in deriving the model.

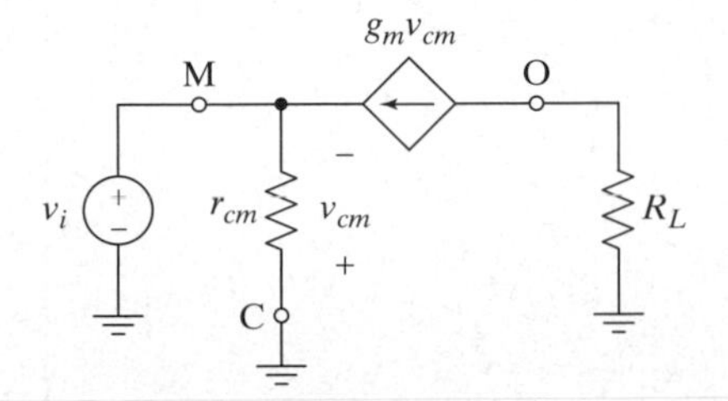

Figure 8–10 A common-control connection.

ASIDE A8.1 THE SPLIT-SOURCE TRANSFORMATION AND SOURCE ABSORPTION THEOREM

Consider the transconductance amplifier shown in Figure A8-1(a). This model is particularly useful when the input is connected between terminals 1 and 2 and the output is taken between terminals 3 and 2, as shown in part (b) of the figure. However, suppose that the input is connected between terminals 1 and 2 as before, but the output is now taken between terminals 3 and 1, as shown in part (c). Although we can certainly analyze this circuit in a straightforward way, we will often find it convenient to have an alternate model available.

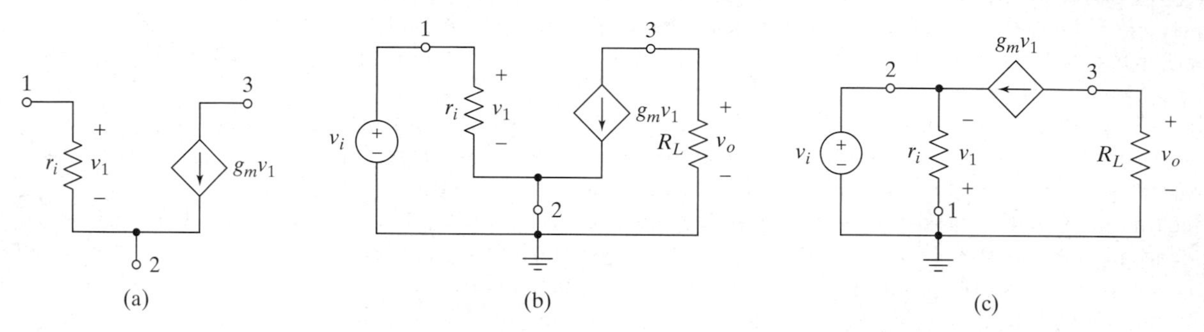

Figure A8-1 (a) A transconductance amplifier. (b) The amplifier connected with terminal 2 common to the input and output. (c) The amplifier connected with terminal 1 common to the input and output.

To derive the new model, we start with the model in part (a) of the figure and put a second controlled-current source in series with the existing source, as shown in Figure A8-2(a). Since these sources have the same value, we have not changed the operation of the circuit in any way. It is like putting two voltage sources of the same value in parallel. Since we are injecting and removing the same current from the node in between the two current sources, we are free to connect it to any other node without affecting that node (i.e., KCL remains unchanged at the node, since we add and subtract the same current). In part (b) of the figure, we connect this node to terminal 1. This operation is called a ***split-source transformation*** and is sometimes a very useful manipulation. Now notice that the current source in parallel with r_i is controlled by the voltage across itself. Therefore, this current source looks like a resistor of value $1/g_m$. If a resistor of value $R = 1/g_m$ is used in place of the current source, as shown in part (c) of the figure, the current through the resistor will be $v_i/R = g_m v_i$. This manipulation is one example of the ***source absorption theorem*** and is another useful circuit manipulation. If we now combine $1/g_m$ in parallel with r_i, we get the model shown in part (d) of the figure, where we have also shown a signal source and load as in Figure A8-1(c).

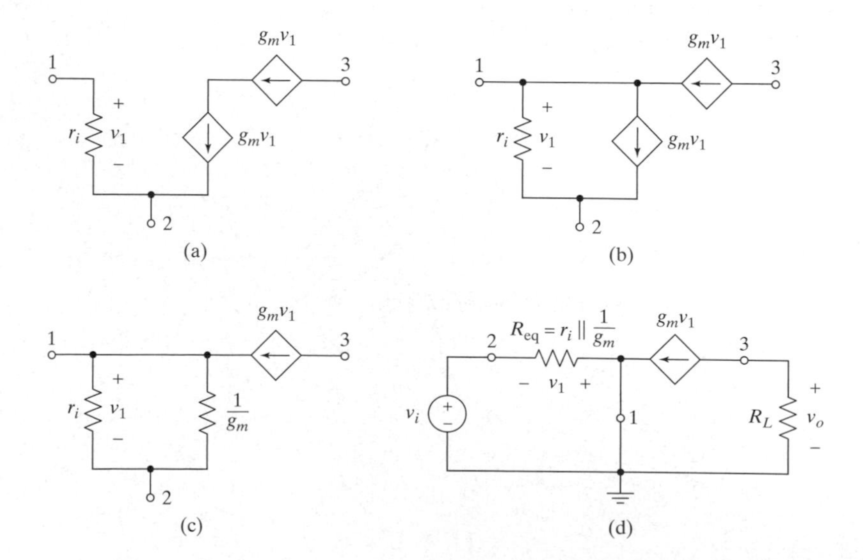

Figure A8-2 (a) The model of Figure A8-1(a) with a second current source added, (b) the middle node connected to terminal 1, (c) the source replaced by $1/g_m$, and (d) the new model used with a source and load as in Figure A8-1(c).

We now return to our generic transistor model and apply the network theorems from the aside. In Figure 8-11(a), we show the generic transistor hybrid-π model with the current source split. In part (b) of the figure, we connect the node in between the two controlled sources to the control terminal of the transistor. We then recognize that one of these controlled sources is controlled by the voltage across itself and can be represented as a resistor. In part (c), we replace that source by its equivalent resistance. In part (d), we combine the two resistors in parallel, redraw the circuit, and change the voltage-controlled current source into a current-controlled current source by recognizing that

$$g_m v_{cm} = g_m r_m i_m, \tag{8.12}$$

where the equivalent resistance has been denoted r_m, since it is the small-signal resistance seen looking in the merge terminal and is given by

$$r_m = \frac{1}{g_m} \Bigg\| r_{cm}. \tag{8.13}$$

Setting (8.12) equal to $a_{ic} i_m$, we see that the common-control current gain is

$$a_{ic} = \frac{i_o}{i_m} = g_m r_m. \tag{8.14}$$

This new model of the transistor is called the ***T model*** because it has the shape of the letter T.

We now need to define under what set of circumstances the hybrid-π and T models are valid. These models assume that the transistor is forward active, but this assumption is correct only for some range of terminal voltages. If the voltages

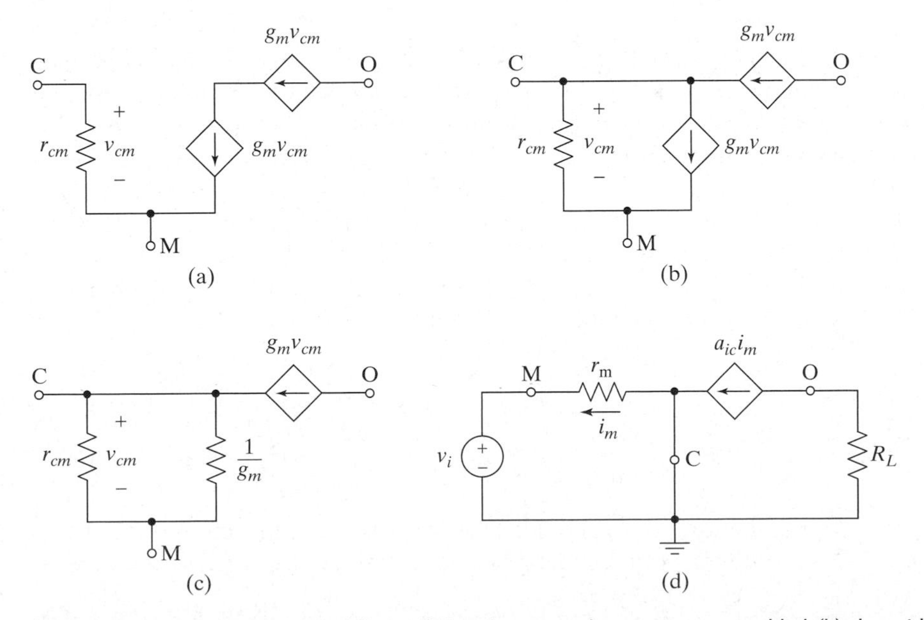

Figure 8–11 (a) The model of Figure 8-9 with a second current source added, (b) the middle node connected to the control terminal, (c) the source replaced by $1/g_m$, and (d) the resulting current-controlled T model with a signal source and load.

are not in this range, then the devices are either in ***cutoff*** (i.e., no current flows), or are in some other region of operation. We assume that when the input voltage of the generic transistor is less than or equal to zero, the device is cut off and no current will flow in any terminal. The model in this region is three open circuited terminals. We also assume that if the voltage from the output terminal to the merge terminal is less than zero, no current can flow in the output. When this voltage is zero, the output and merge terminals are shorted together, and the current is determined by the external circuit.

8.1.5 Bipolar Junction Transistors

In this section, we find two small-signal models for the BJT at low-frequencies. In each case, however, the controlled-current source can be thought of as being either voltage controlled or current controlled. Therefore, we will have four models to choose from when doing our analyses. Since we are interested in small-signal amplifiers in this chapter, we are only concerned about forward-active operation. Because the input of the transistor no longer controls the output when it is saturated or cut off, it doesn't matter whether the input is small-signal or not in these regions. Therefore, the models derived in Chapter 7 for saturation and cutoff are still usable.

When a BJT is operating in the forward-active region, its collector current is given by

$$i_C = I_S e^{v_{BE}/V_T}\left(1 + \frac{v_{CE}}{V_A}\right), \tag{8.15}$$

where the Early effect has been included. By differentiating (8.15) with respect to v_{BE} and evaluating the derivative at the Q point (i.e., $v_{BE} = V_{BE}$ and $v_{CE} = V_{CE}$), we find the ***transconductance***:

$$g_m \equiv \left.\frac{di_C}{dv_{BE}}\right|_{@\text{ Q point}} = \left.\frac{I_S e^{v_{BE}/V_T}(1 + v_{CE}/V_A)}{V_T}\right|_{@\text{ Q point}} = \frac{I_C}{V_T}. \tag{8.16}$$

The transconductance models the variation in collector current caused by changes in the base-emitter voltage. We know from Chapter 6 that for small deviations from the operating point—the Q point—an exponential relationship can be modeled as being linear with reasonable accuracy. Note that (8.15) is identical in form to (8.8) if v_{CE} is constant and the '1' is ignored in (8.8). Therefore, the conclusions drawn about the accuracy of the small-signal approximation for the diode apply equally well to the BJT (*see* Table 6.1).

Similarly, we can differentiate (8.15) with respect to v_{CE} to find the conductance seen from the collector to the emitter. This conductance is called the ***small-signal output conductance***, so we denote it by g_o; we have

$$g_o \equiv \left.\frac{di_C}{dv_{CE}}\right|_{@\text{ Q point}} = \left.\frac{I_S e^{v_{BE}/V_T}}{V_A}\right|_{@\text{ Q point}} \approx \frac{I_C}{V_A}. \tag{8.17}$$

The final approximation in (8.17) holds only if v_{CE} is a small fraction of the Early voltage V_A. It is often more convenient to deal with the ***small-signal output resistance***, r_o, which is the reciprocal of g_o:

$$r_o = \frac{1}{g_o} \approx \frac{V_A}{I_C}. \tag{8.18}$$

This resistance models the change in collector current resulting from changes in the collector-to-emitter voltage.

To find the resistance from base to emitter, we divide (8.15) by beta to get

$$i_B = \frac{I_S e^{v_{BE}/V_T}(1 + v_{CE}/V_A)}{\beta}. \quad \textbf{(8.19)}$$

By differentiating (8.19) and evaluating the derivative at the Q point as before, we find the conductance seen from the base to the emitter *when looking into the base*:[2]

$$g_{be} \equiv \left.\frac{di_B}{dv_{BE}}\right|_{@\text{ Q point}} = \frac{I_B}{V_T} = \frac{I_C}{\beta V_T} = \frac{g_m}{\beta}. \quad \textbf{(8.20)}$$

The reciprocal of this conductance is the ***small-signal resistance from base to emitter*** and is denoted r_π. Taking the reciprocal of (8.20) yields

$$r_\pi = \frac{\beta}{g_m} = \frac{\beta V_T}{I_C} = \frac{V_T}{I_B}. \quad \textbf{(8.21)}$$

This resistor models the change in base current with base-emitter voltage.

The final model resistance that could be included is the ***small-signal resistance from the collector to the base***. This resistor is called r_μ and models the change in base current caused by the collector-base voltage changing. This is a minor effect and can usually be ignored, especially in discrete circuits, for which the impedance levels tend to be lower. The value of r_μ can be shown to always be greater than βr_o and is typically greater than $10\beta r_o$ for modern integrated *npn* transistors [8.1].

It is important to recognize that r_π, r_o, and r_μ are not *real* resistors. They are simply linear *models* that account for the changes in the terminal currents with changes in the terminal voltages. In addition to these small-signal model elements, there are real resistances in series with the BJT terminals. These resistances are caused by the leads to the package, the metal-to-semiconductor contacts, and the bulk semiconductor regions that the currents must traverse before getting to the active regions. If all three of these resistances are included as well (r_c, r_b, and r_{ex}),[3] one possible ***low-frequency small-signal model*** of the BJT is shown in Figure 8-12(a). This model is called the ***hybrid-π model*** and can almost always be simplified to the one shown in Figure 8-12(b) without affecting the accuracy of the analyses. These simplifications are possible because r_{ex} and r_c are small and r_μ is very large. There will be many times when we can further simplify the model to that shown in part (c) of the figure without significant error. The model is called the hybrid-π, because it looks like the Greek letter π (if r_μ is included) and because it is a simplified version of the hybrid-parameter two-port network.

Notice that because the polarities of all voltages and currents are reversed for a *pnp* transistor relative to an *npn*, the hybrid-π model is appropriate for either type of transistor *without modification*. We will often find it convenient to use a current-controlled current source in place of the voltage-controlled current

---

[2] Since the base and emitter currents are different, the resistance (or conductance) seen from base to emitter is different, depending on which terminal you look into. This phenomenon is identical to the impedance reflection discussed in Aside A7.1 and will be discussed again later in this section.

[3] We explain why r_{ex} has the '*x*' in the subscript after (8.26).

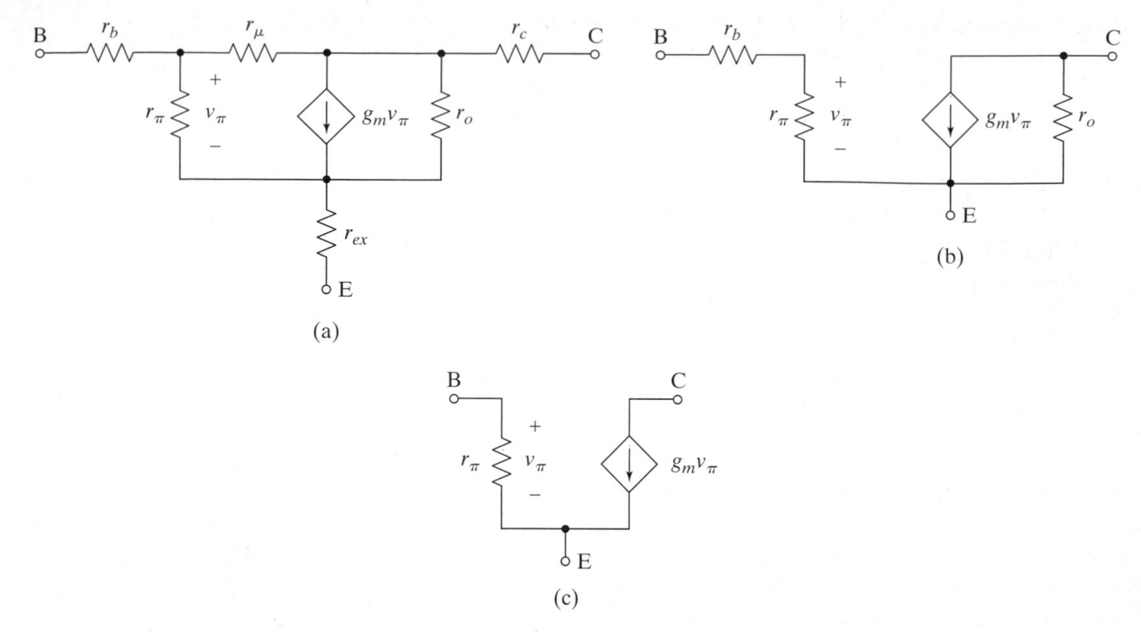

Figure 8–12 The low-frequency hybrid-π model for a BJT (*npn* or *pnp*). (a) The full model, (b) slightly simplified, and (c) the simplest case.

source in the hybrid-π model. The result is the current-controlled hybrid-π model shown in Figure 8-13 for the simplest case. The value of the controlled source is derived from the standard hybrid-π source with the aid of (8.21):

$$g_m v_\pi = g_m i_b r_\pi = \beta i_b. \tag{8.22}$$

The value of β given in (8.22) is not, in general, equal to the DC β used in Chapter 7. When the difference is important, the DC value in forward-active operation is denoted by β_F, as noted in Chapter 7, and the small-signal value in (8.22) is denoted by β_0, where the '0' indicates that this is the small-signal beta at DC (i.e., zero frequency). The reason that β_F and β_0 may differ is that I_C is not a linear function of I_B. The value of β typically rolls off at both low and high currents, as discussed in Chapter 2. β_F is defined as the ratio of I_C to I_B, and β_0 is defined as the slope of the I_C-versus-I_B curve. To see that β_0 is defined as the slope, note that it is equal to the ratio of i_c to i_b and that these quantities are the incremental values (i.e., the changes from the bias point). In other words, we could say that $i_c = \Delta i_C$ and $i_b = \Delta i_B$ and that $\beta_0 = i_c / i_b = \Delta i_C / \Delta i_B$. If we are going to be this careful, then (8.19) is not correct, since it implicitly assumes a constant value for β. The difference between β_F and β_0 is usually quite small—except at very low or very high currents—and can often be ignored. We will not concern ourselves further with the difference. Data sheets and some other books refer to β_F as h_{FE} and β_0 as h_{fe}. This notation reflects the fact that β is the forward gain in a hybrid-parameter two-port representation of the bipolar transistor if the emitter is common to both the input and output.

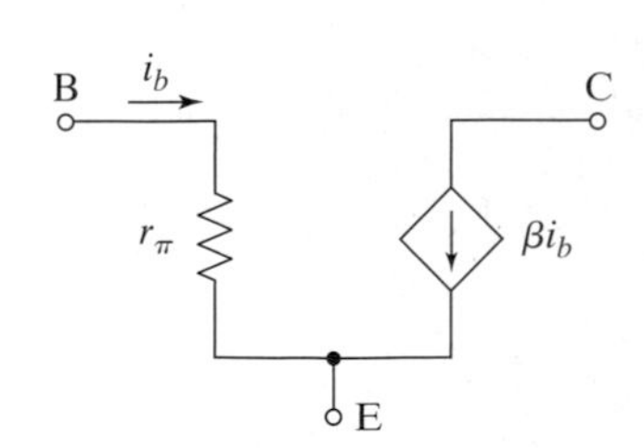

Figure 8–13 A current-controlled version of the simplest hybrid-π model (*npn* or *pnp*).

Another useful small-signal model for the BJT is the ***T model***, so named because in its simplest form it looks like the letter T. It can be derived from the hybrid-π model by circuit manipulations, or directly from the equations for the device. In the T model we consider the emitter, rather than the base, to be the input. The emitter current of the BJT is

$$i_E = i_C + i_B = (\beta + 1) i_B. \tag{8.23}$$

Differentiating with respect to v_{BE} leads to the ***small-signal conductance from emitter to base***

$$g_{eb} \equiv \left.\frac{di_E}{dv_{BE}}\right|_{@\text{ Q point}} = (\beta + 1)\left.\frac{di_B}{dv_{BE}}\right|_{@\text{ Q point}}, \quad \textbf{(8.24)}$$

which, from (8.20), is

$$g_{eb} = (\beta+1)\frac{g_m}{\beta}. \quad \textbf{(8.25)}$$

The ***small-signal emitter resistance*** is defined to be the reciprocal of this conductance and is given by

$$r_e = \left(\frac{1}{\beta+1}\right)\frac{\beta}{g_m} = \left(\frac{\beta}{\beta+1}\right)\frac{V_T}{I_C} = \frac{V_T}{I_E}. \quad \textbf{(8.26)}$$

The physical resistance in series with the emitter was called r_{ex} earlier. The subscript x was used to signify that r_{ex} is the ***extrinsic*** resistance, whereas r_e is modeling the ***intrinsic*** action of the junction. We don't need the x subscript on the other extrinsic resistances, because they can't be confused with any of our small-signal model parameters. We can also express the relationship between r_e and r_π from (8.26):

$$r_e = \left(\frac{1}{\beta+1}\right)\frac{\beta}{g_m} = \frac{r_\pi}{\beta+1}. \quad \textbf{(8.27)}$$

From (8.27), we see that the small-signal resistance seen from base to emitter is very different, depending on whether you look into the emitter or the base. In either case the voltage difference is v_{be}, but the currents differ by a factor of $(\beta+1)$, so the resistances do also. This is exactly the same phenomenon that was described in Aside A7.1 on impedance reflection.

The T model appears in Figure 8-14(a) in its simplest form where it only includes the resistor r_e and the controlled-current source. The controlled-current source no longer connects directly to the emitter, because r_e models the emitter current as a function of v_{be}. The controlled source is now only modeling the collector current, which flows to the base. We have also labeled the controlling voltage v_{be} instead of v_π, because this isn't the hybrid-π model. A more common and more useful form of this model is shown in part (b) of the figure, where we have

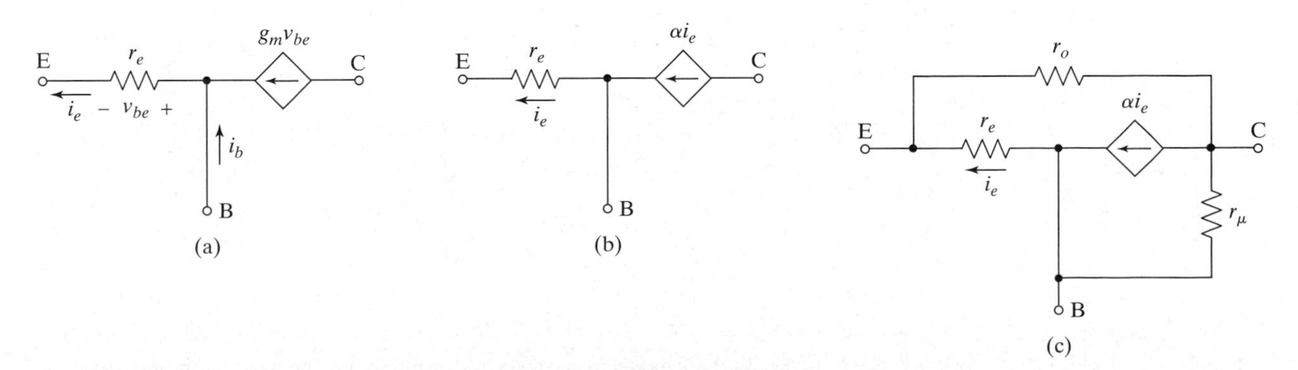

Figure 8–14 The T model for a BJT (*npn* or *pnp*); (a) the simplest topology, (b) the same topology but with a current-controlled source, and (c) with some second-order model elements included.

changed the voltage-controlled current source into a current-controlled current source. This change was accomplished by noting that

$$g_m v_{be} = g_m i_e r_e = g_m i_e \left(\frac{\beta}{\beta+1}\right)\frac{1}{g_m} = \alpha i_e, \tag{8.28}$$

where we have made use of (8.27) and the fact that $\alpha = \beta/(\beta+1)$. The net base current is the difference between the emitter and collector currents:

$$i_b = i_e(1-\alpha) = \frac{i_e}{\beta+1}. \tag{8.29}$$

We can include the other elements in this model, too, if we so desire. For example, Figure 8-14(c) shows the T model with r_o and r_μ. This T model is appropriate for either *npn* or *pnp* transistors, as was the case for the hybrid-π model.

The T model can also be derived directly from the hybrid-π model, as shown in Figure 8-15. Starting with the hybrid-π model in part (a) of the figure, we use the split-source transformation (*see* Aside A8.1) and put a second controlled-current source in series with the existing source as shown in part (b). In part (c) of the figure we connect this node to the base terminal. Now notice that the current source in parallel with r_π is controlled by the voltage across itself. Therefore, from the source absorption theorem, this current source looks like a resistor of value $1/g_m$. If we now combine $1/g_m$ in parallel with r_π, we get

$$r_\pi \| (1/g_m) = \frac{r_\pi(1/g_m)}{r_\pi + \dfrac{1}{g_m}} = \frac{r_\pi}{1 + g_m r_\pi} = \frac{r_\pi}{1+\beta} = r_e, \tag{8.30}$$

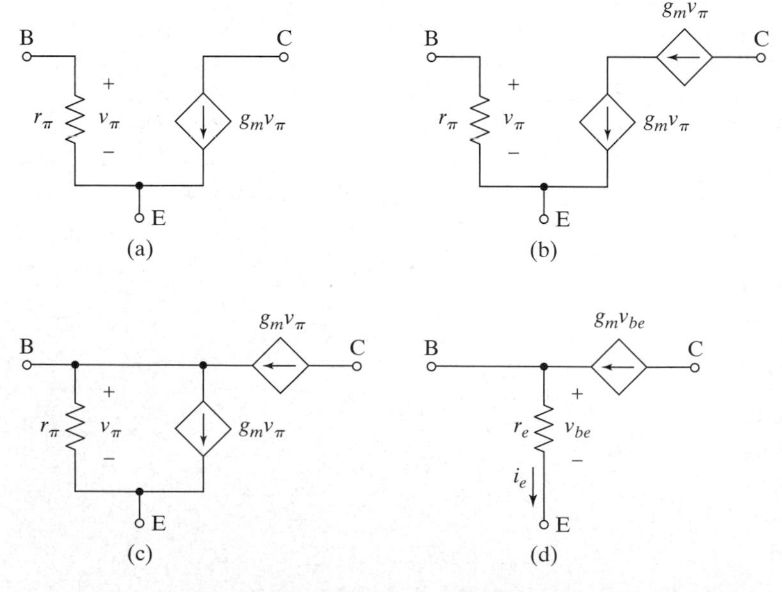

Figure 8–15 Deriving the T model from the hybrid-π model. (a) The hybrid-π model, (b) splitting the source, (c) connecting the middle node to the base, (d) absorbing the controlled source by replacing it by an equivalent resistance, and combining that resistance with r_π to obtain r_e.

where we used the fact that $g_m r_\pi = \beta$, as can be seen from (8.21). We also relabeled v_π as v_{be} again, since the model shown in part (d) of the figure is really the T model, as was first presented in Figure 8-14(a). You need to redraw the circuit a bit to see this.

The T model can also be used as a current-controlled model, as was done for the hybrid-π model. Examining Figure 8-15(d), we see that

$$v_{be} = i_e r_e, \quad \textbf{(8.31)}$$

so we could write the controlled source as

$$g_m v_{be} = g_m r_e i_e = \frac{I_C}{V_T}\frac{V_T}{I_E} i_e = \alpha i_e, \quad \textbf{(8.32)}$$

where α is the common-base current gain encountered in Chapter 7.

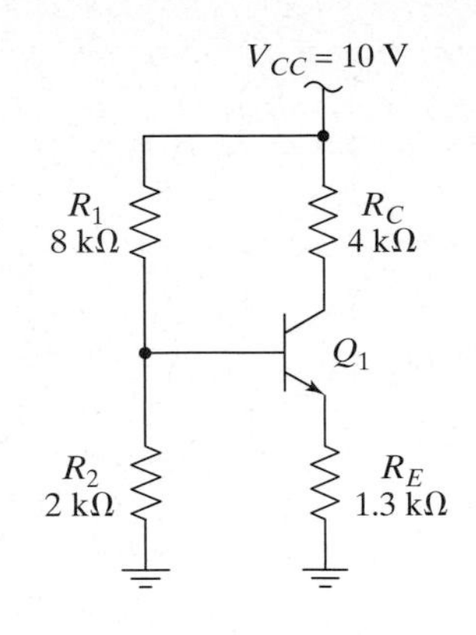

Figure 8–16

Exercise 8.2

Consider the bipolar amplifier shown in Figure 8-16. Determine the small-signal midband AC model for the transistor, and draw the resulting equivalent circuit. Assume that $\beta = 100$ and ignore base-width modulation.

□

Bipolar Transistors in Integrated Circuits Typical integrated BJTs have additional parasitic elements not mentioned in the preceding section. If the devices are junction isolated, which is the most common situation today, there are parasitic diodes associated with the isolation. Figure 8-17 shows simplified cross sections of typical *npn* and *pnp* transistors (*see* Chapter 3 for details). The *npn* shown here is a vertical structure, and the *pnp* is a lateral structure. There are also vertical *pnps*.

In the *npn* transistor, we see that there is a parasitic *pn* junction diode from the isolation to the epitaxial layer. In addition, this diode forms part of a parasitic *pnp* transistor that goes from the base of the *npn* to the epitaxial layer (the *npn* collector) and to the substrate. The substrate junction is rarely modeled as part of a parasitic transistor; rather, SPICE models it using a diode as shown in Figure 8-18(a).

Similarly, the lateral *pnp* transistor has a parasitic diode connecting the base to the substrate. This diode is again a part of two parasitic transistors: one from the *pnp* emitter to the base and then the substrate and the other from the *pnp* collector to the base and the substrate. The parasitic transistors are rarely important, and SPICE models the substrate junction by a diode as shown in Figure 8-18(b).

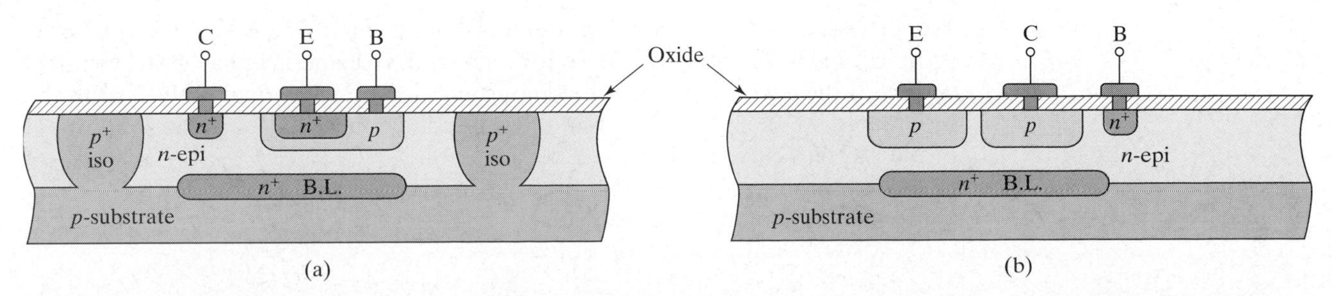

Figure 8–17 (a) Simplified cross section of a typical junction-isolated *npn* transistor. (b) Simplified cross section of a typical junction-isolated lateral *pnp* transistor.

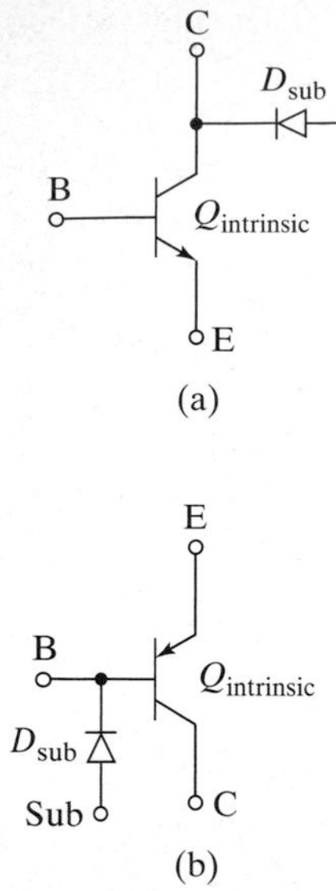

The parasitic substrate diodes rarely have any impact on the small-signal mid-band AC performance of bipolar circuits, but the capacitance of the diodes does affect the high-frequency performance and is frequently included in hand models as discussed in Chapter 9.

Figure 8–18 SPICE models the substrate junction by a diode connected (a) from the collector to the substrate for vertical *npn*'s and (b) from the base to the substrate for lateral *pnp*'s.

8.1.6 MOS Field-Effect Transistors

In this section, we derive two small-signal models for MOSFETs operating in saturation (i.e., forward active). We ignore the body effect until the end of this section. Because the only functional difference between enhancement- and depletion-mode MOSFETs is the value of V_{th}, the small-signal models are the same, and we need to derive them only for enhancement-mode transistors. The model derived for a MOSFET in the linear region of operation in Chapter 7 is also usable for small-signal operation, provided that we ignore the nonlinear term as was done in deriving the linear resistance given by (7.18). The MOSFET model in cutoff was found to be an open circuit and that is still appropriate here as well.

We start our derivation of the small-signal models by remembering that the drain current of an n-channel MOSFET in saturation is given by

$$i_D = \frac{\mu_n C'_{ox}}{2}\frac{W}{L}(v_{GS} - V_{th})^2 = \frac{K'}{2}\frac{W}{L}(v_{GS} - V_{th})^2 = K(v_{GS} - V_{th})^2, \quad \textbf{(8.33)}$$

where K' and K are implicitly defined (as in Chapters 2 and 4) and we have ignored channel-length modulation for now. Differentiating (8.33) with respect to v_{GS} and evaluating the result at the bias point leads to the ***small-signal transconductance***,

$$g_m \equiv \left.\frac{di_D}{dv_{GS}}\right|_{@\text{ Q point}} = 2K(v_{GS} - V_{th})\Big|_{@\text{ Q point}} = 2K(V_{GS} - V_{th}). \quad \textbf{(8.34)}$$

Similarly, for p-channel MOSFET's we get

$$g_m \equiv \left.\frac{di_D}{dv_{SG}}\right|_{@\text{ Q point}} = 2K(v_{SG} + V_{th})\Big|_{@\text{ Q point}} = 2K(V_{SG} + V_{th}). \quad \textbf{(8.35)}$$

Remember that the term $(V_{GS} - V_{th})$ in (8.34) and the term $(V_{SG} + V_{th})$ in (8.35) are both called the *gate overdrive* of the MOSFET and are denoted by V_{ON}. It is also important to remember that K is proportional to W/L. Algebraic manipulation of either (8.34) or (8.35) leads to alternate forms for g_m that are useful for either n- or p-channel devices:

$$g_m = 2\sqrt{KI_D} = \frac{2\,I_D}{(V_{GS} - V_{th})}. \quad \textbf{(8.36)}$$

From (8.34) or (8.35), we see that the transconductance of a MOSFET is linearly proportional to the gate overdrive for a given device and is linearly proportional to W/L if the gate overdrive is held constant (K is proportional to W/L). On the other hand, we see from (8.36) that the transconductance varies as the square root of the DC bias current for a given device.

To find the conductance from drain to source, we need to modify (8.33) to include channel-length modulation:*

* If channel-length modulation significantly affects I_D, then (8.34), (8.35), and (8.36) must also be modified.

$$i_D = K(v_{GS} - V_{th})^2\left(1 + \frac{v_{DS}}{V_A}\right) = K(v_{GS} - V_{th})^2(1 + \lambda v_{DS}). \quad \textbf{(8.37)}$$

Taking the derivative of (8.37) with respect to v_{DS} and evaluating at the Q point leads to the ***small-signal conductance from drain to source***,

$$g_{ds} \equiv \left.\frac{di_D}{dv_{DS}}\right|_{@\text{ Q point}} = \left.\frac{K(v_{GS} - V_{th})^2}{V_A}\right|_{@\text{ Q point}} \approx \frac{I_D}{V_A} = \lambda I_D, \quad \textbf{(8.38)}$$

where the final approximation in (8.38) is valid only if v_{DS} is much less than V_A. The same result is derived for p-channel transistors. The small-signal output resistance is then

$$r_o = \frac{1}{g_{ds}} \approx \frac{V_A}{I_D} = \frac{1}{\lambda I_D}. \quad \textbf{(8.39)}$$

Using (8.34) or (8.36) for n-channel devices and (8.35) or (8.36) for p-channel devices, along with (8.38), we can draw the low-frequency small-signal model for a MOSFET as shown in Figure 8-19(a). Because this model looks like the Greek letter π when the gate-to-drain and gate-to-source capacitances are included and r_o is ignored (we will do this in Chapter 9), and because it is a simplified version of a hybrid-parameter two-port network, we will call it the ***low-frequency small-signal hybrid-π model*** for a MOSFET.

In the case of the MOSFET, the gate current at low frequencies is essentially zero, so there aren't any resistances needed from gate to source or drain to gate. Figure 8-19(b) shows the model with the parasitic resistances in series with the terminals included. These resistances are caused by the metal-to-semiconductor contacts and the bulk regions in series with the active device. Both of these resistors are small and will be ignored unless specifically stated otherwise.

Another model that will be useful is the ***T model*** for the MOSFET. The name of the model comes from the fact that, in its simplest form, it looks like the letter T. To derive this model, consider the source terminal as the input, rather than the gate. In other words, we need to look at the equation for the source current:

$$i_S = i_D = K(v_{GS} - V_{th})^2. \quad \textbf{(8.40)}$$

This equation is only valid at DC, or very low frequencies, where the gate current is negligible. In addition, we have ignored channel-length modulation. We now want to find the effective conductance from source to gate as seen looking into the source. To do this, we differentiate (8.40) with respect to v_{GS} and evaluate the

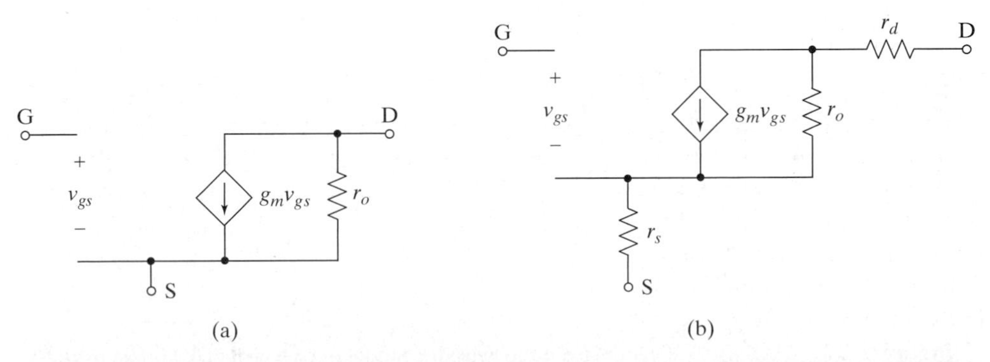

Figure 8–19 (a) The low-frequency small-signal hybrid-π model for a MOSFET (n or p channel). (b) The same model with parasitic resistances in series with the terminals.

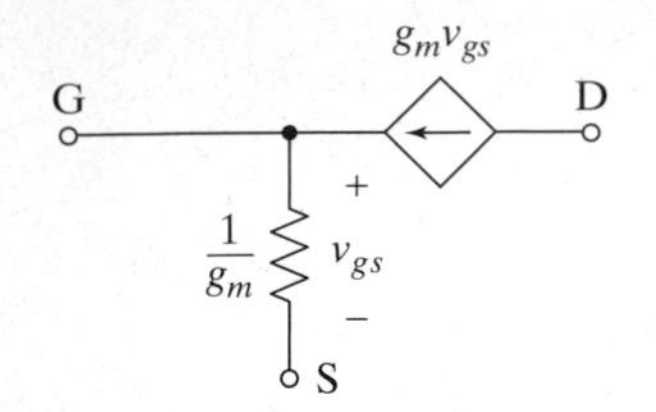

Figure 8–20 The T model of a MOSFET (*n* or *p* channel).

result at the Q point. This evaluation has already been carried out in (8.34) and the result is the transconductance, g_m. The T model is shown in Figure 8-20. Notice that the controlled source now connects from the drain to the gate, instead of from the drain to the source as in the hybrid-π model. Notice also that the gate current is still zero, even though a resistance of $1/g_m$ now shows up from gate to source. To see that i_g is zero, write KCL at the gate:

$$i_g = \frac{v_{gs}}{1/g_m} - g_m v_{gs} = 0. \tag{8.41}$$

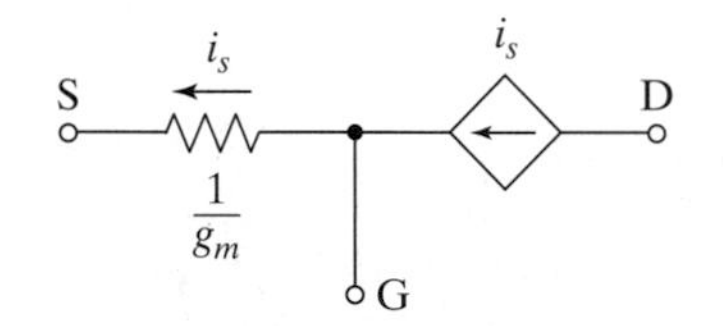

Figure 8–21 A current-controlled version of the MOSFET T model (*n* or *p* channel).

An alternate form of the T-model, as shown in Figure 8-21, is often useful. This form is derived from Figure 8-20 by noting that the source and drain currents are equal.

The T model can also be derived directly from the hybrid-π model as shown in Figure 8-22. Starting with the hybrid-π model in part (a) of the figure, we use the split-source transformation (*see* Aside A8.1) and put a second controlled-current source in series with the existing source as shown in part (b). In part (c) of the figure, we connect this node to the gate terminal. Now notice that the current source from gate to source is controlled by the voltage across itself. Therefore, using the source absorption theorem, we find that this current source looks like a resistor of value $1/g_m$.

Bulk Transconductance One second-order effect that is often important to include in the small-signal MOSFET model is the body effect. This effect was discussed in Chapter 2, and we discovered that the charge in the inversion layer is under the control of both the gate-to-channel voltage as set by v_{GS} and the bulk-to-channel voltage as set by v_{BS}. A DC voltage from bulk to source can be modeled by changing the value of V_{th}, but for a time-varying v_{BS} this effect is modeled by including an additional controlled source as shown in Figure 8-23. (Only the controlled sources are shown, but other elements, like r_o, may be included as well.) The bulk transconductance g_{mb} is the derivative of i_D with respect to v_{BS}. As with all small-signal model parameters, we must evaluate the derivative at the op-

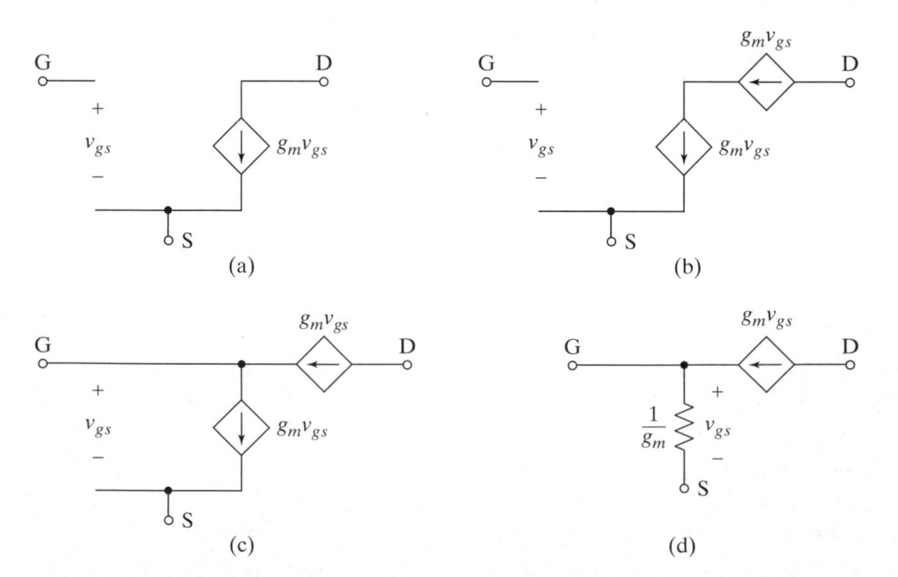

Figure 8–22 Deriving the T model from the hybrid-π model (*n* or *p* channel). (a) The hybrid-π model, (b) splitting the source, (c) connecting the middle node to the gate, and (d) the T model as shown in Figure 8-20.

erating point. In strong inversion the charge density in the channel is a linear function of both v_{GS} and v_{BS}. The change in the inversion layer charge can be found by considering the inversion layer to be the center plate in a series combination of two capacitors, C_{ox} and C_{DR} (C_{ox} is the oxide capacitance and C_{DR} is the depletion-region capacitance). As a result, in Chapter 2 we found that we could relate g_{mb} to g_m by the ratio of these capacitors:

$$\frac{g_{mb}}{g_m} = \frac{C_{DR}}{C_{ox}} = \chi. \tag{8.42}$$

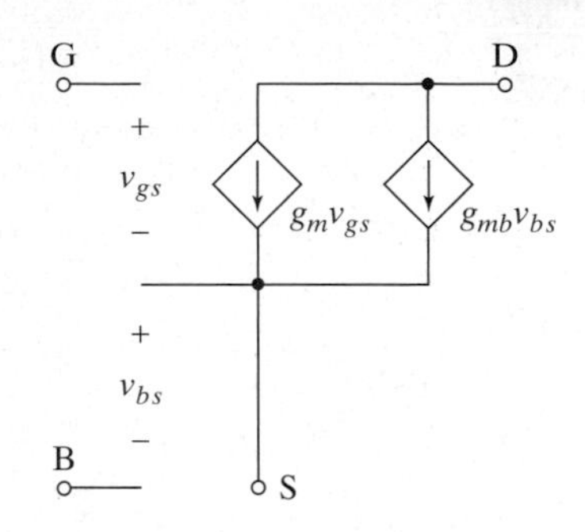

Figure 8–23 Small-signal MOSFET model including the bulk transconductance.

Typical values for this ratio are from 0.1 to 0.3 [8.1]. Unfortunately, SPICE does not use this parameter to specify the bulk transconductance. In Chapter 4, it is shown that

$$\chi = \frac{\gamma}{2\sqrt{2\varphi_f + V_{SB}}}, \tag{8.43}$$

which allows us to calculate χ based on the SPICE parameters **GAMMA** (γ) and **PHI** ($2\varphi_f$) and the DC operating point. In calculating the DC operating point to find V_{SB}, we must allow for the fact that the body effect modifies the value of V_{th} as well.

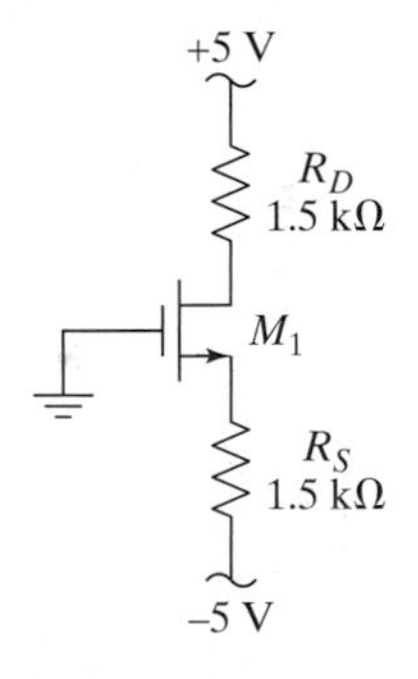

Figure 8–24 A MOS amplifier. (The input and output are not specified.)

Exercise 8.3

Consider the MOS amplifier shown in Figure 8-24, and assume that $K = 2$ mA/V^2 and $V_{th} = 1$ V. Find the small-signal midband AC model for the MOSFET, and draw the resulting equivalent circuit.

MOS Field-Effect Transistors in Integrated Circuits Almost all small-signal MOSFETs are in integrated circuits. The majority of discrete MOSFETs are used as switches or power devices and use specialized fabrication techniques. Therefore, it is extremely rare to need to model a discrete MOSFET for small-signal operation. Nevertheless, the models already presented are reasonable first-order models for discrete or integrated MOSFETs at low frequencies. The main parasitics that have not been discussed and that are of concern in integrated MOSFETs are the parasitic bipolar transistors formed by adjacent MOSFETs in the same substrate. These parasitic bipolar transistors can lead to a destructive latch-up mechanism as discussed in Chapter 2. In addition, there are some parasitics that become important at high frequencies (*see* Chapter 9 for details).

8.1.7 Junction Field-Effect Transistors

The operation of a JFET is identical to that of a depletion-mode MOSFET, except for the fact that the depletion region is controlled by a reverse-biased *pn* junction instead of a metal-oxide-semiconductor capacitor. The drain current equation is usually written differently, however. For operation in the forward-active region, we have

$$i_D = I_{DSS}\left(1 - \frac{v_{GS}}{V_p}\right)^2. \tag{8.44}$$

This equation is identical to the corresponding MOSFET equation if we set $V_p = V_{th}$ and $K = I_{DSS}/V_P^2$. (Remember that V_p is often called $V_{GS(\text{off})}$ on data sheets.) The small-signal transconductance can be found from (8.44) to be

$$g_m \equiv \left.\frac{di_D}{dv_{GS}}\right|_{@\ \text{Q point}} = 2I_{DSS}\left(1 - \frac{V_{GS}}{V_p}\right) = 2\sqrt{I_{DSS}I_D}. \tag{8.45}$$

Using (8.44) for the drain current, we can account for channel-length modulation in the usual way:

$$i_D = I_{DSS}\left(1 - \frac{v_{GS}}{V_p}\right)^2(1 + \lambda v_{DS}). \tag{8.46}$$

The small-signal output conductance is then

$$g_d \equiv \left.\frac{di_D}{dv_{DS}}\right|_{@\ \text{Q point}} = \lambda I_{DSS}\left(1 - \frac{V_{GS}}{V_p}\right)^2 = \lambda I_D = \frac{I_D}{V_A}, \tag{8.47}$$

which leads to the same small-signal output resistance as the MOSFET:

$$r_o = \frac{1}{g_d} = \frac{1}{\lambda I_D} = \frac{V_A}{I_D}. \tag{8.48}$$

The small-signal low-frequency input resistance of a JFET is, for all intents and purposes, infinite, just as it is for a MOSFET. Therefore, using (8.45) and (8.48), the JFET can be modeled using the circuits of Figure 8-19, Figure 8-20, or Figure 8-21. As with the MOSFET, these equations apply equally well for n-channel or p-channel devices. However, unlike MOSFETs, JFETs do not typically suffer from the body effect, since the top and bottom gates are connected together.

Junction Field-Effect Transistors in Integrated Circuits Integrated JFETs are rapidly becoming obsolete, although they are still used in some circuits fabricated in bipolar processes where the JFET can be made without any extra processing steps. The low-frequency small-signal models for integrated JFETs are identical to the models presented for discrete JFETs, except for the inclusion of parasitic junction diodes. Precisely which diodes need to be included and how important they are depends on the specific process used and will not be covered here.

8.1.8 Comparison of Bipolar and Field-Effect Transistors

We now discuss some key differences between bipolar and field-effect transistors for small-signal amplification. We will have more to say after we have studied the high-frequency models in Chapter 9.

The most important difference we can see at this point is that the transconductance of a bipolar transistor is larger than that of a similar-size field-effect transistor operating at the same bias current. The transconductance of a BJT is directly proportional to the bias current I_C, as indicated by (8.16), whereas the transconductance of a FET is only proportional to the square root of I_D, as shown in (8.36) and (8.45). We will see in Section 8.2 that the voltage gain of an amplifier is proportional to g_m, so this difference is significant.

Another major difference between bipolar and field-effect transistors is the input impedance. The FET has a much larger input resistance—essentially infinite for a MOSFET—than the BJT. This is seen clearly by examining the models. The practical significance is seen if you examine the input resistance of MOS and bipolar op amps. In Table 5.1, the input resistances of several op amps are shown. Typical bipolar op amps have input resistances on the order of megohms, typical JFET input op amps have input resistances on the order of teraohms, and typical MOS op amps have input resistances greater than 10 teraohms!

8.2 STAGES WITH VOLTAGE AND CURRENT GAIN

The basic models for our transistors contain three terminals. There are substrate connections in some devices, but they are not essential to the operation of the device. All single-transistor amplifiers directly or indirectly connect one of these three terminals to a reference point (usually ground) and use the other two terminals as the input and output. Since there are three terminals, each of which may be used as the common terminal, there are three possible connections for single-transistor amplifiers.

We begin our study of the small-signal performance of transistor amplifiers with the only connection that is capable of having both voltage and current gain greater than one. Because of this unique ability, this connection is the most frequently used of all three single-transistor amplifier configurations.

8.2.1 A Generic Implementation: The Common-Merge Amplifier

Consider connecting a transistor as shown in the small-signal equivalent circuit in Figure 8-25(a). We will show the DC biasing a bit later. This circuit is called a *common-merge amplifier* because the merge terminal is common to both the input and the output of the circuit (i.e., the merge terminal is connected to ground, whereas the input is applied from control to ground and the output is taken from output to ground). The gain of this circuit can be found by first noting that v_i is given by the voltage divider at the input:

$$v_i = \frac{r_{cm}}{r_{cm} + R_S} v_s. \tag{8.49}$$

The output voltage is then given by

$$v_o = -g_m v_{cm} R_L. \tag{8.50}$$

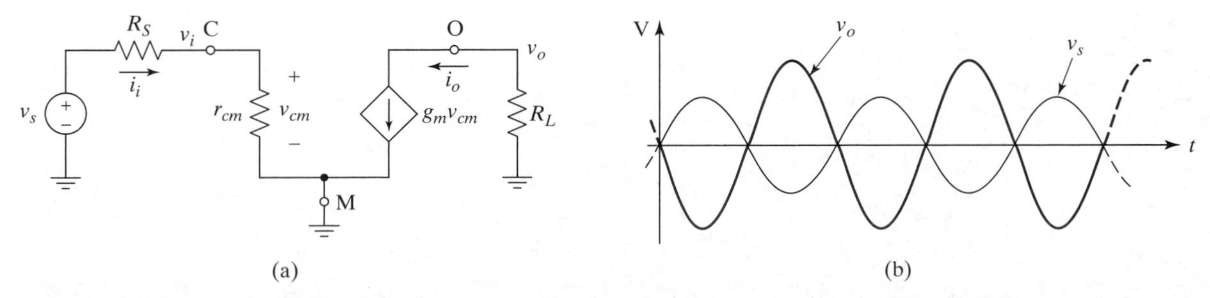

Figure 8–25 (a) The small-signal equivalent circuit of an amplifier using the generic transistor. (b) The small-signal input and output voltages of the circuit.

Noting that $v_{cm} = v_i$ in this circuit and substituting v_i from (8.49) into (8.50); we can derive the overall voltage gain,

$$\frac{v_o}{v_s} = -\left(\frac{r_{cm}}{r_{cm} + R_S}\right) g_m R_L, \tag{8.51}$$

or the voltage gain from v_i to v_o,

$$\frac{v_o}{v_i} = -g_m R_L. \tag{8.52}$$

The gain of this circuit is negative, which means that the output is inverted relative to the input, as shown in Figure 8-25(b). If the component values and bias point are properly chosen, this circuit can produce substantial voltage gain. In addition, notice that the current in the load is i_o and that the current supplied by the source is $i_i = i_c$. Since i_o is greater than i_c, this circuit provides current gain as well as voltage gain. *This is the only single-transistor amplifier connection that can provide both voltage and current gain.*

We can improve our understanding of the operation of this circuit if we examine its operation graphically. To do this, we need to consider the complete amplifier circuit, including biasing. Figure 8-26 shows a complete schematic of this gain stage. The DC supply V_{OO} and R_1, R_2, and R_O are needed only to provide the DC biasing, as discussed in Chapter 7. The two capacitors, C_{IN} and C_{OUT}, are coupling capacitors and are chosen to be large enough to pass the signals of interest with negligible attenuation. Their purpose is to isolate the DC bias voltages from the source and the load. For now, we assume these capacitors are extremely large and therefore have negligible impedance, except at DC.

To examine the operation of this circuit graphically, consider Figure 8-27. This figure shows the forward-active output characteristics of the generic transistor (i.e., a family of i_O versus $v_{OM} = v_O - v_M$ curves for different values of $v_{CM} = v_C - v_M$) along with two load lines and plots of v_{cm} and v_{om} versus time.

For the circuit under consideration, we first plot the DC load line[4] derived from the DC loop equation on the output side, namely,

$$I_O = \frac{V_{OO} - V_{OM}}{R_O}, \tag{8.53}$$

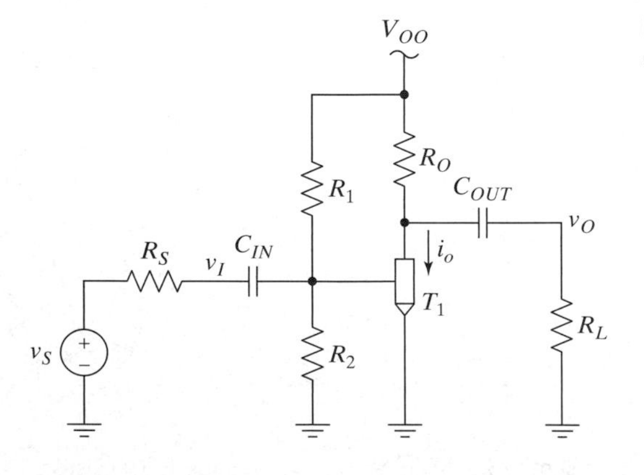

Figure 8–26 A complete amplifier circuit using the generic transistor.

[4] *See* Aside A6.1.

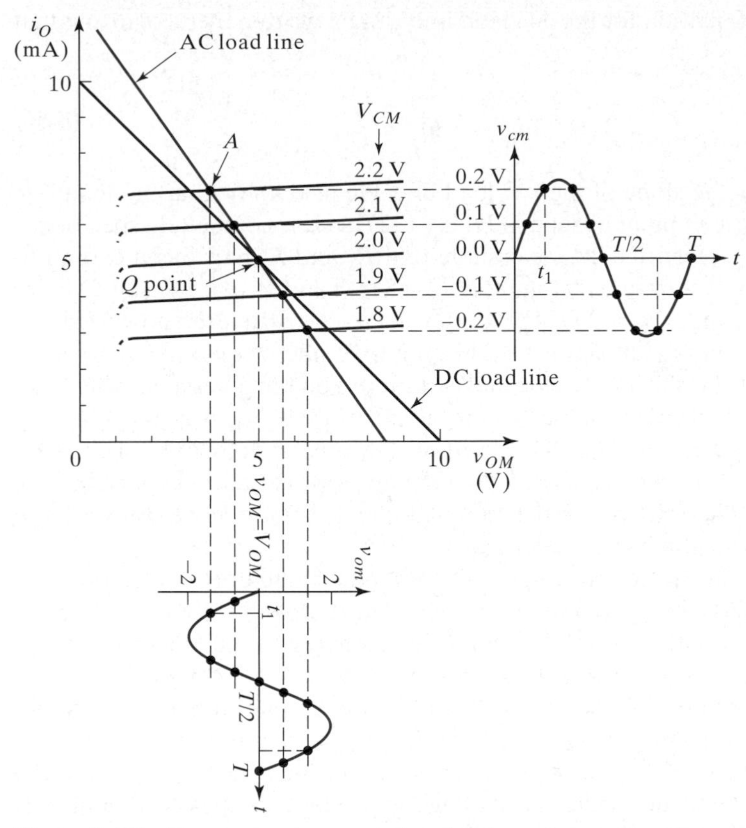

Figure 8–27 The output characteristics of the generic transistor, along with the load lines for the circuit in Figure 8-26 and plots of $v_I = v_{cm}$ and $v_O = v_{om}$ versus time.

where I_O is the DC current into the output terminal of the transistor, as shown in Figure 8-26. The *AC load line* is very much the same, but results from a loop equation written for the output side of the *small-signal midband AC equivalent circuit* shown in Figure 8-28. In constructing this equivalent circuit, we used the fact that it is the **small-signal** equivalent circuit to replace the transistor by its linear small-signal model. We then used the fact that it is the **midband** equivalent circuit to short out the coupling capacitors. Finally, we used the fact that it is the **AC** equivalent circuit to set V_{OO} to zero, since it is a DC supply. The DC Thévenin resistance driving the input of the transistor is given by

$$R_{CC} = R_1 \| R_2. \tag{8.54}$$

The AC load is given by

$$R'_L = R_L \| R_O, \tag{8.55}$$

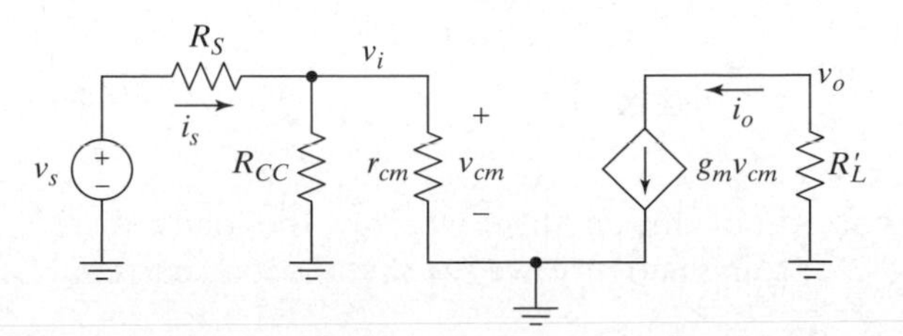

Figure 8–28 The small-signal midband AC equivalent circuit for the amplifier of Figure 8-26.

and the resulting equation for the AC load line can be written by recognizing that $v_o = v_{om}$, so that

$$i_o = \frac{-v_o}{R'_L} = \frac{-v_{om}}{R'_L}. \tag{8.56}$$

This equation sets the slope of the AC load line. We also know that the load line must go through the Q point as seen in Figure 8-27, so it is completely specified.

Returning now to Figure 8-27, we assume that R_1 and R_2 have been chosen to establish the Q point shown in the figure at $I_O = 5$ mA and $V_{OM} = 5$ V. This Q point is achieved by setting $V_{CM} = 2$ V. The output characteristics are shown for v_{CM} equal to the quiescent value and for 0.2 V on either side of the quiescent value. The plot of v_{cm} is drawn off to the side and is centered on the quiescent value; note that because C_{IN} blocks DC and passes the signal, we have $v_{cm} = v_i$, but $V_{CM} \neq V_I$, and similarly, $v_o = v_{om}$, but $V_O \neq V_{OM}$. To find the output voltage for any given value of v_i (say, the value at time t_1), follow the appropriate output characteristic to the left to see where it crosses the AC load line, and then move down on the plot to see the corresponding value of v_{OM}.

For example, at time t_1 the value of v_i is 0.2 V, so the value of v_{CM} is $V_{CM} + 0.2$ V. Following that characteristic to the left, we see that it crosses the AC load line at the point labeled A in the figure. This point represents the simultaneous solution of all of the equations governing the operation of this circuit for this particular value of v_I. Following straight down from point A, we see the corresponding value of v_{OM}. Finally, we recognize that $v_o = v_{om} = v_{OM} - V_{OM}$ and draw the plot of v_o versus time shown. We see from this plot that the output is an inverted version of the input. We also see that the circuit has voltage gain, since the peak value of v_o is about 2 V while the peak value of v_i is about 0.21 V.

Now we continue our analysis of the AC equivalent circuit in Figure 8-28. Because the circuit is unilateral, it is useful to write the overall voltage gain as a product of two simpler transfer functions, as was done in Chapter 1:

$$\frac{v_o}{v_s} = \frac{v_i}{v_s}\frac{v_o}{v_i}. \tag{8.57}$$

If we combine r_{cm} and R_{CC} in parallel, then the first transfer function is a resistive voltage divider. We call this divider the input attenuation factor, as before:

$$\alpha_i = \frac{v_i}{v_s} = \frac{r_{cm} \| R_{CC}}{r_{cm} \| R_{CC} + R_S}. \tag{8.58}$$

The second transfer function is

$$\frac{v_o}{v_i} = \frac{v_o}{v_{cm}} = -g_m R'_L, \tag{8.59}$$

so that the ***overall voltage gain*** for small-signal operation can be expressed as

$$\frac{v_o}{v_s} = -\alpha_i g_m R'_L. \tag{8.60}$$

We can also calculate the current gain of this circuit if we draw the load resistor separately, as shown in Figure 8-29. From the figure, we see that the load current is given by

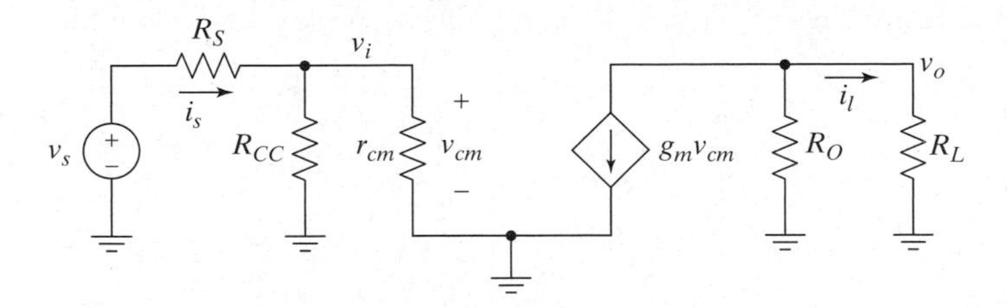

Figure 8–29 The circuit of Figure 8-28 with the load resistor separated.

$$i_l = \frac{-R_O}{R_O + R_L} g_m v_{cm}. \quad \textbf{(8.61)}$$

We can also write

$$i_s = \frac{v_{cm}}{R_{CC} \| r_{cm}} \quad \textbf{(8.62)}$$

and then combine these two equations to write the small-signal ***current gain*** as

$$A_i = \frac{i_l}{i_s} = \frac{-g_m v_{cm}}{v_{cm}/(R_{CC} \| r_{cm})}\left(\frac{R_O}{R_O + R_L}\right) = -g_m(R_{CC} \| r_{cm})\left(\frac{R_O}{R_O + R_L}\right), \quad \textbf{(8.63)}$$

where the minus sign results from defining the load current as positive into the load and again indicates that this stage inverts the signal.

The power gain of the circuit can now be found. The power delivered to the load is $i_l v_o$, and the power delivered by the source is $i_s v_i$ (v_i is used instead of v_s because R_S is part of the Thévenin model for the source, and therefore, we don't want to count power dissipated in R_S as power delivered to our circuit by the source). We can then write the small-signal ***power gain*** of our circuit as

$$A_p = \frac{i_l v_o}{i_s v_i} = \left(\frac{i_l}{i_s}\right)\left(\frac{v_o}{v_i}\right) = A_i A_v, \quad \textbf{(8.64)}$$

where we have implicitly defined the small-signal ***voltage gain*** of the amplifier, A_v. Stating this definition explicitly, we have

$$A_v \equiv \frac{v_o}{v_i}. \quad \textbf{(8.65)}$$

This voltage gain is the typical definition used, and that is why we have assigned a special symbol to it. The *overall* voltage gain given by (8.60) is often of interest, but it is not a directly measurable quantity, since R_S is part of the source. The only voltage we typically have access to in a real system is the voltage to the right of R_S in Figure 8-26, which is v_I. If we used the overall voltage gain as the standard definition, the power gain would not be the product of the voltage and current gains as given in (8.64). Using (8.59) and (8.65) we see that the voltage gain of this circuit is

$$A_v = -g_m R'_L, \quad \textbf{(8.66)}$$

where the minus sign indicates that this is an ***inverting amplifier***, as noted in our discussion of Figure 8-27.

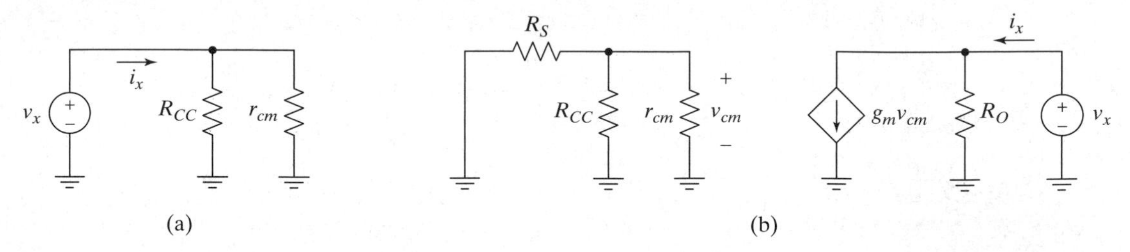

Figure 8–30 (a) The circuit for finding R_i. (b) The circuit for finding R_o.

Finally, it will be useful to find the small-signal midband input and output resistances of this circuit. (We will often drop the modifier "midband" when no confusion is likely.) The ***small-signal input resistance*** of the circuit is the small-signal AC resistance seen by the source.[5] In other words, if all other independent sources are set to zero, and we drive the input by a test voltage v_x and find the resulting current i_x as shown in Figure 8-30(a), the small-signal input resistance is given by

$$R_i = \frac{v_x}{i_x} = R_{CC} \| r_{cm}. \tag{8.67}$$

Similarly, the ***small-signal output resistance*** of the circuit is the small-signal AC resistance seen by the load. To find this resistance, we remove R_L from the equivalent circuit and connect a test source in place of R_L, as shown in Figure 8-30(b). We again set all other independent sources (in this case, v_s) to zero, force v_x, and measure i_x. The small-signal output resistance is given by

$$R_o = \frac{v_x}{i_x} = R_O. \tag{8.68}$$

We can model the small-signal AC performance of this amplifier with one of the unilateral two-port models discussed in Chapter 6. The input resistance we found in (8.67) is the small-signal AC input resistance of the two-port model, and the resistance found in (8.68) is the small-signal AC output resistance of the model. If we are going to use the voltage amplifier model of Figure 6-17(a), the only other parameter that we need is the open-circuit voltage gain. We already have the voltage gain with the load attached; it is given by (8.66). If the output of the amplifier is open circuited, we use R_O instead of R_L', and the open-circuit voltage gain is

$$a_v = -g_m R_O. \tag{8.69}$$

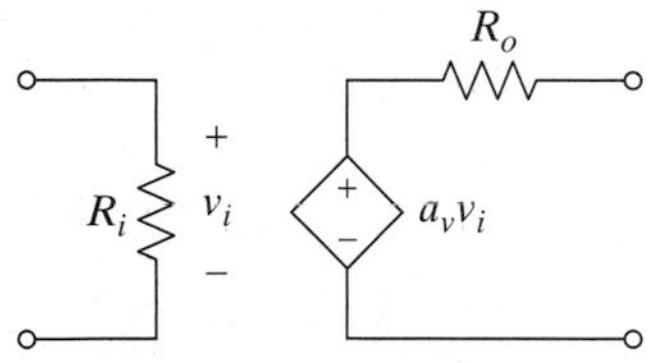

Figure 8–31 The small-signal midband AC unilateral two-port voltage amplifier model for the circuit of Figure 8-26. R_i is given by (8.67), R_o by (8.68), and a_v by (8.69).

The small-signal midband AC unilateral two-port model of the amplifier is shown in Figure 8-31.

Using this model, we can calculate the overall gain of the amplifier with any source and load. Consider the circuit of Figure 8-32. The overall gain of this circuit is found by breaking it up into simpler transfer functions as usual:

$$\frac{v_o}{v_s} = \frac{v_o}{v_i}\frac{v_i}{v_s}. \tag{8.70}$$

[5] This is equivalent to the small-signal midband AC Thévenin resistance seen by the source (i.e., consider the rest of the circuit to be a one-port network connected to the source). It is also called the *driving-point resistance seen by the source.*

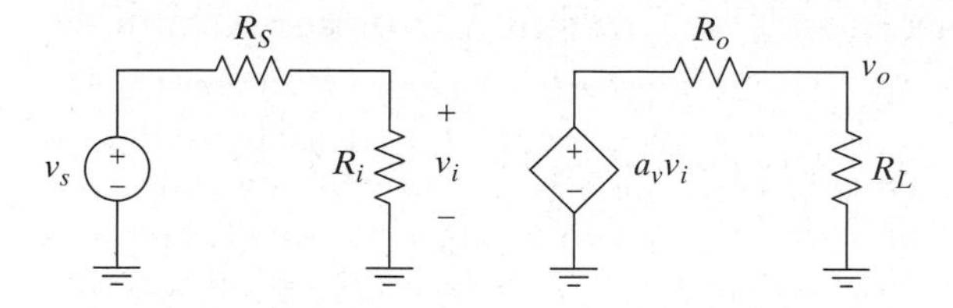

Figure 8–32 The unilateral two-port model connected with a source and load.

The individual transfer functions are given by

$$\frac{v_o}{v_i} = a_v\left(\frac{R_L}{R_L + R_o}\right) = a_v\alpha_o, \quad \textbf{(8.71)}$$

and

$$\frac{v_i}{v_s} = \frac{R_i}{R_i + R_S} = \alpha_i, \quad \textbf{(8.72)}$$

where we have defined the input and output attenuation factors as usual. Putting these equations all together, we get the overall voltage gain:

$$\frac{v_o}{v_s} = \alpha_i a_v \alpha_o = \left(\frac{R_i}{R_i + R_S}\right)a_v\left(\frac{R_L}{R_L + R_o}\right). \quad \textbf{(8.73)}$$

Finally, if we substitute the two-port model parameters for R_i, R_o, and a_v into (8.73), we get

$$\frac{v_o}{v_s} = \left(\frac{R_{CC}\|r_{cm}}{R_{CC}\|r_{cm} + R_S}\right)(-g_m R_O)\left(\frac{R_L}{R_L + R_O}\right). \quad \textbf{(8.74)}$$

A little bit of algebra will convince you that this result is identical to that obtained by direct analysis in (8.60). The advantage of having the unilateral two-port representation for the circuit is fairly clear from (8.73) it leads to a simpler, clearer, and more systematic way of examining the overall performance of the amplifier when connected to a given source and load. This representation also allows us to readily see what happens when the source and/or load are changed.

We are now ready to investigate specific implementations of the amplifier shown in Figure 8-26. In the sections that follow, we investigate bipolar and field-effect transistor implementations.

Exercise 8.4

Consider the circuit in Figure 8-29, and assume that $R_S = 1$ kΩ, $R_L = 5$ kΩ, $R_{CC} =$ 100 kΩ, $r_{cm} = 2$ kΩ, $R_O = 500$ Ω, and $g_m = 20$ mA/V. (a) Find the overall voltage gain of this circuit. (b) Find the elements of the equivalent two-port voltage amplifier model shown in Figure 8-31. (c) Find the overall voltage gain using your model and the circuit in Figure 8-32, and compare your results with part (a).

□

8.2.2 A Bipolar Implementation: The Common-Emitter Amplifier

Figure 8-33 shows an implementation of the amplifier of Figure 8-26 constructed using a BJT. This circuit is called a *common-emitter amplifier*, because the emitter is common to both the input and the output of the circuit (i.e., the emitter is connected to an AC ground, whereas the input is applied from base to ground and the output is taken from collector to ground). The DC bias point analysis of this circuit was covered in Section 7.2.1 and will not be repeated here. The two coupling capacitors C_{IN} and C_{OUT} were discussed when we analyzed Figure 8-26 and serve the same purpose here. The capacitor in parallel with R_E is new, however. This capacitor, C_E in the figure, is called a *bypass capacitor*. It shorts out, or bypasses, R_E for signals while allowing R_E to provide negative feedback for the DC bias point. We will see shortly that if R_E is not bypassed, it significantly lowers the voltage gain of the circuit.

The bias values of the circuit in Figure 8-33 were determined using the rules of thumb presented in Section 7.2.1. First, the value of V_{BB} was set to 2 V by choosing the relative values of R_1 and R_2:

$$V_{BB} = \frac{R_2}{R_1 + R_2} V_{CC} = 2\text{ V}. \tag{8.75}$$

Second, R_E was chosen to yield a collector current of about 2 mA:

$$R_E = \frac{V_E}{I_E} \approx \frac{V_{BB} - 0.7}{I_C} = \frac{1.3}{0.002} = 650\ \Omega. \tag{8.76}$$

Third, R_C was chosen so that V_C is about halfway between V_{CC} and V_E; that is,

$$V_C = \frac{V_{CC} + V_E}{2} = \frac{11.3}{2} = 5.6\text{ V}, \tag{8.77}$$

and

$$R_C = \frac{V_{CC} - V_C}{I_C} = \frac{4.4}{0.002} = 2.2\text{ k}\Omega. \tag{8.78}$$

Finally, assuming that the transistor has $\beta = 100$, we set the current through R_1 and R_2 to be at least times larger than the base current ($I_B = I_C/\beta = 0.002/100 = 20\ \mu\text{A}$); that is,

$$I_{Bias} = \frac{V_{CC}}{R_1 + R_2} \geq 200\ \mu\text{A}, \tag{8.79}$$

so that

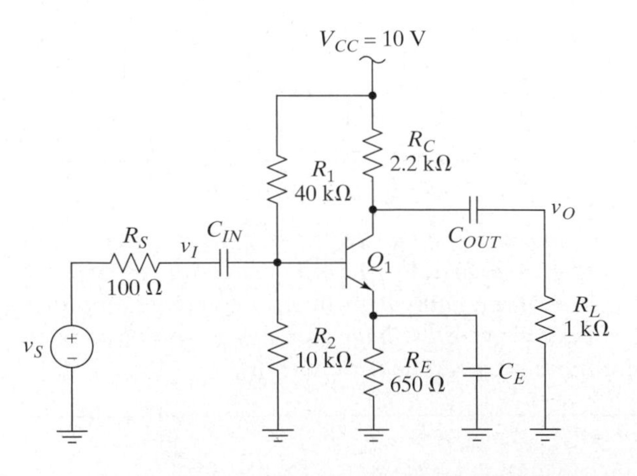

Figure 8–33 A bipolar implementation of an amplifier with both voltage and current gain.

$$R_1 + R_2 \le \frac{10}{200 \times 10^{-6}} = 50\ \text{k}\Omega. \qquad \textbf{(8.80)}$$

Combining (8.75) and (8.80) leads to the given values of R_1 and R_2 as one of the possible solutions. Because of the coupling capacitors, the DC voltage at the load is zero (i.e., $V_O = 0$) and the DC voltage at the input is whatever the DC component of the source is (i.e., V_S).

Now that we understand how the DC bias point was arrived at, although no justification has been given yet for choosing $I_C = 2$ mA, let's examine the small-signal AC performance. Figure 8-34 shows the small-signal midband AC equivalent circuit for this amplifier. In this circuit, the coupling capacitors and bypass capacitor are all shorted out, the DC supply is replaced by a ground (since the AC component of the supply voltage is zero by definition), and the transistor has been replaced by the low-frequency hybrid-π model. We have ignored r_o in the model, since it is much larger than the discrete resistors in parallel with it. It is almost always valid to ignore r_o in discrete circuits, although this is not true for integrated circuits.

The resistor R_{BB} is the DC Thévenin resistance driving the base, so that

$$R_{BB} = R_1 \| R_2 = 8\ \text{k}\Omega, \qquad \textbf{(8.81)}$$

and the AC load is given by

$$R_L' = R_L \| R_C = 687\ \Omega. \qquad \textbf{(8.82)}$$

The transconductance of the transistor is given by (8.16)

$$g_m = \frac{I_C}{V_T} = \frac{0.002}{0.026} = 077\ \text{mA/V}, \qquad \textbf{(8.83)}$$

and r_π is given by (8.21) with our assumed $\beta = 100$:

$$r_\pi = \frac{\beta}{g_m} = \frac{100}{0.077} = 1.3\ \text{k}\Omega. \qquad \textbf{(8.84)}$$

The circuit in Figure 8-34 is identical to the circuit in Figure 8-28 if we simply change the names of R_{CC} and r_{cm} to R_{BB} and r_π, respectively. The overall voltage gain is therefore given by (8.60) with the appropriate substitutions; that is,

$$\frac{v_o}{v_s} = -\alpha_i g_m R_L' \qquad \textbf{(8.85)}$$

where the input attenuation factor is

$$\alpha_i = \frac{R_{BB} \| r_\pi}{R_{BB} \| r_\pi + R_S} = 0.92. \qquad \textbf{(8.86)}$$

Plugging in the given values yields an overall voltage gain of –49. The voltage gain, A_v, is given by (8.66) with the values for this circuit substituted:

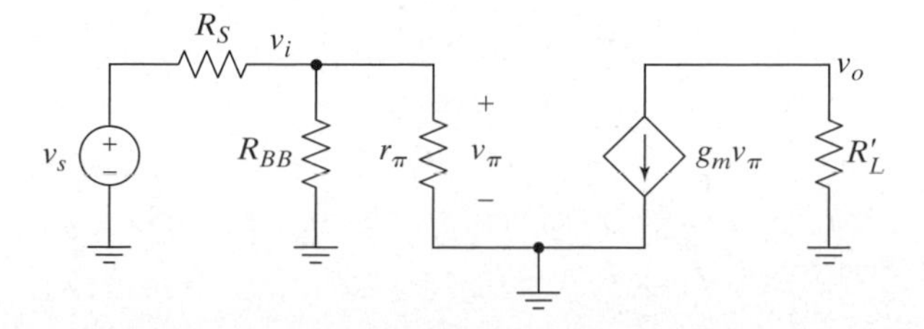

Figure 8–34 The small-signal midband AC equivalent circuit for the amplifier of Figure 8-33.

$$A_v = \frac{v_o}{v_i} = -g_m R'_L = -53. \tag{8.87}$$

The current gain of this circuit is given by (8.63) with appropriate substitutions:

$$A_i = \frac{i_l}{i_s} = -g_m(R_{BB} \| r_\pi)\left(\frac{R_C}{R_C + R_L}\right) = -59. \tag{8.88}$$

The power gain of the circuit is the product of the voltage gain and current gain, as in (8.64). For this circuit, we get

$$A_p = A_v A_i = 3124. \tag{8.89}$$

Example 8.1

Problem:
Run a SPICE simulation to check the gain obtained for the circuit in Figure 8-33.

Solution:
We performed the simulation using the Nsimple transistor model described in Appendix B and 100 μF coupling and bypass capacitors. The PROBE plot of v_O for the AC simulation is shown in Figure 8-35. From this figure, we determine that the overall voltage gain is −44 (the minus sign is not apparent from the figure, only the magnitude), which agrees fairly well with our analytical result of −49.

■

Let's now move on to find a unilateral two-port model for this amplifier. Because the small-signal AC equivalent circuit, Figure 8-34, is the same as that for the generic transistor amplifier (Figure 8-28), we can make use of (8.68) for the output resistance. The output resistance of this circuit is therefore

$$R_o = R_C, \tag{8.90}$$

where we have made the appropriate change in variable. Similarly, the input resistance is given by (8.67) by changing the variables:

$$R_i = R_{BB} \| r_\pi. \tag{8.91}$$

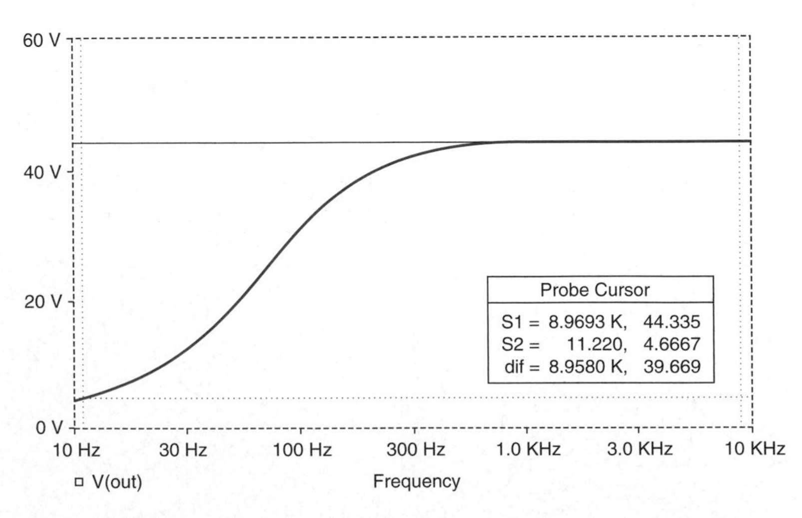

Figure 8–35 The PROBE AC simulation result for v_O. Cursor S1 shows the midband gain.

Finally, the open-circuit voltage gain is given by (8.69) with a change in variable:

$$a_v = \left.\frac{v_o}{v_i}\right|_{\text{open circuit}} = -g_m R_C. \tag{8.92}$$

The output attenuation factor of this amplifier is then given by

$$\alpha_o = \frac{R_L}{R_L + R_C}, \tag{8.93}$$

which is equal to 0.31 for the values used here. The low value for α_o reflects the fact that this circuit has too high an output resistance to be able to effectively drive this load.

There is another major problem with the amplifier circuit presented in Figure 8-33; the gain cannot be set independently of the DC voltage at the collector of the transistor. To see this, rewrite (8.92) as

$$a_v = -g_m R_C = \frac{-I_C}{V_T} R_C = -\frac{V_{R_C}}{V_T} = -\frac{V_{CC} - V_C}{V_T}. \tag{8.94}$$

The only variable in this equation is V_C (assuming that V_{CC} is given), so this severely restricts the designer. Once we have established the DC bias conditions we want for the circuit (based on signal swing or some other criterion), the small-signal AC open-circuit voltage gain is also determined. We can get around this restriction by only bypassing part of R_E, as shown in Figure 8-36.

The small-signal midband AC equivalent circuit for this amplifier is shown in Figure 8-37, where we have again ignored r_o. Before analyzing the circuit, we note that, in general, we will be interested in finding the input and output resistances of our circuit, in addition to the gain. Therefore, we don't want to find the midband AC Thévenin equivalent driving the base of the transistor, since, if we did so, we would lose the clear identity of the source. We will instead start our analysis by noting that the input to the amplifier is just to the right of R_S, labeled v_i in the figure, and we will first compute the voltage gain A_v. We can then find the input resistance and the input attenuation factor. The overall voltage gain is then the product of A_v and α_i.

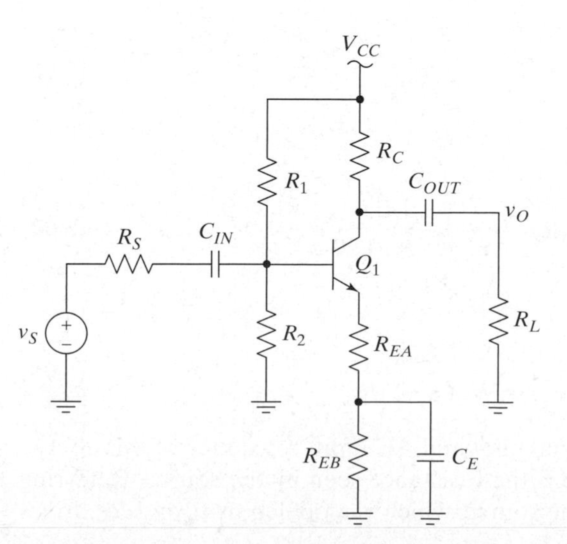

Figure 8–36 A common-emitter amplifier with partially unbypassed emitter resistance.

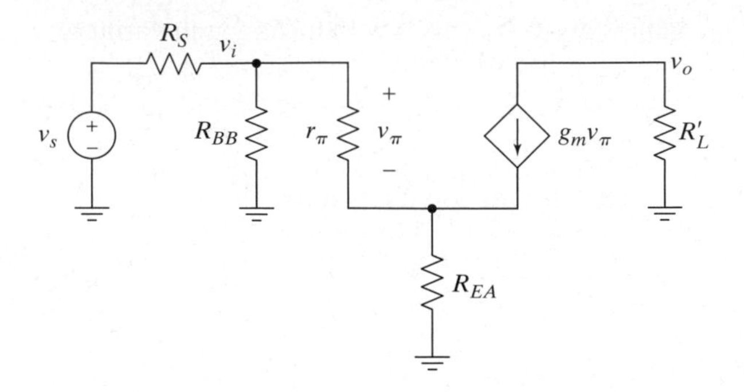

Figure 8–37 The small-signal midband AC equivalent circuit of the amplifier in Figure 8-36.

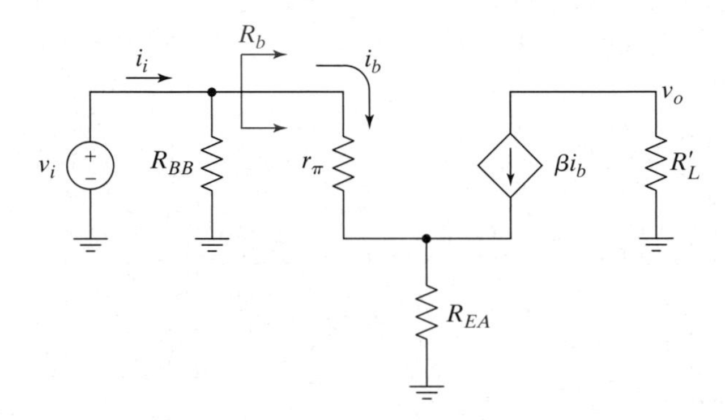

Figure 8–38 The equivalent circuit from Figure 8-37 redrawn.

To make it clear what we are doing, we redraw the circuit as shown in Figure 8-38. The input is now v_i, and we have used the current-controlled hybrid-π model for the transistor because it will simplify the analysis. We write KVL around the base-emitter loop to obtain

$$v_i - i_b r_\pi - (\beta+1) i_b R_{EA} = 0, \tag{8.95}$$

from which we get

$$i_b = \frac{v_i}{r_\pi + (\beta+1)R_{EA}}. \tag{8.96}$$

At the output we find that

$$v_o = -\beta i_b R'_L = \frac{-\beta R'_L v_i}{r_\pi + (\beta+1)R_{EA}}, \tag{8.97}$$

from which we get

$$A_v = \frac{v_o}{v_i} = \frac{-\beta R'_L}{r_\pi + (\beta+1)R_{EA}}. \tag{8.98}$$

We now wish to find the small-signal midband AC input resistance of this circuit. By definition, the input resistance is the resistance seen by the source. Referring back to Figure 8-37, we see that the source, which is modeled by v_s and R_S, drives the node we have labeled v_i. Therefore, the resistance seen by this source is precisely the resistance seen by the v_i generator in Figure 8-38 and is given by the ratio v_i/i_i. By examining the circuit, we see that the input resistance is equal to R_{BB} in parallel with the resistance seen looking in the base of the transistor (from the base to ground), which we call R_b, as shown in the figure. We can find this resistance from (8.96):

$$R_b = \frac{v_i}{i_b} = r_\pi + (\beta+1)R_{EA}. \tag{8.99}$$

This result is a specific example of an extremely important concept called ***impedance reflection***. Impedance reflection is covered in more detail in Aside A8.2. The input resistance can now be written as

$$R_i = R_{BB} \| [r_\pi + (\beta+1)R_{EA}], \tag{8.100}$$

and the input attenuation factor is

$$\alpha_i = \frac{R_i}{R_i + R_S} = \frac{R_{BB} \| [r_\pi + (\beta+1)R_{EA}]}{R_{BB} \| [r_\pi + (\beta+1)R_{EA}] + R_S}. \tag{8.101}$$

Finally, the output resistance can be found by redrawing the circuit from Figure 8-38 to show R_C and R_L separately, as is done in Figure 8-39. We have also set the independent source at the input to zero and have applied a test source where the load was connected so that we can find R_o.

With $v_i = 0$, the base current is also zero, so the controlled-current source has zero value and looks like an open circuit. The resulting small-signal midband AC output resistance is

$$R_o = \frac{v_x}{i_x} = R_C. \tag{8.102}$$

It is an interesting problem to calculate what this resistance is if we include r_o in the transistor model. Although the calculation is more difficult (*see* Problem P8.30), the result shows that the output resistance of the transistor is boosted by the presence of R_{EA} and is even larger than r_o. Therefore, the answer in (8.102) is approximately correct.

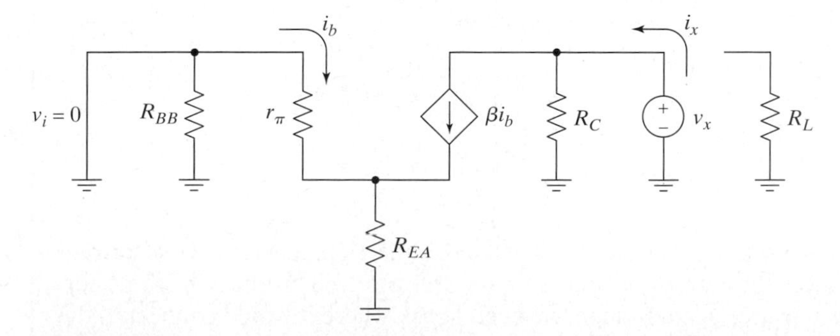

Figure 8–39 The circuit of Figure 8-38 redrawn to find the output resistance.

ASIDE A8.2 IMPEDANCE REFLECTION REVISITED: THE SMALL-SIGNAL MIDBAND AC CASE

We considered the phenomenon of DC impedance reflection in Aside A7.1. In the current Aside, we look at the small-signal midband AC case. The only difference in the equivalent circuits is that the DC base-emitter junction model is a battery of value 0.7 V, whereas the incremental change in v_{BE} is modeled by a resistor in the small-signal case.

Consider the small-signal midband AC equivalent circuit shown in Figure A8-3. We have used the current-controlled version of the hybrid-π model and have included external resistors in series with the emitter and base. The collector is grounded, but where it is connected does not affect our analysis.

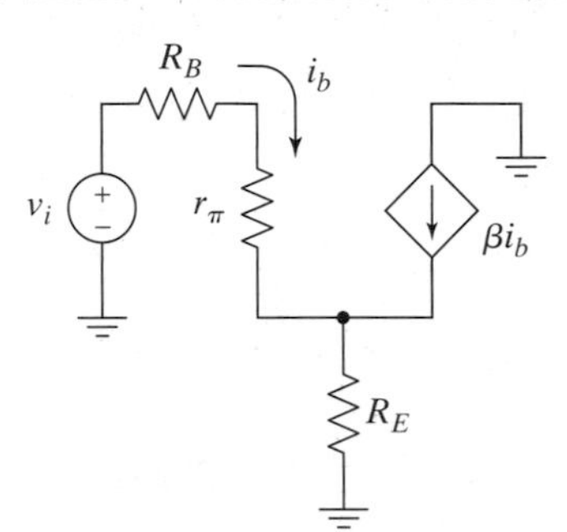

Figure A8-3 A small-signal midband AC equivalent circuit.

Writing KCL at the emitter reveals that the current in R_E is $(\beta+1)i_b$. Then writing KVL around the base-emitter loop leads to

$$v_i - i_b(R_B + r_\pi) - i_b(\beta+1)R_E = 0. \quad \textbf{(A8.1)}$$

The total resistance seen by the input source is therefore

$$R_i = \frac{v_i}{i_b} = R_B + r_\pi + (\beta+1)R_E \quad \textbf{(A8.2)}$$

and we say that R_E is *reflected into the base* by multiplying it by $(\beta+1)$. This is a manifestation of the fact that R_E shows up in the base-emitter loop, but the current through it is $(\beta+1)$ times larger than the current in the base.

Now consider the circuit shown in Figure A8-4. If you write KVL around this loop, you again obtain (A8.1). In fact, if you were given (A8.1) and asked to draw the circuit from which it was derived, you probably would have drawn Figure A8-4, not Figure A8-3. The circuits in Figures A8-3 and A8-4 are equivalent in a limited sense: They produce the same KVL equation and therefore have the same voltage drops across each element, and the current in Figure A8-4 is the same as the current in the base in Figure A8-3. Therefore, Figure A8-4 will yield the same input impedance as Figure A8-3. Note that the current in Figure A8-4 is not the same as in the emitter in Figure A8-3, so the two circuits are not completely equivalent.

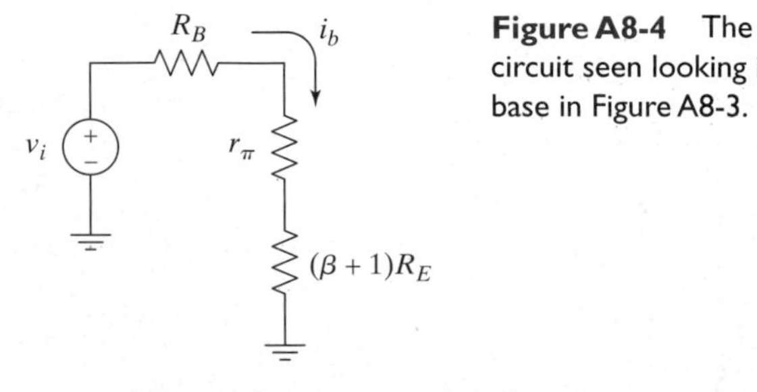

Figure A8-4 The equivalent circuit seen looking into the base in Figure A8-3.

We will frequently find it useful to redraw our schematics using impedance reflection. Since the base current and v_π will be the same as in the original circuit, the controlled source for the collector will produce the same current. If we wish to show the collector current source again though, we need to ask where it would connect to the circuit shown in Figure A8-4. We answer that question by showing an alternate derivation of the circuit shown in

Figure A8-4. In Figure A8-5, we derive the circuit in Figure A8-4 by manipulating the original circuit. Part (a) of the figure is a repeat of Figure A8-3. In parts (b) and (c) of the figure, we make use of the split-source transformation by first splitting the current source into two equal sources in series and then grounding the center connection. To get from part (c) to part (d), you first change the controlled source in parallel with R_E into a voltage-controlled source with v_e as the controlling variable, and then you use source absorption and combine the parallel resistors to arrive at the circuit in (d) (*see* Problem P8.28).

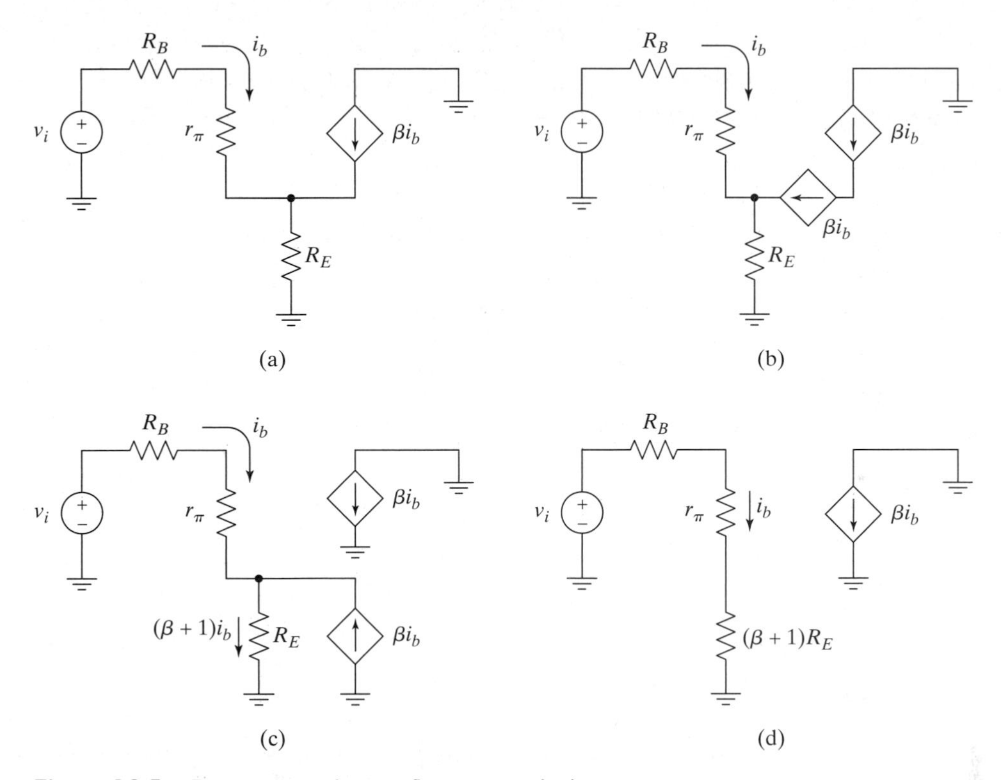

Figure A8-5 Deriving impedance reflection into the base.

A similar result applies when we look into the emitter side of the circuit, as shown in Figure A8-6. In this case, we picture the input source in series with R_E and find the small-signal midband resistance seen looking into the emitter. Notice that having i_i flow into the emitter is not a problem: Since this is only the time-varying component of the emitter current, the *total* emitter current would still flow *out* of the emitter for an *npn* transistor (i.e., since $i_E = I_E + i_e$, it is possible for i_e to be negative while i_E remains strictly positive).

Figure A8-6 A small-signal midband AC equivalent circuit for finding the resistance seen looking into the emitter.

R_B i_b r_π βi_b R_E i_i v_i

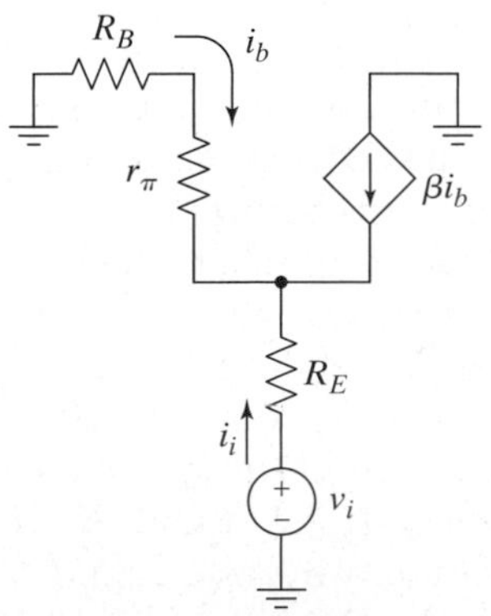

In this circuit, we can again write KVL in the base-emitter loop and use the fact that $i_i = -i_e = -(\beta+1)i_b$ to obtain

$$v_i - i_i R_E - \frac{i_i}{\beta + 1}(r_\pi + R_B) = 0. \quad \textbf{(A8.3)}$$

From this we can find the resistance seen by the source:

$$R_i = \frac{v_i}{i_i} = R_E + \left(\frac{r_\pi + R_B}{\beta + 1}\right). \quad \textbf{(A8.4)}$$

We see that any resistance in the base circuit, which includes r_π since the current through it is i_b, is *reflected into the emitter* by dividing by $(\beta+1)$. We can rewrite (A8.4) by using (8.27) to replace $r_\pi/(\beta+1)$ by r_e:

$$R_i = R_E + r_e + \frac{R_B}{\beta + 1}. \quad \textbf{(A8.5)}$$

This equation makes more intuitive sense if we redraw the circuit in Figure A8-6 using the T model for the transistor, as shown in Figure A8-7.

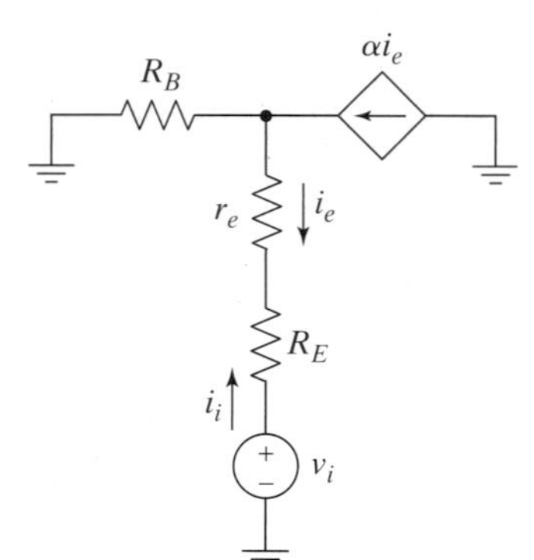

Figure A8-7 The circuit of Figure A8-6 redrawn using the T model.

It is left as an exercise to show that this circuit yields the same input resistance (*see* Problem P8.29).

Before leaving this section, we reexamine (8.98) under the conditions $(\beta + 1)R_{EA} \gg r_\pi$ and $R_L \gg R_C$, which are common in good designs. If these conditions are met, the gain from (8.98) can be approximated by

$$A_v \approx \frac{-R_C}{R_{EA}}. \quad \textbf{(8.103)}$$

In this case, the unbypassed part of the emitter resistance is stabilizing the gain against changes in transistor parameters in the same way that the total DC emitter resistance stabilizes the bias point. We will discover when we study feedback in Chapter 10 that this is the main advantage of negative feedback; it stabilizes the circuit and makes its performance less sensitive to changing parameters. The penalty paid is a reduction in gain from what is achievable without feedback. Because negative feedback is degenerative; that is, it opposes changes in the voltages and currents in the circuit, the unbypassed part of the total external emitter resistance, R_{EA}, is often referred to as the **emitter degeneration** resistor. The circuit with fully bypassed emitter resistance, Figure 8-33, also has degeneration, but only for the DC bias point, and the term is usually applied only with reference to the signal.

The feedback produced by the addition of emitter resistance was explained qualitatively in Chapter 7 for the DC bias point, and the effect of unbypassed emitter resistance on the AC gain is the same. When the AC base voltage increases, the AC collector current in the transistor also increases. This increase in i_c then causes a larger voltage drop across R_{EA}, which, in turn, reduces v_{be}. The reduction in v_{be} is in opposition to the original change, since it decreases the AC collector current. Consequently, the net change in i_c is not as large as it would have been without R_{EA}. This reduction in gain produces the benefit that the gain is less dependent on the small-signal parameters of the transistor than it would be without the degeneration, which is clear, since (8.103) does not depend on the transistor parameters at all. Even when the approximations made in arriving at (8.103) don't hold, however, the gain is less sensitive to device parameters and bias point than it would be without degeneration. Inspection of (8.98) reveals that the gain is less sensitive to changes in β and r_π than it would be without R_{EA}.

Exercise 8.5

Consider the amplifier shown in Figure 8-36, and let $R_1 = 6.8\ k\Omega$, $R_2 = 4.7\ k\Omega$, $R_{EB} = 1\ k\Omega$, $R_C = 2\ k\Omega$, $R_S = 100\ \Omega$, and $R_L = 5\ k\Omega$. Find the change in the gain as β varies from 50 to 200, expressed as a percentage of the value at $\beta = 100$ for (a) $R_{EA} = 30\ \Omega$ and (b) $R_{EA} = 90\ \Omega$. Explain the results.

□

Finally, we illustrate that the gain of this stage can also be found using the T model. Figure 8-40 is a repeat of Figure 8-38, but with the T model used in place of the hybrid-π model.

In this circuit the voltage v_i appears directly across r_e and R_{EA} in series, so we can immediately write

$$i_e = \frac{v_i}{r_e + R_{EA}}. \tag{8.104}$$

The output voltage is then given by

$$v_o = -\alpha i_e R_L', \tag{8.105}$$

so that the voltage gain is

$$A_v = \frac{-\alpha R_L'}{r_e + R_{EA}}, \tag{8.106}$$

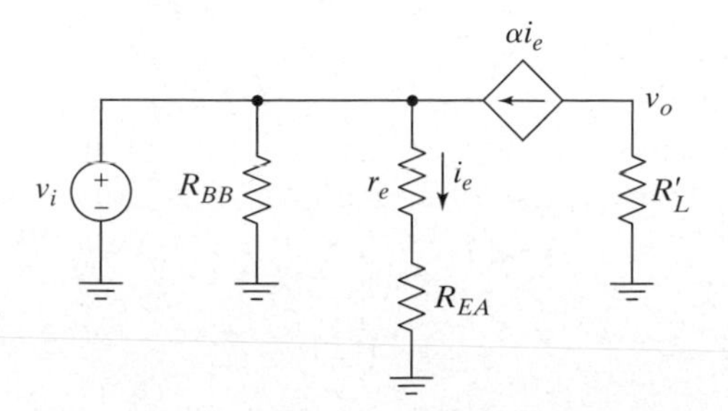

Figure 8–40 The circuit of Figure 8-38 with the T model used in place of the hybrid-π model.

which is the same as (8.98), as you can verify (*see* Problem P8.31). If $R_{EA} \gg r_e$ and $R_L \gg R_C$, then

$$A_v \approx \frac{-R_C}{R_{EA}}, \tag{8.107}$$

as before.

Example 8.2

Problem:
Find the small-signal midband AC values for R_i, R_o, and a_v for the amplifier shown in Figure 8-41. Assume that $\beta = 50$.

Solution:
We begin the analysis by finding the operating point of the transistor. First find V_{BB}:

$$V_{BB} = \frac{R_2}{R_1 + R_2} V_{CC} = 7.3 \text{ V}.$$

If we assume that the base current can be ignored, then $V_B \approx V_{BB}$, and we can say that $V_E \approx 8$ V. Therefore, the emitter current is

$$I_E = \frac{V_{CC} - V_E}{R_E} \approx 2 \text{ mA}.$$

The collector current should be about the same; therefore, the collector voltage is

$$V_C = I_C R_C \approx 4 \text{ V}.$$

We can now check and see that the transistor is, indeed, forward active. Also, the base current is about 40 μA, which is more than 10 times less than the current in R_1 (which is about 480 μA). Consequently, ignoring the base current was acceptable.

Given this bias point, we find

$$g_m = \frac{I_C}{V_T} = 77 \text{ mA/V}$$

at room temperature. We also have

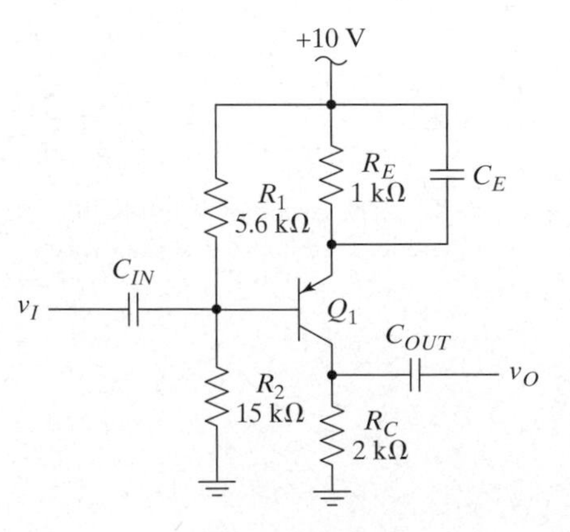

Figure 8–41 A *pnp* common-emitter amplifier.

$$r_\pi = \frac{\beta}{g_m} = 650\ \Omega.$$

The resulting small-signal equivalent circuit is shown in Figure 8-42, where we have ignored r_o and used

$$R_{BB} = R_1 \| R_2 = 4\ \text{k}\Omega.$$

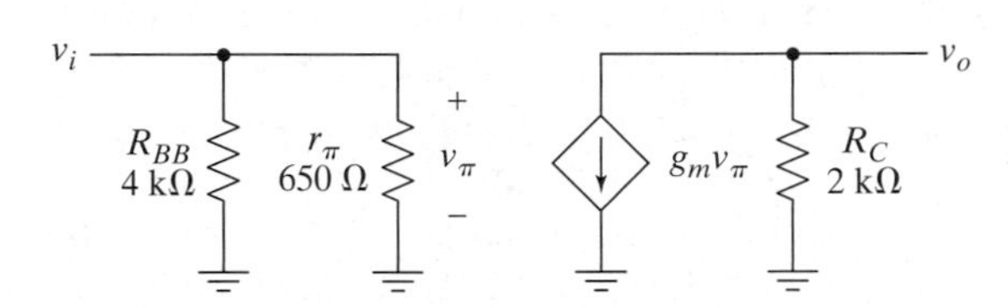

Figure 8–42 The small-signal equivalent circuit for the *pnp* amplifier.

The input resistance of this circuit is

$$R_i = R_{BB} \| r_\pi = 559\ \Omega.$$

The output resistance is (the load is not shown; R_C is part of the amplifier, not the load)

$$R_o \approx R_C = 2\ \text{k}\Omega,$$

and the open-circuit voltage gain is

$$a_v = \frac{v_o}{v_i} = -g_m R_C = -154.$$

■

8.2.3 A MOSFET Implementation: The Common-Source Amplifier

Figure 8-43 shows a MOSFET implementation of the amplifier of Figure 8-26. This circuit is called a *common-source amplifier*, because the source of the transistor is common to both the input and the output (i.e., the source is connected to an AC ground, while the input is applied from gate to ground and the output is

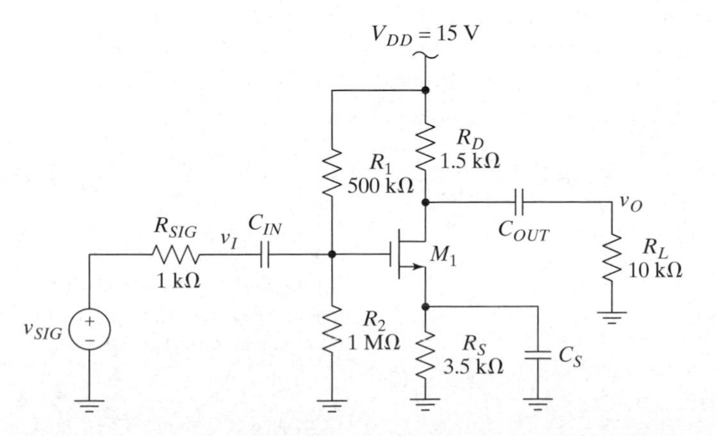

Figure 8–43 A MOSFET implementation of an amplifier with both voltage and current gain.

taken from drain to ground). The signal source is labeled v_{SIG} to avoid confusion with the source of the transistor. The DC bias point analysis of this circuit was covered in Section 7.2.2 and will not be repeated here. The two coupling capacitors C_{IN} and C_{OUT} were discussed when we analyzed Figure 8-26 and serve the same purpose here.

The capacitor in parallel with R_S is new, however. This capacitor, C_S in the figure, is called a **bypass capacitor.** It shorts out, or bypasses, R_S for signals, while allowing R_S to provide negative feedback for the DC bias point. We will see shortly that if R_S is not bypassed, it significantly lowers the voltage gain of the circuit.

The resistors for this circuit have been calculated to provide a reasonably stable bias point with $I_D = 2$ mA and a 3-V symmetric swing capability at the drain, assuming that the FET has $K = 500\ \mu A/V^2$ and $V_{th} = 1$ V. We noted in Section 7.2.2 that we would often not be able to satisfy the rule of thumb for V_S (i.e., that it be at least 10 V for bias stability), and this circuit is a good example. With a 15-V supply, we cannot set V_S at 10 V and still provide the desired drain current (the required V_{GS} is 3 V) with any room for output swing at the drain. Therefore, we compromise and set V_S at 7 V. The gate voltage is

$$V_G = \frac{R_2}{R_1 + R_2} V_{DD} = 10 \text{ V}. \qquad \textbf{(8.108)}$$

With the given drain current, the source voltage is (2 mA)(3.5 kΩ) = 7 V. Therefore, $V_{GS} = 3$ V, and we can check that the drain current is correct. We have

$$I_D = K(V_{GS} - V_{th})^2 = 0.5(3 - 1)^2 \text{ mA} = 2 \text{ mA}, \qquad \textbf{(8.109)}$$

as given. Now consider the drain voltage. If V_D is allowed to drop any lower than $V_G - V_{th}$, the transistor will be in the linear region. Also, V_D cannot go higher than V_{DD}. Therefore, we center V_D in this available range as suggested by our rule of thumb. The resulting drain voltage is

$$V_D = V_{DD} - I_D R_D = 15 - (2 \text{ mA})(1.5 \text{ k}\Omega) = 12 \text{ V}, \qquad \textbf{(8.110)}$$

which is halfway between V_{DD} and $V_G - V_{th}$.

Now that we understand how the DC bias point was arrived at (although no justification has been given yet for choosing $I_D = 2$ mA), let's examine the small-signal AC performance. Figure 8-44 shows the small-signal midband AC equivalent circuit for this amplifier. In this circuit, the coupling capacitors and bypass capacitor are all shorted out, the DC supply is replaced by ground (since the AC component of the supply voltage is zero by definition), and the transistor has been replaced by the low-frequency hybrid-π model. We have ignored r_o in the model, since it is much larger than the discrete resistors in parallel with it. It is almost always valid to ignore r_o in discrete circuits, although it is not for integrated circuits.

The gate-to-source voltage is given by the voltage divider at the input:

$$v_{gs} = \frac{R_{GG}}{R_{GG} + R_{SIG}} v_{sig}. \qquad \textbf{(8.111)}$$

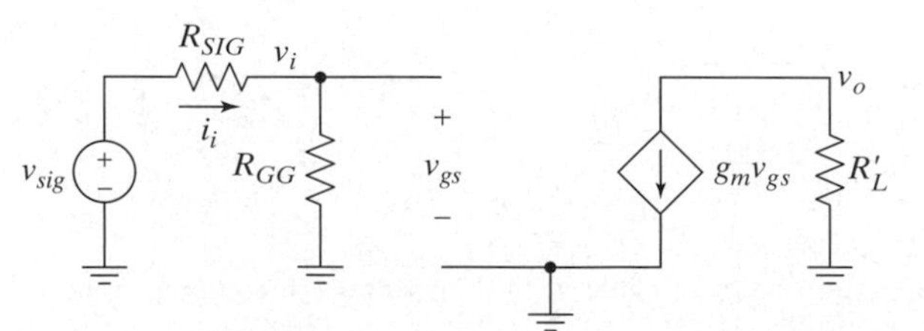

Figure 8–44 The small-signal midband AC equivalent circuit for the amplifier of Figure 8-43.

So the input attenuation factor is

$$\alpha_i = \frac{v_i}{v_{sig}} = \frac{v_{gs}}{v_{sig}} = \frac{R_{GG}}{R_{GG} + R_{SIG}}. \quad \textbf{(8.112)}$$

Because the DC gate current of the MOSFET is zero, the value of R_{GG} can be very large, so it is usually possible to set $v_{gs} \approx v_{sig}$ and therefore, $\alpha_i \approx 1$. For the present example, $R_{GG} = 333\ \text{k}\Omega$ and v_{gs} is about equal to v_{sig}. The output voltage is

$$v_o = -g_m v_{gs} R'_L = \alpha_i(-g_m R'_L) v_{sig}, \quad \textbf{(8.113)}$$

so the overall voltage gain is

$$\frac{v_o}{v_{sig}} = \alpha_i(-g_m R'_L). \quad \textbf{(8.114)}$$

The voltage gain, A_v, is found from (8.113) by recognizing that $v_i = v_{gs}$:

$$A_v = \frac{v_o}{v_i} = --g_m R'_L. \quad \textbf{(8.115)}$$

Using (8.36) we find that $g_m = 2$ mA/V, and plugging this value into (8.115) along with the given resistor values yields a gain of −2.6.

The input resistance of this circuit is defined to be the resistance seen by the signal source. We must remember that R_{SIG} is part of the source, so

$$R_i = \frac{v_i}{i_i} = R_{GG} = 333\ \text{k}\Omega. \quad \textbf{(8.116)}$$

To find the output resistance, we must separate R_L and R_D as shown in Figure 8-45. We also set the independent signal source to zero and placed a test source where the load was. With the signal source set to zero, we have $v_{gs} = 0$, and therefore, the controlled-current source is an open circuit (i.e., the current is zero). The resulting output resistance is

$$R_o = \frac{v_x}{i_x} = R_D = 1.5\ \text{k}\Omega. \quad \textbf{(8.117)}$$

Given the similarity of Figure 8-44 to the corresponding figure for the generic transistor amplifier, Figure 8-28, we can make use of the results from that circuit by changing variables. Using (8.63) and substituting v_{gs} for v_{cm}, R_{GG} for R_{CC}, and R_D for R_O, and realizing that r_{cm} is infinite, we find that the current gain of our common-source amplifier is

$$A_i = \frac{i_l}{i_s} = -g_m R_{GG}\left(\frac{R_D}{R_D + R_L}\right). \quad \textbf{(8.118)}$$

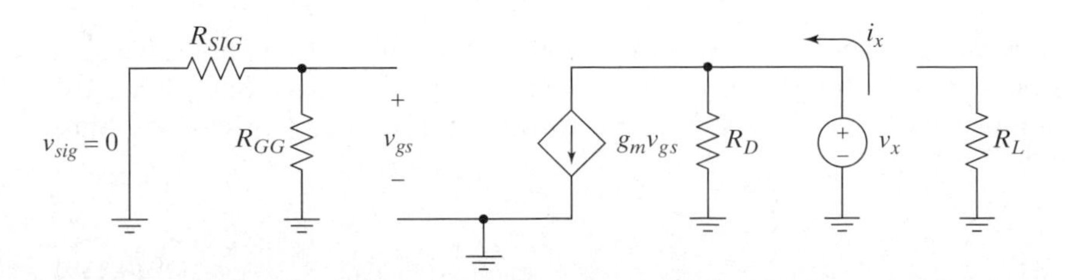

Figure 8–45 The circuit of Figure 8-44 redrawn to find R_o.

Plugging in the values for our circuit, we get $A_i = -87$. This value is not terribly meaningful, however, since we could make it arbitrarily large by increasing R_1 and R_2 without affecting the operation of the circuit. For a discrete circuit, there will be a practical limit set by the available resistors that is not much larger than we have here, but for integrated MOS amplifiers, the input current can be very nearly zero, so the current gain is, for all intents and purposes, infinite at low frequencies. Using (8.118) and (8.115), we can calculate the power gain of this circuit to be about 226.

Example 8.3

Problem:
Use SPICE to verify our analytical result for the gain of the amplifier shown in Figure 8-43.

Solution:
A SPICE simulation was performed using the Nmedium_mos model from Appendix B and 100 μF coupling and bypass capacitors. A plot of v_O for the AC simulation is shown in Figure 8-46. From the figure, we determine that the overall voltage gain is −2.6 (the minus sign is not apparent from the figure, only the magnitude). The result agrees exactly with our calculations, because we used an extremely simple transistor model for the simulation.

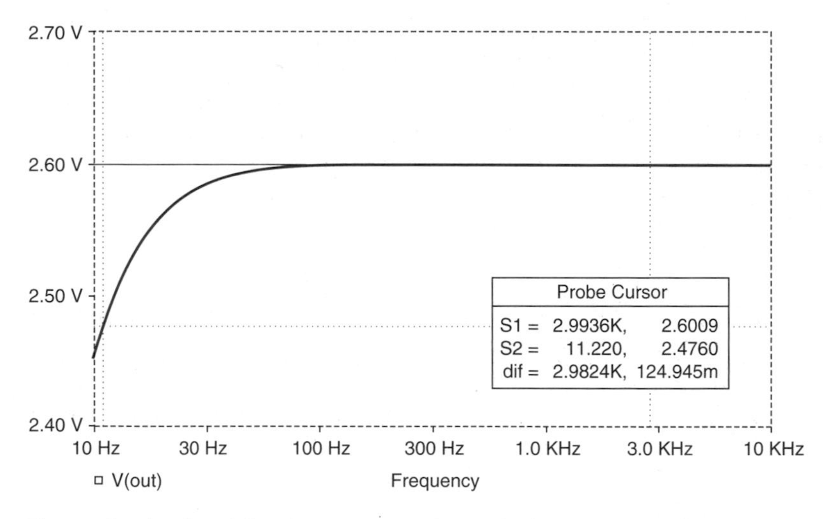

Figure 8–46 The AC simulation result for v_O. Cursor S1 shows the midband gain.

■

Let's now move on to find a unilateral two-port model for this amplifier. We need only one more parameter, the open-circuit voltage gain a_v, to have our model. This gain is found from (8.115) by setting R_L to infinity (i.e., an open circuit). Since $R'_L = R_L \| R_D$, we get

$$a_v = \left.\frac{v_o}{v_i}\right|_{\text{open circuit}} = -g_m R_D = -3. \tag{8.119}$$

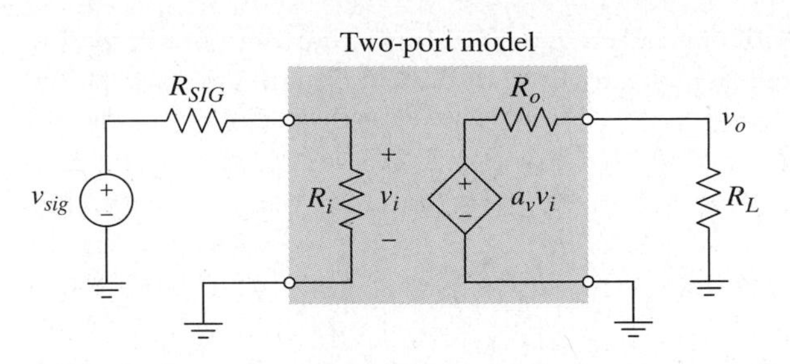

Figure 8–47 The amplifier from Figure 8-43 represented as a unilateral two-port.

Using (8.116), (8.117), and (8.119), we have completely specified a unilateral two-port model of a voltage amplifier. We show this amplifier model with the signal source and load in Figure 8-47 and calculate the overall gain in the familiar way:

$$\frac{v_o}{v_{sig}} = \alpha_i a_v \alpha_o = \left(\frac{R_i}{R_i + R_{SIG}}\right)(-g_m R_D)\left(\frac{R_L}{R_L + R_o}\right). \quad \textbf{(8.120)}$$

You can confirm that (8.120) is the same result as obtained directly in (8.114) (*see* Problem P8.42).

If we rewrite (8.119), we can see a limitation of this circuit, namely, that the open-circuit voltage gain is dependent on the DC bias point. Using (8.33) and (8.34), we find that

$$a_v = -g_m R_D = \frac{-2I_D R_D}{V_{GS} - V_{th}} = \frac{-2I_D R_D}{V_{ON}}. \quad \textbf{(8.121)}$$

This equation indicates that the gain is proportional to the ratio of the voltage drop across R_D to the gate overdrive. In integrated circuit design, we have control over the size of the FET (i.e., the ratio W/L) and can therefore set the gate overdrive to be independent of the bias current. However, in a discrete design, the gate overdrive will be proportional to the square root of I_D. Therefore, the bias voltages and currents cannot be chosen independently of the gain if we use this circuit. We may want to make these choices separately, though; for example, the bias voltages and currents determine the swing limits and power dissipation, which we may not want linked to the gain.

A circuit that offers the designer more freedom to choose the bias point and gain independently is shown in Figure 8-48. In this circuit, we have only bypassed a portion of the source resistor.

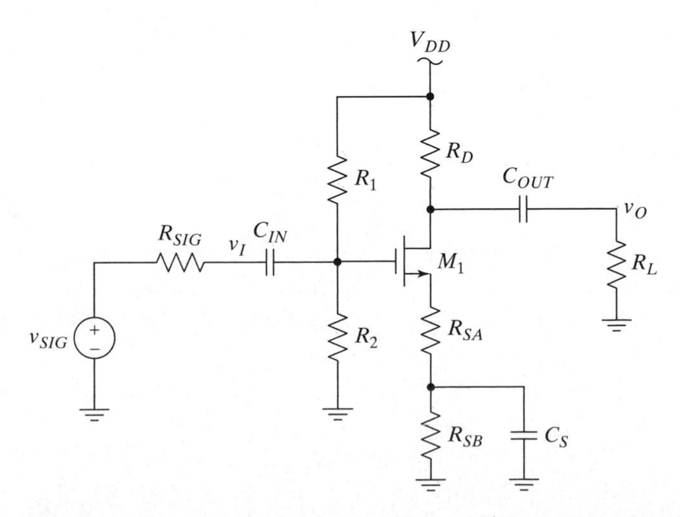

Figure 8–48 A common-source amplifier with partially unbypassed source resistance.

To find the small-signal AC midband performance of this circuit, we first draw the equivalent circuit shown in Figure 8-49. Using this equivalent circuit, we see that the gate voltage is given by

$$v_g = v_i = \frac{R_{GG}}{R_{GG} + R_{SIG}} v_{sig} = \alpha_i v_{sig}. \tag{8.122}$$

The source voltage is

$$v_s = g_m v_{gs} R_{SA}, \tag{8.123}$$

and combining these two equations, we can write

$$v_{gs} = v_g - v_s = \alpha_i v_{sig} - g_m v_{gs} R_{SA}, \tag{8.124}$$

which can be solved to yield

$$v_{gs} = \frac{\alpha_i v_{sig}}{1 + g_m R_{SA}}. \tag{8.125}$$

The output voltage is given by

$$v_o = -g_m v_{gs} R'_L. \tag{8.126}$$

Combining (8.125) and (8.126), we find the overall voltage gain:

$$\frac{v_o}{v_{sig}} = \frac{-\alpha_i g_m R'_L}{1 + g_m R_{SA}}. \tag{8.127}$$

The voltage gain from v_i to v_o is found by setting the input attenuation factor in (8.127) equal to one, since $v_i = v_g$:

$$A_v = \frac{v_o}{v_i} = \frac{-g_m R'_L}{1 + g_m R_{SA}}. \tag{8.128}$$

The input and output resistances of the circuit in Figure 8-49 are the same as for the unbypassed case: $R_i = R_{GG}$ and $R_o = R_D$. The open-circuit voltage gain is given by (8.128), modified by setting R_L to infinity:

$$a_v = \left.\frac{v_o}{v_i}\right|_{\text{open circuit}} = \frac{-g_m R_D}{1 + g_m R_{SA}}. \tag{8.129}$$

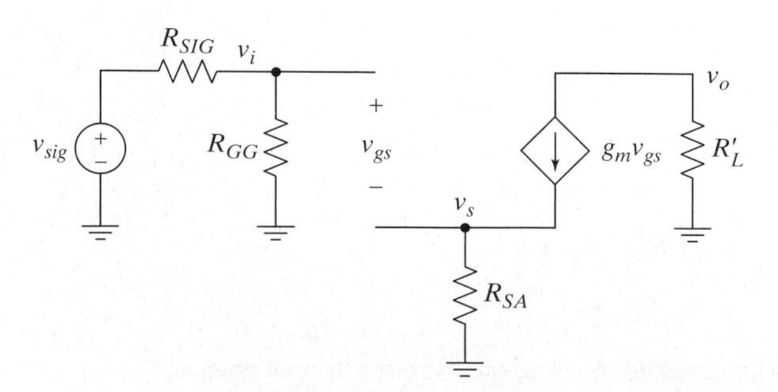

Figure 8–49 The small-signal midband AC equivalent circuit for the amplifier in Figure 8-48.

The unbypassed part of the source resistance represents negative AC feedback; that is, it tends to stabilize the gain just as negative feedback stabilizes the bias point. If the transconductance increases for some reason (e.g., a change in bias point with temperature), the gain of the circuit with all of R_S bypassed will change in proportion to g_m. When there is unbypassed source resistance, however, (8.129) shows that the gain will not increase in direct proportion to an increase in g_m. In fact, if $g_m R_{SA} \gg 1$ (which is hard to achieve with MOSFETs), the gain becomes approximately independent of g_m. Because negative feedback is also called degenerative feedback, R_{SA} is sometimes referred to as a ***degeneration resistor***, or ***source degeneration***. Because we can control the gate overdrive and I_D separately in IC design (by varying W/L), source degeneration is not used very often.

Example 8.4

Problem:

Find the small-signal midband AC values for R_i, R_o, and a_v for the amplifier shown in Figure 8-50. Assume that $V_{th} = -1$ V and $K = 500\ \mu A/V^2$.

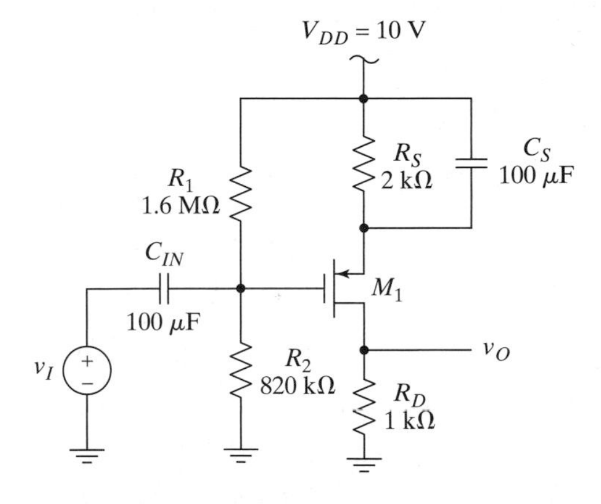

Figure 8–50 A PMOS common-source amplifier.

Solution:

We begin the analysis by finding the operating point of the transistor. The gate voltage is

$$V_G = \frac{R_2}{R_1 + R_2} V_{DD} = 3.4\text{ V}.$$

Writing KVL around the DC gate-source loop, we get

$$V_{DD} - I_D R_S - V_{SG} - V_G = 0.$$

We now plug in the equation for I_D and rewrite to find

$$V_{SG} = \frac{-(2KR_S V_{th} + 1) \pm \sqrt{(2KR_S V_{th} + 1)^2 - 4KR_S(KR_S V_{th}^2 + V_G - V_{DD})}}{2KR_S}$$

$$= 2.92\ V,$$

where we have thrown out the negative solution as being impossible. With this source-to-gate voltage, we have $I_D = 1.84$ mA. The resulting drain voltage is $V_D = I_D R_D = 1.84$ V, and we can see that the device is forward active, as assumed.

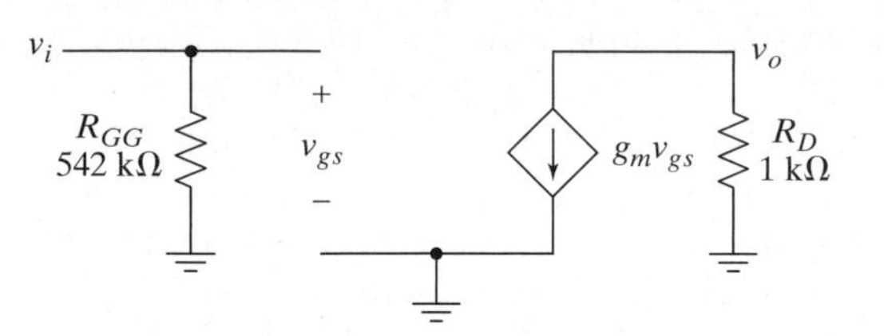

Figure 8–51 The small-signal equivalent circuit for the PMOS amplifier.

Given this bias point, we find that

$$g_m = 2K(V_{SG} + V_{th}) = 1.92 \text{ mA/V}$$

The resulting small-signal equivalent circuit is shown in Figure 8-51, where we have ignored r_o and used

$$R_{GG} = R_1 \| R_2 = 542 \text{ k}\Omega.$$

The input resistance of this circuit is

$$R_i = R_{GG} = 542 \text{ k}\Omega.$$

The output resistance is (the load is not shown, R_D is part of the amplifier, not the load)

$$R_o \approx \text{R}_\text{D} = 1 \text{ k}\Omega,$$

and the open-circuit voltage gain is

$$a_v = \frac{v_o}{v_i} = -g_m R_D = -1.92.$$

A SPICE simulation of this circuit using the Pmedium_mos model from Appendix B confirms both the bias point and the small-signal gain.

8.3 VOLTAGE BUFFERS

In the preceding section, we noted that stages with both voltage and current gain have a relatively high output resistance and are not suitable for driving a low-impedance load directly. In this section, we cover stages that have current gain greater than one, but voltage gain less than or equal to one. These stages have high input resistances, low output resistances, and can therefore be used as output stages for amplifiers of the type shown in the previous section. Their high input resistances will not load down the preceding stages, and their low output resistances will allow them to drive small loads without significant attenuation.

Because their voltage gains are typically about equal to one, they are often called ***voltage followers*** (i.e., the output voltage follows the input voltage). Because they are used to isolate—or buffer—loads from preceding gain stages, these stages are also called ***voltage buffers***.

We illustrate the usefulness of voltage buffers with the aid of Figure 8-52. Part (a) of the figure shows an amplifier driving a load without a voltage buffer, and part (b) of the figure shows the same amplifier and load with a voltage buffer added between them. For part (a), the voltage gain is

$$A_v = \frac{v_o}{v_i} = a_{v1}\alpha_{o1}, \tag{8.130}$$

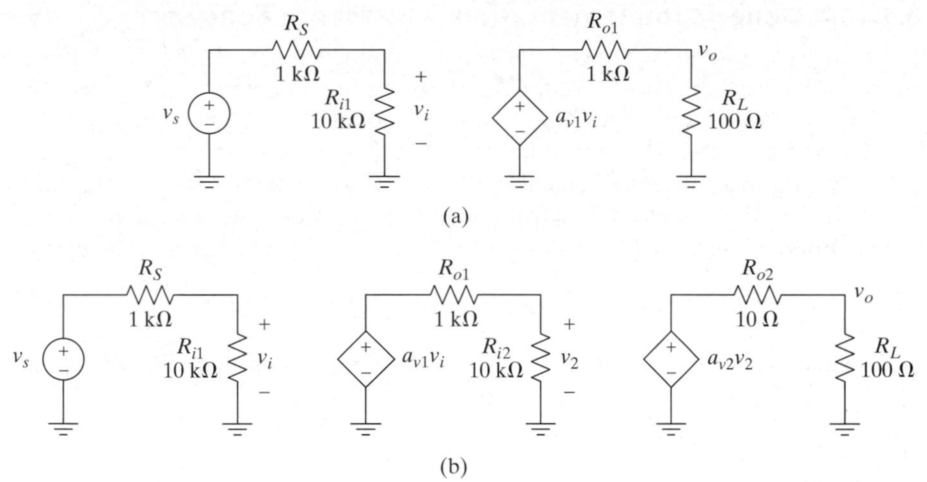

Figure 8–52 (a) An amplifier driving a load and (b) the same amplifier driving the same load through a voltage buffer. $a_{v1} = 20$ and $a_{v2} = 0.9$.

where the output attenuation factor is

$$\alpha_{o1} = \frac{R_L}{R_L + R_{o1}}. \tag{8.131}$$

If the output resistance of this amplifier, R_{o1}, is large—which is often the case if a_v is large—α_{o1} may be small and the overall voltage gain will suffer. Plugging in the given numbers yields $\alpha_{o1} = 0.091$ and $A_v = 1.82$ for this example. If a voltage buffer is inserted between the amplifier and load, as in part (b), however, the voltage gain is

$$A_v = \frac{v_o}{v_i} = a_{v1}\alpha_{12}a_{v2}\alpha_{o2}, \tag{8.132}$$

where

$$\alpha_{12} = \frac{R_{i2}}{R_{i2} + R_{o1}} \tag{8.133}$$

is the attenuation factor coupling the two amplifiers and

$$\alpha_{o2} = \frac{R_L}{R_L + R_{o2}} \tag{8.134}$$

is the new output attenuation factor. If $R_{i2} \gg R_{o1}$ and $R_{o2} \ll R_{o1}$, α_{12} will be nearly equal to one and α_{o2} will be much larger than α_{o1}. In this case, the new gain given by (8.132) will be significantly larger than the original gain given by (8.130), even if a_{v2} is slightly less than one. Plugging in the given numbers, we find that $\alpha_{12} = \alpha_{o2} = 0.91$ and $A_v = 14.9$ in this case.

The voltage buffer functions as does an impedance transformer. By placing the buffer in series with a small load, we can make the load appear as a large impedance to the preceding stage without significantly attenuating the signal.

8.3.1 A Generic Implementation: The Merge Follower

Let's now examine how we can construct a voltage buffer. Consider the circuit shown in Figure 8-53. Part (a) of the figure shows the schematic of a merge follower, wherein the preceding amplifier stage is modeled by a Thévenin equivalent. The DC voltage out of the preceding stage would be used to bias the follower. Part (b) of the figure is the small-signal midband AC equivalent circuit of the buffer. The output of the circuit is taken from the merge terminal, and the transistor's output terminal is connected to an AC ground. The current through R_L is given by

$$i_l = i_i + g_m v_{cm} = i_i(1 + g_m r_{cm}). \tag{8.135}$$

Using this result, we can write KVL around the loop containing v_s and R_L and get

$$v_s = i_i[R_S + r_{cm} + (1 + g_m r_{cm})R_L]. \tag{8.136}$$

The output voltage of the circuit is

$$v_o = i_l R_L = i_i(1 + g_m r_{cm})R_L, \tag{8.137}$$

so combining this fact with (8.136), we get the overall voltage gain:

$$\frac{v_o}{v_s} = \frac{(1 + g_m r_{cm})R_L}{R_S + r_{cm} + (1 + g_m r_{cm})R_L}. \tag{8.138}$$

The voltage gain from v_i to v_o, A_v is given by (8.138) with R_S set equal to zero. The gain in (8.138) can be made very close to one, so the voltage at the merge terminal follows the voltage at the input, which is why the circuit is called a ***merge follower***. Since the transistor's output terminal is tied to ground—which is common to both the input and the output—the circuit could also be called a ***common-output stage***.

The small-signal midband AC input resistance is found from (8.136) by recognizing that v_i is $v_s - i_i R_S$; that is

$$R_i = \frac{v_i}{i_i} = r_{cm} + (1 + g_m r_{cm})R_L = r_{cm} + (1 + a_{im})R_L, \tag{8.139}$$

where we have used the generic transistors common-merge current gain, a_{im}, in the final step. The input resistance is guaranteed to be larger than R_L and is typically many times larger. The output resistance of the amplifier can be found by re-

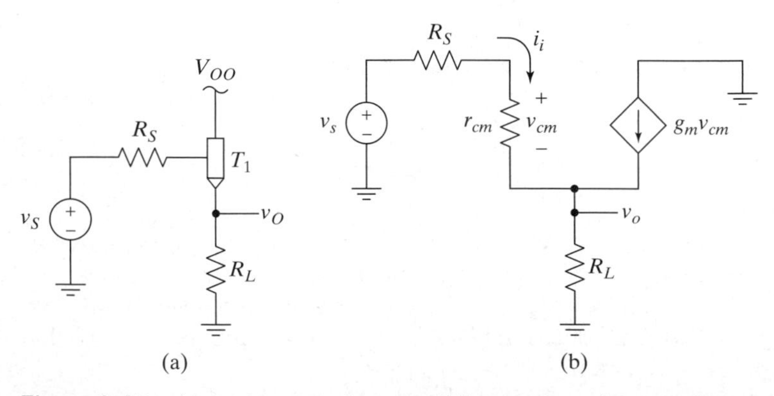

Figure 8–53 (a) A schematic of a merge follower where the preceding amplifier stage is modeled by a Thévenin equivalent. (b) The small-signal midband AC equivalent circuit of the buffer amplifier.

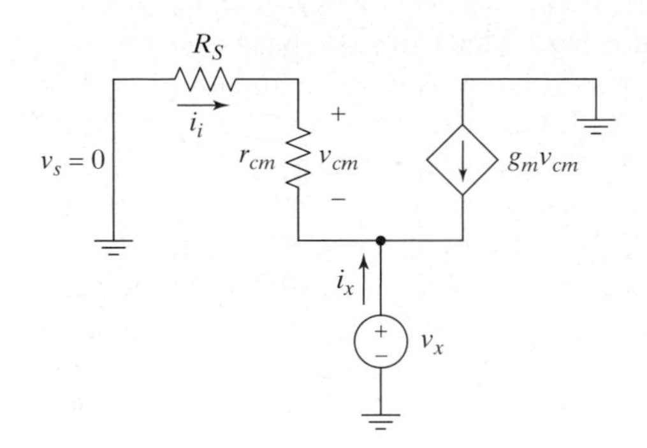

Figure 8–54 The circuit for finding the output resistance of the amplifier in Figure 8-53.

moving the load, setting the independent source v_s to zero, and driving the output with a test source as shown in Figure 8-54. Using KCL at the top of v_x, we see that

$$i_x = -i_i - g_m v_{cm} = -i_i(1 + g_m r_{cm}) = -i_i(1 + a_{im}). \quad \textbf{(8.140)}$$

We can also see directly from the circuit that

$$i_i = \frac{-v_x}{R_S + r_{cm}}. \quad \textbf{(8.141)}$$

Combining (8.140) and (8.141) we find the small-signal midband AC output resistance:

$$R_o = \frac{v_x}{i_x} = \frac{R_S + r_{cm}}{1 + g_m r_{cm}} = \frac{R_S + r_{cm}}{1 + a_{im}}. \quad \textbf{(8.142)}$$

Since a_{im} is large for all real transistors, the output resistance of the voltage buffer will be low.

If the source resistance driving this circuit is the output resistance of a preceding gain stage, we typically have $R_S \gg r_{cm}$, and the output resistance is usually much less than R_S. In other words, the output resistance is much less than the output resistance of the preceding stage, as assumed in the example of Figure 8-52.

8.3.2 A Bipolar Implementation: The Emitter Follower

Consider the circuit shown in Figure 8-55. This is a bipolar implementation of a voltage buffer. The signal input is capacitively coupled to the base, and the output is capacitively coupled from the emitter. If the circuit is designed properly, the change in base-emitter voltage will be small compared to the signal, so the voltage

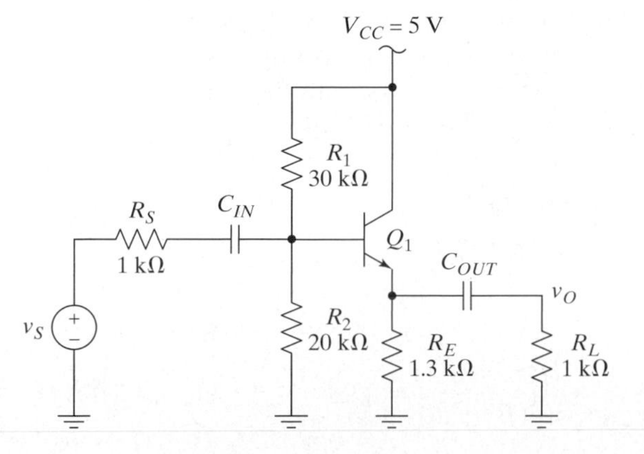

Figure 8–55 An emitter follower.

at the emitter will *follow* the voltage at the base. That is, as the base voltage goes up, the emitter voltage will go up. As the base voltage goes down, the emitter voltage will go down. This is why the circuit is called an **emitter follower**. For the voltage at the emitter to go up, the transistor must supply more current, of course, and this does require that v_{BE} increase. Nevertheless, since the current is an exponential function of the base-emitter voltage, it is possible to make the change in v_{BE} negligible compared to the signal swing (i.e., the small-signal midband AC voltage gain can be close to one). The circuit is also called a **common-collector stage**, because the collector is tied to a signal ground, which is common to both the input and the output. (The input voltage is from base to ground and the output voltage is from emitter to ground.)

The biasing of this emitter follower circuit is very much the same as for the common-emitter stage. The Thévenin voltage driving the base is

$$V_{BB} = \frac{R_2}{R_1 + R_2} V_{CC} = 2 \text{ V}, \tag{8.143}$$

and the Thévenin resistance is

$$R_{BB} = R_1 \| R_2 = 12 \text{ k}\Omega. \tag{8.144}$$

Assuming that the base current can be ignored, the base voltage is about equal to $V_{BB,}$ and the resulting emitter voltage is 1.3 V. The DC emitter current is then $V_E/R_E = 1$ mA, and the collector current is nearly the same. If the β is 100, the corresponding base current is 10 μA. Since the current in R_1 and R_2 is about $V_{CC}/(R_1 + R_2) = 100$ μA, which is 10 times I_B, we were safe in ignoring the base current in our initial calculations.

The small-signal midband AC equivalent circuit for the emitter follower of Figure 8-55 is shown in Figure 8-56(a), where the current-controlled hybrid-π model has been used for the transistor. We have also combined R_L and R_E in parallel and called the result R_L'. The value of r_π is calculated with the bias point just given:

$$r_\pi = \frac{\beta}{g_m} = \frac{\beta V_T}{I_C} = \frac{100(0.026)}{0.001} = 2.6 \text{ k}\Omega. \tag{8.145}$$

The gain of the circuit can be found most easily by redrawing the circuit as shown in part (b) of the figure, wherein we have made use of impedance reflection. Remember that this circuit is equivalent to the circuit in part (a) *only* in the sense that both circuits yield the same value for i_b and have the same loop equation (i.e., the voltage drops are all the same). Since the voltage drops are all the same, we can use this circuit to find the gain. Let's first find the gain from v_i to v_o. By inspection, v_o is given by a resistive voltage divider; that is

$$v_o = \frac{(\beta+1)R_L'}{r_\pi + (\beta+1)R_L'} v_i. \tag{8.146}$$

So the gain is

$$A_v = \frac{v_o}{v_i} = \frac{(\beta+1)R_L'}{r_\pi + (\beta+1)R_L'} = 0.96. \tag{8.147}$$

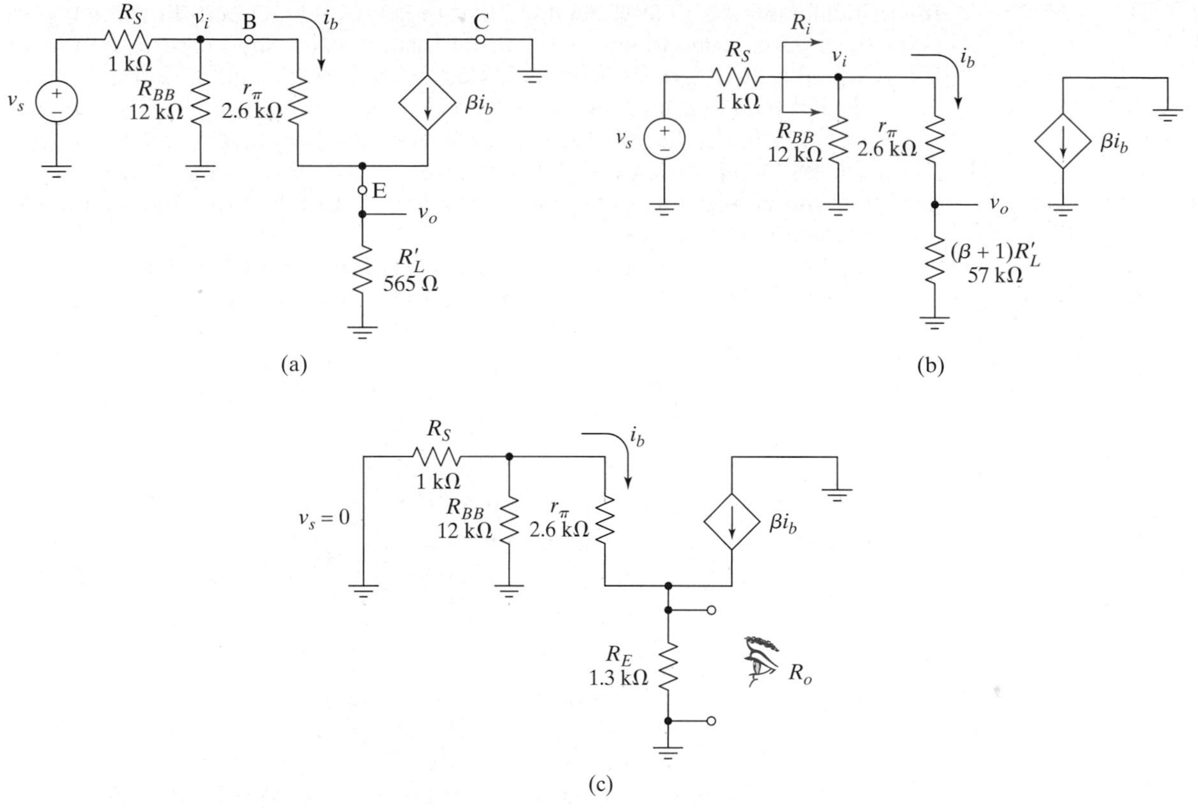

Figure 8–56 (a) The small-signal midband AC equivalent circuit for the emitter follower in Figure 8-55. (b) The circuit redrawn using impedance reflection. (c) Finding the output resistance.

The input resistance can be seen in part (b) of the figure to be given by

$$R_i = R_{BB} \| [r_\pi + (\beta+1)R'_L] = 10\ \text{k}\Omega. \tag{8.148}$$

To find the output resistance of the circuit, we redraw the circuit as shown in part (c) of the figure. Here we have set the independent source v_s to zero and removed the load. We could force a test voltage and measure the resulting current to find R_o (you are encouraged to do so if the following analysis is not clear to you), but we need not do so if we make use of impedance reflection again. In Figure 8-57, we have combined R_S and R_{BB} in parallel (with v_s set to zero) and have reflected them into the emitter along with r_π. The output resistance can be calculated directly by inspection of the figure and yields

$$R_o = R_E \left\| \left(\frac{r_\pi + R_S \| R_{BB}}{\beta+1} \right) = R_E \right\| \left(r_e + \frac{R_S \| R_{BB}}{\beta+1} \right) = 34\ \Omega. \tag{8.149}$$

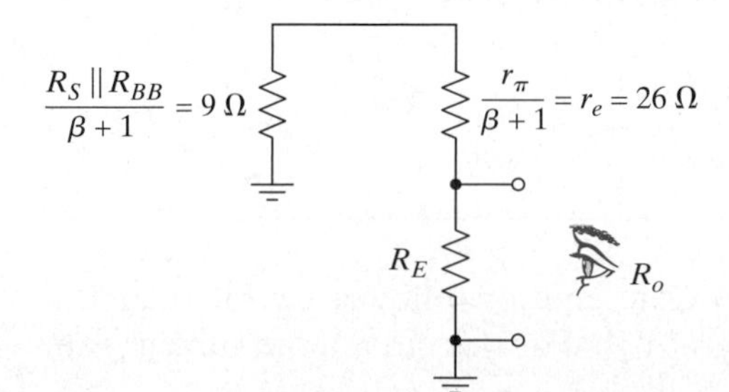

Figure 8–57 The circuit of Figure 8-56 reflected into the emitter.

It is critically important to understand that the equivalent circuit shown in Figure 8-56(b) *cannot* be used to find the output resistance. As stated before, this circuit is equivalent only in terms of the voltages around the loop and the current *on the base side*. Since the output resistance involves looking into the emitter of the transistor, and the current in the emitter is different from the current in the base by a factor of $(\beta+1)$, this circuit will yield the wrong answer by a factor of $(\beta+1)$. Similarly, the equivalent circuit of Figure 8-57 *cannot* be used to find the input resistance of the circuit.

We can now draw a unilateral two-port model of the original circuit as shown in Figure 8-58(a). The open-circuit voltage gain needed for the two-port model is given by (8.147) with R_L set to infinity:

$$a_v = \left.\frac{v_o}{v_i}\right|_{\text{open circuit}} = \frac{(\beta+1)R_E}{r_\pi + (\beta+1)R_E} = 0.98. \tag{8.150}$$

Using this model, we see that the overall voltage gain is

$$\frac{v_o}{v_s} = \left(\frac{R_i}{R_i + R_S}\right)a_v\left(\frac{R_L}{R_L + R_o}\right) = \alpha_i a_v \alpha_o. \tag{8.151}$$

Plugging in the numbers yields $\alpha_i = 0.91$, $\alpha_o = 0.97$ and an overall gain of 0.86. Notice that if the emitter follower where not used, and the source drove the load directly as shown in Figure 8-58(b), the gain would be one-half. Not only would the gain be lower if the source drove the load directly, but the gain would be a strong function of the value of R_L. However, with the emitter follower, the value of R_L can change quite a bit before it causes any significant change in the gain.

The emitter follower can also be analyzed using the T model of the transistor. The results are the same as we obtained using the hybrid-π model, but the gain calculation is a bit simpler.

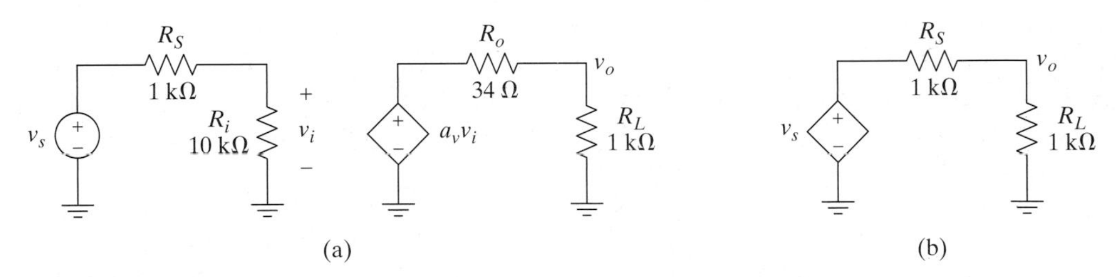

Figure 8–58 (a) The circuit of Figure 8-56(a) redrawn using a unilateral two-port model for the amplifier. (b) The source driving the load directly.

Exercise 8.6

Find the small-signal midband AC values of R_i, R_o, and a_v for the emitter follower circuit in Figure 8-55, using the T model for the transistor.

□

Finally, we calculate the power gain of the emitter follower circuit in Figure 8-55. The power gain of the circuit is $A_p = A_v A_i$. For this circuit, the current gain can be written

$$A_i = \frac{i_o}{i_i} = \frac{v_o/R_L}{v_i/R_i} = \frac{R_i}{R_L}\frac{v_o}{v_i} = \frac{R_i}{R_L}A_v, \quad \textbf{(8.152)}$$

so the power gain is

$$A_p = A_i A_v = \frac{R_i}{R_L}A_v^2, \quad \textbf{(8.153)}$$

which, using the results in (8.147) and (8.148) along with the given load resistor, yields $A_p = 9.2$. We conclude this section by reiterating that the emitter follower has current gain and power gain greater than one, but the voltage gain is always less than one.

Example 8.5

Problem:
Find the small-signal midband AC values for R_i, R_o, and a_v for the emitter follower shown in Figure 8-59. Assume that $\beta = 100$.

Solution:
We begin the analysis by finding the operating point of the transistor. First, we find V_{BB}:

$$V_{BB} = \frac{R_2}{R_1 + R_2}V_{CC} = 4.1 \text{ V}.$$

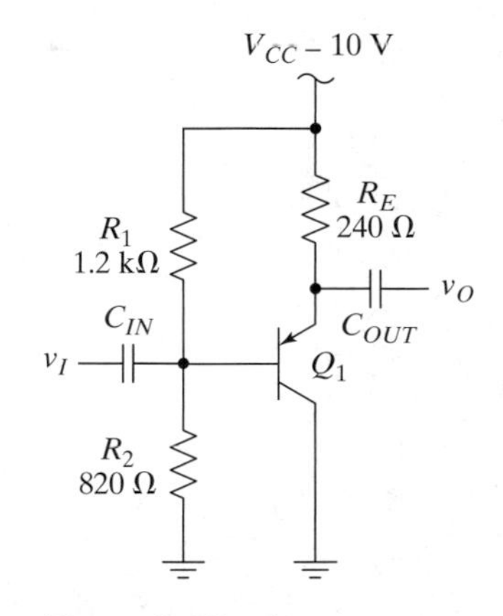

Figure 8–59 A *pnp* emitter follower.

If we assume that the base current can be ignored, $V_B \approx V_{BB}$, and we can say that $V_E \approx 4.8$ V. Therefore, the emitter current is

$$I_E = \frac{V_{CC} - V_E}{R_E} \approx 20 \text{ mA}.$$

We can now check and see that the transistor is forward active. Also, the base current is about 200 μA, which is more than 10 times less than the current in R_1 (which is about 5 mA), so ignoring the base current caused no problems.

Given this bias point, we find that

$$g_m = \frac{I_C}{V_T} = 0.77 \text{ A/V}$$

at room temperature. We also have

$$r_\pi = \frac{\beta}{g_m} = 130\ \Omega$$

and

$$r_e = \frac{V_T}{I_E} = 1.3\ \Omega.$$

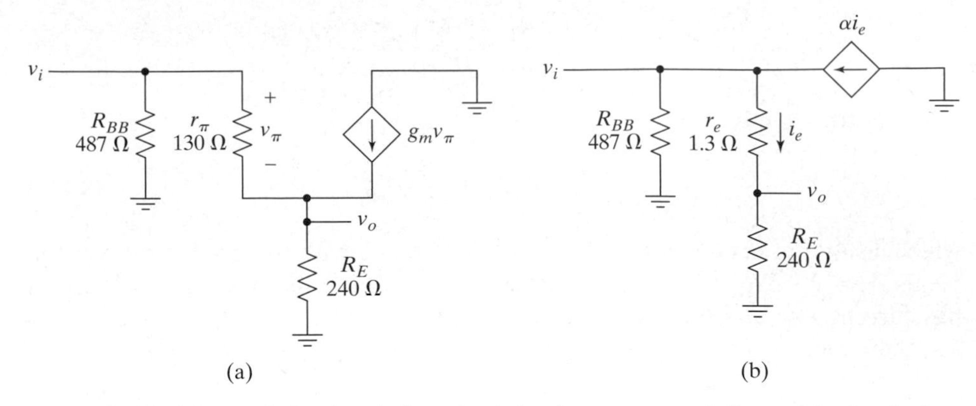

Figure 8–60 The small-signal equivalent circuit for the *pnp* emitter follower (a) using the hybrid-π model and (b) using the T model.

The resulting small-signal equivalent circuits are shown in Figure 8-60, where we have ignored r_o, used

$$R_{BB} = R_1 \| R_2 = 487\ \Omega,$$

and drawn the circuit using the hybrid-π equivalent in part (a) of the figure and the T model in part (b).

Using impedance reflection in Figure 8-60(a), we find that the input resistance of this circuit with no load connected is

$$R_i = R_{BB} \| [r_\pi + (\beta+1)R_E] = 477\ \Omega.$$

If a load is connected, it shows up in parallel with R_E in this equation. The output resistance depends on the source resistance R_S and is again found using impedance reflection:

$$R_o = R_E \left\| \left[\frac{R_S \| R_{BB} + r_\pi}{\beta+1} \right] \right. .$$

If, for example, $R_S = 100\ \Omega$, then $R_o = 2.1\ \Omega$. This very low output resistance is partly a result of the large bias current used in this example. The open-circuit voltage gain is most easily found using the T model in part (b) of the figure and is

$$a_v = \frac{v_o}{v_i} = \frac{R_E}{R_E + r_e} = 0.995$$

if no load is connected. A SPICE simulation of this circuit reveals that I_E is about 5% larger than we predicted, but the small-signal gain is very accurate.

■

8.3.3 A FET Implementation: The Source Follower

Consider the circuit shown in Figure 8-61. This is a MOSFET implementation of a voltage buffer stage. The signal input is capacitively coupled to the gate, and the output is capacitively coupled from the source of the transistor. If the circuit is designed properly, the change in gate-to-source voltage will be small compared with the amplitude of the signal, so the voltage at the source will *follow* the voltage at the gate. That is, as the gate voltage goes up, the source voltage will go up. As the gate voltage goes down, the source voltage will go down. This is why the circuit is called a ***source follower***.

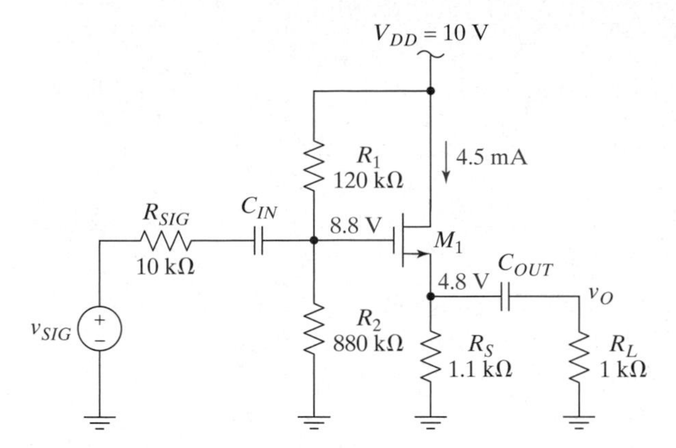

Figure 8–61 A source follower.

For the voltage at the source to go up, the transistor must supply more current, and this does require that v_{GS} increase. Nevertheless, it is possible to make the change in v_{GS} small compared to the signal swing, so that the change in v_O is almost as large (i.e., the small-signal midband AC voltage gain is close to one). The circuit is also called a ***common-drain stage***, because the drain is tied to a signal ground, which is common to both the input and the output. The input voltage is measured from gate to ground, and the output voltage from source to ground.

The biasing of this source follower circuit is much the same as for the common-source stage. The Thévenin voltage driving the gate is

$$V_{GG} = \frac{R_2}{R_1 + R_2} V_{DD} = 8.8 \text{ V}, \tag{8.154}$$

and the Thévenin resistance is

$$R_{GG} = R_1 \| R_2 = 106 \text{ k}\Omega. \tag{8.155}$$

Since the DC gate current is zero, the gate voltage is equal to V_{GG}. This particular circuit was designed to provide a drain current of about 4.5 mA with a transistor that has $K = 500\ \mu\text{A/V}^2$ and $V_{th} = 1$ V. The resulting V_{GS} is 4 V, so the source voltage is $V_S = 4.8$ V.

The small-signal midband AC equivalent circuit for the source follower of Figure 8-61 is shown in Figure 8-62, where the hybrid-π model has been used for the transistor. We have also combined R_L and R_S in parallel and called the result R'_L.

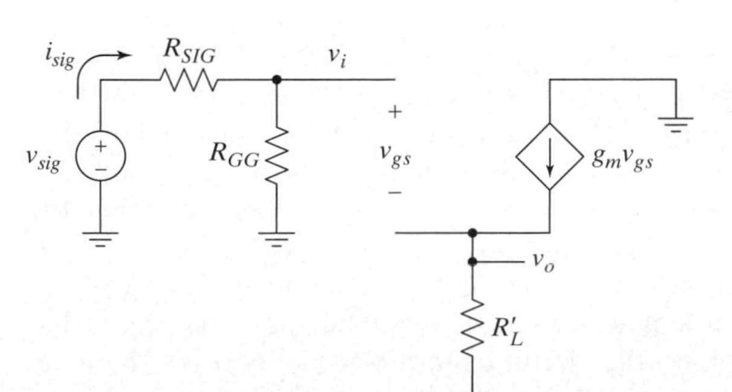

Figure 8–62 The small-signal midband AC equivalent circuit for the source follower in Figure 8-61.

The output voltage is now given by

$$v_o = g_m v_{gs} R'_L. \tag{8.156}$$

We substitute $v_{gs} = v_i - v_o$ into (8.156) and solve to obtain the voltage gain,

$$A_v = \frac{v_o}{v_i} = \frac{g_m R'_L}{1 + g_m R'_L}. \tag{8.157}$$

For this example, the result is $A_v = 0.61$, which is not as close to one as we might like. We will return and see why in a moment. But first we find the input and output resistances of the source follower. The small-signal input resistance can be seen from Figure 8-62 to be equal to R_{GG}:

$$R_i = \frac{v_i}{i_{sig}} = R_{GG} = 106\ \text{k}\Omega. \tag{8.158}$$

To find the output resistance, we have to set the signal source to zero, remove the load, and connect a test source to the terminals where the load was connected as shown in Figure 8-63. Since the gate current is zero, $v_i = 0$, and the gate is effectively connected to ground. Therefore, the controlled current source is controlled by the voltage across itself, and we can use the source absorption theorem to replace the controlled source by its equivalent resistance, $1/g_m$. The output resistance is then R_S in parallel with $1/g_m$, or

$$R_o = R_S \| \frac{1}{g_m} = \frac{R_S}{1 + g_m R_S}, \tag{8.159}$$

which is equal to 256 Ω for the example at hand.

We can now draw a model of the original circuit, making use of our unilateral two-port model as shown in Figure 8-64(a). The input and output resistances for the model are given by (8.158) and (8.159). The open-circuit voltage gain is given by (8.157) with R_L set to infinity (i.e., $R'_L = R_S$); that is,

$$a_v = \left.\frac{v_o}{v_i}\right|_{\text{open circuit}} = \frac{g_m R_S}{1 + g_m R_S}, \tag{8.160}$$

which is 0.77 for this example.

Using this model, we see that the overall voltage gain is

$$\frac{v_o}{v_{sig}} = \left(\frac{R_i}{R_i + R_{SIG}}\right) a_v \left(\frac{R_L}{R_L + R_o}\right) = \alpha_i a_v \alpha_o. \tag{8.161}$$

Plugging in the numbers yields $\alpha_i = 0.91$, $\alpha_o = 0.80$ and an overall gain of 0.56. Notice that if the source follower were not used, and the signal source drove the load directly as shown in part (b) of the figure, the gain would be 0.09. Not only would the gain be lower if the signal source drove the load directly, but the gain would be a strong function of the value of R_L. With the source follower, however, the value of R_L can change quite a bit before it causes a large change in the gain.

We now return to the gain and consider why it is so low. The problem here is that the bias current was not high enough, which causes two problems. The first is that the output resistance is too high, so the circuit cannot effectively drive the load. We can see this by examining the output attenuation factor, which we found to be

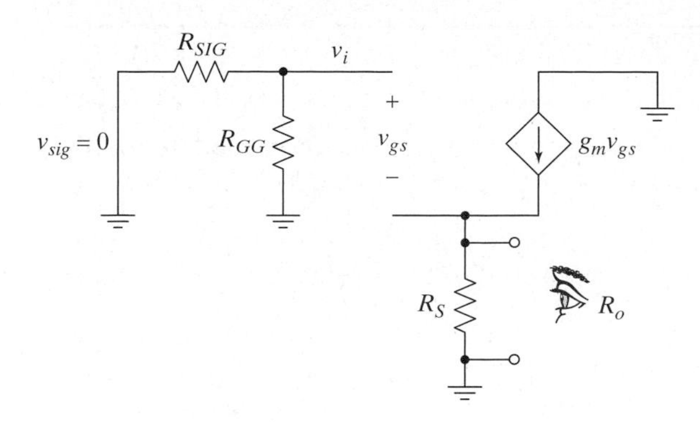

Figure 8–63 Finding the output resistance of the source follower.

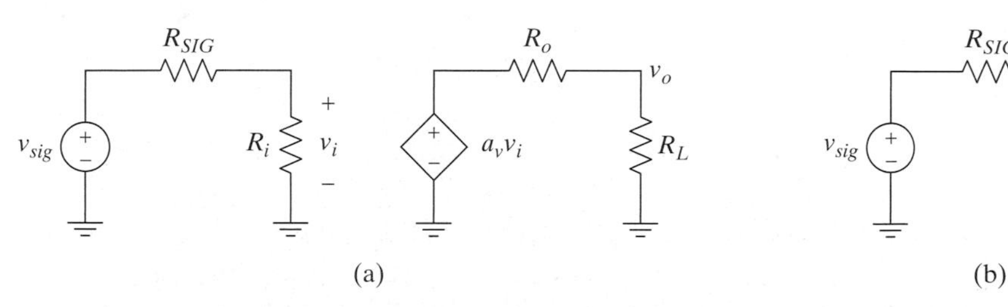

Figure 8–64 (a) The circuit of Figure 8-62 redrawn using a unilateral two-port model for the amplifier. (b) The signal source driving the load directly.

only 0.8. The second problem is that the transconductance is low, so the open-circuit voltage gain is not as high as we would like; we previously found that to be 0.77. Both of these problems can be remedied by increasing the DC drain current.

The source follower can also be analyzed using the T model of the transistor. The results will be the same as were obtained using the hybrid-π model, but the gain calculation is a bit simpler.

Exercise 8.7

Find the small-signal midband AC values of R_i, R_o, and a_v for the source follower circuit in Figure 8-61 using the T model for the transistor.

□

Finally, we calculate the power gain of the source follower circuit in Figure 8-61. The power gain of the circuit is $A_p = A_v A_i$. The current gain can be written

$$A_i = \frac{i_o}{i_i} = \frac{v_o/R_L}{v_i/R_i} = \frac{R_i}{R_L}\frac{v_o}{v_i} = \frac{R_i}{R_L}A_v, \tag{8.162}$$

so the power gain is

$$A_p = A_i A_v = \frac{R_i}{R_L}A_v^2, \tag{8.163}$$

which, using our earlier results, yields $A_p = 33$. This power gain is somewhat arbitrary, since we can raise R_i without affecting the operation of the circuit, as was noted for the common source stage. There is again a practical limit for discrete circuits, but the power gain for integrated circuit source followers can be extremely large at low frequencies. We conclude this section by reiterating that this stage can have current gain and power gain greater than one, but the voltage gain is less than one.

Example 8.6

Problem:

Find the small-signal midband AC values for R_i, R_o, and a_v for the source follower shown in Figure 8-65. Assume that $V_{th} = -1$ V and $K = 2$ mA/V^2.

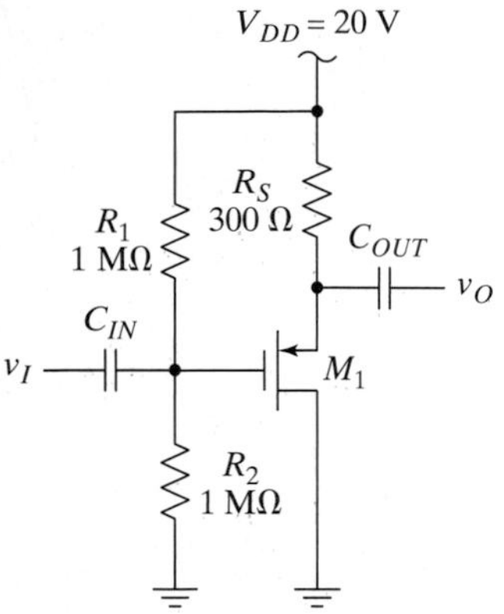

Figure 8–65 A PMOS source follower.

Solution:

We begin the analysis by finding the operating point of the transistor. The gate voltage is

$$V_G = \frac{R_2}{R_1 + R_2} V_{DD} = 10\text{ V}.$$

Writing KVL around the DC gate-source loop, we get

$$V_{DD} - I_D R_S - V_{SG} - V_G = 0.$$

We now plug in the equation for I_D and rewrite to find

$$V_{SG} = \frac{(2KR_SV_{th} + 1) \pm \sqrt{(2KR_SV_{th} + 1)^2 - 4KR_S(KR_SV_{th}^2 + V_G - V_{DD})}}{-2KR_S}$$
$$= 4.13\ V,$$

where we have thrown out the negative solution as being impossible. With this source-to-gate voltage, we have $I_D = 20$ mA, and we can see that the device is forward active as assumed.

Given this bias point, we find that

$$g_m = 2K(V_{SG} + V_{th}) = 12.5\text{ mA/V}$$

The resulting small-signal equivalent circuit is shown in Figure 8-66, where we have ignored r_o and used

$$R_{GG} = R_1 \| R_2 = 500\text{ k}\Omega.$$

The input resistance of this circuit is

$$R_i = R_{GG} = 500\text{ k}\Omega.$$

The output resistance is

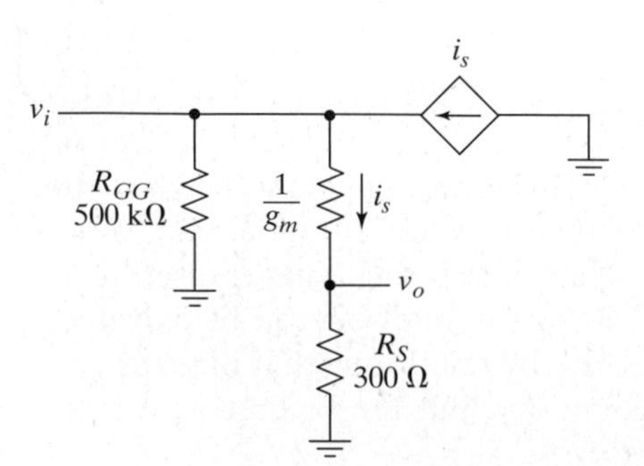

Figure 8–66 The small-signal equivalent circuit for the PMOS source follower.

$$R_o = R_S \| \frac{1}{g_m} = 63\ \Omega,$$

and the open-circuit voltage gain is

$$a_v = \frac{v_o}{v_i} = \frac{R_S}{1/g_m + R_S} = 0.79.$$

A SPICE simulation of this circuit using the Plarge_mos model from Appendix B confirms the bias point and the small-signal gain. ■

8.4 CURRENT BUFFERS

So far we have covered two of the three possible single-transistor amplifier stages. In this section, we cover the third possibility: the ***current buffer***. This circuit is a complement to the voltage buffer covered in Section 8.3. The current buffer has a voltage gain greater than one, a current gain less than or equal to one, a low input resistance, and a large output resistance. In discrete circuit design, this stage is most frequently used to extend the bandwidth of an amplifier; we defer discussion of that application to Chapter 9. For IC design, this stage is frequently used to increase the output impedance of another transistor. This application will be presented in Section 8.5.2.

8.4.1 A Generic Implementation: The Common-Control Amplifier

Consider the circuit shown in Figure 8-67. The generic transistor is being driven by a Norton source this time, since we wish to focus on the operation of the circuit as a current buffer. However, the circuit does have voltage gain. This circuit is called a ***common-control amplifier***, since the control terminal is common to both the input and the output (i.e., the control terminal is tied to ground while the signal is input between the merge terminal and ground, and the output is taken from the output terminal relative to ground).

Now consider the small-signal equivalent circuit shown in Figure 8-68, where we have used the T model for the transistor. If the input impedance of the transistor is smaller than R_S, most of i_s will flow through the transistor rather than R_S. We find that

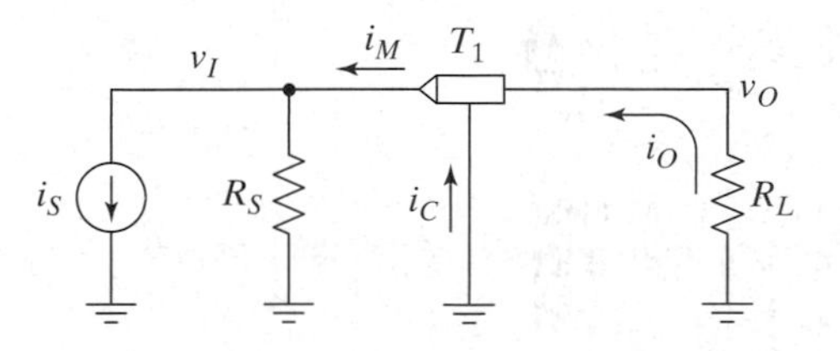

Figure 8–67 A current buffer.

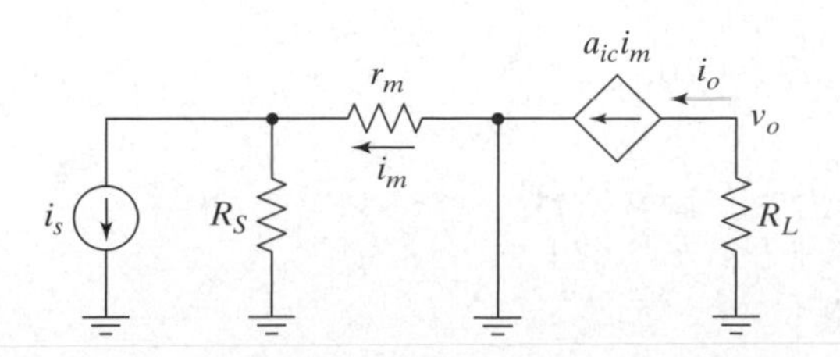

Figure 8–68 The small-signal equivalent circuit for the buffer in Figure 8-67.

$$i_m = \frac{R_S}{r_m + R_S} i_s = \alpha_i i_s \quad \textbf{(8.164)}$$

and note that since r_m is typically small, we get $\alpha_i \approx 1$, as desired (as long as R_S is not too small). We next note that $i_o = a_{ic} i_m$ and that a_{ic} is typically close to one, so that we often have $i_o \approx i_s$.

To see why the transistor is useful in this connection, consider what happens if the signal source in Figure 8-67 is connected directly to the load without the transistor present. In that case, the current in the load would be

$$i_O = \frac{R_S}{R_S + R_L} i_S, \quad \textbf{(8.165)}$$

which could be small if R_L is large. Therefore, if we have a signal source with a large output resistance so that it looks like a current source, and we want to get as much of the signal as possible into our load, it helps to include the transistor as a buffer, as long as the transistor's input impedance is significantly smaller than R_L.

8.4.2 A Bipolar Implementation: The Common-Base Amplifier

A bipolar implementation of a current buffer is shown in Figure 8-69. This circuit is called a ***common-base amplifier***, since the base of the transistor is common to both the input and the output (i.e., the base is tied to ground while the signal is input between the emitter and ground, and the output is taken from the collector relative to ground). To simplify the biasing circuitry, we have assumed that dual supplies are available. As always in this chapter, we assume that the coupling capacitor is large enough that it has negligible impedance at the frequencies of interest.

We find the bias point for this circuit by first noting that the base is at ground, so the emitter voltage is about –0.7 V, and the current is

$$I_C \approx I_E = \frac{V_E - V_{EE}}{R_E} = 1 \text{ mA}, \quad \textbf{(8.166)}$$

where we have assumed that β is large. The DC collector voltage is then

$$V_C = V_{CC} - I_C R_C = 3 \text{ V}, \quad \textbf{(8.167)}$$

and we see that the transistor is forward active, as we have implicitly assumed. We now draw the small-signal midband AC equivalent circuit for the amplifier, as shown in Figure 8-70. We have used the T model here, since it is more convenient. The small-signal emitter resistance in this example is

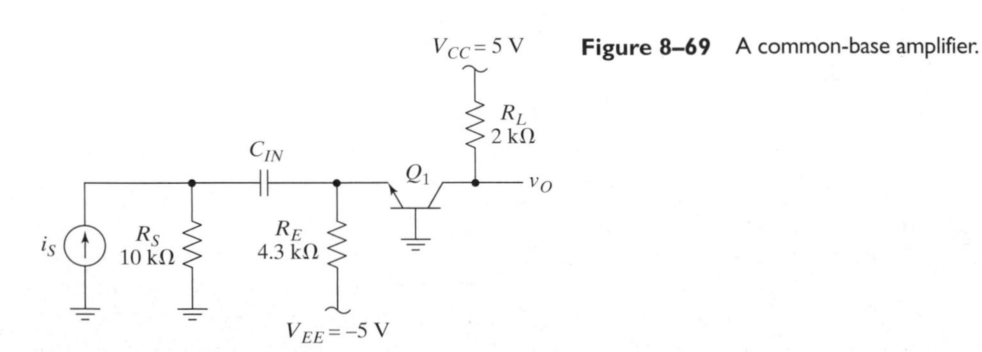

Figure 8–69 A common-base amplifier.

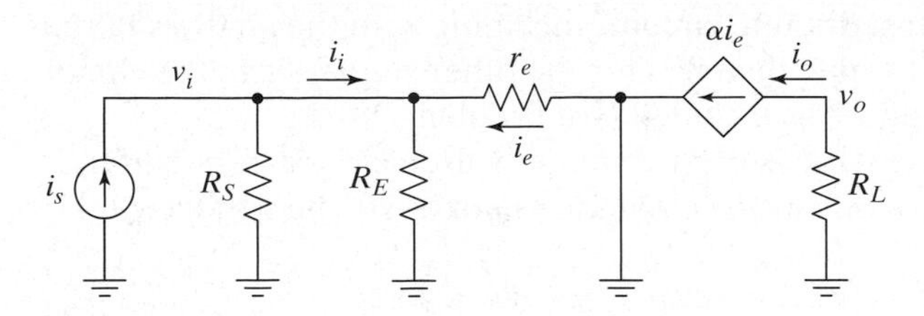

Figure 8–70 The small-signal midband AC equivalent circuit for the amplifier in Figure 8-69.

$$r_e = \frac{V_T}{I_E} = 26\ \Omega, \tag{8.168}$$

assuming room temperature.

The output current for this circuit is

$$i_o = \alpha i_e. \tag{8.169}$$

The emitter current, given by a divider, is

$$i_e = \frac{-R_E}{R_E + r_e} i_i, \tag{8.170}$$

and we can combine these equations to find the current gain:

$$A_i = \frac{i_o}{i_i} = \frac{-\alpha R_E}{R_E + r_e} = a_i. \tag{8.171}$$

In (8.171), we have explicitly shown that the short-circuit current gain, a_i, is the same in this case, since the output attenuation factor is one. Assuming that $\beta = 100$ (so that $\alpha = 0.99$) and using the given values, we find that $A_i = a_i = -0.98$, which is very close to one. The overall current gain of the circuit is

$$\frac{i_o}{i_s} = \alpha_i a_i, \tag{8.172}$$

where the input attenuation factor is

$$\alpha_i = \frac{i_i}{i_s} = \frac{R_S}{R_S + R_i}, \tag{8.173}$$

and the small-signal input resistance is

$$R_i = R_E \| r_e. \tag{8.174}$$

From this example, we find that $R_i = 25.8\ \Omega$, which is very nearly r_e. We, also find that $\alpha_i = 0.997$ and the overall current gain is -0.98. The small-signal output resistance is found by setting the source in Figure 8-70 to zero and looking back into the circuit from the load. With $i_s = 0$, we see that $i_e = 0$, too, so the output resistance is infinite for this transistor model of the circuit. The resistance is infinite in this case only because we have ignored the output resistance of the transistor, which is a reasonable thing to do for most discrete circuits. Including r_o in the analysis of the common-base stage is more trouble than it is for the other stages, since the circuit in Figure 8-70 is not unilateral if r_o is included (*see* Problem P8.72).

Although the common-base stage is most commonly used as a current buffer, it does provide voltage gain as well. For the circuit in Figure 8-70, the voltage gain is

$$A_v = \frac{v_o}{v_i} = \frac{-i_o R_L}{i_i R_i} = -a_i \frac{R_L}{R_i}, \quad \textbf{(8.175)}$$

where we have made use of (8.171). Alternatively, we can see from the figure directly that

$$i_e = -\frac{v_i}{r_e}. \quad \textbf{(8.176)}$$

Using (8.176) and KCL at the collector node, we get

$$\frac{v_o}{v_i} = \frac{\alpha R_L}{r_e}, \quad \textbf{(8.177)}$$

which can be shown to be equal to (8.175). For this example, we get $A_v = 76$. We are also sometimes interested in the transresistance of the common-base stage. For this circuit, we find that

$$\frac{v_o}{i_i} = \frac{v_o}{v_i}\frac{v_i}{i_i} = A_v R_i = \frac{\alpha R_E R_L}{r_e + R_E}. \quad \textbf{(8.178)}$$

Exercise 8.8

Redraw the circuit of Figure 8-70 using the unilateral two-port model for the current amplifier (i.e., the buffer). Derive an equation for the overall current gain and compare your result with (8.172).

8.4.3 A FET Implementation: The Common-gate Amplifier

A FET implementation of a current buffer is shown in Figure 8-71. This circuit is called a ***common-gate amplifier***, since the gate of the transistor is common to both the input and the output (i.e., the gate is tied to ground while the signal is input between the source and ground and the output is taken from the drain relative to ground). To simplify the biasing circuitry, we have assumed that dual supplies are available. As always in this chapter, we assume that the coupling capacitor is large enough that it has negligible impedance at the frequencies of interest.

Considering the DC bias, we write KVL around the gate-source loop to obtain

$$V_{SS} + I_D R_S + V_{GS} = 0. \quad \textbf{(8.179)}$$

We next plug in the square-law equation for I_D in saturation, solve the resulting quadratic for V_{GS}, and throw out the impossible solution. The result for this example (with $K = 500\ \mu A/V^2$ and $V_{th} = 1$ V) is that $V_{GS} = 3.86$ V and $I_D = 4.1$ mA. The drain voltage is

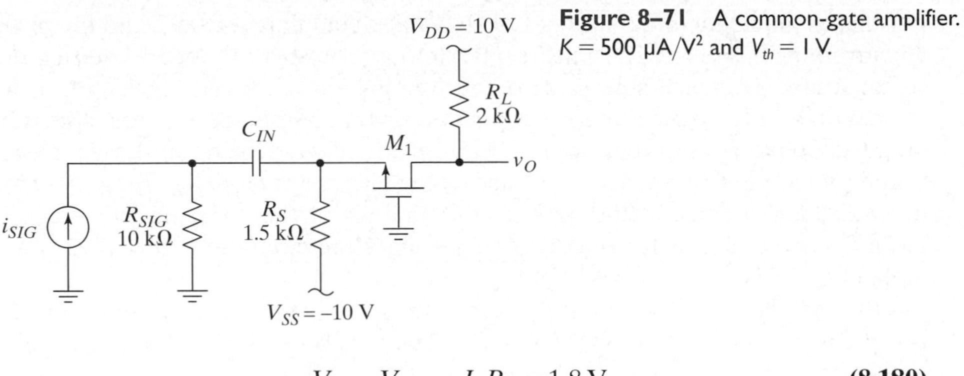

Figure 8–71 A common-gate amplifier. $K = 500\ \mu A/V^2$ and $V_{th} = 1$ V.

$$V_D = V_{DD} - I_D R_D = 1.8\text{ V}, \tag{8.180}$$

and the transistor is forward active, as assumed. Given the DC bias point, we find that $g_m = 2.86$ mA/V and draw the small-signal midband AC equivalent circuit shown in Figure 8-72. We have used the T model in this equivalent circuit to simplify the analysis that follows.

The output current for this circuit is given by

$$i_o = i_s = \frac{-R_S}{R_S + \dfrac{1}{g_m}} i_i = \frac{-g_m R_S}{1 + g_m R_S} i_i, \tag{8.181}$$

and the current gain is then

$$A_i = \frac{i_o}{i_i} = \frac{-g_m R_S}{1 + g_m R_S} = a_i. \tag{8.182}$$

In (8.182) we have indicated that the short-circuit current gain, a_i, is the same as A_i, since the output attenuation factor is one. Using the given values, we find that $A_i = a_i = 0.81$. The overall current gain of the circuit is

$$\frac{i_o}{i_{sig}} = \alpha_i a_i, \tag{8.183}$$

where the input attenuation factor is

$$\alpha_i = \frac{i_i}{i_{sig}} = \frac{R_{SIG}}{R_{SIG} + R_i} \tag{8.184}$$

and the small-signal input resistance is

$$R_i = R_S \| \frac{1}{g_m}. \tag{8.185}$$

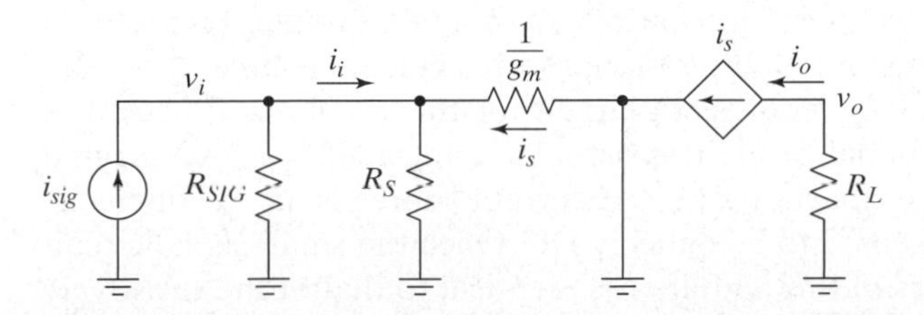

Figure 8–72 The small-signal midband AC equivalent circuit for the amplifier in Figure 8-71.

For this example, we find that $R_i = 284\ \Omega$. We also find that $\alpha_i = 0.97$, and the overall current gain is -0.79. The small-signal output resistance is found by setting the signal source in Figure 8-72 to zero and looking back into the circuit from the load. With $i_{sig} = 0$, we see that $i_s = 0$ too, so the output resistance appears to be infinite. The output resistance is not really infinite; it only is here because we have ignored the output resistance of the transistor, which is a reasonable thing to do for most discrete circuits. Including r_o in the analysis of the common-gate stage is more trouble than it is for the other stages, since the circuit in Figure 8-72 is not unilateral if r_o is included (*see* Problem P8.78).

Although the common-gate stage is most commonly used as a current buffer, it does provide voltage gain as well. For the circuit in Figure 8-72, the voltage gain is

$$A_v = \frac{v_o}{v_i} = \frac{-i_o R_L}{i_i R_i} = a_i \frac{R_L}{R_i}, \tag{8.186}$$

where we have made use of (8.182). Alternatively, we can see from the figure directly that

$$i_s = -g_m v_i. \tag{8.187}$$

Using (8.187) and KCL at the drain node, we get

$$\frac{v_o}{v_i} = g_m R_L, \tag{8.188}$$

which can be shown to be equal to (8.186). For this example, we get $A_v = 5.7$. We are also sometimes interested in the transresistance of the common-gate stage. For this circuit, we find that

$$\frac{v_o}{i_i} = \frac{v_o}{v_i}\frac{v_i}{i_i} = A_v R_i = \frac{g_m^2 R_S R_L}{1 + g_m R_S}. \tag{8.189}$$

Exercise 8.9

Redraw the circuit of Figure 8-72 using the unilateral two-port model for the current amplifier (i.e., the buffer). Derive an equation for the overall current gain and compare your result with (8.183).

□

8.5 INTEGRATED AMPLIFIERS

In Section 7.4, we discussed the reasons that integrated circuit biasing is dramatically different than discrete circuit biasing. One difference is that current mirrors are used frequently when biasing ICs. (We cover the small-signal midband AC performance of current mirrors in Section 8.5.2). As an example, Figure 8-73 shows a generic single-transistor amplifier biased with a current source.

The small-signal midband AC equivalent circuit for this amplifier is shown in Figure 8-74. The merge terminal of the transistor is connected to an AC ground because of the large bypass capacitor. The DC current source is an AC open circuit, since the current is assumed to be entirely DC. Once the small-signal circuit has been drawn, the analysis of this amplifier is identical to that of the resistively biased amplifiers we considered earlier in this chapter. Consequently, our previous results apply.

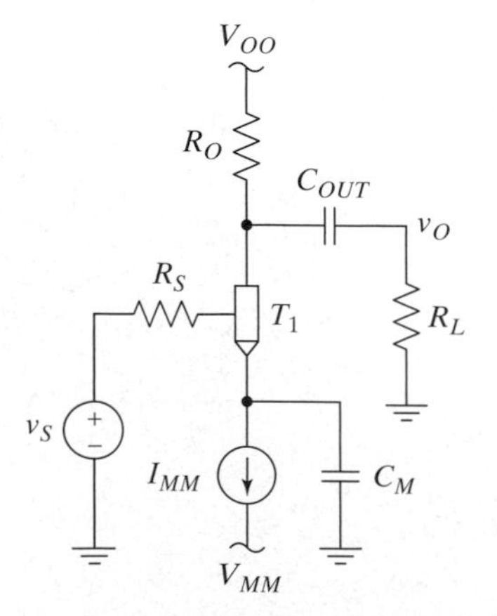

Figure 8–73 A generic single-transistor amplifier with current source biasing.

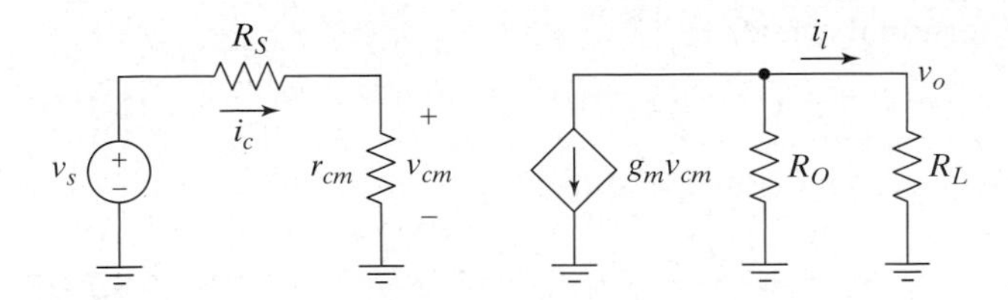

Figure 8–74 The small-signal midband AC equivalent circuit for the amplifier in Figure 8-73.

Although an amplifier on an IC could theoretically be made in the manner shown in Figure 8-73, this is not done in practice. The large bypass capacitor would have to be off chip, which is a serious disadvantage, and there simply isn't any compelling reason to make an amplifier this way. Instead, IC design relies heavily on differential amplifiers, which are discussed in Section 8.6.

Finally, one issue that is absent in discrete amplifiers, but very much a factor in IC design, is the body effect in MOSFETs. This effect is dealt with in Section 8.5.1, which may be skipped without loss of continuity.

8.5.1 The Body Effect in FET Amplifiers

Our previous analyses of MOSFET amplifiers have not included the bulk transconductance caused by the body effect. This simplification is justified in discrete circuit design because the body and source are tied together. However, MOS ICs are fabricated using either p-well, n-well, or twin-well processes (*see* Chapter 3). If the process has only p-type wells, all p-channel transistors will share the n-type substrate as their body. To guarantee that all source-to-bulk and drain-to-bulk junctions are reverse biased, the substrate will be tied to the most positive supply available; call it V_{DD}. The result is that p-channel transistors will exhibit the body effect unless their sources are tied to V_{DD}. Similarly, in an n-well process, all of the n-channel transistors will exhibit the body effect unless their sources are tied to V_{SS}, the most negative supply. Twin-well processes have the advantage that every transistor can have its source and body connected together. In the following sections, we consider the body effect for each of the single-transistor amplifier configurations.

The Body Effect in Common-Source Amplifiers When single-ended common-source amplifiers are made on an IC, they frequently have the source terminal of the transistor tied to an AC ground, so v_{BS} is zero and the bulk transconductance has no effect on the small-signal performance. If there is source degeneration, however, the gain is reduced because of the body effect. Consider, for example, the amplifier shown in Figure 8-75(a). The small-signal midband AC equivalent circuit is shown in part (b) of the figure.

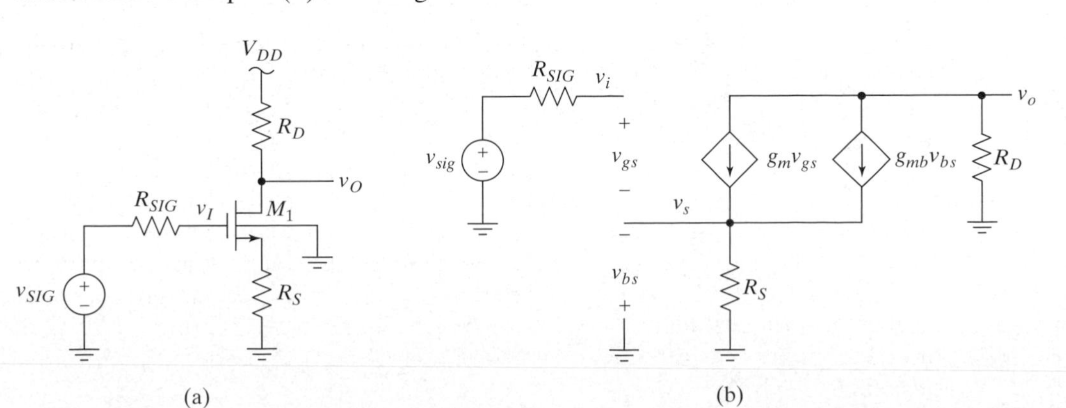

Figure 8–75 (a) A common-source amplifier with source degeneration and (b) the small-signal midband AC equivalent circuit.

Using KCL at the source terminal, we write

$$v_s = (g_m v_{gs} + g_{mb} v_{bs}) R_S. \quad \textbf{(8.190)}$$

We then recognize that $v_{bs} = -v_s$ and solve to obtain

$$v_s = \frac{g_m R_S v_{gs}}{1 + g_{mb} R_S}. \quad \textbf{(8.191)}$$

Since $i_g = 0$, we also know that $v_{gs} = v_{sig} - v_s$, and we can use (8.191) to find

$$v_{gs} = \frac{1 + g_{mb} R_S}{1 + (g_m + g_{mb}) R_S} v_{sig}. \quad \textbf{(8.192)}$$

Finally, we write KCL at the output to get

$$v_o = -(g_m v_{gs} + g_{mb} v_{bs}) R_D \quad \textbf{(8.193)}$$

and combine this with (8.191) and (8.192) to get

$$\frac{v_o}{v_{sig}} = \frac{-g_m R_D (1 + 2g_{mb} R_S)}{1 + (g_m + g_{mb}) R_S}. \quad \textbf{(8.194)}$$

Notice that if $g_{mb} = 0$, this equation reduces to (8.129), as it should. As an example, if $g_{mb} = 0.1g_m$ and $g_m R_S = 1$, the gain with the bulk transconductance included is 12.5% lower than without it.

The Body Effect in Source Followers Consider the source follower shown in Figure 8-76(a). The small-signal midband AC equivalent with bulk transconductance included is shown in part (b) of the figure. We have neglected the output resistance for simplicity, but including it in the analysis is not difficult, since it shows up in parallel with R_L. The bulk node is assumed to be connected to ground, since this is the most negative potential in the circuit. Neither the source nor the gate is connected to ground however, so v_{bs} is not equal to zero or v_{gs}.

Writing KCL at the output node leads to

$$v_o = (g_m v_{gs} + g_{mb} v_{bs}) R_L. \quad \textbf{(8.195)}$$

We substitute $v_{gs} = v_i - v_o$ and $v_{bs} = -v_o$ into (8.195) and solve to get

$$A_v = \frac{v_o}{v_i} = \frac{g_m R_L}{1 + (g_m + g_{mb}) R_L}. \quad \textbf{(8.196)}$$

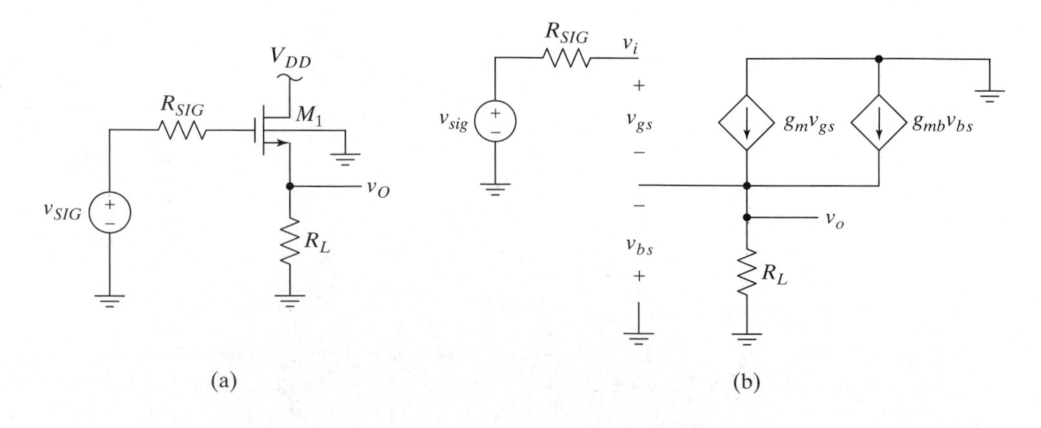

Figure 8–76 (a) A source follower. (b) The small-signal midband AC equivalent circuit of the source follower with the body effect included.

Notice that the maximum gain of the source follower occurs when R_L is very large and is given by

$$A_{v\max} = \frac{g_m}{g_m + g_{mb}} = \frac{1}{1 + \chi}. \quad \textbf{(8.197)}$$

Remembering that χ is typically in the range 0.1 to 0.3, we see that the maximum gain of a source follower with the body effect accounted for is about 0.8 to 0.9. This problem is sometimes a serious one and can be avoided only by using a transistor in a separate well so that its body and source can be tied together.

The body effect also changes the output resistance of the source follower, but in this case, the change is for the better. If we set $v_{sig} = 0$ in Figure 8-76 and look in from the load to find R_o, we see that $v_g = 0$, so $v_{gs} = v_{bs}$. The controlled sources can then be combined in parallel with $g_{meff} = g_m + g_{mb}$. The resulting source is controlled by the voltage across itself and can be replaced by a resistor of value $1/g_{meff}$ by using the source absorption theorem. The resulting output resistance is

$$R_o = R_S \| \frac{1}{g_{meff}} \quad \textbf{(8.198)}$$

and is smaller than the result without the body effect.

The Body Effect in Common-Gate Amplifiers The AC equivalent of a common-gate amplifier is shown in Figure 8-77(a). We assume that the bias current is supplied by the preceding stage, which is represented by the Thévenin signal source (i.e., $I_S = I_{SIG}$). The small-signal midband AC equivalent circuit with the body effect included is shown in part (b) of the figure. We have used the hybrid-π model for the transistor and have ignored r_o for simplicity. The body would be connected to V_{SS} in this circuit, which is an AC ground. Therefore, $v_{gs} = v_{bs}$, and the two controlled sources appear in parallel as shown. Because the generators are in parallel and are controlled by the same voltage, they can be combined into a single controlled source with a transconductance

$$g_{meff} = g_m + g_{mb}. \quad \textbf{(8.199)}$$

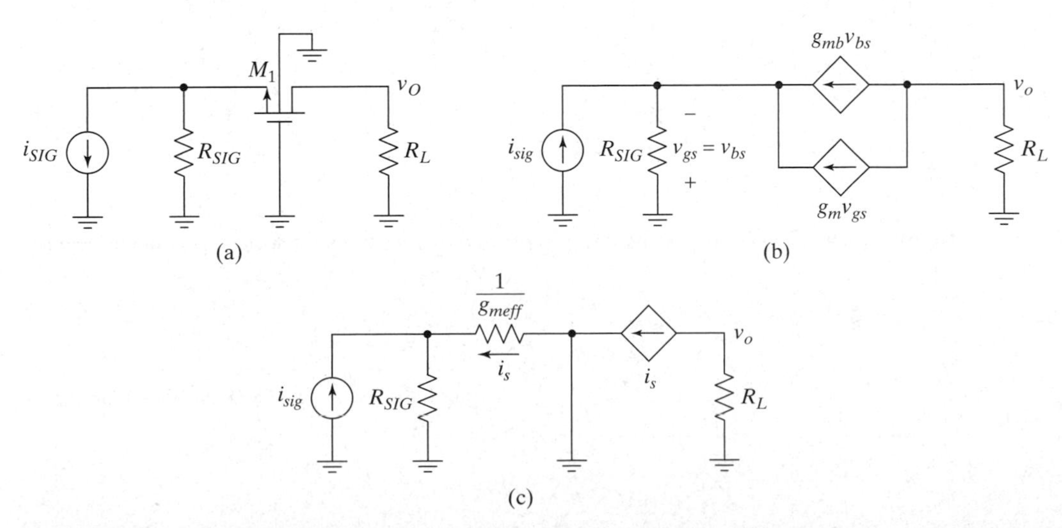

Figure 8–77 (a) A common-gate amplifier. (b) The small-signal midband AC equivalent circuit for the amplifier with the bulk transconductance included. (c) The same circuit redrawn using the T model and the effective transconductance.

Using this effective transconductance, we redraw the small-signal equivalent circuit using the T model for the transistor as shown in part (c) of the figure. The net effect of the bulk transconductance in this case is to increase the effective transconductance of the device, as indicated in (8.199).

8.5.2 Current Mirrors

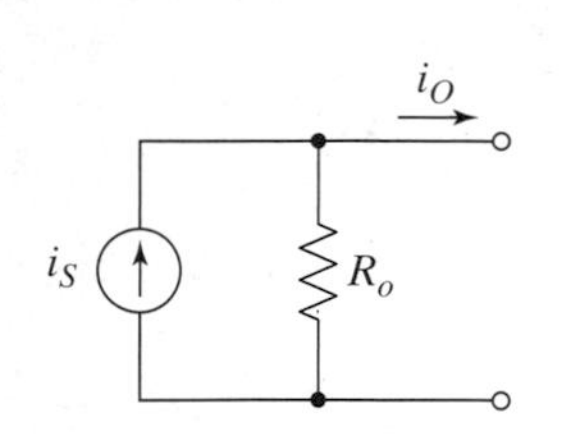

Figure 8–78 The Norton model of a current source.

We now turn our attention to the small-signal midband performance of current mirrors. We will sometimes refer to current mirrors as current sources, since they are frequently used as such. Current sources are typically modeled using a Norton equivalent as shown in Figure 8-78. An ideal current source has an infinite output resistance.

We will adopt the Norton model for the outputs of our current mirrors; therefore, we are interested in finding the small-signal midband AC output resistance of each mirror, in addition to the short-circuit output current. Consider, for example, the generic transistor mirror shown in Figure 8-79(a) and its small-signal equivalent circuit shown in part (b) of the figure. We have not shown the signal source, but it is understood to be a current source. We have also not shown the load, but it is assumed to hold the output voltage of T_2 so that the transistor is forward active.

The first thing we notice is that the controlled source for T_1 is controlled by the voltage across itself and can be replaced by an equivalent resistance of $1/g_{m1}$. The net result is that the diode-connected transistor can be represented by a single resistor whose value is

$$r_{eq1} = r_{o1} \| r_{cm1} \| \frac{1}{g_{m1}} \approx \frac{1}{g_{m1}}. \tag{8.200}$$

Note that the small-signal midband AC model of a diode-connected transistor is just a resistor with a value given by (8.200). Using the simplified model for T_1, we can redraw the equivalent circuit as shown in Figure 8-80.

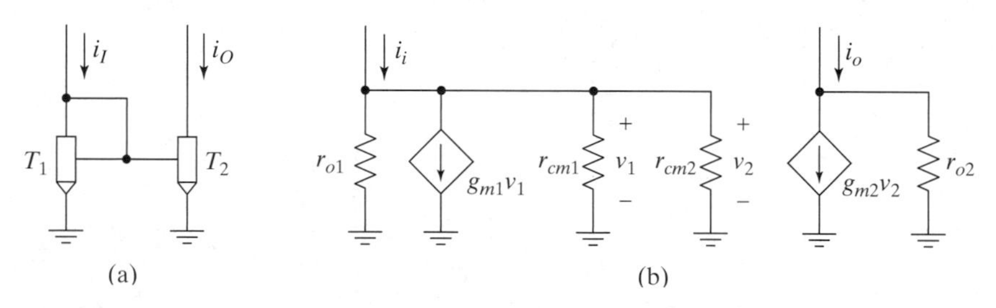

Figure 8–79 (a) A generic transistor current mirror and (b) its small-signal equivalent circuit.

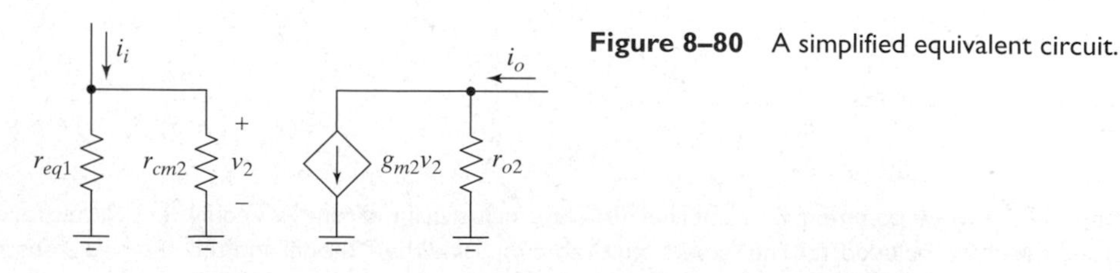

Figure 8–80 A simplified equivalent circuit.

We can now characterize the small-signal performance of this current mirror. Examining the figure reveals that the input resistance is given by

$$R_i = r_{eq1} \| r_{cm2}. \quad \textbf{(8.201)}$$

The short-circuit output current is given by

$$i_{o\text{ s.c.}} = g_{m2}v_2 = g_{m2}(r_{eq1} \| r_{cm2})i_i \quad \textbf{(8.202)}$$

(note that with the output shorted to ground, all of $g_{m2}v_2$ will flow to the output rather than through r_{o2}), so the short-circuit current gain is

$$a_i = \frac{i_{o\text{ s.c.}}}{i_i} = g_{m2}(r_{eq1} \| r_{cm2}). \quad \textbf{(8.203)}$$

We have kept the subscripts in the equations up to this point to allow for the possibility of the transistors being scaled so that $i_O \neq i_I$. Nevertheless, the transistors must be matched for the mirror to function (i.e., they must be identical except for some area scaling). If we consider the case where the transistors are identical, the subscripts can be dropped, assuming that $v_O \approx v_I$ so that the total currents are approximately equal. We also note that $1/g_m \ll r_{cm}$, so with identical transistors, (8.203) can be approximated by

$$a_i = g_m(r_{eq} \| r_{cm}) = g_m\left(r_{cm} \left\| \frac{1}{g_m} \right\| r_{cm}\right) \approx g_m\left(\frac{1}{g_m}\right) = 1, \quad \textbf{(8.204)}$$

as expected. The small-signal current gain of a mirror is not usually of interest. Consequently, we won't normally bother to find it. Nevertheless, it is nice to see that the analysis yields the expected result.

The output resistance of the current mirror is perhaps its most important characteristic and is found by setting the input to zero (i.e., setting $i_i = 0$) and finding the resistance seen looking in from the output terminal. With the input set to zero, we have $v_2 = 0$, so the controlled current source is also zero, and

$$R_o = r_{o2} \quad \textbf{(8.205)}$$

by inspection. The output resistance of a transistor is a function of the bias current and decreases as the bias current increases. Consequently, it is always true that we can increase the output resistance by decreasing the current. Therefore, the only fair way to compare the output resistances of different mirrors is to make the comparison at equal bias currents.

There are many occasions when the output resistance of a simple mirror like the one in Figure 8-79(a) is inadequate. One way of improving the output resistance is to use a **cascode** mirror. The term "cascode" is a holdover from the days of vacuum tubes and refers to a cascade through the cathode (equivalent to the merge terminal of our generic transistor). Therefore, a cascode mirror is two mirrors cascaded together, as shown in Figure 8-81.

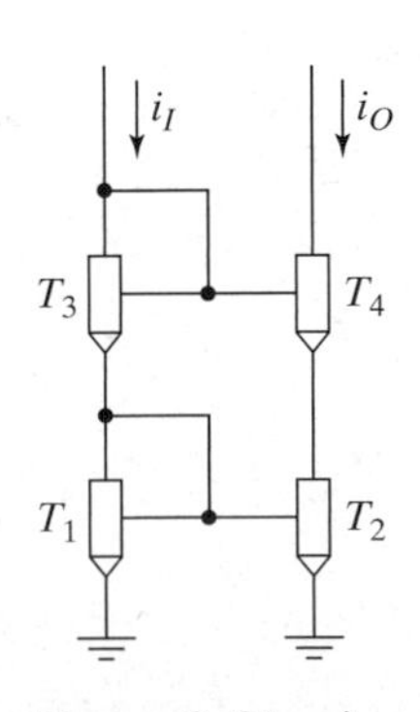

Figure 8-81 A cascode current mirror.

We could certainly analyze this circuit by replacing all of the transistors with appropriate models and then writing all of the necessary equations. The resulting equivalent circuit is complex though, so the brute-force analysis of the circuit would not lend much insight. A much better approach is to recognize that if the control currents of the transistors can be neglected, then $i_{O1} = i_{O3} = i_I$ and $i_{O2} = i_{O4} = i_O$. Assuming that all of the transistors are identical, we know that $i_{O1} \approx i_{O2}$ (the approximation is to ignore the effect of the small difference between v_{O1} and v_{O2} on these currents). Therefore, we can conclude that $i_O \approx i_I$. The areas of the transistors in this mirror can be scaled just the like the areas of the devices in the simple mirror, so at this point, there is no obvious difference between the two circuits.

The input resistance of the cascode mirror is approximately

$$R_i \approx 2r_{eq} \approx \frac{2}{g_m}, \tag{8.206}$$

which is twice the value for the simple mirror, but is still small. The major difference between the cascode and simple current mirrors is the output resistance. A reasonable approximation to the output resistance of the cascode mirror can be found in the following way: First, since R_i is small, and i_i is set equal to zero to find R_o, we can say that the control terminals of T_2 and T_4 are approximately AC grounds. Then, since $v_{cm2} = 0$, its controlled current source is also zero, and we can model it with just its output resistance, r_{o2}. The resulting approximate equivalent circuit is shown in Figure 8-82, where we have retained some of the subscripts to avoid confusion in the analysis (but $r_{o2} = r_{o4}$) and we have driven the output with a test current, i_x, to find R_o.

The output resistance of this circuit is $R_o = v_x/i_x$, so we need to find v_x. We chose to drive the circuit with a test current instead of a test voltage, because it makes the analysis that follows simpler. First note that the current i_a must be equal to i_x. If this isn't clear, write KCL at the top and bottom of r_{o4} to convince yourself. Therefore, we can write KCL at the top of r_{o2} and find

$$v_{cm4} = -i_a(r_{cm4} \| r_{o2}) = -i_x(r_{cm4} \| r_{o2}). \tag{8.207}$$

If we can find the voltage drop across r_{o4}, we can find v_x and be done. Writing KCL at the output node, we find that

$$i_{r_{o4}} = i_x - g_{m4}v_{cm4}, \tag{8.208}$$

and substituting (8.207) into this equation, we find that

$$i_{r_{o4}} = i_x\left[1 + g_{m4}(r_{cm4} \| r_{o2})\right]. \tag{8.209}$$

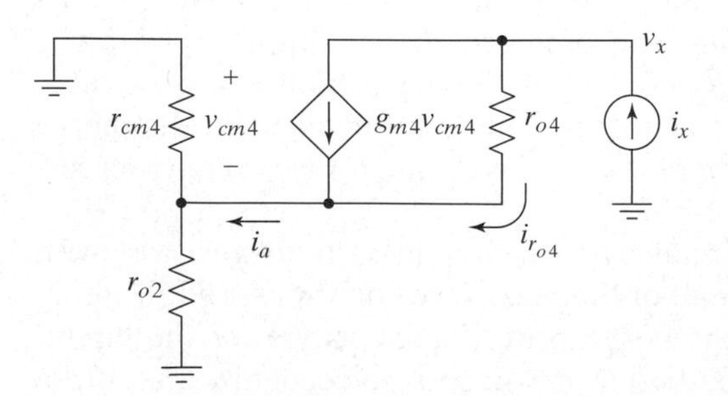

Figure 8–82 An approximate equivalent circuit for finding R_o for the cascode current mirror.

We now use KVL to write

$$v_x = -v_{cm4} + i_{r_{o4}} r_{o4}. \tag{8.210}$$

Finally, substituting (8.207) and (8.209) into (8.210) yields

$$R_o = r_{cm4} \| r_{o2} + \left[1 + g_{m4}(r_{cm4} \| r_{o2}) \right] r_{o4}. \tag{8.211}$$

The output resistance of the cascode connection is therefore much higher than that of the simple mirror, as stated at the outset. The result given in (8.211) is only an approximation, since we assumed that the control terminals of both transistors were AC grounds. A more accurate analysis shows that the output resistance is less than predicted by (8.211).

Another important parameter for a current source is its ***voltage compliance***, or just compliance, which is the range of output voltages over which the current source will function properly. We also refer to the ***headroom*** required by a mirror, which is the minimum supply voltage for which it will function. For the simple source in Figure 8-79, we can see that the source will stop functioning properly if the output voltage is reduced to the point where T_2 leaves the forward-active region of operation. The voltage at which that occurs varies a great deal, depending on the type of transistor. We will examine compliance separately in the sections on bipolar and FET mirrors.

There is often a trade-off between output resistance and compliance, which is why we treat compliance here instead of as part of the DC biasing in Chapter 7. The tradeoff results from the fact that raising the output resistance usually requires adding elements in series with the output leg of the mirror, and each of these elements requires some small voltage drop across it to function.

There are many other variations on the current mirror that could be considered (*see* [8.1] for some of them). In the subsections that follow, we examine bipolar and MOS mirrors separately. Prior to doing so, however, we want to show how a mirror can be used as a load for an amplifier. Because a mirror uses active devices, we call a load made using a mirror an ***active load***.

Consider the generic transistor amplifier shown in Figure 8-83. Transistor T_1 is being used as a single-transistor amplifier with both voltage and current gain, whereas p-type transistors T_2 and T_3 form a current mirror that is used to bias the circuit and simultaneously act as a high-resistance AC load. The resistor sets the current in the p-type mirror, and if T_2 and T_3 are identical, the bias current in T_1 is approximately the same. The DC component of v_I must be the correct value to support the bias current I_{BIAS}.

To draw the small-signal midband AC equivalent circuit for this amplifier, we note that the voltage at the top of R_1 is a DC voltage. Therefore, the AC component of the control current of T_3 is zero. Consequently, the net small-signal model of the current mirror is just the output resistance of T_3 connected to AC ground.

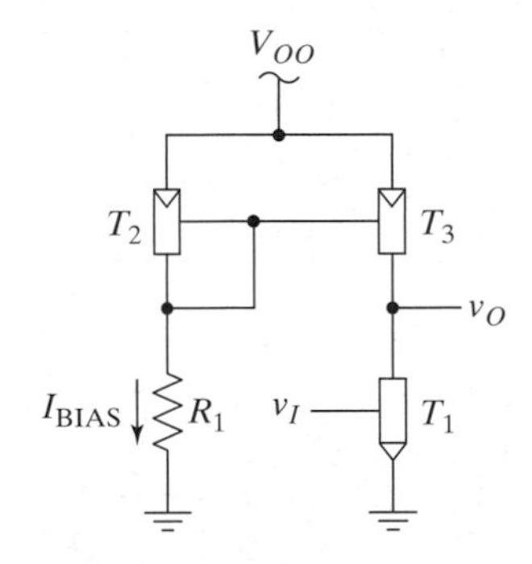

Figure 8–83 An actively loaded generic transistor amplifier.

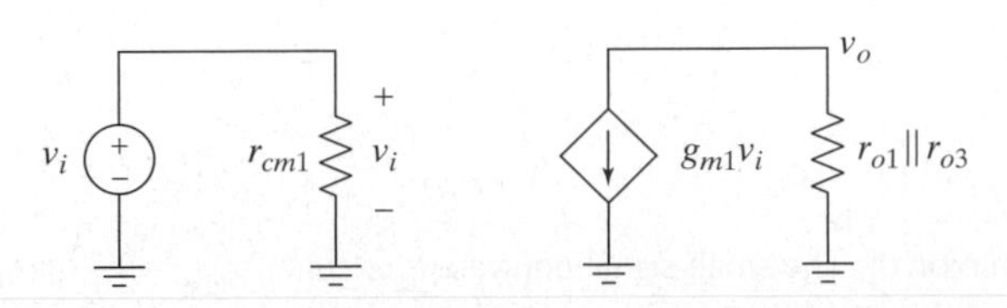

Figure 8–84 The small-signal equivalent circuit for the amplifier in Figure 8-83.

This output resistance then shows up in parallel with the output resistance of T_1. The resulting equivalent circuit is shown in Figure 8-84.

The gain of this circuit is found by inspection to be

$$A_v = \frac{v_o}{v_i} = -g_{m1}(r_{o1} \| r_{o3}). \tag{8.212}$$

Since the output resistances in (8.212) can be quite large, the resulting voltage gain can also be very large. The main advantage of the circuit for use in an IC is that the large gain is obtained without requiring a large resistor. A large resistor has two main disadvantages: First, it takes a large amount of area and is, therefore, expensive, second, there would be a large DC voltage drop across the resistor, which would necessitate using large supply voltages.

Bipolar Current Mirrors In this section we examine the characteristics of some bipolar mirrors. First consider the Widlar mirror first presented in Chapter 7; the circuit is shown in Figure 8-85(a). The small-signal midband AC equivalent is shown in part (b) of the figure, and we have replaced the diode-connected transistor by its equivalent resistance.

In Chapter 7 we showed that the DC output current of this mirror is given by the transcendental equation

$$I_O = \frac{V_T}{R_E} \ln \frac{I_{C1}}{I_O}. \tag{8.213}$$

For the moment, let's assume that the base currents can be safely neglected at DC (so that $I_{C1} \approx I_I$), and that the circuit has been designed to yield $I_O = I_I/N$ (N must be >1). The output resistance of the mirror is found using the same method used for the circuit of Figure 8-82. The result is

$$R_o = R_{eq} + (1 + g_{m2}R_{eq})r_{o2}, \tag{8.214}$$

where

$$R_{eq} = \left(\frac{1}{g_{m1}} + r_{\pi 2}\right) \Big\| R_E. \tag{8.215}$$

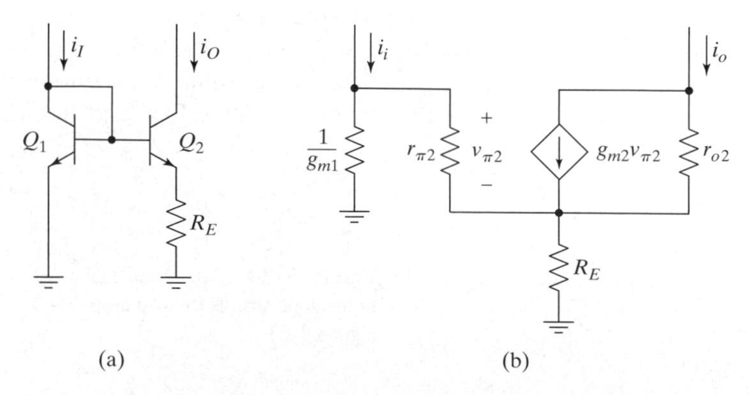

Figure 8–85 (a) A Widlar current mirror. (b) The small-signal equivalent circuit.

We can simplify this result by realizing that

$$r_{\pi 2} = N r_{\pi 1} = \frac{N\beta}{g_{m1}}, \tag{8.216}$$

so that $1/g_{m1}$ can be safely ignored in (8.215). Now we note that

$$r_{\pi 2} = \frac{\beta}{g_{m2}} = \frac{\beta V_T}{I_{C2}}, \tag{8.217}$$

and, if we assume that $I_{C2} \approx I_{E2}$,

$$R_E = \frac{V_{E2}}{I_{C2}}. \tag{8.218}$$

Comparing (8.217) and (8.218), we see that $r_{\pi 2} \gg R_E$ for typical values of V_{E2} and β. (A 60 mV drop across R_E will yield $N \approx 10$ at room temperature, whereas βV_T is probably several volts.) Therefore, using these observations we note that $R_{eq} \approx R_E$ and simplify (8.214) to

$$R_o \approx g_{m2} R_E r_{o2}, \tag{8.219}$$

which can be significantly larger than r_{o2}.

It is interesting to see what the largest possible output resistance is for this bipolar mirror. If we let R_E approach infinity, $R_{eq} \approx r_{\pi 2}$, and we have

$$R_o \approx g_{m2} r_{\pi 2} r_{o2} = \beta r_{o2}, \tag{8.220}$$

which is very large indeed. In fact, this value is large enough that we should go back and see if we need a more complicated model for the transistor. The only model element that we have ignored that could potentially affect our result is r_μ. Remember that r_μ connects from the collector to the base and is typically greater than $10\beta r_o$ in modern transistors (*see* Section 8.1.5). If we include $r_{\mu 2}$ in Figure 8-85(b) and ignore the small resistor $1/g_{m1}$ in series with it, we see that the output resistance of the current mirror is $r_{\mu 2}$ in parallel with what we derived in (8.214). Therefore, the limiting result we obtained in (8.220) is accurate.

The compliance of this mirror is very good; that is, the circuit can operate with very little headroom. Notice from Figure 8-85 that the mirror will function properly as long as v_O is high enough that Q_2 remains forward active. If the output voltage is too low, Q_2 enters saturation, and the mirror ceases to function properly. The voltage at which this occurs is approximately equal to the drop across R_E plus 0.2 V. Therefore, for most mirrors of this type, we only need a few tenths of a volt for the circuit to work. The minimum voltage required at the input is somewhat larger—about 0.7 V—but this is usually less of a problem.

Exercise 8.10

Consider the current mirror shown in Figure 8-86, and assume that all three transistors are identical. Q_3 is a cascode-connected transistor (i.e., it is a common-base connection off of the collector of Q_2) used to increase the output resistance of the mirror. Find the small-signal midband AC output resistance for the mirror. Include $r_{\mu 3}$ in your analysis. Assume that the output is connected to a voltage that keeps Q_3 forward active and that $I_I = 100\ \mu A$, $\beta = 100$, $V_A = 50$ V, and $r_\mu = 10\beta r_o$.

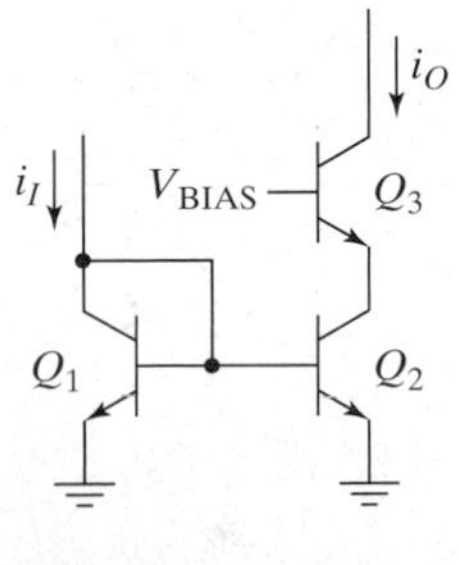

Figure 8–86

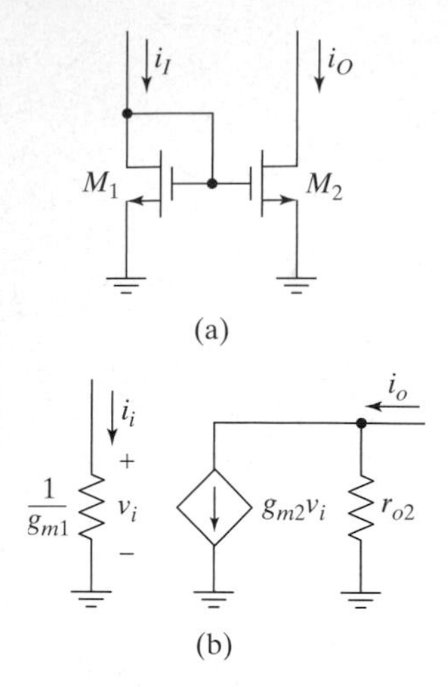

Figure 8–87 (a) A simple MOS current mirror. (b) The small-signal midband AC equivalent circuit.

FET Current Mirrors Consider the MOS current mirror shown in Figure 8-87(a) and the corresponding small-signal midband AC equivalent circuit shown in part (b). We see by inspection that the output resistance of this circuit is equal to r_{o2}. If we desire to increase the output resistance at a fixed bias current, we can accomplish it by increasing the length of the transistor. This will reduce the channel-length modulation effect as noted in Chapter 2 and will therefore decrease λ (or, equivalently, increase V_A). Although this method works to a degree, it quickly requires very large devices, which are expensive. It also does not provide a dramatic increase in output resistance. All other approaches to increasing the output resistance require modifying the circuit topology.

One mirror that provides significantly better output resistance is the ***Wilson mirror*** shown in Figure 8-88(a). A small-signal equivalent circuit for finding the output resistance is given in part (b) of the figure. Transistors M_1 and M_2 make up the basic mirror, and M_3 functions as a common-gate buffer to raise the output resistance.[6] It may at first appear as though the circuit is drawn incorrectly and that M_1 should be diode connected instead of M_2, but this is not the case. Because of the presence of M_3, there is a feedback loop that makes the circuit work properly.

The DC operation of this mirror can be described as follows: Assume that we start with all currents equal to zero and then suddenly apply a DC input current. Initially, the transistors are all off, and the input current will flow into the parasitic capacitance at the input node (i.e., the drain of M_1), which will cause the voltage at that node to ramp upward. The input voltage will continue to rise, and M_2 and M_3 will both turn on, which causes current to flow in the output leg. The output current is then mirrored back to the input node by M_1 and M_2. The input voltage will cease to change only when the drain current of M_1 equals I_I. If all of the transistors are identical, $I_O = I_I$. Other ratios can also be used in the circuit.

Let's now turn our attention to the circuit in part (b) of the figure and calculate the output resistance of this mirror. We have used a single equivalent resistor to model the diode-connected transistor in this circuit. We have chosen to drive the output with a test current instead of a test voltage, because it simplifies the analysis that follows. The output resistance is $R_o = v_x/i_x$, so we need to find v_x. Using KVL and Ohm's law, we write

$$v_x = v_2 + i_{ro3}r_{o3}. \tag{8.221}$$

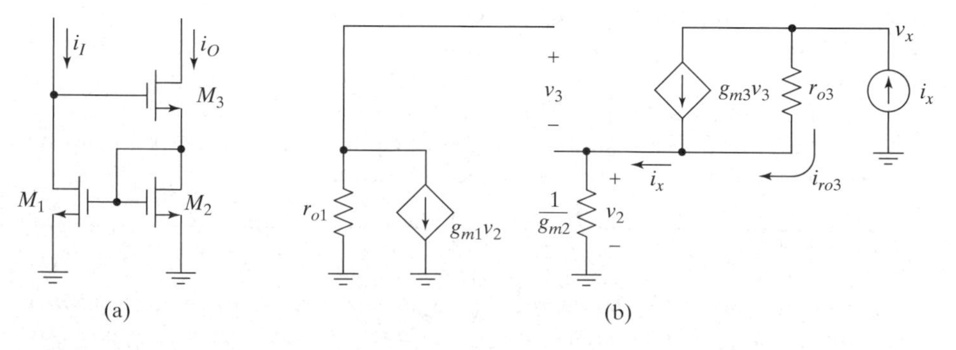

Figure 8–88 A Wilson current mirror.

[6] We do not call this a cascode connection, because M_3 is cascaded with a diode-connected transistor, not a common-source transistor.

The test current i_x also flows into M_2 as shown in the figure, so we can write

$$v_2 = \frac{i_x}{g_{m2}}. \quad \textbf{(8.222)}$$

We again use KVL and Ohm's law to obtain

$$v_3 = -g_{m1}v_2r_{o1} - v_2 = -v_2(1 + g_{m1}r_{o1}). \quad \textbf{(8.223)}$$

KCL at the drain of M_3 leads to

$$i_{ro3} = i_x - g_{m3}v_3, \quad \textbf{(8.224)}$$

and combining these four equations leads to

$$R_o = \frac{v_x}{i_x} = \frac{1}{g_{m2}} + r_{o3}\left[1 + \frac{g_{m3}}{g_{m2}}(1 + g_{m1}r_{o1})\right]. \quad \textbf{(8.225)}$$

When all three devices are identical, this can be simplified to

$$R_o \approx g_m r_o^2, \quad \textbf{(8.226)}$$

which can be extremely large.

The high output resistance of the Wilson mirror is achieved at the expense of a reduced compliance relative to the simple mirror. The compliance is reduced because we now have two devices stacked on the output side. If we continue to assume for simplicity that all the devices are identical, all of the currents and gate-to-source voltages are equal as well. The DC gate-to-source voltage of each MOSFET is then

$$V_{GS} = V_{th} + \sqrt{\frac{I_D}{K}} = V_{th} + V_{ON}. \quad \textbf{(8.227)}$$

The voltage at the source of M_3 is just the V_{GS} of M_2 and is given by (8.227). The voltage at the gate of M_3 is then

$$V_{G3} = 2V_{GS} = 2V_{th} + 2V_{ON}, \quad \textbf{(8.228)}$$

and this transistor will leave saturation when its drain voltage, which is the output, drops to

$$V_{O\min} = V_{G3} - V_{th} = V_{th} + 2V_{ON}. \quad \textbf{(8.229)}$$

When M_3 leaves the saturation region, the mirror stops working properly, so the compliance of the mirror is given by (8.229). A typical value is around 1 to 2 V. The minimum input voltage required is given by (8.228) and is typically around 1.5 to 2.5 V. It is obviously difficult to use mirrors like this in low-voltage circuits, but some very clever alternatives have been developed that retain high output impedance while achieving low minimum working voltages (e.g., the Sooch cascode [8.3]). As a final note, we point out that since $V_{DS1} \neq V_{DS2}$ in this circuit, the output current is not truly equal to the input current for identical devices. This offset can be overcome by adding a diode-connected MOSFET in series with the drain of M_1 [8.1].

8.6 DIFFERENTIAL AMPLIFIERS

The differential amplifier was introduced in Chapter 7, and we examined the DC biasing in detail there. In this section we wish to examine the small-signal mid-band AC performance. We begin with a qualitative description of how the circuit works and then proceed to discuss the small-signal performance quantitatively for both differential and common-mode signals. The large-signal performance of differential amplifiers is considered in detail in Chapter 12, although we will briefly examine the limits here. We assume you are familiar with differential-mode and common-mode signals and notation; if this is not true, you should review Aside A5.1 before proceeding.

We start with a complete analysis of a generic differential pair and then move on to discuss bipolar and MOS implementations. Each of those discussions will be brief and will draw heavily on the results we derive for the generic case.

8.6.1 The Generic Differential Pair

We first present a qualitative description of the operation, followed by a detailed analysis of the small-signal operation. We then present a very useful technique for analyzing differential amplifiers: the half-circuit analysis technique. We close by examining an actively loaded differential pair.

Qualitative Analysis Consider the differential amplifier shown in Figure 8-89. V_{OO} is a positive supply, V_{MM} is a negative supply, and I_{MM} is the tail current for this differential pair. We have assumed that the common-mode input voltage is zero, so we do not bother to use subscripts to indicate that the input shown is differential. Note that the input to T_1 is positive $v_I/2$ and the input to T_2 is negative $v_I/2$, so the differential input is v_I.

When $v_I = 0$, both transistors have equal control voltages. Since their merge terminals are also at the same voltage, the tail current splits equally between the two transistors, and $i_{M1} = i_{M2} = I_{MM}/2$, as found in Chapter 7. We can then write KVL around the loop containing R_1 and R_2 and find

$$v_O = i_{O2}R_2 - i_{O1}R_1, \tag{8.230}$$

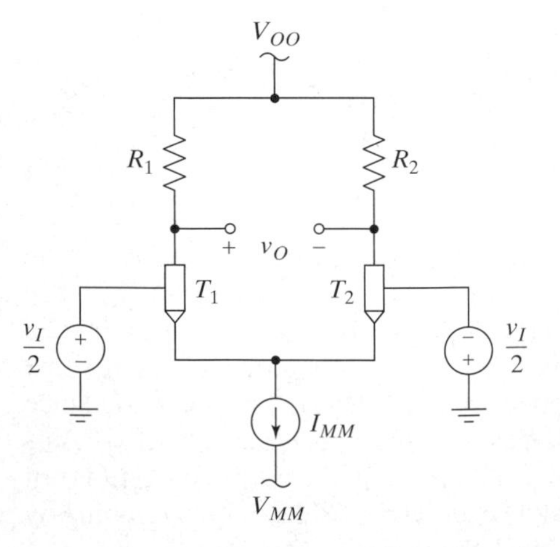

Figure 8–89 A generic differential amplifier with purely differential-mode input.

which, assuming that the control currents of the transistors are negligible and $v_I = 0$, is

$$v_O \approx \frac{I_{MM}}{2}(R_2 - R_1). \tag{8.231}$$

Therefore, so long as $R_1 = R_2$, the differential output voltage will be zero when the differential input voltage is zero. Now let's consider what happens when $v_I \neq 0$. Writing KVL around the input loop, we find that

$$\frac{v_I}{2} - v_{CM1} + v_{CM2} + \frac{v_I}{2} = 0, \tag{8.232}$$

which can be rewritten

$$v_I = v_{CM1} - v_{CM2}. \tag{8.233}$$

Suppose v_I is a small positive voltage; we then see that $v_{CM1} > v_{CM2}$, so we know that $i_{O1} > i_{O2}$, as well. Assuming that the resistors are equal, we see from (8.230) that v_O will be negative in this case. The larger v_I becomes, the larger the magnitude of the output voltage will be. At some point, the input voltage will become large enough that essentially all of the tail current will flow through T_1 alone. At that point, we have $i_{O1} \approx I_{MM}$ and $i_{O2} \approx 0$, so using (8.230), we get

$$v_{O\min} = -I_{MM}R_1. \tag{8.234}$$

Similarly, if we let v_I be very large and negative, we get

$$v_{O\max} = I_{MM}R_2. \tag{8.235}$$

The large-signal transfer characteristic of the differential pair is shown in Figure 8-90.

With the output taken as shown in Figure 8-89, the differential pair is an inverting amplifier, which is apparent in Figure 8-90. Nevertheless, we could simply reverse the polarity of the output in Figure 8-89, and the amplifier would be noninverting.

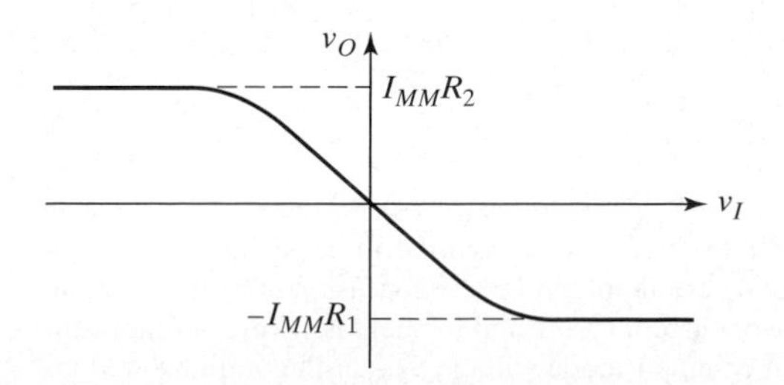

Figure 8–90 The transfer characteristic of the differential pair when driven with a purely differential input.

Now consider the amplifier driven with a purely common-mode signal, as in Figure 8-91. In this case, we can write KVL around the loop containing the control terminals and find that

$$v_{CM1} = v_{CM2}, \tag{8.236}$$

which is independent of the magnitude of the common-mode input.[7] Therefore, we can also state that $i_{O1} = i_{O2}$, which is again independent of the common-mode input voltage, and using (8.230), we can say that $v_O = 0$ so long as the resistors are equal. In other words, if the elements in the circuit are all perfectly matched (i.e., T_1 is the same as T_2 and $R_1 = R_2$), the circuit does not respond to common-mode inputs at all as long as the transistors remain forward active, as was assumed. When all of the elements are perfectly matched in this way, we say the circuit is ***balanced***.

If the circuit is not balanced, even with (8.236) being true, it is possible to have the output currents be a little different or the resistors be a little different. In either case, the differential output would not be zero. So long as i_{O1} and i_{O2} remain fixed, however, v_O will be constant as v_{ICM} changes. This constant nonzero output voltage is called the ***output offset*** of the amplifier.

We can now see an important advantage of differential amplifiers: The input can be DC coupled to the amplifier without affecting the DC biasing. In a single-ended amplifier, the DC input voltage is important to the biasing, and the source cannot be DC coupled unless it is capable of supplying the DC bias current and holds the voltage at an appropriate level. In the case of a differential amplifier, however, as long as the input is within some range of voltages, called the ***common-mode input range***, the amplifier will work. Therefore, DC coupling to differential amplifiers is common (e.g., operational amplifiers).

Detailed Small-Signal Analysis Now that we have a basic understanding of the qualitative operation of the differential pair, we examine the small-signal performance in more detail. Consider the small-signal midband AC equivalent circuit shown in Figure 8-92. This is the equivalent circuit for the amplifier of Figures 8-89 and 8-91, but we have now allowed for both common-mode and differential-mode

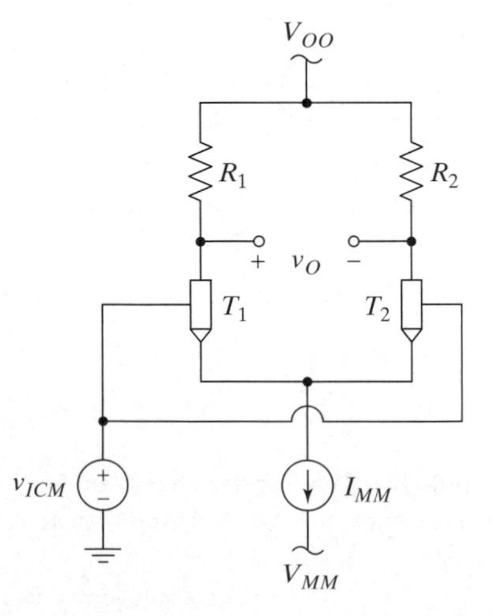

Figure 8–91 A generic differential amplifier with purely common-mode input.

[7] Do not confuse the subscripts CM in (8.236) with the term common-mode; in this equation, the subscripts indicate the voltage from control to merge. There shouldn't be any confusion of which meaning is intended, since, when a common-mode voltage or current is referred to, there is always another subscript along with the CM—for example, the input common-mode voltage, v_{icm}, or the output common-mode voltage, v_{ocm}.

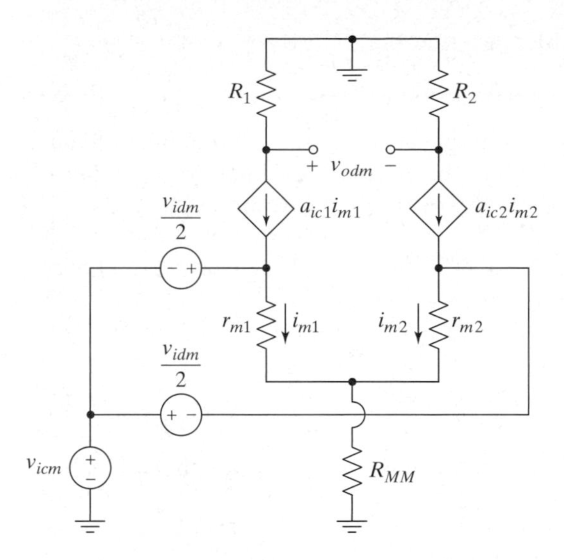

Figure 8–92 The small-signal midband AC equivalent circuit of the generic differential amplifier.

inputs. The DC voltage sources are AC grounds as usual, and we have assumed that the tail current source is not perfect, but has some finite output resistance, R_{MM}. We have used the T model for the transistors because it is somewhat easier to use in this case.

We use superposition to find the total response of this circuit to differential- and common-mode inputs. Let's first analyze the response to a differential input by setting $v_{icm} = 0$. We use superposition again for the two differential sources and find the voltage at the merge node (i.e., at the top of R_{MM}). Using only the source connected to T_1, we find that

$$v_m\Big|_{\text{due to } T_1 \text{ source}} = \frac{r_{m2} \| R_{MM}}{r_{m2} \| R_{MM} + r_{m1}} \frac{v_{idm}}{2}, \tag{8.237}$$

and similarly, using only the source connected to T_2, we find that

$$v_m\Big|_{\text{due to } T_2 \text{ source}} = \frac{-r_{m1} \| R_{MM}}{r_{m1} \| R_{MM} + r_{m2}} \frac{v_{idm}}{2}. \tag{8.238}$$

If the circuit is balanced, the contributions to v_m made by the two differential-mode input sources exactly cancel each other, and we find that $v_m = 0$. We will return to this result later; for now, we just use it in deriving the following equations. Starting at the top of the common-mode input source and writing KVL around the input loop yields

$$\frac{v_{idm}}{2} - i_{m1}r_{m1} + i_{m2}r_{m2} + \frac{v_{idm}}{2} = 0. \tag{8.239}$$

Assuming a balanced circuit so that $v_m = 0$ means that there isn't any current in R_{MM}, writing KCL at the merge node reveals that $i_{m1} = -i_{m2}$. Combining this fact with (8.239) yields

$$i_{m1} = -i_{m2} = \frac{v_{idm}}{2r_m}, \tag{8.240}$$

where we have dropped the subscripts on r_m since the circuit is balanced. This same result is obtained immediately if you consider the current source to be perfect so that R_{MM} is infinite.

Writing KVL around the output loop, we find that

$$v_{odm} = a_{ic2} i_{m2} R_2 - a_{ic1} i_{m1} R_1. \quad \textbf{(8.241)}$$

Assuming that the common-control current gains are about equal to one (which is a very good approximation), using the fact the circuit is assumed to be balanced, letting $R_1 = R_2 = R_O$, and substituting (8.240) into (8.241), we find that

$$v_{odm} \approx \frac{-v_{idm} R_O}{r_m}. \quad \textbf{(8.242)}$$

We now define the ***differential-mode gain*** of the amplifier:

$$A_{dm} \equiv \frac{v_{odm}}{v_{idm}} \approx \frac{-R_O}{r_m}. \quad \textbf{(8.243)}$$

If we choose to use one of the single-ended outputs instead (i.e., either v_{o1} or v_{o2}), the gain will then be

$$\frac{v_{o1}}{v_{idm}} = \frac{-v_{o2}}{v_{idm}} \approx \frac{-R_O}{2r_m}, \quad \textbf{(8.244)}$$

which is one-half the differential-mode gain.

Now let's consider the common-mode input while retaining our assumption of a balanced circuit. Setting the differential-mode input to zero, we see that r_{m1} and r_{m2} appear in parallel. Also, i_{m1} and i_{m2} are equal and are given by one-half of the total common-mode source current:

$$i_{m1} = i_{m2} = \frac{1}{2} \frac{v_{icm}}{r_{m1} \| r_{m2} + R_{MM}} = \frac{v_{icm}}{r_m + 2R_{MM}}. \quad \textbf{(8.245)}$$

We can now write

$$v_{o1} = v_{o2} = -a_{ic} i_m R_O = \frac{-a_{ic} R_O v_{icm}}{r_m + 2R_{MM}}. \quad \textbf{(8.246)}$$

Since the circuit is balanced, the differential output voltage $(v_{o1} - v_{o2})$ will be zero for a purely common-mode input. The equations are messier, but we could follow the same procedure to find v_{o1} and v_{o2} when a common-mode input is applied and the circuit is not balanced. In this case there will be a nonzero differential output (*see* Problem P8.96). Using (8.246), we can find the ***common-mode gain*** of the balanced circuit:

$$A_{cm} = \frac{v_{ocm}}{v_{icm}} = \frac{v_{o1}}{v_{icm}} = \frac{-a_{ic} R_O}{r_m + 2R_{MM}} \approx \frac{-R_O}{2R_{MM}}. \quad \textbf{(8.247)}$$

One very important measure of how well a differential amplifier works is the ***common-mode rejection ratio***, or CMRR.[8] The CMRR is defined as the magnitude of the ratio of the differential-mode gain to the common-mode gain:

$$CMRR \equiv \left| \frac{A_{dm}}{A_{cm}} \right|, \quad \textbf{(8.248)}$$

and the CMRR is frequently expressed in dB:

$$CMRR = 20 \log \left| \frac{A_{dm}}{A_{cm}} \right|. \quad \textbf{(8.249)}$$

[8] Refer to Aside A5.2 for a discussion of the importance of the common-mode rejection.

The question we need to ask is; Which differential- and common-mode gains are we to use in the CMRR? The answer is clear if we have a single-ended output voltage (e.g., an op amp). In that case, we define A_{dm} to be the gain from v_{idm} to v_o, and we define A_{cm} to be the gain from v_{icm} to v_o. As an example, if we use v_{o1} as the output of our circuit, we can find the CMRR by using (8.247) and (8.244). The result is

$$CMRR \approx \frac{R_{MM}}{r_m} \quad \textbf{(8.250)}$$

when the circuit is balanced. The approximations made in the final result in (8.250) will be valid the vast majority of the time. The fact that the CMRR increases as R_{MM} increases explains in part why current sources (i.e., current mirrors) with high output resistance are desirable.

If the output of the circuit is taken differentially, it is harder to decide which gains we should use to find the CMRR. Various books treat this situation differently, so there isn't any universally agreed-upon definition. The important thing is to consider the specifics of the situation you are dealing with. If the output of this first stage is being used differentially, what is the next stage? Is its output still taken differentially? In general, there are four gains required to completely specify an amplifier with differential inputs and outputs. (Such an amplifier is often called a ***fully differential*** amplifier.) One way to specify such an amplifier is to define the gains from each input terminal to each output terminal where all voltages are considered single ended (i.e., referenced to ground), as illustrated in Figure 8-93.

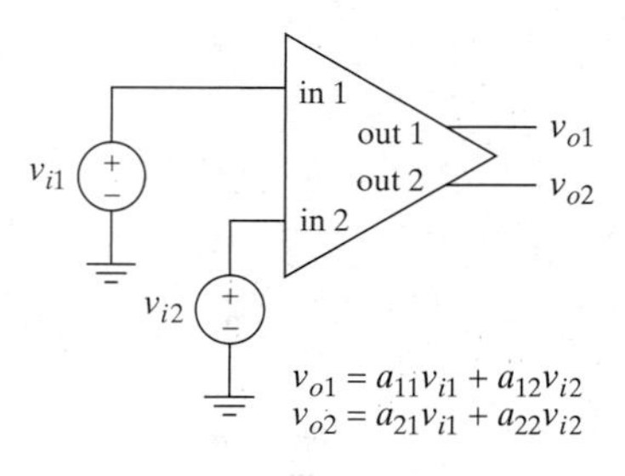

Figure 8–93 One approach to completely specifying a fully differential amplifier.

However, a more common approach is to define the differential-mode gain, A_{dm}, as the ratio of the change in differential output to the change in differential input that caused it and the common-mode gain, A_{cm}, as the ratio of the change in common-mode output to the change in common-mode input that caused it and then to specify the two ***conversion gains***: the common-mode to differential-mode gain, $A_{cm\text{-}dm}$ (the ratio of the change in the differential output to the change in common-mode input that caused it), and the differential-mode to common-mode gain, $A_{dm\text{-}cm}$ (the ratio of the change in the common-mode output to the change in differential input that caused it). This situation is illustrated in Figure 8-94.

If the gains are defined as in Figure 8-94, the CMRR could still be defined as the ratio of A_{dm} to A_{cm}. However, doing so is a bit misleading, because the common-mode to differential-mode conversion gain is likely to cause more trouble than the common-mode gain, since if the output is really used differentially, the next stage probably has some common-mode rejection of its own. A better definition in the case of a purely differential amplifier is to call the CMRR the ratio of A_{dm} to $A_{cm\text{-}dm}$. As noted earlier, in a given application you must consider what the final

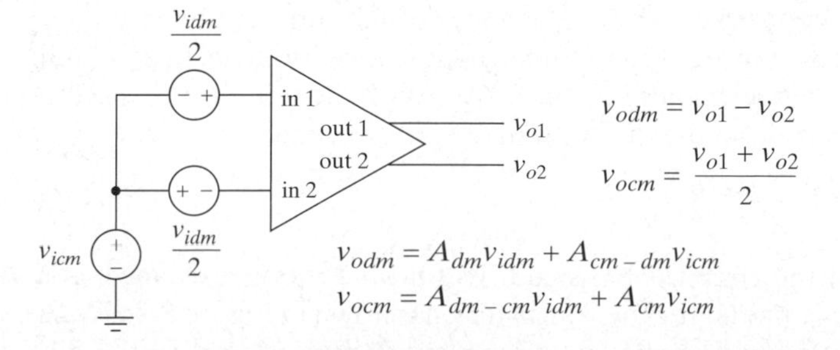

Figure 8–94 The most common approach to completely specifying a fully differential amplifier.

output will be and then see how the common-mode input will affect it. You will generally need to take into account the conversion gain as well as the common-mode gain of the input amplifier. Unless specifically noted otherwise, in this text we will assume a single-ended output, as in (8.250), to avoid confusion.

Example 8.7

Problem:
Consider a differential pair being driven single-ended as shown in Figure 8-95. Find the output voltages, assuming that the amplifier has $A_{dm} = 100, A_{cm} = 2, A_{cm\text{-}dm} = 0.1$, and $A_{dm\text{-}cm} = 0.01$.

Solution:
We need to find the differential- and common-mode components driving the amplifier; then we can use the equations given in Figure 8-94 directly. The AC inputs to the differential pair are v_s and ground. The differential input is therefore $v_{idm} = v_s$, and the common-mode input is $v_{icm} = v_s/2$. The differential output voltage is given by

$$v_{odm} = A_{dm}v_{idm} + A_{cm-dm}v_{icm} = 100v_s + 0.1(v_s/2) = 100.05v_s.$$

The common-mode output voltage is given by

$$v_{ocm} = A_{dm-cm}v_{idm} + A_{cm}v_{icm} = 0.01v_s + 2(v_s/2) = 1.01v_s.$$

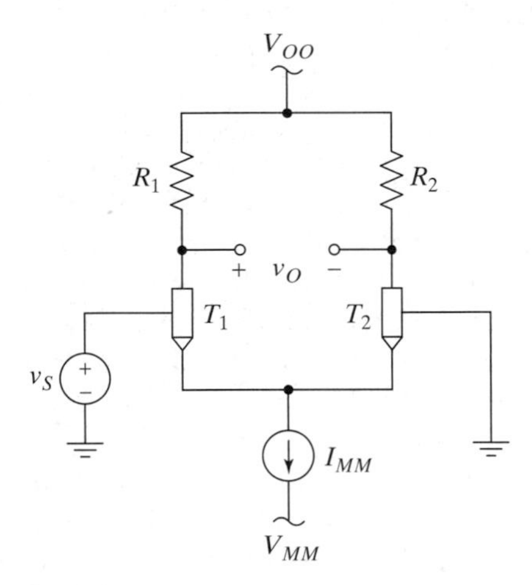

Figure 8–95 A differential pair driven by a single-ended source.

We next consider the input and output resistances of the differential amplifier. The differential output resistance is found by setting the input sources to zero and looking back into the output terminals in Figure 8-92. With the input sources set to zero, i_{m1} and i_{m2} are both zero, so the differential output resistance is

$$R_{od} = R_1 + R_2, \tag{8.251}$$

which is equal to $2R_O$ if the circuit is balanced. To find the input resistance, it is easiest to use the hybrid-π model for the transistors, as shown in Figure 8-96. With a purely differential-mode input as shown, the current in R_{MM} is zero, so the differential input resistance is

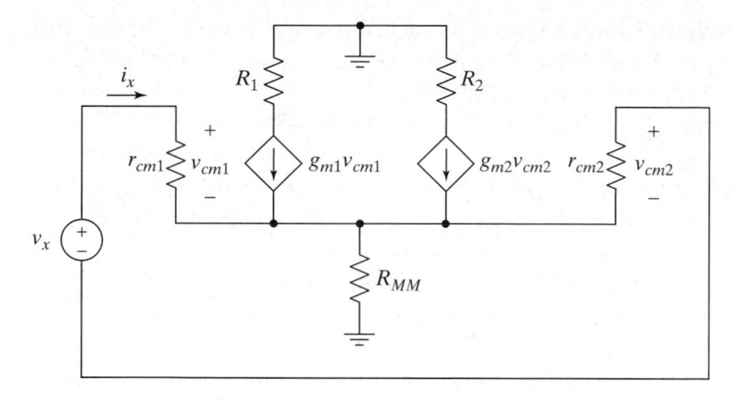

Figure 8–96 Finding the differential input resistance of the circuit.

$$R_{id} = \frac{v_x}{i_x} = r_{cm1} + r_{cm2}, \tag{8.252}$$

which is equal to $2r_{cm}$ when the circuit is balanced.

In the next subsection, we show a simpler way to derive the results we have already obtained by direct analysis. We didn't start with the simpler method, because it is based on some circuit manipulations that may be a bit troubling at first. Nevertheless, the treatment that follows leads to a more elegant way of thinking about differential pairs and is the basis of the intuition that good designers have.

The Half-Circuit Analysis Technique We begin this section by going back and reconsidering the conclusion drawn from (8.237) and (8.238). Based on these equations, we concluded that the voltage at the merge node was zero for a purely differential input. This result can be generalized on the basis of the symmetry of the circuit. Consider the circuit as we have redrawn it in Figure 8-97. Here we have split the current source resistance into two equal resistors ($2R_{MM}$ in parallel with $2R_{MM}$ is equal to R_{MM}, so we haven't changed the circuit), and we have included a resistance for the signal source, R_S, but have also split it in half along with the voltage

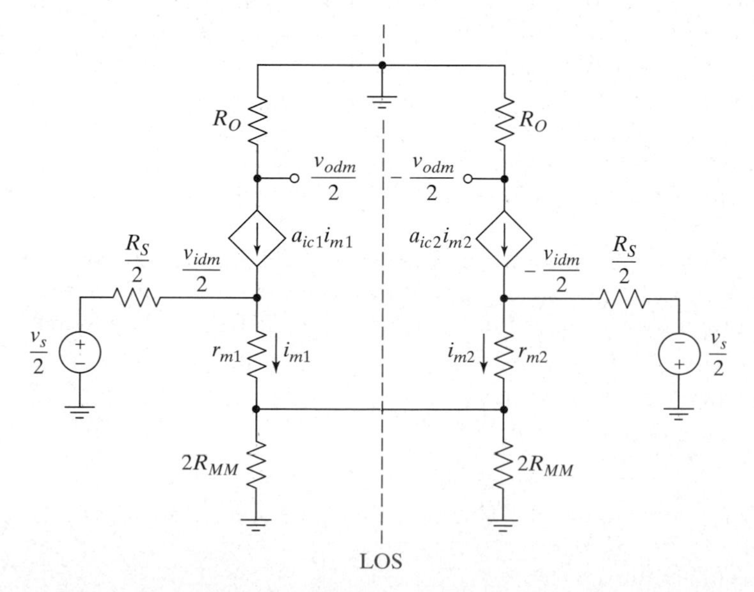

Figure 8–97 LOS for the differential pair.

(i.e., if you put the two Thévenin sources, in series you get a single input with open-circuit voltage v_s and resistance R_S). We have also indicated that a ***line of symmetry*** (LOS) can be drawn down the middle of the circuit. As the name implies, the circuit is completely symmetric about the LOS. With a pure differential input making up two equal but opposite polarity sources as shown, when we use superposition to write equations like (8.237) and (8.238), the resulting contributions will always cancel for any node crossed by the LOS. In other words, any line that crosses the LOS can be connected to a ***differential-mode virtual ground***. It is a virtual ground because the node isn't actually connected to ground, but the node voltage is always zero for differential-mode signals. In this example there is only one such node, the merge node.

The differential output voltage is the difference between the outputs on the left and right sides of the LOS in the figure. Therefore, we have denoted those voltages as $\pm v_{odm}/2$. Now we take advantage of the symmetry of this circuit and draw the ***differential-mode half circuit*** shown in Figure 8-98. Because the circuit is assumed to be perfectly balanced, we have removed unnecessary subscripts. If we analyze this half-circuit, the results will apply equally well to either half of the original circuit, except that all the signs will change, since the two halves are driven differentially. Therefore, the analysis of a differential amplifier has been reduced to that of the familiar single-transistor stage with voltage and current gain.

Examining this circuit, we use Ohm's law to write

$$\frac{v_{odm}}{2} = -a_{ic} i_m R_O. \tag{8.253}$$

Using Ohm's law again, we write

$$i_m = \frac{v_{idm}}{2r_m}. \tag{8.254}$$

Combining these equations, we find that

$$A_{dm} = \frac{v_{odm}}{v_{idm}} = \frac{-a_{ic}R_O}{r_m} \approx \frac{-R_O}{r_m}, \tag{8.255}$$

which agrees with (8.243), derived earlier. We also note that this equation is in agreement with (8.52), derived for the single-transistor amplifier. (R_O here is the same as R_L there.) To see that these equations agree exactly, you must note that $g_m = a_{ic}/r_m$, *see* (8.14). This result is important and warrants some discussion; the differential-mode voltage gain of a differential pair is exactly the same as that of a single-transistor amplifier.

Given that a differential pair is more complicated, and will be noisier (since it has more active devices), the question arises; Why would we use a differential pair? There are several answers to this question. First, the CMRR can help reject

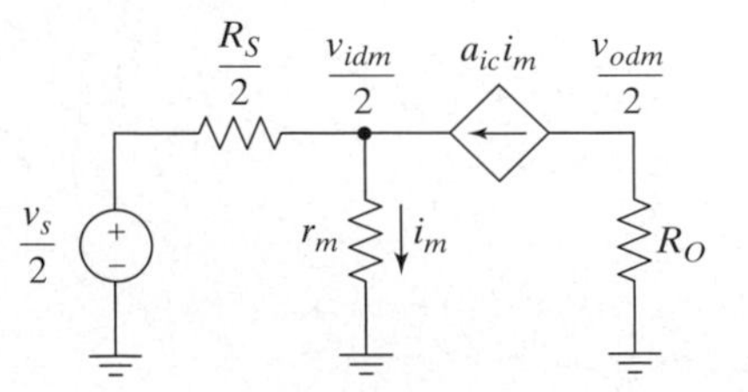

Figure 8–98 The differential-mode half circuit.

noise in some situations, as pointed out in Aside A5.2. Second, the CMRR also allows us to use DC-coupled sources, since their exact DC voltage is not critical. Third, the differential input and DC coupling combine to allow greater flexibility in use (e.g., operational amplifiers). Fourth, we can bias the differential pair without using many resistors. (The resistive load can be replaced by a mirror.) Fifth, the differential-mode virtual ground that appears at the merge terminals allows us to do away with the need for a bypass capacitor. Go back and look at the amplifier shown in Figure 8-73, and recall that we said we would find a way to do away with the bypass capacitor. The differential pair is the circuit that allows us to use current source biasing without a bypass capacitor, while simultaneously retaining the benefit of not needing an input coupling capacitor. We could simply connect the merge terminal to ground in Figure 8-73 to do away with the bypass capacitor, but then the input DC voltage would be needed to bias the transistor *and must be exactly the right value* to produce the desired DC current.

Now return to the circuit in Figure 8-98, and find the differential output resistance. Using the circuit shown in Figure 8-99, where we recognize that since the output voltage is only $v_{odm}/2$ the test source should be $v_x/2$, we find

$$R_{od} = \frac{v_x}{i_x} = 2R_O, \tag{8.256}$$

which again agrees with the earlier result of (8.251) when $R_1 = R_2 = R_O$.

To find the differential input resistance, we switch to using the hybrid-π model again, as shown in Figure 8-100. The result is

$$R_{id} = \frac{v_x}{i_x} = 2r_{cm}, \tag{8.257}$$

in agreement with (8.252) for the balanced case.

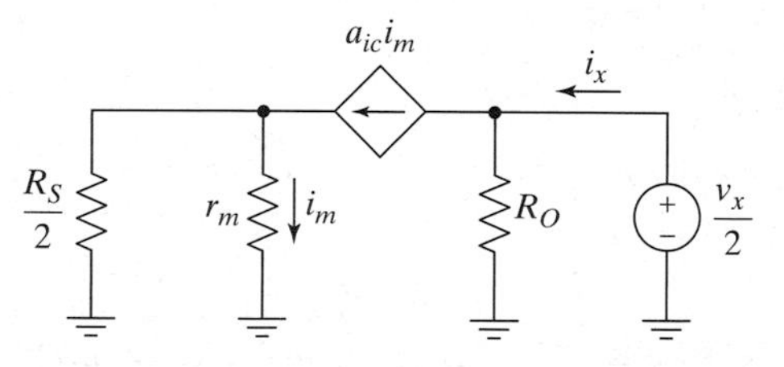

Figure 8–99 Finding the differential output resistance using the half-circuit.

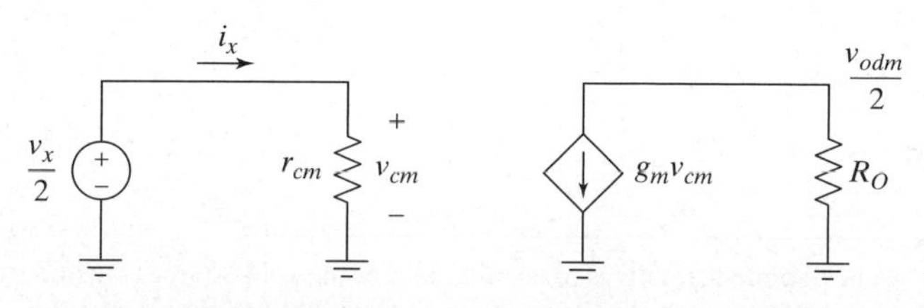

Figure 8–100 Finding the differential input resistance using the half-circuit.

We now turn our attention to a common-mode input as shown in Figure 8-101. The line of symmetry has again been included in the figure. In this case, we have duplicated the common-mode input source.[9] With equal inputs on both sides of the LOS, we conclude that the current through any branch crossing the LOS (e.g., i_x in the figure) will be zero. To see that this statement is true, note that if you break the connection, you get the same voltage on either side; therefore, when the connection is present, no current will flow through it.

Since no current will flow in branches that cross the LOS when a common-mode input is applied to a balanced differential pair, we can break these branches. We then need to analyze only either half of the circuit, and the results will be the same for the other half. We therefore draw the ***common-mode half circuit*** shown in Figure 8-102. Using Ohm's law, we write

$$i_m = \frac{v_{icm}}{r_m + 2R_{MM}}. \tag{8.258}$$

Using Ohm's law again, we write

$$v_{ocm} = -a_{ic}i_m R_O, \tag{8.259}$$

and combining these equations, we find the common-mode gain,

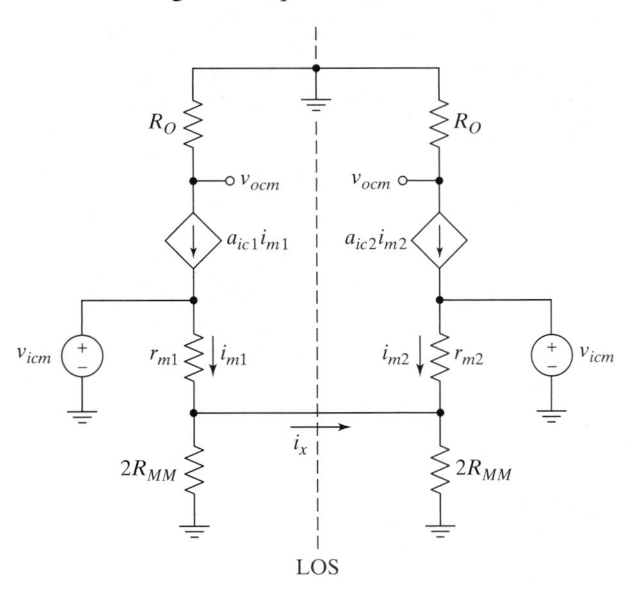

Figure 8–101 The LOS shown with a common-mode input.

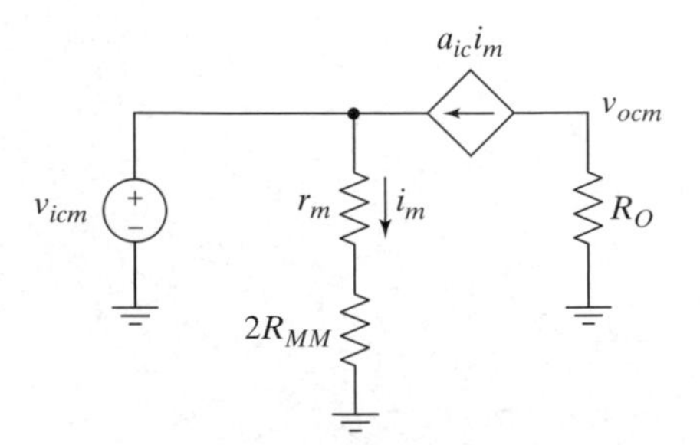

Figure 8–102 The common-mode half-circuit.

[9] In drawing this equivalent, what we have done formally is to use the dual of the split-source transformation covered in Aside A8.1; that is, we put two equal voltage sources in parallel and then break the connection between them.

$$A_{cm} = \frac{v_{ocm}}{v_{icm}} = \frac{-a_{ic}R_O}{r_m + 2R_{MM}} \approx \frac{-R_O}{2R_{MM}}, \tag{8.260}$$

in agreement with (8.247).

The Actively Loaded Differential Pair In Chapter 7 we briefly discussed the biasing of an actively loaded differential pair.[10] The circuit we considered is repeated here as Figure 8-103 for convenience. Our interest in this section is to perform a small-signal midband AC analysis of this amplifier.

The small-signal midband AC equivalent circuit is shown in Figure 8-104, where we have used the T model for all four devices. The subscripts are used only to allow us to uniquely refer to the different currents and to distinguish n-type and

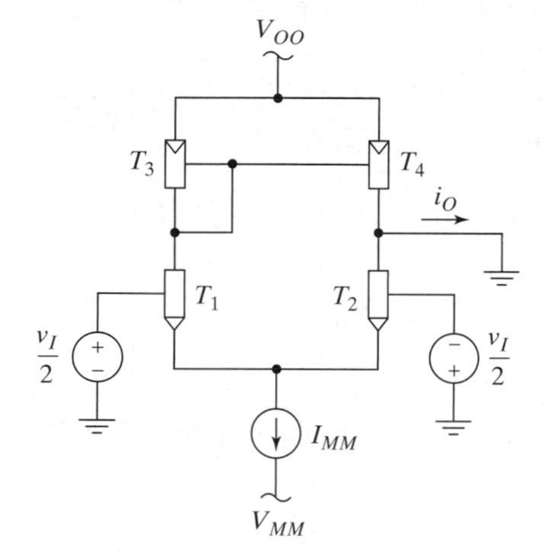

Figure 8–103 An actively loaded differential pair.

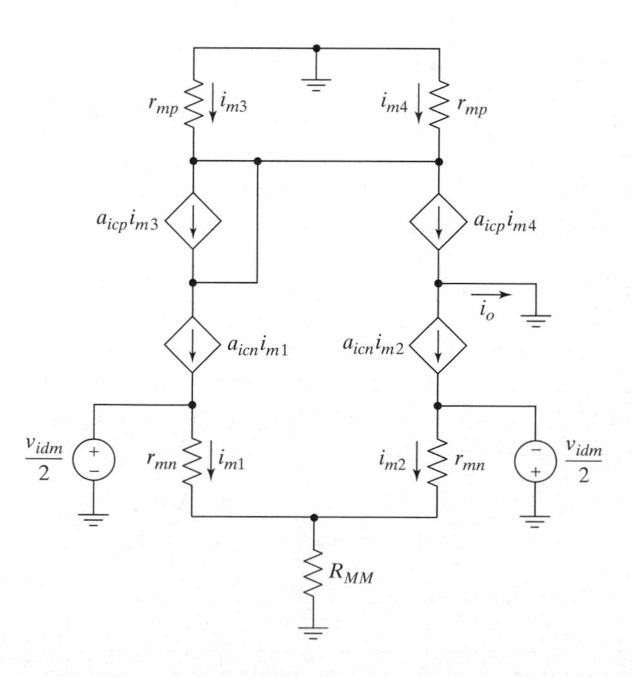

Figure 8–104 The small-signal midband AC equivalent circuit for the actively loaded differential pair.

[10] This type of active load is sometimes called a ***current-mirror load*** to distinguish it from an active load with differential outputs (*see* Problem P8.98).

p-type transistors. The devices are all assumed to be matched, and the DC bias currents in the left and right sides of the pair are assumed equal, so the small-signal model parameters are identical.[11] We also assume that the circuit is driving a short-circuit load. We will remove this restriction later. For now, we restrict ourselves to a differential input as shown.

The bottom half of this circuit has already been analyzed, and we know that

$$i_{m1} = -i_{m2} = \frac{v_{idm}}{2r_{mn}}, \tag{8.261}$$

as in (8.240). We next write KCL at the output of T_1 and find that

$$a_{icn}i_{m1} - i_{m3} - (1 - a_{icp})i_{m4} = 0. \tag{8.262}$$

With perfect matching and ignoring output resistance, $i_{m3} = i_{m4}$. Therefore, we can solve for i_{m4} and get

$$i_{m4} = \frac{a_{icn}i_{m1}}{2 - a_{icp}}. \tag{8.263}$$

Finally, we write KCL at the output node and use (8.261) and (8.263) to find

$$i_o = \left(\frac{a_{icp}a_{icn}}{2 - a_{icp}} + a_{icn}\right)i_{m1} = \left(\frac{a_{icp}a_{icn}}{2 - a_{icp}} + a_{icn}\right)\frac{v_{idm}}{2r_{mn}} \approx \frac{v_{idm}}{r_{mn}}, \tag{8.264}$$

which is about the same as the output current of a single common-merge transistor driven by v_{idm}. If this stage is connected to a load resistor of value R_L, the voltage gain will be

$$\frac{v_o}{v_{idm}} \approx \frac{R_L}{r_m} \approx g_m R_L, \tag{8.265}$$

which is again about the same as a single common-merge transistor. Note that the gain is about twice what would be achieved if we used a single-ended output from a resistively loaded differential pair as in (8.244). We can see intuitively why the active load should have twice the gain: For a resistive load, the current through either load resistor is just the output current of one transistor; for the active load, the output current of T_1 gets mirrored to the output and adds to the output current of T_2 to make i_o approximately twice as large ($i_o \approx i_{m4} - i_{m2} \approx i_{m1} - i_{m2} \approx 2i_{m1}$). Finally, we note that ignoring the small-signal output resistance of the transistors, as we have done here, is not a good approximation if the load resistance becomes large, which is common in ICs. We will address this issue in the sections that follow.

We next want to examine the common-mode gain of this circuit and make use of our earlier results for the differential pair. When the circuit is driven by a purely common-mode input, we have, from (8.245), that

$$i_{m1} = i_{m2} = \frac{v_{icm}}{r_m + 2R_{MM}} \approx \frac{v_{icm}}{2R_{MM}}. \tag{8.266}$$

We now write KCL at the output and again use (8.263) with (8.266) to obtain

[11] We cannot say that this circuit is balanced, since one side of the differential pair is loaded by a diode-connected transistor and the other is not. T_3 and T_4 have different DC voltages at their output terminals and therefore have slightly different currents, because the transistors have finite output resistance. We are not modeling that nonidentity here.

$$i_o = a_{icp} i_{m4} - a_{icn} i_{m2} = a_{icn} i_{m1}\left(\frac{a_{icp}}{2 - a_{icp}} - 1\right) = \frac{a_{icn} v_{icm}}{2R_{MM}}\left(\frac{2a_{icp} - 2}{2 - a_{icp}}\right). \quad \textbf{(8.267)}$$

If we again assume that the circuit drives a load R_L, the common-mode gain is

$$A_{cm} = \frac{v_o}{v_{icm}} = \frac{a_{icn} R_L}{R_{MM}}\left(\frac{a_{icp} - 1}{2 - a_{icp}}\right), \quad \textbf{(8.268)}$$

which is nearly equal to zero, since $a_{ic} \approx 1$. We can see intuitively why the common-mode gain of this circuit should be nearly zero: Since the output currents of T_1 and T_2 are equal, and the mirror serves to mirror the output current of T_1 over to the output, we would expect the difference between i_{o4} and i_{o2} to be small ($i_o \approx 0$); therefore, $v_o \approx 0$, even with a load present. If the mirror formed by T_3 and T_4 were perfect (i.e., $i_4 = i_3$), the common-mode gain would be zero.

The actively loaded differential pair does a much better job of converting a differential signal into a single-ended signal than a resistively loaded differential pair does. As noted, the differential-mode gain is twice what we get with the resistive load, and now we see that the common-mode gain is much lower. Therefore, the CMRR will be significantly better with the active load. As a result of these advantages, the actively loaded differential pair is by far the most common input stage for an operational amplifier.

8.6.2 A Bipolar Implementation: The Emitter-Coupled Pair

We first cover the emitter-coupled pair (ECP) with a resistive load. We then go on to examine the operation of the circuit with an active load.

The Resistively Loaded ECP A resistively loaded emitter-coupled pair is shown in Figure 8-105(a), and its small-signal midband AC equivalent using T models is shown in part (b) of the figure. We assume the tail current source has a small-signal output resistance of 250 kΩ and Q_1 and Q_2 are perfectly matched. We can use the results we

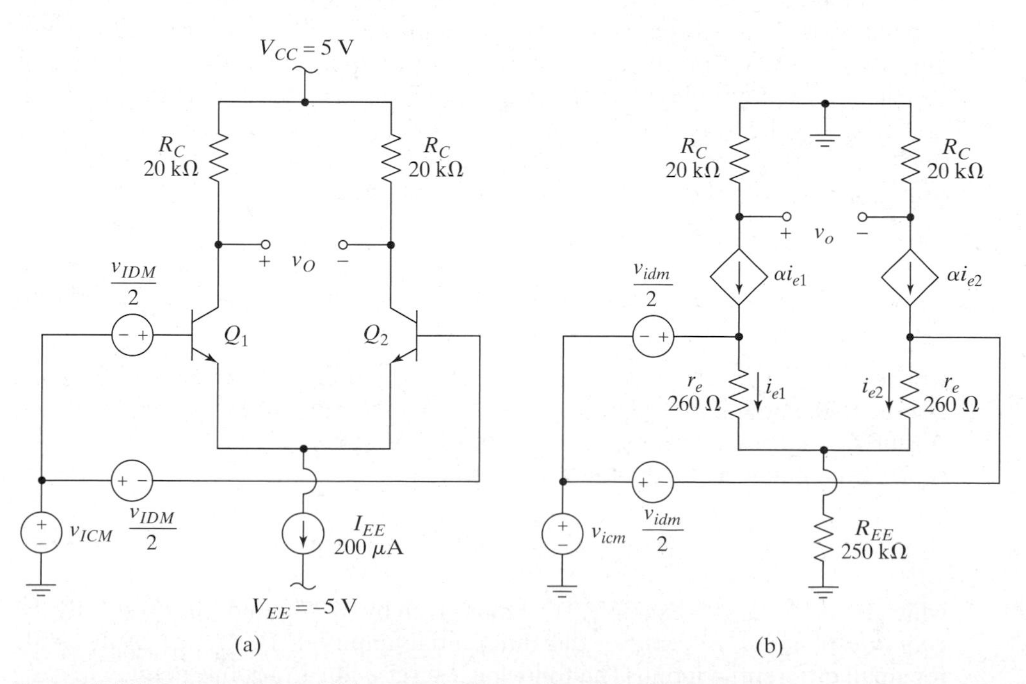

Figure 8–105 (a) An ECP and (b) its small-signal midband AC equivalent circuit.

derived for the generic differential amplifier of Figure 8-92 directly here. Using (8.243), but revising the notation for the bipolar case, we find the differential-mode gain:

$$A_{dm} = \frac{v_{odm}}{v_{idm}} \approx \frac{-R_C}{r_e} = -77. \tag{8.269}$$

The gain to either collector used single ended is one-half of the differential gain.
Using (8.247), we find the common-mode gain to either collector to be

$$A_{cm} = \frac{v_{ocm}}{v_{icm}} \approx \frac{-R_C}{2R_{EE}} = -0.04. \tag{8.270}$$

The CMRR for taking the output single ended is given by (8.250) and is

$$CMRR \approx \frac{R_{EE}}{r_e} = 962, \tag{8.271}$$

or 60 dB. The differential input resistance is given by (8.252) and is

$$R_{id} = 2r_\pi = 52\ \text{k}\Omega, \tag{8.272}$$

assuming that $\beta = 100$. The differential output resistance is given by (8.251) and is

$$R_{od} = 2R_C = 40\ \text{k}\Omega. \tag{8.273}$$

This output resistance was calculated with the output resistance of the transistors ignored. Since the emitters are a differential-mode virtual ground, the output resistances of Q_1 and Q_2 show up in parallel with the load resistors. If $V_A = 50$ V, the output resistances are both $r_o = 500$ kΩ, which is certainly reasonable to ignore here.

As a final point, we note that there is a restricted range of common-mode input voltages over which this circuit can function properly. We have assumed in all of our analyses so far that the transistors are forward active. For that assumption to be true, the common-mode input cannot be too high or too low. If the common-mode input voltage is too high, the transistors will saturate. Assuming as usual that $V_{CEsat} = 0.2$ V and $V_{BE} = 0.7$ V, we see that the transistors will saturate if the common-mode input exceeds a maximum value of

$$V_{ICM\max} = V_C - V_{CEsat} + V_{BE} = V_{CC} - \frac{I_{EE}R_C}{2} - V_{CEsat} + V_{BE}, \tag{8.274}$$

which is 3.5 V for our present example. The common-mode input voltage must also be high enough that Q_1 and Q_2 are on and the tail-current source functions. Assuming that the current source requires 0.2 V of headroom, the minimum acceptable common-mode input for our example is

$$V_{ICM\min} = V_{EE} + V_{mirror,min} + V_{BE}, \tag{8.275}$$

which is −4.1 V for this example. The limits given by (8.274) and (8.275) specify the ***common-mode input range*** of this differential amplifier. This range applies only for small differential inputs. The following exercise illustrates this point.

Exercise 8.11

Drive the differential amplifier in Figure 8-105 with a single-ended source as shown in Figure 8-106. What is the range of input voltages for which the circuit will work properly?

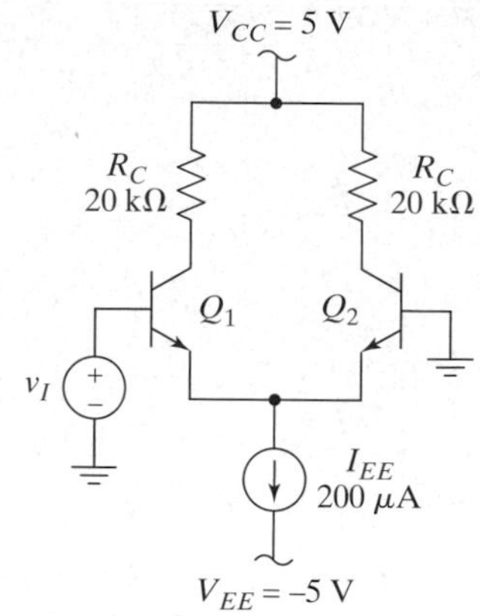

Figure 8–106 The ECP driven single ended.

Now consider what happens when we add resistors in series with the emitters of an ECP, as shown in Figure 8-107(a). The small-signal equivalent circuit is shown in part (b) of the figure. This situation is analogous to using unbypassed emitter resistance in a common-emitter amplifier, and these resistors are again called emitter degeneration resistors. The resistors provide negative feedback, which reduces the gain, but allows it to be less sensitive to the bias point of the circuit.

The analysis of this circuit is identical to that of the circuit in Figure 8-105, except that r_e is replaced by $r_e + R_E$. For example, the differential-mode gain is

$$A_{dm} = \frac{v_{odm}}{v_{idm}} = \frac{-\alpha R_C}{r_e + R_E}, \tag{8.276}$$

which is analogous to (8.269). If $R_E \gg r_e$, the gain of this stage is set by the resistor ratio. An added advantage of degeneration is that it extends the input voltage range over which the circuit behaves approximately linearly. The large-signal performance of the ECP is covered in Chapter 12.

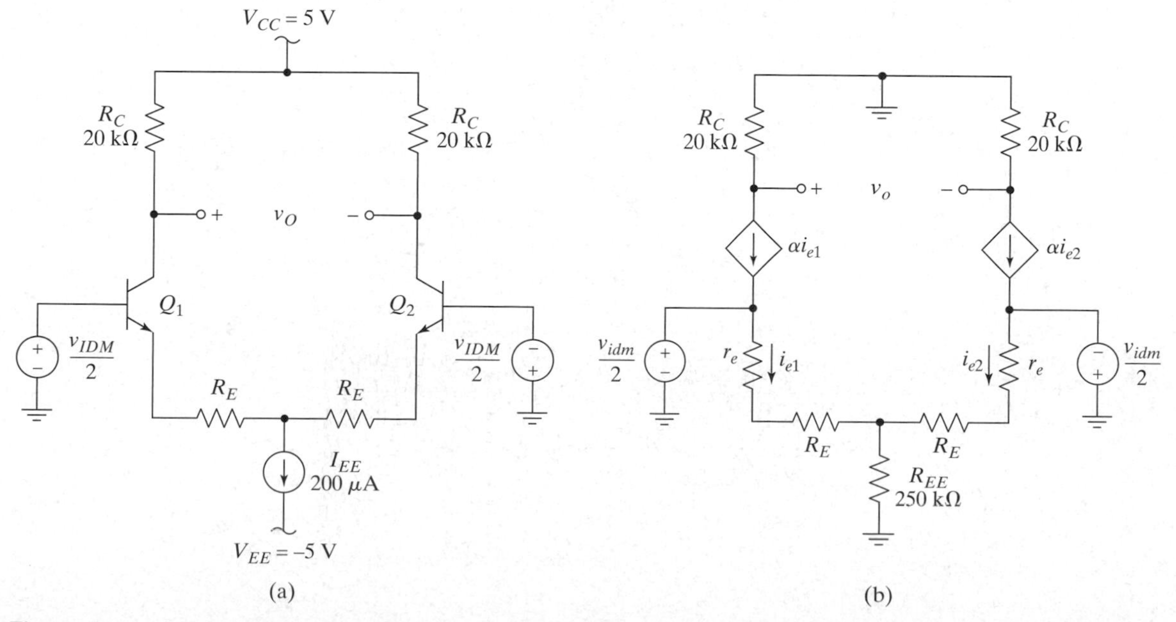

Figure 8–107 (a) The ECP of Figure 8-105 with emitter degeneration added. (b) The small-signal equivalent circuit.

The Actively Loaded ECP An actively loaded ECP is shown in Figure 8-108. If this amplifier drives a short circuit and has a pure differential input, the small-signal midband AC output current is given by (8.264) and is

$$i_o \approx \frac{v_{idm}}{r_e} \approx g_m v_{idm}. \tag{8.277}$$

If the circuit drives a load R_L, the differential voltage gain is

$$A_{dm} \approx g_m R_L, \tag{8.278}$$

as noted in (8.265). The maximum value for this gain is found when the output is open circuited. The effective load resistance is then a result of the output resistances of Q_2 and Q_4, as shown in Figure 8-109(a). We can greatly simplify this circuit if we make a few reasonable approximations and think about the operation of the circuit at a higher level. First, ignore the base currents in the transistors (i.e., set $\alpha = 1$ and $r_\pi = \infty$). Note that since the effective resistance of the diode-connected transistor is small (i.e., $1/g_{m3}$ is small), the AC voltage at the collector of Q_1 will be small. If we also assume that the AC voltage at the emitters of Q_1 and Q_2 is nearly zero (we will return to this assumption later), the voltage across r_{o1} will be small, and the current through it can be neglected. Tying all of this together, we see that the output of the controlled source of Q_4 is given by

$$g_{m4} v_{\pi 4} \approx g_{m4} i_{e1} (1/g_{m3}) = i_{e1}, \tag{8.279}$$

and using the same assumptions, we find that the output of Q_2 is $\alpha_2 i_{e2} \approx i_{e2}$. We now make use of (8.261) and find the total current being forced into the output node:

$$g_{m4} v_{\pi 4} - \alpha_2 i_{e2} \approx i_{e1} - i_{e2} = \frac{v_{idm}}{r_{e1}} \approx g_{m1} v_{idm}. \tag{8.280}$$

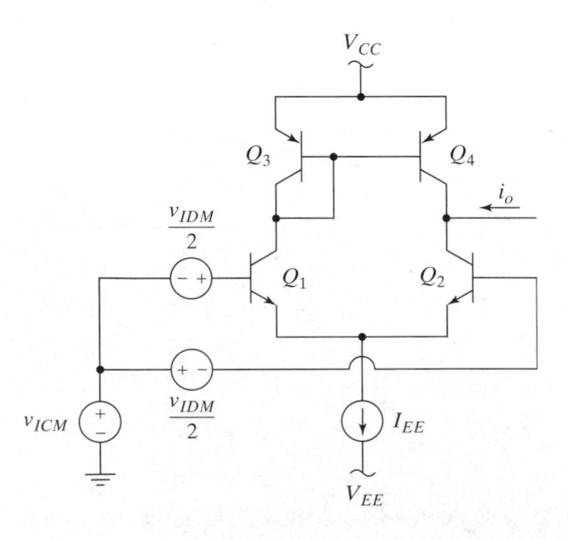

Figure 8–108 An actively loaded ECP.

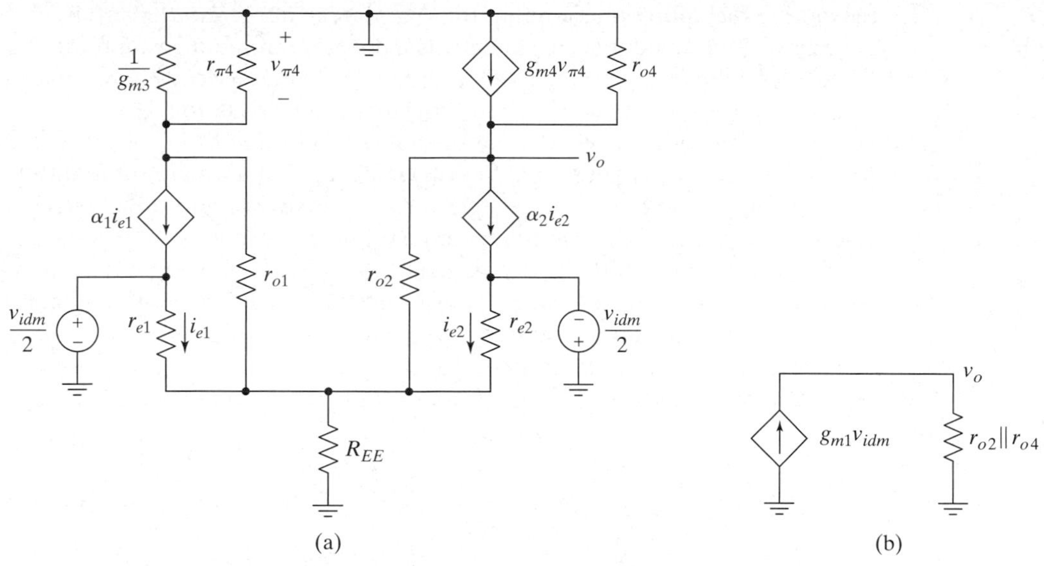

Figure 8–109 (a) The small-signal equivalent circuit for the actively loaded ECP with purely differential inputs. (b) An approximate equivalent circuit at the output node.

In part (b) of the figure, we show an approximate equivalent circuit for the output of this amplifier. The total current into the output node is given by (8.280), and the total resistance to ground is r_{o2} in parallel with r_{o4}, where we have again assumed that the AC voltage at the emitters of Q_1 and Q_2 is zero. The resulting differential gain of the circuit is found from part (b) of the figure to be

$$A_{dm} = \frac{v_o}{v_{idm}} \approx g_{m1}(r_{o2} \| r_{o4}). \quad \textbf{(8.281)}$$

Exercise 8.12

Consider the circuit in Figure 8-108, and assume that $V_{AN} = 75$ V, $V_{AP} = 50$ V, and $I_{EE} = 500$ μA. (a) Find A_{dm}. (b) If a resistive load were used instead, how large would the resistor need to be to achieve the same gain (taking the output single ended)? (c) How large would the supply voltage need to be?

We can now see that the gain given by (8.278) is correct only if $R_L \ll (r_{o2} \| r_{o4})$. In general, the gain with a load attached is given by

$$A_{dm} = \frac{v_o}{v_{idm}} \approx g_{m1}(R_L \| r_{o2} \| r_{o4}). \quad \textbf{(8.282)}$$

Now reexamine the approximation that the AC voltage at the emitters of Q_1 and Q_2 is near zero. We proved that this node was a differential-mode ground for our original differential pair by writing (8.237) and (8.238) and recognizing that these contributions cancel for a balanced circuit. In the present situation, however, the circuit is not completely balanced, because the load presented to Q_1 is different from that presented to Q_2. Therefore, the two collector voltages are quite different, and if we include the transistors' output resistors as we have here, the currents in r_{o1} and r_{o2} will not cancel each other, so the emitters cannot be at ground potential. Nevertheless, the emitters are connected to the two input sources through small resistances (r_e) and the voltage at the emitters will therefore be small *compared with* the large output voltage. As a result, it is reasonable, *on the output node*, to approximate r_{o2} as being connected to ground.

Using (8.268), we find that the common-mode gain of this circuit when driving a load R_L is

$$A_{cm} = \frac{v_o}{v_{icm}} = \frac{\alpha_n R_L}{R_{EE}}\left(\frac{\alpha_p - 1}{2 - \alpha_p}\right) \approx 0, \tag{8.283}$$

where we have used subscripts to denote the *pnp* and *npn* current gains, α_p and α_n, respectively. The common-mode gain of this circuit is much worse if the devices are not perfectly matched. As seen in Chapter 5, good op amps have CMRRs on the order of 100–120 dB.

Exercise 8.13

Consider again Exercise 8.12, and add a separate load resistor to ground with $R_L =$ 25 kΩ. If $R_{EE} = 50$ kΩ, $\beta_n = 150$, and $\beta_p = 25$, (a) find A_{dm}, (b) find A_{cm}, and (c) find the CMRR.

□

8.6.3 A FET Implementation: The Source-Coupled Pair

We first cover the resistively loaded source-coupled pair (SCP). After that, we examine the operation of the SCP with an active load.

The Resistively Loaded SCP A resistively loaded source-coupled pair is shown in Figure 8-110(a), and its small-signal midband AC equivalent using T models is shown in part (b) of the figure. We assume the tail current source has a small-signal output resistance of 250 kΩ and Q_1 and Q_2 are perfectly matched. We can use the results derived for the generic differential amplifier of Figure 8-92 directly here. Using (8.243), but revising the notation for the FET case, we find the differential-mode gain:

$$A_{dm} = \frac{v_{odm}}{v_{idm}} \approx -g_m R_D. \tag{8.284}$$

The gain to either drain used single ended is one-half of the differential gain. If we assume that the transistors have $K = 100\ \mu\text{A/V}^2$ and $V_{th} = 0.5$ V, the differential gain with a single-ended output for this circuit is $A_{dm} = 1.3$.

Using (8.247), we find the common-mode gain to either drain to be

$$A_{cm} = \frac{v_{ocm}}{v_{icm}} \approx \frac{-R_D}{2R_{SS}}, \tag{8.285}$$

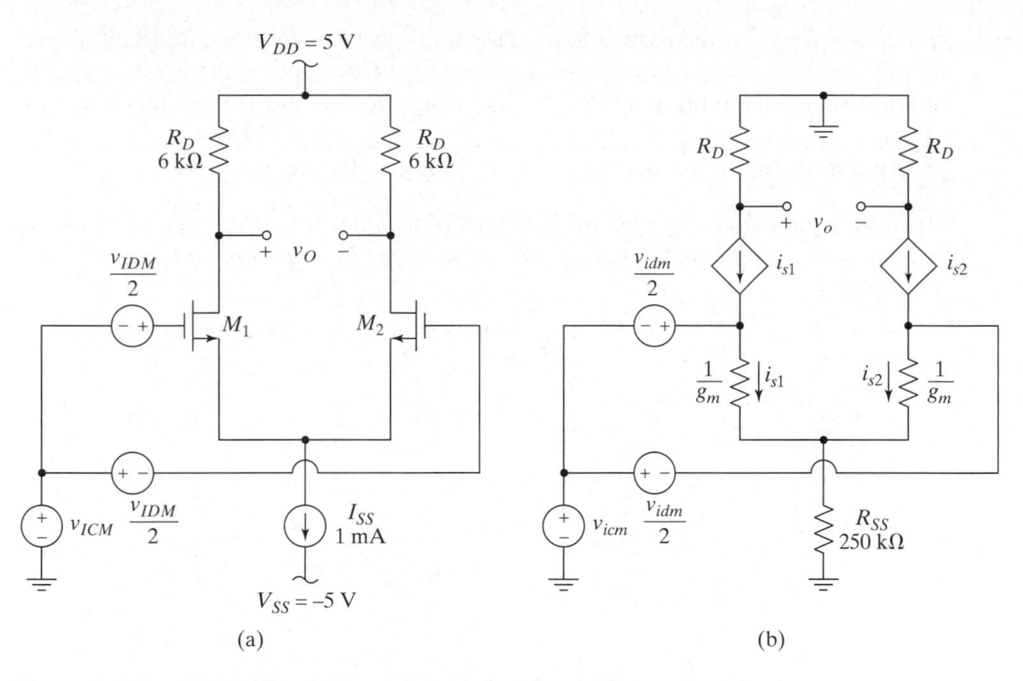

Figure 8–110 (a) A SCP and (b) its small-signal midband AC equivalent circuit.

which is −0.012 in this example. The CMRR for taking the output single ended is given by (8.250) and is

$$CMRR \approx g_m R_{SS}, \tag{8.286}$$

which is 112 (41 dB) in this example. The differential input resistance is infinite, and the differential output resistance, from (8.251), is

$$R_{od} = 2R_D, \tag{8.287}$$

which is 12 kΩ. This output resistance was calculated with the output resistance of the transistors ignored. Since the sources are a differential-mode virtual ground, the output resistances of M_1 and M_2 show up in parallel with the load resistors. If $V_A = 50$ V (i.e., $\lambda = 0.02$), the output resistances are both $r_o = 100$ kΩ, which is certainly reasonable to ignore here.

We note that there is a restricted range of common-mode input voltages over which this circuit can function properly. We have assumed in all of our analyses so far that the transistors are forward active. For that assumption to be true, the common-mode input cannot be too high or too low. If the common-mode input voltage is too high, the transistors will enter the linear region. The transistors enter the linear region if the common-mode input exceeds

$$V_{ICM\max} = V_D + V_{th} = V_{DD} - \frac{I_{SS}R_D}{2} + V_{th}, \tag{8.288}$$

which is 2.5 V for our present example. The common-mode input voltage must also be high enough that M_1 and M_2 are on and the tail-current source functions. Assuming that the current source requires 0.5 V of headroom to function, the minimum acceptable common-mode input for our example is

$$V_{ICM\min} = V_{SS} + V_{mirror,\min} + V_{GS} = V_{SS} + V_{mirror,\min} + V_{th} + \sqrt{\frac{I_D}{K}}, \tag{8.289}$$

which is −1.76 V for the case at hand. The limits given by (8.288) and (8.289) specify the ***common-mode input range*** of this differential amplifier. This range applies only for small differential inputs. The following exercise illustrates this point.

Exercise 8.14

Drive the differential amplifier in Figure 8-110 with a single-ended source as shown in Figure 8-111. What is the range of input voltages for which the circuit will work properly?

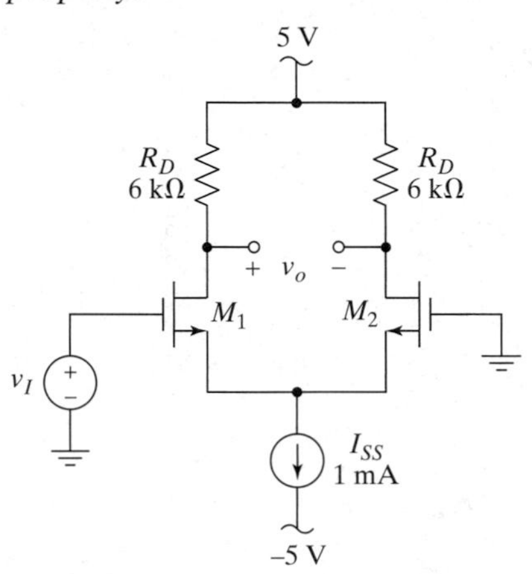

Figure 8–111 The SCP driven single-ended.

The Actively Loaded SCP An actively loaded SCP is shown in Figure 8-112. If this amplifier drives a short circuit and has a purely differential input, the output current, according to (8.264) is

$$i_o \approx g_{mn} v_{idm}, \tag{8.290}$$

where g_{mn} is the transconductance of one of the n-channel devices, which is not necessarily the same as the transconductance of the p-channel devices.

If the circuit drives a load R_L, the differential voltage gain is

$$A_{dm} \approx g_{mn} R_L, \tag{8.291}$$

as in (8.265). The maximum value for this gain is found when the output is open circuited. The effective load resistance is then due to the output resistances of M_2

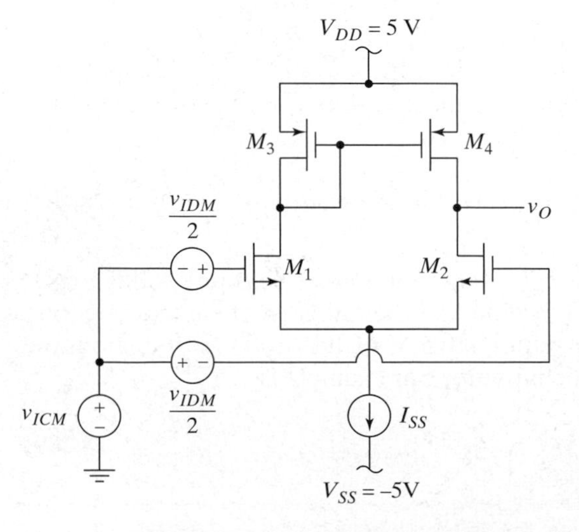

Figure 8–112 An actively loaded SCP.

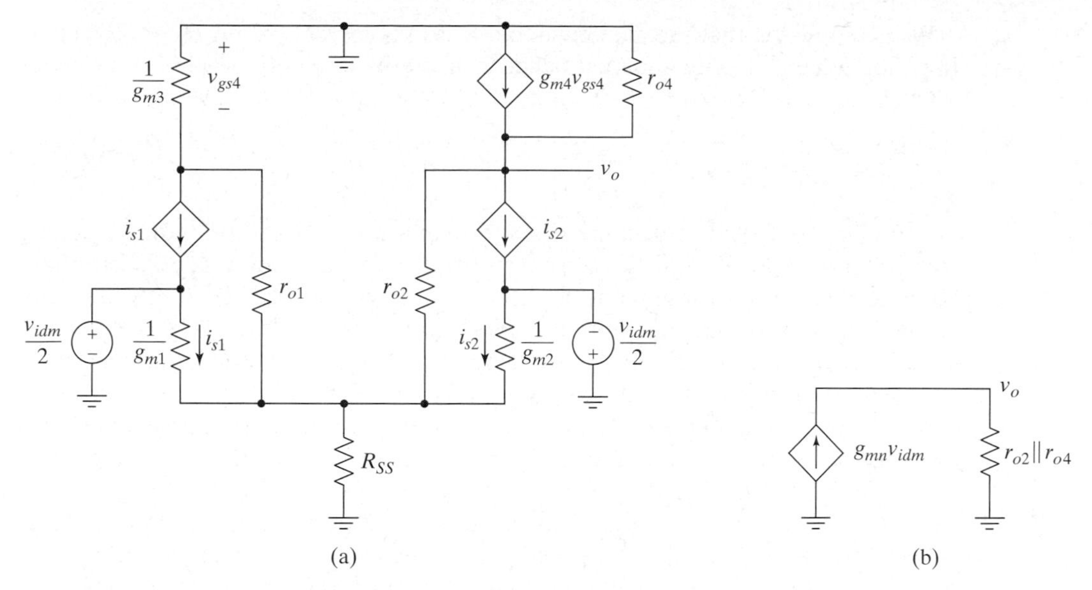

Figure 8–113 (a) The small-signal equivalent circuit for the actively loaded SCP with purely differential inputs. (b) An approximate equivalent circuit at the output node.

and M_4 as shown in Figure 8-113(a). We can greatly simplify this circuit if we make a few reasonable approximations and think about the operation of the circuit at a higher level. First, note that since the effective resistance of the diode-connected transistor is small (i.e., $1/g_{m3}$ is small), the AC voltage at the drain of M_1 will be small. If we also assume that the AC voltage at the sources of M_1 and M_2 is about zero (we will return to this assumption later), the voltage across r_{o1} will be small and the current through it can be neglected. Tying all of this together, we see that the output of the controlled source of M_4 is given by

$$g_{m4}v_{sg4} \approx g_{m4}i_{s1}(1/g_{m3}) = i_{s1}. \tag{8.292}$$

We now make use of (8.261) and find the total current being forced into the output node:

$$g_{m4}v_{sg4} - i_{s2} \approx i_{s1} - i_{s2} = g_{mn}v_{idm}. \tag{8.293}$$

In part (b) of the figure, we show an approximate equivalent circuit for the output of this amplifier. The total current into the output node is given by (8.293), and the total resistance to ground is r_{o2} in parallel with r_{o4}, where we have again assumed that the AC voltage at the sources of M_1 and M_2 is near zero. The resulting differential gain of the circuit is found from part (b) of the figure to be

$$A_{dm} = \frac{v_o}{v_{idm}} \approx g_{mn}(r_{o2}||r_{o4}). \tag{8.294}$$

Exercise 8.15

Consider the circuit in Figure 8-112, and assume that $\lambda_n = 0.0133\ V^{-1}$, $\lambda_p = 0.02\ V^{-1}$, $I_{SS} = 500\ \mu A$, $K_n = 100\ \mu A/V^2$, and $V_{thn} = 0.5\ V$. (a) Find A_{dm}. (b) If a resistive load were used instead, how large would the resistor need to be to achieve the same gain (taking the output single ended)? (c) How large would the supply voltage need to be?

We can now see that the gain given by (8.291) is correct only if $R_L \ll (r_{o2} \| r_{o4})$. In general, the gain with a load attached is given by

$$A_{dm} = \frac{v_o}{v_{idm}} \approx g_{mn}(R_L \| r_{o2} \| r_{o4}). \tag{8.295}$$

We now go back and reexamine the approximation that the AC voltage at the sources of M_1 and M_2 is about zero. We proved that this node was a differential-mode ground for our original differential pair by writing (8.237) and (8.238) and recognizing that these contributions cancel for a balanced circuit. In the present situation, however, the circuit is not completely balanced, because the load presented to M_1 is different from that presented to M_2. Therefore, the two drain voltages are quite different, and if we include the transistors' output resistors as we have here, the currents in r_{o1} and r_{o2} will not cancel each other, so the sources cannot be at ground potential. Nevertheless, the sources of the transistors are connected to the two signal sources through small resistances $(1/g_m)$, and the voltage at the sources will therefore be small *when compared with* the large output voltage. As a result, it is reasonable *on the output node*, to approximate r_{o2} as being connected to ground.

Using (8.268), we find that the common-mode gain of this circuit is zero, since the common-gate current gain of a MOSFET is exactly one at low frequencies. The common-mode gain of this circuit is not zero if the devices are not perfectly matched. As seen in Chapter 5, good op amps have CMRRs on the order of 100–120 dB.

Exercise 8.16

Consider again the circuit in Exercise 8.15. Calculate the CMRR of the amplifier if there is an external load of $R_L = 10$ kΩ. Assume that the (W/L) of M_4 is 5% larger than that of M_3 and the current source output resistance is $R_{SS} = 100$ kΩ.

□

8.7 MULTISTAGE AMPLIFIERS

The single-transistor amplifiers and differential pairs covered in the preceding sections are rarely used by themselves. Most applications require that two or more of these stages be connected together. We call the series connection of two or more amplifier stages a ***cascaded*** or a ***multistage*** amplifier. The design of multistage amplifiers necessarily depends on a thorough understanding of the characteristics of the three different types of single-transistor amplifier, so we summarize the relative performances of these stages in Table 8.1. A more extensive discussion is presented in Chapter 9 after we have covered frequency response. Cascades involving differential stages are discussed separately in Section 8.7.4.

In the sections that follow, we discuss multistage bipolar amplifiers, FET amplifiers, amplifiers that use both bipolar and FET devices, and, finally, amplifiers that use differential pairs. Additional multistage amplifier topologies are considered in Chapters 9 and 10; we focus here on stages that are motivated by small-signal midband AC requirements.

Table 8.1 Relative Values of the Midband AC Characteristics for the Three Single-Transistor Amplifier Stages

Amplifier	R_i	R_o	A_v	A_i
Stages with voltage and current gain	high	high	high	high
Voltage buffers	high	low	<1	high
Current buffers	low	high	high	≤ 1

8.7.1 Multistage Bipolar Amplifiers

In discrete circuit design, the most common multistage amplifier is a common-emitter emitter-follower cascade. The common-emitter stage provides voltage gain and a reasonably high input impedance, while the emitter-follower allows the amplifier to drive small and possibly varying loads without affecting the gain. More advanced output stages are covered in Chapter 12; at this point we are interested only in seeing how we can combine the characteristics of single-transistor amplifiers to achieve some overall design goal. The use of common-base stages in amplifier cascades is presented in Section 8.5.2 and in Chapter 9.

Suppose we are asked to design an amplifier that will have a voltage gain equal to 20 dB, an input resistance of at least 50 kΩ, and an output resistance of no more than 100 Ω when driven from $R_S = 10$ kΩ and while driving $R_L = 1$ kΩ. We can certainly achieve the desired gain and input impedance with a single common-emitter stage like the one in Figure 8-114, but to achieve the output impedance, we would need to use a large bias current, which would make it impossible to meet the input impedance specification, as we now demonstrate with some calculations.

Assume that $V_{CC} = 10$ V and that $\beta = 100$. The output resistance of this amplifier is roughly just R_C. Therefore, to meet the specification for R_o, we must set $R_C = 100$ Ω. The collector current is then $I_C = (V_{CC} - V_C)/R_C$, so the higher we make V_C, the lower the current and power will be. To achieve a reasonable signal swing at the collector of the transistor, however, we can't set V_C too close to V_{CC}. (Swing is covered in detail in Chapter 12. For now, we will approach it intuitively; the output can't swing any farther than the collector of Q_1 can, and the positive swing on

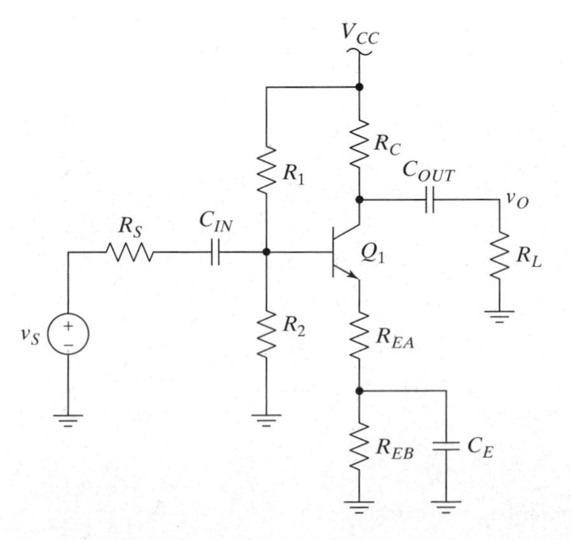

Figure 8-114 A common-emitter amplifier.

v_{C1} is limited by V_{CC}.) A reasonable value for V_C would be 8 V, which implies that $I_C = 20$ mA. The base current is then 0.2 mA, and to satisfy our rule of thumb for a stable design, we need to set $I_{R1} = 2$ mA. But using a current this large in the bias resistors will not allow us to meet the input impedance specification, since the input resistance is given by

$$R_i = R_{BB} \| [r_\pi + (\beta+1)R_{EA}], \tag{8.296}$$

which is always less than R_{BB} alone. For a given current in the bias string, the largest value of R_{BB} is achieved when $R_1 = R_2$, so with $I_{R1} = 2$ mA, the largest possible R_{BB} is 1.25 kΩ, 40 times smaller than we need. Therefore, we cannot make this circuit work even if we cheat on our rule of thumb for I_{R1} and also raise V_C a bit higher; we need a new topology.

We could focus on the input resistance alone and solve the problem by adding an emitter follower in front of the common-emitter stage. However, that still leaves us with a very large bias current in Q_1 and probably a reduced signal swing capability. A better solution is to add the emitter follower after the common-emitter stage. The emitter follower will be able to provide a low output resistance without requiring such a large current and will allow us to achieve our specifications with better signal swing and power dissipation. Since the input impedance of the emitter follower is high (*see* Table 8.1), it will not load down the common-emitter stage. The topology we choose is shown in Figure 8-115.

The input resistance of this amplifier is still given by (8.296). The gain is given by using impedance reflection and defining $R'_C = R_C \| [r_{\pi 2} + (\beta_2 + 1)(R_{E2} \| R_L)]$ and then combining it with (8.106) and (8.150) to obtain (*see* Problem P8.117)

$$A_v = \frac{-\alpha_1 R'_C}{r_{e1} + R_{EA}} \left(\frac{(\beta_2 + 1)(R_{E2} \| R_L)}{(\beta_2 + 1)(R_{E2} \| R_L) + r_{\pi 2}} \right). \tag{8.297}$$

Note that if the gain of the emitter follower is close to one, $R_{EA} \gg r_{e1}$, and $R'_C \approx R_C$, we then have $A_v \approx -R_C/R_{EA}$. In practice, these conditions are frequently met in a good design. The output resistance of this amplifier is found by using impedance reflection:

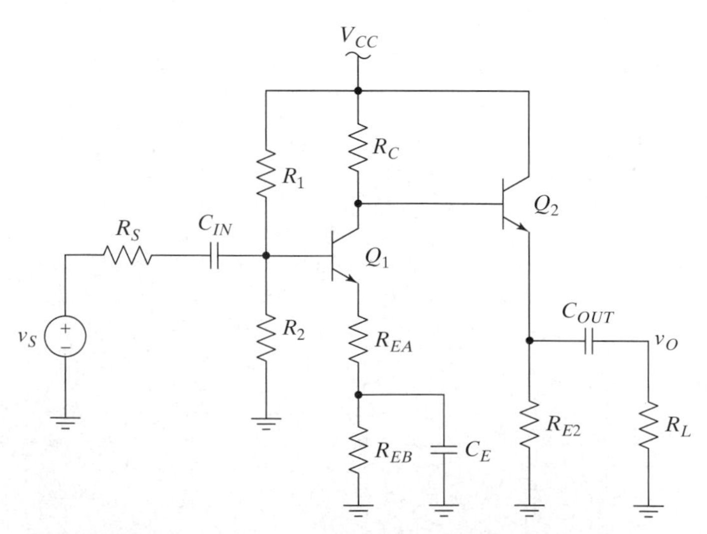

Figure 8–115 A common-emitter, emitter-follower cascade.

$$R_o = R_{E2} \Bigg\| \left[r_{e2} + \frac{R_C}{(\beta_2 + 1)} \right]. \tag{8.298}$$

Now that we have equations for R_i, R_o, and A_v, we can proceed with the design. There isn't a single correct way of approaching a design like this, nor is there an optimum solution,[12] but in this case we have already seen that the required R_i is going to be difficult to achieve, so it makes sense to start there.

To achieve $R_i \geq 50$ kΩ we need R_{BB} to be even larger. Let's do some quick calculations assuming that we want $R_{BB} = 100$ kΩ and see what happens. If we set $V_B = V_{CC}/2$, then $R_1 = R_2$ if we ignore I_B. Therefore, $R_1 = R_2 = 200$ kΩ will achieve our goal. With these values, we get $I_{R1} = 25$ μA. If we then follow our rules of thumb for biasing, we find, $I_{C1} = 250$ μA and $I_{C2} = 2.5$ mA. Now examine the bias voltages: Starting with $V_{B1} = 5$ V, we have $V_{E1} \approx 4.3$. Then we set V_{C1} halfway between $(V_{E1} + V_{CEsat})$ and V_{CC} to maximize swing; thus, $V_{C1} = 7.2$ V. Finally, we have $V_{E2} \approx 6.5$ V. Now we can check whether this bias point can also achieve the desired R_0. We have $r_{e2} = V_T/I_{E2} = 10.4$ Ω, $R_{E2} = V_{E2}/I_{E2} = 2.6$ kΩ, and $R_C = (V_{CC} - V_{C1})/I_{C1} = 11.2$ kΩ. Plugging these into (8.298) with $\beta_2 = 100$ yields $R_o = 116$ Ω. We are close, but not quite there. If we cheat on our rule of thumb just a bit though, and scale the currents up by a factor of twelve each time, instead of ten, we arrive at $I_{C1} = 300$ μA and $I_{C2} = 3.6$ mA. The resistors then turn out to be $r_{e2} = 7.2$ Ω, $R_{E2} = 1.8$ kΩ and $R_C = 9.1$ kΩ, where we have used the nearest 5 % resistor values for R_{E2} and R_C. The resulting R_o is 92 Ω, just within specification. We could cheat on the rule a bit more and get some additional safety margin, but for now we will stick with these choices.

We next find the emitter resistors for Q_1. With $I_{C1} = 300$ μA and $V_{E1} = 4.3$ V, the total emitter resistance needs to be 14.3 kΩ. To find how to split this, we examine the gain. The gain of the emitter follower is the term in parentheses in (8.297) and is equal to 0.99, so we just call it one and ignore that term. Also, $\alpha = 0.99$ for these transistors, so we can ignore it as well. Therefore, we can solve for R_{EA},

$$R_{EA} = \frac{-R_C}{A_v} - r_{e1}, \tag{8.299}$$

which is 823 Ω in this case. We choose the nearest 5% value and set $R_{EA} = 820$ Ω, which also determines $R_{EB} = 13$ kΩ. Now, since we are not setting $I_{R1} \geq 10 I_{B1}$, it makes sense to allow for the base current when we are finding R_1 and R_2. We have $I_{B1} = 3$ μA, and we wanted to set I_{R1} to 25 μA, so we know that $I_{R2} = 22$ μA. With $V_{B1} = 5$ V, we find $R_1 = 200$ kΩ and $R_2 = 227$ kΩ. Using the nearest 5% values we set $R_1 = 200$ kΩ and $R_2 = 220$ kΩ. Using (8.296) we find $R_i = 49$ kΩ, slightly below what we desired. We could fix this by decreasing I_{C1} or I_{R1} or both.

This circuit was simulated in SPICE using the Nsimple model described in Appendix B. The results were[13] $A_v = 18.8$ dB, $R_i = 49$ kΩ, and $R_o = 93$ Ω. In addition, the swing limits were 2.4 V and –2.2 V (the swing limits of this circuit are obtained analytically in Chapter 12), and the circuit dissipates 38 mW.

We see from this example that our knowledge of the individual stages enabled us to pick a cascade that can meet the required design specifications. We also went through the design process and came up with a first-cut design that almost meets the specifications and that can be modified slightly to meet them.

---

[12] We do not have every possible requirement specified (e.g., we don't have an explicit swing requirement), and we don't know the relative importance of the different specifications that may require trade-offs (e.g., R_i and A_v); therefore, no optimum can be defined. This example is typical in that it is almost never possible to find an optimum solution to a design problem.

[13] The SPICE file is called ce_ef.sch on the CD. If you run it, remember to look at v_o/v_i, not v_o/v_s.

8.7.2 Multistage FET Amplifiers

Let's first consider a common-source source-follower cascade. The common-source stage provides voltage gain and a high input impedance, while the source-follower allows the amplifier to drive small and possibly varying loads without affecting the gain too much. More advanced output stages are covered in Chapter 12; at this point, we are interested only in seeing how we can combine the characteristics of single-transistor amplifiers to achieve some overall design goal. The use of common-gate stages in cascades is presented in Section 8.5.2 and in Chapter 9.

Suppose we are asked to design an amplifier that will have a voltage gain equal to 20 dB and an output swing capability[14] of at least ±1 V with ±5-V supplies, $K = 2\ \text{mA/V}^2$, $V_{th} = 1$ V, $R_{SIG} = 100\ \text{k}\Omega$, and $R_L = 100\ \Omega$. We first consider the simplest possible design, a single common-source stage. Figure 8-116 shows a stage using current source biasing.

The small-signal midband AC voltage gain of this stage is

$$A_v = \frac{v_o}{v_i} = -g_m R_L', \tag{8.300}$$

where $R_L' = R_L \| R_D$. Notice that the maximum possible gain for this stage with the given load occurs when $R_L \ll R_D$, so that

$$|A_v|_{\max} = g_m R_L = 2R_L\sqrt{KI_{D\max}}. \tag{8.301}$$

Note also that the maximum possible V_{GS} is 5 V (a practical current source requires some drop across it to work; this is a theoretical maximum), and therefore, the maximum drain current,

$$I_{D\max} = K(V_{GS\max} - V_{th})^2, \tag{8.302}$$

is 32 mA. Plugging this current into (8.301) along with the given values yields a maximum gain of 1.6. We could correct this problem by finding a transistor with a bigger K, but we would still need to use a large current, and such a transistor may not be available. A better solution is to add a source follower as shown in

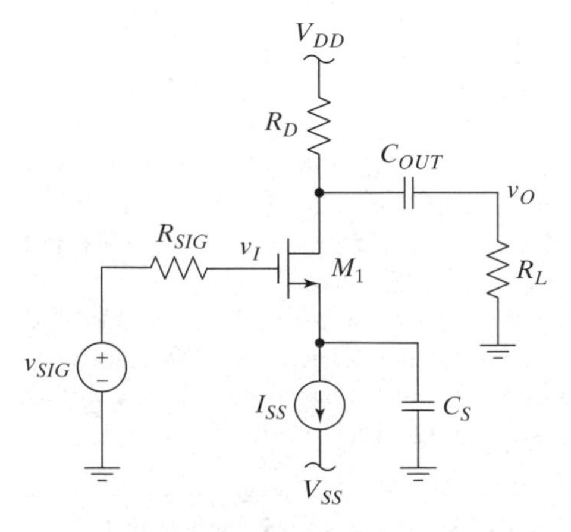

Figure 8–116 A common-source amplifier with current source biasing.

---

[14] Swing is covered in detail in Chapter 12; we will take an intuitive approach here for the sake of this example.

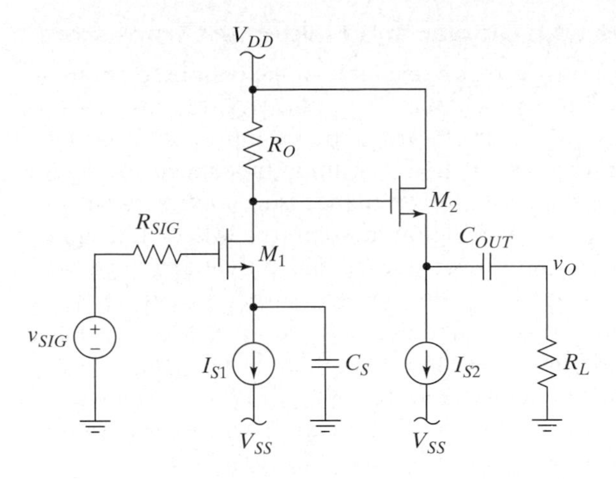

Figure 8–117 A common-source source-follower cascade.

Figure 8-117. We have again used a current source to bias the stage. Since the small-signal midband AC input impedance of the source follower is infinite, it will not load down the common-source stage, and we won't have any problem achieving the desired gain.

To carry out the design of this amplifier, we first list the known constraints. The gain of the follower will be less than one, and we want a swing of ±1 V at the output. Therefore, we require that V_{D1} be able to swing more than ±1V. Since v_{D1} is limited on the high side by V_{DD}, we conclude that $V_{D1} < 4$ V. To be safe, let's set $V_{D1} = 3$ V. Next, we see that the best we can do for the negative swing is to have M_2 cut off (only during the negative transient of v_{D1}), at which time the entire current I_{S2} momentarily flows through R_L. Therefore, we must have $I_{S2} > 10$ mA to achieve $v_{omin} = -1$ V. Allow a safety margin and choose $I_{S2} = 12$ mA. This current requires that $V_{GS2} = 3.45$ V, so with $V_{D1} = 3$ V. we find $V_{S2} = -0.45$ V. Now, the gain of the source follower is

$$\frac{v_o}{v_{d1}} = \frac{R_L}{R_L + 1/g_{m2}}, \tag{8.303}$$

which is 0.5 for this design. We could make this gain higher, but only at the cost of increasing the DC-power consumption, so we will leave it as it is. We will compensate by setting the gain of the common-source stage to 20 instead of 10. The gain of the common-source stage is

$$A_v = -g_m R_D = -2R_D\sqrt{KI_{D1}}. \tag{8.304}$$

We also know that $R_D = (V_{DD} - V_D)/I_{D1}$, so we can plug this into (8.304) and solve to find

$$I_{D1} = \frac{4K(V_{DD} - V_D)^2}{|A_v|^2}, \tag{8.305}$$

which works out to be 80 μA, for a gain of 20. With I_{D1} set to 80 μA, we obtain $R_D = 25$ kΩ.

This circuit was simulated using SPICE and the Nlarge_mos models from Appendix B. The simulated gain is 19.9 dB, and the simulated swing limits are 1.1 V and −1.2 V (see the file cs_sf.sch on the CD).

8.7.3 Multistage Amplifiers with Bipolar and Field-Effect Transistors

If both bipolar and field-effect transistors are available, it is possible to combine them to produce cascaded amplifiers with characteristics superior to those achievable with either type of transistor alone. There are many examples of mixed BJT and FET designs. In general, FETs have much higher input impedance and so are useful as the input stage when very high input impedance is desired, as, for example, in active probes for oscilloscopes and digital multimeters. Also, FETs make better switches for turning off and on power devices, since they can have very small voltage drops from drain to source when they are pushed heavily into the linear region. However, bipolar transistors are easier to bias and use for simple amplifying tasks, especially with small supply voltages, and, most importantly, bipolar transistors have significantly higher gain than MOSFETs of comparable size when using equal bias currents. Bipolar transistors can also have lower output resistance when used as followers, again caused by their higher g_m.

The analysis and design of mixed FET and BJT circuits follow directly from the analysis and design of single-transistor-type circuits, as already covered. The only significant difference is that the designer needs to be aware of the advantages and disadvantages of each type of device.

8.7.4 Multistage Amplifiers with Differential Pairs; Operational Amplifiers

Differential signal processing is routinely used in IC design. Two important advantages to using differential signals are that they can provide significant immunity to interfering signals, since interference is frequently common-mode (*see* Asides A5.1 and A5.2), and they allow us to use DC coupling and therefore avoid coupling and bypass capacitors.

In addition to the use of fully differential amplifiers (i.e., differential inputs and outputs) as a part of larger ICs, differential input stages are used in all operational amplifiers. Operational amplifiers are discussed at length in Chapter 5, in this section we briefly discuss the classic architecture of an operational amplifier and discuss how the different stages combine to achieve the overall performance of the device. The detailed discussion of operational amplifier design rightly belongs to more advanced texts. If you are interested, we recommend [8.1] and [8.3] as excellent places to start.

Figure 8-118 shows a block diagram of the classic operational amplifier structure. The input stage (first stage) has differential inputs and a single-ended output and is used to provide voltage gain, a high input impedance, a differential input, and a single-ended output. The gain stage (second stage) is used to provide additional voltage gain. The level shift circuit is optional, but is used to properly bias the output stage and may be part of the second stage, the output stage, or a separate block. The output stage (third stage) provides a low impedance output and the ability to drive small loads with relatively large signals.

A simplified schematic of the classic architecture is shown in Figure 8-119 using generic transistors. The first stage is an actively loaded differential pair, the second stage is a single-ended common-merge transistor, and the output stage is a

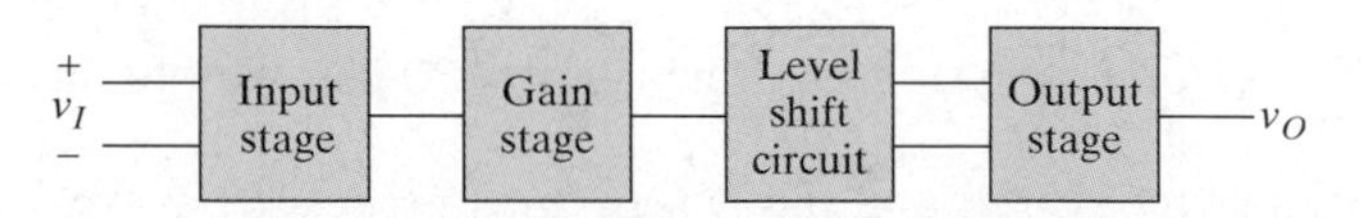

Figure 8–118 The block diagram of a classic operational amplifier.

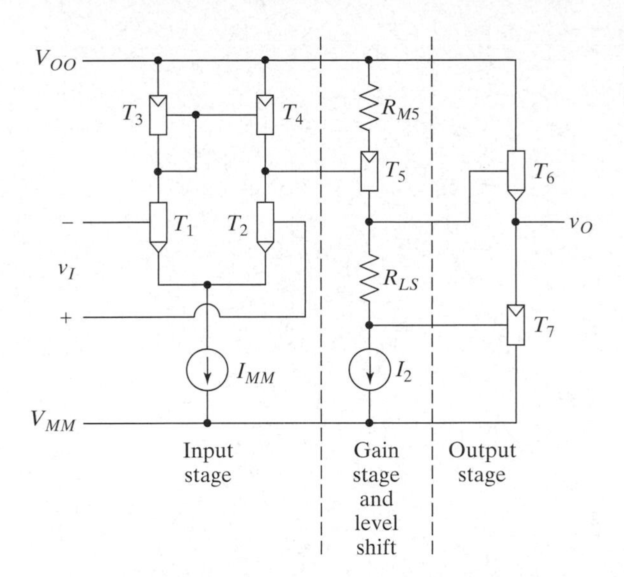

Figure 8–119 A simplified schematic of an operational amplifier.

push–pull voltage buffer (covered in detail in Chapter 12, but briefly described here). The level shift is most conveniently drawn as part of the second stage in this case. Because the current through the resistor R_{LS} is constant (ignoring the small currents into the control inputs of the output stage), the voltage drop across it is constant as well. This voltage drop allows T_6 and T_7 to both be biased on when no input is applied. The buffer is a combination of an n-type and a p-type voltage follower. The n-type transistor provides output current when $v_O \geq 0$, and the p-type transistor provides output current when $v_O \leq 0$.

The current sources shown in the schematic are implemented using current mirrors. I_2 establishes the bias current for T_5 and also acts as an active load, so that the gain of the second stage is quite large. The following two examples show simplified versions of typical BJT and MOSFET operational amplifiers.

Example 8.8

Problem:
Determine the DC biasing and small-signal midband AC open-circuit voltage gain for the simplified operational amplifier shown in Figure 8-120. Assume that all of the transistors have $V_A = 50$ V and $\beta = 100$.

Solution:
We first need to identify the different blocks in the circuit and then break the analysis down into smaller pieces. The first stage of the op amp comprises Q_1, Q_2, Q_3, Q_7, and Q_8. Q_3 provides the tail current and, along with Q_4, Q_5, and Q_9, sets up the biasing. Q_{10} is the common-emitter gain stage (the second stage) and has Q_5 as an active load. Q_6 and Q_{11} form the push–pull output stage.

The DC bias currents are established by R_{BIAS}. Since Q_4 and Q_9 are diode-connected, the bias current is

$$I_{BIAS} = \frac{V_{CC} - V_{EE} - 1.4}{R_{BIAS}}. \quad (8.306)$$

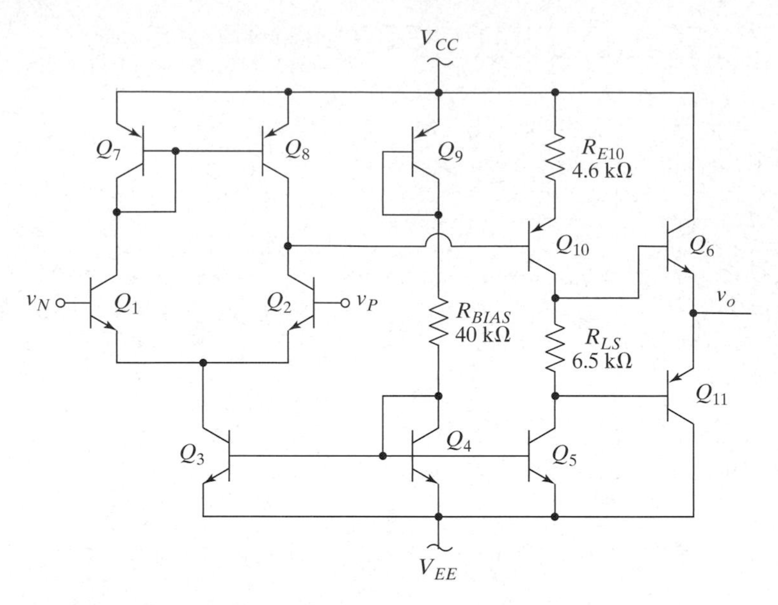

Figure 8–120 A simplified BJT operational amplifier.

With ± 5V supplies, this works out to be 215 μA. This current flows in R_{E10}, so we now know that $V_{E10} = V_{CC} - I_{BIAS}R_{E10} = 4$ V and, therefore, $V_{B10} \approx 3.3$ V. We also know that $r_{o10} = r_{o5} = 233$ kΩ and $g_{m10} = 8.3$ mA/V. The voltage drop across R_{LS} is seen to be 1.4 V, so with no input voltage, Q_6 and Q_{11} are both on.

To find the open-circuit voltage gain we set R_L to infinity, so the input impedance of the output stage is also infinite and the effective load for Q_{10} is $r_{o10} \| r_{o5}$, which is 117 kΩ. The gain of the second stage is, therefore, $A_{v2} = -(r_{o10} \| r_{o5})/(r_{e10} + R_{E10}) = -24.8$.

The bias current splits equally in Q_1 and Q_2, so for all of the transistors in the input stage, we find that $g_m = 4.1$ mA/V and $r_o = 465$ kΩ. The effective load for the first stage is the input impedance of the second stage. Using impedance reflection, we find that this input resistance is $R_{b10} = r_{\pi 10} + (\beta+1)R_{E10} = 477$ kΩ. Using (8.282), we calculate the first stage gain to be $A_{v1} = 641$.

With an open-circuit load, the output stage will have a gain of one; therefore, the overall open-circuit voltage gain is $A_v = A_{v1}A_{v2} = -15{,}900$. The minus sign applies only for inputs at the base of Q_1 (the inverting input of the op amp).

We cannot perform an open-loop simulation of this circuit in SPICE. There is a small systematic input offset voltage caused by the asymmetric load on the first stage and, coupled with the large voltage gain, the output will always be fully saturated one way. Therefore, we simulated the circuit using a standard noninverting op amp circuit with a gain of 10. (We used 9 kΩ and 1 kΩ for the feedback resistor network.) Using this configuration, SPICE[15] finds a bias current of 219.4 μA (we predicted 215 μA), a first-stage gain of 571 (we had 641) a second-stage gain of −34.5 (we had −24.8), and an overall gain of −19,587 (we had −15,900).

■

[15] This circuit is too large to simulate with the student version of PSPICE and is not, therefore, on the CD.

Example 8.9

Problem:
Determine the DC biasing and small-signal midband AC open-circuit voltage gain for the simplified operational amplifier shown in Figure 8-121. Assume that all of the transistors have $K = 100\ \mu A/V^2$, $V_{th} = \pm 0.5$ V as appropriate, and $\lambda = 0.02\ V^{-1}$.

Solution:
We first need to identify the different blocks in the circuit and then break the analysis down into smaller pieces. The first stage of the op amp comprises M_1, M_2, M_3, M_4, and M_9. M_9 provides the tail current and, along with M_8, M_{10}, and M_{11}, sets up the biasing. M_5 is the common-source gain stage (the second stage) and has M_{11} as an active load. M_6 and M_7 form the push–pull output stage.

The DC bias currents are established by R_{BIAS}. Since M_8 and M_{10} are diode-connected and have the same value of K and the same magnitude of V_{th}, the bias current is

$$I_{BIAS} = \frac{V_{DD} - V_{SS} - 2V_{GS10}}{R_{BIAS}}. \quad \textbf{(8.307)}$$

Using $V_{GS10} = V_{th} + \sqrt{I_{BIAS}/K}$ in this equation produces a quadratic that can be solved for V_{GS10}. With $\pm$ 5V supplies, the result is 2.08V and leads to a 250 μA bias current.

We also know that $V_{SG5} = V_{GS10}$ since the devices are matched and have the same current. Therefore, $V_{G5} \approx 3$ V. Using the known current, we also find that $r_{o5} = r_{o11} =$ 200 kΩ and $g_{m5} = 0.316$ mA/V. The DC voltage drop across R_{LS} is seen to be 3 V, so with no input voltage, M_6 and M_7 both have a gate overdrive of 1 V, which leads to a bias current of 100 μA. The effective load for M_5 is $r_{o11} \| r_{o5}$, which is 100 kΩ. (R_{LS} has negligible effect and is ignored.) The gain of the second stage is, therefore, $A_{v2} = -g_{m5}(r_{o11} \| r_{o5}) = -31.6$.

The bias current splits equally in M_1 and M_2, so for all of the transistors in the input stage, we find that $g_m = 0.158$ mA/V and $r_o = 400$ kΩ. Using (8.294), we calculate the first stage gain to be $A_{v1} = 31.6$.

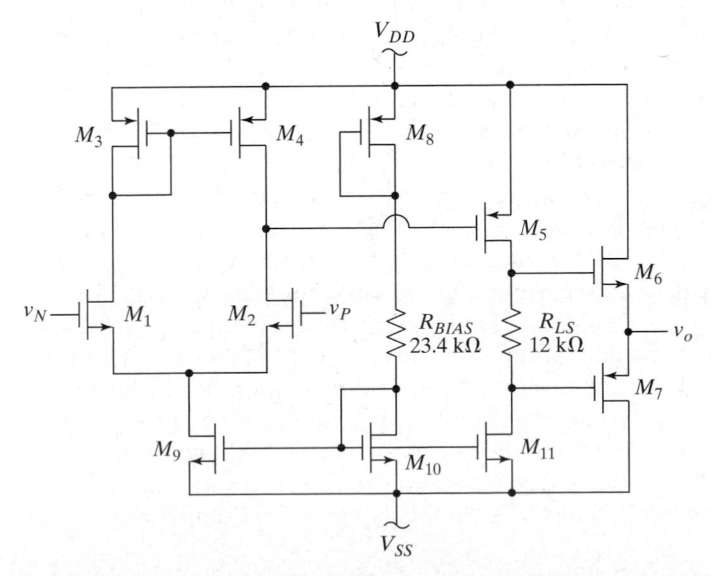

Figure 8–121 A simplified MOSFET operational amplifier.

With an open-circuit load, the output stage will have a gain of one; therefore, the overall open-circuit voltage gain is $A_v = A_{v1}A_{v2} = -999$. The minus sign applies only for inputs at the gate of M_1 (the inverting input of the op amp).

We cannot perform an open-loop simulation of this circuit in SPICE. There is a small systematic input offset voltage caused by the asymmetric load on the first stage, and coupled with the large voltage gain, the output will always be fully saturated one way. Therefore, we simulated the circuit using a standard noninverting op amp circuit with a gain of 10. (We used 90 kΩ and 10 kΩ for the feedback resistor network.) Using this configuration, SPICE[16] finds a bias current of 239 μA (we predicted 250 μA), a first-stage gain of 40 (we had 31.6), a second-stage gain of −28.6 (we had −31.6), and an overall gain of −1073 (we had −999).

■

8.8 COMPARISON OF BJT AND FET AMPLIFIERS

We discussed some of the differences between BJTs and FETs, in Sections 2.10 and 8.1.8. In this section, we focus on the differences between the small-signal midband AC performances of BJT and FET amplifiers and current mirrors. We include current mirrors because they affect the performance of IC amplifiers.

The most obvious differences between BJT and FET amplifiers are that FET amplifiers have much higher input impedances than BJT amplifiers and BJT amplifiers have much higher voltage gains than FET amplifiers (at the same bias current). The difference in voltage gain is a direct result of the difference in transconductance pointed out in Section 8.1.8.

Examining voltage buffers in particular, we see that in addition to the difference in gain caused by the transconductance, the gain of MOS source followers may be further reduced by the body effect (*see* Section 8.5.1). The output resistance of a source follower is also higher than that of an emitter follower (at the same bias current), because of the lower FET transconductance; *see* (8.149) and (8.159).

Turning our attention to current mirrors for the moment, we note that the output resistance of a bipolar mirror is limited. We found in (8.220) that even when the transistor is driven by a perfect current source, the output resistance is finite; and if we somehow try to make R_o larger, r_μ will limit us. Our models for FET transistors do not include a resistor similar to r_μ, they can theoretically achieve almost infinite values of output resistance by cascoding stages. However, there is a limit to the output resistance achievable in MOSFET mirrors due to substrate currents that we have not discussed [8.1]. Nevertheless, FET mirrors are capable of higher output resistance than bipolar mirrors are (*see* Problem P8.136).

In general, the range of input voltages over which FET amplifiers appear approximately linear is wider than the corresponding range for bipolar amplifiers. This difference is caused by the fact that the collector current of a BJT is a function of v_{BE}/V_T, whereas the drain current of a FET is a function of $v_{GS}/(V_{GS} - V_{th})$ and the gate overdrive, $(V_{GS} - V_{th})$, is typically much larger than the thermal voltage, V_T (*see* Chapter 12 for a derivation of the large-signal transfer characteristics and a discussion of the distortion resulting from nonlinear operation). The bipolar linear range can be increased by adding emitter degeneration, but doing so requires adding a resistor. In the case of a FET, the range can be increased by increasing the gate overdrive or by adding source degeneration. The ability to vary the gate overdrive adds an extra dimension of flexibility to the design of FET amplifiers.

---

[16] This circuit is too large to simulate with the student version of PSPICE and is not, therefore, on the CD.

The matching achievable with BJTs of a given size is better than that achievable with FETs of comparable size. One important ramification of this fact is that bipolar differential amplifiers have lower offset voltages than FET differential amplifiers do. In addition, because the gains of BJT amplifiers are higher, the offsets of later stages do not matter as much, whereas in the case of FET amplifiers, the offsets of later stages may be important when referred back to the input.

If we consider low-voltage design, bipolar amplifiers have distinct advantages over FETs. The v_{BE} of a BJT is typically around 0.7 V and does not change much, even for large changes in the collector current. The v_{GS} of FETs, on the other hand, is often much larger (since V_{th} alone is often around 0.7 V) and increases more as the drain current is increased. In addition, $V_{CE\text{sat}}$ is typically around 0.2 V, which is usually lower than the minimum drain-to-source voltage required to keep a FET forward active, $V_{DS\text{sat}}$. For both of these reasons, bipolar amplifiers can function well with supply voltages lower than FET amplifiers require.

We end this brief section with two comments addressing more advanced issues. First, because FETs are square-law devices, whereas BJTs are exponential, FET amplifiers typically have lower distortion (*see* Chapter 12). Second, the noise performances of FET and BJT amplifiers are quite different. When the resistance of the signal source is low, bipolar amplifiers usually have lower noise than FET amplifiers do. On the other hand, when the resistance of the signal source is high, the situation is reversed. As a final comment on noise, we note that MOSFETs have significantly more low-frequency noise than bipolar transistors.

SOLUTIONS TO EXERCISES

8.1 Assuming that the average diode voltage is 0.7 V, the bias analysis yields

$$I_D = \frac{V_I - 0.7}{R_1} = 2.3 \text{ mA}.$$

Using (8.10), we then that find $r_d = 11.3\ \Omega$. The resulting small-signal midband AC equivalent circuit is shown in Figure 8-122. Using this circuit, we find that

$$v_o = \left(\frac{r_d}{r_d + R_1}\right)v_i = 0.22\cos(628t) \text{ mV}.$$

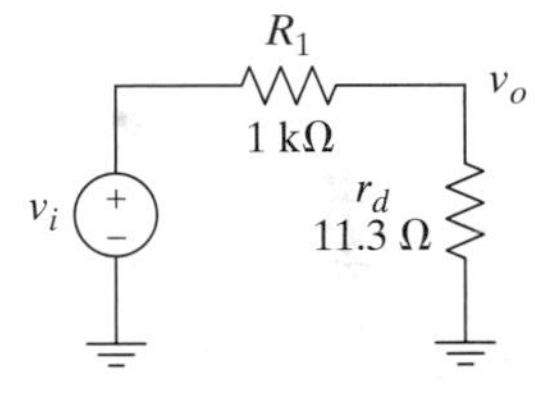

Figure 8–122 The small-signal midband AC equivalent circuit.

Notice that we did not need to know any details about the diode in order to arrive at this estimate (*see* Problem P8.7).

8.2 Assuming as usual that $V_{BE} \approx 0.7$ V and that I_B is much less than the current in R_1, we find that $V_B \approx V_{BB} = 2$ V, so $V_E \approx 1.3$ V, and $I_C \approx 1$ mA. We now check that the transistor is forward active as assumed; since $V_C = 6$ V, it is. We also check that I_B is much less than the current in R_1. (It is.) We now use (8.21) to find $r_\pi = 2.6$ kΩ and (8.16) to get $g_m = 38.5$ mA/V. The resulting equivalent circuit is shown in Figure 8-123.

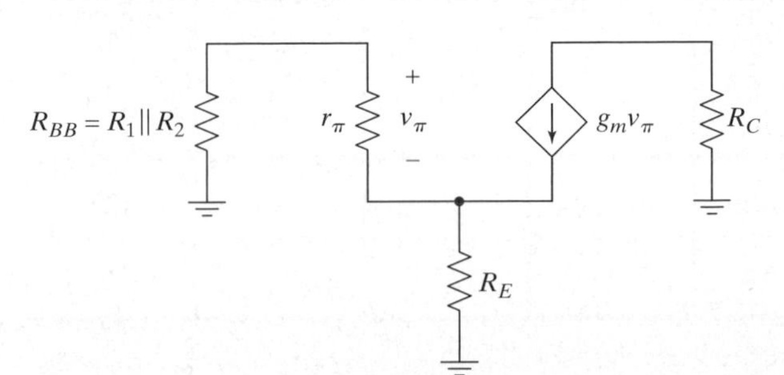

Figure 8–123 The small-signal midband AC equivalent circuit.

8.3 Assuming that the transistor is forward active, we have

$$I_D = K(V_{GS} - V_{th})^2.$$

From the circuit, we have

$$V_{GS} = 5 - I_D R_S.$$

Combining these two equations yields a quadratic for I_D:

$$I_D^2 K R_S^2 - I_D(1 + 8KR_S) + 16K = 0.$$

Solving the quadratic and plugging in the numbers yields $I_D = 2$ mA or 3.56 mA. Since $I_D = 3.56$ mA would require a drop of more than 5 V across R_S, we conclude that $I_D = 2$ mA. Using (8.36), we obtain $g_m = 4$ mA/V. The resulting small-signal midband AC equivalent circuit is shown in Figure 8-124.

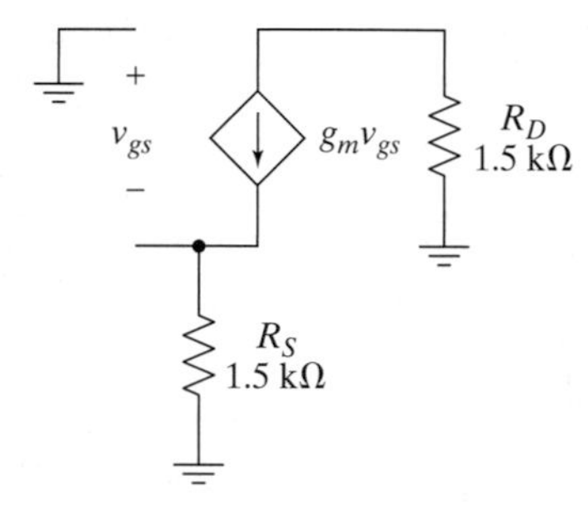

Figure 8–124 The small-signal midband AC equivalent circuit.

8.4 **(a)** Plugging the given values into (8.58) and (8.60), we find that $v_o/v_s = -6$. **(b)** From (8.67), we obtain $R_i = 1.96$ kΩ; from (8.68), we find that $R_o = 500$ Ω; and from (8.69), we obtain $a_v = -10$. **(c)** Using (8.74), we find that $v_o/v_s = -6$, which agrees with our previous result.

8.5 To save on number crunching, we set up an Excel spreadsheet for this problem. The spreadsheet first calculates V_{BB}, R_{BB}, V_E, and R'_L using our standard approximations. Then, for each value of R_{EA}, it finds I_C, and for each value of β, it finds r_π, R_b, R_i, and α_i from (8.99) – (8.101), A_v from (8.98), v_o/v_s, and the percentage differences between the gains at $\beta = 50$ and 200 and the nominal gains with $\beta = 100$. We then ran a SPICE simulation using the Nsimple model described in Appendix B. The results are shown in Table 8.2 and show that the larger R_{EA} yields a more stable gain, as expected. The simulated gains agree quite well with the calculated gains for large values of β, but for lower values they tend to differ by more because we are ignoring the base current in our hand calculations. The percent differences do not agree well, because they are based on the small differences between relatively large numbers; therefore, small errors in the large numbers lead to large errors in the small differences.

Table 8.2

R_{EA}	30 Ω			90 Ω		
β	50	100	200	50	100	200
v_o/v_s by hand	−26.3	−26.9	−27.3	−12.0	−12.2	−12.3
% difference by hand	2.4	0	1.3	1.6	0	0.8
v_o/v_s by SPICE	−25.9	−26.9	−27.4	−11.9	−12.2	−12.4
% difference by SPICE	3.7	0	2	2.2	0	1.2

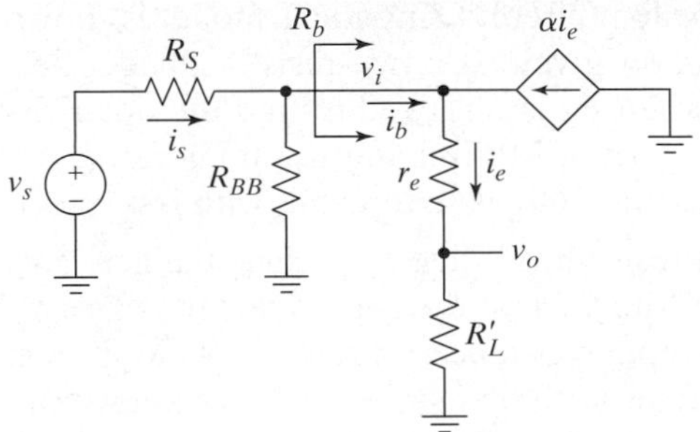

Figure 8–125

8.6 The small-signal midband AC equivalent circuit using the T model is shown in Figure 8-125. The voltage gain is now given by a two-resistor divider network as $A_v = R_L'/(R_L' + r_e)$, and when R_L is removed to find the open-circuit gain, we obtain $a_v = R_E/(R_E + r_e)$. Multiplying the numerator and denominator of this result by $(\beta+1)$ yields (8.150) as before, but with less effort.

The small-signal midband AC input resistance is seen to be R_{BB} in parallel with the resistance looking in the base; call it R_b, as shown in the figure. This resistance is given by

$$R_b = \frac{v_i}{i_b} = \frac{v_i}{(1-\alpha)i_e} = \frac{(r_e + R'_L)}{(1-\alpha)},$$

and using $\alpha = \beta/(\beta+1)$ yields $R_b = (\beta+1)(r_e + R_L') = r_\pi + (\beta+1)R_L'$. Combining this in parallel with R_{BB} yields the same result as (8.148). To find R_o, we need to split R_L' into R_L in parallel with R_E, remove R_L, set v_s to zero, and find the resistance looking in where R_L had been connected. The circuit is shown in Figure 8-126.

By inspection, we see that R_o is equal to R_E in parallel with the resistance seen looking into the emitter, R_e, as shown in the figure. We have $R_e = -v_x/i_e$. Writing KVL around the loop, we obtain $-v_x = (1-\alpha)i_e(R_S \| R_{BB}) + i_e r_e$. Using $\alpha = \beta/(\beta+1)$ and solving, that we find $R_e = r_e + (R_S \| R_{BB})/(\beta+1)$. Putting this in parallel with R_E yields (8.149) again. Finally, we note that the input and output resistances were easier to find using the hybrid-π model and impedance reflection.

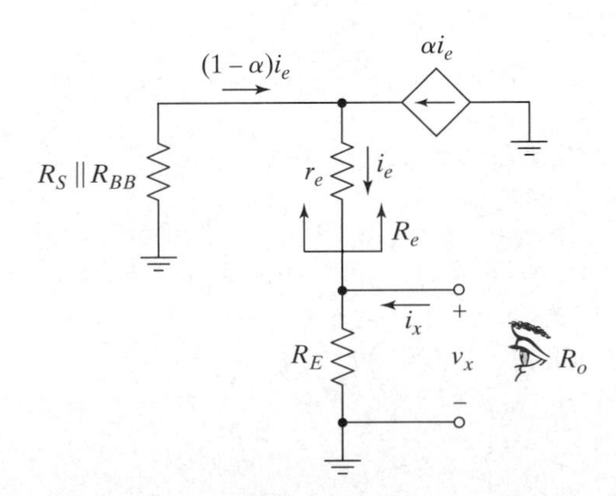

Figure 8–126

8.7 The small-signal midband AC equivalent circuit using the T model is shown in Figure 8-127. The voltage gain is now given by a two-resistor divider network, $A_v = R'_L/(R'_L + 1/g_m)$, and when R_L is removed to find the open-circuit gain, we obtain $a_v = R_S/(R_S + 1/g_m)$. Multiplying the numerator and denominator of this result by g_m yields (8.160) as before, but with less effort.

The small-signal midband AC input resistance is just R_{GG}, since the gate current is zero as noted on the figure. (Write KCL at the gate to see this.) This result agrees with (8.158). To find the output resistance, we remove R_L, look back into the circuit, and set v_{sig} to zero. Now note that since $i_g = 0$, we know that, $v_g = 0$. Therefore, the top of the resistor $1/g_m$ is connected to AC ground and

$$R_o = R_S \| \frac{1}{g_m},$$

as in (8.159).

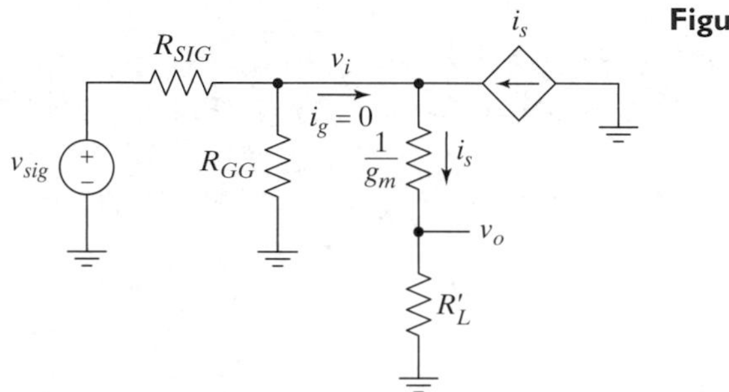

Figure 8–127

8.8 The circuit is shown in Figure 8-128. R_i is given by (8.174), R_o is infinite, and the short-circuit current gain is given by (8.171). Using this circuit, we find that the overall current gain is given by

$$\frac{i_o}{i_s} = \left(\frac{R_S}{R_S + R_i}\right) a_i \left(\frac{R_o}{R_o + R_L}\right) = \alpha_i a_i \alpha_o,$$

which is the same as (8.172) because R_o is infinite and therefore, $\alpha_o = 1$.

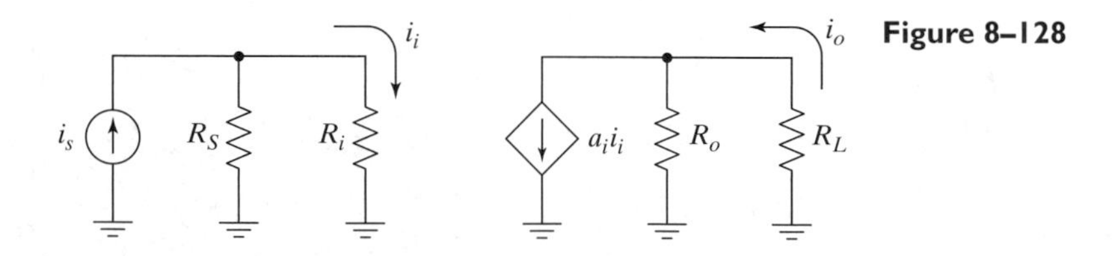

Figure 8–128

8.9 The circuit is shown in Figure 8-129. R_i is given by (8.185), R_o is infinite, and the short-circuit current gain is given by (8.182). Using this circuit, we find that the overall current gain is given by

$$\frac{i_o}{i_{sig}} = \left(\frac{R_{SIG}}{R_{SIG} + R_i}\right) a_i \left(\frac{R_o}{R_o + R_L}\right) = \alpha_i a_i \alpha_o,$$

which is the same as (8.183) since R_o is infinite and, therefore, $\alpha_o = 1$.

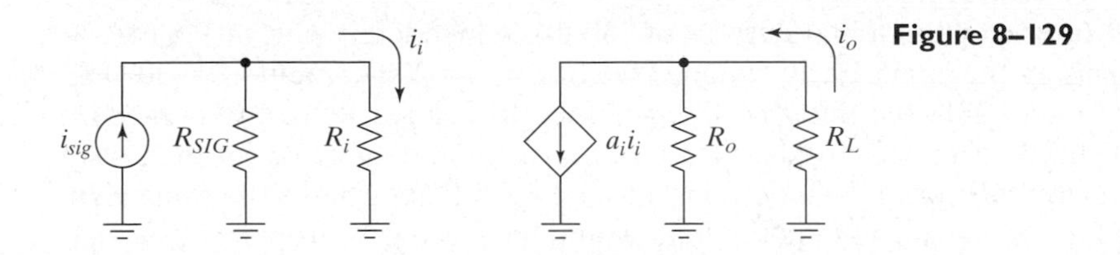

Figure 8–129

8.10 The small-signal midband AC equivalent circuit is shown in Figure 8-130. With a constant input current, the controlled source for Q_2 is a constant value, so it is an open circuit in the AC equivalent, and only r_{o2} shows up.

The output resistance of this circuit is

$$R_o = \frac{v_o}{i_o}.$$

However, before plodding on with the analysis, we note that $r_{\mu 3}$ is connected from v_o to ground and shows up in parallel with some resistance that we will call R_x; that is, $R_o = r_{\mu 3} \| R_x$ where R_x is found using the circuit in Figure 8-130 with $r_{\mu 3}$ removed. The analysis is then identical to that carried out for (8-82), and the result is $R_x = r_{\pi 3} \| r_{o2} + [1 + g_{m3}(r_{\pi 3} \| r_{o2})] r_{o3}$. Plugging in the numbers provided yields $R_o = 44\ \text{M}\Omega$, which is so large that it is likely that some other parasitic would limit it in a real circuit.

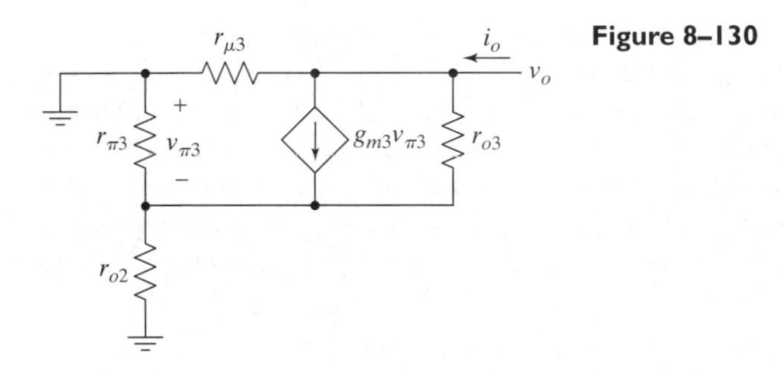

Figure 8–130

8.11 We might think of solving this problem by converting the voltages at the bases of Q_1 and Q_2 into common-mode and differential-mode components and then using the values given by (8.274) and (8.275). Unfortunately, the result will be wrong, as we now illustrate. The common-mode value of v_{B1} and v_{B2} is given by $v_{ICM} = v_I/2$, and the differential value is the same, $v_{IDM} = v_I/2$. However, in deriving the limits given by (8.274) and (8.275), we implicitly assumed that $v_{IDM} \ll v_{ICM}$ since we only considered v_{ICM}. This assumption is obviously not true in the present example.

To find the allowable input range in this example, we note that $v_{B2} = 0$ at all times. Therefore, virtually all of the tail current flows through Q_1 for any v_I greater than a few tenths of a volt. Similarly, virtually all of the tail current will flow through Q_2 for any v_I more than a few tenths of a volt below ground. So the circuit only works as an amplifier for inputs of plus or minus a few tenths of a volt.

8.12 Each transistor has a bias current of 250 μA, which leads to $g_m = 9.6$ mA/V. Also, using the given Early voltages, we have $r_{op} = 200$ kΩ and $r_{on} = 300$ kΩ. **(a)** Using (8.281), we find that $A_{dm} = 1{,}150$. **(b)** If a resistive load is used instead, the resistor must replace the *pnp* transistor and make up for the loss of the current mirrored from Q_1, so it must be 400 kΩ to achieve the same gain. **(c)** With 250 μA through it, the load would have a 100 V drop across it, so if $V_{CC} = 100$ V and the common-mode input was set to zero, the circuit would just work.

8.13 With the given betas, we find that $\alpha_n = 0.993$ and $\alpha_p = 0.962$. Using (8.282), we find that $A_{dm} = 199$, and using (8.283), we find that $A_{cm} = -0.0092$. The resulting CMRR is 21,600, or 87 dB.

8.14 We might think of solving this problem by converting the voltages at the gates of M_1 and M_2 into common-mode and differential-mode components and then using the values given by (8.288) and (8.289). Unfortunately, the result will be wrong, as we now illustrate. The common-mode value of v_{G1} and v_{G2} is given by $v_{ICM} = v_I/2$, and the differential value is the same, $v_{IDM} = v_I/2$. However, in deriving the limits given by (8.288) and (8.289), we implicitly assumed that $v_{IDM} \ll v_{ICM}$ since we only considered v_{ICM}. This assumption is obviously not true in the present example.

To find the allowable input range in this example, we note that $v_{G2} = 0$ at all times. Therefore, the limits on v_I are determined by finding the large-signal transfer characteristic. For large inputs, the tail current will flow almost entirely through one transistor or the other and the output will saturate at a limiting value. We derive the large-signal transfer characteristic of a SCP in Example 12.11.

8.15 Each transistor has a bias current of 250 μA, which leads to $g_m = 0.316$ mA/V. Also, using the given data, we have $r_{op} = 200$ kΩ and $r_{on} = 300$ kΩ. **(a)** Using (8.294), we find that $A_{dm} = 38$. **(b)** If a resistive load is used instead, the resistor must replace the PMOS transistor and make up for the loss of the current mirrored from M_1, so it must be 400 kΩ to achieve the same gain. **(c)** With 250 μA through it, the load would have a 100 V drop across it, so if $V_{DD} = 100$ V and the common-mode input was set to zero, the circuit would just barely work.

8.16 We again assume that we can reasonably ignore the currents in r_{o1} and r_{o2} in finding the drain currents of M_1 and M_2. The common-mode half circuit for finding i_{d1} is shown in Figure 8-131 and yields $i_{d1} = v_{icm}/(1/g_m + 2R_{SS})$. With the given 5% mismatch in the mirror, we can say that the controlled current source of M_4 will have a value equal to 1.05 times i_{d1}. In addition, by symmetry, we know that $i_{d2} = i_{d1}$. Therefore, the output current is $i_o = 1.05i_{d1} - i_{d2} = 0.05i_{d1}$, and the resulting output voltage is R_L times the current. Plugging in the equation for i_{d1} and solving yields the common-mode gain, $A_{cm} = 0.05/(1/g_m + 2R_{SS})$, which evaluates to 0.0087 for this example. Using (8.295), we find that $A_{dm} = 11.2$, so the CMRR = 1,290 (62 dB).

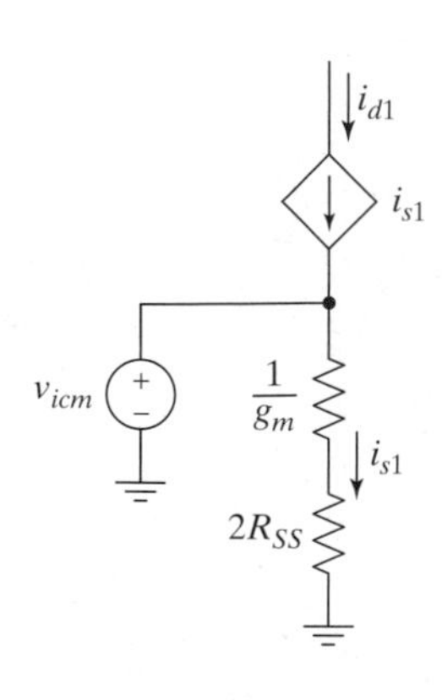

Figure 8–131

CHAPTER SUMMARY

- Each of the transistors studied has both a hybrid-π and a T model, and one model is sometimes more convenient to use than the other.

- It is always incumbent on the designer to pick the simplest model that will do the job. The designer must decide which elements can be safely ignored in a given situation; therefore, the designer must be familiar with the operation of the devices and what specific aspect of the operation each model element represents.
- The designer needs to be able to bounce back and forth between different levels of abstraction when thinking about circuit design. For example, it is sometimes convenient to think about a circuit at the transistor level, and at other times it is more convenient to use a two-port model for an entire amplifier.
- There are three single-transistor amplifier configurations; they correspond to each of the three intrinsic terminals of a transistor being part of both the input and output loop (i.e., one terminal is common to both loops).
- The most frequently used stage is the one that has both voltage and current gain greater than one. It is called a common-emitter or common-source amplifier.
- The voltage buffer stage has a voltage gain less than—but near to—one. It is used to provide a large input impedance and a small output impedance, and allows us to drive small loads without loading down the preceding gain stage. The buffer is called an emitter follower or source follower.
- The current buffer stage has a current gain less than or equal to one. It is used to boost the output impedance or bandwidth of other stages. This buffer is called a common-base or common-gate amplifier.
- Current mirrors can be used as current sources to bias amplifiers and also as active loads. The advantage of an active load is that the effective small-signal resistance is very high, but the DC voltage drop across the load does not have to be that high, and they don't take as much space on an IC as a large resistor would.
- Differential amplifiers are used to provide some immunity to interfering signals and to allow DC coupling and avoid bypass capacitors. Basic differential pairs are called either emitter-coupled pairs or source-coupled pairs.
- Practical amplifiers are almost always made up of a cascade of individual stages. By cascading stages, we can use the different characteristics of each type of stage to produce whatever overall amplifier characteristics we desire.
- Bipolar and field-effect transistors have many similarities, but they also have some key differences that the designer must be aware of.

REFERENCES

[8.1] P.R. Gray, P.J. Hurst, S.H. Lewis and R.G. Meyer, *Analysis and Design of Analog Integrated Circuits*, 4E, New York: John Wiley & Sons, 2001.

[8.2] A.S. Sedra and K.C. Smith, *Microelectronic Circuits*, 4/E, New York: Oxford University Press, 1998.

[8.3] D.A. Johns and K. Martin, *Analog Integrated Circuit Design*. New York, NY: John Wiley & Sons, 1997.

[8.4] D.A. Neamen, *Electronic Circuit Analysis and Design*, Chicago: Irwin, 1996.

PROBLEMS

8.1 Low-Frequency Small-Signal Models for Design

8.1.2 Linear Passive Devices (Rs, Ls, & Cs)

P8.1 Consider two circuits connected to the same 10V supply, as shown in Figure 8-132. The wire connecting the supply to the circuits has 1 Ω of resistance, as shown. Assume that circuit 2 is a digital block and that we can model the current it draws, i_2, as a 10-MHz square wave varying between 100 and 120 mA. Further, assume that circuit 1 is an analog circuit that draws a constant 10 mA. **(a)** What is the AC portion of v_{CC}? **(b)** Now add a capacitor to ground in parallel with the two circuits. How large must this capacitor be to reduce the peak AC on v_{CC} to 1 mV?

S P8.2 Use SPICE to confirm your results from Problem P8.1. Show transient plots of v_{CC} with and without the capacitor present.

P8.3 Figure 8-133 shows a 10V DC power supply connected to a circuit. The wires connecting the two are modeled by a 1-Ω resistor in series with a 10-nH inductor. The circuit is modeled by 100 pF of capacitance to ground. In addition, there is a noise source that has 20 pF of capacitance coupling it to the power supply connection on the circuit. This coupling might be caused, for example, by the power supply line being near a nonshielded cable carrying digital data. Assume the noise source is a 50 MHz, 1 V peak sine wave. **(a)** What is the AC portion of v_{CC}? **(b)** Now add a filter capacitor to ground in parallel with the circuit. How large must this capacitor be to reduce the peak AC on v_{CC} to 1 mV? (*Hint*: assume that the new capacitor will dominate the impedance from v_{CC} to ground, and then check your assumption.)

S P8.4 Use SPICE to confirm your results from Problem P8.3. Show transient plots of v_{CC} with and without the capacitor present.

P8.5 **(a)** Consider the equivalent circuit in Figure 8-4. If $R_S = 1\ \text{k}\Omega$ and $R_i = 100\ \text{k}\Omega$, how large does C_{IN} need to be to set the input pole at 20 Hz? **(b)** Repeat the problem with $R_i = 1\ \text{k}\Omega$.

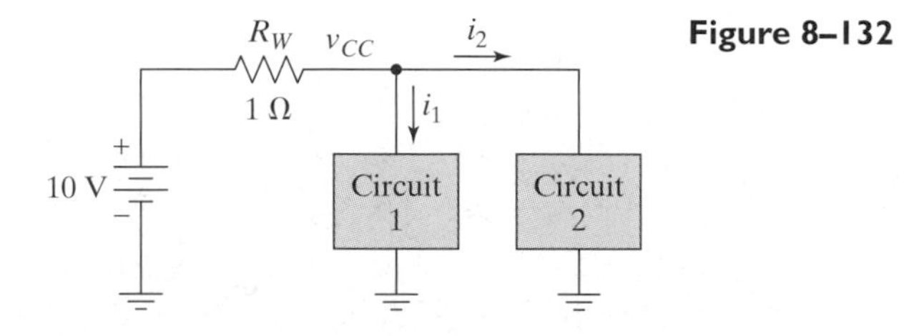

Figure 8–132

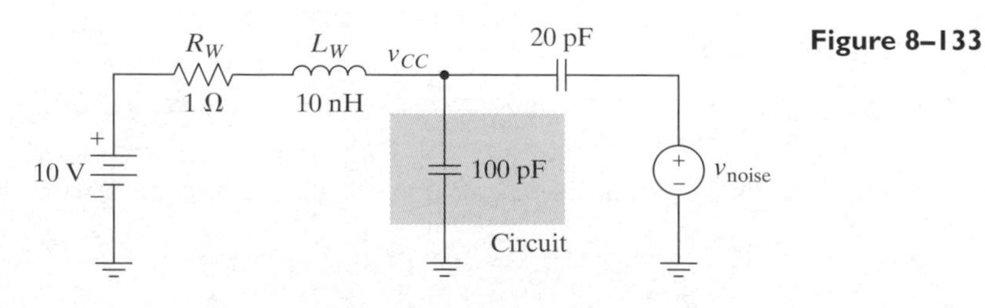

Figure 8–133

P8.6 For the equivalent circuit shown in Figure 8-4, assume that $R_S = 100\ \Omega$, $R_i = 10\ \text{k}\Omega$, $R_o = 10\ \Omega$, $R_L = 100\ \Omega$, $C_{IN} = C_{OUT} = 1\ \mu\text{F}$, and the open-circuit gain of the amplifier, $a(j\omega)$, has a DC gain of 100 and a pole at 5 MHz. Plot the magnitude of the overall gain of the circuit.

8.1.3 Diodes

S P8.7 **(a)** Use SPICE to solve Exercise 8.1, and model the diode by $n = 1$ and $I_S = 10^{-14}$ A. **(b)** Increase and decrease I_S by a factor of 10, and note the change in v_o.

8.1.5 Bipolar Junction Transistors

D P8.8 Consider the circuit shown in Figure 8-134. Determine the resistor values necessary to establish a bias point that yields $g_m = 40$ mA/V and $V_{CE} = 5$ V.

P8.9 Consider the bipolar amplifier shown in Figure 8-135. Determine the small-signal midband AC model for the transistor, and draw the resulting equivalent circuit. Assume that $\beta = 100$ and $V_A = 30$ V.

P8.10 Consider the bipolar amplifier shown in Figure 8-136. Determine the small-signal midband AC model for the transistor, and draw the resulting equivalent circuit. Assume that $\beta = 100$, $V_A = 50$ V, and $r_b = 50\ \Omega$.

S P8.11 Use SPICE to verify your small-signal model in Problem P8.10. Use the Nsimple3 model described in Appendix B.

8.1.6 MOS Field-Effect Transistors

D P8.12 Consider the circuit shown in Figure 8-137, and assume that $K = 2\ \text{mA/V}^2$ and $V_{th} = 1$ V. Find values for the resistors that will establish $g_m = 2$ mA/V and $V_{DS} = 4$ V.

S P8.13 Use SPICE and the Nlarge_mos model presented in Appendix B to verify that your design in Problem P8.12 achieves the desired small-signal transistor parameters.

P8.14 Consider the MOS amplifier shown in Figure 8-137, and assume that $R_1 = 270\ \text{k}\Omega$, $R_2 = 680\ \text{k}\Omega$, $R_S = 20\ \text{k}\Omega$, $R_D = 6.8\ \text{k}\Omega$, $K = 100\ \mu\text{A/V}^2$, and $V_{th} = 0.5$ V. Find the small-signal midband AC model for the MOSFET and draw the resulting equivalent circuit.

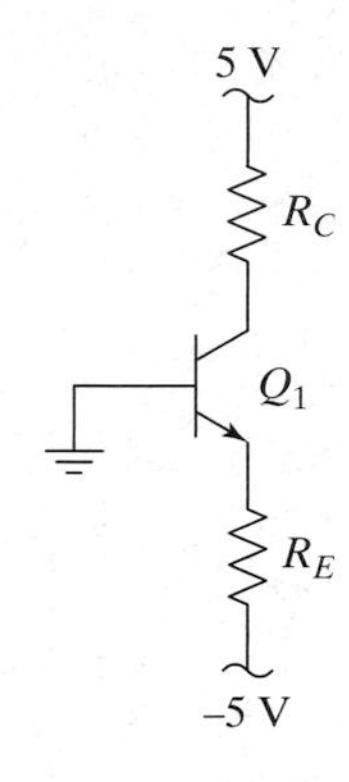

Figure 8–134

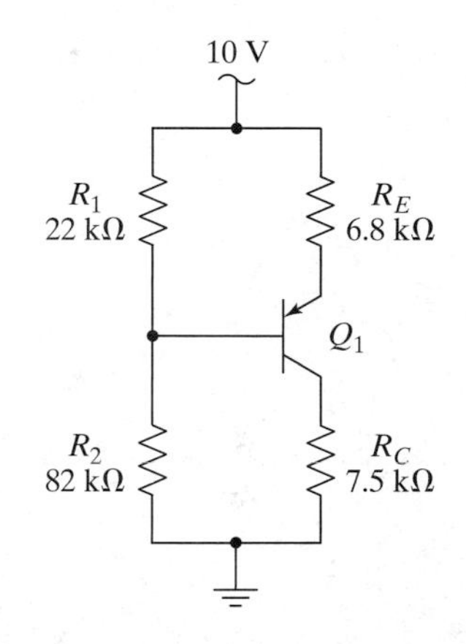

Figure 8–135

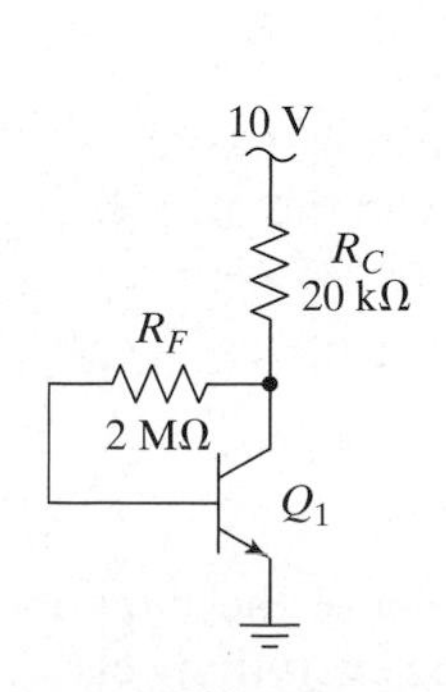

Figure 8–136

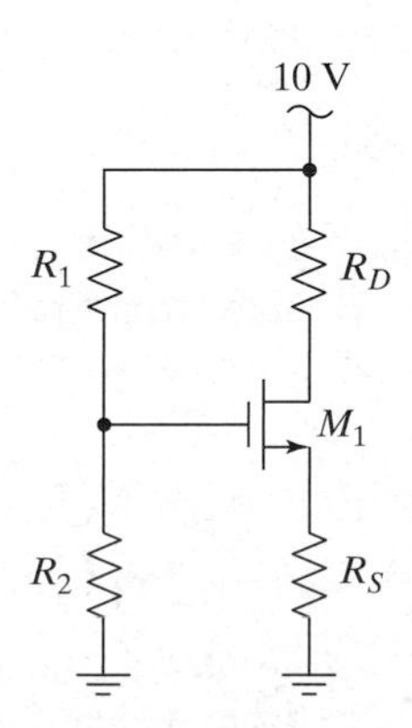

Figure 8–137

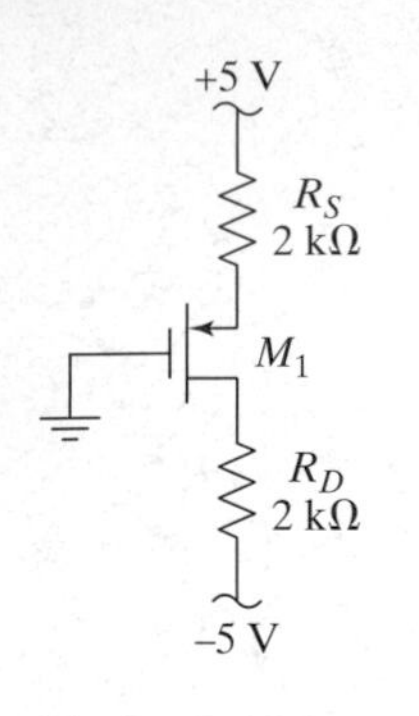

Figure 8–138

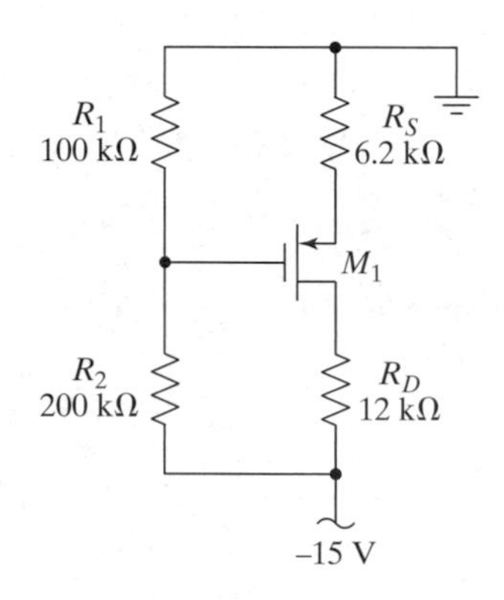

Figure 8–139

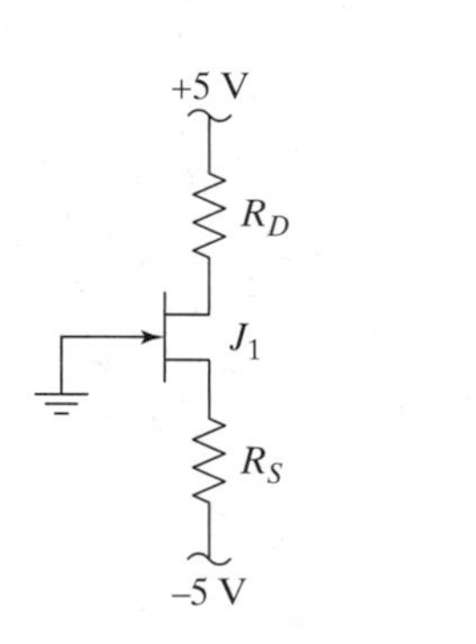

Figure 8–140

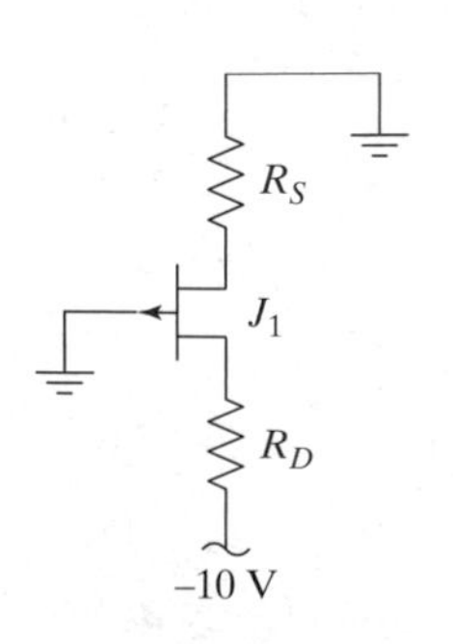

Figure 8–141

P8.15 Consider the MOS amplifier shown in Figure 8-138, and assume that $K = 2\ \text{mA/V}^2$ and $V_{th} = -1$ V. Find the small-signal midband AC model for the MOSFET, and draw the resulting equivalent circuit.

S P8.16 Use SPICE and the Plarge_mos model presented in Appendix B to verify that the small-signal model parameters you found in Problem P8.15 are correct.

P8.17 Consider the MOS amplifier shown in Figure 8-139, and assume that $K = 100\ \mu\text{A/V}^2$ and $V_{th} = -0.5$ V. Find the small-signal midband AC model for the MOSFET, and draw the resulting equivalent circuit.

S P8.18 Use SPICE and the Psimple_mos model presented in Appendix B to verify that the small-signal model parameters you found in Problem P8.17 are correct.

8.1.7 Junction Field-Effect Transistors

P8.19 Consider the JFET amplifier shown in Figure 8-140, and assume that $R_D = 30\ \text{k}\Omega$, $R_S = 46\ \text{k}\Omega$, $I_{DSS} = 250\ \mu\text{A}$, and $V_p = -1$ V. Find the small-signal midband AC model for the JFET, and draw the resulting equivalent circuit.

S P8.20 Use SPICE and the Jn1 model shown in Appendix B to verify your results from Problem P8.19. You must set `beta = 250` Ω, in the model.

D P8.21 Consider the JFET amplifier shown in Figure 8-140 and assume that $I_{DSS} = 250\ \mu\text{A}$ and $V_p = -1$ V. Derive values for the resistors so that you establish that $g_m = 0.25$ mA/V and $V_{DS} = 3$ V.

S P8.22 Use SPICE and the Jn1 model shown in Appendix B to verify your design from Problem P8.21. You must set `beta = 250` Ω, in the model.

P8.23 Consider the JFET amplifier shown in Figure 8-141, and assume that $R_D = 4\ \text{k}\Omega$, $R_S = 5.6\ \text{k}\Omega$, $I_{DSS} = 250\ \mu\text{A}$ and $V_p = 1$ V. Find the small-signal midband AC model for the JFET, and draw the resulting equivalent circuit.

S P8.24 Use SPICE and the Jp1 model shown in Appendix B to verify your results from Problem P8.23. You must set `beta = 250` Ω, in the model.

D P8.25 Consider the JFET amplifier shown in Figure 8-141, and assume that $I_{DSS} = 250\ \mu\text{A}$ and $V_p = 1$ V. Derive values for the resistors so that you establish $g_m = 0.4$ mA/V and $V_{DS} = -4$ V.

S P8.26 Use SPICE and the Jp1 model shown in Appendix B to verify your design from Problem P8.25. You must set `beta = 250` Ω, in the model.

8.2 Stages with Voltage and Current Gain

8.2.1 A Generic Implementation: The Common-Merge Amplifier

P8.27 Consider the graphic analysis shown in Figure 8-27 and the circuit from which it was derived. If R_L is varied while keeping everything else constant, what is the minimum slope possible for the AC load line? What are the corresponding values of R_L and v_o/v_s?

8.2.2 A Bipolar Implementation: The Common-Emitter Amplifier

P8.28 Derive the circuit shown in Figure A8-5(d) from the circuit shown in part (c) of the figure.

P8.29 Show that the circuit in Figure A8-7 yields the input resistance given by (A8.4).

P8.30 Derive the output resistance of the circuit in Figure 8-39, with r_o included in the transistor model.

P8.31 Show that (8.107) is equal to (8.98).

D P8.32 Design the circuit of Figure 8-36 to yield a small-signal midband gain of -10 ± 0.5 with $R_S = 500\ \Omega$, $R_L = 5\ \text{k}\Omega$, and $V_{CC} = 15$ V. Choose the bias point to be stable, minimize power, and yield an input resistance of no less than 5 kΩ and an output resistance of no more than 1 kΩ. Assume that $\beta = 100$, and use only standard 5% resistor values. (*Hint*: You may not be able to strictly stay with the rules of thumb from Chapter 7, but stay as close as possible to them.)

S P8.33 Use SPICE and the Nsimple model from Appendix B to verify your design in Problem P8.32.

D P8.34 Design the circuit of Figure 8-36 to yield a small-signal midband gain of -10 ± 0.5 with $R_S = 100\ \Omega$, $R_L = 10\ \text{k}\Omega$, and $V_{CC} = 10$ V. Choose the bias point to be stable, minimize power, and to yield an input resistance of no less than 1 kΩ and an output resistance of no more than 2 kΩ. Assume that $\beta = 100$, and use only standard 5% resistor values.

S P8.35 Use SPICE and the Nsimple model from Appendix B to verify your design in Problem P8.34.

P8.36 Draw the small-signal midband AC equivalent circuit for the amplifier in Figure 8-142, and calculate the overall voltage gain and power gain. Assume that $\beta = 100$.

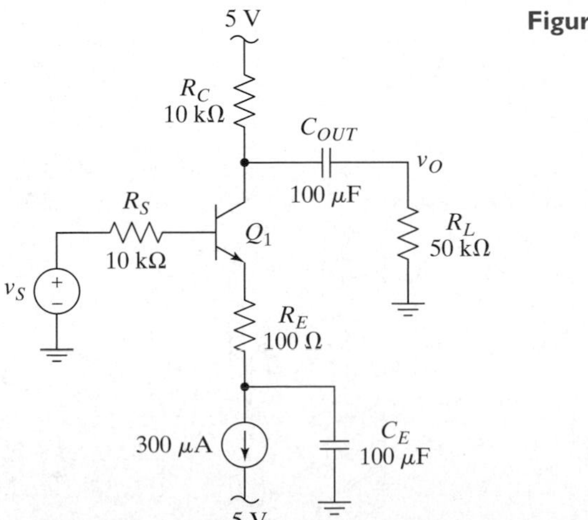

Figure 8–142

P8.37 Draw the small-signal midband AC equivalent circuit for the amplifier in Figure 8-143, and calculate the overall voltage gain and power gain. Assume that $\beta = 100$.

P8.38 Consider the circuit of Figure 8-144. Assume that $\beta = 100$, $V_T = 26$ mV, and $V_A = 100$ V. **(a)** Draw the DC equivalent circuit and determine I_C and I_B. Assume that $V_{BE} = 0.7$ V. **(b)** Draw the small-signal midband AC equivalent circuit. Find r_π, g_m, and r_o for the transistor biased as shown. **(c)** Find v_o/v_s, R_i, and R_o for the amplifier circuit.

P8.39 Consider the circuit of Figure 8-145. Assume that $\beta = 50$ and $V_T = 26$ mV. **(a)** Draw the DC equivalent circuit and determine I_C and I_B. Assume that $V_{BE} = -0.7$ V. **(b)** Draw the small-signal midband AC equivalent circuit. Find r_π and g_m for the transistor biased as shown, ignore r_o. **(c)** Find v_o/v_s, R_i, and R_o for the amplifier circuit.

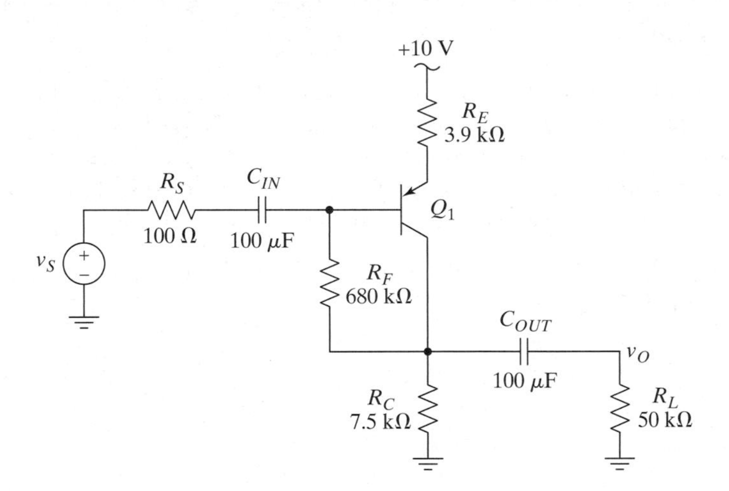

Figure 8–143

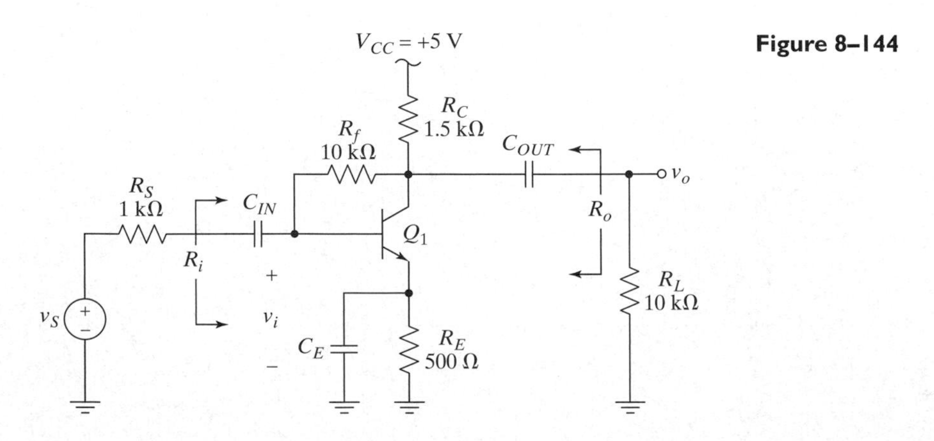

Figure 8–144

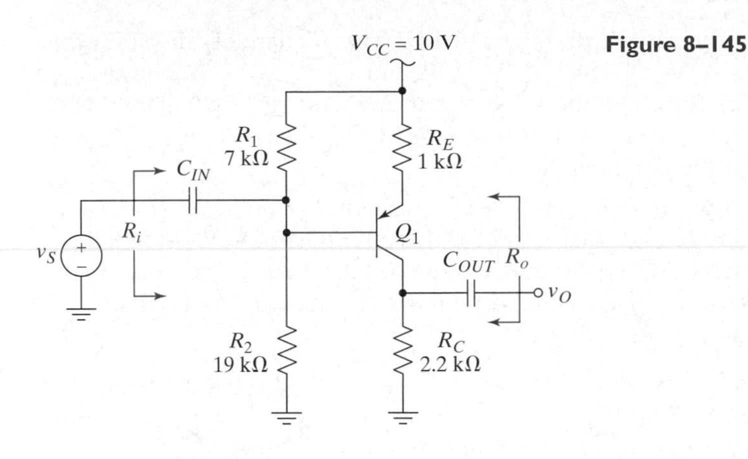

Figure 8–145

P8.40 The schematic of Figure 8-146 is divided into a source block, an amplifier block, and a load block. Assume that $V_T = 26$ mV, $V_{BE} = 0.7$ V, and $\beta = 100$. Assume that all capacitors are AC short circuits at midband frequencies. **(a)** Find I_C, assuming that $V_B = V_{BB}$. **(b)** Find g_m and r_π. **(c)** Draw the unilateral two-port model for the amplifier block using a controlled voltage source. Consider midband AC only. Label R_i, a_v, and R_o clearly. **(d)** Find α_i, the input attenuation factor for the circuit. Find α_o, the output attenuation factor for the circuit. Find $\alpha_i a_v \alpha_o$, the overall voltage gain. **(e)** Suppose the transistor β becomes infinite and all other parameters stay the same. How are α_i, A_v, and α_o each affected?

8.2.3 A MOS Implementation: The Common-Source Amplifier

P8.41 Using the T model for the transistor, analyze the small-signal midband AC voltage gain of the circuit in Figure 8-43. Show that your result is identical to (8.115).

P8.42 Show that (8.120) is equal to (8.114).

P8.43 Using the T model for the transistor, analyze the small-signal midband AC voltage gain of the circuit in Figure 8-48. Show that your result is identical to (8.128).

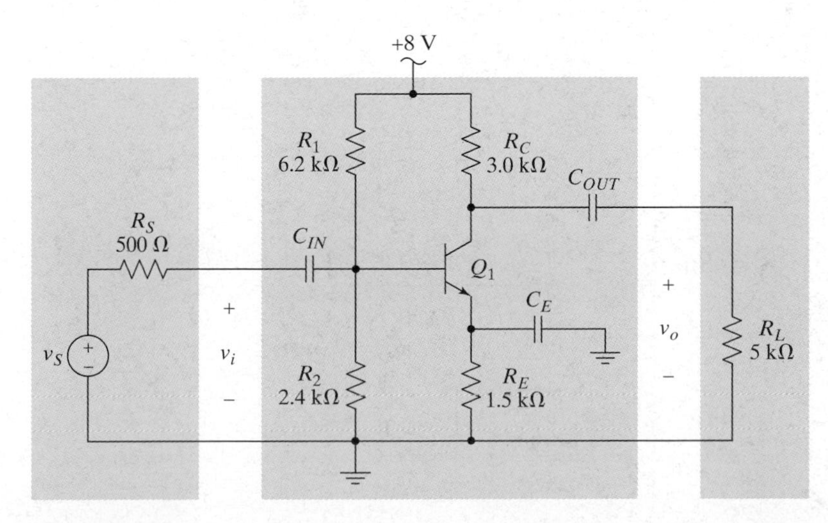

Figure 8–146

P8.44 Consider the circuit of Figure 8-147. The p-channel MOSFET has $K = 500\ \mu A/V^2$, $\lambda = 4.3 \times 10^{-2}\ V^{-1}$, and $V_{th} = -1$ V. **(a)** Find I_D, V_{GS}, V_D, and V_S. **(b)** Draw the small-signal midband AC equivalent circuit. Find g_m and g_{ds} for the transistor biased as shown. **(c)** Find v_o/v_{sig}, R_i, and R_o for the amplifier circuit.

P8.45 Consider the circuit in Figure 8-148. Assume $K = 500\ \mu A/V^2$ and $V_{th} = 1$ V. **(a)** Find the DC bias point for this circuit. **(b)** Draw the small-signal midband AC equivalent circuit. **(c)** Calculate the small-signal midband AC voltage gain. **(d)** Find the small-signal midband AC input and output resistances.

S P8.46 Use SPICE and the Nmedium_mos model in Appendix B to verify your results from Problem P8.45.

P8.47 Consider the circuit in Figure 8-149. Assume $K = 2\ mA/V^2$ and $V_{th} = 1$ V. **(a)** Find the DC bias point for this circuit. **(b)** Draw the small-signal midband AC equivalent circuit. **(c)** Calculate the small-signal midband AC voltage gain. **(d)** Find the small-signal midband AC input and output resistances.

S P8.48 Use SPICE and the Nlarge_mos model in Appendix B to verify your results from Problem P8.47.

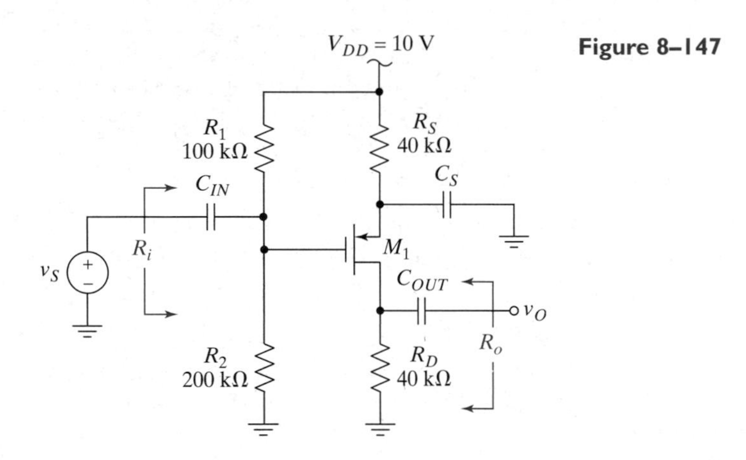

Figure 8–147

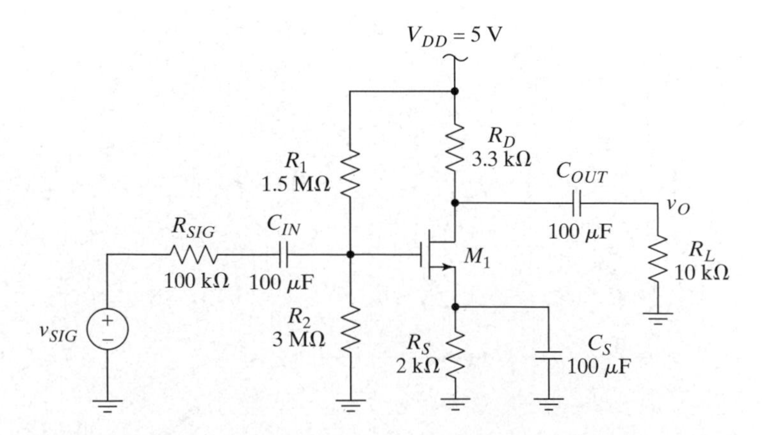

Figure 8–148

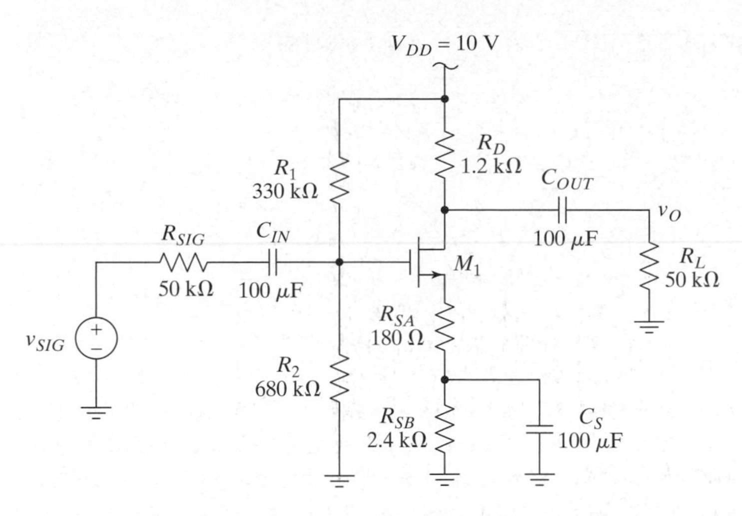

Figure 8–149

S P8.49 **(a)** Draw the PMOS amplifier that is equivalent to the NMOS amplifier in Figure 8-149. Use the Plarge_mos model from Appendix B, and establish the same bias point and gain. **(b)** Use SPICE to confirm that your PMOS amplifier is equivalent in performance to the NMOS amplifier from Problem P8.47.

8.3 Voltage Buffers

8.3.1 A Generic Implementation: The Merge Follower

P8.50 **(a)** Redraw the small-signal midband AC equivalent circuit for the buffer in Figure 8-53(a) using the T model for the generic transistor. **(b)** Using this circuit, derive the overall small-signal midband AC voltage gain, input resistance, and output resistance, and show that your results are the same as (8.138), (8.139), and (8.142).

8.3.2 A Bipolar Implementation: The Emitter Follower

P8.51 Consider the emitter follower shown in Figure 8-150, and assume that $V_S = 2$ V and $\beta = 100$. **(a)** Find the DC bias point and small-signal midband AC model for the transistor. Draw the small-signal midband AC equivalent circuit. **(b)** Find the small-signal midband AC voltage gain v_o/v_i and the overall voltage gain v_o/v_s. **(c)** Find the small-signal midband AC input and output resistances.

S P8.52 Use SPICE and the Nsimple model shown in Appendix B to verify your analyses of the circuit in Problem P8.51.

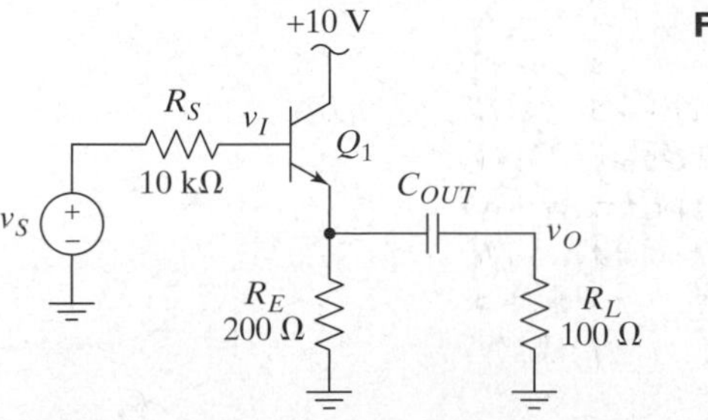

Figure 8–150

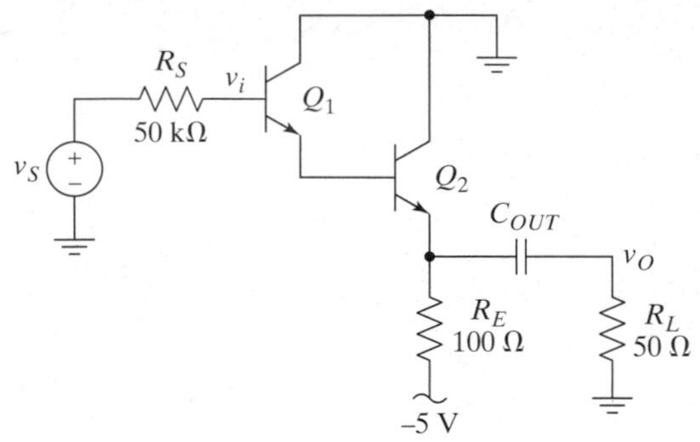

Figure 8–151

D P8.53 Consider the emitter follower shown in Figure 8-150, and assume that $V_S = 2$ V and $\beta = 100$. Find a new value for R_E so that the small-signal midband AC output resistance is $\leq 50\ \Omega$ and the power is minimized.

P8.54 Consider the voltage follower shown in Figure 8-151, and assume that $V_S = 0$, $\beta_1 = 50$, and $\beta_2 = 100$. This connection of two transistors is called a ***Darlington connection***. **(a)** Find the DC bias point and small-signal midband AC models for the transistors. Draw the small-signal midband AC equivalent circuit. **(b)** Find the small-signal midband AC voltage gain v_o/v_i and the overall voltage gain v_o/v_s. **(c)** Find the small-signal midband AC input and output resistances.

P8.55 Consider the emitter follower shown in Figure 8-152, and assume that $\beta = 50$. **(a)** Find the DC bias point and small-signal midband AC model for the transistor. Draw the small-signal midband AC equivalent circuit. **(b)** Find the small-signal midband AC voltage gain v_o/v_i and the overall voltage gain v_o/v_s. **(c)** Find the small-signal midband AC input and output resistances.

S P8.56 Use SPICE and the Psimple model shown in Appendix B to verify your analyses of the circuit in Problem P8.55. You will need to change the value of β in the model to 50.

D P8.57 Reconsider the circuit from Problem P8.55. Redesign the bias network to achieve $R_i \geq 1\ \text{k}\Omega$ and $R_o \leq 20\ \Omega$.

P8.58 Consider the emitter follower shown in Figure 8-153, and assume that $V_S = 0$ V and $\beta = 100$. **(a)** Find the DC bias point and small-signal midband AC model for the transistor. Draw the small-signal midband AC equivalent circuit. **(b)** Find the small-signal midband AC voltage gain v_o/v_i and the overall voltage gain v_o/v_s. **(c)** Find the small-signal midband AC input and output resistances.

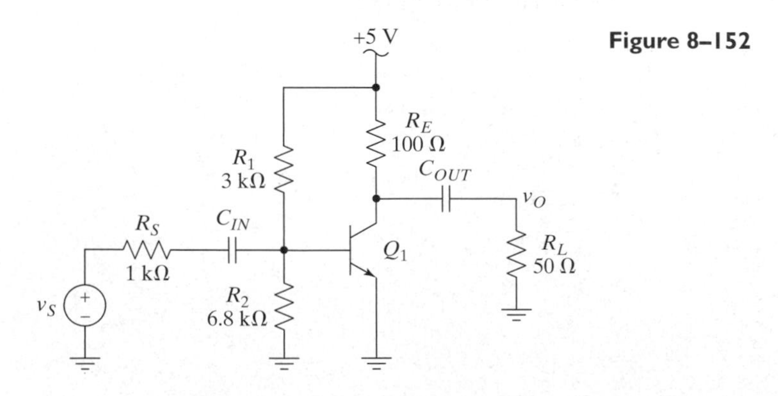

Figure 8–152

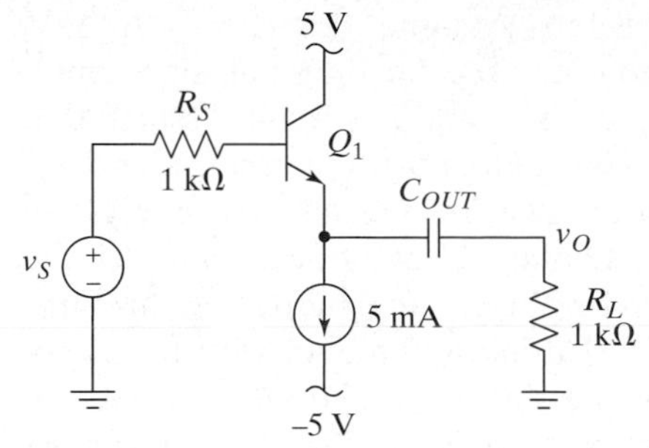

Figure 8–153

P8.59 Refer to the circuit of Figure 8-154. Assume that $V_T = 26$ mV, $V_A = 100$ V, and $\beta = 100$. **(a)** Find the midband AC parameters R_i, A_v, and R_o for the amplifier circuit. *Use the T model for the transistor.* In your analysis at this point, use the model parameters of the T model and R_E and R_{BB}. Later, after calculating the bias point, the numerical values for the model parameters and resistors will be plugged in. Ignore r_o and r_μ at this time. **(b)** Now include the effects of the transistor parameter r_o in the analysis. Which answers to part (a) will change, and what will they change to? Look carefully at the position of r_o in the T model and in the circuit. **(c)** Now include the effects of the transistor parameter r_μ in the analysis. Which answers to part (a) will change, and to what will they change? **(d)** Compare your answers obtained by using a T model with those obtained by using a hybrid-π model for the transistor. They should agree. **(e)** Find I_C for the transistor with the bias conditions shown. Assume that $V_{EB} = 0.7$ V. **(f)** For this amplifier circuit, is it important to include r_o and r_μ in the analysis? Why or why not?

P8.60 Find the value of R_E that yields the largest overall midband gain that can be obtained from the follower in Figure 8-155. Assume $\beta = 150$ and $V_S = 5$ V. What is the corresponding gain?

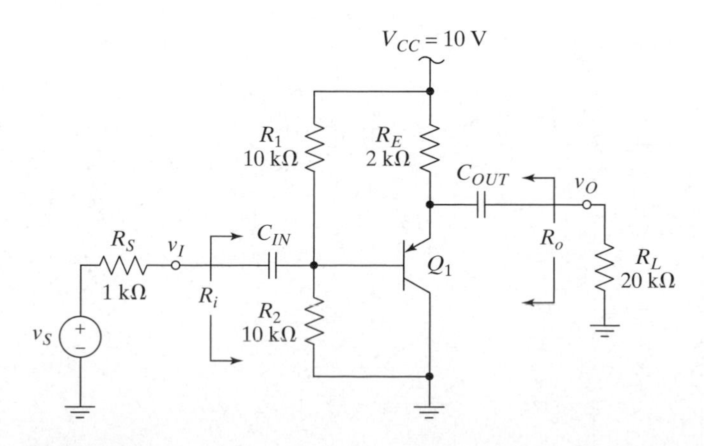

Figure 8–154

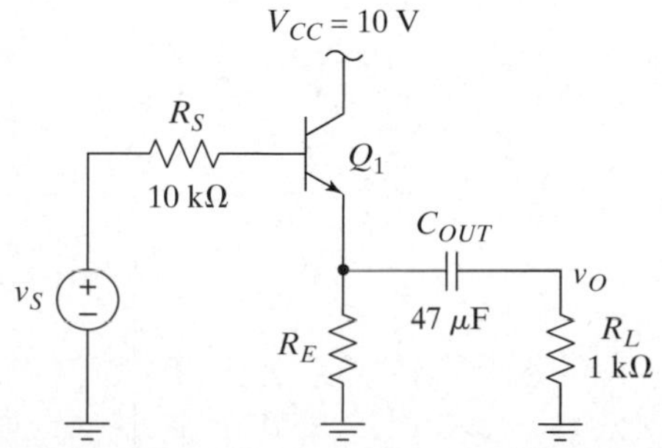

Figure 8–155

P8.61 Refer to the circuit of Figure 8-156. Assume that $V_T = 26$ mV, $V_A = 100$ V, and $\beta = 100$. **(a)** Find equations for the small-signal midband AC parameters R_i, a_v, and R_o for the amplifier circuit. Assume that all capacitors are AC short-circuits at midband frequencies. *Use the T model for the transistor.* Express your answers in terms of α, r_e, R_E, and R_{BB}. Later, after calculating the bias point, the numerical values for the model parameters and resistors will be plugged in. Ignore r_o at this time. **(b)** Compare your answers obtained using a T model with those obtained using a hybrid-π model for the transistor. (You do not need to show the derivation of the answers from the hybrid-π model; just state them.) **(c)** Find I_C for the transistor with the bias conditions shown. Assume that $V_{BE} = 0.7$ V. **(d)** Find the numerical values of R_i, a_v, and R_o. **(e)** Now include the effects of the transistor parameter r_o in the analysis. Which answers to part (a) will change, and what will they change to? Look carefully at the position of r_o in the T model and in the circuit. Comment on the effect of including r_o in the analysis.

8.3.3 A FET Implementation: The Source Follower

P8.62 Consider the source follower shown in Figure 8-157, and assume that $K = 2$ mA/V^2 and $V_{th} = 1$ V. **(a)** Find the DC bias point and small-signal midband AC model for the transistor. Draw the small-signal midband AC equivalent circuit. **(b)** Find the small-signal midband AC voltage gain v_o/v_i and the overall voltage gain v_o/v_{sig}. **(c)** Find the small-signal midband AC input and output resistances.

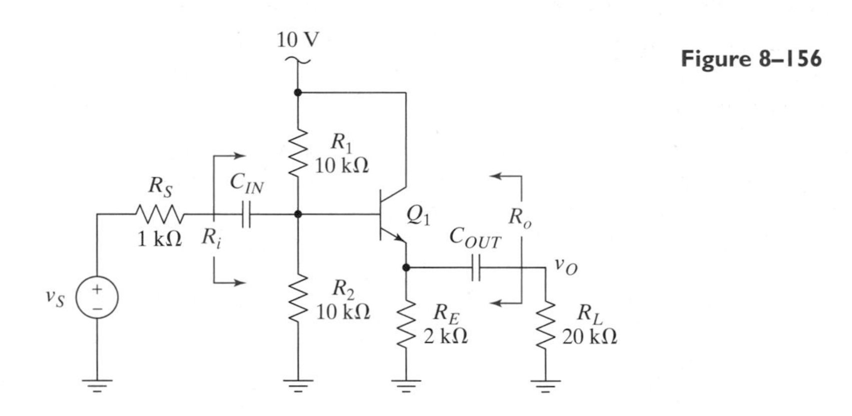

Figure 8–156

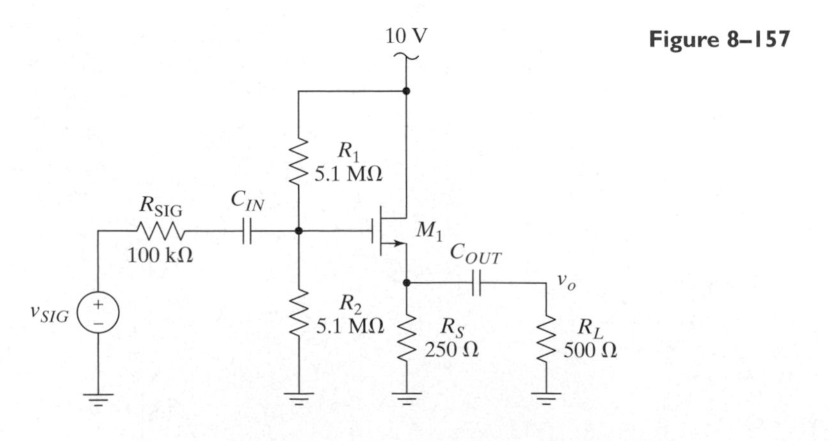

Figure 8–157

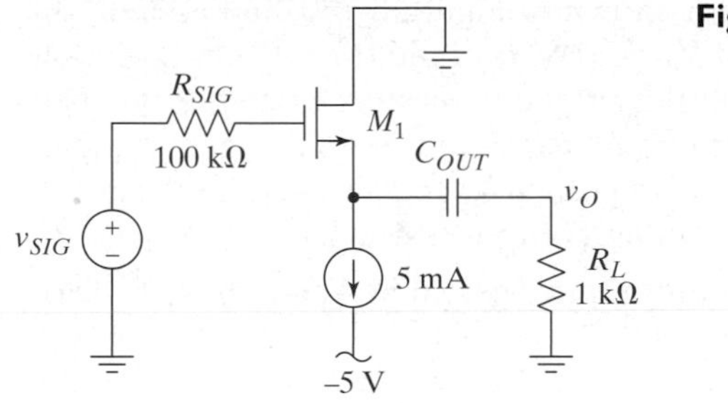

Figure 8–158

S P8.63 Use SPICE and the Nlarge_mos model shown in Appendix B to verify your answers to Problem P8.62.

D P8.64 Consider the source follower shown in Figure 8-157, and assume that $K = 2\ \text{mA/V}^2$ and $V_{th} = 1$ V. Find a new value for R_S so that the small-signal midband AC output resistance is $\leq 50\ \Omega$ and the DC power is minimized. Keep $2\ \text{V} \leq V_S \leq 8$ V.

P8.65 Consider the source follower shown in Figure 8-158, and assume that $V_{SIG} = 0$, $K = 2\ \text{mA/V}^2$, and $V_{th} = 1$ V. **(a)** Find the DC bias point and small-signal midband AC model for the transistor. Draw the small-signal midband AC equivalent circuit. **(b)** Find the small-signal midband AC voltage gain v_o/v_i and the overall voltage gain v_o/v_{sig}. **(c)** Find the small-signal midband AC input and output resistances.

S P8.66 Use SPICE and the Nlarge_mos model shown in Appendix B to verify your answers to Problem P8.65.

P8.67 Consider the source follower shown in Figure 8-159, and assume that $K = 2\ \text{mA/V}^2$ and $V_{th} = -1$ V. **(a)** Find the DC bias point and small-signal midband AC model for the transistor. Draw the small-signal midband AC equivalent circuit. **(b)** Find the small-signal midband AC voltage gain v_o/v_i and the overall voltage gain v_o/v_{sig}. **(c)** Find the small-signal midband AC input and output resistances.

S P8.68 Use SPICE and the Plarge_mos model shown in Appendix B to verify your answers to Problem P8.67.

D P8.69 Consider the source follower shown in Figure 8-159, and assume that $K = 2\ \text{mA/V}^2$ and $V_{th} = -1$V. Find a new value for R_S so that the small-signal midband AC output resistance is $\leq 100\ \Omega$ and the DC power is minimized. Keep $2\ \text{V} \leq V_S \leq 13$ V.

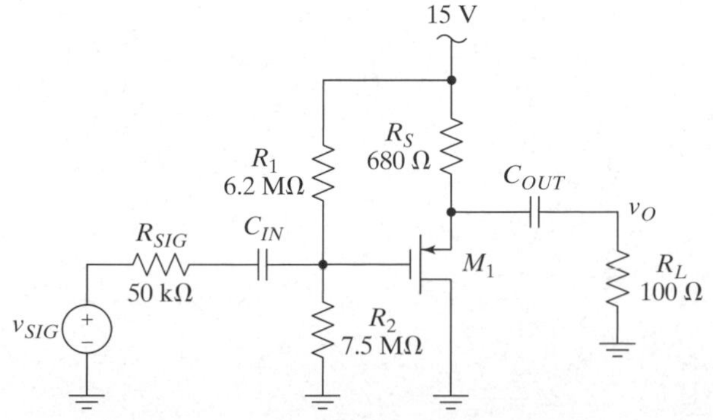

15 V
R_S
680 Ω
R_1
6.2 MΩ
C_{OUT}
v_O
R_{SIG}
C_{IN}
M_1
R_L
100 Ω
50 kΩ
v_{SIG}
R_2
7.5 MΩ

Figure 8–159

P8.70 Consider the source follower shown in Figure 8-160, and assume that $V_{SIG} = 0$, $K = 2$ mA/V^2, and $V_{th} = -1$ V. **(a)** Find the DC bias point and small-signal midband AC model for the transistor. Draw the small-signal midband AC equivalent circuit. **(b)** Find the small-signal midband AC voltage gain v_o/v_i and the overall voltage gain v_o/v_{sig}. **(c)** Find the small-signal midband AC input and output resistances.

S P8.71 Use SPICE and the Plarge_mos model shown in Appendix B to verify your answers to Problem P8.70.

8.4 Current Buffers

8.4.2 A Bipolar Implementation: The Common-Base Amplifier

P8.72 Derive the small-signal midband AC input resistance, output resistance, and short-circuit current gain of the common-base amplifier in Figure 8-69 with the output resistance of the transistor included. (Hint: Find A_v first.)

P8.73 Consider the common-base stage shown in Figure 8-161, and assume that C_B is a large bypass capacitor and $\beta = 100$. We have drawn the circuit in this way to emphasize the similarity in basic structure to the common-emitter stage. **(a)** Find the DC bias point and small-signal midband AC model for the transistor. Draw the small-signal midband AC equivalent circuit. Ignore r_o. **(b)** Find the small-signal midband AC voltage gain v_o/v_i. **(c)** Find the small-signal midband AC transresistance v_o/i_i. **(d)** Find the small-signal midband AC input and output resistances.

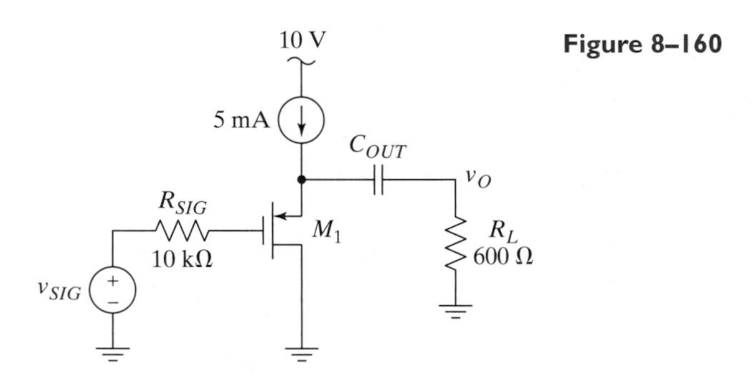

Figure 8–160

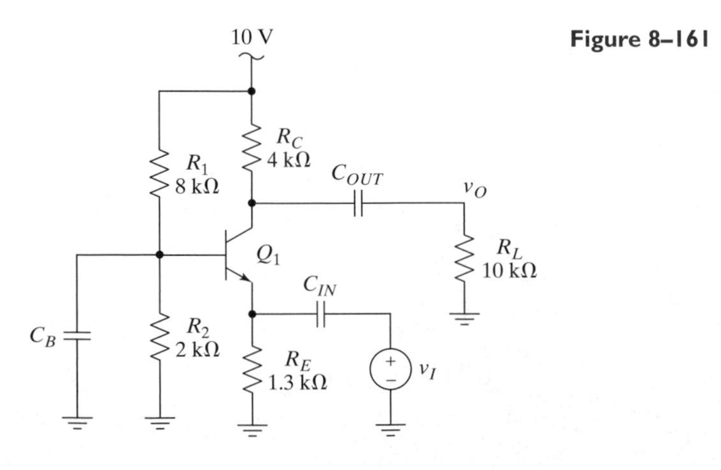

Figure 8–161

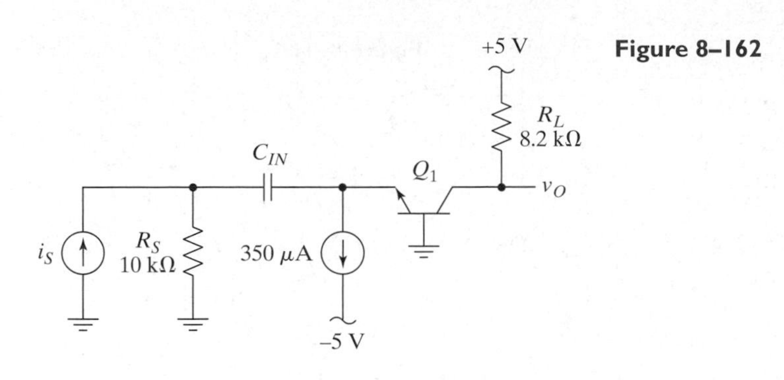

Figure 8–162

P8.74 Consider the common-base stage shown in Figure 8-162, and assume that $\beta = 100$. Ignore r_o. **(a)** Find the DC bias point and small-signal midband AC model for the transistor. Draw the small-signal midband AC equivalent circuit. **(b)** Find the small-signal midband AC current gain i_o/i_i and the overall small-signal midband AC current gain i_o/i_s. **(c)** Find the small-signal midband AC input resistance. **(d)** Find the small-signal midband AC voltage gain v_o/v_i.

S P8.75 Use SPICE and the Nsimple model presented in Appendix B to confirm your answers to Problem P8.74.

P8.76 Consider the common-base stage shown in Figure 8-163, and assume that $\beta = 100$. **(a)** Find the DC bias point and small-signal midband AC model for the transistor. Draw the small-signal midband AC equivalent circuit. Ignore r_o. **(b)** Find the small-signal midband AC current gain i_o/i and the overall small-signal midband AC current gain i_o/i_s. **(c)** Find the small-signal midband AC input resistance. **(d)** Find the small-signal midband AC voltage gain v_o/v_i. **(e)** Find the small-signal midband AC transresistance v_o/i_i.

S P8.77 Use SPICE and the Psimple model presented in Appendix B to confirm your answers to Problem P8.76.

8.4.3 A FET Implementation: The Common-Gate Amplifier

P8.78 Derive equations for the small-signal midband AC input resistance, output resistance, and short-circuit current gain of the common-gate amplifier in Figure 8-71 with the output resistance of the transistor included. (Hint: Find A_v first.)

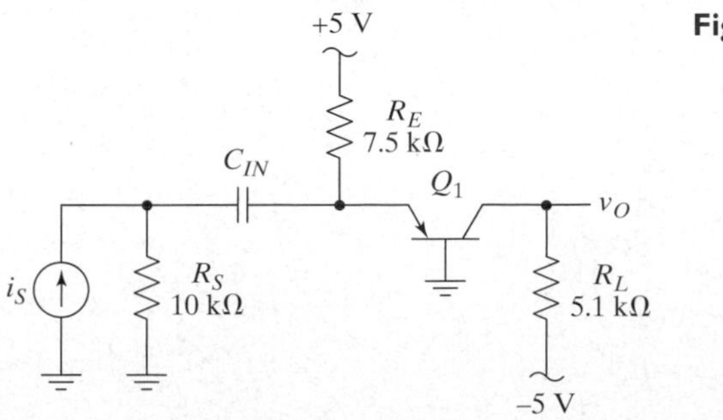

Figure 8–163

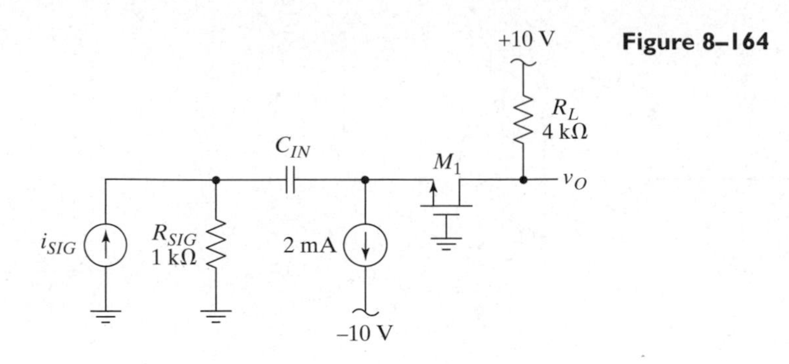

Figure 8–164

P8.79 Consider the common-gate amplifier shown in Figure 8-164, and assume that $K = 500\ \mu\text{A/V}^2$ and $V_{th} = 1$ V. **(a)** Find the DC bias point and small-signal midband AC model for the transistor. Draw the small-signal midband AC equivalent circuit. Ignore r_o. **(b)** Find the small-signal midband AC voltage gain v_o/v_i. **(c)** Find the overall small-signal midband AC current gain i_o/i_{sig}. **(d)** Find the small-signal midband AC input resistance.

S P8.80 Use SPICE and the Nmedium_mos model shown in Appendix B to verify your answers to Problem P8.79.

P8.81 Consider the common-gate amplifier shown in Figure 8-165, and assume that $K = 500\ \mu\text{A/V}^2$ and $V_{th} = -1$ V. **(a)** Find the DC bias point and small-signal midband AC model for the transistor. Draw the small-signal midband AC equivalent circuit. Ignore r_o. **(b)** Find the small-signal midband AC voltage gain v_o/v_i. **(c)** Find the overall small-signal midband AC current gain i_o/i_{sig}. **(d)** Find the small-signal midband AC input resistance.

S P8.82 Use SPICE and the Pmedium_mos model shown in Appendix B to verify your answers to Problem P8.81.

P8.83 Consider the common-gate amplifier shown in Figure 8-166, and assume that $K = 2\ \text{mA/V}^2$ and $V_{th} = 1$ V. **(a)** Find the DC bias point and small-signal midband AC model for the transistor. Draw the small-signal midband AC equivalent circuit. Ignore r_o. **(b)** Find the small-signal midband AC transresistance v_o/i_i. **(c)** Find the small-signal midband AC input resistance.

S P8.84 Use SPICE and the Nlarge_mos model shown in Appendix B to verify your answers to Problem P8.83.

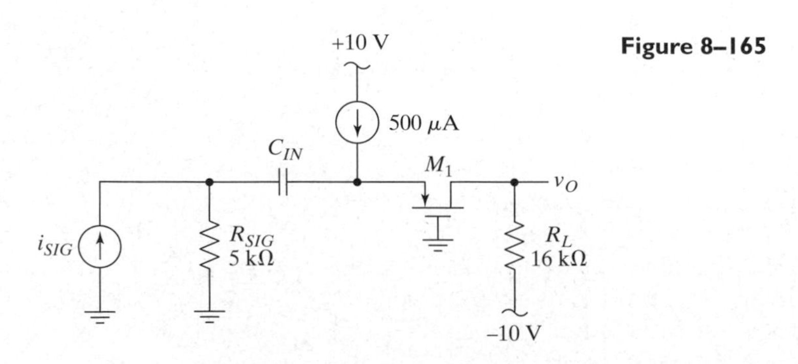

Figure 8–165

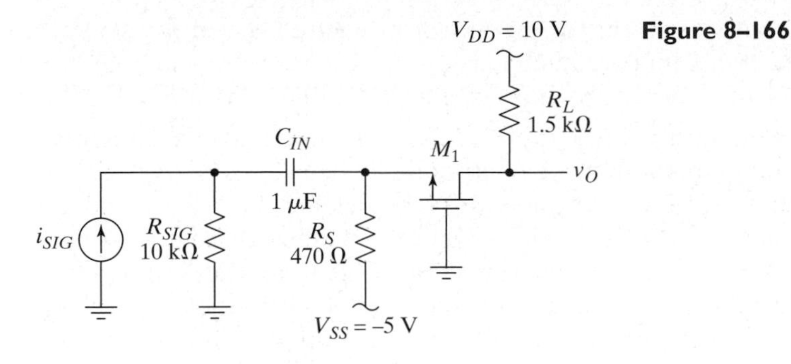

Figure 8–166

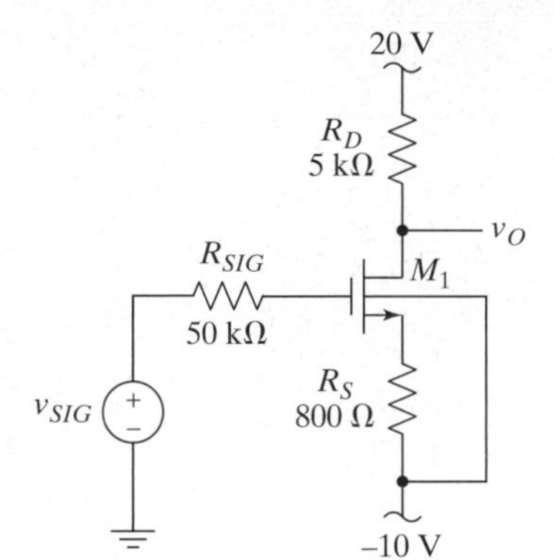

Figure 8–167

8.5 Integrated Amplifiers

8.5.1 The Body Effect in FET Amplifiers

P8.85 Repeat Problem P8.65, but include the bulk transconductance in your analysis. Assume that $\gamma = 0.5$ and the bulk is connected to the -5 V supply. Don't forget to allow for body effect changing V_{th}; use (4.34), assume that $2\varphi_f = 0.6$ V, and the value of V_{th} given in the problem is V_{th0}.

P8.86 Use SPICE and the Nlarge_mos model in Appendix B to verify your solution to Problem P8.85. You will need to edit the model in your circuit to include the parameter **GAMMA**. Set **GAMMA** = 0.5.

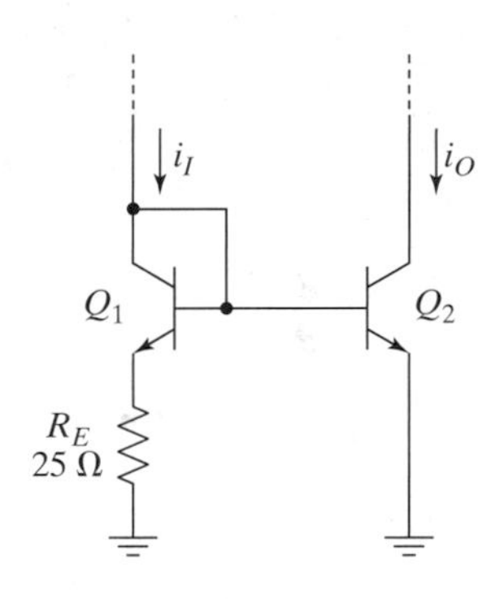

Figure 8–168

P8.87 Consider the common-source amplifier in Figure 8-167, and assume that $V_{SIG} = 0$ V. The transistor has $K = 100\ \mu\text{A/V}^2$, $V_{th0} = 0.5$ V, and $\gamma = 0.5$. **(a)** Find the DC bias point with the body effect included. Use (4.34), and assume that $2\varphi_f = 0.6$ V. **(b)** Draw the resulting small-signal midband AC equivalent circuit, including bulk transconductance but ignoring r_o. **(c)** Derive an equation for the voltage gain.

P8.88 Use SPICE and the Nnominal_mos model in Appendix B to verify your solution to Problem P8.87. You will need to edit the model in your circuit to include the parameter **GAMMA**. Set **GAMMA** = 0.5.

8.5.2 Current Mirrors

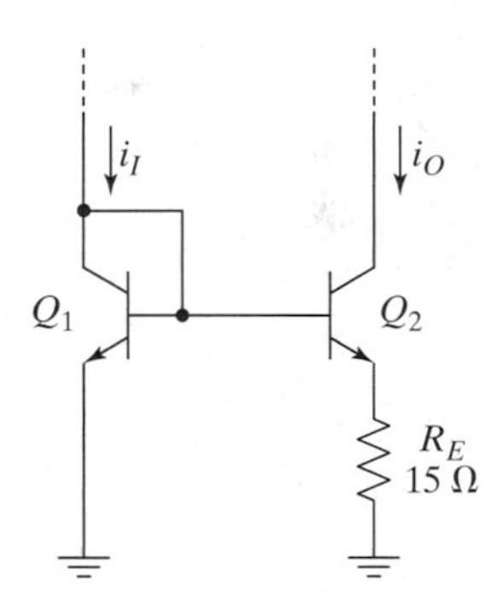

Figure 8–169

P8.89 **(a)** Calculate the small-signal midband AC output resistance of the mirror in Figure 8-168. Assume that the transistors are all kept in forward-active operation and that $I_I = 700\ \mu\text{A}$, $\beta = 100$, and $V_A = 75$ V. **(b)** Estimate the compliance of this mirror (i.e., the minimum output voltage required for the mirror to work properly).

P8.90 **(a)** Calculate the small-signal midband AC output resistance of the mirror in Figure 8-169. Assume that the transistors are all kept in forward-active operation and that $I_I = 100\ \mu\text{A}$, $\beta = 100$, and $V_A = 75$ V. **(b)** Estimate the compliance of this mirror (i.e., the minimum output voltage required for the mirror to work properly).

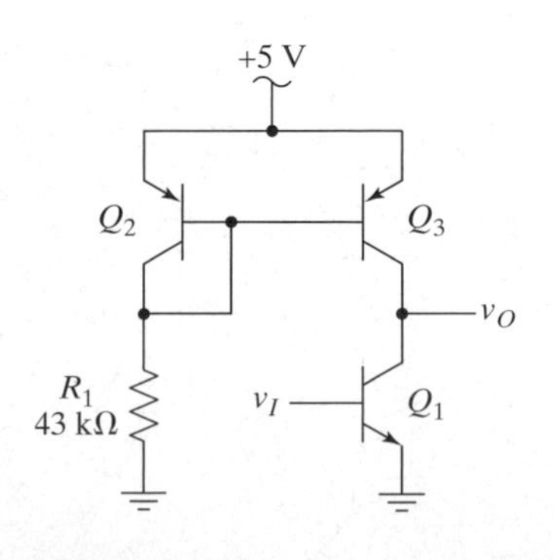

Figure 8–170

P8.91 **(a)** Calculate the voltage gain of the actively loaded common-emitter amplifier in Figure 8-170. Assume that $\beta = 100$ and $V_A = 75$ V for all devices and that V_I is somehow set to whatever value is needed to hold Q_1 and Q_3 in the forward-active region of operation. **(b)** If we desired to achieve the same gain using a resistive load on Q_1 and the same bias current, how large would the resistor need to be? How large would the supply voltage need to be?

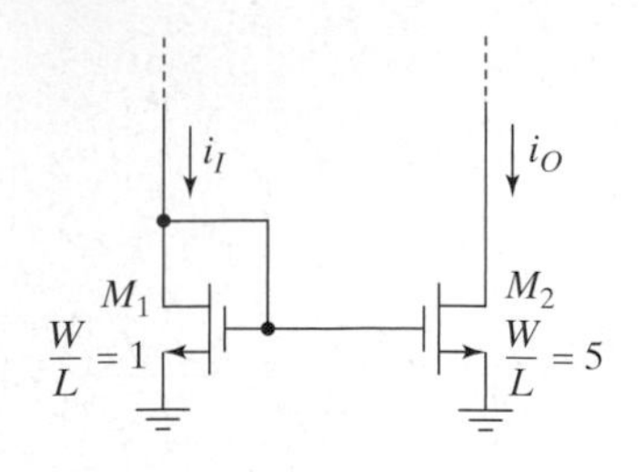

Figure 8–171

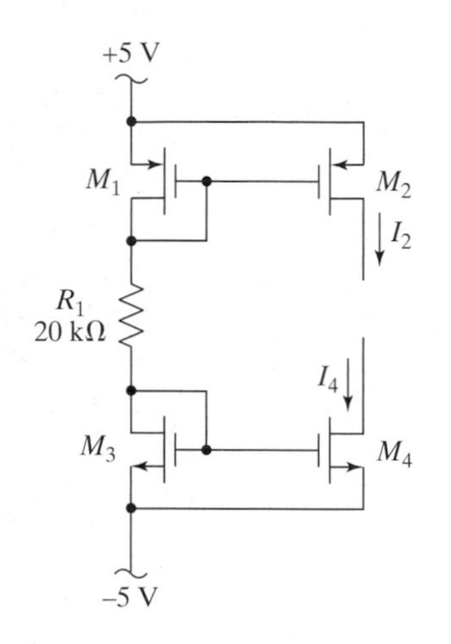

Figure 8–172

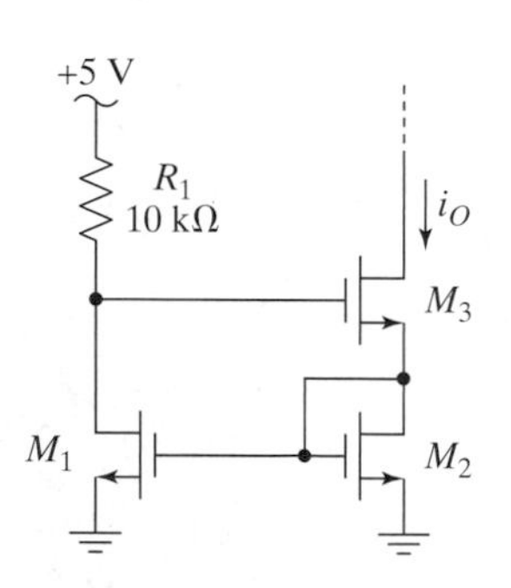

Figure 8–173

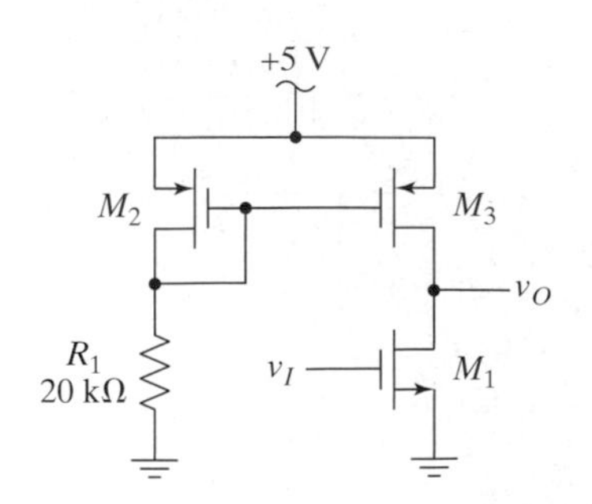

Figure 8–174

P8.92 Find the output current for the mirror in Figure 8-171 if $I_I = 250\ \mu A$ and if the transistors are matched (except for their sizes) and kept in forward-active operation. Ignore channel-length modulation.

P8.93 Consider the mirror in Figure 8-172, and assume that the transistors are all kept in forward-active operation and that $K = 100\ \mu A/V^2$, V_{th} = 0.5 V and $\lambda = 0.02\ V^{-1}$ for the NMOS devices, and $V_{th} = 0.5$ V and $\lambda = 0.03\ V^{-1}$ for the PMOS devices. Ignore the effect of channel-length modulation on the DC currents. **(a)** Calculate the small-signal midband AC output resistance as seen looking into the drain of M_2. **(b)** Calculate the small-signal midband AC output resistance as seen looking into the drain of M_4.

P8.94 **(a)** Calculate the small-signal midband AC output resistance of the mirror in Figure 8-173. Assume that the transistors are all kept in forward-active operation and that $K = 100\ \mu A/V^2$, $V_{th} = 0.5$ V, and $\lambda = 0.02\ V^{-1}$. Ignore the small effect the output voltage has on the DC drain current of M_3. **(b)** Estimate the compliance of this mirror (i.e., the minimum output voltage required for the mirror to work properly).

P8.95 **(a)** Calculate the voltage gain of the actively loaded common-source amplifier in Figure 8-174. Assume that $K = 100\ \mu A/V^2$, $V_{th} = \pm 0.5$ V, and $\lambda = 0.02\ V^{-1}$ for all devices and that V_I is somehow set to whatever value is needed to hold M_1 and M_3 in the forward-active region of operation. **(b)** If we desired to achieve the same gain using a resistive load on M_1 and the same bias current, how large would the resistor need to be? How large would the supply voltage need to be?

8.6 Differential Amplifiers

8.6.1 The Generic Differential Pair

P8.96 Derive an equation for the small-signal differential-mode output voltage of the generic differential pair in Figure 8-92, assuming a purely common-mode input, but allowing for the r_m's of the transistors and the loads to be different.

P8.97 Consider Figure 8-92 and allow for $R_1 \neq R_2$. Derive an expression for the resulting common-mode to differential-mode conversion gain.

P8.98 Consider the amplifier shown in Figure 8-175, and assume that both n-type transistors are perfectly matched and all three p-type transistors are perfectly matched. Assume further that the current sources are perfectly matched so that the DC bias voltages at the output terminals of T_1 and T_2 work out to keep all of the transistors in the forward-active region of operation. (To accomplish this with real circuits requires additional circuitry, called common-mode feedback.) This circuit is also an actively loaded differential pair, but this active load provides a differential output. **(a)** Draw the small-signal midband AC equivalent circuit, and **(b)** find the small-signal midband AC voltage gain.

8.6.2 A Bipolar Implementation: The Emitter-Coupled Pair

P8.99 Consider again the ECP in Figure 8-105(a). If the resistor tied to Q_1's collector is $R_1 = 19$ kΩ, the resistor tied to Q_2's collector is $R_2 = 20$ kΩ, and the current source has an output resistance of $R_{EE} = 50$ kΩ, find the common-mode to differential-mode conversion gain. Assume the transistors are matched and have $\beta = 100$.

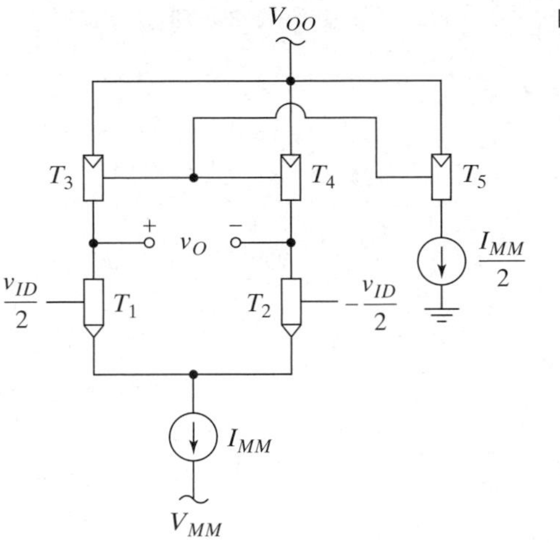

Figure 8–175

P8.100 Use SPICE and the Nsimple model from Appendix B to confirm your answer to Problem P8.99.

P8.101 Consider the circuit shown in Figure 8-176, and assume that the current source has an output resistance $R_{EE} = 100\ \text{k}\Omega$ and that $\beta = 100$. **(a)** Find the differential-mode gain, common-mode gain, and CMRR if the output is taken single-ended from the collector of Q_2 and the DC common-mode input voltage is zero. **(b)** Would you expect the results from part (a) to change if the DC common-mode input voltage was changed to 2 V? Why or why not? **(c)** Find the small-signal midband AC common- and differential-mode input resistances.

S P8.102 Use SPICE and the Nnominal model from Appendix B to confirm the results you obtained in Problem P8.101.

D P8.103 Using the schematic shown in Figure 8-176, design the circuit to have a differential-mode gain of 100 with $I_{EE} = 200\ \mu\text{A}$. What is the maximum allowable DC common-mode input voltage for the resulting circuit?

P8.104 Consider the circuit shown in Figure 8-177, and assume that the current source has an output resistance $R_{EE} = 50\ \text{k}\Omega$ and that $\beta = 100$. Find the small-signal midband AC voltage gain.

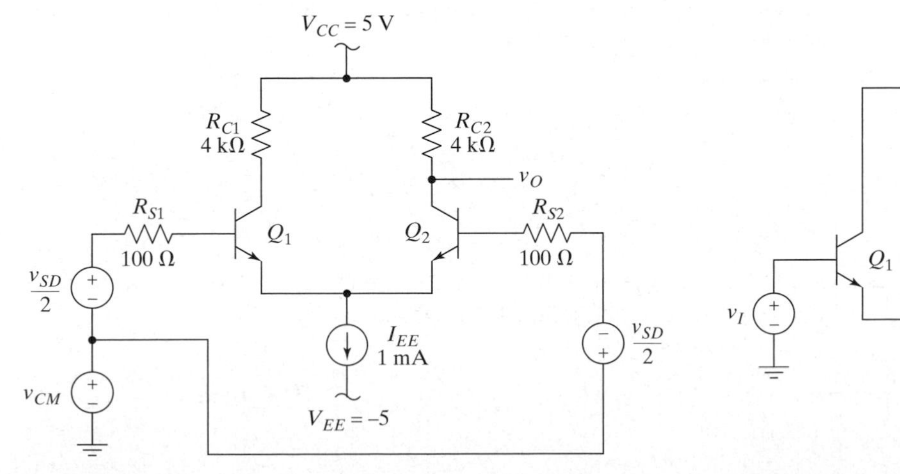

Figure 8–176

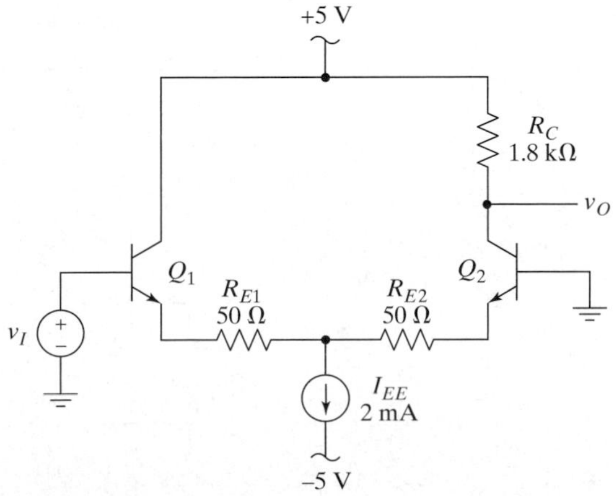

+5 V

R_C 1.8 kΩ

v_O

Q_1 Q_2

R_{E1} 50 Ω

R_{E2} 50 Ω

v_I

I_{EE} 2 mA

−5 V

Figure 8–177

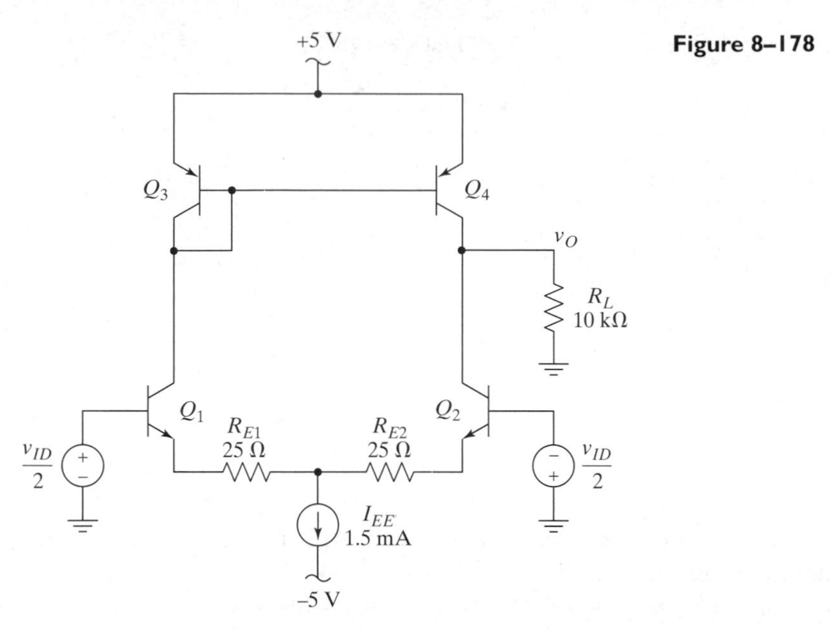

Figure 8–178

P8.105 Consider the circuit in Figure 8-108, and assume that $V_{ICM} = 0$, $V_{AN} = V_{AP} = 50$ V and $I_{EE} = 3$ mA. **(a)** Find A_{dm}. **(b)** If a resistive load were used instead, and the differential output considered, how large would the resistors need to be to achieve the same gain? **(c)** How large would the supply voltage need to be?

S P8.106 Use SPICE along with the Nnominal and Pnominal models from Appendix B to verify your answer to part (a) of Problem P8.105.

P8.107 Consider the amplifier in Figure 8-178. Find the small-signal midband AC differential-mode gain.

P8.108 Consider the amplifier in Figure 8-178. If the current source output resistance is $R_{EE} = 50$ kΩ and the *pnp* $\beta = 25$, but the circuit is ideal in every other way, what is the small-signal common-mode gain?

8.6.3 A FET Implementation: The Source-Coupled Pair

P8.109 Find the differential-mode gain, common-mode gain, and CMRR for the circuit shown in Figure 8-179 if the output is taken single-ended from the drain of M_2 and the DC common-mode input voltage is zero. Assume that the current source has an output resistance $R_{SS} = 100$ kΩ and the transistors have $K = 100\ \mu\text{A/V}^2$ and $V_{th} = 0.5$ V.

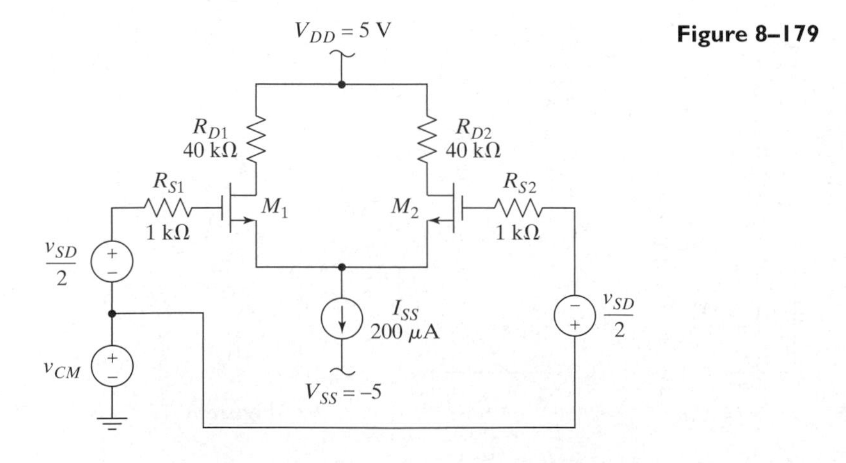

Figure 8–179

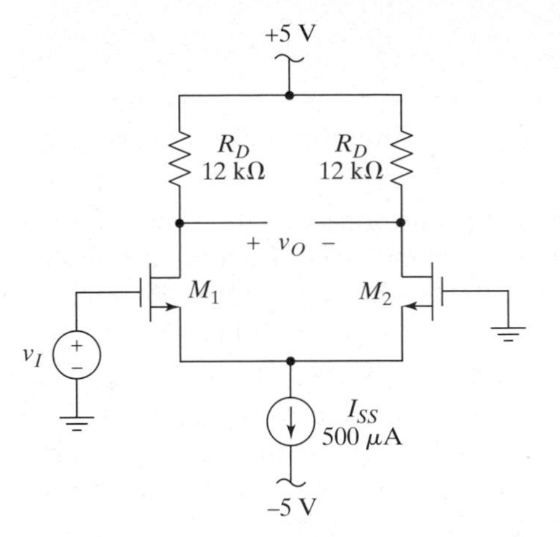

Figure 8–180

S P8.110 Use SPICE and the Nsimple_mos model in Appendix B to verify your results for Problem P8.109.

P8.111 Consider again the circuit of Problem P8.109, but let $R_{D1} = 41$ kΩ. Determine the small-signal midband AC common-mode to differential-mode conversion gain for the circuit if the output is taken differentially; $v_O \equiv v_{D1} - v_{D2}$.

P8.112 Consider the amplifier in Figure 8-180, and assume that $K = 2$ mA/V^2 and $V_{th} = 1$ V. **(a)** Draw the small-signal midband AC equivalent circuit and determine the voltage gain from the single-ended input to the differential output. **(b)** Repeat part (a) with $R_{SS} = 50$ kΩ included.

S P8.113 Use SPICE and the Nlarge_mos model in Appendix B to verify your results for Problem P8.112.

P8.114 Consider the amplifier in Figure 8-181, and assume $V_{CM} = 0$, and that both the NMOS and PMOS transistors have $K = 100$ μA/V^2, $V_{th} = \pm 0.5$ V, and $\lambda = 0.02$ V^{-1}. Find A_{dm}.

P8.115 Use SPICE and the Nsimple2_mos and Psimple2_mos models in Appendix B to verify your answer to Problem P8.114.

P8.116 Consider again the circuit from Problem P8.114. Add a load resistor of 25 kΩ to ground, and assume that the current source output resistance is $R_{SS} = 50$ kΩ and that the (W/L) of M_1 is 5% larger than that of M_2 (but everything else remains the same). **(a)** Calculate A_{dm}. **(b)** Calculate A_{cm} and **(c)** find the CMRR.

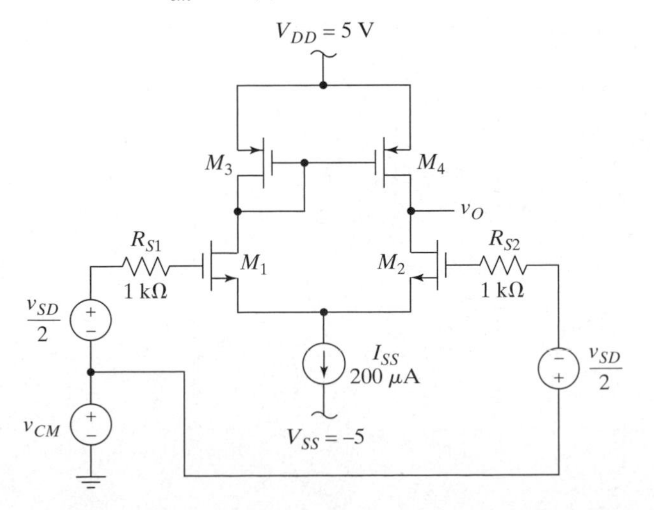

Figure 8–181

8.7 Multistage Amplifiers

8.7.1 Multistage Bipolar Amplifiers

P8.117 Derive (8.296) and show that R_{BB} is maximum when $V_B = V_{CC}/2$. Derive (8.297) and (8.298).

D P8.118 Design an emitter-follower common-emitter cascade to achieve a voltage gain equal to 20 dB, an input resistance of at least 50 kΩ, and an output resistance of no more than 100 Ω when driven from $R_S = 10$ kΩ and while driving $R_L = 1$ kΩ. Assume that $V_{CC} = 10$ V and that $\beta = 100$.

P8.119 Consider the amplifier shown in Figure 8-182. The transistors are 2N2857s, and the model for them is presented in Appendix B. **(a)** Find the DC bias points and small-signal midband AC models for all three transistors, and then draw the resulting equivalent circuit. **(b)** Find the small-signal midband AC voltage gain. **(c)** Find the small-signal midband AC input and output resistances.

S P8.120 Use SPICE and the 2N2857 model presented in Appendix B to confirm your solutions to Problem P8.119.

P8.121 Consider the amplifier shown in Figure 8-183. The transistors are 2N2857s, and the model for them is presented in Appendix B. **(a)** Find the DC bias points and small-signal midband AC models for both transistors. Then draw the resulting equivalent circuit. **(b)** Find the small-signal midband AC voltage gain. **(c)** Find the small-signal midband AC input and output resistances.

S P8.122 Use SPICE and the 2N2857 model presented in Appendix B to confirm your solutions to Problem P8.121.

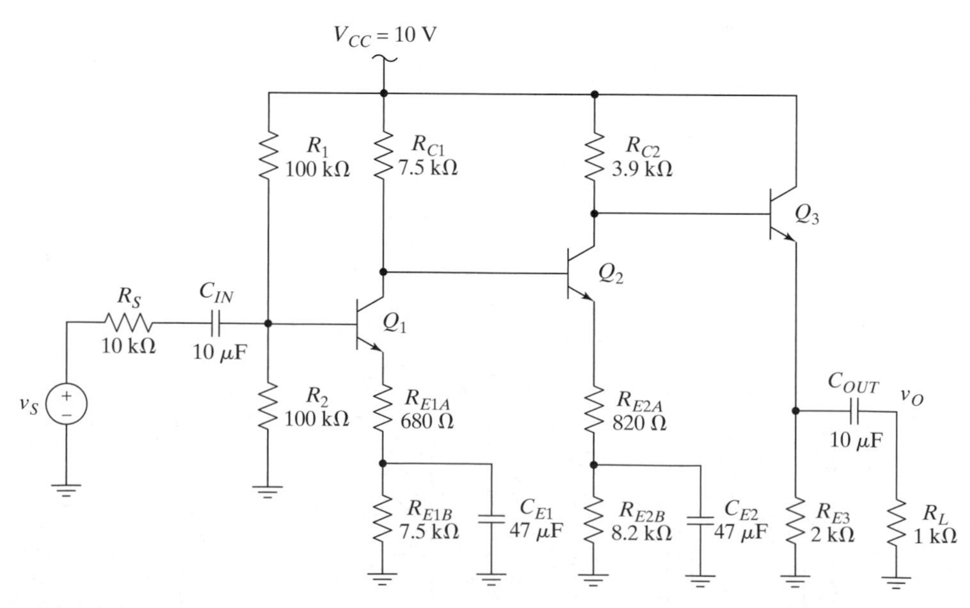

Figure 8–182

P8.123 Consider the amplifier shown in Figure 8-184. Q_1 is a 2N2857 transistor and Q_2 is a 2N3904 transistor. Both models are presented in Appendix B. **(a)** Find the DC bias points and small-signal midband AC models for both transistors, and then draw the resulting equivalent circuit. **(b)** Find the small-signal midband AC voltage gain. **(c)** Find the small-signal midband AC input and output resistances.

S P8.124 Use SPICE and the 2N2857 and 2N3904 models presented in Appendix B to confirm your solutions to Problem P8.123.

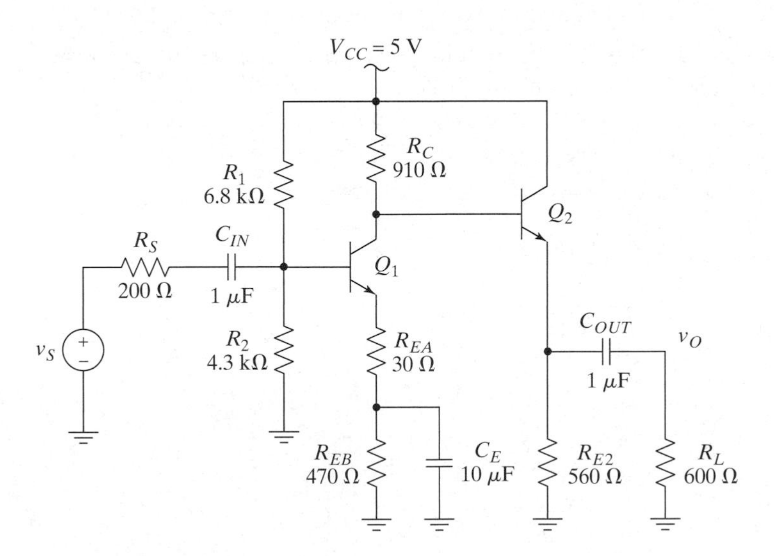

Figure 8–183

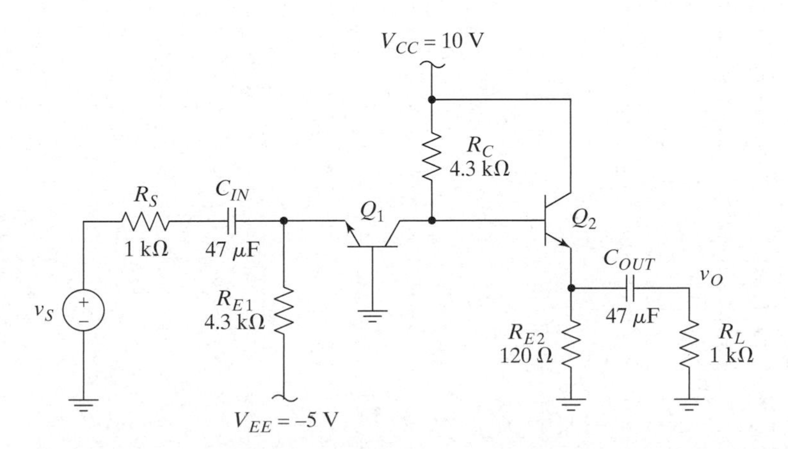

Figure 8–184

P8.125 Consider the circuit of Figure 8-185. The transistors have $\beta = 100$ and $V_T = 26$ mV. **(a)** Consider the case in which the AC-coupled load is connected to v_{O2}. Find v_{o2}/v_s for midband AC. **(b)** Consider the case in which the AC-coupled load is connected to v_{O1} and Q_2 is not present (so the 300 Ω emitter resistor of Q_2 is also not present). Find v_{o1}/v_s for midband AC. **(c)** Explain the function of the emitter-follower circuit formed with Q_2.

P8.126 Consider the amplifier shown in Figure 8-186. The transistors are 2N2222s, and the model for them is presented in Appendix B. **(a)** Find the DC bias points and small-signal midband AC models for both transistors, and then draw the resulting equivalent circuit. (*Hint*: Q_2 is a common-base stage.) **(b)** Find the small-signal midband AC voltage gain. **(c)** Find the small-signal midband AC input and output resistances.

S P8.127 Use SPICE and the 2N2222 model presented in Appendix B to confirm your solutions to Problem P8.126.

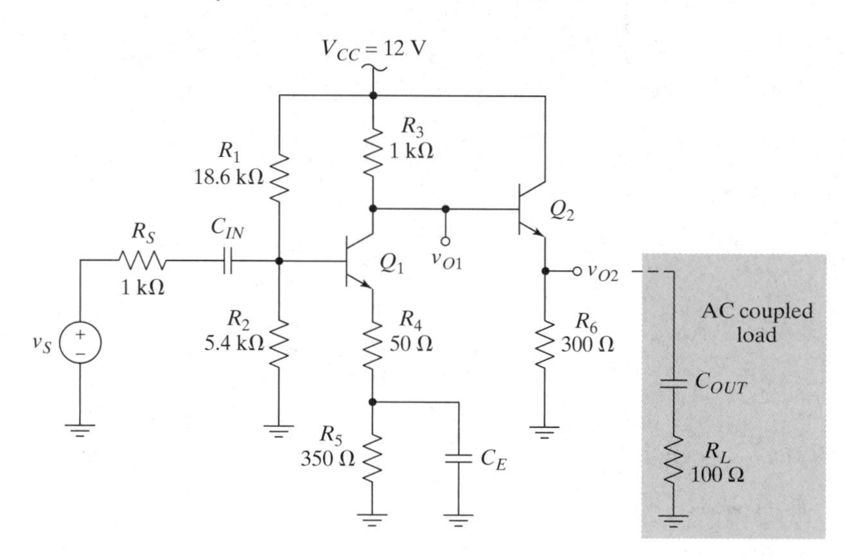

Figure 8–185

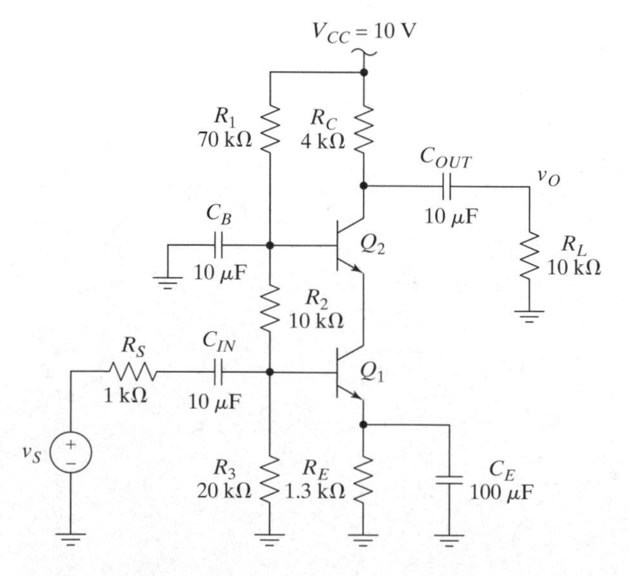

Figure 8–186

8.7.2 Multi-Stage FET Amplifiers

P8.128 Consider the amplifier shown in Figure 8-187. The transistors have $K = 100\ \mu A/V^2$ and $V_{th} = 0.5$ V. **(a)** Find the DC bias points and small-signal midband AC models for both transistors, and then draw the resulting equivalent circuit. Assume $V_S = 0$. **(b)** Find the small-signal midband AC voltage gain. **(c)** Find the small-signal midband AC input and output resistances.

S P8.129 Use SPICE and the Nsimple_mos model presented in Appendix B to confirm your solutions to Problem P8.128.

P8.130 Consider the amplifier shown in Figure 8-188. M_1 and M_2 have $K = 100\ \mu A/V^2$ and $V_{th} = 0.5$ V, and M_3 has $K = 2\ mA/V^2$ and $V_{th} = 1$ V. **(a)** Find the DC bias points and small-signal midband AC models for all three transistors, and then draw the resulting equivalent circuit. **(b)** Find the small-signal midband AC voltage gain. **(c)** Find the small-signal midband AC input and output resistances.

S P8.131 Use SPICE and the Nsimple_mos and Nlarge_mos models presented in Appendix B to confirm your solutions to Problem P8.130.

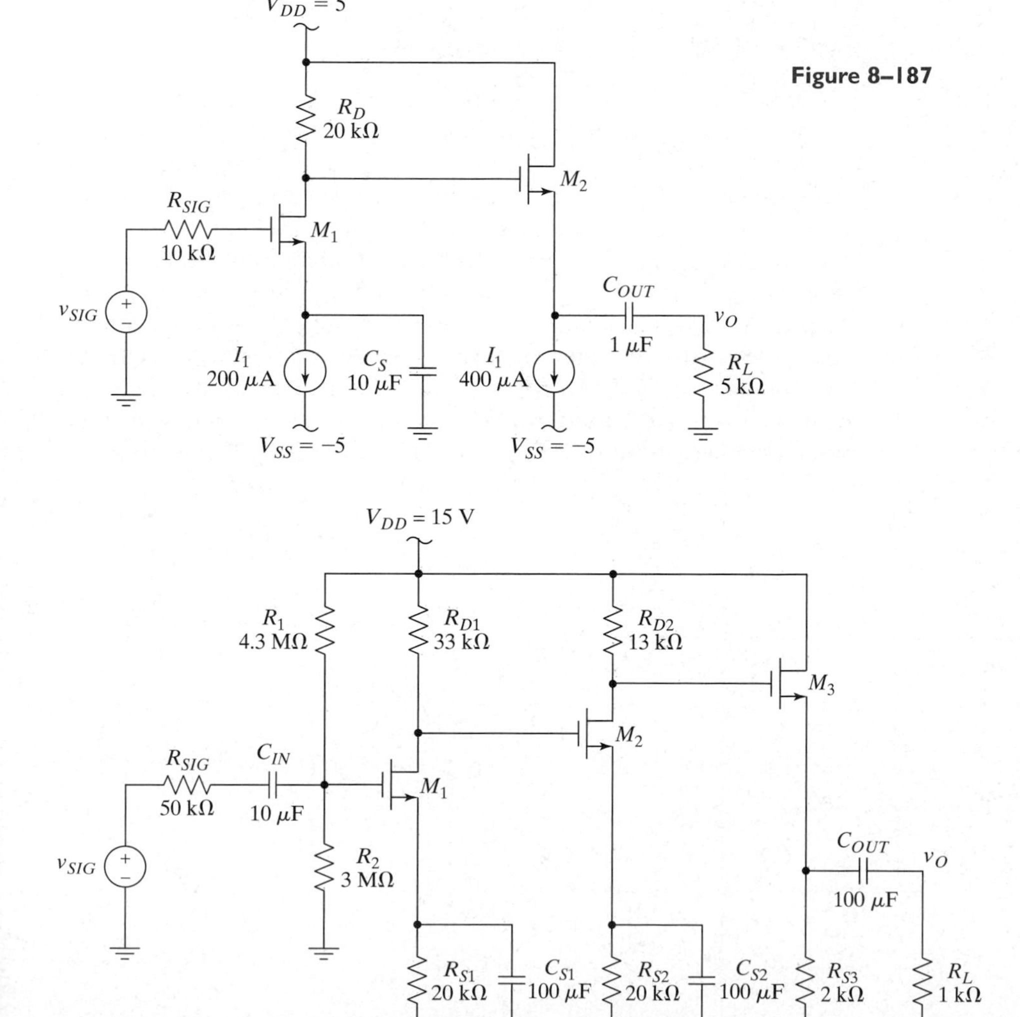

Figure 8–187

Figure 8–188

P8.132 Consider the amplifier shown in Figure 8-189. The transistors have $K = 100\ \mu A/V^2$ and $V_{th} = 0.5$ V. **(a)** Find the DC bias points and small-signal midband AC models for both transistors, and then draw the resulting equivalent circuit. (*Hint*: M_2 is a common-gate stage.) **(b)** Find the small-signal midband AC voltage gain. **(c)** Find the small-signal midband AC input and output resistances.

S P8.133 Use SPICE and the Nsimple_mos model presented in Appendix B to confirm your solutions to Problem P8.132.

8.7.3 Multistage Amplifiers with Bipolar and Field-Effect Transistors

P8.134 Consider the amplifier shown in Figure 8-190. M_1 has $K = 100\ \mu A/V^2$ and $V_{th} = 0.5$ V, and Q_1 and Q_2 have $\beta = 100$. **(a)** Find the DC bias points and small-signal midband AC models for all three transistors, and then draw the resulting equivalent circuit. **(b)** Find the small-signal midband AC voltage gain. **(c)** Find the small-signal midband AC input and output resistances.

S P8.135 Use SPICE and the Nsimple_mos and Nsimple (BJT) models presented in Appendix B to confirm your solutions to Problem P8.134.

8.8 Comparison of BJT and FET Amplifiers

P8.136 This problem examines the limits of cascoding BJTs and MOSFETs to improve output resistance. Consider a common-base BJT stage as in Figure 8-191(a) and a common-gate MOSFET stage as in part (b) of the figure. In both cases, the degeneration resistors (R_E and R_S) are intended to model preceding stages (for example, the output resistance of a transistor that has the cascode added to increase its output resistance) and the bias voltages are AC grounds. **(a)** Derive the output resistance for the BJT circuit as R_E goes to infinity; this is the maximum achievable resistance even when many stages are cascoded. (Be careful! Can you ignore r_μ?) **(b)** Derive the output resistance of the MOSFET circuit as R_S goes to infinity. **(c)** What can you conclude about the relative achievable output resistances of BJT and MOSFET mirrors?

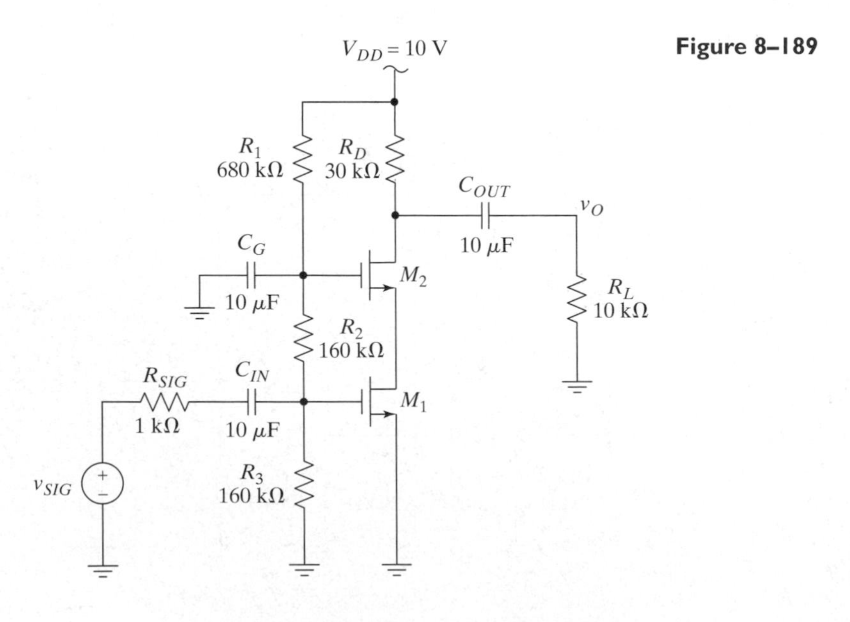

Figure 8–189

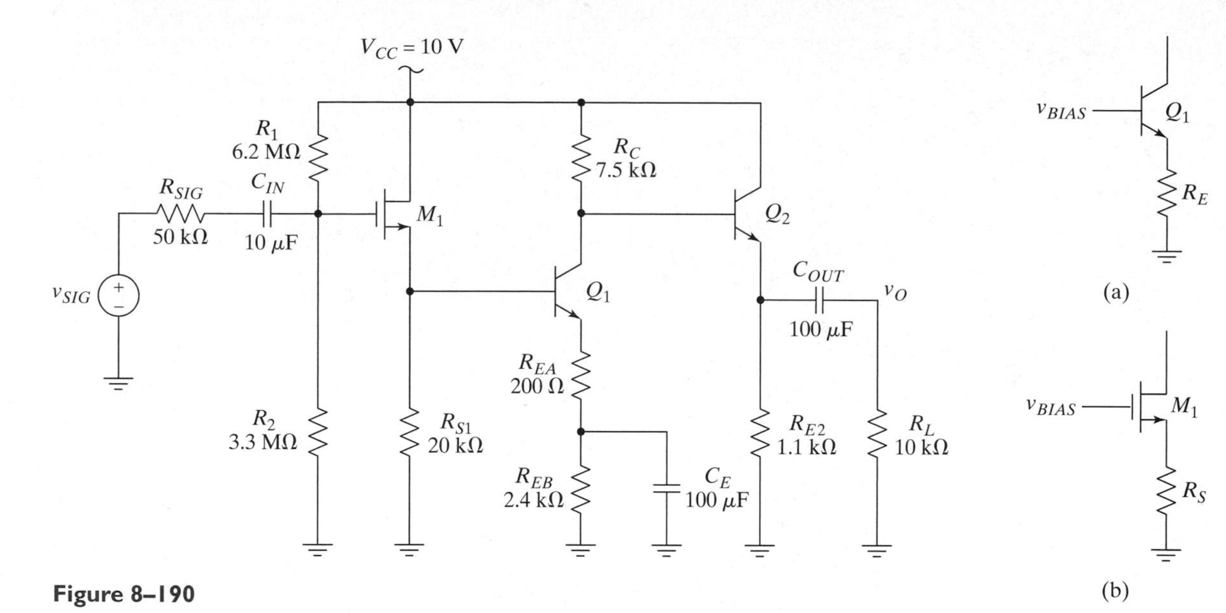

Figure 8–190

Figure 8–191

P8.137 **(a)** Consider an *npn* BJT with $I_C = 1mA$. What is its g_m? **(b)** Now consider an *n*-channel MOSFET with $I_D = 1$ mA. If it has $K' = 5.31 \times 10^{-5}$ A/V^2 (where $K' = \mu_n C'_{OX}$), What W/L ratio is required for it to have the same g_m as the bipolar transistor in part (a)? If the minimum feature size is 2 μm, then the minimum value for L is 2 μm. What is the minimum area of the transistor (in square microns) if the area = WL plus an additional 8μm · W to allow for source and drain contacts? (In a process with a 2 μm minimum feature size, a bipolar transistor can have an area as small as 13 μm × 19 μm = 247 μm^2.)

CHAPTER 9

Amplifier Frequency Response

Frequency response is important in many electronic systems and can also be observed in nature. The sky is blue because high-frequency blue light is scattered more than lower-frequency light. The sunset is red because the higher-frequency light is scattered so much while taking this longer path through the atmosphere that it gets absorbed (i.e., the energy drops to nearly zero).

INTRODUCTION

Preceding chapters have discussed DC biasing and the small-signal midband AC performance of amplifiers. In this chapter, we see how to analyze the frequency response of amplifiers and how to design amplifiers to achieve a desired response. We assume that the desired response is flat within some range of frequencies (the midband, as already noted in Chapter 8). We seek methods for determining that range of frequencies and design techniques to extend the range. Many of the analysis techniques discussed here apply to other linear circuits as well, but amplifiers will be the main topic. Filters, which are used to shape the frequency response of a system to achieve various design goals, are discussed in Chapter 11.

As in the previous two chapters, we begin with a discussion of the models needed for passive and active devices. We then proceed to a discussion of techniques for determining the midband range of a given circuit, called the ***bandwidth***, and finally give some specific circuit examples. We again use the generic transistor to introduce discussions of the different circuit topologies and then present specific results for each type of transistor.

As always in this text, a major goal of the chapter is to develop the techniques necessary to analyze circuits in an intuitive and reasonable manner so that we can use the results for design. The analysis of the complete frequency response (i.e., a knowledge of the frequencies of all poles and zeros) of even a simple amplifier is usually too complex to do by hand, and even if a result is obtained, it is often too complex to be any use in the design process. Therefore, we will concentrate on finding the cutoff frequencies,[1] which determine the bandwidth. If a knowledge of the frequencies of other poles or zeros is desired (e.g., for compensation of a feedback loop, as discussed in Chapter 10), they can sometimes be found by hand analysis; but more often, computer simulation would be used.

In this chapter, we assume that you are familiar with complex impedances and transfer functions. We also assume you are familiar with single-pole transfer functions as covered in Aside A1.6, and AC- and DC-coupled transfer functions, as presented in Sections 6.5.2 and 8.1.2.

9.1 HIGH-FREQUENCY SMALL-SIGNAL MODELS FOR DESIGN

Section 8.1 covered the small-signal midband AC models we use for active and passive devices. In this section, we extend those models for use at higher frequencies. But, not all parasitic elements are explicitly presented here. For example, all diodes and transistors have series inductance in the wires connecting them with the rest of the circuit. These additional parasitic elements are extremely important at very high frequencies. Nevertheless, this section focuses on models suitable for use at moderate frequencies (up to a few hundred MHz at most). Many of the additional parasitic elements can be included in a straightforward way, based on the material presented here. The key issue is deciding what level of modeling is necessary for a given application.

9.1.1 Independent Sources

The models for independent sources have been discussed for DC analysis in Chapter 7 and for AC analysis in Chapter 8. It makes no difference whether the frequencies are high or low for these models; therefore, the models used in Chapter 8 apply equally well here.

9.1.2 Linear Passive Elements (Rs, Ls, & Cs)

In analyzing the frequency response of electronic circuits, it is occasionally necessary to reevaluate the models used for passive components. For example, a wire-wound precision resistor also has significant inductance, which must be taken into account at higher frequencies. In fact, all passive components have ***parasitic*** elements in series or parallel with the desired element (e.g., the inductance of the wire-wound resistor), as shown in Figure 9-1.

At very high frequencies, it may not be possible to use ***lumped*** models for all elements.[2] A lumped model is one that uses a single ideal element to model each

---

[1] These are often called the -3 dB frequencies. They are simply the frequencies at which the specified transfer function has dropped by 3 dB (i.e., $1/\sqrt{2}$) from its midband value.

[2] The boundary is usually taken to be when the dimensions of the circuit exceed one-tenth of the wavelength of the voltage wave. Voltage waves in circuits travel at roughly $1/2$ the speed of light, so the wavelength in meters is approximately $\lambda \approx 150/f$ (in MHz). For a device with a physical length of 1 inch, the boundary between lumped and distributed analysis occurs at about 5.9 GHz. In lumped analysis, we use Kirchhoff's laws, which are only an approximation to Maxwell's equations [9.1].

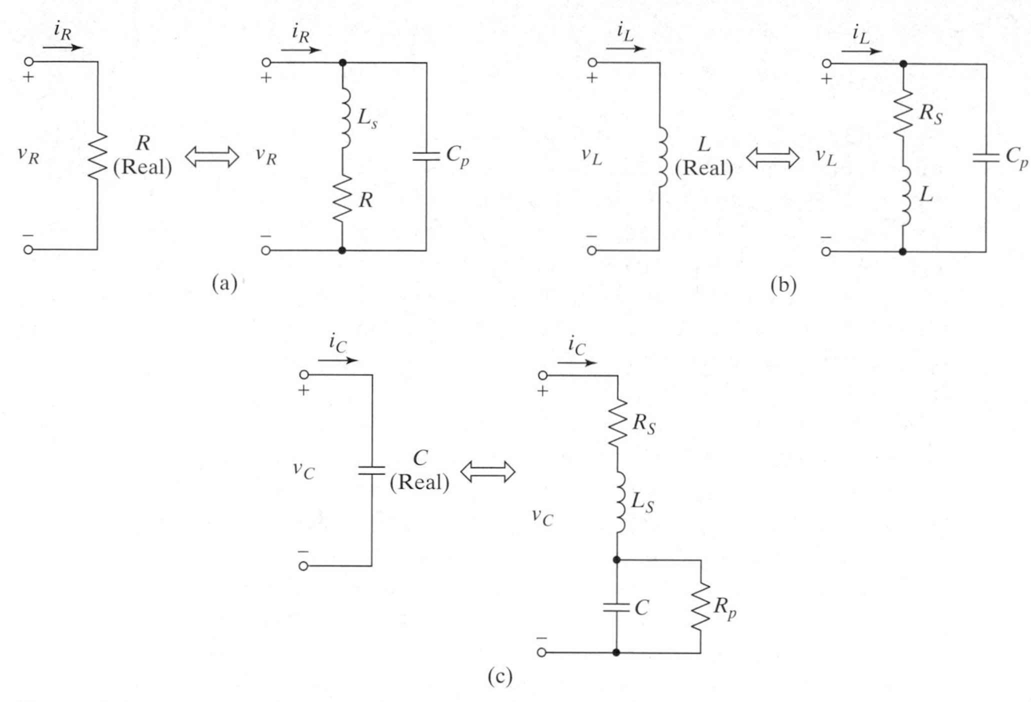

Figure 9–1 One possible model for (a) a real resistor, (b) a real inductor, and (c) a real capacitor. The elements used in the models are ideal resistance, capacitance, and inductance. R_s is the series parasitic resistance (caused by the leads), R_p is the parallel parasitic resistance, and L_s and C_p are the parasitic series inductance and parallel capacitance, respectively.

characteristic (e.g., resistance, inductance) of a device. (The models in Figure 9-1 are lumped.) Lumped models can be quite complex when they are used to approximate a distributed system. For example, a wire-wound resistor could be modeled by a string of R's and L's in series (each RL combination modeling a single turn), with a capacitor in parallel with each series RL combination to model the capacitance from one turn to the next. Models this complicated are not typically employed, however, especially in hand analysis.

When lumped models fail, we can use ***distributed*** models instead. A distributed model is a model that allows the voltages and currents to be functions of physical position as well as time. Other techniques exist for modeling distributed systems in a more elegant way, but they are beyond the scope of this text. Nevertheless, there are parasitic effects that are commonly encountered even at much lower frequencies. In this subsection, we will discuss a few of these effects and present models that can be used for passive devices if the effects are significant. We will discuss discrete and integrated passive elements separately.

Discrete Passive Elements All real resistors have some series inductance and some parallel capacitance, as shown in Figure 9-1(a). With the exception of wire-wound resistors, which are typically avoided except at lower frequencies, these parasitic elements are not usually important at frequencies where lumped models are acceptable.

All real inductors have some series resistance and parallel capacitance, as shown in Figure 9-1(b). The parallel capacitance can usually be ignored, but there are many cases where the series resistance needs to be accounted for. This resistance is usually specified by the manufacturer, often by specifying the ***quality factor***, or Q, of the inductor. The Q is a measure of how much energy is stored in the inductor relative to how much is dissipated. In the context of Figure 9-1(b), if we consider low frequencies where C_p can be ignored, the Q is given by

$$Q = \frac{|v_L|}{|v_R|} = \frac{\omega L}{R_s}. \tag{9.1}$$

This equation allows us to find R_s given the Q, the test frequency, and the inductance. As an example, one commonly available molded inductor of 10 μH has a Q of 50 at a test frequency of 7.9 MHz. Using (9.1), we find that $R_s = 10\ \Omega$.

All real capacitors have some series resistance and inductance and some parallel resistance, as shown in part (c) of Figure 9-1. The parallel resistance is a result of leakage current through the dielectric and can be significant in some applications. Electrolytic capacitors can have especially large leakage currents and are usually used only as power supply filter capacitors. The leakage currents are often specified on the data sheets.

In addition to the leakage current, the series inductance of capacitors can cause problems at surprisingly low frequencies. Because of this inductance, a capacitor will exhibit series resonance at some frequency. For frequencies above this ***self-resonant frequency,*** the capacitor appears mostly inductive, and its impedance *increases* with increasing frequency. The self-resonant frequencies of many popular capacitors are in the 10 kHz to 100 MHz range, with aluminum electrolytic capacitors at the bottom of the range and silver–mica capacitors at the high end. It is common for capacitors used as bypass capacitors to look inductive at the frequencies of interest. However, as long as the overall impedance is low enough, they still serve the purpose of bypassing the elements they are in parallel with.

The impedance of a capacitor at frequencies below the self-resonant frequency can always be modeled as just a resistor in series with a capacitor. (Both element values will, in general, be functions of frequency.) If this two-element model is used, the series resistance is called the ***effective series resistance***, or ESR. The ESR is *not* the same as R_s in Figure 9-1, but it yields a useful simplified model. For example, the power losses in a capacitor can be readily calculated using the ESR. Since the ESR is modeling several different effects, it is a function of frequency, as already noted.

Frequently, the ESR is not given directly, but it is specified by giving the ***dissipation factor*** (DF) of the capacitor. The DF can be defined as

$$\text{DF} = \frac{1}{2\pi}\ \frac{\text{energy dissipated in one cycle}}{\text{peak energy stored}} = \frac{1}{Q}, \tag{9.2}$$

where Q is the quality factor. The DF is sometimes specified as a dimensionless number (<1) and sometimes as a percentage. Using this definition, and assuming that the capacitor is modeled by a series resistance R_s and a capacitance C, we can derive (*see* Problem P9.4)

$$\text{DF} = \omega R_s C, \tag{9.3}$$

The DF is also sometimes given as

$$\text{DF} = \tan\delta = \omega R_s C. \tag{9.4}$$

where δ is defined to be the difference between the phase of the overall capacitor impedance and $-90°$. For an ideal capacitor, both δ and the DF are zero. The DF is also equal to the ratio of the magnitudes of the voltages across the resistance and the capacitance in the two-element model (*see* Problem P9.5):

$$\mathrm{DF} = \frac{|v_R|}{|v_C|} = \omega R_s C. \tag{9.5}$$

Remembering from (9.2) that the DF is $1/Q$, we see that (9.5) agrees with (9.1) for the Q of an inductor.

Example 9.1

Problem:
Find the ESR of a 0.001-μF polyester-film capacitor with a dissipation factor of 1% at 1 kHz.

Solution:
Using (9.3) with DF = 0.01 (i.e., 1%), we find that $R_s = 1.6$ kΩ. This value may seem unbelievably large, but you must compare it with the reactance of the capacitor at the same frequency: $X_C(1\ \text{kHz}) = 1/(2\pi \cdot 1\text{k} \cdot 0.001\mu) = 159\ \text{k}\Omega$. Note that X_C is 100 times larger than the ESR, which is exactly what (9.5) implies for a DF of 1%. ■

Exercise 9.1

Given a 1-μH inductor with a Q of 44 measured at 25 MHz, what is R_S? Ignore C_p during your calculation. □

Exercise 9.2

Measurements on a 0.01-μF capacitor show that the impedance reaches a minimum value of 1.5 Ω at 5 MHz. (a) Ignoring C_p, find values for R_S and L_S. (b) What is the dissipation factor for this capacitor at 1 kHz? □

Integrated Passive Elements Resistors, capacitors, and inductors, as fabricated in integrated circuits, have different high-frequency parasitics than do their discrete counterparts. We will briefly point out the differences in this section for those readers familiar with IC technology.

As noted in Section 8.1, inductors are rarely used in ICs at this time. Nevertheless, they are used, especially for high-frequency circuits. Integrated inductors can be modeled in much the same way as discrete inductors.

Diffused resistors have parasitic *pn* junctions, as discussed in Section 8.1. At high-frequencies, the capacitance of a *pn* junction can be a significant parasitic and should be included in the model used for design. Usually, a simple lumped model is adequate, although there are occasions when a more complicated lumped approximation to a distributed model is required. Thin-film resistors also have parasitic capacitance to the underlying substrate, but it is typically much smaller than for diffused resistors.

Capacitors in integrated circuits are usually made by using parallel-plate structures, where the plates are either conductors (metal or heavily doped polysilicon) or conductive regions in the semiconductor substrate. When a conductive region in the substrate is used as one of the plates of the capacitor, there is a significant parasitic capacitance from this plate to the surrounding substrate regions. These extra parasitic capacitances are often important and must be included in the models used.

ASIDE A9.1 BROADBAND DECOUPLING

In designing electronic circuits, it is frequently necessary to bypass some element or a power supply over a wide range of frequencies. Since any real capacitor exhibits self resonance, this task can be more difficult than you might at first imagine. For example, if a large capacitor on the order of 10 μF is needed to provide an adequate bypass at low frequencies, this capacitor will usually have a self-resonant frequency between 100 kHz and 10 MHz, depending on the type of capacitor chosen. If the self-resonant frequency is 500 kHz, ignoring the parallel capacitance, we determine the series inductance to be $L_s = 1/(\omega_o^2 C)$, which is 10 nH. The impedance of this capacitor will then be equal to 6.4 Ω at 100 MHz. Depending on the application, this impedance may not be small enough.

One technique that is frequently employed in this situation is to put a smaller capacitor with a higher self-resonant frequency in parallel with the large capacitor. The logic is that the small capacitor will take over and provide an acceptable bypass for frequencies above the self-resonant frequency of the larger capacitor. But is this logic correct?

Consider a specific example where the large capacitor is the previously considered 10-μF electrolytic with a self-resonant frequency of 500 kHz and the smaller capacitor is a 0.01-μF polyester-film capacitor with a self-resonant frequency of 20 MHz. We calculate the inductance of the smaller capacitor to be 6.3 nH. The two capacitors in parallel can be modeled as shown in Figure A9-1 if we neglect the series resistance.

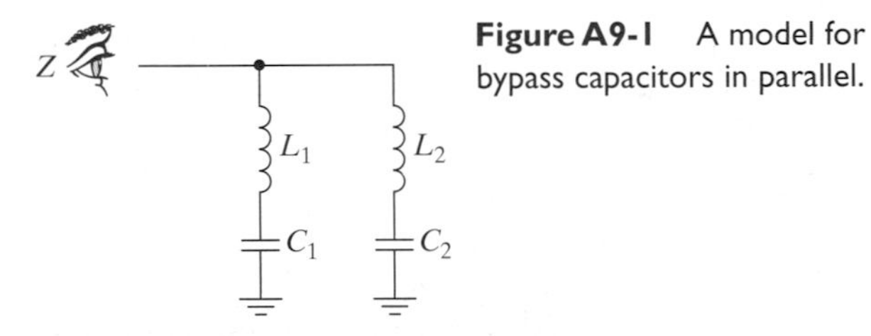

Figure A9-1 A model for two bypass capacitors in parallel.

Using this model, we find that the reactance of the large capacitor is $X_1 = \omega L_1 - 1/\omega C_1$; similarly, the reactance of the smaller capacitor is $X_2 = \omega L_2 - 1/\omega C_2$. The total impedance of the parallel combination is given by

$$Z(j\omega) = \frac{-(\omega L_1 - 1/\omega C_1)(\omega L_2 - 1/\omega C_2)}{j(\omega L_1 - 1/\omega C_1) + j(\omega L_2 - 1/\omega C_2)}.$$

This impedance is plotted, along with the reactances of each capacitor, in Figure A9-2. The figure also indicates the resonant frequencies of the capacitors, and we see that their reactances go through zero at those frequencies, as expected. What may be unexpected is the peak in the total impedance, which is forced to occur *between* the self-resonant frequencies of the two capacitors.

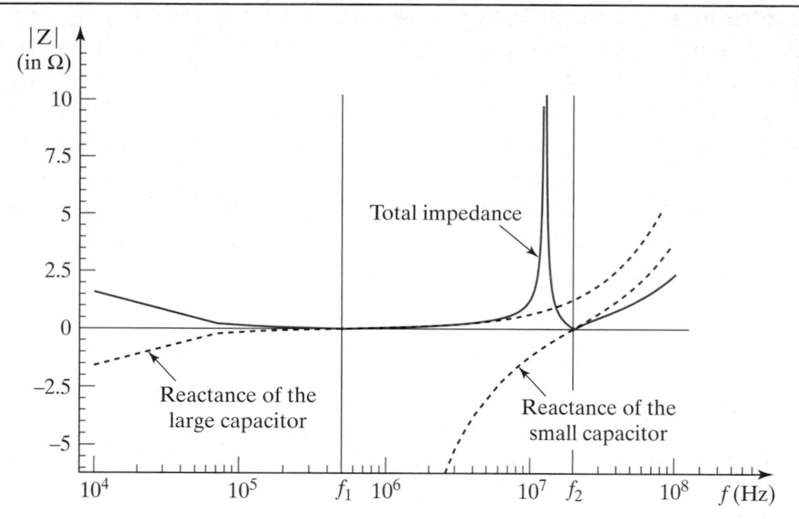

Figure A9-2 The total impedance and individual reactances for the model in Figure A9-1.

The peak in the total impedance can be explained in the following way: For frequencies above the self-resonant frequency of the large capacitor, ω_1, it appears inductive; whereas, for frequencies below the self-resonant frequency of the small capacitor, ω_2, it appears capacitive. Therefore, for frequencies between ω_1 and ω_2, we have, in effect, a parallel LC circuit, and the total impedance peaks when this parallel LC circuit becomes resonant (i.e., the net inductance from the large capacitor cancels the net capacitance of the smaller capacitor).

Putting these two capacitors in parallel does not therefore accomplish the desired goal. It is certainly true, as can be seen by comparing the total impedance with the reactance of the large capacitor, that the parallel combination has a lower impedance for high frequencies; however, we have introduced a pole into the impedance at a frequency between the self-resonant frequencies of the two capacitors. In other words, there is a narrow band of frequencies for which this combination will not be an effective bypass at all. This can have a serious detrimental effect on the performance of an amplifier.

We should not conclude from this example that capacitors should never be put in parallel—only that we need to be careful, and as always, we need to be using the appropriate level of modeling to understand the situation.

9.1.3 Diodes

Both *pn*-junction diodes and Schottky diodes exhibit capacitance because of charge storage in the depletion region, as discussed in Section 2.4.5. The small-signal capacitance caused by the change in depletion-region charge is called the ***junction capacitance*** and is given by

$$C_j \equiv \frac{dQ_{DR}}{dV_D} = \begin{cases} \dfrac{C_{j0}}{(1 - V_D/V_0)^{m_j}} & \text{in reverse bias} \\ 2C_{j0} & \text{in forward bias} \end{cases}, \tag{9.6}$$

where C_{j0} is the zero-bias value of the junction capacitance, V_0 is the barrier potential, m_j is an exponent that is a function of the doping profile in the device (typically between 0.2 and 0.5), and the value given for C_j in forward bias is only an approximation.

In addition to the junction capacitance, *pn*-junction diodes exhibit minority-carrier storage in forward bias, as discussed in Section 2.4.5. This charge storage term leads to the small-signal ***diffusion capacitance***, given by

$$C_d \equiv \frac{dQ_F}{dV_D} = \tau_F \frac{I_D}{V_T}, \tag{9.7}$$

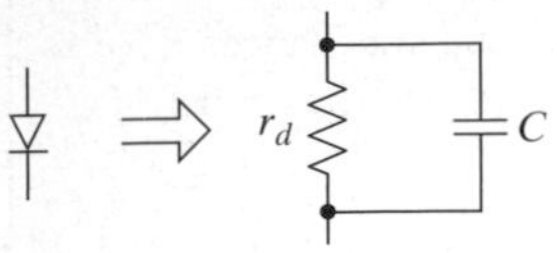

Figure 9–2 A small-signal model for a diode that is valid for high frequencies. r_d is present only in forward bias. The value of C depends on the type of diode and whether it is forward or reverse biased; see text for details.

where Q_F is the minority-carrier charge stored in forward bias and τ_F is the associated time constant, often called the ***forward-transit time***. This equation is valid only in forward bias.

The small-signal model for a diode presented in Chapter 8 can be modified for higher frequencies by adding a capacitor in parallel with the small-signal resistance, as shown in Figure 9-2. In forward bias, the capacitor is given by (9.6) for Schottky diodes and by (9.6) plus (9.7) for *pn*-junction diodes. The small-signal resistance shown is not present in reverse bias; only the capacitor is present, and its value is given by (9.6) alone. The small-signal resistance was found earlier to be $r_d = nV_T/I_D$.

Exercise 9.3

A pn-junction diode has $V_0 = 0.6$ V, $C_{j0} = 2$ pF, $I_S = 10^{-14}$ A, $n = 1$, $m_j = 0.33$, and $\tau_F = 500$ ps. Derive the high-frequency small-signal AC model when (a) $V_D = 0.65$ V and (b) $V_D = -3$ V.

Integrated circuits usually use diode-connected transistors in place of actual diodes, as noted in Section 8.1. Nevertheless, junction diodes do show up as parasitic elements in transistors, as discussed in Section 8.1. When parasitic diodes are present, the models presented in the preceding section are used.

9.1.4 The Generic Transistor

All of the transistors presented in this book have at least two parasitic capacitors that are important to include in the high-frequency model. Figure 9-3 shows those two capacitors added to our models for the generic transistor. The first capacitor is called C_{cm}, and it appears between the control and merge terminals. The other capacitor is called C_{oc} and appears between the output and control terminals. Note that in the current-controlled version of the high-frequency T model the controlling current, i'_m, is *not* the external merge current; it is the current through r_m alone. (The external merge current flows through r_m in parallel with C_{cm}.)

9.1.5 Bipolar Junction Transistors

Bipolar junction transistors have capacitances associated both with *pn* junctions and with the storage of excess minority carriers (mostly in the base) when external currents flow. The details are presented in Section 2.5.5. We are interested only in operation in the forward-active region in this chapter, so the collector-base junction will be reverse biased, and will exhibit a junction capacitance given by (the same as (9.6) in reverse bias except for the subscripts)

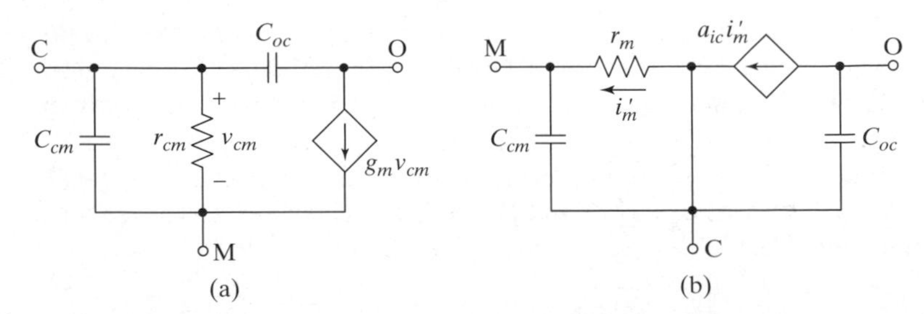

Figure 9–3 (a) The high-frequency hybrid-π model and (b) the high-frequency T model for the generic transistor.

$$C_{\mu} \equiv \frac{C_{\mu 0}}{(1 - V_{BC}/V_0)^{m_{jc}}}. \quad \textbf{(9.8)}$$

C_{μ} is also sometimes called C_{jc}. (It is called C_{μ} because it is in parallel with r_{μ}.) Equation (9.8) is valid for *npn* transistors. For *pnp* transistors, V_{BC} needs to be replaced by V_{CB} (i.e., the voltage should be negative when the junction is reverse biased). The coefficient m_{jc} is typically between about 0.2 and 0.5, and the built-in potential V_0 is typically between 0.5 V and 1 V.

The forward-biased emitter-base junction exhibits both junction capacitance and diffusion capacitance. Because this capacitance appears in parallel with r_{π} in our hybrid-π model, we call it C_{π}. This small-signal capacitance is given by

$$C_{\pi} = C_{je} + C_d \approx 2C_{je0} + g_m \tau_F, \quad \textbf{(9.9)}$$

where the junction capacitance in forward bias is an approximation and the time constant τ_F is frequently called the ***forward-transit time***, or ***base-transit time***. τ_F is not exactly equal to the base-transit time, but we will ignore that detail here.

The high-frequency hybrid-π model for a BJT is shown in Figure 9-4. The two capacitors given by (9.8) and (9.9) have been included in the model, but we have still ignored the extrinsic emitter and collector resistors and r_{μ}. The base resistance and output resistance can frequently be ignored in this model, but r_b affects the high-frequency performance more often than it does the midband performance and must be included in some circumstances. The nonreactive model elements were found in Section 8.1 and are $g_m = I_C/V_T$, $r_{\pi} = \beta_0/g_m$, and $r_o = V_A/I_C$.

In specifying the high-frequency model of a BJT, it is common to give the frequency at which the common-emitter current gain (β) goes to unity, called the ***transition frequency***. Aside A9.2 explains the terminology used and relates the transition frequency to the model parameters.

Finally, we note that we can also add capacitors to the T model of the BJT, as shown in Figure 9-5. We have left r_b out of this model, because it rarely affects the performance of common-base stages. It would be included in series with the base connection if necessary. The other elements in the model were previously found to be $\alpha = \beta/(\beta+1)$ and $r_e = r_{\pi} \| 1/g_m \approx 1/g_m = V_T/I_E$. As noted in Section 9.1.4, the controlling current, i_e' is *not* the external emitter current; it is only the current in r_e. (The external emitter current flows through r_e in parallel with C_{π}.)

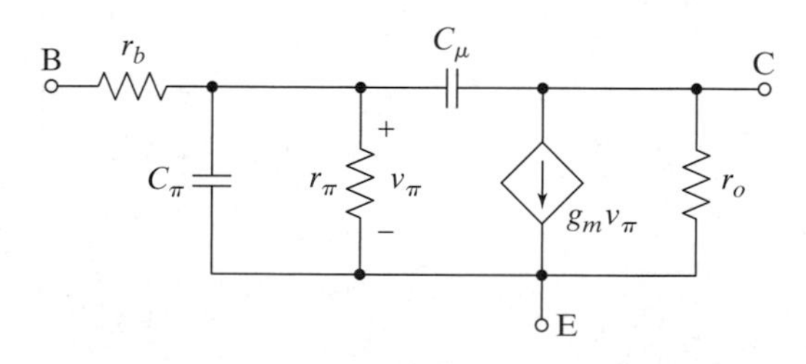

Figure 9–4 The high-frequency hybrid-π model for a BJT.

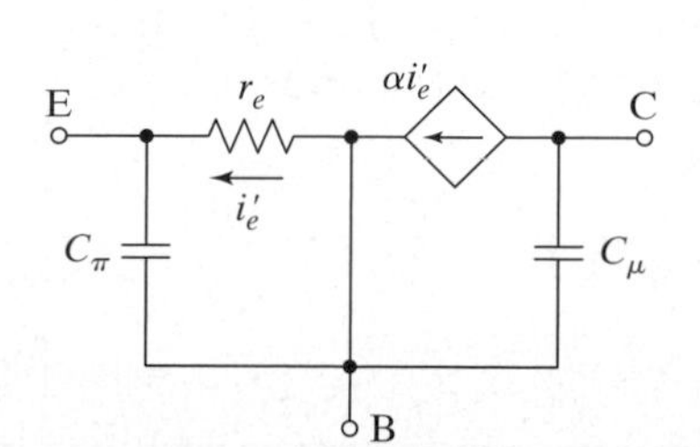

Figure 9–5 The high-frequency T model of a BJT with r_b omitted.

ASIDE A9.2 THE TRANSITION FREQUENCY OF A BJT

We often desire to compare the performance of one bipolar process with another. In addition, we would like to have a parameter that describes the high-frequency capability of a bipolar device in a more intuitive and direct way than is possible by specifying the parameters of the high-frequency hybrid-π model. The most common parameter used for these purposes is the ***transition frequency***, which is denoted f_T or ω_T. The transition frequency is defined to be the frequency at which the short-circuit common-emitter current gain goes to unity. Consider the circuit shown in Figure A9-3, wherein we drive the base of the transistor with a current source and measure the short-circuit output current. The high-frequency hybrid-π model has been used. The output resistance of the transistor has been neglected, since it has a short circuit in parallel with it and will not, therefore, have any current in it.

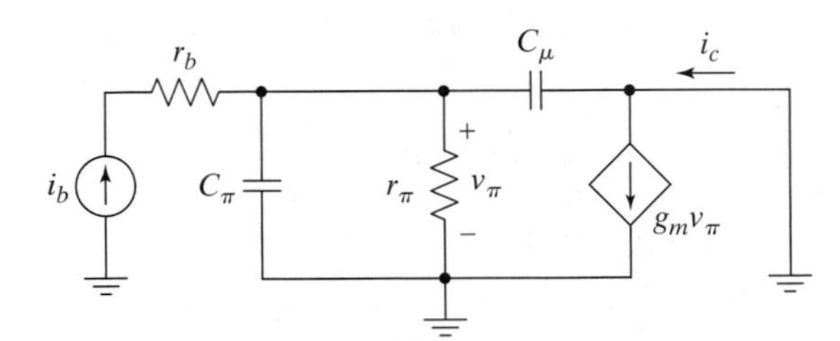

Figure A9-3 A circuit for determining the short-circuit common-emitter current gain.

Notice that C_μ and C_π are in parallel in this circuit, and we can write

$$V_\pi(j\omega) = I_b(j\omega)\left[r_\pi \| (C_\pi + C_\mu)\right] = I_b(j\omega)\frac{r_\pi}{1 + j\omega r_\pi(C_\pi + C_\mu)}. \quad \textbf{(A9.1)}$$

Now, assuming that the current from the controlled source is much larger than the current through C_μ (a reasonable assumption for most frequencies of practical interest; *see* Problem P9.12), we can write

$$I_c(j\omega) \approx g_m V_\pi(j\omega). \quad \textbf{(A9.2)}$$

Finally, we can solve for the current gain:

$$\beta(j\omega) = \frac{I_c(j\omega)}{I_b(j\omega)} = \frac{g_m r_\pi}{1 + j\omega r_\pi(C_\pi + C_\mu)}. \quad \textbf{(A9.3)}$$

This can also be expressed as

$$\beta(j\omega) = \frac{\beta_o}{1 + j\omega/\omega_\beta}, \quad \textbf{(A9.4)}$$

where

$$\beta_0 = g_m r_\pi \quad \textbf{(A9.5)}$$

as before, and we have implicitly defined the −3 dB frequency to be

$$\omega_\beta = \frac{1}{r_\pi(C_\pi + C_\mu)}. \quad \textbf{(A9.6)}$$

Using (A9.4), we can solve for the frequency at which $|\beta(j\omega)| = 1$. This frequency is, in a sense, the largest frequency at which the transistor is useful, and is defined to be the transition frequency. Since the transition occurs at a frequency well above ω_β, we obtain

$$|\beta(j\omega_T)| = 1 = \left|\frac{\beta_0}{1 + j\dfrac{\omega_T}{\omega_\beta}}\right| \approx \frac{\beta_0\omega_\beta}{\omega_T},$$

which yields

$$\omega_T = 2\pi f_T = \beta_o \omega_\beta. \tag{A9.7}$$

Substituting from (A9.5) and (A9.6), we find that

$$\omega_T = \frac{g_m}{C_\pi + C_\mu}. \tag{A9.8}$$

We can rewrite (A9.8) in a useful way by remembering that $C_\pi = g_m \tau_F + C_{je}$:

$$\omega_T = \frac{g_m}{g_m \tau_F + C_{je} + C_\mu}. \tag{A9.9}$$

Using (A9.9), we can see how f_T varies with bias current. (Remember that $g_m = I_C/V_T$.) For large bias currents, the junction capacitance terms are negligible and $\omega_T \approx 1/\tau_F$. For small bias currents, the junction capacitance terms dominate and $\omega_T \approx g_m/(C_{je} + C_\mu)$. In this region, ω_T is approximately proportional to I_C. If the bias current is made too large, the value of τ_F increases so that ω_T drops at very high currents as well. (The increase in τ_F at high currents is briefly discussed in Section 2.5.5.) The net result is that ω_T is a maximum for some current, which depends on the individual transistor. However, it decreases for much smaller or larger bias currents, as shown in Figure A9-4.

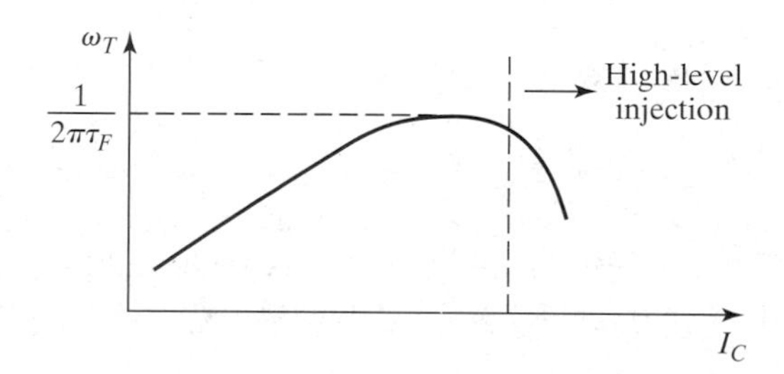

Figure A9-4 The variation of ω_T with collector current.

The data sheets for bipolar transistors usually specify f_T at some current, as well as the junction capacitances C_{je} and C_{jc} (i.e., C_μ), but they don't typically specify τ_F, even though it is the parameter used by SPICE and other simulation programs to model the diffusion capacitance. Therefore, (A9.9) has much practical significance if a SPICE model is needed for a device.

This same notational problem is encountered in the high-frequency hybrid-π model if we use the current-controlled version of the source. At midband, the source is equal to $\beta_0 i_b$, as shown in Chapter 8, but when capacitors are included in the model, we cannot use this same formula, since it would lead to a constant current from the controlled source if i_b were held constant. As shown in Aside A9.2, this is not what happens in the device. If, however, we let the controlled source be given by $\beta_0 i_\pi$, where i_π is defined to be the current in r_π, as shown in Figure 9-6, we obtain the proper results (*see* Problem P9.17).

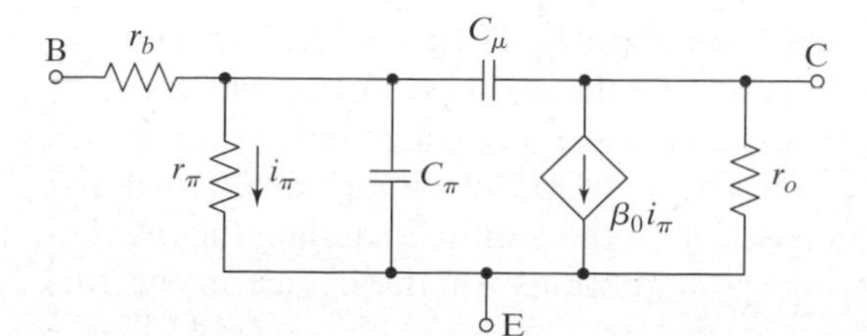

Figure 9–6 The current-controlled version of the high-frequency hybrid-π BJT model.

Exercise 9.4

Consider a BJT with $I_S = 10^{-14}$A, $C_{jc0} = 2$ pF, $C_{je0} = 1$ pF, $V_0 = 0.6$ V, $m_{jC} = 0.33$, $\tau_F = 200$ ps, $r_b = 50\ \Omega$, $V_A = 60$ V, and $\beta_0 = 100$. Find the complete small-signal high-frequency AC hybrid-π and T models when $V_{BE} = 0.67$ V and $V_{CE} = 4$ V.

In addition to the parasitic elements just presented, other elements may be present in integrated bipolar transistors. Section 8.1 discussed the different parasitic junction diodes present in a typical integrated BJT. For high-frequency analysis, each of the parasitic diodes has a capacitance associated with it, which is modeled exactly as shown for a diode in Section 9.1.3.

9.1.6 MOS Field-Effect Transistors

Charge storage in MOSFETs was discussed in Section 2.6.4. The main capacitive element in a MOSFET is the gate-to-channel capacitor formed by the MOS structure. This structure is a parallel-plate capacitor, and the total capacitance is given by

$$C_{g-ch} = \frac{\epsilon_{ox} WL}{t_{ox}}, \tag{9.10}$$

where ϵ_{ox} is the dielectric constant of the oxide, t_{ox} is the thickness of the oxide, and W and L are the width and length of the channel, respectively. The gate-to-channel capacitance leads to capacitance from the gate to the source, C_{gs}, and from the gate to the drain, C_{gd}. When the device is operating in the linear region, C_{g-ch} splits equally between C_{gs} and C_{gd}. Therefore, we have

$$C_{gs} = C_{gd} = \frac{\epsilon_{ox} WL}{2t_{ox}}. \tag{9.11}$$

When the transistor is forward active, however, the channel is pinched off at the drain end, and the gate-to-channel capacitance appears from the gate to the source. In addition, the value is reduced because of the linear reduction in channel charge from the source to the point at which the channel pinches off. A detailed analysis shows that in forward-active operation we get (*see* Section 2.6.4 for details)

$$C_{gs} = \frac{2}{3} \frac{\epsilon_{ox} WL}{t_{ox}} \tag{9.12}$$

and

$$C_{gd} \approx 0. \tag{9.13}$$

In addition to the gate-to-channel capacitance just described, there is a small capacitance due to the gate conductor overlapping the diffused source and drain regions in the device. This capacitance is called the ***overlap*** capacitance, and it adds to each of the capacitances already given. The overlap capacitance is proportional to the width of the device, but is not a function of the length. Including the overlap capacitance, C_{ol}, leads to the following two equations for the device in the forward-active region of operation:

$$C_{gs} = \frac{2}{3}\frac{\epsilon_{ox}WL}{t_{ox}} + C_{ol} \tag{9.14}$$

and

$$C_{gd} = C_{ol}. \tag{9.15}$$

The high-frequency MOSFET models are shown in Figure 9-7. Part (a) of the figure shows the small-signal model in the linear region, where the drain-to-source resistance is given by (see (7.18))

$$R_{ds} = \frac{1}{2K(V_{GS} - V_{th})} \tag{9.16}$$

and the capacitors are given by (9.11) with the overlap capacitance added

$$C_{gs} = C_{gd} = \frac{\epsilon_{ox}WL}{2t_{ox}} + C_{ol}. \tag{9.17}$$

Part (b) of the figure shows the high-frequency hybrid-π model for forward-active operation. The capacitors are given by (9.14) and (9.15), and the transconductance and output resistance are given by the equations derived in Chapter 8:

$$g_m = 2K(V_{GS} - V_{th}) = 2\sqrt{KI_D}, \tag{9.18}$$

and

$$r_o = \frac{1}{g_{ds}} = \frac{V_A}{I_D} = \frac{1}{\lambda I_D}. \tag{9.19}$$

Part (c) of the figure shows the high-frequency T model, which is also valid for forward-active operation, and the same equations apply as for the hybrid-π model. As noted in Section 9.1.4, the controlling current, i'_s, is *not* the external source current; it is only the current in $1/g_m$. (The external source current flows through $1/g_m$ in parallel with C_{gs}.) We left r_o out of the T model, but it can be included.

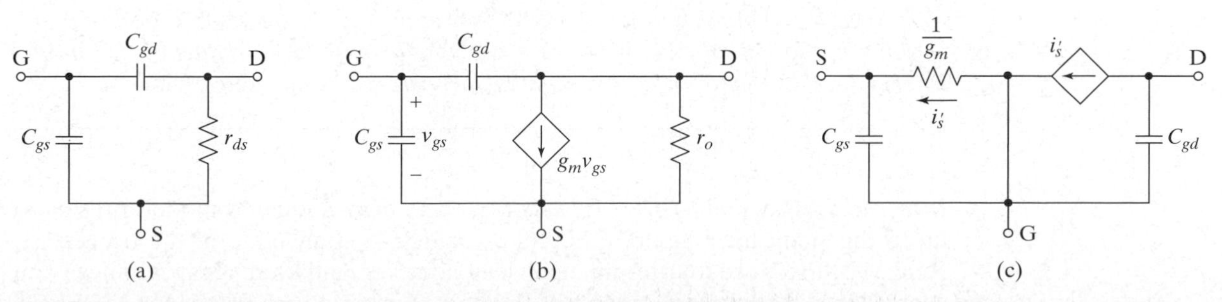

Figure 9–7 The high-frequency small-signal models for a MOSFET: (a) in the linear region, (b) the hybrid-π model for forward-active operation (i.e., saturation), and (c) the T model for forward-active operation.

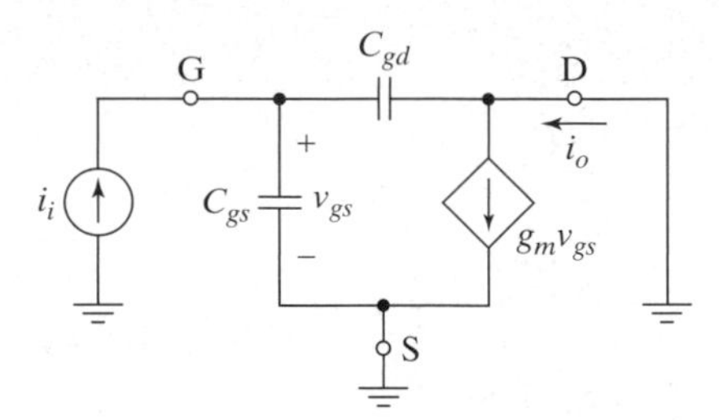

Figure 9–8 Finding f_T for a FET.

A figure of merit that is often used to specify the high-frequency capabilities of a FET is the ***transition frequency***, ω_T, or f_T. The transition frequency is defined as the frequency at which the short-circuit common-source current gain goes to unity. This frequency is, in some sense, the maximum usable frequency of the transistor. Figure 9-8 shows a common-source FET driven by a current source and with the output shorted.

Whether or not we include r_o in this circuit is irrelevant, since it has a short circuit in parallel with it and therefore will not have any current in it. The short-circuit current gain is found from this circuit to be

$$A_i(j\omega) = \frac{I_o(j\omega)}{I_i(j\omega)} = \frac{g_m - j\omega C_{gd}}{j\omega(C_{gs} + C_{gd})} = \frac{g_m}{j\omega(C_{gs} + C_{gd})} - \frac{C_{gd}}{C_{gs} + C_{gd}}, \tag{9.20}$$

where the final term represents the constant fraction of the input current that flows through C_{gd}. If we ignore this usually small term and set the magnitude of (9.20) equal to unity, we find that

$$\omega_T = \frac{g_m}{C_{gs} + C_{gd}}, \tag{9.21}$$

or

$$f_T = \frac{g_m}{2\pi(C_{gs} + C_{gd})}. \tag{9.22}$$

Note from (9.20) that the current gain of a FET goes to infinity as the frequency approaches zero. This is to be expected, since the DC gate current is zero if we ignore the extremely small leakage current present.

Exercise 9.5

Consider an n-channel MOSFET with a 250-Å-thick gate oxide, $W = 5\ \mu m$, $L = 1\ \mu m$, $C_{ol} = 2$ fF, $V_{th} = 0.7$ V, and $\lambda = 0.035\ V^{-1}$. Assume that the electron mobility in the channel is 675 cm^2/Vs. Find the complete small-signal high-frequency AC model when $V_{GS} = 1.5$ V and $V_{DS} = 3$ V. What is the corresponding value of ω_T?

□

Integrated MOS Field-Effect Transistors The high-frequency models presented up to this point have ignored the bulk—or body—connection of the transistors. Real MOSFETs are four-terminal devices, however, and there is capacitance from each of the other three terminals to the bulk. A simple high-frequency MOSFET model that includes the bulk connection is shown in Figure 9-9.

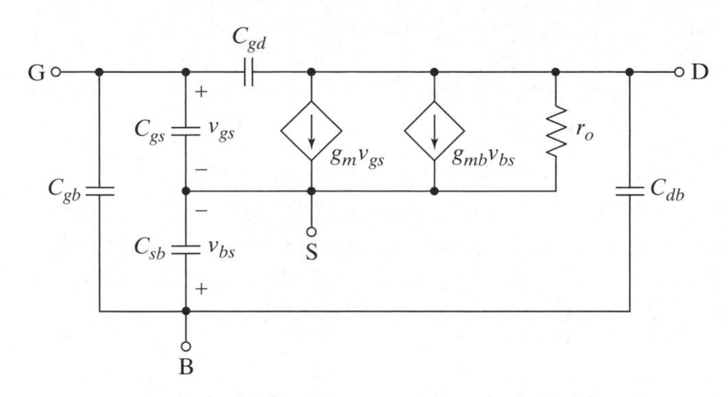

Figure 9–9 A high-frequency MOSFET model, including the bulk connection.

The source-to-bulk and drain-to-bulk capacitances (C_{sb} and C_{db} ,respectively) are a result of the *pn* junctions formed by the source and drain diffusions. These capacitances are voltage dependent, as described for a *pn*-junction diode in Section 9.1.3, but for hand analysis, they are almost always modeled as constants. The gate-to-bulk capacitance is caused by the gate conductor forming a parallel-plate capacitor with the substrate. The gate and substrate act as the plates and the oxide is the dielectric. This capacitor is outside the active area of the transistor; otherwise, it is called the gate-to-channel capacitance. Consequently, it is no different from the parasitic capacitance to the substrate found for every conductor on top of the oxide.

All three of these capacitors are parasitic elements that are not part of the intrinsic device. Since there are other parasitics that must also be included for our analyses to be accurate, we will not model these elements separately and will show only the intrinsic capacitors (C_{gs} and C_{gd}) in our models.

9.1.7 Junction Field-Effect Transistors

JFETs are modeled exactly like depletion-mode MOSFETs, except for the equations used for the device capacitances and the different notation that is sometimes used for the drain current. The notational differences have been dealt with in Section 8.1. The different capacitance equations are presented here.

Because there is a reverse-biased *pn* junction connecting the gate to the channel, the gate-to-channel capacitance is that of a reverse-biased *pn*-junction diode. As noted in Section 2.7.3, the total gate-to-channel capacitance has the form

$$C_{gch} = \frac{C_{gch0}}{\left(1 + \dfrac{V_{GCH}}{V_0}\right)^{m_j}}, \tag{9.23}$$

where the signs are correct for a *p*-channel JFET. For an *n*-channel JFET, V_{GCH} will be negative, and the sign in the denominator sum needs to be changed. When a JFET is in the linear region, the gate-to-channel capacitance splits equally between C_{gs} and C_{gd}, and (9.23) can be used with V_{GCH}, taken to be the average of V_{GS} and V_{GD}. In saturation, C_{gs} is much larger than C_{gd}, and we typically use

$$C_{gs} = \frac{C_{gs0}}{\left(1 + \dfrac{V_{GS}}{V_0}\right)^{m_j}} \tag{9.24}$$

and

$$C_{gd} = \frac{C_{gd0}}{\left(1 + \frac{V_{GD}}{V_0}\right)^{m_j}}. \tag{9.25}$$

The high-frequency small-signal AC models of integrated JFETs are identical to those presented for discrete JFETs, except for the inclusion of parasitic junction diodes. Precisely which parasitic diodes need to be included and how important they are depends on the specific structure used and will not be covered here.

9.2 STAGES WITH VOLTAGE AND CURRENT GAIN

Section 6.5.2 briefly discussed the frequency response of both DC-coupled and AC-coupled amplifiers. The log-magnitude plots for both types of amplifier are repeated in Figure 9-10 for convenience. We are interested in being able to analyze amplifier circuits to determine both the low and high ***cutoff frequencies***, ω_{cL} and ω_{cH}. The cutoff frequencies are the frequencies where the gain drops by 3 dB from its midband value and so are sometimes called the ***−3-dB frequencies***. They are also sometimes called the ***half-power frequencies***, since the power delivered to a load is one-half of the midband value. They are not necessarily equal to any pole frequency, although they *may* be equal or close to one.

An important characteristic of the amplifier's performance is its small-signal ***bandwidth***. The small-signal bandwidth of an amplifier is typically defined as $BW = \omega_{cH} - \omega_{cL}$ for an AC-coupled circuit and as $BW = \omega_{cH}$ for a DC-coupled circuit. We call it the small-signal bandwidth, since we are assuming linear circuits in drawing the responses shown in Figure 9-10.

Small-signal midband AC analysis has already been covered in Chapter 8, so we can use that as a starting point in finding the bandwidth of an amplifier. While performing the small-signal midband AC analysis, we approximated all coupling and bypass capacitors as short circuits and all parasitic capacitors in our devices as open circuits. In other words, we assumed that for frequencies much below ω_{cH}, the capacitors that contribute to the high-frequency roll-off were small enough in value and located in the circuit in such a way that we could open-circuit them without making a significant error. To determine the value of the upper cutoff frequency, however, we need to keep these capacitors in our circuit.

Similarly, when doing midband analysis, we assumed that for frequencies well above ω_{cL}, the capacitors that contribute to the low-frequency roll-off are large enough in value and located in a way that allowed us to approximate them as short circuits without making significant errors. To determine what the lower cutoff frequency is, however, we need to include these capacitors in the analysis.

Aside A9.3 presents a theorem that will be useful in subsequent analyses.

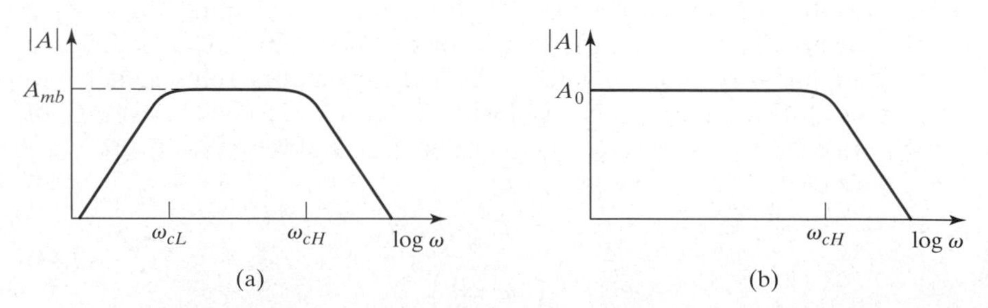

Figure 9–10 The log-magnitude plots of (a) an AC-coupled amplifier and (b) a DC-coupled amplifier.

ASIDE A9.3 MILLER'S THEOREM

Consider the circuit shown in Figure A9-5, where an ideal voltage amplifier has an impedance connected from its output back to its input. This feedback element prevents the overall circuit from appearing to be unilateral and therefore change makes the analysis more difficult and less intuitive.

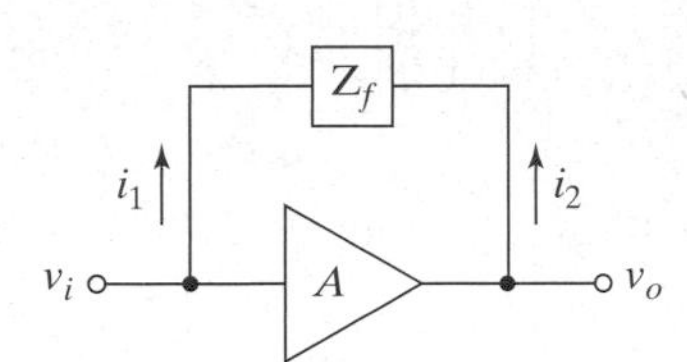

Figure A9-5 An ideal voltage amplifier with a feedback impedance.

However, we can calculate equivalent impedances from each side to ground and derive an equivalent circuit, as shown in Figure A9-6.

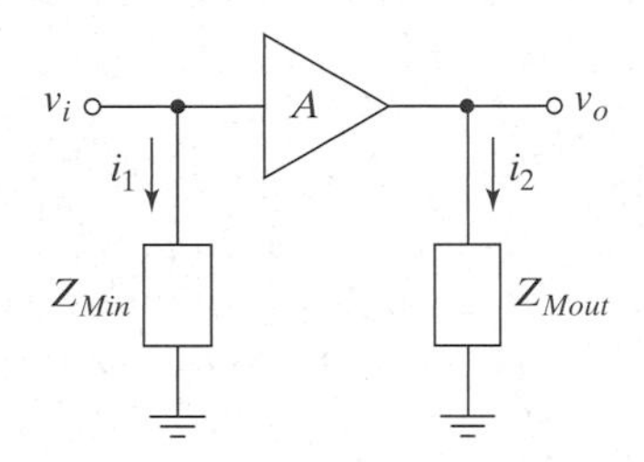

Figure A9-6 An equivalent circuit.

For these two circuits to be equivalent, it must be true that all the voltages and currents remain the same under all conditions. In the circuit in Figure A9-5 we have

$$i_1 = \frac{v_i - v_o}{Z_f} = \frac{v_i(1 - A)}{Z_f} \tag{A9.10}$$

and

$$i_2 = \frac{v_o - v_i}{Z_f} = \frac{v_o(1 - 1/A)}{Z_f}. \tag{A9.11}$$

For the circuit in Figure A9-6, we have

$$i_1 = \frac{v_i}{Z_{Min}} \tag{A9.12}$$

and

$$i_2 = \frac{v_o}{Z_{Mout}}. \tag{A9.13}$$

Equating (A9.10) and (A9.12) allows us to find the ***Miller input impedance***

$$Z_{Min} = \frac{Z_f}{1 - A}. \tag{A9.14}$$

Similarly, equating (A9.11) and (A9.13) allows us to find the ***Miller output impedance***

$$Z_{Mout} = \frac{Z_f}{1 - 1/A}. \tag{A9.15}$$

If we use the Miller input and output impedances from (A9.14) and (A9.15) in Figure A9-6, the circuit is completely equivalent to the one in Figure A9-5, but is obviously unilateral and therefore easier to analyze. This technique is useful only if we assume that $v_o = Av_i$ independent of Z_f. If the gain does depend on Z_f, using this theorem will usually not simplify the problem. In addition, if the gain is a function of frequency, using the theorem usually does not result in any significant simplification. Even when the algebra required to find an answer is not greatly simplified, however, this theorem still presents a powerful way of thinking about some circuits.

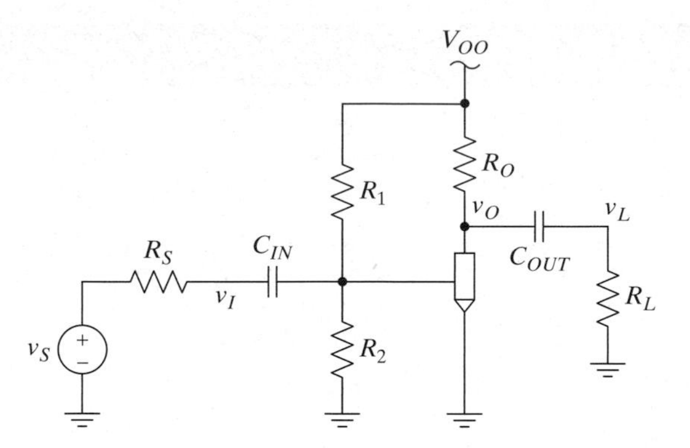

Figure 9–11 A common-merge amplifier.

9.2.1 A Generic Implementation: The Common-Merge Amplifier

Section 8.2 analyzed the midband performance of a common-merge amplifier. The circuit is repeated in Figure 9-11 for convenience. This section examines how to find both the low and high cutoff frequencies of that circuit.

The first step in finding either cutoff frequency is to draw the appropriate equivalent circuit. There are three small-signal equivalent circuits that we may be interested in, depending on what we want to find. All three of them are shown in Figure 9-12 for the common-merge amplifier. Part (a) of the figure shows the ***small-signal low-frequency AC equivalent circuit***, part (b) shows the small-signal midband AC equivalent circuit, and part (c) shows the ***small-signal high-frequency AC equivalent circuit***. Part (d) of the figure shows the frequency response of the amplifier and indicates the frequency range in which each equivalent circuit is valid. Because these are *small-signal* equivalent circuits, they all use a linear model for the transistor. Both the low-frequency and midband equivalent circuits use the low-frequency small-signal model, since they are valid for frequencies below those where the small parasitic capacitances in the transistor will affect the response. The difference between these two equivalents is that the low-frequency equivalent includes the coupling capacitors (and bypass capacitors, if any are present), whereas the midband equivalent is valid only for frequencies that are high enough that these capacitors are accurately modeled as short circuits.

The difference between the midband and high-frequency equivalent circuits is that the high-frequency small-signal model for the transistor is used in the high-frequency equivalent circuit. Because these are all *AC* equivalent circuits, the DC power supplies appear as grounds as they did in Chapter 8.

Once we have these equivalent circuits, we can proceed with our analyses. An introductory treatment is given first, followed by a more advanced method.

Introductory Treatment Using the low-frequency equivalent circuit of Figure 9-12(a), we can derive the low cutoff frequency. We start by writing the overall gain as a product of simpler transfer functions:

$$\frac{v_l}{v_s} = \frac{v_l}{v_{cm}}\frac{v_{cm}}{v_s}. \tag{9.26}$$

We can now write each of these transfer functions by inspection using the methods of Aside A1.6. First, we recognize that v_{cm}/v_s is a single-pole high-pass transfer function. Then we note that the zero is at DC, since v_{cm} only goes to zero at DC where C_{IN} is an open circuit. The gain at high frequencies is given by a resistor divider, and the pole is found by shorting v_s, removing C_{IN}, and looking back into the terminals to find the resistance seen by it. The resulting transfer function is

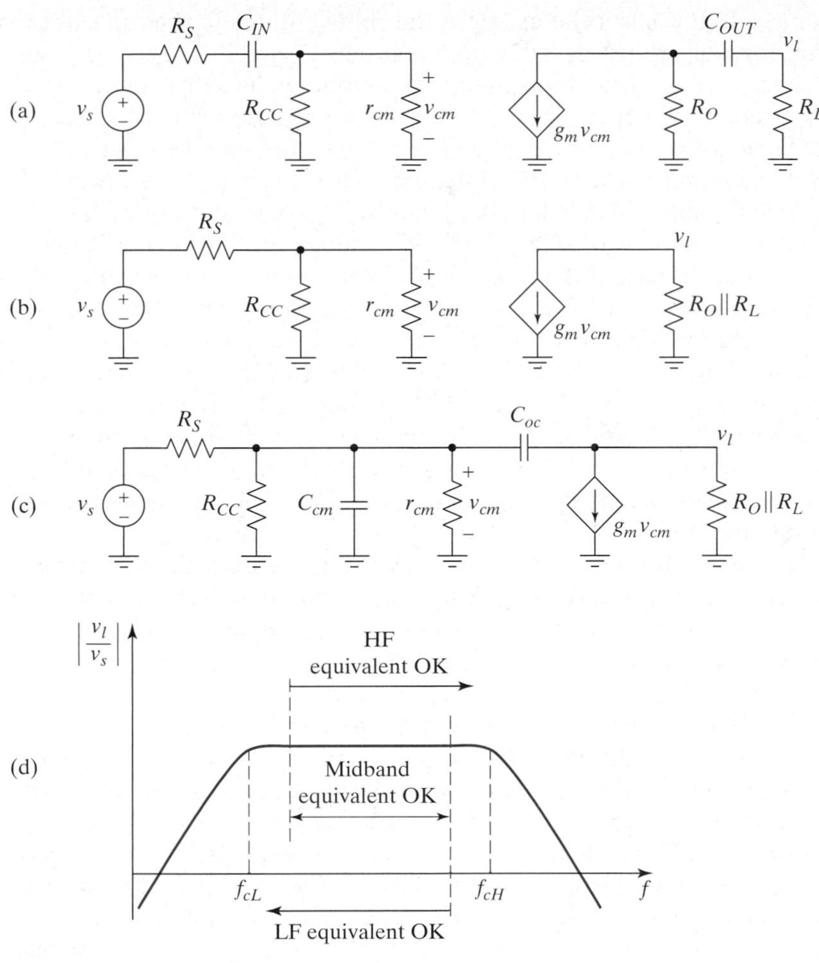

Figure 9–12 The small-signal equivalent circuits for the common-merge amplifier: (a) low-frequency AC, (b) midband AC, and (c) high-frequency AC. (d) The frequency response of the amplifier showing the range of validity for each equivalent circuit.

$$\frac{V_{cm}(j\omega)}{V_s(j\omega)} = \frac{j\omega\left(\dfrac{R_{CC}\|r_{cm}}{R_{CC}\|r_{cm} + R_S}\right)}{j\omega + \dfrac{1}{(R_S + R_{CC}\|r_{cm})C_{IN}}}. \tag{9.27}$$

Moving to the output side, we again have a single-pole high-pass circuit with a DC zero. The transfer function is given by

$$\frac{V_l(j\omega)}{V_{cm}(j\omega)} = \frac{-j\omega[g_m(R_O\|R_L)]}{j\omega + \dfrac{1}{(R_O + R_L)C_{OUT}}}. \tag{9.28}$$

We can combine these equations and write the overall transfer function for the low-frequency equivalent circuit. The result has two poles. To find the low cutoff frequency, we need to set the magnitude of the transfer function equal to $1/\sqrt{2}$

and solve for ω_{cL}. In this particular example, the contribution of each capacitor to the low cutoff is readily seen, since the input and output circuits are separate single-pole circuits. As a practical aside, note that anytime you can calculate a time constant for a capacitor without having to decide what to do with any other capacitor, that capacitor is ***independent*** of all other capacitors, and there is a pole at the reciprocal of its time constant. This statement is true even if you can't see how to split the circuit up into individual units and even if the circuit is not unilateral.

This brute force method will not always yield results useful for design. Without more sophisticated analysis tools, we would probably want to use computer simulation if we had a more complex low-frequency equivalent circuit. For example, even a three-capacitor circuit would require solving three equations in three unknowns (assuming that it can't be separated into single-pole circuits) and would yield a result too complicated to be of much use in design. In general, finding ω_{cL} in the manner just described is not very intuitive, except for the case of simple circuits. It would also involve a lot of algebra if we had a more complicated amplifier. A far more reasonable approximate approach will be presented in the next part of this section.

Let's now find the high cutoff frequency for this circuit. Examination of the circuit in Figure 9-12(c) reveals a difficulty: Because of the presence of C_{oc}, the circuit is not unilateral. Although we could write two nodal equations (at v_{cm} and v_l) and solve them for the transfer function, it will be easier and far more intuitive to first make the circuit unilateral by using Miller's theorem (presented in Aside A9.3). To apply the theorem, we need to find the voltage gain from v_{cm} to v_l (i.e., the gain across C_{oc}). Unfortunately, this gain does depend on C_{oc} to some extent, so the theorem does not simplify the analysis if we apply it exactly. Nevertheless, if the current through C_{oc} is small compared with $g_m v_{cm}$, the gain from v_{cm} to v_l is approximately independent of C_{oc}, and we can expect the theorem to provide results that are approximately correct while simplifying the analysis significantly. Removing C_{oc} and defining $R'_L = R_O \| R_L$, we can write

$$\frac{v_l}{v_{cm}} = -g_m R'_L, \tag{9.29}$$

which is the midband gain. Using this gain along with the formulas in (A9.14) and (A9.15), we can redraw the circuit as shown in Figure 9-13. The overall input capacitance is denoted C_{in} and is equal to C_{cm} in parallel with the Miller input capacitance given by

$$C_{Min} = (1 + g_m R'_L) C_{oc}, \tag{9.30}$$

so that

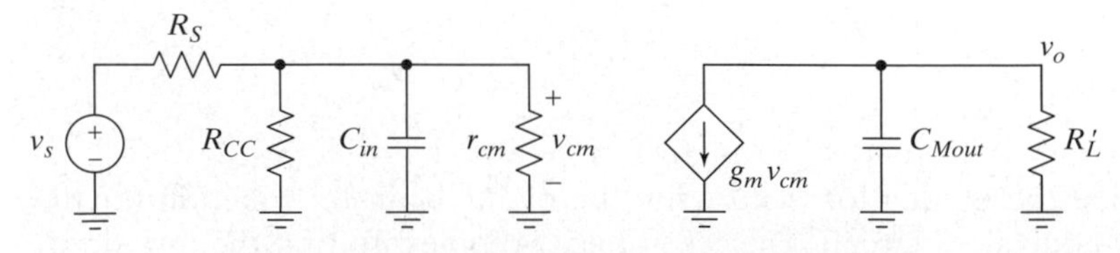

Figure 9–13 The equivalent circuit from Figure 9-12(c) after application of Miller's theorem.

$$C_{in} = C_{cm} + (1 + g_m R'_L)C_{oc}. \quad \textbf{(9.31)}$$

This input capacitance is usually dominated by C_{Min} if the gain of the stage is reasonably large. The Miller output capacitance is then

$$C_{Mout} = (1 + 1/g_m R'_L)C_{oc} \approx C_{oc}. \quad \textbf{(9.32)}$$

We can check to see whether or not the current through C_{oc} is small compared with $g_m v_{cm}$ by examining what fraction of the output current flows through C_{Mout}, which, owing to the Miller theorem, should be about the same as the current through C_{oc}. (It would be exactly the same if we used the frequency-dependent gain instead of the midband gain to find C_{Mout}.) At DC, none of the current flows through this capacitor, and at $\omega = 1/C_{Mout}R'_L$ the current through C_{Mout} is one-half of the output current. (To see this, set the capacitive reactance equal to R'_L). Therefore, for frequencies much below $\omega = 1/C_{Mout}R'_L$, using Miller's theorem with the midband gain will provide results that are reasonable for this circuit.

We can now find the high cutoff frequency of the circuit in Figure 9-13. We need to be careful when interpreting the results we obtain with this equivalent circuit, however. The circuit is unilateral, and each part is a single-pole circuit. Consequently, we are tempted to write the overall transfer function as a product of two separate single-pole transfer functions. Because C_{Min} and C_{Mout} both model the same element, however (C_{oc}), the two single-pole circuits are not independent, and we cannot treat them as such. We will see in the next section how to estimate the high cutoff frequency with both C_{in} and C_{Mout} included, but for now we note that because C_{in} is much larger than C_{Mout}, it tends to dominate the cutoff frequency (i.e., it leads to a lower cutoff frequency).[3] Therefore, the high cutoff frequency can be estimated by calculating the pole frequency caused by C_{in} alone in the circuit of Figure 9-13. The result is

$$\omega_{cH} \approx \frac{1}{(R_S \| R_{CC} \| r_{cm})C_{in}}. \quad \textbf{(9.33)}$$

When we use Miller's theorem with the midband gain and also ignore C_{Mout} while finding ω_{cH}, we say we are making the ***Miller approximation***.[4] We still use C_{Mout} as just outlined to determine whether or not the approximation is valid. If $1/R'_L C_{Mout}$ is much higher than the cutoff frequency calculated using (9.33), then (9.33) is a reasonable approximation (i.e., the midband gain is a reasonable approximation for frequencies near ω_{cH}). Equivalently, we can say that if

$$\frac{1}{\omega_{CH}} \gg \tau_{Mout}, \quad \textbf{(9.34)}$$

---

[3] The relative contribution of each capacitor to the cutoff frequency depends on the equivalent resistance seen by each capacitor, too, but the difference in the capacitors is so large that the statement made here is usually true.

[4] Actually, strictly speaking, we ignore *all* capacitors other than C_{Min} when making the Miller approximation. In this example, C_{cm} is directly in parallel with C_{Min}, though, so it would be silly not to include it. Also, when more capacitors are ignored, we need to figure out some way to check whether or not the approximation is still valid. We will deal with this complication later as the need arises.

then the approximation is acceptable. (We have deliberately avoided plugging in the formulas for our specific example, since we want an equation that is true in general.) In practice, if $1/\omega_{cH}$ is greater than three times τ_{Mout}, the approximation is usually reasonable.

Note that the frequency at which half of the current goes through C_{Mout} is the "pole" frequency of the single-pole circuit on the output side of the equivalent circuit in Figure 9-13. The word *pole* is in quotes because, as noted, there isn't a pole at this frequency, since this capacitor is not independent of C_{in}. In fact, if we find the two poles of the original circuit and compare them with (9.33) and $1/R'_L C_{Mout}$, we see that the dominant pole given by (9.33) is reasonably accurate, but the second pole may be very far away from $1/R'_L C_{Mout}$. Therefore, to avoid confusion, we prefer to state the test for the validity of the Miller approximation in the form given by (9.34).

Example 9.2

Problem:

Use the Miller approximation to find f_{cH} for the circuit in Figure 9-12(c), given that $R_S = 1\ \text{k}\Omega$, $R_{CC} = 20\ \text{k}\Omega$, $C_{cm} = 5$ pF, $r_{cm} = 5\ \text{k}\Omega$, $C_{oc} = 1$ pF, $g_m = 10$ mA/V, and $R'_L = 5\ \text{k}\Omega$. Then use exact analysis to find both pole frequencies. Compare the dominant pole with f_{cH} and the highest pole with $1/R'_L C_{Mout}$.

Solution:

Using (9.31), (9.32), and the given data, we find that $C_{in} = 56$ pF and $C_{Mout} = 1.02$ pF. Using (9.33), we then find $f_{cH} \approx 3.6$ MHz. Finally, we calculate $1/(2\pi R'_L C_{Mout}) = 31.2$ MHz.

Using the pole-zero analysis option in HSPICE[5] reveals that the two pole frequencies for this circuit are 3.2 MHz and 393 MHz. The lower pole agrees quite well with the dominant pole found using the Miller approximation, but note that the higher pole is more than 10 times $1/(2\pi R'_L C_{Mout}) = 31.2$ MHz, which clearly demonstrates that you cannot treat the circuit of Figure 9-13 as two independent single-pole circuits.

■

As previously mentioned, C_{in} is dominated by C_{oc} if the gain is reasonably large. Therefore, C_{oc} determines the bandwidth when the gain is large. We can see intuitively why this should happen if we look at the original circuit (Figure 9-11). Since C_{oc} appears from input to output, when the input voltage increases by an amount Δv_I, the output voltage decreases by an amount $|A_v|\Delta v_I$, where A_v is the voltage gain from input to output. (A_v is negative.) The current through C_{oc} is then larger than would be seen for a capacitor of the same size connected to ground. In fact, the total change in voltage across C_{oc} is $\Delta v_{Coc} = \Delta v_I(1 - A_v)$, so the current increases by this same factor, and the effective capacitance seen to ground is also increased by this same factor.

Knowing that C_{oc} is dominant in determining the high cutoff frequency for this circuit when the gain is large is important from a design point of view. If we wish to extend the bandwidth of the stage, we can either reduce the voltage gain, or reduce C_{oc}, or reduce the resistance seen by C_{Min}. We will also discover in Section 9.6

[5] HSPICE is a product of Cadence Corporation.

that we can significantly improve the bandwidth with the addition of another transistor. The sections on bipolar and field-effect implementations of this circuit show that we can extend the bandwidth of this stage in other ways as well.

Exercise 9.6

Use the Miller approximation to find f_{cH} for the circuit in Figure 9-12(c), given that $R_S = 50\ \Omega$, $R_{CC} = 20\ k\Omega$, $C_{cm} = 5\ pF$, $r_{cm} = 5\ k\Omega$, $C_{oc} = 0.5\ pF$, $g_m = 0.01\ A/V$, and $R'_L = 5\ k\Omega$. Is the Miller approximation valid in this case?

□

An important figure of merit for amplifiers is the **gain-bandwidth product**, or GBW. The GBW is a useful means for comparing amplifiers. The GBW is defined to be the product of the magnitude of the midband gain and the bandwidth of the amplifier. For a DC-coupled amplifier, the GBW is then

$$GBW = |\alpha_i A_v|\omega_{cH}. \tag{9.35}$$

The overall voltage gain has been used in (9.35), since the bandwidth is always calculated including the source resistance R_S and is therefore the bandwidth from v_S to v_O. For AC-coupled amplifiers, it is almost always true that $\omega_{cL} \ll \omega_{cH}$, and (9.35) can still be used without significant error.

For the common-merge amplifier example, we can combine (9.29), (9.33), and the input attenuation factor to find the GBW. The attenuation factor is given by

$$\alpha_i = \frac{R_{CC}\|r_{cm}}{R_{CC}\|r_{cm} + R_S} = \frac{R_i}{R_i + R_S}, \tag{9.36}$$

where R_i is the small-signal midband AC input resistance. After combining these equations, the result is

$$\begin{aligned} \text{GBW} &\approx \left(\frac{R_i}{R_i + R_S}\right)\frac{g_m R'_L}{(R_S\|R_{CC}\|r_{cm})C_{in}} \\ &\approx \left(\frac{R_i}{R_i + R_S}\right)\frac{g_m R'_L}{(R_S\|R_i)C_{in}} = \frac{g_m R'_L}{R_S C_{in}}. \end{aligned} \tag{9.37}$$

Making use of (9.31) and simplifying, we find that

$$\text{GBW} \approx \frac{g_m R'_L}{R_S g_m R'_L C_{oc}} = \frac{1}{R_S C_{oc}}, \tag{9.38}$$

which is a particularly simple result that relates the GBW product to basic quantities. Equations such as this are useful for design, because they clearly show us the limitations of a given topology. The trade-off between gain and bandwidth is important and is discussed further in Aside A9.4.

Exercise 9.7

Consider the amplifier whose small-signal AC equivalent circuit is shown in Figure 9-14. C_C is a large coupling capacitor that contributes a DC zero and a low-frequency pole. C_{in}, on the other hand, contributes a zero at infinite frequency and a high-frequency pole. C_{fb} is, as the notation indicates, a feedback capacitor. Since there is inverting voltage gain across C_{fb}, it will be multiplied by the Miller effect and will contribute to the high-frequency roll-off. (It also contributes a high-frequency zero). (a) Determine approximate values for f_{cL} and f_{cH}. (b) What is the GBW for this amplifier?

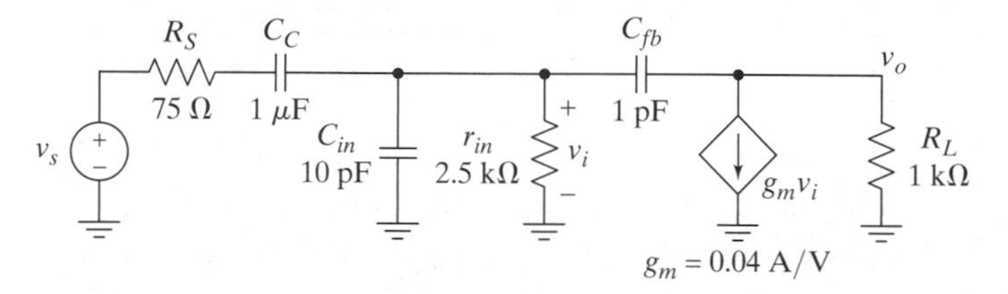

Figure 9–14 Small-signal AC model of an amplifier.

□

Advanced Method: Open- and Short-Circuit Time Constants We just found that estimating the low cutoff frequency with our existing methods is a cumbersome task for all but the simplest amplifier. In addition, the method used up to this point does not often lend insight into the operation of the circuit, nor is it particularly well suited for design. The method used for finding the high cutoff frequency, on the other hand, is quite intuitive and leads to reasonable results and equations suitable for use in the design process. However, not all amplifier stages can be analyzed by applying the Miller approximation. Therefore, we now introduce a new set of tools for finding approximations to the cutoff frequencies: the ***open- and short-circuit time constants.***

The approximations we arrive at based on these time constants have three distinct advantages:

1. The time constants are relatively straightforward to find; we can often do it by inspection.
2. The results are immediately useful for design, since they clearly show us which part of the circuit—if any—dominates the bandwidth.
3. The resulting estimates of the cutoff frequencies are conservative (i.e., ω_{cL} is always lower than we predict and ω_{cH} is always higher than we predict).[6]

On the downside, the techniques only work for circuits without inductors, although this is not a major restriction when dealing with amplifiers. Also, the approximations depend on one pole being dominant for each cutoff frequency, otherwise there may be a relatively large error (around 20% is not uncommon).

We start by examining how to find a reasonable estimate of the low cutoff frequency of a circuit with a bandpass transfer function (e.g., the amplifier in Figure 9-11). When we have finished with ω_{cL}, we will consider the high cutoff frequency, ω_{cH}.

[6] The results will always be conservative for the circuit we analyze, but if we have already used other approximations to simplify the original circuit prior to applying the open- or short-circuit time constants, the resulting bandwidth is not guaranteed to be conservative.

ASIDE A9.4 THE RELATIONSHIP BETWEEN GAIN AND BANDWIDTH IN AMPLIFIERS

Consider a DC-coupled amplifier with a single-pole transfer function given by

$$A(j\omega) = \frac{A(j0)}{1 + \dfrac{j\omega}{\omega_p}}. \tag{A9.16}$$

The gain-bandwidth product of this amplifier is defined to be the product of the low-frequency gain and the pole frequency, namely,

$$\text{GBW} = A(j0)\omega_p, \tag{A9.17}$$

and is an important measure of the performance of the amplifier. A Bode magnitude plot[1] of the transfer function is shown in Figure A9-7 and consists of two asymptotes.

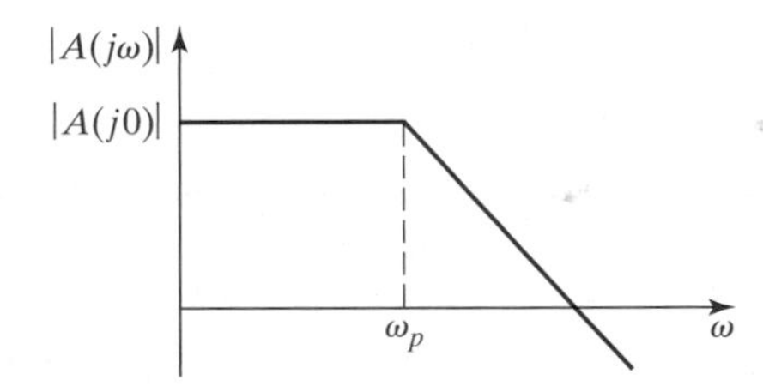

Figure A9-7 The Bode magnitude plot of a single-pole transfer function.

For frequencies well below ω_p, the gain is approximately constant and equal to $A(j0)$. For frequencies well above ω_p, the gain is approximately given by

$$A(j\omega) \approx \frac{\omega_p A(j0)}{j\omega}, \tag{A9.18}$$

so that the magnitude drops in proportion to $1/\omega$.

On the high-frequency asymptote described by (A9.18), the magnitude of the product of the gain and frequency at any given point is

$$|A(j\omega)|\omega = |A(j0)|\omega_p = \text{GBW}. \tag{A9.19}$$

If we can change the amplifier in a way that affects either the gain or the bandwidth, but that keeps the GBW constant, or at least approximately so, we can see from (A9.19) that we can trade gain for bandwidth. For example, if the gain is reduced by a factor of two while keeping GBW constant, then the bandwidth is doubled. If the amplifier has a steeper roll-off of gain versus frequency (i.e., it has more than one pole in the frequency range of interest), the tradeoff won't be one for one, but we can still trade gain for bandwidth. Chapter 10 discusses this issue further.

[1] We can also draw an asymptotic approximation to the phase. In that case, we say that the phase is approximately zero for frequencies less than $1/10$ of of the pole frequency and is approximately –90° for frequencies greater than 10 times the pole frequency. We approximate the phase as linear between these two points, so that it passes through –45° right at the pole frequency. (For a review of Bode plots, *see* Aside AD.2 in Appendix D on the CD.)

To find an approximation to ω_{cL}, we first assume that it is always possible to draw a low-frequency equivalent circuit, as in Figure 9-12(a), so that we are dealing with a high-pass transfer characteristic. If we further assume that one of the poles in the circuit dominates the high-pass cutoff frequency, the problem reduces to finding an approximation for the frequency of this **dominant pole** in the circuit. Since we are dealing with a high-pass characteristic, the dominant pole is the *highest* frequency pole in the circuit. (Don't allow the preceding statement to confuse you: We are still looking for the *low* cutoff frequency of the original circuit, but because we are examining only the low-frequency equivalent, ω_{cL} is approximately the highest frequency pole in this equivalent.) We will examine the accuracy of the approximation when there isn't a dominant pole later. We also must assume that all of the zeros in the transfer function are at frequencies well below the dominant pole, but this assumption usually is true in amplifier circuits.

Given that we have a low-frequency equivalent circuit, we find the short-circuit time constant associated with each capacitor in the circuit and then sum the reciprocals of these time constants to find the approximate cutoff frequency. In other words,

$$\omega_{cL} \approx \sum_{i=1}^{n} \frac{1}{\tau_{is}}, \qquad \textbf{(9.39)}$$

where there are n capacitors in the equivalent circuit. The short-circuit time constant associated with capacitor C_i is

$$\tau_{is} = R_{is}C_i, \qquad \textbf{(9.40)}$$

and the **short-circuit driving-point resistance** seen by capacitor C_i is denoted R_{is}. This resistance is seen when you set all of the independent sources in the circuit to zero, remove C_i, short all of the other capacitors, and look back into the terminals where C_i was connected. A detailed justification for this approximation is given in the companion section on the CD, but for now we can see that the result is reasonable by examining (9.39). If one of the capacitors does cause a dominant pole (i.e., it is substantially higher than all the other poles in the circuit), then for frequencies near this pole, all of the other capacitors will have negligible impedances and can be approximated as short circuits. Making this approximation, we can find the pole frequency of the resulting single-pole circuit by using the method presented in Aside A1.6. The resulting pole frequency will be the reciprocal of the short-circuit time constant for the dominant capacitor. In this circumstance, all of the other short-circuit time constants will be larger, and the sum in (9.39) will be dominated by this one time constant.

Be very careful here: There is a strong tendency to associate a pole with the reciprocal of each time constant, and doing so is *completely wrong*! For example, if we look at a capacitor that does not cause the dominant pole, for frequencies near the pole caused by this capacitor, the dominant capacitor will appear as an *open circuit*, and the pole cannot possibly be given by the reciprocal of the short-circuit time constant. It is possible to find time constants by using combinations of open- and short-circuiting the remaining capacitors and, by so doing, find the nondominant poles as well [9.3], but this method is not often used now that computer programs can find the poles and zeros of a circuit.

An individual pole frequency is equal to the reciprocal of a time constant only for a single-pole network. If the time constant associated with a capacitor can be found without needing to know what to do with other capacitors (i.e., whether to short them out or open-circuit them), the circuit containing the capacitor is a single-pole circuit (although it may be part of a much larger circuit), and there is a pole directly associated with the time constant. Also, note that the short-circuit time-constant method finds an approximation to the *highest* frequency pole in whatever circuit it is applied to; the designer must make sure that it is applied to the low-frequency equivalent circuit if an estimate of ω_{cL} is desired.

The use of short-circuit time constants is best demonstrated by example.

Example 9.3

Problem:

Find an estimate of the low cutoff frequency for the common-merge amplifier with the low-frequency equivalent circuit shown in Figure 9-15.

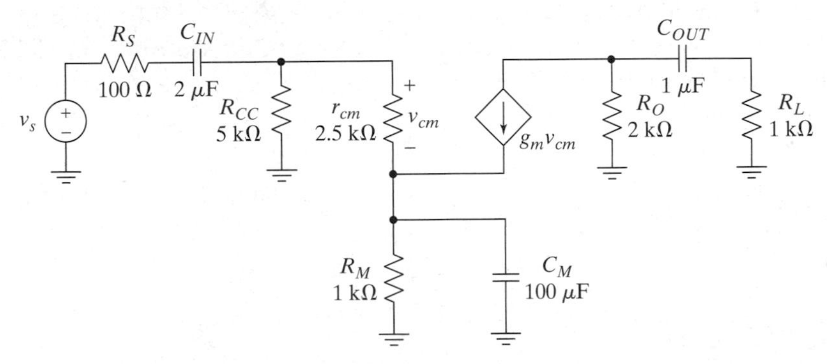

Figure 9–15 The low-frequency equivalent circuit of a common-merge amplifier.

Solution:

This circuit has three capacitors, and they cannot be combined in series or parallel, so we need to calculate three short-circuit time constants and use (9.39) to estimate the low cutoff frequency.

To find the time constant for C_{IN}, we set the independent source v_s to zero, short C_M and C_{OUT}, and find the resistance seen looking back into the terminals where C_{IN} was connected. (What we do with C_{OUT} doesn't matter for this particular circuit, but C_M does.) The procedure is illustrated in Figure 9-16.

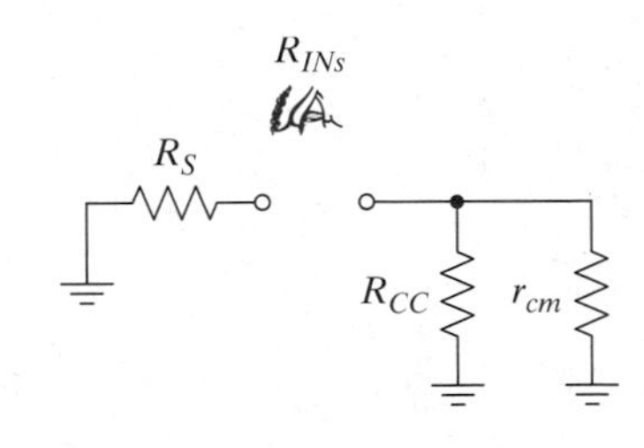

Figure 9–16 Finding the short-circuit driving-point resistance for C_{IN}.

The result is

$$R_{INs} = R_S + R_{CC} \| r_{cm} = 1{,}767\ \Omega,$$

which leads to a short-circuit time constant of

$$\tau_{INs} = R_{INs} C_{IN} = 3.5\ \text{ms}.$$

Finding the short-circuit time constant for C_{OUT} is similar. With C_{IN} and C_M shorted and v_s set to zero, we see that $v_{cm} = 0$, so the controlled current source is an open circuit, and

$$\tau_{OUTs} = R_{OUTs} C_{OUT} = (R_O + R_L) C_{OUT} = 3\ \text{ms}.$$

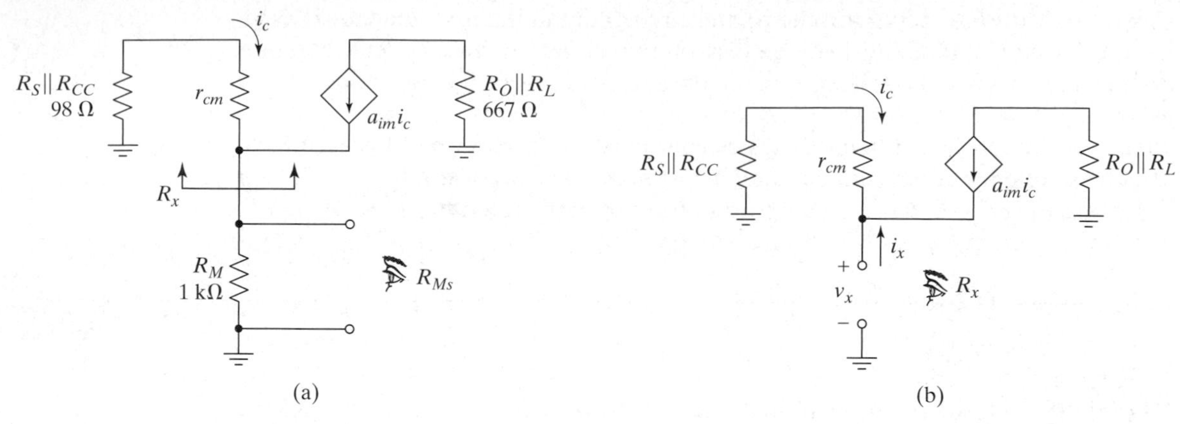

Figure 9–17 Finding the short-circuit driving-point resistance for C_M.

The short-circuit time constant for C_M is found by using the circuit shown in Figure 9-17. We have used the current-controlled version of the transistor model in this case because it simplifies the analysis.

From part (a) of the figure, we see that

$$R_{Ms} = R_M \| R_x \, .$$

We can find R_x as shown in part (b) of the figure. First note that by KCL,

$$i_c(1 + a_{im}) = -i_x \, .$$

Then write KVL to obtain

$$v_x = -i_c(R_S \| R_{CC} + r_{cm}) \, .$$

Finally, combining these two equations results in

$$R_x = \frac{v_x}{i_x} = \frac{R_S \| R_{CC} + r_{cm}}{1 + a_{im}} \, .$$

Assuming that $a_{im} = 99$ for this example yields $R_x = 26\ \Omega$. Finally, we find that $\tau_M = R_{Ms} C_M = 2.5$ ms.

Now using these results along with (9.39) yields

$$\omega_{cL} \approx \frac{1}{\tau_{INs}} + \frac{1}{\tau_{OUTs}} + \frac{1}{\tau_{Ms}} = 1 \text{ krad/s}.$$

■

Let's compare the approximate result found in Example 9.3 with what is obtained by computer simulation. A simulation of the circuit in Figure 9-15 reveals that $\omega_{cL} = 754$ rad/s, which is 25% lower than the estimate (Exam3.sch). While this is a very large error, the result is still useful for three reasons. First, the approximation was conservative, so the design will still work (i.e., we usually don't mind if the bandwidth of an amplifier is a bit larger than we designed it to be). Second, the analysis still shows us how to increase or decrease ω_{cL} and how much a change in any given capacitor or driving-point resistance will affect ω_{cL}. Third, we do not often require an extremely accurate estimate of bandwidth.

Even though our result is useful, the error is larger than we would like. However, the particular example given here has a large error because there isn't a dominant pole. In fact, exact analysis reveals that the poles of this circuit are at 66 rad/s, 333 rad/s, and 612 rad/s. The highest frequency pole is only about twice the second pole frequency; therefore, we expect a rather large error with this technique. When the short-circuit time constants are used to estimate ω_{cL}, the error is typically between 10 and 25%, but the resulting estimate is conservative.[7] The error can be very small if one pole is truly dominant, but that is not usually the case for ω_{cL}.

As a final step, we should check that the zeros in the low-frequency equivalent circuit are all low enough to be safely ignored. We already have one confirmation of this fact, since our analytical result agrees quite well with the simulated result. In practice, this agreement would probably be sufficient. Nevertheless, for completeness, we take a moment to examine the zeros in the circuit.

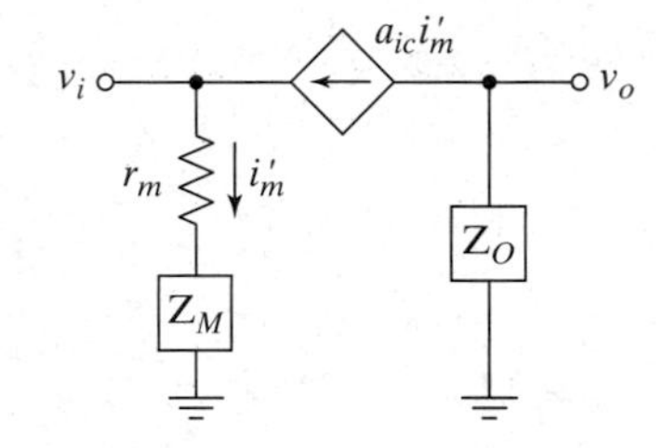

Figure 9–18 The small-signal model of a common-merge stage.

Both C_{IN} and C_{OUT} contribute zeros at DC, because they are in series with the signal and, as the frequency goes to zero, they become open circuits and cause the gain to go to zero. The emitter bypass capacitor is a bit more troubling, but can still be handled in a very intuitive way. Consider the small-signal model of a common-merge stage shown in Figure 9-18. We have used the midband T model and shown the external components in series with the output and merge terminals as general impedances. The implicit assumption is that we are examining frequencies low enough that the transistor's internal capacitances can be ignored. The gain of this circuit is

$$\frac{V_o(j\omega)}{V_i(j\omega)} = \frac{-a_{ic}Z_O(j\omega)}{r_m + Z_M(j\omega)}. \tag{9.41}$$

If we consider the effect of C_M alone in our circuit, the output impedance is purely real and equal to R_O, whereas the merge impedance is R_M in parallel with C_M. At DC, the capacitor is an open circuit and the gain is

$$\frac{V_o(j0)}{V_i(j0)} = \frac{-a_{ic}R_O}{r_m + R_M}. \tag{9.42}$$

As the frequency goes to infinity, the capacitor becomes a short and the gain increases to

$$\frac{V_o(j\infty)}{V_i(j\infty)} = \frac{-a_{ic}R_O}{r_m} = -g_m R_O. \tag{9.43}$$

[7] The results are conservative *for the circuit we analyze*. Of course, we are usually analyzing an approximation to a real circuit, so the results are not *guaranteed* to be conservative in practice, although they usually are if our approximations are acceptable.

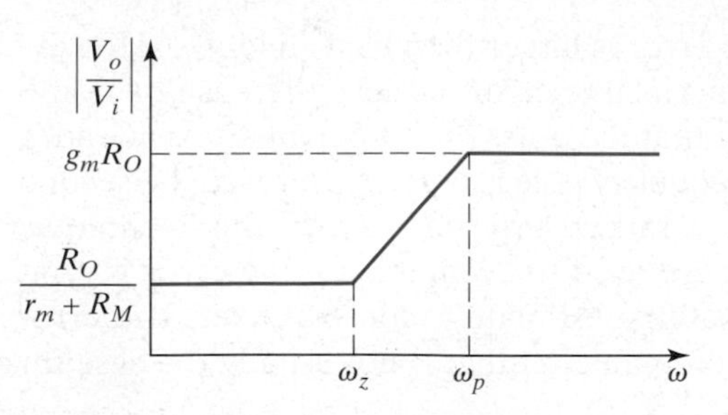

Figure 9–19 A plot of the approximate magnitude versus frequency for the circuit in Figure 9-18 when $Z_O = R_O$ and $Z_M = R_M \| C_M$.

Therefore, we intuitively see that we have a zero followed by a pole. We can find the separation between the pole and zero frequencies by using a plot of the transfer function magnitude versus frequency, as shown in Figure 9-19. Note that this plot uses linear axes and an asymptotic approximation for the magnitude (i.e., the straight-line asymptotes).

The slope on the plot in between the pole and zero is unity, so the ratio of the frequencies is equal to the ratio of the gains:

$$\frac{\omega_p}{\omega_z} = \frac{g_m R_O}{R_O/(r_m + R_M)} = g_m(r_m + R_M) \approx \frac{r_m + R_M}{r_m}. \tag{9.44}$$

If $R_M \gg r_m$, the pole and zero are widely separated, and our approximation should be fine. Note that we can also derive (9.44) directly from (9.41) by plugging in R_O and $R_M \| C_M$ for the impedances. After a bit of algebra, the result is

$$\frac{V_o(j\omega)}{V_i(j\omega)} = \frac{-a_{ic}(R_O/r_m)(j\omega+1/R_M C_M)}{j\omega+1/(R_M \| r_m)C_M}, \tag{9.45}$$

which leads to (9.44) as well.

Exercise 9.8

Consider the small-signal low-frequency AC equivalent circuit shown in Figure 9-20. This circuit represents a common-merge amplifier and is similar to the one analyzed in Example 9.3. Assume that $a_{im} = 99$, and find an estimate of f_{cL} by using short-circuit time constants.

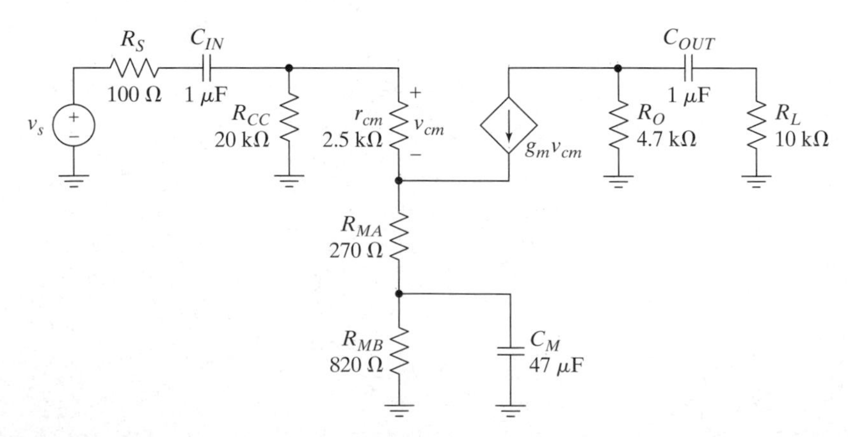

Figure 9–20 The small-signal low-frequency AC equivalent circuit for a common-merge amplifier.

□

Now we can consider how to find an approximation to the high cutoff frequency, ω_{cH}. In this case, assume that it is always possible to draw a high-frequency equivalent circuit as in Figure 9-12(c). The high-frequency equivalent will always have a low-pass transfer characteristic, and if there is a pole that is significantly below all the other poles, then it is dominant and ω_{cH} is approximately equal to that pole frequency. Since we are dealing with a low-pass characteristic, the dominant pole is the *lowest* frequency pole in the circuit. (We are looking for the *high* cutoff frequency of the original circuit, but since we are examining the high-frequency equivalent, ω_{cH} is approximately the lowest pole frequency in this equivalent.) Later we will examine the accuracy of our approximation when there isn't a dominant pole. For the approximation to be reasonable, we must assume that all of the zeros in the transfer function are well above ω_{cH}, which is usually true for amplifiers.

Given a high-frequency equivalent circuit we find the open-circuit time constant associated with each capacitor in the circuit and then take the reciprocal of the sum of these time constants to find the approximate cutoff frequency. In other words,

$$\omega_{cH} \approx \frac{1}{\sum_{i=1}^{n} \tau_{io}}, \tag{9.46}$$

where there are n capacitors in the equivalent circuit, the open-circuit time constant associated with capacitor C_i is

$$\tau_{io} = R_{io} C_i, \tag{9.47}$$

and the **open-circuit driving-point resistance** seen by capacitor C_i is denoted R_{io}. This resistance is the resistance seen when you set all of the independent sources in the circuit to zero, remove C_i, open all of the other capacitors, and look back into the terminals where C_i was connected. A detailed justification for this approximation is given in the companion section on the CD, but we can see that the result is reasonable by examining (9.46). If one of the capacitors does cause a dominant pole (i.e., it is substantially lower than all the other poles in the circuit), then for frequencies near this pole, all of the other capacitors will have very large impedances and can be reasonably approximated by open circuits. Making this approximation, we can find the pole frequency of the resulting single-pole circuit by using the method presented in Aside A1.6. The resulting pole frequency will be the reciprocal of the open-circuit time constant for the dominant capacitor. Since it is the largest time constant by far, it will dominate the sum in (9.46), and the pole will be approximately the same as ω_{cH} given by (9.46).

This result is extremely important. Using (9.46), we can estimate the high cutoff frequency of a circuit far more readily than we can derive the entire transfer function. In addition, the resulting equations are much easier to deal with and allow us to clearly see the contributions of the individual capacitors. We must again resist the temptation to associate a pole with the reciprocal of each time constant. This method is useful only for approximating the cutoff frequency; it will not provide information about nondominant pole frequencies. Finally, we reiterate that the open-circuit time constant method finds an approximation to the *lowest* frequency pole in whatever circuit it is applied to; the designer must make sure that it is applied to the high-frequency equivalent circuit if an estimate of ω_{cH} is desired.

This section concludes with an example of the usage of open-circuit time constants.

Example 9.4

Problem:
Find an estimate of the high cutoff frequency for the common-merge amplifier with the high-frequency equivalent circuit shown in Figure 9-21. This is a repeat of Figure 9-12(c), but with component values included.

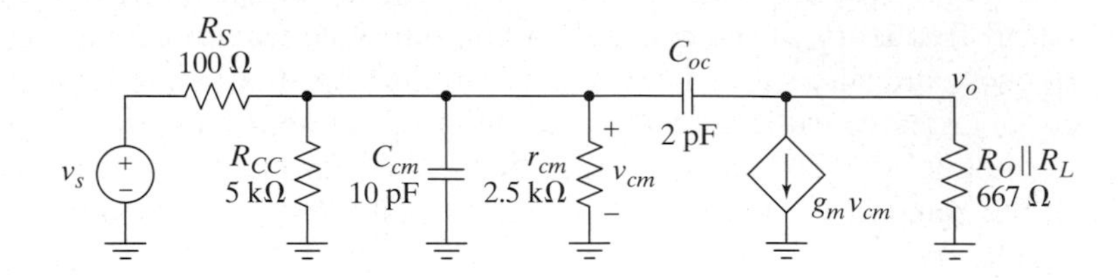

Figure 9–21 A common-merge amplifier high-frequency equivalent circuit.

Solution:
This circuit has two independent capacitors, so we need to calculate two open-circuit time constants and then use (9.46) to estimate the high cutoff frequency.

To find the time constant for C_{cm}, we set the independent source v_s to zero, open C_{oc}, and find the resistance seen looking back into the terminals where C_{cm} was connected. The result is

$$R_{cmo} = R_S \| R_{CC} \| r_{cm} = 94.3\ \Omega,$$

which leads to an open-circuit time constant of

$$\tau_{cmo} = R_{cmo} C_{cm} = 943\ \text{ps}.$$

The open-circuit time constant for C_{oc} is found by using the circuit shown in Figure 9-22.

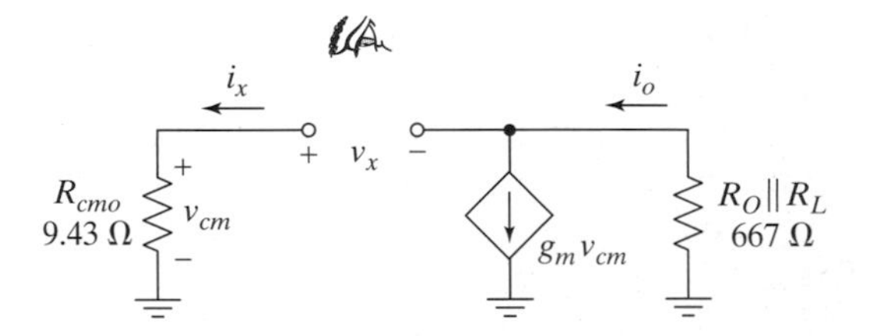

Figure 9–22 Finding the open-circuit driving-point resistance for C_{oc}.

Writing KCL at the output node, we find that

$$i_o = i_x + g_m v_{cm} = i_x(1 + g_m R_{cmo}) .$$

We next write KVL to obtain

$$v_x = i_x R_{cmo} + i_o(R_O \| R_L) = i_x\left[R_{cmo} + (1 + g_m R_{cmo})(R_O \| R_L)\right],$$

and solve to find

$$R_{oco} = \frac{v_x}{i_x} = \left[R_{cmo} + (1 + g_m R_{cmo})(R_O \| R_L)\right]$$

Assuming that $g_m = 0.1$ A/V for this example yields $R_{oco} = 7\ \text{k}\Omega$. Finally, we find that $\tau_{oco} = R_{oco} C_{oc} = 14$ ns.

Now using these results along with (9.46) yields

$$\omega_{cH} \approx \frac{1}{\tau_{cmo} + \tau_{oco}} = 67\ \text{Mrad/s}.$$

■

In this last example, the time constant for C_{oc} is dominant because of the Miller effect. In fact, if you reexamine the equation for R_{oco}, you will note that it can be rewritten as

$$R_{oco} = \left[R_{cmo} + (1 + g_m R_{cmo})(R_O \| R_L)\right] \approx g_m(R_O \| R_L)R_{cmo} = -A_{vmb}R_{cmo}, \quad \textbf{(9.48)}$$

where A_{vmb} is the midband gain across C_{oc}. Multiplying the final term in (9.48) by C_{oc} to get the time constant, we see that the result is equivalent to multiplying C_{oc} by $-A_{vmb}$ to get the Miller input capacitance and then multiplying that by the equivalent resistance seen by C_{Min}. Therefore, the results achieved by this approximation are very nearly the same as those achieved by the Miller approximation if C_{oc} is dominant. If C_{oc} is not dominant, the open-circuit time constant method will usually yield more accurate results.

If we solve for the cutoff frequency in the high-frequency equivalent of Figure 9-21 exactly or use SPICE, we find that $\omega_{cH} = 67$ Mrad/s, as estimated (Exam4.sch). The reason that the estimate is so accurate this time is that there is, in fact, a dominant pole in the circuit. In general, the open-circuit time-constant method yields estimates that are between 0 and 25% lower than the true cutoff frequency. The estimate is again conservative, so even the larger errors are often not a serious problem.

Since our results agree well with SPICE, we could stop here and assume that the zeros must be enough higher than the dominant pole that the approximation is reasonable in this case. We will look at the zeros for completeness, however, rather than relying solely on our agreement with the simulator. Examining Figure 9-21, we note that C_{cm} produces a zero at infinity, since it forces the gain to zero as it shorts out r_{cm}. To see what C_{oc} does to the response, we examine the simplified circuit shown in Figure 9-23.

At DC, C_{oc} is an open circuit and the gain of this circuit is $-g_m R_L$. As the frequency goes to infinity, the gain becomes unity; that is, $v_o = v_i$. The resulting magnitude plot is shown in Figure 9-24, where we have used an asymptotic approximation (i.e., the straight-line asymptotes) and linear scales on both axes. The ratio of the pole and zero frequencies can be found by recognizing that the slope of the middle asymptote on the plot is −1. The result is

$$\frac{\omega_z}{\omega_p} = g_m R_L. \quad \textbf{(9.49)}$$

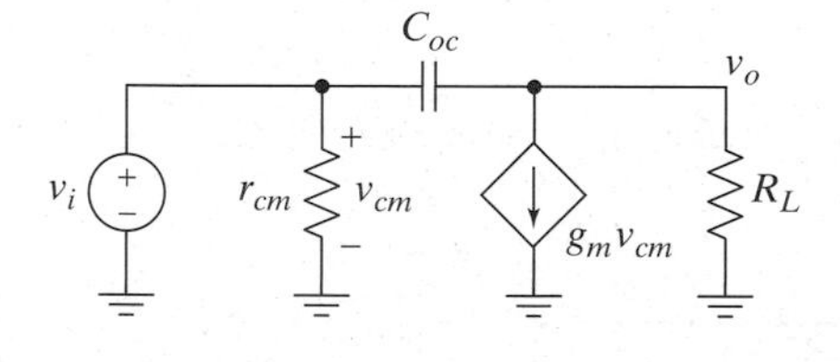

Figure 9–23 A simplified circuit for seeing the zero due to C_{oc}.

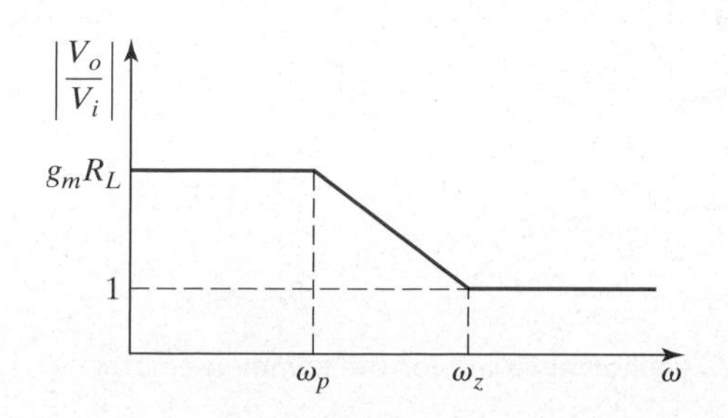

Figure 9–24 The asymptotic magnitude plot for Figure 9-23.

Therefore, so long as the midband gain from control to output is much larger than unity (in practice, four or more is usually good enough), this zero should be well above the pole caused by C_{oc} and should not affect our results.

9.2.2 A Bipolar Implementation: The Common-Emitter Amplifier

We analyze the frequency response of the common-emitter amplifier in two ways. The introductory treatment uses the Miller approximation and standard analysis techniques. The advanced treatment requires an understanding of open-circuit and short-circuit time constants.

Introductory Treatment Consider the common-emitter amplifier shown in Figure 9-25. We already analyzed the DC biasing and small-signal midband AC performance of this circuit in Section 8.2 (see Figure 8-33), where we determined the bias voltages and currents shown in the figure.

We now want to find both the low and high cutoff frequencies for this circuit. To find the low cutoff frequency, we draw the small-signal low-frequency AC equivalent circuit as shown in Figure 9-26.

Writing KVL around the base-emitter loop, we obtain

$$V_b(j\omega) - I_b(j\omega)r_\pi - (\beta+1)I_b(j\omega)Z_E(j\omega) = 0, \tag{9.50}$$

where

$$Z_E(j\omega) = \frac{R_E}{1 + j\omega R_E C_E}. \tag{9.51}$$

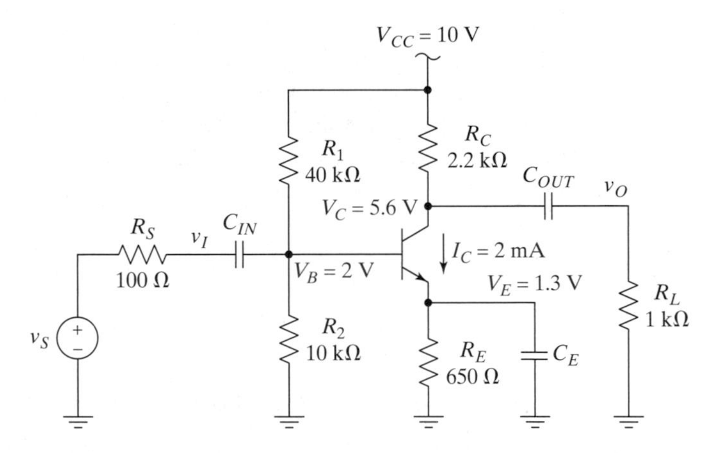

Figure 9–25 A common-emitter amplifier. (This is the same circuit as in Figure 8-33.)

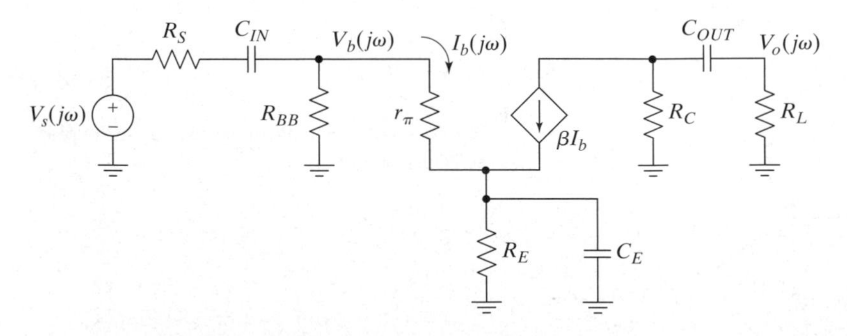

Figure 9–26 The small-signal low-frequency AC equivalent circuit for the common-emitter amplifier of Figure 9-25.

Examining (9.50) and (9.51), we see that we can use impedance reflection with complex impedances just as well as we did with resistances. We can reflect the emitter impedance into the base as shown in Figure 9-27, and the new circuit has the same loop equation from base to ground. Notice that since $Z_C = 1/j\omega C$, we divide C_E by $(\beta + 1)$ instead of multiplying by it.[8]

We can derive the transfer function for the circuit in Figure 9-27 by breaking it into a product of simpler transfer functions as usual:

$$\frac{V_o(j\omega)}{V_s(j\omega)} = \frac{V_o(j\omega)}{V_\pi(j\omega)}\frac{V_\pi(j\omega)}{V_b(j\omega)}\frac{V_b(j\omega)}{V_s(j\omega)}. \tag{9.52}$$

The sequence of transfer functions in (9.52) is not unique, but it will allow us to write each transfer function by inspection. We obtain:

$$\frac{V_b(j\omega)}{V_s(j\omega)} = \frac{Z'_{in}(j\omega)}{Z'_{in}(j\omega) + R_S + \dfrac{1}{j\omega C_{IN}}}, \tag{9.53}$$

where Z'_{in} is the impedance seen looking to the right of C_{IN},

$$Z'_{in}(j\omega) = R_{BB} \| \left[r_\pi + (\beta+1)Z_E(j\omega) \right]. \tag{9.54}$$

Similarly,

$$\frac{V_\pi(j\omega)}{V_b(j\omega)} = \frac{r_\pi}{r_\pi + (\beta+1)Z_E(j\omega)} = \left(\frac{r_\pi}{r_\pi + (\beta+1)R_E}\right)\left(\frac{1 + j\omega\tau_E}{1 + j\omega\tau'_E}\right), \tag{9.55}$$

where we have defined $\tau_E = R_E C_E$ and $\tau'_E = C_E(r_\pi R_E/[r_\pi + (\beta + 1)R_E])$. Now moving to the collector side of the transistor, we recognize that we have a single-pole high-pass circuit with a DC zero and use the method of Aside A1.6 to write

$$\frac{V_o(j\omega)}{V_\pi(j\omega)} = \frac{-j\omega g_m(R_L \| R_C)}{j\omega + \dfrac{1}{(R_C + R_L)C_{OUT}}}. \tag{9.56}$$

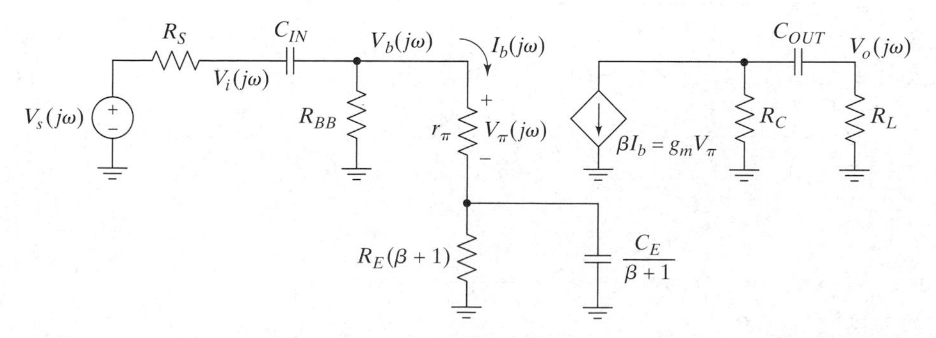

Figure 9–27 The circuit of Figure 9-26 with the emitter impedance reflected into the base.

[8] The β used here is the midband value, β_0, but we ignore the subscript for notational simplicity (a common practice). We do not need to use the high-frequency current-controlled hybrid-π model, because we are not including the transistor's capacitances in this circuit.

Combining (9.53), (9.55), and (9.56) yields the complete transfer function for the small-signal low-frequency AC equivalent circuit. By setting the magnitude of this transfer function equal to $1/\sqrt{2}$, we can solve for ω_{cL}. In practice, we would rarely—if ever—perform this computation by hand. We would use either the approximation techniques presented in the next section or computer simulation. The log magnitude of the resulting transfer function is shown in Figure 9-28. This response is the low-frequency half of the overall AC-coupled response shown in Figure 9-10(a). The high-frequency asymptote is the midband gain of the original circuit. Because we have not included the capacitors that lead to the high-frequency roll-off in the real circuit, the transfer function of this equivalent circuit does not roll off at high-frequencies.

We now turn to finding the high cutoff frequency of the circuit in Figure 9-25. The small-signal high-frequency AC equivalent circuit is shown in Figure 9-29. Because this is a *small-signal* equivalent circuit, we use a linear model for the transistor. Because it is the *high-frequency* equivalent circuit, we short out all of the coupling and bypass capacitors as we did for the midband equivalent circuit. However, we include the parasitic capacitors that are part of the high-frequency hybrid-π model for the transistor. Because it is an *AC* equivalent circuit, the DC power supply appears as a ground. We now include the transistor's intrinsic base resistance r_b, since it can have a significant effect on the high-frequency response when R_S is low. The effective load resistance is given by

$$R'_L = R_L \| R_C. \tag{9.57}$$

The circuit in Figure 9-29 is topologically identical to the high-frequency equivalent for the common-merge amplifier shown in Figure 9-12(c). Therefore, the analysis of the present circuit parallels the earlier development exactly. Using the Miller approximation, we begin by finding the midband gain across C_μ:

$$\frac{v_o}{v_\pi} = -g_m R'_L. \tag{9.58}$$

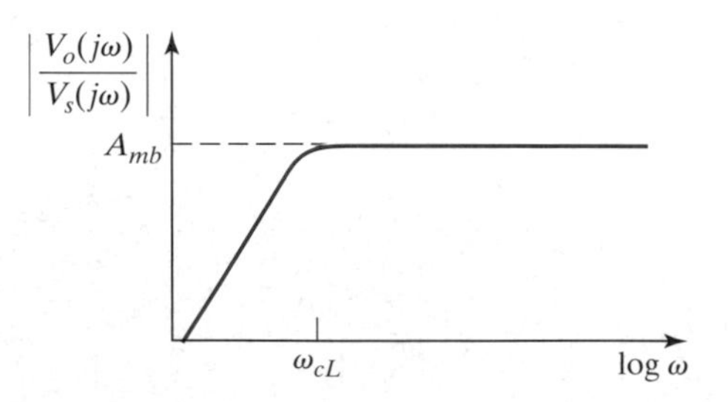

Figure 9–28 The log-magnitude response of the small-signal low-frequency AC equivalent circuit of Figure 9-27.

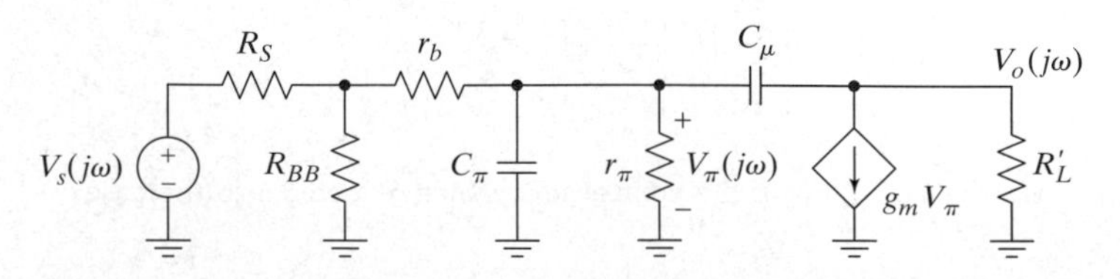

Figure 9–29 The small-signal high-frequency AC equivalent circuit for the amplifier of Figure 9-25.

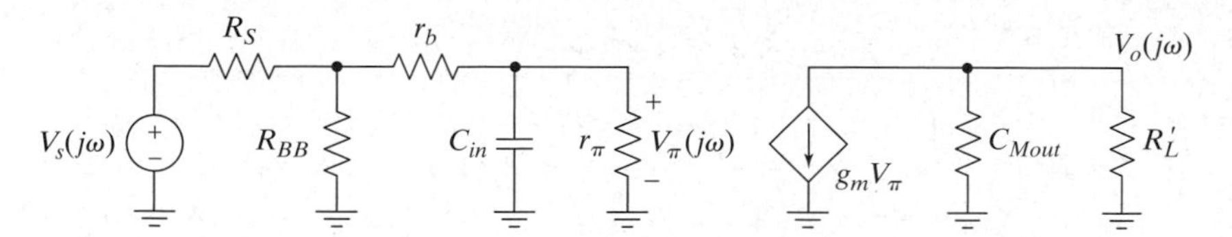

Figure 9–30 The equivalent circuit from Figure 9-29 after application of Miller's theorem.

Using this gain along with the formulas in (A9.14) and (A9.15), we redraw the circuit as shown in Figure 9-30. The overall input capacitance is C_π in parallel with the Miller input capacitance given by

$$C_{Min} = (1 + g_m R_L')C_\mu, \tag{9.59}$$

so that

$$C_{in} = C_\pi + (1 + g_m R_L')C_\mu. \tag{9.60}$$

This input capacitance is usually dominated by C_{Min}, even for stages with modest gain.

The circuit in Figure 9-30 is topologically identical to the common-merge equivalent of Figure 9-13. As noted earlier, we must resist the temptation to treat this circuit as two independent single-pole circuits, since C_{Min} and C_{Mout} are both derived from the same capacitor. Using the Miller approximation, we find ω_{cH} as the pole associated with C_{in}; that is,

$$\omega_{cH} \approx \frac{1}{(R_S' \| r_\pi)C_{in}}, \tag{9.61}$$

where we have defined $R_S' = R_S \| R_{BB} + r_b$.

We can check to see whether or not the Miller approximation is acceptable by using (9.34). The Miller output capacitance is

$$C_{Mout} = (1 + 1/g_m R_L')C_\mu \approx C_\mu, \tag{9.62}$$

and the resistance seen by C_{Mout} is R_L'. We check to see whether or not $1/\omega_{cH}$ is much greater than $R_L' C_{Mout}$. If it is, the approximation is reasonable.

Example 9.5

Problem:

Find the high cutoff frequency for the amplifier in Figure 9-25, assuming that the transistor is a 2N2222 with $\beta = 180$, $r_b = 10\ \Omega$, $\tau_F = 411$ ps, $C_{je0} = 22$ pF, $C_{\mu 0} = 7.3$ pF, $V_0 = 0.75$, and $m_{jc} = 0.34$. (The parameters are all given in Appendix B.)

Solution:

Using the known DC bias point and assuming room temperature, we calculate

$$g_m = \frac{I_C}{V_T} = 77\ \text{mA/V},$$

$$r_\pi = \frac{\beta}{g_m} = 2.3 \text{ k}\Omega,$$

$$C_\mu = \frac{C_{\mu 0}}{\left(1 + \frac{V_{CB}}{V_0}\right)^{m_{jc}}} = 4 \text{ pF},$$

$$C_{je} \approx 2C_{je0} = 44 \text{ pF},$$

$$C_d = g_m \tau_F = 32 \text{ pF},$$

and

$$C_\pi = C_d + C_{je} = 76 \text{ pF}.$$

From (9.57) and the circuit, we see that

$$R_L' = R_L \| R_C = 687 \ \Omega,$$

and therefore,

$$A_v = \frac{v_o}{v_i} = -g_m R_L' = -53,$$

as in Chapter 8. Using this result in (9.31) we find, that

$$C_{in} = C_\pi + (1 + g_m R_L')C_\mu = 292 \text{ pF}.$$

We finally make use of (9.33) to obtain

$$\omega_{cH} \approx \frac{1}{(R_S' \| r_\pi)C_{in}} = 31.5 \text{ Mrad/s},$$

or $f_{cH} \approx 5.0$ MHz.

The last step is to check that the Miller approximation is reasonable for this circuit. To do that, we use (9.32) to find $C_{Mout} \approx C_\mu = 4$ pF and then use (9.34). Since $1/\omega_{cH} = 31.8$ ns is much larger than $R_L' C_{Mout} = 1.5$ ns, the Miller approximation is accurate.

■

A SPICE simulation of the circuit in Example 9.5 reveals that $f_{cH} = 5.4$ MHz, which is 8% higher than our estimate. Considering the simplicity of our result, however, the accuracy is incredibly good. The significance of r_b is examined in Problem P9.45, where we reexamine this circuit with a larger value of r_b.

Note that (9.33) can be further simplified to get

$$\omega_{cH} \approx \frac{1}{(R_S \| R_{BB} \| r_\pi)C_{in}} \approx \frac{1}{R_S g_m R_L' C_\mu}, \tag{9.63}$$

which shows clearly that the bandwidth can be increased by reducing R_S, reducing C_μ, or reducing the midband gain ($g_m R_L'$).

Another way of looking at the performance of this stage is to examine the gain-bandwidth product. Using (9.35) and making use of the fact that $R_i \approx R_{BB} \| r_\pi$

$$\text{GBW} = \left(\frac{R_i}{R_i + R_S}\right)\frac{g_m R_L'}{(R_S \| R_{BB} \| r_\pi)C_{in}} \approx \frac{g_m R_L'}{R_S g_m R_L' C_\mu} = \frac{1}{R_S C_\mu}, \quad \textbf{(9.64)}$$

in complete agreement with (9.38) for the common-merge amplifier. Using (9.64), we can clearly see that decreasing C_μ is beneficial overall, since it increases the GBW. If the decrease in C_μ is achieved by raising V_C, the gain of the stage will be reduced, but since the GBW is increased, the bandwidth increases more than might at first be anticipated. If C_μ is reduced a great deal, then one of the approximations made in (9.64) is no longer valid, namely, we cannot ignore C_π.

Example 9.6

Problem:

Find the high cutoff frequency for the common-emitter amplifier presented in Figure 9-31. This amplifier has only a portion of the emitter resistance bypassed, to provide greater flexibility to the designer as noted in Section 8.2 (*see* Figure 8-36). Assume that a 2N2857 transistor is used, so that $\beta = 100$, $r_b = 10\ \Omega$, $\tau_F = 116$ ps, $C_{je0} = 0.94$ pF, $C_{\mu 0} = 0.89$ pF, $V_0 = 0.75$, and $m_{jc} = 0.3$ (*see* Appendix B for the model).

Solution:

We first solve for the DC bias point of the amplifier and find, that $V_B \approx 1.9$ V, $V_E \approx 1.2$ V, $V_C \approx 2.8$V, and $I_C \approx 2.4$ mA. Using these values, we calculate the small-signal parameters: $r_e = 10.8$, Ω, $r_\pi = 1.1$ kΩ, $g_m = 92$ mA/V, $C_\pi = 12.6$ pF, and $C_\mu = 0.7$ pF.

We next find the midband gain from v_i to v_o, using the T model for the transistor.[9] The result is, see (8.109) for an example analysis

$$A_v \approx \frac{-R_C \| R_L}{r_e + R_{EA}} = -8.9.$$

The small-signal midband AC input resistance is found using impedance reflection,

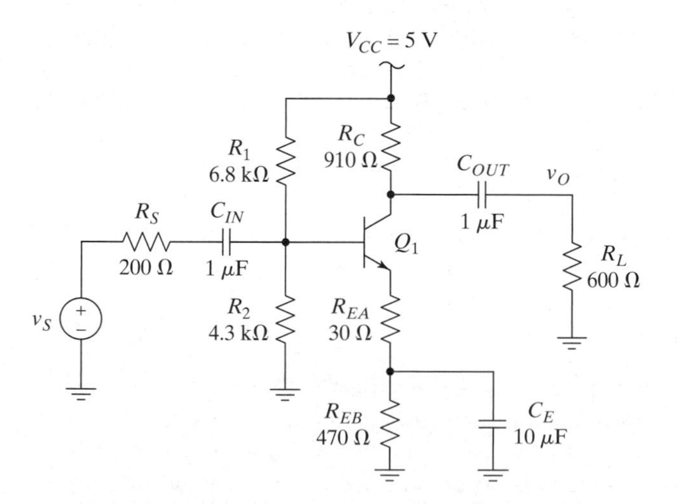

Figure 9–31 A common-emitter amplifier with partially bypassed emitter resistance.

[9] We ignore r_b when finding the midband gain because it complicates the equations, but has a negligible effect on the outcome. We include it in the time constants, however, since its effect may be significant there.

$$R_i = R_{BB} \| \left[r_\pi + (\beta+1)R_{EA} \right] = 1.6 \text{ k}\Omega,$$

and results in an input attenuation factor

$$\alpha_i = \frac{R_i}{R_i + R_S} = 0.89$$

and an overall gain of

$$\frac{v_o}{v_s} = \alpha_i A_v = -7.9.$$

If we assume that the Miller capacitance will dominate the frequency response, we can find an estimate of f_{cH} without too much work. Since the gain from base to collector is A_v, the Miller input capacitance is given by

$$C_{Min} = (1 - A_v)C_\mu = 6.2 \text{ pF}.$$

This capacitor shows up from the base of the transistor to ground and is *not* in parallel with C_π, because of the unbypassed emitter resistance. Assuming for the moment that we can ignore C_π, we forge ahead with the calculation. The resistance seen by C_{Min} is found by defining $R_S' = R_S \| R_{BB} + r_b$ and using impedance reflection and is

$$R_{C_{Min}} = R_S' \| \left[r_\pi + (\beta+1)R_{EA} \right] = 187 \ \Omega,$$

which leads to an estimated pole frequency of

$$f_{cH} \approx \frac{1}{2\pi R_{C_{Min}} C_{Min}} = 137 \text{ MHz}.$$

We can quickly check that $(R_C \| R_L)C_{Mout} = 0.25$ ns is much less than $1/\omega_{cH} = 1.2$ ns, so the current fed forward through C_μ can be safely neglected in this case. However, we have a harder time deciding whether neglecting C_π is justified. If the impedance of C_π is not much larger than r_π at f_{cH}, the gain from the base to v_π will be smaller than in midband, and the Miller approximation is not valid (i.e., the midband gain is not accurate for finding the Miller multiplication of C_μ, and ignoring C_π is not justifiable). In this case, we see that $|1/j\omega_{cH}C_\pi| = 92 \ \Omega$ is actually much smaller than r_π, and the Miller approximation is not valid. Therefore, without the more advanced methods of the next section, we have to either derive the transfer function by hand or compare our results with computer simulations or a breadboard. In this case, the simulated bandwidth is 83 MHz, so our error is about 65%. We have still obtained useful information, however, namely, the fact that the Miller effect is not dominant and that C_π is important in this circuit.

■

The unbypassed emitter resistance in the amplifier in Example 9.6 increases the bandwidth of the circuit while simultaneously reducing the gain. If we redo the calculations of the example with R_{EA} set equal to zero (and R_{EB} increased by the same amount to keep the same bias point), we find that the magnitude of the gain increases to about −33 (A_v) while the calculated bandwidth is reduced to about 40 MHz. Therefore, by increasing R_{EA}, we trade gain for bandwidth. If we look at the gain-bandwidth product in detail, we see that it is not constant as R_{EA} is varied, but does decrease a small amount as R_{EA} increases (*see* Problem P9.44), so we don't get a one-for-one trade-off in this example.

Exercise 9.9

Derive an approximate equation for the gain-bandwidth product for the circuit in Example 9.6. Calculate the value of GBW from your equation and from the results of the example. Do the two answers agree?

Advanced Treatment We now want to find the bandwidth of the common-emitter amplifier in Figure 9-25 by making use of open- and short-circuit time constants. The small-signal low-frequency AC equivalent circuit from Figure 9-26 is repeated in Figure 9-32(a). We can find the short-circuit time constants for C_{IN} and C_{OUT} directly from this figure. With C_E shorted out and v_s set to zero, we remove C_{IN} and look back into the circuit to find the short-circuit driving-point resistance:

$$R_{INs} = R_S + R_{BB} \| r_\pi. \tag{9.65}$$

With C_{IN} and C_E shorted out and v_s set to zero we see that $i_b = 0$, so

$$R_{OUTs} = R_C + R_L. \tag{9.66}$$

To find the short-circuit driving-point resistance for C_E, we first use impedance reflection to redraw the circuit as shown in part (b) of the figure. We have also set v_s to zero and shorted C_{IN}.

The short-circuit driving-point resistance for C_E is then

$$R_{Es} = R_E \left\| \frac{R_S \| R_{BB} + r_\pi}{\beta + 1} \right. . \tag{9.67}$$

Using (9.65) through (9.67), we can find the three short-circuit time constants and plug them into (9.39) to get

$$\omega_{cL} \approx \frac{1}{R_{INs} C_{IN}} + \frac{1}{R_{Es} C_E} + \frac{1}{R_{OUTs} C_{OUT}}. \tag{9.68}$$

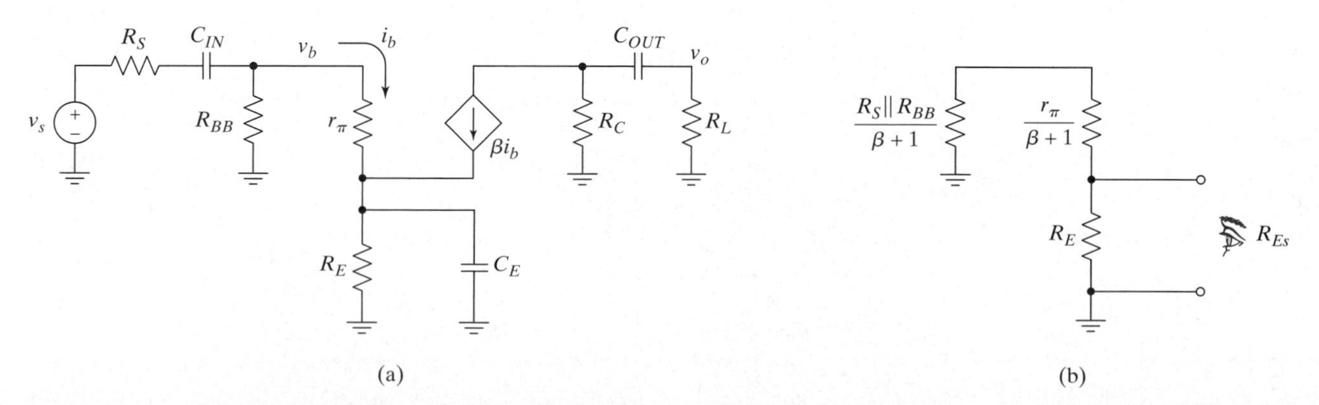

Figure 9–32 (a) The small-signal low-frequency AC equivalent for the common-emitter amplifier of Figure 9-25 and (b) the circuit for finding the short-circuit driving-point resistance seen by C_E.

This is a wonderful equation for design: Each term is readily identified with a particular circuit component, and the individual terms were all derived in a straightforward way. Suppose that we use $C_{IN} = C_{OUT} = 1\ \mu F$ and $C_E = 10\ \mu F$. Plugging in the given component values and assuming a 2N2222 transistor as in Example 9.5, we find that $r_\pi = 2.3\ k\Omega$, $R_{INs} = 1.9\ k\Omega$, $R_{Es} = 13\ \Omega$, and $R_{OUTs} = 3.2\ k\Omega$. The resulting cutoff frequency is $f_{cL} \approx 1.4$ kHz, which is dominated by the time constant associated with C_E. It is common for the time constant associated with the bypass capacitor to dominate, since the short-circuit driving-point resistance for that capacitor is usually much smaller than the other two. Remember that the impedance seen looking into the emitter of a transistor is small. This analytical result is useful for design. Since we know C_E will dominate, we spend our money on a larger capacitor there and use much smaller capacitors for the other two time constants in order to achieve the required bandwidth, which explains why we chose C_E to be 10 times larger than C_{IN} and C_{OUT} for this example.

A SPICE analysis of this circuit yields a low cutoff of 1.2 kHz (Exam5.sch), which is close to our prediction, but lower, so our prediction was conservative, as stated earlier. As a final step, we should check that the zeros in the low-frequency equivalent circuit are all low enough to be safely ignored. We know that they must be, since our simulation results agree so well, but we could also use (9.44) and recognize that when $R_E \gg r_e$, which is certainly true here, our approximation is fine.

Exercise 9.10

Consider the common-emitter amplifier from Figure 9-25. Suppose we want to set $f_{cL} = 1$ kHz. (a) Find the values required for the three capacitors if all three short-circuit time constants are set equal to each other. (b) Now allow τ_{Es} to account for 90% of f_{cL} and τ_{INs} and τ_{OUTs} to each account for 5%, and determine the capacitor values required.

□

Let's now find an approximate value for ω_{cH} using the open-circuit-time constant method. The small-signal high-frequency AC equivalent circuit for the amplifier was shown in Figure 9-29 and is repeated here as Figure 9-33.

Using (9.46), we find that the high cutoff frequency for this circuit will be given by

$$\omega_{cH} \approx \frac{1}{R_{\pi o}C_\pi + R_{\mu o}C_\mu}. \tag{9.69}$$

With the signal source set to zero and C_μ opened, we find by inspection that

$$R_{\pi o} = R'_S \| r_\pi = 104\ \Omega, \tag{9.70}$$

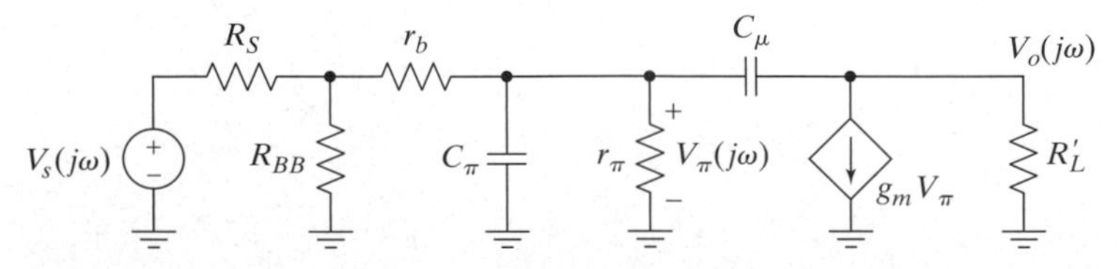

Figure 9–33 The small-signal high-frequency AC equivalent circuit for the amplifier of Figure 9-25.

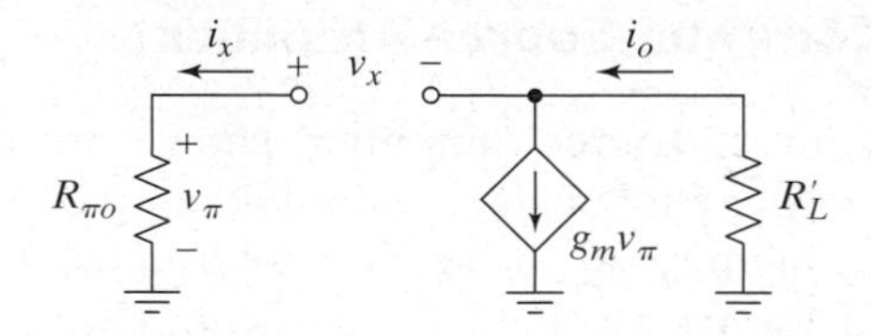

Figure 9–34 The circuit for finding the open-circuit driving-point resistance see by C_μ.

where we have defined $R'_S = R_S \| R_{BB} + r_b$ as before. To find the open-circuit driving-point resistance seen by C_μ, we use the circuit shown in Figure 9-34. This circuit is derived from the one in Figure 9-33 by shorting the signal source and open-circuiting C_π.

Writing KCL at the collector, we find that

$$i_o = i_x + g_m v_\pi = i_x(1 + g_m R_{\pi o}), \tag{9.71}$$

and writing KVL around the outside loop, we get

$$v_x = i_x R_{\pi o} + i_o R'_L. \tag{9.72}$$

Combining these two equations, we find the desired resistance:

$$R_{\mu o} = \frac{v_x}{i_x} = R_{\pi o} + (1 + g_m R_{\pi o})R'_L = 6.3\ \text{k}\Omega. \tag{9.73}$$

Finally, using these numbers along with the capacitances found in Example 9.5, we use (9.69) to find

$$f_{cH} \approx \frac{1}{2\pi(R_{\pi o}C_\pi + R_{\mu o}C_\mu)} = \frac{1}{2\pi(7.9\ \text{ns} + 25.2\ \text{ns})} = 4.8\ \text{MHz}, \tag{9.74}$$

which is dominated by the time constant associated with C_μ, as expected (because of the Miller effect). A SPICE simulation of this circuit shows that $f_{cH} \approx 5.4$ MHz (Exam5.sch). Our prediction is 11% low and conservative. Since our results agree well with SPICE, we could stop here and assume that the zeros must be enough higher than the dominant pole that the approximation is reasonable in this case. Or we could use (9.49) and the gain of our circuit to realize that our approximation is fine.

Exercise 9.11

Redo Example 9.6, using open-circuit time constants instead of the Miller approximation.

□

9.2.3 A FET Implementation: The Common-Source Amplifier

Introductory Treatment Consider the common-source amplifier shown in Figure 9-35. We already analyzed the DC biasing and small-signal midband AC performance of this circuit in Section 8.2 (see Figure 8-43), where we determined the bias voltages and currents shown in the figure. We assumed in that section that $K = 500\ \mu A/V^2$ and $V_{th} = 1$ V, so we will make the same assumptions here. (These parameters are for the Nmedium_mos model in Appendix B.)

We wish to find both the low and high cutoff frequencies of this circuit. To find the low cutoff frequency, we first draw the small-signal low-frequency AC equivalent circuit, as shown in Figure 9-36.

We can derive the transfer function for the circuit in Figure 9-36 by breaking it into a product of simpler transfer functions as usual:

$$\frac{V_o(j\omega)}{V_{sig}(j\omega)} = \frac{V_o(j\omega)}{V_{gs}(j\omega)} \frac{V_{gs}(j\omega)}{V_g(j\omega)} \frac{V_g(j\omega)}{V_{sig}(j\omega)}. \tag{9.75}$$

The sequence of transfer functions in (9.75) is not unique, but it will allow us to write each transfer function by inspection. All three of the intermediate transfer

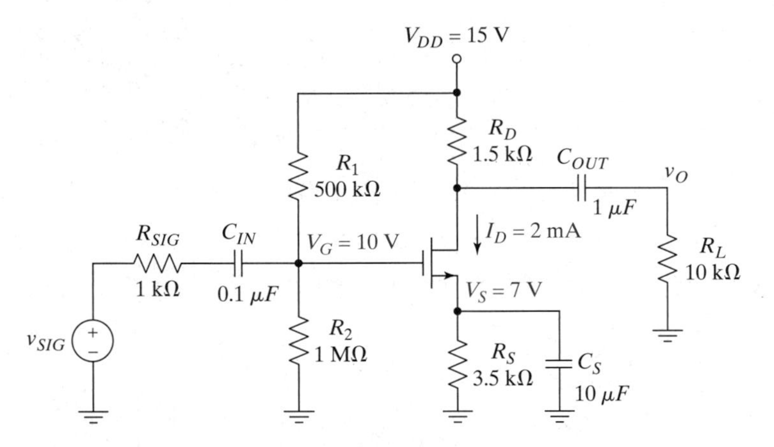

Figure 9–35 A common-source amplifier.

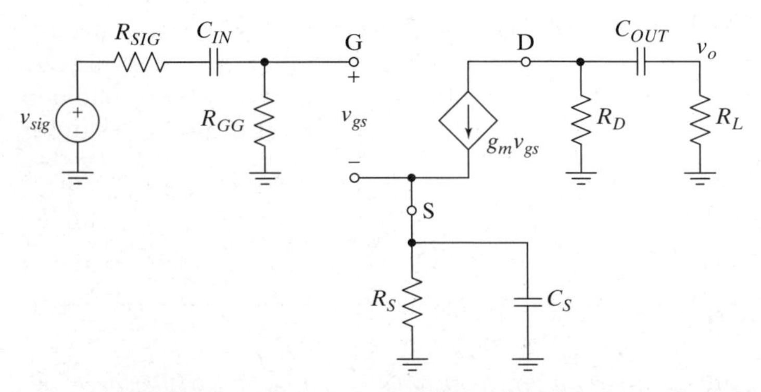

Figure 9–36 The small-signal low-frequency AC equivalent circuit for the common-source amplifier of Figure 9-35.

functions in this equation are single-pole high-pass transfer functions with DC zeros. Therefore, using the methods of Aside A1.6, we find that

$$\frac{V_g(j\omega)}{V_{sig}(j\omega)} = \frac{j\omega\left(\dfrac{R_{GG}}{R_{GG} + R_{SIG}}\right)}{j\omega + \dfrac{1}{(R_{GG} + R_{SIG})C_{IN}}} \tag{9.76}$$

and

$$\frac{V_o(j\omega)}{V_{gs}(j\omega)} = \frac{-j\omega g_m(R_D \| R_L)}{j\omega + \dfrac{1}{(R_D + R_L)C_{OUT}}}. \tag{9.77}$$

To find $V_{gs}(j\omega)/V_g(j\omega)$, we need to find the driving-point resistance for C_S. Doing this requires finding the resistance seen looking back into the source of the FET. Starting from Figure 9-36, we set v_{sig} to zero and arrive at the circuit shown in Figure 9-37(a), where the values of R_{eq1} and R_{eq2} depend on whether we open or short circuit C_{IN} and C_{OUT}. We will see in a moment that neither of these resistors matters, so we defer any further consideration of them. We next recognize that since the gate current is zero, the voltage across R_{eq1} is zero, and this resistor can be removed without affecting the circuit. Using the split-source transformation on the controlled generator, we arrive at the equivalent circuit shown in part (b) of the figure, where R_{eq2} clearly has no effect on R_{Ss}. The current source is now controlled by the voltage across itself and, according to the source absorption theorem, can be replaced by an equivalent resistance of value $1/g_m$. Since R_{eq1} and R_{eq2} never show up in the final result, it doesn't matter what we do with C_{IN} and C_{OUT}. Therefore, we can see that C_S is an independent capacitor, and there is a pole at the reciprocal of the time constant.

The resulting transfer function is

$$\frac{V_{gs}(j\omega)}{V_g(j\omega)} = \frac{j\omega}{j\omega + \dfrac{1}{\left(\dfrac{1}{g_m} \| R_S\right)C_S}}. \tag{9.78}$$

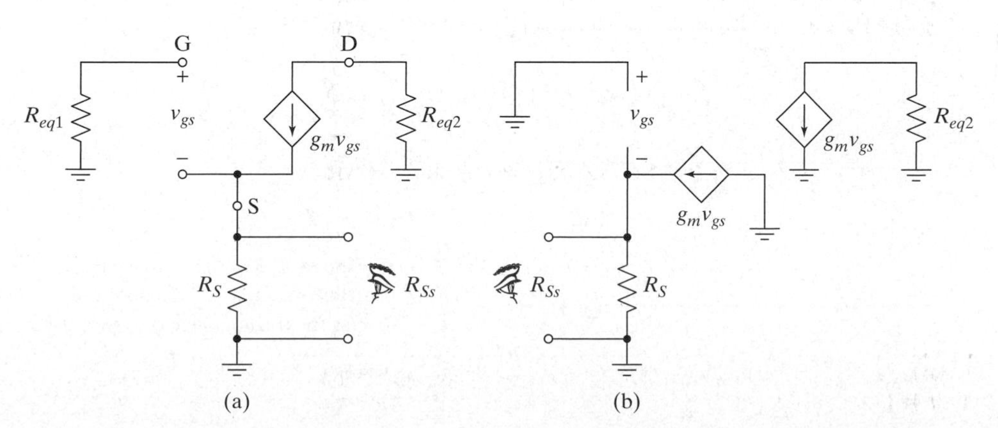

Figure 9–37 Finding R_{Ss} for the circuit in Figure 9-36.

Combining (9.76), (9.77), and (9.78) yields the complete transfer function for the small-signal low-frequency AC equivalent circuit. By setting the magnitude of this transfer function equal to $1/\sqrt{2}$, we can solve for ω_{cL}. In practice, we would rarely—if ever—perform this computation by hand. The log magnitude of the resulting transfer function would look like the one shown in Figure 9-38. This response is the low-frequency half of the overall AC-coupled response shown in Figure 9-10(a). The high-frequency asymptote is the midband gain of the original circuit. Because our equivalent circuit does not include the capacitors that lead to the high-frequency roll-off in the original circuit, the transfer function of this equivalent circuit does not roll off at high-frequencies.

We now turn our attention to finding the high cutoff frequency of the circuit in Figure 9-35. The small-signal high-frequency AC equivalent circuit is shown in Figure 9-39. We reiterate that because this is a *small-signal* equivalent circuit, we use a linear model for the transistor. Because it is the *high-frequency* equivalent circuit, we short out all of the coupling and bypass capacitors, as we did for the midband equivalent circuit. However, we include the parasitic capacitors that are part of the high-frequency hybrid-π model for the transistor. Since this is an *AC* equivalent circuit, the DC power supply appears as a ground. The effective load resistance is given by

$$R'_L = R_L \| R_D. \tag{9.79}$$

The circuit in Figure 9-39 is topologically identical to the high-frequency equivalent for the common-merge amplifier shown in Figure 9-12(c). Therefore, the analysis of the present circuit parallels the earlier development exactly. Assuming that the current through C_{gd} is negligible, we momentarily ignore that capacitor and write

$$\frac{v_o}{v_{gs}} = -g_m R'_L, \tag{9.80}$$

which is the midband gain. Using this gain along with the formulas in (A9.14) and (A9.15), we can redraw the circuit as shown in Figure 9-40. The overall input capacitance is C_{gs} in parallel with the Miller input capacitance given by

$$C_{Min} = (1 + g_m R'_L) C_{gd}, \tag{9.81}$$

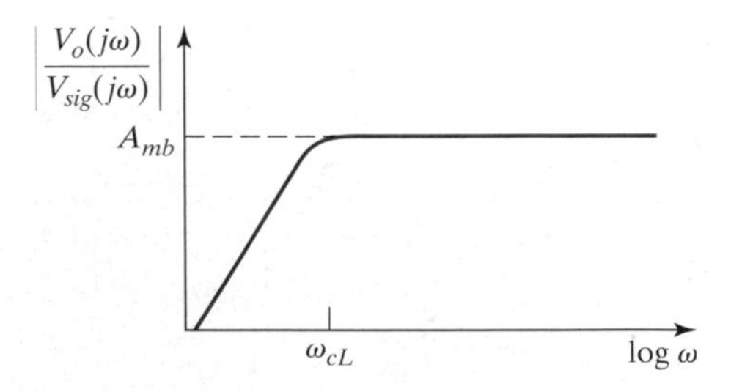

Figure 9–38 The log-magnitude response of the small-signal low-frequency AC equivalent circuit of Figure 9-36.

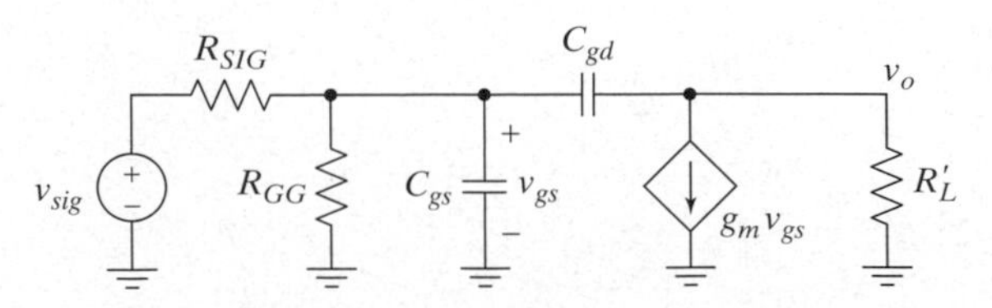

Figure 9–39 The small-signal high-frequency AC equivalent circuit for the amplifier of Figure 9-35.

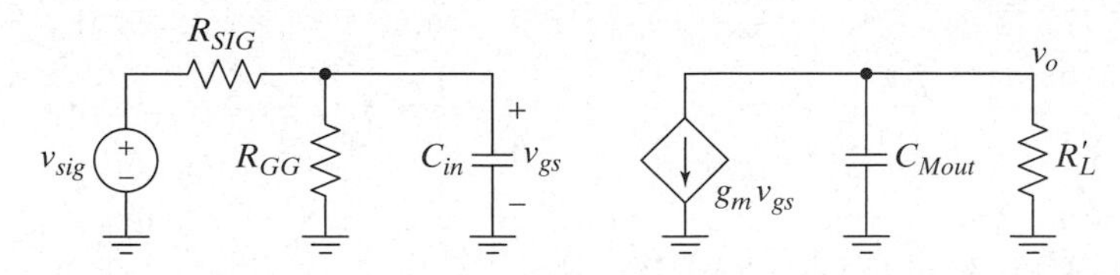

Figure 9–40 The equivalent circuit from Figure 9-39 after application of Miller's theorem.

so that

$$C_{in} = C_{gs} + (1 + g_m R'_L)C_{gd}. \quad \textbf{(9.82)}$$

For MOSFET amplifiers, C_{gd} is much smaller than C_{gs}, and the gain is often low. Therefore, the input capacitance is often not dominated by C_{Min}. Nevertheless, the Miller effect is still important, and ignoring it would certainly lead to erroneous results.

The circuit in Figure 9-40 is again topologically identical to the common-merge equivalent of Figure 9-13. As before, we must resist the temptation to treat this circuit as two independent single-pole circuits, since C_{Min} and C_{Mout} are both derived from the same capacitor. We next recognize that C_{Mout} is used only to determine whether or not the Miller approximation is valid. Using the Miller approximation, we ignore C_{Mout} and find that ω_{cH} is about equal to the pole associated with C_{in}:

$$\omega_{cH} \approx \frac{1}{(R_{SIG} \| R_{GG})C_{in}}. \quad \textbf{(9.83)}$$

We can check to see whether or not the Miller approximation is acceptable by using (9.34). The Miller output capacitance is

$$C_{Mout} = (1 + 1/g_m R'_L)C_{gd} \approx C_{gd}, \quad \textbf{(9.84)}$$

and the resistance seen by C_{Mout} is R'_L. So, we check to see whether or not $1/\omega_{cH}$ is much greater than $R'_L C_{Mout}$. If it is, the approximation is reasonable.

Example 9.7

Problem:
Find the high cutoff frequency for the amplifier in Figure 9-35, assuming that the transistor has $K = 500\ \mu A/V^2$, $V_{th} = 1$ V, and $I_D = 2$ mA, as before. In addition, assume that $C_{gs} = 2.5$ pF and $C_{gd} = 0.25$ pF. (These parameters are for the Nmedium_mos model in Appendix B.)

Solution:
Using the known DC bias point, we calculate

$$g_m = 2\sqrt{KI_D} = 2 \text{ mA/V}.$$

From (9.79) and the circuit, we see that

$$R'_L = R_L \| R_D = 1.3 \text{ k}\Omega.$$

Therefore,

$$A_v = \frac{v_o}{v_i} = -g_m R'_L = -2.6,$$

as in Chapter 8. Using this result in (9.82), we find that

$$C_{in} = C_{gs} + (1 + g_m R'_L)C_{gd} = 3.4 \text{ pF}.$$

We finally make use of (9.83) to obtain

$$\omega_{cH} \approx \frac{1}{(R_{SIG} \| R_{GG})C_{in}} = 295 \text{ Mrad/s},$$

or $f_{cH} \approx 47$ MHz.

The last step is to check that the Miller approximation is reasonable for this circuit. To do that, we use (9.84) to find $C_{Mout} \approx C_{gd} = 0.25$ pF, and we use (9.34) to check that $1/\omega_{cH} = 3.4$ ns is much greater than $R'_L\, C_{Mout} = 0.375$ ns. Since it is, the approximation is valid.

■

A SPICE simulation of the circuit reveals that $f_{cH} = 45.6$ MHz, which is less than 3% lower than our estimate. The main error in our estimate is caused by ignoring the pole on the drain (which we do not know; remember that it is *not* equal to $1/\tau_{Mout}$). Considering the simplicity of our result, however, the accuracy is incredibly good. It is not always this good.

Let's also examine the gain-bandwidth product of this amplifier. Using (9.80), (9.82), and (9.83), we get

$$\begin{aligned} \text{GBW} &= \left(\frac{R_{GG}}{R_{GG} + R_{SIG}}\right) \frac{g_m R'_L}{(R_{SIG} \| R_{GG})[C_{gs} + (1 + g_m R'_L)C_{gd}]} \\ &= \frac{g_m R'_L}{R_{SIG}[C_{gs} + (1 + g_m R'_L)C_{gd}]} = 767 \text{ Mrad/s}, \end{aligned} \tag{9.85}$$

where we have plugged in the numbers from Example 9.7. If the Miller effect dominates the input capacitance, the C_{gs} term in (9.85) can be ignored, as can the '1.' The result is that the midband gain cancels out, and we have

$$\text{GBW} \approx \frac{1}{R_{SIG} C_{gd}}. \tag{9.86}$$

We can clearly see from this equation that decreasing C_{gd} or R_{SIG} is beneficial, since it increases the GBW. This statement is still true for the exact relationship given in (9.85), but depending on the values of the other components, C_{gd} may not have a dominant effect.

Exercise 9.12

Consider the MOS amplifier shown in Figure 9-41. Assume that the source has a DC component $V_{SIG} = 1.5$ V and that the transistor has $V_{th} = 0.5$ V, $K = 100\ \mu A/V^2$, $\lambda = 0\ V^{-1}$, $C_{gs} = 2$ pF and $C_{gd} = 0.5$ pF. (These parameters are for the Nsimple_mos model in Appendix B.) (a) Find the low cutoff frequency. (b) Find the high cutoff frequency. Does the Miller effect dominate? (c) Find the GBW.

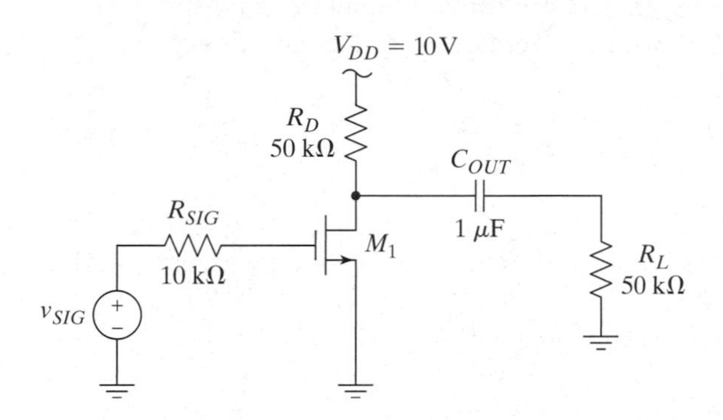

Figure 9–41 A MOS common-source amplifier.

Advanced Treatment We now want to find the bandwidth of the common-source amplifier in Figure 9-35 by making use of open- and short-circuit time constant analysis. The small-signal low-frequency AC equivalent circuit was shown in Figure 9-36. We also calculated the time constants for each of the three capacitors in that section. Since the three capacitors are all independent, we know the three pole frequencies and could calculate the cutoff frequency. Nevertheless, the time constants found there are also what we would get if we found the short-circuit time constants, so we can use the dominant-pole approximation to find a value for ω_{cL}. If we denote those three time constants by the products of their respective short-circuit driving-point resistances and capacitances, then using (9.39) yields

$$\omega_{cL} \approx \frac{1}{R_{INs}C_{IN}} + \frac{1}{R_{Ss}C_S} + \frac{1}{R_{OUTs}C_{OUT}}. \quad \textbf{(9.87)}$$

Examining (9.76), (9.77), and (9.78) we see that

$$\tau_{INs} = R_{INs}C_{IN} = (R_{SIG} + R_{GG})C_{IN} = 33.4 \text{ ms}, \quad \textbf{(9.88)}$$

$$\tau_{Ss} = R_{Ss}C_S = \left(\frac{1}{g_m} \Big\| R_S\right)C_S = 4.4 \text{ ms}, \quad \textbf{(9.89)}$$

and

$$\tau_{OUTs} = R_{OUTs}C_{OUT} = (R_D + R_L)C_{OUT} = 11.5 \text{ ms}. \quad \textbf{(9.90)}$$

Equation (9.87) is wonderful for design; each term is readily identified with a particular circuit component, and the individual terms were all derived in a straightforward way. Plugging in the component values given in Figure 9-35 along with the device parameters from Example 9.7 results in a cutoff frequency of $f_{cL} \approx 55$ Hz, which is dominated by the time constant associated with C_S. A SPICE analysis of this circuit yields a low cutoff of about 40 Hz (Exam7.sch), which is close to, but lower than, our prediction, so our prediction was conservative, as stated earlier. As a final step, we should check that the zeros in the low-frequency equivalent circuit are all low enough to be safely ignored. We know they must be, since our simulation results agree so well, but we could also use (9.44) and recognize that so long as $R_S \gg 1/g_m$, which is certainly true here, our approximation is fine.

Let's now find an approximation to ω_{cH}, using the open-circuit time-constant method. The small-signal high-frequency AC equivalent circuit for the amplifier was shown in Figure 9-39 and is repeated here for convenience as Figure 9-42.

Using (9.46), we find that the high cutoff frequency for this circuit is given by

$$\omega_{cH} \approx \frac{1}{R_{gso}C_{gs} + R_{gdo}C_{gd}}. \qquad \textbf{(9.91)}$$

With the signal source set equal to zero and C_{gd} opened, we find by inspection that

$$R_{gso} = R_{GG} \| R_{SIG} = 997\ \Omega. \qquad \textbf{(9.92)}$$

To find the open-circuit driving-point resistance seen by C_{gd}, we use the circuit shown in Figure 9-43. This circuit is derived from the one in Figure 9-42 by setting the signal source to zero and open-circuiting C_{gs}.

Writing KCL at the drain, we find that

$$i_o = i_x + g_m v_{gs} = i_x(1 + g_m R_{gso}), \qquad \textbf{(9.93)}$$

and writing KVL around the outside loop, we get

$$v_x = i_x R_{gso} + i_o R'_L. \qquad \textbf{(9.94)}$$

Combining these two equations yields the desired resistance:

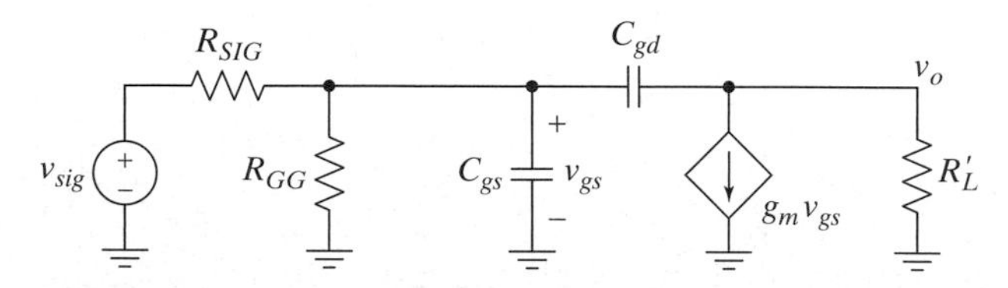

Figure 9–42 The small-signal high-frequency AC equivalent circuit for the amplifier of Figure 9-35.

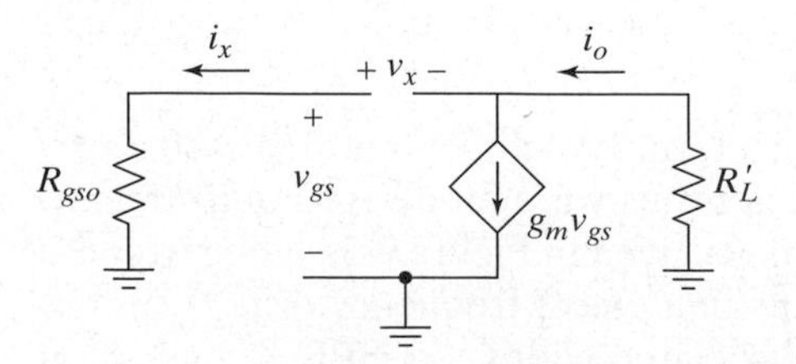

Figure 9–43 The circuit for finding the open-circuit driving-point resistance seen by C_{gd}.

$$R_{gdo} = \frac{v_x}{i_x} = R_{gso} + (1 + g_m R_{gso})R'_L = 4.9\ \text{k}\Omega. \tag{9.95}$$

Finally, using these numbers along with the capacitances given in Example 9.7, we use (9.91) to find

$$f_{cH} \approx \frac{1}{2\pi(R_{gso}C_{gs} + R_{gdo}C_{gd})} = \frac{1}{2\pi(2.5\ \text{ns} + 1.23\ \text{ns})} = 42.8\ \text{MHz}. \tag{9.96}$$

Because of the Miller effect, the time constant associated with C_{gd} has a much larger effect than you might have expected based on the relative values of the capacitors. As noted earlier, a SPICE simulation of this circuit shows that $f_{cH} \approx 45.7$ MHz (Exam7.sch), which is again fairly close to our prediction and shows the prediction to be conservative. Since our results agree well with SPICE, we could stop here and assume that the zeros must be enough higher than the dominant pole that the approximation is reasonable, or we could use (9.49) and the gain of our circuit to check our approximation. (It is fine in this case.)

Exercise 9.13

Consider again the MOS amplifier of Exercise 9.12. Calculate the approximate value of f_{cH}, using open-circuit time constants. Compare your results with the simulated value of 2.7 MHz. Which approximation—open-circuit time constants or Miller—is more accurate? Why?

□

9.2.4 Comparison of Bipolar and FET Implementations

The major differences between bipolar and field-effect transistor amplifiers were presented in Section 8.8. In the current brief section, we want to add a few comments relating to frequency response. Since FETs do not have minority-carrier charge storage as a part of the fundamental operation of the device (i.e., there isn't any diffusion capacitance in FETs), they sometimes have smaller parasitic capacitances than similar-size bipolar transistors at the same bias currents. The gains of FET amplifiers are much lower than those of similar bipolar amplifiers at the same current, however, as has already been pointed out in Section 8.8. Therefore, the GBW of bipolar amplifiers is usually significantly larger than that of comparable FET amplifiers.

If we restrict our discussion to just MOSFET and bipolar transistors (i.e., excluding JFETs and MESFETs), a more esoteric difference shows up. The parasitic capacitances in a bipolar transistor are all voltage dependent, whereas the major capacitances in a MOSFET are not. To be specific, C_π and C_μ are both rather strongly bias dependent, whereas C_{gs} and C_{gd} are due to oxide capacitance and are not, to first order, bias dependent. The major implication of this fact is that MOSFETs have less distortion than bipolar transistors, since bias-dependent capacitors are nonlinear elements.

Because MOSFETs have essentially zero DC gate current and, less importantly, because the gate capacitance is linear, we sometimes deliberately store charge on the gate of a MOSFET and use it as an analog memory element. The major application for this technique is sample-and-hold circuits, which are presented in Chapter 13, although separate capacitors are still used most of the time.

9.3 VOLTAGE BUFFERS

We introduced voltage buffers in Section 8.3 and analyzed their small-signal mid-band AC performance there. In the current section, we investigate their frequency response. We will find that, in general, the high cutoff frequency of voltage buffers significantly exceeds that of stages with both voltage and current gain, so when these amplifiers are cascaded, the buffer usually does not limit the overall bandwidth.

9.3.1 A Generic Implementation: The Merge Follower

Consider the merge follower shown in Figure 9-44. Because these buffers are most frequently biased by the preceding amplifier stage, we have omitted any separate biasing at the control terminal and have assumed that the signal source shown is the Thévenin equivalent for the output of the previous stage. As a result, the source contains a DC bias term in addition to the signal and has a relatively large Thévenin resistance, which is the output resistance of the preceding stage.

Since we are assuming that the merge follower is DC coupled to the preceding stage, the only capacitor present in the low-frequency equivalent circuit is C_{OUT}, as shown in Figure 9-45. This is a single-pole circuit, and we can use the method of Aside A1.6 to write the transfer function. The high-frequency gain is found by shorting C_{OUT}, which leads to a circuit identical to the one in Figure 8-53(b) if we replace R_L by R_L', defined to be the parallel combination of R_L and R_M.

Using (8.138) from Chapter 8 we have

$$\frac{V_o(j\infty)}{V_s(j\infty)} = \frac{(1 + g_m r_{cm})R_L'}{R_S + r_{cm} + (1 + g_m r_{cm})R_L'} = \frac{(1 + a_{im})R_L'}{R_S + r_{cm} + (1 + a_{im})R_L'}. \quad \textbf{(9.97)}$$

To find the pole frequency we set v_s to zero, remove C_{OUT} and look back into the terminals where it was connected. The resistance seen by C_{OUT} is R_L in series with the parallel combination of R_M and the resistance seen looking back into the merge terminal—call it R_m. This resistance was given by (8.142) in Chapter 8 to be,

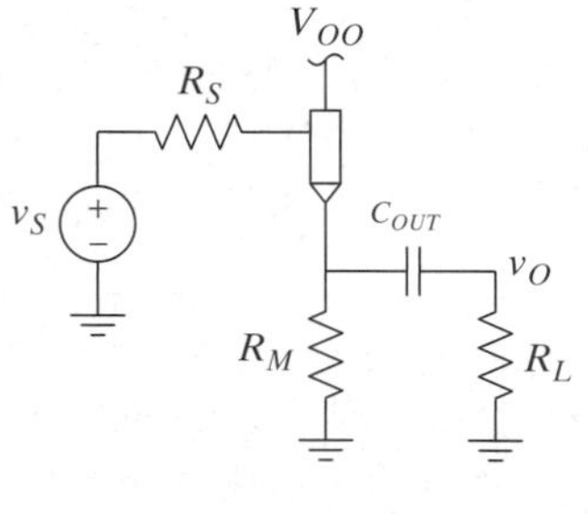

Figure 9–44 A merge follower driven by the Thévenin equivalent of the preceding amplifier stage.

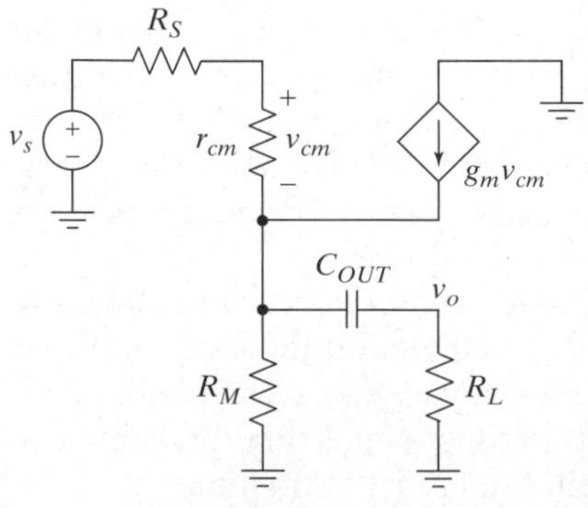

Figure 9–45 The small-signal low-frequency AC equivalent circuit for the buffer in Figure 9-44.

$$R_m = \frac{R_S + r_{cm}}{1 + g_m r_{cm}} = \frac{R_S + r_{cm}}{1 + a_{im}}. \tag{9.98}$$

The overall low-frequency transfer function is then given by

$$\frac{V_o(j\omega)}{V_s(j\omega)} = \frac{V_o(j\infty)}{V_s(j\infty)} \frac{j\omega}{j\omega + \omega_p}, \tag{9.99}$$

where

$$\omega_p = \frac{1}{\left[R_L + \left(R_M \middle\| \frac{R_S + r_{cm}}{1 + a_{im}}\right)\right] C_{OUT}}. \tag{9.100}$$

We know that the zero of this transfer function is at DC, since that is where C_{OUT} becomes an open circuit, and v_o will go to zero. Therefore, ω_{cL} is approximately equal to the pole frequency given by (9.100).

Exercise 9.14

Find f_{cL} for the circuit in Figure 9-44, given that $C_{OUT} = 10\ \mu F$, $R_S = 10\ k\Omega$, $R_M = 360\ \Omega$, $R_L = 1\ k\Omega$, $r_{cm} = 1.5\ k\Omega$, and $g_m = 20\ mA/V$.

We now turn our attention to finding an estimate of the high cutoff frequency for this circuit. An introductory treatment of the subject is given first, followed by a more advanced discussion that uses open-circuit time constants.

Introductory Treatment The small-signal high-frequency AC equivalent for the circuit of Figure 9-44 is shown in Figure 9-46(a). This circuit has two poles in its transfer function. Deriving the exact transfer function would require a fair amount of algebra and would yield a rather complicated result while not lending much insight into the operation of the circuit. We will take a different tack and see if we can arrive at a reasonable and intuitive approximation to the response. We start by noting that C_{oc} shunts the input to ground and will therefore contribute to

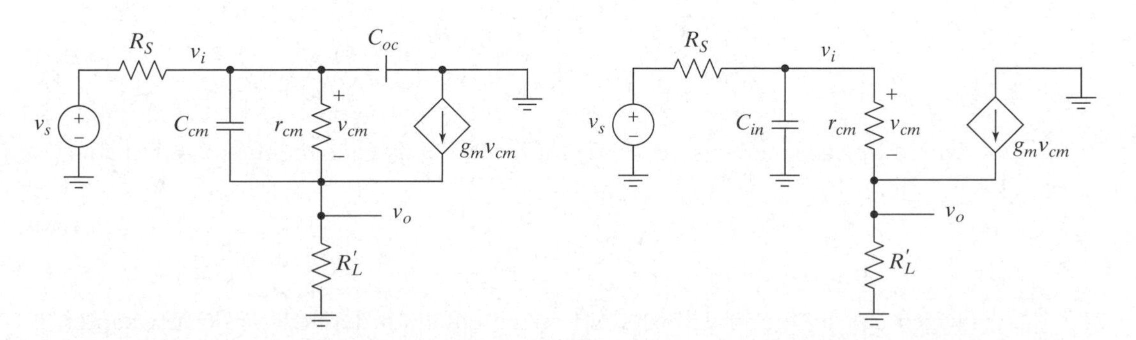

Figure 9–46 (a) The small-signal high-frequency AC equivalent for the buffer in Figure 9-44. (b) After applying Miller's approximation.

a low-pass characteristic with a zero at infinity (i.e., C_{oc} will cause the gain to go to zero only when $\omega = \infty$, which is where the capacitive reactance is zero). We also note that C_{cm} is connected between the input and the output of the amplifier; therefore, we consider using Miller's theorem. Assuming that the gain from input to output is known, we can use (A9.14) and (A9.15) to split C_{cm} up into two components (remember that $Z_C = 1/sC$):

$$C_{Min} = C_{cm}(1 - A_v), \quad \textbf{(9.101)}$$

and

$$C_{Mout} = C_{cm}\left(1 - \frac{1}{A_v}\right). \quad \textbf{(9.102)}$$

If we assume for the moment that we can use the midband gain as a reasonable approximation, the gain we desire is found by using (9.97) with R_S set equal to zero (the midband gain *is* the high-frequency gain of the low-frequency equivalent circuit):

$$A_v = \frac{v_o}{v_i} = \frac{(1 + a_{im})R_L'}{r_{cm} + (1 + a_{im})R_L'} = \frac{R_L'}{r_m + R_L'}. \quad \textbf{(9.103)}$$

This gain is less than, but usually close to, unity. The Miller input capacitance now appears in parallel with C_{oc}, and we denote the combination C_{in} as shown in Figure 9-46(b):

$$C_{in} = C_{oc} + C_{Min}. \quad \textbf{(9.104)}$$

We will discuss the Miller output capacitance later.

We note from (9.101) that if the gain of the follower is close to one, the Miller input capacitance will be much smaller than C_{cm}. This fact makes good sense intuitively: If the gain from v_i to v_o were exactly equal to unity, there wouldn't be any change in the voltage across C_{cm} at all, and the current through it would be zero. Therefore, when the gain is close to unity, we expect very little current in C_{cm}, and the resulting Miller input capacitance is small. The circuit in Figure 9-46(b) has a single-pole low-pass transfer function, which can be written using the method of Aside A1.6. We have

$$\frac{V_o(j\omega)}{V_s(j\omega)} = \left(\frac{(1 + a_{im})R_L'}{R_S + r_{cm} + (1 + a_{im})R_L'}\right)\frac{j\omega}{j\omega + \omega_{cH}}, \quad \textbf{(9.105)}$$

where we have made use of (9.97) for the midband gain and the pole frequency is

$$\omega_{cH} \approx \frac{1}{(R_S \| [r_{cm} + (1 + a_{im})R_L'])C_{in}}, \quad \textbf{(9.106)}$$

which can be found with the aid of the equation derived for R_i in Chapter 8, (8.139). This result, which is based on the Miller approximation as used before, will be a reasonable approximation to the actual frequency response, as long as we are justified in using the midband gain to find the Miller input capacitance. We cannot use (9.34) to check our approximation, since this circuit has a different topology, but we can use a similar method. If using the midband gain is justified, not only is C_{Min} accurate, but ignoring C_{Mout} is also reasonable. If C_{Mout} matters, the gain from v_i to v_o will not be equal to the midband gain when we reach the pole given by (9.106).

The midband gain is calculated by open-circuiting C_{cm}, so the current out of the merge terminal is equal to the current through r_{cm}, plus the current from the controlled source; that is, $i_m = (1 + a_{im})v_{cm}/r_{cm}$ at midband. We can check the approximation by checking to see whether or not the current through C_{cm} is negligible in comparison to the midband value of i_m. The current through C_{cm} is equal to $j\omega C_{cm}V_{cm}$. Setting the magnitude of this current to be much less than the midband value of i_m we find that the current in C_{cm} is negligible as long as

$$|j\omega C_{cm}|V_{cm} \ll (1 + a_{im})\frac{V_{cm}}{r_{cm}} \tag{9.107}$$

and

$$\omega \ll \frac{(1 + a_{im})}{r_{cm}C_{cm}}.$$

In other words, the approximation is valid so long as

$$\omega_{cH} \ll \frac{(1 + a_{im})}{r_{cm}C_{cm}}, \tag{9.108}$$

where ω_{cH} is given by (9.106). For most buffers, the approximation will turn out to be a reasonable one.

Exercise 9.15

Find f_{cH} for the circuit in Figure 9-44, given that $R_S = 10\ k\Omega$, $R_M = 360\ \Omega$, $R_L = 1\ k\Omega$, $r_{cm} = 1.5\ k\Omega$, $g_m = 20\ mA/V$, $C_{cm} = 10\ pF$, and $C_{oc} = 1.5\ pF$.

Advanced Treatment We now want to find an approximation to the high cutoff frequency of the merge follower using the open-circuit time-constant method. We use the equivalent circuit in Figure 9-46(a) to calculate the two open-circuit time constants. The open-circuit driving-point resistance seen by C_{oc} is the same as the resistance seen by C_{in} in our analysis in the preceding section, so using the resistance from (9.106), we have

$$\tau_{oco} = R_{oco}C_{oc} = (R_S \| [r_{cm} + (1 + a_{im})R_L'])C_{oc}. \tag{9.109}$$

We find the resistance seen by C_{cm} by using the circuit shown in Figure 9-47. We arrive at this circuit by setting v_s to zero and open-circuiting C_{oc}. We have drawn the circuit in a somewhat unusual way to allow us to define the resistance R_x as shown.

We now have

$$R_{cmo} = r_{cm} \| R_x. \tag{9.110}$$

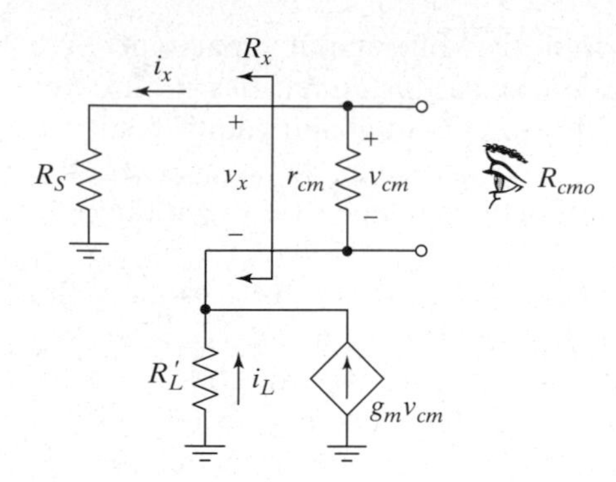

Figure 9–47 Finding the open-circuit driving-point resistance seen by C_{cm}.

We find R_x by first writing KCL at the top of the load:

$$i_L = i_x - g_m v_{cm} = i_x - g_m v_x. \tag{9.111}$$

We then write KVL,

$$v_x = i_x R_S + i_L R'_L, \tag{9.112}$$

and substitute (9.111) into (9.112) to get

$$R_x = \frac{v_x}{i_x} = \frac{R_S + R'_L}{1 + g_m R'_L}. \tag{9.113}$$

Now we have the time constant

$$\tau_{cmo} = R_{cmo} C_{cm} = \left[r_{cm} \middle\| \left(\frac{R_S + R'_L}{1 + g_m R'_L} \right) \right] C_{cm}, \tag{9.114}$$

and we can finally use (9.46) to write

$$\omega_{cH} \approx \frac{1}{\tau_{oco} + \tau_{cmo}} = \frac{1}{(R_S \| [r_{cm} + (1 + a_{im}) R'_L]) C_{oc} + \left[r_{cm} \middle\| \left(\dfrac{R_S + R'_L}{1 + g_m R'_L} \right) \right] C_{cm}}. \tag{9.115}$$

At this point, you can again appreciate the power of this method of analysis. While (9.115) looks imposing at first glance, each term is readily identifiable and is derived in a straightforward and meaningful way. You can readily imagine how this equation might look if it had been derived by using nodal analysis instead. While the result would be similar, the form of the equation would not be nearly as useful for design. If we go through the algebra to compare this result with our previous result from (9.106), which was derived by using the Miller approximation, we find that the term due to C_{oc} is identical, and the term due to C_{cm} differs by $(R_S + R'_L)/R_S$, with (9.115) being the lower approximation of the two. For most sets of reasonable component values, this difference will not be large.

The Miller approach has the advantage of making it intuitively clear why C_{cm} should have a small effect, whereas the open-circuit time-constant method is more conservative and a bit easier computationally. The open-circuit time-constant method also has the advantage of being applicable when the Miller approximation is not valid, or when there are more capacitors involved and the Miller approximation is insufficient.

Exercise 9.16

Repeat Exercise 9.15, using open-circuit time constants. Compare the results.

Finally, we need to check that the zeros in this circuit are all higher than ω_{cH}, or else our approximations will not be valid. We observed at the start of Section 9.3.1 that C_{oc} contributes a zero at infinity, so we are left to consider C_{cm}. First of all, C_{cm} will contribute some additional capacitance to ground from the control node (think of C_{Min} from our earlier analysis), but this will serve only to lower the frequency of the pole associated with C_{oc} (since C_{Min} shows up in parallel with C_{oc}). To investigate the zero caused by C_{cm}, we may consider just the gain from v_c to v_o. Therefore, consider the circuit shown in Figure 9-48, where we have used the T model because it makes the analysis simpler.

Because a perfect voltage source drives the control node, neither the controlled current source nor C_{oc} will have any effect on the transfer function. The transfer function is given by the impedance divider:

$$\frac{V_o(j\omega)}{V_c(j\omega)} = \frac{R'_L}{R'_L + \dfrac{r_m}{1 + j\omega r_m C_{cm}}}. \tag{9.116}$$

This result can be rewritten in our standard form for a high-pass transfer function (from Aside A1.6) with a value of one at infinity (when C_{cm} becomes a short circuit),

$$\frac{V_o(j\omega)}{V_c(j\omega)} = \frac{j\omega + \dfrac{1}{r_m C_{cm}}}{j\omega + \dfrac{r_m + R'_L}{r_m R'_L C_{cm}}} = \frac{j\omega + \dfrac{1}{r_m C_{cm}}}{j\omega + \dfrac{1}{(r_m \| R'_L) C_{cm}}}. \tag{9.117}$$

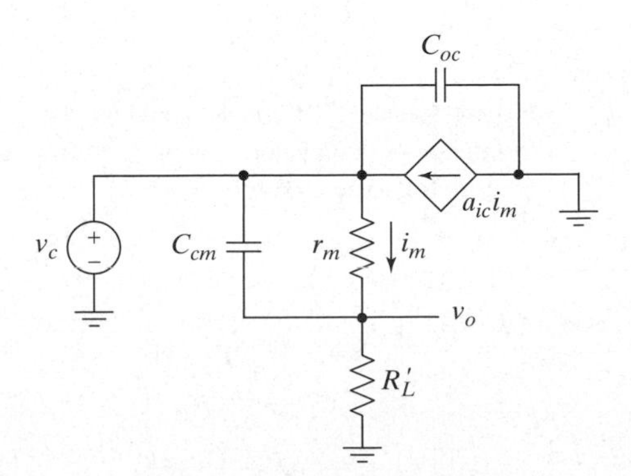

Figure 9–48 Examining the transfer function from v_c to v_o.

Examining this equation, we see two things. First, given that R'_L is usually much larger than r_m, we conclude that the pole and zero will be very close to each other. Second, because r_m is so small, both the pole and the zero are well above ω_{cH}. Therefore, our approximation should be fine in this case.

9.3.2 A Bipolar Implementation: The Emitter Follower

We can apply the results derived for the merge follower directly to the emitter follower. Consider the circuit shown in Figure 9-49, wherein the signal source and resistance represent the Thévenin equivalent of the preceding amplifier stage, so v_S includes a 5 V DC component.

A DC bias point analysis on this amplifier shows that $V_E = 4$ V and $I_C = 27$ mA. Using this bias point and performing a small-signal midband AC analysis yields $g_m = 1.04$ A/V, $r_\pi = 173\ \Omega$, $R_i = 11$ kΩ, $\alpha_i = 0.85$, $A_v = v_o/v_i = 0.98$, and $v_o/v_s = 0.84$. A SPICE simulation produces nearly identical results (Emitfol.sch)

The low cutoff frequency of this circuit can be found with the aid of the equation we derived for the merge follower, (9.100). Changing the variable names as appropriate and plugging in the numbers from our hand analysis of the bias point, we find that

$$f_{cL} \approx \frac{1}{2\pi\left[R_L + \left(R_E \left\| \frac{R_S + r_\pi}{\beta+1}\right.\right)\right]C_{OUT}} = 30 \text{ Hz}. \quad \textbf{(9.118)}$$

A SPICE simulation also yields $f_{cL} = 30$ Hz. The close agreement is to be expected for a single-pole response like this. We now turn our attention to the high cutoff frequency for this circuit.

Introductory Treatment The small-signal high-frequency AC equivalent circuit for the emitter follower in Figure 9-49 is shown in Figure 9-50 and, except for the component names, is identical to the corresponding figure for the merge follower. We have neglected r_b for this circuit. Using the bias point already found for this circuit and the model parameters for the 2N2222, we find that

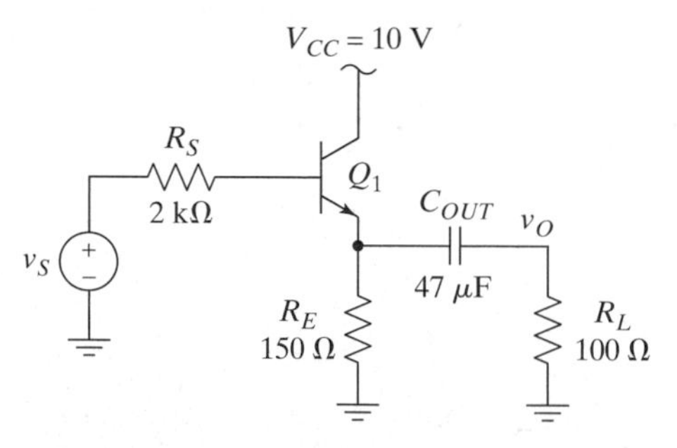

Figure 9–49 An emitter follower circuit. The model parameters for the 2N2222 transistor are given in Appendix B and were already used in Example 9.5.

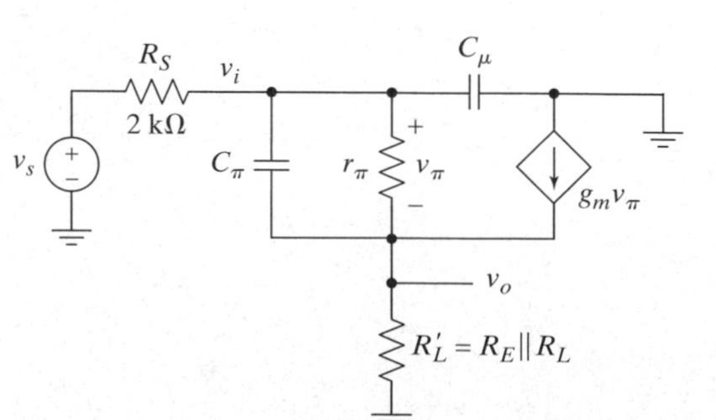

Figure 9–50 The small-signal high-frequency AC equivalent circuit for the emitter follower of Figure 9-49.

$$C_\pi \approx 2C_{je0} + g_m\tau_F = 471 \text{ pF} \quad (9.119)$$

and

$$C_\mu = \frac{C_{\mu 0}}{\left(1 + \dfrac{V_{CB}}{V_0}\right)^{m_{jc}}} = 3.6 \text{ pF.} \quad (9.120)$$

We can use the Miller approximation to estimate the high cutoff frequency for this circuit, as was done for the merge follower in Section 9.3.1. The gain across C_π can be found by using either the T model or impedance reflection and the hybrid-π model. With either method, the result is

$$A_v = \frac{v_o}{v_i} = \frac{R'_L}{r_e + R'_L} = \frac{(\beta+1)R'_L}{r_\pi + (\beta+1)R'_L} = 0.984. \quad (9.121)$$

The Miller input capacitance is

$$C_{Min} = (1 - A_v)C_\pi = 7.4 \text{ pF.} \quad (9.122)$$

The total equivalent capacitance to ground from the base is

$$C_{eq} = C_\mu + C_{Min} = 11.0 \text{ pF,} \quad (9.123)$$

and making use of (9.106), we see that the high cutoff frequency is approximately

$$f_{cH} \approx \frac{1}{2\pi(R_S \| [r_\pi + (\beta+1)R'_L])C_{eq}} = 8.5 \text{ MHz.} \quad (9.124)$$

Using (9.108) to check whether or not the Miller approximation is justified in this example, we find that

$$\omega_{cH} = 54 \text{ Mrad/s} \ll \frac{(\beta+1)}{r_\pi C_\pi} = 2.2 \text{ Grad/s,} \quad (9.125)$$

so the approximation should be quite good. A SPICE simulation of the circuit predicts that $f_{cH} = 8.4$ MHz (Emitfol.sch), very close to our result, as expected. We note that if R_S is low, it is necessary to include r_b in the analysis to obtain accurate results.

Advanced Treatment We now check the high cutoff frequency of the emitter follower, using the method of open-circuit time constants. Since the high-frequency equivalent for the emitter follower given in Figure 9-50 is topologically identical to the high-frequency equivalent for the merge follower given in Figure 9-46, we can use the results derived there directly. The high cutoff frequency is approximately given by

$$f_{cH} \approx \frac{1}{2\pi(\tau_{\mu o} + \tau_{\pi o})}. \quad (9.126)$$

From (9.109), we have

$$\tau_{\mu o} = (R_S \| [r_\pi + (\beta+1)R'_L])C_\mu = 6.1 \text{ ns,} \quad (9.127)$$

and from (9.114),

$$\tau_{\pi o} = \left[r_\pi \middle\| \left(\frac{R_S + R_L'}{1 + g_m R_L'} \right) \right] C_\pi = 12.9 \text{ ns.} \tag{9.128}$$

Using these values in (9.126) yields $f_{cH} \approx 8.4$ MHz, which is the same as predicted by SPICE. Although one of the time constants is more than twice as large as the other in this case, the agreement with SPICE is better than would normally be expected.

High-Frequency Emitter Followers[10] Capacitively loaded emitter followers can have very poorly damped step responses and can even oscillate for certain source impedances. Consider the circuit shown in Figure 9-51. We have represented the load as a parallel *RC* rather than just a capacitor, to yield more general results. Consequently, if the resistor is small enough, the response of the circuit will be well behaved.

The small-signal high-frequency equivalent circuit for this emitter follower is shown in Figure 9-52. The controlled source can be expressed as

$$g_m V_\pi(s) = g_m Z_\pi(s) I_b(s) = \frac{g_m r_\pi}{1 + s\tau_\pi} I_b(s) \tag{9.129}$$

$$= \frac{\beta_o}{1 + s\tau_\pi} I_b(s),$$

where $\tau_\pi = r_\pi C_\pi$. The impedance of the parallel *RC* combination in the emitter circuit is

$$Z_E(s) = \frac{R_E}{1 + s\tau_E}, \tag{9.130}$$

where $\tau_E = R_E C_E$.

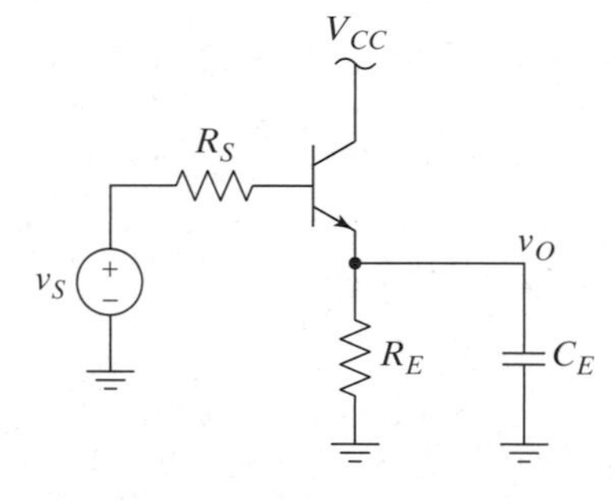

Figure 9–51 An emitter follower driving a capacitive load.

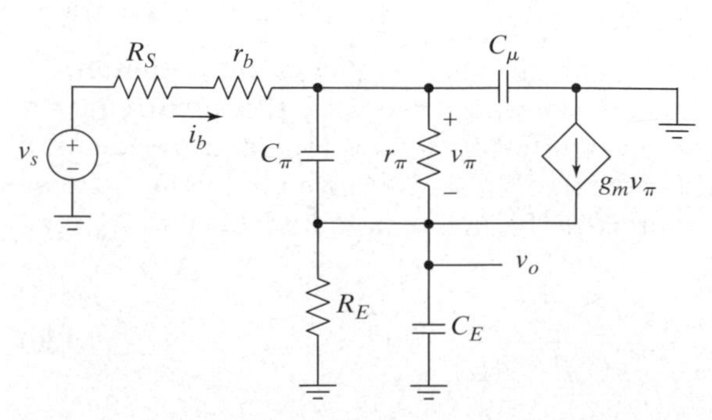

Figure 9–52 The small-signal high-frequency equivalent circuit for the emitter follower of Figure 9-51.

[10] This section is not necessary for any of the material that follows it.

Before proceeding with our analysis, we are going to simplify this circuit on the basis of prior knowledge of the problem that we wish to illustrate. We first note that C_μ forms a low-pass filter with R_S and r_b and tends to help the circuit behave properly. Consequently, we will ignore it. Also, the problem occurs for frequencies well above $1/(r_\pi C_\pi)$, which is slightly greater than ω_β, the –3 dB frequency of the current gain $\beta(j\omega)$. For frequencies significantly greater than $1/(r_\pi C_\pi)$, C_π has effectively shunted r_π, and we can crudely approximate the parallel combination by a short circuit. In addition, for these high frequencies, we can approximate the controlled source of (9.129) by

$$g_m V_\pi(s) \approx \frac{\beta_o}{s\tau_\pi} I_b(s) \approx \frac{I_b(s)}{s\tau_T}, \tag{9.131}$$

where

$$\tau_T = \frac{1}{\omega_T} = \frac{1}{\beta_o \omega_\beta} = \frac{r_\pi(C_\pi + C_\mu)}{\beta_o} \approx \frac{r_\pi C_\pi}{\beta_o}. \tag{9.132}$$

Making these two approximations and defining $R'_S = R_S + r_b$, we arrive at the high-frequency equivalent circuit shown in Figure 9-53(a). It is very important to note that this circuit is a valid approximation only for frequencies well above $1/(r_\pi C_\pi)$.

Writing KVL around the base-emitter loop in part (a) of the figure, we see that we can reflect the emitter impedance into the base as shown in part (b). This new circuit is equivalent in terms of the voltage drops and the current in the base. Using the circuit in part (b) of the figure, we can write

$$\frac{V_o(s)}{V_s(s)} = \frac{Z_E(s)\left(1 + \dfrac{1}{s\tau_T}\right)}{R'_S + Z_E(s)\left(1 + \dfrac{1}{s\tau_T}\right)}, \tag{9.133}$$

which, after some manipulation, can be rewritten as

$$\frac{V_o(s)}{V_s(s)} = \left(\frac{1}{R'_S\tau_T C_E}\right)\frac{(1 + s\tau_T)}{s^2 + s\dfrac{R'_S + R_E}{R'_S\tau_E} + \dfrac{1}{R'_S\tau_T C_E}}. \tag{9.134}$$

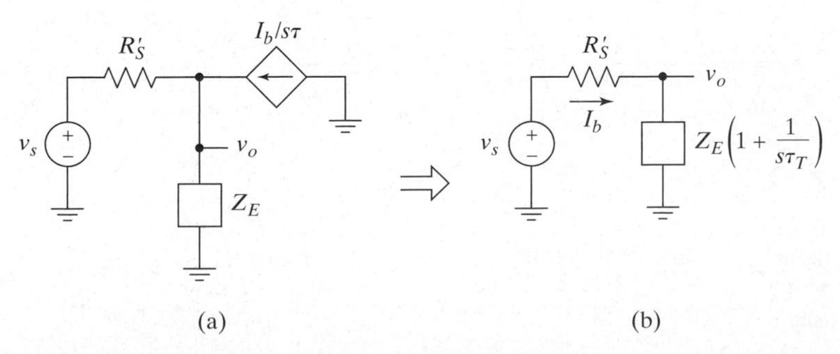

Figure 9–53 (a) The simplified equivalent for frequencies much greater than $1/(r_\pi C_\pi)$. (b) The circuit after reflecting the emitter impedance into the base.

The denominator of this transfer function can be written using the standard form of a second-order transfer function (*see* Section 10.1.5); that is

$$\frac{V_o(s)}{V_s(s)} = \left(\frac{1}{R_S'\tau_T C_E}\right)\frac{(1 + s\tau_T)}{s^2 + 2\zeta\omega_n s + \omega_n^2}, \tag{9.135}$$

where the undamped natural frequency is

$$\omega_n = \sqrt{\frac{1}{R_S'\tau_T C_E}} \tag{9.136}$$

and the damping coefficient is

$$\zeta = \frac{R_S' + R_E}{2}\sqrt{\frac{\tau_T}{R_S' R_E^2 C_E}}. \tag{9.137}$$

From (9.137), it is clear that large values of C_E lead to small damping coefficients and, therefore, to overshoot and ringing in the step response of the emitter follower. In addition, if there is some inductance in series with the source (some parasitic inductance is always present), it is possible for the circuit to have sustained oscillations. Figure 9-54 shows a circuit simulated in PSPICE with several different values for L_S (Ef_ring.sch). The transient simulation results when a step input was applied are shown in Figure 9-55.

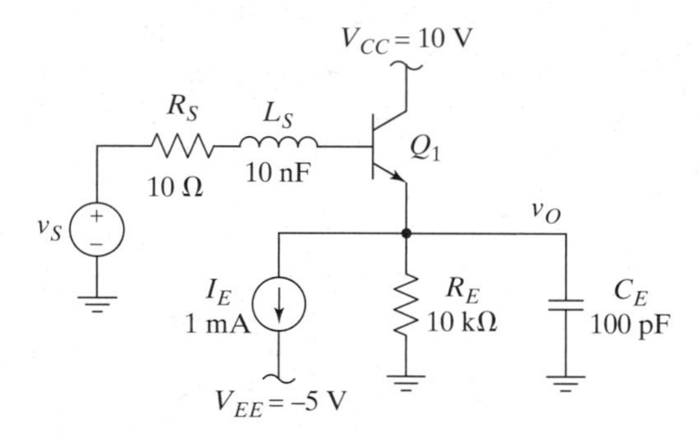

Figure 9–54 An emitter follower with capacitive load and inductive source.

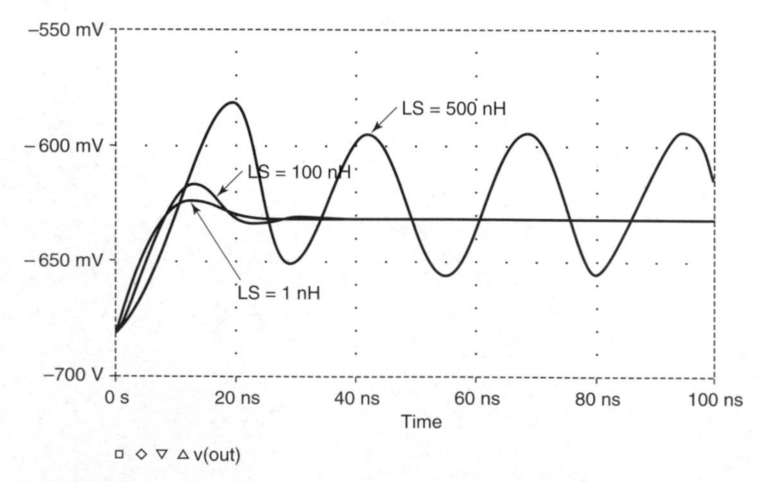

Figure 9–55 The transient simulation results for the circuit in Figure 9-54 when a step input is applied.

With the inductance set to zero in this circuit, (9.137) predicts that $\zeta = 0.28$, which leads to a peak overshoot of 40%, slightly more than is observed in the figure for small inductance. Although the peaking is fairly accurately predicted by this analysis, the ringing for small values of inductance is not quite as bad as would be expected. Nevertheless, the analysis properly predicts the problem and yields useful approximate results.

9.3.3 A FET Implementation: The Source Follower

Consider the circuit shown in Figure 9-56, where the signal source and resistance represent the Thévenin equivalent of a preceding amplifier stage, so v_{SIG} includes a 7-V DC component. Assume that the MOSFET has $K = 2$ mA/V^2, $V_{th} = 1$ V, $C_{gs} = 5$ pF, and $C_{gd} = 0.5$ pF. (These parameters are for the Nlarge_mos model in Appendix B.)

A DC bias point analysis on this amplifier shows that $V_{GS} \approx 3.24$ V, $V_S \approx 3.8$ V, and $I_D \approx 10$ mA. Using this bias point, we find that $g_m \approx 9$ mA/V. The small-signal low-frequency equivalent circuit for the source follower is shown in Figure 9-57. The T model is used for the transistor because it is more convenient for the analyses that we will perform here. (You are encouraged to repeat the analyses with the hybrid-π model and convince yourself of this fact.) The midband gain is found by shorting C_{OUT} and is (the voltage drop across R_{SIG} is zero because $i_g = 0$)

$$\frac{v_o}{v_{sig}} = \frac{R_S \| R_L}{R_S \| R_L + 1/g_m}. \tag{9.138}$$

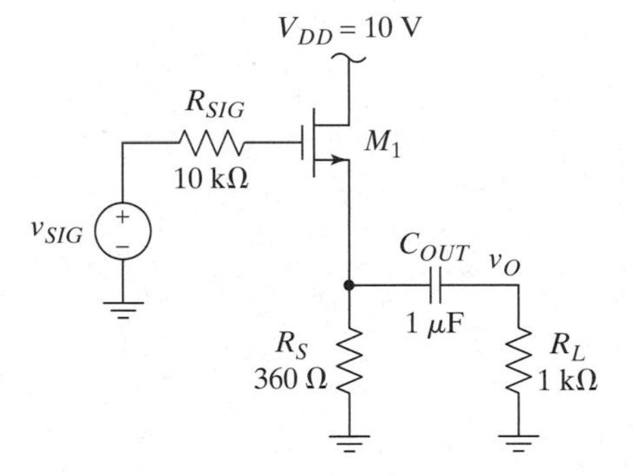

Figure 9–56 A source follower circuit.

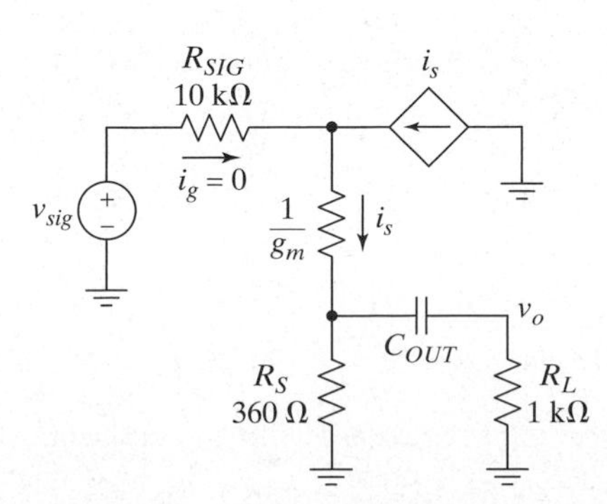

Figure 9–57 The small-signal low-frequency equivalent circuit for the source follower in Figure 9-56.

Plugging in the numbers for our example, we find that $v_o/v_{sig} = 0.7$. A SPICE simulation reveals that the gain is 0.71 (Sf.sch).

Because there is only one capacitor in this equivalent circuit, we can use the method of Aside A1.6 to find the pole frequency. We first set the signal source in Figure 9-57 to zero. Then, removing C_{OUT} and looking to the left, we see a total resistance to ground of $R_S\|(1/g_m)$, since the voltage drop across R_{SIG} is zero. Looking to the right from C_{OUT}, we see R_L to ground. Therefore, the total resistance seen by C_{OUT} is $[R_L + R_S\|(1/g_m)]$. The low cutoff frequency is then found to be

$$f_{cL} \approx \frac{1}{2\pi\left[R_L + R_S\left\|\frac{1}{g_m}\right.\right]C_{OUT}}, \quad \textbf{(9.139)}$$

in perfect analogy with (9.100) for the merge follower. Plugging in the numbers from our analysis of the bias point, we find that $f_{cL} = 147$ Hz. SPICE yields $f_{cL} =$ 146 Hz, which is very close to the predicted value. The close agreement is to be expected for a single-pole response like this.

We now turn our attention to the high cutoff frequency for this circuit.

Introductory Treatment The small-signal high-frequency AC equivalent circuit for this source follower is shown in Figure 9-58, and, except for the component names and the fact that r_{cm} is absent, it is identical to the corresponding figure for the merge follower, Figure 9-46.

We can use the Miller approximation to estimate the high cutoff frequency for this circuit, as was done for the merge follower in Section 9.3.1. The midband gain across C_{gs} is the gain already found in (9.138), since $v_i = v_{sig}$:

$$A_v = \frac{v_o}{v_i} = \frac{v_o}{v_{sig}} = \frac{R_S\|R_L}{R_S\|R_L + 1/g_m}. \quad \textbf{(9.140)}$$

Using this gain, we find that the Miller input capacitance is

$$C_{Min} = (1 - A_v)C_{gs} = 1.5 \text{ pF}. \quad \textbf{(9.141)}$$

The total equivalent capacitance to ground from the gate is then

$$C_{eq} = C_{gd} + C_{Min} = 2 \text{ pF}. \quad \textbf{(9.142)}$$

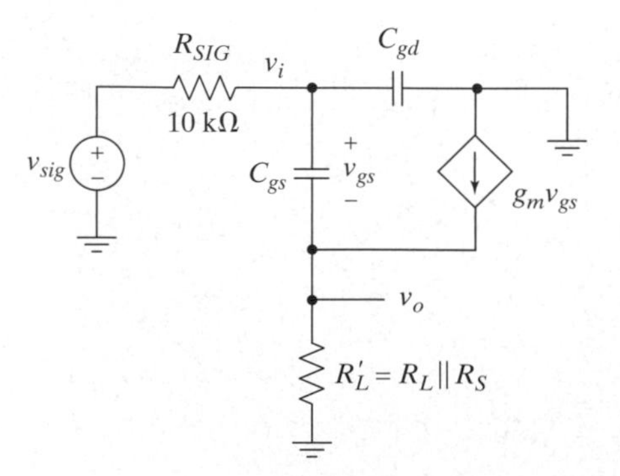

Figure 9–58 The small-signal high-frequency AC equivalent circuit for the source follower of Figure 9-56.

The resistance seen by C_{eq} is R_{SIG}, so the high cutoff frequency is approximately

$$f_{cH} \approx \frac{1}{2\pi R_{SIG} C_{eq}} = 8 \text{ MHz}. \quad \textbf{(9.143)}$$

This result is identical to the one derived for the merge follower in (9.106) if we set r_{cm} and a_{im} equal to infinity. We now check whether or not the Miller approximation is justified in this example by checking to see if the current fed forward through C_{gs} can be ignored in comparison with $g_m v_{gs}$ for frequencies near ω_{cH}. If it can be, then using the midband gain, where C_{gs} is open circuited, to find ω_{cH} is reasonable. The current through C_{gs} is $j\omega C_{gs} V_{gs}(j\omega)$, so this current can be safely ignored for frequencies near ω_{cH} if

$$|j\omega_{cH} C_{gs} V_{gs}(j\omega_{cH})| \ll |g_m V_{gs}(j\omega_{cH})|$$

or

$$\omega_{cH} \ll \frac{g_m}{C_{gs}}. \quad \textbf{(9.144)}$$

Plugging in the known values, we see that the approximation is valid so long as 5 Mrad/s $\ll 2.2 \cdot 10^{13}$ rad/s, which is certainly true. A SPICE simulation of the circuit predicts that $f_{cH} = 8$ MHz, in agreement with our prediction (Sf.sch).

Advanced Treatment We now use the method of open-circuit time constants to find the high cutoff frequency. Our analysis and results are perfectly analogous to those for the merge follower from Section 9.3.1. There are only two capacitors to consider, and the high cutoff frequency is approximately given by

$$f_{cH} \approx \frac{1}{2\pi(\tau_{gdo} + \tau_{gso})}. \quad \textbf{(9.145)}$$

The equivalent resistance seen by C_{gd} is the same as that seen by C_{eq} when one is using the Miller approximation; therefore,

$$\tau_{gdo} = R_{SIG} C_{gd} = 5 \text{ ns}. \quad \textbf{(9.146)}$$

The open-circuit time constant for C_{gs} is the same as that found for C_{cm} in the merge follower, except that $r_{cm} = \infty$ here. Therefore, using (9.114) with $r_{cm} = \infty$ and changing the notation appropriately, we obtain

$$\tau_{gso} = \left(\frac{R_{SIG} + R_L'}{1 + g_m R_L'}\right) C_{gs} = 15.3 \text{ ns}, \quad \textbf{(9.147)}$$

where $R_L' = R_L \| R_S$. Using these values in (9.145) yields $f_{cH} \approx 7.85$ MHz, which is less than 1% lower than the value determined by SPICE. Notice that the time constant due to C_{gs} dominates in determining the bandwidth, even though the Miller effect reduces the effective size of this capacitor. The capacitor still dominates for two reasons: It is much larger than C_{gd} to begin with, and the gain of this source follower is not very close to unity. Consequently, the Miller effect does not reduce the effective value of the capacitor too much.

9.4 CURRENT BUFFERS

We introduced current buffers in Section 8.4 and analyzed their small-signal mid-band AC performance there. In this section, we examine their bandwidth and demonstrate that the bandwidth of a current buffer is much larger than the bandwidth of a stage with both voltage and current gain.

9.4.1 A Generic Implementation: The Common-Control Amplifier

We will not discuss the low cutoff frequency for this stage, since the techniques used for the other stages apply equally well here. There aren't any important conclusions to be reached; in addition, this stage is rarely used in isolation and is most often DC coupled to the surrounding stages.

Since a common-control amplifier functions as a current buffer, as pointed out in Chapter 8, it is usually driven by a relatively high source impedance. Therefore, we use a Norton equivalent to represent the source in Figure 9-59, where we show the small-signal high-frequency AC equivalent circuit for a common-control amplifier. We also use the T model for the transistor, since it is the most convenient model to use here.

Introductory Treatment The circuit in Figure 9-59 is unilateral, so we can analyze it in a straightforward manner. In fact, the input and output sides are each single-pole circuits, and we can write their transfer functions by inspection. We note that, on the input side, we have a low-pass transfer function with a zero at infinity. Therefore, we write

$$\frac{I_m'(j\omega)}{I_s(j\omega)} = \left(\frac{-R_S}{R_S + r_m}\right)\frac{1}{1 + j\omega(R_S \| r_m)C_{cm}}. \tag{9.148}$$

Similarly, on the output side, we have

$$\frac{V_o(j\omega)}{I_m'(j\omega)} = \frac{-a_{ic}R_L}{1 + j\omega R_L C_{oc}}. \tag{9.149}$$

We combine these two equations to find the overall gain of the amplifier;

$$\frac{V_o(j\omega)}{I_s(j\omega)} = \frac{a_{ic}R_L\left(\dfrac{R_S}{R_S + r_m}\right)}{(1 + j\omega R_L C_{oc})(1 + j\omega(R_S \| r_m)C_{cm})}. \tag{9.150}$$

We can set the magnitude of (9.150) equal to $1/\sqrt{2}$ times its DC value and solve to find the high cutoff frequency.

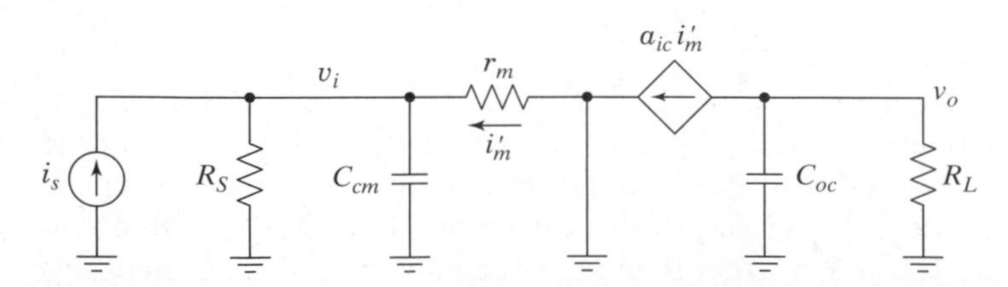

Figure 9–59 The small-signal high-frequency AC equivalent circuit for a common-control amplifier stage.

Advanced Treatment Since the circuit of Figure 9-59 comprises two single-pole circuits, there is little benefit to using the open-circuit time-constant method to approximate the high cutoff frequency. Nevertheless, we point out that the time constants for the capacitors have already been found in writing (9.148) and (9.149); they are

$$\tau_{cmo} = (R_S \| r_m)C_{cm} \tag{9.151}$$

and

$$\tau_{oco} = R_L C_{oc}. \tag{9.152}$$

Therefore, we can write

$$\omega_{cH} \approx \frac{1}{(R_S \| r_m)C_{cm} + R_L C_{oc}}. \tag{9.153}$$

This result will be reasonably accurate if one of the poles is dominant. If the two pole frequencies given by (9.148) and (9.149) are close to each other, however, it would be better to use (9.150) and solve for the cutoff frequency directly.

Note that the time constant in (9.151) is typically much smaller than the time constant for C_{cm} in the common-merge amplifier because r_m is so small. The corresponding time constant for the common-merge amplifier was found in Example 9.4 to be $\tau_{cmo} = R_{cmo}C_{cm} = (R_S \| R_{CC} \| r_{cm})C_{cm}$. Although R_S is typically smaller for that circuit, the overall time constant is usually much larger than (9.151). However, even more significant is the difference in τ_{oco} for the common-merge and common-control circuits. In Example 9.4 we found that, for the common-merge amplifier, $\tau_{oco} = [R_{cmo} + (1 + g_m R_{cmo})(R_O \| R_L)]C_{oc}$. This expression shows that the time constant includes the Miller multiplication of C_{oc} (i.e., the $g_m(R_O \| R_L)C_{oc}$ term), whereas (9.152) does not include this multiplication. Therefore, the bandwidth of the common-control stage is usually much higher than that of the common-merge stage, as originally claimed.

9.4.2 A Bipolar Implementation: The Common-Base Amplifier

Consider the common-base current buffer shown in Figure 9-60, where we have capacitively coupled the signal from a Norton equivalent source to allow us to easily bias the stage. A common-base stage would not usually be used in isolation like this, but doing so here will make the discussion more concrete without complicating the circuit.

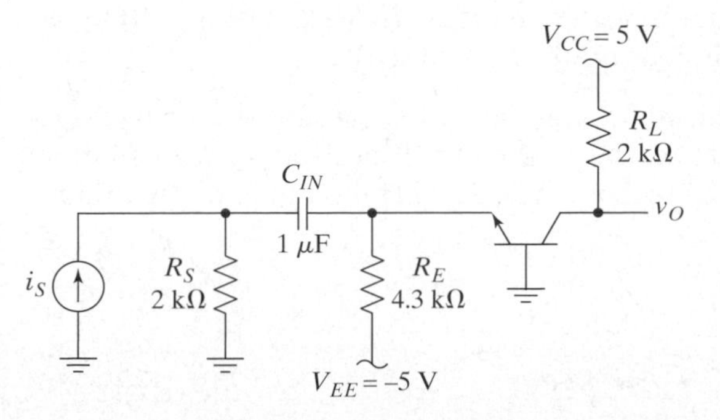

Figure 9–60 A common-base amplifier.

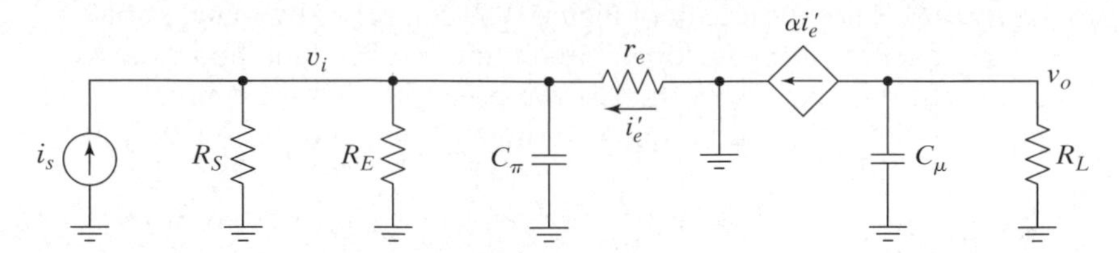

Figure 9–61 The small-signal high-frequency AC equivalent circuit for the amplifier in Figure 9-60.

An analysis of the operating point of this circuit reveals that $I_C \approx 1$ mA, $V_E \approx -0.7$ V, and $V_C \approx 3$ V. Assuming, that the transistor is a 2N2222 and using this operating point, we find that $C_\pi \approx 60$ pF, $C_\mu \approx 4.2$ pF, and $r_e \approx 26\ \Omega$. A small-signal high-frequency AC equivalent circuit is shown in Figure 9-61, where we have used the T model for the transistor. Ignoring r_b, as we have done here, is usually fine for common-base stages, since it rarely has a significant effect on f_{cH}.

Exercise 9.17

Find the low cutoff frequency for the common-base amplifier of Figure 9-60.

Introductory Treatment If we combine R_S and R_E in parallel, the circuit in Figure 9-61 is topologically identical to that in Figure 9-59, and we can make use of the results derived there. We again have two single-pole low-pass transfer functions cascaded, and making use of (9.150), we find that the overall gain is

$$\frac{V_o(j\omega)}{I_s(j\omega)} = \frac{\alpha R_L\left(\dfrac{R_S'}{R_S' + r_e}\right)}{(1 + j\omega R_L C_\mu)(1 + j\omega(R_S' \| r_e)C_\pi)}, \tag{9.154}$$

where we have defined $R_S' = R_S \| R_E$. Using the data for our circuit, we find a midband gain of 1.95 kΩ (the gain is a transresistance and has units of ohms) and poles at 104 MHz and 19 MHz (the lower frequency pole is due to C_μ and dominates because r_e is so small). Using HSPICE, we find that the two poles are at 18.6 MHz and 141.6 MHz. The agreement on the lower pole frequency is quite good, whereas the agreement on the higher pole is fairly poor. The difference is mostly caused by HSPICE using a different capacitance model, which in this case results in a significantly lower value for C_π. Since the higher pole is almost 10 times the lower pole, we expect the high cutoff frequency to be very well approximated by the lower pole; that is, we expect $f_{cH} \approx 19$ MHz. Using SPICE, we find that $f_{cH} = 18.1$ MHz, which agrees quite well (cb.sch).

Advanced Treatment To find an approximation to ω_{cH} using the open-circuit time-constant method, we again make use of the results derived for the common-control amplifier. Using (9.151) through (9.153), suitably modified, we find that

$$\tau_{\pi o} = (R_S' \| r_e)C_\pi \tag{9.155}$$

and

$$\tau_{\mu o} = R_L C_\mu, \tag{9.156}$$

from which we get

$$f_{cH} \approx \frac{1}{2\pi[(R_S'\|r_e)C_\pi + R_L C_\mu]} = 16\text{ MHz}, \quad \textbf{(9.157)}$$

which is about 12% below the value predicted by SPICE (cb.sch). We see again that the estimate is conservative. We can conclude, then, on the basis of our agreement with SPICE, that the zeros are all well above f_{cH}, or we can examine Figure 9-61 and see that both zeros are at infinity, since that is where the capacitors become short circuits.

9.4.3 A FET Implementation: The Common-Gate Amplifier

Consider the common-gate current buffer shown in Figure 9-62, where we have capacitively coupled the signal from a Norton equivalent source to allow us to easily bias the stage. A common-gate stage would not usually be used in isolation like this, but doing so here will make the discussion more concrete without complicating the circuit.

Assume that the MOSFET has $K = 2\text{ mA/V}^2$, $V_{th} = 1$ V, $C_{gs} = 5$ pF, and $C_{gd} = 0.5$ pF. A DC bias point analysis of this amplifier shows that $V_{GS} \approx 2.6$ V and $I_D \approx 5$ mA. Using this bias point and performing a small-signal midband AC analysis yields $g_m \approx 6.4$ mA/V and a midband gain of 1.1 kΩ. (The gain is a transresistance and has units of ohms.) A SPICE simulation finds that the gain is 1.11 kΩ (cg.sch). A small-signal high-frequency AC equivalent circuit is shown in Figure 9-63, where we have used the T model for the transistor.

Exercise 9.18

Find the low cutoff frequency for the common-gate amplifier of Figure 9-62.

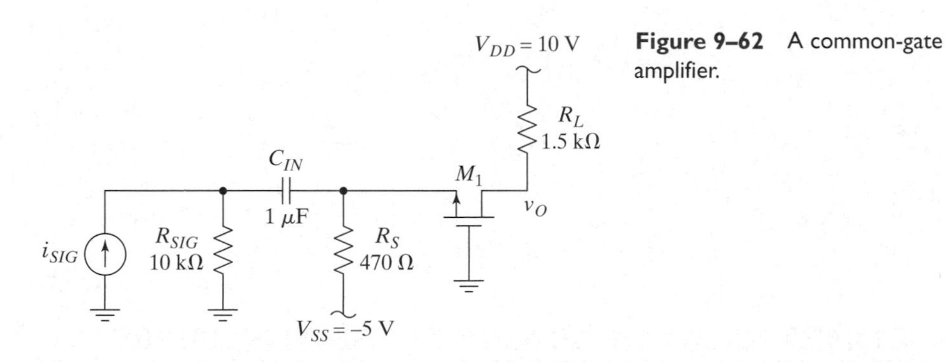

Figure 9–62 A common-gate amplifier.

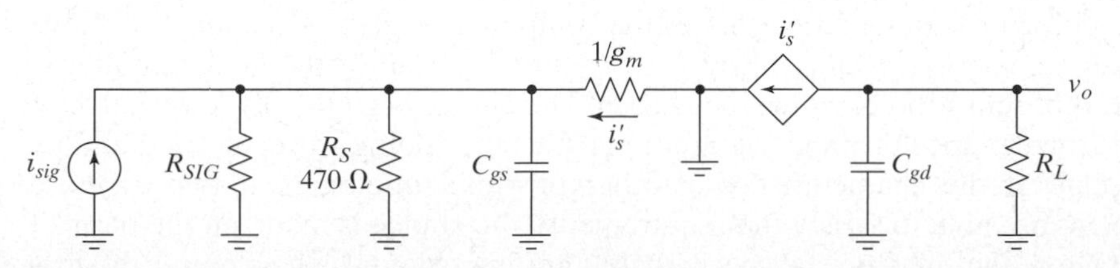

Figure 9–63 The small-signal high-frequency AC equivalent circuit for the amplifier in Figure 9-62.

Introductory Treatment If we combine R_{SIG} and R_S in parallel, the circuit in Figure 9-63 is topologically identical to that in Figure 9-59, and we can make use of the results derived there. We again have two single-pole low-pass transfer functions cascaded, and making use of (9.150) we find that the overall gain is

$$\frac{V_o(j\omega)}{I_{sig}(j\omega)} = \frac{R_L\left(\dfrac{R'_{SIG}}{R'_{SIG} + 1/g_m}\right)}{(1 + j\omega R_L C_{gd})(1 + j\omega(R'_{SIG}\|1/g_m)C_{gs})}, \tag{9.158}$$

where we have defined $R'_{SIG} = R_{SIG}\|R_S$. Using the data for our circuit, we find poles at 212 MHz and 274 MHz. The lower frequency pole is due to C_{gd}, and we note that neither pole dominates. Using SPICE, we find that the two poles are at 212 MHz and 272 MHz, which agrees very well with our prediction. Since both poles are similar in value, we expect the high cutoff frequency to be lower than either pole. Setting the magnitude of (9.158) equal to $1/\sqrt{2}$, we find that $f_{cH} \approx 153$ MHz. Using SPICE, we find that $f_{cH} = 154$ MHz, which agrees quite well.

Advanced Treatment To find an approximation to ω_{cH}, using the open-circuit time constant method, we again make use of the results derived for the common-control amplifier. Using (9.151) through (9.153) suitably modified, we find that

$$\tau_{gso} = \left(R'_{SIG}\bigg\|\frac{1}{g_m}\right)C_{gs} \tag{9.159}$$

and

$$\tau_{gdo} = R_L C_{gd}, \tag{9.160}$$

from which we get

$$f_{cH} \approx \frac{1}{2\pi\left[\left(R'_{SIG}\bigg\|\dfrac{1}{g_m}\right)C_{gs} + R_L C_{gd}\right]} = 120 \text{ MHz}, \tag{9.161}$$

which is about 22% below the value predicted by SPICE. The error is large in this case, since neither of the poles is dominant. We can check the frequencies of the zeros and see that they are all well above f_{cH} by examining Figure 9-63. Both zeros are at infinity, since that is where the capacitors become short circuits.

9.5 COMPARISON OF SINGLE-STAGE AMPLIFIERS

Transistor-level circuit design begins by selecting a topology suitable for the task at hand (e.g., a common-merge or common-control amplifier stage). After selecting a topology, the designer calculates the component values necessary to achieve the design goals. If the chosen topology cannot meet all of the goals simultaneously, a different topology must be chosen. The choice is guided by knowledge of what particular specifications were not met by the topology already tried and by knowledge of the characteristics of other possible topologies. If two or more topologies are able to satisfy the requirements, the choice is made on the basis of other factors that may not be specified, but are important, such as power dissipation, cost, and reliability.

To choose appropriate topologies and to know how to modify them when that is needed, a good designer must understand the characteristics of many common circuits. In this section, we compare the different single-stage amplifiers presented in the preceding sections. The differences between bipolar and FET amplifiers with voltage and current gain were discussed in Section 9.2.4. The main differences in the buffer stages are in their midband performance—not their frequency response—and those differences were discussed in Section 8.8. Since we are interested in practical results and conclusions here rather than analysis techniques, there is no reason to include the generic transistor in our discussion; therefore, we present the results for bipolar and FET amplifiers in the sections that follow.

9.5.1 Bipolar Amplifiers

Consider the general bipolar amplifier shown in Figure 9-64. Table 9.1 shows approximate equations for the midband voltage and current gains, the midband input and output resistances, and the high cutoff frequency of this circuit for all three configurations. It is assumed that both the source and load are capacitively coupled to the amplifier and that R_E is not bypassed. If all or part of R_E is bypassed, the unbypassed part should be used for the small-signal quantities given here. It is also assumed that the transistor's output resistance and base resistance can be safely ignored.

For the common-emitter configuration, the signal is applied to the base (node 1) relative to ground, and the output is taken from the collector (node 2) relative to ground. For the emitter follower, the input is applied to the base (node 1), but the output is taken from the emitter (node 3). In addition, it is assumed that R_C is removed (i.e., set to zero) for the emitter follower. For the common-base configuration, the input is applied to the emitter (node 3), the output is taken from the collector (node 2), and it is assumed that the base is an AC ground (i.e., node 1 is tied to ground). In each case, the source and load resistors are included in parallel with the appropriate elements. In some cases, two equivalent expressions are given in the table: one that is derived using thc hybrid-π model and one using the T model.

There are a number of things that we should note from the equations in this table. First of all, the common-emitter stage is the only stage that has both voltage and current gains greater than unity. Second, the magnitudes of the voltage gains of the common-emitter and common-base stages are equal if R_E is bypassed in the common-emitter stage. Both the common-emitter and emitter follower stages have large input resistance, whereas the common-base stage input resistance is low. On the other hand, the output resistances of both the common-emitter and common-

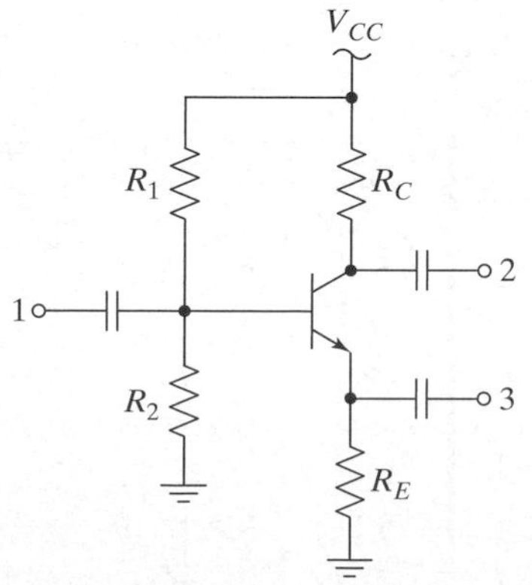

Figure 9–64 A general bipolar single-transistor amplifier.

Table 9.1 Approximate Equations Describing the Small-Signal Performance of the Three Single-Transistor Bipolar Amplifiers ($R_{BB} \equiv R_1 \| R_2$)

	Common emitter	Common collector (emitter follower)	Common base
$A_v = (v_o/v_i)$	$\dfrac{-\alpha R_C \| R_L}{r_e + R_E} = \dfrac{-g_m r_\pi (R_C \| R_L)}{r_\pi + (\beta + 1)R_E}$	$\dfrac{R_E \| R_L}{R_E \| R_L + r_e} = \dfrac{(\beta + 1)(R_E \| R_L)}{(\beta + 1)(R_E \| R_L) + r_\pi}$	$\dfrac{\alpha (R_C \| R_L)}{r_e}$
$A_i = (i_o/i_i)$	$\left(\dfrac{-R_C}{R_C + R_L}\right)\beta\left(\dfrac{R_{BB}}{R_{BB} + r_\pi + (\beta + 1)R_E}\right)$	$\left(\dfrac{R_E}{R_E + R_L}\right)(\beta + 1)\left(\dfrac{R_{BB}}{R_{BB} + r_\pi + (\beta + 1)(R_E \| R_L)}\right)$	$\left(\dfrac{R_C}{R_C + R_L}\right)\alpha\left(\dfrac{R_E}{R_E + r_e}\right)$
R_i	$R_{BB} \Big\| \left[r_\pi + (\beta + 1)R_E\right]$	$R_{BB} \Big\| \left[r_\pi + (\beta + 1)(R_E \| R_L)\right]$	$R_E \| r_e$
R_o	R_C	$R_E \Big\| \dfrac{R_{BB} \| R_S + r_\pi}{\beta + 1}$	R_C
f_{cH}	$\dfrac{1}{2\pi\left\{\left(r_\pi \Big\| \dfrac{R_S \| R_{BB} + R_E}{1 + g_m R_E}\right)C_\pi + \left[R_S \| R_i + \left(1 + \dfrac{\beta(R_{BB} \| R_S)}{R_{BB} \| R_S + r_\pi + (\beta + 1)R_E}\right)\left(R_C \| R_L\right)\right]C_\mu\right\}}$	$\dfrac{1}{2\pi\left[\left(r_\pi \Big\| \dfrac{R_S \| R_{BB} + R_E \| R_L}{1 + g_m(R_E \| R_L)}\right)C_\pi + \left(R_S \| R_i\right)C_\mu\right]}$	$\dfrac{1}{2\pi\left[\left(r_\pi \Big\| \dfrac{R_E \| R_S}{1 + g_m(R_E \| R_S)}\right)C_\pi + \left(R_C \| R_L\right)C_\mu\right]} = \dfrac{1}{2\pi[(r_e \| R_E \| R_S)C_\pi + (R_C \| R_L)C_\mu]}$

Table 9.2 Relative Values of the Midband AC Characteristics and High Cutoff Frequencies for the Three Single-Transistor Bipolar Amplifier Stages

Amplifier	R_i	R_o	A_v	A_i	f_{cH}
Common emitter	high	high	high	high	low
Common base	low	high	high	<1	high
Common collector (emitter follower)	high	low	<1	high	high

base stages are high, whereas the emitter follower has a low output resistance. Finally, the cutoff frequencies of the emitter follower and common-base stage are much higher than that of the common-emitter amplifier, predominantly because $R_C = 0$ for the emitter follower and R_{BB} is bypassed in the common-base stage.

Also, notice that the equations indicate that the emitter follower is not unilateral, whereas the other two amplifiers are, at least with the simplified models used to derive these results. This is evident because the output resistance of the emitter follower is a function of R_S, and the input resistance is a function of R_L. Neither statement is true of the other two topologies.

The approximate high cutoff frequencies quoted in this table were found using open-circuit time constants. The two time constants were found once for the general circuit, and we then made the appropriate substitutions to include R_S and R_L and to delete R_{BB} and R_C, where appropriate (*see* Problem P9.109). A simpler listing of the relative characteristics of these amplifiers is shown in Table 9.2.

9.5.2 FET Amplifiers

Consider the general FET amplifier shown in Figure 9-65. Table 9.3 shows approximate equations for the midband voltage and current gains, the midband input and output resistances, and the high cutoff frequencies of this circuit for all three configurations. It is assumed that both the source and load are capacitively coupled to the amplifier. It is also assumed that the transistor's output resistance and the body effect can be safely ignored.

For the common-source configuration, the signal is applied to the gate (node 1) relative to ground, and the output is taken from the drain (node 2) relative to ground. For the source follower, the input is again applied to the gate (node 1), but the output is taken from the source (node 3). In addition, it is assumed that R_D is removed (i.e., set to zero) for the source follower. For the common-gate configuration, the input is applied to the source, the output is taken from the drain, and it is assumed that the gate is an AC ground (*i.e.*, node 1 is connected to ground). In each case, the signal source and load resistors are included in parallel with the appropriate elements. In some cases, two equivalent expressions are given in the table: one is derived using the hybrid-π model and one is derived using the T model.

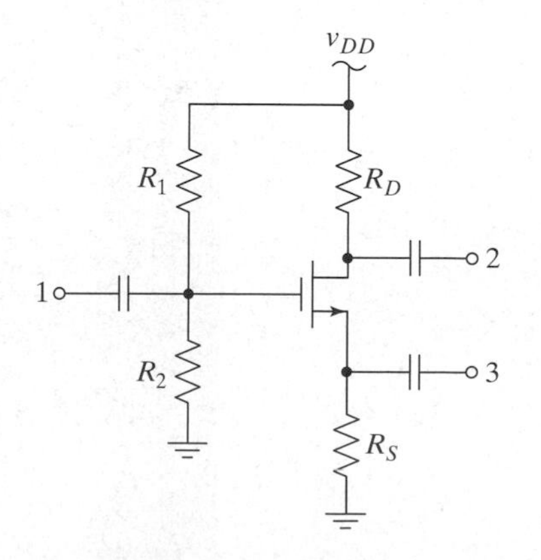

Figure 9–65 A general FET single-transistor amplifier.

There are a number of things that we should note from the equations in this table. First of all, the common-source stage is the only stage that has both voltage and current gains greater than unity. Second, the magnitudes of the voltage gains of the common-source and common-gate stages are equal. Both the common-source and source follower stages have large input resistance, whereas the common-gate stage input resistance is low. On the other hand, the output resistances

Table 9.3 Approximate Equations Describing the Small-Signal Performance of the Three Single-Transistor FET Amplifiers

	Common source	Common drain (source follower)	Common gate
$A_v = (v_o/v_i)$	$\dfrac{-g_m(R_D \Vert R_L)}{1 + g_m R_S}$	$\dfrac{R_S \Vert R_L}{R_S \Vert R_L + \dfrac{1}{g_m}} = \dfrac{g_m(R_S \Vert R_L)}{1 + g_m(R_S \Vert R_L)}$	$g_m(R_D \Vert R_L)$
$A_i = (i_o/i_i)$	$\left(\dfrac{R_D}{R_D + R_L}\right)\left(\dfrac{-g_m R_{GG}}{1 + g_m R_S}\right)$	$\left(\dfrac{R_S}{R_S + R_L}\right)\dfrac{g_m R_{GG}}{1 + g_m(R_S \Vert R_L)}$	$\left(\dfrac{R_D}{R_D + R_L}\right)\left(\dfrac{R_S}{R_S + \dfrac{1}{g_m}}\right)$
R_i	R_{GG}	R_{GG}	$R_S \Big\Vert \dfrac{1}{g_m}$
R_o	R_D	$\dfrac{1}{g_m} \Big\Vert R_S$	R_D
f_{cH}	$\dfrac{1}{2\pi\left\{\left(\dfrac{R_{SIG} \Vert R_{GG} + R_S}{1 + g_m R_S}\right)C_{gs} + \left[R_{SIG} \Vert R_{GG} + \left(1 + \dfrac{g_m(R_{SIG} \Vert R_{GG})}{1 + g_m R_S}\right)(R_D \Vert R_L)\right]C_{gd}\right\}}$	$\dfrac{1}{2\pi\left[\left(\dfrac{R_{SIG} \Vert R_{GG} + R_S \Vert R_L}{1 + g_m(R_S \Vert R_L)}\right)C_{gs} + (R_{SIG} \Vert R_{GG})C_{gd}\right]}$	$\dfrac{1}{2\pi\left[\left(\dfrac{R_S \Vert R_{SIG}}{1 + g_m(R_S \Vert R_{SIG})}\right)C_{gs} + (R_D \Vert R_L)C_{gd}\right]} = \dfrac{1}{2\pi\left[\left(R_{SIG} \Big\Vert \dfrac{1}{g_m} \Big\Vert R_S\right)C_{gs} + (R_D \Vert R_L)C_{gd}\right]}$

Table 9.4 Relative Values of the Midband AC Characteristics and High Cutoff Frequencies for the Three Single-Transistor MOSFET Amplifier Stages

Amplifier	R_i	R_o	A_v	A_i	f_{cH}
Common source	∞	high	high	∞	low
Common gate	low	high	high	<1	high
Common drain (source follower)	∞	low	<1	∞	high

of both the common-source and common-gate stages are high, whereas the source follower has a low output resistance. Finally, the cutoff frequencies of the source follower and common-gate stage are much higher than that of the common-source amplifier, predominantly because $R_D = 0$ for the source follower and R_{GG} is bypassed in the common-gate stage.

The approximate high cutoff frequencies quoted in this table were found by using open-circuit time constants. The two time constants were found once for the general circuit, and we then made the appropriate substitutions to include R_{SIG} and R_L and to delete R_{GG}, R_S, and R_D where appropriate (*see* Problem P9.115). A simpler listing of the relative characteristics of these amplifiers is shown in Table 9.4.

9.6 MULTISTAGE AMPLIFIERS

Many applications cannot be handled with single-transistor amplifiers, but require the use of a ***cascaded***, or ***multistage***, amplifier. In this section, we analyze the frequency responses and gain-bandwidth products of several multistage amplifiers. We depend on a thorough understanding of the characteristics of the three different types of single-transistor amplifier, so we summarize them in Table 9.5, which applies equally well to MOSFET or bipolar amplifiers and so is stated in terms of the generic transistor.

Table 9.5 Relative Values of the Midband AC Characteristics and High Cutoff Frequencies for the Three Generic Single-Transistor Amplifier Stages

Amplifier	R_i	R_o	A_v	A_i	f_{cH}
Common merge (voltage and current gain)	high	high	high	high	low
Common control (current buffer)	low	high	high	<1	high
Common output or merge follower (voltage buffer)	high	low	<1	high	high

Although we often cascade different types of amplifier stages to achieve a combination of characteristics that is unachievable by any one stage, it is also beneficial to cascade identical stages. For example, suppose that we need to design an amplifier with an overall voltage gain A_T. Furthermore, suppose that a common-merge amplifier is acceptable in terms of the terminal impedances, but we need to achieve the highest bandwidth possible. Do we want to design a single common-merge amplifier with gain A_T, or do we want to cascade more than one common-merge stage? Consider the simple amplifier analyzed in Section 9.2.1 and repeated in Figure 9-66 for convenience. The midband AC gain of this amplifier is

$$A_v = \frac{v_o}{v_i} = -g_m R'_L \qquad \textbf{(9.162)}$$

as found before, and $R'_L = R_L \| R_O$. If we assume that the bandwidth is dominated by the Miller effect, we then find from (9.33) that the high cutoff frequency is about

$$\omega_{cH} \approx \frac{1}{(R_S \| R_{CC} \| r_{cm})C_{in}}, \qquad \textbf{(9.163)}$$

where

$$C_{in} = C_{cm} + (1 + g_m R'_L)C_{oc}, \qquad \textbf{(9.164)}$$

as in (9.31). Now suppose that we cascade several of these stages. In this case, the effective load resistance for each stage will be the input impedance of the next stage, and the effective source resistance of each stage will be the R_O of the preceding stage. Depending on the supply voltage available, the number of stages cascaded, and other details, we may or may not be able to dispense with the intermediate coupling capacitors and biasing resistors, but these will not affect the following analysis in a fundamental way, so we will ignore them.

Suppose that we design our cascade so that each individual stage is identical and so that $R_O \ll r_{cm}$. In this case, (9.163) can be simplified to[11]

$$\omega_{cH} \approx \frac{1}{R_O C_{in}}, \qquad \textbf{(9.165)}$$

and, in addition, R'_L is approximately just R_O, so that the midband AC gain is approximately $A_v \approx -g_m R_O$, and the input attenuation factor of each stage is nearly unity. As a first-order approximation, we can also say that

$$C_{in} \approx g_m R_O C_{oc} = |A_v| C_{oc}, \qquad \textbf{(9.166)}$$

so that the gain-bandwidth product of each stage is approximately

$$GBW = |\alpha_i A_v \omega_{cH}| \approx \frac{g_m R_O}{g_m R_O^2 C_{oc}} = \frac{1}{R_O C_{oc}}. \qquad \textbf{(9.167)}$$

[11]For simplicity we are ignoring the first stage, which is driven by R_S, not R_O. If $R_O \ll R_S$, the first stage may have to have lower gain to avoiding limiting the overall BW.

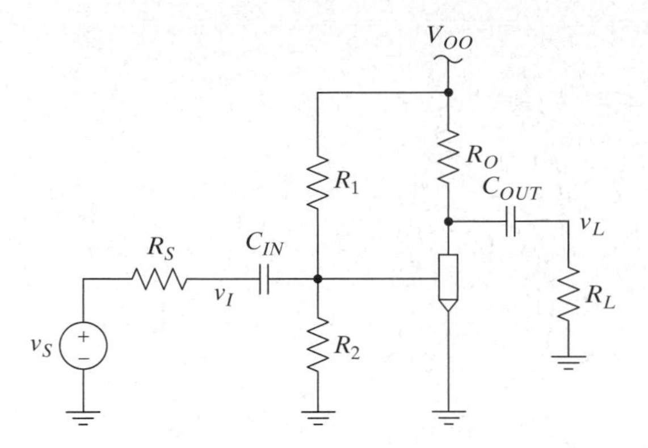

Figure 9–66 A common-merge amplifier.

This equation is in complete agreement with our previous result, (9.38), since the effective source resistance of each stage is R_O. From (9.167), we see that if we change the gain of each stage by varying g_m (by changing the DC bias point), while keeping R_O and C_{oc} fixed, we will keep the GBW approximately constant. In other words, for increasing g_m, the gain will increase and the bandwidth will decrease in direct proportion; this is a manifestation of the gain-bandwidth trade-off discussed in Aside A9.4.

This analysis has now led to an interesting question; if we can vary the gain of a stage while keeping the GBW constant, and we want to achieve an overall gain of A_T, how many stages should we cascade to maximize the resulting cutoff frequency? With the simplifying assumptions made here, the optimum number of stages to cascade for an overall gain of 100 turns out to be nine (*see* Problem P9.120). We would not use that many stages, however, because the cost would be too high. The bandwidth increases significantly for each additional stage up to about three stages; after that, the increases are smaller—but still significant—so you do sometimes see two or three stages cascaded together.

We now begin our examination of combinations of different stages. We start with an extremely common multistage amplifier: the common-merge merge-follower cascade. We then cover the very important and commonly used cascode amplifier (i.e., the common-merge common-control cascade). In each case, we cover the bipolar and MOSFET equivalents immediately after discussing the generic amplifier.

There are many different kinds of cascaded amplifiers, but the selection presented here covers the most common combinations, and the analyses of other cascades are handled in a similar fashion. Some other examples are given in the problems at the end of the chapter. In all of the following sections, we focus entirely on the high cutoff frequency, since finding the low cutoff frequency is no different for the cascades than it is for the individual stages, and the cascades have no significant effect on the achievable values of f_{cL}.

9.6.1 The Common-Merge Merge-Follower Cascade

This amplifier was introduced in Section 8.6, where it was shown that it combines the characteristics of the two single-transistor stages to obtain high input resistance, high voltage gain, low output resistance, and a reasonable ability to drive small loads. From Table 9.5 and our discussion in Section 9.3, we expect that the

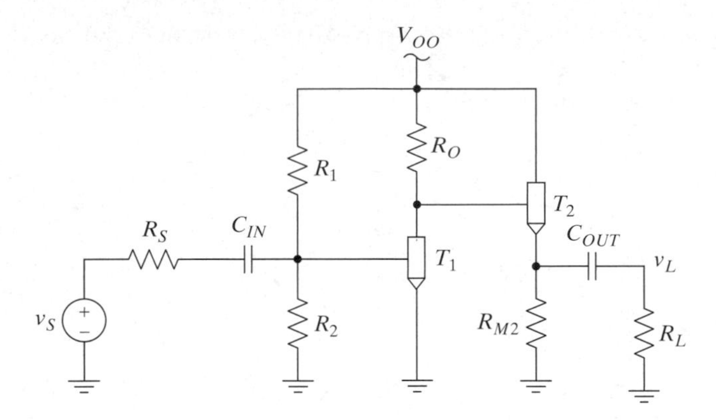

Figure 9–67 A common-merge merge-follower cascade.

overall bandwidth of this cascade will be limited by the common-merge stage. We can prove this by examining the circuit shown in Figure 9-67. We begin with an introductory treatment of the bandwidth followed by a more advanced treatment that requires a knowledge of open-circuit time constants.

Introductory Treatment An approximation to the overall bandwidth of this cascade can be obtained by assuming that the Miller effect will dominate the bandwidth of the common-merge stage. If that is true, the Miller output capacitance of the common-merge stage is neglected, and we calculate the bandwidth of the first stage on the basis of the single-pole circuit on the input side. If the Miller approximation is not valid, we would have to simultaneously solve the equations to determine the transfer function and then solve for the bandwidth if we wanted an exact answer. In practice, we would use open-circuit time constants to find an approximate value as shown in the following section. We will proceed on the basis of the assumption that the Miller effect dominates.

The common-merge stage of this amplifier sees an effective midband load resistance equal to its own output resistor, R_O, in parallel with the input resistance of T_2. Calling this effective load resistance R_O', we have

$$R_O' = R_O \| [r_{cm2} + (1 + a_{im2})(R_{M2} \| R_L)]. \tag{9.168}$$

Using this load and assuming that the Miller effect dominates the frequency response of the common-merge stage, we can use (9.33) to write an equation for the cutoff frequency of the first stage. We have

$$\omega_{cH1} \approx \frac{1}{(R_S \| R_{CC} \| r_{cm1})C_{in}}, \tag{9.169}$$

where

$$C_{in} = C_{cm1} + (1 + g_{m1}R_O')C_{oc1}, \tag{9.170}$$

as in (9.31). We can check the accuracy of the Miller approximation by using (9.34) as before, although that is not sufficient by itself: We also need to check the cutoff frequency of the merge follower stage, since the dominant pole for that stage occurs at its input, which is the output of the common-merge stage. In other words, if the merge follower has a significant effect, it causes the gain from the source to the output node of T_1 to roll off, and using the midband gain for the Miller effect is not a reasonable approximation. We will discuss this issue more a bit later.

The merge-follower sees an effective "source" resistance equal to the output resistance of the preceding common-merge stage. Using this fact along with (9.106), we find that the cutoff frequency of the second stage (i.e., the merge follower) is approximately

$$\omega_{cH2} \approx \frac{1}{\{R_O \| [r_{cm2} + (1 + a_{im2})(R_{M2} \| R_L)]\} C_{in2}} = \frac{1}{R'_O C_{in2}}, \quad \textbf{(9.171)}$$

where, making use of (9.101), (9.103), and (9.104) we have

$$C_{in2} = C_{oc2} + \left(\frac{r_{m2}}{r_{m2} + R_{M2} \| R_L}\right) C_{cm2}. \quad \textbf{(9.172)}$$

For most, but not all, sets of practical values, $\omega_{cH1} \ll \omega_{cH2}$, and the overall bandwidth of the circuit is approximately given by (9.169) alone. When ω_{cH2} is less than or near to ω_{cH1}, the gain from the input to the output of T_1 is less than the midband value by the time C_{oc1} starts to affect the response. Therefore, the Miller multiplication of C_{oc1} is less than assumed in (9.170). This error in the multiplication would tend to make the estimated cutoff frequency in (9.169) too low, but since we also neglect C_{Mout} in the Miller approximation, which tends to make our estimate higher, it is difficult to predict the end result. All we can say is that the estimate of ω_{cH1} is not likely to be accurate. If ω_{cH2} is much less than ω_{cH1}, ω_{cH2} should be a reasonable approximation to the overall cutoff frequency of the cascade.

When ω_{cH1} is dominant, the value of ω_{cH2} is meaningless, except to help confirm the validity of the approximation. This is true because, while the current fed forward through C_{oc1} may be neglected for frequencies near ω_{cH1}, there is no guarantee that it can be safely neglected at much higher frequencies; therefore, our approximation for ω_{cH2} may not be accurate.

Exercise 9.19

Find approximate values for ω_{cH1} and ω_{cH2} for the amplifier in Figure 9-67, given the following data: $R_1 = 47\ k\Omega$, $R_2 = 22\ k\Omega$, $R_O = 4.7\ k\Omega$, $R_{M2} = 330\ \Omega$, $R_S = 75\ \Omega$, $R_L = 75\ \Omega$, $g_{m1} = 20\ mA/V$, $r_{cm1} = 2\ k\Omega$, $r_{cm2} = 1\ k\Omega$, $r_{m2} = 25\ \Omega$, $a_{im2} = 99$, $C_{cm1} = C_{cm2} = 10\ pF$, and $C_{oc1} = C_{oc2} = 1\ pF$. Does the Miller effect dominate? What is the approximate value of the cutoff frequency for the cascade?

Advanced Treatment To use open-circuit time constants to find the high cutoff frequency of the circuit in Figure 9-67, we first draw the high-frequency small-signal equivalent circuit shown in Figure 9-68, where we have defined $R_{CC} = R_1 \| R_2$ and $R'_L = R_L \| R_{M2}$, as usual. We then find the four open-circuit time constants and approximate the cutoff frequency by

$$\omega_{cH} \approx \frac{1}{\tau_{cm1o} + \tau_{oc1o} + \tau_{cm2o} + \tau_{oc2o}}. \quad \textbf{(9.173)}$$

By inspection, we find that

$$\tau_{cm1o} = R_{cm1o} C_{cm1} = (R_S \| R_{CC} \| r_{cm1}) C_{cm1}. \quad \textbf{(9.174)}$$

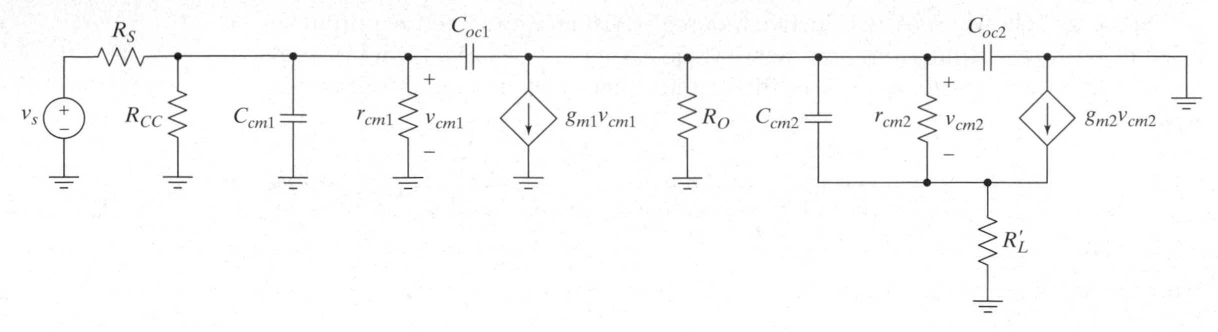

Figure 9–68 The high-frequency small-signal equivalent circuit for the multistage amplifier in Figure 9-67.

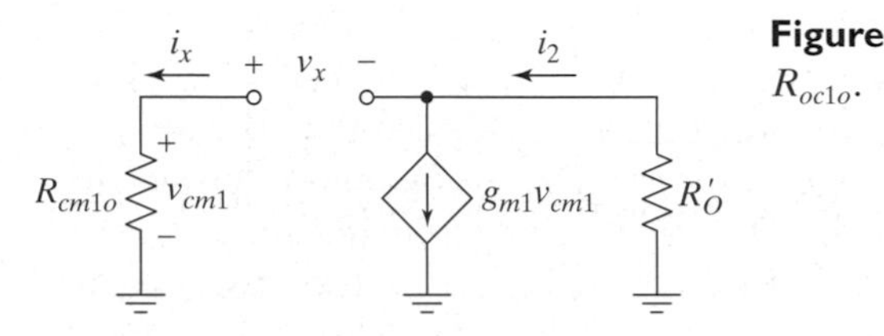

Figure 9–69 The equivalent circuit for finding R_{oc1o}.

To find R_{oc1o}, the driving-point resistance seen by C_{oc1}, we use the equivalent circuit shown in Figure 9-69, where we have combined R_O in parallel with the input resistance of T_2 and called the result R'_O, as in (9.168).

By KVL, we can write

$$v_x = i_x R_{cm1o} + i_2 R'_O, \tag{9.175}$$

and by KCL, we can write

$$i_2 = i_x + g_{m1}v_{cm1} = i_x(1 + g_{m1}R_{cm1o}). \tag{9.176}$$

Combining these equations, we find that

$$R_{oc1o} = \frac{v_x}{i_x} = R_{cm1o} + (1 + g_{m1}R_{cm1o})R'_O, \tag{9.177}$$

and therefore,

$$\tau_{oc1o} = R_{oc1o}C_{oc1} = [R_{cm1o} + (1 + g_{m1}R_{cm1o})R'_O]C_{oc1}. \tag{9.178}$$

To find the time constants for T_2, we first look at Figure 9-68 and note that with the signal source set equal to zero and the other capacitors open circuited, v_{cm1} is zero. Therefore, $g_{m1}v_{cm1} = 0$, and we arrive at the equivalent circuit of Figure 9-70, which is topologically identical to 9-46(a). The open-circuit time constants for that circuit were found in Section 9.3.1. Using (9.109) and (9.114), we have

$$\tau_{oc2o} = (R_O \| [r_{cm2} + (1 + a_{im2})R'_L])C_{oc2} = R'_O C_{oc2} \tag{9.179}$$

and

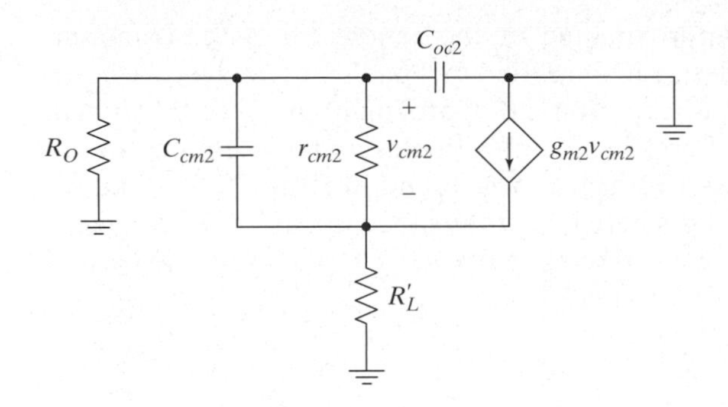

Figure 9–70 Equivalent circuit for finding the time constants for T_2.

$$\tau_{cm2o} = \left[r_{cm2} \middle\| \left(\frac{R_O + R'_L}{1 + g_{m2}R'_L} \right) \right] C_{cm2}. \tag{9.180}$$

We can now substitute (9.174), (9.178), (9.179), and (9.180) into (9.173) to find the overall bandwidth. The open-circuit time-constant method will yield an accurate result if any one of the four time constants is larger than the sum of the other three by a significant amount (say, a factor of three or more).

Exercise 9.20

Repeat Exercise 9.19, using the open-circuit time-constant approach. You now need to know that $g_{m2} = 40$ mA/V. Which time constant dominates? Do the results agree with those obtained before? Why or why not?

9.6.2 A Bipolar Implementation: The Common-Emitter Emitter-follower Cascade

A common-emitter emitter-follower cascade is shown in Figure 9-71. The small-signal midband AC performance of this amplifier was analyzed in Section 8.6. In the current section, we calculate approximate values for the high cutoff frequency.

Introductory Treatment Since the bandwidth of a common-emitter stage is usually much lower than that of an emitter follower, the bandwidth of the amplifier in

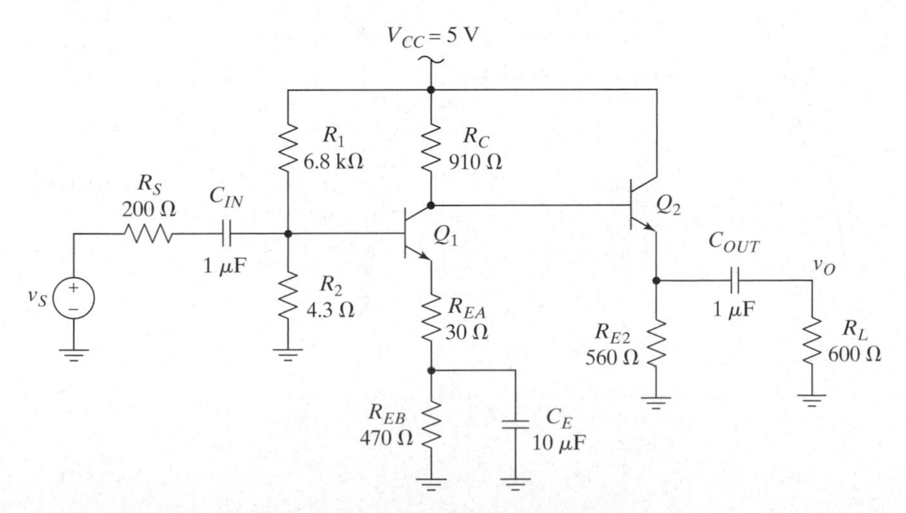

Figure 9–71 A common-emitter emitter-follower cascade

Figure 9-71 is frequently well approximated by the bandwidth of the common-emitter stage alone. For a well-designed amplifier, the loading caused by the emitter follower can often be ignored for a first-cut approximation as well, although we will include it in our analysis here. The bandwidth of a common-emitter amplifier with some unbypassed emitter resistance was found in Example 9.6, and we can use the results from that analysis here by defining the effective collector resistance to be $R'_C = R_C \| R_{i2}$, where R_{i2} is the resistance seen looking into the base of Q_2. Using impedance reflection, we can write

$$R'_C = R_C \| [r_{\pi 2} + (\beta+1)R'_L], \quad \textbf{(9.181)}$$

where $R'_L = R_{E2} \| R_L$. Then, making use of the results from Example 9.6, we have the midband gain of the first stage,

$$A_{v1} = \frac{-R'_C}{r_{e1} + R_{EA}}, \quad \textbf{(9.182)}$$

the Miller input capacitance,

$$C_{Min} = (1 - A_{v1})C_{\mu 1} \quad \textbf{(9.183)}$$

and the resulting cutoff frequency

$$f_{cH} \approx \frac{1}{2\pi(R'_S \| [r_{\pi 1} + (\beta+1)R_{EA}])C_{Min}}, \quad \textbf{(9.184)}$$

where $R'_S = R_S \| R_{BB} + r_{b1}$. The Miller approximation is valid in this case only if the time constant due to C_{Mout} is much less than $1/\omega_{cH}$ *and* the impedance of C_π is greater than r_π at ω_{cH}. As seen in Example 9.6, neglecting C_π is not always reasonable. In addition, $C_{\mu 2}$ shows up in parallel with C_{Mout} in this cascade, so the time constant due to C_{Mout} is larger than it was for the common-emitter stage alone. Therefore, to properly approximate the bandwidth of this cascade, we often need the open-circuit time-constant method presented in the next section.

The simulated voltage gains from v_s to the collector of Q_1 and from v_s to v_o for the circuit in Figure 9-71 are shown in Figure 9-72 (Ce_ef.sch). We see from this result that the overall voltage gain is not much lower than the gain to the collector of Q_1, so the emitter-follower gain is close to 1. We also see that the bandwidth from v_s to v_o is

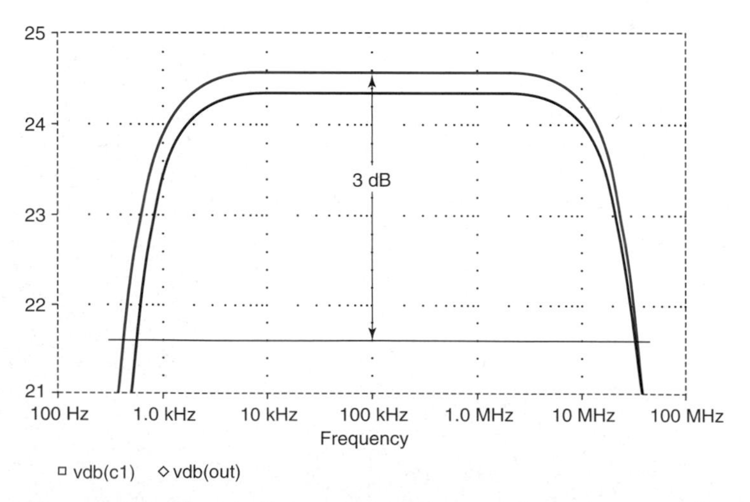

Figure 9–72 The gains from v_s to the collector of Q_1 and from v_s to v_o for the circuit in Figure 9-71.

nearly the same as from v_s to the collector of Q_1, so the transfer function of the emitter follower has not degraded the overall bandwidth. The plot also shows that the bandwidth of the overall circuit is about 36 MHz. Compared with the simulated bandwidth of 82 MHz from Example 9.6, it would appear at first glance as if the input capacitance of the emitter follower has significantly reduced the overall bandwidth.

However, the reduction in bandwidth is caused by the fact that the overall gain is now higher. For the original circuit, the gain was simulated to be −7.08. With the emitter follower, the simulated gain is −16.4. The difference is a result of the emitter follower buffering the load from the collector of Q_1, so the effective collector load for Q_1 is larger, and its gain is therefore larger. If the Miller effect were dominant, the results of Exercise 9.9 would indicate that the gain-bandwidth product is given by GBW $\approx 1/R_S C_\mu$, which is independent of the collector load. Therefore, increasing the gain from 7.08 to 16.4 by varying the collector load would yield an expected bandwidth of 35 MHz, very close to what we get. In other words, the GBW of the cascade is about the same as the GBW of the common emitter alone, and the emitter follower has not hurt us in this respect. A detailed analysis would be significantly more complicated, but even the crude approximations used here yield useful insights.

Example 9.8

Problem:

Find an approximation to f_{cH} for the circuit in Figure 9-71, using the Miller approximation. As in Example 9.6, assume that a 2N2857 transistor is used, so that $\beta = 100$, $r_b = 10\ \Omega$, $\tau_F = 116$ ps, $C_{je0} = 0.94$ pF, $C_{\mu 0} = 0.89$ pF, $V_0 = 0.75$, and $m_{jc} = 0.3$.

Solution:

A DC bias analysis reveals that $V_{B1} \approx 1.9$ V, $V_{E1} \approx 1.2$ V, $V_{C1} \approx 2.8$ V, $V_{E2} \approx 2.1$ V, $I_{C1} \approx 2.4$ mA, and $I_{C2} \approx 3.8$ mA. Using these values, we calculate the small-signal parameters: $r_{e1} = 10.8\ \Omega$, $r_{\pi 1} = 1.1$ kΩ, $g_{m1} = 92$ mA/V, $C_{\pi 1} = 12.6$ pF, $C_{\mu 1} = 0.7$ pF, $r_{e2} = 6.8\ \Omega$, $r_{\pi 2} = 680\ \Omega$, $g_{m2} = 150$ mA/V, $C_{\pi 2} = 18.8$ pF, and $C_{\mu 2} = 0.59$ pF.

Plugging the values into (9.184) yields $f_{cH} \approx 54$ MHz, which is 50% higher than the simulated value of 36 MHz (Ce_ef.sch). The approximation is very poor in this case, because the Miller effect is not dominant. The time constant for C_{Mout} in parallel with $C_{\mu 2}$ is $R'_C(C_{Mout} + C_{\mu 2}) = 0.65$ ns, much smaller than $1/\omega_{cH} = 2.9$ ns, but the impedance of $C_{\pi 1}$ at ω_{cH} is 223 Ω, which is smaller than $r_{\pi 1}$, so using the midband gain in the Miller approximation and ignoring $C_{\pi 1}$ is not accurate.

■

Advanced Treatment The solution to Exercise 9.11 derived the open-circuit time constants for a common-emitter amplifier with partially unbypassed emitter resistance. We will use the results from that exercise here, but we encourage you to derive the equations yourself before looking there for help.

For the amplifier of Figure 9-71, there are four open-circuit time constants to calculate: $\tau_{\pi 1o}$, $\tau_{\mu 1o}$, $\tau_{\pi 2o}$, and $\tau_{\mu 2o}$. The overall high cutoff frequency is then given by

$$f_{cH} \approx \frac{1}{2\pi(\tau_{\pi 1o} + \tau_{\mu 1o} + \tau_{\pi 2o} + \tau_{\mu 2o})}. \tag{9.185}$$

Using the results of Exercise 9.11, we have

$$\tau_{\pi 1o} = \left(r_{\pi 1} \middle\| \frac{R'_S + R_{EA}}{1 + g_{m1} R_{EA}} \right) C_{\pi 1}, \tag{9.186}$$

where $R'_S = R_S \| R_{BB} + r_{b1}$. Similarly, we obtain

$$\tau_{\mu 1o} = \left[R_{eq} + R'_C \left(1 + \frac{g_{m1} r_{\pi 1} R_{eq}}{r_{\pi 1} + (\beta+1) R_{EA}} \right) \right] C_{\mu 1}, \quad \textbf{(9.187)}$$

where $R_{eq} = R'_S \| [r_{\pi 1} + (\beta+1) R_{EA}]$ and R'_C is R_C in parallel with the input resistance of Q_2 as given by (9.181). $C_{\mu 2}$ is connected from the base of Q_2 to AC ground, so the time constant is found by inspection to be

$$\tau_{\mu 2o} = R'_C C_{\mu 2}. \quad \textbf{(9.188)}$$

The time constant for $C_{\pi 2}$ is found in a manner completely analogous to $\tau_{\pi 1o}$. The result is

$$\tau_{\pi 2o} = \left(\frac{R_C + r_b + R_{E2} \| R_L}{1 + g_{m2}(R_{E2} \| R_L)} \right) C_{\pi 2}. \quad \textbf{(9.189)}$$

Example 9.9

Problem:
Find an approximation to f_{cH} for the circuit in Figure 9-71, using open-circuit time constants. As in Example 9.8, assume that a 2N2857 transistor is used, so that $\beta = 100$, $r_b = 10\ \Omega$, $\tau_F = 116$ ps, $C_{je0} = 0.94$ pF, $C_{\mu 0} = 0.89$ pF, $V_0 = 0.75$, and $m_{jc} = 0.3$.

Solution:
As in Example 9.8, a DC bias analysis reveals that $V_{B1} \approx 1.9$ V, $V_{E1} \approx 1.2$ V, $V_{C1} \approx 2.8$ V, $V_{E2} \approx 2.1$ V, $I_{C1} \approx 2.4$ mA, and $I_{C2} \approx 3.8$ mA. Using these values, we again obtain $r_{e1} = 10.8\ \Omega$, $r_{\pi 1} = 1.1$ kΩ, $g_{m1} = 92$ mA/V, $C_{\pi 1} = 12.6$ pF, $C_{\mu 1} = 0.7$ pF, $r_{e2} = 6.8\ \Omega$, $r_{\pi 2} = 680\ \Omega$, $g_{m2} = 150$ mA/V, $C_{\pi 2} = 18.8$ pF, and $C_{\mu 2} = 0.59$ pF.

Plugging these values into (9.186) through (9.189), we find that $\tau_{\pi 1o} = 0.72$ ns, $\tau_{\mu 1o} = 3.58$ ns, $\tau_{\mu 2o} = 0.52$ ns, and $\tau_{\pi 2o} = 0.51$ ns. The resulting cutoff frequency is found with (9.185) to be $f_{cH} \approx 30$ MHz. The simulated value is 36 MHz (Ce_ef.sch), so our approximation is 20% low. The error is caused by the fact that the dominant time constant, $\tau_{\mu 1o}$, is not much larger than the sum of the others and because our DC bias analysis and resulting AC model parameters do not exactly agree with SPICE. Nevertheless, the value is a reasonable estimate and is conservative.

■

Exercise 9.21

Recalculate the results for Examples 9.8 and 9.9, assuming that the source resistance is changed to $R_S = 1$ kΩ. Is the Miller effect a better or worse approximation in this case? Why?

□

9.6.3 A FET Implementation: The Common-Source Source-Follower Cascade

A common-source source-follower cascade is shown in Figure 9-73. The small-signal midband AC performance of this amplifier was analyzed in Section 8.6. In this section, we calculate approximate values for the high cutoff frequency.

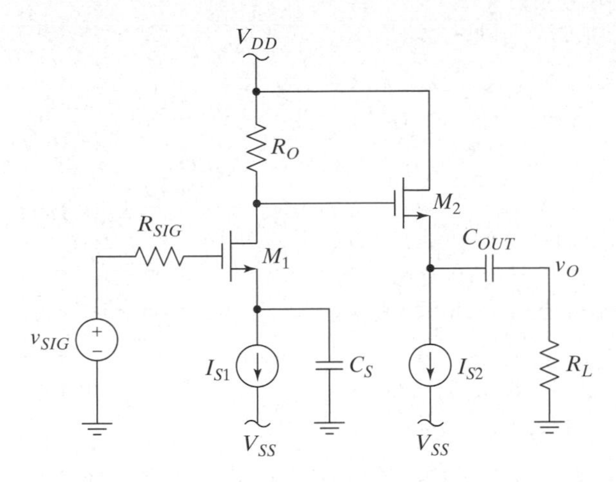

Figure 9–73 A common-source source-follower amplifier.

Introductory Treatment If we assume that the bandwidth of the source follower in Figure 9-73 is much greater than that of the common-source amplifier, we can estimate f_{cH} by analyzing the high-frequency small-signal AC equivalent circuit shown in Figure 9-74. We have used the T model for M_2 in this figure, since that is usually more convenient for a source follower stage. If we further assume that the Miller effect dominates, we can use the midband gain to bring C_{gd1} down as C_{Min}, which will appear in parallel with C_{gs1}, as shown in Figure 9-75. Since the input current of the source follower is zero for midband frequencies, the gain is $-g_m R_D$ and the Miller input capacitance is

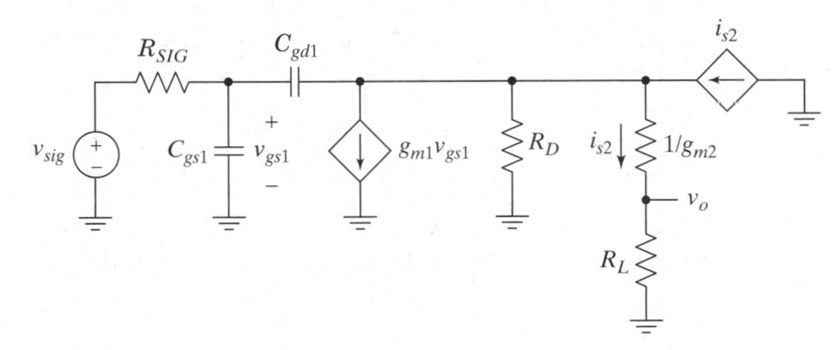

Figure 9–74 The high-frequency small-signal equivalent circuit for the amplifier in Figure 9-73, assuming that the common-source stage dominates the bandwidth.

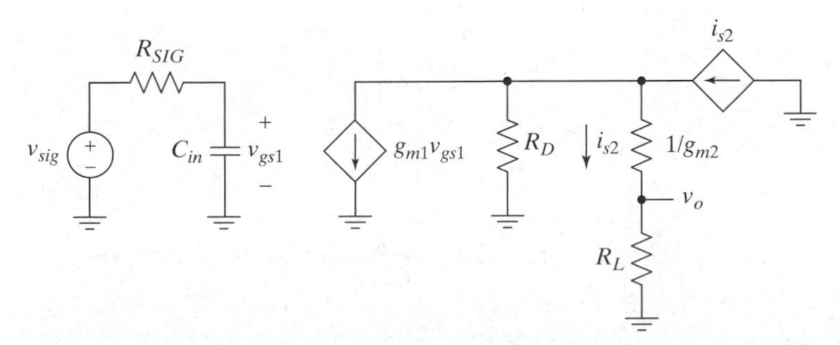

Figure 9–75 The high-frequency small-signal equivalent circuit from Figure 9-74 after application of the Miller approximation.

$$C_{Min} = C_{gd1}(1 + g_{m1}R_D). \tag{9.190}$$

The resulting total input capacitance is $C_{in} = C_{Min} \| C_{gs1}$, the resistance seen by this capacitor is R_{SIG}, and the bandwidth is

$$f_{cH} \approx \frac{1}{2\pi R_{SIG} C_{in}}. \tag{9.191}$$

We then check the Miller approximation. Since C_{Mout} shows up in parallel with C_{gd2}, the approximation is valid so long as

$$R_D(C_{Mout} + C_{gd2}) \ll R_{SIG}C_{in}. \tag{9.192}$$

If the Miller approximation is not valid, or if the bandwidth of the source follower is not much larger than the common-source stage, we need to use more advanced techniques to find the bandwidth of this cascade.

Exercise 9.22

Find f_{cH} for the circuit in Figure 9-73, assuming that the Miller effect dominates and the bandwidth of the source follower is very high. Assume that $R_{SIG} = 10\ k\Omega$, $R_D = 40\ k\Omega$, $R_L = 10\ k\Omega$, $I_{SS1} = 100\ \mu A$, $I_{SS2} = 200\ \mu A$, and the transistor has $V_{th} = 0.5\ V$, $K = 100\ \mu A/V^2$, $\lambda = 0\ V^{-1}$, $C_{gs} = 2\ pF$, and $C_{gd} = 0.5\ pF$.

Advanced Treatment The complete high-frequency small-signal AC equivalent circuit of the amplifier shown in Figure 9-73 is given in Figure 9-76. The high cut-off frequency of this amplifier is approximately given by

$$f_{cH} \approx \frac{1}{2\pi(\tau_{gs1o} + \tau_{gd1o} + \tau_{gs2o} + \tau_{gd2o})}. \tag{9.193}$$

With the signal source set to zero and the other capacitors opened, we find by inspection that

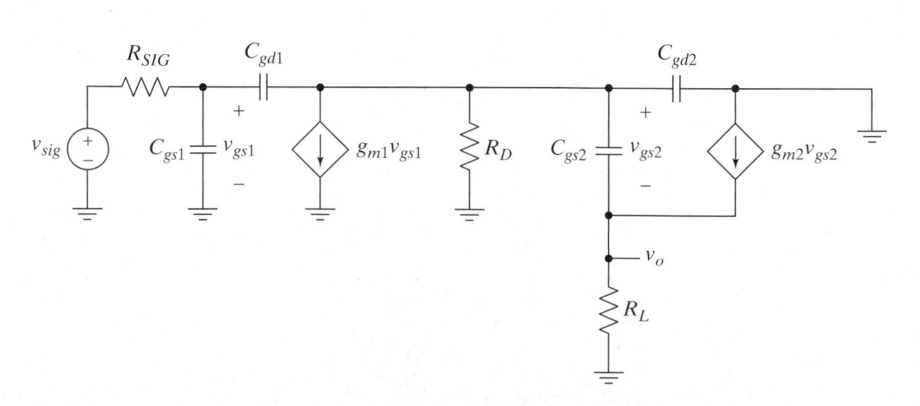

Figure 9–76 The complete high-frequency small-signal AC equivalent circuit for the amplifier in Figure 9-73.

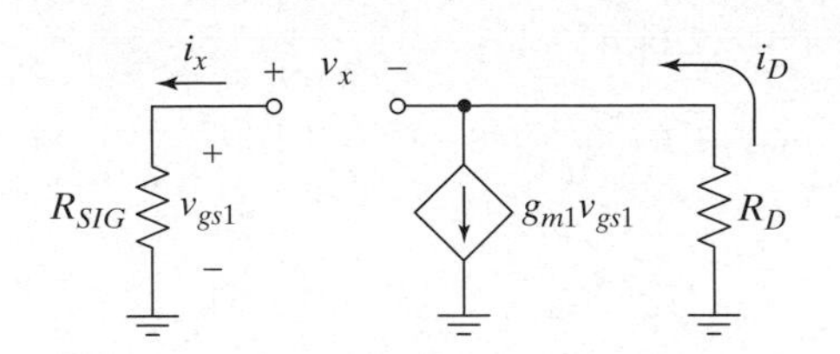

Figure 9–77 Equivalent circuit for finding R_{gd1o}.

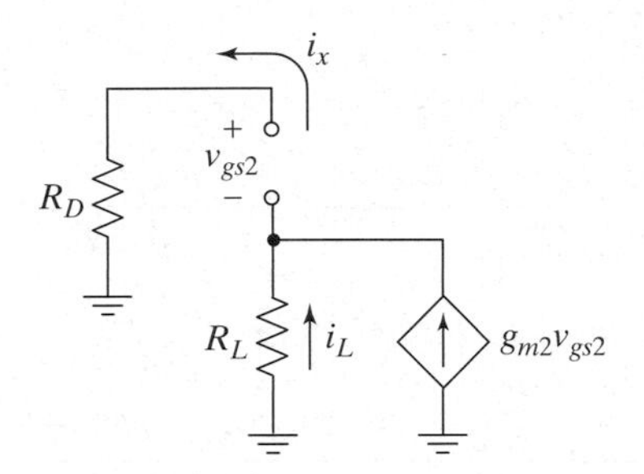

Figure 9–78 Equivalent circuit for finding τ_{gs2o}.

$$\tau_{gs1o} = R_{SIG}C_{gs1}. \tag{9.194}$$

Similarly, with $v_{sig} = 0$ and the capacitors opened, we see that $v_{gs1} = 0$, so the $g_{m1}v_{gs1}$ source is zero, and

$$\tau_{gd2o} = R_D C_{gd2}. \tag{9.195}$$

To find τ_{gd1o}, we use the equivalent circuit shown in Figure 9-77 and use KVL to write

$$v_x = i_x R_{SIG} + i_D R_D. \tag{9.196}$$

We then use KCL and Ohm's law to obtain

$$i_D = i_x + g_{m1}v_{gs1} = i_x(1 + g_{m1}R_{SIG}). \tag{9.197}$$

Substituting (9.197) into (9.196) and solving yields

$$R_{gd1o} = \frac{v_x}{i_x} = R_{SIG} + (1 + g_{m1}R_{SIG})R_D, \tag{9.198}$$

so that we have

$$\tau_{gd1o} = [R_{SIG} + (1 + g_{m1}R_{SIG})R_D]C_{gd1}. \tag{9.199}$$

To find τ_{gs2o}, we use the equivalent circuit shown in Figure 9-78. In drawing this circuit, we realize that $v_{gs1} = 0$, since $v_{sig} = 0$. Therefore, the $g_{m1}v_{gs1}$ source is zero. All we have looking to the left from the gate of M_2 is R_D. Using this figure, we know that $R_{gs2o} = v_{gs2}/i_x$, and we can use KVL to write

$$v_{gs2} = i_x R_D + i_L R_L. \tag{9.200}$$

We then use KCL to obtain

$$i_L = i_x - g_{m2}v_{gs2}, \tag{9.201}$$

and we substitute this into (9.200), and solve to obtain

$$\tau_{gs2o} = R_{gs2o}C_{gs2} = \left(\frac{R_D + R_L}{1 + g_{m2}R_L}\right)C_{gs2}. \quad \textbf{(9.202)}$$

Using (9.202), (9.199), (9.195), and (9.194) in (9.193), we can find the high cutoff frequency.

Exercise 9.23

Repeat Exercise 9.22, using open-circuit time constants to approximate f_{cH}. Compare your results.

□

9.6.4 The Cascode Amplifier: A Common-Merge Common-Control Cascade

Recall from Section 9.2.1 that the high cutoff frequency of a common-merge amplifier is frequently limited by the Miller effect. The problem stems from the large inverting gain from one side of C_{oc} to the other; this inverting gain makes C_{oc} appear as a much larger effective capacitance to ground at the control node. We can increase the bandwidth of the stage while retaining almost all of the other characteristics by inserting a current buffer in series with the output as shown in Figure 9-79. We have used the small-signal model for the transistor and have shown the current buffer as a three-terminal block to illustrate the basic idea.

If the current gain of the buffer is nearly unity (i.e., $i_l \approx i_o$) and the output resistance of the buffer is much larger than R_L, then the overall amplifier yields the same midband gain as the common-merge stage would by itself. If the input resistance of the buffer is much smaller than R_L, however, which it usually is, the voltage gain across the common-merge stage alone will be much less than it would be without the current buffer. Therefore, the Miller multiplication will yield a much smaller C_{Min}, and the bandwidth of the common-merge stage will be significantly improved.

We can use a common-control stage as the current buffer. The cascade of a common-merge and a common-control amplifier is called a ***cascode*** amplifier. The name is a holdover from the days of vacuum tubes and refers to a cascade through the cathode (equivalent to the merge terminal of our generic transistor).

Since the circuit of Figure 9-79 requires cascading a common-merge stage with a common-control stage (as the current buffer), the bandwidth of the cascode cannot be significantly larger than the common-merge stage alone unless the

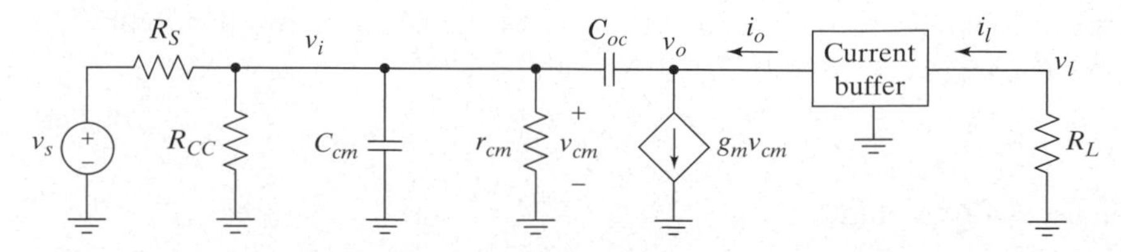

Figure 9–79 Increasing the bandwidth of a common-merge amplifier by adding a current buffer.

bandwidth of the common-control stage alone is itself significantly higher than that of the common-merge stage alone. Fortunately, as we have already seen, this condition is usually satisfied.

Consider the generic cascode amplifier shown in Figure 9-80. We see from this schematic that the biasing of the cascode amplifier is similar to that of a common-merge stage. We have simply split the top resistor in the bias string in half and inserted T_2 in series with the output of T_1. To make the midband input resistance of T_2 small, we need the bypass capacitor C_C from the control terminal to ground.

The high-frequency small-signal AC equivalent circuit of the cascode is shown in Figure 9-81. We have used the T model for T_2, since that proves to be most convenient. In this circuit, the Thévenin resistance driving the control of T_1 is given by $R_{CC} = R_2 \| R_3$, since the top of R_2 is an AC ground. In addition, we have defined $R'_L = R_L \| R_O$ as usual.

Introductory Treatment Since the cascode connection is specifically intended to prevent the Miller multiplication of C_{oc1} from dominating the frequency response, we cannot use the Miller approximation *by itself* and expect to obtain reasonable results. Nevertheless, we can use the approximation to help estimate the bandwidth of the circuit. Note that the midband gain from control to output of T_1 is approximately $-g_{m1}r_{m2} \approx -1$, so the Miller input capacitance is about equal to $2C_{oc1}$ and is in parallel with C_{cm1}. The bandwidth of the transfer function from v_s to v_{c1} is therefore approximately

$$f_{cH1} \approx \frac{1}{2\pi(R_S \| R_{CC} \| r_{cm1})(C_{cm1} + 2C_{oc1})}. \quad \textbf{(9.203)}$$

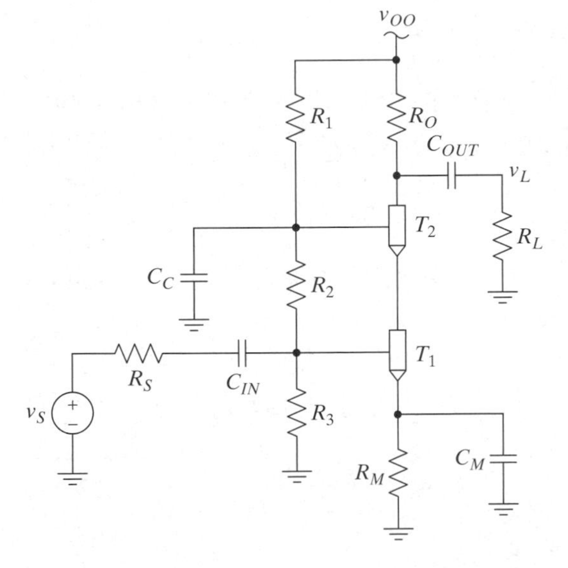

Figure 9–80 A generic cascode amplifier.

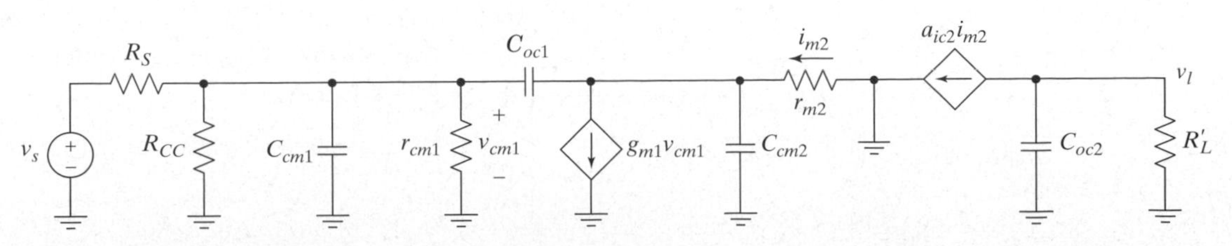

Figure 9–81 The high-frequency small-signal AC equivalent circuit of the cascode amplifier from Figure 9-80.

Since the output of T_1 is connected by r_{m2} to ground, and r_{m2} is usually small, we can reasonably ignore the capacitance from this node to ground (C_{Mout1} in parallel with C_{cm2}). The output side of T_2 is a single-pole circuit with

$$f_{cH2} \approx \frac{1}{2\pi R_L' C_{oc2}}. \tag{9.204}$$

The cascade of these transfer functions has an overall cutoff frequency found by setting the magnitude of the product equal to $1/\sqrt{2}$ times the DC gain. Ignoring the DC gain, we can say that

$$\frac{1}{\sqrt{2}} \approx \left| \left(\frac{1}{1 + j\omega_{cH}(R_S \| R_{CC} \| r_{cm1})(C_{cm1} + 2C_{oc1})} \right) \left(\frac{1}{1 + j\omega_{cH} R_L' C_{oc2}} \right) \right| \tag{9.205}$$

and solve to find ω_{cH}.

Advanced Treatment The high cutoff frequency of the circuit in Figure 9-81 is given by

$$f_{cH} \approx \frac{1}{2\pi(\tau_{cm1o} + \tau_{oc1o} + \tau_{cm2o} + \tau_{oc2o})}. \tag{9.206}$$

We find by inspection that

$$\tau_{cm1o} = (R_S \| R_{CC} \| r_{cm1}) C_{cm1}, \tag{9.207}$$

$$\tau_{cm2o} = r_{m2} C_{cm2}, \tag{9.208}$$

and

$$\tau_{oc2o} = R_L' C_{oc2}. \tag{9.209}$$

To find τ_{oc1o}, we use the circuit shown in Figure 9-82. Writing KVL, we find that

$$v_x = i_x(R_S \| R_{CC} \| r_{cm1}) + i_{m2} r_{m2}. \tag{9.210}$$

Using KCL, we obtain

$$i_{m2} = i_x + g_{m1} v_{cm1} = i_x[1 + g_{m1}(R_S \| R_{CC} \| r_{cm1})], \tag{9.211}$$

now substitute, and solve to get

$$\tau_{oc1o} = R_{oc1o} C_{oc1} = \left[(R_S \| R_{CC} \| r_{cm1}) + [1 + g_{m1}(R_S \| R_{CC} \| r_{cm1})] r_{m2} \right] C_{oc1}. \tag{9.212}$$

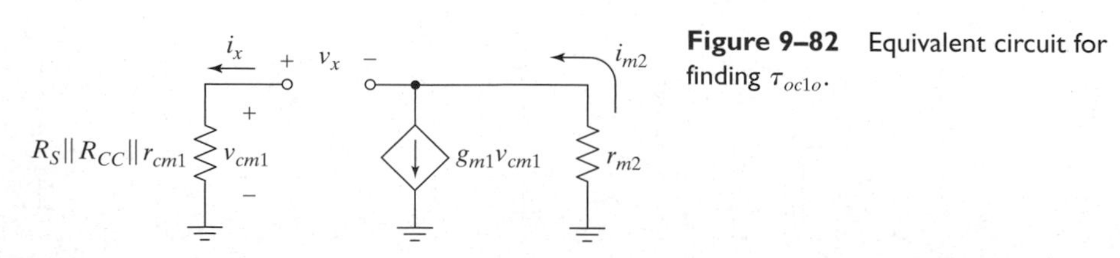

Figure 9–82 Equivalent circuit for finding τ_{oc1o}.

Plugging (9.212), (9.209), (9.208), and (9.207) into (9.206) yields an approximate value for the cutoff frequency.

9.6.5 The Bipolar Cascode Amplifier: A Common-Emitter Common-Base Cascade

A bipolar cascode amplifier is shown in Figure 9-83. An analysis of the DC bias point reveals that $V_{B1} = 2$ V, $V_{B2} = 3$ V, $V_{C2} = 6$ V, and $I_{C1} = 1$ mA. The corresponding small-signal model parameters are $r_{\pi 1} = 4.7$ kΩ, $r_{e2} = 26$ Ω, $g_{m1} = 38$ mA/V, $C_{\pi 1} = C_{\pi 2} = 60$ pF, $C_{\mu 1} = 6.6$ pF, and $C_{\mu 2} = 4.5$ pF, where we have used the 2N2222 model parameters from Appendix B.

The high-frequency small-signal AC equivalent circuit is shown in Figure 9-84, where we have used the T model for the common-base transistor. We have ignored r_{b2}, since it usually does not have a significant impact on the frequency response of the common-base stage (*see* Problem P9.142). In this circuit, the Thévenin resistance driving the base is $R_{BB} = R_2 \| R_3$, and the effective load is $R_L' = R_L \| R_C$. Open circuiting all of the capacitors and ignoring r_{b1}, we find the midband gain;

$$A_v = \frac{v_o}{v_i} = -g_{m1}\alpha_2 R_L' \approx -g_{m1}R_L', \tag{9.213}$$

which is the same as for the equivalently loaded common-emitter stage alone.

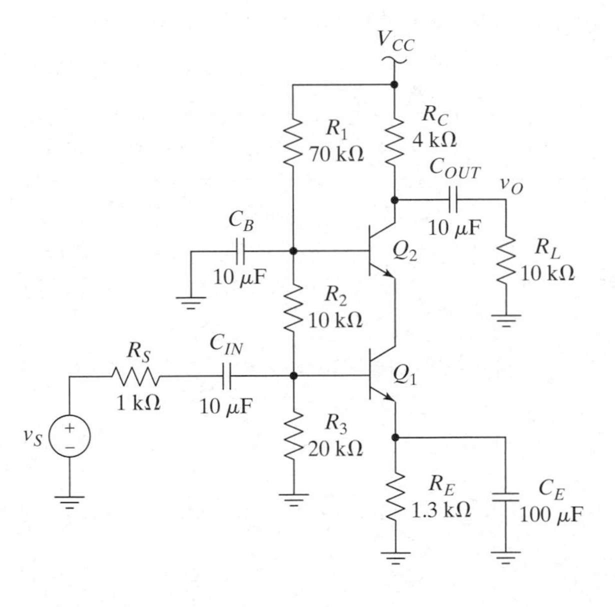

Figure 9–83 A bipolar cascode amplifier.

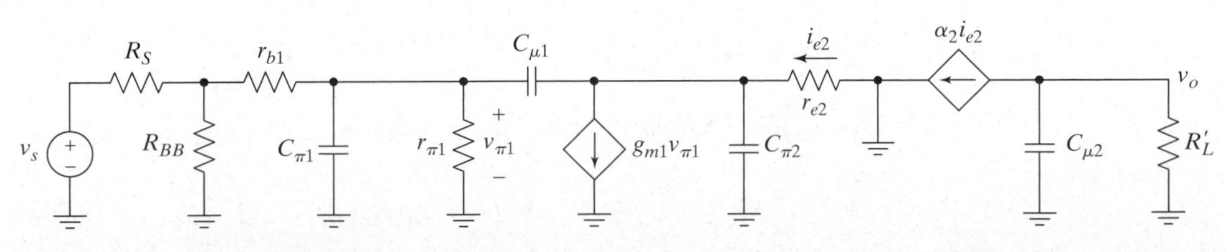

Figure 9–84 The high-frequency small-signal AC equivalent circuit of the cascode amplifier in Figure 9-83.

Introductory Treatment As noted in Section 9.6.4, the gain across $C_{\mu 1}$ is −1. To be explicit, we see from Figure 9-84 that at midband,

$$v_{c1} = -g_{m1} r_{e2} v_{\pi 1}, \quad \textbf{(9.214)}$$

so the gain across $C_{\mu 1}$ is

$$\frac{v_{c1}}{v_{\pi 1}} = -g_{m1} r_{e2} = \frac{-I_{C1}}{V_T} \frac{V_T}{I_{E2}} = -1, \quad \textbf{(9.215)}$$

where the final equality follows because $I_{C1} = I_{E2}$. Therefore, we can use (9.203) with the appropriate substitutions made:

$$f_{cH1} \approx \frac{1}{2\pi (R_S \| R_{BB} \| r_{\pi 1})(C_{\pi 1} + 2C_{\mu 1})} = 3 \text{ MHz}. \quad \textbf{(9.216)}$$

Similarly, using (9.204), we have

$$f_{cH2} \approx \frac{1}{2\pi R_L' C_{\mu 2}} = 12.4 \text{ MHz}. \quad \textbf{(9.217)}$$

Using these values in (9.205) and solving yields (see Problem P9.143) $f_{cH} \approx$ 2.8 MHz, about 12% lower than the value obtained by simulation, which is 3.2 MHz (Exam10.sch).

Advanced Treatment Using the equivalent circuit shown in Figure 9-84, we can derive four open-circuit time constants for the cascode amplifier. The overall cutoff frequency will then be

$$f_{cH} \approx \frac{1}{2\pi (\tau_{\pi 1o} + \tau_{\mu 1o} + \tau_{\pi 2o} + \tau_{\mu 2o})}. \quad \textbf{(9.218)}$$

Setting v_s to zero and open circuiting the other capacitors, we see from Figure 9-84 that

$$\tau_{\pi 1o} = R_{\pi 1o} C_{\pi 1} = [(R_S \| R_{BB} + r_{b1}) \| r_{\pi 1}] C_{\pi 1}. \quad \textbf{(9.219)}$$

To find $\tau_{\mu 1o}$, we use the equivalent circuit in Figure 9-85. Using KVL, we write

$$v_x = i_x R_{\pi 1o} + i_{e2} r_{e2}, \quad \textbf{(9.220)}$$

and, using KCL, we write

$$i_{e2} = i_x + g_{m1} v_{\pi 1} = i_x (1 + g_{m1} R_{\pi 1o}). \quad \textbf{(9.221)}$$

Substituting (9.221) into (9.220) and solving, we obtain

$$R_{\mu 1o} = \frac{v_x}{i_x} = R_{\pi 1o} + (1 + g_{m1} R_{\pi 1o}) r_{e2} \quad \textbf{(9.222)}$$

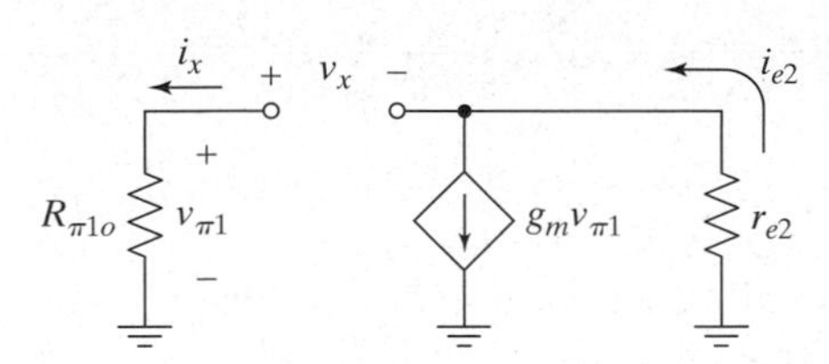

Figure 9–85 Circuit for finding $R_{\mu 1o}$.

and

$$\tau_{\mu 1o} = R_{\mu 1o} C_{\mu 1} = [R_{\pi 1o} + (1 + g_{m1} R_{\pi 1o}) r_{e2}] C_{\mu 1}. \quad \textbf{(9.223)}$$

Also, with $v_s = 0$ and the capacitors open circuited, $v_{\pi 1} = 0$, and so

$$\tau_{\pi 2o} = r_{e2} C_{\pi 2}. \quad \textbf{(9.224)}$$

Finally, $v_s = 0$ sets $v_{\pi 1} = 0$, which in turn sets $i_{e2} = 0$, so by inspection, we have

$$\tau_{\mu 2o} = R'_L C_{\mu 2}. \quad \textbf{(9.225)}$$

Using (9.219), (9.223), (9.224), and (9.225) in (9.218) yields the high cutoff frequency for the circuit.

Example 9.10

Problem:
Find the approximate high cutoff frequency for the cascode in Figure 9-83, using open-circuit time constants. Use the 2N2222 transistor model from Appendix B.

Solution:
We noted in the text that a DC bias analysis reveals that $V_{B1} = 2$ V, $V_{B2} = 3$ V, $V_{C2} =$ 6 V, and $I_{C1} = 1$ mA, and the corresponding small-signal model parameters are $r_{\pi 1} = 4.7\ \text{k}\Omega$, $r_{e2} = 26\ \Omega$, $g_{m1} = 38$ mA/V, $C_{\pi 1} = C_{\pi 2} = 60$ pF, $C_{\mu 1} = 6.6$ pF, and $C_{\mu 2} = 4.5$ pF. In addition, r_b is 10 Ω for both transistors.

Using (9.219), (9.223), (9.224), (9.225), and the given values, we find that $\tau_{\pi 1o} =$ 44.5 ns, $\tau_{\mu 1o} = 9.9$ ns, $\tau_{\pi 2o} = 1.6$ ns, and $\tau_{\mu 2o} = 12.9$ ns. Plugging into (9.218) yields $f_{cH} \approx 2.3$ MHz, which is 28% below the simulated value of 3.2 MHz.

The error is relatively large in this example because none of the time constants is dominant, which is often true in a cascode amplifier. Nevertheless, the relative contributions of the different time constants are close to correct, and the overall answer is again conservative.

■

9.6.6 The MOSFET Cascode Amplifier: A Common-Source Common-Gate Cascade

A MOSFET cascode amplifier is shown in Figure 9-86. The transistors in this circuit have $K = 400\ \mu A/V^2$, $V_{th} = 0.5$ V, $C_{gs} = 8$ pF, and $C_{gd} = 2$ pF. (They are Nsimple_mos devices as in Appendix B, but with their width increased by a factor of 4.) A DC bias analysis of this amplifier reveals that $V_{G1} = 1.6$ V, $V_{G2} = 3.2$ V, $I_{D1} = I_{D2} = 484\ \mu A$, and $V_{D2} = 5.2$ V. The corresponding small-signal model parameters are $g_{m1} = g_{m2} = 880\ \mu A/V^2$.

The high-frequency small-signal AC equivalent circuit for this amplifier is shown in Figure 9-87. We have used the T model for the common-gate transistor, since it is most convenient. If we open circuit all of the capacitors, we find that the small-signal midband gain is

$$A_v = \frac{v_o}{v_i} = -g_m(R_L \| R_D) = -4.4. \tag{9.226}$$

Introductory Treatment As noted in Section 9.6.4, the gain across C_{gd1} is −1. To be explicit, we see from Figure 9-87 that at midband,

$$v_{d1} = \frac{-g_{m1}}{g_{m2}} v_{gs1}, \tag{9.227}$$

so the gain across C_{gd1} is

$$\frac{v_{d1}}{v_{g1}} = \frac{-g_{m1}}{g_{m2}} = -1, \tag{9.228}$$

where the final equality follows because the transistors have equal drain currents. We can therefore use (9.203) with the appropriate substitutions made:

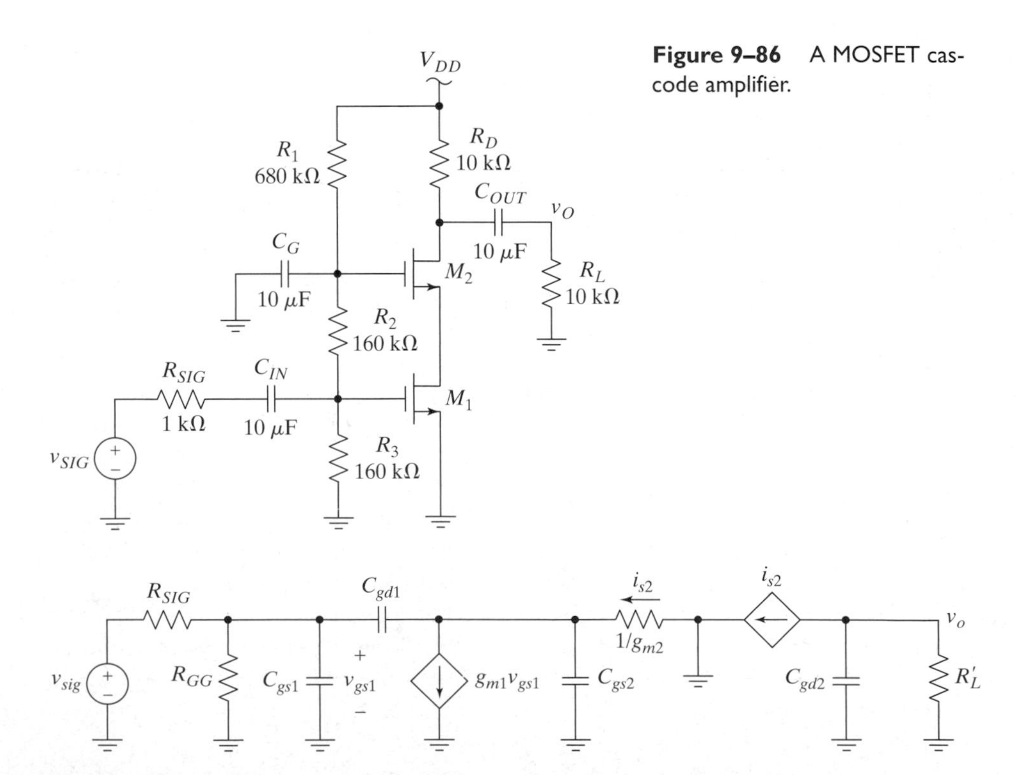

Figure 9–86 A MOSFET cascode amplifier.

Figure 9–87 The high-frequency small-signal AC equivalent circuit for the amplifier in Figure 9-86.

$$f_{cH1} \approx \frac{1}{2\pi(R_{SIG}\|R_{GG})(C_{gs1} + 2C_{gd1})} = 13.4 \text{ MHz}. \quad \textbf{(9.229)}$$

Similarly, using (9.204), we have

$$f_{cH2} \approx \frac{1}{2\pi R'_L C_{gd2}} = 15.9 \text{ MHz}. \quad \textbf{(9.230)}$$

Using these values in (9.205) and solving yields (*see* Problem P9.151) $f_{cH} \approx$ 9.3 MHz, about 32% higher than the value obtained by simulation, which is 7.1 MHz (Exam11.sch).

Advanced Treatment Using the equivalent circuit shown in Figure 9-87, we can derive four open-circuit time constants for the cascode amplifier. The overall cut-off frequency will then be

$$f_{cH} \approx \frac{1}{2\pi(\tau_{gs1o} + \tau_{gd1o} + \tau_{gs2o} + \tau_{gd2o})}. \quad \textbf{(9.231)}$$

With v_{sig} set equal to zero and the capacitors open-circuited, we find that

$$\tau_{gs1o} = R_{gs1o}C_{gs1} = (R_{SIG}\|R_{GG})C_{gs1}. \quad \textbf{(9.232)}$$

To find the driving-point resistance seen by C_{gd1}, we can use the circuit shown in Figure 9-88. We first use KVL to write

$$v_x = i_x R_{gs1o} + \frac{i_{s2}}{g_{m2}} \quad \textbf{(9.233)}$$

and KCL to write

$$i_{s2} = i_x + g_{m1}v_{gs1} = i_x(1 + g_{m1}R_{gs1o}). \quad \textbf{(9.234)}$$

We then combine these two equations and solve to obtain

$$R_{gd1o} = \frac{v_x}{i_x} = R_{gs1o} + \frac{(1 + g_{m1}R_{gs1o})}{g_{m2}}, \quad \textbf{(9.235)}$$

and finally

$$\tau_{gd1o} = \left[R_{gs1o} + \frac{(1 + g_{m1}R_{gs1o})}{g_{m2}}\right]C_{gd1}. \quad \textbf{(9.236)}$$

We next note that with v_{sig} set to zero and the capacitors opened, v_{gs1} is also zero, so

$$\tau_{gs2o} = \frac{C_{gs2}}{g_{m2}}. \quad \textbf{(9.237)}$$

With v_{sig} set to zero, we again see that v_{gs1} is zero; therefore, so is i_{s2}, and we find that

$$\tau_{gd2o} = R'_L C_{gd2}. \quad \textbf{(9.238)}$$

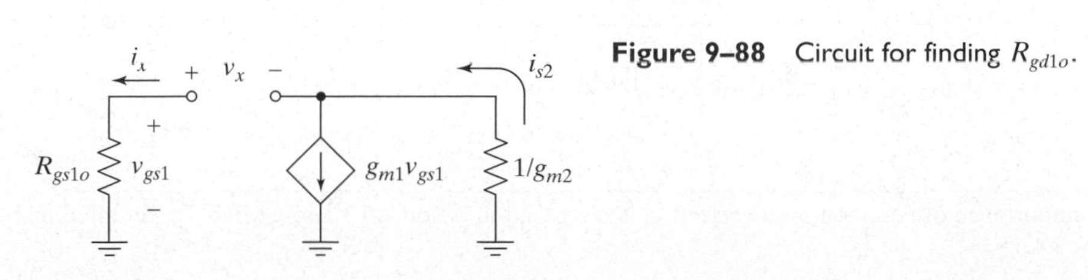

Figure 9–88 Circuit for finding R_{gd1o}.

Example 9.11

Problem:
Find the approximate high cutoff frequency for the cascode in Figure 9-86, using open-circuit time constants. The transistors are Nsimple_mos as in Appendix B, but with the width multiplied by a factor of four.

Solution:
We noted in the text that a DC bias analysis yields $V_{G1} = 1.6$ V, $V_{G2} = 3.2$ V, $I_{D1} = I_{D2} = 484$ μA, and $V_{D2} = 5.2$ V. The corresponding small-signal model parameters are $g_{m1} = g_{m2} = 880\ \mu\text{A/V}^2$. The capacitors are given in the text as $C_{gs} = 8$ pF and $C_{gd} = 2$ pF.

Using (9.232), (9.236), (9.237), (9.238), and the given values, we find that $\tau_{gs1o} = 7.9$ ns, $\tau_{gd1o} = 6.2$ ns, $\tau_{gs2o} = 9.1$ ns, and $\tau_{gd2o} = 12.9$ ns. Plugging into (9.231) yields $f_{cH} \approx 4.4$ MHz, which is 38% below the simulated value of 7.1 MHz.

The error is relatively large in this example, because none of the time constants are dominant, which is often true in a cascode amplifier. Nevertheless, the relative contributions of the different time constants are close to correct, and the overall answer is again conservative.

■

9.7 DIFFERENTIAL AMPLIFIERS

The small-signal midband AC performance of differential amplifiers was analyzed in Section 8.6. In the current section, we examine the frequency response of differential stages. Because differential amplifiers are almost always DC coupled, we will not be concerned with the low cutoff frequency in this section.

When a differential input is applied, resistively loaded differential amplifiers can be analyzed using the half-circuit concept presented in Section 8.6. In that case, the frequency response is identical to the single-transistor stage with both voltage and current gain (i.e., the common-merge, common-emitter, and common-source stages). When the input is a combination of differential-mode and common-mode signals, we are often interested in how the common-mode rejection ratio[12] (CMRR) varies with frequency, so we will consider that issue here. Finally, we will examine the frequency response of actively loaded differential stages.

9.7.1 The Generic Differential Pair

Consider the resistively loaded generic differential pair shown in Figure 9-89. This is similar to Figure 8-89, and we analyzed its midband operation in Section 8.6.

The high-frequency differential-mode half circuit for the amplifier in Figure 9-89 is shown in Figure 9-90 and is topologically identical to the high-frequency equivalent of the common-merge stage analyzed in Section 9.2.1. We can find an approximation to f_{cH} for this circuit by using either the Miller approximation or the more advanced method of open-circuit time constants. Using the Miller approximation, we find that (*see* Problem P9.161)

$$f_{cH} \approx \frac{1}{2\pi\left(\frac{R_S}{2}\middle\|r_{cm}\right)\left[(1 + g_m R_O)C_{oc} + C_{cm}\right]}, \tag{9.239}$$

[12] The importance of common-mode rejection is discussed in Aside A5.2 and CMRR is presented in Section 5.4.6.

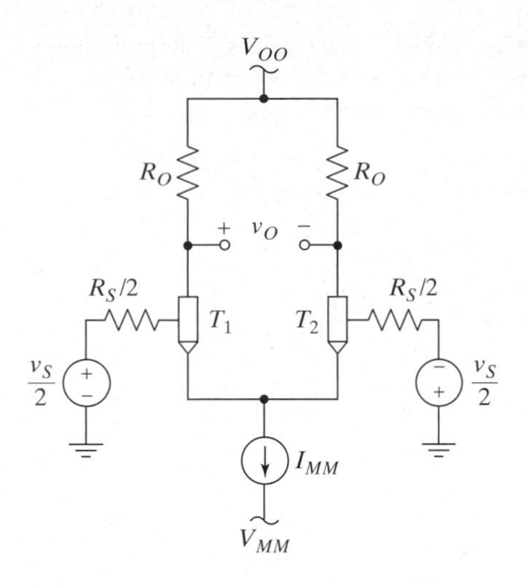

Figure 9–89 A resistively loaded generic differential pair.

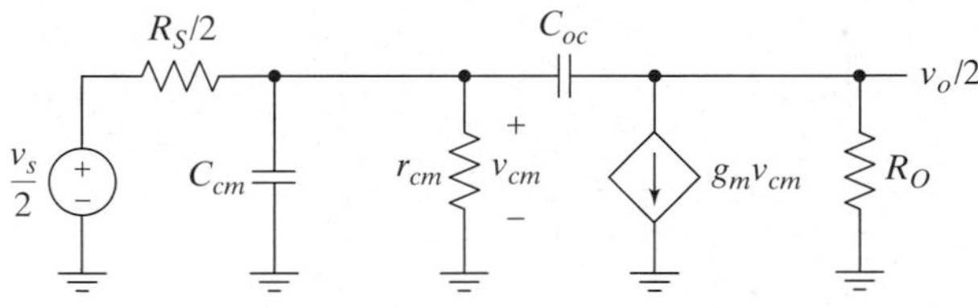

Figure 9–90 The high-frequency differential-mode half circuit for the amplifier in Figure 9-89.

and using open-circuit time constants, we get (*see* Problem P9.162)

$$f_{cH} \approx \frac{1}{2\pi(R_{cmo}C_{cm} + [R_{cmo} + (1 + g_m R_{cmo})R_O]C_{oc})}, \tag{9.240}$$

where $R_{cmo} = (R_S/2)\|r_{cm}$. The two answers are almost the same if the Miller multiplication of C_{oc} dominates.

Now consider the amplifier driven by a combination of common-mode and differential-mode signals, as shown in Figure 9-91.[13] The output is assumed to be taken single ended from the output of T_2 as shown. The common-mode gain of this circuit depends critically on how we model the current source. If we use an ideal model for I_{MM}, the merge terminal in the common-mode half-circuit equivalent is floating (i.e., it isn't connected to anything). As a result, the currents in the transistors will be zero and the common-mode AC output voltage will be zero, which implies that the common-mode gain is zero. However, the situation is quite different for a more realistic model of the current source.

Figure 9-92 shows the common-mode half circuit for this amplifier if we model the current source with a small-signal AC output resistance R_{MM}. To make the circuit symmetric, we must split R_{MM} into two equal resistors in parallel: hence, the resistance in the half-circuit equivalent is $2R_{MM}$. The low-frequency gain of this circuit is

$$A_{cm0} = \left(\frac{r_{cm} + (1 + a_{im})2R_{MM}}{r_{cm} + (1 + a_{im})2R_{MM} + R_S/2}\right)\left(\frac{-g_m r_{cm} R_O}{r_{cm} + (1 + a_{im})2R_{MM}}\right)$$

$$\approx \frac{-R_O}{2R_{MM}}, \tag{9.241}$$

[13] When used alone, the subscript *cm* always means conrol-to-merge, as in v_{cm}. But when used with other characters in this section (e.g., v_{scm}), it refers to the common-mode value of the quantity.

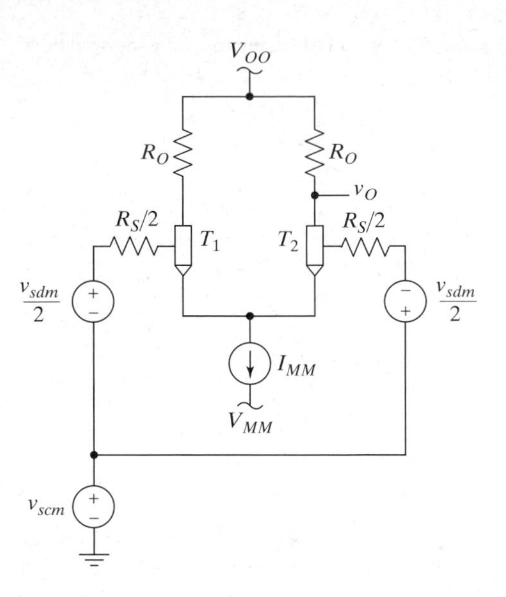

Figure 9–91 The generic differential amplifier driven by a combination of differential- and common-mode signals.

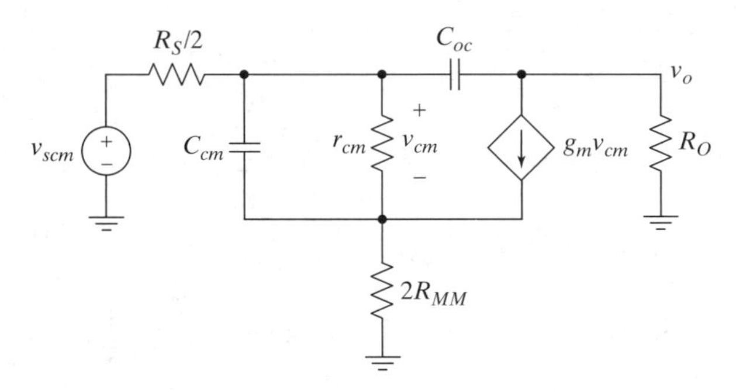

Figure 9–92 The common-mode half circuit for the amplifier in Figure 9-91.

where the first term in parentheses is the attenuation caused by R_S, the second term is the gain from the control to the output, and the approximation follows by remembering that $a_{im} = g_m r_{cm}$ and assuming that $a_{im} \gg 1$. This low-frequency gain is usually much less than unity, since R_{MM} is typically quite large.

As the frequency is increased, the two capacitors affect the circuit in opposite directions. Because it shunts r_{cm}, C_{cm} will cause the gain to roll off, but because the gain is typically low to begin with, this is a minor effect. The major effect is due to C_{oc}, which will couple the signal directly to the output and cause an increase in the gain for typical component values. As the frequency approaches infinity and the capacitors are nearly short circuits, the gain is given by the two-resistor voltage divider:

$$A_{cm}(\infty) \approx \frac{R_O \| 2R_{MM}}{R_O \| 2R_{MM} + R_S/2}. \quad \textbf{(9.242)}$$

In this case, the common-mode gain would have the general form shown in Figure 9-93.

A more realistic model for the current source would also include some capacitance, as shown in Figure 9-94. To make the circuit symmetric so that we can draw

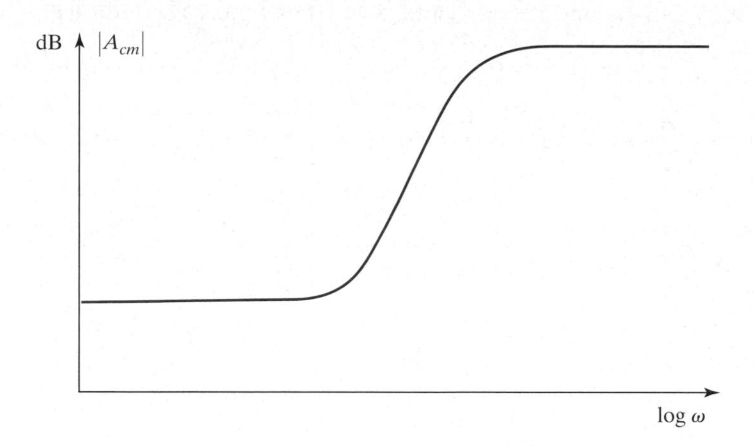

Figure 9–93 The common-mode gain for the circuit of Figure 9-92.

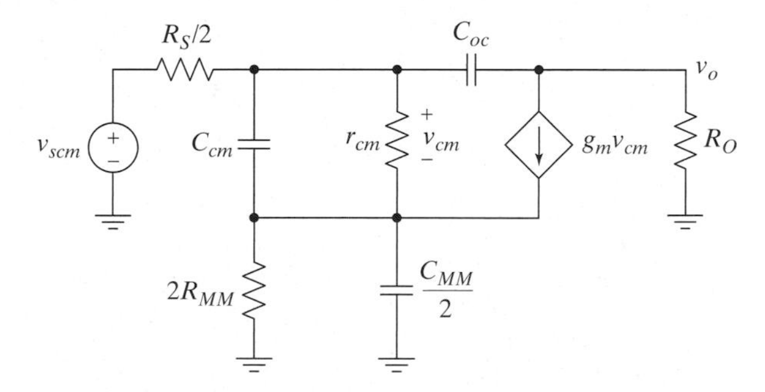

Figure 9–94 A more realistic common-mode half circuit for the amplifier in Figure 9-91.

a half-circuit equivalent, we split the capacitance into two equal capacitors in parallel; hence, the capacitor is $C_{MM}/2$. The analysis of this circuit is more complicated, and the detailed response depends on the values, which may vary over a wide range, but it is still true that the common-mode gain increases for a range of frequencies. The qualitative difference is that the common-mode gain eventually rolls off.

In this circuit, there are two possible causes of the gain increasing with frequency: First, C_{oc} couples the signal to the output, bypassing the transistor. Second, as $C_{MM}/2$ shunts $2R_{MM}$, the input divider in (9.241) will increase and cause the gain to increase. Which of these mechanisms will dominate depends on the component values. The roll-off in common-mode gain at high frequencies is caused by $C_{MM}/2$ and C_{cm} forming a low-pass filter with $R_S/2$ (and any series resistance in the control terminal of the transistor). To picture this situation, we note that for very high frequencies, r_{cm} and $2R_{MM}$ can be neglected, since their resistance will be much larger than the reactance of the capacitors in parallel with them. In this case, we see that C_{cm} and $C_{MM}/2$ are in series with each other and form an RC low-pass filter with $R_S/2$. The general form for the common-mode gain is then

$$A_{cm}(j\omega) = A_{cm0}\frac{(1 + j\omega/\omega_z)}{(1 + j\omega/\omega_p)}, \tag{9.243}$$

where A_{cm0} is given by (9.241) and the zero and pole frequencies depend on the details of the circuit (*see* Problems P9.163 and P9.164 for examples).

The CMRR for the amplifier in Figure 9-91 is defined to be the ratio of the differential-mode gain (from v_{sdm} to v_o) to the common-mode gain (from v_{scm} to v_o). Making use of our previous results and assuming that the Miller effect dominates the differential gain, we obtain

$$\text{CMRR} = \frac{A_{dm0}}{A_{cm0}} \frac{(1 + j\omega/\omega_p)}{(1 + j\omega/\omega_{cHd})(1 + j\omega/\omega_z)} \tag{9.244}$$

where A_{cm0} is given by (9.241), the zero and the pole have not been given in detail, ω_{cHd} is found from (9.239) to be

$$\omega_{cHd} \approx \frac{1}{\left(\frac{R_S}{2}\middle\| r_{cm}\right)\left[(1 + g_m R_O)C_{oc} + C_{cm}\right]} \tag{9.245}$$

or from (9.240) to be

$$\omega_{cHd} \approx \frac{1}{R_{cmo}C_{cm} + \left[R_{cmo} + (1 + g_m R_{cmo})R_O\right]C_{oc}}, \tag{9.246}$$

and

$$A_{dm0} = \frac{g_m R_O r_{cm}}{2(r_{cm} + R_S/2)}. \tag{9.247}$$

Note that (9.239) and (9.240) were derived for the gain from a differential input to a *differential* output, but since the gain rolls off equally at the output of each transistor, the bandwidth is the same when the output is taken single-ended; hence, (9.245) and (9.246) are correct.

Example 9.12

Problem:
Plot the common-mode gain, differential-mode gain, and CMRR as a function of frequency for the differential amplifier in Figure 9-91. Compare the results with those predicted by (9.245), (9.246), (9.247), (9.241), (9.242), and Problem P9.163. Assume that $R_O = 8\ \text{k}\Omega$, $R_S = 1\text{k}\Omega$, $R_{MM} = 100\ \text{k}\Omega$, $C_{MM} = 4$ pF, $r_{cm} = 10\ \text{k}\Omega$, $g_m = 10$ mA/V, $C_{oc} =$ 1.2 pF, and $C_{cm} = 14$ pF.

Solution:
Plugging into the equations cited, we obtain $A_{dm0} = 31.6$ dB, f_{cHd}(Miller) = 3 MHz, f_{cHd}(open-circuit τ) = 2.5 MHz, and $A_{cm0} = -0.04$, $A_{cm}(\infty) = 0.94$, and the pole and zero frequencies for the CM gain are $f_z = 0.8$ MHz and $f_p = 19$ MHz.

Using SPICE with the given values, we obtain the plots shown in Figure 9-95. From these plots, we find that $A_{dm0} = 31.6$ dB, which agrees with our calculation; $f_{cHd} = 2.6$ MHz, which agrees well with the value calculated by open-circuit time constants and, to a lesser degree, with the value calculated using the Miller

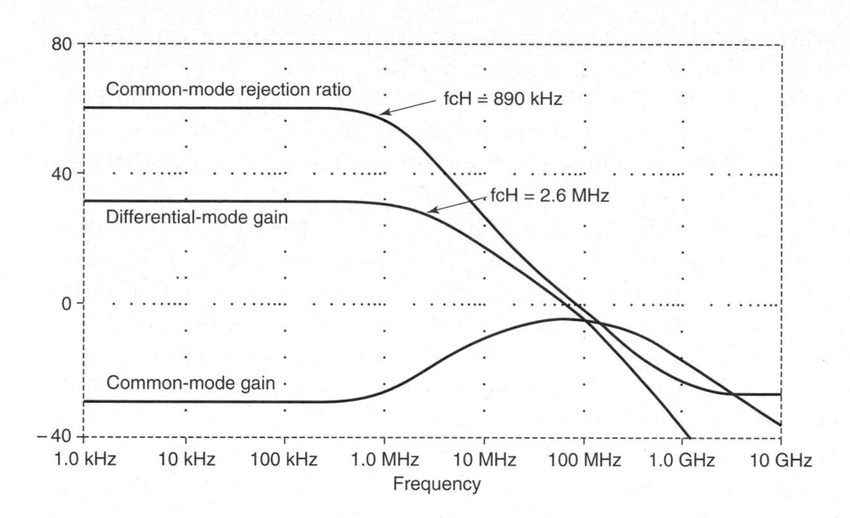

Figure 9–95

approximation; $A_{cm0} = -0.04$, which is the same as our calculation; and $f_z = 1$ MHz, which agrees well with our calculation. The values of f_p and $A_{cm}(\infty)$ cannot be seen on the plot, because of the high-frequency roll-off in the common-mode gain, which is unaccounted for in the hand analysis. Nevertheless, it appears from the plot that if the high-frequency roll-off were removed, the calculated values would be close.

As illustrated by this example, the bandwidth of the CMRR is typically lower than that of the differential-mode gain. Note that if the value of R_{MM} were larger, as is often the case, then the common-mode gain would start to increase at a lower frequency and the bandwidth of the CMRR would be even lower.

The Generic Differential Pair with an Active Load The generic differential pair with an active load is shown in Figure 9-96. A complete analysis of the frequency response of this stage is quite tedious and not usually necessary for design. We will instead perform a simple approximate analysis that yields useful results.

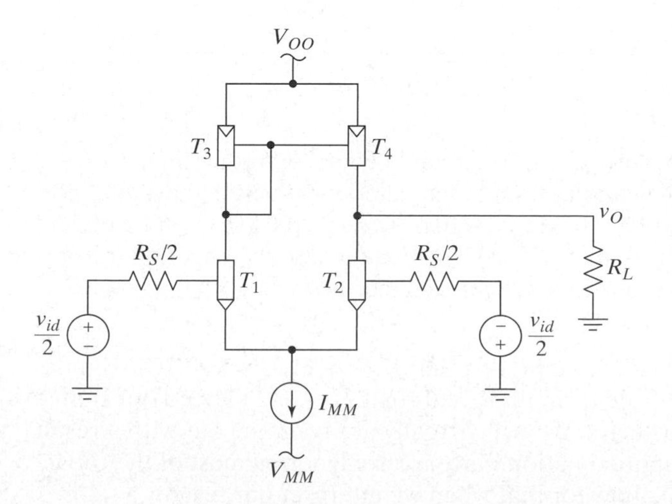

Figure 9–96 An actively loaded generic differential pair.

Since T_3 is diode-connected, it represents a small AC impedance and prevents the voltage at the output of T_1 from changing much. As a result, capacitance at this node does not significantly affect the frequency response. Also, if the output resistances of the transistors are large, the fact that the outputs of T_1 and T_2 are not tied to a symmetric load has a negligible effect on the differential-mode symmetry at the merge terminals of T_1 and T_2, and we can still approximate that node as a differential-mode ground. Since the node is approximately constant at ground potential, capacitance at this node does not affect the frequency response. There are only two other locations to examine in this circuit: the output node and the differential input attenuation factor (i.e., the transfer function from the differential source to the control terminals of T_1 and T_2). To simplify the discussion, let's first assume that R_S is small enough that the differential input attenuation factor has a much higher cutoff frequency than that of the rest of the circuit.

The total capacitance at the output node comprises C_{oc4} connected to the output of T_1, C_{oc2} connected to the control terminal of T_2, and whatever parasitic capacitance may be present, $C_{\text{parasitic}}$. We have already concluded that the output of T_1 is nearly an AC ground, so we can say that C_{oc4} approximately connects to ground. If the voltage gain of the amplifier is large, the AC voltage at the control of T_2 will be much smaller than at the output, and we can also say that C_{oc2} approximately connects to AC ground. Given these assumptions, the total capacitance from the output node to ground is

$$C_o \approx C_{oc2} + C_{oc4} + C_{\text{parasitic}}. \tag{9.248}$$

The total small-signal AC resistance to ground is given by

$$R_o \approx r_{o2} \| r_{o4} \| R_L, \tag{9.249}$$

where we have again assumed that the merge terminals of T_1 and T_2 are approximately a differential-mode ground. Given these values and still ignoring the input attenuation factor, we can approximate the high cutoff frequency of this circuit as

$$f_{cH} \approx \frac{1}{2\pi R_o C_o}. \tag{9.250}$$

This approximation is crude, but it is sometimes useful, as the following example will illustrate.

Example 9.13

Problem:
Use (9.250) to predict the bandwidth of the circuit in Figure 9-96, and then compare the results with a SPICE simulation. Assume that all four transistors have identical small-signal parameters with $r_{cm} = 10\ \text{k}\Omega$, $g_m = 10\ \text{mA/V}$, $r_o = 200\ \text{k}\Omega$, $C_{oc} = 1.2\ \text{pF}$, and $C_{cm} = 14\ \text{pF}$. In addition, assume that $R_S = 0\ \Omega$ (so that the input attenuation factor is unity), and R_L is infinite (i.e., no external load is connected).

Solution:
Using (9.250) and the given data, we predict that $f_{cH} = 663\ \text{kHz}$. A SPICE simulation of the small-signal equivalent circuit reveals that $f_{cH} = 442\ \text{kHz}$, so our simple result gets us in the right ballpark. We will virtually always come out with a result that is too high using this approximation, since we are ignoring most of the capacitors. Consequently, we can allow for that when we interpret our results.

If we include some source resistance, our approximation is not as good. For example, using $R_S = 100\ \Omega$ yields $f_{cH} = 380$ kHz. The approximation becomes much worse as R_S increases; this effect is due to the input attenuation factor, which is discussed next.

■

The differential amplifier does have some effective input capacitance to ground from each input (i.e., from the control terminal of T_1 to ground and from the control terminal of T_2 to ground). Therefore, when R_S is nonzero, the input attenuation factor will be a low-pass transfer function and, depending on the values, may limit the bandwidth.

Even though we ignored the asymmetry in the load in the preceding analysis, it does have a significant effect on the performance of this type of differential amplifier when the input attenuation factor is significant. Because the output terminal of T_1 is nearly an AC ground, there isn't any Miller multiplication of C_{oc1}, whereas, the gain across C_{oc2} is quite large and leads to a large Miller input capacitance on that side. As a result, the signal amplitude at the control terminal of T_2 rolls off at a lower frequency than it does at the control terminal of T_1.

If we drive the circuit asymmetrically, as shown in Figure 9-97, we get some interesting results. When we connect the signal source to the control of T_1 and connect the control of T_2 to ground, we do away with the Miller multiplication and achieve a higher bandwidth. In essence, the circuit looks like a merge follower driving a common-control amplifier, but it has the benefit of having the merge follower output current mirrored to the load to boost the DC gain. If we modify the SPICE circuit used in Example 9.13 and use $R_S = 1\ \text{k}\Omega$ to emphasize the point, we find that driving the control terminal of T_1 (part (a) of the figure) leads to a bandwidth of 440 kHz, while driving the control terminal of T_2 (part (b) of the figure) yields a bandwidth of 110 kHz. This large difference in bandwidths can be important in applications employing negative feedback, since the bandwidth from the overall input is different from the bandwidth seen by the feedback signal.

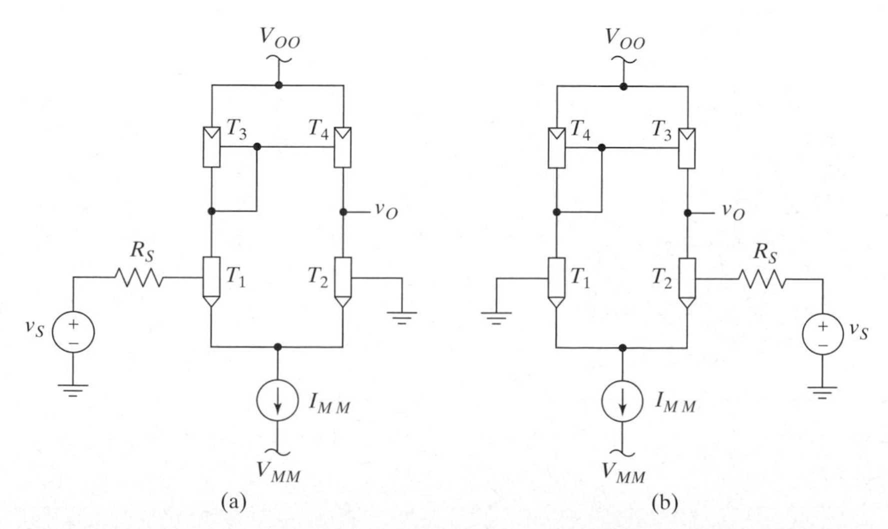

Figure 9–97 Driving the differential pair single ended. (a) Driving T_1 and (b) driving T_2.

9.7.2 A Bipolar Implementation: The Emitter-Coupled Pair

A bipolar differential pair, or emitter-coupled pair, with resistive loads is shown in Figure 9-98. Whether we take the output differentially or single-ended from either of the collectors to ground, the high cutoff frequency for a differential-mode input can be approximated by (9.239) or (9.240), with the appropriate changes in notation and the inclusion of r_b; the Miller approximation yields

$$f_{cH} \approx \frac{1}{2\pi\left[\left(\frac{R_S}{2} + r_b\right)\middle\|r_\pi\right]\left[(1 + g_m R_C)C_\mu + C_\pi\right]}, \tag{9.251}$$

while the open-circuit time, constant method leads to

$$f_{cH} \approx \frac{1}{2\pi(R_{\pi o}C_\pi + [R_{\pi o} + (1 + g_m R_{\pi o})R_C]C_\mu)}, \tag{9.252}$$

where $R_{\pi o} = r_\pi \| (r_b + R_S/2)$.

If we examine the frequency response for a common-mode input, we find from (9.243) that the gain is approximately given by

$$A_{cm}(j\omega) = A_{cm0}\frac{(1 + j\omega/\omega_z)}{(1 + j\omega/\omega_p)}, \tag{9.253}$$

where the DC common-mode gain is

$$A_{cm0} = \left(\frac{r_\pi + (1 + \beta)2R_{EE}}{r_\pi + (1 + \beta)2R_{EE} + R_S/2}\right)\left(\frac{-g_m r_\pi R_C}{r_\pi + (1 + \beta)2R_{EE}}\right) \approx \frac{-R_C}{2R_{EE}}, \tag{9.254}$$

as in (9.241). The pole and zero in (9.253) are again dependent on the details (*see* Problems P9.168 and P9.169).

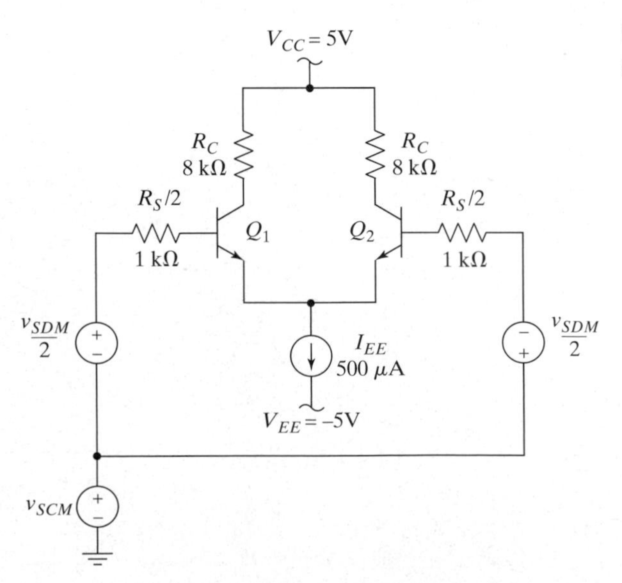

Figure 9–98 An emitter-coupled pair with resistive loads.

Consider using a single-ended output from the circuit in Figure 9-98, say, $v_O = v_{C2}$. The CMRR is then defined to be the ratio of the differential-mode gain (from v_{sdm} to v_o) to the common-mode gain (from v_{scm} to v_o). Making use of our previous results and assuming that the Miller effect dominates the differential gain, we again obtain

$$\text{CMRR} = \frac{A_{dm0}}{A_{cm0}} \frac{(1 + j\omega/\omega_p)}{(1 + j\omega/\omega_{cHd})(1 + j\omega/\omega_z)}, \tag{9.255}$$

where A_{cm0} is now given by (9.254), the zero and the pole have not been given in detail, ω_{cHd} is found from (9.251) to be

$$\omega_{cHd} \approx \frac{1}{\left(\frac{R_S}{2}\middle\|r_\pi\right)\left[(1 + g_m R_C)C_\mu + C_\pi\right]}, \tag{9.256}$$

or from (9.252) to be

$$\omega_{cHd} \approx \frac{1}{R_{\pi o}C_\pi + \left[R_{\pi o} + (1 + g_m R_{\pi o})R_C\right]C_\mu}, \tag{9.257}$$

and

$$A_{dm0} = \frac{g_m R_C r_\pi}{2(r_\pi + R_S/2)}. \tag{9.258}$$

Example 9.14

Problem:
Find the differential-mode gain, common-mode gain, and CMRR as a function of frequency for the circuit shown in Figure 9-98 if the output is taken single-ended from the collector of Q_2 and the DC common-mode input voltage is zero. Assume that the current source has an output resistance $R_{EE} = 200$ kΩ and a capacitance $C_{EE} = 10$ pF. Assume that the transistors have $\beta = 100$, $r_b = 62.5\ \Omega$, $\tau_F = 1$ ns, $C_{je0} = 0.45$ pF, and $C_{jc0} = 2$ pF. (These parameters are for the nominal model given in Appendix B.) Furthermore, assume that the impedance of the current source dominates the common-mode frequency response, so the results of Problem P9.169 apply.

Solution:
With $I_C = I_{EE}/2 = 250\ \mu$A for each transistor, we calculate $g_m = 9.6$ mA/V, $r_\pi = 10.4$ kΩ, $r_e = 104\ \Omega$, $C_\pi = 10.5$ pF, and $C_\mu = 1.2$ pF. From the midband AC analysis, we obtain

$$\alpha_{idm} = \frac{v_{idm}}{v_{sdm}} = \frac{2r_\pi}{2r_\pi + R_S} = 0.91$$

for the differential input attenuation factor and, therefore,

$$A_{dm0} = \frac{v_o}{v_{sdm}} = \frac{\alpha_{idm}(-g_m R_C)}{2} = -35$$

for the gain. The common-mode input resistance is extremely large, so we can ignore the input attenuation factor and write

$$A_{cm0} = \frac{v_o}{v_{scm}} = \frac{-R_C}{r_e + 2R_{EE}} = -0.02.$$

The corresponding DC CMRR is 65 dB. According to the Miller approximation, the high cutoff frequency for the differential-mode gain is given by (9.251) and is $f_{cHd} = 1.6$ MHz. Using open-circuit time constants and (9.252), we obtain $f_{cHd} =$ 1.6 MHz again.

Using the result presented in Problem P9.169, we find that the common-mode gain has a zero at 80 kHz and a pole at 306 MHz.

The results of a SPICE simulation of this circuit are shown in Figure 9-99. From this output, we can measure the zero in the common-mode gain,which we find occurs at 104 kHz, fairly close to our prediction. We do not predict the roll-off in the common-mode gain when we only include C_{EE} in the analysis, however. From this plot, we also see that $A_{cm0} = -0.02$, as predicted, and $A_{dm0} = -36$, close to our prediction. The simulation also reveals that $f_{cHd} = 1.4$ MHz, so our estimate was 14% high.

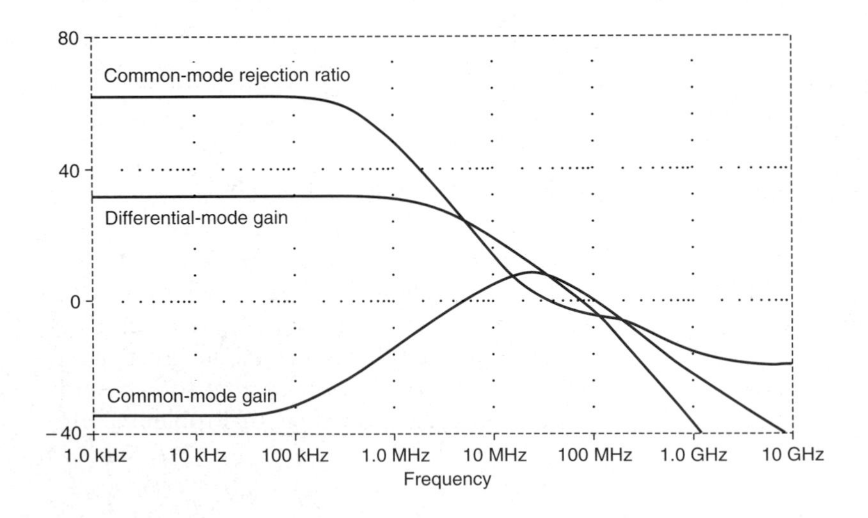

Figure 9–99 SPICE output.

The Emitter-Coupled Pair with Active Load An emitter-coupled pair with active load is shown in Figure 9-100. As with the generic differential stage with active load, we can find a simple approximation to the differential-mode high cutoff frequency by making some simplifying approximations. We first note that even with the asymmetrical load, the emitters of Q_1 and Q_2 will be an approximate differential-mode ground. We then note that the diode-connected Q_3 provides a low-resistance path to AC ground, so capacitance at that node doesn't matter. Therefore, if we ignore the input attenuation factor, the frequency response of the stage can be approximated by using the single time constant at the output. If we allow for an external load resistance and capacitance (both to ground), we then have

$$f_{cH} \approx \frac{1}{2\pi(r_{o2}\|r_{o4}\|R_L)(C_{\mu2} + C_{\mu4} + C_L)}. \tag{9.259}$$

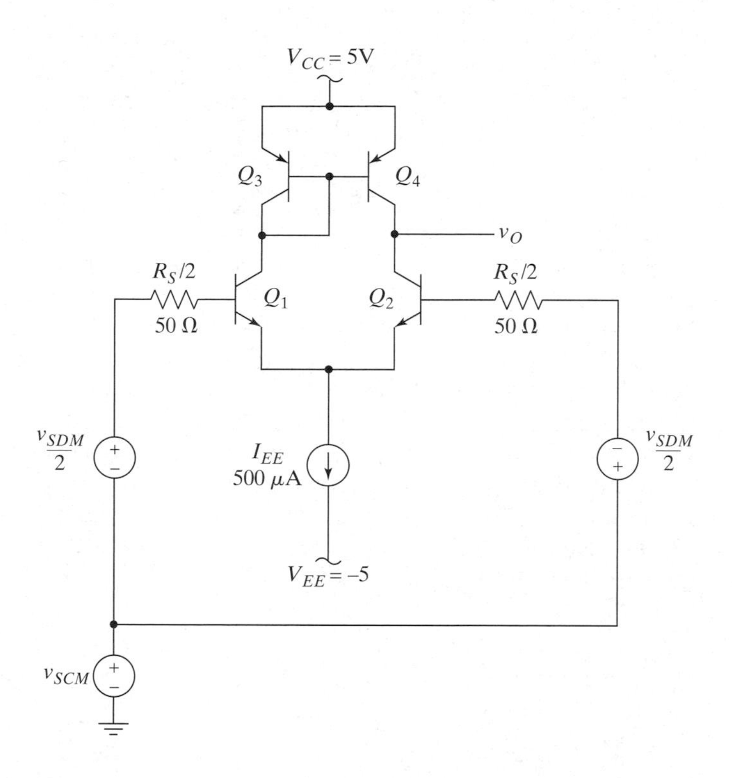

Figure 9–100 An emitter-coupled pair with active load.

Example 9.15

Problem:

Find the differential-mode high cutoff frequency for the circuit in Figure 9-100, and compare it with the value obtained from SPICE. Assume that both the *npn* and *pnp* transistors have $\beta = 100$, $r_b = 62.5\ \Omega$, $V_A = 50$ V, $\tau_F = 1$ ns, $C_{je0} = 0.45$ pF, and $C_{jc0} =$ 2 pF (these parameters are for the nominal and nominalp models in Appendix B with V_A added) that $R_S = 100\ \Omega$, and that there is an external load capacitance (perhaps due to driving the next stage) of $C_L = 10$ pF.

Solution:
With $I_C = I_{EE}/2 = 250\ \mu A$ for each transistor, we calculate $g_m = 9.6\ mA/V$, $r_\pi = 10.4\ k\Omega$, $r_e = 104\ \Omega$, $r_o = 200\ k\Omega$, $C_\pi = 10.5\ pF$, $C_{\mu 2} = 1.1\ pF$, and $C_{\mu 4} = 2\ pF$. Using (9.259), we predict that $f_{cH} = 121\ kHz$. The SPICE simulation result is shown in Figure 9-101 and reveals that $f_{cH} = 97\ kHz$. The approximation becomes significantly less accurate as C_L decreases or R_S increases, since the input attenuation factor rolls off at frequencies at or near the f_{cH} predicted from the output time constant alone.

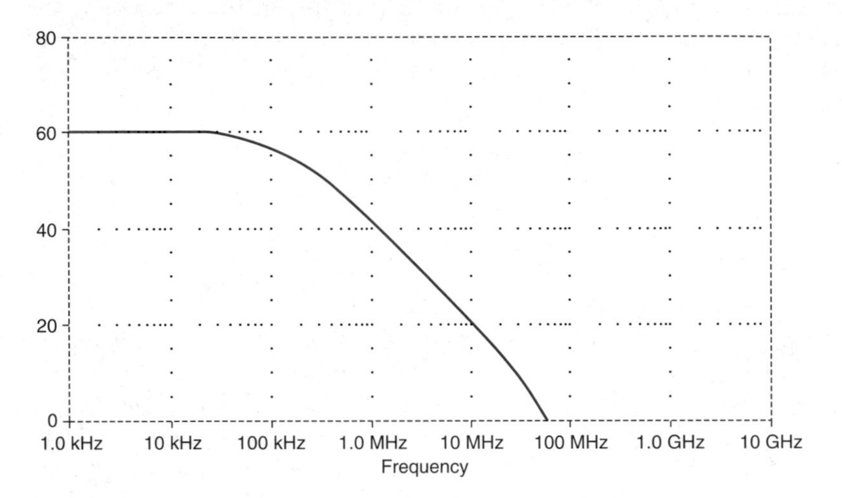

Figure 9–101 SPICE output.

As pointed out in Section 9.7.1, when this circuit is driven single ended, the bandwidths can be significantly different, depending on which side the input is connected to. Figure 9-102 shows the simulated responses when the circuit is driven from a 1 kΩ source connected to one of the inputs, whereas the other input is grounded. Note that the difference in bandwidths would be smaller if a load capacitance were included. If the load capacitance is large enough, the gain rolls off due to the load rather than the Miller effect, and single-ended bandwidths are nearly the same.

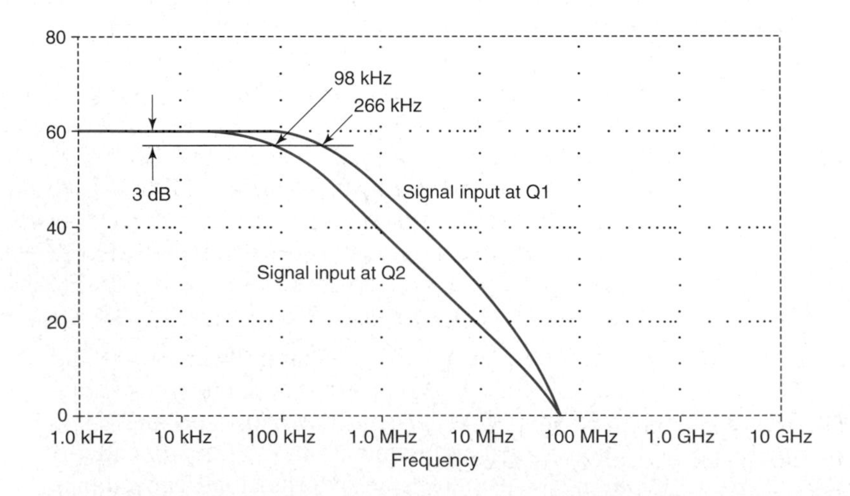

Figure 9–102 The responses of the circuit in Figure 9-100 when driven single ended. In each case, the signal source is tied to one input (with $R_S = 1\ k\Omega$), whereas the opposite input is grounded.

9.7.3 A FET Implementation: The Source-Coupled Pair

A MOSFET source-coupled pair is shown in Figure 9-103. Whether we take the output differentially or single ended from either of the drains to ground, the high cutoff frequency for a differential-mode input can be approximated by (9.239) or (9.240) with the appropriate changes in notation and the removal of r_{cm}; the Miller approximation yields

$$f_{cH} \approx \frac{1}{2\pi\left(\frac{R_{SIG}}{2}\right)\left[(1 + g_m R_D)C_{gd} + C_{gs}\right]}, \tag{9.260}$$

whereas the open-circuit time-constant method leads to

$$f_{cH} \approx \frac{1}{2\pi\left(\frac{R_{SIG}}{2}C_{gs} + \left[\frac{R_{SIG}}{2} + \left(1 + \frac{g_m R_{SIG}}{2}\right)R_D\right]C_{gd}\right)}. \tag{9.261}$$

If we examine the frequency response for a common-mode input, we find from (9.243) that the gain is approximately given by

$$A_{cm}(j\omega) = A_{cm0}\frac{(1 + j\omega/\omega_z)}{(1 + j\omega/\omega_p)}, \tag{9.262}$$

where the DC common-mode gain is (R_{SS} models the current source)

$$A_{cm0} = \left(\frac{-R_D}{1/g_m + 2R_{SS}}\right) \approx \frac{-R_D}{2R_{SS}}, \tag{9.263}$$

as in (9.241). The pole and zero in (9.262) are again dependent on the details (*see* Problems P9.175 and P9.176).

Consider using a single-ended output from the circuit in Figure 9-103, say, $v_O = v_{D2}$. The CMRR is then defined to be the ratio of the differential-mode gain (from v_{sdm} to v_o) to the common-mode gain (from v_{scm} to v_o). Making use of our previous results and assuming that the Miller effect dominates the differential gain, we again obtain

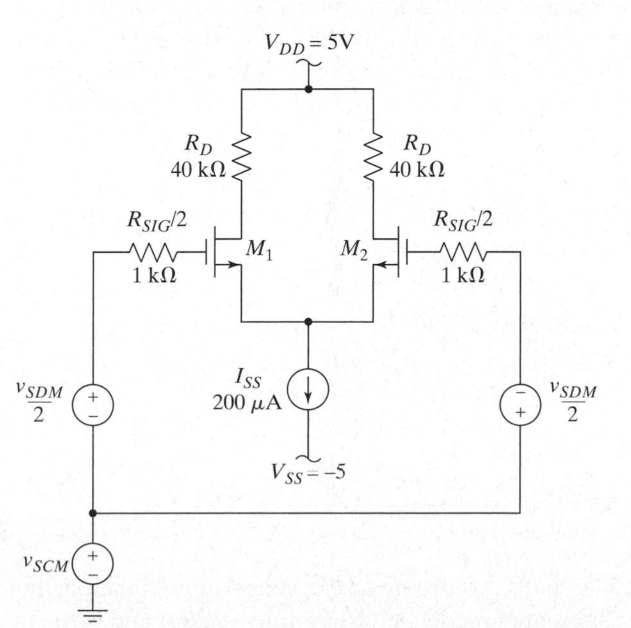

Figure 9–103 A MOSFET source-coupled pair with resistive loads.

$$\text{CMRR} = \frac{A_{dm0}}{A_{cm0}} \frac{(1 + j\omega/\omega_p)}{(1 + j\omega/\omega_{cHd})(1 + j\omega/\omega_z)} \tag{9.264}$$

where A_{cm0} is now given by (9.263), the zero and the pole have not been given in detail; ω_{cHd} is found from (9.260) to be

$$\omega_{cHd} \approx \frac{1}{\left(\frac{R_{SIG}}{2}\right)\left[(1 + g_m R_D)C_{gd} + C_{gs}\right]} \tag{9.265}$$

or, from (9.261) to be

$$\omega_{cHd} \approx \frac{1}{\frac{R_{SIG}}{2}C_{gs} + \left[\frac{R_{SIG}}{2} + \left(1 + \frac{g_m R_{SIG}}{2}\right)R_D\right]C_{gd}}, \tag{9.266}$$

and

$$A_{dm0} = \frac{g_m R_D}{2}. \tag{9.267}$$

Example 9.16

Problem:
Find the differential-mode gain, common-mode gain, and CMRR as a function of frequency for the circuit shown in Figure 9-103 if the output is taken single ended from the drain of M_2 and the DC common-mode input voltage is zero. Assume that the current source has an output resistance $R_{SS} = 100\ \text{k}\Omega$ and a capacitance $C_{SS} = 10$ pF. Assume that the transistors have $K = 100\ \mu\text{A/V}^2$, $C_{gs} = 2$ pF, $C_{gd} = 0.5$ pF, and $V_{th} = 0.5$ V. (These parameters are for the Nsimple_mos model in Appendix B.) Furthermore, assume that the impedance of the current source dominates the common-mode frequency response, so the results of Problem P9.176 apply.

Solution:
With $I_D = I_{SS}/2 = 100\ \mu$A and $K = 100\ \mu\text{A/V}^2$ for each transistor, we calculate $g_m = 200\ \mu$A/V. From the midband AC analysis, we obtain

$$A_{dm0} = \frac{v_o}{v_{sdm}} = \frac{(-g_m R_D)}{2} = -4$$

for the gain. We also have

$$A_{cm0} = \frac{v_o}{v_{scm}} = \frac{-R_D}{2R_{SS}} = -0.2.$$

The corresponding DC CMRR is 26 dB. According to the Miller approximation, the high cutoff frequency for the differential-mode gain is given by (9.260) and is $f_{cHd} = 24.5$ MHz. Using open-circuit time constants, we utilize (9.261) to obtain $f_{cHd} = 6$ MHz.

These results indicate that the Miller approximation is not valid for this circuit, and we expect the simulated bandwidth to be much closer to the open-circuit time-constant result.

Using the result presented in Problem P9.176, we find that the common-mode gain has a zero at 159 kHz and a pole at 6.4 MHz.

The results of a SPICE simulation of this circuit are shown in Figure 9-104. From this output, we can measure the DC CMRR, which is 27 dB, close to our prediction of 26 dB. The zero in the common-mode gain occurs at 176 kHz, again, close to our prediction. We do not predict the high-frequency roll-off in the common-mode gain when we only include C_{SS} in the analysis, but the first pole is close to our prediction. From this plot, we also see that $A_{cm0} = -0.2$, as predicted, and $A_{dm0} = -4.3$, close to our prediction. The simulation also reveals that $f_{cHd} =$ 6.3 MHz, so our estimate based on open-circuit time constants is only 5% low.

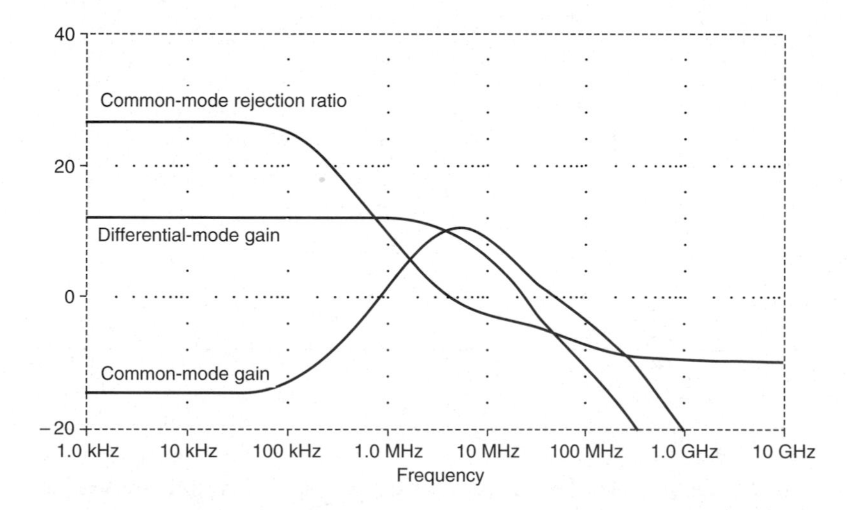

Figure 9–104 SPICE output.

■

The SCP with an Active Load A MOSFET source-coupled pair with active load is shown in Figure 9-105. As with the generic differential stage with active load, we can find a simple approximation to the differential-mode high cutoff frequency by making some simplifying approximations. We first note that even with the asymmetrical load, the sources of M_1 and M_2 will be an approximate differential-mode ground. We then note that the diode-connected M_3 provides a low-resistance path to AC ground. Therefore, if we ignore the input attenuation factor, the frequency response of the stage can be approximated by using the single time constant at the output. If we allow for an external load resistance and capacitance (both to ground), we have

$$f_{cH} \approx \frac{1}{2\pi(r_{o2} \| r_{o4} \| R_L)(C_{gd2} + C_{gd4} + C_L)}. \quad (9.268)$$

Example 9.17

Problem:

Find the high cutoff frequency for the circuit in Figure 9-105, and compare it with the value obtained from SPICE. Assume that both the NMOS and PMOS transistors have $K = 100\ \mu A/V^2$, $\lambda = 0.02\ V^{-1}$, $C_{gs} = 2$ pF, $C_{gd} = 0.5$ pF, and $V_{th} = 0.5$ V (these parameters are for the Nsimple2_mos and Psimple2_mos models in Appendix B), that $R_{SIG} = 2\ k\Omega$, and that there is an external load capacitance (perhaps due to driving the next stage) of $C_L = 5$ pF, but $R_L = \infty$.

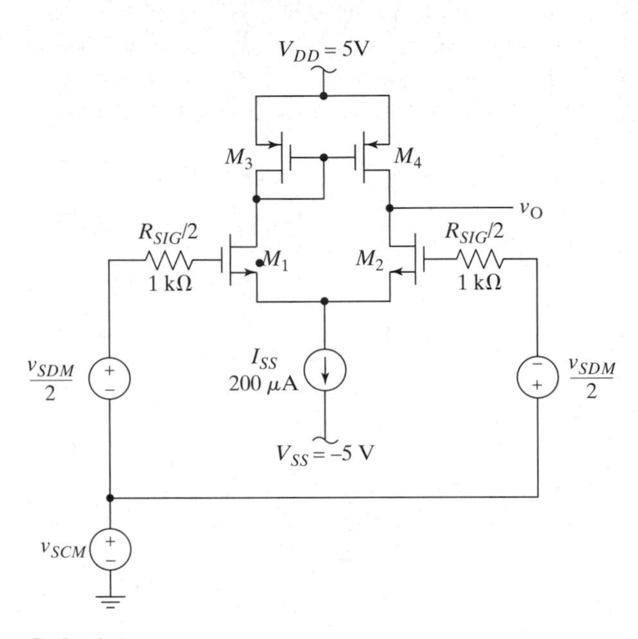

Figure 9–105 A MOSFET source-coupled pair with active load.

Solution:

With $I_D = I_{SS}/2 = 100\ \mu A$ for each transistor, we calculate $g_m = 200\ \mu A/V$ and $r_o = 500\ k\Omega$. Using (9.268), we predict that $f_{cH} = 106$ kHz. The SPICE simulation result is shown in Figure 9-106 and reveals that $f_{cH} = 90$ kHz. The approximation becomes less accurate as C_L decreases or R_{SIG} increases, since the input attenuation factor rolls off at frequencies at or near the f_{cH} predicted from the output time constant alone.

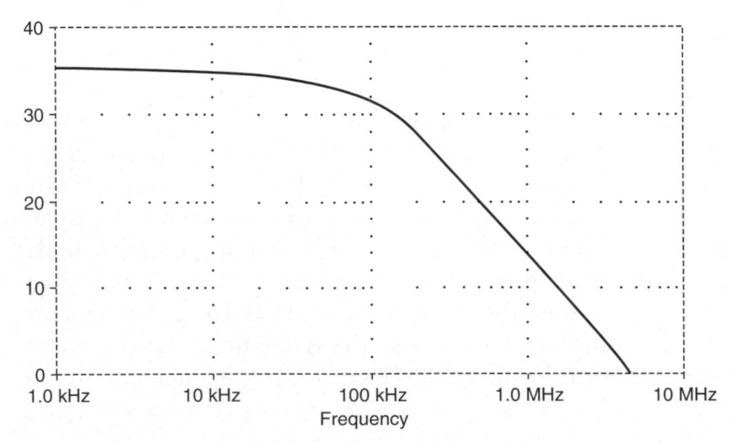

Figure 9–106 SPICE output.

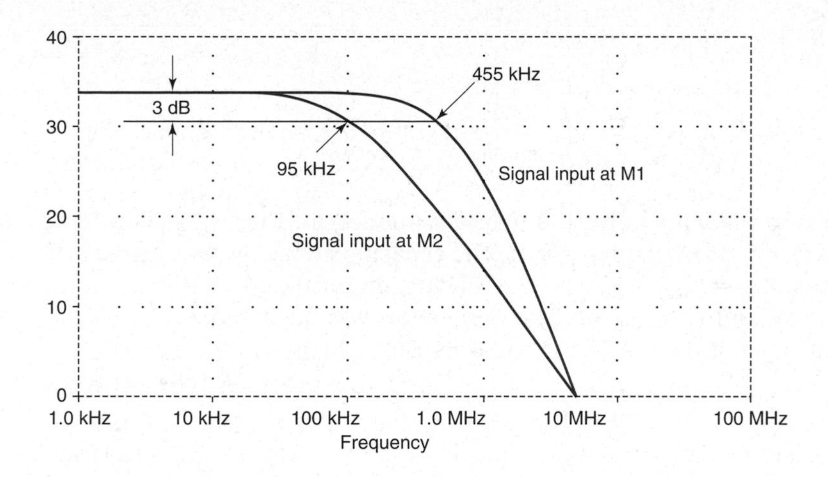

Figure 9–107 The responses of the circuit in Figure 9-105 when driven single ended. In each case, the signal source is tied to one input (with $R_{SIG} = 100\ k\Omega$), whereas the opposite input is grounded.

As pointed out in Section 9.7.1, when this circuit is driven single ended, the bandwidths can be significantly different, depending on which side the input is connected to. Figure 9-107 shows the simulated responses when the circuit is driven from a 100 kΩ source connected to one of the inputs, whereas the other input is grounded. Note that the difference in bandwidths would be smaller if a load capacitance were included. If the load capacitance is large enough, the gain rolls off due to the load rather than the Miller effect and the single-ended bandwidths are nearly the same.

SOLUTIONS TO EXERCISES

9.1 From (9.1), we find that $R_S = \omega L/Q = 2\pi\,(25\ \text{M})(10\ \mu)/44 = 3.6\ \Omega$.

9.2 Ignoring C_p, we find that the impedance of the equivalent circuit in Figure 9-1(c) is

$$Z = \sqrt{R_s^2 + (\omega L_s - 1/\omega C)^2}.$$

This impedance has a minimum value of R_s at

$$\omega_o = 1/\sqrt{L_s C};$$

therefore,

$$R_s = 1.5\ \Omega \text{ and } L_s = 1/[(2\pi)^2(5\ M)^2(0.01\ \mu)] = 0.1\ \mu\text{H}.$$

The dissipation factor is found using (9.4) and is 9.4×10^{-5}.

9.3 (a) In this case the diode is forward biased. The DC current is found to be $I_D = 720\ \mu\text{A}$ and $r_d = 36\ \Omega$. Using (9.6), we find that $C_j \approx 4$ pF, and from (9.7), we obtain $C_d = 13.8$ pF. Therefore, the total capacitance is 17.8 pF. (b) The diode is now reverse biased, so r_d is not present in the model, and the capacitance is given by (9.6) alone. Plugging in the values given, we find that $C_j = 1.1$ pF.

9.4 We first find

$$I_C = I_S e^{V_{BE}/V_T}\left(1 + \frac{V_{CB}}{V_A}\right) = 1.66 \text{ mA}.$$

The midband parameters for the hybrid-π model are then $g_m = 64$ mA/V, $r_\pi = 1.6$ kΩ, $r_o = 36$ kΩ, and $r_b = 50$ Ω. The capacitances are given by (9.8) and (9.9), using $V_{BC} = V_{BE} - V_{CE} = -3.3$ V. Plugging in the given values yields $C_\pi = 14.8$ pF and $C_\mu = 1.1$ pF. The parameters for the T model are all the same, and, in addition, $r_e = 15.5$ Ω and $\alpha = 0.99$.

9.5 With the given data, we find $C'_{ox} = \epsilon_{ox}/t_{ox} = (8.854 \times 10^{-14})(3.9)/(250 \times 10^{-8}) =$ 138 nF/cm^2 and $K = (\mu_n C'_{ox}/2)(W/L) = (675)(1.38 \times 10^{-7})(0.5)(5) = 233$ μA/V^2. Therefore, the DC current is $I_D = K(V_{GS} - V_{th})^2(1 + \lambda V_{DS}) = 165$ μA. Using this current, we find that $g_m = 3.9$ mA/V and $r_o = 173$ kΩ. Using (9.14) and (9.15), since the device is in saturation, we find that $C_{gs} = 6.6$ fF and $C_{gd} = 2$ pF. Finally, using (9.21), we obtain $\omega_T = 453$ Grad/s, or $f_T = 72$ GHz.

9.6 Using (9.31), (9.32), and the given data, we find that $C_{in} = 30.5$ pF and $C_{Mout} =$ 0.51 pF. Using (9.33), we then find that $f_{cH} \approx 106$ MHz. To check the validity of the Miller approximation, we first calculate $R'_L C_{Mout} = 2.6$ ns and then compare this with the input time constant:

$$1/\omega_{CH} \approx (R_S \| R_{CC} \| r_{cm})C_{in} = 1.5 \text{ ns}.$$

According to (9.34), then, the Miller approximation is definitely *not* acceptable in this case. In fact, a SPICE simulation shows that f_{cH} is slightly over 41 MHz, well off from our estimate.

9.7 (a) The low-frequency small-signal AC equivalent circuit is formed by open-circuiting C_{in} and C_{fb} in Figure 9-14. Using the method of Aside A1.6, we then find that $f_{cL} \approx 1/(2\pi C_C(R_S + r_{in})) = 62$ Hz. Simulation shows that $f_{cL} = 62$ Hz, as expected. The high-frequency small-signal AC equivalent circuit is found by short-circuiting C_C in Figure 9-14. We then make use of the Miller approximation to find the total input capacitance, which is C_{in} in parallel with C_{Min}, or

$$C'_{in} = C_{in} + C_{fb}(1 + g_m R_L) = 51 \text{ pF}$$

and the Miller output capacitance,

$$C_{Mout} = C_{fb}(1 + 1/g_m R_L) = 1.03 \text{ pF}.$$

According to the Miller approximation, we have $f_{cH} \approx 1/[2\pi(R_S \| r_{in})C'_{in}] =$ 43 MHz. We then check the approximation by using (9.34): $1/\omega_{CH} =$ 3.7 ns $> R_L C_{Mout} = 1.03$ ns. SPICE simulation shows that the actual cutoff frequency is 35 MHz. (b) The midband gain of this amplifier is

$$v_o/v_s = (-g_m R_L r_{in})/(r_{in} + R_S) = -38.8,$$

so the calculated GBW is

$$\text{GBW} = (38.8)(35\text{ M}) = 1.4 \times 10^9.$$

9.8 We start by calculating the short-circuit driving-point resistance seen by C_{IN}. Using impedance reflection, we find that the result is $R_{INs} = R_S + R_{CC} \| [r_{cm} + (a_{im} + 1) R_{MA}] = 12\text{ k}\Omega$. The resulting time constant is $\tau_{INs} = R_{INs} C_{IN} = 12$ ms. We next find the resistance seen by C_M. The result is

$$R_{Ms} = R_{MB} \left\| \left[R_{MA} + \frac{R_S \| R_{CC} + r_{cm}}{a_{im} + 1} \right] \right. = 217\ \Omega.$$

The resulting time constant is $\tau_{Ms} = R_{Ms} C_M = 10.2$ ms. The resistance seen by C_{OUT} is $R_{OUTs} = R_O + R_L = 14.7\text{ k}\Omega$, so the time constant is $\tau_{OUTs} = R_{OUTs} C_{OUT} = 14.7$ ms. Using (9.39), we find that $\omega_{cL} \approx 249$ rad/s, or $f_{cL} \approx 40$ Hz. SPICE simulation that reveals that $f_{cL} = 27$ Hz, so our result is off by nearly 50%, but it is a conservative estimate. The large error is due to the fact that none of the reciprocal time constants dominate.

9.9 Ignoring r_b in Example 9.6 and using (9.35), we have

$$\text{GBW} = |\alpha_i A_v| \omega_{cH} = \frac{|A_v|}{R_S(1 - A_v)C_\mu}.$$

If the input attenuation factor is close to unity and A_v is negative and much larger than unity,

$$\text{GBW} \approx \frac{|A_v|}{R_S |A_v| C_\mu}$$

and we obtain (9.64) again:

$$\text{GBW} \approx \frac{1}{R_S C_\mu}.$$

Using the given values, we get GBW = $6.25 \times 10^9\ \text{sec}^{-1}$. Using the simulated gain and bandwidth from the example we obtain, GBW = $3.7 \times 10^9\ \text{sec}^{-1}$. The corresponding error is quite large, but our simple equation did yield the correct order of magnitude and provides a useful insight.

9.10 (a) Using (9.68), we find that each time constant should be $3/\omega_{cL} = 477\ \mu$s. Using the values for the short-circuit resistances, we find that $C_{IN} = 0.25\ \mu$F, $C_{OUT} = 0.15\ \mu$F, and $C_E = 36.7\ \mu$F. (b) We now set $\tau_{Es} = 1/(.90\omega_{cL}) = 177\ \mu$s and $\tau_{INs} = \tau_{OUTs} = 1/(0.05\omega_{cL}) = 3.2$ ms, which leads to $C_{IN} = 1.7\ \mu$F, $C_{OUT} = 1\ \mu$F, and $C_E = 13.6\ \mu$F. Note that the cost of these capacitors would likely be dominated by the large emitter bypass capacitor. Consequently, the second solution would be less expensive to implement; also, the differences can be much larger in some circuits.

9.11 We need to find two open-circuit time constants; $\tau_{\pi o}$ and $\tau_{\mu o}$. To find $\tau_{\pi o}$, we first use the split-source transformation on the controlled current source to arrive at the equivalent circuit shown in Figure 9-108(a), where we have defined

$$R_S' = R_S \| R_{BB} + r_b.$$

We then remove r_π and recognize that

$$R_{\pi o} = r_\pi \| R_x,$$

where R_x is found from the circuit in Figure 9-108(b).Using KVL, we write

$$v_x = i_x R_S' + i_A R_{EA}$$

and, using KCL, we write

$$i_A = i_x - g_m v_x.$$

Combining these equations and solving, we find that

$$R_x = \frac{v_x}{i_x} = \frac{R_S' + R_{EA}}{1 + g_m R_{EA}}.$$

Plugging in the given numbers, we find

$$R_{\pi o} = 57\ \Omega,$$

therefore, $\tau_{\pi o} = R_{\pi o} C_\pi = 0.72$ ns. To find $\tau_{\mu o}$, we again use the split-source transformation, followed by impedance reflection, to draw the circuit as shown in Figure 9-109.

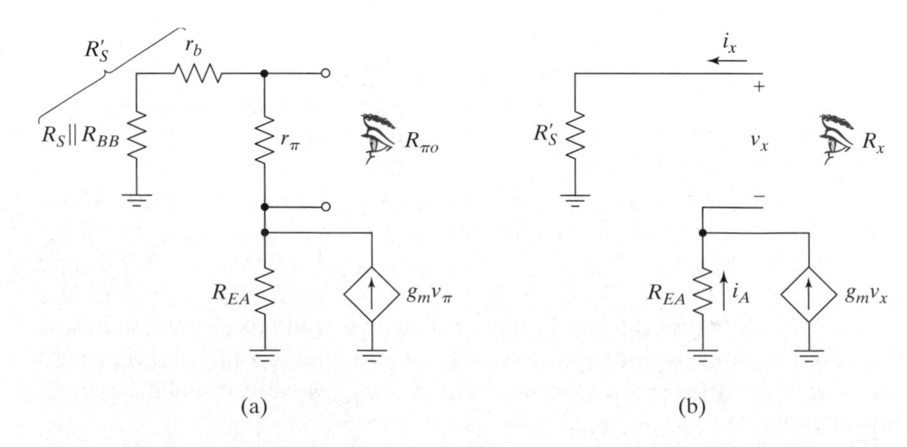

Figure 9–108

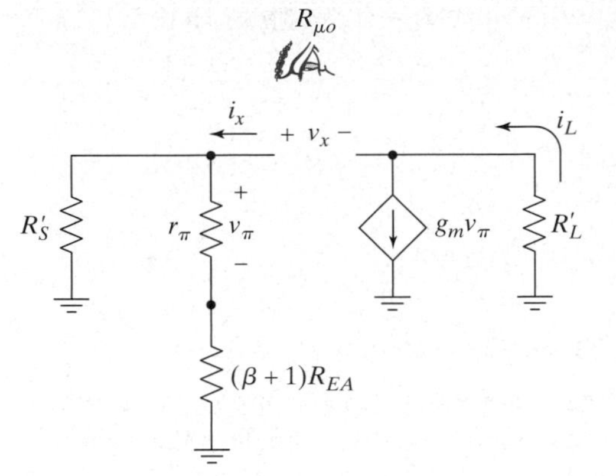

Figure 9–109

We start the analysis of this circuit by defining

$$R_{eq} = R'_S \| [r_\pi + (\beta + 1)R_{EA}].$$

We then use KVL to write

$$v_x = i_x R_{eq} + i_L R'_L$$

and KCL to write

$$i_L = i_x + g_m v_\pi.$$

Using a voltage divider, we have

$$v_\pi = i_x R_{eq} \frac{r_\pi}{r_\pi + (\beta + 1)R_{EA}}.$$

Substituting this relation into the equation we got using KCL and then substituting that into the loop equation yields

$$R_{\mu o} = \frac{v_x}{i_x} = R_{eq} + R'_L\left(1 + \frac{g_m r_\pi R_{eq}}{r_\pi + (\beta + 1)R_{EA}}\right).$$

Using the given values, we arrive at

$$\tau_{\mu o} = R_{\mu o} C_\pi = 1.5 \text{ ns}.$$

Finally, using (9.69), we find that $f_{cH} \approx 70$ MHz. This result is 15% lower than the simulated value of 83 MHz (Exam6.sch). The inaccuracy is because the largest time constant is not dominant, although it is about twice as big as the other. In other words, there isn't a dominant pole. Nevertheless, the result is more accurate than was obtained using the Miller approximation, and very importantly, this result is a conservative estimate of the bandwidth.

9.12 We begin by finding the DC bias point. Since $I_G = 0$, we know that

$$V_{GS} = V_{SIG} = 1.5\text{ V}.$$

Therefore, $I_D = K(V_{GS} - V_{th})^2 = 100\ \mu\text{A}$, $V_D = V_{DD} - I_D R_D = 5$ V, and

$$g_m = 2\sqrt{KI_D} = 200\ \mu\text{A/V}.$$

The small-signal equivalent circuit is shown in Figure 9-110.

(a) To find the low cutoff frequency, we open-circuit C_{gs} and C_{gd} to produce the low-frequency equivalent circuit. We then have a single-pole high-pass transfer function, and using the methods of Aside A1.6, we find that

$$f_{cL} = \frac{1}{2\pi C_{OUT}(R_D + R_L)} = 16\text{ Hz}.$$

(b) To find the high cutoff frequency, we short-circuit C_{OUT} to produce the high-frequency equivalent circuit. We next apply the Miller approximation. The midband gain from gate to drain is $A_v = v_o/v_g = -g_m(R_L \| R_D) = -5$. Using this, we find $C_{Min} = 6\ C_{gd} = 3$ pF. Combining this in parallel with C_{gs}, we have a total input capacitance of 5 pF. The resistance seen by this capacitance is $R_{SIG} = 10\ \text{k}\Omega$, so

$$f_{cH} \approx \frac{1}{2\pi C_{in} R_{SIG}} = 3.2\text{ MHz}.$$

We check the approximation using (9.34). $C_{Mout} = 1.2\ C_{gd} = 0.6$ pF, and it sees a resistance of $R_L \| R_D = 25\ \text{k}\Omega$, so we find that $1/\omega_{cH} = 50$ ns is only slightly more than three times the value of $C_{Mout}(R_L \| R_D) = 15$ ns. In this case, we expect the Miller approximation to be reasonable, but not extremely close. SPICE simulation of this circuit reveals that $A_v = -5$, $f_{cL} = 16$ Hz, and $f_{cH} =$ 2.7 MHz. We expect the gain and f_{cL} to be accurate, and they are. The value for f_{cH} is 19% higher, because the Miller approximation is not great in this instance.

(c) GBW $= A_v \omega_{cH} = 1 \times 10^8\ \text{sec}^{-1}$. (The sign of the gain is ignored and $v_o/v_{sig} =$ A_v for this circuit in midband, since $i_g = 0$.)

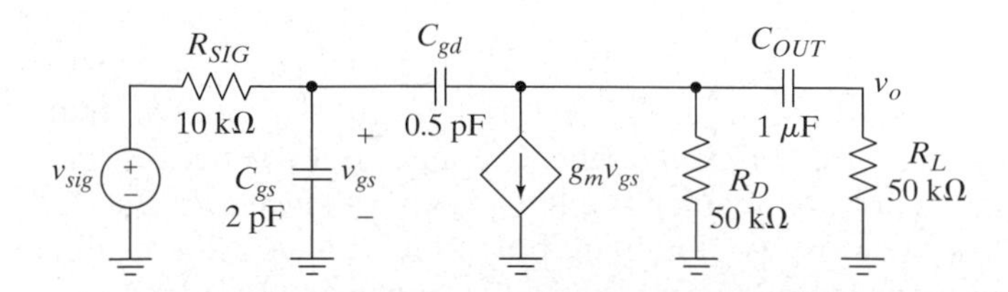

Figure 9–110 The small-signal equivalent circuit for the amplifier in Figure 9-41.

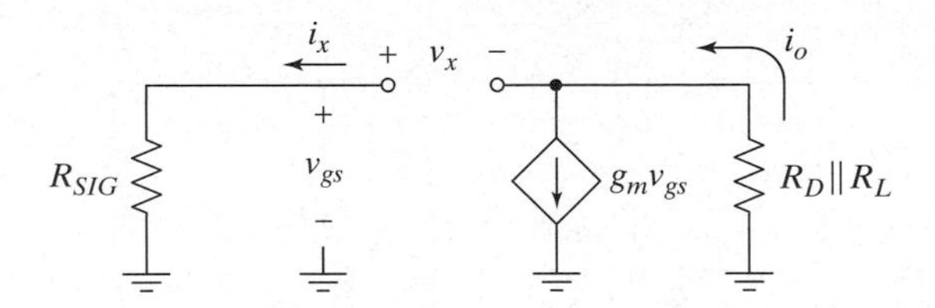

Figure 9–111 Circuit for finding the open-circuit driving-point resistance seen by C_{gd}.

9.13 The high-frequency equivalent circuit is found from the circuit in Figure 9-110 by short circuiting C_{OUT}. The open-circuit driving-point resistance seen by C_{gs} is then found to be R_{SIG} by inspection. To find the open-circuit resistance seen by C_{gd}, we use Figure 9-111. Using KVL, we write

$$v_x = i_x R_{SIG} + i_o(R_D \| R_L).$$

Using KCL, we write

$$i_o = i_x + g_m v_{gs} = i_x(1 + g_m R_{SIG}).$$

Combining these results, we find that

$$R_{gdo} = \frac{v_x}{i_x} = R_{SIG} + (R_D \| R_L)(1 + g_m R_{SIG}).$$

Finally, we use (9.46) to obtain

$$f_{cH} \approx \left(\frac{1}{2\pi}\right)\frac{1}{R_{SIG}C_{gs} + [R_{SIG} + (R_D \| R_L)(1 + g_m R_{SIG})]C_{gd}} = 2.5 \text{ MHz}.$$

This answer is 7.4% lower than the simulated bandwidth. The value obtained using open-circuit time constants is closer to the correct value because the Miller effect does not completely dominate the response. In addition, the open-circuit time constant approximation is conservative.

9.14 For the generic transistor, $a_{im} = g_m r_{cm} = 30$. Now, using (9.100) we obtain $f_{cL} \approx \omega_p/2\pi = 13.5$ Hz. SPICE simulation confirms that $f_{cL} = 13.5$ Hz.

9.15 $a_{im} = 30$, as in Exercise 9.14, and $r_m = r_{cm}/(1 + a_{im}) = 48.4\ \Omega$. So, using (9.101), (9.103), and (9.104), we find $A_v = 0.845$ and $C_{in} = 3$ pF. Then, using (9.106), we obtain $f_{cH} = 10.8$ MHz. Using (9.108), we see that

$$\omega_{cH} = 67.7 \text{ Mrad/s} \ll 2.07 \text{ Grad/s};$$

the approximation is valid. SPICE simulation yields $f_{cH} = 10.6$ MHz, in close agreement with our result.

9.16 Using (9.109), we find, $\tau_{oco} = 7.4$ ns, and from (9.114), we obtain $\tau_{cmo} = 7.8$ ns. Then, making use of (9.115), we arrive at $f_{cH} = 10.5$ MHz. It is actually surprising that this result is so close to the simulated value, since there isn't a dominant time constant! We would not generally expect such agreement.

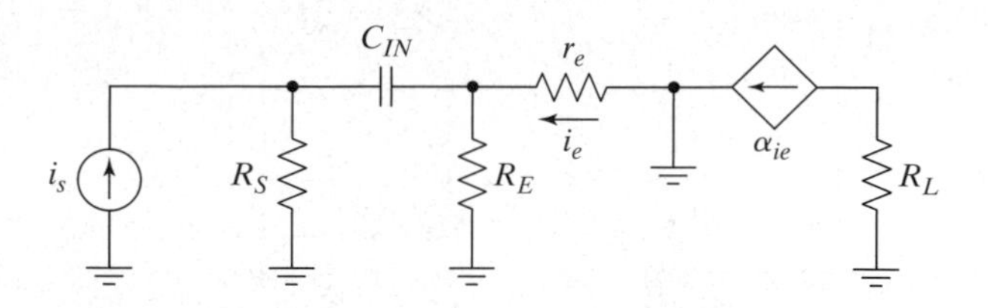

Figure 9–112 The small-signal low-frequency AC equivalent circuit for the common-base amplifier in Figure 9-60.

9.17 The small-signal low-frequency AC equivalent circuit is shown in Figure 9-112. This is a single-pole circuit, and we can use the method of Aside A1.6 to find the pole. The resistance seen by C_{IN} is found by setting $i_s = 0$ (i.e., open circuit the source) and is $(R_S + R_E \| r_e)$, so the cutoff frequency is

$$f_{cL} = \frac{1}{2\pi(R_S + R_E \| r_e)C_{IN}} = 79 \text{ Hz}.$$

9.18 The small-signal low-frequency AC equivalent circuit is shown in Figure 9-113. This is a single-pole circuit, and we can use the method of Aside A1.6 to find the pole. The resistance seen by C_{IN} is found by setting $i_{sig} = 0$ (i.e., open-circuit the source) and is

$$\left(R_{SIG} + R_S \middle\| \frac{1}{g_m}\right),$$

so the cutoff frequency is

$$f_{cL} = \frac{1}{2\pi\left(R_{SIG} + R_S \middle\| \frac{1}{g_m}\right)C_{IN}} = 16 \text{ Hz}.$$

9.19 We first find $R_{CC} = R_1 \| R_2 = 15\text{ k}\Omega$. Then, we use (9.168) to find $R'_O = 2.83\text{ k}\Omega$ and (9.170) to get $C_{in} = 67.6$ pF. Plugging the values into (9.169), we obtain $f_{cH1} \approx 33$ MHz. We next use (9.172) to find $C_{in2} = 3.9$ pF and plug into (9.172) to find $f_{cH2} \approx 14.4$ MHz. In this example, f_{cH2} is nearly three times less than f_{cH1}; therefore, the Miller approximation is not valid, and f_{cH1} is not accurate. The overall bandwidth in the example is limited by the merge follower and is approximately equal to f_{cH2}. This is unusual, but it does occur. The reason it happened can be seen by examining (9.169); the resistance seen by C_{in} is approximately just R_S, since it is so low and dominates the parallel combination. In addition, the gain of the merge follower is much less than unity (about 0.3), so the Miller input capacitance of the merge follower is larger than would normally be expected. (Remember that since the merge follower has a noninverting voltage gain, the Miller effect actually *decreases* the equivalent input capacitance.)

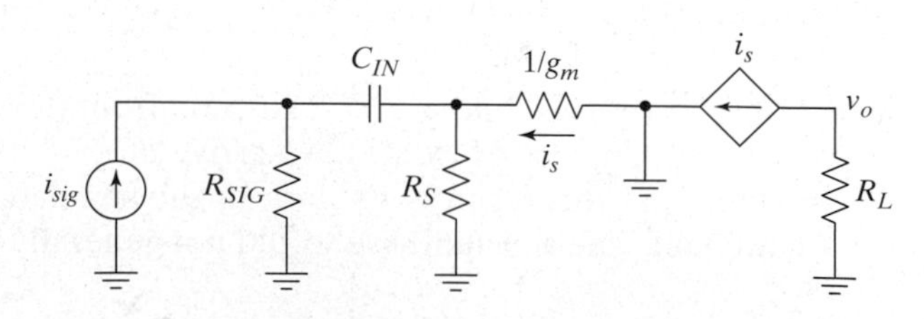

Figure 9–113 The small-signal low-frequency AC equivalent circuit for the common-gate amplifier in Figure 9-62.

9.20 Using (9.174), (9.178), (9.179), and (9.180), we find that $\tau_{cm1o} = 0.72$ ns, $\tau_{oc1o} = 7.0$ ns, $\tau_{oc2o} = 2.8$ ns, and $\tau_{cm2o} = 8.2$ ns, respectively. Plugging into (9.173), we find that $f_{cH} \approx 8.5$ MHz. None of the four time constants are dominant, so the open-circuit time constant-method may be well below the correct value. The result obtained here is significantly less than the one from Exercise 9.19. The large difference is due to the different approximations made in the two solutions. Since there isn't a dominant time constant, the approximations made in the earlier solution are actually better in this case; a SPICE simulation of the AC equivalent circuit reveals a cutoff frequency of 12.3 MHz. This is not far off from the value obtained in Exercise 9.19, but the value obtained here is 31% lower. The fact that the Miller approximation came closer to the correct answer is purely coincidental in this case, since the approximation is not valid as pointed out in the solution to 9.19.

9.21 With $R_S = 1$ kΩ, we obtain $R'_S = 735$ Ω, $R_{eq} = 624$ Ω, and all other parameters the same as in Examples 9.8 and 9.9. Plugging into (9.184), we obtain $f_{cH} \approx$ 16.1 MHz. Using (9.186) through (9.189), we find that $\tau_{\pi 1o} = 2.16$ ns, $\tau_{\mu 1o} =$ 10.5 ns, $\tau_{\mu 2o} = 0.52$ ns, and $\tau_{\pi 2o} = 0.51$ ns. The resulting cutoff frequency is found with (9.185) to be $f_{cH} \approx 11.6$ MHz. The simulated bandwidth in this case is 14.3 MHz. The Miller approximation is more accurate now, since the Miller effect is dominant.

9.22 We first calculate $g_{m1} = 2\sqrt{KI_{D1}} = 200$ μA/V. Then, using (9.190), we find that $C_{Min} = 4.5$ pF and $C_{in} = C_{Min} \| C_{gs1} = 6.5$ pF. Plugging into (9.191), we obtain $f_{cH} \approx 2.4$ MHz. We check the Miller approximation using (9.192). The Miller output capacitance is

$$C_{Mout} = C_{gd1}\left(1 + \frac{1}{g_{m1}R_D}\right) = 0.56 \text{ pF},$$

so the approximation is valid if $R_D(C_{Mout} + C_{gd2}) = 42.5$ ns is much less than $R_{SIG}C_{in} = 65$ ns. Obviously, the approximation is on shaky ground in this example. A SPICE simulation reveals that $f_{cH} = 1.4$ MHz; so our approximation was over 70% higher!

9.23 We first calculate $g_{m2} = 2\sqrt{KI_{D2}} = 283$ μA/V. Then, using g_{m1} from Exercise 9.22, along with (9.202), (9.199), (9.195), and (9.194), we find that $\tau_{gs2o} =$ 26.1 ns, $\tau_{gd1o} = 65$ ns, $\tau_{gd2o} = 20$ ns, and $\tau_{gs1o} = 20$ ns, respectively. Plugging these values into (9.193), we obtain $f_{cH} \approx 1.2$ MHz. This answer is only 14% below the simulated value of 1.4 MHz (Exer22.sch). The error is caused by the fact that τ_{gd1o} is not larger than the sum of the other time constants.

CHAPTER SUMMARY

- We need to carefully model the passive and active devices in our circuits when we are analyzing their frequency response. If the models are inadequate, the results will be, too.
- A common parameter for specifying the high-frequency capability of a transistor is the transition frequency, f_T.

- When a capacitor (or any impedance) appears from the output to the input of a voltage amplifier, the effective impedance to ground is different; this is called the Miller effect.
- A common situation in amplifiers is for a feedback capacitor across an inverting gain stage to appear as a much larger capacitor to ground at the input and thereby limit the upper cutoff frequency.
- Open- and short-circuit time constants can be used to approximate the upper and lower cutoff frequencies of amplifiers. The approximations are quite accurate if there is a dominant pole and are useful for design, even when a dominant pole does not exist.
- Single-transistor stages with both voltage and current gain usually have significantly lower high cutoff frequencies than either current buffers or voltage buffers. Therefore, buffer stages can usually be cascaded with stages having both voltage and current gain without lowering the overall bandwidth.
- Sometimes, cascading stages can reduce the Miller multiplication of a feedback capacitor (the cascode connection) or can reduce the driving-point resistance seen by some capacitor (e.g., a voltage buffer driving a stage with voltage and current gain) and can thereby increase the bandwidth of the stage.
- The bandwidths of differential stages depend on whether they are driven single ended or differentially.
- The CMRR usually has a lower cutoff frequency than the differential gain of a differential amplifier.

REFERENCES

[9.1] T. H. Lee, *The Design of CMOS Radio-Frequency Integrated Circuits*. Cambridge, UK: Cambridge University Press, 1998.

[9.2] E. Kreyszig, *Advanced Engineering Mathematics*. 8/E. New York: John Wiley & Sons, 1998. (Also, any good book on linear algebra will have a discussion of Cramer's rule.)

[9.3] A. Davis & E. Moustakas, "Decomposition Analysis of Active Networks," *International Journal of Electronics*, 46, pp. 449–456, May 1979 *or* "Decomposition Analysis of Active Networks,"Asilomar Conference on Circuits, Systems, and Computing, Pacific Grove, California, November 1978.

PROBLEMS

9.1 High-Frequency Small-Signal Models for Design

9.1.2 Linear Passive Elements (Rs, Ls, & Cs)

P9.1 A 10 μH inductor is specified as having a Q of 50 at 10 MHz. Ignoring C_p, what is the value of R_s?

P9.2 What is the Q of an inductor that has $L = 0.15$ μH and $R_s = 0.3\ \Omega$; **(a)** at 10 KHz, and **(b)** at 10 MHz?

P9.3 Suppose a 10 μF capacitor is used as a bypass capacitor. It is known that the capacitor is self-resonant at 1 MHz. **(a)** Ignoring any series resistance, plot the impedance of this capacitor as a function of frequency for frequencies from 1 Hz to 10 MHz. **(b)** If the capacitor is used to bypass a 470-Ω resistor, and it must have an impedance less than one-tenth the value of the resistor to be an effective bypass, over what range of frequencies can this capacitor be used?

P9.4 Derive (9.3) from (9.2) and the description given in the text.

P9.5 Prove that (9.5) is correct by drawing a phasor diagram for the voltages in a series connection of a resistor and a capacitor.

P9.6 Consider again the capacitor from Example 9.1. At low frequencies, the energy loss in a capacitor is due mostly to the work required to polarize the dielectric. Therefore, the model for the capacitor can be more intuitively represented as a single resistor in parallel with an ideal capacitor. Calculate the value of this parallel resistance so that this model is equivalent to the series model at 1 kHz.

SP9.7 Consider a resistor divider made by using two identical 10-kΩ resistors and driven by an ideal voltage source. The output voltage at the center of the divider is ideally equal to one-half of the input. Now suppose the resistors are on an integrated circuit and each resistor has 1 pF of parasitic capacitance to ground, distributed along the length of the resistor. Model each resistor by three equal capacitors and two equal resistors, as shown in Figure 9-114. Given this model, use SPICE to generate a plot of the transfer function for the two-resistor divider circuit. What is the −3-dB frequency?

9.1.3 Diodes

P9.8 Suppose a *pn*-junction diode has $V_D = 680$ mV and a total small-signal capacitance of 6 pF when $I_D = 500$ μA and $V_T = 26$ mV. Furthermore, suppose the capacitance is measured to be 4 pF when $I_D = 100$ μA. Derive the complete high-frequency model when $V_D = 600$ mV.

P9.9 Consider a *pn*-junction diode with $I_S = 10^{-15}$ A, $n = 1$, $\tau_F = 200$ ps, $C_{j0} = 1.5$ pF, $m_j = 0.3$, $V_0 = 0.62$ V, and $V_T = 26$ mV. Determine the complete small-signal model when **(a)** $V_D = -2$ V and **(b)** $V_D = 0.65$ V.

P9.10 Consider a Schottky diode with $I_S = 5 \times 10^{-9}$ A, $n = 1$, $C_{j0} = 0.5$ pF, $m_j = 0.3$, $V_0 = 0.58$ V, and $V_T = 26$ mV. Determine the complete small-signal model when **(a)** $V_D = -3$ V and **(b)** $V_D = 0.3$ V.

P9.11 Consider a *pn*-junction diode with $I_S = 3 \times 10^{-15}$ A, $n = 1$, $\tau_F = 0.8$ ns, $C_{j0} = 3$ pF, $m_j = 0.35$, $V_0 = 0.7$ V, and $V_T = 26$ mV. Determine the complete small-signal model when **(a)** $V_D = 0.6$ V and **(b)** $V_D = 0.7$ V.

9.1.5 Bipolar Junction Transistors

P9.12 Derive the equation for $\beta(j\omega)$, including the current through C_μ (*see* Aside A9.2). Show that the equation is the same as (A9.4) except for the addition of a zero at a frequency greater than ω_T.

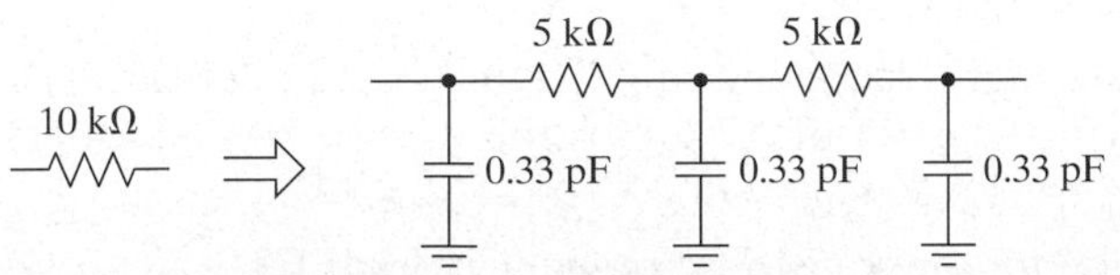

Figure 9–114 A lumped-element model for an integrated resistor.

P9.13 Consider an *npn* BJT with $I_S = 3 \times 10^{-14}$A, $C_{jc0} = 3$ pF, $C_{je0} = 1.2$ pF, $V_0 = 0.65$ V, $m_{jc} = 0.33$, $\tau_F = 400$ ps, $r_b = 75\ \Omega$, $V_A = 40$ V, and $\beta_0 = 80$. **(a)** Find the complete small-signal high-frequency AC hybrid-π and T models when $V_{BE} = 0.6$ V and $V_{CE} = 3$ V. **(b)** What is the corresponding value of ω_T?

P9.14 For a BJT with $C_{je0} = 3$ pF and $\tau_F = 1$ ns, at what value of I_C does the diffusion capacitance equal the base-emitter junction capacitance?

P9.15 If you need to come up with a simplified model for a BJT that is valid only for frequencies well above ω_β, would you need to include r_π in your model? Why or why not?

P9.16 If you are modeling a BJT in a situation where the base is driven by a nearly ideal current source, is including r_b in the model important? How about if you are driving the transistor from a nearly ideal voltage source?

P9.17 Prove that the current-controlled model shown in Figure 9-6 yields the same $\beta(j\omega)$ as the voltage-controlled model used in Aside A9.2.

9.1.6 MOS Field-Effect Transistors

P9.18 Consider a *p*-channel MOSFET with a 100 Å (1 Å $= 10^{-10}$ meter) thick gate oxide, $W = 1$ μm, $L = 0.5$ μm, $C_{ol} = 0.5$ fF, $V_{th} = -0.9$ V, $K' = 21$ μA/V^2, and $\lambda = 0.08$ V^{-1}. **(a)** Find the complete small-signal high-frequency AC model when $V_{GS} = -2$ V and $V_{DS} = -3$ V. What is the corresponding value of ω_T? **(b)** Find the complete small-signal high-frequency AC model when $V_{GS} = -2$ V and $V_{DS} = -0.5$ V.

P9.19 If the width of a MOSFET is doubled, but everything else is kept the same, what happens to the value of ω_T if **(a)** V_{GS} is held constant and **(b)** I_D is held constant?

P9.20 How does the ω_T of a MOSFET vary with I_D? Is there a practical limit to the maximum value that can be attained?

P9.21 Consider an *n*-channel MOSFET with an 80 Å (1 Å $= 10^{-10}$ meter) thick gate oxide, $W = 0.8$ μm, $L = 0.4$ μm, $C_{ol} = 0.2$ fF, $V_{th} = 0.6$ V, and $\lambda = 0.35$ V^{-1}. Assume that the electron mobility in the channel is 675 cm^2/Vs. Find the complete small-signal high-frequency AC model when $V_{GS} = 1$ V and $V_{DS} = 2$ V. What is the corresponding value of ω_T?

P9.22 Use the split-source transformation, and derive the MOSFET high-frequency T model given in Figure 9-7(c) directly from the hybrid-π model in part (b) of the figure.

P9.23 Prove that the current-controlled model shown in Figure 9-7(c) yields the same $A_i(j\omega)$ as the voltage-controlled model used in the text; see Equation (9.20).

9.1.7 Junction Field-Effect Transistors

P9.24 Find the complete high-frequency small-signal AC model for a *p*-channel JFET with $C_{gs0} = 2$ pF, $C_{gd0} = 0.1$ pF, $V_0 = 0.65$ V, $m_j = 0.33$, $I_{DSS} = 250$ μA, $V_p = 3$ V, $V_{GS} = 1.8$ V, and $V_{DS} = -3.2$ V.

P9.25 Find the complete high-frequency small-signal AC model for an *n*-channel JFET with $C_{gs0} = 3$ pF, $C_{gd0} = 0.2$ pF, $V_0 = 0.7$ V, $m_j = 0.33$, $I_{DSS} = 150$ μA, $V_p = -1.5$ V, $V_{GS} = -0.7$ V, and $V_{DS} = 3$ V.

P9.26 Derive an equation for the transition frequency of a JFET.

9.2 Stages with Voltage and Current Gain

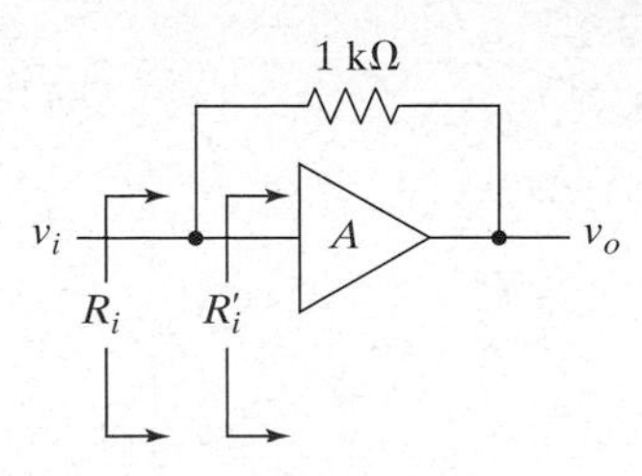

Figure 9–115

P9.27 Consider the circuit of Figure 9-115. Assume initially that R_i', the input impedance of the amplifier itself, is infinite. The voltage gain of the amplifier is A. **(a)** Find R_i, the impedance seen looking in from v_i, if $A = 0.5$. Give a numerical answer. **(b)** Find R_i if $A = -0.5$. **(c)** Find R_i if $A = -10$. **(d)** In the case that $R_i' = 10\ \text{k}\Omega$ and $A = 0.5$, What is R_i?

P9.28 Find the current out of the DC source in the circuit shown in Figure 9-116. The amplifier shown is a perfect gain block with a voltage gain of four. Use the Miller theorem. Explain the result you get.

P9.29 Repeat Problem P9.28, but let the amplifier have a low-pass transfer function with a DC gain of 4 and a pole at 1 kHz. What is the input current for frequencies well above 1 kHz? For DC?

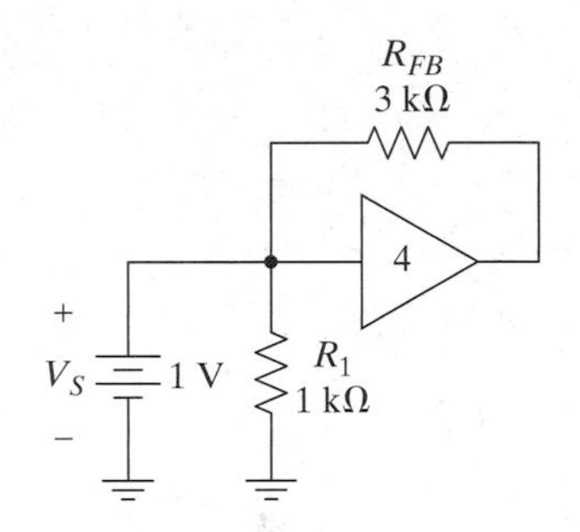

Figure 9–116

P9.30 Consider the circuit shown in Figure 9-117. What is the input impedance of this circuit? The amplifier shown is a perfect gain block with a voltage gain of four. (*Hint:* try using the Miller theorem.) Explain the result you get.

P9.31 Repeat Problem P9.30, but let the amplifier have a low-pass transfer function with a DC gain of 4 and a pole at 1 kHz. What is the input impedance for frequencies well above 1 kHz? For DC?

9.2.1 A Generic Implementation: The Common-Merge Amplifier

Introductory Treatment

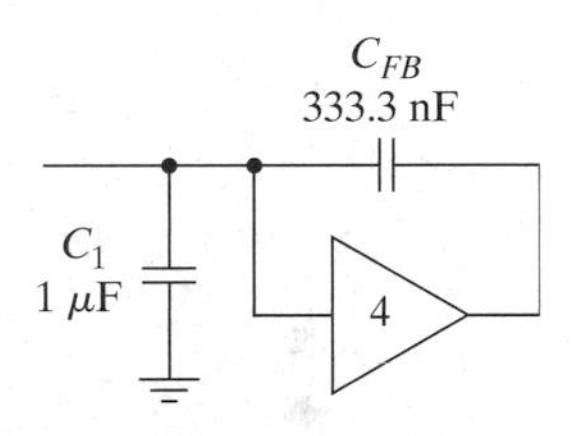

Figure 9–117

P9.32 Consider the common-merge amplifier shown in Figure 9-118. Assume that $g_m = 40$ mA/V and $r_{cm} = 4\ \text{k}\Omega$. **(a)** Find the values of C_{IN} and C_{OUT} necessary for the low-frequency equivalent circuit to have two poles at 100 Hz. **(b)** Find an approximate value for the midband gain. **(c)** Find an approximate value for f_{cH}. Assume that $C_{cm} = 6$ pF and $C_{oc} = 2$ pF.

P9.33 Consider the amplifier in Figure 9-118. Derive an approximate equation for the gain-bandwidth product. How does the GBW vary with R_S? With g_m?

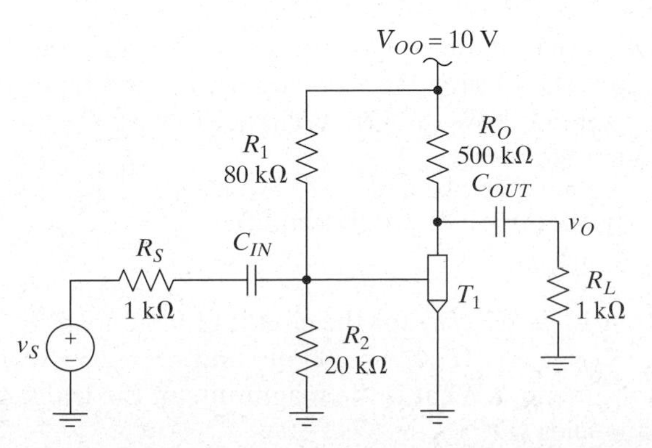

Figure 9–118 A common-merge amplifier.

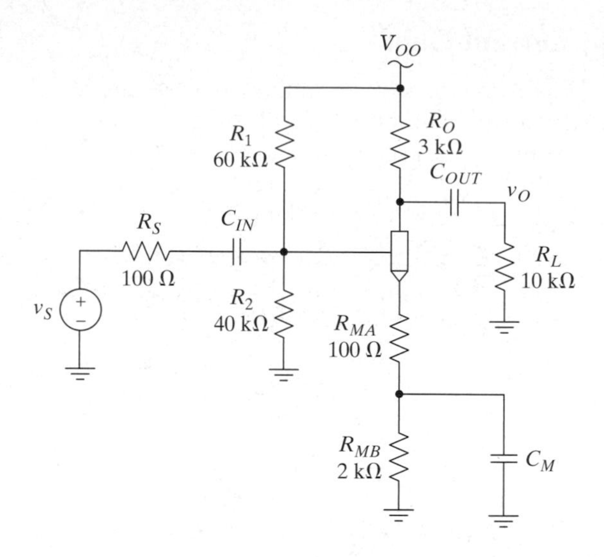

Figure 9–119 A common-merge amplifier.

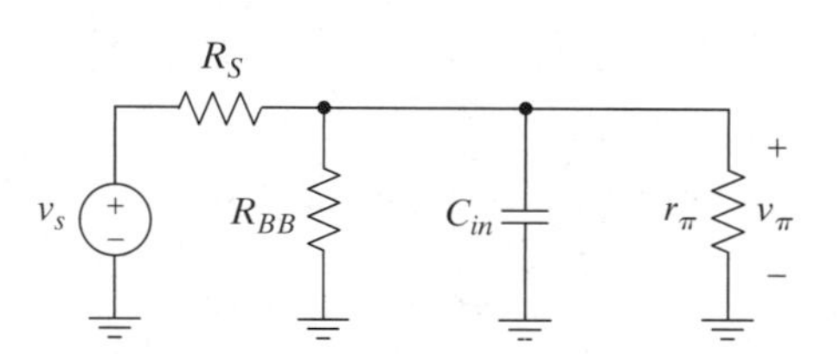

Figure 9–120

Advanced Method: Open- and Short-Circuit Time Constants

DP9.34 Consider the common-merge amplifier of Figure 9-119. Find values for C_{IN}, C_{OUT}, and C_M so that $f_{cL} \approx 10$ Hz. Pick the capacitors so that all three time constants contribute equally to f_{cL}. Assume $r_{cm} = 3$ kΩ and $a_{im} = 99$.

P9.35 Draw the high-frequency small-signal AC equivalent circuit for the common-merge amplifier in Figure 9-119. Find an approximation to f_{cH}, assuming that $g_m = 40$ mA/V, $r_{cm} = 3$ kΩ, $C_{cm} = 3.5$ pF, and $C_{oc} = 1.2$ pF.

P9.36 Derive an equation for the gain-bandwidth product of the common-merge amplifier in Figure 9-119. Using the data supplied in the figure and in Problem P9.35, examine how the GBW changes when R_{MA} is changed by ±25%. Explain the result.

9.2.2 A Bipolar Implementation: The Common-Emitter Amplifier

Introductory Treatment

P9.37 Find the transfer function $V_\pi(j\omega)/V_s(j\omega)$ for the circuit of Figure 9-120. Given that $R_S = 5$ kΩ, $R_{BB} = 50$ kΩ, $C_{in} = 200$ pF, and $r_\pi = 2.5$ kΩ, what is the high cutoff frequency? What is the magnitude of the transfer function at low frequencies (DC)?

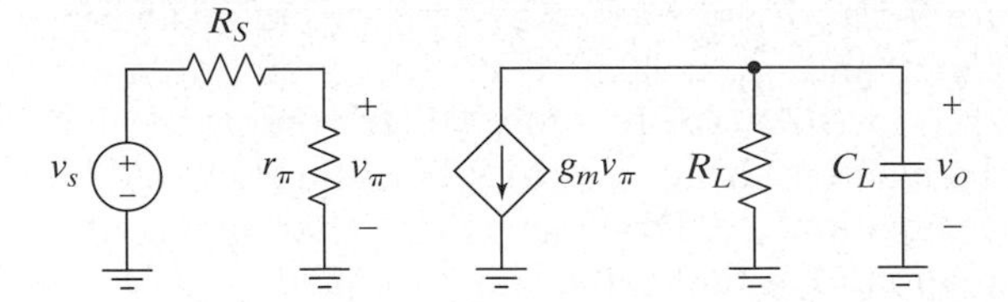

Figure 9–121

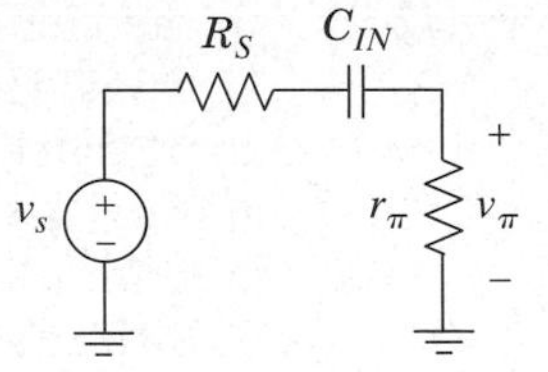

Figure 9–122

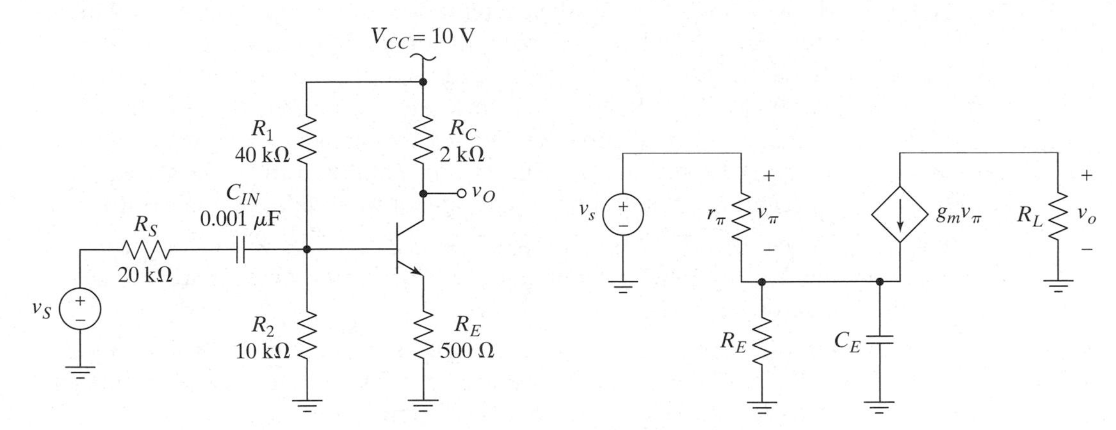

Figure 9–123

Figure 9–124

P9.38 Find the transfer function $V_o(j\omega)/V_s(j\omega)$ for the circuit of Figure 9-121. Given that $R_S = 1\,\text{k}\Omega$, $r_\pi = 2.5\,\text{k}\Omega$, $g_m = 40\,\text{mA/V}$, $R_L = 5\,\text{k}\Omega$, and $C_L = 1000\,\text{pF}$, what is the high cutoff frequency? What is the magnitude of the transfer function at low frequencies (DC)?

P9.39 Find the transfer function $V_\pi(j\omega)/V_s(j\omega)$ for the circuit of Figure 9-122. Given that $R_S = 1\,\text{k}\Omega$, $r_\pi = 2.5\,\text{k}\Omega$, and $C_{IN} = 0.01\,\mu\text{F}$, what is the pole frequency? What is the magnitude of the transfer function at high frequencies?

P9.40 Consider the circuit of Figure 9-123. The transistor has $\beta = 100$. You may assume that $V_B = V_{BB}$ and $V_{BE} = 0.7$ V. **(a)** Find the magnitude of $V_o(j\omega)/V_s(j\omega)$ at midband AC, using the small-signal model. **(b)** Find the low cutoff frequency.

P9.41 Consider the small-signal low-frequency equivalent Figure 9-124, where $r_\pi = 2.5\,\text{k}\Omega$, $g_m = 40\,\text{mA/V}$, $R_E = 1.3\,\text{k}\Omega$, $C_E = 30\,\mu\text{F}$, and $R_L = 5\,\text{k}\Omega$. **(a)** Use impedance reflection to model the impedance seen looking into the base. The circuit is shown in Figure 9-125, where R_E' and C_E' represent the reflected values of R_E and C_E respectively. State the values of R_E' and C_E', and express the impedance seen looking into the base as a function of r_π, R_E' and C_E'. **(b)** Find the transfer function $V_\pi(j\omega)/V_s(j\omega)$ in terms of r_π, R_E' and C_E'. (Use the circuit of Figure 9-125.) Calculate the pole and zero frequencies of this transfer function. **(c)** Calculate the pole of $V_\pi(j\omega)/V_s(j\omega)$, using the method presented in the text in Aside A1.6, and compare your results with part (b). **(d)** Calculate the magnitude of $V_o(j\omega)/V_s(j\omega)$ at midband AC. Calculate the pole and zero frequencies of $V_o(j\omega)/V_s(j\omega)$.

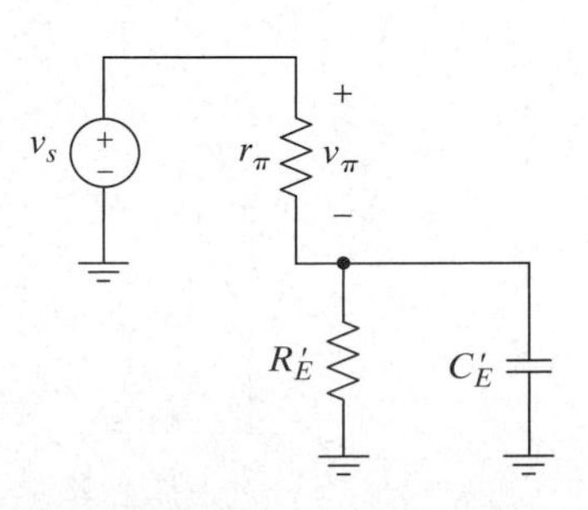

Figure 9–125

P9.42 In the circuit of Figure 9-126, $R_S = 1\ \text{k}\Omega$, $C_\pi = 0.4\ \text{pF}$, $r_\pi = 2.5\ \text{k}\Omega$, $C_\mu = 5.6\ \text{fF}$, $R_L = 2\ \text{k}\Omega$, and $g_m = 40\ \text{mA/V}$. The prefix "femto," abbreviated by "f," stands for 10^{-15}. **(a)** To apply Miller's theorem, what approximation must be made in this case? **(b)** What gain A should be used for Miller's approximation? **(c)** Draw the circuit after application of Miller's approximation. Evaluate the high-frequency pole of $V_\pi(j\omega)/V_s(j\omega)$. **(d)** Evaluate the time constant due to the output Miller capacitance, C_{Mout}. What does this value imply about the validity of the Miller approximation in this case?

P9.43 Consider the circuit of Figure 9-127. **(a)** Draw the high-frequency small-signal AC equivalent circuit before the application of Miller's theorem to C_μ. You may combine R_1 and R_2 into one resistor, R_{BB}. **(b)** What gain A should be used in the application of Miller's approximation to C_μ? Express A in terms of g_m, r_π, and the labels of the circuit elements in the figure. **(c)** Draw the small-signal equivalent circuit after the application of Miller's approximation to C_μ. Draw and label C_{Min} and C_{Mout}, and show equations for their values in terms of C_μ.

P9.44 Derive the GBW for the circuit in Example 9.6 without approximating C_{Min} by $-A_v C_\mu$. Show that the GBW does increase slightly as R_{EA} is reduced.

P9.45 Find the approximate value of ω_{cH} for the common-emitter amplifier of Figure 9-25 if a 50 Ω base resistance is included in the small-signal transistor model. Use the data in Example 9.5.

P9.46 Derive an equation for the GBW of the amplifier in Figure 9-31 with r_b included. Use the Miller approximation for the bandwidth, and assume that $R_{BB} \gg R_S$. Does r_b increase or decrease the GBW? (*Hint:* Consider the Thévenin equivalent to the left of r_b.)

Figure 9–126

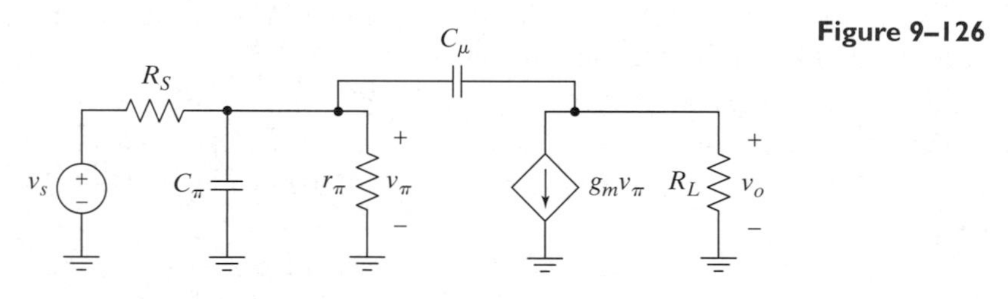

Figure 9–127

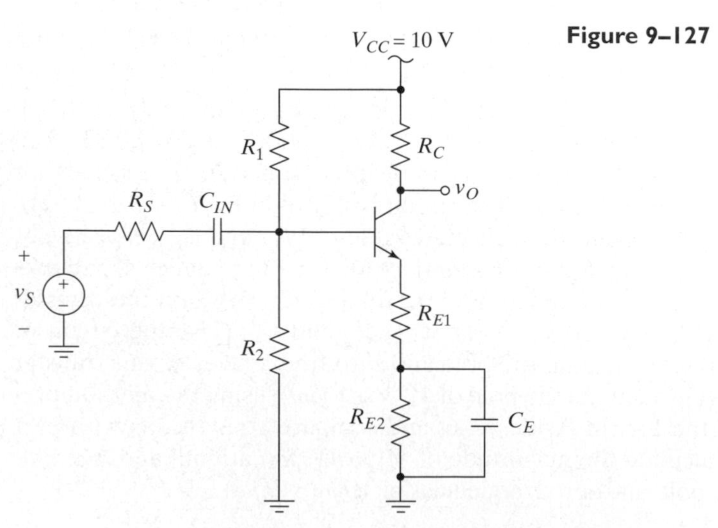

Advanced Treatment

P9.47 Consider the circuit shown in Figure 9-128. Using the 2N2222 model presented in Appendix B and the values given, **(a)** calculate the DC bias point. **(b)** Draw the low-frequency small-signal AC equivalent circuit, and find an approximate value for f_{cL}. **(c)** Draw the high-frequency small-signal AC equivalent circuit, and find an approximate value for f_{cH}.

S P9.48 Confirm the results of Problem P9.47, using SPICE.

D P9.49 Consider the circuit shown in Figure 9-128. Find new values for R_1, R_2, R_{EA}, R_{EB}, and R_C so that the overall midband gain is −50, the bias point is stable, and f_{cH} is as high as possible. (*Hint:* Consider what will improve the GBW, and start from the design shown in the figure.) You do not need to allow for any significant swing capability, but you do need to keep $V_B \geq 2$ V for bias stability. Compare the GBW of your design with that of Problem P9.47. Is it better?

P9.50 Find approximate equations for f_{cL} and f_{cH} for the circuit shown in Figure 9-129. Assume that the current through R_B is negligible compared to $g_m v_\pi$. This problem is quite difficult.

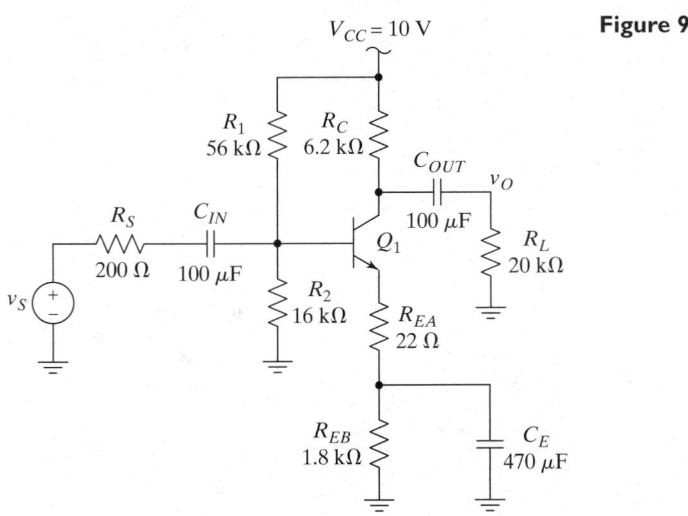

Figure 9–128

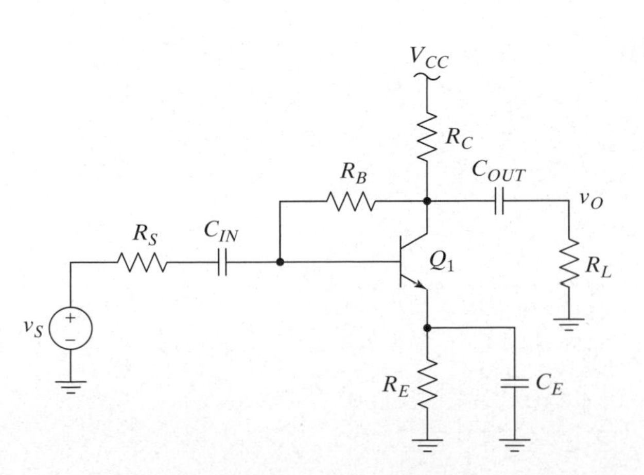

Figure 9–129

P9.51 Find approximate equations for f_{cL} and f_{cH} for the circuit shown in Figure 9-130. Assume that the current through R_B is negligible compared to $g_m v_\pi$. This problem is quite difficult.

P9.52 Consider the circuit shown in Figure 9-131. Using the 2N3906 model presented in Appendix B, along with the values given on the schematic, **(a)** calculate the DC bias point, **(b)** draw the low-frequency small-signal AC equivalent circuit and find an approximate value for f_{cL}, and **(c)** draw the high-frequency small-signal AC equivalent circuit and find an approximate value for f_{cH}.

S P9.53 Confirm the results of Problem P9.52, using SPICE.

D P9.54 Consider the circuit shown in Figure 9-131. Find new values for R_1, R_2, R_{EA}, R_{EB} , and R_C so that the overall midband gain is −50, the bias point is stable, and f_{cH} is as high as possible. (*Hint:* Consider what will improve the GBW, and start from the design shown in the figure.) You do not need to allow for any significant swing capability, but you do need to keep $V_B \leq 8$ V for bias stability (i.e., V_{BB} in the small-signal equivalent must be greater than or equal to 2 V). Compare the GBW of your design with that of Problem P9.52. Is your design better?

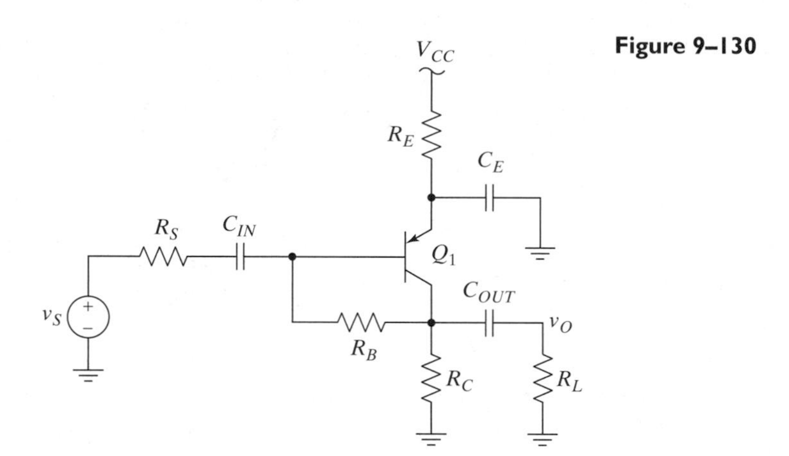

Figure 9–130

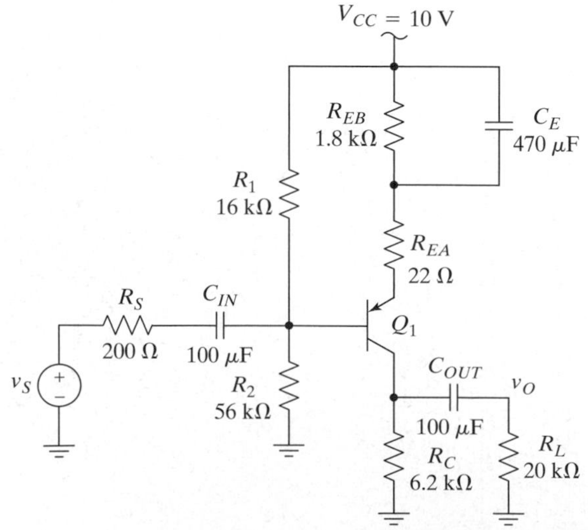

Figure 9–131

9.2.3 A FET Implementation: The Common-Source Amplifier

Introductory Treatment

P9.55 Find the transfer function, $V_o(j\omega)/V_{sig}(j\omega)$, for the circuit shown in Figure 9-132 for low and midband frequencies. Ignore the output resistance of the transistor.

P9.56 Use the Miller approximation to find an equation for f_{cH} for the amplifier in Figure 9-132. How would you check whether or not the approximation is valid in this case?

P9.57 Find the transfer function, $V_o(j\omega)/V_{sig}(j\omega)$, for the circuit shown in Figure 9-133 for low and midband frequencies. Ignore the output resistance of the transistor.

P9.58 Use the Miller approximation to find an equation for f_{cH} for the amplifier in Figure 9-133. How would you check whether or not the approximation is valid in this case?

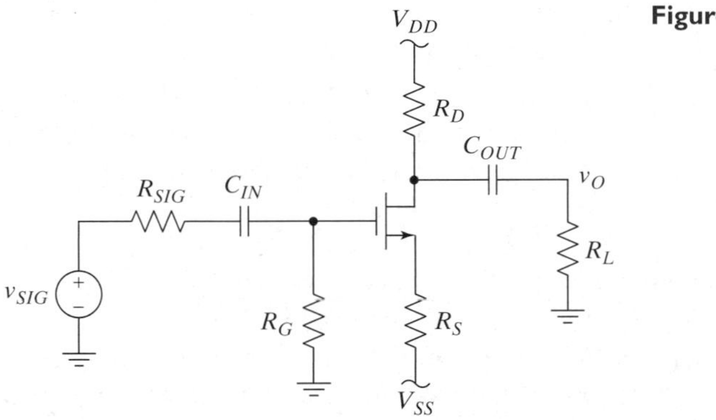

Figure 9–132

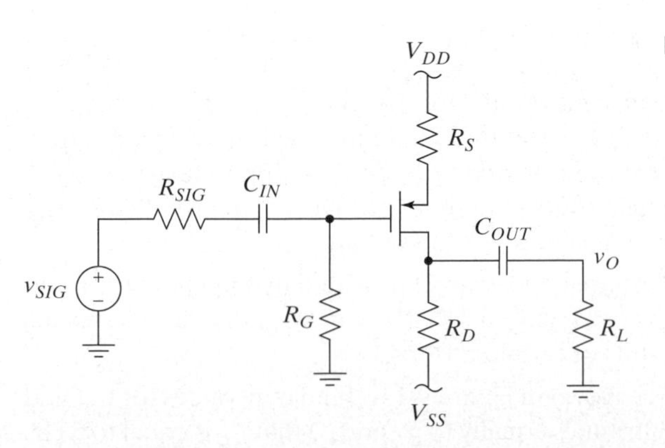

Figure 9–133

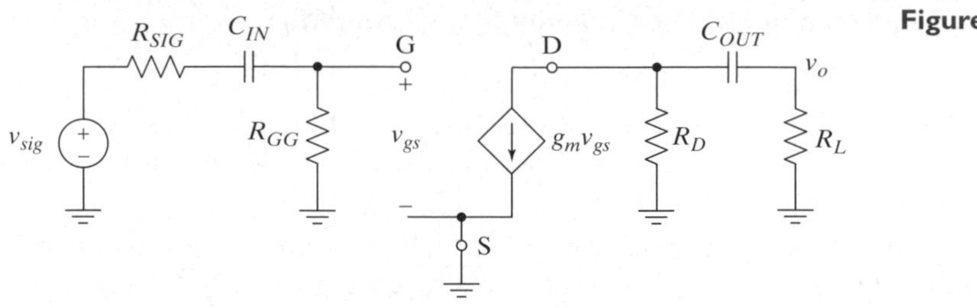

Figure 9–135

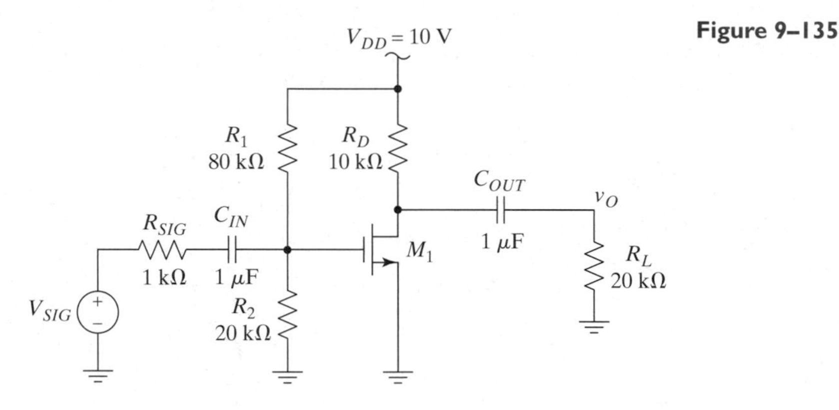

P9.59 Derive the transfer function, $V_o(j\omega)/V_{sig}(j\omega)$, for the low-frequency small-signal equivalent circuit shown in Figure 9-134.

P9.60 Consider the amplifier shown in Figure 9-135. The transistor is modeled by the Nsimple_mos model found in Appendix B. **(a)** Find the DC bias point and the full small-signal model for the transistor. **(b)** Draw the low-frequency equivalent circuit and find an approximation to f_{cL}. **(c)** Draw the high-frequency equivalent circuit and find an approximation to f_{cH}, assuming that the Miller effect dominates. **(d)** Determine whether or not the Miller approximation is valid for this circuit.

S P9.61 Use SPICE and the Nsimple_mos model provided in Appendix B to simulate the circuit shown in Figure 9-135. How do your simulation results compare with the results of Problem P9.60?

D P9.62 Consider the amplifier shown in Figure 9-135. Find new values for C_{IN} and C_{OUT} so that each contribute equally to f_{cL} and so that f_{cL} is equal to 5 Hz.

D P9.63 Consider the amplifier shown in Figure 9-135. Find new values for R_1, R_2, and R_D so that the midband overall gain is −2 and $f_{cL} = 10$ Hz. Assume $K = 100\ \mu\text{A/V}^2$.

P9.64 Consider the amplifier shown in Figure 9-136. The transistor is modeled by the Psimple_mos model found in Appendix B. **(a)** Find the DC bias point and the full small-signal model for the transistor. **(b)** Draw the low-frequency equivalent circuit and find an approximation to f_{cL}. **(c)** Draw the high-frequency equivalent circuit and find an approximation to f_{cH}, assuming that the Miller effect dominates. **(d)** Determine whether or not the Miller approximation is valid for this circuit.

S P9.65 Use SPICE and the Psimple_mos model provided in Appendix B to simulate the circuit shown in Figure 9-136. How do your simulation results compare with the results of Problem P9.64?

D P9.66 Consider the amplifier shown in Figure 9-136. Find new values for C_{IN} and C_{OUT} so that each contributes equally to f_{cL} and so that f_{cL} is equal to 5 Hz.

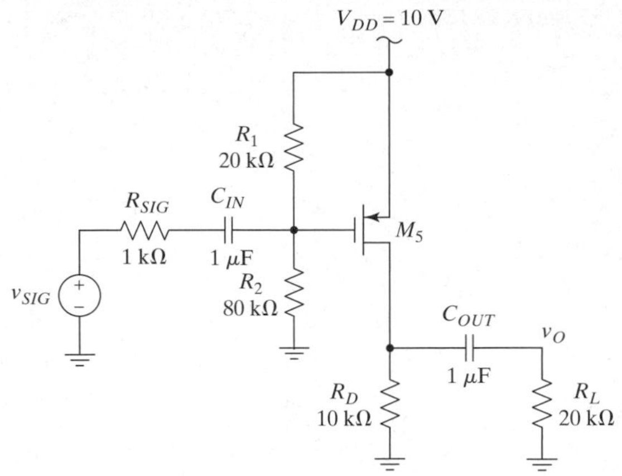

Figure 9–136

D P9.67 Consider the amplifier shown in Figure 9-136. Find new values for R_1, R_2, and R_D so that the midband gain is -2 and $f_{cL} = 10$ Hz. Assume $K = 100\ \mu A/V^2$.

Advanced Treatment

P9.68 Consider the amplifier shown in Figure 9-135. The transistor is modeled by the Nsimple_mos model found in Appendix B. **(a)** Find the DC bias point and the full small-signal model for the transistor. **(b)** Draw the low-frequency equivalent circuit and find an approximation to f_{cL}, using short-circuit time constants. **(c)** Draw the high-frequency equivalent circuit and find an approximation to f_{cH}, using open-circuit time constants. **(d)** If you previously did Problem P9.60, compare the result derived in part (c) of that problem with part (c) of this problem. Which answer is more accurate? Why?

P9.69 Consider the amplifier shown in Figure 9-136. The transistor is modeled by the Psimple_mos model found in Appendix B. **(a)** Find the DC bias point and the full small-signal model for the transistor. **(b)** Draw the low-frequency equivalent circuit and find an approximation to f_{cL}, using short-circuit time constants. **(c)** Draw the high-frequency equivalent circuit and find an approximation to f_{cH}, using open-circuit time constants. **(d)** If you previously did Problem P9.64, compare the result derived in part (c) of that problem with part (c) of this problem. Which answer is more accurate? Why?

P9.70 Find approximate equations for f_{cH} and f_{cL} for the amplifier shown in Figure 9-137.

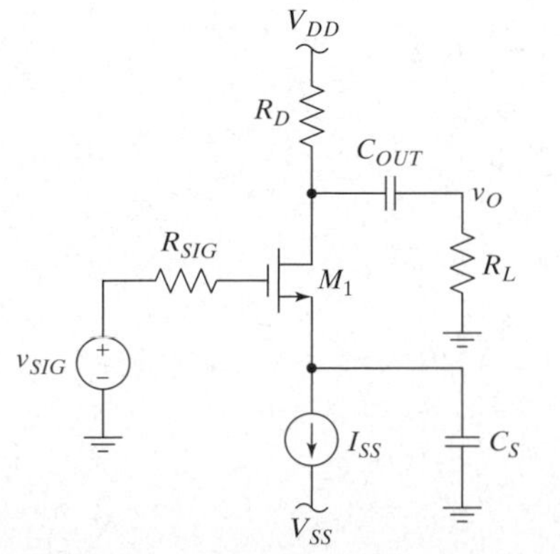

Figure 9–137

Figure 9–138

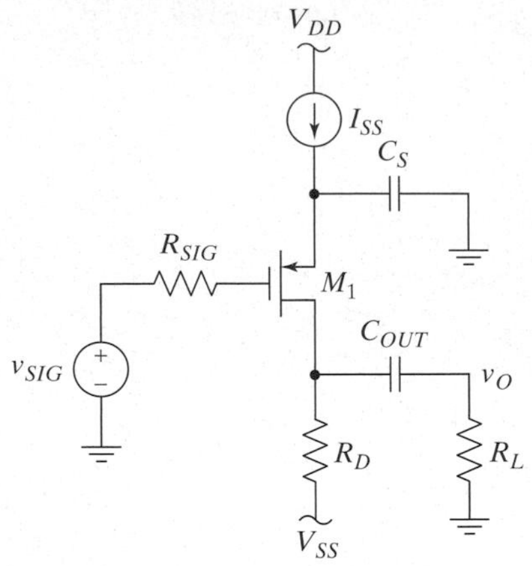

P9.71 Find approximate equations for f_{cH} and f_{cL} for the amplifier shown in Figure 9-138.

9.3 Voltage Buffers

9.3.1 A Generic Implementation: The Merge Follower

Introductory Treatment

P9.72 Find an approximate equation for f_{cH} for the merge follower shown in Figure 9-139.

Advanced Treatment

P9.73 Find an equation for f_{cH} for the merge follower in Figure 9-139 by using the open-circuit time-constant method.

P9.74 Compare (9.115) and (9.106), and show that the terms due to C_{oc} are identical in each equation and the terms due to C_{in} differ by $(R_S + R_L')/R_S$, with ω_{cH} in (9.115) being the lower approximation of the two.

9.3.2 A Bipolar Implementation: The Emitter Follower

Introductory Treatment

P9.75 Find an approximate value for f_{cH} for the emitter follower in Figure 9-140. Use the model given in Appendix B for the 2N3904 transistor. The source has a DC component of $V_S = 5$ V. Is the Miller approximation valid?

S P9.76 Confirm the results of Problem P9.75 by running a SPICE simulation of the circuit in Figure 9-140. Use the 2N3904 model from Appendix B for the transistor.

Figure 9–139

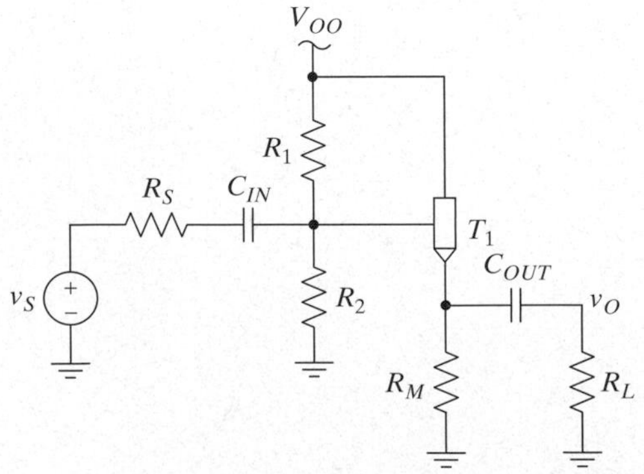

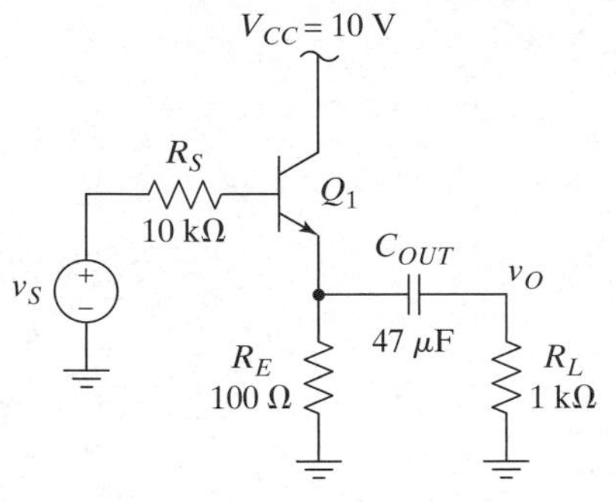

Figure 9–140

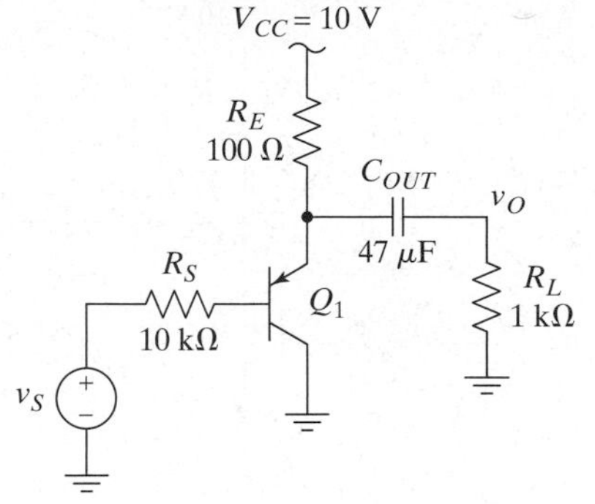

Figure 9–141

D P9.77 Change the value of R_E in Figure 9-140 so that the overall midband gain is 0.8. Assume $V_S = 5$ V. What is the new value of f_{cH}? Use the 2N3904 transistor model from Appendix B.

P9.78 Assuming that $V_S = 5$ V, what, approximately, is the maximum bandwidth obtainable from the circuit in Figure 9-140? Assume that only R_E can be varied and use the 2N3904 transistor model from Appendix B.

P9.79 Find an approximate value for f_{cH} for the emitter follower in Figure 9-141. Use the model given in Appendix B for the 2N3906 transistor. The source has a DC component of $V_S = 5$ V.

S P9.80 Confirm the results of Problem P9.79 by running a SPICE simulation of the circuit in Figure 9-141.

Advanced Treatment

P9.81 Repeat Problem P9.75, using open-circuit time constants.

P9.82 Repeat Problem P9.79, using open-circuit time constants.

9.3.3 A FET Implementation: The Source Follower

Introductory Treatment

P9.83 Find an approximate value for f_{cH} for the source follower in Figure 9-142. Use the model given in Appendix B for the Nlarge_mos transistor. The source has a DC component of $V_{SIG} = 5$ V. Is the Miller approximation valid?

S P9.84 Confirm the results of Problem P9.83 by running a SPICE simulation of the circuit in Figure 9-142. Use the Nlarge_mos transistor model from Appendix B.

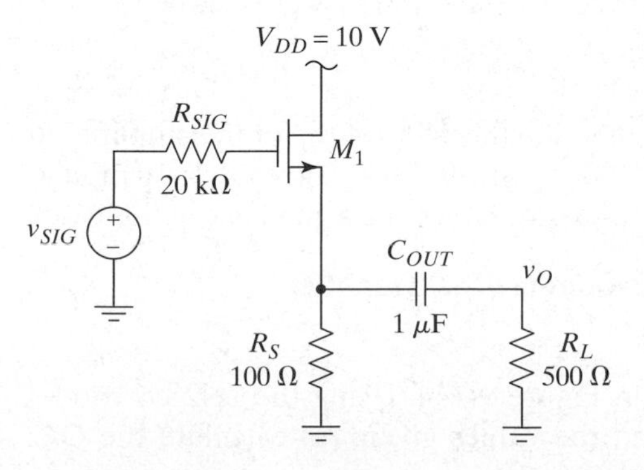

Figure 9–142

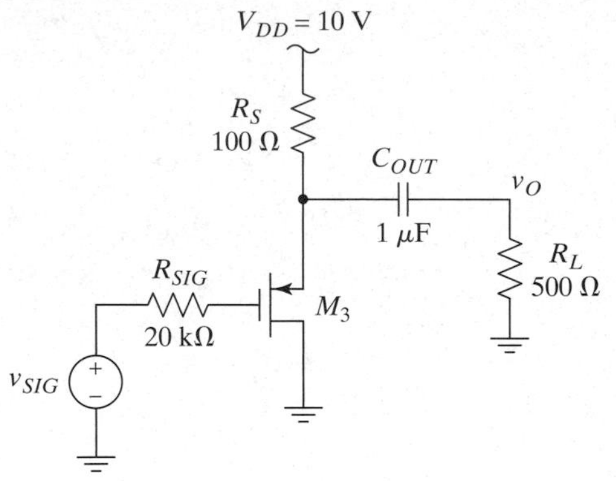

Figure 9–143

D P9.85 Change the value of R_S in Figure 9-142 so that the midband gain is 0.6. Assume $V_{SIG} = 5$ V. What is the new value of f_{cH}? Use the Nlarge_mos transistor model from Appendix B.

P9.86 Assuming that $V_{SIG} = 5$ V, what, approximately, is the maximum bandwidth obtainable from the circuit in Figure 9-142? Assume that only R_S can be varied and use the Nlarge_mos transistor model from Appendix B.

P9.87 Find an approximate value for f_{cH} for the source follower in Figure 9-143. Use the model given in Appendix B for the Plarge_mos transistor. The source has a DC component of $V_{SIG} = 5$ V.

S P9.88 Confirm the results of Problem P9.87 by running a SPICE simulation of the circuit in Figure 9-143.

Advanced Treatment

P9.89 Repeat Problem P9.83, using open-circuit time constants.

P9.90 Repeat Problem P9.87, using open-circuit time constants.

9.4 Current Buffers

9.4.1 A Generic Implementation: The Common-Control Amplifier

Introductory Treatment

P9.91 Using (9.150), find the value of f_{cH} for the amplifier in Figure 9-59 if $R_S = 10$ kΩ, $r_m = 50$ Ω, $a_{ic} = 0.98$, $R_L = 10$ kΩ, $C_{cm} = 8$ pF, and $C_{oc} = 1$ pF.

P9.92 Using (9.150), find the value of f_{cH} for the amplifier in Figure 9-59 if $R_S = 20$ kΩ, $r_m = 100$ Ω, $a_{ic} = 0.99$, $R_L = 10$ kΩ, $C_{cm} = 5$ pF, and $C_{oc} = 1.5$ pF.

P9.93 Consider the amplifier in Figure 9-59. If $R_S = 10$ kΩ, $r_m = 100$ Ω, $a_{ic} = 0.99$, $R_L = 20$ kΩ, and $C_{oc} = 1.5$ pF, how large would C_{cm} have to be for the two poles to have the same frequency? Explain this answer intuitively.

Advanced Treatment

P9.94 Using open-circuit time constants, find the value of f_{cH} for the amplifier in Figure 9-59 if $R_S = 5$ kΩ, $r_m = 50$ Ω, $a_{ic} = 0.99$, $R_L = 10$ kΩ, $C_{cm} = 10$ pF, and $C_{oc} = 2$ pF. Do you expect the approximation to be a good one in this case?

9.4.2 A Bipolar Implementation: The Common-Base Amplifier

Introductory Treatment

P9.95 Consider the circuit shown in Figure 9-144. Using the 2N2857 model presented in Appendix B and the values given, **(a)** calculate the DC, bias point. **(b)** Draw the low-frequency small-signal AC equivalent circuit and find an approximate value for f_{cL}. **(c)** Draw the high-frequency small-signal AC equivalent circuit and find an approximate value for f_{cH}, using (9.154). The poles are the same whether the Thévenin or Norton equivalent is used for the source.

S P9.96 Confirm the results of Problem P9.95, using SPICE.

P9.97 **(a)** Find an approximate value for f_{cL} for the circuit shown in Figure 9-145. **(b)** Determine a new value for C_{IN} that would cause both coupling capacitors to contribute equally to f_{cL}.

P9.98 Consider the circuit shown in Figure 9-145. Using the 2N3906 model presented in Appendix B and the values given, **(a)** calculate the DC bias point. **(b)** Draw the low-frequency small-signal AC equivalent circuit and find an approximate value for f_{cL}. **(c)** Draw the high-frequency small-signal AC equivalent circuit and find an approximate value for f_{cH}, using (9.154). The pole frequencies do not depend on whether we use a Thévenin or Norton model for the source.

S P9.99 Confirm the results of Problem P9.98 using SPICE.

Advanced Treatment

P9.100 Repeat Problem P9.95, using short- and open-circuit time-constants. Are the approximations reasonable in this case?

P9.101 Repeat Problem P9.98, using short- and open-circuit time-constants. Are the approximations reasonable in this case?

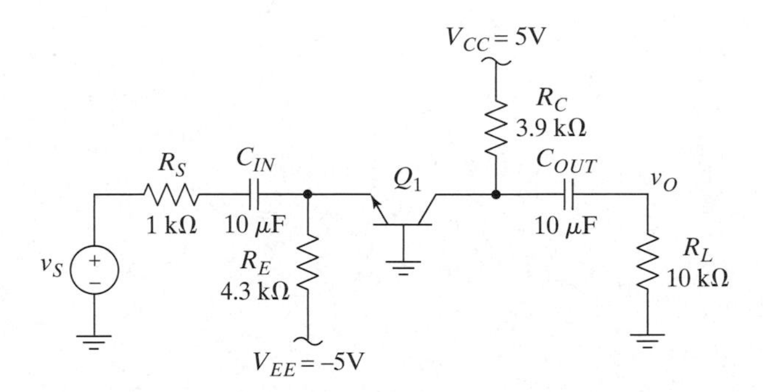

Figure 9–144

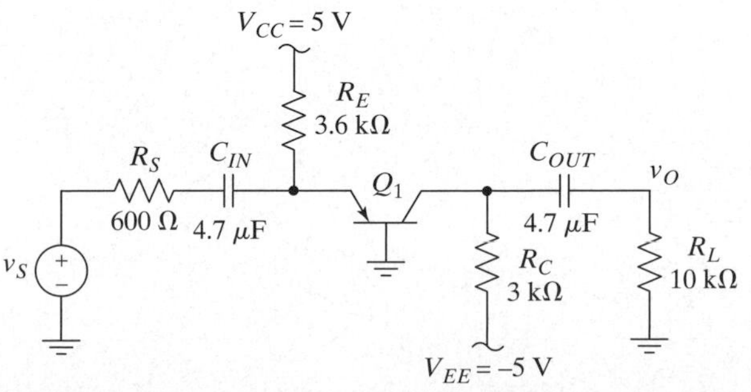

Figure 9–145

9.4.3 A FET Implementation: The Common-Gate Amplifier

Introductory Treatment

P9.102 Consider the circuit shown in Figure 9-146. Using the Nlarge_mos model presented in Appendix B and the values given, **(a)** calculate the DC bias point. **(b)** Draw the low-frequency small-signal AC equivalent circuit and find an approximate value for f_{cL}. **(c)** Draw the high-frequency small-signal AC equivalent circuit and find an approximate value for f_{cH} using (9.158).

S P9.103 Confirm the results of Problem P9.102, using SPICE.

P9.104 Consider the circuit shown in Figure 9-147. Using the Plarge_mos model presented in Appendix B and the values given, **(a)** calculate the DC bias point. **(b)** Draw the low-frequency small-signal AC equivalent circuit and find an approximate value for f_{cL}. **(c)** Draw the high-frequency small-signal AC equivalent circuit and find an approximate value for f_{cH}, using (9.158).

S P9.105 Confirm the results of Problem P9.104, using SPICE.

Advanced Treatment

P9.106 Repeat Problem P9.102, using short- and open-circuit time-constants. Are the approximations reasonable in this case?

P9.107 Repeat Problem P9.104 using short- and open-circuit time-constants . Are the approximations reasonable in this case?

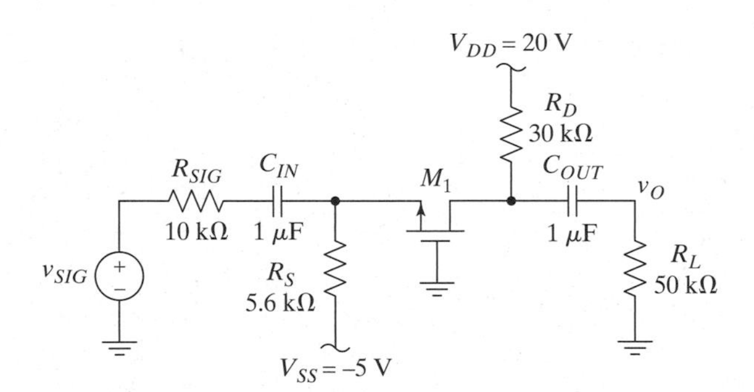

Figure 9–146

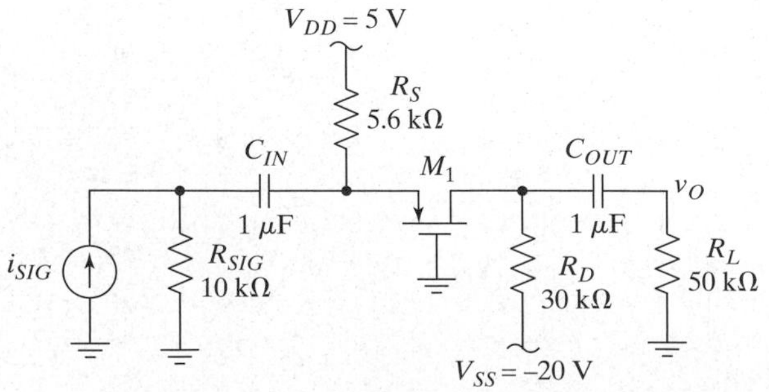

Figure 9–147

9.5 Comparison of Single-Stage Amplifiers

9.5.1 Bipolar Amplifiers

P9.108 Derive all of the results given in Table 9.1, except for the high cutoff frequencies.

P9.109 Derive all of the high cutoff frequencies given in Table 9.1.

D P9.110 If we are to design an amplifier that will be driven by a Norton equivalent source and we want the input attenuation factor, which is a current divider in this case, to be as close to unity as possible, what is the best single-transistor topology to choose? Why?

D P9.111 If we want a single-transistor amplifier that will provide a voltage gain of 10 (magnitude) and a good input attenuation factor when driven from a Thévenin source, what type of stage should we choose? Why?

P9.112 Can we make an amplifier with a noninverting voltage gain of 10 and a large input impedance, using a single bipolar transistor? Explain your answer.

P9.113 Is it possible to make a single-transistor bipolar amplifier with both voltage and current gains near unity and a low input impedance? Explain your answer.

9.5.2 FET Amplifiers

P9.114 Derive all of the results given in Table 9.3, except for the high cutoff frequencies.

P9.115 Derive all of the high cutoff frequencies given in Table 9.3.

D P9.116 If we are to design an amplifier that will be driven by a Norton equivalent source and we want the input attenuation factor, which is a current divider in this case, to be as close to unity as possible, what is the best single-transistor topology to choose? Why?

D P9.117 If we want a single-transistor amplifier that will provide a voltage gain of 10 (magnitude) and a good input attenuation factor when driven from a Thévenin source, what type of stage should we choose? Why?

P9.118 Can we make an amplifier with a noninverting voltage gain of 10 and a large input impedance, using a single field-effect transistor? Explain your answer.

P9.119 Is it possible to make a single-transistor amplifier with both voltage and current gains near unity and a low input impedance? Explain your answer.

9.6 Multistage Amplifiers

P9.120 Consider cascading N identical common-merge amplifiers together. Assume that each stage has a single-pole low-pass response and that the GBW product remains constant as the gain and bandwidth are varied (for each individual stage). Derive an equation for the high cutoff frequency of the cascade in terms of the GBW of the individual stages, the desired overall low frequency gain A_{T0}, and the number of stages N. If $A_{T0} = 100$, which N yields the largest high cutoff frequency? What is ω_{cH} as a fraction of GBW?

9.6.1 The Common-Merge Merge-Follower Cascade

Introductory Treatment

P9.121 Find approximate values for ω_{cH1} and ω_{cH2} for the amplifier in Figure 9-67, given the following data: $R_1 = 47\ \text{k}\Omega$, $R_2 = 22\ \text{k}\Omega$, $R_O = 4.7\ \text{k}\Omega$, $R_{M2} = 330\ \Omega$, $R_S = 1\ \text{k}\Omega$, $R_L = 10\ \text{k}\Omega$, $g_{m1} = 20$ mA/V, $r_{cm1} = 2\ \text{k}\Omega$, $r_{cm2} = 1\ \text{k}\Omega$, $r_{m2} = 25\ \Omega$, $a_{im2} = 99$, $C_{cm1} = C_{cm2} = 10$ pF, and $C_{oc1} = C_{oc2} = 1$ pF. Does the Miller effect dominate? What is the approximate value of the cutoff frequency for the cascade?

S P9.122 Use SPICE to verify the results derived in Problem P9.121.

Advanced Treatment

P9.123 Find f_{cH} for the circuit in Problem P9.121, using open-circuit time constants. Assume $g_{m2} = 40$ mA/V. Are any of the time constants dominant?

9.6.2 A Bipolar Implementation: The Common-Emitter Emitter-Follower Cascade

Introductory Treatment

P9.124 Use the Miller approximation to find an approximate value for f_{cH} for the cascaded amplifier in Figure 9-148. Assume that the emitter follower does not affect the frequency response. Use the model given for a 2N2857 in Appendix B. Check the Miller approximation (taking the input capacitance of the emitter follower into account).

S P9.125 Check the results of Problem P9.124, using SPICE.

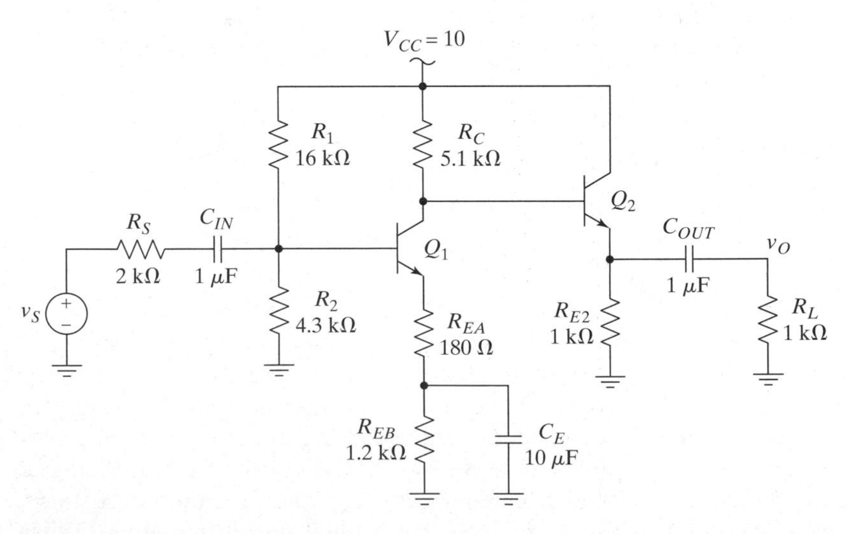

Figure 9–148

D P9.126 Using the topology shown in Figure 9-148 with 2N2857 transistors and a 5-V supply, design an amplifier to achieve an overall midband voltage gain of 20 dB with $R_S = 1\text{ k}\Omega$ and $R_L = 100\ \Omega$. Obey the rules of thumb for bias design and use $C_{IN} = 68\ \mu\text{F}$, $C_E = 10\ \mu\text{F}$, and $C_{OUT} = 4.7\ \mu\text{F}$. Calculate the approximate f_{cH} for your design. Is there anything you can do to increase f_{cH} while still meeting all of the specifications?

S P9.127 Confirm the results of Problem P9.126, using SPICE.

D P9.128 Repeat Problem P9.124 with R_S changed to 10 kΩ. Now add an emitter follower in front of the common-emitter stage, and make whatever changes to the circuit are necessary to keep the midband gain and the bias currents of Q_1 and Q_2 about the same as they were before. What is f_{cH} for the new amplifier? Explain why the emitter follower has improved the bandwidth.

S P9.129 Confirm the results of Problem P9.128, using SPICE.

Advanced Treatment

P9.130 Find an approximation for f_{cH} for the circuit in Figure 9-148, using open-circuit time constants. Do not ignore the emitter follower. Use the model given for a 2N2857 in Appendix B.

S P9.131 Check the results of Problem P9.130 using SPICE.

P9.132 Find an approximation for f_{cH} for the circuit in Figure 9-149 using open-circuit time constants. Do not ignore the emitter follower. Use the 2N2857 model for Q_1, and the 2N3904 model for Q_2, both from Appendix B.

S P9.133 Check the results of Problem P9.132, using SPICE.

D P9.134 Change the values of R_{E1}, R_C and R_{E2} as necessary to maximize the value of f_{cH} in the circuit shown in Figure 9-149 while keeping an overall midband voltage gain of $v_o/v_s = 5$. Use the 2N2857 model for Q_1 and the 2N3904 for Q_2. Both models are in Appendix B. What aspect of the circuit's performance did you sacrifice to increase the bandwidth?

S P9.135 Confirm the results of Problem P9.134, using SPICE.

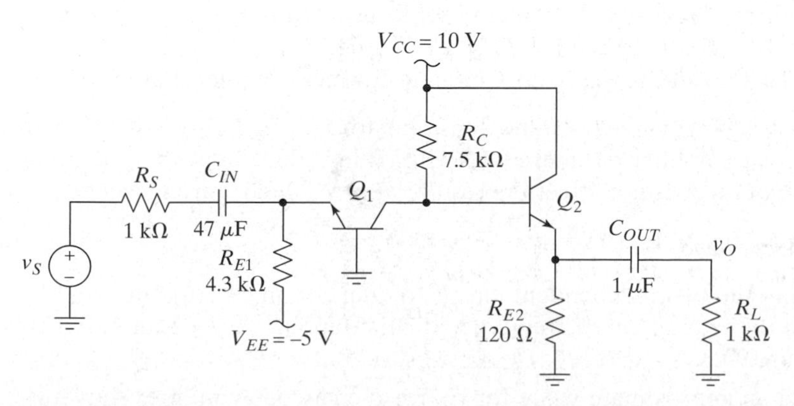

Figure 9–149

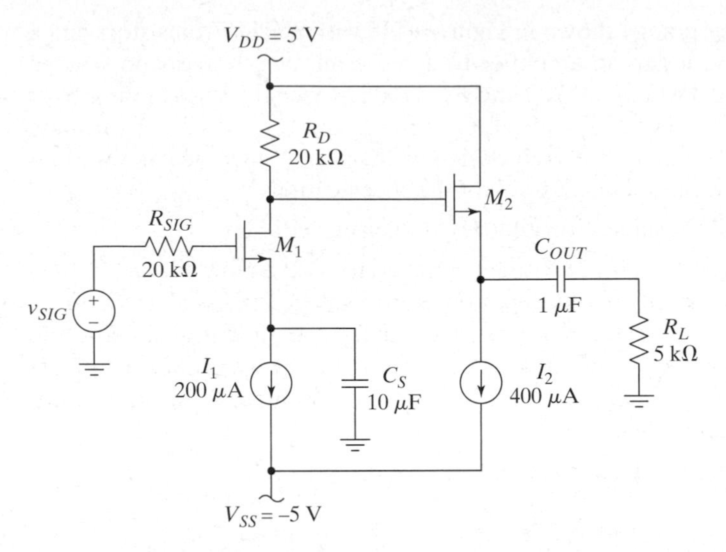

Figure 9–150

9.6.3 A FET Implementation: The Common-Source Source-Follower Cascade

Introductory Treatment

P9.136 Use the Miller approximation to find an approximate value for f_{cH} for the cascaded amplifier in Figure 9-150. Assume that the source follower does not affect the frequency response. Use the model given for Nsimple_mos in Appendix B. Check the Miller approximation (taking the input capacitance of the source follower into account).

S P9.137 Check the results of Problem P9.136, using SPICE.

Advanced Treatment

P9.138 Find an approximate value for f_{cH} for the amplifier in Figure 9-150, making use of the method of open-circuit time constants. Do not ignore the source follower. Use the model given for Nsimple_mos in Appendix B.

S P9.139 Check the results of Problem P9.138, using SPICE.

D P9.140 Use the circuit shown in Figure 9-150, and change R_D and the two bias currents as necessary to achieve a midband overall voltage gain of −5. What is the value of f_{cH} for your design? Can you increase f_{cH} by making some change to your bias point?

S P9.141 Confirm the results of Problem P9.140, using SPICE.

9.6.5 The Bipolar Cascode Amplifier: A Common-Emitter Common-Base Cascade

P9.142 Derive the open-circuit time constants for $C_{\pi 2}$ and $C_{\mu 2}$ in the bipolar cascode amplifier of Figure 9-83 with r_{b2} included. Show that the presence of r_{b2} does not affect the results when r_{o1} and r_{o2} are neglected.

Introductory Treatment

P9.143 Using the two independent cutoff frequencies in (9.216) and (9.217), find the overall cutoff frequency for the bipolar cascode amplifier in Figure 9-83.

P9.144 Find an approximate value for f_{cH} for the cascode amplifier shown in Figure 9-151. Use the 2N2222 model given in Appendix B.

S P9.145 Confirm the results of Problem P9.144, using SPICE.

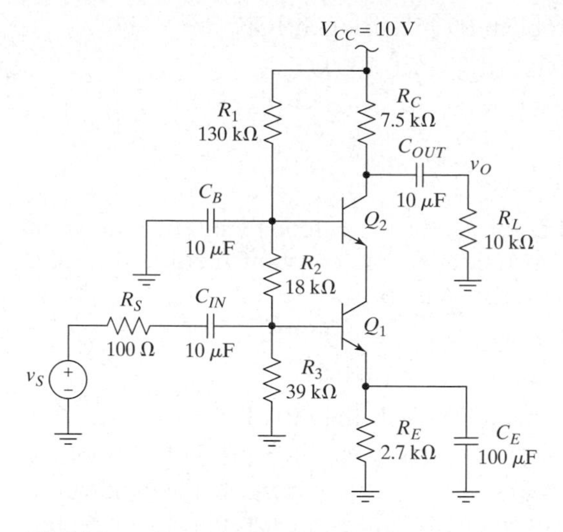

Figure 9–151

P9.146 Repeat Problem P9.144 with R_S set to 10 kΩ. What dominates the bandwidth in this case?

Advanced Treatment

P9.147 Find an approximate value for f_{cH} for the cascode in Figure 9-151, using the open-circuit time, constant method. Use the 2N2222 model given in Appendix B and ignore r_b in Q_2.

S P9.148 Confirm the results of Problem P9.147, using SPICE.

P9.149 Repeat Problem P9.147 with R_S set to 10 kΩ. What dominates the bandwidth in this case?

D P9.150 Consider, the circuit from Problem P9.149. Change the topology of the amplifier in a way that will significantly increase the bandwidth. (*Hint:* consider how to reduce the driving-point resistance seen by the dominant time constant.)

9.6.6 The MOSFET Cascode Amplifier: A Common-Source Common-Gate Cascade

Introductory Treatment

P9.151 Using the two independent cutoff frequencies in (9.229) and (9.230), find the overall cutoff frequency for the MOSFET cascode amplifier.

P9.152 Find an approximate value for f_{cH} for the cascode amplifier shown in Figure 9-152. Use the Nlarge_mos model given in Appendix B.

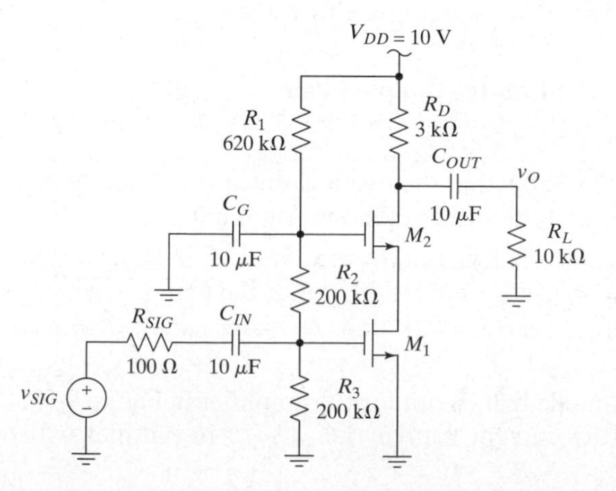

Figure 9–152

S P9.153 Confirm the results of Problem P9.152, using SPICE.

P9.154 Repeat Problem P9.152 with R_{SIG} set to 100 kΩ.

S P9.155 Confirm the results of Problem P9.154, using SPICE.

Advanced Treatment

P9.156 Find an approximate value for f_{cH} for the cascode amplifier shown in Figure 9-152, using the open-circuit time-constant method. Use the Nlarge_mos model given in Appendix B.

S P9.157 Confirm the results of Problem P9.156, using SPICE.

P9.158 Repeat Problem P9.156 with R_{SIG} set to 100 kΩ.

S P9.159 Confirm the results of Problem P9.158, using SPICE.

D P9.160 Consider again the circuit from Problem P9.158. Change the topology of the amplifier in a way that will significantly increase the bandwidth. (Hint: consider how to reduce the driving-point resistance seen by the dominant time constant.)

9.7 Differential Amplifiers

9.7.1 The Generic Differential Pair

P9.161 Derive Equation (9.239).

P9.162 Derive Equation (9.240). Show that the result is the the same as Equation (9.239) if the Miller effect is dominant.

P9.163 Consider the common-mode half circuit in Figure 9-92. If C_{oc} dominates the frequency response, show that the gain is given by (9.243) with $\omega_p \approx 1/[(R_O + R_S/2)C_{oc}]$ and $\omega_z \approx |A_{cm0}/A_{cm}(\infty)|\omega_p$. (Hint: recall that $R_{MM} \gg r_{cm}$.)

P9.164 Consider the common-mode half circuit in Figure 9-94. If the impedance of the current source (i.e., $C_{MM}/2$ in parallel with $2R_{MM}$) dominates the frequency response and R_S is small, show that the common-mode gain is $V_o(j\omega)/V_{scm}(j\omega) \approx -(R_O/2R_{MM})(1 + j\omega R_{MM}C_{MM})/(1 + j\omega C_{MM}/2g_m)$.

P9.165 (a) Plot the CMRR for the generic differential amplifier in Figure 9-91 if $R_O = 8$ kΩ, $R_S = 1$kΩ, $R_{MM} = 100$ kΩ, $r_{cm} = 10$ kΩ, $g_m = 10$ mA/V, $C_{oc} = 1.2$ pF, and $C_{cm} = 16$ pF. You will need the equations given in Problem 9-61, in addition to some of the equations in the text. (b) What happens to the CMRR if R_S is reduced to 100 Ω? (c) What happens to the CMRR if R_{MM} is increased to 1 MΩ?

9.7.2 A Bipolar Implementation: The Emitter-Coupled Pair

P9.166 Derive Equation (9.251).

P9.167 Derive Equation (9.252). Show that the result is the approximately the same as Equation (9.251) if the Miller effect is dominant.

P9.168 Consider the common-mode half circuit for the amplifier in Figure 9-98. If C_μ dominates the frequency response, show that the gain is given by (9.253) with $\omega_p \approx 1/[(R_C + R_S/2)C_\mu]$ and $\omega_z \approx |A_{cm0}/A_{cm}(\infty)|\omega_p$. (Hint: recall that $R_{EE} \gg r_\pi$.)

P9.169 Consider the common-mode half circuit for the amplifier in Figure 9-98. If the impedance of the current source (i.e., $C_{EE}/2$ in parallel with

$2R_{EE}$) dominates the frequency response and R_S is small, show that the common-mode gain is

$V_o(j\omega)/V_{scm}(j\omega) \approx -(R_C/2R_{EE})(1 + j\omega R_{EE}C_{EE})/(1 + j\omega C_{EE}/2g_m)$

S P9.170 Use SPICE with the nominal transistor model from Appendix B and the circuit shown in Figure 9-98. Model the current source with a perfect source in parallel with $C_{EE} = 5$ pF and $R_{EE} = 250$ kΩ. **(a)** Plot the CMRR. **(b)** What happens to the CMRR if R_S is reduced to 200 Ω? (c) What happens to the CMRR if R_{EE} is increased to 1 MΩ?

P9.171 **(a)** Find the differential-mode gain, common-mode gain, and CMRR as a function of frequency for the circuit shown in Figure 9-153 if the output is taken single ended from the collector of Q_2 and the DC common-mode input voltage is zero. Use the nominal model from Appendix B, but set `xtf` = 0 in the model. Assume that the current source has an output resistance $R_{EE} = 100$ kΩ and a capacitance $C_{EE} = 10$ pF and that the impedance of the current source dominates the common-mode frequency response, so the results of Problem P9.169 apply. **(b)** Would you expect the results from part (a) to change if the DC common-mode input voltage were changed to 2 V? Why or why not?

S P9.172 Use SPICE to confirm the results you obtained in Problem P9.171.

The Emitter-Coupled Pair with Active Load

P9.173 Consider the circuit in Figure 9-100. Let the circuit drive a load consisting of $R_L = 10$ kΩ in parallel with $C_L = 5$ pF. Use $R_{EE} = 100$ kΩ in parallel with $C_{EE} = 10$ pF as the small-signal model of the current source. Assume that Q_4 is 10% larger in emitter area than Q_3. Ignore internal device capacitances and derive an approximate transfer function for the common-mode gain as a function of frequency.

S P9.174 Use SPICE and the nominal and nominalp models in Appendix B to confirm your results from P9.173. Because mismatch in the mirrors is critical to the CM gain, set `BF` = 10,000 in the nominalp models in order to remove beta error and obtain agreement with your hand analysis.

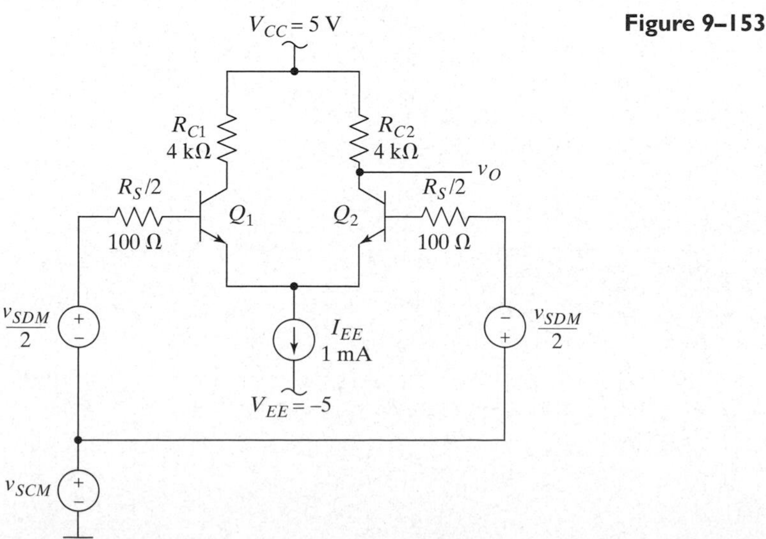

Figure 9–153

9.7.3 A FET Implementation: The Source-Coupled Pair

P9.175 Consider the common-mode half circuit for the amplifier in Figure 9-103. If C_{gd} dominates the frequency response, show that the gain is given by (9.262) with $\omega_p \approx 1/[(R_D + R_S/2)C_{gd}]$ and $\omega_z \approx |A_{cm0}/A_{cm}(\infty)|\omega_p$.

P9.176 Consider the common-mode half circuit for the amplifier in Figure 9-103. If the impedance of the current source (i.e., $C_{SS}/2$ in parallel with $2R_{SS}$) dominates the frequency response, show that the common-mode gain is $V_o(j\omega)/V_{scm}(j\omega) \approx -(R_D/2R_{SS})(1 + j\omega R_{SS}C_{SS})/(1 + j\omega C_{SS}/2g_m)$.

S P9.177 Find the differential-mode gain, common-mode gain and CMRR as a function of frequency for the circuit shown in Figure 9-103 if the output is taken differentially, the resistor connected to the drain of M_2 is $R_{D2} = 41$ kΩ, and the DC common-mode input voltage is zero. Use the Nsimple_mos model from Appendix B. Assume the current source has a small-signal output resistance $R_{SS} = 100$ kΩ and a capacitance $C_{SS} =$ 10 pF and that the impedance of the current source dominates the common-mode frequency response so the results of Problem P9.176 apply.

P9.178 Use SPICE and the Nsimple_mos model from Appendix B to confirm your results from P9.177.

The SCP with an Active Load

P9.179 Consider the circuit in Figure 9-105. Let the circuit drive a load consisting of $R_L = 20$ kΩ in parallel with $C_L = 5$ pF. Use $R_{SS} = 100$ kΩ in parallel with $C_{SS} = 5$ pF as the small-signal model of the curent source. Assume that M_4 is 5% larger in width than M_3. Ignore internal device capacitances and derive an approximate transfer function for the common-mode gain as a function of frequency.

S P9.180 Use SPICE and the Nsimple_mos and Psimple_mos models in Appendix B to confirm your results from P9.179. You can produce the 5% mismatch by giving M_4 a multiplier of $M = 1.05$.

CHAPTER 10

Feedback

Negative feedback is ubiquitous in nature and in everyday life. A skier, for example, constantly compares visual and tactile information to a mental image of his or her desired position. Any error is used to adjust his or her body position in order to reduce the error; which is precisely the function of a negative feedback loop.

INTRODUCTION

Feedback is present in almost every electronic system, in all biological systems, and in many mechanical systems as well. There are two different kinds of feedback; ***negative feedback*** and ***positive feedback***.

Negative feedback is also called ***degenerative feedback***. In a system with negative feedback, the output is fed back to the input in a way that tends to decrease the output; hence, negative feedback is inherently stable. Negative feedback is commonly used in amplifiers, filters, and many other types of electronic circuits. We encounter negative feedback in a number of places in this text. For example, feedback is used to stabilize the operating point of transistor amplifiers and to set the gain of amplifiers constructed using op amps. In the first part of this chapter, we undertake a thorough treatment of negative feedback, with particular emphasis on amplifiers. The material is first presented in a way that emphasizes the fundamental properties of feedback amplifiers. Later sections present progressively more detailed refinements and additions, which are necessary for dealing with practical implementations.

Positive feedback is also called ***regenerative feedback***. In a system with positive feedback, the output is fed back to the input in a way that tends to produce a greater output (i.e., it regenerates the output). Positive feedback is unstable and is avoided in designing amplifiers.[1] However, we will see that it is used in oscillators (oscillators are unstable in the sense that they produce an output with no input), and it is also useful for some other circuits (e.g., comparators with hysteresis). Oscillators are covered in Section 10.2.

10.1 NEGATIVE FEEDBACK

Negative feedback was invented by Harold S. Black of Bell Laboratories in 1927. He was working to produce an amplifier with stable gain, since the vacuum tube amplifiers used in the telephone system at that time had to be manually adjusted on a regular basis. The problem was—and still is—that without feedback, it is very difficult to produce an amplifier whose gain does not change with temperature, supply voltage, and other variables. In this section, we present the analysis of negative feedback amplifiers, starting with a block-diagram level approach and progressing to more and more sophisticated analysis techniques and models. Aside A10.1 provides a brief summary of the interesting history of this important invention.

10.1.1 Ideal Block Diagram Analysis

Since the output of an amplifier is supposed to be the same as the input—only larger—Black reasoned that if he took a fraction of the output and compared it with the input, the difference should ideally be zero. If this difference is not zero, it represents an error, and we should be able to change the output in such a way that we drive the error towards zero. Figure 10-1 presents a block diagram of a system with negative feedback. Each of the blocks is assumed to be unilateral, and we can write

$$S_o = aS_e = a(S_i - S_f) = a(S_i - bS_o), \tag{10.1}$$

where S_o is the output signal, S_i is the input signal, S_f is the feedback signal, S_e is the error, a is the ***open-loop gain*** of the ***forward amplifier***, and b is the ***feedback factor*** (i.e., the gain of the feedback block). Algebraic manipulation of (10.1) yields the ***closed-loop gain***,

$$A_f = \frac{S_o}{S_i} = \frac{a}{1 + ab}, \tag{10.2}$$

where the subscript f reminds us that this is the gain *with feedback*.

We can now see the main advantage of feedback. If $ab \gg 1$, the transfer function becomes

$$A_f = \frac{a}{1 + ab} \approx \frac{1}{b}, \tag{10.3}$$

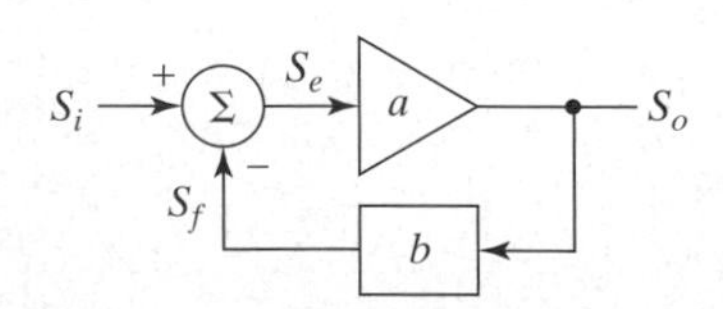

Figure 10–1 Block diagram of a negative-feedback amplifier.

[1] It isn't *always* unstable; the gain around the loop must be larger than unity, as we will see later.

ASIDE A10.1 HISTORY: THE INVENTION OF NEGATIVE FEEDBACK

Harold Black was born in Massachusetts in 1898 and went to work for the Western Electric Company's West Street Laboratories in 1921, which became part of Bell Laboratories in 1925. His starting salary was $32 a week. One of his first assignments was to work on the design of amplifiers used to drive long phone lines. Black recognized the need to include many such amplifiers—called repeaters—in long telephone lines to make up for the drop in strength as the signal travels down the line. The problem is that when many such amplifiers are connected in series, the distortion of the amplifiers grows to unacceptable levels, and gain errors accumulate as well. Therefore, Black began to investigate distortion and stability in vacuum tube amplifiers [10.4, 10.5].

After working on this problem for a number of years, in August 1927 Black conceived of the negative-feedback amplifier while riding to work on the Lackawanna Ferry. He wrote the first description of the system on a page of *The New York Times*. A key development along the way occurred when Black restated the problem in a new way: He viewed the output of an imperfect amplifier as consisting of both the desired output, which is an amplified replica of the input, and unwanted "distortion" (the term is used here to refer to all unwanted changes, whether true distortion or gain changes). This led him to think of ways of removing that distortion rather than thinking of ways of preventing its introduction. He realized that by comparing a scaled version of the output with the input, he could generate an "error" that could be used to drive the output towards the proper value. This example is a great demonstration of the fact that carefully restating a difficult problem will sometimes lead to a solution.

After field tests in 1930 and 1931 proved that his circuit worked, it became widely used by Bell Telephone. However, getting the circuit patented was more difficult. Black applied for a patent in 1928, but the patent office regarded the invention with skepticism, and it wasn't until nine years later, in 1937, that the patent was issued.

Thirty years after the invention of negative feedback, M.J. Kelly, the president of Bell Laboratories, called it one of the two most significant inventions of the previous 50 years. (The other invention was the audio vacuum tube developed by Lee De Forest.) Black received a number of awards for the invention, including becoming a fellow of the Institute of Radio Engineers (the IRE was a precursor of the Institute of Electrical and Electronics Engineers—IEEE) and being inducted into the National Inventors Hall of Fame in 1981. He died in December 1983.

so that the closed-loop gain, A_f, is independent of the open-loop gain, a, and is entirely determined by the feedback factor. Since the closed-loop gain is greater than unity for an amplifier, the feedback factor is less than unity and can be set by very stable passive components (e.g., a resistor divider). Therefore, we have achieved the goal of making an amplifier with stable gain. The price we pay for this stability is a loss in gain: If $ab \gg 1$, then we also know that $a \gg 1/b \approx A_f$, so the closed-loop gain is much smaller than the open-loop gain. This fact does not represent a significant problem, however, since it is relatively easy to make amplifiers with very large, but poorly controlled, gains.

We can see that the system shown here uses *negative* feedback, since the signal fed back from the output is subtracted from the input. Therefore, if something in the loop changes in a way that would increase S_o (e.g., if a increases due to a change in temperature), the increase is subtracted from the input, which reduces the error signal driving the forward amplifier, so that the increase in S_o is reduced. If the feedback signal were added to the input instead, the error would tend to grow until something in the loop limited the growth; that would be positive feedback. In this section, we will focus on amplifiers; therefore, we restrict our attention to negative feedback. We can quantify how well the feedback opposes a change in the gain by calculating how much the closed-loop gain changes for a given change in the open-loop gain.

Example 10.1

Problem:

Assume that the forward amplifier in the block diagram of Figure 10-1 has a gain of 1000. Find the value of b needed to set the closed-loop gain equal to 20 by using (10.3). What is the exact value of the resulting closed-loop gain? By how much does the closed-loop gain change if a decreases by 25%?

Solution:

Using (10.3), we find that the desired $b = 1/20 = 0.05$. Using (10.2), we find that the exact gain is $A_f = 19.6$. If a decreases by 25%, we have $a = 750$, and the resulting closed-loop gain is $A_f = 19.5$. The change in closed-loop gain is less than 1% for a 25% change in the open-loop gain.

■

Example 10.1 demonstrates clearly how insensitive the closed-loop gain is to changes in the open-loop gain. Aside A10.2 presents an alternative way of determining this sensitivity.

At the beginning of the section, we stated that the error would be driven toward zero by the application of negative feedback. From Figure 10-1, we write

$$S_e = S_i - S_f = S_i - bS_o = S_i(1 - A_f b), \tag{10.4}$$

and substituting (10.2) for A_f yields

$$S_e = \frac{S_i}{1 + ab}. \tag{10.5}$$

Therefore, if the quantity ab is large, the error will in fact be small. As ab approaches infinity, the error approaches zero for all finite inputs. Hence, we have confirmed our initial intuitive understanding of the operation of negative feedback.

The product of the open-loop gain and the feedback factor, ab, has shown up several times in our equations already and is obviously of significance. Notice from Figure 10-1 that if you start at the input of the forward amplifier and work your way around the loop, you find that the gain around the loop is equal to $-ab$. The minus sign is a result of the fact that this is a negative feedback connection. The quantity ab is called the ***loop gain***, which is sometimes denoted by L.

Exercise 10.1

We desire to build a negative-feedback amplifier with a closed-loop gain of 10 that will remain accurate to within ±1%, whereas the open-loop gain changes by ±50%. Assume that the feedback factor remains constant. What nominal values of a and b are required? (Use the minimum possible value for a.)

□

10.1.2 Ideal Analysis and the Characteristics of Negative Feedback

In the preceding section, we discovered that negative feedback stabilizes the closed-loop gain with respect to changes in the open-loop gain. This is the primary

ASIDE A10.2 NORMALIZED GAIN SENSITIVITY

We can quantify the sensitivity of the closed-loop gain to changes in open-loop gain by using the normalized small-signal sensitivity (fully explained in Aside CD-A7.1). Applying the definition of sensitivity to the equation $A_f = a/(1+ab)$, we find that the normalized sensitivity of the closed-loop gain, A_f, to changes in the open-loop gain, a, is given by

$$S_a^{A_f} \equiv \lim_{\Delta a \to 0} \frac{\Delta A_f/A_f}{\Delta a/a} = \frac{a}{A_f}\frac{dA_f}{da} = \frac{1}{1+ab}. \tag{A10.1}$$

For example, if $a = 1{,}000$ and $b = 0.05$, $S_a^{A_f} = 1/51 = 0.02$. In other words, if a changes by 1%, A_f will change by only 0.02%. The sensitivity is a good figure of merit for comparing different systems, but we need to be careful in using the sensitivity to calculate the magnitude of an expected change. Since the sensitivity is based on using a derivative, and A_f is not a linear function of a, the results are generally accurate only for small changes.

advantage of feedback, but other aspects of the open-loop amplifier's performance are also altered by the application of feedback. In this section, we will examine some of these other changes. When negative feedback is applied to an amplifier, the following changes occur:

1. The overall gain is reduced.
2. The input and output impedances are changed.
3. The bandwidth is increased.[2]
4. Nonlinear distortion is reduced.

To discuss these changes in more detail, it is necessary to analyze a negative-feedback amplifier with a model more detailed than the block diagram of Figure 10-1. Accordingly, consider the amplifier whose unilateral two-port model is shown in Figure 10-2(a), and apply negative feedback to the amplifier as shown in Figure 10-2(b). The feedback network is modeled by a single

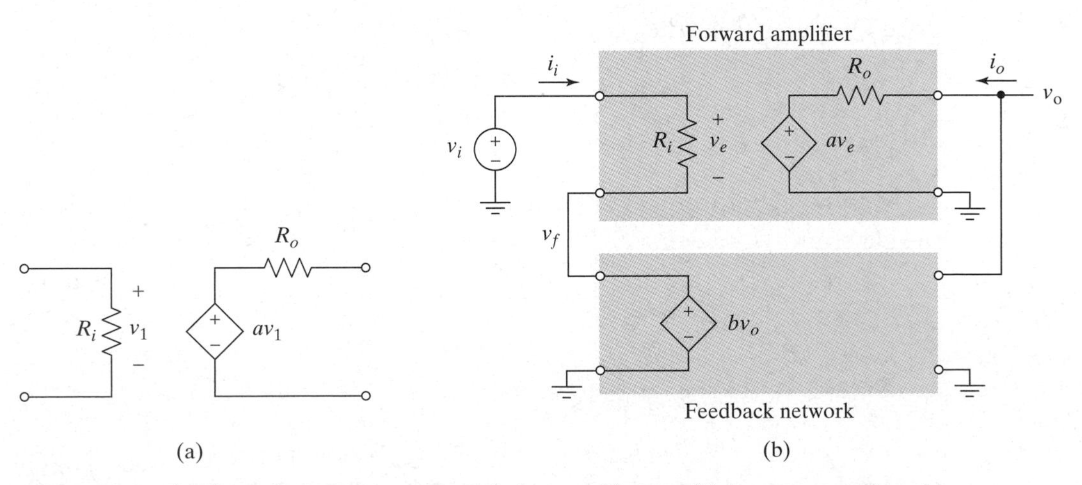

Figure 10-2 (a) The unilateral two-port model of an amplifier and (b) the same amplifier with negative feedback applied.

[2] Do conclude that feedback yields the largest bandwidth possible (*see* the detailed discussion later in this section).

controlled source in this case. In other words, the feedback network is ideal in the sense that it has infinite input resistance and zero output resistance. (The input of the feedback network is connected to v_o, and the output of the feedback network is v_f.) We have not shown a load, so we are implicitly assuming that R_L is infinite. Also, note that the input voltage of the amplifier, v_1, becomes the error voltage of the feedback amplifier.

Reduction in Gain Examining 10-2b, we write

$$v_o = av_e = a(v_i - bv_o), \tag{10.6}$$

from which we again derive

$$A_f = \frac{v_o}{v_i} = \frac{a}{1 + ab} = \frac{a}{1 + L}. \tag{10.7}$$

If the loop gain is large (i.e., $L \gg 1$), then $A_f \approx 1/b$, as before. This stabilization of the overall gain is achieved at the expense of a reduction in gain by the factor $1 + L$ (i.e., the closed-loop gain, A_f, is equal to the open-loop gain, a, divided by $1 + L$).

Modification of Input and Output Resistances The application of negative feedback also changes the input and output resistances of the amplifier. From Figure 10-2(a), we see that the original input and output resistances are R_i and R_o, respectively. If we denote the ***input resistance with feedback*** as R_{if}, then, from part (b) of the figure and the definition of input resistance, we have

$$R_{if} = \frac{v_i}{i_i}. \tag{10.8}$$

Using Ohm's law, we write

$$i_i = \frac{v_i - bv_o}{R_i}, \tag{10.9}$$

and substituting for v_o from (10.7), we get

$$i_i = \frac{v_i}{R_i}(1 - A_f b). \tag{10.10}$$

We now use (10.10) and (10.8) to write

$$R_{if} = \frac{R_i}{1 - A_f b}, \tag{10.11}$$

and we substitute for A_f using (10.7) to arrive at the final result:

$$R_{if} = R_i(1 + ab) = R_i(1 + L). \tag{10.12}$$

In this case, the input resistance has been increased by the factor $1 + L$, the same factor by which the gain was decreased.

Now let's examine the ***output resistance with feedback***, R_{of}. From the circuit and the definition, we have

$$R_{of} = \frac{v_o}{i_o} \tag{10.13}$$

where we must remember to set the independent source to zero (i.e., set $v_i = 0$). Following a procedure analogous to the one just used to find R_{if}, we write (note that with $v_i = 0$ we have $v_e = -bv_o$)

$$i_o = \frac{v_o - av_e}{R_o} = \frac{v_o(1 + ab)}{R_o} \tag{10.14}$$

and combine (10.13) and (10.14) to get

$$R_{of} = \frac{R_o}{1 + ab} = \frac{R_o}{1 + L}. \tag{10.15}$$

Therefore, the output resistance has been reduced by the now familiar factor $1 + L$.

In general, the input and output resistances are either increased or decreased by the factor $1 + L$, depending on the type of feedback connection. For the circuit shown in Figure 10-2(b), the forward amplifier and feedback amplifier are in series at the input and in parallel—or shunt—at the output.

To see that the input is a series connection, start at the input to the overall amplifier, v_i, and trace a path towards the amplifier. There is no option in tracing this path. You must go through the forward amplifier and the feedback network and then return to the source through ground; therefore, it is a series connection. On the other hand, starting from the output of the overall amplifier, v_o, and tracing a path toward the amplifier requires making a choice. We can either go through the output port of the forward amplifier to ground, or we can go through the input port of the feedback network to ground and then return to the load, which is not shown here since it is assumed to be infinite. In other words, the forward amplifier and feedback network are connected in parallel, or shunt, at the output.

We can be more mathematical in stating what happens when we trace a path from the input source or the output. In the case of the input, we can write a single loop equation that shows how the error signal is related to the input and the feedback:

$$v_e = v_i - v_f. \tag{10.16}$$

On the output side, we need to write a nodal equation instead of a loop equation. The currents in the forward amplifier and feedback network are different; only the voltage, v_o, is the same for both. Therefore, the feedback network has no direct way to measure the output current, and v_o must be the variable that is being fed back to the input (i.e., $v_f = bv_o$).

Feedback amplifiers are classified by the types of connections used at the input and output. The amplifier of Figure 10-2 is called a ***series–shunt*** amplifier, because the input connection is series and the output connection is shunt. There are four possible feedback connections, as illustrated in Aside A10.3.

ASIDE A10.3 THE FOUR FEEDBACK TOPOLOGIES

Figure A10-1 shows the four connections possible for negative-feedback amplifiers. Note that the two-port models chosen for the forward amplifier and the feedback network differ, depending on the type of connection. It is most convenient to use a Thévenin model for the feedback network output when the input of the amplifier is a series connection, as in parts (a) and (b) of the figure, since voltages are summed. For parts (a) and (b) of the figure, we obtain

$$v_e = v_i - v_f, \tag{A10.2}$$

where, for part (a), $v_f = bv_o$, and for part (b) $v_f = bi_o$. Similarly, it is most convenient to use a Norton model for the feedback network output when the amplifier has a shunt input connection as in parts (c) and (d) of the figure. For the shunt input connection, we have

$$i_e = i_i - i_f, \tag{A10.3}$$

where, for part (c), $i_f = bv_o$, and for part (d) $v_f = bi_o$.

When the output connection is shunt, as in parts (a) and (c) of the figure, the output voltage is the variable being fed back, and the controlled source in the feedback network is therefore controlled by v_o. The output of the forward amplifier can be reasonably modeled as either a Norton or Thévenin equivalent (although it will

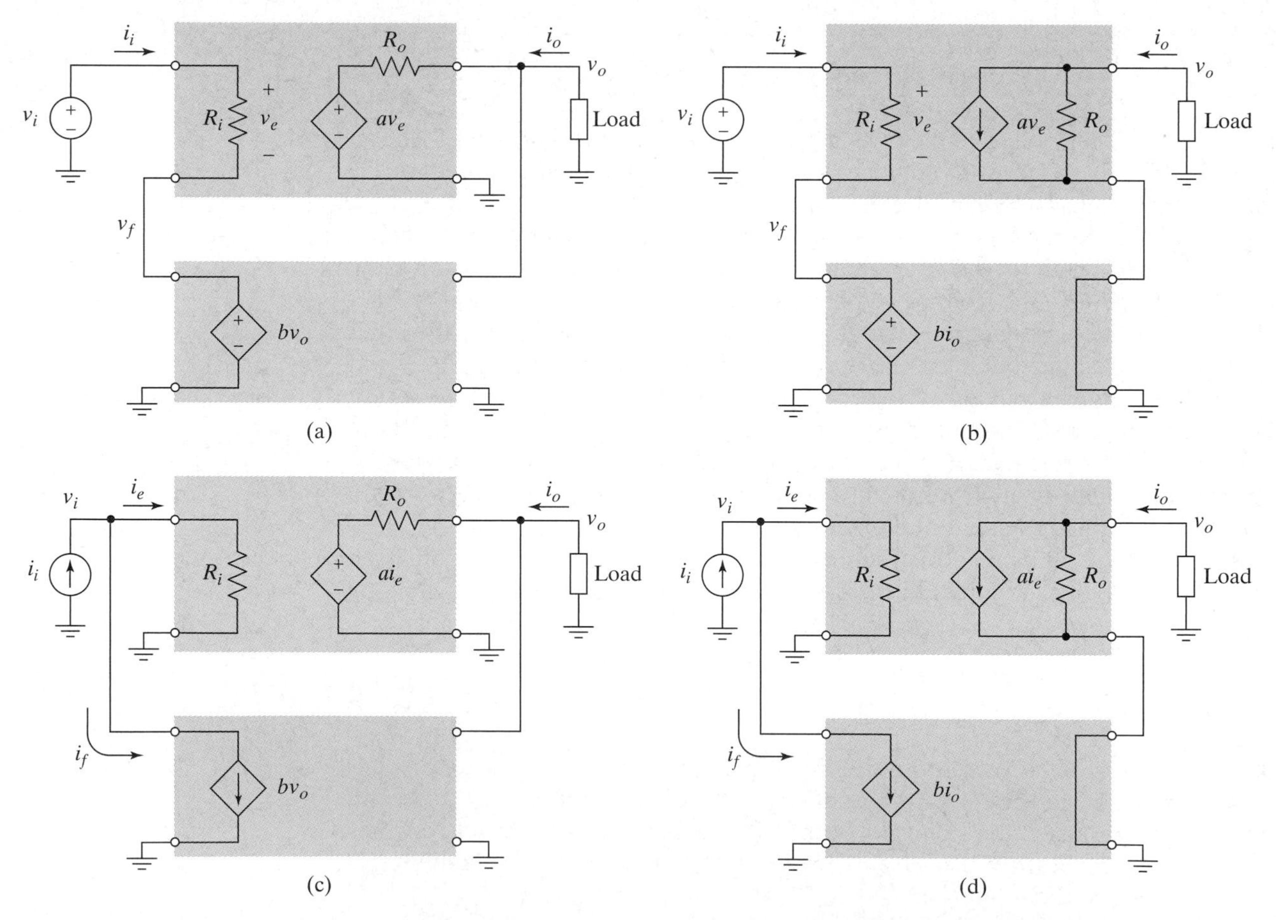

Figure A10-1 The four feedback topologies: (a) series–shunt, (b) series–series, (c) shunt–shunt, and (d) shunt–series. The loads may be ideal in some cases (i.e., an open circuit when v_o is the output, as in (a) and (c), and a short circuit when i_o is the output, as in (b) and (d)).

matter which is used when the feedback network is allowed to be bilateral; *see* Section 10.1.4). We use a Thévenin equivalent here, since the output variable is voltage, and it is usually most convenient to think of the forward amplifier as having a voltage output. Similarly, when the output connection is series, as in parts (b) and (d) of the figure, the output current is fed back and controls the source in the feedback network. Consequently, we use a Norton equivalent for the forward amplifier output.

To determine which connection is being used in a given circuit, we first identify the forward amplifier and the feedback network (realizing that some components may be outside the loop and not part of either network). The choice is not always unique: Some elements may be considered to be part of the forward amplifier or the feedback network. What is important is to clearly identify the feedback variable, input variable, and output variable.

Having identified these variables, we then check the type of connection, which will also show whether or not our determination was correct. We start by tracing a path from the signal source towards the amplifier. If we must go through the forward amplifier to get to the feedback network, they are in series. In that case, if we chose v_i and v_f as the input and feedback variables, we were correct and can write an equation like (A10.2) and identify v_e as the input to the forward amplifier. If, on the other hand, we can choose whether to go into the feedback network or the forward amplifier while tracing the path, the connection is shunt, and we should have chosen i_i and i_f as the variables. In either case, if we made a mistake, we can now correct it.

We next go through a similar process at the output. We start from the load and trace a path towards the amplifier. If we must go through the forward amplifier to get to the feedback network, they are in series. If we can choose which one to go into, they are in shunt. If they are in series, then we must use i_o as the output variable. Choosing i_o for the series connection is dictated by the fact that the output voltage does not appear across the feedback network, while i_o does flow through it (*see* parts (b) and (d) of the figure). Therefore, v_o is not available to the feedback network, but i_o is. Similarly, if the output is a shunt connection, we must use v_o as the output variable, since i_o is not available to the feedback network (i.e., it does not flow through the feedback network).

Returning to the calculation of the input and output resistances, we consider the four topologies shown in Figure A10-1. For both of the circuits with a series input connection (i.e., parts (a) and (b) of the figure) we write

$$i_i = \frac{v_i - bx_o}{R_i}, \tag{10.17}$$

where x_o is used to denote either v_o or i_o as appropriate. In either case we also know that (assume that the loads are ideal for now)

$$x_o = av_e = a(v_i - bx_o), \tag{10.18}$$

which yields

$$A_f = \frac{x_o}{v_i} = \frac{a}{1 + ab}. \tag{10.19}$$

Substituting (10.19) into (10.17) and solving yields

$$R_{if} = \frac{v_i}{i_i} = R_i(1 + ab) = R_i(1 + L), \tag{10.20}$$

as before. Similarly, for both of the circuits with a shunt input connection (i.e., parts (b) and (d) of the figure), we write

$$v_i = i_e R_i = (i_i - bx_o)R_i. \tag{10.21}$$

For both circuits, it is true that (assume the loads are ideal again)

$$x_o = ai_e = a(i_i - bx_o), \tag{10.22}$$

so we have

$$A_f = \frac{x_o}{i_i} = \frac{a}{1 + ab}. \tag{10.23}$$

Substituting (10.23) into (10.21) and solving yields the final result:

$$R_{if} = \frac{v_i}{i_i} = \frac{R_i}{1 + ab} = \frac{R_i}{1 + L}. \tag{10.24}$$

In other words, the input resistance of the forward amplifier is decreased by the factor $1 + L$ for a shunt connection. Applying an analysis similar to that shown in (10.13) through (10.15) yields the result that the output resistance with feedback is increased by the factor $1 + L$ for a series output connection and is decreased by the same factor for a shunt output connection.

Exercise 10.2

Derive the output resistances with feedback for all four circuits in Figure A10-1.

One way of remembering whether the input and output resistances will be increased or decreased by the feedback is to think about resistors. The total resistance always increases if a resistor is added in series and always decreases if a resistor is added in parallel. These facts remind us that the input or output resistance with feedback increases for a series connection and decreases for a shunt connection.

Exercise 10.3

A simple model for an operational amplifier is shown in Figure 10-3(a). Using this model, examine the circuits in parts (b) and (c) of the figure, and determine what type of feedback connection is used in each case. Identify the variable being fed back, and then write the input summation equation for each circuit.

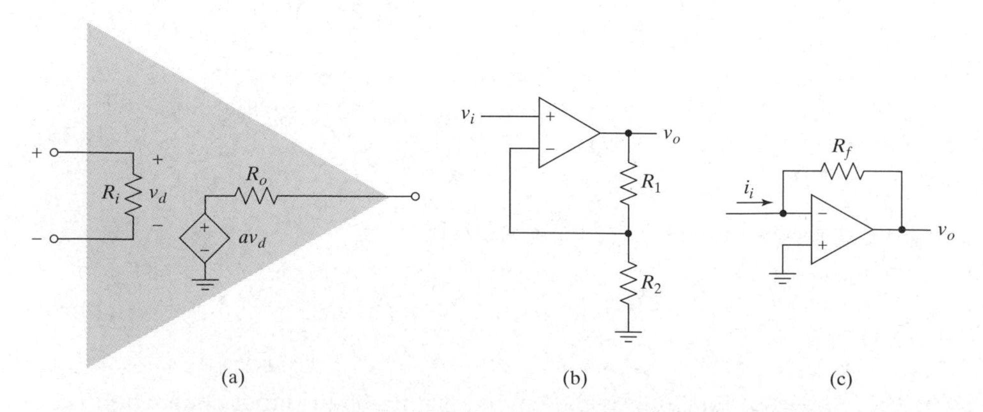

Figure 10–3 (a) An op amp model. (b) and (c) show feedback amplifiers.

Increase in Bandwidth Let's now return to considering the series–shunt amplifier of Figure 10-2. Assume that the gain of the forward amplifier is a function of frequency and is adequately described by a single-pole low-pass transfer function:

$$a(j\omega) = \frac{a(j0)}{1 + j\omega/\omega_p} = \frac{a_0}{1 + j\omega/\omega_p}. \quad \textbf{(10.25)}$$

Plugging (10.25) into the equation for the closed-loop gain, (10.7), yields

$$A_f(j\omega) = \frac{a(j\omega)}{1 + a(j\omega)b} = \frac{A_{f0}}{1 + j\omega/\omega_{pf}}, \quad \textbf{(10.26)}$$

where

$$A_{f0} = A_f(j0) = \frac{a_0}{1 + a_0 b} = \frac{a_0}{1 + L(j0)} = \frac{a_0}{1 + L_0}, \quad \textbf{(10.27)}$$

$L(j0) = L_0$ is the DC loop gain, and

$$\omega_{pf} = \omega_p(1 + a_0 b) = \omega_p(1 + L_0). \quad \textbf{(10.28)}$$

The DC gain given by (10.27) agrees with our previous results where we implicitly ignored the frequency response. The closed-loop pole given by (10.28) is $(1 + L_0)$ times the open-loop pole frequency, which shows that the application of negative feedback extends the bandwidth of the amplifier.

Increasing the bandwidth of an amplifier is often a desirable effect, and negative feedback does achieve this goal. Nevertheless, be careful not to draw a common, but incorrect, conclusion. While the bandwidth of the forward amplifier will certainly be increased when negative feedback is applied, it does *not* follow that using feedback will achieve the highest possible bandwidth.

For example, suppose we want to design an amplifier with a gain of 10 and the largest bandwidth possible. For a first try, we design an amplifier with a gain of 1000 and use feedback to reduce the gain to 10. Denote the bandwidth of the open-loop amplifier with a gain of 1000 by ω_{p1000}, and assume that a single-pole transfer function like (10.25) is an adequate model. When the loop is closed to bring the gain down to 10, the bandwidth will be extended by the same factor that the gain was reduced by, namely, a factor of 100. Denoting this closed-loop bandwidth by ω_{pf}, we have $\omega_{pf} = 100\omega_{p1000}$.

We don't want to compare the bandwidth of the open-loop amplifier with the closed-loop bandwidth, because the gain of this open-loop amplifier is much higher. Instead, we need to design a second open-loop amplifier that has a gain of 10 so that we compare amplifiers with equal gain. Denote the bandwidth of this second amplifier by ω_{p10}. We then compare the closed-loop bandwidth of the first amplifier, ω_{pf}, to the open-loop bandwidth of the second, ω_{p10}.

It usually turns out that the bandwidth of the second, lower gain, amplifier is higher than the closed-loop bandwidth of the first amplifier; that is, $\omega_{p10} > \omega_{pf}$. While this may seem counterintuitive, it is a result of the fact that, to design the open-loop amplifier with a gain of 1000, you would have to sacrifice a lot of bandwidth to get a gain so high. You don't necessarily recover all of this bandwidth when you use negative feedback to bring the gain back down. The closed-loop amplifier would have one significant advantage over the second open-loop amplifier in this example, however; the gain of the closed-loop amplifier would be stable, while the open-loop gain of 10

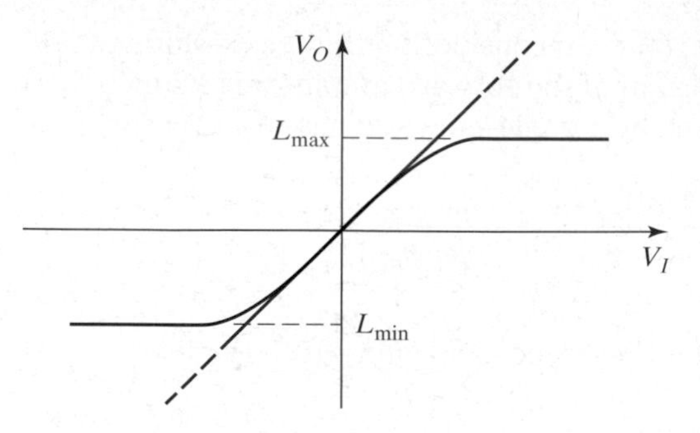

Figure 10–4 The transfer characteristic of an ideal linear amplifier (in black) and a more realistic transfer characteristic showing the upper and lower swing limits and nonlinearity (in blue).

would be likely to be sensitive to temperature and other parameters. In general, the highest bandwidth amplifiers do not use negative feedback. The penalty paid is that it is more difficult to make these circuits have predictable and stable performance.

Exercise 10.4

Suppose we have an open-loop amplifier with $a_0 = 50{,}000$ and $f_p = 10$ Hz. If we connect this amplifier in a unity-gain feedback connection, what it the closed-loop bandwidth f_{pf}?

Reduction in Nonlinear Distortion Real amplifiers are never completely linear. At the very least, the transfer characteristic will deviate from a straight line as it approaches the maximum limits of its output swing (L_{max} and L_{min}), as shown in Figure 10-4. The small-signal gain of the amplifier at any particular value of V_I is the slope of the V_O-versus-V_I transfer characteristic at that point. Therefore, when the transfer characteristic is nonlinear, the gain of the amplifier changes as V_I changes. Since negative feedback stabilizes the closed-loop gain with respect to changes in the open-loop gain, it is also effective at reducing the effects of nonlinearity in the forward amplifier. In other words, since feedback helps to make the gain more constant, and the gain is the slope of the transfer characteristic, the characteristic of the closed-loop amplifier will be closer to a straight line.

Example 10.2

Problem:

Consider the nonlinear amplifier whose large-signal transfer characteristic is illustrated in Figure 10-5. The horizontal axis is labeled v_E, since we are going to use this amplifier as the forward amplifier in a feedback loop. How much more linear is the transfer characteristic if the amplifier is connected in a negative-feedback loop with a feedback factor of 1/10? Sketch the resulting large-signal closed-loop transfer characteristic.

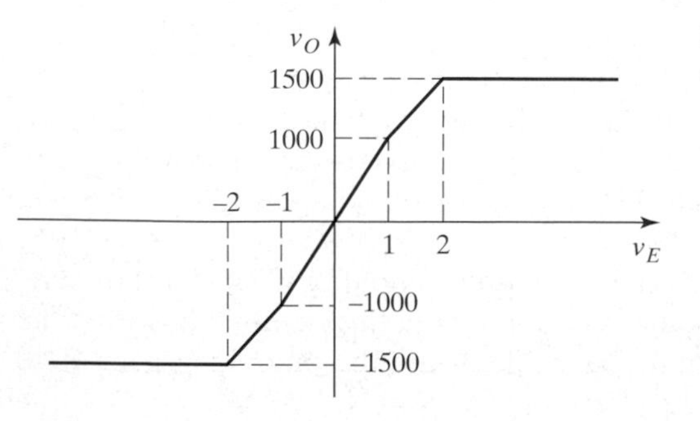

Figure 10–5 A nonlinear transfer characteristic.

Solution:

For the open-loop amplifier, the input *is* v_E, so $v_I = v_E$. The maximum gain of the open-loop amplifier occurs for values of v_I between -1 V and 1 V and is 1000. For $1 < |v_I| < 2$, the gain is 500, and for $|v_I| > 2$, the gain is zero. The change in gain seen between $v_I = 0.5$ V and $v_I = 1.5$ V is 50% of the maximum value.

When this amplifier is connected in a negative-feedback loop with a feedback factor of 1/10, however, v_E is no longer equal to v_I. Nevertheless, the forward amplifier itself remains unchanged, so we know what v_E must be for a given v_O. For example, to achieve $v_O = 1500$ V requires $v_E = 2$ V, but $v_E = v_I - bv_O$, so $v_I = v_E + bv_O = 2 + 150 = 152$ V. Similarly, the other breakpoint in the open-loop transfer characteristic occurs at $v_O = 1000$ V, which requires $v_E = 1$ V and $v_I = 101$ V. The resulting closed-loop transfer characteristic is shown in Figure 10-6. Note that for $|v_O| < 1000$ V, the closed-loop gain is $1000/(1 + 1000/10) = 9.9$, and the closed-loop gain for $1000 < v_O < 1500$ is $500/(1 + 500/10) = 9.8$, which represents only a 1% reduction in gain. When the gain of the open-loop amplifier is zero (for $|v_E| > 2$ V), the gain of the closed-loop amplifier is also zero.

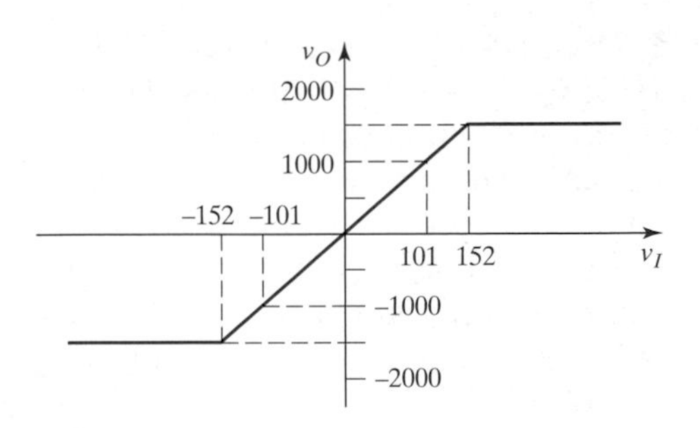

Figure 10–6 The closed-loop large-signal transfer characteristic.

Exercise 10.5

Suppose we have an amplifier with a nonlinear gain described by $a = 250$ for $|v_{IN}| > 1$ V and $a = 1{,}000$ for $|v_{IN}| \le 1$ V. What is the maximum closed-loop gain we can achieve while keeping the variation in gain less than 1% of the value at $v_{IN} = 0$?

The Effect of Negative Feedback on Noise[3] In some situations, negative feedback can be used to reduce noise in electronic circuits, and in other cases, feedback may be part of an overall solution to improve the ***signal-to-noise ratio*** (SNR). However, feedback alone cannot improve SNR, as is sometimes erroneously reported. In this section, we will first consider a circuit in which feedback reduces the noise on a DC current. We will then move on to consider the SNR in an amplifier with feedback.

Consider, for example, the current source shown in Figure 10-7. The current source is constructed using a generic transistor, a resistor, and an ideal operational

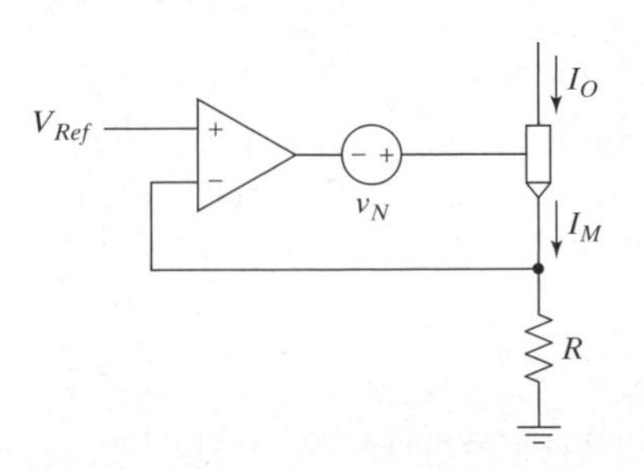

Figure 10–7 A current source with noise in the transistor included. Note that the output of the transistor is shown open, but it must connect through some circuit to the positive power supply.

[3]This section may be skipped without loss of continuity.

amplifier. The noise in the transistor is modeled by a noise source in series with the transistor's input. The current source takes a DC reference voltage as an input and produces a DC current as its output.

Because of the infinite gain of the op amp and the negative feedback, there is a virtual short between the op amp inputs; therefore, the voltage at the top of the resistor is equal to V_{Ref}. Since the input current to the op amp is zero, the current out of the merge terminal of the transistor is

$$I_M = \frac{V_{Ref}}{R}, \tag{10.29}$$

and, assuming the current gain of the transistor is nearly one (i.e., the control current is nearly zero) the output current is

$$I_O \approx \frac{V_{Ref}}{R}. \tag{10.30}$$

The noise voltage source did not enter into the equation because it is contained in the feedback loop and its effect is canceled. One way to see why the noise is canceled is to realize that the noise added to the op amp output could equivalently be produced by a time-varying gain for the op amp. We know that the variation in closed-loop gain is equal to the variation in open-loop gain divided by $(1 + L)$. Therefore, with the open-loop gain of the op amp approaching infinity, L approaches infinity, and the equivalent variation in closed-loop gain goes to zero. For finite op amp gain, the effect of the noise is reduced by the gain of the op amp, but not completely canceled (*see* Exercise 10.6).

Exercise 10.6

Consider the circuit in Figure 10-7, and assume that $R = 1\ k\Omega$, $V_{Ref} = 2$ V, and the gain of the op amp is 100,000. If the noise at the input to the transistor is 10 μV rms, what is the rms noise on i_O?

□

We now consider the case of a noisy amplifier, as shown in Figure 10-8. We have again modeled the noise by a single input source, and we write by inspection that

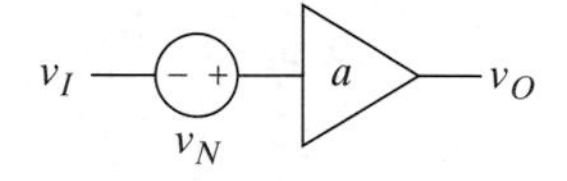

Figure 10–8 A noisy amplifier modeled by a noiseless amplifier, a, in series with an input noise source, v_N.

$$v_O = a(v_I + v_N). \tag{10.31}$$

The SNR at the output of this amplifier is

$$\text{SNR}_o = \frac{av_I}{av_N} = \frac{v_I}{v_N}. \tag{10.32}$$

If this SNR is not good enough for the given application, we may design a special low-noise preamplifier and place it in front of our amplifier as shown in Figure 10-9. Since the preamplifier will be the first amplifier in a chain, it does not need to deal with re-

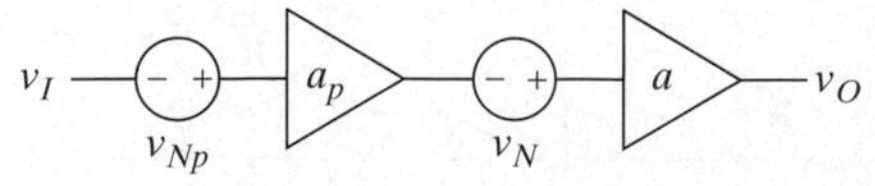

Figure 10–9 Cascading the original amplifier with a low-noise preamplifier to increase the SNR.

ceiving large signals or driving low-impedance loads. Therefore, we can optimize the design to minimize the noise that it contributes. The output is now given by

$$v_O = a_p a(v_I + v_{NP}) + a v_N, \tag{10.33}$$

and the output SNR is

$$\mathrm{SNR}_o = \frac{a_p a v_I}{a_p a v_{Np} + a v_N} = \frac{a_p v_I}{a_p v_{Np} + v_N}. \tag{10.34}$$

Neglecting the noise from the preamplifier, we see from (10.34) that the SNR due to the input noise of our original amplifier alone is

$$\mathrm{SNR}_o\Big|_{\text{due to } v_N \text{ alone}} = \frac{a_p v_I}{v_N}, \tag{10.35}$$

which is larger than the SNR given by (10.32) by the factor a_p, the preamp gain. In other words, the contribution to the SNR by our original amplifier has been reduced. Assuming we do a good enough job of optimizing our preamplifier for low noise, the overall SNR will be improved.

A problem with this solution is that the overall gain of the amplifier is now much larger; it is $a_p a$ instead of just a. Depending on the signal levels, this extra gain may cause the signal at the output to be large enough that the amplifier saturates. If we desire to use the low-noise preamplifier to increase our SNR, but we don't want to change the overall gain, we can use negative feedback as shown in Figure 10-10. The overall gain of this closed-loop system can be equal to the original gain a if the feedback factor is chosen properly.

We can manipulate the block diagram of Figure 10-10 to simplify the analysis.[4] First notice that we can pull the noise source for our original amplifier to the left and combine it with the preamplifier noise. To do this and keep the overall noise the same, we must divide our original noise source by the gain of the preamplifier, a_p. The resulting diagram is shown in Figure 10-11(a). Next, we recognize that the input to the preamplifier is $v_I + v_{Np} + v_N/a_p - bv_O$, and it doesn't matter what order we sum these terms in. Therefore, we can pull the noise source to the left of the

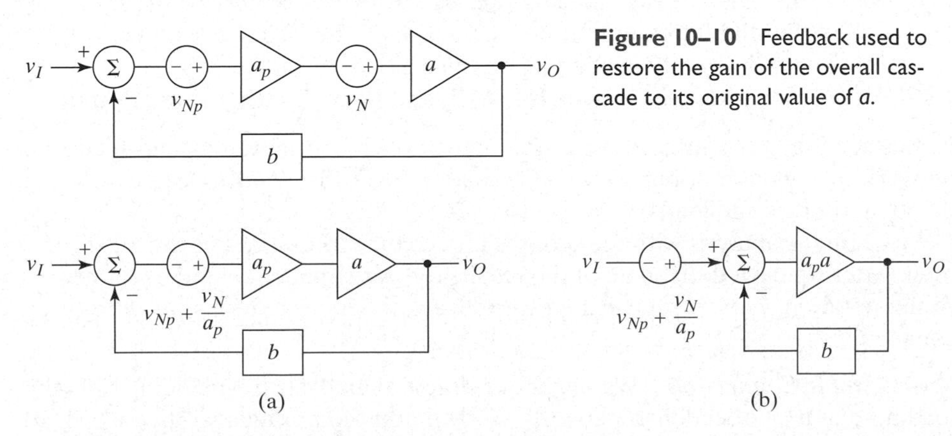

Figure 10–10 Feedback used to restore the gain of the overall cascade to its original value of a.

Figure 10–11 Manipulating the block diagram of Figure 10-10 to simplify the analysis.

[4] For the sake of simplicity in our illustration we are adding the noise voltages directly. To handle independent noise sources correctly, we add mean square values instead.

input summer as shown in part (b) of the figure. Now we use our standard feedback analysis to obtain

$$v_O = \frac{a_p a}{1 + a_p ab}\left(v_I + v_{Np} + \frac{v_N}{a_p}\right) = A_f\left(v_I + v_{Np} + \frac{v_N}{a_p}\right), \tag{10.36}$$

and we can use this result to write the SNR at the output:

$$\text{SNR}_o = \frac{A_f v_I}{A_f\left(v_{Np} + \dfrac{v_N}{a_p}\right)} = \frac{a_p v_I}{a_p v_{Np} + v_N}. \tag{10.37}$$

This is the same as the SNR for the system without feedback, as given by (10.34). In other words, the feedback did not affect the SNR at all. Practical feedback networks introduce more noise and actually degrade the SNR, but the degradation is usually small.

Exercise 10.7

Consider the circuits shown in Figures 10-8 and 10-10. Assume that $a = 20$, $v_N = 500\ \mu V$ rms, $a_p = 50$, and $v_{Np} = 10\ \mu V$ rms. Find the value of b required to keep the gain equal to 20 for the cascaded amplifier, and find the SNRs for both circuits, assuming that $v_I = 25$ mV rms. For simplicity, sum the noise voltages just like you would regular signals.

□

10.1.3 First-Order Practical Analysis

In this section, we present the analysis of practical feedback amplifiers. In the previous sections we idealized the system in three important ways:

1. We used unilateral models for both the forward amplifier and the feedback network.
2. We did not include any loading due to the feedback network (i.e., we assumed that the input and output resistances of the feedback network were either zero or infinity depending on the connection).
3. We used ideal sources and loads (i.e., R_S and R_L were either zero or infinite).

In this section, we continue to assume that the forward amplifier and feedback network are unilateral, but we include loading from the feedback network and nonideal sources and loads.

We begin the analysis with the series–shunt connection, since it is the most common. After we have derived all of the results for this connection and explained all of the notation, we will repeat the procedure with less detail for the other three connections.

Series–Shunt Connection We begin our practical analysis by including only the loading due to the feedback network. We will add the nonideal source and load later. We will discover that by changing our model slightly, we can make use of our previous results, even with the feedback network loading included.

A series–shunt feedback amplifier is shown in Figure 10-12, where we have used unilateral two-port models for the forward amplifier and the feedback network.

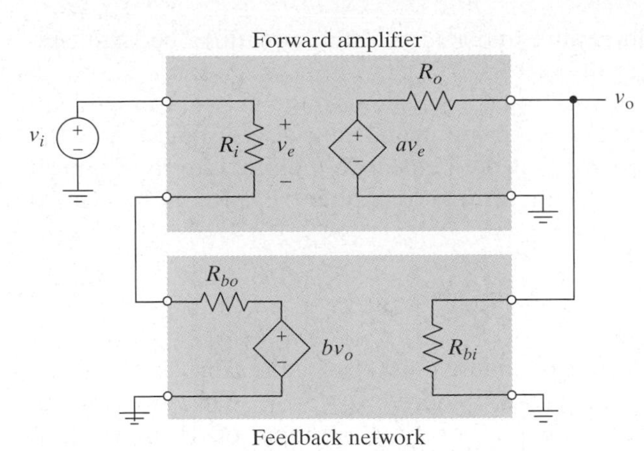

Figure 10–12 A series–shunt feedback amplifier with feedback network loading included.

We used a Thévenin equivalent for the feedback network output, since we want to be able sum voltages around the amplifier's input loop. We also used a Thévenin equivalent for the output of the forward amplifier, since it is the output voltage that is being fed back (as opposed to the current). The resistors R_{bi} and R_{bo} denote the ***input and output resistance of the feedback network***, respectively.

There are a number of different ways in which we could analyze this amplifier. Our goals here are to develop an approach that is general enough to allow us to apply it to many practical examples, is systematic enough to help us avoid mistakes, and that will lend insight so that we can make the transition from analysis to design. With these goals in mind, we begin the analysis by simplifying the circuit.

In Figure 10-13(a), we have repeated the amplifier from Figure 10-12 and have identified a two-port network that includes the forward amplifier and the loading of the feedback network. In part (b) of the figure, we have absorbed the loading of the feedback network into the forward amplifier to form a new "primed" circuit. The purpose of making this transformation is to simplify the analysis by doing it in stages. We also benefit from the simplification, because it allows us to more clearly see how the feedback works. Finally, if we develop a consistent style of analysis,

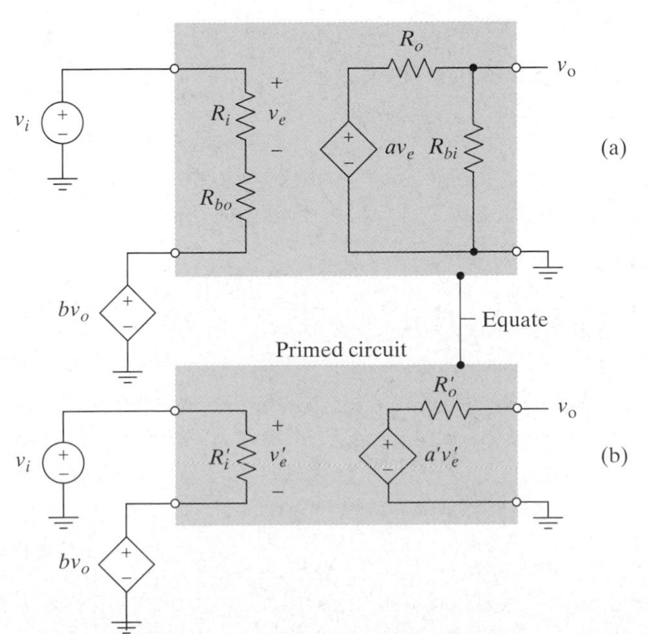

Figure 10–13 (a) The feedback amplifier of Figure 10-12 with the two-port identified and (b) the simplified equivalent feedback amplifier.

we will be able to reuse old results, and many of the equations become predictable; this will help guard against errors.

For the primed two-port network to be equivalent to the network in part (a), the two networks must have the same input and output resistances and the same gains. By inspection, we see that the resistance looking into the input of the two-port in part (a) is $R_i + R_{bo}$; therefore, the input resistance of the primed circuit is

$$R_i' = R_i + R_{bo}. \tag{10.38}$$

The input port voltage in part (a) of the figure must also be the same as it is in part (b) if the two-port networks are equivalent; therefore, $v_e' = v_i - bv_o$. We find the output resistance of a two-port network by setting the input of the network to zero. In this case, we set v_e' to zero in part (b) of the figure and $v_e' = v_i - bv_o$ equal to zero in part (a). In both cases the result is that the controlled voltage source has zero value, which means that it looks like a short circuit. Therefore, we can see by inspection that

$$R_o' = R_o \| R_{bi}. \tag{10.39}$$

Using the fact that $v_e' = v_i - bv_o$ again and examining the two-resistor divider network at the input of the two-port network in part (a) we see that

$$v_e = \frac{R_i}{R_i + R_{bo}} v_e', \tag{10.40}$$

which will be needed when we compare the gains. The output voltage of the network in part (a) of the figure is

$$v_o = \frac{R_{bi}}{R_{bi} + R_o} a v_e, \tag{10.41}$$

and the output voltage of the network in part (b) is

$$v_o = a' v_e'. \tag{10.42}$$

Equating (10.41) and (10.42) and making use of (10.40), we find that

$$a' = a\left(\frac{R_i}{R_i + R_{bo}}\right)\left(\frac{R_{bi}}{R_{bi} + R_o}\right). \tag{10.43}$$

Because Figure 10-13(b) is topologically equivalent to Figure 10-2(b), we can make use of the results derived earlier by replacing R_i, a, and R_o with R_i', a', and R_o', respectively. From (10.7), we get

$$A_f = \frac{a'}{1 + a'b} = \frac{a'}{1 + L}. \tag{10.44}$$

We do not need a prime on L, because we have defined the loop gain in exactly the same way we did before; only the notation for the forward amplifier elements has changed. Using (10.12) and (10.15), we obtain

$$R_{if} = R_i'(1 + L) \tag{10.45}$$

and

$$R_{of} = \frac{R_o'}{1 + L}. \tag{10.46}$$

We see from this example that including the feedback network loading does not change our previous results. The only difference in the analysis is that we added the steps necessary to find the primed network. Example 10.3 shows how to apply this analysis technique. After this example, we go on to consider nonideal source and load connections.

Example 10.3

Problem:
Consider the generic transistor amplifier shown in Figure 10-14. Find the small-signal midband voltage gain, input resistance, and output resistance for this amplifier.

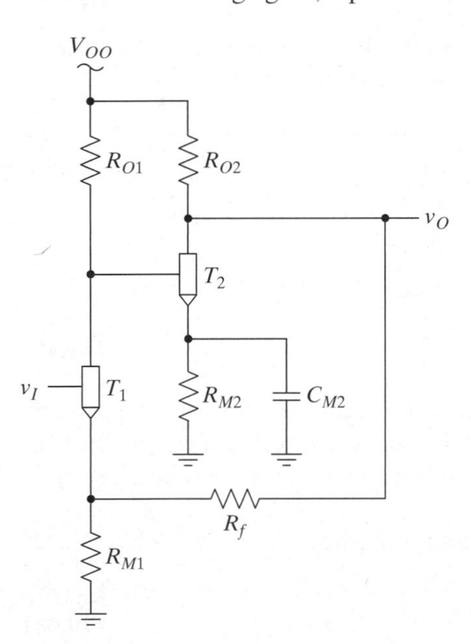

Figure 10–14 A transistor amplifier with feedback.

Solution:
To solve this problem, we first need to identify the forward amplifier, feedback network, and type of feedback connection (*see* Aside A10.3). Then we can draw small-signal models and see how to apply our two-port analysis to the circuit.

We note by inspection that the forward amplifier comprises T_1, T_2, R_{O1}, R_{M2}, C_{M2}, and R_{O2}, and the feedback network comprises R_{M1} and R_f. This identification is not always obvious and, in some instances, is not even unique. We will show how we find it for a number of different examples as we progress through this chapter. Nevertheless, if it isn't clear to you in a given circuit, don't let that stop you. Make your best guess and move on; if you have chosen incorrectly, it will become evident later, since something will not work correctly in the analysis.

Given this identification, we see that the feedback connection is a series connection at the input, since, starting from v_i, you must go through T_1 to get to the feedback network. However, the output connection is shunt, since, from v_o, you can return to ground through either the forward amplifier or the feedback network. This circuit is, therefore, a series–shunt amplifier.

We are now in a position to show how to apply the two-port analysis technique to this example. Unfortunately, as is often the case, there is a complication. Since the current into the control terminal of T_1 is not equal to the current out of the merge terminal, the input to our forward amplifier is not a port. In fact, the forward amplifier is really a three-port network! The control and merge terminals of T_1 both form ports with ground, and the output is a port (also relative to ground). Nevertheless, if we forge ahead with a small-signal model, we will see how to solve this problem. The feedback network we identified in Figure 10-14 is a two-port network, and we derive the parameters for the unilateral two-port model of this network before drawing the small-signal circuit for the entire amplifier.

Because we have a series–shunt feedback connection, it is most convenient to use a Thévenin equivalent for the output of the feedback network, as shown in Figure 10-15. Since we are approximating the network as being unilateral, the input side does not contain a controlled source.

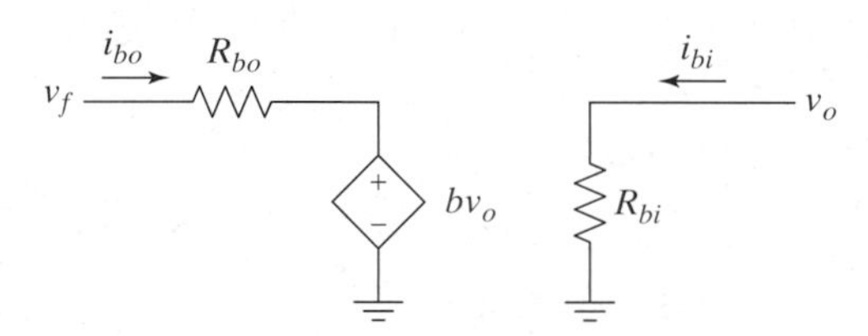

Figure 10–15 The unilateral two-port model for the feedback network.

We first write an equation for the output voltage of the feedback network:

$$v_f = R_{bo} i_{bo} + b v_o. \tag{10.47}$$

WARNING! It is extremely important to write this equation for the proper variable; if we incorrectly write the equation for the feedback network output current when it is the voltage that is fed back, the analysis that follows will be extremely awkward.

From (10.47), we can see how to find R_{bo} and b. We have

$$R_{bo} = \left. \frac{v_f}{i_{bo}} \right|_{v_o=0} \tag{10.48}$$

and

$$b = \left. \frac{v_f}{v_o} \right|_{i_{bo}=0}. \tag{10.49}$$

Applying these definitions to the circuit in part (a) of the figure, we set v_o to zero (i.e., short that point to ground) and look in from the left to find

$$R_{bo} = R_{M1} \| R_f, \tag{10.50}$$

and we set i_{bo} to zero (i.e., we open the connection from R_{M1} to T_1) and calculate the v_f that results from a given v_o to find

$$b = \frac{R_{M1}}{R_{M1} + R_f}. \tag{10.51}$$

Now focus attention on the input of the feedback network. From the figure, we write

$$R_{bi} = \frac{v_o}{i_{bi}}. \tag{10.52}$$

When we try to apply this equation to the real circuit, however, we find that it matters whether we set v_f or i_{bo} to zero. In other words, the network is not unilateral as we have assumed; if it were, the input resistance would be independent of what was connected to the output port, as is true for the model in Figure 10-15. So we know that the unilateral model is only an approximation, but we still need to find R_{bi}.

To figure out how to find R_{bi}, we need to recognize that the equations for a two-port network have two independent variables and two dependent variables, and the independent variables are the same for each port. From (10.47), we see that i_{bo} and v_o are the independent variables in this two-port model. Therefore, the equation at the other port must be

$$i_{bi} = a_r i_{bo} + \left(\frac{1}{R_{bi}}\right) v_o, \tag{10.53}$$

where a_r is the reverse current gain of the network and is the term that we are ignoring when we use the unilateral model. From (10.53) we see that

$$R_{bi} = \left.\frac{v_o}{i_{bi}}\right|_{i_{bo}=0}. \tag{10.54}$$

A useful mnemonic for remembering what to do to find R_{bi} is the following: For a <u>sh</u>unt input connection, <u>sh</u>ort the feedback network output to find R_{bi} and for a <u>se</u>ries input connection, <u>se</u>ver (or open) the feedback network output to find R_{bi}, [10.3].

Applying (10.54) to the original circuit in Figure 10-14, we set i_{bo} to zero by breaking the connection going up to T_1 from R_{M1} and find

$$R_{bi} = R_f + R_{M1}. \tag{10.55}$$

We will address the issue of whether or not ignoring the reverse gain of the feedback network (i.e., assuming that it is unilateral) is accurate in Section 10.1.4; for now, we just note that the approximation is often acceptable, since the signal-fed forward through the feedback network is typically much smaller than the signal fed forward through the forward amplifier.

Using the two-port model just derived for the feedback network, the small-signal midband AC equivalent circuit of our amplifier is shown in Figure 10-16(a), where we have defined $R_{i2} \equiv R_{O1} \| r_{cm2}$.

As noted before, we can see from the circuit that the forward amplifier is not a two-port (since i_{c1} is not the same as the current into the feedback network output, i_{bo}). Nevertheless, we can absorb the feedback network loading and derive the prime network, as was done before. Figure 10-16(a) has the prime network outlined. Part (b) of the figure shows the resulting model, using a two-port model for the prime network. This figure is identical to Figure 10-13(b), so all of the results derived for that circuit apply equally well here.

WARNING! The circuit in part (b) of Figure 10-16 is not completely equivalent to the circuit in part (a) of the figure. In part (b), the forward amplifier is modeled by a two-port network, and the input current is therefore also the current into the feedback generator, whereas in part (a) of the figure, the input current, i_{c1}, is not equal to the current into the feedback generator, i_{bo}.

Nevertheless, since we are not interested at this point in the current in the feedback generator, and the voltage from the generator does not depend on that current, this partial equivalence is good enough. We see from our derivation that the voltages are the same in both input loops, since we equate KVL, and the current is the same at the input to the amplifier, so the results will be useful for finding the gain and input resistance.

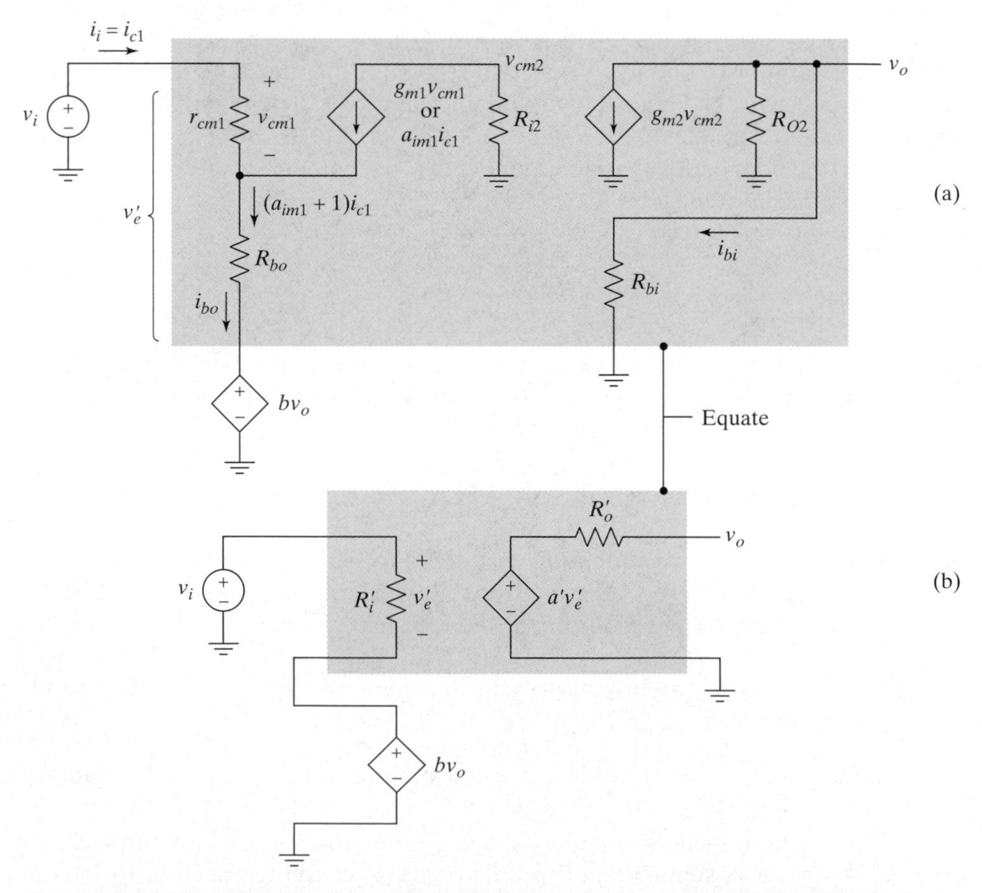

Figure 10–16 Simplifying the series–shunt amplifier by using the prime network. (a) Shows the small-signal equivalent circuit for the amplifier with the two-port model for the feedback network, and (b) shows the forward amplifier and feedback network loading replaced by the prime network.

To finish the example, we must find values for a', R_i', and R_o' for the circuit in Figure 10-16. We have noted on the schematic in part (a) of the figure that the current out of the merge terminal of T_1 is $(a_{im1}+1)i_{c1}$. Using this fact and noting that the voltage across the input port is v_e', we use KVL to write

$$v_e' = i_{c1}r_{cm1} + (a_{im1} + 1)i_{c1}R_{bo} \tag{10.56}$$

for the circuit in part (a) of the figure. For the circuit in part (b) of the figure, we get

$$v_e' = i_{c1}R_i', \tag{10.57}$$

and setting these two results equal, we obtain

$$R_i' = r_{cm1} + (a_{im1} + 1)R_{bo}. \tag{10.58}$$

We next wish to find the gain, a'. We can see from the circuit in part (b) of the figure that

$$a' = \frac{v_o}{v_e'}. \tag{10.59}$$

For the circuit in part (a) of the figure, we obtain

$$\frac{v_o}{v_e'} = \frac{v_o}{v_{cm2}}\frac{v_{cm2}}{v_{cm1}}\frac{v_{cm1}}{v_e'}. \tag{10.60}$$

For the first term on the right-hand side of (10.60), we get

$$\frac{v_o}{v_{cm2}} = -g_{m2}(R_{O2}\|R_{bi}) \tag{10.61}$$

by inspection. Similarly, the second term is

$$\frac{v_{cm2}}{v_{cm1}} = -g_{m1}R_{i2}. \tag{10.62}$$

The last term in (10.60) can be found with the aid of (10.57) and (10.58):

$$\frac{v_{cm1}}{v_e'} = \frac{i_{c1}r_{cm1}}{v_e'} = \frac{r_{cm1}}{r_{cm1} + (a_{im1} + 1)R_{bo}}. \tag{10.63}$$

Finally, we combine these results to obtain

$$a' = -g_{m2}(R_{O2}\|R_{bi})(-g_{m1}R_{i2})\left(\frac{r_{cm1}}{r_{cm1} + (a_{im1} + 1)R_{bo}}\right). \tag{10.64}$$

The last element of the prime network, R_o', is found by setting v_e' equal to zero in both circuits and looking into the output terminals. By inspection, we find that

$$R_o' = R_{O2}\|R_{bi}. \tag{10.65}$$

These results can now be plugged into (10.44), (10.45), and (10.46) to find A_f, R_{if}, and R_{of}. ■

Exercise 10.8

Consider the operational amplifier circuit shown in Figure 10-17, where a unilateral two-port model has been used for the op amp (enclosed in the triangle). (a) Identify the feedback network, verify that this is a series–shunt connection, and specify what quantity is equivalent to v_e. (b) What quantity is equal to the feedback voltage, v_f? (c) Find equations for the elements of the unilateral two-port model for the feedback network. (d) Find equations for the parameters of the primed network. (e) Give the equations for A_f, R_{of}, and R_{if}. (f) If the gain of the op amp, a, goes to infinity, show that the closed-loop gain goes to $1 + R_1/R_2$, as is expected from ideal op amp analysis. What are the corresponding values of R_{if} and R_{of}?

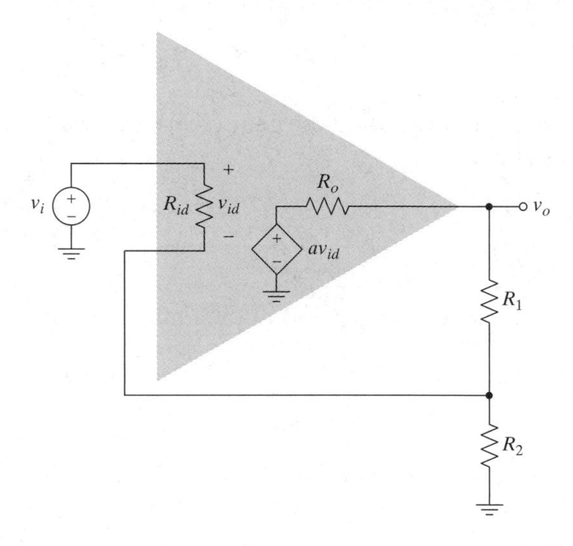

Figure 10–17 A series–shunt amplifier.

We are now ready to add a nonideal source and load to the series–shunt amplifier as shown in Figure 10-18(a). The prime network for this circuit is identical to the one we found in Figure 10-13(b). Therefore, we can redraw the circuit as shown in Figure 10-18(b).

The only difference between the simplified feedback amplifier of Figure 10-18(b) and the ideal amplifier dealt with in Section 10.1.1 is the addition of

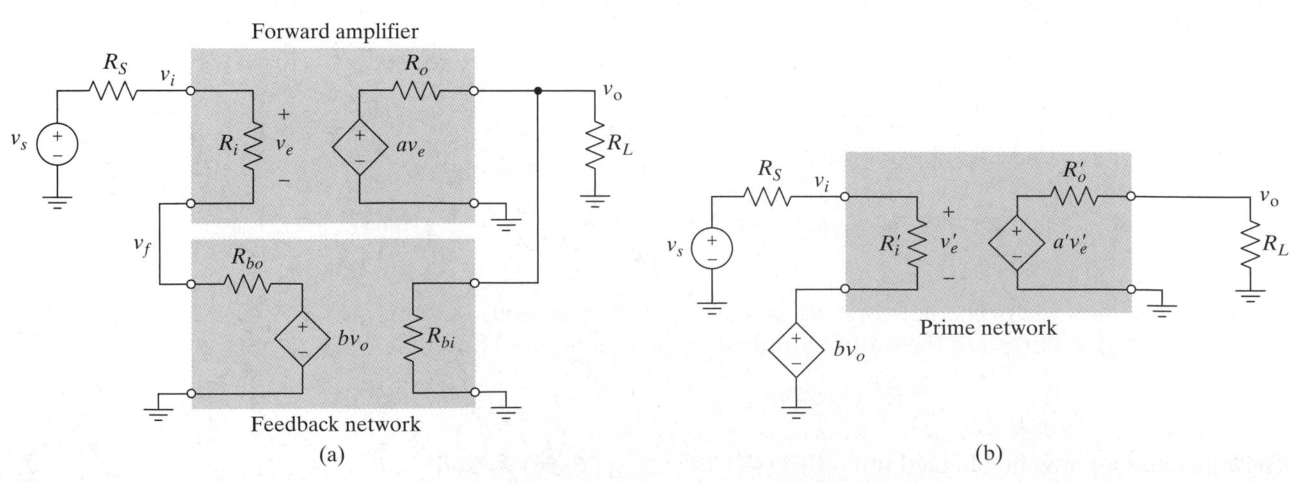

Figure 10–18 (a) The series–shunt amplifier with nonideal source and load added. (b) The amplifier after absorbing the feedback network loading into the primed network.

R_S and R_L. Because of the nonzero source resistance, we can consider two different voltage gains for this amplifier: the overall voltage gain v_o/v_s and the voltage gain v_o/v_i.[5] We expect to see an output attenuation factor in both of these gains, and we expect to see an input attenuation factor in the overall gain. With these preliminary observations made, we begin by considering the open-loop voltage gain of the circuit.

We find the open-loop gain of the amplifier by setting the feedback factor equal to zero. In other words, we set the controlled source that represents the feedback to zero. The open-loop voltage gain of the circuit in Figure 10-18(b) is given by

$$A_i \equiv \left.\frac{v_o}{v_i}\right|_{b=0} = a'\left(\frac{R_L}{R_L + R_o'}\right) = a'\alpha_o', \tag{10.66}$$

where we have defined an output attenuation factor in the usual way.[6] Notice that we added a subscript 'i' to indicate that this gain was from v_i, not v_s. To find the closed-loop gain, we use KVL in the output loop to write

$$v_o = \alpha_o' a' v_e' = \alpha_o' a'(v_i - bv_o), \tag{10.67}$$

which leads to

$$v_o = \frac{\alpha_o' a' v_i}{1 + \alpha_o' a' b}. \tag{10.68}$$

Using (10.68) and (10.66), we find the closed-loop voltage gain:

$$A_{if} = \frac{v_o}{v_i} = \frac{\alpha_o' a'}{1 + \alpha_o' a' b} = \frac{A_i}{1 + A_i b}. \tag{10.69}$$

As expected, this gain contains an output attenuation factor but not an input attenuation factor. We can compare (10.69) with the result obtained for the ideal case, (10.2), and notice that the form of the equation is identical. The only difference is the open-loop gain. The open-loop gain used here, as given by (10.66), includes the effects of the feedback network loading (absorbed into the prime network) and an output attenuation factor.

We can repeat the analysis just given to find the overall voltage gain. We again begin by finding the open-loop gain:

$$A_s \equiv \left.\frac{v_o}{v_s}\right|_{b=0} = \frac{v_e'}{v_s}\frac{v_o}{v_e'} = \left(\frac{R_i'}{R_i' + R_S}\right)a'\left(\frac{R_L}{R_L + R_o'}\right) = \alpha_i' a' \alpha_o', \tag{10.70}$$

Here, the input and output attenuation factors have again been defined in the usual way.[7] Notice that we added a subscript 's' to indicate that this gain was from v_s, not v_i. We next write KVL at the output to obtain

---

[5] Remember from Section 8.2.1 that we define the voltage gain to be v_o/v_i, so that the power gain will be the product of the voltage and current gains. Nevertheless, it is often useful to consider the overall voltage gain as well.

[6] This is *not* the output attenuation factor of the feedback amplifier, however, since the output resistance *with feedback* is not equal to R_o'; that is why we have used the prime superscript on the attenuation factor. Rather, it is the output attenuation factor of the open-loop amplifier with feedback loading.

[7] The input attenuation factor is again *not* the attenuation factor of the closed-loop amplifier, but only of the open-loop amplifier with feedback loading, hence the prime superscript.

$$v_o = \alpha_o' a' v_e', \tag{10.71}$$

and then write KVL at the input to get

$$v_e' = \alpha_i'(v_s - b v_o). \tag{10.72}$$

Combining (10.71) and (10.72) yields

$$v_o = \frac{\alpha_i' a' \alpha_o' v_s}{1 + \alpha_i' a' \alpha' b}, \tag{10.73}$$

and finally, using (10.73) and (10.70), we get the overall gain with feedback:

$$A_{sf} = \frac{v_o}{v_s} = \frac{A_s}{1 + A_s b}. \tag{10.74}$$

This equation again has the familiar form. Examination of Figure 10-18(b) reveals that A_{sf} and A_{if} should differ only by the input attenuation factor for the closed-loop amplifier. In other words, we should have

$$A_{sf} = \alpha_{if} A_{if}, \tag{10.75}$$

where

$$\alpha_{if} = \frac{R_{if}}{R_{if} + R_S}. \tag{10.76}$$

We leave it as an exercise for you to show that (10.75) is correct (*see* Problem P10.15).

At this point, the notation we use deserves some attention, since it has necessarily gotten a bit complicated. Because you may want to refer back to this notation again, we explain it in Aside A10.4.

Now that we have equations for the voltage gain of our practical amplifier, we move on to examine the input and output resistances with feedback. For the circuit in Figure 10-18(b), the input resistance with feedback is defined by

$$R_{if} = \frac{v_i}{i_i}, \tag{10.77}$$

as shown in Figure 10-19.

The input current in this circuit is given by

$$i_i = \frac{v_i - b v_o}{R_i'} = \frac{v_i(1 - A_{if} b)}{R_i'}, \tag{10.78}$$

where we have made use of (10.69). Plugging (10.78) into (10.77) yields

$$R_{if} = \frac{R_i'}{1 - A_{if} b} = \frac{R_i'}{1 - \dfrac{A_i b}{1 + A_i b}} = R_i'(1 + A_i b). \tag{10.79}$$

Since A_i is the open-loop gain from v_i to v_o, the product $A_i b$ is a loop gain. Making use of our notational scheme, we can call this the loop gain seen when the input is taken to be v_i and write

$$R_{if} = R_i'(1 + A_i b) = R_i'(1 + L_i). \tag{10.80}$$

We expect the input resistance to be increased by a factor $(1 + L)$ for a series input connection, as was derived in (10.12), but when we include R_S and R_L, we have to ask

ASIDE A10.4 NOTATION USED IN PRACTICAL FEEDBACK ANALYSIS

For practical feedback analysis, we must deal with loading due to the feedback network and nonideal source and load. In addition, we need a way of keeping track of open-loop and closed-loop parameters and a way of distinguishing between overall gain (i.e., from v_s or i_s) and normal voltage or current gain (i.e., from v_i or i_i). Because we use lowercase variables for transistor and two-port parameters, we use uppercase variables for gains of feedback networks. In addition, we use the subscript i to represent gains determined from v_i or i_i, and we use the subscript s to denote gains determined from v_s or i_s. The subscript f is used when we refer to closed-loop parameters ($f \Rightarrow$ with feedback) and is absent for open-loop parameters. The prime superscript is used on all quantities that uniquely refer to the primed network—that is, the network formed by absorbing feedback loading into the forward amplifier.

For example, A_i and A_{if} denote the open- and closed-loop gains, respectively, from v_i (or i_i) to the output, while A_s and A_{sf} denote the open- and closed-loop gains respectively, from v_s (or i_s) to the output. Similarly, the closed-loop input and output resistances are denoted by R_{if} and R_{of}, respectively. Also, a', R_i', R_o', α_i', and α_o' denote the no-load gain, input resistance, output resistance, input attenuation factor, and output attenuation factor respectively, of the open-loop prime network.

Finally, when dealing with functions of frequency, we refer to the DC values by adding a subscript '0'; for example, $A_i(j0) = A_{i0}$.

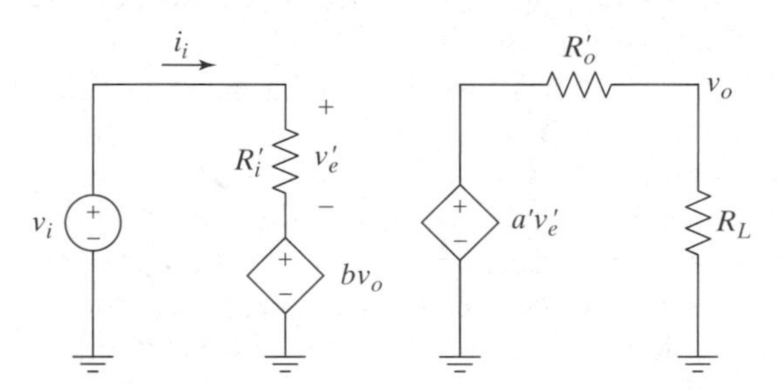

Figure 10–19 Determining the input resistance with feedback of the practical series–shunt amplifier.

ourselves what we mean by *loop gain*. To find the input resistance of the amplifier, we must, by definition, remove the source and look into the amplifier to see what the resistance is. In other words, we know that R_{if} cannot depend on R_S, which is part of our model for the source. Therefore, we cannot calculate the loop gain with R_S in place (i.e., it cannot include both attenuation factors, as it does in (10.74), where R_S is included in the analysis). So it makes sense that the loop gain used in (10.80) is L_i, the loop gain without the divider due to R_S.

Let's now move on to find the output resistance with feedback, R_{of}. In this case, we remove R_L, set v_s to zero, and apply a test source as shown in Figure 10-20. Therefore, we know that R_S will show up in the equation, but R_L cannot.

The output resistance with feedback is given by

$$R_{of} = \frac{v_x}{i_x} \tag{10.81}$$

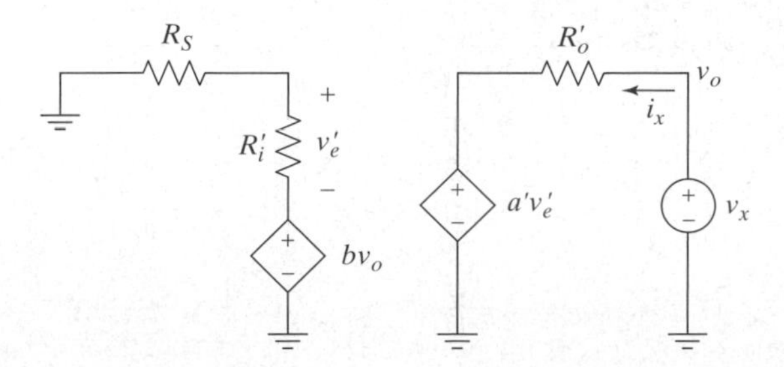

Figure 10–20 Circuit for finding the output resistance of the series–shunt feedback amplifier of Figure 10-18(b).

The current from our test source is given by

$$i_x = \frac{v_x - a'v'_e}{R'_o}. \tag{10.82}$$

With $v_s = 0$, we have $v'_e = -\alpha'_i b v_o$ and, using the fact that $v_x = v_o$, we get

$$i_x = \frac{v_x\left[1 + a'b\left(\dfrac{R'_i}{R'_i + R_S}\right)\right]}{R'_o}. \tag{10.83}$$

Plugging (10.83) into (10.81) and making use of the definition of α'_i given implicitly in (10.70) leads to

$$R_{of} = \frac{R'_o}{1 + \alpha'_i a'b} = \frac{R'_o}{1 + L_o}, \tag{10.84}$$

where we have implicitly defined the loop gain seen when an input is applied at v_o and have denoted it by L_o. We have not seen this particular loop gain before, but it makes good sense; since we are now driving the network from v_o (with R_L removed), we expect to see α'_i in the equation, but not α'_o.

The facts that the input resistance of a feedback amplifier is a function of the load, and the output resistance is a function of the source, are sometimes bothersome to students. These facts come about because the feedback amplifier is not a unilateral two-port network, which is what we are used to dealing with. In fact, from Figure 10-18(b), we see that by removing the source and load, we have the two-port model for the series–shunt feedback amplifier shown in Figure 10-21. That this network is not unilateral is evident from the fact that both controlled sources are present.

We complete our consideration of the practical series–shunt feedback connection by examining the effect of feedback on the bandwidth of the amplifier. We know from the ideal analysis that the bandwidth will be improved by a factor $(1 + L)$ for a single-pole open-loop response, but on the basis of our analyses in this section, we recognize that there is some question as to which loop gain should be used. We answer this question by using direct analysis. First, assume that the forward amplifier gain is given by

$$a(j\omega) = \frac{a_0}{1 + j\omega/\omega_p}. \tag{10.85}$$

Now, consider the gain from v_s to v_o. Plugging (10.85) into (10.43) and the result into (10.70) yields

Figure 10–21 The full two-port model of the series–shunt feedback amplifier.

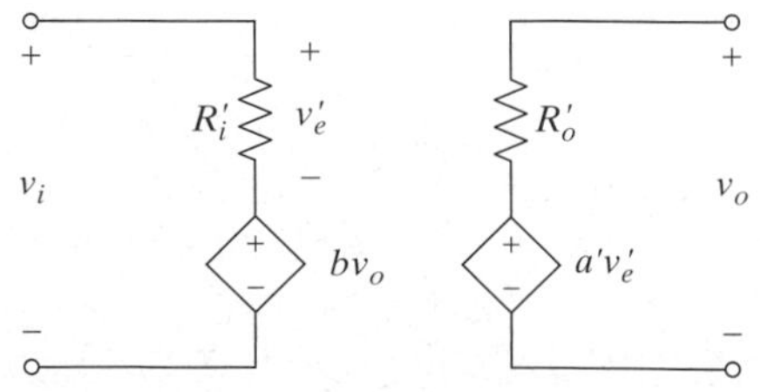

$$A_s(j\omega) = \frac{\alpha_i'\alpha_o'a_0\left(\dfrac{R_i}{R_i + R_{bo}}\right)\left(\dfrac{R_{bi}}{R_{bi} + R_o}\right)}{1 + j\omega/\omega_p} = \frac{\alpha_i'a_0'\alpha_o'}{1 + j\omega/\omega_p}. \quad \textbf{(10.86)}$$

Plugging (10.86) into (10.74) yields

$$A_{sf}(j\omega) = \frac{A_{sf0}}{1 + j\omega/\omega_{sf}}, \quad \textbf{(10.87)}$$

where

$$A_{sf0} = \frac{\alpha_i'a_0'\alpha_o'}{1 + \alpha_i'a_0'\alpha_o'b} = \frac{\alpha_i'a_0'\alpha_o'}{1 + L_s} \quad \textbf{(10.88)}$$

and

$$\omega_{sf} = \omega_p(1 + \alpha_i'a_0'\alpha_o'b) = \omega_p(1 + L_s). \quad \textbf{(10.89)}$$

In writing these equations we implicitly defined L_s to be consistent with our notation; it is the loop gain seen when we consider v_s to be the input, so that R_S and R_L are both included. We note that the closed-loop gain in (10.88) and bandwidth in (10.89) have been modified by the same term, $(1 + L_s)$.

We now consider the gain from v_i to v_o and start by substituting (10.85) into (10.43) and that result into (10.66) to find

$$A_i(j\omega) = \frac{\alpha_o'a_0\left(\dfrac{R_i}{R_i + R_{bo}}\right)\left(\dfrac{R_{bi}}{R_{bi} + R_o}\right)}{1 + j\omega/\omega_p} = \frac{a_0'\alpha_o'}{1 + j\omega/\omega_p}. \quad \textbf{(10.90)}$$

Plugging (10.90) into (10.69) yields

$$A_{if}(j\omega) = \frac{A_{if0}}{1 + j\omega/\omega_{if}}, \quad \textbf{(10.91)}$$

where

$$A_{if0} = \frac{a_0'\alpha_o'}{1 + a_0'\alpha_o'b} = \frac{a_0'\alpha_o'}{1 + L_i} \quad \textbf{(10.92)}$$

and

$$\omega_{if} = \omega_p(1 + a_0'\alpha_o'b) = \omega_p(1 + L_i). \quad \textbf{(10.93)}$$

We again note that the closed-loop gain and bandwidth have been modified by the same term, $(1 + L_i)$, where L_i was defined in (10.80).

It is interesting that the closed-loop bandwidths in (10.89) and (10.93) are not the same. The difference is caused by whether we consider v_s or v_i to be the input to the amplifier. (R_S can be present in both cases, but for v_o/v_i, we don't need to know v_s.)

Since the input attenuation factor is guaranteed to be less than unity, we know that ω_{if} is larger than ω_{sf}. This result makes sense once we examine the input attenuation factor of the closed-loop amplifier. The input attenuation factor for the amplifier with feedback is

$$\alpha_{if} = \frac{R_{if}}{R_{if} + R_S}. \tag{10.94}$$

where $R_{if} = R_i'(1 + L_i)$. As the input frequency is increased, the gain of the forward amplifier rolls off, which causes the loop gain, L_i, to roll off as well. Therefore, the value of R_{if} is not constant with frequency *even if* R_i' is constant. It would be more accurate to call R_{if} an input impedance rather than a resistance. Since the magnitude of Z_{if} drops with increasing frequency, it appears capacitive. This effective input capacitance forms a low-pass filter with R_S and causes the input attenuation factor to roll off with frequency. When the overall gain from v_s to v_o is examined, this roll off in input attenuation factor is included, but it is not included when the gain from v_i to v_o is examined; therefore, $\omega_{if} > \omega_{sf}$.

It is an interesting and sometimes unfortunate characteristic of feedback amplifiers that a frequency-dependent gain causes purely real input and output resistances to become frequency-dependent impedances (*see* Problems P10.11 and P10.12). For DC-coupled amplifiers, the loop gain has a low-pass transfer characteristic. At low frequencies, the loop gain is its largest and is nearly independent of frequency, so the input and output impedances are resistive and are given by R_{if} and R_{of}. As the frequency increases, the loop gain drops and eventually becomes so small that the feedback loop is essentially inoperative. At that point, the input and output impedances again appear resistive and are approximately equal to R_i' and R_o'. Because of these facts, shunt connections look inductive (the impedance increases with frequency), whereas series connections look capacitive (the impedance drops with increasing frequency). When calculating $a(j\omega)$ for a forward amplifier, if ω_p depends on R_S, you can't use the same ω_p for both ω_{if} and ω_{sf}. You must set $R_S = 0$ when finding the ω_p used in the equation for ω_{if} (see the companion section on the CD for an explanation).

Example 10.4

Problem:

(a) Find A_{if}, A_{sf}, R_{if}, and R_{of} for a series–shunt amplifier with the following specifications. The forward amplifier has $R_i = 1$ kΩ, $R_o = 1$ kΩ, and $a = 250$ V/V. The feedback network has $R_{bi} = 5$ kΩ, $R_{bo} = 100$ Ω, and $b = 0.05$ V/V. The source has a Thévenin resistance $R_S = 1$ kΩ and the load is $R_L = 1$ kΩ.

(b) Now assume that the forward amplifier has the same DC gain just given, but also has a single pole at 10 kHz. Find values for f_{if} and f_{sf}.

Solution:

(a) The resulting schematic is identical to Figure 10-12, and we begin the analysis by finding the prime network as shown in Figure 10-13(b). Using (10.38), we find that $R_i' = 1.1$ kΩ; from (10.39), we get $R_o' = 833$ Ω; and from (10.43), we obtain $a' = 189.4$ V/V. Now we use (10.66) to find $A_i = 103.3$ V/V and (10.70) to get $A_s = 54.1$ V/V. With these open-loop gains, we use (10.69) and (10.74) to find $A_{if} = 16.8$ V/V and $A_{sf} = 14.6$ V/V, respectively. The input and output resistances are found by using (10.80) and (10.84) and are $R_{if} = 6.8$ kΩ and $R_{of} = 140$ Ω, respectively.

(b) With the given pole frequency and the previously obtained results for the DC gains, we use (10.89) to obtain $f_{sf} = 37$ kHz and (10.93) to find $f_{if} = 61.6$ kHz.

■

Exercise 10.9

Use SPICE to confirm the results of Example 10.4.

Exercise 10.10

Consider the circuit shown in Figure 10-22, where an op amp has been modeled by a differential-mode input resistance,[8] gain, and output resistance. Find A_{if}, A_{sf}, R_{if}, and R_{of} for this circuit. Refer back to Exercise 10.8.

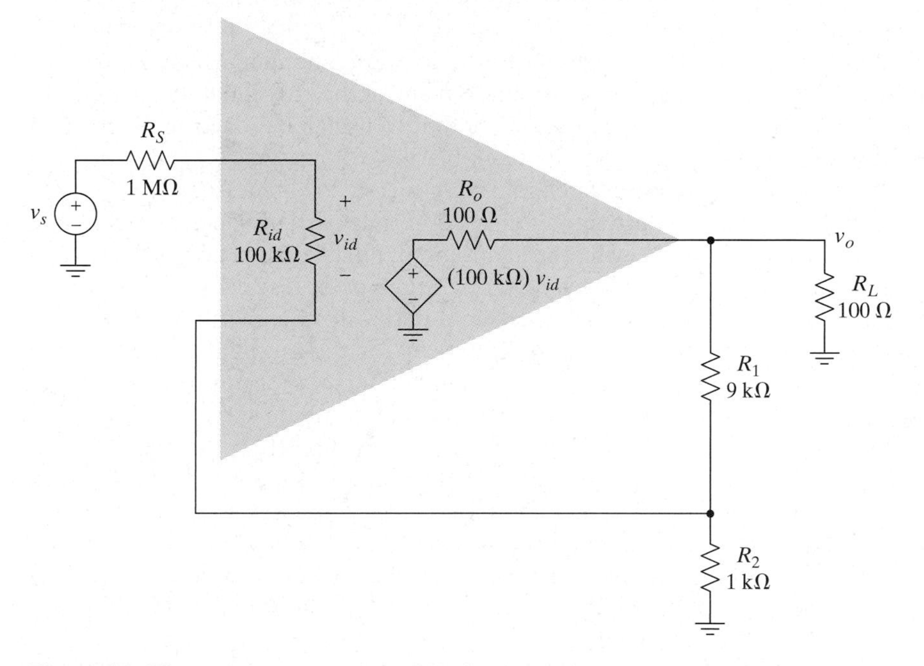

Figure 10–22 An op amp circuit with feedback.

Exercise 10.11

Use SPICE to confirm the results you obtained in Exercise 10.10.

Exercise 10.12

Assume that the open-loop gain of the op amp in Exercise 10.10 has a single-pole low-pass response with $f_p = 10$ Hz. What are the bandwidths f_{if} and f_{sf}?

[8] If you are not familiar with the term *differential-mode*, don't worry. It is the input resistance between the two inputs, rather than the resistance to ground from either input, which is a common-mode resistance. The topic of differential-mode and common-mode signals is covered in Chapter 5 (*see* Asides A5.1 and A5.2).

Exercise 10.13

Use a macro model like the one shown in Figure 4-20 for the op amp, and confirm your results from Exercise 10.12 by using SPICE.

We have now completed our consideration of the practical series–shunt amplifier. Our analyses have all been based on two-port models, however, and before we go on to consider other topologies, we need to examine how to apply our methods to circuits that are not initially modeled by two-port networks.

Consider the amplifier shown in Figure 10-23. This amplifier is a cascade of two common-merge stages and a merge follower with ***global feedback***. We call the feedback global, because it encompasses the entire forward amplifier, which is a cascade of individual stages. If we also have feedback in any of the individual stages, we refer to that as ***local feedback***. The choice of whether to use global or local feedback is complicated by many considerations (e.g., biasing voltages and bandwidth), but global feedback does provide greater gain stability (*see* Problem P10.43).

The topology of the forward amplifier used here is partially determined by the kind of feedback used, as we will see in a moment. First, let's see how we can identify the forward amplifier, feedback network, and type of feedback connection.

We start at the source and trace the signal path into the amplifier. Reviewing Aside A10.3 may be useful. Ignoring R_1 and R_2 for the moment, since they are only for biasing purposes, we see that the signal enters on the control terminal of T_1. We also note that the input to the second common-merge stage, T_2, is connected to the output terminal of T_1. Therefore, the connection from the merge terminal of T_1 to the overall amplifier output is the feedback path, not the forward signal path. In other words, T_1 is a common-merge amplifier stage as originally claimed, but it has the overall amplifier output fed back to its merge terminal. Ignoring the feedback for the moment, we finish the forward amplifier by tracing the rest of the signal path. We thereby determine that the forward amplifier is a

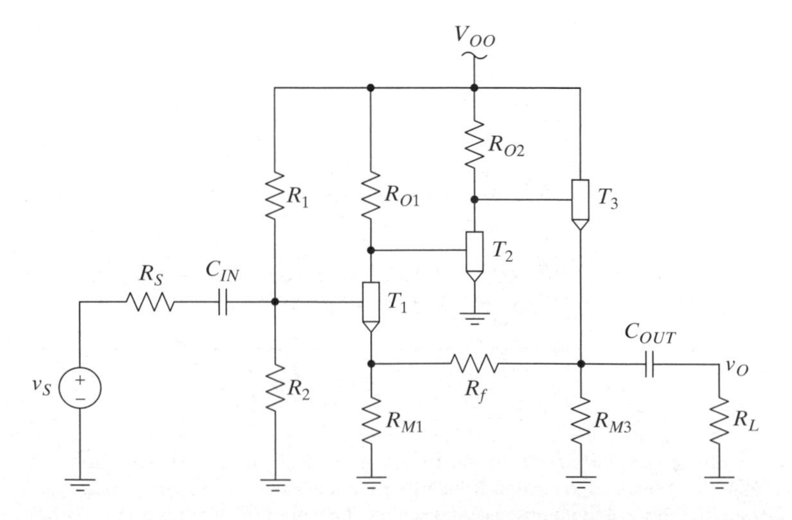

Figure 10–23 A practical series–shunt feedback amplifier using generic transistors.

cascade of two common-merge stages (T_1 and T_2) followed by a merge follower (T_3), as stated at the outset of the discussion.

We now focus our attention on the feedback network. There isn't a unique choice for what we consider to be part of the feedback network in this case. We could say that R_{M1} and R_{M3} are part of the forward amplifier and that the feedback network is made up of R_f alone, but this would not turn out to be a convenient choice (although we do refer to R_f as the feedback resistor). If we consider R_{M1} to be part of the forward amplifier, we see that the first common-merge stage has local feedback, which is caused by R_{M1} not being bypassed. The global feedback somehow combines with the local feedback. If, on the other hand, we consider R_{M1} to be part of the feedback network, this complication is removed. Whether or not we choose to consider R_{M3} part of the feedback network is not very important, but we will call it part of the feedback network for our analysis.

Now that the feedback network has been identified, we need to confirm that this is indeed a series–shunt connection and that the feedback is negative. To determine the type of connection, it is best to picture the forward amplifier and the feedback network as blocks, as shown in Figure 10-24. We then start from the signal source and trace a path towards the amplifier. In this case, we must go through the forward amplifier to get to the feedback network, so the two are in series at the input. Therefore, we know that it is voltages that will be summed. Recognizing that the input voltage to the forward amplifier is v_{cm1}, which is the error voltage for the feedback loop, we have

$$v_e = v_{cm1} = v_i - v_f, \tag{10.95}$$

where v_f is the feedback network output voltage. If we start at the load and trace a path back into the amplifier, we see that we can go into either the forward

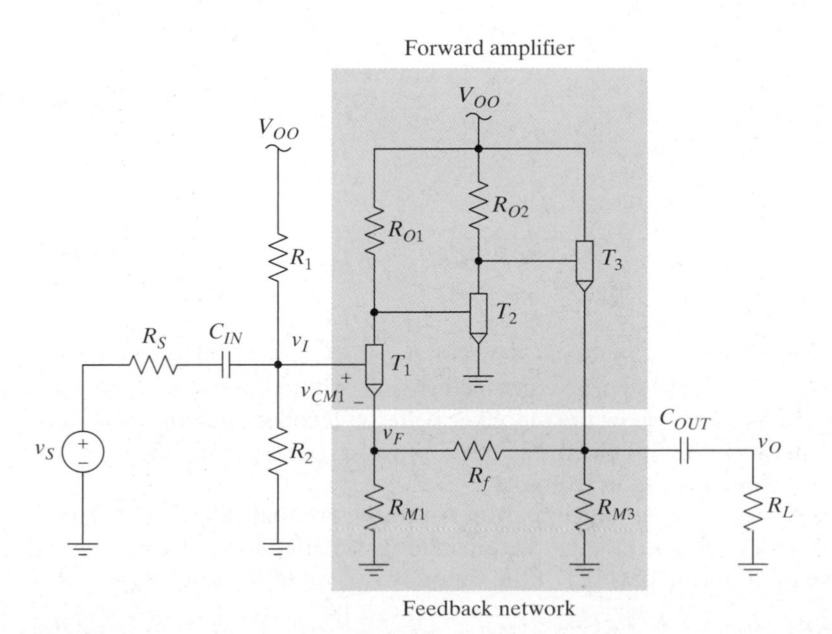

Figure 10–24 The amplifier from Figure 10-23 with the forward amplifier and feedback networks shown as blocks.

amplifier or the feedback network, so the two are in shunt, and it is the output voltage that will be the input to the feedback network. Finally, we note that although R_1 and R_2 are part of the biasing for the forward amplifier, they are outside of the feedback loop.

The final thing we need to check before proceeding with the analysis is whether or not this feedback is negative. We start by asking what happens if the error, $v_e = v_{cm1}$, is increased a little bit. The output current of T_1 will then increase, which will decrease the input voltage to the second common-merge stage. The second common-merge stage again inverts the signal, so the input voltage to the merge follower will increase. The output voltage will also increase, since the merge follower does not invert the signal. Finally, when v_o increases, the voltage fed back to the merge terminal of T_1 will increase. The increase in v_{m1} tends to *decrease* the value of v_{cm1}, so the feedback has reduced the magnitude of the original disturbance and is in fact negative. If the forward amplifier had only one common-merge stage, the feedback would be positive instead of negative. Consequently, we see that the type of feedback affected our choice of topology for the forward amplifier.

The circuit in Figure 10-23 is similar to the one we analyzed in Example 10.3, so we can draw on the results obtained there to reduce the amount of work we need to do here. First note that we again have a forward amplifier that is not a two-port network; therefore, we will find the two-port network for the feedback and then draw the small-signal midband AC equivalent circuit for the overall amplifier. After doing that, we will absorb the feedback loading into the forward amplifier to form the prime network and proceed.

The feedback network identified in Figure 10-24 is the same as the one for the circuit in Example 10.3, except for the addition of R_{M3}. Using the procedure described in the example, we find the elements of the unilateral two-port model for the feedback network:

$$R_{bo} = \left.\frac{v_f}{i_{bo}}\right|_{v_o=0} = R_{M1} \| R_f, \tag{10.96}$$

$$b = \left.\frac{v_f}{v_o}\right|_{i_{bo}=0} = \frac{R_{M1}}{R_{M1} + R_f}, \tag{10.97}$$

and

$$R_{bi} = \left.\frac{v_o}{i_{bi}}\right|_{i_{bo}=0} = R_{M3} \| (R_f + R_{M1}). \tag{10.98}$$

Using these elements, we draw the small-signal midband AC equivalent circuit for the amplifier in Figure 10-25(a). We have combined R_1 and R_2 into the bias resistor R_{CC}. In part (b) of the figure, we show the amplifier with the feedback loading absorbed into the prime network. We find the prime network parameters by equating the two-port networks, as demonstrated in Example 10.3.

We have noted on the schematic in part (a) of the figure that the current out of the merge terminal of T_1 is $(a_{im1}+1)i_{c1}$. Utilizing this fact and noting that the voltage across the input port in part (b) of the figure is v_e', we use KVL to write

$$v_e' = i_{c1} r_{cm1} + (a_{im1} + 1) i_{c1} R_{bo} \tag{10.99}$$

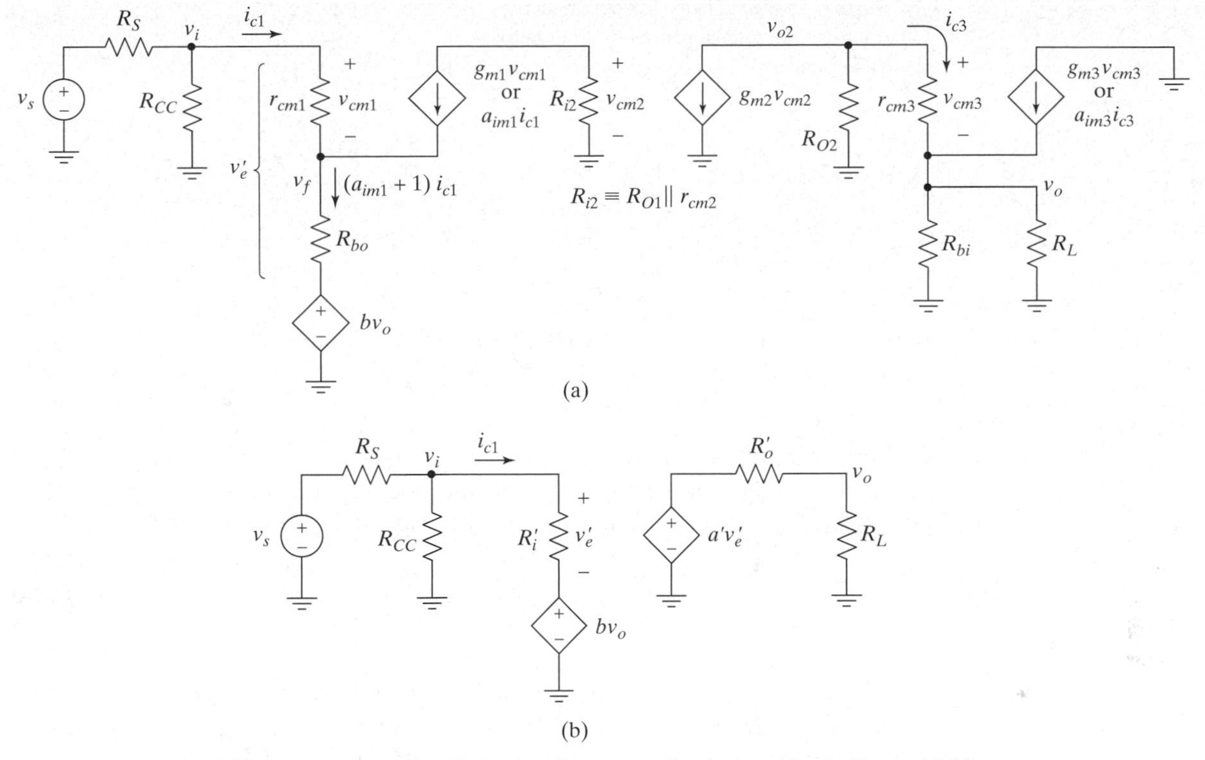

Figure 10–25 (a) The small-signal midband AC equivalent circuit for the amplifier in Figure 10-23. (b) The simplified circuit using the prime network.

for the circuit in part (a) of the figure. For the circuit in part (b) , we get

$$v_e' = i_{c1}R_i' \tag{10.100}$$

and setting these two results equal we obtain

$$R_i' = r_{cm1} + (a_{im1} + 1)R_{bo}. \tag{10.101}$$

We next turn our attention to finding the voltage gain of the prime network. If we drive both two-port networks with an input voltage v_e' and examine the open-circuit output voltages, the circuit in part (b) of the figure yields

$$\frac{v_o}{v_e'} = a'. \tag{10.102}$$

For the circuit in part (a), we write

$$\frac{v_o}{v_e'} = \frac{v_o}{v_{o2}}\frac{v_{o2}}{v_{cm2}}\frac{v_{cm2}}{v_{cm1}}\frac{v_{cm1}}{v_e'}. \tag{10.103}$$

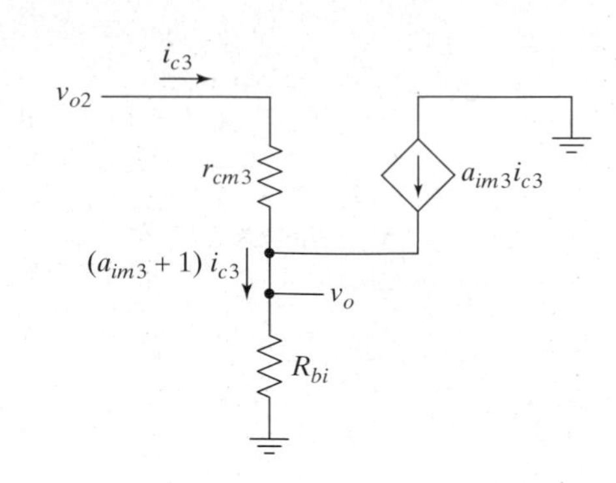

Figure 10–26 Figure for obtaining the open-circuit voltage gain of the merge follower stage.

The first term on the right-hand side of (10.103) can be found with the aid of Figure 10-26. (Remember that we want the open-circuit output voltage, so R_L is removed.) Writing KVL from v_{o2} to ground and solving, we obtain

$$\frac{v_o}{v_{o2}} = \frac{(a_{im3} + 1)R_{bi}}{(a_{im3} + 1)R_{bi} + r_{cm3}}. \tag{10.104}$$

The second term on the right-hand side of (10.103), v_{o2}/v_{cm2}, can also be found with some help from Figure 10-26. The equivalent resistance seen looking into the control terminal of T_3 is equal to v_{o2}/i_{c3}. Using KVL, we obtain

$$R_{i3} \equiv \frac{v_{o2}}{i_{c3}} = r_{cm3} + (a_{im3} + 1)R_{bi}. \tag{10.105}$$

This resistor shows up in parallel with R_{O2} in the circuit of Figure 10-25. Therefore, we write

$$v_{o2} = -g_{m2}v_{cm2}(R_{O2}\|R_{i3}), \tag{10.106}$$

from which we obtain

$$\frac{v_{o2}}{v_{cm2}} = -g_{m2}(R_{O2}\|R_{i3}). \tag{10.107}$$

To find the third term on the right-hand side of (10.103), we use the voltage-controlled source for T_1. We then define $R_{i2} = R_{O1}\|r_{cm2}$ and write

$$\frac{v_{cm2}}{v_{cm1}} = -g_{m1}R_{i2}. \tag{10.108}$$

Finally, the last term on the right-hand side of (10.103) can be found by writing KVL from v_i to ground, as was done in deriving R_i'. We obtain

$$\frac{v_{cm1}}{v_e'} = \frac{r_{cm1}}{r_{cm1} + (a_{im1} + 1)R_{bo}}. \tag{10.109}$$

We can combine (10.109), (10.108), (10.107), (10.104), and (10.103) to obtain an equation for v_o/v_e' for the circuit in Figure 10-25(a). Setting the result equal to (10.102), which was obtained for part (b) of the figure, we have

$$a' = \left(\frac{r_{cm1}}{r_{cm1} + (a_{im1} + 1)R_{bo}}\right)(-g_{m1}R_{i2})(-g_{m2}(R_{O2}\|R_{i3}))\left(\frac{(a_{im3} + 1)R_{bi}}{(a_{im3} + 1)R_{bi} + r_{cm3}}\right). \quad \textbf{(10.110)}$$

This equation is rather imposing, but the origin of every term is clear, and it was not too difficult to derive. It again demonstrates the importance of knowing how to do analysis for the purpose of design, as opposed to analysis for the sake of getting the answer. There is only one element left to determine in the primed circuit model: R_o'. From the circuit in Figure 10-25(b), we see that when the input, v_e', is set to zero, the controlled source, $a'v_e'$, will also be zero, and the output resistance of the two-port network will be R_o'. To find the corresponding output resistance of the network in part (a) of the figure, we make use of Figure 10-27. In part (a) of the figure, we note that with $v_e' = 0$, we have $v_{cm1} = 0$. Therefore, $v_{cm2} = 0$, and so is the current source $g_{m2}v_{cm2}$. Looking in from the output to find the resistance as shown in part (a), we see that

$$R_o' = R_{bi}\|R_x \quad \textbf{(10.111)}$$

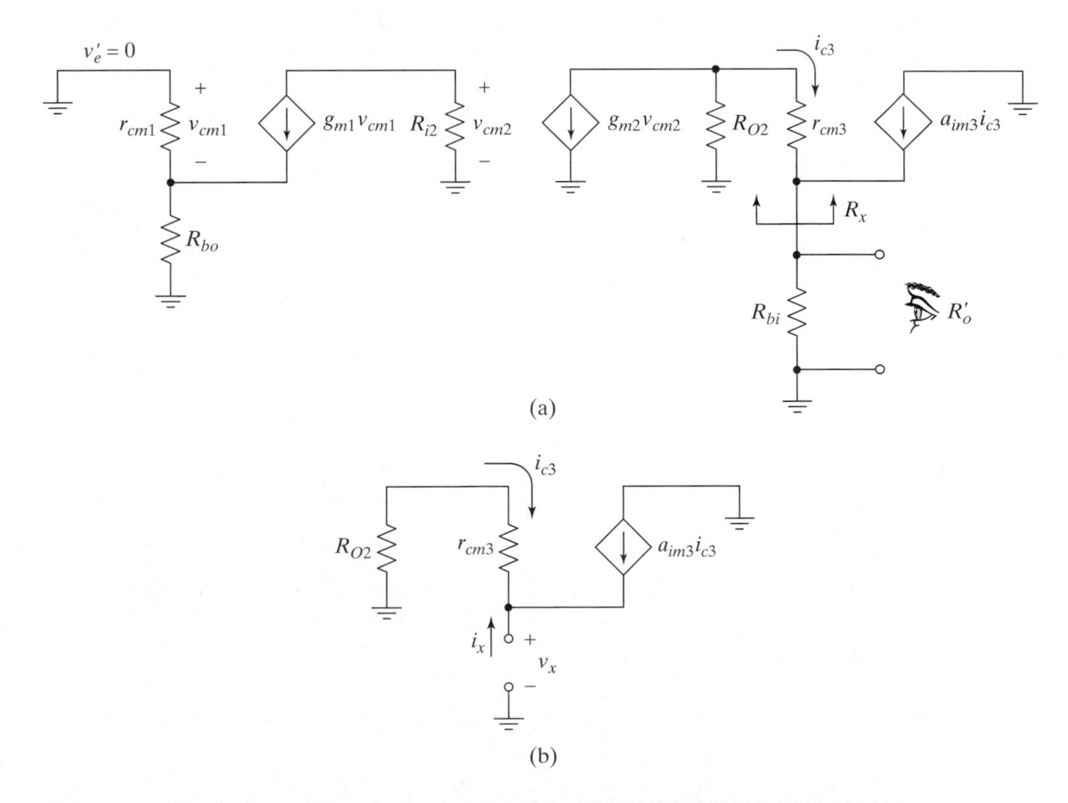

Figure 10–27 Finding the output resistance of the prime network from Figure 10-25(a).

where R_x is the resistance seen looking into the merge terminal of T_3 and is found from part (b) of the figure to be

$$R_x = \frac{v_x}{i_x}. \tag{10.112}$$

Using KCL, we recognize that

$$i_x = -i_{c3}(a_{im3} + 1). \tag{10.113}$$

Using KVL, we write

$$v_x = i_{c3}(R_{O2} + r_{cm3}), \tag{10.114}$$

and combining these equations, we obtain[9]

$$R_x = \frac{R_{O2} + r_{cm3}}{a_{im3} + 1}. \tag{10.115}$$

We now have all of the elements describing Figure 10-25(b). The only difficulty remaining in applying our previous analyses to this circuit is the presence of R_{CC}. This resistor has no effect on the parameters that depend on the gain from v_i to v_o, since it appears from v_i to ground. In other words, A_i, A_{if}, and ω_{if} remain unchanged from the values calculated for Figure 10-13(b). However, parameters that depend on the gain from v_s to v_o are affected by R_{CC}. This effect can be included by finding the Thévenin equivalent for the source with R_{CC} included, as shown in Figure 10-28. The modified source is described by (the prime used here has nothing to do with the prime network; it simply denotes the fact that we have modified the original source)

$$v_s' = v_s\left(\frac{R_{CC}}{R_S + R_{CC}}\right) \tag{10.116}$$

and

$$R_S' = R_S \| R_{CC}. \tag{10.117}$$

Using the Thévenin source, we can modify our previous analysis to accommodate R_{CC}. The new gain A_s is given by

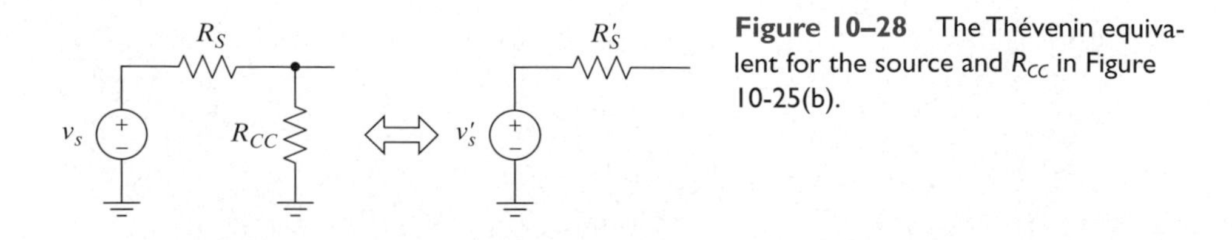

Figure 10–28 The Thévenin equivalent for the source and R_{CC} in Figure 10-25(b).

---

[9] Readers familiar with the bipolar transistor will recognize this resistance as being R_{O2} and r_{cm3} reflected into the merge terminal, in perfect analogy with reflecting into the emitter of a BJT.

$$A_s' \equiv \left.\frac{v_o}{v_s'}\right|_{b=0} = \frac{v_e'}{v_s'}\frac{v_o}{v_e'} = \left(\frac{R_i'}{R_i' + R_S'}\right)a'\left(\frac{R_L}{R_L + R_o'}\right)$$
$$= \alpha_i' a' \alpha_o' \quad \textbf{(10.118)}$$

in direct analogy with (10.70). We have used a prime on A_s because the source is not the original source, but the Thévenin source of Figure 10-28. The new input attenuation factor for the prime network is defined in the same way we defined it before, namely,

$$\alpha_i' \equiv \frac{v_e'}{v_s'} = \left(\frac{R_i'}{R_i' + R_S'}\right), \quad \textbf{(10.119)}$$

but now R_S' is used in place of R_S. Using (10.118), we have

$$A_{sf}' = \frac{v_o}{v_s'} = \frac{A_s'}{1 + A_s' b}, \quad \textbf{(10.120)}$$

which has the familiar form. In order to obtain the gain from the original source, v_s, to the output, we write

$$A_{sf} = \frac{v_o}{v_s} = \frac{v_o}{v_s'}\frac{v_s'}{v_s} = A_{sf}'\frac{v_s'}{v_s} = A_{sf}'\left(\frac{R_{CC}}{R_{CC} + R_S}\right), \quad \textbf{(10.121)}$$

where we have made use of (10.116).

The output impedance with feedback is the same as we had in (10.84); that is

$$R_{of} = \frac{R_o'}{1 + \alpha_i' a' b} = \frac{R_o'}{1 + L_o}. \quad \textbf{(10.122)}$$

The value of ω_{sf} is given by

$$\omega_{sf} = \omega_p(1 + \alpha_i' a_0' \alpha_o' b) = \omega_p(1 + L_s). \quad \textbf{(10.123)}$$

The fact that we have used the Thévenin equivalent source does not affect the value of ω_{sf}, since no reactive elements were involved; therefore, we don't need a prime on ω_{sf}. Similarly, because using the Thévenin equivalent does not change the resistance seen looking toward the source from the circuit, α_i' is the same, and therefore, so is L_s.

Finally, the input resistance with feedback of our present circuit is R_{CC} in parallel with the input resistance seen from the Thévenin source, which is given by (10.79) from our analysis of Figure 10-13(b). The result for the present circuit is

$$R_{if} = R_{CC} \| [R_i'(1 + A_i b)], \quad \textbf{(10.124)}$$

where

$$A_i \equiv \left.\frac{v_o}{v_i}\right|_{b=0} = a'\left(\frac{R_L}{R_L + R_o'}\right) = a'\alpha_o', \quad \textbf{(10.125)}$$

as before. The value of A_{if} is the same as we had previously, as is the bandwidth from v_i to v_o.

We have now completed the application of our analysis technique to a practical transistor amplifier. The process has been tedious and the notation somewhat complex, but the procedures, taken one at a time, are not too difficult. The key points to remember when analyzing feedback amplifiers are to be very careful with notation, to simplify the circuit in stages, as we have done, and to then end up with a circuit that will allow you to make use of previously derived results as much as possible.

Exercise 10.14

Consider the circuit shown in Figure 10-29, where an op amp has been modeled by differential-mode and common-mode input resistances, gain, and output resistance. This is the same circuit as was used in Exercise 10.10, but we have now added the common-mode input resistances, R_{C1} and R_{C2}, to the model. Find A_{sf}, R_{if}, and f_{sf} for this circuit.

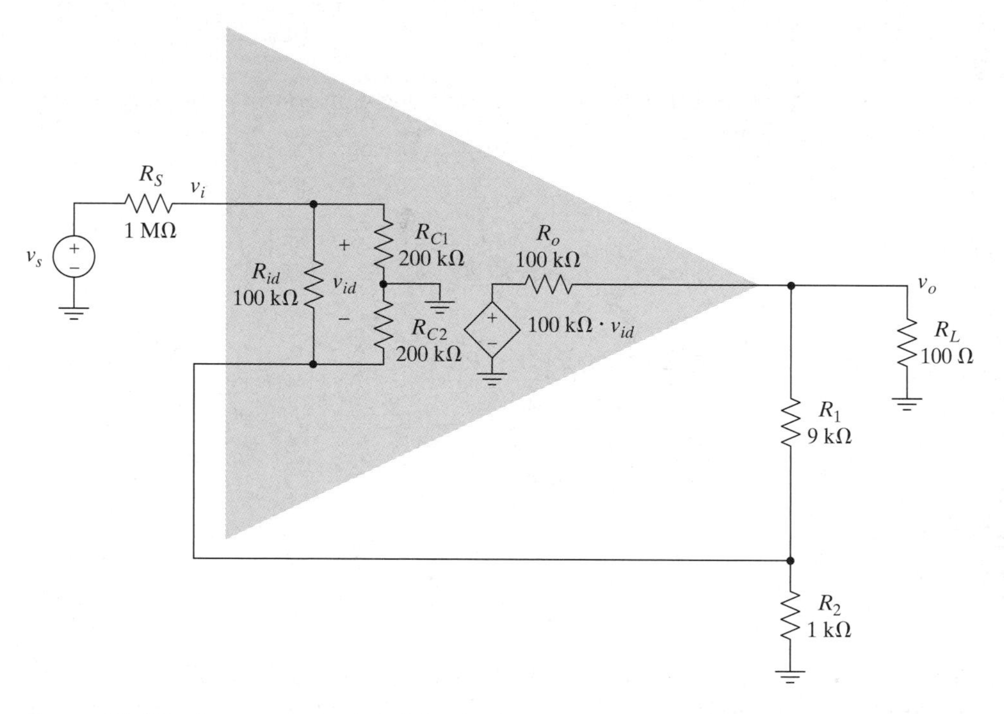

Figure 10–29 An op amp circuit with feedback.

Exercise 10.15

Use SPICE to confirm the results of Exercise 10.14.

Series–Series Connection

Figure 10-30(a) shows a series–series negative feedback amplifier that uses unilateral two-port networks for both the forward amplifier and the feedback network.

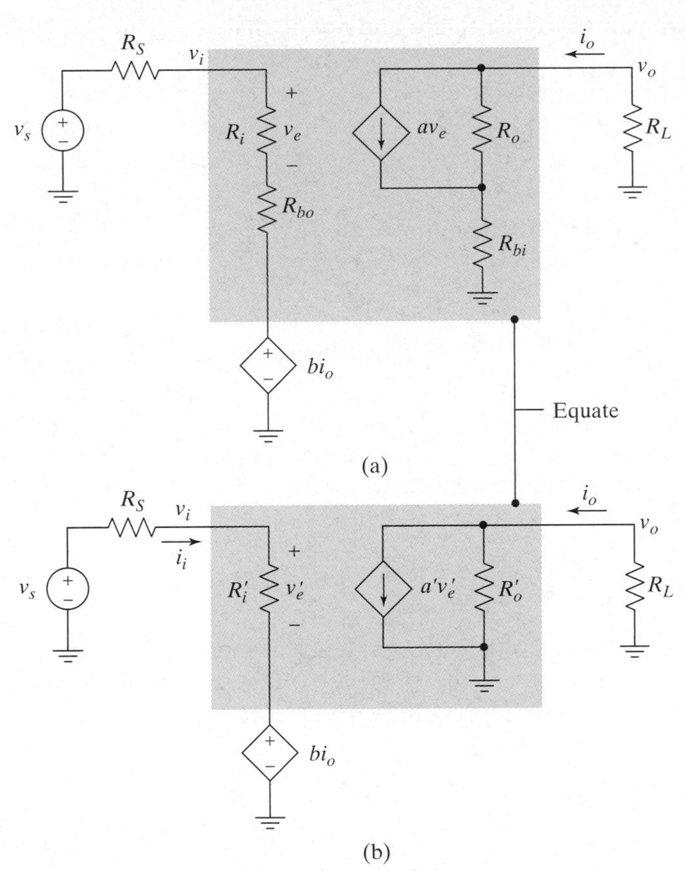

Figure 10–30 (a) A series–series negative feedback amplifier and (b) the feedback network loading absorbed into the forward amplifier to form the primed network.

Thévenin models have been used for the signal source and the feedback network output because of the series connection at the input. A Norton equivalent has been used for the amplifier output, since it is the output current that is being fed back. In part (b) of the figure, we have absorbed the feedback network loading into the forward amplifier to create the primed network.

Equating the two two-port networks as shown in the figure, we find by inspection that

$$R_i' = R_i + R_{bo}, \tag{10.126}$$

which is the same as we had for the series–shunt connection. On the output side we compare short-circuit currents (i.e., with $R_L = 0$) in this case. The short-circuit output current for the network in part (a) of the figure is

$$i_{o\ \text{s.c.}} = av_e\left(\frac{R_o}{R_o + R_{bi}}\right). \tag{10.127}$$

For the primed circuit in part (b), we obtain

$$i_{o\ \text{s.c.}} = a'v_e'. \tag{10.128}$$

Equating these two currents, we find that

$$a' = a\left(\frac{R_o}{R_o + R_{bi}}\right)\frac{v_e}{v_e'}. \tag{10.129}$$

Examining the circuit in part (a) of the figure and noting that $v_e' = (v_i - bi_o)$, we see that

$$v_e = \left(\frac{R_i}{R_i + R_{bo}}\right)v_e'. \tag{10.130}$$

Combining this result with (10.129), we obtain

$$a' = a\left(\frac{R_o}{R_o + R_{bi}}\right)\left(\frac{R_i}{R_i + R_{bo}}\right). \tag{10.131}$$

Finally, we note that with $v_e' = 0$, the controlled current sources in both networks have zero value, which implies that they look like an open circuit ($i = 0$ for an open circuit). So, looking into the output ports of the networks when $v_e' = 0$ and equating the resistances seen, we get

$$R_o' = R_o + R_{bi}. \tag{10.132}$$

The circuit in part (b) of the figure is the same as the ideal series–series amplifier shown in Figure A10-1(b), except for the presence of the nonideal load and source. We begin our analysis by considering the open-loop gain from v_i to i_o. With $b = 0$, we have $v_i = v_e'$ and

$$A_i = \left.\frac{i_o}{v_i}\right|_{b=0} = a'\alpha_o', \tag{10.133}$$

where

$$\alpha_o' = \frac{R_o'}{R_o' + R_L}. \tag{10.134}$$

With feedback included (i.e., $b \neq 0$), we write

$$i_o = a'\alpha_o' v_e' = a'\alpha_o'(v_i - bi_o), \tag{10.135}$$

which can be solved to yield

$$A_{if} = \frac{i_o}{v_i} = \frac{a'\alpha_o'}{1 + a'\alpha_o' b} = \frac{A_i}{1 + A_i b}. \tag{10.136}$$

For the overall gain, we have

$$A_s = \left.\frac{i_o}{v_s}\right|_{b=0} = \alpha_i' a' \alpha_o', \tag{10.137}$$

where

$$\alpha_i' = \frac{R_i'}{R_i' + R_S}. \tag{10.138}$$

Including the feedback, we have

$$i_o = a'\alpha_o' v_e' = \alpha_i' a' \alpha_o' (v_s - b i_o), \quad \textbf{(10.139)}$$

which can be solved to obtain

$$A_{sf} = \frac{i_o}{v_s} = \frac{\alpha_i' a' \alpha_o'}{1 + \alpha_i' a' \alpha_o' b} = \frac{A_s}{1 + A_s b}. \quad \textbf{(10.140)}$$

We now calculate the input resistance with feedback by using Figure 10-30(b). We have

$$R_{if} = \frac{v_i}{i_i}, \quad \textbf{(10.141)}$$

and, using Ohm's law, we obtain

$$i_i = \frac{v_i - b i_o}{R_i'} = \frac{v_i(1 - bA_{if})}{R_i'}. \quad \textbf{(10.142)}$$

Substituting for A_{if} from (10.136) yields

$$R_{if} = \frac{v_i}{i_i} = \frac{R_i'}{1 - \dfrac{A_i b}{1 + A_i b}} = R_i'(1 + A_i b) = R_i'(1 + L_i), \quad \textbf{(10.143)}$$

as for the series–shunt amplifier. The output resistance with feedback is found by removing the load resistance in Figure 10-30(b), setting the independent signal source to zero, and looking in from the output to find

$$R_{of} = \frac{v_o}{i_o}. \quad \textbf{(10.144)}$$

Using KCL, we find the current into R_o', namely,

$$i_{R_o'} = i_o - a' v_e', \quad \textbf{(10.145)}$$

and then use this current and Ohm's law to write

$$v_o = R_o'(i_o - a' v_e'). \quad \textbf{(10.146)}$$

We next examine the input of the circuit (remembering that $v_s = 0$) and write

$$v_e' = -\alpha_i' b i_o, \quad \textbf{(10.147)}$$

where α_i' is given in (10.138). Using (10.147) in (10.146), we obtain

$$R_{of} = \frac{v_o}{i_o} = R_o'(1 + \alpha_i' a' b) = R_o'(1 + L_o). \quad \textbf{(10.148)}$$

In other words, the output resistance with feedback is increased by one plus the loop gain, as expected. The loop gain used is the loop gain seen in driving the loop from the output, so R_L is not present in the answer.

If the forward amplifier has a single-pole low-pass transfer function, the analysis carried out for the series–shunt amplifier in (10.85) through (10.93) applies equally well here, with very minor variations in a few terms. First, we again assume that the forward amplifier gain is given by

$$a(j\omega) = \frac{a_0}{1 + j\omega/\omega_p}. \quad \textbf{(10.149)}$$

Now, consider the gain from v_s to v_o first. Plugging (10.149) into (10.131) and the result into (10.137) yields

$$A_s(j\omega) = \frac{\alpha_i' a_0 \alpha_o' \left(\dfrac{R_i}{R_i + R_{bo}}\right)\left(\dfrac{R_o}{R_{bi} + R_o}\right)}{1 + j\omega/\omega_p} = \frac{\alpha_i' a_0' \alpha_o'}{1 + j\omega/\omega_p}. \quad \textbf{(10.150)}$$

Plugging (10.150) into (10.140) yields

$$A_{sf}(j\omega) = \frac{A_{sf0}}{1 + j\omega/\omega_{sf}}, \quad \textbf{(10.151)}$$

where

$$A_{sf0} = \frac{\alpha_i' a_0' \alpha_o'}{1 + \alpha_i' a_0' \alpha_o' b} = \frac{\alpha_i' a_0' \alpha_o'}{1 + L_s} \quad \textbf{(10.152)}$$

and

$$\omega_{sf} = \omega_p(1 + \alpha_i' a_0' \alpha_o' b) = \omega_p(1 + L_s). \quad \textbf{(10.153)}$$

Equations (10.151) through (10.153) are identical to the ones arrived at earlier, so the systematic approach has paid off. We can have greater confidence in our analyses when we arrive at familiar results. In fact, it is generally true that once we have absorbed loading and arrived at the primed network for a given feedback connection, we are able to apply previously derived results without further work.

We now consider the gain from v_i to v_o and start by substituting (10.149) into (10.131) and the result into (10.133) to find

$$A_i(j\omega) = \frac{a_0 \alpha_o' \left(\dfrac{R_i}{R_i + R_{bo}}\right)\left(\dfrac{R_o}{R_{bi} + R_o}\right)}{1 + j\omega/\omega_p} = \frac{a_0' \alpha_o'}{1 + j\omega/\omega_p}. \quad \textbf{(10.154)}$$

Plugging (10.154) into (10.136) yields

$$A_{if}(j\omega) = \frac{A_{if0}}{1 + j\omega/\omega_{if}}, \quad \textbf{(10.155)}$$

where

$$A_{if0} = \frac{a_0'\alpha_o'}{1 + a_0'\alpha_o'b} = \frac{a_0'\alpha_o'}{1 + L_i} \tag{10.156}$$

and

$$\omega_{if} = \omega_p(1 + a_0'\alpha_o'b) = \omega_p(1 + L_i). \tag{10.157}$$

We again note that the closed-loop gain and the bandwidth have been modified by the same term.

The comments made earlier in comparing the closed-loop bandwidths in (10.89) and (10.93) apply equally well to (10.153) and (10.157).

We finish this section with a transistor-level example. The amplifier shown in Figure 10-31 is sometimes called a ***series–series triple***. As was the case with the series–shunt amplifier example, the complexity of this circuit is partially determined by the need to make the feedback negative. In fact, a comparison of this figure with the series–shunt amplifier in Figure 10-23 shows that the feedback loop is the same; therefore, we know that the feedback is again negative. The only difference is that T_3 is not used as a merge follower. In fact, T_3 serves as a ***phase splitter***. The signal at the merge terminal of T_3 is in phase with the signal on its control terminal, whereas the signal at the output terminal is 180° out of phase with the control input (i.e., it is inverted); therefore, the signals at the merge terminal and the output terminal are out of phase with each other—hence the name *phase splitter*.

There are some difficulties in applying our standard analysis methods to this circuit. We have already encountered and overcome the difficulties at the input,[10] but there is an additional complication at the output. The current being sensed by

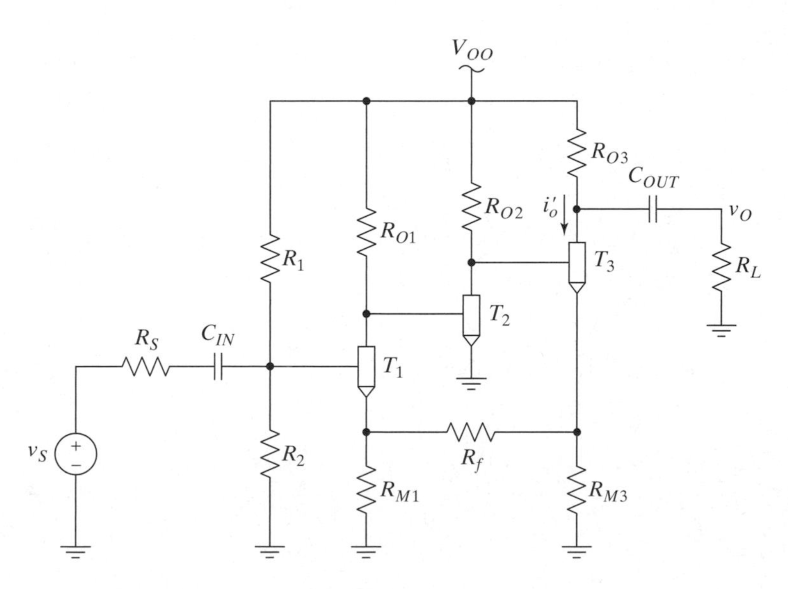

Figure 10–31 A series–series triple.

[10] First, the fact that the forward amplifier is not a two-port network, and second, that R_1 and R_2 are outside of the feedback loop.

the feedback network, which comprises R_{M1}, R_{M3}, and R_f, is not the same as the output current of T_3. Furthermore, the output current of T_3 is not the output current of the amplifier (i.e., the load current). In the analysis that follows, we make use of the results previously obtained for the series–shunt connection and modify or extend them as necessary. It is possible to analyze this amplifier as a series–shunt feedback amplifier by pretending that v_{M3} is the output and then converting the results at the end. However, that approach is awkward, so we will not pursue it here.

We first model the feedback network as a unilateral two-port network. The model is shown in Figure 10-32(b). The current into the feedback network is not equal to i'_o, but is nearly equal to i'_o if the current gain of the transistor is large. We will assume for now that $a_{im3} \gg 1$, so that this figure is approximately correct. The equation at the output port of the feedback network is

$$v_f = i_{bo}R_{bo} + bi'_o, \tag{10.158}$$

which leads to

$$R_{bo} = \left.\frac{v_f}{i_{bo}}\right|_{i'_o=0} = R_{M1} \| (R_f + R_{M3}), \tag{10.159}$$

where we have applied the formula to the circuit in Figure 10-32(a) by inspection. Using (10.158) again, we find that (*see* Problem P10.44)

$$b = \left.\frac{v_f}{i'_o}\right|_{i_{bo}=0} = \left[R_{M3} \| (R_f + R_{M1})\right]\left(\frac{R_{M1}}{R_{M1} + R_f}\right). \tag{10.160}$$

To find the input resistance of the feedback network, we again note that the independent variables used in the equation at the input port must be the same as those used at the output port—that is, in (10.158). Therefore, the equation at the input port is

$$v_{bi} = a_r i_{bo} + R_{bi} i'_o, \tag{10.161}$$

where a_r is the reverse gain of the feedback network and is the term we are ignoring when we use the unilateral model. From (10.161), we find that

$$R_{bi} = \left.\frac{v_{bi}}{i'_o}\right|_{i_{bo}=0} = R_{M3} \| (R_f + R_{M1}). \tag{10.162}$$

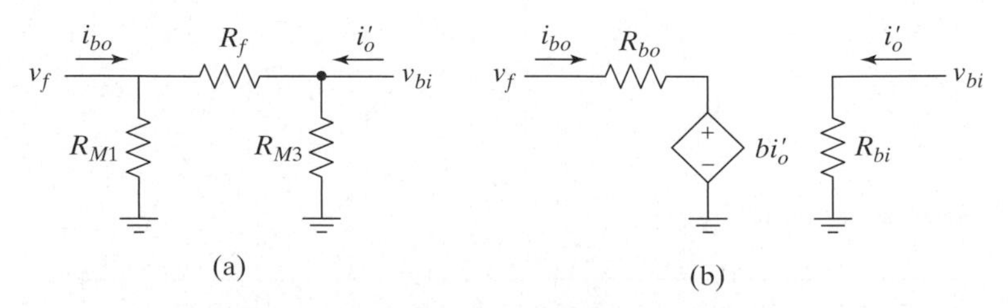

Figure 10–32 (a) The original feedback network and (b) the unilateral two-port model of the feedback network for the amplifier in Figure 10-31.

It does matter what we do with the output port of the feedback network when we find R_{bi} (i.e., whether we set $i_{bo} = 0$ or $v_f = 0$); therefore, we know that the feedback network is not truly unilateral and our use of a unilateral two-port model is only an approximation.

Figure 10-33(a) shows the resulting small-signal midband AC model of the amplifier. In drawing this figure, we have defined

$$R_{i2} = R_{O1} \| r_{cm2} \tag{10.163}$$

and

$$R_{CC} = R_1 \| R_2 \tag{10.164}$$

as before. We have also defined

$$R'_L = R_{O3} \| R_L \tag{10.165}$$

and i'_o for convenience. We will put the results in terms of v_o later.

Part (b) of the figure uses the prime network to replace the forward amplifier with feedback network loading. The input resistance of the prime network is the same as we found for the series–shunt amplifier:

$$R'_i = r_{cm1} + (a_{im1} + 1)R_{bo}. \tag{10.166}$$

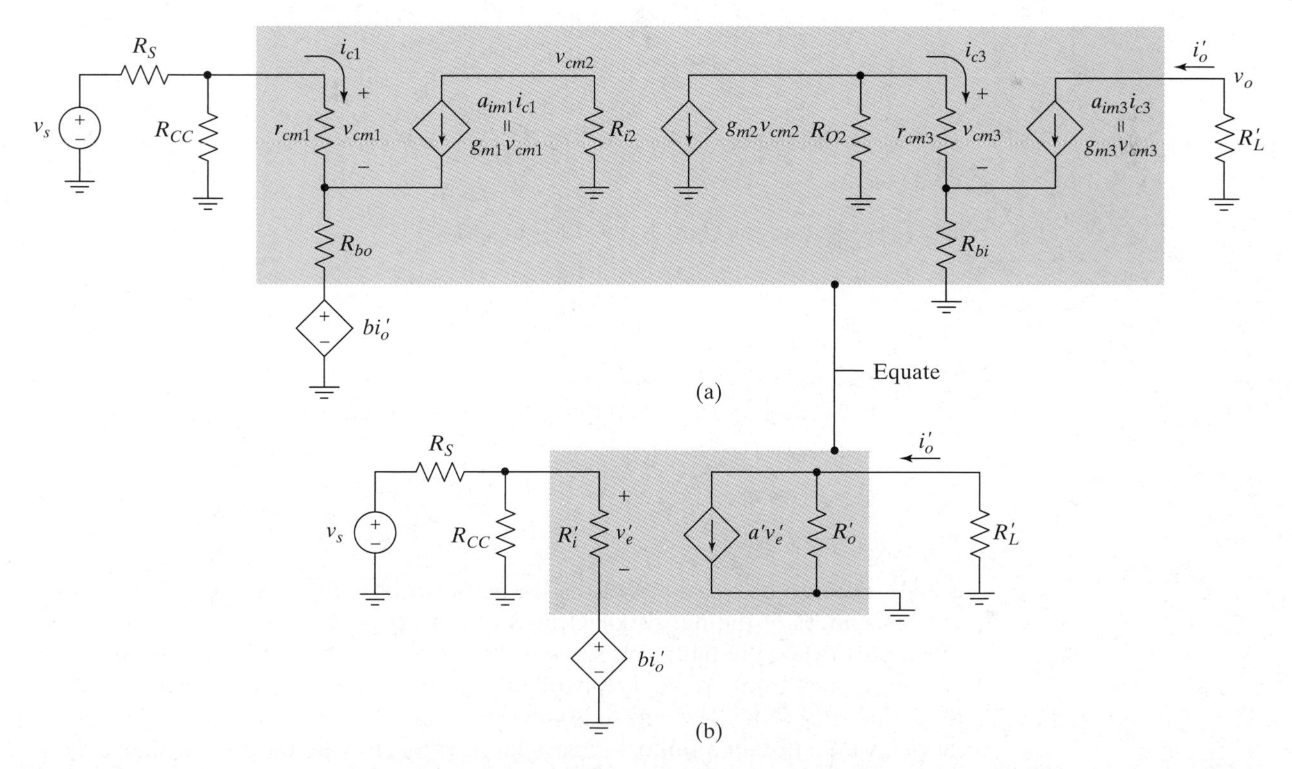

Figure 10–33 (a) The small-signal midband AC model of the amplifier in Figure 10-31 with the feedback network modeled by a unilateral two-port network. (b) The forward amplifier and feedback network loading combined into the prime network.

The gain of the forward amplifier is found by comparing the short-circuit output currents in parts (a) and (b) of the figure. For part (a), we obtain

$$\frac{i'_{o\ s.c.}}{v'_e} = \frac{i'_{o\ s.c.}}{i_{c3}} \frac{i_{c3}}{v_{cm2}} \frac{v_{cm2}}{v_{cm1}} \frac{v_{cm1}}{v'_e}, \quad \textbf{(10.167)}$$

where the last two ratios on the right-hand side were found for the series–shunt amplifier and apply equally well here; thus,

$$\frac{v_{cm2}}{v_{cm1}} = -g_{m1}R_{i2} \quad \textbf{(10.168)}$$

and

$$\frac{v_{cm1}}{v'_e} = \frac{r_{cm1}}{r_{cm1} + (a_{im1} + 1)R_{bo}}. \quad \textbf{(10.169)}$$

The second ratio on the right-hand side of (10.167) is found by using a current divider and is

$$i_{c3} = \frac{-R_{O2}}{R_{O2} + R_{i3}} g_{m2}v_{cm2} \quad \textbf{(10.170)}$$

where R_{i3} is the resistance seen looking into the control terminal of T_3 as before (but now R_L is not in parallel with R_{bi}) that is,

$$R_{i3} = r_{cm3} + (a_{im3} + 1)R_{bi}. \quad \textbf{(10.171)}$$

Therefore,

$$\frac{i_{c3}}{v_{cm2}} = \frac{-R_{O2}}{R_{O2} + R_{i3}} g_{m2}. \quad \textbf{(10.172)}$$

The first term on the right-hand side of (10.167) is

$$\frac{i'_{o\ s.c.}}{i_{c3}} = a_{im3} \quad \textbf{(10.173)}$$

by inspection. For part (b) of the circuit, we obtain $i'_{o\ s.c.}/v'_e = a'$. Equating this with (10.167) and making the appropriate substitutions, we get

$$a' = a_{im3}\left(\frac{-g_{m2}R_{O2}}{R_{O2} + R_{i3}}\right)(-g_{m1}R_{i2})\left(\frac{r_{cm1}}{r_{cm1} + (a_{im1} + 1)R_{bo}}\right). \quad \textbf{(10.174)}$$

Finally, the output resistance of the prime network is found by equating the output resistances of the networks in parts (a) and (b), when v'_e is set equal to zero. From part (a) of the figure we see that the output resistance is infinite, since the current comes from an ideal controlled source. In other words, v_o will not affect the value of i'_o at all. If we include the small-signal output resistance of T_3 in our model, we do obtain a finite, but very large value, for the output resistance. Nevertheless, given our present model, we see that the output resistance from part (b) of the figure is R'_o, and in this example, R'_o is infinite.

WARNING! If we did include the output resistance of T_3 in our equivalent circuit while finding R_o', we could *not* then use our standard analysis for the output resistance with feedback. Approximating the current into the feedback network as being equal to i_o' is reasonable for finding the gain of this circuit, but is not effective for finding the output resistance. (In fact, the result can be off by orders of magnitude [10.1].) This is a common problem in series output connections and deserves a brief digression for explanation, so we have discussed the issue further in Aside A10.5.

Now that we have completely specified the prime network in Figure 10-33(b), we can solve for the gains, bandwidths, and input and output resistances of the closed-loop amplifier. We can make use of most of the results we derived earlier in this section, since Figure 10-33(b) is topologically identical to Figure 10-30(b), with the exception of R_{CC}. Since R_{CC} is connected from v_i to ground, it cannot affect the gain or bandwidth from v_i to i_o', and we obtain the same results as before (except that we must use R_L' in place of R_L); A_i, A_{if}, and ω_{if} are given by (10.133), (10.136), and (10.157), respectively. The transfer function from v_s to i_o' *is* affected by the presence of R_{CC}, however. We handle this situation by finding the Thévenin

ASIDE A10.5 OUTPUT RESISTANCE OF SERIES–SERIES TRIPLE

Notice in Figure 10-33(a) that the current going to the left at the bottom of the controlled source for T_3 is i_o' and will remain so even if r_{o3} is included in the model as shown in Figure A10-2. Note also in this figure that the rest of the circuit has been modeled by a loop gain, $\widetilde{L}$, driven by the voltage v_{m3}. (The basic loop is the same as the series–shunt triple, so we are free to consider this voltage to be the feedback variable.) Examining this figure reveals that r_{o3} is outside the feedback loop and will not, therefore, have its value modified by $(1 + L_o)$.

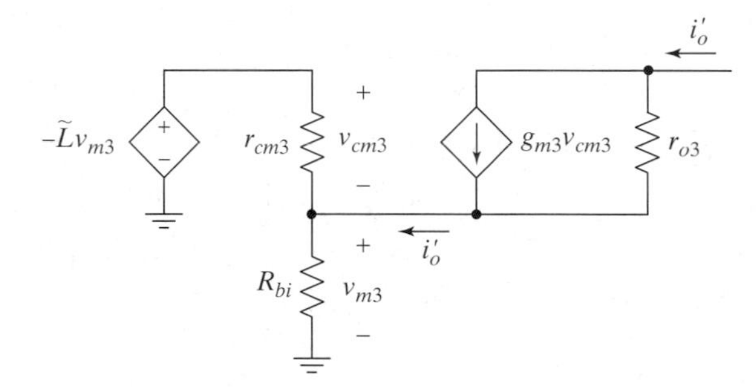

Figure A10-2 Examining the effect of r_{o3} on R_o'.

Using the circuit in Figure A10-2, we can derive the resistance seen looking into the output of T_3. The result is (*see* Problem P10.52)

$$R_{o3} = \left(R_{bi} \,\middle\|\, \frac{r_{cm3}}{1+\widetilde{L}}\right) + \left[1 + g_{m3}(1+\widetilde{L})\left(R_{bi} \,\middle\|\, \frac{r_{cm3}}{1+\widetilde{L}}\right)\right] r_{o3}, \qquad \textbf{(A10.4)}$$

which simplifies to

$$R_{o3} \approx (1 + g_{m3}r_{cm3})r_{o3} = (a_{im3} + 2)r_{o3} \qquad \textbf{(A10.5)}$$

when $\widetilde{L}$ is large. Since the common-merge current gain, a_{im}, is typically quite large, the output resistance is very large.

equivalent for the source with R_{CC} just as we did for the series–shunt example in the previous section (*see* Figure 10-28 and the related equations). Making use of that model, we can modify our previous result for A_s to

$$A'_s = \left.\frac{i'_o}{v'_s}\right|_{b=0} = \frac{v'_e}{v'_s}\frac{i'_o}{v'_e} = \left(\frac{R'_i}{R'_i + R'_S}\right)a'\alpha'_o$$

$$= \alpha'_i a'\alpha'_o, \tag{10.175}$$

where we now have

$$\alpha'_i \equiv \frac{v'_e}{v'_s} = \frac{R'_i}{R'_i + R'_S} \tag{10.176}$$

and

$$\alpha'_o = \frac{i'_o}{i'_{o\ s.c.}} = \frac{i'_o}{a'v'_e} = \frac{R'_o}{R'_o + R'_L}. \tag{10.177}$$

Using (10.175), we have the closed-loop gain;

$$A'_{sf} = \frac{A'_s}{1 + A'_s b}. \tag{10.178}$$

Because (10.178) has the same form as (10.140), and the bandwidth from v'_s to v_o is the same as from v_s to v_o,[11] the closed-loop bandwidth is again given by (10.153), but with α'_i and α'_o given by (10.176) and (10.177).

To find R_{of}, consider Figure 10-34, which shows the output portion of Figure 10-33(a) with R'_L separated into R_{O3} and R_L, with r_{o3} included. The resistance seen from R'_L is R_{o3}. This resistance is infinite if r_{o3} is not included, but it is still very large even when this resistor is included, as noted in Aside A10.5. From Figure 10-34, we see that R_{of} is equal to R_{o3} in parallel with R_{O3}. Since R_{o3} is infinite if r_{o3} is not included, we obtain $R_{of} = R_{O3}$ in that case. It is not worth including r_{o3} in the analysis, since we know that even though R_{o3} will not turn out to be infinite, it will be quite large, and we will still have R_{of} approximately equal to R_{O3}.

The input resistance of the series–series triple is the same as we obtained for the series–shunt amplifier. R_{CC} shows up in parallel with the resistance seen looking into the forward amplifier (with feedback applied), so we obtain

$$R_{if} = R_{CC}\|[R'_i(1 + A_i b)], \tag{10.179}$$

where

$$A_i = a'\left(\frac{R'_o}{R'_o + R'_L}\right) = a'\alpha'_o \tag{10.180}$$

as before.

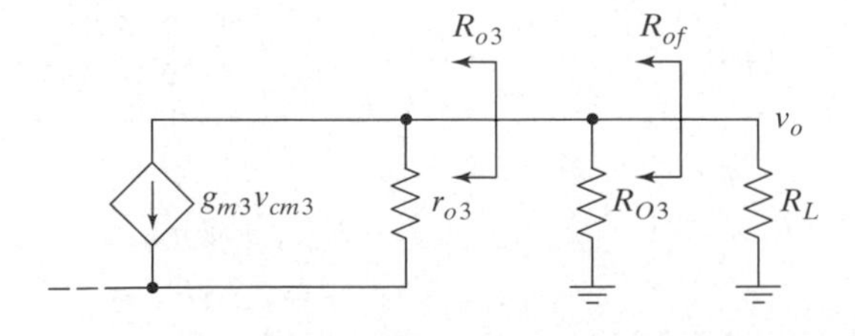

Figure 10–34 The output of the network in Figure 10-33(a) with R'_L separated into R_{O3} and R_L.

[11] This statement is true because the Thévenin equivalent is purely resistive.

It is very important to note that since this amplifier is a series–series connection, the output *must* be considered to be i_o' when we are doing the feedback analysis. If v_o is the output we are really interested in, we *must* wait until the final step to convert from i_o' to v_o. Examining Figure 10-33(b) again, we see that $v_o = i_o' R_L'$. Therefore, if we want the gain from v_s to v_o, we have

$$\frac{v_o}{v_s} = \frac{v_o}{i_o'}\frac{i_o'}{v_s'}\frac{v_s'}{v_s} = \frac{v_o}{i_o'}A_{sf}'\frac{v_s'}{v_s} = R_L'A_{sf}'\left(\frac{R_{CC}}{R_{CC}+R_S}\right). \quad \textbf{(10.181)}$$

The gain from v_i to v_o is also found by multiplying A_{if} by R_L'. The units of A_{sf} are conductance (i.e., A/V), so multiplying it by a resistance produces the necessary dimensionless voltage gain (V/V).

Exercise 10.16

Consider the bipolar series–series triple in Figures 10-35 and 10-36. (a) Find the value of A_{sf}. (b) Find the values of R_{if} and R_{of}. Assume that $\beta = 100$.

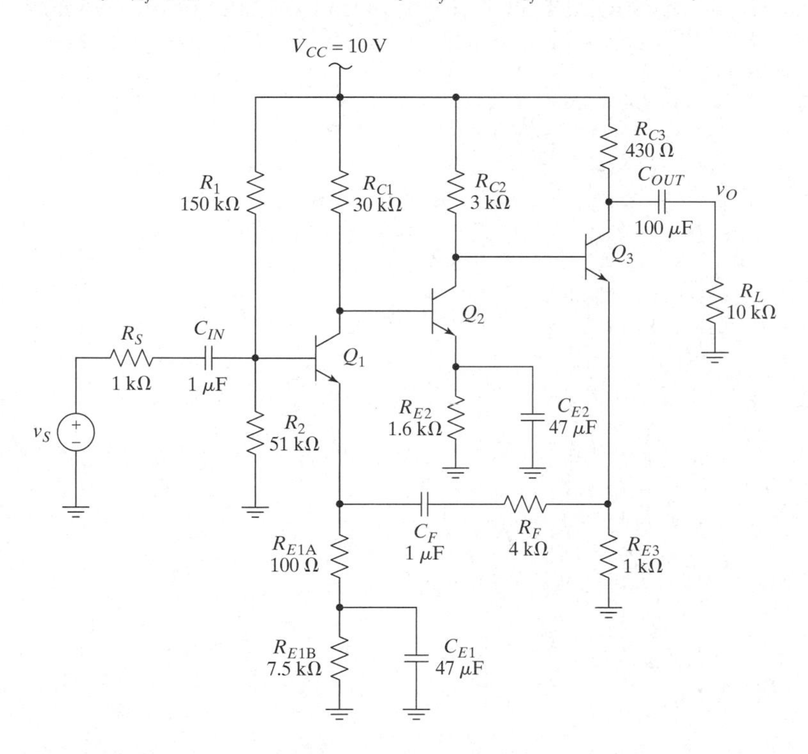

Figure 10–35 A bipolar series–series triple.

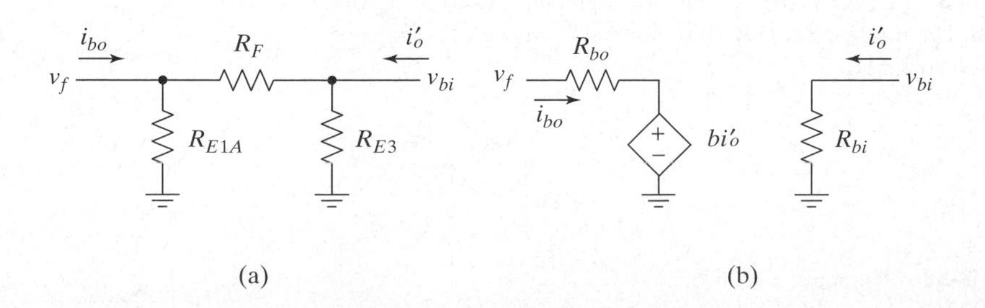

Figure 10–36 (a) The small-signal midband equivalent circuit for the feedback network in Figure 10-35. (b) The unilateral two-port model of the feedback network.

Shunt–Shunt Connection A practical shunt–shunt amplifier is shown in Figure 10-37(a), using two-port models for the forward amplifier and the feedback network. The primed circuit formed by absorbing the feedback network loading into the forward amplifier is shown in part (b) of the figure.

We can redraw the input circuit of Figure 10-37(a) as shown in Figure 10-38. Equating the two-port networks as indicated in Figure 10-37, but using Figure 10-38 to see the input to the first circuit more clearly, we find by inspection that

$$R_i' = \frac{v_i}{i_e'} = R_i \| R_{bo}. \tag{10.182}$$

Using Figure 10-38 again, we see that

$$i_e = \left(\frac{R_{bo}}{R_{bo} + R_i}\right) i_e'. \tag{10.183}$$

We find the gain a' by equating the open-circuit output voltages of the two networks in Figure 10-37. For the circuit in part (a) of the figure, the open-circuit output is (i.e., R_L is removed),

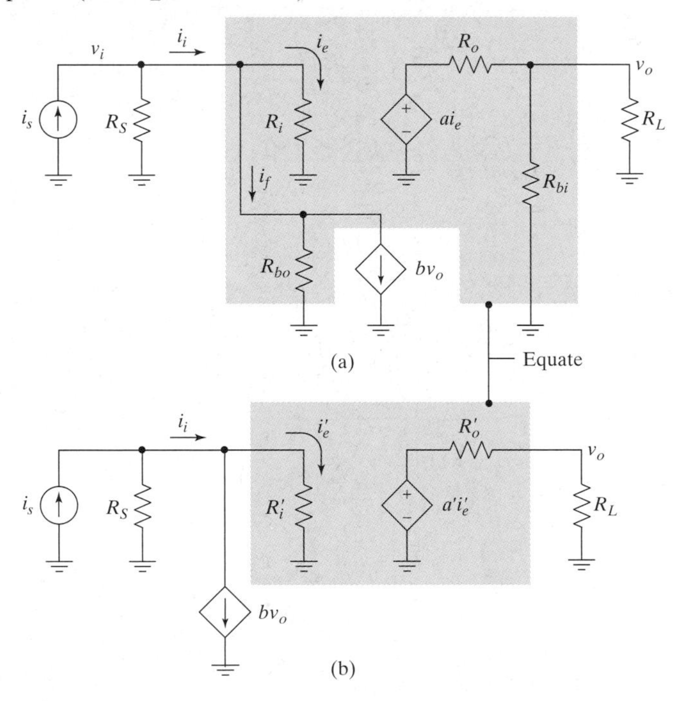

Figure 10–37 (a) A practical shunt–shunt amplifier using two-port models. (b) The model after absorbing feedback network loading into the forward amplifier.

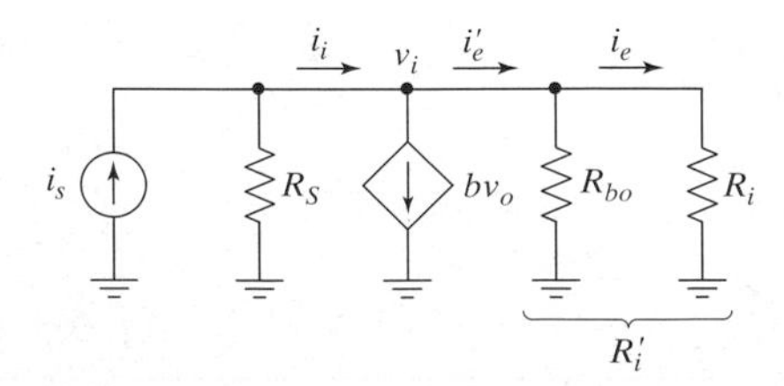

Figure 10–38 The input circuit of Figure 10-37(a) redrawn to make it easier to see R_i' and how the currents divide.

$$v_{o\ o.c.} = \left(\frac{R_{bi}}{R_{bi} + R_o}\right) a i_e. \qquad \textbf{(10.184)}$$

The open-circuit output of the network in part (b) is $a' i'_e$, so equating these outputs yields

$$a' = a\left(\frac{R_{bi}}{R_{bi} + R_o}\right)\frac{i_e}{i'_e} \qquad \textbf{(10.185)}$$

and substituting (10.183) into (10.185) we obtain

$$a' = a\left(\frac{R_{bi}}{R_{bi} + R_o}\right)\left(\frac{R_{bo}}{R_{bo} + R_i}\right). \qquad \textbf{(10.186)}$$

We now find the output resistance of each network. In either case, when i'_e is set to zero, the current-controlled voltage source is also zero. Therefore, looking back into the circuit in part (a), we find an output resistance of $R_o \| R_{bi}$. For the circuit in part (b), the output resistance is R'_o. Setting these equal to each other, we arrive at

$$R'_o = R_o \| R_{bi}. \qquad \textbf{(10.187)}$$

We now know all of the parameters in the primed circuit of Figure 10-37(b), and we can proceed with the analysis. When the loop is opened (i.e., $b = 0$), i'_e is equal to i_i, and the open-loop gain from i_i to v_o is

$$A_i = \left.\frac{v_o}{i_i}\right|_{b=0} = a'\left(\frac{R_L}{R_L + R'_o}\right) = a'\alpha'_o, \qquad \textbf{(10.188)}$$

where we have implicitly defined α'_o in the usual way. With feedback included, we have

$$v_o = a'\alpha'_o i'_e = a'\alpha'_o(i_i - bv_o), \qquad \textbf{(10.189)}$$

which yields

$$A_{if} = \frac{v_o}{i_i} = \frac{a'\alpha'_o}{1 + a'\alpha'_o b} = \frac{A_i}{1 + A_i b}, \qquad \textbf{(10.190)}$$

as expected. The overall open-loop gain is

$$A_s = \left.\frac{v_o}{i_s}\right|_{b=0} = \left(\frac{R_S}{R_S + R'_i}\right) a' \left(\frac{R_L}{R_L + R'_o}\right) = \alpha'_i a' \alpha'_o, \qquad \textbf{(10.191)}$$

with α'_i defined as usual. With the feedback included, we have

$$v_o = a'\alpha'_o i'_e = a'\alpha'_o\alpha'_i(i_s - bv_o), \qquad \textbf{(10.192)}$$

from which we obtain the expected result:

$$A_{sf} = \frac{v_o}{i_s} = \frac{\alpha'_i a'\alpha'_o}{1 + \alpha'_i a'\alpha'_o b} = \frac{A_s}{1 + A_s b}. \qquad \textbf{(10.193)}$$

We now focus on finding the terminal resistances for the closed-loop amplifier. The input resistance with feedback can be found by using KCL at the input node along with Ohm's law to write

$$v_i = (i_i - bv_o)R_i', \tag{10.194}$$

which, when combined with (10.190), yields

$$v_i = i_i(1 - A_{if}b)R_i' = i_i\left(1 - \frac{A_i b}{1 + A_i b}\right)R_i'. \tag{10.195}$$

Finally, we obtain

$$R_{if} = \frac{v_i}{i_i} = \frac{R_i'}{1 + A_i b} = \frac{R_i'}{1 + L_i}. \tag{10.196}$$

To find the output resistance, we look at Figure 10-37(b), and with i_s set equal to zero, we have

$$i_e' = -\alpha_i' b v_o. \tag{10.197}$$

Writing KVL around the output loop and making use of (10.197), we obtain

$$i_o = \frac{v_o - a' i_e'}{R_o'} = \frac{v_o(1 + \alpha_i' a' b)}{R_o'}, \tag{10.198}$$

which yields

$$R_{of} = \frac{v_o}{i_o} = \frac{R_o'}{1 + \alpha_i' a' b} = \frac{R_o'}{1 + L_o}. \tag{10.199}$$

If the forward amplifier has a single-pole low-pass transfer function given by

$$a(j\omega) = \frac{a_0}{1 + j\omega/\omega_p}, \tag{10.200}$$

then we can substitute this into (10.188) and the result into (10.190) to obtain

$$A_{if}(j\omega) = \frac{A_{if0}}{1 + j\omega/\omega_{if}}, \tag{10.201}$$

where

$$A_{if0} = \frac{a_0' \alpha_o'}{1 + a_0' \alpha_o' b}, \tag{10.202}$$

$$\omega_{if} = \omega_p(1 + a_0' \alpha_o' b), \tag{10.203}$$

and

$$a_0' = a_0\left(\frac{R_{bi}}{R_{bi} + R_o}\right)\left(\frac{R_{bo}}{R_{bo} + R_i}\right). \tag{10.204}$$

Exercise 10.17

Derive the closed-loop bandwidth for the transfer function from i_s to v_o for the shunt–shunt amplifier in Figure 10-37, given that the forward gain is described by (10.200).

Example 10.5

Problem:

Consider the shunt–shunt transistor amplifier shown in Figure 10-39. Identify the feedback network and forward amplifier. Then draw the small-signal midband AC equivalent circuit, using the appropriate unilateral two-port model for the feedback network. Derive the primed circuit, redraw the schematic, and use the results of this section to find equations for v_o/v_i, R_{if}, and R_{of}.

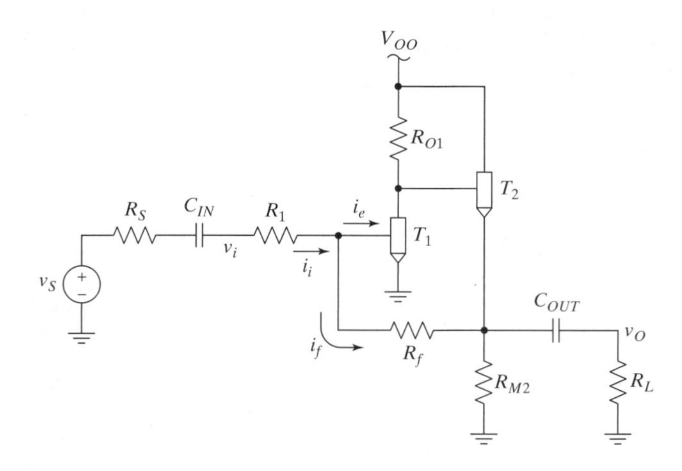

Figure 10–39 A shunt–shunt amplifier using generic transistors.

Solution:

The forward amplifier is composed of T_1, T_2, and R_{O1}. The feedback network comprises R_f and R_{M2}. C_{IN}, C_{OUT}, and R_1 are outside of the feedback loop. The two-port model for the feedback network is shown in Figure 10-40(b), and from that figure we write

$$i_f = \left(\frac{1}{R_{bo}}\right)v_{bo} + bv_o. \tag{10.205}$$

Remembering that the same independent variables must appear in the other port equation, we write

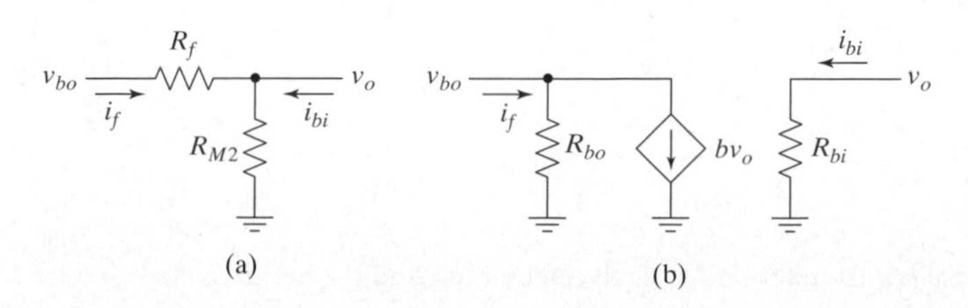

Figure 10–40 (a) The feedback network from Figure 10-39. (b) The unilateral two-port model for the feedback network.

$$i_{bi} = a_r v_{bo} + \left(\frac{1}{R_{bi}}\right) v_o, \qquad \textbf{(10.206)}$$

where a_r is the reverse gain we will ignore. From (10.205), (10.206), and the circuit in Figure 10-40(a) we find

$$b = \left.\frac{i_f}{v_o}\right|_{v_{bo}=0} = \frac{-1}{R_f}, \qquad \textbf{(10.207)}$$

$$R_{bo} = \left.\frac{v_{bo}}{i_f}\right|_{v_o=0} = R_f, \qquad \textbf{(10.208)}$$

and

$$R_{bi} = \left.\frac{v_o}{i_{bi}}\right|_{v_{bo}=0} = R_f \| R_{M2}. \qquad \textbf{(10.209)}$$

Using these results, we draw the small-signal midband AC equivalent circuit shown in Figure 10-41(a). In this figure, we have also combined R_1 with the source and have used a Norton equivalent because the input connection is shunt. (Therefore, it is currents that are summed.) We have

$$i_s' = \frac{v_s}{R_S + R_1} \qquad \textbf{(10.210)}$$

and

$$R_S' = R_S + R_1. \qquad \textbf{(10.211)}$$

The primed circuit is used to simplify the amplifier and is shown in Figure 10-41(b). By direct analysis, we obtain

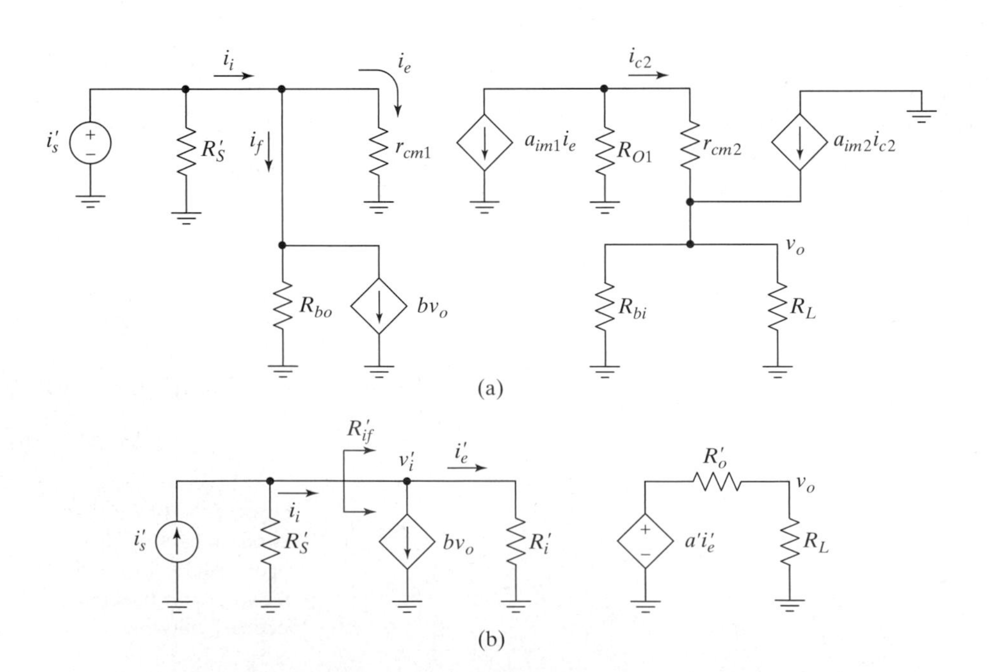

Figure 10–41 (a) The small-signal midband AC equivalent circuit and (b) the simplified circuit.

$$R_i' = r_{cm1} \| R_{bo}, \tag{10.212}$$

$$R_o' = R_{bi} \left\| \left(\frac{r_{cm2} + R_{O1}}{1 + a_{im2}} \right) \right., \tag{10.213}$$

and

$$a' = \frac{v_{o\ o.c.}}{i_e'} = \frac{v_{o\ o.c.}}{i_{c2}} \frac{i_{c2}}{i_e} \frac{i_e}{i_e'}$$

$$= (1 + a_{im2})R_{bi} \left(\frac{-a_{im1}R_{O1}}{R_{O1} + r_{cm2} + (1 + a_{im2})R_{bi}} \right) \left(\frac{R_{bo}}{R_{bo} + r_{cm1}} \right). \tag{10.214}$$

For the input resistance we have

$$R_{if}' = \frac{v_i'}{i_i} = \frac{R_i'}{1 + a'\alpha_o' b} = \frac{R_i'}{1 + L_i}, \tag{10.215}$$

where

$$\alpha_o' = \frac{R_L}{R_L + R_o'}. \tag{10.216}$$

We finally get

$$R_{if} = \frac{v_i}{i_i} = R_{if}' + R_1. \tag{10.217}$$

For R_{of}, we obtain

$$R_{of} = \frac{R_o'}{1 + \alpha_i' a' b} = \frac{R_o'}{1 + L_o}, \tag{10.218}$$

where

$$\alpha_i' = \frac{R_S'}{R_S' + R_i'}. \tag{10.219}$$

For the open-loop gain, we find

$$A_i = \left. \frac{v_o}{i_i} \right|_{b=0} = a'\alpha_o', \tag{10.220}$$

which leads to

$$A_{if} = \frac{v_o}{i_i} = \frac{a'\alpha_o'}{1 + a'\alpha_o' b} = \frac{A_i}{1 + L_i} \tag{10.221}$$

for the closed-loop gain. The gain we desired, however, is the voltage gain. For that we find

$$\frac{v_o}{v_i} = \frac{v_o}{i_i} \frac{i_i}{v_i} = \frac{A_{if}}{R_{if}}. \tag{10.222}$$

It is interesting to note that if $L_i \gg 1$, $v_o/v_i \approx -R_f/R_1$, $R_{if} \approx R_1$, and R_{of} is very small. ■

Exercise 10.18

Consider the bipolar amplifier in Figure 10-42; this circuit is called a Bell Telephone doublet and is a simple wideband amplifier that can be made to have well-controlled input and output impedances. (You can add a resistor in series with the output to set the output resistance.) Find equations and values for v_o/v_i, R_{if}, and R_{of}. Verify your results with SPICE, using the nominal transistor model from Appendix B. Assume that $\beta = 100$.

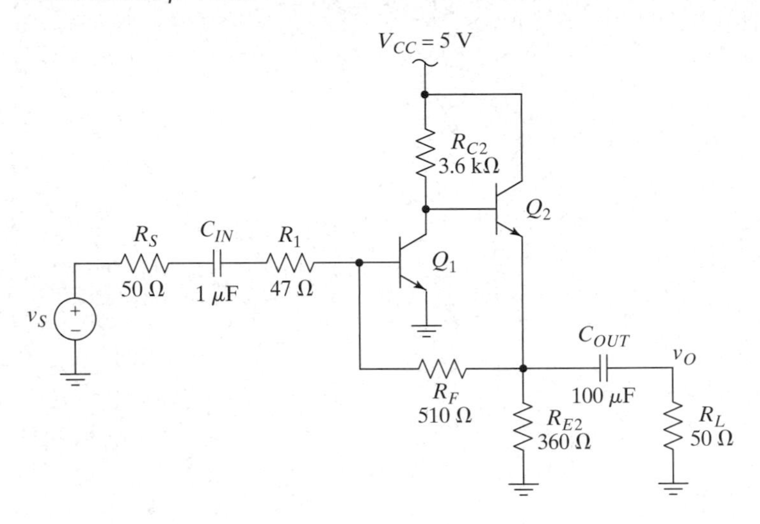

Figure 10–42 A bipolar shunt–shunt amplifier.

Shunt–Series Connection

Figure 10-43(a) shows a shunt–series amplifier modeled with two-port networks. Part (b) of the figure shows the circuit after absorbing the feedback network loading into the primed network. For this connection, we derive (*see* Problem P10.59)

$$R_i' = R_i \| R_{bo}, \tag{10.223}$$

$$a' = \frac{i_{o\,\text{s.c.}}}{i_e'} = a\left(\frac{R_o}{R_o + R_{bi}}\right)\left(\frac{R_{bo}}{R_{bo} + R_i}\right), \tag{10.224}$$

where $i_{o\,\text{s.c.}}$ is the short-circuit output current and

$$R_o' = R_o + R_{bi}. \tag{10.225}$$

In addition, by following the previous analyses, we find

$$A_i = \left.\frac{i_o}{i_i}\right|_{b=0} = a'\alpha_o' \tag{10.226}$$

and

$$A_s = \left.\frac{i_o}{i_s}\right|_{b=0} = \alpha_i' a' \alpha_o', \tag{10.227}$$

where

$$\alpha_i' = \frac{R_S}{R_S + R_i'} \tag{10.228}$$

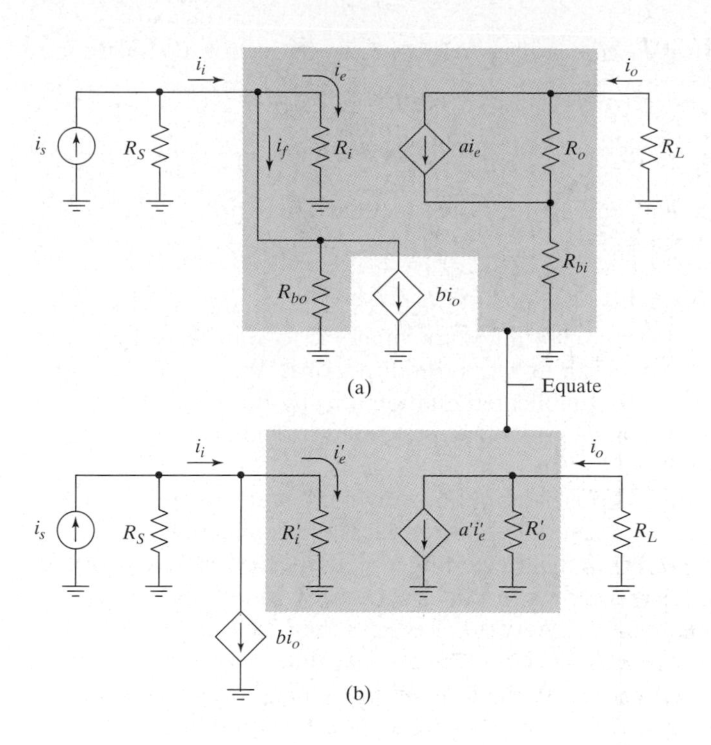

Figure 10–43 (a) A shunt–series amplifier. (b) The primed network.

and

$$\alpha'_o = \frac{R'_o}{R'_o + R_L}. \tag{10.229}$$

The closed-loop gains are

$$A_{if} = \frac{i_o}{i_i} = \frac{A_i}{1 + A_i b} \tag{10.230}$$

and

$$A_{sf} = \frac{i_o}{i_s} = \frac{A_s}{1 + A_s b}. \tag{10.231}$$

The input and output resistances are, predictably,

$$R_{if} = \frac{R'_i}{1 + A_i b} = \frac{R'_i}{1 + L_i} \tag{10.232}$$

and

$$R_{of} = R'_o(1 + \alpha'_i a' b) = R'_o(1 + L_o). \tag{10.233}$$

If the forward amplifier has a single-pole low-pass response, each closed-loop bandwidth is extended by the same factor that the corresponding gain is decreased.

10.1.4 Advanced Analysis

In this section, we briefly consider two additional advanced topics in the analysis of feedback amplifiers. First, we examine how to handle the analyses when the forward amplifier and/or feedback network cannot reasonably be modeled as unilateral networks. Second, we discuss alternative methods for obtaining the loop gain and a common misconception about the equivalence of the loop gain and the return ratio of a controlled source.

Full Two-Port Analysis of Feedback Amplifiers Prior to Section 10.1.3, we idealized feedback amplifiers in three important ways:

1. We used unilateral models for both the forward amplifier and the feedback network.
2. We did not include any loading due to the feedback network.
3. We used ideal sources and loads.

In Section 10.1.3 we removed the last two of these restrictions, and in this section, we want to remove the first one and allow for bilateral forward amplifiers and feedback networks (i.e., signals can flow in both directions). We will find that the analyses are only slightly more complicated mathematically, but they become less intuitive. In addition, the methods previously presented provide acceptably accurate answers a vast majority of the time.

Consider the shunt–shunt feedback amplifier shown in Figure 10-44. We choose to examine a shunt–shunt connection first because it is the connection most likely to have its performance significantly affected by a bilateral feedback network. In Figure 10-44, we have used the full two-port network representations for both the forward amplifier and the feedback network. Because the blocks are now bilateral, the inputs and outputs of each block are not obvious unless you look at the notation used for the voltages present in the loop. In this schematic, a_r and a are the reverse and forward transconductances of the forward amplifier, respectively, whereas b and b_r are the feedback factor and reverse gain of the feedback network, respectively.

Even though this is a shunt–shunt amplifier, and we are therefore interested in v_o, we have used Norton models for the output of the forward amplifier and the input of the feedback network. We did this because these two ports are in parallel and we will be able to add the current sources in parallel. Similarly, we used a Norton model for the input port of the forward amplifier because we will be able to combine it with the feedback generator. We also used the input voltage as the controlling variable for both the forward amplifier and the reverse transmission in the feedback network. Using the voltage seems strange, since a shunt connection at the input dictates that current is what is fed back, but since it is the voltage that is the same across both networks in a shunt connection, using it as the controlling variable allows us to combine the two-port networks, as we will now demonstrate.

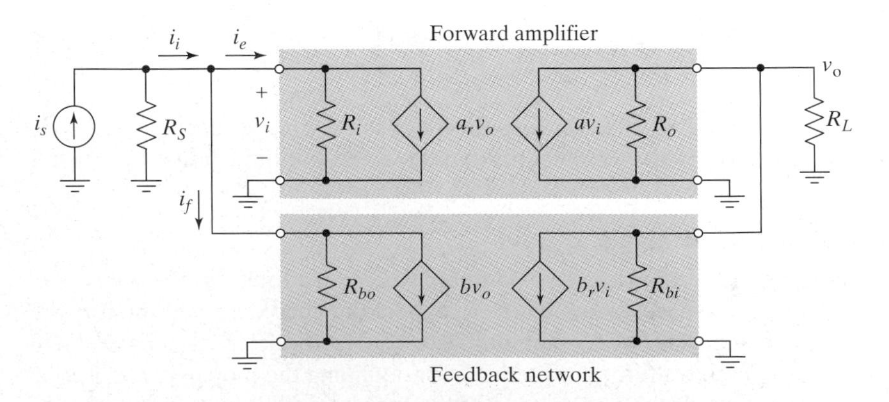

Figure 10–44 A shunt–shunt feedback amplifier.

Figure 10-45 shows the [illegible] of the elements combined. In par[illegible]

(10.234)

and

(10.235)

as before. Then, we combi[illegible] at the input, since they are both [illegible] as the feedback generator, and w[illegible]

(10.236)

We have also combined t[illegible] nd have shown them as the source [illegible] is

(10.237)

At this point, we can see t[illegible] feedback connection. If we convert [illegible] equivalent circuit and change the con[illegible] nt–shunt configuration as shown in [illegible] ply. Comparing Figures 10-45 and [illegible] equal to each other, yielding

(10.238)

and then recognize that v_i [illegible]

(10.239)

In practical shunt–shunt am[illegible] gain of the forward amplifier will hav[illegible] ain of the feedback network can so[illegible] mple will show. In addition, at high f[illegible] r falls, the reverse gain of the feedba[illegible]

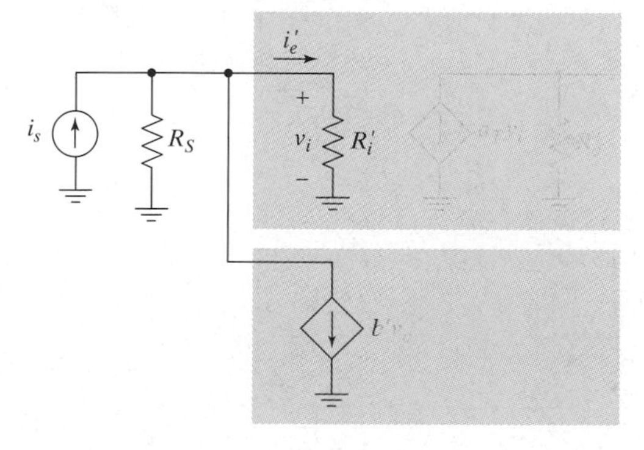

Figure 10–45 [illegible] nplifier from Figure 10-44 [illegible] of the ele-ments combined.

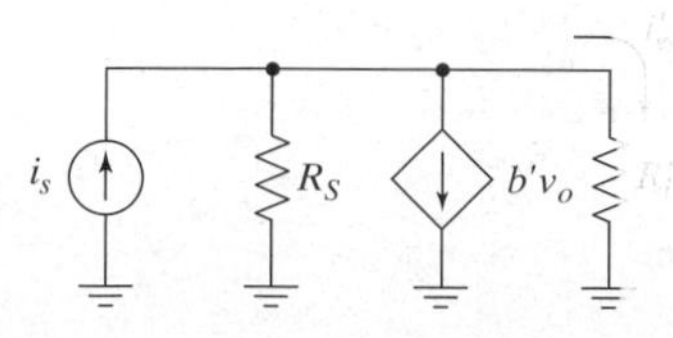

Figure 10–46 [illegible] unt–shunt amplifier [illegible] form.

Example 10.6

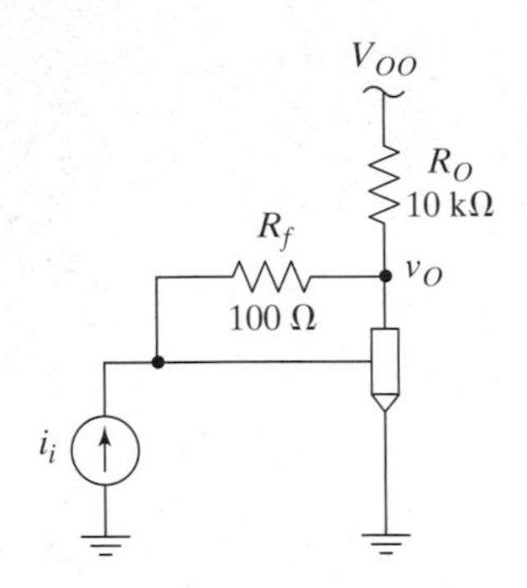

Figure 10–47 A generic shunt–shunt amplifier.

Problem:
Consider the generic amplifier shown in Figure 10-47 where $I_I = 0$. This amplifier uses shunt–shunt feedback, which is usually called local shunt feedback in this configuration. Assume that $g_m = 19.2$ mA/V and $r_{cm} = 5.2$ kΩ. Find the gain, v_o/i_i, (a) using the unilateral model for the feedback network and (b) using a bilateral model for the feedback network. Compare the results. You may assume that the reverse gain of the forward amplifier is negligible.

Solution:
(a) The feedback network in this circuit is just R_f, so if we use the unilateral model shown in Figure 10-48(b), we find that

$$R_{bo} = \left.\frac{v_i}{i_f}\right|_{v_o=0} = R_f,$$

$$R_{bi} = \left.\frac{v_o}{i_{bi}}\right|_{v_i=0} = R_f,$$

and

$$b = \left.\frac{i_f}{v_o}\right|_{v_i=0} = \frac{-1}{R_f}.$$

Using this model for the feedback network and the model for the transistor, we draw the small-signal midband AC model for the amplifier, as shown in Figure 10-49(a). We

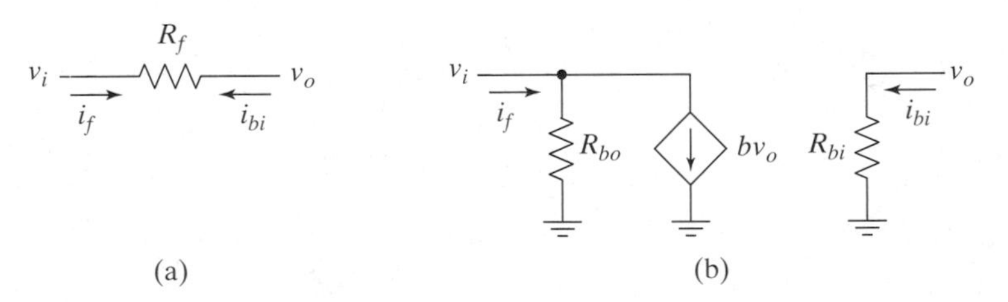

Figure 10–48 (a) The feedback network and (b) the unilateral two-port model.

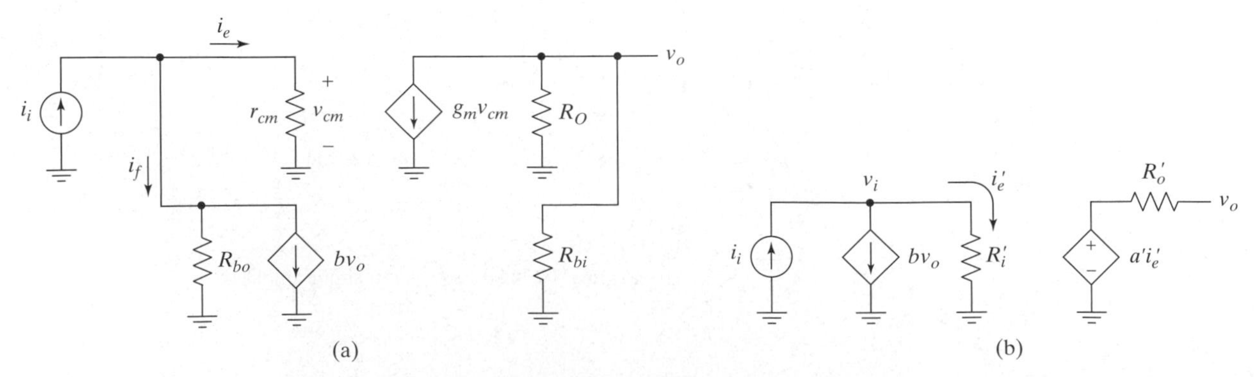

Figure 10–49 (a) The small-signal midband AC equivalent circuit for the amplifier. (b) The circuit after absorbing feedback network loading and conversion of the output to a Thévenin equivalent.

then absorb the feedback network loading and redraw the output circuit, using a Thévenin equivalent as shown in part (b) of the figure. Equating the networks we find

$$R_i' = r_{cm} \| R_{bo},$$
$$R_o' = R_O \| R_{bi}$$

and

$$a' = -g_m R_i' R_o'.$$

Using this figure, we now find that

$$A_i = \left. \frac{v_o}{i_i} \right|_{b=0} = a'$$

and

$$A_{if} = \frac{v_o}{i_i} = \frac{A_i}{1 + A_i b} = \frac{a'}{1 + a'b}.$$

Plugging in the values for this problem yields $A_{if} = -65.2\ \Omega$.

(b) We now use the bilateral two-port model for the feedback network, as shown in Figure 10-50. The component values are the same as those found before, except that we now also have

$$b_r = \left. \frac{i_{bi}}{v_i} \right|_{v_o=0} = \frac{-1}{R_f}.$$

It is not always true that $b = b_r$, as found here.

Using this model for the feedback network and absorbing the feedback loading, we arrive at the model shown in Figure 10-51(a), where

$$a_T = g_m + b_r = g_m - 1/R_f$$

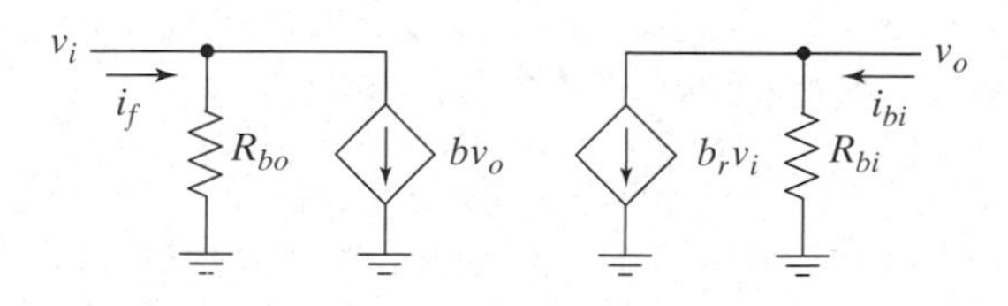

Figure 10–50 The bilateral two-port model for the feedback network.

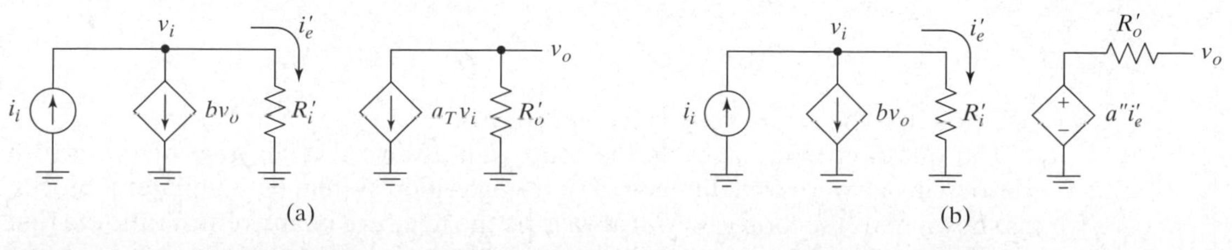

Figure 10–51 (a) The amplifier with feedback loading and reverse gain absorbed. (b) Converting the output to a Thévenin equivalent.

and all of the other components are the same as before. In part (b) of the figure, we have converted the output circuit to a Thévenin equivalent. Equating the two circuits, we find that

$$a'' = -a_T R_i' R_o'.$$

Using this figure, we now obtain

$$A_i = \left.\frac{v_o}{i_i}\right|_{b=0} = a''$$

and

$$A_{if} = \frac{v_o}{i_i} = \frac{A_i}{1 + A_i b} = \frac{a''}{1 + a''b}.$$

Plugging in the values for this problem yields $A_{if} = -47.3\ \Omega$. The answer we got using the unilateral model for the feedback network was 38% higher. A SPICE simulation of this circuit reveals that $A_{if} = -47.6\ \Omega$, very close to the value obtained here. (The difference is a result of slight differences in the model parameter values, since a real transistor model was used.)

■

The other three feedback topologies can also be analyzed with bilateral models for the forward amplifier and feedback network. By using the appropriate choice of two-port network, the bilateral models can always be converted into our standard form, as just illustrated for the shunt–shunt case. A complete treatment of feedback amplifiers using bilateral networks is presented in [10.1].

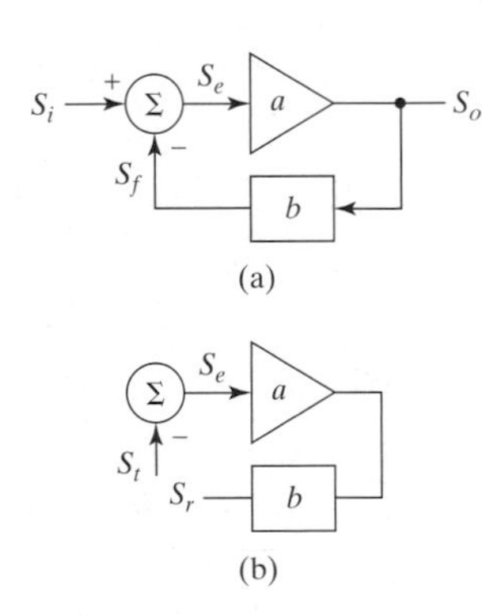

Figure 10–52 (a) Block diagram of negative feedback. (b) Breaking the loop to find the loop gain.

Alternative Methods for Finding the Loop Gain We have repeatedly seen in this chapter that the loop gain is an important quantity in determining the performance of feedback amplifiers. In addition, as we show in Section 10.1.5, the loop gain is the critical factor in determining the stability of a system with negative feedback. Because the loop gain is so important, textbooks often present methods for directly computing the loop gain of a circuit. Unfortunately, there is a serious danger inherent in these methods, which is why we have deferred discussing them to this point. The purpose of this section is to briefly address the issue of how we can directly determine the loop gain of an amplifier.

Consider again the block diagram representation of a negative-feedback amplifier shown in Figure 10-1 and repeated in Figure 10-52(a) for convenience. If we break the loop at any point, as illustrated in part (b) of the figure, set the input, S_i, to zero, inject a test signal, S_t, and examine the signal returned by the loop, S_r, we can define the ***return ratio***,

$$RR = \frac{S_r}{S_t} = -ab. \tag{10.240}$$

We find that the return ratio is the negative of the loop gain in this case.

The question then arises, Is the loop gain always the negative of the return ratio? The answer is a definite no. For a single-loop system with unilateral blocks, as shown here, the loop gain will always be the negative of any return ratio we find (we can break the loop in different places), as long as we maintain the same loading when we break the loop. If the amplifier has more than one feedback loop, which is common, or has bilateral blocks, which is also common, the loop gain is not necessarily the negative of the return ratio. Example 10.7 demonstrates that the loop gain and return ratio can be significantly different.

Example 10.7

Problem:
Consider again the generic shunt–shunt amplifier from Example 10.6. In that example, we found that $A_{if} = a''/(1 + a''b)$, which leads to $L = a''b = (g_m - 1/R_f)(r_{cm}\|R_f)(R_O\|R_f)(1/R_f)$. Now compute a return ratio directly from the small-signal midband AC equivalent circuit without using two-port models, and compare the value obtained with L.

Solution:
The small-signal midband AC equivalent circuit for the amplifier is shown in Figure 10-53(a), and the loop is broken at the output of the controlled current source in part (b) of the figure. Since the loop gain is dimensionless and the circuit is a single-loop amplifier, it doesn't matter whether the return ratio is a current gain or a voltage gain; we will get the same answer (*see* Problem P10.71). We can now write

$$i_r = -g_m v_{cm} = -g_m\left(\frac{r_{cm}}{r_{cm} + R_f}\right)v_o$$
$$= \frac{-g_m r_{cm}}{r_{cm} + R_f}\left[R_O\|(R_f + r_{cm})\right]i_t,$$

from which we obtain

$$RR \equiv \frac{i_r}{i_t} = \frac{-g_m r_{cm}}{r_{cm} + R_f}\left[R_O\|(R_f + r_{cm})\right].$$

Plugging in the values given in Example 10.6, we find that $L = 0.89$ (using the bilateral model for the feedback network) and $RR = -65.2$—a significant difference, indeed!

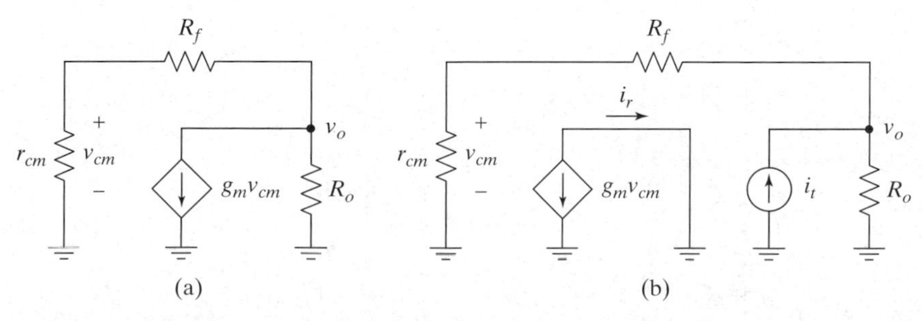

Figure 10–53 (a) The generic shunt–shunt amplifier small-signal midband AC equivalent circuit. (b) Breaking the loop to calculate the return ratio for the controlled source.

The easiest way to avoid the problem encountered in Example 10.7 is to go through the analyses presented in this chapter and find L from the final equivalent circuit, e.g., like the one in Figure 10-51(b). In reality, the work required to compute a return ratio from the original circuit is almost as great as finding the primed network anyway, so the small amount of extra work required to follow the procedure presented here is usually well worth it. It is also possible to do the analysis of feedback amplifiers by using the return ratio concept. A good summary of the relationship between the two analysis techniques is presented in [10.9].

We close this section with a brief discussion of how to use SPICE to determine the loop gain of an amplifier. The simulation of closed-loop systems is often frustrating for students studying circuit design, but it doesn't need to be. The problem is that the loop often needs to remain closed to establish the DC biasing, but we need to open the loop to determine the forward gain, feedback factor, or loop gain. Consider the generic series–shunt amplifier from Figure 10-23, which is repeated in Figure 10-54 for convenience.

If we want to simulate the open-loop gain or the loop gain of this amplifier, we have to find a way to break the loop for signals while maintaining the DC connections so that the biasing is correct. Two ways to accomplish this task are briefly described in the next few paragraphs.

First, we can run a simulation to establish the DC operating point and then use that operating point and some independent sources as shown in Figure 10-55 to break the loop for signals while maintaining both the DC biasing and the AC loading caused by the feedback network. Note in the figure that the independent voltage source called V_{M3} is set equal to the DC voltage appearing at the merge terminal of T_3 in the original circuit. By placing this source where it is, we guarantee that the DC biasing at the merge terminal of T_1 remains the same. Since the small-signal midband AC output resistance of the feedback network is determined by looking into the feedback network with the v_o shorted to ground, using this DC voltage source also establishes an AC ground on the right side of R_f, so that the AC loading presented to the merge terminal of T_1 is the same as in the original circuit. The DC current source I_{M1} is set equal to the DC merge current in T_1 in the original circuit and guarantees that the DC biasing at the merge terminal of T_3 remains the same. In addition, since the small-signal midband AC input

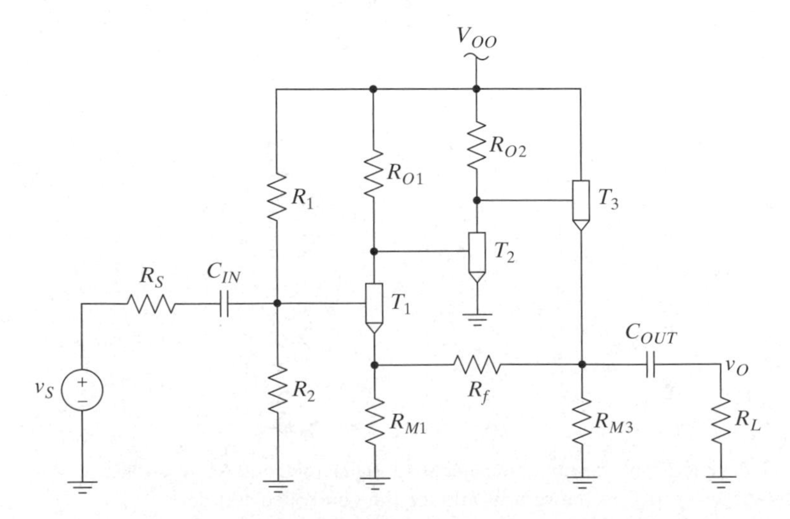

Figure 10–54 A generic series–shunt amplifier.

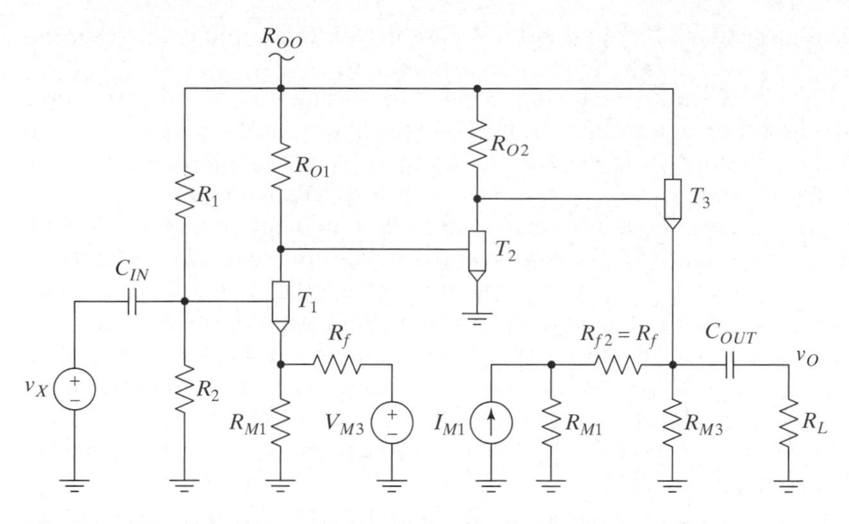

Figure 10–55 Breaking the loop to simulate a series–shunt amplifier.

resistance of the feedback network is found by open-circuiting the feedback network output, and the DC current source is an AC open circuit, the AC loading at the merge terminal of T_3 is the same as in the original network. We use this circuit to find the gain from the control terminal of T_1 to the output (not from the signal source to the output, since the input divider includes R_1 and R_2, which are outside the feedback loop). We use this forward gain as A_i and use the calculated value of b to determine the loop gain. (b is frequency independent here and is easy to find. Consequently, we don't need to simulate it; $b = R_{M1}/(R_{M1} + R_f)$).

A second method for simulating the loop gain is described in detail in [10.10] and is illustrated here by means of the shunt–shunt amplifier shown in Figure 10-47. This method simulates the loop gain directly, rather than just the open-loop forward gain. The procedure has three steps and uses three separate copies of the circuit in SPICE, as shown in Figure 10-56.

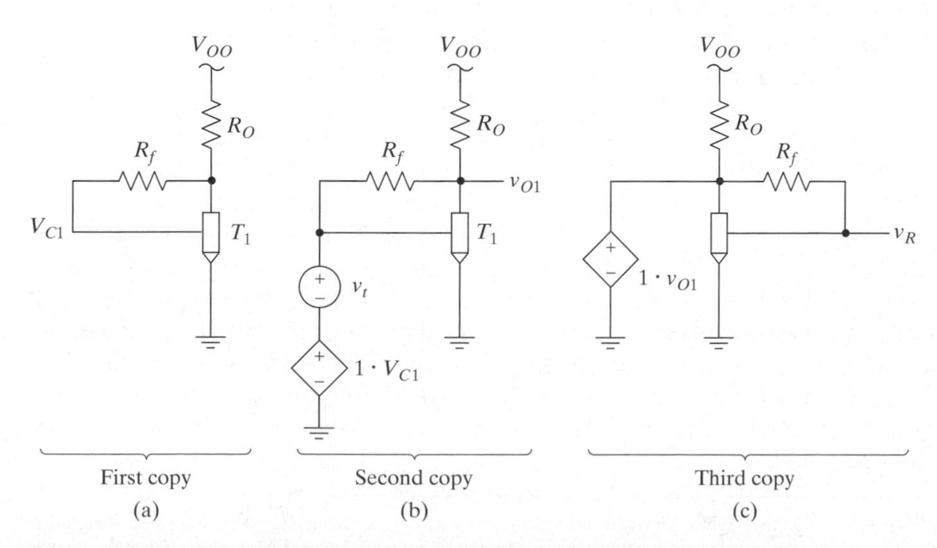

Figure 10–56 Using SPICE to simulate the loop gain of a shunt–shunt amplifier.

First, set the signal source driving the first copy of the amplifier to zero; this copy will be used to establish the DC operating point.

Second, drive the input to the second copy of the circuit with an AC test source and a controlled source, to establish the DC bias. The controlled source is controlled by the first copy of the circuit. If the input is a shunt connection, these sources should be voltage sources in series as shown here. If the input is a series connection, the sources should be current sources in parallel. In addition to supplying the test input and the DC biasing, these sources disable the feedback in the second copy of the circuit. For the example shown here, the feedback is a current. Since the voltage at the input node is set by the voltage sources, the magnitude of the feedback current has no effect. In other words, the feedback is shorted out.

The third step is to take the amplifier output from the second copy of the circuit and use it to drive the output of the third copy. This forces the AC signal and the DC biasing to be correct at the output. We set the original signal source to zero in this third copy of the circuit (i.e., a short for a voltage source and an open for a current source), and we then measure the signal returned. In this case, the test input was a voltage, and we measure the voltage returned. For a series input connection, we would have forced a current, and we would measure the current through the shorted signal source as the return signal. The return ratio is given by $RR = v_r/v_t$ here; for a series input connection, it would be $RR = i_r/i_t$.

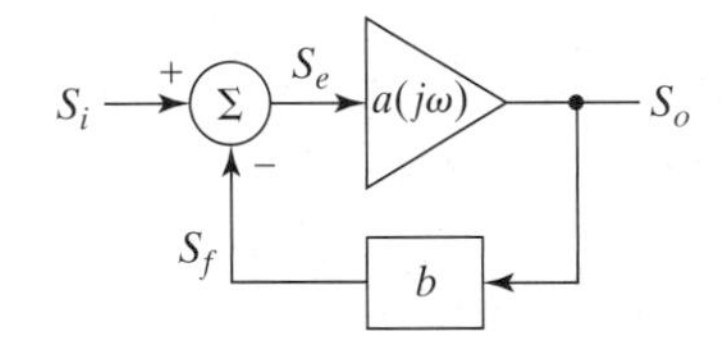

Figure 10–57 An ideal feedback amplifier.

10.1.5 Stability of Systems with Feedback

Up to this point, we have ignored the phase shifts that might accumulate as a signal propagates around a feedback loop. These phase shifts pose a potentially serious problem, however. For example, consider the ideal feedback amplifier shown in Figure 10-57, where the forward amplifier's gain is a function of frequency and the feedback factor is constant.

Given the frequency-dependent gain of the forward amplifier, the transfer function for the closed-loop amplifier is

$$A_f(j\omega) = \frac{S_o(j\omega)}{S_i(j\omega)} = \frac{a(j\omega)}{1 + a(j\omega)b}. \tag{10.241}$$

Suppose that the total phase shift seen in the forward amplifier is exactly 180° for a sinusoidal signal at frequency ω_{180}; in other words, a sinusoid at this frequency is inverted at the output of the forward amplifier. In this case, the forward gain at ω_{180} is actually negative. If the magnitude of the loop gain is exactly unity at the frequency ω_{180}, the denominator in (10.241) is equal to zero and the closed-loop gain goes to infinity.

The practical implication of having the gain go to infinity is that an output can exist even when the input is zero. If a signal at the frequency ω_{180} somehow gets started in the loop in Figure 10-57, it will sustain itself, even with S_i equal to zero, since the gain around the loop is exactly unity and the 180° phase shift in $a(j\omega)$ cancels the minus sign at the summer. When a circuit can produce a signal without any input (other than the DC biasing providing the power), we say that it is ***oscillating***. We do sometimes want to design oscillators, but if we are trying to design an amplifier, oscillations need to be avoided. Any amplifier that can oscillate is called ***unstable***.[12]

---

[12] We are now implicitly using the word *stable* in a new way. We previously discussed how to design amplifiers with stable operating points. In that context, the word *stable* meant that the operating point did not change with temperature or other variables. In the present context, the word *stable* implies that a circuit cannot have a sustained oscillation.

One frequently asked question is, If we know in advance that our input signal will never contain a component at the frequency ω_{180}, is the potential oscillation still a problem? The short answer is a very definite *yes*. In all real electronic systems there is always noise present, so this self-sustaining oscillation may get started whether or not we deliberately introduce an input at that frequency. The terminology used is that the noise *excites* the instability. In addition to noise, transient signals, such as occur when equipment is turned on, also may excite the oscillation. Therefore, it is never reasonable or safe to assume that an unstable amplifier will not oscillate.

What if the magnitude of the loop gain is greater than unity at the frequency ω_{180}? From (10.241), we see that the closed-loop gain no longer goes to infinity, so is the amplifier still unstable? The answer is usually—but not always—yes. If the phase of the loop gain is a monotonic function of frequency (i.e., it either continues to get more negative or stays the same as the frequency increases), a loop with gain greater than unity at ω_{180} will be unstable. If the phase is not monotonic, we need to know more before we can say whether or not the loop is unstable. We address this issue again after we have introduced gain and phase margins.

To see why the answer is usually yes, consider the loop in Figure 10-57, and assume that the phase is monotonic and the loop gain is greater than unity at ω_{180}. The resulting system has poles in the right-half of the complex plane, and, as a result, the natural response contains an exponentially growing function of time. Now imagine that a small sine wave at frequency ω_{180} has just been introduced into the loop. The sine wave will grow exponentially with time; but in practice there will be some nonlinearity in the loop that will limit the magnitude of the signal. At this point the loop will continue to oscillate, but with the amplitude of the oscillation limited by the nonlinearity. An example is shown in Figure 10-58; part (a) shows the nonlinearity of the forward amplifier and part (b) shows the resulting waveform.

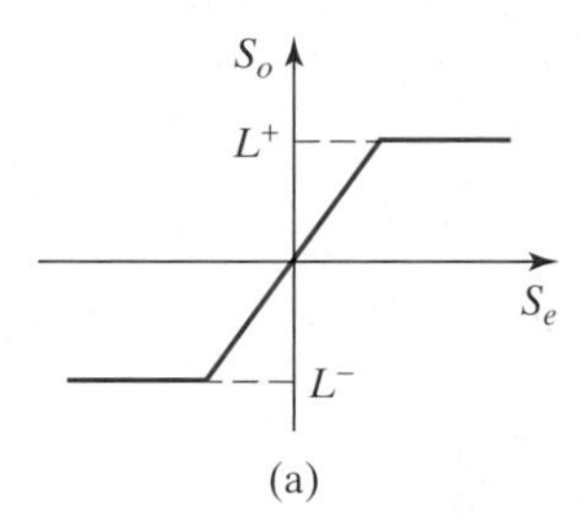

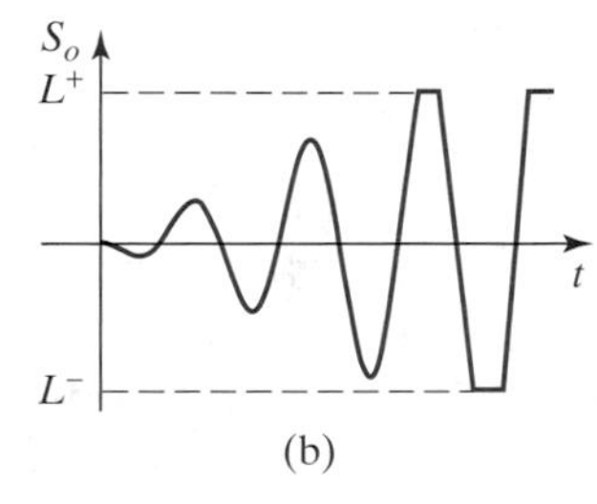

Figure 10–58 (a) A possible transfer characteristic for the forward amplifier in Figure 10-57. (b) The resulting output if the loop gain is slightly greater than one at the frequency ω_{180}, and an oscillation is started at $t = 0$.

Finally, what if the loop gain is slightly less than unity at the frequency ω_{180}? Does this pose a problem? In this case the system is ***unconditionally stable***, meaning that with no input applied, there cannot be any steady-state output. Nevertheless, if something excites the loop at the frequency ω_{180} (e.g., a sudden change in input or supply voltage), there may be a very slowly decaying oscillation. Although it will eventually decay away, it may take a long time if the loop gain is very close to unity. In this case, our amplifier is still not usable. Systems that exhibit decaying oscillatory responses are stable, but we still refer to their ***relative stability***. We will see in the next section how we quantify the degree of relative stability in a given amplifier. But first, let's summarize the conditions necessary for instability to occur.

If the following two conditions are met, a negative-feedback amplifier may be unstable:[13]

> The total phase shift around the loop is 180° at frequency ω_{180} (i.e., $\angle L(j\omega_{180}) = 180°$).
>
> The magnitude of the loop gain at ω_{180} is greater than or equal to unity (i.e., $|L(j\omega)| \geq 1$).

[13] These rules apply only when the forward amplifier and feedback network are both *minimum-phase* networks (i.e., networks that only have poles in the left-half of the complex plane and zeros in either the left-half of the plane or on the imaginary axis). All the amplifiers we have studied and most feedback networks of practical importance are minimum phase or approximately minimum phase.

These conditions together constitute a simplified version of what is called the ***Nyquist stability criterion***. Remember that if the conditions are met, an amplifier may oscillate at this frequency, regardless of whether or not an input signal is applied at that frequency.

Gain and Phase Margins In this section, we present quantitative measures for how close an amplifier is to being unstable. In other words, we want to know how much margin for error we have in the design; if something changes a bit from our desired value, will the amplifier still be stable? There are two measures commonly used: the ***gain margin*** and the ***phase margin***.

The gain margin is a measure of how much additional loop gain would be required to cause the amplifier to be unstable and is positive for a stable amplifier. Since the amplifier is unstable if the loop gain is unity (i.e., 0 dB) at ω_{180}, the gain margin (in dB) is defined by

$$GM = -|L(j\omega_{180})|_{\text{in dB}}. \tag{10.242}$$

Note that the gain margin is not defined if the phase of the loop gain never reaches $-180°$, so first- and second-order systems do not have a gain margin (i.e., these systems are unconditionally stable—the gain can have any value).

The phase margin is similarly defined; it is the extra phase shift that would need to be added to the loop to make the total phase shift equal $-180°$ at the frequency where the magnitude of the loop gain is unity. The phase margin is positive for a stable amplifier. We call the frequency where the loop gain is unity the ***unity-gain frequency*** and denote it by ω_u. (*Note*: ω_u is *not* where the closed-loop or open-loop gain goes to one, but where the *loop* gain goes to unity.) Since the phase of the loop gain is negative, the difference between it and $-180°$, which is the phase margin, is

$$PM = \angle L(j\omega_u) + 180°. \tag{10.243}$$

We can make use of Bode plots to find the gain and phase margins of an amplifier, as shown in Figure 10-59. In this figure, we have plotted the asymptotic approximation[14] to the magnitude of the open-loop gain of a forward amplifier with the transfer function

$$a(j\omega) = \frac{10^5}{(1 + j\omega/10^2)(1 + j\omega/10^5)(1 + j\omega/10^7)}. \tag{10.244}$$

We have also plotted the reciprocal of the magnitude of the feedback factor, $1/|b|$, and have assumed that it is independent of frequency and equal to 50 dB (i.e., $|b| = -50$ dB). The phase plot is also an asymptotic approximation, and the phase is the sum of the phases of $a(j\omega)$ and b, which is equal to the phase of $a(j\omega)$ in this case. The magnitude of the loop gain expressed in dB is given by

$$\begin{aligned} |L(j\omega)|_{\text{in dB}} &= 20\log(|a(j\omega)| \, \| \, b(j\omega)|) \\ &= 20\log|a(j\omega)| + 20\log|b(j\omega)| \\ &= 20\log|a(j\omega)| - 20\log\left(\frac{1}{|b(j\omega)|}\right), \end{aligned} \tag{10.245}$$

[14] If you are unfamiliar with Bode plots, see Aside D.1 in Appendix D on the CD.

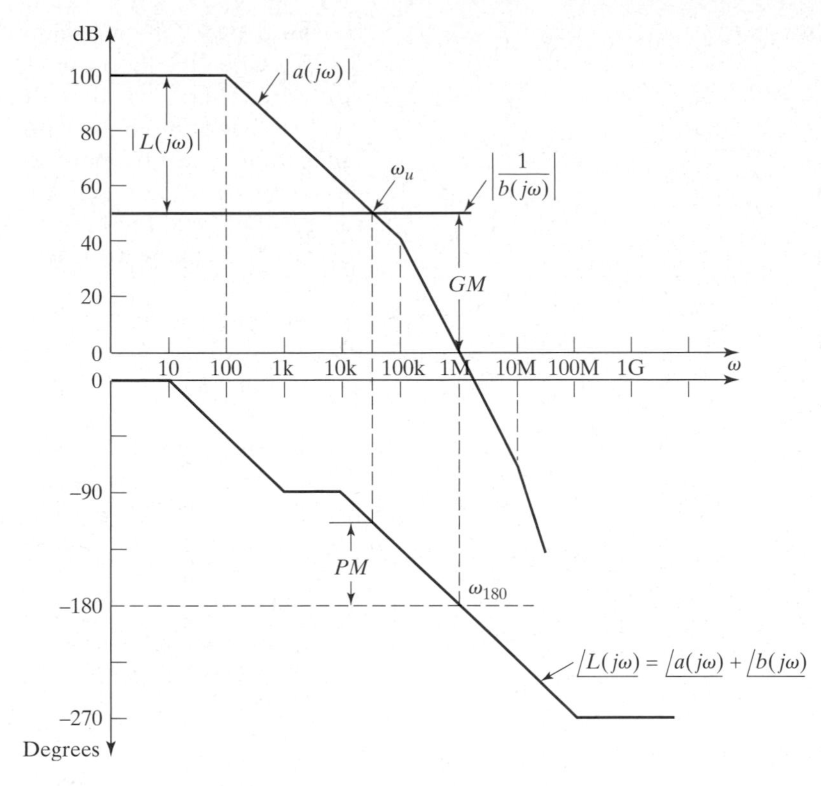

Figure 10–59 The Bode plot for a feedback amplifier with $a(j\omega)$ given by (10.244) and $b =$ −50 dB.

so it is the difference between the plot of the open-loop gain magnitude and the magnitude of the reciprocal of the feedback factor as shown in the figure. Therefore, the frequency, ω_u, which is where the loop gain goes to unity (i.e., 0 dB), is given by the intersection of the two magnitude plots as shown. The phase margin is then the difference between the phase of the loop gain and −180°, as indicated. The phase margin is positive in this case, so the system is stable. The gain margin is found from the plot by locating ω_{180}, the frequency where $\angle L(j\omega) = -180^\circ$, and then measuring the loop gain at that frequency as shown. In this case the loop gain is negative, so the gain margin is positive, indicating a stable system.

If the loop gain has several poles and zeros within some frequency band, then it is possible for the phase to be nonmonotonic. If the phase is nonmonotonic, there may be more than one gain and phase margin to check. In this case, you may need to use a more sophisticated technique like a Nyquist plot [10.2], since *it is possible for the system to be stable* even if one or more of the gain and phase margin pairs indicate otherwise. However, you are safe if all of the gain and phase margin pairs indicate that the system is stable.

If we consider a low-pass transfer function with monotonic phase, as we have here, there will only be one phase margin and one gain margin, and they will both be either positive or negative. It is also possible for a forward amplifier with a bandpass characteristic to be unstable for frequencies below the low cutoff, but this situation is not usually seen.

Notice [illegible] Figure 10-59, the gain and phase ma[illegible]r that when the loop gain is large, the [illegible]tely $1/b$. Therefore, we have the most tr[illegible]ve-feedback amplifiers with very *low* clo[illegible]nterintuitive at first, but it is completely [illegible]uce the closed-loop gain, we must increas[illegible]nce it is the signal fed back to the input tha[illegible]ems are more likely with larger feedback [illegible]

Since it is [illegible]lems, and we wouldn't usually use an ampli[illegible] internally compensated operational amplifi[illegible] ***unity-gain stable***—that is, stable for any closed-loop ga[illegible]

The Bode p[illegible]and gain margins works equally well when b is [illegible]iculty encountered is to remember that the n[illegible]n b leads to a zero in $1/b$. Conversely, you m[illegible]f b, not $1/b$, to the phase of a to produce the ph[illegible]

We finish this [illegible]sign values for the gain and phase margins are in t[illegible]5°, respectively. These choices lead to reasonably w[illegible]de sufficient room for component variation under [illegible]

Exercise 10.19

What is the minimum [illegible]hich the amplifier shown in Figure 10-59 is stable?

□

Closed-Loop Pole Locations and Transient Response In this section, we take a closer look at systems that may have poor relative stability. The transient response and frequency response are directly related for an LTI system, but the practical consequences of having a system that has poor relative stability are sometimes easier to spot in the time domain. Therefore, we will examine the transient response of these systems. Our focus will be on the step response, since this is a common practice in both circuit and control system design and yields a great deal of insight and information about the system.

Since the phase of the loop gain must be −180° for a negative-feedback amplifier to be unstable, it is not possible for systems with only one or two poles to be unstable. In fact, systems with only one pole are not only unconditionally stable, but always have a phase margin of at least 90°, which translates into their having very well behaved responses. The step response of a system with only one pole never overshoots the final value, but simply approaches it with the well-known RC step response, as shown for a single-pole RC low-pass filter in Figure 10-60. The capacitor current in part (a) of the figure is

$$i(t) = C\frac{dv_o(t)}{dt}. \tag{10.246}$$

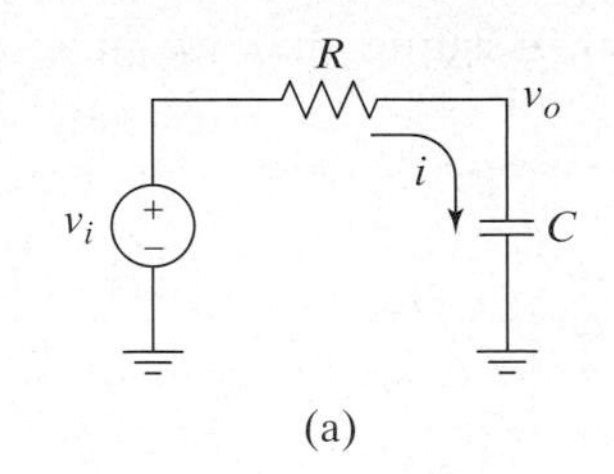

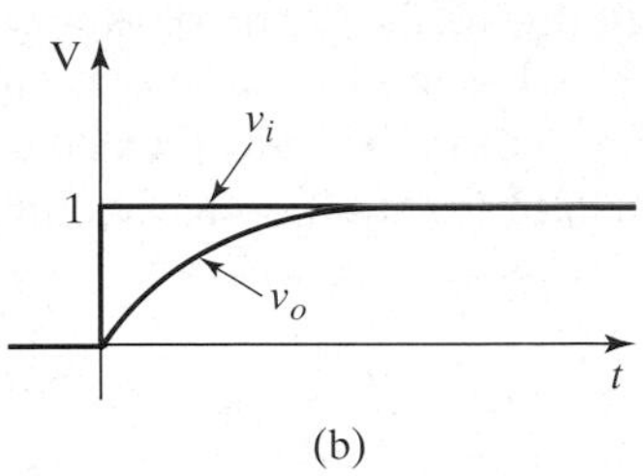

Figure 10–60 (a) An RC low-pass filter and (b) the step response.

This current can also be expressed as

$$i(t) = \frac{v_i(t) - v_o(t)}{R}, \tag{10.247}$$

and combining these results, we arrive at the differential equation

$$v_i(t) = v_o(t) + RC\frac{dv_o(t)}{dt}. \tag{10.248}$$

If the input voltage is a unit step function as shown in part (b) of the figure, the solution to this differential equation for $t \geq 0$ is

$$v_o(t) = 1 - e^{-t/RC}. \tag{10.249}$$

This output is also shown in the figure; note that there isn't any overshoot or ringing present. The amount of time it takes the output to settle to within 5% of its final value is found by setting (10.249) equal to 0.95 and solving for t. The result is

$$t_{5\%} = -RC\ln(0.05) = 3RC. \tag{10.250}$$

Although the response is well behaved, the settling time given by (10.250) is longer than can be obtained with a second-order system.

If an amplifier has a single-pole low-pass response, its step response will also be given by (10.249), and the time constant (RC in the equation) will be the reciprocal of the pole frequency. If we connect this single-pole amplifier in a negative-feedback loop, we already know that the pole frequency will be increased by one plus the loop gain. This increase in bandwidth is equivalent to reducing the time constant in the step response by a factor of one plus the loop gain. Therefore, the closed-loop system will be faster, but it will still behave in the same manner; in other words, the form of the step response would still be given by (10.249).

We now move on to consider second-order systems. Assume that we have a voltage amplifier with a transfer function given by

$$\frac{V_o(s)}{V_e(s)} = a(s) = \frac{a_0\omega_n^2}{s^2 + 2\zeta\omega_n s + \omega_n^2}. \tag{10.251}$$

The form used for the denominator here is common in the field of control theory [10.2]. The frequency ω_n is the ***undamped natural frequency***, and the coefficient ζ is called the ***damping coefficient***. If $\zeta = 0$, the amplifier poles are a complex conjugate pair on the imaginary axis, and the step response of the amplifier (i.e., the output that is produced by a unit step at the input) is given by

$$v_o(t) = 1 - \cos\omega_n t \qquad \text{for } t \geq 0. \tag{10.252}$$

This equation shows that with no damping (i.e., with $\zeta = 0$), the step response oscillates forever at the undamped natural frequency (hence the name). If $0 < \zeta < 1$, the system is ***underdamped***. In that case, the poles are a conjugate pair in the left half of the complex plane, and the step response is given by

$$v_o(t) = 1 - \frac{e^{-\zeta\omega_n t}}{\sqrt{1 - \zeta^2}}\sin\left(\omega_d t + \tan^{-1}\frac{\sqrt{1 - \zeta^2}}{\zeta}\right) \text{ for } t \geq 0, \tag{10.253}$$

where

$$\omega_d = \omega_n\sqrt{1 - \zeta^2} \tag{10.254}$$

is called the ***damped natural frequency***. In this case, the step response has ringing at the frequency ω_d, which is less than ω_n, but the ringing does eventually decay away. The smaller ζ is, the longer the ringing takes to decay.

If $\zeta = 1$, the system is ***critically damped***, the poles are identical and real, and the step response is

$$v_o(t) = 1 - e^{-\omega_n t}(1 + \omega_n t) \qquad \text{for } t \geq 0, \tag{10.255}$$

which does not exhibit any overshoot or ringing.

For values of ζ greater than unity, the system is ***overdamped*** and the poles are distinct and real. As the value of ζ increases, the separation between the poles increases. When ζ is very nearly unity, the overdamped system can be approximated by (10.255), while for values of ζ larger than about 1.5, the system can be approximated by the single-pole step response of (10.249) with RC replaced by $[\omega_n(\zeta - \sqrt{\zeta^2 - 1})]^{-1}$.

The locations of the poles in the complex plane are shown in Figure 10-61(a) for different values of ζ, and sample step responses are shown in part (b) of the figure. The settling time and rise time of a second-order system can be significantly better than that of a single-pole system, so this type of response is sometimes desirable. We must avoid letting the overshoot and ringing get too severe, however. Therefore, it is important to know what happens to the response when the amplifier described by (10.251) is used in a closed-loop situation.

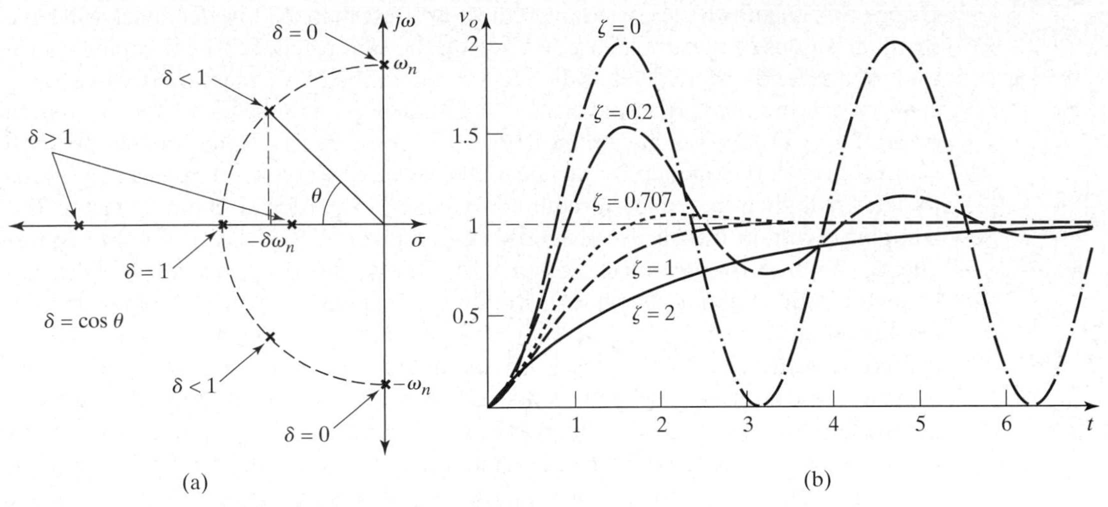

Figure 10–61 (a) The positions of the poles of the second-order system described by (10.251) as a function of ζ. (b) Sample step responses.

When the amplifier described by (10.251) is connected in a negative-feedback loop with a constant feedback factor b, the closed-loop gain is

$$A_f(s) = \frac{\dfrac{a_0\omega_n^2}{s^2 + 2\zeta\omega_n s + \omega_n^2}}{1 + \dfrac{a_0\omega_n^2 b}{s^2 + 2\zeta\omega_n s + \omega_n^2}} = \frac{a_0\omega_n^2}{s^2 + 2\zeta\omega_n s + \omega_n^2 + a_0\omega_n^2 b}, \quad \textbf{(10.256)}$$

and writing this in the standard form, we obtain

$$A_f(s) = \frac{A_{f0}\omega_{nf}^2}{s^2 + 2\zeta_f\omega_{nf}s + \omega_{nf}^2}, \quad \textbf{(10.257)}$$

where

$$A_{f0} = \frac{a_0}{1 + a_0 b}, \quad \textbf{(10.258)}$$

as expected,

$$\omega_{nf} = \omega_n\sqrt{1 + a_0 b}, \quad \textbf{(10.259)}$$

and

$$\zeta_f = \frac{\zeta}{\sqrt{1 + a_0 b}}. \quad \textbf{(10.260)}$$

Using (10.259) and (10.260), we can predict the effect that negative feedback will have on the positions of the amplifier poles and on the step response. The damping coefficient of a second-order system is directly related to the phase margin discussed earlier. A plot of the phase margin versus the closed-loop damping factor for the system described by (10.257) is shown in Figure 10-62 for different values of the DC loop gain, a_0b. The curves are almost identical for reasonably large loop gains and allow us to determine the phase margin required to achieve a particular damping factor. The damping factor, in turn, is usually chosen to achieve a desired step response or frequency response. A common choice for ζ_f is 0.707, which yields a reasonably fast step response with minimal overshoot, virtually no ringing, and a good phase margin (around 65°).

Predicting the effect of feedback on amplifiers with more than two poles is extremely difficult in general, and computer-aided methods would typically be used. Some electronic systems can be approximated as either first- or second-order systems in order to gain some reasonable insight into their operation and stability, but because the phase of higher order poles can rapidly accumulate and degrade the phase margin, it is best to use computer-aided techniques for the final analysis of most systems.

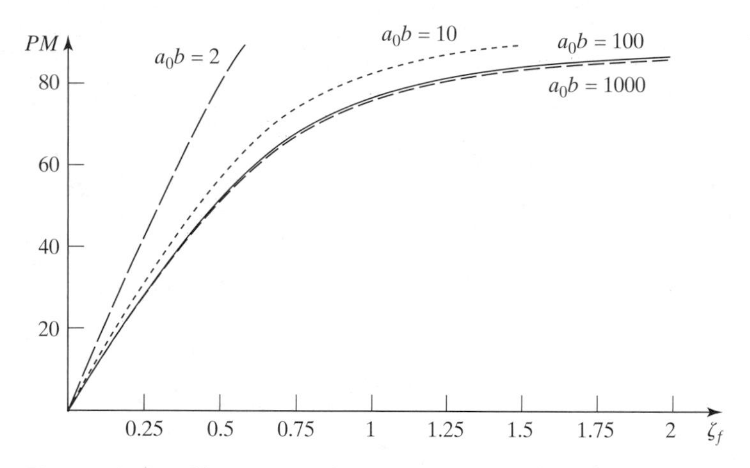

Figure 10–62 Phase margin (in degrees) versus closed-loop damping factor for the system described by (10.257).

Example 10.8

Problem:
Use SPICE to find the closed-loop step response and phase margin for a feedback amplifier described by (10.257) if the forward amplifier has $a_0 = 2$, $\zeta\zeta = 0.707$, and $\omega_n = 2$ rad/s, and the feedback factor is equal to unity (i.e., $b = 1$).

Solution:
With PSPICE, the **ELAPLACE** block can be used to describe the forward amplifier transfer function as in (10.251). The resulting schematic is shown in Figure 10-63. (*Note*: The summer block cannot be changed to subtract the feedback term; therefore, a gain of −1 is inserted into the feedback path.)

The transient simulation result is shown in Figure 10-64.

Using (10.259) and (10.260) with the given data we calculate $\zeta_n = 0.4$ and $\omega_{fn} = 3.46$ rad/s, which corresponds to $f_{fn} = 0.55$ Hz. We then plot the step response using (10.253), together with these closed-loop values. The resulting plot is shown in Figure 10-65 and agrees with the SPICE result.

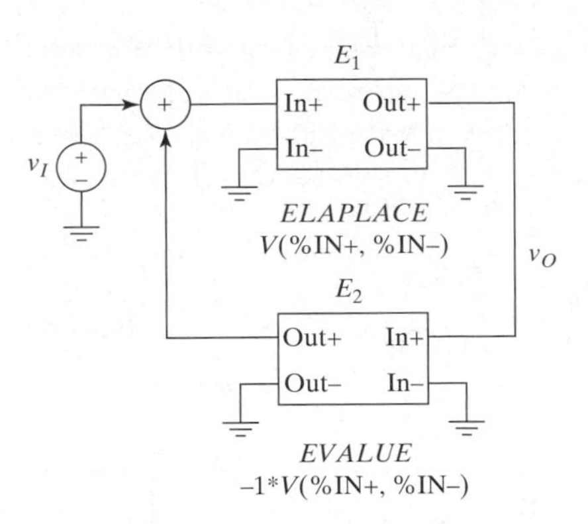

Figure 10–63 PSPICE schematic.

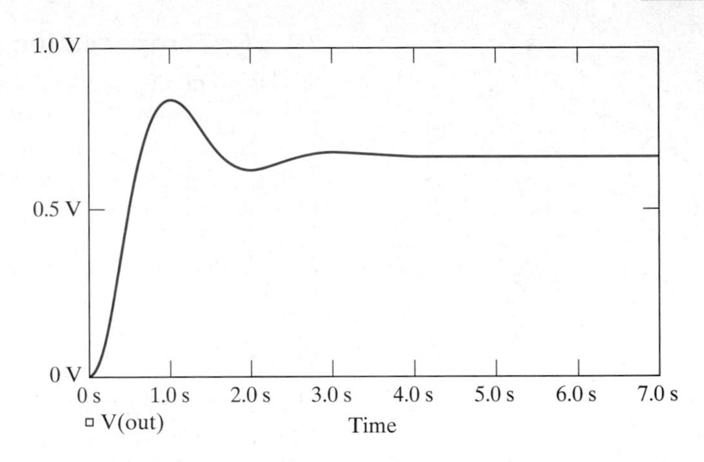

Figure 10–64 Transient simulation result.

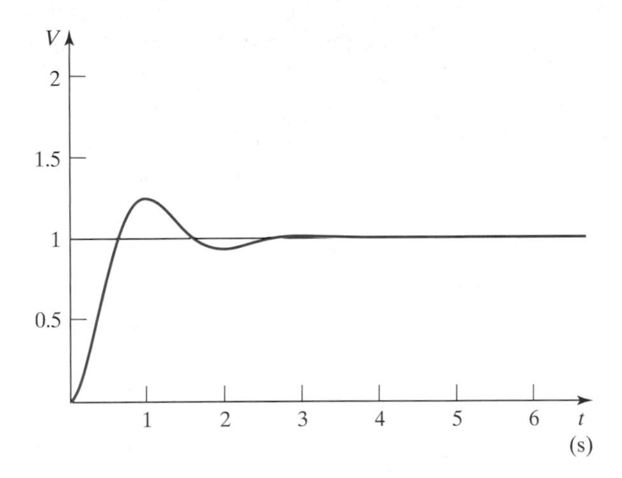

Figure 10–65 Calculated step response.

The PM was found by using SPICE to generate the Bode plots. The PSPICE schematic is shown in Figure 10-66 (Exam8b.sch) and the resulting Bode plots are given in Figure 10-67. The phase margin is seen to be $180 - 111 = 69°$, and, using Figure 10-62 with the DC loop gain of 2, we find that this corresponds to $\zeta_n = 0.4$, as predicted by (10.260).

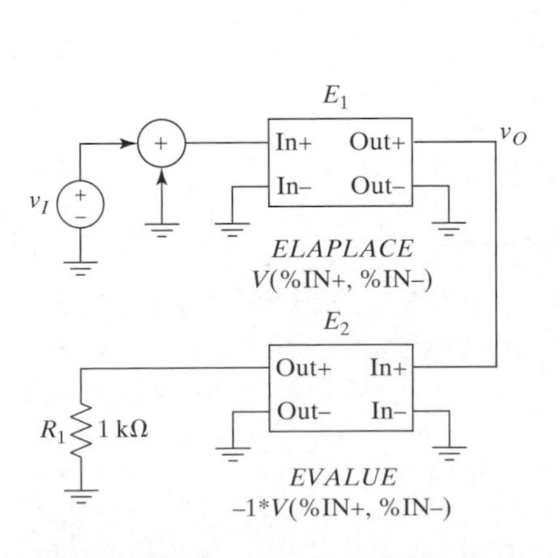

Figure 10–66 PSPICE schematic to find loop gain and phase.

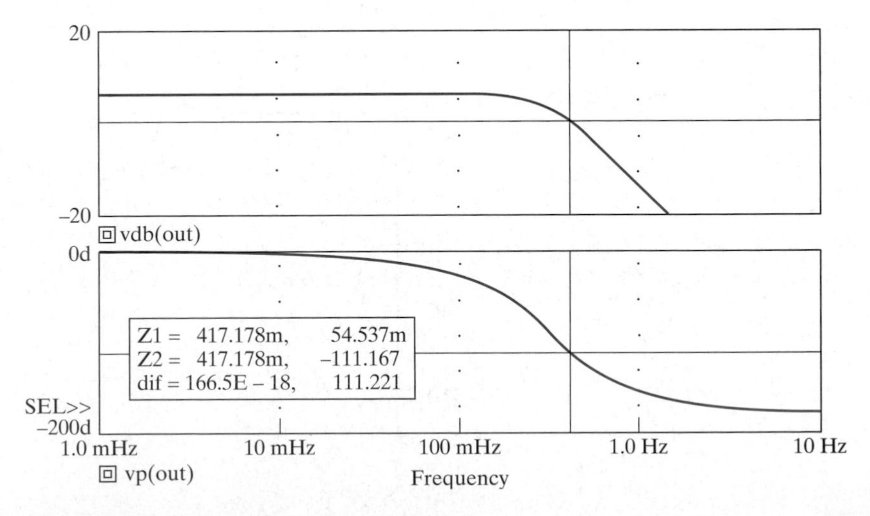

Figure 10–67 Bode plots for loop gain.

10.1.6 Compensation

In the preceding section, we saw how to predict the response of closed-loop amplifiers that may have poor relative stability. In many instances, it is necessary to ***compensate*** a feedback amplifier either to guarantee that it is stable or to achieve a desired phase margin. For example, consider a negative-feedback amplifier using a forward amplifier with response

$$a(j\omega) = \frac{10^6}{(1 + j\omega/10^4)(1 + j\omega/10^8)^2}. \tag{10.261}$$

The Bode magnitude and phase plots for this amplifier are shown in Figure 10-68. If we use the amplifier in a negative-feedback configuration with a constant feedback factor, the phase of the loop gain will be equal to the phase of $a(j\omega)$, which is shown in the figure. This phase passes through −180° at the frequency $\omega_{180} = 100$ Mrad/s, and the magnitude of the open-loop gain is 40 dB at that point. Therefore, if we desire a closed-loop gain of less than 40 dB, the loop will be unstable.

Suppose we want to construct an amplifier with a closed-loop gain of 20 dB and a phase margin of 45° and we want to use this forward amplifier. To achieve a

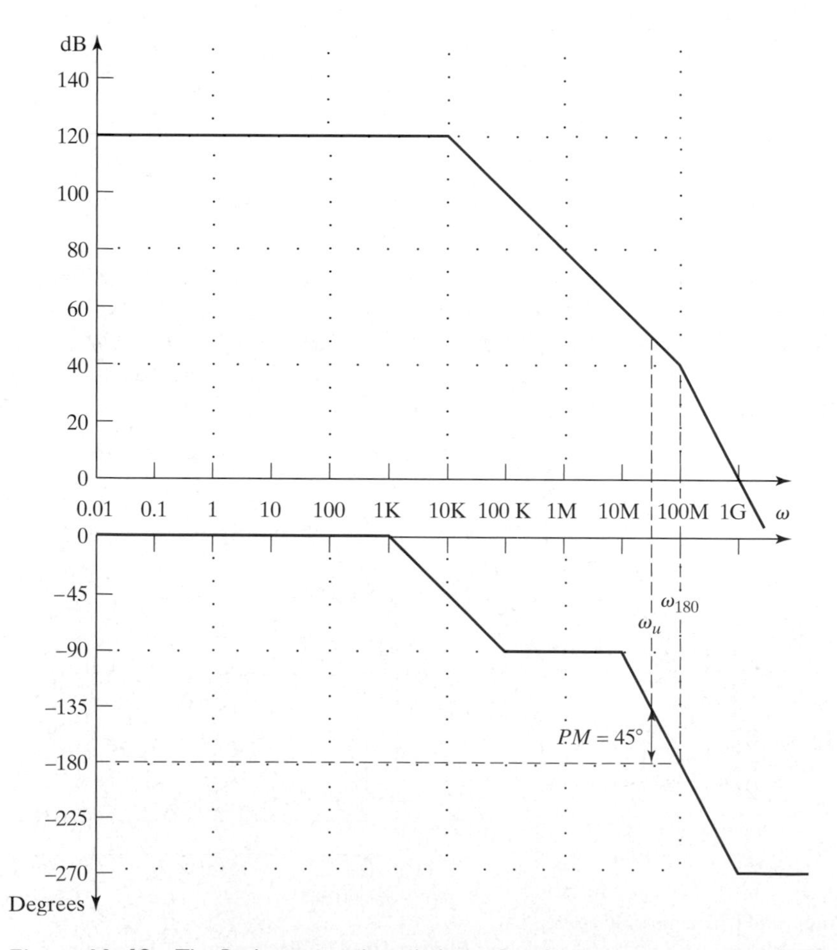

Figure 10–68 The Bode magnitude and phase plots for the open-loop gain given by (10.261).

phase margin of 45° for a loop containing the forward amplifier described by (10.261) and a constant feedback factor, the loop gain must go to unity at the frequency where the phase of $a(j\omega)$ is $-135°$. From Figure 10-68, we see that we need to set $\omega_u = 32$ Mrad/s. We also see from the figure that $|a(j\omega_u)| = 50$ dB, so if we desire a closed-loop gain of 20 dB, the open-loop gain is 30 dB higher than it should be to achieve the desired phase margin. One way we could solve this problem is to simply lower the open-loop gain by 30 dB. If we do that, the phase margin for a closed-loop gain of 20 dB will be 45°, as desired. The gain margin for the compensated amplifier is then 10 dB, as shown in Figure 10-69.

Lowering the open-loop gain does achieve the desired phase margin, but we pay a penalty for achieving the stability in this way. Recall from Section 10.1.1 that both the magnitude of the error and the sensitivity of the closed-loop gain to changes in the open-loop gain depend on the factor $(1 + L)$. Therefore, reducing the open-loop gain sacrifices some of the advantages of negative feedback.

We now present two other methods of compensation. Neither of these methods reduces the DC loop gain, but they do sacrifice bandwidth. The first method is to add a dominant pole into the loop. Adding a dominant pole is usually straightforward in practice and can always compensate a loop sufficiently to

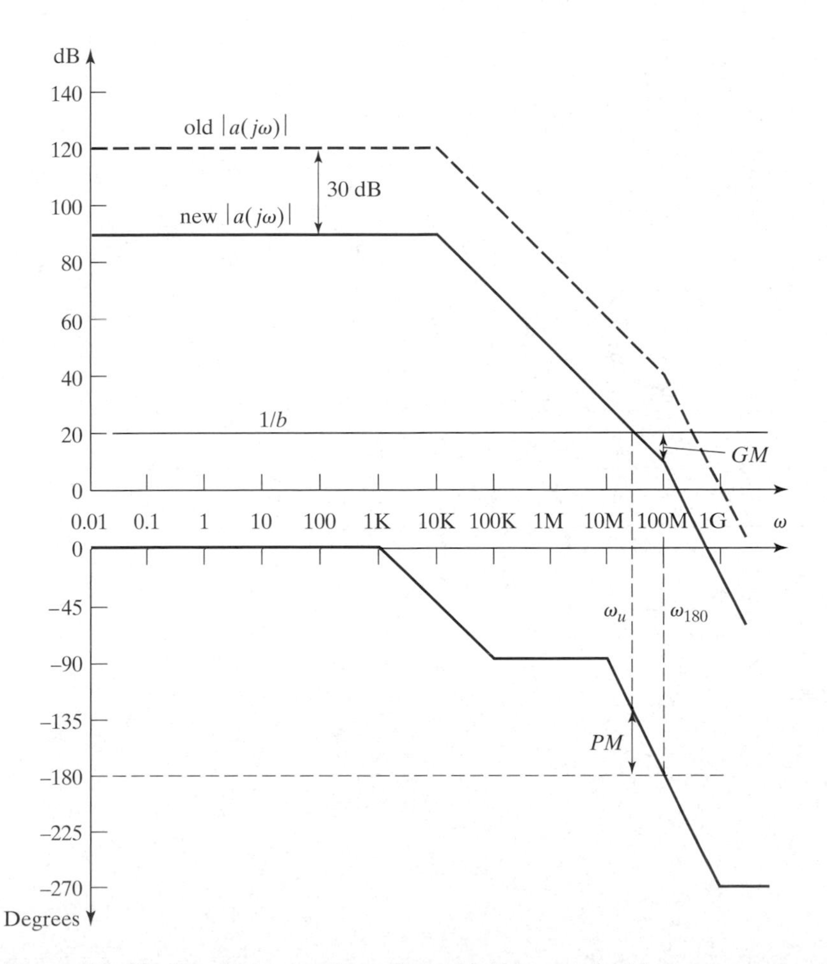

Figure 10–69 The Bode plots from Figure 10-68 repeated, with a new open-loop gain 30 dB lower and a $1/b$ of 20 dB also shown. The resulting gain and phase margins are shown on the plots as well.

produce a stable system, but we will discover that the new pole can be at a frequency much lower than the original dominant pole in the loop, so the reduction in bandwidth can be large.

Consider again the transfer function of (10.261), but assume that we have added a dominant pole; that is, we have added a pole that is lower in frequency than all of the existing poles. In this case, the pole we have at 10 krad/s in the original transfer function becomes our second pole. Assuming that the other poles are separated from this one by at least one order of magnitude, the phase of the loop gain will be −135° at 10 krad/s. (We have picked up −90° from the new dominant pole, plus −45° from the pole at 10 krad/s.) Therefore, we want to set the loop gain to unity at 10 krad/s, as shown in Figure 10-70. Once we have determined the unity-gain frequency, we know that there is only one pole to the left of it, so the slope of the Bode magnitude plot to the left of the crossover is −20 dB/decade. Therefore, we can draw a line going through 0 dB at ω_u with a slope of −20 dB/decade; where this line intersects the original DC gain determines the frequency required for our new dominant pole. The construction of the Bode plot and the resulting dominant pole frequency are shown in the figure. In this case, we find that we must add a pole at 0.1 rad/s, quite a drop from the original 10 krad/s.

The final method of compensation we present in this introductory treatment is to reduce the frequency of the existing dominant pole in the loop. This method

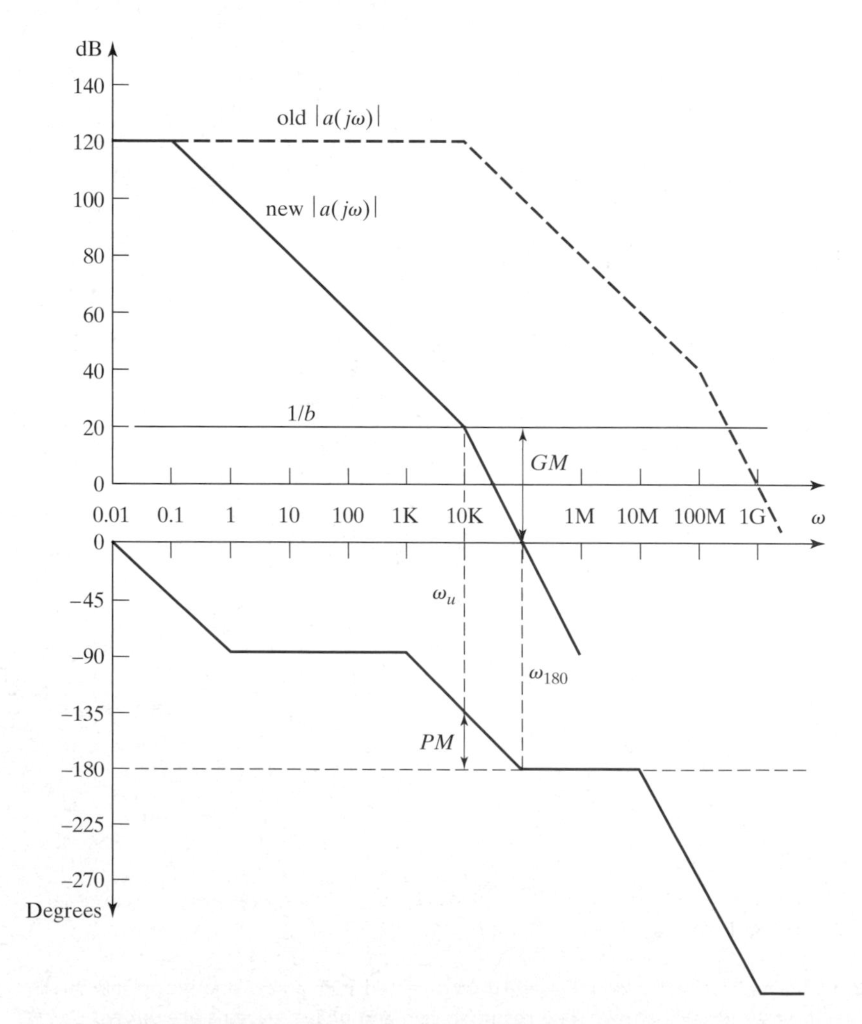

Figure 10–70 Finding the frequency required for a new dominant pole if the system is to achieve a closed-loop gain of 20 dB with a 45° phase margin.

can be used only when we know which capacitance in our circuit sets the dominant pole and we have some way of adding to this capacitance. In this case, we again find the unity-gain frequency required to achieve the desired phase margin and then draw a line with a -20 dB/decade slope intersecting $1/b$ at ω_u. Where this line intersects the DC gain defines the new frequency required for our dominant pole. The construction technique is illustrated for our present example and a 45° phase margin in Figure 10-71. In this case, we end up having to reduce our dominant pole frequency to about 320 rad/s. Comparing that frequency with the frequency of an added dominant pole, which was found to be 0.1 rad/s, we see that the reduction in bandwidth is considerably less severe with this technique.

The next example provides one illustration of how it is sometimes possible to move the frequency of a dominant pole. The illustration also shows that we can sometimes reduce the pole magnitude significantly without requiring large capacitors. Since capacitors require a large area to implement on an integrated circuit, finding ways to compensate amplifiers with small capacitors is important in IC design. The method demonstrated by this example is the method most often used to internally compensate IC operational amplifiers.[15]

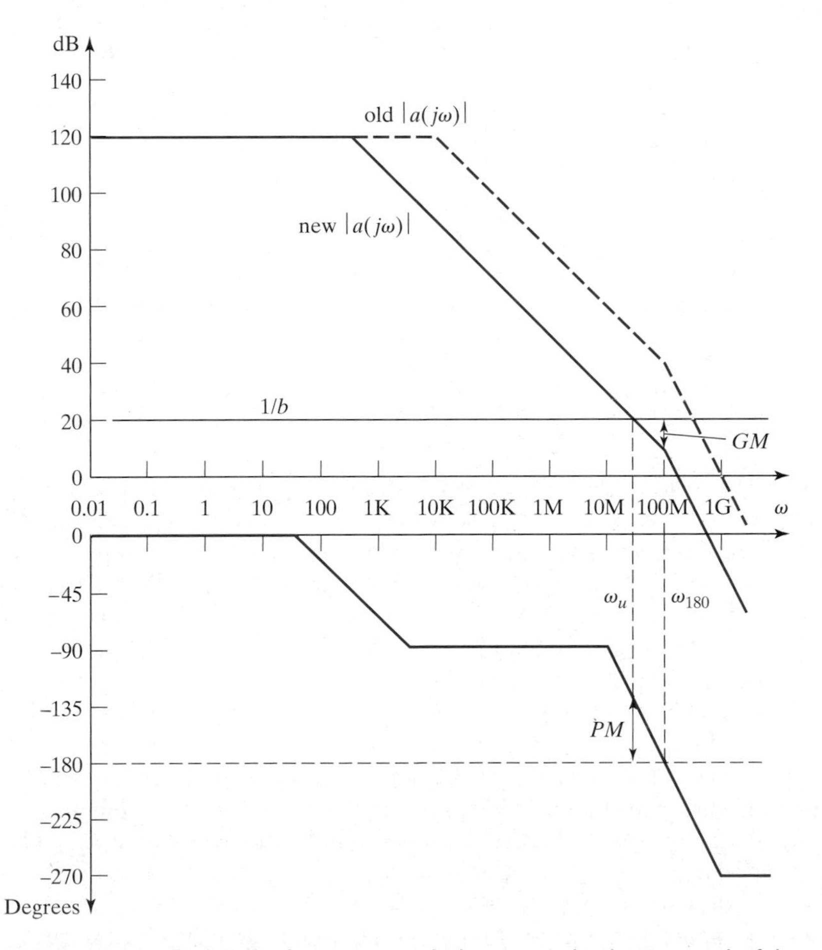

Figure 10–71 Finding the frequency to which to move the dominant pole if the system is to achieve a closed-loop gain of 20 dB with a 45° phase margin.

[15] If you are familiar with the internal structure of operational amplifiers from Section 8.7.4, you know that the second stage is usually a single-ended stage with voltage and current gain, so the example that follows applies directly.

Example 10.9

Problem:
All single-transistor amplifier stages that have both voltage and current gain (e.g., common emitter, common source) can be represented by a small-signal midband AC equivalent circuit like the one shown in Figure 10-72. The bandwidth of these stages is frequently dominated by the Miller effect. If this amplifier is part of the forward amplifier in a feedback loop, and we desire to lower the dominant pole frequency, what can we do?

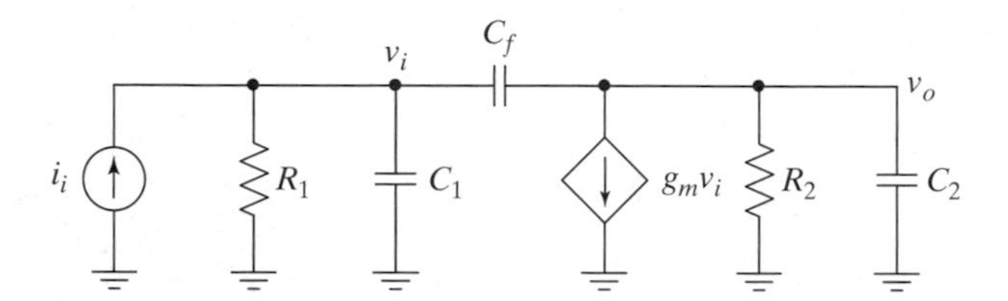

Figure 10–72

Solution:
The Miller input capacitance of this stage is

$$C_{Min} = C_f(1 - A_v) \approx -A_v C_f \approx g_m R_2 C_f,$$

where we have made the usual approximations and have used the midband gain for the Miller approximation. The resistance seen by C_{Min} is R_1, so if the bandwidth of this stage is dominated by the Miller effect, the bandwidth is approximately given by

$$\omega_{cH} \approx \frac{1}{g_m R_1 R_2 C_f}.$$

The capacitance we have called C_f here is the internal capacitance of the transistor and is usually quite small. If we want to lower the dominant frequency, examination of the equation quickly reveals that we can add an external compensation capacitor, C_c, in parallel with C_f and achieve our desired goal. If we do this, we also gain the benefit that the compensation capacitor is multiplied by the Miller effect, so we can use a smaller capacitor than you might initially suspect. This technique of compensation is frequently referred to as ***Miller compensation***.

When Miller compensation is used, we often gain a second benefit: In addition to the dominant pole frequency being lowered, the second-highest pole frequency is frequently increased. Examination of the methodology we use for this kind of compensation, as shown in Figure 10-71, reveals that if the second highest pole frequency moves up, we don't sacrifice as much bandwidth when compensating the amplifier. The fact that the second pole moves up in frequency while the dominant pole moves down is frequently called ***pole splitting***.

To see how pole splitting works, we first write the nodal equations for the circuit in Figure 10-72, using phasors and complex admittances. We have

$$V_o(Y_2 + Y_f) - V_i Y_f = -g_m V_i$$

$$-V_o Y_f + V_i(Y_1 + Y_f) = I_i,$$

where Y_1 is the admittance of R_1 in parallel with C_1, Y_2 is the admittance of R_2 in parallel with C_2, and Y_f is the admittance of C_f in parallel with C_c. Letting $C_f' = C_f \| C_C$ and solving these equations, we find that

$$\frac{V_o(s)}{I_i(s)} = \frac{R_1R_2(sC_f' - g_m)}{s^2R_1R_2(C_1C_2 + C_f'(C_1 + C_2)) + s(R_1(C_1 + C_f') + R_2(C_2 + C_f') + g_mR_1R_2C_f') + 1}.$$

We can express this transfer function as

$$\frac{V_o(s)}{I_i(s)} = \frac{N(s)}{(s + p_1)(s + p_2)} = \frac{N(s)/p_1p_2}{\dfrac{s^2}{p_1p_2} + s\left(\dfrac{1}{p_1} + \dfrac{1}{p_2}\right) + 1},$$

and, by approximating p_1 by the pole frequency found using the Miller approximation and assuming that it dominates (i.e., that $p_2 \gg p_1$), we find that

$$p_2 \approx \frac{g_mC_f'}{C_1C_2 + C_f'(C_1 + C_2)}.$$

This equation shows that as C_C is increased, p_2 will increase. The model we have used here is a bit simplistic. The poles *do not always* split when a more realistic model is used for the transistor, but they usually do.

■

Exercise 10.20

Suppose you have an amplifier with a transfer function given by

$$a(j\omega) = \frac{10^5}{(1 + j\omega/10^3)(1 + j\omega/10^5)(1 + j\omega/10^6)} \qquad \textbf{(10.262)}$$

and you apply negative feedback to the amplifier, using a feedback network that is not a function of frequency. (a) What is the lowest closed-loop gain for which you can maintain a PM of at least 45° without using any compensation? (b) If you desire to have a closed-loop gain of 30 dB with a 45° PM, and you compensate the loop by moving the dominant pole frequency down, what new pole frequency should be used?

□

In addition to the techniques presented here, there are more advanced compensation techniques presented in books on control theory. We will briefly illustrate two of the most common advanced techniques here, but we refer you to books on control theory for the details (e.g., [10.2]).

Two commonly used more advanced techniques are illustrated in Figure 10-73. In both cases, the amplifier from Exercise 10.20 is compensated, so you can compare the results with those achieved in the exercise.

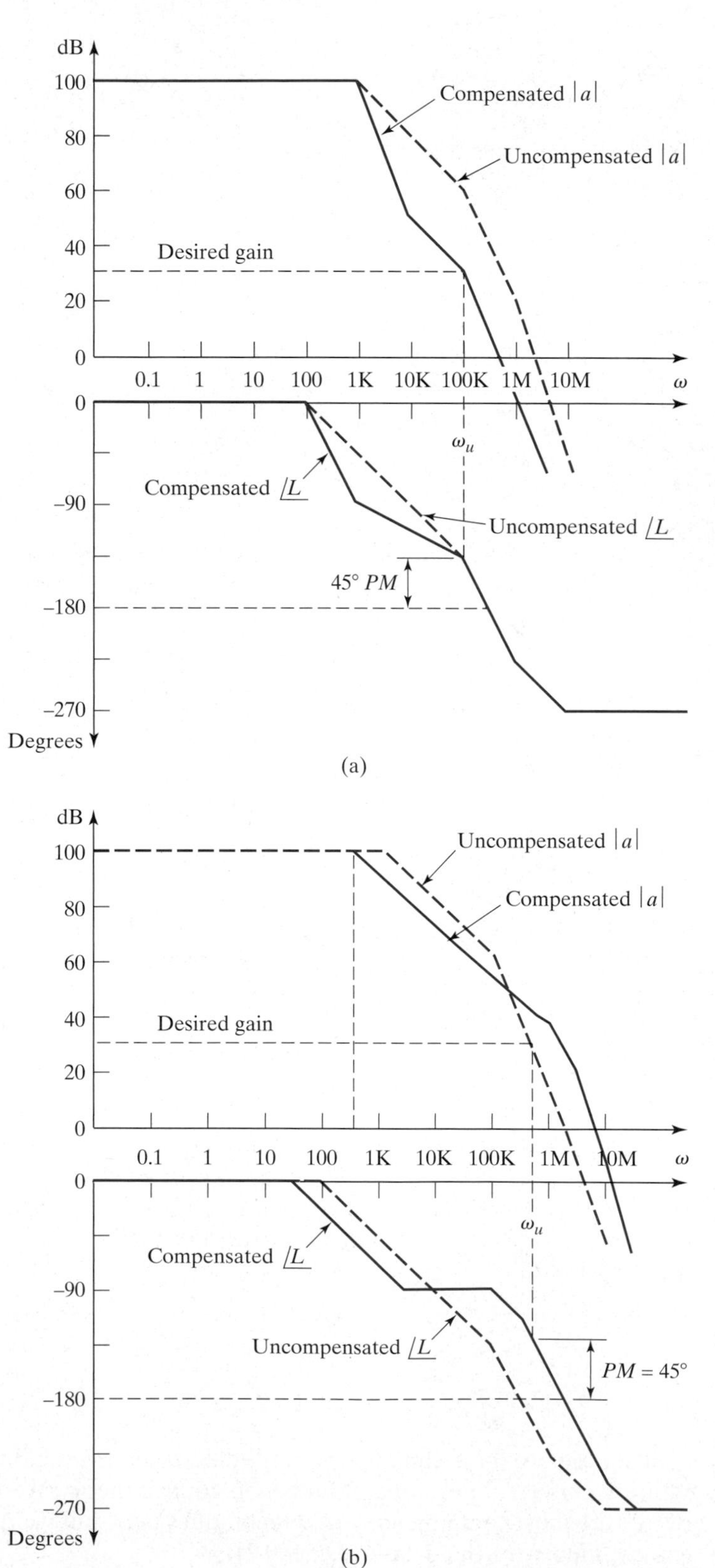

Figure 10–73 (a) Lag–lead compensation. (b) Lead–lag compensation.

In the first technique, called *lag–lead compensation*, a pole–zero pair is placed in the loop as shown in part (a) of the figure. The spacing between the pole and the zero is set so that it brings the magnitude response down sufficiently to achieve the desired phase margin. The frequencies of the pole–zero pair are located far enough below the next pole that the phase shift caused by the pair does not push the phase too close to −180°. (The maximum phase shift depends on the spacing between the pole and zero.) In this example, the pole was placed at 1 krad/s and the zero at 10 krad/s. The net result is that the loop gain has a cutoff frequency nearly equal to that of the uncompensated open-loop amplifier.

The *lead–lag compensation* technique is illustrated in Figure 10-73(b). In this technique, a zero–pole pair is used to produce a phase bulge that causes the unity-gain frequency to be pushed higher, where the loop gain is lower. In this example, we have also had to lower the dominant pole a small amount to achieve the 45° phase margin with 30-dB closed-loop gain. The compensation zero was set at 100 krad/s, and the pole was set at 32 Mrad/s, which causes the unity-gain crossing to occur at 562 krad/s. (The zero and pole frequencies are frequently set so that the zero crossing is halfway between them on a logarithmic plot, although this is not optimal). There is a danger in placing the compensation zero at the same frequency as one of the open-loop poles, as we have in this example. In practice, the pole–zero cancellation will not be exact, as a result of slight component variations, and the resulting closely spaced pole–zero pair may adversely affect the transient response.

10.2 POSITIVE FEEDBACK AND OSCILLATORS

Oscillators are an important class of circuits and are used in almost every electronic system. For example, oscillators are employed to produce the sinusoidal signals that are used as the carriers[16] for radio and television broadcasts and to produce the square waves used as clocks in computers and other synchronous digital systems.

An oscillator is a circuit designed to provide a periodic output with no input signal applied; DC power is required. Oscillators depend on positive feedback to keep the oscillation going. A positive feedback loop is shown in Figure 10-74. Although we will not use an input signal, one is shown to compare the positive feedback loop with the negative feedback loop studied in the previous section and to indicate how the oscillations begin.

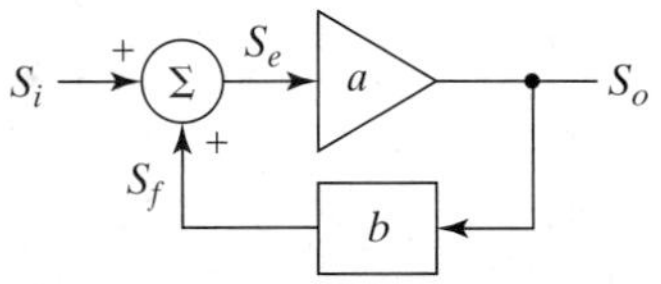

S_i + Σ S_e a S_o + S_f b

Figure 10–74 A positive feedback loop.

The output signal in this figure is given by

$$S_o = aS_e = a(S_i + bS_o), \tag{10.263}$$

which leads to

$$\frac{S_o}{S_i} = \frac{a}{1 - ab}. \tag{10.264}$$

The only difference between this result and the result derived for the negative-feedback amplifier in (10.2) is the sign in front of the loop gain. If the input signal is zero, the only way we can have a finite output from this circuit is for the transfer function given by (10.264) to equal infinity. The transfer function will be infinite at

---

[16] We call the signal a *carrier* because the information to be transmitted is used to *modulate*, or change, some characteristic of the wave (e.g., frequency modulation). The modulated sine wave can then be thought of as carrying the signal.

a frequency ω_o if $L(j\omega_o) = a(j\omega_o)b(j\omega_o) = 1$. This condition is known as the ***Barkhausen criterion*** for oscillation. Equivalently, we can say that the circuit can oscillate at ω_o if the following two conditions are met:

1. The magnitude of the loop gain is unity at wo (i.e., $|L(j\omega_o)| = 1$).
2. The phase of the loop gain is zero or some multiple of 360° at wo (i.e., $\angle L(j\omega_o) = 0$ or $n360°$ for integer n).

An alternative view of a positive feedback loop for an oscillator is shown in Figure 10-75. In this figure, we explicitly recognize that no input signal will be present. We can still identify the loop gain as $L = ab$, and we can write

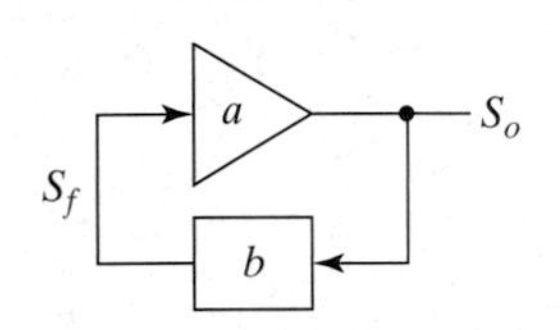

Figure 10–75 An alternative view of a positive feedback loop for an oscillator.

$$S_o = aS_f = abS_o. \tag{10.265}$$

There are only two solutions possible for this equation: the trivial case when S_o is zero, and the useful case where $ab = 1$, which is the Barkhausen criterion.

If the Barkhausen criterion is satisfied exactly, the circuit can provide a constant-amplitude periodic output. The circuit oscillates at the frequency where the loop gain is unity. If we make passive elements the dominant frequency dependence in the loop, then the frequency of oscillation can be made predictable and stable with respect to changes in temperature, supply voltage, and other factors. Since we have difficulty predicting or controlling the exact gain and frequency response of electronic amplifiers, having the key performance specifications depend on accurate and stable passive components is desirable. This observation is similar to the one made for negative-feedback amplifiers wherein we noted that the closed-loop gain could be stable if it depended on the passive feedback network rather than the forward amplifier.

There are two general categories of oscillator that we will discuss: ***sinusoidal oscillators*** and ***nonsinusoidal oscillators***. Although no oscillator produces a perfect sine wave, and nonsinusoidal oscillators can be used with filters to produce sine waves, the categories are useful because the methods used to produce oscillation are different. Sinusoidal oscillators are sometimes called ***linear*** or ***small-signal*** oscillators, although neither of these terms is completely accurate. Nonsinusoidal oscillators are sometimes called ***nonlinear*** oscillators.

10.2.1 Sinusoidal Oscillators

For sinusoidal oscillators we use combinations of resistors, capacitors, inductors, and crystals to form frequency-selective networks. These networks are then used in positive feedback loops with amplifiers to produce oscillators. One question we need to answer is, Do we know that the oscillations will start without some effort on our part? The answer can be found by returning to the loop shown in Figure 10-74. We note that there will always be noise and other small disturbances present in our circuit, so if the Barkhausen criterion is met, a small input signal will be present in the loop and the oscillations will begin. If the loop gain is exactly one, however, the oscillations may be very small, and they will never grow. More importantly, it is impossible to produce real circuits that have a loop gain of exactly unity. If the gain drops to slightly less than unity, the oscillations will decay away with time. Therefore, all real oscillators use a loop gain greater than unity.

With a loop gain greater than unity, the oscillations will grow with time until the growth is stopped by a nonlinearity as discussed in Section 10.1.5. Therefore, all real oscillators are nonlinear circuits. Nevertheless, we can still make profitable use of linear circuit analysis techniques to determine whether or not oscillations will start and at what frequency they will occur. The kind of nonlinearity seen in the loop is important, however, since it will determine how much distortion is produced (*see* Chapter 12 for a discussion of nonlinear distortion).

There are two main design issues we need to address to produce stable low-distortion oscillators: We need to carefully control the frequency at which the loop phase is equal to 0°, and we need to limit the magnitude of the oscillations in the loop without causing too much distortion.

We begin by examining oscillators that use RC feedback networks. We go on to look at oscillators that use LC networks and crystals.

RC Oscillators Consider the circuit in Figure 10-76 as a first example. This circuit is usually called a **Wien–Bridge** oscillator, or, occasionally (and more accurately), a **Wien-type** oscillator [10.6]. We assume that the gain of the forward amplifier, a, is known and is constant for frequencies near the oscillation frequency. We will come back to this assumption later. The feedback factor for the circuit is given by

$$b(j\omega) = \frac{V_f(j\omega)}{V_o(j\omega)} = \frac{Z_p}{Z_p + Z_s} = \frac{1}{3 + j\omega\tau + \dfrac{1}{j\omega\tau}}, \qquad \textbf{(10.266)}$$

where $\tau = RC$, Z_p is the parallel combination of R and C, and Z_s is the series combination of R and C. The Barkhausen criterion for oscillation is then

$$L(j\omega) = \frac{a}{3 + j\left(\omega\tau - \dfrac{1}{\omega\tau}\right)} = 1. \qquad \textbf{(10.267)}$$

The loop will oscillate when the phase of this loop gain is zero, which occurs at the frequency where the imaginary term in the denominator goes to zero:

$$\omega_o = \frac{1}{RC}. \qquad \textbf{(10.268)}$$

If the gain of the forward amplifier is equal to three at this frequency, the magnitude of the loop gain is equal to unity, and the loop will oscillate.

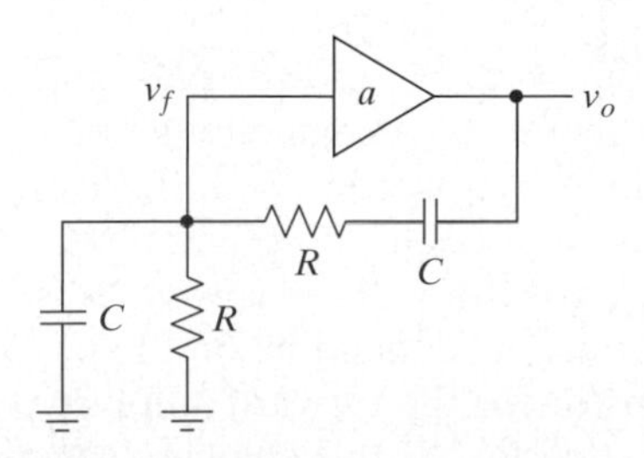

Figure 10–76 A Wien–Bridge oscillator.

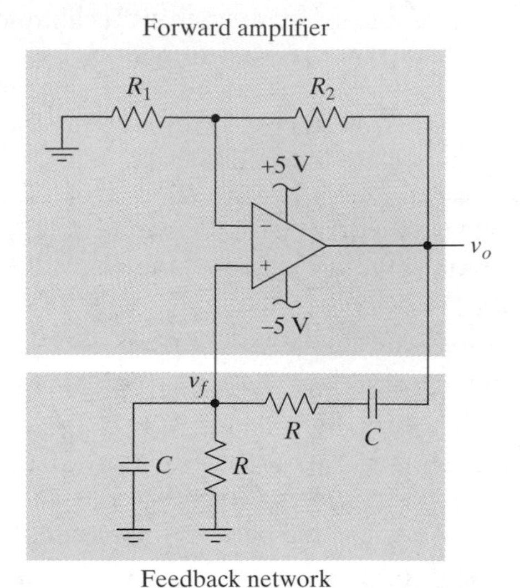

Figure 10–77 A practical Wien–Bridge oscillator. The power supply connections to the op amp are shown for clarity in the discussion about the output swing being limited.

To make this oscillator practical, we need to be sure that the oscillations will be present, even with reasonable parameter variations caused by changing temperature, supply voltage, component values, and so on. To do this, we use a loop gain larger than unity, which is accomplished by setting the gain of the forward amplifier greater than three by a sufficient margin to allow for all variations. Doing this will guarantee that oscillations start and grow. The next step is to limit the magnitude of the oscillations. One way to limit the amplitude is to allow the amplifier to reach its swing limits.

For example, consider the circuit shown in Figure 10-77, and assume that the output of the operational amplifier can swing fully from the positive supply to the negative supply. The gain of this forward amplifier is $a = v_o/v_f = (1 + R_2/R_1)$. Note that the "forward amplifier" here uses negative feedback. If we set $R_2 = 3R_1$ we have a gain of four, and the circuit will oscillate at the frequency $\omega_o = 1/RC$. The oscillations will grow until the operational amplifier output limits the swing.

Using the operational amplifier to limit the magnitude of the oscillations will work, but has three potential problems. First, it may produce an oscillation that is larger in magnitude than desired. Second, it may limit the frequency of oscillation, because the operational amplifier may not be able to recover fast enough from having the output saturate. And third, it may produce more distortion than is acceptable. In general, sharp nonlinearities (often called *hard* nonlinearities) produce more distortion, while smoother nonlinearities (often called *soft* nonlinearities) produce less distortion.

One way to produce lower magnitude oscillations in this circuit, while preventing the operational amplifier output from saturating, is to use diode limiters, as shown in Figure 10-78.

Resistor R_{D1} and diode D_1 limit the negative output swing, and resistor R_{D2} and diode D_2 limit the positive output swing. The resistors must be small enough that the loop gain is reduced below unity when either diode is forward biased. This circuit functions by allowing the gain of the forward amplifier to be greater than three, so that the loop gain is greater than unity, until the voltage across R_2 is about ± 0.7 Volt. At this point, either D_1 or D_2 will turn on, depending on the polarity of the output, and the effective feedback resistor for the forward amplifier will approximately be R_2 in parallel with either R_{D1} or R_{D2}. If this change drops

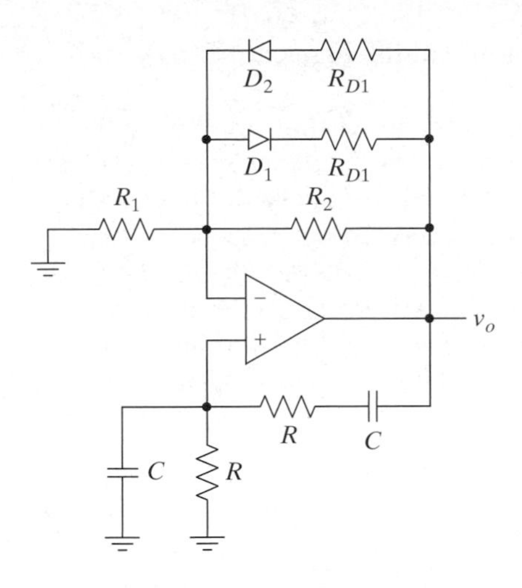

Figure 10–78 The Wien–Bridge oscillator of Figure 10-77 with diode limiters added. Following our standard practice, the op amp supplies are not shown.

the gain of the forward amplifier to less than three, the oscillations will stop growing. In essence, the loop oscillates with a magnitude such that the average loop gain is exactly unity.

Other, more advanced methods can also be used to control the loop gain in an oscillator. For example, the circuit shown in Figure 10-79 uses an incandescent bulb as a nonlinear resistor. The circuit is identical to the Wien-Bridge oscillator in Figure 10-77, except that R_1 has been replaced by the bulb. When the bulb is first turned on, the loop gain is set to be larger than unity so that oscillations begin and grow. As the current in the bulb increases, however, the bulb heats up and its resistance increases, which brings the gain back down. A properly designed circuit of this type can have very low harmonic distortion (.003% can be achieved).[17]

The Wien–Bridge oscillator implemented using an operational amplifier cannot usually be used at very high frequencies because the gain of the operational amplifier needs to be approximately constant at the frequency of oscillation (i.e., the phase shift in the op amp should be nearly zero). This requirement typically limits

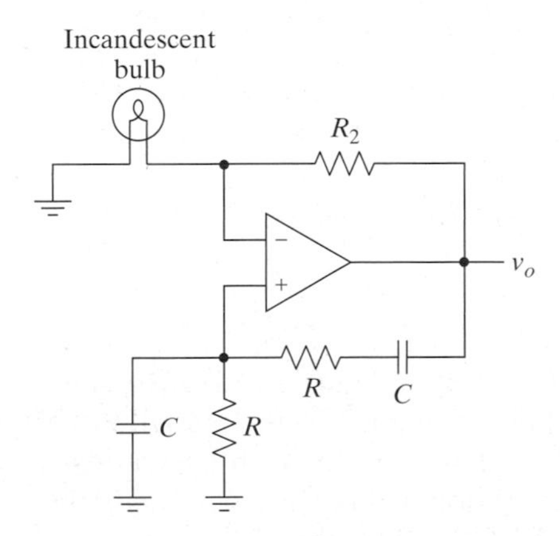

Figure 10–79 A low-distortion Wien–Bridge oscillator.

[17] An interesting aside, showing how important low distortion can be, is that the Hewlett-Packard company was founded when Hewlett and Packard discovered how to use an incandescent bulb to produce the lowest distortion audio oscillator available at the time.

operation to frequencies less than about one-tenth of the high cutoff frequency of the forward amplifier.

The Wien–Bridge oscillator, along with other oscillators that use *RC* networks in the feedback path, has another problem as well: The feedback network is not very selective (i.e., the gain-versus- frequency characteristic does not have a sharp peak). In fact, the gain of the feedback network described by (10.266) drops only by about 10% for frequencies an octave away from ω_o. The practical implication of this fact is that distortion (caused by the nonlinearity used to limit the magnitude of the oscillations) is not filtered out very effectively. The nonlinearity used to control the gain is therefore very important in determining the distortion. This issue is addressed by Exercise 10.21.

Exercise 10.21

Design the Wien–Bridge oscillator shown in Figure 10-77 to oscillate at 10 kHz and have a loop gain of 4/3. Assume that the op amp is ideal and the oscillations will be limited by the op amp. Then simulate the circuit,[18] using the op amp model shown in Figure 10-80. Measure the actual frequency of oscillation. Using that frequency, have SPICE print out the Fourier components. (The Fourier outputs are very sensitive to the fundamental frequency given, so don't just use 10 kHz.) What is the total harmonic distortion (i.e., the RMS sum of the first eight harmonics)? Now add diode limiters as shown in Figure 10-78. Pick the resistor values so that the loop gain is reduced to 2.5/3 when the limiters are active. Simulate the circuit again, and check the distortion, using the D1N4148 model from Appendix B. How much lower is the distortion than in the original circuit? Can you explain why the frequency of operation is closer to 10 kHz when the diode limiters are present?

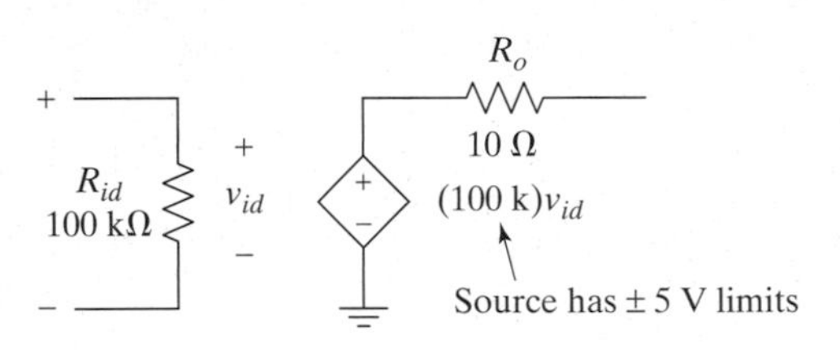

Figure 10–80 The op amp macro model for use in Exercise 10.21. The controlled source has swing limits at ±5 V.

We close this section with a brief treatment of one more RC oscillator: the phase shift oscillator. A phase shift oscillator is shown in Figure 10-81. We assume that the forward amplifier is ideal and has a gain of $-a$; that is, the magnitude of the gain is a and there is a 180° phase shift. If the phase shift through the feedback network is set to 180°, the Barkhausen phase criterion will be met. Analyzing this feedback network reveals that the phase shift is 180° at a frequency

$$\omega_o = \sqrt{6}/RC \tag{10.269}$$

[18] You will need to provide some means of starting the oscillations in SPICE. You can use an initial condition on one of the capacitors, or you can insert a voltage pulse somewhere in the loop.

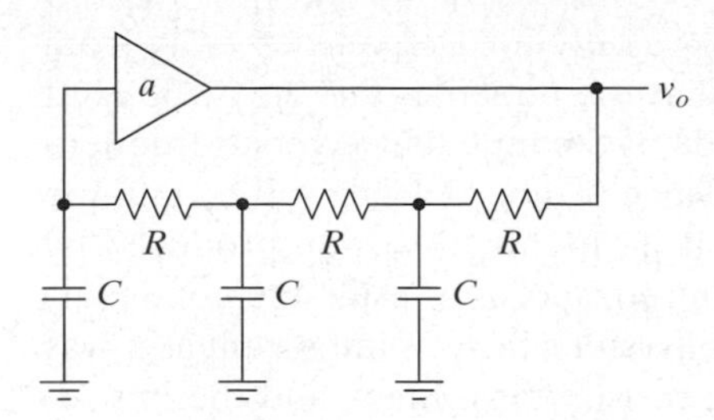

Figure 10–81 A phase shift oscillator.

(*see* Problem P10.97). At that frequency, the gain of the feedback network is 1/29, so if we set $a = 29$, we meet the Barkhausen magnitude criterion.

Exercise 10.22

Use SPICE to simulate a phase shift oscillator like the one in Figure 10-81. Use $R = 1\ k\Omega$, $C = 390\ nF$, and $a = -40$. Use an ideal voltage-controlled voltage source for the amplifier, but include swing limits at ± 1 V. What frequency will this circuit oscillate at? Will the oscillations start? Will they grow? What will be the final amplitude?

LC Oscillators In this section, we examine sinusoidal oscillators that use a parallel resonant circuit, often called a ***tank circuit***,[19] as shown in Figure 10-82(a). The impedance of this tank is

$$Z_t(s) = \frac{1}{Y(s)} = \frac{1}{sC + 1/sL} = \frac{sL}{1 + s^2LC}. \tag{10.270}$$

The impedance of the tank is infinite at the resonant frequency given by $\omega_o = \sqrt{1/LC}$. In other words, $i_l = -i_c$ at ω_o, and the external current into the tank is zero. One advantage of *LC* oscillators based on a resonant tank is that the tank

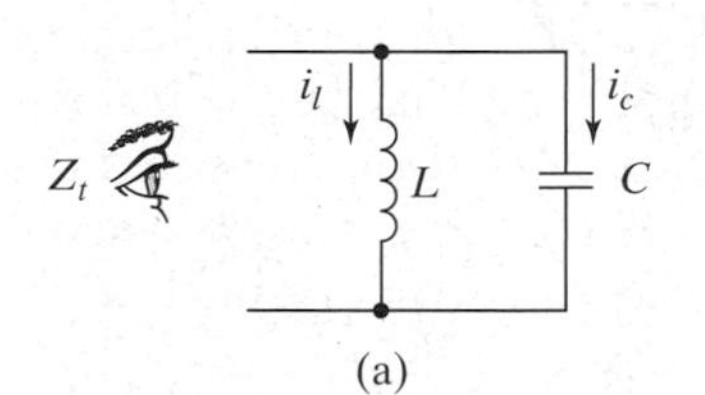

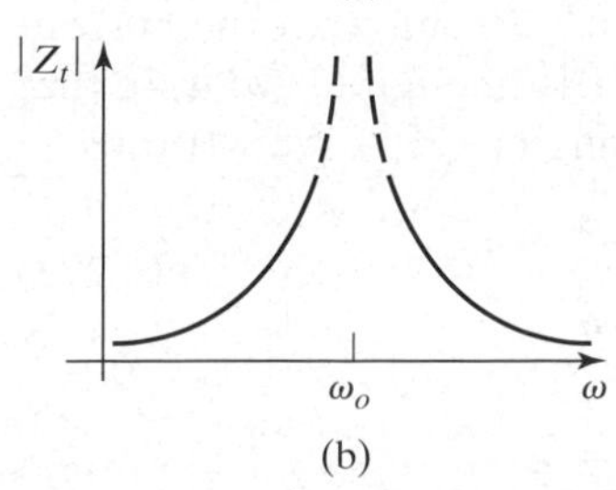

Figure 10–82 (a) A tank circuit and (b) the magnitude of the impedance versus frequency.

[19] The circuit stores energy like a tank stores water.

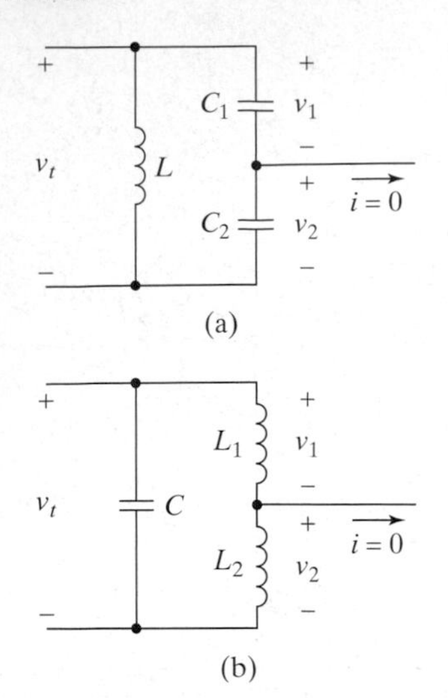

Figure 10–83 The tank (a) with a capacitive divider and (b) with an inductive divider.

is very selective; that is, the impedance drops rapidly for frequencies away from ω_o. Evaluating (10.270) shows, for example, that the impedance is only 0.33 Ω at $2\omega_o$. Part (b) of the figure shows the magnitude of the impedance versus frequency; the selectivity is apparent from this plot. Since this impedance will dictate the loop gain, the extreme selectivity implies that harmonic distortion produced by the nonlinearity in the loop will be heavily filtered, and, as a result, the nonlinearity used to limit the amplitude of the oscillations isn't nearly as important as it was for *RC* oscillators. All real tank circuits have resistance in them, and the impedance is not infinite at ω_o. Nevertheless, even practical tank circuits can be extremely selective.

The two oscillators that we will examine modify the tank circuit by using either a capacitive voltage divider, as shown in Figure 10-83(a), or an inductive voltage divider, as shown in part (b) of the figure. If the current out of the center tap on the divider is zero, then for the capacitive divider, we have

$$\frac{V_1(j\omega)}{V_2(j\omega)} = \frac{i_c Z_{C1}}{i_c Z_{C2}} = \frac{1/j\omega C_1}{1/j\omega C_2} = \frac{C_2}{C_1}. \tag{10.271}$$

Similarly, for the inductive divider, we have

$$\frac{V_1(j\omega)}{V_2(j\omega)} = \frac{i_l Z_{L1}}{i_l Z_{L2}} = \frac{j\omega L_1}{j\omega L_2} = \frac{L_1}{L_2}. \tag{10.272}$$

Using the capacitive divider on the tank, we form a ***Colpitts oscillator***, as shown in Figure 10-84. This circuit uses a single-transistor common-merge amplifier to provide the voltage gain and a resonant tank with a capacitive divider to form the feedback network. The inductor labeled ***RF choke*** is a large inductor used to provide a DC short to the power supply while ensuring that the top of the tank circuit is not connected to an AC ground. The resistor divider formed by R_1 and R_2 is used to establish the DC voltage at the transistor's control terminal and, combined with R_M, sets the DC bias current in the transistor. The two capacitors C_B and C_M are large bypass capacitors. The feedback voltage appears at the top of the tank (i.e., across C_2) and is connected to the control terminal of T_1 through C_B.

The small-signal model for the oscillator is shown in Figure 10-85, where we have assumed that the frequency of oscillation is in the midband of the transistor amplifier and have therefore ignored the parasitic capacitances, shorted the bypass capacitors, and opened the RF choke. We have included the small-signal output resistance of the transistor in our model, since it may significantly affect the gain.

To use (10.271) to find the feedback factor, we must assume that the current into $R_2 \| r_{cm}$ is very small (i.e., $i_i \approx 0$). Assuming that i_i is negligible, we note that v_o appears across C_1 and v_f appears across C_2, so, using (10.271), we write

$$v_f = -v_o\left(\frac{C_1}{C_2}\right). \tag{10.273}$$

Therefore, the feedback factor is

$$b = \frac{v_f}{v_o} = \frac{-C_1}{C_2}. \tag{10.274}$$

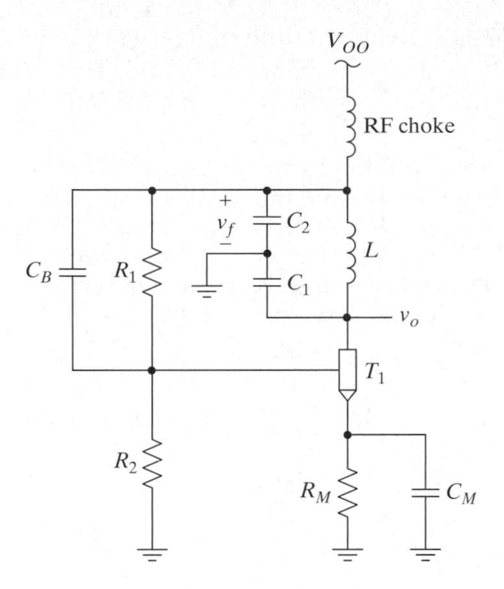

Figure 10–84 A Colpitts oscillator.

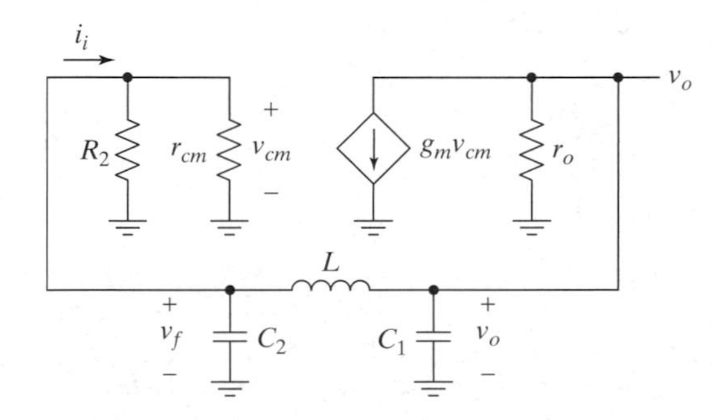

Figure 10–85 The small-signal mid-band model for the Colpitts oscillator.

Analysis of the forward amplifier at ω_o where the feedback looks like an open circuit yields a gain of

$$a = \frac{v_o}{v_f} = -g_m r_o. \tag{10.275}$$

So long as the loop gain ab is larger than unity, the circuit will oscillate at $\omega_o = \sqrt{1/LC_t}$, where C_t is the series combination of C_1 and C_2.

It is not uncommon for the assumption that $i_i \approx 0$ to be wrong. In that case, we can still find the feedback factor by examining the transfer function of the third-order low-pass filter comprising C_1, C_2, L, and $R_2 \| r_{cm}$ (*see* Problem P10.102). Alternatively, if $C_2 = \alpha C_1$ and $R_2 \| r_{cm} \gg \sqrt{L(1+\alpha)/\alpha C_1}$ we can show that the impedance seen looking into the feedback network at the resonant frequency ω_o is approximately equal to $\alpha^2(R_2 \| r_{cm})$ (*see* Problem P10.103). Therefore, the effective load presented to the transistor is $R_{eff} \approx R_2 \| r_{cm} \| r_o$, and the magnitude of the loop gain turns out to be $|L(j\omega_o)| \approx g_m(r_o \| R_2 \| r_{cm})/\alpha$ (see Problem P10.104).

With the loop gain larger than unity, the oscillations will grow until they are limited by the nonlinearity of the transistor. At that point, a linear analysis using the circuit in Figure 10-85 is no longer appropriate. Nevertheless, we know that the circuit will oscillate, and we know the frequency of oscillation. (A detailed discussion of the nonlinear behavior of oscillators is beyond the scope of this text. If you are interested in a more advanced treatment, we refer you to [10.6] or [10.7].)

The circuit in Figure 10-84 also has the advantage that the signal can be coupled out of the circuit using a transformer, as shown in Figure 10-86. This kind of coupling provides electrical isolation as well as the ability to match the impedance of a load.

A ***Hartley oscillator*** is similar to a Colpitts; the difference is that the inductor in the tank circuit is divided as shown in Figure 10-83(b) and the feedback is derived from the tapped inductor.

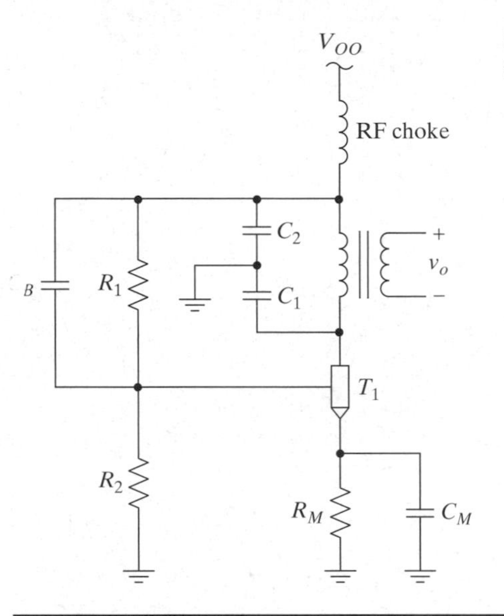

Figure 10–86 A Colpitts oscillator with transformer coupling at the output.

Exercise 10.23

Consider the Colpitts oscillator shown in Figure 10-87. Determine the frequency of oscillation. Simulate the circuit and verify that it works. Examine a transient plot of the collector current and comment on what you see. You will need a voltage pulse in series with the base to get the oscillations started. (This is necessary only in SPICE—not in a real circuit.) Also, you may need to specify a smaller-than-default step size in the transient simulation if the oscillations don't begin, or if they die away.

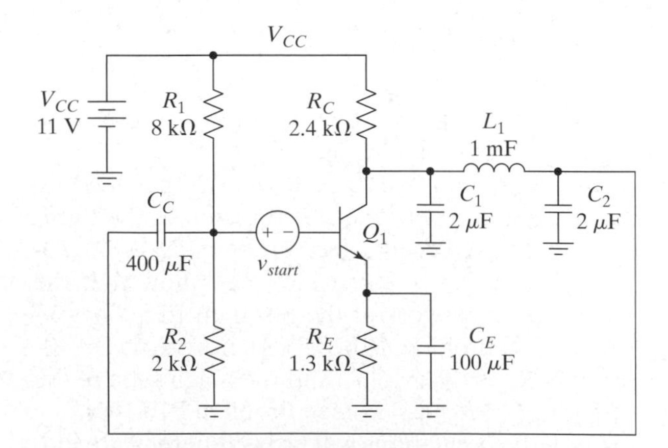

Figure 10–87 A Colpitts oscillator. The voltage source in series with the base is a small pulse used to initiate oscillation in SPICE.

Crystal Oscillators Oscillators based on RLC feedback networks can have very good frequency stability if the components are carefully chosen. However, many applications require that the frequency of oscillation be more stable than it is possible to achieve with even the best RLC circuits. When very good frequency stability is required, quartz crystals are used as the frequency-selective component. A quartz crystal is a piece of quartz (i.e., glass, a crystal of SiO_2) that has been specially cut and has had metal contacts plated on two surfaces. Quartz is *piezoelectric*, which means that physical strain induces a voltage and vice versa. As a result, an applied electric field causes mechanical vibrations, and, depending on the cut, the crystal resonates at a particular frequency. The schematic symbol and equivalent circuit of a crystal are shown in Figure 10-88.

Table 10.1 gives values for the electrical models of several common crystals. These parameters are usually specified, but if they aren't, C_o, C_1, Q, and ω_o are specified, and L is determined from $L \approx 1/C_1\omega_o^2$, whereas R_S is determined from $R_S \approx \omega_o L/Q$. A Colpitts crystal oscillator circuit is shown in Figure 10-89 where the crystal is used in place of the inductor of an LC tank. The crystal is, by itself, a resonant circuit; the only reason for the external capacitors is to enable us to make a voltage divider and generate a 180° phase shift, as was done for the Colpitts oscillator in the previous section. The coupling capacitor, C_C, is used to block DC current in the crystal.

Table 10.1 Typical Component Values for Common Cuts of Quartz Oscillator Crystals (Courtesy of RCA)

Frequency	32 kHz	280 kHz	525 kHz	2 MHz
R_S	40 kΩ	1820 Ω	1400 Ω	82 Ω
L	4.8 kH	25.9 H	12.7 H	0.52 H
C_1	4.91 fF	12.6 fF	7.24 fF	12.2 fF
C_o	580 pF	450 pF	475 pF	350 pF
Q	25,000	25,000	30,000	80,000

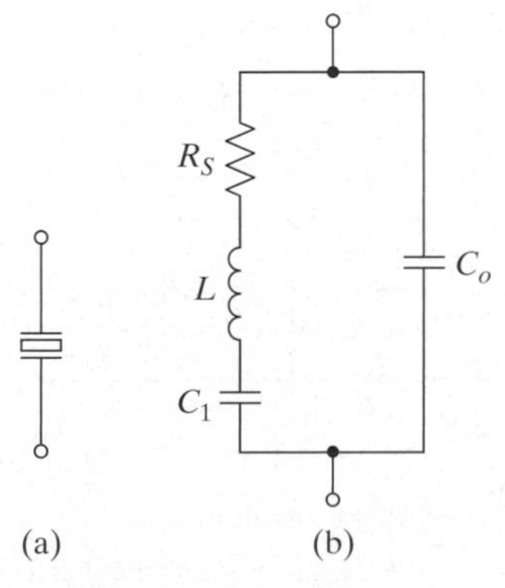

Figure 10–88 (a) The schematic symbol for a quartz crystal and (b) the equivalent circuit.

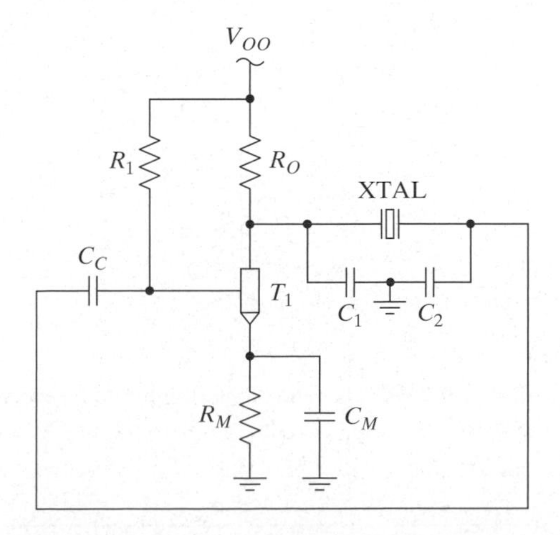

Figure 10–89 A Colpitts crystal oscillator.

10.2.2 Nonsinusoidal Oscillators

The most common type of nonsinusoidal oscillator is the ***relaxation oscillator***. All relaxation oscillators use some form of limit cycle; for example, the voltage on a capacitor can be ramped up and down between two limits. The cycle proceeds by first ramping the voltage up until some threshold, or limit, is reached. At that point, the circuit changes so that the voltage ramps back down toward the lower limit. When the lower limit is reached, the circuit returns to its original state, and the voltage starts ramping back up again.

A block diagram of a relaxation oscillator is shown in Figure 10-90, where we have chosen to use current sources to charge and discharge the capacitor. The amplifier symbol with a hysteresis loop inside it is a comparator with hysteresis (covered in Section 5.3.1). This comparator has two thresholds, an upper and a lower one. The output of the comparator controls the position of the switch. If the switch is in the position shown, the capacitor voltage ramps up linearly, since the current is constant; the voltage is

$$v_C(t) = \frac{1}{C}\int_0^t i_C(\lambda)d\lambda + v_C(0). \tag{10.276}$$

When the capacitor voltage reaches the upper threshold, the output of the comparator changes. The change in comparator output also closes the switch and moves the comparator threshold to the lower value. With the switch closed, the capacitor voltage ramps down with time, since the total current is now out of the capacitor. When the input reaches the lower threshold, the output of the comparator changes again, causing the switch to open and the threshold to move back to the upper value. The resulting waveforms are shown in the figure.

Also as shown in the figure, the output of the comparator is a square wave, and in this example, it has a 50% duty cycle.[20] During either half cycle, the slope of the capacitor voltage is found by differentiating (10.276); the magnitude of the result is

$$\left|\frac{dv_C(t)}{dt}\right| = \frac{I}{C}. \tag{10.277}$$

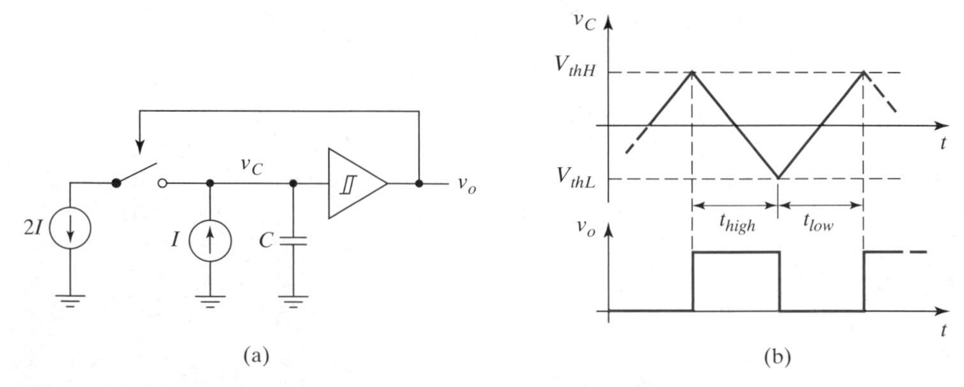

Figure 10–90 (a) The block diagram of a relaxation oscillator and (b) the resulting waveforms.

[20] *Duty cycle* refers to the ratio of the time a binary digital waveform is high to the total period. The term also gets used to describe the symmetry of other waveforms, like triangle waves.

The total change in capacitor voltage during either half period is $V_{thH} - V_{thL}$, so the period is

$$T = \frac{2C(V_{thH} - V_{thL})}{I}. \quad \textbf{(10.278)}$$

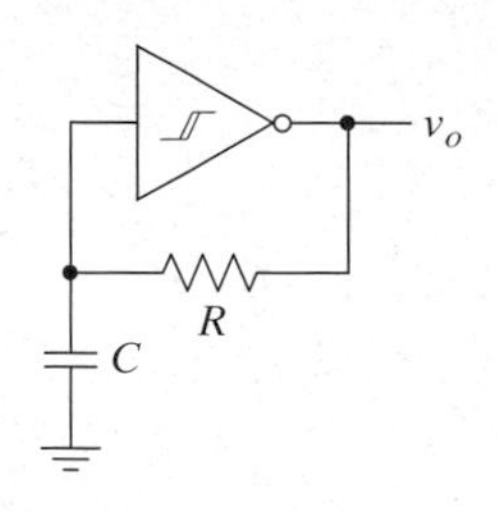

Figure 10–91 A relaxation oscillator using *RC* charging.

It is also possible to charge the capacitor through a resistor, as shown in Figure 10-91. In this figure, the comparator output is always at the positive or negative limit. When it is at the positive limit, the capacitor charges up toward that limit with an exponential rise and a time constant *RC*. When the upper threshold of the comparator is reached, the comparator's output goes low. (It is an inverting comparator, as denoted by the bubble on the output.) The capacitor now charges down toward the lower swing limit. When the comparator's lower threshold is reached, its output goes high and the process begins all over again.

Notice that the voltage on the capacitor in our original example, Figure 10-90, is a nice sawtooth waveform. Notice also that by varying the currents during t_{high}, and t_{low}, we can vary the duty cycle of the waveform. This type of circuit is often used in *function generators*, which typically provide sine, square, and triangle wave outputs with variable duty cycles and periods. The *RC*-type relaxation oscillator can also vary the duty cycle if we make some provision for different resistors to be used in the charge and discharge phases. Because this type of circuit is so versatile, and because it is easily integrated due to the lack of inductors, this type of oscillator has become very popular for many applications. We will now present a brief introduction to the most popular commercial oscillator circuit based on a relaxation oscillator.

Exercise 10.24

Consider the relaxation oscillator shown in Figure 10-91. Assume that the comparator is perfect, that the thresholds are at ±1 V, and that the output swing limits are ±5 V. Given, R = 10 kΩ and C = 0.01 μF that (a) sketch the output voltage and the capacitor voltage as a function of time, assuming that v_O just switched high at t = 0, and (b) determine the frequency of oscillation.

□

The 555 Timer The 555 Timer is the most popular commercial timer available.[21] The circuit can be used to produce a well-controlled delay, an oscillator, or devices satisfying any of a large number of other functions. In the terminology used in digital circuits, the 555 can function as either a *monostable multivibrator*, also known as a *one-shot*, or an *astable multivibrator*. A one-shot is a circuit that provides a single pulse of known width at its output whenever it is triggered. An astable multivibrator is an oscillator that typically produces a square wave instead of a sine wave or other function.

As an oscillator, the circuit is capable of producing waveforms with different duty-cycles and can also be used as a *voltage-controlled oscillator* (VCO). The frequency of oscillation of a VCO is controlled by an input voltage, and this

[21] As of 1997, the 555 timer had been in continuous production by a number of different companies for 25 years without an update to the original design. For 23 of those years, the 555 was the best-selling IC of all, analog or digital [10.8]. It was redesigned and improved in 1997 and continues to be in heavy use today.

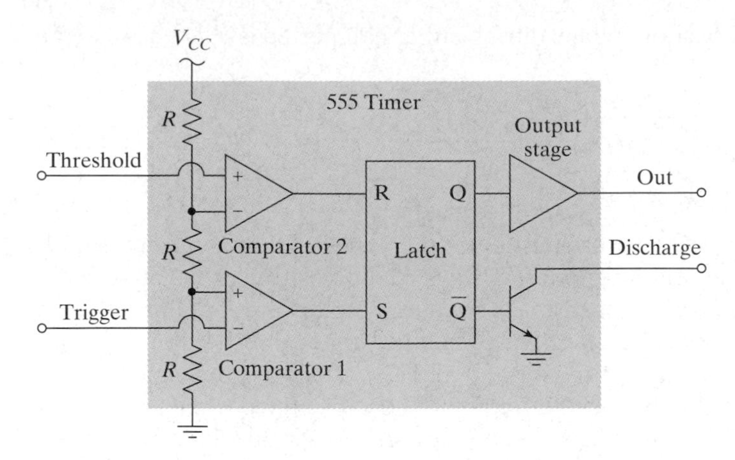

Figure 10–92 The block diagram of a 555 Timer IC.

functional block is useful in a number of applications. A simplified block diagram of a 555 Timer is shown in Figure 10-92. The circuit contains two comparators, a three-resistor divider network to establish the comparator thresholds, a set–reset latch, an output stage, and a discharge transistor.

The resistor divider establishes a voltage equal to $V_{CC}/3$ at the positive input of comparator 1 and a voltage of $2V_{CC}/3$ at the negative input of comparator 2. When the SR latch is in the reset state, the output voltage is low, and the voltage at the base of the discharge transistor is high, which turns the transistor on hard, pulling current into its collector. The transistor will typically saturate and pull the voltage at its collector nearly to ground. When the SR latch is set, the output voltage is high, and the base of the discharge transistor is pulled low so that the transistor is turned off. The state of the latch is controlled by the threshold and trigger inputs.

The 555 can be set up to operate as a monostable multivibrator, or one-shot, as shown in Figure 10-93. In this configuration, it produces a single output pulse when it is triggered. The accuracy of the pulse width is on the order of 1% and has a 50-ppm/°C temperature coefficient. Therefore, the width of this pulse is suitable for use as a delay (i.e., to delay the time of some event from the occurrence of the trigger event).

The circuit operation can be described by assuming that we start with the SR latch reset, the discharge transistor on, the capacitor voltage (i.e., the threshold voltage) low, and the trigger high. In this state, the discharge transistor is pulling enough current through R_1 to hold the capacitor voltage low (ideally zero), and the output of comparator 2 is therefore low. With the trigger input high, the output of comparator 1 is low as well, so the latch holds its previous state (reset in this case). When the trigger input is taken low, the output of comparator 1 goes high, the latch is set, and the output voltage switches high, as shown in Figure 10-94. The discharge transistor also turns off at this time, and the capacitor voltage charges up through R_1. For proper operation, the trigger input must return high prior to the end of the output pulse. The capacitor voltage during the time the output is high is given by

$$v_C(t) = V_{CC}(1 - e^{-t/R_1C}) \tag{10.279}$$

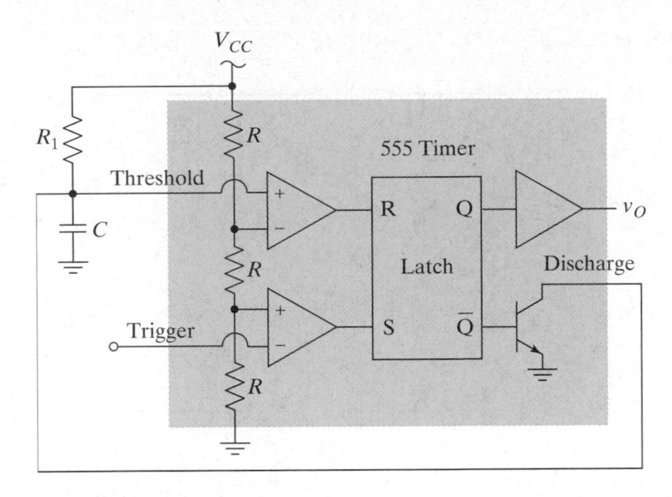

Figure 10–93 The 555 Timer connected as a one shot.

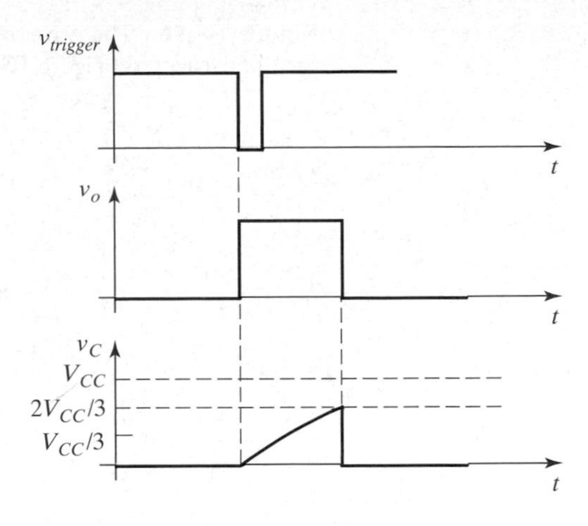

Figure 10–94 The waveforms for the one shot of Figure 10-93.

and is shown in the figure. The voltage charges up toward V_{CC}. However, when it reaches $2V_{CC}/3$, the output of comparator 2 goes high and resets the latch.[22] When the latch is reset, the output goes low, the discharge transistor quickly pulls the capacitor voltage back down to zero, and the process is ready to begin again. The pulse width is found by setting (10.279) equal to $2V_{CC}/3$ and solving for t:

$$t_{pw} = -R_1 C \ln(1/3) = 1.1 R_1 C. \tag{10.280}$$

Exercise 10.25

Design a 555 one-shot circuit to produce a pulse that is 2.5 ms wide when it is triggered.

Figure 10-95 shows the 555 used as an oscillator, or astable multivibrator. The accuracy of the frequency of oscillation is about 2%, and the temperature coefficient (i.e., the drift of the frequency with temperature) is about 100 ppm/°C. We

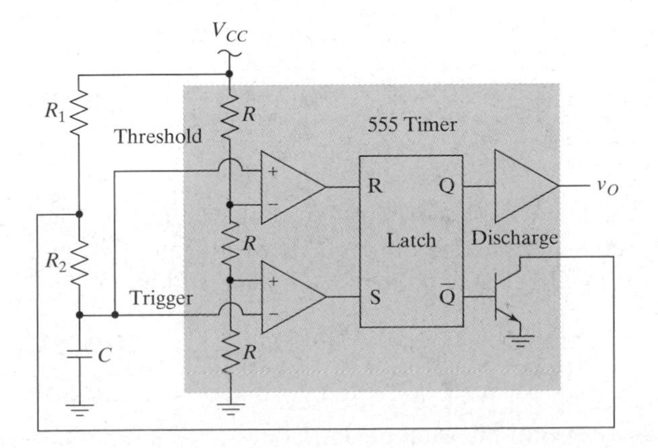

Figure 10–95 The 555 connected as an oscillator.

[22] If the trigger input is still high at this time, then the output of comparator 1 will also be high. The latch would then be simultaneously receiving set and reset signals, and the results are unpredictable.

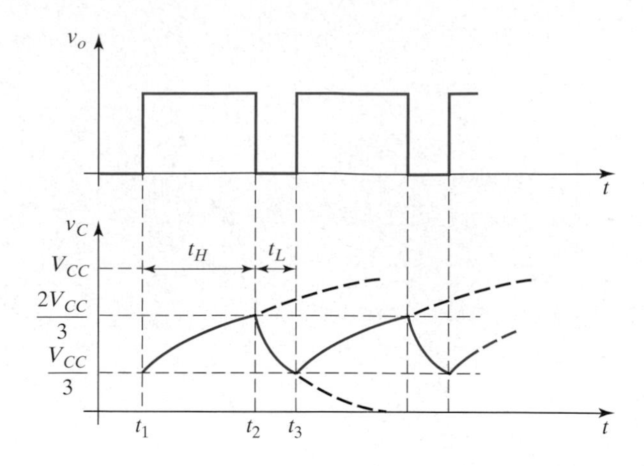

Figure 10–96 The waveforms for the oscillator in Figure 10-95.

start our analysis of this mode by assuming that the output has just switched high and the capacitor voltage is charging up toward V_{CC} from $V_{CC}/3$, as shown at time t_1 in Figure 10-96. If the capacitor voltage were allowed to charge all the way to V_{CC}, the total change in voltage would be $2V_{CC}/3$. Therefore, the capacitor voltage for $t_1 \leq t \leq t_2$ is

$$v_C(t) = \frac{V_{CC}}{3} + \frac{2V_{CC}}{3}\left(1 - e^{-(t-t_1)/(R_1+R_2)C}\right). \quad \textbf{(10.281)}$$

When the capacitor voltage reaches $2V_{CC}/3$, the output of comparator 2 goes high and resets the latch. As a result, the output goes low and the discharge transistor turns on, pulling the voltage at the node between R_1 and R_2 nearly to ground. We denote this time as t_2 in the figure. The time the output stays high is found by setting (10.281) equal to $2V_{CC}/3$:

$$t_H = t_2 - t_1 = -(R_1 + R_2)C\ln(1/2) = 0.7(R_1 + R_2)C. \quad \textbf{(10.282)}$$

With the top of R_2 near ground potential, the capacitor voltage discharges toward ground from the $2V_{CC}/3$ starting point. Assuming that the top of R_2 is at ground, we find that v_C is described by

$$v_C(t) = \frac{2V_{CC}}{3}e^{-(t-t_2)/R_2C}. \quad \textbf{(10.283)}$$

When the voltage reaches $V_{CC}/3$, denoted by t_3 in the figure, the output of comparator 1 goes high, setting the latch and starting the cycle all over again. The time the output is low is found by setting (10.283) equal to $V_{CC}/3$; the result is

$$t_L = t_3 - t_2 = -R_2C\ln(1/2) = 0.7R_2C. \quad \textbf{(10.284)}$$

Combining (10.282) and (10.284), we find the total period

$$T = t_H + t_L = 0.7(2R_2 + R_1)C, \quad \textbf{(10.285)}$$

and the frequency of oscillation is

$$f = \frac{1}{T} = \frac{1}{0.7(2R_2 + R_1)C}. \quad \textbf{(10.286)}$$

In this example, we showed a duty cycle of about 75% (i.e., v_O is high about 75% of the time). The duty cycle can be varied by changing the values of R_1 and R_2, but notice from (10.282) and (10.284) that t_H is always greater than t_L. A different circuit can overcome this limitation, as illustrated in Problem P10.114.

Exercise 10.26

Design a 555 oscillator to produce an output frequency of 25 kHz with a 67% duty cycle.

□

SOLUTIONS TO EXERCISES

10.1 We want $A_f = a/(1 + ab) = 10 \pm 1\%$ when $a = a_o \pm 50\%$ and b is constant. Using these data, we write

$$\frac{1.5a_o}{1 + 1.5a_ob} = 10.1$$

and

$$\frac{0.5a_o}{1 + 0.5a_ob} = 9.9.$$

Solving these two equations simultaneously yields $a_o = 666.4$ and $b = 0.098$. This *is* the minimum value of a that will work. Any larger value will yield a variation in A_f of less than 1%.

10.2 To find R_{of} we set either i_i or v_i to zero, remove the load, force either i_o or v_o, and measure the other. Then $R_{of} = v_o/i_o$. First consider the shunt output connections, parts (a) and (c) of Figure A10-1. In both cases, we have

$$i_o = \frac{v_o - ax_e}{R_o}$$

and $x_e = -bv_o$ where x_e is v_e in part (a) and i_e in part (c). Substituting x_e into the equation for i_o and solving yields

$$R_{of} = \frac{R_o}{1 + ab} = \frac{R_o}{1 + L}$$

in both cases. Now consider the series output connections in parts (b) and (d) of the figure. We have $v_o = R_o(i_o - ax_e)$ and $x_e = -bi_o$, where x_e is v_e in part (b) and i_e in part (d). Substituting x_e into the equation for v_o and solving yields $R_{of} = R_o(1 + ab) = R_o(1 + L)$ in both cases.

10.3 For the circuit in part (b), we tentatively identify the feedback network as comprising R_1 and R_2 together. The forward amplifier is the operational amplifier. If we trace a path from the input source towards the amplifier, we see that we must go through the forward amplifier to get to the feedback network, so the input connection is a series one. On the output side, if we start from where a load would be connected and trace a path towards the amplifier, we see that we can choose whether to go into the forward amplifier or the feedback network, so they are in shunt. Therefore, this is a series–shunt connection. The variable being fed back is v_o and the input summation is $v_e = v_d = v_i - v_f$, where v_f is the voltage at the top of R_2.

For the circuit in part (c), we identify the feedback network as R_f and the forward amplifier as the operational amplifier. Tracing a path from the input,we see the two are in shunt, and, similarly, starting from the output, we see that this is a shunt connection. Therefore, this is a shunt–shunt amplifier. The feedback variable is i_o and the input summation is $i_e = i_i - i_f$, where i_e is the current into the negative input terminal on the amplifier and i_f is the current going from left to right through R_f.

10.4 To achieve unity gain, we set $b = 1$. Therefore, $L_0 = 50{,}000$, and using (10.28), we find that $f_{pf} = 500$ kHz.

10.5 Using the given data, we write

$$A_{f\max} = \frac{1000}{1 + 1000b}$$

and

$$0.99A_{f\max} = \frac{250}{1 + 250b}.$$

Simultaneously solving these equations leads to $b = 0.29$ and $A_{f\max} = 3.4$.

10.6 The voltage driving the control terminal of the transistor is $v_C = a(V_{Ref} - v_M) + v_N$, where v_M is the voltage at the merge terminal and a is the op amp gain. Notice that we can rewrite this equation as

$$v_C = a\left(V_{Ref} + \frac{v_N}{a} - v_M\right),$$

so if we divide the noise by a, we can move it to the op amp input, sum it with V_{Ref}, and have an equivalent circuit. We say we have *referred the noise to the input* when we do this. The effective input is now

$$v_{I\mathrm{eff}} = V_{Ref} + \frac{v_N}{a},$$

and the output current is given by (10.30) if we use this effective input in place of V_{Ref}. Plugging in, we see that the noise in i_O is given by v_N/aR and is 0.1 pA rms. Note that if the 10-μV rms noise had been imposed across R directly (by using a unity-gain buffer to drive the control terminal on the transistor, but no feedback), the noise current would be 10 nA rms, 100,000 times worse than is achieved with feedback.

10.7 The gain of the circuit in Figure 10-10 is $A_f = a_p a/(1 + a_p ab)$. Setting this gain equal to 20 and solving for b yields $b = 0.049$. The SNR of the circuit in Figure 10-8 is given by (10.32); plugging in the given values, we find that SNR = 50, or 34 dB. The SNR for the circuit in Figure 10-10 is given by (10.37); plugging in the values, we find that SNR = 1250, or 62 dB. The improvement is caused by the low-noise preamplifier, but the feedback was required to keep the gain set to 20.

10.8 (a) The op amp is the forward amplifier and the feedback network is composed of R_1 and R_2. Tracing a path from v_i, we have to go through the forward amplifier to get to the feedback network, so the input connection is a series one. Also, tracing a path from v_o, we can return to ground through the forward amplifier or the feedback network, so the output connection is a shunt one. The error voltage is the differential input to the op amp: $v_e = v_{id}$. (b) The feedback voltage is the voltage at the top of R_2. (c) Using the formulas from Example 10.3 yields

$$b = \left.\frac{v_f}{v_o}\right|_{i_{bo}=0} = \frac{R_2}{R_1 + R_2},$$

$$R_{bi} = \left.\frac{v_o}{i_{bi}}\right|_{i_{bo}=0} = R_1 + R_2,$$

and

$$R_{bo} = \left.\frac{v_f}{i_{bo}}\right|_{v_o=0} = R_1 \| R_2.$$

(d) Again making use of the results in Example 10.3, we have

$$R_i' = R_{id} + R_{bo}, R_o' = R_o \| R_{bi} \quad \text{and} \quad a' = \left(\frac{R_{id}}{R_{id} + R_{bo}}\right) a \left(\frac{R_{bi}}{R_{bi} + R_o}\right).$$

(e) Using (10.44), (10.45), and (10.46) along with the previous results, we have

$$A_f = \frac{a'}{1 + a'b} = \frac{\left(\frac{R_{id}}{R_{id} + R_{bo}}\right) a \left(\frac{R_{bi}}{R_{bi} + R_o}\right)}{1 + \left(\frac{R_{id}}{R_{id} + R_{bo}}\right) a \left(\frac{R_{bi}}{R_{bi} + R_o}\right)\left(\frac{R_2}{R_1 + R_2}\right)},$$

$$R_{of} = \frac{R_o'}{1 + a'b} = \frac{R_o \| R_{bi}}{1 + \left(\frac{R_{id}}{R_{id} + R_{bo}}\right) a \left(\frac{R_{bi}}{R_{bi} + R_o}\right)\left(\frac{R_2}{R_1 + R_2}\right)},$$

and

$$R_{if} = R_i'(1 + a'b) = (R_{id} + R_{bo})\left[1 + \left(\frac{R_{id}}{R_{id} + R_{bo}}\right) a \left(\frac{R_{bi}}{R_{bi} + R_o}\right)\left(\frac{R_2}{R_1 + R_2}\right)\right].$$

(f) From the result in part (e), as a goes to infinity we have

$$\lim_{a \to \infty} A_f = \frac{1}{b} = \frac{R_1 + R_2}{R_2} = 1 + \frac{R_1}{R_2}.$$

Since a' goes to infinity as a does, we also get $R_{if} = \infty$ and $R_{of} = 0$.

10.9 The schematic for the PSPICE file is shown in Figure 10-97. The input impedance is shown in Figure 10-98 and indicates that $Z_{if} = 6.8$ kΩ, purely real, for low frequencies, as expected. The impedance becomes capacitive for a while and then settles out to being equal to R_i' at high frequencies. The output impedance was determined by grounding the left side of R_S, removing R_L, and driving the output with a test source. The impedance is shown in Figure 10-99 and at low frequencies is equal to R_{of}, as calculated in the example. At high frequencies, the output impedance looks inductive for a while and eventually approaches R_o' as the loop gain goes to zero. (When the loop gain is very small, the feedback is essentially not present, so we expect that $R_{if} \approx R_i'$ and $R_{of} \approx R_o'$ at very high frequencies.) The gains, A_{if} and A_{sf}, are shown in Figure 10-100 and agree exactly with the calculations in the example. The cutoff frequencies, f_{sf} and f_{if}, also agree with our calculations.

10.10 The basic analysis of the circuit is performed in Exercise 10.8. The difference here is that we have added a source and a load. Referring to the previous results and plugging in the values given here, we have $R_{bi} = 10$ kΩ, $R_{bo} =$ 900 Ω, $R_i' = 100.9$ kΩ, $R_o' = 99$ Ω, $a' = 98.1$ k, and $b = 0.1$. We now use (10.66) to find that $\alpha_o' = 0.503$ and $A_i = 49.3$ k. Using (10.69), we have $A_{if} = 10.0$. We next use (10.70) to find that $\alpha_i' = 0.092$ and $A_s = 4.78$ k. (10.74) then yields

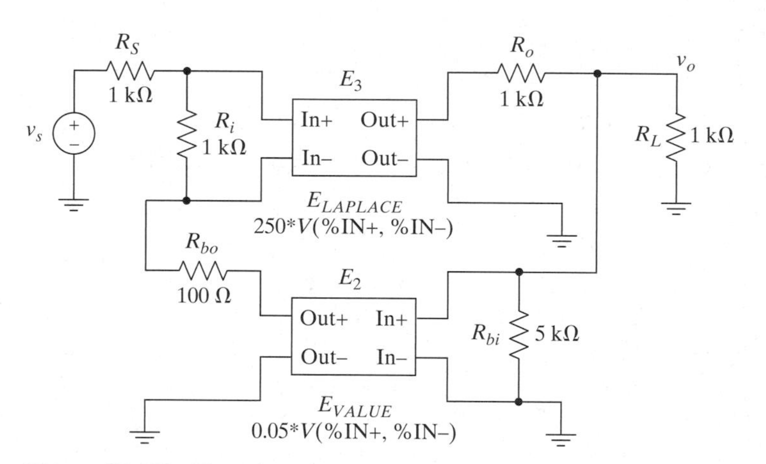

Figure 10–97 The schematic for simulation.

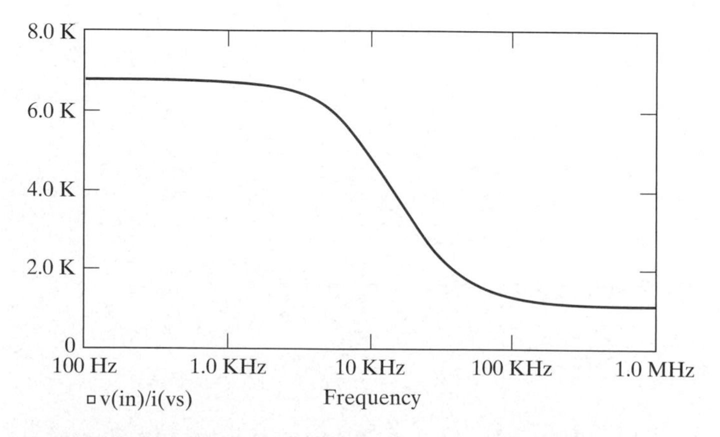

Figure 10–98 The input impedance.

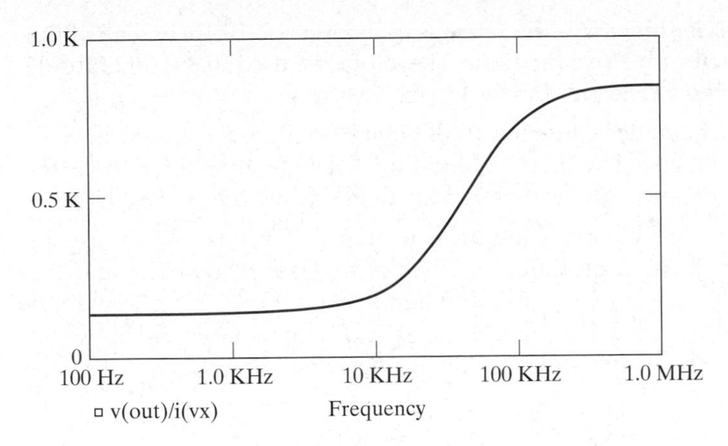

Figure 10–99 The output impedance.

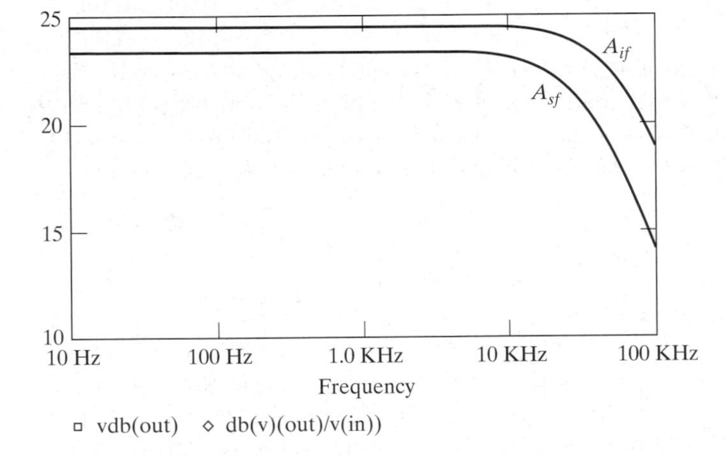

Figure 10–100 The gains.

$A_{sf} = 9.98$. Finally, using (10.80) and (10.84), we find that $R_{if} = 498$ MΩ and $R_{of} = 0.11$ Ω. In a practical circuit, other factors (e.g., common-mode input resistance) would limit the value of R_{if} to something substantially less than the value predicted here.

10.11 The PSPICE schematic is shown in Figure 10-101, and the gains, A_{if} and A_{sf}, are found by simulation to be 10 and 9.98, as predicted. The values of R_{if} and R_{of} are also found to be 498 MΩ and 0.11 Ω, as predicted.

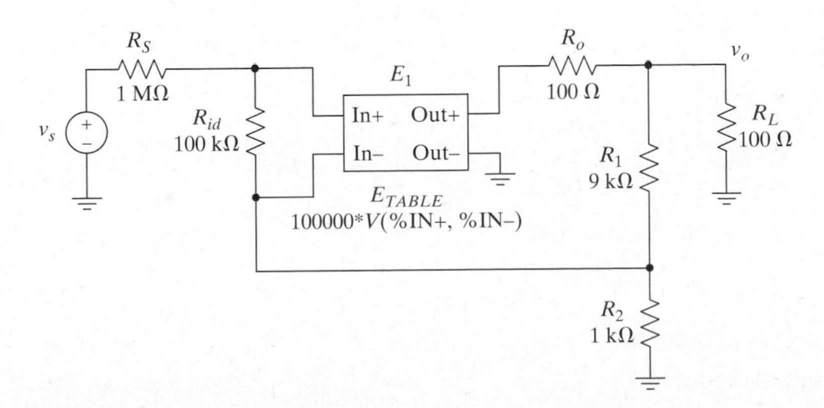

Figure 10–101 The PSPICE schematic.

10.12 The value of a_0' in this case is the same as a' in Exercise 10.10, and all of the attenuation factors are also the same. Therefore, we use (10.89) and (10.93) to obtain $f_{sf} = 4.55$ kHz and $f_{if} = 49.4$ kHz.

10.13 The PSPICE schematic is the same as in Figure 10-101, except that the controlled source is an ELAPLACE block that includes the single-pole response. The simulation shows that our calculated bandwidths are correct.

10.14 The analysis of this circuit is identical to that of Exercise 10.10, except that the value of R_{bo} is modified because of the presence of R_{C2} and the input circuit is altered by R_{C1}. We can handle R_{C1} by pulling it to the left and absorbing it into a Thévenin equivalent for the source, as shown in Figure 10-102. The result is that we have a new source given by

$$v_s' = v_s\left(\frac{R_{C1}}{R_{C1} + R_S}\right) = 0.167v_s$$

and $R_S' = R_S \| R_{C1} = 167$ kΩ.

The new value of R_{bo} is $R_{bo} = R_1 \| R_2 \| R_{C2} = 896\ \Omega$ and is so close to that obtained without R_{C2} that there isn't any noticeable effect on the gain, bandwidth, or input resistance. Therefore, the only change to our previous results is to allow for the presence of R_{C1}. Using (10.121) suitably modified, we have

$$A_{sf} = A_{sf}'\left(\frac{R_{C1}}{R_{C1} + R_S}\right)$$

where A_{sf}' is the same as A_{sf} in Exercise 10.10, except that we need to use R_S' in place of R_S. Using R_S' we calculate $\alpha_i' = 0.377$, $A_s' = 18.6$ kΩ, and, therefore, $A_{sf}' = 9.99$. Finally, this leads to $A_{sf} = 1.67$. The value of R_{if} is now given by an equation identical in form to (10.124): $R_{if} = R_{C1} \| [R_i'(1 + A_i b)] = 200$ kΩ. The bandwidth is given by (10.123), where all of the terms are the same as in Exercise 10.10, except for α_i'. Using our new value of $\alpha_i' = 0.377$ and all other values from Exercise 10.10, we find that $f_{sf} = 18.6$ kHz.

10.15 The PSPICE schematic is shown in Figure 10-103. The simulation results agree with our calculations.

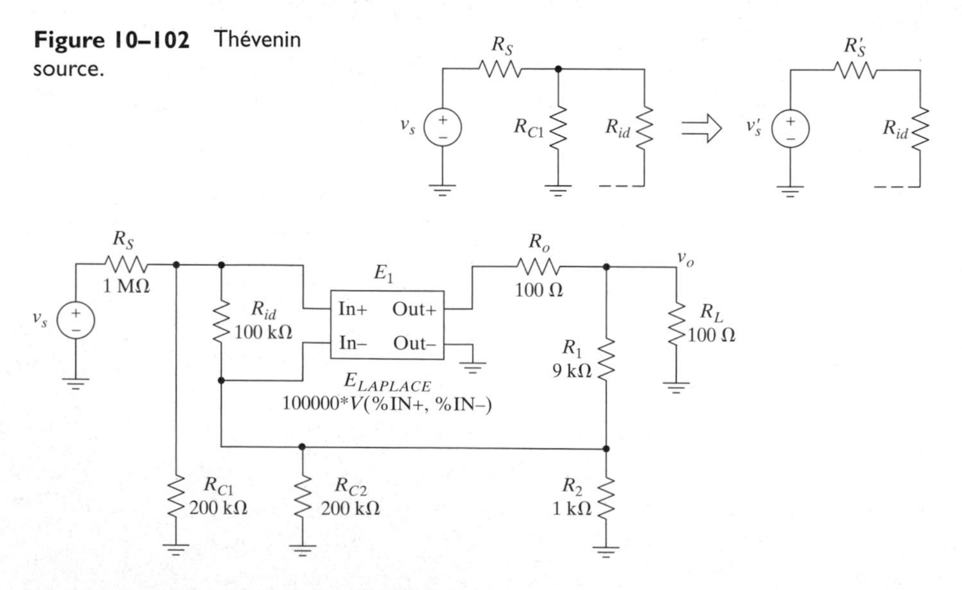

Figure 10–102 Thévenin source.

Figure 10–103 The PSPICE schematic.

10.16 (a) We first perform a DC bias point analysis and find (note that C_F causes the global feedback to be present only for signals) that $I_{C1} = 240\ \mu A$, $I_{C2} = 1.3$ mA, and $I_{C3} = 5.5$ mA. Using these values, we calculate $g_{m1} = 9.23$ mA/V, $g_{m2} = 50$ mA/V, $g_{m3} = 212$ mA/V, $r_{\pi 1} = 10.9\ k\Omega$, $r_{\pi 2} = 2\ k\Omega$, and $r_{\pi 3} = 472\ \Omega$. Following the analysis of the generic series–series triple, we then find the following small-signal midband AC quantities: $R_{BB} = R_1 \| R_2 = 38.1\ k\Omega$, $R_S' = R_S \| R_{BB} = 974\ \Omega$, $R_L' = R_L \| R_{C3} = 412\ \Omega$, $R_{i2} = R_{C1} \| r_{\pi 2} = 1.88\ k\Omega$, $R_{bi} = R_{E3} \| (R_{E1A} + R_F) = 804\ \Omega$, $R_{i3} = r_{\pi 3} + (\beta + 1)R_{bi} = 81.7\ k\Omega$ $R_{bo} = R_{E1A} \| (R_F + R_{E3}) = 98\ \Omega$, $R_i' = r_{\pi 1} + (\beta+1)R_{bo} = 20.8\ k\Omega$, R_o' is infinite (ignoring r_o of Q_3),

$$b = R_{bi}\left(\frac{R_{E1A}}{R_{E1A} + R_F}\right) = 19.6\ \Omega,$$

$$\alpha_i' = \frac{R_i'}{R_i' + R_S'} = 0.954,$$

$$\alpha_o' = \frac{R_o'}{R_o' + R_L'} = 1,$$

$$a' = \beta_3\left(\frac{-g_{m2}R_{C2}}{R_{C2} + R_{i3}}\right)(-g_{m1}R_{i2})\left(\frac{r_{\pi 1}}{r_{\pi 1} + (\beta+1)R_{bo}}\right) = 1.6\ \text{A/V},$$

$$A_s' = \alpha_i' a' \alpha_o' = 1.54\ \text{A/V},$$

$$A_{sf}' = \frac{A_s'}{1 + A_s' b} = 49.5\ \text{mA/V},$$

and

$$\frac{v_o}{v_s} = R_L' A_{sf}'\left(\frac{R_{BB}}{R_{BB} + R_S}\right) = 19.9.$$

A SPICE simulation reveals that $v_o/v_s = 19.6$, so our result is quite close (the error is due to our approximate DC bias point calculation.) (b) As in the generic transistor example, we predict $R_{of} = R_{C3} = 430\ \Omega$. Then, using (10.179) and (10.180), we find that $R_{if} = R_{BB} \| [R_i'(1 + a'\alpha_o' b)] = 36\ k\Omega$. The SPICE simulation reveals that $R_{if} = 36\ k\Omega$ and $R_{of} = 430\ \Omega$.

10.17 Substitute (10.200) into (10.186) and then into (10.191) to obtain

$$A_s(j\omega) = \frac{\alpha_i' a_0' \alpha_o'}{1 + j\omega/\omega_p}.$$

Then substitute this equation into (10.193) and do some algebra to find

$$A_{sf}(j\omega) = \frac{\alpha_i' a_0' \alpha_o'}{1 + \alpha_i' a_0' \alpha_o' b + j\omega/\omega_p} = \frac{A_{sf0}}{1 + j\omega/\omega_{sf}},$$

where

$$A_{sf0} = \frac{\alpha_i' a_0' \alpha_o'}{1 + \alpha_i' a_0' \alpha_o' b}$$

and $\omega_{sf} = \omega_p(1 + \alpha_i' a_0' \alpha_o' b) = \omega_p(1 + L_s)$.

10.18 We start with a DC bias point analysis. Assuming that we can neglect the DC drop across R_f (we'll check this in a moment), we have $V_{E2} \approx 0.7$ V, which leads to $I_{E2} \approx 2$ mA. Also, we can say that $V_{C1} \approx 1.4$ V, so $I_{C1} \approx 1$ mA. With $\beta = 100$, we then have $I_{b1} \approx 10\ \mu A$ and $I_{B1}R_f = 0.5$ mV, which is certainly negligible. This bias point leads to $r_{\pi 1} = 2.6\ k\Omega$ and $r_{\pi 2} = 1.3\ k\Omega$. We next identify the feedback network as comprising R_f and R_{E2} and the forward amp as Q_1, Q_2, and R_C. C_{IN}, C_{OUT}, and R_1 are outside of the feedback loop. The feedback network parameters are calculated as in Example 10.5; we get, $R_{bi} = R_f \| R_{E2} = 211\ \Omega$, $R_{bo} = R_f = 510\ \Omega$, and $b = -1/R_f = -1.96$ mA/V. The results from the example apply exactly by changing the notation as appropriate for the BJT. We find that $a' = -48.2\ \text{k}\Omega$, $R_i' = 426\ \Omega$, $R_o' = 39.8\ \Omega$, $R_S' = 97\ \Omega$, $\alpha_i' = 0.185$, $\alpha_o' = 0.56$, $R_{if}' = 7.9\ \Omega$, $R_{if} = 55\ \Omega$, $R_{of} = 2.2\ \Omega$, $A_i = 27\ \text{k}\Omega$, $A_{if} = 501\ \Omega$, and $v_o/v_i = 9.1$. A SPICE simulation of this circuit reveals that $v_o/v_i = 9$, $R_{if} = 55\ \Omega$, and $R_{of} = 2.3\ \Omega$.

10.19 The minimum stable closed-loop gain is equal to the open-loop gain at ω_{180} and is 0 dB. Note that if $1/b$ is 0 dB, then ω_u is equal to ω_{180}, and both the *GM* and *PM* are zero.

10.20 The Bode plots for the amplifier are shown in Figure 10-104. (a) To have a 45° phase margin without compensation, we need ω_u to be 100 krad/s, as shown on the plot. For the loop gain to be 0 dB at this frequency, we need $1/b$ to be 60 dB, which is also the lowest closed-loop gain we can achieve. (b) From the plot, we see that in order for the magnitude of a to intersect $1/b$ at 30 dB at $\omega_u =$ 100 krad/s, which is required to achieve a 45° phase margin, the dominant pole must be moved down to 32 rad/s (halfway between 10 and 100 on a logarithmic scale).

10.21 Using (10.268) and setting $R = 1\ \text{k}\Omega$, we find that $C = 15.9$ nF. Using $R_1 = 1\ \text{k}\Omega$ as well, we find that $R_2 = 3\ \text{k}\Omega$ to get a loop gain of 4/3. When the diode limiters are added, we find that the series resistors must be 6.7 kΩ so that either one of them in parallel with R_2 will be 2.5 kΩ. The resulting SPICE circuit is shown at the top of pg 767 (an initial condition of 0.1 V was added to C_1 to get the oscillations to start in SPICE):

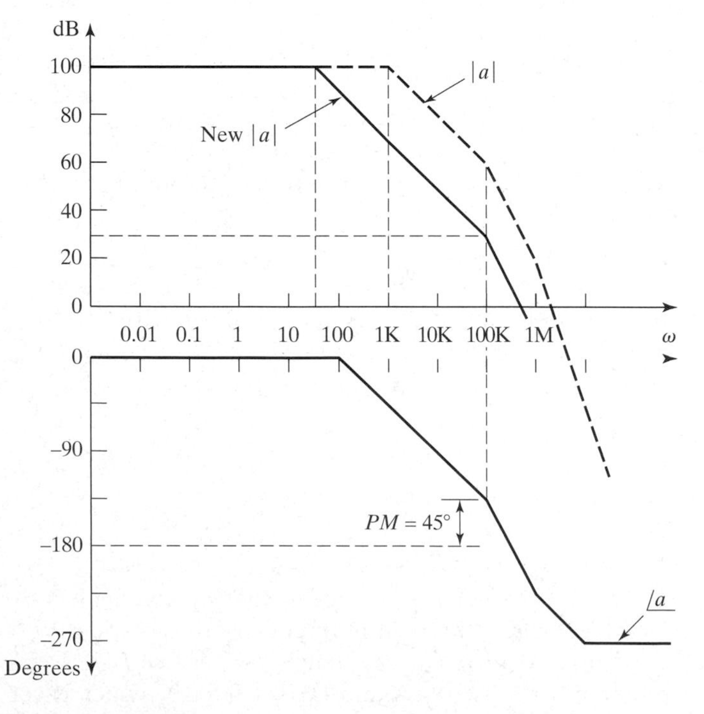

Figure 10–104

PSPICE schematic for exercise 10.21

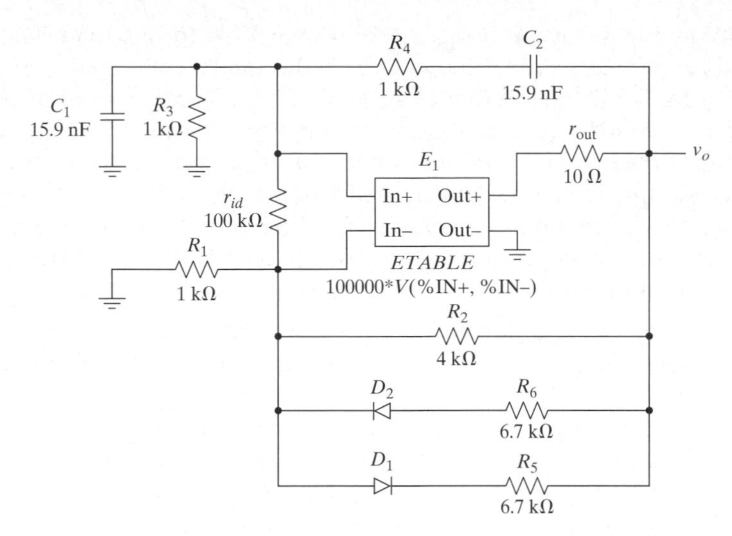

Transient simulations were run on this circuit, both with and without the diode limiters in place (see Exer21a.sch and Exer21b.sch). The Fourier components for the circuit without the diodes are as follows:

```
FOURIER COMPONENTS OF TRANSIENT RESPONSE V(out)

DC COMPONENT = 2.766734E-02
```

HARMONIC NO	FREQUENCY (HZ)	FOURIER COMPONENT	NORMALIZED COMPONENT	PHASE (DEG)	NORMALIZED PHASE (DEG)
1	7.700E+03	5.868E+00	1.000E+00	-9.044E+01	0.000E+00
2	1.540E+04	5.491E-01	9.357E-02	2.717E+00	9.315E+01
3	2.310E+04	1.088E+00	1.854E-01	8.136E+01	1.718E+02
4	3.080E+04	5.752E-01	9.802E-02	-1.712E+02	-8.074E+01
5	3.850E+04	3.227E-01	5.500E-02	-1.210E+02	-3.056E+01
6	4.620E+04	2.648E-01	4.513E-02	-3.758E+00	8.668E+01
7	5.390E+04	1.651E-01	2.813E-02	7.048E+01	1.609E+02
8	6.160E+04	1.280E-01	2.082E-02	1.207E+02	2.111E+02
9	6.930E+04	1.006E-01	1.715E-02	-1.076E+02	-1.717E+01

```
TOTAL HARMONIC DISTORTION = 2.436198E + 01 PERCENT
```

The Fourier components for the circuit with the diodes areas follows:

```
FOURIER COMPONENTS OF TRANSIENT RESPONSE V(out)

DC COMPONENT = -5.945318E-02
```

HARMONIC NO	FREQUENCY (HZ)	FOURIER COMPONENT	NORMALIZED COMPONENT	PHASE (DEG)	NORMALIZED PHASE (DEG)
1	9.200E+03	5.696E+00	1.000E+00	5.277E+01	0.000E+00
2	1.840E+04	1.005E-01	1.763E-02	6.378E+01	1.101E+01
3	2.760E+04	8.687E+01	1.525E-01	1.362E+02	8.344E+01
4	3.680E+04	5.251E-02	9.217E-03	-1.307E+02	-1.834E+02
5	4.600E+04	3.236E-01	5.681E-02	-1.690E+02	-2.218E+02
6	5.520E+04	3.245E-02	5.697E-03	8.552E+01	3.275E+01
7	6.440E+04	5.721E-02	1.004E-02	-1.107E+02	-1.635E+02
8	7.360E+04	3.098E-02	5.438E-03	-1.026E+02	-1.554E+02
9	8.280E+04	5.961E-02	1.046E-02	-8.147E+01	-1.342E+02

```
TOTAL HARMONIC DISTORTION = 1.647825E + 01 PERCENT
```

The total harmonic distortion dropped from over 24% to less than 17% with the addition of the diode limiters. With the diode limiters, the frequency of oscillation is reasonably close to the value predicted. Without the diode limiters, the frequency of oscillation is lower because the output voltage swings from −5 V to +5 V and, when it is at one limit or the other, the circuit is not functioning as a linear circuit. In fact, the incremental loop gain is zero, since the op amp is not amplifying at all when its output is saturated. Under these conditions, the frequency depends on RC charging and discharging, not on the phase shift through the network.

10.22 Using (10.269), we find that the circuit will oscillate at 1 kHz. We also see that the gain is more than sufficient to start oscillations, so the oscillations will grow until they are limited to ±1 V by the amplifier. The PSPICE schematic is shown in Figure 10-105 and includes a voltage source (which is a pulse) to start the oscillations. The output is shown in Figure 10-106 and shows that the frequency is 1 kHz, as expected, and the magnitude grows until limited, as predicted.

10.23 The small-signal midband AC equivalent circuit is shown in Figure 10-107, where we have defined $R_i = R_1 \| R_2 \| r_\pi$ and $R'_C = R_C \| r_o$. A bias point analysis reveals that $I_C = 1$ mA, so using the 2N2222 model presented in Appendix B, we find that $r_o = 74$ kΩ, $r_\pi = 4.7$ kΩ, and $g_m = 38.5$ mA/V. This figure is topologically identical to Figure 10-85, so the results derived there apply. In particular, we find that the frequency of oscillation is

$$f_o = \frac{1}{2\pi\sqrt{LC_t}} = 5 \text{ kHz}.$$

The schematic used for simulation is shown in the exercise (Figure10-87). A sample output is shown in Figure 10-108. This figure shows both the output voltage and the collector current of the transistor. Notice that the transistor is on only for brief periods of time, but even though the current does not look anything at all like a sine wave, the selectivity of the tank circuit causes the output to be a reasonable sine wave. (The collector current is periodic with a period of 5 kHz, so that its Fourier series contains a component at 5 kHz, which is what the tank circuit selects.) Note that we used a simulation step size of 1 μs. If you use the default, the simulation does not work correctly. Also, if you view the output from zero to about 6 or 7 ms, you will see the oscillations build up and their amplitude stabilize.

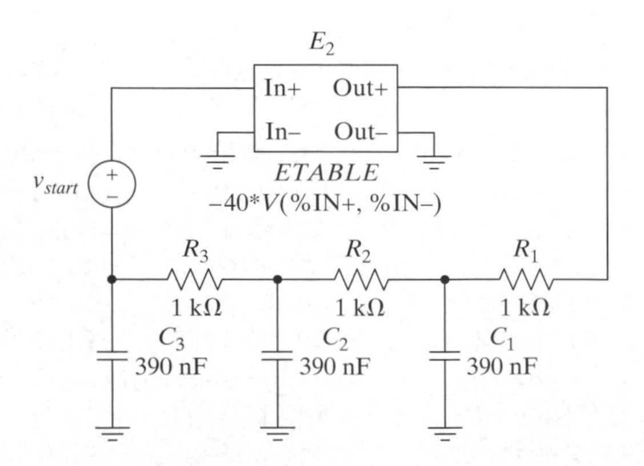

Figure 10–105

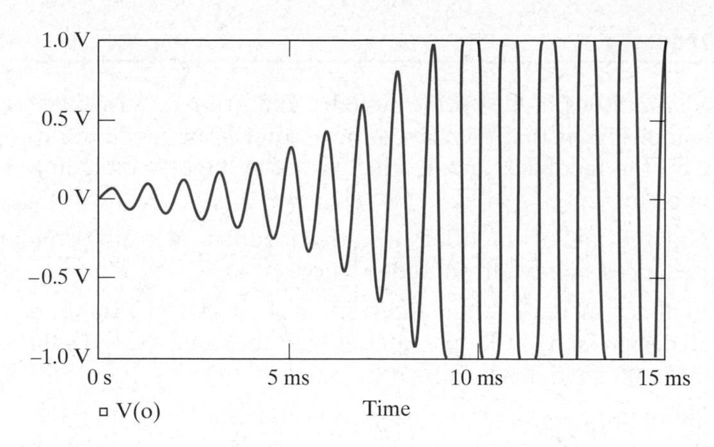

Figure 10–106

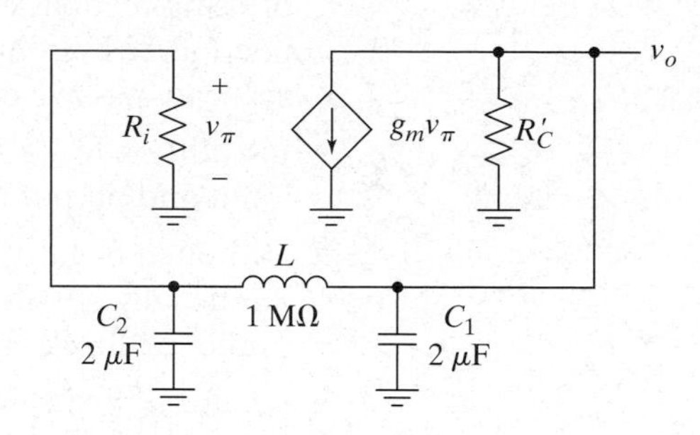

Figure 10–107

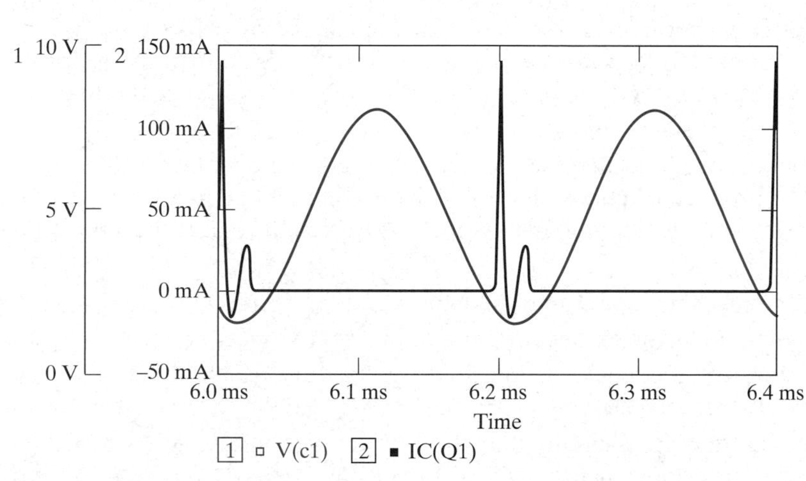

Figure 10–108

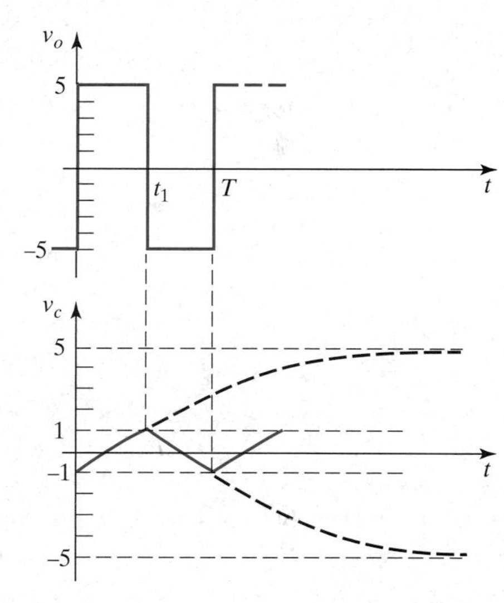

Figure 10–109

10.24 The output and capacitor voltages are shown in Figure 10-109. For $0 \le t \le t_1$, the output is high, and the capacitor voltage is $v_C(t) = -1 + 6(1 - e^{t/RC})$. Setting this equation equal to 1 V (the upper threshold), we find that $t_1 = 40.5$ μs. By symmetry, we know that the period is $T = 2t_1 = 81.1$ μs, which implies that the frequency of oscillation is 12.3 kHz.

10.25 We use the circuit shown in Figure 10-93. From (10.280), we find that, to obtain a 2.5-ms pulse width, we need $R_1C = 2.27$ ms. Therefore, if we use $C = 0.1$ μF and $R_1 = 22$ kΩ, we will be close.

10.26 Use the circuit in Figure 10-95. To achieve a 67% duty cycle, we need to set $t_H/(t_L + t_H) = 0.67$. In other words, we need $t_H = 2t_L$. Using (10.282) and (10.284), we see that we can achieve this goal by setting $R_1 = R_2$. So, using (10.286), we find that one possible solution is $C = 0.0015$ μF and $R_1 = R_2 = 13$ kΩ.

CHAPTER SUMMARY

- Negative feedback is ubiquitous in electronic systems. The primary advantage of using feedback in amplifiers is that the closed-loop gain can be made to depend almost entirely on the feedback factor, which is set by highly stable and accurate passive components.
- In addition to stabilizing the gain of amplifiers, negative feedback also alters the input and output impedances, bandwidth, and distortion.
- Although negative feedback increases the bandwidth of the forward amplifier, it is usually possible to design an open-loop amplifier with the same overall gain as the feedback amplifier and a higher bandwidth.
- Feedback by itself cannot improve the signal-to-noise ratio of an amplifier, but it can be used with a low-noise preamplifier as a part of a solution. It can reduce the amount of noise present in DC current sources.
- There are four different feedback topologies, depending on whether voltages or currents are summed at the input and whether the output voltage or current is fed back.
- Series input connections increase the input impedance by a factor $(1 + L_i)$, and shunt input connections decrease the input impedance by a factor $(1 + L_i)$, where L_i is the loop gain with the source removed, but the load included.
- Series output connections increase the output impedance by a factor $(1 + L_o)$, and shunt output connections decrease the output impedance by a factor $(1 + L_o)$, where L_o is the loop gain with the load removed, but the source included.
- We analyze practical feedback amplifiers by
  1 modeling the feedback network as a unilateral two-port network,
  2 absorbing the feedback loading into the forward amplifier to form the primed network, and then
  3 calculating the open-circuit gain A_i or A_s as appropriate. We know then that the closed-loop gains will be given by $A_{if} = A_i/(1 + A_i b)$ and $A_{sf} = A_s/(1 + A_s b)$.
- When the forward amplifier or feedback network cannot reasonably be approximated as unilateral, we can use full two-port models to represent them and can then manipulate the circuit to have the same form as that which we analyzed for unilateral systems.
- There are several techniques available for simulating the loop gain of negative-feedback amplifiers, but we must be careful, since the loop gain is *not* generally equal to the negative of the return ratio around the loop, as is sometimes erroneously stated.
- Negative-feedback amplifiers can become unstable if the loop accumulates an additional 180° of phase shift at a frequency where the loop gain is greater than or equal to unity. In this context, the word unstable does not mean that the gain, or some other parameter, changes with time; rather, it means that the circuit can oscillate.
- We quantify how close a given amplifier is to being unstable by specifying the gain and phase margins. The gain margin is the amount of extra loop gain required to cause the loop to be unstable, and the phase margin is the additional phase shift that will cause the loop to be unstable.

- We can compensate an amplifier to cause it to be stable for a given closed-loop gain by
  1. reducing the open-loop gain,
  2. decreasing the dominant pole frequency,
  3. introducing a new dominant pole, or
  4. using a more advanced technique like lead–lag compensation.
- Positive feedback is used to produce oscillators, which are an extremely important class of circuits. Oscillators can produce nearly sinusoidal signals, square waves, triangle waves, and other waveforms. They are used, for example, to generate the carrier signals for radio and television broadcasts and to generate the clock waveforms employed in computers and other modern synchronous digital systems.
- Practical oscillators use loop gains greater than unity to guarantee that the oscillations will start. The oscillations then grow until some nonlinearity in the loop limits their amplitude or changes the loop gain. As a result of this nonlinearity, distortion is produced and must be removed if a pure sine wave is desired.
- Sinusoidal oscillators use the phase shift through *RC* networks or the frequency selectivity of *LC* circuits to determine the frequency of oscillation of a positive feedback loop. *LC* oscillators tend to have lower distortion, due to the frequency selectivity inherent in the resonant circuit.
- Nonsinusoidal oscillators use limit cycles to produce oscillations. The most popular example of this type of oscillator is the common 555 IC Timer circuit.

REFERENCES

[10.1] R. C. Jaeger, *Microelectronic Circuit Design*. New York: McGraw-Hill, 1997.

[10.2] K. Ogata, *Modern Control Engineering*, Upper Saddle River, NJ 3/E. Prentice Hall, 1996.

[10.3] A.S. Sedra and K.C. Smith, *Microelectronic Circuits*, F/E: New York. Oxford University Press, 1998.

[10.4] H.S. Black, "Inventing the negative feedback amplifier," *IEEE Spectrum*, Dec., 1977.

[10.5] J.E. Brittain, "Harold S. Black and the negative feedback amplifier," *Proceedings of the IEEE*, Vol. 85, No. 8, Aug., 1997, p. 1335.

[10.6] D.O. Pederson and K. Mayaram, *Analog Integrated Circuits for Communication*. Kluwer Academic Publishers, 1991.

[10.7] T.H. Lee, *The Design of CMOS Radio-Frequency Integrated Circuits*. Cambridge University Press, 1998.

[10.8] H.R. Camenzind, "Redesigning the old 555," *IEEE Spectrum*, Sept. 1997, p. 80.

[10.9] P.J. Hurst, "A comparison of two approaches to feedback circuit analysis," *IEEE Trans. on Education*, Vol. 35, No. 3, August, 1992, pp 253–261 (The return ratio analysis presented in this paper is topologically identical to the procedure presented in A.M. Davis, "A general method for analyzing feedback amplifiers," *IEEE Trans. on Education*, Vol. E-24, No. 4, November, 1981, pp 291–293.)

[10.10] P.J. Hurst, "Exact simulation of feedback circuit parameters," *IEEE Trans. on Circuits and Systems*, Vol. 38, No. 11, November, 1991, pp. 1382–1389.

PROBLEMS

10.1 Negative Feedback

10.1.1 Ideal Block Diagram Analysis

P10.1 Suppose we want to construct an amplifier with a closed-loop gain of 50. We want the gain to vary by no more than ± 5% while the open-loop gain varies by ± 50%. Determine values for *a* and *b* that will satisfy these constraints.

P10.2 Repeat Problem P10.1, but allow *b* to vary by ± 2% as well.

P10.3 Suppose you are designing a feedback system to control some industrial process. You must keep the error less than 1% of the input amplitude, and you need a closed-loop gain of 10. Determine the required values of *a* and *b*.

10.1.2 Ideal Analysis and the Characteristics of Negative Feedback

P10.4 We desire to construct a negative-feedback amplifier having a closed-loop gain of 20, using a forward amplifier with the large-signal transfer characteristic shown in Figure 10-110. **(a)** What value of *b* must we use to set $A_f = 20$ for S_I near zero? **(b)** Draw the large-signal transfer characteristic for the closed-loop amplifier, and label all breakpoint values. **(c)** What is the small-signal closed-loop gain for $S_I = 2$? How would you find this gain from the transfer characteristic you drew in part (b)?

P10.5 Consider a negative-feedback amplifier constructed by using a forward amplifier with the large-signal transfer characteristic shown in Figure 10-110 and having $R_i = 25$ kΩ and $R_o = 20$ Ω. The closed-loop amplifier has $b = 0.066$ and uses a shunt–series topology. What are the small-signal input and output resistances, R_{if} and R_{of}, for S_I near zero?

P10.6 A negative-feedback amplifier has a forward amplifier with a single-pole low-pass transfer function with a DC gain of 100,000 and a pole at 5 Hz. **(a)** What is the bandwidth of the closed-loop amplifier if the closed-loop gain is 20? **(b)** What is the bandwidth of the closed-loop amplifier if the closed-loop gain is unity? (This is called the unity-gain bandwidth.)

P10.7 An amplifier with a single-pole low-pass transfer function with a DC gain of 300 and a pole at 5 kHz also has $R_i = 10$ kΩ and $R_o = 50$ Ω. We want to use this amplifier in a negative-feedback connection to produce a closed-loop amplifier with a gain of 10. We want to achieve the

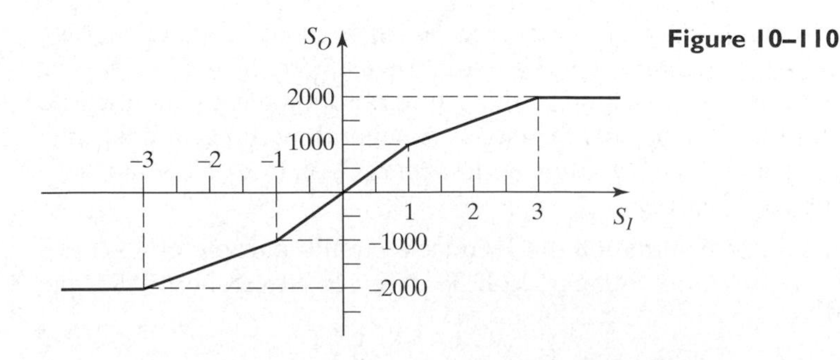

Figure 10–110

largest input resistance and smallest output resistance possible. **(a)** What type of feedback connection should we use? **(b)** What value of b is required? **(c)** What are the resulting values of R_{if} and R_{of}? **(d)** What is the resulting closed-loop bandwidth?

P10.8 An amplifier with a single-pole low-pass transfer function with a DC gain of 700 and a pole at 1 kHz also has $R_i = 1\ \text{k}\Omega$ and $R_o = 100\ \Omega$. We want to use this amplifier in a negative-feedback connection to produce a closed-loop amplifier with a gain of 20. We want to achieve the smallest input and output resistances possible. **(a)** What type of feedback connection should we use? **(b)** What value of b is required? **(c)** What are the resulting values of R_{if} and R_{of}? **(d)** What is the resulting closed-loop bandwidth?

P10.9 An amplifier with a single-pole low-pass transfer function with a DC gain of 500 and a pole at 2 kHz also has $R_i = 20\ \text{k}\Omega$ and $R_o = 10\ \text{k}\Omega$. We want to use this amplifier in a negative-feedback connection to produce a closed-loop amplifier with a gain of 15. We want to achieve the largest input and output resistances possible. **(a)** What type of feedback connection should we use? **(b)** What value of b is required? **(c)** What are the resulting values of R_{if} and R_{of}? **(d)** What is the resulting closed-loop bandwidth?

P10.10 An amplifier with a single-pole low-pass transfer function with a DC gain of 400 and a pole at 1 kHz also has $R_i = 1\ \text{k}\Omega$ and $R_o = 750\ \Omega$. We want to use this amplifier in a negative-feedback connection to produce a closed-loop amplifier with a gain of 10. We want to achieve the smallest input resistance and largest output resistance possible. **(a)** What type of feedback connection should we use? **(b)** What value of b is required? **(c)** What are the resulting values of R_{if} and R_{of}? **(d)** What is the resulting closed-loop bandwidth?

10.1.3 First-Order Practical Analysis

P10.11 Consider the input impedance of a feedback amplifier where the forward amplifier has a single-pole low-pass response as given in (10.85) and the feedback is independent of frequency. Assume that $R_i' = 50\ \text{k}\Omega$, $\omega_p = 1$ krad/s, and $L_{i0} = 200$. **(a)** Assume that the input of the feedback amplifier is a series connection. Show that, in this case, the input impedance is equivalent to the circuit illustrated in Figure 10-111(a). Find values for all three components. **(b)** Assume that the input of the feedback amplifier is a shunt connection. Show that, in this case, the input impedance is equivalent to the circuit illustrated in Figure 10-111(b). Find values for all three components.

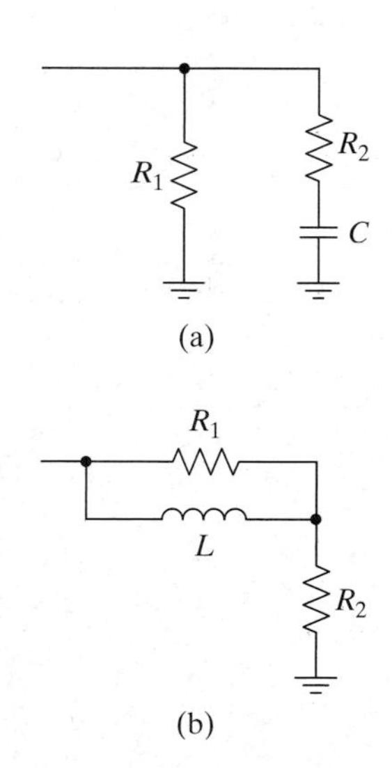

Figure 10–111

P10.12 Consider the output impedance of a feedback amplifier where the forward amplifier has a single-pole low-pass response as given in (10.85) and the feedback is independent of frequency. Assume that $R_o' = 500\ \Omega$, $\omega_p = 1$ krad/s, and $L_{i0} = 200$. **(a)** Assume that the output of the feedback amplifier is a series connection. Derive the output impedance as a function of frequency and draw an equivalent RLC circuit. Provide values for each of the components in the equivalent circuit. **(b)** Repeat (a), assuming that the output is a shunt connection.

P10.13 Consider the small-signal equivalent circuit for the feedback amplifier shown in Figure 10-112, where the forward amplifier is modeled as having a single-pole transfer function with a pole frequency $\omega_p = 1/R_1C$. Show, by direct analysis of the time constant associated with C, that the closed-loop pole frequency is given by $\omega_{pf} = \omega_p(1 + Ab)$, as expected.

P10.14 For each of the circuits shown in Figure 10-113, indicate what kind of feedback connection is used (e.g., series–shunt) and whether the feedback is positive or negative.

Series–Shunt Connection

P10.15 Show that (10.75) is correct.

P10.16 Consider the series–shunt amplifier shown in Figure 10-114. The transistors are 2N2857's, and the model parameters are given in Appendix B. **(a)** Identify the forward amplifier and the feedback network. **(b)** Draw the small-signal midband AC model for the amplifier, using a unilateral two-port model for the feedback network. Calculate the parameters of the feedback network model. **(c)** Determine the midband values of A_{if} and A_{sf}. **(d)** Determine the midband values of R_{if} and R_{of}.

S P10.17 Use SPICE and the Q2N2857 model presented in Appendix B to confirm your results for Problem P10.16. Explain how you find each parameter, and provide the raw output as well as the final answer.

P10.18 Continue with Problem P10.16. Find a single-pole approximation to the forward amplifier's frequency response, and use it to determine values for f_{if} and f_{sf}.

S P10.19 Use SPICE and the Q2N2857 model presented in Appendix B to confirm your results for Problem P10.18. Hand in a SPICE plot of the frequency response showing the values of f_{if} and f_{sf}.

P10.20 Change the design of the circuit in Figure 10-114 to produce an amplifier with an overall closed-loop voltage gain of about 5.

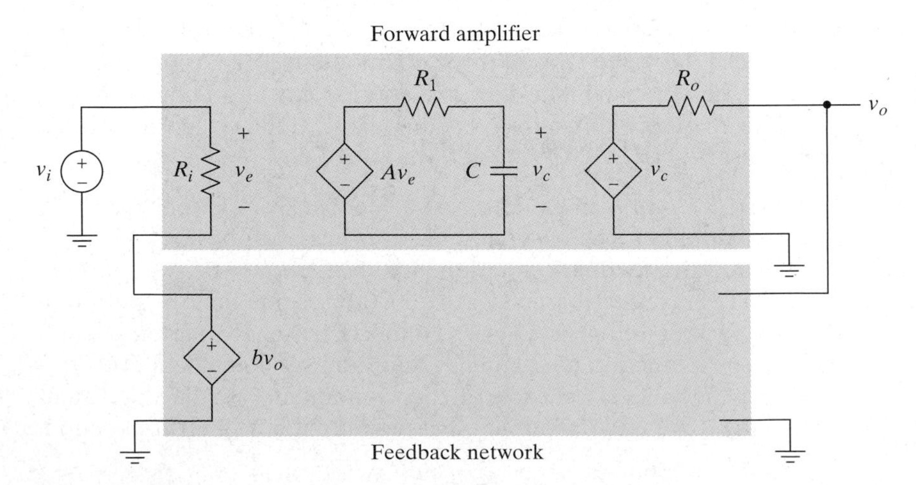

Figure 10–112

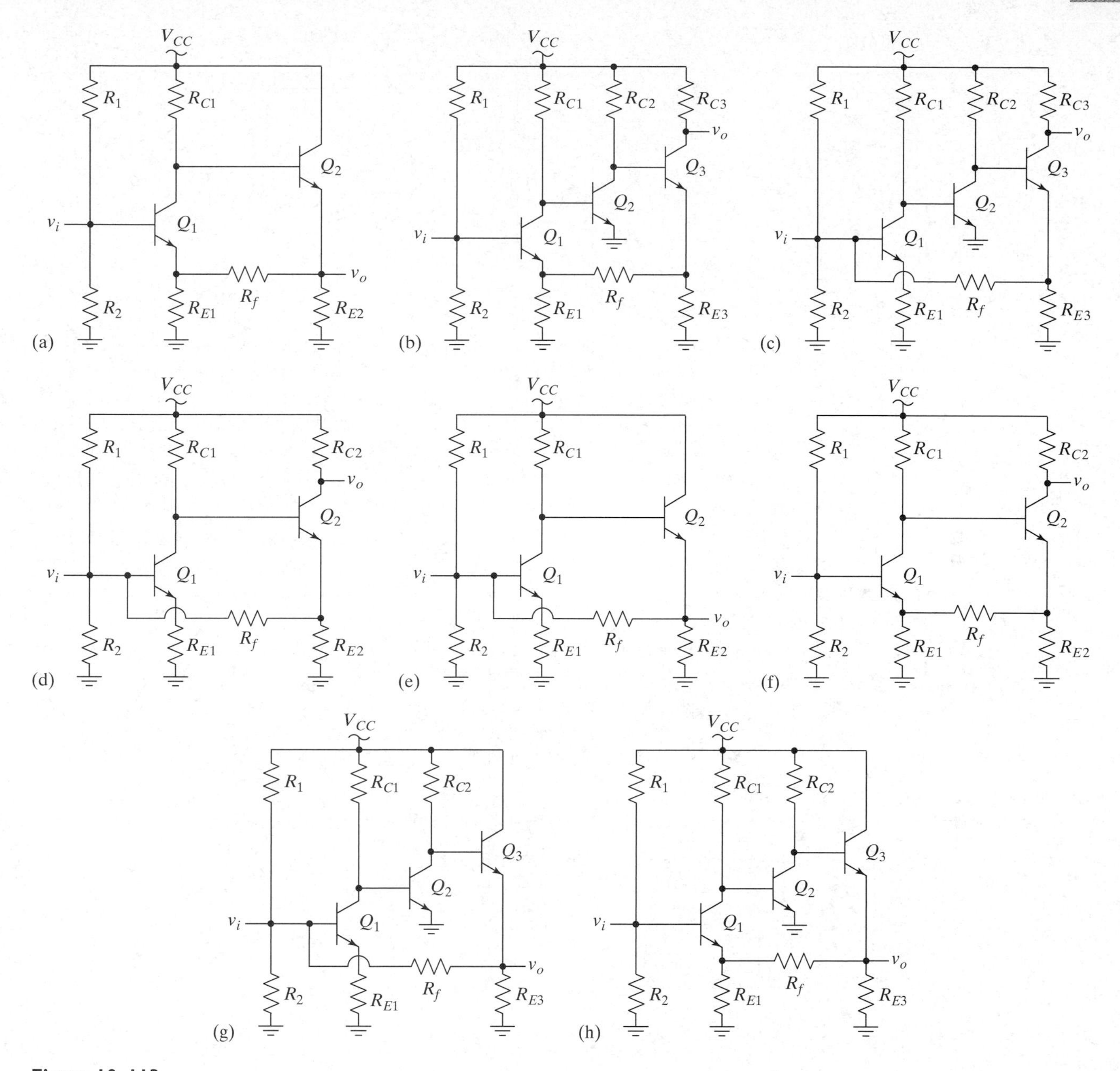

Figure 10–113

P10.21 Consider the series–shunt amplifier shown in Figure 10-115 and assume $V_S = 0$ V. The transistors are 2N2857's, and the model parameters are given in Appendix B. **(a)** Identify the forward amplifier and the feedback network. **(b)** Draw the small-signal midband AC model for the amplifier, using a unilateral two-port model for the feedback network. Calculate the parameters of the feedback network model. **(c)** Determine the midband values of A_{if} and A_{sf}. **(d)** Determine the midband values of R_{if} and R_{of}.

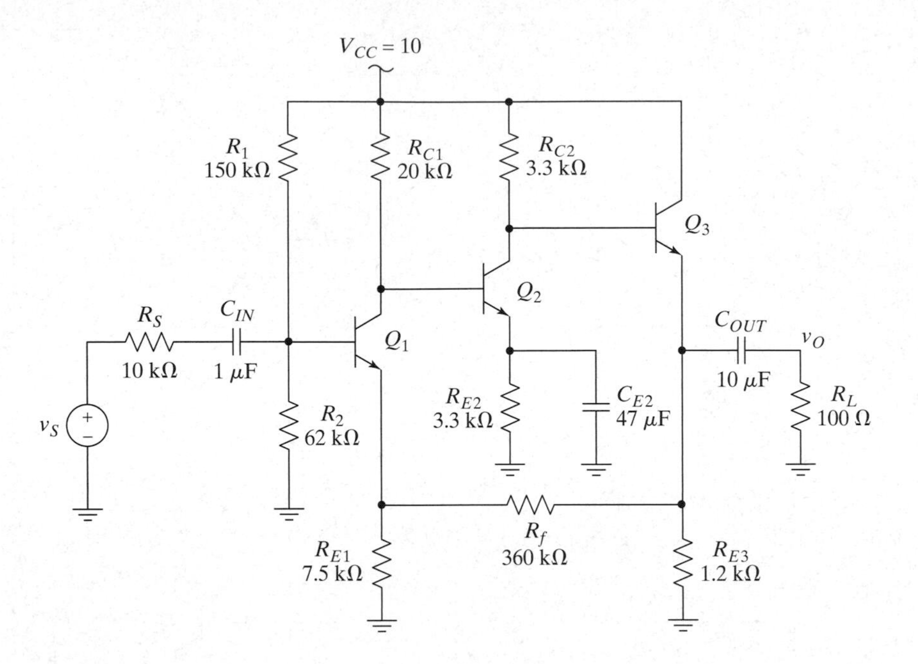

Figure 10–114

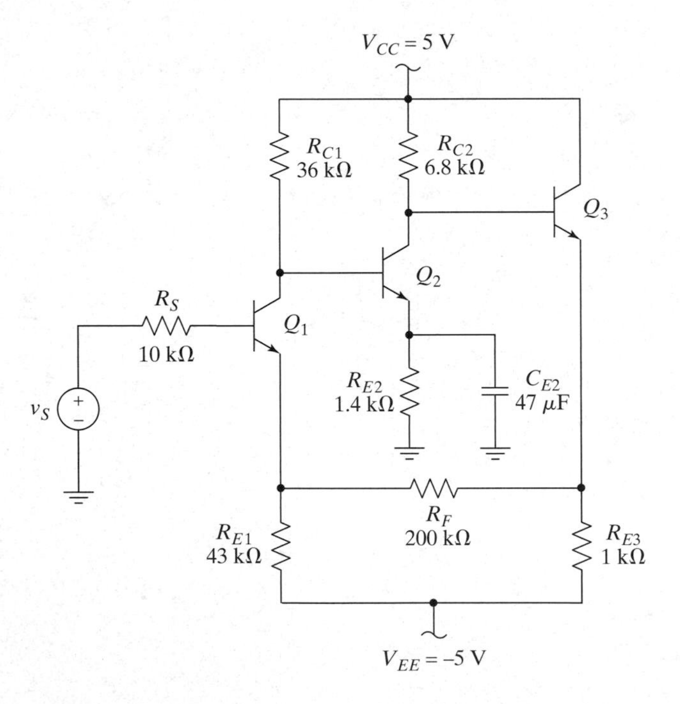

Figure 10–115

S P10.22 Use SPICE and the Q2N2857 model presented in Appendix B to confirm your results for Problem P10.21. Explain how you find each parameter, and provide the raw output as well as the final answer.

P10.23 Continue with Problem P10.21. Find a single-pole approximation to the forward amplifier's frequency response, and use it to determine values for f_{if} and f_{sf}.

S P10.24 Use SPICE and the Q2N2857 model presented in Appendix B to confirm your results for Problem P10.23. Hand in a SPICE plot of the frequency response showing the values of f_{if} and f_{sf}.

P10.25 Consider the op amp circuit shown in Figure 10-116. Assume that the DC gain of the op amp is 200,000 and that the circuit has a single pole at 3 Hz. **(a)** Find $A_{sf}(j\omega)$. **(b)** Find the DC values of R_{if} and R_{of}.

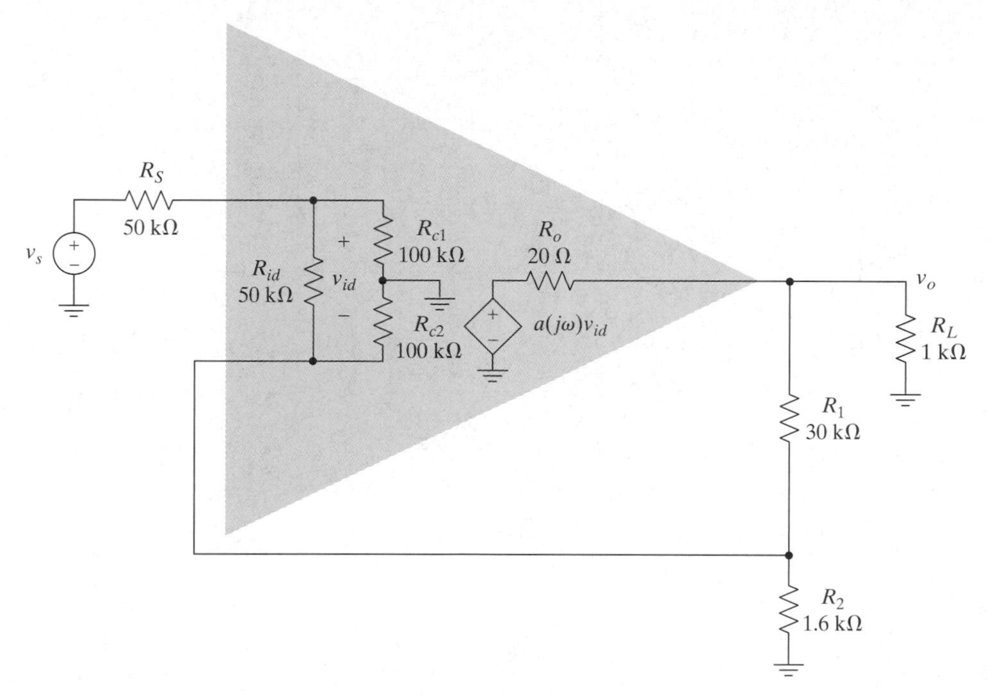

Figure 10–116

S P10.26 Use SPICE to confirm your results for Problem P10.25. Explain how you find each parameter, and provide the raw output as well as the final answer.

P10.27 Consider the series–shunt amplifier shown in Figure 10-117. Use the transistor model parameters given in Appendix B. M_1 and M_2 are both Nsimple-mos and M_3 is Nlarge-mos. **(a)** Identify the forward amplifier and the feedback network. **(b)** Draw the small-signal midband AC model for the amplifier, using a unilateral two-port model for the feedback network. Calculate the parameters of the feedback network model. **(c)** Determine the midband values of A_{if} and A_{sf}. **(d)** Determine the midband values of R_{if} and R_{of}.

S P10.28 Use SPICE and the Nsimple_mos and Nlarge_mos models presented in Appendix B to confirm your results to Problem P10.27. Explain how you find each parameter, and provide the raw output as well as the final answer.

P10.29 Continue with Problem P10.27. Find a single-pole approximation to the forward amplifier's frequency response, and use it to determine values for f_{if} and f_{sf}. Note: Since the forward amplifier's input impedance is purely capacitive, use $R_{SIG} = 0$ to find f_{if}.

S P10.30 Use SPICE and the Nsimple_mos and Nlarge_mos models presented in Appendix B to confirm your results for Problem P10.29. Hand in a SPICE plot of the frequency response showing the values of f_{if} and f_{sf}.

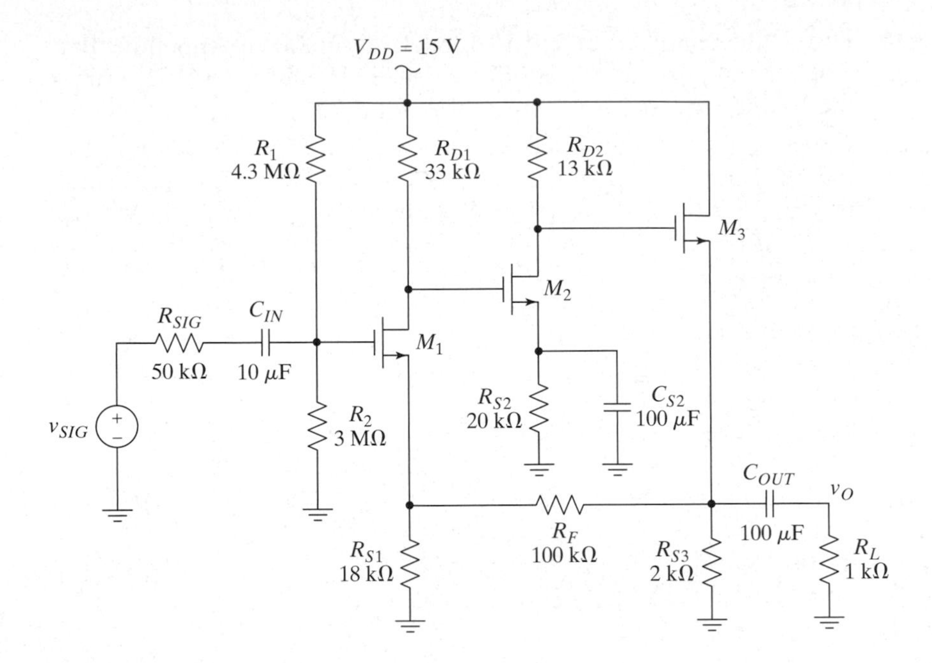

Figure 10–117

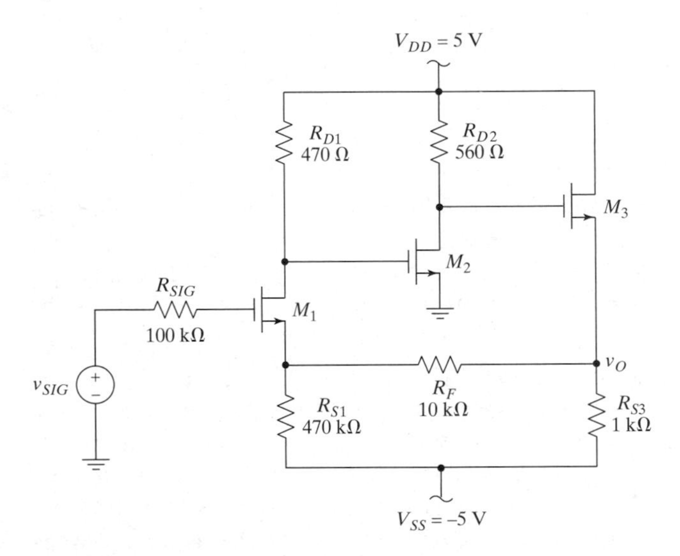

Figure 10–118

P10.31 Consider the series–shunt amplifier shown in Figure 10-118. Use the Nlarge_mos transistor model parameters given in Appendix B. **(a)** Identify the forward amplifier and the feedback network. **(b)** Draw the small-signal midband AC model for the amplifier, using a unilateral two-port model for the feedback network. Calculate the parameters of the feedback network model. **(c)** Determine the midband values of A_{if} and A_{sf}. **(d)** Determine the midband values of R_{if} and R_{of}.

S P10.32 Use SPICE and the Nlarge_mos model presented in Appendix B to confirm your results for Problem P10.31. Explain how you find each parameter, and provide the raw output as well as the final answer.

P10.33 Continue with Problem P10.31. Find a single-pole approximation to the forward amplifier's frequency response, and use it to determine values for f_{if} and f_{sf}. Note: Since the forward amplifier's input impedance is purely capacitive, use $R_{SIG} = 0$ to find f_{if}.

S P10.34 Use SPICE and the Nlarge_mos model presented in Appendix B to confirm your results for Problem P10.33. Hand in a SPICE plot of the frequency response showing the values of f_{if} and f_{sf}.

P10.35 Consider the emitter-coupled pair emitter-follower cascade shown in Figure 10-119. This circuit is, in essence, a very simple operational amplifier. Use the nominal transistor model given in Appendix B. **(a)** Determine the DC biasing, and draw the small-signal midband AC equivalent circuit. Use a unilateral two-port model for the feedback network formed by R_1 and R_2. You may find it most convenient to think of the circuit as an emitter-follower (Q_1) cascaded with a common-base stage (Q_2) and then another emitter-follower (Q_3). **(b)** Determine the closed-loop gain A_{if}. **(c)** Determine the input resistance R_{if}. **(d)** Determine the output resistance R_{of}. (The load is not shown; it would be connected from v_O to ground.)

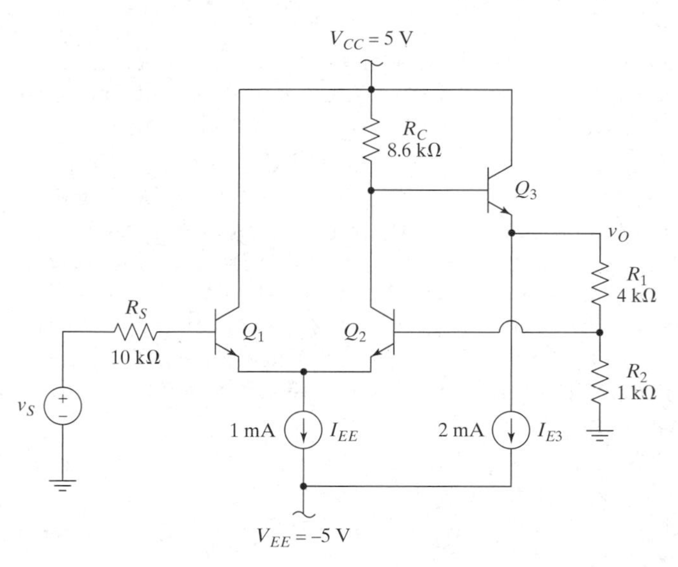

Figure 10–119

S P10.36 Use SPICE and the nominal model presented in Appendix B to confirm your results for Problem P10.35. Explain how you find each parameter, and provide the raw output as well as the final answer.

P10.37 Consider the source-coupled pair source-follower cascade shown in Figure 10-120. This circuit is, in essence, a very simple operational amplifier. Use the Nlarge-mos transistor model given in Appendix B. **(a)** Determine the DC biasing and draw the small-signal midband AC equivalent circuit. Use a unilateral two-port model for the feedback network formed by R_1 and R_2. You may find it most convenient to think of the circuit as a source-follower (M_1) cascaded with a common-gate stage (M_2) and then another source-follower (M_3). **(b)** Determine the closed-loop gain A_{if}. **(c)** Determine the input resistance R_{if}. **(d)** Determine the output resistance R_{of}. (The load is not shown; it would be connected from v_O to ground.)

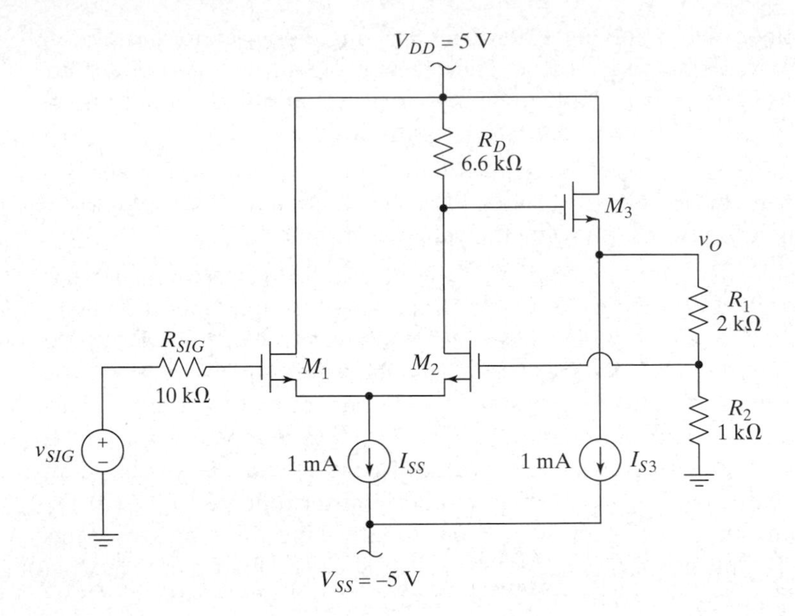

Figure 10–120

S P10.38 Use SPICE and the Nlarge_mos model presented in Appendix B to confirm your results for Problem P10.37. Explain how you find each parameter, and provide the raw output as well as the final answer.

P10.39 Consider the emitter-coupled pair emitter-follower cascade shown in Figure 10-121. This circuit is, in essence, a very simple operational amplifier. Use the nominal and nominal P transistor models given in Appendix B. **(a)** Determine the DC biasing and draw the small-signal midband AC equivalent circuit. Use a unilateral two-port model for the feedback network formed by R_1 and R_2. **(b)** Determine the closed-loop gain, A_{if}. **(c)** Determine the input resistance R_{if}. **(d)** Determine the output resistance R_{of}. (The load is not shown; it would be connected from v_O to ground.)

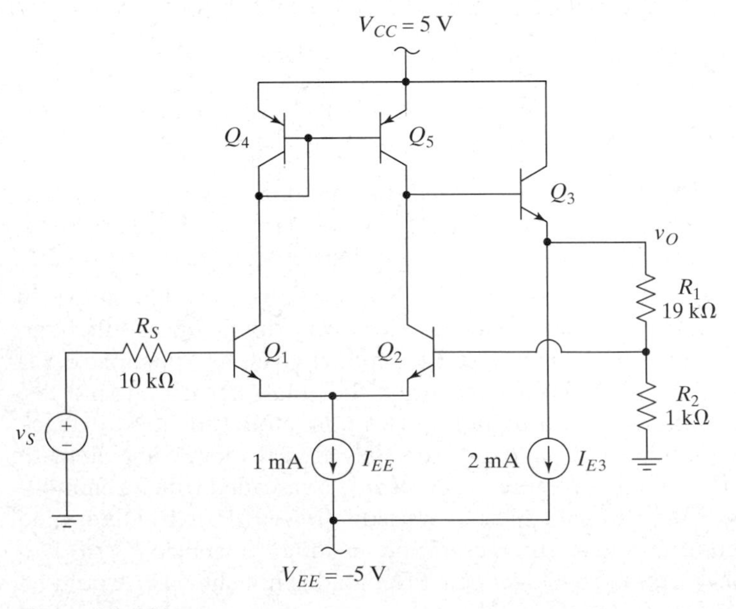

Figure 10–121

S P10.40 Use SPICE and the nominal and nominalp models presented in Appendix B to confirm your results for Problem P10.39. Explain how you find each parameter, and provide the raw output as well as the final answer.

P10.41 Consider the source-coupled pair source-follower cascade shown in Figure 10-122. This circuit is, in essence, a very simple operational amplifier. Use the Nlarge_mos and Plarge _mos models from Appendix B, but set $\lambda = 0.02$ in the models for M_2 and M_5. **(a)** Determine the DC biasing and draw the small-signal midband AC equivalent circuit. Use a unilateral two-port model for the feedback network formed by R_1 and R_2. **(b)** Determine the closed-loop gain A_{if}. **(c)** Determine the input resistance R_{if}. **(d)** Determine the output resistance R_{of}. (The load is not shown; it would be connected from v_O to ground.)

S P10.42 Use SPICE and the Nlarge_mos and Plarge_mos models presented in Appendix B to confirm your results for Problem P10.41. Explain how you find each parameter, and provide the raw output as well as the final answer.

P10.43 This problem illustrates that global feedback is more effective than local feedback in terms of stabilizing the closed-loop gain with respect to the open-loop gain. Consider the block diagrams shown in Figure 10-123, and assume that a is the same in each circuit, that all of the blocks are unilateral, and that the two feedback factors are chosen so that the overall closed-loop gains are the same (i.e., v_o/v_i is the same for each circuit). If A_1 is the nominal closed-loop gain of the circuit in part (a) of the figure and A_2 is the nominal closed-loop gain of the circuit in part (b), show that

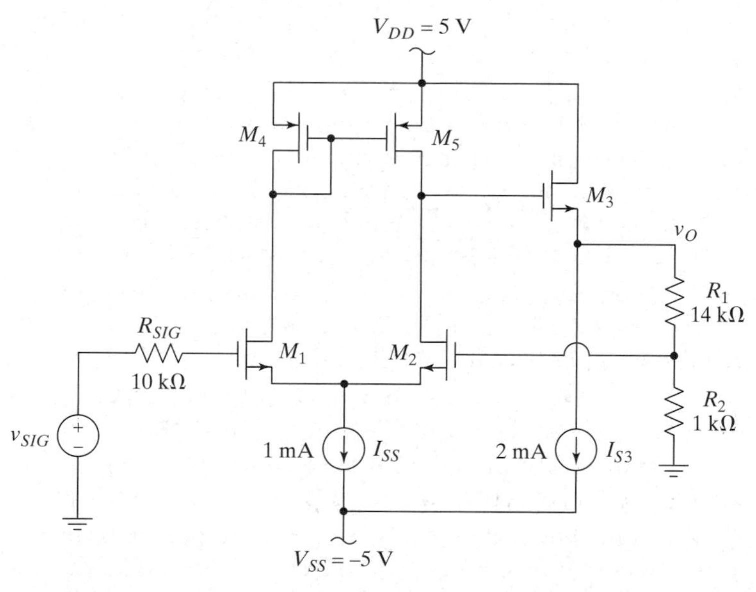

Figure 10–122

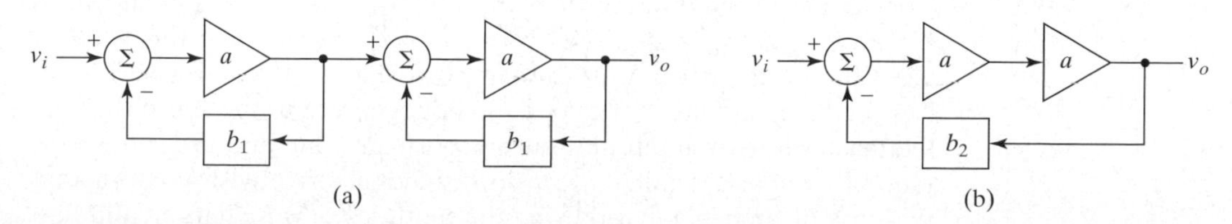

Figure 10–123

$$S_a^{A_2} = \frac{S_a^{A_1}}{1 + ab_1};$$

in other words, the circuit with global feedback is less sensitive to changes in a. (*see* Aside A10.2)

Series–Series Connection

P10.44 Derive (10.160).

P10.45 Consider the transconductance amplifier shown in Figure 10-124. In order to work properly, the output voltage must be high enough to keep the transistor in saturation. The input source has a DC component of 1 V, in addition to a small AC signal. This amplifier uses series–series feedback. **(a)** Identify the feedback network and derive a unilateral two-port model for it. **(b)** Model the operational amplifier as having a differential input resistance of 100 kΩ, an open-circuit voltage gain of 50,000, and an output resistance of 100 Ω (the model is the same as in Exercise 10.10, except for the gain), and use a small-signal midband AC model for the MOSFET; assume that $K = 100\ \mu A/V^2$, $V_{th} = 0.5$ V, and $\lambda = 0.02$. Draw the resulting equivalent circuit. **(c)** Determine A_{if}. **(d)** Determine R_{if}. **(e)** Determine R_{of}. (*Note*: Since the common-gate current gain of a MOSFET is unity for low frequencies, the forward amplifier *is* a two-port, and our formulas apply *at low frequencies*. The forward amplifier is not, however, a two-port at high frequencies.)

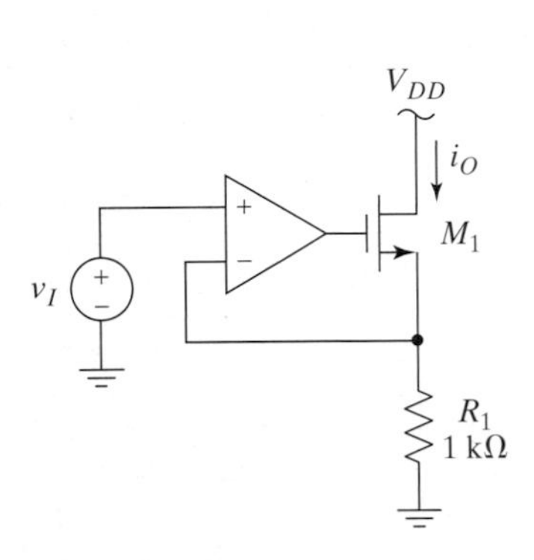

Figure 10–124

S P10.46 Use SPICE and the Nsimple2_mos model in Appendix B to verify your solutions to Problem P10.45. Explain how you find each parameter and provide the raw output as well as the final answer.

P10.47 Consider the series–series triple in Figure 10-125, and assume that $\beta = 100$. **(a)** Identify the forward amplifier and the feedback network. **(b)** Draw the small-signal midband AC model for the amplifier, using a unilateral two-port model for the feedback network. Calculate the parameters of the feedback network model. **(c)** Determine the midband values of A_{if} and A_{sf}. **(d)** Determine the midband values of R_{if} and R_{of}.

S P10.48 Use SPICE and the nominal model in Appendix B to verify your solutions to Problem P10.47. Explain how you find each parameter, and provide the raw output as well as the final answer.

P10.49 Continue with Problem P10.47. Find a single-pole approximation to the forward amplifier's frequency response, and use it to determine values for f_{if} and f_{sf}. Use the nominal transistor model presented in Appendix B.

S P10.50 Use SPICE and the nominal model presented in Appendix B to confirm your answers to Problem P10.49. Hand in a SPICE plot of the frequency response showing the values of f_{if} and f_{sf}.

P10.51 Consider the common-emitter amplifier with degeneration shown in Figure 10-126. We have previously analyzed circuits like this without using the techniques of this chapter; however, this circuit is a series–series feedback amplifier. The circuit is degenerate in the sense that the feedback network input and output ports are one and the same. This type of feedback is called *local series* feedback. We wish to show that this circuit can be analyzed using the methods of this chapter, and that the results agree with those obtained previously. **(a)** Draw the

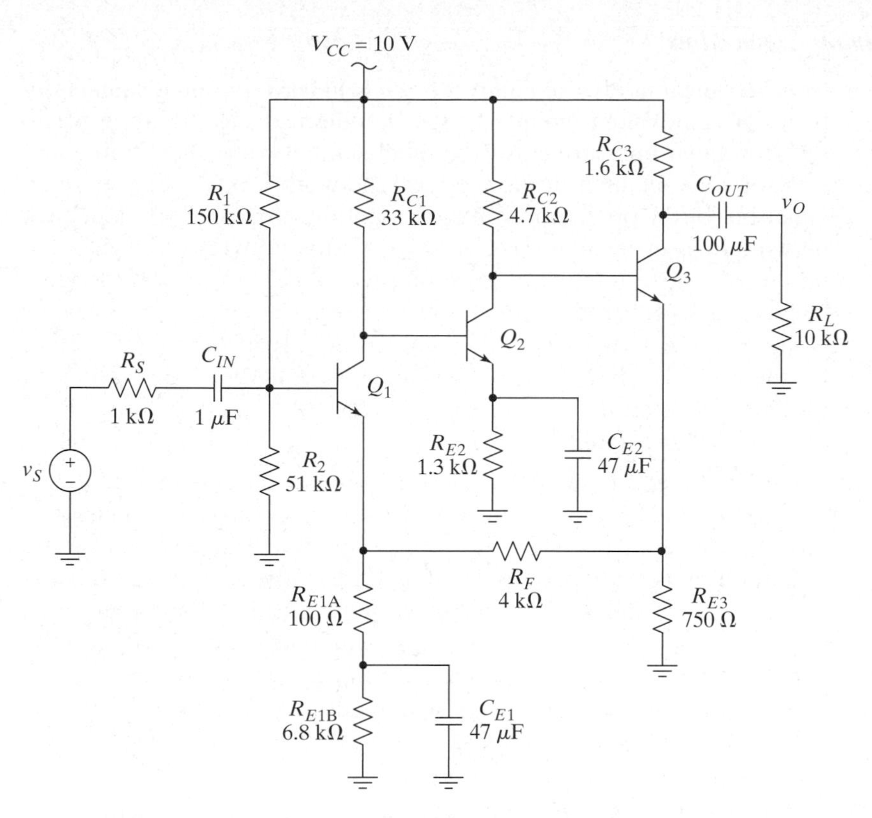

Figure 10–125

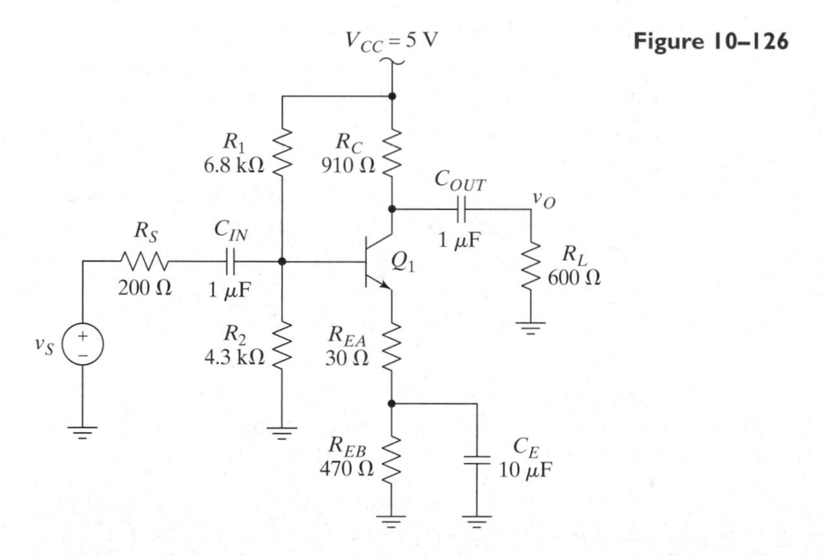

Figure 10–126

small-signal midband AC equivalent circuit for this amplifier. Use a split-source transformation to separate the input and output loops, and use a controlled source to model the feedback. Identify the output variable being fed back and the feedback variable that is being summed at the input. **(b)** Find A_i. **(c)** Find A_{if}. **(d)** Find R_{if}. **(e)** Compare your results with those obtained by direct analysis.

P10.52 Consider the equivalent circuit shown in Figure A10-2. **(a)** Starting from Figure 10-33(a), derive an equation for $\widetilde{L}$. **(b)** Derive (A10.4).

Shunt–shunt Connection

P10.53 Consider the amplifier in Figure 10-127. **(a)** Draw the small-signal midband AC equivalent circuit for the amplifier, using the appropriate unilateral two-port model for the feedback network. Derive an equation for each element in the feedback network model. **(b)** Derive an equation for the open-loop voltage gain of the amplifier with feedback network loading included (i.e., $A_i = v_o/i_i$ with $b = 0$). **(c)** Write an equation for the open-loop input resistance with feedback network loading included, R_i'. **(d)** Write an equation for the open-loop output resistance with feedback network loading included, R_o' . **(e)** Derive an equation for the overall closed-loop voltage gain, v_o/v_s. **(f)** Write an equation for the input resistance with feedback, R_{if}. **(g)** Write an equation for the output resistance with feedback, R_{of}.

P10.54 Consider the amplifier shown in Figure 10-128, and assume that $K = 500\ \mu A/V^2$ and $V_{th} = 1$ V for both devices. **(a)** Draw the small-signal midband AC equivalent circuit for the amplifier, using the appropriate unilateral two-port model for the feedback network. Derive an equation for each element in the feedback network model and determine the values of the elements. **(b)** Derive equations for the closed-loop *voltage* gains, v_o/v_i and v_o/v_{sig}, and determine their values. **(c)** Write an equation for the input resistance with feedback, R_{if}, and determine its value. **(d)** Write an equation for the output resistance with feedback, R_{of}, and determine its value.

S P10.55 Use SPICE and the Nmedium_mos model presented in Appendix B to verify your results for Problem P10.54. Explain how you find each parameter, and provide the raw output as well as the final answer.

P10.56 Consider the amplifier shown in Figure 10-129. Assume that the operational amplifier has $R_{id} = 100\ k\Omega$, $R_o = 50\ \Omega$, and an open-circuit voltage gain of 100,000. **(a)** Redraw the circuit, using the op amp model and the appropriate unilateral two-port model for the feedback network. Derive equations for, and determine the value of, all the elements in the two-port model. **(b)** Derive an equation for the closed-loop *voltage* gain, v_o/v_i, and determine its value. **(c)** Write an equation for the input resistance with feedback, R_{if}, and determine its value. **(d)** Write an equation for the output resistance with feedback, R_{of}, and determine its value.

S P10.57 Use SPICE to verify your results for Problem P10.56. Explain how you find each parameter, and provide the raw output as well as the final answer.

P10.58 Consider the shunt–shunt circuit shown in Figure 10-130. Assume that the gain of the amplifier has the correct sign to make the feedback negative. **(a)** What must the units of b be if we are to analyze this circuit with our standard feedback techniques? **(b)** Circle the feedback network, and find an equation for $b(j\omega)$ as a function of ω and the components given. **(c)** Assuming that the amplifier is ideal (i.e., $a = \infty$, $R_i = \infty$, and $R_o = 0$), derive an equation for the closed-loop gain as a function of ω and the components given.

Shunt–Series Connection

P10.59 Derive (10.223), (10.224), and (10.225).

Figure 10–127

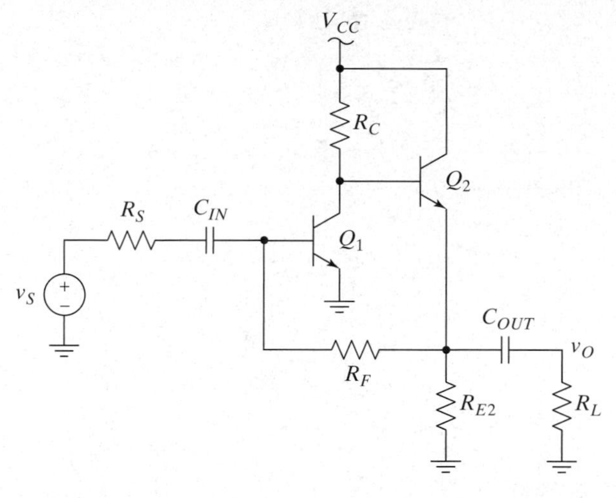

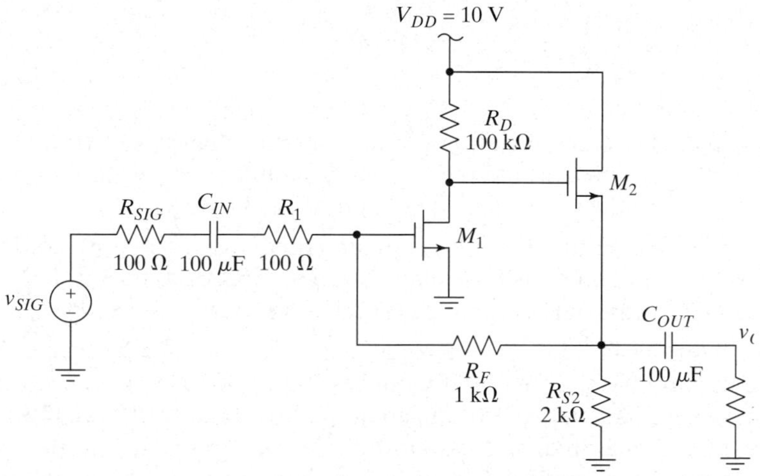

$V_{DD} = 10$ V

R_D 100 kΩ

M_2

R_{SIG} C_{IN} R_1

100 Ω 100 μF 100 Ω

M_1

v_{SIG}

C_{OUT}

v_O

R_F 1 kΩ

R_{S2} 2 kΩ

100 μF

Figure 10–128

Figure 10–129

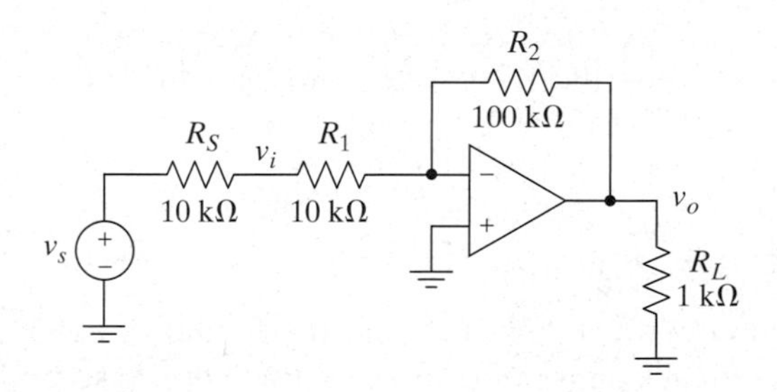

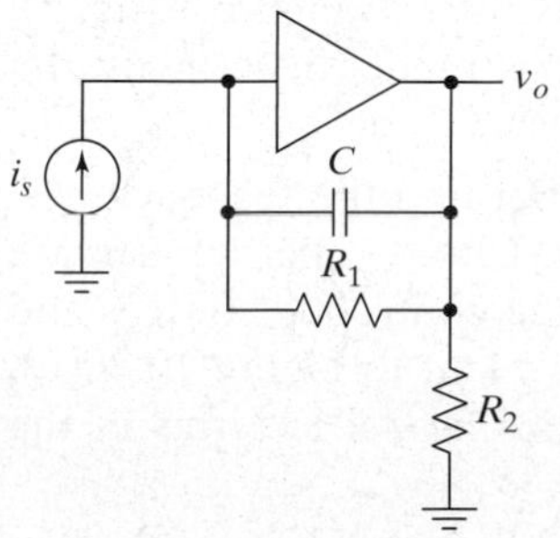

v_o

i_s

C

R_1

R_2

Figure 10–130

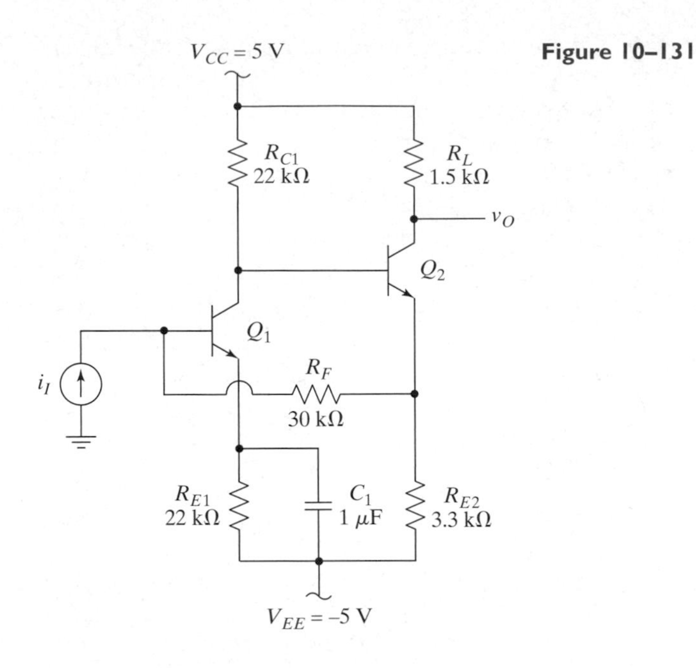

Figure 10–131

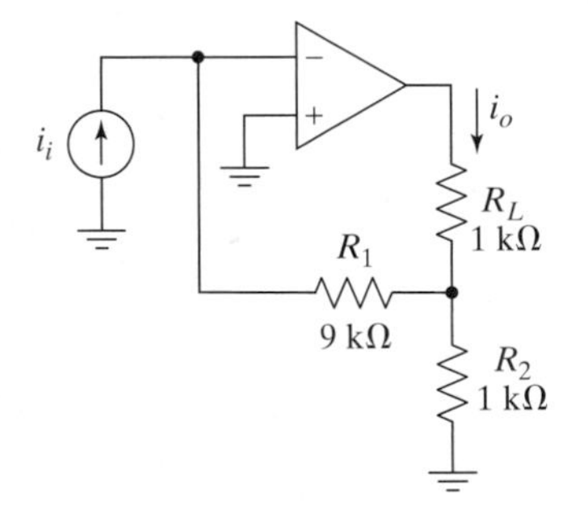

Figure 10–132

P10.60 Consider the shunt–series amplifier shown in Figure 10-131, and assume that $\beta = 100$. (a) Find the small-signal midband AC current gain, i_o/i_i. (b) Find the small-signal midband AC value of R_{if}.

S P10.61 Use SPICE and the Nsimple model in Appendix B to verify your results for Problem P10.60. Explain how you find each parameter, and provide the raw output as well as the final answer.

P10.62 Consider the shunt–series circuit shown in Figure 10-132. Assume that the op amp has a differential input resistance of 100 kΩ, an open-circuit voltage gain of 100,000, and an output resistance of 100 Ω. **(a)** Find the small-signal midband AC closed-loop current gain, i_o/i_i. **(b)** Find the small-signal midband AC values of R_{if} and R_{of}. **(c)** Show that the closed-loop current gain is approximately equal to $-(1 + R_1/R_2)$.

S P10.63 Use SPICE to confirm your answers to Problem P10.62. Explain how you find each parameter, and provide the raw output as well as the final answer.

10.1.4 Advanced Analysis

Full Two-Port Analysis of Feedback Amplifiers

P10.64 Use bilateral models for the forward amplifier and the feedback network of a series–shunt feedback amplifier. Choose the appropriate two-port models so that you can combine the controlled sources and reduce the circuit to a topology that is equivalent to Figure 10-18(b). Give equations for all of the elements of this model in terms of the original two-port networks.

P10.65 Use bilateral models for the forward amplifier and the feedback network of a series–series feedback amplifier. Choose the appropriate two-port models so that you can combine the controlled sources and reduce the circuit to a topology that is equivalent to Figure 10-30(b). Give equations for all of the elements of this model in terms of the original two-port networks.

P10.66 Use bilateral models for the forward amplifier and the feedback network of a shunt–series feedback amplifier. Choose the appropriate two-port models so that you can combine the controlled sources and reduce the circuit to a topology that is equivalent to Figure 10-43(b). Give equations for all of the elements of this model in terms of the original two-port networks.

P10.67 Consider the local shunt amplifier shown in Figure 10-133 where $I_I = 0$. Assume that $\beta = 100$. **(a)** Find the gain, v_o/i_i, using a unilateral approximation for the feedback network. **(b)** Repeat part (a), using a bilateral model for the feedback network. Compare your results. In both cases, assume that the forward amplifier is unilateral.

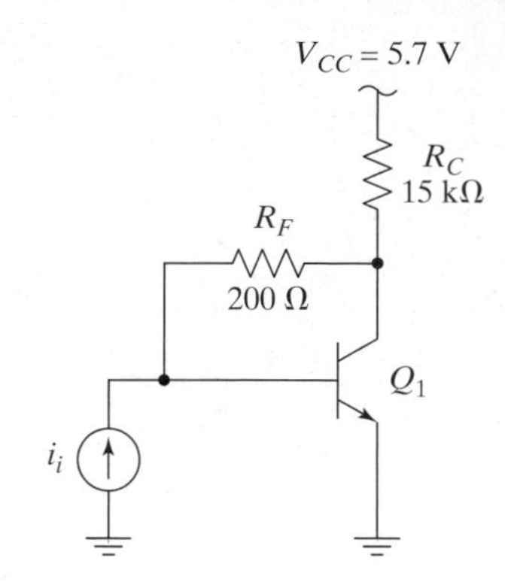

Figure 10–133

S P10.68 Use SPICE and the Nsimple model presented in Appendix B to simulate the performance of the amplifier in Figure 10-133. Which answers from Problem P10.67 are closest to being correct? Why? Explain how you find each parameter, and provide the raw output as well as the final answer.

P10.69 Consider the local shunt amplifier shown in Figure 10-134 where $I_I = 0$. Assume that $K = 500\ \mu A/V^2$ and $V_{th} = 1$ V. **(a)** Find the gain, v_o/i_i, using a unilateral approximation for the feedback network. **(b)** Repeat part (a), using a bilateral model for the feedback network. Compare your results. In both cases, assume that the forward amplifier is unilateral.

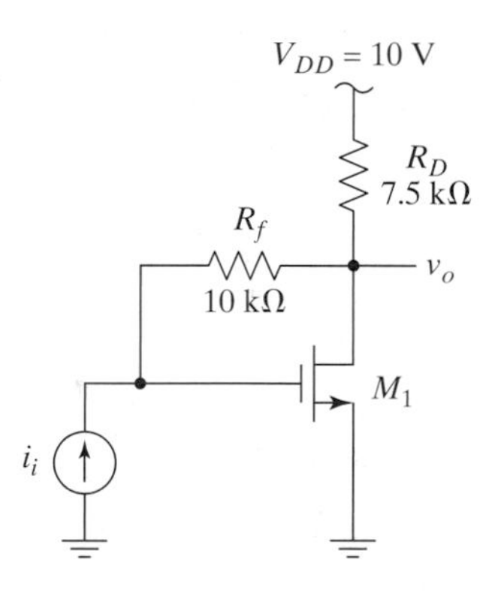

Figure 10–134

S P10.70 Use SPICE and the Nmedium_mos model presented in Appendix B to simulate the performance of the amplifier in Figure 10-134. Which answers from Problem P10.69 are closest to being correct? Why? Explain how you find each parameter, and provide the raw output as well as the final answer.

Alternative Methods for Finding the Loop Gain

P10.71 The amplifier in Example 10.7 is a single-loop amplifier, and the return ratio will be the same no matter where the loop is broken, so long as the loading is maintained. Calculate the return ratio by inserting a test voltage source to drive v_o and then looking at the voltage returned. (Be sure to load the controlled current source with the same resistance it had before you broke the loop to insert the test source.) Show that the return ratio is the same as that calculated in Example 10.7.

S P10.72 Consider again the series–shunt amplifier from Problem P10.16. Employing the method presented in Figure 10-55, use SPICE to simulate the open-loop gain of this circuit. Then use the calculated value of b, along with this forward gain, to determine A_{if}, and compare the value you get with what you obtain from a simulation of the original circuit. Hand in either a SPICE schematic or a net list plus the raw output to show your results.

S P10.73 Consider again the shunt–shunt amplifier from Problem P10.67. Employing the method presented in Figure 10-56, use SPICE to simulate the loop gain of this circuit. Calculate the loop gain with the methods of previous sections, and compare the result with the simulation. Hand in either a SPICE schematic or, net list plus the raw output to show your results.

10.1.5 Stability of Systems with Feedback

S P10.74 Use SPICE to simulate an unstable feedback loop and produce a picture similar to Figure 10-58(b). You can use resistors, capacitors, and gain blocks, or you can use analog behavioral modeling elements (e.g., ELAPLACE). Use a DC loop gain of −10 and a total of four identical poles at 500 kHz. Put in a block (e.g., ETABLE) to limit the amplitude to ±1 V. You may need to use initial conditions or a small voltage pulse to start the oscillation.

P10.75 Repeat Problem P10.74, but vary the DC loop gain from −1 to −10. What do you observe? Explain the results.

Gain and Phase Margins

P10.76 The Bode plot in Figure 10-135 shows the gain and phase of the forward amplifier in a negative feedback loop. The feedback factor, b, is constant. **(a)** What is the minimum closed-loop gain, A_f, for which this loop is stable? **(b)** What is the value of closed-loop gain, A_f, for which the phase margin will be 45°?

P10.77 Repeat Problem P10.76 for the Bode plot in Figure 10-136.

P10.78 Repeat Problem P10.76 for the Bode plot in Figure 10-137.

P10.79 Repeat Problem P10.76 for the Bode plot in Figure 10-138.

P10.80 Consider the feedback amplifier shown in Figure 10-139. Assume that the op amp is ideal, except that the gain is given by

$$A(j\omega) = \frac{V_o(j\omega)}{V_e(j\omega)} = \frac{10^4}{(1 + j2\pi f/(2\pi \times 10^4))(1 + j2\pi f/(2\pi \times 10^6))}.$$

(a) Draw the magnitude and phase of A. **(b)** Draw the appropriate unilateral two-port model and the find the value of $b(j\omega)$. **(c)** Assuming that the pole due to the capacitor in the feedback network occurs at 1 kHz and that the DC feedback factor is 0.01, find the resulting gain and phase margins. Is the circuit stable with this feedback?

P10.81 Consider a negative-feedback amplifier where the forward amplifier is described by

$$a(j\omega) = \frac{10^5}{\left(1 + \dfrac{j\omega}{2\pi \times 10^4}\right)\left(1 + \dfrac{j\omega}{2\pi \times 10^6}\right)\left(1 + \dfrac{j\omega}{2\pi \times 10^8}\right)}.$$

Draw the Bode magnitude and phase plots for the amplifier. (a) Is this amplifier unity-gain stable? (b) What is the phase margin if the amplifier is connected in negative feedback with $b = 1/100$?

P10.82 Consider again the amplifier from Problem P10.81. If the feedback network is described by

$$b(j\omega) = \frac{0.01}{1 + \dfrac{j\omega}{2\pi \times 10^2}},$$

(a) draw the Bode plots to show the loop gain and phase. **(b)** Find the *PM*. **(c)** Find the *GM*.

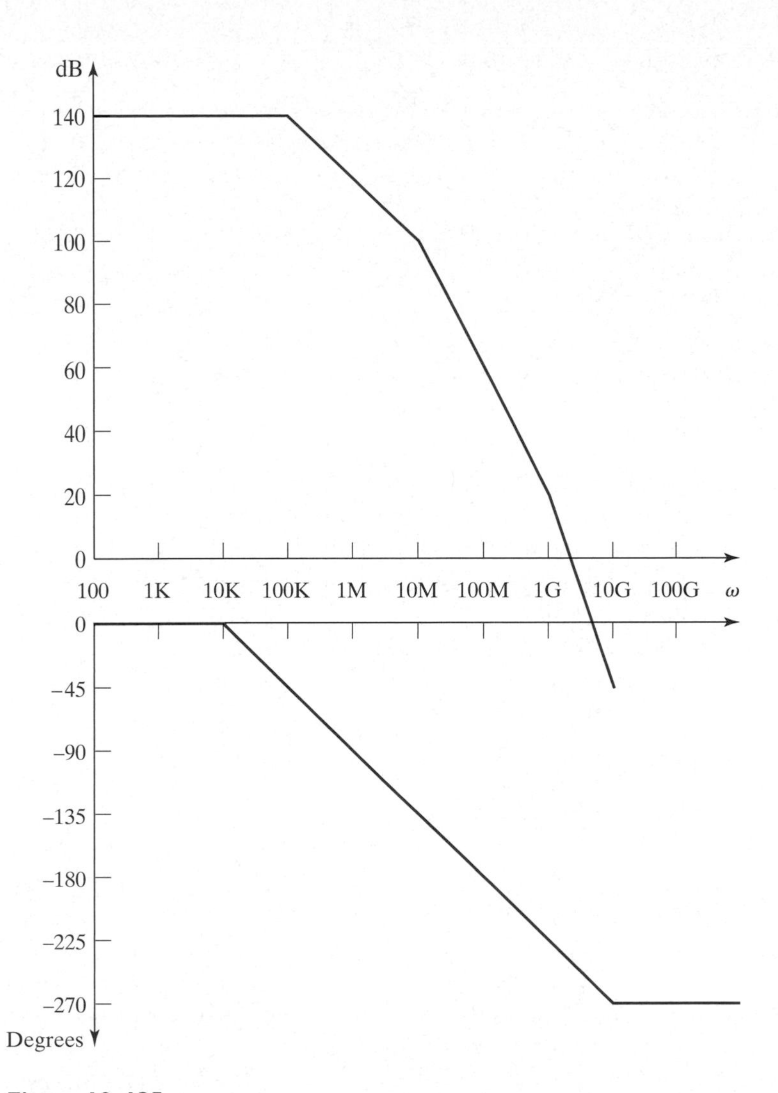

Figure 10–135

Closed-Loop Pole Locations and Transient Response

P10.83 Suppose we have a forward amplifier described by a second-order transfer function as in (10.251) and that $\omega_n = 6.28$ MRad/s, $\zeta = 1.2$, and $a_0 = 50$. **(a)** What is the largest value of b that can be used with this amplifier and still keep $\zeta_f \geq 0.707$? **(b)** What is the approximate phase margin that corresponds to this value of b?

P10.84 Suppose we have a forward amplifier described by a second-order transfer function as in (10.251) and that $\omega_n = 10$ MRad/s, $\zeta = 1$, and $a_0 = 100$. If the amplifier is used in a feedback loop with $b = 0.24$, approximately what is the peak overshoot in the closed-loop step response?

10.1.6 Compensation

P10.85 The Bode plot in Figure 10-135 shows the gain and phase of the forward amplifier in a negative feedback loop. The feedback factor, b, is constant. We desire a closed-loop gain of 40 dB and a phase margin of 45°. **(a)** If you compensate this system by adding a new dominant pole without changing any of the existing pole frequencies, what must the

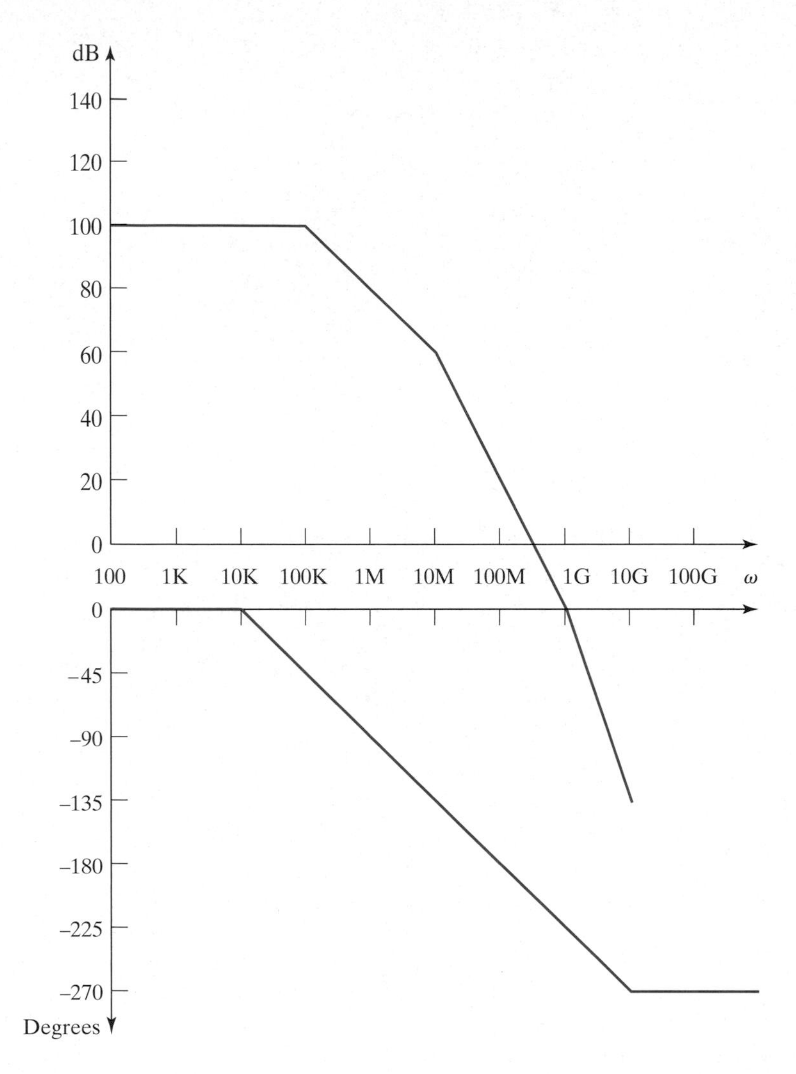

Figure 10–136

new pole frequency be? **(b)** If you compensate the loop by moving the existing dominant pole down in frequency without changing any of the higher poles, what must its new frequency be?

P10.86 The Bode plot in Figure 10-135 shows the gain and phase of the forward amplifier in a negative feedback loop. The feedback factor, b, is constant. **(a)** If we desire to reduce the DC gain of this amplifier so that it is unity-gain stable with a 45° phase margin, what must the new open-loop DC gain be? **(b)** If we desire to reduce the frequency of the dominant pole so that the amplifier is unity-gain stable with a 45° phase margin, what must the new pole frequency be? Assume that the other poles remain unchanged. **(c)** If we desire to add a new dominant pole so that the amplifier is unity-gain stable with a 45° phase margin, what must the frequency of this new pole be? Assume that the other poles remain unchanged.

P10.87 The Bode plot in Figure 10-136 shows the gain and phase of the forward amplifier in a negative feedback loop. The feedback factor, b, is constant. We desire a closed-loop gain of 20 dB and a phase margin of 45°. **(a)** If you compensate this system by adding a new dominant pole without changing any of the existing pole frequencies, what must the

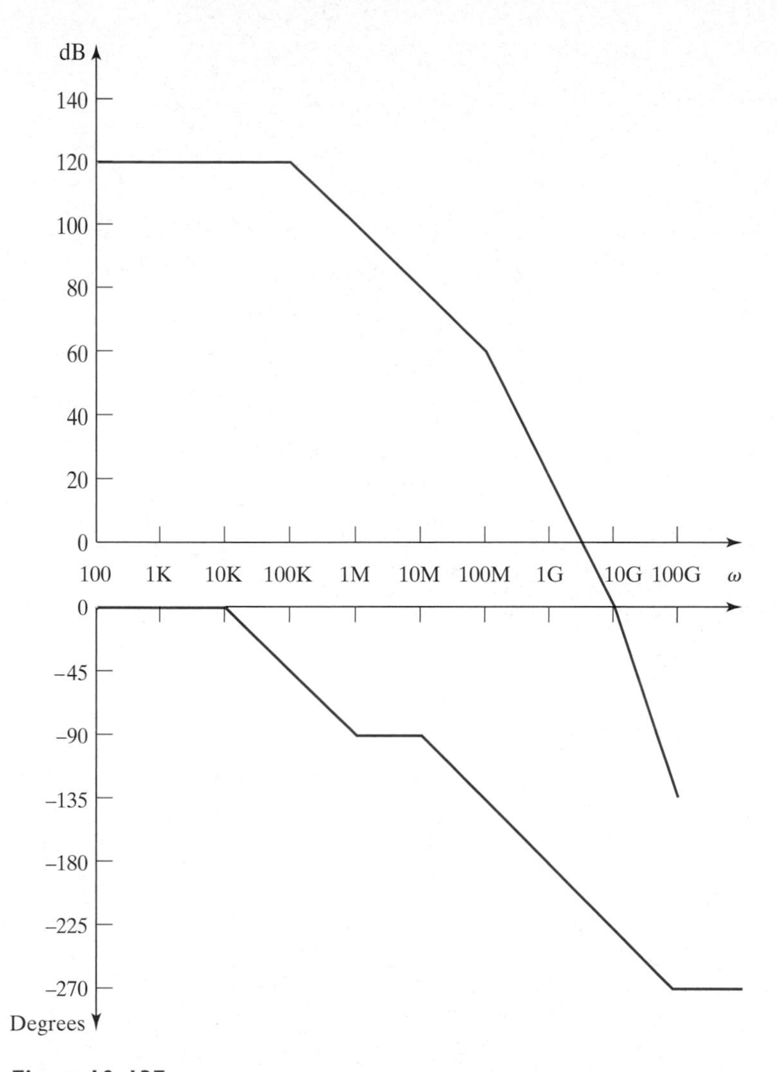

Figure 10–137

new pole frequency be? **(b)** If you compensate the loop by moving the existing dominant pole down in frequency without changing any of the higher poles, what must the new frequency be?

P10.88 The Bode plot in Figure 10-136 shows the gain and phase of the forward amplifier in a negative feedback loop. The feedback factor, b, is constant. **(a)** If we desire to reduce the DC gain of this amplifier so that it is unity-gain stable with a 45° phase margin, what must the new open-loop DC gain be? **(b)** If we desire to reduce the frequency of the dominant pole so that this amplifier is unity-gain stable with a 45° phase margin, what must the new pole frequency be? Assume that the other poles remain unchanged. **(c)** If we desire to add a new dominant pole so that the amplifier is unity-gain stable with a 45° phase margin, what must the frequency of the new pole be? Assume that the other poles remain unchanged.

P10.89 The Bode plot in Figure 10-137 shows the gain and phase of the forward amplifier in a negative feedback loop. The feedback factor, b, is constant. We desire a closed-loop gain of 40 dB and a phase margin of 45°. **(a)** If you compensate this system by adding a new dominant pole

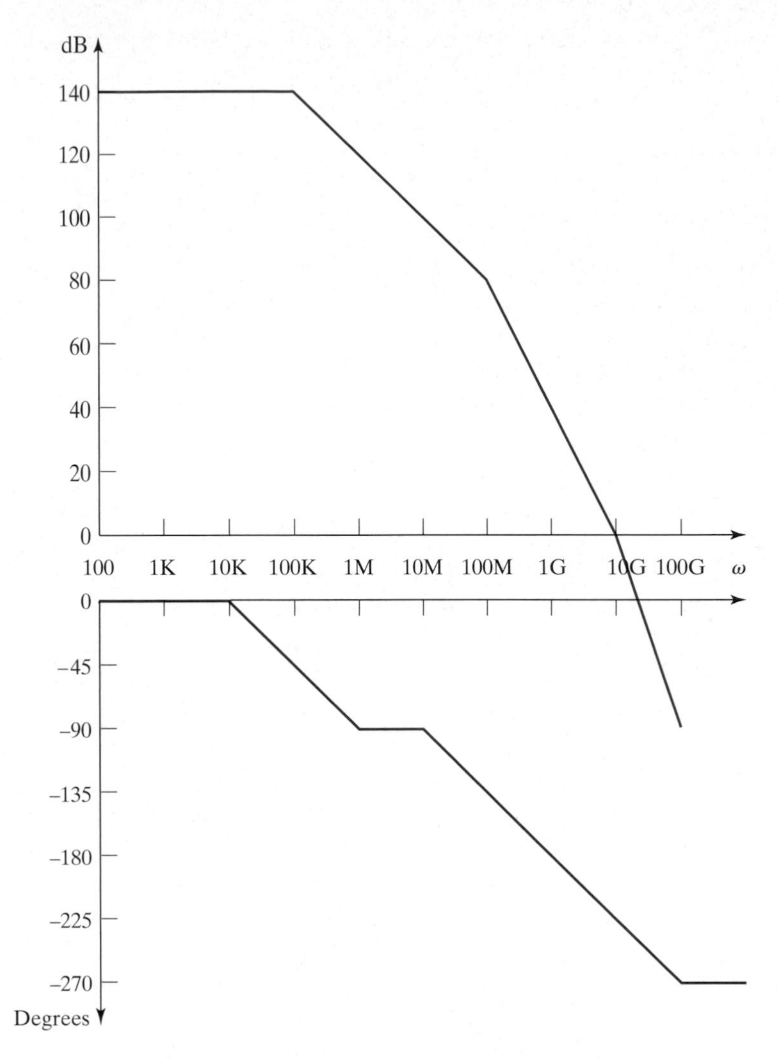

Figure 10–138

without changing any of the existing pole frequencies, what must the new pole frequency be? **(b)** If you compensate the loop by moving the existing dominant pole down in frequency without changing any of the higher poles, what must the new frequency be?

P10.90 The Bode plot in Figure 10-137 shows the gain and phase of the forward amplifier in a negative feedback loop. The feedback factor, b, is constant. **(a)** If we desire to reduce the DC gain of this amplifier so that it is unity-gain stable with a 45° phase margin, what must the new open-loop DC gain be? **(b)** If we desire to reduce the frequency of the dominant pole so that this amplifier is unity-gain stable with a 45° phase margin, what must the new pole frequency be? Assume that the other poles remain unchanged. **(c)** If we desire to add a new dominant pole so that this amplifier is unity-gain stable with a 45° phase margin, what must the frequency of the new pole be? Assume that the other poles remain unchanged.

P10.91 The Bode plot in Figure 10-138 shows the gain and phase of the forward amplifier in a negative feedback loop. The feedback factor, b, is constant. We desire a closed-loop gain of 40 dB and a phase margin of 45°. **(a)** If you compensate this system by adding a new dominant pole without changing any of the existing pole frequencies, what must the

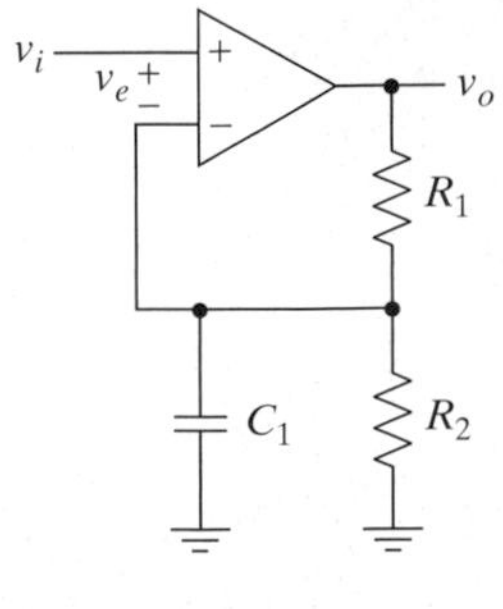

Figure 10–139

new pole frequency be? **(b)** If you compensate the loop by moving the existing dominant pole down in frequency without changing any of the higher poles, what must the new frequency be?

P10.92 The Bode plot in Figure 10-138 shows the gain and phase of the forward amplifier in a negative feedback loop. The feedback factor, b, is constant. **(a)** If we desire to reduce the DC gain of this amplifier so that it is unity-gain stable with a 45° phase margin, what must the new open-loop DC gain be? **(b)** If we desire to reduce the frequency of the dominant pole so that this amplifier is unity-gain stable with a 45° phase margin, what must the new pole frequency be? Assume that the other poles remain unchanged. **(c)** If we desire to add a new dominant pole so that this amplifier is unity-gain stable with a 45° phase margin, what must the frequency of the new pole be? Assume that the other poles remain unchanged.

P10.93 Consider the amplifier shown in Figure 10-140. Using the nominal model from Appendix B, calculate the bandwidth of the amplifier. Then determine the value of compensation capacitor required (from base to collector) to lower the bandwidth to 300 kHz.

S P10.94 Use SPICE and the nominal model from Appendix B to confirm your results for Problem P10.93, edit the transistor model to set `itf = 10m`. Is the Miller effect dominant in this circuit? Show plots of the frequency response before and after adding the compensation capacitor.

P10.95 Consider the amplifier shown in Figure 10-141 and assume $V_{SIG} = 1.5$ V. Using the Nsimple_mos model from Appendix B, calculate the bandwidth of the amplifier. Then determine the value of compensation capacitor required (from gate to drain) to lower the bandwidth to 700 kHz.

S P10.96 Use SPICE and the Nsimple_mos model from Appendix B to confirm your results for Problem P10.95. Is the Miller effect dominant in this circuit? Show plots of the frequency response before and after adding the compensation capacitor.

10.2.1 Sinusoidal Oscillators

RC Oscillators

P10.97 Derive the transfer function for the feedback network of the oscillator in Figure 10-81, and show that it has a 180° phase shift at a frequency $\omega_o = \sqrt{6}/RC$ and that the gain at that frequency is 1/29.

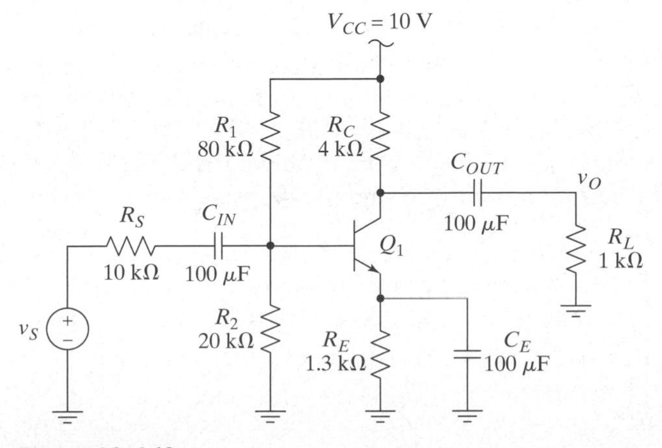

Figure 10–140

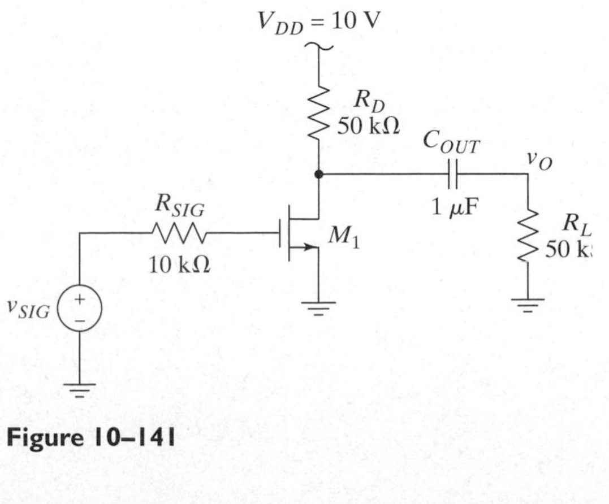

Figure 10–141

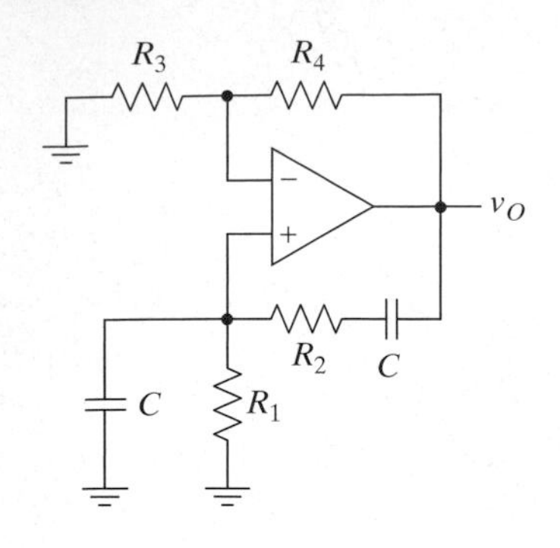

Figure 10–142

D P10.98 Design a phase shift oscillator using two op amps, some resistors, and some capacitors. Set the frequency of oscillation to 5 kHz and allow the amplitude to be limited by the op amps.

S P10.99 Use a simple op amp model and SPICE to verify your design in Problem P10.98. Run a transient simulation and show the output. (Be careful to make some provision for starting the oscillations. You may also need to set the maximum transient step size.)

P10.100 Consider the Wien–Bridge oscillator shown in Figure 10-142. **(a)** Derive an equation for the frequency of oscillation, assuming that the op amp is ideal. **(b)** Derive the minimum value of R_4 necessary for oscillation to occur. Your answer should be in terms of R_1, R_2, R_3, and C.

P10.101 Reverse the locations of the series and parallel *RC* networks in Figure 10-142, and repeat Problem P10.100 for the new circuit.

LC Oscillators

P10.102 Calculate the true loop gain for the small-signal midband AC equivalent circuit of the Colpitts oscillator given in Figure 10-85. Since this is a single loop and we have not used a unilateral two-port network for the feedback, the loop gain is the gain found by disconnecting the controlled current source from the circuit, injecting a test current where the controlled source was, and then calculating the short-circuit output current of the controlled source, which is the response to this test current. Solve for the frequency where the phase of the loop gain is zero. State the conditions under which this frequency is reasonably approximated by $\omega_o = \sqrt{1/LC_t}$, where C_t is the total capacitance of C_1 in series with C_2.

P10.103 Consider the circuit shown in Figure 10-143. Show that if $C_2 = \alpha C_1$,

$$Z_{eff}(j\omega_o) = \alpha R_i\left[\alpha - j\frac{1}{R_i}\sqrt{\frac{L(1+\alpha)}{\alpha C_1}}\right],$$

where $\omega_o = \sqrt{(C_1 + C_2)/LC_1C_2} = 1/\sqrt{LC_t}$. Therefore, when

$$R_i \gg \sqrt{\frac{L(1+\alpha)}{\alpha C_1}},$$

we have $Z_{eff}(j\omega_o) \approx \alpha^2 R_i$.

P10.104 Use the schematic in Figure 10-144 to show that the magnitude of the loop gain of the Colpitts oscillator is approximately given by $|L(j\omega_o)| \approx g_m(r_o \| \alpha^2 R_2 \| \alpha^2 r_{cm})/\alpha$ when $R_2 \| r_{cm} \gg \sqrt{L(1+\alpha)/\alpha C_1}$. Assume that $\alpha < 1$, and use the result from Problem P10.103. *See* Problem 10.102 for a discription of how to find the loop again.

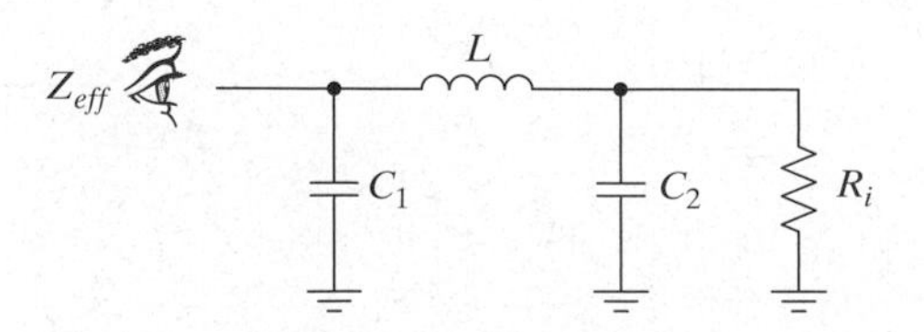

Figure 10–143

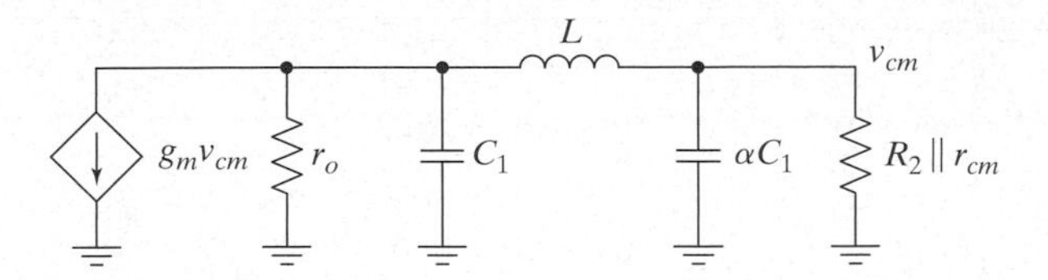

Figure 10–144

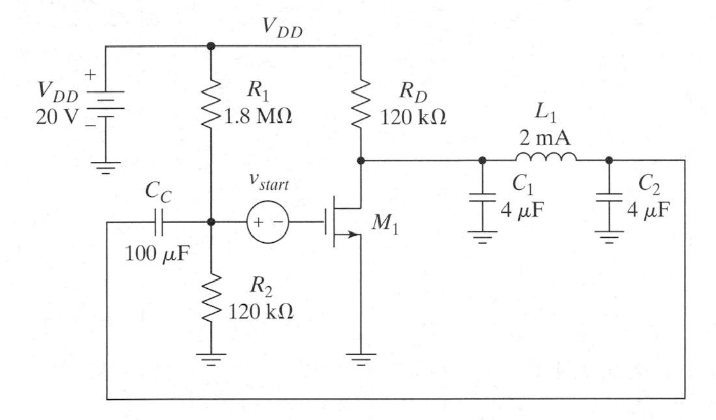

Figure 10–145

P10.105 Consider ,the Colpitts oscillator shown in Figure 10-145. The transistor has $K = 2$ mA/V^2 and $V_{th} = 1$ V. **(a)** Calculate the approximate forward gain of the amplifier at the resonant frequency of the feedback network. State your assumptions. **(b)** Calculate the approximate feedback factor at resonance. State your assumptions. **(c)** At what frequency will this circuit oscillate if the loop gain is large enough? **(d)** Is the loop gain large enough to cause oscillations?

S P10.106 Use SPICE and the Nlarge_mos model from Appendix B to perform a transient simulation on the circuit shown in Figure 10-145. Set the maximum step size in the transient simulation to 10 μs (step ceiling in SPICE), the print step to 10 μs, and the end time to 150 ms. (This simulation may take a while.) Insert a voltage pulse source in series with the gate as shown to start the oscillations; set the pulse to go from 0 V to 1 V immediately and to have a pulse width of 200 μs. Does the frequency of oscillation agree with what you predicted in Problem P10.105?

P10.107 Consider the Colpitts oscillator shown in Figure 10-146. Note that the feedback is connected to the emitter, so the transistor is being used in the common-base configuration (but the base is not an AC ground). **(a)** Explain how you can demonstrate that this circuit has positive feedback. **(b)** What frequency of oscillation would a simplistic linear analysis predict for this circuit? **(c)** Approximately how large would you expect the oscillation at the collector to be? (Hint: consider the DC voltage at the collector and assume a symmetric oscillation.)

S P10.108 Use the 2N2222 model from Appendix B and SPICE to simulate the circuit shown in Figure 10-146 and confirm your analysis from P10.107. Perform a transient simulation for 6 μs with a 0.01-μs maximum step size. Include a voltage pulse generator in series with the base to initiate oscillation. (Use a 0.1-V pulse with 10-ns rise and fall times and a 500-ns pulse width). Hand in a plot of the output voltages at the collector and emitter. Why is the oscillation at the emitter more sinusoidal?

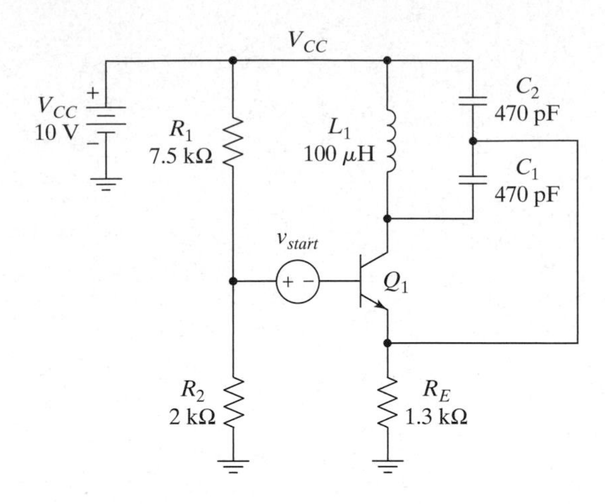

Figure 10–146

P10.109 Consider the Hartley oscillator shown in Figure 10-147. **(a)** Explain how you can demonstrate that this circuit has positive feedback. **(b)** What frequency of oscillation would a simplistic linear analysis predict for this circuit?

S P10.110 Use the 2N2222 model from Appendix B and SPICE to simulate the circuit shown in Figure 10-147. Perform a transient simulation for 50 μs with a 0.01 μs maximum step size. Include a voltage pulse generator in series with the base to initiate oscillation. (Use a 0.1-V pulse with 10 ns rise and fall times and a 500-ns pulse width.) Hand in a plot of the output voltage. Were your predictions in P10.109 accurate?

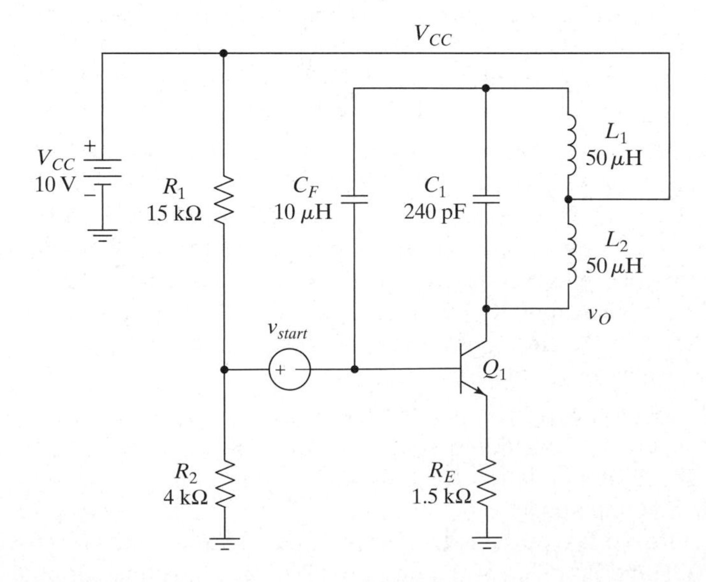

Figure 10–147

10.2.2 Nonsinusoidal Oscillators

P10.111 Consider the relaxation oscillator shown in Figure 10-148. Assume that the diodes, and comparator are perfect, that the comparator thresholds are at ±1 V, and that the output swing limits are ±5 V. **(a)** Sketch the output voltage and the capacitor voltage as a function of time, assuming that v_O just switched high at $t = 0$, and **(b)** determine the frequency of oscillation.

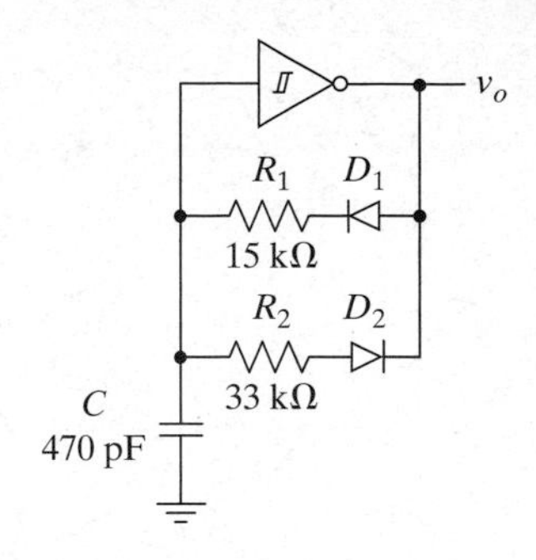

Figure 10–148

The 555 Timer

P10.112 Consider the 555 oscillator shown in Figure 10-149. **(a)** Draw the waveforms for the output voltage and the capacitor voltage for at least one period. **(b)** Determine the frequency of oscillation.

S P10.113 Use the 555D model in the evaluation version of PSPICE, and simulate the circuit in Figure 10-149 to verify your answers to Problem P10.112. You will need to add a load resistor to ground from the output, and you need to tie the reset pin to V_{CC} and add a 0.01-μF decoupling capacitor from the control pin to ground. Hand in a plot of the output voltage and capacitor voltage as a function of time.

P10.114 Consider the 555 oscillator circuit shown in Figure 10-150. This circuit allows the output to have a duty cycle less than 50%. Show that the output is high for $t_H = 0.7R_1C$ and low for

$$t_L = (R_1 \| R_2)C\ln\left(\frac{R_2 - 2R_1}{2R_2 - R_1}\right).$$

P10.115 The voltage at the negative input of comparator 2 in a 555 Timer is nominally equal to $2V_{CC}/3$ and is called the *control voltage*. This node is made available externally and, for many applications, has a bypass capacitor connected to ground to ensure that the reference voltages to the comparators are relatively free from noise. This pin can also be used to change the internal thresholds, however, which allows the 555 timer to be used as a pulse-width modulator, as shown

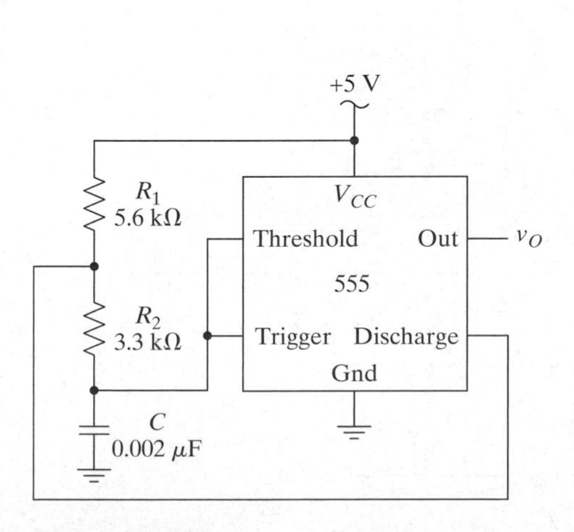

Figure 10–149

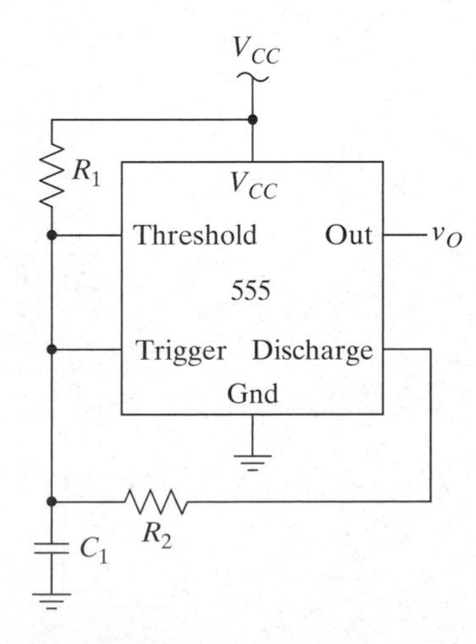

Figure 10–150

in Figure 10-151. A pulse-width modulator uses a periodic trigger input which can be generated by another 555, and produces pulses whose widths depend on the control voltage. Derive an equation for the pulse width as a function of the control voltage. The range of allowable control voltages is restricted, but for now assume that the 555 continues to work properly.

P10.116 The control voltage input of a 555 Timer is described in Problem P10.115. In the current problem, we illustrate how to use this voltage to produce a frequency modulator. A frequency modulator is an oscillator with an electronically variable frequency of oscillation. Consider the oscillator shown in Figure 10-152. Derive an equation for the frequency of oscillation as a function of the control voltage. Assume that the 555 works properly as the control voltage is varied.

D P10.117 If $V_{CC} = 5$ V, the control voltage of a 555 timer must be approximately in the range $1 \leq V_C \leq 4.3$ for the circuit to function properly. Suppose we desire to make a pulse-width modulation circuit that will allow a modulating signal to vary from 0 to 5 V. We use the circuit shown in Figure 10-153, which is driven by a periodic trigger signal as in Figure 10-151. Calculate values for R_1 and R_2 so that the control voltage will be 2.43 V when $v_M = 0$ and 4.23 V when $v_M = 5$ V. The internal resistors in the 555 bias string are all equal to 5 kΩ.

D P10.118 Design a pulse-width modulator that produces one pulse every 3 ms and varies the positive pulse width from 0.5 ms to 1.5 ms as an input voltage is varied from 0 V to 5 V. Use a single 5-V power supply, two 555 Timers, and some resistors and capacitors. Use the first 555 to periodically trigger the second 555. The second 555 should be configured as a voltage-controlled one-shot. Consider the circuit shown in Problem P10.117.

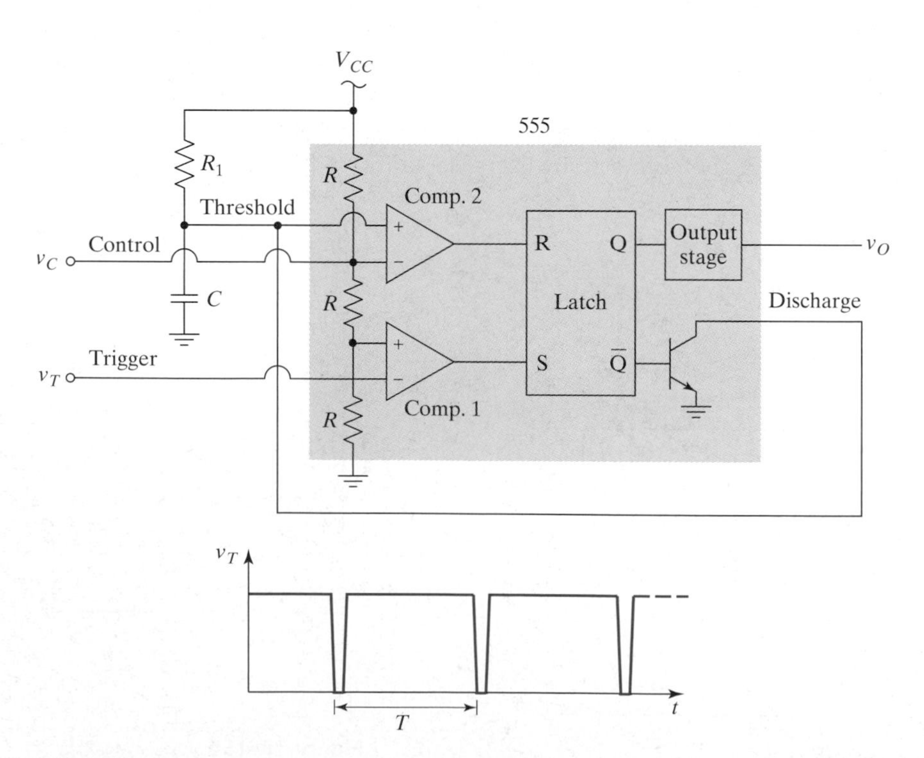

Figure 10–151

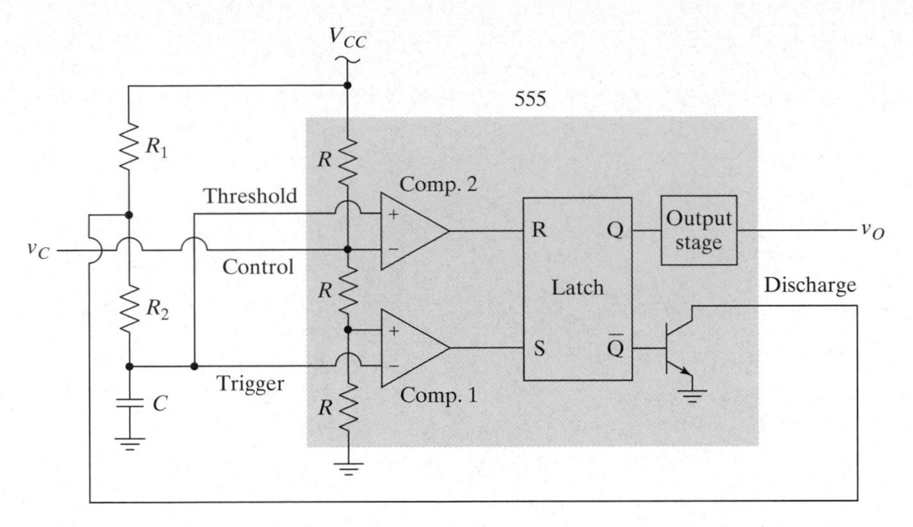

Figure 10–152

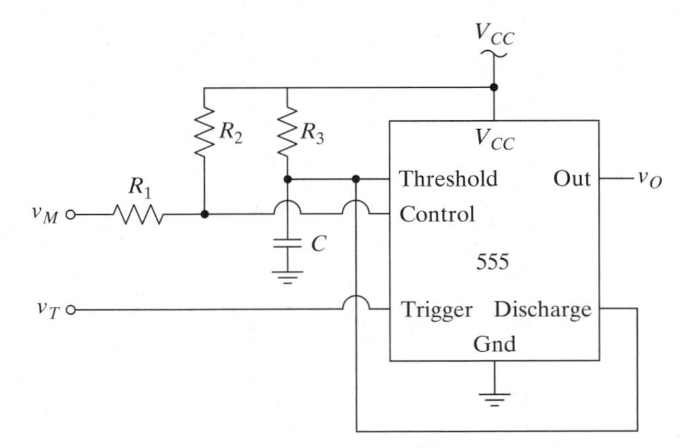

Figure 10–153

CHAPTER 11

Filters and Tuned Amplifiers

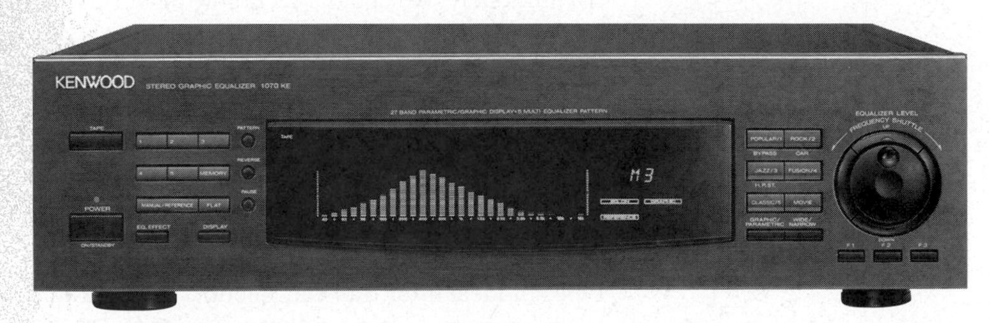

Graphic equalizers like this one are banks of bandpass filters that allow the listener to tailor the sound to a particular environment. (Photo courtesy of Kenwood USA Corporation. Used by permission.)

INTRODUCTION

There are many applications in electronics where a circuit must be able to pass signals within a specified range of frequencies while rejecting signals outside of that range. In this chapter, we explore circuits used for this purpose. When the circuits do not provide any gain or any function other than rejecting certain frequencies, they are called *filters*. Filters may be passive or active. Sometimes, amplifiers are built to include filtering as a natural part of their operation; such circuits are called *tuned amplifiers* and are also covered here. Finally, we briefly introduce the *phase-locked loop*, a very important building block in communication systems.

11.1 FILTERS

All circuit designers need to understand some basic principles and notation relating to filters. The purpose of this section is to provide you with that background. We will not pursue filter theory in detail; instead, we refer you to one of the many good references on this subject if you need more information (e.g., [11.2]). Classic *RLC* filter theory is one of the few engineering disciplines wherein all of the necessary results have been worked out and incorporated into handbooks and computer programs. There is still a lot of research being done on filter design, but the focus is on *implementations* rather than basic theory.

This section focuses on filter transfer functions and how we can produce circuits that yield a given desired transfer function. We present a few examples of passive *RLC* filters here. Active *RC* filters using op amps are presented in Section 5.2, but we present a brief discussion of their utility, with reference to other types of filters, here. We end the section with a brief introduction to discrete-time filtering techniques.

11.1.1 Ideal Transfer Functions

In this section, we examine ideal filter transfer functions in both the frequency domain and the time domain. We then state, without proof, some important general conclusions that are supported by the examples given.

Frequency-Domain Description In the simplest case, there are five different types of ideal filter transfer function, as illustrated in Figure 11-1: ***low-pass***, ***high-pass***, ***bandpass***, ***bandstop*** (or ***notch***), and ***allpass***.

The ideal filters described by these transfer functions pass all signals in the ***passband*** with a gain of unity and completely reject all signals in the ***stopband*** (i.e., the gain in the stopband is zero). The cutoff frequencies are typically defined by the points where the magnitude drops to $1/\sqrt{2}$ (i.e., it drops by 3 dB). In real filters, the magnitude cannot change discontinuously (as illustrated in Figure 11-1), so there is some ***transition band*** between the passband and the stopband.

The allpass filter is unique in that it does not have a stopband. An ideal allpass filter affects only the phase of the signal. This type of filter is useful, for example, in equalizing a communication channel to avoid distortion caused by a nonlinear phase (*see* Aside 12.1).

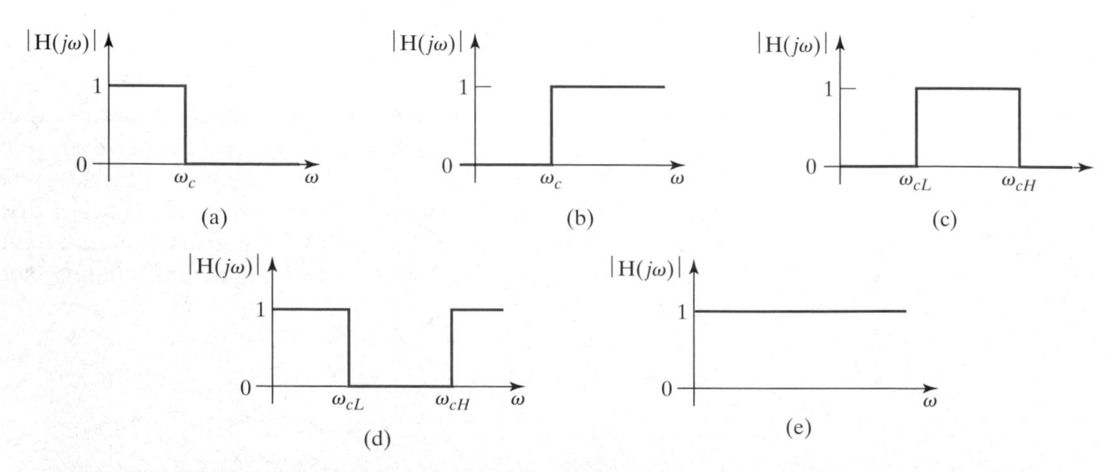

Figure 11–1 Ideal filter transfer functions: (a) low-pass, (b) high-pass, (c) bandpass, (d) bandstop (or notch), and (e) allpass.

Time-Domain Description Consider an ideal low-pass filter with a cutoff frequency of ω_c and a linear phase with a slope of $-t_p$ (i.e., the filter has a constant time delay of t_p seconds), as shown in Figure 11-2(b). This transfer function is described by

$$H(j\omega) = \begin{cases} e^{-j\omega t_p} & -\omega_c \le \omega < \omega_c \\ 0 & \text{elsewhere} \end{cases} \tag{11.1}$$

and has an impulse response given by

$$h(t) = \frac{1}{2\pi}\int_{-\omega_c}^{\omega_c} e^{j\omega(t-t_p)}d\omega = \frac{1}{\pi}\int_0^{\omega_c}\cos[\omega(t-t_p)]d\omega = \frac{\omega_c}{\pi}\,\frac{\sin[\omega_c(t-t_p)]}{\omega_c(t-t_p)}. \tag{11.2}$$

The function $\sin x/x$ is frequently encountered in communication theory and is called the *sinc* function $(\text{sinc}\,(x) = \sin x/x)$ [11.1]. This impulse response is shown in Figure 11-2(c) and is noncausal (i.e., the response appears at the output prior to the arrival of the impulse at the input).[1] This noncausal response is caused by our filter having zero gain in the stopband and is one manifestation of the fact that the filter is not realizable.

The step response is the integral of the impulse response:

$$s(t) = \int_{-\infty}^{t} h(\lambda)d\lambda = \frac{\omega_c}{\pi}\int_{-\infty}^{t}\frac{\sin[\omega_c(\lambda-t_p)]}{\omega_c(\lambda-t_p)}d\lambda. \tag{11.3}$$

If we let $x = \omega_c(\lambda-t_p)$, then $d\lambda = dx/\omega_c$, and we have

$$s(t) = \frac{1}{\pi}\int_{-\infty}^{\omega_c(t-t_p)}\frac{\sin x}{x}dx = \frac{1}{2} + \frac{1}{\pi}\int_0^{\omega_c(t-t_p)}\frac{\sin x}{x}dx, \tag{11.4}$$

and, defining the sine integral as $Si(t) = \int_0^t \frac{\sin x}{x}dx$, we have

$$s(t) = \frac{1}{2} + \frac{Si[\omega_c(t-t_p)]}{\pi}. \tag{11.5}$$

The sine integral cannot be evaluated in closed form, but the step response given by (11.5) is plotted in Figure 11-2(d). We again see the anticipatory, or noncausal, nature of the filter, since the output rings[2] prior to the arrival of the input step. We also notice that the response has overshoot and ringing, which may not be acceptable in some situations.

This example has illustrated a few important points that are true of all filters:

1. Any transfer function that is strictly bandlimited (i.e., that has zero gain in some region) is noncausal and therefore not physically realizable.
2. Networks with a linear phase have symmetric impulse and step responses.
3. Filters with abrupt changes in gain at the cutoff frequency will exhibit overshoot and ringing in their step response. A smooth transition from the passband to the stopband is required to keep the transient response well behaved.

---

[1] Noncausal responses are sometimes called anticipatory, since the output anticipates the arrival of the input.

[2] To "ring" simply means to oscillate back and forth and is usually said of a decaying oscillation.

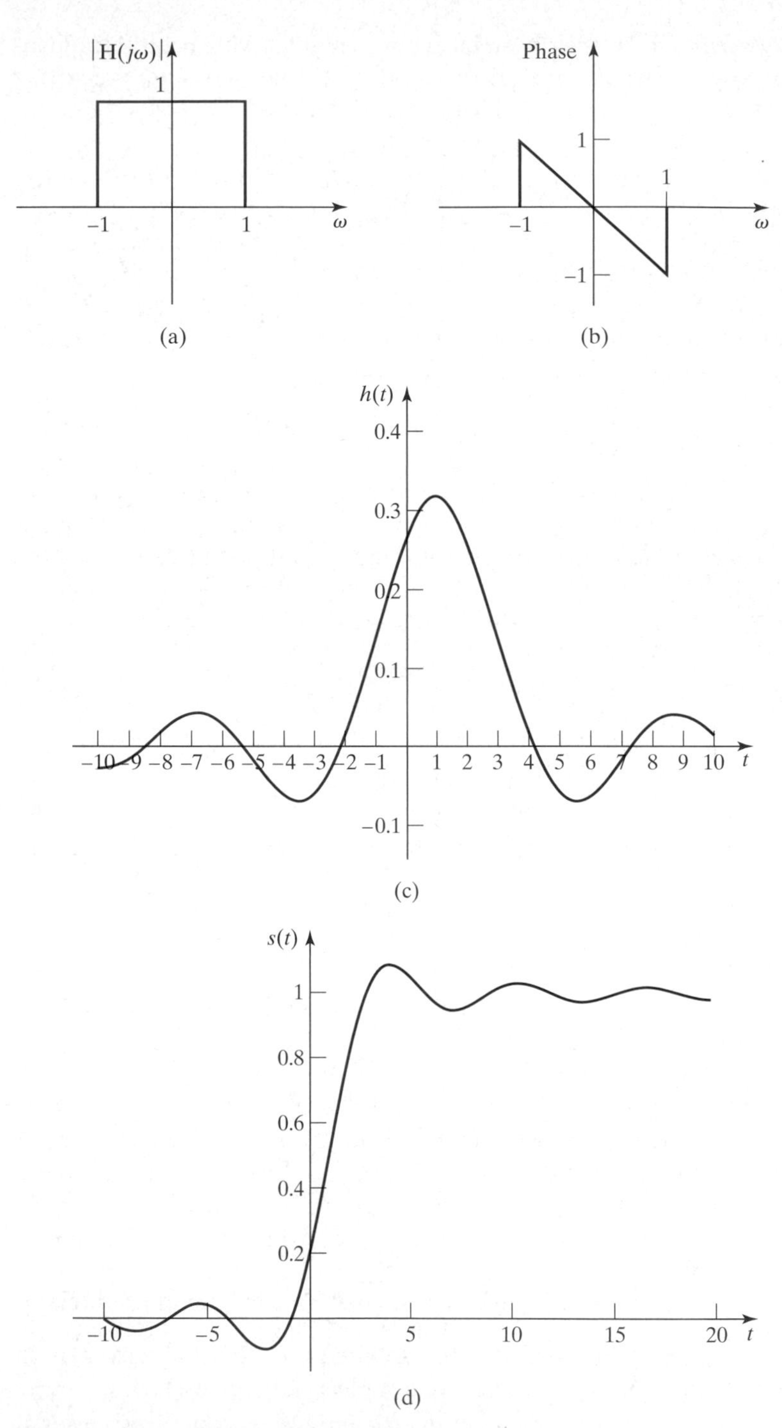

Figure 11–2 (a) The Bode magnitude plot and (b) phase plot for an ideal low-pass filter with cutoff frequency $\omega_c = 1$ rad/s and delay $t_p = 1$ s. (c) The impulse response of the filter and (d) the step response (*see* LPF.nb on CD).

11.1.2 Practical Transfer Functions

The ideal transfer functions presented in Figure 11-1 are not physically realizable. In addition, it is not possible to independently choose the magnitude response in the frequency domain and the transient response in the time domain. Therefore, we must frequently make trade-offs in choosing the filter for a given application.

Since ideal transfer functions are not realizable, and perhaps not even desirable given their transient responses, we now investigate how practical filters approximate ideal characteristics. We demonstrate approximation methods with a single example: the Taylor approximation [11.2, pp, 96–99]. We then present other approximations, but without explanation. We show only a low-pass filter, although the results can be extended to other types of filters as shown in Section 11.1.4. In addition, we deal with normalized component values and frequencies, for simplicity. The component values and frequencies can be scaled for a given application, as shown in Section 11.1.3. We end this section with a brief discussion of practical bandpass and allpass filters.

The Taylor Magnitude Approximation: Butterworth Filters Suppose we have a desired transfer function, $H_d(j\omega)$, and we wish to approximate it using some passive network with a transfer function $H(j\omega)$. In this section, we will restrict our attention to the low-pass filter transfer function shown in Figure 11-2(a, b). The question then is, How can we determine the parameters of a passive network so that its response comes as close as possible to the desired response?

First, we choose the particular characteristic we wish to optimize. For example, say that we want the square of the magnitude of the transfer function to be as close as possible to the square of the magnitude of the desired transfer function. In the Taylor approximation technique, we use Taylor series expansions of both $|H(j\omega)|^2$ and $|H_d(j\omega)|^2$ and equate coefficients of like powers of ω. (The Taylor series expands a function about a certain point; for a low-pass filter, we typically choose to expand around $\omega = 0$.)

We will be able to equate only the first N terms in these two series, where N depends on the complexity of the network. Since the coefficients in the Taylor series expansions are found by differentiating the functions, this procedure guarantees that the first $N - 1$ derivatives of our approximation are the same as those of the desired function. Therefore, if $N = 3$ in our example, the value of the function itself, the slope of the function, and the rate of change of the slope are all the same as for the desired magnitude function in Figure 11-2(a) at $\omega = 0$. We call this particular approximation the ***maximally flat*** magnitude response.

It can be shown that for rational functions (the magnitude and phase of real transfer functions are always rational), the maximally flat criterion can be satisfied by equating coefficients of like powers of s (or ω) in the numerator and denominator [11.2]. Using this technique, we can find maximally flat magnitude responses and phase responses.

One magnitude response that satisfies the maximally flat criterion is the ***Butterworth*** response, given by

$$|H(j\omega)| = \frac{1}{\sqrt{1 + \omega^{2N}}}. \tag{11.6}$$

This equation is normalized; that is, $|H(j0)| = 1$ and $|H(j1)| = 1/\sqrt{2}$ for all values of N. The poles of these normalized transfer functions are equally spaced around the unit circle in the left half of the s-plane and are symmetric about the real axis, as shown in Figure 11-3 for $N = 4$. The normalized Butterworth magnitude responses are plotted in Figure 11-4 for $N = 1, 2, 3,$ and 4.

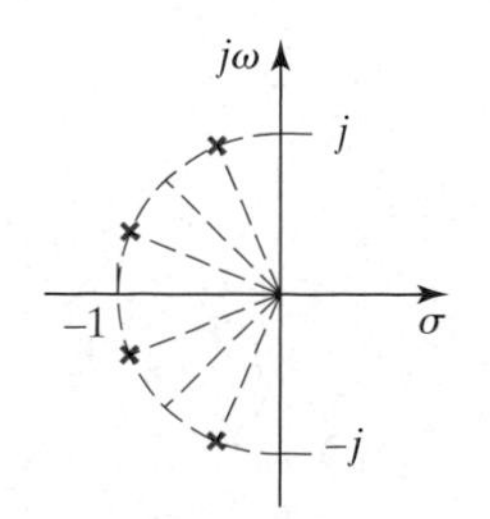

Figure 11–3 The pole locations of a Butterworth transfer function with $N = 4$. The poles are equally spaced around the left half of a unit circle and are symmetric about the real axis.

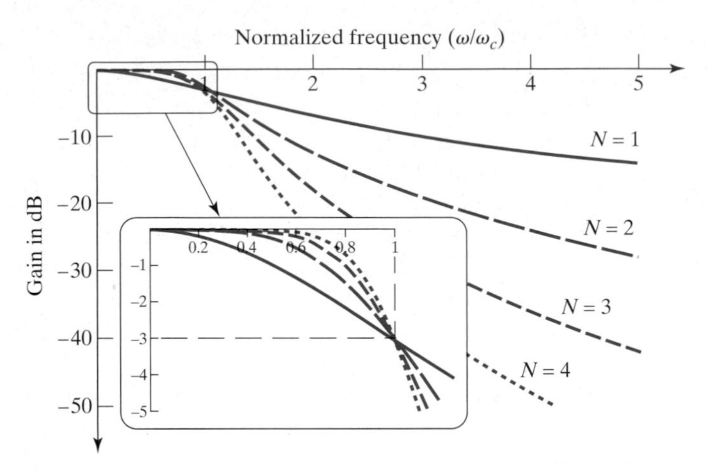

Figure 11–4 The Butterworth magnitude responses for $N = 1, 2, 3,$ and 4 (see Butter.nb on CD).

The Taylor approximation emphasizes making the approximation as good as possible at one point: the point about which we derived our Taylor series (typically $\omega = 0$ for low-pass filters). As a result, the approximation is fairly poor for frequencies approaching the cutoff frequency. In fact, the Butterworth response is reasonable both in magnitude and phase only for frequencies less than about half of the cutoff frequency; after that, the magnitude starts to roll off and the phase is nonlinear. In addition, the step response of Butterworth filters shows significant overshoot and ringing and is not usually acceptable in pulse transmission systems. The advantages of the Butterworth response are its simplicity and the insensitivity of the response to component variations [11.2].

Example 11.1

Practical filters are frequently specified by a template that the transfer function must fit within, as illustrated for a low-pass transfer function in Figure 11-5. The filter is assumed here to have a maximum gain of 0 dB, which is consistent with a passive filter. The ***maximum passband ripple***, as shown, specifies how much the gain can vary within the passband. For this low-pass filter, the passband extends from DC to ω_p, the passband edge. In addition, a ***minimum stopband attenuation*** is specified. The filter must provide at least this much attenuation for any signal in the stopband, which begins at the stopband edge, ω_s. The region between the passband edge and the stopband edge is the transition band. The selectivity of the filter is determined by how steep the transfer function must roll off in the transition band.

The example shown here has a gradual transition from the passband to the stopband. Very steep transitions from the passband to the stopband require high-order filters and will necessarily have rapidly varying phase as well.[3] This rapidly varying phase will not be linear with frequency and will lead to distortion, as noted in Aside 12.1. There is, therefore, a trade-off between the selectivity of the circuit and the transmission distortion.

[3] Cascading similar filters does not significantly sharpen the transition in the vicinity of ω_c(called the "knee"), but it does increase the slope of the response for $\omega \gg \omega_c$ (*see* Problem P11.5).

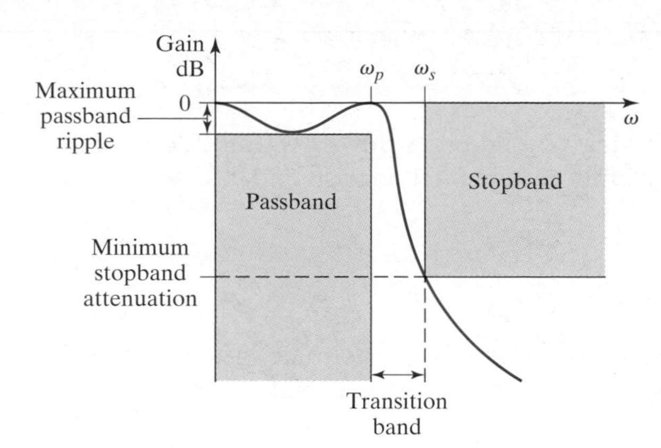

Figure 11–5 Specifications for a low-pass filter. A transfer function that meets these specifications is also shown.

Example 11.2

Problem:

Consider the low-pass filter shown in Figure 11-6. Assume that $R = 1\ \Omega$ (the normalized case). Determine values for L and C that will yield a Butterworth response with a cutoff frequency of 1 rad/s.

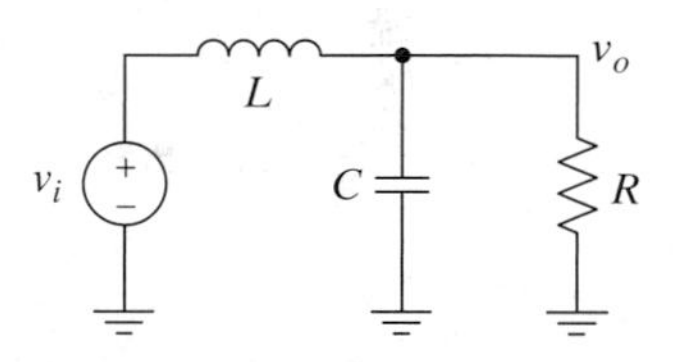

Figure 11–6 A low-pass filter.

Solution:

The square of the magnitude of the transfer function can be written as

$$|H(j\omega)|^2 = \frac{1}{1 + L(L - 2C)\omega^2 + L^2C^2\omega^4}.$$

Equating coefficients of like powers of ω in the numerator and denominator, we find that $1 = 1$ and $L = 2C$. We have enough constraints and don't need to go on to the ω^4 term. We also set the cutoff frequency to unity, so that

$$|H(j1)|^2 = \frac{1}{2} = \frac{1}{1 + L(L - 2C) + L^2C^2}.$$

Plugging in $L = 2C$ leads to $C = \sqrt{1/2} = 0.707$ F, and then $L = 1.414$ H. The SPICE output is shown in Figure 11-7. The response is maximally flat, as expected, and has a cutoff frequency of 0.159 Hz (1 rad/s), as desired.

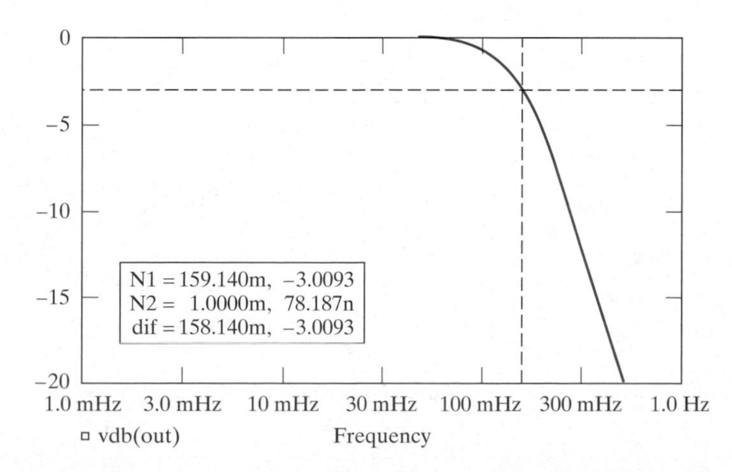

Figure 11–7 SPICE output.

Example 11.3

Problem:

Consider the following second-order low-pass transfer function, given in a *standard form* frequently used in filter analysis and design:

$$H(s) = \frac{1}{s^2 + s\dfrac{\omega_o}{Q} + \omega_o^2}.$$

Determine the pole locations for this transfer function. Find the value of Q necessary to obtain a Butterworth response, and plot the magnitude and phase responses for $Q = 1, 0.707$, and 0.3.

Solution:

The poles are given by

$$p_{1,2} = \frac{-\omega_o}{2Q}\left(1 \pm \sqrt{1 - 4Q^2}\right).$$

For values of Q less than or equal to 1/2, the poles are real. For Q greater than 1/2, the poles are complex conjugates and are given by

$$p_{1,2} = \frac{-\omega_o}{2Q}\left(1 \pm j\sqrt{4Q^2 - 1}\right).$$

To have a Butterworth response, the poles need to be on radials that form ±45° angles with the negative real axis. Therefore, the real and imaginary parts are equal, and we must have $Q = 1/\sqrt{2} = 0.707$. The poles are then at $p_{1,2} = -\omega_o/\sqrt{2}(1 \pm j)$ and lie on a circle of radius ω_o, meaning that the filter BW is ω_o.

The magnitude and phase responses are plotted in Figure 11-8 for $Q = 1, 0.707$ and 0.3. Note that for purely real poles ($Q = 0.3$), the transfer function approximates a single-pole transfer function at low frequencies, due to the lower frequency pole dominating. The plot for $Q = 0.707$ is the maximally flat response; any larger value of Q will lead to peaking in the magnitude response, as demonstrated in the plot for $Q = 1$.

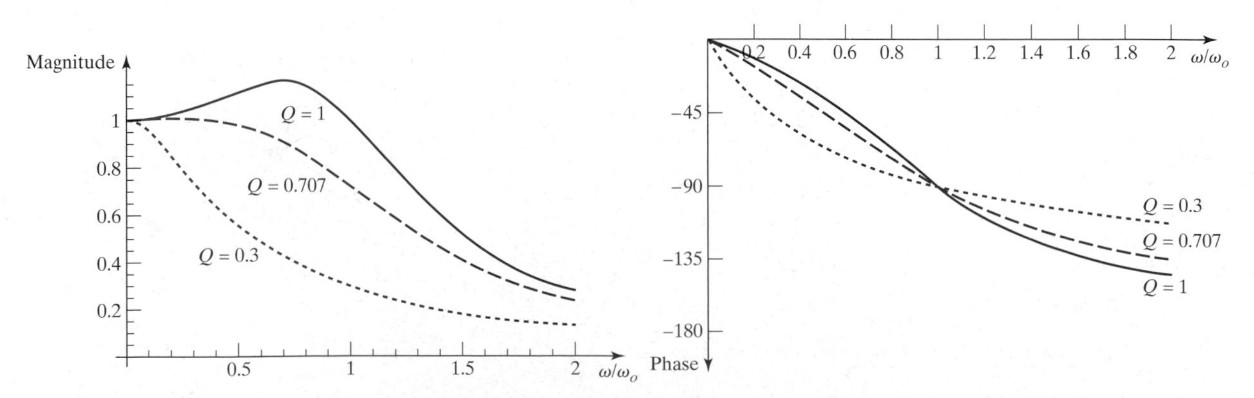

Figure 11–8 Magnitude and phase of the standard second-order transfer function for $Q = 1, 0.707$, and 0.3 (see Exam3.nb on CD).

Exercise 11.1

Consider the low-pass filter shown in Figure 11-6. Assume that R = 1 Ω (the normalized case). Determine values for L and C that will yield a Butterworth response with a cutoff frequency of 3 rad/s.

The Equiripple Magnitude Approximation: Chebyshev Filters Another common approximation is based on assigning equal significance to every point in the passband. In this case, the maximum error in the approximation is minimized when all peak errors are equal across the passband. The net result is called an ***equiripple*** or ***Chebyshev*** approximation [11.2].

The normalized Chebyshev transfer functions are given by

$$|H(j\omega)| = \frac{1}{\sqrt{1 + \varepsilon C_N^2(\omega)}}, \quad \textbf{(11.7)}$$

where the ripple magnitude is set by ε and the first four Chebyshev polynomials are[4]

$$C_N(\omega) = \begin{cases} 1 & N = 0 \\ \omega & N = 1 \\ 2\omega^2 - 1 & N = 2. \\ 4\omega^3 - 3\omega & N = 3 \\ 8\omega^4 - 8\omega^2 + 1 & N = 4 \end{cases} \quad \textbf{(11.8)}$$

The normalization used here is different than in the Butterworth case. Instead of using a −3-dB cutoff frequency equal to unity, the Chebyshev filters use $\omega = 1$ as the edge of the ripple band (i.e., the point were the ripple can first exceed the specified error). The peak-to-peak passband ripple in dB is given by

$$\text{ripple}_{p-p} = 10\log(1 + \varepsilon^2). \quad \textbf{(11.9)}$$

The number of ripple peaks (maximum or minimum) in the passband is equal to N. The magnitude responses for $N = 3$ and $N = 4$ with $\varepsilon = 0.509$ (a 1-dB ripple) are shown in Figure 11-9. This figure clearly shows that $\omega = 1$ is *not* the −3-dB cutoff, but is instead the edge of the ripple band. The cutoff frequency is given by

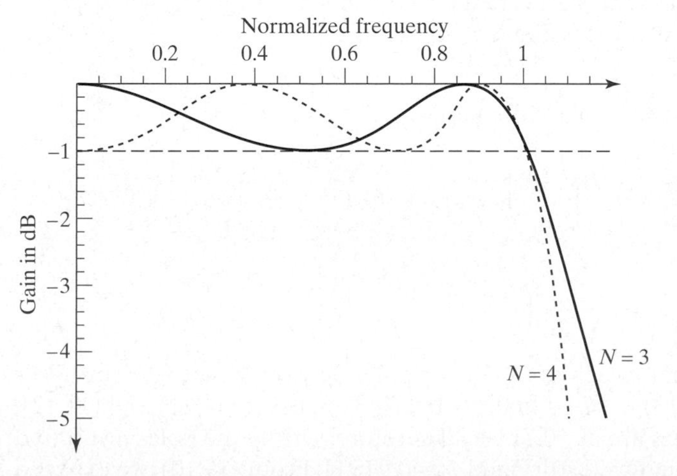

Figure 11–9 The Chebyshev magnitude responses for $N = 3$ and $N = 4$ with $\varepsilon = 0.509$ (a 1-dB ripple). The frequency is normalized to the edge of the ripple band, instead of the cutoff frequency (see Cheby.nb on CD).

[4] These polynomials are found using a recursive relationship shown in [11.2].

$$\omega_c = \cosh\left[\frac{1}{N}\cosh^{-1}\left(\frac{1}{\varepsilon}\right)\right]. \quad \textbf{(11.10)}$$

Finally, note that for odd values of N, the DC gain is on a peak, whereas for even values of N, the DC gain is at a minimum.

The poles of a Chebyshev transfer function are symmetric about the real axis, are equally spaced, and lie on the left half of an ellipse whose major axis is coincident with the $j\omega$ axis and that is centered at the origin. For normalized filters, this ellipse is inside of the unit circle. The real and imaginary parts of the kth pole for the filter are given by [11.2]

$$\sigma_k = -\sinh(\xi)\sin\left(\frac{(2k-1)\pi}{2N}\right) \quad \textbf{(11.11)}$$

and

$$\omega_k = \cosh(\xi)\cos\left(\frac{(2k-1)\pi}{2N}\right) \quad \textbf{(11.12)}$$

for $k = 1, 2, ..., N$, where

$$\xi = \frac{1}{N}\sinh^{-1}(1/\varepsilon). \quad \textbf{(11.13)}$$

Considering the error over the entire passband has led to a response that is more uniform across the passband, although the Chebyshev response approaches that of a Butterworth as the ripple goes to zero. The Chebyshev response rolls off more rapidly than a Butterworth of comparable complexity. The phase is again nonlinear, especially near the cutoff frequency, so distortion may result. The step response of a Chebyshev filter again exhibits overshoot and ringing and is usually not suitable for pulse transmission systems.

Example 11.4

Problem:

Consider the low-pass filter shown in Figure 11-6. Assume that $R = 1\ \Omega$ (the normalized case). Determine values for L and C that will yield a Chebyshev response with a ripple band edge of 1 rad/s and a 0.5-dB ripple.

Solution:

The transfer function in the Laplace domain is

$$H(s) = \frac{1}{1 + sL + s^2LC},$$

which has roots at

$$s_{1,2} = \frac{1}{2C}\left(-1 \pm j\sqrt{\frac{4C}{L} - 1}\right).$$

Using (11.9), we find that to achieve 0.5 dB ripple, we need $\varepsilon = 0.349$. This filter has $N = 2$, so, using (11.13), we find that $\xi = 0.887$. Now, using (11.11) and (11.12), we find that the two poles are at $-0.713 \pm j$. Therefore, setting the poles just found to these values, we obtain $C = 0.701$ F and $L = 0.945$ H. From (11.10), we expect a cutoff frequency of 0.221 Hz.

Using these values in SPICE, we obtain the magnitude response shown in Figure 11-10. This response has a peak ripple of 0.488 dB, a ripple band edge of 0.158 Hz (0.99 rad/s), and a cutoff frequency of 0.229 Hz (1.44 rad/s). All values are very close to the predictions.

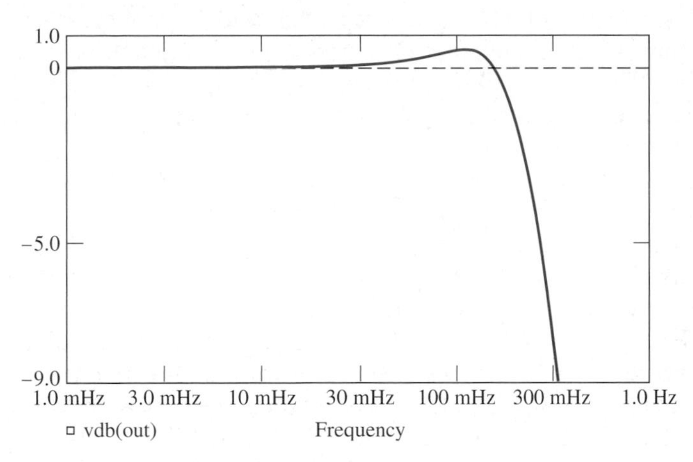

Figure 11–10 SPICE output.

Exercise 11.2

Consider the low-pass filter shown in Figure 11-6. Assume that $R = 1\ \Omega$ (the normalized case). Determine values for L and C that will yield a Chebyshev response with a ripple band edge of 1 rad/s and a 1-dB ripple.

Phase and Time-Domain Approximations: Bessel Filters The Taylor and equiripple approximations can be applied to the phase of a transfer function as well as to the magnitude. In fact, since the phase needs to be linear to avoid distortion, we often deal with the negative of the derivative of the phase, called the *group delay*. It can be shown that for narrowband modulated signals, the delay seen by the modulating signal in traveling through a system is approximately equal to the group delay [11.3].

If the Taylor approximation is used to produce a maximally flat group delay, the result is known as a *Bessel* filter [11.2]. It is also possible to use an equiripple approximation to constant group delay. Bessel filters have less overshoot and ringing than Butterworth or Chebyshev filters and so perform better in pulse transmission systems, but they do have slower transitions from the passband to the stopband, as we would expect, given the better behaved phase response. The poles of a Bessel transfer function, like those of the Chebyshev, are symmetric about the real axis, are equally spaced, and lie on the left half of an ellipse whose major axis is coincident with the $j\omega$ axis and that is centered at the origin. Unlike the Chebyshev filter, though, this ellipse is outside the unit circle for the normalized transfer function.

It is also possible to work directly in the time domain and come up with filters to approximate a given step response. These filters are particularly useful for pulse transmission systems. We will not cover such filters here; a good treatment is available in [11.2] and in other filter textbooks.

Practical Bandpass and Allpass Filters We now briefly examine a practical bandpass filter. Figure 11-11(a) shows the magnitude and phase of an ideal bandpass filter, and the characteristics of a typical practical bandpass realization are shown in part (b) of the figure. The bandwidth, $\omega_{-3\text{dB}} = \omega_{cH} - \omega_{cL}$, is often much less than the center frequency, ω_o. When it is, we can frequently simplify the analyses as shown in Section 11.1.4.

A measure of the effectiveness of the filter is provided by the slope of the skirt of the magnitude response. Quantitatively, this measure is called the *selectivity* and is given either by the ratio of the frequency where the gain is −60 dB to the frequency where it is −6 dB or by the ratio of the frequency where the gain is −30 dB to the frequency where it is −3 dB, ω_{-30}/ω_{cH} in the figure.

We end this section with a brief discussion of allpass filters. Allpass filters are used to adjust the phase of a transfer function without affecting the magnitude response. An allpass filter results from a pole–zero diagram where the zeros in the right-half plane are mirror images of the poles in the left-half plane. For example, consider the circuit shown in Figure 11-12(a). The transfer function of this circuit is

$$\frac{V_o(j\omega)}{V_i(j\omega)} = \frac{1 - j\omega RC}{1 + j\omega RC}. \tag{11.14}$$

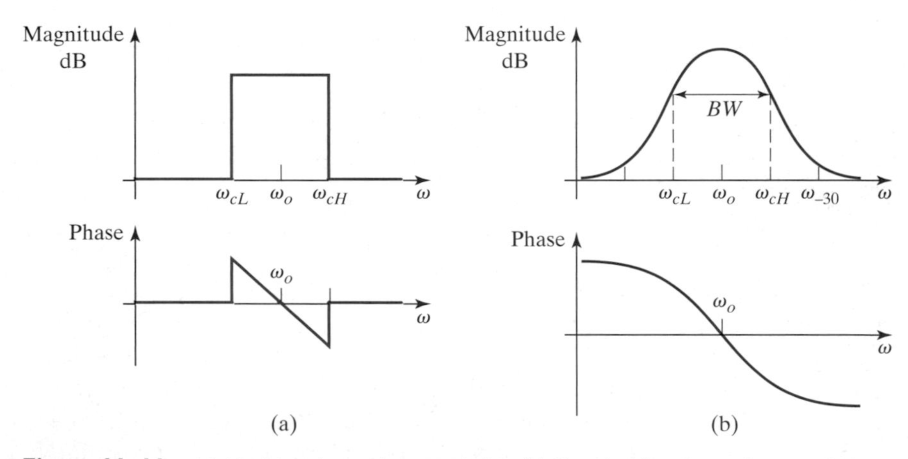

Figure 11–11 (a) Ideal bandpass characteristics. (b) Practical bandpass characteristics.

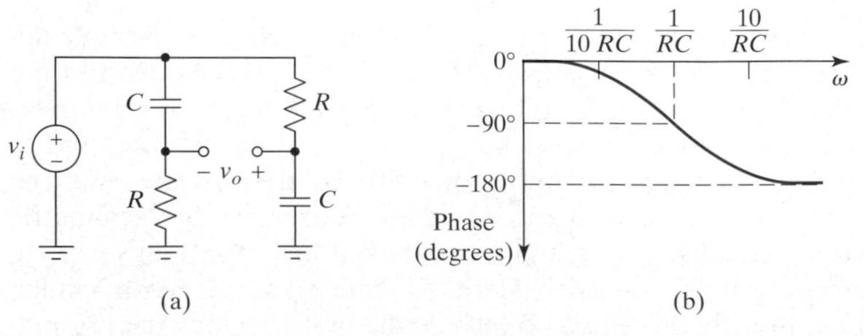

Figure 11–12 A single-pole, single-zero allpass filter.

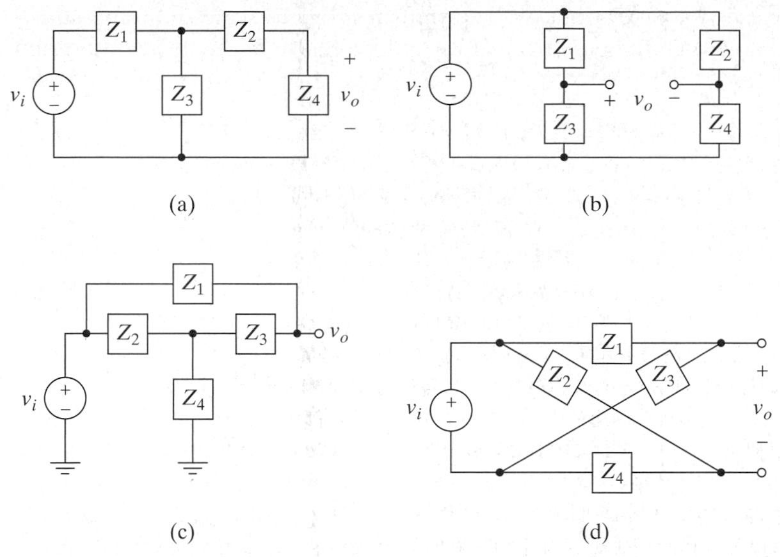

Figure 11–13 Network topologies: (a) A ladder network, (b) a bridge network, (c) a bridged-T network, and (d) a lattice network.

The magnitude of this transfer function is unity for all frequencies, and the phase is plotted in part (b) of the figure. There is one real pole in the left-half plane (at $s = -1/RC$) and one real zero in the right-half plane (at $s = 1/RC$).

Far more complicated allpass transfer functions are possible, but they cannot be achieved by ***ladder networks***, which have the topology shown in Figure 11-13(a). Ladder networks are simple to analyze, but they are always ***minimum phase***. A minimum-phase network is a network whose poles are strictly in the left-half plane and whose zeros are either in the left-half plane or on the real axis. No poles or zeros are allowed in the right-half plane. Since allpass filters have zeros in the right-half plane, they are examples of nonminimum-phase networks. The topologies shown in parts (b)–(d) of Figure 11-13 can all produce nonminimum-phase networks.

11.1.3 Normalization (Frequency and Component Value Scaling)

It is convenient to work with normalized transfer functions and component values, but real filters are not often made with 1-Ω resistors, 1-F capacitors, or 1-H inductors and do not typically have cutoff frequencies of 1 rad/s. Therefore, we need to be able to scale the component values and operating frequencies of networks that we design. Consider, for example, a single-pole low-pass filter, as shown in Figure 11-14.

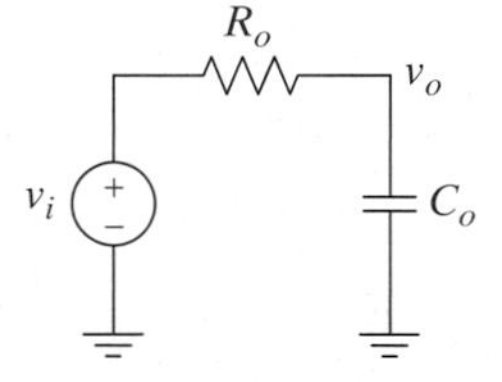

Figure 11–14 A low-pass filter.

This filter has a pole at $\omega_p = 1/(R_oC_o)$. If we set $R_o = 1\ \Omega$ and $C_o = 1$ F, then $\omega_o = 1$ rad/s. Now, suppose that we wish to use a more realistic value for the resistor, so we scale it up by a factor a: $R = aR_o$. To keep the same filter response, we need to simultaneously scale the capacitor down by a factor a: $C = C_o/a$. The resulting filter has the same pole frequency. What we have done is to keep the impedances the same at a given frequency. If an inductor is included in the circuit, we need to scale it as well: $L = aL_o$.

Suppose we also want to scale the pole frequency of this filter to a new value b times larger than ω_o. We can guarantee that the response stays the same in every way, except for being shifted up in frequency, if we set the impedances to be the same at the new pole frequency as they were at the normalized pole frequency. This requires using $C_{new} = C_{old}/b$ and $L_{new} = L_{old}/b$. The resistor does not change, since its impedance is independent of frequency.

Combining these two operations, we see that in order to scale the impedances up by a factor a and scale the pole frequency up by a factor b, the new component values are given by

$$R = aR_o,$$

$$L = \frac{aL_o}{b},$$

and

$$C = \frac{C_o}{ab}, \tag{11.15}$$

where R_o, C_o, and L_o are the normalized component values.

It is also sometimes useful to know how to scale the poles and zeros in the complex plane. For example, consider the Butterworth poles in Figure 11-3. Suppose we want to know what happens to these poles if we desire a filter with a cutoff frequency of 10 krad/s instead of 1 rad/s. Since the shapes of both the magnitude and phase responses depend on the geometry of the pole and zero positions in the complex plane, we only need to linearly scale the real and imaginary axes. Therefore, instead of having the poles lie on a circle of unit radius, the new poles lie on a circle with a radius of 10,000. If zeros are present, they scale in exactly the same way.

If the transfer functions are written in the proper forms, then the absolute magnitude of the transfer function will not be affected by this scaling. For example, if the Butterworth transfer function from Figure 11-3 is expressed as

$$H(j\omega) = \frac{1}{\left(1 - \frac{j\omega}{p_1}\right)\left(1 - \frac{j\omega}{p_1^*}\right)\left(1 - \frac{j\omega}{p_2}\right)\left(1 - \frac{j\omega}{p_2^*}\right)}, \tag{11.16}$$

where $p_1 = -0.924 + j0.393$ and $p_2 = -0.393 + j0.924$, then scaling the poles will not affect the DC gain, since (11.16) is equal to unity at $\omega = 0$, no matter what the poles are. While this form works for low-pass transfer functions, care must be exercised in choosing the form to use, or the constant multiplier, when scaling bandpass, high-pass, and bandstop transfer functions (*see* Example 11.5).

Exercise 11.3

Starting from the normalized low-pass filter design in Figure 11-14 with $R_o = 1\ \Omega$ and $C_o = 1$ F, determine the resistor value necessary to produce a filter with a cutoff frequency of 100 kHz while using a capacitor of 1000 pF.

□

11.1.4 The Narrowband Approximation

Many of the applications for frequency-selective networks are in communication systems where the information is transmitted by modulating some parameter (e.g., amplitude or frequency) of a high-frequency carrier wave, as mentioned in Chapter 1. The frequency-domain representations of these signals have sidebands on either side of the carrier frequency, as we demonstrate for amplitude modulation in Aside A11.1.

ASIDE A11.1 AMPLITUDE MODULATION

We illustrate in this aside that communication systems can frequently be treated as narrowband bandpass systems. Consider a communication channel that uses *amplitude modulation* (e.g., AM broadcast radio). The input signal is a voltage representing the message to be transmitted, $v_m(t)$. This signal can be expressed as a Fourier series[1] and is assumed to be band limited to a fairly low frequency (e.g., music is typically band limited to about 20 kHz).

A circuit called a *modulator* translates this band of frequencies, called the *baseband*, up to a new band of frequencies centered about the *carrier* frequency, ω_c, without changing relative magnitudes or phases of the individual components. The resulting AM signal is given by

$$v(t) = [V_c + v_m(t)] \cos\omega_c t, \qquad \textbf{(A11.1)}$$

where V_c is the amplitude of the carrier signal. A sample message signal and the resulting modulated waveform are shown in Figure A11-1.

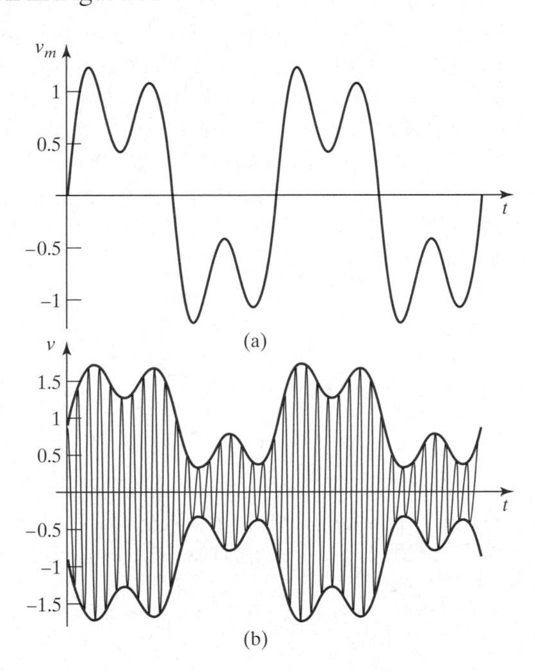

Figure A11-1 (a) A message signal and (b) the resulting amplitude-modulated signal (see AM_sig.nb on the CD).

The magnitude of $v_m(t)$ is always adjusted to be less than V_c, so that the envelope of the modulated signal never drops to zero. To illustrate, consider the special case of a single frequency. The message signal has the form $v_m(t) = V_m \cos(\omega_m t + \theta_m)$. In this case, using (A11.1), we obtain

$$\begin{aligned} v(t) &= [V_c + V_m \cos(\omega_m t + \theta_m)]\cos\omega_c t \\ &= V_c[1 + m\cos(\omega_m t + \theta_m)]\cos\omega_c t, \end{aligned} \qquad \textbf{(A11.2)}$$

where $m = V_m/V_c$ is called the *modulation index* and is always less than unity. In addition, it is always true that $\omega_c \gg \omega_m$. We may rewrite (A11.2) using the trigonometric identity for the product of two cosines as

$$\begin{aligned} v(t) = V_c\cos\omega_c t &+ \frac{m}{2} V_c\cos[(\omega_c + \omega_m)t + \theta_m] \\ &+ \frac{m}{2} V_c\cos[(\omega_c - \omega_m)t - \theta_m]. \end{aligned} \qquad \textbf{(A11.3)}$$

[1] To be represented as a Fourier series, the wave must be periodic. But, for any finite-length message, we can always form a periodic waveform by imagining the message to be repeated over and over again.

Note that (A11.3) has three separate frequencies: the carrier and two components spaced symmetrically about the carrier.

In the general case when the message signal is more complex, there are many components (or a continuous spectrum) on each side of the carrier. These components above and below the carrier frequency are called the ***upper*** and ***lower sidebands*** of the AM signal, respectively.

To receive the AM signal, the receiver must filter the signal to get rid of other signals that may have frequencies close to that of the desired signal. These other frequencies would interfere with the latter's detection. This is one example where a combination of filters and tuned amplifiers can be useful.

If the total bandwidth of the system is less than one-tenth of the carrier frequency, we call the system a ***narrowband*** system. There are some significant simplifications that can be used in the analysis and design of narrowband systems, as we now demonstrate.

We wish to derive a method for transforming a narrowband bandpass system into an equivalent low-pass system. One simple way to build a bandpass filter is to drive a parallel *RLC* tank circuit with a current source (e.g., the output current of a transistor) and to use the voltage developed across the tank circuit as the output, as shown in Figure 11-15(a). In this case, the transfer function is the impedance of the tank circuit. The impedance is equal to the reciprocal of the admittance; consequently, if we want to examine the poles of the transfer function, we need to look at the zeros of the admittance. Using a subscript B, for bandpass, on the complex variable s, we write the admittance of the tank circuit as

$$Y(s_B) = G + s_B C + \frac{1}{s_B L} = G + C\left(s_B + \frac{\omega_o^2}{s_B}\right), \tag{11.17}$$

where the resonant frequency of the tank is

$$\omega_o = \frac{1}{\sqrt{LC}}. \tag{11.18}$$

Now consider the admittance of a single-pole low-pass filter as shown in part (b) of the figure. Using the subscript L, for low pass, on the complex variable s, we see that the admittance is

$$Y(s_L) = G + s_L C. \tag{11.19}$$

Equating (11.18) and (11.19), we can derive a ***low-pass-to-bandpass transformation***[5]

$$s_L = s_B + \frac{\omega_o^2}{s_B}, \tag{11.20}$$

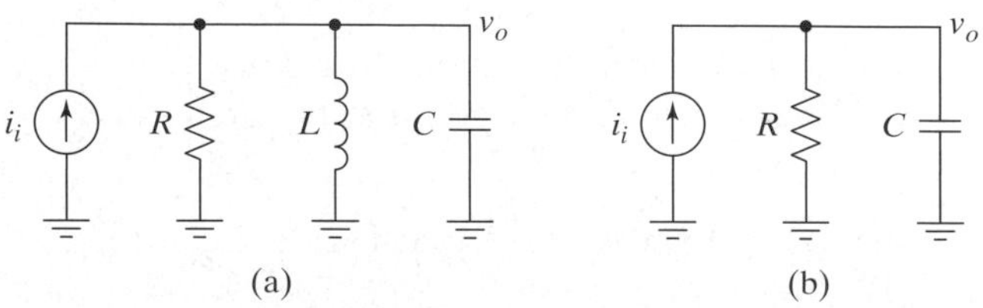

(a) (b)

Figure 11–15 (a) An *RLC* bandpass filter and (b) a single-pole low-pass filter.

[5] Other transformations that are useful are low-pass to highpass, $S_H = 1/S_L$ and low-pass to bandstop (notch), $S_L = S_N/(S_N^2 + \omega_o^2)$.

or

$$s_B = \frac{s_L}{2} \pm \sqrt{\left(\frac{s_L}{2}\right)^2 - \omega_o^2}. \tag{11.21}$$

The bandpass filter has two complex-conjugate poles corresponding to the one pole for the low-pass filter, as we would expect. If the bandpass filter is a narrowband system, we can say that $|s_L/2| \ll \omega_o$ as we will see in a moment. Therefore, the term $(s_L/2)^2$ in the radical in (11.21) can be neglected, and we arrive at

$$s_B \approx \frac{s_L}{2} \pm j\omega_o \tag{11.22}$$

for the narrowband case. In other words, the bandpass pole locations are equal to the low-pass pole location scaled by 1/2 and translated by $\pm j\omega_o$.

As an example, if a low-pass filter has the pole shown in Figure 11-16(a), the narrowband low-pass-to-bandpass transformation yields the poles shown in part (b) of the figure. From part (b) of this figure we see that $|p_L/2| \ll \omega_o$, so for frequencies near the passband of the low-pass filter, our earlier approximation that $|s_L/2| \ll \omega_o$ is seen to be true.

We next demonstrate that for frequency response calculations near the center frequency in a narrowband system, the zeros and conjugate pole can also be ignored. Figure 11-17 shows the poles and zero of a narrowband bandpass transfer function given by

$$H(s) = \frac{Ks}{(s-p)(s-p^*)}. \tag{11.23}$$

The magnitude of this transfer function is

$$|H(j\omega)| = \frac{K\omega}{|j\omega-p|\,|j\omega-p^*|}. \tag{11.24}$$

Examining Figure 11-17, we see that, for frequencies greater than the lower cutoff, $\omega > \omega_{cL}$, the vector from the zero to a point in the passband is approximately one-half the length of the vector from p^* to the passband. In other words, the following expression is approximately constant:

$$\frac{\omega}{|j\omega-p^*|} \approx \frac{1}{2}. \tag{11.25}$$

So long as this transfer function satisfies the narrowband approximation, the error in (11.25) is less than 5%. Using (11.25) in (11.24), we can write

$$|H(j\omega)| \approx \frac{K/2}{|j\omega-p|} \tag{11.26}$$

for $\omega > \omega_{cL}$. Therefore, zeros at the origin or on the real axis, poles on the real axis, and the conjugate pole can be ignored in frequency response calculations in or above the bandpass region of a narrowband system. As a result, we focus on the poles and zeros in the vicinity of $j\omega_o$ when designing narrowband systems. This simplification allows us to use the methods we examined for low-pass systems, such as the Butterworth approximation, when designing narrowband bandpass systems. Example 11.5 illustrates the point.

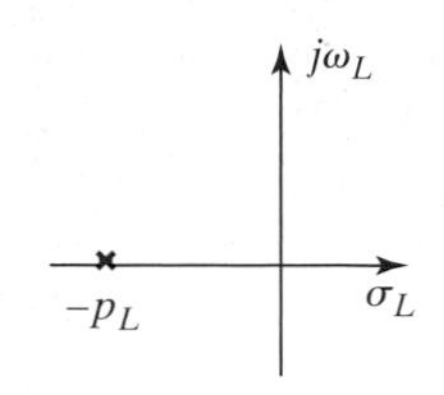

(a)

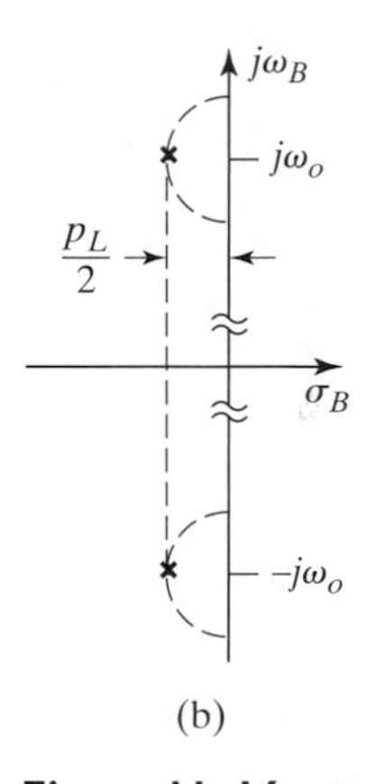

(b)

Figure 11–16 (a) Low-pass pole location. (b) Corresponding bandpass pole locations found using the narrowband low-pass-to-bandpass transformation.

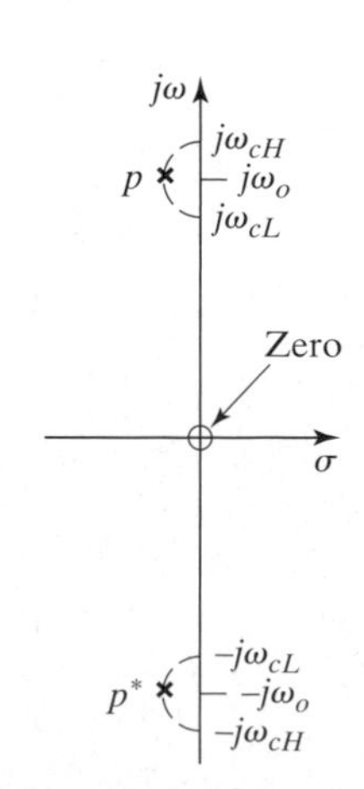

Figure 11–17 Poles and zero for the narrowband transfer function (11.23).

Example 11.5

Problem:
Determine the transfer function of an eighth-order Butterworth bandpass filter with a center frequency of 10 krad/s and a bandwidth of 100 rad/s.

Solution:
Note that when we transform a low-pass transfer function to a bandpass transfer function, we double the number of poles, as seen by (11.22). Therefore, to produce an eight-pole bandpass filter, we start with a four-pole low-pass filter. Using the pole positions shown in Figure 11-3 and scaling them to achieve a bandwidth of 100 rad/s, we find that the low-pass poles are at

$$p_{L1} = -92.4 + j38.3, p_{L2} = -38.3 + j92.4$$

and their conjugates. We then use (11.22) and arrive at the bandpass poles

$$p_{B1} = -46.2 + j10{,}019.2, \; p_{B2} = -19.2 + j10{,}046.2$$
$$p_{B3} = -46.2 + j9980.8, \; p_{B4} = -19.2 + j9953.8$$

and their conjugates. We then have to determine the constant multiplier necessary to set the magnitude of the transfer function equal to unity at ω_o. The resulting transfer function is

$$H(j\omega) = \frac{1.00247*10^{24}}{\left[\begin{array}{c}(j\omega - p_{B1})(j\omega - p_{B1}^*)(j\omega - p_{B2})(j\omega - p_{B2}^*) \\ (j\omega - p_{B3})(j\omega - p_{B3}^*)(j\omega - p_{B4})(j\omega - p_{B4}^*)\end{array}\right]},$$

where the constant multiplier was determined numerically. The resulting magnitude response in the passband is plotted in Figure 11-18.

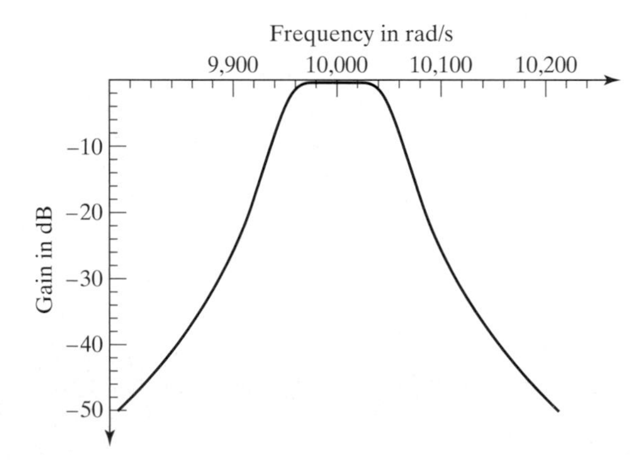

Figure 11–18 The Butterworth bandpass magnitude response (see Butter.nb on CD).

11.1.5 Integrated Filters and Simulated Inductors

Because integrated circuits are relatively inexpensive to produce, and because having fewer components on a printed circuit board generally leads to lower cost, higher reliability, smaller size, and lower power, it is often important to integrate as much as possible. Therefore, a lot of work has been put into producing integrated filters to replace passive *RLC* filters. Because large-value capacitors and resistors consume large areas on an IC, they are extremely expensive and are avoided.

Integrated inductors are, at this point in time, restricted to specialized applications and very high frequencies, because they have poor Q, have small values, and are not always compatible with standard IC fabrication. In this section, we briefly discuss the approaches taken to produce integrated filters, including simulating inductors.

If we restrict ourselves to using only resistors, capacitors, and active components to design filters, it may appear as if we have an impossible job, since passive *RC* circuits cannot produce complex poles, which are necessary for all but the simplest of filters. It turns out, however, that passive *RC* circuits can have complex zeros; so, by using these circuits in the feedback path of an amplifier, we can generate complex poles.

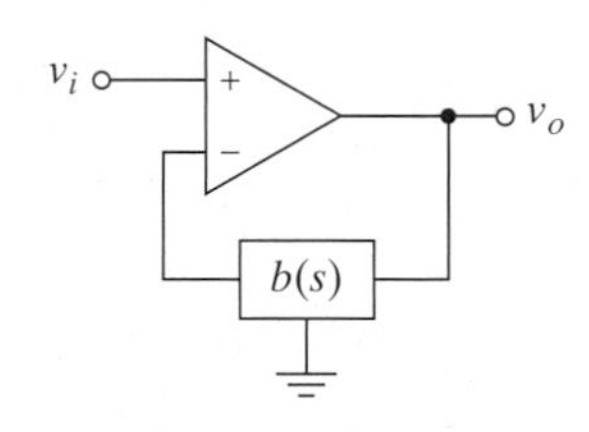

Figure 11–19 A feedback amplifier with a frequency-dependent feedback network.

Consider the feedback amplifier shown in Figure 11-19. Assume that the forward amplifier is ideal and that the feedback network is a function of frequency, as indicated.

The transfer function of this amplifier is given by

$$\frac{V_o(s)}{V_i(s)} = \frac{a}{1 + ab(s)}, \qquad \textbf{(11.27)}$$

and, if we write the feedback function as $b(s) = N_b(s)/D_b(s)$, we see that, for large loop gain (i.e., $ab \gg 1$), we have

$$\frac{V_o(s)}{V_i(s)} \approx \frac{1}{b(s)} = \frac{D_b(s)}{N_b(s)}. \qquad \textbf{(11.28)}$$

We see that the poles of the closed-loop transfer function are approximately given by the zeros of $b(s)$. Since a purely passive *RC* network can have complex zeros, using it in a feedback loop allows us to produce a filter with complex poles. (Another approach to achieving this goal is illustrated in Problem P11.17.)

Using the principle illustrated in (11.28) is more complicated than it may appear, since the zeros of the closed-loop transfer function determine the overall response. For example, if $b(s)$ has a quadratic numerator, it will also have a quadratic denominator (the number of finite poles is always equal to or greater than the number of finite zeros), and the resulting closed-loop transfer function will have two finite zeros as well as the two finite complex poles. If the zeros are both below the frequency of the complex poles, the response is a high-pass filter. (To see this, picture the Bode magnitude plot in a very general way: The zeros cause the magnitude to increase with frequency, and then the poles cause it to level off.) If one zero is below the pole frequency and one is above it, the result is a bandpass filter (the magnitude response rises due to the low-frequency zero, rolls off due to the presence of two poles, and then flattens due to the high-frequency zero), and if both zeros are above the pole frequency, the result is a low-pass response (the poles cause the response to roll off, and the zeros cause it to flatten out at very high frequencies).

To obtain the desired response, the feedback network usually needs to be transformed in some way and the signal injected in a different location than is shown in Figure 11-19. Injecting the signal somewhere else in the loop does not affect the pole locations of the closed-loop response, but it does allow us to choose the zeros. (A complete treatment of this topic goes beyond the objectives of this chapter; an introduction is provided in [11.11]. An example of a transformed network is given in Problems P11.18 and P11.19.)

Active *RC* filters can also have voltage gains greater than unity, as well as low output resistances and high input resistances. Low output resistance and high input resistance allow us to achieve power gain, cascade stages without significant interaction, and add impedance matching elements if necessary. Active *RC* filters were discussed in Section 5.2 in conjunction with op amp circuits, although the concept of viewing them as passive *RC* feedback networks with ideal forward amplifiers was not presented. The resistors and capacitors can be integrated with the amplifier, so these filters may be fully integrated.

Another useful technique for producing filters without inductors is to use a ***generalized impedance converter*** (GIC) to simulate an inductor using only resistors, capacitors, and op amps. A GIC is shown in Figure 11-20 and can be analyzed without too much difficulty if we assume the op amps to be ideal.

Each op amp is connected with negative feedback, so we can conclude that it forces its inputs to be at the same voltage (i.e., there is a virtual short). Using this fact and starting from the input source, we see that the node between Z_2 and Z_3 is equal to V_i, and so is the voltage at the top of Z_5. Now, because ideal op amps have zero input current, we can see that $I_i = I_1$, $I_2 = I_3$, and $I_4 = I_5$. Therefore,

$$I_4 = \frac{V_3 - V_i}{Z_4} = I_5 = \frac{V_i}{Z_5}, \tag{11.29}$$

which leads to

$$V_3 = \left(1 + \frac{Z_4}{Z_5}\right)V_i. \tag{11.30}$$

Similarly,

$$I_2 = \frac{V_2 - V_i}{Z_2} = I_3 = \frac{V_i - V_3}{Z_3}, \tag{11.31}$$

which leads to

$$V_2 = \frac{V_i(Z_2 + Z_3) - V_3 Z_2}{Z_3}. \tag{11.32}$$

Finally, we write

$$I_i = I_1 = \frac{V_i - V_2}{Z_1} \tag{11.33}$$

and substitute (11.30) into (11.32) and the result into (11.33) to obtain

$$Z_i = \frac{V_i}{I_i} = \frac{Z_1 Z_3}{Z_2 Z_4} Z_5. \tag{11.34}$$

This equation describes the GIC. Note that if Z_4 is a capacitor and all of the other impedances are resistors, then we have

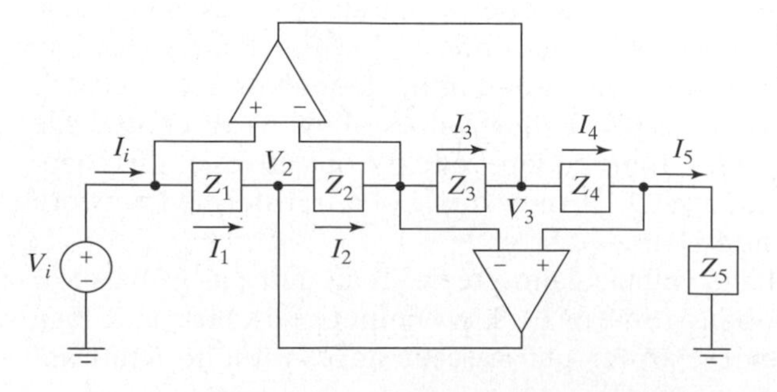

Figure 11–20 A generalized impedance converter.

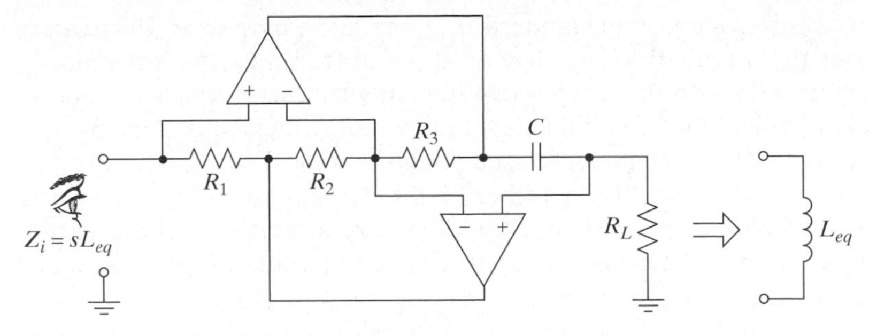

Figure 11–21 The GIC used to simulate an inductor.

$$Z_i = s\left(\frac{R_1 R_3 R_5 C}{R_2}\right) = sL_{eq}, \tag{11.35}$$

where we have defined the equivalent input inductance, L_{eq}. Other conversions are also possible. Figure 11-21 shows the GIC used to simulate an inductor.

Exercise 11.4

What is the equivalent inductance for the GIC circuit in Figure 11-21 if $R_1 = R_2 = R_3 = R_5 = 1\ k\Omega$ and $C = 470\ pF$?

11.1.6 Discrete-Time Filters

Not all filters operate on continuous-time analog signals. There are also digital filters that operate on binary "words" and discrete-time analog filters that operate on sampled, or discrete-time, analog signals. (Aside A1.7 shows different types of signals.) In this section, we first briefly introduce a common discrete-time analog filtering technique, ***switched-capacitor filters***. We then present the transversal ***finite impulse response*** filter, which can be implemented as a continuous-time analog, discrete-time analog, or digital filter.

One fundamental characteristic of discrete-time circuits which must be remembered is that they operate only on samples of the signal. Therefore, as discussed in Aside A13.1, we must be sure that the sample rate is greater than twice the highest frequency contained in the signals in order to avoid aliasing. If we allow aliasing, the sampling operation necessarily loses information; that is, we cannot uniquely reconstruct the original signal from its samples. If you are not familiar with discrete-time system analysis, you can still gain a qualitative idea of how the filters in this section work, but you may not be able to understand the transfer functions presented. Any good book on discrete-time linear systems will provide the necessary background (for example, [11.7]).

We will need to add to our system of notation in this section to allow for the discrete nature of the signals. We use $v[n]$ to refer to a voltage that is a function of discrete time. A system clock is always used, and the period of the clock, T_c, is usually our reference. Consequently, this notation denotes the value at time nT_c.

Switched-Capacitor Filters For a number of different reasons, it is often not practical to fully integrate *RLC* or active *RC* filters. First, at very low frequencies, the resistor, capacitor and inductor values required are too large. Second, to construct predictable filters, we need to control *RC* products and/or *L*/*R* ratios, neither of which can be done well in standard IC fabrication processes. Third, there are applications that require us to change a filter cutoff frequency electronically, which is difficult to do with these types of filters. Fourth, these filters are not always compatible with digital CMOS integrated circuits, which constitute the majority of all integrated circuits manufactured. Finally, many systems naturally deal with periodic samples of a waveform, rather than the continuous-time waveform. For all of these reasons, switched-capacitor filters are attractive and are used a great deal. In addition to filtering, switched-capacitor circuits can perform many other signal-processing functions as well.

In Section 5.2, we analyzed the op amp integrator circuit, which is repeated here as Figure 11-22. Since the capacitor provides negative feedback and the op amp has infinite gain, the inverting input is driven to zero volts and is called a virtual ground. The resistor then functions as a voltage-to-current converter, the current flows into the capacitor, and the output voltage is given by

$$v_o(t) = \frac{-1}{RC}\int_0^t v_i(\lambda)d\lambda + v_o(0). \tag{11.36}$$

A *switched-capacitor* version of the integrator is shown in Figure 11-23. Assume that both switches are initially open and that both capacitors are fully discharged. Then, if $SW1$ closes while $SW2$ remains open, the input voltage appears across C_1, and the charge on this capacitor is $q_1[0] = C_1 v_i[0]$.

The next step is to open $SW1$ and then close $SW2$ (we *must* be sure to get $SW1$ open *before* $SW2$ closes; this is called a *break-before-make* switching operation). The op amp will then work to drive the voltage at the inverting input to ground, which requires moving charge off of C_1. If we assume that the op amp input current is negligible, this charge can be transferred only to C_2, and charge is conserved during the transfer, so the charge on C_2 at this time is $q_2[1] = -q_1[0] = -C_1 v_i[0]$. (We have defined the polarity of the charge and voltage on C_2 as shown in the figure.) The corresponding output voltage is

$$v_o[1] = v_2[1] = \frac{q_2[1]}{C_2} = -\frac{C_1}{C_2} v_i[0]. \tag{11.37}$$

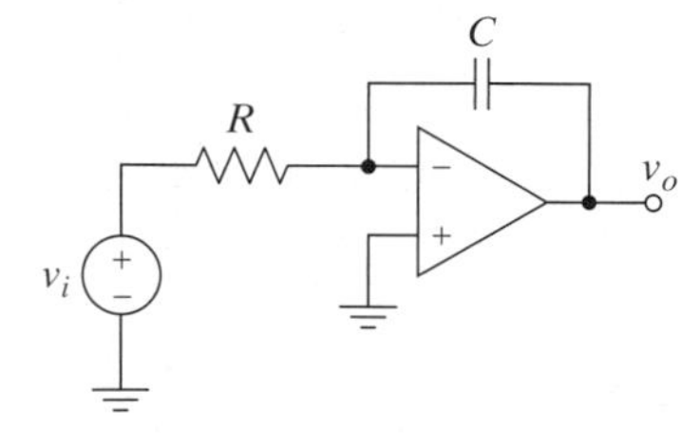

Figure 11–22 An *RC* integrator.

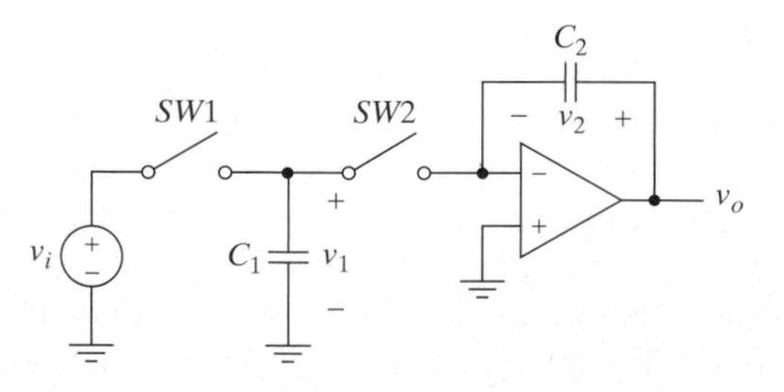

Figure 11–23 A switched-capacitor integrator.

We then open $SW2$ and start the whole process over again. The charge on C_2 accumulates, so that after n cycles we have

$$v_o[n] = -\frac{C_1}{C_2}\sum_{k=0}^{n-1} v_i[k], \tag{11.38}$$

which is a discrete-time integration. We can rewrite (11.38) as

$$v_o[n] = v_o[n-1] - \frac{C_1}{C_2} v_i[n-1]. \tag{11.39}$$

If you are familiar with z transforms, you will notice that we can write this as

$$V_o(z) = z^{-1}\left[V_o(z) - \frac{C_1}{C_2} V_i(z)\right], \tag{11.40}$$

from which we can derive the z-domain transfer function:

$$\frac{V_o(z)}{V_i(z)} = -\left(\frac{C_1}{C_2}\right)\frac{z^{-1}}{1 - z^{-1}}. \tag{11.41}$$

Notice that the transfer function given implicitly by (11.38) or explicitly by (11.41) depends on a ratio of capacitor values. In integrated circuits, these ratios can be controlled to about ±0.1%, so the degree of control we have is much greater than we would have if we used an active RC integrator. The RC product in (11.36) cannot typically be controlled to better than about ±30% on an IC.

An alternative view of this circuit is obtained by recognizing that the two switches and the capacitor in Figure 11-23 serve to convert the input voltage into an average current, just as the resistor in Figure 11-22 does. Specifically, if there is one switching cycle every T_c seconds, $q = C_1 v_i$ coulombs of charge are moved every T_c seconds. (Assume that v_i is constant over this period.) Therefore, the average current is

$$\bar{i} = \frac{\Delta q}{T_c} = \frac{C_1}{T_c} v_i, \tag{11.42}$$

whereas in Figure 11-22 the current is

$$i = \frac{v_i}{R}. \tag{11.43}$$

Equating these two currents, we see that the switches and the capacitor together (i.e., the switched capacitor) have an equivalent resistance of

$$R_{eq} = \frac{T_c}{C_1} = \frac{1}{f_c C_1}. \tag{11.44}$$

This equation is important because it shows us that the equivalent time constant of the switched-capacitor integrator is

$$\tau_{eq} = R_{eq} C_2 = \frac{C_2}{f_c C_1}, \tag{11.45}$$

which depends on the capacitor ratio and the clock frequency. In other words, we can vary the time constant by varying the clock frequency. This dependence on clock frequency is implicitly present in (11.39) and (11.41), since the time between samples depends on the clock.

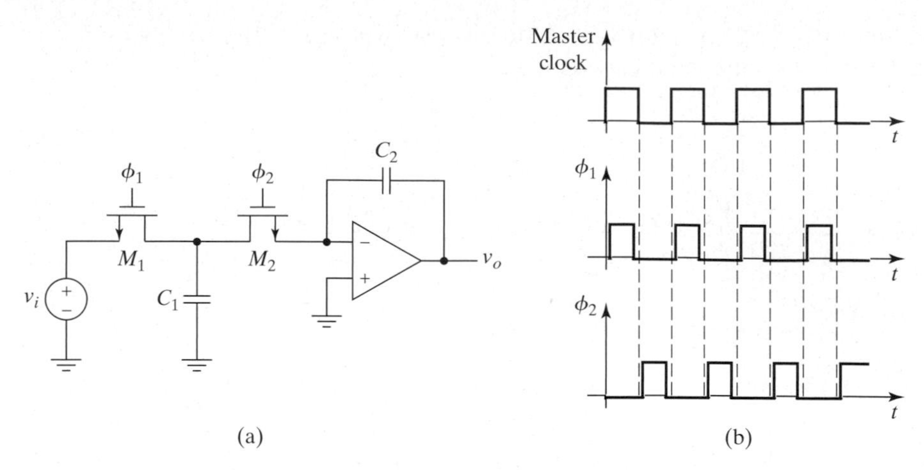

Figure 11–24 (a) The switched-capacitor integrator. (b) The nonoverlapping clocks.

The switches in switched-capacitor circuits are implemented using MOSFETs, as shown in Figure 11-24(a), which presents the integrator again. The clock waveforms used to drive the switches are shown in part (b) of the figure. The important aspect of these clocks is that they are ***nonoverlapping***; that is, they are never high at the same time. Nonoverlapping clocks guarantee that the two switches will never be on at the same time.

There are many practical issues to be dealt with in the design of switched-capacitor circuits. For example, problems arise due to

1. Parasitic capacitances present in the circuit,
2. Op amp offset voltages,
3. The finite on resistance of the MOSFET switches,
4. Charge injection from the MOSFET switches,
5. Variation in MOSFET on resistance with channel voltage,
6. and many others [11.5].

Many circuit configurations have been developed that are insensitive to parasitic capacitances and capable of canceling op amp offset voltages and so on.

We conclude this section with a brief presentation of a switched-capacitor amplifier and a low-pass filter.

Example 11.6

Problem:
Design an amplifier, using switched-capacitor circuits.

Solution:
If we take the switched-capacitor integrator from Figure 11-24(a) and add a switch across C_2 as shown in Figure 11-25, this will cause C_2 to be discharged every clock cycle. Therefore, the charge will not accumulate and we will have (refer to (11.39))

$$v_o[n] = -\frac{C_1}{C_2} v_i[n-1],$$

which is a gain of C_1/C_2 and a delay of one clock period. The output is valid only on ϕ_2, since, during ϕ_1, the switch across C_2 is on and the output voltage is zero.

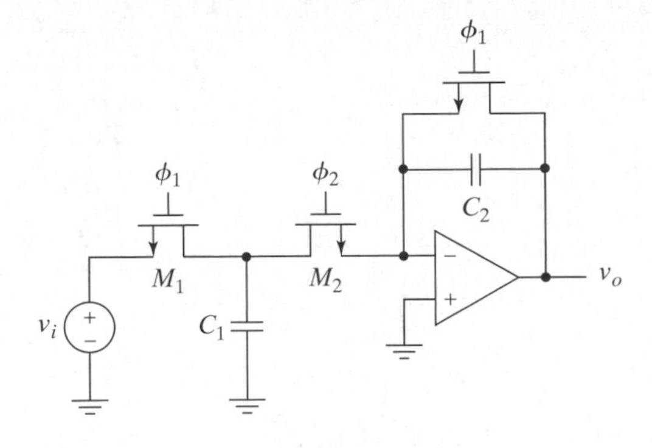

Figure 11-25 A switched-capacitor gain stage.

Example 11.7

Problem:
Design a switched-capacitor equivalent to the continuous-time single-pole low-pass filter shown in Figure 11-26. This filter has a DC gain of unity and a cutoff frequency of about 20 kHz. It has been designed to make the capacitor value small. That way, the switched-capacitor equivalent will not require large capacitors either. Assume that 1-MHz nonoverlapping clocks are available.

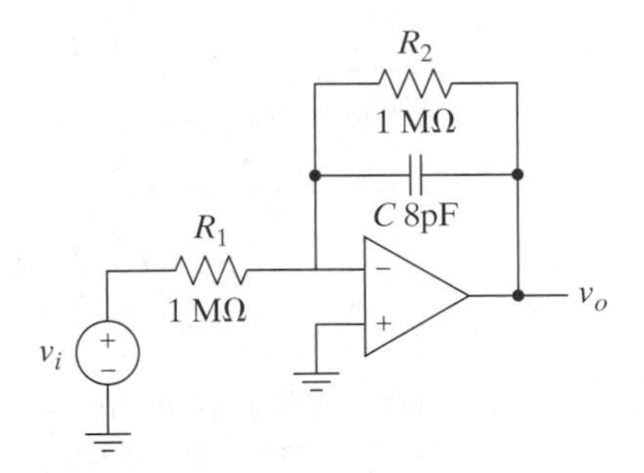

Figure 11-26 An active *RC* low-pass filter.

Solution:
The equivalent switched-capacitor circuit is shown in Figure 11-27. We arrived at this circuit by replacing each of the resistors in Figure 11-26 with a capacitor and two switches. The equivalent resistance for each of these switched capacitors is given by (11.44), so with a 1-MHz clock, we get a 1-MΩ equivalent resistance with a capacitor of 1 pF as shown. This switching scheme is a simplification of what would be used in practice, but it does work and allows us to keep the example simple.

A SPICE simulation of this circuit using voltage-controlled switches yields the results shown in Figure 11-28 for a 20-kHz input. Since we have to use transient simulations of the circuit, we cannot easily plot the frequency response. Nevertheless, by examining the phase shift and magnitude reduction at 20 kHz, we see that the pole frequency is close to 20 kHz. Note that this filter is inverting, so there will be a 180° phase shift for low frequencies. Note also that the fact that the output is a sampled-and-held signal is evident from the plot.

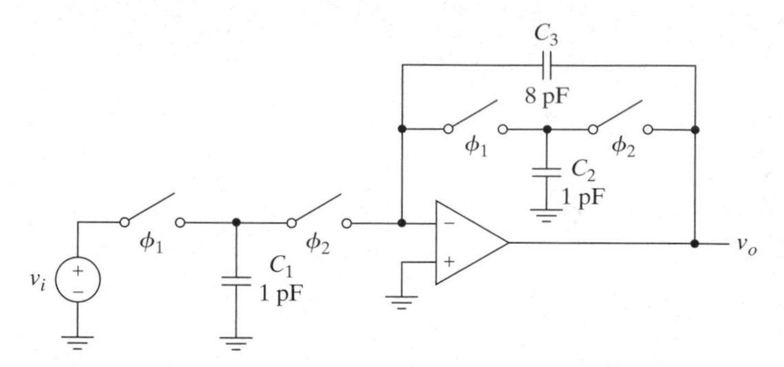

Figure 11–27 The switched-capacitor filter that is equivalent to the low-pass filter in Figure 11-26.

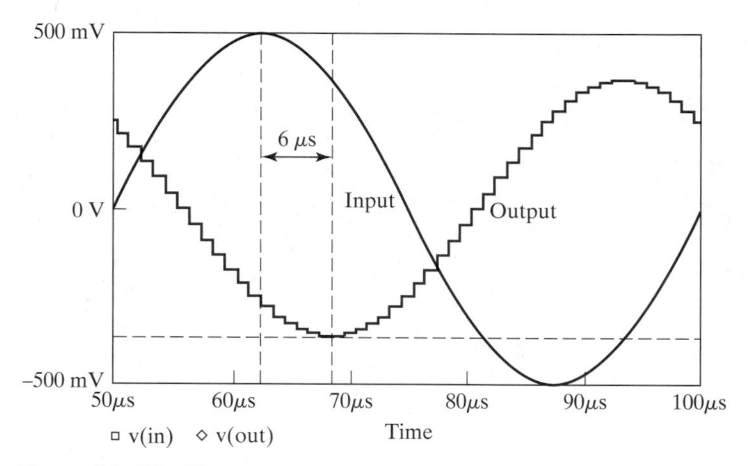

Figure 11–28 The simulated input and output waveforms for a 20-kHz input The 6 μs delay shown corresponds to nearly 45° of phase at 20 kHz, which shows that the pole frequency is near 20 kHz.

Finite Impulse Response Filters All of the filters discussed up to this point have impulse responses that last forever (although they may asymptotically go to zero as $t \to \infty$); such filters are called *infinite impulse response* (IIR) filters. In this section, we discuss filters that have impulse responses that are nonzero only for a finite time. These are called *finite impulse response* (FIR) filters and have many applications in modern communication systems.

Two significant advantages of FIR filters are that they can be made to have a perfectly linear phase, which is a requirement for distortionless transmission (*see* Aside 12.1), and they are guaranteed to be stable. The guaranteed stability is most important in adaptive filters, since it greatly simplifies the adaptation algorithm.[6] The most frequently used adaptation algorithm is the least-mean squares algorithm, which is well described in the classic text by Widrow and Stearns [11.6]. The stability of FIR filters is the result of the fact that they do not have any nonzero poles; that is, they are all-zero filters.

[6] You can readily imagine that adapting the poles of a continuous-time bandpass filter could be problematic, since you would have to find a way to guarantee that they never entered the right-half plane.

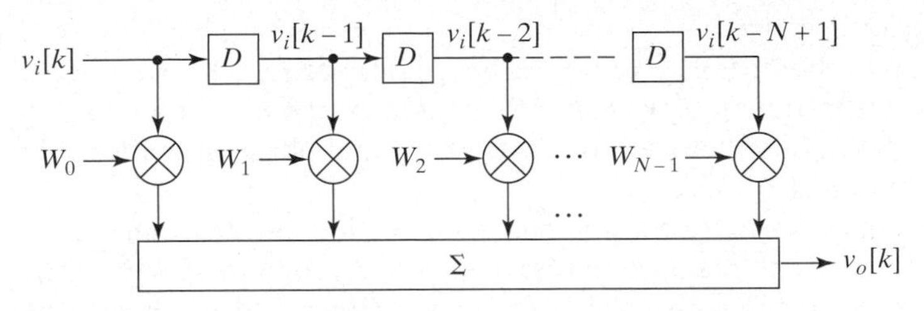

Figure 11–29 A transversal, or tapped-delay line, FIR filter.

One way to make an FIR filter is with a tapped-delay line as shown in Figure 11-29, where the blocks with a D in them represent delays of one clock period (i.e., z^{-1} in z-transform notation and e^{-sT_c} in Laplace transform notation) and the circles with crosses in them are multipliers. This filter has N taps and the W_js are called the tap weights. The delay line provides a series of past inputs for the filter to operate on, and the output is the weighted sum of these values. The equation for such a filter is

$$v_o[k] = \sum_{j=0}^{N-1} W_j v_i[k-j]. \tag{11.46}$$

This structure is sometimes called a ***transversal filter***. Transversal filters can be constructed using continuous-time delay elements, but most frequently employ discrete-time delays, as assumed here. They can be entirely digital, in which case the delay line is a sequence of shift registers, or they can be analog, in which case the delay line is a sequence of track-and-hold circuits.

The transfer function of the filter shown in Figure 11-29 can be written by inspection. In the z-domain, we have

$$H(z) = \frac{V_o(z)}{V_i(z)} = \sum_{k=0}^{N-1} W_k z^{-k}, \tag{11.47}$$

and in the Laplace domain, we have

$$H(s) = \frac{V_o(s)}{V_i(s)} = \sum_{k=0}^{N-1} W_k e^{-skT_c}. \tag{11.48}$$

These transfer functions do not have any poles (except at zero), so they are called *all-zero transfer functions*. The absence of poles is what makes the FIR structure unconditionally stable, as noted earlier, but it also limits what the filter can do. Nevertheless, these filters are commonly used in modems, disc drive read channels, Ethernet receivers, and other communication systems. The transfer functions are meaningful only for frequencies below $f_c/2$, since this is a discrete-time system. Example 11.8 illustrates the point.

Example 11.8

Problem:

Find the magnitude and phase response of a five-tap FIR filter with tap weights equal to 0.05, 0.2, 0.5, 0.2, and 0.05. Assume that a 44-kHz sample rate is used (i.e., a 44-kHz clock is used to sample the signals, and the delays in the filter are all equal to one period of this clock, or 22.7 μs).

Solution:
The transfer function of the filter is given by either (11.47) or (11.48). To plot the magnitude of this transfer function, we can either replace z by $e^{j\omega T_c}$ in (11.47) or replace s by $j\omega$ in (11.48). In either case, the magnitude and phase of the transfer function are plotted in Figure 11-30.

The magnitude response is that of a low-pass filter with a cutoff frequency of about 6.5 kHz. Notice that the magnitude response is valid only to 22 kHz; from there on to 44 kHz we see the negative frequency image. (We could also have plotted the magnitude response from −22 kHz to 22 kHz.) The DC gain is unity, which is the sum of the tap weights. The phase of this filter is linear, which is a result of the fact that the tap weights are symmetric about the center tap. Any FIR filter with symmetric tap weights will have a linear phase.

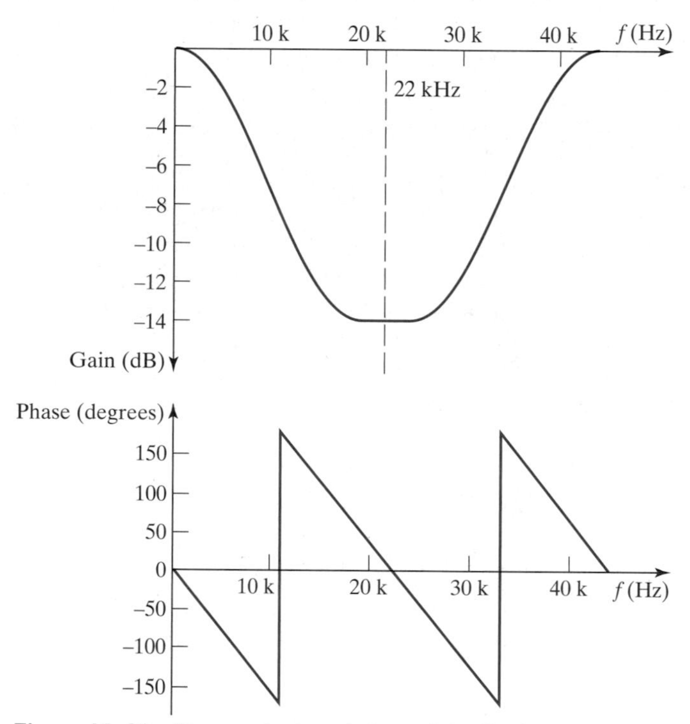

Figure 11–30 The magnitude and phase of the filter's transfer function (see Exam8.nb on the CD).

Example 11.9

Problem:
Use SPICE to simulate a continuous-time version of the FIR filter in Example 11.8. Simulating a continuous-time version is much simpler in SPICE than to simulate a discrete-time version.

Solution:
The SPICE schematic is shown in Figure 11-31. We have used ideal transmission lines to implement the delays. Tapped transmission lines are sometimes used to build discrete FIR filters, and they are a convenient way to simulate an ideal delay in SPICE. The multipliers have current outputs so that the summation can be accomplished by wiring all of the multiplier outputs together.

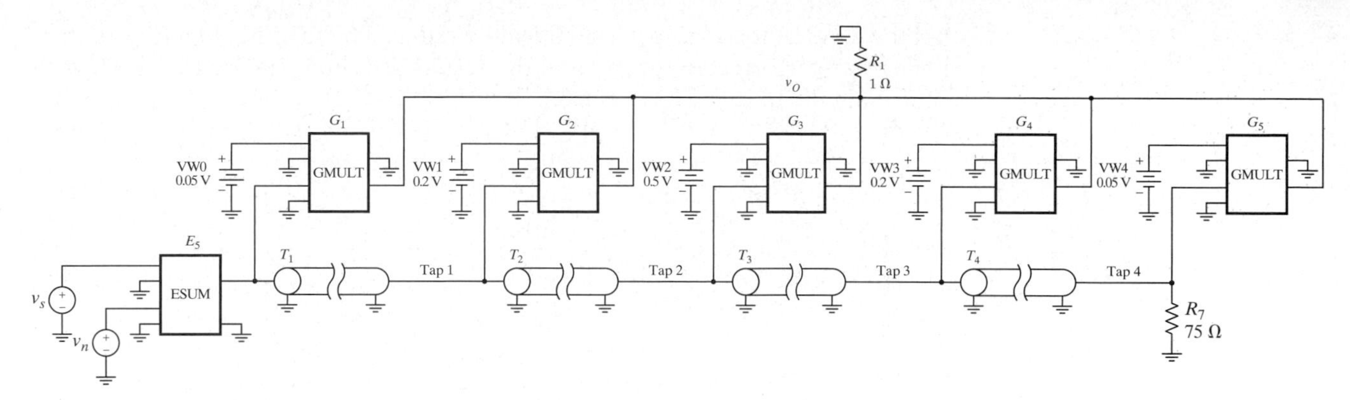

Figure 11–31 A SPICE circuit for simulating the five-tap FIR filter from Example 11.8 (*see* Exam8.sch on the CD).

The magnitude and phase of the simulated transfer function for the filter are shown in Figure 11-32 and agree with the calculated values in Example 11.8. Note that even though this filter is a continuous-time system, the output is calculated by using delayed versions of the input, and since the delays are all multiples of 22.7 μs, the transfer function is still meaningful only for frequencies less than 22 kHz. If we plot it for frequencies greater than 44 kHz, the pattern shown here repeats, as expected.

To demonstrate this filter's performance in the time domain, we used a 1-V peak 1-kHz signal with a 0.5-V peak 20-kHz noise added to it. Figure 11-33 shows the input and output waveforms in the time domain. It is evident from this figure

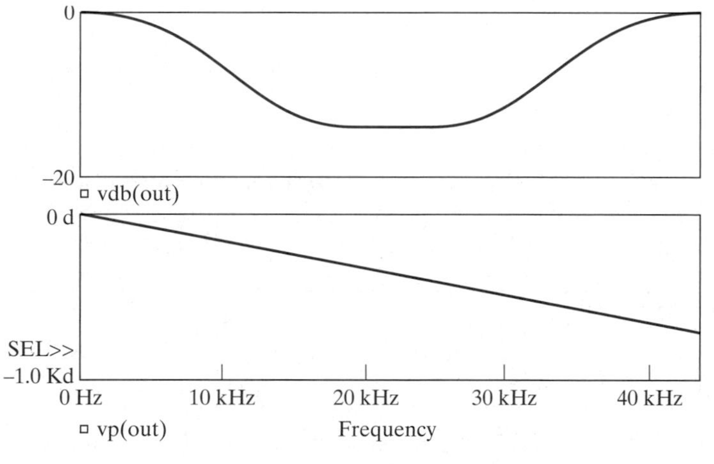

Figure 11–32 The simulated response.

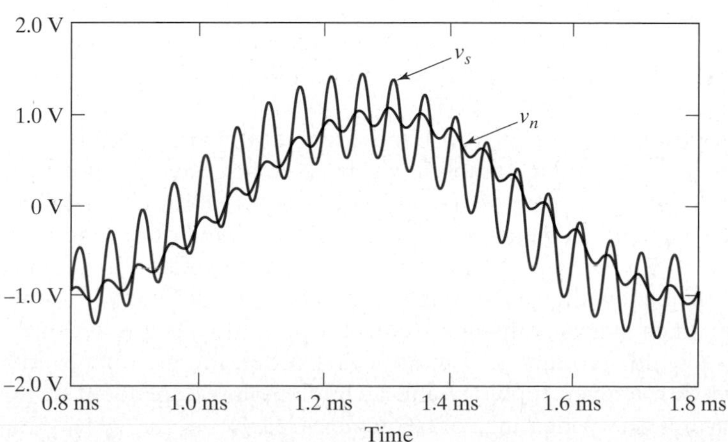

Figure 11–33 Transient simulation output.

that the 20-kHz noise has been significantly reduced by the filter. An FFT of these signals was performed in probe, and the results are shown in Figure 11-34, again confirming the filter's effectiveness.

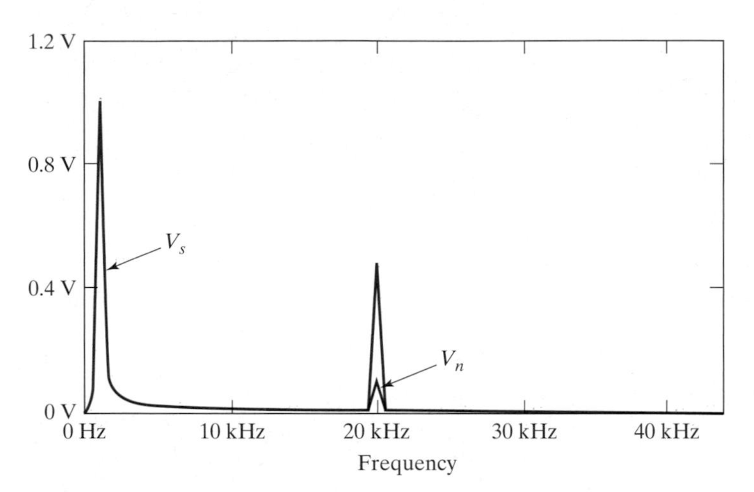

Figure 11–34 An FFT of the waveforms in Figure 11-33.

Exercise 11.5

Use MATLAB to produce the samples out of the filter described in Example 11.8 when it is driven by a 1-Vpeak 1-kHz sine wave signal with a 0.5-Vpeak 20-kHz noise term added. Use an ideal brick wall reconstruction filter on these samples, and then plot the result. Does it look like Figure 11-33? It should.

11.2 TUNED AMPLIFIERS

In this section, we consider the analysis and design of ***tuned amplifiers***. Unlike the amplifiers seen elsewhere in this book, the amplifiers we consider here are designed to amplify a narrow band of frequencies centered around some frequency f_o. Tuned amplifiers are frequently used in communication applications, such as radio and television.

One of the major problems associated with the design of tuned amplifiers is the potential for oscillation. While many amplifiers with feedback have the potential to oscillate, as we show in Chapter 10, some tuned amplifiers are particularly prone to do so, even when no feedback is deliberately used. A major advantage of tuned amplifiers is that it turns out to be much easier to achieve amplification at high frequencies if the bandwidth required is small. In fact, the difficulty in amplifier design is not so much in achieving high-frequency operation as it is in achieving a wide gain-bandwidth product. In fact, to a first order, the gain-bandwidth product achievable with a given topology and technology is independent of the center frequency over a very wide range of frequencies [11.4].

11.2.1 Single-Tuned Amplifiers

A single-tuned amplifier has only one tuned circuit. In this section, we investigate the design of single-tuned amplifiers, discuss the stability of various designs, and examine how to overcome the relatively low Q of practical inductors by using transformers.

Single-Tuned Amplifier Design A single transistor amplifier can use an RLC tank circuit at the output as illustrated in Figure 11-35 with a generic transistor; note that these circuits do not show the biasing details.

The small-signal AC equivalent circuit for the amplifier in Figure 11-35(a) is shown in Figure 11-36. The parasitic capacitance of the transistor that shows up between the control and output terminals is included in parallel with the tank capacitance, since the source resistance is zero in this example, so that $C_{eq} = C + C_{oc}$ (neglecting current fed forward through C_{OC}). In a similar fashion, the output resistance of the transistor can be absorbed in parallel with R to obtain R_{eq}. The equivalent circuit corresponding to the amplifier in Figure 11-35(b) would be the same, except for the input resistance and the direction of the controlled current source.

The voltage gain of this circuit is given by

$$A_v = \frac{V_o}{V_i} = \frac{-g_m}{G_{eq} + sC_{eq} + 1/sL}$$

$$= -\frac{g_m}{C_{eq}} \frac{s}{s^2 + s/R_{eq}C_{eq} + 1/LC_{eq}}, \tag{11.49}$$

where $G_{eq} = 1/R_{eq}$. The magnitude of the gain is

$$|A_v(j\omega)| = \frac{g_m}{C_{eq}} \frac{\omega}{\sqrt{(\omega/R_{eq}C_{eq})^2 + (1/LC_{eq} - \omega^2)^2}}. \tag{11.50}$$

The magnitude reaches its maximum value at the resonant frequency of the equivalent tank circuit. This frequency is also called the center frequency of the tuned amplifier and is

$$\omega_o = \frac{1}{\sqrt{LC_{eq}}}. \tag{11.51}$$

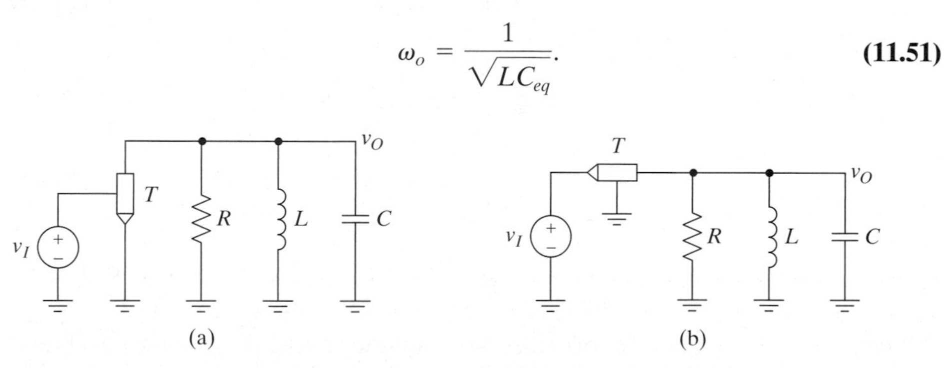

Figure 11–35 Two single-stage single-tuned amplifiers: (a) a stage with voltage and current gain (common merge) and (b) a stage with voltage gain (common control).

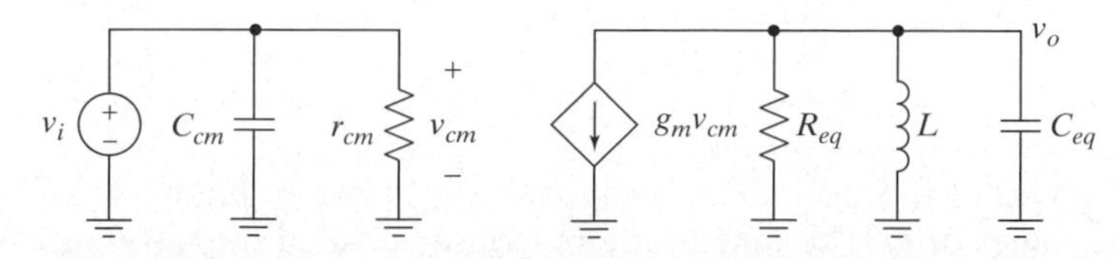

Figure 11–36 The small-signal AC equivalent circuit for the amplifier in Figure 11-35(a).

The magnitude of the gain at ω_o is

$$|A_v(j\omega_o)| = g_m R_{eq}, \tag{11.52}$$

since the impedance of the LC parallel combination goes to infinity at ω_o. The cutoff frequencies are found by setting the square of (11.50) equal to one half of (11.52). The solutions are

$$\omega_{cH} = \frac{1}{2R_{eq}C_{eq}} + \sqrt{\omega_o^2 + \frac{1}{4R_{eq}^2C_{eq}^2}} \tag{11.53}$$

and

$$\omega_{cL} = \frac{-1}{2R_{eq}C_{eq}} + \sqrt{\omega_o^2 + \frac{1}{4R_{eq}^2C_{eq}^2}}. \tag{11.54}$$

The bandwidth is given by the difference

$$\text{BW} = \omega_{cH} - \omega_{cL} = \frac{1}{R_{eq}C_{eq}}, \tag{11.55}$$

and, using (11.53), (11.54), and (11.55), we can solve for ω_o in terms of the cutoff frequencies. The result is

$$\omega_o = \sqrt{\omega_{cH}\omega_{cL}}, \tag{11.56}$$

so ω_o is the geometric mean of the cutoff frequencies. We also note that the phase of $A_v(j\omega)$ is zero at ω_o and is $\pm 45°$ at the cutoff frequencies. If the amplifier satisfies the narrowband approximation, then we can show that

$$\omega_o \approx \frac{\omega_{cL} + \omega_{cH}}{2}. \tag{11.57}$$

The ***quality factor*** of a tuned amplifier is defined as

$$Q = \frac{\omega_o}{\text{BW}} \tag{11.58}$$

and is a measure of how selective the circuit is; for a given ω_o, the Q is higher for smaller bandwidths (i.e., more selective circuits).

We can write a completely normalized version of the transfer function in (11.49) by making use of (11.51), (11.52), (11.55), and (11.58). The result is

$$\frac{A_v(j\omega)}{A_v(j\omega_o)} = \frac{1}{1 + jQ\left(\dfrac{\omega}{\omega_o} - \dfrac{\omega_o}{\omega}\right)}. \tag{11.59}$$

The magnitude and phase of this normalized response are shown in Figure 11-37 for several different values of Q. The magnitude response is called the ***universal resonance curve***.

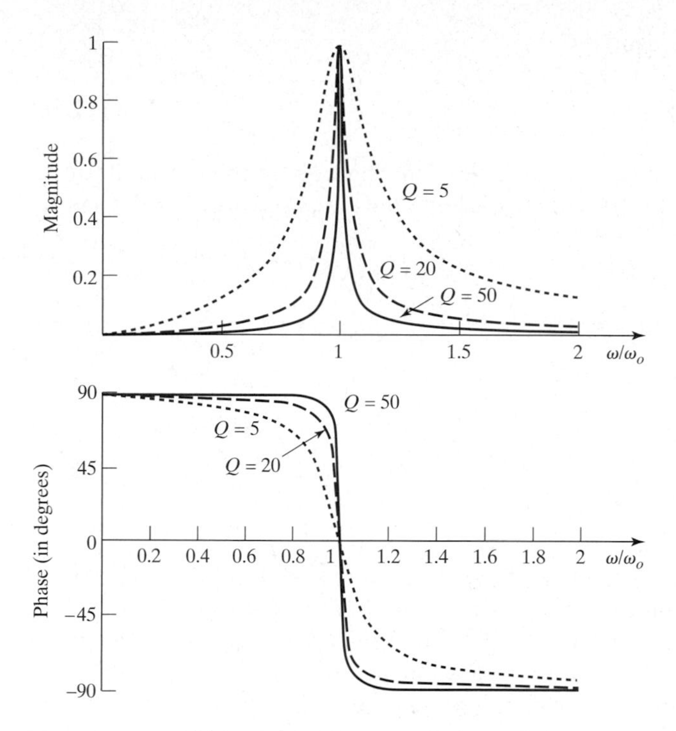

Figure 11–37 Universal resonance curve for a single-tuned amplifier (*see* universal.nb on CD).

It is also possible to put the tuned circuit at the input of a single-transistor amplifier if it is driven from a relatively large impedance. A generic circuit is shown in Figure 11-38, where the Norton model of the source has been used, the source resistance is absorbed into the tank circuit, and the biasing circuitry is not shown.

We can draw the small-signal AC equivalent circuit for this amplifier as shown in Figure 11-39. We have absorbed the transistor's output resistance in parallel with R_O to obtain $R_o' = r_o \| R_O$ and we have absorbed the transistor's input resistance and capacitance into the tank so that $R_{eq} = r_{cm} \| R$ and $C_{eq} = C + C_{in}$, where C_{in} is the total input capacitance of the common-merge stage. Using the Miller effect, we obtain $C_{in} = C_{cm} + C_{oc}(1 + g_m R_o')$.

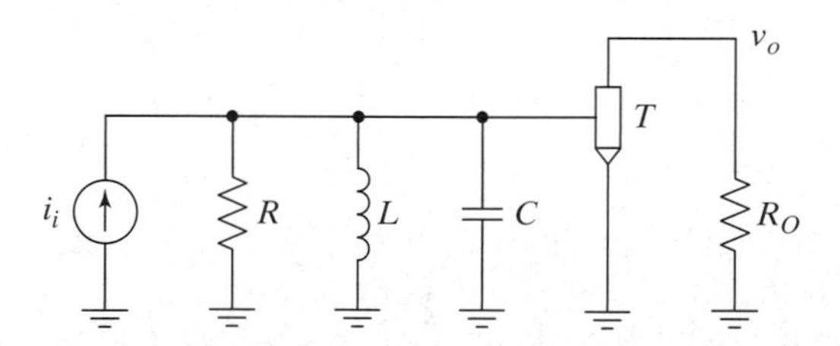

Figure 11–38 A single-tuned amplifier with the tank circuit at the input.

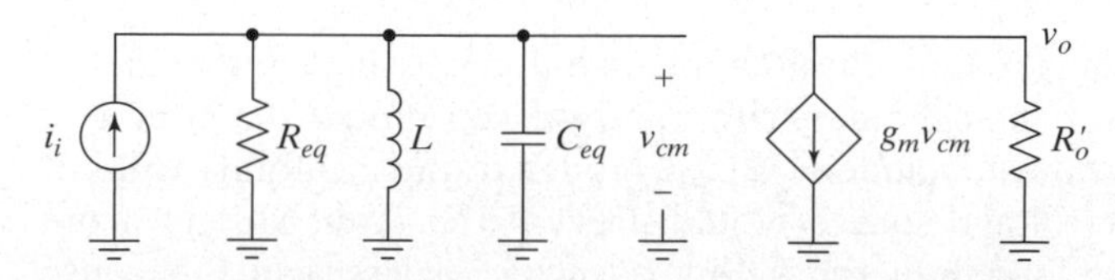

Figure 11–39 The small-signal AC equivalent circuit for Figure 11-38.

The gain from i_i to v_o is given by

$$A(j\omega) = \frac{-g_m R_o'}{C_{eq}} \frac{s}{s^2 + s/R_{eq}C_{eq} + 1/LC_{eq}}. \tag{11.60}$$

This equation has the same form as (11.49), differing only in the constant multiplier. Therefore, the results presented earlier apply equally well to the corresponding circuit.

Example 11.10

Problem:
Design the amplifier in Figure 11-35(a) to have a midband gain of −10, a center frequency of 1 MHz, and a bandwidth of 10 kHz. Assume that $g_m = 4$ mA/V, $r_o = 20$ kΩ, $C_{cm} = 40$ pF, and $C_{oc} = 10$ pF.

Solution:
Using (11.52) and the gain specification, we determine that $R = 2.9$ kΩ. From the bandwidth specification and (11.55), we determine that $C = 6.4$ nF. Finally, using (11.51), we find that L must be 3.97 μH to obtain the desired center frequency.

A SPICE simulation of this solution, using the small-signal AC equivalent circuit in Figure 11-36 and the values calculated here, results in the transfer function shown in Figure 11-40, which is very close to the desired performance; we obtain $f_o = 998$ kHz, $A_{vo} = -10.12$, and BW = 10 kHz.

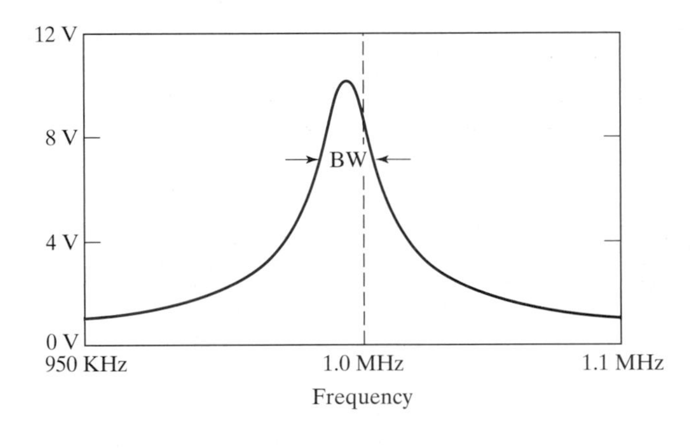

Figure 11–40 SPICE output (see Exam10.sch on the CD).

Single-Tuned Amplifier Stability Although the two circuits presented in the previous section function the same in the ideal case, the amplifier with the tuned circuit at the output is prone to being unstable (i.e., it can have a natural response that does not decay away). The reason for this instability is that the input impedance of the transistor can become negative for frequencies below ω_o. This negative input resistance is caused by the Miller effect and the reactive load, as we now demonstrate.

For frequencies below ω_o, the tank circuit appears inductive, since more current flows through the inductor than the capacitor, and we can model the circuit as shown in Figure 11-41. We must include R_S in the model, or the instability will not show up, since a perfect voltage source would prevent the input voltage from changing. The value of the load inductor is now L_{eq}, which is less than L because the capacitor in parallel with it has canceled part of the inductance. This equivalent inductance is frequency dependent, since the capacitive and inductive reac-

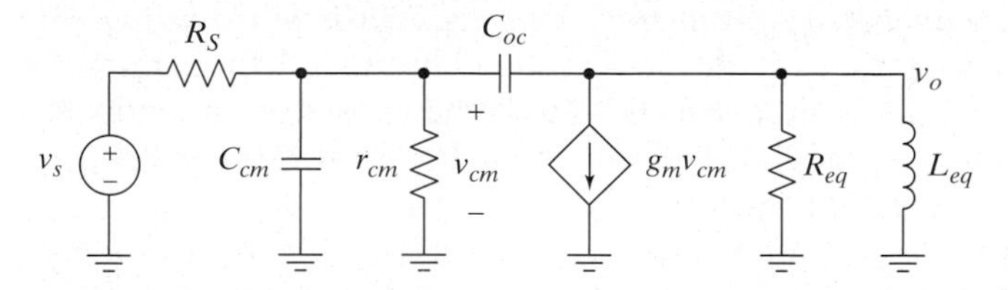

Figure 11–41 The small-signal AC equivalent circuit of the tuned amplifier in Figure 11-36 for frequencies below ω_o.

tances vary differently with frequency. The equivalent inductance is found by setting the admittance of the circuit equal to that of the original parallel LC combination at a given frequency:

$$\frac{1}{j\omega L_{eq}} = Y(j\omega) = \frac{1}{j\omega L} + j\omega C_{eq}. \quad \textbf{(11.61)}$$

This equation can be solved to yield $L_{eq} = L/(1 - \omega^2 LC_{eq})$.

We temporarily ignore r_{cm} and C_{cm} and focus on the circuit to the right of them. Using the Miller theorem, we find that the equivalent input impedance of the transistor is given by

$$Z_{in}(j\omega) = \frac{Z_{C_{oc}}(j\omega)}{1 - A_v(j\omega)} = \frac{1}{j\omega C_{oc}(1 - A_v(j\omega))}. \quad \textbf{(11.62)}$$

Assuming that the current through C_{oc} is negligible in comparison with $g_m v_{cm}$, we can apply the Miller approximation and say that the gain is

$$A_v(j\omega) = \frac{V_o(j\omega)}{V_{cm}(j\omega)} \approx -g_m Z_{eq}(j\omega) = \frac{-j\omega g_m L_{eq}}{1 + j\omega L_{eq}/R_{eq}}. \quad \textbf{(11.63)}$$

Using (11.63) in (11.62), we find that

$$Z_{in}(j\omega) = \frac{-g_m R_{eq}^2}{\omega^2 L_{eq} C_{oc}(1 + g_m R_{eq})^2 + R_{eq}^2 C_{oc}/L_{eq}} - j\frac{R_{eq}^2 + \omega^2 L_{eq}^2(1 + g_m R_{eq})}{\omega^3 L_{eq}^2 C_{oc}(1 + g_m R_{eq})^2 + \omega R_{eq}^2 C_{oc}}. \quad \textbf{(11.64)}$$

The negative real component of Z_{in} is what can cause instability. Now include r_{cm} and C_{cm} again, and focus on the equivalent circuit at the amplifier input. The resulting circuit is shown in Figure 11-42, where

$$R_1 = r_{cm} \| \mathrm{Re}[Z_{in}(j\omega)] \quad \textbf{(11.65)}$$

and

$$X_1 = \frac{1}{j\omega C_{cm}} \Bigg\| j\,\mathrm{Im}[Z_{in}(j\omega)]. \quad \textbf{(11.66)}$$

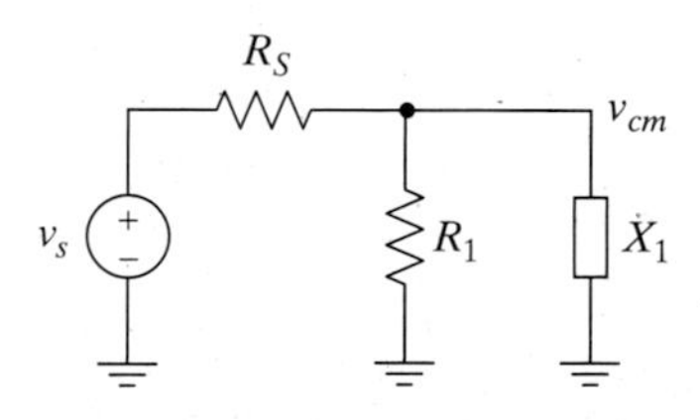

Figure 11–42 The equivalent input network for the amplifier of Figure 11-41.

We have drawn the input reactance with a box, because it may be capacitive or inductive, depending on the component values and the frequency.

If the total input reactance is capacitive, the gain from v_s to v_{cm} has a pole at $s = -1/(R_1 \| R_S)C_1$ in the complex plane. If, on the other hand, the input reactance is inductive, then there is a pole at $s = -(R_1 \| R_S)/L_1$. In either case, if $R_1 \| R_S = r_{cm} \| R_S \| \mathrm{Re}[Z_{in}(j\omega)]$ is negative, the pole is in the right half plane and the circuit is unstable (i.e., it's natural response contains an exponentially increasing component).

Inductor Quality, Impedance Transformation, and Transformer Coupling In the design of tuned amplifiers, it is often true that the Q of the inductors may not be high enough to achieve the desired selectivity and the capacitor and inductor values required may not be practical. We can often use transformers to help achieve a practical solution in such cases.

As shown in Section 9.1.2, the Q of an inductor is given by $Q \approx \omega L/R_s$ for frequencies below the self-resonant frequency of the inductor. We can transform this series resistance into an equivalent parallel resistance as shown in Figure 11-43. The equivalent parallel resistance, R_p, is determined by equating the admittance of the series circuit,

$$Y(j\omega) = \frac{1}{R_s + j\omega L} = \frac{R_s - j\omega L}{R_s^2 + \omega^2 L^2}, \tag{11.67}$$

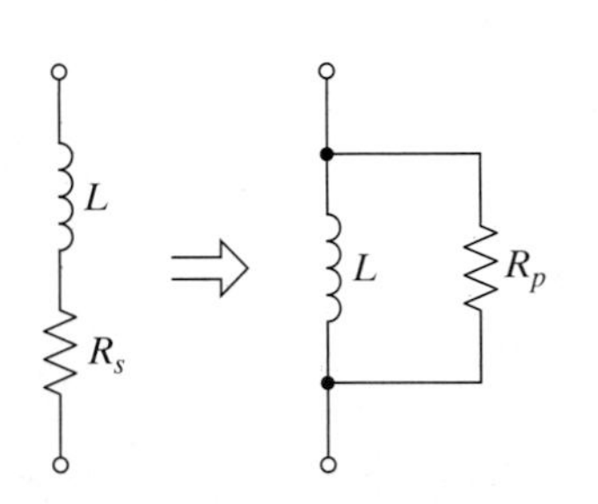

Figure 11–43 An inductor with series resistance converted to an equivalent parallel circuit.

to that of the parallel circuit,

$$Y(j\omega) = \frac{1}{R_p} + \frac{1}{j\omega L}. \tag{11.68}$$

If we have a high-Q coil, then $\omega L \gg R_s$ for frequencies below the self-resonant frequency, and we can rewrite (11.67) and set it equal to (11.68) to obtain

$$Y(j\omega) \approx \frac{R_s}{\omega^2 L^2} - j\frac{1}{\omega L} = \frac{1}{R_p} + \frac{1}{j\omega L}, \tag{11.69}$$

from which we get

$$R_p = \frac{\omega^2 L^2}{R_s} = \omega L Q. \tag{11.70}$$

If an inductor is used in a parallel RLC tank circuit and the Q of the inductor is not high enough, this parallel resistor may be too small to allow the overall tank circuit Q to be as high as desired, and the tuned amplifier's selectivity will be decreased.

Exercise 11.6

For the circuit shown in Figure 11-35(a), assume that $L = 10\ \mu H$ with $Q = 200$ at $\omega = 10^7$ rad/s. Also, assume that the transistor's parameters are $g_m = 5$ mA/V, $r_o = 20$ kΩ, $C_{cm} = 10$ pF, and $C_{oc} = 2$ pF. (a) Determine the values of R and C required to produce a bandwidth of 200 krad/s and a center frequency of $\omega_o = 10^7$ rad/s. (b) What is the voltage gain at ω_o?

We can overcome inductors with low Qs by using impedance-transforming networks. There are many types of impedance-transforming networks, and the choice of which one to use depends on the frequency range and the impedance levels. One such network, which is particularly attractive in tuned amplifiers, is the **autotransformer** shown in Figure 11-44(a). The double vertical lines next to the transformer winding signify that this is an ideal autotransformer.[7]

[7] By "ideal," we mean that it is lossless and has a unity coefficient of coupling between the coils. It is not reasonable to let the inductance go to infinity as is done for an ideal two-winding transformer, since the inductance of the autotransformer is often important, as we will see. A good discussion of basic transformer principles and equivalent circuits is given in [11.1].

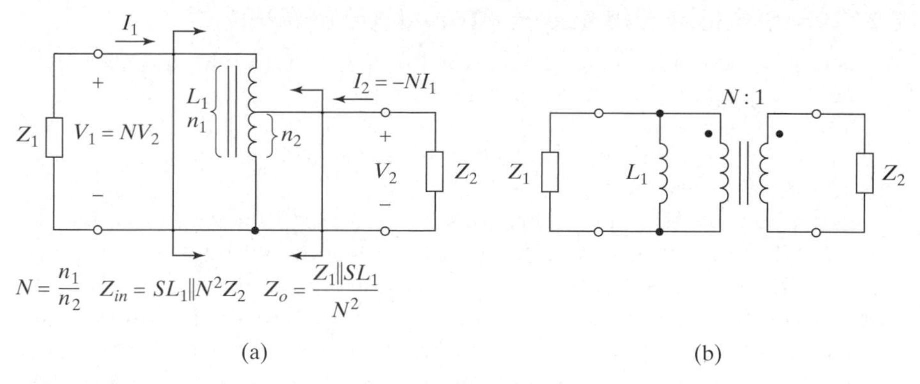

Figure 11–44 (a) Autotransformer with impedance transforming relations. (b) Equivalent model of (a).

An autotransformer, as the name implies, has a single continuous winding instead of separate primary and secondary windings. This single winding is tapped in one or more places, so that we still get transformer action. The coils are typically wound on a ferrite core, and a coefficient of coupling very close to unity can be achieved. The voltage–current relationship of the ideal autotransformer is given by

$$-\frac{I_2}{I_1} = \frac{V_1}{V_2} = \frac{n_1}{n_2} \equiv N > 1, \tag{11.71}$$

where we have defined the turns ratio N.

As shown in the figure, the autotransformer is a two-port network, and the impedance seen looking into port 1 is equal to L_1 in parallel with N^2 times the impedance connected to port 2. Similarly, the impedance looking into port 2 is equal to L_1 in parallel with the impedance connected to port 1, all divided by N^2.

A model for the ideal autotransformer is shown in part (b) of the figure, where L_1 is the total inductance and N is the turns ratio (*see* Problem P11.30 for a derivation of this model). The double vertical bars signify an ideal transformer (i.e., no losses and infinite inductances, but with a turns ratio of N). Note also that when the transformer is used as shown here, the voltage gain is reduced by a factor of $1/N$, and the current gain is increased by a factor of N. For this reason, we usually consider power gain in tuned amplifiers, which is unaffected in a lossless transformer.

Exercise 11.7

Consider the circuit shown in Figure 11-45, and assume that $L_1 = 100\ \mu H$, $N = 10$, $R_1 = 100\ k\Omega$, and $C_1 = 1\ nF$. Determine the impedance seen looking into port 2.

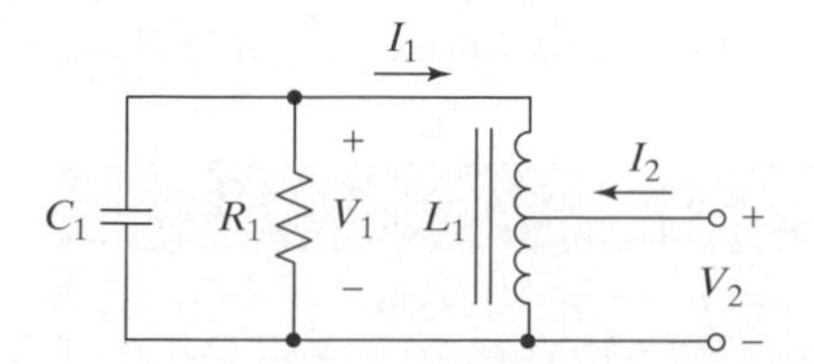

Figure 11–45

11.2.2 Synchronous and Stagger Tuned Amplifiers

In this section, we consider tuned amplifiers that have more than one tuned circuit. If each tuned circuit has the same center frequency and bandwidth, then the system is said to be **synchronously tuned**. If the tuned circuits are different and arranged so that the overall effect is to produce a desired narrowband response, the amplifier is said to be **stagger tuned**.

The overall bandwidth of a synchronously tuned amplifier will, of course, be lower than that of one of the tuned circuits alone. To find out how much smaller, we start with the normalized universal resonance curve described by (11.59). Suppose we cascade N identical tuned circuits, and suppose they do not interact with each other (or suppose interaction was allowed for when making them all the same). The cutoff frequencies can be determined by setting the square of the magnitude of the transfer function equal to one-half:

$$\frac{1}{2} = \left(\frac{1}{1 + Q^2 \left(\frac{\omega}{\omega_o} - \frac{\omega_o}{\omega} \right)^2} \right)^N . \tag{11.72}$$

This equation leads to a quadratic equation for ω, and the two solutions are given by

$$\omega_{1,2} = \frac{\omega_o}{2Q} \sqrt{2^{1/N} - 1} \left(1 \pm \sqrt{1 + \frac{4Q^2}{2^{1/N} - 1}} \right). \tag{11.73}$$

Since the transfer function we started with is double sided (i.e., there are peaks at $\pm\, \omega_o$), there are really four cutoff frequencies given by the positive and negative values of the two solutions found in (11.73). Taking the positive sign inside the parentheses in (11.73) leads to the positive value of ω_{cH}, and taking the negative sign leads to $-\omega_{cL}$. Therefore, the bandwidth is

$$\begin{aligned} \mathrm{BW}_N &= \omega_{cH} - \omega_{cL} \\ &= \frac{\omega_o}{2Q} \sqrt{2^{1/N} - 1} \left[\left(1 + \sqrt{1 + \frac{4Q^2}{2^{1/N} - 1}} \right) + \left(1 - \sqrt{1 + \frac{4Q^2}{2^{1/N} - 1}} \right) \right] \\ &= \frac{\omega_o}{Q} \sqrt{2^{1/N} - 1} = \mathrm{BW} \sqrt{2^{1/N} - 1}, \end{aligned} \tag{11.74}$$

where BW is the bandwidth of a single stage as given by (11.58). It is interesting to note that this is the same relationship we obtain for the bandwidth reduction in cascaded identical single-pole low-pass transfer functions (*see* P9.120). The amount by which the bandwidth is reduced for a cascade of N identical single-pole stages is sometimes called the **shrinkage factor**, $S_N = \sqrt{2^{1/N} - 1}$. The shrinkage factor is given in Table 11.1 for $N = 1$ to 10.

The shrinkage factor can be avoided by the use of *stagger tuning*. In stagger tuning, the individual tuned circuits are not the same, but rather are arranged to achieve the desired overall response. For example, a maximally flat magnitude response (i.e., Butterworth) can be obtained as illustrated in Figure 11-46. Each individual tuned circuit is designed to realize a DC zero and one pair of the

Table 11.1 Bandwidth Shrinkage Factor

N	2	3	4	5	6	7	8	9	10
S_N	0.64	0.51	0.43	0.39	0.35	0.32	0.30	0.28	0.27

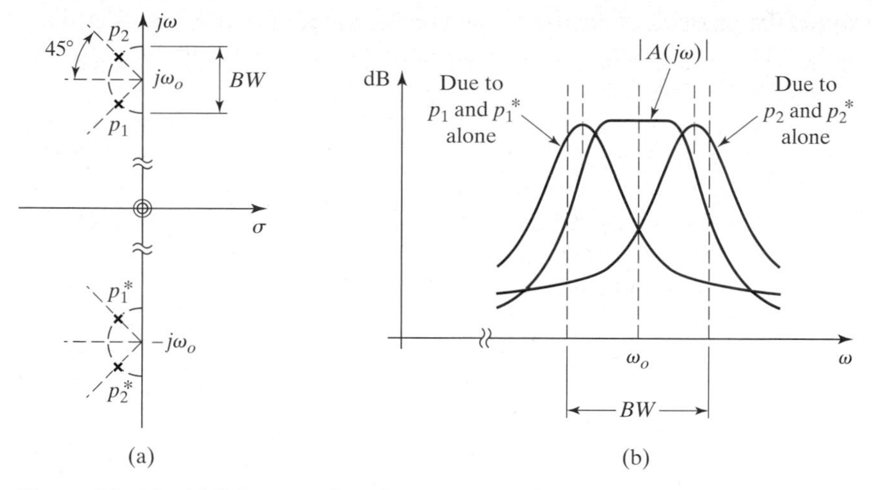

Figure 11–46 (a) Pole–zero plot of a stagger-tuned maximally flat magnitude design using two single-tuned stages. (b) Magnitude responses for the individual tuned circuits and the overall stagger-tuned design (see Stagger.nb on the CD).

complex-conjugate poles required for the response. Other responses can also be obtained. Consequently, stagger tuning is often employed when good selectivity is required.

Amplifiers that have multiple tuned circuits are difficult to design and are very sensitive to component variations if the tuned circuits interact. Therefore, it is important to use amplifier topologies that can effectively isolate the tuned circuits from each other. One circuit that works well with multiple tuned circuits is the cascode amplifier introduced in Chapter 9. The cascode was shown to have a large bandwidth precisely because it isolated the load from the common-merge transistor, which had the effect of nearly eliminating the Miller multiplication of C_{oc}. This same isolation allows the cascode to have tuned circuits at both the input and the output and to have these circuits operate independently.

A generic tuned cascode amplifier is shown in Figure 11-47(a) exclusive of biasing. The small-signal AC equivalent is shown in part (b) of the figure, where we

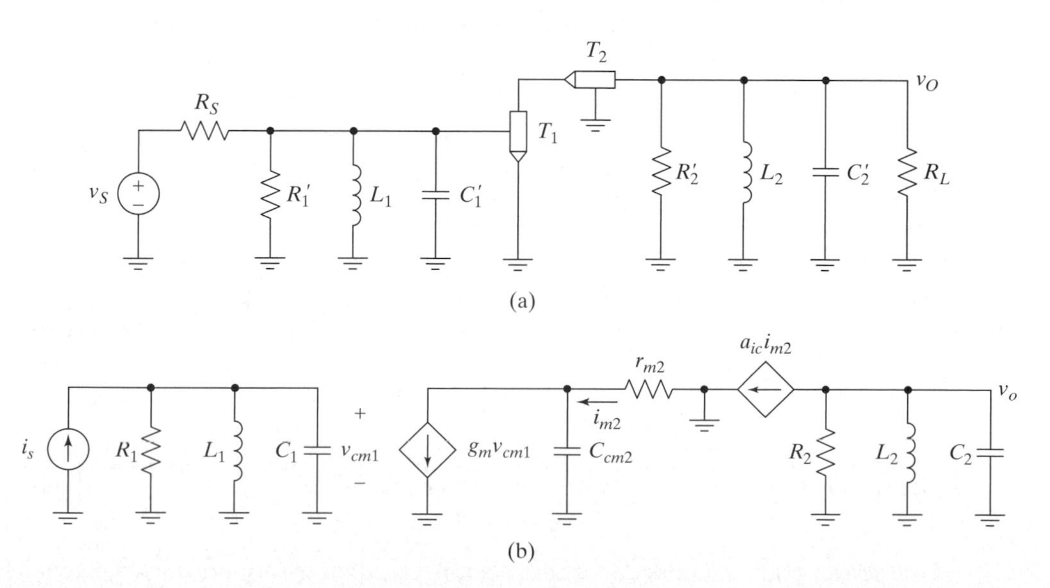

Figure 11–47 (a) A generic cascode amplifier exclusive of biasing. (b) The small-signal AC equivalent circuit.

have absorbed the parasitic capacitances and resistances of the transistors into the tank circuits wherever possible. We have also switched to a Norton model for the source and have absorbed R_S into the tank circuit. We have

$$R_1 = R_S \| r_{cm1} \| R_1', \tag{11.75}$$

$$C_1 = C_1' + C_{cm1} + C_{Min}, \tag{11.76}$$

where C_{Min} is the Miller input capacitance ($C_{Min} \approx 2C_{oc1}$),

$$R_2 \approx r_{o2} \| R_2' \| R_L\ , \tag{11.77}$$

and

$$C_2 = C_2' + C_{oc2}\ . \tag{11.78}$$

The one capacitor we cannot absorb into one of the tank circuits is C_{cm2}, but since it is in parallel with a very small resistance, r_{m2}, it can be ignored for frequencies at which this circuit is useful. Ignoring C_{cm2}, we find that the gain of the amplifier is

$$\begin{aligned} A_v = \frac{V_o}{V_s} &= \frac{V_o}{I_{m2}}\frac{I_{m2}}{V_{cm1}}\frac{V_{cm1}}{I_s}\frac{I_s}{V_s} \\ &= \left(\frac{-\ a_{ic}s/C_2}{s^2 + s/R_2C_2 + 1/L_2C_2}\right)g_m\left(\frac{s/C_1}{s^2 + s/R_1C_1 + 1/L_1C_1}\right)\frac{1}{R_S} \\ &= \frac{-g_m a_{ic}}{R_S C_1 C_2}\left(\frac{s}{s^2 + s/R_2C_2 + 1/L_2C_2}\right)\left(\frac{s}{s^2 + s/R_1C_1 + 1/L_1C_1}\right). \end{aligned} \tag{11.79}$$

In the subsections that follow, we consider the synchronous and stagger tuning of this circuit.

Synchronously Tuned Design If the circuit in Figure 11-47 is synchronously tuned, we set the center frequencies and bandwidths of both tuned circuits to the same values. Because there are two tuned circuits cascaded in the transfer function, the overall bandwidth will be reduced by the shrinkage factor of 0.64, as shown in Table 11.1. To achieve an overall bandwidth of BW, we set the individual bandwidths equal to BW/0.64 = 1.56BW. Therefore, we would set

$$1.56\text{BW} = \frac{1}{R_1C_1} = \frac{1}{R_2C_2} \tag{11.80}$$

and

$$\omega_o^2 = \frac{1}{L_1C_1} = \frac{1}{L_2C_2}. \tag{11.81}$$

From (11.79), the voltage gain at ω_o is

$$A_{vo} = \frac{-g_m a_{ic} R_1 R_2}{R_S}. \tag{11.82}$$

If the source and load impedances are also specified, transformers may be needed at the input and output to simultaneously satisfy all constraints.

Example 11.11

Problem:
Design the amplifier in Figure 11-47 to achieve $f_o = 10.7$ MHz and BW = 200 kHz. (These are the values used in the IF section of broadcast-band FM radios.) Assume that for both transistors, $r_{cm} = 8.6\ k\Omega$, $g_m = 18$ mA/V, $C_{cm} = 43$ pF, $C_{oc} = 5$ pF, and $r_o = 100\ k\Omega$. Also assume $R_L = 10\ k\Omega$ and $R_S = 1\ k\Omega$.

Solution:
From the required BW and (11.80), we determine that $R_1C_1 = R_2C_2 = 510$ ns. From (11.75), (11.76), and the given device data, we find that $R_1 = 896\ \Omega \| R_1'$ and $C_1' = C_1 - 53$ pF . Using standard values, we see that one possible choice is $R_1' = 4.7\ k\Omega$ and $C_1' = 620$ pF. Similarly, we use (11.77) and (11.78) to find that $R_2' = 10\ k\Omega$ and $C_2' = 100$ pF will work.

We next use (11.81), the desired center frequency and the capacitor values just found to determine that $L_1 = 0.329\ \mu H$ and $L_2 = 2.11\ \mu H$.

Note that we have had to use more precise inductors, since the standard values will not produce the results requested (*see* Examll.sch on the CD). ■

Exercise 11.8

Repeat Example 11.11, but set both inductors equal to 1 µH. What capacitor values are required now? Use standard values. □

Stagger-Tuned Design In stagger tuning, each circuit resonates at a different frequency. Suppose, for example, we wish to produce a Butterworth response as shown in Figure 11-46. We start by rewriting (11.79) in factored form:

$$A_v(s) = \frac{-g_m a_{ic}}{R_S C_1 C_2}\left(\frac{s}{(s - p_2)(s - p_2^*)}\right)\left(\frac{s}{(s - p_1)(s - p_1^*)}\right). \qquad \textbf{(11.83)}$$

The poles in (11.83) need to be placed in the locations shown in Figure (11-46)(a). We start with the equivalent low-pass response. To achieve a Butterworth response with a bandwidth equal to BW, we set the poles on a circle of radius BW and at 45° angles with respect to the real and imaginary axes. Therefore, we derive the two low-pass poles

$$p_{L1,2} = \frac{\text{BW}}{\sqrt{2}}(-1 \pm j). \qquad \textbf{(11.84)}$$

Using the narrowband transformation given by (11.22), we can then derive the four bandpass poles as

$$p_{B1-4} = -\frac{\text{BW}}{2\sqrt{2}} \pm j\left(\omega_o \pm \frac{\text{BW}}{2\sqrt{2}}\right). \qquad \textbf{(11.85)}$$

Using Figure 11-46, we can now assign

$$p_1, p_1^* = -\frac{\text{BW}}{2\sqrt{2}} \pm j\left(\omega_o - \frac{\text{BW}}{2\sqrt{2}}\right) \tag{11.86}$$

and

$$p_2, p_2^* = -\frac{\text{BW}}{2\sqrt{2}} \pm j\left(\omega_o + \frac{\text{BW}}{2\sqrt{2}}\right). \tag{11.87}$$

We can plug each set of poles into the corresponding term in (11.83) and expand to get the quadratic equations; the results are

$$s^2 + s\frac{\text{BW}}{\sqrt{2}} + \omega_o^2\left(1 - \frac{\text{BW}}{\omega_o\sqrt{2}}\right) + \frac{\text{BW}^2}{4} \tag{11.88}$$

for the polynomial from p_1 and p_1^*, and

$$s^2 + s\frac{\text{BW}}{\sqrt{2}} + \omega_o^2\left(1 + \frac{\text{BW}}{\omega_o\sqrt{2}}\right) + \frac{\text{BW}^2}{4} \tag{11.89}$$

for the polynomial from p_2 and p_2^*. The design procedure now is to equate coefficients of like powers of s in (11.88) and (11.89) with the corresponding denominators in (11.79) to arrive at the required component values. Following this procedure, we find that

$$R_1C_1 = R_2C_2 = \frac{\sqrt{2}}{\text{BW}}, \tag{11.90}$$

$$\frac{1}{L_1C_1} = \frac{\text{BW}^2}{4} + \omega_o^2\left(1 - \frac{\text{BW}}{\omega_o\sqrt{2}}\right), \tag{11.91}$$

and

$$\frac{1}{L_2C_2} = \frac{\text{BW}^2}{4} + \omega_o^2\left(1 + \frac{\text{BW}}{\omega_o\sqrt{2}}\right). \tag{11.92}$$

Finally, the midband gain is found by using the narrowband approximation. Using (11.23) and (11.26) in (11.83), we obtain

$$|A_v(j\omega)| \approx \frac{g_m a_{ic}}{4R_SC_1C_2}\frac{1}{|j\omega - p_1||j\omega - p_2|}. \tag{11.93}$$

Then, from Figure 11-46(a), we see that

$$|j\omega_o - p_1| = |j\omega_o - p_2| = \frac{\text{BW}}{2}, \tag{11.94}$$

so

$$|A_v(j\omega_o)| \approx \frac{g_m a_{ic}}{R_SC_1C_2\text{BW}^2}. \tag{11.95}$$

Finally, using (11.90) in (11.95), we can write

$$|A_v(j\omega_o)| \approx \frac{g_m a_{ic} R_1 R_2}{2R_S}. \tag{11.96}$$

We note from (11.82), (11.90), and (11.95) that, for equal bandwidths and resistors, the midband gain of the synchronously-tuned stage is larger than that of the stagger-tuned stage by a factor of 2:

$$\frac{A_v(j\omega_o)_{\text{synch.}}}{A_v(j\omega_o)_{\text{stagger}}} = \frac{\dfrac{g_m a_{ic} R_1 R_2}{R_S}}{\dfrac{g_m a_{ic}}{R_S C_1 C_2 \text{BW}^2}} = \frac{R_1 R_2}{\dfrac{R_1 R_2 \text{BW}^2}{2\text{BW}^2}} = 2. \tag{11.97}$$

If, in addition to the ω_o and BW specifications, the source and load terminations are also specified, the use of transformers at the input and output may be required.

We end by pointing out that it is also common to use a differential pair to implement an amplifier with two tuned circuits. The differential pair can be thought of as a voltage buffer (e.g., merge follower) followed by a gain stage (e.g., common merge) and provides isolation between the two tuned circuits, just as the cascode connection does. The differential stage is somewhat easier to bias, however, and it does not require as large a supply voltage. We demonstrate this circuit in Problems P11.42 and P11.46.

Example 11.12

Problem:
Design the amplifier in Figure 11-47 to achieve $f_o = 10$ MHz and BW $= 1$ MHz with stagger tuning and a Butterworth response. Assume that for both transistors, $r_{cm} = 8.6\ \text{k}\Omega$, $g_m = 18$ mA/V, $C_{cm} = 43$ pF, $C_{oc} = 5$ pF, and $r_o = 100\ \text{k}\Omega$. Also assume $R_L = 10\ \text{k}\Omega$ and $R_S = 1\ \text{k}\Omega$.

Solution:
From the required BW and (11.90), we determine that $R_1C_1 = R_2C_2 = 225$ ns. From (11.75), (11.76), and the given device data, we find that $R_1 = 896\ \Omega \| R_1'$ and $C_1' = C_1 - 53$ pF. Using standard values, we see that one possible choice is $R_1' = 510\ \Omega$ and $C_1' = 620$ pF. Similarly, we use (11.77) and (11.78) to find that $R_2' = 2.7\ \text{k}\Omega$ and $C_2' = 100$ pF will work.

We next use (11.91) and the desired center frequency to determine that $L_1 = 0.351\ \mu\text{H}$. Similarly, (11.92) leads to $L_2 = 1.97\ \mu\text{H}$. (Exact values have again been used.) (*See* Exam12.sch on the CD.) ■

Exercise 11.9

Repeat Example 11.12, but set both inductors equal to 1 μH. What resistor and capacitor values are required now? Use standard values. □

11.3 PHASE-LOCKED LOOPS

Phase-Locked Loops (PLLs) are extremely important in many applications. For example, they are used in frequency synthesis, FM demodulation, and clock recovery in communication systems. In this section, we present a first-order introduction to these intriguing systems. If you are interested in knowing more, we strongly recommend the introductory treatment in [11.9], or, for more advanced treatments, the tutorial at the start of [11.10] is very good, as is the material in [11.4] and [11.8].

The PLL is a negative feedback loop that forces the phase of the periodic signal at the output of a ***voltage-controlled oscillator*** (VCO) to be approximately equal to the phase of the input signal. Because frequency is the derivative of phase with respect to time, the only way the phases of two signals can be equal at all times is for the frequencies to be equal as well. Therefore, a PLL also guarantees that the frequency of the VCO is equal to the frequency of the input. PLLs can also be made using square waves for the signals. For simplicity, we will focus on sinusoidal PLLs.

The block diagram of a basic PLL is shown in Figure 11-48. The ***phase detector*** generates an error that is, ideally, proportional to the difference in phase between the loop input and the VCO output; that is,

$$v_E(t) = K_D[\phi_I(t) - \phi_{VCO}(t)], \tag{11.98}$$

where K_D is the phase detector gain in volts per radian, $\phi_I(t)$ is the total instantaneous input phase, and $\phi_{VCO}(t)$ is the total instantaneous phase of the VCO output. The ***loop filter*** attenuates rapid variations in the error, so the loop responds only to slower variations, and, although virtually always used, the loop filter is not absolutely essential to the operation of the loop. The VCO has a periodic output with an instantaneous frequency given by

$$\omega_{VCO}(t) = \omega_{FR} + K_o v_C(t), \tag{11.99}$$

where ω_{FR} is called the ***free-running frequency***, K_o is the VCO gain in radians per second per volt, and $v_C(t)$ is the control voltage into the VCO.

Let's now consider the operation of the PLL, assuming that it is ***locked***—in other words, assuming that the phase and frequency of the VCO output are equal to the input phase and frequency, respectively. If the input frequency and phase change slowly enough and stay within the PLL's range, the PLL will track these changes, as shown in Figure 11-49. This figure shows the input waveform (which, since the loop is locked, is the same as the VCO output) changing frequency, and the resulting control voltage to the VCO. The control voltage plot could also be a plot of the input frequency versus time if the vertical scale were changed appropriately. Notice that this figure illustrates one possible use of a PLL; if we use the control voltage as the output, the PLL demodulates a frequency-modulated signal.

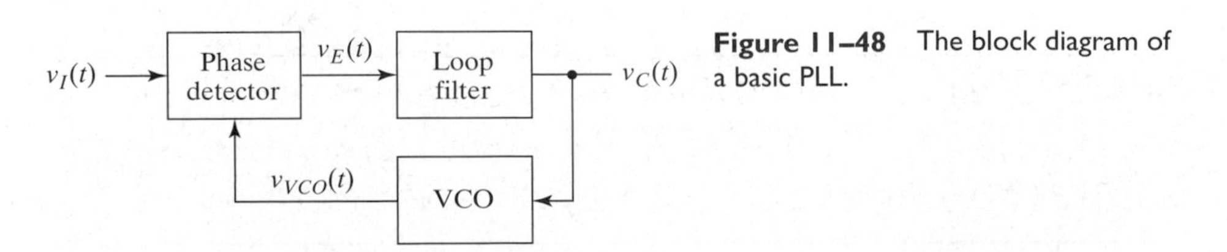

Figure 11–48 The block diagram of a basic PLL.

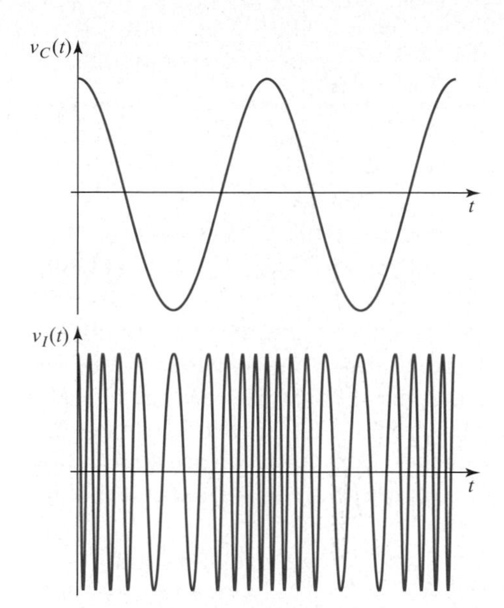

Figure 11–49 The VCO control voltage and loop input voltage for a PLL when tracking changes in the input frequency (see PLL.nb on the CD).

Now let's take a more detailed look at the operation of the loop when it is locked. Assume that the input is given by

$$v_I(t) = \cos[\phi_I(t)] = \cos[\omega_o t + \theta_I(t)] \tag{11.100}$$

and that the VCO output is

$$v_{VCO}(t) = \cos[\phi_{VCO}(t)] = \cos[\omega_o t + \theta_{VCO}(t)], \tag{11.101}$$

where we have explicitly shown that the phase of the signal is the entire argument to the cosine. What we often call the phase, θ, is really just the phase offset, whereas the frequency is the rate of change of the total phase. In this example, the frequency is equal to ω_o if the phase offset is constant. Using (11.100) and (11.101) in (11.98), we have

$$v_E(t) = K_D[\phi_I(t) - \phi_{VCO}(t)] = K_D[\theta_I(t) - \theta_{VCO}(t)], \tag{11.102}$$

or, in the Laplace domain,

$$V_e(s) = K_D[\Theta_i(s) - \Theta_{vco}(s)]. \tag{11.103}$$

Now, using (11.99) along with (11.101), we find that

$$\omega_{VCO}(t) = \omega_{FR} + K_o v_C(t) = \frac{d}{dt}\left[\omega_o t + \theta_{VCO}(t)\right]. \tag{11.104}$$

Writing the control voltage in (11.104) in terms of its DC and AC components leads to

$$\omega_{FR} + K_o[V_C + v_c(t)] = \omega_o + \frac{d\theta_{VCO}(t)}{dt}. \tag{11.105}$$

Examining (11.105), we see that

$$\omega_o = \omega_{FR} + K_o V_C \tag{11.106}$$

and

$$\frac{d\theta_{VCO}(t)}{dt} = K_o v_c(t), \quad \textbf{(11.107)}$$

or, in the Laplace domain,

$$s\Theta_{vco}(s) = K_o V_c(s). \quad \textbf{(11.108)}$$

In writing (11.108) and using it in the following analyses, we are linearizing the system by ignoring the DC component of the control voltage, which is needed only to bring the loop into lock and handle the initial offset between the average input frequency and the free-running frequency of the VCO. The loop can be represented as a linear system only when it is locked. In essence, we are subtracting the $\omega_o t$ terms from (11.100) and (11.101) and are using $\theta_I(t)$, and $\theta_{VCO}(t)$ as the signals of interest.

Assuming that the loop filter transfer function is $F(s)$, we also know that

$$V_c(s) = V_e(s)F(s). \quad \textbf{(11.109)}$$

We can combine (11.103), (11.108), and (11.109) to derive the transfer function from the input phase to the control voltage:

$$\frac{V_c(s)}{\Theta_i(s)} = \frac{sF(s)K_D}{s + K_o K_D F(s)}. \quad \textbf{(11.110)}$$

This transfer function would be of interest, for example, if we used the PLL to demodulate a phase-modulated signal. If, on the other hand, we use the PLL as an FM demodulator, the transfer function needs to be in terms of the input frequency. Since frequency is the derivative of phase, we have

$$\Omega_i(s) = s\Theta_i(s), \quad \textbf{(11.111)}$$

which, combined with (11.110), leads to

$$\frac{V_c(s)}{\Omega_i(s)} = \frac{V_c(s)}{s\Theta_i(s)} = \frac{K_D F(s)}{s + K_o K_D F(s)}. \quad \textbf{(11.112)}$$

This transfer function deserves a bit of explanation. $\Omega_i(s)$ is the frequency-domain representation (i.e., the Laplace transform) of $\omega_i(t)$, the input frequency as a function of time. Since the frequency of the input may be changing with time, we can speak of the frequency at which the input frequency is changing.

We now draw a Laplace-domain block diagram for the PLL in the locked condition, as shown in Figure 11-50. The phase detector is represented as the combination of a summer and a gain, whereas the VCO is represented as the combination of a gain block and an integrator. The input may be considered to be either the frequency, Ω_i, or the phase, Θ_i, of the input signal. A direct analysis of the feedback loop (*see* Section 10.1) yields (11.110) and (11.112), as you can verify. We emphasize at this point that this block diagram is valid only when the loop is locked. The operation of a PLL while it is acquiring lock is highly nonlinear (*see* [11.9]) and cannot be described by Laplace transforms.

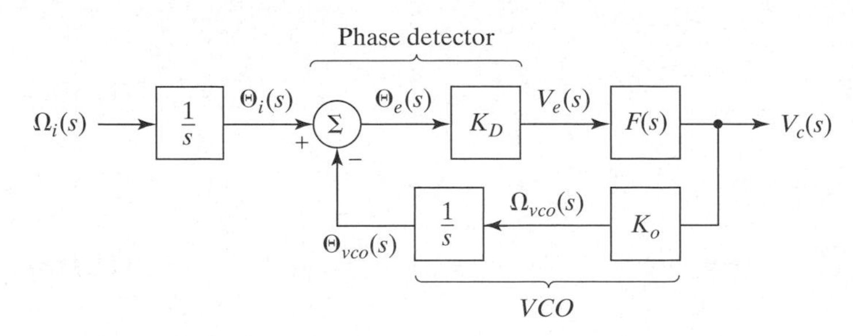

Figure 11–50 A Laplace-domain block diagram for the PLL when locked.

Very narrow bandwidth PLLs are sometimes used to reduce the jitter in a signal. Jitter is the cycle-to-cycle variation in the period, and if the frequency at which the period changes is higher than the cutoff frequency of the PLL's transfer function, the PLL can attenuate this high-frequency jitter. The result is a purer sinusoidal signal.[8] In this case, the VCO output is used as the output of the PLL, and the transfer function of interest is

$$\frac{\Omega_{vco}(s)}{\Omega_i(s)} = \frac{K_o K_D F(s)}{s + K_o K_D F(s)}, \tag{11.113}$$

as can be determined from the block diagram in Figure 11-50.

It is often of importance to know the range of frequencies over which a PLL can acquire lock, called the *capture range* or *acquisition range*, and the range of frequencies over which it can stay locked once lock is acquired, which is called the *lock range*. The capture range is always smaller than the lock range. Determining the capture range is difficult, since it is a nonlinear problem. A good introductory treatment is presented in [11.9], but we will not pursue it here.

The lock range, however, can be found by examining the steady-state error signal out of the phase detector. Unless the input frequency to the PLL is exactly equal to the free-running frequency of the VCO, the control voltage will have to be nonzero for the loop to be locked. Therefore, we can see whether or not the phase detector and loop filter are capable of supplying the required control voltage. Real phase detectors have periodic transfer functions; one example is shown in Figure 11-51. Note that the maximum output from this phase detector is πK_D. Therefore, if the steady-state voltage out of the phase detector necessary to keep the loop locked exceeds πK_D, the loop will not be able to maintain lock.

Assume that we start with a loop in the locked condition and then introduce a step change in the input frequency. If the magnitude of the step is $\Delta\omega$, we have

$$\Omega_i(s) = \frac{\Delta\omega}{s} \tag{11.114}$$

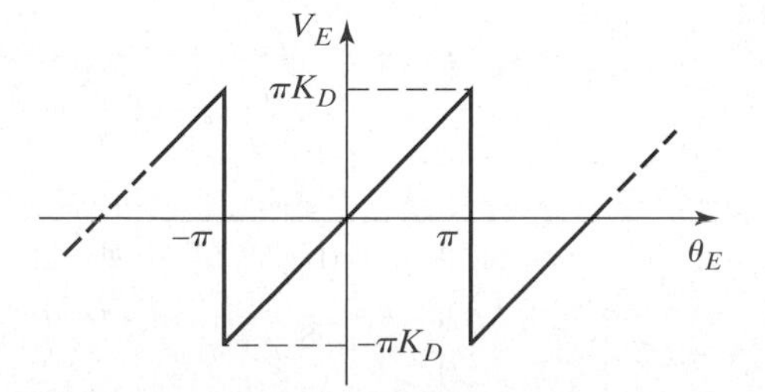

Figure 11–51 One possible phase detector transfer function.

[8] This statement assumes that the jitter due to the VCO alone is smaller than that due to the input signal, which is true only in some situations.

and

$$\Theta_i(s) = \frac{\Delta\omega}{s^2}. \tag{11.115}$$

Using the final-value theorem of Laplace transforms, we have

$$\lim_{t\to\infty} v_E(t) = \lim_{s\to 0} sV_e(s) = \lim_{s\to 0} s\frac{V_e(s)}{\Theta_i(s)}\Theta_i(s). \tag{11.116}$$

Substituting (11.109) into (11.108) and using that result in (11.103), we derive

$$\frac{V_e(s)}{\Theta_i(s)} = \frac{sK_D}{s + K_oK_DF(s)}. \tag{11.117}$$

We can use (11.115) and (11.117) in (11.116) to find the steady-state error resulting from a step change in frequency, but the result will depend on the loop filter we use. Therefore, we consider three different loop filters in the sections that follow.

11.3.1 First-Order PLLs

If we don't use a filter in the PLL at all, $F(s) = 1$. Using (11.115) and (11.117) in (11.116) in this case leads to

$$v_E(\infty) = \lim_{s\to 0}\frac{\Delta\omega K_D}{s + K_oK_D} = \frac{\Delta\omega}{K_o}. \tag{11.118}$$

Setting this equal to the maximum output of the phase detector as given in Figure 11-51, we find the lock range:

$$\text{Lock Range} = \Delta\omega_{\max} = \pi K_oK_D. \tag{11.119}$$

In deriving (11.119) we have implicitly assumed that the VCO range exceeds the lock range, which is usually, but not always, true. If the VCO range is smaller, then it limits the lock range. We have also assumed that the loop can respond rapidly enough to the change in input frequency that it continues to function in a linear manner. This assumption is almost certainly false if we really make a step change in the input frequency, so the lock range found here is valid only for very slow changes in the input frequency.

Now notice that with $F(s) = 1$, the transfer function for an FM demodulator, given by (11.112), can be written for real frequencies as

$$\frac{V_c(j\omega)}{\Omega_i(j\omega)} = \frac{K_D}{j\omega + K_oK_D} = \frac{1/K_o}{1 + \dfrac{j\omega}{K_oK_D}}, \tag{11.120}$$

which is a single-pole low-pass filter with a pole frequency of

$$\omega_p = K_oK_D. \tag{11.121}$$

Note that we can similarly determine that the transfer function in (11.113) becomes

$$\frac{\Omega_{vco}(j\omega)}{\Omega_i(j\omega)} = \frac{1}{1 + j\omega/\omega_p}. \tag{11.122}$$

We now make several comments about these results. First, examining (11.120) shows why we call this a first-order PLL: The denominator polynomial is a first-order polynomial. Second, (11.120) shows that an FM demodulator rejects high-frequency changes in the input frequency, as noted earlier, and (11.122) similarly shows that a PLL can reduce high-frequency variations in the frequency of a signal (i.e., high-frequency jitter). Third, and most important, (11.119) and (11.121) show that the lock range and bandwidth of a first-order PLL are proportional to each other. Therefore, *in a first-order PLL, it is not possible to simultaneously achieve a wide lock range and a low bandwidth*, even though both would frequently be desirable.

11.3.2 Second-Order PLLs

Now suppose that we use a filter described by

$$F(s) = \frac{1 + s/\omega_z}{1 + s/\omega_p} \quad \textbf{(11.123)}$$

in the Laplace domain or by

$$F(j\omega) = \frac{1 + j\omega/\omega_z}{1 + j\omega/\omega_p} \quad \textbf{(11.124)}$$

in the Fourier domain. In this case, we can prove that the lock range is the same as we had for the first-order loop if the phase detector is again described by Figure 11-51 (*see* Problem P11.50):

$$\text{Lock Range} = \Delta\omega_{\max} = \pi K_o K_D. \quad \textbf{(11.125)}$$

We can also show that (*see* Problem P11.51)

$$\frac{V_c(s)}{\Omega_i(s)} = \frac{K_D \omega_p (1 + s/\omega_z)}{s^2 + s\omega_p(1 + K_o K_D/\omega_z) + \omega_p K_o K_D}, \quad \textbf{(11.126)}$$

which can be written with the denominator in the standard form we introduced in Section 10.1.5;, namely,

$$\frac{V_c(s)}{\Omega_i(s)} = \frac{K_D \omega_p (1 + s/\omega_z)}{s^2 + 2\zeta\omega_n s + \omega_n^2}, \quad \textbf{(11.127)}$$

where the undamped natural frequency is

$$\omega_n = \sqrt{\omega_p K_o K_D} \quad \textbf{(11.128)}$$

and the damping coefficient is

$$\zeta = \frac{1}{2}\left(1 + \frac{K_o K_D}{\omega_z}\right)\sqrt{\frac{\omega_p}{K_o K_D}}. \quad \textbf{(11.129)}$$

Similarly,

$$\frac{\Omega_{vco}(s)}{\Omega_i(s)} = \frac{K_D K_o \omega_p (1 + s/\omega_z)}{s^2 + 2\zeta\omega_n s + \omega_n^2}, \quad \textbf{(11.130)}$$

where (11.128) and (11.129) still apply. Equations (11.127) and (11.130) show why we call this a second-order PLL: The denominator polynomial is second-order.

Although the bandwidth of the transfer functions given by (11.126) and (11.130) depends on the value of ζ, it is usually within about $\pm 30\%$ of ω_n. To achieve a small bandwidth and a large lock range simultaneously, we need to have a low pole frequency in the filter ($\omega_p \ll K_o K_D$) and a zero sufficiently below the product $K_o K_D$ (i.e., $\omega_z \ll K_o K_D$). If the zero is not this low, the value of the damping coefficient, as given by (11.129), will be very low because $\omega_p \ll K_o K_D$.

11.3.3 Type-II PLLs

All of the PLLs considered up to this point in time have a single integration in their loop gain (i.e., a single term of the form $1/s$, rather than $1/(s - p)$). They are, therefore, called *Type-I* PLLs. All PLLs are at least Type I, since the VCO itself contains an integration, which is inherent in going from frequency to phase. If a second integration is added into the loop, the loop becomes a *Type-II* loop. The advantage of a Type-II loop can be understood intuitively in the following way.

If the loop filter has an integrator, the steady-state error out of the phase detector will go to zero for any fixed offset between the VCO free-running frequency and the PLL input frequency. (Otherwise the output of the integrator would continue to change and the circuit would not be at steady state.) Therefore, the lock range will no longer be determined by the maximum output of the phase detector. Rather, the VCO will now determine the lock range.

The major advantage gained by a Type-II PLL is that the loop bandwidth and the lock range are no longer intimately linked. Therefore, a low bandwidth and a large lock range can be simultaneously achieved with a well-behaved loop. A Type-II PLL is very similar to a second-order Type-I loop with a very low pole frequency in the loop filter, so satisfactory performance can often be achieved by placing a zero at a frequency sufficiently below $K_o K_D$ as discussed in the previous section.

Example 11.13

PLLs can be simulated using the analog behavioral modeling capabilities in SPICE. For example, a PSPICE schematic for a simple PLL is shown in Figure 11-52. The input source and the VCO are both modeled as EVALUE elements so that their instantaneous frequencies can be changed on the basis of a modulation input. The phase detector is a multiplier, and the loop filter is a single-pole low-pass filter implemented using an ELAPLACE block.

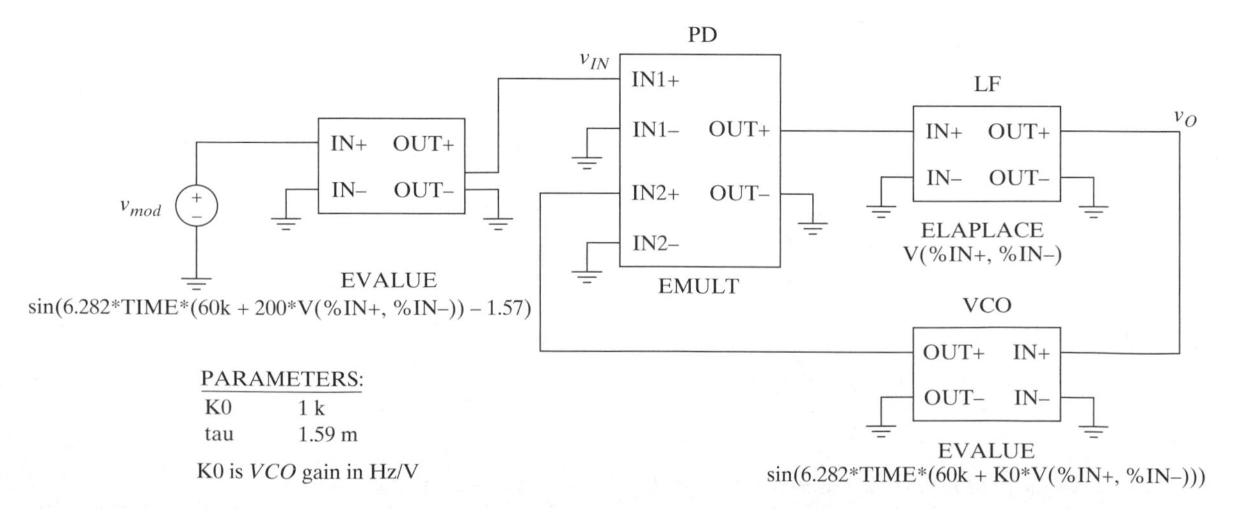

Figure 11–52 PSPICE schematic for an analog behavioral model of a PLL (see Exam13.sch on the CD).

With the modulation input pulsing the input frequency by 20 Hz, the control voltage of the VCO is shown in Figure 11-53. We can see that the PLL does follow the step changes in input frequency.

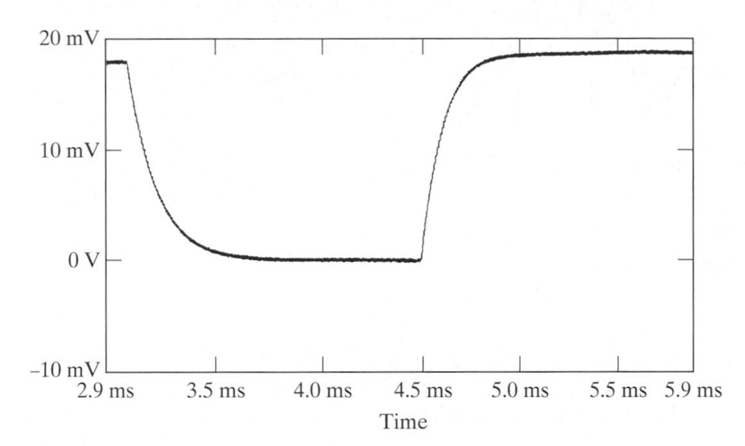

Figure 11–53 VCO control voltage when the input frequency is stepped back and forth by 20 Hz.

PLLs are available as ICs (e.g., the LM565) and are also commonly used as a component in larger ICs. The phase detectors in these loops may be true multipliers, or, for digital signals, they may be just a flip-flop. The VCOs can be implemented using a number of different oscillators. For example, you could use a Colpitts oscillator as presented in Chapter 10 and use a varactor diode as an electronically variable capacitor to change the frequency of oscillation. In a loop using digital waveforms, the 555 timer discussed in Chapter 10 can be used as the VCO.

SOLUTIONS TO EXERCISES

11.1 As in Example 11.2, we find that we need $L = 2C$. Then, to set the cutoff frequency to 3 rad/s we require, that

$$|H(j3)|^2 = \frac{1}{2} = \frac{1}{1 + 9L(L - 2C) + 81L^2C^2}.$$

Plugging in $L = 2C$ leads to $C = 0.24$ F and $L = 0.48$ H.

11.2 Using (11.9) and the desired ripple of 1 dB, we find $\varepsilon = 0.51$. Then, with $N = 2$ we use (11.13) to find $\xi = 0.714$. Next, we use (11.11) and (11.12) to find the poles are at $s_{1,2} = -0.549 + j0.895$. Equating these poles with the equation for the poles of our circuit as given in Example 11.4 leads to $C = 0.911$ F and $L = 0.996$ H. Running a SPICE simulation with these values confirms the peak ripple to be 1 dB.

11.3 The desired cutoff frequency is 6.28×10^5 times larger than the normalized circuit, so $b = 6.28 \times 10^5$. Using this value of b, the given value of C, the fact that $C_o = 1$ F and (11.15), we find that $a = 1590$. Therefore, $R = 1.6$ kΩ.

11.4 Plugging the given values into (11.35) yields 0.47 mH.

11.5 The MATLAB program is on the CD (*see* Exer11_5.m). The output is shown in Figure 11-54, and does look like Figure 11-33.

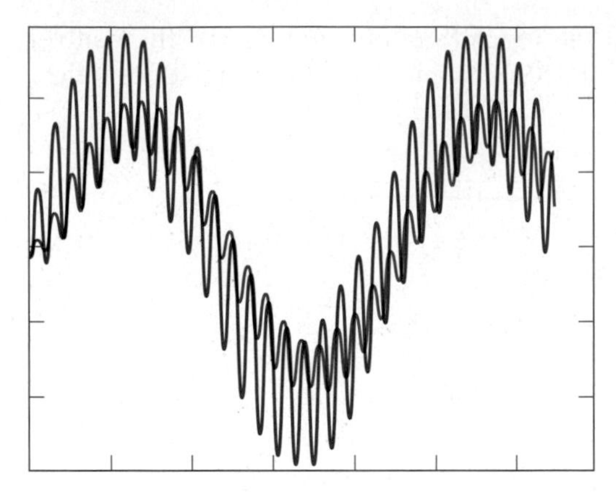

Figure 11–54 See Exer11_5.m on the CD.

11.6 (a) We first use (11.70) at ω_o to find $R_p = 20$ kΩ. From (11.51), the desired ω_o and the given L, we obtain $C_{eq} = 1$ nF, which is so much larger than the transistor's parasitic capacitors, that we also have $C = 1$ nF. Using this capacitance in (11.55) along with the requested bandwidth, we obtain $R_{eq} = 5$ kΩ. We also know that $R_{eq} = r_o \| R \| R_p$, so we obtain $R = 10$ kΩ. (b) Plugging the known values into (11.52), we find that $A_v = -25$.

11.7 The total impedance seen looking into the output will be the parallel combination of L_1, R_1, and C_1 transformed. As shown in Figure 11-44 and described in the text, we transform the impedance by dividing by N^2. We can do this one component at a time. Since the impedances of resistors and inductors are proportional to their values, dividing the impedances by N^2 is equivalent to dividing the values by N^2. Therefore, we obtain $R_{eq} = 1$ kΩ and $L_{eq} = 1$ μH. For the capacitor, on the other hand, the impedance varies as $1/C$; therefore, the equivalent capacitance is N^2 times higher: $C_{eq} = 0.1$ μF.

11.8 Following the procedure in the example, we obtain $C_1' = 160$ pF and $C_2' = 220$ pF.

11.9 Following the procedure in the example, we obtain $C_1' = 220$ pF, $R_1' = 11$ kΩ, $C_2' = 240$ pF, and $R_2' = 11$ kΩ.

CHAPTER SUMMARY

- Frequency-selective circuits are frequently required in electronic communication systems and other applications.
- We cannot choose the frequency-domain and time-domain performance of filters independently. Filter responses that roll off quickly in the frequency domain (i.e., are highly selective) have large overshoots and ringing in their step response.
- Different approximation techniques are used for making realizable filters that come as close as possible to an ideal filter in some particular characteristic. For example, a Butterworth response results from coming as close as possible to the ideal magnitude-squared response in terms of the Taylor series expansions. The result is that the first N derivatives are the same, which leads to a maximally flat response.
- Keeping the square of the magnitude as close to ideal as possible over the entire passband leads to the Chebyshev, or equiripple, response characteristic.
- Filters are often designed by using normalized values, which are then scaled to achieve the actual design.

- Many bandpass systems have a passband much smaller than the center frequency and can be more easily analyzed by using the narrowband approximation.
- Integrated filters can be built in a number of different ways, but usually the problem is to overcome the limitations of integrated inductors and the poor absolute tolerances of passive components on an IC.
- Inductors can be simulated on an IC by using resistors, capacitors, and op amps. One way to do this is to use a generalized impedance converter.
- Many filters operate on discrete samples of a signal rather than the continuous-time signal itself.
- Switched-capacitor filters are a popular type of discrete-time filter that replace large resistors with switches and capacitors. One major advantage is that filter responses depend on capacitor ratios. In addition, filters can be tuned to some extent by varying the clock speed.
- FIR filters do not have poles and are therefore unconditionally stable and easier to adapt. They can be implemented using continuous-time analog, discrete-time analog, or digital circuits.
- Tuned amplifiers combine amplification and filtering. They must be carefully designed to avoid stability problems.
- Tuned amplifiers with multiple tuned circuits can be synchronously tuned or stagger tuned, depending on the desired overall transfer function.
- Phase-locked loops are extremely useful feedback systems that provide a high degree of selectivity while tracking slow changes in frequency. They can be used for FM demodulation and many other applications.

REFERENCES

[11.1] A.M. Davis, *Linear Circuit Analysis*. Boston, MA: PWS Publishing Company, 1998.

[11.2] H.J. Blinchikoff and A.I. Zverev, *Filtering in the Time and Frequency Domains*. New York: John Wiley & Sons, 1976.

[11.3] S. Haykin, *An Introduction to Analog and Digital Communications*. New York: John Wiley & Sons, 1989.

[11.4] T.H. Lee, *The Design of CMOS Radio-Frequency Integrated Circuits*. Cambridge, U.K.: Cambridge University Press, 1998.

[11.5] R. Gregorian and G.C. Temes, *Analog MOS Integrated Circuits for Signal Processing*. New York: John Wiley & Sons, 1986.

[11.6] B. Widrow and S.D. Stearns, *Adaptive Signal Processing*. Englewood Cliffs, NJ: Prentice-Hall Inc., 1985.

[11.7] B.P. Lathi, *Linear Systems and Signals*. Carmichael, CA: Berkeley-Cambridge Press, 1992.

[11.8] J.W.M Bergmans, *Digital Baseband Transmission and Recording*. Dordrecht, The Netherlands: Kluwer Academic Publishers, 1996.

[11.9] P.R. Gray, P.J. Hurst, S.H. Lewis and R.G. Meyer, *Analysis and Design of Analog Integrated Circuits*, 4E, New York: John Wiley & Sons, 2001.

[11.10] B. Razavi, Ed., *Monolithic Phase-Locked Loops and Clock Recovery Circuits*. Piscataway, NJ: IEEE Press, 1996.

[11.11] A.S. Sedra and K.C. Smith, *Microelectronic Circuits*, 4/E. New York: Oxford University Press, 1998.

PROBLEMS

11.1 FILTERS

11.1.2 Practical Transfer Functions

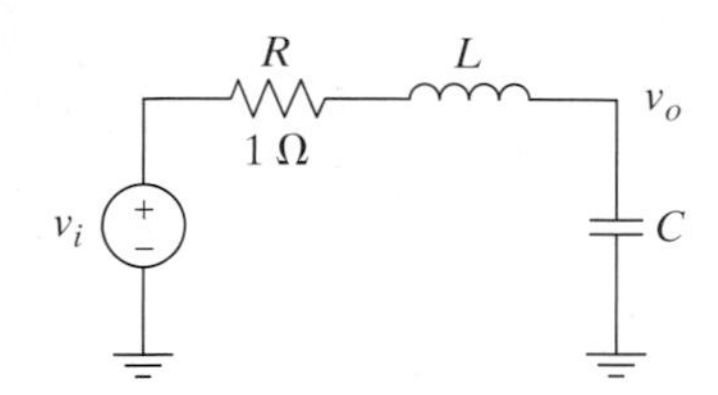

Figure 11–55

D P11.1 Consider the low-pass filter shown in Figure 11-55. Assume that $R = 1\ \Omega$ (the normalized case). Determine values for L and C that will yield a Butterworth response with a cutoff frequency of 1 rad/s.

S P11.2 Use SPICE to confirm your results from Problem P11.1. Hand in the plot of the magnitude response.

D P11.3 Consider the low-pass filter shown in Figure 11-6. Assume that $R = 1\ \Omega$ (the normalized case). Determine values for L and C that will yield a Chebyshev response with a ripple band edge of 1 rad/s and a 0.75-dB ripple.

S P11.4 Use SPICE to confirm your results from Problem P11.3. Hand in the plot of the magnitude response.

S P11.5 Use SPICE, MATLAB, or some other program to plot the overall transfer functions that result from cascading 2, 10, and 20 identical single-pole low-pass filters. Use perfect voltage buffers in between the filters so that they don't load each other down. Scale the filters so that the −3-dB frequency of the cascade is the same for all three situations. Based on these results, do you think you would get a sharper transition if you cascaded more of the same stages?

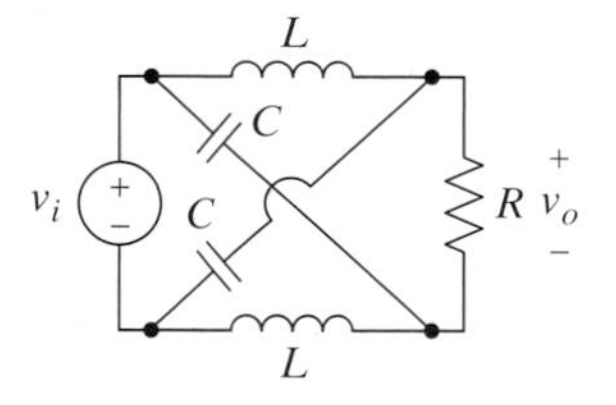

Figure 11–56

P11.6 Derive the transfer function for the allpass network shown in Figure 11-56, assuming that $R = 1\ \text{k}\Omega$, $L = 15.9$ mH, and $C = 15.9$ nF. Prove that the magnitude is constant and plot the phase versus frequency. This is a first-order lattice-type allpass network.

11.1.3 Normalization (Frequency and Component Value Scaling)

D P11.7 Determine the component values necessary for the Butterworth filter in Example 11.2 to have a cutoff frequency of 25 kHz and a resistor of 3 kΩ.

S P11.8 Use SPICE to confirm your results from Problem P11.7. Hand in the plot of the magnitude response.

D P11.9 Determine the component values necessary for the Chebyshev filter in Example 11.4 to have a cutoff frequency of 25 kHz and a resistor of 3 kΩ. Don't forget the difference between the cutoff frequency and the ripple band edge. (*see* [11.10]).

S P11.10 Use SPICE to confirm your results from Problem P11.9. Hand in the plot of the magnitude response.

D P11.11 Consider the low-pass filter shown in Figure 11-55. Assume that $R = 10\ \text{k}\Omega$. Determine values for L and C that will yield a Butterworth response with a cutoff frequency of 10 kHz.

S P11.12 Use SPICE to confirm your results from Problem P11.11. Hand in the plot of the magnitude response.

D P11.13 Consider the low-pass filter shown in Figure 11-6. Assume that $R = 4.7\ \text{k}\Omega$. Determine values for L and C that will yield a Chebyshev response with a cutoff frequency of 50 kHz and a 1-dB ripple. Don't forget the difference between the cutoff frequency and the ripple band edge (*see* [11.10]).

S P11.14 Use SPICE to confirm your results from Problem P11.13. Hand in the plot of the magnitude response.

11.1.4 The Narrowband Approximation

P11.15 Determine the transfer function of a fourth-order Butterworth band-pass filter with a center frequency of 5 krad/s and a bandwidth of 100 rad/s.

P11.16 Determine the transfer function of a sixth-order equiripple band pass filter with a maximum peak-to-peak ripple of 0.2 dB, a center frequency of 10 MHz, and a 200-kHz bandwidth.

11.1.5 Integrated Filters and Simulated Inductors

P11.17 Consider a closed-loop amplifier with a forward gain given by

$$a(s) = \frac{1 + s\tau}{s\tau(1 + s2\tau)}.$$

Show that the closed-loop response has complex-conjugate poles when $b(s) = 1$.

P11.18 Consider the circuit shown in Figure 11-57, where the amplifier is assumed to be ideal and have a voltage gain of unity. This circuit has positive feedback, but it is stable because the loop gain is less than unity. The closed-loop gain is given by $A(s) = 1/(1 - b(s))$. Recall that the denominators of all possible transfer functions in a given network are the same (since the denominator describes the natural response of the network and does not depend on where we inject a signal). Therefore, no matter where we inject the signal, the denominator of the resulting transfer function will be $1 - b(s)$. **(a)** Show that the poles of the transfer function are given by the solutions of $s^2 + s/R_pC_1 + 1/\tau_1\tau_2 = 0$, where $R_p \equiv R_1 \| R_2$, $\tau_1 \equiv R_1C_1$, and $\tau_2 \equiv R_2C_2$. **(b)** Use the form shown in Example 11.3 for the denominator polynomial, and show that $\omega_o = \sqrt{1/\tau_1\tau_2}$ and $Q = R_pC_1\omega_o$. **(c)** Let $R_1 = R_2 = R$ and $C_2 = \alpha^2C_1$, where α is a constant less than unity. Show that $\omega_o = 1/\alpha\tau_1$ and $Q = 1/2\alpha$.

P11.19 One example of how to use the circuit shown in Figure 11-57 is given in Figure 11-58(a). The signal is being injected at the bottom of R_1. Note that when $v_s = 0$, the loop topology is identical to that of Figure 11-57. An equivalent circuit for this filter is shown in Figure 11-58(b). By direct analysis of this circuit, show that the overall transfer function is $V_o(s)/V_s(s) = \omega_o^2/(s^2 + s\omega_o/Q + \omega_o^2)$, where $\omega_o = \sqrt{1/\tau_1\tau_2}$ and $Q = R_pC_1\omega_o$, as in Problem P11.18. Note that the numerator of the transfer function is a constant. Therefore, the zeros are at infinity and the circuit is a low-pass filter.

S P11.20 Use SPICE to simulate the circuit shown in Figure 11-58(b) with $R =$ 1 kΩ, $C_1 = 100$ pF, and $\alpha = 0.025$. Show a plot of the magnitude response near ω_o, using a linear frequency scale, and verify that the results given in Problem P11.19 are correct.

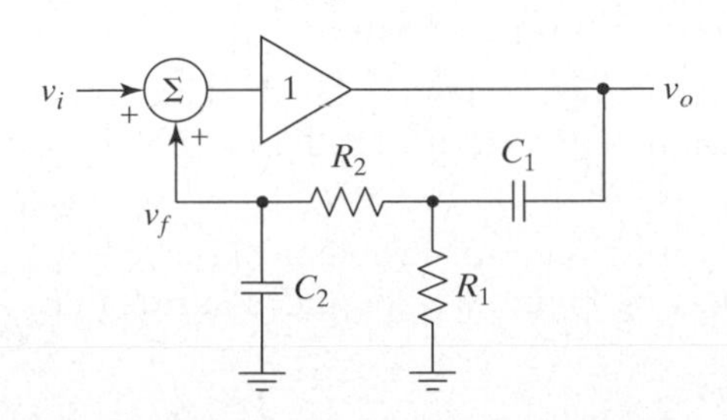

Figure 11–57

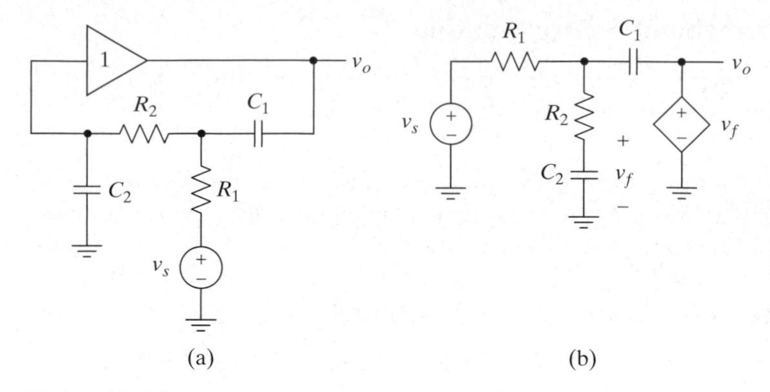

Figure 11–58

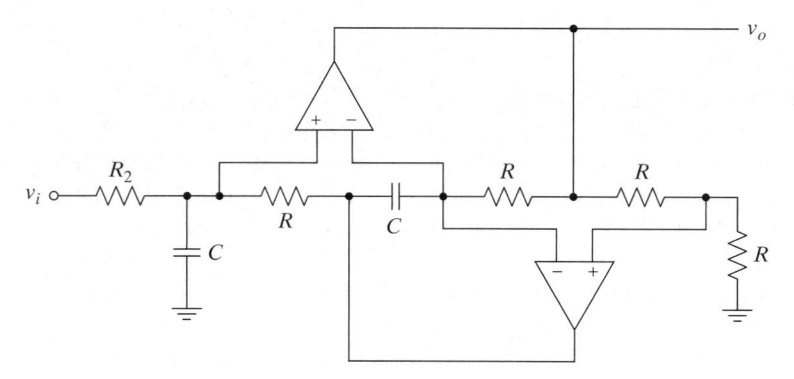

Figure 11–59

D P11.21 Consider the circuit shown in Figure 11-59. **(a)** Show that the transfer function is

$$\frac{V_o(s)}{V_i(s)} = \frac{2s(\omega_o/Q)}{s^2 + s(\omega_o/Q) + \omega_o^2}$$

where $\omega_o = 1/RC$ and $R_2 = QR$. (*Hint*: Find the impedance looking into the GIC first.) **(b)** Design the circuit for a $Q = 50$ and $\omega_o = 100$ krad/s with $R = 1$ kΩ.

11.1.6 Discrete-Time Filters

Switched-Capacitor Filters

P11.22 Consider the low-pass filter shown in Figure 11-27. **(a)** What happens to the −3-dB frequency of this filter if the clock frequency changes from 1 MHz to 900 kHz? **(b)** From 1 MHz to 1.1 MHz?

D P11.23 Using the low-pass filter topology shown in Figure 11-27, change the capacitor values to achieve a design with a cutoff frequency of 50 kHz with a 500-kHz clock.

S P11.24 Perform a SPICE simulation of your circuit for Problem P11.23. Show a transient output similar to that of Figure 11-28, and confirm that your design works as expected.

Finite Impulse Response Filters

P11.25 Consider an ideal FIR filter with seven taps. The weights are given by [0.05, 0.1, 0.2, 0.3, 0.2, 0.1, 0.05]. **(a)** Generate a plot of the magnitude and phase response of this filter for a 100-kHz clock rate. What is the cutoff frequency? **(b)** What does the cutoff frequency become if the clock rate is changed to 200 kHz?

11.2 TUNED AMPLIFIERS

11.2.1 Single-Tuned Amplifiers

D P11.26 Consider the amplifier shown in Figure 11-60. Determine the values of C_1 and R that are required to obtain a center frequency of 10.7 MHz and a BW of 200 kHz if $L_1 = 1\ \mu\text{H}$.

S P11.27 Use SPICE and the Nnominal model from Appendix B to simulate your circuit from Problem P11.26. Show a plot of the magnitude of the output in dB versus frequency, using a linear frequency scale. Show that the center frequency and bandwidth are correct.

D P11.28 Consider the amplifier shown in Figure 11-61. Determine the values of C_1 and R that are required to obtain a center frequency of 10.7 MHz and a BW of 200 kHz if $L_1 = 1\ \mu\text{H}$.

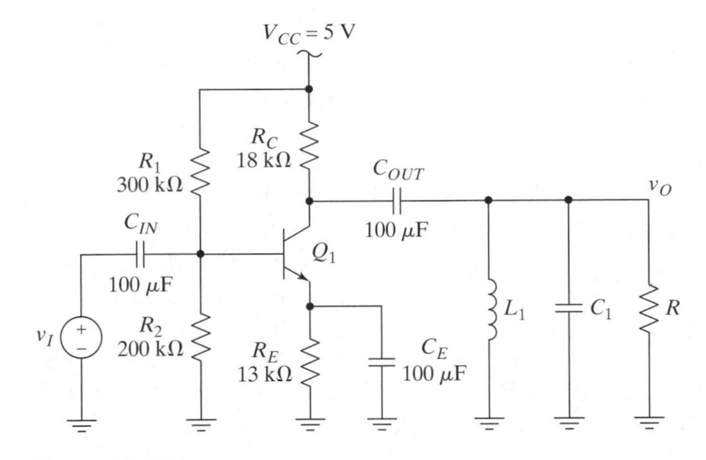

Figure 11–60

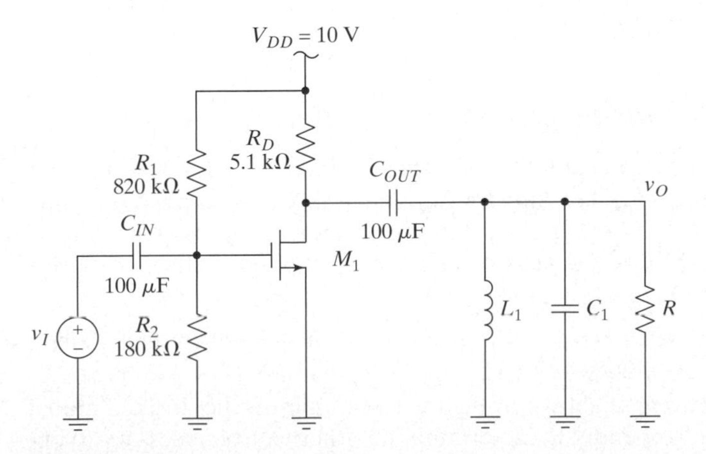

Figure 11–61

S P11.29 Use SPICE and the Nlarge_mos model from Appendix B to simulate your circuit from Problem P11.28. Show a plot of the magnitude of the output in dB versus frequency, using a linear frequency scale. Show that the center frequency and bandwidth are correct.

P11.30 This problem works through the derivation of the model presented for the autotransformer in Figure 11-44(b). Consider the circuits shown in Figure 11-62. The autotransformer is shown in part (a) of the figure, and we define $N = n_1/n_2$ as before. In part (b) of the figure, we model the autotransformer by using a normal two-winding transformer with the windings connected in series. The equations that describe this transformer in the Laplace domain are (*see* [11.1])

$$V_1'(s) = sL_1'I_1 + sMI_2'$$

and

$$V_2(s) = sMI_1 + sL_2I_2' ,$$

where the mutual inductance is $M = \sqrt{L_1'L_2}$, since the coupling coefficient is assumed to be unity. The turns ratio for this transformer is $N' = n_1'/n_2 = \sqrt{L_1'/L_2}$, and since $n_1 = n_1' + n_2$, we see that $N = N' + 1$. Finally, $L_1 = L_1' + L_2 + 2M$. **(a)** Using these equations and the circuit shown, derive an equation for $V_1(s)$ in part (b) of the figure in terms of L_1, I_1, I_2, and N. **(b)** Now use the circuit in part (c) of the figure along with the relations for the ideal two-winding transformer ($V_2 = V_1/N$ and $I_2 = -NI_1'$) to derive another expression for $V_1(s)$ in terms of L_1, I_1, I_2, and N. **(c)** Show that the two expressions found in parts (a) and (b) are equivalent. This result proves that the model in part (c) of the figure is equivalent to the circuit in part (a). **(d)** Derive the equation shown for Z_{in} in Figure 11-44(a).

P11.31 Suppose we wish to design a single-tuned amplifier with the small-signal equivalent shown in Figure 11-63(a) and we want to achieve a center frequency of 70 MHz with a BW of 500 kHz. **(a)** If we use an inductor with $L = 0.05$ μH and a Q of 100 at 70 MHz, what is the value of capacitance required? **(b)** What is the minimum BW achievable (i.e., let R go to infinity)? **(c)** Now suppose we use an autotransformer with $L = 0.2$ μH, a Q of 200 at 70 MHz, and $N = 2$ as shown in Figure 11-63(b). What value of capacitance is required now? **(d)** What is the minimum BW achievable now? **(e)** Determine the value of R that will result in a BW of 500 kHz.

11.2.2 Synchronous and Stagger-tuned Amplifiers

D P11.32 Consider the cascode amplifier shown in Figure 11-47, and assume that $R_S = 100\ \Omega$, $R_L = 10\ \text{k}\Omega$, and, for both devices, $C_{cm} = 25$ pF, $C_{oc} = 5$ pF, $r_{cm} = 10\ \text{k}\Omega$, $r_o = 50\ \text{k}\Omega$, and $g_m = 20$ mA/V. Determine the other values necessary to obtain a synchronously tuned amplifier with $f_o = 10$ kHz and BW = 500 Hz. Look at (11.75) through (11.78).

D P11.33 Repeat Problem P11.32, but with a stagger-tuned design achieving a Butterworth response. Look at (11.75) through (11.78).

D P11.34 Consider the circuit shown in Figure 11-64 and use the 2N2222 model presented in Appendix B. Determine the other values necessary to obtain a synchronously tuned amplifier with $f_o = 10$ kHz and BW = 500 Hz.

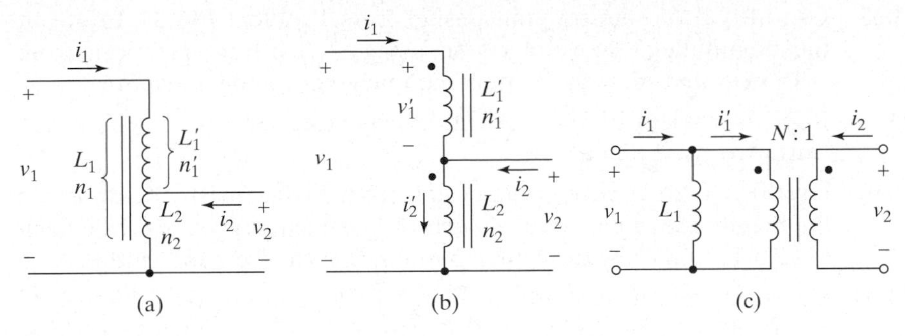

Figure 11–62

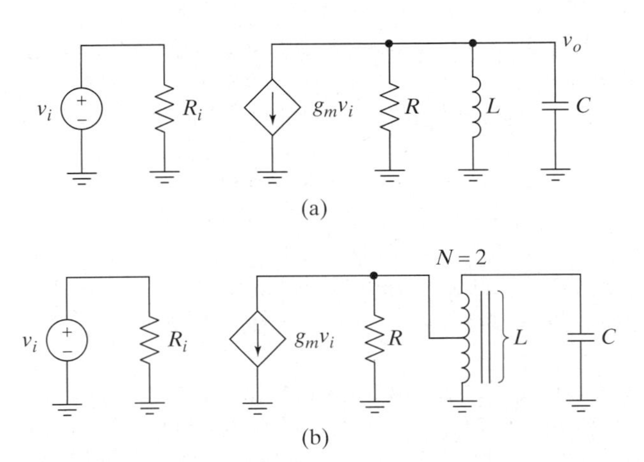

Figure 11–63

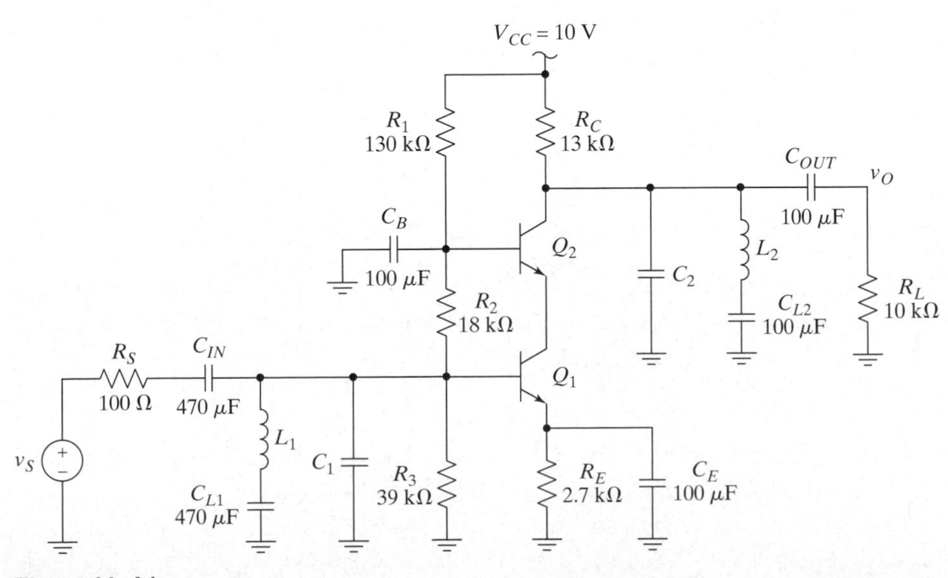

Figure 11–64

S P11.35 Use SPICE to confirm your design from Problem P11.34. Print out the magnitude of the voltage gain, using a linear frequency scale from 5–15 kHz, and show the center frequency, gain, and bandwidth.

D P11.36 Repeat Problem P11.34, but with a stagger-tuned design achieving a Butterworth response.

S P11.37 Use SPICE to confirm your design from Problem P11.36. Print out the magnitude of the voltage gain, using a linear frequency scale from 5–15 kHz, and show the center frequency, gain, and bandwidth.

D P11.38 Consider the circuit shown in Figure 11-65 and use the Nlarge_mos model presented in Appendix B. Determine the other values necessary to obtain a synchronously tuned amplifier with $f_o = 10$ kHz and BW = 500 Hz.

S P11.39 Use SPICE to confirm your design from Problem P11.38. Print out the magnitude of the voltage gain, using a linear frequency scale from 5–15 kHz, and show the center frequency, gain, and bandwidth.

D P11.40 Repeat Problem P11.38, but with a stagger-tuned design achieving a Butterworth response.

S P11.41 Use SPICE to confirm your design from Problem P11.40. Print out the magnitude of the voltage gain, using a linear frequency scale from 5–15 kHz, and show the center frequency, gain, and bandwidth.

D P11.42 Consider the emitter-coupled pair tuned amplifier shown in Figure 11-66. Use the nominal *npn* model from Appendix B, and assume that $R_S = 100$. Design the circuit to achieve a midband voltage gain of 20 and a synchronously tuned response with f_o = 10.7 MHz and BW = 200 kHz.

S P11.43 Use SPICE to confirm your design from Problem P11.42. Print out the magnitude of the voltage gain, using a linear frequency scale from 10.4–11 MHz, and show the center frequency, gain, and bandwidth.

D P11.44 Repeat Problem P11.42, but with a stagger-tuned design achieving a Butterworth response.

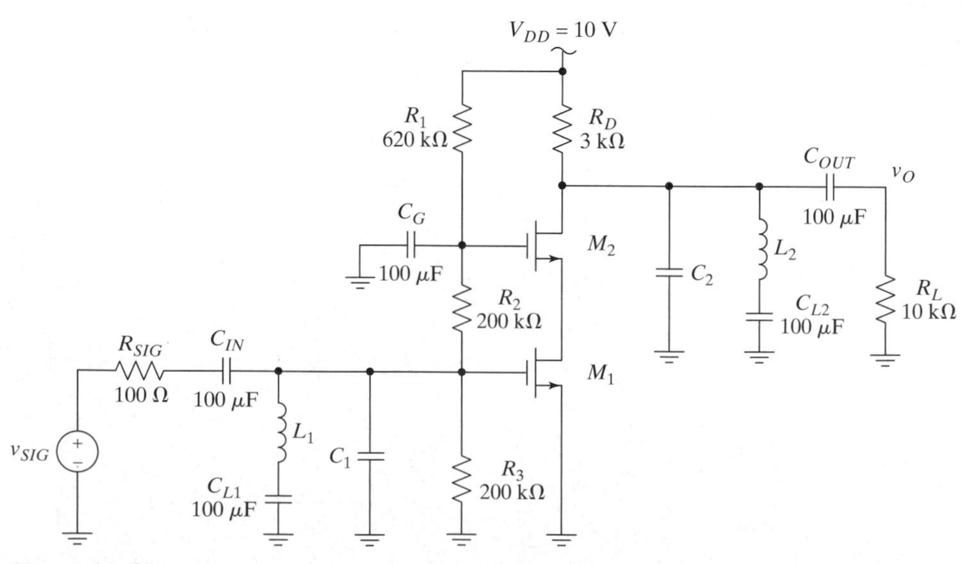

Figure 11–65

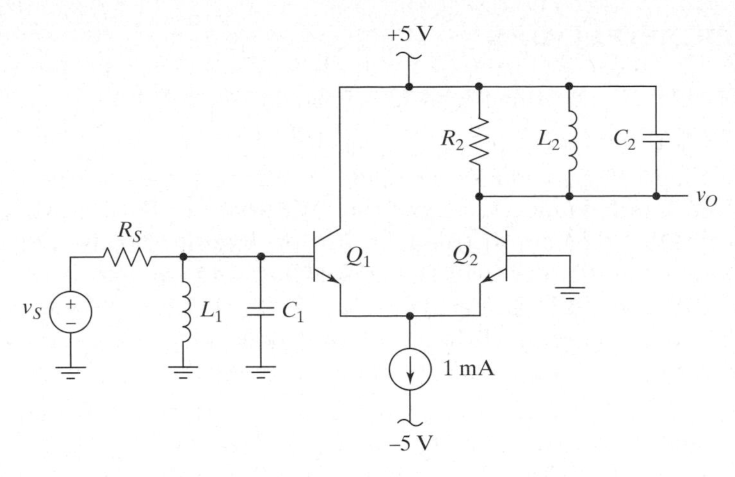

Figure 11–66

S P11.45 Use SPICE to confirm your design from Problem P11.44. Print out the magnitude of the voltage gain, using a linear frequency scale from 10.4–11 MHz, and show the center frequency, gain, and bandwidth.

D P11.46 Consider the source-coupled pair tuned amplifier shown in Figure 11-67. Use the Nlarge_mos model from Appendix B, and assume that $R_{SIG} = 100\ \Omega$. Design the circuit to achieve a midband voltage gain of 10 and a synchronously tuned response with $f_o = 10.7$ MHz and BW = 200 kHz.

S P11.47 Use SPICE to confirm your design from Problem P11.46. Print out the magnitude of the voltage gain, using a linear frequency scale from 10.4–11 MHz, and show the center frequency, gain, and bandwidth.

D P11.48 Repeat Problem P11.46, but with a stagger-tuned design achieving a Butterworth response.

S P11.49 Use SPICE to confirm your design from Problem P11.48. Print out the magnitude of the voltage gain, using a linear frequency scale from 10.4–11 MHz, and show the center frequency, gain, and bandwidth.

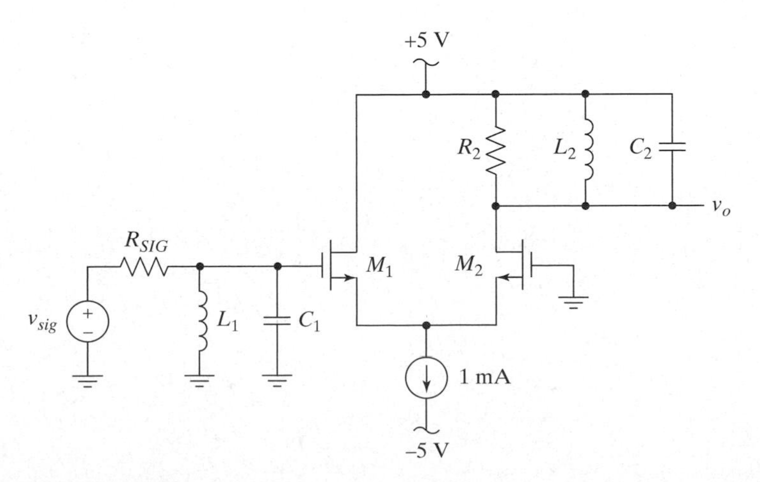

Figure 11–67

11.3 PHASE-LOCKED LOOPS

P11.50 Derive (11.125). What assumptions do you need to make?

P11.51 Derive (11.126) and (11.127).

P11.52 Consider using a PLL as an FM demodulator. We are, therefore, interested in the transfer function $V_c(s)/\Omega_i(s)$. We desire to have a lock range of 1 MHz and a cutoff frequency for the transfer function of 50 kHz. **(a)** Will a first-order PLL work? If not, why not? **(b)** If we use a second-order PLL with a filter described by (11.124), to what frequencies must the pole and zero be set to obtain $\zeta = 0.707$ and $\omega_n = 2\pi \times 50$ kHz? (This is a Butterworth response.)

P11.53 Suppose a first-order PLL is used to reduce the jitter on a signal. **(a)** What is the transfer function $\Theta_{vco}(j\omega)/\Theta_i(j\omega)$? **(b)** If the lock range is set to 100 kHz, what is the corresponding −3-dB frequency?

CHAPTER 12

Low-Frequency Large-Signal AC Analysis

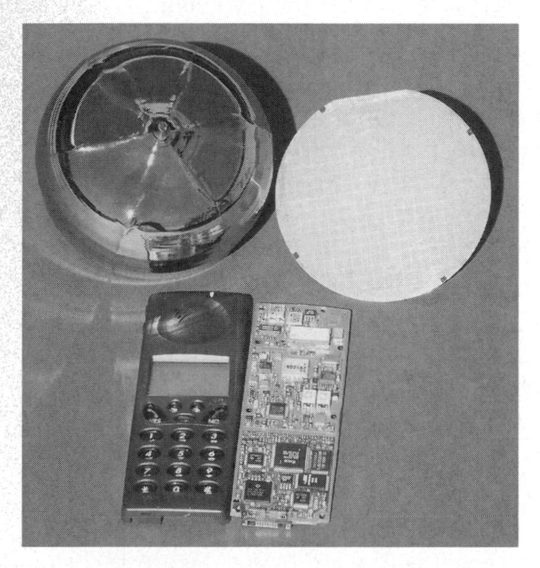

A cell phone next to a silicon wafer and the top end of a silicon ingot. The power amplifier is a critical component in cell phones since it dissipates the most power by far. Analyzing power amplifiers requires the techniques covered in this chapter.

INTRODUCTION

In Chapters 8 through 11, we assumed that the signals were small enough that small-signal models provided acceptably accurate results. However, there are many circuits that operate with large signals and that must therefore be analyzed using large-signal models. We covered low-frequency large-signal models in Chapter 7 and used them for DC bias point analyses there. In this chapter, we use those models for circuits with large low-to-mid-frequency AC signals present. In addition, in Section 12.2, we develop a large-signal midband model for coupling and bypass capacitors.

We will not explicitly cover high-frequency large-signal analysis here. Simultaneously modeling both the nonlinear and frequency-dependent characteristics of circuits requires sophisticated mathematical tools like the Volterra series or numerical methods like harmonic balance. Such methods are beyond the scope of this text. If you are interested, one good treatment can be found in [12.1].

We begin by examining some common nonlinear diode circuits in Section 12.1. We then examine the swing limits and distortion of transistor amplifiers in Section 12.2. We close by covering output stages in more detail in Section 12.3. Some nonlinear op amp circuits are also presented in Chapter 5.

12.1 DIODE CIRCUITS

The most commonly used characteristic of diodes by far is their ability to rectify a signal. In other words, they allow significant current flow only in one direction. This capability is used most commonly to convert AC signals into DC signals, as covered in the first part of this section. After rectifiers, we discuss diode limiting, clamping, and multiplying circuits. We finish the section with a brief treatment of diode logic circuits.

12.1.1 Diode Rectifiers

We often desire to ***rectify*** an AC waveform—that is, to convert it to DC. For example, electrical power is more efficiently transmitted using AC, but most electronic circuits require DC power supplies to operate. Therefore, electronic equipment that is powered by the AC line almost always contains a ***power supply***, which serves to convert the incoming AC power to one or more DC supply voltages.

Half-Wave Rectifier A ***half-wave rectifier*** is shown in Figure 12-1. The circuit allows only half of the sinusoidal input voltage through to the output, since the diode will only allow current to flow in one direction. When drawing the output voltage in this figure, we assumed that the peak input voltage, V_m, is much greater than the voltage necessary to turn the diode on. To use this circuit to produce a DC voltage, we add a filter to remove the AC component from the output.

Example 12.1

Problem:
Consider the half-wave rectifier in Figure 12-1. Draw the input and output waveforms if $V_m = 1$ V.

Solution:
With the input voltage this low, we cannot ignore the voltage drop across the diode when it is forward biased. Figure 12-2 shows the circuit redrawn using our large-signal diode models for each region of operation.

During the time the diode is reverse biased, we see clearly from Figure 12-2(a) that $v_O = 0$, since there cannot be any current flow in the resistor. For $v_I > 0.7$ V, the diode turns on, and from part (b) of the figure we see that $v_O(t) \approx v_I(t) - 0.7$ V. The resulting waveforms are shown in Figure 12-3.

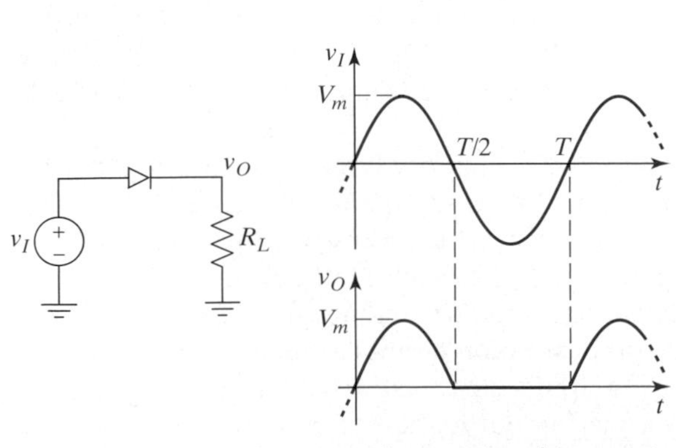

Figure 12–1 A half-wave rectifier circuit and its associated waveforms.

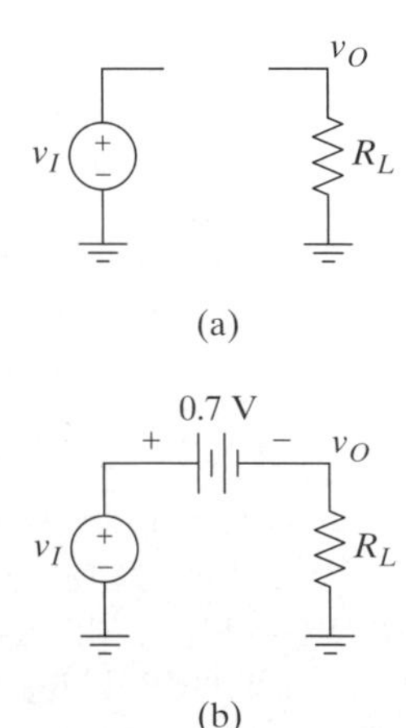

Figure 12–2 The rectifier of Figure 12-1 modeled (a) when $v_I \le 0.7$ and the diode is reverse biased and (b) when $v_I > 0.7$ and the diode is forward biased.

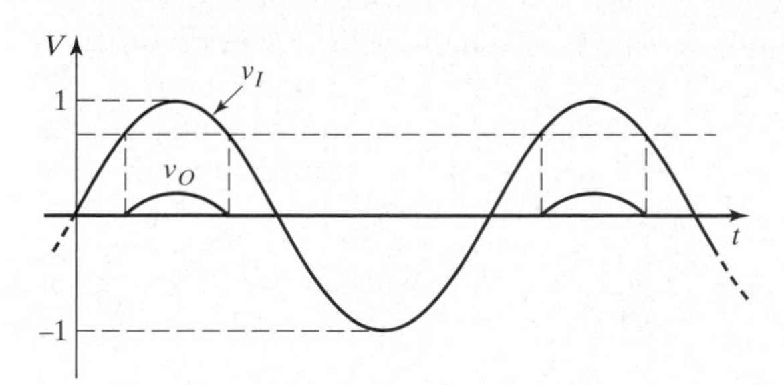

Figure 12–3 The waveforms for the rectifier with V_m = 1 V.

Example 12.1 illustrates an important point about rectifying smaller AC voltages: The peak output amplitude drops quickly when the forward voltage across the diode becomes significant relative to the peak magnitude of the input. In fact, if the input magnitude is much below 0.7 V for a silicon diode rectifier, the output will be negligible.[1]

From Figure 12-3, we see that the diode is conducting for less than half of each period of the input sine wave. The time the diode is on during each period is called the ***conduction time***. Frequently, the ***conduction angle*** is specified instead; this is defined as the length of time the diode is conducting, expressed as a proportion of the input phase (360° per cycle). For a half-wave rectifier, we would like the conduction angle to be nearly 180°.

A half-wave rectifier with a single capacitor to filter the output is shown in Figure 12-4, along with the input and output voltages when $V_m = 12$ V. We can see from the figure that the output voltage is approximately 0.7 V lower than the input voltage at the positive peak of the sine wave. The ***ripple voltage***, V_r, is defined as the maximum deviation in the output, as shown in the figure.

When the capacitor has been charged up to its peak voltage, the capacitor prevents the output from dropping fast enough to follow the input. Therefore, the diode turns off. At this time, the output follows an RC discharge given by the familiar equation (set $t = 0$ at the peak and assume that the peak output is $V_p = V_m - 0.7$ V)

$$v_O(t) = V_p e^{-t/RC}. \qquad (12.1)$$

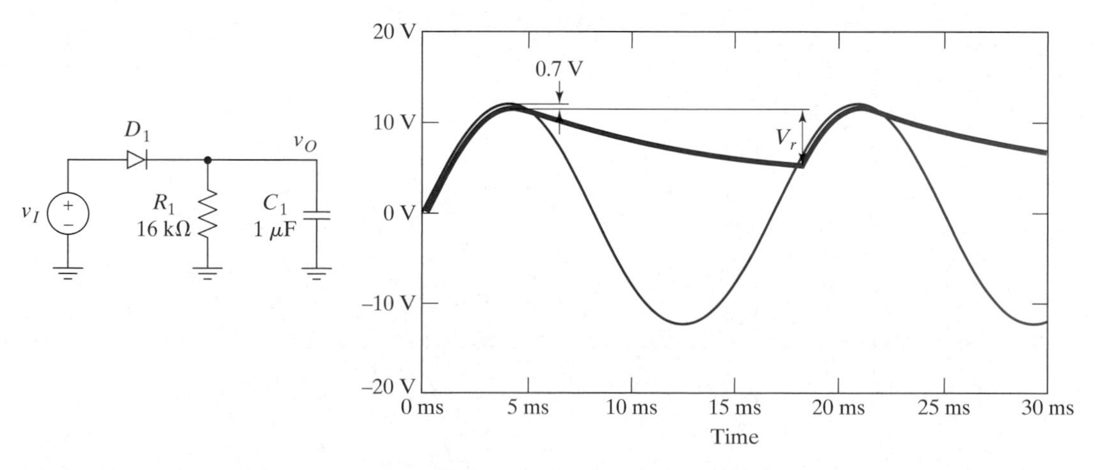

Figure 12–4 (a) A half-wave rectifier with a single capacitor. (b) The input and output voltages. (See Fig4.sch on the CD.)

[1] This problem can be overcome with an op amp, so long as the frequency isn't too high. See the superdiode circuit shown in Chapter 5.

When the input comes back up on the next positive half-cycle, the diode turns on again and pulls the output back up to V_p. If $RC \gg T$, then the droop is only the beginning of an RC discharge and can be approximated by the straight line given by $v_O(t) \approx V_p(1 - t/RC)$. For small ripple, we can say that the output follows this line for nearly T seconds. The resulting ripple is (*see* Problem P12.2)

$$V_r \approx \frac{V_p T}{RC}. \tag{12.2}$$

In other words, when $RC \gg T$, we have an almost constant output voltage equal to V_p. It is also possible to develop a negative output voltage by using a half-wave rectifier, as shown in Problem P12.3.

Exercise 12.1

How large must the capacitor in Figure 12-4(a) be if we desire to have less than 5% ripple with a 60-Hz input? Assume that $V_p \approx V_m$.

To choose a suitable diode for a given rectifier application, we need to know what the peak current in the diode is. Figure 12-5 is a repeat of Figure 12-4(b), but with the filter capacitor equal to 21 μF and with a plot of the diode current included. After the first half-cycle of the input, the diode carries current only during a small portion of the positive half-cycle, but the peak current can be larger than you might expect. In this example, $V_p \approx 11.3$ V. If a constant 11.3 V were applied to the load resistor, the source current would be 706 μA. However, the peak diode current is about 21 mA. This large current is required because the diode must supply all of the current needed by the load for an entire period T during only a short time (the conduction time of the diode).

If we assume that the capacitor discharge is approximately a straight line, as was done in deriving (12.2), we can estimate the conduction time of the diode for this circuit. Note that we can flip the time axis as shown in Figure 12-6 to make the math easier. Using part (b) of the figure, we see that

$$V_p \cos \omega t_c = V_p - V_r \ . \tag{12.3}$$

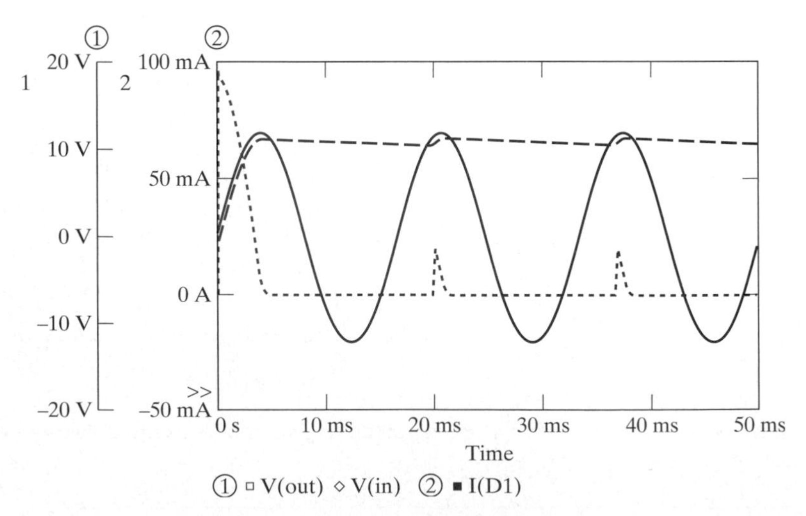

Figure 12–5 The waveforms for the rectifier in Figure 12-4(a), including the diode current (see Fig5.sch on the CD).

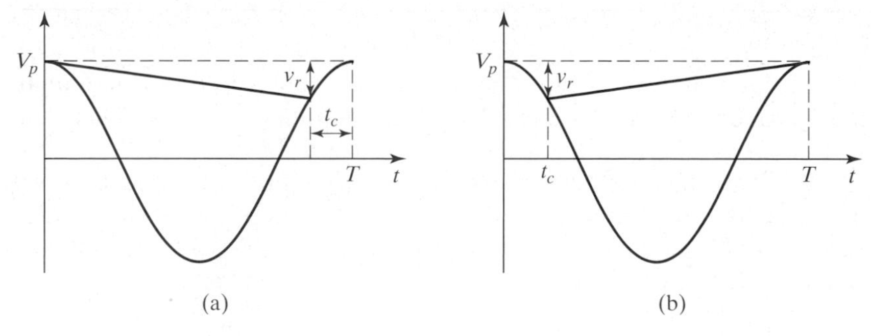

Figure 12–6 (a) An idealization of the output waveform and (b) a time-reversed version to use in finding t_c.

Assuming that the product ωt_c is small, we can approximate the cosine by the first two terms of a Taylor series ($\cos x \approx 1 - x^2/2$) and solve to obtain

$$t_c \approx \frac{1}{\omega}\sqrt{\frac{2V_r}{V_p}}. \qquad \textbf{(12.4)}$$

From Figure 12-5, we see that we can approximate the diode current pulses as right triangles with a base width of t_c and a height equal to the peak diode current, I_p. Assuming that the DC output voltage is approximately V_p (i.e., the ripple is small), we have an average output current of $I_O \approx V_p/R_L$. The total charge drawn by the load each period must equal the total charge supplied by the diode current pulse. Therefore,

$$\frac{t_c I_p}{2} \approx \frac{TV_p}{R_L}, \qquad \textbf{(12.5)}$$

which leads to

$$I_p \approx \frac{2V_p T}{R_L\, t_c} = I_O \frac{2T}{t_c}. \qquad \textbf{(12.6)}$$

Plugging in the values for our example, we find that $t_c \approx 0.84$ ms and $I_p \approx 28$ mA. These values agree reasonably well with the simulation results (*see* Fig5.sch on the CD). In addition to this peak current, we see from Figure 12-5 that the transient current through the diode when the power supply is first turned on can be much larger. The exact value depends on where during the cycle the circuit is first connected. The diode we choose to use must be capable of handling this large ***surge current***.

We also need to know the ***peak inverse voltage*** (PIV) that the rectifier diode will be subjected to. Assuming that $V_p \approx V_m$ and the ripple is small, we see from Figure 12-5 that the PIV is approximately twice V_m. The reverse-breakdown voltage of the diode must be greater than the PIV.

Finally, we need to consider the power dissipated in the diode. Real diodes have some small series resistance. Since the diode current can have a large peak value, the power dissipation of the small series resistor can dominate, and we need to include this resistance in our model (*see* Problem P12.4). We can reduce the power dissipated in the diode by using a smaller filter capacitor, since this leads to a longer conduction time and therefore a lower peak current. The trade-off is that the ripple is increased. An alternative approach is to use the full-wave rectifier presented next.

Example 12.2

The purpose of this example is to show how a rectifier can be used to demodulate an amplitude-modulated (AM) signal. Amplitude modulation is introduced in Aside A11.1.

Figure 12-7 shows a half-wave rectifier circuit, an AM input, and the resulting simulated output. The time constant of the rectifier is chosen to be small enough that the output can follow changes in the envelope of the input, but large enough that the ripple caused by the high-frequency carrier is not too large.

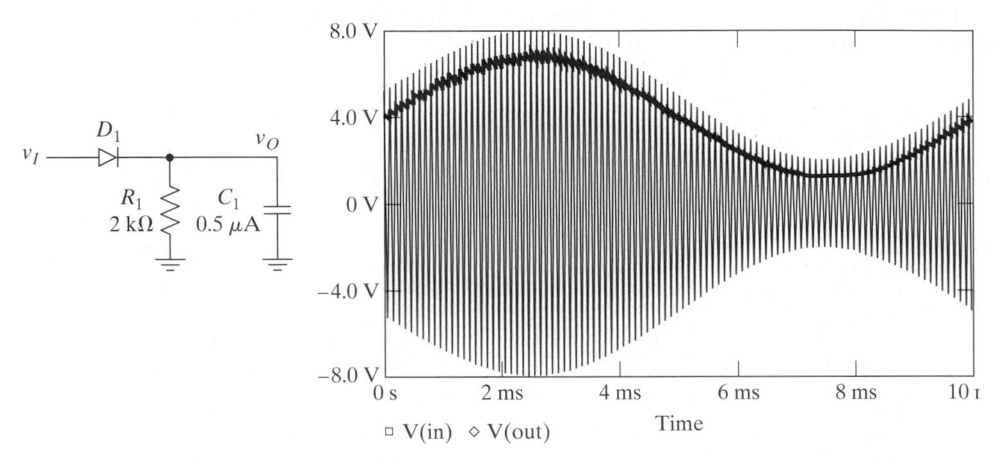

Figure 12–7 (a) An AM demodulator. (b) The input and output waveforms for a single-frequency modulating signal (f_{mod} = 100 Hz and f_c = 10 kHz).

It is interesting to note that the ripple in the output is larger on the positive half-cycle of the modulation. Since the output is always discharging to ground, the larger the output voltage is, the steeper the initial discharge is. Therefore, the ripple is always worse for larger outputs.

■

Full-Wave Rectifiers Full-wave rectifiers are more complex than half-wave rectifiers, but they achieve lower ripple with the same filter capacitance and have lower peak currents and lower average power dissipation in the diodes. A full-wave rectifier circuit is shown in Figure 12-8. The circuit uses a transformer with a center-tapped secondary and the center tap tied to ground as shown.[2] Because the secondary is one continuous winding, the voltages induced in the two halves are in phase; that is, when v_1 is positive, v_2 is as well. Using a transformer has the added

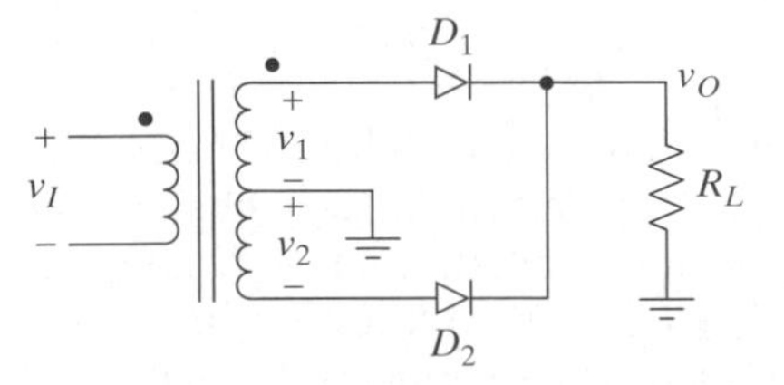

Figure 12–8 A full-wave rectifier circuit.

[2] For those not familiar with transformers, they are covered in most basic *RLC* circuit textbooks. One particularly good treatment is given in [12.2].

advantage of providing isolation between the input and output. This isolation is useful both for safety and, sometimes, for a situation in which the "grounds" available to the power source and the load are not the same.

During the positive half-cycle of the input voltage, both v_1 and v_2 are positive, so D_1 is forward biased and D_2 is reverse biased. Therefore, the load current flows through D_1, and v_O follows the input, except for the small offset voltage caused by the forward drop across D_1. During the negative half-cycle of the input, both v_1 and v_2 are negative, so D_2 is forward biased and D_1 is reverse biased. In this case, the current flows through D_2, and v_O follows the *negative* of the input, with a small offset caused by the forward drop across D_2. Input and output waveforms are shown in Figure 12-9 for a 60-Hz, 12-V peak input.

This figure shows that the circuit now provides an output pulse on both the positive and negative half-cycles of the input. The ripple frequency is now twice the input frequency (120 Hz for the waveform in Figure 12-8), and it is easier to filter it to achieve a given ripple magnitude. To illustrate this point, we simulated the full-wave rectifier of Figure 12-8, using the same component values as the half-wave rectifier in Figure 12-4; $R_1 = 16\ \text{k}\Omega$ and $C_1 = 1\ \mu\text{F}$. We plot the output voltages for both of these circuits, along with the unfiltered full-wave rectified output in Figure 12-10 (*see* Fig10.sch on the CD).

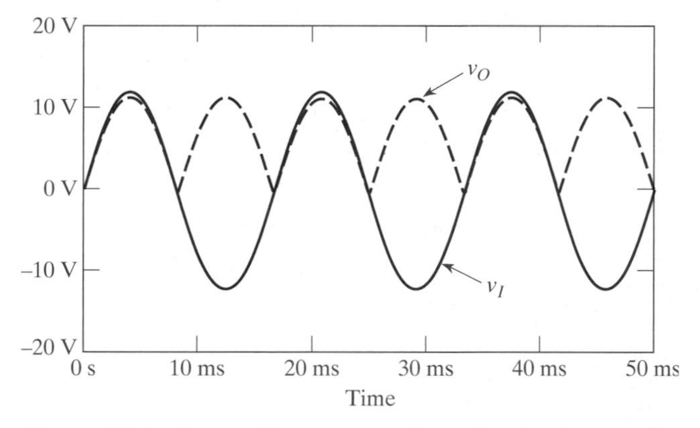

Figure 12–9 Input and output waveforms for the circuit in Figure 12-8. It is assumed that the secondary has twice as many turns as the primary, so that each half yields a voltage equal to v_I.

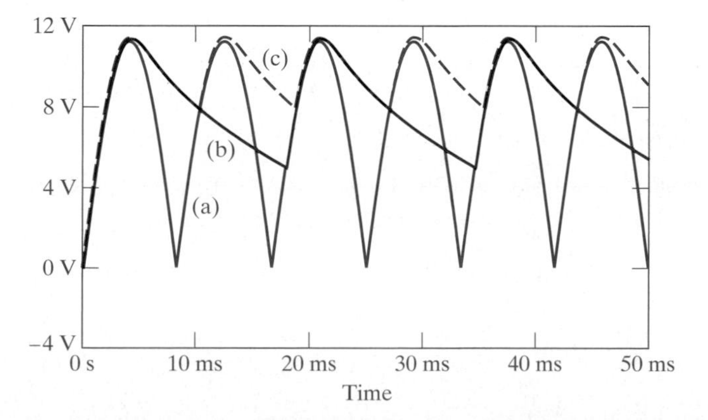

Figure 12–10 (a) The unfiltered full-wave rectified output, (b) the filtered half-wave rectified output, and (c) the filtered full-wave rectified output (see Fig10.sch on the CD).

Notice that the filtered half-wave rectified output skips every other peak in the full-wave rectified output. Since the filters are the same, however, the filtered outputs coincide for the initial part of the droop. If the ripple is small (smaller than shown here), the output voltage decays almost linearly, as assumed in deriving (12.2). For the small-ripple case, the full-wave rectified output has about one-half the ripple of the corresponding half-wave rectified output; that is,

$$V_r \approx \frac{V_p T}{2RC}. \tag{12.7}$$

Because the full-wave rectified and filtered ripple is smaller, the conduction time of each diode is also smaller (roughly half) than the diode in the half-wave rectifier. The simulated diode currents in the two circuits are plotted in Figure 12-11 along with the output voltages; in this case, both circuits have $R_1 = 16\ \text{k}\Omega$ and $C_1 = 21\ \mu\text{F}$, as determined in Exercise 12.1, to produce 5% ripple in the half-wave circuit (*see* Fig12-11.sch on the CD). Note that the initial current surge is the same in both circuits, but after steadystate has been achieved, the current spikes are about half as large in the full-wave circuit.

In summary, the full-wave rectifier can achieve the same ripple as a half-wave rectifier with a capacitor only half as large. In addition, the diode currents in steady state are about half as large (for the same filter, which means about half as much ripple). However, the PIV seen by the diodes in the full-wave rectifier is the same as in the half-wave circuit.

A negative-output full-wave rectifier is shown in Figure 12-12. The operation is identical to the circuit in Figure 12-8, except that the diodes conduct on the opposite half-cycles and the resulting output voltage is negative.

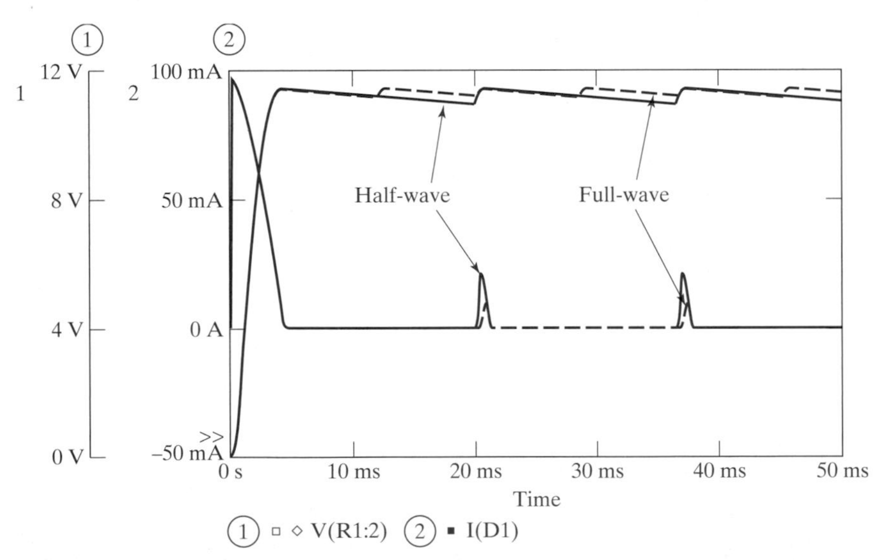

Figure 12–11 Full-wave and half-wave rectified and filtered outputs and the corresponding diode currents. (Only the current in D_1 is shown for the full-wave rectifier–see Fig11.sch on the CD.)

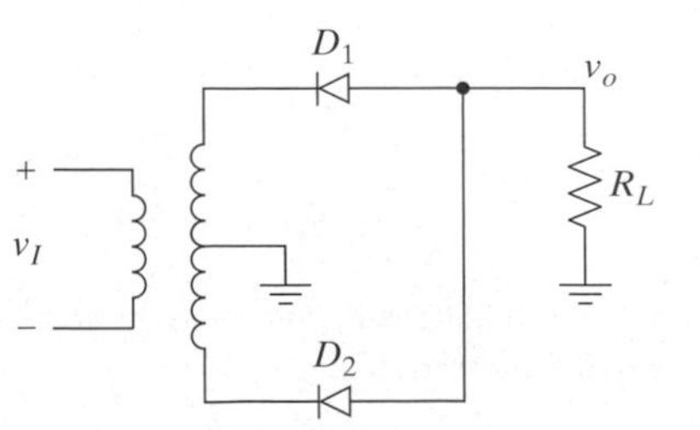

Figure 12–12 A negative-output full-wave rectifier.

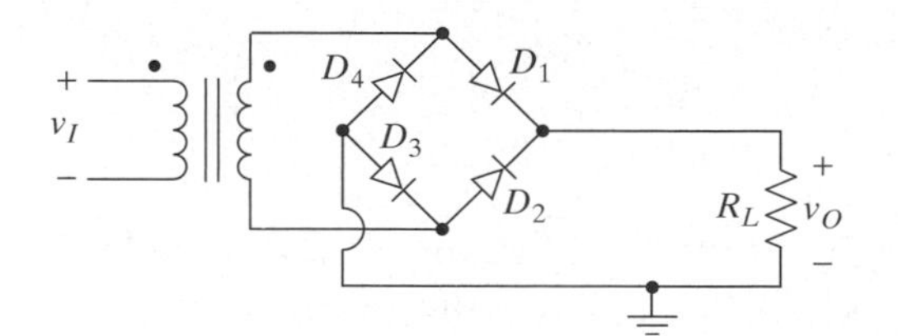

Figure 12–13 A full-wave bridge rectifier circuit.

Another common type of full-wave rectifier uses a diode bridge circuit as shown in Figure 12-13. This circuit has the advantages of not requiring a center-tapped transformer and reducing the PIV seen by the diodes by a factor of two.

During the positive half-cycle of the input, D_1 and D_3 are conducting while D_2 and D_4 are off. During the other half-cycle, D_2 and D_4 are conducting while D_1 and D_3 are off. In either case, the current flows through the load in the same direction, producing a positive full-wave rectified version of the input. A negative output voltage can be obtained simply by moving the ground to the top of R_L and taking the output from the bottom of it. The output waveform has the same shape as in Figure 12-9, but there are now two diode drops to account for rather than one, so the output voltage is a bit lower. The conduction times of the diodes and the currents are the same as in the full-wave rectifier of Figure 12-8, but the PIV seen by each diode is only half as large (*see* Problem P12.6).

12.1.2 Limiting, Clamping, and Multiplying Circuits

Diodes are frequently used to limit and clamp waveforms to certain levels. In addition, they can be used in voltage multipliers. In this section, we explore some of the most common applications.

Limiting Circuits A diode *limiter*, or *clipper*, circuit is shown in Figure 12-14. When the output voltage is less than V_r plus the forward drop on the diode, the diode is off and $v_O = v_I$. When v_I is large enough to turn the diode on, however, the diode will conduct current, which causes a voltage to develop across R_1. Since the diode current can increase rapidly with small increases in the voltage across it, and since any increase in the current causes an increase in the drop across R_1, the output voltage does not change much once the diode turns on. As a result, the output is limited, or clipped, as shown in the figure. (We assumed that the diode drop was negligible when we drew the output waveform.)

Other clipping circuits and some applications are examined in the problems.

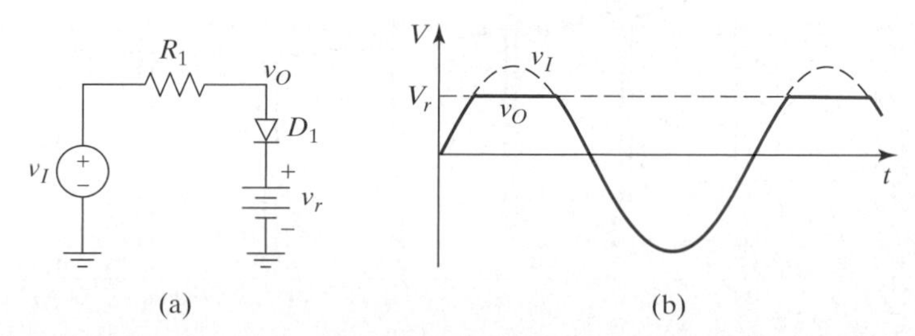

Figure 12–14 (a) A diode limiter circuit and (b) the associated waveforms. (We assume that $V_r >> v_D$.)

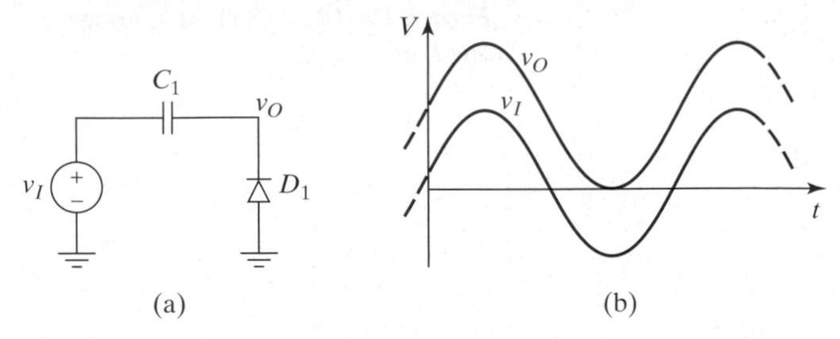

Figure 12–15 (a) A clamping circuit and (b) the input and output waveforms.

Clamping Circuits A diode *clamp*, or level restoring, circuit is shown in Figure 12-15 along with the input and output waveforms. For simplicity, assume that the diode is ideal for the discussion that follows, and assume also that the capacitor is initially uncharged. When the input goes negative, the capacitor will charge until the negative peak is reached. As the input amplitude decreases from this peak, the diode turns off and the current stops. Once the capacitor is charged to the peak negative value of the input, the diode will be off all of the time, and the output voltage will be the same as the input, but shifted so that the output never goes negative. The output is clamped to always being a positive voltage. If the diode is not ideal, the minimum output voltage is negative and equal to one diode drop, whereas the maximum positive output voltage is twice the peak input minus one diode drop (*see* Problem P12.12).

This circuit is also sometimes called a ***DC restore*** circuit. The name arises because of a problem called ***baseline wander***. Baseline wander occurs, for example, in wireline communication systems (e.g., in Ethernet over twisted pairs) where transformers are used to couple the signals to the line. Since a transformer cannot pass DC, if the signal transmitted over the line contains a DC component, the received signal will not be a faithful reproduction of what was sent, and we may have trouble detecting the data. We can encode the data to ensure that no DC component exists, but there will often still be a slowly varying nonzero average value. Since transformers severely attenuate low frequencies, the resulting slowly time-varying average value is not transmitted, and as a result, the received signal exhibits baseline wander.

This situation is illustrated in Figure 12-16. Here a binary signal (−10 or +10) is sent through the transformer. We assume that prior to the time shown, the signal was rapidly alternating between +10 and −10 and had zero average value. But starting with the time shown, the signal contains more +10 values than −10 values

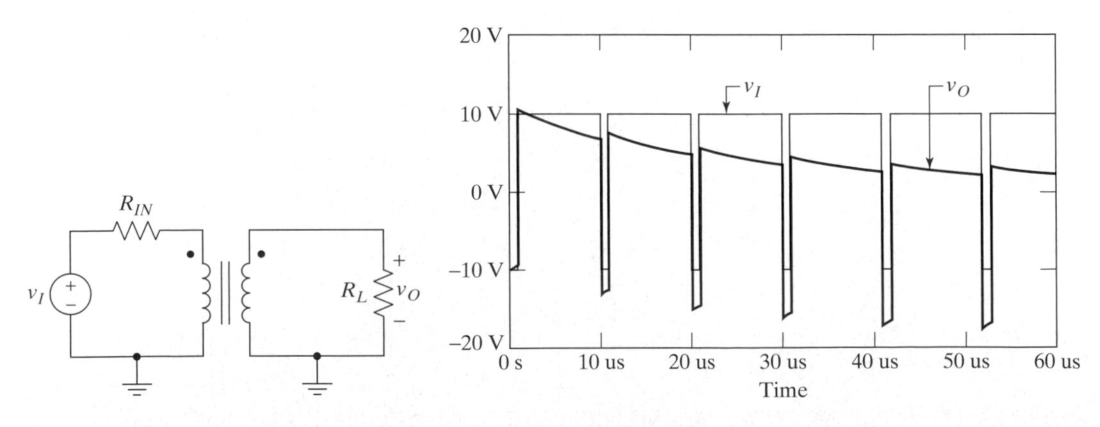

Figure 12–16 (a) A signal being sent through a transformer. (b) Associated waveforms.

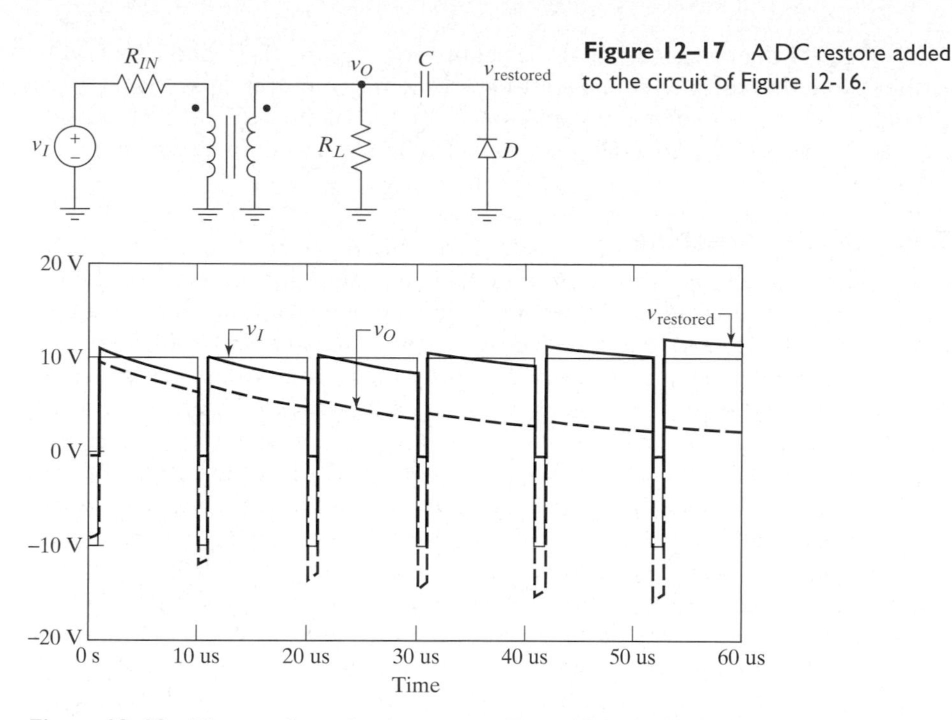

Figure 12–17 A DC restore added to the circuit of Figure 12-16.

Figure 12–18 The waveforms for the circuit in Figure 12-17 (see DC_Wander.sch on the CD).

and therefore has a slowly increasing positive average value. This slowly varying average cannot pass through the transformer, so the waveform at the output is shifted down as shown to establish an average value of zero.

If we pass this output through a DC restore circuit as shown in Figure 12-17, the restored output appears as shown in Figure 12-18 (*see* DC_Wander.sch on the CD). Notice that although the droop on each individual pedestal still shows up, the long-term drift in the average value is substantially removed.

Multiplying Circuits Diodes can also be used to produce DC output voltages that are higher than the peak AC input. Notice in Figure 12-15 that the peak value of the output voltage is twice that of the input. Therefore, if we follow a clamp circuit by a rectifier, we can double the input voltage. One such ***voltage-doubler*** circuit is shown in Figure 12-19, along with the associated waveforms for a 10-V peak sinusoidal input (*see* Doubler.sch on the CD).

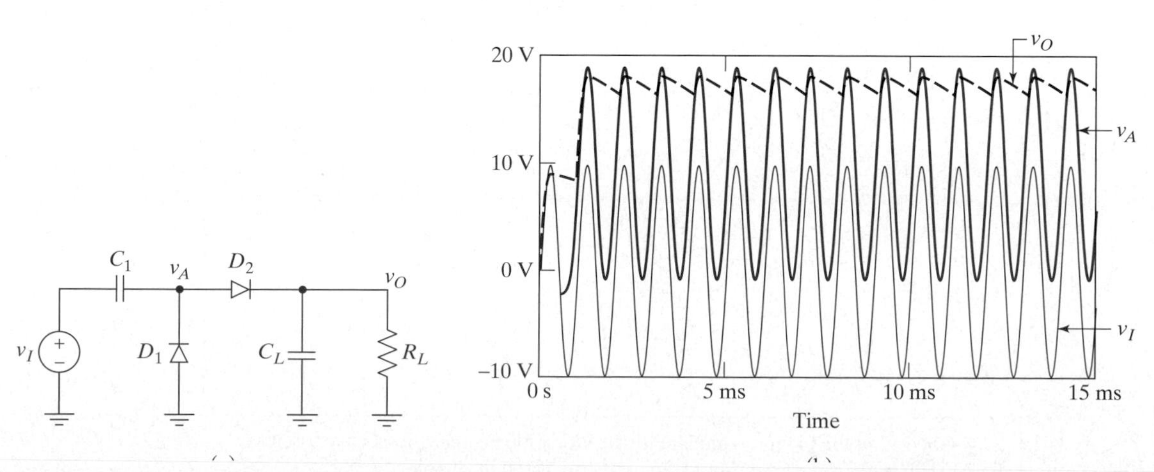

Figure 12–19 A voltage-doubler circuit and its associated waveforms (see Doubler.sch on the CD).

+5 V
R_1
v_A
v_B
v_O
(a)

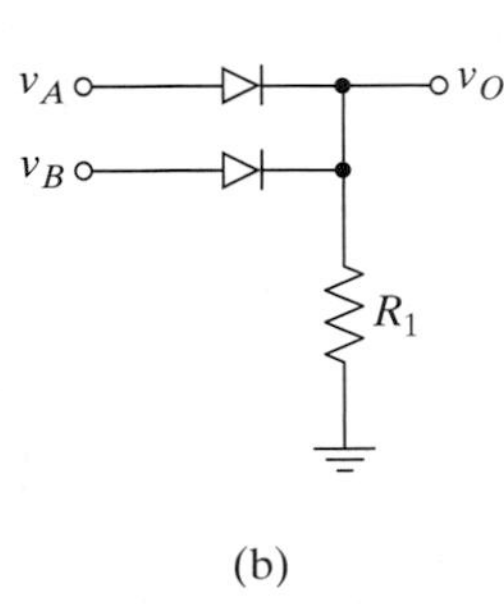

(b)

Figure 12–20 (a) AND circuit. (b) OR circuit.

This circuit is the combination of the clamp from Figure 12-15 and a half-wave rectifier. Note that after an initial start-up transient, the output approaches twice the peak input minus two diode drops—one from the clamp and one from the rectifier. This idea can also be extended to quadruple the input voltage, as shown in Problem P12.15.

12.1.3 Diode Switching

Since diodes conduct significant currents only in one direction, they can be used as switches in many applications. For example, it is possible to perform simple logic functions with diodes and to use them as switches in a sample-and-hold operation.

Two circuits that implement logic functions using diodes are shown in Figure 12-20. The circuit in part (a) of the figure implements a logical AND.[3] In other words, the output voltage is high (5 V) only when both inputs are high (5 V); otherwise, the output is low (0 V). If either of the inputs is low (0 V), then the output will also be low (about 0.7 V). Since the output low level is higher than the input low level, we could not cascade circuits like this. Therefore, it is not usable as a general-purpose logic gate. Nevertheless, it is sometimes useful to perform a single logical operation with circuits similar to this, especially when the voltage levels involved will not allow the use of standard logic families without level shifting. The circuit in part (b) of the figure implements a logical OR function; the output is high (about 4.3 V) if either of the inputs is high (5 V) and is low (0 V) only if both inputs are low (0 V).

An example of using diodes to produce a sample-and-hold circuit[4] is shown in Figure 12-21. The current sources in this circuit can be switched off and on. We will assume that the two current sources and all four diodes are perfectly matched. When the current sources are on, the output voltage is equal to the input voltage and tracks any changes in it (so long as v_I doesn't change too fast). We can see this by considering what happens if v_O is not equal to v_I. Suppose, for example, that $v_I > v_O$; in this case, the voltages across D_3 and D_2 are greater than the voltages across D_4 and D_1. Consequently, there is a net current into the hold capacitor, which raises its voltage. The net current into the capacitor goes to zero only when the voltages across all four diodes are equal, which requires that $v_O = v_I$.

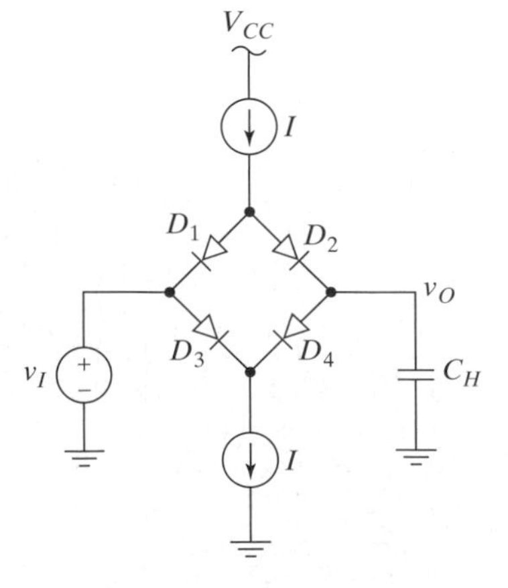

Figure 12–21 A diode-bridge sample-and-hold circuit.

[3] See Chapter 14 for a complete description of logic operations.
[4] The use of sample-and-hold circuits is discussed in Chapter 13.

12.2 AMPLIFIERS

In this section we investigate the large-signal performance of amplifiers, excluding special-purpose output stages. We see how to calculate the swing limits and efficiency of bipolar amplifiers in Section 12.2.1 and of FET amplifiers in Section 12.2.2. In Section 12.2.3, we show how to calculate harmonic distortion.

We begin with a brief generic discussion of large-signal midband analysis techniques.[5] Consider Figure 12-22, which shows a generic transistor amplifier (minus biasing details) with both voltage and current gain.

To calculate the maximum swing available at the load, we need to develop a large-signal midband model for the coupling capacitor. Because there can't be any DC current through the capacitor, the DC voltage on the load must be zero (i.e., $V_L = 0$). The DC voltage at the output of the transistor is V_O, so the DC voltage across the coupling capacitor is $V_{COUT} = V_O$. If the voltage at the output of the transistor has a sinusoidal signal added to the DC bias value, the voltages appear as shown in Figure 12-23. Because the capacitor is so large, the circuit is not capable of moving enough charge onto or off of it to change its voltage noticeably during one period of the signal. (Remember that $Q = CV$, so for a large C, you need to move a lot of charge to see any appreciable change in voltage.)[6] In other words, the voltage across the capacitor is approximately constant, as shown in the figure; that is $v_{COUT}(t) \approx V_{COUT} = V_O$.

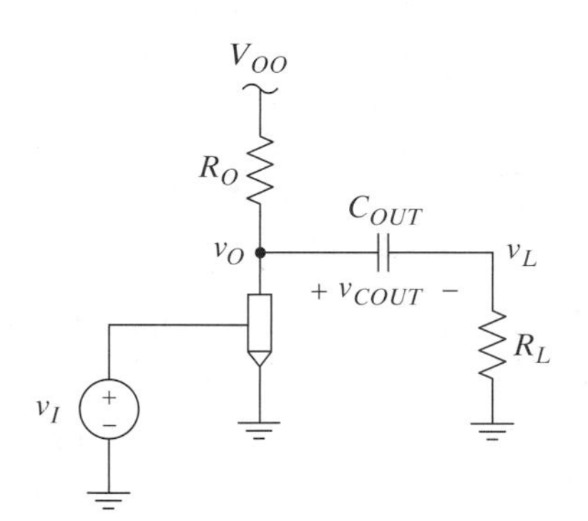

Figure 12–22 A generic transistor amplifier with both voltage and current gain.

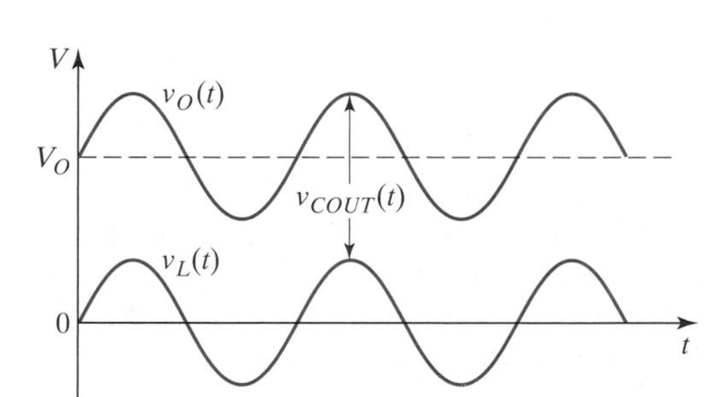

Figure 12–23 The voltages for the circuit in Figure 12-22 if a sinusoidal signal is used.

---

[5] We deliberately do not say midband AC analysis, because we are not going to limit ourselves to just the AC component of the signal. Rather, for large-signal analysis, we are concerned with the total instantaneous values of the voltages and currents. Therefore, we cannot set the DC power supplies to zero.

[6] An analogy may be useful here. Think of a large water reservoir: If you add water to it or pump water out of it using a fire hose, it will take a very long time to change the level of the reservoir. If you oscillate between pumping water out of and into the reservoir every hour, you will never cause a noticeable change in the level.

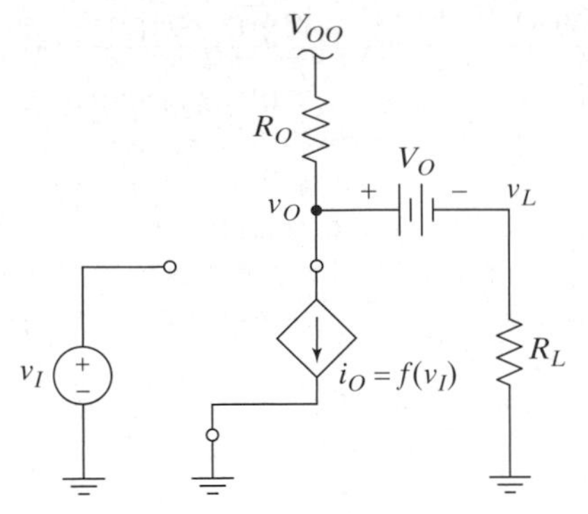

Figure 12–24 The large-signal equivalent circuit for Figure 12-22.

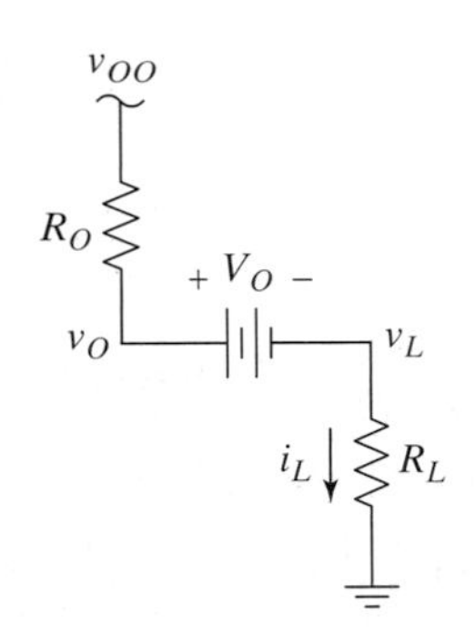

Figure 12–25 The instantaneous large-signal equivalent circuit for the amplifier in Figure 12-24 when the transistor output current is zero.

The fact that $v_{COUT}(t)$ is constant suggests a reasonable large-signal midband model for the coupling capacitor—that is, a battery. In Figure 12-24, we have replaced the capacitor by its model and have also replaced the generic transistor by its low-frequency large-signal model, a voltage-controlled current source.

Using this equivalent circuit, we can determine the maximum swing available at the output. The maximum instantaneous positive output voltage will occur when the transistor is turned off completely so that $i_O(t) = 0$. In this case, the output circuit appears as shown in Figure 12-25. *It is extremely important to note that this equivalent circuit is valid only for a single instant in time.* It is also important to notice that it is a large-signal midband equivalent circuit, so it necessarily includes the DC power supply. In other words, we are not only interested in the variations from the bias point, which require an AC equivalent circuit and would allow us to ground the DC power supply, but we are interested in the total instantaneous values of the voltages and currents.

We find $v_O(t)$ by writing KVL (we also drop the explicit reference to time, since the notation alone indicates which quantities are functions of time):

$$V_{OO} - i_L R_O - V_O - i_L R_L = 0 \ . \tag{12.8}$$

This leads to

$$v_{L\max} = \left(\frac{R_L}{R_L + R_O}\right)(V_{OO} - V_O). \tag{12.9}$$

The minimum output voltage (i.e., the maximum negative output in this case) will occur when the transistor pulls so much current through R_O that v_O drops far enough to cause the transistor to leave the forward-active region of operation. At that point the model for the transistor needs to be replaced by a model that depends on what type of transistor we are dealing with. Assume for now that v_O can drop to ground: therefore, $v_{L\min} = V_{COUT} = -V_O$.

In general, the positive and negative limits won't have the same magnitude. If we want the output to be an unclipped sine wave, we have to hold the peak value of the sine wave to the smaller of the two limits. We call the resulting maximum swing the **maximum symmetric swing** of the amplifier. The maximum symmetric swing available from this amplifier is given by

$$\begin{aligned} v_{L\max\_sym} &= \min\{v_{L\max}, |v_{L\min}|\} \\ &= \min\left\{\left(\frac{R_L}{R_L + R_O}\right)(V_{OO} - V_O), V_O\right\}. \end{aligned} \tag{12.10}$$

For amplifiers that need to deliver large amounts of power, it is often important to know how efficient they are in converting the DC power supplied to them into signal power in the load. In general, an amplifier is supplied with power by the signal source v_S and by the DC power supply[7] V_{OO}; call these powers P_S and P_{supply}, respectively. (These are the average values of those powers.) The power delivered by the signal is usually insignificant compared with the power supplied by the DC power supply, so it gets ignored in calculating the efficiency. The amplifier supplies signal power to the load, and we denote the average value by P_L. The *efficiency* of the amplifier, expressed as a percentage, is defined to be

$$\eta = \frac{P_L}{P_{\text{supply}}} \times 100\% \tag{12.11}$$

The total signal power in the load depends on what signal we provide, so there isn't a unique answer for the efficiency of the amplifier. Nevertheless, it is reasonable to see what the efficiency is if we drive the load as hard as possible (without causing clipping of the waveform); in other words, we use a sinusoidal signal with a peak value of $v_{L\text{max\_sym}}$. The average power delivered to a load resistor R_L by a sine wave with a peak value of $v_{L\text{max\_sym}}$, is

$$P_L = \frac{v^2_{L\text{max\_sym}}}{R_L T} \int_0^T \cos^2\left(\frac{2\pi}{T}t\right)dt = \frac{v^2_{L\text{max\_sym}}}{2R_L}. \tag{12.12}$$

Using this power for P_L in (12.11) yields the maximum efficiency.

12.2.1 Signal Swing in Bipolar Amplifiers

We will first investigate the three single-transistor amplifier configurations and will then discuss swing in cascaded amplifiers.

Common-Emitter Amplifiers Consider the common-emitter amplifier in Figure 12-26. The small-signal performance of this amplifier was analyzed in Section 8.2.2. We are interested here in determining the maximum symmetric swing and efficiency of the circuit.

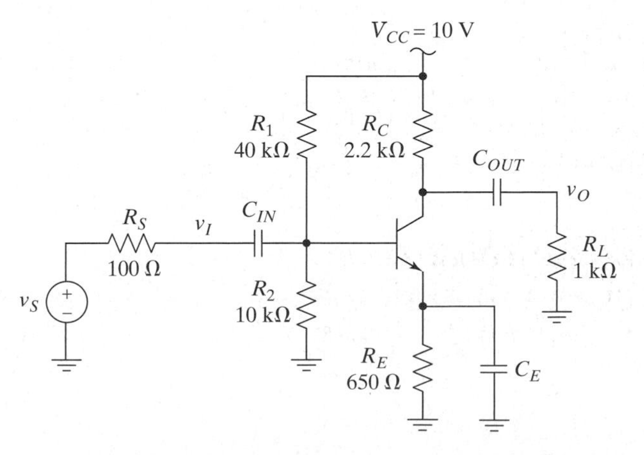

Figure 12–26 A bipolar implementation of an amplifier with both voltage and current gain.

[7] There may be more than one power supply, in which case we sum their powers.

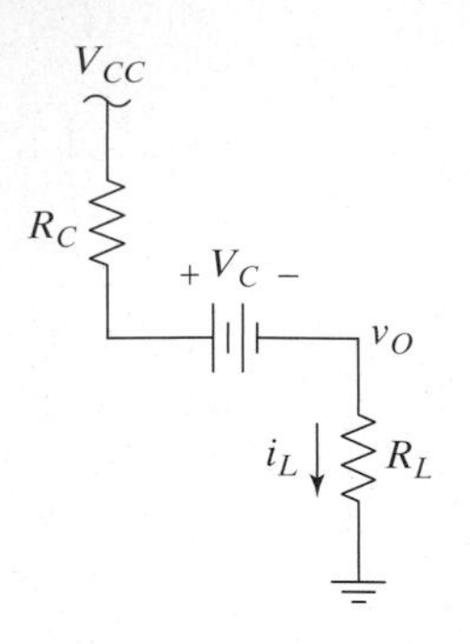

Figure 12–27 The output of the circuit in Figure 12-26 when the transistor is cut off.

The maximum output voltage for this circuit will occur when the transistor in Figure 12-26 is cut off. At that instant of time, all of the current in the load flows through R_C, and we have a resistive voltage divider formed by R_C and R_L, as shown in Figure 12-27. The total voltage across these two resistors is $(V_{CC} - V_C)$, since the DC voltage across the capacitor is equal to V_C. The resulting maximum instantaneous positive output voltage is

$$v_{L\max} = \left(\frac{R_L}{R_L + R_C}\right)(V_{CC} - V_C). \quad \textbf{(12.13)}$$

Plugging in the numbers, we get $v_{L\max} = (1/3.2)(4.4) = 1.38$ V. The positive swing capability is low because R_L is smaller than R_C and the voltage divider formed by these two resistors shows up in the equation. We will return to this point later. An alternative representation of (12.13) is sometimes useful. Note that the DC voltage across R_C is $V_{R_C} = (V_{CC} - V_C) = I_C R_C$; substituting this into (12.13) yields

$$v_{L\max} = I_C\left(\frac{R_C R_L}{R_L + R_C}\right) = I_C(R_C \| R_L). \quad \textbf{(12.14)}$$

This equation shows that, for a given R_C and R_L, the maximum swing is proportional to the DC bias current in the transistor.

The minimum output voltage occurs when the transistor saturates heavily. When that happens, the collector voltage will be equal to the emitter voltage plus $V_{CE\text{sat}}$. Using $V_{CE\text{sat}} = 0.2$ V as our model, we find that

$$v_{C\min} = V_E + V_{CE\text{sat}} = 1.3 + 0.2 = 1.5 \text{ V}. \quad \textbf{(12.15)}$$

The output voltage at this time will be

$$v_{L\min} = v_{C\min} - V_C = 1.5 - 5.6 = -4.1 \text{ V}. \quad \textbf{(12.16)}$$

Since the peak negative value is larger than the peak positive value, the positive value limits the maximum symmetric swing at v_L to be $v_{L\max\_sym} = 1.38$ V.

At this point, we can reexamine our choice of DC collector voltage. In Chapter 7, we indicated that choosing V_C halfway between V_{CC} and V_E is reasonable; we are now in a position to see why. If we set $R_L \gg R_C$ in the present example, then (12.13) leads to $v_{L\max} \approx V_{CC} - V_C$. To obtain maximum symmetric swing, we would like the positive and negative swing limits to be the same; therefore, using (12.13) and (12.16) and assuming that $V_{CE\text{sat}} = 0$ for simplicity, we set

$$\begin{aligned} v_{L\max} &= -v_{L\min} \\ V_{CC} - V_C &= V_C - V_E \end{aligned} \quad \textbf{(12.17)}$$

and solve to obtain

$$V_C\Big|_{\text{max. symmetric swing}} = \frac{V_{CC} + V_E}{2}. \quad \textbf{(12.18)}$$

For a good design, the condition $R_L \gg R_C$ is often met, so (12.18) is a reasonable way to choose V_C if no other specification requires a different choice. In the present example, however, R_L is not much larger than R_C, and this result is not correct. We can solve for the value of V_C that will produce maximum symmetric swing when R_L and V_{CEsat} are accounted for, but the solution depends on R_C, which is itself a function of V_C and I_C. Therefore, we either need to solve iteratively or hold the value of R_C constant (*see* Problem P12.20).

We now turn our attention to the efficiency of this amplifier. To calculate the efficiency, we need to determine the total power delivered by the DC power supply (V_{CC}). By summing currents at V_{CC} in Figure 12-26, we see that the DC current drawn from the supply is I_C plus the current through R_1. We know that $I_C = 2$ mA. We also know that the base current of the transistor is negligible compared with the current in R_1, so the current through R_1 and R_2 can be approximated by disconnecting the base of the transistor. The result is that $I_{R1} \approx V_{CC}/(R_1 + R_2) =$ 200 μA. Therefore, the total power delivered by the supply is $P_{supply} = (2.2\text{ mA})(10\text{ V}) =$ 22 mW.

From (12.12), we find the average power delivered to the load using a sinusoid with a peak value equal to the maximum symmetric swing (1.4 V peak):

$$P_L = \frac{(1.4)^2}{R_L T}\int_0^T \cos^2\left(\frac{2\pi}{T}t\right)dt = \frac{(1.4)^2}{2R_L} = 1.0\text{ mW}. \quad \textbf{(12.19)}$$

Using this signal power in the load, along with the supply power of 22 mW, we find the maximum efficiency possible with an undistorted sine wave to be

$$\eta_{max} = \frac{0.001}{0.022} \times 100\% = 5\%. \quad \textbf{(12.20)}$$

This efficiency is very low and points out that this amplifier is not a good choice for an output stage driving a load resistance as small as is used here. There are three problems with this stage in terms of efficiency. First, some of the available swing is dropped across R_E in order to obtain a stable bias point. Second, the divider between R_C and R_L causes a drop in available output swing. Third, the transistor is on all of the time and dissipates significant power. We can improve the efficiency of the stage somewhat with a different design, but it will never be very good. We investigate more efficient output stages in Section 12.3.

Exercise 12.2

By examining (12.13), we might conclude that we can improve the swing of this circuit by decreasing R_C, but this exercise will demonstrate that that is not the case unless we also change the gain. Calculate the value of I_C required for the circuit in Figure 12-26 to keep the same voltage gain (i.e., $A_v = -53$) with R_C reduced to 1 kΩ. What is the maximum swing possible with this stage now. (Assume that we keep $V_E = 1.3$ V for bias stability.)

□

Example 12.3

Problem:
Run a SPICE simulation to check the swing results obtained for the circuit in Figure 12-26.

Solution:
We performed the simulation using the Nsimple transistor model described in Appendix B and 100-μF coupling and bypass capacitors. The PROBE plot of v_O for the transient simulation is shown in Figure 12-28. From this figure, we determine that the swing limits are $v_{L\min} = -4.72$ V and $v_{L\max} = 1.25$ V, which agree fairly well with our approximations. The simulated minimum limit is lower for two reasons: V_C is higher in the simulation, and the transistor is heavily saturated in the simulation, with V_{CEsat} much less than 0.2 V.

Common-Base Amplifiers A common-base amplifier is shown in Figure 12-29. This is the same amplifier examined in Section 8.4.2, except that we have shown a separate load here while keeping the AC load (i.e., $R_C \| R_L$) the same.

We leave it as an exercise to draw out the large-signal equivalent circuits for finding the maximum and minimum swing limits for this circuit. The results are the same as we obtained for the common-emitter amplifier in (12.13) and (12.16).

Common-Collector Amplifiers (Emitter Followers) An emitter follower is shown in Figure 12-30. We analyzed the small-signal midband performance of this stage in Section 8.3.2. We found there that the DC bias current is $I_C = 1$ mA and that $V_E = 1.3$ V.

The large-signal midband model for the circuit of Figure 12-30 is shown in Figure 12-31, where we have assumed that the transistor is forward active and have used the fact that the large-signal midband model of the coupling capacitor is a battery. We have also simplified the circuit by driving the base of the transistor directly from a source v_B. The voltage across the coupling capacitor is V_E, since V_O must be zero.

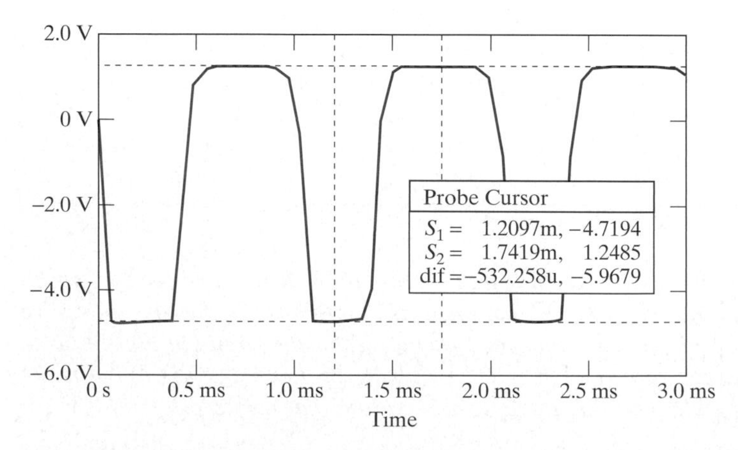

Figure 12–28 The PROBE transient simulation result for v_{OUT}. Cursor S_1 is set to show the negative swing limit, and S_2 shows the positive limit.

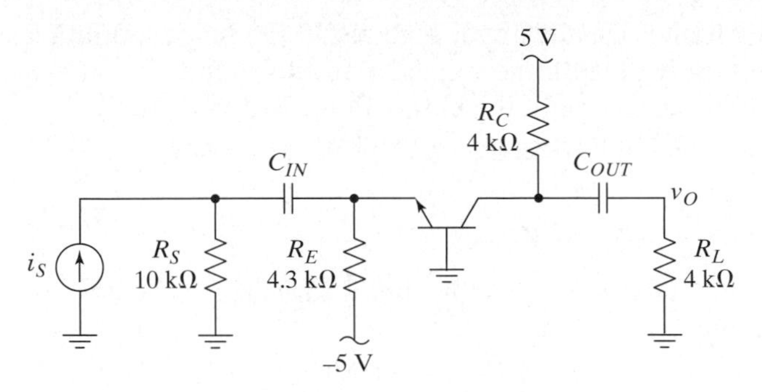

Figure 12–29 A common-base amplifier.

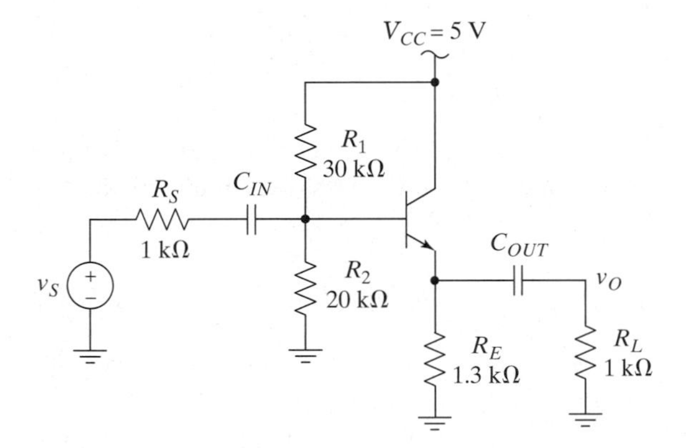

Figure 12–30 An emitter follower.

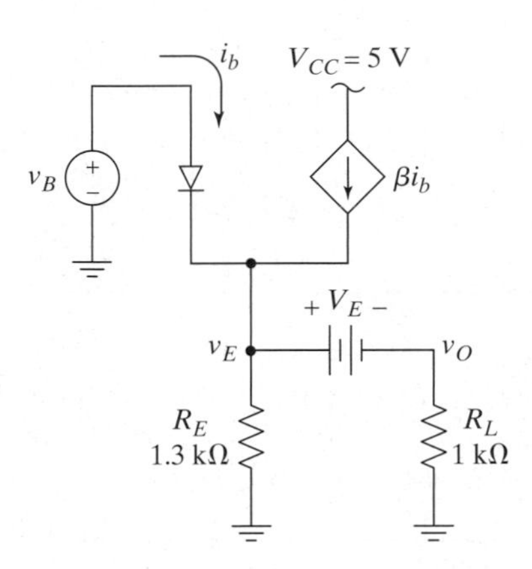

Figure 12–31 The large-signal equivalent circuit for the emitter follower.

The maximum instantaneous output voltage does not appear to have any limit, given the circuit in Figure 12-31, but that is not really the case. If v_B is taken too high, the base-collector junction of the transistor will become forward biased, and the transistor will saturate. The upper limit is therefore found by assuming that the transistor is saturated. When that occurs, we have

$$v_E = V_{CC} - V_{CEsat}, \tag{12.21}$$

which is equal to 4.8 V in this example. With this value of emitter voltage, the output voltage is

$$v_{Omax} = v_E - V_E = V_{CC} - V_{CEsat} - V_E, \tag{12.22}$$

which is 3.5 V in our example (assuming, as usual, that $V_{CEsat} = 0.2$ V).

The practical upper swing limit is usually not as high as (12.22), since saturating the transistor requires the base voltage to go higher than the supply, V_{CC}. If the stage driving the base uses the same supply, the base voltage is usually limited by V_{CC}. When this is the case, the maximum output is given by

$$\begin{aligned} v_{O\max} &= v_{B\max} - v_{BE} - V_E \\ &\approx V_{CC} - 0.7 - V_E. \end{aligned} \tag{12.23}$$

The lower output limit occurs when v_B goes so low that it turns the transistor off. With the transistor cut off, the current being injected into v_E is zero, so the node voltage is as low as possible. The equivalent circuit is shown in Figure 12-32. In this case, the two resistors are in series with each other and with the coupling capacitor, which acts like a battery. The minimum (maximum negative) output voltage is given by the resulting resistor divider:

$$v_{O\min} = \frac{-R_L}{R_L + R_E} V_E. \tag{12.24}$$

The result is -0.57 V for our example. An alternative representation of (12.24) that is often useful is found by substituting $V_E = I_E R_E$ into the equation. The result is

$$v_{O\min} = I_E \frac{-R_L R_E}{R_L + R_E} = -I_E(R_L \| R_E). \tag{12.25}$$

This result shows that for a given set of resistor values, the minimum achievable swing is proportional to the DC bias current in the transistor.

Using the maximum symmetric swing, we can calculate the maximum efficiency of the circuit. The total supply current is $I_C + I_{BIAS} = 1.1$ mA, so the power delivered by the supply is $P_{\text{supply}} = 5.5$ mW. The maximum average load power with a 0.57-V peak sine wave is 1.62 μW, and the efficiency is $\eta = 0.03\%$, which is extremely low. We can improve that efficiency significantly by choosing a different bias point, as shown in Exercise 12.3, but more sophisticated output stages are needed if high efficiency is desired. These stages are discussed in Section 12.3.

Exercise 12.3

Assume that (12.23) is accurate, and use it along with (12.24) to determine the value of V_E that will maximize the symmetric swing capability of the emitter follower in Figure 12-30. Determine the maximum efficiency with this new value of V_E.

Cascaded Amplifiers In this section, we use examples to combine and extend the results of the previous sections to examine the swing limits of cascaded amplifiers.

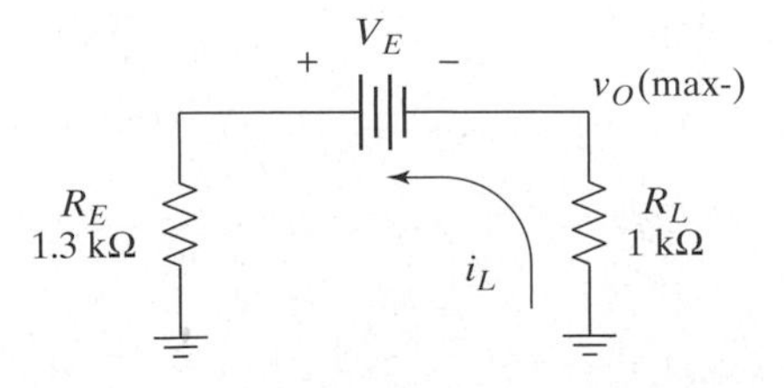

Figure 12–32 The equivalent circuit when the transistor is cutoff.

Example 12.4

Problem:
Consider the common-emitter emitter-follower cascade shown in Figure 12-33; note that the DC bias voltages and currents are shown on the schematic. This amplifier was designed in Section 8.7.1. Determine the swing limits for the amplifier.

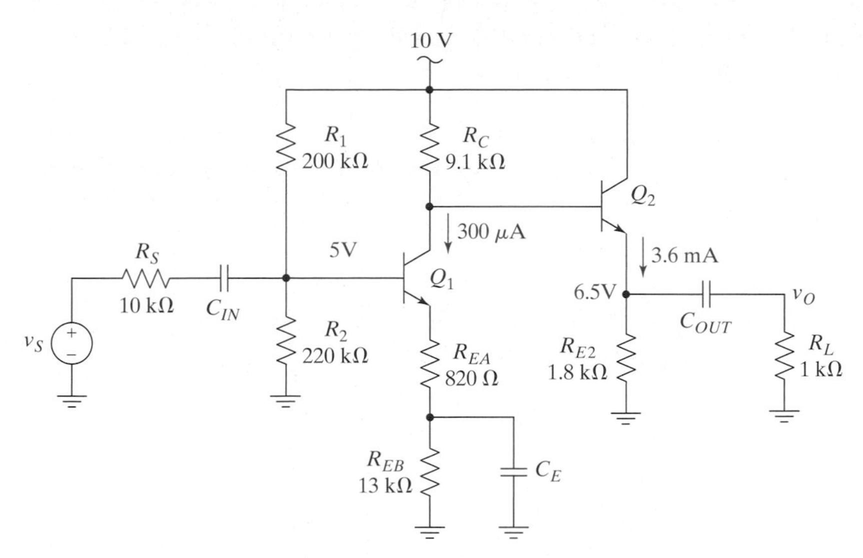

Figure 12–33 A common-emitter emitter-follower cascade.

Solution:
The swing may be limited by either stage, so we must determine the swing limitations of each stage. If the base of Q_2 can swing far enough, the emitter follower will limit the maximum and minimum output voltages to the values given by (12.23) and (12.24), respectively. For this example, the limits imposed by the emitter follower are $v_{O\text{max}}(\text{EF}) = 2.8$ V and $v_{O\text{min}}(\text{EF}) = -2.3$ V.

The common-emitter stage is not exactly like the one we analyzed before, so we need to derive new equations for this circuit. If the input swings low enough to cause Q_1 to cut off, the maximum output voltage from the circuit can be found by using the equivalent circuit shown in Figure 12-34(a). This circuit can be rearranged by making use of impedance reflection as shown in part (b) of the figure to facilitate writing the node equation for $v_{E2\text{max}}$.

The node equation for $v_{E2\text{max}}$ is ($R'_C = R_C/(\beta+1)$, as shown in the figure)

$$v_{E2\text{max}}\left(\frac{1}{R'_C}+\frac{1}{R_{E2}}+\frac{1}{R_L}\right)-\frac{V_{CC}-0.7}{R'_C}-\frac{V_{E2}}{R_L}=0, \tag{12.26}$$

and we solve this equation to obtain

$$v_{E2\text{max}}=\left(\frac{V_{CC}-0.7}{R'_C}+\frac{V_{E2}}{R_L}\right)(R'_C\|R_{E2}\|R_L). \tag{12.27}$$

Finally, the maximum instantaneous output, as dictated by the common-emitter stage, is

$$v_{O\text{max}}(CE) = v_{E2\text{max}} - V_{E2}. \tag{12.28}$$

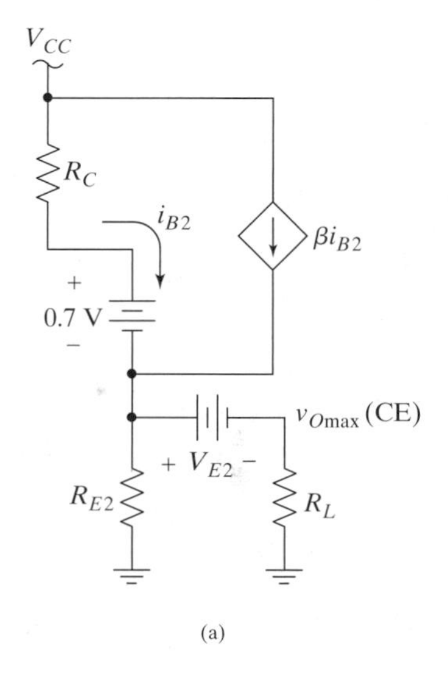

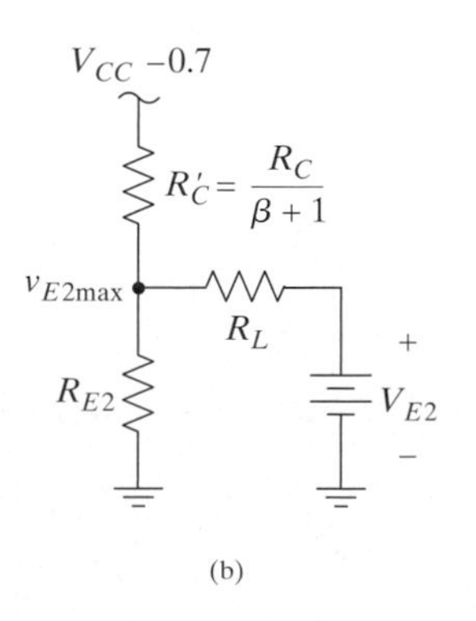

Figure 12–34 (a) Large-signal midband equivalent circuit when Q1 is cutoff. (b) The circuit rearranged to write the node equation for v_E2.

For the example at hand, this evaluates to 2.2 V. Since this limit is smaller than that set by the emitter follower, it indicates that the collector of Q_1 can't swing far enough for the emitter follower to limit the swing, and the maximum output of the overall circuit is 2.2 V.

If the input swings high enough to saturate Q_1, the resulting equivalent circuit can be drawn as shown in Figure 12-35 if we assume that the base current of Q_1 is negligible compared with its collector current. Although this approximation is not great if the transistor is heavily saturated, it is certainly acceptable most of the time.

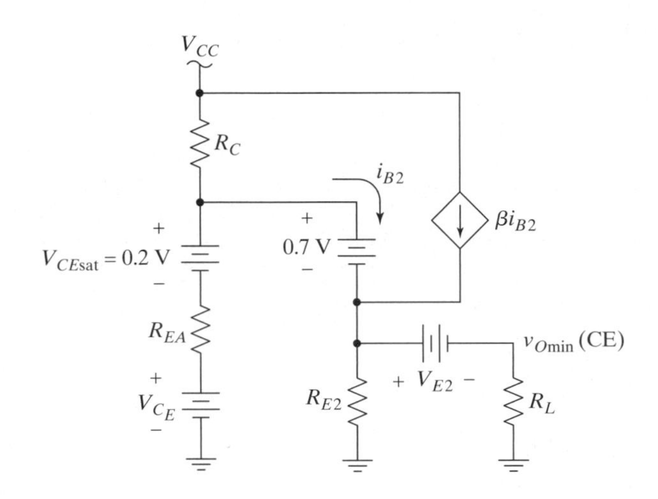

Figure 12–35 Large-signal midband equivalent circuit when Q_1 is saturated and the base current of Q_1 is ignored.

This circuit can be solved in the same way as Figure 12-34(b). (Find a Thévenin equivalent for everything to the left of the base of Q_2 and the circuit is topologically identical to the previous one.) The result is

$$v_{O\min}(\text{CE}) = \left(\frac{V_{Th} - 0.7}{R_{Th}} + \frac{V_{E2}}{R_L}\right)(R_{Th}\|R_{E2}\|R_L) - V_{E2}, \tag{12.29}$$

where

$$V_{Th} = V_{C_E} + 0.2 + (V_{CC} - V_{C_E} - 0.2)\frac{R_{EA}}{R_{EA} + R_C} \tag{12.30}$$

and

$$R_{Th} = R_{EA}\|R_C. \tag{12.31}$$

Plugging in, we find that $v_{O\min}(\text{CE}) = -2.6$ V for this example. Since the limit imposed by the emitter follower is smaller, it dominates, and $v_{O\min} = -2.3$ V.

A SPICE simulation of this circuit using the Nsimple models from Appendix B reveals that the limits are 2.4 and −2.2 V, reasonably close to the values of 2.2 and −2.3 V determined here.

■

The procedure illustrated in the preceding example is common to all multistage amplifiers—namely, you must examine the swing limits of each stage (while it is connected to the adjacent stages) and see which ones limit the swing. If there is more than one stage with a voltage gain larger than unity, however, it would be unusual for any but the last stage to limit the swing since the signal swing is smaller in the early stages. For example, if two common-emitter stages are cascaded, and each stage has a gain of 10, then when the output signal is 1 V peak to peak, the signal at the output of the first stage is only 0.1 V peak to peak. Therefore, for the first stage to limit the overall swing capability, it would have to have a swing limit less than one-tenth that of the final stage.

The signal swing in a cascaded amplifier with global feedback is examined in the next example. If you have not yet covered feedback (Chapter 10), you may skip this example without any loss in continuity.

Example 12.5

Problem:

Consider the series–series triple shown in Figure 12-36. The small-signal performance of this amplifier was analyzed in Exercise 10.16; we now want to find the swing limits for the circuit. The DC bias point was also found in Exercise 10.16; we had $I_{C1} = 240\ \mu A$, $I_{C2} = 1.3$ mA, and $I_{C3} = 5.5$ mA, from which we obtain $V_{C1} = 2.8$ V, $V_{E1} = 1.8$ V, $V_{C2} = 6.1$ V, $V_{E2} = 2.1$ V, $V_{E3} = 5.4$ V, and $V_{C3} = 7.6$ V.

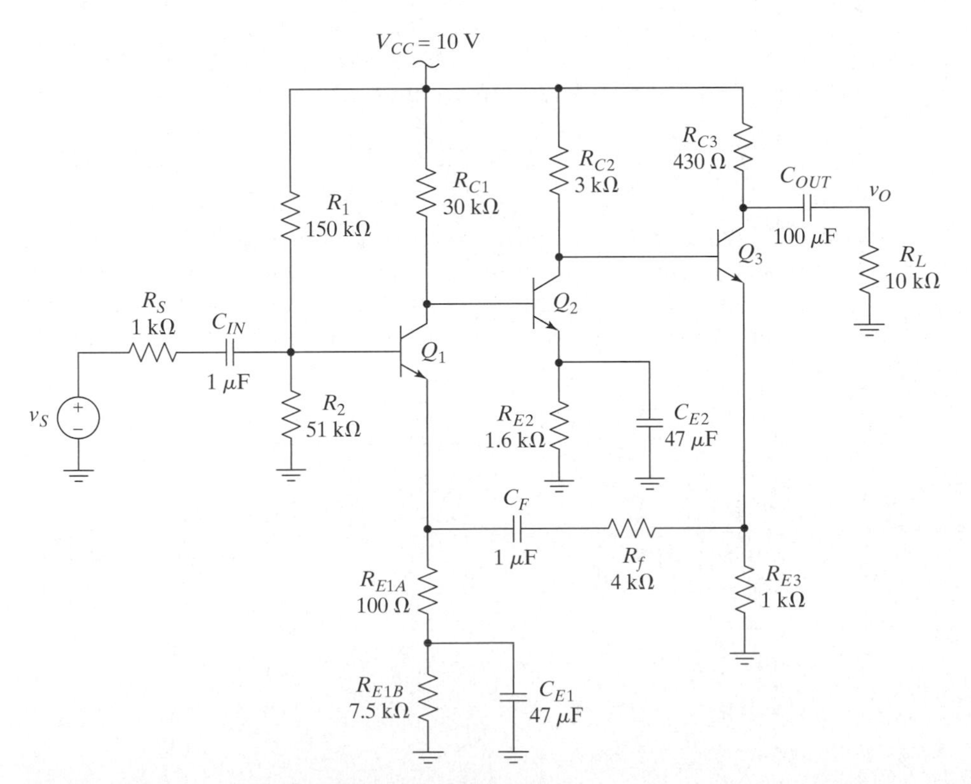

Figure 12–36 A series–series triple.

Solution:

We first note that the gain of each individual stage in the forward amplifier is unaffected by the global feedback. The overall gain is reduced because the input to the forward amplifier (the error signal) is reduced, not because the gain of the forward amplifier is reduced. The small-signal midband AC voltage gain of the final stage in the forward amplifier is

$$A_{v3} = \frac{v_{c3}}{v_{c2}} = -g_{m3}R'_L\left(\frac{r_{\pi3}}{r_{\pi3} + (\beta+1)R_{bi}}\right), \quad \textbf{(12.32)}$$

where $R'_L = R_L \| R_{C3} = 412\ \Omega$ and the feedback network input resistance is $R_{bi} = R_{E3}\|(R_{E1A} + R_F) = 804\ \Omega$. Plugging in $g_{m3} = 212$ mA/V, $r_{\pi3} = 472\ \Omega$, and $\beta = 100$, we find that $A_{v3} = -0.5$. Therefore, the signal at the collector of Q_2 will be twice as large as at the output. Similarly, the gain of the second stage is

$$A_{v2} = \frac{v_{c2}}{v_{c1}} = -g_{m2}(R_{C2}\|R_{i3}), \quad \textbf{(12.33)}$$

where $R_{i3} = r_{\pi3} + (\beta+1)\,R_{bi} = 81.7$ kΩ, and $g_{m2} = 50$ mA/V. This equation yields $A_{v2} = -145$, so the swing at the collector of Q_1 will be 145 times smaller than at the collector of Q_2. Therefore, we only need to examine the swing limits at the collectors of Q_2 and Q_3.

Examining the last stage first, we see that

$$v_{O\max}(Q_3) = (V_{CC} - V_{C3})\frac{R_L}{R_L + R_{C3}} = 2.3\text{ V}, \quad \textbf{(12.34)}$$

but we must remember that for the output to swing that far, the collector of Q_1 would need to swing down twice as far—that is, by −4.6 V. The minimum voltage at the collector of Q_2 occurs when Q_2 is saturated. The resulting equivalent circuit is shown in Figure 12-37. We see from this figure that

$$v_{C2\min} = V_{E2} + 0.2 = 2.3\text{ V}. \quad \textbf{(12.35)}$$

Since $V_{C2} = 6.1$ V, the maximum negative swing at that point is $6.1 - 2.3 = 3.8$ V. Therefore, the maximum positive swing at the collector of Q_3 is equal to $-3.8\,(A_{v3}) = 1.9$ V and is limited by the swing at the collector of Q_2. The maximum instantaneous output is equal to the maximum swing at the collector of Q_3, so $v_{O\max} = 1.9$ V.

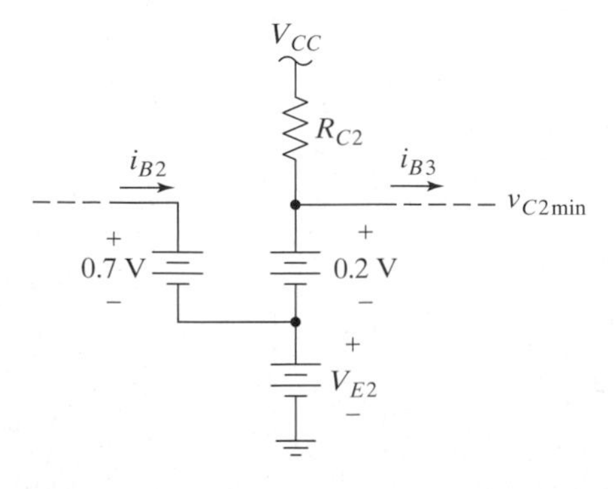

Figure 12–37 The equivalent circuit for finding $v_{C2\min}$.

The minimum voltage at the output is found when Q_3 saturates (if it can—we'll check that later). The equivalent circuit is shown in Figure 12-38(a). Since the overall voltage gain of the closed-loop amplifier is nearly 20, we can crudely approximate the input voltage as a constant, which leads to saying that v_{E1} is a constant. Therefore, the voltage on the right side of C_F is constant, and since the DC current through R_f must be zero, the DC voltage on either side of it must be the same and equal to $V_{E3} = 5.5$ V, which explains why we have a 5.5-V source in the figure. We have not shown what drives the base of Q_3, because we are going to ignore the base current in this approximate analysis. In part (b) of the figure, we used a Thévenin equivalent for the circuit in the emitter, ignored the base current (assuming that it is small compared with i_{C3}, even though the transistor is saturated), and labeled some currents.

The Thévenin equivalent is given by

$$V_{Th} = \frac{5.5R_{E3}}{R_{E3} + R_f} = 1.1 \text{ V and } R_{Th} = R_f \| R_{E3} = 800\ \Omega. \quad \textbf{(12.36)}$$

Writing KVL from the ground up through the transistor to V_{CC}, we find that

$$V_{CC} = V_{Th} + i_E R_{Th} + 0.2 + i_3 R_C. \quad \textbf{(12.37)}$$

Now we write KCL at the collector node to see that

$$i_E = i_3 - i_L. \quad \textbf{(12.38)}$$

We find the load current by writing KVL from ground to V_{CC} going through the load and solving to obtain

$$i_L = \frac{V_{CC} - V_{C3} - i_3 R_C}{R_L}. \quad \textbf{(12.39)}$$

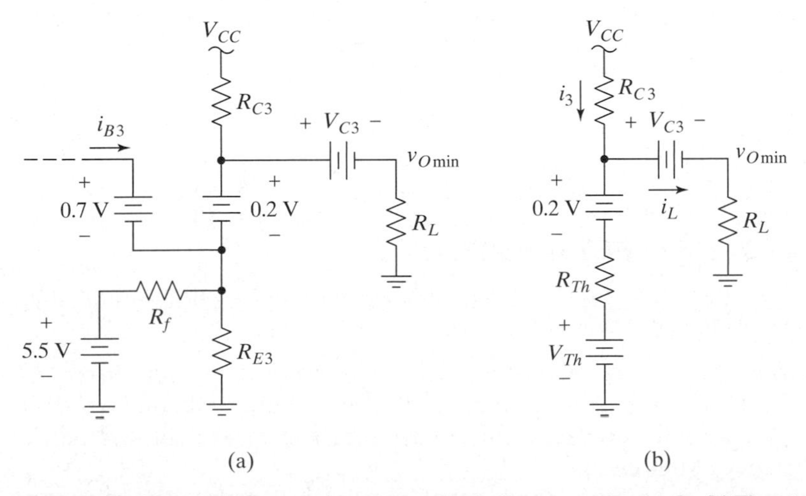

Figure 12–38 (a) The equivalent circuit for finding $v_{O\min}$ due to Q_3. (b) The circuit after some simplifications.

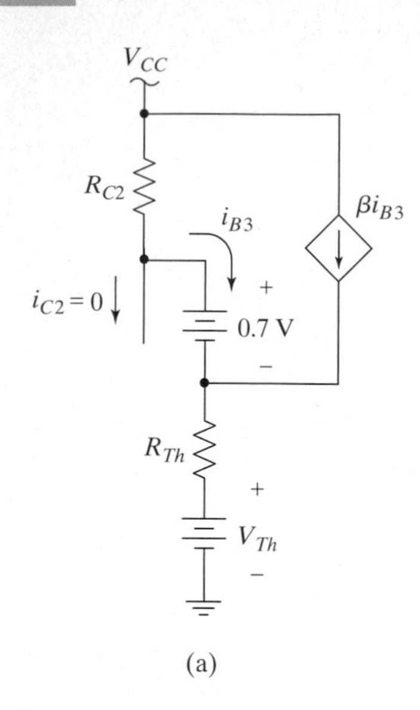

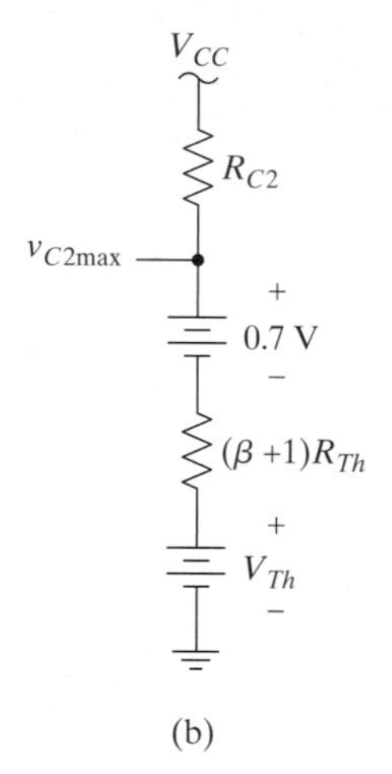

Figure 12–39 (a) The equivalent circuit for finding v_{C2max}. (b) The circuit after using impedance reflection.

We can now plug (12.39) into (12.38), substitute the result into (12.37), and solve for the current:

$$i_3 = \frac{V_{CC} - V_{Th} - 0.2 + (V_{CC} - V_{C3})R_{Th}/R_L}{R_{Th} + R_C(1 + R_{Th}/R_L)}. \quad \textbf{(12.40)}$$

Plugging in the values for this example yields $i_3 = 7$ mA. Then, from Figure 12-38(b), we see that the minimum output, as determined by Q_3, is

$$v_{Omin}(Q_3) = V_{CC} - i_3 R_C - V_{C3} = -0.6 \text{ V}. \quad \textbf{(12.41)}$$

Finally, we need to determine whether or not the collector voltage of Q_2 can swing high enough to saturate Q_3. If it can't, then the minimum output will be determined by Q_2 instead of (12.41).

The maximum voltage at the collector of Q_2 occurs when Q_2 is cut off. The resulting equivalent circuit is shown in Figure 12-39(a), where we have again used the Thévenin equivalent for the circuit in the emitter of Q_3. Using impedance reflection, we can redraw the circuit as shown in part (b) of the figure to find v_{C2max}.

From the circuit in part (b) of the figure, we obtain

$$v_{C2max} = 0.7 + V_{Th} + (V_{CC} - 0.7 - V_{Th})\frac{(\beta+1)R_{Th}}{(\beta+1)R_{Th} + R_{C2}}, \quad \textbf{(12.42)}$$

and, plugging in, we find $v_{C2max} = 9.7$ V. Since $V_{C2} = 6.1$ V, the positive swing is 3.6 V, which is more than large enough to support the minimum output given by (12.41). Therefore, the minimum output voltage is $v_{Omin} = -0.6$ V.

A SPICE simulation of this circuit yields $v_{Omax} = 2.1$ V and $v_{Omin} = -0.5$ V, which agree well with our calculated values of 1.9 V and −0.6 V.

■

12.2.2 Signal Swing in FET Amplifiers

We will first investigate the three single-transistor amplifier configurations and will then discuss swing in cascaded amplifiers.

Common-Source Amplifiers Consider the common-source amplifier in Figure 12-40. The small-signal performance of this amplifier was analyzed in Section 8.2.3. We are interested here in determining the maximum symmetric swing and efficiency of this circuit.

The maximum positive output voltage for the circuit will occur when the transistor in Figure 12-40 is cut off. At that point, all of the current in the load flows through R_D, and we have a resistive voltage divider formed by R_D and R_L, as

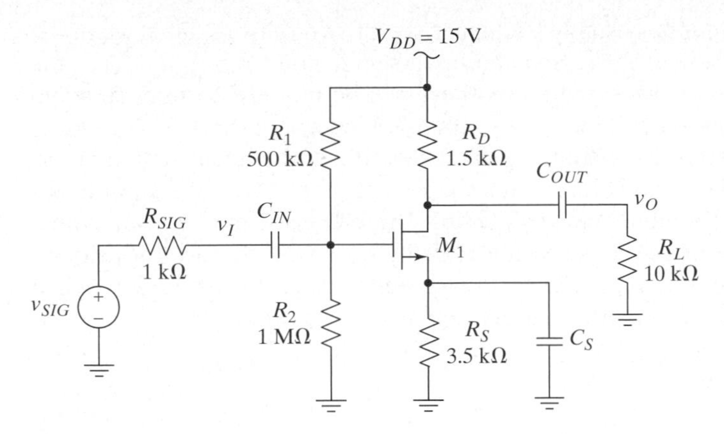

Figure 12–40 A MOSFET implementation of an amplifier with both voltage and current gain. (I_D = 2 mA, V_D = 12 V, and V_S = 7 V.)

shown in Figure 12-41. The total voltage across these two resistors is $V_{DD} - V_D$, since the DC voltage across the capacitor is equal to V_D. The resulting maximum positive output voltage is

$$v_{L\max} = \left(\frac{R_L}{R_L + R_D}\right)(V_{DD} - V_D). \tag{12.43}$$

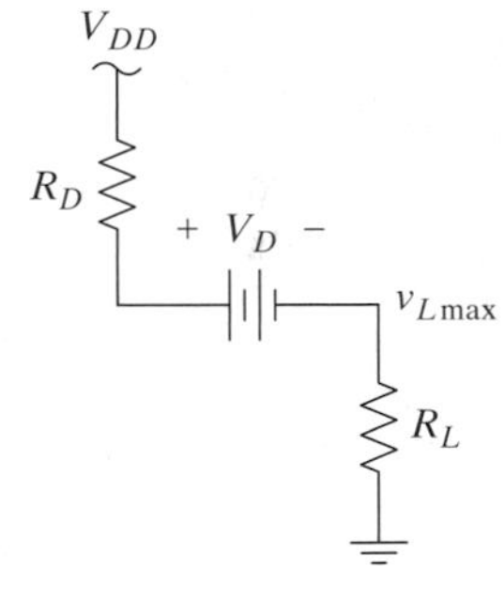

Figure 12–41 The output of the circuit in Figure 12-40 when the transistor is cut off.

Plugging in the numbers, we get $v_{L\max} = (10/11.5)(3) = 2.6$ V. An alternative representation of (12.43) is sometimes useful. Note that the DC voltage across R_D is $V_{R_D} = (V_{DD} - V_D) = I_D R_D$; substituting this into (12.43) yields

$$v_{L\max} = I_D\left(\frac{R_D R_L}{R_L + R_D}\right) = I_D(R_D \| R_L). \tag{12.44}$$

This equation shows that for a given R_D and R_L, the maximum swing is proportional to the DC bias current in the transistor.

The minimum theoretical output voltage occurs when the transistor enters the linear region of operation and has a drain-to-source resistance that is negligible compared with R_D. At that point, the drain and source voltages are about equal, and we find that

$$v_{D\min} \approx V_S = 7\text{ V}. \tag{12.45}$$

In practice, making the FET resistance negligible would often require an unreasonably large input voltage. The practical limit is set by examining the circuit when v_I is at its maximum positive value (*see* Problem P12.27). Nevertheless, the absolute minimum output voltage is

$$v_{L\min} = v_{D\min} - V_D, \tag{12.46}$$

which, from (12.45) yields $v_{L\min} = v_{D\min} - V_D \approx V_S - V_D = -5$ V for this example.

Since the peak negative value is larger than the peak positive value in this example, the positive value limits the maximum symmetric swing at v_L to be $v_{L\max\_sym} = 2.6$ V.

At this point, we can reexamine our choice of DC drain voltage. In Section 8.2.3, we stated that choosing V_D halfway between V_{DD} and $V_G - V_{th}$ was reasonable for achieving maximum swing. We are now in a position to examine this issue more carefully. The absolute minimum output voltage for this stage is given by (12.46), but the practical minimum is usually less negative, for two reasons. First, when the transistor has entered the linear region, the small-signal gain from v_{gs} to v_o (i.e., the slope of v_O versus v_{GS}) changes; therefore, there will be more distortion. To avoid this distortion, we may not be able to use signals which swing that far. Second, as just noted, if we include the resistance of the transistor in the linear region in our analysis, we see that the input voltage has to be very high to cause the output to approach the absolute minimum given by (12.46). Therefore, we can come up with a conservative estimate of the lower swing limit by not allowing the transistor to enter the linear region. Since the transistor enters the linear region when $V_D = V_G - V_{th}$, the corresponding conservative minimum output is

$$\begin{aligned} v_{L\min} &= v_{D\min} - V_D = V_G - V_{th} - V_D \\ &= V_S + V_{GS} - V_{th} - V_D \\ &= V_S + V_{ON} - V_D. \end{aligned} \quad \textbf{(12.47)}$$

The final form of the answer in this equation shows that using a large DC bias voltage at the source terminal or a large gate overdrive reduces the negative swing magnitude.

Depending on the allowable distortion or the available input swing, the real minimum output lies somewhere between the values given by (12.46) and (12.47). At the other end of the range, the transistor cuts off fairly abruptly, so we can use the point at which the transistor enters cutoff as the limit.

If we set $R_L \gg R_D$ in the present example, (12.43) leads to $v_{L\max} \approx V_{DD} - V_D$. To obtain maximum symmetric swing, we would like the positive and negative swing limits to have the same magnitude; therefore, using (12.43) and (12.47) to be conservative, we set (note that $V_G = V_{GS} + V_S$)

$$\begin{aligned} v_{L\max} &= -v_{L\min} \\ V_{DD} - V_D &= V_D - (V_G - V_{th}) \end{aligned} \quad \textbf{(12.48)}$$

and solve to obtain

$$V_D \Big|_{\text{max. symmetric swing}} = \frac{V_{DD} + (V_G - V_{th})}{2}. \quad \textbf{(12.49)}$$

For a good design, the condition $R_L \gg R_D$ is usually met, so (12.49) is a reasonable way to choose V_D if no other specification requires a different choice. Frequently, however, the value of V_D will be dictated by other constraints (e.g., DC coupling to the next stage may require a certain value for biasing purposes).

We can also solve for the value of V_D that will produce maximum symmetric swing when R_L is not much larger than R_D, but the solution depends on R_D, which is itself a function of V_D and I_D. Therefore, we need to either solve iteratively or fix the value of R_D (*see* Problem P12.28).

Let's now conclude our examination of the common-source circuit in Figure 12-40 by calculating the efficiency of the amplifier. To do that, we need to determine the total power delivered by the DC power supply (V_{DD}). The current through R_1 can be neglected, so the DC current drawn from the supply is approximately $I_D = 2$ mA. Therefore, the total power delivered by the supply is $P_{\text{supply}} = (2 \text{ mA})(15 \text{ V}) =$ 30 mW.

The maximum symmetric swing for this amplifier was found to be 2.6V peak; the average power delivered to the load using a sinusoid with this peak value is

$$P_L = \frac{(2.6)^2}{R_L T}\int_0^T \cos^2\left(\frac{2\pi}{T}t\right)dt = \frac{(2.6)^2}{2R_L} = 338\ \mu\text{W}. \quad \textbf{(12.50)}$$

Using this signal power in the load along with the supply power of 30 mW, we find the maximum efficiency possible with an undistorted sine wave to be

$$\eta_{max} = \frac{0.338}{30} \times 100\% = 1.1\% \quad \textbf{(12.51)}$$

This efficiency is very low and points out that this amplifier is not a good choice for an output stage driving a load resistance as small as that used here. There are three problems with this stage in terms of efficiency. First, some of the available swing is dropped across R_S to obtain a stable bias point. Second, the divider between R_D and R_L causes a drop in available output swing, which would be much worse if R_L were lower. Third, the transistor is on all of the time and dissipates significant power.

Exercise 12.4

Use the circuit of Figure 12-40, and assume that $V_{DD} = 5$ V, $R_{SIG} = 100$ kΩ, $R_L = 10$ kΩ, $K = 2$ mA/V², and $V_{th} = 1$ V. (a) Design the amplifier to allow a swing of ±1 V and to minimize power consumption. (b) What is the resulting efficiency with an output sine wave of 1V peak?

Example 12.6

Problem:
Use SPICE to verify our analytical results for the gain and swing of the amplifier shown in Figure 12-40.

Solution:
A SPICE simulation was performed using the Nmedium_mos model from Appendix B. Plots of v_O for the AC and transient simulations are shown in Figures 12-42 and 12-43, respectively. From these figures, we determine that the overall voltage

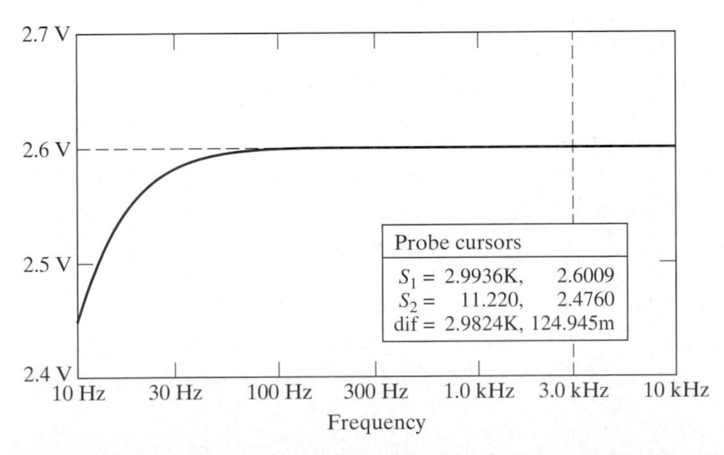

Figure 12–42 The AC simulation result for v_{OUT}. Cursor S_1 shows the midband gain.

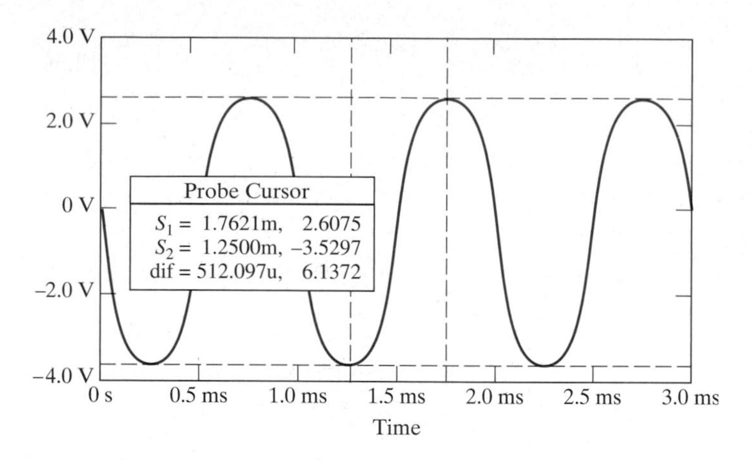

Figure 12–43 The transient simulation result for v_O. Cursor S_l shows the positive limit and S_2 shows the negative limit.

gain is −2.6 (the minus sign is not apparent from the figure; only the magnitude is), and the swing limits are $v_{\text{Lmin}} = -3.5$ V and $v_{\text{Lmax}} = 2.6$ V with a 2-V peak sine wave input. Except for the negative swing limit, the results agree exactly with calculations, because we used an extremely simple model for the transistor. The disagreement on the negative swing limit is due to the peak input voltage amplitude. As noted in the text, the input would need to be very large to make the resistance of the transistor negligible in the linear region of operation. The negative swing limit does approach the absolute minimum limit as the peak input is increased; for example, if the peak input is set to 5 V, the limit is −4.2 V. ■

Common-Gate Amplifiers A common-gate amplifier is shown in Figure 12-44. This is the same amplifier examined in Section 8.4.3, except that we have shown a separate load here while keeping the AC load (i.e., $R_D \| R_L$) the same.

We leave it as an exercise to draw out the large-signal equivalent circuits for finding the maximum and minimum swing limits for this circuit. The results are the same as we obtained for the common-source amplifier in (12.43) and (12.46).

Common-Drain Amplifiers (Source Followers) A source follower is shown in Figure 12-45. The transistor has $K = 500\ \mu\text{A/V}^2$ and $V_{th} = 0.8$ V. We analyzed the small-signal midband performance of this stage in Section 8.3.3. We found there that the DC bias current is $I_D = 4.5$ mA and that $V_S = 5$ V.

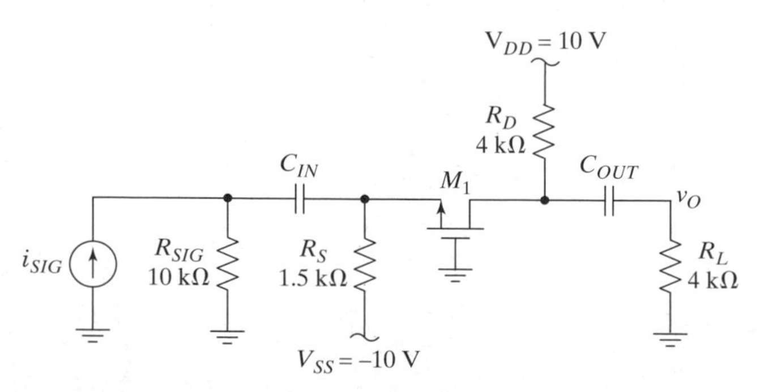

Figure 12–44 A common-gate amplifier.

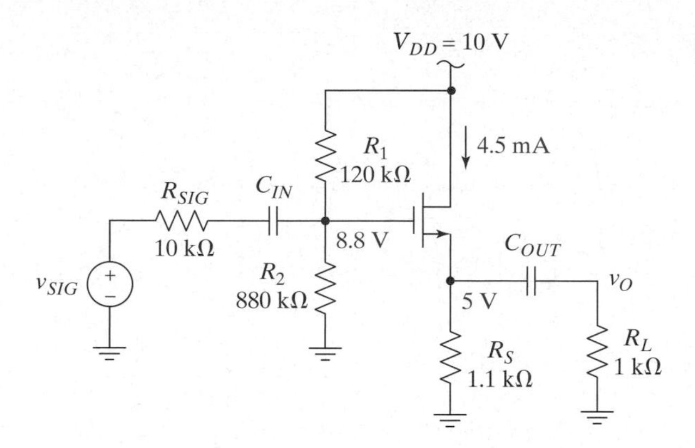

Figure 12–45 A source follower.

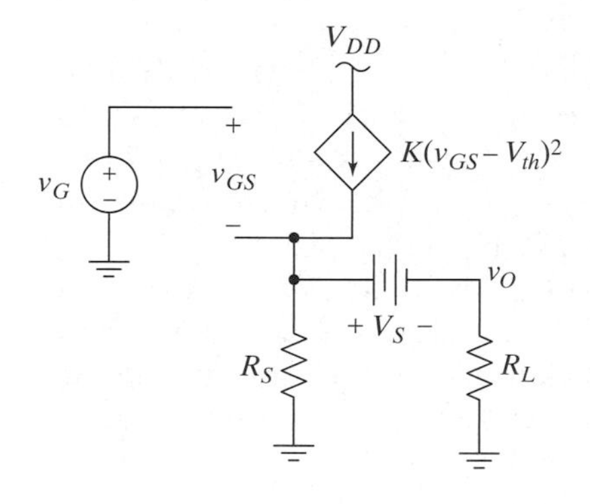

Figure 12–46 The large-signal midband equivalent circuit for the source follower when the transistor is forward active.

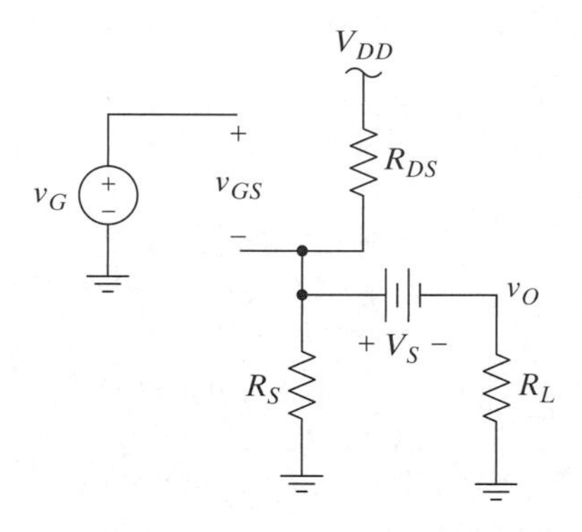

Figure 12–47 The large-signal midband equivalent circuit for the source follower when the transistor is in the linear region.

The large-signal model for the circuit of Figure 12-45 is shown in Figure 12-46, where we have assumed that the transistor is forward active and have used the fact that the large-signal midband model of the coupling capacitor is a battery. We have also simplified the circuit by driving the gate of the transistor directly from a source v_G. The voltage across the coupling capacitor is V_S, since V_O must be zero.

The maximum output voltage does not appear to have any limit, given the circuit in Figure 12-46, but that is not the case. If v_G is taken too high, the transistor will enter the linear region of operation. The upper limit is therefore found by assuming that the transistor is in the linear region and redrawing the circuit as shown in Figure 12-47, where

$$R_{DS} \approx \frac{1}{2K(V_{GS} - V_{th})} = \frac{1}{2KV_{ON}}, \tag{12.52}$$

for small V_{DS}, as in (7.18).

Using this figure, we see that

$$v_{O\max} = v_{S\max} - V_S, \tag{12.53}$$

and we can find $v_{S\max}$ by writing KCL at the source node and solving. The result is

$$v_{S\max} = \left(\frac{V_S}{R_L} + \frac{V_{DD}}{R_{DS}}\right)(R_{DS}\|R_S\|R_L). \tag{12.54}$$

Unfortunately, R_{DS} is a function of v_S, so we either need to plug (12.52) into (12.54) and solve again, or iterate to find the solution. It turns out that iterating is fairly quick here. For the present example, let's assume that we use a 5-V peak sine wave for the signal so that $v_{G\max} = 5 + V_G = 13.8$ V, and let's guess that $v_{S\max} = 9$ V (i.e., the gain of the source follower is assumed to be less than unity, since $\Delta v_G = 5$ V and we set $\Delta v_S = 4$ V). That leads to $v_{GS\max} = 4.8$ V and $R_{DS} = 250\ \Omega$. Plugging the latter value into (12.54) yields $v_{S\max} = 7.6$ V. Therefore, let's try $v_{S\max} = 8$ V as our second guess. We now find that $v_{GS\max} = 5.8$ V and $R_{DS} = 200\ \Omega$, which yields $v_{S\max} = 7.96$ V. Therefore, for this example, $v_{S\max} = 8$ V and $v_{O\max} = 3$ V. The current required to supply R_S and R_L at this higher voltage is provided by the transistor.

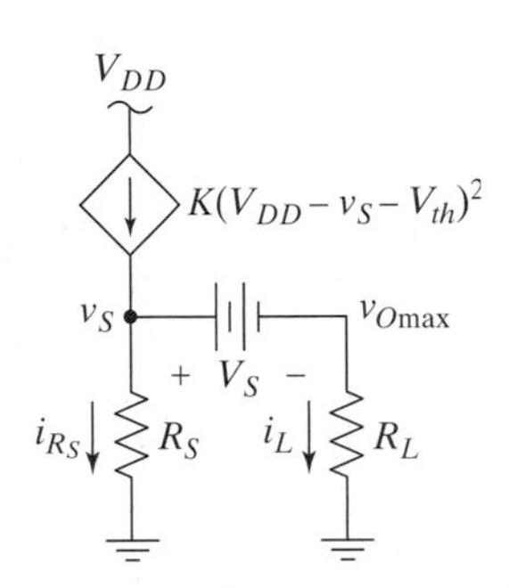

Figure 12–48 The large-signal equivalent circuit for finding $v_{O\max}$ when $v_G \leq V_{DD}$.

The practical maximum output voltage is usually lower than we just found, because the gate voltage is usually limited by V_{DD} (e.g., if the source follower is driven by a preceding stage without the coupling capacitor). In this case, the equivalent circuit is shown in Figure 12-48. The transistor will be forward active, and $v_{GS} = V_{DD} - v_S$, so the drain current is given by $i_D = K(V_{DD} - v_S - V_{th})^2$ as shown. We can solve this circuit by writing KCL at the source:

$$K(V_{DD} - v_S - V_{th})^2 = i_{R_S} + i_L = \frac{v_S}{R_S} + \frac{v_S - V_S}{R_L}. \tag{12.55}$$

From (12.55), we can write a quadratic equation for v_S and solve. The result is

$$v_S = V_{DD} - V_{th} + \frac{1}{2KR_p} - \frac{1}{2K}\sqrt{\frac{4K(V_{DD} - V_{th})}{R_p} + \frac{1}{R_p^2} - \frac{4KV_S}{R_L}}, \tag{12.56}$$

where $R_p \equiv R_S\|R_L$ and we have taken the solution that is consistent with the circuit. Plugging in the values for our example, we find the maximum value of v_S to be 5.75 V. Using (12.53), we find that $v_{O\max} = 0.75$ V, much lower than we obtained when the gate voltage was allowed to exceed the supply.

The lower output limit occurs when v_G goes so low that it turns the transistor off completely. With the transistor cut off, the current out of the source terminal is zero, so the node voltage is as low as possible. The equivalent circuit in this case is shown in Figure 12-49; the two resistors are in series with each other and the coupling capacitor, which acts like a battery. The maximum negative output voltage is given by the resulting resistive divider,

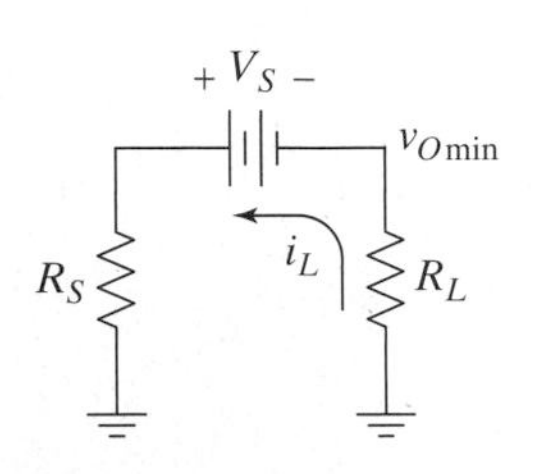

Figure 12–49 The equivalent circuit when the transistor is cut off.

$$v_{O\min} = \frac{-R_L}{R_L + R_S} V_S, \tag{12.57}$$

and the result is –2.4 V for our example. An alternative representation of (12.57) that is often useful is found by substituting $V_S = I_S R_S$ into the equation. The result is

$$v_{O\min} = I_S \frac{-R_L R_S}{R_L + R_S} = -I_S(R_L\|R_S). \tag{12.58}$$

This result shows that for a given set of resistor values, the minimum achievable swing is proportional to the DC bias current in the transistor.

If we allow the gate voltage to exceed the supply, than $v_{O\max} = 3$ V and $v_{O\min} = -2.4$ V for this circuit, and the maximum symmetric swing is limited by $v_{O\min}$ to ±2.4 V. Using this value, we can calculate the maximum efficiency of the circuit. The total supply current is $I_D + I_{BIAS} \approx I_D = 4.5$ mA, so the power delivered by the supply is $P_{supply} = 45$ mW. The maximum average load power with a 2.4-V peak sine wave is 2.9 mW, and the efficiency is $\eta = 6.4$ %. This efficiency is quite low. If the gate voltage is limited to the supply, the maximum symmetric swing is limited by the maximum output to ±0.75 V and the efficiency is only 0.6%, which is very low. We can improve that efficiency significantly by choosing a different bias point, as shown in Exercise 12.5, but more sophisticated output stages are needed if high efficiency is desired. These stages are discussed in Section 12.3.

Exercise 12.5

Consider the source follower in Figure 12-45 and assume that the maximum source voltage is given by (12.56). Plugging this into (12.53) yields the maximum output voltage. Use this result along with (12.57) to determine the value of V_S that will maximize the symmetric swing capability of the circuit. (You cannot solve the equations analytically, so use a spreadsheet or other program to iteratively find the result.) Determine the maximum efficiency with this new value of V_S.

Cascaded Amplifiers In this section, we use an example to combine and extend our previous results to examine the swing limits of cascaded amplifiers.

Example 12.7

Problem:

Consider the common-source source-follower cascade with current source biasing shown in Figure 12-50. Note that the DC bias voltages and currents are given on the schematic. This amplifier was designed in Section 8.7.2. Determine the swing limits for the amplifier.

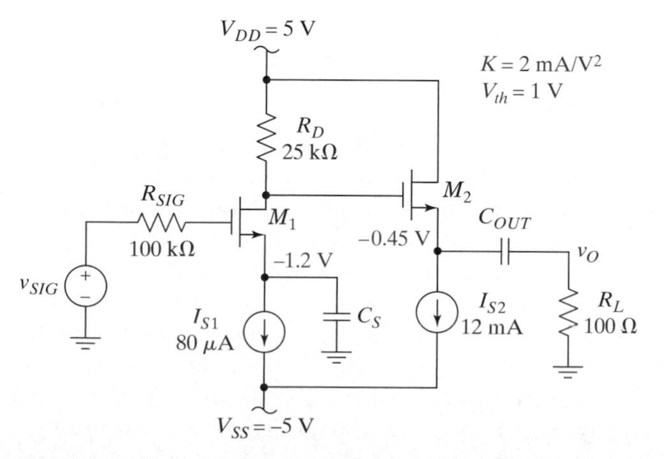

Figure 12–50 A common-source source-follower cascade.

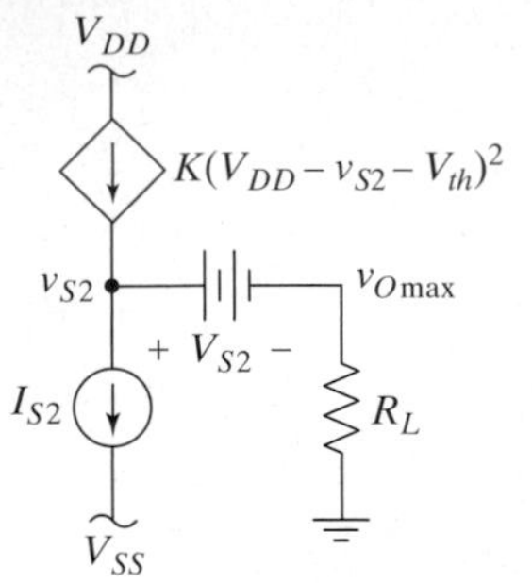

Figure 12–51 The large-signal midband equivalent circuit for finding $v_{o\max}$.

Solution:
The maximum output voltage for this circuit will occur when M_1 is cut off. At that instant, the gate voltage of M_2 will be equal to V_{DD} and M_2 will be in saturation. The large-signal midband equivalent circuit is shown in Figure 12-51.

Writing KCL at the source of M_2, we have

$$K(V_{DD} - v_{S2} - V_{th})^2 = I_{S2} + \frac{v_{O\max}}{R_L}. \tag{12.59}$$

We also note that $v_{S2} = v_{O\max} + V_{S2}$ and rewrite the equation as

$$K(V_{DD} - V_{S2} - V_{th} - v_{O\max})^2 = I_{S2} + \frac{v_{O\max}}{R_L}. \tag{12.60}$$

Defining $V_1 = V_{DD} - V_{S2} - V_{th}$ for simplicity, we expand the square in (12.60) and solve the resulting quadratic to obtain

$$v_{O\max} = V_1 + \frac{1}{2KR_L} \pm \sqrt{\frac{V_1}{KR_L} + \frac{1}{4K^2R_L^2} + \frac{I_{S2}}{K}}. \tag{12.61}$$

Plugging in the values for this example, we find that $v_{O\max}$ is either 12.82 V or 1.08 V. Since the maximum change in v_{D1} is 2 V (i.e., from the DC bias value up to V_{DD}), there is no way that $v_{O\max}$ can be larger than that. Therefore, $v_{O\max} = 1.08$ V.

The minimum output voltage occurs when M_2 is cut off (if that can happen) and is given by

$$v_{O\min} = -I_{S2}R_L = -1.2 \text{ volts}, \tag{12.62}$$

since all of I_{S2} will flow through the load. The source voltage of M_2 is $v_{S2} = v_O + V_{S2}$, so if the output is at the value given by (12.62), we have $v_{S2} = -1.65$ V. As long as the gate of M_2 is less than -0.65 V, it will be cut off. Since the drain of M_1 can swing down to -1 V before it leaves saturation, it can certainly swing low enough to cut M_2 off, and the result we have is correct.

A SPICE simulation of this circuit using the Nlarge_mos models from Appendix B reveals that the limits are 1.1 V and -1.2 V, as determined here.

■

The procedure illustrated in the preceding example is common to all multistage amplifiers—namely, you must examine the swing limits of each stage (while connected to the adjacent stages) and see which ones limit the swing. For example, if the drain of M_1 had not been able to swing low enough to cut off M_2, we would have had to determine $v_{O\min}$ basis of the on the minimum value of v_{D1}.

If stages with voltage gains larger than unity are cascaded together, it is unusual for any but the last stage to limit the swing since the signal swing is smaller in the early stages. For example, if two common-source stages are cascaded and each stage has a gain of 10, then when the output signal is 1 V peak to peak, the signal at the output of the first stage is only 0.1 V peak to peak. Therefore, for the first stage to limit the overall swing capability, it would have to have a swing limit less than one-tenth that of the final stage.

12.2.3 Distortion in Amplifiers

In the preceding sections, we discussed how to determine the maximum swing limits of amplifiers. Knowing these limits is useful, but we can rarely allow an amplifier to operate near them, since the signal will be badly distorted. In fact, we are often interested in finding how much distortion is produced by a given amplifier with a given signal amplitude. In this section, we briefly introduce this topic. More advanced discussions can be found in [12.1, 12.4, 12.6, and 12.7].

When a nonlinear circuit is driven by a sinusoidal source, it produces components at the output that are not present in the source, but are harmonically related to it. Although linear circuits can cause dispersion, which distorts the shape of an input waveform, they cannot produce new frequency components. Therefore, harmonic distortion is the hallmark of a nonlinear circuit. In this section, we are interested primarily in examining distortion produced by nonlinear circuits; the topic of dispersion caused by linear networks is briefly addressed in Aside A12.1 on distortionless transmission.

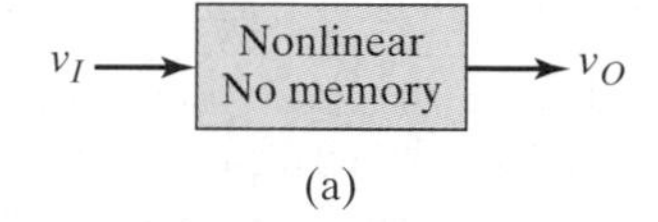

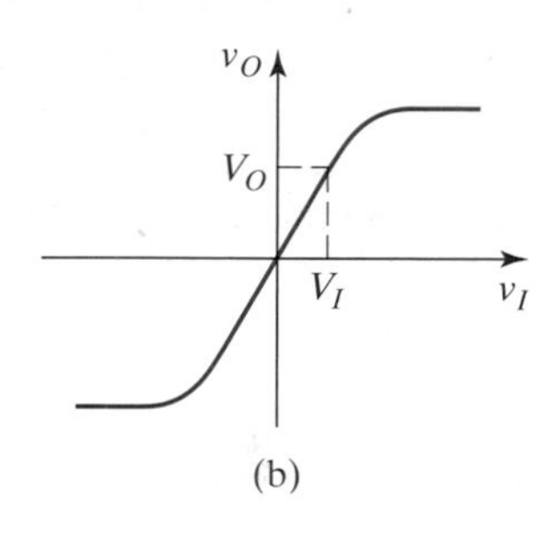

Figure 12–52 (a) A nonlinear circuit without memory and (b) the circuit's DC transfer characteristic.

We will restrict our attention to ***memoryless nonlinearities***— that is, circuits which do not contain nonlinear energy storage elements (e.g., nonlinear capacitors). Since all solid-state devices contain nonlinear capacitances, this restriction means that we can only find distortion for midband frequencies where the capacitors can all be ignored.[8] We will also restrict ourselves to differentiable nonlinearities, but this restriction does not limit the practical utility of the method.

Suppose we have a nonlinear system as shown in Figure 12-52. Suppose further that the nonlinearity is differentiable and memoryless. In this case, we can use a power series expansion (i.e., a Taylor series) to represent the output. The Taylor series expansion of a function $f(x)$ about the point a is given by

$$f(x) = f(a) + f'(a)(x - a) + \frac{f''(a)(x - a)^2}{2!} + \cdots$$

$$+ \frac{f^{(n-1)}(a)(x - a)^{n-1}}{(n - 1)!} + R_n, \tag{12.63}$$

where R_n is the remainder and $f(x)$ must have n derivatives. Applying this expansion to the nonlinearity represented in part (b) of the figure, we obtain

$$v_O = V_O + \left.\frac{dv_O}{dv_I}\right|_{@V_I}(v_I - V_I) + \left.\frac{d^2v_O}{dv_I^2}\right|_{@V_I}(v_I - V_I)^2 + \cdots . \tag{12.64}$$

We now note that the AC component of the input is $v_i = v_I - V_I$, and we are only interested in the AC component of the output $v_o = v_O - V_O$; therefore, we rewrite (12.64) as

$$v_o = v_i\left.\frac{dv_O}{dv_I}\right|_{@V_I} + v_i^2\left.\frac{d^2v_O}{dv_I^2}\right|_{@V_I} + \cdots, \tag{12.65}$$

which can be written as a general power series

$$v_o = \sum_{n=1}^{N} a_n v_i^n. \tag{12.66}$$

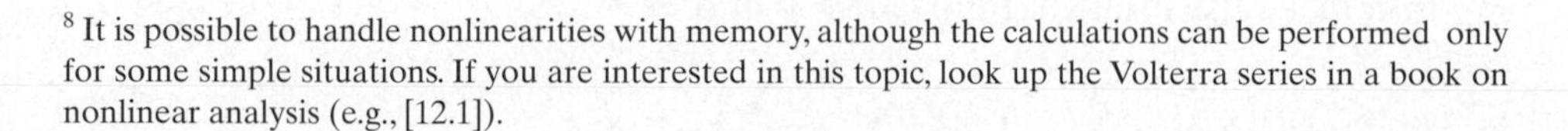

[8] It is possible to handle nonlinearities with memory, although the calculations can be performed only for some simple situations. If you are interested in this topic, look up the Volterra series in a book on nonlinear analysis (e.g., [12.1]).

ASIDE A12.1 DISTORTIONLESS TRANSMISSION

There are many systems where we desire the output of the system to be a faithful replica of the input—for example, communication systems like the telephone and sound reproduction systems like stereo amplifiers. Being a bit more precise, we can say that if $x(t)$ is the input to a system and $y(t)$ is the output, ***distortionless transmission*** requires that

$$y(t) = Gx(t - \tau) \qquad \textbf{(A12.1)}$$

where G is a constant gain and τ is a constant delay. In other words, the output should be a scaled and delayed version of the input.

Suppose the input to the system is given by

$$x(t) = \cos(\omega_1 t + \theta_1) + \cos(\omega_2 t + \theta_2). \qquad \textbf{(A12.2)}$$

If the system transmits this signal without distortion, then the output must be

$$y(t) = G\cos(\omega_1(t - \tau) + \theta_1) + G\cos(\omega_2(t - \tau) + \theta_2). \qquad \textbf{(A12.3)}$$

We can now see what is required for distortionless transmission. First, the gain must be constant for all terms in the input, so it is clear from (A12.3) that the gain must be independent of frequency. Second, each component must be delayed by the same amount of time; consequently, examining the components in (A12.3) shows that the phase shift caused by the network must be a linear function of frequency with a slope equal to $-\tau$. To make this point explicit, note that (A12.3) can be rewritten as

$$y(t) = G\cos(\omega_1 t + \varphi_1 + \theta_1) + G\cos(\omega_2 t + \varphi_2 + \theta_2), \qquad \textbf{(A12.4)}$$

where $\varphi_1 = -\omega_1\tau$ is the phase shift caused by the network for the component at frequency ω_1 and $\varphi_2 = -\omega_2\tau$ is the phase shift caused by the network for the component at frequency ω_2. In general, the phase shift for a component at frequency ω_x must be $\varphi_x = -\omega_x\tau$ if that component is going to have a time delay of τ seconds in traveling through the network.

Therefore, if the output of a system is to be a faithful replica of the input, the magnitude of the transfer function must be constant, and the phase must be a linear function of frequency (and pass through the origin) for frequencies less than the maximum frequency of interest. The distortion caused by having frequency-dependent gain or nonlinear phase is more properly referred to as ***dispersion***, since it leads to a pulse being dispersed (i.e., spread out in time) as it travels through the system. Frequency-dependent gain can cause dispersion because of the filtering process (e.g., a pulse is broadened as it passes through a low-pass filter) and nonlinear phase results in dispersion because the different frequency components making up the pulse arrive at the output at different times. Therefore, LTI systems can distort the waveform due to dispersion; they cannot, however, produce harmonic distortion.

If the input to this system is a single-frequency sinusoid

$$v_i = \hat{V}\cos\omega_1 t, \qquad \textbf{(12.67)}$$

then the output is

$$v_o = a_1\hat{V}\cos\omega_1 t + a_2\hat{V}^2\cos^2\omega_1 t + \cdots. \qquad \textbf{(12.68)}$$

We now make use of the trigonometric identities

$$\cos^2 A = \frac{1}{2}(1 + \cos 2A) \qquad \textbf{(12.69A)}$$

and

$$\cos^3 A = \frac{1}{4}(3\cos A + \cos 3A) \qquad \textbf{(12.69B)}$$

to obtain

$$v_o = \begin{cases} a_1 \hat{V} \cos \omega_1 t & \Rightarrow \text{fundamental} \\ + \dfrac{a_2 \hat{V}^2}{2}(1 + \cos 2\omega_1 t) & \Rightarrow \text{DC} + 2^{\text{nd}} \text{ harmonic.} \\ + \dfrac{a_3 \hat{V}^3}{4}(3 \cos \omega_1 t + \cos 3\omega_1 t) & \Rightarrow \text{fundamental} + 3^{\text{rd}} \text{ harmonic} \\ + \vdots & \end{cases} \qquad \textbf{(12.70)}$$

Notice that the first term in the series, which is the linear term, produces the desired output at the fundamental: $a_1 \cos \omega_1 t$. This term is precisely what we would get from a small-signal analysis of the circuit. The second-order term produces a DC component and a term at the second harmonic, the third-order term produces terms at the fundamental and the third harmonic, and so on. Generally, all even-order terms in the series will produce a DC level shift and even-order distortion for harmonic frequencies up to and including the order of the term. All odd-order terms in the series will produce a term at the fundamental and odd-order distortion for all harmonics up to and including the order of the term.

We can regroup the terms in (12.70) so that all of the terms at each frequency are together:

$$v_o = \begin{cases} \dfrac{a_2 \hat{V}^2}{2} + \cdots & \Rightarrow \text{DC} \\ + \left(a_1 \hat{V} + \dfrac{3a_3 \hat{V}^3}{4} + \cdots\right) \cos \omega_1 t & \Rightarrow \text{fundamental} \\ + \left(\dfrac{a_2 \hat{V}^2}{2} + \cdots\right) \cos 2\omega_1 t & \Rightarrow 2^{\text{nd}} \text{ harmonic} \\ + \left(\dfrac{a_3 \hat{V}^3}{4} + \cdots\right) \cos 3\omega_1 t & \Rightarrow 3^{\text{rd}} \text{ harmonic} \\ + \vdots & \end{cases} \qquad \textbf{(12.71)}$$

From this equation, we see that the output of a nonlinear system will generally contain a ***DC level shift*** and ***harmonic distortion***, in addition to the desired output at the fundamental. Also, the magnitude of the fundamental term is different than is predicted by linear analysis. (This effect is called ***gain compression***.)

To simplify the calculations, it is common to ignore all but the first term contributing to each harmonic. In other words, only the first term in parentheses for each harmonic in (12.71) is kept. We call this the ***small-distortion approximation***. So long as $\hat{V}$ is small enough, this approximation will be reasonably accurate.

We now define the ***n^{th}-order fractional harmonic distortion*** as

$$HD_n = \frac{\text{amplitude of the } n^{\text{th}} \text{ harmonic component}}{\text{amplitude of the fundamental component}}. \quad \textbf{(12.72)}$$

Using the small-distortion approximation along with (12.71) and the definition in (12.72), we find the second-harmonic distortion to be

$$HD_2 = \frac{|a_2|\hat{V}^2/2}{|a_1|\hat{V}} = \frac{1}{2}\left|\frac{a_2}{a_1}\right|\hat{V}. \quad \textbf{(12.73)}$$

We can check whether or not the small-distortion approximation is valid by varying $\hat{V}$ and seeing if HD_2 changes proportionally. Similarly, we find the third-harmonic distortion

$$HD_3 = \frac{|a_3|\hat{V}^3/4}{|a_1|\hat{V}} = \frac{1}{4}\left|\frac{a_3}{a_1}\right|\hat{V}^2. \quad \textbf{(12.74)}$$

The higher order terms are all found in a similar manner. Finally, we define the ***total harmonic distortion*** as

$$THD = \sqrt{\sum_{n=2}^{N} |HD_n|^2}. \quad \textbf{(12.75)}$$

When the input to a nonlinear system contains more than one frequency, these terms combine to produce terms at sum and difference frequencies. The terms are called ***intermodulation distortion*** (*IM*). The magnitudes of the various *IM* terms can be directly related to the magnitudes of the corresponding *HD* terms, so reducing *HD* is guaranteed to reduce *IM* distortion also. *IM* distortion can be a significant problem, especially in radio receivers, because the distortion terms can be at frequencies very close to the original signals. In fact, the third-order *IM* from two strong radio stations can exceed the signal of a weak station nearby. We leave it to the references to discuss this topic. In the remainder of this section, we present a few examples of how to calculate distortion. In each case, we compare the hand calculations with SPICE simulations.

Simulating distortion with SPICE can be problematic, so a brief discussion is warranted. In some versions of SPICE, you can obtain distortion results based on the small-distortion approximation by using the **.DISTO** command as a part of the AC analysis mode. This command is covered in [12.3] and will not be discussed here, since PSPICE does not include it. An alternative method, which is useful even when the small-distortion approximation is not valid, is to use the Fourier analysis option in the transient simulation mode. In using this option, however, you must be sure to simulate the circuit for a long enough time that steady-state operation has been achieved. You must also be sure that the maximum simulation step size is small enough and that the option **RELTOL**, which sets the tolerance for variations from one time step to the next, is small enough. In general, you need to adjust these parameters until they no longer significantly affect the output. Very small high-order distortion terms are likely to be highly sensitive to these parameters and are therefore probably not accurate. Nevertheless, the main distortion products are usually fairly accurate.

Example 12.8

Problem:
Consider the simplified common-emitter amplifier shown in Figure 12-53. Calculate the values of HD_2, HD_3, HD_4, and HD_5.

Solution:
The output voltage of this amplifier is given by

$$v_O = V_{CC} - i_C R_C, \tag{12.76}$$

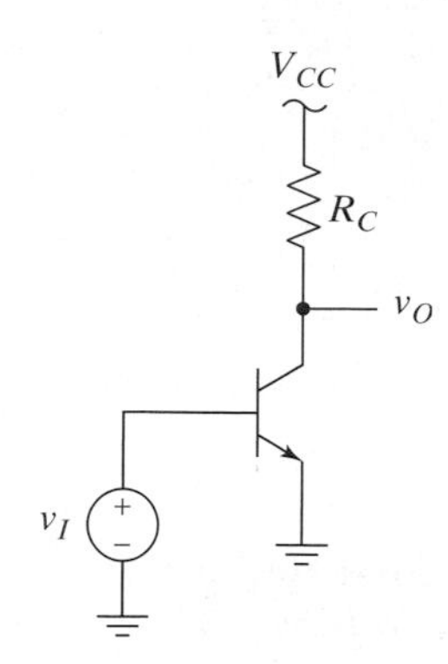

Figure 12–53 A simplified common-emitter amplifier.

and the collector current is

$$\begin{aligned} i_C &= I_S e^{v_I/V_T} = I_S e^{(V_I+v_i)/V_T} = I_S e^{V_I/V_T} e^{v_i/V_T} \\ &= I_C e^{v_i/V_T}. \end{aligned} \tag{12.77}$$

The Taylor series expansion for an exponential is

$$e^x = 1 + x + \frac{x^2}{2!} + \frac{x^3}{3!} + \cdots. \tag{12.78}$$

Combining these three equations leads to

$$v_O = V_{CC} - I_C R_C\left[1 + \frac{v_i}{V_T} + \frac{1}{2}\left(\frac{v_i}{V_T}\right)^2 + \frac{1}{6}\left(\frac{v_i}{V_T}\right)^3 + \cdots\right], \tag{12.79}$$

and examining only the AC component, we finally arrive at

$$v_o = -I_C R_C\left[\frac{v_i}{V_T} + \frac{1}{2}\left(\frac{v_i}{V_T}\right)^2 + \frac{1}{6}\left(\frac{v_i}{V_T}\right)^3 + \cdots\right]. \tag{12.80}$$

Equating (12.80) with the general power series in (12.66), we find that

$$a_1 = -g_m R_C, \tag{12.81}$$

as expected (since this is the small-signal gain)

$$a_2 = \frac{a_1}{2V_T}, \tag{12.82}$$

and, in general,

$$a_k = \frac{a_1}{k!V_T^{k-1}} \text{ for all } k \geq 1. \tag{12.83}$$

Assuming that $v_i = \hat{V}\cos\omega t$ and using (12.83) and (12.72), subject to the small-distortion approximation, we obtain

$$HD_k = \frac{a_k}{a_1}\left(\frac{\hat{V}}{2}\right)^{k-1} = \frac{1}{k!2^{k-1}}\left(\frac{\hat{V}}{V_T}\right)^{k-1}. \tag{12.84}$$

If $\hat{V} = 10$ mV, then at room temperature we obtain $HD_2 = 0.096$, $HD_3 = 6.2 \times 10^{-3}$, $HD_4 = 3.0 \times 10^{-4}$, and $HD_5 = 1.1 \times 10^{-5}$. Since these terms are dropping off so rapidly, we can get a reasonable estimate of the THD by using just the four terms. The result is $THD \approx 9.6\%$. This is a very large value for THD. In practical amplifiers, we use feedback to reduce the nonlinearity and hence the distortion.

A SPICE simulation of this circuit predicts that $HD_2 = 0.090$, $HD_3 = 5.0 \times 10^{-3}$, $HD_4 = 1.8 \times 10^{-4}$, $HD_5 = 1 \times 10^{-5}$, and $THD = 9\%$. These numbers are reasonably close to the calculated values.

Example 12.9

Problem:

Consider the simplified common-source amplifier shown in Figure 12-54. Calculate the values of HD_2, HD_3, HD_4, and HD_5.

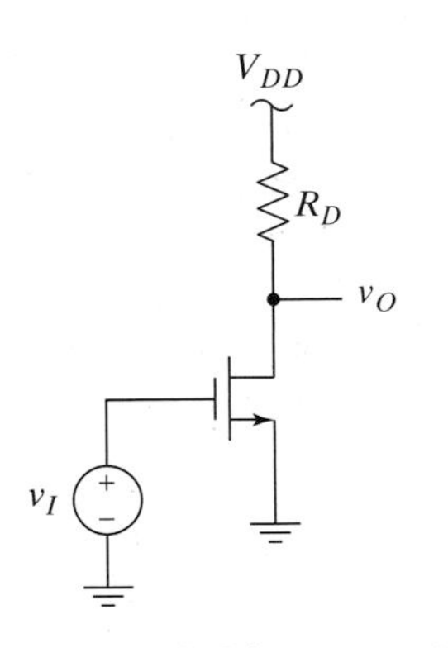

Figure 12–54 A simplified common-source amplifier.

Solution:

The output voltage of this amplifier is

$$v_O = V_{DD} - i_D R_D, \tag{12.85}$$

and the drain current is

$$i_D = K(v_{GS} - V_{th})^2. \tag{12.86}$$

Using the fact that $v_{GS} = V_{GS} + v_{gs}$ we can, after some algebra, rewrite (12.86) as

$$i_D = K[(V_{GS} - V_{th})^2 + 2(V_{GS} - V_{th})v_{gs} + v_{gs}^2]. \tag{12.87}$$

We then plug (12.87) into (12.85) to obtain

$$v_O = V_{DD} - KR_D[(V_{GS} - V_{th})^2 + 2(V_{GS} - V_{th})v_{gs} + v_{gs}^2]. \tag{12.88}$$

We are interested only in the AC component of the output, so we examine

$$\begin{aligned} v_o &= v_O - V_O = v_O - [V_{DD} - KR_D(V_{GS} - V_{th})^2] \\ &= -KR_D[v_{gs}^2 + 2(V_{GS} - V_{th})v_{gs}]. \end{aligned} \tag{12.89}$$

Comparing (12.89) with the general power series in (12.66), we find that

$$a_1 = -2K(V_{GS} - V_{th})R_D = -g_m R_D, \tag{12.90}$$

as expected (since this is the small-signal gain) and

$$a_2 = -KR_D. \tag{12.91}$$

All other terms are zero.

Assuming that $v_i = \hat{V} \cos \omega t$ and using (12.90), (12.91), and (12.72), subject to the small-distortion approximation, we obtain

$$HD_2 = \frac{\hat{V}}{4(V_{GS} - V_{th})}, \tag{12.92}$$

and all other terms are zero. If $\hat{V} = 10$ mV and the gate overdrive is $(V_{GS} - V_{th}) = 0.5$ V, then we obtain $HD_2 = 0.005$. Since there is only one nonzero term,[9] the THD is also 0.005, or 0.5%, which is moderately large. In practical amplifiers, we use feedback to reduce the nonlinearity and hence, the distortion. A SPICE simulation of this circuit using the Nlarge_mos model from Appendix B predicts that $HD_2 = 0.005$, $HD_3 = 1.8 \times 10^{-5}$, $HD_4 = 1.3 \times 10^{-5}$, $HD_5 = 1.1 \times 10^{-5}$, and $THD = 0.5\%$. The second-order distortion agrees quite well, but the magnitudes of the small higher order terms are very sensitive to the simulation parameters (e.g., the step size).

■

Notice that the distortion for the MOS amplifier in Example 12.9 is much smaller than that of the bipolar amplifier in Example 12.8, even though the input has the same amplitude. In general, MOS amplifiers exhibit greater linearity than do bipolar amplifiers, but this can be deceiving. Since the gain of the FET amplifier is much lower, it is often more reasonable to compare the distortion when the output amplitudes are the same (*see* Problem P12.34).

Example 12.10

Problem:
Consider the simplified differential amplifier shown in Figure 12-55, and assume that the two transistors are perfectly matched and at the same temperature. Calculate the values of HD_2, HD_3, HD_4, and HD_5.

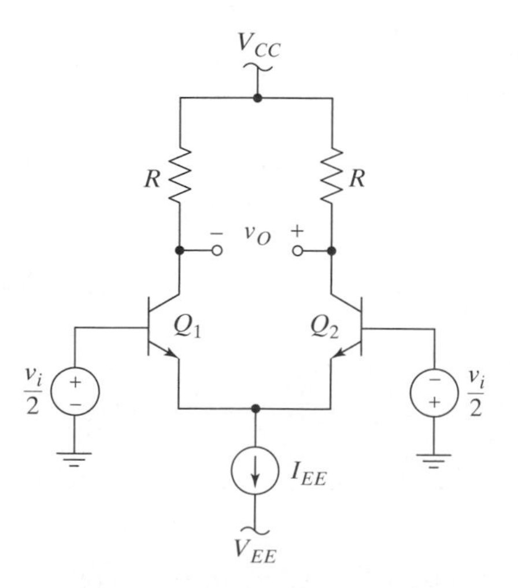

Figure 12–55 A simplified differential amplifier.

---

[9]Real MOSFETs do not obey the square-law relationship perfectly and do have other terms in the power series, especially for very small devices. Nevertheless, the other distortion terms can be quite small, particularly for very large devices.

Solution:
Writing KVL around the base-emitter loop yields

$$\frac{v_i}{2} - v_{BE1} + v_{BE2} + \frac{v_i}{2} = 0, \tag{12.93}$$

which can be manipulated to yield

$$\begin{aligned} v_i = v_{BE1} - v_{BE2} &= V_T\left(\ln\frac{i_{C1}}{I_S} - \ln\frac{i_{C2}}{I_S}\right) \\ &= \mathrm{V_T} \ln\frac{i_{C1}}{i_{C2}}. \end{aligned} \tag{12.94}$$

or

$$\frac{i_{C1}}{i_{C2}} = e^{v_i/V_T}. \tag{12.95}$$

Summing currents at the emitters yields

$$i_{C1} + i_{C2} = \alpha I_{EE}, \tag{12.96}$$

and using this equation along with (12.95), we obtain

$$\alpha I_{EE} = i_{C1}\left(1 + \frac{i_{C2}}{i_{C1}}\right) = i_{C1}(1 + e^{-v_i/V_T}). \tag{12.97}$$

We can now solve (12.97) for i_{C1}:

$$i_{C1} = \frac{\alpha I_{EE}}{1 + e^{-v_i/V_T}} = \frac{\alpha I_{EE} e^{v_i/2V_T}}{e^{v_i/2V_T} + e^{-v_i/2V_T}}. \tag{12.98}$$

Similarly, we obtain

$$i_{C2} = \frac{\alpha I_{EE} e^{-v_i/2V_T}}{e^{v_i/2V_T} + e^{-v_i/2V_T}}. \tag{12.99}$$

We now write KVL around the output loop and use (12.98) and (12.99) to write

$$v_O = R(i_{C1} - i_{C2}) = \alpha I_{EE} R\left(\frac{e^{v_i/2V_T} - e^{-v_i/2V_T}}{e^{v_i/2V_T} + e^{-v_i/2V_T}}\right), \tag{12.100}$$

which can be written as

$$v_O = \alpha I_{EE} R \tanh\left(\frac{v_i}{2V_T}\right). \tag{12.101}$$

Equation (12.101) is the large-signal midband transfer characteristic of the differential pair and is plotted in Figure 12-56.

We can then use the Taylor series expansion of the hyperbolic tangent function to write (note that the DC differential output voltage is zero, so the total instantaneous value is equal to the AC value)

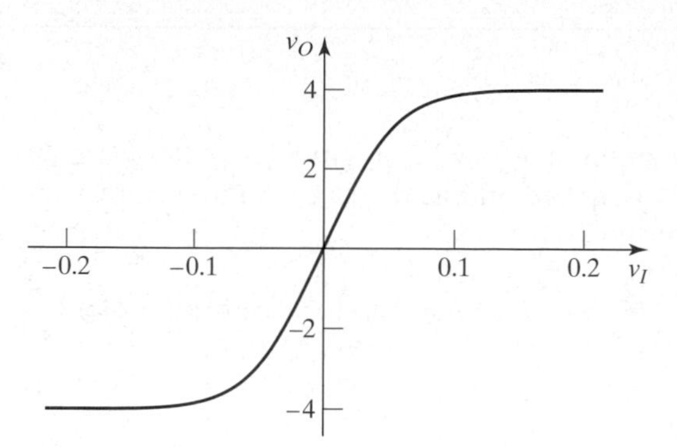

Figure 12–56 The large-signal mid-band transfer characteristic of the differential pair in Figure 12-55, plotted for the case $I_{EE} = 100\ \mu A$, $R = 40\ k\Omega$, $\alpha = 0.99$, and $V_T = 0.026$ V.

$$v_O = v_o = \alpha I_{EE}R\left[\frac{v_i}{2V_T} - \frac{1}{3}\left(\frac{v_i}{2V_T}\right)^3 + \frac{2}{15}\left(\frac{v_i}{2V_T}\right)^5 - \cdots\right]. \quad \textbf{(12.102)}$$

The first thing we notice about this expression is that the differential circuit has canceled all even-order terms in the output. This phenomenon occurs because the even-order terms in the collector currents have the same signs and don't contribute to any differential output voltage (e.g., $x^2 = (-x)^2$). Equating (12.102) with our standard power series in (12.66), we find the coefficients

$$a_1 = \frac{\alpha I_{EE}R}{2V_T} = g_m R, \quad \textbf{(12.103)}$$

$$a_3 = \frac{-\alpha I_{EE}R}{24V_T^3} = \frac{-a_1}{12V_T^2}, \quad \textbf{(12.104)}$$

and

$$a_5 = \frac{a_1}{120V_T^4}, \quad \textbf{(12.105)}$$

and all even terms are equal to zero. Notice that the first coefficient, (12.103), is the small-signal AC gain, as expected.

If the input is $v_i = \hat{V}\cos \omega t$, then using (12.72), we find,

$$HD_3 = \frac{1}{12V_T^2}\left(\frac{\hat{V}}{2}\right)^2 = \frac{1}{48}\left(\frac{\hat{V}}{V_T}\right)^2 \quad \textbf{(12.106)}$$

and

$$HD_5 = \frac{1}{120V_T^4}\left(\frac{\hat{V}}{2}\right)^4 = \frac{1}{1920}\left(\frac{\hat{V}}{V_T}\right)^4. \quad \textbf{(12.107)}$$

Comparing these values with those found in Example 12.8 for the common-emitter amplifier, we see that the third-order distortion is one-half the value it is in the common-emitter stage, but the fifth-order distortion is the same. If we use $\hat{V} = 10$ mV as before, at room temperature we obtain $HD_3 = 3.1 \times 10^{-3}$ and $HD_5 = 1.1 \times 10^{-5}$; all even-order terms are zero.

A SPICE simulation of this circuit predicts that $HD_2 = 3.7 \times 10^{-5}$, $HD_3 = 2.9 \times 10^{-3}$, $HD_4 = 9.5 \times 10^{-6}$, $HD_5 = 2.4 \times 10^{-6}$, and $THD = 0.28\%$. The main term is reasonably close to the calculated value and the others are negligible in comparison.

Example 12.11

Problem:
Consider the simplified differential amplifier shown in Figure 12-57, and assume that the two transistors are perfectly matched and at the same temperature. Calculate the values of HD_2, HD_3, HD_4, and HD_5.

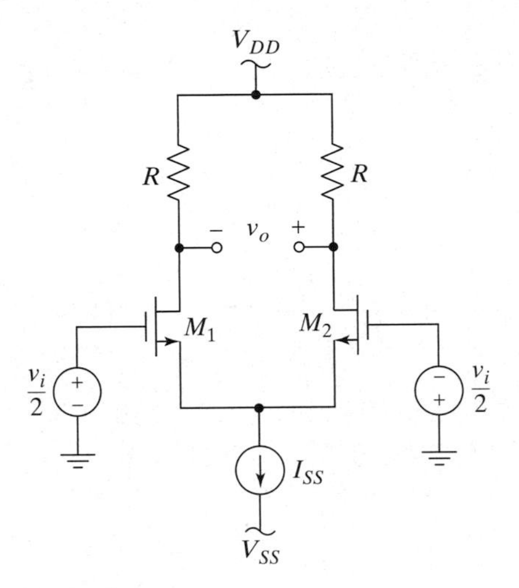

Figure 12–57 A simplified differential amplifier.

Solution:
Writing KVL around the gate-source loop yields

$$\frac{v_i}{2} - v_{GS1} + v_{GS2} + \frac{v_i}{2} = 0, \tag{12.108}$$

from which we obtain

$$v_i = v_{GS1} - v_{GS2} = \sqrt{\frac{i_{D1}}{K}} - \sqrt{\frac{i_{D2}}{K}}. \tag{12.109}$$

KCL at the source node yields

$$i_{D1} + i_{D2} = I_{SS}, \tag{12.110}$$

and, substituting (12.110) into (12.109), we obtain

$$v_i = \sqrt{\frac{i_{D1}}{K}} - \sqrt{\frac{I_{SS} - i_{D1}}{K}}, \tag{12.111}$$

which is valid for $i_{D1} \leq I_{SS}$, which corresponds to $v_i \leq \sqrt{I_{SS}/K}$. Squaring both sides of (12.111) and manipulating, we find that

$$i_{D1}^2 - I_{SS}i_{D1} + \frac{1}{4}(I_{SS} - Kv_i^2) = 0, \tag{12.112}$$

which has two solutions:

$$i_{D1} = \frac{I_{SS}}{2} \pm \frac{v_i}{2}\sqrt{2KI_{SS} - K^2v_i^2}. \tag{12.113}$$

Because of the symmetry in the circuit, the positive solution is i_{D1} and the negative solution is i_{D2}.

Writing KVL in the drain loop yields

$$v_O = (i_{D1} - i_{D2})R, \tag{12.114}$$

which, after substitution from (12.113), becomes

$$v_O = v_i R\sqrt{2KI_{SS}}\sqrt{1 - \frac{Kv_i^2}{2I_{SS}}}. \tag{12.115}$$

Equation (12.115) is valid only for $|v_i| \le \sqrt{I_{SS}/K}$. This equation is plotted in Figure 12-58.

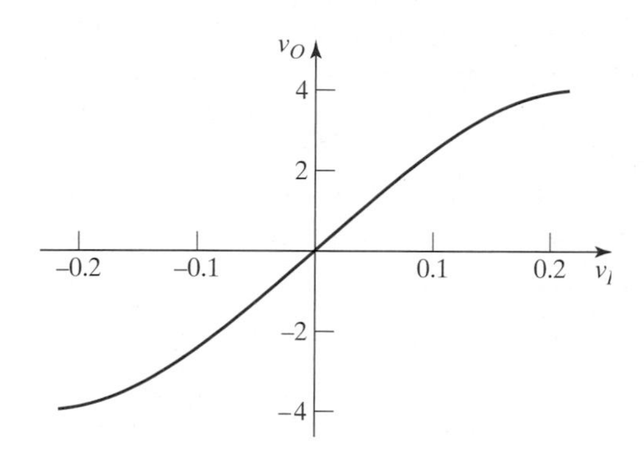

Figure 12–58 The large-signal midband transfer characteristic of the differential pair in Figure 12-57, plotted for the case $I_{SS} =$ 100 μA, $R = 40$ kΩ, and $K = 2$ mA/V². The plot is made for the full range over which the equation is valid ($-0.22 \le v_i \le 0.22$).

Since (12.115) is valid only for $|v_i| \le \sqrt{I_{SS}/K}$, we also know that $(Kv_i^2/2I_{SS}) \le 1/2$. Therefore, we can use the expansion

$$\sqrt{1 + x} = 1 + \frac{1}{2}x - \frac{1}{2\cdot 4}x^2 + \frac{1\cdot 3}{2\cdot 4\cdot 6}x^3 - \cdots, \tag{12.116}$$

which is valid for $-1 \le x \le 1$, to simplify the equation. The result is

$$v_O = R\sqrt{2KI_{SS}}\left(v_i - \frac{K}{4I_{SS}}v_i^3 - \frac{K^2}{32I_{SS}^2}v_i^5 - \cdots\right). \tag{12.117}$$

The first thing we notice about this expression is that the differential circuit has canceled all even-order terms in the output. This phenomenon occurs because the even-order terms in the drain currents have the same signs and don't contribute to any differential output voltage (e.g., $x^2 = (-x)^2$). Equating (12.117) with our standard power series in (12.66), we find the coefficients

$$a_1 = R\sqrt{2KI_{SS}} = 2R\sqrt{KI_D} = g_m R, \tag{12.118}$$

$$a_3 = -\frac{KR\sqrt{2KI_{SS}}}{4I_{SS}} = \frac{-Ka_1}{4I_{SS}}, \tag{12.119}$$

and

$$a_5 = \frac{-a_1 K^2}{32 I_{SS}^2}, \tag{12.120}$$

and all even terms are equal to zero. Notice that the first coefficient, (12.118), is the small-signal AC gain, as expected.

If the input is $v_i = \hat{V}\cos \omega t$, then using (12.72), we find that all even-order terms are zero,

$$HD_3 = \frac{K}{16I_{SS}}\hat{V}^2, \tag{12.121}$$

and

$$HD_5 = \frac{K^2}{512 I_{SS}^2} \hat{V}^4 . \tag{12.122}$$

Comparing these values with those found in Example 12.9 for the common-source amplifier, we see that distortion in the differential pair is radically different. In fact, it is somewhat surprising at first glance that we have odd-order distortion, even though we used a strictly square-law model for the FETs. We can explain the presence of these odd-order terms by first noting that a square-law characteristic does produce a second harmonic term at the sources of the transistors. The v_{GS} of both transistors then includes a difference between the fundamental at the gate and the second harmonic at the source. When this difference passes through the square-law characteristic, a third harmonic is produced. The second and third harmonics then mix and produce a fifth harmonic, and so on.

If we calculate the distortion for this circuit using $\hat{V} = 10$ mV, $K = 2$ mA/V^2, and $I_{SS} = 100$ μA, we find that all even-order terms are zero, $HD_3 = 1.3 \times 10^{-4}$, $HD_5 = 7.8 \times 10^{-9}$, and $THD = 0.013\%$.

A SPICE simulation of this circuit using $\hat{V} = 10$ mV, the Nlarge_mos model from Appendix B, $R = 40$ kΩ, and $I_{SS} = 100$ μA predicts $HD_2 = 3.5 \times 10^{-5}$, $HD_3 = 1.1 \times 10^{-4}$, $HD_4 = 1.4 \times 10^{-5}$, $HD_5 = 9.4 \times 10^{-6}$, and $THD = 0.011\%$. The main term is reasonably close to the calculated value and the other terms are negligible in comparison.

■

12.3 OUTPUT STAGES

Voltage buffers were introduced in Section 8.3, where we showed that they have a high input resistance and a low output resistance and serve to isolate a low-impedance load from the preceding gain stage. Because they are the final stage in any cascade of amplifiers, they must handle the largest signal swing. We presented methods of calculating swing limits for bipolar amplifiers in Section 12.2.1 and for FET amplifiers in Section 12.2.2. We then showed how to calculate distortion in Section 12.2.3. In Sections 12.2.1 and 12.2.2, we found that single-transistor buffers have very low efficiency when driving small loads. In this section, we present special-purpose output stages designed to drive small loads with significantly higher efficiency. Since it requires large power to drive a small resistance load with large swings, output stages used for this purpose are often called ***power amplifiers***. Since output stages must frequently dissipate significant power, we finish this section by examining the important topics of thermal modeling and heat sinks.

12.3.1 Class A Output Stages: Classification and Efficiency of Output Stages

At the beginning of Section 12.2, we defined the efficiency of an amplifier to be

$$\eta = \frac{P_L}{P_{\text{supply}}} \times 100\%, \tag{12.123}$$

which is (12.11) repeated for convenience, and both powers are averages. The efficiency is maximized by achieving the largest possible peak amplitude for the output while simultaneously minimizing the power consumed from the supply.

Assuming a sinusoidal signal with peak amplitude $\hat{V}_o$, we find that the power in the load is $P_L = \hat{V}_o^2/2R_L$, as shown in (12.12).

For example, consider a simple one-transistor buffer, as in Figure 12-59. The maximum symmetric swing would usually be limited by the negative limit and is given by

$$v_{L\max\_sym} = -v_{L\min} = V_M \frac{R_L}{R_L + R_M}. \quad \textbf{(12.124)}$$

Assuming that the DC supply and load are fixed, the maximum efficiency for this circuit is 8.3% and is obtained by choosing $R_M = R_L$ and $V_M = 2V_{OO}/3$ (*see* Problem P12.38). The corresponding maximum symmetric swing is $V_{OO}/3$.

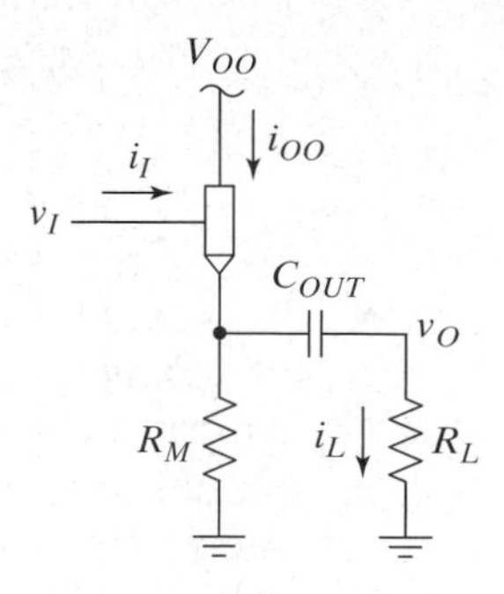

Figure 12–59 A generic one-transistor voltage buffer: the merge follower.

Exercise 12.6

Assume that the output stage in Figure 12-59 has $V_{OO} = 10$ V and $R_L = 100\ \Omega$. (a) What DC bias current should we have in the transistor to achieve maximum efficiency? (b) What is the resulting $v_{O\max\_sym}$?

The poor efficiency of this circuit is the result of the fact that the current drawn from the supply exceeds the current needed to drive the load by a significant amount. If we assume that i_I can be neglected, then when $v_O = 0$, $i_{OO} = V_M/R_M$, since $i_L = 0$. With a sinusoidal output voltage, the total current drawn from the supply will be sinusoidal, as shown in Figure 12-60, and will have an average value of $I_{OO} = V_M/R_M$ and a peak value of $\hat{i}_{OO}(\text{pk}) = 2V_M/R_M$. With the circuit biased for maximum efficiency, the minimum supply current is zero, since the transistor just cuts off at the peak negative output voltage. If the current does not go to zero at that point, the circuit is wasting some power and the efficiency is lower.

We can improve on the efficiency of the circuit in Figure 12-59 by achieving a larger maximum symmetric swing with the same supply power. One circuit that accomplishes this is shown in Figure 12-61. This circuit uses two power supplies, and the DC bias current is I_{MM}, so the power provided by the supply is $P_{\text{supply}} = I_{MM}(V_{OO} - V_{MM})$. If the supplies are symmetric, $P_{\text{supply}} = 2I_{MM}V_{OO}$. Assuming that the current source is ideal and that the transistor will work for any

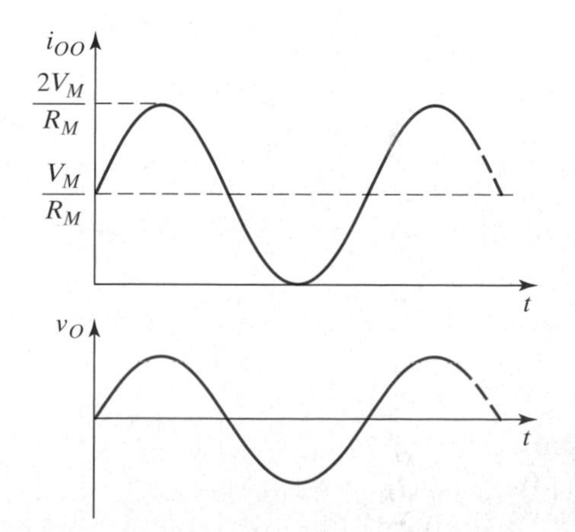

Figure 12–60 The supply current and output voltage for the circuit in Figure 12-59 when it is biased for maximum efficiency.

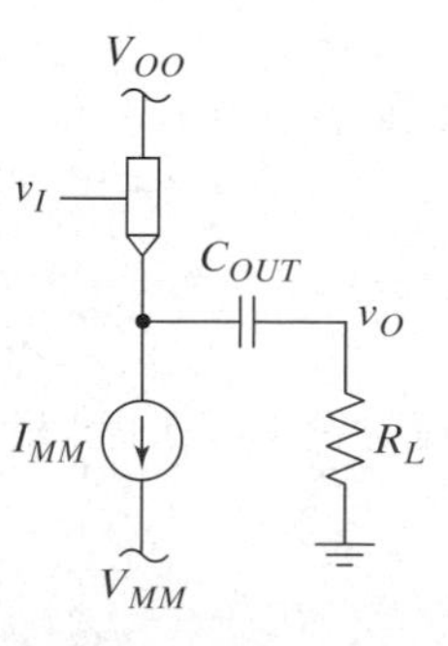

Figure 12–61 A more efficient class A output stage.

$V_{OM} \geq 0$, the maximum symmetric swing is V_{OO} and is achieved when $I_{MM} = V_{OO}/R_L$. The resulting efficiency is

$$\eta_{max} = \frac{P_{L\max}}{P_{supply}} \times 100\% = \frac{V_{OO}^2/2R_L}{2V_{OO}I_{MM}} \times 100\% = 25\%, \qquad \textbf{(12.125)}$$

which is the maximum efficiency attainable with a class A output stage (*see* Aside A12.2). We will define precisely what a class A stage is later in this section. In designing a class A output stage, it is also important to note that the power dissipated by the transistor can be quite high; in fact, it is highest when there isn't any signal present (*see* Problem P12.40). In addition, you must pay careful attention to the peak current and voltage in the device. Some general comments about high-power transistors, thermal modeling, and heat sinks are given in Section 12.3.5.

ASIDE A12.2 MAXIMUM THEORETICAL EFFICIENCY OF A CLASS A AMPLIFIER

Two Class A amplifiers are shown in Figure A12-1: an inverting amplifier and a voltage buffer. Also shown in the figure is a family of output curves for the transistor, along with the load line.

The output of the voltage buffer can be expressed as $v_L = V_{OO} - v_{OM}$, and the output of the common-merge amplifier is $v_L = v_{OM}$. In either case, if we assume that the transistor works for any $v_{OM} \geq 0$, then the maximum output swing capability of $\pm V_{OO}/2$is achieved when we choose $V_{OM} = V_{OO}/2$ for the DC bias point, as shown in the figure. With this bias point, we have $I_O = V_{OO}/2R_L$, and the peak load current is $\hat{i}_L = V_{OO}/R_L$.

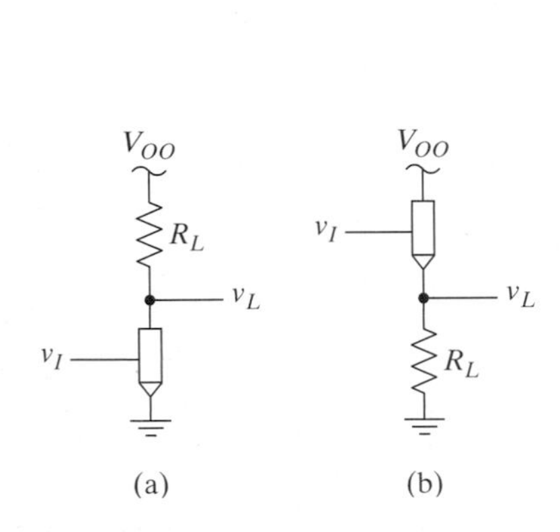

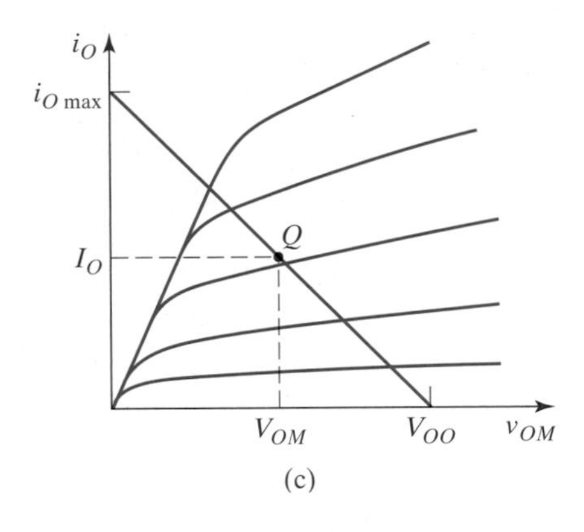

Figure A 12-1 (a) A Class A inverting amplifier (common merge). (b) A Class A voltage buffer and (c) the family of output curves for the transistor and a load line that applies to either circuit.

Assuming a sinusoidal signal, the maximum RMS output voltage is $V_l = \hat{v}_L/\sqrt{2}$ and the maximum RMS output current is $I_l = \hat{i}_L/\sqrt{2}$. The average power delivered to the load is then

$$P_L = I_l V_l = \frac{\hat{i}_L \hat{v}_L}{2} \leq \frac{(i_{O\max}/2)(V_{OO}/2)}{2} = \frac{I_O V_{OO}}{4}. \qquad \textbf{(A12.5)}$$

The average supply power is just $P_{supply} = I_O V_{OO}$, so the maximum theoretical efficiency of a Class A amplifier is

$$\eta_{max} = \frac{I_O V_{OO}/4}{I_O V_{OO}} \times 100\% = 25\%. \qquad \textbf{(A12.6)}$$

This efficiency can be doubled to 50% if an inductor or transformer is used, as shown in Example 12.12.

Exercise 12.7

Assume that the output stage in Figure 12-61 has $V_{OO} = -V_{MM}$ and $R_L = 100\ \Omega$. (a) What value of V_{OO} will set the DC power equal to that in the circuit in Exercise 12.6? (b) What value of I_{MM} should be used for maximum efficiency? (c) What is the resulting v_{Omax_sym}? (d) Show that the difference in efficiency is entirely owing to the increased swing capability.

The transistor in Figure 12-61 conducts current all of the time. (Even in the ideal circuit, it goes to zero only for an infinitesimal time at the negative peak excursion; in any practical circuit, it would not go to zero at all, or the waveform would be distorted.) We say that the **conduction angle**, θ_c, for the device is 360°; in other words, it conducts current for 360° out of every cycle of the waveform.

Output stages are classified by the conduction angle of the transistor or transistors that constitute the stage. In a **class A** output stage, $\theta_c = 360°$, so the device is always on. In a **class B** output stage, $\theta_c = 180°$, so each transistor is on for half of each cycle, and we usually use two transistors in a push–pull arrangement as will be shown later. In a **class AB** output stage, $180° < \theta_c < 360°$, and two devices are again used in a push–pull arrangement. In a **class C** output stage, $\theta_c < 180°$. The transistor currents, i_O, for the different output stages are illustrated in Figure 12-62.

The operation of more efficient output stages is covered in the sections that follow. We now present an example of how to double the efficiency and swing of a Class A output stage by using a transformer.

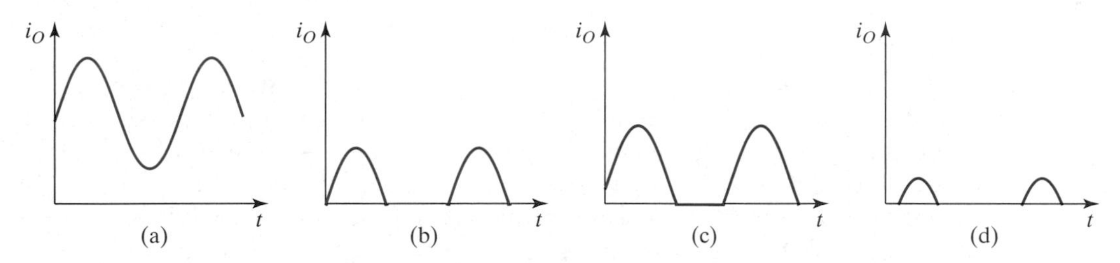

Figure 12–62 The transistor output currents in (a) a class A output stage, (b) a class B output stage, (c) a class AB output stage, and (d) a class C output stage.

Example 12.12

Problem:

Consider the transformer-coupled output stage shown in Figure 12-63. Assume that the transformer is ideal, that the input can go above V_{OO} and below ground, and that the transistor will work for any $v_{OM} \geq 0$. Determine the maximum output swing possible, the bias current needed to achieve maximum efficiency, and the maximum efficiency of the circuit. Assume that the input current can be safely ignored.

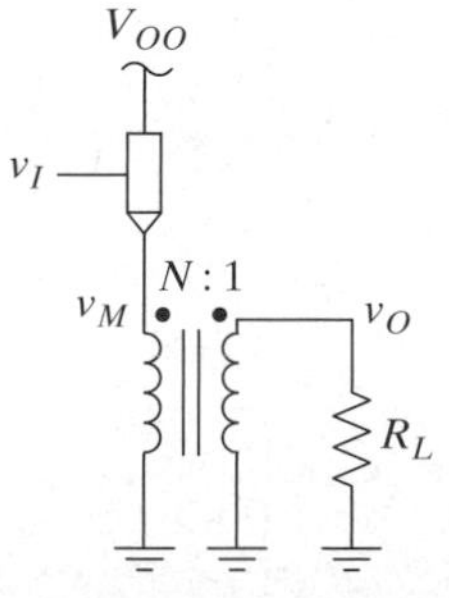

Figure 12–63 A transformer-coupled output stage.

Solution:
Using the ideal transformer relations,[10] we know that $v_O = v_M/N$ and the effective load resistance seen by the transistor is N^2R_L. Note that the primary of the transformer acts as a DC short, so the DC bias voltage at the merge terminal of the transistor is $V_M = 0$, and the DC bias current is determined by the DC component of the input voltage. The transformer is not a short for AC signals, however, so the merge terminal still follows variations in the input voltage, as shown in Figure 12-64 where a full-scale output swing is assumed (*see* Problem P12.41).

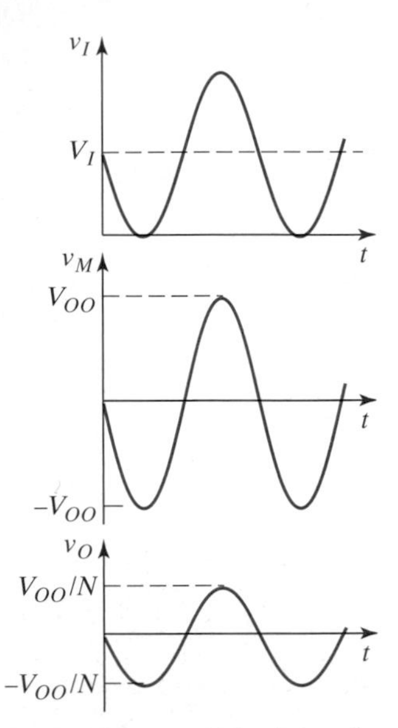

Figure 12–64 Waveforms for the circuit of Figure 12-63 with a full-scale output.

The maximum merge voltage in the ideal case is V_{OO}, and since the average must be zero, the minimum merge voltage will be $-V_{OO}$. The peak current required of the transistor is then $\hat{i}_M = V_{OO}/N^2R_L$, and maximum efficiency is obtained when the average current is equal to the peak current (since the transistor current will then just go to zero on the negative peaks). Therefore, the total supply power is

$$P_{\text{supply}} = \frac{V_{OO}^2}{N^2R_L}. \tag{12.126}$$

The voltage swing on the load is $1/N$ times as large as the swing at the merge terminal, so the average load power is

$$P_L = \frac{V_{OO}^2}{2N^2R_L}, \tag{12.127}$$

and the maximum efficiency is

$$\eta = \frac{P_L}{P_{\text{supply}}} \times 100\% = 50\%. \tag{12.128}$$

twice that of a Class A output stage without the transformer. The transformer also provides impedance translation if $N > 1$, so the output impedance of the stage is reduced, and the input impedance is increased (because the effective load is larger, which reflects into the input). These benefits are obtained at the expense of swing, which is reduced by N. The maximum theoretical efficiency is independent of N. Finally, we note that the same increase in swing can be obtained by using just an inductor, rather than a transformer (*see* [12.4] for examples), and that the same technique can be used in an inverting Class A gain stage (i.e., a common-merge stage).

■

12.3.2 Class B and AB Output Stages

This section introduces the most common output stages used in integrated circuit design: the Class B *push–pull output stage* and the Class AB push–pull output stage. At the end of the section, we present an aside that derives the maximum theoretical efficiency of these stages and examines the power dissipated in the transistors as a function of the power delivered to the load.

Because the operation of the two stages depends on the large-signal operation of the transistors, we will describe the operation of the bipolar and FET stages sepa-

[10] For those not familiar with transformers, one good treatment is presented in [12.2].

rately. Nevertheless, there are a few comments we can make that apply equally to both. A generic Class B, or push–pull, output stage is shown in Figure-12-65(a) and the static transfer characteristic is shown in part (b) of the figure.

When the input voltage is near zero, both transistors are cut off and the output is zero. As v_I becomes sufficiently positive, the *n*-type transistor will turn on and act as a voltage buffer, whereas the *p*-type transistor is off. The reverse statement is true for negative values of the input. The transfer characteristic clearly shows the dead band that results for inputs that are too small to significantly bias either transistor. The key word in the previous sentence is "significantly"—a class B output with ideal transistors does not have this dead band because the transistors turn on abruptly.

Because the DC bias current in this circuit is zero, there isn't any power dissipation when the input is zero, and the overall efficiency is high—theoretically, as high as 78.5% (*see* Aside A12.3). In addition, since transistors are used to pull the output both high and low, the driving capability of the circuit is more symmetric than a Class A output stage. The major disadvantage of this stage is the ***crossover distortion*** that results from the shape of the static transfer characteristic near the origin. In Figure 12-66, we show a sinusoidal input and the corresponding output waveform. The crossover distortion is clearly visible as shoulders in the output waveform.

There are two main techniques for reducing crossover distortion to an acceptable level. The first is to use negative feedback, as discussed in Chapter 10 and illustrated in Example 12.13, and the second is to modify the stage to make it a Class AB stage, which we illustrate after that example.

Figure 12–65 A generic Class B output stage.

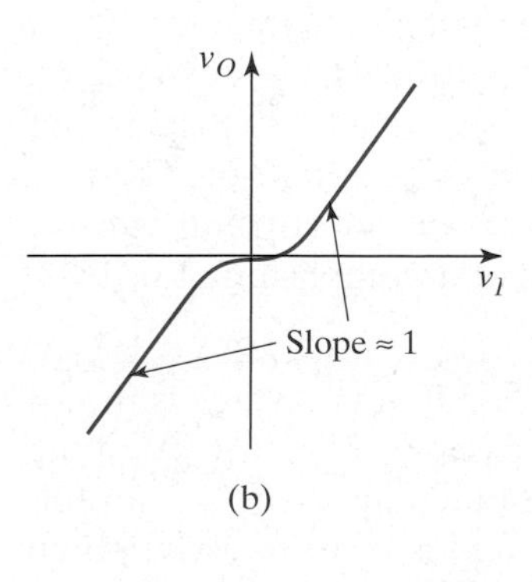

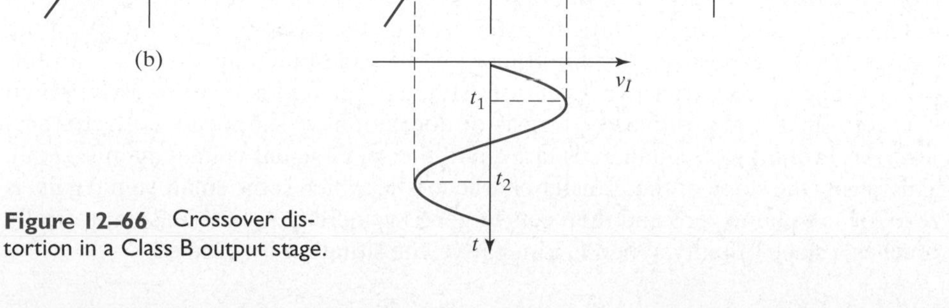

Figure 12–66 Crossover distortion in a Class B output stage.

Example 12.13

Problem:
Consider the negative-feedback amplifier with a Class B output stage shown in Figure 12-67(a). Assume that the op amp is ideal, except that it has a finite gain equal to A; that is, $v_A = A(v_I - v_O)$. For simplicity, we approximate the transfer characteristic of the Class B output stage as shown in part (b) of the figure; it has a slope of zero for $-V_1 \leq v_A \leq V_1$ and a slope of unity elsewhere. Determine the static transfer characteristic of the closed-loop amplifier.

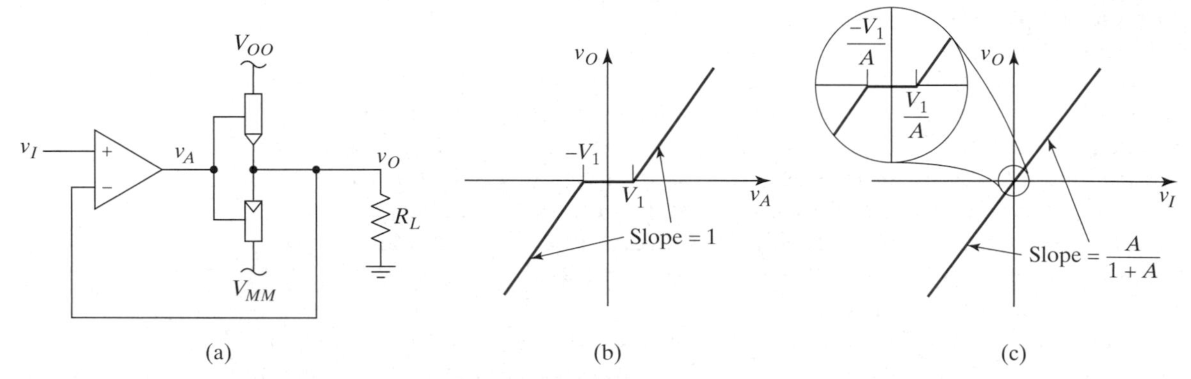

Figure 12–67 (a) A negative-feedback amplifier with a Class B output stage. (b) The static transfer characteristic of the output stage. (c) The static transfer characteristic of the closed-loop amplifier.

Solution:
Start by considering the case where $v_I = v_A = v_O = 0$. Then, as the input is increased, v_A increases, too; however, initially, v_O does not change. When the input reaches $v_I = V_1/A$, however, the output starts to follow further changes in the input. The slope of the closed-loop gain in this region is $A_f = A/(1 + A)$, which is very nearly unity for large values of op amp gain. Note that the size of the dead band in the transfer characteristic has been reduced by the op amp gain, as shown in part (c) of the figure.

■

We can also reduce crossover distortion by adding some circuitry to bias the transistors so that they have a small current flowing in them even when the signal is absent. The result is a cross between a true Class B stage and a Class A stage and is called a Class AB stage. The efficiency can be close to that of a Class B stage, but will certainly be lower. A conceptual schematic is shown in Figure 12-68. The batteries will provide some bias for the transistors so that the output currents will not be zero, even when the input is zero. Several methods of accomplishing this biasing are illustrated in the problems.

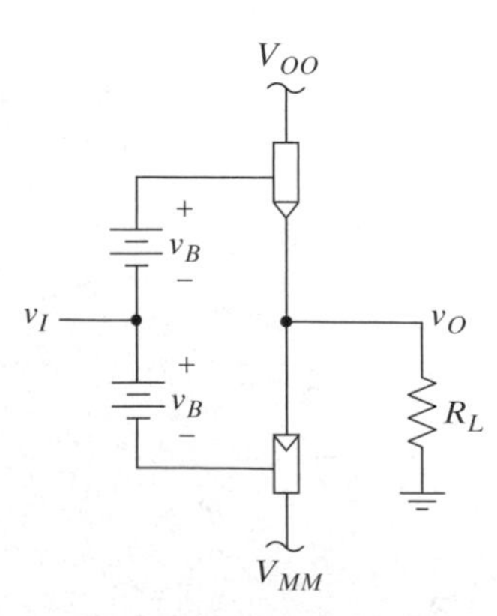

Figure 12–68 Conceptual schematic of a Class AB output stage.

Bipolar Class B and AB Output Stages A bipolar Class B output stage is shown in Figure 12-69(a), and the static transfer characteristic is shown in part (b) of the figure. When v_I is positive, the *npn* transistor is on and functions as an emitter follower, while the *pnp* transistor is cutoff. Although the *npn* is forward active when v_I is very small and positive, the transistor does not have significant collector current. As a result, g_m is small, r_π is large, and the small-signal voltage gain is small. This is why the slope of the transfer characteristic, which is the small-signal gain, is zero for v_I equal to zero and then gets larger as v_I increases, until it asymptotically reaches a slope of unity. When v_I is negative, the situation is reversed.

ASIDE A12.3 MAXIMUM EFFICIENCY OF CLASS B OUTPUT STAGES AND TRANSISTOR POWER DISSIPATION

A generic Class B output stage is shown in Figure A12-2, where we have assumed symmetric supplies. The output current in the n-type transistor is $i_N = v_O/R_L$ for $v_I \geq 0$ and is zero for $v_I < 0$. Similarly, $i_P = -v_O/R_L$ for $v_I \leq 0$ and zero for $v_I > 0$. The voltage across the n-type transistor is $v_N = V_{OO} - v_O$, and across the p-type transistor we have $v_P = v_O + V_{OO}$.

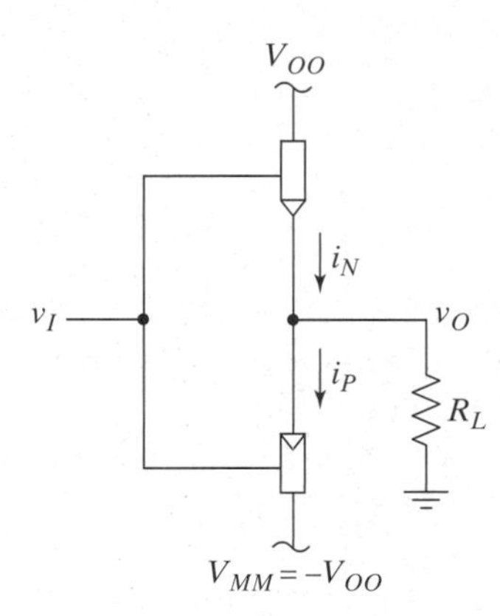

Figure A12-2 A generic Class B output stage.

Assuming a sinusoidal input and ideal transistors, and ignoring crossover distortion, we can express the output voltage as $v_O = \hat{V}_o \sin\omega t$. The average positive supply current is then

$$I_{OO} = \frac{1}{T}\int_0^T i_N(t)dt. \qquad \textbf{(A12.7)}$$

Recognizing that the input is positive for $0 \leq t \leq \pi/\omega$, we substitute for i_N and obtain

$$I_{OO} = \frac{\hat{V}_o}{R_L T}\int_0^{\pi/\omega} \sin(\omega t)\, dt = \frac{\hat{V}_o}{\pi R_L}. \qquad \textbf{(A12.8)}$$

Because of the symmetry of the circuit, the average current from the negative supply is the same, so the overall supply power is $P_{supply} = 2I_{OO}V_{OO}$. The average power in the load is $P_L = \hat{V}_o^2/2R_L$, so the efficiency of the stage is

$$\eta = \frac{\hat{V}_o^2/2R_L}{2V_{OO}(\hat{V}_o/\pi R_L)} \times 100\%. \qquad \textbf{(A12.9)}$$

The maximum efficiency is obtained when $\hat{V}_o = V_{OO}$ and is

$$\eta_{\max} = \frac{\pi}{4} \times 100\% = 78.5\%. \qquad \textbf{(A12.10)}$$

This efficiency is not possible with practical transistors (since the output cannot swing from V_{OO} to V_{MM}), so it is only an upper limit.

We are also interested in examining the maximum power dissipated in either of the transistors in this stage. (Owing to the symmetry, it is the same for either transistor.) The instantaneous power for one cycle of the input for the n-type transistor is

$$p_N = v_N i_N = \begin{cases} \dfrac{V_{OO}V_o}{R_L}\sin\omega t - \dfrac{\hat{V}_o^2}{R_L}\sin^2\omega t\,; & 0 \leq t \leq \dfrac{\pi}{\omega} \\ 0\,; & \dfrac{\pi}{\omega} < t \leq \dfrac{2\pi}{\omega} \end{cases}, \qquad \textbf{(A12.11)}$$

and the average power in the transistor is then

$$P_N = \frac{\hat{V}_o}{TR_L}\left[V_{OO}\int_0^{\pi/\omega}\sin(\omega t)dt - \hat{V}_o\int_0^{\pi/\omega}\sin^2(\omega t)dt\right]$$

$$= \frac{\hat{V}_o}{2\pi R_L}\left(2V_{OO} - \frac{\pi\hat{V}_o}{2}\right). \qquad \textbf{(A12.12)}$$

We can express this equation in terms of the average power in the load, $P_L = \hat{V}_o^2/2R_L$, and the maximum power in the load, $P_{max} = V_{OO}^2/2R_L$. After some algebra, the result is

$$P_N = \frac{2}{\pi}\sqrt{P_L P_{max}} - \frac{P_L}{2}. \qquad \textbf{(A12.13)}$$

This equation is plotted in Figure A12-3, where we have normalized both axes to the maximum power in the load. The maximum power dissipation in the transistor occurs when the power being delivered to the load is only 41% of the maximum, and the efficiency at that point is only 50%. Consequently, the stage works best when driven near the maximum swing limit (but, of course, the distortion is worse as you approach the swing limit). This plot illustrates an important advantage of Class B output stages: The maximum power dissipated in either transistor is only 20% of the maximum power that can be delivered to the load. Contrast this with a Class A stage, where the transistor has to dissipate twice the maximum load power, and you see that the transistors in a class B stage do not need to be nearly as large as those in a Class A stage.

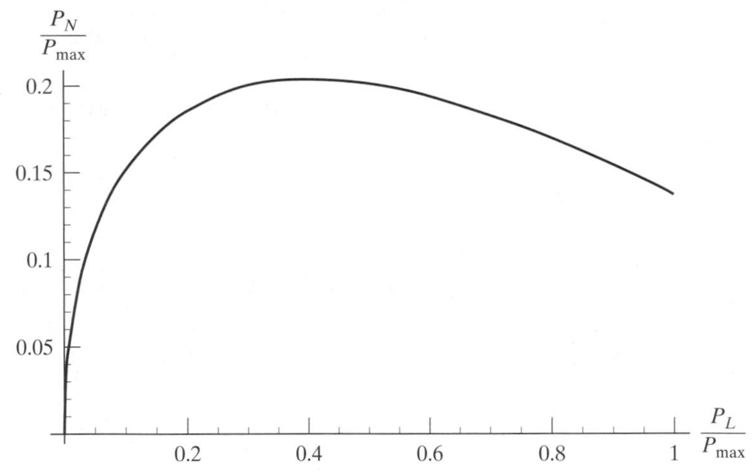

Figure A12-3 The power dissipated in either transistor of the push–pull output stage as a function of the power delivered to the load.

Exercise 12.8

Consider the output stage in Figure 12-69(a) at room temperature. If $R_L = 50\ \Omega$ and $I_S = 10^{-14}$ A for both transistors, find the low-frequency small-signal voltage gain when (a) $V_I = 100$ mV, (b) $V_I = 500$ mV, and (c) $V_I = 1$ V. (d) Relate the answers you get to the characteristic in Figure 12-69(b).

FET Class B and AB Output Stages A MOSFET Class B output stage is shown in Figure 12-70(a), and the static transfer characteristic is shown in part (b) of the figure. When v_I is positive and greater than the threshold voltage, the *n*-channel transistor is on and functioning as a source follower while the *p*-channel transistor is cutoff. When v_I is negative, the situation is reversed. For $V_{thp} < v_I < V_{thn}$, both transistors are cut off, and the output is zero. This description of the operation assumed our simple square-law model for the MOSFET and ignored subthreshold conduction, but these approximations do not significantly affect the results.

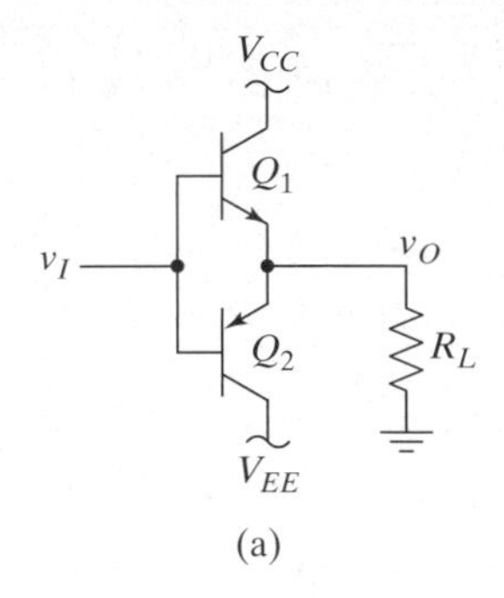

(a)

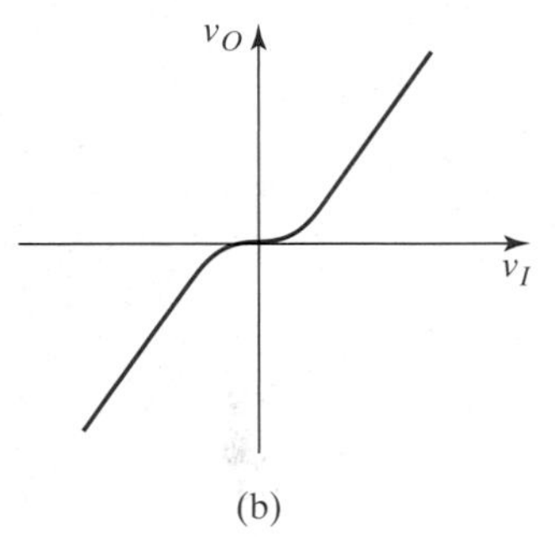

(b)

Figure 12–69 (a) A Class B bipolar output stage. (b) The static transfer characteristic.

Exercise 12.9

Consider the output stage in Figure 12-70(a) with $R_L = 50\ \Omega$. Assume that $K = 2\ mA/V^2$ for both transistors, $V_{thn} = 1$ V, and $V_{thp} = -1$ V. Find the low-frequency small-signal voltage gain when (a) $V_I = 500$ mV, (b) $V_I = 2$ V, and (c) $V_I = 7$ V. (d) Relate the answers you get to the characteristic in Figure 12-70(b).

12.3.3 More Advanced Output Stages

A large number of different output stages have been developed over the years for special-purpose applications where high efficiency is important. For example, in battery-operated transmitters (e.g., walkie-talkies, cellular telephones), it is important to be efficient, since the power drain on the battery is significant even if all of the power gets to the antenna.[11] In audio power amplifiers, efficient output stages are desirable because the power supply required and cooling requirements can be significantly reduced.[12]

Two common approaches to making efficient output stages are to have the transistors be on only for a small fraction of the cycle (e.g., Class C amplifiers) and to change the supply voltage used by the final output stage as needed (e.g., Class G amplifiers). We very briefly introduce each of these approaches here.

Class C stages are very specialized, are highly nonlinear, and are used mostly in radio-frequency circuits (*see* [12.4] for a good treatment). We have encountered class C operation elsewhere in the text, but not in an output stage. For example, the transistor amplifier in the Colpitts oscillator examined in Exercise 10.23 operates as a class C amplifier. The large amount of distortion produced by Class C operation is filtered out by the resonant tank circuit used for feedback. Class C amplifiers cannot typically be used without tuned circuits, because of the harmonic distortion they produce, so we will not discuss them further here. Nevertheless, there is a similar type of amplifier, sometimes called a **switching amplifier**, that works on the same principle of keeping the transistor off a large portion of the time (i.e., using a very small conduction angle).

The advantage of a switching amplifier over a Class C amplifier is that the switching is done at a frequency much higher than that of the signal.(The "switching" in a Class C amplifier is at the signal frequency.) Using high-frequency switching makes it much easier to filter the unwanted components out of the output.

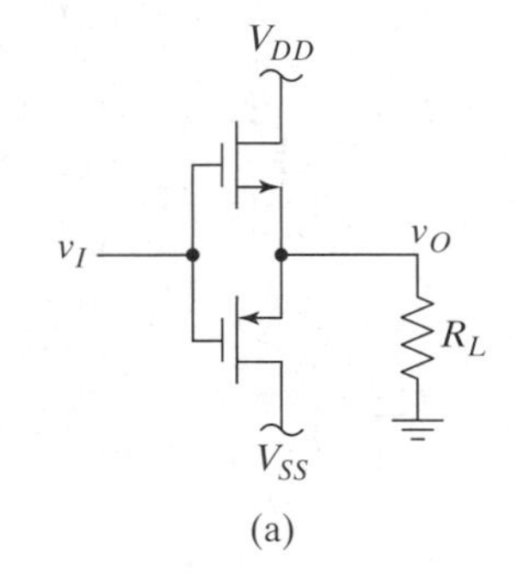

(a)

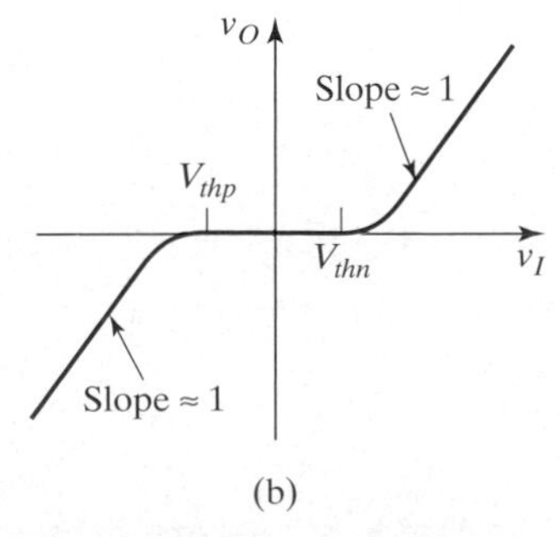

(b)

Figure 12–70 (a) A Class B MOSFET output stage. (b) The static transfer characteristic.

[11] Unfortunately, some modulation formats require that the power amplifier be linear, which does not allow many types of efficient output stages to be used (e.g., Class C). Overcoming this trade-off is an active area of research at the present time.

[12] For high-end audio equipment, Class A amplifiers are preferred by some because they have lower distortion than some other, more efficient, amplifiers.

As an example of a switching amplifier, consider the operation of the circuit shown in Figure 12-71 along with the waveforms. The input signal—a sine wave in this example—is first multiplied by a high-frequency pulse stream with a very small duty cycle. The resulting amplitude-modulated stream of pulses is applied to a standard Class B output stage. The transistors in this stage are on only for brief periods of time, so the efficiency is extremely high. At the output, the waveform is passed through a low-pass filter to remove the pulses and recover the original waveform. In the example shown here, the filter is just a single-pole *RC*, so the output waveform has noticeable high-frequency content.

A second approach to produce high efficiency output stages is to vary the power supply as needed. A conceptual push–pull amplifier is shown in Figure 12-72. The circuit operates in the following way: For small positive inputs, T_1 is the only device on, and it supplies the current to R_L. For larger positive inputs, the circuit labeled X_1 turns on and supplies more current to the load. We assume that $V_2 > V_1$, so that when T_1 is the only device on, the power dissipated in the circuit is lower than it would be if T_1 tied directly to V_2. Nevertheless, by adding the extra circuitry, we can still pull the load to a voltage higher than V_1. The operation for negative inputs is the same. Example 12.14 demonstrates the idea.

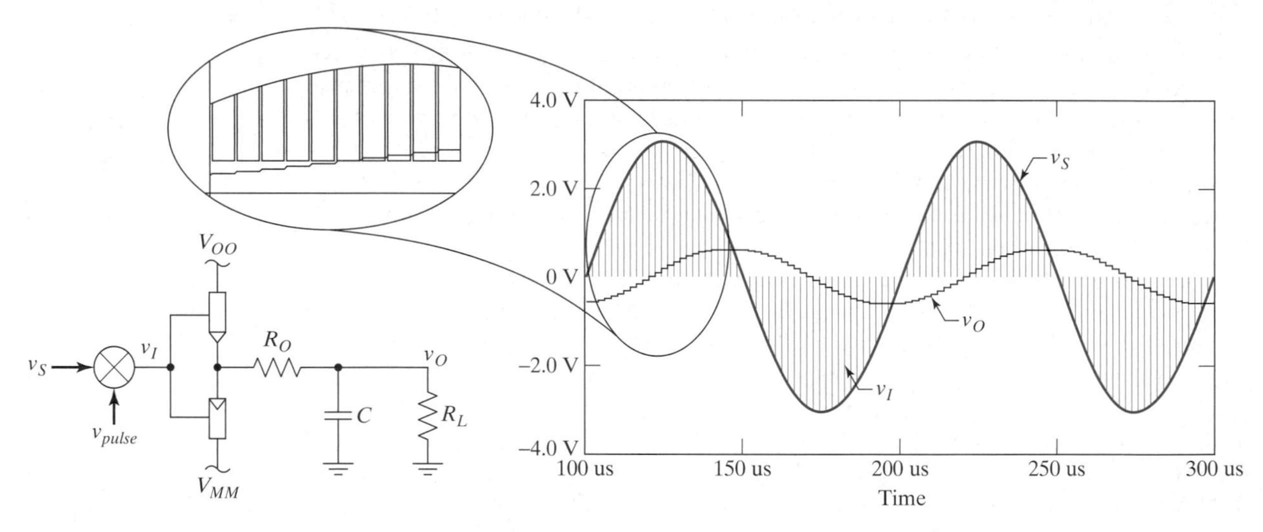

Figure 12–71 (a) A switching power amplifier (the circle with a cross through it represents a multiplier) and (b) the waveforms (see swit_amp.sch on the CD).

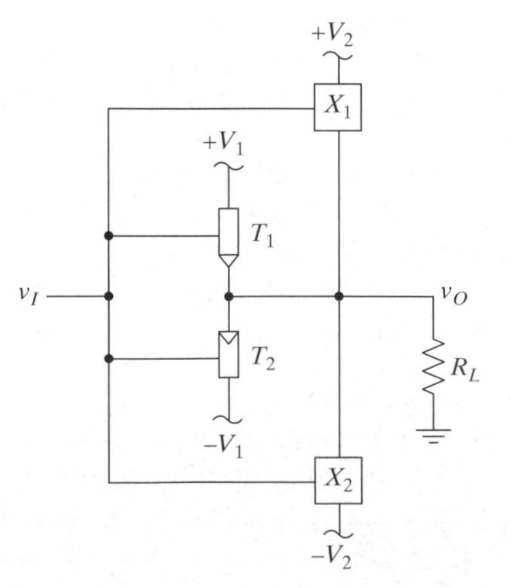

Figure 12–72 A conceptual high-efficiency output stage.

Example 12.14

Problem:
Examine the operation of the Class G output stage shown in Figure 12-73. Assume that the input voltage has a 4-V DC component.

Solution:
For small input voltages, Q_2 is off, and current is supplied to the load through Q_1 and D_1. While this is true, the power dissipated in the circuit is smaller than it would be if the current were supplied by the 10-V supply, so the efficiency is higher. The collector of Q_1 is at about 4.3 V at this time, and the base of Q_2 is about 0.7 V higher than v_I. So, once v_I gets to about 4.3 V, Q_2 starts to turn on and pulls the collector of Q_1 up above 4.3, thus turning off D_1.

When v_I is large enough to turn Q_2 on, the current is supplied by the 10-V supply and the output can be pulled higher than 5 V, which would have been impossible with the 5-V supply alone. Therefore, this circuit essentially has two separate supply voltages and uses the higher voltage only when necessary, thereby increasing the efficiency. Figure 12-74 shows the output waveform and the voltage at the collector of Q_1 for an input with a 4-V DC component and a 3-V peak sine wave at 10 kHz. Note that the collector of Q_1, which acts as the supply voltage, goes up when needed.

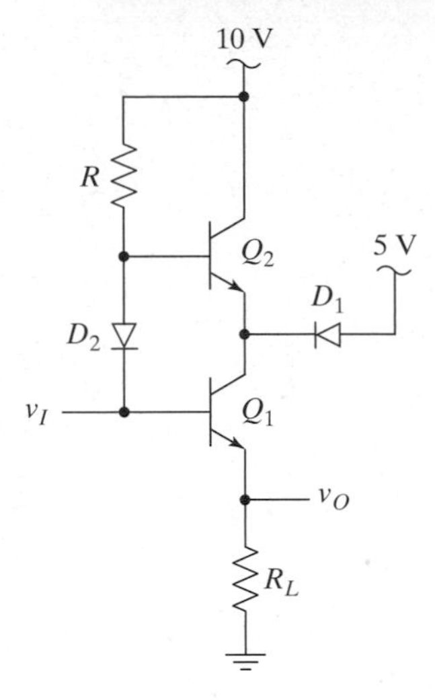

Figure 12–73 A Class G output stage.

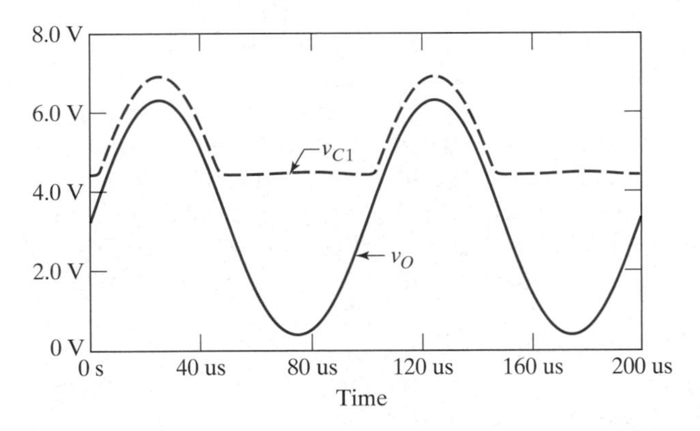

Figure 12–74 The waveforms of the circuit of Figure 12-73 (see Exam14.sch).

12.3.4 Power Transistors, Thermal Modeling, and Heat Sinks

The voltages, currents, and power can all be quite large in an output stage, so we need to consider all three carefully during the design process. In Chapter 7 we discussed the safe operating area of a transistor, and we must be especially mindful of that when designing output stages. With that said, the power dissipated by the transistors in an output stage sometimes requires special attention. If the devices dissipate too much power, they will be destroyed as heat builds up. It is sometimes necessary to provide heat sinks and sometimes even to provide fans for extra cooling. There are even some very high-power circuits that use liquid cooling! In this section, we discuss how to model the heat flow and how to determine whether or not a heat sink or other precautions are necessary. Before we discuss thermal modeling, however, we want to point out some very important characteristics of bipolar and field-effect transistors when used in power amplifiers. A good detailed treatment of power transistors and applications is given in [12.5].

Bipolar Power Transistors When the base-emitter junction of a bipolar transistor heats up, the current will increase if the bias voltage is kept constant. This extra current will then cause more power dissipation and will further increase the temperature of the junction. The situation can easily get out of hand and lead to what is called ***thermal runaway***.

The problem of thermal runaway implies that it is not generally possible to put power bipolar transistors in parallel to increase their power-handling capability. If the transistors are not perfectly matched, and one of them starts to draw slightly more current, it will heat up more than the others. The increase in temperature will cause a further increase in the current. Eventually, the one device will take most or all of the current until it burns up. This problem may also show up if an individual transistor has nonuniform current flow across the base-emitter junction area. The region with a higher current density may heat up locally, draw more current, heat up more, and eventually destroy the device. This problem is called ***second breakdown***.

Another problem that is often troubling with power BJTs is the base current. For example, suppose the collector current is 10 A and the beta is 100, the base current is then equal to 100 mA, still a very large value. This problem implies that driving power BJTs is not always an easy task. Finally, the turnoff time of a power BJT can be significant owing to the minority-carrier charge storage in the base.

Power MOSFETs To use a MOSFET to switch or control very large currents, we would like to minimize the channel resistance. This implies using very short channel lengths and very large widths. Unfortunately, with a conventional short-channel MOSFET, the drain-to-source breakdown voltage tends to be small due to punch-through, as discussed in Chapter 2. As the drain voltage is increased, the drain-to-bulk depletion region gets wider. Because the doping of the drain is higher than the doping of the body, the depletion region moves mostly in the body. If it moves far enough, it will touch the source-to-body depletion region and cause punch-through.

Power MOSFETs are constructed differently to avoid punch through and optimize high-current performance. One common type of structure is shown in cross section in Figure 12-75. This structure is called a ***vertical diffused MOSFET*** (VDMOS) and has only three terminals.

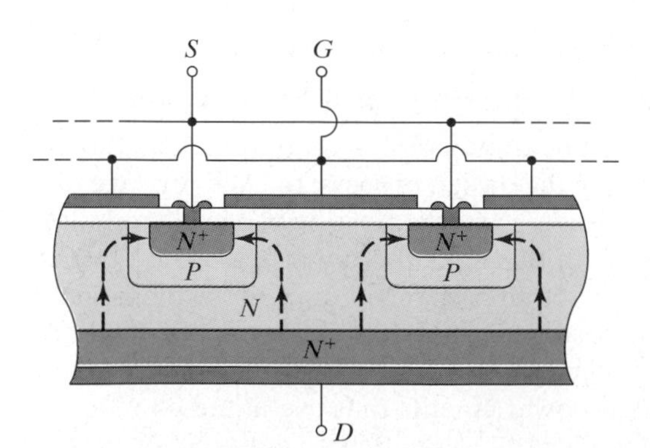

Figure 12–75 An n-channel vertical diffused MOSFET transistor. There are many small source islands connected in parallel. The drain current flows as indicated by the dashed lines.

When a gate voltage greater than the threshold is applied, and an inversion layer is formed at the surface of the p-type body regions, the drain current in this device flows vertically from the drain to the surface and then across to the source. As the drain voltage is increased, the depletion region between the drain and the body moves mostly into the lightly doped n-type region, and punch-through is avoided.

One interesting characteristic of power MOSFETs is that for reasonable currents, the drain current is approximately proportional to the gate overdrive (i.e., $v_{GS} - V_{th}$) rather than to the square of the overdrive. Therefore, the transconductance reaches a maximum value and cannot be increased further by increasing the gate overdrive.

The threshold voltage of a MOSFET decreases with increasing temperature, which causes an increase in drain current at a fixed v_{GS}, whereas the coefficient K decreases with increasing temperature (owing to decreasing mobility), causing a decrease in drain current. There exists a bias point at which these two effects cancel and the drain current does not change with temperature. Nevertheless, for large currents, the current decreases with increasing temperature, so power MOSFETs are inherently stable with respect to temperature (i.e., an increase in power will increase the temperature, which will tend to decrease the current and therefore the power). This implies, for example, that power MOSFETs can be put in parallel to handle larger currents.

Power MOSFETs have essentially infinite DC current gain, so the static current of the drive circuits is zero, which makes power MOSFETs very easy to drive. They also do not have minority-carrier storage, so they turn on and off quickly. One problem with driving power MOSFETs, however, is that they have large gate areas and hence large input capacitance. (Several nF are not uncommon.) The input capacitance requires large dynamic input current if the gate voltage is to be changed rapidly.

Thermal Modeling and Heat Sinks Because power transistors dissipate large amounts of power, their internal temperatures will usually be significantly higher than the surrounding ambient temperature. Manufacturers will guarantee the device specifications only so long as the internal temperature (called the junction temperature, with reference to a pn junction inside the device) does not exceed some limit, usually 125°C or 150°C. In addition, the failure rate for semiconductor devices doubles for every 10–15°C rise in temperature. For both of these reasons, it is important to know how to calculate the junction temperature of a power transistor (and other circuits as well!).

Heat can be transferred by conduction, convection, or radiation. Let's consider only conduction for the moment. Imagine that a temperature difference, ΔT, exists between the ends of a bar of material with length l and cross-sectional area A. There will be a net flow of energy from the hotter end to the cooler one. The energy flow per unit time is the power transferred and is given by

$$P_{cond} = \frac{\lambda A \Delta T}{l}, \tag{12.129}$$

where λ is the thermal conductivity of the material (usually in W/m/°C) [12.5].

We see from (12.129) that the driving force for heat flow is a difference in temperature. We can use an electrical analogy here and say that temperature is analo-

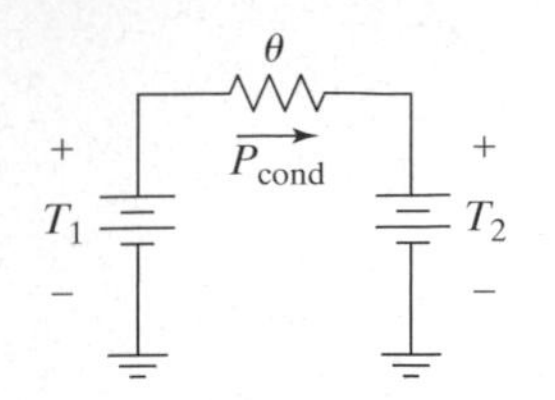

Figure 12–76 An electrical analog for (12.129).

gous to voltage and heat flow is analogous to current. If we define the thermal resistance to be

$$\theta = \frac{\Delta T}{P}, \tag{12.130}$$

then we can represent the situation described by (12.129) with the circuit shown in Figure 12-76, and in this case, $\theta = \lambda l/A$. The units for thermal resistance are °C/W.

Continuing with the analogy, if a junction is dissipating P_j watts of power, we can represent that dissipation as a current source (i.e., there is a certain amount of energy that must be removed per unit time). If we represent the ambient temperature as ground (zero volts) and if we know the thermal resistance from the junction to the ambient, θ_{JA}, we can then find the junction temperature (relative to ambient) by using the equivalent circuit shown in Figure 12-77.

The resulting junction temperature is

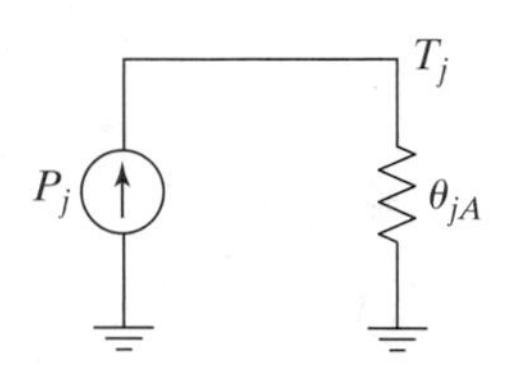

Figure 12–77 Equivalent circuit for finding the temperature of a junction dissipating P_j watts.

$$T_j = T_A + P_j\theta_{JA}, \tag{12.131}$$

where T_A is the ambient temperature. If the junction temperature must be kept below T_{max}, then we must keep the power dissipation below

$$P_{jmax} = \frac{T_{jmax} - T_A}{\theta_{JA}}. \tag{12.132}$$

In addition, there is always an absolute maximum power that can be dissipated independent of the junction temperature, P_{max}. Plotting (12.132) subject to the maximum power restriction leads to the ***power-derating curve*** shown in Figure 12-78.This curve shows how the maximum allowable power dissipation in the transistor varies with ambient temperature. The maximum power of the transistor is specified at some temperature T_{Ao}, and for any temperature above that value, the derating begins. Typically, T_{Ao} is near 25°C, but not always. As noted earlier, T_{jmax} is usually 150°C. A similar plot can be made showing the maximum power dissipation versus the case temperature of the transistor.

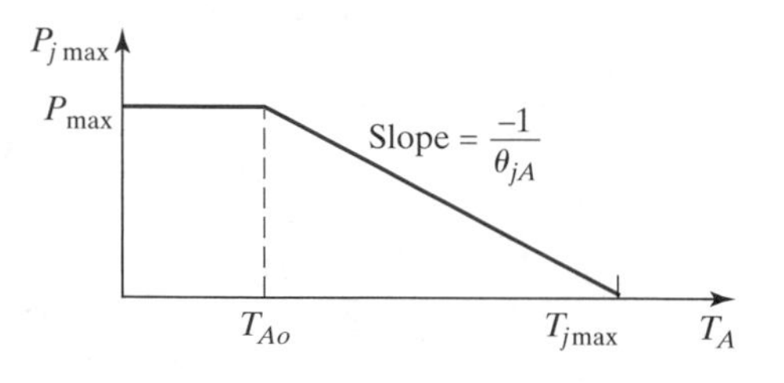

Figure 12–78 A transistor power derating curve.

Example 12.15

Problem:
Determine the maximum dissipation allowed at an ambient temperature of 100°C for a 2-W power transistor rated at 30°C. Assume that $\theta_{JA} = 60°C/W$.

Solution:
We first find $T_{jmax} = T_{Ao} + P_{max}\theta_{JA} = 150°C$. We then use (12.132) to determine that $P_{jmax} = 0.83$ W.

■

If we want to dissipate more power at a given ambient temperature than is allowed by (12.132), we can lower the value of θ_{JA} by using a heat sink. The total thermal resistance has three components: the junction-to-case resistance, θ_{JC}; the case-to-sink resistance, θ_{CS}; and the sink-to-ambient resistance, θ_{SA}. The total is

$$\theta_{JA} = \theta_{JC} + \theta_{CS} + \theta_{SA}. \quad \textbf{(12.133)}$$

The value of θ_{JC} is fixed by the type of packaging used, but the values of θ_{CS} and θ_{SA} depend on how we mount the case and what kind of heat sink, if any, is used. If we just let the case sit in still air, $\theta_{CA} = \theta_{CS} + \theta_{SA}$ will have the largest value possible. If we provide a heat sink, a fan, or both, we can lower the value substantially by improving the rate at which heat is removed by conduction, convection, and radiation.

Typical values of these resistances are $\theta_{JA} \approx 50°\text{C/W}$ (without any heat sink or forced airflow), $\theta_{JC} \approx 1°\text{C/W}$, $\theta_{CS} \approx 0.5°\text{C/W}$, and $\theta_{SA} \approx 0.5$ to $10°\text{C/W}$, depending on the type of heat sink.

Exercise 12.10

A transistor is rated as being able to dissipate 1 W safely at 25°C and has a maximum junction temperature of 125°C. (a) What is θ_{JA}? (b) How much power can this transistor dissipate at 50°C?

SOLUTIONS TO EXERCISES

12.1 Using (12.2) along with the assumption that $V_p \approx V_m$ and the desired ripple magnitude of 5%, we find that $C = T/(0.05R)$. Plugging in the given values leads to $C = 21\ \mu\text{F}$.

12.2 With $R_C = 1\ \text{k}\Omega$ we find that $R'_L = R_L \| R_C = 500\ \Omega$ and using $A_v = -g_m R'_L$ we see that we must set $g_m = 106$ mA/V, which requires that $I_C = 2.76$ mA at room temperature and results in $V_C = 7.24$ V. Using this new design and (12.13), we find $v_{L\max} = 1.4$ V, and from (12.15) and (12.16) we get $v_{L\min} = -5.7$ V, so $v_{L\max\_sym} =$ 1.4 V as before. It is interesting to see why the positive swing didn't change, even though we reduced R_C. It is a direct result of our desiring to keep the gain constant. Note that

$$\begin{aligned} g_m(R_C \| R_L) &= g_m \frac{R_C R_L}{R_C + R_L} = \frac{I_C R_C}{V_T} \frac{R_L}{R_C + R_L} \\ &= \frac{(V_{CC} - V_C)}{V_T} \frac{R_L}{R_C + R_L} = \frac{v_{L\max}}{V_T}, \end{aligned}$$

so, for this topology, if we keep the gain constant, we also keep the swing constant. Problem P12.20 explores how to design this circuit for maximum swing.

12.3 To attain the maximum symmetric swing, we set $v_{O\max} = -v_{O\min}$. Using (12.23) and (12.24) leads to

$$V_E = (V_{CC} - 0.7) \Big/ \left(1 + \frac{R_L}{R_L + R_E}\right) = 3.0\ \text{V},$$

which produces $v_{Omax\_sym} = 1.3$ V. Therefore, the bias current is now $I_C \approx I_E = V_E/R_E = 2.3$ mA. Assuming that $I_{R1} = 100$ μA, the total supply current is 2.4 mA, and the DC power is 12 mW. With a 1.3-V peak sine wave at the output, the signal power in the load is 845 μW, so the efficiency is 7%.

12.4 **(a)** We start by noting that $R_o \approx R_D$. To determine what value to choose for V_D, we need to know how our choice will affect the power dissipation. The power will be proportional to I_D, which is given by $I_D = (V_{DD} - V_D)/R_D$. Therefore, we solve (12.43) for the value of R_D required to meet the swing specification for a given choice of V_D. The result is

$$R_D = R_L\left(\frac{V_{DD} - V_D}{v_{L\max}} - 1\right).$$

We can plug this result into the equation for I_D and have both R_D and I_D as a function of V_D. If we examine the derivative of I_D with respect to V_D, we see that it is positive. Therefore, increasing V_D also increases I_D, and we can conclude that we should choose the minimum value of V_D possible in order to minimize the power dissipation. The absolute minimum value for V_D would result if we let V_S go to ground (i.e., short out R_S), but this would make the bias point highly dependent on the exact values of V_{th} and K. If we calculate the value of I_D for various values of V_D we see that the current is very close to the minimum value when $V_D = 2$ V, which leads to $R_D = 20$ kΩ, $I_D = 150$ μA, and $V_{GS} = 1.27$ V. To meet the minimum swing requirement, we use (12.47) to find that $V_G - V_{th} = 1$ V. Therefore, we set $V_G = 2$ V. This leads to $V_S = V_G - V_{GS} = 0.73$ V and $R_S = V_S/I_D = 4.87$ kΩ. We achieve the desired gate voltage with $R_1 = 3$ MΩ and $R_2 = 2$ MΩ. The resulting circuit will not have the absolute lowest possible power, nor will it have a very stable operating point. However, it is a reasonable compromise. **(b)** The power consumed by the circuit is almost entirely due to I_D and is 0.75 mW. The signal power for a 1-V peak sine wave in a 10-kΩ load is 50 μW, so the efficiency is 7%. A SPICE simulation of this circuit using the Nlarge_mos model from Appendix B yields swing limits of +1.0 and −1.3 V, close to our design goals.

12.5 The maximum symmetric swing is achieved when $v_{O\max} = -v_{O\min}$. We define the difference in swing amplitudes, called "delta swing," to be $v_{O\max} + v_{O\min}$. We then use the solver function in an Excel spreadsheet to vary V_S and find that value which sets delta swing to zero. With the values given in this circuit, we find $V_S = 3.74$ V. This leads to $I_D = 3.4$ mA, a swing of ± 1.78 V, and an efficiency of 4.7%.

12.6 The efficiency is maximum when $V_M = 2V_{OO}/3 = 6.67$ V and $R_M = R_L = 100$ Ω. **(a)** $I_M = V_M/R_M = 66.7$ mA, and **(b)** using (12.124), we have $v_{L\max\_sym} = V_M/2 =$ 3.33 V.

12.7 **(a)** The power in Exercise 12.6 is $P_{supply} = I_M V_{OO} = 666.7$ mW. The power in this circuit is $P_{supply} = 2I_{MM}V_{OO}$, and $I_{MM} = V_{OO}/R_L$ for maximum symmetric swing, so plugging in, we find that $V_{OO} = 5.77$ V. **(b)** $I_{MM} = 57.7$ mA. **(c)** $v_{Omax\_sym} = 5.77$ V. **(d)** The maximum swing in the previous circuit was 3.33 V.

Since the supply power and load resistance are the same in both circuits, the ratio of the efficiencies is equal to the ratio of the squares of the maximum swings, which is $(5.77/3.33)^2 = 3$. The efficiency of the previous circuit was 8.3%, and the efficiency of this circuit is 25%; note that 25 = 3(8.3).

12.8 In all three cases, the *pnp* transistor is cut off and the *npn* acts as an emitter follower by itself. The DC current in the transistor is $I_C = I_S e^{V_{BE}/V_T}$, and we can find V_{BE} by iterating; $V_{BE} = V_I - I_C R_L$. The low-frequency small-signal voltage gain is $A_v = R_L/(R_L + r_e)$, and $r_e = V_T/I_E$. Assuming that $I_E = I_C$ and plugging in the numbers, we obtain **(a)** $I_C = 0.47$ pA, $r_e = 55.5$ GΩ and $A_v = 9 \times 10^{-10}$, **(b)** $I_C = 2.2$ μA, $r_e = 11.6$ kΩ, and $A_v = 0.004$, and **(c)** $I_C = 5.9$ mA, $r_e = 4.4$ Ω, and $A_v = 0.92$. **(d)** The low-frequency small-signal gain is equal to the slope of the characteristic at each point. So, for $V_I = 100$ mV, the gain is nearly zero, for $V_I = 500$ mV, it is much larger, but still small, and for $V_I = 1$ V, the gain is nearly equal to unity as expected.

12.9 In all three cases, the *p*-channel transistor is cut off and the *n*-channel acts as a source follower by itself. The DC current in the transistor is $I_D = K(V_{GS} - V_{th})^2$, and $V_{GS} = V_I - I_D R_L$. We can combine these two equations to obtain the quadratic $V_{GS}^2 + V_{GS}(1/KR_L - 2V_{th}) + V_{th}^2 - V_I/KR_L = 0$, which can then be solved for V_{GS}. The low-frequency small-signal voltage gain is $A_v = R_L/(R_L + 1/g_m)$, and $g_m = 2\sqrt{KI_D}$.

(a) The transistor is off, so the gain is zero. **(b)** $I_D = 1.7$ mA, $1/g_m = 1.1$ kΩ, and $A_v = 0.04$, and **(c)** $I_D = 35.6$ mA, $1/g_m = 59.2$ Ω, and $A_v = 0.46$. **(d)** The low-frequency small-signal gain is equal to the slope of the characteristic at each point. So, for $V_I = 500$ mV, the gain is zero, for $V_I = 2$ V, it is much larger, but still small, and for $V_I = 7$ V, the gain is much higher. The gain in this case does not approach unity unless v_I gets much bigger, because the value of R_L is so low.

12.10 **(a)** Using (12.132) and the data provided, we find that $\theta_{JA} = 100°\text{C/W}$. **(b)** Using (12.132) again with the thermal resistance found in (a), we find that $P_{max}(50°\text{C}) = 0.75$ W.

CHAPTER SUMMARY

- Diodes are frequently used to rectify AC voltages. The applications of rectifiers include AC-to-DC power conversion and amplitude demodulation.
- Half-wave rectifiers allow only the positive (or negative) half of the input signal to pass through to the output.
- Full-wave rectifiers allow positive input signals to pass through to the output and also invert the sign of the negative portion of the input. Full-wave rectifiers are more complex than half-wave rectifiers, but they achieve lower ripple and have less stringent requirements for the diodes.
- Transformers are often used to achieve electrical isolation and to simplify rectifier design.
- Diodes can also be used to perform level restoring, level clamping, and simple switching functions.
- The swing limits of amplifiers are calculated using large-signal midband models for coupling and bypass capacitors and the large-signal low-frequency models presented in Chapter 7.
- We are interested in finding the maximum symmetric swing and efficiency of amplifiers that deal with large signals.
- All amplifiers have nonlinear distortion. We can approximate the magnitude of the distortion by using Taylor series expansions for the nonlinearities in the transistors.

- Due to nonlinearities, output tones appear at harmonics of the fundamental frequencies applied to an amplifier. These tones are referred to as harmonic distortion.
- Total harmonic distortion is the RMS sum of all of the harmonic distortion terms.
- When more than one frequency is present at the input of an amplifier, there will be components at the output at frequencies that are linear combinations of the input frequencies. These components are referred to as intermodulation distortion.
- When an output stage is used to provide significant amounts of power to a load, the efficiency of the output stage becomes a major concern. There are several classes of output stage defined by what percentage of the time the transistors are on.
- Class A output stages have the transistors on all of the time. They have the lowest distortion and lowest efficiency of any amplifier.
- Class B output stages have each transistor on for one-half of the input cycle (assuming a sine wave input). They are more efficient than Class A stages, but have more distortion, particularly crossover distortion.
- Class AB amplifiers have each transistor on for more than half of a cycle, but less than a full cycle. They significantly reduce the crossover distortion of Class B amplifiers, but have lower efficiency.
- Other classes of output stage are possible that either use multiple supply voltages or keep the output transistors on for very brief periods of time to achieve high efficiency.
- Transistors can get hot and be damaged as a result of power dissipation. We can model heat flow using electrical analogs and can then determine how hot the junction of a transistor will get.
- Heat sinks can be used to reduce the thermal resistance seen from a transistor's junction to the ambient. Lowering this resistance will allow the transistor to maintain a lower junction temperature.

REFERENCES

[12.1] S. A. Maas, *Nonlinear Microwave Circuits*. Norwood, MA: Artech House Inc., 1988.

[12.2] A.M. Davis, *Linear Circuit Analysis*. Boston: PWS Publishing Company, 1998.

[12.3] A. Vladimirescu, *The SPICE Book*. New York: John Wiley&Sons, Inc., 1994.

[12.4] T.H. Lee, *The Design of CMOS Radio-Frequency Integrated Circuits.* Cambridge, U.K.: Cambridge University Press, 1998.

[12.5] N. Mohan, T.M. Undeland, and W.P. Robbins, *Power Electronics*, 2nd Edition. New York: John Wiley&Sons, Inc., 1995.

[12.6] B. Razavi, *RF Microelectronics*. Englewood Cliffs, NJ: Prentice Hall, 1997.

[12.7] D. A. Johns and K. Martin, *Analog Integrated Circuit Design*. New York: John Wiley&Sons, Inc., 1997.

PROBLEMS

12.1 DIODE CIRCUITS

12.1.1 Diode Rectifiers

P12.1 Derive an equation for the conduction angle of the half-wave rectifier shown in Figure 12-1 as a function of the peak input voltage, V_m. Assume that the diode turns on abruptly when the forward voltage is V_o.

P12.2 Use the Taylor series expansion of the exponential in (12.1), and assume that $RC \gg t$ to derive the straight-line approximate for $v_O(t)$ given in the text and the resulting ripple given by (12.2).

P12.3 Consider the half-wave rectifier circuit shown in Figure 12-79. Assume that $v_I(t) = 20 \cos(628t)$ and that the diode is ideal except for having a 0.7-V forward drop. **(a)** Sketch the output waveform if the capacitor is removed, and explain when the diode is on or off. **(b)** Sketch the output when the capacitor is present and the ripple is about 10%. **(c)** What value of RC time constant is required to set the ripple to 10%?

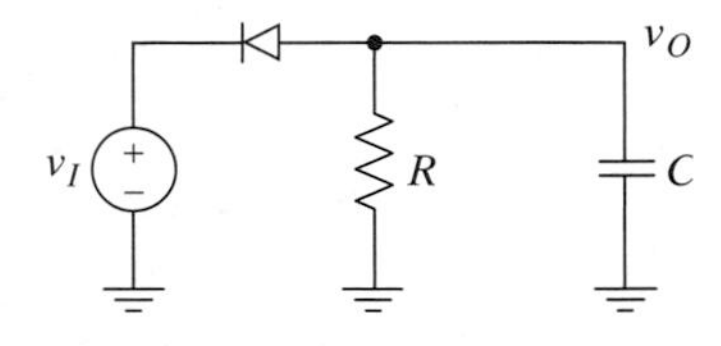

Figure 12–79

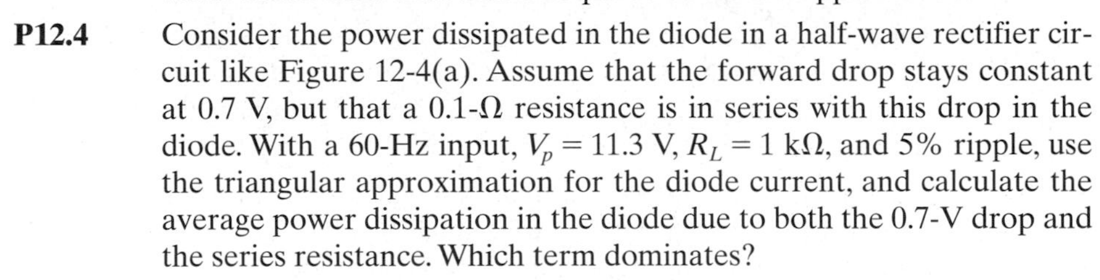

P12.4 Consider the power dissipated in the diode in a half-wave rectifier circuit like Figure 12-4(a). Assume that the forward drop stays constant at 0.7 V, but that a 0.1-Ω resistance is in series with this drop in the diode. With a 60-Hz input, $V_p = 11.3$ V, $R_L = 1$ kΩ, and 5% ripple, use the triangular approximation for the diode current, and calculate the average power dissipation in the diode due to both the 0.7-V drop and the series resistance. Which term dominates?

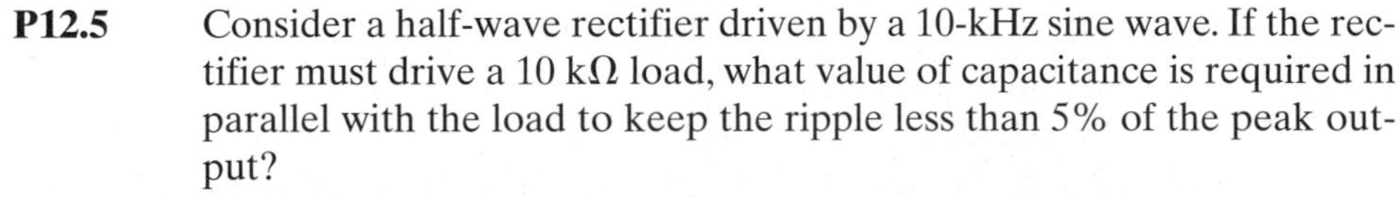

P12.5 Consider a half-wave rectifier driven by a 10-kHz sine wave. If the rectifier must drive a 10 kΩ load, what value of capacitance is required in parallel with the load to keep the ripple less than 5% of the peak output?

P12.6 Draw the waveforms for the full-wave bridge rectifier circuit in Figure 12-13, and determine the PIV seen by the diodes.

D P12.7 Using a transformer with a turns ratio of N, design a full-wave rectifier that will produce 5 V DC at its output with no more than 2% ripple when driving a 1-kΩ load. Assume that the AC input is the standard 120-Vrms 60-Hz line voltage. Determine N and the value of filter capacitor needed.

12.1.2 Limiting, Clamping, and Multiplying Circuits

P12.8 A simple equivalent model of a practical circuit that limits the output voltage swing in both directions is shown in Figure 12-80. Such circuits can be used, for example, to strip away amplitude variations prior to demodulating a frequency-modulated signal. Some frequency demodu-

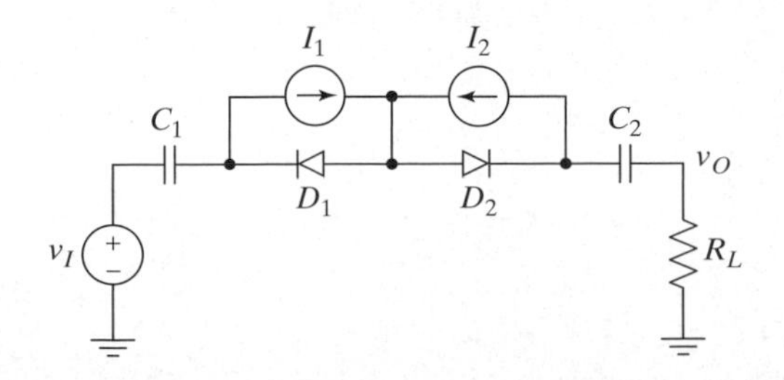

Figure 12–80 A diode limiter that limits the output voltage in both directions.

lators are sensitive to the amplitude of the signal as well as the frequency, so we need to make the amplitude constant prior to demodulating the signal. Assuming that the diodes are ideal, that $I_1 = I_2 = 1$ mA, that C_1 and C_2 are equal and very large, and that $R_L = 2.2$ kΩ, **(a)** determine the output clip levels of this circuit. **(b)** For what range of input voltages will the output not be clipped? **(c)** How do you think the operation of the circuit is affected by nonideal diodes?

P12.9 Consider the circuit shown in Figure 12-81. Assume that the diodes are ideal. Plot the transfer characteristic of this circuit, V_O/V_I, for $-2\text{ V} \leq v_I \leq 2$ V.

P12.10 Consider the circuit shown in Figure 12-81 with a 100-Hz trianglular wave input with a peak value of 2 V. Sketch the output waveform that would result. Notice that if you filtered this waveform, it would be a reasonable approximation to a sine wave. This circuit is called a trianglular-to-sine wave-converter.

D P12.11 Design a circuit similar to Figure 12-81, but using six different clip levels: ±1 V, ±1.5 V, and ±2 V. Assume ideal diodes. Set the circuit up so that for inputs less than 1 V in magnitude, the gain is unity. For inputs between 1 V and 1.5 V in magnitude, the gain should be 0.75, for inputs between 1.5 and 2, the gain should be 0.5, and for inputs larger than 2 V, the gain should be zero.

P12.12 Draw the steady-state input and output waveforms for the clamp circuit in Figure 12-15, but including a nonideal diode. Assume that the diode has a constant 0.7-V drop when forward biased and use a 10-V peak input sine wave. What are the maximum and minimum output voltages?

S P12.13 Use a 1-μF capacitor, a 1-kHz, 10-V peak sine wave input, and the D1N4002 diode model in PSPICE, and simulate the circuit in Figure 12-15. Show the output waveform and verify your answers to Problem P12.12.

P12.14 Sketch the steady-state output waveform, and determine the output voltage for the circuit shown in Figure 12-82. Assume that the diodes have constant forward drops equal to 0.7 V and that the input is a 10-V peak sine wave.

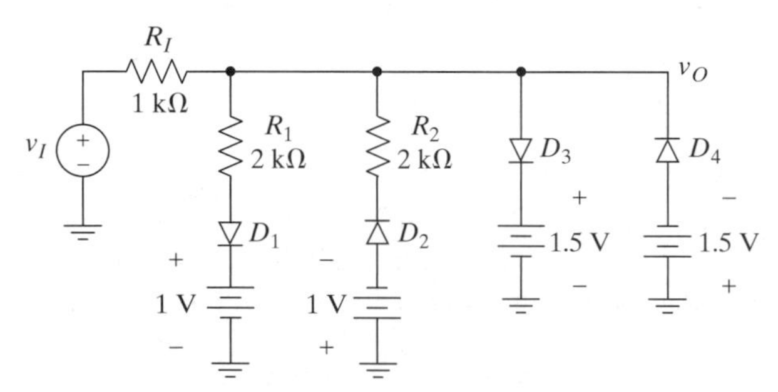

Figure 12–81

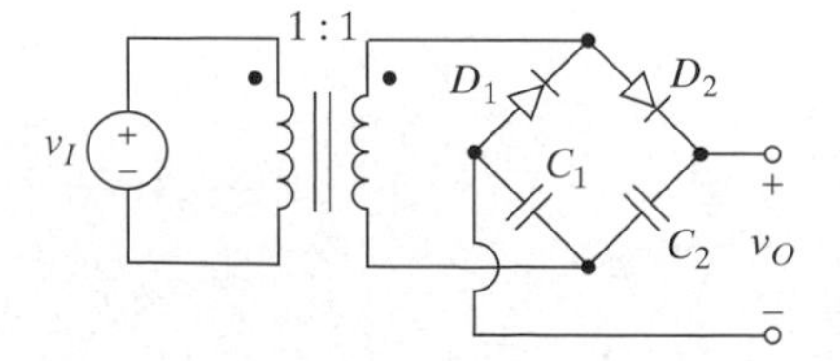

Figure 12–82

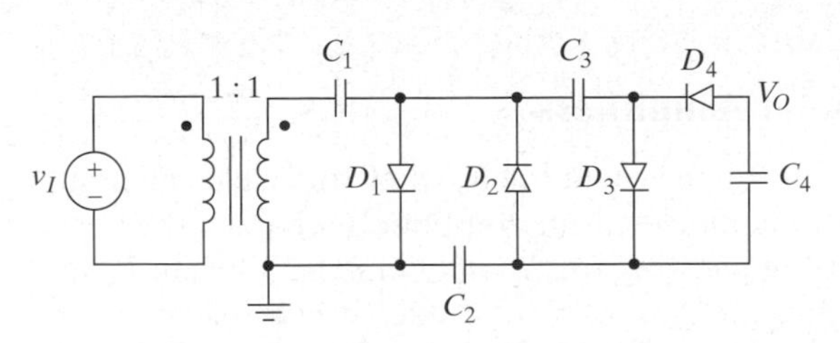

Figure 12–83 A voltage quadrupler.

P12.15 Assume that the circuit in Figure 12-83 is driven by a 10-V peak sine wave and that all of the diodes are ideal. **(a)** Determine the steady-state voltages across each capacitor. **(b)** What is the voltage V_O?

12.1.3 Diode Switching

P12.16 Consider the circuit shown in Figure 12-84. Assume that the diodes are ideal, except that they have a 0.7-V drop when forward biased. Assume further that all three inputs are either 0 V or 5 V. Consider all eight possible combinations of the three binary inputs, and calculate the output voltage in each case. Assuming that any output voltage of 1 V or less is considered a logical low (false), what is the logical description of the operation of this circuit?

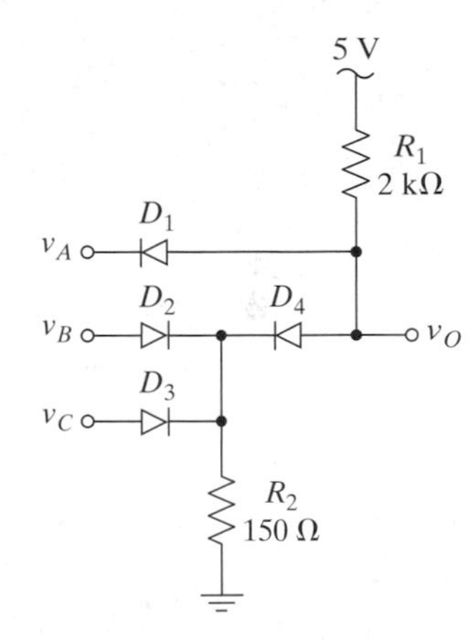

Figure 12–84

P12.17 Consider the circuit shown in Figure 12-85. Assume that the diodes are ideal and perfectly matched, except that they have a 0.7-V drop when forward biased. What is the value of v_O, expressed as a function of the three input voltages?

P12.18 Consider the circuit shown in Figure 12-85. Assume that the diodes are perfectly matched and have $I_S = 10^{-14}$ amps and $n = 1$. **(a)** If $v_1 = 1$ V, $v_2 = 2$ V, and $v_3 = 3$ V, what is v_O? **(b)** Assuming that the three input voltages always differ by at least 60 mV, derive an approximate expression for v_O as a function of the input voltages. (*Hint*: Consider what currents can reasonably be ignored.) **(c)** If $v_1 = 1$ V, $v_2 = 1$ V, and $v_3 = 3$ V, what is v_O? **(d)** What is the maximum error you will ever see if you use the expression derived in part (b)?

P12.19 Consider the sample-and-hold circuit shown in Figure 12-21. Assume that the diodes are ideal and perfectly matched, except that they have a 0.7-V drop when forward biased. Assume also that $V_{CC} = 5$ V, $I = 1$ mA, and $C_H = 1$ pF. **(a)** If v_I has been equal to 1 V for a long time and then abruptly switches to 2 V at $t = 0$, draw the resulting output voltage for $t \geq 0$. **(b)** How long will it take the output voltage to reach 2 V? **(c)** If the input is a 1-V peak sine wave, what is the maximum frequency that can be used if we want to maintain $v_O = v_I$ at all times?

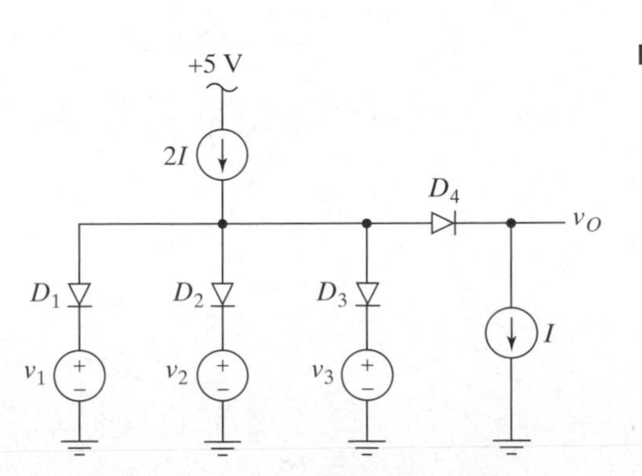

Figure 12–85

12.2 AMPLIFIERS

12.2.1 Signal Swing in Bipolar Amplifiers

P12.20 **(a)** Derive an equation for the value of R_C that will yield maximum symmetric swing for a common-emitter amplifier by redoing the analysis that led to (12.18), but allowing for R_L not being larger than R_C and also including V_{CEsat}. Assume that everything except V_C is held constant. **(b)** If $V_{CC} = 10$ V, $R_L = R_C = 1$ kΩ, $V_E = 1.3$ V, and $V_{CEsat} = 0.2$ V, as in Exercise 12.2, what value of V_C leads to maximum symmetric swing? **(c)** What is the corresponding v_{Lmax_sym}? **(d)** What is the resulting I_C? **(e)** What is the corresponding gain?

P12.21 Consider the common-emitter amplifier shown in Figure 12-86. **(a)** Show that the maximum output voltage is

$$v_{O\max} = (V_{CC} - V_C)\frac{R_L}{R_L + R_C}.$$

(b) What region of operation is the transistor in when the output approaches the value given by the equation in (a)? **(c)** Show that the minimum output voltage is

$$v_{O\min} \approx \left(\frac{V_{CC}}{R_C} + \frac{V_{C_E} + 0.2}{R_{E2}} + \frac{V_C}{R_L}\right)(R_C \| R_{EA} \| R_L) - V_C$$

where we have assumed that $V_{CEsat} = 0.2$ V and that the base current is negligible compared with the collector current, even when the transistor is saturated.

P12.22 Find the maximum symmetric swing obtainable from the amplifier in Figure 12-87.

S P12.23 Use the Psimple model from Appendix B together with SPICE to confirm your result in Problem P12.22. Use a 10-kHz input sinusoid with a peak amplitude just large enough to cause the output to be clipped on both halves. Hand in your transient plot of v_L.

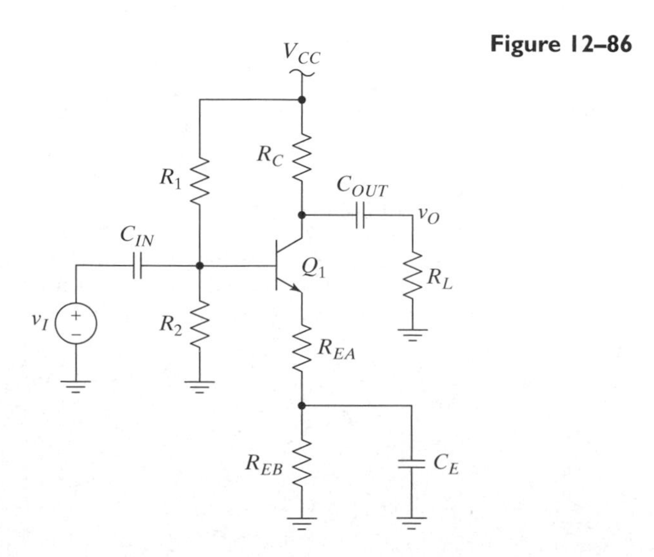

Figure 12–86

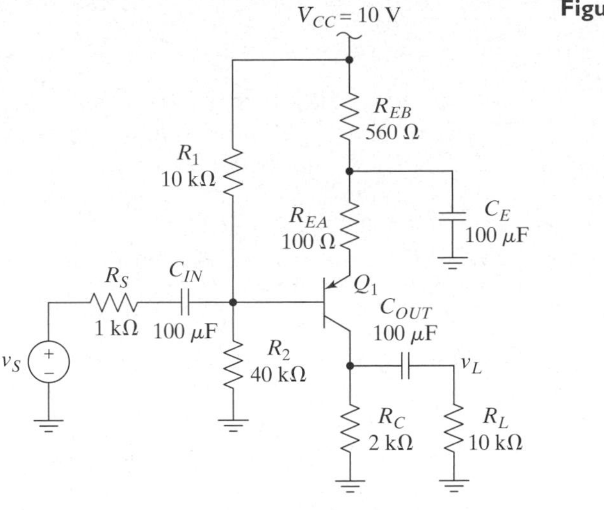

Figure 12–87

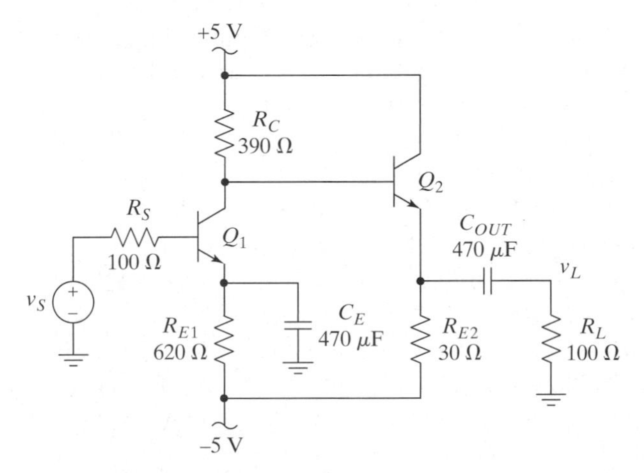

Figure 12–88

P12.24 Find the maximum symmetric swing obtainable from the amplifier in Figure 12-88. Assume that $V_S = 0$, $\beta = 100$, and the product $I_{B1}R_S$ is negligible.

S P12.25 Use the Nsimple model from Appendix B together with SPICE to confirm your result in Problem P12.24. Use a 10-kHz input sinusoid with a peak amplitude just large enough to cause the output to be clipped on both halves. Hand in your transient plot of v_L.

P12.26 Consider a common-emitter emitter-follower cascade, as shown in Figure 12-89. Assume that $V_{CC} = 5$ V. Suppose we need to design the amplifier so that the output voltage can swing at least ±1 V and the values of R_L and R_S are fixed, but we are going to determine everything else. **(a)** What is the absolute maximum DC bias voltage that can be used at the collector of Q_1? Why? **(b)** What is the absolute minimum DC bias voltage that can be used at the collector of Q_1? Why? **(c)** Considering just the minimum output voltage, do we want to choose a larger or smaller value for V_{C1} to minimize the DC bias current required in Q_2? Assume that Q_2 can be cut off and that we design the circuit to just meet the swing requirement for the chosen value of V_{C1}.

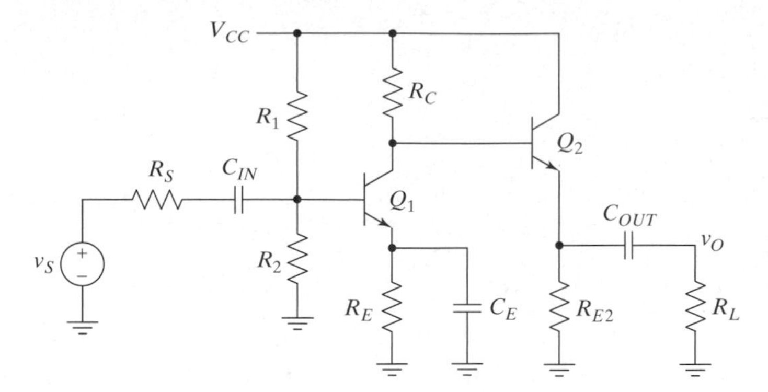

Figure 12–89

12.2.2 Signal Swing in FET Amplifiers

P12.27 Equation (12.45) gives an absolute minimum output voltage for the common-source amplifier, assuming that R_{DS} is negligible. Use the approximation for R_{DS} given in (7.18), and derive an equation for $v_{L\min}$ as a function of v_I for the circuit in Figure 12-40.

P12.28 Redo the analysis that led to (12.49), but allow for R_L not being much larger than R_D. Assume that everything except V_D is held constant.

P12.29 Determine the small-signal midband AC gain and the midband swing limits for the amplifier shown in Figure 12-90 when v_{SIG} is a 1-V peak sine wave. The transistor has $K = 2\ \text{mA/V}^2$ and $V_{th} = 1$ V.

S P12.30 Use SPICE and the Nlarge_mos model from Appendix B to confirm your results from Problem 12-25. Hand in plots of the small-signal AC gain and a transient simulation showing the output when the input is driven by a 1-V peak sine wave at 1 kHz.

P12.31 Determine the small-signal midband AC gain and the midband swing limits for the amplifier shown in Figure 12-91. The transistors have $K = 2\ \text{mA/V}^2$ and $V_{th} = 1$ V.

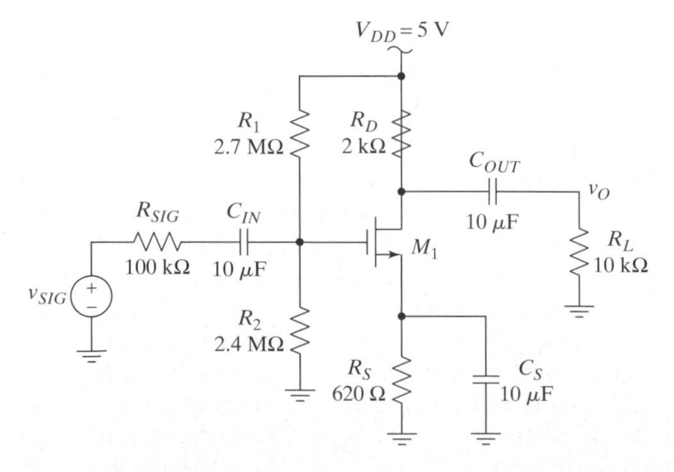

Figure 12–90

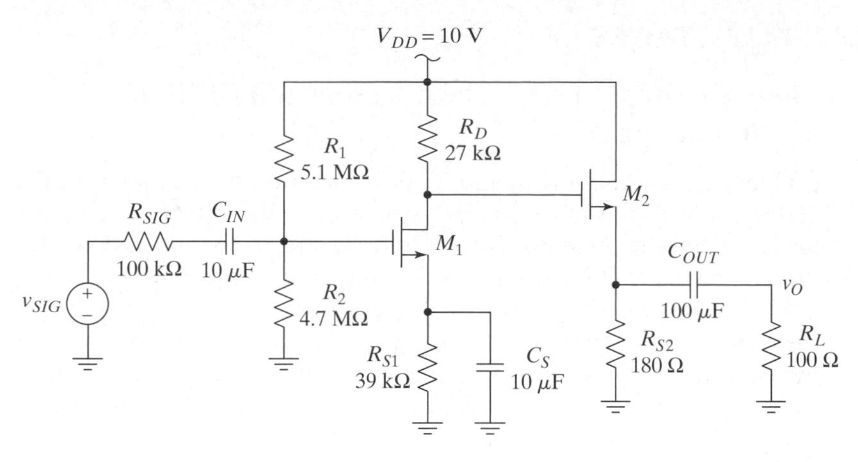

Figure 12–91

S P12.32 Use SPICE and the Nlarge_mos model from Appendix B to confirm your results from Problem 12-23. Hand in plots of the small-signal AC gain and a transient simulation showing the output when the input is driven by a 0.25-V peak sine wave at 1 kHz.

D P12.33 Redesign the amplifier in Example 12.7 so that the output can swing ± 2 V while keeping the same small-signal gain and transistors (challenging!).

12.2.3 Distortion in Amplifiers

P12.34 Recalculate the distortion of the common-source amplifier in Example 12.9 with an input amplitude set large enough that the fundamental component in the output is equal to the fundamental component at the output of the common-emitter amplifier in Example 12.8. Assume $I_C = 1$ mA, $I_D = 0.5$ mA, $R_C = 2$ kΩ, $R_D = 5$ kΩ, and the gate overdrive is 0.5 V. How do the *THD*s compare in this case?

P12.35 Consider what happens to the distortion of a common-emitter amplifier as the source resistance changes. **(a)** Suppose, for simplicity, that the common-emitter current gain of the amplifier (i.e., β) is constant. If the amplifier is driven by an ideal current source instead of an ideal voltage source as in Example 12.8, what would the resulting *THD* be? **(b)** Based on your result in part (a), what do you think happens to the *THD* as the magnitude of R_S is increased?

P12.36 Find HD_2 and HD_3 for the circuit shown in Figure 12-92 if the current gain is given by (I_{BQ} is the quiescent base current—i.e., with no AC input)

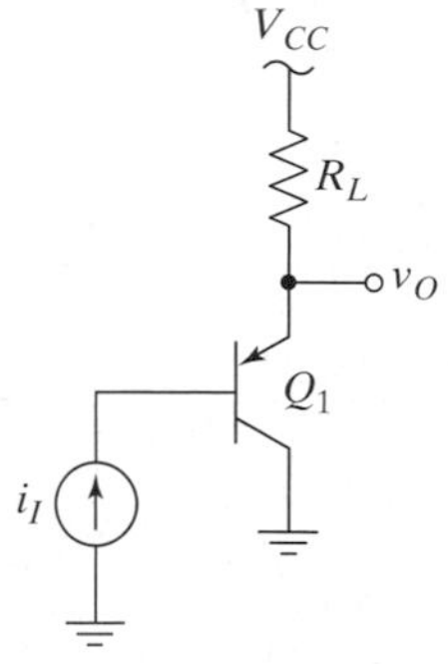

$$\beta = \frac{i_C}{i_B} \approx 150\left[1 + 0.02\left(\frac{i_b}{I_{BQ}}\right) - 0.1\left(\frac{i_b}{I_{BQ}}\right)^2\right].$$

Be very careful with the notation! Use $i_I = 10^{-4}(1 + 0.05\cos\omega t)$.

V_{CC}
R_L
v_O
Q_1
i_I

Figure 12–92

P12.37 If we desire to use a common-source MOSFET amplifier at a fixed bias current, do we need to make the device's W/L larger or smaller to reduce the value of HD_2?

12.3 OUTPUT STAGES

12.3.1 Class A Output Stages: Classification and Efficiency of Output Stages

P12.38 **(a)** Derive the values of R_M and V_M that maximize the efficiency of the merge follower in Figure 12-59. Assume that the supply voltage and load resistance are fixed, and use a sinusoidal signal with a peak amplitude set by the maximum symmetric swing limit. Do not forget that V_M must be low enough to allow for the positive swing, and assume that the transistor can operate with any $V_{OM} \geq 0$ and that the input voltage can exceed V_{OO}. (*Hint*: First find the optimum value of R_M, assuming a fixed V_M. Consider what V_M must be.) **(b)** What is the corresponding v_{Omax_sym}?

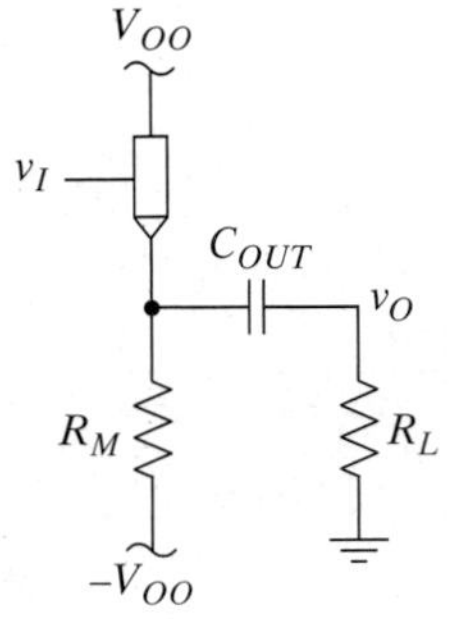

Figure 12–93

P12.39 Consider the circuit shown in Figure 12-93. **(a)** What value of R_M will lead to the highest efficiency? Assume that the supply voltage and load resistance are fixed, and use a sinusoidal signal with a peak amplitude set by the maximum symmetric swing limit. Do not forget that V_M must be low enough to allow for the positive swing, and assume that the transistor can operate with any $V_{OM} \geq 0$ and that the input voltage can exceed V_{OO}. (*Hint*: First find the optimum value of R_M, assuming a fixed V_M.) **(b)** What value of V_M will maximize the efficiency? **(c)** What is the resulting v_{Omax_sym}? **(d)** What is the resulting maximum efficiency?

P12.40 Consider the ideal Class A output stage in Figure 12-61. Assume that $V_{OO} = -V_{MM} = 5$ V, $R_L = 100\ \Omega$, $V_{OM} = 5$ V, $I_{MM} = 50$ mA, and the output is a 5-V peak sine wave at 10 kHz. **(a)** Draw the waveforms for i_L, i_M, and v_{OM} for at least one full cycle. Label the peak values clearly. **(b)** Derive an equation for the instantaneous power $p_T = i_M v_{OM}$ dissipated in the transistor, assuming that the input current can be neglected. Make a plot of the function. **(c)** What is the peak power dissipated in the transistor? Show that it is equal to twice the maximum power in the load. **(d)** What is the average power dissipated in the transistor? **(e)** What signal level leads to the most power being dissipated in the transistor?

S P12.41 Consider the transformer-coupled output stage in Example 12.12. Use a controlled current source and an ideal transformer to model the circuit in SPICE as shown in Figure 12-94. The transformer model is called XFRM_LINEAR in SPICE. Set the coupling parameter to 1, set L1 to 100, and L2 to 1 (this models a transformer with a turns ration of 10). Set the transconductance to be 1, use $R_L = 10\ \Omega$, $V_{OO} = 5$ V, and use

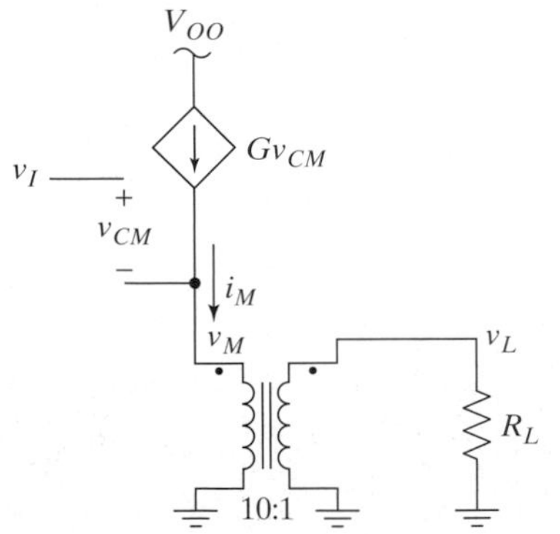

Figure 12–94

a 5 V peak 10 kHz sinusoidal input source with a 100 mV DC component. Run a transient simulation and provide plots for v_L, v_M, and i_M. Explain the operation.

12.3.2 Class B and AB Output Stages

P12.42 Starting from (A12.12), show that the maximum power dissipation in the transistor occurs for $\hat{V}_o = 2V_{OO}/\pi$ and that the corresponding efficiency is 50%.

P12.43 What is the efficiency of an ideal Class B output stage if the input is a square wave and the output swing is as large as possible?

Bipolar Class B and AB Output Stages

P12.44 Consider the push–pull output stage shown in Figure 12-69(a). Sketch a family of three static transfer characteristics—one for a small value of R_L, one for a large value of R_L, and one for a value of R_L in the middle of the two extremes. Explain why you think the curves will look the way you have sketched them.

S P12.45 Use SPICE and the Nsimple and Psimple models from Appendix B to verify your answer to Problem P12.44. Sweep the input from –2 V to 2 V. Hand in a plot of the transfer characteristic, showing all three values of R_L.

P12.46 Consider the Class AB output stage shown in Figure 12-95. We wish to explore some of the practical limitations of this kind of biasing. Assume that the two transistors are perfectly matched and that the base currents and Early effect can be safely neglected. Assume also that both transistors have $I_S = 10^{-14}$ A, that $V_I = 0$, and that the circuit is at room temperature. **(a)** Derive an equation for I_{C1} as a function of I_R, R, I_S, and V_T. **(b)** If we want to set $I_{C1} = 1$ µA, and we use $I_R = 10$ µA, how large do the resistors need to be? **(c)** If I_R varies by ±10%, how much does I_{C1} vary?

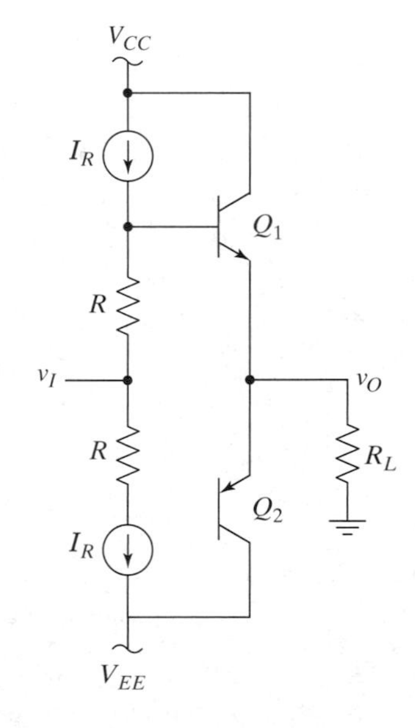

Figure 12–95

P12.47 Consider the Class AB output stage shown in Figure 12-95, but replace the two resistors with diodes (placed so that they are forward biased by the current). Assume that the two transistors are perfectly matched and that the base currents and Early effect can be safely neglected. Assume also that both transistors and the two diodes have $I_S = 10^{-14}$ A, that $V_I = 0$, and that the circuit is at room temperature. Derive an equation for I_{C1} as a function of I_R. (*Hint*: You must consider the large-signal forward-biased models for the diodes and transistors.)

P12.48 Consider the Class AB output stage shown in Figure 12-96. Assume that all three transistors are perfectly matched, that $V_O = 0$ V, and that the circuit is at room temperature. The circuit comprising Q_3, R_1, and R_2 is called a $\boldsymbol{V_{BE}}$ **multiplier**. If we ignore the base current of Q_3, then the currents in R_1 and R_2 are the same and the voltage across the multiplier is $V_{CE3} = V_{BE3}(1 + R_1/R_2)$. If we also ignore the base currents of Q_1 and Q_2, and assume $I_{C3} \gg I_{R1}$, then $I_{C3} = I_R$. (a) Show that

$$I_{C1} = I_S\left(\frac{I_R}{I_S}\right)^{\frac{R_1+R_2}{2R_2}}$$

(b) If $I_S = 10^{-14}$ A and $I_R = 10$ µA, and if we want $I_{CI} = 1$ µA, find values for R_1 and R_2 that will work.

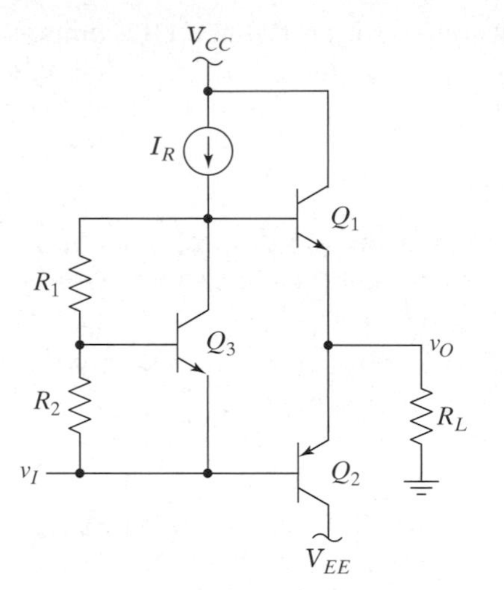

Figure 12–96

FET Class B and AB Output Stages

P12.49 Consider the Class B output stage in Figure 12-70(a). Assume that the supplies are ±10 V, $V_{thn} = 1$ V, $V_{thp} = -1$ V, and $K = 2$ mA/V^2 for both transistors. If the input is a 5-V peak sine wave and $R_L = 1$ kΩ, **(a)** what is the peak output amplitude? **(b)** What is the efficiency? (Approximate the output as a sine wave.)

P12.50 Consider the Class AB output stage shown in Figure 12-97. Assume that $R_L = 1$ kΩ, $V_{thn} = 1$ V, $V_{thp} = -1$ V, and $K = 2$ mA/V^2 for both transistors. **(a)** If we desire to have the small-signal voltage gain be ≥ 0.9 for v_I near zero, how large do we need to make V_B? **(b)** Assuming that the gates of the transistors cannot exceed the supply voltages, what is the maximum output swing corresponding to your solution in part (a)? **(c)** If the output is sinusoidal with the full swing found in part (b), what is the efficiency of this stage?

P12.51 Follow the same procedure as in Problem P12.50, but vary the small-signal gain obtained for v_I near zero. Plot the efficiency of the stage versus this small-signal gain for gains from 0.1 to 0.9.

D P12.52 Consider the CMOS Class AB output stage shown in Figure 12-98. Assume that $I_B = 50$ μA, $R_L = 10$ kΩ, $V_{DD} = -V_{SS} = 5$ V, $V_{thn} = 1$ V, $V_{thp} = -1$ V, and $K' = 20$ μA/V^2 for all transistors. **(a)** Determine (W/L)s for the transistors that will yield a small-signal midband AC voltage gain of 0.8 when $V_O = 0$ V. **(b)** Using your solution to part (a), and assuming that the gate of M_1 can't be pulled any higher than V_{DD} (because the current source stops functioning) and that $V_{SS} \leq v_I \leq V_{DD}$, determine the swing limits for this output stage. **(c)** With the largest sinusoidal output possible, and using your results from part (b), determine the efficiency of this stage. **(d)** What could you do to increase the efficiency? Would increasing the efficiency hurt the performance in any way?

12.3.3 More Advanced Output Stages

P12.53 Consider the switching amplifier circuit shown in Figure 12-99. Assume that the pulses go from 0 V to 1 V and have a pulse width of 200 ns with

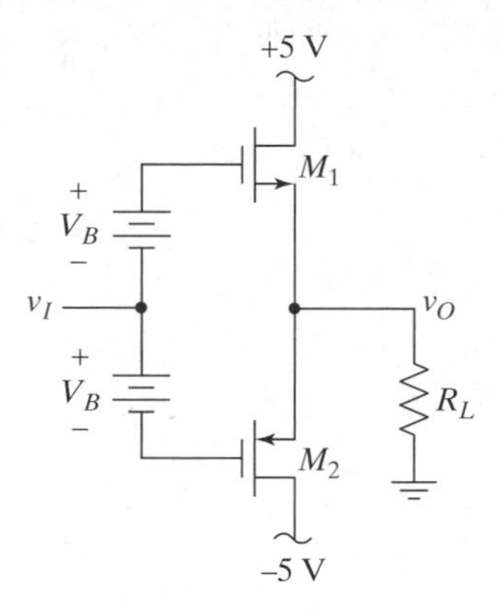

Figure 12–97

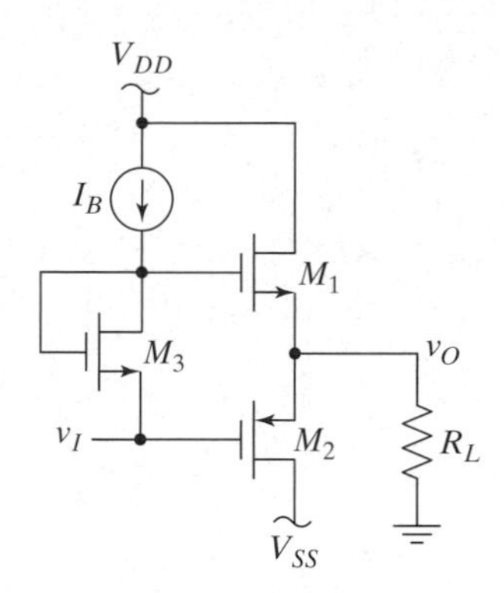

Figure 12–98

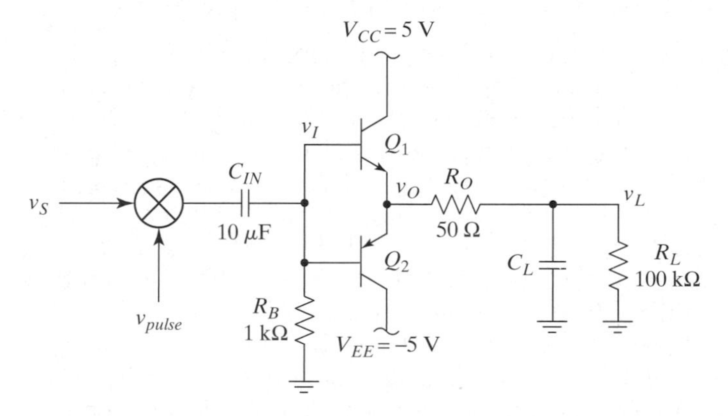

Figure 12–99

a period of 2 μs. R_B is used in this circuit to keep the input voltage centered on zero. **(a)** Calculate the value of C_L required to have a 20-kHz cutoff frequency for the output low pass. **(b)** Draw v_I, v_O and v_L for one full cycle of v_S with $v_S = 3 \sin(2\pi \cdot 100\ \text{k} \cdot t)$.

S P12.54 Use SPICE and the Nsimple and Psimple models presented in Appendix B to simulate the circuit shown in Figure 12-99. Set up the pulse to go from 0 V to 1 V and have a pulse width of 200 ns with a period of 2 μs, and use a 3-V peak 10-kHz sinusoidal input. Hand in a plot of the output waveform confirming that your results for Problem P12.53 are correct.

P12.55 Consider the switching amplifier circuit shown in Figure 12-100. Assume that the pulses go from 0 V to 1 V and have a pulse width of 200 ns with a period of 2 μs. R_G is used in this circuit to keep the input voltage centered on zero. **(a)** Calculate the value of C_L required to have a 20-kHz cutoff frequency for the output low pass. **(b)** Draw v_I, v_O and v_L for one full cycle of v_S with $v_S = 5 \sin(2\pi \cdot 100\ \text{k} \cdot t)$.

S P12.56 Use SPICE and the Nlarge_mos and Plarge_mos models presented in Appendix B to simulate the circuit shown in Figure 12-100. Set up the pulse to go from 0 V to 1 V and have a pulse width of 200 ns with a period of 2 μs, and use a 5-V peak 10-kHz sinusoidal input. Edit the models to only include K_P and V_{t0}, plus set $K_P = 80$ m. Hand in a plot of the output waveform confirming that your results for Problem P12.55 are correct.

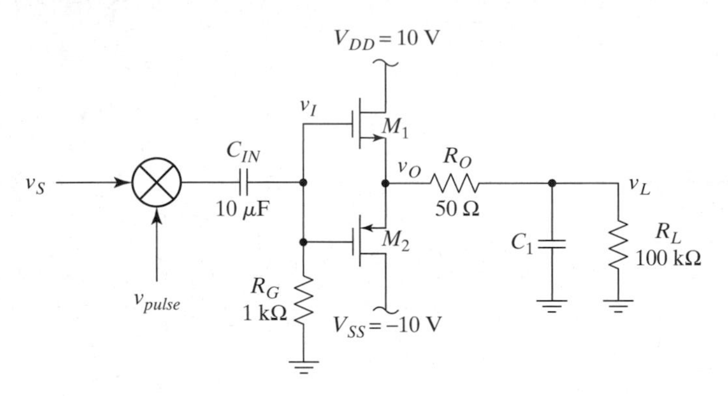

Figure 12–100

12.3.4 Power Transistors, Thermal Modeling, and Heat Sinks

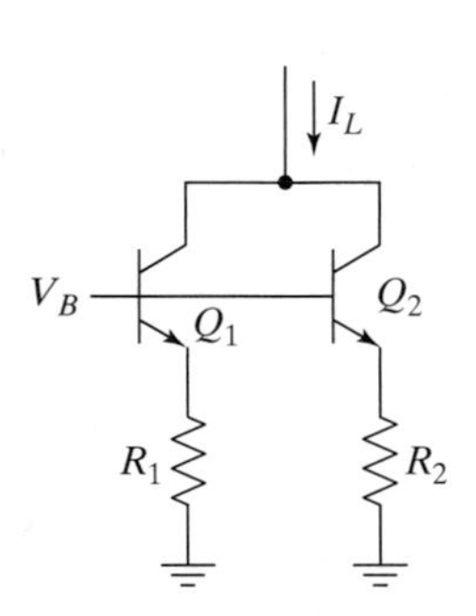

Figure 12–101

P12.57 Suppose we want to put two BJTs in parallel to handle a large load current as shown in Figure 12-101. Assume that $R_1 = R_2 = 0$ for this problem. **(a)** If $I_{S1} = 1.3I_{S2}$ and $I_L = 10$ A, what are I_{C1} and I_{C2}? **(b)** Assuming that $V_{CE} = 0.2$ V and ignoring the base currents, determine the powers dissipated by both transistors, given your answers to (a). **(c)** Assuming that the transistors do not have heat sinks, that $\theta_{JA} = 50°\text{C/W}$, and $T_A = 25°\text{C}$, find the junction temperatures of both transistors. **(d)** Assume that $I_{S2} = 10^{-15}$ A at 25°C and that $I_S(T) = I_S(25) \times 10^{(T-25)/15}$ for both transistors. If the junction temperatures are at the values you calculated in part (c), what would the ratio I_{C1}/I_{C2} be? (Take into account the original 30% difference *and* the temperature difference.) **(e)** Given your answer to part (d), what will happen if the circuit is left on for a long time?

P12.58 Suppose we want to put two BJTs in parallel to handle a large load current, as shown in Figure 12-101. If $I_{S1} = 1.3I_{S2}$ and $I_L = 2$ A, how large do we need to make R_1 and R_2 in order to guarantee that I_{C1} and I_{C2} are within 5% of each other? Assume that the resistors are perfectly matched.

P12.59 A class A power amplifier is rated to provide 10 W of maximum power to a load and has an efficiency of 20%. The maximum allowable junction temperature is 150°C, the ambient temperature is 25°C, and $\theta_{JC} \approx \theta_{CS} = 0.5°\text{C/W}$. Determine the maximum allowable thermal resistance of the heat sink.

P12.60 A power transistor is rated to handle 120 W maximum and has $\theta_{JC} = 0.5°\text{C/W}$. If the transistor is operated with a heat sink with $\theta_{SA} = 0.5°\text{C/W}$, determine the maximum power that can be dissipated at an ambient temperature of 50°C if the maximum junction temperature is 200°C and $\theta_{CS} = 0.2°\text{C/W}$.

P12.61 Determine the maximum power that can be handled by a 100-W power transistor with $T_{j\max} = 200°\text{C}$ and $T_A = 40°\text{C}$ **(a)** when no heat sink is used and $\theta_{JA} = 50°\text{C/W}$ and **(b)** when a sink is used and $\theta_{JC} = \theta_{CS} = 0.5°\text{C/W}$ and $\theta_{SA} = 1°\text{C/W}$.

CHAPTER 13

Data Converters

Music reproduction with CD players is a common example of the many uses of analog-to-digital and digital-to-analog conversion.

INTRODUCTION

Almost all modern electronic systems use a combination of analog and digital circuits so that signals are processed partly in the *analog domain* and partly in the *digital domain*. Each domain has advantages and disadvantages that a circuit designer must understand. In this chapter, we start with a brief discussion of the differences between *analog signal processing* (ASP) and *digital signal processing* (DSP) as background for the main topic of the chapter: *data conversion circuits*.

Data converters are circuits that convert analog signals into digital signals or vice versa and are an important branch of modern electronic circuit design. The coverage here is intended to provide an introduction to those who go on in the field of circuit design and to allow others to be informed users of data converters. For those who want more detailed information, we refer you to [13.1], [13.2], and [13.5] and the many other references available on data conversion circuits.

13.1 OVERVIEW

We begin by providing some motivation for our study of data converters. To that end, we present a brief comparison of analog and digital signal processing and the usage of data converters in that context. We then go on to present the transfer functions and specifications of data converters.

13.1.1 Analog and Digital Signal Processing and Applications of Data Converters

Many of the signals processed in electronic circuits are derived from quantities like pressure, temperature, electric field strength, and light intensity, which are analog in nature; that is, they can assume any value within some continuous range (ignoring the discrete nature of light and charge). Nevertheless, there are often advantages to processing these signals in the digital domain. For example, digital communication and storage are superior to analog techniques in many ways.

Consider an analog communication channel as shown in Figure 13-1. The original transmitted signal is $x(t)$, and the received signal is

$$y(t) = x(t) + n(t), \tag{13.1}$$

where $n(t)$ is the noise added by the channel. The added noise is not desired, but is an unavoidable characteristic of all real channels. In a system like this, there isn't any way to completely remove the added noise at the receiver, so the ***fidelity*** of the signal has been reduced (i.e., the signal received is not a completely faithful copy of what was sent).

However, if the signal to be transmitted is first converted to a stream of binary digits (bits) as illustrated in Figure 13-2, these bits can be transmitted instead of the orig-

Transmitter → $x(t)$ → Channel → $y(t)$ → Receiver

Figure 13–1 An analog communication system.

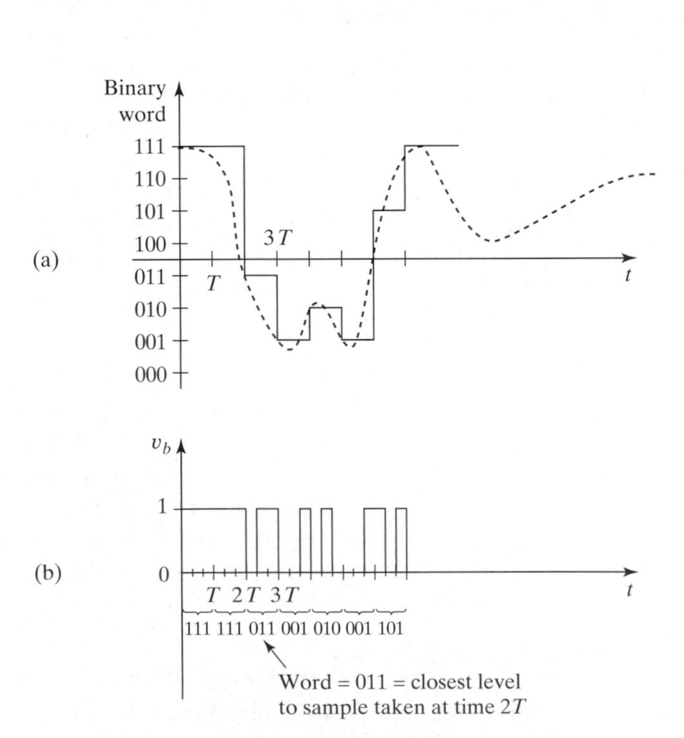

Figure 13–2 (a) A continuous-time analog signal is periodically sampled at multiples of the sample period T, and each sample is rounded or quantized to the nearest of eight discrete values. (b) The resulting three-bit binary words are each represented as a sequence of three bits, with each bit consuming one-third of a period T.

inal analog signal. The continuous-time analog signal shown in part (a) of the figure is first periodically *sampled* at multiples of the sample period T. Each sample is then rounded, or *quantized*, to the nearest discrete value. In the example shown here, there are eight discrete values corresponding to a three-bit binary word. The difference between the actual value of the signal at the sample time and the nearest discrete level is called the *quantization error*. The resulting binary words can each be represented as a sequence of bits as shown in part (b) of the figure.

Once we have a binary data stream to deal with, there are many different ways we can transmit or store the data. For example, suppose we simply transmit the waveform in part (b) of Figure 13-2 directly by sending one bit every $T/3$ seconds, representing a binary one by a voltage of one volt and a binary zero by a voltage of zero. When noise is added to this waveform, the resulting received signal might appear as shown in Figure 13-3. The original binary data can be recovered from this signal with an error rate that approaches zero, as long as we don't exceed the information-carrying capacity of a given channel.[1] For example, suppose we use the received signal shown in the figure as the input to a comparator with a threshold at one-half volt. Then, except for errors caused by extremely rare large noise spikes, the output of the comparator will be a perfect copy of the original noiseless signal. The probability of any one bit in the detected stream being wrong is called the *bit-error rate* (BER) and is extremely low in modern digital transmission and storage systems (in the range of 10^{-2} to 10^{-8} without coding). In addition, we can further improve performance by using *coding* to allow us to recover from some errors.

Although we have glossed over many important details in this example, it does serve to illustrate the fundamental advantage of digital communication and data storage: We can, at least in principle, recover the original data with perfect fidelity. In the case of analog communication and storage, we can recover only an approximation to the original data.

At this point, you may recognize a slight deception in the example just given. Although we can recover the original binary data exactly, the original binary data do not exactly correspond to the original analog signal. In other words, if we started with the signal shown in Figure 13-2(a), our received binary data would not allow us to reconstruct the signal exactly, because our initial

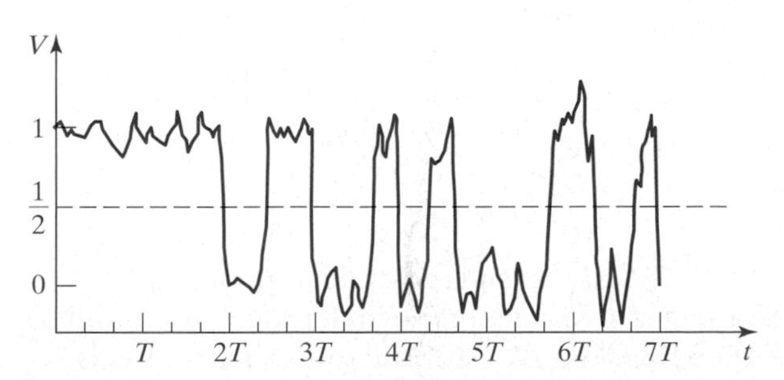

Figure 13–3 An example of what the received signal might look like if the binary signal from Figure 13-2(b) was transmitted over a channel like the one in Figure 13-1.

[1] The field of information theory deals in part with calculating the maximum amount of information that can be transmitted over a given channel in a given time with a specified fidelity. A good treatment of this topic can be found in [13.4].

quantization necessarily introduced errors. By careful design, however, we can make ***analog-to-digital converters*** (ADCs) that provide highly accurate representations of the original analog waveform and allow us to transmit and store signals with significantly greater fidelity than is possible using analog techniques. If the quantization errors in a sampled system are viewed as noise (called ***quantization noise***), it can be shown that the peak signal-to-noise ratio (SNR) that can be attained for an N-bit converter is (*see* Problem P13.1)

$$SNR_{\text{peak}} = 6N + 1.76 \text{ dB}, \tag{13.2}$$

where this SNR applies only when the input is a full-scale sine wave. The SNR will be lower for any smaller signal, since the quantization noise remains the same.

Example 13.1

Probably the best known example of digital storage of analog data is the compact disc introduced for music in 1983. The signal to be recorded is first sampled at a rate of 44.1 kHz, and each sample is then quantized to a 16-bit word. These 16-bit words are coded to allow for error recovery and are stored on the disc.

Problem:
(a) Ignoring coding, so that each sample is represented by the original 16-bit word alone, how many bits are required to represent one minute of recorded music? (b) If the maximum voltage allowed is 10 V (when the sample is 1111111111111111), how small is the smallest voltage (0000000000000001)? (c) What is the peak SNR we can obtain with this converter?

Solution:
(a) Sixty seconds at 44.1 ksamples/s is 2.65×10^6 samples. At 16 bits per sample this requires 4.23×10^7 bits (i.e., 42 Mbits or 5.3 MBytes). (b) $2^{16} = 65{,}536$ different levels, one of which is zero. Therefore, the code for 65,535 = 10 V, and the code for 1 is equal to 10/65,535 = 153 μV. (c) Using (13.2), we find that the peak SNR is 98 dB. For comparison, a decent consumer-grade analog tape recorder achieves an SNR of around 60 dB.

■

In addition to the advantages of digital transmission and storage, digital signal processing has other advantages over analog signal processing. Digital computations (e.g., multiplication) can be carried out to arbitrary precision (32-bit processing is common), whereas analog multiplication and other similar operations are limited to about 0.1–1% precision, which corresponds to about 7 bits. Also, digital circuit design is more systematic than analog circuit design. Consequently, the computer-aided design (CAD) techniques are more advanced and more effective. As a result of these differences in CAD tools, the design time is usually much shorter for digital circuits than for analog circuits, and the probability of a design working as expected the first time is much higher. It is also much easier to develop tests for digital circuits and they have far more sophisticated built-in test capabilities—a major advantage in this age when the overall cost of testing can be over 50% of the cost of producing a complex integrated circuit. Digital circuits can also take advantage of newer processes more easily. Using modern scalable design rules, it is possible to take an existing design and scale it down to the smaller dimensions of a new process without making significant design changes. The result is a design that is predictably faster and consumes less power. Although analog circuit performance usually improves with smaller geometries as well, it does not scale in a simple or predictable manner and the design usually needs to be redone to take advantage of the better process.

With all of the advantages just listed, you might be wondering why anything is still done with analog circuits. The demise of analog circuits for all but a few niche areas has been predicted many times over the past 25 years and has not yet come to pass. It is not likely to happen in the foreseeable future either, because analog signal processing does possess a few key advantages. If a given function can be reasonably implemented with analog circuits (i.e., the accuracy and complexity of computation required do not exceed the capabilities of analog circuits), then the function can usually be implemented in analog circuits that are smaller, operate at higher speeds, and consume less power than their digital counterparts. All three of these advantages are significant, especially in high-volume markets, where the higher development costs of analog circuits can be amortized, or where higher speed or lower power are required for the application.

Analog circuits are also required in many applications to provide the inputs and outputs that interface with sensors, actuators, and other devices. A typical DSP system includes the functions shown in the block diagram in Figure 13-4. Systems that contain a mixture of analog and digital circuits are called ***mixed-signal circuits***. A large percentage of all signal processing is presently done using mixed-signal chips.

The input low-pass filter (LPF) is used to prevent a problem called ***aliasing***, which is explained in Aside A13.1. This filter is usually called an ***antialiasing filter***. The signal is then sampled by the ***sample-and-hold*** (S/H)[2] circuit, which is often included in the ADC. The ADC is used to convert each sample of the analog input into an N-bit digital word, which is then processed by the DSP block. After processing, the digital outputs are sent to a ***digital-to-analog converter*** (DAC). The output of the DAC must be low-pass filtered to smooth out the steps in the waveform and produce the desired analog output. This last filter is sometimes called a ***reconstruction filter***, since it reconstructs the continuous-time waveform from its sampled-and-held values. To illustrate this point, suppose we feed a DAC

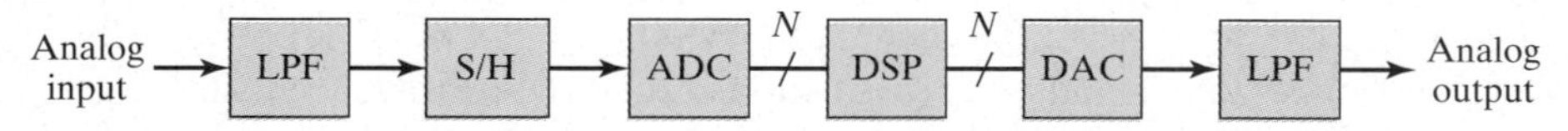

Figure 13–4 A typical DSP system. (The lines with a slash through them and the N next to them signify N lines in parallel.)

---

[2] Sometimes called a sample-and-hold amplifier (SHA).

ASIDE A13.1 SAMPLE RATE AND ALIASING

If a continuous-time analog signal is sampled fast enough, we can completely reconstruct the original waveform from its samples without loss of information. We wish to understand this process better and see how fast we must sample the signal. The sampling process can be viewed mathematically as a multiplication by an infinite sequence of impulses, as shown in Figure A13-1(a). The signal input is $x(t)$, and the infinite sequence of impulses is denoted by $s(t)$, the sampling signal, which is

$$s(t) = \sum_{n=-\infty}^{\infty} \delta(t - nT). \tag{A13.1}$$

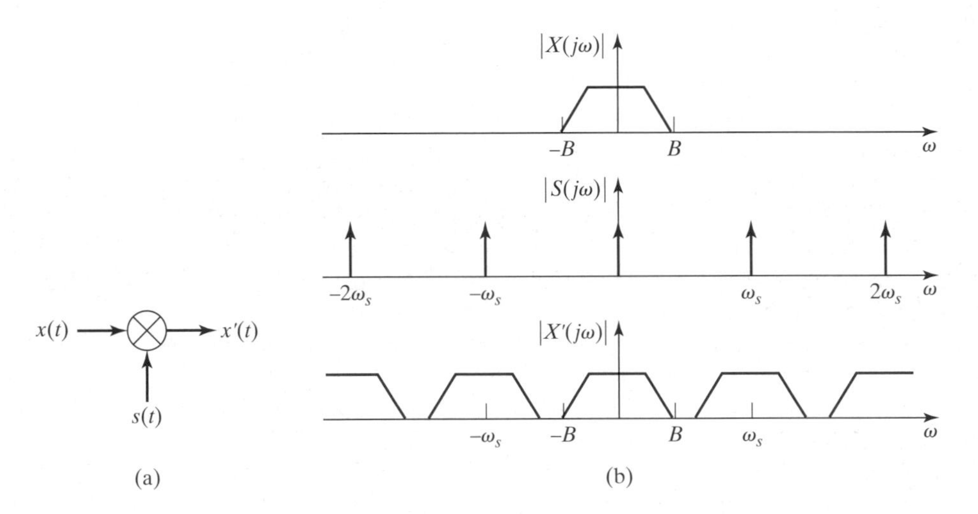

Figure A13-1 (a) Sampling shown as a multiplication process. (b) The resulting spectra.

The Fourier transform of $s(t)$ is again an infinite sequence of impulses, given by

$$S(j\omega) = \frac{2\pi}{T}\sum_{k=-\infty}^{\infty} \delta(\omega - k\omega_s), \tag{A13.2}$$

where ω_s is the sampling frequency ($\omega_s = 2\pi/T$). Since multiplication in the time domain is equivalent to convolution in the frequency domain, the Fourier transform of the output of the multiplier is given by

$$X'(j\omega) = \frac{1}{T}\sum_{k=-\infty}^{\infty} X(j\omega - kj\omega_s). \tag{A13.3}$$

In part (b) of the figure, we show the corresponding spectra. If, as is shown in the figure, the spectrum of $X(j\omega)$ is bandlimited to a frequency B, and if $\omega_s > 2B$, then the sampled spectrum given by (A13.3) is as shown at the bottom. The spectrum of $X(j\omega)$ is repeated every $k\omega_s$. From this last spectrum, we see that the original signal can be recovered by using a low-pass filter, so long as the condition $\omega_s > 2B$ is met. If this condition is not met, the copies of the spectrum of $X(j\omega)$ overlap, and we can no longer find a unique reconstruction of the original waveform. This overlap causes a phenomenon known as ***aliasing***, which is most easily illustrated in the time domain.

The name *aliasing* comes from the fact that when a signal is not sampled at a frequency at least twice as high as the signal itself, the samples do not uniquely describe that signal, so it has an alias at another frequency. In Figure A13-2, we show a 1-kHz sine wave and a 6-kHz sine wave, both sampled at a 5-kHz rate. Notice that the samples fall right on top of each other, so there isn't any way to tell these two signals apart given the samples alone. In other words, the sine wave at 6 kHz is an alias, or alternate, for the sine wave at 1 kHz. The problem is the result of the fact that the 6-kHz sine wave is not sampled at a rate greater than 12 kHz.

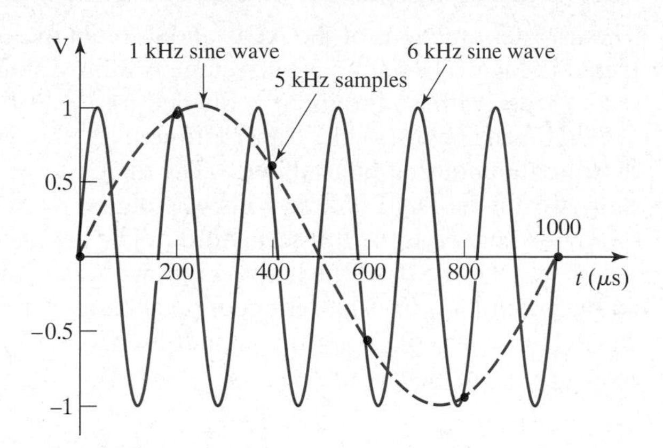

Figure A13-2 An illustration of aliasing (see aliasing.nb on CD).

The minimum sampling frequency, $\omega_s > 2B$, is called the *Nyquist frequency*. We must be sure to sample a signal at greater than the Nyquist frequency, or we will not be able to uniquely reconstruct the signal from the samples. Therefore, a low-pass filter is used in front of a sampler to remove high-frequency components from the input; this is the so-called antialiasing filter.

with the digital words representing samples of a sine wave. The output for a 4-bit DAC is shown in Figure 13-5(a). This waveform clearly approximates the desired sine wave, but it contains a number of steps that are not present in the actual sine wave and also has obvious quantization errors. These steps can be smoothed out by using a low-pass filter. Part (b) of the figure shows the result of low-pass filtering the waveform in part (a). With smaller steps (i.e., more resolution in the DAC) and a better filter, the resulting sine wave can be quite good.

Because the DSP block deals with discrete samples of the analog input, it is important to know how fast we must sample the input waveform in order to fully describe it. Aside A13.1 discusses the rate at which we must sample a given waveform to avoid losing information.

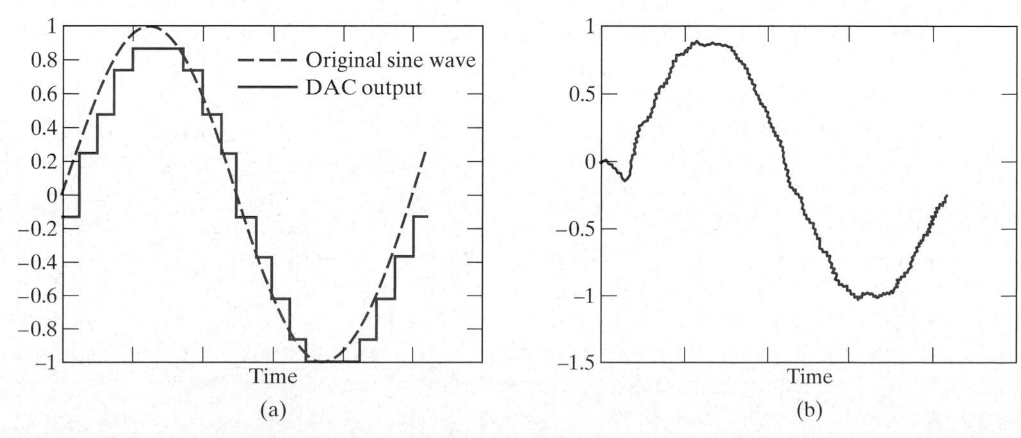

Figure 13–5 (a) The raw output of a DAC approximating a sine wave. (b) The DAC output after passing through a low-pass filter.

13.1.2 Data Converter Transfer Functions and Specifications

The transfer functions of the ADC and DAC in the system of Figure 13-4 are illustrated in Figure 13-6 for a 3-bit system. A 3-bit system has $2^3 = 8$ total discrete values to work with, so the input is divided into $2^3 - 1 = 7$ ranges as shown. For the DAC, shown in part (a), the transfer function is discrete; that is, there are only eight discrete input/output pairs allowed. The digital word 000 corresponds to an analog output from the DAC of zero volts. The digital word 001 corresponds to an output of $FS/8$, where FS is the full-scale output. The quantity $FS/8$ is the value of the ***least significant bit*** (LSB) of the binary input word. The digital word 010 corresponds to an output of $FS/4$. This pattern goes on as shown in the figure until the final digital word, 111, which corresponds to an output of $7/8(FS)$. In general, for an N-bit DAC with a full-scale output of FS,

$$\text{LSB} = \frac{FS}{2^N}. \tag{13.3}$$

Since the input is continuous, so is the ADC transfer function shown in part (b) of the figure. For inputs between zero and $FS/16$, which is one-half of the LSB, the ADC output will be 000. For inputs between $FS/16$ and $3/16(FS)$, the output is 001, and so on. If the ADC and the DAC are connected back to back (i.e., the DSP block in Figure 13-4 is replaced by wires), the output will equal the input only for the points indicated as solid dots on the ADC transfer function. However, notice that the quantization error will never be larger than (1/2) LSB. The ***resolution*** of the ADC is the smallest difference that can be resolved; in other words, it is the minimum change in V_I that will guarantee the output changes. It is $FS/2^N$ for an N-bit converter. For the DAC, the resolution is the same and represents the minimum change in V_O possible. Sometimes, the resolution of a converter is specified simply by giving the number of bits.

Real data converters are not perfect, of course. Figure 13-7 shows a nonideal transfer characteristic for a 3-bit DAC. Notice that the output voltages resulting from the different binary inputs do not lie on the ideal line. The ideal change in output voltage between each adjacent pair of outputs is one LSB. The real change will, in general, be different, and the difference is called the ***differential nonlinearity***(DNL), as shown on the plot. The DNL can be specified independently for each input word, but usually only the maximum DNL is specified. Another measure of nonlinearity is the ***integral nonlinearity*** (INL). The INL is the difference between each sample and a straight line that passes through the two endpoints (shown as a dashed line in the figure). The INL is also sometimes specified with respect to the best straight-line fit to the data (in a mean-squared-error sense),

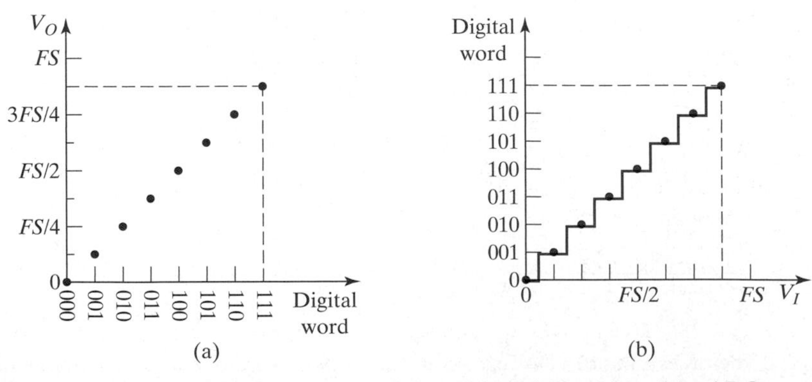

Figure 13–6 The transfer functions of (a) a 3-bit DAC and (b) a 3-bit ADC.

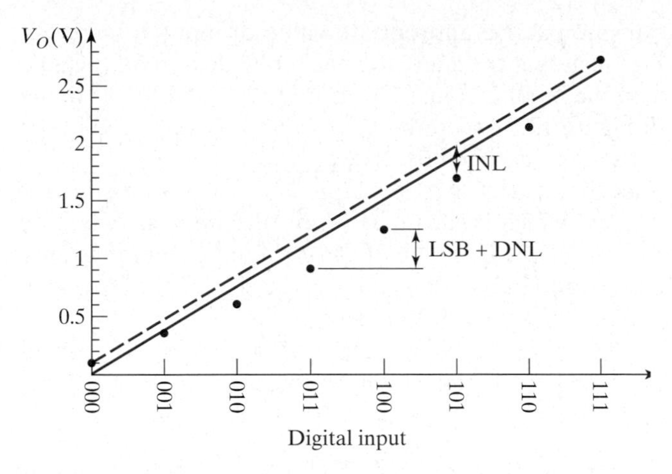

Figure 13–7 A nonideal 3-bit DAC. The points would ideally lie on the solid straight line shown. INL is measured with respect to the dashed line fitted to the endpoints of the DAC output (see dac_err.nb on CD).

which leads to lower values, but may be more appropriate if the gain and offset of the DAC output are adjusted in some way. The gain is the slope of the line, and the offset is where it intercepts the vertical axis.

DNL affects the quantization step size and therefore would increase the quantization noise of the system, whereas INL represents a deviation from linearity and is therefore related to the amount of distortion produced by the converter. Both DNL and INL are specified as a fraction of an LSB.

The errors in ADCs are similarly defined. The DNL is the difference between the actual width of each step and the ideal width of 1 LSB, as illustrated in Figure 13-8. The INL is the difference between a transition and the straight line passed through the tops of the first and last transitions.[3] A ***missing code*** occurs if the digital output skips over one possible value as the input is changed. Missing-code errors are unique to ADCs, but can be corrected in the digital domain, since

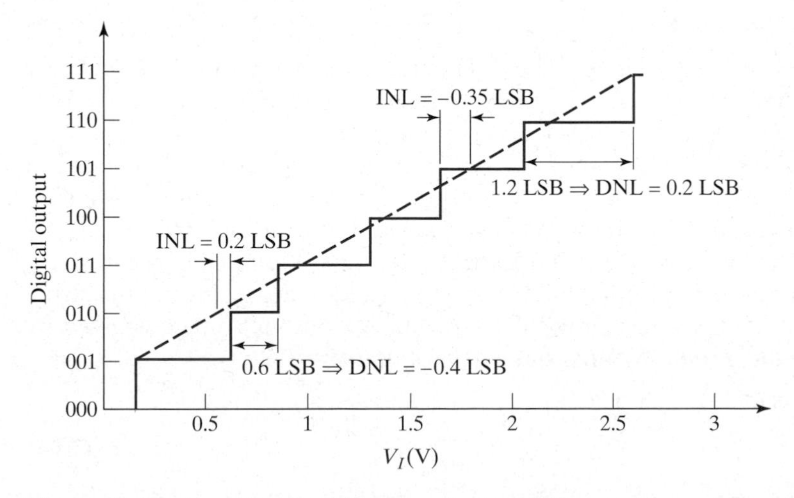

Figure 13–8 A nonideal 3-bit ADC showing how DNL and INL are defined (see adc_err.nb on CD).

[3] It is again possible to use a best fit line instead of using the endpoints, but the endpoints yield a conservative value.

the output does make a transition at the appropriate value of input voltage. On the other hand, if the DNL is such that the stair step has too wide a tread (i.e., the output does not change when the input increases by more than one LSB, as in the last full step on the right in Figure 13-8), then the error cannot be repaired in the digital domain because information has been irretrievably lost.

If the errors in either a DAC or an ADC become large enough, it is possible for the output to be ***nonmonotonic***, which means that when the input increases by one LSB, the output actually decreases. This type of error is fatal in some systems. For example, if the converter is used in a feedback loop, the sign of the loop gain changes in that small region and an oscillation can occur.

The rest of this chapter focuses on the operation and design of ADCs and DACs. Because DACs are somewhat simpler and are sometimes used in the design of an ADC, they are covered first.

Exercise 13.1

Suppose that an analysis of the noise present in a given converter shows that the smallest usable LSB is 10 μV. What supply voltage would be required to produce a 16-bit converter? How about a 20-bit converter? A 30-bit converter?

□

13.2 DIGITAL-TO-ANALOG CONVERTERS

In this section, we examine the design of DACs. All DACs must be able to produce different output voltages or currents based on a digital input. Therefore, there must be some method for precisely dividing a reference input into a set of voltages or currents. We first treat DACs made using resistors to divide the reference and then cover DACs made using capacitive dividers.

13.2.1 Resistive DACs

One straightforward way to make a DAC is to use a resistor string to produce all of the output voltages necessary and then use a set of switches to select which output to use, as shown for a 3-bit DAC in Figure 13-9. The seven switches are controlled by the three binary inputs as shown, and the amplifier at the output is a buffer to guarantee that the output does not load down the resistor string (i.e., does not draw substantial current from one node in the string and thereby change the voltage). The resistor string divides the reference, V_r, into eight steps (2^N, where $N = 3$ in this example), so that the voltage across each resistor in the string corresponds to one LSB. For an N-bit converter, the output voltage is given by (assuming that the buffer amplifier has a gain of unity)

$$V_O = \frac{V_r}{2^N}\sum_{i=0}^{N-1} 2^{N-1-i} b_i. \qquad \textbf{(13.4)}$$

This architecture has the advantage that the output is guaranteed to be monotonic, but it uses a large number of resistors and switches and is impractical for converters with more than a few bits of resolution.

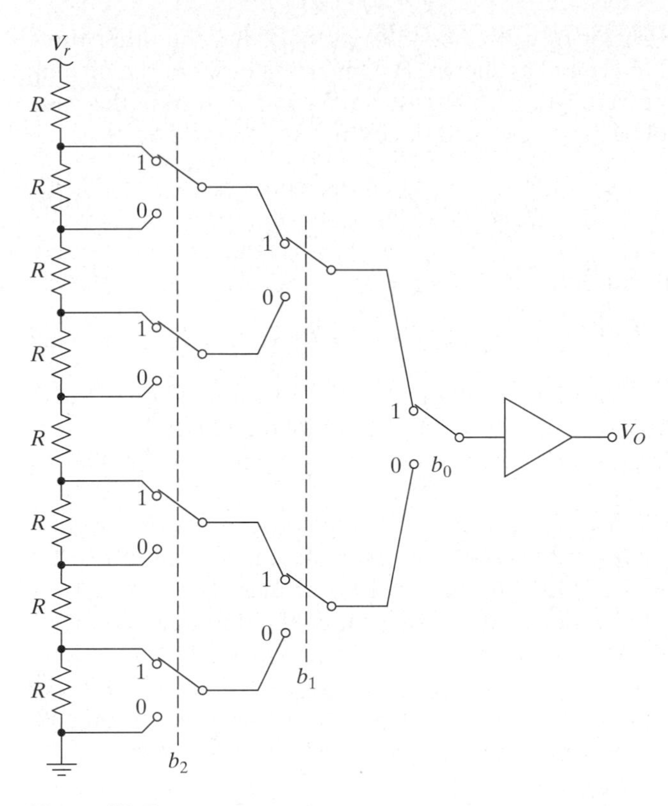

Figure 13–9 A resistor-string DAC.

Exercise 13.2

How many resistors and switches would be required to implement an 8-bit DAC, using the architecture shown in Figure 13-9?

A DAC can also be implemented using a ***binary-weighted array*** of resistors and an op amp, as illustrated in Figure 13-10 for a 4-bit DAC. The binary input word is used to control the switches in this circuit, which could be implemented by metal-oxide semiconductor field-effect transistors (MOSFETs). If the input word is

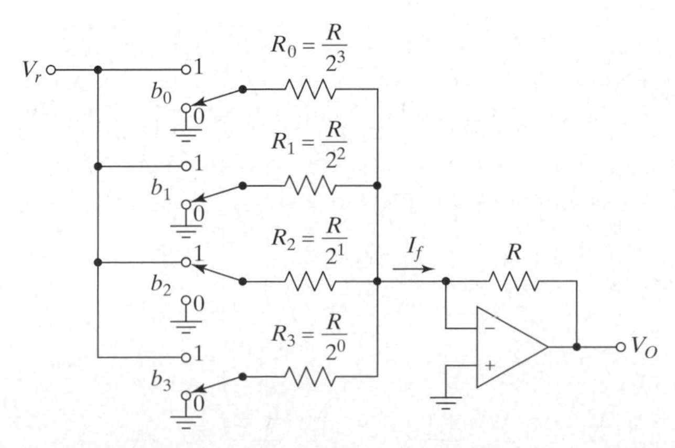

Figure 13–10 A DAC constructed using a binary-weighted array of resistors.

0010, for example, the switch for b_2 will be connected to V_r, and the other three switches will be connected to ground as shown. The inverting input of the op amp is a virtual ground, so the currents through R_0, R_1, and R_3 will be zero in this case. The current through R_2 will be $I_2 = V_r/R_2$, and the output voltage will be

$$V_O = \frac{-R}{R_2}V_r = -2V_r. \quad \mathbf{(13.5)}$$

In general, the output voltage will be

$$V_O = -V_r(2^3 b_0 + 2^2 b_1 + 2^1 b_2 + 2^0 b_3), \quad \mathbf{(13.6)}$$

where b_X will be 1 or 0, corresponding to whether bit X in the input word is 1 or 0. The full-scale reference voltage for this converter, as defined in the transfer functions in Figure 13-6, is $2^4 V_{LSB} = -16V_r$. The maximum output voltage is smaller by one LSB; it is $V_{O\,max} = -15V_r$.

One serious problem with this circuit is the wide range of resistors required for a practical converter. For a 16-bit converter, for example, the spread in resistor values is $2^{15} = 32{,}768$. If $R = 100$ kΩ, then $R_{15} = 100$ kΩ and $R_0 = 3.0518$ Ω. A change of only 0.0031% in the value of R_{15} would change the output by one LSB. This accuracy is way beyond what can be maintained in practical circuits. Another problem with this architecture is that the current drawn from the reference changes with the input word. If the reference source has a nonzero source impedance, then this changing current will produce a changing reference voltage.

Exercise 13.3

(a) How many resistors and switches would be required to implement an 8-bit DAC, using the architecture shown in Figure 13-10? (b) What would the ratio of the largest to the smallest resistor be?

Exercise 13.4

Consider an 8-bit DAC, using the architecture shown in Figure 13-10. If $R_7 = 100$ kΩ, how much can it change before the output changes by one LSB?

One way to avoid the spread in resistor values is to make use of an **R–2R Ladder**. An R–$2R$ ladder is shown in Figure 13-11, and, as the name implies, uses only two values of resistor: R and $2R$. The input to the ladder is either the current $2I$ or the

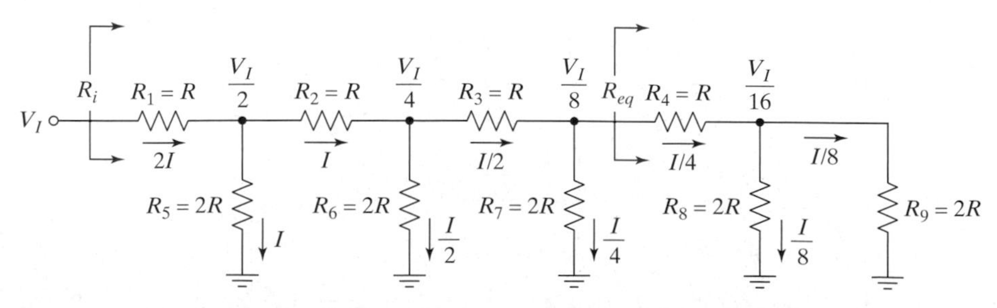

Figure 13–11 An R–$2R$ ladder.

voltage V_I. Starting from the right side of the ladder, we see that there are two resistors to ground (R_8 and R_9), and both are equal to $2R$. Therefore, the current flowing to the right through R_4 divides equally between the two resistors. Also, the two combine in parallel to form an equivalent resistance to ground of R. This equivalent resistance is in series with R_4, which is also equal to R; therefore, the voltage at the left of R_4 is divided equally, with half across R_4 and half across R_8 and R_9. Also, the total resistance of R_4 in series with R_8 and R_9 is $2R$ (shown as R_{eq} on the schematic), which now shows up in parallel with R_7 to ground. Therefore, the current going to the right through R_3 divides equally, with half going through R_7 and half going into R_{eq}. Also, the voltage at the left of R_3 is divided equally, half across R_3 and half across R_7 in parallel with R_{eq}. We continue to work our way back to the left in this fashion, with the currents splitting in half at each rung of the ladder as shown. At each node, the current splits equally, and the voltage is one-half of the value on the next node to the left, as shown in the figure. The voltage is also halved each time you move to the right one rung on the ladder, so this ladder has produced a binary-weighted array of both currents and voltages.

Exercise 13.5

Consider the R–2R ladder, of Figure 13-11. Assume that all of the resistors are ideal, except for R_5. How much can the value of this resistor vary before the current through it changes by $I/2^{15}$? How does this compare with the accuracy required for a 16-bit converter using the binary-weighted array of resistors?

A DAC can be made using the R–$2R$ ladder, as illustrated for a 4-bit DAC in Figure 13-12. This schematic is only conceptual, but we will show one way to implement the switches in Example 13.2. For now, we note that, as before, the switch positions depend on the corresponding bit in the input word, and the currents are summed by using the virtual ground at the input of the op amp. The output voltage in this case is given by

$$V_O = -R_f \sum_{i=0}^{3} \frac{b_i I_r}{2^i}. \tag{13.7}$$

For this circuit to work properly, we must ensure that the on resistances of the switches are negligible in comparison with $2R$ and that there aren't any significant offset voltages (i.e., the bottoms of the $2R$ resistors in the ladder must be at ground potential for the ladder to work properly).

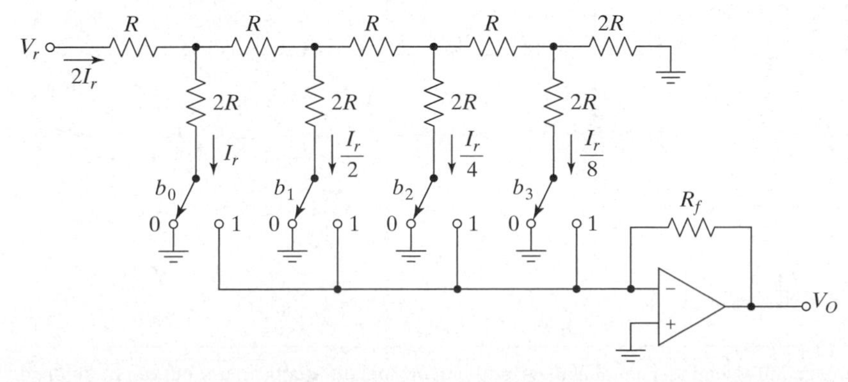

Figure 13–12 A conceptual schematic for a 4-bit DAC.

Example 13.2

A practical implementation of the DAC shown in Figure 13-12 is given in Figure 13-13, using generic transistors.[4] This schematic assumes that the circuit is fabricated as an integrated circuit so that the transistors can be matched. The op amp on the left is used in a feedback loop with T_{Ref} as a current source. Due to the virtual short circuit at the op amp inputs, the voltage at the merge terminal of T_{Ref} is V_{ref}, and the current is (V_{MM} is a negative voltage)

$$I = \frac{V_{ref} - V_{MM}}{2R}. \tag{13.8}$$

The current in T_0 is the same as the current in T_{Ref}, and the other currents are binary weighted due to the R–$2R$ ladder. In order for the ladder to work properly, however, the voltages at the merge terminals of these transistors must all be the same. Because the voltages on the control terminals of T_0 through T_3 are all the same, while the currents in the transistors are different, the voltages on the merge terminals would not be the same if the transistors were all identical, and as a result, the ladder would not work properly. This potential problem is overcome by scaling the sizes of the these transistors in a binary fashion so that the voltages on the merge terminals are the same. This solution is not practical for DACs with more than a few bits of resolution, however, due to the large transistors required. Other methods are used that are beyond the scope of the present treatment.

Another factor that could prevent the output currents of these transistors from achieving the desired binary weights is the output resistance of the transistors.

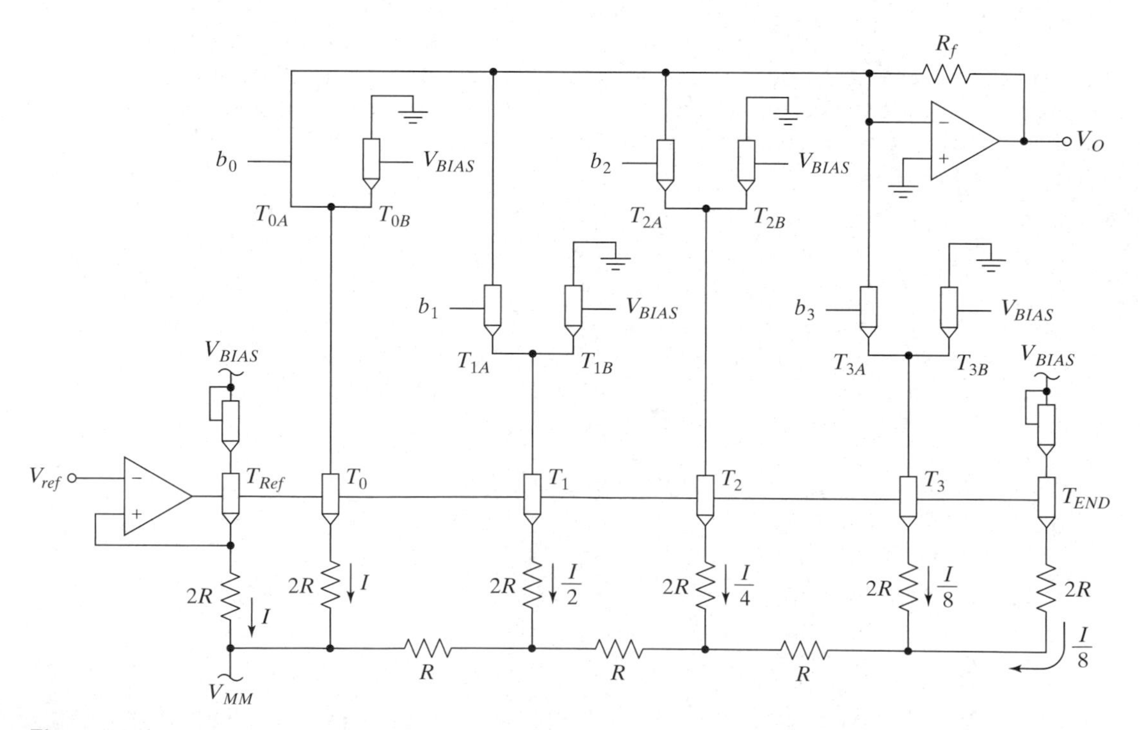

Figure 13–13 A practical 4-bit DAC.

[4] This example requires an understanding of current mirrors and differential pairs, but can be skipped without loss of continuity.

This problem is overcome by keeping the voltages on the output terminals all about the same. The differential-pair current switches above the ladder transistors keep the output voltages of the ladder transistors at $V_{BIAS} - V_{CM}$. (The V_{CM}s of the current switches can again be made uniform by scaling the transistors.) The output terminals of T_{Ref} and T_{End} are both connected to V_{BIAS} through diode-connected transistors to maintain their output voltages at about the same level.

Finally, the currents from the ladder transistor outputs are all applied to the tails of the differential pairs, which are controlled by the digital input word. For example, when b_0 is greater than V_{BIAS}, which represents the most significant bit in the input word being a one, T_{0A} takes essentially all of the current through T_0 and passes it to the inverting input of the op amp. If $b_0 < V_{BIAS}$, on the other hand, the current from T_0 is pulled out of ground through T_{0B}. All of the other differential pairs switch their currents in the same way. The op amp sums the currents at its inverting input and forces the sum to flow through R_f, which produces the output voltage

$$V_O = IR_f \sum_{i=0}^{3} \frac{b_i}{2^i}. \tag{13.9}$$

13.2.2 Capacitive DACs

DACs can be made using only capacitors, switches, and op amps. A 3-bit example using a binary-weighted array of capacitors is shown in Figure 13-14. The circuit works in the following way: When SW2 is in the reset position as shown, the op amp is connected as a unity-gain buffer, and both the output and the inverting input are at ground potential. With SW1 also in the reset position, the total input capacitance will be charged up to V_r. The total input capacitance is determined by the digital input word. Each switch is closed if the corresponding bit is one and open if it is zero, so that the total input capacitance is

$$C_i = \sum_{n=0}^{2} b_n 2^{2-n} C. \tag{13.10}$$

Therefore, the total charge on this capacitor is

$$Q_i = V_r \sum_{n=0}^{2} b_n 2^{2-n} C. \tag{13.11}$$

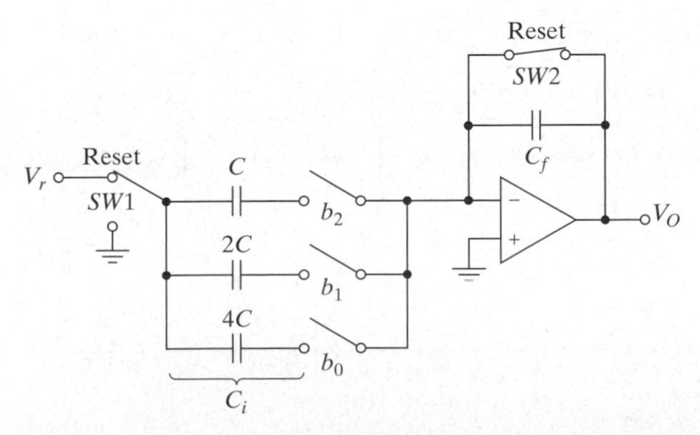

Figure 13–14 A capacitive DAC using a binary-weighted array of capacitors.

After C_i has been charged to V_r, we open SW2 and connect SW1 to ground. The op amp will now work to keep the inverting input at ground, but to do so, it must change its output voltage so that the feedback capacitor absorbs all of the charge on C_i. The resulting output voltage is

$$V_O = \frac{Q_i}{C_f} = \frac{V_r \sum_{n=0}^{2} b_n 2^{2-n} C}{C_f}, \tag{13.12}$$

and, if we set $C_f = C$, we get

$$V_O = V_r \sum_{n=0}^{2} b_n 2^{2-n}, \tag{13.13}$$

which is precisely what we want for a DAC.

This DAC has the same major problem as the binary-weighted resistive DAC shown in Figure 13-10, namely, for more than just a few bits of resolution, the spread in capacitor values required is too large to be practical. We can again overcome the problem by using different architectures, but they will not be pursued here.

13.3 ANALOG-TO-DIGITAL CONVERTERS

A time-varying input must be sampled and held prior to being converted into its digital equivalent. This sample-and-hold operation is inherent in some ADCs, but not all. Therefore, we will briefly discuss S/H circuits here before going on to talk about ADC design.

13.3.1 Sample-and-Hold Circuit

A S/H circuit that samples the input every T seconds is shown conceptually in Figure 13-15(a). The switch could be implemented by one or two MOSFETs or by more complex circuits using diode bridges or op amps. Part (b) of the figure shows the ideal input and output waveforms of the S/H for a sine wave input.

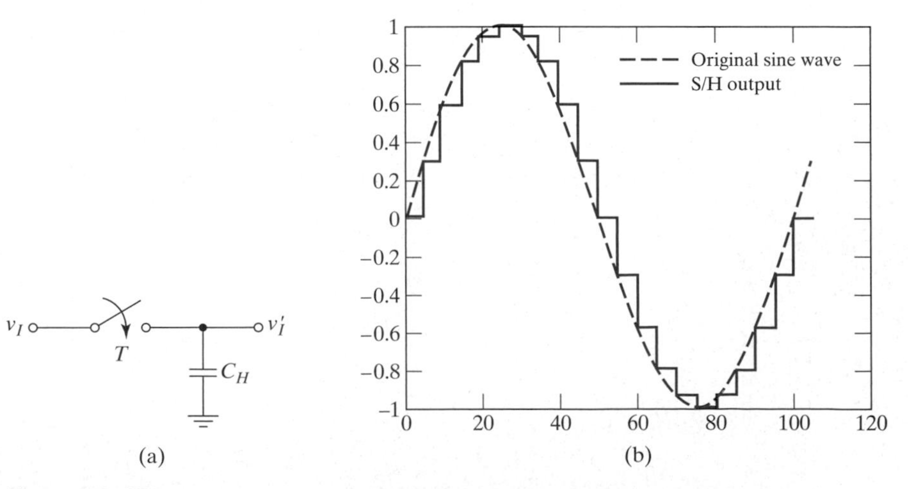

Figure 13–15 (a) A conceptual S/H schematic and (b) v_I and v_I'.

In the ideal case, the S/H closes the switch for an instant to sample the input voltage and then opens the switch to hold the value for T seconds. In reality, the switch always has some resistance and must be held closed for a while, called the ***acquisition time***, to allow the voltage on the capacitor to charge up to v_I. When the switch opens, the voltage on the capacitor is held for the duration of the period T.

There are, of course, errors that must be allowed for in practical S/H circuits. For example, when the switch opens, there is usually a small amount of charge transferred to or from the hold capacitor by the circuits comprising the switch. This charge transfer leads to a small change in the held voltage; the change is called the ***hold step*** of the S/H. In addition, some charge leaks off of the hold capacitor during the hold time, so the voltage held on the capacitor will ***droop*** with time.

In the rest of this section we assume the ADCs are preceded by an ideal S/H unless noted otherwise. As a result, we treat the inputs to the ADCs as constants during each conversion cycle. The field of ADC design is a very active one, and there are many modern approaches to building ADCs that are not presented here. This section is only a basic primer to the field.

13.3.2 The Successive Approximation ADC

An ADC can be made by placing a DAC in a negative feedback loop, as shown in Figure 13-16. The digital input word to the DAC is varied until the DAC output is approximately equal to the input voltage, v_I. A binary search pattern is used so that one bit of the digital output is determined on each clock cycle. For example, during the first clock cycle, the circuit determines whether the input is greater than or less than one-half of the reference. Assuming that the input is greater than half of the reference, the second cycle would again divide the range in half and would test whether the input was greater than or less than three-fourths of the reference. This process continues with each new cycle dividing the range in half again and determining one more bit of the digital output word.

The comparison is made by the op amp shown in the figure. The ***successive approximation register*** (SAR) uses the output of the op amp to change the digital input to the DAC. As briefly described earlier, the conversion starts with the MSB and progresses toward the LSB one bit at a time. For example, for a 3-bit conversion, the SAR would first output the binary word '100'. If v_I was larger than the resulting DAC output, $v_{DAC} = V_r/2$, then the MSB is equal to one and the next word out of the SAR would be '110,' which yields $v_{DAC} = 3V_r/4$. However, if $v_I < v_{DAC}$ for the first comparison, the MSB is zero, and the next word would be '010', which yields $v_{DAC} = V_r/4$. Following this pattern, an N-bit conversion requires N clock cycles. The output of the ADC is the digital input to the DAC, as shown. This kind of ADC is called a ***successive approximation*** ADC.

Successive approximation ADCs are simple, can be reasonably accurate, and are medium speed, but they do require an accurate DAC. We next present ADCs that are either slower and simpler or faster and more complicated than the successive approximation type.

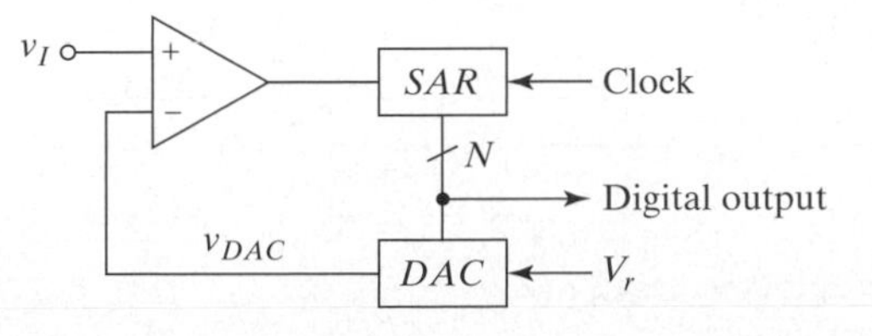

Figure 13–16 A successive approximation ADC.

Exercise 13.6

What is the conversion time of a 16-bit successive approximation ADC if a 1-MHz clock is used?

13.3.3 The Dual-Slope ADC

Another type of ADC is the ***dual-slope*** converter shown in Figure 13-17(a). This converter is slow, but it can achieve high resolution and linearity with relatively simple circuits. For the conversion to be accurate, the only part of the converter that must be absolutely accurate is the reference voltage, V_r. Other errors in the circuits, such as offset voltages, gain errors, and random variations in the clock period, tend to cancel or average out and do not affect the accuracy or linearity very much. As a result of this robustness, the dual-slope converter is popular for applications where accuracy is required, but speed is not essential—for example, digital multimeters.

The conversion is accomplished by first integrating the input for a fixed time and then measuring how long it takes to integrate back to zero using a fixed slope (proportional to $-V_r$). The time it takes to integrate back to zero will provide a measure of the ratio of v_I to V_r, as we will now show.

The converter consists of an integrator block, a comparator, some control circuits, a switch, a clock, a counter, and a reference. With the switch in the position shown in the figure, the input voltage is integrated. Assuming that the integrator was reset at the start of the cycle (i.e., its output was set equal to zero), the output voltage will ramp up with a slope proportional to v_I, as shown in part (b) of the figure. This integration is performed for a fixed number of clock cycles, 2^N for an N-bit converter. At the end of this integration time, the integrator output has reached a voltage proportional to v_I and the number of clock periods, namely,

$$v_x(t_1) = K2^N T_{CL} v_I, \tag{13.14}$$

where T_{CL} is the clock period and K is a constant that depends on the integrator circuitry. At time t_1, the control block starts the counter and switches the integrator input to $-V_r$. The integrator output voltage now ramps down with a fixed slope determined by V_r. When the voltage reaches zero, as determined by the comparator, the counter is stopped. The time it takes the integrator output to reach zero is proportional to how large the starting value was, given by (13.14), and how fast the ramp down was, which is proportional to V_r:

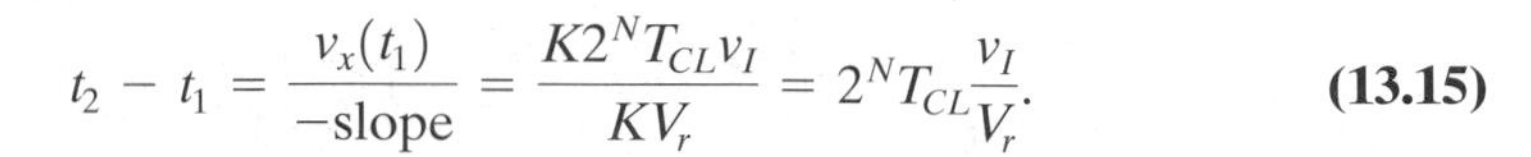

$$t_2 - t_1 = \frac{v_x(t_1)}{-\text{slope}} = \frac{K2^N T_{CL} v_I}{KV_r} = 2^N T_{CL}\frac{v_I}{V_r}. \tag{13.15}$$

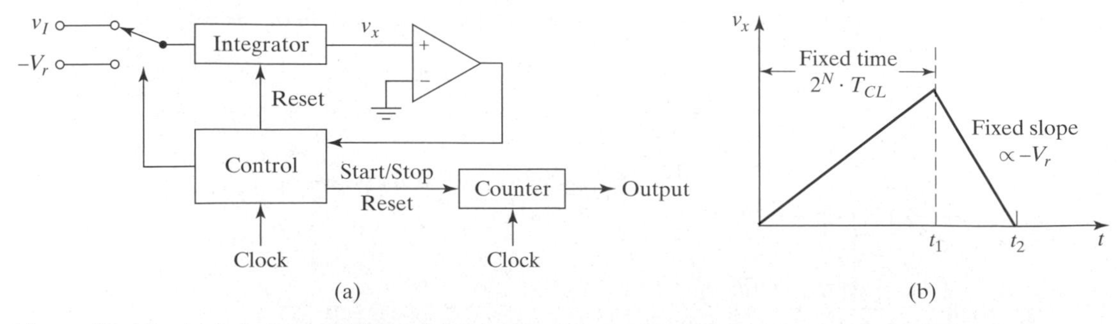

Figure 13–17 (a) A dual-slope ADC and (b) the integrator output voltage.

The count in the counter at time t_2 is the smallest integer greater than $(t_2 - t_1)/T_{CL}$, so we have

$$\text{count} = \left\lceil 2^N\left(\frac{v_I}{V_r}\right)\right\rceil, \quad \textbf{(13.16)}$$

where $\lceil x \rceil$, the ceiling operator, returns the smallest integer greater than x. This count is the output of the ADC. The proportionality factor in (13.14) cancels out when (13.16) is derived, because it affects the slope of v_I the same during both the ramp up and the ramp down. When the conversion is finished, the control block resets the counter and integrator, switches the integrator input to v_I, and starts the process over again.

The dual-slope ADC does not require a S/H circuit. If v_I is changing during the conversion, the peak integrator output as given in (13.14) will simply be proportional to the average of the input during that first portion of the conversion cycle. Therefore, any sinusoidal input with an integral number of periods during time t_1 will average out to zero, and the ADC will not work. This limitation turns out to be an advantage in some applications, however. For example, for a DC voltmeter, if the time t_1 is set to 1/60 of a second, then 60-Hz power-line noise will be rejected. This feature is called ***normal-mode rejection***.

Exercise 13.7

Consider a 10-bit dual-slope ADC with a clock period of 16.3 μs. What is the longest time a conversion can take?

□

13.3.4 The Flash ADC

The fastest kind of ADC is called, appropriately, a ***flash*** converter. A 2-bit flash ADC is shown in Figure 13-18. This circuit uses a resistor string to divide the reference into three equally spaced levels and then simultaneously compares the input with all three levels. The outputs of the comparators can then be examined to determine the magnitude of the input. The converter is called a flash converter because the comparison with all possible levels is made at one time. If v_I is equal to $0.6V_r$ for example, the bottom two comparators in this example would have high outputs, and the top comparator would have a low output. The decoding circuit can then determine from these outputs that v_I is between $V_r/2$ and $3V_r/4$.

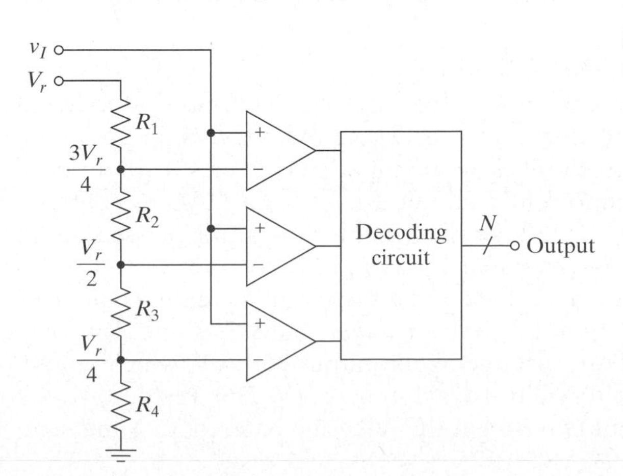

Figure 13–18 A 2-bit flash converter.

Flash converters are the fastest converters made, but require a huge amount of circuitry and power for high resolution. For an N-bit flash converter, we need to generate $2^N - 1$ reference levels, which requires 2^N identical resistors, and we need $2^N - 1$ comparators. The major problem with this converter is the large number of comparators required. For example, for a 10-bit converter, we need 1,023 comparators, and for a 16-bit converter, we need 65,535 comparators. If each of these comparators dissipates 5 mW, the 16-bit converter requires 328 watts of power. For lower resolution ADCs, however, the flash architecture is used when high speed is the primary requirement.

The outputs of the comparators in a flash converter form what is called a *thermometer code*. The name is derived from the fact that all of the outputs corresponding to comparison levels below v_I should be high, whereas all of the outputs corresponding to comparison levels above v_I should be low. In this ideal case, all the decoding logic does is find the two comparators where the outputs switch from high to low (between the second and third comparators in our example) and then output the appropriate digital word.

For high resolution flash ADCs, the difference between two adjacent comparator thresholds becomes small, and there exists the possibility that the thermometer code will have what are called *bubbles* in it. (Bubbles are also called *sparkles*.) For example, the outputs of the comparators, in sequence from the bottom to the top, might be [1111 · · · 111**1010**000 · · · 000]. The one zero surrounded by ones in this sequence is called a bubble (in boldface). The decoding circuits must be able to deal with this situation and make a reasonable choice as to what the output should be. A simple choice is to pick the last transition from one to zero (i.e., the transition in which no ones exist to the right).

Since the comparators in a flash ADC sample the input, a S/H is not absolutely necessary. Nevertheless, a S/H is frequently used to avoid the errors caused when the sampling times of all the comparators are not exactly the same. Also note that the converter in Figure 13-18 makes output code transitions when $V_I = nLSB$, for integer n, which is different than we show in Figure 13-6(b). Either method can be used in practice.

Exercise 13.8

If a 12-bit flash ADC is constructed using comparators that consume 10 mW of power each, how much power will the converter require? Assume that the comparators dominate the power.

13.3.5 Other ADC Architectures

There are many other ADC architectures that have not been presented here. In particular, pipelined converters work by breaking the conversion up into a sequence of subconversions. Each block in the pipeline converts its input into a one- (or more) bit digital representation and then uses a DAC to subtract the result from the input. What is left is called the *residue* and is passed on to the next stage in the pipeline. For example, if the input is 2 V and the full-scale reference is 3 V, the first stage would decide that the input is greater than one-half of the reference, so the output bit for that stage would be a one. This one would be fed into a one-bit DAC, producing an output of 1.5 V, which would be subtracted from the input to yield a residue of 0.5 V. This residue would then be amplified by a factor of two (to get the full scale back to 3 V) and sent into the next stage.

The operation of the pipelined converter is therefore similar to a successive approximation converter, but with each successive approximation accomplished at a different stage of the pipeline. The net result is faster conversion than a successive approximation converter at the expense of some extra circuitry and with some latency (i.e., the output at one time is due to the input sample at some previous time, since the sample has to propagate down the pipeline).

There are other more sophisticated approaches to making ADCs with various combinations of speed, resolution, power dissipation, and cost. For example, oversampled delta sigma converters and folding and interpolating converters are presently quite popular for many applications. If you are interested in these more sophisticated techniques, we recommend looking in [13.1] and [13.2] as a good starting point.

SOLUTIONS TO EXERCISES

13.1 Using (13.3), we see that the required full-scale voltage is 2^N times the LSB. The supply voltage must be at least equal to the full-scale output. Therefore, for 16 bits, we need a supply greater than 0.66 V. For 20 bits, we need at least 10.5 V, and for 30 bits, we would need 10,737 V!

13.2 We would need $2^8 = 256$ resistors and $2^8 - 1 = 255$ switches.

13.3 **(a)** We need one resistor for each bit, plus the feedback resistor, for a total of nine resistors. We need one switch per bit, or eight switches. **(b)** The largest resistor would be $2^7 = 128$ times the value of the smallest one.

13.4 The current corresponding to an LSB is $I_{\text{LSB}} = V_r/100\text{ k}\Omega$. The nominal value of R_7 is $R_7 = (100\text{ k}\Omega)/2^7 = 781.25\Omega$. Assume that the value of R_7 decreases so that it is equal to $(1 - x)781.25\Omega$. We set the change in the current through R_7 equal to the LSB current and solve for x:

$$\frac{1}{390.6(1 - x)} - \frac{1}{390.6} = \frac{1}{100{,}000}.$$

The result is $x = 0.0078$, or, in other words, R_7 can change only by 0.78%.

13.5 All of the resistors to the right of R_5 are equivalent to a resistor of value $2R$. Therefore, the current in R_5 is given by $I_5 = 2I[2R/(2R + R_5)]$, where we have assumed that the current in R_1 is $2I$. Setting $I_5 = I(1 \pm 1/2^{15})$ leads to $R_5 = 2R(1 \pm 0.00006)$. In other words, R_5 must be accurate to within $\pm$0.006%. This accuracy requirement is four times less stringent than the 0.0015% required for the binary-weighted array. In addition, we don't have to try and produce resistors that differ by a factor of 32,768!

13.6 We require one clock period per bit, or 16 clock periods. At 1 MHz, this is 16 μs.

13.7 If the input is equal to the full-scale voltage, the ramp up and ramp down will both take $2^{10} = 1{,}024$ clock cycles. Therefore, a full conversion would take 2,048 periods, or 33.3 ms.

13.8 The converter will have $2^{12} - 1 = 4{,}095$ comparators, and at 10 mW each, the total power will be 41 W.

CHAPTER SUMMARY

- There are many applications in which digital signal processing has significant advantages over analog signal processing.
- Data converters are a major component of many digital signal-processing systems.
- ADCs must sample the continuous-time input at a rate at least twice as high as the highest frequency in the signal if the samples are to be a unique representation of the input.
- Data converters can suffer from a number of different errors, including differential nonlinearity, integral nonlinearity, nonmonotonicity, and missing codes.
- DACs can be constructed by using resistive ladders and switches, with R–$2R$ ladders and switches, and with capacitor arrays and switches.
- ADCs can be constructed by using DACs in a feedback loop, but the performance of the resulting ADC is limited by the performance of the DAC.
- Dual-slope ADCs are slow, but can have high resolution with relatively simple circuitry.
- Flash ADCs are the fastest converters, but they consume a large amount of power and are impractical for high-resolution converters.

REFERENCES

[13.1] R. van de Plassche, *Integrated Analog-to-Digital and Digital-to-Analog Converters*. Boston: Kluwer Academic Publishers, 1994.

[13.2] B. Razavi, *Principles of Data Conversion System Design*. New York, IEEE Press, 1995.

[13.3] E.A. Lee and D.G. Messerschmitt, *Digital Communication*: 2/E. Boston: Kluwer Academic Publishers, 1994.

[13.4] R.G. Gallager, *Information Theory and Reliable Communication*. New York: John Wiley and Sons, 1968. (This is a classic reference in the field—old and difficult, but well worthwhile.)

[13.5] Analog Devices, *Analog-Digital Conversion Handbook*, 3/E. Englewood Cliffs, NJ: Prentice-Hall, 1986 (a good reference for the specifications of converters, even though much of the data is old).

PROBLEMS

13.1 Introduction

13.1.1 Analog and Digital Signal Processing and Applications of Data Converters

P13.1 Assume that the quantization error in a sampled system is a uniformly distributed random variable in the range from $-\text{LSB}/2$ to $\text{LSB}/2$. The RMS value of the quantization noise is the square root of the variance of this uniform random variable. **(a)** Find the RMS value of the quantization noise for an N-bit converter with a full-scale output of FS. **(b)** Find the RMS value of a sine wave with a peak-to-peak value equal to FS. **(c)** Take the ratio of the RMS signal to the RMS noise, and express the result in decibels. You should obtain (13.2).

P13.2 Suppose a given piece of music has a maximum dynamic range of 40 dB (i.e., the ratio of the signal power in the loudest passage to the signal power in the quietest passage is 10,000). Further, assume that you would like to record this music with an SNR of at least 80 dB (even in the quietest section). Assuming that the signal level in the loudest section of the music is set equal to the full-scale input of an ADC, how many bits must the ADC have to achieve the desired SNR?

P13.3 Consider sampling music at a 44.1-kHz rate. If we desire to have a −3-dB bandwidth of 20 kHz and to have the original spectrum be at least 60 dB down before it crosses the first copy (i.e., the spectrum shifted up to 44.1 kHz), how many poles must we use in the anti-aliasing filter?

13.1.2 Data Converter Transfer Functions and Specifications

P13.4 The eight output voltages of a 3-bit DAC with a 5-V full scale are measured to be 0.1, 0.74, 1.4, 2.04, 2.64, 3.25, 3.85, and 4.5 V, in order. **(a)** What is the LSB in volts? **(b)** What are the ideal output voltages? **(c)** What is the maximum DNL (expressed as a fraction of an LSB)? **(d)** What is the maximum INL if the endpoint line is used (expressed as a fraction of an LSB)?

P13.5 The output code transitions for a 3-bit ADC with a 5-V full scale are measured to occur for input voltages of 0.3, 0.91, 1.51, 2.18, 2.7, 3.41, and 4.12 V in order. (There aren't any missing codes.) **(a)** What is the LSB in volts? **(b)** What are the ideal code transition voltages? **(c)** What is the maximum DNL (expressed as a fraction of an LSB)? **(d)** What is the maximum INL if the first and last transitions are used to define the straight line (expressed as a fraction of an LSB)?

P13.6 The output voltages for a 3-bit DAC are given in Table 13.1. Fill in the ideal values, DNL, the values that fall on the straight line between the two endpoints, and the resulting INL. V_r is the full-scale reference voltage.

Table 13.1

Digital Input	Ideal v_O/V_r	Real v_O/V_r	DNL (LSB)	End-point Line (v_O/V_r)	INL (LSB)
000		0.020			
001		0.103			
010		0.212			
011		0.331			
100		0.458			
101		0.582			
110		0.720			
111		0.910			

P13.7 Complete Problem P13.6 and then add two new columns to the table: the points corresponding to the best straight-line fit to the data (in a mean-squared-error sense) and the INL resulting from using this line instead of the endpoint line.

P13.8 Complete Problem P13.6 and then add three new columns to the table: calibrated output, calibrated DNL, and calibrated INL. For the calibrated output, apply a gain and an offset to the DAC output so that the two endpoints have the ideal values. Then calculate the DNL and INL for these new calibrated output values.

13.2 Digital-to-Analog Converters

13.2.1 Resistive DACs

P13.9 Consider the 2-bit digital-to-analog converter shown in Figure 13-19. Assume that the op amp is ideal. **(a)** Fill in the values of V_O in the truth table (i.e., give the voltages). For this part, you should assume that the NMOS transistors are perfect switches (i.e., $V_{DSn} = 0$ when $b_n = 1$ and $I_{Dn} = 0$ when $b_n = 0$). **(b)** Assuming that the value of R is known, find an equation for the minimum W/L required to guarantee that the error in V_O is less than 1/2 LSB when the digital input is $b_1 = 1$ and $b_0 = 0$. Assume that the gate voltages of the MOSFETs are 5 V when the corresponding bit is 1 and 0 V when the bit is 0. Assume that $V_{th} = 1$ V and that the values of μ_n and C'_{ox} are known (but leave them as variables in your equation). You may safely assume that the V_{DS} of any FET operating in the linear region is very small.

P13.10 Consider the 2-bit DAC in Figure 13-20. **(a)** Find the ideal output voltages for all input codes if $V_{FS} = -V_r = -3$ V. **(b)** If $R = 2$ kΩ and the op amp and switches are ideal, find the value of R_f needed to achieve the values given in (a). **(c)** Suppose the op amp has a +100-mV offset voltage (model it as a +100-mV source from ground to the noninverting input of the op amp) and that the switches have 250 Ω of resistance when they are on. Find the output voltages for all input codes for this case. (*Hint*: Find the Thévein equivalent driving the op amp for each code.) **(d)** Find the DNL for each input code. **(e)** Find the INL for each input code, using the endpoint straight line. [*Note*: The DNL and INL are small in this case even though the errors are relatively large, because the DAC has so few bits (and, hence, a large LSB).]

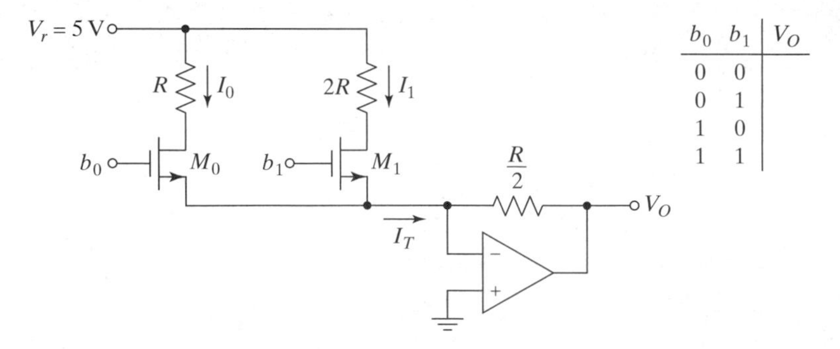

Figure 13–19

P13.11 Consider a 2-bit resistor-string DAC. **(a)** What are the ideal output voltages as fractions of the reference voltage? **(b)** Find the minimum and maximum possible output voltages as fractions of the reference if all of the resistors have a ±5% tolerance. **(c)** Find the worst case DNL for each code for the outputs found in (b).

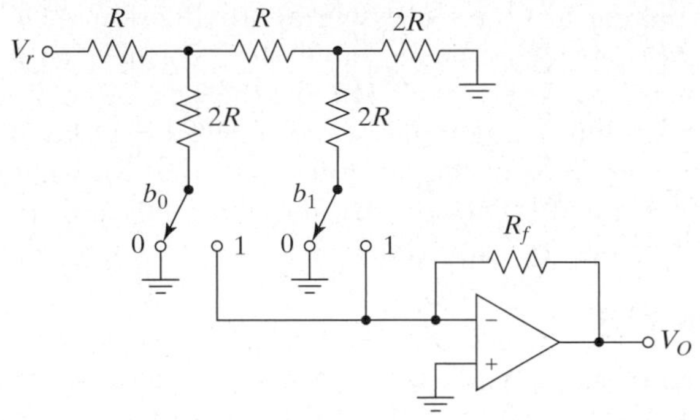

Figure 13–20

P13.12 Consider the 4-bit DAC in Example 13.2, and assume that $IR_f = 5$ V. Assume that the R–$2R$ ladder and the bottom current sources that utilize it are perfect, so that the tail currents of the differential pairs have their ideal values. Suppose that the voltages used to switch the differential pairs do not cause them to switch fully and, therefore, 5% of the current flows in the transistor that is supposed to be off and 95% flows in the transistor that is supposed to be on. **(a)** Derive the output voltage for each input word. **(b)** Derive the DNL for each input word. **(c)** Derive the INL for each input word, using the endpoint straight line.

13.2.2 Capacitive DACs

P13.13 Consider a 10-bit capacitive DAC made using the architecture shown in Figure 13-14. **(a)** If the minimum capacitor is set to 0.2 pF, how much total capacitance is required for this DAC? **(b)** How large an error can be tolerated on the largest capacitor if the DNL is to be kept below 1/2 LSB?

13.3 Analog-to-Digital Converters

13.3.1 Sample-and-Hold Circuit

P13.14 Consider a S/H as in Figure 13-15. **(a)** If the resistance of the switch is 10 Ω when the switch is on, the hold capacitor is 1 pF, and we desire to acquire the value to within 0.1% accuracy, what is the acquisition time of the S/H? **(b)** If a buffer amplifier is connected to the hold capacitor, how much current can it draw if the droop must be less than 0.1% in 100 ns for a held voltage of 1 V?

13.3.2 The Successive Approximation ADC

P13.15 **(a)** Draw v_{DAC} as a function of time during the conversion cycle of a 4-bit successive approximation ADC with $v_I = 2.46$ V, $V_r = 3$ V, and a 1-MHz clock. **(b)** What is the final digital output? **(c)** How long did the conversion take? **(d)** How long would the conversion take for $v_I = 0.56$ V?

P13.16 Consider making a 3-bit successive approximation ADC, using the DAC described by the data in Table 13.1. **(a)** What are the code transition points for the resulting ADC (i.e., for what values of v_I does the output code change)? **(b)** Find the DNL and INL of the resulting ADC for every output code.

P13.17 Consider making a 3-bit successive approximation ADC, using the DAC described by the data in Table 13.1. Further, assume that the comparator has a 50-mV offset. (Model this as a 50-mV source in series with the noninverting input of the comparator.) **(a)** What are the code transition points for the resulting ADC (i.e., for what values of v_I does the output code change)? **(b)** Find the DNL and INL of the resulting ADC for every output code.

13.3.3 The Dual-Slope ADC

P13.18 Consider a dual-slope ADC, as shown in Figure 13-17. Assume that the comparator has an offset voltage, V_{OFF}, but is otherwise perfect. Also assume that when the integrator is reset, the circuit forces v_x to zero as determined by the comparator. Show how this offset will affect the results derived in (13.14) through (13.16).

P13.19 Suppose you want to make a 16-bit dual-slope ADC with 60-Hz normal-mode rejection. **(a)** What is the fastest clock frequency you can use? **(b)** What input voltage yields the longest conversion time? **(c)** What is the longest time a conversion can take with this clock frequency?

13.3.4 The Flash ADC

P13.20 Suppose you are designing a flash ADC, using a 3-V reference. If the comparators will have offset voltages of 0.5 mV or less, but everything else is perfect, what is the maximum resolution you can design for and avoid bubbles in your thermometer code? (Offset voltages can have either sign.)

P13.21 Use a combination of NAND and NOR gates to design a thermometer-to-BCD converter for a 3-bit flash ADC that implements the following rule: The highest point in the thermometer code that has two adjacent ones is valid. Also, if only the bottom comparator output is high, the output should be decimal one. BCD is binary-coded decimal and simply refers to the representation where the MSB is on the left and the LSB is on the right; therefore, decimal zero is 000, decimal 3 is 011, decimal 4 is 100, and so on. As examples of the rule, if the seven comparator outputs are (bottom comparator listed on the left, top on the right) [XX11000], where Xs mean you don't care what the value is, the BCD output should be 100. Similarly, [XXX1100] should yield 101, and so on. The exception is that [1000000] yields 001.

CHAPTER 14

Gate-Level Digital Circuits

Digital signals can be transmitted and received with greater fidelity than analog signals can, which is a major reason for the dominance of digtial data transmission.

INTRODUCTION

In this chapter, we begin our discussion of purely digital circuits and systems. A digital ***logic gate*** is a circuit that implements some logical function. Digital circuits can be described at several levels. In this chapter, we present digital circuits at the gate level. Chapter 15 presents the transistor-level design of digital circuits. Most digital circuit design is done at the gate level or higher; but, since this book focuses on transistor-level circuit design, this chapter only introduces logic gates as a way of motivating and enabling the transistor-level design shown in Chapter 15. There are a number of excellent texts that cover digital circuit design at the gate level or above—for example, [14.1] and [14.2].

The vast majority of circuits designed and built in the world today are digital. Why are digital circuits and systems so common? The main reasons are:

1 Digital signals can be transmitted, received, stored, and retrieved with essentially no degradation in signal quality;[1]
2 Digital circuits and systems lend themselves to a more systematic design approach that can be highly automated at the gate level and above;
3 Digital systems are much easier to test than analog systems;
4 Digital systems can have high-resolution long-term memory; and
5 Digital systems can readily be made programmable.

Although digital circuits can use any number of discrete levels, the vast majority of digital circuits and systems use ***binary logic***. Binary logic uses only two levels. If more than two levels are used, we call it ***multivalued logic***. The two levels in binary logic are commonly referred to as ***0*** and ***1***, ***off*** and ***on***, or ***low*** and ***high***, respectively. In this text, we will restrict ourselves to binary systems.

To design linear circuits using nonlinear devices, we need to restrict our attention to small signals where the devices can be acceptably approximated as linear (*see* Chapter 6). However, nonlinearities are essential to the construction of digital circuits. To make a decision (e.g., that something is a 0 or a 1), we must have a nonlinearity. A truly linear system is not capable of making a decision; the output may be closer to one level than another, but the circuit itself does not decide or indicate that it is.

In this chapter, we first cover some basic background on digital circuits, number systems, and logic gates. We then discuss some very common applications of logic gates. We close with a brief section discussing the important topic of reflections on transmission lines.

14.1 BACKGROUND AND BINARY LOGIC

In this section, we first consider some fundamental characteristics of digital circuits. Then we discuss number systems and present several commonly encountered systems. Finally, we discuss the basic logical operations that all digital systems use and the algebra necessary to manipulate logical operators.

14.1.1 Fundamental Characteristics of Digital Circuits

The fundamental requirement of digital electronic circuits is that the electrical variables (current or voltages) that represent information be discrete as opposed to continuous. We will restrict our attention to binary discrete variables. Each ***binary digit*** (***bit***) assumes one of two possible states, usually denoted as zero and one. The physical states of the circuit have a permissible range, but the range for each state is nonoverlapping, so that, ideally, the circuit is always unambiguously in one state or the other. The digital-circuit designer must ensure that the signal is clearly within one of the two allowed regions and never in the forbidden region, as shown in Figure 14-1. Because the two states are made widely separated, digital circuits are tolerant of much larger component variations than are linear circuits and are, therefore, typically more reliable.

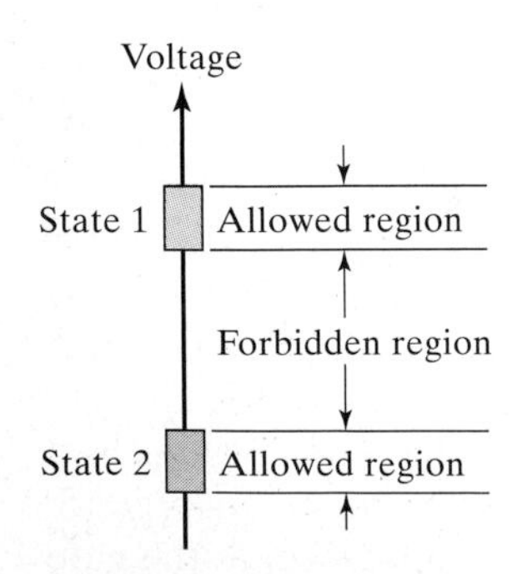

Figure 14–1 Node voltage levels showing allowed and forbidden ranges of values.

[1] This topic is discussed further in the introduction to Chapter 13 on data converters.

14.1.2 Number Systems

To understand the operation of digital circuits, we must become familiar with the binary number system and its use in logic circuits. The role played by the two digits 0 and 1 in a binary system is the same as that played by the 10 digits (0, 1, 2,...,9) in the decimal system. In a decimal system, we use the ***base***, or ***radix***, 10. For example, the digits in the number 1984 (one thousand nine hundred eighty-four) have the following positions and meaning:

$$(1984)_{10} = 1 \cdot 10^3 + 9 \cdot 10^2 + 8 \cdot 10^1 + 4 \cdot 10^0 \quad \textbf{(14.1)}$$

Similarly, the number 64.35 is

$$(64.35)_{10} = 6 \cdot 10^1 + 4 \cdot 10^0 + 3 \cdot 10^{-1} + 5 \cdot 10^{-2}. \quad \textbf{(14.2)}$$

Note that we use numbers from 0 to 9 multiplied by powers of 10, depending on their positions. A number N to any base r can be expressed as

$$(N)_r = a_n r^n + a_{n-1} r^{n-1} + \cdots + a_0 r^0 + a_{-1} r^{-1} + \cdots, \quad \textbf{(14.3)}$$

where $n = 1,2,3,...,$ $r =$ base (e.g., for decimal, $r = 10$; for binary, $r = 2$; for octal, $r = 8$) and $a_n =$ digits with values between 0 and $r - 1$. A radix point is used to separate digits that represent positive powers of the radix from digits that represent negative powers of the radix. Following this pattern for a binary number, we see that the binary number 1001 can be broken down as

$$(1001)_2 = 1 \cdot 2^3 + 0 \cdot 2^2 + 0 \cdot 2^1 + 1 \cdot 2^0 = (9)_{10}, \quad \textbf{(14.4)}$$

and similarly, 101.01 is

$$(101.01)_2 = 1 \cdot 2^2 + 0 \cdot 2^1 + 1 \cdot 2^0 + 0 \cdot 2^{-1} + 1 \cdot 2^{-2} = (5.25)_{10}. \quad \textbf{(14.5)}$$

The bit furthest to the right in the binary number is called the ***least significant bit*** (***LSB***), and the leftmost bit is called the ***most significant bit*** (***MSB***).

Two other bases that are commonly encountered are ***octal*** and ***hexadecimal***. In the octal system, the basis is 8 and the digits used are 0, 1, ..., 7. In the hexadecimal system the base is 16 and the digits used are 0, 1, ..., 9, A, B, C, D, E, F. Table 14.1 shows how some numbers are represented in the decimal, binary, octal, and hexadecimal bases.

To convert a binary number to octal we group the binary digits into sets of three, starting with the LSB (the final group may have fewer than 3 bits):

$$1110111100 = \underbrace{001}_{1} + \underbrace{110}_{6} + \underbrace{111}_{7} + \underbrace{100}_{4} = (1674)_8. \quad \textbf{(14.6)}$$

If we group the binary digits into sets of four, starting with the LSB, the number can be converted to the hexadecimal system. For example,

$$1111101101 = \underbrace{0011}_{3} + \underbrace{1110}_{E} + \underbrace{1101}_{D} = (3ED)_{16}. \quad \textbf{(14.7)}$$

Four bits taken together are called a ***nibble***, and eight bits taken together are referred to as a ***byte***. Whatever the standard-length binary number is in a given system is called a ***word***. Common digital systems use word lengths of 16, 32, or 64 bits (i.e., 2, 4, or 8 bytes).

Table 14.1 Equivalent Representations for the Decimal Numbers 0–31 Using Decimal, Binary, Octal, and Hexadecimal Bases

Decimal	Binary	Octal	Hexadecimal	Decimal	Binary	Octal	Hexadecimal
0	00000	00	00	16	10000	20	10
1	00001	01	01	17	10001	21	11
2	00010	02	02	18	10010	22	12
3	00011	03	03	19	10011	23	13
4	00100	04	04	20	10100	24	14
5	00101	05	05	21	10101	25	15
6	00110	06	06	22	10110	26	16
7	00111	07	07	23	10111	27	17
8	01000	10	08	24	11000	30	18
9	01001	11	09	25	11001	31	19
10	01010	12	0A	26	11010	32	1A
11	01011	13	0B	27	11011	33	1B
12	01100	14	0C	28	11100	34	1C
13	01101	15	0D	29	11101	35	1D
14	01110	16	0E	30	11110	36	1E
15	01111	17	0F	31	11111	37	1F

Another system sometimes encountered is the ***binary-coded decimal*** (***BCD***) system. In this system, each decimal digit is represented by four binary digits (b_3, b_2, b_1, b_0). In the BCD system, each decimal digit is converted individually, as opposed to converting the whole number.[2] For example, the decimal number 17 is given by

$$(17)_{10} = (0001),(0111) \tag{14.8}$$

in BCD, but its binary representation is given by

$$(17)_{10} = (10001)_2.$$

Another common code is the ***American Standard Code for Information Interchange*** (***ASCII***). ASCII is a 7-bit code wherein each 7-bit binary code word maps into a single character. For example, the capital letter A is represented by 1000001, a question mark is represented by 0111111, and so on. The ASCII code is used for storing and manipulating alphanumeric data on a computer, but is not important for simple digital logic, so we will not consider it further here.

Exercise 14.1

Convert the decimal number 832 into (a) binary, (b) octal, (c) hexadecimal, and (d) BCD format.

[2] Only the digits 0–9 are coded. The codes (1010) through (1111) are not used.

14.1.3 Binary Logic Gates

The three basic logic operations in digital circuits are designated OR, AND, and NOT. More complex logic operations are defined in terms of these three basic operations. The output voltage (or current) of a gate represents either a logical 1 or 0. If V(1) > V(0), we have positive logic, and if V(1) < V(0), we have negative logic. To avoid confusion, we shall henceforth use positive logic, as is most often done.

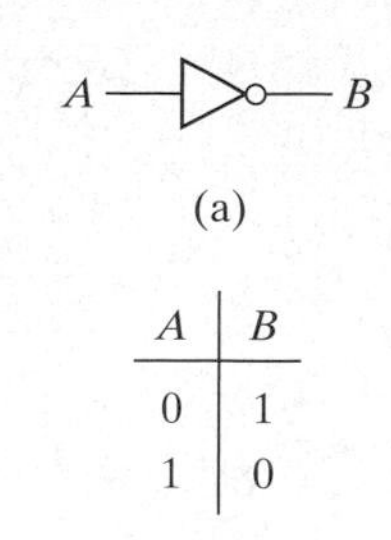

A	B
0	1
1	0

(b)

Figure 14–2 (a) Inverter, or NOT gate. (b) Truth table.

The Inverter (NOT Gate) The ***inverter***, or ***NOT*** gate, is shown symbolically in Figure 14-2(a) and performs the NOT or inversion operation (since the variables are binary, if B is not equal to A, it must be the inverse of A;—hence the terminology). Inversion is also sometimes called complementation. The NOT operation is denoted by

$$B = \overline{A}, \tag{14.9}$$

and we say that B is the inverse, or complement, of A. Sometimes, a prime is used instead of the overbar to indicate inversion of a variable (e.g., $B = A'$ for (14.9)). The circle at the output of the triangle amplifier symbol in Figure 14-2(a) signifies inversion. If it were absent, we would have a buffer (i.e., a noninverting unity-gain amplifier).

The table in Figure 14-2(b), which is called a ***truth table***, completely describes the operation of the inverter. The truth table is a listing of the values of the dependent variable in terms of all possible combinations of values of the independent variables. The truth table is used to prove logic equations and may also be viewed as defining what is meant by a logical operation, such as that shown in (14.9).

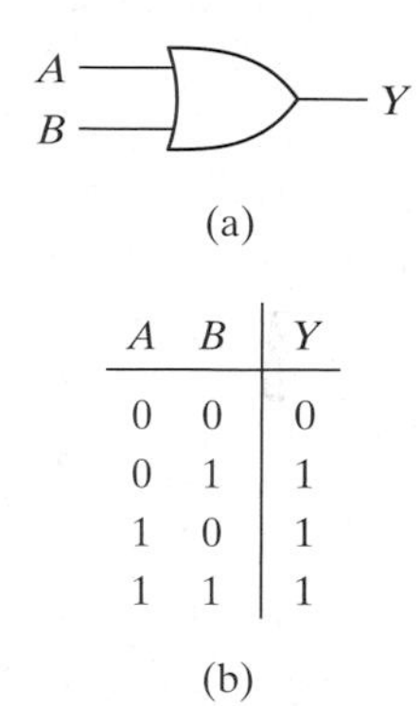

A	B	Y
0	0	0
0	1	1
1	0	1
1	1	1

(b)

Figure 14–3 (a) OR gate. (b) Truth table.

The OR Gate The ***OR*** gate, shown symbolically in Figure 14-3(a) (for two inputs), performs the logical OR operation. The function of the OR gate is expressed mathematically by

$$Y = A + B, \tag{14.10}$$

where "+" is used to denote the logical OR operation and does not have the same meaning as in ordinary algebra. The truth table for the OR operation is shown in part (b) of the figure. The output is a logical 1 if A or B or both are logical 1.

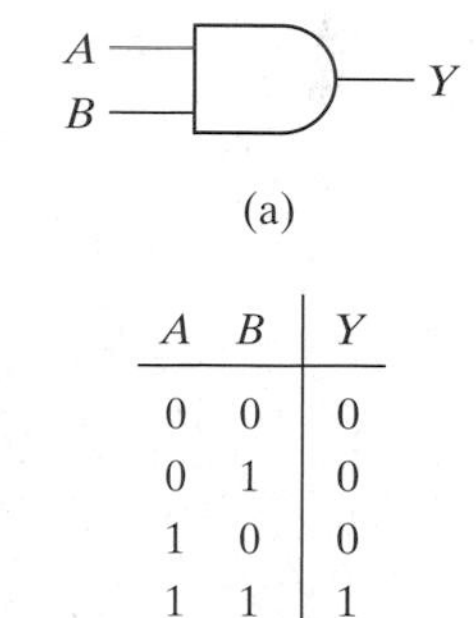

A	B	Y
0	0	0
0	1	0
1	0	0
1	1	1

(b)

Figure 14–4 (a) AND gate. (b) Truth table.

The AND Gate The ***AND*** gate, shown symbolically in Figure 14-4(a) (for two inputs), performs the logical AND operation. The function of the AND gate is expressed mathematically by

$$Y = A \cdot B, \tag{14.11}$$

where "·" indicates the logical AND operation and is not the same as multiplication. The AND symbol is sometimes omitted so that AB is equivalent to $A \cdot B$. The truth table for the AND operation is shown in part (b) of the figure. The output is a logical 1 if A and B are both logical 1.

Exercise 14.2

Fill out the truth table for the circuit shown in Figure 14-5.

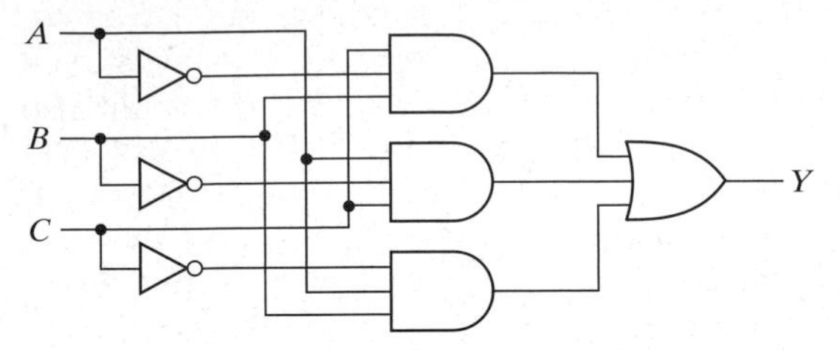

Figure 14–5 A logic circuit.

Boolean Algebra and De Morgan's Theorems The algebra of discrete variables is called ***Boolean algebra***. When the discrete variables are binary, Boolean algebra[3] is sometimes called ***switching algebra***, but we will use the more common terminology.

The order of precedence for Boolean operations is first, NOT; second, AND; and third, OR. The order can be modified or made explicit by using parentheses, as in normal algebra. For example, the expression $\overline{A}B + C$ is evaluated as ([NOT A] AND B) OR C, while the expression $\overline{A}(B + C)$ is evaluated as (NOT A) AND (B OR C).

The commutative law holds for both AND and OR functions:

$$AB = BC \qquad \text{and} \qquad A + B = B + A. \tag{14.12}$$

The associative law also holds for both AND and OR operations:

$$(AB)C = A(BC) = ABC \tag{14.13a}$$

and

$$(A + B) + C = A + (B + C) = A + B + C. \tag{14.13b}$$

Finally, the distributive law holds both ways: AND distributes over OR; that is,

$$A(B + C) = AB + AC; \tag{14.14a}$$

and, unlike regular algebra, OR also distributes over AND:

$$A + BC = (A + B)(A + C). \tag{14.14b}$$

Finally, the next two identities, known as ***De Morgan's Theorems***, are very useful in dealing with logic circuits. The theorems are

$$\overline{A + B} = \overline{A} \cdot \overline{B}, \tag{14.15}$$

and

$$\overline{A \cdot B} = \overline{A} + \overline{B}. \tag{14.16}$$

[3] Boolean algebra was developed by George Boole in 1847 and was first applied to switching networks by Claude Shannon in 1939.

De Morgan's theorems show that the complement of a Boolean function is obtained if we change each OR symbol to an AND, change each AND symbol to an OR, and complement each variable. With the help of these theorems, we can also show that every logic function can be implemented with either ANDs and NOTs or ORs and NOTs (*see* Problems P14.15 and P14.16). The theorems can be verified by filling out truth tables. Also, they apply to more then two variables.

It can be shown that any Boolean function can be implemented using NOT, AND, and OR functions [14.2]. Such a set of functions is referred to as a ***functionally complete set.*** Either the AND function or the OR function can be removed, and the set is still functionally complete. The NOT function, however, is essential. Therefore, it is far more common to use NAND and NOR gates, which are just AND and OR gates with their outputs inverted, as shown in the next few sections. Using NAND or NOR gates allows us to also implement the NOT function. We will also see in Chapter 15 that it is more straightforward to implement NAND and NOR gates at the transistor level than it is to implement AND and OR gates at that level.

Exercise 14.3

Make a truth table with columns for A, B, $A + B$, $\overline{A}$, $\overline{B}$, and $\overline{A} \cdot \overline{B}$. Fill the table out for all combinations of A and B, and, by so doing, demonstrate that (14.15) is true.

The NOR and NAND Gates The ***NOR*** gate is simply an OR gate followed by an inversion and can be implemented using an OR gate and an inverter, as shown in Figure 14-6(a). For a true NOR gate, the inversion is schematically indicated by the circle on the output of the gate, as shown in part (b) of the figure. If a separate inverter is shown on a schematic, as in part (a), it should mean that the gate has physically been implemented by combining an OR gate and an inverter. The function of the NOR gate is expressed mathematically by

$$Y = \overline{A + B}. \tag{14.17}$$

The truth table for the NOR gate is given in part (c) of the figure. The NOR gate by itself can implement a functionally complete set of logic functions (*see* Problem P14.13).

The ***NAND*** gate is simply an AND gate followed by an inversion, as shown in Figure 14-7(a). The function of the NAND gate is expressed mathematically by

$$Y = \overline{A \cdot B}. \tag{14.18}$$

The truth table for the NAND gate is given in part (b) of the figure. The NAND gate by itself can implement a functionally complete set of logic functions (*see* Problem P14.14).

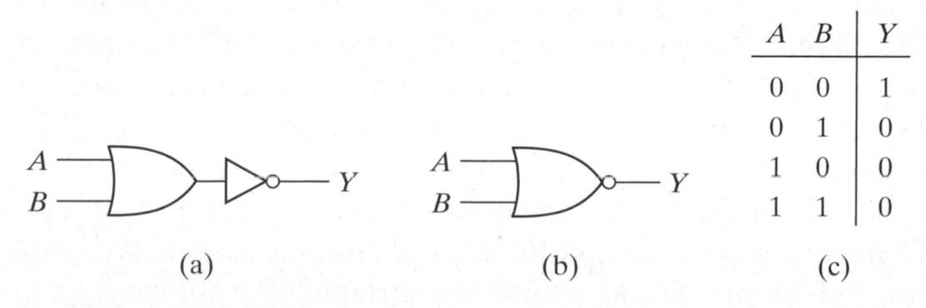

A	B	Y
0	0	1
0	1	0
1	0	0
1	1	0

(c)

Figure 14–6 (a) NOR gate constructed using an OR gate and an inverter, (b) the NOR schematic symbol, and (c) truth table.

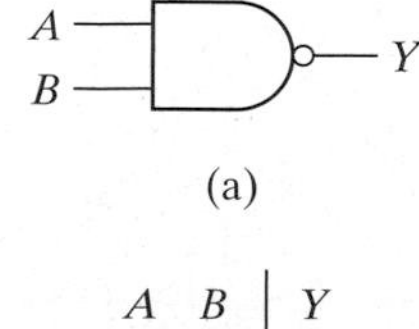

A	B	Y
0	0	1
0	1	1
1	0	1
1	1	0

(b)

Figure 14–7 (a) NAND gate. (b) Truth table.

Combinational Logic Circuits We can combine single logic gates to perform complex functions. If the systems that we build do not contain any memory, then the output can be a function only of the present inputs. Such systems are called ***combinational logic***. We will see in the next section how to add memory to a logic circuit.

There are many techniques available for systematically designing a combinational logic circuit to perform a given function described by either a Boolean expression or a truth table. These techniques are taught in courses on logic design and will not be pursued here. We will show only circuits that are simple enough to design without these techniques.

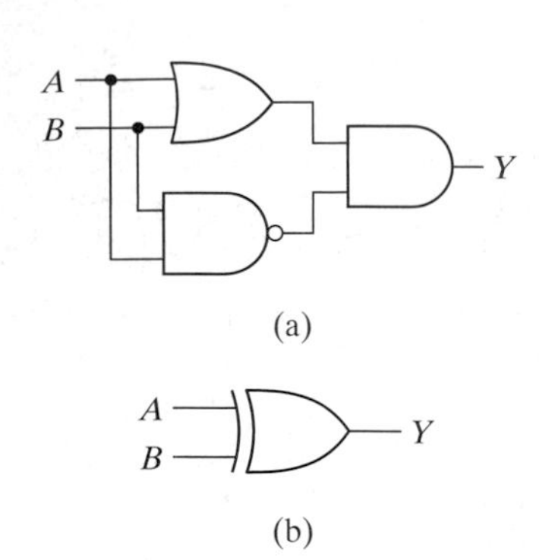

Figure 14–8 (a) An exclusive OR circuit and (b) the schematic symbol.

Exercise 14.4

*Fill in a truth table for the circuit shown in Figure 14-8(a). This is called an **exclusive OR** function. The typical schematic symbol is shown in part (b) of the figure.*

14.2 FLIP-FLOPS

In Section 14.1, we considered the operation of basic logic gates and showed that they can be combined to produce combinational logic circuits. Since combinational logic circuits don't have any memory, there are many useful functions that they cannot perform. In this section, we show how to design ***flip-flops***, which operate as one-bit memory cells. Flip-flops are also called ***latches***. Logic circuits constructed using flip-flops can have the present output be a function of both the past and present inputs. Such circuits are called ***sequential logic circuits***.

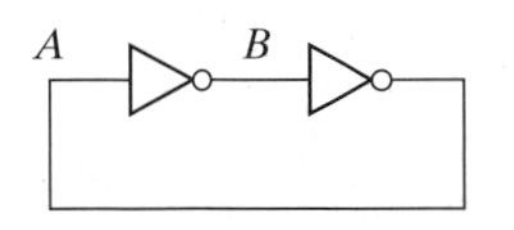

Figure 14–9 A bistable circuit.

All flip-flops are based on the same principle: Positive feedback is used to produce a circuit that is ***bistable***. A bistable circuit is one that has two stable operating points. Which operating point the circuit is in is called the ***state*** of the circuit. If the state can be sensed and changed, then the circuit can function as a one-bit memory element.

The simplest bistable circuit is constructed using two inverters in a loop as shown in Figure 14-9. This circuit only has two nodes, A and B. Because of the inverters, if A is high, B must be low and vice versa; hence, the circuit has two stable states.

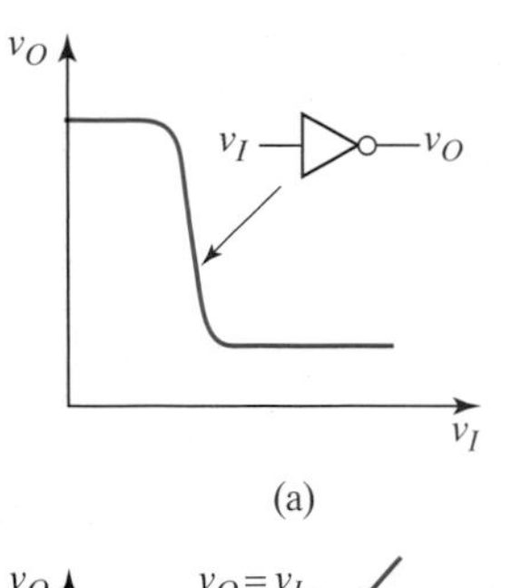

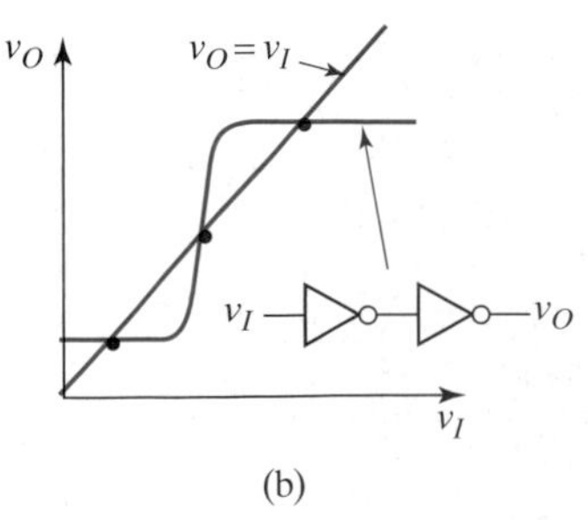

Figure 14–10 (a) One inverter and its transfer characteristic. (b) The transfer characteristic for two inverters in series and the load line for the circuit.

The operation of the bistable circuit can also be viewed using a plot of the transfer characteristic of the two inverters in series, as shown in Figure 14-10. Part (a) of the figure shows the static transfer characteristic of one of the inverters. When the input voltage is below the threshold (a logical ZERO), the output voltage is high (a logical ONE). When the input voltage is greater than the threshold, the output voltage is low. In part (b) of the figure, we show the transfer characteristic that results from putting both inverters in series. Any solution of the equations for this circuit must also lie on this characteristic. Because of the external connection, the input and output voltages of the series connection of the two inverters must be the same. Therefore, we draw a line with a slope of unity on the plot as well. This line is called the ***load line***, because it represents the external load connection for the two inverters in series. Any solution of the equations for this circuit must also lie on the load line. Therefore, when the equations are simultaneously solved, the only possible operating points are found where the straight line intersects the transfer characteristic. There are three intersections on the plot, but only two of them are stable, as we will now demonstrate.

The point where the load line intersects the middle of the transfer characteristic is not stable. To see that this statement is true, suppose for the moment that the circuit is at this point. If the input voltage increases at all (due to noise or some change in the circuit), the output voltage of the inverters must also increase. But the output *is* the input, so as it increases, it causes further increases in the output, and the original change is magnified. This positive feedback will quickly drive the circuit to the top operating point shown. At that point, the input and output of the two-inverter chain are high and the midpoint (v_B in Figure 14-9) is low, so the circuit is stable and can remain in this state forever. If we started at the midpoint and let the input voltage decrease a bit, we would end up at the lower operating point, which is again stable.

In the sections that follow, we show how we can move this bistable circuit from one operating point to the other. The internal positive feedback will then hold the circuit at that state until we deliberately change it; hence, the circuit has memory.

14.2.1 The Set-Reset Flip-Flop

A set-reset (SR) flip-flop is shown in Figure 14-11(a). A table describing the function of the circuit is shown in part (b) of the figure, and the schematic symbol is shown in part (c). This ***function table*** is similar to a truth table, but it describes a dynamic situation, not a static one. [4] The output is the output at some discrete time, denoted by Q_n, and the table includes an entry for the previous state of the flip- flop (Q_{n-1}). Although the circuit is drawn differently, the two NOR gates are in series, just like the inverters in Figure 14-10(b). The configuration shown here is usually described as cross coupled. The flip-flop has two outputs that are complements of each other. We usually consider the Q output to be the state of the flip-flop.

The circuit operates in the following way: If both inputs (S and R) are zero, the previous state is retained. Suppose, for example, that Q_{n-1} is high (i.e., ONE). Then the output of the bottom NOR, which is $\overline{Q}_{n-1}$, will be low (i.e., ZERO), independently of what S is. In this case, both inputs to the top NOR are low, so its output is high, as originally assumed. Now suppose that Q_{n-1} is low. In this case, both inputs to the bottom NOR are low, so $\overline{Q}_{n-1}$ is high. Therefore, the output of the top NOR, Q_{n-1}, will be low, as assumed.

Now consider what happens when the set input, S, goes high while R remains low. The output of the bottom NOR, $\overline{Q}_{n-1}$, will now go low, *independent of what the previous state of the circuit was*. With R low as well, this guarantees that Q_n will go high (i.e., the flip-flop has been "set"). Note that S does not have to stay high. Once the flip-flop is set, the S input can go low again, and the state will be re-

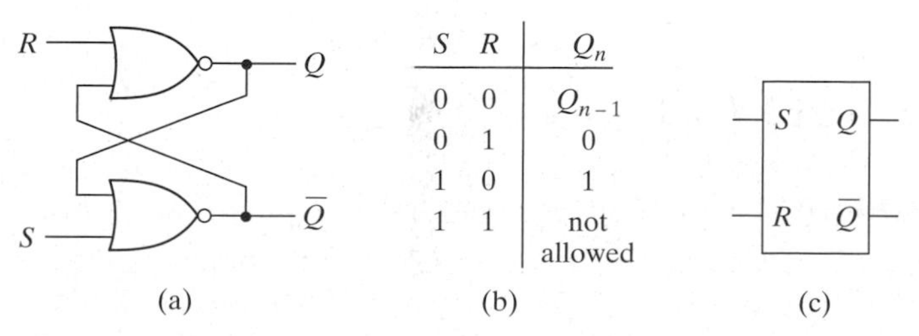

S	R	Q_n
0	0	Q_{n-1}
0	1	0
1	0	1
1	1	not allowed

(b)

Figure 14–11 (a) An SR flip-flop, (b) a table describing the circuits function, and (c) the schematic symbol.

[4] The function table is also sometimes called a characteristic table or a next-state table.

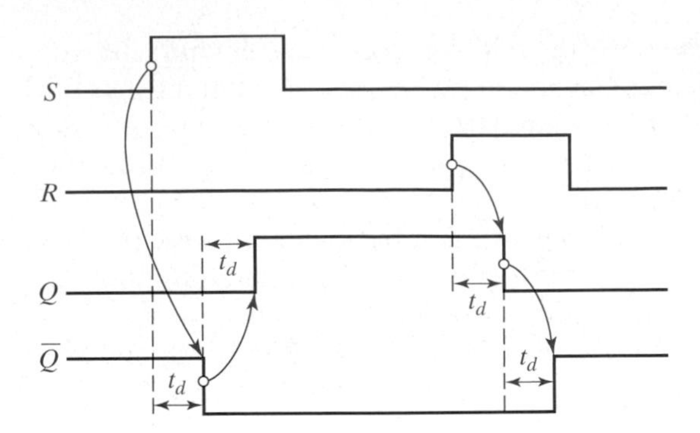

Figure 14–12 A timing diagram for the SR flip-flop. The arrows indicate which transition causes the following change.

tained. This sequence of events is illustrated in Figure 14-12. The figure shows that there is some delay through each gate, so it takes a time t_d for the change at the gate input to affect its output.

The operation of the reset input is similar. If R goes high while S is kept low, the output of the top NOR, Q_n, will go low (i.e., the flip-flop is "reset"). With Q_n and S both low, the bottom NOR output will be high. The reset input can go low again, and this new state will be retained. This sequence is also illustrated in Figure 14-12.

Finally, we note that both inputs should not be allowed to go high at the same time. If this happens, both NOR outputs go low, so Q and $\overline{Q}$ are not complements anymore. Also, if both inputs are high and then go low at exactly the same time, we can't predict what the resulting output state will be, since both outputs will try to go high, which is a condition that cannot be sustained. Which output will actually stay high depends on mismatches in the NOR gates and cannot be predicted.

Exercise 14.5

A clocked SR flip-flop is shown in Figure 14-13. The clock input is driven by a periodic square wave with a period much greater than twice the gate delay. Show that the outputs of the flip-flop can change only when the clock input is high. Therefore, S and R matter only when the clock input is high.

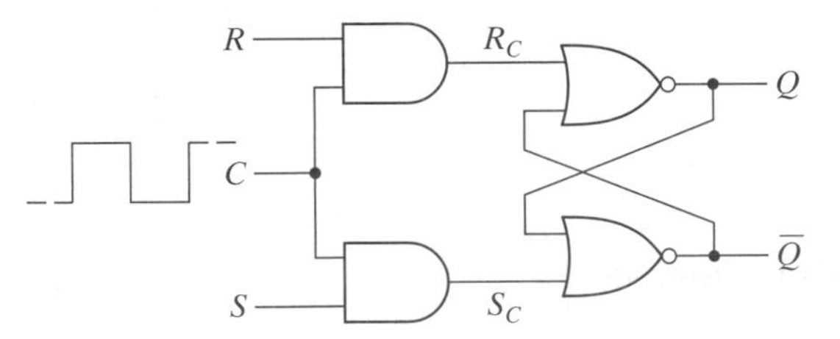

Figure 14–13 A clocked SR flip-flop.

14.2.2 The JK Flip-Flop

The fact that the output of an SR flip-flop is undefined if both inputs go high is troublesome in many applications. The JK flip flop avoids this problem and is more flexible in its operation. The JK flip-flop is a clocked flip-flop; that is, it requires a separate clock input to operate. This clock signal is usually a square wave with a fixed period. Logic circuits that require a clock and that only allow output transitions to occur in synchrony with the clock are called ***synchronous-logic circuits***. The clock can be generated using an astable multivibrator, as discussed in Section 10.2.2.

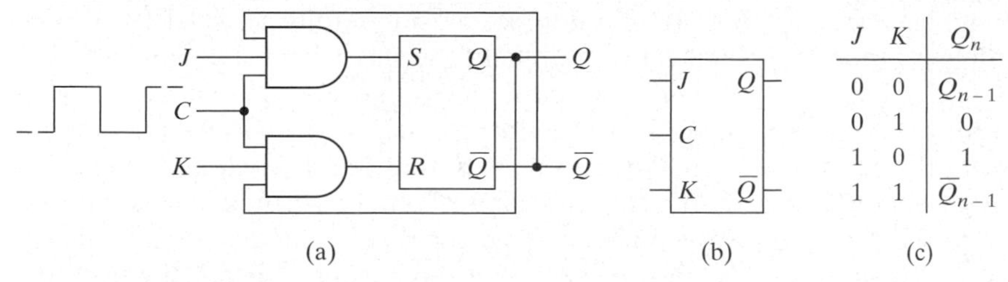

Figure 14–14 (a) A JK flip-flop made using an SR flip-flop. (b) The Schematic symbol for a JK flip-flop and (c) the function table. (The flip-flop only changes state when the clock is high.)

A JK flip-flop is shown in Figure 14-14(a); the schematic symbol is shown in part (b) of the figure, and the function table is shown in part (c). The AND gates serve to ***enable*** the inputs to the SR flip-flop. That is, only when the clock is high are the J and K inputs able to affect the SR flip-flop. In addition to needing the clock to be high, the J input affects S only if the SR flip-flop is currently reset, and the K input affects R only if the flip-flop is currently set. Therefore, we see that when both J and K are low, S and R will be low, and the flip-flop will hold its present state just like the SR flip-flop. When J is high and the flip-flop is currently reset (i.e., $\overline{Q}_{n-1}$ is high), the flip-flop will be set when the clock goes high, independently of what K is. If K is high and the flip-flop is currently set (i.e., Q_{n-1} is high), the flip-flop will reset when the clock goes high, independently of what J is. It follows that if both J and K are high, the flip-flop will toggle its state when the clock goes high. When operated in the toggle mode, a JK flip-flop is sometimes called a T flip-flop.

The JK flip-flop as shown in Figure 14-14 has a major problem: It will work only if the clock pulse width (i.e., the time the clock is high) is short compared with the propagation delay of the gate. To understand this limitation, consider what happens when J and K are both high and Q_{n-1} is low. In this case, the output of the flip flop will toggle when the clock goes high, as indicated in the function table. But, if the output toggles and the clock is still high, the output will toggle again. This process will repeat until either the clock goes low or J or K changes. In order to avoid this problem, we use ***master–slave*** JK flip-flops.

A master-slave JK flip-flop is shown in Figure 14-15. The master flip-flop is enabled when the clock is high, so the data are latched into the master during that portion of the clock cycle. During that time, $\overline{C}$ is low and the slave is disabled and holds the previous value. Then, when the clock goes low, $\overline{C}$ goes high and enables the slave. The data from the master are then transferred to the slave and show up at the output. Since the master and slave flip-flops are never enabled at the same time, the output will not continue to toggle if the clock is held in any one state for too long. The clock does have to remain in each state long enough to allow for the propagation delay through one of the flip-flops.

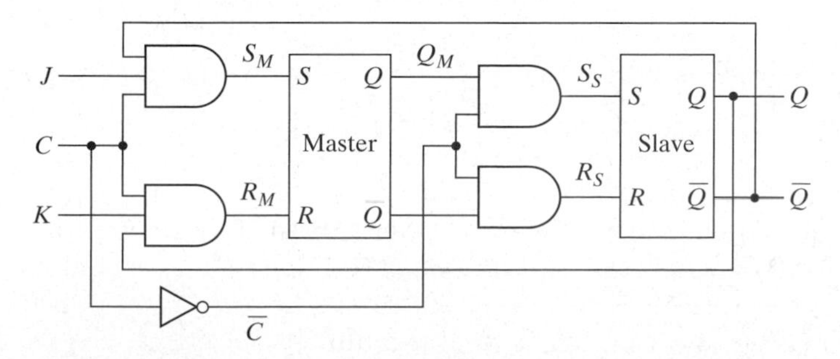

Figure 14–15 A master–slave JK flip-flop.

In designing a master–slave JK flip-flop, we must carefully consider the propagation delays of the individual gates to prevent the slave from changing before it should. For example, in the figure, the data on S_M and R_M can change one gate delay after the clock goes high. The slave clock, which is $\overline{C}$, goes low one inverter delay after the clock goes high. We must be sure that the slave clock changes before the output of the master flip-flop can change; otherwise, the data will pass on through to the slave and we will not have accomplished our purpose. Similarly, when the clock goes low, we must be sure that the master is disabled before the slave outputs can change.

The JK flip-flop just described is a ***level-triggered*** flip-flop; that is, the master is enabled when the clock level is high, and the slave is enabled when the clock level is low. The problem with level-triggered JK flip-flops is that they are sensitive to glitches on the inputs at certain points in the operation. For example, suppose that the previous state of the flip-flop was $Q = 0$ and that we are now ready for the next clock cycle. Suppose further that $J = 0$ and $K = 1$, so we are resetting the flip-flop again; in other words, we don't want the state to change. In this case, while the clock is high, both S_M and R_M are low, so the master flip-flop output should not change. However, if a positive glitch occurs on the J input prior to the clock going low, it can pass through to S_M and set the master flip-flop. Since Q is low, the AND gate driving R_M is disabled, so we don't have any opportunity for the flip-flop to be reset. As a result, when the clock goes low, this error will be passed on to the slave. A similar situation exists if we are trying to set the flip-flop when it is already set. A positive glitch on the K input can cause an erroneous reset. This problem is sometimes called ***ones catching***, since the flip-flop has captured an erroneous ONE. We could make the problem far less likely to occur if we used a clock with a very short positive pulse, but a much better solution is to use an ***edge-triggered*** JK flip-flop.

An edge-triggered JK flip-flop is shown in Figure 14-16(a), and the schematic symbol is shown in part (b) of the figure. The triangle inside the block in part (b) indicates that the flip-flop is ***edge triggered,*** as explained in a moment, and the bubble indicates that it is negative edge triggered (i.e., the input is latched on the negative-going edge of the clock).

To understand how this circuit operates, we need to first examine the input gate structure. Consider, for example, the situation where $Q = 0$ and we want to set the flip-flop, so $J = 1$. Part of the input structure is shown in Figure 14-17(a) for this case, and the corresponding waveforms are shown in part (b) of the figure.

The bubbles at the input of the second gate invert the inputs so that the AND is true when both inputs are low. Because $Q = 0$, we know that $\overline{Q} = 1$. Now, with $J = 1$, the output of the NAND gate, J_C, will be the inverse of the clock, delayed by one gate delay. Therefore, when the clock goes low, J_C will go high one gate delay later, as shown. During that gate delay, both inputs to the second gate are low, so the AND is true and S goes high. In other words, the negative edge of the clock has

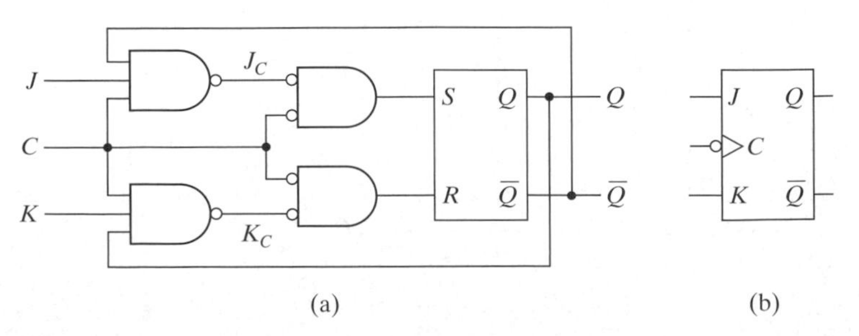

Figure 14–16 (a) An edge-triggered JK flip-flop and (b) the schematic symbol for it.

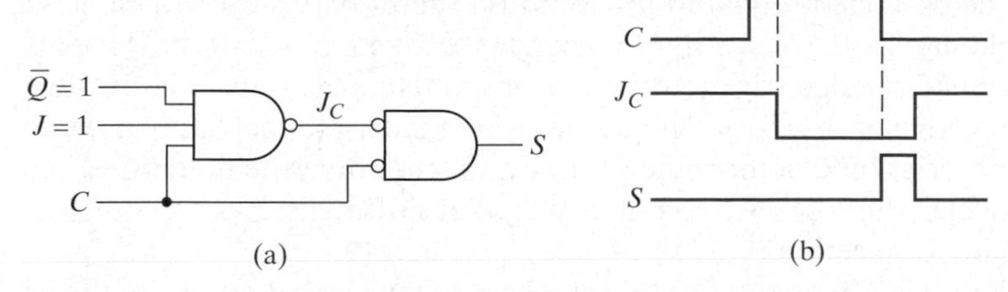

Figure 14–17 (a) A part of the input circuit in Figure 14-16 when $Q = 0$. (b) The resulting waveforms.

produced a narrow pulse on the S line as a result of the J input being high. Similarly, if the K input is high and $Q = 1$, a negative clock edge will produce a narrow pulse on the R line. In this way, the SR flip-flop is set or reset only on the negative clock edge. As long as the J and K inputs are held constant for some short time prior to the clock edge (called the ***setup time***) and are held constant for some short time after the clock edge (called the ***hold time***), the circuit is insensitive to glitches on the inputs. It is also possible to make positive edge-triggered circuits (*see* Problem P14.26).

Exercise 14.6

Fill in a function table for the circuit shown in Figure 14-18. What function does this circuit perform?

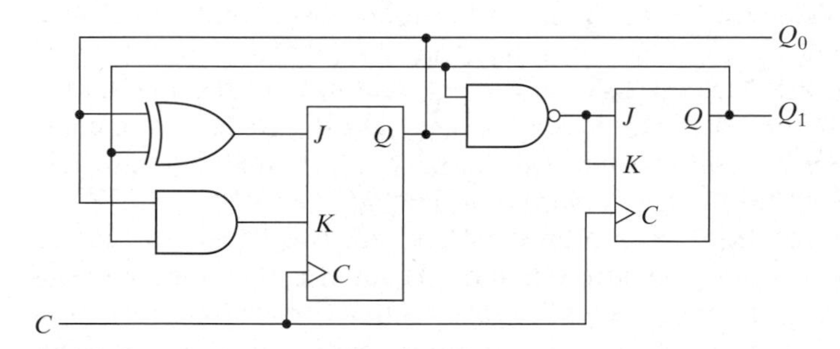

Figure 14–18

14.2.3 The D Flip-Flop

A D flip-flop is shown in Figure 14-19(a), and its schematic symbol is shown in part (b) of the figure. This flip-flop implements a digital delay; that is, the output at

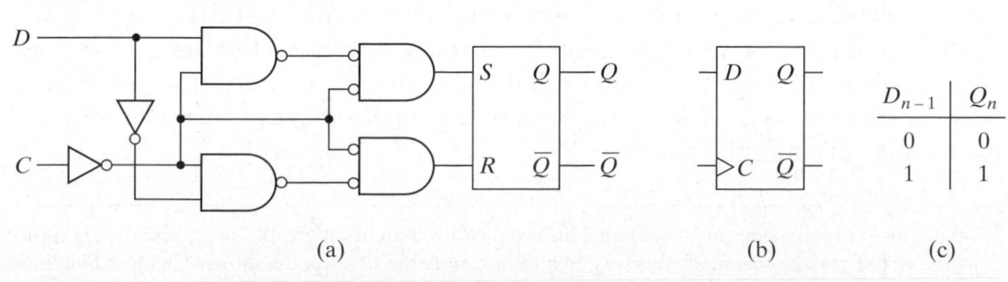

D_{n-1}	Q_n
0	0
1	1

Figure 14–19 (a) A D flip-flop, (b) its schematic symbol, and (c) the function table.

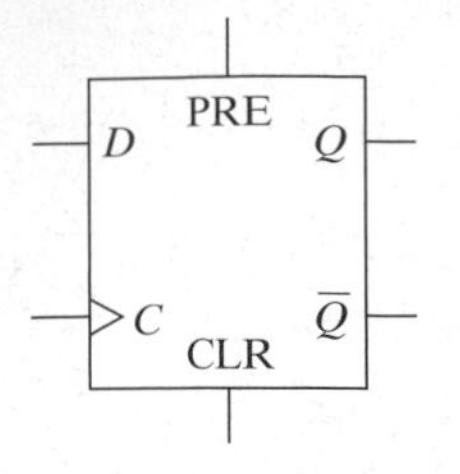

Figure 14–20 A D flip flop with preset and clear inputs.

the end of each clock cycle is equal to the input on the previous cycle, as seen in the function table in part (c) of the figure—hence the name D flip-flop. This particular circuit is positive-edge triggered, so the output changes state slightly after the positive-going edge of the clock. The output is insensitive to the value of the D input, except for a brief time before (the setup time) and after (the hold time) the positive clock edge. D flip-flops are commonly used in shift registers and counters, as discussed in the next section.

Clocked flip-flops also frequently have ***asynchronous*** clear and preset inputs, as shown for a D flip-flop in Figure 14-20. The preset input will set the flip-flop so that $Q = 1$ at any time, regardless of the state of the clock; that is what is meant by being asynchronous. In similar fashion, the clear input will clear the flip-flop so that $Q = 0$ at any time.

14.3 SHIFT REGISTERS AND COUNTERS

In this section, we briefly consider the two most common applications of flip-flops: shift registers and counters. A ***shift register*** is an array of flip-flops that can store or delay binary data. There are many applications of shift registers and several different types of shift registers. The simplest shift register can be constructed from a sequence of D flip-flops, as shown in Figure 14-21(a). The binary data are sequentially presented to the circuit, and each bit is shifted one place to the right on each clock edge, as shown in part (b) of the figure, where we have assumed that all the flip-flops started out in state zero. The delay through a gate is assumed to be so small that it doesn't show up in this figure, but you must remember that since each flip-flop is clocked by the same clock, we are assuming that the next flip-flop in the line will latch the output of the preceding flip-flop as it was just *prior* to the arrival of the clock edge. At any given point in time, the state of the shift register is simply the set of stored values $[Q_0Q_1Q_2Q_3]$. So, for example, at time t_1 in the figure, the state of this 4-bit register is 1011. Note that this state is, in fact, a record of the four most recent input bits at time t_1.

Another application of flip-flops is counters. There are two types of counters: ***ripple counters*** and ***synchronous counters***. In a ripple counter, the input pulses to be counted ripple through a sequence of flip-flops so that each flip-flop is driven by the preceding flip-flop in the line. In a synchronous counter, all of the flip-flops have the same clock input, so that they all change state at the same time.

A 3-bit ripple counter implemented with JK flip-flops is shown in Figure 14-22(a).[5] The flip-flops have their J and K inputs tied high so that they are in the toggle mode. Except for the first flip-flop, each flip-flop is driven by the Q output of the preceding one. Since the output of FF_0 toggles on each positive clock edge, its output is a square wave at half the clock frequency, as shown in part (b) of the figure. The output of FF_0 then drives FF_1, so that Q_1 is a square wave at one-fourth the frequency of the input. This pattern continues and if we consider the output state to be $[Q_2Q_1Q_0]$, the state counts down from 111 (decimal 8) to 000 as shown. In this case, when the final count is reached the counter simply starts over. It is also possible to make counters that count up or that can count either up or down, depending on another input. Also, note that the input to be counted does not have to be a periodic clock as shown here, it can be a nonperiodic sequence of pulses, and the circuit will count the pulses.

[5] The inputs to the flip-flops are shown in a different order than in Figure 14-16(b), because it is more convenient to draw the schematic this way. In addition, only the Q output is shown. Changes like these are common in drawing schematics when the inputs and outputs are all labeled.

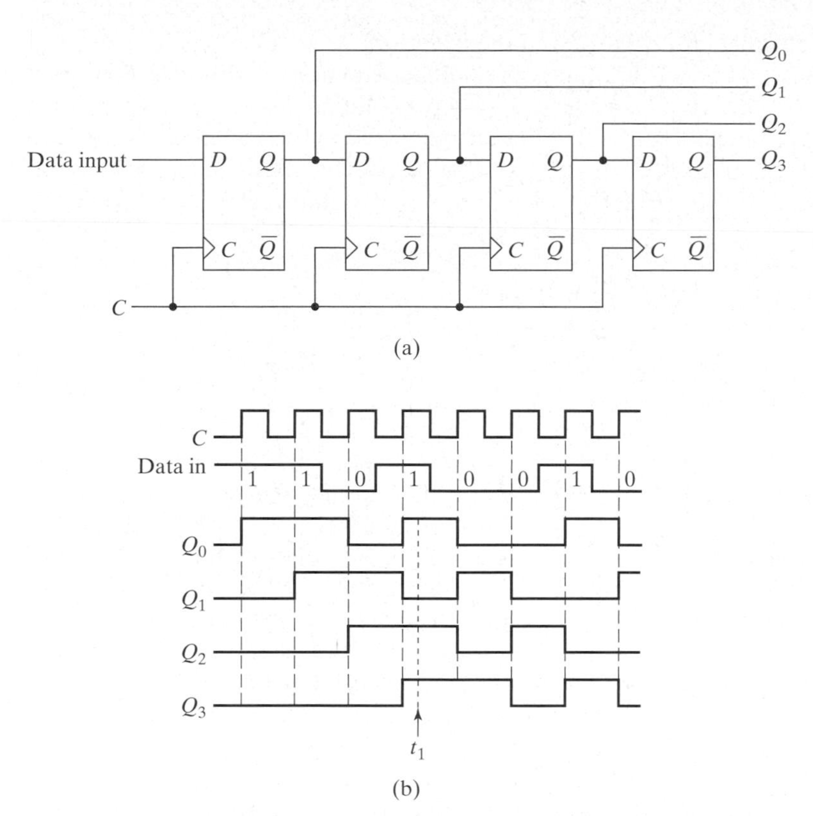

Figure 14–21 A 4-bit shift register made using D flip-flops.

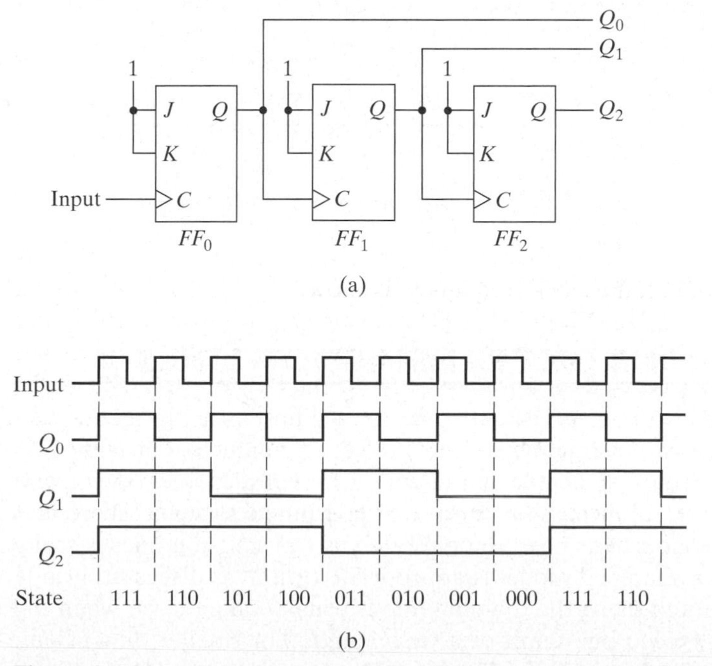

Figure 14–22 (a) A 3-bit ripple counter. (b) The associated waveforms.

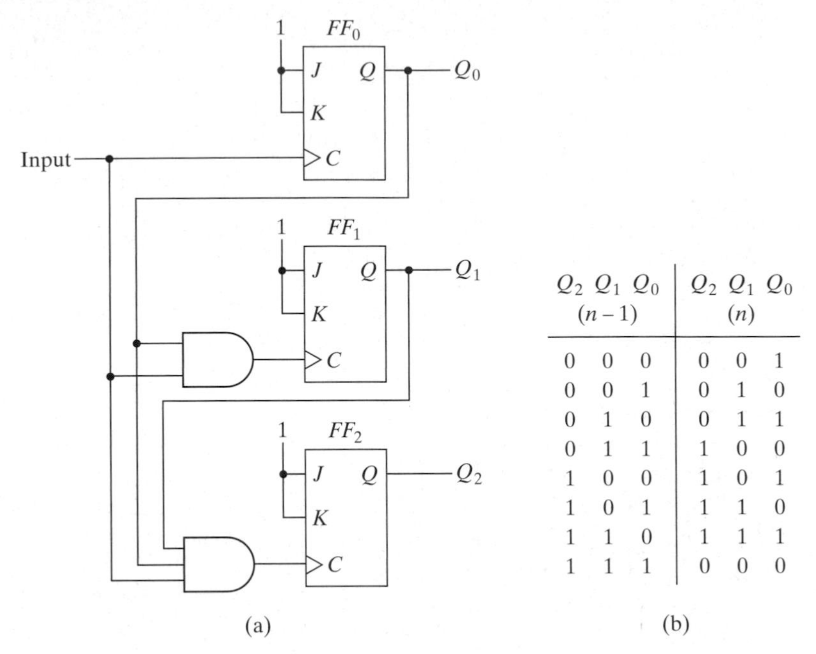

Q_2 Q_1 Q_0 $(n-1)$			Q_2 Q_1 Q_0 (n)		
0	0	0	0	0	1
0	0	1	0	1	0
0	1	0	0	1	1
0	1	1	1	0	0
1	0	0	1	0	1
1	0	1	1	1	0
1	1	0	1	1	1
1	1	1	0	0	0

Figure 14–23 (a) A 3-bit synchronous binary counter and (b) its function table.

A 3-bit synchronous binary counter is shown in Figure 14-23(a), and its function table is shown in part (b) of the figure. On the first clock pulse, the AND gates in front of FF_1 and FF_2 will not allow the clock pulse through, so only FF_0 changes state, and the output goes from 000 to 001, as shown in the function table. Then, on the second pulse, the AND gate in front of FF_1 allows the clock through, so both FF_0 and FF_1 change states, and the next output state is 010, as it should be. You can follow through the rest of the function table and convince yourself that the output goes from 000 to 111 in order and then starts over again.

Exercise 14.7

Show how to make a 4-bit ripple counter using only positive edge-triggered D flip-flops.

□

14.4 REFLECTIONS ON TRANSMISSION LINES

We close this chapter with an important topic for digital design: reflections on transmission lines. A transmission line is simply a pair of conductors providing an electrical circuit between two points, as illustrated for an idealized source and load in Figure 14-24(a). Transmission lines are distributed systems. Therefore, whether or not interconnections need to be treated as transmission lines is really a question of whether a lumped model is appropriate or not, as discussed briefly in Section 9.1.2. As noted there, the dividing line is usually taken to be when the physical dimensions exceed one-tenth of a wavelength. For smaller dimensions, we can usually use lumped models and Kirchhoff's laws as acceptable approxima-

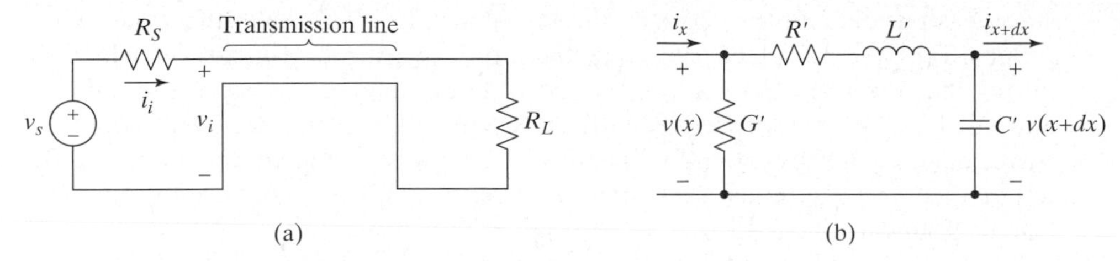

Figure 14–24 A transmission line connecting a source and load.

tions. For larger dimensions, we need to use distributed models and Maxwell's equations. (Kirchhoff's laws are approximations to Maxwell's equations that assume infinitely fast propagation of signals [14.3].)

The primary characteristics of a transmission line are its resistance per unit length, R', capacitance per unit length, C', inductance per unit length, L', and conductance per unit length, G'. These parameters are illustrated in the lumped model in part (b) of Figure 14-24. If this model represents a differential element of the line of length dx, then the model becomes exact as $dx \rightarrow 0$.

The secondary characteristics of the transmission line are more commonly used and are calculated from the primary characteristics. The most important single parameter is the ***characteristic impedance***,

$$Z_0 = \sqrt{\frac{R' + j\omega L'}{G' + j\omega C'}}. \tag{14.19}$$

At every point in the line, the ratio of the voltage to the current is equal to Z_0. Note that if the resistance and conductance of the line can be neglected, which implies that the line is an ideal ***lossless transmission line***, the characteristic impedance is purely resistive. We will assume lossless lines from this point on.

Most transmission lines encountered in digital systems are either coaxial cables, twisted pairs, traces on a printed-circuit board, or metal lines on an integrated circuit. Coaxial cables usually have impedances of 50 Ω or 75 Ω, and the waves travel at about two-thirds the speed of light in a vacuum. Twisted pairs typically have impedances near 100 Ω and the waves travel a bit faster than one-half the speed of light in a vacuum (around 1 ns of delay for 7 in of transmission line, or 144 ps/in). Traces on PCBs typically have impedances around 100–150, and the speed of propagation is close to that of twisted pairs. Notice that if we use transition times of 1 ns or more in our circuits and keep the trace lengths down to about 1 in or so, reflections are not a big problem. This explains why reflections are not a big problem on ICs yet: The time delay for a signal traversing the IC is too small. As the speed of logic circuits continues to increase, however, reflections will likely become a concern there, too.

When a transmission line is driven by an AC voltage, this ***incident*** voltage travels down the line as a wave, along with its associated current wave. When the waves reach the end of the line, the ratio of voltage to current must equal the load resistance (or impedance). This boundary condition produces some interesting and important results.

Consider, for example, what happens if the load impedance is infinite. In this case, the current must go to zero at the end of the line. The only way this can happen is if a ***reflected wave*** is launched from the load toward the source. The voltage and current at any point in the cable will then be the superposition of the incident waves and the reflected waves. The incident and reflected voltages will add at each point in the line, and the reflected current will subtract from the incident current since it is traveling in the opposite direction. For the total current to be zero, the reflected current must have the same magnitude as the incident current, which requires that the reflected voltage also be of equal magnitude. Since the voltages add, the load voltage will be twice the input voltage. This reflected wave will then travel back to the source and will add to the voltage at the source. If the source impedance does not match the cable (i.e., if $R_S \neq Z_0$), there will be another reflection back toward the load.

We illustrate the reflections seen when driving an open-circuit load (i.e., $R_L = \infty$) in Figure 14-25(a). We assume a transit delay of τ_d seconds for the cable and use a single pulse for the source. For simplicity, we assume that the source impedance matches the line. When the source voltage first changes, the waves have not had sufficient time to travel down the line, so the voltage at the input to the line is determined by a voltage divider with the line impedance and R_S; hence, the initial input voltage step is one-half as high as the source. When this wave arrives at the open-circuited end of the line, the reflected voltage is equal to the incident voltage and adds to it. Therefore, the load voltage is equal to unity, as shown. When the reflected voltage wave arrives back at the input one delay later, the input voltage is also equal to unity. Since the source is assumed to match the line, no further reflections occur.

When the transmission line drives a short circuit (i.e., $R_L = 0$), the voltage at the end of the line must be zero, which requires that the reflected voltage wave be the opposite of the incident wave. This situation is illustrated in part (b) of Figure 14-25. In this case, when the reflection gets back to the input, it cancels the input completely. This situation is useful for generating short pulses: If you drive a shorted transmission line with a square wave, you will get a pulse of width $2\tau_d$ at the input. The pulse will be negative for a negative transition at the input, as shown.

If the load matches the line, then there isn't any reflection at all, as shown in part (c) of the figure. Finally, we note that the ratio of the reflected voltage to the incident voltage is called the ***reflection coefficient*** and is given by

$$\Gamma = \frac{V_r}{V_i} = \frac{Z_L - Z_0}{Z_L + Z_0}, \tag{14.20}$$

where we have allowed for the general case of a complex load impedance.

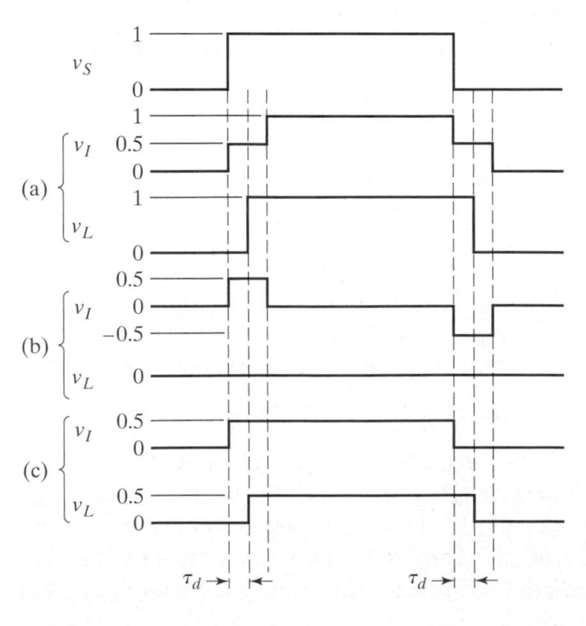

Figure 14–25 Waveforms for the circuit of Figure 14-24, assuming that $R_S = Z_0$ (a) when R_L is infinite, (b) when R_L is zero, and (c) when $R_L = Z_0$.

Now that we have a basic understanding of how and why reflections occur, we need to understand when they are a potential problem and how to deal with them. We assume that any mismatch either will be slight or will be due to a large load impedance (e.g., the input of some other gate); therefore, we ignore the possibility of a shorted line.

In general, reflections are not a problem when the delay of the line is much shorter than the rise and fall times on the input. In this case, the reflections are never seen, because they are over with by the time the input finishes changing. If the delay of the line is significant relative to the transition times, however, then we need to be sure that the lines are terminated in an impedance that is at least close to the characteristic impedance of the line, or else the digital waveforms will not be clean steps. It is important to note that the line does not need to be terminated at both ends, although doing so will help in the case of a mismatch. We can see that a proper termination at just one end will work by examining Figure 14-25 again. Note that although the input voltage does not make clean steps, the load voltage does. Since this load voltage would be the input to the next gate, the reflections back to the source are not a problem, because they have been absorbed by the source.

If R_S is not equal to Z_0, however, then we get a reflection back toward the load again. This situation is illustrated in Figure 14-26, which is a SPICE output for a circuit like that in Figure 14-24(a), but with $R_S = 75\ \Omega$, $Z_0 = 100\ \Omega$, and $R_L = 500\ \Omega$ (tranline.sch). Note that the reflection now bounces back and forth, and we have two reflection coefficients: At the load we have $\Gamma_L = 0.667$, and at the source we have $\Gamma_S = -0.143$.

A convenient way of keeping track of multiple reflections is to use a ***bounce diagram,*** as shown in Figure 14-27. Bounce diagrams are also called ***lattice diagrams.*** When the source voltage steps to 1 V at $t = 0$, v_I also steps, but only to $v_I = v_S Z_0/(Z_0 + R_S) = 0.57$ V. Then, one propagation delay later (2 ns in the figure), v_L steps to the sum of the original incident wave and the first reflection, v_{R1}:

$$v_L(\tau_d) = v_I(0) + v_{R1} = (1 + \Gamma_L)v_I(0) = 0.95 \text{ V}. \qquad \textbf{(14.21)}$$

As given implicitly in (14.21), the first reflection is $v_{R1} = \Gamma_L v_I(0) = 0.38$ V. This reflected wave comes back to the input and, at time $2\tau_d$, adds to the input and launches a second reflection heading back toward the load. This second reflection has amplitude $v_{R2} = \Gamma_S v_{R1} = -0.054$ V. The input voltage at the source end at this time is equal to the sum of all of the waves: the original one (0.57 V), the first

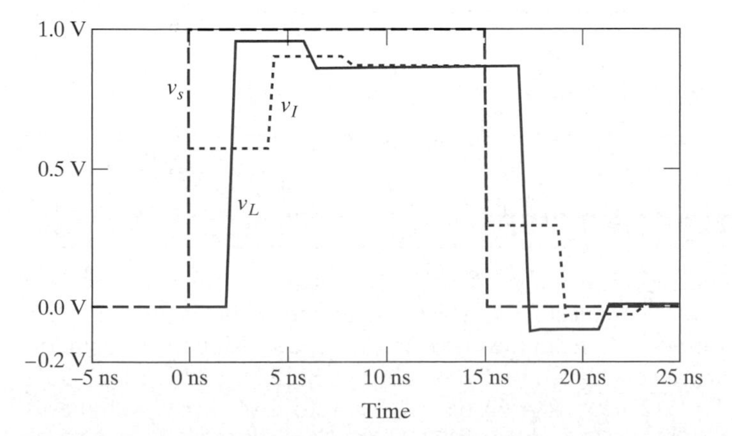

Figure 14–26 SPICE output for a pulse source in the circuit in Figure 14-24(a) with $R_S = 75\ \Omega$, $Z_0 = 100\ \Omega$, $R_L = 500\ \Omega$, and $\tau_d = 2$ ns (from tranline.sch).

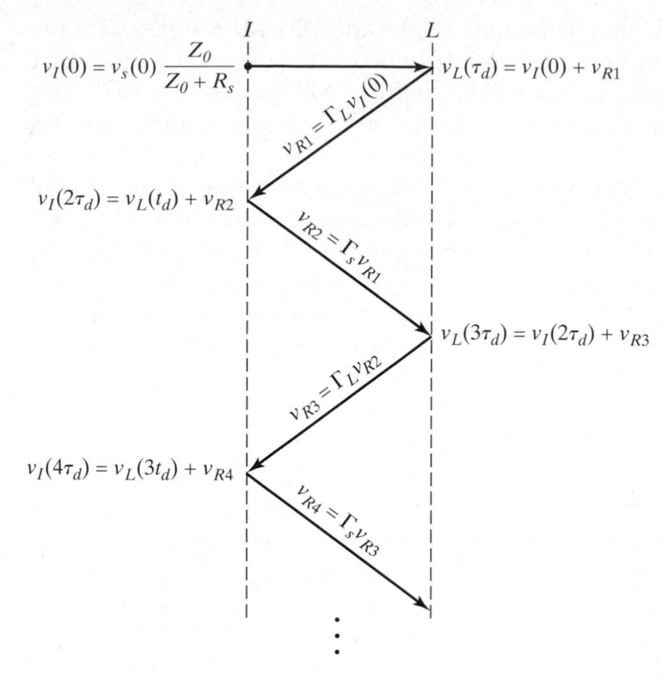

Figure 14–27 A bounce diagram.

reflection (0.38 V) and the second reflection (–0.054 V), for a total of 0.896 V. In other words, the input at this time is given by

$$v_I(2\tau_d) = v_L(\tau_d) + v_{R2} = v_I(0)[1 + \Gamma_L(1 + \Gamma_S)]. \tag{14.22}$$

When the second reflection arrives back at the load, it generates a third reflection given by $v_{R3} = \Gamma_L v_{R2} = -0.036$ V, and the output voltage is given by

$$v_L(3\tau_d) = v_I(2\tau_d) + v_{R3} = v_I(0)\{1 + \Gamma_L[1 + \Gamma_S(1 + \Gamma_L)]\}, \tag{14.23}$$

which equals 0.86 V. This pattern continues as shown in Figure 14-27. Note that the reflection coefficients are always less than unity, so the reflections do decay away with time.

Exercise 14.8

What is the magnitude of the sixth reflection for the situation shown in Figure 14-26?

SOLUTIONS TO EXERCISES

14.1 **(a)** One way to perform the conversion simply repeats the following process over and over: Find the smallest N for which the number to be converted is less than 2^N. Then subtract 2^{N-1} from the number and repeat, using the remainder. Keep track of the powers of two that you subtract. So, $832 < 2^{10} = 1024$. $832 - 2^9 = 320$. $320 < 2^9 = 512$. $320 - 2^8 = 64$. $64 = 2^6$. Therefore, the binary number is given by $2^9 + 2^8 + 2^6$, or $832 = (1101000000)_2$. **(b)** The procedure is similar to that used in (a), but with one extra step: We need to divide at each stage to see how many times 8^{N-1} will go into the number. So, $832 < 8^4 = 4096$. $832 \div 8^3 = 1$ + remainder of 320. $320 < 8^3 = 512$. $320 \div 8^2 = 5$ with remainder of zero. Therefore, $832 = (1500)_8$. **(c)** Similarly, $832 < 16^3 = 4096$. $832 \div 16^2 = 3$ + remainder of 64. $64 < 16^2$. $64 \div 16 = 4$ with no remainder. Therefore, $832 = (340)_{16}$. **(d)** For BCD, we convert each digit to a 4-bit binary number individually. You can do this by inspection or use the process outlined in (a). Either way, the result is 832 = (1000),(0011),(0010).

14.2 The complete truth table is as follows:

A	B	C	Y
0	0	0	0
0	0	1	0
0	1	0	0
0	1	1	1
1	0	0	0
1	0	1	1
1	1	0	1
1	1	1	0

14.3 The complete truth table is as follows (note by comparing the third and sixth columns that (14.15) is true):

A	B	$A + B$	$\overline{A}$	$\overline{B}$	$\overline{A} \cdot \overline{B}$
0	0	0	1	1	1
0	1	1	1	0	0
1	0	1	0	1	0
1	1	1	0	0	0

14.4 The truth table is as follows:

A	B	Y
0	0	0
0	1	1
1	0	1
1	1	0

14.5 When the clock input is low, the outputs of the two AND gates, S_C and R_C, are low, independent of S and R. Therefore, the outputs of the flip-flop cannot change—*see* the table in Figure 14-11(b). When the clock input is high, $S_C = S$, $R_C = R$, and the flip-flop works exactly as the circuit in Figure 14-11 does.

14.6 The function table follows. Note, if the circuit starts in state 00, it goes to state 01 and then never gets to state 00 again. Instead, it counts from 1 to 3 (in decimal) over and over again and either output has a frequency one-third that of the clock.

$Q_o(n-1)$	$Q_1(n-1)$	$Q_o(n)$	$Q_1(n)$
0	0	0	1
0	1	1	0
1	0	1	1
1	1	0	1

14.7 A four-bit ripple counter that uses only positive edge-triggered D flip-flops is shown in Figure 14-28. Q_0 is the LSB in the output word.

14.8 Following the pattern shown in the bounce diagram of Figure 14-27, the sixth reflection will have amplitude $v_{R6} = \Gamma_S v_{R5} = \Gamma_S \Gamma_L v_{R4} = \Gamma_S^3 \Gamma_L^3 v_I(0) = -0.5$ mV.

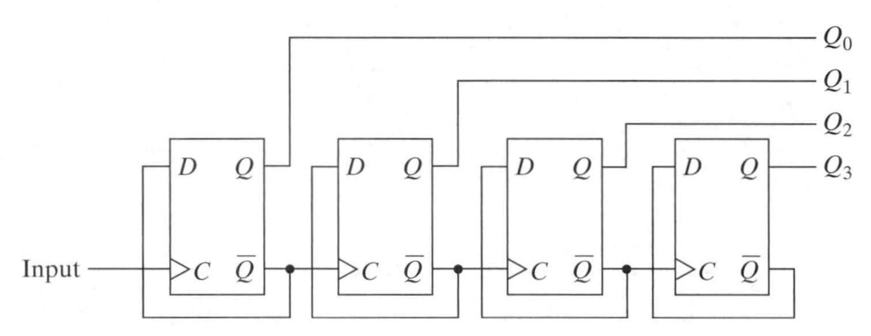

Figure 14–28

CHAPTER SUMMARY

- Logic gates perform basic logical functions like AND, OR, and NOT.
- Most logic circuits use binary inputs and outputs.
- Boolean algebra is used to manipulate logical expressions.
- A functionally complete set of Boolean functions is a set that allows any Boolean function to be implemented. Either AND and NOT or OR and NOT together make up functionally complete sets.
- Flip-flops, or latches, are bistable circuits that can store one binary digit.
- Flip-flops can be level triggered or edge triggered.
- Flip-flops can be combined to make shift registers and counters.

REFERENCES

[14.1] J.F. Wakerly, *Digital Design Principles and Practices*. Englewood Cliffs, NJ: Prentice Hall, 1990.

[14.2] C.H. Roth, Jr., *Fundamentals of Logic Design*. 4/E, Boston, Massachusetts: PWS Publishing Company, 1995.

[14.3] T.H. Lee, *The Design of CMOS Radio-Frequency Integrated Circuits*. Cambridge, UK: Cambridge University Press, 1998.

PROBLEMS

14.1 Introduction

P14.1 Design a circuit that will implement the exclusive-OR function (EXOR) described by the truth table in Table 14.2, using only AND, OR, and NOT gates.

P14.2 Design a circuit that will implement the exclusive OR function (EXOR) described by the truth table in Table 14.2 using only NAND and NOR gates.

P14.3 Convert the decimal number 1,235 to **(a)** binary, **(b)** octal, **(c)** hexadecimal, and **(d)** BCD.

P14.4 Convert the decimal number 1,679 to **(a)** binary, **(b)** octal, **(c)** hexadecimal, and **(d)** BCD.

P14.5 Convert the binary number 1010111011 to **(a)** decimal, **(b)** octal, **(c)** hexadecimal, and **(d)** BCD.

P14.6 Convert the binary number 1101100101 to **(a)** decimal, **(b)** octal, **(c)** hexadecimal, and **(d)** BCD.

P14.7 Convert the octal number 573 to **(a)** binary, **(b)** decimal, **(c)** hexadecimal, and **(d)** BCD.

P14.8 Convert the octal number 362 to **(a)** binary, **(b)** decimal, **(c)** hexadecimal, and **(d)** BCD.

P14.9 Convert the hexadecimal number A2D5 to **(a)** binary, **(b)** octal, **(c)** decimal, and **(d)** BCD.

P14.10 Convert the hexadecimal number 2FCB3 to **(a)** binary, **(b)** octal, **(c)** decimal, and **(d)** BCD.

P14.11 Convert the BCD number 0110,1001,1000,1001 to **(a)** binary, **(b)** octal, **(c)** hexadecimal, and **(d)** decimal.

P14.12 Convert the BCD number 1001,0111,1000,1001 to **(a)** binary, **(b)** octal, **(c)** hexadecimal, and **(d)** decimal.

P14.13 Prove that a NOR gate, by itself, can implement a functionally complete set of logic functions. Do this by showing how to use a two-input NOR gate to implement a NOT gate, an AND gate, and an OR gate.

P14.14 Prove that a NAND gate, by itself, can implement a functionally complete set of logic functions. Do this by showing how to use a two-input NAND gate to implement a NOT gate, an AND gate, and an OR gate.

Table 14.2 Truth Table for an Exclusive-OR Function

A	B	Y
0	0	0
0	1	1
1	0	1
1	1	0

P14.15 Use De Morgan's theorems to demonstrate that the AND and NOT functions constitute a functionally complete set of logic functions (i.e., show that you can produce an OR function from a NOT function and an AND function).

P14.16 Use De Morgan's theorems to demonstrate that the OR and NOT functions comprise a functionally complete set of logic functions. (i.e., show that you can produce an AND function from a NOT and an OR function)

P14.17 Use truth tables to verify that (14.15) and (14.16) are true for three variables.

P14.18 Use a truth table to prove that the OR function is distributive over the AND function.

P14.19 Use a truth table to prove that the AND function is distributive over the OR function.

D P14.20 Design a circuit using only NOR gates that will perform the same logic function as the circuit in Figure 14-5. Confirm that your circuit works properly by filling out a truth table for it.

D P14.21 Design a circuit that will implement the function $Y = A(B + C) + D$ using only NOR gates. Confirm that your circuit works properly by filling out a truth table for it.

14.2 Flip-Flops

P14.22 Draw a timing diagram and fill in a function table for the circuit shown in Figure 14-29. What is it?

P14.23 Derive a function table for the circuit shown in Figure 14-30.

P14.24 Draw a timing diagram and fill in a function table for the circuit shown in Figure 14-31. What is this circuit?

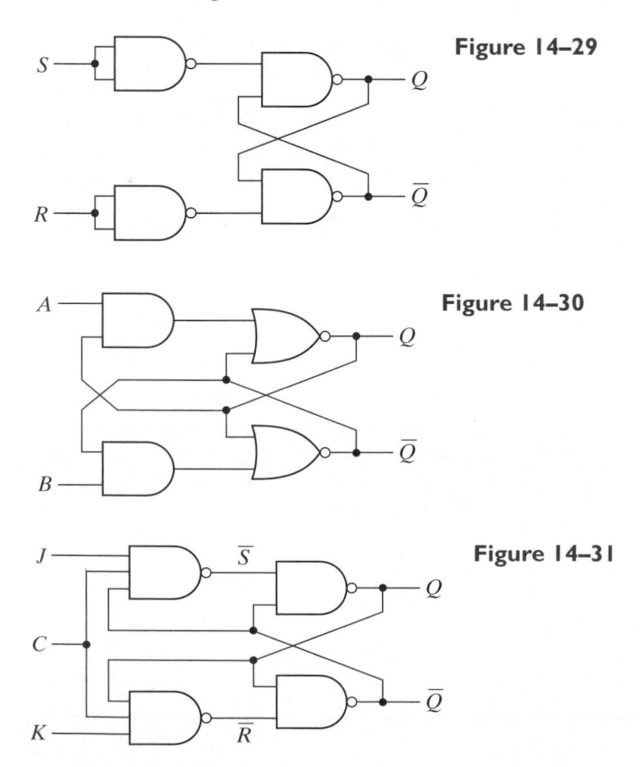

Figure 14–29

Figure 14–30

Figure 14–31

P14.25 Consider a level-triggered JK flip-flop, as in Figure 14-15, and let it be driven by the inputs shown in Figure 14-32. Further, assume that the initial state of the flip-flop is $Q = 0$. Draw S_M, R_M, Q_M, $\overline{C}$, S_S, and Q as a function of time, along with the input waveforms shown. Is the output of the flip-flop after the clock cycle shown correct?

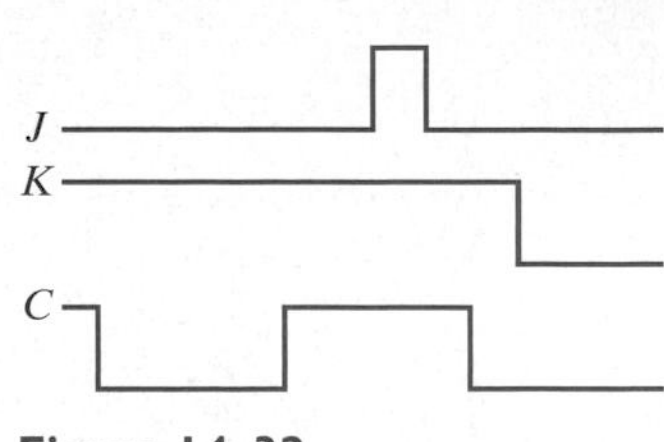

Figure 14–32

P14.26 Consider the circuit shown in Figure 14-33. Assuming that the clock is a square wave with a period significantly greater than twice a gate delay, that each gate has the same delay, and that $\overline{J}$ is low, draw waveforms for C, J_C, and S for one full period of the clock.

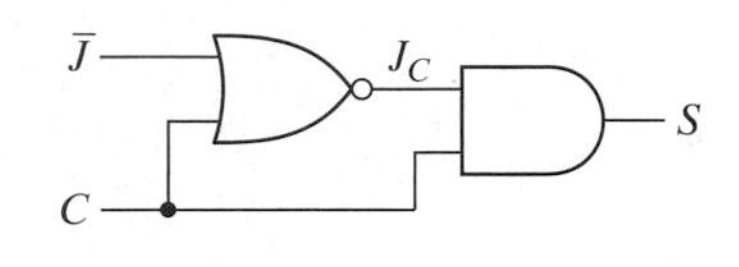

Figure 14–33

P14.27 Consider the D flip-flop shown in Figure 14-19(a). Draw a timing diagram showing the output switching from zero to one and then back to zero. Show the D input, the clock, the set and reset inputs to the SR flip-flop, and the Q output. Assume that all the gates and the SR flip-flop have the same delay and that the clock period is significantly greater than two gate delays.

P14.28 Show how to implement a D flip-flop using a JK flip-flop and one or more logic gates.

14.3 Shift Registers and Counters

D P14.29 Design a 3-bit counter that will count up (i.e., from 000 to 111), using positive edge-triggered JK flip-flops.

D P14.30 Design a three-bit ripple counter that will count up (i.e., from 000 to 111) or down (i.e., from 111 to 000), using positive edge-triggered JK flip-flops and some logic gates. There should be a single binary input, U/D, that determines whether the counter counts up or down, ($U/D = 1$ implies count up; $U/D = 0$ implies count down.)

P14.31 Fill in a function table for the circuit in Figure 14-34, assuming that it starts in state 0000. Write the state as $[Q_3Q_2Q_1Q_0]$. What is this circuit?

14.4 Reflections on Transmission Lines

P14.32 Draw a bounce diagram for a lossless transmission line with $Z_0 = 50\ \Omega$ and $\tau_d = 0.5$ ns when the line is driven by a pulse of amplitude 1 and has instantaneous transitions. Assume that $R_S = 40\ \Omega$ and $R_L = 10\ \text{k}\Omega$. How long does it take for the load voltage to be within 1% of its final value?

P14.33 Assume that a lossless transmission line with $Z_0 = 120\ \Omega$ and $\tau_d = 1$ ns is driven by a source with $R_S = 3\ \Omega$. If the source voltage makes an abrupt step of 1 V, how close to 120 Ω does R_L need to be in order to guarantee that the output is within 1% of its final value in 3 ns?

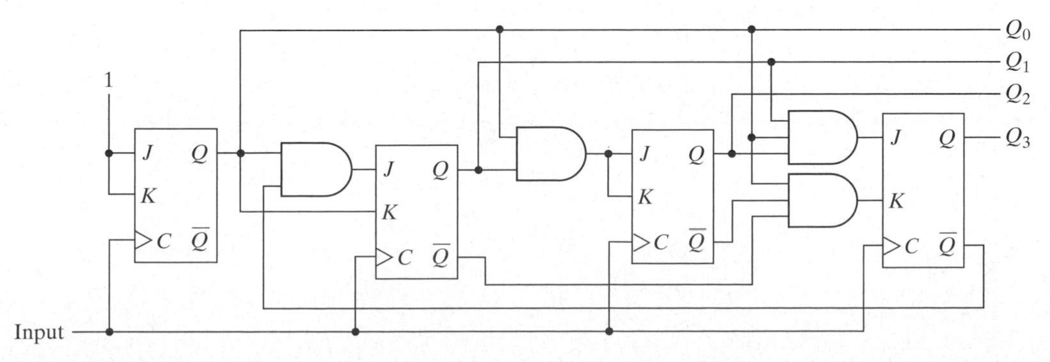

Figure 14–34

CHAPTER 15

Transistor-Level Digital Circuits

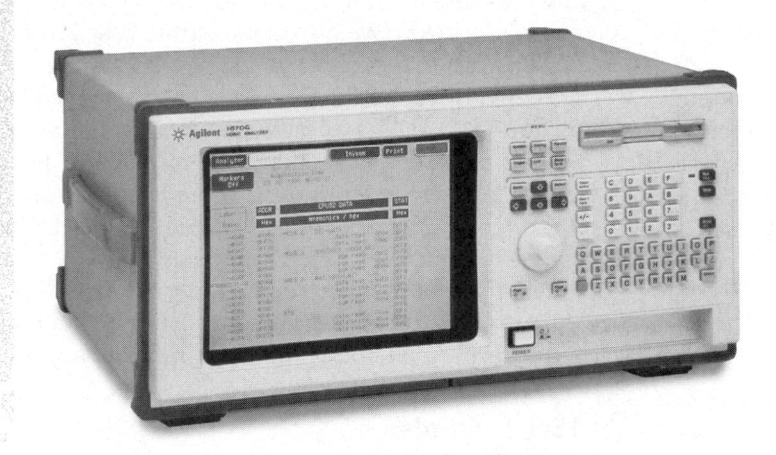

Logic analyzers are used to examine the detailed operation of complex logic circuits. They allow the designer to trigger on different events and see many signals represented simultaneously. (Photo courtesy of Agilent Technologies. Reprinted with permission.)

INTRODUCTION

In the previous chapter, we introduced gate-level digital circuits. In this chapter, we introduce the design of transistor-level digital circuits. We begin with a discussion of the models we use for the components that comprise these circuits. We then progress to a discussion of how to characterize logic gates in terms of their ***static performance*** (i.e., without any signals changing) and their ***dynamic performance*** (i.e., with changing signals). We then go on to discuss the analysis and design of logic circuits in different technologies and different logic families.

A logic family is a set of logic circuits (e.g., gates, flip-flops) implemented using a particular technology (e.g., CMOS, TTL) so that the different circuits can be connected together to perform more complex functions. If gates of different families are connected together, they may or may not be compatible in terms of input and output voltage levels, input and output currents, and other factors. Some of the logic families presented in this section are suitable only for integrated circuit design (e.g., NMOS); that is, they are not available as separate packages containing AND gates, inverters, and other basic functions. Other families presented are typically not used in the design of custom integrated circuits (e.g., TTL), but are available only as ***small-scale integrated circuits*** (SSIs) like AND gates, or ***medium-scale integrated circuits*** (MSIs) like counters.[1]

15.1 DEVICE MODELING FOR DIGITAL DESIGN

In this section, we present models for the various devices encountered in digital circuits. We are concerned at this point with models suitable for use in hand analysis and design. More detailed models, as would typically be used for computer-aided simulation and design, are discussed in Chapter 4. In each of the sections that follow, we first deal with DC voltages and currents (i.e., the static situation) and then discuss how to model the devices during switching transients (i.e., the dynamic situation). We also demonstrate how large-signal transient analyses can be performed.

The modeling of independent sources and linear passive devices (Rs, Ls & Cs) for the DC analysis of logic circuits is the same as it is for analog circuits, so the models presented in Section 7.1 are used here. For transient analyses of digital circuits, the independent sources are also modeled as in Section 7.1.1, but the total instantaneous values are used instead of the DC values. Linear passive elements can be modeled with the DC models for most transient situations, and we will do that in this chapter. You should be aware, however, that for very fast transients, extreme accuracy, or special circumstances, it may be necessary to include some of the parasitic elements from the high-frequency models presented in Section 9.1.2.

15.1.1 Diodes

In this section, we discuss how to model diodes for digital circuit analysis. We start with the models for DC, or static, analysis, and then discuss the models used for analyzing switching transients.

DC Models (Static Models) The low-frequency models for the diode presented in Section 7.1.3 apply equally well to digital circuit analysis. You should be familiar with the material in that section before proceeding with this chapter. The models we will most frequently use here are the constant-voltage model for the forward-biased diode and the open-circuit model for the diode when cut off.

Dynamic Models and Switching Transients in Diodes[2] In this section, we consider how to model the behavior of diodes when they switch from off to on or from on to off. Our goal is to understand what model parameters affect the switching times and to be able to make reasonable estimates with hand analyses. These analyses are necessarily approximations because we are dealing with large-signal transient phenomena to which linearized models are not applicable.

The depletion region of a diode changes width with the applied voltage, and, therefore, the charge stored in this depletion region also changes, as discussed in Section 2.4.5. This charge storage mechanism leads to the ***junction capacitance***. This capacitance is nonlinear, however, and for digital circuits, we must often deal with large transients where we cannot reasonably approximate the junction capacitance as a constant for the entire signal swing. The junction capacitance of a diode is given by

[1] Circuits with fewer than 10 gates on a chip are usually called SSI circuits. MSI chips have between 10 and 100 gates per chip, large-scale integrated (LSI) circuits have from about 100 to 10,000 gates per chip, and very-large scale integrated (VLSI) circuits have over 10,000 gates per chip.

[2] A more detailed discussion of switching transients is available in [15.1] and [15.2].

$$C_j = \begin{cases} \dfrac{C_{j0}}{\left(1 - \dfrac{V_D}{V_0}\right)^{m_j}}; & \text{in reverse bias} \\ 2C_{j0}; & \text{in forward bias,} \end{cases} \qquad \textbf{(15.1)}$$

where C_{j0} is the zero bias value of the junction capacitance, V_0 is the built-in potential of the junction, $2C_{j0}$ is approximate, and m_j is an exponent that is a function of the doping profile in the device and is typically between 0.33 and 0.5.

In addition to possessing a junction capacitance, a forward-biased *pn*-junction diode stores charge in the form of excess minority carriers, caused by diffusion across the junction. This charge storage mechanism is the ***diffusion capacitance***. Schottky diodes are majority-carrier devices and do not have diffusion capacitance. Diffusion capacitance is again a nonlinear function of the applied voltage, but we are able to express the total charge stored in terms of the current in the diode and a time constant, as was done in (2.61). In that equation, we assumed an n^+-p diode, so the charge storage was caused by the minority electrons on the *p* side of the junction. If we allow for the more general case, we can denote the total diffusion charge stored in a forward-biased diode as Q_F and write

$$Q_F = I_D \tau_F, \qquad \textbf{(15.2)}$$

where τ_F is a characteristic lifetime associated with the diode. It is very important to note that (15.2) describes a *steady-state* situation. The current that flows by diffusion is determined by the slopes of the minority-carrier densities, and it is quite possible for the total charge present to be greater or less than predicted by (15.2) if a transient situation is considered.

We now turn our attention to a specific circuit and consider a diode driven by a voltage source in series with a resistor, as shown in Figure 15-1(a). If the input voltage is zero for $t < 0$ and then switches to a positive value V_F, as shown in part (b) of the figure, we wish to investigate how the diode responds to this sudden change. Because the voltage across the diode is zero initially and cannot change

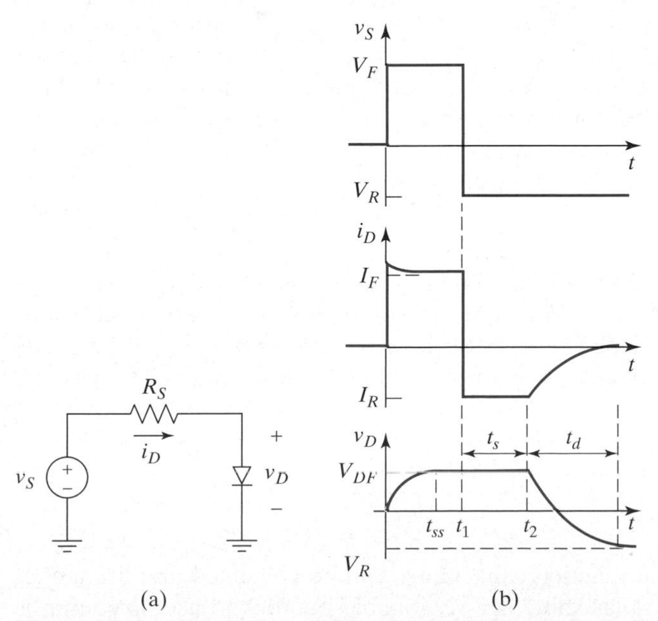

Figure 15–1 (a) A diode switching circuit. (b) The waveforms for the circuit.

without changing the charge in the depletion region, the diode acts like a capacitor and will not allow the voltage across it to change instantaneously (unless an impulse of current is supplied). Therefore, the diode voltage at $t = 0^+$ is still zero, as shown in part (b)The initial current is therefore given by

$$i_D(0^+) = \frac{V_F - v_D(0^+)}{R_S} = \frac{V_F}{R_S} \quad \textbf{(15.3)}$$

and flows by diffusion across the depletion region. The carriers do, of course, drift through the bulk regions to arrive at the depletion region, but the voltage drops across the bulk regions are negligibly small because they conduct current very easily. We will, therefore, ignore these voltage drops, as is typically done in analyzing *pn*-junction behavior.

Once the current starts to flow, the capacitance of the junction charges up and the diode voltage increases, as shown in part (b) of the figure. As the diode voltage increases, the current decreases as predicted by the form of (15.3) and shown in the figure. Once the bulk of the charge has been provided to the junction, the diode current settles down to very close to its steady-state value. It still takes longer, however, for the diode to reach steady state, denoted by t_{ss}, since the excess minority-carrier charge stored in the diode continues to build up (*see* Aside A15.1); this typically takes on the order of a few minority-carrier transit times.

Now assume that the diode has reached steady-state and the voltage steps down to a negative value, as shown at time t_1 in the figure. We know that the diode will eventually become reverse biased and the current will be nearly zero, but initially, this is not true. Because of the excess minority-carrier charge stored in the device, the current in the reverse direction will, in fact, be significant and approximately constant for a time called the ***storage time*** of the diode, t_s. This initial reverse current is (V_R and I_R are negative)

$$I_R = \frac{V_R - v_D(t_1)}{R_S}. \quad \textbf{(15.4)}$$

When the storage time is over, the excess minority carriers have been removed from the device, and all that remains is to discharge the junction capacitance, which is a normal *RC* discharge except that the capacitance is changing during the discharge. The time required to discharge, the depletion capacitance is called the ***delay time***, t_d, and the total time required to turn the diode off is called the ***reverse-recovery time*** of the diode, given by

$$t_{rr} = t_s + t_d. \quad \textbf{(15.5)}$$

The storage time depends on the total excess minority-carrier charge stored and the magnitude of the current used to remove this charge (the reverse current). The excess charge stored depends on both the minority-carrier transit time and the steady-state forward current just prior to turning the diode off, I_F. The relation between t_s and τ_T is

$$t_s = \tau_T \ln\left(1 - \frac{I_F}{I_R}\right). \quad \textbf{(15.6)}$$

If we assume the change in junction capacitance can be neglected and use a constant value of C_j (the zero-bias value is a reasonable number to use, since that is what the capacitance is at the end of the time), the delay time can be found from the exponential RC decay equation (replace the diode with the model in Figure 9-2 for reverse bias, and solve the circuit — *see* Problem P15.1):

$$v_D(t) = V_R + (V_{DF} - V_R)e^{-(t-t_2)/R_SC_j}; \quad t \geq t_2 \tag{15.7}$$

Setting $v_D(t_d) = V_R + 0.1(V_{DF} - V_R)$ to find the end of the delay time yields

$$t_d = R_SC_j\ln 10 = 2.3R_SC_j. \tag{15.8}$$

We demonstrate the switching of a diode with an example. The example is followed by an aside that discusses the switching transient in more detail. Finally, we point out that the diode goes through several distinct regions of operation during the switching transient and that we have, in essence, been using a different model in each region. Breaking the transient excursion up into distinct regions is typical of large-signal transient analysis and is why having a reasonable understanding of the operation of the devices is important.

Example 15.1

Problem:
Use SPICE to model the circuit in Figure 15-2, and determine how long it takes the diode to turn on and off. Assume that the input source switches from zero to 3 V and then, after the diode reaches steady state, switches down to −3 V. Use the D1N914 diode model in PSPICE. The minority-carrier transit time for this diode model is $\tau_T = 11.54$ ns.

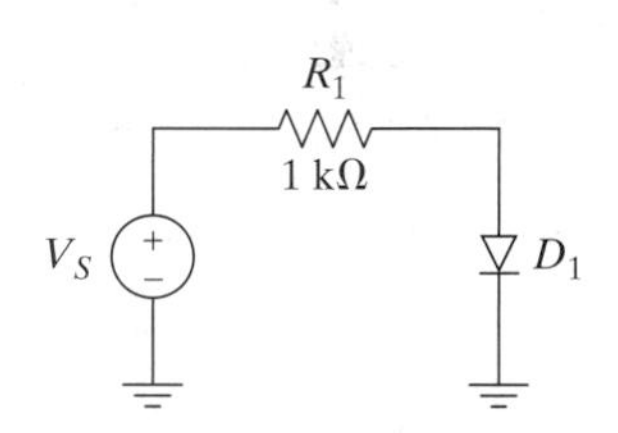

Figure 15–2 A diode circuit.

Solution:
The SPICE output is shown in Figure 15-3 (Exam1.sch). Note that the initial current is 3 mA, as predicted by (15.3), but then it decreases to the positive steady-state value as the diode voltage builds up. After the diode has reached steady state, the input voltage drops. The current waveform shows that the initial reverse current is near −3.7 mA, as is predicted by (15.4) with $v_D(t_1) \approx 0.7$ V. The current stays at this level for a time close to that predicted by (15.6) $t_s \approx 6.8$ ns. The reverse recovery time from the simulation is about 12 ns. If we use (15.5), (15.6), and (15.8) along with the zero-bias junction capacitance of $C_j = 4$ pF, we obtain $t_{rr} =$ 16 ns, which is an acceptable approximation.

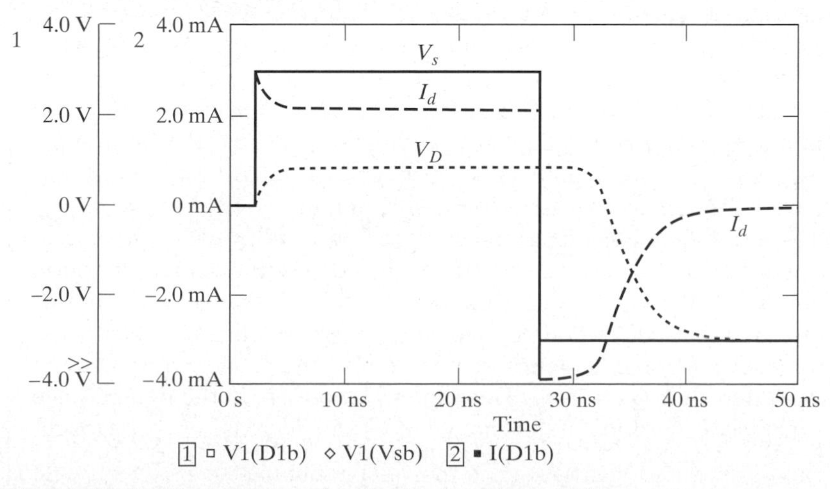

Figure 15–3 The waveforms for the circuit in Figure 15-2.

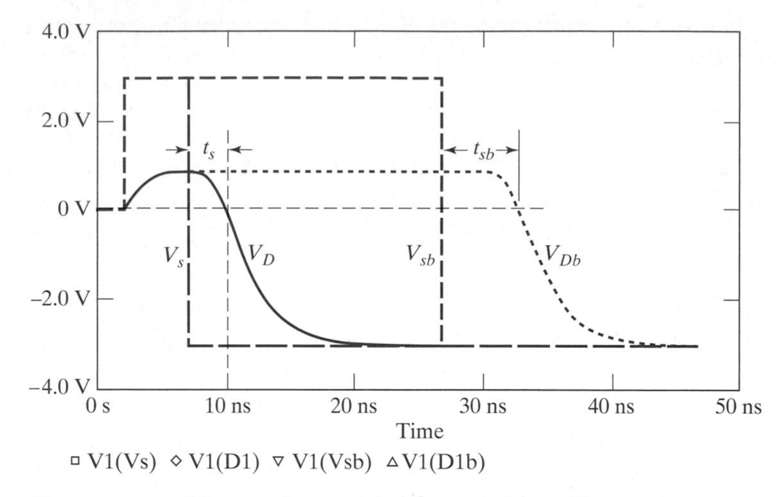

Figure 15–4 The waveforms with different pulse widths; t_s is the storage time with a 5-ns wide pulse, and t_{sb} is the storage time with the original 25-ns wide pulse.

For illustrative purposes, we now use SPICE to show what happens if the diode is not allowed to reach steady state prior to being turned off. If we don't leave the pulse high long enough, the excess minority-carrier density will not have sufficient time to reach steady state. Therefore, when the diode is turned off, there isn't as much charge to remove, and the storage time is significantly reduced. Figure 15-4 shows what happens if the pulse width is reduced to 5 ns, compared with our original 25 ns. Note that it takes the diode about $2\tau_T$ to reach steady state, or about 24 ns in this case.

■

Exercise 15.1

Consider the circuit in Figure 15-2 again. If the resistor is 4.7 kΩ and the diode has $\tau_T = 10$ ns and $C_j = 5$ pF, find the reverse recovery time if the input has been at 5 V and switches to -2 V.

□

ASIDE A15.1 DIODE SWITCHING - DETAILED VIEW

We wish to consider the diode switching illustrated in Figure A15-1. In particular, we want to examine what happens to the excess minority-carrier profiles in the diode. Let's restrict our attention to one type of carrier for simplicity—say, the electrons on the p side of the junction. The diffusion current is proportional to the slope of the minority-carrier concentration at the edge of the depletion region, so at $t = 0^+$ the slope must change as indicated in part (c) of the figure. It doesn't require moving much charge to bring about this change, so it can occur almost instantaneously. The minority-carrier concentration at the edge of the depletion region is a function of the bias voltage across the diode; so, as the current flows and provides the necessary charge, the voltage across the diode and the minority-carrier density at the edge of the depletion region increase together as shown in parts (b) and (c) of the figure. Since the diode is being forward biased in this case, the depletion region also gets narrower as indicated, although the change is very small and is greatly exaggerated in the figure. A steady state is reached when the depletion region reaches the proper width and the diffusion charge has built up to the steady-state value, $Q_F = I_D\tau_F$. This time is denoted t_{ss} in the figure. Since the current in this circuit decreases as v_D increases, the slope of the change in v_D will decrease as it approaches V_{DF} as shown.

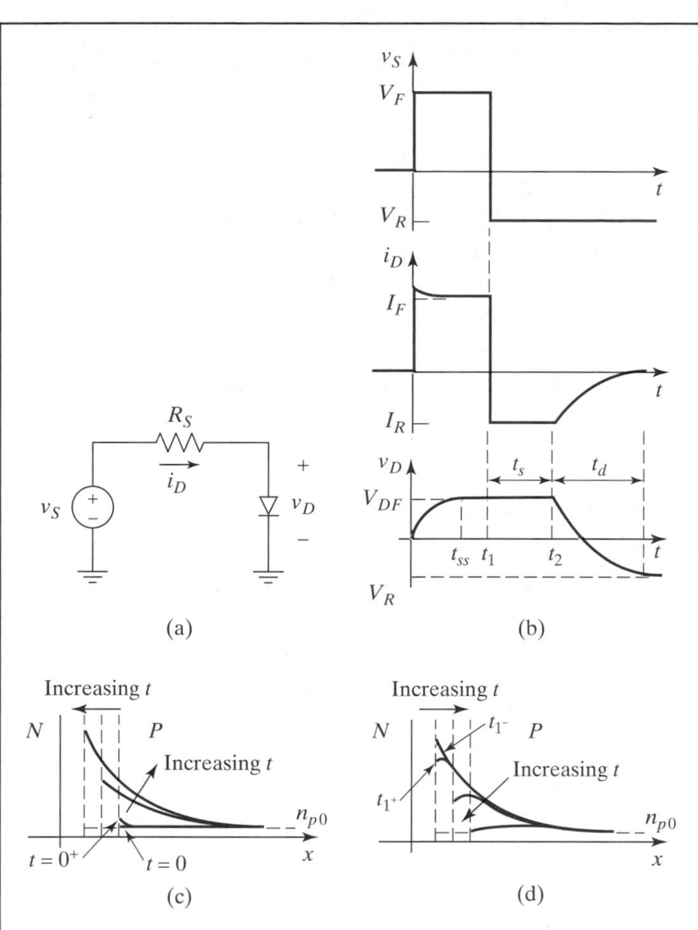

Figure A15-1 (a) A diode switching circuit. (b) The waveforms for the circuit. (c) The minority-carrier densities on the p side of the junction as a function of time while the diode is turning on; the dashed line is the edge of the depletion region. (d) The minority-carrier densities on the p side of the junction as a function of time while the diode is turning off; the dashed line is the edge of the depletion region.

Now consider what happens when the input voltage changes abruptly to a negative value, V_R, at time t_1, as shown in the figure. The diode voltage again cannot change instantaneously, but the current does change instantaneously as shown. At this point in time, an interesting phenomenon occurs: The slopes of the minority-carrier concentrations change rapidly to support the new current, as shown in part (d) of the figure. (Only the p side of the junction is shown for simplicity.) Since it does not require moving very much charge to change the profile as shown at time t_{1^+}, this change can take place very rapidly. As time progresses, the external current stays constant, and the minority carrier profile changes as shown in part (d) of the figure, while keeping the slope at the edge of the depletion region constant. The minority-carrier profile relaxes toward the equilibrium value because of diffusion of carriers away from the peak concentration (in both directions) and because of recombination. During this time, the current across the depletion region is composed mostly of electrons diffusing to the depletion region from the peak concentration on the P side and then drifting across under the influence of the field. Because of the excess minority carriers stored (electrons on the P side in this example) this drift current can be supplied for some time. Also, the depletion region does get wider during this time, but the increase in negative charge on the P side of the depletion region is compensated for by the decrease in excess minority carriers, which are also negative, so the diode voltage does not change appreciably.[1] The key point is that the diode does not start to become reverse biased until all of the excess minority carriers have been removed from the edge of the depletion region. When the excess minority-carrier concentration at the edge of the depletion region gets to zcro, the junction starts to become reverse biased. The time it takes to bring the excess minority-carrier concentration at the edge of the depletion region to zero is called

[1] You may be bothered by this statement, since the voltage across the depletion region is determined by integrating the charge density twice and we have always ignored the mobile charge when performing these integrations (i.e., we make the depletion approximation). Nevertheless, the mobile charge should be taken into account in a proper analysis.

the *storage time* of the diode and is denoted by t_s in the figure. The time required to remove the rest of the minority carrier charge and to change the charge in the depletion region to the reverse bias value is called the *delay time* and is denoted t_d in the figure. (It is also sometimes called the transition time.)

Once the excess minority-carrier density at the edge of the depletion region gets to zero, the slope of the density will start to decrease as more carriers diffuse away from the peak and recombine. Therefore, the external current decreases during this time as shown in part (b) of the figure. The sum of t_s and t_d is called the *reverse recovery time* of the diode. This time is denoted t_{rr} and is frequently specified on data sheets for discrete diodes.

The reverse recovery time of Schottky diodes is much smaller than that of *pn*-junction diodes, because Schottky diodes are majority-carrier devices and do not have any excess minority carriers to remove. Therefore, only the junction capacitance needs to be charged, and the storage time is eliminated.

15.1.2 Bipolar Junction Transistors

In this section, we discuss the models used for bipolar transistors in digital circuits. We first cover the models used for DC, or static, analysis and then go on to examine the models used for transient analysis.

DC Models (Static Models) The low-frequency BJT models presented in Section 7.1.4 apply equally well to digital circuit analysis and will not be repeated here. You should be familiar with the material in that section before proceeding with this chapter. We do need to make a few additional comments about the device when operating in saturation, however.

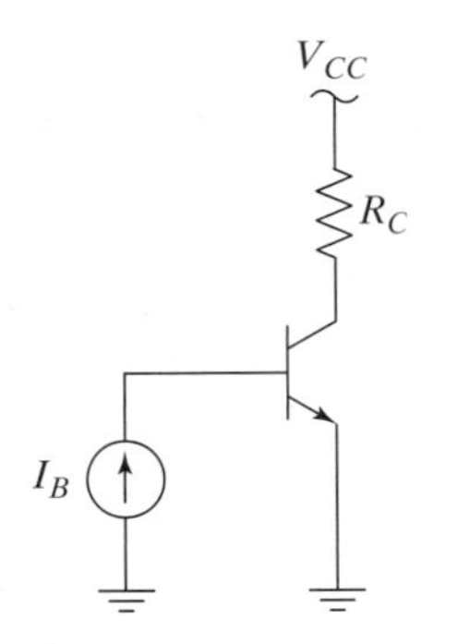

Figure 15–5 A BJT driven with a base current.

Consider the circuit shown in Figure 15-5, and assume that I_B is large enough to drive the transistor into saturation. We can calculate the collector current for this case as

$$I_{C\text{sat}} = \frac{V_{CC} - V_{CE\text{sat}}}{R_C}, \tag{15.9}$$

and, as noted in Chapter 7, we will use $V_{CE\text{sat}} = 0.2$ V as a reasonable approximation. The exact value of $V_{CE\text{sat}}$ is a function of how hard we drive the transistor and is derived in Example 4.4.

The base current necessary to just saturate the transistor is $I_B = I_{C\text{sat}}/\beta$. Any base current larger than this value will drive the transistor further into saturation. Since we sometimes want to ensure that a transistor is saturated under all conditions, it is useful to have a measure of how "hard" we are saturating the transistor (i.e., how much base current we are providing above and beyond what is required to just cause saturation). We can develop such a measure by first defining the ***forced beta*** as

$$\beta_{\text{forced}} = \frac{I_{C\text{sat}}}{I_B}. \tag{15.10}$$

When the transistor is just on the edge of saturation we have $\beta_{\text{forced},} = \beta_F$. As we drive the transistor harder and harder (i.e., we increase I_B), we see from (15.10) that β_{forced} drops. Therefore, we can define the ***base overdrive factor*** as

$$\text{base overdrive factor} = \frac{\beta_F}{\beta_{\text{forced}}}, \tag{15.11}$$

or alternatively, by finding the minimum base current that would be required to support $I_{C\text{sat}}$ (i.e., $I_{B\min} = I_{C\text{sat}}/\beta_F$), we find (*see* Problem P15.9)

$$\text{base overdrive factor} = \frac{I_B}{I_{B\min}}. \qquad (15.12)$$

We will see later that the base overdrive factor can be used in design to ensure that the transistor remains saturated at all times.

Example 15.2

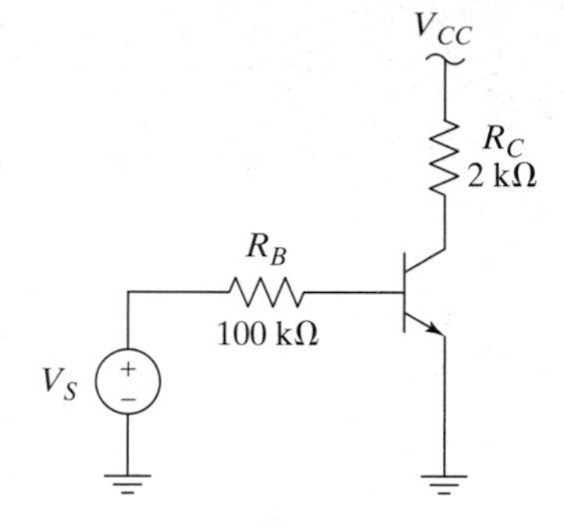

Figure 15–6 A BJT circuit.

Problem:

Consider the circuit shown in Figure 15-6, and assume that $\beta_F = 100$ and $V_{CC} = 5$ V. Determine the values of I_B, I_C, and V_C for (a) $V_S = 1$ V and (b) $V_S = 2$ V. (c) How large must R_C be to saturate the transistor with a forced beta of 20 when $V_S = 2$ V?

Solution:

Assume that the transistor is forward active, and write KVL around the base loop to find

$$I_B = \frac{V_S - 0.7}{R_B}.$$

We then find $I_C = \beta_F I_B$ and $V_C = V_{CC} - I_C R_C$. Plugging in the numbers for part (a), we obtain $I_B = 3$ µA, $I_C = 300$ µA, and $V_C = 4.4$ V. Note that if the 0.7, V approximation for V_{BE} is off by only 0.1 V, the values of I_B and I_C will change by 33%! Therefore, this model is not a good one to use with V_S this low. In a case like this, we would need to use the nonlinear equation for V_{BE} and solve iteratively. For part (b), we obtain $I_B = 13$ µA, $I_C = 1.3$ mA, and $V_C = 2.4$ V. In this case, the answers do not change a lot if V_{BE} changes by ±0.1 V. (c) Given that $V_S = 2$ V, we just found that $I_B = 13$ µA. To have a forced beta of 20, we must have $I_{C\text{sat}} = 20\,(13\ \mu\text{A}) = 260$ µA. Using (15.9), we find that we must have $R_C = 18.5$ kΩ. ■

Exercise 15.2

For the circuit considered in Figure 15-6, what value of V_S is required to achieve a forced beta of 10? □

Dynamic Models and Switching Transients in BJTs[3] Bipolar transistors have capacitances associated with each junction that have a significant impact on their dynamic behavior. In particular, each junction has junction capacitance associated with the depletion region, and if the junction is forward biased, it also has diffusion capacitance associated with the excess minority-carrier profile. The junction capacitances are given by

$$C_{jx} = \begin{cases} \dfrac{C_{jx0}}{\left(1 - \dfrac{V_{Bx}}{V_0}\right)^{m_{jx}}}; & \text{in reverse bias} \\ 2C_{jx0}; & \text{in forward bias,} \end{cases} \qquad (15.13)$$

[3] A more detailed discussion of switching transients is available in [15.1] and [15.2].

where x is either c for the collector-base junction or b for the emitter-base junction, C_{jx0} is the zero-bias value of the junction capacitance, V_0 is the built-in potential of the junction, and m_{jx} is the junction exponent, as shown in Section 2.5.5.

The junctions also have diffusion capacitance when they are forward biased. The diffusion capacitance is caused by excess minority-carrier charge stored mostly in the base[4]. In Section 2.5.5, we showed that the forward-active charge storage term is given by

$$Q_F = I_C \tau_F . \tag{15.14}$$

This equation is valid only for steady-state operation, but it allows us to calculate the amount of excess charge required to support a given collector current. When the transistor is either reverse active or saturated, the collector-base junction also produces excess minority carriers, and we have a reverse charge term

$$Q_R = I_R \tau_R \tag{15.15}$$

where we have implicitly defined I_R as the reverse transistor's collector current. These two charge storage terms can be added when the transistor is saturated.

Since most digital circuit design is done for integrated circuits rather than discrete circuits, we also note that BJTs on an IC have parasitic capacitances to the substrate. Exactly where these capacitors connect and how significant they are depends on the process used and sometimes also on whether an *npn* or a *pnp* transistor is being used. The most significant of these parasitic capacitances for the circuits we will consider is the collector-to-substrate capacitor on an *npn* transistor. Most processes still use junction isolation (*see* Chapter 3), so this capacitor is the capacitance of a reverse-biased *pn*-junction diode.

We now consider what happens when a BJT is switched on and off. Consider the circuit shown in Figure 15-7(a), where the input voltage is initially negative and then switches to a positive value at $t = 0$, as shown in part (b) of the figure. We assume that V_P is large enough to cause the transistor to saturate. We first describe the switching transient without using equations and then, in Aside A15.2, derive approximate equations.

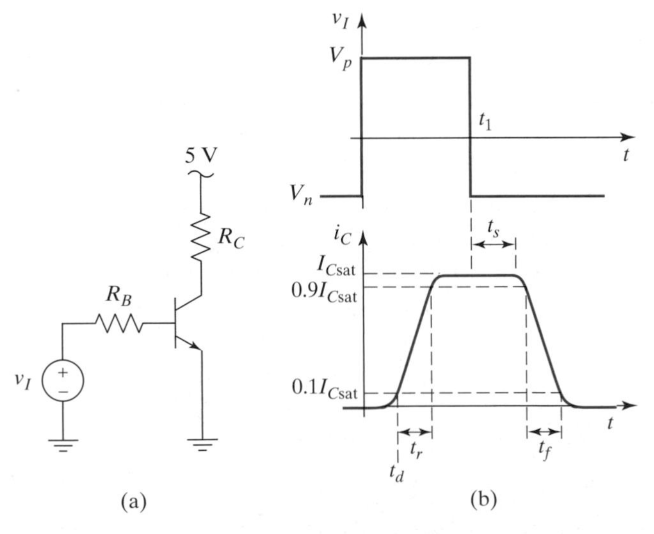

Figure 15–7 (a) A BJT switching circuit. (b) The waveforms for the circuit.

[4] There is also a small amount of minority-carrier charge storage in the emitter caused by reverse injection.

When the input voltage is negative, the transistor is cut off and no external current flows. When v_I switches to V_p, the transistor does not turn on instantly, because the base-emitter junction capacitance must be charged to forward bias the junction before any current can flow. Even after the junction becomes forward biased, there is a small delay while the carriers injected into the base from the emitter traverse the base and arrive at the collector. Finally, it takes some time for the current to build to an appreciable level, say, 10% of I_{Csat}. We define the ***delay time*** as the time required for the collector current to rise to 10% of its final saturation value and denote it by t_d, as shown in the figure.

After the transistor has turned on, the collector current increases as the base current supplies the charge necessary to build up the excess minority-carrier profile in the base and to charge the depletion capacitances of the base-emitter and base-collector junctions. We define the ***rise time*** to be the time required for the collector current to increase from 10% to 90% of its final value and denote it by t_r. This definition of rise time is a general definition that we will use in other circumstances as well.

During the final part of the transition, while i_C is increasing to the full saturation value I_{Csat}, the collector-base junction becomes more and more forward biased. Because the base-collector junction is forward biased in saturation, the minority-carrier density no longer goes to zero at the edge of the base-collector depletion region, and there is excess charge stored in the base. The situation is shown in Figure 15-8 and can be viewed as a superposition of the forward- and reverse-active regions of operation, as is also shown in the figure. Both junctions are forward biased, and the minority carrier density at each edge of the base is set by the forward bias on that junction. The collector current is the diffusion current in the base, which is proportional to the slope of the minority-carrier profile. This picture allows you to see why the collector current stops increasing in saturation: As the collector-base junction becomes more forward biased, the minority carrier density at the collector edge of the base increases, which prevents the slope of the total minority-carrier profile in the base from increasing.

The transistor is not in steady-state operation until all of the excess charge has been supplied and the charge profiles have stopped changing. If the transistor is switched back off before reaching steady state, the time required to turn the transistor off will be shorter than normal.

Now consider what happens when the input voltage switches back to a negative value, as shown at time t_1 in the figure. We assume that t_1 is large enough that the transistor has attained steady-state operation after turning on. When the base current first becomes negative and starts pulling charge out of the base, the collector current does not change. This initially surprising result is due to the excess charge

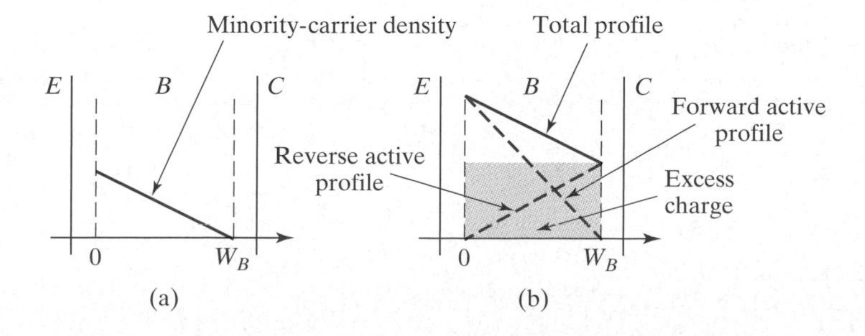

Figure 15–8 (a) The minority-carrier profile in a BJT in forward- active operation. (b) The profile in saturation, showing that it is the superposition of the forward- and reverse-active profiles.

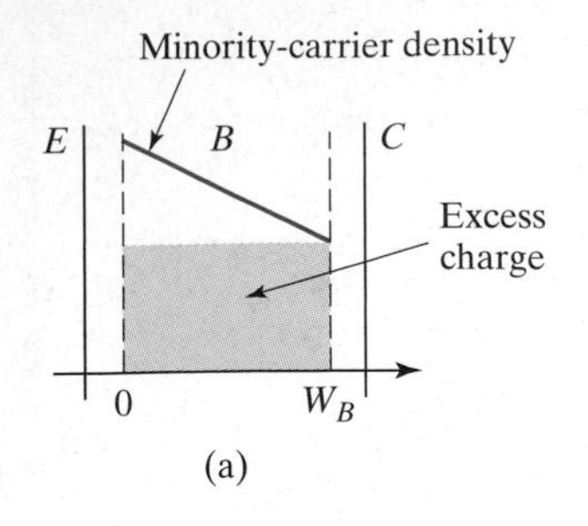

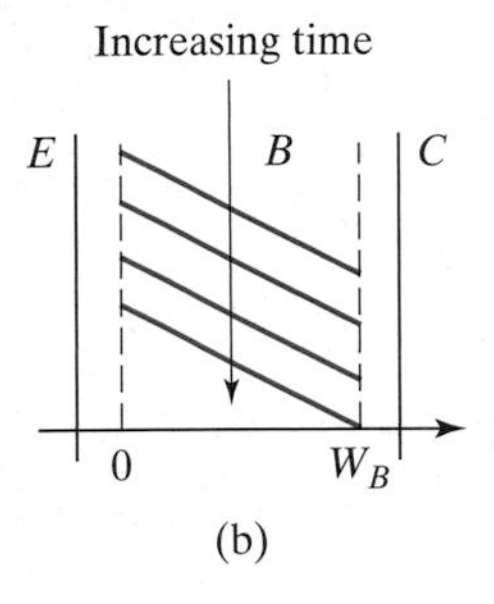

Figure 15–9 (a) The minority-carrier profile in a BJT in saturation. (b) The profile as the transistor is taken out of saturation.

stored in the base during saturation. Figure 15-9(a) shows the minority-carrier profile in the base of a transistor in saturation. As the base current starts pulling charge out of the base, the profile changes as shown in part (b) of the figure. Since the collector current is proportional to the slope of this density, the current does not start to change until all of the excess charge has been removed. The time required to remove the excess charge is called the ***storage time*** of the transistor and is denoted by t_s in Figure 15-7. It is also sometimes called the *saturation time* of the transistor.

The storage time of the transistor increases as the base overdrive factor increases, since extra base overdrive translates into more excess base charge. The storage time is reduced if we pull a larger current out of the base to turn the device off.

The ***fall time*** of the collector current is specified as the time it takes the current to fall from 90% of the saturation value to 10% of the saturation value (analogous to the rise time) and is denoted t_f, as shown in the figure. The fall time is a function of how hard we drive the transistor to turn it off.

If the ***reverse overdrive factor***[5] is reasonably large (say, > 5), we can ignore the time it takes the collector current to drop to 90% of its saturation value. Therefore, the storage time and the fall time completely specify how long it takes the transistor to turn off.

The storage time and fall time are usually the dominant times in switching a BJT into and out of saturation, because the reverse base current is typically small and we depend on recombination to do much of the work. If the transistor is driven hard into saturation (i.e., it has a small forced beta), the storage time usually dominates. We can improve this switching speed significantly by adding a capacitor in parallel with R_B in Figure 15-7(a) (*see* Problem P15.7). The capacitor will store charge that can be used to rapidly remove charge from the base of the transistor while it is turned off. This solution is not usually practical in ICs, however, because the required capacitor is too large. The best way to increase the switching speed of a BJT is to not allow it to saturate. We will explore two ways of accomplishing this later in the chapter: the Schottky clamped transistor and emitter-coupled logic.

ASIDE A15.2 BIPOLAR TRANSISTOR SWITCHING TIMES

The purpose of this aside is to derive approximate equations for the different times involved in switching a BJT into and out of saturation. We will break the analysis up into different regions determined by the physics of the device and will then handle each region separately. The situation we wish to examine is illustrated in Figure A15-2.

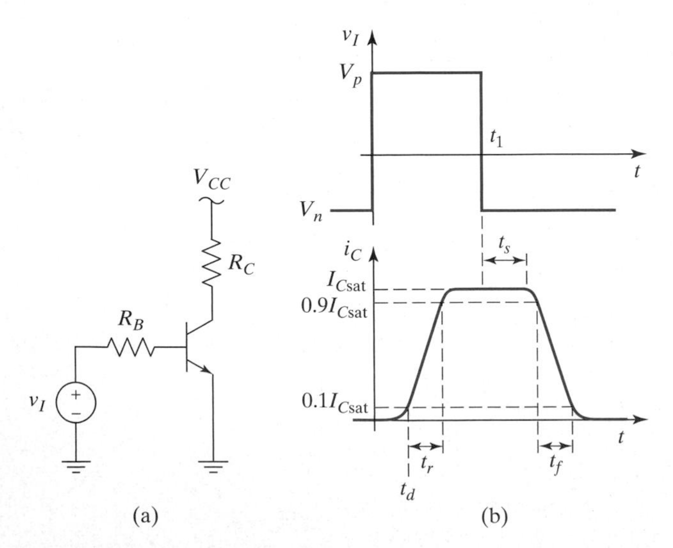

Figure A15-2 (a) A BJT switching circuit. (b) The waveforms for the circuit.

Assuming that v_I has been negative for some time, the transistor will be cut off just prior to $t = 0$. Since the base-emitter junction has capacitance, the voltage across it cannot change instantaneously unless we provide an impulse of current. Therefore, we must first determine how long it takes to charge the base-emitter and base-collector depletion regions so that the transistor is just barely on ($I_C = 0.1I_{Csat}$); this is the delay time, t_d. The collector-base junction (CBJ) is reverse biased this entire time, and we can approximate it as a constant capacitor equal to the average of the values at $t = 0$ and $t = t_d$; call this $\overline{C}_{jc}(t_d)$. The emitter-base junction (EBJ) changes from being reverse biased to being forward biased, but we can again approximate it with an average value during the period, namely, $\overline{C}_{je}(t_d)$. Using these two average capacitances in parallel and allowing for base resistance in the transistor, we find that the circuit is a series RC with $R = R_B + r_b$ and $C = \overline{C}_{jc}(t_d) + \overline{C}_{je}(t_d)$. The resulting step response yields the following expression for v_{BE}, which is valid during t_d:

$$v_{BE}(t) = V_p - (V_p - V_n)e^{-t/\tau_d}. \tag{A15.1}$$

Here

$$\tau_d = (R_B + r_b)(\overline{C}_{jc}(t_d) + \overline{C}_{je}(t_d)). \tag{A15.2a}$$

The average capacitances are found using (15.13) and the junction voltages at the beginning and end of the delay time. Since $v_C(t_d) \approx v_C(0) = V_{CC}$, we have;

$$\overline{C}_{jc}(t_d) = \frac{C_{jc0}}{2}\left(\frac{1}{\left(1 - \dfrac{V_n - V_{CC}}{V_0}\right)^{m_{jc}}} + \frac{1}{\left(1 - \dfrac{0.7 - V_{CC}}{V_0}\right)^{m_{jc}}}\right) \tag{A15.2b}$$

and

$$\overline{C}_{je}(t_d) = \frac{C_{je0}}{2}\left(2 + \frac{1}{\left(1 - \dfrac{V_n}{V_0}\right)^{m_{jc}}}\right). \tag{A15.2c}$$

We can now choose the value of v_{BE} that corresponds to $I_C = 0.1I_{Csat}$ and solve to find t_d. For example, if $I_C = 0.1I_{Csat}$ with $v_{BE} = 0.7$ V, we set (A15.1) equal to 0.7 and solve for t_d:

$$t_d = \tau_d \ln\left(\frac{V_p - V_n}{V_p - 0.7}\right). \tag{A15.3}$$

The exact value we use for v_{BE} does not change the result for t_d much, so long as V_p is much larger. Therefore, although we can't say that the collector current will actually be 10% of the saturation value when $v_{BE} = 0.7$, the value we obtain for t_d is still a reasonable estimate.

We now move on to consider the rise time of the collector current. The key approximation we make here is to say that the base current will be approximately constant during this time. Specifically,

$$\bar{i}_B(t_r) = \frac{V_p - \bar{v}_{BE}}{R_B} \approx I_{BF} = \frac{V_p - 0.7}{R_B}, \tag{A15.4}$$

where I_{BF} is the forward base current when the device reaches steady state. During the rise time the transistor is forward active, so ignoring the small change in the EBJ depletion region charge, the total charge we must supply is the sum of the change in the diffusion charge, $\Delta Q_d = \Delta I_C\tau_F = 0.8I_{Csat}\tau_F$, and the change in the CBJ depletion charge. We can approximate the change in CBJ depletion charge by using an average value for the CB junction capacitance, $\overline{C}_{jc}(t_r)$, and the relation $Q = CV$. The result is

$$\begin{aligned}\Delta Q_{jc}(t_r) &= \overline{C}_{jc}(t_r)\Delta V_{CB} = \overline{C}_{jc}(t_r)\Delta I_C R_C \\ &= 0.8R_C\overline{C}_{jc}(t_r)I_{Csat},\end{aligned} \tag{A15.5}$$

where the bias on C_{jc} is approximately $0.7 - V_{CC}$ at the start of the rise time and is near zero at the end of the rise time (since the transistor is nearly saturated) so that

$$\overline{C}_{jc}(t_r) = \frac{C_{jc0}}{2}\left(1 + \frac{1}{\left(1 - \frac{0.7 - V_{CC}}{V_0}\right)^{m_{jc}}}\right). \tag{A15.6}$$

Assuming that this total change in charge is supplied by the average base current in (A15.4) during the time t_r, and ignoring recombination in the base during this time, we find that

$$t_r \approx \frac{0.8I_{Csat}}{\bar{i}_B(t_r)}(\tau_F + R_C\overline{C}_{jc}(t_r)), \tag{A15.7}$$

or, recognizing this base current as the forced base current, we can express it in terms of the forced beta:

$$t_r \approx 0.8\beta_{forced}(\tau_F + R_C\overline{C}_{jc}(t_r)). \tag{A15.8}$$

This equation shows us that if we drive the transistor hard (i.e., with a large i_B), the rise time will be shorter, which makes good intuitive sense. If recombination in the base is included, the resulting rise time is longer, but the difference is usually small. The amount of recombination increases as the total charge in the base increases, but we can use the value at the midpoint, where $I_C = I_{Csat}/2$, as an average. If the transistor were conducting this current in steady state, the base current would be $I_B = I_{Csat}/2\beta_F$. If we assume that the current is entirely due to recombination, the actual current available to charge the transistor is $I_{BF} - I_{Csat}/2\beta_F$. If we use this base current in (A15.7), we obtain a slightly more accurate value for t_r.

Now let's examine the storage time for the transistor. The total minority-carrier profile in the base appears as shown in Figure A15-3 when the transistor is saturated. In this figure, we have divided the total profile up into two pieces: the saturation charge, Q_S, and the forward-active charge, Q_{FA}. The triangle representing the forward-active charge has a slope proportional to I_{Csat}, but the saturation charge does not contribute to the external current flow. If the saturation charge is removed and the forward-active charge moves down (without changing the slope), the external current stays the same.

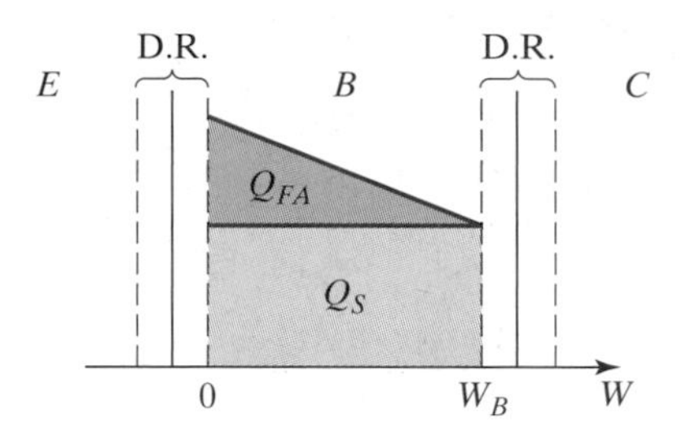

Figure A15-3 The minority-carrier profile in the base of a saturated transistor. (D.R. = depletion region.)

Since the saturation charge represents the excess charge in the base, in steady-state operation it is proportional to the excess base current:

$$I_{Bx} = I_{BF} - I_{Csat}/\beta_F. \tag{A15.9}$$

So we have

$$Q_S = I_{Bx}\tau_S, \tag{A15.10}$$

where τ_S is called the ***saturation time constant***. This time constant is analogous to τ_F and can be thought of as the average lifetime for excess base charge. Since the excess base charge is due to a superposition of the forward- and reverse-active regions of operation, the saturation time constant is a function of both the forward and reverse transit times. A detailed analysis reveals that [15.2]

$$\tau_S = \frac{\alpha_F(\tau_F + \alpha_R\tau_R)}{1 - \alpha_F\alpha_R}. \quad \textbf{(A15.11)}$$

The saturation time constant is sometimes specified on transistor data sheets for discrete transistors, but is usually determined empirically for integrated transistors. Equation (A15.11) underestimates the value of τ_S for most transistors, since it ignores reverse injection into the collector in the reverse transistor, which is quite large when the collector doping level is lower than the base, as is usually the case. Using (A15.9) through (A15.11) (or a measured value for τ_S), we can find the total excess charge stored when the transistor is saturated.

Now that we know how much charge must be removed during the storage time, we need to determine the rate at which it is removed. There are two mechanisms at work in removing this charge: recombination and the reverse base current. Immediately after the input voltage has dropped as shown in Figure A15-2(b), the base voltage has not yet changed, and the reverse base current is approximately equal to

$$I_{BR} = \frac{V_n - 0.7}{R_B}. \quad \textbf{(A15.12)}$$

In addition, we know that it would take a base current equal to I_{BF} to support the excess charge in the base; in other words, recombination removes charge at a rate equivalent to a current of magnitude I_{BF} flowing *out* of the base. Therefore, at time t_1^+, charge is being removed at a rate that is equivalent to a base current of (remember that I_{BR} is negative)

$$I_{Beq.}(t_1^+) = I_{BR} - I_{BF}. \quad \textbf{(A15.13)}$$

As soon as any excess charge is removed, however, the equivalent base current is reduced in magnitude. In fact, the rate at which charge is removed continues to decrease as the charge is removed. A detailed analysis reveals that $Q_S(t)$ is exponential for $t \geq t_1$ [15.2]:

$$Q_S(t) = \tau_S\left[\left(I_{BR} - \frac{I_{Csat}}{\beta_F}\right) + (I_{BF} - I_{BR})e^{-(t-t_1)/\tau_S}\right]. \quad \textbf{(A15.14)}$$

Note that the initial slope of (A15.14) is equal to $I_{BR} - I_{BF}$, in agreement with (A15.13). (A15.14) is plotted in Figure A15-4. The dashed line indicates that (A15.14) would predict negative saturation charge for times greater than t_s. However, the equation does not apply once the charge has gone to zero. If fact, we find t_s by setting (A15.14) equal to zero when $(t - t_1) = t_s$:

$$t_s = \tau_S \ln\left(\frac{I_{BF} - I_{BR}}{\frac{I_{Csat}}{\beta_F} - I_{BR}}\right). \quad \textbf{(A15.15)}$$

Note that in finding t_s we have not ignored the recombination of charge in the base. In fact, if $V_n = 0$, the reverse base current is very small and recombination is likely to dominate the storage time.

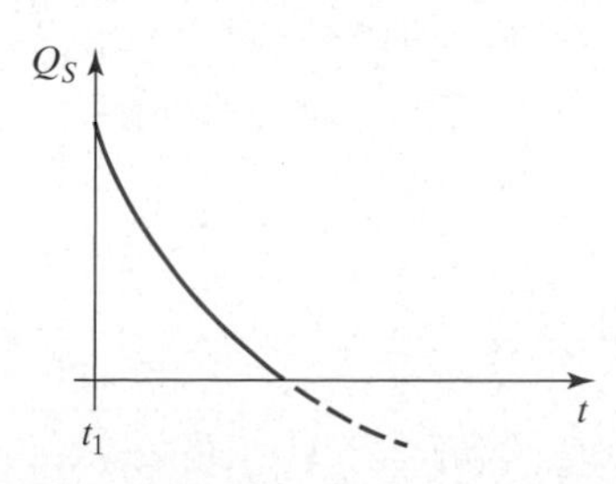

Figure A15-4 The saturation charge as a function of time for $t \geq t_1$.

Now let's examine the fall time, t_d. The total charge that must be moved during the fall time is equal to that moved during the rise time. The equivalent base current available to remove the charge is made up of the reverse base current and the average of the recombination current: $I_{Beq} = I_{BR} - I_{Csat}/2\beta_F$. Note that we cannot ignore recombination in the fall time, since the reverse current may be very small. Combining this equivalent current and the change in charge, we find that (the base current is negative—hence the change in sign)

$$t_f \approx \frac{-0.8 I_{Csat}}{I_{BR} - I_{Csat}/2\beta_F}\left(\tau_F + R_C\overline{C}_{jc}(t_r)\right). \quad \textbf{(A15.16)}$$

Finally, we note that there is again a final recovery time for the transistor to shut off completely. This time is analogous to the delay time at the beginning of the transistor turning on, except that now v_{BE} only asymptotically approaches the final value of V_n, so we can't solve the exponential equation by setting $v_{BE} = V_n$. We do know that the time constant is again given by (A15.2), though, and we can approximate this recovery time as being two or three time constants. (Remember that an exponential gets 95% of the way to its final value in three time constants.)

Example 15.3

Problem:

Find all of the times in Figure A15-2 if $V_{CC} = 5$ V, $V_p = 5$ V, $V_n = -1$ V, $R_C = 1$ kΩ, $R_B = 10$ kΩ, and the transistor has $\beta_F = 100$, $\beta_R = 1$, $r_b = 62.5$ Ω, $\tau_F = 1$ ns, $\tau_R = 50$ ns, $C_{je0} = 0.45$ pF, $C_{jc0} = 2$ pF, $m_{je} = m_{jc} = 0.33$, and $V_0 = 0.75$ V.

Solution:

Using (A15.2) along with the given data, we find that $\overline{C}_{je}(t_d) = 1$ pF, $\overline{C}_{jc}(t_d) = 0.17$ pF, and $\tau_d = 1.9$ ns. Therefore, using (A15.3) we find that $t_d \approx 4$ ns.

Next, we calculate $I_{Csat} = 4.8$ mA and use (A15.4) to obtain $I_{BF} = 0.43$ mA and (A15.6) to find $\overline{C}_{jc}(t_r) = 1.5$ pF. We plug these numbers and the given data into (A15.7) to find $t_r \approx 23$ ns.

The storage time is found by using (A15.11) to yield $\tau_S = 51$ ns, (A15.12) to yield $I_{BR} = -0.17$ mA, and (A15.15) to yield $t_s = 52$ ns.

The fall time is found using (A15.16): $t_f \approx 49$ ns.

A SPICE simulation of this circuit using the given model parameters yields the output shown in Figure 15-10. From this output, we determine that $t_d = 8$ ns, $t_r = 24$ ns, $t_s = 50$ ns, and $t_f = 48$ ns. These values are reasonably close to our calculated estimates.

[5] The reverse overdrive factor is the ratio of the magnitude of the reverse base current to I_{Bmin}, in perfect analogy to (15.12).

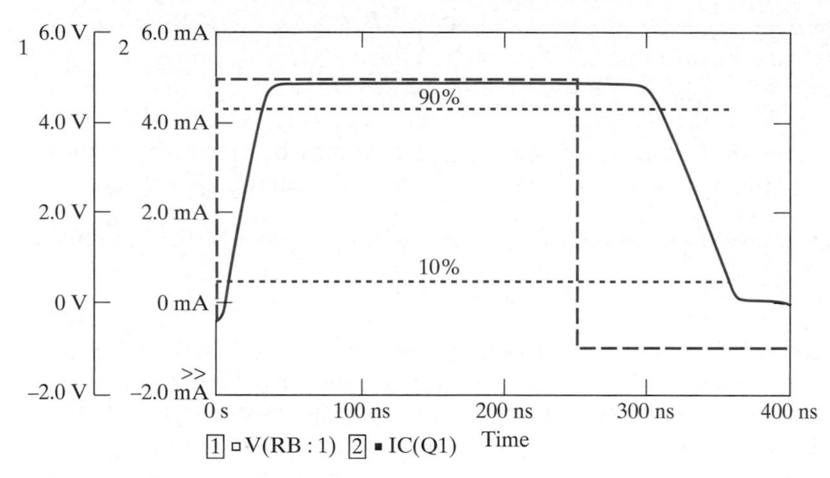

Figure 15–10 SPICE output

■

Exercise 15.3

Consider a BJT switching out of saturation with a circuit like that shown in Figure 15-7(a). Would you expect the storage time to be significantly reduced if the value of R_B was decreased while keeping everything else the same?

□

15.1.3 MOS Field-Effect Transistors

In this section, we examine models used for MOS transistors in digital circuits. We begin with the DC, or static, model and then discuss the model used for transient analyses.

DC Models (Static Models) The low-frequency MOSFET models presented in Section 7.1.5 apply equally well to digital circuit analysis and will not be repeated here. You should be familiar with the material in that section before proceeding with this chapter. However, we do need to make one addition to those models: For the analyses presented in this chapter, we often need to include the channel-length modulation effect in the large-signal model for the MOSFET in saturation.

Note that the drain current of a MOSFET in saturation can be written as

$$I_D = K(V_{GS} - V_{th})^2(1 + \lambda V_{DS}) = K(V_{GS} - V_{th})^2 + \lambda K(V_{GS} - V_{th})^2 V_{DS}$$
$$= K(V_{GS} - V_{th})^2 + \frac{V_{DS}}{R_{DS}}. \quad \textbf{(15.16)}$$

This equation shows that a saturated MOSFET looks like a controlled current source in parallel with a resistance given by

$$R_{DS} = \frac{1}{\lambda K(V_{GS} - V_{th})^2}, \quad \textbf{(15.17)}$$

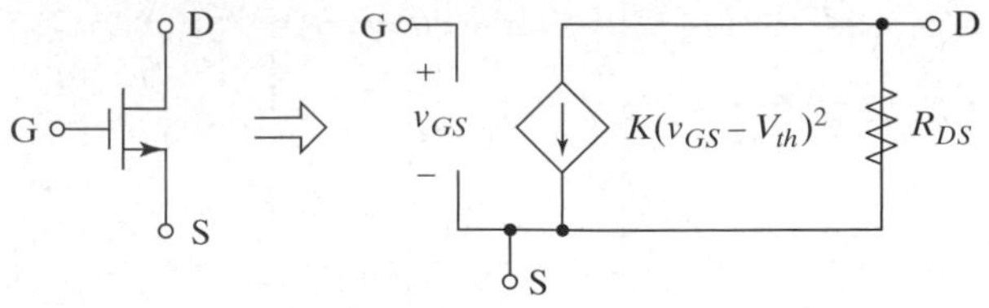

Figure 15–11 A model for a MOSFET in saturation, including the effect of channel-length modulation.

as shown in Figure 15-11. If you are familiar with small-signal analysis for linear circuits, you may recognize that this is the small-signal output resistance of the MOSFET. The only reason a small-signal model parameter will work for digital circuit analysis, even though the swing is definitely not small, is that the model used for the MOSFET output resistance is a linear model (i.e., the factor $(1 + \lambda V_{DS})$ modifying the basic drain current is linear in V_{DS}).

Switching Transients in MOSFETs (Dynamic Models)[6] The AC model for a MOSFET is presented in Section 9.1.6, so you should be familiar with that section before proceeding with the MOS circuits in this chapter. The most important distinguishing characteristic of MOSFETs in terms of their dynamic behavior in digital circuits is that they are majority-carrier devices and do not, therefore, exhibit minority-carrier charge storage. The most important capacitances to consider are the gate-to-channel capacitance and the source- and drain-to-substrate capacitances. In addition, there is a small parasitic capacitance from the gate to the substrate, which we will ignore here. The gate-to-channel capacitance is linear, so except for the fact that its connections to the source and drain are different in the saturation and linear regions, it is straightforward to deal with. The junction capacitances are usually dealt with in hand analysis by using an average value. We demonstrate how to do transient analysis in MOSFET circuits as a part of our analysis of CMOS inverters in Section 15.3.

15.2 SPECIFICATION OF LOGIC GATES

In this section, we first present definitions relating to the static performance of logic gates and then go on to definitions relating to the dynamic performance. Because the simplest possible logic gate is an inverter, we use the inverter whenever possible to define the parameters.

15.2.1 Static Specifications

Logic Swing and Noise Margins An inverter is shown in Figure 15-12(a), and its static, or DC, transfer characteristic is shown in part (b) of the figure. The static transfer characteristic is also sometimes called the voltage-transfer characteristic (VTC). Because we want the outputs of a logic gate to be unambiguously high or

[6] A more detailed discussion of switching transients is available in [15.1] and [15.2].

low, we want to avoid operating on the part of the characteristic where V_O changes significantly with changing v_I. We therefore define our logic levels and valid input ranges by finding the two points on the transfer characteristic where the slope is -1, as shown in the figure[7].

With these definitions, the output high level, V_{OH}, is guaranteed to be the minimum high level this gate will output so long as the input is a valid low. A valid low input is defined as any input voltage less than or equal to V_{IL}. Similarly, the output low level, V_{OL}, is guaranteed to be the maximum low level the gate will output so long as the input is a valid high. A valid high input is defined as any input voltage greater than or equal to V_{IH}. We do not indicate in our notation that these values are maximums and minimums, because we want to reserve that notation for the maximum and minimum values that will be seen over process, temperature, and supply variations.

Using the logic levels as defined we can further define the ***logic swing*** to be

$$\text{Logic swing} = V_{OH} - V_{OL}, \qquad (15.18)$$

and the ***transition region*** to be

$$\text{Transition region} = V_{IH} - V_{IL}. \qquad (15.19)$$

We would like our logic swing to be large enough that it is always clear whether a given signal is high or low, so ignoring power dissipation and other factors, we would like the logic swing to be large. On the other hand, we would like our transition region to be as small as possible, since it is the range of inputs for which the output is neither a valid high nor a valid low. Manufacturers usually specify the minimum values of V_{OH} and V_{IH} and the maximum values of V_{OL} and V_{IL}, which specify the minimum logic swing and transition region.

Now consider what happens when one inverter (or some other logic gate) drives another, as shown in Figure 15-13(a). In this case, we will assume that the two inverters are nominally identical and therefore have the same specifications (e.g., V_{OH}).

Consider, for example, the situation when V_{OA} is high. We would like to know that gate B will always properly interpret this output. If we can guarantee that $V_{IH} < V_{OH}$

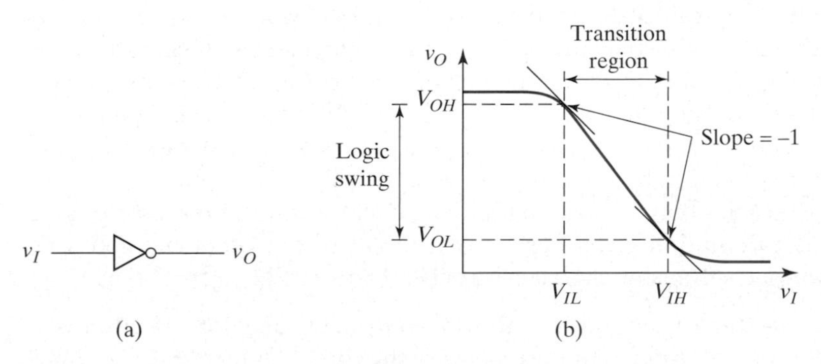

Figure 15–12 (a) An inverter. (b) The DC transfer characteristic.

---

[7] If the magnitude of the incremental gain of the inverter (dV_O/dV_I) is greater than unity, a disturbance at the input will be amplified and may cause an error after a string of similar logic gates.

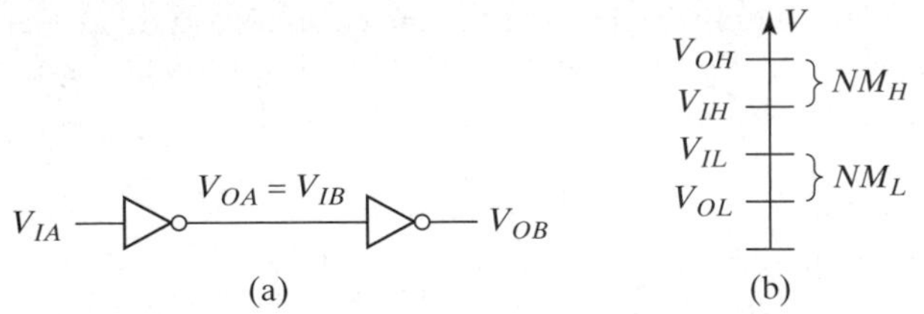

Figure 15–13 (a) Two gates in series. (b) A one-dimensional plot of the input and output voltage specifications.

under all circumstances, then the gates will function properly for high levels. Similarly, if $V_{IL} > V_{OL}$ under all circumstances, the gates will function properly for low levels, too. We can picture the situation graphically by showing the input and output levels on a one-dimensional plot, as in part (b) of the figure.

Noise and other disturbances are always present in real circuits, so we want to guarantee that these logic gates work together properly even in the presence of noise. In other words, we need to be sure that even if some noise is added to or subtracted from the output voltage of gate A, gate B will still properly interpret the signal. When the output of A is high, we can subtract noise from this output, and B will still properly interpret the signal, so long as the noise is smaller than the difference between V_{OH} and V_{IH}. Therefore, we define the ***high noise margin*** as

$$NM_H = V_{OH} - V_{IH}. \tag{15.20}$$

Similarly, the difference between V_{IL} and V_{OL} is a measure of the amount of noise that can be added to V_{OL} before it gets misinterpreted. Therefore, we define the ***low noise margin*** as

$$NM_L = V_{IL} - V_{OL}. \tag{15.21}$$

In general, larger noise margins are desirable, although a trade-off is necessary. Large noise margins require a large logic swing, which yields longer transition times (i.e., slower operation) and higher power dissipation.

Noise margins can be defined differently than we have done here. One common definition is to still use (15.20) and (15.21), but to use the nominal values of V_{OH} and V_{OL} instead of the values at the unity-slope points as we have done. The definition used here has two distinct advantages: It has a logical basis (that the incremental gain of the gate should be less than unity when being driven properly), and it yields conservative values for the noise margins. Both of these techniques fail to work properly if the transfer characteristic of the gate is highly asymmetric [15.4], and other methods should be used in that case. We will only examine gates with transfer characteristics that are symmetric or very close to it. In the case of symmetric gates, the definition given here is equivalent to the other methods preferred for asymmetric transfer characteristics [15.4 and 15.5].

Fan-In and Fan-Out In digital circuits, the output of one gate is often connected to the input of more than one gate as shown in Figure 15-14. ***Fan-out*** refers to the maximum number of gates that can be driven by a gate, while remaining within the guaranteed specifications. The manufacturer usually specifies the fan-out number N. Similarly, logic gates often have more than one input. In the simplest case, the number of inputs is called the ***fan-in***. There are cases, however, where more than one gate might drive a single input. In this case, fan-in refers to the maximum number of gates that can drive a gate while

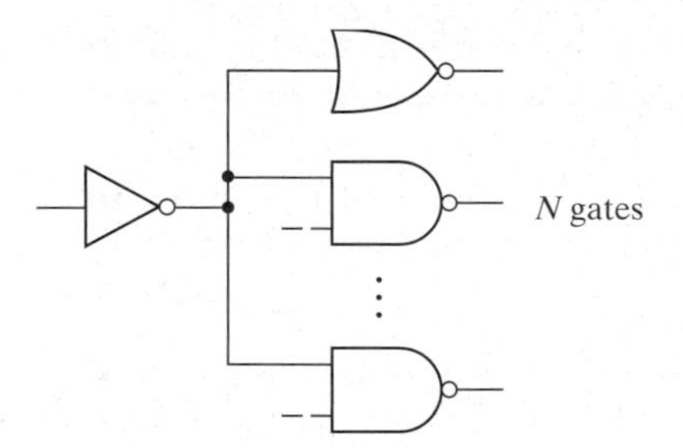

Figure 15–14 Illustration of fan out.

remaining within the guaranteed specifications. We will discuss this issue more when we get to specific examples.

Static Power Dissipation Some logic gates have currents that flow even when the inputs and outputs are constant. This current flow leads to power dissipation (which is wasted as heat) and is not desirable. For an inverter, the power dissipated when the output is high will not generally be the same as the power dissipated when the output is low. We define the ***average static power dissipation*** of an inverter as the average of the dissipation when V_O is high and when V_O is low. We calculate this power for each of the different kinds of logic gates we consider.

15.2.2 Dynamic Specifications

Propagation Delay and Transition Times When the input to an inverter changes, the output will change in response, as shown in Figure 15-15. We assume that the nominal input and output levels are the same and denote them by V_{OH} and V_{OL}.[8] As indicated in the figure, the output will not respond immediately to the change in input. It takes some time for the change to propagate to the output. The standard measure for this ***propagation delay*** is to take the difference between the times when the input and output waveforms pass through their midpoints, as shown on the figure. There are two different propagation delays in general: the delay when the output is going from low to high, denoted t_{PLH}, and the delay when the output is going from high to low, denoted t_{PHL}. The propagation delay is often quoted as the average of these two times:

$$t_P = \frac{t_{PHL} + t_{PLH}}{2}. \tag{15.22}$$

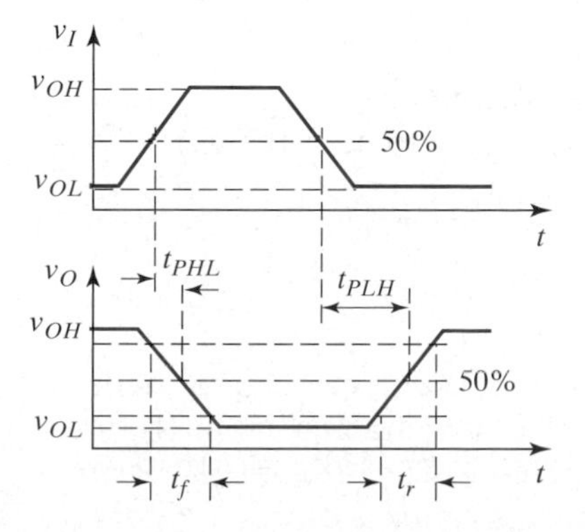

Figure 15–15 Illustration of propagation delay and transition times.

[8] We will use V_{OH} and V_{OL} to denote the nominal values except when calculating noise margins.

The outputs of logic gates cannot switch from one state to another infinitely fast. Therefore, in addition to the propagation delay, we specify the ***transition times***, as illustrated in Figure 15-15. There are two transition times: the ***rise time*** and the ***fall time***, respectively denoted t_r and t_f in the figure. These transition times are usually specified as the time required for the voltage to change from 10% of the total swing to 90% of the total swing.

Dynamic Power Dissipation In addition to the static power dissipation discussed previously, logic gates dissipate power during the transitions from one state to another. This power is a function of the rate at which these transitions occur and is called ***dynamic power dissipation***. Typically, the average power dissipated in resistive elements (if there are any) during the transition times is nearly equal to the average of the static power dissipated in the high and low output states. This fact, combined with the fact that the transition times are frequently a small percentage of the overall clock period, implies that the power dissipated in resistors is included in the static power calculations. Therefore, we concentrate here on the power dissipated while charging and discharging a capacitive load,[9] as shown in Figure 15-16(a). In this figure, we do not show the details of the circuitry driving the capacitive load (indicated as a box with an X through it), since we don't need to know them. It is sufficient for our present purposes simply to require that the circuit not contain any energy storage elements.

Consider a positive output transition first (i.e., from low to high). In this case, we can use the circuit shown in part (b) of the figure. Any paths to ground in the circuit other than those through the load capacitor are assumed to have been taken into account when we calculated the static power, so in this case all of the current through the circuit goes into the load. (This is the best case possible, any other paths for current to flow to ground would waste power.) The capacitor voltage changes from V_{OL} to V_{OH}, and all of the current flowing into the capacitor must go through the circuit. The energy supplied by V_{DD} during this positive transition, denoted $E_{V_{DD}}(L \rightarrow H)$, is given by the integral of the instantaneous power, $P(t) = V_{DD} i(t)$:

$$E_{V_{DD}}(L \rightarrow H) = V_{DD} \int_0^{t_{LH}} i(t)dt. \qquad \textbf{(15.23)}$$

here, t_{LH} is the time required for the positive transition, and we assume that it starts at $t = 0$. Since we are not allowing any energy storage in X, and the current

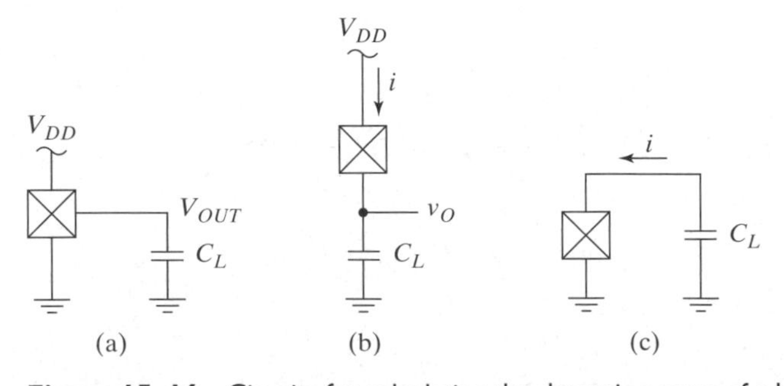

Figure 15–16 Circuits for calculating the dynamic power of a logic gate with a capacitive load. (a) The gate (the block marked X) cannot contain any energy storage elements. (b) The circuit for considering the low-to-high transition. (c) The circuit for considering the high-to-low transition.

---

[9] We ignore inductors, since they are not used in any of the digital circuits we will study.

all flows into C_L, we know that the integral in (15.23) equals the total charge delivered to the load capacitor. In other words,

$$\int_0^{t_{LH}} i(t)dt = \Delta Q_{C_L} = C_L \Delta V_O = C_L(V_{OH} - V_{OL}). \tag{15.24}$$

Using (15.24) in (15.23), we find the total energy supplied by V_{DD} during the positive transition:

$$E_{V_{DD}}(L \rightarrow H) = C_L V_{DD}(V_{OH} - V_{OL}). \tag{15.25}$$

The total energy supplied to the capacitor during this time is given by (remember that the energy stored on a capacitor is $CV^2/2$)

$$E_{C_L}(L \rightarrow H) = \frac{C_L}{2}(V_{OH}^2 - V_{OL}^2). \tag{15.26}$$

Finally, the energy dissipated in the circuit X during the positive transition is the difference between the energy supplied by V_{DD} and the energy delivered to C_L:

$$\begin{aligned} E_X(L \rightarrow H) &= E_{V_{DD}}(L \rightarrow H) - E_{C_L}(L \rightarrow H) \\ &= C_L V_{DD}(V_{OH} - V_{OL}) - \frac{C_L}{2}(V_{OH}^2 - V_{OL}^2). \end{aligned} \tag{15.27}$$

We can now find the energy dissipated in X during a negative transition (i.e., from high to low) by using the circuit in part (c) of Figure 15-16. We again assume that any extraneous paths for current have been accounted for in calculating the static power, so that all of the current flows through the load. Since the current through the circuit is same as the current through the load and the voltage across the circuit is the same as the voltage across the load, the total energy dissipated in the circuit is the same as the energy supplied by the load:

$$E_X(H \rightarrow L) = E_{C_L}(H \rightarrow L) = \frac{C_L}{2}(V_{OH}^2 - V_{OL}^2). \tag{15.28}$$

If the output of the gate alternates between high and low periodically, we can find the average power consumed by the gate by first finding the energy dissipated in the gate during one cycle and then dividing this energy by the time per cycle. The total energy dissipated in the circuit during one cycle (i.e., the output goes from low to high and then from high to low) is the sum of the terms in (15.27) and (15.28):

$$\begin{aligned} E_X(\text{one cycle}) &= E_X(L \rightarrow H) + E_X(H \rightarrow L) \\ &= C_L V_{DD}(V_{OH} - V_{OL}). \end{aligned} \tag{15.29}$$

If this process is repeated every T seconds (i.e., the switching frequency is $f_{SW} = 1/T$), the average dynamic power consumed by the gate is

$$P_{\text{dynamic}} = \frac{E_X(\text{one cycle})}{T} = f_{SW} E_X(\text{one cycle}). \tag{15.30}$$

If we examine the common situation where V_{OH} is equal to the positive supply rail (i.e., V_{DD}) and V_{OL} is ground, we find from (15.29) that

$$E_X(\text{one cycle}) = C_L V_{DD}^2, \tag{15.31}$$

and combining this result with (15.30) we get the classic result for dynamic power dissipation:

$$P_{\text{dynamic}} = f_{SW} C_L V_{DD}^2. \tag{15.32}$$

Power–Delay Product A common *figure of merit* for comparing logic gates is the *power–delay product* (PDP), given by multiplying the total power (i.e., the sum of the static and dynamic powers) by the propagation delay:

$$\text{PDP} \equiv P_{\text{total}} t_p = (P_{\text{static}} + P_{\text{dynamic}}) t_p. \tag{15.33}$$

A figure of merit (FOM) is a specification that can be used to compare the relative merits of different approaches to solving some design problem or the relative performance of different designs. The FOM is often expressed as a function of two or more specifications involved in a trade-off. In this case, the trade-off is between power and speed. Since achieving higher speed is desirable, but usually requires increased power consumption, which is not desirable, the product of power and delay is a good FOM to use.

The PDP has the units of energy. Since we would like our logic to be fast and low power, the lower the PDP is, the better. Typical PDPs for modern logic gates are on the order of tens of femtojoules (femto $= 10^{-15}$), although they can be considerably higher and do depend on the external loads and the speed of operation of the circuit.

15.3 MOS DIGITAL CIRCUITS

The first commercially successful logic family was made using bipolar transistors.[10] The early forms of bipolar logic suffered from large power dissipation, however, so MOS alternatives were introduced. The first MOS logic to enjoy widespread commercial success used NMOS circuits exclusively (e.g., the first microprocessor, the Intel 4004, introduced in 1972). But NMOS gates consume significant amounts of static power, as we will demonstrate. During the 1980s, CMOS became the dominant technology for digital and mixed-signal circuits. There are a number of reasons for this dominance, including the facts that CMOS logic has lower power dissipation, rail-to-rail logic swings (i.e., the output can swing from the positive supply voltage, or "rail" to the negative supply voltage), and small-size gates. In addition, MOS transistors make good switches and have high input resistance; both of these characteristics are useful for analog circuits.

We begin this section with a brief discussion of an NMOS inverter. We then move on to discuss CMOS logic gates in some detail and close with an introduction to MOS memories. We will almost always use a single supply in drawing our schematics, so we connect the gates between V_{DD} and ground. It should be understood that ground can be replaced with a negative power supply, which is called V_{SS} in MOS circuits, since it usually connects to the source terminals of NMOS devices. We will only cover so-called *static logic* gates in this chapter. This simply means that when a given input is presented to the gate, the output will be available some short time later without the need to switch the operating mode of the

[10] TTL, introduced in the early 1960s.

gate. *Dynamic logic*, on the other hand, has different precharge and evaluation phases, and the output is not always valid. Dynamic logic will not be addressed here. A good presentation can be found in [15.7].

15.3.1 NMOS

We can construct an MOS inverter, using an enhancement-mode transistor and a load resistor as shown in Figure 15-17(a). The drain characteristics of the transistor are shown in part (b) of the figure, along with the ***load line*** of the circuit, which we derive next.

We are interested in determining the DC transfer characteristic of this gate, and we assume that the output is taken from the drain of the transistor. We first note that $V_{GS} = V_I$ and $V_O = V_{DS}$. We then write KVL for the drain loop as[11]

$$V_{DD} - I_D R_D - V_{DS} = 0 \quad \textbf{(15.34)}$$

and rearrange this equation to obtain an equation for the drain current as a function of V_{DS}:

$$I_D = \frac{V_{DD}}{R_D} - \frac{V_{DS}}{R_D}. \quad \textbf{(15.35)}$$

This equation describes the restriction placed on the values of I_D and V_{DS} by the circuit external to the transistor. Since this circuit can be thought of as the load presented to the transistor, we call the plot of (15.35) the load line. The transistor also restricts the values of I_D and V_{DS}, as indicated by the drain characteristics for the device. The transfer characteristic for this circuit is obtained by varying $V_I = V_{GS}$ and then finding the value of $V_O = V_{DS}$ corresponding to each value of V_I. For each value of V_I, we are restricted by the transistor to a characteristic curve like one of those shown in part (b) of the figure. (There are an infinite number, of course; we show only those for a few discrete values of V_{GS}.) The intersection of

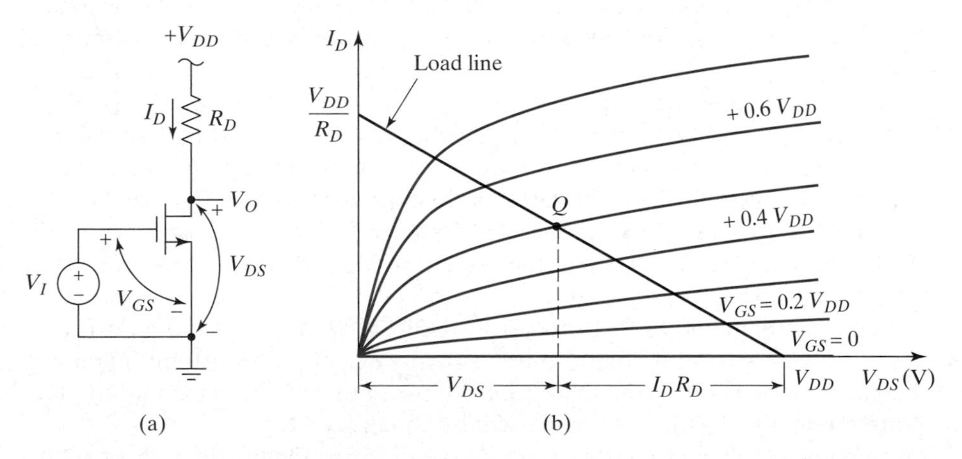

Figure 15–17 (a) An NMOS inverter circuit. (b) The drain characteristics of the transistor and the load line of the circuit.

[11] To see that a "loop" exists, remember that V_{DD} is a DC supply connected from ground to the point shown in the schematic. Therefore, current flows up from ground, through the V_{DD} source, and then down through R_D and the transistor to return to ground.

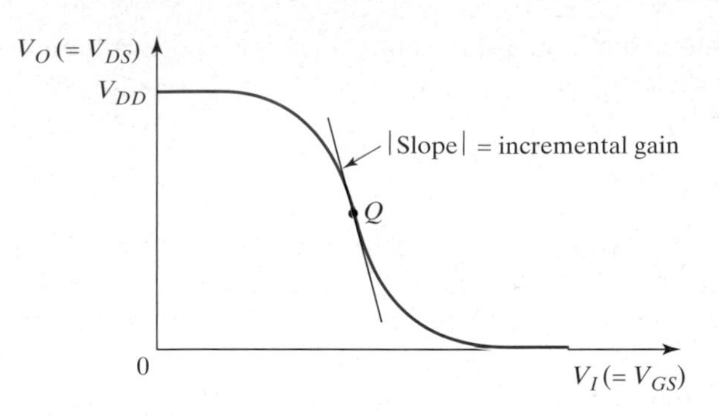

Figure 15–18 The DC transfer characteristic of the circuit in Figure 15-17.

this curve with the load line represents the simultaneous solution of the equations describing the circuit and the transistor. If we sweep V_I and find V_O in this manner, we can plot the transfer characteristic shown in Figure 15-18.

We note that the slope of the transfer characteristic in the linear region near Q, which is the incremental, or small-signal, gain of the device, is larger for larger values of R_D. We have used the letter 'Q' for this point because if the circuit were used as an amplifier[12], it would be the operating point when the signal was zero; in other words, it would be the ***quiescent*** operating point.

Although it might be reasonable to implement the circuit shown in Figure 15-17 with discrete devices, most transistor-level logic design is done for integrated circuits. Resistors in integrated circuits consume large areas (sometimes as much as several thousand transistors!) and are usually avoided if they can be replaced by transistors. There are other problems with this gate as well. First, it consumes static power when the output is low, even if the load is only a capacitor. Second, there is a trade-off between this static power consumption and the propagation delay that practically prevents the gate from being very fast (*see* Problem P15.17).

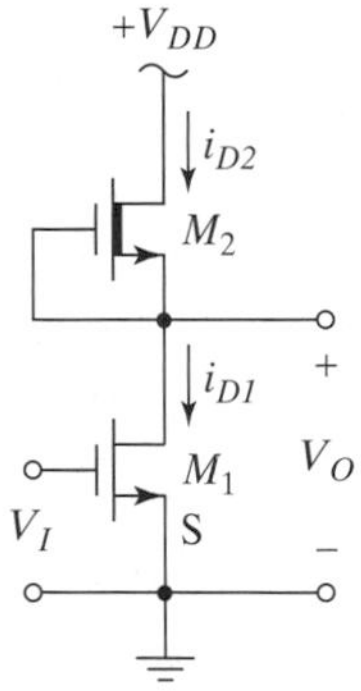

An NMOS inverter circuit that does not use a resistive load is shown in Figure 15-19. A depletion-mode NMOS transistor is used as the load in this gate instead of a resistor. Most MOS processes do not have depletion-mode devices available, so this circuit is not common anymore. Nevertheless, it is worth briefly examining how it works.

Figure 15–19 An NMOS inverter with a depletion-type load.

The DC transfer characteristic of the inverter can be constructed graphically from the characteristics of the driver M_1 (an enhancement-type NMOS transistor) and the load M_2 (a depletion-type NMOS transistor). In parts (a) and (b) of Figure 15-20, we show the drain characteristics of the enhancement-type and depletion-type transistors, respectively. The W/L ratio of the load transistor is usually made smaller than the W/L ratio of the drive transistor in order to obtain a steep transfer characteristic (i.e., high gain).

From the circuit in Figure 15-19, we note that $V_I = V_{GS1}$, $V_{GS2} = 0$ and $V_O = V_{DS1} = V_{DD} - V_{DS2}$. We further note that if there isn't any current drawn from the output of the gate (for example, if the load is purely capacitive and we examine the DC response), then the two drain currents will be equal: $I_{D1} = I_{D2}$.

We can now graphically construct the DC transfer characteristic of this gate by using the $V_{GS} = 0$ curve for the depletion-mode device, and the family of curves for the enhancement-mode device, as shown in Figure 15-21(a). Notice that $V_{DS1} = V_{DD} - V_{DS2}$, so the curve for M_2 is flipped horizontally and moved to the right by V_{DD}. For any value of V_I,we use the corresponding curve from the family

[12] Those of you familiar with amplifier design will recognize that this circuit is a common-source amplifier.

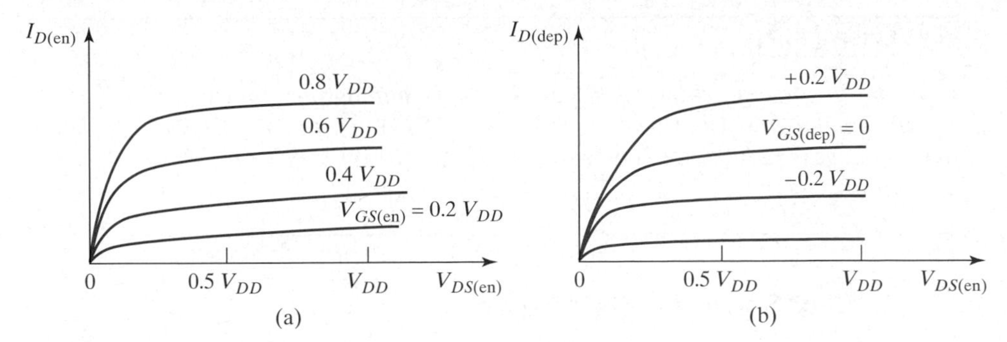

Figure 15–20 Drain characteristics of (a) the enhancement-mode NMOS transistor and (b) the depletion-mode NMOS transistor.

of curves for M_1 and find V_O as the intersection of that curve and the characteristic for M_2. In essence, the characteristic curve for M_2 with $V_{GS} = 0$ is the load line for M_1. The resulting transfer characteristic is shown in part (b) of the figure. Note especially that when the output of the gate is low (point *D* on the curve), there is a constant current flowing. In other words, the static power dissipation of the gate is not zero. We will see in the next section that the static power dissipation of CMOS gates is zero. This fact, along with smaller propagation delays and other advantages, has led to CMOS becoming the dominant technology in use today.

It is also possible to construct NMOS gates with enhancement-mode loads. NMOS logic has the advantage of using a relatively simple process that results in small transistors and, therefore, circuits that consume less area. Because the circuits use less area and are simpler to process, they are less expensive than circuits fabricated in CMOS. As noted before, however, NMOS circuits have poorer performance than CMOS, especially when driving off-chip loads. Because CMOS has become the dominant technology in use today, we will not pursue NMOS further. If you are interested in a more in-depth treatment of NMOS, we refer you to [15.6] or [15.2].

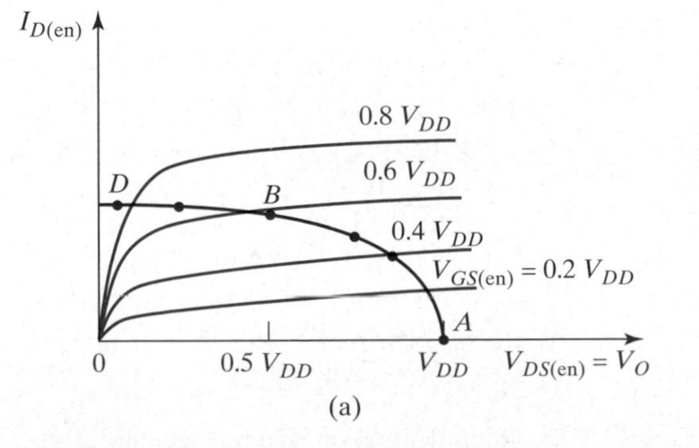

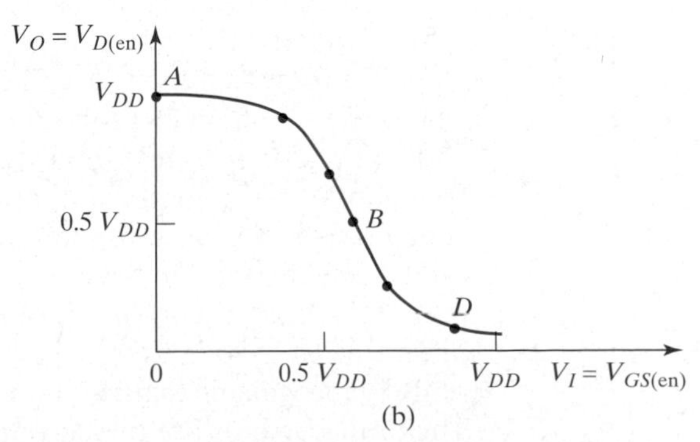

Figure 15–21 (a) The drain characteristics of M_1 with the load characteristic for M_2 (i.e., the curve at $V_{GS2} = 0$). (b) The resulting DC transfer characteristic of the gate.

Exercise 15.4

Consider the inverter shown in Figure 15-17(a), and assume that $V_{DD} = 3$ V, $R_D =$ 150 kΩ and the transistor has $V_{th} = 0.8$ V, $K = 25\ \mu A/V^2$, and $\lambda = 0.05\ V^{-1}$. (a) What input voltage is required to achieve $V_{DS} = 1.5$ V? (b) What is the incremental gain at this point?

15.3.2 CMOS Inverter (NOT gate)

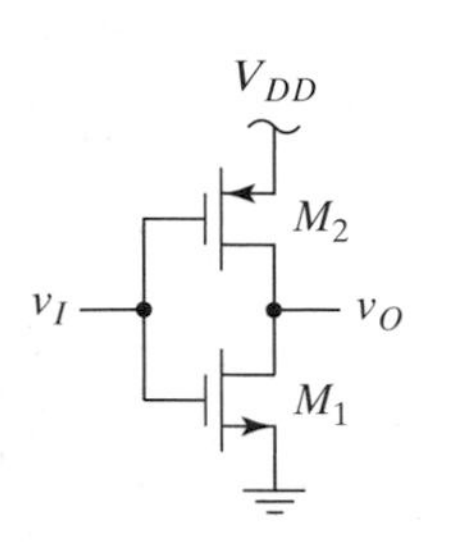

Figure 15–22 A CMOS inverter.

Static Transfer Characteristic We start by deriving the static transfer characteristic of the CMOS inverter shown in Figure 15-22. To simplify the analysis and concentrate our attention on the essential features, we initially assume that the devices are perfectly matched, so that $V_{thN} = |V_{thP}| = V_{th}$, $K_N = K_P = K$, and $\lambda_N = \lambda_P = \lambda$. This assumption is not necessary to the operation of the circuit, although the performance will be closer to ideal if the circuit meets these conditions. We also assume that V_{th} is less than $V_{DD}/2$. This second assumption is essential to the analysis, but it is always true. Note that because the surface mobility of electrons is about 2.5 times the surface mobility of holes in silicon, the PMOS transistors need to be about 2.5 times the size of the NMOS transistors in order to achieve $K_N = K_P$ as desired; in other words,

$$\left(\frac{W}{L}\right)_{\text{PMOS}} \approx 2.5\left(\frac{W}{L}\right)_{\text{NMOS}}. \tag{15.36}$$

Finally, we have also ignored the substrate connections in the figure. It is implicitly assumed that the body of the NMOS device is connected to ground and the body of the PMOS device is connected to V_{DD} so that the bulk-to-source voltages for both devices are zero. This assumption is virtually always true for CMOS inverters, but we will have to reconsider the substrate connections when we examine more complex logic gates. We begin our analysis by providing a simple description of the circuit operation and then follow with a detailed derivation of the DC transfer characteristic.

If the input voltage is equal to V_{DD}, the NMOS transistor will be on (either in saturation or in the linear region) and the PMOS transistor will be off. So long as there isn't any DC current in the load, the NMOS transistor will pull the output all the way down to ground and the steady-state low output voltage will be zero. Similarly, when the input voltage is zero, the PMOS transistor will be on, the NMOS will be off, and the output voltage will be equal to V_{DD}.

The static (or DC) transfer characteristic of this circuit is found as the locus of all points for which $I_{DN} = I_{DP}$. In other words, each point of the transfer characteristic is the simultaneous solution of two equations, one relating I_{DN} to V_I and V_O, and one relating I_{DP} to V_I and V_O. We must be careful to use the equations corresponding to the regions of operation the devices are in at a given point on the characteristic. Because of the symmetry in the circuit, it is reasonable to assume (and we will soon confirm) that the transfer characteristic will go through the point $V_I = V_{DD}/2$, $V_O = V_{DD}/2$. For these voltages we see that both transistors will be in saturation so long as $V_{th} < V_{DD}/2$. We can write equations for the two drain currents by recognizing that $V_{GSN} = V_I$, $V_{DSN} = V_O$, $V_{SGP} = V_{DD} - V_I$, and $V_{SDP} = V_{DD} - V_O$. The drain currents are then given by

$$I_{D1} = K(V_I - V_{th})^2(1 + \lambda V_O) \tag{15.37}$$

and

$$I_{D2} = K(V_{DD} - V_I - V_{th})^2(1 + \lambda[V_{DD} - V_O]) \quad \textbf{(15.38)}$$

Setting these two equations equal to each other and substituting $V_I = V_{DD}/2$ leads to $V_O = V_{DD}/2$, as expected. We can draw the DC transfer characteristic of the gate starting from this middle point. To determine the slope of the characteristic as it passes through the midpoint we use the equivalent circuit shown in Figure 15-23, where the transistors have been replaced by the large-signal model presented in Figure 15-11.

Examining this circuit, we write

$$\begin{aligned} V_O &= K(R_{DS1}\|R_{DS2})\left[(V_{DD} - V_I - V_{th})^2 - (V_I - V_{th})^2\right] \\ &= K(R_{DS1}\|R_{DS2})(V_{DD} - 2V_{th})(V_{DD} - 2V_I). \end{aligned} \quad \textbf{(15.39)}$$

This equation is valid so long as both transistors remain in saturation. The equation shows that the output voltage is a linear function of the input. If we consider the ideal case, λ goes to zero (i.e., there isn't any channel-length modulation) and the drain resistance goes to infinity. In this ideal case, the slope of V_O versus V_I is infinite. We use this result to draw the middle portion of the DC transfer characteristic shown in Figure 15-24.

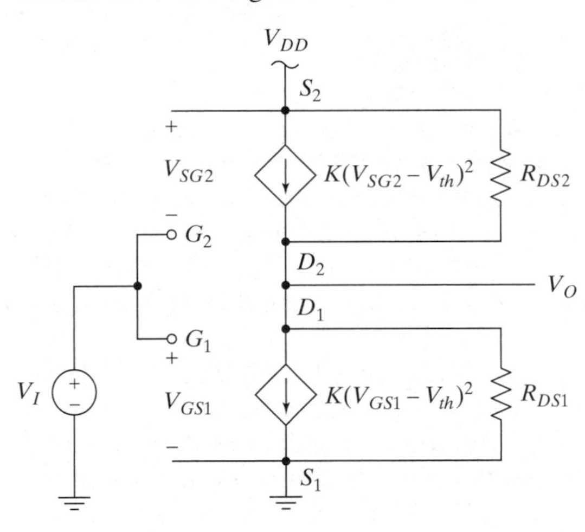

Figure 15–23 The CMOS inverter modeled for the region where both transistors are in saturation ($K_N = K_P = K$, $V_{thN} = |V_{thP}| = V_{th}$, and $\lambda_N = \lambda_P = \lambda$).

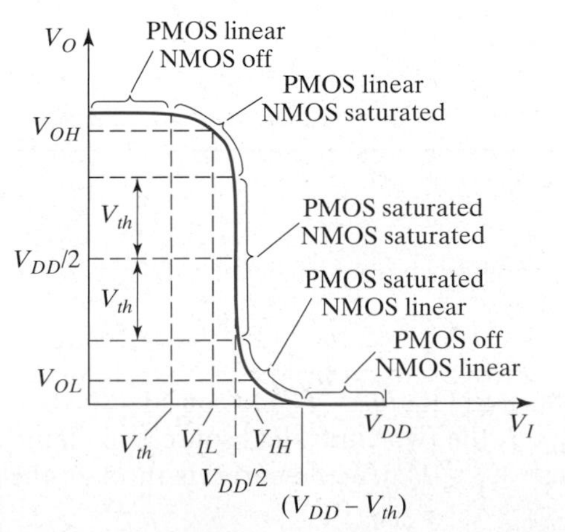

Figure 15–24 The DC transfer characteristic of an ideal symmetric CMOS inverter.

The next task in generating the DC transfer characteristic is to determine how far V_O can change before either one of the transistors leaves the saturation region. Consider the NMOS transistor first. When the drain voltage falls to a threshold below the gate, then an inversion layer will form at the drain end of the channel. In other words, the device will no longer be saturated and will enter the linear region. When this happens, the NMOS transistor can be modeled by a resistor. The resistor is nonlinear, but the current through it is given by (7.17) with the appropriate substitutions ($V_{GS1} = V_I$ and $V_{DS1} = V_O$):

$$I_D = K\left[2(V_I - V_{th})V_O - V_O^2\right]. \tag{15.40}$$

The effective drain-to-source resistance of the NMOS transistor in this region is given by the ratio of V_{DS} ($= V_O$) to I_D:

$$R_{DS1} = \frac{V_O}{I_D} = \frac{1}{K\left[2(V_I - V_{th}) - V_O\right]}. \tag{15.41}$$

With the NMOS transistor in Figure 15-23 replaced by the resistor given by (15.41), we can write

$$V_O = K(V_{GS2} - V_{th})^2(R_{DS1}\|R_{DS2}) = K(R_{DS1}\|R_{DS2})(V_{DD} - V_I - V_{th})^2. \tag{15.42}$$

From this equation, we can see that as V_I increases, V_O will decrease. As V_O decreases, the value of R_{DS1} decreases, and as a result, the slope of V_O versus V_I decreases. This decreasing slope leads to the curved part of the characteristic shown in Figure 15-24 for the region where the NMOS transistor is in the linear region and the PMOS transistor is saturated. Finally, when $V_I \geq V_{DD} - V_{th}$, the PMOS transistor turns off and V_O is reduced to zero. No current can flow in this state, so the static power dissipation is zero.

Because of the assumed symmetry of the circuit, a completely analogous discussion applies to the other half of the transfer characteristic. When V_O gets above $V_{DD}/2$ by V_{th}, the PMOS transistor enters the linear region and the gain drops. When V_I is less than V_{th}, the NMOS transistor turns off completely and the output voltage is equal to the positive supply, V_{DD}. No current can flow in this situation, so the static power dissipation of the inverter is zero for both $V_O = V_{DD}$ and $V_O = 0$.

Example 15.4

Problem:
Determine what happens to the static transfer characteristic of the inverter if $K_P \neq K_N$.

Solution:
Before considering the case at hand, go back to Figure 15-24 and consider the balanced case for a moment. When $V_I = V_{DD}/2$, the two controlled sources in Figure 15-23 have identical currents, and because $V_O = V_{DD}/2$, the currents through the resistors are also equal.

Now, if $K_P > K_N$, then the PMOS transistor has greater current drive capability than the NMOS transistor has. Therefore, when both devices are in saturation, the currents from the controlled sources will be equal when $V_I > V_{DD}/2$. The currents through the resistors will still be equal when $V_O = V_{DD}/2$, so the midpoint of the transfer characteristic will occur when $V_I > V_{DD}/2$ and $V_O = V_{DD}/2$. This will result in the entire transfer characteristic shifting to the right, as shown in Figure 15-25.

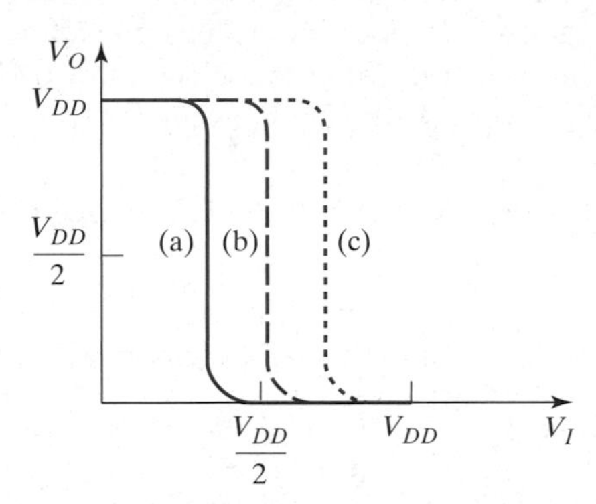

Figure 15–25 Inverter transfer characteristics for (a) $K_P < K_N$, (b) $K_P = K_N$, and (c) $K_P > K_N$ (see Exam4.sch for a simulation).

Similarly, if $K_P < K_N$, the midpoint of the transfer characteristic will occur for $V_I < V_{DD}/2$ and $V_O = V_{DD}/2$, and the whole characteristic will shift to the left as shown.

We can view an inverter as a comparator with a built-in threshold or switching point. In other words, the symmetric inverter shown as curve (b) in Figure 15-25 functions like an inverting comparator with a threshold of $V_{DD}/2$. Note that by varying the size of the transistors in an inverter, we can change the switching point of the inverter, which is sometimes useful.

Noise Margins We now turn our attention to calculating the noise margins, which requires that we first find the values of V_{OH}, V_{OL}, V_{IH}, and V_{IL}. To find the values of V_{OL} and V_{IH}, we first equate the drain current of the NMOS transistor operating in the linear region as given by (15.40) to the drain current of the PMOS transistor in saturation as given by (15.38):

$$2(V_I - V_{th})V_O - V_O^2 = (V_{DD} - V_I - V_{th})^2. \tag{15.43}$$

We have assumed that $\lambda = 0$ in deriving (15.43), since we are still considering the ideal inverter. If we differentiate both sides of (15.43) with respect to V_I, we get

$$2(V_I - V_{th})\frac{dV_O}{dV_I} + 2V_O - 2V_O\frac{dV_O}{dV_I} = -2(V_{DD} - V_I - V_{th}). \tag{15.44}$$

We now set $dV_O/dV_I = -1$, $V_I = V_{IH}$, and $V_O = V_{OL}$ and solve to obtain

$$V_{OL} = V_{IH} - V_{DD}/2. \tag{15.45}$$

Substituting this back into (15.43) in place of V_O, using V_{IH} in place of V_I, and solving yields (*see* Problem P15.21)

$$V_{IH} = \frac{5V_{DD}^2 - 12V_{DD}V_{th} + 4V_{th}^2}{8(V_{DD} - 2V_{th})}. \tag{15.46}$$

The values of V_{IL} and V_{OH} can be solved for in a similar manner, but for the symmetric case we can say that $V_{OH} = V_{DD} - V_{OL}$ and $V_{IL} = V_{DD} - V_{IH}$. Using these values, we can find the noise margins of the ideal symmetric CMOS gate: $NM_H = V_{OH} - V_{IH}$ and $NM_L = V_{IL} - V_{OL}$.

We have shown how to calculate the logic levels and noise margins given a single transfer characteristic. Manufacturers typically specify minimum and maximum values obtained from many individual transfer characteristics in order to take into account process variations. In addition, values are often given at different temperatures, although the DC transfer characteristics of CMOS logic gates are not very sensitive to temperature, so the changes are small.

Example 15.5

Problem:
Find the values of V_{OH}, V_{OL}, V_{IH}, V_{IL}, and the noise margins for an ideal symmetric CMOS inverter with $V_{DD} = 5$ V and $V_{th} = 1$ V.

Solution:
Using (15.46), we obtain $V_{IH} = 2.88$ V and then, using (15.45), we find $V_{OL} = 0.38$ V. By symmetry, we have $V_{OH} = V_{DD} - V_{OL} = 4.62$ V and $V_{IL} = V_{DD} - V_{IH} = 2.12$ V. Using these values in (15.20) and (15.21), we obtain $NM_H = NM_L = 1.74$ V, which is quite good.

■

Exercise 15.5

Find the values of V_{OH}, V_{OL}, V_{IH}, V_{IL}, and the noise margins for an ideal symmetric CMOS inverter with $V_{DD} = 3$ V and $V_{th} = 0.8$ V.

□

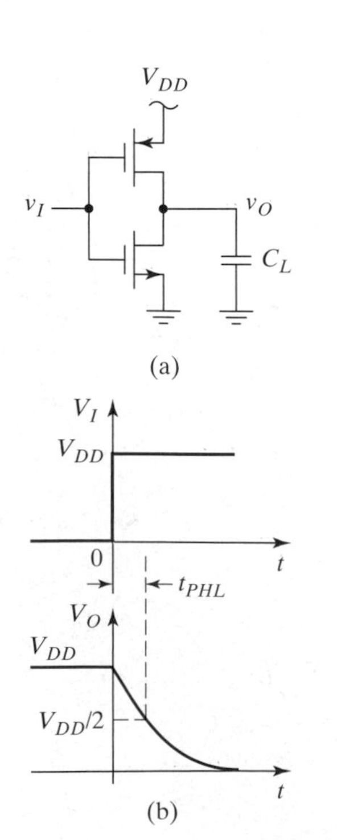

Figure 15–26 (a) A CMOS inverter with capacitive load. (b) The input and output waveforms for calculating t_{PHL}.

Propagation Delay (Switching Speed) We next consider the propagation delay of a CMOS inverter. Because the input resistance of a MOSFET is essentially infinite, the only load that is usually present with CMOS logic circuits is the input capacitance of the next gate (or gates). A simple model for the situation is shown in Figure 15-26(a). If the inverter is driving other CMOS gates, the load capacitor is not a constant capacitor, but depends on the output voltage. Nevertheless, we usually use an approximate average value for hand computations.

Because we will still assume the gate to be perfectly symmetric, the propagation delay from low to high will be the same as from high to low. Therefore, we only consider t_{PHL} as shown in part (b) of the figure. The input voltage is assumed to switch instantly from zero to V_{DD} at time $t = 0$. We further assume that the gate was in the previous state long enough to reach steady state, so that $v_O(0-) = V_{DD}$, as shown.

Just before the input switches (at $t = 0-$), the PMOS transistor is on (in the linear region) and the NMOS transistor is off. When the input switches to V_{DD}, the PMOS transistor will turn off almost immediately and the NMOS transistor will turn on almost immediately.

A perfectly reasonable estimate of the propagation delay can be calculated by hand using ***charge-control analysis***. In performing charge-control analysis, we use a piecewise linear approximation to the overall transfer characteristic. We then examine the period corresponding to each linear segment separately. For each of these linear segments, we calculate the total change in charge and then approximate the currents as constant over that period. We then approximate the time using $\Delta t = \Delta Q/\bar{I}$. The key to obtaining accurate answers with this technique is to choose the linear regions so that neither the capacitances being charged nor the currents doing the charging change too much during the interval.

For the case at hand, we note that for $t > 0$, the equivalent circuit looks like Figure 15-27, where the PMOS transistor is not shown because it is off (i.e., it is an open circuit) and the drain current of the NMOS transistor discharges the capacitor. As the output voltage drops, the NMOS transistor changes from being saturated to operating in the linear region. Although the drain current certainly changes during this time, a reasonable first-order approximation to its average value can be found by averaging the values at the beginning and end of the period:

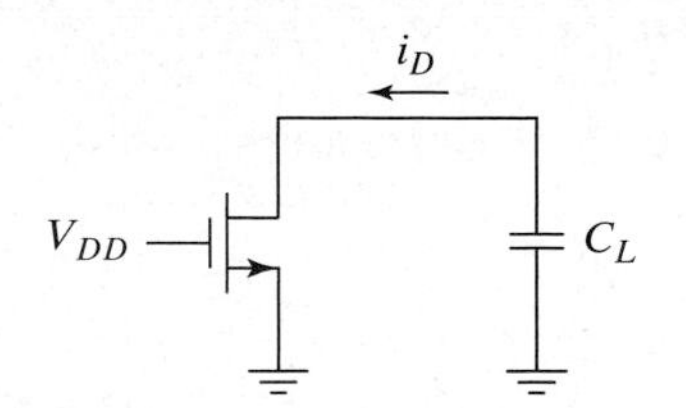

Figure 15–27 The equivalent circuit for the inverter of Figure 15-26(a) for $t > 0$.

$$\bar{I}_{DN} \approx \frac{i_{DN}(0+) + i_{DN}(t_{PHL})}{2}. \tag{15.47}$$

Since $V_I = V_O = V_{DD}$ at t=0+, we know that the transistor is in saturation, and we have

$$i_{DN}(0+) = K(V_{DD} - V_{th})^2, \tag{15.48}$$

where we have ignored channel-length modulation for simplicity. At the end of the propagation delay, the drain (or output) voltage has dropped to $V_{DD}/2$ by definition, so the transistor is in the linear region, and its drain current is given by

$$i_{DN}(t_{PHL}) = K\left[2(V_{DD} - V_{th})\frac{V_{DD}}{2} - \frac{V_{DD}^2}{4}\right]. \tag{15.49}$$

We can use (15.48) and (15.49) in (15.47) to estimate the average current flowing during t_{PHL}.

The total change in the capacitor voltage during t_{PHL} is

$$\Delta V_{C_L}(\text{during } t_{PHL}) = \frac{V_{DD}}{2}, \tag{15.50}$$

so the total change in the charge on the capacitor is

$$\Delta Q_{C_L}(\text{during } t_{PHL}) = \frac{C_L V_{DD}}{2}. \tag{15.51}$$

Since this charge is all removed by the drain current of the NMOS transistor, we can estimate t_{PHL} as

$$t_{PHL} \approx \frac{\Delta Q_{C_L}(\text{during } t_{PHL})}{\bar{I}_{DN}} = \frac{C_L V_{DD}}{2\bar{I}_{DN}}. \tag{15.52}$$

Because of the symmetry we have assumed, we can say that

$$t_{PLH} \approx \frac{C_L V_{DD}}{2\bar{I}_{DP}}. \tag{15.53}$$

Notice from (15.48) and (15.49) that $\bar{I}_{DN}$ is proportional to K. Also, recall that K is proportional to W/L. Therefore, we can conclude that the propagation delays are inversely proportional to the respective transistor sizes:

$$t_{PHL} \propto \frac{1}{\left(\frac{W}{L}\right)_N} \text{ and } t_{PLH} \propto \frac{1}{\left(\frac{W}{L}\right)_P}. \quad \textbf{(15.54)}$$

Further, recall from (15.36) that the PMOS device must be approximately 2.5 times larger than the NMOS device if the gate is to be symmetric (e.g., have $t_{PHL} = t_{PLH}$). Therefore, a minimum-size symmetric inverter requires approximately the area of 3.5 minimum-size devices.

The sizes of transistors used in CMOS inverters is determined by the load they must drive and the speed required. We see from (15.53) that if C_L increases, the transistors must become proportionally larger to keep the speed constant. However, if the size of an inverter is increased, the inverter or gate prior to it will be driving a larger load (since larger transistors have larger input capacitance). In light of this observation, an interesting situation to examine is the optimum way to increase the size of inverters when a large load must be driven (e.g., to drive signals off of an IC). We explore this question in Aside A15.3.

Finally, we note that the rise and fall times for a CMOS inverter can also be found by applying the charge-control method presented here. They are usually in the neighborhood of twice the propagation delays.

Example 15.6

Problem:

Find the propagation delays for an ideal symmetric CMOS inverter with $V_{DD} = 5$ V, $V_{th} = 0.8$ V, $C_L = 0.2$ pF, and $K = 25$ μA/V^2.

Solution:

We first use (15.48) and (15.49) in (15.47) to find $\bar{I}_{DN} = 405$ μA. We then use (15.52) to obtain $t_{PHL} \approx 1.23$ ns. By symmetry, t_{PLH} is the same.

A SPICE simulation predicts that $t_{PHL} = 1.19$ ns, very close to our approximation.

■

Exercise 15.6

Suppose the NMOS transistor in Example 15.6 has $W = 1.5$ μm and $L = 0.75$ μm and the PMOS transistor has $W = 3.75$ μm and $L = 0.75$ μm. What must the new dimensions be if we want the gate to be just as fast when driving $C_L = 1$ pF?

□

Dynamic Power Dissipation We derived the dynamic power dissipation of an inverter in Section 15.2.2. In that section, we assumed that all of the current through the inverter was used to charge or discharge the load capacitance. However, when both transistors in a CMOS inverter are on, there is a direct path for current from V_{DD} to ground through the transistors. The current that flows in this path leads to extra power dissipation that is not accounted for in our previous analysis. This power depends on how long both transistors are on during the transition, and the added dynamic power dissipation is, therefore, a function of the

ASIDE A15.3 OPTIMUM SCALING OF CMOS INVERTERS

An interesting problem arises when a large load capacitance must be driven by a CMOS logic circuit (e.g., when driving a signal off of an IC). If we use a minimum-size inverter to drive the large C_L, the propagation delay of that inverter will be very long, since it is proportional to C_L as derived in (15.53). If we decide to use a large inverter to drive C_L, then this inverter will have a large input capacitance, and the minimum-size gate driving it will have a long delay. Therefore, we conclude that we want to use a string of inverters to drive the load and gradually increase the sizes of the inverters in the string, as shown pictorially in Figure A15-5.

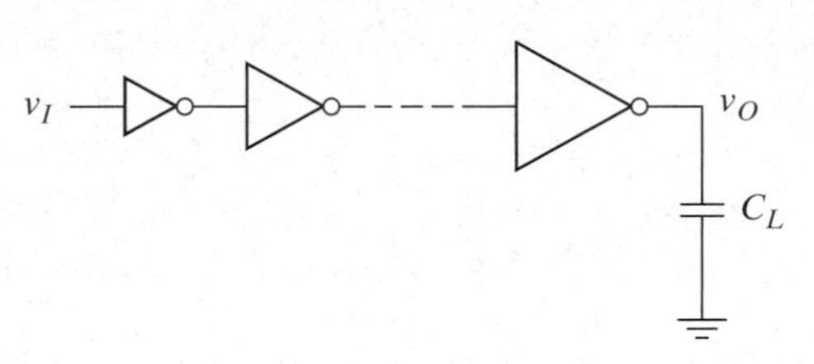

Figure A15-5 A chain of inverters driving a large capacitive load.

The question then is this: What scaling factor should we use if we desire to minimize the overall propagation delay of the chain of inverters? Suppose that the first inverter in the chain is a minimum-size symmetric inverter. Its input capacitance is C_1, and its average output current to charge or discharge the next stage is $\bar{I}_1$. We assume that each inverter is scaled up in size by a factor S. The input capacitance to the second inverter is then $C_2 = SC_1$, and its output current will be $\bar{I}_2 = S\bar{I}_1$. The propagation delay of the first inverter is given by (15.53) as

$$t_p = \frac{C_2 V_{DD}}{2\bar{I}_1} = S\frac{C_1 V_{DD}}{2\bar{I}_1} = St_0, \tag{A15.17}$$

where we have defined the delay of one minimum-size symmetric inverter driving another minimum-size symmetric inverter as t_0. Similarly, for the ith inverter in the chain, we have

$$t_p = \frac{C_{i+1} V_{DD}}{2\bar{I}_i} = \frac{S^i C_1 V_{DD}}{2S^{i-1}\bar{I}_1} = St_0, \tag{A15.18}$$

so that for N inverters, we obtain a total propagation delay of

$$t_p[N] = NSt_0. \tag{A15.19}$$

For a chain of N inverters, the scale factor must satisfy the constraint

$$S^N C_1 = C_L, \tag{A15.20}$$

which leads to (take the natural logarithm of both sides)

$$N = \frac{\ln(C_L/C_1)}{\ln S}. \tag{A15.21}$$

To find the optimum scaling factor we plug (A15.21) back into (A15.19) and differentiate $t_p[N]$ with respect to S. We then set the derivative equal to zero to find the minimum and obtain

$$S_{opt.} = e = 2.72. \tag{A15.22}$$

Substituting this back into (A15.21) gives us the optimal number of inverters in the chain:

$$N_{opt.} = \ln\left(\frac{C_L}{C_1}\right). \tag{A15.23}$$

The number given by (A15.23) is not an integer, so this optimum solution cannot be used exactly, but it gives us a target to shoot for. This solution is also a simplification of the real problem, since the total capacitance driven by each inverter includes some capacitance due to its own transistors and some wiring capacitance, and neither of these terms scale as assumed here. A more accurate analysis reveals that the scaling factor should be larger than we found in (A15.22), typically near 3.6 for modern processes [15.7].

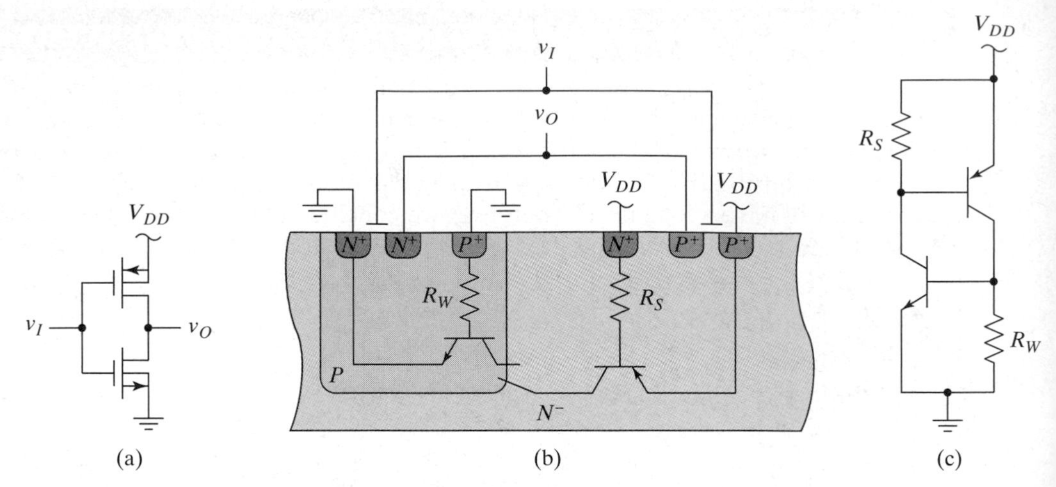

Figure 15–28 (a) A CMOS inverter, (b) a simplified layout of the inverter, and (c) the parasitic SCR.

input transition times. Manufacturers will sometimes specify a ***power dissipation capacitance*** that is added to the load capacitance for power calculations. This fictitious capacitance accounts for the transient current through the transistors, while allowing us to use the same equation, (15.32), to determine the dynamic power dissipation of a gate.

Latchup A CMOS inverter is shown in Figure 15-28(a), including the bulk connections of the transistors. A simplified drawing of the layout is shown in part (b) of the figure, assuming a *p*-well process. There is a parasitic *npnp* structure formed by the NMOS source, *p*-well, substrate, and PMOS source, respectively. This four-layer structure looks like a silicon-controlled rectifier (SCR) and, as presented in Section 2.9, it can be modeled by two bipolar transistors as shown in part (c) of the figure. The operation of this circuit, without the resistors, is discussed in Section 2.9. We show there that when a current gets started in the structure, the collector current of the *pnp* transistor supplies the base current of the *npn* transistor and vice versa. Once in this state, called latchup, the circuit will conduct a current from V_{DD} to ground that is limited only by series resistance and that will often destroy the device if not interrupted externally. Latchup is prevented in modern CMOS circuits by layout rules designed to minimize the gain of the parasitic bipolar transistors. For example, the well resistance, R_W, and the substrate resistance, R_S, shunt the base-emitter junctions as shown in the figure. If these resistances are small, it is much harder for the devices to turn on.

15.3.3 CMOS NOR and NAND Gates

In this section, we consider the design of single CMOS logic gates. We begin by looking at a two-input NOR gate. The output of a NOR gate should be low if either of the inputs is high. We can accomplish this task by putting two NMOS transistors in parallel, with their sources connected to ground, their drains tied together, and their gates used as the inputs, as shown in Figure 15-29. This circuit is not complete, however; it needs some kind of pull-up, as noted on the schematic. The pull-up could be a resistor, as in NMOS logic, but as noted in Section 15.3.1, resistors consume a lot of area and are avoided in integrated circuit design. In addition, if a resistor is used, the gate is not symmetric, is slow in pulling the output high, and has static power dissipation.

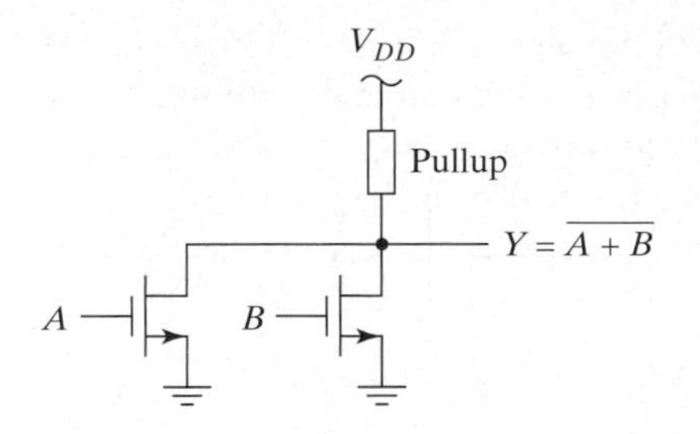

Figure 15–29 A NOR gate using NMOS transistors and an unspecified pull-up circuit.

A much better way to implement the pull-up circuit is to use PMOS devices. From De Morgan's theorems, we know that the NOR function can be rewritten as

$$\overline{Y} = A + B \Rightarrow Y = \overline{A} \times \overline{B}. \tag{15.55}$$

In other words, if both inputs are low, the output should be high. Since both inputs need to be low, this implies that we should stack two PMOS transistors in series for the pullup. If both of their gates go low, they both turn on and provide a current path to pull the output high. The resulting CMOS two-input NOR gate is shown in Figure 15-30.

The DC transfer characteristic from either input to the output of the NOR gate could be derived in much the same way as we obtained the transfer characteristic for the inverter. The transfer characteristic from input A to the output would depend on the voltage present on B and vice versa. If we want the NOR gate to be symmetric, so that t_{PHL} and t_{PLH} are the same and the switching threshold is near $V_{DD}/2$, then we must choose the sizes of the transistors correctly. As we pointed out in our discussion of the CMOS inverter, the surface mobility of electrons is about 2.5 times the surface mobility of holes in silicon, so the PMOS transistors need to be about 2.5 times the size of the NMOS transistors to account for this fact. In the NOR gate, we have the added problem that the PMOS transistors appear in series.

When two or more devices are in series, we need to increase their sizes (i.e., W/L) to maintain the same propagation delay. A rough, but effective, approximation is to say that two devices in series appear like a device with twice the length; therefore, in order to keep K constant, we should double the width as well. This approach would be exactly correct if the devices were both in the linear region, had precisely the same V_{SG}'s, and had small V_{SD}'s, since they would then function

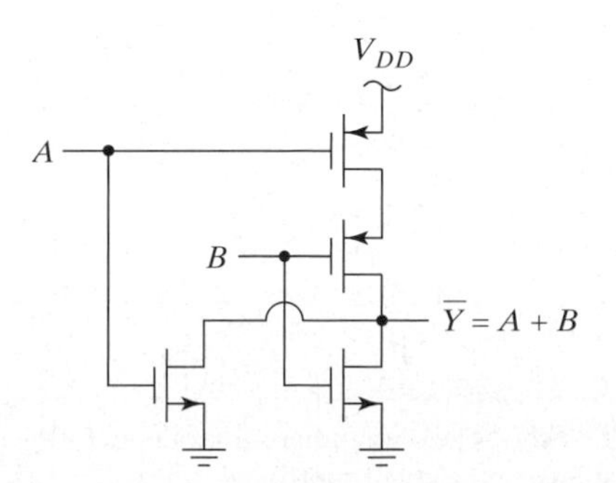

Figure 15–30 A two-input CMOS NOR gate.

as resistors in series.[13] Unfortunately, the PMOS devices in Figure 15-30 are not in the linear region for the entire transition time, nor do they have exactly the same V_{SG}, nor is V_{SD} small for both of them during the entire transition. Nevertheless, this rough approximation is reasonably effective for determining the sizes of these series devices. Therefore, for an N-input NOR gate, the transistors need to be sized according to

$$\left(\frac{W}{L}\right)_{\text{PMOS}} \approx 2.5N\left(\frac{W}{L}\right)_{\text{NMOS}}. \tag{15.56}$$

So, if the NMOS transistors are minimum size, the PMOS transistors for our two-input NOR gate will be 5 times minimum size, and the overall gate will consume approximately the equivalent of twelve minimum size transistors. You might think that we could reduce the sizes of the NMOS transistors, since they are in parallel in this circuit, but there are two reasons we would not typically do so. First, if the devices are minimum size devices with $(W/L) = 1$, to reduce (W/L) would actually require increasing the area of the device (area $= WL$). Second, there will be times when only one input goes high, so we probably (though not always) want the propagation delay to be equal to that of an inverter, even under this worst-case condition.

Another problem that arises with the series transistors is that they may suffer from the body effect. If the CMOS process is a p-well process, then all of the PMOS transistors are fabricated in the n-type substrate, and all of their bulk connections go to the positive supply, as shown in Figure 15-31. This means that only one of the PMOS transistors will have $V_{SB} = 0$. The net result of the body effect will be to decrease the effectiveness of the PMOS transistors at pulling the output high and to change the switching threshold of the gate to be slightly lower than $V_{DD}/2$. These problems become more severe as we stack more transistors in series. Because of the drop in performance and the increased size of gates with many series transistors, it is often worthwhile to redesign a complex combinatorial logic circuit to use several smaller gates rather than a single gate with many inputs.

We now turn our attention to the CMOS NAND gate shown in Figure 15-32. We now need both inputs to be high in order to pull the output low; hence, we have two NMOS transistors in series to pull the output down. We can again use De Morgan's theorems to rewrite the equation as $Y = \overline{A \cdot B} = \overline{A} + \overline{B}$, which shows that the output should be pulled up if either input is low; hence, we have two PMOS devices in parallel to pull the output up.

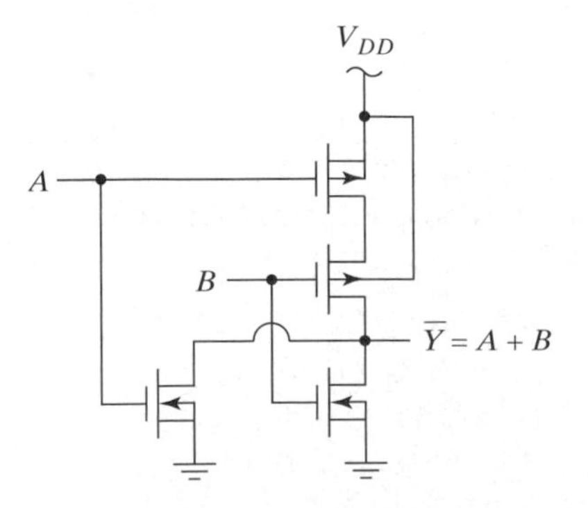

Figure 15–31 A two-input CMOS NOR gate showing the bulk connections. A p-well process has been assumed.

[13] In other words, we are using a simple switch model for the FETs. They are either OFF (an open circuit), or ON. When ON they appear as a switch with a resistance inversely proportional to W/L.

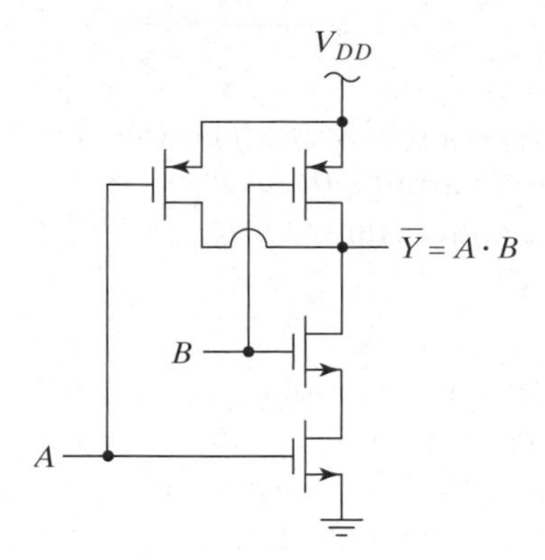

Figure 15–32 A two-input CMOS NAND gate.

The transistor sizes are determined using the same method as we applied to the NOR gate. If we want the gate to have the same propagation delays as a minimum-size symmetric inverter, then the two NMOS devices should have $W = 2W_{min}$. The PMOS transistors, on the other hand, can be the same size as they would be in the minimum-size symmetric inverter, namely, 2.5 times the minimum-size transistor. For an N-input NAND gate, the transistors should be sized so that

$$\left(\frac{W}{L}\right)_{PMOS} \approx \frac{2.5}{N}\left(\frac{W}{L}\right)_{NMOS}. \quad \textbf{(15.57)}$$

Returning to our two-input NAND example, if the NMOS transistors are made twice the minimum size to achieve the same speed as a minimum-size symmetric inverter, then the PMOS transistors will be 2.5 times the minimum size, and the overall gate will take the area of about nine minimum-size transistors. Comparing this with our two-input NOR gate, we find that the two-input NAND gate, uses only 75% as much area. Therefore, if all else is equal, we prefer NAND gates over NOR gates in IC design.

The NAND gate also suffers from the body effect if the n-channel devices cannot be placed in separate wells (i.e., if the process is an n-well process), but in this case it would cause the switching threshold to be slightly greater than $V_{DD}/2$. The stacked devices do again cause a degradation in performance, as discussed for the NOR circuit. Therefore, we avoid making NAND gates with too many inputs.

Example 15.7

Problem:
If we build a two-input NOR gate using minimum-size transistors for all four transistors, approximately how much longer will t_{PHL} and t_{PLH} be than they are for a minimum size symmetric inverter?

Solution:
Using the same logic as was used to arrive at (15.56), we can say that using minimum-size transistors in place of the properly scaled devices would increase the propagation delay by 2(2.5) = 5 times. This factor applies only to t_{PLH}, since it is the PMOS devices that are in series. For t_{PHL}, the worst-case value (when only one input goes high) is the same as t_{PHL} for the minimum size symmetric inverter and the best-case value (if both inputs go high at once) is roughly half as large.

■

Exercise 15.7

Determine the transistor sizes we should use in a three-input NOR gate if we want the propagation delays to be the same as for a minimum-size symmetric inverter.

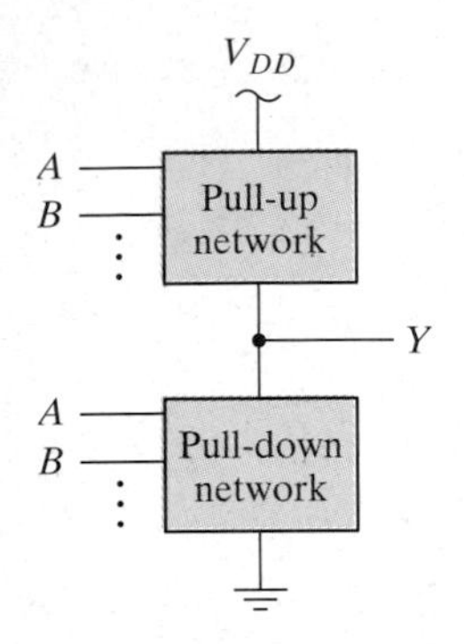

Figure 15–33 A general CMOS logic gate. Both the pull-up and the pull-down networks have the same inputs.

15.3.4 More Complex CMOS Gates

In deriving the two-input NOR and NAND structures in the previous section, we illustrated a general principle that is used in designing more complex CMOS gates; namely, the pull-up and pull-down networks must have complementary Boolean functions. Consider the general logic gate illustrated in Figure 15-33. The pull-up network is made up of PMOS devices and is derived by looking at the equation for Y; in other words, what combination of inputs should cause the output to be high? The pull-down network, on the other hand, is made up of NMOS devices and is derived by looking at the equation for $\overline{Y}$; in other words, what combination of inputs should cause the output to be low?

If the two networks do not implement complementary Boolean functions, then there will be at least one input word for which the output is not defined, either because it isn't being pulled up or down, or because it is simultaneously being pulled both ways. We illustrate this principle with an example. Suppose we want to design a gate that will implement the function

$$Y = \overline{A} \cdot (\overline{B} + \overline{C} \cdot \overline{D}). \tag{15.58}$$

This equation naturally describes the PMOS pull-up network shown in Figure 15-34. To derive the pull-down network, we complement (15.58):

$$\overline{Y} = \overline{\overline{A} \cdot (\overline{B} + \overline{C} \cdot \overline{D})} = A + B \cdot (C + D). \tag{15.59}$$

Examination of the pull-down network in the figure shows that it implements (15.59) directly.

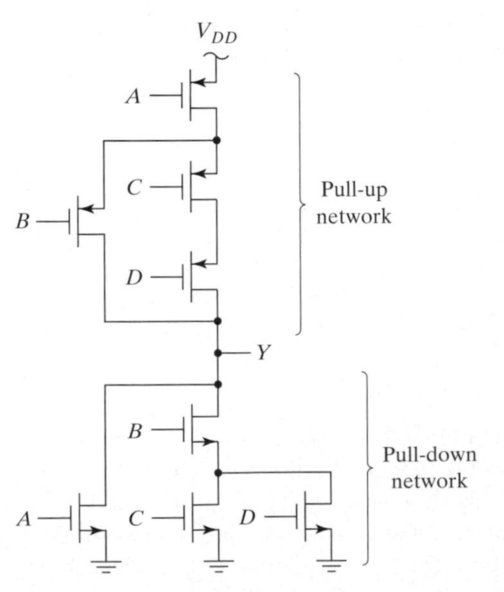

Figure 15–34 A CMOS gate to implement (15.58).

Notice that we can also derive the one network from its complement by using a series–parallel transformation: Every time we see transistors or groups of transistors in series in one network, there must be transistors or groups of transistors in parallel in the complementary network. For example, in Figure 15-34, we see that the transistors driven by C and D are in series in the pull-up network, but in parallel in the pull-down network. Similarly, the transistor driven by B is in parallel with the two series devices C and D in the pull-up network, but is in series with the corresponding two parallel devices in the pull-down network.

As a final note about the design of CMOS logic gates, we may not always want to design the gates to have worst-case propagation delays equal to those of an inverter. If we carefully examine the critical path in a complex combinatorial logic network, it may be possible to save significantly on area by choosing the transistor sizes to optimize the critical path, which may mean that some propagation delays in the network are much longer than others. This kind of custom design of digital circuits is sometimes done for special applications, but it complicates the design process considerably and makes digital design a bit more like analog design.

One simple observation that is often important is that the delays from the different inputs of a gate to the output are not the same. For example, consider a two-input NAND gate as in Figure 15-32. If B is already high and A then switches from low to high, the value of t_{PHL} is longer than it would be if we reversed the inputs. In other words, the B input to this gate is the fastest. This statement is true, since there will always be parasitic capacitance at the nodes in between series elements; therefore, the element closest to the output will be the fastest.

Exercise 15.8

Using only NMOS and PMOS transistors, design a circuit that performs the logical operation $\overline{Y} = (A + B) \cdot C$.

□

15.3.5 Other Types of CMOS Logic

The CMOS logic gates studied thus far suffer from a few important problems:

1. A gate with N inputs requires $2N$ transistors, since the pull-up and pull-down networks each require one transistor per input variable.
2. The gate becomes large and slow when multiple inputs are used. Although we have shown how to scale devices to overcome the speed limitation to some extent, this method does not completely overcome the problem, and NAND and NOR gates with a fan-in greater than four should be avoided [15.7].
3. The propagation delay of a gate is proportional to the load capacitance it drives, which makes it difficult for these gates to drive some loads. A string of progressively larger inverters can be used, as discussed in Aside A15.3, but doing so is still slow and requires more area.

Circuit designers have come up with different types of CMOS gates to overcome these problems. While no other approach has become as widely used as standard CMOS logic, they are important and are used quite a bit in custom IC design. We will only briefly introduce a few techniques here and refer you to [15.7] for a very good detailed treatment.

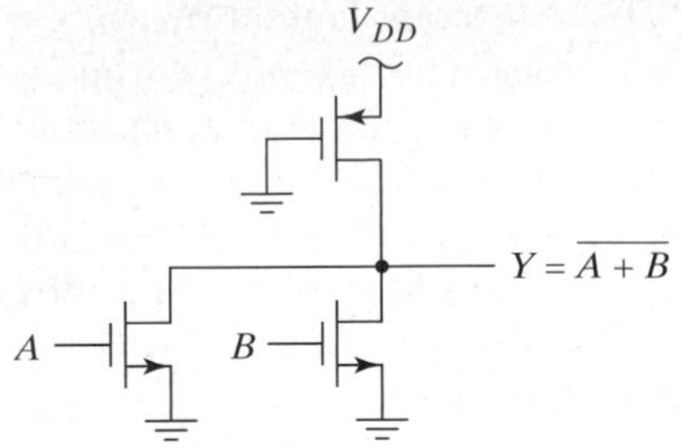

Figure 15–35 A pseudo-NMOS NOR gate.

One way of reducing the number of transistors required by a CMOS gate is to simplify either the pull-up or pull-down network. Since either network implements the full logic function, only one of them is necessary for basic functionality. Figure 15-35 illustrates the idea. In this figure, we have used the standard NMOS pull-down network, but have replaced the PMOS pull-up network with a single PMOS load device. This gate is called a ***pseudo-NMOS*** gate, since it resembles a depletion-load NMOS gate and requires only $N + 1$ transistors for N inputs. The disadvantages are that the output no longer swings from rail to rail (so the noise margins are smaller) and the gate has static power dissipation.

There is a trade-off between the speed and power of the pseudo-NMOS gate. To keep the static power low, we would like the PMOS device to be small so that the current it supplies when the output is low is small. On the other hand, we would like the transistor to be large to keep t_{PLH} small. Pseudo-NMOS gates are not widely used, because the static power dissipation, which is in the neighborhood of 1 mW per gate, becomes prohibitive for very many gates. The static power of the gate is zero when the output is high, however. Consequently, these gates are attractive in applications with a large fan-in and where the majority of outputs are high (so that the static power is zero)—for example, in memory address decoders [15.7].

Another very useful alternative to standard CMOS logic is called ***pass-transistor logic***. A pass transistor is simply a switch; in its simplest form, it can be implemented with a single MOS transistor. In pass transistor logic, two transistors in series must both be on for a signal to get through, so they implement a logical AND operation. Similarly, two transistors in parallel implement a logical OR, since a signal can get through if either one of them is on. In a sense, we have already seen pass-transistor logic; it is implicitly used in the design of standard CMOS gates. The pull-up networks are PMOS pass-transistor logic and the pull-down networks are made up of NMOS pass transistors.

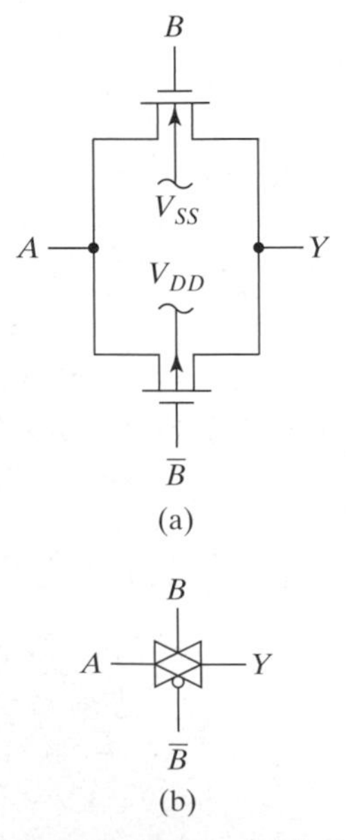

Figure 15–36 (a) A CMOS transmission gate and (b) the schematic symbol for it.

In true pass-transistor logic, we do not have pull-up and pull-down networks, so it is beneficial to use a combination of a PMOS and an NMOS transistor in what is called a CMOS transmission gate as illustrated in Figure 15-36. We have used the four-terminal schematic symbol for the MOSFETs in this figure for two reasons. First, the devices are used as symmetric switches, so we decide which terminal is the "source" on the basis of the applied voltages. Second, we want to emphasize that the body terminals are not tied to either the source or the drain terminal, since we can't be certain which one is acting as the source. Also, note that the NMOS body is tied to the most negative potential in the circuit, which is typically called V_{SS} in MOS circuits and may or may not be equal to ground. The properties of the transmission gate are explored further in Aside A15.4.

ASIDE A15.4 THE CMOS TRANSMISSION GATE

A single MOS transistor can function as a switch, as shown in Figure A15-6(a) for an n-channel device. We have used the four-terminal schematic symbol for the transistor for two reasons. First, the device is used as a symmetric switch, so we decide which terminal is the "source" on the basis of the applied voltages. Second, we want to emphasize that the body terminal is not tied to either the source or drain terminal, since we can't be certain which one is acting as the source. Also, note that the body is tied to the most negative potential in the circuit, which is typically called V_{SS} (although it may be ground in some circuits). This switch has a serious limitation, however, If the gate is pulled up to V_{DD} to turn the switch on, the input must be lower than $V_{DD} - V_{th}$, or the switch will turn off. In addition, the resistance of the switch is a strong function of the input voltage, which is a serious problem if we desire to switch analog signals. The CMOS transmission gate shown in part (b) of the figure overcomes these problems.

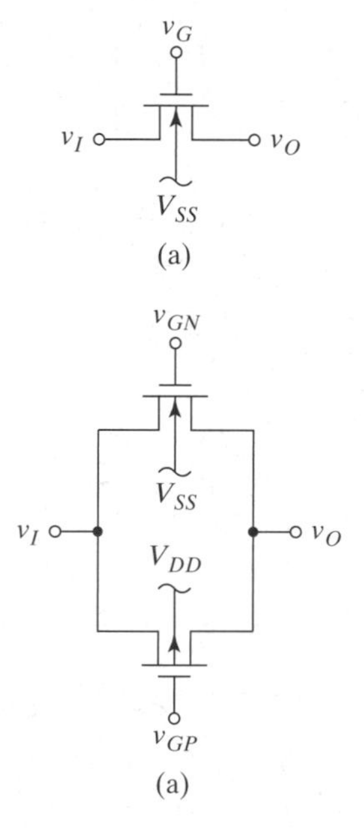

Figure A15-6 (a) A single NMOS switch and (b) a CMOS transmission gate.

If we assume that $V_{thN} = -V_{thP} = V_{th}$, then when $v_{GN} = V_{DD}$ and $v_{GP} = V_{SS}$ and $V_{SS} + V_{th} \leq v_I \leq V_{DD} - V_{th}$, both transistors in the transmission gate are on and the resistance is smaller than with either one alone. Also, the resistance is fairly constant, since the resistance of the PMOS device decreases as v_I increases while the resistance of the NMOS device increases, with increasing v_I. For $V_{DD} - V_{th} < v_I \leq V_{DD}$, only the PMOS device is on, and for $V_{SS} \leq v_I < V_{SS} + V_{th}$, only the NMOS device is on, but in both cases, the transistor that is on has a very large gate-to-channel voltage and, therefore, a low on resistance.

This gate sets $Y = A$ whenever B is high. When B is low, A and Y are not connected. A common application of pass-transistor logic is demonstrated by the four-to-one ***multiplexer*** circuit shown in Figure 15-37. The function of this circuit is to connect one of the four inputs (I_0,I_1,I_2,I_3) to the output. Which input is connected is determined by the two-bit binary select signal (S_1,S_0). If the inputs are binary, we can represent the operation of this network by the Boolean expression

$$Y = I_0 \cdot \overline{S_0} \cdot \overline{S_1} + I_1 \cdot \overline{S_0} \cdot S_1 + I_2 \cdot S_0 \cdot \overline{S_1} + I_3 \cdot S_0 \cdot S_1. \quad \textbf{(15.60)}$$

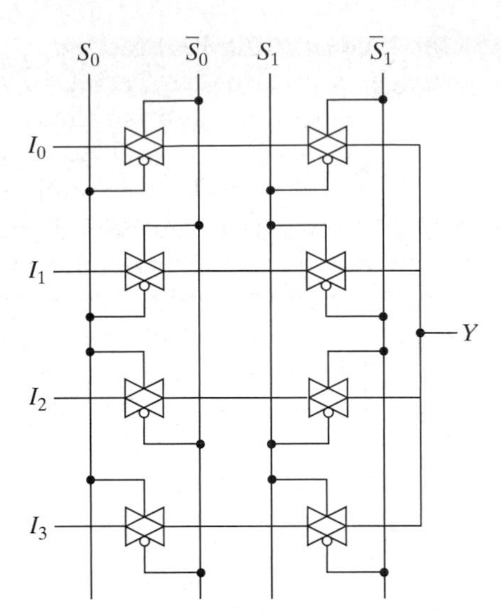

Figure 15–37 A four-to-one multiplexer circuit.

A major use of this network is in switching analog signals. Since the transmission gates represent a low impedance when on and a high impedance when off, the inputs and output in this circuit can be any voltage between 0 and V_{DD} and the circuit will function properly.

Exercise 15.9

Draw a two-input pseudo-NMOS NAND gate and describe its operation.

15.3.6 MOS Memory

Many digital systems use some kind of memory. Memory is used to store operating systems, programs, data, and other information in computers and many other digital and mixed-signal systems. Memory can be generally divided into systems with movable parts (e.g., disk drives) and systems without movable parts (e.g., semiconductor memory). Memories with moving parts have a lower cost per bit and are well suited for use as mass storage. They are not typically used as the primary memory in a computer because the data must be accessed serially and because they have much longer access times than semiconductor memories. In this section, we are interested only in giving a brief introduction to semiconductor memory made with MOS devices.

One kind of semiconductor memory allows the user to both write data to and read data from any location at any time and is called ***random-access memory*** (RAM). Another kind of memory does not allow the user to write data in normal operation and is therefore called ***read-only memory*** (ROM). The data stored in most ROMs can also be accessed randomly, so the term RAM, which is applied exclusively to memory that we can both write to and read from, is used to mean more than it says. RAM is *volatile* memory, since it loses its data when power is removed.

Read-only memories may be programmed during manufacture, in which case they cannot be programmed by the user at all, or they may be programmable. ***Programmable ROMs*** (PROMs) may require that they be programmed by means of special equipment prior to their use, or they may be able to be programmed while in the circuit. PROMs are further classified on the basis of whether or not they can be erased once thry are programmed. One kind of PROM is erased by exposure to ultraviolet light and is called UV-erasable PROM or just ***erasable PROM*** (EPROM), while another kind can be erased electrically and is called EEPROM (***electrically erasable PROM***). ROM is *nonvolatile* memory, since it retains its data when power is removed. This aspect of ROMs makes them useful for the permanent storage of instructions for microprocessor-controlled appliances, games, and similar devices.

Modern semiconductor memories are the largest integrated circuits manufactured, and memories with 1 billion bits of storage (1 Gb) and over 2 billion transistors will soon be commercially available. Memory design is a specialized area of circuit design, and we will provide only a brief introduction here. Nevertheless, this treatment will show you applications for a number of concepts and techniques presented elsewhere in the book.

Memory Architecture The architecture used for a memory has a huge impact on the achievable performance. Consider, for example, a 64-Mb memory chip.[14] If we organized the chip as a one-dimensional array of memory cells, and if each cell was approximately square (which is close to correct), the chip would be 64 million times as large in one dimension as it is in the other! This is clearly not practical.

One practical way to organize a memory is shown in Figure 15-38. The memory is logically arranged as a three-dimensional array, but is physically laid out as a two-dimensional array of two-dimensional arrays. The appropriate subarray is chosen by using a ***block address***, and then the appropriate bit in the subarray is chosen by using a combination of ***row*** and ***column*** addresses. The largest single subarray usually contains no more than 256 kbits of storage, because the performance suffers if these subarrays get too large.

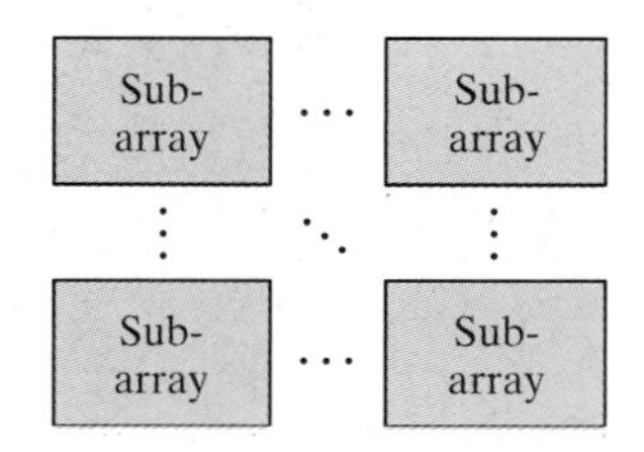

Figure 15–38 The architecture of a large semiconductor memory.

A single subarray from the memory in Figure 15-38 is illustrated in Figure 15-39, where we have assumed the memory to be a RAM. From this figure, we see that each subarray is fed an M-bit row address, an N-bit column address, an L-bit data word, and a command to tell the array whether to read or write the data at the specified address (R/W). The individual storage cells are connected in rows called ***word lines*** and in columns called ***bit lines***. The terminology evokes the image of a

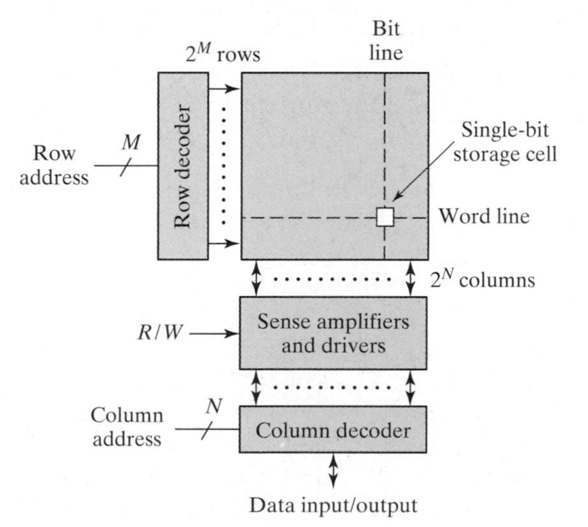

Figure 15–39 The architecture of one of the subarrays in Figure 15-38.

[14] The sizes of memories are almost always specified by rounding a base-10 number to a number that is a power of 2 times 1,000 or 1,000,000 (or, in the future, probably 1,000,000,000). But the size is understood to actually be a power of 2. Therefore, 1 kb really means 2^{10} = 1,024 bits, 64 kb really means 2^{16} = 65,536 bits, 1 Mb really means 2^{20} = 1,048,576 bits, and 64 Mb really means 2^{26} = 67,108,864 bits.

binary word being stored in each row and the column picking out a particular bit in the word, although there isn't any reason why the data in a particular subarray row needs to correspond to one word in the system using the memory.

The individual memory cells in this array may be made in a number of different ways, but there are two major categories: The RAM cells may be *static* or *dynamic*. ***Static RAM*** (SRAM) will retain its memory so long as the power is connected. The individual cells are constructed using flip-flops or some other bistable circuit. Static RAM has the advantages that the data can be read without destroying what is stored and the operation is simple. On the other hand, ***dynamic RAM*** (DRAM) is based on storing charge on a capacitor. Since the charge will eventually leak off, dynamic memory cells need to be refreshed periodically (typically every few ms)—hence the name dynamic. While DRAMs have the added complication of refresh circuitry, the basic cells are simpler, so the overall density of the memory is much higher. In the sections that follow, we discuss the design of the different basic memory cells and the peripheral circuits.

Exercise 15.10

Suppose a 32-Mb memory is made by using 64-kb subarrays. (a) How many subarrays are needed? (b) How many bits will we require for the block address? (c) For the row and column addresses? (Assume the that subarrays are square.)

Static RAM Cells An SRAM memory cell can be constructed with six transistors, as shown in Figure 15-40. The circuit is a bistable latch made from coupled inverters, as discussed in Section 14.2, along with two access transistors (M_1 and M_2) to allow us to set or reset the cell from the bit line (BL). We have shown the body contacts for the access transistors to emphasize that they are being used as pass gates. The word line (WL) is used to turn on the access transistors. In this way, all of the cells along a given word line in the memory have their access transistors turned on, but we read or write data only from the cell connected to the bit lines we have selected.

Notice that if the bit stored in this cell is a ONE ($D = 1$), we have $\overline{D} = 0$, and the inverters are in one of the two stable states. If the word line is low ($WL = 0$), the two access transistors are off and the cell simply retains the stored value. In

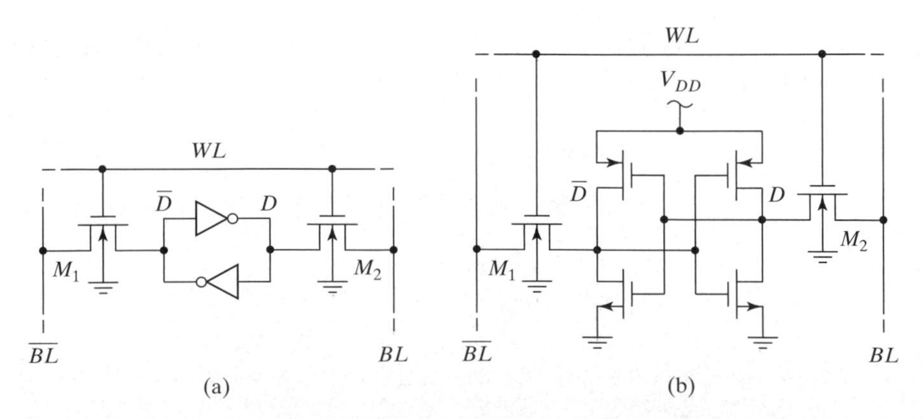

Figure 15–40 The six-transistor SRAM cell. (a) The circuit with the latch shown as coupled inverters and (b) the transistor-level schematic.

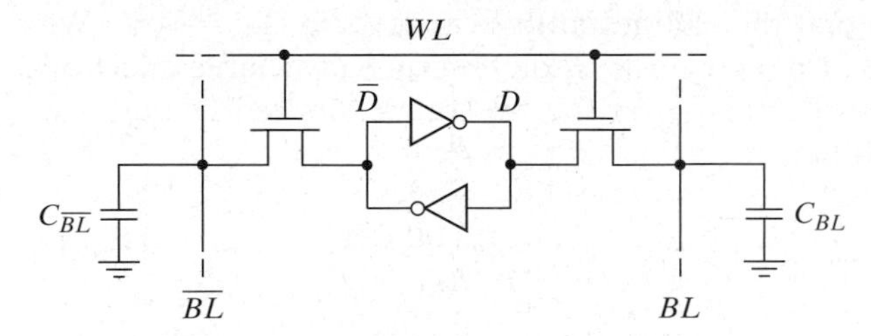

Figure 15–41 The six-transistor SRAM cell showing the bit-line capacitance.

order to read what is stored in the cell, we connect BL and $\overline{BL}$ to high-impedance inputs (e.g., a logic gate input) and set the word line high ($WL = 1$). Then M_1 and M_2 will both be on, and the memory cell will force $BL = D = 1$ and $\overline{BL} = \overline{D} = 0$. To write information into this location, we must connect BL and $\overline{BL}$ to low-impedance sources capable of forcing the voltages on these lines to the desired values and then set $WL = 1$.

When the cell is connected in a large memory, there will be significant parasitic capacitance to ground from both bit lines, as shown in Figure 15-41. The presence of this parasitic capacitance could be a problem when the cell is read. Suppose, for example, that just prior to reading the gate, $\overline{BL}$ is low, BL is high, D is low, and $\overline{D}$ is high. Then when we set WL high to read the gate, the existing levels on the bit lines may change the state of the cell if the parasitic capacitances are large enough. For this reason and others, both bit lines are usually precharged to the same value (V_{DD} and $V_{DD}/2$ are common) prior to a read operation.

Exercise 15.11

(a) Ignoring all of the decoders and other peripheral circuits, how many transistors are there in a 1-Mb SRAM chip if the six-transistor cell is used? (b) Assuming that each transistor requires a 1 μm × 1 μm area and that the wiring and contacts will consume 95% of the total area (which is not unreasonable!), how large must the IC be? (Assume that it is square, and give the length of a side in mm)?

□

Dynamic ROM Cells A one-transistor DRAM cell is shown in Figure 15-42 and consists of one transistor and the storage capacitor, C_S. This is the dominant memory cell in use today. There is again a large parasitic capacitor on the bit line as shown. To read the value of this cell, we first precharge the bit line to a voltage V_{pre} that stores a charge $Q_{BL} = V_{pre}C_{BL}$ on the parasitic capacitor. Assume that the

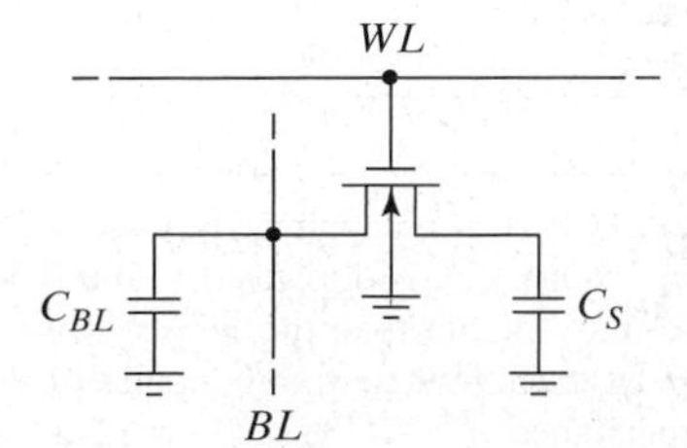

Figure 15–42 A one-transistor DRAM memory cell.

initial voltage on C_S is V_S so that the charge on this capacitor is $Q_S = V_S C_S$. We then set the word line high to turn on the transistor. Ignoring any charge injection from the switch, we find that the charge on the two capacitors redistributes to yield a final voltage on the bit line of

$$V_{BL} = \frac{Q_S + Q_{BL}}{C_S + C_{BL}} = \frac{V_S C_S + V_{pre} C_{BL}}{C_S + C_{BL}}. \quad \mathbf{(15.61)}$$

The total change in the bit-line voltage caused by reading the cell is

$$\Delta V_{BL} = V_{BL} - V_{pre} = \frac{C_S}{C_S + C_{BL}}(V_S - V_{pre}). \quad \mathbf{(15.62)}$$

Since C_{BL} is usually 10 to 100 times larger than C_S, the change in bit-line voltage is quite small and must be amplified by a sense amplifier. It is important to keep C_S as small as possible, because that minimizes the size of the memory cell, which allows larger memories to be built on one IC. Larger memory arrays also cause C_{BL} to increase, of course, so this problem is exacerbated even further. Also, note that reading the data in the storage cell destroys the data. Therefore, along with the need to periodically refresh the contents of a DRAM cell, we must also rewrite the contents after a read. Finally, we point out that the one-transistor DRAM cell provides only a single-ended output (that is, we only get D), whereas the SRAM cell we examined provides a differential output (D and $\overline{D}$). The differential output is twice as large (assuming the same single-ended values) and, therefore, easier to sense.

To write a bit into the cell, we set the bit line to either V_{DD} or 0 and set the word line high momentarily. Note that when BL is high, the voltage on the storage capacitor will not rise all the way to V_{DD}, since a single NMOS transistor has been used as the pass gate. Ignoring the body effect, we find that the stored voltage will be only $V_S(\text{high}) = V_{DD} - V_{th}$. This decrease in the stored voltage can be quite significant; for example, with a 3-V supply and a 0.8-V threshold, there is a 27% reduction in the high value stored. Because of this problem, modern DMOS memories typically **bootstrap**[15] the word line to a voltage higher than V_{DD}. Also, note that which side of the pass gate acts as the source of the transistor depends on the value of the bit line voltage and the stored voltage. There are many details and variations left out of the present discussion, and we refer you to [15.7] for a more thorough treatment.

Exercise 15.12

Consider writing a '1' into the DRAM cell in Figure 15-42. If $V_S = 0$ V prior to the write, describe the operation of the pass transistor during the write (i.e., which end acts as the source? What region of operation is the device in?).

---

[15] The word *bootstrap* is used to refer to any of a number of techniques whereby one voltage is made to differ from another voltage by a fixed amount. For example, if a capacitor is charged to, say, 1 V and is then connected to V_{DD} using switches, we can obtain a voltage about 1 V higher than the supply. This capacitor could then drive the word line higher than the supply. How much higher would depend on the size of the bootstrapping capacitor relative to the word line capacitance.

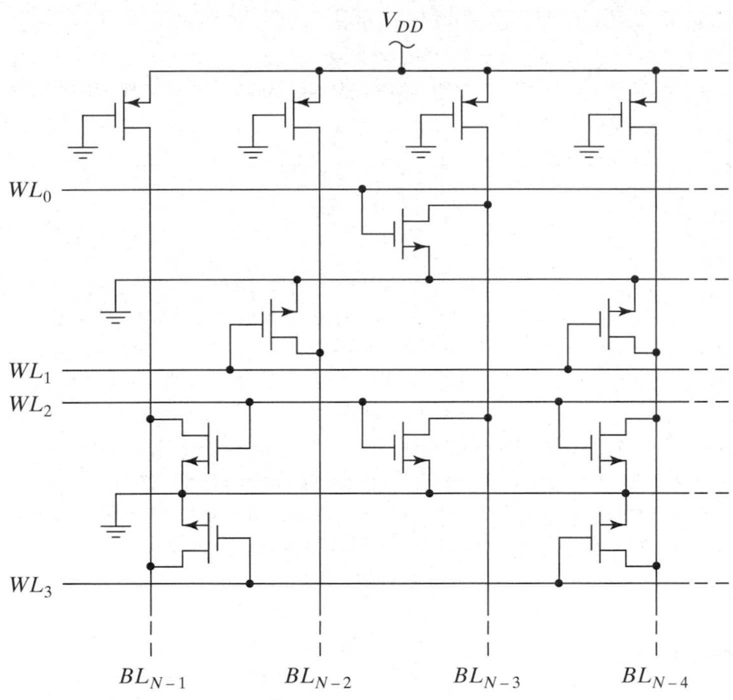

Figure 15–43 The upper-left 16 bits of a NOR ROM array.

ROM There are a number of different ways to make ROM cells; we will cover only one of them here. For those interested, a good in-depth treatment is given in [15.7]. Figure 15-43 shows the upper-left 16 bits of a NOR ROM array with *N* columns. The vertical bit lines each have a PMOS pull-up to V_{DD}, and the voltage on a bit line will, therefore, be V_{DD} unless it is pulled down by an NMOS transistor connected to ground. The NMOS devices are located at the intersections of the word lines and the bit lines. If the word line connected to a transistor's gate is high, then that transistor will pull the bit line low. If you look at the structure of any one bit line, you will see that it is a pseudo-NMOS NOR gate—hence the name NOR ROM.

The transistors in this memory cell are sized differently than they would be for a pseudo-NMOS NOR gate. The PMOS device is made larger in the memory array to decrease the time required to pull the bit line high, which can have substantial capacitance in larger memories ($C_{BL} > 1$ pF). The penalty paid for making this device larger is that the bit-line voltage corresponding to a zero is larger, and, therefore, the noise margin of the NOR gate is smaller. We can live with a lower noise margin in a memory, however, because the true logic level will be regenerated by the sense amplifier.

Commercial ROMs are laid out with a transistor at the crossing of every bit line and word line. Data are then programmed into the array by somehow determining which transistors are connected or active in the circuit. One approach is to customize a contact layer so that only some transistors actually contact the word line and bit line. A second approach is to implant transistors that are always off (for the NOR ROM). Either one of these approaches requires the manufacturer to customize the memory and therefore introduces a delay in product development.

PROMs were developed to allow customers to program their own memory chips. PROMs are programmed by storing charge on a floating gate transistor to change its threshold voltage [15.7] and will not be covered here.

Exercise 15.13

What are the four 4-bit words that are stored in the portion of the ROM shown in Figure 15-43?

□

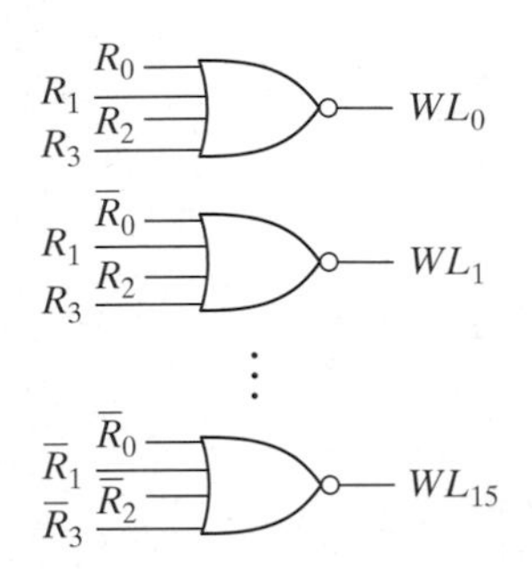

Figure 15–44 A row decoder for a 4-bit address word.

Decoders The memory-array architecture illustrated in Figure 15-39 shows two types of decoders: row decoders to produce the word lines and column decoders to select the appropriate bit line(s). In all of the memories we have discussed, the word line for the selected storage cell must go high to enable either a read or a write operation, while all other word lines must remain low. This decoding operation can be implemented using a NOR gate for each word line, as illustrated for a 4-bit address word in Figure 15-44. In drawing this figure, we have assumed that the complement of each address bit is available. Note that we want WL_1 to be high when the row-address input, $[R_3, R_2, R_1, R_0]$, is [0001]. In other words,

$$WL_1 = R_0 \cdot \overline{R_1} \cdot \overline{R_2} \cdot \overline{R_3} = \overline{\overline{R_o} + R_1 + R_2 + R_3}, \tag{15.63}$$

which is implemented by the NOR gate shown in the figure.

This decoder obviously becomes a problem for larger address words, owing to the large fan-in of the NOR gates. The problem is not as bad as it sounds, however, since the NOR gates can all be implemented using pseudo-NMOS NOR logic, in which case the decoder looks identical to a NOR ROM array. Nevertheless, as we noted in discussing complex CMOS logic gates, it is better to do a two-step decode to reduce the fan-in of the gates. Considering again the 4-bit decoder in Figure 15-44, note that the expression for the fifth word line can be rewritten as

$$\begin{aligned} WL_5 &= R_o \cdot \overline{R_1} \cdot R_2 \cdot \overline{R_3} = \overline{\overline{R_o} + R_1 + \overline{R_2} + R_3} \\ &= \overline{\overline{R_o \cdot \overline{R_1}} + \overline{R_2 \cdot \overline{R_3}}} \end{aligned} \tag{15.64}$$

where the first expression is the natural one to write, the second leads to a single NOR gate decoder as in Figure 15-44, and the final expression allows us to decode the word line in two steps, as shown in Figure 15-45. If we follow this same method, we can perform a two-step decoding for the entire row-address word.

The column decoder in Figure 15-39 is essentially a 2^N-to-1 multiplexer. We showed how to build a four-to-one multiplexer in Figure 15-37, and the column multiplexer could be built that way. However, there would 2^N lines, and each line would have N transmission gates on it requiring two transistors; therefore, this circuit would require $N2^{N+1}$ transistors. As an example, for $N = 10$, this works out to 20,480 transistors. The method would also put N transmission gates in series on

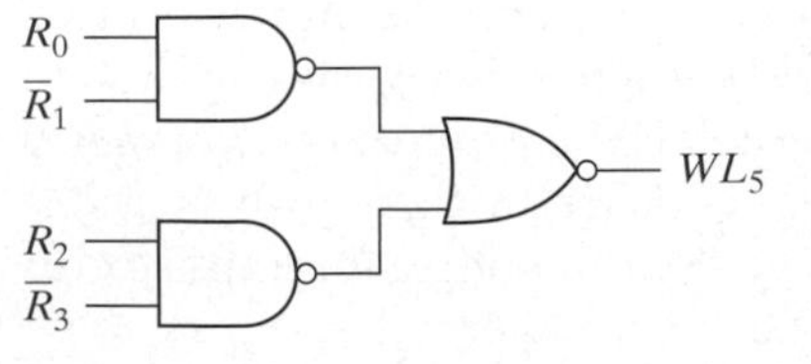

Figure 15–45 A two-step decoder for the fifth word line in Figure 15-44.

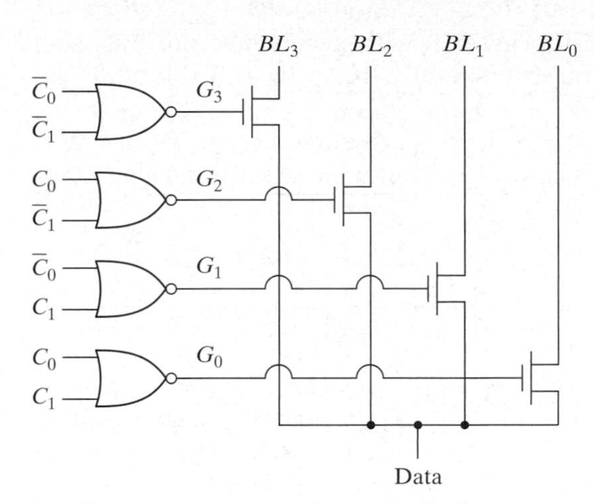

Figure 15–46 A four-to-one bit line decoder. (The transistors are all NMOS.)

each bit line, which would add capacitance and resistance to the lines. In this application, the full transmission gates are not required, so a better choice would be the multiplexer shown in Figure 15-46. This circuit uses single NMOS pass transistors in series with each bit line (the bulk connections have not been shown, for simplicity) and a NOR decoder (just like the row decoders already discussed) to drive the pass transistor gates based on the column-address word $[C_0,C_1]$. We have again assumed that the complements of the column address are available. This decoder places only one pass transistor in series with each bit line.

This bit line decoder requires 2^N pass transistors and 2^N lines in the NOR decoder, with N NMOS transistors each and one PMOS pull-up. Therefore, the NOR decoder requires $(N + 1)2^N$ transistors, and the whole bit line decoder requires $(N + 2)2^N$ transistors. For $N = 10$, this works out to be 12,288 transistors.

Another possible implementation is the *tree decoder* shown in Figure 15-47. This decoder does not require an address decoder in front of it and uses only $2^N + 2^{N-1} + \cdots 2^2 + 2 = 2(2^N - 1)$ transistors. For $N = 10$, this is only 2,046 transistors, less than one-tenth as many as in a transmission gate multiplexer. The

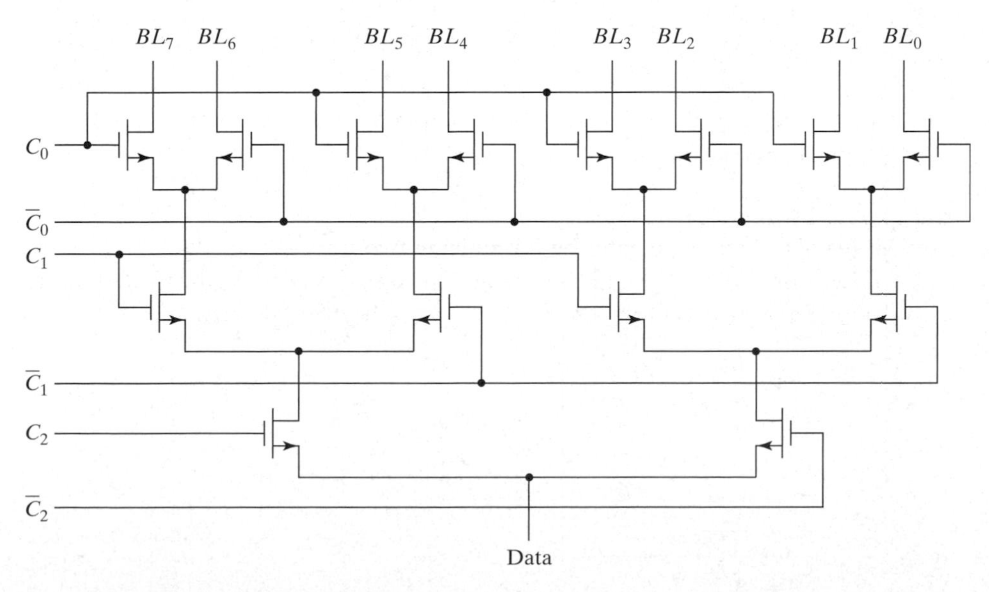

Figure 15–47 An eight-to-one tree decoder.

tree decoder also adds N pass transistors in series with each bit line, however, so it is slow. In addition, the series pass transistors reduce the swing available at the decoder output (*see* Problem P15.51).

It is possible to combine the decoder structures presented here to get decoders that use a reasonable number of transistors without unnecessarily slowing down the circuit.

Exercise 15.14

Draw the transistor-level schematic for the NOR decoder shown at the gate level in Figure 15-46. Use the NOR decoder structure shown as a NOR ROM array in Figure 15-43.

Sense Amplifiers Many sense amplifiers are typical analog amplifiers, as covered elsewhere in this text. For SRAMs and ROMs, it is common for the swing on the sense amplifier to be limited in some way so that they don't experience a full logic swing. Using a smaller swing saves power, since the amplifiers don't charge and discharge the large bit line capacitances as much. In one-transistor DRAMs, on the other hand, we need to be sure that a full logic swing is restored by the sense amplifier, because doing so will ensure that we replace the value that was held on the storage capacitor. (Remember that the read process on a one-transistor DRAM cell is destructive.) The most common circuit used for the sense amplifier in DRAM memories is the cross-coupled inverter. This circuit is the latch that we have already considered in Section 14.2 and earlier in this section when we examined the six-transistor SRAM cell. In this application, however, there is an additional set of switches (M_1 and M_2) to connect the inverters to the power supplies, as shown in Figure 15-48.

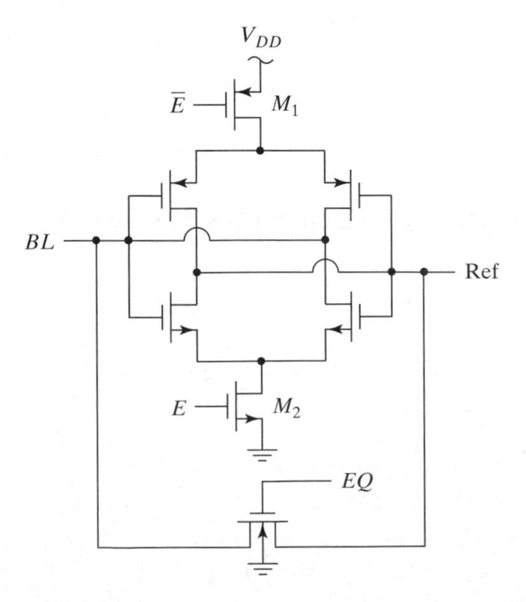

Figure 15–48 A DRAM sense amplifier.

Notice that the latch has a differential input, so we need to supply it with a reference voltage that is midway between the high and low outputs of the cell we are sensing. Also, the latch is on only when the enable input is high ($E = 1$). We begin the read process by turning on the equalizing switch by setting EQ high, and enabling the latch while the word line signal is still low (so that none of the DRAM cells are connected); this causes the cross-coupled inverters to be biased at the unstable midpoint of their transfer characteristic (*see* Figure 14-10). Next, the equalizing switch is turned off, the latch is disabled, and when that is done, the word line is pulled high. When the storage cell is connected to the bit line, the voltage will be pulled above or below the reference, thereby establishing an imbalance at the inputs to the latch. We then enable the latch again, and the positive feedback in the cross-coupled latch will rapidly cause the bit line to swing all the way in the direction of the imbalance. The level stored on the DRAM cell is thereby restored as we sense it.

There are many details of the timing of the read and write operations that we have overlooked in this brief overview of memory circuits. We refer the interested reader to [15.7] for a more detailed treatment.

15.4 BIPOLAR LOGIC CIRCUITS

Bipolar logic circuits can be divided into two major categories: *saturating logic* and *nonsaturating logic*. Saturating logic circuits have one or more of the transistors in saturation for some particular logic state. Nonsaturating logic does not allow the transistors to saturate. Saturating logic has the advantage that the voltage across the transistor (from collector to emitter) is relatively stable with respect to changes in device parameters, temperature, and supply voltage. But it is much slower than nonsaturating logic, since it takes time to remove the excess charge stored in the base of a saturated transistor (*see* Section 15.1.2 and Aside A15.2). Saturating logic is essentially obsolete at this time, but is still occasionally used.

The most common example of saturating logic is the old 7400 series *transistor–transistor logic* (TTL). More modern TTL logic families have been made nonsaturating through the use of *Schottky clamps*, as we will discuss. Even the newer forms of TTL logic are rapidly being replaced by the newest CMOS logic families (e.g., 74HCXX), which offer comparable speeds at lower power.

The most common example of nonsaturating bipolar logic circuits is *emitter-coupled logic* (ECL). ECL has been available for over 30 years and continues to evolve, but it is still used in applications requiring very high speeds (100–500 MHz), and variants of ECL are used in custom digital circuits at speeds well above 500 MHz.

15.4.1 Transistor–Transistor Logic (TTL)

Because TTL circuits are essentially obsolete, our treatment will be brief. Nevertheless, there is a great deal that can be learned about the operation of bipolar transistors in switching applications by studying TTL gates. It is also useful to see how TTL gates evolved out of earlier logic circuits, so we begin with a brief look at the development of TTL logic. We then cover the operation of standard TTL logic and finally discuss how the gates can be modified to prevent saturation. For those interested in more detailed treatments of TTL, we suggest [15.8].

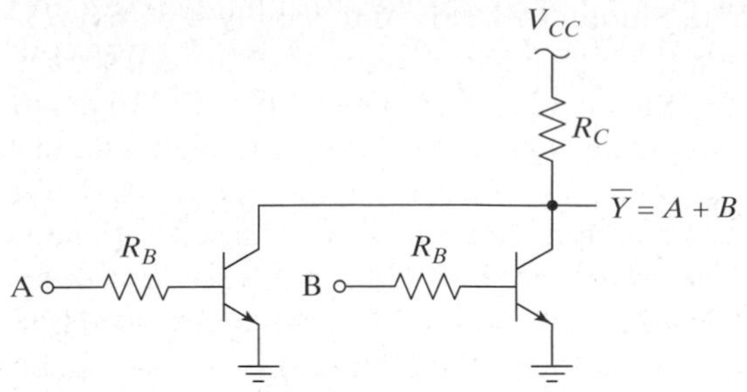

Figure 15–49 An RTL NOR gate.

Development of TTL It is interesting and useful to briefly go over the development of TTL logic circuits. Let's consider the NOR function as an example. A two-input NOR requires that the output be low if either of two inputs is high. We can achieve this function by using two *npn* transistors in parallel with their collectors tied together at the output, their emitters both grounded, and their bases acting as the inputs. The output can then be connected to V_{CC} through a ***pull-up resistor*** as shown in Figure 15-49. The resulting gate is called a ***resistor–transistor logic*** (RTL) gate. Notice that if we remove one of the two transistors along with its R_B, we have an RTL inverter, which is equivalent to a common-emitter amplifier. So we see that the simplest bipolar inverter is really an overdriven amplifier.

This gate will function with appropriate choices for the values of the resistors, but it suffers from several problems. The most notable problems are low noise margins and large power-delay products (PDPs). In discrete design, these problems may not be severe for some applications, and circuits similar to this may be the most convenient way of implementing a one-of-a-kind design on a discrete test board. Nevertheless, the performance of this kind of gate is unacceptable for any complex logic function.

An alternative bipolar gate topology is shown in Figure 15-50 and is an example of ***diode–transistor logic*** (DTL). This circuit implements a two-input NAND gate. (If we remove one of the two input diodes, we have a DTL inverter.) The voltage V_{BB} can be either ground or some negative voltage (typically, −2 V). If either of the inputs is low, it pulls the voltage at the bottom of the 2-kΩ resistor low, and no base current can be provided to the BJT, so it is off. With the BJT off, the collector resistor pulls the output high. For the output to go low, both inputs must be high. The base current of the BJT is then provided through the 2-kΩ resistor, and the transistor's collector current pulls the output low.

This gate suffers from many problems also, the worst being that it is slow and has a large PDP. The slow speed is caused by using a resistive pull-up at the output (as opposed to using another transistor as an ***active pull-up***) and by the time required to remove the excess base charge when turning the BJT off. The turnoff

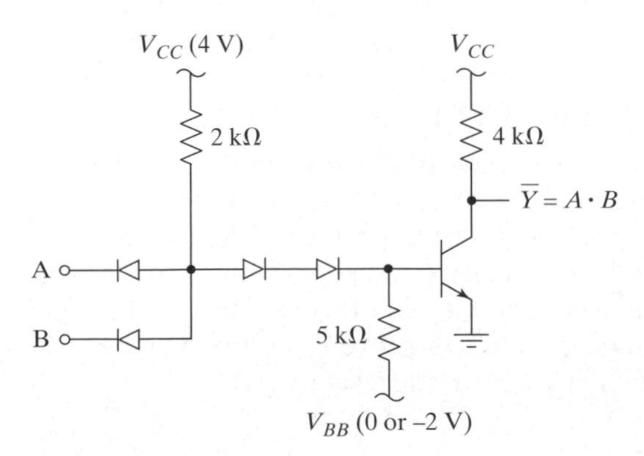

Figure 15–50 A DTL NAND gate.

time of the BJT can be improved by making V_{BB} negative, but then two supplies are required. The speed could also be improved by using a smaller resistor connected to the base, but then the static power dissipation will go up.

Exercise 15.15

Consider three RTL inverters in series. Ground the input to the first inverter, and assume that $V_{CC} = 5$ V, $R_C = 5$ kΩ, and $R_B = 38$ kΩ for all three. What are the nominal values of V_{OL} and V_{OH} for these inverters with a fan-in of unity and a fan-out of unity?

Standard TTL The classic TTL NAND gate shown in Figure 15-51 overcomes the major problems of the DTL NAND gate. The input stage serves the same purpose as the two input diodes in the DTL gate. The multi-emitter transistor can be viewed as two transistors with their bases and collectors tied together. The phase-splitter stage has two outputs, one at its emitter and one at its collector, that are 180° out of phase with each other (i.e., one is high while the other is low and vice versa). The output stage is called a ***totem-pole*** output and provides both active pull-up and pull-down transistors. If Q_1 is made with a single emitter, the circuit is a TTL inverter.

We now analyze the gate when the output is high. This will occur if either one of the two inputs is low, as shown in Figure 15-52. The voltages present at the different nodes in the circuit are also indicated on the figure. We assume that the low input is 0.2 V and the high input is 3.6 V (typical for TTL). With input B low, the corresponding base-emitter junction of Q_1 is forward biased. The base voltage will be about 0.7 V above the emitter, or about 0.9 V as shown. The base current will then be

$$I_B = \frac{5 - 0.9}{4\ \text{k}} \approx 1\ \text{mA}. \tag{15.65}$$

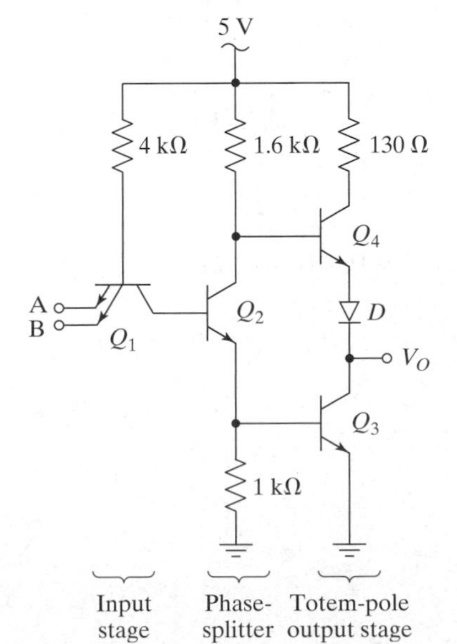

Figure 15–51 A basic two-input TTL NAND gate.

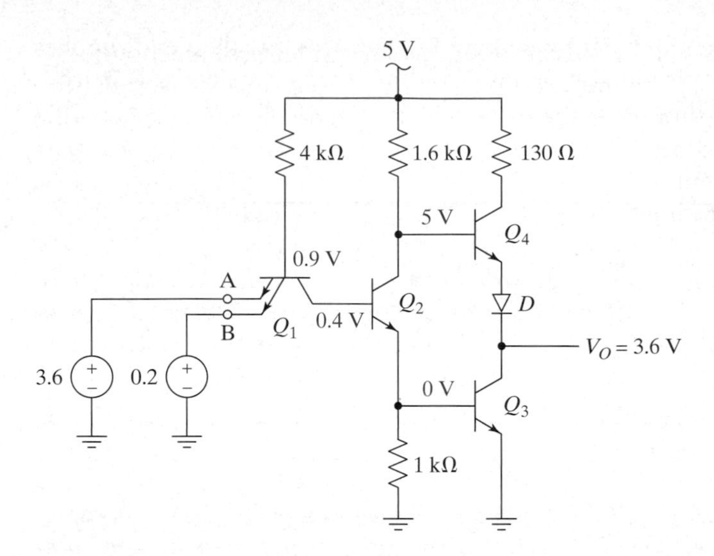

Figure 15–52 The TTL NAND gate from Figure 15-51 when one input is low.

This base current is sufficient to support a large collector current, but the collector current would have to flow *out of* the base of Q_2. Therefore, the collector current will essentially be zero and Q_1 will be heavily saturated. The voltage at the collector of Q_1 is then about 0.4 V as shown. With the base of Q_2 at about 0.4 V, we can conclude that both Q_2 and Q_3 must be off. (To turn them on would require $V_{B2} = V_{BE2} + V_{BE3} \approx 1.4$ V.) With Q_2 cutoff, the voltage at the base of Q_4 is pulled high to nearly 5 V. In this state both the diode and Q_4 will be turned on. With no load connected, the only current through them will be the leakage current of Q_3, but they will be on, nonetheless. The voltage at the output is therefore about 3.6 V as shown.

If the low input now switches back high, the gate will switch toward the steady-state solution shown in Figure 15-53. In this case, we need to notice that there are three *pn* junctions from the base of Q_1 to ground: the base-collector junction of Q_1 and the base-emitter junctions of Q_2 and Q_3. Therefore, the base of Q_1 cannot rise much above 2.1 V, and the bases of Q_2 and Q_3 are at about 1.4 V and 0.7 V, respec-

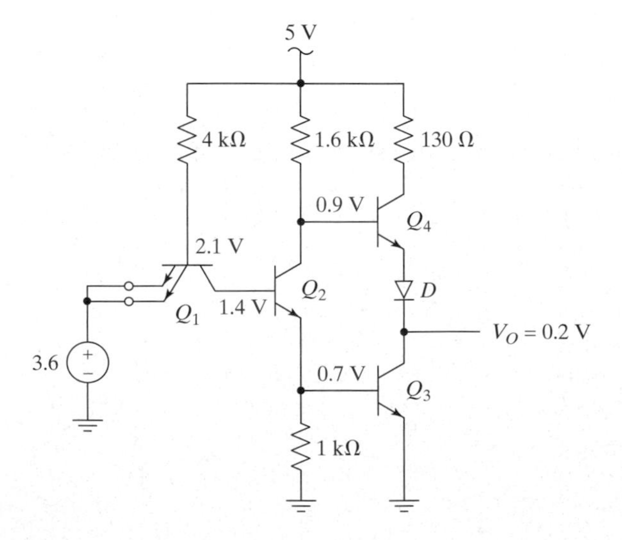

Figure 15–53 The TTL NAND gate from Figure 15-51 when both inputs are high.

tively. With both inputs at 3.6 V, we can see that both emitter-base junctions of Q_1 must be reverse biased. Since the base-collector junction is forward biased, however, the transistor is operating in the reverse-active region. The reverse beta (β_R) is typically quite low. In fact, in TTL circuits it is deliberately made much less than unity, so the input current is not large. (This increases the fan-out of the gate, since each gate can supply the input currents to several gates.) The base current of Q_1 is now given by

$$I_{B1} = \frac{5 - 2.1}{4\ \text{k}} \approx 0.75\ \text{mA}. \quad \textbf{(15.66)}$$

Because of the small value of β_R, the collector current of Q_1 is nearly equal to the base current given by (15.66). The collector current of Q_1 is also the base current of Q_2, and because this base current is so large, Q_2 is driven hard into saturation, and its collector voltage is only about 0.9 V as shown ($V_{C2} = V_{BE3} + V_{CEsat}$). This voltage appears at the base of Q_4 and is insufficient to turn Q_4 and the diode in series with it on, so they are both off. The emitter current of Q_2 is quite large; specifically,

$$I_{E2} = I_{B2} + I_{C2} \approx 0.75\ \text{mA} + \frac{5 - 0.9}{1.6\ \text{k}} = 3.3\ \text{mA}, \quad \textbf{(15.67)}$$

but the current through the 1-kΩ resistor is limited to

$$I_{R1k} = \frac{V_{BE3}}{1\ \text{k}} \approx 0.7\ \text{mA}. \quad \textbf{(15.68)}$$

The rest of I_{E2} is forced into Q_3 as base current, so $I_{B3} \approx 2.6$ mA. This is more than enough current to saturate Q_3 (provided that the fan-out specification of the gate is not exceeded), so the output voltage is about 0.2 V as shown.

Several comments can now be made about the performance of the standard TTL gate shown. Most important, the gate is very slow because some of the transistors are allowed to saturate in each state. It takes time to remove the excess base charge from these transistors, and that slows down the switching time, as seen in Section 15.1.2 and Aside A15.2. The charge stored in the base of Q_3 is the dominant term slowing down the gate. The saturation of Q_2 does not affect the speed nearly as much, because when either of the inputs goes low, Q_1 turns on, and its collector current pulls the charge out of the base of Q_2 rapidly. The saturation of Q_1 also does not affect the speed much, since it switches from saturation to reverse active, rather than cutoff, and the excess base charge does not need to be removed.

The charge stored in the base of Q_3 could be removed faster if the 1-kΩ resistor were made smaller or if it connected to a negative supply instead of ground. Making the resistor smaller increases the power consumption of the gate, however, and using a negative supply complicates the use of the gate. We will see a better solution to this problem later.

Another major problem with this TTL gate is that a large current spike occurs during the time the output is changing from low to high. This spike is caused by the fact that Q_2 can turn off quickly, but Q_3 cannot. When Q_2 turns off, its collector voltage goes high quickly and turns Q_4 and the diode on. When they first turn on, however, Q_3 has not yet had sufficient time to turn off. The result is that there is a current path from V_{CC} to ground through the 130-Ω resistor, Q_4, the diode, and Q_3. The 130-Ω is there to limit the magnitude of this current, but the current can be large even with this resistor present (on the order of tens of mA). This current

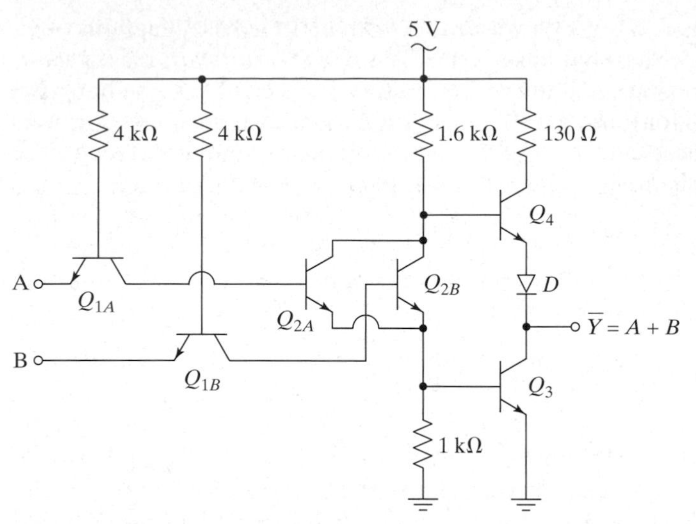

Figure 15–54 A 2-input TTL NOR gate.

spike is troublesome because all power supply connections have some series resistance and inductance and the current therefore leads to high-frequency voltage noise on the power supply.

A TTL NOR gate is shown in Figure 15-54. In this circuit, there are two phase-splitter transistors: Q_{2A} and Q_{2B}. If either of them turns on, the output will be pulled low. The rest of the circuit functions the same as the NAND gate.

Commercial TTL gates also usually have a diode connected from each input to ground (anode on ground, cathode on the input), so that you can't have a large negative voltage spike at the input (which might be due, for example, to ringing on a transmission line). These clamps are added because any large negative input could destroy Q_1.

Example 15.8

In this example, we simulate the behavior of a TTL NAND gate to examine the switching transient and the current spike that occurs.

We use the circuit of Figure 15-53, but tie the B input high (3.6 V) and step the A input from 3.6 V to 0.2 V and then back again. The transistor model used is not for any particular TTL process; it is just a rough estimate of what the proper model might be:

```
.model NTTL npn(is=1e-16 bf=50 br=.2 rb=75 vaf=50 tf=.5n
tr=50n cje=.5p cjc=1p ccs=1.5p)
```

The results of the simulation are shown in Figure 15-55. The resulting propagation delays are approximately correct for a typical standard TTL logic gate. The negative glitch seen in v_O at the start of t_{PLH} is caused by the negative input step capacitively coupling through to the output.

A careful analysis of the switching times is quite complicated, but referring back to Aside A15.2, we can understand what is going on. When v_A goes low, Q_1 is taken from the reverse-active region into saturation, which can happen quickly, since the base current of Q_1 is so large. Q_1 then pulls enough current out of the base of Q_2 that Q_2 is taken out of saturation quickly as well. The bulk of the delay is caused by Q_3 coming out of saturation. Using (A15.11) and the given device data, we determine that $\tau_S = 15$ ns for these transistors. Then, we use the SPICE simulation to find that the forward base current in Q_3 is 2.2 mA when it is saturated, and we use (A15.12) to find that the reverse base current is -0.7 mA. Plugging these values into (A15.15) with the realization that $I_{Csat} = 0$ for this no-load simulation, we predict that $t_s = 15$ ns, not too far off from the simulated value.

When v_A goes high, the output does not change right away, because we have to provide for the delay time involved in taking Q_1 out of saturation and for turning Q_2 and Q_3 on. These times are all shorter than the time required to take Q_3 out of saturation, so t_{PHL} is substantially smaller than t_{PLH}.

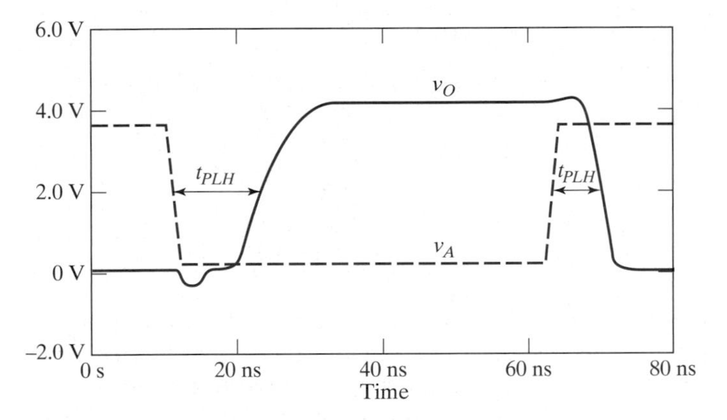

Figure 15–55 Transient simulation results for a standard TTL NAND gate. The propagation delays are $t_{PLH} \approx 12.5$ ns and $t_{PHL} \approx 7$ ns.

Figure 15-56 again shows the input and output voltages, but this time the supply current, i_{CC}, is also shown. The supply current momentarily spikes to a value of nearly 16 mA during t_{PLH}, since Q_3 is not off yet.

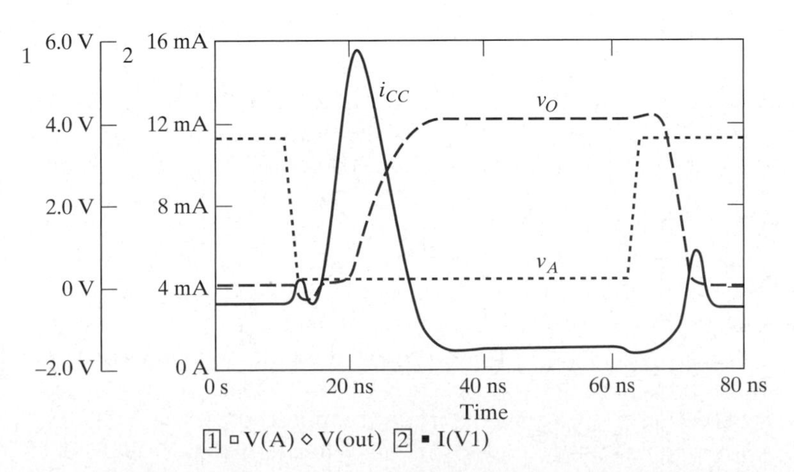

Figure 15–56 Transient output showing the current spike for the TTL NAND gate.

Exercise 15.16

A DM7404 IC has six TTL inverters in one package. Suppose the total parasitic inductance in series with the V_{CC} lead is 40 nH (package inductance and capacitance vary widely; see [15.7]), and assume that each inverter has the current waveform shown in Figure 15-57 when switching from low to high. (This is an approximation to the current simulated in Example 15.8.) If all six inverters happen to switch at the same time, what does the internal V_{CC} look like? Assume that $V_{CC} = 5$ V external to the package and that the parasitic series resistance in the line is zero.

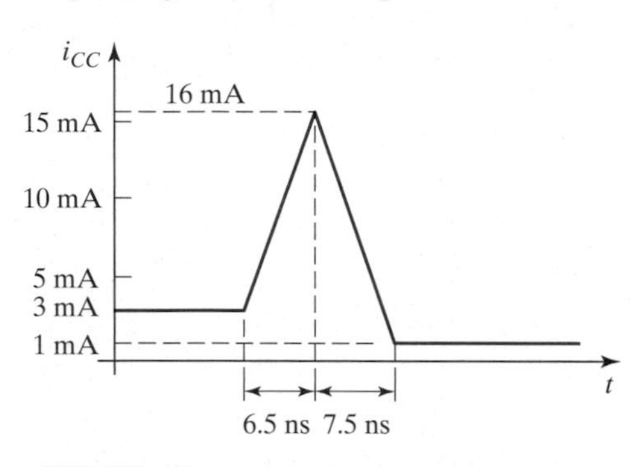

Figure 15–57 An approximation to the current in a TTL gate when the output switches from high to low.

Schottky TTL One common enhancement to the standard TTL gate is to use Schottky diodes to clamp the base-collector junctions of Q_1, Q_2, and Q_3 to prevent them from saturating. A ***Schottky-clamped transistor*** is shown in Figure 15-58. The Schottky diode turns on with a forward bias of only about 0.3 V or 0.4 V, so it diverts the extra base current without allowing the base-collector junction of the BJT to be significantly forward biased. Since the Schottky diode does not suffer from minority-carrier charge storage, it can turn off very fast.

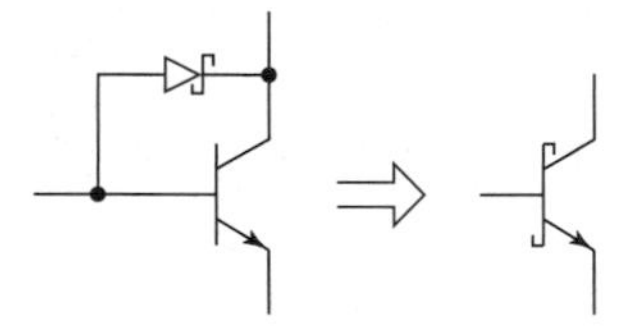

Figure 15–58 A Schottky-clamped transistor.

A simplified schematic for a Schottky-TTL (STTL) NAND gate is shown in Figure 15-59. Many other forms of Schottky-clamped TTL gates have also been developed that consume less power while still achieving high-speed operation. More comprehensive treatments of these variations of TTL can be found in [15.6] and [15.8], but will not be covered here, since TTL is obsolete.

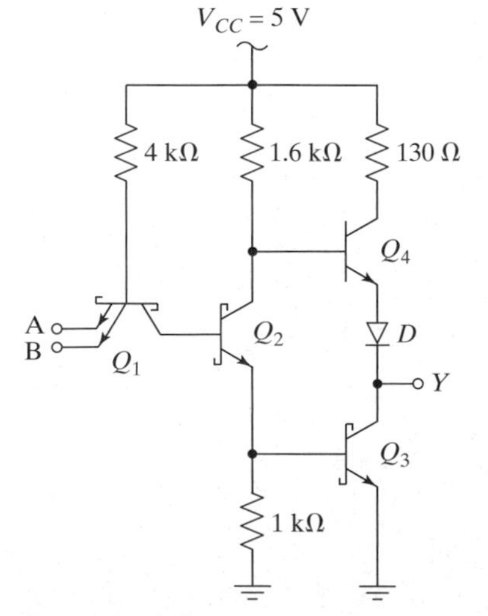

Figure 15–59 A simplified Schottky-TTL NAND gate.

15.4.2 Emitter-Coupled Logic

Emitter-coupled logic (ECL) is the fastest form of logic. The basic principle is that currents are switched instead of voltages. Switching currents can be faster because the devices remain forward active and the voltage swings are very small, so the circuits aren't slowed down by having to charge and discharge node capacitances over large ranges. Several different ECL families have been produced over the years, and, in addition, variations on the basic gate are used in very high-speed custom designs. In fact, if just the basic core of the ECL gate is used (i.e., without the emitter-follower output buffers), the resulting logic gate is called *current-mode logic* (CML). CML can be used when the gate does not have to supply much current to a load (e.g., for logic that is strictly internal to an integrated circuit) and is often used in custom designs. It can be used with MOSFETs as well.

We will illustrate the operation of ECL gates with one specific example: the 10K logic family. After a brief general discussion of the gate operation we examine the static and dynamic performance in more detail. We will then close with a look at the biasing circuitry.

Basic Operation A 10K family ECL OR–NOR gate is shown in Figure 15-60 with external load resistors connected to V_{EE}. We will discuss the load resistors more later. The bias voltage V_{BB} (which is nominally -1.29 V for 10K ECL) is derived internally, but for now we will ignore the details of how it is generated. Note that standard 10K ECL voltage levels are specified for operation from a single *negative* supply (-5.2 V). The circuit itself has no way of knowing where the "real" ground is, of course, so you can run ECL gates off of a single positive supply. However, ground is used for a good reason, as we will shall soon see.

The main advantage of ECL is its speed, but there are other advantages as well. For one thing, ECL gates provide differential outputs. Having differential outputs available sometimes helps simplify the implementation of a given logic function. Differential outputs also help reduce noise on the power supply caused by a changing supply current. If both outputs are equally loaded, then while one is switching from low to high, the other is switching from high to low. If everything is perfectly symmetric, the sum of the two currents (in Q_3 and Q_4) will ideally be constant (*see* Problem P15.59); therefore, the drop across any series resistance or inductance in the supply lead will also be constant. If the outputs are not symmetrically loaded (as in a logic gate without differential outputs), this cancellation does not occur.

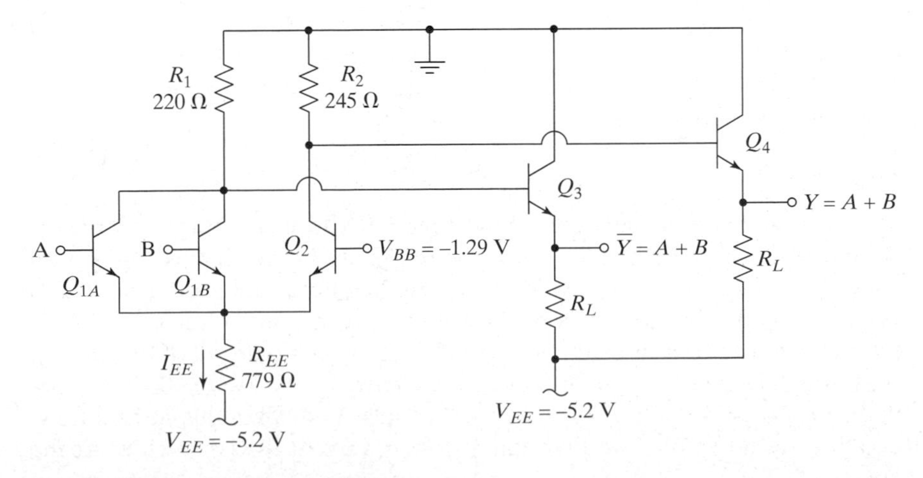

Figure 15–60 An ECL OR–NOR gate.

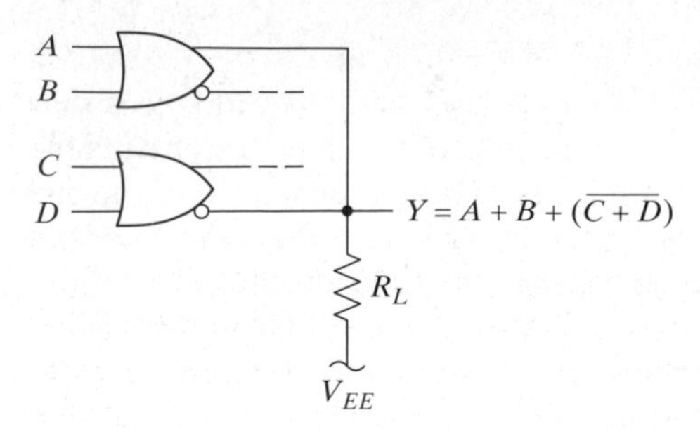

Figure 15–61 A wired-OR connection of two ECL outputs.

Finally, we note that the outputs of more than one gate can be tied together directly as shown in Figure 15-61. Since the emitter follower of either gate can pull the output high, the output in this case is seen to be $Y = A + B + \overline{(C + D)}$. In other words, we can implement an OR function just by tying two outputs together (and using the NOR outputs, we can implement an AND function as well); this kind of connection is called either a wired-OR or emitter dotting (because tying the two emitters together on a schematic requires a connection dot).

Now let's return to the circuit in Figure 15-60. If A and B are both low (substantially below -1.29; typically, -1.77), then Q_{1A} and Q_{1B} are essentially off, and virtually all of the current I_{EE} flows through Q_2. This current is given by

$$I_{EE2} = \frac{V_{BB} - V_{BE2} - V_{EE}}{779} \approx \frac{3.2}{779} = 4.1 \text{ mA}, \tag{15.69}$$

where we have added the subscript '2' to I_{EE} to indicate that this is the value when Q_2 takes almost all of the current. Assuming that β is large, we have $I_C \approx I_E$, and we can write

$$V_{C2} = V_{B4} = -I_C R_2 = -1 \text{ V}. \tag{15.70}$$

Since Q_{1A} and Q_{1B} have essentially no current through them, we also get

$$V_{C1} = V_{B3} \approx 0. \tag{15.71}$$

Using (15.70) and (15.71), we find that

$$V_Y = V_{B4} - V_{BE} \approx -1.7 \text{ V} \tag{15.72}$$

and

$$V_{\overline{Y}} = V_{B3} - V_{BE} \approx -0.7 \text{ V}. \tag{15.73}$$

The typical outputs are actually somewhat lower, because the output current is usually large enough that the base-emitter voltages are larger than 0.7 V. Typical values are $V_{OH} = -0.88$ and $V_{OL} = -1.77$ V. (When the output is high, the current in R_L is larger, so the V_{BE} of the emitter-follower transistor is larger, too).

If either A or B (or both) go high (to -0.88 V), then essentially all of the current will flow through Q_{1A} and/or Q_{1B}. In this case, we have $V_{E1} = -0.88 - 0.7 = -1.58$, and, therefore, $I_{EE1} = (-1.58 - V_{EE})/R_{EE} = 4.65$ mA. With all of this current flowing through Q_{1A} and/or Q_{1B}, we find that $V_{B3} = -(4.65 \text{ mA})R_1 = -1$ V. At the same time, $I_{C2} \approx 0$, so $V_{B4} \approx 0$ V. Therefore, we have $Y \approx -0.7$ V and $\overline{Y} \approx -1.7$ V. (The typical values are again -0.88 V and -1.77 V, respectively, due to larger V_{BE}.) From this brief description of the operation, we can see that the Y output implements the OR function and the $\overline{Y}$ output implements the NOR function.

We can now explain why ECL gates use ground and V_{EE} as their supplies instead of V_{CC} and ground. Notice from the preceding analyses that the output voltages depend on the most positive supply voltage (ground), the collector resistors, and the value of I_{EE}. Noise on the negative supply, V_{EE}, will not affect the outputs very much, but noise on ground will. Since ground is usually the least noisy point in a system, it helps to have the output voltages referenced to it rather than some positive supply.

Let's now consider again the loads connected to the gate. Since ECL gates are usually used for high-speed logic, the traces connecting gates on a printed-circuit board (PCB) are dealt with as transmission lines (*see* Section 14.4). The loads that are used are usually designed to match the impedance of the line being driven by the gate in order to minimize reflections. The load is typically connected near the input to the gate being driven, so that the line is properly terminated at the receiving end, as shown in Figure 15-62, where we have assumed that the lines have a characteristic impedance of 150 Ω. It is also common to save power by using two resistors on each line to form a Thévenin equivalent load that terminates the line to -2 V instead of -5.2 V (*see* Problem P15.60).

In commercial ECL gates, the two emitter-follower output transistors have their own ground pin separate from the rest of the circuit. Since the loads and logic swings are not perfectly symmetric, there can be relatively large changes in the sum of these two currents during the switching transient. By using a separate package pin, these transients are isolated from the main part of the logic gate. In addition, commercial gates have large-value pull-down resistors (50 kΩ) connected from each input to V_{EE}. These resistors guarantee that unused inputs are valid logic low.

Finally, we note that the NAND and AND functions can be implemented in ECL by using two other techniques: stacking the emitter-coupled pairs and collector dotting. The ECP at the core of the ECL gate can be stacked vertically with other ECPs to implement the NAND function, as illustrated in Figure 15-63. This kind of stacking is also sometimes called *series gating*. The input voltage ranges are not the same for all inputs, but this complication can be overcome with relatively simple level-shifting circuits. In this example, both inputs must be high to steer I_{EE} through R_1 and cause the output to go low.

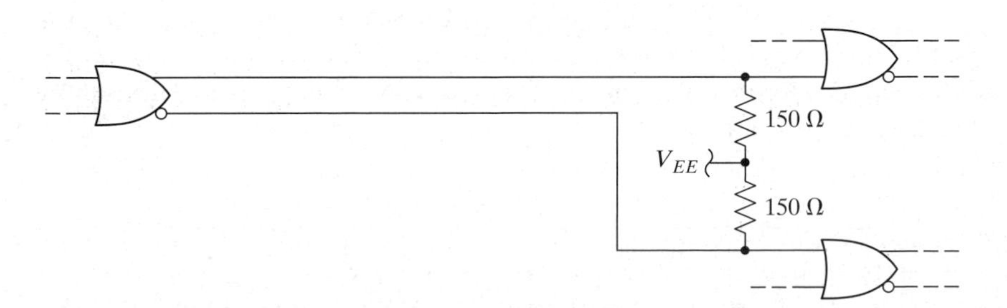

Figure 15–62 A typical ECL load connection.

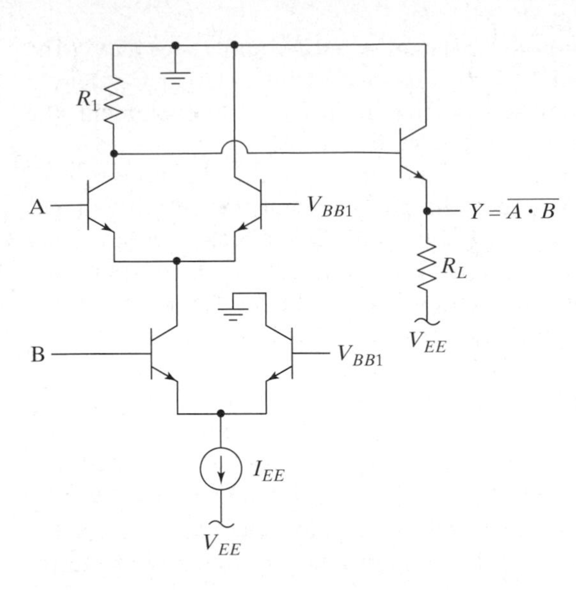

Figure 15–63 An illustration of series gating.

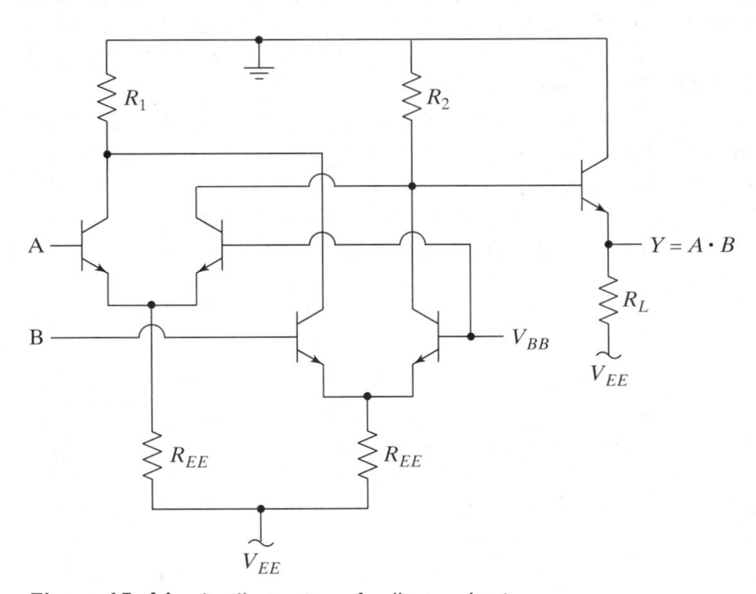

Figure 15–64 An illustration of collector dotting.

Collector dotting is illustrated in Figure 15-64 and produces the AND function. This connection is called *collector "dotting,"* because we have tied the collectors of two ECL inverters together, which requires a connection "dot" in the schematic. In this example, if both A and B are high, no current will flow through R_2, and the output will be high. Note that the output at the bottom of R_1 is *not* the complement of the actual output in this circuit, though, so we don't have a differential output available.

Exercise 15.17

Use collector dotting to create an ECL gate that implements the function $Y = (A + B) \cdot C$.

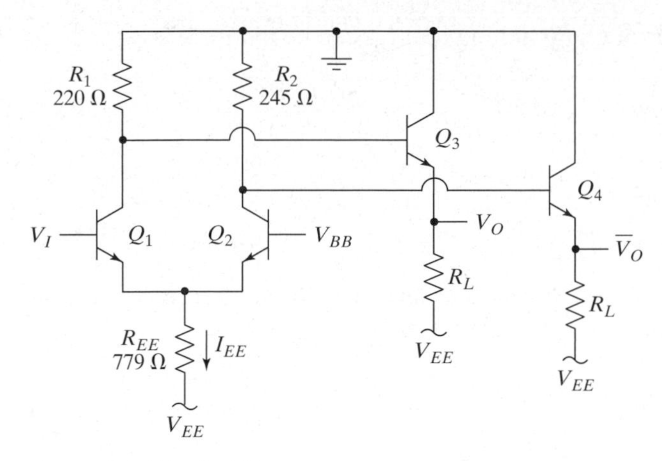

Figure 15–65 An ECL inverter.

Static Transfer Characteristic and Noise Margins We now derive the static, or DC, large-signal transfer characteristic of an ECL inverter. Consider the inverter in Figure 15-65, and assume that the transistors are perfectly matched (so that $I_{S1} = I_{S2} = I_S$) and at the same temperature (so $V_{T1} = V_{T2} = V_T$). Writing KVL around the base-emitter loop yields

$$V_I - V_{BE1} + V_{BE2} - V_{BB} = 0, \qquad \textbf{(15.74)}$$

which can be manipulated to give

$$\begin{aligned} \Delta V_{BE} &= V_I - V_{BB} = V_{BE1} - V_{BE2} \\ &= V_T\left(\ln\frac{I_{C1}}{I_S} - \ln\frac{I_{C2}}{I_S}\right) = \mathrm{V_T}\ln\frac{I_{C1}}{I_{C2}}, \end{aligned} \qquad \textbf{(15.75)}$$

or

$$\frac{I_{C1}}{I_{C2}} = e^{\Delta V_{BE}/V_T}. \qquad \textbf{(15.76)}$$

Summing currents at the emitters yields

$$I_{C1} + I_{C2} = \alpha I_{EE}, \qquad \textbf{(15.77)}$$

where we approximated I_{EE} as being constant. Using this equation along with (15.76), we obtain

$$\alpha I_{EE} = I_{C1}\left(1 + \frac{I_{C2}}{I_{C1}}\right) = I_{C1}(1 + e^{-\Delta V_{BE}/V_T}). \qquad \textbf{(15.78)}$$

We now solve (15.78) for I_{C1}:

$$I_{C1} = \frac{\alpha I_{EE}}{1 + e^{-\Delta V_{BE}/V_T}}. \qquad \textbf{(15.79)}$$

The voltage at the collector of Q_1 (assuming that I_{B3} can be ignored) is

$$V_{C1} = -I_{C1}R_1 = \frac{-\alpha I_{EE}R_1}{1 + e^{-\Delta V_{BE}/V_T}} = \frac{-\alpha I_{EE}R_1}{1 + e^{(V_{BB}-V_I)/V_T}}. \tag{15.80}$$

The output voltage is

$$V_O = V_{C1} - V_{BE3}. \tag{15.81}$$

We find V_{OL}, V_{IL}, V_{OH}, and V_{IH} by taking the derivative of (15.81) and setting it equal to -1. Approximating V_{BE3} as a constant, we have

$$\frac{dV_O}{dV_I} = \frac{dV_{C1}}{dV_I} = \frac{-\alpha I_{EE}R_1 e^{(V_{BB}-V_I)/V_T}}{V_T(1 + e^{(V_{BB}-V_I)/V_T})^2} = -1, \tag{15.82}$$

and solving (15.82) for V_I, we find that

$$V_I = V_{BB} - V_T\ln\left(\frac{\alpha I_{EE}R_1 - 2V_T \pm \sqrt{(\alpha I_{EE}R_1)^2 - 4V_T\alpha I_{EE}R_1}}{2V_T}\right). \tag{15.83}$$

Figure 15-66 is a plot of (15.81) made using $I_{EE} = 4.6$ mA, $R_1 = 220\ \Omega$, $\alpha = 0.98$, and $V_{BE3} = 0.8$ V. (Remember, we said that V_{BE} is larger because of the large current used.) From this plot, we see that the transfer characteristic is symmetric about $V_I = V_{BB}$, as we expect. The negative solution in (15.83) is V_{IH}. When $V_I = V_{IH}$, the output is nearly at its minimum value, and $I_{C1} \approx I_{EE}$. Therefore,

$$V_{IH} \approx V_{BB} + V_T\ln\left(\frac{I_{EE}R_1}{V_T} - 1\right). \tag{15.84}$$

Plugging in the numbers yields $V_{IH} = -1.20$ V. Plugging this value back into (15.80) and then (15.81) yields $V_{OL} = -1.77$ V. We also obtain $V_{IL} = -1.38$ V and $V_{OH} = -0.83$ V. The corresponding noise margins are

$$NM_H = V_{OH} - V_{IH} = 0.37\text{ V} \tag{15.85}$$

and

$$NM_L = V_{IL} - V_{OL} = 0.39\text{ V}. \tag{15.86}$$

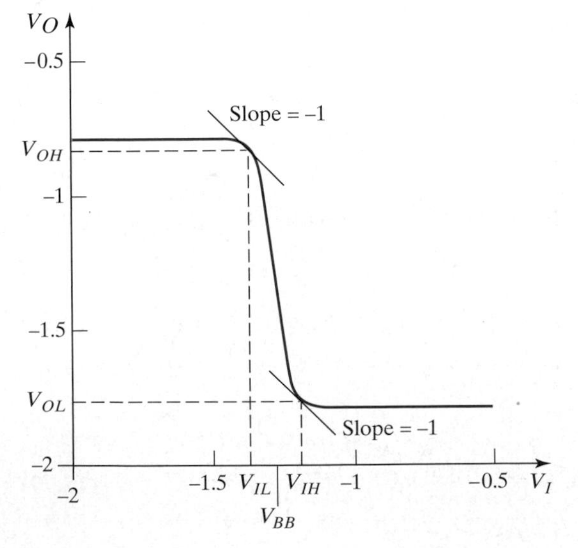

Figure 15–66 A plot of (15.81) made using I_{EE} = 4.6 mA, R_1 = 220 Ω, α = 0.98, and V_{BE3} = 0.8 V.

These values are strong functions of the value assumed for V_{BE3}. The other output ($\overline{V_O}$) has the same values as this one because, when Q_1 is on fully, the value of I_{EE} is about 4.6 mA and $I_{EE}R_1 = 1$ V, and when Q_2 is on fully, I_{EE} is about 4.1 mA and $I_{EE}R_2 = 1$ V.

Example 15.9

Problem:
Use SPICE to generate a DC transfer characteristic like the one shown in Figure 15-66. Explain any differences. Use the Nsimple model from Appendix B and 1-kΩ load resistors.

Solution:
The SPICE output is shown in Figure 15-67. This plot is very close to the one in Figure 15-66 for $-3 \leq V_I \leq -1.1$, but deviates quite significantly for more positive values of V_I. In fact, we see two different regions of behavior. For $-1.1 \leq V_I \leq -0.5$, the output continues to drop as V_I is increased. The analysis we used to derive (15.80) assumed that the value of I_{EE} was constant, but in this circuit I_{EE} is set by the emitter voltage of the ECP. As V_I increases, the emitter voltage increases, and so does I_{EE}. The net result is that V_{C1} and V_O decrease. The slope of this segment is approximately equal to $-R_1/R_{EE}$ (*see* Problem P15.61). The region where V_O increases is caused by Q_1 saturating. With the base-collector junction of Q_1 saturated, the collector will follow V_I, but approximately 0.7 V below it. The slope in this region approaches unity.

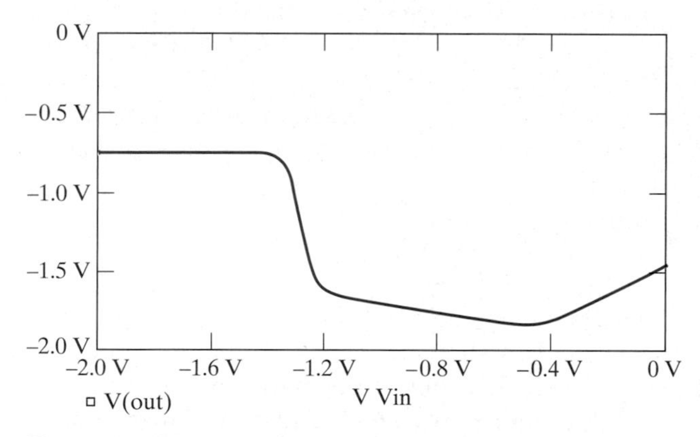

Figure 15–67 Simulated transfer characteristic.

■

Exercise 15.18

If we assumed that the V_{BE} of Q_3 was 0.7 V instead of 0.8 V in Figure 15-65, what would the resulting noise margins be? What can you conclude from this result?

□

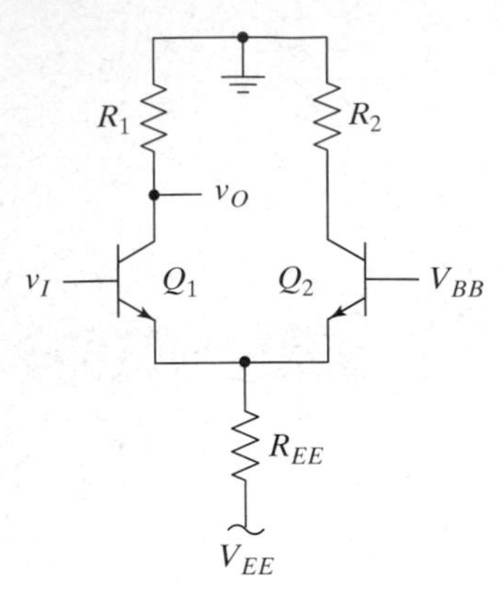

Figure 15–68 The core of an ECL inverter.

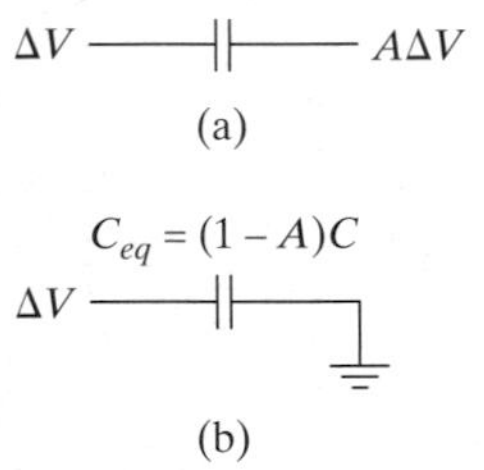

Figure 15–69 (a) A capacitor with the voltage changes on each side. (b) A capacitor to ground that requires the same charge, given the change ΔV.

Propagation Delay and Fan-Out To accurately predict the propagation delay of an ECL gate can be quite complicated and would usually be done using SPICE. Nevertheless, it is important for design purposes to have an understanding of what determines the switching speed. Therefore, we will investigate a couple of limiting cases analytically, and in so doing, we will illustrate the charge-control analysis method further. A more detailed analysis of the switching of an ECL gate can be found in [15.7].

Because the emitter followers at the output of an ECL gate can supply a large amount of current and the input does not require much current, the static fan-out of an ECL gate would be quite large (around 100 or more, depending on how much you allow the swing to be degraded). The fan-out in practical circuits is much lower, however, and is actually limited by the acceptable propagation delay. As we shall soon see, the delay of an ECL gate can be dominated by the load capacitance it needs to drive. Therefore, the larger the fan-out, the slower the gate. In practice, the fan-out is usually no more than 10 gates, although it can be significantly lower, depending on parasitic capacitance and the speed required for the application.

The core of an ECL gate (i.e., without the emitter followers) is shown in Figure 15-68. (This is a CML gate, as noted earlier.) We will analyze this circuit first and then discuss what happens when the emitter followers and external loads are included. We consider only the switching of the inverting output (i.e., the collector of Q_1), although the analysis for the switching at the collector of Q_2 is nearly identical.

We examine two possible limiting cases for the switching speed of this gate. In the first case, the speed is limited by the transistor because of charge being supplied to the depletion regions and the active base through the internal base resistance. The second case is when the external load resistor charging up the total capacitance at the collector node limits the speed. In both cases, we will use average values for the junction capacitors, since they change value during the switching transient. We will also modify the values of the capacitors to account for the voltage swings on both sides of each capacitor.

For example, consider the capacitor shown in Figure 15-69(a), where the voltage swing on one side is ΔV and the swing on the other side is $A\Delta V$. The total change in charge on the capacitor is then

$$\Delta Q = C(1 - A)\Delta V, \qquad \textbf{(15.87)}$$

so that looking in from the left side, we see an effective capacitor of value $C_{eff} = (1 - A)C$ connected to ground as shown in part (b) of the figure. This *effective capacitance* can be larger or smaller than C, depending on the sign of A. For example, if $A = 1$, there isn't any change of voltage across the capacitor, and the effective capacitance is zero. If $A = -1$, the voltage swing across the capacitor is $2\Delta V$, and the capacitor is equivalent to a capacitance of $2C$ connected to ground. This effective capacitance is the same as the Miller capacitance presented in Aside A9.3 and frequently used in analog circuit design.

Returning to our example and examining the case when the transistor limits the speed, we can crudely approximate the circuit as shown in Figure 15-70. If there is an external source resistance as well, it can be summed with r_b. The capacitor C_{beq} is the equivalent capacitance seen from the base to ground and is composed of three terms: the average effective base-emitter junction capacitance, the average effective collector-base junction capacitance, and the average diffusion capacitance. The voltage v_b' is the internal base voltage, and the collector current is determined by it with no further delay.

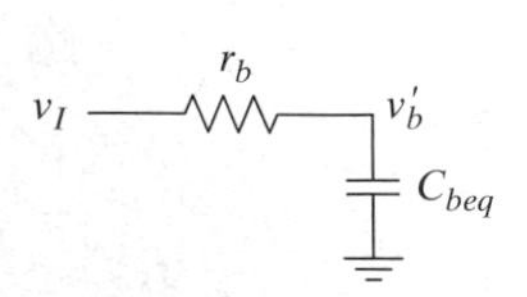

Figure 15–70 A model for finding the propagation delay when the speed is limited by the transistor.

The average effective base-emitter junction capacitance is found by first approximating the capacitance of the forward-biased junction with a constant value of $C_{je} \approx 2C_{je0}$, as in our standard model. We then look at the swing on both sides of this capacitor to find its effective value. On the base side, we will have a swing approximately equal to the change in the input voltage. Assume that the input will swing an amount equal to

$$\Delta V_I = V_{OH} - V_{OL}. \tag{15.88}$$

The emitter voltage will be determined by either Q_1 or Q_2, depending on which base voltage is higher. The emitter will approximately be the higher of the two base voltages minus V_{BE}. Hence, the swing at the emitter will be

$$\Delta V_E = (V_{OH} - 0.7) - (V_{BB} - 0.7), \tag{15.89}$$

where we have assumed that V_{BE} is a constant 0.7 V for both Q_1 and Q_2. The effective value of C_{je} will then be

$$C_{je-eff} \approx \left(1 - \frac{\Delta V_E}{\Delta V_I}\right) 2C_{je0}. \tag{15.90}$$

Similarly, the effective average collector-base junction capacitance is

$$C_{jc-eff} \approx \left(1 - \frac{\Delta V_C}{\Delta V_I}\right) \overline{C_{jc}}, \tag{15.91}$$

where, assuming that the differential pair switches fully, we have

$$\Delta V_C = 0 - I_{EE}R_1 \tag{15.92}$$

and

$$\overline{C_{jc}} = \frac{C_{jc}(V_I = V_{OH}) + C_{jc}(V_I = V_{OL})}{2}. \tag{15.93}$$

We use (15.13) to find the values of C_{jc} at each extreme of the signal swing. When $V_I = V_{OH}$, we have $V_{BC} = V_{OH} + I_{EE}R_1$, and when $V_I = V_{OL}$, we have $V_{BC} = V_{OL}$. Finally, the average diffusion capacitance is equal to the total change in the diffusion charge divided by the change in input voltage:

$$\overline{C_D} = \frac{\Delta Q_D}{\Delta V_I} = \frac{\tau_F \Delta I_C}{\Delta V_I}. \tag{15.94}$$

Putting this all together, we have

$$C_{beq} \approx \left(1 - \frac{\Delta V_E}{\Delta V_I}\right) 2C_{je0} + \left(1 - \frac{\Delta V_C}{\Delta V_I}\right) \overline{C_{jc}} + \frac{\tau_F \Delta I_C}{\Delta V_I}, \tag{15.95}$$

with $\overline{C_{jc}}$ given by (15.93). Using Figure 15-70, and assuming that v_I switches from V_{OL} to V_{OH}, we can write

$$v_b'(t) = V_{OL} + (V_{OH} - V_{OL})(1 - e^{-t/r_b C_{beq}}). \tag{15.96}$$

Setting this voltage equal to the midpoint of the swing, $(V_{OH} - V_{OL})/2$, we solve for the propagation delay (because of symmetry, $t_{pLH} = t_{pHL} = t_p$):

$$t_p = 0.69 r_b C_{beq}. \tag{15.97}$$

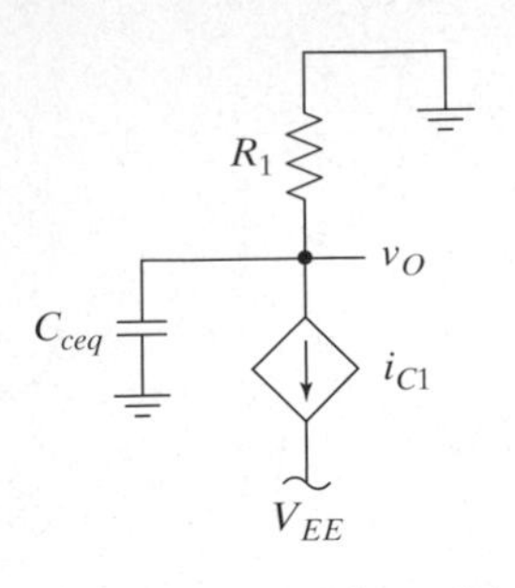

Figure 15–71 A model for finding the propagation delay when the speed is limited by the load.

Now consider the case where the speed is limited by the external load. In this case, we can model the circuit as shown in Figure 15-71. We assume that the collector current source in this case switches instantaneously between zero and I_{EE} as v_I switches between V_{OL} and V_{OH}, respectively. We also lump all of the capacitance at the collector node into C_{ceq}.

The equivalent capacitance at the collector node, C_{ceq}, has three contributing terms: the effective average collector-base junction capacitance, the capacitance of any load connected to the collector, and the collector-to-substrate capacitance of the transistor. The effective average collector-base junction capacitance is given by (15.91), but with $\Delta V_I/\Delta V_C$ instead of $\Delta V_C/\Delta V_I$ since we are now looking at the capacitor from the other side. The load capacitance depends on what we drive with the output (e.g., an emitter follower), the collector-to-substrate capacitor is usually a reverse-biased *pn*-junction diode, and we will need to use an average value over the swing.

If v_I switches from V_{OL} to V_{OH} at $t = 0$, i_{C1} switches from zero to I_{EE}, and using Figure 15-71, we can derive (for $t \geq 0$)

$$v_O(t) = -I_{EE} R_1 (1 - e^{-t/R_1 C_{ceq}}). \tag{15.98}$$

Similarly, as v_I switches from V_{OH} to V_{OL} at $t = 0$, i_{C1} switches from I_{EE} to zero and we can derive

$$v_O(t) = -I_{EE} R_1 e^{-t/R_1 C_{ceq}}. \tag{15.99}$$

Setting either one of these equations equal to the midpoint of the swing, $v_O = -I_{EE}R_1/2$, yields the expression for the propagation delay:

$$t_p = 0.69 R_1 C_{ceq}. \tag{15.100}$$

Example 15.10 illustrates the use of these analytical results and compares them with simulation results.

Example 15.10

Problem:

Determine the propagation delay of the ECL core in Figure 15-68 under three sets of circumstances;

1 with $r_b = 50\ \Omega$ and $R_1 = 50\ \Omega$, so that the transistor limits the speed,
2 with $r_b = 0$ and $R_1 = 220\ \Omega$, so that the load limits the speed, and
3 with $r_b = 50\ \Omega$ and $R_1 = 220\ \Omega$.

Assume that the input switches between $V_{OL} = -0.7$ V and $V_{OH} = -1.7$ V, that $V_{EE} = -5.2$ V, and that $R_{EE} = 779\ \Omega$. Compare the results with SPICE simulations. In all cases, use an external resistor to model r_b, and use the Nsimple model from Appendix B and edit the model to set $I_S = 10^{-14}$ A, $\beta_F = 100$, $\beta_R = 0.5$, $\tau_F = 0.5$ ns, $\tau_R = 5$ ns, $C_{jeo} = 2$ pF, and $C_{jco} = 1$ pF. We will need to use the SPICE defaults VJC = 0.75 V and MJC = 0.33.

Solution:

Our analyses indicate that the swings are symmetric, so consider only the case where v_I switches from V_{OH} to V_{OL}. We then have $\Delta V_I = -1$ V. Using (15.89), we find that $\Delta V_E = -0.6$ V in this case. Also, when $v_I = V_{OL}, i_{C1} \approx 0$, and when $v_I = V_{OH}$, we have $i_{C1} \approx (V_{OH} - 0.7 - V_{EE})/R_{EE} = 4.9$ mA. Therefore, $\Delta I_C = -4.9$ mA in (15.94) and $\overline{C_D} = 2.4$ pF.

Now, in case 1, $I_{EE}R_1 = 0.24$ V, so $\Delta V_C = 0.24$ V. Also, using (15.13) along with $V_{BC} = -0.5$ when $v_I = V_{OH}$ and $V_{BC} = -1.7$ when $v_I = V_{OL}$ yields $\overline{C_{jc}} \approx 0.76$ pF. Plugging into (15.91) yields an effective value of 0.93 pF. Using (15.90) yields $C_{je-eff} = 1.6$ pF. Finally, we have $C_{beq} = 2.4$ pF $+ 0.93$ pF $+ 1.6$ pF $= 5$ pF, and, using (15.97), we arrive at $t_P \approx 0.17$ ns.

In case 2, we find that $I_{EE}R_1 = 1$ V, so $\Delta V_C = 1$ V. Using (15.13) along with $V_{BC} = 0.3$ when $v_I = V_{OH}$ and $V_{BC} = -1.7$ when $v_I = V_{OL}$ yields[16] $\overline{C_{jc}} \approx 0.94$ pF and using (15.91) yields an effective value of 1.9 pF. Using (15.100) yields $t_p \approx 0.29$ ns.

The simplest way to consider case 3 is to say that the base of the transistor must charge up first to change the current and the current will then change the collector voltage. This logic would imply that we should sum the two times given in cases 1 and 2, which yields $t_p \approx 0.46$ ns. In reality, both nodes can charge simultaneously, so the time will not be as long as the sum.

A SPICE simulation of the system was performed for all three cases. The results are summarized in Figure 15-72, which shows both the high-to-low and low-to-high output transitions for all of the cases.

The simulation shows that the input step couples to the output directly. This fact is most noticeable for the case 2 simulation, since r_b is zero. Coupling does really occur, although it would be smaller than we see here if we included some load capacitance. We ignore this initial step and use the midpoint of the two steady-state values to determine the propagation delay (e.g., -0.5 V for case 2).

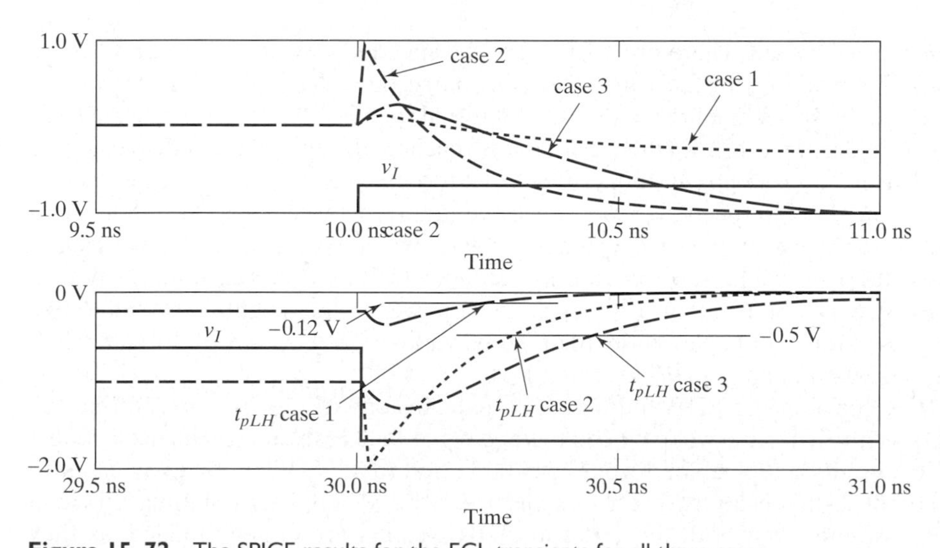

Figure 15–72 The SPICE results for the ECL transients for all three cases.

[16] We used the reverse-biased equation in (15.13) when $V_{BC} = 0.3$, since the circuit is only mildly forward biased.

From this simulation, we determine that in case 1 the propagation delays are $t_{PLH} = 0.22$ ns and $t_{PHL} = 0.36$ ns, for an average value of 0.29 ns, a fair amount longer than our prediction. The error is caused mostly by the fact that the voltage on the collector is not driven by a low impedance. The collector-base capacitance has to be charged through r_b and R_1 until the internal collector current of the transistor becomes significant. Since the transistor is already on at the start of t_{PLH}, that value is closer to our prediction.

For case 2, the simulation predicts that $t_{PLH} = 0.28$ ns and $t_{PHL} = 0.24$ ns, for an average of 0.26 ns, reasonably close to our prediction. For case 3, the simulation yields $t_{PLH} = 0.44$ ns and $t_{PHL} = 0.47$ ns, which is close to the sum of our predictions. If our prediction for case 1 had been more accurate, our sum would have been too large, as expected.

■

The speed of most ECL gates is limited by the collector load (case 2 in Example 15.10), which leads to a definite speed–power trade-off. If we desire to keep the logic swing and noise margins fixed, we must keep the product $I_{EE}R_C$ fixed. The propagation delay of the gate is proportional to R_C, however, as shown by (15.100). If we are designing CML circuits on an IC, the power is dominated by $I_{EE}V_{EE}$. Therefore, if we want a fixed swing of magnitude ΔV, we can write the PDP of CML as

$$\begin{aligned} \text{PDP}_{CML} &\approx (I_{EE}V_{EE})(R_C C_{ceq}) \\ &= \frac{I_{EE}V_{EE}\Delta V C_{ceq}}{I_{EE}} = V_{EE}\Delta V C_{ceq}, \end{aligned} \tag{15.101}$$

which is a constant. Therefore, to increase the speed of the gate, we must decrease R_C and increase I_{EE} by the same factor. Of course, there is a limit to how far we can increase the speed; the limit is set by the internal speed of the transistors, which was case 1 in Example 15.10. When that limit is reached, the speed does not go up with further increases in power. In fact, if you put too much current through the transistors in an effort to speed them up, τ_F will eventually increase owing to high-level injection and the circuit will actually slow down. (Increasing τ_F increases the capacitance—*see* (15.94).) As seen in Example 15.10, the propagation delay of a CML gate can be less than 1 ns. Assuming a power in the neighborhood of 9 mW ($I_{EE}V_{EE}$) and a propagation delay of 0.5 ns yields a PDP of 4.5 pJ. In fact, CML gates now often have PDPs less than 1 pJ.

Now consider the PDP of a true ECL gate, instead of a CML gate. To find the PDP of an ECL gate, we first need to know what load resistors will be used. In discrete circuits, a typical load would be 100 Ω tied to a −2-V supply (You can also accomplish this using two resistors and the −5.2-V supply.) Assuming this load and assuming that both the OR and NOR outputs are equally loaded, as they should be for maximum noise immunity, we can calculate the power dissipated by an ECL gate. The ECP inside the gate consumes about 9 mW, owing to I_{EE} flowing from ground to V_{EE}. The load that is high consumes $(V_{OH} - (-2))/R_L \approx (-.88 + 2)/100 = 11.2$ mA, and the load that is low consumes about $(-1.77 + 2)/100 = 2.3$ mA, for a total power of (13.5 mA)(2) = 27 mW in the loads.

The total static power for the gate is then about 36 mW. The dynamic power is usually negligible in comparison with this static power so we assume an average power per gate of 36 mW. The typical propagation delay of a discrete ECL gate (10K series) is about 1.5 ns, so the PDP for this load configuration is 54 pJ. We can readily see that having to drive large off-chip loads is very expensive in terms of power.

With emitter followers added, the speed may be limited either by the ECL core analyzed here or by the time constant associated with the loads on the emitter followers. If the collector of Q_1 drops too quickly, for example, the emitter follower connected to it will cut-off momentarily, and the load voltage will have to decay with the time constant set by the external load. If the load is able to easily follow the changes at the collector of Q_1, the ECL core circuit will limit the speed. There is a middle region, of course, where both circuits affect the speed, and the total delay is longer than either one alone, but less than the sum.

Exercise 15.19

Consider again case 2 in Example 15.10. If there is a 2-pF load connected to the gate, what would you predict the propagation delay to be?

Biasing Circuits Because the noise margins of ECL gates are so small and because the gates consume large amounts of power and heat up significantly, it is important that the magnitude of the voltage swing not change appreciably with temperature and that it remain properly centered about V_{BB}. The bias circuit that generates V_{BB} ensures that this happens, as we now demonstrate.

Consider the bias circuit shown in Figure 15-73 along with the rest of an ECL gate. The diodes would be implemented as diode-connected BJTs[17], so we call the voltages across them V_{BE}. We begin with a qualitative description of the circuit's operation and then proceed with an analysis. If the current through R_3 and R_4 is reasonably constant over temperature, the voltage across R_4 will be constant. The base voltage of Q_5 is then equal to a constant plus $2V_{BE}$. When the bias circuit is connected to the ECL gate and the input is less than V_{BB}, Q_2 is on and the emitter voltage of Q_2 will be $2V_{BE}$ below the base of Q_5. In other words, the voltage across R_{EE} should be about equal to the voltage across R_4 and should be relatively constant. The voltage across R_2 at this time, which is ΔV, will be R_2/R_{EE} times the voltage across R_{EE} and will also be constant. Therefore, this biasing circuit achieves one of the two goals we set; namely, that the swing should be relatively constant over temperature. It also turns out that the circuit centers the swing on V_{BB}, as we will demonstrate by analysis in a moment. Finally, we note that Q_5 provides a low-impedance drive for V_{BB}, so that a single bias circuit can be used for multiple gates.

The output high voltage of the ECL gate is $V_{OH} \approx -V_{BE3}$, since the collector voltage of Q_1 (or Q_2) is essentially ground when it is high. Similarly, the output low voltage is $V_{OL} = -\Delta V - V_{BE3}$, where $\Delta V = I_{EE1}R_1 = I_{EE2}R_2$. To keep the swing centered on V_{BB}, we would like to have

[17] Diode-connected transistors are covered in Section 8.1.3.

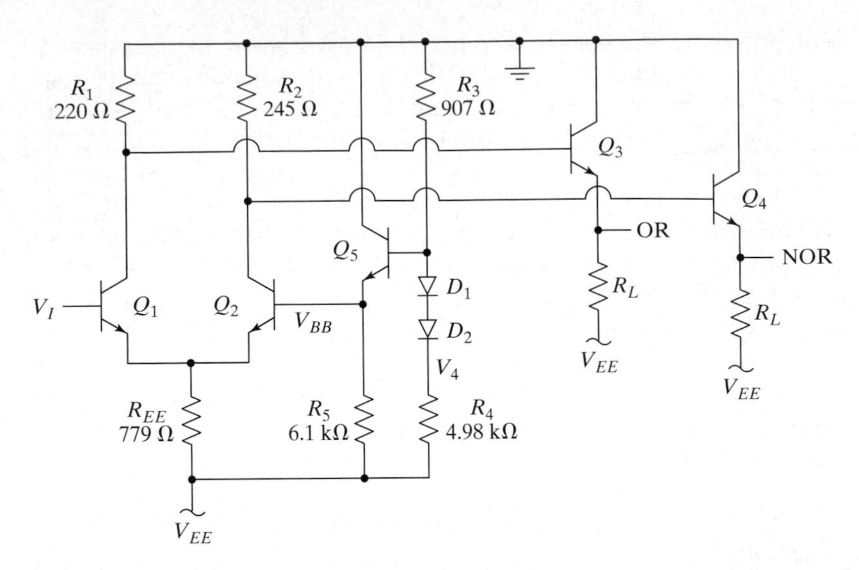

Figure 15–73 The bias circuit for 10K family ECL circuits shown with the rest of the gate.

$$V_{BB} = \frac{V_{OL} + V_{OH}}{2} \approx \frac{-\Delta V - 2V_{BE}}{2} \tag{15.102}$$

at all times. If the input to an ECL gate is driven by another ECL gate with output equal to V_{OH}, then the voltage at the emitter of Q_1 is $V_{E1} = V_{OH} - V_{BE} = -2V_{BE}$. This voltage produces a current in R_{EE} equal to $I_{EE1} = (-2V_{BE} - V_{EE})/R_{EE}$, which then flows almost entirely through Q_1, yielding a drop across R_1 equal to

$$\Delta V = (-2V_{BE} - V_{EE})\frac{R_1}{R_{EE}}. \tag{15.103}$$

Plugging this value for the swing back into (15.102) yields

$$V_{BB} = +V_{EE}\frac{R_1}{2R_{EE}} - V_{BE}\left(1 - \frac{R_1}{R_{EE}}\right) \tag{15.104}$$

for our desired V_{BB}.

Now examine the bias circuit itself to derive an equation for V_{BB}. Ignoring the base current of Q_5 we see that the voltage at the top of R_4 is

$$V_4 = \left(\frac{R_4}{R_4 + R_3}\right)(-V_{EE} - 2V_{BE}) + V_{EE}. \tag{15.105}$$

The voltage at the base of Q_5 is $2V_{BE}$ higher than this, so the emitter of Q_5, which is V_{BB}, is only V_{BE} higher; rearranging the equation a bit, we arrive at

$$V_{BB} = V_{EE}\left(1 - \frac{R_4}{R_3 + R_4}\right) + V_{BE}\left(1 - \frac{2R_4}{R_3 + R_4}\right). \tag{15.106}$$

Note that this is exactly the form we desire. Equating (15.104) and (15.106), we find that we should set

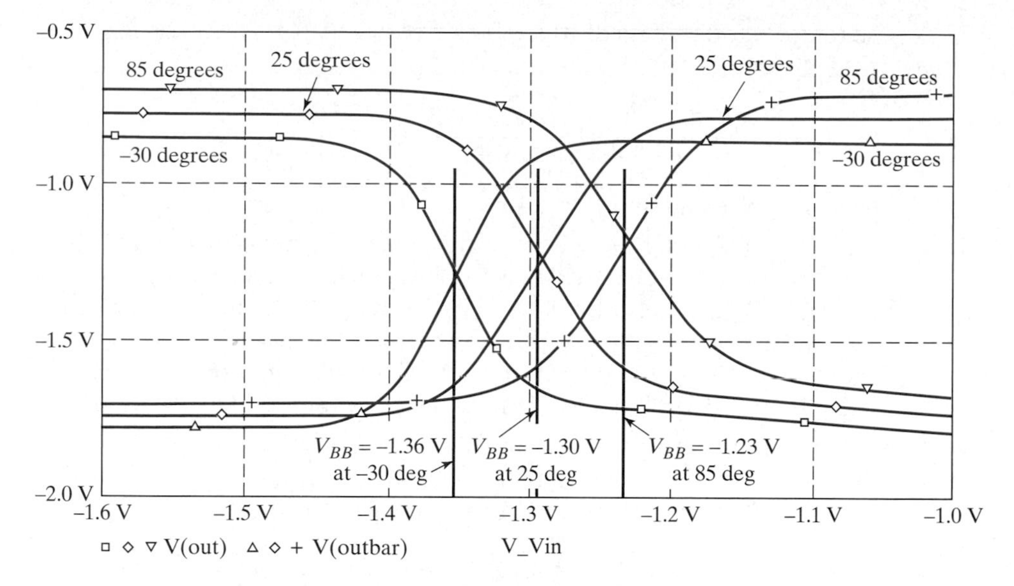

Figure 15–74 A SPICE simulation of the ECL circuit in Figure 15-73.

$$\frac{R_1}{2R_{EE}} = \left(1 - \frac{R_4}{R_3 + R_4}\right), \tag{15.107}$$

which is approximately true for the ECL gate. Notice that satisfying this condition results in (15.102) being true at any temperature, so we have achieved the desired goal. We must point out that this analysis has been *very* crude! The most significant errors are that not all of the V_{BE}'s of all of the transistors are the same (although their changes with temperature are, approximately), the current through R_4 does change with temperature, and the change in the swing is different if we consider the case when I_{EE} flows through Q_2 (since the base of Q_2 will not be equal to V_{OH}). Nevertheless, the analysis illustrates some general principles used in designing temperature-independent bias circuits and gives an approximately correct understanding of the bias circuit examined.

We close by presenting the results of a SPICE simulation on the circuit shown in Figure 15-73 (Ecl_l0k.sch). The DC transfer characteristics for both the OR and the NOR outputs are shown in Figure 15-74 at three different temperatures. We have also shown the simulated value of V_{BB} at each temperature, and you can see that it is the switching threshold (i.e., the outputs cross), as expected.

SOLUTIONS TO EXERCISES

15.1 The forward current in the diode will be $I_F = 4.3/5\text{k} = 0.86$ mA, and the initial reverse current will be $I_R = -2.7/5\text{k} = -0.54$ mA. Therefore, the storage is found from (15.6) to be $t_s = 7$ ns. The delay time is found from (15.8) to be $t_d = 54$ ns. Therefore, the reverse recovery time is $t_{rr} = 61$ ns.

15.2 The collector saturation current is

$$I_{C\text{sat}} = \frac{V_{CC} - v_{CE\text{sat}}}{R_C} = \frac{5 - 0.2}{2\text{ k}} = 2.4 \text{ mA}.$$

Therefore, a forced beta of 10 requires that $I_B = 0.24$ mA, but the base current is given by

$$I_B = \frac{V_S - 0.7}{R_B},$$

so we can solve and find $V_S = 24.7$ V.

15.3 No, you would not expect a large difference in the storage time. Even though a smaller value of R_B leads to a larger base overdrive and more excess base charge, it also leads to a larger reverse base current to remove that charge faster. You would, however, expect smaller rise and fall times. Figure 15-75 shows the output of a SPICE simulation identical to that in Example 15.3, except that two values of R_B are used: 1 kΩ and 10 kΩ.

15.4 (a) The drain current of the transistor is $I_D = K(V_I - V_{th})^2(1 + \lambda V_O)$, where we have made use of the facts that $V_{GS} = V_I$ and $V_{DS} = V_O$. Setting the expression for I_D equal to (15.35), we solve for the input voltage and find that

$$V_I = V_{th} + \sqrt{\frac{V_{DD} - V_O}{KR_D(1 + \lambda V_O)}}$$

and, plugging in the given numbers, we obtain $V_I = 1.41$ V. (b) The incremental, or small-signal, gain is the slope of the V_O-versus-V_I characteristic evaluated at the Q point. In other words,

$$\text{Gain} = \left.\frac{dV_O}{dV_I}\right|_{@Q}.$$

Using (15.34), we can write $V_O = V_{DD} - I_D R_D$, so we have

$$\text{Gain} = -R_D\left.\frac{dI_D}{dV_I}\right|_{@Q} = -2R_D K(1 + \lambda V_O)(V_I - V_{th}),$$

which yields -4.9 for this case.

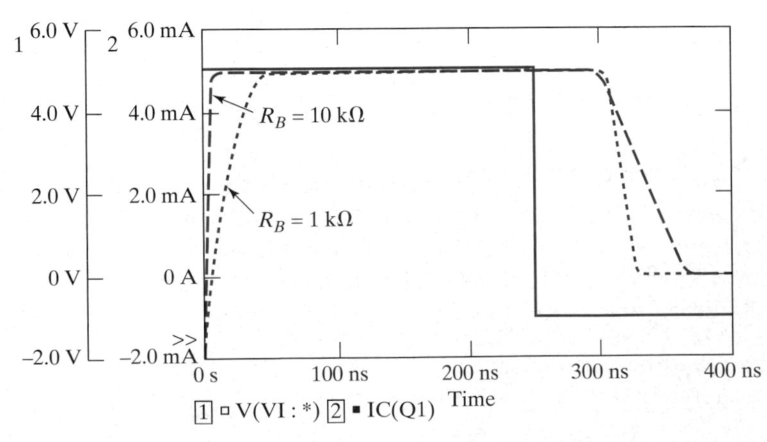

Figure 15–75 SPICE output.

15.5 Following the solution of Example 15.5 exactly, we obtain $V_{IH} = 1.67$ V, $V_{OL} = 0.18$, $V_{IL} = 1.33$, $V_{OH} = 2.82$ V, and $NM_H = NM_L = 1.15$ V. The noise margins are seen to still be quite reasonable, even with a much lower supply voltage.

15.6 The new load capacitor is five times larger than the one in the example; therefore, we must move five times as much charge in the same time. This will require five times as much current, which implies that we need to increase the W/L of both transistors by five times. Therefore, we now have $W = 7.5$ μm and $L = 0.75$ μm for the NMOS and $W = 18.75$ μm and $L = 0.75$ μm for the PMOS.

15.7 The worst-case pull down is a single NMOS transistor; therefore, the NMOS devices can be minimum size. Using (15.56), we determine that the PMOS devices should each be 7.5 times the minimum size. Therefore, the overall gate will consume the area of 24.5 minimum-size transistors.

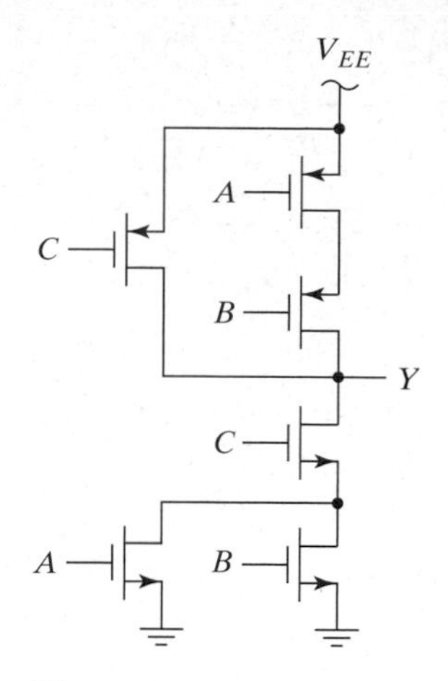

Figure 15–76

15.8 The Boolean expression given describes the pull-down network, since it tells us what must be true to cause the output to be low. Therefore, the pull-down network is as shown in Figure 15-76. Using the series–parallel transformation, we then derive the pull-up network shown. To check ourselves, we use De Morgan's theorems to derive $Y = \overline{(A + B) \cdot C} = \overline{A} \cdot \overline{B} + \overline{C}$, which does describe the pull-up network shown.

15.9 A two-input pseudo-NMOS NAND gate is shown in Figure 15-77. When both inputs are high, the NMOS transistors are on and the output is low. For the output to go low enough to achieve reasonable noise margins, the transistors must be sized so that the NMOS devices can overpower the PMOS device, which tends to pull the output high. If either input goes low, the corresponding NMOS transistor will turn off and the PMOS device will pull the output high.

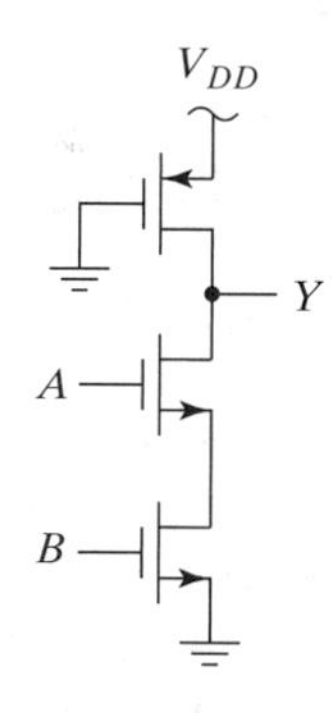

Figure 15–77

15.10 **(a)** Remember that 32 Mb = 2^5 Mb, where 1 Mb really means 2^{20} = 1,048,576 bits, and 64 kb = 2^6 kb where 1 kb really means 2^{10} = 1,024 bits. Therefore, 32 Mb = 2^{15} kb = $2^9 \times 64$ kb. There are 2^9 (i.e., 512) 64-kb subarrays in a 32-Mb memory. **(b)** Since there are 2^9 subarrays, we need a 9-bit block address. **(c)** A 64-kb subarray requires 8-bit word and column addresses (2^{16} = 65,536).

15.11 **(a)** There are six transistors per bit, and 1 Mb really means 2^{20} = 1,048,576 bits, so there are 6,291,456 transistors. **(b)** The transistors will consume 6,291,456 μm^2 = 6.291 mm^2. The wiring consumes 95% of the total area, so the total area is 125.8 mm^2, which requires a square 11.2 mm on a side.

15.12 With $V_{BL} = V_{DD}$, when the word line goes high ($WL = V_{DD}$), there isn't any inversion layer at the left end of the pass transistor (the bit-line side), but with $V_S = 0$ V, there is an inversion layer at the right end (the storage cell side). Therefore, the pass transistor is in saturation, with the right end acting as the source and the current flowing from the bit line onto the storage capacitor. As the storage capacitor charges up, the V_{GS} of the pass transistor is reduced, so that the current is reduced as well. The net result is an exponential like charge-up for the storage capacitor to $V_{DD} - V_{th}$.

15.13 Word 0 is 1101, word 1 is 1010, word 2 is 0100, and word 3 is 0110.

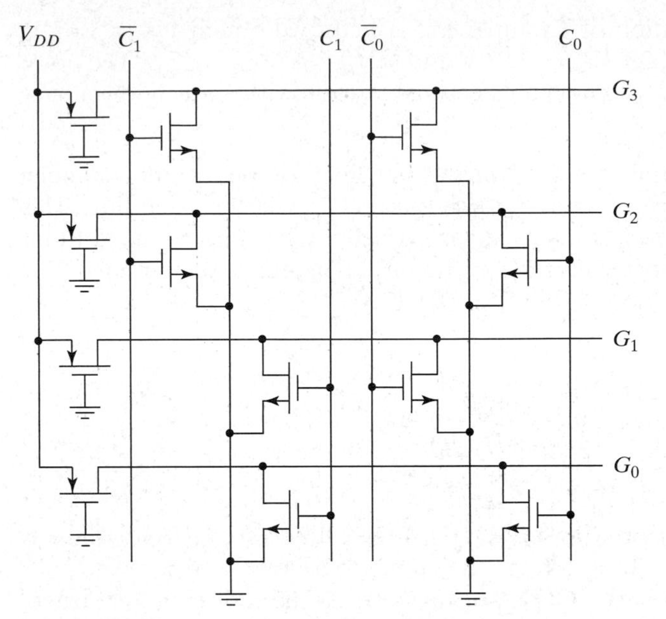

Figure 15–78

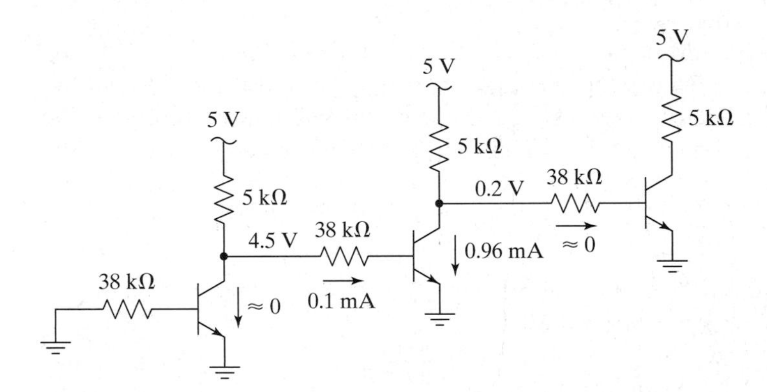

Figure 15–79

15.14 The NOR decoder is shown in Figure 15-78.

15.15 The circuit is shown in Figure 15-79 with the voltages and currents labeled (assuming that $V_{BE} = 0.7$ V and $V_{CEsat} = 0.2$ V). From this figure, we see that $V_{OH} = 4.5$ V and $V_{OL} = 0.2$ V.

15.16 Since the series resistance is zero, the internal supply voltage will be the same as the external supply voltage whenever the current is constant. When the current is changing, however, there will be a drop across the inductance equal to $v_L = Ldi_{CC}/dt$. From Figure 15-57, we see that the slope of the current waveform is 2 mA/ns when rising and −2 mA/ns when falling. With $L =$ 40 nH, this leads to a constant voltage across the inductor of ± 80 mV during the current spike. If all six inverters switched at the same time, we would have a drop of ± 480 mV. Therefore, the internal value of V_{CC} would change from 5 V, to 4.52 V, to 5.48 V, and then back to 5 V during the transition.

15.17 The solution is shown in Figure 15-80.

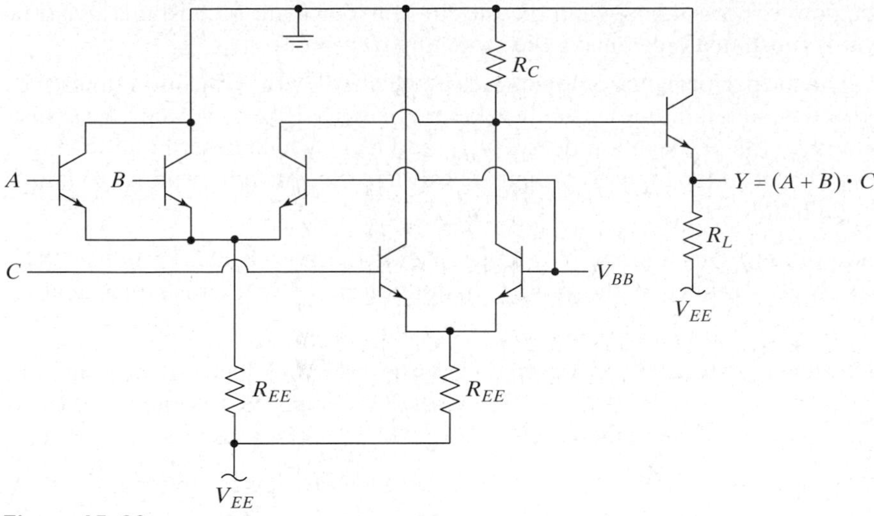

Figure 15–80

15.18 The value for V_{IH} is still the same: $V_{IH} = -1.20$ V. Plugging this value back into (15.80) and then (15.81) while using $V_{BE3} = 0.8$ V yields $V_{OL} = -1.67$ V. By symmetry, we obtain $V_{IL} = -1.38$ V and $V_{OH} = -0.73$ V. The corresponding noise margins are $NM_H = 0.47$ V and $NM_L = 0.29$ V. We conclude that the value of V_{BE3} determines how well the output swing is centered about the switching threshold. Therefore, if the load is changed significantly (so that I_L changes, thereby changing V_{BE3}), one of the noise margins will be decreased.

15.19 In the example, we determined that the average effective collector-base junction capacitance was 1.9 pF. The load capacitance simply adds to this value, yielding a total of 3.9 pF. Using this capacitance in (15.100) yields $t_p \approx$ 0.59 ns. A SPICE simulation indicates an average value of $t_p \approx 0.6$ ns.

CHAPTER SUMMARY

- The excess minority-carrier charge storage in the base of a BJT slows the device down when it is being switched from cutoff to saturation.
- MOSFETs do not store minority carriers and tend therefore to switch faster than BJTs.
- Switching transients can frequently be analyzed using charge control techniques. In other words; break the transient up into small swings, determine how much charge must be moved during each period of time and approximately what current is available to move the charge. The time spent in each region is the total charge that must be moved divided by the current available to move it.
- The static transfer characteristic of logic gates is specified by the logic swing, the transition region and the points where the slope of the transfer characteristic is -1: V_{IL}, V_{OH}, and V_{IH}, V_{OL}.
- The fan-in and fan-out of a logic gate specify the number of inputs and the number of gates, respectively, that can be connected to the output.

- Static power is consumed when the output of a logic gate is constant. Dynamic power is consumed only during the switching transients.
- The dynamic performance of logic gates is specified by the transition times (i.e., the rise time and fall time), usually taken to be from 10% to 90% of the transition, and by the propagation delays (t_{pLH} and t_{pHL}), which describe how long it takes a change at the input to propagate through the gate and produce a change at the output.
- An important figure of merit for logic gates is the power-delay product, PDP, equal to the product of the power dissipation and the average propagation delay.
- NMOS logic gates can be constructed by using resistive loads or by using enhancement- or depletion-mode MOS devices as the load. NMOS logic uses a simple process, but has performance inferior to CMOS and is rarely used anymore.
- CMOS logic uses complementary NMOS and PMOS devices to form the pull-down and pull-up networks, respectively.
- CMOS logic has zero static power dissipation, has rail-to-rail logic swings, is reasonably fast, and can implement complex functions with a minimum of circuitry. It is the dominant logic family in use at this time.
- Other types of MOS logic, such as pseudo-NMOS and pass-transistor logic, are useful in special applications.
- CMOS processes allow several options for building semiconductor memories. These memories may be read only (ROM) or may allow you to read and write data (RAM).
- RAM may be either static (SRAM) or dynamic (DRAM). Static memories use more transistors, but dynamic memories require periodic refreshing.
- Bipolar logic can be divided into saturating and nonsaturating logic. Saturating logic is essentially obsolete at this time, but is used occasionally in special custom designs where the simplicity and stable output levels are advantageous and the speed is not a concern.
- Transistor–transistor logic (TTL) is the most common form of saturating logic. It has been replaced by CMOS logic in almost all applications.
- Emitter-coupled logic is the fastest form of logic because the core logic function is implemented by current-mode logic wherein the devices are always forward active and currents are switched from one device to another without requiring large voltage swings.

REFERENCES

[15.1] J. Millman & H. Taub, *Pulse, Digital, and Switching Waveforms.* New York: McGraw-Hill, 1965. (This is an old text, but still contains some of the best descriptions of how to analyze switching circuits.)

[15.2] D.A. Hodges and H.G. Jackson, *Analysis and Design of Digital Integrated Circuits,* 2 Ed. New York: McGraw-Hill, 1988.

[15.3] J.F. Wakerly, *Digital Design Principles and Practices.* Englewood Cliffs, NJ: Prentice Hall, 1990.

[15.4] J. R. Hauser, "Noise margin criteria for digital logic circuits," *IEEE T. on Education,* Vol. 36, No. 4, 1993, pp. 363–368.

[15.5] J.B. Foley & J.A.R. Bannister, "Analyzing ECL's noise margin," *IEEE Circuits & Devices Mag.,* May 1994, pp. 32–37.

[15.6] R. C. Jaeger, *Microelectronic Circuit Design.* New York: McGraw-Hill, 1997.

[15.7] J. M. Rabaey, *Digital Integrated Circuits.* Upper Saddle River, NJ: Prentice Hall, 1996.

[15.8] T.A. DeMassa & Z. Ciccone, *Digital Integrated Circuits.* New York: John Wiley & Sons, 1996.

PROBLEMS

15.1 DEVICE MODELING FOR DIGITAL DESIGN

15.1.1 Diodes

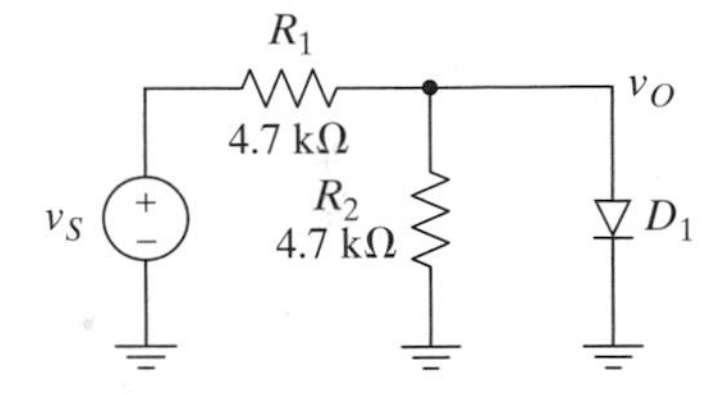

Figure 15–81

P15.1 Derive (15.7) and (15.8).

P15.2 Consider the circuit shown in Figure 15-81. The input source is a 2-MHz square wave with ± 2 V amplitude. Assume that the diode has a minority-carrier transit time of 10 ns and a zero bias junction capacitance of 5 pF. **(a)** What is the storage time of the diode? **(b)** What is the delay time? **(c)** Draw the source voltage and output voltage for one full period of operation of the circuit.

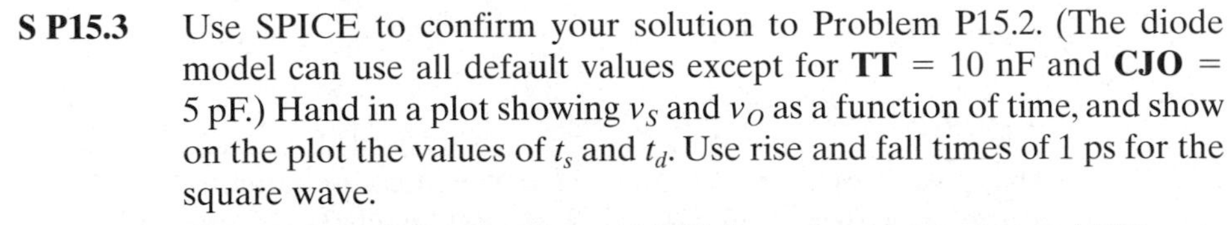

S P15.3 Use SPICE to confirm your solution to Problem P15.2. (The diode model can use all default values except for **TT** = 10 nF and **CJO** = 5 pF.) Hand in a plot showing v_S and v_O as a function of time, and show on the plot the values of t_s and t_d. Use rise and fall times of 1 ps for the square wave.

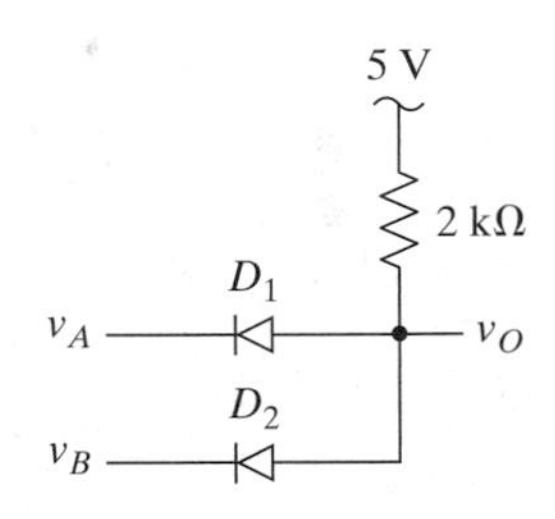

Figure 15–82

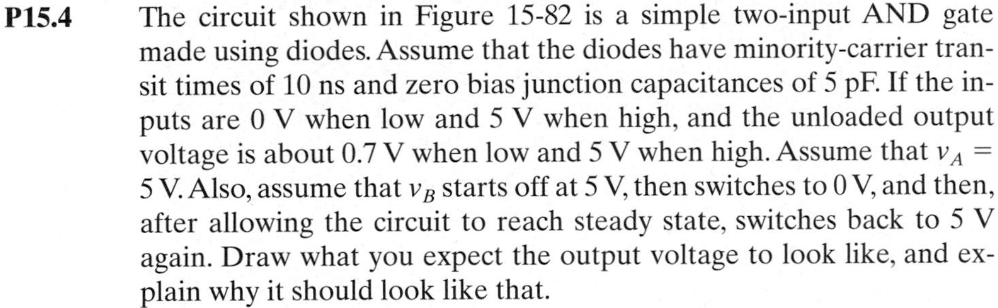

P15.4 The circuit shown in Figure 15-82 is a simple two-input AND gate made using diodes. Assume that the diodes have minority-carrier transit times of 10 ns and zero bias junction capacitances of 5 pF. If the inputs are 0 V when low and 5 V when high, and the unloaded output voltage is about 0.7 V when low and 5 V when high. Assume that v_A = 5 V. Also, assume that v_B starts off at 5 V, then switches to 0 V, and then, after allowing the circuit to reach steady state, switches back to 5 V again. Draw what you expect the output voltage to look like, and explain why it should look like that.

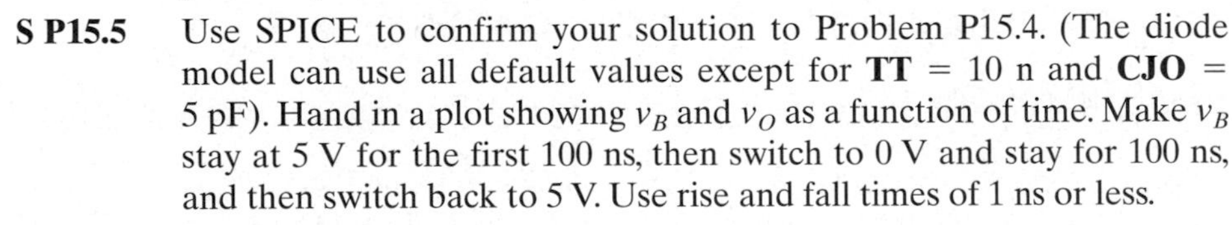

S P15.5 Use SPICE to confirm your solution to Problem P15.4. (The diode model can use all default values except for **TT** = 10 n and **CJO** = 5 pF). Hand in a plot showing v_B and v_O as a function of time. Make v_B stay at 5 V for the first 100 ns, then switch to 0 V and stay for 100 ns, and then switch back to 5 V. Use rise and fall times of 1 ns or less.

15.1.2 Bipolar Junction Transistors

P15.6 Consider the circuit from Example 15.2. If $V_S = V_{CC} = 5$ V and $R_C = 2$ kΩ, what must R_B be in order to obtain a forced beta of 10?

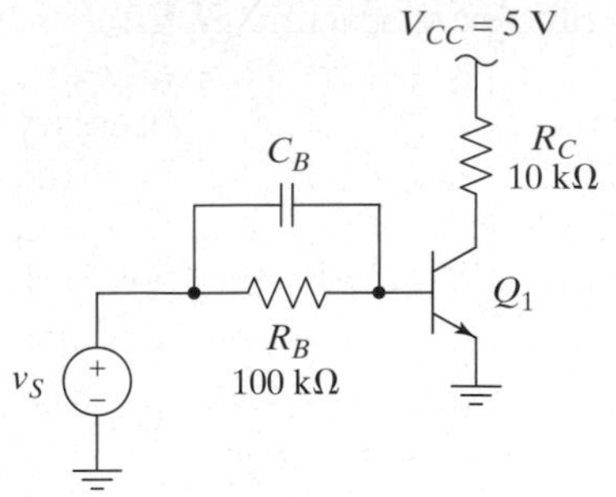

Figure 15–83

P15.7 Consider the circuit shown in Figure 15-83. Assume that the transistor has $\beta_F = 100$, $\beta_R = 1$, $r_b = 62.5\ \Omega$, $\tau_F = 1$ ns, $\tau_R = 50$ ns, $C_{je0} = 0.45$ pF, $C_{jc0} = 2$ pF, $m_{je} = m_{jc} = 0.33$, and $V_0 = 0.75$ V, as in Example 15.3. The capacitor in parallel with R_B has been added to speed up the transistor switching in and out of saturation. **(a)** Assuming that v_S has been equal to 5 V long enough for the circuit to reach steady-state operation, calculate the value of C_B needed for the charge stored on it to equal the excess base charge. **(b)** If the capacitor is set to the value you calculated in (a), what do you think will happen to the collector current if the source voltage abruptly changes to 0 V? Why?

S P15.8 Use SPICE to confirm your answer to Problem P15.7. (Use the transistor model parameters listed in the problem.) Hand in a transient plot of the collector current and source voltage for two cases: one with no speedup capacitor present and one with the speedup capacitor you calculated in Problem P15.7.

P15.9 Prove that (15.11) and (15.12) are equivalent.

15.1.3 MOS Field-Effect Transistors

P15.10 Determine the drain-to-source resistance of an n-channel MOSFET in saturation with $K = 10\ \mu\text{A/V}^2$, $\lambda = 0.015$, $V_{th} = 0.7$ V, and **(a)** $V_{GS} = 1$ V and **(b)** $V_{GS} = 2$ V.

P15.11 Determine the complete dynamic model of an n-channel MOSFET operating in saturation with $W = 0.5\ \mu\text{m}$, $L = 0.25\ \mu\text{m}$, $t_{ox} = 200$ Å (Å = angstrom; 1 angstrom $= 10^{-10}$ meter), $V_{th} = 0.6$ V, and $V_{GS} = 1.2$ V. Ignore the parasitic junction capacitances and the substrate connection. Assume $\mu_n = 650\ \text{cm}^2/\text{vs}$ in the channel.

15.2 SPECIFICATION OF LOGIC GATES

P15.12 Consider an inverter whose static transfer characteristic is described by

$$V_O = 2.5\cos\left(\frac{\pi V_I}{5}\right) + 2.5, \text{ for } 0 < V_I < 5\text{ V}.$$

(a) Sketch the transfer characteristic. **(b)** Find the values of V_{IH}, V_{IL}, V_{OH}, and V_{OL} (not the typical values, but the values necessary to find the noise margins). **(c)** Find the noise margins. **(d)** Find the magnitude of the logic swing. **(e)** Find the magnitude of the transition region.

P15.13 An inverter has the measured static transfer characteristic given in Table 15.1. **(a)** Sketch the transfer characteristic. **(b)** Find the values of V_{IH}, V_{IL}, V_{OH}, and V_{OL} (not the typical values, but the values necessary to find the noise margins). **(c)** Find the noise margins. **(d)** Find the magnitude of the logic swing. **(e)** Find the magnitude of the transition region.

P15.14 What is the dynamic power dissipated by a digital circuit with 100,000 gates if each gate drives 0.5 pF of load capacitance, the clock is 10 MHz, each output toggles once every other clock cycle on average, and **(a)** $V_{DD} = 5$ V, **(b)** $V_{DD} = 3$ V, and **(c)** $V_{DD} = 1$ V.

15.3 MOS DIGITAL CIRCUITS

15.3.1 NMOS

P15.15 Consider the NMOS inverter shown in Figure 15-84, and assume that $R_D = 50$ kΩ and the transistor has $V_{th} = 0.8$ V and $K = 25$ μA/V². **(a)** Find the nominal values of V_{OH}, V_{OL}, V_{IH}, and V_{IL} for this gate. **(b)** Determine the noise margins.

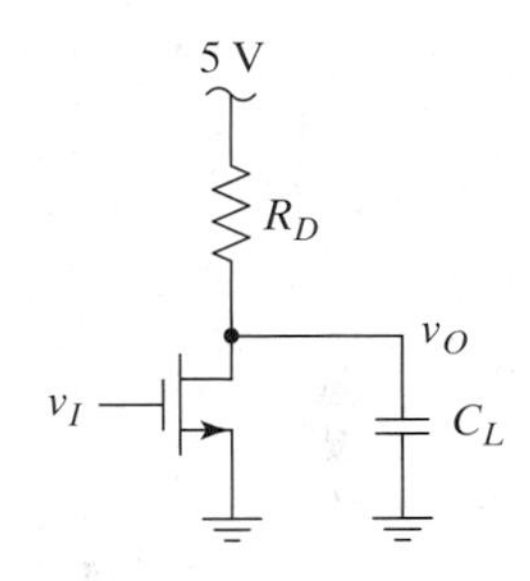

Figure 15–84

S P15.16 Use SPICE to confirm your answer to Problem P15.15. Use the level-1 MOS model with the parameters given in the problem, and hand in a DC sweep showing V_O versus V_I with the values of V_{OH}, V_{OL}, V_{IH}, and V_{IL} marked on the plot.

P15.17 Consider the NMOS inverter shown in Figure 15-84, and assume that the transistor has $V_{th} = 0.8$ V and $K = 25$ μA/V². If $R_D = 50$ kΩ and $C_L = 0.2$ pF, **(a)** derive an equation for t_{PLH} and find its value, assuming that the input has been 5 V for a long time and then abruptly switches to 0 V. Assume further that the transistor turns off immediately, and ignore parasitic capacitance in the transistor. **(b)** Derive an equation for t_{PHL} and find its value, assuming that the input has been 0 V for a long time and then abruptly switches to 5 V. Break your calculation up into two regions: when M_1 is in saturation (ignore channel-length modulation) and when M_1 is in the linear region. In the linear region, approximate the drain current by the average of its value when the transistor just enters the region and its value at t_{PHL}. **(c)** Derive an equation for the average static power consumed by this gate. Note that you cannot reduce the propagation times without increasing the power.

S P15.18 Use SPICE to confirm your answer to Problem P15.17. Use the level-1 MOS model with the parameters given in the problem, and hand in a transient output showing the propagation delays.

Table 15.1

V_I	0	0.4	0.6	0.8	1.0	1.2	1.4	1.5	1.6	1.8	2.0	2.2	2.4	2.6	3.0
V_O	3.0	2.98	2.95	2.9	2.78	2.5	2.0	1.5	1.0	0.5	0.22	0.1	0.05	0.02	0.0

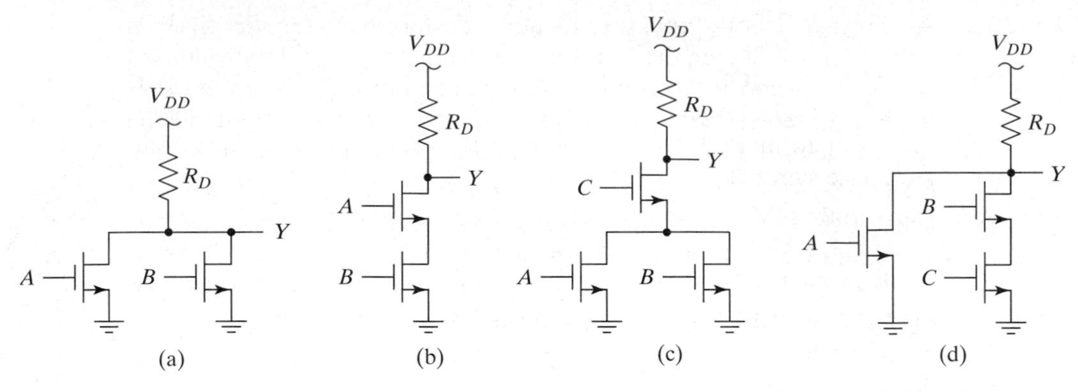

Figure 15–85

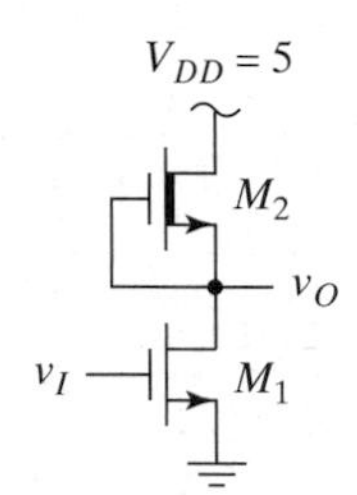

Figure 15–86

P15.19 Determine the Boolean expressions for the NMOS gates shown in Figure 15-85. (Assume that the device sizes and resistor values are chosen properly.)

P15.20 Consider the depletion-load NMOS inverter in Figure 15-86. Assume that $V_{th1} = 0.8$ V, $V_{th2} = -2.5$ V, $K_1 = 22\ \mu A/V^2$, $K_2 = 12\ \mu A/V^2$, and $\lambda = 0.05$ for both devices. **(a)** Find V_O when $V_I = 0$ V. **(b)** Find V_O when $V_I = 2.5$ V. **(c)** Find V_O when $V_I = 5$ V.

15.3.2 CMOS Inverter (NOT gate)

P15.21 Derive (15.46).

P15.22 Find the values of V_{OH}, V_{OL}, V_{IH}, V_{IL}, and the noise margins for a CMOS inverter with $V_{DD} = 5$ V, $V_{thN} = -V_{thP} = 0.8$ V, $\lambda_N = \lambda_P = 0$, $K_N = 25\ \mu A/V^2$, and $K_P = 36\ \mu A/V^2$.

15.3.3 CMOS NOR and NAND Gates

P15.23 **(a)** Determine the sizes of all of the transistors in a four-input NAND gate if we desire to have the propagation delays be equal to those of a minimum-size symmetric inverter. **(b)** How large is the resulting gate in terms of equivalent minimum-size transistors? **(c)** Repeat (a) for a four-input NOR gate. (d) Repeat (b) for the four-input NOR gate, and compare the overall sizes of the NOR and NAND gates.

P15.24 If we build a three-input NOR gate using minimum-size transistors for all transistors, approximately how much longer will t_{PHL} and t_{PLH} be than they are for a minimum-size symmetric inverter? Give both the best- and worst-case values where appropriate.

P15.25 If we build a three-input NAND gate using minimum-size transistors for all transistors, approximately how much longer will t_{PHL} and t_{PLH} be than they are for a minimum-size symmetric inverter? Give both the best- and worst-case values where appropriate.

D P15.26 Determine the sizes of all of the transistors in a three-input NOR gate if we allow the worst case propagation delays to be twice as long as those for a minimum-size symmetric inverter.

D P15.27 Determine the sizes of all of the transistors in a three-input NAND gate if we desire to have t_{PHL} be the same as for a minimum-size symmetric inverter and allow the worst-case t_{PLH} to be three times as long as for a minimum-size symmetric inverter.

15.3.4 More Complex CMOS Gates

D P15.28 Using only NMOS and PMOS transistors, design a circuit that performs the logical operation $\overline{Y} = A + B \cdot C$. Show the transistor sizes necessary to make the propagation delays equal to those of a minimum-size symmetric inverter.

D P15.29 Using only NMOS and PMOS transistors, design a circuit that performs the logical operation $Y = \overline{A} + \overline{B} + \overline{C}$ if only A, B, and C are available as inputs.

P15.30 Using only NMOS and PMOS transistors, design a circuit that performs the logical operation $\overline{Y} = A + \overline{B} \cdot (C + D)$.

D P15.31 **(a)** Using only NMOS and PMOS transistors, design a circuit that performs the logical operation $\overline{Y} = A + B \cdot C$. **(b)** If minimum-size transistors are used throughout, determine the worst-case values of the propagation times compared with those for a minimum-size symmetric inverter.

P15.32 Derive a Boolean expression for the output (Y) of the logic gate shown in Figure 15-87 in terms of the inputs (A, B, and C).

P15.33 Consider the logic gate shown in Figure 15-87. If minimum-size transistors are used throughout, determine the worst-case values of the propagation times compared with those for a minimum-size symmetric inverter.

P15.34 Consider the logic gate shown in Figure 15-87. Determine the transistor sizes necessary to make the worst-case propagation delays equal to those of a minimum-size symmetric inverter.

D P15.35 Using only NMOS and PMOS transistors, design a two-input exclusive-NOR gate.

P15.36 Suppose that we wanted to design a CMOS-like gate using only enhancement-mode NMOS transistors. What disadvantage would the NMOS transistors have, relative to PMOS devices, when used in the pull-up network?

D P15.37 Design a two-input AND gate using a CMOS structure. Can it be done in one stage?

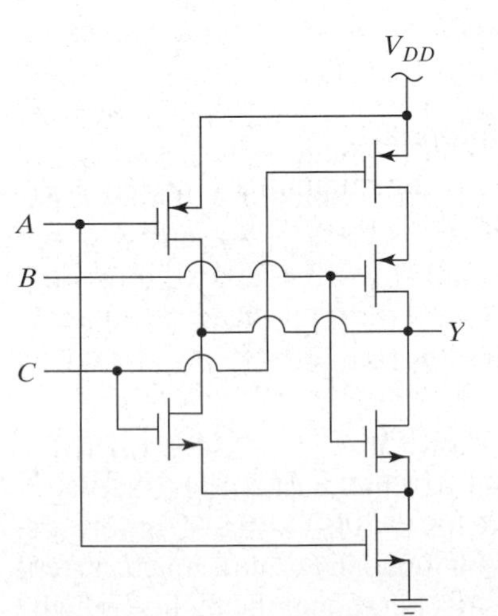

Figure 15–87

15.3.5 Other Types of CMOS Logic

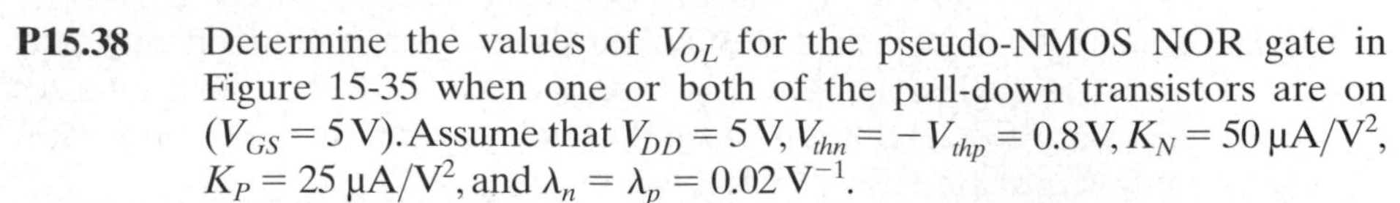

P15.38 Determine the values of V_{OL} for the pseudo-NMOS NOR gate in Figure 15-35 when one or both of the pull-down transistors are on ($V_{GS} = 5$ V). Assume that $V_{DD} = 5$ V, $V_{thn} = -V_{thp} = 0.8$ V, $K_N = 50\ \mu\text{A/V}^2$, $K_P = 25\ \mu\text{A/V}^2$, and $\lambda_n = \lambda_p = 0.02\ \text{V}^{-1}$.

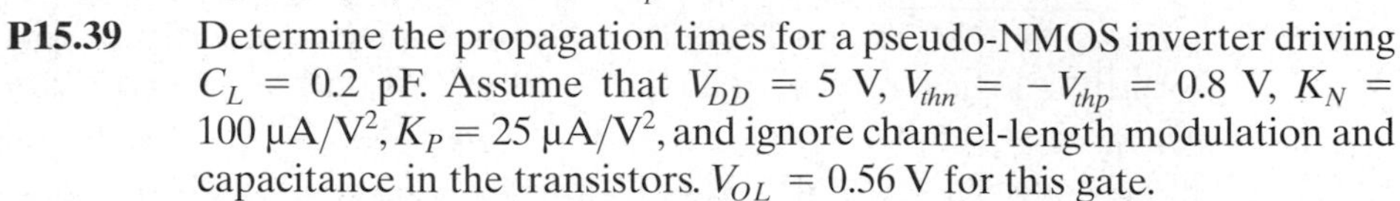

P15.39 Determine the propagation times for a pseudo-NMOS inverter driving $C_L = 0.2$ pF. Assume that $V_{DD} = 5$ V, $V_{thn} = -V_{thp} = 0.8$ V, $K_N = 100\ \mu\text{A/V}^2$, $K_P = 25\ \mu\text{A/V}^2$, and ignore channel-length modulation and capacitance in the transistors. $V_{OL} = 0.56$ V for this gate.

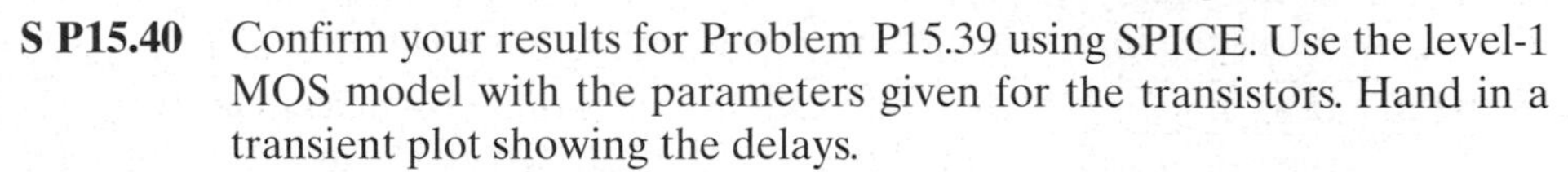

S P15.40 Confirm your results for Problem P15.39 using SPICE. Use the level-1 MOS model with the parameters given for the transistors. Hand in a transient plot showing the delays.

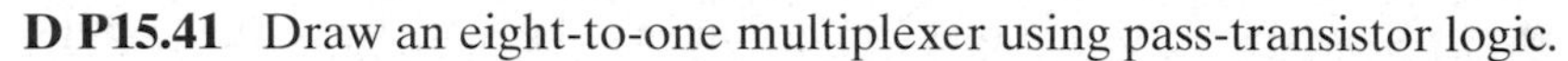

D P15.41 Draw an eight-to-one multiplexer using pass-transistor logic.

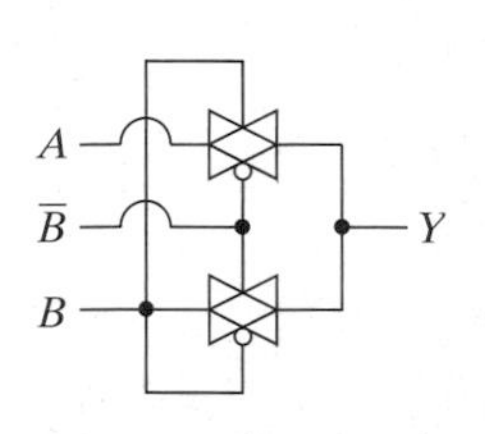

Figure 15–88

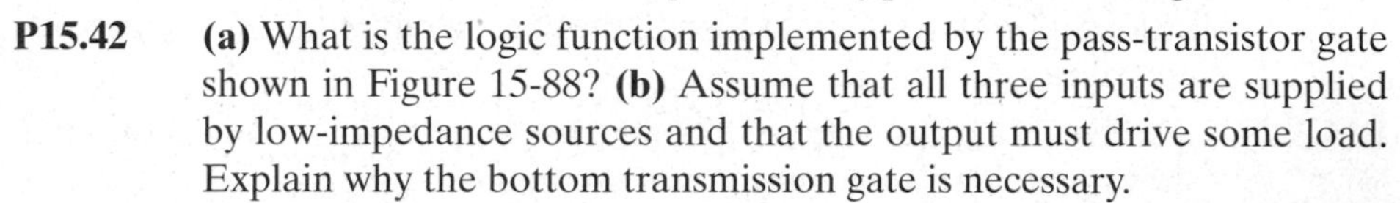

P15.42 **(a)** What is the logic function implemented by the pass-transistor gate shown in Figure 15-88? **(b)** Assume that all three inputs are supplied by low-impedance sources and that the output must drive some load. Explain why the bottom transmission gate is necessary.

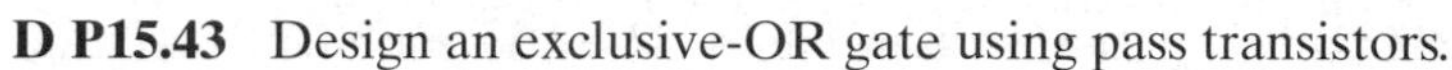

D P15.43 Design an exclusive-OR gate using pass transistors.

15.3.6 MOS Memory

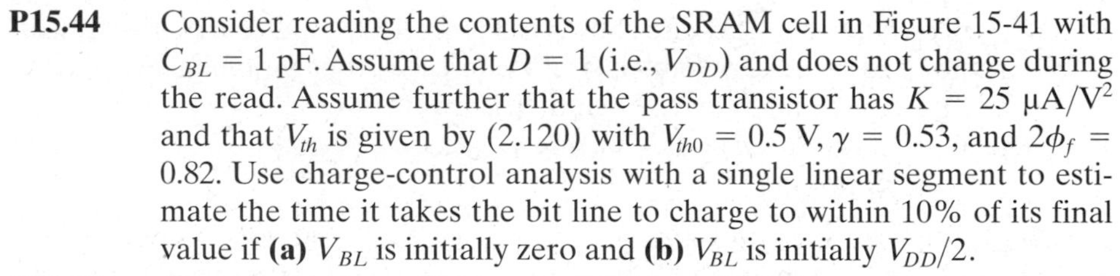

P15.44 Consider reading the contents of the SRAM cell in Figure 15-41 with $C_{BL} = 1$ pF. Assume that $D = 1$ (i.e., V_{DD}) and does not change during the read. Assume further that the pass transistor has $K = 25\ \mu\text{A/V}^2$ and that V_{th} is given by (2.120) with $V_{th0} = 0.5$ V, $\gamma = 0.53$, and $2\phi_f = 0.82$. Use charge-control analysis with a single linear segment to estimate the time it takes the bit line to charge to within 10% of its final value if **(a)** V_{BL} is initially zero and **(b)** V_{BL} is initially $V_{DD}/2$.

P15.45 Consider writing a '0' into the DRAM cell in Figure 15-42. If $V_S = V_{DD}$ prior to the write, describe the operation of the pass transistor during the write (i.e., which end acts as the source? What region of operation is the device in?).

P15.46 Consider writing a '1' into the DRAM cell in Figure 15-42 with $V_S = 0$ V prior to the write, as in Exercise 15.12. Use charge-control analysis to derive an approximate equation for the time required to charge C_S in terms of the transistor model parameters and component values.

P15.47 Assume that the pass gate in a one-transistor DRAM cell can be effectively modeled as a constant resistance when it is on during a read operation. **(a)** Derive an equation for the bit-line voltage as a function of time assuming that the pass transistor is turned on at $t = 0$, that it has an on resistance of R_{on}, and that $v_S(0^+) = V_S$, V_{pre}, C_S, and C_{BL} are known. **(b)** If $C_S = 20$ fF, $C_{BL} = 1$ pF, and $R_{on} = 2\ \text{k}\Omega$, how long will it take the bit-line voltage to change by 90% of the total change it will see?

P15.48 What are the four 8-bit words that are stored in the ROM in Figure 15-89?

P15.49 Consider a ROM with the structure shown in Figure 15-89. **(a)** If the NMOS transistors all have $K_N = 25\ \mu\text{A/V}^2$, what must K_P be to set $V_{OL} = 0.33V_{DD}$? Assume that $V_{DD} = 3$ V and $V_{thn} = -V_{thp} = 0.8$ V. Ignore the body effect. **(b)** Using the value of K_P you found in (a), how much current is consumed by each bit line during a read operation if the bit line is high?

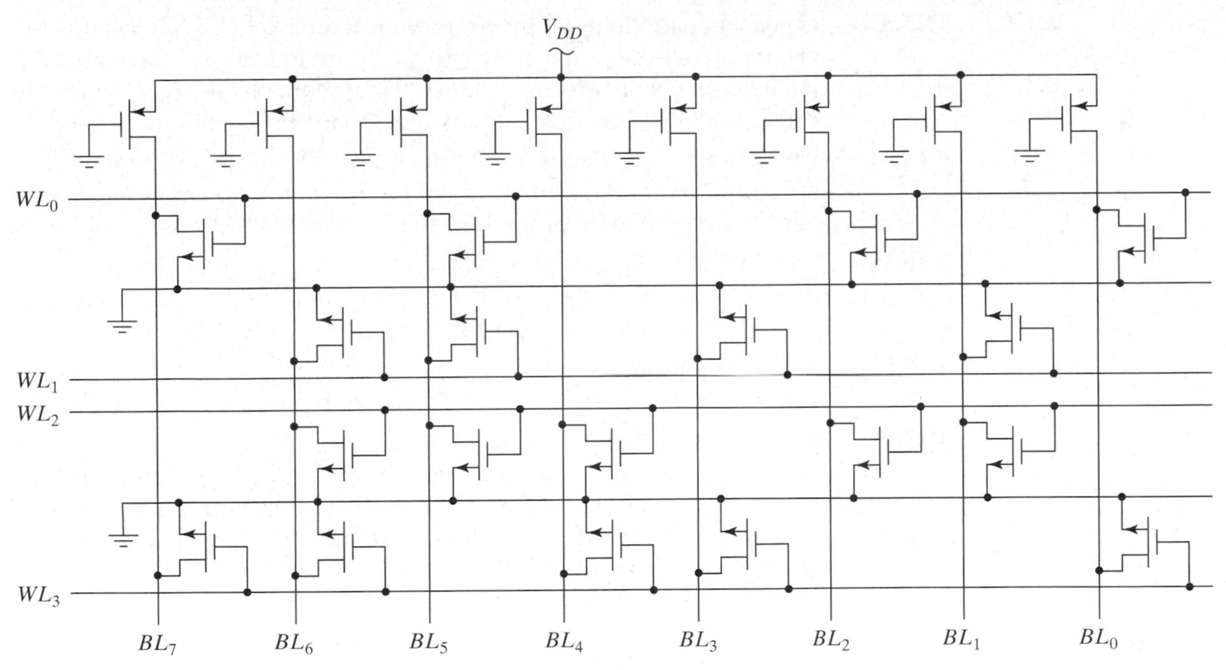

Figure 15–89

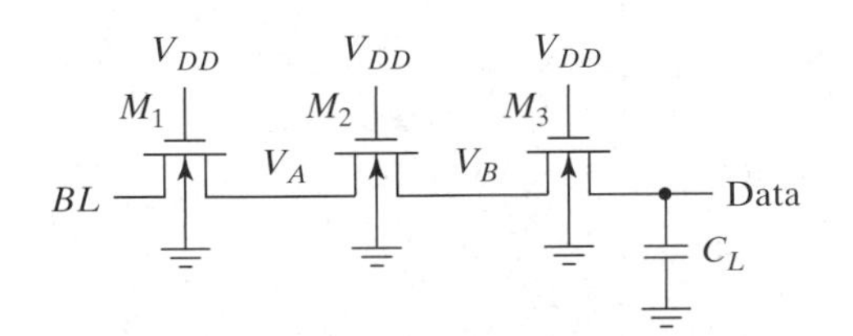

Figure 15–90

(c) If it is low? **(d)** Assuming that half of all bits are '1' on average, what is the average power consumed during a read operation on a 64 kb ROM configured as a square array (i.e., an equal number of word lines and bit lines)?

P15.50 Develop the Boolean expressions and draw the resulting schematic for a two-step word-line decoder for a 4-bit word address.

P15.51 Consider an eight-to-one tree-type column decoder as shown in Figure 15-47. Figure 15-90 shows one path through the decoder when it is selected. (C_L is the input capacitance of the output buffer, which is not shown.) Assume that the transistors have $K = 25\ \mu\text{A/V}^2$ and that V_{th} is given by (2.120) with $V_{th0} = 0.5$ V, $\gamma = 0.53$, and $2\phi_f = 0.82$. Suppose we are reading data out of a cell and the sense amplifier has driven the bit line to $V_{BL} = V_{DD} = 3$ V. Further assume that $V_A = V_B = V_L = 0$ initially. **(a)** What is the voltage V_A? **(b)** What is the voltage V_B? **(c)** What is the final output voltage? **(d)** Repeat (a)–(c) if $V_{BL} = 0$ V.

15.4 BIPOLAR LOGIC GATES

15.4.1 Transistor–Transistor Logic (TTL)

P15.52 Consider again the RTL inverters from Exercise 15.15. If we desire to keep the nominal $V_{OH} > 3.5$ V, what is the maximum fan-out of the inverter?

P15.53 Consider again the RTL inverters from Exercise 15.15. **(a)** What is the maximum fan-out of the inverter if we desire to keep the forced beta of the inverters being driven by it less than or equal to 20? **(b)** What is the corresponding value of V_{OH} (i.e., when the maximum fan-out is used)?

P15.54 Derive a Boolean expression for the output (Y) of the logic gate shown in Figure 15-91 in terms of the inputs (A, B, and C). Assume that all resistors have been chosen properly so that the circuit functions as expected.

P15.55 Consider the TTL gate shown in Figure 15-92. **(a)** What would happen if the 1-kΩ resistor (at the base of Q_3) was connected to -2 V instead of ground? **(b)** Explain why the 130-Ω resistor is included in the circuit.

P15.56 Based on the simulation results presented in Example 15.8, what is the PDP of the two-input TTL NAND gate with no load and $f_{sw} = 1$ MHz?

P15.57 A SPICE simulation of a two-input TTL NAND gate shows that the base current of Q_3 is 2.2 mA when the output is low. Determine the maximum fan-out of this gate if it is driving gates identical to itself and we desire to keep the forced beta of Q_3 less than 10 (to keep V_{OL} low enough).

Figure 15–91

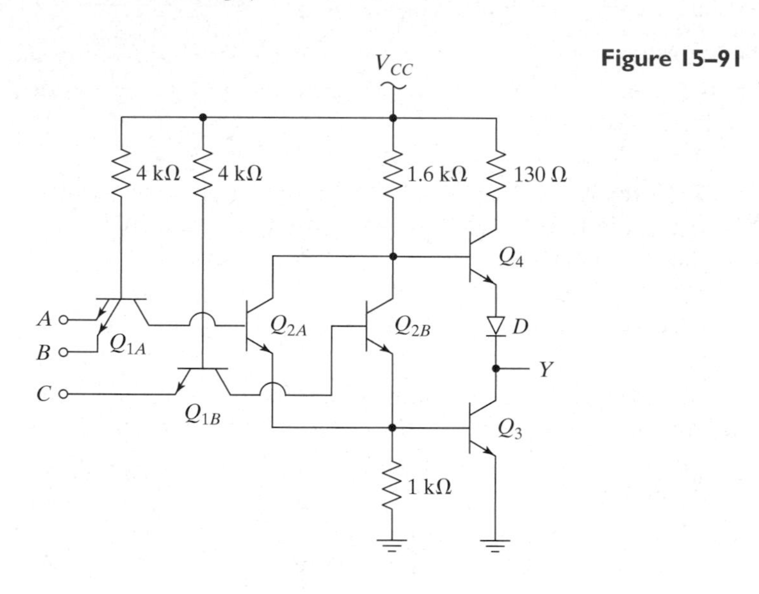

Figure 15–92

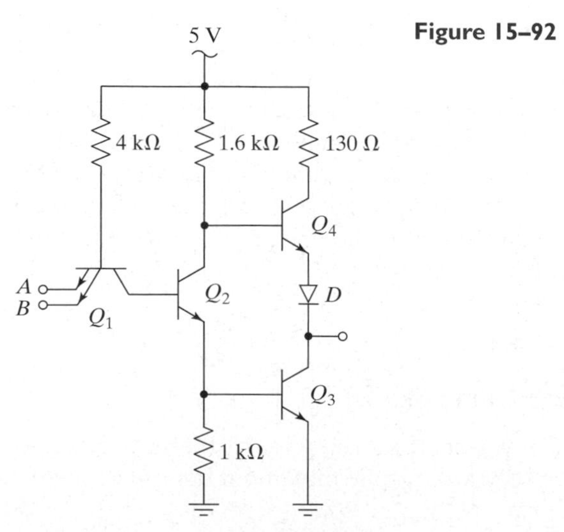

S P15.58 **(a)** Use SPICE to run a DC sweep of the TTL NAND gate shown in Example 15.8 to find the static transfer characteristic (v_O versus v_A while $v_B = 3.6$ V). **(b)** On your plot of the static transfer characteristic, indicate the values of V_{IL}, V_{OL}, V_{IH}, and V_{OH}, and indicate how you got them. **(c)** What are the noise margins of this gate?

15.4.2 Emitter-Coupled Logic

P15.59 Consider driving two identical RC loads with symmetric, linear voltage ramps as shown in Figure 15-93, parts (a) and (b). **(a)** Show that the sum of the two drive currents ($i_T = i_P + i_N$) is constant, and determine what it is. **(b)** Would the current be constant if the two voltages were still symmetric, but not linear, as in part (c) of the figure?

P15.60 **(a)** Design a load for an ECL gate using two resistors, one connected to ground and one to -5.2 V. Choose the resistor values so that the Thévenin equivalent load is 100 Ω tied to -2 V. **(b)** Compute the static power consumed by this load when the level on the line is a logic 1 and a logic 0, and calculate the average static power used. **(c)** Calculate the average static power used by a single 100-Ω resistor connected to -5.2 V. How much power do you save by using the Thévenin equivalent load?

P15.61 Consider the simulated ECL inverter transfer characteristic in Example 15.9. Show that the slope of the characteristic for $-1.1 \leq V_I \leq -0.5$ is approximately equal to $-R_1/R_{EE}$.

P15.62 Consider the custom-designed ECL inverter shown in Figure 15-94. **(a)** Sketch the static transfer characteristic for this gate. **(b)** Derive equations for V_{OH}, V_{OL}, V_{IH}, and V_{IL} (not the typical values, but the values needed for determining noise margins).

S P15.63 Use the Nsimple model from Appendix B to simulate the circuit in Figure 15-94, and confirm your results from Problem P15.62. Hand in a printout of the static transfer characteristic, and mark on the plot the points where the slope equals -1.

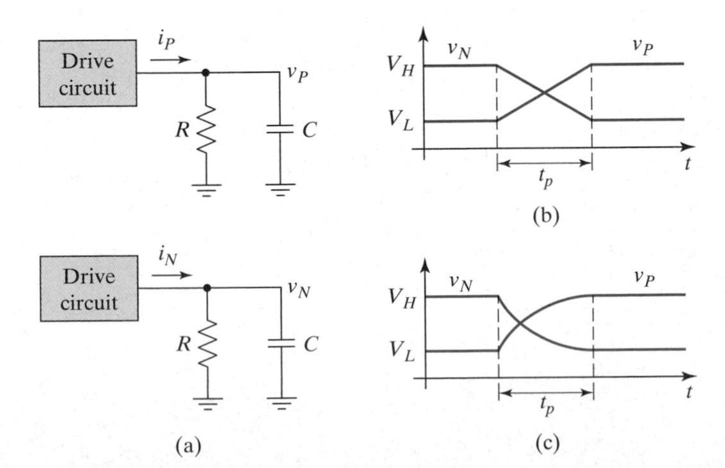

Figure 15–93

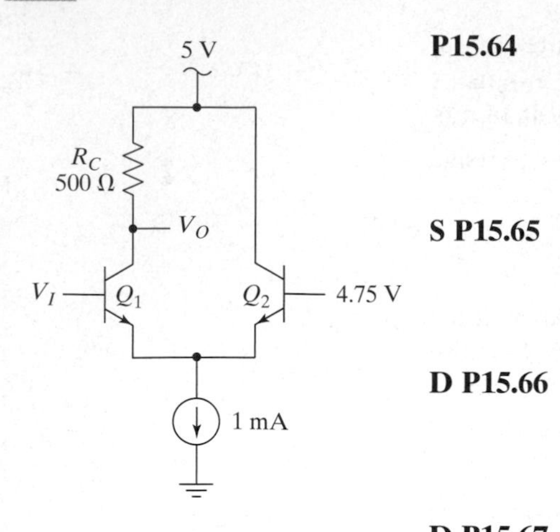

Figure 15–94

P15.64 Consider the custom-designed ECL inverter shown in Figure 15-94, and assume that the transistors have $I_S = 10^{-14}$ A, $\beta_F = 100$, $r_b = 20\ \Omega$, $\tau_F = 0.5$ ns, $C_{jeo} = 2$ pF, and $C_{jco} = 1$ pF. Find an approximate value for the propagation delay of this gate. Is the speed limited by the load time constant or the base time constant?

S P15.65 Use the Nsimple model from Appendix B and edit it to have the parameters given in P15.64. Simulate the circuit in Figure 15-94, and confirm your results from Problem P15.64. Hand in a printout of the output voltage from the transient simulation.

D P15.66 Design a CML gate to operate from a single +3-V supply, implement the function $Y = A + B + C$, and have a logic swing of 1 V and a static power of 30 μW. You may assume that a constant bias voltage is available for V_{BB}, but you must specify its voltage.

D P15.67 Use collector dotting to design a CML gate that implements the logic function $Y = (A + B) \cdot (C + D)$. Use a single 3.3-V supply, and design for a logic swing of 1 V and a static power of 100 μW. You may assume that a constant bias voltage is available for V_{BB}, but you must specify its voltage.

P15.68 Consider the CML logic gate shown in Figure 15-95. **(a)** What logic function is implemented by this gate? **(b)** What is the typical value of V_{OH}? **(c)** of V_{OL}? **(d)** What is the purpose of $Q_5 - Q_{10}$ and $R_1 - R_3$?

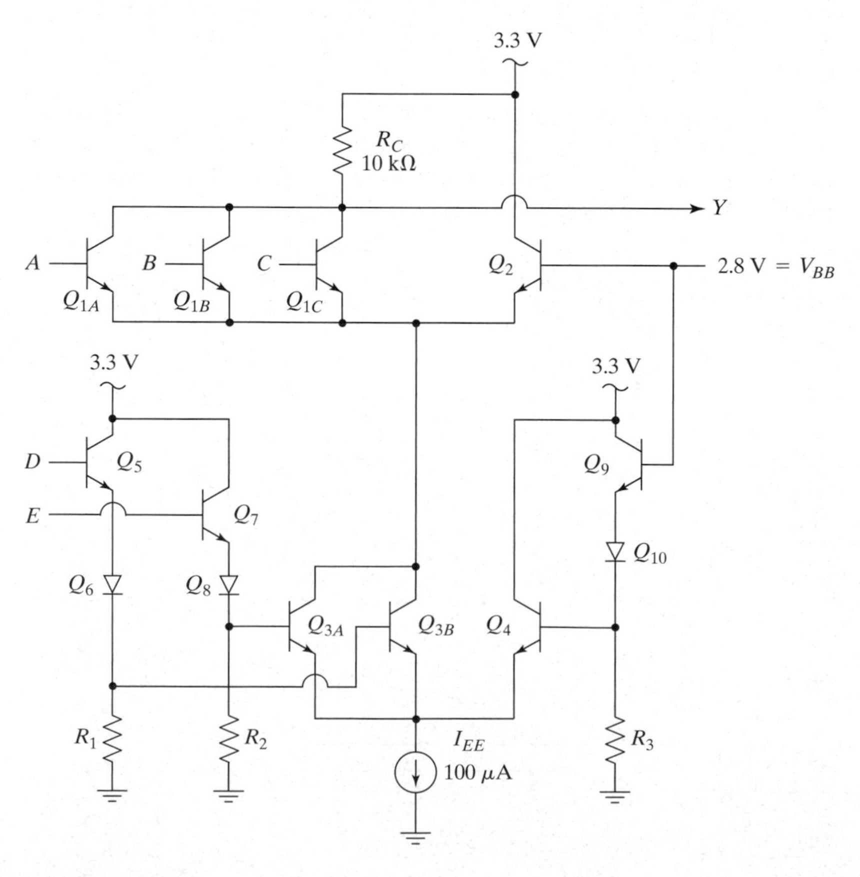

Figure 15–95

D P15.69 Use the wired-OR capability of standard 10K series ECL gates to design a circuit that will implement the function $Y = \overline{A} \cdot (\overline{B} + \overline{C})$, using only two two-input OR/NOR gates with A and B as the available inputs.

D P15.70 Use the wired-OR capability of standard 10K series ECL gates to design a two-input exclusive-OR gate using only two-input OR/NOR gates.

P15.71 Consider the modified ECL inverter shown in Figure 15-96. **(a)** What is the nominal value of V_{OH}? **(b)** Determine the value of R_1 needed to produce a logic swing that is symmetric about V_{BB}.

D P15.72 Consider the custom ECL gate shown in Figure 15-97. Determine the values needed for R_1 and R_2 for this gate to have equal logic swings on both outputs. Make sure that V_{BB} is centered between V_{OH} and V_{OL}.

S P15.73 Use the Nsimple model from Appendix B and SPICE to plot the static transfer characteristic for your design in Problem P15.72.

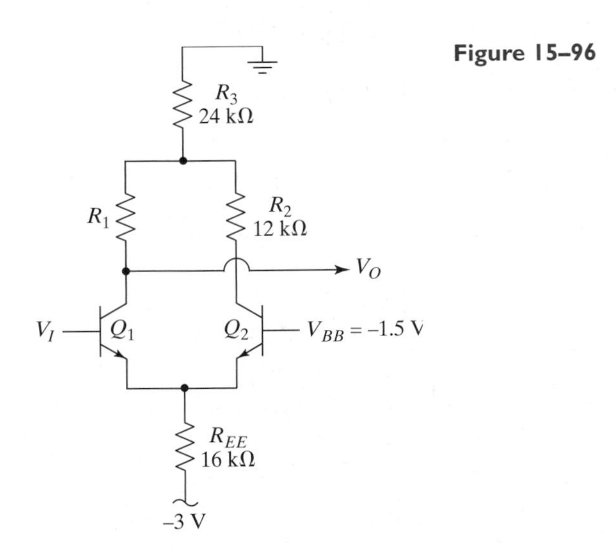

Figure 15–96

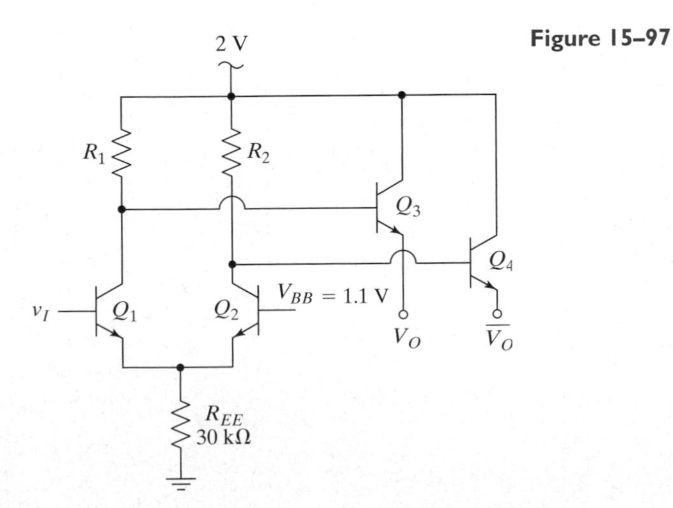

Figure 15–97

APPENDIX A
SPICE Reference

INTRODUCTION

SPICE is the most widely used circuit simulation program available. This appendix is a brief reference for the command and element-input syntax used by SPICE. Although the majority of the commands and element statements are the same in most versions of SPICE, the two versions covered here are PSPICE and HSPICE. We will point out differences where they exist between these versions. The general structure and proper use of SPICE is covered in Chapter 4 and should be understood prior to using the program. The example device models used througout the text are presented in Appendix B.

A.1 RUNNING SPICE

To run SPICE, we first need to create an input file. Unless you are using a schematic entry program, there isn't any required file extension for this input file. Nevertheless, it is good practice to pick an extension and be consistent in its use. The extensions `.cir` and `.ckt` are both common. The details of the input file syntax are given in Section A.2.

When the input file is finished you must run the simulator. How you do this varies from platform to platform. In the MicroSim Schematics program, you run PSPICE directly from a menu or a button. However you call the simulator, when it is finished, it will produce an output file with the same name as your input file but with the extension `.out`. This file contains any error messages SPICE produced, a copy of the input file, and other information depending on what was requested. It usually contains at least a printout of the DC node voltages, element currents, voltages and power. It will also contain the results of any `.print` or `.plot` statements that were included in the input file. If nonlinear devices are present and an AC simulation was performed, it typically also contains the models evaluated at the operating point.

In addition to the output file, SPICE may produce one or more files for the post processor to use. For PSPICE there is only one file and it has the extension `.dat`. For HSPICE, the files have different extensions depending on the type of simulation data they contain; for example, a DC simulation post-processor data file will have the extension `.dc` followed by a number (in case more than one DC simulation was performed).

A.2 THE INPUT FILE

The input file consists of the netlist for the circuit to be simulated, program control statements that affect the way SPICE runs, and simulation commands that tell SPICE what to do with the circuit.[1] These three types of statements are discussed in the following subsections.

The first line of the input file must be the title line and the last line must be the `.end` line (in a packaged set of programs like MicroSim Schematics and PSPICE, you don't need to worry about these details; they are handled automatically when you invoke the simulator from within the schematic capture program). In addition, if `.alter` statements are used, they must be at the end of the file (this statement is only available in HSPICE). Except for the items noted above, the order in which the elements are specified and the order in which the program control and simulation commands are given does not matter. Nevertheless, it is good practice to have a consistent, organized format for your input files, as shown in Chapter 4.

The following sections present an abridged summary of the syntax for the SPICE input file. The SPICE manuals or a complete reference should be consulted for the many options and commands omitted here. The notational conventions used in this summary are shown in Table A.1.

Component values (and other numbers) can be entered in SPICE using scientific notation or suffixes. For example, the number 3.6×10^4 can be entered as `3.6E4` or as `36k`. SPICE does not expect you to include units, and it is a good idea not to, since they can be confused with the suffixes used. For example, if you enter a capacitor value as 1F in SPICE, it gets interpreted as 1 femtofarad, not 1 farad! Some of the suffixes used in SPICE are; `T` $= 10^{12}$, `G` $= 10^9$, `Meg` $= 10^6$, `K` $= 10^3$, `M` $= 10^{-3}$, `U` $= 10^{-6}$, `N` $= 10^{-9}$, `P` $= 10^{-12}$, `F` $= 10^{-15}$. HSPICE also uses `X` for `Meg`.

Nodes in SPICE may be numbered or, in some versions, named. The ground node is special in SPICE and each circuit must have a ground node. It must always be numbered `0` or, if names are allowed, be called `gnd`.

A.2.1 The Netlist

The netlist tells SPICE the topology of the circuit to be simulated and the values or models to be used for each of the components. The simplest components are the passive two-terminal devices; resistors, capacitors and inductors.

Table A.1

A summary of the notation used for specifying SPICE commands.

`xxx, yyy, zzz`	Arbitrary alphanumeric strings.
`{xxx}`	Items enclosed in braces `{}`, are optional. Nested sets of braces indicate that all preceding items must be included. For example; in `{a {b {c {d}}}}`, if you wish to include item `c`, you must also include items `a` and `b`, but you can leave out item `d`.
`[A,B,C]`	A list of items in square brackets implies that any one of the items in the list may be used.

[1] The word netlist is sometimes used to refer to the entire input file, including simulation commands and program control statements.

Resistors, Capacitors and Inductors The general input format for resistors, capacitors and inductors in SPICE is:

```
[R,C,L]xxx node1 node2 value
```

The names must begin with `R`, `L`, or `C`, but the following characters can be numbers and/or alphabetic characters. The value may be given in any numeric format. In addition, a model may be specified in place of the value and temperature coefficients and other options are available, consult the SPICE manual for details on these options. Examples:

```
R3      in      gnd     3.3k
Cin     in      b1      100u
L1      vcc     col     1m
```

Independent Sources The general format for independent sources in SPICE is:

```
[V,I]xxx +node -node {{DC value} {AC magnitude, phase}
{tranfcn}}
```

Voltage source names must begin with `V` and current source names must begin with `I`. Current flows from the `+node` *through the source* to the `-node`. The source may have separate DC, AC, and transient values. `tranfcn` is the function used to describe the transient value of the source. The most commonly used functions are described in the following sections.

Transient Functions; PULSE

```
PULSE(v1 v2 {td {tr {tf {pw {per}}}}})
```

Parameter	Description	Default Value
`v1`	initial value of the pulse	none
`v2`	the value the pulse changes to	none
`td`	delay prior to the pulse starting to change to `v2`	0
`tr`	rise time of the edge from `v1` to `v2`	*Tstep*[2]
`tf`	fall time of the edge from `v2` to `v1` (if the pulse goes back down)	*Tstep*
`pw`	pulse width (i.e., how long it stays at `v2`)	*Tstop* (*Tstep*)[3]
`per`	period for the pulse repeating (if it does)	*Tstop* (*Tstep*)

Example:

```
vin     in gnd pulse(.5 1 1u 2u 3u 5u 12u)
```

as shown in Figure A-1.

Transient Functions; SINUSOID

```
SIN(vo pk {freq {td {df {phase}}}})
```

---

[2] *Tstep* is the step size and *Tstop* is the stop time as specified in the `.tran` statement.

[3] HSPICE defaults are given in parenthesis when different from PSPICE.

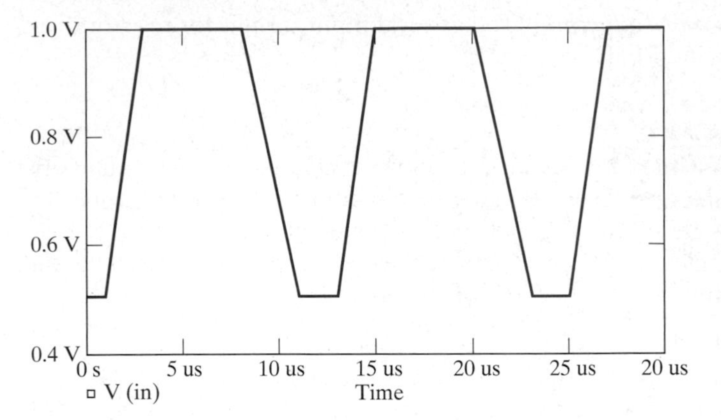

Figure A–1 The pulse produced by the example shown.

Parameter	Description	Default Value
`vo`	offset value	none
`pk`	peak amplitude of the sine wave	none
`freq`	frequency of the sine wave	1/*Tstop*
`td`	delay prior to the sine wave starting	0
`df`	damping factor for an exponentially decaying sine wave	0
`phase`	phase of the sine wave	0

The equation used for the sine wave, $s(t)$, is

$$s(t) = \mathbf{vo} + \mathbf{pk}\sin\left(2\pi\left(\mathbf{freq}(t - \mathbf{td}) + \frac{\mathbf{phase}}{360}\right)\right)e^{-(t-\mathbf{td})\mathbf{df}}.$$

Example:

```
vin      in      gnd      sin(0 1 1k 100u 500 45)
```

as shown in Figure A-2.

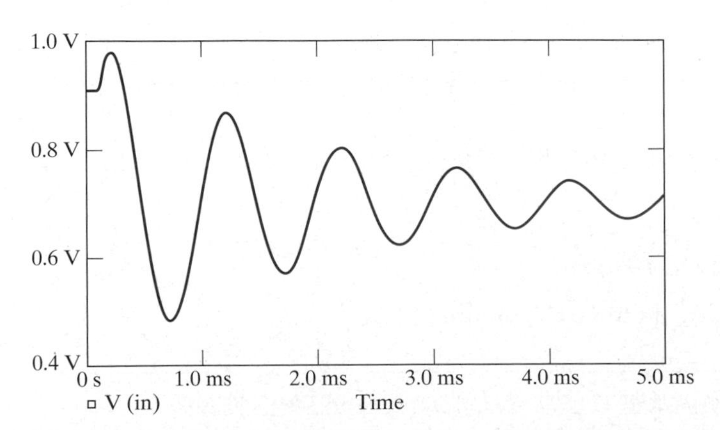

Figure A–2 The sine wave produced by the example shown.

Transient Functions; EXPONENTIAL

```
EXP(v1 v2 {td1 {tau1 {td2 {tau2}}})
```

Parameter	Description	Default Value
v1	initial value of the waveform (value 1)	none
v2	the value the waveform changes to	none
td1	delay prior to the waveform starting to change to **v2**	0
tau1	time constant for exponential change from **v1** to **v2**	*Tstep*[4]
td2	delay prior to the waveform starting to change back to **v1**	*Tstep*
tau2	time constant for exponential change from **v2** to **v1**	*Tstop* (*Tstep*)[5]

The equation used for the exponential, $s(t)$, is

$$s(t) = \begin{cases} \mathbf{v1}; & 0 \le t \le \mathbf{td1} \\ \mathbf{v1} + (\mathbf{v2} - \mathbf{v1})[1 - e^{-(t-\mathbf{td1})/\mathbf{tau1}}]; & \mathbf{td1} < t \le \mathbf{td2} \\ \mathbf{v1} + (\mathbf{v2} - \mathbf{v1})[1 - e^{-(\mathbf{td2}-\mathbf{td1})/\mathbf{tau1}}]e^{(t-\mathbf{td2})/\mathbf{tau2}}; & \mathbf{td2} < t \le Tstop \end{cases}$$

Example:

```
vin      in      gnd      exp(.25 .5 1u 2u 7u 1u)
```

as shown in Figure A-3.

Transient Functions; PIECEWISE LINEAR

```
PWL(t1 v1 {t2 v2} .....)
```

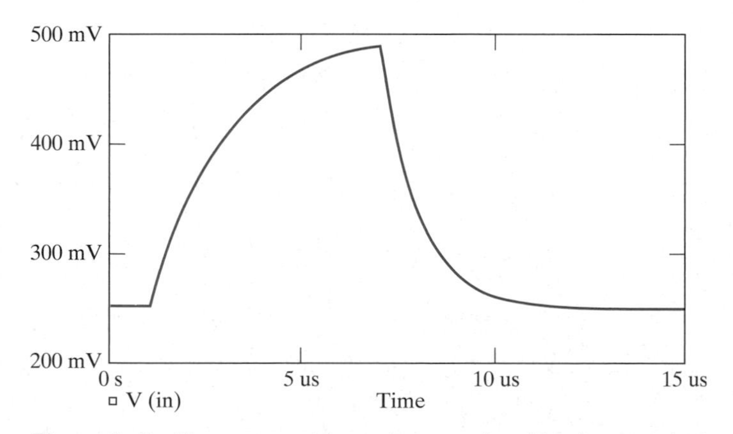

Figure A-3 The exponential waveform produced by the example shown.

[4] *Tstep* is the step size and *Tstop* is the stop time as specified in the **.tran** statement.

[5] HSPICE defaults are given in parenthesis when different from PSPICE.

Parameter	Description	Default Value
t1	time at which waveform equals v1 (value 1)	none
v1	the value of the waveform at t1	none
t2	time at which waveform equals v2	none
v2	the value of the waveform at t2	none

Note: The waveform starts at v1 at $t = 0$ if t1 > 0.
Example:

```
vin    in    gnd    dc 1 pwl(1u 1.5 3u -.5 7u 0 9u 1)
```

as shown in Figure A-4.

Transient Functions; SINGLE-FREQUENCY FREQUENCY MODULATION

```
SFFM(off pk {fc {mod {fs}}})
```

Parameter	Description	Default Value
off	DC offset	none
pk	peak value of the modulated sine wave	none
fc	frequency of the modulated sine wave (the "carrier")	1/*Tstop*[6]
mod	modulation index	0
fs	frequency of the modulating signal	1/*Tstop*

The equation used for the single-frequency FM signal, $s(t)$, is

$$s(t) = \mathtt{off} + \mathtt{pk}\sin[2\pi\mathtt{fc}t + \mathtt{mod}\sin(2\pi\mathtt{fs}t)].$$

Example:

```
vin    in    gnd    sffm(0 1 30k 15 1k)
```

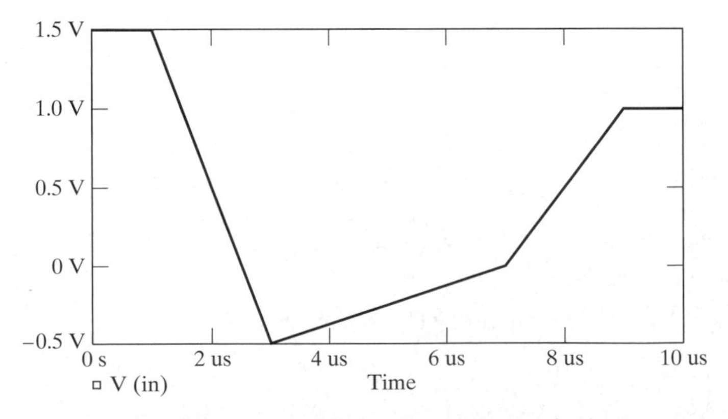

Figure A–4 The waveform produced by the example shown.

[6] *Tstep* is the step size and *Tstop* is the stop time as specified in the .tran statement.

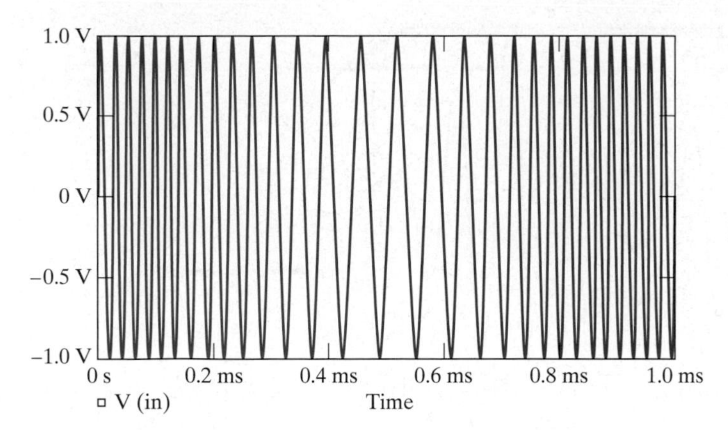

Figure A–5 The waveform produced by the example shown.

as shown in Figure A-5.

Transient Functions; AMPLITUDE MODULATION This function is not available in PSPICE. The syntax shown below is for HSPICE.

```
AM(sa oc fs fc td)
```

Parameter	Description	Default Value
`sa`	signal amplitude	0
`oc`	DC offset (gets multiplied by `sa`)	0
`fs`	modulating signal frequency	1/*Tstop*[7]
`fc`	frequency of the sine wave being modulated (the "carrier")	0
`td`	delay before the start of the waveform	0

The equation for the amplitude modulated signal, $s(t)$, is

$$s(t) = \mathbf{sa}(\mathbf{oc} + \sin[2\pi\mathbf{fs}(t - \mathbf{td})])\sin[2\pi\mathbf{fc}(t - \mathbf{td})]$$

Example:

```
vin    in    gnd    am(5 0 1k 10k 0)
```

as shown in Figure A-6.

Dependent Sources In addition to independent voltage and current sources, SPICE includes all four dependent sources. The voltage-controlled voltage source (VCVS) element name must begin with **E**. The voltage-controlled current source (VCCS) element name must begin with **G**. The current-controlled voltage source (CCVS) element name must begin with **H**. The current-controlled current source (CCCS) element name must begin with **F**. The general format for these elements in SPICE is:

```
[E,G,H,F]xxx +out -out +cntrl -cntrl gain
```

[7] *Tstop* is the stop time as specified in the `.tran` statement.

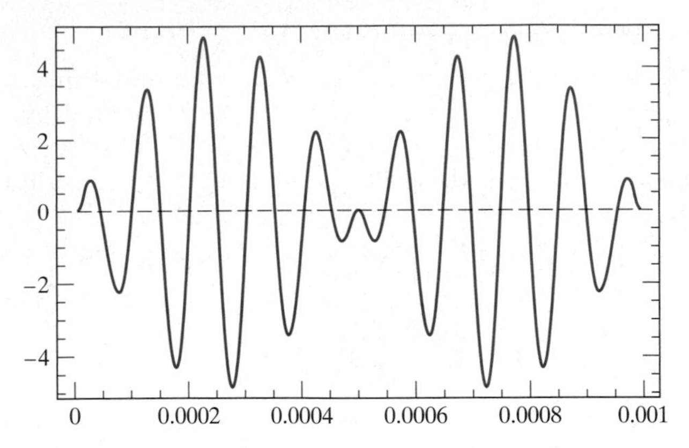

Figure A–6 The waveform produced by the example shown.

The source output is connected between `+out` and `-out`. Current flows from `+out` *through the source* to `-out`. The value of the source is controlled by the signal present at the control nodes; `+cntrl` and `-cntrl`. The signal at the control nodes is multiplied by the gain to obtain the output. In addition to the linear function shown here, the output can be calculated based on a polynomial function of one or more inputs, a lookup table or other expressions. You should consult the program manual for these other options.

Example:

```
g12 2 4 6 7 0.6
```

The source described by this example produces a current from node 2 to node 4 that is equal to 0.6 times the voltage from node 6 to node 7.

The most recent versions of SPICE replace some of the controlled sources with analog behavioral modeling elements. In both HSPICE and PSPICE the **H** and **F** elements are obsolete, although still supported. They are replaced by the newer **G** and **E** sources which are, respectively, a controlled current source and a controlled voltage source. The controlling inputs may now be either a voltage, a current, or a combination. The newer versions of SPICE include many more sophisticated behavioral modeling capabilities; see [A.2] and the manuals for details. Exercise 4.5 provides an example of one form of the new elements.

Diodes The diode is the only two-terminal semiconductor device modeled in SPICE. Diode names must begin with the letter **D**. The general input format for diodes is:

```
Dxxx anode cathode model_name {area}
```

anode is the node name or number connected to the anode of the diode (i.e., the positive side when forward biased) and **cathode** is the node number or name connected to the cathode. **model_name** is the name of the model to be used for this particular diode. The **area** multiplier changes some of the model parameters to account for devices with the same structure, but different sizes. Setting **area** equal to two is roughly the same as putting two diodes in parallel (it isn't exactly the same because the parasitics are different for two discrete devices in parallel than they are for one device with twice the area). The model statement has the format:

```
.MODEL model_name D(list of parameters)
```

The parameters may be listed in any order and none of them are essential since SPICE does have usable default values. The parameters may be separated by commas or by spaces. Be warned, however, that the default values will NOT typically produce accurate results. They are only useful if you want to test what will happen with a generic, and nearly ideal, diode. See Chapter 4 for details on the diode model parameters. Table A.2 is a reproduction of Table 4.1 and summarizes the basic model parameters.

Example:

```
d1    in    out    dsimple
.model dsimple D(IS=1e-15 N=1.3)
```

This example will model an almost ideal diode; it will not breakdown and does not have any capacitance or series resistance.

Transistors The SPICE input syntax for all three-terminal semiconductor devices is similar, but there are some differences. Junction field-effect transistor names begin with `J`, MOSFET names begin with `M` and bipolar junction transistor names begin with `Q`. MESFET's are also modeled in SPICE; in PSPICE a new element, starting with the letter `B`, is used, while in HSPICE the JFET element is used and the model is simply changed. We will not explicitly cover MESFETs here. The general forms of the input statements are:

```
Jxxx nd ng ns model_name {area}
Mxxx nd ng ns sub model_name {L=value} {W=value}
Qxxx nc nb ne {sub} model_name {area}
```

Table A.2 SPICE Diode model parameters.[1]

Name	Parameter	Units	Default	Example
BV	Breakdown voltage	V	infinite	60
CJO[2]	Zero-bias junction capacitance	F	0	1E-12
EG	Bandgap	eV	1.11	1.11 Si 0.69 Schottky 0.67 Ge
IBV	Current at BV	A	$10^{-10}(10^{-3})$	1E-4
IS	Saturation current	A	10^{-14}	5E-15
M	Grading coefficient	none	0.5	0.33
N	Ideality factor	none	1	1.5
RS	Series resistance	Ω	0	5
TT	Transit time	s	0	1E-9
VJ	Built-in potential	V	1 (.8)	0.65

[1] The defaults are for PSPICE and HSPICE; when they are different, the HSPICE values are given in parentheses.

[2] The last character in the parameter name is the letter 'O,' not the number zero.

nd is the drain node, **ng** is the gate node and **ns** is the source node for both types of field-effect transistor. **sub** is the substrate or bulk node for both MOSFETs and BJT's. For the BJT; **nc** is the collector node, **nb** is the base node and **ne** is the emitter node. In all cases **model_name** is the name of the model to be used for the device and **area** is an optional parameter to specify the area of the transistor. Chapter 4 explains how **area** affects some of the model parameters. For the MOSFET, the width and length of the device can also be specified on the element line as shown. They may also be given default values in the **.options** statement, or they may be given values in the model. For all three devices the model statement has the general form

```
.MODEL model_name [type](parameter list)
```

where **type** is either **NJF** or **PJF** for an n-channel or p-channel JFET respectively. For the MOSFET, **type** is **NMOS** or **PMOS** for n-channel or p-channel devices. For the bipolar transistor **type** is either **NPN** or **PNP**.

The parameters may be listed in any order and none of them are essential since SPICE does have usable default values. Be warned, however, that the default values will NOT produce accurate results for a given circuit. They are only useful if you want to test what happens with a generic, and nearly ideal, transistor. See Chapter 4 for details on the model parameters. Tables A.3 and A.4 are repeated from Chapter 4 and summarize the model parameters.

A.2.2 Simulation Commands

The different modes of analysis are discussed in Section 4.2. The purpose of this section is to provide a brief summary of the most commonly used SPICE simulation commands.

.AC The general syntax for the AC simulation command is:

```
.AC [LIN, DEC, OCT} #_pts fstart fstop
```

This command tells SPICE to perform a small-signal AC analysis and to sweep the frequency from **fstart** to **fstop**. The sweep can either be a linear sweep, or a logarithmic sweep. The parameter **LIN** indicates a linear sweep; SPICE then uses a total of **#_pts** values for the frequency, evenly spaced between **fstart** and **fstop**. The parameter **DEC** (or **OCT**) tells SPICE to perform a sweep where the different frequencies are evenly spaced on a logarithmic scale using **#_pts** values per decade (or octave).

Examples:

```
.ac lin 100 1k 10k
```

This example performs a sweep from $f = 1$ kHz to $f = 10$ kHz with 100 equally spaced values.

```
.ac dec 10 100 1meg
```

This example performs a sweep from $f = 100$ Hz to $f = 1$ MHz with 10 values per decade. The values are evenly spaced on a logarithmic scale.

.DC The different syntaxes for the DC sweep simulation command are:

```
.DC {LIN} var start stop incr {nested sweep}
.DC {DEC,OCT} var start stop #_pts {nested sweep}
.DC var LIST value... {nested sweep}
```

Table A.3 SPICE BJT model parameters.[1]

Name	Parameter	Units	Default	Example
`BF`	Forward beta	none	100	180
`BR`	Reverse beta	none	1	5
`CJC`	Zero-bias base-collector junction capacitance	F	0	1E-12
`CJE`	Zero-bias base-emitter junction capacitance	F	0	2E-12
`IKF`	Knee current for high-current forward beta rolloff	A	infinite	5E-2
`IKR`	Knee current for high-current reverse beta rolloff	A	infinite	5E-2
`IS`	Saturation current	A	10^{-16}	3E-16
`ISC`	Base-collector junction leakage saturation current	A	0	1E-16
`ISE`	Base-emitter junction leakage saturation current	A	0	1E-16
`ITF`	TF high-current parameter	A	0	N/A
`MJC`	Base-collector grading coefficient	none	0.33	0.5
`MJE`	Base-emitter grading coefficient	none	0.33	0.33
`NC`	Base-collector junction leakage emission coefficient	none	2	2
`NE`	Base-emitter junction leakage emission coefficient	none	1.5	1.5
`NF`	Forward emission coefficient	none	1	1.2
`NR`	Reverse emission coefficient	none	1	2
`RB`	Base resistance	Ω	0	50
`RC`	Collector resistance	Ω	0	100
`RE`	Emitter resistance	Ω	0	1
`TF`	Forward transit time	s	0	1E-9
`TR`	Reverse transit time	s	0	5E-8
`VAF`	Forward Early voltage	V	infinite	60
`VAR`	Reverse Early voltage	V	infinite	120
`VJC`	Base-collector built-in potential	V	.75	0.65
`VJE`	Base-emitter built-in potential	V	.75	0.65
`VTF`	TF base-collector voltage dependence coefficient	V	infinite	N/A
`XTF`	TF bias dependence coefficient.	none	0	N/A

[1] The defaults are for PSPICE and HSPICE.

Table A.4 SPICE Level-1 MOSFET model parameters.[1]

Name	Parameter	Units	Default
`CBD`	Drain-to-bulk diode zero-bias capacitance	F	0
`CBS`	Source-to-bulk diode zero-bias capacitance	F	0
`CGBO`[2]	Gate-to-bulk overlap capacitance per unit length	F/m	0
`CGDO`[2]	Gate-to-drain overlap capacitance per unit width	F/m	0
`CGSO`[2]	Gate-to-source overlap capacitance per unit width	F/m	0
`GAMMA`	Bulk threshold parameter	$V^{0.5}$	calculated
`IS`	Source-to-bulk and drain-to-bulk diode saturation current	A	10^{-14}
`KP`	transconductance parameter	A/V^2	2×10^{-5} (*n*-ch same, *p*-ch 8.6×10^{-6})
`L`	Drawn channel length (usually on element line)	m	`DEFL` (can be set by `DEFL` in `.OPTIONS`)
`LAMBDA`	Channel-length modulation parameter	V^{-1}	0
`LD`	Lateral diffusion length	m	0
`MJ`	Source-to-bulk and drain-to-bulk diode grading coefficient	none	0.5
`N`	Source-to-bulk and drain-to-bulk diode emission coefficient	none	1
`NSUB`	Substrate doping level	$1/cm^3$	None (1×10^{15})
`PHI`	Surface potential ($2\phi_f$)	V	0.6 (calculated)
`RD`	Series drain resistance	Ω	0
`RS`	Series source resistance	Ω	0
`TOX`	Oxide thickness	m	1×10^{-7}
`TPG`	Type of gate material ($0 \Rightarrow$ Al, $^{+}1 \Rightarrow$ Poly with doping opposite polarity as substrate, $-1 \Rightarrow$ Poly with doping same polarity as substrate)	none	+1
`TT`	Source-to-bulk and drain-to-bulk diode transit time	s	0
`UO`[2]	Surface mobility	$cm^2/V{\cdot}s$	600
`VTO`	Zero-bias threshold voltage	V	0(calculated)
`W`	Drain channel width (usually on element line)	m	`DEFW` (can be set by `DEFW` in `.OPTIONS`)

[1] The defaults are for PSPICE and HSPICE; the HSPICE values are shown in parentheses when different. Examples are not given because the values of some parameters vary tremendously from one device to another and for the other parameters the defaults are good examples.

[2] The last character in the parameter name is the letter 'O,' not the number zero.

These commands tell SPICE to perform a DC analysis while sweeping the value of the variable **var**. There are four options for the parameter **var**:

1) Setting it to the name of an independent source (e.g., **vin**) causes the DC value of the source to be swept.
2) Entering **model_type model_name(parameter_name)** in place of **var** causes the parameter **parameter_name** in the model named **model_name** of type **model_type** to be varied.
3) The temperature may be varied by using **TEMP** in place of var.
4) A global parameter that has been set using a **.param** command can be varied. In all cases, the variable will be swept from **start** to **stop**. If a linear sweep is requested (by specifying **LIN** or by leaving it blank) the variable will be incremented by the amount **incr** each step. If a logarithmic sweep is requested (specified by either **DEC** or **OCT**), the variable is swept from **start** to **stop** with exactly **#_pts** values evenly spaced on each decade (or octave). Specifying **LIST** causes SPICE to sequentially set the variable to each of the values given in the list. For each type of DC sweep a nested sweep can be specified; if it is, the first sweep is performed for each value of the variable in the nested sweep.

Examples:

```
.dc lin vcc 10 15 1
```

This example causes the supply voltage **vcc** to vary from 10 to 15 volts in 1 volt steps.

```
.dc dec rload 10 1k 3
```

This example causes the global parameter **rload** to vary from 10 to 1000 with 3 equally spaced values (on a logarithmic scale) for each decade.

```
.dc TEMP list 0 25 50 75 100
```

This example causes the simulation temperature to be set to 0°C, 25°C, 50°C, 75°C, and 100°C.

.FOUR The general syntax for the Fourier simulation command is:

```
.FOUR freq {last harmonic} output .....
```

The Fourier analysis requires that a **.tran** analysis has been performed. It tells SPICE to decompose the transient output into Fourier components based on the fundamental sine wave at frequency **freq**. By default, nine harmonics are included, but the limit may be set by **last harmonic**. The Fourier decomposition is performed for each current or voltage in the list of outputs, **output ....**

Example:

```
.four 1k 5 v(3)
```

This command tells SPICE to provide the first five harmonics of the Fourier decomposition of the voltage at node three in terms of the fundamental at 1 kHz.

.FUNC The general syntax for the function simulation command is different for HSPICE and PSPICE. For PSPICE the command syntax is:

```
.FUNC fname({arg...}) {body}
```

The name of the function is **fname** and the optional list of arguments is given in parentheses following the name. The body of the function is NOT optional; in this case the curly brackets are a part of the PSPICE input line (some versions of PSPICE do not require the curly brackets). For HSPICE the syntax is:

```
.FUNC fname({arg...})=body
```

Example:

```
.func avg(x,y)  {(x+y)/2} ← this is for PSPICE
.func avg(x,y)=(x+y)/2 ← this is for HSPICE
```

This example finds the arithmetic average of **x** and **y**. You could, for example, print of plot the average of two node voltages.

.OP The general syntax for the operating point simulation command is:

```
.OP
```

This command causes SPICE to perform a DC operating point analysis and print the detailed results to the output file. This analysis is automatically performed, but if the **.OP** statement is not included, the printout is less detailed.

.PARAM The general syntax for the parameter definition simulation command is different for PSPICE and HSPICE. For PSPICE the syntax is:

```
.PARAM name=[value, {expression}]
```

For HSPICE the syntax is:

```
.PARAM name=[value, 'expression']
```

The name of the parameter is **name**, and it may either be given a fixed value (**value**), or it may be given an expression. In PSPICE the expression is enclosed in curly braces and in HSPICE it is enclosed in single quotes.

Examples:

```
.param load=1k ← works for either HSPICE or PSPICE
.param r_out={load/2} ← PSPICE
.param r_out='load/2' ← HSPICE
```

The first example sets the parameter **load** to 1,000. The second example sets the parameter **r_out** to **load** divided by two. This example demonstrates that one parameter can be used in defining another. These parameters can then be used in the netlist wherever a numeric value would typically be given. For example:

```
r4 3 4 r_out
rL 4 0 load
```

These two statements combined with the previous examples set the value of R_L to 1 kΩ and the value of R_4 to 500 Ω. Some versions of PSPICE require the parameter name to be enclosed in curly brackets; for example, the second example above would become

```
rL 4 0 {load}
```

If you are using the MicroSim Schematics program to generate your input file, you must add a **PARAM** pseudocomponent to your schematic before you can use a global parameter, and the parameter name must be enclosed in curly braces when used for the value of any component. See the manual and examples on the CD for complete information.

.SENS The general syntax for the sensitivity simulation command is:

```
.SENS variable
```

This command causes SPICE to perform a small-signal DC sensitivity analysis to find the sensitivity of `variable` with respect to all component values and model parameters in the circuit.

Example:

```
.sens v(out)
```

This example would cause the sensitivities of the voltage at node 'out' to all component values and model parameters to be determined and printed. **WARNING!** This command can produce a very large output file, since it computes the sensitivity to *every* model parameter for *every* device.

.TF The general syntax for the transfer function simulation command is:

```
.TF out_var in_source
```

This command causes SPICE to find and print the DC transfer function from the input source `in_source` to the output variable `out_var`. In addition, SPICE finds and prints the input and output resistances. **WARNING!** This analysis is a linearized DC analysis. Not only are all nonlinear elements replaced by linear models as in a `.AC` command, but all capacitors are open circuited as in a `.DC` command. Therefore, when coupling or bypass capacitors are present, unexpected and meaningless results may be obtained.

.TRAN The general syntax for the transient simulation command is:

```
.TRAN step stop
```

This command causes SPICE to perform a transient simulation for time running from $t = 0$ to $t =$ `stop`. The step size used by the program is adjusted automatically to bring about convergence, but if a `.print` or a `.plot` statement is used the time step for printing or plotting will be `step`. There are several options for this command, but they are different for PSPICE and HSPICE and will not be covered here.

A.2.3 Program Control Statements

This section provides a short summary of the most commonly encountered statements for controlling the input, output and operation of the program. These statements do not directly provide information about the netlist or the types of simulation to be performed, but they do control the way some of those statements are interpreted and how SPICE presents the results.

.OPTIONS The general syntax for the options command is:

```
.OPTIONS option{=value} ...
```

This statement can set many variables that influence the way SPICE interprets the netlist, presents the output, and performs the actual simulation. Some of the options are set by simply including their name in the list, others require you to specify a numeric value. A listing of the most common options is given in Table A.5, but the SPICE manuals should be consulted for a complete list if you are using SPICE extensively. All of the parameters default to reasonable values for most circuits. The parameters that are either off or on default to off.

.PLOT The general syntax for the plot command is:

```
.PLOT [AC,DC,TRAN] var... {(lower_lim,upper_lim)}
```

This command tells SPICE to produce a line-printer plot as a part of the output file. The plot may be a **DC** plot, in which case the x-axis is whatever DC parameter was swept with a **.dc** command, or it may be an **AC** plot, in which case the x-axis is frequency, or it may be a transient (**TRAN**) plot, in which case the x-axis is time. The variables to be plotted are listed next (**var**). All variables listed on one **.plot** statement will be printed on the same plot. The first variable listed will have its values printed as well as plotted. More than one **.plot** statement may be included. The upper and lower limits for the y-axis may be specified if desired.

The magnitude and phase of an AC voltage or current can be specified; for example, for the node voltage at node 2 use **vm(2)** and **vp(2)**. Similarly, use **vdb(2)** to get the value in dB. The voltage from node 2 to node 3 can be specified as **v(2,3)**.

Table A.5 Commonly encountered options

option	Meaning
ABSTOL = X	Sets the absolute tolerance for branch currents in DC and transient simulations.
CHGTOL = X	Sets the charge error tolerance (SPICE conserves charge from one iteration to the next).
DEFL = X	Sets the default channel length for MOS devices.
DEFW = W	Sets the default channel width for MOS devices.
GMIN = X	Sets the minimum conductance SPICE allows between any two nodes in the circuit. (It helps SPICE solve the equations to not have an infinite resistance between any two nodes. In some high impedance circuits this can cause problems and GMIN should be reset).
NOBIAS	Suppresses printing of the DC node voltage table.
NOMOD	Suppresses printing of the model parameters .
NOPAGE	Suppresses page breaks and the printing of the header on each page of output.
POST	Tells HSPICE to store simulation data for use by the post processor (PSPICE uses the **.probe** command instead).
RELTOL = X	Sets the relative tolerance for the change in voltage from one iteration to the next.
TNOM = X	Sets the nominal simulation temperature (default; 27°C).
VNTOL = X	Sets the absolute tolerance for node voltages in DC and transient simulations.

Examples:

```
.plot ac v(out) v(3)
```

This example will cause the small-signal AC voltages at nodes 'out' and '3' to be plotted versus frequency.

```
.plot tran i(r1) (0,20m)
```

This example will plot the current through resistor R_1 as a function of time with the y-axis limits set to 0 and 20 mA.

.PRINT The general syntax for the plot command is:

```
.PRINT [AC,DC,TRAN] var...
```

This command tells SPICE to print a table of the values of the variables listed (`var`) as a part of the output file. The table may be for a `DC` analysis, in which case the rows are whatever DC parameter was swept with a `.dc` command. It may be for an `AC` analysis, in which case the rows are frequency. It may be for a transient (`TRAN`) analysis, in which case the rows are time. More than one `.print` statement may be included.

The magnitude and phase of an AC voltage or current can be specified; for example, for the node voltage at node 2 use `vm(2)` and `vp(2)`. Similarly, use `vdb(2)` to get the value in dB. The voltage from node 2 to node 3 can be specified as `v(2,3)`.

Example:

```
.print dc vm(3,4)
```

This example will cause the magnitude of the voltage appearing between nodes 3 and 4 to be printed for each value of the variable swept by the `.dc` analysis statement.

.PROBE The general syntax for the probe command is:

```
.PROBE
```

This command is only used in PSPICE and tells the simulator to produce a data file for the post processor. The same function is implemented in HSPICE by using the '`.options post`' statement.

.NODESET The general syntax for the nodeset command is:

```
.NODESET V(node {,node})=value ...
```

This command causes SPICE to use the specified initial value when solving the DC bias point or the first point of a transient or DC sweep. It does not permanently set the voltage to that value; it just gives SPICE an initial value to use. It is useful, for example, to ensure that a bistable circuit like a flip-flop starts in a particular state.

Example:

```
.nodeset v(2)=1 v(3,4)=2.5
```

This example sets the initial value for the voltage at node 2 to 1 volt and the initial value for the voltage from node 3 to node 4 to 2.5 volts.

REFERENCES

[A.1] A. Vladimirescu, *The SPICE Book*, New York: John Wiley & Sons, 1994

[A.2] B. Al-Hashimi, *The Art of Simulation using PSpice, Analog and Digital*, New York: CRC Press, 1995

[A.3] P.W. Tuinenga, *SPICE, A Guide to Circuit Simulation & Analysis Using PSpice*. 3E., Englewood Cliffs, NJ: Prentice Hall, 1995

[A.4] D. Foty, *MOSFET Modeling with SPICE, Principles and Practice*, Upper Saddle River, NJ: Prentice Hall, 1997

[A.5] G. Massobrio & P. Antognetti, *Semiconductor Device Modeling with SPICE*. 2E., New York: McGraw-Hill, Inc., 1993

APPENDIX B
Example Device Models

INTRODUCTION

This appendix provides a listing of the device model parameters used in examples, exercises, and problems throughout this book. Many of the models are taken directly from the Microsim model library included with the student edition of PSPICE, and some are made up to be representative examples of generic devices. In some cases, the values used in hand analysis differ somewhat from the parameters used by SPICE. These differences are explained here and the different values are given.

B.1 DEVICE DATA

In the following sections the models for several different devices are presented. Chapter 4 describes the most important of the model parameters used here, you should consult the references if you want more detail. The final section in this appendix provides a printout of the model libraries contained on the CD.

1N4148 The 1N4148 SPICE model is included in the evaluation library with PSPICE (Eval.lib).

```
.model D1N4148     D(Is=2.682n N=1.836 Rs=.5664 Ikf=44.17m
Xti=3 Eg=1.11 Cjo=4p M=.3333 Vj=.5 Fc=.5 Isr=1.565n Nr=2
Bv=100 Ibv=100u Tt=11.54n)
```

2N2222 The 2N2222 SPICE model is included in the evaluation library with PSPICE (Eval.lib).

```
.model Q2N2222     NPN(Is=14.34f Xti=3 Eg=1.11 Vaf=74.03
Bf=255.9 Ne=1.307 Ise=14.34f Ikf=.2847 Xtb=1.5 Br=6.092 Nc=2
Isc=0 Ikr=0 Rc=1 Cjc=7.306p Mjc=.3416 Vjc=.75 Fc=.5 Cje=22.01p
Mje=.377 Vje=.75 Tr=46.91n Tf=411.1p Itf=.6 Vtf=1.7 Xtf=3
Rb=10)
```

As explained in Chapter 4, the value of β depends on more than just the parameter **BF**. A reasonable compromise value to use for hand analysis of the 2N2222 is $\beta = 180$.

2N2857 The 2N2857 SPICE model is not included in newer versions of PSPICE. Nevertheless, the model from an old version is:

```
.model Q2N2857    NPN(Is=69.28E-18 Xti=3 Eg=1.11 Vaf=100 Bf=288
Ne=1.167 Ise=69.28E-18 Ikf=21.59m Xtb=1.5 Br=1.219 Nc=2 Isc=0
Ikr=0 Rc=4 Cjc=893.1f Mjc=.3017 Vjc=.75 Fc=.5 Cje=939.8f
Mje=.3453 Vje=.75 Tr=1.607n Tf=115.7p Itf=.27 Vtf=10 Xtf=30
Rb=10)
```

As explained in Chapter 4, the value of β depends on more than just the parameter **BF**. A reasonable compromise value to use for hand analysis of the 2N2857 is $\beta = 100$.

2N3904 The 2N3904 SPICE model is included in the evaluation library with PSPICE (Eval.lib).

```
.model Q2N3904    NPN(Is=6.734f Xti=3 Eg=1.11 Vaf=74.03
Bf=416.4 Ne=1.259 Ise=6.734f Ikf=66.78m Xtb=1.5 Br=.7371 Nc=2
Isc=0 Ikr=0 Rc=1 Cjc=3.638p Mjc=.3085 Vjc=.75 Fc=.5 Cje=4.493p
Mje=.2593 Vje=.75 Tr=239.5n Tf=301.2p Itf=.4 Vtf=4 Xtf=2
Rb=10)
```

As explained in Chapter 4, the value of β depends on more than just the parameter **BF**. A reasonable compromise value to use for hand analysis of the 2N3904 is $\beta = 150$.

2N3906 The 2N3906 SPICE model is included in the evaluation library with PSPICE (Eval.lib).

```
.model Q2N3906    PNP(Is=1.41f Xti=3 Eg=1.11 Vaf=18.7 Bf=180.7
Ne=1.5 Ise=0 Ikf=80m Xtb=1.5 Br=4.977 Nc=2 Isc=0 Ikr=0 Rc=2.5
Cjc=9.728p Mjc=.5776 Vjc=.75 Fc=.5 Cje=8.063p Mje=.3677
Vje=.75 Tr=33.42n Tf=179.3p Itf=.4 Vtf=4 Xtf=6 Rb=10)
```

As explained in Chapter 4, the value of β depends on more than just the parameter **BF**. A reasonable compromise value to use for hand analysis of the 2N3906 is $\beta = 150$.

B.2 MODEL LIBRARIES FROM THE CD

B.1.1 Book.lib

This library contains a resistor model and BJT models. Comments are on lines beginning with asterisks and should be read; they provide useful information about the models following the comment.

```
* Device model library #1 for INTRODUCTION TO ELECTRONIC CIRCUIT DESIGN
* by R. Spencer and M. Ghausi
*
* Resistor models **********************************
*
.model Rcarbon res(r=1 dev 10% tc1=2m)
*
* Bipolar transistor models **********************************
*
.model Nperfect npn(is=1e-14 bf=1g vaf=1g)
.model Pperfect pnp(is=1e-14 bf=1g vaf=1g)
.model Nsimple npn(is=1e-14 bf=100)
.model Psimple pnp(is=1e-14 bf=100)
```

```
.model Nsimple2 npn(is=5e-14 dev 1% lot 98% bf=125 dev 1% lot 60%)
.model Nsimple2hi npn(is=1e-13 bf=200)
.model Psimple2 pnp(is=5e-14 dev 1% lot 98% bf=125 dev 1% lot 60%)
.model Nsimple3 npn(is=1e-14 bf=100 rb=50 vaf=50)
.model Psimple3 pnp(is=1e-14 bf=100 rb=50 vaf=50)
.model NTTL npn(is=1e-16 bf=50 br=.2 rb=75 vaf=50 tf=.5n tr=50n
cje=.5p cjc=1p ccs=1.5p)
*
.model Nsmall npn(is=1e-15 bf=100)
*
* The following three models are meant to represent different
* fabrication runs for the same transistor. The logic is that the
* emitter was diffused deeper on the fast model. This led to lower tf,
* greater beta and larger IS due to the thinner base, but it also led to
* increased rb for the same reason. The cje is decreased because the
* doping levels would be lower at the deeper junction.
* Note that the transistor called "nominal" below is called Nnominal in
* Microsim's Schematics program.
*
.model nominal npn(is=3.3e-15 bf=100 rb=62.5 tf=1n cje=.45p cjc=2p
itf=1m xtf=5 vtf=10)
*
.model nfast npn(is=1e-14 bf=200 rb=75 tf=.5n cje=.4p cjc=2p itf=1m
xtf=5 vtf=10)
*
.model nslow npn(is=1e-15 bf=50 rb=50 tf=1.5n cje=.5p cjc=2p itf=1m
xtf=5 vtf=10)
*
* This pnp model is the pnp complement to the nominal npn model given
* above and is called Pnominal in Microsim's Schematics program.
*
.model nominalp pnp(is=3.3e-15 bf=100 rb=62.5 tf=1n cje=.45p cjc=2p
itf=1m xtf=5 vtf=10)
*
* This model came from an old PSPICE library
*
.model Q2N2857 NPN(Is=69.28E-18 Xti=3 Eg=1.11 Vaf=100 Bf=288 Ne=1.167
+  Ise=69.28E-18 Ikf=21.59m Xtb=1.5 Br=1.219 Nc=2 Isc=0 Ikr=0 Rc=4
+  Cjc=893.1f Mjc=.3017 Vjc=.75 Fc=.5 Cje=939.8f Mje=.3453 Vje=.75
+  Tr=1.607n Tf=115.7p Itf=.27 Vtf=10 Xtf=30 Rb=10)
*  National  pid=42  case=TO72
*  88-09-07 bam     creation
```

B.1.2 Book2.lib

This library contains FET models. Comments are on lines beginning with asterisks and should be read; they provide useful information about the models following the comment. In all of the following MOSFET models, the values of `w`, `l`, `tox` and `cgdo` have been chosen to yield the specified values for K, Cgs and Cgd. If you want to change the value of K in a model without affecting any of the capacitances, the simplest thing to do is to add the parameter **`KP`**. It will override the internal calculation of K as noted in Chapter 4. Note that $K = (KP/2)*(W/L)$ if you provide KP.

```
* Device model library #2 for INTRODUCTION TO ELECTRONIC CIRCUIT DESIGN
* by R. Spencer and M. Ghausi
*
* JFET models ************************************
*
.model Jn1 NJF(beta=500u vto=-1)
.model Jp1 PJF(beta=500u vto=-1)
*
* MOSFET models **********************************
*
.model Nbig NMOS(level=1 w=1u l=1u kp=1m vto=.7 tox=100n lambda=0)
*
* The following MOS models yield devices with K=100 uA/V^2, Cgs=2p,
* Cgd=.5p and Vth=.5
*
.model Nsimple_mos NMOS(level=1 w=289u l=30u vto=.5 tox=100n lambda=0
cgdo=1.73n)
.model Psimple_mos PMOS(level=1 w=289u l=30u vto=-.5 tox=100n lambda=0
cgdo=1.73n)
*
* The following MOS models are the same as Nsimple except that they
* include output resistance
*
.model Nsimple2_mos NMOS(level=1 w=289u l=30u vto=.5 tox=100n lambda=.02
cgdo=1.73n)
.model Psimple2_mos PMOS(level=1 w=289u l=30u vto=-.5 tox=100n lambda=.02
cgdo=1.73n)
*
* The following devices are the same as Nsimple except they are
* depletion-mode.
*
.model Nsimpled_mos NMOS(level=1 w=289u l=30u vto=-.5 tox=100n lambda=0
cgdo=1.73n)
.model Psimpled_mos PMOS(level=1 w=289u l=30u vto=.5 tox=100n lambda=0
cgdo=1.73n)
*
* The following MOS models yield devices with K=2 mA/V^2, Cgs=5p,
* Cgd=.5p and Vth=1
*
.model Nlarge_mos NMOS(level=1 w=2043u l=10.6u vto=1 tox=100n lambda=0
cgdo=245p)
```

```
.model Plarge_mos PMOS(level=1 w=2043u l=10.6u vto=-1 tox=100n lambda=0
cgdo=245p)
*
* The following MOS models yield devices with K=500 uA/V^2, Cgs=2.5p,
* Cgd=.25p and Vth=1
*
.model Nmedium_mos NMOS(level=1 w=722u l=15u vto=1 tox=100n lambda=0
cgdo=346p)
.model Pmedium_mos PMOS(level=1 w=722u l=15u vto=-1 tox=100n lambda=0
cgdo=346p)
*
```

REFERENCES

[B.1] D.P. Foty, *MOSFET Modeling with SPICE - Principles and Practice*. Upper Saddle River, NJ. Prentice Hall, 1997

[B.2] G. Massobrio and P. Antognetti, *Semiconductor Device Modeling with SPICE*, 2E. New York: McGraw-Hill, 1993

[B.3] Y.P. Tsividis, *Operation and Modeling of the MOS Transistor*, New York: McGraw-Hill, 1987

Answers to Selected Problems

Chapter 1

P1.7 $\dfrac{v_O}{v_S} = -G_m(R_1\|R_2)\left(\dfrac{R_2}{R_1 + R_2}\right)$

P1.10 $\dfrac{i_O}{i_S} = (R_1\|R_2)\left(\dfrac{R_4\|R_L}{R_4\|R_L + R_3}\right)\dfrac{A}{R_L}$

P1.15 $v_{Th} = \dfrac{15}{8} + e^{-t}\,\mathrm{V}$, $R_{Th} = \dfrac{3}{8}\Omega$, and $i_N = 5 + (8/3)e^{-t}\,\mathrm{A}$

P1.18 $\dfrac{V_o(j\omega)}{V_s(j\omega)} = \dfrac{\dfrac{R_i}{R_i + R_S}}{1 + j\omega\dfrac{R_S R_i C_i}{R_i + R_S}}\left(\dfrac{G_m(R_o\|R_L)}{1 + j\omega\dfrac{R_L R_o C_L}{R_L + R_o}}\right)$

P1.23 $\dfrac{V_o(j\omega)}{I_s(j\omega)} = \dfrac{R_2}{1 + j\omega R_2 C}$

P1.29 13 bits

Chapter 2

P2.2 **(a)** $p_n \approx 1.05 \times 10^4\ \mathrm{cm}^{-3}$, **(b)** $2 \times 10^6\ \mathrm{cm}^{-3}$, and **(c)** 2.5 million atoms

P2.5 3.43 kΩ

P2.7 **(a)** 4.2 μA, **(b)** 24.5 μA, and **(c)** 201 μA

P2.12 24°

P2.17 $V_0 = 0.879$ V and $x_{D.R.} = 0.34\ \mu$m

P2.23 −28.3 V

P2.27 28.6 mV

P2.32 $I_S = 3.92 \times 10^{-15}$ A

P2.39 **(a)** saturation, 50 μA, and **(b)** saturation, 2 μA

P2.44 **(a)** $K = 466\ \mu\mathrm{A/V}$, **(b)** linear region and, **(c)** $I_D = 746\ \mu$A

P2.48 333 kΩ

P2.52 **(a)** $V_p = -2.4$ V and **(b)** $I_{DSS} = 12.2\ \mu$A

P2.56 $V_{th} = V_p$ and $K = I_{DSS}/V_p^2$

Chapter 3

None

Chapter 4

P4.7 $I_S = 5.2 \times 10^{-14}$ A

P4.11 $n = 1.3027$ and $I_S = 6.0148 \times 10^{-13}$ A

P4.15 $I_S = 9.72 \times 10^{-15}$ A

Chapter 5

P5.1 **(a)** $V_1 = V_2 = 1$ V, $V_3 = 0$ V, **(d)** $V_2 = 0.1$ V, $V_3 = -1.8$ V
(f) $V_1 = V_2 = 0$ V, $V_3 = -5$ V, **(i)** $V_1 = V_2 = 1.05$ V, $V_3 = 0.1$ V

P5.6 -21.27

P5.11 **(a)** $R_1 = R_3 = 25\,\text{k}\Omega$ and $R_2 = R_4 = 750\,\text{k}\Omega$

(b) $R_{i1} = 25\,\text{k}\Omega$ and $R_{i2} = 800\,\text{k}\Omega$

P5.15 **(a)** $I_O = 0.5$ mA and **(b)** $V_O \geq 0.59$ V

P5.19 $t = 0.25$ s

P5.26 The two stable states have $v_O = 5$ V and $v_O = -10$ V

P5.30 **(a)** $|v_o/v_i| = 1$ and **(b)** $v_O = -0.5$ V

P5.35 $v_O = I_B R_2$

P5.40 **(a)** In order to detect the input for sure, we must have $v_I > 5.5$ mV.

(b) $50\text{ mV} \leq v_O \leq 105\text{ mV}$

P5.45 **(a)** $A_f(\max) = 10$, and **(b)** $f_{-3dB} = 200$ kHz

P5.51 $a = 182{,}000$ V/V and $V_{off} = -3.2$ mV

Chapter 6

P6.3 $v_o(t) = \dfrac{2}{\pi} \cos \omega_0 t$

P6.6 The numerical solution is $V_D = 688$ mV and $I_D = 0.31$ mA.

P6.9 $R_1 = 71\Omega$

P6.12 **(a)** $I_R = 4.94$ mA and $V_R = 5.06$ V

(b) The model is a resistor of value $1.2\,\text{k}\Omega$.

(c) $v_o/v_i = 0.545$

P6.16 $V_0 = 604$ mV

P6.20 **(a)** $g_m \equiv \left.\dfrac{di_o}{dt}\right|_{@\text{ Q point}} = a_1 + 2a_2V_1$

(b) a_1 has units of $1/\Omega$ or A/V and a_2 has units of A/V^2.

P6.25 $r_o = 500\Omega$, $r_i = 20\,\text{k}\Omega$, and $k = 2.5$

Chapter 7

P7.2 $V_O = 1.57$ V

P7.7 **(a)** $I_C = 0.2$ mA, **(b)** $I_B = 0.96\,\mu$A, **(c)** $I_C = 0.25$ mA, and **(d)** $I_C = 0.39$ mA

P7.11 **(a)** Q_1 is forward active and **(b)** $V_B = 1$ V.

P7.15 $R = 18.6\,\text{k}\Omega$

P7.19 **(a)** $W/L = 53$, **(b)** $I_D = 479\,\mu$A, and **(c)** $I_D = 356\,\mu$A

P7.24 $I_D = 259\,\mu$A

P7.29 $R_D = 3\,\text{k}\Omega$

P7.33 One solution is $R_1 = 160\,\text{k}\Omega$, $R_2 = 40\,\text{k}\Omega$, $R_E = 2.6\,\text{k}\Omega$, and $R_C = 9.4\,\text{k}\Omega$.

P7.36 One solution is $R_1 = 15\,\text{k}\Omega$, $R_2 = 60\,\text{k}\Omega$, $R_E = 1.3\,\text{k}\Omega$, and $R_C = 4.35\,\text{k}\Omega$.

P7.43 V_{BB} min = 3.41 V

P7.49 $R_S = 75.8\text{ k}\Omega$ and $R_D = 64.2\text{ k}\Omega$

P7.57 $R_S = 500\Omega$ and $R_D = 50\text{ k}\Omega$

P7.64 $R_1 = 226\text{ k}\Omega$, $R_2 = 774\text{ k}\Omega$, $R_S = 10\text{ k}\Omega$, and $R_D = 10\text{ k}\Omega$

P7.72 Q_1 is forward active and has $I_{C1} = 1$ mA and $V_{CE1} = 3.7$ V.
Q_2 is forward active and has $I_{C2} = 10$ mA and $V_{CE2} = 5.7$ V.

P7.79 M_1 is forward active and has $I_{D1} = 132\ \mu\text{A}$, $V_{GS1} = 1.65$ V, and $V_{DS1} = 6.4$ V.
M_2 is forward active and has $I_{D2} = 2$ mA, $V_{GS2} = 4.98$ V, and $V_{DS1} = 6$ V.

P7.82 **(a)** $I_1 = 1$ mA and $I_2 = 1.1$ mA, **(b)** $I_1 = 1$ mA and $I_2 = 1.4$ mA

P7.86 $I_O \approx 219$ mA

P7.92 **(a)** $I_3 = 1$ mA and $I_4 = 2$ mA, **(b)** $I_3 = 1$ mA and $I_4 = 2$ mA, **(c)** $I_3 = 0.6$ mA and $I_4 = 1.2$ mA, **(d)** Q_1 serves to reduce the error associated with the finite beta of the transistors.

P7.99 $I_1 = I_2 = I_3 = I_5 = 467\ \mu\text{A}$

Chapter 8

P8.1 **(a)** a ± 10 mV 10 MHz square wave, **(b)** $C = 475$ nF

P8.8 $R_C = 673\ \Omega$ and $R_E = 4.13\text{ k}\Omega$

P8.12 $R_1 = 15\text{ M}\Omega$, $R_2 = 5\text{ M}\Omega$, $R_S = 2\text{ k}\Omega$, and $R_D = 10\text{ k}\Omega$

P8.19 The model consists of only one element, the controlled current source that supplies the drain current, $i_d = g_m v_{gs}$, where $g_m = 340\ \mu\text{A/V}$.

P8.25 $R_S = 1.25\text{ k}\Omega$ and $R_D = 36.2\text{ k}\Omega$

P8.36 The small-signal model includes a controlled source with $g_m = 11.4$ mA/V and an input resistance $r_\pi = 8.75\text{ k}\Omega$. The resulting overall voltage gain is $v_o/v_s = -28.9$ and the power gain is $A_p = 737$.

P8.45 **(a)** $I_D = 615\ \mu\text{A}$, $V_D = 3$ V, $V_S = 1.23$ V, and $V_G = 3.33$ V,
(c) $A_v = -2.75$, **(d)** $R_i = 1\text{ M}\Omega$ and $R_o = 3.3\text{ k}\Omega$

P8.55 **(a)** $I_C = 5.6$ mA, $g_m = 215$ mA/V, and $r_\pi = 232\Omega$
(b) $A_v = v_o/v_i = 0.88$ and $v_o/v_s = 0.44$, **(c)** $R_i = 1\text{ k}\Omega$ and $R_o = 15\Omega$

P8.61 **(a)** $A_v = \dfrac{R'_L}{R'_L + r_e}$, $R_i = R_{BB} \| \left[r_\pi + (\beta + 1)R'_L\right]$, and $R_o = R_E \left\| \dfrac{r_\pi + R_S \| R_{BB}}{\beta + 1}\right.$

where $R'_L \equiv R_L \| R_E$. **(c)** $I_C = 2.1$ mA, **(d)** $A_v = 0.99$, $R_i = 4.9\text{ k}\Omega$, and $R_o = 20\Omega$. **(e)** Adding r_o does not have any noticeable effect.

P8.62 **(a)** $I_D = 8$ mA and $g_m = 8\text{ mA/V}^2$ in the small-signal model

(b) $A_v = \dfrac{v_o}{v_i} = 0.57$ and $\dfrac{v_o}{v_{sig}} = 0.55$, **(c)** $R_i = 2.55\text{ M}\Omega$ and $R_o = 83\Omega$

P8.69 One solution is $R_1 = 4.8\text{ M}\Omega$, $R_2 = 10.2\text{ M}\Omega$, and $R_S = 330\Omega$. Only the values of R_1 and R_2 can change.

P8.73 **(a)** $I_C = 1$ mA and $r_e = 26\ \Omega$, **(b)** $A_v = \dfrac{\alpha(R_L \| R_C)}{r_e} = 109$,

(c) $\dfrac{v_o}{i_i} = \dfrac{\alpha(R_L \| R_C)R_E}{R_E + r_e} = 2.8\text{ k}\Omega$,

(d) $R_i = R_E \| r_e = 25.5\ \Omega$ and $R_o = R_C = 4\text{ k}\Omega$

P8.79 **(a)** $I_D = 2$ mA and $g_m = 2$ mA/V, **(b)** $A_v = g_m R_L = 8$,

(c) $\dfrac{i_o}{i_{sig}} = \dfrac{g_m R_{SIG}}{1 + g_m R_{SIG}} = 0.67$, **(d)** $R_i = 1/g_m = 500\ \Omega$

P8.85 **(a)** $I_D = 5\text{ mA}$, $V_{SB} = 2\text{ V}$, $V_{th} = 1.42\text{ V}$, $g_m = 6.3\text{ mA/V}$, and $g_{mb} = 0.98\text{ mA/V}$, **(b)** $\frac{v_o}{v_i} = \frac{v_o}{v_{sig}} = \frac{g_m R_L}{1 + (g_m + g_{mb})R_L} = 0.76$,
(c) $R_i = \infty$ and $R_o = \frac{1}{g_m + g_{mb}} = 137\ \Omega$

P8.90 **(a)**
$I_O = 95\ \mu\text{A}$, $R_o = R_E \| (r_{\pi2} + r_{e1}) + r_{o2}\left[1 + \frac{g_{m2} r_{\pi2} R_E \| (r_{\pi2} + r_{e1})}{r_{\pi2} + r_{e1}}\right] = 833\text{ k}\Omega$
(b) $V_{O\min} \approx 0.2\text{ V}$

P8.93 **(a)** $R_o = 118\text{ k}\Omega$ and **(b)** $R_o = 177\text{ k}\Omega$

P8.97 $A_{cm-dm} = \frac{v_{odm}}{v_{icm}} = \frac{a_{ic}}{r_m}(R_2 - R_1)\left(1 - \frac{R_{MM}}{R_{MM} + r_m/2}\right)$

P8.103 $R_{C2} = 52\text{ k}\Omega$ and $V_{ICM} < 0.3\text{ V}$

P8.104 $A_v = 11.7$

P8.108 $A_{cm} = -3.6 \times 10^{-3}$

P8.111 $A_{cm-dm} = -0.005$

P8.121 **(a)** $I_{C1} = 2.4\text{ mA}$, $I_{C2} = 3.7\text{ mA}$, **(b)** $v_o/v_i = -21.5$, **(c)** $R_i = 1.6\text{ k}\Omega$,
(d) $R_o = 15\ \Omega$

P8.128 **(a)** $I_{D1} = 200\ \mu\text{A}$, $I_{D2} = 400\ \mu\text{A}$, **(b)** $A_v = -3.8$, **(c)** $R_i = \infty$, $R_o = 2.5\text{ k}\Omega$

Chapter 9

P9.2 **(a)** $Q = 3.14 \times 10^{-2}$, **(b)** $Q = 31.4$

P9.6 $R_p = 15.8\text{ M}\Omega$

P9.8 $r_d = 1.13\text{ k}\Omega$ and $C = 3.6\text{ pF}$

P9.13 **(a)** $r_b = 75\ \Omega$, $r_\pi = 6.6\text{ k}\Omega$, $g_m = 12.1\text{ mA/V}$, $r_o = 127\text{ k}\Omega$, $C_\pi \approx 7.2\text{ pF}$,
$C_\mu = 1.8\text{ pF}$, $\alpha = 0.988$, and $r_e = 83\ \Omega$. **(b)** $\omega_T = 1.34\text{ Grad/s}$

P9.16 The base resistance is not important when the base is driven by an ideal current source because it does not affect the base current. It is important when the base is driven by an ideal voltage source because it can significantly affect the value of v_π, especially at high frequencies.

P9.19 **(a)** ω_T is independent of W when V_{GS} is held constant. ω_T is proportional to $1/\sqrt{W}$ when I_D is held constant.

P9.21 $C_{gs} = 1.14\text{ fF}$, $C_{gd} = 0.2\text{ fF}$, $g_m = 403\ \mu\text{A/V}$, $r_o = 35.4\text{ k}\Omega$, and $\omega_T = 300\text{ Grad/s}$.

P9.25 $C_{gs} = 2.39\text{ pF}$, $C_{gd} = 109\text{ fF}$, and $g_m = 107\ \mu\text{A/V}$.

P9.28 $I_S = 0$. The amplifier supplies exactly the current required by R_1 for *any* value of V_S.

P9.33 $GBW \approx \frac{1}{2\pi R_S C_{oc}}$

P9.35 $f_{cH} \approx 31.4\text{ MHz}$

P9.40 **(a)** At midband, $V_o(j\omega)/V_s(j\omega) = -1$, **(b)** $f_{cL} = 5.9\text{ kHz}$

P9.45 $\omega_{cH} \approx 24.5\text{ Mrad/s}$

P9.47 **(a)** $I_C \approx 840\ \mu\text{A}$, **(b)** $g_m = 32\text{ mA/V}$, $r_\pi = 5.6\text{ k}\Omega$, and $r_o = 88\text{ k}\Omega$. $f_{cL} \approx 6.8\text{ Hz}$
(c) $f_{cH} \approx 1.5\text{ MHz}$

P9.55

$$\frac{V_o(j\omega)}{V_{sig}(j\omega)} = \frac{-g_m(R_D \| R_L)\left(\frac{R_G}{R_G + R_{SIG}}\right)}{1 + g_m R_S}\left(\frac{j\omega}{j\omega + \frac{1}{(R_G + R_{SIG})C_{IN}}}\right)\left(\frac{j\omega}{j\omega + \frac{1}{(R_D + R_L)C_{OUT}}}\right)$$

P9.58 $f_{cH} \approx \dfrac{1}{2\pi\left(R_{SIG}\|R_G\right)C_{gd}\left(1 + \dfrac{g_m\left(R_D\|R_L\right)}{1 + g_m R_S}\right)}$, the Miller approximation is valid if

$$\left(R_D\|R_L\right)C_{gd}\left(1 + \frac{1 + g_m R_S}{g_m\left(R_D\|R_L\right)}\right) \ll \left(R_{SIG}\|R_G\right)C_{gd}\left(1 + \frac{g_m\left(R_D\|R_L\right)}{1 + g_m R_S}\right)$$

P9.64 **(a)** $I_D = 225\ \mu\text{A}$, **(b)** $f_{cL} \approx 14.7$ Hz, **(c)** $f_{cH} \approx 44.3$ MHz, **(d)** The approximation is not valid since $\tau_{Mout} = 4.6$ ns, which is on the same order of magnitude as $1/\omega_{cH} \approx 3.6$ ns.

P9.69 **(a)** $I_D = 225\ \mu\text{A}$, **(b)** $f_{cL} \approx 15$ Hz, **(c)** $f_{cH} \approx 24$ MHz, **(d)** In P9.64(c) we obtained $f_{cH} \approx 44.3$ MHz. The answer given here is more accurate since the Miller approximation is not valid.

P9.73 $f_{cH} \approx \dfrac{1}{2\pi\left\{\left(R_S\|R_{CC}\|\left[r_{cm} + \left(a_{im} + 1\right)R_L'\right]\right)C_{oc} + \left[r_{cm}\left\|\left(\dfrac{R_S + R_L'}{1 + g_m R_L'}\right)\right.\right]C_{cm}\right\}}$

where $R_L' = R_M\|R_L$ and $R_{CC} = R_1\|R_2$

P9.75 $f_{cH} \approx 5.1$ MHz, the approximation is valid.

P9.81 $f_{cH} \approx 5.3$ MHz

P9.87 $f_{cH} \approx 2.5$ MHz

P9.90 $f_{cH} \approx 2.5$ MHz

P9.92 $f_{cH} \approx 10.6$ MHz

P9.95 **(a)** $I_C = 1$ mA **(b)** $f_{cL} \approx 640$ Hz **(c)** $f_{cH} \approx 83$ MHz

P9.100 **(a)** $I_C = 1$ mA **(b)** $f_{cL} \approx 17$ Hz **(c)** $f_{cH} \approx 77$ MHz

P9.102 **(a)** $I_D = 615\ \mu\text{A}$ **(b)** $f_{cL} \approx 15.5$ Hz **(c)** $f_{cH} \approx 16.2$ MHz

P9.106 **(a)** $I_D = 615\ \mu\text{A}$ **(b)** $f_{cL} \approx 17.3$ Hz **(c)** $f_{cH} \approx 14$ MHz

P9.110 We want a small input resistance to maximize the input attenuation factor, and this requires a common-base amplifier.

P9.117 To make the input attenuation factor in this case close to one we need a large input resistance. Since we also want a voltage gain greater than one, we must choose a common-source amplifier.

P9.121 $f_{cH1} \approx 2.67$ MHz and $f_{cH2} \approx 22.4$ MHz. Therefore, the Miller approximation should be reasonable and the cutoff frequency is about 2.7 MHz.

P9.124 $f_{cH} \approx 10.1$ MHz, τ_{Mout} is about 3 times smaller than $1/\omega_{cH}$ and $Z_{C_{\pi 1}}(j\omega_{cH}) = 2.5\ \text{k}\Omega$ which is almost equal to $r_{\pi 1}$, therefore, the Miller approximation is not very good in this case.

P9.130 $f_{cH} \approx 6.9$ MHz

P9.136 $f_{cH} \approx 1.5$ MHz, the Miller approximation shouldn't be too bad since (9.192) yields $R_D\left(C_{Mout} + C_{gd2}\right) = 21.8\ \text{ns} \ll R_{SIG}C_{in} = 107$ ns.

P9.138 We find $\tau_{gs1o} = 40$ ns, $\tau_{gd1o} = 76.6$ ns, $\tau_{gs2o} = 16.7$ ns, and $\tau_{gd2o} = 10$ ns, which together yield $f_{cH} \approx 1.1$ MHz. Since no time constant is clearly dominant, the answer may not be extremely accurate.

P9.144 We first find the DC bias point, which yields $I_C \approx 0.52$ mA. Using (9.216) we obtain $f_{cH1} \approx 24.8$ MHz, and using (9.217) we obtain $f_{cH2} \approx 8.84$ MHz. Finally, making use of (9.205) we find $f_{cH} \approx 8$ MHz.

P9.149 We find $\tau_{\pi 1o} = 178$ ns, $\tau_{\mu 1o} = 46.2$ ns, $\tau_{\pi 2o} = 2.6$ ns, and $\tau_{\mu 2o} = 18$ ns, which together yield $f_{cH} \approx 650$ kHz.

P9.152 We first find the DC bias point, which yields $I_D \approx 1.85$ mA. Using (9.229) we obtain $f_{cH1} \approx 266$ MHz, and using (9.230) we obtain $f_{cH2} \approx 138$ MHz. Finally, making use of (9.205) we find $f_{cH} \approx 114$ MHz. The approximation is not very good in this case though, because the time constant we are ignoring, which is at the drain of M_1, would yield

$$f_{cH} \approx g_{m2} \Big/ \left[2\pi\left(C_{gs2} + C_{gd1} \right) \right] = 111 \text{ MHz all by itself.}$$

P9.158 We find $\tau_{gs1o} = 250$ ns, $\tau_{gd1o} = 50.1$ ns, $\tau_{gs2o} = 1.3$ ns, and $\tau_{gd2o} = 1.16$ ns, which together yield $f_{cH} \approx 526$ kHz.

P9.171 **(a)** The DC bias current is $I_C = 0.5$ mA, which yields $g_m = 19.2$ mA/V and $r_\pi = 5.2$ kΩ. Using the model parameters from Appendix B we also find $r_b = 62.5\ \Omega$, $C_\mu = 1.2$ pF, and $C_\pi \approx 20.1$ pF. The DM DC gain is then one-half the value for the circuit with a differential output; $A_{dm0} = v_o/v_{sdm} = -g_m R_C/2 = -38.5$. The DM bandwidth is given by (9.251) and is 8.9 MHz. Using (9.254) we calculate $A_{cm0} \approx -R_C/2R_{EE} = -0.02$. From the results given in P9.169 we find $\omega_z = 1/R_{EE}C_{EE} = 1$ Mrad/s, or $f_z = 159$ kHz, and $\omega_p = 2g_m/C_{EE} = 3.85$ Grad/s, or $f_p = 612$ MHz. Using (9.255) we see that the CMRR has a DC gain of 1930, a zero at 612 MHz, one pole at 159 kHz, and a second pole at 8.9 MHz. **(b)** We only expect a very minor change due to V_{CB} now being 1 V, which yields $C_\mu = 1.5$ pF and $f_{cHd} \approx 7.4$ MHz.

P9.177 We have $I_D = 100\ \mu$A, $g_m = 200\ \mu$A/V, $C_{gs} = 5$ pF, and $C_{gd} = 0.5$ pF. We can use (9.267) to find the gain from v_{sdm} to each drain separately and then combine the results: $v_{d1}/v_{sdm} = -g_m R_{D1}/2 = -4$ and $v_{d2}/v_{sdm} = g_m R_{D2}/2 = 4.1$ so $A_{dm0} = -4 - 4.1 = -8.1$. Similarly, use (9.263) to obtain: $v_{d1}/v_{scm} \approx -R_{D1}/2R_{SS} = -0.2$ and $v_{d2}/v_{scm} \approx -R_{D2}/2R_{SS} = -0.205$ so $A_{cm0} \approx -0.2 - (-0.205) = 0.005$ and the DC CMRR is $CMRR_0 = A_{dm0}/A_{cm0} \approx 1620$ or 64.2 dB. The small change in R_{D2} will not significantly affect the bandwidths, so the results from Example 9.16 apply. Namely, the DM gain has a pole at about 6 MHz, and the CM gain has a zero at about 159 kHz and one pole at about 6.4 MHz.

Chapter 10

P10.1 $a = 665$ and $b = 0.018$

P10.6 **(a)** $f_{pf} = 25$ kHz **(b)** $f_{pf} = 500$ kHz

P10.10 **(a)** Shunt-series, **(b)** $b = 0.0975$, **(c)** $R_{if} = 25\ \Omega$ and $R_{of} = 30$ kΩ, **(d)** $f_{pf} = 40$ kHz

P10.14 **(a)** series-shunt, PFB **(b)** series-series, NFB **(c)** shunt-series, PFB **(d)** shunt-series, NFB **(e)** shunt-shunt, NFB **(f)** series-series, PFB **(g)** shunt-shunt, PFB **(h)** series-shunt, NFB

P10.20 Set $R_f = 52$ kΩ and keep everything else the same.

P10.27 **(a)** The FB network comprises R_{S1} and R_f and *may* include R_{S3}. The forward amplifier consists of M_1, R_{D1}, M_2, R_{D2}, R_{S2}, C_{S2}, and M_3. In addition, if R_{S2} was not included in the FBN, then it must be included here. R_1 and R_2 are outside the feedback loop. **(b)** see solution **(c)** $A_{if} = 3.15$ and $A_{sf} = 3.06$ (note these values are not accurate enough to warrant such precision, we show the extra digits to show their relative magnitudes) **(d)** $R_{if} = 1.77$ MΩ and $R_{of} = 74\ \Omega$

P10.33 $f_{sf} \approx 800$ kHz and $f_{if} \approx 30$ MHz

P10.39 **(a)** $I_{C1} = I_{C2} = 500\ \mu$A, $I_{C3} = 2$ mA, $V_o \approx 0$ V **(b)** $A_{if} = 20$ **(c)** $R_{if} = 20.6$ MΩ **(d)** $R_{of} = 21\ \Omega$

P10.45 **(a)** $R_{bi} = R_{bo} = b = R_1 = 1$ kΩ, **(b)** $I_D \approx 1$ mA, $g_m \approx 632\ \mu$A/V, and $r_o \approx 50$ kΩ, **(c)** $A_{if} = 1.0$ mA/V, **(d)** $R_{if} = 1.9$ GΩ, **(e)** $R_{of} = 1.6$ GΩ

P10.54 **(a)** $R_{bi} = R_{S2} \| R_F = 667\ \Omega$, $R_{bo} = R_F = 1\ \text{k}\Omega$, $b = -1/R_F = -1\ \text{mA/V}$,

(b) $a' = \left(\dfrac{-g_{m1} R_D R_{bo} R_{bi}}{R_{bi} + 1/g_{m2}} \right) = -15.8\ \text{kV/A}$, $R_o' = R_{bi} \| 1/g_{m2} = 375\ \Omega$,

$\alpha_o' = \dfrac{R_L}{R_L + R_o'} = 0.73$, $A_i = a'\alpha_o' = -11.5\ \text{kV/A}$, $A_{if} = \dfrac{A_i}{1 + A_i b} = -920\ \text{V/A}$,

$\dfrac{v_o}{v_i} = \dfrac{A_{if}}{R_1 + R_{if}'} = -5.1$, $\dfrac{v_o}{v_{sig}} = \dfrac{A_{if}}{R_{SIG} + R_1 + R_{if}'} = -3.3$,

(c) $R_{if}' = \dfrac{R_i'}{1 + A_i b} = 80\ \Omega$, $R_{if} = R_1 + R_{if}' = 180\ \Omega$,

(d) $\alpha_i' = \dfrac{R_1 + R_{SIG}}{R_1 + R_{SIG} + R_i'} = 0.167$, $R_{of} = \dfrac{R_o'}{1 + \alpha_i' a' b} = 103\ \Omega$

P10.60 **(a)** $i_o/i_i \approx 9.9$ **(b)** $R_{if} = 212\ \Omega$

P10.67 The unilateral model predicts $A_{if} = -200\ \Omega$, and the bilateral model yields $A_{if} = -120\ \Omega$.

P10.76 **(a)** $A_{f\min 0} = 60\ \text{dB}$ **(b)** $A_{f0} = 100\ \text{dB}$

P10.81 **(a)** The amplifier is not unity-gain stable **(b)** $PM = 22.5°$

P10.86 **(a)** The DC gain must be reduced by 100 dB. **(b)** The first pole must move down to 1 rad/s. **(c)** The new dominant pole needs to be at 0.01 rad/s.

P10.91 **(a)** We must set $\omega_u = 100$ krad/s and $\omega_{pnew} = 1$ rad/s. **(b)** We must set $\omega_u = 100$ Mrad/s and $\omega_{p1}' = 1$ krad/s.

P10.95 The DC bias point is $I_D = 100\ \mu\text{A}$ which yields $g_m = 200\ \mu\text{A/V}$. Also, $C_{gs} = 2$ pF and $C_{gd} = 0.5$ pF. Using open-circuit time constants we find $f_{cH} \approx 2.5$ MHz without compensation. Setting $f_{cH} = 700$ kHz yields $C_C = 1.94$ pF.

P10.100 **(a)** $\omega_o = 1/\left(C\sqrt{R_1 R_2}\right)$ **(b)** $R_4 \geq R_3\left(1 + R_2/R_1\right)$

P10.105 **(a)** Assume the current into the tank is zero and the frequency of oscillation is in the amplifier's midband. Then $v_o/v_{fb} = -120$. **(b)** Assuming we can ignore i_{fb} we have; $v_{fb}/v_o \approx -C_1/C_2 = -1$. **(c)** $f_o = 1/(2\pi\sqrt{LC_T}) = 2.5$ kHz where $C_T = C_1C_2/(C_1 + C_2)$. **(d)** Yes! $|L(j\omega_o)| = |ab| = 120$.

P10.115 $PW = -R_1 C \ln\left(1 - v_C/V_{CC}\right)$

Chapter 11

P11.1 $L = 0.707$ H and $C = 1.414$ F

P11.7 $L = 27$ mH and $C = 1.5$ nF

P11.13 $L = 18.1$ mH and $C = 751$ pF

P11.22 **(a)** $\omega_{-3\text{dB}}$ changes to 112.5 krad/s and **(b)** $\omega_{-3\text{dB}}$ changes to 137.5 krad/s

P11.26 Using the nearest standard values we find $R = 4.7\ \text{k}\Omega$ and $C = 220$ pF

P11.38 $L_1 = 124\ \mu\text{H}$, $L_2 = 2.9$ mH, $C_1 = 2.043\ \mu\text{F}$, and $C_2 = 88.42$ nF

P11.44 $R_2 = 4.24\ \text{k}\Omega$, $L_1 = 19.92$ nH, $L_2 = 821.3$ nH, $C_1 = 11.25$ nF, and $C_2 = 265.8$ pF

P11.53 **(a)** $\Theta_{vco}(j\omega)/\Theta_i(j\omega) = 1/\left(+ j\omega/\omega_p\right)$ where $\omega_p = K_o K_D$, **(b)** with $LR = \pi K_o K_D = 2\pi(100\ \text{kHz})$ we have $\omega_p = 200$ krad/s.

Chapter 12

P12.1 $\theta_c = \left[1 - \dfrac{2}{\pi}\sin^{-1}\left(\dfrac{V_o}{V_m}\right)\right]180°$, where θ_c is in degrees and the arcsine must be in radians.

P12.5 $C = 0.2\ \mu F$

P12.9 We find $v_O = v_I$: $-1\ V \le v_I \le 1\ V$, $v_O = 1/3 + (2/3)v_I$: $1\ V \le v_I \le 1.75\ V$, $v_O = -1/3 + (2/3)v_I$: $-1.75\ V \le v_I \le -1\ V$, and $v_O = \pm 1.5\ V : |v_I| \ge 1.75\ V$.

P12.15 In steady state; $V_{C1} = 10\ V$ (positive on left), $V_{C2} = 20\ V$ (positive on left), $V_{C3} = 20\ V$ (positive on left), $V_{C4} = 20\ V$ (positive on bottom), and $V_O = -40\ V$.

P12.19 **(a)** $v_O = 1 + (10^9)t : 0 \le t \le 1\ ns$, $v_O = 2\ V : t \ge 1\ ns$ **(b)** 1 ns, **(c)** $f_{max} = 159\ MHz$

P12.22 $v_{Lmax} \approx 4.1\ V$ and $v_{Lmin} \approx -3.3\ V$, therefore $v_{Lmax\_sym} \approx 3.3\ V$.

P12.29 First, the DC bias point is: $I_D = 1\ mA$, $V_D = 3\ V$, $V_S = 0.62\ V$. The small-signal midband gain is $A_v = v_o/v_i = -4.7$. The swing limits are $v_{Lmax} \approx 1.7\ V$ and $v_{Lmin} \approx -2.2\ V$.

P12.34 To achieve equal output amplitudes at the fundamental, we need $\hat{V}_{MOS} = 7.69\hat{V}_{BJT}$. Using $\hat{V}_{BJT} = 10\ mV$ we find THD $\approx 9.6\%$ as in Example 12.8. Then, using $\hat{V}_{MOS} = 76.9\ mV$ we obtain THD $\approx 3.8\%$. Although the MOSFET still has lower THD, the difference is much smaller when compared with equal output amplitudes.

P12.40 **(a)** $i_L(t) = 50\sin(2\pi ft)\ mA$, $i_M(t) = 50[1 + \sin(2\pi ft)]\ mA$, and $v_{OM}(t) = 5[1 - \sin(2\pi ft)]\ V$ **(b)** $P_T(t) = 125[1 + \cos(4\pi ft)]\ mW$, **(c)** $P_T(\text{peak}) = 250\ mW$ and $P_{Lmax} = \hat{v}_O^2/2R_L = 125\ mW$ **(d)** $P_{Tavg} = 125\ mW$, **(e)** the peak power is consumed when the signal is zero.

P12.43 The maximum theoretical efficiency in this case is 100%.

P12.47 $I_{C1} = I_{C2} = I_R$

P12.50 **(a)** $V_B = 2.1\ V$ **(b)** $v_{Lmax\_sym} = 2.9\ V$ **(c)** $\eta = 29\%$

P12.57 **(a)** $I_{C1} = 5.65\ A$ and $I_{C2} = 4.35\ A$ **(b)** $P_1 = 1.13\ W$ and $P_2 = 0.87\ W$ **(c)** $T_{j1} = 81.5°$ and $T_{j2} = 68.5°$ **(d)** $I_{C1}/I_{C2} = 2.72$ (and the currents are huge if we assume V_{BE} stays fixed – obviously we would not want to do that) **(e)** Eventually, Q_1 will end up with all of the current.

Chapter 13

P13.2 The minimum resolution required is 20 bits

P13.6

Digital input	ideal vo/vr	real vo/vr	DNL	Line	INL
000	0.000	0.050		0.050	0.000
001	0.125	0.075	−0.100	0.177	0.102
010	0.250	0.150	−0.050	0.304	0.154
011	0.375	0.255	−0.020	0.431	0.176
100	0.500	0.370	−0.010	0.559	0.189
101	0.625	0.580	0.085	0.686	0.106
110	0.750	0.700	−0.005	0.813	0.113
111	0.875	0.940	0.115	0.940	0.000

P13.10 **(a)** For $b_0b_1 = \{00, 01, 10, 11\}$ we have $V_O = \{0\ V, -0.75\ V, -1.5\ V, -2.25\ V\}$ **(b)** $R_f = 4\ k\Omega$ **(c)** For $b_ob_1 = \{00, 01, 10, 11\}$ we now have $V_O = \{0.1\ V, -0.564\ V, -1.267\ V, -1.951\ V\}$ **(d)** The corresponding values of DNL expressed as a fraction of an LSB are $\{0, 0.115, 0.063, 0.088\}$ **(d)** The corresponding values of INL expressed as a fraction of an LSB are $\{0, 0.0262, -0.0004, 0\}$

P13.15 **(a)** v_{DAC} takes on the values $\{1.5\text{ V}, 2.25\text{ V}, 2.625\text{ V}, 2.438\text{ V}\}$ in sequence. It holds each value for 1 microsecond. **(b)** 1110 **(c)** 4 microseconds **(d)** 4 microseconds.

P13.19 **(a)** $f_{CL\max} = 3.93\text{ MHz}$ **(b)** $V_I = V_r$ results in the longest conversion **(c)** The longest conversion takes 33.3 ms with the fastest clock possible.

Chapter 14

P14.2 The truth table shows that $y = \overline{A}\cdot B + A\cdot\overline{B}$, applying DeMorgan's theorem we arrive at $y = \overline{\overline{(\overline{A}\cdot B)}\cdot\overline{(A\cdot\overline{B})}}$. Using NOR gates as inverters we can implement this Boolean expression directly as shown in the following figure.

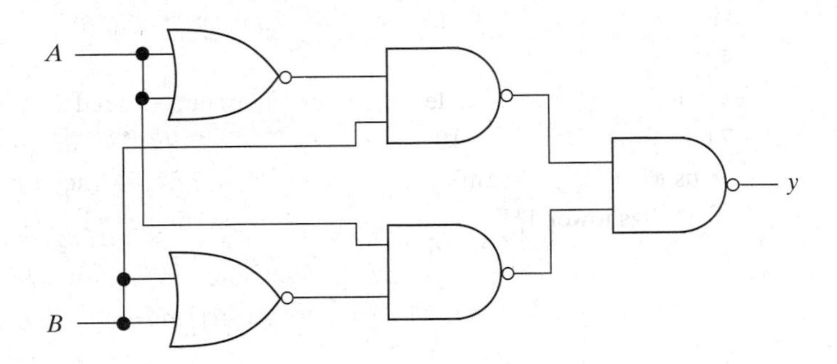

P14.6 **(a)**

$1\times 2^9 + 1\times 2^8 + 0\times 2^7 + 1\times 2^6 + 1\times 2^5 + 0\times 2^4 + 0\times 2^3 + 1\times 2^2 + 0\times 2^1 + 1\times 2^0 = (869)_{10}$

(b) $\underbrace{001}_{1} + \underbrace{101}_{5} + \underbrace{100}_{4} + \underbrace{101}_{5} = (1548)_8$,

(c) $\underbrace{0011}_{3} + \underbrace{0110}_{6} + \underbrace{0101}_{5} = (365)_{16}$, **(d)** $(869)_{10} = (1000), (0110), (1001)$

P14.12 **(a)** It is most convenient to convert a BCD number to decimal before proceeding: $\underbrace{(1001)}_{9}, \underbrace{(0111)}_{7}, \underbrace{(1000)}_{8}, \underbrace{(1001)}_{9} = (9789)_{10}$. Now, note that:
$2^{13} < 9789 < 2^{14}$, and $9789 - 2^{13} = 1597$; $2^{10} < 1597 < 2^{11}$, and $1597 - 2^{10} = 573$; $2^9 < 573 < 2^{10}$, and $573 - 2^9 = 61$; $2^5 < 61 < 2^6$, and $61 - 2^5 = 29$; $2^4 < 29 < 2^5$, and $29 - 2^4 = 13$; $2^3 < 13 < 2^4$, and $13 - 2^3 = 5$; $2^2 < 5 < 2^3$, and $5 - 2^2 = 1 = 2^0$; therefore, $(9789)_{10} = (10011000111101)_2$.

(b) We have: $8^4 < 9789 < 8^5$, and $9789 - 2(8^4) = 1597$; $8^3 < 1597 < 8^4$, and $1597 - 3(8^3) = 61$; $8^1 < 61 < 8^2$, and $61 - 7(8^1) = 5$; $5 = 5(8^0)$; therefore, $(9789)_{10} = (23075)_8$. **(c)** We have: $16^3 < 9789 < 16^4$, and $9789 - 2(16^3) = 1597$; $16^2 < 1597 < 16^3$, and $1597 - 3(16^2) = 61$; $16^1 < 61 < 16^2$, and $61 - 3(16^1) = 13$; $16^0 < 13 < 16^1$, and $13 = 13(16^0)$; therefore, $(9789)_{10} = (233D)_{16}$, **(d)** done in part (a).

P14.16 $A\cdot B = \overline{\overline{A} + \overline{B}}$, which yields the following circuit

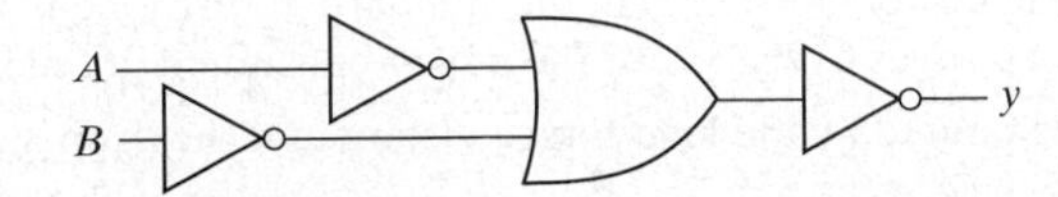

P14.19 Prove that $A(B + C) = AB + AC$. Filling in the truth table below proves this:

Inputs		
A B C	A(B + C)	AB + AC
0 0 0	0	0
0 0 1	0	0
0 1 0	0	0
0 1 1	0	0
1 0 0	0	0
1 0 1	1	1
1 1 0	1	1
1 1 1	1	1

P14.24 The circuit is a JK flip flop.

P14.31 The circuit counts from $(0000)_2 = (0)_{10}$ to $(1001)_2 = (9)_{10}$ and then starts over again.

P14.32 Using the given data and (14.20) we obtain $\Gamma_S = -0.1111$ and $\Gamma_L = 0.99005$. We also find $v_I(0) = v_S(0)\, Z_0/(Z_0 + R_S) = 0.556$ V. Since we are assuming a lossless line and the input step is 1 V, the final value of the load voltage is $v_L(\infty) = R_L/(R_S + R_L) = 0.996$ V. Using the formulas given in Figure 14-27 the first few reflections are; $v_{R1} = 0.550$, $v_{R2} = -0.0611$, $v_{R3} = -0.0605$, and $v_{R4} = 0.006723$. The successive voltages at the load end are; $v_L(0.5\text{ n}) = 1.1056$ V, $v_L(1.5\text{ n}) = 0.9840$ V, and $v_L(2.5\text{ n}) = 0.9973$ V. Therefore, the load is within 1% of its final value after 2.5 ns.

Chapter 15

P15.2 **(a)** $I_F = 128\,\mu$A, $I_R = -723\,\mu$A, and $t_s = 1.63$ ns. **(b)** $t_d = 27$ ns. **(c)** $t_{rr} = 28.6$ ns.

P15.6 $R_B = 17.9\text{ k}\Omega$

P15.10 **(a)** $R_{DS} = 74\text{ M}\Omega$, **(b)** $R_{DS} = 3.9\text{ M}\Omega$

P15.15 **(a)** $V_{IL} = 1.2$ V, $V_{IH} = 2.7$ V, $V_{OL} = 1.15$ V, and $V_{OH} = 4.8$ V **(b)** $NM_L = 0.05$ V, and $NM_H = 2.1$ V

P15.20 **(a)** $V_O = 5$ V, **(b)** $V_O = 1.16$ V, **(c)** $V_O = 0.33$ V.

P15.25 There are three NMOS in series, so $t_{pHL} \approx 3t_{pHL}$ (min. inv.). Worst case is one minimum size PMOS turning on, so $t_{pLH} \approx 2.5t_{pLH}$ (min. inv.). Best case is three minimum size PMOS turning on, so $t_{pLH} \approx (3/2.5)t_{pLH}$ (min. inv.).

P15.32 $Y = \overline{A} + \overline{B} \cdot \overline{C}$ or $\overline{Y} = A \cdot (B + C)$

P15.38 When only one pull-down transistor is on, $V_{OL} = 1.34$ V. When both pull-down transistors are on, $V_{OL} = 0.62$ V.

P15.42 **(a)** $Y = A \cdot B$, **(b)** Without the lower gate, there isn't a current path to pull the load low when $B = 0$.

P15.47 **(a)** $v_{BL}(t) = V_{final} - (V_{final} - V_{pre})e^{-t/\tau}$, where $\tau = R_{on}(1/C_S + 1/C_{BL})^{-1}$ and V_{final} is given by (15.61). **(b)** $t = 90$ ps.

P15.52 The maximum fan-out is 4.

P15.57 The maximum fan-out is 21.

P15.60 **(a)** The load consists of $R_1 = 162\Omega$ to ground and $R_2 = 260\ \Omega$ to -5.2 V. **(b)** The average static power consumed in the Thevenin load is 70.6 mW. **(c)** The average static power consumed in the standard 100 Ω load is 153 mW. The Thevenin load saves over 53% of the power consumed in the load!

P15.64 The speed is limited by the load time constant and the propagation delay is about 1 ns.

P15.71 **(a)** $V_{OH}(\text{nom}) = -1.2$ V, **(b)** $R_1 = 2.2\text{ k}\Omega$.

Index

A

C

E

F

G

H

M

T

U

V

W

Z